5th International School and Conference "Saint Petersburg OPEN" 2018

Optoelectronics, Photonics, Engineering and Nanostructures

Journal of Physics: Conference Series Volume 1124

St. Petersburg, Russia
2-5 April 2018

Part 1 of 2

ISBN: 978-1-5108-8099-3
ISSN: 1742-6588

Printed from e-media with permission by:

Curran Associates, Inc.
57 Morehouse Lane
Red Hook, NY 12571

Some format issues inherent in the e-media version may also appear in this print version.

This work is licensed under a Creative Commons Attribution 3.0 International Licence.
Licence details: http://creativecommons.org/licenses/by/3.0/.

No changes have been made to the content of these proceedings. There may be changes to pagination
and minor adjustments for aesthetics.

Printed with permission by Curran Associates, Inc. (2026)

For permission requests, please contact the Institute of Physics
at the address below.

Institute of Physics
Dirac House, Temple Back
Bristol BS1 6BE UK

Phone: 44 1 17 929 7481
Fax: 44 1 17 920 0979

techtracking@iop.org

Additional copies of this publication are available from:

Curran Associates, Inc.
57 Morehouse Lane
Red Hook, NY 12571 USA
Phone: 845-758-0400
Fax: 845-758-2633
Email: curran@proceedings.com
Web: www.proceedings.com

5th International School and Conference "Saint Petersburg OPEN" 2018

Optoelectronics, Photonics, Engineering and Nanostructures

Journal of Physics: Conference Series Volume 1124

St. Petersburg, Russia
2-5 April 2018

Part 1 of 2

TABLE OF CONTENTS

VOLUME 1

Preface

Peer Review Statement

CRYSTAL GROWTH AND STRUCTURAL PROPERTIES OF SEMICONDUCTOR MATERIALS AND NANOSTRUCTURES

Analytical–Monte Carlo Model of the Growth of in Nanostructures During Droplet Epitaxy on the Triangle-Patterned GaAs Substrates 1
S V Balakirev, M S Solodovnik, I A Mikhaylin, M M Eremenko, O A Ageev

Investigation on Structural Transitions of Graphenes into Diamond Polymorphs at High Pressure 6
V A Greshnyakov, E A Belenkov

Investigation of Titanium Oxide Film on Sapphire Substrate for Gas Sensor 11
Y V Klunnikova, S P Malyukov, A V Sayenko, M I Tolstunov

Nanostructural Features of Anodic Zirconia Synthesized using Different Temperature Modes 15
I A Petrenyov, R V Kamalov, A S Vokhmintsev, N A Martemyanov, I A Weinstein

Structural Properties of Multilayer Heterostructure for Quantum-Cascade Lasers Grown by MBE Growth 20
R R Reznik, N V Kryzhanovskaya, A E Zhukov, A I Khrebtov, Yu B Samsonenko, S V Morozov, G E Cirlin

Spiral Growth of a Crystal Due to Chemical Reaction 24
A V Redkov, S A Kukushkin, A V Osipov

Formation of Hexagonal Silicon Regions in Silicon 30
A A Nikolskaya, D S Korolev, A N Mikhaylov, A I Belov, A A Sushkov, D A Pavlov, D I Tetelbaum

Microstructure and Composition of Zirconium Coatings Obtained by Electro-Spark Alloying 35
V Koshuro, M Fomina, M Fedoseev

Investigation of Deposition Conditions on the Structural Properties of μc-Si:H... 39
A V Uvarov, A S Gudovskikh, V V Fedorov

Structure of Fluorographene and Its Polymorphous Varieties 43
M E Belenkov, V M Chernov, E A Belenkov

Carbon Materials Formed by Polymerization of C_{20} and C_{24} Fullerites 49
M I Tingaev, E A Belenkov

Metal Oxide (Ta-TaOx)-Coatings Obtained by Magnetron Sputtering and Heat Treatment with High-Frequency Currents 53
M Fomina, A Voyko, A Shumilin, V Papshev, A Zakharevich, A Skaptsov, A Fomin

Synthesis of Bismuth Nanowires for Thermoelectric Applications 57
A A Lozovenko, A A Poznyak, G G Gorokh

High Speed Near-Infrared Range Sensor Based on InP/InGaAs Heterostructures 63
K J Smirnov, V V Davydov, S F Glagolev, G V Tushavin

Features of a Zone Thermal Crystallization Semiconductor Thin Films Grown from a Discrete
Liquid Source 68
S N Chebotarev, A V Varnavskaya, I V Gavrus, V N Lozovskii, L M Goncharova, N A
Yakovenko

Comprehensive Study of the Optical and Physical Properties of InGaN Nanocolumns Grown on
Si(111) and por-Si/Si(111) Substrates by Plasma-Assisted Molecular Beam Epitaxy 72
D Zolotukhin, P Seredin, A Lenshin, D Goloshchapov, A Mizerov, E Nikitina

Features of the Mechanism of Gas Sensitivity of the Zinc Oxide Nanorods Arrays to Carbon
Monoxide 77
V. V. Petrov, A. P. Starnikova, Kh. A. Abdullin, D. P. Makarenko

Droplet Epitaxy of In/AlGaAs Nanostructures on the As-Stabilized Surface 81
S V Balakirev, M M Eremenko, I A Mikhaylin, V S Klimin, M S Solodovnik

Investigation of the Electrode Material Influence on the Titanium Oxide Nanosize Structures
Memristor Effect 87
V I Avilov, A S Kolomiytsev, R V Tominov, N I Alyabyeva, E M Bykova

Properties of Semiconductor GaAs Nanoparticles, Synthesized by Combination of Mechanical
Milling Methods and Chemical Etching 91
N Yu Yashina, A J K Al-Alwani, O Yu Tsvetkova, A S Kolesnikova, E G Glukhovskoy, V P
Sevostyanov

Research of Influence Al on Luminescence and Dark Current-Voltage Characteristics of InAs/GaAs
Heterostructures 96
A. S. Pashchenko, E. M. Danilina, N. M. Bogatov

Doping of GaP Layers Grown by Molecular-Beam Epitaxy on Silicon Substrates 100
A A Lazarenko, M S Sobolev, E V Pirogov, E V Nikitina

Multiscale Simulation of Surface Characteristics of Field Emitter Tip 103
K Nikiforov

Plasma-Arc Sputtering for Synthesis of Pt/CeO$_2$ and Pd/CeO$_2$ Nanosystems 107
E A Derevyannikova, T Yu Kardash, A I Stadnichenko, E M Slavinskaya, A V Zaikovskii, S A
Novopashin, A I Boronin

Study of the Geometrical Parameters of in Nanostructures During Droplet Epitaxy on the As-
Stabilized GaAs(001) Surface 112
S V Balakirev, M M Eremenko, I A Mikhaylin, M S Solodovnik

Preparation of Flat Surfaces on Silicon Carbide Substrate using Electron Beam Processing 118
E Yu Gusev, S P Avdeev

Transformation of the Defect Structure of InGaAs and InAlAs Metamorphic Buffer Layers
Depending on Indium Concentration 122
L A Snigirev, A A Sitnikova, D A Kirilenko, N A Bert

The Features of Moiré Pattern on Electron-Microscope Images of Free-Standing Quantum Dots Containing Dislocations 127

L A Sokura, V N Nevedomskiy, N A Bert

Development of Local Catalytic Centers Positioning Technology for Carbon Nanotubes Growth 133

O I Il'In, N N Rudyk, M V Kuzhelev, A A Fedotov

Investigation of InAs/InGaP Nano-Heterostructures Grown by MOVPE for Intermediate Band Solar Cells 137

N A Kalyuzhnyy, S A Mintairov, A M Nadtochiy, V N Nevedomskiy, R A Salii

Suppression of Miscibility Gaps in Vapor-Liquid-Solid InGaAs and InGaN Nanowires 142

T Jean, V G Dubrovskii

Size Effects in LiNbO3 Thin Films Fabricated by Pulsed Laser Deposition 149

Z Vakulov, M Ivonin, E G Zamburg, V S Klimin, D P Volik, D A Golosov, S M Zavadskiy, A P Dostanko, A V Miakonkikh, I E Clemente, K V Rudenko, O A Ageev

GaN Nanostructures Grown on Si(111) by PA-MBE via Droplet Epitaxy: SEM and HRTEM Study 154

G A Sapunov, V V Fedorov, A D Bolshakov, A M Mozharov, L N Dvoreckaia, A Kirilenko, A A Sitnikova, I S Mukhin

Effect of Substrate Temperature on the Properties of Plasma Deposited Silicon Oxide Thin Films 158

E Yu Gusev, J Y Jityaeva, S P Avdeev, O A Ageev

Study of the Effect of Carbon-Containing Gas Pressure on the Geometric Parameters of an Array of Carbon Nanostructures 162

V S Klimin, A A Rezvan, O A Ageev

Influence of Elastic Stresses on the Vapor-Solid-Solid Growth Mechanism of Au-Catalyzed GaAs Nanowires 166

A A Koryakin, S A Kukushkin

Structural Investigation of Light-Emitting A3B5 Structures Grown on Ge/Si(100) Substrate 171

A V Rykov, M V Dorokhin, P S Vergeles, V A Kovalskiy, E B Yakimov, M V Ved', N V Baidus, A V Zdoroveyshchev, V G Shengurov, S A Denisov

The Effect of AlGaAs/GaAs Laser Heterostructure Ionic Surface Cleaning on the Structure and Properties of AlN Films Deposited by the Method of Reactive Ion-Plasma Sputtering 176

E V Fomin, K A Surnin, A D Bondarev, I P Soshnikov, K P Kotliar, S I Pavlov, A V Nashchekin

Theoretical Analysis of the Length Distributions of Ga-Catalyzed GaAs Nanowires 181

Y Berdnikov, V G Dubrovskii, N V Sibirev, A A Koryakin, A. S. Sokolovskii, T Tauchnitz, H Schneider, M Helm, E Dimakis

Genetic Algorithm Application for Solving X-Ray Diffraction Inverse Problem 187

D V Sivkov, V I Punegov

Stress Evolution in AlN Layers Grown on c-Al2O3 by Plasma-Assisted Molecular Beam Epitaxy at Metal-Rich Conditions 192

O A Koshelev, D V Nechaev, S V Ivanov, V N Jmerik

Fabrication Method of the Patterned Mask for Controllable Growth of Low-Dimensional Semiconductor Nanostructures 197

L N Dvoretckaia, A M Mozharov, V V Fedorov, A D Bolshakov, I S Mukhin

MBE Growth of GaAs Nanowires with Modulated Crystal Structure 202
    I V Ilkiv, K P Kotlyar, D A Kirilenko, S P Lebedev, A A Lebedev, P A Alekseev, A D Bouravleuv, G E Cirlin

Nucleation of CdSe Thin Films: The Kinetic Model 206
    A A Koryakin, S A Kukushkin, A V Redkov

NANOBIOTECHNOLOGY, BIOPHYSICS AND BIOPHOTONICS

Energy Transfer in Upconversion Nanoparticles – Phthalocyanine Hybrid Complexes 212
    D A Gvozdev, E P Lukashev, V Z Paschenko

Solid-State Modeling of Human Tracheobronchial Tree for 23 Generations of Airways 217
    V V Makevnina

Peculiarities of Noncontact Cardiac Signal Registration 221
    V A Simon, V A Gerasimov, D K Kostrin, L M Selivanov, A A Uhov

Compact Nuclear Magnetic Spectrometer for Non-Destructive Condition Testing of Biological Objects 227
    N S Myazin, V V Davydov

Influence of Electrostatic Interactions on Cell-Penetrating Peptide-Small Interfering RNA Complex Formation and Intracellular Delivery Efficiency 233
    A Svirina, I Terterov, V Klimenko, S Shmakov, N Knyazev, A Emelyanov, V Vysochinskaya, A Bogdanov

Nanoparticle Tracking Analysis of Extracellular Vesicles Reveals Two Populations of Exosomes 239
    A V Selenina, V A Kulichkova, A N Tomilin, A S Tsimokha

Diagnostic Potential of the Oral Fluid for the Observation People with Multiple Dental Caries by Means of FTIR 243
    P V Seredin, D L Goloshchapov, Y A Plotnikova, Y A Ippolitov, J Vongsvivut

Possible Applications of Bacterial Cellulose in the Manufacture of Electrical Insulating Paper 248
    N Zhuravleva, A Reznik, D Kiesewetter, A Stolpner, A Khripunov

Magnetic Nanoparticles for Medical Application with a Coating Deposited with Various Methods 252
    V N Zorin, E B Naumisheva, V N Postnov, N V Evreinova, K G Gareev, D V Korolev

A Studying of Subphase Temperature and Dissolved Ascorbic Acid Concentration Influence on the Process of Langmuir Monolayer Formation 257
    M M Qassime, V A Goryacheva, A J Al-Alwani, T N Lugovitskaya, A B Shipovskaya, E G Glukhovskoy

Investigation Optical Properties and Functional Surface Characteristics of Nanoparticles Based on Porous Silicon for Applications in Biomedicine 262
    A O Belorus, A I Pastukhov, A S Lenshin, P V Seredin, Yu M Spivak, V A Moshnikov

Fluorescent Method of Bacterial Contamination Control on Meat Surface 267
    M V Mekhrengin, I K Meshkovskii, L A Kaftyreva, V I Guryev, D A Pogorelaya

Study of Electric Properties of Self-Assembled Films of Albumin During Their Dehydration 272
    M A Baranov, A P Alekseenko, E N Velichko

Controlled Modification of Hyaluronic Acid for Photoinduced Reactions in Tissue Engineering.................. 277
A. V. Sochilina, A. G. Savelyev, P. A. Demina, N. V. Ierusalimsky, D. A. Khochenkov, R. A. Akasov, N. V. Sholina, E. V. Khaydukov, A. N. Generalova

Microchip Device with Dry-Stored Reagents for Loop Mediated Isothermal Amplification........................... 283
A Tupik, G Rudnitskaya, A Bulyanitsa, T Lukashenko, D Varlamov, A Evstrapov

Investigation of the Color Effect of the Test Object on the Pupil Response....................................... 288
D Bobrova, N Bikberdina, M Boronenko

Artificial Neural Network Prediction of Lysozyme Solubility for Protein Crystallization............................. 292
A S Sokolovskii, N A Besedina, Yu Berdnikov, V G Dubrovskii

Effective Application of Independent Component Analysis to the Task of Gradual EMG Control 296
M O Shamshin, V A Makarov, S A Lobov

Use of Electrophoretic Light Scattering for Investigation the Parameters of Macromolecules....................... 301
E A Savchenko, E K Nepomnyashchaya, E N Velichko, A N Skvortsov

Fluorescent Detection and Analysis of Single Molecules.. 306
O B Kuznetsova, E A Savchenko, E T Aksenov

Non-Invasive Research of Biological Objects by the Method of Laser Polarimetry..311
M V Putintseva, E T Aksenov, C C Korikov, E N Velichko

In Vitro Model of CNS Neuronal Pathway Recovery using Microfluidic Chips.................................... 315
O Antipova, Y Pigareva, V Kolpakov, A Gladkov, A Bukatin, I Mukhina, V Kazantsev, A Pimashkin

Investigation of Na+/K+-ATPase Role in Cancer Cells' Functioning 320
S A Perkov, A K Emelyanov, S V Shmakov, A A Bogdanov

Accuracy Increase of Blood Viscoelasticity Measurement of the Microfluidic Wheatstone Bridge............... 323
N D Shipulya, S A Konakov

Singlet Oxygen Generation Mechanism in the Presence of Excited Nanoporous Silicon............................. 328
D M Samosvat, O P Chikalova-Luzina, V S Khromov, A G Zegrya, G G Zegrya

Formation and Manipulation of Polyacrylamide Spheroids Doped with Magnetic Nanoparticles in Microfluidic Chip 333
D V Nozdriukhin, N A Filatov, A S Bukatin

Polylactide Film Deposition onto Titanium Surface from Different Solutions 337
A V Romashkin, N S Struchkov, Yu A Polikarpov, V A Petukhov, D D Levin, V K Nevolin

Comparison of Step and Flow-Focusing Emulsification Methods for Water-In-Oil Monodisperse Drops in Microfluidic Chips.. 343
N A Filatov, D V Nozdriukhin, A A Evstrapov, A S Bukatin

Spectroscopic Investigation of Vegetable Pectin - Fluorescent Protein EGFP Interaction in Tumor Cells *in Vitro*.. 348
V V Danilov, V V Klimenko, S V Shmakov, A I Khrebtov, A S Kulagina

LASERS, SOLAR CELLS AND OTHER OPTOELECTRONIC DEVICES

Modulation P-Doping as the Way to Attain Multi-State Lasing in Short-Cavity InAs/InGaAs Quantum Dot Lasers 352
V V Korenev, A V Savelyev, V G Dubrovskii, S Breuer, M V Maximov, A E Zhukov

Injection Microdisk Lasers Based on Multilayers of InGaAs/GaAs Quantum Well-Dot Structures 356
N V Kryzhanovskaya, E I Moiseev, K P Kotlyar, Yu A Guseva, Yu M Shernyakov, S A Scherbak, M V Maximov, A E Zhukov

Optical Properties of $In_{0.8}Ga_{0.2}As$ Quantum Dots in GaAs Photovoltaic Convertors 360
R A Salii, V V Evstropov, S A Mintairov, M A Mintairov, M Z Shvarts, N A Kalyuzhnyy

Features of Direct Digital Synthesis Applications for Microwave Excitation Signal Formation in Quantum Frequency Standard on the Atoms of Cesium 365
A A Petrov, V V Davydov, D V Zalyotov, V E Shabanov, D V Shapovalov

Effect of Carrier Localization on Performance of Coupled Large Optical Cavity Diode Lasers 372
A A Serin, A S Payusov, Yu M Shernyakov, N A Kalyuzhnyy, S A Mintairov, N Yu Gordeev, F I Zubov, M V Maximov, A E Zhukov

Development of Technological Installation for Obtaining Functional Thin Film Elements Used in Solar Elements of Space Equipment 378
T O Zinchenko, V I Kondrashin, E A Pecherskaya, P E Golubkov, K O Nikolaev, F A Abdullin

Spectral Approach to Li–ion Batteries Degradation Rate Estimation 382
E A Yusupova, V G Malyshkin, V V Davydov

$CH_3NH_3PbI_3$ Crystal Growth, Structure and Composition 387
V E Anikeeva, O I Semenova, O E Tereshchenko

Isolation and Optoelectronic Characterization of Si Solar Cells Microstructure Defects 392
A Gajdos, P Skarvada, R Macku, N Papez, L Skvarenina, D Sobola

Investigation of the Effect of Chemical Pre-Treatment on Uniformity of the Silicon Wafer Texturing for Manufacturing a Solar Cell 398
D Kudryashov, A Gudovskikh, A Rodin, N Pinaev

The Optical Method for Condition Control of Flowing Medium 403
N M Grebenikova, K J Smirnov, V V Davydov, V. Y. Rud'

Study of P-Type Contact Topography Influence on Characteristics of Microdisk and Microring Lasers 408
A S Novikov, E I Moiseev, N V Kryzhanovskaya, B. I. Afinogenov, Y A Guseva, K Kotlyar, A A Lipovskii, A G Nasibulin, M V Maximov, V G Tikhomirov, A E Zhukov

The Ways to Improve the Energy Conversion Efficiency in Erbium-Doped Gd_2O_3 Nanoparticles 413
Yu A Kuznetsova, A F Zatsepin, M A Mashkovtsev, V N Rychkov, S P Vyatkina

Theoretical Model for Spatial Separation of Dominating Recombination Region in a-Si:H/c-Si Structures 417
A. D. Maslov, E. V. Bezuglaya

Surface Morphology After Reactive Ion Etching of Silicon and Gallium Arsenide Based Solar Cells 421
N Papež, D Sobola, A Gajdoš, L Škvarenina, R Macku, M Eliáš, A Nebojsa, R Motúz

Reducing of Thermal Resistance of Edge-Emitting Lasers Based on Coupled Waveguides.............. 427
A A Serin, A S Payusov, Yu M Shernyakov, N A Kalyuzhnyy, S A Mintairov, M V Maximov, N Yu Gordeev, A E Zhukov

All-Oxide Heterojunction Solar Cells Formed by Magnetron Sputtering 431
D Kudryashov, A Gudovskikh, A Monastyrenko

Influence of Au Surface Properties on Photon Emission from Localized Metal-Insulator-Metal Tunnel Contact 436
V A Shkoldin, D V Permyakov, M V Zhukov, A A Vasiliev, T A Mamaeva, A O Golubok, A V Uskov, A D Bolshakov, A K Samusev, S Mukhin

Diagnostics of Inhomogeneities of Parameters of Light-Emitting Heterostructures by the Photoelectric Method.. 440
I V Frolov, O A Radaev, V A Sergeev, S V Vasin

Influence of Coating Layers on Characteristics of Microdisk Lasers with InAs/InGaAs Quantum Dots Active Region.. 444
I Y Agafonov, N V Kryzhanovskaya, E I Moiseev, A S Dragunova, M V Fetisova, K P Kotlyar, S A Scherbak, I V Reduto, Yu A Guseva, A A Lipovskii, M V Maximov, A E Zhukov

Influence of Hydrogen Plasma Passivation on Electrical and Spectral Characteristics of GaN Nanowires / Si Solar Cells... 449
K U Shugurov, A M Mozharov, V V Fedorov, A D Bolshakov, G A Sapunov, I S Mukhin

Mixed Halide Perovskite Light Emitting Solar Cell.. 453
D Gets, A Ishteev, T Liashenko, D Saranin, S Makarov, A Zakhidov

Non-Radiative Coupled Donor-Acceptor Pair Recombination in Nanostructures Based on Nitrides and Related Phenomena 458
N A Talnishnikh, E I Shabunina, N M Shmidt, A E Chernyakov, D S Arteev, A A Zybin

Formation of Nanoscale Structures on the Surface of Gallium Arsenide by Local Anodic Oxidation and Plasma Chemical Etching.. 462
V S Klimin, M S Solodovnik, S A Lisitsyn, A A Rezvan, S V Balakirev

Quantum Autonomous Magnetic Field Sensor... 467
S E Logunov, A Yu Koshkin, V V Davydov

Stabilization TiO$_2$ Anatase by F-Ion Doping for Solar Panel Producing........................... 472
S. N. Fedorov, R. V. Kurtenkov, V. V. Vasiliev

Alpha-Factor in Multimode Lasers... 476
I K Boikov, A V Savelyev

Heterointerfaces in the Bottom Tunnel Part of GaInP/GaAs/Ge Solar Cells........................... 481
M A Mintairov, V V Evstropov, S A Mintairov, M Z Shvarts, N A Kalyuzhnyy

Growth and Optical Characterization of 7.5 μm Quantum-Cascade Laser Heterostructures Grown by MBE 486
A V Babichev, A G Gladyshev, E S Kolodeznyi, A S Kurochkin, G S Sokolovskii, V E Bougrov, L Ya Karachinsky, I I Novikov, V V Dudelev, V N Nevedomsky, S O Slipchenko, A V Lutetskiy, A N Sofronov, D A Firsov, L E Vorobjev, N A Pikhtin, A Yu Egorov

Modeling of the Internal Quantum Yield for Solar Cell with Quantum Dots........................... 491
A. Panchak, N Kalyuzhnyy, S. Mintairov, M. Shvarts

Effect of Temperature on Dry Etching of III-V Structures 495
  I A Morozov, A S Gudovskikh, D A Kudryashov, K P Kotlyar, K Y Shubina

Chromaticity Coordinates Temperature Dependence for Blue Laser Diodes for Solid State Laser
Lighting 499
  A Aluyev, Yu Akhmerov, N Kourova, V Logachev, M Mezhenny, A Chelny, A Savchuk, V Murashev, O Rabinovich, S Didenko

Optical Coupling Upset in Multijunction Solar Cells with Built-In Bragg Reflectors 503
  S A Levina, E D Filimonov, V M Emelyanov, M Z Shvarts

Quantum Efficiency Measurement of Subcells in Multi-Junction Solar Cells Based on III-V/Si 507
  A I Baranov, A S Gudovskikh, D A Kudryashov, I A Morozov, A M Mozharov, K S Zelentsov

Temperature Annealing Effect on ITO Film 511
  K P Kotlyar, D A Kudryashov, I P Soshnikov, A V Uvarov, R R Reznik

Silicon Doping of GaP Layers Grown by Time-Modulated PECVD 515
  A I Baranov, I A Morozov, A V Uvarov, G E Yakovlev, J-P Kleider

1.06 μm Wavelength Photodetectors with Metamorphic Buffer Layers Grown on GaAs Substrates.............. 520
  I V Samartsev, S M Nekorkin, B N Zvonkov, N V Dikareva, A V Zdoroveyshchev, A V Rykov, N V Baidus

Sunlight Simulation in Investigating Characteristics of Small-Size Sunlight Concentrators and
Multijunction Solar Cells 524
  E D Filimonov, S A Levina, M Z Shvarts

A Novel Approach to Characterization of Bottom Sub-Cell in Multijunction Solar Cell using
Photoluminescence. 528
  K S Zelentsov, A S Gudovskikh, D A Kudryashov, N V Kryzhanovskaya, N A Kalyuzhnyy, S A Mintairov

Numerical Simulation of the Carbon Nanotubes Transport Layer Influence on Performance of
GaAs Solar Cell.. 532
  S A Raudik, A M Mozharov, D M Mitin, A D Bolshakov, P M Rajanna, A G Nasibulin, I S Mukhin

Investigating the Output Characteristics of Ring Laser in Laser Dynamic Goniometer 537
  N. A. Eno, P. A. Pavlov, J. T. Bagshaw

Investigation of the Electrical and Optical Properties of Nickel Oxide Films Produced by RF
Magnetron Sputtering Method... 542
  A Aglikov, A Mozharov, D Kudryashov, D Sosnin, A Vasilev, A Bolshakov, S Makarov, I Mukhin

Lateral Mode Engineering in Diode Lasers Based on Coupled Ridges... 547
  A S Payusov, A A Serin, Yu M Shernyakov, D A Rybalko, M M Kulagina, M V Maximov, N Yu Gordeev

NANOPHOTONICS, SPECTROSCOPY, MICROCAVITIES, OPTICS, PLASMONICS

Au-SiO$_2$-Si Hybrid Plasmonic Waveguide Micro-Ring Resonator Sensor 551
  M A Butt, S A Degtyarev, S N Khonina

Magnetron Sputtered MoS$_2$: Optical and Structural Analysis 557
    A. I. Belikov, Kyaw Z. Phyo

Effect of Defocusing and Apodization on the Resolution of Incoherent Optical System Beyond the
Rayleigh Limit.. 563
    Andra N. K. Reddy, M. A. Butt, S. N. Khonina

Increasing the Resolution of the Aberrated Optical System Based on Quadratic Amplitude
Apodization 568
    P A Khorin, S A Fomchenkov

Optical Properties and Energy Band Parameters of Luminescent CaMoO$_4$:Bi Ceramics 574
    *R A Parulin, I V Timoshenko, Yu A Kuznetsova, A F Zatsepin, E S Buyanova, Z A
Mikhaylovskaya, M S I Koubisy*

Polarization Plane Rotation for Higher Order Modes in Twisted Optical Fibers with Discrete
Rotationally Symmetric Core 578
    C N Alexeyev, M C Alexeyeva, B P Lapin, D V Vikulin, M A Yavorsky

Room Temperature Lasing from Microdisk Laser in Aqueous Medium 582
    *M V Fetisova, N V Kryzhanovskaya, I V Reduto, E I Moiseev, S A Blokhin, K P Kotlyar, S A
Scherbak, A A Lipovskii, A A Kornev, A S Bukatin, M V Maximov, A E Zhukov*

Self-Assembly of Lines of Microscopic Photonic Crystals.. 586
    M S Ashurov, A A Ezhov, T A Kazakova, S O Klimonsky

Broadband Nanoantenna 589
    D Poletaev, B Sokolenko, V Voytitsky, A Nudga

Novel Highly Efficient Blue-Emitting Branched Oligoarylsilanes 593
    M S Skorotetcky, O V Borshchev, N M Surin, S A Ponomarenko

VOLUME 2

Nonlinear Pulses in Dispersion-Managed Fiber-Optic Systems in Presence of High Losses.......................... 597
    *V Neskorniuk, A Lukashchuk, I Gabitov, A Chipouline, M Malekizandi, G Ovchinnikov, F
Küppers*

Study of Optical Pumping Influence on Carbon Nanotubes Permittivity in THz Frequency Range 602
    P Demchenko, D Gomon, I Anoshkin, D Lioubtchenko, M Khodzitsky

Numerical Study of Optical Properties of Sphere-Gap-Cone Hybrid Nanoantenna.. 607
    Y Sun, S. V. Makarov, D A Zuev

Active and Passive Phase Stabilization for the All-Fiber Michelson Interferometer.. 611
    M. S. Elezov, M. L. Scherbatenko, D. V. Sych, G. N. Goltsman

Binary Diffractive Optics for 3D-Demultiplexing of OAM Beams 616
    A P Porfirev, S A Fomchenkov, G E Gridin, S N Khonina

The Study of Photoluminescence Properties of AlGaAs/GaAs Heterostructure After Ga+ Focused
Ion Beam Etching 621
    G V Voznyuk, I V Levitskii, M I Mitrofanov, M N Mizerov, D N Nikolaev, V P Evtikhiev

Optical Limiting in Nanodiamond Suspension: Shortening of the Laser Pulses 625
    R Yu Krivenkov, K G Mikheev, T N Mogileva, N Nunn, O A Shenderova, G M Mikheev

New Ultra High-Speed All-Optical Coherent D-Trigger 630
S N Bagayev, V S Egorov, V G Nikolaev, I B Mekhov, I A Chekhonin, M A Chekhonin

Comparison of Optical Spectral Devices in the Framework of System Approach 636
V Kazakov, O Moskaletz, M Vaganov

Four-Step Fabrication of SERS-Active Microfluidic Channels 642
E S Babich, E S Gangrskaia, I V Reduto, A A Lipovskii

Second Harmonic Generation and Spontaneous Parametric Down-Conversion in Mie
Nanoresonators 647
M I Petrov, A A Nikolaeva, K S Frizyuk, N A Olekhno

Spatially Modulated Channel Waveguide Elements Optically Written in Photorefractive Lithium
Niobate 651
A D Bezpaly, A S Perin, V M Shandarov

Plasmonic Amplification of Terahertz Radiation in a Double-Layer Graphene Nanoribbon Array 655
I M Moiseenko, O V Polischuk, M Y Morozov, V V Popov

SERS Induced by Two Coupled Monolayers of Au Plasmonic Nanoparticles 659
*V Kaydashev, P Zolotukhin, A Belanova, A Anokhin, A Volodin, A Chernyshev, N Lyanguzov, E
Kaidashev*

Superconducting Nanowire Single-Photon Detector on Lithium Niobate 664
E Smirnov, A Golikov, P Zolotov, V Kovalyuk, M Lobino, B Voronov, A Korneev, G Goltsman

Study of Charge Relaxation in Poled Silicate Glasses 668
D V Raskhodchikov, I V Reshetov, D K Tagantsev, A A Lipovskii, V P Kaasik

Initialization of the Bell States of Two Qubits by Unipolar Pulses 673
M. V. Denisenko, N. V. Klenov, A. M. Satanin

Second Harmonic Generation by Metal Core - Dielectric Shell Spherical Nanoparticles: Spatial vs.
Plasmon Resonances 678
S A Scherbak, A A Lipovskii, E S Babich

Laser-Deposited Hybrid Au-Ag@C Nanoparticles as Efficient SERS & Adsorption Material 684
A A Vasileva, D V Pankin, I E Kolesnikov, S S Rzhevskiy, A A Manshina

Influence of Sputtering Parameters on the Main Characteristics of Ultra-Thin Vanadium Nitride
Films 689
*P I Zolotov, A V Divochiy, Yu B Vakhtomin, A V Lubenchenko, P V Morozov, I V Shurkaeva, K
V Smirnov*

Dielectric Surrounding Decimates Eigenmodes of Microdisk Optical Resonators 695
A V Raskhodchikov, S A Scherbak, N V Kryzhanovskaya, A E Zhukov, A A Lipovskii

Electron Diffusivity Measurements of VN Superconducting Single-Photon Detectors 699
N R Romanov, P I Zolotov, Yu B Vakhtomin, A V Divochiy, K V Smirnov

Phase Diffraction Patterns Optically Induced by Laser Bessel-Like Beams in Photorefractive
Lithium Niobate with a Doped Surface Layer 705
I Trushnikov, A Inyushov, A Perin

Focusing of High-Order Vortex Laser Beams by Binary Axicon with Different Numerical Aperture 710
D A Savelyev, S A Fomchenkov

NEXAFS C1s-Spectra Apparatus Distortion Study 714
A E Mingaleva, O V Petrova, D V Sivkov, N N Shomysov, S V Nekipelov, R N Skandakov, V N Sivkov

Raman Lidar with for Geoecological Monitoring 718
J Ruzankina, V Elizarov, L Konopel'Ko, A Zhevlakov, A Grishkanich

Soft X-Ray Photoemission Study of Ba Adsorption on the Ceramic Multiferroic $BiFeO_3$ 722
P A Dementev, G V Benemanskaya, S N Timoshnev

UV-Induced Refractive Index Changes Due to Silver Molecular Clusters in Photo-Thermo-Refractive Glass 727
V V Gorbyak, A I Sidorov

Solution of the Multiple Light Scattering Problem in Turbid Suspensions via Cross-Correlation Processing 731
Z A Zabalueva, E K Nepomnyashchaya, C C Korikov

Quality Factor Enhancement in Gold-Based Tamm Plasmon Cavity by Sub-Wavelength Structuration 736
A R Gubaydullin, K M Morozov, M A Kaliteevski

Development of Diffractive Optical Elements with Low Surface Roughness by Direct Laser Writing 739
S A Fomchenkov, A P Porfirev

Separation of Inhomogeneous and Homogeneous Broadening Manifestations in InGaAs/GaAs Quantum Wells by Time-Resolved Four-Wave Mixing 743
I A Solovev, Yu V Kapitonov, B V Stroganov, Yu P Efimov, S A Eliseev, S V Poltavtsev

Magneto-Resonance Methods of the Relaxation Rate Measuring for the Proton-Containing Flowing Fluids Composition Studying 747
L I Fatkhutdinova, A N Mamonkina, S V Ermak, V V Semyonov

Transmission of Thermal Imaging by using Infrared Bundle Based on Silver Halide Solid Solution 752
A S Shmygalev, D D Salimgareev, A S Korsakov, B P Zhilkin, V I Terekhov

On-Chip Single-Photon Spectrometer for Visible and Infrared Wavelength Range 756
V Kovalyuk, O Kahl, S Ferrari, A Vetter, G Lewes-Malandrakis, C Nebel, A Korneev, G Goltsman, W Pernice

On-Chip Controlled Placement of Nanodiamonds with a Nitrogen-Vacancy Color Centers (NV) 761
S Komrakova, J Javadzade, V Vorobyov, S Bolshedvorskii, V Soshenko, A Akimov, V Kovalyuk, A Korneev, G Goltsman

Experimental Optimisation of O-Ring Resonator Q-Factor for On-Chip Spontaneous Four Wave Mixing 765
P An, V Kovalyuk, A Golikov, E Zubkova, S Ferrari, A Korneev, W Pernice, G Goltsman

Optimization of Contra-Directional Coupler Based on Silicon Nitride Bragg Rib Waveguide 769
E Zubkova, P An, V Kovalyuk, A Korneev, S Ferrari, W Pernice, G Goltsman

Investigation of Influence Incoherent Background Illumination on the Nonlinear Response of a Lithium Niobate Crystal Sample at Low Light Intensity 774
A V Pustozerov, A S Perin

Bolometric Effect for Detection of sub-THz Radiation with Devices Based on Carbon Nanotubes 779
    M V Moskotin, I A Gayduchenko, G N Goltsman, N Titova, B M Voronov, G F Fedorov, F Pyatkov, F Hennrich

Silicon Nitride Nanophotonic Circuit for On-Chip Spontaneous Four-Wave Mixing 784
    A Golikov, V Kovalyuk, P An, E Zubkova, S Ferrari, W Pernice, A Korneev, G Goltsman

Optimization of On-Chip Photonic Delay Lines for Telecom Wavelengths 788
    A Prokhodtsov, P An, V Kovalyuk, E Zubkova, A Golikov, A Korneev, S Ferrari, W Pernice, G Goltsman

Graphene-Based Tunability of Chiral Metasurface in Terahertz Frequency Range 793
    M S Masyukov, A V Vozianova, A N Grebenchukov, M K Khodzitsky

Graphene-Layer and Graphene-Nanoribbon FETs as THz Detectors 797
    Y E Matyushkin, I A Gayduchenko, M V Moskotin, G N Goltsman, G E Fedorov, M G Rybin, E D Obraztsova

Purcell Effect in GaN-Based Waveguiding Structures 803
    K M Morozov, K A Ivanov, A R Gubaydullin, M A Kaliteevski

Strong Coupling Between Excitons in Transition Metal Dichalcogenides and Optical Bound States
in the Continuum 808
    S. K. Sychev, K. L. Koshelev, Z. F. Sadrieva, A. A. Bogdanov, I. V. Iorsh

Experimental Observation of Symmetry Protected Bound State in the Radiation Continuum in the
Periodic Array of Ceramic Disks 813
    A A Bogdanov, M Balezin, P V Kapitanova, Z Sadrieva, M Belyakov, E A Nenasheva, A F Sadreev

High-Q States and Strong Mode Coupling in High-Index Dielectric Resonators. 817
    S. A. Gladyshev, A. A. Bogdanov, P. V. Kapitanova, M. V. Rybin, K. L. Koshelev, Z. F. Sadrieva, K. B. Samusev, Y. S. Kivshar, M. F. Limonov

RIE for Structuring E-Field Processed Glasses 823
    I Reduto, D Raskhodchikov, E Gangrskaia, V Kaasik, Yu Svirko, A Lipovskii

Photodecomposition of Organic/Inorganic Composite Materials Based on Polyvinylpyrrolidone 827
    A S Kulagina, S K Evstropiev, K V Dukelskii, N A Volkova, K S Evstropyev, N V Nikonorov

Generation of the Trapping Light Structures Based on Vector Fields 832
    N Shostka, O Karakchieva, B V Sokolenko

Tamm Magnetophotonic Structures with Bi-Substituted Iron Garnet Layers at Oblique Incidence 838
    T Mikhailova, S Tomilin, S Lyashko, A Shaposhnikov, A Prokopov, A Karavainikov, A Bokova, V Berzhansky

Epitaxial Films of Garnet Ferrite with Anisotropy "easy Plane" for Magneto-Optical Eddy Current
Flaw Detection 844
    N Lugovskoy, V Berzhansky, D Glechik, A Prokopov

SPINTRONICS, ELECTRO- AND MAGNETOOPTICS

Electrically Controlled Spin Polarization in Suspended GaAs Quantum Point Contacts 848
    D A Pokhabov, A G Pogosov, E Yu Zhdanov, A A Shevyrin, A K Bakarov, A A Shklyaev

The Study of Mechanical Resonances of the Phase Electro-Optic Modulator Based on $LiNbO_3$ for Noise Reduction of Fiber-Optic Gyroscope 852
M A Smolovik, D A Pogorelaya, A A Vlasov, A S Aleynik, V E Strigalev

Magneto-Optical Properties of Metaphosphate and Borate Glasses 856
Dmitrii I Sobolev, Anastasiia N Babkina, Nikolay V Nikonorov

Investigation of the Magnetic Properties of Manganese Silicide Grown on i-GaAs Substrate by Pulsed Laser Deposition 860
Y Kuznetsov, M Dorokhin, A Kudrin, V Lesnikov, E Demidov, V Karzanov

On the Mechanism of Spin-Polarized Injection in (Ga,Mn)As/n+GaAs/InGaAs Zener Tunnel Diode 865
M Ved, M Dorokhin, E Malysheva, A Zdoroveyshchev, Yu Danilov, A Parafin, Yu Kuznetsov

Morphology of Garnet Films for Thermo-Magnetic Recording 870
V Berzhansky, Y Danishevskaya, A Nedviga, M Bektemirova

ELECTRIC, MAGNETIC AND MICROWAVE DEVICES

A 0.3-0.7 THz Flux-Flow Oscillator Integrated with the Slot Antenna and Elliptical Lens 874
N V Kinev, K I Rudakov, A M Baryshev, V P Koshelets

The Development of the Bistable Micromechanical Actuator for Optical Relay 878
Y. B. Enns, E. N. Pyatishev, A. N. Kazakin

Calculation of High-Frequency Conductivity and Hall Constant of a Thin Conductive Layer in the View of Equal Specularity Coefficients of Its Surfaces 884
O V Savenko, I A Kuznetsova, A A Yushkanov

Electrical and Photoelectric Characteristics of Gallium Oxide Films Obtained by RF Magnetron Sputtering 890
A Tsymbalov, J Petrova

Effect of Metal Modifiers on the Characteristics of Resistive Hydrogen Sensors Based on Thin Films of Tin Dioxide 895
A V Almaev, N K Maksimova, E Yu Sevastyanov, E V Chernikov, T A Davydova, T E Smirnova

Discrete Diffraction in Network of Magnonic Crystals 900
E N Beginin, A Yu Sharaevskaya

The Study of Optical Properties of Graphene Intercalated with Ferric Chloride for Application in Terahertz Photonics 905
A D Zaitsev, A N Grebenchukov, P S Demchenko, E T Alonso, M F Craciun, S Russo, A Baldycheva, M K Khodzitsky

Lifetime Testing of a MEMS Switch with Pt-Pt Contact 910
I V Uvarov, A N Kupriyanov

Simulation of Electrical Conductivity of Silicon Diodes with Bismuth Implanted-Ion Profiles 915
S M Loganchuk, S N Chebotarev, D A Arustamyan, A A A Mohamed, L Touel, N M Bogatov

Flexoelectrical Nanogenerator Design using Aligned Carbon Nanotubes 919
M V Il'Ina, A A Konshin, E G Solomin

How to Take Fractional-Order Derivative Experimentally? 923
D D Stupin, A I Lihachev, A V Nashchekin

Ti/4H-SiC Schottky Diode Breakdown Voltage with Different Thickness of 4H-SiC Epitaxial Layer............ 928
    S V Sedykh, S B Rybalka, A Yu Drakin, A A Demidov, N S Ponomaryova, O A Shishkina

The Influence of the Deep Level Type on a Switching Time Delay of GaAs Avalanche S-Diodes 933
    V V Kopyev, T E Smirnova, V L Oleinik, I A Prudaev

Accelerated Degradation HEMT Based on AlGaN / SiC 937
    A S Evseenkov, V G Tikhomirov

Software Complex for Calculating the Initial Section of the Current-Voltage Characteristics of a
Resonant-Tunneling Diode with the Possibility of Computer Statistical Experiment...................................... 942
    K V Cherkasov, S A Meshkov, M O Makeev

Development of the Sheet Electron Beam Focusing System Based on Thermionic and Field
Emission Cathodes 948
    A V Danilushkin, A A Burtsev, K. V. Shumikhin, G. V. Sakhadzhi

Increase of Accuracy of Capacitance Parameters Measurements of Power Semiconductor Modules
on Base IGBT and FRD.. 952
    D A Knyaginin, A Yu Drakin, S B Rybalka, A A Demidov

Influence of Microwave Electromagnetic Field on the Structure of Polymers.. 956
    E Vasinkina, S Kalganova, V Alekseev, Yu Kadykova, S Arzamastsev, A Dolzhikova

Investigation of the Influence of Parameters of Nanoscale Profiling of the Surface of GaAs
Structures by a Combination of Local Anodic Oxidation and Plasma Chemical Etching Methods 960
    V S Klimin, A A Rezvan, I N Kots, N A Naidenko

Research of using Plasma Methods for Formation Field Emitters Based on Carbon Nanoscale
Structures... 964
    V S Klimin, A A Rezvan, O A Ageev

Numerical Simulation of Induction Heating of a Carburizing Container with a Titanium Sample.................. 968
    A Voyko, M Fomina, A Shumilin, I Rodionov, S Kalganova, I Artyukhov, A Fomin

Measurements in Microwave Electrotechnology 972
    S Kalganova, E Vasinkina, V Alekseev, V Lavrentyev, S Trigorly, T Dunaeva

Effect of Metallic Nanoantennas on the Efficiency of the Surface Plasmon-Polariton Generation via
Excitation of Electromagnetic Waves in a Tunnel Junction.. 976
    L N Dvoretckaia, A M Mozharov, A V Uskov, A D Bolshakov, A O Golubok, I S Mukhin

OTHER ASPECTS OF NANOTECHNOLOGY

Metal-Assisted Photoenhanced Wet Chemical Etching of GaN Epitaxial Layers 980
    K Yu Shubina, D V Mokhov, T N Berezovskaya, A M Mizerov, E V Nikitina, A D Bouravleuv

Investigation of Luminescence Quantum Yields of Carbon Dots Synthesized from Ethylene Glycol,
Citric Acid and Berries 984
    M N Egorova, A E Tomskaya, A N Kapitonov, S A Smagulova, A A Alekseev

Formation of a Microheterophase State from Planar Nanoparticles of Graphene at the Oil-Water
Interface... 990
    Yu V Pakharukov, F K Shabiev, R F Safargaliev

The Nanosecond Studies of Granular Carbon Nanostructures Based on High Temperature
Superconductors 994
    M P Faradzheva, A V Prikhodko, O I Konkov, S P Faradzhev

Thermoelectric Peltier Micromodules Processed by Thin-Film Technology 998
    E Bakulin, S Dzyubanenko, S Konakov, A Korotkov, V Loboda, A Yugay

The Influence of Photo-Stimulated Adsorption of Polyelectrolyte Molecules on Electro-Physical
Characteristics of Structures Based on Single Crystal Silicon Substrates 1003
    A V Kozlowski, S V Stetsyura, I V Malyar

Electromechanical Bending Microactuator as Optical Shutter 1008
    R. Kleimanov, Y. Enns, E. Pyatishev, I. Komarevtsev

Application of XRD Methods for the Pilot Studies of New Functional Materials for Photonics 1013
    M. Dermeneva, D. Muravijova, M. Mynbaeva, V. Bougrov, M. Yagovkina

The Formation of Arachidic Acid Langmuir Monolayers on the NiCl$_2$ Solution 1018
    A Chumakov, Ammar J Al-Alwani, A Ermakov, O Shinkarenko, N Begletsova, E Glukhovskoy,
    S Santer

Chemical Analysis of Thin-Film's Colour Generation During Surface Laser Oxidation of TiN-
Coating 1022
    O S Yulmetova, M S Tutova, R F Yulmetova

Studies of the Formation of Copper Nanoparticles Monolayers on the Water Subphase 1026
    N N Begletsova, E I Selifonova, R K Chernova, A S Chumakov, V P Sevostyanov, E G
    Glukhovskoy

Multicaloric Effect in Barium Titanate Nanotube 1032
    I A Starkov, I L Mylnikov, A S Starkov

Scanning Probe Microscopy of AlGaAs/GaAs Diode After Partial Electrical Breakdown 1038
    A O Mikhaylov, P A Alekseev, A A Podoskin, S O Slipchenko, M S Dunaevskiy

Methods of Applying the Reliability Theory for the Analysis of Micro-Arc Oxidation Process 1043
    P E Golubkov, E A Pecherskaya, Y V Shepeleva, A V Martynov, T O Zinchenko, D V
    Artamonov

Structure of "Chromium Steel-Base – Ti-Coating" and Its Production by the Contact Welding 1049
    I Egorov, A Shchelkunov, A Fomin, I Rodionov

Structural Transformations on the Surface of 1.3343 Tool Steel and 12Cr18Ni10Ti Stainless Steel
After Induction Heat Treatment and Quenching 1053
    P Palkanov, A Fomin, I Rodionov

Formation of Wear-Resistant Oxide-Carbide Coatings on Titanium by Electrospark Alloying 1057
    S Mezentsov, V Koshuro, A Fomin, I Rodionov

Modified Multicapillary Glass Structure for SERS Application 1061
    N E Markina, A V Markin, A M Zakharevich, Y S Skibina, I Yu Goryacheva

Carbide Coatings Obtained by Electro-Spark Alloying and Finishing 1066
    V Koshuro

High Robustness of Epitaxial 4H-SiC Graphene to Oxidation Processes 1070
    V. S. Prudkovskiy, K. P. Katin, M. M. Maslov, P. Puech, R. Yakimova, G. Deligeorgis

Ohmic Contacts to N-Type 4H- and 6H-SiC 1074
    V I Egorkin, A V Nezhentsev, V E Zemlyakov, V A Gudkov, V I Garmash

Investigation of the Metal Surfaces Destruction Due to Electrochemical Corrosion and Cavitation,
Methods of Protection with the Use of Polymer Composite Coatings Based on CNT 1078
    A A Goshev, M K Eseev, S N Kapustin

Composition-Dependent Conductivity of $In_xGa_{1-x}As$ Nanowires 1083
    V Sharov, P Alekseev, M Dunaevskiy, R Reznik, G Cirlin

MBE Formation of Self-Catalyzed GaAs Nanowires using ZnO Nanosized Films 1087
    M S Solodovnik, O G Karenkikh, S V Balakirev, S I Petrov, R V Ryzhuk, A N Alexeev, O A Ageev

Cathodic Transformation of Bactericidal Silver-Containing Bioceramic Coatings for Implants 1092
    E Poshivalova, I Rodionov

Synthesis and Properties of a Polyamine-Cumulene/Carbon Nanotubes for Removing Harmful
Substances from Aqueous Solutions 1096
    E A Neskoromnaya, A V Melezhik, O V Alekhina, I V Burakova, E S Mkrtchyan, A E Burakov

Nano-Sized Al-Ni Energetic Powder Material for Heat Release Element of Thermoelectric Device 1102
    S Yu Nemtseva, E A Lebedev, Yu P Shaman, P I Lazarenko, R M Ryazanov, S A Gavrilov, D G Gromov

Investigation of Local Charge Accumulation in Yttria Stabilized Zirconia Films with Au
Nanoparticles by Scanning Kelvin Probe Microscopy 1106
    D O Filatov, O N Gorshkov, A N Mikhaylov, D S Korolev, M N Koriazhkina, M A Ryabova, I N Antonov, M E Shenina, D A Pavlov, M S Dunaevskiy

Adsorption of the Methylene Blue Dye on Carbon Nanocomposites Under Dynamic Conditions: A
Kinetic Study 1110
    I V Burakova, A V Babkin, E A Neskoromnaya, A E Burakov, D A Kurnosov, A G Tkachev

Kinetics of the Adsorption of Synthetic Dyes on a Polyhydroquinone/Graphene Carbon
Nanocomposite 1115
    A V Babkin, A V Melezhik, D A Kurnosov, E S Mkrtchyan, I V Burakova, A E Burakov, E V Galunin

Kelvin Probe Microscopy of $MoSe_2$ Monolayers on Graphene 1121
    B R Borodin, M S Dunaevskiy, F A Benimetskiy, S P Lebedev, A A Lebedev, P A Alekseev

Photosensitive Sulphide Heterostructures Obtained by using Successive Ionic Layer Adsorption
and Reaction on Planar and Profiled Substrates 1126
    N Bogomazova, G Gorokh, A Zakhlebayeva, A Pligovka, A Murashkevich, T Galkovsky

Computer Simulation of Aerosol Nanoparticles Focusing and Deposition Process for Functional
Microstructure Fabrication 1132
    N V Protas, A A Efimov, V K Zemlyanoy, V V Ivanov

Study of the Possibility of using Arachidic Acid as a Monocrystalline Substrate for the Formation
of Monolayers of Aromatic Hydrocarbons 1136
    O A Shinkarenko, A S Chumakov, M V Pozharov, A S Kolesnikova, A J K Al-Alwani, O Yu Tsvetkova, V P Sevostyanov, E G Glukhovskoy

Masking Coating Formation by the Focused Ion Beams Method for Plasma Chemical Treatment 1141
    I N Kots, V S Klimin, V V Polyakova, A A Rezvan, Z E Vakulov, O A Ageev

Size Effect on Memristive Properties of Nanocrystalline ZnO Film for Resistive Synaptic Devices 1146
  N A Shandyba, I V Panchenko, R V Tominov, V A Smirnov, M I Pelipenko, E G Zamburg, Y H
  Chu

Computer Simulation of Femtosecond Pulsed Laser Ablation of Aluminium and Copper 1151
  R V Davydov, V I Antonov

Nanometer-Scale Oxidation of Silicon Surface by ICP Plasma 1156
  I Clemente, A Miakonkikh, S Averkin, K Rudenko

Polarization Photosensitivity of n-P-CdSiAs$_2$ Photodiode 1160
  R V Davydov, V Yu Rud', Yu V Rud', E I Terukov

Spatial Point Pattern Analysis of the Local Current Distribution on the Surface of Multi-Tip Field
Emitters 1165
  S V Filippov, S Carapezzi, E O Popov, A G Kolosko

Investigation of Sintering of Silver Lines on a Heated Plastic Substrate in the Dry Aerosol Jet
Printing 1171
  A A Efimov, K N Minkov, P V Arsenov, N V Protas, V V Ivanov

Structure and Hardness of the Zirconium Surface After Laser Modification 1175
  V Proskuryakov, S Mezentzov, I Rodionov

The Electrochemical Deposition of Silicon - Carbon Thin Films from Organic Solution............................1179
  M N Grigoryev, T N Myasoedova, T S Mikhailova

Investigation of Interaction of Molecules of Inorganic Gases with Surface of Copper-Containing
Polyacrylonitrile 1183
  M M Avilova, V V Petrov

Optical Studies of InAs/GaAs Monolayer Bragg Superlattices..1188
  K A Ivanov, A R Gubaidullin, G Pozina, E V Nikitina, M A Kaliteevski

Multiple Frequency Bloch Oscillations in Natural Superlattices..1193
  K A Ivanov, E I Girshova, E D Kolykhalova, M A Kaliteevski

Enhancement of Spontaneous Emission Probability in a Disordered Media..1199
  K M Morozov, K A Ivanov, A R Gubaydullin, M A Kaliteevski

Room Temperature Lasing in Injection Microdisks with InGaAsN/GaAs Quantum Well Active
Region 1205
  E I Moiseev, M V Maximov, A M Nadtochiy, N V Kryzhanovskaya, D A Sannikov, T Yagafarov,
  M Kulagina, T Niemi, R Isoaho, M Guina, A E Zhukov

Author Index

5th International School and Conference
Saint Petersburg OPEN 2018

Optoelectronics, Photonics, Engineering and Nanostructures

Saint Petersburg, April 2-5, 2018

Acknowledgements

Russian Foundation for Basic Research

RSF
Russian Science Foundation

SPIE. STUDENT CHAPTER
SAINT PETERSBURG
ACADEMIC UNIVERSITY
RUSSIAN ACADEMY
OF SCIENCES

OSA®
The Optical Society

IOP Institute of Physics

Content from this work may be used under the terms of the Creative Commons Attribution 3.0 licence. Any further distribution of this work must maintain attribution to the author(s) and the title of the work, journal citation and DOI.
Published under licence by IOP Publishing Ltd

Editorial Preface

Dear Colleagues,

The 5th International School and Conference "Saint Petersburg OPEN 2018" on Optoelectronics, Photonics, Engineering and Nanostructures was held on April 2 - 5, 2018 at St Petersburg Academic University (St Petersburg, Russia). The School and Conference included a series of invited talks given by leading professors with the aim to introduce young scientists with actual problems and major advances in physics and technology. The invited speakers were

Maurice Skolnick *(University of Sheffield, UK)*

Stefan Breuer *(Technische Universität Darmstadt, Germany)*

Huiyun Liu *(University College London, UK)*

Manfred Bayer *(Technische Universität Dortmund, Germany)*

Andrey Akimov *(University of Nottingham, UK)*

Victor Chaly *(JSC "Svetlana Growth", Russia)*

Artur Dideikin *(Ioffe Institute, Russia)*

Pavel Belov *(ITMO University, Russia)*

Alexey Skvortsov *(Peter the Great St Petersburg Polytechnic University, Russia)*

During the poster session all undergraduate and graduate students attending the conference presented their works. Sufficiently large number of participants with 287 student attendees from 49 cities all over the world allowed the Conference to provide a fertile ground for fruitful discussions between the young scientists as well as to become a perfect platform for the valuable discussions between student authors and highly experienced researchers. The best student papers, which were selected by the Program Committee and by the invited speakers basing on the theses and their poster presentation, were awarded with diplomas of the conference and prizes.

This year "**Saint Petersburg OPEN 2018**" was organized by **St Petersburg Academic University** in cooperation with **Peter the Great St. Petersburg Polytechnic University**. The School and Conference was supported by **Russian Foundation for Basic Research** (*Project #18-32-10009*), **Russian Science Foundation** and **SPIE**. **Saint Petersburg OPEN 2018** successfully continues the tradition of annual schools and seminars for youth on topical problems of physics and technology that have been organized by academician Zhores Alferov at the Academic University since 2009. We invite all the students and young scientists to attend "**Saint Petersburg OPEN**" in 2019!

More details at *http://spbopen.spbau.com/*

**The head of the Program Committee,
rector of St Petersburg Academic University
Nobel Prize Winner academic *Zhores Alferov* is making opening remarks**

**The head of the Organizing Committee
corresponding member of RAS, Prof. *Alexey Zhukov***

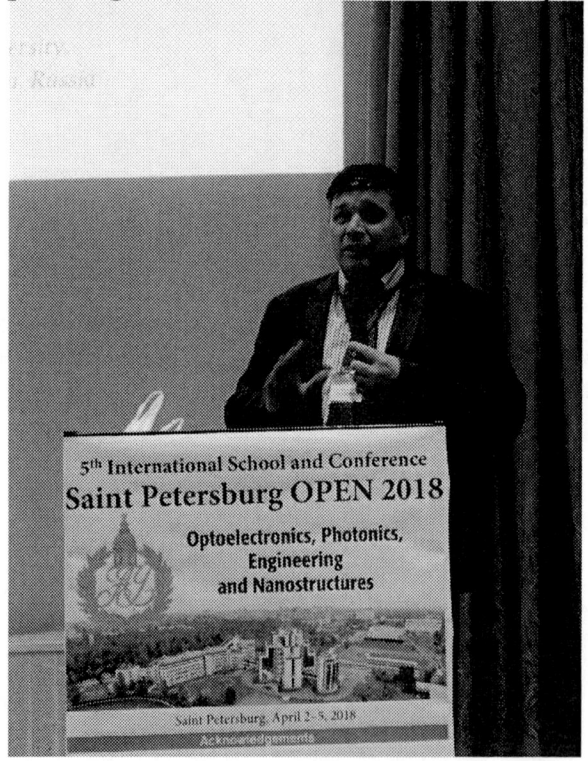

The invited speakers

Prof. Maurice Skolnick

PD Dr. Stefan Breuer

Prof. Huiyun Liu

Prof. Manfred Bayer

Prof. Andrey Akimov

Dr. Victor Chaly

Dr. Artur Dideikin

Prof. Pavel Belov

Prof. Alexey Skvortsov

Poster Session

Elevator speech session

Awards ceremony
Prof. Alexey Zhukov and Prof. Andrey Lipovskii with students

Awarding ceremony of "Saint Petersburg OPEN 2018"

Peer review statement

All papers published in this volume of *Journal of Physics: Conference Series* have been peer reviewed through processes administered by the proceedings Editors. Reviews were conducted by expert referees to the professional and scientific standards expected of a proceedings journal published by IOP Publishing.

Content from this work may be used under the terms of the Creative Commons Attribution 3.0 licence. Any further distribution of this work must maintain attribution to the author(s) and the title of the work, journal citation and DOI.
Published under licence by IOP Publishing Ltd

Analytical–Monte Carlo model of the growth of In nanostructures during droplet epitaxy on the triangle-patterned GaAs substrates

S V Balakirev[1], M S Solodovnik[1,2], I A Mikhaylin[2], M M Eremenko[2], O A Ageev[2]

[1]Department of Nanotechnologies and Microsystems, Southern Federal University, Taganrog 347922, Russia
[2]Research and Education Center "Nanotechnologies", Southern Federal University, Taganrog 347922, Russia

Abstract. Kinetic Monte Carlo simulations combined with the analytical nucleation theory are used to study droplet epitaxy of indium on the GaAs substrates patterned with triangular-shaped nanoholes. The simulation results demonstrate that the growth on the Ga-stabilized surface provides much better selectivity than the formation of nanostructures on the As-stabilized surface because of the stronger bonding with As atoms of the substrate. We estimate technological parameters which allow obtaining required characteristics of nanostructures with precise positioning and complete filling of holes. The distance between nanostructures is considered to be controlled by the growth temperature at an appropriate interhole distance. At the same time, dimensions of nanostructures are outlined by the geometrical parameters of holes and can be adjusted by the deposition thickness.

1. Introduction

Control over the interaction between active elements of nanodevices is an outstanding problem in quantum science and engineering. Great prospects for use in novel devices are associated with nanostructures based on GaAs and related materials [1-8]. Semiconductor quantum dots fabricated by droplet epitaxy are particularly attractive for the realization of quantum computers and photonic integrated circuits, as they can be controllably positioned and varied in geometrical characteristics. During the Stranski-Krastanov growth on patterned GaAs substrates, InAs quantum dots can nucleate outside the holes or form a multiple quantum dot in a single hole [9,10]. At the same time, the droplet epitaxy technique allows formation of a single quantum dot with controllable parameters within each hole [11,12]. However, a detailed investigation of the nucleation and growth processes of nanostructures during droplet epitaxy on patterned substrates is still necessary. To fabricate quantum dot arrays with required geometrical parameters in given positions, appropriate technological parameters should be chosen. Since it is inexpedient to carry out a large number of experiments, it is appropriate to use simulations.

In this work, kinetic Monte Carlo model based on the analytical nucleation theory is used to describe the processes of droplet epitaxy of indium on the nanopatterned GaAs substrates. Previously, we validated the model using the experimental data in a wide range of growth temperatures. The studies demonstrated nonmonotonic dynamics of the adatom supersaturation and island critical size as well as formation of the wetting layer with thickness depending on the substrate temperature [13].

Content from this work may be used under the terms of the Creative Commons Attribution 3.0 licence. Any further distribution of this work must maintain attribution to the author(s) and the title of the work, journal citation and DOI.
Published under licence by IOP Publishing Ltd

Productivity of the analytical approach and versatility of the Monte Carlo method make it possible to consider complicated phenomena during nucleation and growth of islands [14-19].

2. Description of the model

The model is based on the kinetic Monte Carlo simulations which use the results of analytical calculations within the thermodynamic theory of nucleation. The key points of the model agree with those for the flat substrate which were used previously to describe the processes of nucleation and growth during In/GaAs(001) droplet epitaxy [13]. In this study, we develop the model to discuss the growth mechanisms on patterned substrates. Critical regions are formed throughout the simulation area in such a way that the area of the region above the substrate remains the same. Thus, nucleation may occur in any point and the nucleation probability is determined on microscopic level. Lennard–Jones potential is used to calculate the binding energy between atoms of the crystal. A type and an order of events are defined by waiting times which are taken from calculations by the Arrhenius equation.

3. Results and discussion

The results of our simulations on patterned substrates demonstrated a difference between growth processes on the Ga and As-stabilized surfaces. Due to the fact that the bonds of metal atoms (In and Ga) with each other is weaker than those with As atoms, the mobility of adatoms on the Ga-stabilized surface is higher [17]. At the same time, adatoms on the As-stabilized surface stick to the substrate much stronger because they are bound directly to the As atoms of the substrate. As a result, the wetting layer on the As-stabilized surface consists in several monolayers (ML) (Fig. 1b) whereas it is about 1 ML on the Ga-stabilized surface at the growth temperature $T = 100°C$ (Fig. 1a). Our previous study demonstrated that the formation of the wetting layer on the Ga-stabilized surface may be even prevented completely when $T = 150°C$ [17].

This behavior gives an opportunity of better control of the growth characteristics with the technological parameters. The As-stabilized surface can be used when the wetting layer is required for a device structure. However, we consider a situation when the formation of the wetting layer should be suppressed. As Fig. 1 shows, the Ga-stabilized surface is appropriate for this condition. In this case, the deposited material is located in the holes rather than on the flat areas of the substrate (Fig. 1a).

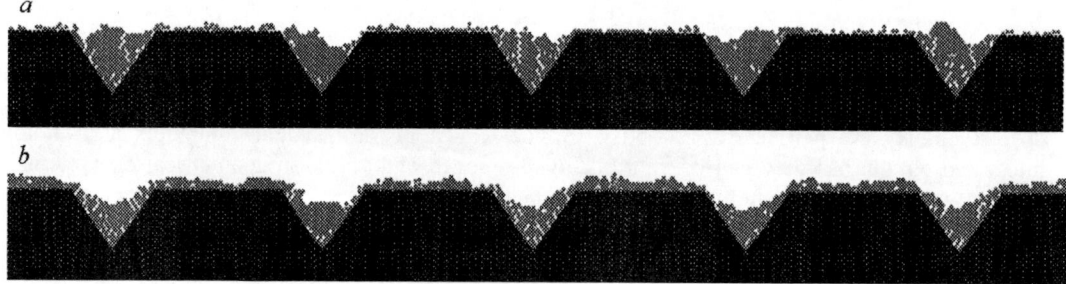

Figure 1. Morphology of the array of In/GaAs(001) nanostructures after deposition of 5 ML of indium on the *a*) Ga-stabilized and *b*) As-stabilized nanopatterned substrate with $d = 10$ nm, $h = 6$ nm, $r = 20$ nm at $T = 100°C$ (length of the simulation area $L = 100$ nm).

The diffusion length of In adatoms on the GaAs substrate is quite large as compared with the diffusion length of Ga adatoms [20-21]. Consequently, a sufficient distance between holes should be provided to prevent nucleation and growth of nanostructures outside the holes. The results of our simulations show that the deposited material can be collected only within the triangular holes (Fig. 2a, 3a, 3b). To ensure the best selectivity of the growth, an appropriate combination of the parameters of patterning and the growth conditions should be selected. When the substrate temperature $T = 100$ °C, the interhole distance $r = 75$ nm is sufficient to localize In nanostructures within the holes (Fig. 2a). The deposition of indium with thickness $H = 5$ ML allows partial filling of the holes. In certain cases, the deposited material exceeds the borders of the hole.

Figure 2. Morphology of the array of In/GaAs(001) nanostructures after deposition of 5 ML of indium on the Ga-stabilized nanopatterned substrate with $d = 16$ nm, $h = 12$ nm and a) $r = 50$ nm, b) $r = 100$ nm at $T = 100$°C ($L = 200$ nm).

As the interhole distance increases up to 150 nm, the nucleation and growth of droplets on the flat part of the substrate is probable (Fig. 2b). Although the diffusion length of In adatoms on the GaAs surface is large, the growth at a low temperatures $T = 100$°C may result in undesirable formation of nanostructures outside the prepared nucleation centers if they are located wide apart. Fig. 3a shows that the growth temperature should be increased up to 200°C to achieve better selectivity on the substrate with the same parameters of patterning. In this case, the deposited material is distributed between the holes whereas it is consumed by droplets formed on the flat surface at a lower temperature (Fig. 2b).

Figure 3. Morphology of the array of In/GaAs(001) nanostructures after deposition of a) 5 ML, b) 4 ML of indium on the Ga-stabilized nanopatterned substrate with $d = 16$ nm, $h = 12$ nm, $r = 100$ nm at $T = 200$°C ($L = 200$ nm).

One of the most advantages of the droplet epitaxy method is a possibility to control the size of nanostructures independently of their surface density [22-26]. Growth on patterned substrates makes it possible to control positions of nanostructures as well as their geometrical characteristics. As Fig. 3b demonstrates, preventing of exceeding the level of the substrate by nanostructures can be achieved along with complete filling of holes. A decrease of the deposition thickness from 5 to 4 ML at $T = 200$°C leads to a decrease in the height of nanostructures relative to the hole bottom from 17 nm to the hole depth h which is equal to 12 nm. It should be noted that the maximum diameter of a nanostructure remains the same and equal to the hole diameter $d = 16$ nm as well as the surface density of nanostructures does not change.

4. Conclusion

Thus, the precise selective growth of nanostructures can be realized during droplet epitaxy of In on triangle-patterned GaAs substrates. However, appropriate technological parameters at the stages of both substrate preparation and indium deposition must be chosen to accomplish the growth with required characteristics. The distance between nanostructures can be controlled by the growth temperature on the substrate patterned with appropriate interhole distance. The deposition thickness allows controlling the degree of filling of holes and dimensions of nanostructures within the holes of appropriate depth and diameter.

Acknowledgments

This work was supported by the Russian Science Foundation Grant No. 15-19-10006. The results were obtained using the equipment of the Research and Education Center and Center for Collective Use "Nanotechnologies" of Southern Federal University.

References

[1] Ageev O A, Smirnov V A, Solodovnik M S, Rukomoikin A V and Avilov V I 2012 *Semiconductors* **46** 1616–1621

[2] Klimin V S, Solodovnik M S, Smirnov V A, Eskov A V, Tominov R V and Ageev O A 2016 *Proceedings of SPIE* **10224** 102241Z

[3] Balakirev S V, Solodovnik M S, Eremenko M M, Mikhaylin I A and Ageev O A 2017 *J. Phys.: Conf. Ser.* **917** 032034

[4] Ageev O A, Solodovnik M S, Balakirev S V and Eremenko M M 2016 *Phys. Solid State* **58** 1045–1052

[5] Moon S, Kim K, Kim Y, Heo J and Lee J 2016 *Sci. Rep.* **6** 30107

[6] Ageev O A, Klimin V S, Solodovnik M S, Eskov A V and Krasnoborodko S Y 2016 *J. Phys.: Conf. Ser.* **741** 012178

[7] Avilov V I, Ageev O A, Smirnov V A, Solodovnik M S and Tsukanova O G 2015 *Nanotech. in Russia* **10** 214-219

[8] Ageev O A, Solodovnik M S, Balakirev S V, Eremenko M M and Mikhaylin I A 2016 *J. Phys.: Conf. Ser.* **741** 012012

[9] Atkinson P, Schmidt O G, Bremner S P and Ritchie D A 2008 *C. R. Phys.* **9** 788–803

[10] Helfrich M, Hu D, Hendrickson J, Gehl M, Rülke D, Gröger R, Litvinov D, Linden S, Wegener M, Gerthsen D, Schimmel T, Hetterich M, Kalt H, Khitrova G, Gibbs H and Schaadt D 2011 *J. Cryst. Growth* **323** 187–90

[11] Solodovnik M S, Balakirev S V, Eremenko M M, Mikhaylin I A, Avilov V I, Lisitsyn S A and Ageev O A 2017 *J. Phys. Conf. Ser.* **917** 032037

[12] Kim J S, Kawabe M and Koguchi N 2006 *Appl. Phys. Lett.* **88** 072107

[13] Balakirev S V, Solodovnik M S and Ageev O A 2018 *Phys. Status Solidi B* **255** 1700360

[14] Ageev O A, Solodovnik M S, Balakirev S V and Mikhaylin I A 2016 *Tech. Phys.* **61** 971

[15] Landau D P and Binder K 2015 *A Guide to Monte Carlo Simulations in Statistical Physics* (Cambridge: Cambridge University Press) p 519

[16] Ageev O A, Solodovnik M S, Balakirev S V, Mikhaylin I A and Eremenko M M 2017 *J. Cryst. Growth* **457** 46

[17] Balakirev S V, Solodovnik M S and Ageev O A 2017 *J. Phys. Conf. Ser.* **917** 032033

[18] Ageev O A, Solodovnik M S, Balakirev S V and Mikhaylin I A 2016 *J. Phys. Conf. Ser.* **681** 012036

[19] Ageev O A, Solodovnik M S, Balakirev S V and Eremenko M M 2016 *J. Vac. Sci. Technol. B* **34** 041804

[20] Grandjean N and Massies J 1993 *J. Cryst. Growth* **134** 51–62

[21] Palma A, Semprini E, Talamo A and Tomassini N 1996 **37** 135–138

[22] Lee J H, Wang Z M and Salamo G J 2007 *J. Phys. Condens. Matt.* **19** 176223

[23] Kim J S and Koguchi N 2004 *Appl. Phys. Lett.* **85** 5893–5895

[24] Bietti S, Bocquel J, Adorno S, Mano T, Keizer J G, Koenraad P M and Sanguinetti S 2015 *Phys. Rev. B* **92** 075425

[25] Alén B, Fuster D, Muñoz-Matutano G, Alonso-González P, Canet-Ferrer J, Martínez-Pastor J, Fernández-Martínez I, Royo M, Climente J I, González Y, Briones F, Hernández D, Molina S I and González L 2017 *AIP Conference Proceedings Series* **1399** 421–422

[26] Bietti S and Sanguinetti S 2012 *Nanoscale Res. Lett.* **7** 551

SPBOPEN 2018   IOP Publishing

Investigation on structural transitions of graphenes into diamond polymorphs at high pressure

V A Greshnyakov[1] and E A Belenkov[1]

[1] Physics Department, Chelyabinsk State University, Chelyabinsk 454001, Russia

Abstract. Modeling of structural transformations of graphenes under strong compression is performed by the density functional theory. It is found that the phase transitions of L_6 graphene layers in diamond polymorphs should occur at pressures of $\sim$ 57-71 GPa, while some diamond polymorphs can be obtained of L_{4-8} graphene layers at 32-52 GPa. Also in this article, powder X-ray diffraction patterns of the «L_{4-8} graphite → LA10» phase transition were calculated.

1. Introduction

Carbon nanostructures are used to create selective adsorbents and catalysts, nano-composite materials, nanoelectronic devices and medical materials [1-3]. Another method of practical use of nanostructures is the production of new high-strength materials with a diamond-like structure. To date, the possibility of the existence of thirty-five diamond-like phases, of which only three phases have been experimentally obtained, is theoretically established [4-6]. These diamond-like phases have different properties, and various carbon materials with the diamond-like structure need to be synthesized for specific practical applications. Therefore, the conditions under which it is possible to synthesize new diamond-like phases must be investigated. Diamond-like phases with the structure different from that of cubic diamond can be obtained by compressing fullerenes, nanotubes, or graphene layers [4-6]. In this paper, the phase transitions of L_6 and L_{4-8} graphenes to diamond polymorphs were calculated under compression.

2. Methods

The investigation of phase transitions in carbon compounds was carried out using the technique described in [7-9]. Simulation of direct phase transitions of graphites into diamond polymorphs was performed under uniaxial compression (Fig. 1). Calculations of the structures and total energies of the carbon phases were carried out by the density functional theory (DFT) method for the local density approximation (LDA) [10] and the generalized gradient approximation (GGA) [11]. Norm-conserving pseudopotentials are employed in conjunction with plane-wave basis sets of a cutoff energy of 60 Ryd and $12 \times 12 \times 12$ k-point grids. First, we used the DFT-calculations to geometrically optimize the phase structure and determine the unit cell parameters. Then, successive changes in the structure of the carbon compounds were calculated in the process of phase transformations. For the calculation of X-ray diffraction patterns of the phase transitions, the molybdenum $K_{\alpha1}$ wavelength was used.

3. Results

Figures 2-4 show dependences of the total energy difference (ΔE_{total}) on the atomic volume (V) characterizing direct and reverse phase transitions between polymorphic modifications of carbon.

Content from this work may be used under the terms of the Creative Commons Attribution 3.0 licence. Any further distribution of this work must maintain attribution to the author(s) and the title of the work, journal citation and DOI.
Published under licence by IOP Publishing Ltd

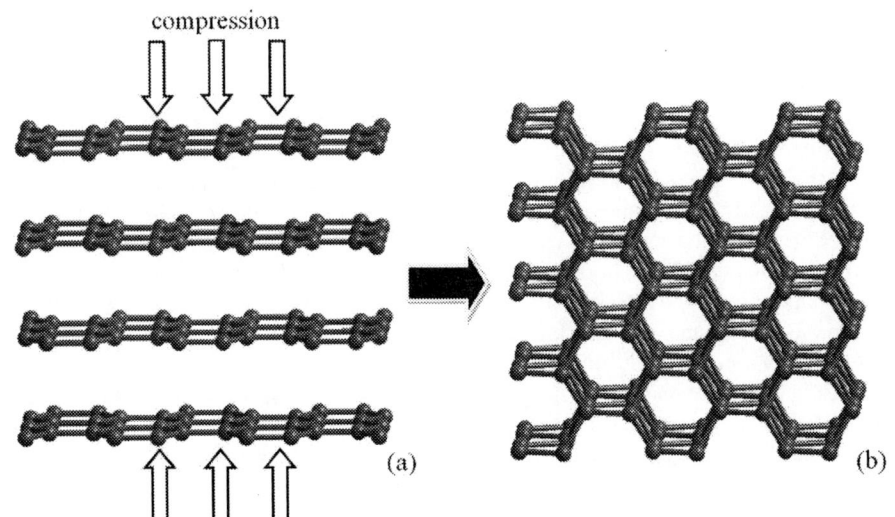

Figure 1. Structure transformation of graphites (a) to diamond polymorphs (b).

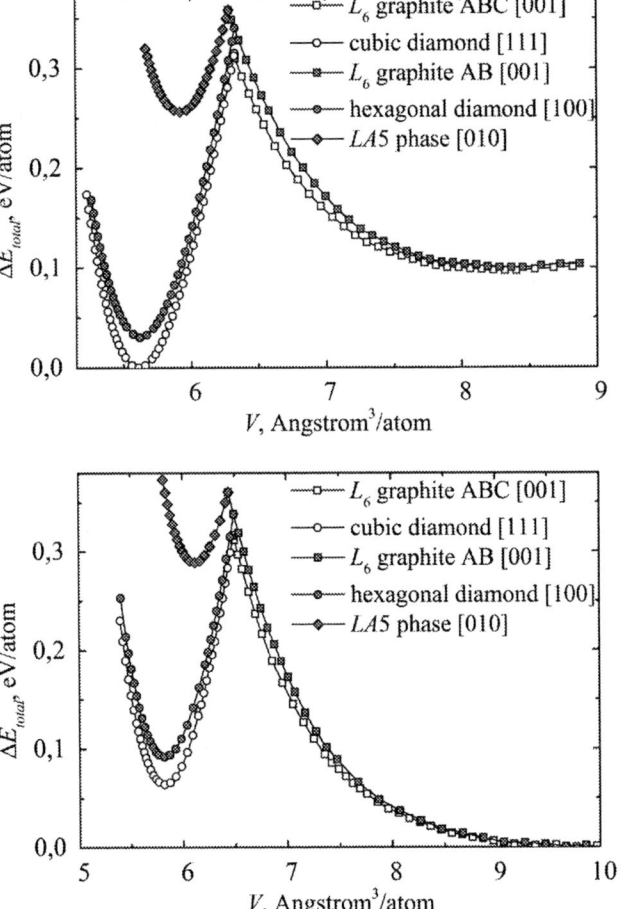

Figure 2. Dependences of the total energy on the volume for L_6 graphites and diamond polymorphs calculated by the DFT-LDA (a) and DFT-GGA (b) methods.

Transformation of L_6 graphites with ABC and AB layer packings into cubic diamond, hexagonal diamond (LA2), and LA5 phase will occur at pressures of 59-64 GPa for the DFT-LDA method and 68-71 GPa for the DFT-GGA method as a result of overcoming energy barriers of 0.216-0.258 and 0.316-0.360 eV/atom, respectively (Fig. 2). The minimum potential barrier is observed for the phase transition of rhombohedral L_6 graphite ABC to cubic diamond, and the maximum barrier height corresponds to the transition of orthorhombic L_6 graphite AB to diamond-like LA5 phase.

The phase transition of L_6 graphites with AA layer packings into hexagonal diamond, LA3, LA5, $P2_1/m$, Pmma, and C2/m phases will occur at pressures of 57-64 GPa for the DFT-LDA method and 66-70 GPa for the DFT-GGA method as a result of overcoming energy barriers of 0.275-0.315 and 0.391-0.431 eV/atom, respectively (Fig. 3). The minimum potential barrier height is observed for the phase transition of hexagonal L_6 graphite AA to hexagonal diamond, and the maximum barrier corresponds to the phase transition this graphite to diamond-like $C2/m$ phase. The formation of LA6 and LA7 phases in the direct phase transition from L_6 graphite AA, apparently, is impossible (Fig. 3).

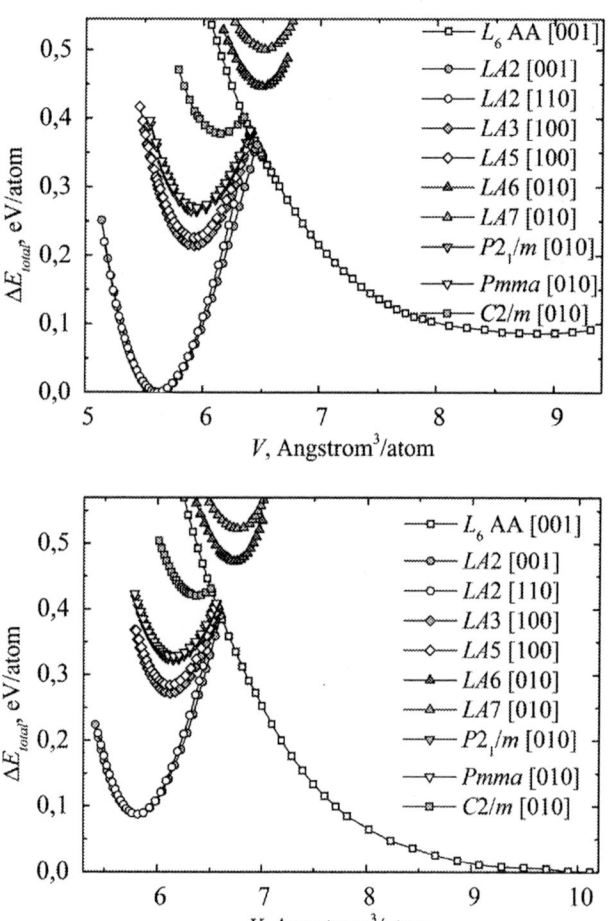

Figure 3. Dependences of the total energy on the volume for L_6 graphite AA and diamond polymorphs calculated by the DFT-LDA (a) and DFT-GGA (b) methods.

Note that cubic and hexagonal diamonds are formed from L_6 graphites at minimum pressures. During the phase transitions, the density increases jumpwise by 9-14 %, and the enthalpy decreases by 0.01-0.81 eV/atom.

Structural transformations of tetragonal L_{4-8} graphites with AB, ABCD, and AA packing into LA3, LA5, LA6, LA7, and LA10 diamond polymorphs can occur at pressures of 32-50 GPa (DFT-LDA) or

40-52 GPa (DFT-GGA) (Fig. 4). Significantly lower values of the pressures at which LA6, LA7, and LA10 phases are formed can be explained by the relatively low energy barriers separating L_{4-8} graphites and these phases. The structural transitions of L_{4-8} graphites into diamond polymorphs are first-order exothermic phase transitions, as a result of which the density will increase by 11-19 % and the energy will be released (~ 0.32-1.04 eV/atom).

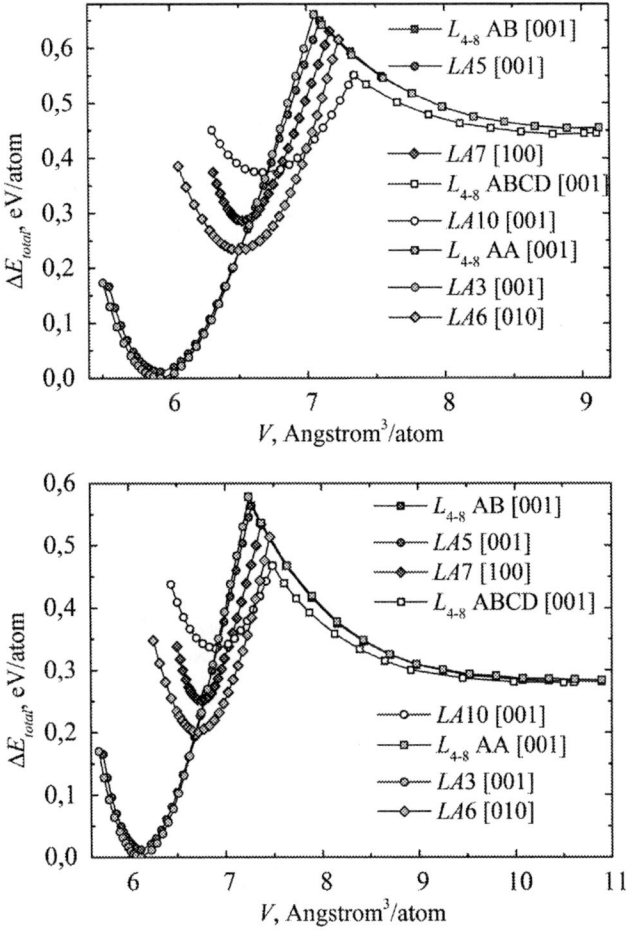

Figure 4. Dependences of the total energy on the volume for L_{4-8} graphites and diamond polymorphs calculated by the DFT-LDA (a) and DFT-GGA (b) methods.

In phase transitions of graphenes into diamond polymorps, the diffraction pattern changes significantly and is accompanied by the disappearance of most of the original maxima and the appearance of a multitude of new maxima. An example of the calculated X-ray diffraction patterns of such a phase transition occurring at the minimum pressure (32 GPa) is shown in Fig. 5.

4. Conclusions

In this paper, the calculations of direct phase transitions of L_6 and L_{4-8} graphene layers in polymorphic varieties of diamond are performed by the DFT method. It is established that the synthesis of cubic and hexagonal diamonds is most likely from graphite on the basis of L_6 layers at pressures from 57 to 71 GPa. Diamond polymorphs with the most strained structures can be obtained from tetragonal L_{4-8} graphene layers at pressures from 32 to 52 GPa. The LA10 phase identification can be carried out using the theoretically calculated X-ray diffraction pattern.

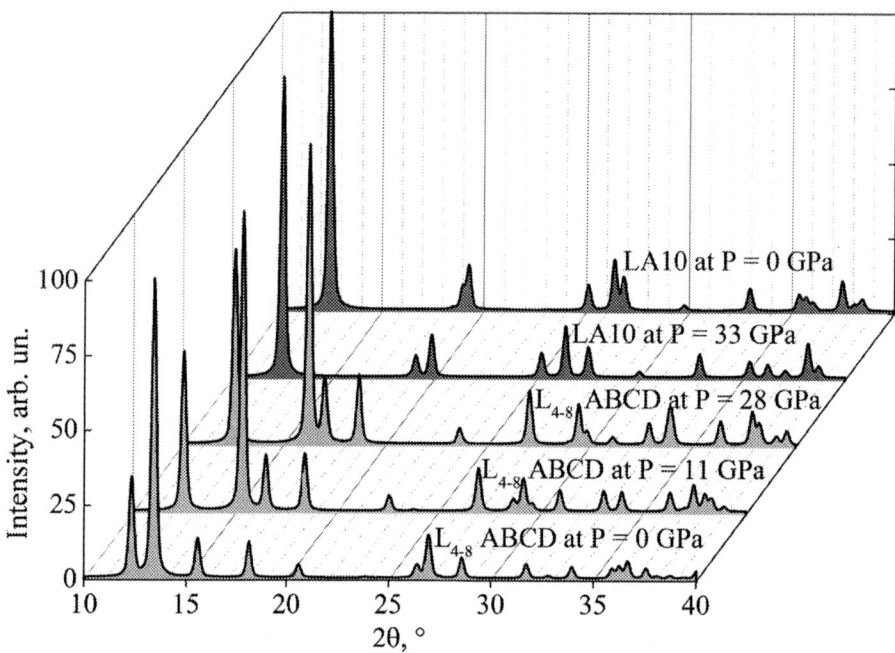

Figure 5. Theoretical powder X-ray diffraction patterns for the phase transition of tetragonal L_{4-8} graphite to LA10 diamond polymorph.

References

[1]   Dmitriev S V, Baimova J A, Korznikova E A and Chetverikov A P 2018 Nonlinear excitations in graphene and other carbon nano-polymorphs *Nonlinear Systems* vol 2 (Nonlinear Phenomena in Biology, Optics and Condensed Matter) ed J F R Archilla, F Palmero *et al* (Switzerland: Springer International Publishing AG) chapter 3 pp 175-195

[2]   Grishakov K S, Katin K P and Maslov M M 2018 *Diam. Relat. Mater.* **84** 112

[3]   Brazhe R A and Savin A F 2017 *Russian Microelectronics* **46** 506

[4]   Belenkov E A and Greshnyakov V A 2015 *Phys. Solid State* **57** 205

[5]   Belenkov E A and Greshnyakov V A 2015 *Phys. Solid State* **57** 1253

[6]   Belenkov E A and Greshnyakov V A 2015 *Phys. Solid State* **57** 2331

[7]   Greshnyakov V A and Belenkov E A 2017 *J. Exp. Theor. Phys.* **124** 265

[8]   Greshnyakov V A and Belenkov E A 2017 *Letters on Materials* **7** 318

[9]   Greshnyakov V A and Belenkov E A 2016 *Tech. Phys.* **61** 1462

[10]   Perdew J P, Zunger A 1981 *Phys. Rev. B* **23** 5048

[11]   Perdew J P, Burke K and Ernzerhof M 1996 *Phys. Rev. Lett.* **77** 3865

SPBOPEN 2018　　　　　　　　　　　　　　　　　　　　　　　　　　IOP Publishing

Investigation of Titanium Oxide Film on Sapphire Substrate for Gas Sensor

Y V Klunnikova[1], S P Malyukov[1], A V Sayenko[1], M I Tolstunov[2]

[1] Institute of Nanotechnology, Electronics and Electronic Equipment Engineering, Southern Federal University, Taganrog 347928, Russia
[2] Institute of Chemistry, Southern Federal University, Rostov-on-Don 344006, Russia

Abstract. We study the surface structure of titanium oxide (TiO_2) film on sapphire substrate obtained by laser annealing (wavelength of 1064 nm). The film was studied by powder x-ray diffractometry. The obtained TiO_2 films can be used as a sensitive material for gas sensors.

1. Introduction

There is a wide range of gas sensor designs due to their importance in various fields of technology (mainly for prevention of fire and explosions, when working with explosive and poisonous gases, and to monitor the environmental situation in large cities).

The principal inorganic materials with gas sensitive properties used in gas sensors are metal oxides which have semiconductor properties and a bandgap width about 3-4 eV. The oxides of tin (SnO_2), tungsten (WO_3), zinc (ZnO), indium (In_2O_3), copper (Cu_2O, CuO), titanium (TiO_2), iron (Fe_2O_3) and their combinations are the most widely used among gas sensitive materials. One of the important features of gas sensors based on titanium oxide (TiO_2) is the possibility of using them at high temperatures due to the high chemical stability of the film [1-4].

2. Investigation of titanium oxide films on sapphire substrate

The film-forming solution of tetraethoxytitanium ($Ti(OC_2H_5)_4$) was deposited on a sapphire substrate with a thickness of 500 μm by centrifugation (centrifuge SPIN NXG-P1, rotor rotation speed of 2000-3000 rpm, application time of 30 s).

The use of sapphire substrate allows subsequent laser annealing of the gas sensitive material. Sapphire substrates cause high adhesion strength to the gas sensitive material and have a high melting point, high chemical and radiation resistance, high hardness and transparency, which leads to the quality and stability improvement of the gas sensitive material [5-8]. After film pre-drying in the oven at 100-120 ° C for 15-20 minutes (the solvent and hydrolysis products are removed from the film) laser annealing is carried out using the radiation of a pulsed solid-state Nd:YAG laser with a wavelength of 1064 nm (temperature on film surface of 500-600 ° C, laser beam scanning rate of 1-10 mm/s) needed to modify the crystalline and defective structure and improve the quality and stability of the gas sensitive material. The use of laser annealing makes it possible to shorten the technological time of obtaining a gas sensitive material in comparison with existing methods (for example, with annealing in a muffle furnace) [9-10].

The main benefits of gas sensor development on sapphire substrate with TiO_2 film are lowering of working temperature, high selectivity to detectable gases, and increased stability in time. Reaching these goals calls for employing modern tendencies of gas sensor construction as gas sensing materials.

Content from this work may be used under the terms of the Creative Commons Attribution 3.0 licence. Any further distribution of this work must maintain attribution to the author(s) and the title of the work, journal citation and DOI.
Published under licence by IOP Publishing Ltd

Improving sensing characteristics and lowering working temperatures is possible in obtaining thin (50-300 nm) gas sensing films that have nanostructure with unique properties.

Figure 1 shows the technological route of titanium dioxide gas sensitive material formation.

Contents, structure and properties of solution and derivative films are determined by both chemical processes in the solution and physical-chemical processes, which occur in subsequent operation and depend on:

- chemical composition of source components in the solution;
- solution preparation order;
- solution aging time;
- methods and patterns of film infliction;
- methods of thermal treatment (drying and annealing).

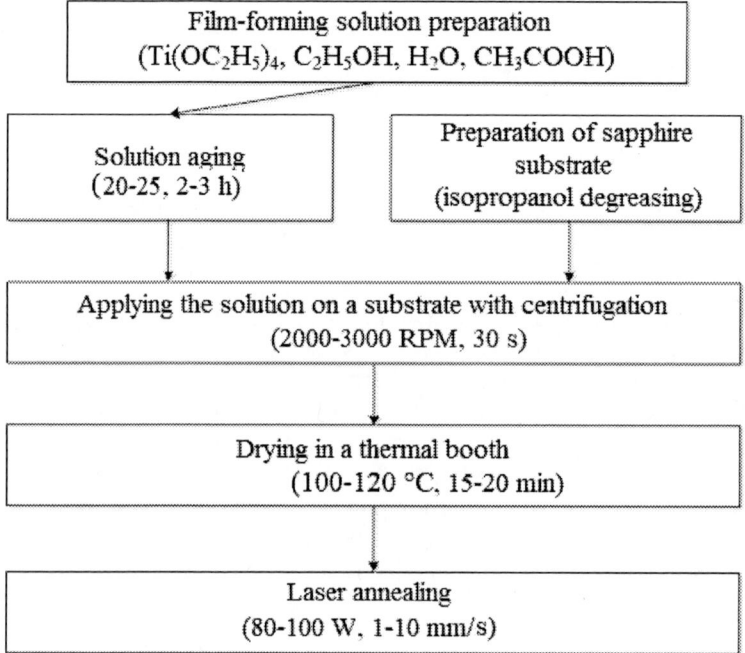

Figure 1. The technological route of titanium dioxide gas sensitive material formation

The proposed method [10] assumes obtaining a thin film gas sensing material based on titanium dioxide (TiO_2) on a $20 \times 20 \times 0.5$ mm sapphire substrate. A thin (100-250 nm) TiO_2 film with platinum electrodes (50-200 nm) is inflicted onto the top surface, while a film (300-400 nm) resistive Ni heater is formed on the bottom side.

The topology and 3D visualization of top and bottom surfaces of the sensitive material of gas sensor are shown in Figure 2.

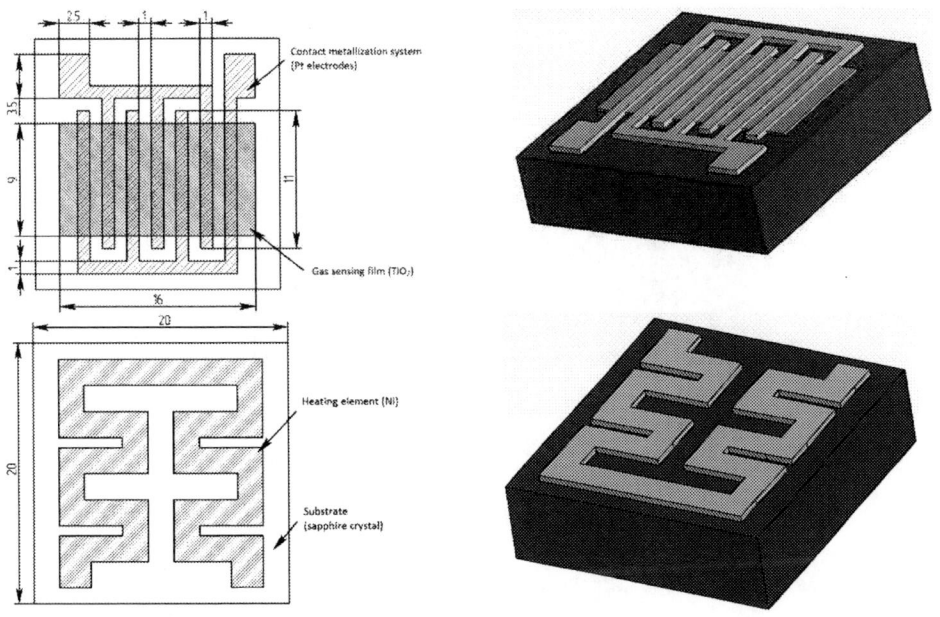

Figure 2. Topology and 3D visualization of the top and the bottom sides of the gas sensing element

The phase composition of the thin-film structure was investigated by powder X-ray diffractometry [11]. We used ARLX'TRA, Thermo ARL diffractometer to perform X-ray phase analysis of obtained thin films. Qualitative analysis of the phase composition was performed using an open database (card index) COD (Crystallographhy Open Database) and Match program. The X-ray roentgenogram of the obtained film, reflexes of the standardized roentgenogram and Miller indices are shown in Figure 3. We have chosen the x-ray roentgenogram for titanium oxide with the structure of rutile (card No. 99-207-1134) for the reference. We can see from the obtained data that the reflexes of the standard sample coincide with the reflexes of the resulting film. Therefore, the material has a phase composition similar to the rutile modification of titanium oxide.

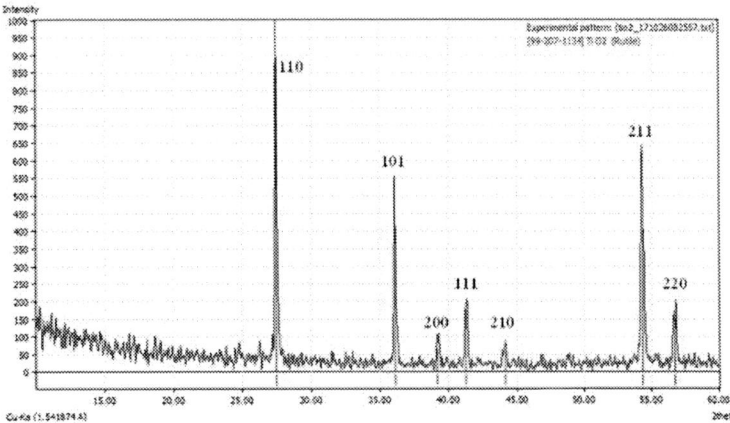

Figure 3. The X-ray roentgenogram of titanium oxide film obtained by laser annealing

By varying the composition and structure, as well as the porosity and the thickness of the gas sensitive layer, we can control the sensitivity and selectivity of the sensor to various external components. The surface morphology of TiO_2 film on sapphire substrate (scanning electron microscopy) is shown in figure 4. The average crystallite size is $15 - 25$ nm. The dimensions of crystallites have a direct impact on the width of the forbidden zone of a titanium oxide thin film.

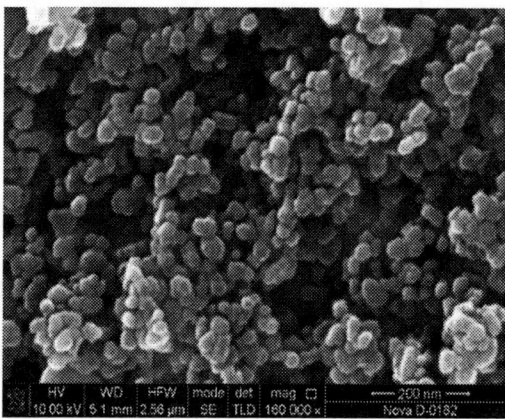

Figure 4. The surface morphology of TiO_2 film on sapphire substrate

3. Investigation results

We have obtained the gas-sensitive material from a film-forming solution of tetraethoxytitanium on the sapphire substrate by laser annealing. We have determined that laser radiation allows to modify the crystalline and defective structure of materials, improve the quality of the oxide, reproducibility of the film parameters and their stability, as well as increase the production capacity of gas sensitive elements. The proposed method for the titanium oxide (TiO_2) film obtaining can be used to produce gas sensors for fire alarms and detecting concentrations of hazardous, toxic and harmful substances [2, 4, 10].

Acknowledgments

This research is supported by the Federal Target Program No. 14.587.21.0025, the project ID is RFMEFI58716X0025.

References

[1]   Gaskov A M, Rumyanceva M N 2000 *Inorganic Materials* **3** 369
[2]   Buslov V, Kozhevnikov V, Kulikov D, Rembeza S, Russkih D 2008 *Modern Electronics* **7** 22
[3]   Horoshih V M, Belous V A 2009 *Semiconductors* **3** 213
[4]   Min-Hyun S, Masayoshi Y, Tetsuya K, Jeung-Soo H, Noboru Y, Kengo S 2009 *Procedia Chemistry* **1** 192
[5]   Dobrovinskaya E R, Lytvynov L A, Pishchik V 2009 *Sapphire. Material, Manufacturing, Applications* (New York: Springer)
[6]   Bogdanov Yu, Kochemasov V, Hasyanova E 2014 *Printed Wiring* **1** 204
[7]   Akselrod M S, Bruni F J 2012 *J. Cryst. Growth* **360** 134
[8]   Cherednichenko D I, Malyukov S P, Klunnikova Yu V 2013 *Sapphire: Stucture, Technology and Applications* (USA: Nova Science Publishers)
[9]   Malyukov S P, Klunnikova Yu V, Saenko A V, Kulikova I V 2014 *J. Phys.: Conf. Ser.*
[10]   Malyukov S P, Klunnikova Yu V, Sayenko 2017 *Panent* 2625096
[11]   Kovba L M, Trunov V K 1976 *X-ray Phase Analysis* (Moscow: Isdatelstvo MGU)

Nanostructural features of anodic zirconia synthesized using different temperature modes

I A Petrenyov[1], R V Kamalov[1], A S Vokhmintsev[1], N A Martemyanov[1], I A Weinstein[1,2]

[1]NANOTECH Centre, Ural Federal University, Yekaterinburg 620002, Russia
[2]Institute of Solid State Chemistry, Russian Academy of Sciences, Yekaterinburg 620990, Russia

Abstract. Nanotubular and nanoporous structures of ZrO_2 were synthesized by potentiostatic anodization with varying the temperatures of anode in the range of $T_A = 0 - 90$ °C and electrolyte in the range of $T_{El} = 20 - 50$ °C. It was shown that difference between T_A and T_{El} had significant influence on growth rate and morphology type of zirconia nanostructures. Optimal parameters of thermal modes for nanotubular ZrO_2 synthesis were discussed.

Introduction

Nanocrystalline ZrO_2 is a promising material for the development of optoelectronic devices, lasers, solar cells, memristors [1-3]. Among various methods of synthesis, for instance, self-propagation high-temperature synthesis, hydrothermal method, laser ablation, etc., anodization of metallic zirconium found wide application as a simple method for obtaining self-oriented and highly ordered arrays of nanotubes [4, 5]. Nanotubular structures of different metal oxides are grown in organic electrolytes containing F-ions [2, 6]. Growth rate of nanostructured anodic oxide layer is governed by kinetic processes of metal oxidation and dissolution of reaction products and depends on temperature [7]. It is known that temperature factor affects morphology and defectiveness of ZrO_2 structure [2], which, in turn, specifies optical and luminescent properties of the material [5]. At the same time, there is no information on the growth features of ZrO_2 nanostructures under conditions of a temperature difference (ΔT) between the anode (T_A) and the electrolyte (T_{El}). Therefore, the aim of this work is to study the effect of ΔT on the morphological characteristics of anodic zirconia.

Samples and experimental methods

Synthesis routine was performed in a cell with two-circuit thermostatic control (Figure 1) at a constant voltage of 60 V. First circuit maintained the temperature of the anode within the range $10 - 50$ °C; second circuit maintained the temperature of the electrolyte within the range of $25 - 50$ °C. The value of $\Delta T = T_A - T_{El}$ was ranged from -25 to 50 °C. A 120 μm thick zirconium foil (99.9 %) served as the anode. The cathode was a stainless steel plate. Zr foil was pretreated in a solution of acids containing $HF/HNO_3/H_2O$ (1:6:20), washed with distilled water and dried in air. ZrO_2 was synthesized by two-stage anodization. The first stage was conducted during 2 hours in the electrolyte (ethylene glycol with adding 0.5 wt. % NH_4F and 5 wt. % H_2O). After removing the primary oxide layer second anodization was carried out under the same conditions.

Content from this work may be used under the terms of the Creative Commons Attribution 3.0 licence. Any further distribution of this work must maintain attribution to the author(s) and the title of the work, journal citation and DOI.
Published under licence by IOP Publishing Ltd

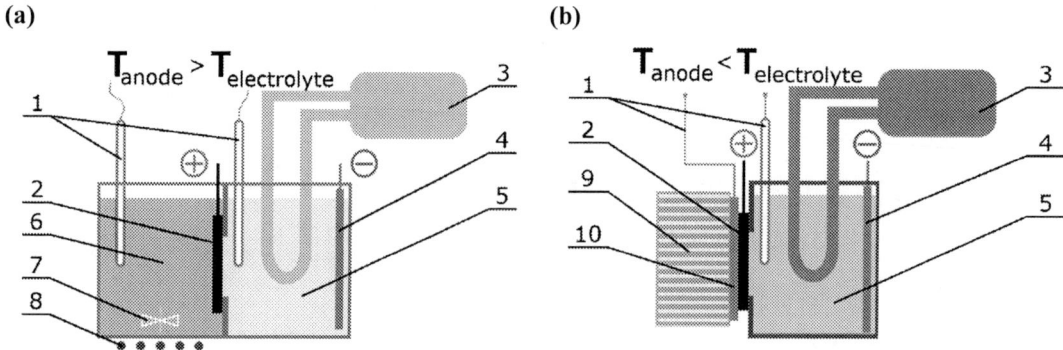

Figure 1 (a, b). Scheme of electrochemical cell for Zr anodization with **(a)** $\Delta T > 0$; **(b)** $\Delta T < 0$; 1 – thermocouples, 2 – anode (Zr foil), 3 – cooling/heating system, 4 – cathode (stainless steel), 5 – electrolyte, 6 – distilled water tank, 7 – stirrer, 8 – heater, 9 – radiator, 10 – thermoelectric cooler.

Morphological parameters of the synthesized samples of nanostructural ZrO_2 were studied using the Carl Zeiss SIGMA VP scanning electron microscope (SEM). SEM images were processed using SIAMS 700 software, inner (d_{in}) diameters of nanotubes and nanoporous, outer (d_{out}) diameter of nanotubes, oxide layer thickness (h) were evaluated.

Results and discussions

Figure 2 shows the SEM images of anodic ZrO_2, synthesized at different temperature modes. Reports show that the morphological changes of zirconia layer occurred under the variation of ΔT. For instance, Figure 2a,b shows nanotubular structure corresponding to $\Delta T = -15\ °C$. The synthesized oxide layer has $h = 3\ \mu m$ and consists of self-ordered nanotubes with $d_{in} = 35\ nm$ and $d_{out} = 55\ nm$. It is noted that the nt-ZrO_2 structure is formed when $T_A < T_{El}$. When $\Delta T = 0\ °C$ the oxide layer is characterized with random nanoporous structure with $h \approx 6\ \mu m$ and $d_{np} \approx 25\ nm$, see Figure 2c. Any regularity of the structure practically disappears (Figure 2d) and the oxide layer becomes more disordered at $\Delta T > 15\ °C$. These structures are referred to the nanoporous type. The oxide layers are thicker about 2-3 fold than the nanotubular ones. Estimations of the morphological characteristics for all synthesized structures are given in the Table 1.

Table 1. The parameters of synthesized nanostructures of anodic ZrO_2.

| № | Temperature, °C | | | Geometric parameters | | | Structure type |
	ΔT	T_A	T_{El}	$h, \pm 0.2\ \mu m$	d_{in}, $\pm 5\ nm$	d_{out}, $\pm 10\ nm$	
1		0	20	3.4	27	75	
2	< 0	10	25	5.1	35	55	nanotubular
3		20	25	5.2	30	45	
4	0	25	25	5.9	25	45	nanotubular
5		50	50	6.2	25	–	nanoporous
6		50	35	8.2	–	–	
7	> 0	70	40	9.1	–	–	nanoporous
8		90	40	9.9	–	–	

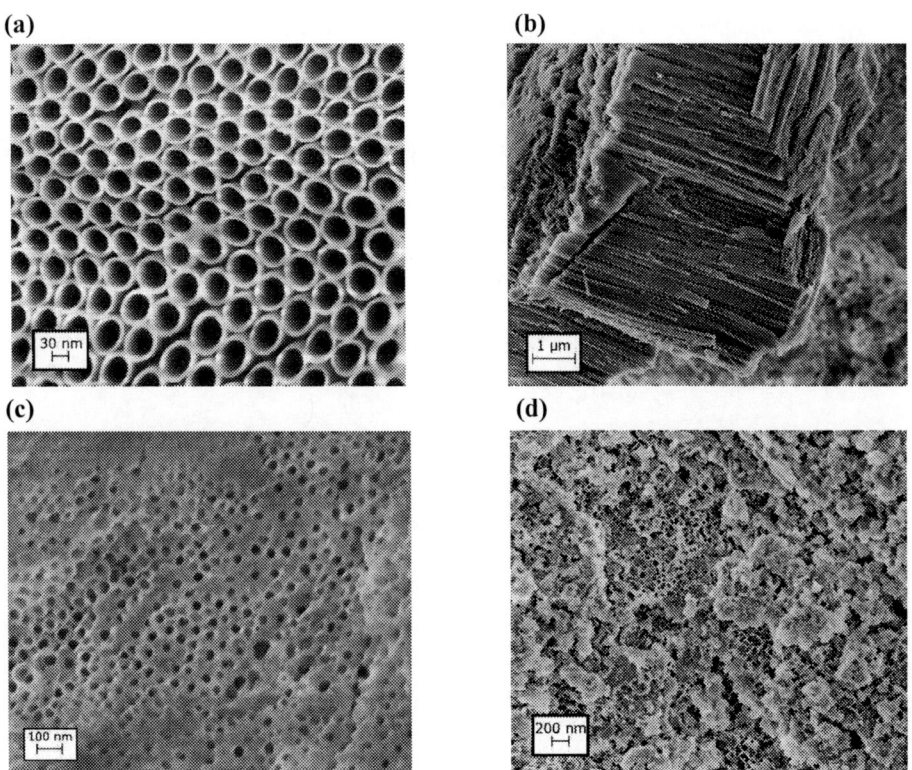

Figure 2 (a, b, c, d). SEM images of the surface **(a, c, d)** and lateral view **(b)** of zirconia layer synthesized at **(a, b)** $\Delta T = -15$ °C, **(c)** at $\Delta T = 0$ °C, **(d)** $\Delta T = 50$ °C.

In order to discuss the influence of the temperature factor (ΔT) on the morphological features of the nanostructured oxide layer formation, major anode processes should be considered. The formation of the ZrO_2 nanostructure is connected to the ratio of the rates of the two competing processes: the anodic oxidation of Zr metal (υ_{ox}) at the metal/oxide interface and the dissolution of oxidation products (υ_{etch}) at the oxide/electrolyte interface [4, 8, 9] (see Fig. 3).

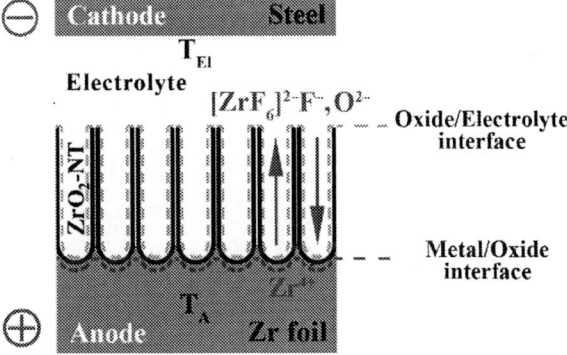

Figure 3. Schematic of the processes occurring during anodization of Zr foil in fluoride containing electrolyte at oxide/electrolyte and metal/oxide interfaces.

On one side, metallic zirconium is oxidized (1a) at the metal/oxide interface. Water molecules are decomposed on the anode surface under the high voltage bias and H^+ and O^{2-} ions (1b) are generated. Oxygen ions interact with the metal to form the zirconium oxide (1c). On the other side, the dissolution process due to the formation of a soluble complex $[ZrF_6]^{2-}$ ion is governed by the migration of Zr^{4+} and F^- ions in the oxide layer under the applied electric field (2a) and (2b) and by chemical etching of the oxide due to the presence of F^- ions in the electrolyte (2c). Described processes can be represented by the following reactions:

$$Zr \rightarrow Zr^{4+} + 4e^- \tag{1a}$$

$$2H_2O \rightarrow 4H^+ + O^{2-} \tag{1b}$$

$$Zr + 2H_2O \rightarrow ZrO_2 + 4H^+ + 4e^- \tag{1c}$$

$$Zr^{4+} + 6F^- \rightarrow [ZrF_6]^{2-} \tag{2a}$$

$$ZrO_2 + 6F^- + 4H^+ \rightarrow [ZrF_6]^{2-} + 2H_2O \tag{2b}$$

$$ZrO_2 + 6HF \rightarrow [ZrF_6]^{2-} + 2H_2O + 2H^+ \tag{2c}$$

On the basis of the described mechanism, the rates of the reactions will be determined by the transport of Zr^{4+}, F^- и O^{2-} ions. Since the temperature influences generation and mobility of ions significantly, a change of T_A and/or T_{El} values leads to a shift of the equilibrium between the two competing processes. Therefore, when $\uparrow\Delta T$ ($T_A > T_{El}$) $\rightarrow \upsilon_{ox}\uparrow$ and $\upsilon_{etch}\downarrow$. In this case, the oxidation process dominates and leads to the increase in the growth rate of the oxide layer. In turn, when $\downarrow\Delta T$ ($T_A < T_{El}$) $\rightarrow \upsilon_{ox}\downarrow$, a $\upsilon_{etch}\uparrow$, the oxidation process slows down and the dissolution is intensified.

By analyzing the data given in the Table 1, we can summarize the following. The h thickness of the oxide layer increases from 3 to 10 μm with an increase in ΔT from -20 to 50 °C. This fact indicates the shift of anode process in the favor of oxidation. Thus, the higher the anode temperature, the thicker oxide layer can be obtained. However, in this case, the structure becomes more defective and disordered (Fig 2c).

The nanotubular structure of anodic zirconia was formed in the interval of -20 °C ≤ ΔT ≤ 0°C. The inner tube diameter of $d_{in} = 30$ nm remains constant within the error. This fact corresponds well with the works [10, 11], since d_{in} value mainly depends from the applied voltage [12] and composition of electrolyte [9], that are unchanged in our experiment. The outer diameter d_{out} and the wall thickness $(d_{out} - d_{in})/2$ increase with the lowering of T_A due to suppression of the dissolution process, because of the strong dependence of the solubility of anodization product ions on temperature; this is in agreement with the studies [2, 12]. It should be mentioned that electrolyte temperature remained practically constant when $\Delta T < 0$ °C what gives a reason to conclude that only the anode temperature influenced the morphology of the oxide layer. In this case, the migration of ions in the solid phase is probably most crucial for dissolution process in comparison with wet chemical etching. Therefore, anode temperature is a key factor to control the morphology of zirconium oxide layer.

Conclusion

The electrochemical installation with two conditioning circuits was designed. Nanotubular and nanoporous structures of zirconia were synthesized by potentiostatic anodization of Zr foil in the fluoride containing electrolyte. It was shown that the morphology parameters of the grown oxide layer significantly depend on the temperature of anode and electrolyte. Nanotubular type structure was observed at anode temperature range from 0 to 20 °C and at electrolyte temperature of 25 °C, nanoporous zirconia with different ordering degree was grown at $0 \leq \Delta T <15$ °C (partially ordered) and at $\Delta T \geq 15\ °C$ (disordered). Anodizing rate or the thickness of the oxide layer were increased by ΔT raise. It is shown that the decreasing of T_A leads to rise of the outer diameter of nanotubes. The mechanisms of the nanostructured anodic zirconia growth were discussed with allowing for processes, occurring during anodization at oxide/electrolyte and metal/oxide interfaces.

Acknowledgments

Act 211 Government of the Russian Federation, contract № 02.A03.21.0006, supported the study. R.V.K. thanks RFBR research project № 18-33-01072 for support. A.S.V. and I.A.W. thank Minobrnauki initiative research project № 16.5186.2017/8.9 for support.

References

[1] Wang X, Zhao J, Du P, Guo L, Xu X, Tang C 2012, *Mater. Res. Bull.*, **47**, 11, 3916
[2] Hosseini M G, Daneshvari-Esfahlan V, Ordikhani-Seyedlar R 2015, *Corros. Eng. Sci. Technol.*, **50**, 7, 533
[3] Vokhmintsev A S, Kamalov R V, Kozhevina A V, Petrenyov I A, Weinstein I A, Martemyanov N A 2018, *Proc. 2018 USBEREIT*, **1**
[4] Amer A W, Mohamed S M, Hafez A M, AlQaradawi S Y, Aljaber A S, Allam N K 2014, *RSC Adv.*, **4**, 68, 36336
[5] Kozhevina A V, Vokhmintsev A S, Kamalov R V, Martemyanov N A, Chukin A V, Weinstein I A 2017, *J. Phys. Conf. Ser.*, **917**
[6] Kamalov R, Vokhmintsev A, Dorosheva I, Kravets N, Weinstein I 2016, *Adv. Sci. Lett.*, **22**, 688
[7] Asgari V, Noormohammadi M, Ramazani A, Kashi M A 2017, *J. Phys. D: Appl. Phys.*, **50**, 375501
[8] Ghicov A, Schmuki P 2009, *Chem. Commun.*, 20, 2791
[9] Lockman Z, Ismail S, Razak K A, Lee L S 2011, *IOP Conf. Ser. Mater. Sci. Eng.*, **18**, 5, 52004
[10] Ismail S, Ahmad Z A, Berenov A, Lockman Z 2011, *Corros. Sci.*, **53**, 4, 1156
[11] Berger S, Jakubka F, Schmuki P 2008, *Electrochem. commun.*, **10**, 12, 1916
[12] Mor G K, Shankar K, Paulose M, Varghese O K, Grimes C A 2005, *Nano Lett.*, **5**, 1, 191

SPBOPEN 2018   IOP Publishing

IOP Conf. Series: Journal of Physics: Conf. Series **1124** (2018) 022005   doi:10.1088/1742-6596/1124/2/022005

Structural properties of multilayer heterostructure for quantum-cascade lasers grown by MBE growth

R R Reznik[1-3], N V Kryzhanovskaya[1], A E Zhukov[1, 4], A I Khrebtov[3], Yu B Samsonenko[3], S V Morozov[5] and G E Cirlin[1, 4]

[1]St Petersburg Academic University, Khlopina 8/3, 194021, St-Petersburg, Russia
[2]Institute for Analytical Instrumentation RAS, Rizhsky 26, 190103, St-Petersburg, Russia
[3]ITMO University, Kronverkskiy pr. 49, 197101, St. Petersburg, Russia
[4]Saint-Petersburg Science Center, Russian Academy of Sciences, Universitetskaya nab. 5, 199034, St. Petersburg, Russia
[5]Institute for Physics of Microstructures of the Russian Academy of Sciences, Academicheskaya 7, 603087, Afonino, Russia

Abstract. GaAs/AlGaAs heterostructure with 226 quantum cascades was synthesized by molecular-beam epitaxy. Structural and optical properties studies have shown high level of homogeneity and quality of epitaxial structure.

1. Introduction

Sources of the terahertz (THz) frequency range are in demand for various applications for both commercial and special purposes, including spectroscopy, determination of trace amounts of various substances, development of systems forming images of objects concealed from ordinary optical systems, wide-bandwidth communication systems, etc. One of the main problems is the lack of compact chromatic sources of radiation of sufficiently high-power for this spectral range. The quantum-cascade laser (QCL) is at present believed to be the most promising candidate for the development of THz systems for different purposes [1, 2]. The first QCLs for the THz range were fabricated in the early 2000s [3, 4]. To date THz QCLs of this kind have been developed, which operate up to 200 K [5] and have an emission power of several tens of mW at liquid-nitrogen temperature. Also, roomtemperature laser sources of terahertz radiation have been developed [6] on the basis of the intraband amplification of the differential frequency of two QCLs of the mid-infrared (IR) spectral range [7].

In this study, we have been investigated the possibility of molecular-beam epitaxy (MBE) synthesis of multiple-period multilayer heterostructure for QCLs fabrication in frequency range around 3 THz., and structural properties of grown heterostructure were studied.

Despite the apparent scientific and practical importance of QCLs, this technology started to be developed in Russia only a few years ago. At that time, there were reports about QCLs grown by

Content from this work may be used under the terms of the Creative Commons Attribution 3.0 licence. Any further distribution of this work must maintain attribution to the author(s) and the title of the work, journal citation and DOI.
Published under licence by IOP Publishing Ltd

molecular beam epitaxy (MBE), with emission at a wavelength of about 5–6 μm [8–10], and those synthesized by epitaxy from metal-organic compounds, intended for a wavelength of ~10 μm [11]. At the same time, no development of THz QCLs in Russia has been reported.

2. Experiments.

The epitaxial structure was synthesized by molecular-beam epitaxy using a Riber Compact 21 MBE machine. Growth was performed on semi-insulating GaAs (100) substrates under arsenic-stabilized conditions. Figure 1 shows the schematic diagram of the grown structure. Particular attention was paid to accurately setting the growth rates and to maintaining their stability during whole the growth run. The growth rates were carefully calibrated on a separate sample directly prior to growth of the multiple-period multilayer heterostructures. The GaAs and AlAs growth rates were set at 0.425 and 0.075 monolayers per second (ML/s). Special high-speed shutters were used to obtain high quality interfaces. In our case, the actuation time of the shutters of the aluminium and gallium sources did not exceed 0.15 seconds. A 200-nm-thick $Al_{0.8}Ga_{0.2}As$ stop-layer was deposited onto a GaAs buffer layer. The active region contained 226 periods. Each cascade contains a GaAs/AlGaAs double quantum well (QW), between whose levels a QCL transition occurs, and a wider QW serving as the injector/extractor of electrons. The active region was bound from above and below by GaAs:Si contact layers (5×10^{18} cm^{-3}) with thicknesses of 60 nm. The middle part of the injector/extractor layers also had n-type doping to a concentration of $~2 \times 10^{16}$ cm^{-3}. When heterostructures are used for the emission in the THz range are synthesized by the MBE with total thicknesses exeeding10 μm, the GaAs growth rate may, nevertheless, decrease due to exhaustion of the gallium source in the course of prolonged deposition. Such change in the growth rate was taken into account during the experiment.

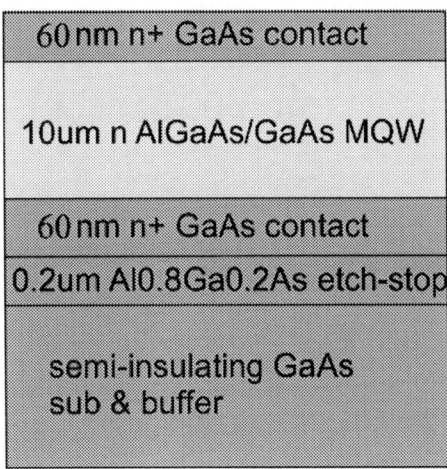

Figure 1. The schematic diagram of the grown heterostructure.

3. Results and Discussion.

The structural properties of the sample were studied by high-resolution X-ray diffraction method (HR-XRD), atomic force microscopy (AFM) and transmission electron microscopy (TEM). For HR-XRD method the D8 DISCOVER Bruker AXS diffractometer (radiation wavelength λ =0.15406 nm) with a primary beam half-width of <12 arcsec in the Ω–2θ scanning mode was used. Figure 2 shows the rocking curve around the symmetric GaAs (004) reflection. The full width at half-maximum (FWHM) of the satellite peaks due to the periodic repetition of QCL cascades is 15–19 arcsec. It is noteworthy that the full width of the superstructure peaks in the model spectrum, found with allowance for bending of the structure under elastic stresses, is 22.4 arcsec. This means that both the possible effects due to the inaccurately of the cascade thickness within the whole structure and the roughness of the heterointerfaces can be neglected, which confirms that the technological parameters for the case of synthesis of the active region are chosen correctly. The calculated rocking curve for the model structure providing the excellent agreement with the experimental data which is shown in Figure 2.

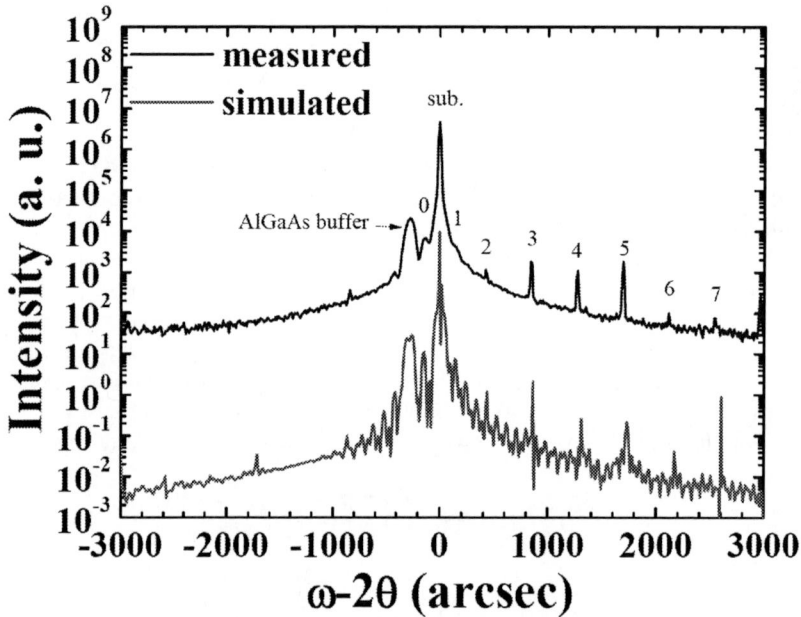

Figure 2. X-ray rocking curve of the QCL structure near the GaAs (004) reflection and the simulated curve.

Figure 3 shows typical AFM (a) and TEM (b) images of grown heterostructure. According to the AFM measurements it seen that average surface roughness is 2 Å, only showing perfect smoothness of the grown structure. In turn, the TEM studies results indicate very abrupt interface between the layers within the whole structure (figure 3b). Both of these facts attest to the high crystallographic and epitaxial quality of the grown structure.

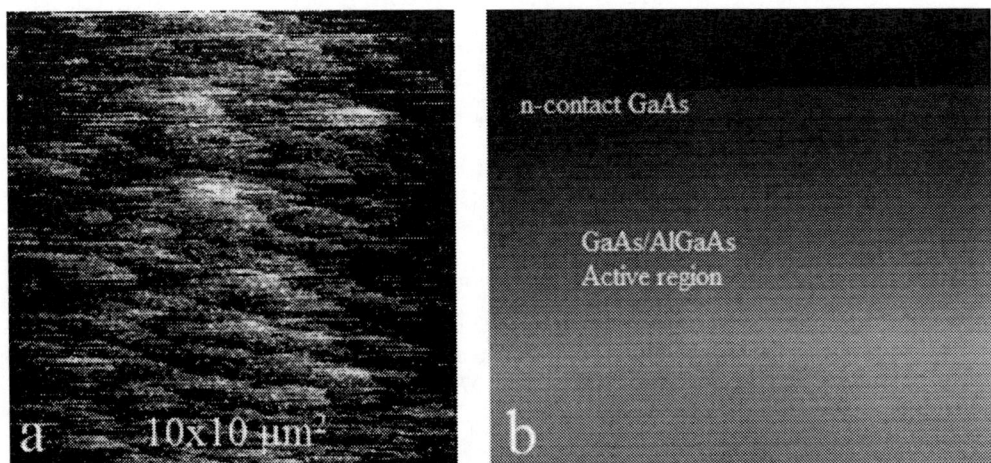

Figure 3. (a) - AFM image of grown sample surface: Terrace length-0.7~0.96 um, RMS roughness-2 Å.; (b) - TEM image of grown structure.

Thus, the method of molecular-beam epitaxy was used to synthesize multiperiod (226 cascades, ~10 μm) GaAs/AlGaAs epitaxial structures intended for use in the fabrication of quantum-cascade lasers in the terahertz range. Structural properties studies have shown high level of homogeneity and quality of the epitaxial structure. The angular width of the superstructure peaks in the X-ray rocking curve does not exceed 20 arcsec, total surface roughness is as less as 2 Å.

Acknowledgments
The study was supported by the Program of Basic Research of the Presidium of the Russian Academy of Sciences "Nanostructures: physics, chemistry, biology, and fundamentals of technology." The samples were grown under support of Russian Science Foundation (project No 18-19-00493).

References
[1] Kazarinov R F, Suris R A 1971 *Sov. Phys. Semicond.* **5** 707
[2] Faist J, Capasso F, Sivco D L, Hutchinson A. L, Cho A Y 1994 *Science* **264** 553
[3] Koehler R, Tredicucci A, et al 2002 *Nature* **417**
[4] Rochat M, Ajili L, Willenberg H, et al 2002 *Appl. Phys. Lett.* **81** 1381
[5] Fathololoumi S, Dupont E, Chan C W I, et al 2012 *Opt. Express* **20** 3866
[6] Lu Q Y, Bandyopadhyay N, Slivken S, et al 2014 *Appl. Phys. Lett.* **104** 221105
[7] M. A. Belkin, Q. J. Wang, C. Pflugl, et al 2009 *IEEE J. Sel. Top. Quantum Electron.* **15** 952
[8] Mamutin V V, Ustinov V M, Boetthcher J, Kuenzel H 2010 *Semiconductors* **44** 962
[9] A. Yu. Egorov, P. N. Brunkov, E. V. Nikitina, et al 2014 *Semiconductors* **48** 1600
[10] A. Yu. Egorov, A. V. Babichev, L. Ya. Karachinskii, et al *Semiconductors* **49** 1527

SPBOPEN 2018   IOP Publishing

IOP Conf. Series: Journal of Physics: Conf. Series **1124** (2018) 022006   doi:10.1088/1742-6596/1124/2/022006

Spiral growth of a crystal due to chemical reaction

A V Redkov[1,3], S A Kukushkin[1-3] and A V Osipov[1,3]

[1]Institute of Problems of Mechanical Engineering RAS, St. Petersburg 199178, Russia
[2]Peter the Great St. Petersburg Polytechnic University, St. Petersburg 195251, Russia
[3]ITMO University, St. Petersburg 197101, Russia
E-mail: avredkov@gmail.com

Abstract. The article is devoted to the study of crystal growth due to chemical reactions in multicomponent media via terrace-kink-step mechanism. The theory of Burton, Cabrera and Frank is developed and applied to multicomponent case. Equations are obtained for speed of movement of single step, group of steps at certain interstep distance. Growth spirals caused by screw dislocations are considered. Growth rate of the crystal is found. Impact of partial pressures of reaction products on diffusion length of adatoms and growth rate is analyzed.

1. Introduction

Nowadays crystals, thin films and methods for their growth are in great demand in industry and there is a significant amount of work devoted to theoretical aspects of crystal growth via different mechanisms: crystallization, nucleation, sublimation and many others. One of the mechanisms playing key role during the growth at low supersaturation is growth due to terrace-kink-step mechanism, at which atoms adsorbed to the surface diffuse one-by-one towards steps and kinks and then incorporate into crystalline lattice. This mechanism was described in the classical works of Burton, Cabrera, and Frank (BCF) and related generalizations in numerous reviews and books [1-2]. A theory has been proposed for describing the growth of single-component crystals via this mechanism. The theory made it possible to predict the dependence of the crystal growth rate on various growth conditions, and to explain why the crystals grow even at very low supersaturations (<1%), when nucleation of nanoislands on the surface is impossible, and therefore growth by the mechanism of nucleation is suppressed. It should be noted that the theory was refined and modified many times, and attempts were made to apply it to multicomponent crystals growing due to chemical reactions, in particular, gallium nitride [3]. However, in general, the theory was developed mainly with respect to single-component crystals growing from its own saturated vapors [1], atomic fluxes or solutions, and as far as we know, there is no generalization of the theory for multicomponent systems with chemical reactions, which would allow one to fully describe the growth of arbitrary single-component or multicomponent crystal from multicomponent media due to chemical reaction. In fact, many modern methods of thin films growth involving chemical reactions are widely used in industry, in particular, semiconductor (thin films of A^{III}-B^{V}, A^{II}-B^{IV} semiconductors). HVPE (hydride vapor phase epitaxy), MOCVD (metal-organic chemical vapor deposition) and MBE (molecular beam epitaxy) should be mentioned, the growth during which occurs at the presence of flows of various chemical components towards the surface. The processes and surface structures, which are

---

Content from this work may be used under the terms of the Creative Commons Attribution 3.0 licence. Any further distribution of this work must maintain attribution to the author(s) and the title of the work, journal citation and DOI.
Published under licence by IOP Publishing Ltd

observed during such multicomponent growth [4], a very similar to ones observed during the growth in single-component system. One can notice steps, terraces, growth spirals. That's why we suppose that all of the mechanisms described by BCF theory are also valid for the growth of multicomponent crystals. The main goal of this work is to extend mathematical apparatus of this theory on multicomponent systems.

To extend existing BCF theory a series of particular problems should be understood and solved. Namely these problems are: 1) growth of single-component crystal due to chemical reaction from multicomponent media (eg. Si epitaxy in atmosphere of SiH_4 and H_2) It should be noted that the formation of molecular crystals also may belong to this case; 2) growth of multicomponent crystal from the atmosphere of vapors of its individual components (eg. growth of GaN via MBE, which involve individual fluxes of Ga and N). After solving these problems one could combine solutions to describe growth of arbitrary multicomponent crystal when the individual components are being delivered to its surface due to chemical reaction (eg. growth of GaN via HVPE, when Ga and N arrive in the form of reagents GaCl and NH_3, which react near the surface).

This work is the first in a series of papers devoted to extension of BCF theory to multicomponent systems and it considers the first abovementioned problem, namely, growth of single-component crystal due to chemical reaction in multicomponent media via terrace-step-kink and spiral mechanisms.

2. Growth of single-component crystal due to chemical reaction

Let us consider a crystal growing from a multicomponent medium as a result of a chemical reaction. In general, the reaction can be written as:

$$\nu_1 A_{1(g)} + \nu_2 A_{2(g)} + \cdots + \nu_N A_{N(g)} \rightarrow S_{(s)} + \nu'_1 B_{1(g)} + \cdots$$

$$+ \nu'_2 B_{2(g)} + \cdots + \nu'_M B_{M(g)} \quad (1)$$

where $A_{i(g)}$ are the initial chemical reagents in liquid or gas state, $B_{i(g)}$ are reaction products, $S_{(s)}$ is a solid crystalline phase; ν_i are the stoichiometric coefficients. As a result of this reaction, $S_{(s)}$ adatoms are delivered to the surface of the crystal. We believe that the growth of a crystal having terraces-steps-kinks is caused by the following factors: the exchange between the vapor and the adsorption layer due to chemical reaction, diffusion of the atoms of the adsorption layer in the direction towards the steps, their subsequent diffusion along steps to kinks, and incorporation into crystalline lattice which ensures movement of the step. The mathematical formulation of the problem is completely analogous to that presented in the works of Burton, Cabrera, and Frank [1], except for the fact that in this paper we consider the growth of a crystal in a multicomponent system. The main difference is that in a multicomponent system different substances are present in the gas phase and on the surface. For simplicity, we assume that the average distance between kinks is significantly less than the diffusion length of the adatoms, i.e. concentration of kinks at the step is sufficiently high and step can be considered as continuous drain, due to reasoning similar to those given in [1].

In general, the crystal $S_{(s)}$ can consist of atoms of one type, for example, silicon (Si), or from atoms of several types, such as gallium nitride (GaN). We will consider the first case, the second will be considered in subsequent papers. It is also believed that only adatoms in adsorbed state participate in the reverse chemical reaction, since much more energy is needed to remove atoms incorporated into lattice (and thus having a large number of bonds). We assume that the reaction of formation of the solid phase occurs directly on the surface, and there is no reaction in the gas phase, which is true for many growth processes.

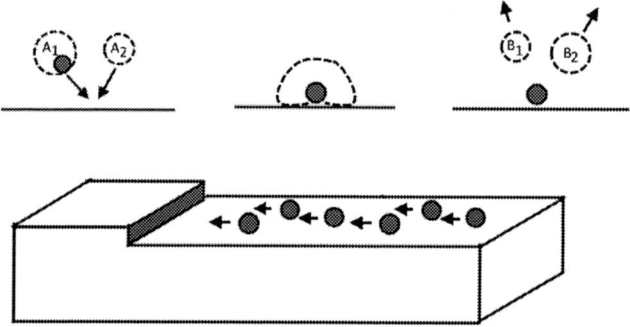

Figure 1. Mechanism of crystal growth due to chemical reaction. Flux of chemical reagents and formation of $S_{(s)}$ adatoms on the surface (upper insets), and their subsequent diffusion and incorporation in steps (lower inset). Evaporation is the process inverse to upper figure.

For comprehensive description of the dependence of the growth rate of the crystal on the conditions, it is necessary to solve a number of individual tasks, as was done in the classical work of Barton, Cabrera, and Frank [1], namely, to find the speed of movement of a single step, the speed of a series of parallel steps, and the speed of movement of a spiral step formed due to screw dislocation.

Firstly, we should find the speed of single step. Using the principle of detailed equilibrium, it is easy to find in a general form the fluxes of atoms moving from the gas phase to the adsorbed state on the surface $j_{v\downarrow}$ and back to gas $j_{v\uparrow}$ and their difference j_v:

$$j_v = j_{v\downarrow} - j_{v\uparrow} = k_d \prod_1^N C_{A_i}^{v_i} - k_r n_s \prod_1^M C_{B_i}^{v_i'} =$$

$$= \left(\frac{K}{K_{eq}} - \frac{n_s}{n_{s0}} \right) k_r n_{s0} \prod_1^M C_{B_i}^{v_i'} \tag{2}$$

where k_d is the rate constant of the direct chemical reaction, $k_r n_{s0}$ is the rate constant of the reverse reaction, $K = \prod_1^N C_{A_i}^{v_i} / \prod_1^M C_{B_i}^{v_i'}$, $K_{eq} = k_r n_{s0} / k_d$ is the equilibrium constant of chemical reaction (1), C_{B_i} and C_{A_i} are concentrations of reagents, n_s and n_{s0} are the true and equilibrium concentrations of adsorbed atoms, respectively. Comparing formula (2) and the equation for the flux from the gas phase obtained in [1] for crystal growth from own vapors, it is not difficult to see that the role of supersaturation in the multicomponent case is played by the ratio $\frac{K}{K_{eq}}$, and the role of the evaporation time of adatoms from the surface is $1/k_r \prod_1^M C_{B_i}^{v_i'}$. It should be noted that in general case during the growth direct evaporation of adatoms from the surface is also possible and expression (2) on the right-hand side should have the term $-n_s/\tau_s$, where τ_s is the true lifetime of adatoms in the absence of a chemical reaction. However, we believe that $\tau_s \gg 1/k_r \prod_1^M C_{B_i}^{v_i'}$, evaporation is extremely slow compared to the rate of reverse chemical reaction, and therefore we neglect it. In further papers we will consider effects which may arise due to competition of fluxes caused by own vapors and by chemical reaction.

To find the speed of movement of an individual step, we introduce the following variables, using the notation of [1]:

$$\sigma = \alpha - 1$$

$$\sigma_s = \alpha_s - 1$$

where σ is the supersaturation in the gas phase, $\alpha = \frac{K}{K_{eq}}$, σ_s is the supersaturation of the adatoms on the surface, $\alpha_s = n_s/n_{s0}$. In general, n_s depends on the distance to the step, whereas K and the concentrations of the components are assumed to be independent from this distance. Further mathematical calculations are very close to that ones obtained in [1]. To find the diffusion flux of adatoms over the surface j_s by analogy with [1] one can write the following equation:

$$j_s = -D \, grad \, n_s = D n_{s0} \, grad \, \varphi \qquad (3)$$

where $\varphi = \sigma - \sigma_s$ is determined by supersaturation in the gas phase and on the surface. The flux of the adatoms from the gas to the surface j_v, using this notation, can be written as follows:

$$j_v = (\alpha - \alpha_s) \, k_r n_{s0} \prod_1^M C_{B_i}^{v_i'} = \varphi \, k_r n_{s0} \prod_1^M C_{B_i}^{v_i'} \qquad (4)$$

If the supersaturation is sufficiently small, then the growth rate of the crystal and, correspondingly, the movement of the steps can also be neglected. In this case, when solving the problem of diffusion of adatoms towards the step, it can be considered immobile, so that the solution becomes stationary. Under these conditions, the value of φ must satisfy the continuity condition:

$$div \, j_s = j_v$$

Substituting j_s (3) and j_v (4), as it was done in [1], we obtain equation for function φ:

$$\lambda_s^2 \nabla^2 \varphi = \varphi \qquad (5)$$

where the mean free path of the adatom of crystalline phase is introduced $\lambda_s^2 = \dfrac{D}{k_r \prod_1^M C_{B_i}^{v_i'}}$. Note that in our case, unlike [1] it is not a constant and depends on the rate of reverse chemical reaction and concentrations of reaction products.

The coordinate system is chosen so that the step lies on the x-axis. The y coordinate determines the distance from the step. As boundary conditions at infinity, we choose $\varphi|_{y\to\infty} = 0$ ($\sigma = \sigma_s$). We will slightly complicate the task for generality and assume that directly at the step the flux of the component, will depend not only on diffusion, but also on the incorporation rate β of adatom into step:

$$-D\frac{d\,n_s}{dx} = \beta V(n_s - n_{s0})\Big|_{y=0}$$

where V is the volume of the adatom. Then by making a replacement one can obtain:

$$D\frac{d\varphi}{dx} = \beta V(\sigma - \varphi)|_{y=0}$$

The solution of this diffusion problem, as it can be easily shown, is $\varphi = C\exp(\pm\frac{y}{\lambda_s})$, where $C = \sigma(1 + \frac{D}{\lambda_s \beta V})^{-1}$. The plus sign refers to the lower half-plane (y <0), the minus sign is for the upper half-plane (y> 0). The flux of adatoms to the step can be calculated using equation (1):

$$j_s|_{y=0} = \frac{D\beta V n_{s0}}{\beta V \lambda_s + D}\,\sigma \tag{6}$$

The speed of movement of the single step ϑ_∞, in this case, according to [1] will be equal to:

$$\vartheta_\infty = 2j_s/n_0 = 2\left(\frac{D\beta V}{\lambda_s \beta V + D}\right)\sigma$$

The factor 2 in (6) is due to the fact that the flux of adatoms arrives from both sides of the step. In extreme cases the expression coincides with the one obtained by Burton, Cabrera and Frank [1].

Let us calculate the speed of movement of group of steps with interstep distance of y_0. The first step is at position $-y_0/2$, the second one is at $y_0/2$, etc. In this case we should solve equation (5) with the following boundary conditions $\varphi'(0) = 0$, $D\frac{d\varphi}{dx} = \beta V(\sigma - \varphi)|_{y=\frac{y_0}{2}}$

The solution is:

$$\varphi = \sigma\,\frac{\beta V \lambda_s ch(\frac{y}{\lambda_s})}{D\,sh\left(\frac{y_0}{2\lambda_s}\right) + \beta V \lambda_s ch(\frac{y_0}{2\lambda_s})}$$

Now we can calculate the flux towards each step and speed of movement of group of steps ϑ_∞^{gr} :

$$\vartheta_\infty^{gr} = \frac{2j_{si}}{n_0} = 2\sigma\,\frac{D\,\beta V}{D\,th\left(\frac{y_0}{2\lambda_s}\right) + \beta V \lambda_s}\,th(\frac{y_0}{2\lambda_s})$$

Next step is the calculation of vertical growth rate R of the spiral step caused by exit of screw dislocation (and therefore growth rate of the crystal). By substituting ϑ_∞^{gr} to equation for R and taking into account distance between steps $y_0 = 4\pi\rho_c$ [1], one can obtain:

$$R = \frac{n_0 V \vartheta_\infty^{gr}}{4\pi\rho_c} = \frac{n_0 V}{2\pi\rho_c}\,\frac{D\,\beta V\,th(\frac{4\pi\rho_c}{2\lambda_s})}{D\,th\left(\frac{4\pi\rho_c}{2\lambda_s}\right) + \beta V \lambda_s}\left(\frac{K}{K_{eq}} - 1\right) \tag{7}$$

where $n_0 V$ – is a height of a step, ρ_c is a half of 2D-critical nucleus size.

One can see, that in case of diffusion growth limit ($\beta V \lambda_s \gg D\,th\left(\frac{4\pi\rho_c}{2\lambda_s}\right)$) growth rate is determined by expression:

$$R = \frac{n_0 V}{2\pi\rho_c} \frac{D}{\lambda_s}\, th\left(\frac{4\pi\rho_c}{2\lambda_s}\right)\left(\frac{K}{K_{eq}} - 1\right) \qquad (8)$$

In case of incorporation limit ($\beta V \lambda_s \ll D\,th\left(\frac{y_0}{2\lambda_s}\right)$) the growth rate is:

$$R = \frac{n_0 V^2}{2\pi\rho_c} \beta\left(\frac{K}{K_{eq}} - 1\right) \qquad (9)$$

and not dependent on interstep distance as expected. One should note that equations (7-9) are correct in case that partial pressure of own vapors is small enough to neglect flux of adatoms from vapor phase, and lifetime of adatoms is determined not by evaporation, but only by reverse chemical reaction rate.

3. Conclusions

The growth of a crystal from multicomponent media due to chemical reaction is considered. The model of growth is proposed based on classical Burton, Cabrera and Frank theory. The model is very similar to classical one, and the main difference is that the ratio of K/K_{eq} plays a role of supersaturation, and the lifetime of adatoms is not constant, but is determined by the rate of reverse chemical reaction and concentration of products of the reaction, which can be pumped out of the growth chamber. It opens up new possibilities in controlling of crystal growth mechanism because one can vary both supersaturation and diffusion length whereas during growth from own vapors one can vary only supersaturation.

References
[1]   Burton W K, Cabrera N, Frank F C 1951 *Phil. Trans. Roy. Soc.* **243** 299
[2]   Chernov A A 1961 *Physics-Uspekhi* **4** 116
[3]   Cadoret R 1999 *Journal of Crystal Growth* **205** 123.
[4]   S. S. Sharofidinov, A. V. Redkov, A. V. Osipov and S. A. Kukushkin, Journal of Physics: Conference Series, 2017, 917, 032028

Formation of hexagonal silicon regions in silicon

A A Nikolskaya, D S Korolev, A N Mikhaylov, A I Belov, A A Sushkov, D A Pavlov and D I Tetelbaum

Lobachevsky University, 23/3 Gagarin prospect, Nizhny Novgorod, 603950, Russia

E-mail: alena.nikolskaya.1994@mail.ru

Abstract. The synthesis of hexagonal phase of silicon (9R polytype) by ion implantation has been studied. Transmission electron microscopy reveals the 9R regions in a subsurface layer of silicon substrate after ion implantation and subsequent heat treatment. The results are discussed from the viewpoint of the effect of mechanical stresses on the hexagonal phase formation.

1. Introduction

Silicon is still the main material of semiconductor electronics. However, due to its indirect band structure, diamond-like silicon has a significant drawback – low efficiency of light emission, which prevents its use in new generations of devices based on optical and electro-optical phenomena. The main ways of solving this problem are the formation of regions of another phase with better luminescent properties (Ge, iron silicide, GaAs, GaN, silicon quantum dots in a dielectric matrix, etc.) on the silicon substrate, doping of silicon with a rare-earth impurity (Er), or a change of the silicon structure by external impact [1].

Despite a great amount of research carried out within the mentioned approaches, none of them have still found wide practical application. Another way is to create nanoregions with hexagonal structure within the ordinary silicon. The change of the symmetry type (transition to a new syngony) creates the prerequisites for a radical change in the band structure and improvement of the luminescence efficiency. Indeed, it has been shown theoretically and experimentally [2, 3] that hexagonal silicon has higher photoluminescence intensity than the usual diamond-like phase. However, there is still no method for synthesis of bulk crystals of hexagonal silicon, which are stable under normal conditions. Therefore, the existing methods for creating thin layers of this modification are of little use for practical purposes. Thus, the development of methods compatible with silicon technology for the production of sufficiently stable inclusions of hexagonal silicon and the study of their properties are relevant both from a fundamental point of view and in connection with the problem of overcoming the pointed disadvantages of silicon. Ion-beam treatment method is most compatible with the traditional methods of silicon technology. In [4], a method for synthesis of the hexagonal (2H) phase of Si by ion implantation in cubic silicon using the high-current irradiation with As^+ ions under conditions of poor thermal contact of the sample with the holder for the realization of heating by ion beam was suggested. However, this method is not quite appropriate for practical usage.

In this work, the formation of hexagonal 9R-Si phase in the subsurface region of silicon after implantation of gallium and nitrogen ions followed by annealing has been for the first time revealed by the transmission electron microscopy (TEM).

Content from this work may be used under the terms of the Creative Commons Attribution 3.0 licence. Any further distribution of this work must maintain attribution to the author(s) and the title of the work, journal citation and DOI.

Published under licence by IOP Publishing Ltd

2. Experimental

The silicon samples were cut from (100) n-Si wafers with the thickness of 380 μm. First, the sample was implanted by N_2^+ (20 keV, $2.6 \cdot 10^{17}$ at/cm^2) with 1100 °C post-implantation annealing. Then N_2^+ ($5 \cdot 10^{16}$ at/cm^2) and Ga^+ ($5 \cdot 10^{16}$ at/cm^2) ions were co-implanted with energies of 40 keV and 80 keV, respectively. The final annealing was performed at 800 °C (30 min) in a dry nitrogen atmosphere.

The {110} cross-sections of the synthesized samples were studied by TEM using the JEOL JEM-2100F microscope.

3. Results and discussion

A cross-sectional transmission electron microscopy image of the silicon sample is shown in Figure 1a. The preliminary 20 keV nitrogen implantation followed by annealing at 1100 °C was used to synthesize a layer of silicon nitride, which prevents the out-diffusion of implanted gallium from the sample during annealing [5]. The corresponding layers are seen on the microscopic image. The topmost layer is the epoxy which used for preparation. Beneath of it, a layer with lighter contrast is seen, which is an ion-synthesized silicon nitride. Inside the latter (in its lower part) dark inclusions are visible, which are identified as inclusions of a gallium-containing phase. The silicon nitride film is separated from the modified silicon substrate by a wavy boundary. Below it, a relatively thick (> 100 nm) layer with a non-uniform contrast is located. The main attention was devoted to studying the structure of this layer. Note that the boundary between this layer and an unmodified substrate is at a depth substantially exceeding the projected ranges of Ga^+ and N^+ ions (R_P = 60 nm). The analysis of the images shows that, within this layer, there are some regions with the structure different from diamond-like silicon. An image obtained with atomic resolution (Figure 1b) allows us to identify the structure of the regions. Simulation of the diffraction pattern by Fourier transform is shown in Figure 1c.

The interpretation of such structure, which often arises in the case of heteroepitaxial growth of silicon on sapphire, is described in some detail in [6]. In Figure 1a, upon closer examination, an ordered arrays of twinning defects are observed in silicon structure, which on the (110) cross section are represented by two series of (111) planes with an angle of 70° 32′ to each other. There are four such series in the diamond structure, but only two of them can be seen on the (110) cross section.

The alternation of the (111) layers in the structure of diamond-like Si can be written as …ABCABCABC… (or …ccc…). In the presence of a twinning boundary, the alternation of layers looks like: …ABCABĊBACBA… (or …cchcc…), where Ċ points out the position of the twin boundary (stacking fault). If another twin boundary arises after a certain number of layers, the order of alternation of close-packed layers becomes the original one, and then a twin lamel (twin pair) appears. It is assumed that, at the rearrangement of a three-layer diamond packing within the regions, the lattice of which remains conjugate to the "defect-free" region of silicon substrate, an energy-optimal number of built-in twin pairs should be multiplied by three.

The main polytypes of diamond, containing simultaneously fragments of hexagonal and cubic packages are described in [7]. The primitive unit cell of one of these polytypes – 9R includes three periods in one direction, each of which in turn includes one fragment of the cubic packing …ABCABCABC … and two fragments of the hexagonal packing (hhc). The stacking sequence in the 9R phase can be described as … ABACACBCB …. The degree of hexagonality of this polytype is 2/3. The change in the periodicity of the packing in the direction perpendicular to the close-packed planes is confirmed by the fact that in the simulated diffraction pattern (Figure 1c) between the central reflection and the reflection, which corresponds to the interplanar distance d_{111} = 0.313 nm in diamond-like 3C-Si, there are two additional reflections (tripling the packing period). Since in this case a structure with hexagonal syngony is considered [8], these reflections can be assigned with four-digit indices, the respective values at interplanar distances (d) of which are d_{0003} = 0.943 nm; d_{0006} = 0.471 nm; d_{0009} = 0.313 nm (Figure 1c).

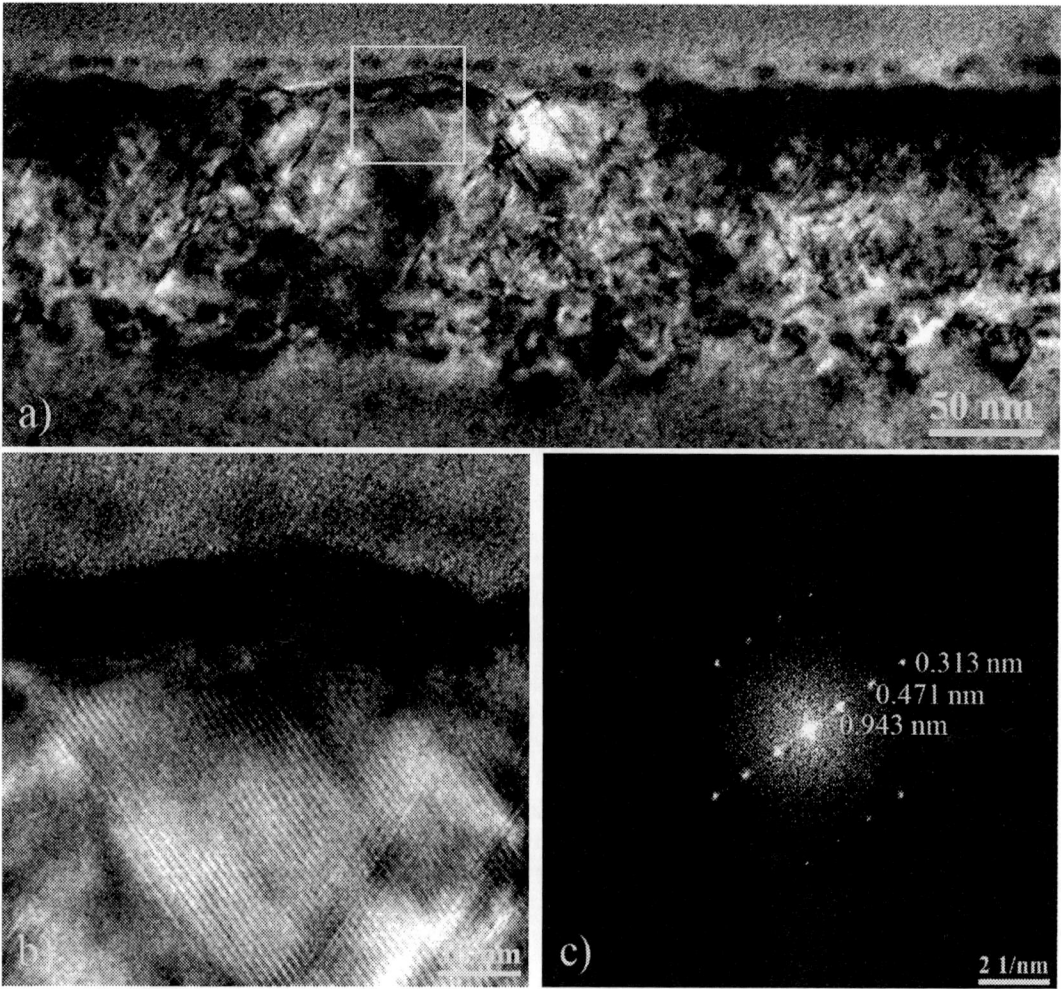

Figure 1(a,b,c). (a) High-resolution cross-sectional TEM image of silicon with pre-synthesized silicon nitride layer irradiated with Ga$^+$ and N$_2^+$ ions and annealed at 800°C; **(b)** enlarged image of the indicated square region; **(c)** Fourier-transform pattern for this image.

Figure 2 shows an image of the "hexagonalized" region, in which two orientations of the 9R phase are clearly visible with the angle between the respective arrays of planes of 70° 32′. This is also revealed on the picture of microdiffraction (Figure 2b), in which two additional reflections are observed along with the main reflections of 3C-Si.

The possible mechanism for the formation of the 9R-Si phase is as follows. First, a layer of silicon nitride is formed on the surface of substrate by ion-beam synthesis and high-temperature annealing (1100 °C). This layer, as shown by the XPS method [5], prevents gallium from leaving the sample due to out-diffusion. It is known that silicon nitride, in comparison with silicon, has a higher limit of plastic deformation. With ion implantation of nitrogen and gallium, a high level of elastic stress arises in this and underlying implanted layer, which exert a strong elastic influence on the underlying layers of the substrate. This creates a high level of stress in it. Then, annealing is carried out at a temperature of 800 °C. It is known that at such temperatures silicon becomes a plastic material, however, the mobility of dislocations (which plastic deformation is usually associated with) at such temperatures is very low. Therefore, the relaxation of elastic deformation occurs not by sliding dislocations, but by shifting the atomic planes (111), leading to multiple twinning, and, as described above, "hexagonalization" with the formation of the 9R-Si phase. Since the elastic deformation decreases with depth, the formation of the 9R phase is limited by a boundary layer with a thickness of ~ 100-200 nm (see Figure 1a). In the absence of a Si_3N_4 film, the stress needed for the formation of hexagonal phase is achieved only locally, so the transformation occurs, but the volume fraction of the 9R phase is low.

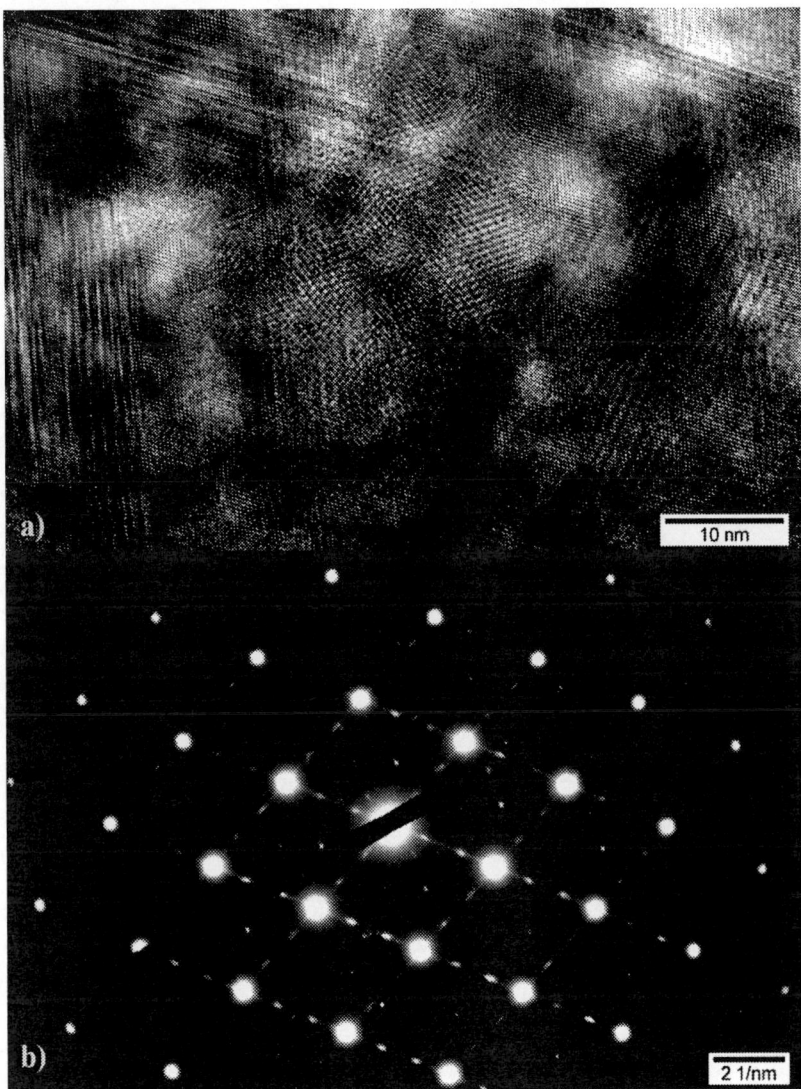

Figure 2(a,b). (a) High-resolution TEM image of a silicon region, which covers several adjoining hexagonalized regions with different orientations; **(b)** picture of microdiffraction for this region.

4. Conclusions

The work demonstrates for the first time the formation of hexagonal phase of 9R-Si upon ion implantation in silicon with thermal treatment. It is shown that a continuous layer of silicon nitride is formed by implantation of nitrogen ions in silicon and post-implantation annealing. During the subsequent double implantation of gallium and nitrogen ions, the hexagonal phase formation process occurs in the silicon at the interface with the ion-synthesized Si_3N_4 film. A mechanism for the formation of hexagonal regions is proposed that includes the relaxation of mechanical stresses induced by ion implantation.

Acknowledgments

The work is supported by the Ministry of Education and Science of the Russian Federation (State Assignment № 16.2737.2017/4.6). One of the authors A.A. Nikolskaya acknowledges the support in the framework of UMNIK program.

References

[1]   Pavesi, L., Guillot, G. (Eds.) (2006). *Optical Interconnects. The Silicon Approach*. Berlin, Heidelberg: Springer-Verlag. 389 p.

[2]   Shmyt'ko I M, Izotov A N, Afonikova N S, Vieira S, Rubio G 1988 *Phys. Sol. State* **40** 687

[3]   Bandet J, Despax B, Caumont M 2002 *J. Phys. D: Appl. Phys* **35** 234

[4]   Tan T Y, Föll H 1981 *Phil. Mag. A* **44** 127

[5]   Korolev D S, Mikhaylov A N, Belov A I, Konakov A A, Vasiliev V K, Nikolitchev D E, Surodin S I, Tetelbaum D I 2017 *Int. J. Nanotech.* **14** 637

[6]   Pavlov D A, Pirogov A V, Krivulin N O, Bobrov A I 2015 *Semiconductors* **49** 95

[7]   Wen B, Zhao J, Bucknum M J, Yao P, Li T 2008 *Diamond & Related Materials* **17** 356

[8]   Korolev D S, Nikolskaya A A, Krivulin N O, Belov A I, Mikhaylov A N, Pavlov D A, Tetelbaum D I, Sobolev N A, Kumar M 2017 *Tech. Phys. Let.* **43** 767

SPBOPEN 2018

Microstructure and composition of zirconium coatings obtained by electro-spark alloying

V Koshuro[1], M Fomina[1], M Fedoseev[1]

[1] Yuri Gagarin State Technical University of Saratov, Saratov 410054, Russia

Abstract. The study focuses on electro-spark alloying of titanium surface and formation of a porous zirconium coating. The coating morphology was formed as a result of melt drop transfer, heat treatment, and partially oxidation. The study establishes the influence of technological regimes of alloying on the surface morphological and chemical composition.

1. Introduction

Some zirconium alloys have found application in medicine along with titanium [1,2]. Much attention is paid to the surface morphology of materials used for the production of intraosseous implants. For example, the surface of dental implants should have the required open porosity to ensure the osseointegration [3-9].

Electro-spark alloying (ESA) is an effective method for the production of porous coatings. This method ensures the formation of highly porous coatings on the surface of biocompatible metals, e.g. tantalum coatings on titanium [10-12]. In this paper, the morphology of the surface and the elemental composition of the coatings produced by ESA with zirconium on the surface of titanium at DC and AC operating current were studied.

2. Methodology

Samples of VT1-0 commercially pure (cp) titanium were subjected to ESA with zirconium grade "E110" (Zr – balance, Nb – 0.9–1.1 wt.%). ESA was conducted in the pulse mode at the following parameters: pulse duration within 10–100 ms, current amplitude of 0.8–2.5 A, and pulse energy ranging within 0.01–0.1 J [2]. In this work, the effect of 3 direct current values was studied: 0.8, 1.5 and 2.5 A. The morphology and elemental composition of the coatings formed at the constant current were compared to the analogous parameters for the layers obtained at AC operating current $I = 1.75$ A.

The surface morphology of the samples was studied using scanning electron microscopy (SEM). SEM combined with energy-dispersive X-ray analysis (EDX) of chemical composition of samples was performed on "MIRA II LMU" with "INCA PentaFETx3" detector. Changes in titanium, zirconium and oxygen concentrations (at.%) were studied. Hardness of the coatings was evaluated by microindentation using "PMT-3M" (at the load of 100 gf).

3. Results

The resulting zirconium coatings on titanium were characterised by heterogeneous surface morphology, and their microstructure resembled the morphology of sprayed coatings [5]. When a DC operating current equaled 0.8 A the zirconium coating consisted of large microparticles that were unevenly distributed over the surface (Figure 1a). Analysis of the chemical composition of the coating surface showed the presence of zirconium – about 24.57 at.%, titanium – about 49.03 at.% and oxygen

Content from this work may be used under the terms of the Creative Commons Attribution 3.0 licence. Any further distribution of this work must maintain attribution to the author(s) and the title of the work, journal citation and DOI.

Published under licence by IOP Publishing Ltd

– 26.09 at.%. Thus, an intensive transfer of the alloying element was observed, as well as the simultaneous oxidation of the base (Ti) and coating (Zr) materials. At the operating current of 1.5 A, a more homogeneous transfer of zirconium micro-drops of a smaller size was observed (Figure 1b).

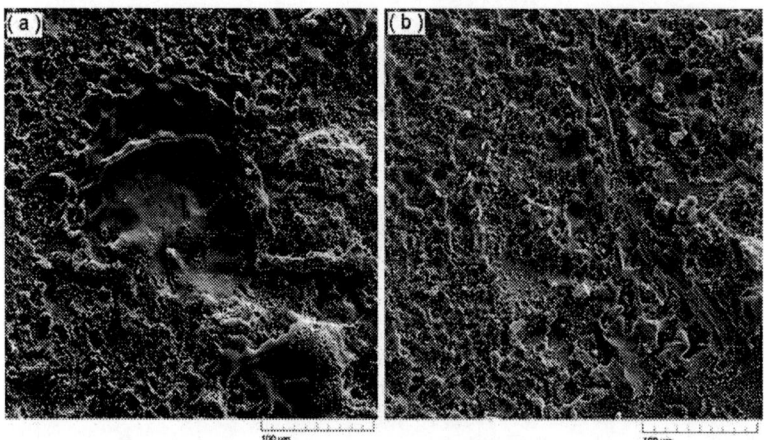

Figure 1(a, b). Morphology of the zirconium coating after ESA at 0.8 A (DC mode) **(a)**; after ESA at 1.5 A (DC mode) **(b)**.

The amount of zirconium in the surface did not exceed 16.21 at.%, whereas the amount of oxygen was much less – about 5.16 at.%. Areas with a high oxygen content of up to 25 at.% were locally observed.

The increase in direct current during ESA to 2.5 A led to a decrease in the surface heterogeneity. Cracks appeared on the coating (Figure 2a). The transfer of zirconium micro-drops decreased sharply, the amount of zirconium did not exceed 2.33 at.% (Figure 2a). The amount of oxygen reached a maximum of about 37.17 at.%. Thus, the reduction in the amount of zirconium can be explained by the accelerated oxidation of the electrode-tool and titanium base materials, as well as by the formation of microparticles with low adhesion.

The surface of the coating formed by ESA with AC operating current $I = 1.75$ A had cracks and a smoother relief (Figure 2b). The content of zirconium was 21.93 at.%, oxygen – about 39.7 at.%, titanium – balance. While analyzing the chemical composition of the coating, it can be assumed that during ESA under AC operating current the oxidation of the coating material occurred. Thus, when using AC mode for ESA, the metal transfer was carried out more intensively than with a constant current.

There were differences in the morphology of the coatings formed by ESA with DC and AC operating current. Metal drops transferred during crystallization on the titanium base acquired an irregular shape in the course of ESA at the constant current (Figure 3a). The average size of the structural elements (particles) was 8.0 μm. The open pore size was 8.1 μm with a total porosity of 54%.

When ESA with the alternating current was conducted, zirconium formed spherical structural elements with the size of 7.5 μm (Figure 3b). The open pore size was significantly less compared to that when the treatment by direct current was performed and it amounted to 4.7 μm. The coatings formed due to ESA with the alternating current were characterized by a low porosity of about 48%.

It should also be noted the formation of nano-sized structural elements (particles) in the form of individual crystals with a size of not less than 60 nm was observed (Figure 3b).

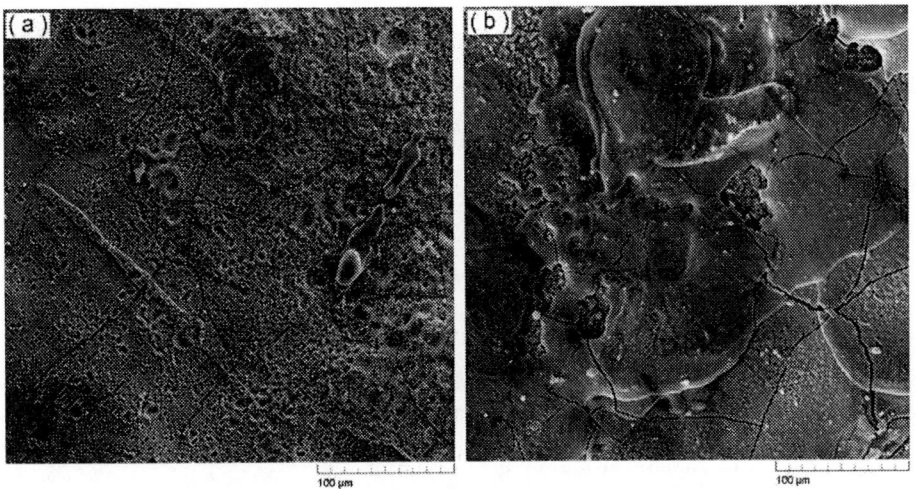

Figure 2(a, b). Morphology of the zirconia coating after ESA at 2.5 A (DC mode) **(a)**; after ESA at 1.75 A (AC mode) **(b)**.

Preliminary data on hardness showed that as a result of ESA, the coatings with a hardness of 10.4±2.0 GPa were formed at the current values I = 2.5 A (DC mode) and 1.75 A (AC mode). High hardness values supported the assumption of the possible oxidation of the base and coating materials in the course of ESA. Probably, nano-sized structural elements formed after ESA at 1.75 A (AC mode) can be oxide crystals phase (Figure 3b).

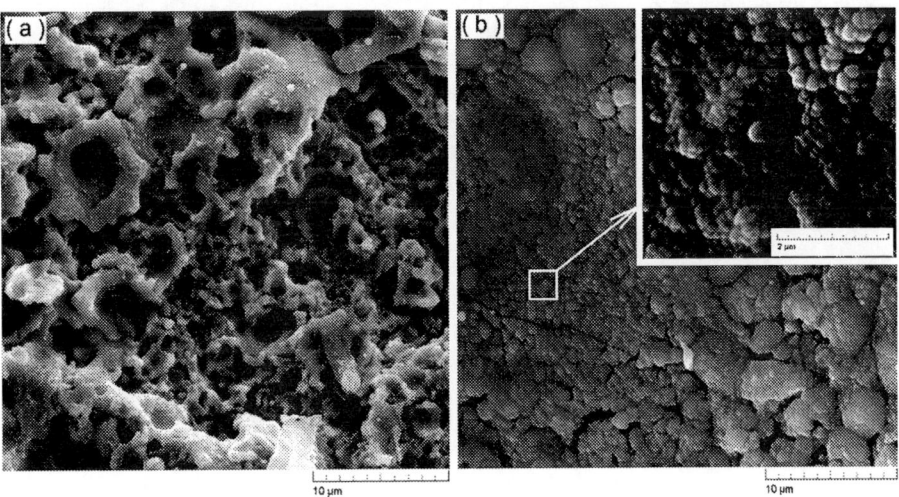

Figure 3(a, b). Morphology of the zirconium coating after ESA at 1.5 A (DC mode) **(a)**; after ESA at 1.75 A (AC mode) **(b)**.

4. Conclusion

The surface morphology of titanium samples after ESA with zirconium was characterized by the formation of the coatings of Ti-Zr-O system. These functional coatings were formed of micro-sized drops (not exceeding 30–50 µm). The most pronounced structure was formed at the operating current of 1.5 A, the concentration of zirconium was about 16.21 at.% and oxygen – at least 5.16 at.%.

The enhanced transfer of the electrode metal occurred when the average operating current of 1.75 A (AC mode) was applied. The resulting coating consisted of zirconium – 39.7 at.%, oxygen and titanium – 21.93 at.%.

Acknowledgments

The research was supported by the Russian Science Foundation (project No. 17-79-10106).

References

[1] Altuna P, Lucas-Taulé E, Gargallo-Albiol J, Figueras-Álvarez O, Hernández-Alfaro F and Nart J 2016 *Int. J. Oral. Maxillofac. Surg.* **45** 842
[2] Nochovnaya N A, Panin P V, Alekseev E B and Bokov K A 2017 *Met. Sci. Heat. Treat.* **58** 520
[3] Shi J, Yang J, Li Z, Zhu L, Li L and Wang X 2017 *J. Alloys. Compd.* **728** 1043
[4] Kalita V I, Komlev D I and Radyuk A A 2016 *Inorg. Mater. Appl. Res.* **7** 536
[5] Fomin A, Fomina M, Koshuro V, Rodionov I, Zakharevich A and Skaptsov A 2017 *Ceram. Int.* **43** 11197
[6] Fomin A, Dorozhkin S, Fomina M, Koshuro V, Rodionov I, Zakharevich A, Petrova N and Skaptsov A 2016 *Ceram. Int.* **42** 10838
[7] Fomin A A, Steinhauer A B, Rodionov I V, Petrova N V, Zakharevich A M, Skaptsov A A and Gribov A N 2013 *Biomed. Eng.* **47(3)** 138
[8] Fomin A A, Steinhauer A B, Rodionov I V, Fomina M A, Zakharevich A M, Skaptsov A A, Gribov A N and Karsakova Ya D 2014 *J. Frict. Wear.* **35(1)** 32
[9] Fomin A A, Fomina M A, Rodionov I V, Koshuro V A, Poshivalova E Yu, Shchelkunov A Yu, Skaptsov A A, Zakharevich A M and Atkin V S 2015 *Tech. Phys. Lett.* **41(9)** 909
[10] Fomin A A, Fomina M A, Koshuro V A, Rodionov I V, Voiko A V, Zakharevich A M, Aman A, Oseev A, Hirsch S and Majcherek S 2016 *Tech. Phys. Lett.* **42** 932
[11] Koshuro V A and Fomin A A 2014 *Proc. 10th Int. Vacuum Electron Sources Conf. and 2nd Int. Conf. on Emission Electronics (Saint-Petersburg) (Saint-Petersburg: IEEE)* p 145
[12] Koshuro V, Fomin A, Fomina M, Rodionov I, Brzhozovskii B, Martynov V, Zakharevich A, Aman A, Oseev A, Majcherek S and Hirsch S 2016 *J. Phys. Conf. Ser.* **741(1)** 012197

SPBOPEN 2018

Investigation of deposition conditions on the structural properties of μc-Si:H

A V Uvarov[1], A S Gudovskikh[1,2], V V Fedorov[1]

[1]St. Petersburg Academic University, St. Petersburg 194021, Russia
[2]St. Petersburg Electrotechnical University "LETI", St. Petersburg 197376, Russia

Abstract. This article is concerned with Raman study of μc-Si:H obtained by PECVD under different deposition conditions. The analysis of crystal fracture and bonded hydrogen content in the layer was carried out. It is shown that as the amorphous phase increases, the content of Si-H bonds prevails. Optimal conditions for deposition of microcrystalline silicon with a low content of hydrogen bonds were defined.

1. Introduction

Today microcrystalline silicon finds application in the field of solar cells, thin-film transistors and optical sensors. One of the promising ways of using microcrystalline silicon is a creation of semiconductor superlattices with quantum-size effect of ultrathin films [1]. However, when such structures are fabricated by plasma-chemical deposition (PECVD), the specific features appears concerning the structure of microcrystalline silicon. High degrees of dilution of silane (SiH_4) in hydrogen are used to obtain microcrystalline silicon, instead of amorphous silicon. This leads to the fact that the concentration of atomic and molecular hydrogen in the resulting layer is quite high. During the formation of subsequent layers the weakly bound hydrogen under the influence of temperature forms a molecule and comes out from the layer. In the formation of multilayer structures, this process of effusion can lead to delamination and destruction of the deposited layers and the formation of cavities in a multilayer structure [2]. The actual problem is production of μc-Si:H with a low content of excess hydrogen in the layer. Thus the estimation of the influence of the deposition conditions on the structural properties and concentration of hydrogen in the μc-Si:H layer is an important issue.

2. Experimental

The layers of microcrystalline silicon were obtained by plasma-enhanced chemical vapor deposition at the temperature of 380°C on fused silica substrates. During deposition process the dilution of silane in hydrogen varied in the range of 1% to 4% and the power of capacitively-coupled 13.56MHz RF plasma varied from 20W (44 mW/cm^2) to 60W (132 mW/cm^2) respectively. The structural properties and concentration of bound hydrogen in the μc-Si:H layers were studied by Raman spectroscopy in the range of 400 – 550 cm^{-1}. The radiation intensity of the Raman spectrometer at a wavelength of 532 nm was adjusted in such a way that there was no local heating of the silicon layer. Maximum intensity of the latitude optical mode (LO) of crystalline silicon in the Raman spectrum corresponds to 520 cm^{-1}, whereas for amorphous silicon it is 480 cm^{-1}. The crystallinity fraction (X_c) of the obtained layers was estimated from the formula in accordance with [3],

$$X_c = \frac{I_{520}}{I_{520}+I_{480}} \tag{1}$$

Content from this work may be used under the terms of the Creative Commons Attribution 3.0 licence. Any further distribution of this work must maintain attribution to the author(s) and the title of the work, journal citation and DOI.
Published under licence by IOP Publishing Ltd

where I_{480} is the Raman intensity of the maximum at a wavenumber of 480 cm^{-1}, I_{520} is the intensity of the LO peak at 520 cm^{-1}. The intensity of each maximum was obtained by deconvolution of the original spectrum. It should be noted that the LO peak for microcrystalline silicon could be shifted to the lower wavenumbers relative to the value for crystalline silicon at 520 cm^{-1} due to decreasing of crystallite size lower 100Å [4]. The presence of bonded hydrogen in microcrystalline silicon was indicated by peaks in the Raman spectrum at 2000 cm^{-1} and 2100 cm^{-1} corresponding to the resonance frequency of Si-H and Si-H$_2$ bonds [5]. Content of n(Si-H) and n(Si-H$_2$) bonds in the layer can be roughly estimated using formulas:

$$n(Si - H) = \frac{I_{2000}}{I_{520}+I_{480}} \qquad (2)$$

$$n(Si - H_2) = \frac{I_{2100}}{I_{520}+I_{480}} \qquad (3)$$

where I_{2000} is the Raman intensity of the maximum at a wavenumber of 2000 cm^{-1}, I_{2100} is the intensity of the maximum at 2100 cm^{-1}.

3. Results

Silicon deposition processes were performed with variations of the pressure, plasma power and the degree of SiH$_4$ dilution in hydrogen. Deposition time in all processes was set to 20 min. Resulting Raman intensity strictly depends from focusing of incident laser which was manually adjusted and should not be compared from sample to sample. But the relative intensity of the maxima on each spectrum can be considered as a reliable indicator of the material properties. For clarity, the intensity of the Raman spectrum in the range of 1900 - 2200 on all graphs was increased by a factor of 10.

3.1 Deposition pressure

The variation in value of deposition pressure was carried out at a constant power of 60 W and a dilution degree of 2%. Raman spectra (figure 1) demonstrate that at deposition pressure above 500 mTorr the saturation of crystalline fraction appears, which is indicated by a more pronounced peak at 520 cm^{-1}.

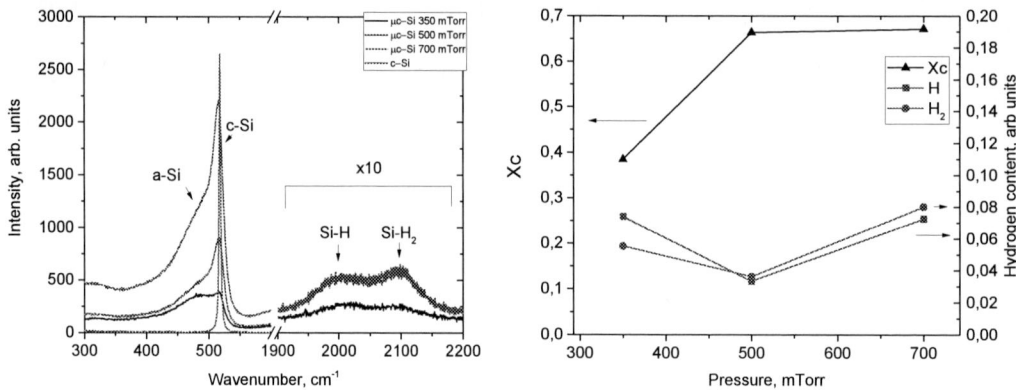

Figure 1. Raman spectra of microcrystalline silicon obtained at various pressures (a). Hydrogen content and crystal fracture in dependence of deposition pressure (b).

Total concentration of hydrogen slightly varies with increasing pressure and has a minimum at 500 mTorr. The intensity of the maximum at 2100 cm^{-1} is higher compared to that at 2000 cm^{-1} in μc-Si:H layers with higher crystallinity.

3.2 Plasma RF power

Raman study of dependence of the plasma RF power on the structure of layers and hydrogen content was carried out at constant pressure of 700 mTorr and SiH$_4$/H$_2$ dilution of 2% (figure 2).

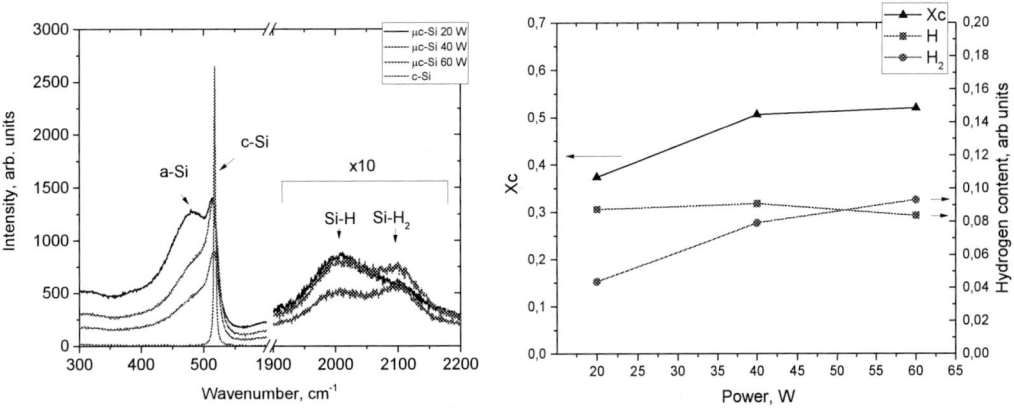

Figure 2. Raman spectra of microcrystalline silicon obtained at various RF power (a). Hydrogen content and crystal fracture in dependence of RF power (b).

It can be said with certainty that the crystal fraction increases with the power of the plasma, which is confirmed by a decrease in intensity of 480 cm^{-1}. Saturation of the influence of power on the crystal fraction occurs at 40 W (88 mW/cm^2). It can also be noted that decrease of amorphous phase fraction indicated by Raman intensity drop at 480 cm^{-1} is followed by rise of Si-H$_2$ bond concentration. The increase of Si-H$_2$ bond concentration with fraction of μc-Si:H is associated with a fact that Si-H$_2$ bonds are mostly located at the μc-Si grain boundaries. .

3.3 SiH$_4$/H$_2$ dilution
The layers were obtained at different SiH$_4$/H$_2$ dilution with constant RF power of 20 W and constant pressure of 700 mTorr (figure 3).

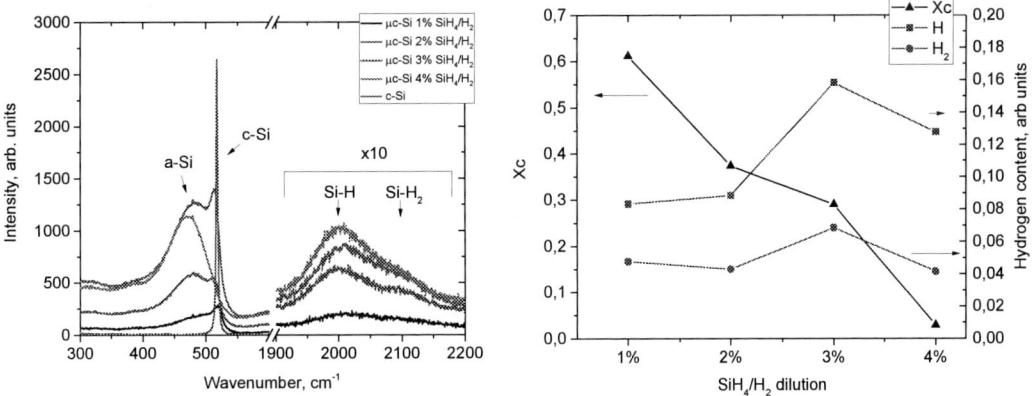

Figure 3. Raman spectra of microcrystalline silicon obtained at various SiH$_4$/H$_2$ dilution (a). Hydrogen content and crystal fracture in dependence of SiH$_4$/H$_2$ dilution (b).

The crystalline fraction strictly depends on the concentration of SiH$_4$ during the deposition process and increases with higher dilution. At SiH$_4$ concentration of 4% or more, the layer is completely amorphous, as indicated by the absence of a maximum at 520 cm^{-1}. Increase of crystalline fraction is clearly followed by decrease of the bonded hydrogen content.

4. Conclusions
We can note a strict dependence of the crystalline fraction of the layer on all three variable parameters. The increase in pressure and deposition power leads to an increase of crystalline fraction. It can be

explained in terms of higher ion energy of Si, and Si-H, Si-H$_2$ radicals for increased specific power of RF plasma discharge. By varying the SiH$_4$ concentration from 1% to 4%, the structure of the layer changes from microcrystalline to completely amorphous followed by increase of the hydrogen content. The increase in pressure does not give an explicit dependence of the hydrogen content in the layer; however, we can note a higher content of Si-H$_2$ bonds at higher pressures. Thus, the optimum parameters for the production of microcrystalline silicon by PECVD with a low content of bonded hydrogen are increase values of pressure and RF power with a lower concentration of SiH$_4$ in hydrogen.

Acknowledgements
This work was supported by the Russian Science Foundation (№ 17-19-01482)

References
[1] K. Jarolimek, R. A. de Groot, G. A. de Wijs, and M. Zeman 2014 Phys. Rev. B **90** 125430
[2] A. S. Gudovskikh, A. V. Uvarov, I. A. Morozov, A. I. Baranov, D. A. Kudryashov, K. S. Zelentsov, A. S. Bukatin, and K. P. Kotlyar 2018 J. Vac. Sci. Technol. A **36** 02D408
[3] A.Voutsas, M.Hatalis, J. Boyce, and A. Chiang 1995 J. Appl. Phys. **78** 6999–7006.
[4] Z. Iqbal and S. Vepiek 1982 J. Phys. C: Solid State Phys. **15** 377-392.
[5] V. A. Volodin, and D. I. Koshelev 2013 J. Raman Spectrosc. **44** 1760-1764

Structure of fluorographene and its polymorphous varieties

M E Belenkov[1], V M Chernov[1] and E A Belenkov[1]

[1]Department of Physics, Chelyabinsk State University, Chelyabinsk 454001, Russia

Abstract. X-ray powder diffraction analysis was used to study structure of fluorographene. In order to interpret experimental results, calculations of crystal structure and electronic properties of fluorographene polymorphs were performed by the method of density functional theory with generalized gradient approximation. Sublimation energies of polymorphic varieties of fluorographene layers differ insignificantly (range from 14.08 to 14.32 eV/CF). Electronic band gaps of fluorographene layers vary from 3.04 to 4.19 eV. It was established that the structure of experimentally synthesized fluorographene corresponds to first type of polymorphic structure.

1. Introduction

Fluorographene is obtained as result of chemical adsorption of fluorine on the surface of graphene layers. First experimentally synthesized fluorographene was obtained in 2010 [1-3]. The fluorographene layers are composed of carbon atoms in the sp^3 hybridization state, each of which forms covalent bonds with three adjacent carbon atoms and one fluorine atom. Fluorographene is a promising material for electronics, since it has semiconductor properties, unlike graphene, which exhibits metallic properties [4-6]. In addition, fluorographene has several advantages over graphane, oxidized graphene and chlorographene. These compounds are obtained by chemical adsorption of hydrogen, oxygen and chlorine atoms on the graphene layers, respectively [7-9]. Graphane is destroyed at the temperature just above room temperature [7,10]. Oxidation of graphene layers occurs unevenly, which causes a significant heterogeneity of properties of the oxidized graphene layers [11]. Chlorographene layers can be sources of toxic gaseous compounds. Fluorographene does not have these drawbacks which peculiar for graphane, oxidized graphene and chlorographene. It is possible to produce graphene layers from fluorographene [3]. A theoretical analysis performed earlier for graphane layers showed that there are different ways of attaching non-carbon atoms to graphene layers, so existence of the five basic structural variants of graphane is possible [12, 13]. Obviously, that in the case of fluorographene, the existence of similar polymorphic varieties is possible. In this paper, structure of the experimentally synthesized fluorographene was studied and theoretical calculations of geometrically optimized structure and electronic properties of five structural fluorographene varieties were performed.

2. Methods

Fluorographene was synthesized as a result of interaction between highly ordered pyrolytic graphite with carbon content percentage of 99.4% having particle size ~ 10 μm, with fluorine. The reaction was carried out for 48 hours at temperature 600 °C and fluorine pressure 1 atm. X-ray powder diffraction analysis of fluorographene was performed by X-ray diffractometer using Cu kα X-rays.

Theoretical analysis shows that there are only five different ways of attaching fluorine atoms on both sides of the graphene layer, so that the structural positions of all carbon atoms are equivalent. Number of other structural varieties of fluorographene layers with nonequivalent structural positions

---

Content from this work may be used under the terms of the Creative Commons Attribution 3.0 licence. Any further distribution of this work must maintain attribution to the author(s) and the title of the work, journal citation and DOI.

Published under licence by IOP Publishing Ltd

of carbon atoms can be much larger, but the most thermodynamically stable layers should be primarily those with equivalent positions of atoms, therefore only layers with equivalent structural positions of atoms were considered further. Basic structural types of fluorographene layers are similar to the five basic structural types of graphane described in [12,13]. The first of these types has a structure similar to the structure of graphene-hydrogen layers described in works [7,14]. Three structural types - type 2, type 3 and type 4 were considered earlier for graphane in work [15]. The possibility of the existence of the fifth basic type of graphane was first studied in article [12].

Fluorographene layers of five basic types, which were theoretically constructed from graphene layers by the addition of fluorine atoms to them, were then geometrically optimized by the methods of the density functional theory (DFT) [16] with the generalized gradient approximation (GGA) [17]. Calculations of electronic properties and geometrically optimized structures were performed using the Quantum ESPRESSO software package [18]. Calculations were performed for bulk structures, namely for stacks of fluorographene layers with interplanar distance of 10 Å. This distance ensures the absence of influence of adjacent layers on each other, so calculated structure of the layers and their properties correspond to values distinctive for isolated layers. Densities of electronic states were calculated using k-point mesh: $12 \times 12 \times 12$. The sublimation energy (E_{sub}) of the fluorographene was calculated as the difference between the total specific energy of the fluorographene layer per the CF molecular group and the energies of the isolated fluorine and carbon atoms.

3. Results and Discussion

Five highly broadened diffraction maxima (broadening ~ 2°) are observed on the X-ray powder diffraction patterns of fluorographene (Figure 1). This indicates the small size of the coherent scattering regions (Table 1) in the sample and makes it difficult to identify the crystal structure. Theoretical calculations of the geometrically optimized structure were performed for the five main structural varieties of fluorographene in order to interpret the X-ray powder diffraction data.

Images of structure of the fluorographene layers and their elementary cells geometrically optimized by the DFT-GGA method are shown in Figure 2. Calculated values of the structural parameters of fluorographene layers are given in Table 2. The crystal lattice of a layer of the first structural type refers to hexagonal system (Hex), the unit cell contains two fluorine atoms and two carbon atoms. The remaining structural fluorographene forms have crystal lattices corresponding to the orthorhombic system (Ort), their elementary cells contain 8 or 16 atoms (Table 2, Figure 2). The layer density of fluorographene layers varies in the range from 1.557 to 1.821 mg / m2, which is approximately twice the density of graphene layers 0.74 mg / m2 (Table 2).

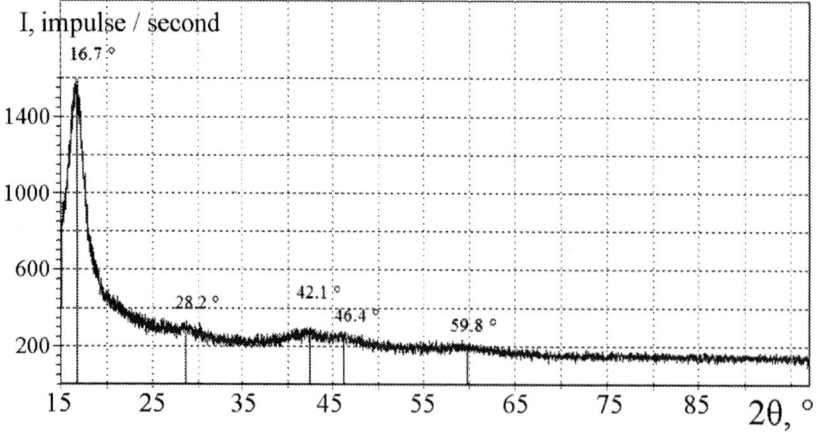

Figure 1. Powder X-ray diffraction patterns of fluorographene.

Table 1. Structural parameters of fluorographene according to the results of X-ray diffraction analysis (d - interplanar spacing, L - average size of coherent scattering regions, θ - diffraction angle, β - integral width of diffraction maximum).

№	2θ, °	θ, °	d, Å	β, °	β_f, °	L, Å
1	16.70	8.35	5.30	1.90	1.70	52
2	28.20	14.10	3.16	2.00	1.80	51
3	42.10	21.05	2.14	2.10	1.90	50
4	46.40	23.20	1.95	2.10	1.90	51
5	59.80	29.90	1.54	2.20	2.00	51

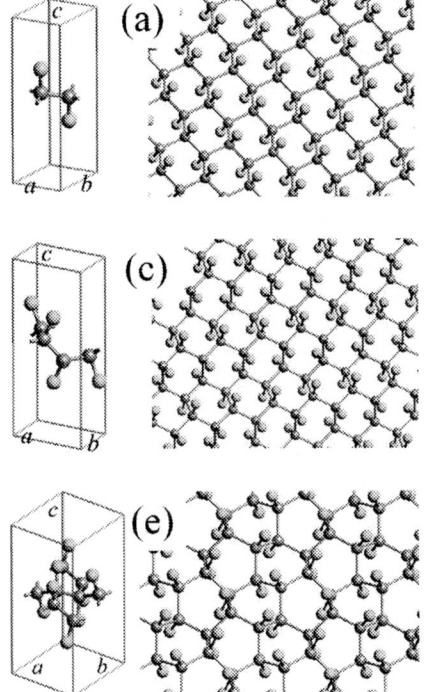

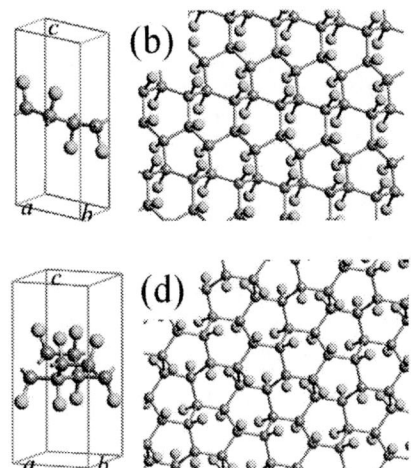

Figure 2 (a, b, c, d, e). Unit cells and geometrically optimized structure of polymorphic varieties of fluorographene layers: (a) type 1; (b) type 2; (c) type 3; (a) type 4; (a) type 5.

Table 2. The structural parameters and properties of the main types of fluorographene layers (T1–T5), as well as hexagonal (L_6) graphene (a, b - lengths of vectors of elementary translations, E_{sub} - sublimation energy, Δ - band gaps, ρ - layer density, N - number of atoms per unit cell).

Type of layer	T1	T2	T3	T4	T5	L_6
Crystal system	Hex	Ort	Ort	Ort	Ort	Hex
a, Å	2.602	2.581	2.607	4.877	5.056	2.491
b, Å		4.570	4.214	4.575	4.620	
γ, °	120	90	90	90	90	120
E_{total}, eV/u.c.	-1601.27	-3202.04	-3202.53	-6403.18	-6404.19	-314.64
E_{total}, eV/(CH)	-800.64	-800.51	-800.63	-800.40	-800.52	-
E_{sub}, eV	14.32	14.19	14.31	14.08	14.20	-
Δ, eV	3.321	3.390	3.557	4.195	3.044	0.00
ρ, mg/m^2	1.706	1.696	1.821	1.793	1.557	0.74
N, at.	4	8	8	16	16	2

In addition, band structure of the fluorographene layers, density of electronic states, and the sublimation energies were calculated. Figures 3 and 4 show these results for the five fluorographene polymorphs. Sublimation energies for different polytypes differ insignificantly: minimal energy is equal to 14.08 eV/CF (type 4), a maximal energy - 14.32 eV/CF (type 1). This indicates the possibility of a stable existence of any of the major structural varieties of fluorographene. Sublimation energies of the fluorographene layers are 25% higher than the Esub layers of graphane [12,13], so fluorographene should indeed be more resistant to thermal effects, as observed experimentally [2,7,10]. Band gaps at the Fermi energy level for the polytypes are in the range from 3.044 to 4.195 eV. This indicates that all polymorphic varieties of fluorographene layers should belong to wide-band semiconductors. Comparison of calculated values of structural parameters and X-ray powder diffraction data showed that the best fit is observed for fluorographene polymorph of the type 1.

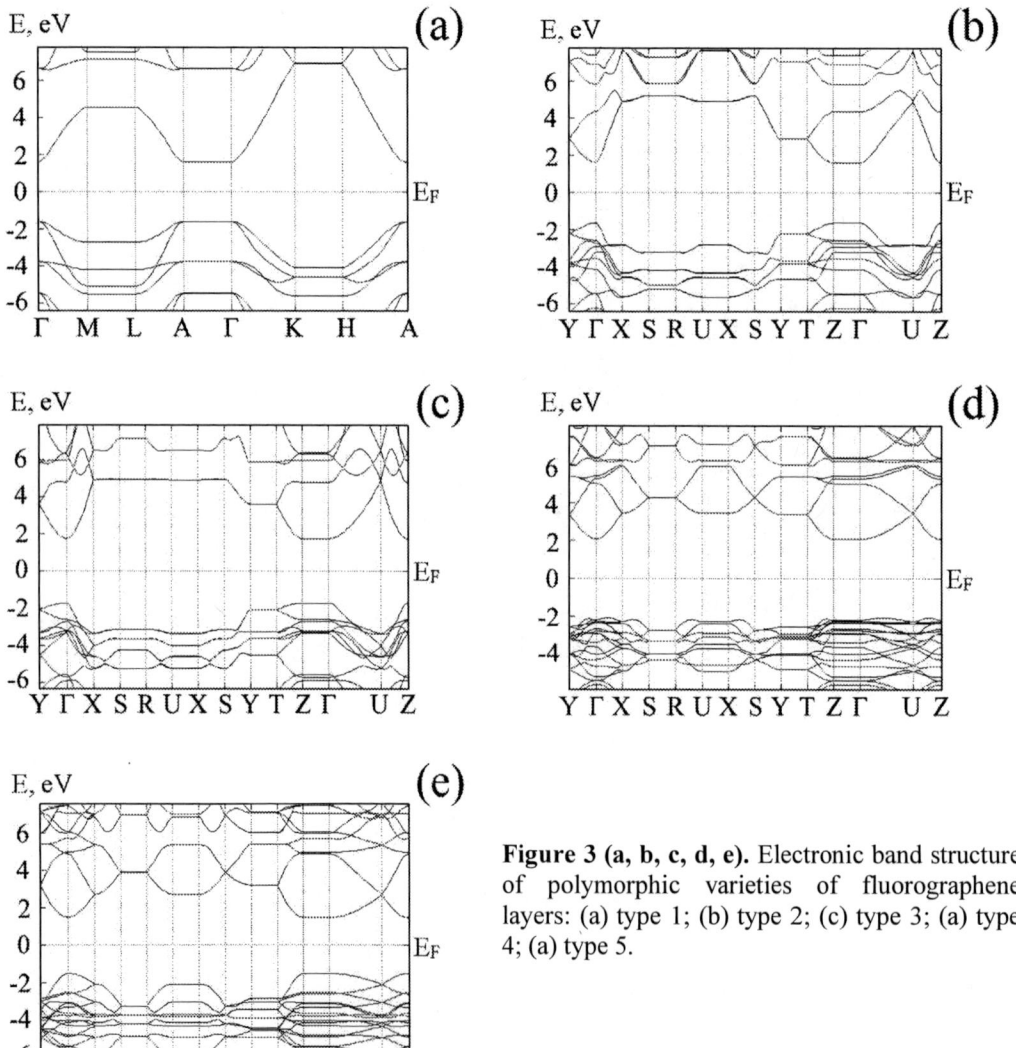

Figure 3 (a, b, c, d, e). Electronic band structure of polymorphic varieties of fluorographene layers: (a) type 1; (b) type 2; (c) type 3; (a) type 4; (a) type 5.

Fluorographene can find wide practical use in electronics, as a basis for designing nanoelectronic devices. Nanoelectronic devices can be obtained by local carbonization of semiconductor fluorographene layers and as a result of that sections of the layers where fluorine atoms were removed will become fragments of the graphene layer and their properties will be metallic. Thus, it is possible to obtain nanoelectronic devices based on fluorographene layers of any of the five basic structural types, and the properties of the resulting structures will differ significantly due to differences in the electronic properties of the original polymorphs of fluorographene. Therefore some structural varieties of fluorographene layers may be more preferable than others for specific practical applications.

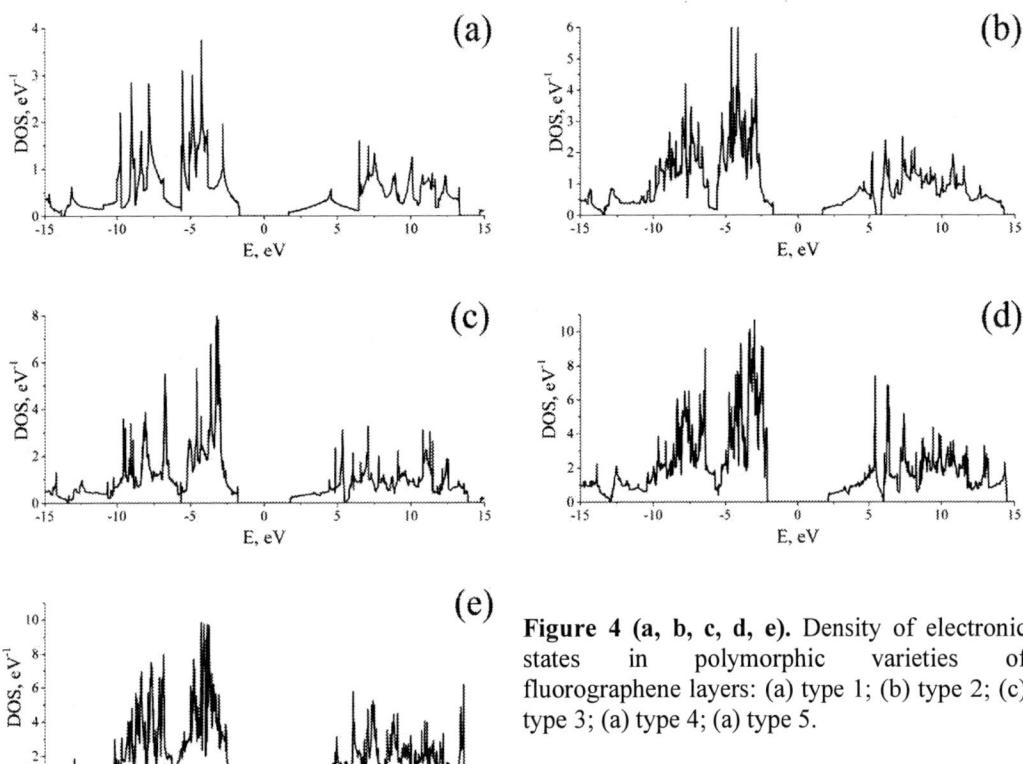

Figure 4 (a, b, c, d, e). Density of electronic states in polymorphic varieties of fluorographene layers: (a) type 1; (b) type 2; (c) type 3; (a) type 4; (a) type 5.

References
[1]   Robinson J T, Burgess J S, Junkermeier C E, Badescu S C, Reinecke T L, Perkins F K, Zalalutdniov M K, Baldwin J W, Culbertson J C, Sheehan P E, Snow E S 2010 *Nano Lett.* **10** 3001
[2]   Nair R R, Ren W Jalil, R, Riaz I, Kravets V G, Britnell L, Blake P, Schedin F, Mayorov A S, Yuan S, Katsnelson M I, Cheng H M, Strupinski W, Bulusheva L G, Okotrub A V, Grigorieva I V, Grigorenko A N, Novoselov K S, Geim A K 2010 *Small* **6** 2877
[3]   Zboril R, Karlicky F, Bourlinos A B, Steriotis T A, Stubos A K, Georgakilas V, Safarova K, Jancík D, Trapalis C, Otyepka M 2010 *Small* **6** 2885
[4]   Novoselov K S, Geim A K, Morozov S V, Jiang D, Zhang Y, Dubonos S V, Grigorieva I V, Firsov A A 2004 *Science* **306** 666
[5]   Novoselov K S 2011 *Phys. Usp.* **54** 1299
[6]   Belenkov E A, Kochengin A E 2015 *Physics of the Solid State* **57** 2126
[7]   Elias D C, Nair R R, Mohiuddin T M G, Morozov S V, Blake P, Halsall M P, Ferrari A C,

Boukhvalov D W, Katsnelson M I, Geim A K, Novoselov K S 2009 *Science.* **323** 610

[8] Chen D, Feng H, Li J 2012 *Chem. Rev.* **112** 6027

[9] Li B, Zhou L, Wu D, Peng W, Yan K, Zhou Y, Liu Z 2011 *ACS Nano* **5** 5957

[10] Openov L A, Podlivaev A I 2010 *Technical Phys. Lett.* **36** 31

[11] Sorokin P B, Chernozatonskii L A 2013 *Phys. Usp.* **56** 105

[12] Belenkova T E, Chernov V M, Belenkov E A 2016 *Radioelectronics. Nanosystems. Information technologies* **8** 49

[13] Belenkova T E, Greshnyakov V A, Chernov V M, Belenkov E A 2017 *Journal of Physics: Conference Series* **917** 032015

[14] Sofo J O, Chaudhari A S, Barber G D 2007 *Phys. Rev. B.* **75** 153401

[15] Wen X D, Hand L, Labet V, Yang T, Hoffmann R, Ashcroft N, Oganov A R, Lyakhov A O 2011 *PNAS* **108** 6833

[16] Koch W A, Holthausen M C 2001 *Chemist's guide to density functional theory. 2nd edition.* (Wiley-VCH Verlag GmbH) p 293

[17] Perdew J P, Chevary J A, Vosko S H, Jackson K A, Pederson M R, Singh D J, Fiolhais C 1992 *Phys. Rev. B* **46** 6671

[18] Giannozzi P, Baroni S, Bonini N, Calandra M, Car R, Cavazzoni C, Ceresoli D, Chiarotti G L, Cococcioni M, Dabo I, Corso A D, Gironcoli S, Fabris S, Fratesi G, Gebauer R, Gerstmann U, Gougoussis C, Kokalj A, Lazzeri M, Martin-Samos L, Marzari N, Mauri F, Mazzarello R, Paolini S, Pasquarello A, Paulatto L, Sbraccia C, Scandolo S, Sclauzero G, Seitsonen A P, Smogunov A, Umari P, Wentzcovitch R M 2009 *J. Phys.: Condens. Matter* **21** 395502

SPBOPEN 2018 IOP Publishing

IOP Conf. Series: Journal of Physics: Conf. Series **1124** (2018) 022011 doi:10.1088/1742-6596/1124/2/022011

Carbon materials formed by polymerization of C_{20} and C_{24} fullerites

M I Tingaev[1], E A Belenkov[1]

[1]Department of Physics, Chelyabinsk State University, Chelyabinsk 454001, Russia

Abstract. In this work the calculations of new hybrid sp^2+sp^3 carbon phases using density-functional theory in generalized gradient approximation were performed. The structure of phases was formed from C_{20} and C_{24} fullerites during their polymerization. As a result of calculations, it was found two new stable hybrid carbon phases, consist from sp^2+sp^3 hybridized carbon atoms. This C_{20} and C_{24} phases have band gap 0 and 1.76 eV, sublimation energy 6.92 end 6.89 eV/atom corresponded.

1. Introduction

Fullerites are molecular crystals formed from fullerenes [1]. In these crystals, fullerenes are connected by van der Waals bonds, so the mechanical properties of fullerites are low, which does not allow to use them as constructional materials. However, under high pressures effect on fullerites, fullerene molecules polymerize and new carbon materials with hard structure and high mechanical characteristics are formed [2,3]. The structure of such materials depends on type of fullerene which was taken as precursor, as well as the pressure that determines the polymerization degree. At pressures $\sim$ 1-8 GPa (T $\sim$ 150-900°C), one-dimensional and two-dimensional polymerized fullerite phases are formed [2,4]. At higher pressures, three-dimensional polymerized sp^2+sp^3 phases are formed in which the proportion of sp^3 hybridized atoms increases with increasing pressure [3,5]. In this work, we performed theoretical calculations of new carbon materials formed by polymerization of C_{20} and C_{24} fullerites.

2. Methods

Structure of sp^2+sp^3 hybrid carbon phases can be obtained by partial cross-linking of fullerenes, nanotubes or graphene layers by the method described in [6,7]. In this work the structure of new hybrid carbon phase were obtained by partial cross-linking of C_{20} and C_{24} fullerites. In structure fullerites, the fullerenes were in 6 coordinated states. Partial cross-linking means that some of sp^2 atoms form an additional covalent bond. Atoms that did not form an additional covalent bond are sp^2 hybridized, and the other ones are atoms with sp^3 atom hybridization. As a result, 6 carbon atoms in each fullerene transformed from the sp^2 state to the sp^3 hybridized state (figure 1). Thus, the ratio sp^2/sp^3 of atoms in the phase obtained from C_{20} fullerite was 7:3, and in the phase from C_{24} fullerite - 1:3. Partially cross-linked structures were geometrically optimized by using density functional theory in the generalized gradient approximation. The calculations used k-point mesh: $12\times12\times12$. To limit the dimension of the set of basic functions, the cutoff energy value was set to 950 eV. Thus, we calculated the final structures of phases, their electronic band structures and the density of electronic states.

Content from this work may be used under the terms of the Creative Commons Attribution 3.0 licence. Any further distribution of this work must maintain attribution to the author(s) and the title of the work, journal citation and DOI.
Published under licence by IOP Publishing Ltd

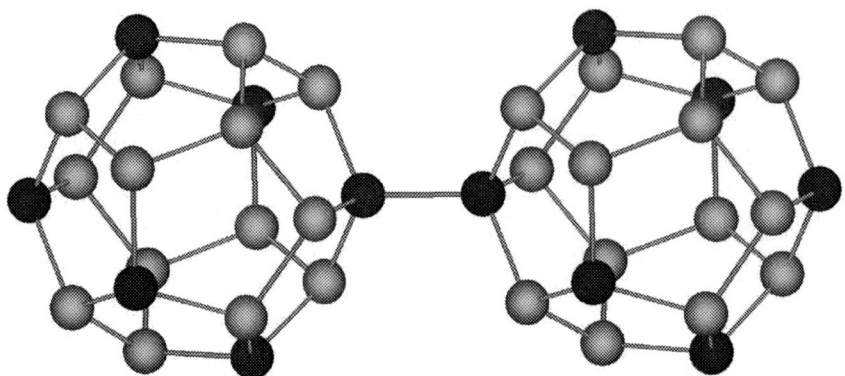

Figure 1. Illustration of partial polymerization of neighboring fullerenes C_{20} in the formation of a hybrid phase (dark atoms are sp^3 hybridized carbon atoms, gray atoms are in the state of sp^2 hybridization).

3. Results

Unit cells and the crystal structure of sp^2+sp^3 phases after geometric optimization are shown in figure 2. The calculated values of the lengths of interatomic bonds vary in the range from 1.378 to 1.618 Å. Different lengths of covalent bonds due to the fact that they are formed by different number of electronic pairs. There are three types of interatomic bonds in structure of phases: bonds between sp^2 and sp^2 atoms, bonds between sp^3 and sp^2 atoms, bonds between sp^3 and sp^3 atoms. The minimal length of links between sp^2 hybridized atoms. The longest value of bonds is observed between sp^3 hybridized atoms. The lengths of bonds in the hybrid phases differ from the values this characteristic of graphite and diamond.

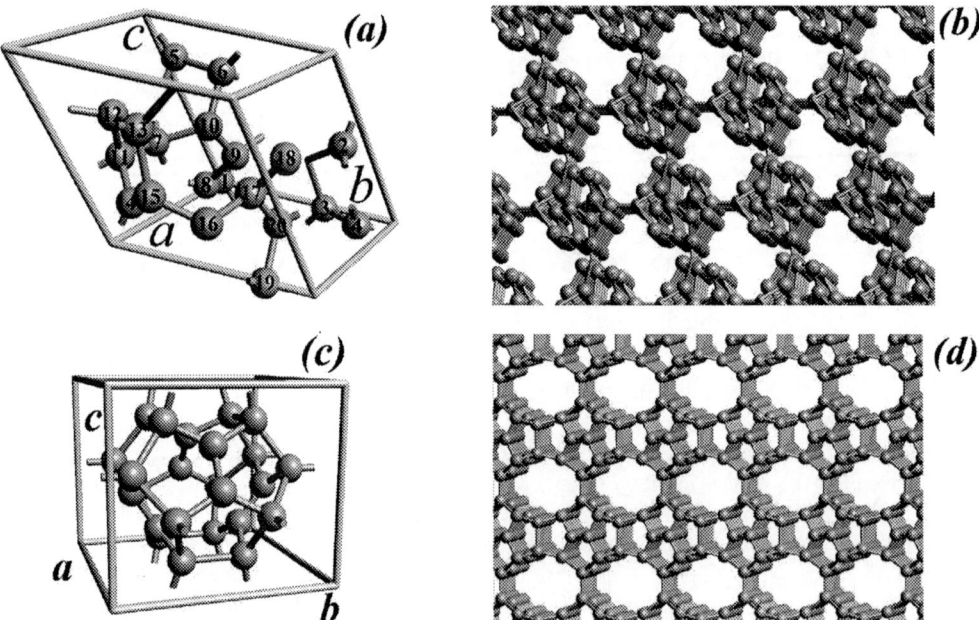

Figure 2 (a, b, c, d). The unit cells (a,c) and the crystal structures (b,d) of sp^2+sp^3 carbon phases obtained by polymerization of: C_{20} fullerite (a,b) and C_{24} fullerite (c,d).

The angles between the bonds in the hybrid phases differ from the characteristic values in structures of cubic diamond and graphite. Thus, by four-coordinated atoms, the angles between bonds vary in the range from 90° and 128.19°, which significantly differs from the angle 109.47° characteristic of the diamond. The angles between bonds by three-coordinated atoms vary from 108.42° to 123.24°, which also differs from the angle of 120° typical for graphite. Such deviations of the angles between interatomic bonds indicate the presence of significant stresses in the structure of the hybrid phase.

The crystal lattice of the phase obtained from C_{20} fullerite refers to the triclinic crystal system, and the phase formed from C_{24} fullerite to orthorhombic crystal system. The unit cells of these phases contain 20 and 24 atoms, respectively. The density of the hybrid phase obtained from C_{20} fullerite is 2.844 g/cm^3, for the second phase it is 2.459 g/cm^3. Densities of these phases are smaller than the density of cubic diamond, but higher than the density of graphite.

Table 1. Parameters of unit cells and structure characteristics of hybrid sp^2+sp^3 carbon phases, graphite (G) and cubic diamond (D) (E_{tot} - total energy, ΔE_{tot} - total difference energy, Esub - sublimation energy, ρ - density, and ΔE_g - band gap, $C_{(20)}a$ and $C_{(24)}a$ - hybrid sp^2+sp^3 carbon phases obtained by polymerization C_{20} fullerite and C_{24} fullerite, $T_{(8,0)}a_{A4c}$ and $T_{(8,0)}a_{A4d}$ - hybrid sp^2+sp^3 carbon phases obtained by polymerization bundles of (8,0) nanotubes).

Phase	$C_{(20)}a$	$C_{(24)}a$	$T_{(8,0)}a_{A4c}$ [7]	$T_{(8,0)}a_{A4d}$ [7]	G	D
a (Å)	5.999	6.13	8.950	6.399	2.487	3.596
b (Å)	5.840	6.108	4.380	4.253	2.487	3.596
c (Å)	5.832	5.198	12.114	6.399	7.333	3.596
α (°)	118.18	90	90.05	70.60	90	90
β (°)	69.89	90	46.19	83.72	90	90
γ (°)	110.01	90	104.36	70.59	120	90
N (atom)	20	24	32	16	4	2
sp^2:sp^3	7:3	1:3	1:1	1:1	–	0
E_{tot} (eV)	-3129.54	-3753.49	-5021.7	-2511.9	-629.35	-314.55
E_{tot} (eV/atom)	-156.477	-156. 395	-156.93	-156.99	-157.34	-157.28
ΔE_{tot} (eV/atom)	0.799	0.881	0.41	0.35	0	0.06
E_{sub} (eV/atom)	6.981	6.899	7.37	7.43	7.78	7.72
ΔE_g (eV)	0	1.758	0.126	0.418	0	5.6
ρ (g/cm^3)	2.844	2.459	1.982	2.059	2.2	3.5

It was also found that the sublimation energies of the hybrid carbon phases (6.92 and 6.89 eV) are lower than the sublimation energies for graphite (7.78 eV), diamond (7.73 eV) and hybrid sp^2+sp^3 carbon phases obtained by polymerization bundles of (8,0) nanotubes [7].

The band gap (ΔE_g) at the Fermi energy level for the $C_{24}a$ phase is 1.76 eV, ΔE_g for the $C_{20}a$ phase equals 0 eV (figures 3,4). Thus, the hybrid phase obtained from C_{24} fullerite should be a semiconductor, and $C_{20}a$ phase should have metallic properties. The hybrid sp^2+sp^3 phases formed from fullerenes and nanotubes have different ratios of atoms in sp^3 and sp^2 hybridization states. Formally, the more sp^2 atoms in a phase, the closer the properties of such a phase should be to graphite. If there is a high proportion of sp^3 atoms, then the properties should be close to those of the diamond. However, for electronic properties this is not correct - the maximum fraction of sp^3 atoms is observed in the $C_{20}a$ phase, while the width of the band gap is zero, similar as in graphite.

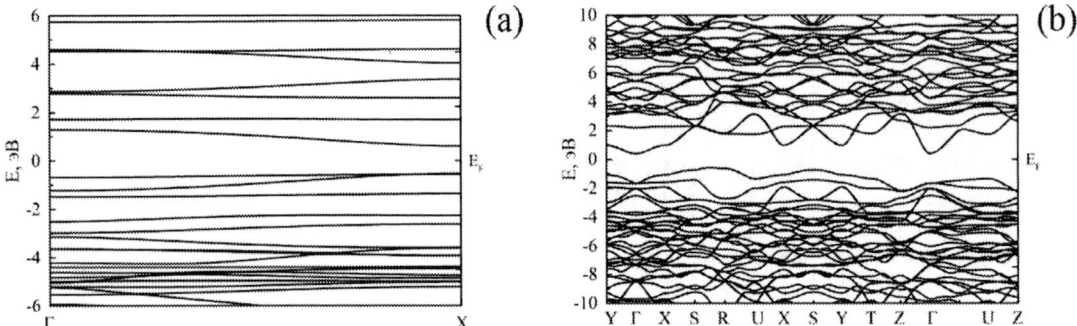

Figure 3. Band structure of sp^2+sp^3 carbon phases obtained by polymerization: C_{20} fullerite (a) and C_{24} fullerite (b).

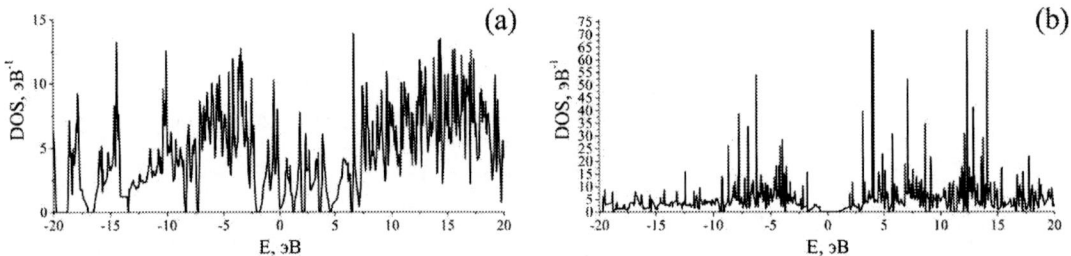

Figure 4. Density of electronic states of sp^2+sp^3 carbon phases obtained by polymerization: C_{20} fullerite (a) and C_{24} fullerite (b).

4. Conclusion

Thus, the structure and properties of two new hybrid sp^2+sp^3 carbon phases, which can be obtained by polymerization of C_{20} and C_{24} fullerites, have been calculated. The value of the sublimation energies of these phases falls within the range typical for carbon materials that stably exist under normal conditions. Hybrid sp^2+sp^3 phases have a three-dimensional polymerized structure and can be used as structural materials. C_{24}a phase has semiconductor properties and it could be used for the development of electronic devices.

References

[1] Duclos S J, Brister K, Haddon R C, Kortan A R, Thiel F A 1991 *Nature* **351** 380
[2] Nunez-Regueiro M, Monceau P, Rassat A, Bernier P, Zahab A 1991 *Nature* **354** 289
[3] Brazhkin V V, Lyapin A G, Antonov Yu V, Popova S V, Klyuev Yu A, Naletov A M, Mel'nik N N 1995 *JETP Lett.* **62** 350
[4] Iwasa Y, Arima T, Fleming R M, Siegrist T, Zhou O, Haddon R C, Rothberg L J, Lyons K B, Carter H L Jr, Hebard A F, Tycko R, Dabbagh G, Krajewski J J, Thomas G A, Yagi T 1994 *Science* **264** 1570
[5] Blank V, Pivovarov G, Lvova N, Gogolinsky K, Reshetov V 1995 *Phys. Lett. A* **205** 208
[6] Belenkov E A, Tingaev M I 2015 *Letters on Materials* **5** 15
[7] Tingaev M I, Belenkov E A 2017 *J. Phys. Conf. Series* **917** 032013

SPBOPEN 2018   IOP Publishing

IOP Conf. Series: Journal of Physics: Conf. Series **1124** (2018) 022012   doi:10.1088/1742-6596/1124/2/022012

Metal oxide (Ta-TaOx)-coatings obtained by magnetron sputtering and heat treatment with high-frequency currents

M Fomina[1], A Voyko[1], A Shumilin[1], V Papshev[1,2], A Zakharevich[3], A Skaptsov[3], A Fomin[1]

[1] Yuri Gagarin State Technical University of Saratov, Saratov 410054, Russia
[2] TAV Dental, Shlomi 2283202, Israel
[3] Saratov State University, Saratov 410012, Russia

Abstract. The paper presents the results of obtaining tantalum films with a thickness of 200–900 nm on titanium using DC magnetron sputtering. As a result, a high-strength "Ti-base – Ta-coating" structure with a hardness of at least 2040±150 HV (20.01±1.47 GPa) was produced. The subsequent modification with high-frequency currents provided an increase in hardness to 3694±443 HV (36.23±4.35 GPa), which was associated with the formation of a thin oxide layer of TaO_x with a tantalum concentration of 74.1–75.2 wt.% and oxygen of about 12.5–13.9 wt.%.

1. Introduction

To improve the functional characteristics of titanium products, including those for medical application, their surface is subjected to a modifying treatment, e.g. plasma spraying and oxidation [1]. Numerous studies showed that some biocompatible metals (titanium, zirconium, tantalum) thermally treated in an oxygen-containing atmosphere had increased hardness and wear resistance [2,3]. It is known that tantalum has the highest biomechanical compatibility among metals and alloys used for medical purposes [4]. This provides its use in the manufacture of individual high-loaded components of implants, as well as functional coatings. In this paper, the possibility of creating a "Ti-base – (Ta-TaO_x)-coating" structure due to magnetron sputtering of tantalum on a titanium base and heat treatment with high-frequency currents (HFC) was studied.

2. Methodology

Experimental samples were made in the form of 2 mm thick cp-titanium "Grade 2" disks (D = 14 mm) with central holes. "Ti-base – Ta-coating" structure was produced by magnetron sputtering (DC-mode). After obtaining the layered structure, the heat treatment with HFC in an oxygen-containing atmosphere was conducted. The temperature varied from 600 to 1200 °C at a high-temperature exposure duration of 30 to 300 s.

The surface morphology of the samples was studied using optical and scanning electron microscopy (SEM) with energy-dispersive X-ray analysis (EDX). To test hardness, the Vickers method (at 10–50 gf) was used.

3. Results

The macrostructure of titanium samples after the tantalum film was formed (through a rectangular mask) was characterized by the presence of a rectangular section with a dark boundary (showed the

---

Content from this work may be used under the terms of the Creative Commons Attribution 3.0 licence. Any further distribution of this work must maintain attribution to the author(s) and the title of the work, journal citation and DOI.

Published under licence by IOP Publishing Ltd

difference in height between the base and the film). After heat treatment, titanium and tantalum had characteristic differences in color. The film of an oxidized TaO_x tantalum treated with HFC at the temperature of 600–800 °C had a dark gray color (Figure 1a). Thermal treatment under these conditions did not lead to a change in the morphology parameters. The increase in the treatment time to $t = 300$ s at a temperature $T = 800\pm15$ °C promoted the formation of microcracks and partial peeling of the metal oxide film.

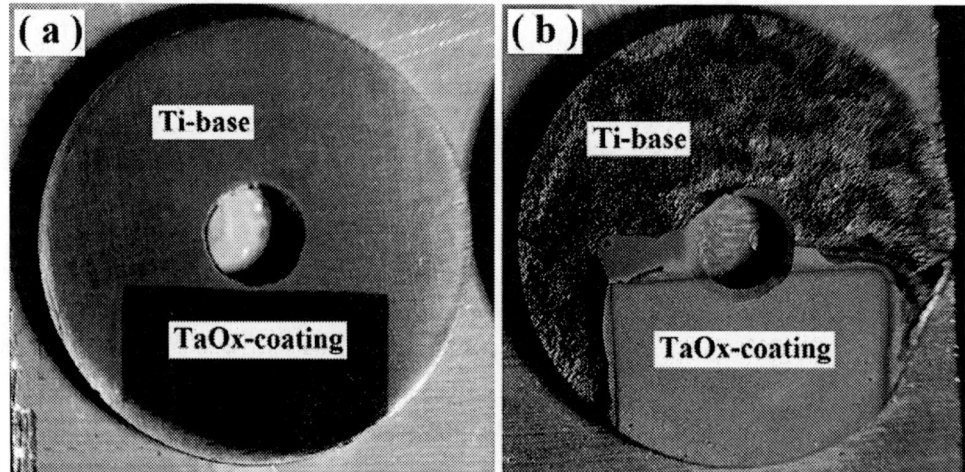

Figure 1(a, b). Titanium disc with TaO_x coating obtained by magnetron sputtering of tantalum (0.6 μm film) and subsequent heat treatment with HFC at the temperature $T = 600$ °C and $t = 30$ s **(a)**; a coated sample produced at the temperature above $T = 1000$ °C **(b)**.

At an increase in the treatment duration from 30 to 300 s and the treatment temperature of 1000–1200 °C, there was a more intense oxidation of a titanium base. The resulting scale of titanium spontaneously separated from the surface, but the oxide layer of TaO_x was retained (Figure 1b). The areas of a titanium sample, where no tantalum film was observed, were intensively oxidized with the formation of a thick oxide coating of rutile [2]. Thus, the tantalum film had a protective role during the high-temperature oxidation.

The concentration of tantalum after magnetron sputtering reached 94.7–97.5 wt.% (Figure 2a). With a film thickness of about 0.8–0.9 μm, a maximum tantalum concentration of about 98.5–99.0 wt.% was achieved. The presence of titanium in the film was associated with the peculiarity of the EDX method, when the pear-shaped area of characteristic X-radiation (at the accelerating voltage of 20 kV) was about 1 μm.

The surface morphology of titanium samples remained practically unchanged after the deposition of a tantalum layer with a thickness of up to 0.8–0.9 μm. The microchannels remained visible after turning and finishing (Figure 2b). The similar character of the morphology was also observed at a low temperature of heat treatment with HFC. The treatment at a temperature of about 600 °C did not lead to any noticeable changes in the morphology, including the nanoscale range. However, with an increase in the temperature to 800 °C and treatment time to 300 s, nanocrystals of tantalum oxide were formed on the microprotrusions. The dimensional parameters of nanograins of tantalum oxide were difficult to establish.

SPBOPEN 2018   IOP Publishing
IOP Conf. Series: Journal of Physics: Conf. Series **1124** (2018) 022012   doi:10.1088/1742-6596/1124/2/022012

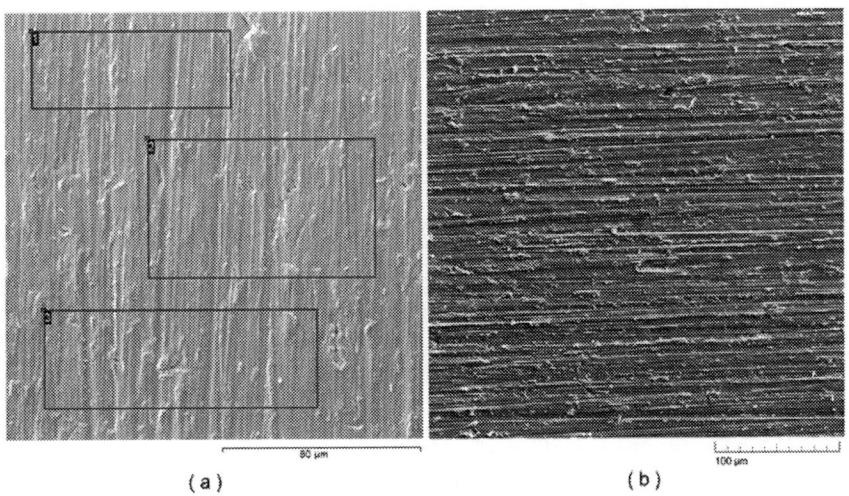

Figure 2(a, b). Morphology of a titanium sample with a TaO_x coating obtained by magnetron sputtering of tantalum (0.6 μm film) and subsequent heat treatment with HFC at the temperature $T = 600$ °C and $t = 30$ s (rectangles show the areas for EDX) **(a)**; a coated sample produced at the temperature above $T = 800$ °C and $t = 30$ s **(b)**.

More distinct grain boundaries were noted at a treatment temperature of about 1000 °C or more. The application of a high-temperature range led to the formation of numerous defects and noticeable cracking (Figure 3a).

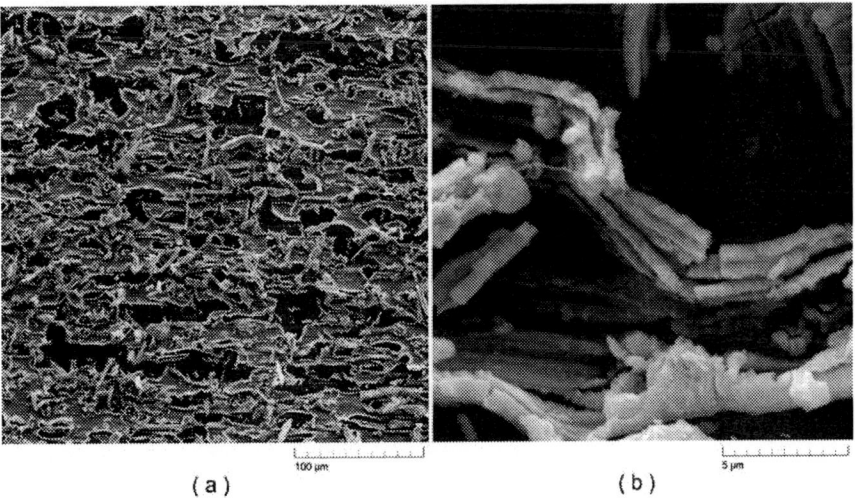

Figure 3(a, b). Morphology of a titanium sample with a TaO_x coating obtained by magnetron sputtering of tantalum and subsequent heat treatment with HFC at the temperature $T = 800$ °C and $t = 300$ s **(a)**; metal oxide coating fragments **(b)**.

55

On the fragments of the oxide film, small crystals were observed (Figure 3b). However, this coating structure can not provide high mechanical characteristics and wear resistance, which is due to low adhesion and cracking.

The concentration of tantalum after oxidation by HFC decreased from 94.7–97.5 wt.% to 74.1-75.2 wt.%, while the oxygen concentration reached 12.5–13.9 wt.%. At an oxygen concentration in the metal oxide coating of not more than 13–14 wt.%, no noticeable cracking and failure occurred. Such coatings had high mechanical characteristics, in particular hardness.

The appearance of the TaO_x oxide layer was accompanied by an increase in Vickers hardness from 2040±150 HV (20.01±1.47 GPa) to 3694±443 HV (36.23±4.35 GPa). In the course of high-temperature treatment, the above-described cracking occurred and the hardness of the coating decreased noticeably to 550–770 HV (5.39–7.55 GPa), and with a prolonged treatment it approached the hardness of a moderately strengthened commercially pure titanium – about 300–400 HV.

4. Conclusion

Thus, after the deposition of tantalum on the titanium surface and further heat treatment with HFC, the "Ti-base – (Ta-TaO_x)-coating" structure was formed. The hardness of metal oxide coatings reached 20–36 GPa, which qualitatively can ensure high wear resistance, provided that the necessary characteristics of the surface morphology are met.

The proposed approach for the modification of the surface of titanium products can ensure an improvement in the quality of medical devices, including implants [2]. The coatings obtained can also find application in the manufacture of electronic components (capacitors), MEMS and sensitive sensors of a wide range of applications [5-10].

Acknowledgments

The research was supported by the Russian Science Foundation (project No. 18-79-10040).

References

[1] Koshuro V A, Nechaev G G and Lyasnikova A V 2014 *Tech. Phys.* **59** 1570
[2] Fomin A, Dorozhkin S, Fomina M, Koshuro V, Rodionov I, Zakharevich A Petrova N and Skaptsov A 2016 *Ceram. Int.* **42** 10838
[3] Koshuro V A, Fomin A, Voyko A, Rodionov I, Zakharevich A, Skaptsov A and Fomin A 2018 *Compos. Struct.* https://doi.org/10.1016/j.compstruct.2018.01.055 (in press)
[4] Rudnev V S, Medkov M A, Lukiyanchuk I V, Steblevskaya N I, Kilin K N and Belobeletskaya M V 2014 *Surf. Coat. Technol.* **258** 1232
[5] Oseev A, Mukhin N, Lucklum R, Zubtsov M, Schmidt M-P, Steinmann U, Fomin A, Kozyrev A and Hirsch S 2018 *Sens. Actuators. B Chem.* **257** 469
[6] Schmidt M-P, Oseev A, Engel C, Brose A, Aman A and Hirsch S 2014 *Procedia Eng.* **87** 88
[7] Aman A, Majcherek S, Schmidt M-P and Hirsch S 2014 *Procedia Eng.* **87** 124
[8] Aman A, Majcherek S, Hirsch S and Schmidt B 2015 *J. Appl. Phys.* **118** 164105
[9] Majcherek S, Aman A and Fochtmann J 2016 *J. Micromech. Microeng.* **26** 025013
[10] Aman S, Aman A and Morgner W 2013 *Compos. Sci. Technol.* **84** 58

Synthesis of bismuth nanowires for thermoelectric applications

A A Lozovenko, A A Poznyak, G G Gorokh

Department of Micro- and Nanoelectronics, Belarusian State University of Informatics and Radioelectronics, Minsk, 220013, Republic of Belarus

Abstract. The processes of synthesis ordered nanowire arrays of bismuth in anodic alumina template ware developed. The sequence of technological operations for the formation of porous templates for the electrochemical deposition of nanowires was described. Optimum electrochemical conditions of reproducible synthesis uniform nanowire arrays were determined. The microstructure and composition of formed structures was studied. The developed techniques are effective for creating perspective nanostructures used in thermoelectric devices.

1. Introduction

At present, alternative sources of energy, such as solar energy, biofuel, wind power, as well as thermal converters, become more relevant. Especially promising is the development of thermoelectric converters of unused industrial heat directly into electrical energy. From a practical point of view, the most important parameter determining the properties of the thermoelectric material is thermoelectric figure of merit $ZT = \sigma S^2 T/k$ where σ, S, and k are the electrical conductivity, the Seebeck coefficient, and the thermal conductivity, respectively. $ZT \geq 3$ - is necessary condition in which it becomes possible to replace the mechanical generators with thermoelectric generators. Nanostructured thermoelectric materials, such as materials with superlattices, systems with quantum wells and dots, quantum wires and nanocomposites can provide high ZT [1-5]. Mathematical model shows, that bismuth nanowires with diameter 5 nm and dopand concentration 10^{18} cm^{-2} provides ZT up to 6.3 [6]. One of the factors affecting the increase in ZT is the additional scattering of phonons at the boundaries of the nanowires, which leads to a decrease in the thermal conductivity [7]. Also, using of porous alumina as a template, that has low thermal conductivity [8], contributes to an increase the thermoelectric figure of merit. In this work we present the technique of creating of nanostructured thermoelectric material based on bismuth nanowires synthesized in porous anodic alumina (PAA) templates.

2. Experimental

The matrices of bismuth nanowires were formed by electrochemical deposition in templates of porous anodic alumina (PAA). The sequence of matrix preparation operations is shown on figure 1. Templates 30 μm thick and with pores diameter of 70 nm were obtained by two-step anodizing of Al foil (99.99%) of 100 μm thickness. A Keysight N5751A power supply was used as the anodizing unit. First stage anodization was carried out in two-electrode electrochemical cell in 0.4 M/dm^{-3} solution oxalic acid at constant applied potential of 55 V (Figure 1, a). As the cathode we used an aluminum plate located parallel to the anodized foil at distance of 10 cm. The electrolyte was intensively agitated by magnetic stirrer RH 2 (IKA) and temperature of electrolyte was kept equal to 10 °C using thermostat. The first stage of anodizing was carried out to a depth of 5 μm. The thickness of the

---

Content from this work may be used under the terms of the Creative Commons Attribution 3.0 licence. Any further distribution of this work must maintain attribution to the author(s) and the title of the work, journal citation and DOI.

Published under licence by IOP Publishing Ltd

formed anodic alumina layer was controlled coulometrically. Further 5 μm thick sacrificial alumina layer, formed during first anodization stage, was selectively dissolved in an aqueous solution containing 1.12 M/dm^3 H$_3$PO$_4$ and 0.6 M/dm^3 CrO$_3$ at 75 ± 1 °C for 25 min (Figure 1, b). The second anodization of Al in 0.4 M/dm^3 oxalic acid at 55 V leads to the formation of ordered porous anodic alumina template with hexagonal arrangement of pores (Figure 1, c). A copper layer with thickness of 1 μm was deposited by magnetron sputtering onto one side of the AAO template to serve as cathode for electrochemical deposition of bismuth nanowires (Figure 1, d). Further, a porous anodic alumina layer, a remaining aluminum layer and a barrier oxide layer were successively selectively dissolved (Figure 1, e). A porous alumina layer was selectively dissolved in solution 1.12 M/dm^3 H$_3$PO$_4$ and 0.6 M/dm^3 CrO$_3$ at 75 ± 1 °C for 25 min. Selective removal of remaining aluminum layer was carried out in 0.01 M/dm^3 CuCl$_2$ and 2 M/dm^3 HCl solution at 15 °C during 20 min. The barrier oxide layer on the bottom of PAA templates was removed in 0.5 M/dm^3 solution of orthophosphoric acid during 10 minutes at a temperature of 50 °C with stirring (Figure 1, f). Pore widening operation was carried out in 2 M/dm^3 solution of H$_2$SO$_4$ with stirring during 15 minutes at temperature of 50 °C (Figure 1, g). Pore widening is necessary to increase porosity and provide the required pore diameter.

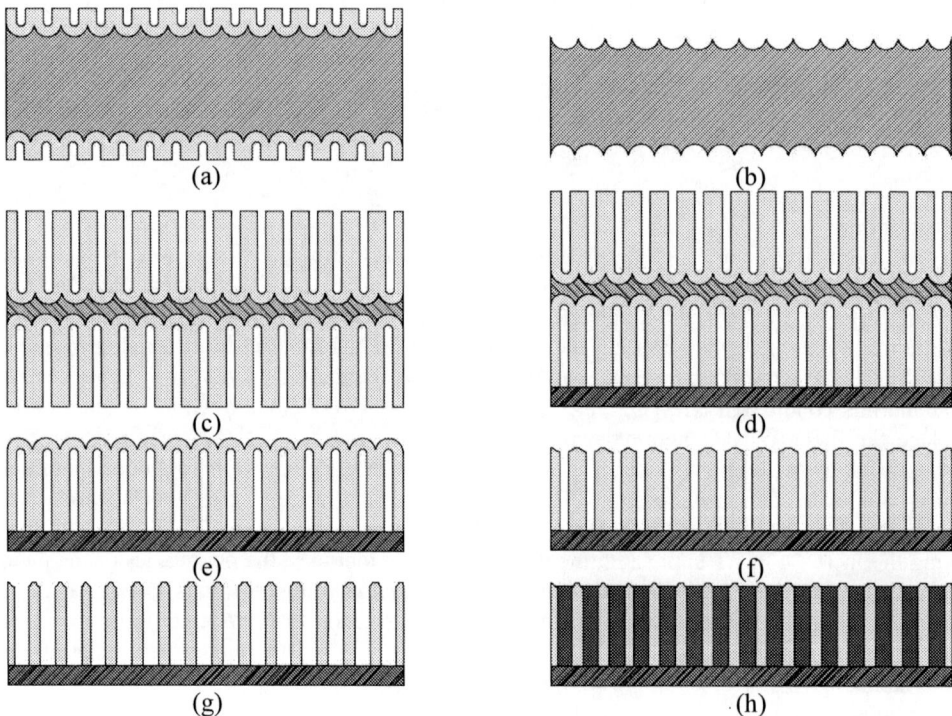

Figure 1. The main phases of the matrices of bismuth nanowires formation process: (a) first stage anodizing of aluminium foil; (b) selective dissolution of sacrificial porous oxide layer; (c) second stage anodizing of structured aluminium foil; (d) magnetron sputtering of copper layer on porous surface; (e) selective dissolution of porous alumina layer and remaining aluminum layer; (f) barrier oxide layer dissolution; (g) pore widening operation; (h) electrochemical deposition of bismuth.

Electrochemical deposition of bismuth into the prepared matrix was carried out in solution consisting of 0.13 M/dm^3 BiCl$_3$, 1.2 M/dm^3 NaCl and 1 M/dm^3 HCl onto copper sublayer, treated with a 0.05 M/dm^3 sulfuric acid during 1 minute at room temperature (Figure 1, h). Cathodic polarization curve was determined during deposition on the copper layer. The curve was determined in the voltage range 0 ... -1.2 volts, with a scanning speed of 0.005 V/s and a step of 0.0025 V, in a three-electrode cell with a silver chloride reference electrode. Carbon electrode uses as anode. PGSTAT302N power supply, connected to a personal computer with the Nova 2.0 software package, was used for the determination of cathodic polarization curves. Deposition into PAA was carried out in the galvanostatic mode at current densities of 10 mA/cm^2, 20 mA/cm^2 and 30 mA/cm^2, which were chosen after analysis of polarization curves. Deposition time was 20 minutes and during electrodeposition the electrolyte was constantly agitated by magnetic stirrer. SEM images of the PAA templates and Bi/PAA nanocomposites were collected using scanning electron microscope of high resolution S-4800 f. Hitachi. A gold layer 5 nm thick was sputtered on the surface of the samples before SEM analysis. An electron probe X-ray spectral microanalysis of electrochemically deposited Bi nanowires in the pores of AOA membranes was carried out on a scanning electron microscope equipped with a special AN 10000 detector from Princeton Gamma-Tech, Inc. This detector registered the characteristic X-ray radiation from all elements that fell under the action of the primary electron beam of the microscope. The spot from the primary ray had a characteristic size of 2 μm. The penetration depth of the beam was from 0.1 μm to several microns.

3. Results and discussion

Structure of PAA templates on different stages of preparation is shown on SEM images (Figure 2). At first stage of anodizing pores nucleate chaotically, mainly on defects in the structure of aluminium (Figure 2, a). During oxidation pores are self-organized and arranged in an ordered hexagonal structure. At second stage of anodizing pores nucleate on pre-patterned metal surface, obtained after the selective dissolution of sacrificial alumina layer. This allow to form porous films with ordered structure (Figure 2, b). After selective dissolution of a porous anodic alumina layer and remaining aluminum layer we obtain a freestanding oxide template with barrier oxide layer and copper layer on another side (Figure 2, c, d). After the local etching of the barrier oxide layer, a pore diameter and porosity was 45 nm and 10% respectively (Figure 2, e, f). Mathematical modelling of thermal conductivity of porous membranes shows, that an increase in porosity of up to 30% leads to a decrease in the thermal conductivity by 30% [8], which can positively affect on the figure of merit of the thermoelectric composite. Thus, the diameter of channels in PAA was widened by etching the received templates, and resulting diameter of pores was 70±2 nm, and porosity about 30% (Figure 2, g, h).

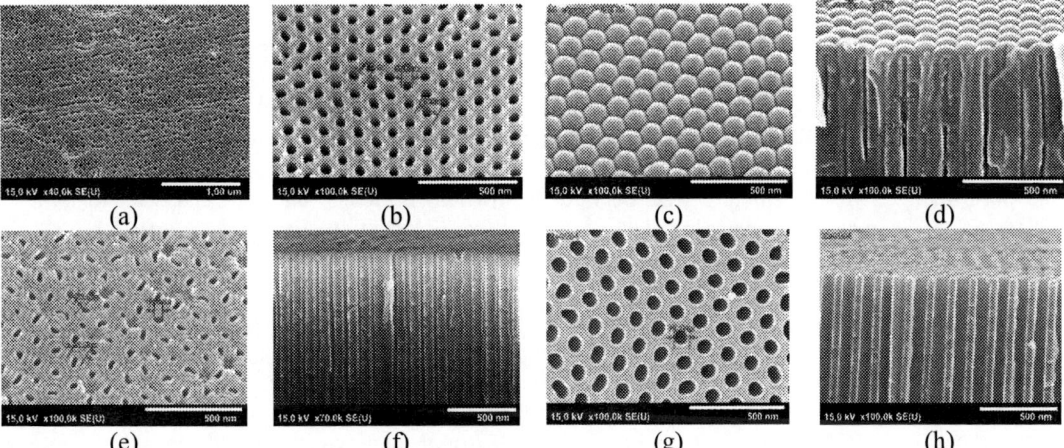

Figure 2. SEM images of anodic alumina templates: (a) surface of alumina foil after first stage of anodizing; (b) ordered surface of porous anodic alumina template after second stage of anodizing; (c, d) barrier oxide layer of porous template; (e, f) surface and cross section of template after barrier oxide layer removal; (g, h) surface and cross section of template with enlarged pores.

The analysis of the polarization curves made it possible to determine the optimal electrical conditions of electrochemical deposition of bismuth nanowires. The polarization curve has three characteristic areas (fig. 3). Area I corresponds to the voltage range at which electrochemical deposition of metal occurs on the copper surface. The maximum current density in the voltage range -0.24 ... -0.31 V (area II) is explained by the limiting ion (Sb⁻) diffusion current I_d. When the limiting diffusion current is reached, it sharply deteriorates the quality of the deposited metal. Therefore, the operating range of the metal deposition current densities below 90% of I_d, at voltage range -0,16…-0,24 V and current densities of 5…24 mA/cm^2. Further polarization (area III) leads to an increase the current density due to hydrogen evolution on copper layer, which prevents the formation of bismuth nanowire in the PAA template.

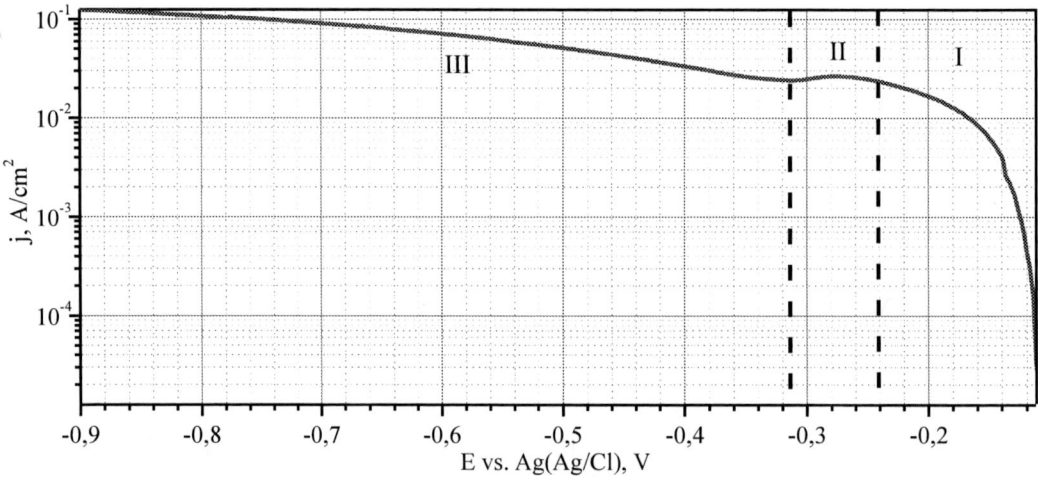

Figure 3. Cathodic polarization curve of porous electrode, based on 30 μm thick PAA template with ordered porous structure and copper sublayer, in solution consisting of 0.13 M/dm^3 BiCl$_3$, 1.2 M/dm^3 NaCl and 1 M/dm^3 HCl. Potential scan rate is 0.005 V per second. The current was normalized on geometric area of the sample (0.8 cm^2)

SEM images of templates with deposited bismuth ware analysed (Figure 4). In sample of nanocomposite with bismuth nanowires formed at current density 10 mA/cm^2 (Figure 4, a) PAA template is filled at 9.84 μm. In this electrochemical condition of formation the growth rate of nanowires is 0.492 μm/min. In each pore a nanowire was synthesized with a diameter corresponding to the pore diameter. PAA template with bismuth nanowires formed at current density 20 mA/cm^2 is filled at 14.1 μm, that corresponds to growth rate of 0.705 μm/min (Figure 4, b). The disproportionality of the deposition rate and the current density can be related to a change in the percentage of current efficiency with the current density increase. In sample formed at current density 30 mA/cm^2 (at voltage above -0.24 V) nanowires inside the porous template are not detected (Figure 4, c). This may be due to the intensive evolution of hydrogen on the copper sublayer, which prevents the formation of bismuth nanowires in the pores and deposition occurs on the surface of the membrane. Thus, the most reproducible and uniform electrochemical deposition is carried out at current densities of 10 ... 20 mA/cm^2 with formation bismuth nanowires at each pore with diameters corresponding to pore diameters of 70 nm.

SPBOPEN 2018   IOP Publishing

IOP Conf. Series: Journal of Physics: Conf. Series **1124** (2018) 022013   doi:10.1088/1742-6596/1124/2/022013

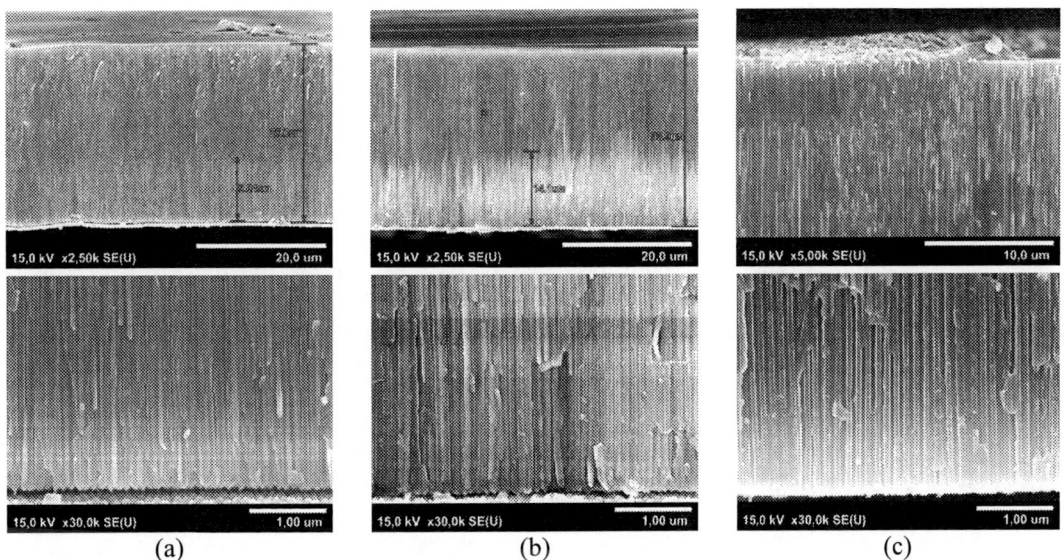

(a)   (b)   (c)

Figure 4. SEM microphotographs of PAA cross-section with bismuth nanowires formed at different current densities: (a) PAA template with Bi nanowire array 9.84 μm length, deposited at 10 mA/cm^2; (b) (a) PAA template with Bi nanowire array 14,1 μm length, deposited at 20 mA/cm^2; (c) PAA template with layer of bismuth on surface of template, deposited at 30 mA/cm^2.

Figure 5 shows the data of investigations of Bi nanowires composition in the porous PAA template as spectra of electron probe x-ray spectral microanalysis. On the spectrum there are lines corresponding to the elemental composition of the original matrix: the line with maximum of 1.62 keV corresponds to aluminium in the PAA structure, with maximum of 0.51 keV – oxygen. The electrochemically deposited nanowires in the pores is reflected by several lines in the spectrum that corresponds with bismuth in various forms (1.87 keV, 2.52 keV, 2.57 keV and 2.74 keV) with a maximum band of 2.42 keV.

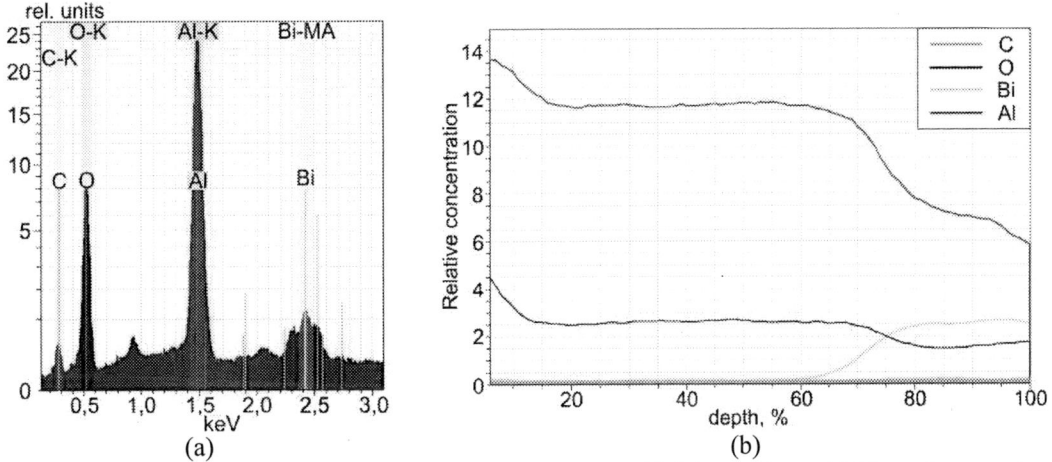

(a)   (b)

Figure 5. Spectrum of energy dispersive X-ray spectroscopy of PAA template with bismuth nanowires deposited at current density 20 mA/cm^2 (a) and distribution of elements in depth of nanocomposite (b).

4. Conclusion

The technique of nanoporous templates formation for the electrochemical synthesis of nanowires from semiconductors and semimetals with large aspect ratio of diameter to length has been developed. This method allows by varying the formation conditions to control the pore sizes and their scaling controllably. Nanoporous templates were used to obtain arrays of bismuth nanowires by electrochemical deposition from chloride solutions. As the result of electrochemical synthesis, the bismuth nanowires are formed in each pore, and their diameters correspond to the pore sizes, and length is determined by the duration of deposition. The developed methods make it possible to reproduce nanowires of semimetals with the required physicochemical properties, which opens the prospect for the creation of wide range of thermoelectric devices, such as thermo-generators, microcoolers, as well as devices operating on quantum effects with low production costs. Further work will be aimed at studying the thermoelectric properties of obtained nanocomposites.

Acknowledgments

The authors gratefully acknowledge the contributions of Reseach & Design Center of JSC «INTEGRAL» and Zhigulin Dmitriy, for providing images from a scanning electron microscope and energy dispersive X-ray spectroscopy investigation.

References

[1] Boukai A., Xu K., Heath J. R., 2006 Adv. Mater., **18**(7) 864
[2] Ali A., Chen Y., Vasiraju V., Vaddiraju S., 2017 Nanotechnology, **28**(28), 282001.
[3] Gorokh G. G., Lozovenko A.A., Bulat L.P., 2017 Semiconductors, **51**(7), 850
[4] Gorokh G. G., Obuhov I. A., Lozovenko A. A., 2015 TKEA, **1** 3.
[5] Obuhov I. A., Gorokh G. G., Lozovenko A. A., Smirnova E., 2017 Nanoindustry, **6**(77), 96.
[6] Lin Y. M., Sun X., Dresselhaus M. S., 2000. Physical Review B, **62**(7), 4610.
[7] Chen G, Narayanaswamy A., Dames C., 2004. Superlattices And Microstructures, **35** 161.
[8] Belahurau Ya. A., Shukevich Ya. I., Barkalin V. V., Khatko V.V., Taratyn I.A., 2012. Nano- and microsystems technology, **1** 18.

SPBOPEN 2018   IOP Publishing

High speed near-infrared range sensor based on InP/InGaAs heterostructures

K J Smirnov[1,2], V V Davydov[3], S F Glagolev[2], G V Tushavin[1,4]

[1] OJSC "NRI "Electron", Saint Petersburg, Russia
[2] Bonch-Bruevich Saint – Petersburg State University of Telecommunications, Saint Petersburg 193232, Russia
[3] Peter the Great Saint – Petersburg Polytechnic University, Saint Petersburg 195251, Russia
[4] Saint Petersburg National Research University of Information Technologies, Mechanics and Optics, Saint Petersburg 197101, Russia

Abstract. Technology of creation effective photocathode based on the InP/InGaAs heterostructures is given. The results of an experimental study of pin-diode, which was used as the receiver of photoelectrons, are presented. Arrangement of the vacuum photoelectronic device with InP/InGaAs photocathode is proposed.

1. Introduction

At present time, problem of detection of low-level short duration signals at the wavelengths from 0,9 to 1,6 µm is extremely actual. The existing photodetectors do not meet the requirements of sensitivity, processing speed and resolution. One of the most preferred solution of this problem is to create vacuum photoelectronic device with photocathode based on the InP/InGaAs heterostructures [1-2]. The purposes of the work was to obtain effective InP/InGaAs photocathode with quantum efficiency on the level of 3-5% and to implement the results of photocathode structure researches in real photodetector which works in near infrared range (NIR).

2. Construction of vacuum photoelectronic device

Photosensitive device is implemented on the basis of the hybrid technology. The photocathode structure and solid-state element, as the receiver of the photoelectrons, are contained in one vacuum value. Applying of this technology simultaneously allows to register low-level signals at the wavelengths from 0,9 to 1,6 µm, to get processing speed of a device on the level of a few nanoseconds and ensure a low level of internal noise as compared to devices with non-vacuum parts. The proposed sensor construction consists of three main parts: device framework with inputs for supplying voltage and outputs, glass with fixed photocathode structure, plate with the receiver of the photoelectrons. Proximity-focusing technology was used to provide a high resolution.

The body and the all internal elements of the detector produced from the vacuum outgassing materials. Non-evaporable getters made from the titanium plates were also included in the device structure. These getters could improve the vacuum level in the working device, which is extremely important for the detector lifetime. The processes of assembling sensor and activation of photocathode structure need ultra-high vacuum conditions. All steps of sensor creation were conducted in the ultra-high vacuum installation, which could reach the level of $\cdot 10^{-9}$ Pa [3].

Content from this work may be used under the terms of the Creative Commons Attribution 3.0 licence. Any further distribution of this work must maintain attribution to the author(s) and the title of the work, journal citation and DOI.
Published under licence by IOP Publishing Ltd

3. Photocathode structure

The photocathode, which operates in the NIR, also known as the transferred electron (TE) photocathode, can move the long-wavelength threshold of photoemission by increasing the internal energy of the photoelectrons. This structure has been realized on the basis of heteropair InP-InGaAs. Titanium grid-electrode (Schottky-barrier) was inflicted on the heterostructure surface. After that photocathode structure was purified by 2-step method (chemical cleaning and vacuum annealing) [4-5]. Atomic clean InP/InGaAs surface activated with Cs and O_2 allows to form effective state of negative electron affinity on photocathode surface [1-2,6]. In the real device backside bombarded technology is implemented. Photocathode structure is irradiated from the InP substrate to the emitter layer. The dependence between the quantum efficiency (QE) of the proposed InP/InGaAs heterostructure which works in transmission mod and λ is represented in figure 1. The photocathode demonstrates stable QE on the level of 3-4% in NIR. Such values of efficiency are enough to create sensor on the basis of developed heterostructures.

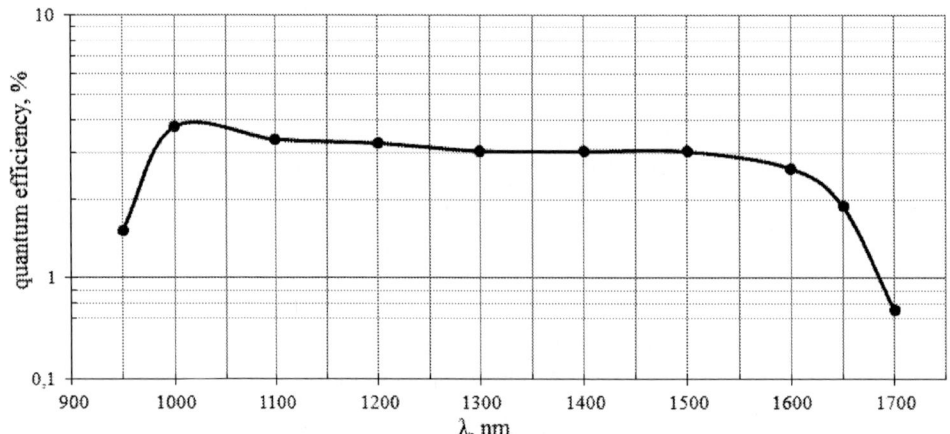

Figure 1. Spectral characteristics of InP/InGaAs heterostructure with bias voltage U_s= 3.4 V.

4. Receiver of photoelectrons

In the device layout with InP/InGaAs photocathode the line of silicon pin-diodes with the number of elements equal to 12 was selected as the receiver of the photoelectrons and the amplifying element. It is a vertical structure on the basis of pure high-resistance silicon obtained by zone melting methods. Special features of creation of pin-diode lines ensure them to work in the reverse bias mode. In this mode a full depletion of the high-resistance base occurs, which ensures drift transfer of the charge carriers. It allows to obtain a time of the signal record at the level of a few nanoseconds. Figure 2 shows the line of pin-diodes on the technical testing adapter and appearance of the device layout.

Our experiments showed that the impulse response front of the photodiode is equal to 1.1 ns. and the impulse response duration is equal to 1.9 ns. Amplification is achieved due to the occurrence in the solid-state element (diode structure) of a large number of charge carriers by the influence of high energy (several keV) electrons emitting from the photocathode.

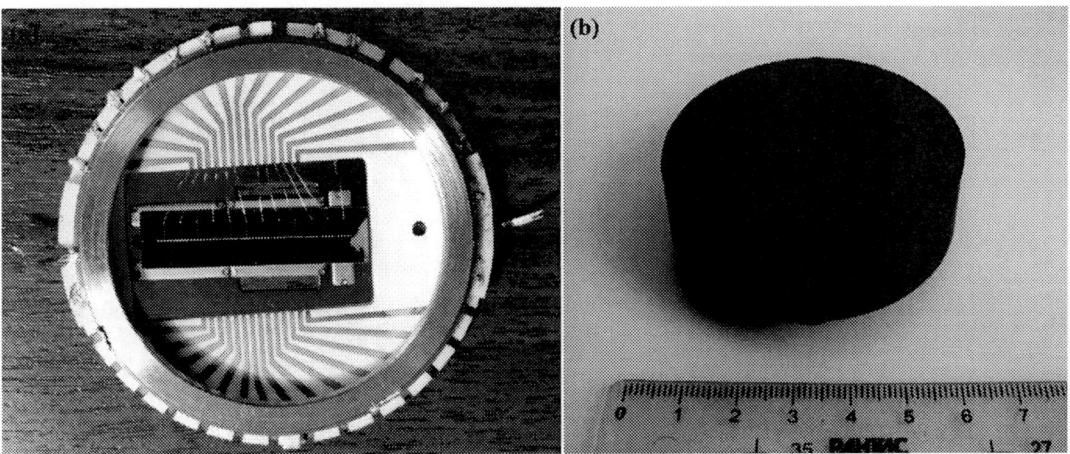

Figure 2(a, b). (a) Line of pin-diodes on technical testing adapter; **(b)** Device layout.

Figure 3 shows the dependence of the amplification factor (M) on the energy of photoelectrons (on the voltage between photocathode and pin-diode) for test diode lines 1-3. M is determined by the ratio of the electrons generated in the Si-structure of the diode line to the electrons which were emitted by the InP/InGaAs heterostructure.

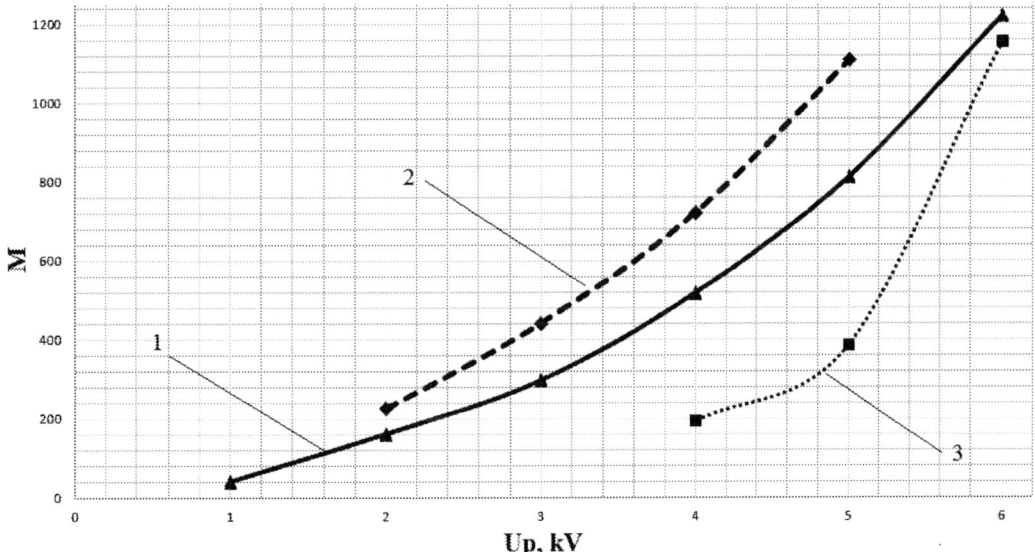

Figure 3. Amplification factor (M) depending on the energy of the photoelectrons Up for test diode lines 1-3.

The results of M obtained on the investigated diode lines reaches 10^3. The diode lines which were used in the real device layouts fulfilled 3 main parameters: processing speed, amplification, dark current of all diode line elements on the level of a few nA.

5. Results

The sensor with the InP/InGaAs photocathode and the pin-diode line was created. Spectral characteristic obtained on the output of one diode line element with bias voltage $U_s = 3.4$ V is shown in figure 4. Obtained results of the device sensitivity on the level of more than 2 A/W in NIR could be significantly increased with the help of various methods of optimizing elements of photosensitive sensor.

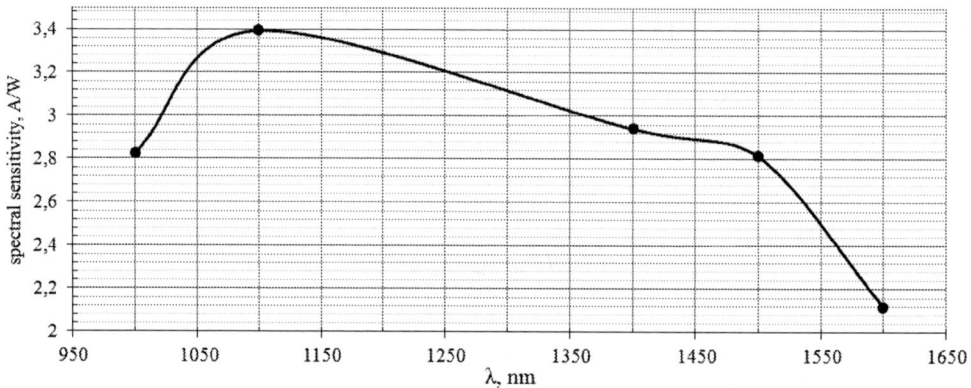

Figure 4. Spectral characteristic of experimental sensor layout with supply voltage U_s= 3.4 V.

Besides the level of sensitivity and processing speed of sensor another important parameter of device exists. That factor is an equitability which demonstrates the level of sensitivity obtained on the different elements of the diode line compared to each other. Figure 5 shows sensitivity of fabricated sensor for all elements of diode line simultaneously. The unevenness of device reaches the value of more than 50% which is highly undesirable. It happens mainly because of the unevenness of the photocathode. The main reasons of it are no ideal conditions of chemical cleaning of the photocathode structure and large active area of the photocathode. It creates the additional difficulties when performing technological operations like vacuum annealing and photocathode activation.

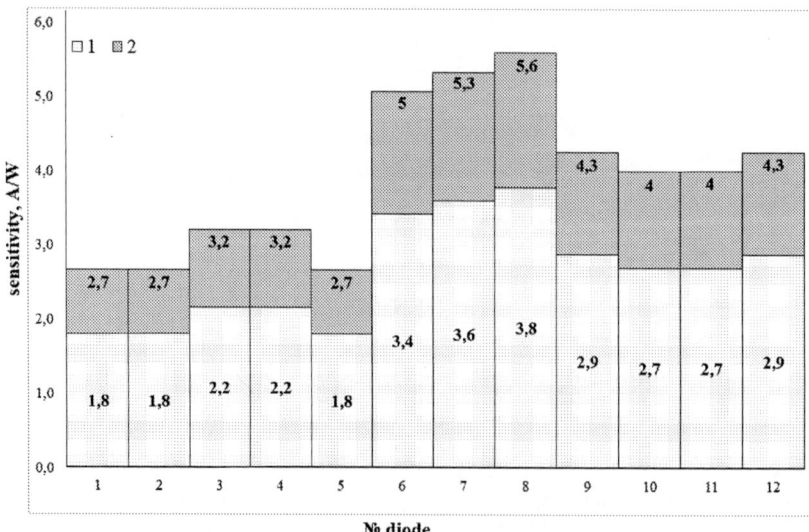

Figure 5. Sensitivity on wavelength λ=1,1 μm for sensor with applied bias voltage 1. U_s= 3.4 V, 2. U_s= 3.6 V.

6. Conclusion

Our sensor with the developed InP/InGaAs photocathode provides the processing speed on the level of a few ns. with sensitivity of the several A/W. Avalanche InGaAs photodiodes (APD) as compared to proposed device got sensitivity on the level below 1 A/W. Developed device could be used for many applications, for example, real-time detection of the reflected laser beam in poor visibility conditions, high-speed location systems [7]. High sensitivity and processing speed of the developed sensor could be also applied in refractometry, nondestructive testing, fiber optics [8-13].

References

[1] Aebi V W, Sykora D F, Jurkovic M J, Costello K A 2011 *Proc. of SPIE*, v. 8033 80330T

[2] Smirnov K J, Medzakovskiy V I, Davydov V V, Vysoczky M G, Glagolev S F 2017 *J. of Phys: Conference Series* **917(6)** 062019

[3] Myazin N S, Smirnov K J, Davydov V V, Logunov S E 2017 *J. of Phys: Conference Series* **929(1)** 012080

[4] Sun Y, Liu Z, Machuca F, Pianetta P, Spicer W 2005 *SLAC-PUB* 11018

[5] Sun Y, Liu Z, Pianetta P 2007 *SLAC-PUB* 12710

[6] Chanlek N, Herbert J, Jones R, Jones L, Middleman K, Militsyn B 2014 *J. Phys. D: Appl. Phys.* **47** 055110

[7] Kovsh A, Zhukov A, Livshits D, Krestnikov I US 2009 Patent N7561607

[8] Myazin N S, Logunov S E, Davydov V V, Rud' V Y, Grebenikova N M, Yushkova V V 2017 *J. of Phys: Conference Series* **929(1)** 012064

[9] Ermolaev A N, Krishpents G P, Davydov V V, Vysoczkiy M G 2016 *J. of Phys: Conference Series* **741(1)** 012171

[10] Davydov V V, Kruzhalov S V, Vologdin V A 2017 *J. of Optical Technology (A Translation of Opticheskii Zhurnal)* **84(8)** p 568-573

[11] Logunov S E, Davydov V V, Vysoczky M G, Koshkin A Y, Rud V Y 2017 *J of Phys: Conference Series* **917(5)** 052028

[12] Davydov V V, Kruzhalov S V, Grebenikova N M, Smirnov K J 2018 *Measurement Techniques* **61(4)** p 365-372

[13] Grebenikova N M, Smirnov K J, Artemiev V V, Davydov V V, Kruzhalov S V 2018 *J. of Phys: Conference Series* **1038(1)** 012089

SPBOPEN 2018   IOP Publishing

Features of a zone thermal crystallization semiconductor thin films grown from a discrete liquid source

S N Chebotarev[1], A V Varnavskaya[1], I V Gavrus[1], V N Lozovskii[1], L M Goncharova[1], N A Yakovenko[2]

[1]Department of Physics and Electronics, Platov South-Russian State Polytechnic University (NPI), Novocherkassk 346421, Russia
[1]Department of Optoelectronics, Kuban State University, Krasnodar, 350040, Russia

Abstract. We studied the effect of the growth cell geometry on the temperature fields and the uniformity of semiconductor layers grown by thermal crystallization method from a discrete liquid source. Calculation results show that the radius of local sources practically does not change the temperature field of the substrate. When the radius of the local evaporators was changed from 2 to 6.5 mm, the observed increase in the maximum temperature did not exceed 2 K. However, increasing the distance between the substrate and the temperature screen fundamentally changes the temperature of the substrate. An increase in the distance from 2 to 25 mm caused a decrease in the substrate temperature by 20 K. It is shown that to achieve uniformity of better than 90% need to use a hexagonal arranged system of round local sources with the radius of $r = 0.6$ cm.

1. Introduction

Thin films play a significant role in the modern technology of electronic semiconductor devices [1]. Methods for obtaining thin-film structures are quite diverse. Among the recognized technological flagships, which include molecular-beam epitaxy [2] and vapor phase epitaxy, there are a number of other growth techniques such as ion-beam crystallization [3,4], liquid-phase epitaxy and zone crystallization from a discrete source [5]. The last technological variant has certain advantages, consisting in protecting the growth zone from the penetration of background impurities as well as the method allows controlling the growth rate due to the temperature variation. The purpose of this paper is to investigate the effect of the geometry of the growth cell on the temperature fields and the uniformity of the growing semiconductor layers.

2. Calculation, experiment and results

The main structural elements of the growth process are a heater, a growth agent source, a substrate and a heat shield. In the experiments we used a graphite source of diameter D, having hexagonally ordered circular local evaporators of radius r. In the local evaporators, a growth agent was loaded, maintained in the liquid phase. In our case, this substance was germanium. The substrate was a 300-mm silicon

---

Content from this work may be used under the terms of the Creative Commons Attribution 3.0 licence. Any further distribution of this work must maintain attribution to the author(s) and the title of the work, journal citation and DOI.
Published under licence by IOP Publishing Ltd

wafer with a diameter of 100 mm. The distance between the source and substrate h was 3 mm. T_1 and T_2 are temperature of the evaporator and the substrate respectively. The thermal regime was controlled by the position of the screen, distant from the substrate by a distance l. The calculation of the temperature fields was carried out numerically, using the surface to surface radiation method [6]. Figure 1 shows an example of calculating the temperature fields of each element of the growth cell, where the cellular structure of the discrete source is also clearly visible.

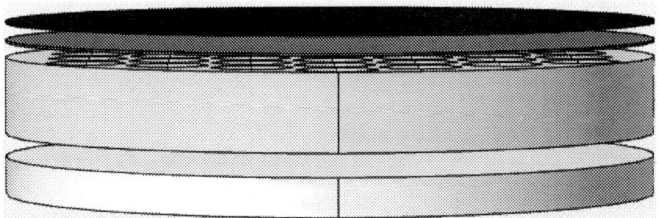

Figure 1. Distribution of the temperature field

Two heat equation were used to calculate the temperature fields. We used the heat equation (1), which takes into account the heat transfer from the heater to the source of the growth substance, as well as the Stefan-Boltzmann equation (2):

$$pC_p \frac{\partial T}{\partial t} + \nabla \cdot \left(-k\nabla T \right) = Q \tag{1}$$

$$-n \cdot \left(-k\nabla T \right) = h \cdot \left(T_{\text{inf}} - T \right) + \frac{\varepsilon}{1-\varepsilon}\left(J_0 - \sigma T^4 \right) \tag{2}$$

Here ρ is the density; k is the thermal conductivity; Q is the volume heat source; n is the surface normal vector; T_{inf} equals the temperature of the convection cooling gas; ε is the surface emissivity; J_0 is the expression for surface radiosity; and σ is the Stefan–Boltzmann constant.

A curve was constructed showing the temperature changes of the source of growth material along the diameter. The radius of the cell was assumed to be 6 mm.

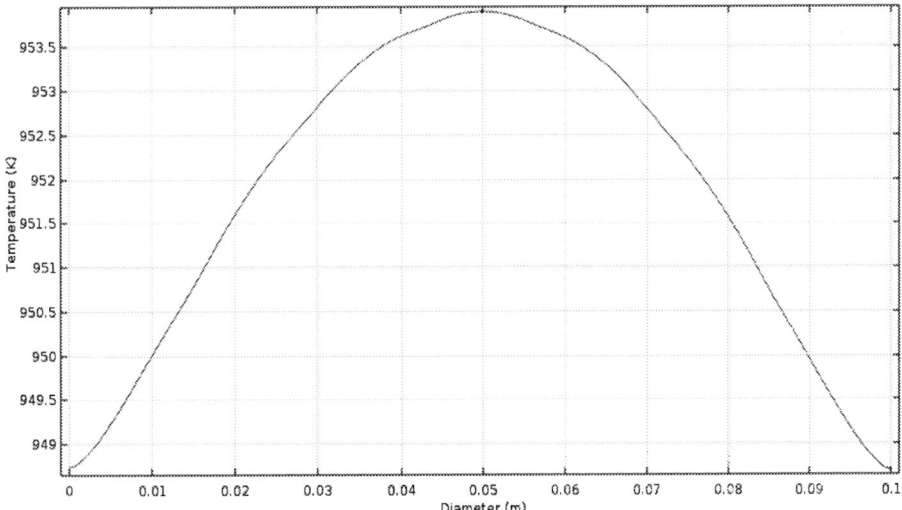

Figure 2. Temperature distribution at source

It can be seen from the graph that the temperature difference between the edge of the source and its center is about 5 K, and also that the curve does not show any bends in the center of the cells with Ge. The graphs show the dependence of the maximum temperature at the center of the substrate on the radius of the local evaporators r (red line) and the distance between the substrate and the screen l (blue line). Figure 3b shows an example of the calculated distribution of the relative thickness of the deposited layer along a line passing through the center of the substrate. The layer thickness was measured by SEM using the nanoscale marks [7].

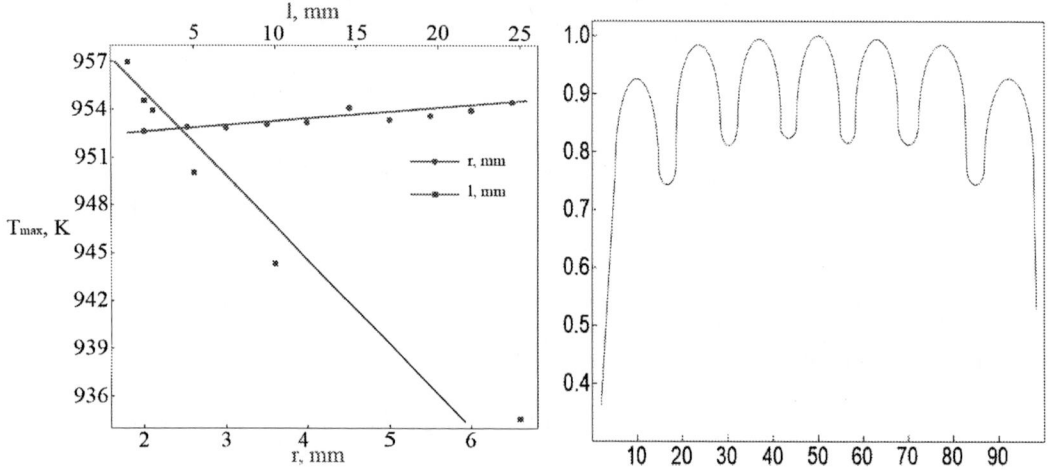

Figure 3 (a, b). (a) Temperature dependence on the radius r and the distance l;
(b) The dependence of non-uniformity thickness on radius of the local sources r

Based on the experimental data for different cell radii, the temperature distribution curves at the source were plotted and the homogeneity values of the layer were calculated using the formula:

$$\delta = \frac{n_{max} - n_{min}}{n_{max}} \qquad (3)$$

Figure 4 shows the calculated and experimental dependence of the homogeneity of the δ layer on the radius r. The obtained dependence practically decreases linearly with increasing radius. The experimental data reflected the crosses support the conclusions of the theory.

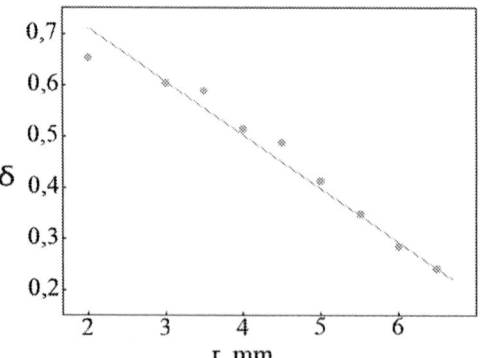

Figure 4. Dependence of the homogeneity δ of the layer on the radius r of the cell

3. Conclusions

The influence of the geometry of the growth zone on the temperature field of the substrate and the uniformity of the deposited layer was studied. It is shown that the dimensions of the local evaporators relatively only slightly change the temperature field of the substrate, but determine the homogeneity of the layer decisively. The method uses the system of geometrical arranged local sources filled with liquid germanium. It is shown that to achieve uniformity of better than 90% for a hexagonal arranged system of round local sources with the radius of $r = 0.6$ cm.

Acknowledgements

The authors would like to thank for financially supporting this work the Russian Foundation for Basic Research [grant number 17-08-01206].

References

[1] Kryzhanovskaya N V, Polubavkina Yu S, Scherbak S A, Moiseev E I., Zhurikhina V V, Zubov F I, Lipovskii A A, Kulagina M M, Troshkov S I, Zadiranov Yu M, Maximov M V, Zhukov A E 2017 *J. Appl. Phys.* **121** 043104

[2] Nikiforov A I, Timofeev V F, Pchelyakov O P 2015 *Appl. Surf. Sci.* **354** 450

[3] Chebotarev S N, Pashchenko A S, Lunin L S, Zhivotova E N, Erimeev G A, Lunina M L 2017 *Beilstein J. Nanotech.* **8** 12

[4] Chebotarev S N, Pashchenko A S, Lunin L S, Irkha V A 2016 *Nanotech. in Russia.* **11** 435

[5] Yatsenko A N, Chebotarev S N, Lozovskii V N, Mohamed A A, Erimeev G A, Goncharova L M, Varnavskaya A A 2017 *J. Phys: Conf. Ser.* **917** 032008

[6] Colomer G., Costa M., Consul R, Oliva A 2004 *Int. J. of Heat and Mass Transf.,* **47** 257

[7] Lozovskii V N, Chebotarev S N, Irkha V A, Valov G V 2010 *Tech. Phys. Lett.* **36** 737

SPBOPEN 2018

IOP Conf. Series: Journal of Physics: Conf. Series **1124** (2018) 022016

IOP Publishing

doi:10.1088/1742-6596/1124/2/022016

Comprehensive study of the optical and physical properties of InGaN nanocolumns grown on Si(111) and por-Si/Si(111) substrates by plasma-assisted molecular beam epitaxy

Zolotukhin D*[1], Seredin P[1], Lenshin A[1], Goloshchapov D[1,3], Mizerov A[2], Nikitina E[2]

[1]Voronezh State University, Universitetskaya pl. 1, Voronezh, 394018, Russia
[2] St. Petersburg Academic University, Khlopina 8/3, 194021 St. Petersburg, Russia
[3]Voronezh State Technical University, Voronezh, Russia

Abstract. We report on a successful growth of self-organized $In_{0.3}Ga_{0.7}N$ nanorods by low-temperature plasma-assisted molecular beam epitaxy on a Si(111) substrate with and without preformed thin porous Si layer (*por-Si*). XPS study of both samples confirmed In content predicted according to the growth conditions as well as the analysis of PL spectra. Moreover, InGaN nanorods grown on *por-Si* layer showed ~20% higher PL intensity.

1. Introduction

III-nitride semiconductors are of great interest for applications in high temperature/power electronic devices as well as photonic devices in a wide spectral range from ultraviolet (the GaN bandgap is 3.45 eV [1]) to near-infrared (the InN bandgap is ~0.7 eV [2]). However, due to the lack of the native GaN substrates, III-N based structures conventionally grow on heteroepitaxial substrates with a high thermal expansion coefficient and lattice mismatches leading to high threading dislocation generation. Many dislocation reduction techniques such as lateral overgrowth, low-temperature buffer layer, multiple intermediate layers have been already developed. However it is still difficult to reduce the dislocation density to be less than 10^6 cm^{-2}. It is well known that large nanorod free surface helps to suppress propagation of the threading dislocations and stress generation [3], thus InGaN nanorods structure seems to be suitable for fabrication light-emitting devices with a high efficiency.

In this paper we report on fabrication and comprehensive study of the InGaN nanorods structure grown on a compliant Si-substrate by a low temperature plasma-assisted MBE (LT PA MBE) with up to 20% higher PL intensity than ones formed on traditional Si(111) substrate.

2. Experimental

All Si substrates were pre-treated using the Shiraki method [4]. Then a 20 nm-thick por-Si layer was formed on several substrates using the original method of selective etching. After that all the substrates was annealed at 820°C for an hour until the 7×7 surface reconstruction was observed by RHEED. All of the growth processes were performed by using Veeco Gen 200 PA MBE setup. For comparison nanorods were grown in the one growth process on conventional Si(111) substrate and substrate with a por-Si layer (sample **a** and **b**, respectively) without nitridation step. LT InGaN nanorods were formed at F_{III}/F_N=1 flux ratio, whereas fluxes was F_{Ga}=0.04 ML/s, F_{In}=0.02 ML/s and F_N=0.06 ML/s and the

Content from this work may be used under the terms of the Creative Commons Attribution 3.0 licence. Any further distribution of this work must maintain attribution to the author(s) and the title of the work, journal citation and DOI.
Published under licence by IOP Publishing Ltd

substrate temperature of $T_s=400°C$. RHEED was used for *in situ* control of the surface morphology of the layers. The surface morphology also was investigated *ex situ* by atomic force microscope (AFM), scanning electron microscopy (SEM) and with optical microscope (OM). Crystalline quality was studied by XRD analysis, mobility and carrier concentration was determined by analysis of the Hall measurements data. Ternary alloy composition was confirmed by XPS and PL spectra analysis.

3. Results

The lateral sizes distribution histogram obtained from the AFM study of the substrate and samples surfaces (Figure 1) shows that the use of a por-Si layer has a crucial influence on the distribution of the nanorods diameters.

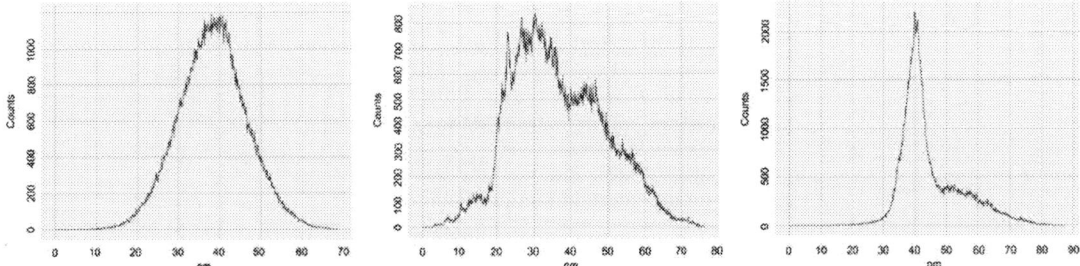

Figure 1. The lateral sizes distribution histogram obtained from AFM study of por-Si layer (left), InGaN/Si(111) layer (centre) and InGaN/por-Si/Si(111) layer (right).

In addition, cross section SEM images of all heterostructures (Figure 2) illustrate rather planar morphology for all of the interfaces. One can also observe a more pronounced nanocolumnar structure in case of the sample **b**.

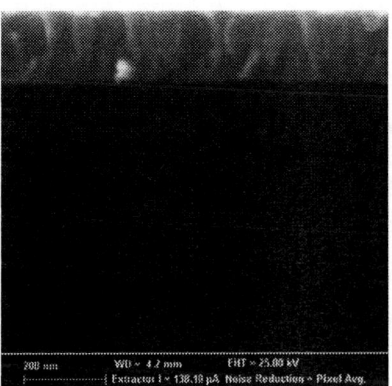 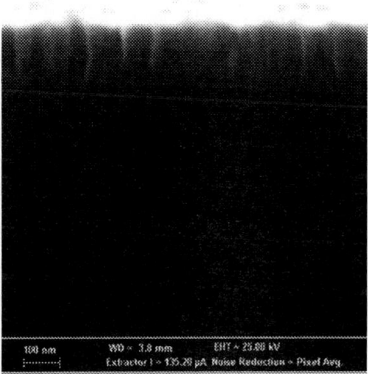

Figure 2 Cross-section SEM images of the GaN/Si(111) (left) and GaN/por-Si/Si (right) nanorods.

XPS spectra of both samples are listed at figure 3.

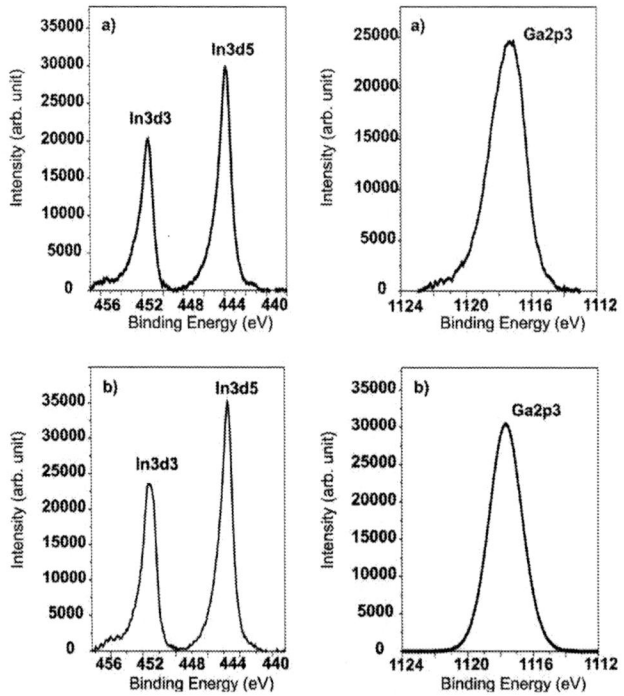

Figure 3 XPS spectra of the InGaN/Si(111) (**a**) and InGaN/por-Si/Si (**b**) nanorods.

PL spectres of the samples which are listed at Figure 4 revealed ~20% higher PL peak intensity for sample grown on *por-Si* layer.

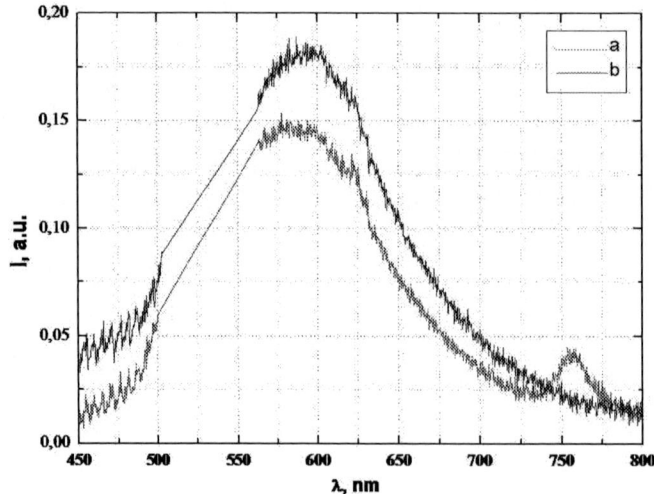

Figure 4 PL spectra of the InGaN/Si(111) (**a**) and InGaN/por-Si/Si (**b**) nanorods.

Table 1. Data obtained from XRD analysis

Sample	c, Å	a, Å	ε_{xx}	ε_{zz}	ρ_{screw}, cm^{-1}	ρ_{edge}, cm^{-1}
a	5.4005	3.0178	-0.087	$7.06 \cdot 10^{-3}$	$3.15 \cdot 10^{10}$	$1.97 \cdot 10^{9}$
b	5.4156	3.0253	-0.086	$7.04 \cdot 10^{-3}$	$3.10 \cdot 10^{10}$	$1.60 \cdot 10^{9}$

4. Discussion

The threading dislocations reduce the emission efficiency, increase the leakage current, and reduce the life time of III-N based devices. Usage of nanorods-like structure is the one of the possible ways to create energy efficient LED structures with high light extraction surface due to presence of the large free nanorods surface, which acts as dislocation suppression instrument due to their inclination. AFM and SEM study of the surface morphology confirmed nanorods-like structure of the samples and increased average distance between them. Thus, since the average distance between nanorods was increased – light extraction coefficient should become higher. This fact can be confirmed by PL spectra analysis of both samples. As can be seen from figure 4 the maximum PL intensity of sample **b** grown on *por-Si* layer is by ~20% higher than that one for sample **a** grown on Si(111) substrate. Also PL maximum peaks intensity positions are 585 nm and 574 nm for sample **a** and **b**, respectively. We believe that this peak position shift is related with ternary alloy composition fluctuation as well as with the difference in the elastic strain in the InGaN layers. As it was shown in *K. O'Donnell* work [5] PL peak position is related indirectly with InGaN bandgap and it can be evaluated as

$$E_{PL}(x) = -1.54 + 1.45 E_g(x) \tag{1}$$

The $E_g(x)$ dependence for $In_xGa_{1-x}N$ alloys

$$E_g(x) = 3.51 - 2.75x \tag{2}$$

Using equations (3) and (4) we can calculate In composition in $In_xGa_{1-x}N$ ternary alloy as $x_a \approx x_b \approx 0.33\text{-}0.34$ which correlates well with growth conditions. Moreover, as it was shown in [6] In content can be evaluated from XPS spectra of the samples as

$$X_{In} = \frac{I_{In3d5}/F_{In3D5}}{(I_{In3d5}/F_{In3D5} + I_{Ga2p3}/F_{Ga2p3})} \tag{3}$$

Whereas I_i – integrated peak intensity and subscript denotes peak type and F – correlation coefficient (F_{Ga2p3}=2.75 and F_{In3d5}=4.53). Using formula (3) we can calculate In content as x_a=0.34 and x_b=0.32. It should be noted that basing on XPS analysis we can assume that the free surface of nanorods is oxidized, that is confirmed by asymmetrical form of the In3d5 peaks.

As can be seen from the data obtained from XRD analysis (table 1) InGaN layer grown on *por-Si* layer have slightly better crystalline quality.

The van der Pauw–Hall-effect measurements were performed at room temperature. Both samples revealed p-type of conductivity with carrier concentration of $3 \cdot 10^{19}$ cm^{-3} and $3.64 \cdot 10^{19}$ cm^{-3} and carrier mobility of 104.7 cm^2/(V·s) and 103.4 cm^2/(V·s) for samples **a** and **b**, respectively.

Summarizing this discussion, InGaN nanorod-like layers with reasonable crystalline quality and p-type conductivity were formed by plasma-assisted molecular beam epitaxy on compliant *por-Si* and traditional Si(111) substrates. PL and XPS study independently confirmed x≈0.32-0.34 In content in both samples. PL spectra also revealed up to 20% higher peak intensity growth for InGaN layer formed on *por-Si* layer. AFM measurement of nanorods diameter distribution showed that more than 75% of nanorods grown on the compliant substrate have the same diameters ~ 40 nm contrary to nanorods grown on Si(111) substrates which diameters are statistically distributed in range of 20 – 60 nm. Thus, usage of compliant Si substrates with nano-porous Si layer seem to be a suitable way to integrate III-N technology with existing Si technology.

Acknowledgments

The research as part of the development of nanoheterostructures as well as monitoring and investigation of the fundamental properties of samples is conducted under support of the grant of President of Russian Federation MD-188.2017.2 and grant of the Russian Ministry of Education as part of the state's task for higher-education institutions for scientific research 2017-2019 11.4718.2017/BCH. This growth experiments were supported financially by the Ministry of Education and Science of the Russian Federation № 16.9789.2017/BCH.

References

[1]   V. Bougrov *et al* 2001 *Properties of Advanced SemiconductorMaterials GaN, AlN, InN, BN, SiC, SiGe*, Eds. Levinshtein M.E., Rumyantsev S.L., Shur M.S., John Wiley & Sons, Inc., New York, 2001, 1-30.
[2]   K. M. Yu *et al.* 2005 Appl. Phys. Lett. **86**, 071910
[3]   Kishino K, Ishizawa S 2015 Nanotechnology **26**, 225602
[4]   Lenshin A *et al.* 2014 Technical Physics **59**, No. 2, 224
[5]   O'Donnell KP *et al.* 2001 J. Phys Cond. Mat. **13**, 6977
[6]   Fang Z *et al.* J Appl Phys 2014 **115**, 043514

Features of the mechanism of gas sensitivity of the zinc oxide nanorods arrays to carbon monoxide

V.V. Petrov[1], A.P. Starnikova[1], Kh.A. Abdullin[2], D.P. Makarenko[3]

[1] Southern Federal University, Research and Education and Centre "Microsystem technics and multisensor monitoring systems", Taganrog, 347922, Russia
[2] Al Farabi Kazakh National University, National Nanotechnology Laboratory of Open Type, Almaty, 050040, Kazakhstan
[3] JSC "VNIIHOLODMASH" , Moskow 127410, Russia

Abstract. In the paper electrophysical properties of ZnO nanorod arrays are investigated and features of the mechanism of gas sensitivity of the sensors to carbon monoxide (II) and humidity are examined. The ZnO nanorod arrays were grown on p-type silicon substrates. Metal contacts formed over the array of nanorods. Sensory structures were heated to a temperature of 100-200 ° C and subjected to carbon monoxide (II) at a concentration of 100-1000 ppm in nitrogen atmosphere and humidity at 50-75% in air. The results of the research showed that the properties of sensor structures based on ZnO nanorod arrays grown on silicon p-type substrates with respect to carbon monoxide (II) and humidity differ from the properties of chemoresistive sensors based on zinc oxide films.

1. Introduction

The problem of air pollution has become global due to industrial development. Therefore, the development of new methods and instruments for monitoring the air environment based on modern technologies of micro- and nanoelectronic sensors is an actual problem. Currently, inorganic oxide materials and nanostructures based on them are being intensively studied. Metal oxides (SnO_2, TiO_2, In_2O_3, etc.) are used to create various electronic devices. In particular, ZnO is a promising material as a sensitive element of gas sensors [1, 2].

2. Experiment

Arrays of ZnO nanorods were synthesized on p-type silicon substrates by the hydrothermal method, The ZnO nanorod arrays were synthesized on p-type silicon substrates by the hydrothermal method. A thin seed layer of zinc oxide ZnO was previously deposited by the sol-gel method on the surface of thoroughly cleaned Si substrates. The sol was prepared by dissolving zinc acetate in ethanol. Uniform distribution of the sol on the surface of the substrate was achieved by centrifugation at a rotation speed of about 2000 rpm, followed by drying at 130 ° C and annealing at 350 ° C [3]. Hydrothermal synthesis of ZnO nanorods was carried out in an aqueous solution of zinc nitrate and hexamethylenetetramine ($C_6H_{12}N_4$) in a glass beaker in which a substrate with a seed layer of ZnO was placed in a vertical position on a fluoroplastic holder. Hydrothermal treatment was carried out in a temperature range of 90-97 ° C for 3 hours with vigorous stirring. Samples were washed with deionized water and dried. The formed ZnO nanostructures of predominantly vertical orientation have an average transverse dimension of about 30-40 nm (Fig. 1).

Content from this work may be used under the terms of the Creative Commons Attribution 3.0 licence. Any further distribution of this work must maintain attribution to the author(s) and the title of the work, journal citation and DOI.
Published under licence by IOP Publishing Ltd

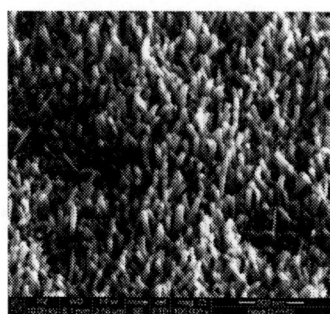

Figure 1.SEM images of the as-grown ZnO nanorod layer on a Si substrate

Then a V-Cu-Ni contact metallization of 0.3-0.4 μm thick was deposited on the surface of the samples. The resistance of the obtained gas sensor samples was from hundreds of ohms to several kOhm. Electrophysical and gas sensitive properties were measured using an automated installation for determining the parameters of gas sensors at the Center for Collective Use "Microsystem Technics and Integral Sensors". The current-voltage characteristics of the created structure with contact metallization on a glass substrate are shown in Fig. 2 (a). In Fig. 2 (b) the temperature dependence of the resistance of the structure in the temperature range from room temperature to 250 ° C is presented. In addition, the dependence of the resistance of the created structure on humidity was measured. Humidity in the range 50 - 75% was established by saturated solutions of salts [4] - Fig. 2 (c).

(a)   (b)

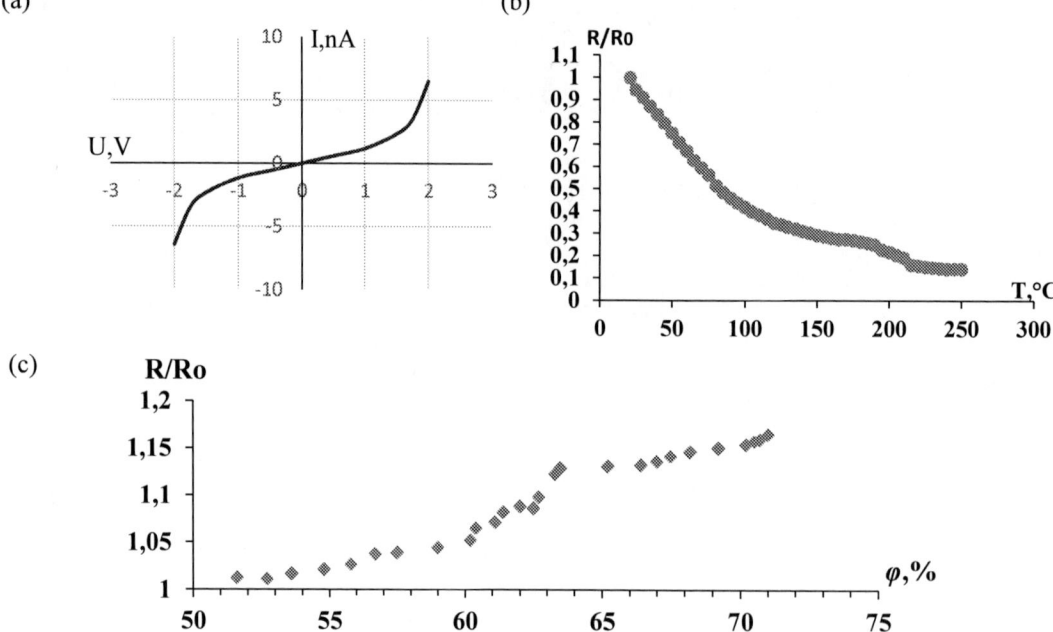

(c)

Figure 2(a, b, c). (a) Volt-ampere characteristics; The dependence of the normalized resistance of the formed structure on the basis of the ZnO nanorods array on the temperature (b) and on the humidity (c)

Measurements of gas sensitivity properties with respect to CO concentration of 1000 ppm at operating temperatures of 100 - 200∘C were carried out on an automated gas calibration bench of the [2]. The air (or nitrogen) and CO injection was carried out in the same scheme with a cycle that was repeated several times: the gas inlet was purged with air (or nitrogen). The gas flow rate was 0.3 l/min.

The dynamics of the sensor response at gas concentrations of 100, 500 and 1000 ppm and at temperatures of 100 and the answer of the sensor 200∘C is shown in Fig. 3(a, b, c).

(a)   (b)

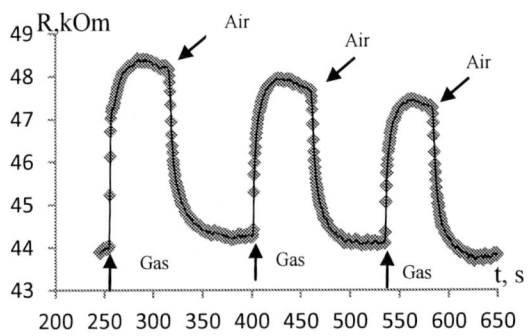 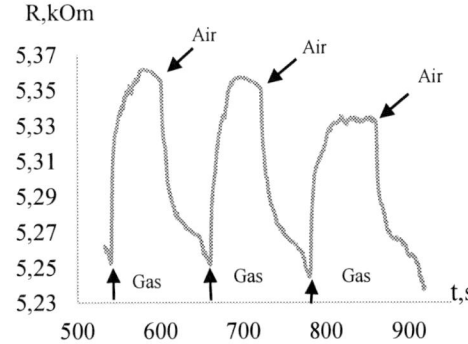

(c)

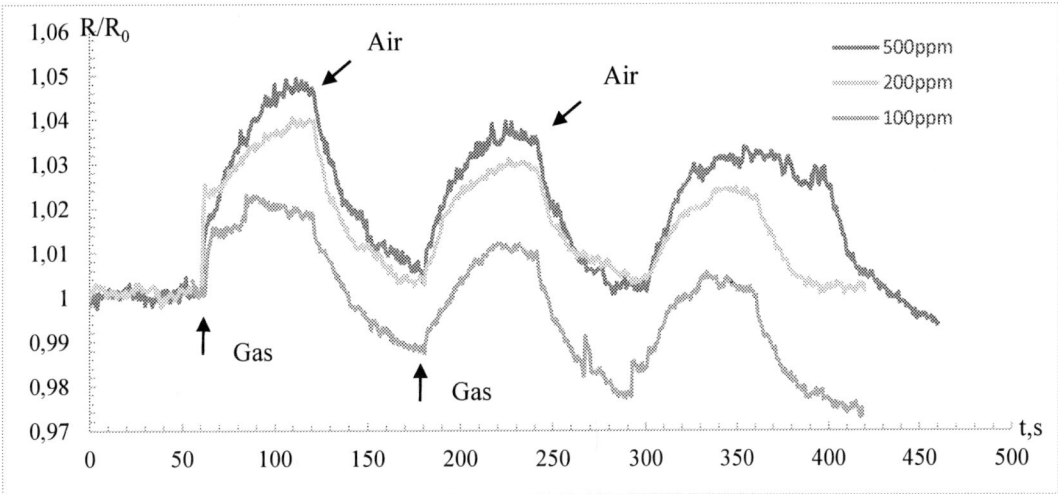

Figure 3(a, b, c). The response of the sensor to CO (1000 ppm) at 200°C (a); The response of the sensor to CO (1000 ppm) at 100°C (b) and normalized response to CO 100, 200, 500 of ppm at 200 ° C (c)

3. Results and discussion

In Fig. 1 the morphology of samples of ZnO nanorod arrays grown on a silicon substrate is shown. As can be seen from the figure, the ZnO nanorods have an average transverse dimension of about 30-40 nm and a length of up to 1 μm. Studies of electrophysical properties have shown that V-Cu-Ni contact metallization forms a contact close to an ohmic one (Figure 2 (a)), and the temperature dependence of the resistance shows a semiconductor character. The obtained ZnO samples are n-type conductivity semiconductor, but formed sensory structures based on them under the influence of moisture and CO molecules show a response not characteristic for a gas-sensitive material of n-type conductivity.

As is known at temperatures below 200oC on the surface of zinc oxide, adsorbed oxygen dissociates into atomic oxygen, from which oxygen ions O2- and O- [5] are formed. When carbon monoxide molecules interact with oxygen ions, the following reactions occur:

$$2CO + O_2^- \rightarrow 2CO_2 + e^-, (1)$$

$$CO + O^- \rightarrow CO_2 + e^-, (2)$$

When water molecules interact with the surface of ZnO and atomic oxygen, the following reactions occur on its surface [6]:

$$H_2O + O_o + 2Zn_{xn} \leftrightarrow 2(OH - Zn) + VO^{..} + 2e^-, \quad (3)$$

Where Oo is the oxygen atom at the oxygen site and VO$^{..}$ is the vacancy created at the oxygen site.
It is seen that when reactions (1) - (3) proceed to the surface layer of ZnO, electrons will be generated whose concentration and, accordingly, the conductivity of the near-surface layer will increase and the resistance decrease. However, we observe an increase in resistance when exposed to molecules of carbon monoxide (II) and moisture (Fig. 3). An explanation of this behavior of the sensory structure can be given under the assumption of the formation of two back-to-back p-n junctions. In this case, the measuring current goes from the metal contact to the ZnO nanorods, then to the silicon substrate, since its resistance is less than that of the ZnO nanorods, and, further, to the second metal contact through the ZnO nanorods. In this structure, one of the p-n-transition is always in reverse bias state; its voltage-current characteristic has the form shown in Figure 2(a). Under the influence of water molecules or CO molecules on the surface of ZnO nanorods in the n-semiconductor, the concentration of charge carriers increases. This leads to a decrease in the leakage current of the reverse biased p-n-junction. If the leakage current of the p-n junction falls, the resistance of the entire sensor structure will increase.
Thus, studies of the electrophysical properties and gas sensitivity characteristics of ZnO nanorod arrays formed on p-type silicon have shown that the obtained sensor structure is two back-to-back connected p-n junctions formed by n-type zinc oxide nanorods and p-type silicon substrate. The response of such a sensory structure is manifested in an increase in resistance when exposed to carbon monoxide (II) molecules and water.

References
[1] Myasoedova TN, Yalovega GE, Shmatko VA, Funik AO, Petrov VV 2016 *Sensors and Actuators B*. **230** 167
[2] Petrov V.V., Kalazhokov Zamir.Kh., Kalazhokov Zaur.Kh., Nadda M.Z., Kalazhokov Kh.Kh. 2016 Proc. 13 Int. Sci.-Tech. Conf. on Actual Problems Of Electronic Instrument Engineering (APEIE) vol 1 P4 (Novosibirsk: Novosibirsk State Technical University) p 21.
[3] Abdullin KhA, Bakranov NB, Ismailov DV, Kalkozova JK, Kumekov SE, Podrezova LV, Cicero G 2014 *Semiconductors* **48** No 4, 471
[4] Changa S.-P., Changa S.-.J., Lua C.-Y., Lib M.-Ju, Hsub C.-L., Chiouc Yu-Z., Hsuehd T.-J., Chen I-C. 2010 *Superlattices and Microstructures* **47** 772–778.
[5] Takata M., Tsubone D., Yanagida H. 1976 *Journal of The American Ceramic Society.* **59** 1-2, p.4
[6] W.P. Tai, J.H. Oh, 2002 *J.Mater. Sci., Mater. Electron.* **13** p.391–394.

Droplet epitaxy of In/AlGaAs nanostructures on the As-stabilized surface

S V Balakirev[1], M M Eremenko[2], I A Mikhaylin[2], V S Klimin[1], M S Solodovnik[1,2]

[1]Department of Nanotechnologies and Microsystems, Southern Federal University, Taganrog 347922, Russia
[2]Research and Education Center "Nanotechnologies", Southern Federal University, Taganrog 347922, Russia

Abstract. The article presents the results of experimental studies of the regimes of formation of self-organizing In/AlGaAs nanostructures by the method of droplet epitaxy under As-stabilization conditions at different Al content in the surface layer. Dependences of the influence of the growth temperature, surface composition, and deposition thickness on the geometric characteristics of the In nanodroplet arrays such as density, size and dispersion. The possibility of controlling the parameters of nanostructures array by changing the Al content in the surface layer. An unusual dynamics of the change in the critical thickness of the formation of In nanodroplers is revealed with a change in the composition of the surface.

1. Introduction

The existing methods for the formation of A3B5 quantum dots by molecular beam epitaxy (MBE) are based on the Stransky-Krastanov mechanism – the elastic relaxation of mechanical stresses in lattice-mismatched systems, the growth of which is accomplished by the simultaneous deposition of components of III and V groups [1–3]. The disadvantages of this approach are the need for mismatch of crystal lattices, as well as the interdependence between the density and size of quantum dots formed by such a mechanism [1]. Moreover, the mismatch magnitude should lie within a certain range of values [1]. The combination of these factors dramatically narrows the range of combinations of materials and the ability to control the geometric characteristics of quantum dots [4, 5].

At the same time, droplet epitaxy [6, 7], based on the separate deposition of the components, lacks these drawbacks. Droplet epitaxy allows not only independently control the density and size of quantum dots [8, 9], but also to form quantum dots in any A3B5 systems [7, 10-13]. Moreover, the method is promising for the integration of low-dimensional A3B5 systems with silicon technology, as well as for the creation of different types of hybrid systems of great interest for optoelectronics and THz devices [14-20].

The use of multi-stage droplet epitaxy techniques [21-23] allows to effectively minimize the influence of negative factors associated with pre-surface surface treatment in the case of using modified substrates [24–29], which also favorably distinguishes this method from other A3B5 quantum dot synthesis techniques.

However, today there is no unambiguous representation not only of the microscopic processes and behavior of individual atoms with growth in the drip epitaxy regime in multicomponent systems, but there are also practically no studies on growth on surfaces of variable composition, which is important for controlling the parameters of the quantum dots formed.

Content from this work may be used under the terms of the Creative Commons Attribution 3.0 licence. Any further distribution of this work must maintain attribution to the author(s) and the title of the work, journal citation and DOI.
Published under licence by IOP Publishing Ltd

In the present work, we carry out experimental studies of the formation of the self-assembled In nanostructures on the planar surfaces of AlGaAs epitaxial layers on the surfaces of epitaxial layers of variable composition.

2. Experiment

Experimental studies of the droplet epitaxy of In/AlGaAs nanostructures were carried out using the SemiTEq STE 35 MBE system. GaAs(001) wafers of the epi-ready class were used as substrates. After the procedure for thermal desorption of the oxide and smoothing of the GaAs surface in the As flux, a 400 nm thick GaAs buffer was grown at standard conditions [30, 31]: substrate temperature $T = 580°C$, effective growth rate $V = 1$ ML/s and V/III flux ratio $J = 4$. Then, under the same conditions, an AlGaAs layer 10 nm thick was grown, the mole fraction of Al in which varied depending on the sample stepwise and amounted to 0.25, 0.5, and 1. Then all the sources were simultaneously closed, and the substrate temperature was lowered to the values at which the studies of self-organizing nanostructures formation were carried out. The cooling modes were selected in such a way as to ensure that the surface has the As-stabilized structure, observed during the formation of the functional layer by RHEED.

After stabilizing the temperature of the sample at a predetermined value, In deposition was carried out. The substrate temperature T in this case varied from 150 to 300°C in order to avoid desorption of adatoms and activation of exchange processes that made it difficult to interpret the experimental data. The growth rate varied in the range 0.25-0.5 ML/s. The deposition thickness of In (H) varied in the range 0.5-3 ML with step 0.25 ML. After completion of growth, the substrate was rapidly cooled in order to suppress undesirable material redistribution over the surface due to surface diffusion.

3. Results and discussion

The results of experimental studies of the In/AlGaAs system showed a significant difference from the results obtained for Ga/AlGaAs [32],which is due to much greater mobility and chemical activity of In adatoms on the surface.Because of the particular features of the system under consideration, the greatest interest is in the range of small effective thicknesses, which makes it possible in the future to ensure the formation of quantum dots with sizes providing acceptable structural perfection and energy characteristics. The high mobility of the In adatoms results in a substantial increase in the dimensions of the nanostructures in comparison with Ga. Thus, at an equivalent deposition thickness of 3 ML, the structure diameter was 3.67 nm for Ga and 20.21 nm for In at $T = 150°C$. Increasing the temperature to $T = 300°C$, the structure size increased to 6.7 nm for Ga and up to 98.41 nm for In. It should be especially noted that the dispersion of the sizes of the nanostructures In is much lower than for the Ga – on average, it decreases from 30-40% at $T = 150°C$ to 5-15% at $T = 300°C$, which is also due to higher values of the surface diffusion of adatoms [33–37].

Figures 1 and 2a show the dependence of the average size and density, respectively, for the two extreme cases – for In/GaAs and In/AlAs. As it follows from an analysis of the dependences obtained, an increase in the Al content in the epitaxial layer significantly changes the kinetics of the growth processes. At growth temperatures up to 200°C, an increase in the Al fraction leads to an increase in the density In droplets from $2.2 \cdot 10^9$ cm^{-2} to $2.1 \cdot 10^{10}$ cm^{-2} while approximately preserving their sizes about 20 nm. Increase in growth temperature to 300°C leads to an increase in the size of the structures with an increase in the Al mole fraction, while maintaining the difference in density by an order of magnitude. Thus, at deposition thickness of 3 ML the average In droplet diameter increases from 98 nm for GaAs surface to 177 nm for AlAs surface. And at the same time the droplet density decreases from $1.8 \cdot 10^8$ cm^{-2} до $3.7 \cdot 10^7$ cm^{-2}, respectively.

In general, the range of variation in the density of droplet arrays also tends to expand as the Al content in the surface epitaxial layer increases.

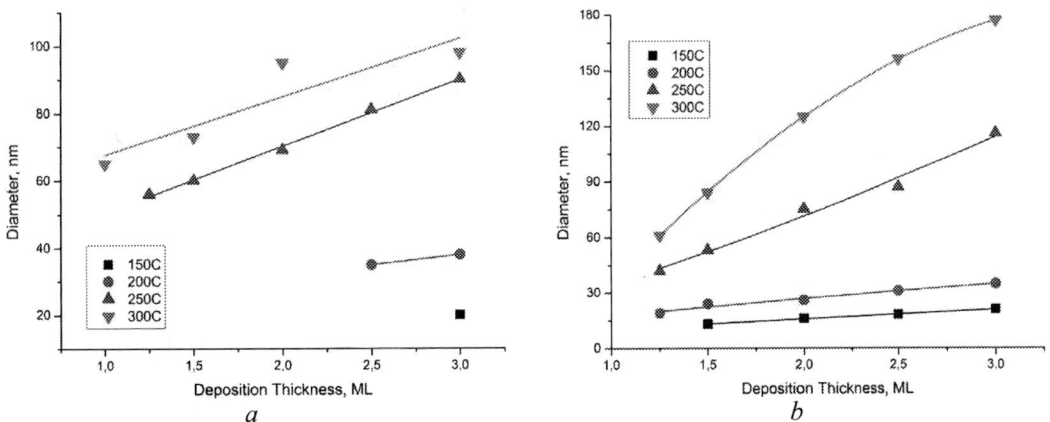

Figure 1. Dependence of the average size of In/GaAs (*a*) and In/AlAs (*b*) droplets on the effective deposition thickness at different growth temperatures.

Particular attention should be paid to the nature of the change in the critical thickness and, as a consequence, the thickness of the wetting layer with increasing Al content in the surface layer. In a case of GaAs surface (Al fraction is 0%) it has a pronounced tendency to decrease until it completely disappears when going into the incomplete condensation mode as it show at Figure 2*b*. At the same time in a case of AlAs surface (Al fraction is 100%) the thickness of the wetting layer is practically unchanged and is 1.0-1.25 ML over the entire temperature range under consideration. This unusual behavior of the system is due, apparently, to an increase in the chemical interaction between In adatoms and the surface when the composition of the functional layer changes. To give an unambiguous explanation of the processes and mechanisms underlying these phenomena will allow their further study with the use of the theoretical approaches developed by us [38, 39].

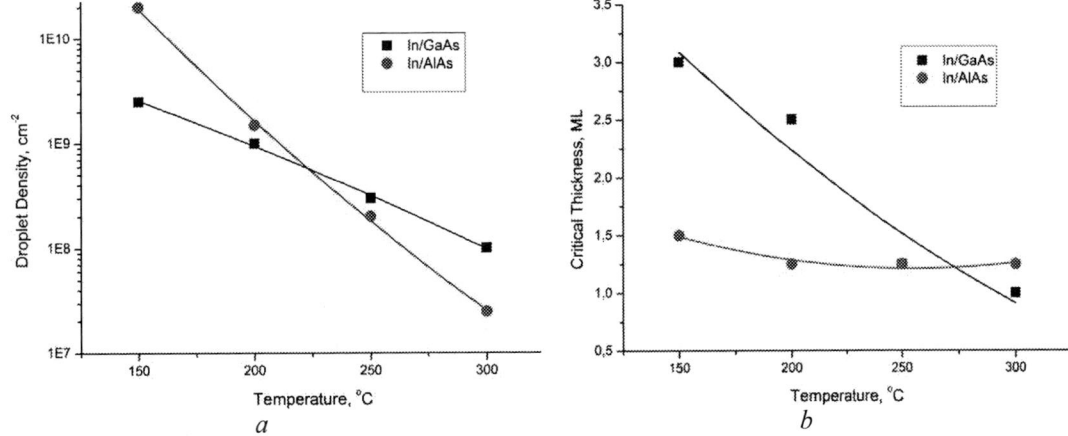

Figure 2. The temperature dependences of the density (*a*) and the critical thickness of the formation (*b*) of an ensemble of In/(Ga,Al)As nanostructures.

Statistical processing of the experimental data obtained by us showed that ensembles of self-organizing nanostructures significantly change their structural characteristics during the transition of formation from the subcritical thickness of deposition to supercritical (see Figure 3).

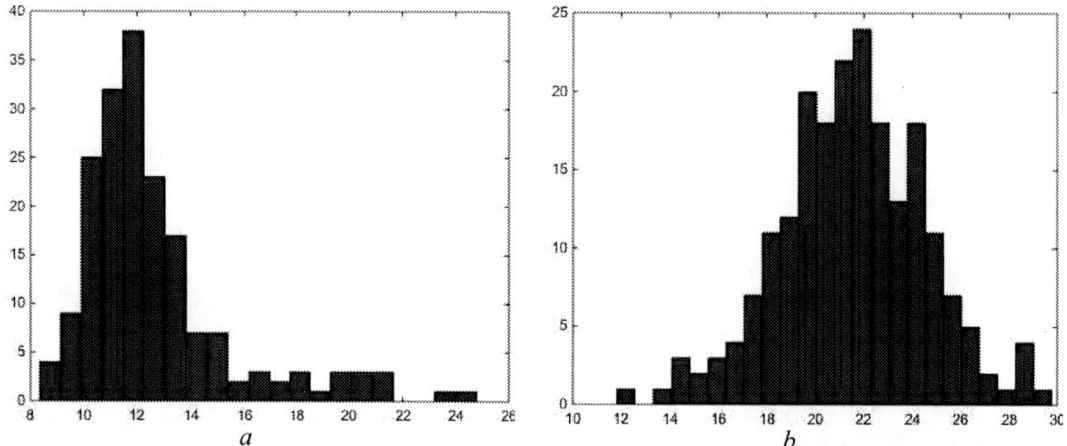

Figure 3. Histograms of the size distribution of an ensemble of In/AlAs nanostructures formed in different deposition modes: *a*) subcritical ($T = 150°C$, $H = 1.5$ ML) and *b*) supercritical ($T = 150°C$, $H = 3.0$ ML).

As can be seen from the Figure 3, in the subcritical region, the geometric parameters of the ensemble have a pronounced bimodal distribution, which, as we move to the supercritical region, changes to a unimodal distribution with a sequential decrease in the dispersion to the previously designated 5-15% as the substrate temperature increases.

It should be separately noted that an increase in the exposure time of structures, subject to the preservation of thermodynamic conditions on the surface, does not significantly change the nature of the distribution.

4. Conclusion

In summary, we carried out the experimental studies of the formation of the self-assembled In nanodroplets on the planar surfaces of AlGaAs epitaxial layers under As-stabilization conditions at different Al content in the surface layer. It is shown that an increase in the Al content in the epitaxial layer significantly changes the kinetics of the processes on the surface. This leads to an expansion of the range of possible densities of the obtained In nanodroplets without a significant change in their sizes. In addition, it was shown that with the increase of the Al fraction on the surface, the critical thickness of the formation of the In nanostructures is reduced and stabilized at a value of 1 ML. Also shown is the change in the distribution of the geometric characteristics of the nanostructures during a transition from subcritical to supercritical deposition modes.

Acknowledgements

This work was supported by the Russian Science Foundation Grant No. 15-19-10006. The results were obtained using the equipment of the Research and Education Center and Center for Collective Use "Nanotechnologies" of Southern Federal University.

References

[1]   Dubrovskii V G 2014 *Nucleation theory and growth of nanostructures* (Berlin: Springer)

[2] Heidemeyer H, Müller C, Schmidt O 2004 *J. Cryst. Growth.* **261** 444–449
[3] Helfrich M, Hu D Z, Hendrickson J, Gehl M, Rülke D, Gröger R, Litvinov D, Linden S, Wegener M, Gerthsen D, Schimmel T, Hetterich M, Kalt H, Khitrova G, Gibbs H M, Schaadt D M 2011 *J. Cryst. Growth.* **323** 187–190
[4] Martin-Sanchez J, Gonzalez Y, Gonzalez L, Tello M, Garcia R, Granados D, Garcia J M, Briones F 2005 *J. Cryst. Growth.* **284** 313–318
[5] Schramboeck M, Andrews A M, Roch T, Schrenk W, Strasser G 2007 *Microelectronic Engineer.* **84** 1443–1445
[6] Kim J S, Koguchi N 2004 *Appl. Phys. Lett.* **85** 5893– 5895
[7] Heyn C, Stemmann A, Schramm A, Welsch H, Hansen W, Nemcsics A 2007 *Phys. Rev. B* **76** 075317
[8] Lee J, Wang Z M, Salamo G 2007 *J. Phys. Condens. Matt.* **19** 176223
[9] Sablon K A, Lee J H, Wang Z M, Shultz J H, Salamo G J 2008 *Appl. Phys. Lett.* **92** 203106
[10] Egawa T, Ogawa A. Jimbo T, Umeno M 1998 *Jap. J. Appl. Phys.* **37** 1552
[11] Kazi Z I, Egawa T, Jimbo T, Umeno M 1999 *IEEE Photonics Technol. Lett.* **11** 1563–1565
[12] Mano T, Kuroda T, Mitsuishi K, Yamagiwa M, Guo X, Furuya K, Sakoda K, Koguchi N 2007 *J. Cryst. Growth.* **301** 740–743
[13] Wu J, Li Zh, Shao D, Manasreh M O, Kunets V P, Wang Z M, Salamo G J, Weaver B D 2009 *Appl. Phys. Lett.* **94** 171102
[14] Mano T, Watanabe K, Tsukamoto S, Fujioka H, Oshima M, Koguchi N 1999 *Jap. J. Appl. Phys.* **38** L1009
[15] Mano T, Watanabe K, Tsukamoto S, Fujioka H, Oshima M, Koguchi N 2000 *J. Cryst. Growth.* **209** 504–508
[16] Wang Z M, Holmes K, Mazur Y I, Ramsey K A, Salamo G J 2006 *Nanosc. Res. Lett.* **1** 57–61
[17] Lee J, Wang Z M, Hirono Y, Salamo G 2011 *IEEE Transac. Nanotech.* **10** 600–605
[18] Mano T, Watanabe K, Tsukamoto S, Koguchi N 2000 *Appl. Phys. Lett.* **76** 3543–3545
[19] Somaschini C, Bietti S, Koguchi N, Sanguinetti S 2009 *Nano Lett.* **9** 3419–3424
[20] Somaschini C, Bietti S, Sanguinetti S, Koguchi N, Fedorov A 2010 *Nanotech.* **21** 125601
[21] Stemmann A, Heyn C, Koppen T, Kipp T, Hansen W 2008 *Appl. Phys. Lett.* **93** 123108
[22] Koppen A, Koppen T, Grave M, Wildfang S, Mendach S, Hansen W, Heyn C 2009 *J. Appl. Phys.* **106** 064315
[23] Vasilenko M A, Neizvestny I G, Shwartz N L 2015 *Comput. Mater. Sc.* **102** 286–292
[24] Ageev O A, Smirnov V A, Solodovnik M S, Rukomoikin A V, Avilov V I 2012 *Semiconductors* **46** 1616–1621
[25] Avilov V I, Ageev O A, Smirnov V A, Solodovnik M S, Tsukanova O G 2015 *Nanotech. in Russia* **10** 214–219
[26] Klimin V S , Tominov R V, Eskov A V, Krasnoborodko S Y, Ageev O A 2017 *J. Phys.: Conf. Ser.* **917** 092005
[27] Tominov R V, Bespoludin V V, Klimin V S , Smirnov V A, Ageev O A 2017 *IOP Conf. Ser.: Mater. Sci. Eng.* **256** 012023
[28] Klimin V S, Solodovnik M S, Smirnov V A, Eskov A V, Tominov R V, Ageev O A 2016 *Proceedings of SPIE* **10224** 102241Z
[29] Ageev O A, Klimin V S, Solodovnik M S, Eskov A V, Krasnoborodko S Y 2016 *J. Phys.: Conf. Ser.* **741** 012178
[30] Ageev O A, Solodovnik M S, Balakirev S V and Eremenko M M 2016 *Phys. Solid State* **58** 1045–1052
[31] Ageev O A, Solodovnik M S, Balakirev S V, Eremenko M M and Mikhaylin I A 2016 *J. Phys.: Conf. Ser.* **741** 012012
[32] Solodovnik M S, Balakirev S V, Eremenko M M, Mikhaylin I A, Avilov V I, Lisitsyn S A, Ageev O A 2017 *J. Phys. Conf. Ser.* **917** 032037
[33] Ageev O A, Solodovnik M S, Balakirev S V, Mikhaylin I A, Eremenko M M 2017 *J. Cryst.*

Growth **457** 46–51

[34] Ageev O A, Solodovnik M S, Balakirev S V, Eremenko M M 2016 *J. Vac. Sci. Technol. B* **34** 041804

[35] Balakirev S V, Solodovnik M S, Eremenko M M, Mikhaylin I A, Ageev O A 2017 *J. Phys. Conf. Ser.* **917** 032034

[36] Ageev O A, Solodovnik M S, Balakirev S V, Mikhaylin I A 2016 *Technical Physics* **61** 971–977

[37] Ageev O A, Solodovnik M S, Balakirev S V, Mikhaylin I A 2016 *J. Phys. Conf. Ser.* **681** 012036

[38] Balakirev S V, Solodovnik M S, Ageev O A 2018 *Phys. Status Solidi B* **255** 1700360

[39] Balakirev S V, Solodovnik M S, Ageev O A 2017 *J. Phys. Conf. Ser.* **917** 032033

SPBOPEN 2018   IOP Publishing

Investigation of the electrode material influence on the titanium oxide nanosize structures memristor effect

V I Avilov[1], A S Kolomiytsev[1], R V Tominov[1], N I Alyabyeva[2], E M Bykova[3]

[1]Southern Federal University, Department of Nanotechnologies and Microsystems Technology, Taganrog, 347922, Russia

[2]Université Paris-Sud, Espace Technologique, Saint-Aubin, 91405, France

[3]Taganrog Scientific Research Institute of Communications, Taganrog, 347900, Russia

Abstract. The paper presents the investigation results of the titanium oxide nanosized structure memristor effect with various upper electrodes. It was shown that titanium oxide nanosized structures obtained by the local anodic oxidation method exhibit a memristor effect without carrying out an additional electroforming operation, and also the material of the upper electrode affects the structure memristor effect characteristics.

1. Introduction

Nanoscale structures demonstrating the memristor effect have broad prospects for using in the development and creation of resistive memory elements RRAM and synaptronics due to their ability to switch between high-resistance state (HRS) and low-resistance state (LRS) when switching voltage pulses are applied [1-3]. At the same time, the memristor structures based on titanium oxide are most preferable, since they have a high read/write speed, low power consumption, high reproducibility of parameters and stability of characteristics [4-5]. Analysis of modern methods of lithography has shown that classical technological processes based on the use of optical lithography have serious limitations in the production of memristor structures with dimensions less than 10 nm, in addition, the structures thus formed require an additional electroforming operation to implement the memristor switching mechanisms [6] . The method of local anodic oxidation (LAO) is one of the most promising methods for the oxides production, since it allows the electroforming-free memristor oxide nanosized structures (ONS) formation with high spatial resolution [7-13]. In the formation of memristor structure, in addition to the formation of titanium oxide, an important task is the upper contact electrodes precise formation, since the contact electrodes material has a significant effect on the memristor ONS characteristics. Precise contact formation can be realized by local ion-stimulated deposition methods by a focused ion beam from the gas phase [14-16]. Thus, it is important and relevant to study the upper electrode material influence on the memristor effect parameters on the basis of titanium oxide nanosized structures obtained using probe nanolithography methods.

2. Experiment

Content from this work may be used under the terms of the Creative Commons Attribution 3.0 licence. Any further distribution of this work must maintain attribution to the author(s) and the title of the work, journal citation and DOI.

Published under licence by IOP Publishing Ltd

In the work, experimental studies were carried out on a thin titanium film 20 nm thick, formed by magnetron sputtering on the surface of the Si/SiO_2 structure. Titanium film local anodic oxidation was carried out on a raster graphic template using a scanning probe microscope (SPM) Solver P47 Pro (NT MDT, Russia). In the course of experimental studies, a titanium ONS with lateral dimensions of 2×2 μm and a thickness of 2.9±0.2 nm was formed (Figure. 1).

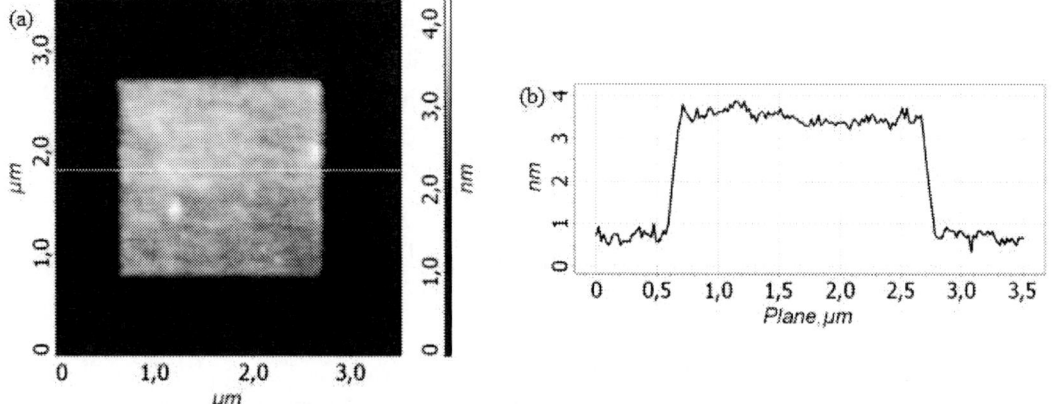

Figure 1(a, b). (a) AFM-image of titanium ONS; **(b)** profile along the line.

After that, on the formed oxide structure, the current-voltage characteristic were measured in the current mode of AFM spectroscopy, using various cantilevers with platinum, nitride-titanium and carbon coating as the upper contact electrode. In this case, the thin titanium film acted as the lower contact electrode (Figure. 2).

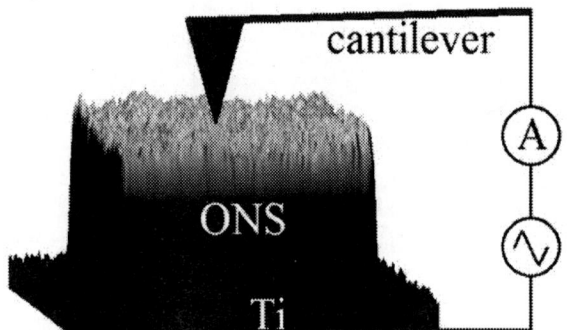

Figure 2. Scheme of measuring current-voltage characteristic

3. Results and discussion

Obtained dependences analysis showed that the formed titanium ONS exhibits a memristor effect without additional electroforming operation, in addition, it was shown that the use of cantilevers with different coatings as the upper electrode has a significant effect on the memristor effect parameters (Figure. 3).

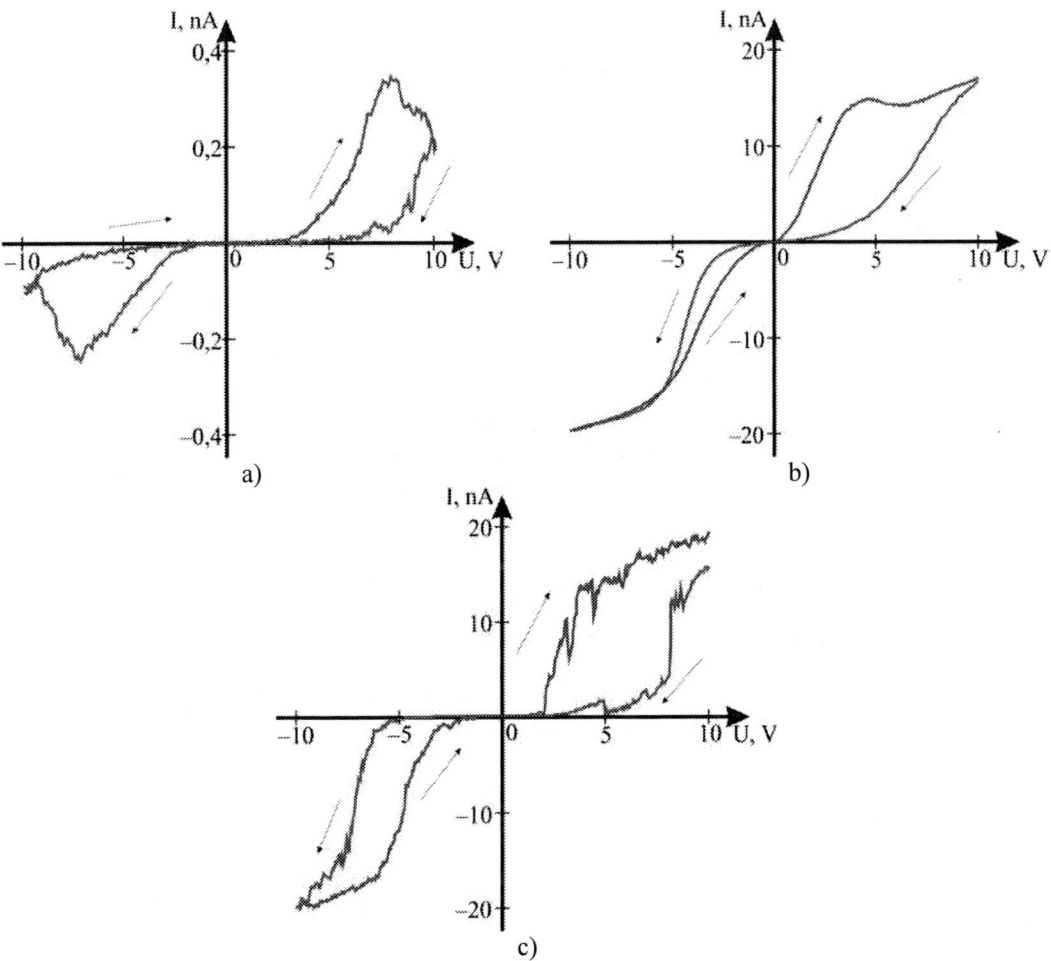

Figure 3(a, b, c). Current-voltage characteristic of titanium ONS obtained with cantilever coated by **(a)** platinum; **(b)** titanium nitride; **(c)** carbon.

Thus, when using a cantilever with Pt coating, a symmetrical I-V characteristic was obtained (Figure. 3a) with low current values, with the switching voltages to HRS (U_{res}) and to LRS (U_{set}) being 7.5 and -7.5 V, respectively. At the same time, the resistance of the ONS of titanium in the HRS is 327 GΩ, and in the LRS is 22 GΩ, and the resistance ratio in these states is 15.

When using a cantilever with TiN coating, an asymmetric I-V characteristic was obtained (Figure. 3b), however, in the negative voltage range, the memristor effect is negligible. The switching voltages U_{res} and U_{set} were 4.5 and -5 V, respectively, the resistance of the structure in the HRS is 4.65 GΩ, and in the LRS is 0.39 GΩ and the resistance ratio in these states is 12.

When using a cantilever with a carbon coating, an asymmetric -V characteristic is also observed (Figure 3c), with switching voltages U_{res} and U_{set} equal to 7.5 and -8 V, respectively, where the current through the ONS is absent when applying voltages less than $\pm$ 5 V in case of HRS and when applying voltages less than $\pm$ 2 V in case of LRS. The resistance of the structure in the HRS state is 1.74 GΩ, and in the LRS state it is 0.3 GΩ, while the resistance ratio in the high-resistance state to the low resistance is 6.

4. Conclusion

Thus, the work showed the influence of the upper contact electrode material on the memristor effect of oxide nanoscale titanium structures. It is shown that the greatest ratio of resistances in the HRS and

LRS is observed when cantilevers with platinum coating are used as electrodes, and the lowest switching voltages are used with cantilevers with nitride-titanium coating. The obtained results can be used to develop technological processes for the formation of RRAM elements based on oxide nanoscale structures, as well as synaptronic elements in the development of architectures based on neuromorphic systems.

References

[1] Ting-Chang C, Kuan-Chang C, Tsung-Ming T, Tian-Jian C and Simon M. 2016 Resistance random access memory *Materials Today* **19** N5 254-64

[2] Cong Y, Jiaji W, Gang H, Jieqiong Z, Tengfei D, Pin H and HaoW 2016 Physical Mechanism and Performance Factors of Metal Oxide Based Resistive Switching Memory: A Review *Journal of Materials Science & Technology* **32** 1-11

[3] Fei X, Jiqian Z, Tingting F, Shoufang H, Maosheng W 2018 Physical Synchronous dynamics in neural system coupled with memristive synapse *Nonlinear Dynamics* **92** 1395-1402

[4] Pan F, Gao S, Chen C, Song C and Zeng F 2014 Recent progress in resistive random access memories: Materials, switching mechanisms, and performance *Materials Science and Engineering* R **83** 1-59

[5] Sieu D and Ramanathan S 2011 Adaptive oxide electronics: A review *J. Appl. Phys.* **110** 071101

[6] Amer, S., Hasan, M.S., Rose, G.S. 2018 Analysis and Modeling of Electroforming in Transition Metal Oxide-Based Memristors and Its Impact on Crossbar Array Density *IEEE Electron Device Letters* **39** 813909

[7] Chen A 2016 A review of emerging non-volatile memory (NVM) technologies and Applications *Solid-State Electronics* **125** 25–38

[8] Avilov, V.I., Polupanov, N.V., Tominov, R.V., Smirnov, V.A., Ageev, O.A. 2017 Scanning probe nanolithography of resistive memory element based on titanium oxide memristor structures *IOP Conference Series: Materials Science and Engineering* **256** 012001

[9] Linggang Z, Jian Z, Zhonglu G and Zhimei S 2015 An overview of materials issues in resistive random access memory *J Materiomics* **1** 285–95

[10] Avilov, V.I., Ageev, O.A., Jityaev, I.L., Kolomiytsev, A.S., Smirnov, V.A. 2016 Investigation of memristor effect on the titanium nanowires fabricated by focused ion beam *Proceedings of SPIE - The International Society for Optical Engineering* **10224** 102240T

[11] Siles P F, Archanjo B S, Baptista D L, Pimentel V L and Joshua J 2011 Nanoscale lateral switchable rectifiers fabricated by local anodic oxidation *J. Appl. Phys.*, **110** 024511

[12] Avilov, V.I., Ageev, O.A., Blinov, Y.F., Konoplev, B.G., Polyakov, V.V., Smirnov, V.A., Tsukanova, O.G. 2015 Simulation of the formation of nanosize oxide structures by local anode oxidation of the metal surface *Technical Physics* **60** 717-723

[13] Avilov, V.I., Ageev, O.A., Smirnov, V.A., Solodovnik, M.S., Tsukanova, O.G. 2015 Studying the modes of nanodimensional surface profiling of Gallium Arsenide epitaxial structures by local anodic oxidation surface *Nanotechnologies in Russia* **10** 214-219

[14] Ageev, O. A.; Vnukova, A. V.; Gromov, A. L.; Il'in, O. I.; Kolomiytsev, A. S.; Konoplev, B. G.; Lisitsyn, S. A. 2014 Analysis of modes of nanoscale profiling during ion-stimulated deposition of W and Pt using the method of focused ion beams *Nanotechnologies Russ* **9** 145–150

[15] Ageev, O.A.; Alekseev, A.M.; Vnukova, A.V.; Gromov, A.L.; Kolomiytsev, A. S.; Konoplev, B. G.; Lisitsyn, S. A. 2014 Studying the resolving power of nanosized profiling using focused ion beams *Nanotechnologies Russ* **9** 26–30

[16] Ageev O.A., Kolomiytsev A.S., Bykov A.V., Smirnov V.A., Kots I.N. 2015 Fabrication of advanced probes for atomic force microscopy using focused ion beam *Microelectronics Reliability* **55** 2131–2134

SPBOPEN 2018

IOP Publishing

IOP Conf. Series: Journal of Physics: Conf. Series **1124** (2018) 022020   doi:10.1088/1742-6596/1124/2/022020

Properties of semiconductor GaAs nanoparticles, synthesized by combination of mechanical milling methods and chemical etching

N Yu Yashina[1,2], A J K Al-Alwani[1,3], O Yu Tsvetkova[2], A S Kolesnikova[2], E G Glukhovskoy[1,2], V P Sevostyanov[4]

[1] Department of Nano- and Biomedical Technologies, Saratov State University, Saratov 410012, Russia

[2] Education and Research Institute of Nanostructures and Biosystems, Saratov State University, Saratov 410012, Russia

[3] Babilon University, Babel, Iraq

[4] Scientific Research Institute of Technology of Organic, Inorganic Chemistry and Biotechnology, Saratov 410012, Russia

Abstract. The results of realization of the method for obtaining A3B5 nanoparticles with the help of ball-milling and additional liquid-phase etching of the single-crystal material are presented in the article. The conditions for liquid chemical etching of powders of A3B5-type semiconductors (GaAs), which based on the classical technique for polishing the semiconductor surface with a peroxide-ammonia mixture were studied. The dependence of the etching rate on the time and the use of surfactants (cetyltrimethylammonium bromide) were investigated to control the etch rate and nanoparticle sizes. The size distribution and chemical composition of the obtained nanoparticles were studied using scanning electron microscopy (BSE, EDX). This research has shown that nanoparticles A3B5, obtained by this method, have sizes from 2-5 nm, suitable for the synthesis of semiconductor quantum dots.

1. Introduction

Present concept of nanotechnology include materials, processes and various kinds of objects with a characteristic nano-range from 1 to 100 nm, where the size is the main factor in the creation of materials and other systems with specific properties for further applications [1,2]. Typical nano-objects such as quantum dots attract special attention due to their properties. The fabrication of GaAs quantum dots (QDs) is important in applications for optical devices, electronic and other devices [3]. Synthesize of A3B5 not possible by direct chemical methods, but they can be grown as monocrystals from melts of equimolecular amounts of components according to the method Ya. Czochralski [4,5]. The A2B6-systems (in contrast to A3B5-materials) are easily obtained in the form of colloidal solutions (solutions of quantum dots ZnS, CdS, CdSe, etc.) [6,7]. This allows them to be used as independent units in other technological processes, to be modified their composition, structure, and other properties. It can easy to build up inorganic shells to the nanoparticles A2B6 in the composition of colloidal solutions and can set other properties for them: chemical properties, different width of the bandgap, can be oxidized, passivation their surface, cover with layers of organic stabilizer, etc. [8-12].

Content from this work may be used under the terms of the Creative Commons Attribution 3.0 licence. Any further distribution of this work must maintain attribution to the author(s) and the title of the work, journal citation and DOI.

Published under licence by IOP Publishing Ltd

At a certain stage, it is possible to apply chemical or electrochemical etching processes to reducing the size of the nanoparticles by analogy with massive materials**Error! Reference source not found.**. However, the information of the chemical etching of powders and reduction of their sizes up to the nanoscale is practically absent.

In accordance with all this, the aim of this work was to study the synthesis of nanoparticles of A3B5 semiconductors by liquid chemical etching. The most popular narrow-gap semiconductors of this series, gallium arsenide (GaAs) is in high demand in the scientific and applied fields.

2. Experimental

The starting powder of GaAs was obtained by grinding single-crystal plates in a planetary ball mill of type Pulverisette (Fritsch, Germany). Grinding was carried out in two stages when the speed of rotation of the grinding bowls 600 rpm. Grinding time at each process was 1 h. As the grinding ball was used hard metal tungsten carbide (WC) with diameters of 10 and 1 mm on first and second stages, respectively.

After dispersing, the particles sizes were examined by scanning electron microscopy (SEM) Tescan Mira II LMU. SEM-analysis of the chemical elemental composition of the powders studied and showed the content of gallium and arsenic about 49% each, as well as traces of oxygen to 2%, possibly due to partial oxidation of the substance during grinding.

The granulometric analysis was performed with the help of Malvern Zetasizer NanoZS.

For etching GaAs powder has taken as a base the oxidative peroxide-ammonia mixture which is well known in the electronics industry. In our study, we implemented 2 options based on this mixture.The components in each option were presented in the following amounts and proportions:

- Option#1: consisting of deionized water, an aqueous solution of ammonia and hydrogen peroxide ($H_2O:aqNH_4OH:H_2O_2$) in volume ratio30:6:9; 15 ml mixture was used for etching per 1 g of dry GaAs powder;

- Option#2: the etching mixture was prepared like Option#1, but using 0.001 M aqueous solution of cationogenic surfactant [$(C_{16}H_{33})N(CH_3)_3$]Br – cetyltrimethylammonium bromide (CTAB)instead of deionized water;15 ml mixture – per 1 g of dry GaAs powder.

3. Results and discussion

Option#1:

When a GaAs sample is contacted with etching mixture, a rapid exothermic reaction is observed in the first few seconds of which the temperature of the reaction mixture was raised to T=70 °C. The first 2-3 minutes the temperature was kept quite high. After that, the temperature began to gradually decrease. Then the temperature solution gradually decreased to room temperature after 30 minutes.

During the experiment, samples were taken for measuring particle size distribution and composition at different time periods of 1, 2, 3, 5, 30 and 90 min from the beginning of etching. And each aliquot was placed in eppendorf centrifuge tubes, where pre-filled deionized water to quench the chemical reaction (stop bath). Within 5 minutes, the particles in the solution had a size of 2-5 nm. After 90 min from the start of etching the selected sample has predominant size of ~84 nm, indicating the agglomeration process. Fig.1 shows the nanoparticles size distribution.

The elemental composition of the sample of gallium arsenide after etching was examined on a scanning electron microscope. EDX analysis showed that the elemental composition of GaAs nanoparticles after etching in the peroxide-ammonia corresponded almost to the initial. The dried GaAs solution residue contained (weight %): Ga – 27, As – 27; O – 39; N – 7. In fact, gallium arsenide in this chemical reaction is a catalyst for the decomposition of hydrogen peroxide. As a result of this decomposition, atomic oxygen is formed, the resulting oxygen reacts with ammonia and can oxidize it to ammonium nitrate and it explains the presence of nitrogen in results of EDX analysis. At the same time, the reactor contained gallium arsenide particles (as the black sediment) that did not react. When the components interact, their concentration in the solution changes, which affected the final result, i.e. the size distribution of the nanoparticles and their concentration.

It is well known that the smaller particle size the higher tendency to agglomeration, which is undesirable for obtaining for example quantum dots etc.

To avoid such a negative effect of agglomeration, stabilizers are used, which are applied to the nanoparticles and preventing them from sticking together. In this study, the surfactant of cationogenic cetyltrimethyl ammonium bromide (CTAB) – $[(C_{16}H_{33})N(CH_3)_3]Br$ was used as a stabiliser, which forms stable micelles in an aqueous medium.

Etching solution in Option#2 was prepared on the base of 0.001 M aqueous solution of CTAB (see section of «Materials and methods»). At this suitable concentration the system of «H_2O-CTAB» micelles formed. Such micelle system was used for stabilization of micro- and nanoparticles of gallium arsenide. The granulometric composition of etched GaAs powders in solution #2 during 5 and 30min within 1-4 nm. After 90min the nanoparticles with predominance sustainable size 3 nm (Fig. 1).

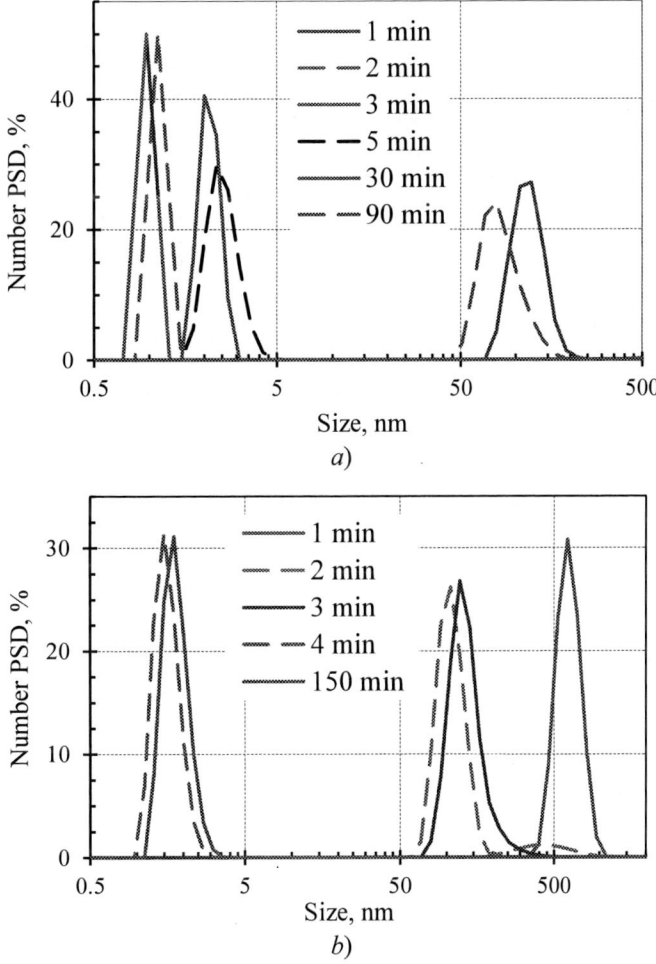

a)

b)

Figure 1. Dependence of the nanoparticles size distribution on etching time: (*a*) – Option#1 GaAs; (*b*) –Option#2 GaAs.

Fig.2 shows the temperature and acidity changes in time during the synthesis of nanoparticles. Since the consumption of aqueous ammonia in the system becomes pH = 9.5 and remains stable over time, indicating an attenuation reaction.

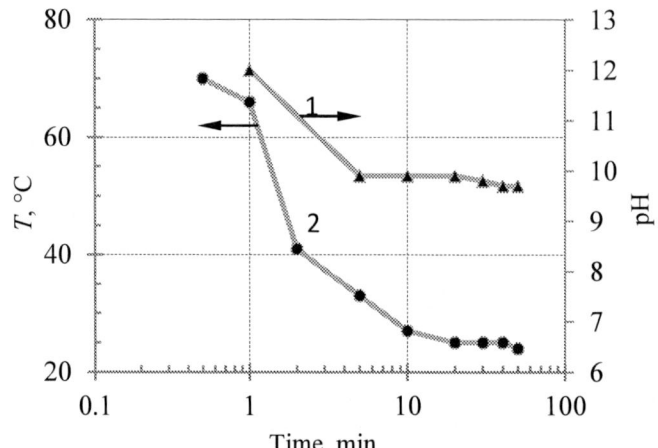

Figure 2. Change in acidity (1) and temperature (2) from the etching time: (1) – pH for GaAs; (2) – T for GaAs

4. Conclusions

Reproducible and accurate synthesis of A3B5 nanosize powder is possible by using of liquid chemical etching with help of peroxide-ammonia solution. In this method the starting powder of A3B5 which that obtained by grinding can be used. The described method allows to obtained nanoparticles with sizes from 2-5 nm (for GaAs). When using water-soluble surfactants, it is possible to reduce the process of particle agglomeration and slow down the rate of etching. The typical etching reaction time is 5-10 minutes. The obtained nanoparticles with 2-5 nm are suitable for the using as the semiconductor quantum dots.

Acknowledgments

This work was supported by grants from the Russian Foundation for Basic Research Projects No 17-07-00407-a, 17-07-00139 and Presidential scholarship SP-2502.2016.1.

References

[1]   Bera D, Qian L, Tseng T-K and Holloway P H 2010 *Quantum Dots and Their Multimodal Applications: A Revie,*Materials, **3** 2260-2345

[2]   Ganeev R A, Ryasnyanskiy A I and Usmanov T 2007 *Optical and nonlinear optical characteristics of the Ge and GaAs nanoparticle suspensions prepared by laser ablation*, Optics Communications, **272** 242-246

[3]   Seravalli L, Frigeri P, Nasi L, Trevisi G and Bocchi C 2010 *Metamorphic quantum dots: Quite different nanostructures*, Journal of Applied Physics, **108** 064324

[4]   Czochralski J 1918 *Einneues Verfahren zurMessung der Kristallisationsgeschwindigkeit der Metalle*, Zeitschriftfür Phys. Chemie, **92** 219-221

[5]   Brodie I and Muray J J 1992 *The physics of micro/nano-fabrication*, Springer

[6]   Qin H, Meng R, Wang N and Peng X 2017 *Photoluminescence intermittency and photo-bleaching of single colloidal quantum dot*, Adv Mater, **29** 1606923

[7]   Rudko G, Fediv V, Davydenko I et al 2016 *Synthesis of capped $A^{II}B^{VI}$ nanoparticles for fluorescent biomarker*, Nanoscale Research Letters, **11** 1-6

[8]   Krupko E V, Grodzyuk G Ya, Khalavka Yu B et al 2011 *Effect of the composition of the reaction mixture on the preparation of L-cysteine-stabilized CdS nanoparticles and their optical properties*, Theoretical and Experimental Chemistry, **47** 101-107

[9]   Chang J and Waclawik E R 2014 *Colloidal semiconductor nanocrystals: controlled synthesis and surface chemistry in organic media*, RSC Advances, **4** 23505-23527

[10] Pal U, Bautista-Hernández A, Koshizaki N et al 2001 *Synthesis of GaAs nanoparticles embedded in SiO$_2$ matrix by radio frequency co-sputtering technique*, ScriptaMaterialia, **44** 1841-1846

[11] Levchenko I V, Stratiychuk I B, Tomashyk V N et al 2017 *Chemical interaction of InAs, InSb, GaAs, and GaSb crystal surfaces with (NH$_4$)$_2$Cr$_2$O$_7$–HBr–citric acid etching solutions*, Inorganic Materials, **53** 1109-1114

[13] Bioud Y A, Boucherif A, Belarouci A et al 2016 *Chemical Composition of Nanoporous Layer Formed by Electrochemical Etching of p-Type GaAs*, Nanoscale Res. Lett., **11** 1-8

Research of influence Al on luminescence and dark current-voltage characteristics of InAs/GaAs heterostructures

A.S. Pashchenko[1], E.M. Danilina[1], N.M. Bogatov[2]

[1]Federal research centre the Southern Scientific Centre RAS, 41, Chekhov st., Rostov-on-Don, 344006, Russia;

[2]Kuban State University, 149, Stavropolskay st., Krasnodar, 350040, Russia

Abstract. A study was made of $Al_xGa_{1-x}As$ wide-bandgap potential barriers influence on photoluminescence and dark current-voltage characteristics of InAs/GaAs heterostructures obtained by an ion-beam deposition method. It was established that employing $Al_xGa_{1-x}As$ barriers cause a shift of InAs quantum dots ground-state photoluminescence emission peak in high-energy band (blue shift), increase in intensity and decrease of full width a half at maximum. In addition, the measurements of dark current-voltage characteristics show that increase of Al content in barrier leads to decrease of bias voltage (to 0.48 V) of altering charge carriers transfer mechanism from thermoelectron emission to tunnelling assisted by external electric field. It was established that the use of the $Al_{0.4}Ga_{0.6}As$ potential barrier can produce minimal dark current (10^{-8} A) in InAs/GaAs heterostructures with quantum dots.

1. Introduction

The quantum dots (QDs) heterostructures based on AIIIBV materials [1-3], Ge/Si [4] et al are of great interest for the researchers. This is due to next generation high-efficient photosensitive optoelectronic devices development. The major problem for existing infrared photodetectors (HgCdTe, II-type superlattices, GaSb) is necessity their cooling to increase the detectivity. One way to solve this problem is to grow semiconductor heterostructures with QDs. The photogenerated charge carriers confinement in QD produces a decrease of thermoelectron emission and dark current. Therefore, a study of charge carrier transfer mechanism and photoluminescence of InAs/GaAs heterostructures with wide-bandgap potential barrier are of special importance.

2. Experimental details

The paper studies InAs/GaAs heterostructures growed by ion-beam deposition method [5]. Deposition was carried out on GaAs semi-insulating substrate with a crystallographic orientation (100). The calibration functions of InAs and GaAs sputtering yield, beam energy, oblique angle and charge density were shown in works [5-6].

The polycrystalline targets were used as sources of growth material for AlGaAs potential barrier. Three types of samples are developed: 1) with GaAs barrier; 2) with $Al_{0.2}Ga_{0.8}As$ barrier; 3) with $Al_{0.4}Ga_{0.6}As$ barrier. As a first step n$^+$-GaAs buffer layer was formed at the temperature of 883 K and the pressure in the growth chamber of $3.7 \cdot 10^{-7}$ Pa. The accelerating beam voltage was 450 V at the current density of $3.2 \cdot 10^{-4}$ A/cm^2. At the next step an i-GaAs barrier layer was deposited at the same conditions. Then the temperature was down to 808 K after 15 s pause. The InAs QDs were formed at the beam voltage of 250 V and the ion current density of $4.5 \cdot 10^{-6}$ A/cm^2. The QDs

Content from this work may be used under the terms of the Creative Commons Attribution 3.0 licence. Any further distribution of this work must maintain attribution to the author(s) and the title of the work, journal citation and DOI.

Published under licence by IOP Publishing Ltd

coating by the AlGaAs barrier layer was accompanied an increase in temperature to 823 K. Further, a n⁺-GaAs front layer was grown on it.

An investigation of luminescent properties was carried out at temperature of 90 K in spectral range of 0.9 – 1.3 eV. The source of optical radiation was an injection laser with a wavelength of 402 nm and power 8.5 mW. The photoluminescent signal registration was carried out by the MDR-23 monochromator and a photodetector device with photodiode PDG-3600. The effect of exciting laser radiation on the photoluminescence spectra of heterostructures was eliminated by an optical filter Y-1.4x. Figure 1 shows optical scheme for measuring photoluminescence.

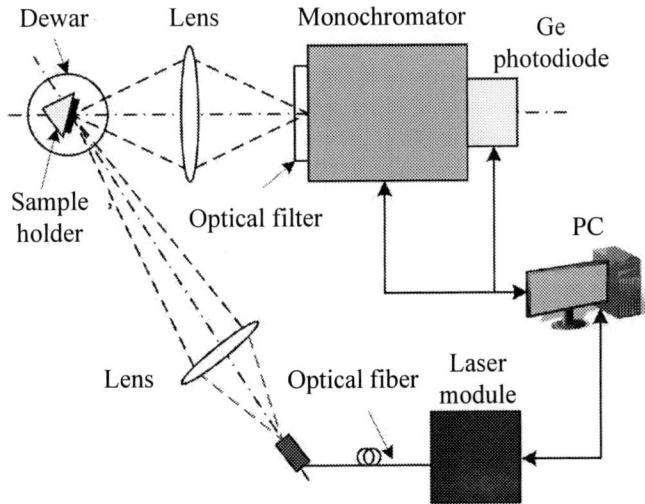

Figure 1. Optical measuring scheme of experimental samples photoluminescence

The measurements of dark current-voltage characteristics were made of using picoammeter Keithley 6485.

3. Results and discussion

The photoluminescent properties of grown heterostructures were studies. In a steady state mode, the radiation and recombination rate came to equilibrium and stood reliant on the photogenerated charge carrier lifetime at the QD energy level or energy level formed by the GaAs wetting layer (WL) during photoluminescence excitation in the sample. As a result, QDs ground-state (GS) and WL were photoluminescent sources in formed samples.

Figure 2 shows photoluminescence spectrum for three types of samples with the different barriers: GaAs; $Al_{0.2}Ga_{0.8}As$; $Al_{0.4}Ga_{0.6}As$. The existence of peaks from WL suggests QDs grew in the Stransky-Krastanov mode [7].

3.1 Photoluminescence

Photoluminescence spectrum of sample with GaAs barrier is characterized by GS peaks with energy 1.15 eV. The large width of GS photoluminescence emission peak is a resulting from InAs QDs size dispersion. The QDs locate in close proximity (surface density varies from 10^{10} to 10^{12} cm⁻²). The electron wave-function overlap on discrete energy level arises in QDs array. It leads to a formation of the energy subband in quantum well. The emission and absorption spectra depend up the subband location. The smaller the dimensional variation of QD and mechanical stresses influence the smaller the width of the energy subband and photoluminescence spectrum. The InAs QDs are in electrostatic field of GaAs barrier at the thickness of the barrier layer of 10 nm. This field provides for an excited electron emission from QDs level to the i-GaAs heterostructures matrix by two main mechanisms: overbarrier emission and tunnelling. The potential barrier height for electrons is 0.52 eV. The second emission peak is shifted to a blue region of spectrum (1.26 eV). This peak characterizes emission transition in WL.

The AlGaAs solid solutions are a wide-bandgap semiconductor. The energy gap width depends up a composition and varies from 1.4 to 2.1 eV. The wide-bandgap materials usage makes possible the functional characteristics improvement [8; 9]. Application of these potential barriers alters an elastic strain distribution in QDs layer and gives rise to shift to high-energy region.

The measurement results for $Al_{0.2}Ga_{0.8}As$ barrier with energy gap width of 1.71 eV are presented on figure 2. The GS peaks of the sample are shifted on ≈ 0.39 eV compared to GaAs barrier. At the same time, WL peaks aren't shifted because InAs QDs were covered from above. A high intensity of photoluminescence spectrum of sample with $Al_{0.4}Ga_{0.6}As$ barrier is accounted by photogenerated charge carrier localization in quantum well formed by InAs QD. $Al_xGa_{1-x}As$ potential barrier height for $x = 0.2$ at.% is 0.65 eV and for $x = 0.4$ at.% is 0.77 eV.

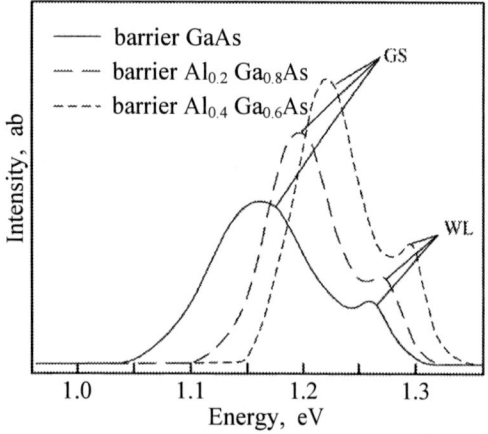

Figure 2. Measurement of photoluminescence spectrum with a different potential barriers $Al_xGa_{1-x}As$

Figure 3. Dark current-voltage characteristics of InAs/GaAs heterostructures with a different potential barriers

3.2 Dark current-voltage characteristics

The dark current-voltage characteristics of samples were measured for charge carrier transport investigate. The results are presented on figure 3. It was established that in the $0 - 0.5$ V range charge carrier transport mechanism from QD to conduction band of heterostructure is a thermoelectron emission. The current reaches saturation for all types of samples. In addition, increase of bias voltage is appeared in the $0.5 - 1.5$ V range. At the voltage over 1.5 V transport mechanism alters from thermoelectron emission to tunnelling assisted by an external electric field. It is accounted for by the shift band scheme in the direction of field. This is valid for the samples with $Al_xGa_{1-x}As$ barriers. Along with that, the width of potential barrier decreases, as a result, probability of charge carrier tunnelling through AlGaAs barrier increases essentially. The main transport mechanism from InAs QD becomes tunnelling. Figure 3 is shown that $Al_{0.4}Ga_{0.6}As$ barrier causes decrease of bias voltage as high as 0.48 V. It was established that the usage of the $Al_{0.4}Ga_{0.6}As$ potential barrier can produce minimal dark current (10^{-8} A) in InAs/GaAs heterostructures with QDs.

4. Conclusion

Thus the employing of $Al_xGa_{1-x}As$ wide-bandgap potential barriers leads to a blue shift of photoluminescence emission peak, an intensity increase and a decrease of full width a half at maximum. This effect is accounted for by an increase of the charge carriers localization in a quantum well under an increase of $Al_xGa_{1-x}As$ potential barrier height.

The measurements of dark current-voltage characteristics of the samples show that an increase of Al content in barrier leads to decrease of voltage at which alters charge carriers transport from thermoelectron emission to tunnelling assisted by an external electric field. It was established that the usage of the $Al_{0.4}Ga_{0.6}As$ potential barrier can produce minimal dark current (10^{-8} A) in InAs/GaAs heterostructures with QDs. That results can be employed when developing of effective infrared photodetector.

Acknowledgments

This work was carried out within the framework of the state assignment of the SSC RAS for 2018 (the project registration number 01201354240), and also with financial support by the Russian Foundation for Basic Research [grant numbers 16-38-60127 mol_a_dk, 17-08-01206].

References

[1] Hirakawa K, Lee S-W, Lelong P, Fujimoto S, Hirotani K, Sakaki H 2002. *J. Microelectronic Engineering.* **63(1)** 185

[2] Brunkov P N, Kovsh R, Ustinov V M, Musikhin Y G, Ledentsov N N, Konnikov S G, Polimeni A, Patanè A, Main P C, Eaves L, Kapteyn C M 1999. *Journal of Electronic Materials.* **28(5)** 486

[3] Chakrabarti S, Stiff-Roberts D, Su X.H., Bhattacharya P, Ariyawansa G, Perera G U 2005. *Journal of Physics D. Applied Physics.* **38(13)** 2135

[4] Yakimov A I, Dvurechenskii A V, Proskuryakov Y Y, Nikiforov A I, Pchelyakov O P, Teys S A, Gutakovskii AK 1999. *Applied Physics Letters.* **75(10)** 1413

[5] Chebotarev S N, Pashchenko A S, Williamson A, Lunin L S, Irkha V A, Gamidov V A 2015. *Technical Physics Letters.* **41(7)** 661

[6] Chebotarev S N, Pashchenko A S, Lunin L S, Zhivotova E N, Erimeev G A, Lunina M L 2017. *Beilstein Journal of Nanothechnology.* **2017 (8)** 12

[7] Trukhanov E M, Teys SA 2016. Bulletin of the Russian Academy of Sciences: Physics. **80(6)** 641

[8] Lin S Y 2001. *Applied Physics Letters.* **78(18)** 2784

[9] Chen Z H, Baklenov O, Kim E T, Mukhametzhanov I, Tie J, Ye Z, Madhukar A, Campbell J C 2001. *Journal of Applied Physics.* **89(8)** 4558

Doping of GaP layers grown by molecular-beam epitaxy on silicon substrates

A A Lazarenko, M S Sobolev, E V Pirogov, E V Nikitina

Nanoelectronic Lab, St. Petersburg Academic University, St. Petersburg 194021, Russia

Abstract We investigated the possibility of doping GaP layers with silicon and beryllium to produce contact layers to create light-emitting devices on a silicon substrate. In the process of epitaxial growth, it was possible to solve the problem of the appearance of antiphase regions and germinating dislocations which can capture the carriers. As a result, we obtain GaP layers of n- and p-type with a high degree of doping of good structural quality by molecular-beam epitaxy, suitable for making contacts for LEDs on a Si substrate.

1. Introduction

Nowadays, silicon is the most common semiconductor material in the production of semiconductor electronics. The implementation of light emitting devices based on silicon will dramatically increase the functionality of silicon microelectronics and open up entirely new areas of its application.

Due to the fact that silicon has an indirect structure of electronic zones, the implementation of effective light-emitting sources directly on silicon is extremely difficult task.

This is due to the fact that radiative transitions in direct-band semiconductors are a first-order process and the transition probability is high. In indirect-gap semiconductors, radiative recombination appears as a second-order process, so the probability of radiative transitions is much lower. In addition, in indirect-gap semiconductors, as the degree of excitation increases, the losses associated with absorption of radiation on injected free carriers increase more rapidly than amplification.

Dilute nitrides alloys (III-V-N group material) are of wide interest all over the world. On their basis it is possible to create light-emitting devices on a silicon substrate, since they have a direct structure of the energy bands and they can be lattice-matched with silicon. However, the creation of many widely used electronic devices (diodes, transistors, solar cells, etc.) is impossible without the introduction of doping impurities and the production of layers with specified properties to create a contact layer.

Addition of CBr4, Be or Zn in GaP leads to the appearance of p-GaP, while the addition of S, Te, and Si - to n-GaP. However, doping of III-V layers grown on silicon substrates is a difficult task because of the high concentration of defects in the layer caused by the appearance of antiphase regions and stacking faults [1].

2. Experiment

In this work, doped layers of n-GaP and p-GaP grown by molecular-beam epitaxy (MBE) on silicon substrates were investigated. Epitaxial growth was carried out using the MBE VEECO GEN III. Doping of grown materials is provided by single-zone Si and Be sources with low thermal inertia.

To avoid the problem of the formation of antiphase regions during the growth of III-V compounds on silicon (IV material), vicinal silicon substrates (001) disoriented by 4° in the [110] direction were used. In addition, a special method of growing the initial layer (migration-enhanced epitaxy, MEE) was used [2].

The carrier concentration and mobility in n-GaP/Si p-GaP/Si layers was determined by Hall measurements.

3. Results and discussion

There are some difficulties associated with the growth of initial GaP layer on silicon because of the mismatch between the lattice constants and the different number of valence electrons. In our previous works [3] it was shown that it is possible to create $GaP_{1-x}N_x$ layers on silicon substrates of good structural quality by molecular-beam epitaxy. It was shown [3] that the thickness of the occurrence of defects is 50-100 nm, and the density of dislocation outcrops is ~2×10^8 cm^{-2}, which suggests that the GaP buffer layer is of high structural quality.

However, such a number of dislocations on the heterointerface of the Si substrate and the GaP buffer layer do not make it possible to create contact through the substrate. In addition, during the epitaxial growth, Ga and P materials are diffused into the Si substrate [4].

The solution of the problem is the creation of intra-cavity contacts (figure 1), in which with the special etching and passivation of heterostructure layers, the electrodes to the doped p- and n-type layers are provided from above through the active layers.

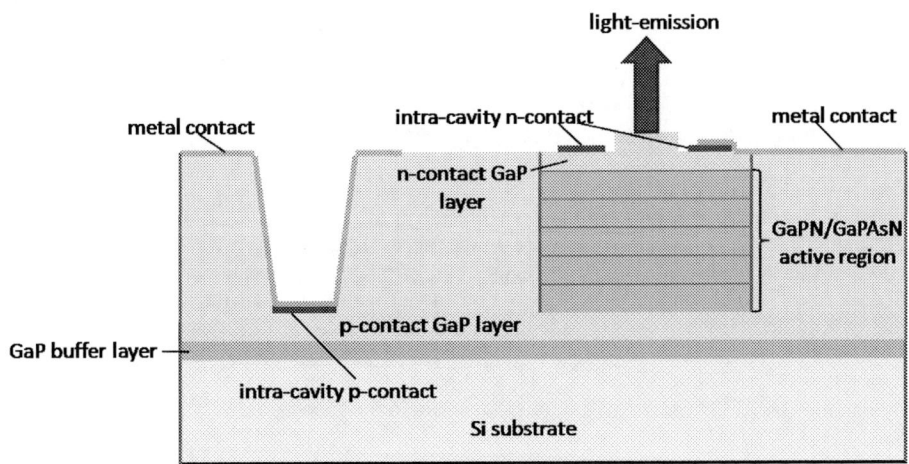

Figure 1. Schematic representation of the intra-cavity contacts to the Si/GaP/GaP(As)N LED heterostructure

Thus, the problem arises of creating heavily doped GaP layers of both p-type and n-type.
Using of bulk doping of GaP by Si did not allow obtaining a high concentration of carriers in the layer. This is due to the peculiarities of the effusion source of silicon and, as a consequence, to the low flux of silicon during the growth of the heterostructure. Therefore, it was decided to use a delta-doping (figure 2).

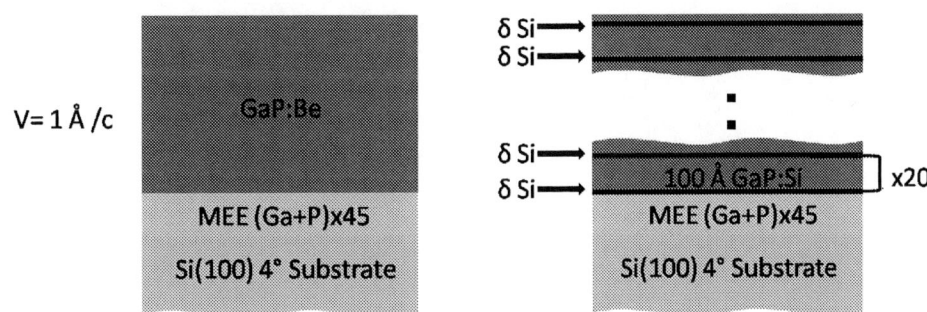

Figure 2. Schematic representation of the doping layers a) GaP:Be and b) GaP:Si

The samples were consisted of an initial layer, grown by the MEE (45 periods) and a doped GaP layer. Doping with beryllium was carried out evenly over the entire thickness of the layer. A periodic structure (20 periods) consisting of a silicon layer (δSi) and 100 Å GaP:Si was used to obtain the heavily doped GaP:Si layer. The growth rate was 1 Å/s, the total thickness of GaP layers was 0.8 µm.
The concentration and mobility of charge carriers at room temperature were determined (table 1).

Table 1. The concentration and mobility of charge carriers at room temperature.

Sample	Concentration (cm^{-2})	Mobility cm^2/(V·s)
GaP:Be	$1,0 \cdot 10^{20}$	7
GaP:Si	$6,0 \cdot 10^{18}$	13
pure GaP (holes) [5]	-	≤ 150
pure GaP (electrons) [5]	-	≤ 250

The concentration of free charge carriers in the material can decrease due to the large number of defects in the material. However, in our case it was possible to achieve high values (carrier concentration), which indicates a good structural quality of the grown layer.

For comparison, table 1 shows the mobility of electrons and holes in undoped GaP. It is seen that the carrier mobility in a pure material is very small, but in the obtained GaP:Si and GaP:Be layers it decreases approximately 20 times. This is due, first of all, to a high degree of doping, as well as to a fairly low mobility of carriers in pure material. In addition, carrier mobility is reduced due to scattering by roughness and defects that are present at the Si/GaP heterointerface. It should be noted that for the creation of light-emitting devices, the mobility of charge carriers in contact layers is not a key parameter.

4. Conclusion
We investigated the possibility of doping GaP layers with silicon and beryllium to produce contact layers to create light-emitting devices on a silicon substrate. In the process of epitaxial growth, it was possible to solve the problem of the appearance of antiphase regions and germinating dislocations which can capture the carriers. As a result, we obtain GaP layers of n- and p-type with a high degree of doping of good structural quality by molecular-beam epitaxy, suitable for making contacts for LEDs on a Si substrate.

Acknowledgments
This work is supported by RFBR Grant № 17-08-00840 and № 16-58-150006.

References
[1]   Ohlsson Et Al., Applied Physics Letters, V. **80** (24), 4546-4548, (2002)
[2]   Y. Takagi, H. Yonezu, K. Samonji, T. Tuji, N. Ohshima, J. Crystal Growth 187 **42** (1998)
[3]   Kryzhanovskaya, N.V., Polubavkina, Y.S., Nevedomskiy, V.N. et al, Semiconductors, Volume **51**, Issue 2, pp 267–27 (2017)
[4]   Sobolev, M.S., Lazarenko, A.A., Nikitina, E.V. et al. Semiconductors, Volume **49**, Issue 4, pp 559–562 (2015)
[5]   http://www.matprop.ru/GaP_electric#Transport date 31.03.2018

Multiscale simulation of surface characteristics of field emitter tip

K Nikiforov

Saint Petersburg State University, 7/9 Universitetskaya nab., St. Petersburg, 199034 Russia

Abstract. Paper presents multi-scale modelling of the field emitter tip surface characteristics. At micro-scale an approximation of emitter shape is obtained, at meso-scale crystallographic faces are constructed for application of a semi-empirical regression model of work function distribution, on the nano-scale surface atoms coordinates are calculated that serves as base data for all the above mentioned levels of detail. Electric field distribution at all scale levels is calculated.

1. Introduction

An important problem of using computational models for simulation the properties and characteristics of emission systems is accounting for multi-scale nature of the physical phenomena that occur during the process of field electron emission. One of the principal problems of multi-scale modeling is the need to conjoin a number of different models describing the behavior and properties of complex systems with different levels of detail.

A combination of classic and quantum approach in emission system modeling at different scales is an extraordinarily complex and currently topical problem. The solution for the problem of constructing a multi-scale models that would unite conceptually different algorithms for describing nanostructured system behavior on different levels in the hierarchy — i.e. on nano-, meso-, micro- and macro-scale — allows to conjoin the formulations for 1D-, 2D- and 3D-problems. The study gives the examples of such models and systems.

2. Simulation of structural, crystallographic, work function and electric field characteristics of field emitter tip surface

Multi-scale modeling is not only modeling on different scales (from subatomic to macro-level), but also on different levels, as the results for one level of scaling can be used like the input data for the next one. The object of study is a metal single crystal emitter tip made out of metal wires of diameters about $0.1-0.15$ mm by the use of anode electric etching. The field emission phenomena occur with electric field strength of the order of 10^9-10^{11} V/m. Usually such fields are observed on surfaces of emitters with tip curvature radii of about $10-1000$ nm. For example Figure 1 shows possible shapes approximating emitter as equipotential surfaces of the electric field [1] generated by charged cone with a sphere at the top (dotted line). As can be seen, the emitter apex shape is close to a hemisphere.

One of the problems that require multi-scale approach is the problem of computing the electric field above the emission surface. This is caused by the direct influence on the result by both nano- and micro- (atom packing density distribution, work function values, local surface curvature radius etc. [2−7]) and macroparameters (geometry of the electrode systems which defines the distribution of the macroscopic electric field).

Content from this work may be used under the terms of the Creative Commons Attribution 3.0 licence. Any further distribution of this work must maintain attribution to the author(s) and the title of the work, journal citation and DOI.

Published under licence by IOP Publishing Ltd

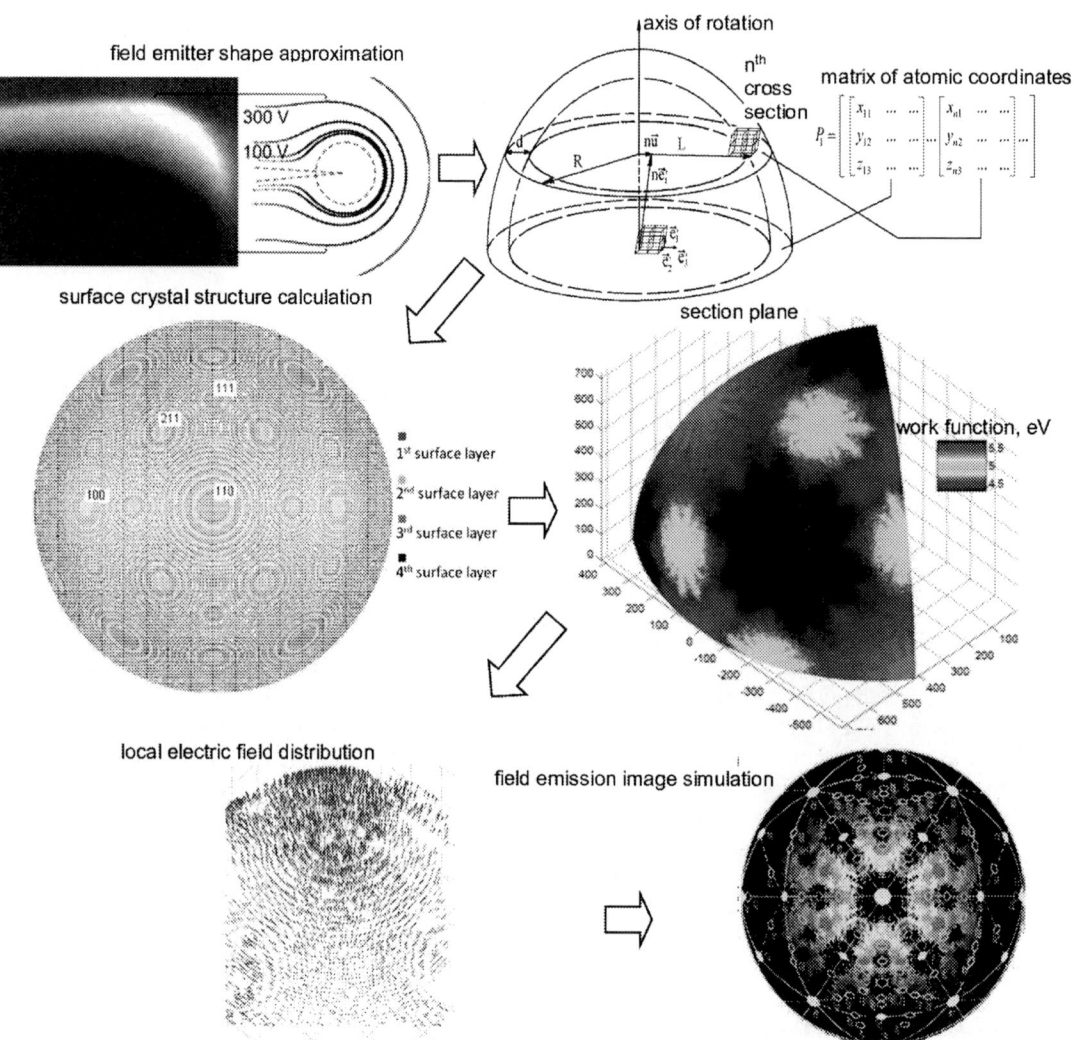

Figure 1. Models hierarchy chart: shape approximation of emitter; construction of a set of coordinates of surface atoms, localization of crystallographic faces (the model structure of the apex surface, where the surface layers of the atoms of different depths of 0.2, 0.4, 0.6 and 0.8 lattice parameters and Miller indices for major crystal faces are represented); calculation of work function map, local electric field vectors over surface atoms (calculated by the finite-differences method) and simulated pattern of field electron emission.

The number of the detalization levels of different scales during modeling depends on the complexity of shape and structure of the emitter. Those levels can also be categorized by the three principal functions that are simultaneously performed by the electric field in field emission systems: generation, acceleration and transportation. Field electron current is generated in process of emission on the nanoscale: under the influence of the field the surface threshold turns into a potential barrier which can be tunneled through with a non-zero probability, as described by the quantum theory. The electric field strength which causes the width of the potential barrier to be in order of nanometers, is formed due to the presense of nanostructure details of the surface and due to amplification of the larger scale field. Hence, this level of modeling is donned nanoscale. The most important problem of modeling on nanoscale is defining the exact maximum distance to the surface on which the nanoscale defects have any significant impact, and computing the field enhancement coefficient.

On the microscale the field distribution is defined by microscale geometric parameters of field emission systems. The non-uniform distribution of surface work function values causes patch field effect due to contact potential difference established between areas having different work function values.

On the macroscale the field emission electrode systems usually corresponds to a calculation area of complex shape which includes the boundaries of the emitter with large surface curvature and small size, the fact that leads to a rather large range of the characteristic sizes in the same geometric configuration. Moreover, the exponential dependence of the field emission current density [8] requires increased precision for taking into account the emitter's boundary conditions:

$$j(F) = e^3 / \left(8\pi h \Phi t^2(y_0)\right) F^2 \exp\left(8\pi (2m)^{1/2} / (3heF) \Phi^{3/2} v(y_0)\right),$$

$$y_0 = e\left(eF/4\pi\varepsilon_0\right)^{1/2} / \Phi$$

where m and e are electron mass and charge, h is Planck's constant, Φ is the work function, F is the electric field strength, $t(y_0)$ and $v(y_0)$ are Nordheim's elliptic functions, which can be approximated as [9]

$$v(y) \approx 1 + y^2(\ln y - 3)/3, \qquad t(y) \approx 1 + y^2(1 - \ln y)/9.$$

The microscale electric field's distribution near the emitter surface can be calculated analytically by the sphere-on-cone model [1]. Within the framework of this model one equipotential surface (Figure 1) of the electric field created by the charged orthogonal cone with the sphere at the top is taken as the emitter and the other one, as the anode. The electric field strength is given by the expression

$$F_{macro} = V_R \left(nr^n + (n+1)a^{2n+1}r^{-n-1} \right)^2 \left(P_n(\cos\theta) \right)^2 / R^{2n} r^2 + \left(a^{2n+1}r^{-n-1} - r^n \right)^2 (n+1)^2 \left(\cos\theta P_n(\cos\theta) - P_{n+1}(\cos\theta) \right)^2 / R^{2n} r^2 \sin^2\theta \right)^{1/2},$$

where V_R and R are the anode voltage and anode distance; the shape of the anode is approximated by

$$r \approx R / P_n(\cos\theta)^{1/n},$$

a is the sphere radius; P_n is the Legendre function of the first kind of nth order given by the expression

$$P_n(\cos\theta) = \frac{2}{\pi} \int_0^\theta \cos\left(n + \frac{1}{2}\right) x \Big/ \left(2(\cos x - \cos\theta)\right)^{1/2} dx$$

and n is determined from the condition $P_n(\cos\theta_0) = 0$. On the surface of cone and sphere there is the condition $V = 0$.

In this approach one has to compute local values of the work function for various crystallographic planes of the emitter tip, using either ab initio calculation methods or empirical models [10, 11]. Figure 1 shows the hierarchy chart of models and their relations useful for understanding and development of the theory and relating it to experimental data.

3. Conclusion

This study presents the mathematical atomistic model of characteristics of emitter surface that is the study object of atom probe tomography, field emission electron/ion microscopy and field desorption microscopy, where crystallographic, work function and electric field characteristics of field emitter tip surface are of great importance [12].

In order to achieve a more in-depth understanding of the existing monocrystalline emitters, and to develop them further it is necessary to study further, considering both macroscopic processes (electric field distribution in the interelectrode gap; formation of the spatial charge) and microscopic parameters (formation and deformation of crystallographic faces, distribution of atomic packing density and of work function). Such a combined approach to the processes with temporal and spatial scales being that different is only possible using multiscale computer modeling.

Acknowledgments

Scientific research were performed at the Research park of St. Petersburg State University Computing Center.

References
[1] Dyke W P, Trolan J K, Dolan W W and Barnes G. 1953 *J. Appl. Phy.* **24** 570
[2] Nikiforov K A and Egorov N V 2015 Simulation of specimen structure in atom probe tomography and field electron microscopy *2015 International Conference on Mechanics – Seventh Polyakhov's Reading (Saint Petersburg)* (Piscataway: IEEE) 7106762
[3] Nikiforov K A, Egorov N V and Lunkovskiy M N 2015 Modelling of field emitter surface structure *Journal of Physics: Conference Series* **643** 012010
[4] Nikiforov K and Krasnova A 2014 Model of field electron emitter surface structure *2014 Tenth International Vacuum Electron Sources Conference (IVESC)* (Saint Petersburg) (Piscataway: IEEE) 6892018
[5] Nikiforov K A, Egorov N V, Shen C-C 2009 *J. Surf. Invest. X-Ray, Synchrotron and Neutron Techniques* **3** 833
[6] Nikiforov K A and Lunkovskiy M N 2016 Algorithm for computation of coordinates of atoms on field emitter tip surface, *2016 Young Researchers in Vacuum Micro/Nano Electronics (VMNE-YR), (Saint Petersburg)* (Piscataway: IEEE) 7880413
[7] Nikiforov K A 2016 Modelling of crystallographic faces of field emitter tip *2016 Young Researchers in Vacuum Micro/Nano Electronics (VMNE-YR) (Saint Petersburg)* (Piscataway: IEEE) 7880412
[8] Egorov N and Sheshin E 2017 *Field Emission Electronics* (Cham: Springer)
[9] Forbes R G 2006 *Applied Physics Letters* **89** 113
[10] Egorov N V, Antonov A Y and Gribkova I M 2014 Statistical test of a single semiempirical work function model *J. Surf. Invest. X-Ray, Synchrotron and Neutron Techniques* **8** 138
[11] Antonov A Yu, Varayun' M I, Gribkova I M and Pigul' E Yu 2015 Mathematical modelling of the work function distribution on a monocrystalline cathode surface *2015 International Conference "Stability and Control Processes" in Memory of V.I. Zubov (SCP) (Saint Petersburg)* (Piscataway: IEEE) 7342101
[12] Gault B 2012 *Atom Probe Microscopy* (Heidelberg: Springer)

Plasma-arc sputtering for synthesis of Pt/CeO$_2$ and Pd/CeO$_2$ nanosystems

E A Derevyannikova[1], T Yu Kardash[1,2], A I Stadnichenko[1,2], E M Slavinskaya[1,2], A V Zaikovskii[3], S A Novopashin[3] A I Boronin[1,2]

[1] Boreskov Institute of Catalysis SB RAS, Novosibirsk 630090, Russia
[2] Novosibirsk State University, Novosibirsk 630090, Russia
[3] Kutateladze Institute of Thermophysics SB RAS, Novosibirsk 630090, Russia

Abstract. Plasma-arc (PA) sputtering method was applied for synthesis of Pt/CeO$_2$ and Pd/CeO$_2$ nanoparticles. PA represents a joint sputtering of a graphite electrode with various metals (Ce with Pt or Pd). This novel approach allows obtaining highly dispersed metals and ceria on the first stage of synthesis. The thermal treatment of an initial composite in air at 600°C -1000°C results in the burnt of a carbon shell and formation of CeO$_2$ with a fluorite-type structure. XPS and TEM analysis showed that Pd and Pt species are stabilized in the ionic form due to strong interaction with ceria. Obtained nanoparticles have a high potential of application as catalysts for efficient low temperature CO oxidation.

1. Introduction

Nanosized systems built from Pt, Pd and ceria are used as active phases for various oxidation catalytic reactions and application, such as three-way catalysts for combustion of exhaust gases [1, 2]. Currently, for the preparation of Pd/CeO$_2$ and Pt/CeO$_2$ systems are used chemical methods, such as impregnation [3, 4], precipitation [5], microemulsification [6] and solution combustion [7] methods. Nevertheless, techniques able to produce structures with unusual morphology, phase composition or elements oxidation states are highly in demand. Alternative techniques include microwave discharge [8], aerosol synthesis [9], hydrothermal synthesis [10], pulsed laser ablation [11, 12] and plasma-arc processing [13]. The application of an electric arc for synthesis of nanomaterials relates to the pioneering work [14] on fullerene synthesis. Then, the same technology was used for synthesis of carbon nanotubes [15]. An electric arc in buffer inert gas at decreased pressure was used for the synthesis of metal-carbon nanoparticles. Therefore, application of novel synthetic method for ceria nanoparticles is very attractive from fundamental point of view. In current study, plasma-arc technology is used as efficient method to synthesize of Pt/CeO$_2$ and Pd/CeO$_2$ nanosystems.

2. Experimental/methods

The first step in the catalyst preparation was the synthesis of the PdCeC and PtCeC composites. The experiments were carried out in a direct current electric arc, which had a current of 100 A, in a buffer gas (helium) at 25 Torr. The spray electrode (anode) was a graphite rod 70 mm in length and 7 mm in diameter. A hole was drilled in the center of the electrode for the insertion of a cerium rod 2.8 mm in diameter. Pd or Pt foil was mounted to achieve a Pd/Ce and Pt/Ce weight ratio of 3% and 5%, respectively. Monatomic spray products diffused in the buffer gas from the hot zone of the arc, which resulted in the cooling and heterogeneous condensation of the spray products. Then, the synthesized composite materials were calcined in air at temperatures of 600°C – 1000°C to obtain Pd/CeO_2 and Pt/CeO_2 samples.

The chemical composition was determined using the X-ray fluorescence method. The specific surface of the samples S_{BET} was determined by BET method, using argon thermal desorption with a Sorbtometr-M adsorption analyzer.

X-ray photoelectron spectroscopy (XPS) was performed using an ES-300 (KRATOS Analytical) photoelectron spectrometer equipped with MgKα (hν = 1253.6 eV) radiation sources. The X-ray source was used in power equal to 70 watts, thus no sample reduction took place during the measurements. The spectrometer was calibrated using the $Au4f_{7/2}$ (84.0 eV) and $Cu2p_{3/2}$ (932.7 eV) lines of pure metallic surfaces of gold and copper. An in house homemade software package, XPS-Calc, was used for the mathematical treatment of the XPS spectra. The Shirley model and the Gauss–Lorentz functions were used for background subtraction and curve fitting, respectively.

X-ray diffraction (XRD) patterns were recorded on a Bruker D8 Advance instrument using CuKα radiation and the Bragg-Brentano focusing geometry. A LynxEye (Bruker) multi strip detector was used for intensity measurements. Data acquisition was performed in the 2θ range of 20–85°, at a 0.05° step and with a counting time of 3 sec. Rietveld refinement for quantitative analysis was carried out using the software package Topas V.4.3. The lengths of coherent scattering domain (CSD) were calculated using LVol-IB values (i.e. volume weighted mean column lengths based on integral breadth).

Transmission electron microscopy (TEM) investigation was performed using JEM-2010 (JEOL Ltd., Japan) and JEM-2200FS (JEOL Ltd., Japan) electron microscopes operated at 200 kV to obtain HRTEM images. STEM HAADF mode was employed together with EDX spectroscopy. The samples for the TEM study were prepared on perforated carbon film mounted on a copper grid.

3. Result and discussion

Electron microscopy (TEM) and energy dispersive X-ray analysis (EDX) of the as-synthesized PdCeC and PtCeC materials indicated that it consisted mainly of amorphous carbon particles 5–50 nm in size with inclusions of crystal nanoparticles 2–10 nm in size that contained cerium and palladium or platinum. Figure 1a shows the typical image of the as-prepared composite in the PtCeC system. The analysis of the electron diffraction data shows that the structure of the nanocrystals corresponds to the Ce_2O_3 structure (ICDD PDF-2 #00-023-1048). The similar results were obtained for the plasma-arc obtained PdCeC composite [16].

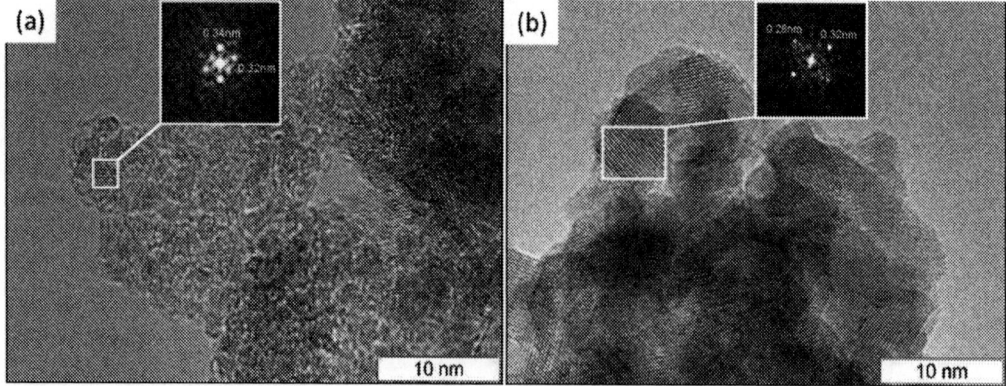

Figure 1. TEM data for the PtCeC composite **(a)** and Pt/CeO_2 nanocomposite **(b)** obtained by calcination of the initial composite at 600°C.

Based on the TGA measurements, we have chosen the temperatures above 600°C as calcination temperatures to remove carbon from initial composite samples to obtain oxide nanoparticles [16, 17]. TEM and XRD analysis (Figure 1b and Figure 2) showed a formation of CeO_2 (ICDD PDF-2 #00-001-

0112) with a fluorite-type nanoparticles after calcination of PdCeC and PtCeC materials at 600°C in air for 4 hours. The XRD pattern for the Pt/CeO$_2$ sample (Figure 2a) shows only diffraction peaks of the fluorite phase. However, for the Pd/CeO$_2$ sample (Figure 2b), the graphite phase is remained after calcination at 600°C and disappears only after calcination at 800°C. After calcination at 800°C Pt metal phase (ICDD PDF-2 #00-016-4610) appears for the Pt/CeO$_2$ nanoparticles, whereas for the Pd/CeO$_2$ the metallic phase is not formed even after calcination at 900°C. The analysis of the diffraction data was performed using the Rietveld fitting. The Table 1 shows the obtained structural parameters and the CSD sizes of the fluorite phase. The main difference between Pd/CeO$_2$ and Pt/CeO$_2$ samples is the size of the CSD of the fluorite phase. The CSD size of the Pd/CeO$_2$ samples is 17 nm after calcination at 600°C, whereas 5 nm particles are observed for the Pt/CeO$_2$ sample at the same calcination temperature.

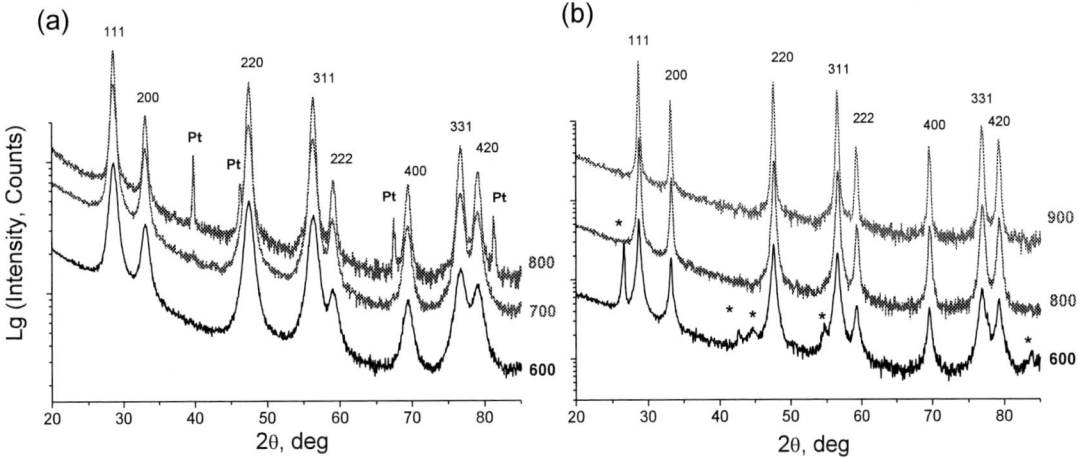

Figure 2. X-ray diffraction data for the Pt/CeO$_2$ **(a)** and Pd/CeO$_2$ **(b)** nanoparticles obtained after calcination of the Pt(Pd)CeC composite in air. * - graphite phase.

Table 1. Phase composition and structural parameters of the Pd/CeO$_2$ and Pt/CeO$_2$ nanoparticles.

Sample	T_{calc}, °C	Phase composition	S_{BET}, m^2/g	CeO$_2$ lattice parameter, Å	CSD size, nm
Pd/CeO$_2$	600	CeO$_2$ fluorite +Graphite	14.5	5.411 (1)	17 (1)
	800	CeO$_2$ fluorite	8.7	5.412 (1)	27 (1)
	900	CeO$_2$ fluorite	3.0	5.412 (1)	51 (5)
Pt/CeO$_2$	600	CeO$_2$ fluorite	110	5.415 (1)	6.6 (1)
	700	CeO$_2$ fluorite	71	5.415 (4)	8.6 (1)
	800	CeO$_2$ fluorite + Pt metal	48	5.413 (3)	12.7 (1)

According to XPS data surfaces of all produced samples are presented by ceria, oxygen, platinum/palladium and carbon. No contaminations were detected. The concentration of carbon is relatively high. In case of PtCeC composites the carbon is presented by common form – carbon black with C-C, C-H bonds (E_b(C1s)=284.8 eV) and carbonates (E_b(C1s)=284.6 eV). The rise of annealing temperature leads to decrease of carbon concentration from 34%at. at 600°C down to 20%at. at 800-900°C. In case of PdCeC composite the amount of carbon is much more – 57%at. after annealing at 600°C and 32%at. after 800°C. In PdCeC sample carbon is presented by sp^2 hybrid state (E_b(C1s)=284.6 eV + E_b(π-π*)=290-292 eV), sp^3 structures (E_b(C1s)=286.0 eV) and carbonates (E_b(C1s)=288-289 eV).

Figure 3 shows the XPS data for the Pd/CeO$_2$ and Pt/CeO$_2$ samples calcined at 600°C. The curves fitting of the Pd3d and Pt4f lines revealed the presence of different ionic states. In the both Pt/CeO$_2$ and Pd/CeO$_2$ samples metallic components of platinum (E$_b$(Pt4f$_{7/2}$)=71.1eV) and palladium (E$_b$(Pd3d$_{5/2}$)=335.2 eV) were not detected. The analysis of the Pt4f line (Figure 3a) indicates on two components with E$_b$(Pt4f$_{7/2}$)=72.7 eV and 74.6 eV, that could be assigned unambiguously to ionic species Pt^{2+} and Pt^{4+}, respectively. The observed binding energies are somewhat higher one assigned usually to platinum in PtO and PtO$_2$ [18]. Such Pt ionic species could also be associated with surface oxide clusters of Pt-CeO$_2$ solid solution [12]. The quota of Pt^{4+}state is not more than 10%. According to XPS data this state totally disappears after calcination at 700°C and higher. At the same time Pt^{2+} state is stable under samples annealing.

For the Pd/CeO$_2$ sample (Figure 3b), both detected doublets are assigned to the Pd^{2+} states, as Pd^{4+} has E$_b$(Pd3d$_{5/2}$)=338.1±0.1 eV and is known to be unstable. Thus the first doublet with E$_b$(Pd3d$_{5/2}$)= 337.9 eV corresponds to Pd^{2+} ions in the Pd$_x$Ce$_{1-x}$O$_{2-\delta}$ solid solution [19]. The second component with E$_b$(Pd3d$_{5/2}$) = 336.7 eV corresponds to PdO surface species [20]. Similarly, the metallic platinum (E$_b$(Pt4f$_{7/2}$) = 71.1 eV) was not detected for the Pt/CeO$_2$ sample. It is clearly seen that the main state of palladium is Pd$_x$Ce$_{1-x}$O$_{2-\delta}$ solid solution (close to 80% of all observed palladium).

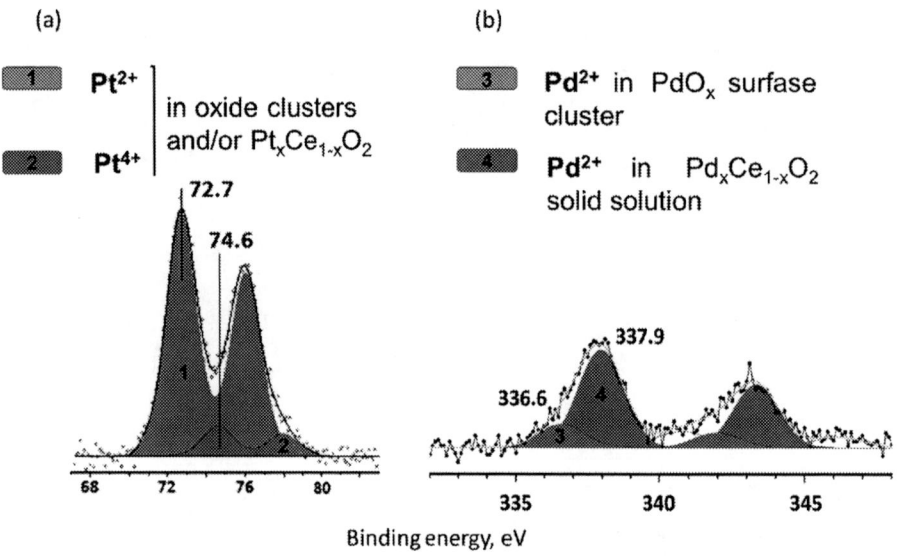

Figure 3. XPS data for the Pt/CeO$_2$ **(a)** and Pd/CeO$_2$ **(b)** samples calcined at 600°C

The obtained data demonstrates that plasma-arc sputtering of the highly defective metal particles with subsequent thermal treatment allows obtaining ceria nanoparticles with ionic species of noble metals – platinum and palladium. It should be noted, that platinum and palladium particles effectively dissolve in ceria, forming solid solution, single-ion or subnanosized cluster on ceria surface layers. The stabilization of the Pt and Pd ions on the surface of ceria is known in the literature. However, the plasma-arc synthesis due to simultaneous sputtering of ceria with noble metals is highly efficient due to formation of defective ceria particles. The obtained composites have a great potential in heterogeneous catalysis for oxidation reactions including efficient low-temperature oxidation reactions [12, 16, 17]. The performed catalytic testing of the calcined at 600°C samples showed their capability to oxidize CO at T<100°C. Thus, it could be concluded that PtCeO$_2$ and PdCeO$_2$ nanocomposites have low-temperature oxidation activity, very attracting for catalytic application.

4. Conclusions

The plasma-arc sputtering method was adapted for the synthesis of composite palladium–ceria and platinum-ceria materials. The Pd/CeO$_2$ and Pt/CeO$_2$ nanosystems were synthesized in two steps: step 1 – direct synthesis of carbon composite PdCeC and PtCeC in the plasma- arc chamber, and step 2 – calcination of the composite with carbon burnout in air at 600°C - 900°C. The obtained Pt(Pd)-ceria nanoparticles efficiently stabilize Pt and Pd ionic species, which makes the obtained samples highly attractive in oxidation catalysis.

Acknowledgments

This work was conducted within the framework of budget project for Boreskov Institute of Catalysis. The reported study was partially funded by RFBR and Novosibirsk region according to the research project № 17-43-540738 and № 17-03-00754.

References

[1] Colussi S, Gayen A, Camellone M F, Boaro M, Llorca J, Fabris S, Trovarelli A 2009 *Angew Chem Inter Edit* **48** 8481

[2] Gandhi H S, Graham G W, McCabe R W 2003 *J. Catal.* **216** 433

[3] Trevenin P O, Alcalde A, Pettersson L J, Jaras S G, Fierro J L G 2003 *J.Catal.* **215** 78

[4] Craciun R, Daniell W, Knozinger H 2002 *App Catal A: Gereral* **230** 153

[5] Shen W J, Matsumurs Y 2000 *J. Mol. Catal. A: Chemical* **153** 165

[6] Kurnatowska M, Kepinski L, Mista W 2012 *App Catal B: Enviromental* **117** 135

[7] Hegde M S, Madras G, Patil K C 2009 *Acc Chem Res* **42** 704

[8] Glaspell G, Hassan H, Elzatahry A, Abdalsayed V, El-Shall M 2008 *Topics in Catalysis* **47** 22

[9] Strobel R, Alfons A, Pratsinis S E 2006 *Advanced Powder Technology* **17** 457

[10] Wu Z, Li M, Howe J, Meyer H M, Overbury S H 2010 *Langmuir* **26** 16595

[11] Dong W, Reichenberger S, Chu S, Weide P, Ruland H, Barcikowski S, Wagener P, Muhler M 2015 *J Catal* **330** 497

[12] Slavinskaya E M, Stadnichenko A I , Muravyov V V, Kardash T Y, Derevyannikova E A, Zaikovskii V I, Stonkus O A, Lapin I N, Svetlichnyi V A, Boronin A. I. 2018 *ChemCatChem* **10** 2232

[13] Hinokuma S, Murakami K, Uemura K, Matsuda M, Ikeue K, Tsukahara N, MacHida M 2009 *Topics in Catalysis* **52** 2108

[14] Kroto H W, Heath J R, O'Brien S C, Curl R F, Smalley R E 1985 *Nature* **318** 162

[15] Kratschmer W, Lamb L D, Fostiopoulos K, Hoffman D R 1990 *Nature* **347** 354

[16] Gulyaev R V, Slavinskaya E M, Novopashin S A, Smovzh D V, Zaikovskii A V, Osadchii D Yu, Bulavchenko O A, Korenev S V, Boronin A I 2014 *Appl. Catalysis B* **147** 132

[17] Kardash T Yu , Slavinskaya E M, Gulyaev R V, Zaikovskii A V, Novopashin S A, Boronin A I 2017 *Topics in Catalysis* **60** 898

[18] Svintsitskiy D A, Kibis L S, Stadnichenko A I, Koscheev S V, Zaikovskii V I, Boronin A I 2015 *ChemPhysChem* **16** 3318

[19] Gulyaev R V, Kardash T Y, Malykhin S E , Stonkus, O. A. Ivanova, A. S. Boronin, A. I. 2014 *Phys Chem Chem Phys* **16** 13523

[20] Ivanova AS, Slavinskaya EM, Gulyaev RV, Zaikovskii VI, Stonkus O.A., Danilova I.G., Plyasova L.M., Polukhina I.A., Boronin A.I. 2010 *Appl Catal B* **97** 57

SPBOPEN 2018   IOP Publishing

Study of the geometrical parameters of In nanostructures during droplet epitaxy on the As-stabilized GaAs(001) surface

S V Balakirev[1], M M Eremenko[2], I A Mikhaylin[2], M S Solodovnik[1,2]

[1]Department of Nanotechnologies and Microsystems, Southern Federal University, Taganrog 347922, Russia
[2]Research and Education Center "Nanotechnologies", Southern Federal University, Taganrog 347922, Russia

Abstract. The droplet epitaxy of indium on the As-stabilized GaAs(001) surface is investigated using theoretical model and experimental studies. The model is developed on the basis of a combination of the nucleation thermodynamic theory and kinetic Monte Carlo algorithm. The surface density of droplets is observed to decrease with increasing temperature whereas both the droplet diameter and height increases because of the intensive attachment of In adatoms to larger, i.e. more stable, islands. At the same time, the droplet aspect ratio demonstrates a nonmonotonic temperature dependence. Although a general tendency for the aspect ratio with increasing temperature is falling, there is a temperature at which the aspect ratio is maximal both for simulations and experiments. This is caused by the temperature balancing between the wetting of the substrate and the formation of islands with near-spherical shape. The simulation results are in good agreement with obtained experimental data in a wide range of substrate temperatures.

1. Introduction

The use of nanostructures opens wide prospects for the realization of nanolasers, photonic integrated circuits, quantum computers and other novel devices. III–V semiconductor materials are of particular interest due to their optoelectronic properties and technological capabilities [1-6]. However, precise control of the geometrical characteristics of nanostructures is necessary to ensure required device parameters. The Stranski-Krastanov growth mode is commonly used to obtain an array of self-organized quantum dots in a lattice-mismatched material system, such as Ge/Si [7,8] or InAs/GaAs [9,10]. However, the impossibility to fabricate quantum dots in lattice-matched systems and to exercise an independent control of their size and surface density imposes a number of shortcomings on the Stranski-Krastanov growth method as compared with droplet epitaxial formation of quantum dots [11-17]. The latter method has been actively used for last three decades to fabricate nanostructures with different shape and size, including InAs/GaAs quantum dots of segmental shape [18-20]. The influence of the substrate temperature on the characteristics of nanostructures has been investigated during both stages of droplet epitaxy: formation of metallic droplets [16,21] and their crystallization in the flux of group V molecules [20,22]. Nevertheless, mechanisms of nucleation and growth of metallic nanostructures still remain unclear. One of important issues is a balance between the formation of an island and the substrate wetting.

In the present work, we carry out a complex investigation of geometrical parameters during In/GaAs(001) droplet epitaxy on the As-stabilized surface. We use the combination of analytical theory and kinetic Monte Carlo method to compare the experimental results with the simulation and to

Content from this work may be used under the terms of the Creative Commons Attribution 3.0 licence. Any further distribution of this work must maintain attribution to the author(s) and the title of the work, journal citation and DOI.
Published under licence by IOP Publishing Ltd

analyze the microscopic processes. This approach allows combining productivity of the analytical model and versatility of Monte Carlo simulations [23-27].

2. Description of the model

The model used in the current work was previously developed and applied to the droplet epitaxial growth on the Ga- and As-stabilized GaAs surfaces [11,28]. It is based on the (1+1)-dimensional approach with calculation of interatomic binding energies using the expression of the Lennard–Jones potential. Deposition, desorption and diffusion of atoms through six nearest sites are implemented with the probability defined by the Arrhenius equation. The probability determines a waiting time for a certain event and put into a queue in the general time scale.

We previously validated the model by the experiments in a wide range of growth temperatures [11]. The studies demonstrated nonmonotonic dynamics of the adatom supersaturation and island critical size as well as formation of the wetting layer with thickness depending on the substrate temperature. In this work, we use the As-stabilized GaAs substrate with the (001) orientation to study the dependence of the droplet geometrical parameters, such as surface density, curvature radius, height and aspect ratio, on the growth temperature. A linear size of the simulation area equal to 200 nm is considered to be sufficient for the low-temperature growth (below 300 °C) to estimate the characteristics of nanostructures.

3. Experiment

The samples were grown on the GaAs(001) epi-ready substrates using the the SemiTEq STE 35 molecular beam epitaxy system with solid state sources. First, we carried out a standard procedure of the oxide removal and deposited a 400-nm thick GaAs buffer at a substrate temperature of 580°C. Then we closed the arsenic source and cooled the substrate down to the growth temperature at which the droplet epitaxy studies were carried out. The substrate temperature was varied from 150 to 350°C in 50°C increments. We preliminarily calibrated the indium growth rate by the growth of InGaAs ternary compounds. A growth rate of 0.5 ML/s and an equivalent deposition thickness of 4 ML were used for all samples. After the In droplets had been formed, we quenched the samples and transferred to the SEM characterization.

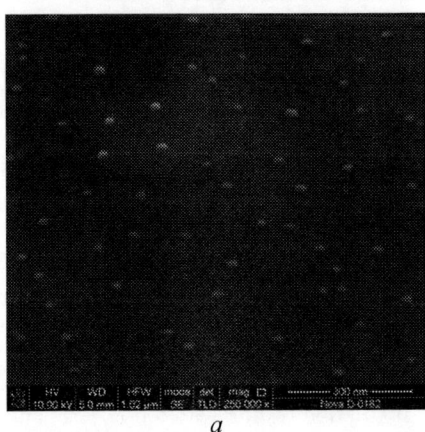
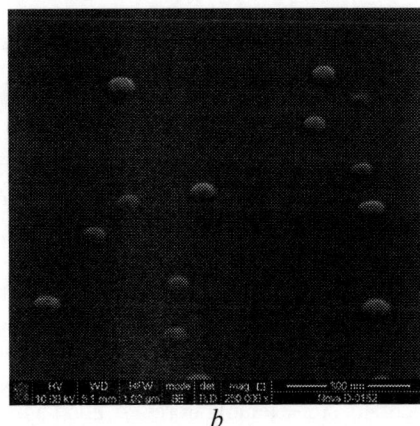

a *b*

Figure 1. SEM images of the droplet arrays after deposition of 4 ML of indium at $T = 150$°C (*a*) and $T = 250$°C (*b*).

The results of experimental studies demonstrated that the droplets have a shape of spherical segments with an acute contact angle (Figure 1). The droplet arrays are not spatially ordered, but the size distribution is quite uniform. Tiny droplets are practically not observed which is due to the fact that small islands of subcritical size are not stable and tend to decompose rapidly according to the classical nucleation theory [29,30]. During the period while the substrate is being quenched the processes of Ostwald ripening and island stabilization occur.

The experiments show that the surface density of droplets is low, as compared with other material systems, such as Ga/GaAs or Ga/AlGaAs [12,31-33]. This is due to the higher mobility of In adatoms and smaller diffusion length of Ga adatoms [34]. The high diffusivity of In adatoms on the GaAs substrate is an advantage of this system which allows achieving a sufficient distance between nanostructures and eliminate undesirable quantum-mechanical interactions.

Figure 1 demonstrates that an increase of the substrate temperature leads to an increase in the average droplet size and to a decrease in the surface density of droplets. This phenomena is not unexpected since it was observed in a number of growth systems, including droplet epitaxy [16,21]. The reason of this behavior is the decomposition of unstable nuclei and surface diffusion of released atoms towards the most stable ones as a result of the mobilization connected with an increase in the growth temperature.

4. Simulation results and discussion

The simulation was carried out for the experimental temperature range with the same deposition thickness and annealing during 5 seconds. Morphology of droplets at the substrate temperature $T = 150\ ^{\circ}C$ and $T = 250\ ^{\circ}C$ is presented in Figure 2. The islands nucleate randomly on the surface of the substrate, grow and take a shape of non-ideal spherical segment. The wetting layer with thickness exceeding 1 monolayer is formed on the surface. In contrast to the Ga-stabilized surface, on which formation of the wetting layer is suppressed almost completely [28], the wetting layer is formed on the As-stabilized surface and its thickness decreases with increasing temperature, which is confirmed by our previous study [11]. This behavior is explained by the fact that binding energy of metallic atoms (In and Ga atoms) between each other is lower than the covalent bond between In and As atoms. The larger barrier is needed to be overcome to start the nucleation. Consequently, the wetting layer of larger thickness remains on the surface.

Figure 2. Morphology of In droplets after deposition of 4 monolayers of indium on the As-stabilized GaAs(001) surface at $v = 0.5$ ML/s and *a*) $T = 150\ ^{\circ}C$, *b*) $T = 250\ ^{\circ}C$. The simulation area is 300 nm long.

One can see that the islands vary in geometrical parameters depending on the growth temperature. The surface density of droplets decreases twice with the temperature increasing from 150 °C to 250 °C. At the same time, an increase in the droplet diameter is also observed. Figure 3 shows the quantitative difference between the geometrical characteristics of In nanostructures obtained at different substrate temperatures. A good agreement of the simulation results with the experiments in a wide range of temperatures is observed. A slight discrepancy is explained by the specifics of the experiments.

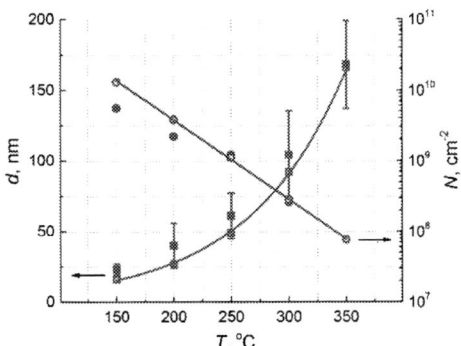 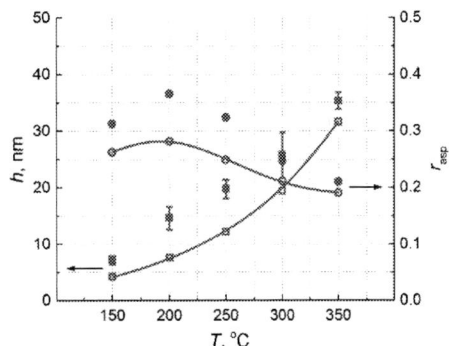

Figure 3. Temperature dependence of the droplet diameter and density ($H = 4$ ML, $v = 0.5$ ML/s). Filled figures – experiments, open figures – simulation.

Figure 4. Temperature dependence of the droplet height and aspect ratio ($H = 4$ ML, $v = 0.5$ ML/s). Filled figures – experiments, open figures – simulation.

Figure 4 demonstrates that the droplet height increases with increasing temperature which is also connected with the intensification of the surface diffusion. However, the average aspect ratio of a droplet expressed as a ratio of the droplet diameter to the droplet height does not change monotonically. It is known that the wetting of the substrate by the liquid phase becomes better with increasing temperature [35], but we reveal that there is a maximum value of the aspect ratio which does not correspond to the minimal value of the temperature. Although there is no precise coincidence of the simulation results with the experiments, there is a common regularity concerning the temperature dependence of the aspect ratio. Our model takes into account the fact that at low temperatures adatoms are not as mobile as at high temperatures. They have a larger energetic barrier to detach from the substrate and form a droplet with larger contact angle. The presence of the peak in the temperature dependence of the aspect ratio is a consequence of the temperature balance between the substrate wetting and the mutual attraction between In atoms.

5. Conclusion
In summary, we carried out the experimental and theoretical study of geometrical parameters of In nanostructures during droplet epitaxial growth on the As-stabilized GaAs(001) surface. The analytical–Monte Carlo model we used in the present study demonstrated a good agreement with the experiments in a wide range of growth conditions. It was observed that the surface density of droplets decreases with increasing temperature whereas the average droplet diameter and height increases. We revealed that the temperature dependence of the droplet aspect ratio has a maximum value at $T = 200$ °C which is explained by the fact that the contact angle cannot increase continuously at low temperatures. The location of the peak is determined by the balance between the substrate wetting and the detachment of adatoms from the surface to form a droplet with a large contact angle.

Acknowledgements
This work was supported by the Russian Science Foundation Grant No. 15-19-10006 and by Grant of the President of the Russian Federation No. MK-2629.2017.8. The results were obtained using the equipment of the Research and Education Center and Center for Collective Use "Nanotechnologies" of Southern Federal University.

References
[1]   Ageev O A, Smirnov V A, Solodovnik M S, Rukomoikin A V and Avilov V I 2012 *Semiconductors* **46** 1616–1621

[2] Klimin V S, Solodovnik M S, Smirnov V A, Eskov A V, Tominov R V and Ageev O A 2016 *Proceedings of SPIE* **10224** 102241Z

[3] Mokkapati S and Jagadish C 2009 *Mater. Today* **12** 22–32

[4] Ageev O A, Klimin V S, Solodovnik M S, Eskov A V and Krasnoborodko S Y 2016 *J. Phys.: Conf. Ser.* **741** 012178

[5] Avilov V I, Ageev O A, Smirnov V A, Solodovnik M S and Tsukanova O G 2015 *Nanotech. in Russia* **10** 214-219

[6] Ageev O A, Solodovnik M S, Balakirev S V, Eremenko M M and Mikhaylin I A 2016 *J. Phys.: Conf. Ser.* **741** 012012

[7] Brehm M and Grydlik M 2017 *Nanotechnology* **28** 392001

[8] Santalla S N, Kanyinda-Malu C and Cruz R M 2003 *J. Cryst. Growth* **253** 190–197

[9] Grundmann M, Stier O and Bimberg D 1995 *Phys. Rev. B* **52** 11969

[10] Yamaguchi K, Yujobo K and Kaizu T 2000 *Jpn. J. Appl. Phys.* **39** L1245–L1248

[11] Balakirev S V, Solodovnik M S and Ageev O A 2018 *Phys. Status Solidi B* **255** 1700360

[12] Solodovnik M S, Balakirev S V, Eremenko M M, Mikhaylin I A, Avilov V I, Lisitsyn S A and Ageev O A 2017 *J. Phys. Conf. Ser.* **917** 032037

[13] Somaschini C, Bietti S, Koguchi N and Sanguinetti S 2010 *Appl. Phys. Lett.* **97** 203109

[14] Somaschini C, Bietti S, Koguchi N and Sanguinetti S 2009 *Nano Lett.* **9** 3419–24

[15] Mano T, Watanabe K, Tsukamoto S, Fujioka H, Oshima M and Koguchi N 1999 *Jpn. J. Appl. Phys.* **38** L1009–11

[16] Lee J H, Wang Z M and Salamo G J 2007 *J. Phys. Condens. Matt.* **19** 176223

[17] Lee J H, Wang Z M, Kim E S, Kim N Y, Park S H and Salamo G J 2009 *Nanoscale Res. Lett.* **5** 308–14

[18] Mano T, Watanabe K, Tsukamoto S, Imanaka Y, Takamasu T, Fujioka H, Kido G, Oshima M and Koguchi N 2000 *Jpn. J. Appl. Phys.* **39** 4580–3

[19] Lee J, Wang Z and Salamo G 2009 *IEEE Trans. Nanotechnol.* **8** 431–436

[20] Noda T, Mano T and Sakaki H 2011 *Crystal Growth & Design* **11** 726–8

[21] Kim J S and Koguchi N 2004 *Appl. Phys. Lett.* **85** 5893–5895

[22] Noda T and Mano T 2008 *Appl. Surf. Sci.* **254** 7777

[23] Ageev O A, Solodovnik M S, Balakirev S V and Eremenko M M 2016 *J. Vac. Sci. Technol. B* **34** 041804

[24] Balakirev S V, Solodovnik M S, Eremenko M M, Mikhaylin I A and Ageev O A 2017 *J. Phys. Conf. Ser.* **917** 032034

[25] Ageev O A, Solodovnik M S, Balakirev S V and Mikhaylin I A 2016 *Technical Physics* **61** 971–977

[26] Reyes K, Smereka P, Nothern D, Millunchick J M, Bietti S, Somaschini C, Sanguinetti S and Frigeri C 2013 *Phys. Rev. B* **87** 165406

[27] Ageev O A, Solodovnik M S, Balakirev S V and Mikhaylin I A 2016 *J. Phys. Conf. Ser.* **681** 012036

[28] Balakirev S V, Solodovnik M S and Ageev O A 2017 *J. Phys. Conf. Ser.* **917** 032033

[29] Dubrovskii V G 2014 *Nucleation Theory and Growth of Nanostructures* (Berlin: Springer) p 601

[30] Kelton K and Greer A L 2010 *Nucleation in Condensed Matter: Applications in Materials and Biology* (Amsterdam: Elsevier) p 756.

[31] Ageev O A, Solodovnik M S, Balakirev S V and Eremenko M M 2016 *Phys. Solid State* **58** 1045

[32] Somaschini C, Bietti S, Sanguinetti S, Koguchi N and Fedorov A 2010 *Nanotechnology* **21** 125601

[33] Lee C D, Park C, Lee H J, Park S, Lee K S, Park C, Noh S and Koguchi N 1998 *Jpn. J. Appl. Phys.* **37** 7158.

[34] Ageev O A, Solodovnik M S, Balakirev S V, Mikhaylin I A and Eremenko M M 2017 *J. Cryst.*

Growth **457** 46

[35] Zhang H, Chen Y, Zhou G, Tang C and Wang Z 2012 *Nanoscale Res. Lett.* **7** 600

Preparation of flat surfaces on silicon carbide substrate using electron beam processing

E Yu Gusev, S P Avdeev

Southern Federal University, Institute of Nanotechnology, Electronics and Equipment Engineering, Research and Educational Centre "Nanotechnologies", Taganrog 347900, Russia

Abstract. Electron beam processing of 6H-SiC{0001} surfaces is studied. Initial substrates contain polishing scratches on (0001) face and rough surface of (000-1) face specially developed by KOH etching which are eliminated by electron beam processing. Processing is carried out in vacuum of 0.1 mTorr, energy about of 69 kJ, and background temperature of 1100 K in presence of 1-30 μm thick silicon film on the substrates. Atomic force microscopy images show near atomically flat surfaces which roughness decreases in 22-23 times for (000-1) and 1.5-2.0 times for (0001), respectively. Root-mean square value of processed surfaces are 0.27-0.30 nm for 5×5 μm^2 scan area. The steps (height of 0.5-1.0 nm) between the terraces (length of 500 nm) correspond to a height of 2-4 Si-C bilayers. The obtained changing of roughness factor also indicates a decrease of real surface area of the faces during electron beam processing.

1. Introduction

The key parameter to achieve quality epilayers and device structures on silicon carbide (SiC) substrate is quality of the surface prior to epitaxy. Commercially available SiC wafers have a large number of scratches and a residual subsurface defects or damaged layer [1-4]. This layer induced by mechanical impacts of abrasive particles during grinding and rough polishing is not eliminated even by chemical-mechanical polishing (CMP). Prior to epitaxial growth and device fabrication it is necessary to remove damaged layer and decrease surface roughness. It is known that finishing of silicon carbide is very difficult owing to its high hardness and chemical resistance. In recent years, several preparation techniques based on CMP are proposed to eliminate the damaged layer [2-4]. These are methods included surface modification and soft abrasive polishing. However, this approach does not exclude mechanical action and, accordingly, the residual layer is not completely removed. Moreover SiC polishing processes cannot be the same for each polytype, orientation and doping levels and has to be tuned to reach surface quality [1]. These drawbacks result in necessity to develop additional damage-free finishing processing technique for SiC wafers.

More promising, not only for epitaxy but also for graphene obtaining, non-mechanical approach to polish SiC is thermal surface modification [5-7]. And the thermal source can be vacuum annealing [7], laser and low-energy electron beams [5,6].The perspective of the latter has been shown by previously received results [5,6,8]. After upgrade of the equipment, refining of production tools aimed at expanding the temperature range of processing, it became possible to carry out experimental studies of SiC polishing.

In this work we investigated surface preparation technique of 6H-SiC substrates by electron beam processing.

Content from this work may be used under the terms of the Creative Commons Attribution 3.0 licence. Any further distribution of this work must maintain attribution to the author(s) and the title of the work, journal citation and DOI.
Published under licence by IOP Publishing Ltd

2. Experiments

The 6H-SiC substrates used for the study are Lely crystals doped to density of about (Nd-Na) 10^{17} cm^{-3}. We use {0001} faces as samples: as-polished (0001) face and chemical etched (000-1) face. The latter samples were etched in molten potassium hydroxide (KOH) for 35 minutes to obtain "initial" increased surface roughness. Then all samples were cleaned according to standard RCA cleaning procedure and coated by 1-30 µm thick silicon layers in order to remove damage layer by electron beam activated dissolution. The technique is based on SiC dissolution mechanism in a pure Si melt [9] and presented with experimental methodic and processing conditions in [6,10]. We also gives physical-mathematical model designed to investigation of temperature distribution and melt lifetime in the structure during processing in [11]. The electron beam processing was carried out in vacuum of 0.1 mTorr, energy about of 69 kJ, and background temperature of 1100 K. After processing Si layer was removed in HNO$_3$:HF (3:1) solution. Measurements on the surfaces were made using Solver P47 Pro (NT-MDT, Russia) atomic force microscope in tapping mode.

3. Results and discussion

The surface morphology and profiles of {0001} faces for as-received or initial and electron beam processed samples are shown in Fig.1. Fig.2. The values of its roughness parameters: peak-to-peak (Ry), ten point height (Rz), average height (Ha), average roughness (Ra), root-mean-square (Rq) roughness, surface skewness (Rsk) and coefficient of kurtosis (Rk) are summarized in Table 1.

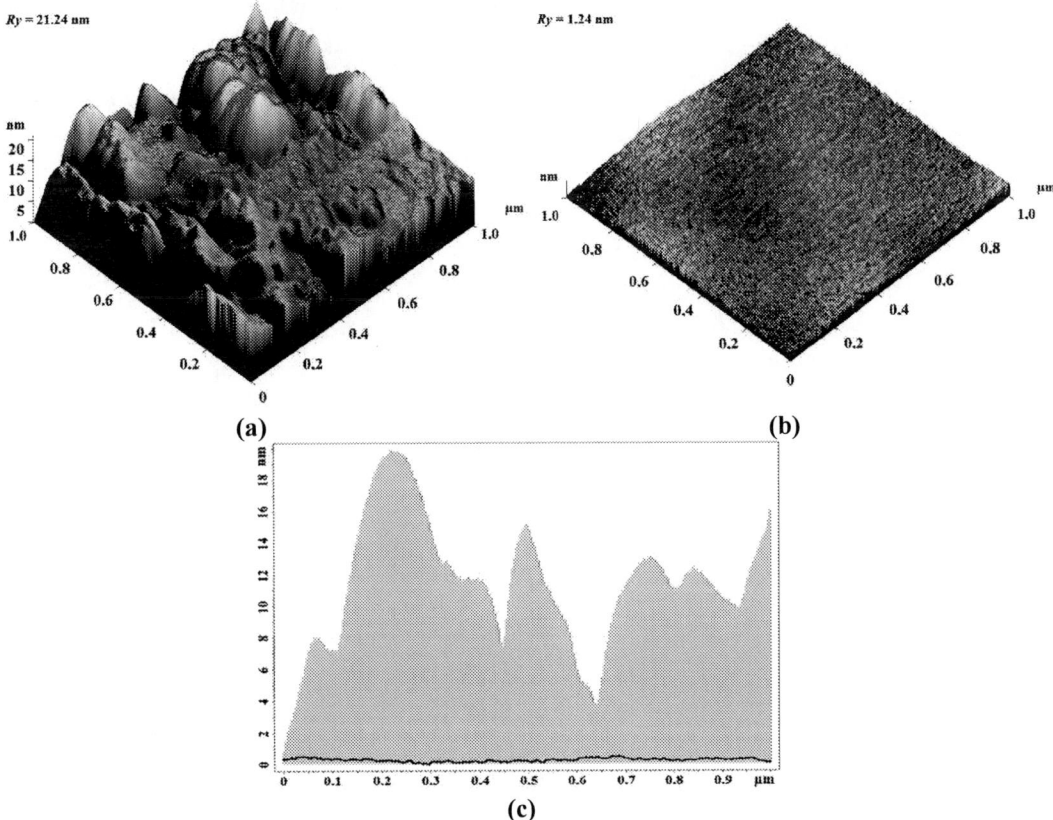

Figure 1(a-c). AFM images (a,b) of the (000-1) 6H-SiC substrate (a) KOH etched, (b) electron beam processed, and the corresponding profiles (c) to (a) – shading in gray, (b) – black line

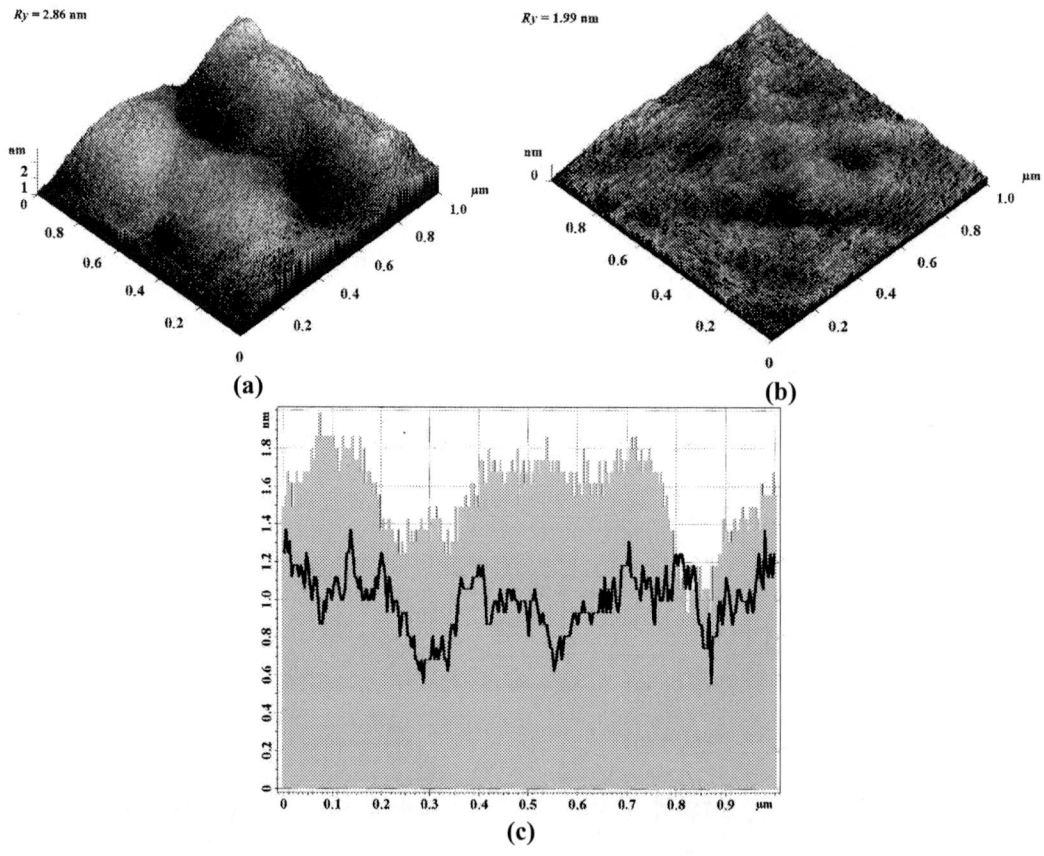

Figure 2(a-c). AFM images (a,b) of the (0001) 6H-SiC substrate (a) as-recieved, (b) electron beam processed, and the corresponding profiles (c) to (a) – shading in gray, (b) – black line

Table 1. Effect of electron beam processing on surface roughness of {0001} 6H-SiC substrate

Parameter	(000-1) face		(0001) face	
	KOH etched	Processed	As-received	Processed
Ry, nm	52.67±1.88	2.27±0.24	4.61±0.52	2.04±0.21
Rz, nm	26.02±0.89	1.14±0.11	2.35±0.28	1.05±0.12
Ha, nm	16.63±1.38	1.19±0.19	2.02±0.55	1.01±0.10
Ra, nm	5.09±0.60	0.21±0.02	0.39±0.09	0.23±0.06
Rq, nm	6.67±0.60	0.27±0.02	0.53±0.12	0.30±0.06
Rsk, -	0.91±0.17	−0.02±0.08	0.91±0.41	0.27±0.05
Rk, -	2.34±1.63	0.21±0.09	3.83±0.50	0.63±0.41
D, -	<2.37>	<2.13>	<2.29>	<2.21>
Fr, -	<7.78>	<1.95>	<4.99>	<3.2>

As seen in figures and table, the roughness values of initial and as-received samples decreased by processing in about of 22-23 times for (000-1) and 1.5-2 times for (0001), respectively. The RMS values of both type processed samples are 0.27-0.30 nm for area of 5×5 μm^2. Surface skewness and kurtosis were changed from nonzero (asymmetric distribution) for initial surfaces to zero value (symmetric narrow distribution), which corresponds to normal and leptokurtic distribution. This behavior confirms the surface profile became more uniform. The steps (height of 0.5-1.0 nm) which can be seen between the terraces (length of 500 nm) correspond to a height of 2-4 Si-C bilayers [12]. The obtained changing of fractal dimension (D) and roughness factor (Fr, see Table 1) also indicates a decrease of real surface area of the faces during electron beam processing.

Processing conditions were chosen on the basis of the model [11] so as to ensure the proper thickness of the molten silicon, its temperature and dissolution time. The melt thickness at a fixed temperature determines the limit of the soluble thickness of the substrate (damaged layer). In our experiment, the thickness of the removed layer correlates with the average height (Ha) value and was from 1 to 15 nm.

4. Conclusion
In this paper electron beam processing of the (0001) and (000-1) faces of 6H-SiC has been presented. Surface roughness of the samples have been decreased in 1.5-20 times to 0.27-0.30 nm for RMS values. The steps of 0.5-1.0 nm in height (2-4 Si-C bilayers) and terraces up to 500 nm in length were found. The obtained changing of calculated roughness factor showed a significant decrease of real surface area of the faces during processing that indicates dissolution processes of substrate material.

Thus, electron beam processing of silicon carbide covered by silicon initiating a dissolution processes, has been allowed to produce formation of a clean, near-atomically flat surfaces of the substrates with removal of the damaged layer. This technique could be considered as a superfinishing damage-free polishing complementary to standard chemical-mechanical process.

Acknowledgment
The equipment of the Research and Educational Centre "Nanotechnologies" of Southern Federal University was used for the study.

References
[1] Monnoye S, Turover D, Vicente P 2003 in: Choyke W J, Matsunami H, Pensl G (Eds.) 2003 Silicon Carbide: recent major advances, Springer-Verlag, Berlin-Heidelberg 699-710
[2] Friedrichs P, Kimoto T, Ley L. Pensl G (Eds.) 2010 Silicon carbide: Volume 1: Growth, defects. And novel application, Wiley-VCH Verlag GmbH & Co. KGaA Weinheim
[3] Deng H, Liu N, Endo K, Yamamura K 2018 *Applied Surface Science* **434** 40-48
[4] Yang X, Sun R, Ohkubo Y, Kawai K, Arima K, Endo K, Yamamura K 2018 *Electrochimica Acta* **271** 666-676
[5] Agueev O A, Avdeev S P, Svetlichnyi A M, Konakova R V, Milenin V V, Lytvyn P M, Lytvyn O S, Okhrimenko O B, Soloviev S I, Sudarshan T S 2004 *Materials Science Forum* **483-485** 725-728
[6] Ageev O A, Avdeev S P, Gusev E Yu, Konoplev B G, Svetlichnyi A M, Cherednichenko D I 2010 Russian patent Registration No 2389109
[7] Lebedev S P, Petrov V N, Kotousova I S, Lavrent'ev A A, Dement'ev P A, Lebedev A A, Titkov A N 2011 *Materials Science Forum* **679-680** 437-440
[8] Avdecv S P, Kravchenko A A, Gusev E Yu, Petrov S N 2007 *Proceedings of SPIE - The International Society for Optical Engineering* **6636** 66360K
[9] Syväjärvi M, Yakimova R, Janzén E 1997 *Diamond and Related Materials* **6** 1266-1268
[10] Agueev O A, Avdeev S P, Cherednichenko D I, Gusev E Yu 2009 *Silicon carbide and related materials, ISSCRM-2009, NovGU, Great Novgorod* 138-141
[11] Ageev O A, Avdeev S P, Gusev E Yu, Klovo A G 2009 *Silicon carbide and related materials, ISSCRM-2009, NovGU, Great Novgorod* 142-146
[12] Powell J A, Larkin D J, Abel P B 1995 *Journal of Electronic Materials* **24, 4** 295-301

SPBOPEN 2018   IOP Publishing

Transformation of the defect structure of InGaAs and InAlAs metamorphic buffer layers depending on indium concentration

L A Snigirev[1,2], A A Sitnikova[2], D A Kirilenko[2], N A Bert[2]

[1] Saint Petersburg Electrotechnical University "LETI", Saint Petersburg 197376, Russia

[2] Ioffe Institute, Saint Petersburg 194021, Russia

Abstract. A study of samples grown by molecular beam epitaxy using the technology of $In_xAl_{1-x}As$ or $In_xGa_{1-x}As$ metamorphic buffer layer formation on GaAs (001) substrate was performed by transmission electron microscopy. The 1 μm thick buffer layers with square-root dependence of In content on buffer layer thickness were studied in plan-view and cross-section geometries. The results demonstrate cascade step-wise relaxation of misfit strain along buffer layer thickness with the reduction of dislocation density to less than 10^{-6} cm^{-2} at the subsurface region. An inhomogeneous distribution of dislocations observed on plan-view TEM images is explained by overlap of split-level dislocation networks with different periods.

1. Introduction

Metamorphic buffer layers (MBL) have made it possible to obtain device structures with a high content of In on GaAs substrates that are cheaper and more available than InP substrates generally used for this purpose. The transition from the GaAs substrate to the active layer is conventionally accomplished by gradual change in the lattice constant along the MBL through layer composition alteration. Misfit strain arising in MBL because of the lattice mismatch relaxes by formation of crystalline structure defects, mainly dislocations. The specificity of such MBL is that the defects can generally remain inside the transition layer that allows one to achieve a low defect density in the overgrown active area. Sublinearly graded buffer layers were found to provide lowest misfit dislocation density and less residual strain as compared with step-wise or linearly graded ones [1].

The purpose of this work was to elucidate misfit dislocation network features in $In_xAl_{1-x}As$ and $In_xGa_{1-x}As$ MBL with square-root and linear In distribution.

2. Experiment

Samples were grown by molecular-beam epitaxy (MBE) on GaAs (001) substrates which were annealed before the MBL growth. A growth rate of 0.6 ML per second was kept constant during the whole growth process. Deposition of MBL was finished with a thin region with a constant indium fraction slightly reduced relative to the maximum. The thickness of the MBL was approximately 1 μm.

For samples A, B and C containing $In_xAl_{1-x}As$ MBL two-stage growth was used: during the initial stage (up to 0.2 μm) the substrate temperature continuously lowered from 375 °C to 330 °C, at the last stage (after 0.2 μm) the growth was performed at a constant temperature of 330 °C. The fraction of indium was increasing according to the root dependence, starting from x=0.05 and ending with the maximum fraction of 0.8-0.9. The root dependence was realized by approximation with several linear segments of which the number was from 7 to 21 for different samples.

Content from this work may be used under the terms of the Creative Commons Attribution 3.0 licence. Any further distribution of this work must maintain attribution to the author(s) and the title of the work, journal citation and DOI.

Published under licence by IOP Publishing Ltd

Samples D and E with $In_xGa_{1-x}As$ MBL were grown at the constant temperature from the range 380-400°C, growth rate being (0.6 - 0.8) ML per second. The In content was increasing linearly to a maximum fraction of 0.4. The details of the growth procedure can be found in [2,3].

Since transmission electron microscopy (TEM) is a suitable space-resolution tool to characterize the linear structural defects it has been used for the study of misfit dislocation array generated in InGaAs MBL. Cross-section and plane-view specimens for TEM study were prepared by a conventional technique including preliminary mechanical thinning and final ion milling. The specimens were investigated employing Philips EM420 and Jeol JEM-2100F microscopes using diffraction contrast imaging and selected area electron diffraction (SADP).

3. Results and discussion
Investigation of the samples with MBL was carried out by TEM in both cross-section and plan-view geometries.

3.1. Cross section
Fig. 1 demonstarates cross-sectional TEM images of the samples A, B and C with MBL of approximately equal thickness (about 1 μm). The samples slightly differ in the maximum In content (the maximum concentration of indium is 81%, 85% and 87% for A, B, C, respectively) and in number of linear segments approximating the root dependence of In content. The MBL extends from the GaAs substrate to the interface with devise structure which is represented as a thin dark line in the upper part of the images in Fig. 1. One more thin line of the dark contrast closer to the surface represents quantum well. The dislocations represented on the images as thin lines of dark contrast are predominantly contained within the bottom region of the MBL while the upper part of the MBL is almost free of defects. The multistage dislocation network at the MBL bottom region is known to necessarily form to relieve the mismatch strain arising due to growing In content. What stands out is nonuniform distribution of dislocations over the MBL thickness within the bottom region: they are concentrated at curved stripes separated by almost dislocation-free intervals.

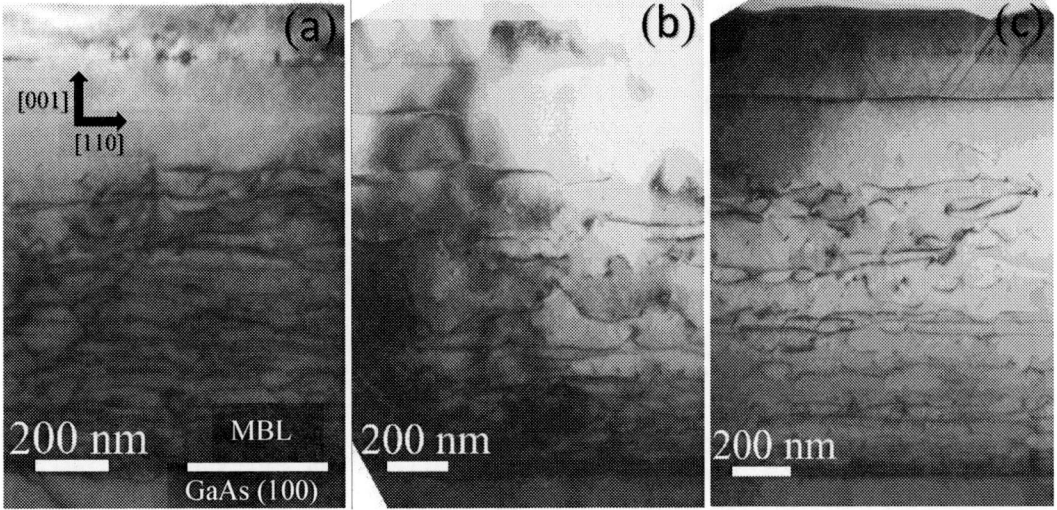

Figure 1 (a, b, c). 220 bright-field (110) cross-section TEM images of samples A (a), B (b), C (c)

To monitor these alterations, the dislocation density was measured along the MBL thickness. The dislocation density was determined as total dislocation length divided by a selected volume, the TEM lamella thickness being measured by means of convergent beam diffraction. In Fig. 2 the measured density of dislocations is plotted against the distance from the substrate-MBL interface for the samples A, B and C. As it can be seen, the dislocation density along the MBL thickness has local minima and maxima, while it decreases almost to zero at the upper part of MBL. As it also seen, the average dislocation density in the sample A is higher than that in the samples B and C.

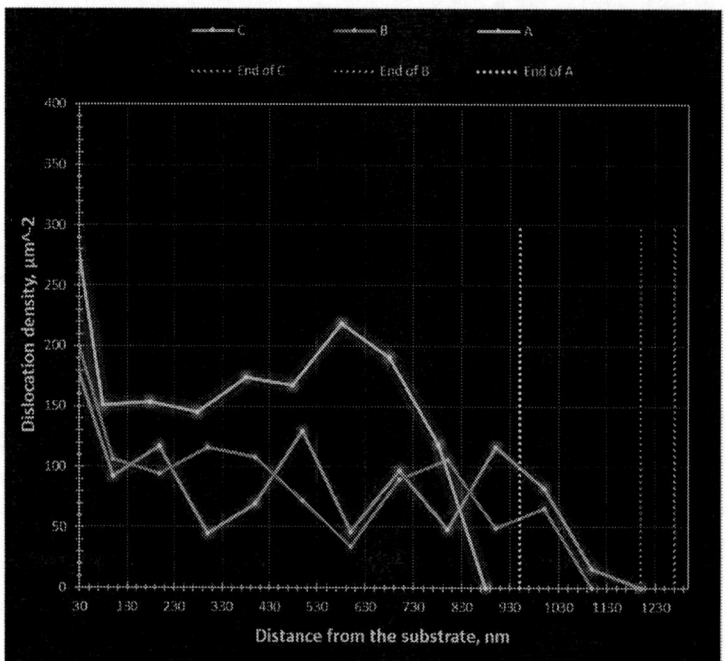

Figure 2. Distribution of Dislocation Density over the MBL thickness

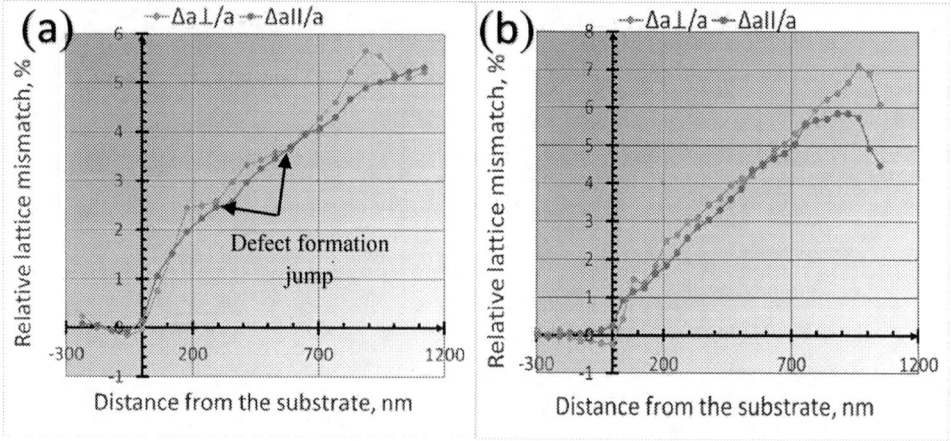

Figure 3 (a, b). Variation of relative lattice mismatch over the MBL thickness for samples A (a) and C (b): $\Delta a^{\perp}/a_{sub}$ (blue) and $\Delta a^{\parallel}/a_{sub}$ (red)

In addition, the relative lattice mismatch along ($\Delta a^{\perp}/a_{sub}$) and normal ($\Delta a^{\parallel}/a_{sub}$) to the growth direction was determined in dependence on the distance from the substrate-MBL interface by means of measurements of corresponding reciprocal lattice vector length on SADP with the reference to GaAs substrate interplanar distances [4]. The results of relative lattice mismatch measurements for the samples A and C are shown in Figure 3.

The graphs in Fig. 3 show that the value of $\Delta a^{\perp}/a_{sub}$ at some regions becomes greater than $\Delta a^{\parallel}/a_{sub}$, which indicates that the lattice there becomes elastically strained, i.e generation of misfit dislocations is essentially suppressed, and elastic energy accumulates during the growth of this parts of MBL. When the elastic energy reaches a critical value, the intensive generation of mismatch dislocations happens that is reflected in that the components of relative lattice mismatch in both directions ($\Delta a^{\perp}/a_{sub}$ and $\Delta a^{\parallel}/a_{sub}$) become equal to each other. These points are marked in Figure 3a as "jumps" in the dislocation formation process.

Thus, the nonuniformity of the contrast on the cross section TEM image is explained by cascade generation of mismatch dislocations due to stepwise relaxation of strain.

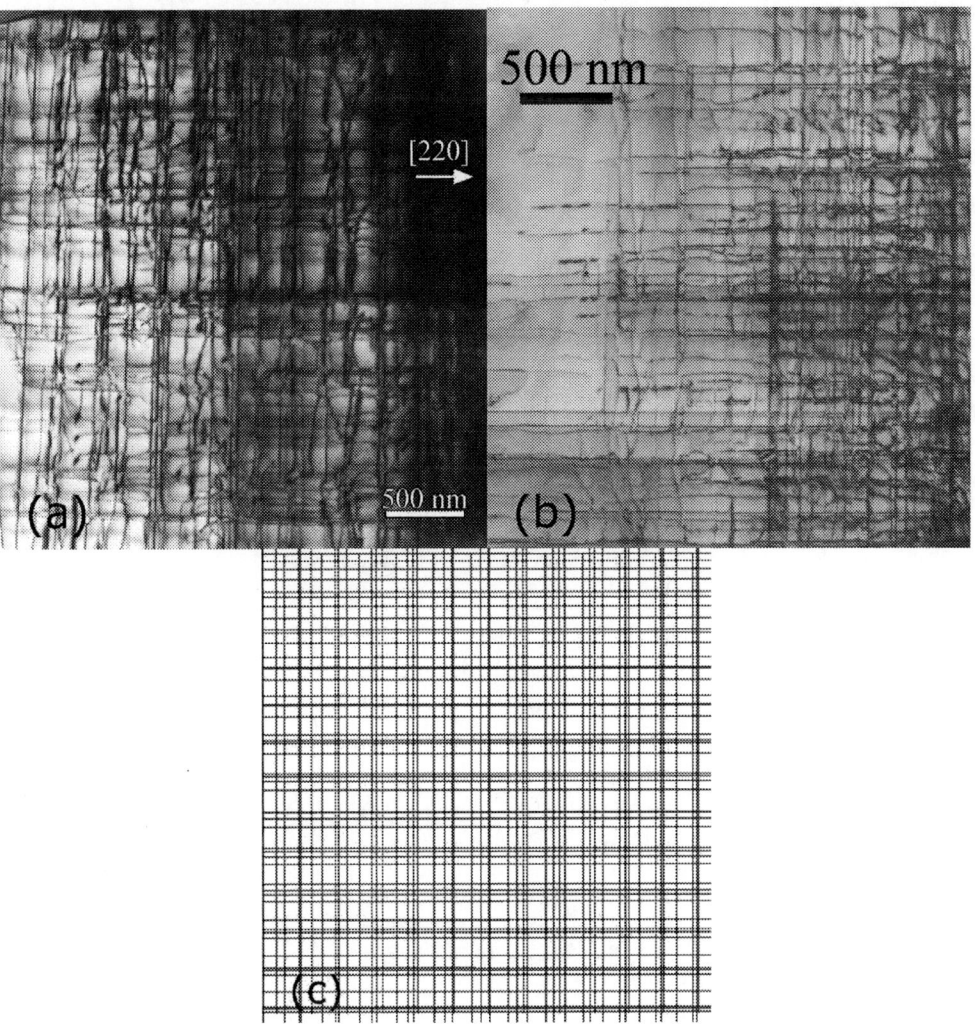

Figure 4 (a, b, c). 220 bright-field plane-view TEM images of the samples D **(a)** and E **(b)**, and modeled moiré fringes **(c)**.

3.2. Planar images

TEM plan-view images of samples D and E represented in Fig. 4 show dense misfit dislocation network. The network is seen to be irregular with alternating dislocation groupings and rarefactions. Such irregularity cannot be explained by the variation in the plastic relaxation extent (i.e. misfit strain variation) over the MBL growth surface. To elucidate the reason of dislocation groupings and rarefactions appearance on the plan-view TEM images we performed a simple simulation of projection of four overlaid square lattices with different periods. The result of the simulation is presented in Fig. 4c and seems to be similar to experimental TEM images. We can conclude that the observed irregularity in dislocation distribution on plan-view TEM images is apparent and occur as the result of translational moiré effect due to the presence of multistage dislocation network. The appearance of apparent dislocation groupings and rarefactions on the plan-view images of MBL allows us to conclude that the multistage dislocation network has somewhat dissimilar periods at different levels.

4. Conclusion

TEM study of 1 μm thick $In_xAl_{1-x}As$ and $In_xGa_{1-x}As$ MBLs grown on GaAs (001) substrate with square-root and linear profile of In content was performed to characterize the misfit dislocation network. The obtained results demonstrate cascade step-wise generation of misfit dislocations with increasing MBL thickness yielding the reduction of dislocation density to less than 10^{-6} cm^{-2} at the MBL subsurface region. An inhomogeneous distribution of dislocations observed on plan-view TEM images is explained by overlap of split-level dislocation networks with different periods.

Acknowledgments

We are grateful to V.A. Soloviev for the provided samples. TEM studies were carried out using equipment of the Federal Joint Research Center "Material science and characterization in advanced technology" supported by the Ministry of Education and Science of the Russian Federation (id RFMEFI62117X0018).

References

1. Kujofsa T Ayers J E. 2014 *J. Vac. Sci. Technol. B*, **32**, 031205
2. Soloviev V A et al. 2018 *Semiconductors*, **52**, no. 1, pp. 120–125
3. Sorokin S V et al. 2016 *J. Cryst. Growth*, **455**, pp. 83–89
4. Baidakova M V et al. 2016 *Tech. Phys. Lett.*, vol. 42, no. 5, pp. 464–467

SPBOPEN 2018

IOP Publishing

IOP Conf. Series: Journal of Physics: Conf. Series **1124** (2018) 022028   doi:10.1088/1742-6596/1124/2/022028

The features of moiré pattern on electron-microscope images of free-standing quantum dots containing dislocations

L A Sokura, V N Nevedomskiy, N A Bert

Ioffe Institute, Saint Petersburg 194021, Russia

Abstract. Simulation of moiré pattern arising on plan-view transmission electron-microscope image of nanoisland (quantum dot) containing dislocations was performed on the base of finite-element method. The presence of dislocations with a screw component of Burgers vector in the nanoisland is shown to be energetically unfavourable. On the contrary, edge dislocations allow relieving mismatch stress most effectively. The simulation results provide a means to determine the type of dislocation inside quantum dot by the specific features of the moiré pattern on transmission electron microscope image.

1. Introduction

Transmission electron microscopy (TEM) is one of the most common methods for studying microstructure of materials. The TEM image is known to be a projection of electron transparent volume of TEM specimen on registration plane along the incidence beam, therefore the image of an object within the area under study may overlaps with the image of another one. When two overlapping objects are crystalline and rotated relative to each other or lattice mismatched rotational or translational moiré pattern may appear on the TEM image. The translational moiré pattern is often observed on plan-view TEM images of free-standing quantum dots which, from standpoint of microstructure, are crystalline nanoislands (NI) on the surface of crystalline substrate with different lattice constant. If NI is coherent to the substrate the translational moiré pattern consists of periodic parallel fringes which appear due to elastic relaxation in the upper part of NI [1, 2]. However, when NI contains a defect it causes crystal lattice distortions which result in the significant modification of moiré pattern that is demonstrated in Figure 1.

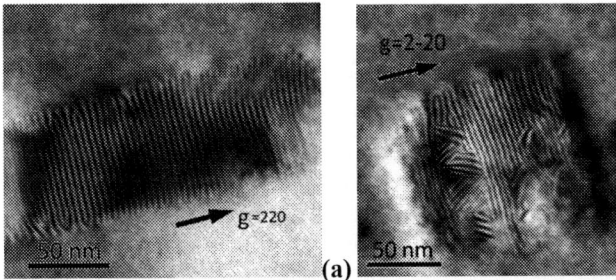

Figure 1 (a, b). Plan-view TEM images of InSb/InAs nanoislands representing **(a)** moiré fringes and **(b)** complicated moiré pattern obtained under two-beam conditions with diffraction vector g=220 and 2-20, accordingly.

Content from this work may be used under the terms of the Creative Commons Attribution 3.0 licence. Any further distribution of this work must maintain attribution to the author(s) and the title of the work, journal citation and DOI.

Published under licence by IOP Publishing Ltd

Contrast from dislocations on the plan-view TEM image appears to be overlaid with moiré pattern that hinders visualization and analysis of the dislocation type [3].

The purpose of this work is to study the impact of different dislocation types in NI on the appearance of the moiré pattern.

2. Experimental method

In order to determine the type of defect contained in the sample from plan-view TEM image, we carried out a simulation of moiré patterns. Moiré patterns received from simulation were used then for comparison with experimental ones. The modeling of images consisted of two stages:

1) Calculation of the displacement field according to the elasticity theory (isotropic case);

2) Calculation of TEM images using the Howie-Whelan equations on the base of calculated displacement field.

The nanoisland-substrate model was based on the results of TEM studies of InSb/InAs quantum dots (Figure 2, a) [3]. The faces of the island are formed by {111} planes. Experimental images have shown the presence of misfit dislocations at NI/substrate interface (Figure 2, a). Also NI contained a stacking fault.

To simplify the calculations, the shape of the nanoisland in NI-substrate model was taken as a truncated cone with an angle of 55° at the base (Figure 2, b). The simulated NI had a height of 7 nm and base diameter of 40 nm.

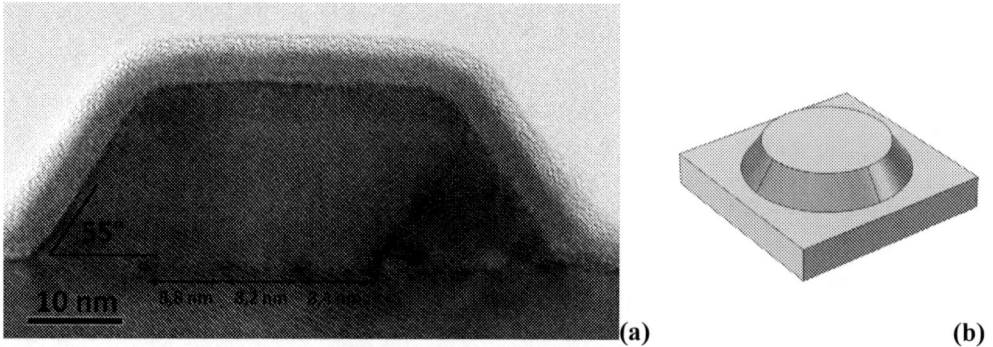

Figure 2 (a, b). (a) Cross-section HREM image of an InSb/InAs nanoisland containing dislocations and stacking fault; **(b)** the geometry of the nanoisland-substrate model.

The displacement field was calculated for a NI containing various configurations of dislocations:
- without dislocations;
- one Shockley dislocation (screw, Burgers vector b=a/6*[121]);
- one Frank dislocation (edge, Burgers vector b=a/3*[111]);
- two Frank dislocations (configurations with parallel and perpendicular Burgers vectors);
- one perfect 60°-dislocation (mixed, b=a/2*[101]);
- two perfect 60°-dislocations (configurations with parallel and perpendicular Burgers vectors);
- one perfect 90°-dislocation (edge, b=a/2*[101]);
- two perfect 90°-dislocations (configuration with parallel and perpendicular Burgers vectors).

Using the finite-element method we calculated general displacements and elastic stress fields in NI, which arise due to the NI and substrate lattice mismatch and the presence of structural defects. For the calculation, conventional equations of the elasticity theory for a homogeneous isotropic material were solved, namely differential equilibrium equations (1), Hooke's law (2), and the Cauchy equation (3):

$$\frac{\partial \sigma_{ij}}{\partial x_j} = 0$$

(1)

$$\sigma = E\varepsilon \tag{2}$$

$$\varepsilon_{ij} = \frac{1}{2}\left(\frac{\partial u_i}{\partial x_j} + \frac{\partial u_j}{\partial x_i}\right) \tag{3}$$

where σ_{ij} – are the components of the stress tensor ($i, j = x, y, z$); ε_{ij} – are the components of the elastic strain tensor; u_i – are the components of the elastic displacements vector; E – is the elastic modulus (or Young's modulus, for InAs – 5.09 Pa, for InSb – 4.28 Pa).

The initial strain ε_0 in the NI was expressed as the compression of the NI lattice (a_{QD}) with respect to the lattice of the substrate (a_{sub}):

$$\varepsilon_0 = \frac{a_{QD} - a_{sub}}{a_{sub}}\begin{pmatrix} 1 & 0 & 0 \\ 0 & 1 & 0 \\ 0 & 0 & 1 \end{pmatrix} \tag{4}$$

For the InSb/InAs system lattice mismatch is 6 %.

To take into account the contribution of the displacement field caused by dislocations, a thin material layer was introduced into the NI model with an initial strain determined by the direction and magnitude of the Burgers vector.

By solving equations of the elasticity theory the total elastic energy of the system was calculated for each configuration by:

$$W_s = \sum_{i,j} \frac{1}{2}c_{ij}(\varepsilon_i - \varepsilon_{i0})(\varepsilon_j - \varepsilon_{j0}) \tag{5}$$

where c_{ij} – are the components of the elastic modulus tensor.

The elastic energy depended on the type of defects introduced and their position in the NI. For each configuration, the dislocation position in the NI corresponding to the elastic energy minimum was determined (for example, for the perfect 90°-dislocation in Figure 3). These dislocation positions were used later to simulate the TEM images.

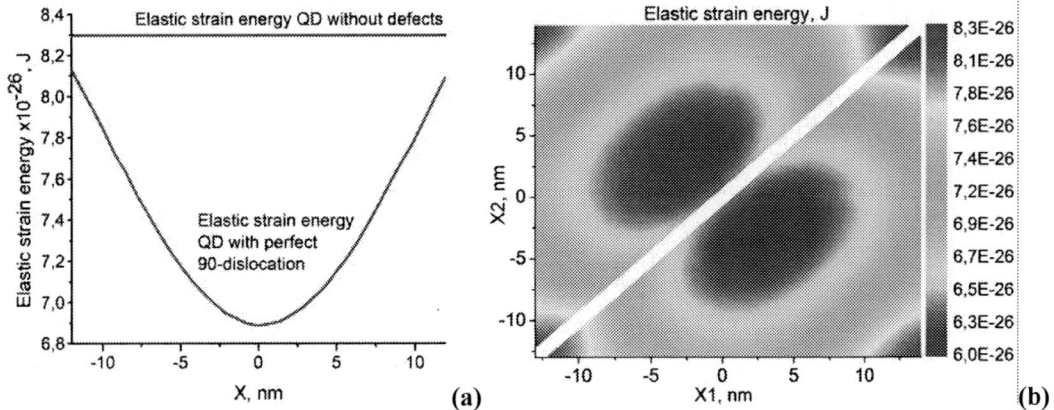

Figure 3 (a, b). Total elastic energy of a nanoisland-substrate system with one **(a)** and two **(b)** complete 90°-dislocations. The axes are the distance from the dislocation to the island center (x, nm) and the elastic energy of the system ($*10^{-26}$ J).

Based on the calculated displacement field, a simulation of the TEM images of a coherent NI and NI containing various type dislocations was carried out. To describe electron diffraction in a crystal, we used the dynamic Howie-Whelan approximation, which contains two differential equations for the wave functions of the transmitted (Φ_o) and diffracted (Φ_g) beams [2]:

$$\frac{d\Phi_o(z)}{dz} = \frac{\pi i}{\xi_g} \Phi_g(z) \exp(2\pi i \vec{g}\vec{u})$$

$$\frac{d\Phi_g(z)}{dz} = \frac{\pi i}{\xi_g} \Phi_o(z) \exp(-2\pi i \vec{g}\vec{u}) + 2\pi i \vec{s} \Phi_g(z)$$

(7)

where $\vec{g}$ – is the diffraction vector, the diffraction vector g is perpendicular to the direction of incidence electron beam z, ξ_g – is the extinction length, $\vec{s}$ – is the parameter of deviation (in our model $s = 0$), $\vec{u}$ – is the displacement vector obtained from the solution of the elasticity theory equations.

3. Experimental results

The calculation showed that after the elastic stress relaxation, the total elastic energy of the system decreases by 10 times, and after the dislocations introduction, by another 6-27%, depending on their type and quantity (Figure 4). Of the considered models, NI with two 90°-dislocations had the lowest energy. The minimum energy was obtained for symmetrically located dislocations relative to the island center and a distance between them was 8.5 nm. It correlates well with the experimental data – the distance between dislocations founded from cross-section HREM image of an InSb/InAs nanoisland was 8.2-8.8 nm (Figure 2, a).

It was also found that the introduction of the edge dislocation makes it possible to relieve the stresses in the system most effectively, and the introduction of the screw dislocation, on the contrary, leads to increase in the system energy for any arrangement in the island.

As a result of TEM image simulation, it was found that the introduction of dislocations in the NI leads to different distortions in the moiré pattern (Figure 4). It can be seen that moiré pattern of NI with dislocations undergoes considerable changes when rotating the diffraction vector (g). TEM image loses symmetry under the conditions g = 400 and g = 0-40. It is also seen that the introduction of dislocations in NI leads to appearance of additional moiré fringes in the image with the diffraction vector g = 220. At g = 2-20 moiré pattern for NI with edge dislocations does not differ from the images of coherent nanoisland, unlike the dislocations with screw component of the Burgers vector.

The introduction of dislocations in NI leads to bending of the moiré fringes at g = 400 and g = 0-40 for edge dislocations and at g = 2-20 for screw dislocations. The number of bend fringes is determined by the number of dislocation.

The obtained results showed that moiré patterns obtained under various diffraction conditions allow to identify the dislocation type in NI. For example, the perfect 60°-dislocation presence can be determined from the moiré fringes at g = 2-20. But to distinguish the perfect 90°-dislocation from the Frank partial dislocation occur very difficult from the moiré pattern.

The simulated images of NI with dislocations correlate well with experimental ones and explain the observed features of the moiré pattern (Figure 5).

Nanoisland	System energy, $*10^{-26}$ J	g = 220	g = 400	g = 2-20	g = 0-40
Coherent NI	8,30				
NI with Frank particle dislocation	7,24				
NI with two Frank particle dislocations with parallel Burgers vectors	6,53				
NI with two Frank particle dislocations with perpendicular Burgers vectors	6,51				
NI with perfect 60°-dislocation	7,83				
NI with two perfect 60°-dislocation with parallel Burgers vectors	7,66				
NI with two perfect 60°-dislocation with perpendicular Burgers vectors	7,46				
NI with perfect 90°-dislocation	6,89				
NI with two perfect 90°-dislocation	6,05				

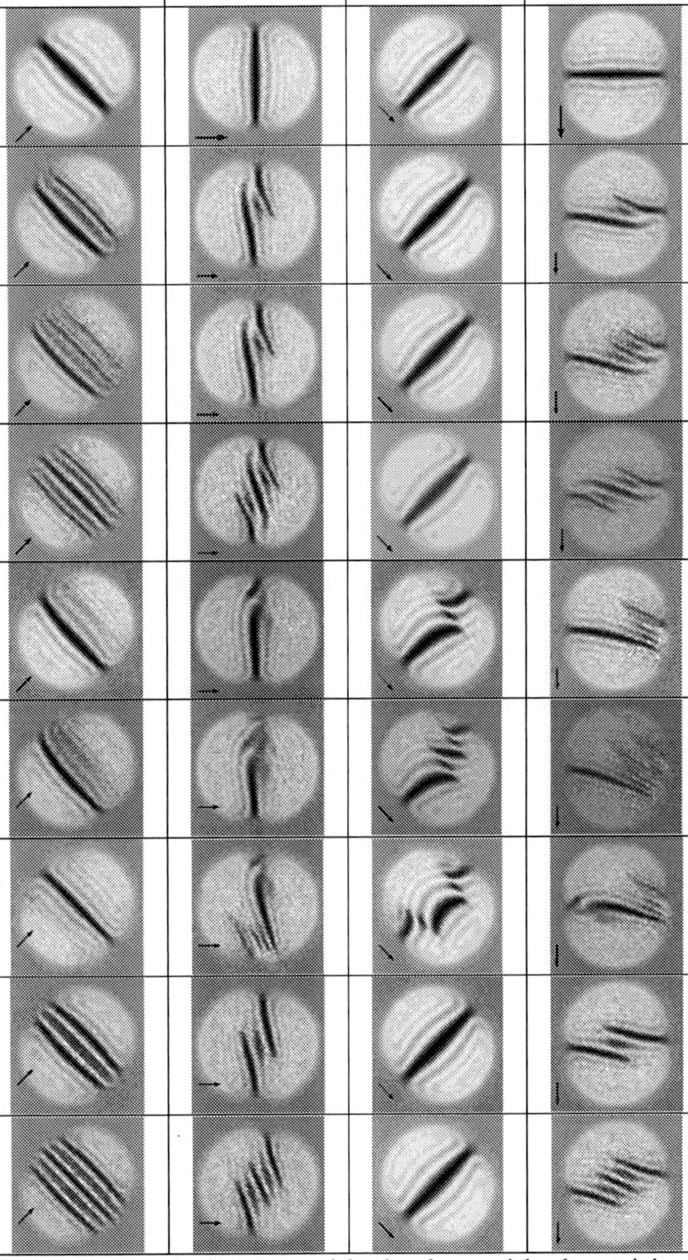

Figure 4. Simulated electron microscope images of coherent nanoisland and a nanoisland containing partial Frank dislocations, perfect 60°- and 90°-dislocations. The images are calculated for different directions of the diffraction vector g.

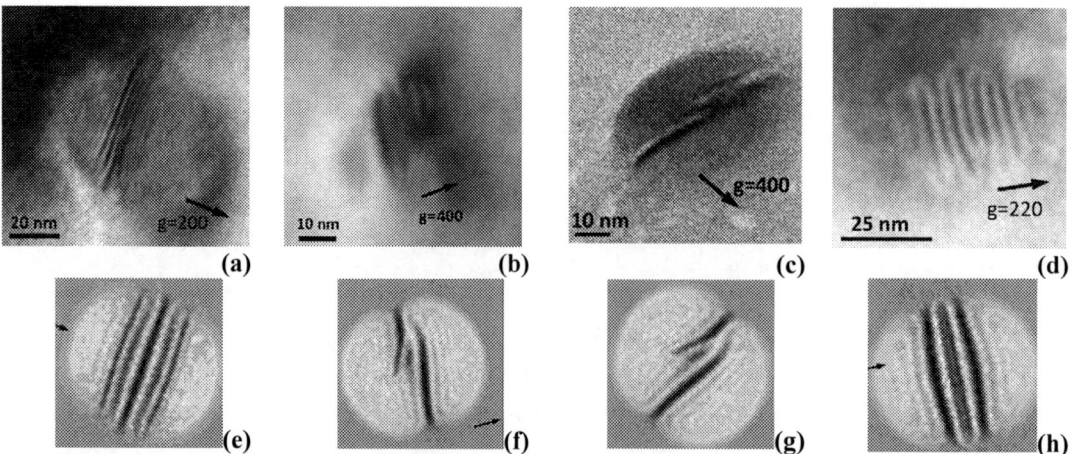

Figure 5 (a–h). Comparison of experimental (a – d) and simulated (e – h) TEM images from InSb nanoislands in the presence of a different type dislocations: with two Frank particle dislocations (a, e), with one Frank particle dislocation (b – c, f – g) and with perfect 90°-dislocation (d, h).

4. Conclusion

A model of NI with dislocations at the nanoisland/substrate interface is constructed. From the calculation of the total elastic energy minimum is obtained that the distance between dislocations in the NI is in good agreement with found experimentally. It confirms the consistency of the constructed model and the calculations performed.

It is shown that the presence of dislocations with a screw component of the Burgers vector in the NI is unlikely, since this is energetically unfavourable. Edge dislocations, on the contrary, allow relieving the system stresses most effectively and, therefore, happen in real objects more often.

The simulation results allow determining the type of dislocation inside quantum dot by moiré pattern on TEM image of NI in plan-view geometry. For example, the experimentally observed features of the moiré pattern from the InSb quantum dot are possible to explain by the presence of Frank particle dislocations and perfect 90°-dislocations in it.

Acknowledgments

TEM characterization was performed using equipment owned by the Federal Joint Research Center "Material science and characterization in advanced technology" with financial support by Ministry of Education and Science of the Russian Federation (id RFMEFI62117X0018).

References

[1] Bert N.A., Freidin A.B., Kolesnikova A.L., et al. On strain state and pseudo-moire TEM contrast of InSb quantum dots coherently grown on InAs surface // Phys. Status Solidi A. Appl. Mat. 2010. Vol. 207. Pp. 2323–2326;
[2] D.B. Williams and C.B. Carter (1996): Transmission Electron Microscopy, Plenum Press, NY;
[3] N A Bert, V N Nevedomskiy, L A Sokura, "Features of microstructure of InSb quantum dots on InAs substrate", Journal of Physics: Conference Series 586 (2015) 012004.

Development of local catalytic centers positioning technology for carbon nanotubes growth

O I Il'in[1], N N Rudyk[1], M V Kuzhelev[2] and A A Fedotov[1]

[1] Southern Federal University, Institute of Nanotechnologies, Electronics and Electronic Equipment Engineering, Taganrog, 347922, Russia
[2] «Rehm RUS» LLC, Moscow, 127018, Russia

Abstract. The paper presents the results of developing positioning catalytic centers technology for the local growth of carbon nanotubes using non-standard techniques of nanometer resolution lithography explosive. It is shown that using of ZnO as a sacrificial layer is promising for the formation of nanoscale catalytic centers, as well as actuality of using focused ion beam as a lithographic tool. An array of local catalytic centers with a diameter of 35 nm, completely repeating the given topology, suitable for the growth of carbon nanotubes is obtained. The obtained results can be used for developing technological processes formation of nano- and microelectronic structures, nano- and microsystems based on single or local CNTs arrays.

1. Introduction

Carbon nanotubes (CNTs) are promising material, which is associated with progress in the creation of a new element base for vacuum micro- and nanoelectronics [1-3]. Methods that make it possible to growth a single CNTs in places defined by the design of the device are the most interesting in device applications. [4, 5]. A solution of this problem is the using of lithographic techniques to create and position the CNT growth catalytic center (CC) with dimensions less than 50 nm. The most promising method with same resolution is electron-beam lithography. However, high cost, impossibility of providing high-temperature processes with the resist layers, complexity of equipment and low productivity are the main factors holding back this method of lithography. Development of local catalytic centers positioning technology for carbon nanotubes growth without described problems, is an actual task.

A promising method that allows to modify the surface of a substrate with a nanometer resolution is the method of focused ion beams (FIB) [6]. However, direct ion-beam etching of the catalytic Ni film leads to appearance of radiation defects and partial doping of the catalyst by ions of source (Ga^+).

The aim of the work is developing of local catalytic centers positioning technology for carbon nanotubes growth, excluding direct interaction between FIB and the catalytic layer.

2. Experiments and methods

Si (100) was used as substrates. ZnO sacrificial layer (Fig. 1) sputtering was performed by pulsed laser deposition (PLD) system Peeoner 100 (Neocera, USA) from a ZnO target (Kurt J. Lesker Co). The deposition thickness was 60 nm. Positioning of catalytic centers on the substrate was performed by local etching of array of "points" by FIB method by the gallium liquid metal ion source on Nova

---

Content from this work may be used under the terms of the Creative Commons Attribution 3.0 licence. Any further distribution of this work must maintain attribution to the author(s) and the title of the work, journal citation and DOI.

Published under licence by IOP Publishing Ltd

Nanolab 600 (FEI, Netherlands). The beam current was 0.1 pA, the accelerating voltage - 30 kV. At each "point" of the array, the dwell time of the beam was increased. On the FIB-profiled samples, Ni catalytic layer were formed by magnetron sputtering (in Ar atmosphere, pressure 0.5 Pa) on the Auto500 (BOC Edwards, UK) from the Ni target 99.995% (Kurt J. Lesker Co) at 150 °C temperature. Ni thickness was 15 nm. For selective etching of ZnO and formation of CNT growth local CCs, the samples were chemically etched in 40% aqueous ammonia solution for 10 sec at a temperature of 24 °C. CNT growth was provide by plasma chemical vapor deposition (PECVD) in an atmosphere of acetylene (C2H2) and ammonia (NH3) on samples with local CC by using multifunctional nanotechnology complex NANOFAB (NT-MDT, Russia). The growth temperature was 680 °C, the growth time was 15 min, the acetylene and ammonia flows are 70 and 210 cm^3/min, respectively. Analysis of the samples was provided by using atomic force (AFM) and scanning electron microscopy (SEM) on the Ntegra scanning probe microscope system (NT-MDT, Russia) and Nova Nanolab 600, respectively.

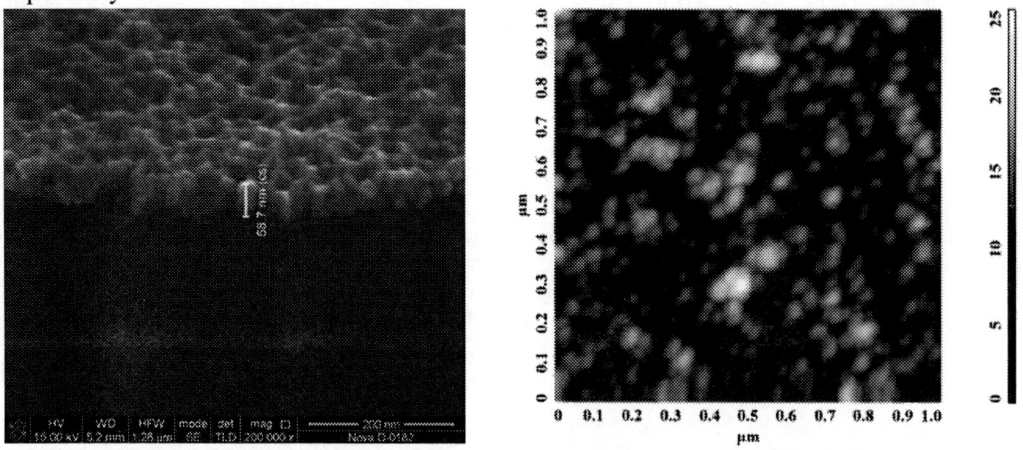

Figure 1. SEM (a) and AFM (b) images of ZnO layer.

3. Results and discussion

Analysis of obtained SEM and AFM data has shown that the PLD method allows the create of sacrificial material films with the necessary physico-chemical and morphological properties. Precise control of the of thickness, roughness, structure, adhesion and chemical composition parameters of the formed sacrificial layer, allows to create of thin films of material with the required parameters necessary for further technological operations. It is found that obtained ZnO films have a nanocrystalline structure, the grain size was 12 ± 3 nm, surface roughness 8 ± 1 nm.

AFM and SEM samples research after FIB-etching have shown (Figure 2, a) that increasing of beam dwell time results to increment a depth and diameter of etching. From the SEM images obtained (Figure 2, b), it is seen that NH_4OH exposure, induce a selective dissolution of the ZnO film and removal of the ZnO/Ni structure occurs, as a result on the substrate are formed of catalytic Ni regions with a given geometry. Thus, this operation allowed to remove the metal located on the sacrificial layer and save the catalytic centers on the substrate in the FIB-profiled areas.

SPBOPEN 2018   IOP Publishing

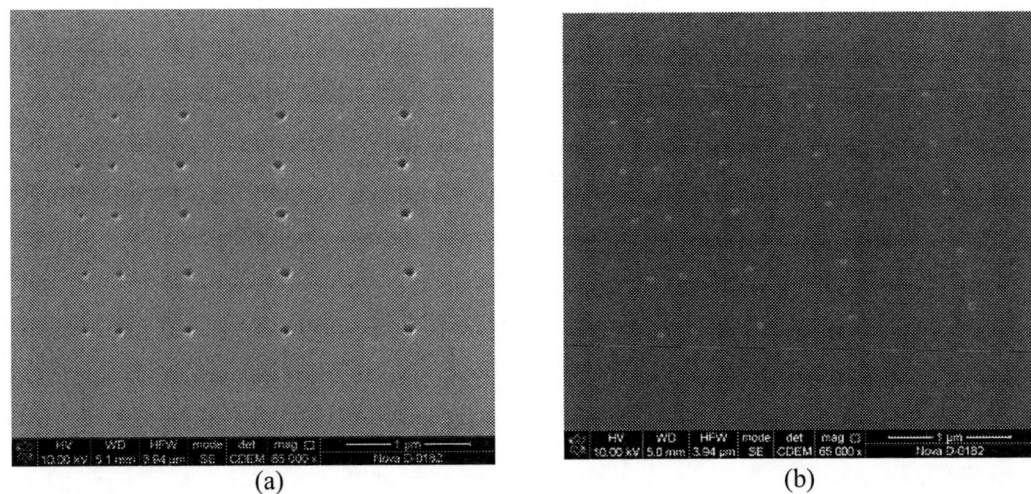

(a)   (b)

Figure 2. SEM image of the structure: ZnO/Si after FIB etching (a); Ni/ZnO/Si after selective etching in NH_4OH (b)

The use of an etchant of non-aggressive composition allows us to apply the developed technique for creating devices for nanoelectronics in the formation of catalytic regions on multilayer structures with a complex topology that are critical to the effects of chemically aggressive solutions. At the same time, the developed technology makes it possible to form a CC with a diameter of ~35 nm (Figure 3, a). However, the FIB method imposes high demands on thickness uniformity, as well as the need to refine the ion-beam etching modes of the sacrificial layer, without affecting the substrate material.

A similar experiment with Ni deposition by the PLD method did not make it possible to obtain local catalytic centers. This can be attributed to the fact that the Ni particles (atoms, ions, clusters) deposited by the PLD method have a higher energy compared to the magnetron sputtering, which leads to their penetration into the substrate structure with increasing Ni layer adhesion, violating the further formation of catalytic centers during annealing films in CNTs synthesis process. In this case, films formed by the magnetron sputtering have adhesion, which satisfies the catalytic centers formation conditions.

Later, CNT was grown on the formed local catalytic centers by the PECVD. The analysis of the SEM images of the samples after growth (Figure 3, b, c) showed that the formation of nanotubes occurs in predetermined areas formed at the topology formation stage by FIB.

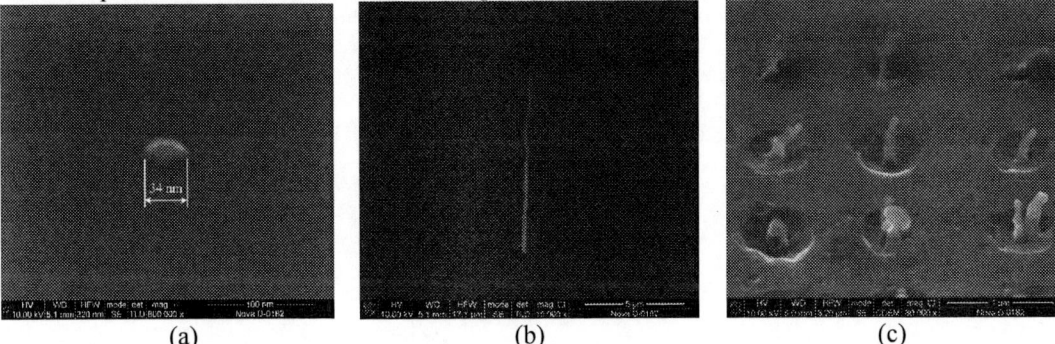

(a)   (b)   (c)

Figure 3. SEM image of a single catalytic center (a); a single CNT (b); an array of local CNTs (c)

It has been established that with the increase in the diameter of catalytic areas over 250 nm, the formation of 2 and more CNTs is observed under the presented modes. Thus, the developed

135

technology can be used to create new promising elements of nanoelectronics, micro- and nanosystems [7].

4. Conclusion

As a result of this work, a technique for nanolithography, suitable for the formation of nanoscale catalytic centers and growth of CNTs on them, was developed.

It is shown that the PLD method allows to create films of the ZnO sacrificial layer with the necessary physicochemical and morphological properties. Precise control of the thickness, roughness, structure and adhesion parameters of the formed sacrificial layer allows to create of the thin films material suitable for creation of the single CNTs and arrays based on them. Although that the FIP method imposes high demands on uniformity of thickness, as well as the necessary to refine the sacrificial layer ion-beam etching modes, shown the perspective of using FIB-technology for creating catalytic areas and CC with a diameter from 30 nm suitable for the single CNTs growth. In this case, the use of ZnO as sacrificial layer has a part of advantages associated with the possibility of carrying out high-temperature technological processes, due to the temperature-resistant sacrificial layer; a large set of selective etchants; ensuring the necessary adhesion of the sacrificial layer material (low adhesion) and functional material (high adhesion) to the substrate. Also the developed technology allows to avoid the operations connected with resist soft baking, development and hard baking.

References

[1]   Ageev O A, Blinov Y F, Il'ina M V., Il'in O I, Smirnov V A and Tsukanova O G 2016 Study of adhesion of vertically aligned carbon nanotubes to a substrate by atomic-force microscopy Phys. Solid State 58 309–14.
[2]   Ageev O A, Il'in O I, Rubashkina M V., Smirnov V A, Fedotov A A and Tsukanova O G 2015 Determination of the electrical resistivity of vertically aligned carbon nanotubes by scanning probe microscopy Tech. Phys. 60 1044–50.
[3]   Ageev O A, Blinov Y F, Ilina M V, Ilin O I and Smirnov V A 2016 Modeling and experimental study of resistive switching in vertically aligned carbon nanotubes J. Phys. Conf. Ser. 741 12168.
[4]   Il'ina M V., Il'in O I, Blinov Y F, Smirnov V A, Kolomiytsev A S, Fedotov A A, Konoplev B G and Ageev O A 2017 Memristive switching mechanism of vertically aligned carbon nanotubes Carbon N. Y. 123 514–24.
[5]   Peng L-M, Zhang Z and Wang S 2014 Carbon nanotube electronics: recent advances Mater. Today 17 433–42.
[6]   Ageev O A, Kolomiytsev A S, Bykov A V, Smirnov V A and Kots I N 2015 Fabrication of advanced probes for atomic force microscopy using focused ion beam MR 55 2131–4.
[7]   Il'ina M, Il'in O, Blinov Y, Konshin A, Konoplev B and Ageev O 2018 Piezoelectric Response of Multi-Walled Carbon Nanotubes *Materials (Basel).* **11** 638.

SPBOPEN 2018

Investigation of InAs/InGaP nano-heterostructures grown by MOVPE for intermediate band solar cells

N A Kalyuzhnyy[1], S A Mintairov[1], A M Nadtochiy[1,2], V N Nevedomskiy[1], R A Salii[1]

[1] Ioffe Institute, St.-Petersburg, Russia, Polytechnicheskaya str. 26
[2] St Petersburg Academic University, St Petersburg, Russia, Hlopina str. 8/3

E-mail: nickk@mail.ioffe.ru

Abstract. InAs/GaInPnano-heterostructures obtained by embedding 1.3 - 2.1 monolayers of InAs in the GaInP matrix lattice-matched to a GaAs substrate have been grown by the MOVPE and investigated. The formation of large nano-objects with lateral size of 150 – 200 nm, which did not depend on the InAs thickness, has been observed by TEM. The nano-objects had aheight of about 10 nm, and their shape can be described as a discontinuous quantum well. Photoluminescence (PL) analysis has shown that the position of the PL peak is practically independent on the InAs thickness and is about 940 nm. It has been supposed that the mechanism of nano-object formation differs from quantum dot deposition in the Stranski-Krastanov growth mode. Probably they are formed due to intensive As-P intermixing, and InAs acts only as nucleus for the further InGaAsP nano-object formation.

1. Introduction

Semiconductor solar cells (SCs) based on AIIIBV heterostructures are ones of the most efficient solutions among the renewable energy sources. Multijunction (MJ) solar cells, which are consisted of several series-connected p–n junctions, allow reducing the fundamental losses such as incomplete absorption and carrier thermalization inherent to single-junction SCs with a fixed bandgap. Another approach, which has been proposed theoretically [1] for reducing the fundamental losses, is a SC with intermediate band (IB). The approach promises to grow equivalent of a triple-junction SC in a single-junction SC due to creating bandgap-energy electron-hole pairs using two sub-bandgap photons that use a set of energy band located within the semiconductor energy bandgap as steppingstone for the electron-hole generation process. InAs quantum dots (QDs) were proposed [2] as a way for the realization of the IB solar cell concept, since confined states around the QD have energy levels located within the matrix bandgap that can be used as an intermediate band.

Unfortunately, practical realization of this concept is not yet in sight, since the majority of developments has been carried out in the field of IB SCs still based on the InAs/GaAs QD system, which is not well appropriate for this purpose due to insufficient depth of the InAs QD electronic level in matrix (<0.2 eV). In spite of that the system of InAs QDs in GaAs matrix has been studied for a variety of applications, including telecommunication lasers on GaAs substrates [3, 4] and solar cells with an expanded spectral range [5]. In particular, InAs/GaAs QD system has demonstrated the long-wavelength contribution to the quantum efficiency of a single-junction GaAs SC [6-8] that can be useful for MJ SCs with improved photogenerated current balance [9].

Content from this work may be used under the terms of the Creative Commons Attribution 3.0 licence. Any further distribution of this work must maintain attribution to the author(s) and the title of the work, journal citation and DOI.
Published under licence by IOP Publishing Ltd

Another candidate for the IB SC is InAs QDs in the InGaP matrix [10].According to theoretical calculations the main requirement for IB SCs is to have the band gap of a matrix of ~ 1.95 eV [1], and the depth of a QD electronic level larger than 0.4 eV [9]. The $In_{0.49}Ga_{0.51}P$ alloy lattice-matched to GaAs substrate has a band gap up to 1.91 eV, depending on the ordering degree in the solid solution [11, 12]. Therefore, the InAs QD electronic level in the $In_{0.49}Ga_{0.51}P$ matrix is deeper than in GaAs one. Moreover, $In_{0.49}Ga_{0.51}P$ is the material of the top subcell in GaInP/Ga(In)As/Ge triple-junction SCs. Therefore, the InAs/GaInP QD technology can be applied for both the multi-junction SCs and single-junction IB SCs.

Today, the main epitaxial method to grow a wide range of semiconductor devices (such as lasers, light emitting diodes and solar cells) is the metal organic vapor phase epitaxy (MOVPE). This method allows creating SC structures of any complexity and also demonstrates the possibility of nano-dimensional object growth such as QDs.

The purpose of the work is the development of MOVPE epitaxial technology of InAs/GaInP nano-heterostructures and the study of their optical and physical properties.

2. Experimental procedure

All structures have been grown using the MOVPE technique by means of R&D installation with a horizontal reactor at low pressure (100 mbar). Metal-organic compounds were used as sources of Group III elements (trimethylgallium (TMGa), trimethylaluminum (TMAl)and trimethylindium (TMIn)). Arsine (AsH_3) and phosphine (PH_3) were used as a sources of the Group V elements (As and P). Experiments were carried out on (100) GaAs substrates misoriented to the (111) direction by 6°. The structures (figure 1) were grown lattice-matched to a substrate and included the InAs material with the thickness varied within the range of 1.3 - 2.1 monolayers (ML). The InAs material was deposited on a 250 nm-thick GaInP layer at low temperature (520°C) and then was heated under arsine flow and covered by a 250 nm-thick GaInP layer. AlInP barriers were used to prevent photo-generated carriers' losses in photoluminescence (PL) measurements.

To obtain the PL spectra, a 532 nm YAG:Nd laser of 350 mW power was used. All measurements were carried out by means of a cooled Ge photodetector using a standard lock-in amplifier.

The cross-section of the investigated structures was examined by TEM. The samples were prepared for the TEM study using the conventional procedure of preliminary mechanical thinning followed by final ion beam milling (3.5 keVAr+ ions). The TEM study was carried out with a JEOL JEM 2100F microscopes at 200 kV accelerating voltage.

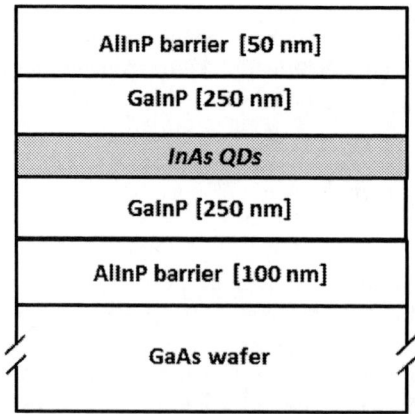

Figure 1. The schematic of experimental structures.

3. Results and discussion

Figure 2 shows cross-section TEM images for the structures with InAs thickness of 1.5 ML. The large InAs nano-objects connected by a thin layer can be seen. The distance between the objects is 20-150 nm (figure 2a), their lateral size is in the range of 150 – 200 nm and height is about 10 nm (figure 2b). TEM images show that InAs objects formation in the GaInP matrix occurs not by the classic Stranski-Krastanov (SK) growth mode, and it can be structurally described as a discontinuous quantum well according to the TEM cross-section image. We shall call them "quantum pancakes" (QPs). A wetting layer is not observed in comparison with InAs/GaAs QDs. Such a QP nano-object formation could be explained by intensive As-P intermixing after/during InAs deposition on the GaInP surface. As a result of the intermixing, the quaternary InGaAsP solid solution is formed instead of InAs quantum dots. TEM investigations of InAs/GaInP nano-heterostructures have shown that the lateral size of the nano-objects is practically independent on InAs thicknesses in the examined range. This also points to the mechanism, where deposited InAs material acts only as nucleus for the further InGaAsP QP growth.

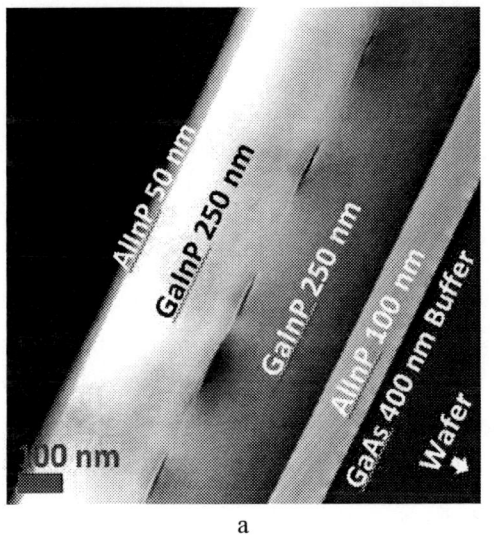
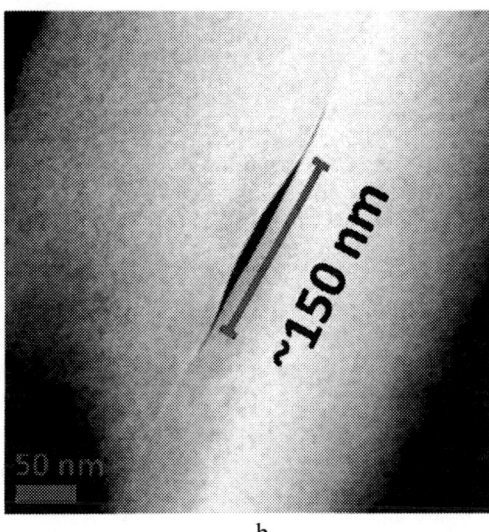

a   b

Figure 2. TEM cross-section images at different resolutions of the experimental structure with a 1.55 ML-thick InAs layer.

Figure 3 shows that for the layers with different InAs thicknesses, the PL peak position remains unchanged at ~940 nm. This fact, together with TEM observation, demonstrates the relatively constant physical parameters of obtained objects – their density, the lateral size and height in the investigated range of InAs thicknesses. Such a behavior is strongly different in comparison with that of InAs QDs in the GaAs matrix, where the PL peak wavelength increases with the increase of InAs thickness, and a sharp maximum of the PL intensity is revealed in the range of 1.6-1.8 InAs MLs [6]. Moreover, the InAs/GaAs QDs grown by MOVPE usually demonstrate the multi-modal distribution of size and height [13], in contrast with the QP observed here.

Based on these results, the relative PL integral intensity versus InAs thickness was calculated and normalized to the highest PL spectra (figure 4). The optimal InAs thickness values were determined to be in the range 1.45 - 1.58 ML at which the maximum PL intensity of QDs was observed, which a little bit less than for InAs/GaAs QDs.

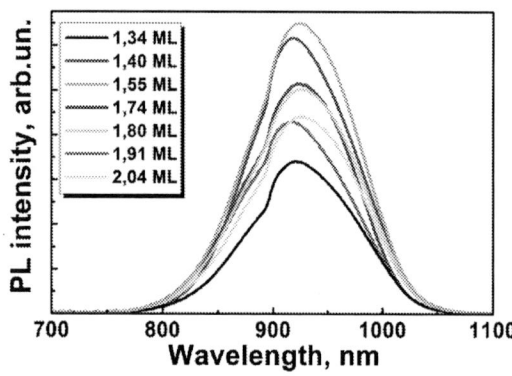

Figure 3. PL spectra of structures with different InAs thicknesses.

Figure 4. The dependence of PL intensity on the InAs thickness.

4. Conclusions

In the paper, the InAs/GaInP nanoheterostructures have been grown by the MOVPE, and their optical and physical properties have been investigated. The obtained data allows concluding that embedding InAs into the GaInP matrix during MOVPE leads to considerable differences in the process of nano-object formation in comparison with classic InAs/GaAs system, where InAs QDs in Stranski-Krastanov growth mode are formed. Instead, we obtained massive, separately located nano-objects with height of 10 nm that has been characterized by means of TEM. Such InAs/GaInP heterostructures have demonstrated large lateral size (150-200 nm) of nano-objects and an independence of the PL peak position on the InAs thickness. All these results point to the mechanism, where deposited InAs material acts only as nucleus for the large nano-object formation, probably due to intensive As-P intermixing.

We believe that developed InAs/GaInP nano-heterostructures are promising for creating IB SCs due to heterostructure high PL intensity.

Acknowledgements

The work has been supported by the Russian Foundation for Basic Research (grant No. 16-08-01004). TEM measurements were carried out on the equipment of the Federal Joint Research Centre «Material science and characterization in advanced technology» (Ioffe Institute, St.Petersburg, Russia).

References

[1] Luque A and Marti A 1997 *Phys. Rev. Lett.* **78** pp 5014–5017

[2] Luque A, Martí A, Stanley C, López N, Cuadra L, Zhou D, et al. 2004 *Journal of Applied Physics* **96** p 903

[3] Salhi A, Fortunato L, Martiradonna L, Todaro M T, Cingolani R,Passaseo A, Vittorio M. De 2007 *Semiconductor Science and Technology* **22** pp 396–398

[4] Kageyama T, Nishi K, Yamaguchi M, Mochida R, Maeda Y, Takemasa K, Tanaka Y, Yamamoto T, Sugawara M, Arakawa Y 2011 *2011 Conference on Lasers and Electro-Optics Europe and 12th European Quantum Electronics Conference* (22-26 May 2011)

[5] Blokhin S A, Sakharov A V, Nadtochy A M, Pauysov A S, Maximov M V, Ledentsov N N, Kovsh A R, Mikhrin S S, Lantratov V M, Mintairov S A, Kaluzhniy N A and Shvarts M Z 2009 *Semiconductors* **43** pp 514–518

[6] Kalyuzhnyy N A, Mintairov S A, Salii R A, Nadtochiy A M, Payusov A S, Brunkov P N, Nevedomsky V N, Shvarts M Z, Martí A, Andreev V M and Luque A 2016 *Progress in Photovoltaics* **24** pp 1261–1271

[7] Bailey C G, Forbes D V, Polly S J, Bittner Z S, Dai Y, Mackos C, Raffaelle R P and Hubbard S

M 2012 *Journal on Photovoltaics*, **2** pp 269–275

[8] Ramiro I, Martí A, Antolín E and Luque A 2014 *Journal of Photovoltaics* **4** pp 736–748

[9] Mintairov M A, Evstropov V V, Shvarts M Z, Mintairov S A, Salii R A and Kalyuzhnyy N A 2016 *AIP Conf. Proc.* (St. Petersburg, Russia) **1748**

[10] Ramiro I, Villa J, Lam P, Hatch S, Wu J, Lopez E, Antolin E, Liu H, Marti A and Luque A 2015 *Journal on Photovoltaics* **5** pp 840–845

[11] Zunger A and Mahajan S *Handbook on Semiconductors* (Elsevier Scince B.V.) pp 1399–1514

[12] Kalyuzhnyy N A, Mintairov S A, Mintairov M Aand Lantratov V M 2009 *Proceedings of the 24nd European Photovoltaic Solar Energy Conference* (Hamburg, Germany) pp 538-544

[13] Salii R A, Mintairov S A, Brunkov P N, Nadtochiy A N, Payusov A S, Kalyuzhnyy N A 2015 *Semiconductors* **49** pp 1111–1118

SPBOPEN 2018   IOP Publishing

Suppression of miscibility gaps in vapor-liquid-solid InGaAs and InGaN nanowires

T Jean[1,2] and V G Dubrovskii[2]

[1]Université Clermont Auvergne, CNRS, SIGMA Clermont, Institut Pascal, F-63000 Clermont-Ferrand, France
[2]ITMO University, Kronverkskiy pr. 49, 197101 St. Petersburg, Russia

Abstract. Miscibility gaps in ternary III-V and III-N alloys often prevent their compositional tuning required for new generation electronic and optoelectronic devices. Here, we show how the miscibility gaps are suppressed on kinetic grounds at high enough supersaturations in liquid droplets catalyzing the vapor-liquid-solid growth of ternary III-V and III-N nanowires. We give two examples for highly mismatched InGaAs and InGaN material systems in terms of the compositional diagrams describing the solid composition as a function of the indium content in the liquid or vapor phase.

Introduction

III-V and III-N nanowires (NWs) are widely considered as promising building blocks for fundamental nanoscience and nanotechnology [1-3]. NWs allow for almost unlimited bottom-up design and coherent growth on lattice-mismatched substrates without forming misfit dislocations [4-6]. Useful NW structures in most cases comprise ternary III-V NWs and heterostructures within such NWs. Immiscible III-V alloys such as InGaAs and InGaN have enabled new classes of optoelectronic devices, but it is admittedly challenging to tune their operating wavelength. This requires an increased In content x in $In_xGa_{1-x}As$ or $In_xGa_{1-x}N$, prevented by the indium segregation within the miscibility gaps. Beyond the advantage of dislocation-free growth on dissimilar substrates, NWs may help to extend the compositional range of such alloys and ultimately to fabricate high quality photonic heterostructures with tunable and well-controlled compositions [7-13]. However, the mechanisms behind the formation of ternary III-V NWs are very different from the thin film case and their comprehensive understanding is beyond reach to this end (see Ref. [3] for a detailed review).
Consequently, here we develop a model that attributes the wide compositional range of VLS III-V NWs to the purely kinetic growth regime without macroscopic nucleation. This requires a high degree of supersaturation in liquid droplets catalyzing the VLS growth, which is accessible in epitaxial techniques with high material inputs such as metal organic vapor phase epitaxy (MOVPE) or hydride vapor phase epitaxy (HVPE).

Model

The composition of ternary III-V NWs grown by the VLS method can be described using four different approaches – (i) equilibrium model [14,15]; (ii) binary nucleation model [15-17], (iii) regular

---

Content from this work may be used under the terms of the Creative Commons Attribution 3.0 licence. Any further distribution of this work must maintain attribution to the author(s) and the title of the work, journal citation and DOI.
Published under licence by IOP Publishing Ltd

growth model [7,18] and (iv) kinetic growth model [19,20]. Following the kinetic approach of Refs. [19,20], we define the composition of a ternary $A_xB_{1-x}D=(AD)_x(BD)_{1-x}$ NW as $x = i/(i+j)$, where i is the number of AD pairs, j is the number of BD pairs, and $i + j$ is the total number of III-V pairs in a growing island. Putting dx/dt to zero yields the natural result [19]

$$x = \frac{di/dt}{di/dt + dj/dt},$$ (1)

showing that the kinetically controlled composition is given by the ratio of the growth rate of AD fraction in the island over the total growth rate of the island.

For the growth rates of AD and BD fractions in the island, we take the usual approximations justified earlier in Ref. [21]

$$\frac{di}{dt} = K_{AD}\left(e^{\mu_A^L + \mu_D^L} - e^{\mu_{AD}^S}\right), \quad \frac{dj}{dt} = K_{BD}\left(e^{\mu_B^L + \mu_D^L} - e^{\mu_{BD}^S}\right).$$ (2)

Here, μ^L are the chemical potentials for atoms A, B and D in liquid and μ^S are the chemical potentials of AD or BD pairs in solid, all measured in thermal units of k_BT. The crystallization rates K_{AD} and K_{BD} summarize kinetic growth effects associated with different diffusivities of atoms in the liquid phase and crystallization rates of AD and BD pairs, respectively. Both growth rates in Eq. (2) equal zero when $\mu_A^L + \mu_D^L = \mu_{AD}^S$ and $\mu_B^L + \mu_D^L = \mu_{BD}^S$, by definition of thermodynamic equilibrium in a ternary liquid-solid system [14,15]. Using Eqs. (2) in Eq. (1), it is easy to obtain

$$\frac{1-x}{x} = \frac{K_{BD}}{K_{AD}} e^{(\mu_B^L - \mu_A^L)}\left(\frac{1 - e^{-\Delta\mu_{BD}}}{1 - e^{-\Delta\mu_{AD}}}\right).$$ (3)

The chemical potential differences $\Delta\mu_{AD}$ and $\Delta\mu_{BD}$ depend on the solid composition x and liquid composition $y = c_A/(c_A + c_B)$ (with c_A and c_B being the atomic concentrations of atoms A and B in liquid) according to

$$\Delta\mu_{AD} = \mu_A^L(y) + \mu_D^L(y) - \mu_{AD}^0 - \ln x - \omega_s(1-x)^2,$$
$$\Delta\mu_{BD} = \mu_B^L(y) + \mu_D^L(y) - \mu_{BD}^0 - \ln(1-x) - \omega_s x^2.$$ (4)

Here, μ_{AD}^0 and μ_{BD}^0 are the chemical potentials of pure binary AD and BD solids and ω_s is the interaction constant in thermal units. The solid state is described within the regular solution model [17], although it is not critical. The factor in the right hand side of Eq. (3) depends only on y, while all characteristics of the solid state enter the exponential terms $\exp(-\Delta\mu_{AD})$ and $\exp(-\Delta\mu_{BD})$. The kinetic factor K_{BD}/K_{AD} should not depend on y for ternary materials based on the group III intermixing for arsenides, phosphides and nitrides, because both crystallization rates are proportional to the concentration of highly volatile arsenic or nitrogen atoms, which cancels in the ratio of K_{BD} over K_{AD}. Any geometrical effect associated with the size dependence of the crystallization rates is proportional to a certain power of the island size $i + j$ depending on the island growth mechanism [22], and cancels in the K_{BD}/K_{AD} ratio.

Despite its simplicity, Eq. (3) clearly shows the main effect - for small $\Delta\mu_{AD}$ and $\Delta\mu_{BD}$, the compositional diagram $x(y)$ contains the miscibility gap described by the quadratic interaction terms in Eqs. (4) at $\omega_s > 2$, while at $\Delta\mu_{AD} \gg 1$ and $\Delta\mu_{BD} \gg 1$ the dependence on x in the right hand side disappears. Of course, the miscibility gaps can be modified by strain induced by the lattice mismatch (see, for example, Refs. [23] and [24]), the effect which is not taken into account in Eqs. (4). However, Eq. (3) shows that the miscibility gap is fully suppressed on kinetic grounds at high enough supersaturations in the liquid phase even if it is present in bulk thermodynamics. Such high degrees of supersaturation in the catalyst droplets are locally accessible in MOVPE or HVPE VLS techniques with high material inputs [25].

Equation (3) is considerably simplified in the case of ternaries based on the indium (A=In, D=V) and gallium (B=Ga, D=V) intermix due to the known high stability of indium in the liquid phase such that the indium content in the catalyst droplet is very close to unity [16,17]. In the following, we will additionally assume that $\varepsilon = (K_{InV} / K_{GaV})\exp(\mu_{InV}^0 - \mu_{GaV}^0) << 1$. This strong inequality is consistent with $y \cong 1$, because indium-rich liquid requires that the crystallization rate of In-V pairs is much lower than that of Ga-V pairs. Using Eqs. (4), we can re-arrange Eq. (3) as

$$\zeta = \varepsilon\left(\frac{1-x}{x}\right) + \alpha(1-x)\left[e^{\omega_s x^2} - \varepsilon e^{\omega_s (1-x)^2}\right],$$
(5)

with

$$\zeta = e^{-[\mu_{In}^L - \mu_{Ga}^L - (\mu_{In}^0 - \mu_{Ga}^0)]},$$
(6)

$\varepsilon = (K_{InV} / K_{GaV})\exp(\mu_{InV}^0 - \mu_{GaV}^0)$ and $\alpha = \exp[-(\mu_{In}^L + \mu_V^L - \mu_{InV}^0)]$. Clearly, the α parameter changes from unity in the equilibrium case to zero at very high supersaturations in liquid.

Chemical potentials of the indium and gallium atoms in the liquid phase are given by [17]

$$\mu_{In}^L = \mu_{In}^0 + \ln c_{In} + \varphi_{In} = \mu_{In}^0 + \ln[y(c_{In} + c_{Ga})] + \varphi_{In},$$

$$\mu_{Ga}^L = \mu_{Ga}^0 + \ln c_{Ga} + \varphi_{Ga} = \mu_{Ga}^0 + \ln[(1-y)(c_{In} + c_{Ga})] + \varphi_{In},$$
(7)

where the φ terms sum up the terms describing interactions of indium and gallium with all other atoms. At $y \cong 1$, we can neglect the y dependence of all the terms in Eqs. (7) except for the leading logarithmic dependence $\ln(1 - y)$ for gallium. Taking the difference, we obtain

$$1 - y = Q\left[\varepsilon\left(\frac{1-x}{x}\right) + \alpha(1-x)e^{\omega_s x^2} - \alpha\varepsilon(1-x)e^{w_s(1-x)^2}\right],$$
(8)

with $Q = \exp[\mu_{In}^0 - \mu_{Ga}^0 - (\mu_{InV}^0 - \mu_{GaV}^0) + \varphi_{In} - \varphi_{Ga}]$. At $\varepsilon << 1$, the last term in the brackets can be neglected with respect to the second one. Using the notations

$$a = Q\varepsilon, \quad \delta = \alpha / \varepsilon,$$
(9)

Eq. (8) takes the final form

$$y = 1 - a\left[\frac{1-x}{x} + \delta(1-x)e^{\omega_s x^2}\right].$$
(10)

This solid-liquid compositional diagram is controlled by three parameters. The thermodynamic parameter a should be much smaller than unity to ensure that the droplet is indium-rich. The kinetic δ factor describes the relative importance of macroscopic nucleation with respect to the growth kinetics and tends to zero in the purely kinetic growth regime when $\mu_{In}^L + \mu_V^L >> \mu_{InV}^0$.

Equation (10) is our final result for the $x(y)$ dependence. We now want to find the relationship between the liquid composition y and the indium fraction in vapor in HVPE growth of InGaN NWs at high temperatures, as in Ref. [20]. We consider stationary solutions of the kinetic equations for the total numbers of the indium (N_{In}) and gallium (N_{Ga}) atoms in liquid $dN_{In} / dt = V_{In} - N_{In} / \tau_{In}^L - di / dt$, $dN_{Ga} / dt = V_{Ga} - N_{Ga} / \tau_{Ga}^L - dj / dt$, with V_{In} and V_{Ga} as the total atomic influxes of indium and gallium into the droplet, and τ_{In}^L and τ_{Ga}^L as the mean lifetimes of the indium and gallium atoms in liquid. Neglecting the di / dt term for indium corresponds to the case where N_{In} is controlled by the indium desorption rather than its consumption due to the VLS growth of a NW. This explains the strong inequality $N_{In} >> N_{Ga}$, which is required for $y \cong 1$. Thus, we obtain the stationary solution in the form $N_{In} = V_{In}\tau_{In}^L$. As for the gallium atoms, nothing can be said about N_{Ga} from the stationary solution to the kinetic equation. However, since gallium desorbs much less than indium, we can simply assume a temperature independent $N_{Ga} = const$ in the first approximation. Next, we notice that

$N_{Ga}/N_{In} = (1-y)/y \cong 1-y$ and $V_{Ga}/V_{In} = (1-z)/z$, where $z = V_{In}/(V_{In}+V_{Ga})$ is by definition the indium content in vapor. These considerations yield $1-y = (\tau_*/\tau_{In}^L)(1-z)/z$, with $\tau_* = N_{Ga}/V_{Ga}$. Comparing this to Eq. (5), we find the $x(z)$ dependence explicitly in the form

$$z = [1 + bF(x)]^{-1}, \quad F(x) = \frac{1-x}{x} + \delta(1-x)e^{\omega_s x^2}, \tag{11}$$

with $b = a(\tau_{In}^L/\tau_*)$.

Results and discussion

Let us now see how the obtained expressions change the compositional diagrams of VLS InGaAs and InGaN NWs. We first note that the earlier results of Refs. [16] and [17], obtained within the frame of binary nucleation theory with a saddle point of the island formation energy, predict the presence of the miscibility gaps in ternary VLS NWs whenever $\omega_s > 2$. This is due to the main assumption on the nucleation-limited composition, determined by the composition of the critical island (or nucleus). Our model is different and takes into account that the NW composition may change in the follow-up stage where the island grows to fill the monolayer slice of a NW. This growth can be driven by either thermodynamics or kinetics. In particular, Figure 1 shows the solid-liquid $[x(y)]$ compositional diagram for VLS InGaAs NWs grown at 450 °C. According to the data of Ref. [17], the ω_s value at this temperature equals 2.375, while the a value should be very small to ensure that y is close to unity. Choosing $a = 0.0005$, we obtain the compositional diagrams shown in Fig. 1. As expected from Eq. (10), large δ correspond to the thermodynamically controlled NW compositions with the miscibility gaps, although slightly modified with respect to purely thermodynamic values of Refs. [16], [17]. Decreasing δ leads to the kinetically limited VLS growth regime where the miscibility gap disappears.

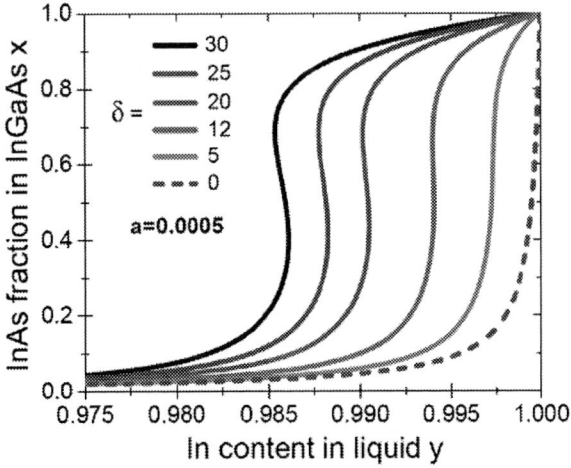

Figure 1. Solid-liquid compositional diagram for InGaAs NWs grown at 450°C ($\omega_s = 2.375$). The small $a = 0.0005$ describes a high stability of indium in liquid such that any appreciable fraction of InAs in solid requires very high indium content in liquid (more than 0.97), as in Refs. [16] and [17]. Large $\delta > 20$ yield the miscibility gaps in solid InGaAs, while smaller δ correspond to the kinetically controlled VLS growth in which all solid compositions become possible. The dashed line corresponds to $\delta = 0$.

These results qualitatively explain the experimental data of Refs. [7] and [28], reporting stationary compositional tuning of gold-catalyzed MOVPE InGaAs NWs throughout the entire compositional range. Most probably, InGaAs NWs described in Refs. [7,28] grew at high supersaturations via regular crystallization without macroscopic nucleation. In this case, the entire range of solid compositions becomes accessible, as described by our model. We note, however, that the solid composition remains very sensitive to the indium content in liquid even at small δ and requires more indium-rich droplets to access any appreciable InAs content in NWs.

Figures 2 and 3 show the solid-liquid $[x(y)]$ and solid-vapor $[x(z)]$ compositional diagrams for self-catalyzed InGaN NWs, obtained from Eqs. (10) and (11), respectively. The ω_s value was set to 3.33 at the growth temperature of 645°C according to the data of Ref. [26], corresponding to the experimental conditions of Ref. [20]. This value yields the miscibility gap of unstrained $In_xGa_{1-x}N$ between $x = 0.12$ and $x = 0.88$, described by the wavy regions of both diagrams in the thermodynamically controlled regime at $\delta = 10$. Increasing the chemical potential difference of InN pairs in liquid and solid decreases the δ values and the miscibility gap starts to shrink, as we saw earlier in Fig. 1 for InGaAs NWs. The gap is completely suppressed at $\delta \cong 1.5$ for these plausible parameters. The dashed curves at $\delta = 0$ correspond to the purely kinetic VLS growth regime. This regime is described by the one-parametric equation for the solid-vapor composition

$$x = \frac{bz}{1+(b-1)z},\tag{12}$$

where the miscibility gap disappears. For any b, it becomes possible to cover the entire compositional range of InGaN by simply changing the indium content in vapor. The results of Ref. [20] demonstrate how the solid composition can be tuned even at a fixed z by changing the growth temperature in HVPE process. Indeed, for high enough temperatures typically employed in HVPE, indium desorbs more than gallium and this changes the temperature-dependent b value in favor of more gallium-rich droplets and consequently NWs at higher temperatures. This effect can be described by a simple Arrhenius-like dependence of the factor τ_{In}^L / τ_*, as discussed in more detail in Ref. [20].

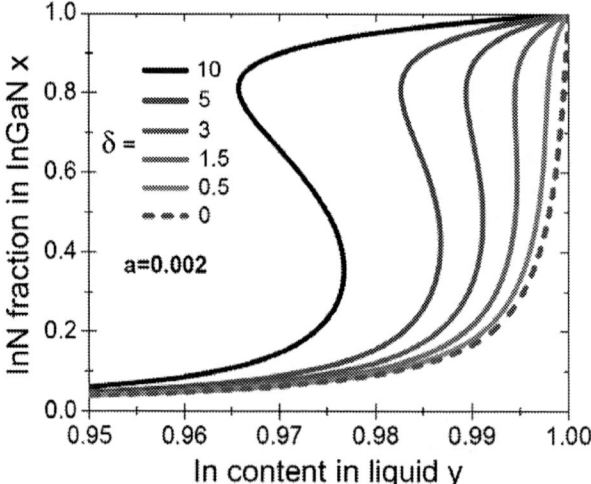

Figure 2. Solid-vapor compositional diagram for self-catalyzed InGaN NWs grown at 645°C (corresponding to $\omega_s = 3.33$), at $a = 0.002$.

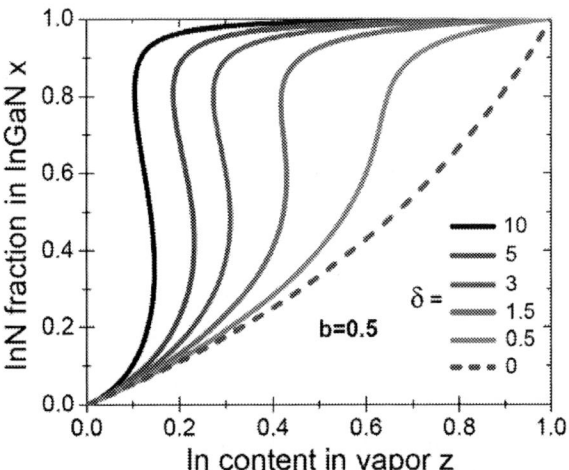

Figure 3. Solid-vapor compositional diagram for self-catalyzed InGaN NWs grown at 645°C (corresponding to $\omega_s = 3.33$). The b value in was set to 0.5, corresponding to an InN fraction of about 0.4 for the vapor composition $z = 0.6$ in the purely kinetic growth regime (shown by dashed lines at δ =0). The miscibility gap, described by the wavy regions of the diagrams, is present for large enough δ but shrinks to zero for δ smaller than 1.5 for these parameters.

In conclusion, we have shown that the miscibility gap of any ternary material, if present in the bulk thermodynamics at a growth temperature, is completely suppressed under high enough supersaturations of a metastable mother phase. This fundamental property is not specific for NWs. However, ternary VLS NWs give an example of the growth system where supersaturations in a nano-sized catalyst can be extremely high compared to the standard thin film case. For InGaAs and InGaN NWs catalyzed by indium-rich droplets, it is possible to cover the entire compositional range of InV fractions in solid, which is not always accessible in the conventional growth techniques.

Acknowledgement

VGD gratefully acknowledges financial support received from the Ministry of Education and Science of the Russian Federation under grant 14.613.21.0055 (project ID: RFMEFI61316X0055).

References

[1] Zhang A, Zheng G, Lieber C M 2016 *Nanowires: building blocks for nanoscience and nanotechnology*, Springer, 2016
[2] Dubrovskii V G 2015 *Theory of VLS growth of compound semiconductors*, 93 ed Anna Fontcuberta I Morral, Shadi A Dayeh and Chennupati Jagadish (Burlington: Academic Press) pp. 1-78
[3] Dubrovskii V G 2017 *J. Phys. D: Appl. Phys.* **50** 453001.
[4] Glas F 2015 *Strain in nanowires and nanowire heterosructures* Semiconductors and Semimetals 93 ed Anna Fontcuberta I Morral, Shadi A Dayeh and Chennupati Jagadish (Burlington: Academic Press) pp.79-123
[5] Ng K W, Ko W S, Tran T T D, Chen R, Nazarenko M V, Lu F, Dubrovskii V G, Kamp M, Forchel A, Chang-Hasnain C J 2013 *ACS Nano* **7** 100
[6] Zhang X, Dubrovskii V G, Sibirev N V, Ren X 2011 *Cryst. Growth Des.* **11** 5441
[7] Ameruddin A S, Caroff P, Tan H H, Jagadish C, Dubrovskii V G 2015 *Nanoscale* **7** 16266
[8] Messing M E, Wong-Leung J, Joyce H J, Zanolli Z, Gao Q, Tan H H, Wallenberg L R, Johansson J, Jagadish C 2011 Growth of straight InAs on GaAs nanowire heterostructures *Nano Lett.* **11** 3899

[9] Dick K A, Bolinsson J, Borg B M, Johansson J 2012 *Nano Lett.* **12**, 3200

[10] Zannier V, Ercolani D, Gomes U P, David J, Gemmi M, Dubrovskii V G, Sorba L 2016 *Nano Lett.* **16** 7183

[11] Qian F, Li Y, Gradecak S, Lieber C M 2004 *Nano Lett.* **4** 20

[12] Sekiguchi H, Kishino K, Kikuchi A 2010 *Appl. Phys. Lett.* **96** 231104

[13] Kuykendall T, Ulrich P, Aloni S, Yang P 2007 *Nat. Mater.* **6** 951

[14] Priante G, Glas F, Patriarche G, Pantzas K, Oehler F, Harmand J C 2016 *Nano Lett.* **16**, 1917.

[15] Glas F 2017 *Cryst. Growth Des.* **17** 4785

[16] Ghasemi M, Johansson J 2017 *Cryst. Growth Des.* **17** 1630

[17] Dubrovskii V G, Koryakin A A, Sibirev N V 2017 *Mater. Des.* **132** 400

[18] Dubrovskii V G 2015 *Cryst. Growth Des.* **15** 5738

[19] Johansson J, Ghasemi M 2017 *Phys. Rev. Materials* **1** 040401(R)

[20] Roche E, André Y, Avit G, Bougerol C, Castelluci D, Réveret F, Gil E, Médard F, Leymarie J, Jean T, Dubrovskii V G, Trassoudaine A 2018 *Nano Lett.* (submitted)

[21] Dubrovskii V G, Grecenkov J 2015 *Cryst. Growth Des.* **15** 340

[22] Dubrovskii V G 2009 *J. Chem. Phys.* **131** 164514

[23] Glas F 1987 J. Appl. Phys. **62** 3201

[24] Karpov S Yu 1998 *MRS Internet J. Nitride Semicond. Res.* **3** 16

[25] Gil E, Dubrovskii V G, Avit G, André Y, Leroux C, Lekhal K, Grecenkov J, Trassoudaine A, Castelluci D, Monier G, Ramdani M R, Robert-Goumet C, Bideux L, Harmand J C, Glas F 2014 *Nano Lett.* **14** 3938

[26] Adhikari J, Kofke D 2004 *J. Appl. Phys.* **95** 6129

[27] Jung C S, Kim H S, Jung G B, Gong K J, Cho Y J, Jang S Y, Kim C H, Lee C, Park J 2011 *J. Phys. Chem.* **115** 7843

SPBOPEN 2018
IOP Publishing

Size effects in LiNbO3 thin films fabricated by pulsed laser deposition

Z Vakulov[1], M Ivonin[1], E G Zamburg[2], V S Klimin[1], D P Volik[3], D A Golosov[4], S M Zavadskiy[4], A P Dostanko[4], A V Miakonkikh[5], I E Clemente[5], K V Rudenko[5], O A Ageev[1]

[1]Research and Education Centre «Nanotechnology», Southern Federal University, Taganrog 347922, Russian Federation
[2]Department of Electrical & Computer Engineering, National University of Singapore, Singapore 119077, Singapore
[3]Department of Radio Engineering Electronics, Southern Federal University, Taganrog 347922, Russian Federation
[4]Department of Research and Development, Belarusian State University of Informatics and Radioelectronics, Minsk 220013, Belarus
[5]Institute of Physics and Technology RAS, Moscow 117218, Russian Federation

Abstract. This paper shows the results of the study of size effects of electro-physical parameters in nanostructured $LiNbO_3$ thin films fabricated by pulsed laser deposition. Obtained results shown that with an increase in the number of laser pulses from 50 000 to 150 000, the maximum value of the amplitudes (psi) increases from 36 to 47, and phase shift (delta) varies from 160 to 175. It was found that with increasing of the film thickness from 45.3 to 192.9 nm, charge carrier concentration in the films increasing from $4.1 \cdot 10^{12}$ to $8.75 \cdot 10^{12}$ cm^{-3}. An increase in the grain diameter from 118 to 172 nm causes a gradual increase in charge carriers mobility from 125.714 to 505.841 $cm^2/V \cdot s$.

1. Introduction

Currently, lithium niobate ($LiNbO_3$) bulk crystals are widely used in acousto-optic [1] and piezoelectric devices as well as waveguide structures with low losses [2, 3]. Due to its physical properties $LiNbO_3$ can be considered as the backbone of modern integral electronics. Thus, the formation of $LiNbO_3$ thin films on a silicon substrate will allow combining optical and electronic components on the same chip [4].

$LiNbO_3$ thin films can be fabricated by the following growth techniques: molecular-beam epitaxy [5], sol-gel method [6], magnetron sputtering [7], and pulsed laser deposition (PLD) [8, 9]. The latter is the most suitable for complex oxides since it has many technological parameters, as well as the ability of preserving the stoichiometric composition of the target [10]. Moreover, PLD allows obtaining films in a wide range of electro-physical and optical properties, which in turn depend on the morphological parameters of the growing films. Therefore, the purpose of this paper is studying the influence of size effects in $LiNbO_3$ films on their electro-physical and optical properties.

2. Experiment

In order to fabricate $LiNbO_3$ films nanotechnological facility NANOFAB NTK-9 (NT-MDT, Russia), comprising PLD module Pioneer 180 (Neocera Co., USA) was used. The $LiNbO_3$ target was ablated

Content from this work may be used under the terms of the Creative Commons Attribution 3.0 licence. Any further distribution of this work must maintain attribution to the author(s) and the title of the work, journal citation and DOI.
Published under licence by IOP Publishing Ltd

by excimer KrF laser (λ=248 nm) (Coherent Inc., USA) with energy density on target surface about 1.5 J/cm^2. The number of laser pulses varied from 50 000 to 200 000 at repetition rate of 10 Hz. The argon pressure in the growth chamber varied from 10^{-3} to 10^0 Torr. Films with a thickness from 45 to 180 nm were deposited at a temperature from 300 to 600 °C on Si and SiO$_2$ substrates.

The morphology of obtained films was studied by atomic force microscopy (AFM) in semi-contact mode using Ntegra probe nanolaboratory (NT -MDT, Russia). Electro-physical parameters of the films have been studied by measuring the Hall electromotive force at Ecopia HMS-3000 system (Ecopia Co., Korea). The dependences of the optical characteristics were studied on spectral ellipsometer M-2000X (Woollam J.A. Co, USA) under beam angle of 65° in the wavelength range from 240 to 1000 nm with a 10-nm pitch. The size of the spot was about 2×5 mm. Since the ellipsometry is indirect technique, the values of the optical constants can be estimated by applying the model to recalculate measured values of the amplitudes (psi) and the phase shift (delta) into the optical parameters of samples. Elemental composition of obtained films was studied by energy dispersive X-ray (EDX) spectroscopy using Nova Nanolab 600 scanning electron microscope (SEM) (FEI Co., The Netherlands).

3. Results and discussion

Figure 1 shows AFM images of LiNbO$_3$ films fabricated under different amounts of laser pulses on Si substrates.

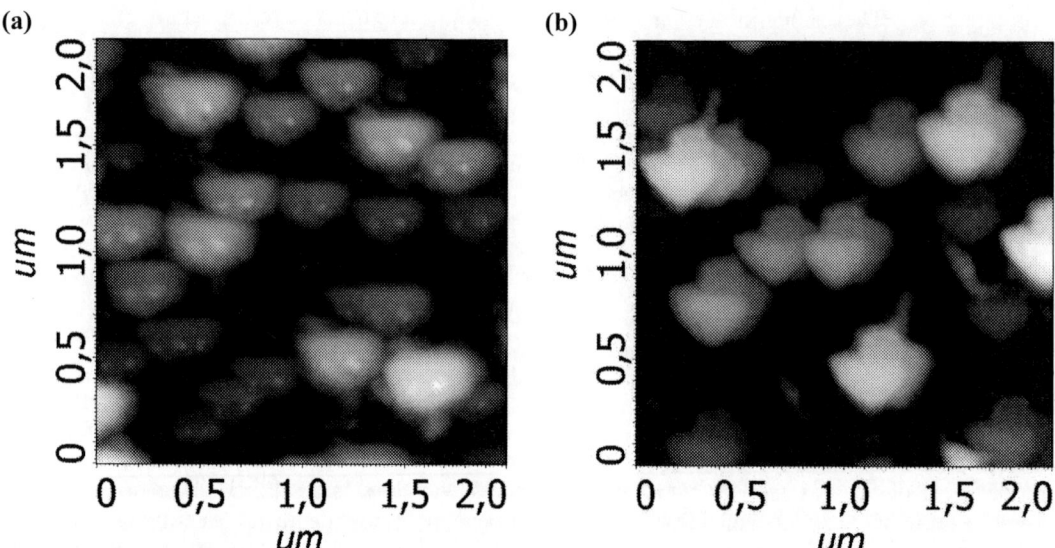

Figure 1 (a, b). AFM images of nanostructured LiNbO$_3$ films on Si fabricated under different number of laser pulses: 50 000 **(a)** and 100 000 **(b)**.

With increasing number of pulses from 50 000 to 100 000, the roughness of nanostructured LiNbO$_3$ films increased from 7.764 to 8.20 nm, which may be associated with a longer thermal effect on the film obtained under 100 000 pulses compare to the film obtained under 50 000 pulses. Wherein, in both cases formation of large droplets on the film surface with the diameter of 180 − 220 nm were observed.

Figure 2 shows cross-section of LiNbO$_3$ thin film on SiO$_2$ and the dependence of LiNbO$_3$ films roughness on Ar pressure fabricated on Si and SiO$_2$ substrates. The SiO$_2$ layer was grown on the Si surface by chemical vapor deposition method. The thickness of the layer was about 100 nm.

(a)   (b)

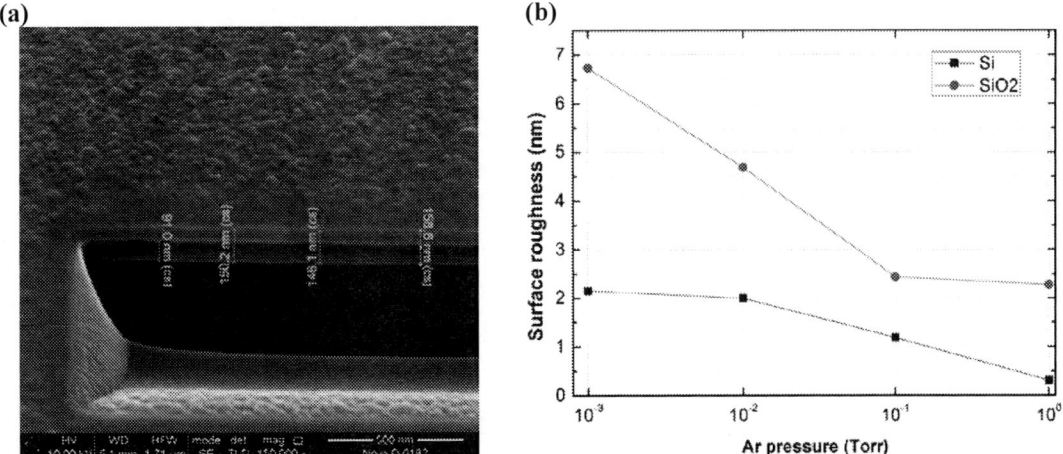

Figure 2 (a, b). Cross-section of $LiNbO_3$ thin film on SiO_2 **(a)**, and the dependence of surface roughness on Ar pressure **(b)**.

For $LiNbO_3$ films fabricated over a wide range of growth conditions on Si substrates, the formation of large droplets is typical as Figure 3 (a) shows.

(a)   (b)

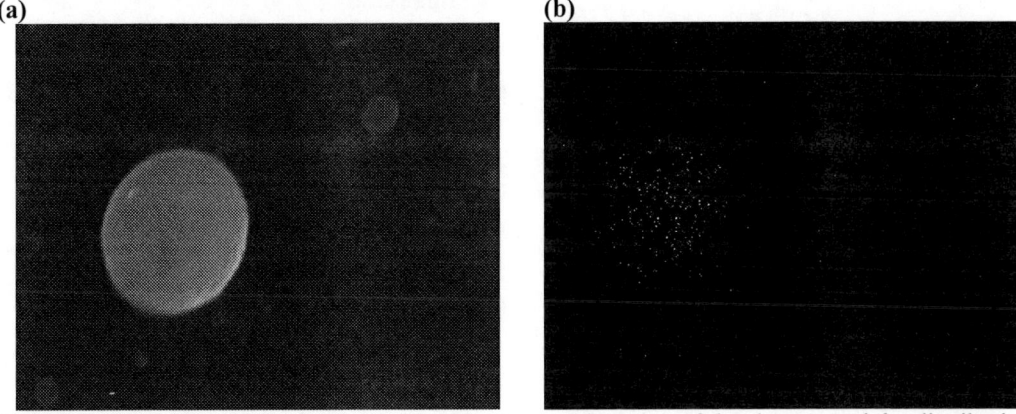

Figure 3 (a, b). Elemental mapping: electron micrograph region of droplet **(a)**, and the distribution of niobium element **(b)**.

The results of EDX study shows the formation of niobium oxide clusters on the $LiNbO_3$ film surface fabricated on Si substrate under an O_2 pressure of 10^{-2} Torr. Using of SiO_2 as substrates and addition of Ar on the film forming step are allowed to reduce the number of the droplets as Figure 2 (a) shows.

Figure 4 shows the dependencies of the electro-physical parameters of nanostructured $LiNbO_3$ films on grain size and films thickness. It was found that increase in the film thickness from 45.3 to 192.9 nm results in increasing of charge carrier concentration from $4.1 \cdot 10^{12}$ to $8.75 \cdot 10^{12}$ cm^{-3}. Increasing in grain size from 118 to 172 nm causes a gradual increase in the charge carrier mobility from 125.714 to 505.841 $cm^2/V \cdot s$.

(a)

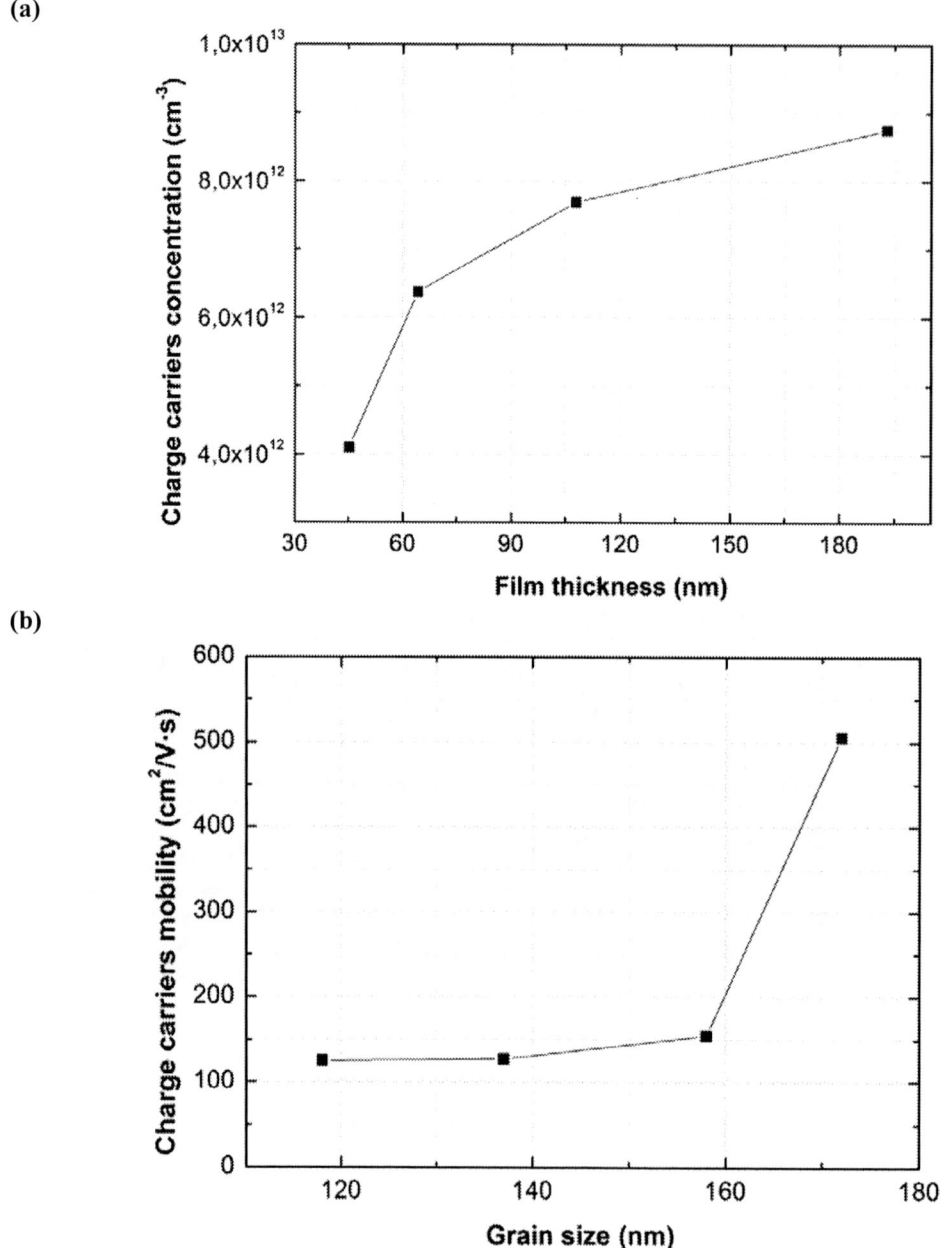

(b)

Figure 4 (a, b). Dependences of charge carriers concentration on LiNbO$_3$ films thickness **(a)** and mobility of charge carriers on the grain size **(b)**.

The increase in the mobility of charge carriers can be associated with enhancing of the crystal structure of the films, as well as decreasing of LiNbO$_3$ film roughness due to increasing of grain size. This fact is confirmed by the results of AFM and SEM studies.

Figure 5 shows the spectral dependences of psi and delta for films with different thicknesses. It has been shown that increases in the number of laser pulses from 50 000 to 150 000 results in changing of psi maximum value from 36 to 47, and delta from 160 to 175.

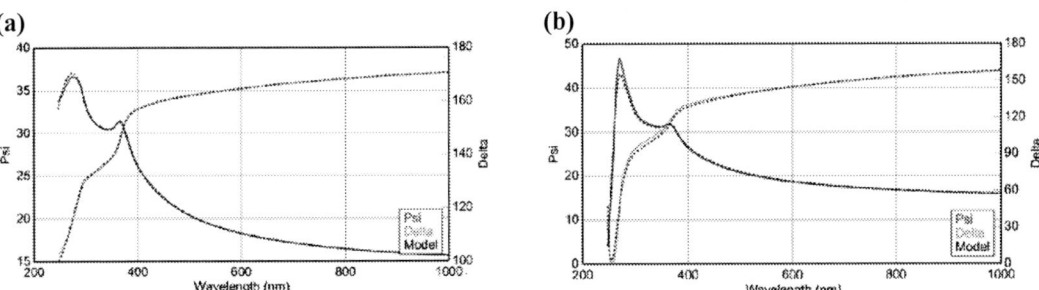

Figure 5 (a, b). Dependences of amplitude (psi) and phase shift (delta) on the wavelength for samples fabricated under different number of laser pulses: 50 000 **(a)** and 150 000 **(b)**.

The refractive index value (n) calculated on the basis of the measured psi and delta in the visible part of the spectrum was 2.01 – 1.97, while the value of extinction coefficient (k) did not exceed 0.05.

4. Conclusion

Obtained results shows that increasing in the number of laser pulses from 50 000 to 150 000 results in increasing the maximum value of psi from 36 to 47, and delta from 160 to 175. It was established that increasing of film thickness from 45.3 to 192.9 nm results in increasing of charge carriers concentration from $4.1 \cdot 10^{12}$ to $8.75 \cdot 10^{12}$ cm^{-3}. Increasing in grain size from 118 to 172 nm causes gradual increasing in charge carriers mobility from 125.714 to 505.841 cm^2/V·s. The increasing in charge carriers mobility can be associated with enhancing of crystal quality of films, as well as a decreasing in film roughness and the number of grains due to an increase of their diameter. This fact is confirmed by the results of AFM and SEM studies.

Obtained results can be used as a physical and technological basis for the development and manufacture of integrated acousto-optic devices, as well as sensitive elements of sensors using various effects of surface acoustic waves.

Acknowledgments

This work was financially supported by the Russian Foundation for Basic Research № 16-57-00028-Bel_a, and proposal № 18-57-00011 Bel_a. The work was done on the equipment of Research and Education Centre «Nanotechnology» and Collective Use Centre «Nanotechnology», Southern Federal University.

References

[1] Shibaev S S, Volik D P, Pomazanov A V 2015 *Technical Physics* **60** 568
[2] Wang P, Qi J, Liu Z, Liao Y, Chu W, Cheng Y 2017 *Nature. Scientific reports* **5** 1274
[3] Bazzan M, Sada C 2016 *Appl. Phys. Rev.* **2** 040603
[4] Liu Y, Li J, Zhang Z, Kong Y 2017 *Ferroelectr.* **520** 34
[5] Tellekamp M, Shank J, Goorsky M, Doolittle W 2016 *J. Electron. Mater.* **45** 6292
[6] Rubešová K, Mikolášová D, Hlásek T, Jakeš V, Nekvindová P, Bouša D, Oswald J 2016 *J. Lumin.* **176** 260
[7] Zhukov R, Bykov A, Kiselev D, Malinkovich M, Parkhomenko Yu 2014 *J. Alloys Compd.* **586** S336
[8] Wang X, Ye Z, Wu G, Cao L, Zhao B 2005 *Mater. Lett.* **59** 2994
[9] Kilburger S, Chety R, Millon E, Bin P Di, Bin C Di, Boulle A, Guinebretière R 2007 *Appl. Surf. Sci.* **253** 8263
[10] Tominov R V, Zamburg E G, Khakhulin D, Klimin V S, Smirnov V A, Chu Y H, Ageev, O A 2017 *J. Phys.: Conf. Ser.* **917** 032023

SPBOPEN 2018   IOP Publishing

GaN nanostructures grown on Si(111) by PA-MBE via droplet epitaxy: SEM and HRTEM study

G A Sapunov[1], V V Fedorov[1], A D Bolshakov[1], A M Mozharov[1],
L N Dvoreckaia[1], D A Kirilenko[2,3], A A Sitnikova[3], I S Mukhin[1,2]

[1]St.Petersburg Academic University, 194021, St. Petersburg, Russia
[2]ITMO University, 197101, St. Petersburg, Russia
[3]Ioffe Physical-Technical Institute of the Russian Academy of Sciences, 194021 St. Petersburg, Russia

Abstract. In this report, we demonstrate that the use of GaN seeding layer prepared by droplet epitaxy technique using plasma assisted molecular beam epitaxy (PA-MBE) can lead to the epitaxial stabilization of GaN nanostructures with different lattice orientations. From the high-resolution transmission electron microscopy (HRTEM) studies, it is shown that {111} faceted zinc-blende (ZB) GaN islands act as the fast nucleation sites for the growth of GaN nanowires (NWs) with hexagonal wurtzite (WZ) structure. Further GaN nanostructure morphology depends on lattice orientation of the seeding island. It will be shown that further intergrowth and coalescence of GaN NWs lead to changes in the mutual orientation.

1. Introduction

Growth of group-III nitride nanostructures, such as GaN NWs, attracts a lot of attention due to their importance both in fundamental research and in development of future optoelectronic nano-scaled devices [1]. Small contact area with the substrate, commonly observed for the epitaxial nanostructures, can provide efficient lateral strain relaxation and defect-free NW growth, as it has been demonstrated for GaN NWs [2]. Formation of GaN NWs on Si complies the self-induced growth mechanism. This technique does not involve the substrate surface patterning and catalyst particles use which makes it attractive both from technological and further device application points of view. In this study the Ga-droplets deposition on the silicon surface prior to the NWs growth was used [3,4]. We demonstrate that this technique can lead to synthesis of inclined WZ GaN NWs, formed on {111}-facets of GaN ZB nanoislands.

2. Experimental

GaN nanostructures were grown by PA-MBE using Veeco GEN-III MBE machine equipped with an inductively coupled RF-plasma Riber nitrogen source and group-III material solid source. Silicon (111) p-type wafers with a 4° miscut oriented towards <1-1 0> were cleaned using Shiraki method and used as substrates. Ga droplets were formed using Ga deposition on the reconstructed Si (111)-(7x7) substrate surface at 200°C. The seeding layers had an equivalent thickness in the range of 0.6-2.5 ML controlled in-situ with RHEED by monitoring the Ga-($\sqrt{3}$x$\sqrt{3}$)R30° surface reconstruction, corresponding to the 0.3ML of Ga [4]. In the following the droplet layer was exposed under the flux of activated nitrogen and annealed up to GaN growth temperature (790-810°C). Further

---

Content from this work may be used under the terms of the Creative Commons Attribution 3.0 licence. Any further distribution of this work must maintain attribution to the author(s) and the title of the work, journal citation and DOI.
Published under licence by IOP Publishing Ltd

GaN growth was carried out under nitrogen rich growth conditions generally leading to the formation of vertical NWs [5].

Structure and morphology of the synthesized nanoheterostructures were studied with scanning electron microscopy (Zeiss SUPRA 25-30-63) and high-resolution transmission electron microscopy (HRTEM) using a JEOL 2100F microscope (200kV). The samples for HRTEM studies were prepared by standard methods involving ion sputtering at the last stage.

3. Results

Surface morphology of GaN nanostructures grown using 0.6 ML thick seeding layer is presented on SEM image (Fig.1). Cross section view is presented on TEM image (Fig.2 a). Analysis of SEM and TEM images shows the presence of nanostructures with different morphology, namely tripods, vertical and inclined NWs. Closer view of vertical and inclined NWs is presented on (Fig.2 b, c).

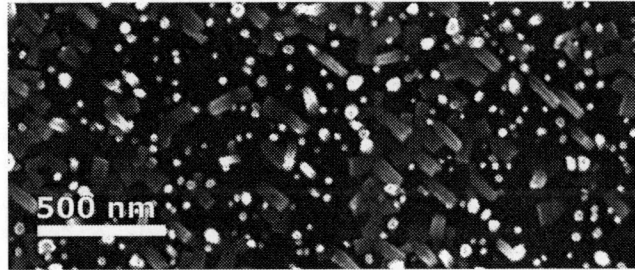

Figure 1. SEM images of the GaN nanostructures, top view.

HRTEM study of the synthesized structures (see Fig. 2 b) demonstrates the presence of a nanometer-sized (~5nm) nucleation core at the base part of inclined NWs [6]. Close-up view of its central part is presented on Fig.3. Notably, there is no any nucleation particles at the base of vertical NWs (Fig.2 c).

Figure 2. HRTEM images of GaN nanostructures, cross section view a) distant view b) pair of inclined NWs protruding from the common seeding particle c) vertical NW

Fast Fourier transform (FFT) analysis of TEM image was used to investigate crystal structure and lattice orientation of GaN nanostructure is presented on the Fig.3. FFT analysis demonstrates that the crystal structure of the nanostructure core is different from its NW-shaped branches: FFT pattern of the core corresponds to the cubic stacking sequence of the close packed planes, whereas pattern of branches corresponds to the hexagonal structure [7], [8]. FFT images from inclined NWs and their core correspondingly matched with schematic model of reciprocal space nodes (represented with dashed circles) of the hexagonal [11-20] and cubic [1-10] axis zones.

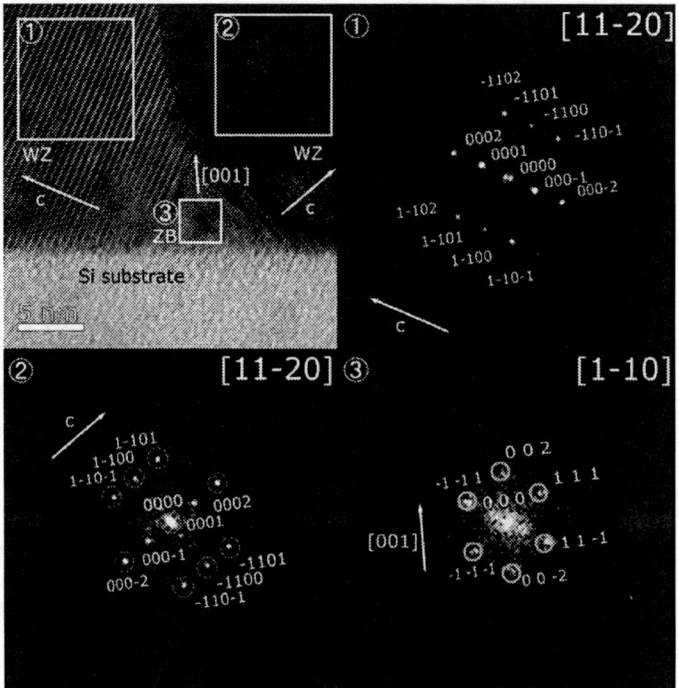

Figure 3. Fast Fourier transform of TEM image of the inclined NWs and common core matched with schematic model of reciprocal space nodes

It can be seen, that ZB core {001} crystalline plane is oriented parallel to the substrate surface (Fig. 3). Apparently, {111} ZB planes serve as nucleation sites for WZ NWs. Therefore, geometrically expected mutual angle between c-axis of coalescing NWs is 109.47°, corresponding to the angle between [111] and [-1-1 1] axis of ZB origin. However, FFT analysis shows that the angle equals ~116°, meaning the formation of energetically favorable grain boundary plane (-1103) instead of the geometrically expected (-3308), which requires a change in the mutual orientation, although there is no visible lattice distortion on FFT image of ZB core.

4. Conclusions

In this work growth of GaN NWs on Si (111) substrate by PA-MBE using droplet epitaxy technique was studied. It was found that Ga droplet nitridation leads to formation of epitaxially oriented ZB GaN nanoislands with {001} planes aligned parallel to Si(111). It was shown that during further GaN growth, {111}-facets of GaN nanoislands play a role of nucleation sites for inclined GaN nanocolumns. It was also found that coalescence of the inclined NWs can lead to change in the mutual orientation of NWs and core.

Acknowledgments

This work was carried out with the support of the Russian Foundation for Basic Research (18-32-00899 mol_a) and Russian Science Foundation (grant 18-72-00231). TEM characterizations were performed using equipment of the Federal Joint Research Center "Material science and characterization in advanced technology" supported by the Ministry of Education and Science of the Russian Federation (id RFMEFI62117X0018).

References

[1] M. Mandl et al., Phys. status solidi - Rapid Res. Lett., vol. 7, no. 10, pp. 800–814, Oct. 2013.

[2] S. Zhao et al., Quantum Electron., vol. 44, pp. 14–68, Nov. 2015.

[3] A. Mozharov et al., Phys. Status Solidi - Rapid Res. Lett., vol. 9, no. 9, pp. 507–510, 2015.

[4] A. Kawazu and H. Sakama 1988 *Phys. Rev. B* vol. **37** no. 5 pp. 2704–2706

[5] V. Consonni Phys. status solidi - Rapid Res. Lett., vol. 7, no. 10, pp. 699–712, 2013.

[6] V. Fedorov et al., Journal of Physics: Conf. Series 917 (2017) 032040

[7] J. Jo, Y. Tchoe, G.-C. Yi, and M. Kim, "Real-Time Characterization Using in situ RHEED Transmission Mode and TEM for Investigation of the Growth Behaviour of Nanomaterials," Sci. Rep., vol. 8, no. 1, p. 1694, Dec. 2018.

[8] C. Bayram et al., "Cubic Phase GaN on Nano-grooved Si (100) via Maskless Selective Area Epitaxy," Adv. Funct. Mater., vol. 24, no. 28, pp. 4492–4496, Jul. 2014.

[9] V. Fedorov et al., "Droplet epitaxy mediated growth of GaN nanostructures on Si (111) via plasma-assisted molecular beam epitaxy", CrystEngComm, 2018, Accepted Manuscript.

Effect of substrate temperature on the properties of plasma deposited silicon oxide thin films

E Yu Gusev, J Y Jityaeva, S P Avdeev, O A Ageev

Southern Federal University, Institute of Nanotechnologies, Electronics and Equipment Engineering, Research and Educational Centre "Nanotechnologies" Taganrog 347922, Russia

Abstract. Silicon oxide (SiO_x) films were obtained from $SiH_4/NO_2/Ar$ by plasma-enhanced chemical vapors deposition (PECVD) technique. Effect of substrate temperature on the properties of the deposited films have been presented. The substrate temperature of the deposition process varied from 150 to 450°C at fixed working pressure of 1000 mTorr and RF power of 10 W. The films were investigated by atomic force and scanning electron microscopies as well as laser ellipsometry. The grain size, root-mean-square roughness and refractive index of the films were in the range of 20-250 nm, 0.2-2.4 nm and 1.5-2.0, respectively. The refractive indices and stoichiometry of SiO_x films are discussed using the Lorentz-Lorenz formula.

1. Introduction

The decrease Fresnel reflection and increase a fraction of a transmitted solar radiation is an actual scientific and technical problem for optics and solar power engineering [1,2]. It is solved by texturing surface and forming antireflective coatings [1-7]. They contribute to increasing efficiency and reducing costs of solar cells and improve light transmission and contrast in optics [5,6]. However, the refractive index and the film thickness must correspond to the theoretical values calculated to achieve the maximum effect.

The antireflective coatings on basic of silicon oxide films are the preferable for amorphous hydrogenated and polycrystalline silicon solar cells. Silicon oxide is relatively inexpensive and possessing both antireflection and passivating properties that reducing recombination losses [5,8].

Plasma enhanced chemical vapor deposition method (PECVD) allows to obtain uniform oxide films in a single technological cycle with Si:H, that it possible to effectively minimize the influence of negative factors and does not lead to degradation of the electrophysical and optical properties of the materials solar cells [5]. Any more, it allows to form layers at a temperature below 300°C, which is especially important in the manufacture of solar cells on glass and flexible substrates.

The purpose of the study is to determine the effect of deposition temperature on optical properties silicon oxide thin films obtained by PECVD.

2. Experimental details

The silicon oxide (SiO_x) thin films were prepared from $Ar:SiH_4:N_2O$ gas mixture by plasma enhanced chemical vapor deposition technique (PlasmaLab 100) [9,10]. Silicon oxide thin films were deposited on peeled a n-type silicon wafers (100). Total gas flow, gas ratio, RF power and total pressure were constant to 980 sccm and 161.5:8.5:710 sccm ($Ar:SiH_4:N_2O$), 10 W, 1000 mTorr, respectively.

Content from this work may be used under the terms of the Creative Commons Attribution 3.0 licence. Any further distribution of this work must maintain attribution to the author(s) and the title of the work, journal citation and DOI.
Published under licence by IOP Publishing Ltd

Temperature was varied in the range of 150-450°C. Film thickness were controlled by profilometry (AlphaStep D-100) of the steps formed by photolithography (MJB4). Root mean squared (RMS) roughness was determined by atomic force microscopy (NTEGRA Probe Nanolaboratory) and scanning electron microscopy (Nova NanoLab600) with increase by a factor of 250 000 [11,12]. The refractive index was determined by laser ellipsometry (LEF-3M) on $\lambda=632.8$ nm and then film stoichiometry was calculated in accordance to [13].

3. Results and discussion

A series of samples with thin (less than 50 nm) SiO_x films was fabricated. Surface morphology of the samples was studied (Fig.1). The grain size and RMS roughness decreased from 200-250 nm to 20-40 nm and from 2.4 to 0.2 nm, respectively (Fig.2,a), the film became denser. The deposition rate increased from 38 to 44 nm/min.

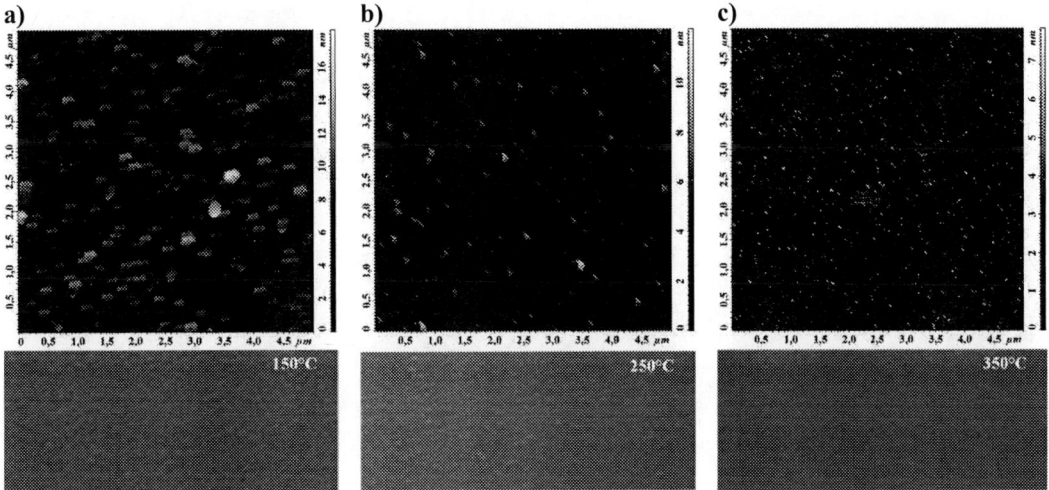

Figure 1(a-c). AFM and SEM images of SiO_x films deposited at various temperatures: **(a)** 150°C, **(b)** 250°C, **(c)** 350°C

Then, ellipsometric measurements of the films were carried out. The obtained dependence of the refractive index on temperature (Fig.2,b) has specifics.

With a decrease in the grain size and roughness with increasing temperature, we should expect changes in the properties of the film material towards a transition to the properties of a bulk material, i.e. tendency of the refractive index to the value of 1.45-1.48 (stoichiometric silicon dioxide) [14]. However, the obtained the index values exceed the level and their difference increases with temperature. It can be assumed that despite the constancy of gas mixture composition and flow, pressure and power, the temperature affects the structure and composition of SiO_x films [15,16].

The change in stoichiometry of SiO_x film can be expressed as

$$Si_{(1-1/2\,x)} + SiO_{2\ x/2} \leftrightarrow SiO_x,$$

and, thus, as the temperature increases, the transition from films with a large value of x to smaller occurs through the SiO phase ($n = 1.85$ [17]), and it shifted toward the silicon [18], nanocrystalline silicon precipitation. The dependence of the refractive index on the composition was constructed on the basis of the Lorentz-Lorenz formula [13]:

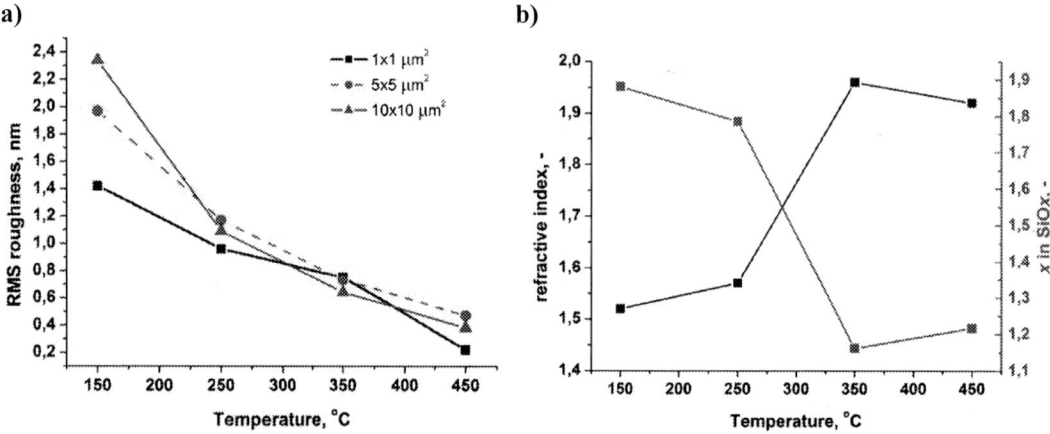

Figure 2(a,b). Roughness **(a)**, refractive index and x **(b)** of SiO$_x$ films vs. deposition temperature

$$n^2 = \left(\frac{n_{Si}^2 X_{Si}}{n_{Si}^2 + 2} + \frac{n_{SiO_2}^2 X_{SiO_2}}{n_{SiO_2}^2 + 2} \right) \bigg/ \left(\frac{X_{Si}}{n_{Si}^2 + 2} + \frac{X_{SiO_2}}{n_{SiO_2}^2 + 2} \right),$$

where n is the refractive index of the film, X_{Si}, X_{SiO_2} and n_{Si}, n_{SiO_2}, are the volume fractions and refractive indices of components Si and SiO$_2$, respectively.

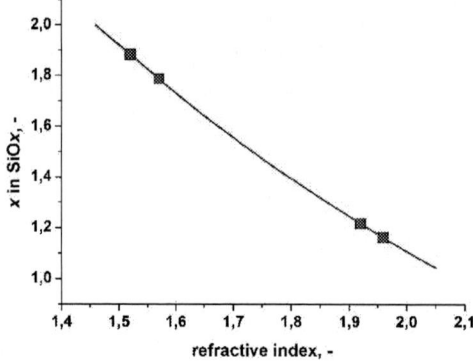

Figure 3. Value of x in SiO$_x$ vs. refractive index

According to the assumption, using the Lorentz-Lorenz formula, the values of x in SiO$_x$ (Fig.3) were calculated and the effect of temperature was determined (Fig. 2,b, blue colour). However, for the adequacy of a stoichiometric calculation and subsequent analysis of the obtained results, verification of the assumption is required. Investigations of the elemental composition and structure of films should be carried out.

4. Conclusion

Thin silicon oxide films of less than 50 nm thick were fabricated. The influence of deposition temperature on morphology (grain size and roughness) and refractive index in the range of 150-450°C has been studied. Atomic force microscopy, scanning electron microscopy and laser ellipsometry were used for measuring. The grain size and roughness decreased from 200-250 nm to 20-40 nm and from 2.4 to 0.2 nm with a rise of temperature. At the same time, the refractive index (1.52-1.96) exceeded the characteristic value for stoichiometric oxide and increased with temperature. It was assumed that temperature affects the structure and composition of SiO_x films and tends to saturate with silicon at high temperatures. The values of x in SiO_x have been found to decrease with temperature from 1.9 to 1.2.

The obtained non-stoichiometric films of silicon oxide can be considered for use in silicon solar cells as passivation and antireflection layers. Future directions of research are mentioned.

Acknowledgment

This work was financially supported by Southern Federal University (grant VnGr-07/2017-02). The equipment of the Research and Educational Centre "Nanotechnologies" of Southern Federal University was used for the study.

References

[1] Chu S, Cui Y, Liu N 2016 *Nature materials* **16** 16-22
[2] Green M A, Hishikawa Y, Warta W, Dunlop E D, Levi D H, Hohl-Ebinger J, Ho-Baillie A W Y 2017 *Prog Photovolt Res Appl* **25** 668-676
[3] Klimin V S, Tominov R V, Eskov A V, Krasnoborodko S Y, Ageev O A 2017 *J. Phys.: Conf. Ser.* **917** 092005
[4] Klimin V S, Solodovnik M S, Smirnov V A, Eskov A V, Tominov R V, Ageev O A 2016 *Proc. of SPIE* **10224** 102241Z
[5] Raut H K, Ganesh V A, Nair A S, Ramakrishna S 2011 *Energy & Environment Science* **4** 3779
[6] Yang B, Lee M 2013 *Applied Surface Science* **284** 565-568
[7] Scubert M F, Mont F M, Chhajed S, Poxson D J, Kim J K, Shubert E F 2008 *Optics Express* **16** 5290-5298
[8] Park S P, Park H, Kim D, Nam J, Yang J Y 2017 *Current Applied Physics* **17** 517-521
[9] Gusev E Yu, Jityaeva J Y, Bykov Al V 2017 *J. Phys.: Conf. Ser.* **929** 012055
[10] Gusev E Y, Jityaeva J Y, Geldash A A, Ageev O A 2017 *J. Phys.: Conf. Ser.* **917** 032029
[11] Ageev O A, Alyabieva N I, Konoplev B G, Smirnov V A, Tkachuk V V 2014 *Advanced Materials Research* **894** 374-378
[12] Ageev O A, Ilin O I, Rubashkina M V, Smirnov V A, Fedotov A A, Tsukanova O G 2015 *Technical Physics* **60** 1044-1050
[13] Jacobsson R 1975 *Physics of thin films* **8** 51-98
[14] http://www.ioffe.ru/SVA/NSM/nk/#Oxides]
[15] Fazio M A, Perani M, Brinkmann N, Terheiden B, Cavalcoli 2017 *Journal of Alloys and Compounds* **725** 163-170
[16] Guha S, Yang J, Yan B 2013 *Solar Energy Materials and Solar Cells* **119** 1-11
[17] Rinnert H Vergnat M, Bumeau A 2001 *Journal of Applied Physics* **89** 237-243.
[18] Green M A, Keevers J 1995 *Progress in photovoltaics: Research and application* **3** 189-192

SPBOPEN 2018   IOP Publishing

Study of the effect of carbon-containing gas pressure on the geometric parameters of an array of carbon nanostructures

V S Klimin[12], A A Rezvan[1] and O A Ageev[2]

[1]Department of Nanotechnology and Microsystems, Southern Federal University, Taganrog 347922, Russia
[2]Research and Education Center "Nanotechnologies", Southern Federal University, Taganrog 347922, Russia

Abstract. In this article, the formation of the production of high-efficiency photovoltaic devices, solar cells is considered. Increased efficiency is achieved through the use of carbon nanomaterial. The use of carbon nanotubes increases the efficiency of solar photovoltaic due to efficient charge transfer in the device. Effective use of carbon nanomaterial in these devices is achieved through the creation of an orderly network of cross-walled carbon nanotubes. The work on the formation of such arrays is the chosen method of plasma treatment with chemical vapor deposition. The result shows that at an optimal pressure of the carbon-containing gas, vertically oriented carbon pipes grow. At pressures below, amorphous carbon grows on the surface. In particular, it has been established that the array of vertically oriented carbon nanostructures with a diameter of (30-70) nm, height (5-7) μm and a density of 3.7×10^5 cm^{-2} can be grown on a nickel-vanadium-silicon structure.

1. Introduction

At the present time, the component base of nanoelectronics is actively developing and materials used in such a base are being expanded [1-6]. Also, there are a sufficiently large number of methods for obtaining nanostructures that allow obtaining memory elements of a nanoscale device with a long operating time of devices [7-10]. One of the promising materials is carbon nanostructures. The increased interest in the study of carbon nanostructures is due, on the one hand, to their unique physicochemical properties, due to which they are an attractive object of fundamental research, and on the other hand by broad prospects of applied use [11-16]. The use of such materials will reduce the load and improve the mass-dimensional characteristics of devices and increase the strength and life of the devices [17].

To implement practical developments using carbon nanostructures, selective synthesis of nanotubes with controlled parameters and properties is required. One of the most promising methods of this synthesis is chemical vapor deposition initiated by plasma [18]. The chemical vapor deposition initiated by plasma method allows obtaining arrays of carbon nanostructures oriented perpendicular to the substrate. A feature of this method is the use of catalytic centers [19-20]. The purpose of this work is to optimize the regimes for obtaining arrays of carbon nanostructures by chemical vapor deposition, which affect the geometric parameters. Geometric parameters affect the electro physical parameters of carbon nanostructures. One of these parameters is the power of the plasma source.

2. Description of the method

Content from this work may be used under the terms of the Creative Commons Attribution 3.0 licence. Any further distribution of this work must maintain attribution to the author(s) and the title of the work, journal citation and DOI.
Published under licence by IOP Publishing Ltd

Experimental studies of the influence of carbon-containing gas pressure on the geometric parameters of arrays of carbon nanostructures were carried out with the growth of carbon nanostructures by the method of plasma-chemical deposition from the gas phase. Arrays of carbon nanostructures were formed on the structures of nickel-chromium-silicon, nickel-vanadium-silicon, nickel-titanium-silicon, and nickel-aluminum-silicon. Nickel was used as the material for the formation of catalytic centers. Also, the silicon substrate and sublayer materials were used in the experiments, the sublayer materials served to provide a lower contact to the carbon nanostructures and as a diffusion barrier for the absence of interaction between nickel and silicon. Catalytic centers were formed by heating a thin nickel film for 20 minutes. The thickness of the nickel film was 20 nm, the thickness of the sublayer material was 50 nm. At the stage of formation of catalytic centers, the temperature was 650 °C, the flow of gaseous ammonia N_{NH3} = 15 cm3 / min., argon flow N_{Ar} = 40 cm3 / min. After the catalytic centers were formed on the substrate and the parameters of the catalytic centers were controlled by the methods of AFM and SEM, the influence of the pressure of a carbon-containing gas on the geometric parameters of carbon nanostructures was studied. Three groups of samples were formed at different pressures of the working atmosphere of gases in the reactor at 1.5, 3, and 4.5 Torr. The temperature of formation of the array of carbon nanostructures was 750 ° C, the flow of gaseous ammonia N_{NH3} = 210 cm3 / min, the flow of argon N_{C2H2} = 70 cm3 / min. The power of the plasma source was 240 W. Measurement of the geometric parameters of carbon nanostructures was carried out using scanning electron microscopy and scanning probe microscopy. The arrays of carbon nanostructures were oriented perpendicular to the substrate.

3. Results

As a result of experimental studies, arrays of carbon nanostructures were obtained. For modes of obtaining carbon nanostructures with a working gas atmosphere pressure of 1.5 Torr, an amorphous carbon film with no structure was formed on the substrate the film thickness was 2 micrometers (Figure 1.).

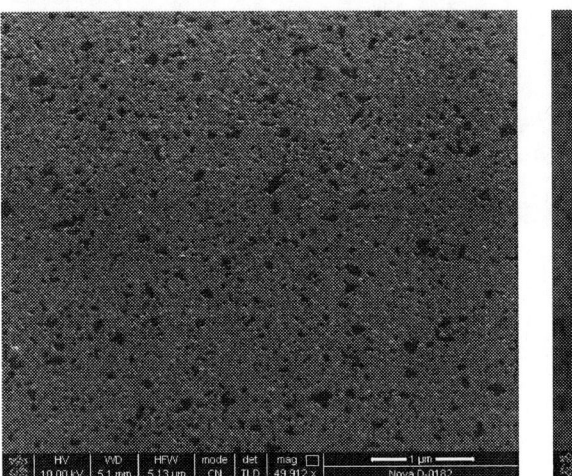

Figure 1. SEM images of amorphous carbon film obtained at pressures of carbon-containing gas 1,5 Torr.

At the remaining pressure values of the carbon-containing gas, the carbon nanoscale structures shown in Figure 2.

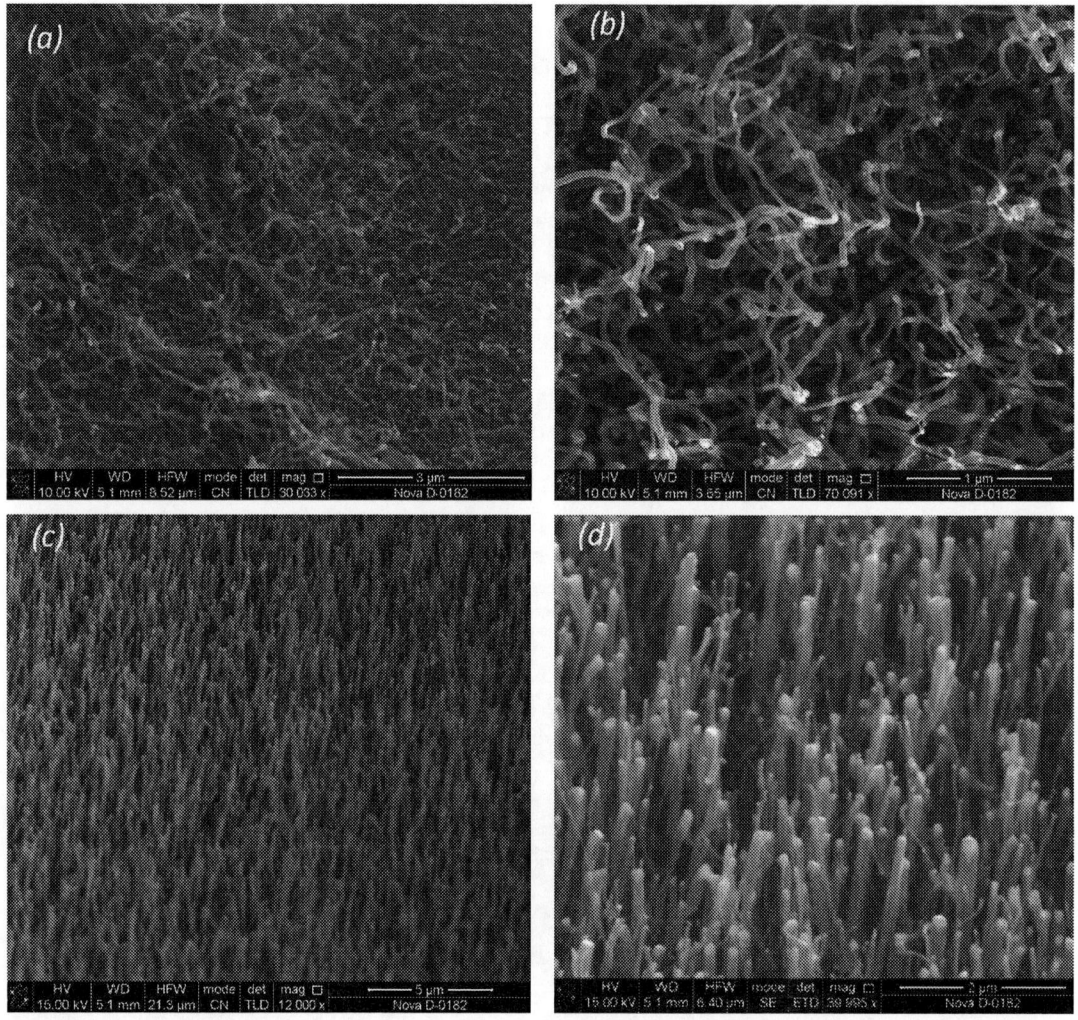

Figure 2 (a, b, c, d). SEM images of CNT arrays obtained at different pressures of carbon-containing gas: **(a, b)** 3 Pa; **(c, d)** 4,5 Pa.

From the obtained SEM images it is possible to determine the geometric parameters of the carbon nanotubes entering the array. Figure 2 .a, b. shows the array of carbon nanostructures formed at a pressure of carbon-containing gas of 3 Torr. This array is formed on a nickel-titanium-silicon structure. The height of carbon structures in this array is 1 μm, and the diameter is 10-30 nm, the distance between the tubes in the array is the largest among the scans presented, and the tubes in this array have the smallest dispersion of geometric parameters. Figure 2 (c, d) shows an array of carbon nanostructures formed at a carbon-containing gas pressure of 4.5 Torr. This array is formed on a nickel-titanium-silicon structure. The height of carbon structures in this array is 3 μm, the diameter of nanotubes is 20-50 nm. Tubes in this array are characterized by a large aspect ratio, but they are assembled into bundles of 5-10 pieces.

4. Discussion and Conclusions

In the work, experimental studies of the formation of arrays of carbon nanostructures were carried out. The influence of the pressure of a carbon-containing gas on the geometric and electro physical parameters of carbon nanostructures is shown. General patterns of the influence of technological parameters (pressure, sublayer material, plasma power and growth time) on the geometric parameters of carbon nanotubes in the array are determined. In particular, it has been established that the array of vertically oriented carbon nanostructures with a diameter of (30-70) nm, height (5-7) μm and a density of 3.7×10^5 cm^{-2} can be grown on a nickel-vanadium-silicon structure at gas flow rates $N_{C2H2} = 70$ cm3 / min and $N_{NH3} = 210$ cm3 / min, pressure 4.5 Torr, 750 ° C, plasma power W = 2 W. Such results allow selecting the modes of the technological process, for the formation of carbon nanostructures with the required parameters.

Acknowledgments

This work was supported by the Russian Foundation for Basic Research Project № 16-29-14023 ofi_m. The results were obtained using the equipment of Common Use Center and Education and Research Center "Nanotechnologies" of Southern Federal University.

References

[1] Nagase M, Nakamatsu K, Matsui S, Namatsu H 2005 *Japanese J. of Appl. Phys.* 44(7) 5409

[2] Klimin V S, Il'Ina M V, Il'In O I, Rudyk N N, Ageev O A 2017 *J. Phys.: Conf. Ser.* **917** 092023

[3] Rouhrig M, Thiel M, Worgull M and Houlscher H 2012 *Small* **8** 3009

[4] Ageev O A, Klimin V S, Solodovnik M S, Eskov A V, Krasnoborodko S Y 2016 *J. Phys.: Conf. Ser.* **741** 012178

[5] Morgan C G, Kratzer P and Scheffler M 1999 *Phys. Rev. Lett.* **82** 4886

[6] Tominov R V, Bespoludin V V, Klimin V S, Smirnov V A, Ageev O A 2017 *IOP Conf. Ser.: Mater. Sci. Eng.* **256** 012023

[7] Murdick D A, Wadley H N G and Zhou X W 2007 *Phys. Rev.* B **75** 125318

[8] Shiraishi K and Ito T 1998 *Phys. Rev.* B **57** 6301

[9] Klimin V S, Solodovnik M S, Smirnov V A, Eskov A V, Tominov R V, Ageev O A 2016 *Proc. of SPIE* **10224** 102241Z-1

[10] Amrani A, Djafari Rouhani M and Mraoufel A 2011 *Appl. Nanosci.* **1** 59

[11] Klimin V S, Tominov R V, Eskov A V, Krasnoborodko S Y, Ageev O A 2017 *J. Phys.: Conf. Ser.* **917** 092005

[12] Kangawa Y, Ito T, Taguchi A, Shiraishi K, Irisawa T and Ohachi T 2002 *Appl. Surf. Sci.* **190** 517

[13] Tominov R V, Zamburg E G, Khakhulin D, Klimin V S, Smirnov V A, Chu Y H, Ageev, O.A. 2017 *J. Phys.: Conf. Ser.* **917** 032023

[14] Daweritz L and Ploog K 1994 *Semicond. Sci. Tech.* **9** 123

[15] Il'Ina M V, Y F. Blinov, Il'In O I, Klimin V S, Ageev O A 2016 *Proc. of SPIE* **10224** 102240U-1

[16] Foxon C T and Joyce B A 1977 *Surf. Sci.* **64** 293

[17] Rudyk N N, Il'In O I, Il'Ina M V, Fedotov A A, Klimin V S, Ageev O A 2017 *J. Phys.: Conf. Ser.* **917** 082008

[18] Tok E S, Neave J H, Zhang J, Joyce B A and Jones T S 1997 *Surf. Sci.* **374** 397

[19] Kley A, Ruggerone P and Scheffler M 1997 *Phys. Rev. Lett.* **79** 5278

[20] Autumn K, Liang Y A, Hsieh S T, Zesch W, Chan W P, Kenny Th W, Fearing R and Full R J 2000 *Nature* **405** 681

SPBOPEN 2018

Influence of elastic stresses on the vapor-solid-solid growth mechanism of Au-catalyzed GaAs nanowires

A A Koryakin[1,2,3], S A Kukushkin[2]

[1]St. Petersburg Academic University, Khlopina 8/3, St.Petersburg 194021, Russia
[2]ITMO University, Kronverkskiy pr. 49, St. Petersburg 197101, Russia
[3]SPbD JSCC RAS - Branch of SRISA, Saint Petersburg 194021, Russia

Abstract. Influence of elastic stresses on the vapor-solid-solid growth mechanism of Au-catalyzed GaAs nanowires is studied. The elastic energy of the triangle-shaped island formed on the catalyst-nanowire interface is calculated. It is shown that the nucleation time of the triangle-shaped island is several times higher than that of the disk-shaped island in the case of nucleation of semi-coherent islands. The contribution to the free energy of island formation from the formation of island side walls is higher for the triangle-shaped island, although, its contribution from the elastic energy is lower when islands nucleate at the catalyst free surface.

1. Introduction
Nanowires (NWs) of III-V semiconductors are widely considered as building blocks for novel devices [1]. The III-V NWs are often synthesized at present by catalytic epitaxial methods. As a catalyst material, various metals (e.g. gold, silver) or group III elements (e.g. indium, gallium) can be used. If the NW growth proceeds at a temperature below the estimated melting point of catalyst particle it is often difficult to determine its phase state [2, 3]. However, most of the theoretical works that study the catalytic NW growth consider only the vapor-liquid-solid growth mechanism. Relatively few works (e.g. [2-4]) are dedicated to the vapor-solid-solid (VSS) growth mechanism.

This report continues the investigation of Au-catalyzed GaAs NW growing in the direction [111] via the VSS mechanism started in [5]. It was shown [5] that the VSS growth can proceed by the As diffusion along the catalyst-NW interface, whereas the volume diffusion flux is insufficient to feed the growing monolayer. In this paper, the estimation of the nucleation time in the case of nucleation of the triangle-shaped island at the catalyst-NW interface is performed.

2. Elastic energy of the triangle-shaped island
The growing NW facet is fed by the material fluxes coming from the vapor phase or side walls. The elements of group III and V dissolved into the catalyst particle that is considered as a solid solution in the case of VSS growth reach the catalyst-NW interface and form islands of monolayer height [6]. Here, we assume that the growth is limited by the As flux because the As concentration in catalyst is usually of the order of one percent or less whereas the concentration of Ga is of the order of tens of percent. Therefore, the As fluxes only should be taken into account. In figure 1, the NW top is schematically shown with two possible ways of As diffusion, interface and volume diffusion [5]. As a result of the lateral growth of island, the monolayer forms and the vertical NW growth proceeds [6].

Consider the free energy of island formation in the following form [7]

$$\Delta F(i) = \alpha \, i^{1/2} - (\Delta\mu - w)i, \tag{1}$$

Content from this work may be used under the terms of the Creative Commons Attribution 3.0 licence. Any further distribution of this work must maintain attribution to the author(s) and the title of the work, journal citation and DOI.
Published under licence by IOP Publishing Ltd

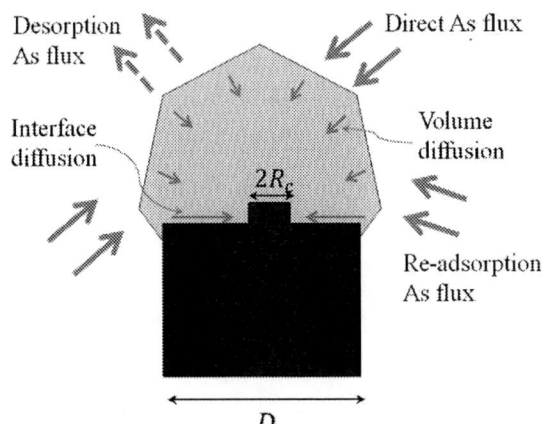

Figure 1. Schematic illustration of the As material fluxes coming into the catalyst particle during the NW growth.

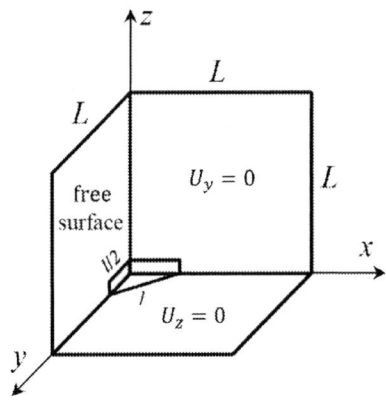

Figure 2. Model geometry of the system used for the calculation of elastic energy of triangle-shaped island formation.

where $\Delta\mu$ is the difference of the chemical potential per the GaAs pair between the catalyst particle and GaAs crystal; i is the number of GaAs pair in the island; $\alpha = 2 \cdot 3^{-1/4}(2\gamma + \gamma_{SV})(h\Omega)^{1/2}$ characterizes the contribution from the formation of island side walls in the case of triangle-shaped island of the monolayer height h and with the side $l = R/3^{1/2}$, R is the island radius; γ is the energy of interface of island side walls and catalyst, γ_{SV} is the energy of interface of island side walls and vapor phase; Ω is the volume of the GaAs pair in solid phase; w is the elastic energy per GaAs pair.

In the work [5], the dependence of w on i was found numerically. The function $w(i)$ can be approximated by the equation [5, 8]:

$$w(i) = w_0 + b(\ln i + d)/i^{1/2}, \tag{2}$$

w_0, b and d are constants (w_0 and b are expressed in meV below). In present work, the nucleation of the triangle-shaped islands is studied. We consider the formation of the triangle-shaped island at the catalyst free surface to estimate the maximum relaxation of elastic stresses caused by the difference of atomic densities of GaAs and catalyst material (figure 2).

The calculation of elastic stresses is performed by the finite element method [5, 9]. To imitate the phase transition, the effective coefficient of thermal expansion is introduced [9]. For the catalyst material and GaAs island, this coefficient equals to 0 and ε, respectively, where ε is the free deformation tensor that corresponds to the reconstruction of Ga sublattice in catalyst to the Ga sublattice in GaAs crystal. To find the minimal elastic stresses in the system, we perform the calculation for the different initial deformations of the island such that [5] $\varepsilon_{zz} = 3\varepsilon - 2\varepsilon_{xx}$, where ε_{xx} changes from 0 to 1.5ε, $\varepsilon_{yy} = \varepsilon_{xx}$, $\varepsilon_{ij} = 0$ for $i \neq j$, ε is the lattice mismatch between the Au-Ga catalyst and GaAs crystal. For simplicity, we also assume that $\varepsilon_{ij} > 0$ owing to the fact that $\varepsilon > 0$. We use the following simplified geometry for the estimation. The catalyst is considered in the form of the cuboid with the size $Lx2Lx2L$, where $L = 50$ nm. Taking into account the symmetry of the system, the calculation can be realized for $1/4$ part of the system that is schematically shown in figure 2. The boundary conditions for the equation of the elasticity theory are: $U_y(y = 0) = 0$, $U_z(z = 0) = 0$, $\sigma_{ij} = 0$ at the surface $x = 0$, $x = L$, $y = L$, $z = L$, where U_i is the deformation vector, σ_{ij} is the stress tensor. The results of the calculation by the finite element method for the case of nucleation in the

AuGa catalyst ($\varepsilon = 0.092$) and AuGa$_2$ catalyst ($\varepsilon = 0.172$) are presented in figure 3. The elastic coefficients of Au-Ga and GaAs can be found in the paper [5]. For comparison, the elastic energies obtained in [5] for the disk-shaped islands are also plotted on figure 3.

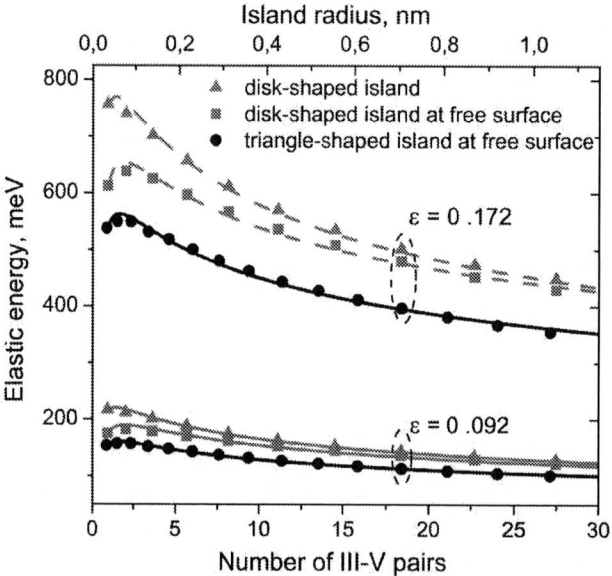

Figure 3. Dependence of the minimum elastic energy of island formation at $\varepsilon = 0.092$ (AuGa) and $\varepsilon = 0.172$ (AuGa$_2$). The results are shown for the case of formation of the triangle-shaped island at the free catalyst surface (circle symbols), disk-shaped island formed far from the free catalyst surface (triangle symbols) [5] and disk-shaped island at the free catalyst surface (square symbols) [5]. The results of fitting by the equation (2) are shown as solid lines.

The coefficients of the fitting function (3) in the case of the triangle-shaped island equal to: $w_0 = 20.33$, $b = 90.01$, $d = 1.507$ ($\varepsilon = 0.092$) and $w_0 = 71.04$, $b = 314.6$, $d = 1.507$ ($\varepsilon = 0.172$). Thus, it is seen from the figure 2 that the elastic energy per one III-V pair is several tens of percent lower in the case of nucleation of the triangle-shaped island compared to the formation of the disk-shaped island.

3. Results and discussion

Let us calculate the nucleation time of the triangle-shaped island using for a rough estimation the following diffusion equation that accounts for the As flux along the catalyst-NW interface [5]:

$$\frac{\partial C_{As}}{\partial t} = D_{As}\Delta C_{As}, \tag{3}$$

where $\Delta = 1/r \ \partial/\partial r \ (r \ \partial \cdot/\partial r)$, r is the polar radius, D_{As} is the diffusion coefficient on the catalyst-NW interface, C_{As} is the As concentration. The boundary conditions and initial condition for the equation (3) are as follows [5]:

$$j_{As,r}(D/2,t) = -(j_0 - k_0 C_{As}^2(D/2,t)), \tag{4}$$

$$j_{As,r}(0,t) = 0,$$

$$C_{As}(r,0) = 0.$$

Here $j_{As,r}$ is the As radial flux along the interface; j_0 is the As flux per unit of the length of the perimeter of the NW top facet; the flux j_0 includes the direct As flux and the re-adsorption As flux (see figure 1); $k_0 \sim k_{As}h$, where k_{As} is the coefficient that characterizes the As evaporation flux; D is the NW diameter. The time required for the first island formation can be estimated by means of equating the nucleation time τ_N and growth time [5]:

$$\tau_N\big(C_{As}(r,t)\big) = t, \tag{5}$$

where $\tau_N \sim 1/\pi(D/2)^2 I$ [6], $I = N_0 W^+(i_c) Z \exp(-\Delta F(i_c)/kT)$ is the nucleation intensity, N_0 is the number of adsorption sites on the catalyst-NW interface; $W^+(i_c)$ is the rate at which the GaAs pair attach to the island, Z is the Zeldovich factor, i_c is the critical size of the island, k is the Boltzmann constant; T is the growth temperature. To estimate the order of τ_N, we use the formula $W^+(i_c) = 2\pi N_0 D_{As} C_{As}^\infty / \ln \lambda_{As}/R_c$ obtained for the case of the disk-shaped island [5], where C_{As}^∞ is the equilibrium As concentration at $R = \infty$, R_c is the critical island radius, $\lambda_{As} \sim R$ is the diffusion length of As species at the catalyst-NW interface, and expression for Z and $\Delta F(i_c)$ derived in [5]. The results of self-consistent solution of the equations (3)-(5) show that the nucleation time of the semi-coherent triangle-shaped island at the free catalyst surface (when $\gamma \sim 0.3$ Jm^{-2} and we assume $\gamma_{SV} = \gamma$ for a simple consideration), $\tau_N \sim 0.01$ s, is several times higher than that of the disk-shaped island. Therefore, the nucleation of the semi-coherent disk-shaped island is energetically more favorable process than the nucleation of the semi-coherent triangle-shaped island. This result can be explained by the fact that the surface energy term to the nucleation barrier is higher for the triangle-shaped island because it has a longer length of the perimeter but the elastic energy term is lower. However, the nucleation of the coherent triangle-shaped island ($\gamma \sim 0.01$ Jm^{-2}) proceeds at a higher rate than the nucleation of the coherent disk-shaped island, although, the orders of magnitude of nucleation time is the same in both cases ($\tau_N \sim 10^{-4}$ s). This is associated with the decrease of the role of the side wall term to the free energy. The parameters of the model used in the computation are as follows [5]: $D_{As} \sim 10^{-12}$ $m^2 s^{-1}$, $T = 420\,^\circ C$, $R = 50$ nm, $k_{As} = 3.9 \cdot 10^{21}$ $m^{-2} s^{-1}$. The chemical potential difference $\Delta\mu$ and desorption coefficient k_{As} are calculated by means of the method presented in [10].

Acknowledgments
The reported study was funded by RFBR according to the research project № 18-32-00559.

References
[1] Dasgupta N P, Sun J, Liu C, Brittman S, Andrews S C, Lim J, Gao H, Yan R, Yang P 2014 *Advanced Materials* **26** 2137-2184
[2] Persson A I, Larsson M W, Stenstrom S, Ohlsson B J, Samuelson L, Wallenberg L R 2004 *Nature Mater.* **3** 677-681
[3] Dick K A, Deppert K, Martensson T, Mandl B, Samuelson L, Seifert W 2005 *Nano Lett.* **5** 761-764
[4] Krogstrup P, Yamasaki J, Sorensen C B *et al.* 2009 *Nano Lett.* **9** 3689-3693
[5] Koryakin A A, Kukushkin S A, Sibirev N V 2018 *Semiconductors* (submitted)
[6] Dubrovskii V G, Sibirev N V 2004 *Phys. Rev. E* **70** 031604
[7] Christian J W 2002. The Theory of Transformations in Metals and Alloys (Amsterdam: Pergamon).
[8] Khachaturyan A G 2008 Theory of structural transformations in solids (New York: Dover)

[9]   Chen C R, Li S X, Zhang Q 1999 *Mater. Sci. Eng. A* **272** 398-409

[10]   Glas F, Ramdani M R, Patriarche G, Harmand J C 2013 *Phys. Rev. B* **88** 195304

Structural investigation of light-emitting A3B5 structures grown on Ge/Si(100) substrate

A V Rykov[1], M V Dorokhin[1], P S Vergeles[2], V A Kovalskiy[2], E B Yakimov[2], M V Ved'[1], N V Baidus[1], A V Zdoroveyshchev[1], V G Shengurov[1], S A Denisov[1]

[1]Research Institute of Physics and Technology, Lobachevsky State University, Nizhni Novgorod, 603950, Russia
[2]Institute of Microelectronics Technology, Russian Academy of Sciences, Chernogolovka, 142432, Russia

Abstract. In this paper we have investigated light-emitting diodes based on GaAs/InGaAs heterostructures grown on a Ge/Si(100) substrates. Ge layers were deposited by the "hot wire" method, and A3B5 layers were grown by the low pressure MOCVD. Structures were investigated by the methods of electroluminescence spectroscopy and electron beam induced currentimaging in a scanning electron microscope. Technological ways to improve crystalline quality of active region of light-emitting structures grown on Ge/Si(100) substrate were shown.

1. Introduction

There is an increasing demand for Si substrates as the basis of optoelectronics devices, alternatively to GaAs [1]. Growing A3B5 semiconductor layers of proper crystalline quality on the Si substrate opens new prospects for the creation of such microwave devices, light-emitting diodes (LEDs), photodetectors, solar cells, etc. GaAs layers can be grown directly on the Si substrate, but due to the difference in the lattice parameter and the thermal expansion coefficient of GaAs and Si, layers contain structural defects of high density. A more prominent way is to use different buffer layers introduced between the Si substrate and the GaAs film. These layers should have intermediate values of the lattice parameter and thermal expansion coefficient. For instance, growth of Ge layers on Si and following growth of GaAs layers is a widespread technique.

Structures we produced are the basis for creation of hybrid A3B5/Si LEDs with ferromagnetic contact.We used ferromagnetic CoPt contact for the purpose of injecting spin-polarized carriers into active region resulting in emission of partially circularly polarized lightat the room temperature [2].The main problem of those hybrid diodes is low intensity of luminescence due to high density of defects. Two types of defects influence luminescence the most: threading dislocations and antiphase boundaries (APBs). Antiphase domain is a region where Ga and As sublattices are swapped forming a boundary propagating through structure [3]. We focused on technological ways to increase electroluminescence intensity by reducing significantly the density of APBs in the present paper.

Electron beam induced current (EBIC) in a scanning electron microscope has been exploited to investigate the structure of electrically active defects in our samples. This technique combines spatial visualization and high resolution detection of electrically active defects in a structure with the

---

Content from this work may be used under the terms of the Creative Commons Attribution 3.0 licence. Any further distribution of this work must maintain attribution to the author(s) and the title of the work, journal citation and DOI.
Published under licence by IOP Publishing Ltd

possibility for calculation of minority carrier diffusion length. This parameter characterizes quality of a semiconductor material and determines the main characteristics of many semiconductor devices [4].

2. Fabrication of structures

Epitaxial structures were grown on Si substrates with (100) orientation. At the first stage, intermediate layers were deposited on the substrate by the method of molecular beam epitaxy (MBE). First, 200 nm thick Si layer was grown at 800°C using a sublimation source. Second, the substrate temperature was decreased to 350°C and germane (GeH$_4$) was injected into the growth chamber at pressure of $4 \cdot 10^{-4}$ Torr. Germane pyrolytically decomposes under these conditions in the region of the tantalum wire heated up to 1350°C temperature [5]. Atomic Ge precipitated on a substrate as a result of the pyrolytic decomposition of the germanes, and 750 nm thick Ge layer was grown.

At the second stage, the A3B5 structure was grown at pressure of 50 mbar using the AIX 200RF MOCVD system. Precursors were used as follows: trimethylgallium for Ga, trimethylaluminum for Al, arsine for As, silane for n-type doping of buffer, and carbon tetrachloride for p-type delta-doping of cap layer. The technique and sequence of growth of A3B5 buffer layers coincided with those described in [6] where layers of the laser structure were formed after the growth of transition layers of AlAs/GaAs and thick n$^+$-GaAs buffer. In the present work, the n-GaAs layer ($n = 5 \cdot 10^{17}$ cm^{-3}) and the InGaAs quantum well (QW) were grown on top of heavily doped n$^+$-buffer. Two structures were fabricated and investigated with the different modes of a buffer layer growth. In the first structure the buffer layer thickness was 2 μm and the growth temperature was 700°C (structure A). For the second structure fabrication, the buffer layer growth temperature was increased to 750°C and the buffer layer thickness was increased to 3 μm (structure B). In addition a set of 5 shallow InGaAs QWs were grown in the beginning of the structure B buffer layer fabrication to create a dislocation barrier.

3. Results and discussion

The photoluminescence (PL) spectra at room temperature (not shown) reveal a peak at 965 nm wavelength corresponding to recombination in the quantum well. At the same time, PL intensity is by one order of magnitude lower than intensity obtained on the control structure grown on a GaAs substrate. This situation is typical for A3B5 structures grown on Si substrate due to high density of electrically active defects in active region.Top CoPt contacts were deposited on sample surface [7] and the Ohmic contact was alloyed into the buffer layer away from CoPt contact for the electroluminescence (EL) study. The EL spectra are shown in figure 2. EL study shows that increase in doping level and in thickness of buffer layer as well as inclusion of InGaAs dislocation filters leads to EL intensity increase by more than an order of magnitude.

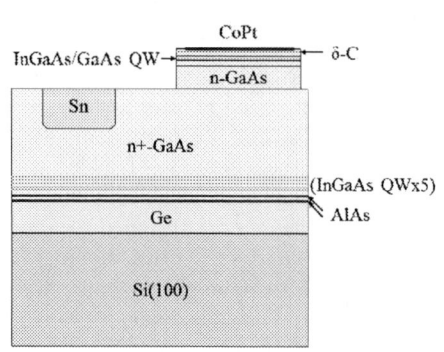

Figure 1. Scheme of investigated LEDs.

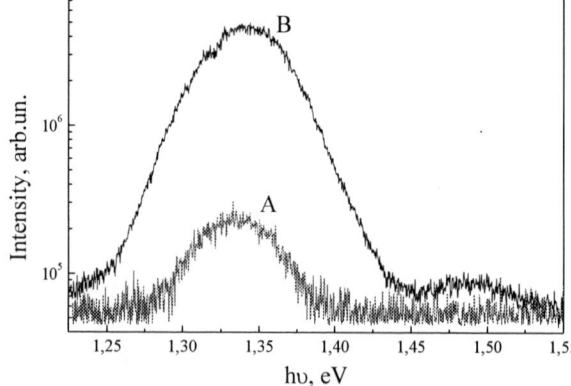

Figure 2. EL spectra of A3B5/Ge/Si LEDs with CoPt contact at 77K.

Structural investigations of LED structures with InGaAs/GaAs QWs were carried out by EBIC technique. Details of EBIC technique were described elsewhere [4]. Figure 3 shows EBIC micrograph of the LED structure A obtained at accelerating voltage of 35 kV. It can be noted that the images show a high density (about 10^8 cm^{-2}) of defects with dark contrast, associated with threading dislocations. APBs consist of Ga-Ga and As-As bonds resulting in creation of large amount of non-radiative recombination centers and the whole antiphase boundary might result in closed loops with dark contrast on EBIC picture (figure 3). Shape of these loops remains the same as the accelerating voltage increases from 5 kV to 35 kV. This fact confirms association of the loops with APBs, which propagate from the initially formed transition layer to the surface.

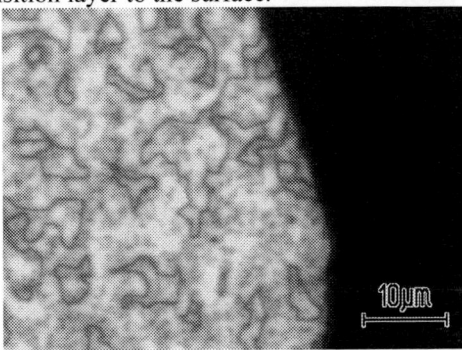

Figure 3. EBIC micrigraph of LED structure A.

At the same time, no loops presumably associated with APBs were observed for structure with thicker n$^+$-GaAs buffer and additional dislocation filter layers (figure 4). Absence of APBs might be a result of combination of several factors. The level of doping was raised in order to increase luminescence intensity. Greater level of doping was obtained by increasing mole fraction of silane in growth chamber as well as increasing temperature of substrate from 700°C up to 750°C. Self-annihilation of antiphase boundaries in GaAs epilayers on Ge substrates with 6° offcut was reported in [8]. In our case, industrially compatible Si(100) substrates without additional offcut were used, but it is reasonable to assume that APBs could self-annihilate at the same conditions: higher temperature and thicker A3B5 layer. Overall thickness of A3B5 structure with thicker buffer reaches 3.5 μm considering all transition layers and dislocation filters. In our case, EBIC signal was collected from the depth of 0.2-0.3 μm (space charge region (SCR) + minority carrier diffusion length). It is safe to say that if self-annihilation occurs at some point, it takes place beyond the signal collection depth of the EBIC technique.

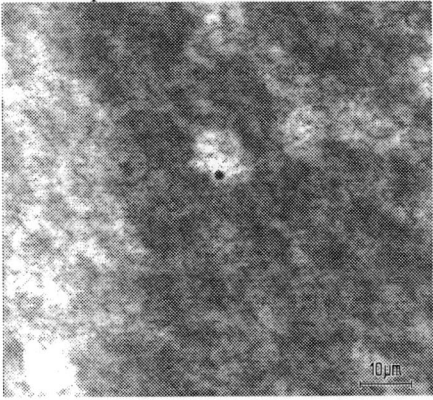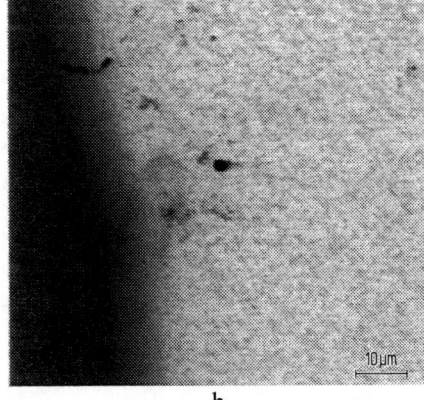

a   b

Figure 4. EBIC micrographs of the same spot of LED structure B obtained at an accelerating voltage of (a) 35 kV and (b) 5 kV

The comparison of the images obtained at different accelerating voltages for structure B shows that they differ substantially. Large-scale inhomogeneities of current collection (bright regions in figure 4(a)) are revealed for EBIC images at high accelerating voltages, while such inhomogeneities are absent in EBIC micrographs at low energies of the primary electron beam (figure 4(b)). In some cases, these bright regions reach a size of 10×10 μm^2. At the same time, micrographs in the secondary electron imaging (SEI) mode have not revealed any defects on the surface, which allows us to state that the heterogeneity of the EBIC signal is not related to the surface.

To obtain additional information on nature of inhomogeneities revealed, dependencies of the EBIC collection efficiency on the accelerating voltage were measured both for the signal averaged directly over light regions and for the signal averaged over a large area. The latter procedure made it possible to minimize the contribution from the light regions at high accelerating voltages. Figure 5 shows these dependencies. It can be seen that for accelerating voltages greater than 10 keV, the curve for the light region lies higher than the curve obtained by averaging over a large area. This means that there is actually more current flowing in the light regions, as compared with the rest of the structure area at high accelerating voltages of the primary electron beam.

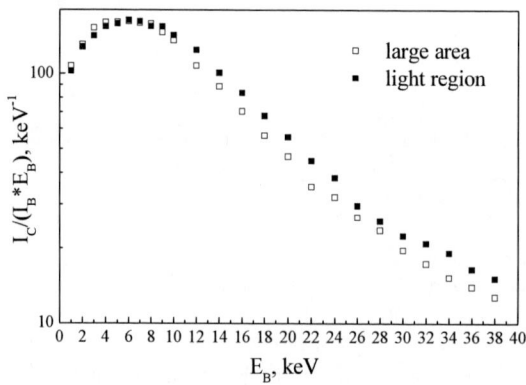

Figure 5. Normalized dependencies of the EBIC collection efficiency on accelerating voltage: hollow squares - a curve for a signal averaged over a large area, filled squares - for a light region.

According to the calculation of the band diagram of such a structure, the InGaAs QW is deep in the SCR, whose width is ~ 0.1 μm at zero bias voltage, and, hence, cannot affect the EBIC collection efficiency. At the same time, it has been established from the simulation of the obtained dependences of the EBIC collection efficiency that the diffusion length of the minority charge carriers is comparable with the width of the SCR and varies in the range 0.1-0.2 μm. Thus, proceeding from the analysis, it can be assumed that the appearance of regions with an increased current flow may be associated with fluctuations in the SCR width caused by the heterogeneity of the δ-layer of carbon. This assumption is also confirmed by the fact, that described inhomogeneities were not revealed in EBIC images of structure A, which does not include carbon δ-layer.

In conclusion we have grown and investigated InGaAs/GaAs LED structure on Ge/Si(100) substrate. Although the EL intensity is relatively low compared to LED structures on native substrate, there is room for improvement. A combination of changes in LED growth technology led to obtaining a LED structure with APB-free active region. This effect in combination with variations in doping of structure led to significant increase in EL intensity of the LED.

Acknowledgments

This study was supported by the Russian Foundation for Basic Research (projects 18-32-00636 (MOCVD growth and EBIC imaging), 16-07-01102, and 17-37-80008) and Ministry of Education and Science of Russian Federation (projects 16.7443.2017/BCh and SP-2450.2018.5).

References

[1] Yu. B. Bolkhovityanov, O. P. Pchelyakov 2009 *Open Nanoscience Journal* **3** 20

[2] A V Rykov, M V Dorokhin, P S Vergeles, N V Baidus, V A Kovalskiy, E B Yakimov, O A Soltanovich 2018 *Journal of Physics: Conf. Series* **993** 012014

[3] S. M. Ting, E. A. Fitzgerald 2000 *Journal of Applied Physics* **87** 2618

[4] N. M. Shmidt, P. S. Vergeles, E. B. Yakimov 2007 *Semiconductors* **41**(4) 491-494

[5] V. G. Shengurov, S. A. Denisov, V. Yu. Chalkov, Yu. N. Buzynin, M. N. Drozdov, A. N. Buzynin, P. A. Yunin 2015 *Technical Physics Letters* **41** (1) 36–39

[6] V. Ya. Aleshkin, N. V. Baidus, A. A. Dubinov, Z. F. Krasilnik, S. M. Nekorkin, A. V. Novikov, A. V. Rykov, D. V. Yurasov, A. N. Yablonskiy 2017 *Semiconductors* **51**(5) 663-666

[7] A. V. Zdoroveyshchev, M. V. Dorokhin, P. B. Demina, A. V. Kudrin , O. V. Vikhrova, M. V. Ved', Yu. A. Danilov, I. V. Erofeeva, R. N. Krjukov, D. E. Nikolichev 2015 *Semiconductors* **49** 1601

[8] M. K. Hudait, S. B. Krupanidhi 2001 *Journal of Applied Physics* **89** 5972

SPBOPEN 2018   IOP Publishing

The effect of AlGaAs/GaAs laser heterostructure ionic surface cleaning on the structure and properties of AlN films deposited by the method of reactive ion-plasma sputtering

E V Fomin[1,2], K A Surnin[1,2], A D Bondarev[2], I P Soshnikov[2-4], K P Kotliar[3], S I Pavlov[2], A V Nashchekin[2]

[1] Saint Petersburg Electrotechnical University "LETI", Saint-Petersburg 197376, Russia
[2] Ioffe Institute of RAS, Saint-Petersburg 194021, Russia
[3] Academic University, Saint-Petersburg 194021, Russia
[4] Institute for Analytical Instrumentation of RAS, Saint-Petersburg 198095, Russia

Abstract: The paper presents studies of the structural and optical properties of AlN thin films grown on an AlGaAs/GaAs semiconductor laser heterostructure with the use of preliminary ionic surface cleaning and without it. After cleavage of the heterostructure, a natural surface oxide forms on the facet. In the paper, the ionic etching regime providing the removal of the surface oxide layer without significant defects in the semiconductor heterostructure is determined.

Introduction

The most important characteristic of modern semiconductor lasers, along with power, is their reliability and durability. Usually, degradation of laser heterostructures occurs due to catastrophic optical mirrors damage (COMD) [1][2]. The main cause of COMD are surface defects in the active region of the facet cavity, the presence of which results in the absorption of the laser's own radiation and the appearance of nonradiative recombination. These processes leads to local heating with the breaking of chemical bonds and the generation of new defects, which, in turn, leads to the destruction of the structure. On a certain output power threshold, the laser mirror gets irreversible damage and the laser fails.

To weaken the effect of COMD, passivation of the surface of the facets is used. For example, a coating is applied to the surface of the resonator, which weakens nonradiative recombination and increases the threshold power before COMD [2-5][7]. Passivation may be carried out chemically [2-4][6]. The advantage of this method is that the facets are not exposed to high-energy particles. One more way is to cleave the laser structure in a vacuum and apply coating without extraction it to air, which is a technically complex and expensive [4][5]. Another option is to carry out cleaning of the heterostructure facets surface by ionic etching in an inert gas atmosphere before applying the passivation coating. A similar work has already been presented [7], however, in the case of ZnSe, as the authors notice, it is necessary to additionally deposit antireflective or reflective coatings over the passivation layer, which makes the process more complex. The negative side of this method is that in the process of etching it is possible to damage the structure itself. Therefore, the development of methods for forming passivation coatings of AlGaAs/GaAs heterostructures with the possibility of removing a defective layer is an actual and important problem in the technology of high-power semiconductor lasers.

In the paper we investigate the possibility of using thin AlN films as passivation coatings for high-power AlGaAs/GaAs lasers using preliminary cleaning of the facets of the heterostructures prior to passivation layer deposition to remove a layer of surface oxide that is formed naturally after cleavage of the structure in air and in further negatively affects the laser lifetime.

Content from this work may be used under the terms of the Creative Commons Attribution 3.0 licence. Any further distribution of this work must maintain attribution to the author(s) and the title of the work, journal citation and DOI.
Published under licence by IOP Publishing Ltd

To carry out the cleaning, the method of ionic etching of the facets of the heterostructures was chosen. The cleaning procedure is followed by the deposition of AlN passivation coating at a single process, without extraction of the structure to air. This method allows us to clean the surface from impurities and, in contrast to chemical passivation, from the layer of native oxide, Al_2O_3, which allows it to be used for processing the aluminum-containing heterostructures [4][6].

Aluminum nitride has been chosen as a passivation coating because it has a number of features: high strength characteristics, chemical resistance, high thermal conductivity and decomposition temperature, wide variation of the refractive index. In contrast to ZnSe in [7], AlN due to its optical properties, physical and chemical resistance, can simultaneously serve as a passivating and antireflective coating for mirrors of high-power semiconductor lasers, which greatly simplifies the process and reduces its cost.

Experimental

To evaluate the effect of ionic cleaning the samples of AlGaAs/GaAs laser heterostructures obtained using MOCVD technology [8] were taken. The AlN coatings was deposited on the facets of the heterostructures after cleavage and exposition in the air for a predetermined time. In addition, control samples of AlN films deposited on an oriented GaAs (001) substrate were investigated. Deposition of AlN films was carried out by the method of reactive ion-plasma sputtering in the "UB-744" triode-type vacuum system by sputtering a target made of pure Al (99.999%) in N_2 atmosphere. As a working gas for ionic cleaning, Ar was used. The purity of working gases is not worse than 99.99%. The residual pressure before the process was not higher than 1.0x10-6 Torr. The pressure of working gases during ionic cleaning and during the deposition process was maintained at 2.0-3.0x10^{-3} Torr.

The thickness and refractive index of all films were evaluated on the LEF-3M-1 ellipsometer at a wavelength of 633 nm. The structure of the films was studied with scanning electron microscopy using the Supra 25 microscope. Quantitative analysis of SEM images was carried out using the DIAna TEM program [9].

To estimate the results of ionic etching, a special series of thin film samples was prepared with a variation in the time of formation of the oxide layer from 30 to 120 minutes, with and without the ionic cleaning procedure. The accelerating voltage on the sample holder during the ionic cleaning process was +30 V.

Results and discussion

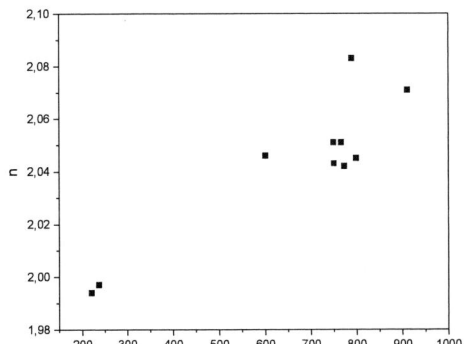

Fig. 1. Dependence of the refractive index on the thickness of the film

The thickness of the films measured by ellipsometry was from 220 to 240 Å for samples prepared for estimation of the results of ionic etching. The refractive index for these samples ranged from 1.994 to 1.997. Typical thickness values for control samples of films grown in working processes are 650-800 Å, and the refractive index is 2.00-2.08 (Fig. 1). For films thicker than 700 Å, an additional thickness estimation with the SEM images was carried out. The results of thickness measurements with SEM and ellipsometry are in good agreement. Taking into account the measured control samples, a tendency to increase the refractive index while increasing film thickness was found. The increase in the refractive index while increas of film thickness can be explained by the fact that at the beginning of the process aluminum nitride grows in the amorphous phase because of large mismatch of the lattice parameters with GaAs, gradually converting to microcrystalline phase, as the thickness of the films grows. This effect is especially noticeable in the presence of residual oxygen in the chamber, as was previously observed in [10].

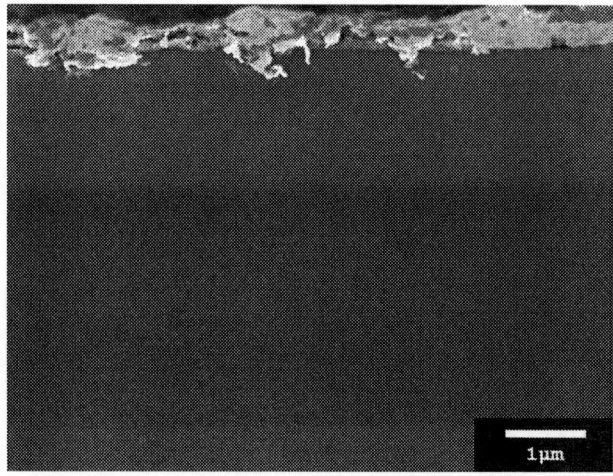

Fig. 2. SEM image of an AlN film deposited on an AlGaAs structure, 120 min exposure to air with etching

Fig. 3. Distribution of the relative intensity of secondary electron emission for different samples

The SEM image of the surface of the AlGaAs laser structure after ionic cleaning and deposition of the AlN film is shown in Fig 2. Figure 3 shows the results of a quantitative analysis of the intensity of secondary electron emission from samples prepared under different conditions. The changes in the intensity of the SEE reflects the features of the chemical composition of the various regions of the heterostructure. One can see transitions between layers and the presence of a quantum well, expressed by a local increase in intensity at around 2500 nm. It has been found that for samples that was not undergo ionic cleaning, the contrast of the images increases, and the resolution decreases with increasing of oxidation time from 30 to 120 min before deposition of AlN. For samples obtained with the use of ionic cleaning, a weak dependence of the image quality on the oxidation time is observed. This allows us to conclude that the surface condition of the samples is comparable after cleaning, regardless of the time of formation of the oxide layer. The results obtained can be explained by the processes of destruction of the oxide layer and sputtering of oxygen when bombarded with argon ions. At the same time the refractive index of films deposited after the ionic cleaning procedure is usually higher than for films deposited on an oxidized surface (2.0-2.08 versus 1.98-1.99).

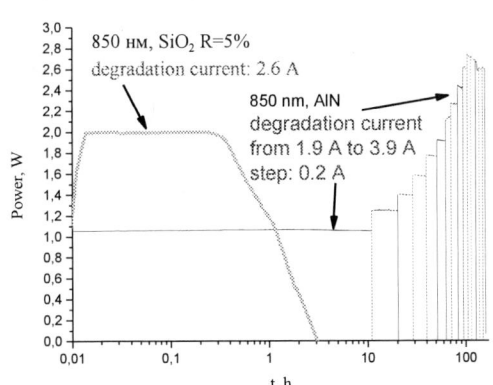

Fig. 4. Comparison of the degradation time of lasers with mirrors of SiO₂ and AlN

A study to estimate the degradation time of laser structures revealed a significant increase in the lifetime of the structure when an AlN coating was used for passivation of the facets instead of the SiO_2 that was used previously (Fig. 4). With a pump current of 2.6 A, the laser with a SiO_2 mirror starts to degrade in less than an hour and quickly fails. At the same time, laser with an AlN mirror was able to work for more than 150 hours with a gradual increase in pump current, including more than 100 hours at currents exceeding 2.5 A. The onset of gradual degradation for laser passivated with AlN is noticeable at pump currents above 3.5 A. While working with a pumping current of 3.9 A, a power drop from 2.73 to 2.6 W was observed in 25 hours for AlN-passivated laser.

Conclusion

The results of the studies shows the advisability of using aluminum nitride films as a passivation coating for the high-power semiconductor lasers facets. Incorporation of the stage of preliminary ionic surface cleaning into the deposition technology also seems reasonable due to the improvement in the quality of deposited coatings and the increase in the reliability and longevity of lasers. At the same time, more detailed studies of the possible damage to the structure in the process of ionic etching are needed, and further improvement of the deposition technology is required to effectively control the quality of the coatings.

Acknowledgment

We wish to thank the head of Laboratory of semiconductor luminescence, Pikhtin N.A. and the deputy director of the Center of Physics of Nanoheterostructures, Bert N.A. for all-round support. This work was partially supported by grant from the RFBR (18-07-01364 A).

Reference

1. J. Souto et al., Catastrophic optical damage of high power InGaAs/AlGaAs laser diodes / Microelectronics Reliability, Sep. 2016, v. 64, pp. 627-630

2. G. Beister et al., Non-radiative current in InGaAs/AlGaAs laser diodes as a measure of facet stability / Solid-State Electronics, 1998, Vol. 42, No. 11, pp. 1939-1945

3. V.N. Bessolov et al. Sulfur passivation of InGaAs/AlGaAs SQW laser facets / Materials Science and Engineering 544 (1997) 350-382

4. P. Ressel et al., Novel Passivation Process for the Mirror Facets of Al-Free Active-Region High-Power Semiconductor Diode Lasers / Photonics Technology Letters, May 2005, vol. 17, no. 5, pp. 962-964

5. N. Chang et al., ZnSe for Mirror Passivation of High Power GaAs Based Lasers / Electronics Letters, 15th Aug. 1996, v. 32, n. 17, pp. 1595-1596

6. A. V. Ankudinov et al., Nanorelief of an oxidized cleaved surface of a grid of alternating $Ga_{0.7}Al_{0.3}As$ and GaAs heterolayers / Semiconductors, May 1999, v. 33, № 5, pp. 555-558

7. X. Shu et al., ZnSe by electron-beam evaporation used for facet passivation of high power laser diodes / Solid-State Electronics 49 (2005) 2016–2017

8. D. A. Vinokurov et al., 850-nm Diode Lasers with Different Ways of Compensating for Internal Mechanical Stresses in an AlGaAs:P/GaAs Heterostructure / Semiconductors, 2013, v. 47, №. 8, pp. 1075-1078

9. Soshnikov I.P., Gorbenko O.M., Ledentsov N.N., Golubok A.O. / Semiconductors, 2001, v. 35, № 3, pp. 347-352

10. Ya. V. Lubyanskiy et al., Oxygen Nitrogen Mixture Effect on Aluminum Nitride Synthesis by Reactive Ion Plasma Deposition / Semiconductors, 2018, v. 52, No 2, pp. 184-188

Theoretical analysis of the length distributions of Ga-catalyzed GaAs nanowires

Y Berdnikov[1], V G Dubrovskii [1], N V Sibirev[1], A A Koryakin[1,2], A. S. Sokolovskii[1], T Tauchnitz[3,4], H Schneider[3], M Helm[3,4], and E Dimakis[3]

[1] ITMO University, Kronverkskiy pr. 49, 197101, St. Petersburg, Russia
[2]St. Petersburg Academic University, Khlopina 8/3, 194021 St. Petersburg, Russia
[3]Institute of Ion Beam Physics and Materials Research, Helmholtz-Zentrum Dresden-Rossendorf, 01328 Dresden, Germany
[4]Technische Universität Dresden, 01062 Dresden, Germany

Abstract. In this work, we analyze the length distributions of self-catalyzed GaAs nanowires grown on in-situ processed SiOx/Si(111) substrates. We propose the model that accounts for (i) gradual emerging of Ga droplets and nanowires, (ii) fluctuation-induced broadening of the length distributions, and (iii) the effect of nucleation antibunching in individual nanowires on the length distribution shapes. Both theoretical and experimental results show the broadest length distributions on unprocessed substrate, narrowing in samples with processed oxide surface and sub-Poissonian length distribution in samples with an optimized substrate preparation.

Introduction

Gold-free self-catalyzed growth of III-V nanowires (NWs) by the vapor-liquid-solid (VLS) mechanism [1-3] is proven effective for monolithic integration of III-V materials with silicon electronic platform, opening new perspectives in the design of novel optoelectronic devices. Certain applications require a high level of size uniformity within the array of NWs in terms of their length and diameters [4]. It has recently been shown that the diameter homogeneity of self-catalyzed III-V NWs can be accessed in the so-called "self-focusing" growth regime [3,5]. Despite the fact that narrow NW length distributions (LDs) are more difficult to achieve, markedly sub-Poissonian LDs of self-catalyzed GaAs NWs were reported in Ref. [6]. However, this paper considers the statistics of NW lengths only in one specific growth protocol. In this work, we study theoretically and experimentally the LDs within the ensembles self-catalyzed GaAs NWs grown by different procedures yielding different NW densities, based on the experimental data of Ref. [7].

Description of experiments

During the sample preparation [7], *in situ* surface modification procedure (SMP) of native SiOx/Si(111) substrates was used for some samples that precedes the GaAs NW growth and decouples the Ga/SiOx interaction from the subsequent Ga-assisted NW nucleation. The SMP consisted of the following three steps: (1) thermal annealing of the substrate, (2) Ga deposition and droplet formation, and (3) second thermal annealing of the substrate for complete evaporation of the Ga droplets. The subsequent growth of GaAs NWs was performed at a substrate temperature T of 615 °C, with the Ga flux fixed at 0.16 ML/s (as given by the equivalent growth rate on planar GaAs(001) substrate) and the V/III ratio of 11. The substrate was first exposed to the As4 flux and the NW growth was initiated 10 min later by opening the Ga shutter. After 15 min of growth, the As4 and Ga beams were interrupted simultaneously and the

---

Content from this work may be used under the terms of the Creative Commons Attribution 3.0 licence. Any further distribution of this work must maintain attribution to the author(s) and the title of the work, journal citation and DOI.
Published under licence by IOP Publishing Ltd

substrate temperature was ramped down to 300 °C at a rate of 50 °C/min. We compare the distributions over the NW lengths measured for three samples prepared differently with or without SMP.

Model

In the growth process of Ref. [7], no Ga droplet pre-deposition is employed and hence the droplets always emerge concomitantly with NWs. The formation of new droplets enables the NW nucleation at the time-dependent rate $J(t)$. Within the model for the kinetics of Ga adatoms on $SiO_x/Si(111)$ developed in Ref. [2], the nucleation rate of NWs emerging from the newly formed droplets is given by

$$J(t) = J_0 g^2 \left[\left(\frac{t}{\Delta t} \right)^{\frac{3}{2}} \right]. \qquad (1)$$

Here, J_0 is a constant, $g(y) = e^{-y} \int_0^y dx\, e^x x^{-1/3}$ and Δt is the characteristic nucleation time of NWs catalyzed by the randomly emerging Ga droplets. Without the effect of kinetic fluctuations [8], the NW LD can be obtained in the form [2]

$$f(L, L_{Max}) = A g^2 \left[\left(\frac{L_{max}-L}{\Delta L} \right)^{\frac{3}{2}} \right], \qquad (2)$$

with A as the normalization constant and L_{max} as the maximum NW length in the ensemble after a given growth time. The width of this LD is controlled by the parameter $\Delta L = V\Delta t$ which includes the length-independent instantaneous growth rate of NWs V (limited by the As flux [1-3,5-7,9-11]).

Equation (2) does not account for kinetic broadening of the LD described by the second derivative in the continuum Fokker-Planck equation [8]. According to the general growth theory [8,11,12], more precise LD is obtained by convoluting the NW nucleation rate with Green's function that describes time evolution of the LD starting from the delta-like peak at $t = 0$. According to Refs. [10,11], Green's function in the case of length-independent axial growth rate of NWs is given by

$$F(s,t) = \frac{1+\varepsilon\rho(s)}{\sqrt{2\pi D(t)}} \exp \left[-\frac{(\rho(s)-z(t))^2}{2D(t)} \right], \qquad (3)$$

Here, s is NW length measured in the numbers of monolayers (MLs) ($s = L/h$, where L is the NW length in nm and $h = 0.326$ nm is the height of the GaAs ML in the growth direction), $\rho(s) = (e^{\varepsilon s} - 1)/\varepsilon$ is the invariant size, $z(t) = (e^{\varepsilon Vt} - 1)/\varepsilon$ is the most representative invariant size [8] and $D(t) = z(t) + \varepsilon z^2(t)/2$ is the variance of the LD. The ε parameter determines the strength of nucleation antibunching caused by the instantaneous depletion of the droplet with its As in each ML growth cycle [10-14]. As shown in Ref. [13], the probability to nucleate new ML decreases by the factor $\exp(-\varepsilon)$ immediately after the formation of preceding ML. This effect leads to a sub-Poissonian Green's function, while Poisson distribution corresponds to the VLS growth without antibunching (at $\varepsilon = 0$) [14]. The resulting LD is now obtained numerically from

$$f(s,t) = B \int_0^t d\tau\, F(s,\tau) J(t-\tau), \qquad (4)$$

with $F(s,\tau)$ given by Eq. (3), $J(t)$ defined in Eq. (1) and B as the normalization constant. In this way, we are able to account for three effects: (i) random nucleation of Ga droplets and subsequently NWs, (ii) fluctuation-induced broadening of the LDs and (iii) possible effect of nucleation antibunching in individual NWs on the LD shapes. The maximum NW length L_{max} and the growth rate V in Eqs. (2) and (4) can be measured directly. Therefore, the shape and width of the NW LD in our model is controlled by the two parameters, the characteristic nucleation time Δt and the antibunching parameter ε which enters the $\rho(s)$, $z(t)$ and $D(t)$ values in Green's function given by Eq. (3). Clearly, the Δt is responsible for random nucleation of NWs from Ga droplets with the universal nucleation rate given by

Eq. (2), while the ε describes the narrowing effect induced by nucleation antibunching with respect to the Poissonian case with $\varepsilon = 0$.

Results and discussion

We now compare the experimental NW LDs for samples 1, 2 and 3 and their fits by our model expressions. For sample 1, which was obtained without SMP (as in the case of self-catalyzed GaAs NWs described in Ref. [2]), we observe very broad and asymmetric LD shown in Fig. 1 (a), similarly to Ref. [2]. The deterministic curve given by Eq. (2) yields the fit which is almost identical to that obtained from Eq. (4) that includes kinetic fluctuations. The characteristic nucleation time equals 35 s. The NW nucleation step is quite long, leading to a long left tail of the LD. We can thus conclude that the main source of the LD width in the regime of simultaneous Ga and As deposition is the initial incubation time required to form the Ga droplets in the oxide holes [1], while the follow-up fluctuation-induced Poissonian broadening is almost negligible. This is confirmed by the fact that the observed LD features a pronounced asymmetry toward the long left tail corresponding to the NWs that have emerged later, without any noticeable transition to the symmetric Poissonian shape. Standard deviation of this LD, $\sigma = \sqrt{D}$ (both σ and D are given for the dimensionless LDs, i.e., measured in MLs and squared MLs, respectively) equals 1376 MLs, which is 16 times larger than the Poissonian standard deviation $\sigma_p = \sqrt{\langle s \rangle} = 88$ MLs for the measured mean length of 7789 MLs.

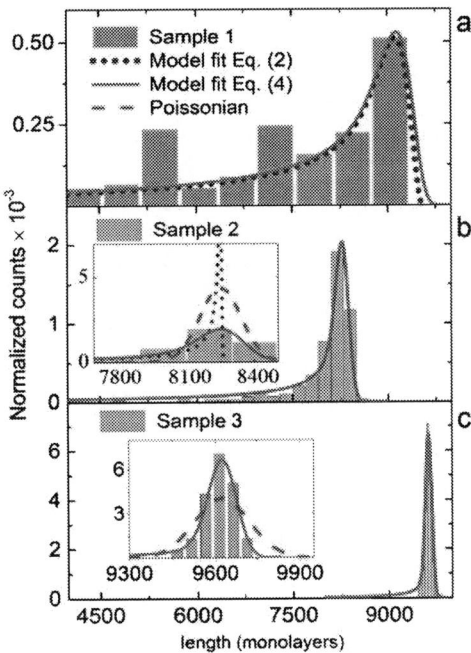

Figure 1. Measured LDs for samples 1 (a), 2 (b) and 3 (c) (histograms), fitted by Eqs. (2) and (4) in (a) and (b) and Eq. (4) in (c). The inserts in (b) and (c) show the zoomed views of the central parts of the histograms in comparison with the fits by Eqs. (2) and (4) and Poisson distribution for the same mean length (solid magenta lines). Standard deviations of the LDs are 1376 MLs in (a), 784 MLs in (b) and 60 MLs in (c). The LD shown in (c) is 1.6 times narrower than Poisson LD, whose standard deviation equals 98 MLs.

All three steps of the SMP were performed during the substrate preparation of sample 2. In this case, we observe more narrow LD shown in Fig. 1 (b), whose standard deviation $\sigma = 428$ MLs is more than two times smaller than in sample 1 for a similar mean length. However, this standard deviation remains much larger than the Poissonian value of 89 MLs for $\langle s \rangle = 7956$ MLs. The deterministic LD given by Eq. (2) yields a much narrower LD than observed experimentally. Therefore, in this case the width of the resulting LD should be due to a combined effect of the NW nucleation randomness and kinetic fluctuations, best fitted by Eq. (4) with $\Delta t = 1.3$ s. The extracted value of the antibunching parameter ε stays within the interval of the fitting error, showing the effective absence of nucleation antibunching in this case.

Surprisingly narrow sub-Poissonian LD was obtained in sample 3 after the optimization of the SMP. Figure 3 (c) clearly demonstrates that this LD is noticeably narrower than Poissonian, with $\sigma = 60$ MLs against $\sigma_p = 98$ MLs for the mean length $\langle s \rangle = 9612$ MLs. The model fit gives a very short nucleation time Δt on the order of 0.01 s, showing that all droplets emerge simultaneously after the optimized SMP and almost instantaneously give rise to NWs. The LD shape is almost symmetric for such a short nucleation delay. The best fit for Green's function gives the value of ε around 0.0002, corresponding to a weak nucleation antibunching. It leads, however, to saturation of standard deviation of Green's function at the value of $1/\sqrt{2\varepsilon} \cong 50$ MLs with respect to infinitely increasing width in the Poissonian case [15].

Table 1. Statistical parameters of different NW LDs

Sample	Growth temperature T (°C)	Mean length $\langle s \rangle$ (MLs)	Standard deviation σ (MLs)
Sample 1, this work	615	7789	1376
Sample 2, this work	615	7956	428
Sample 3, this work	615	9612	60
Sample 1 of Ref. [2]	604	13365	1006
Sample 2 of Ref. [2]	624	10172	2196
Sample 3 of Ref. [2]	643	7733	2248

Table 1 summarizes the statistical parameters of our NW LDs along with the corresponding data for the three samples of Ref. [2] that were grown at different temperatures by simultaneously depositing Ga and As after the annealing step, as in our sample 1. Figure 2 shows that standard deviations for self-catalyzed GaAs NWs grown without SMP follow the clear trend to narrow up at lower temperatures due to a shorter nucleation time [2]. However, all these LDs are very broad and asymmetric, with a typical shape shown in Fig. 1 (a) and well described by Eq. (2). Their $\sigma/\sqrt{\langle s \rangle}$ ratios, showing the dimensionless LD widths with respect to Poisson distribution, are on the order of ten or even more. Introducing the complete SMP of SiOx/Si(111) substrates before depositing GaAs improves the length homogeneity within the NW ensembles. The narrowing effect is rather small for sample 2 but large for sample 3, where the resulting LD is narrower than Poissonian.

In Ref. [6], markedly sub-Poissonian LDs were systematically obtained within a range of the growth times for Ga-catalyzed GaAs NWs grown on SiOx/Si(111) substrates via a lithography-free procedure. Sub-Poissonian LDs reported in Ref. [6] were attributed to a combination of relatively fast nucleation of NWs and nucleation antibunching. Additionally, the collective effect of shadowing of the arsenic flux by neighboring NWs in the directional MBE deposition technique was shown to broaden the LDs of long enough NWs. Here, we only report the LDs corresponding to a fixed moment of time (15 min). Their time evolution will be considered elsewhere. However, we strongly believe that sub-Poissonian narrowing described here for sample 3 has the same origin as in Ref. [6]. As regards the differences in SMP for samples 2 and 3 that lead to the striking improvement in the length homogeneity, the full picture requires a detailed study of the initial nucleation step from differently prepared Ga droplets. At the moment, we note that (i) the resulting droplet/NW density in sample 3 is 30 times smaller than in sample 2 and (ii) the mean NW length in sample 3 is 1.3 times larger than in sample 2. The nucleation of Ga droplets on unpatterned SiOx/Si(111) substrates is a complex process that includes the Ga precipitation into the thermally annealed oxide openings [2]. Lower surface density of available holes makes it easier to collect the Ga adatoms in the absence of competition for the gallium diffusion flux, which could be the first explanation of the faster nucleation in sample 3 than in sample 2. Longer NWs in sample 3 under very similar deposition conditions should be due to a smaller shadowing in the lower density array, which improves the length homogeneity according to Ref. [6]. Finally, the NWs of sample 3 are slightly thinner, and nucleation antibunching is known to be stronger for smaller diameter NWs [11,13,14].

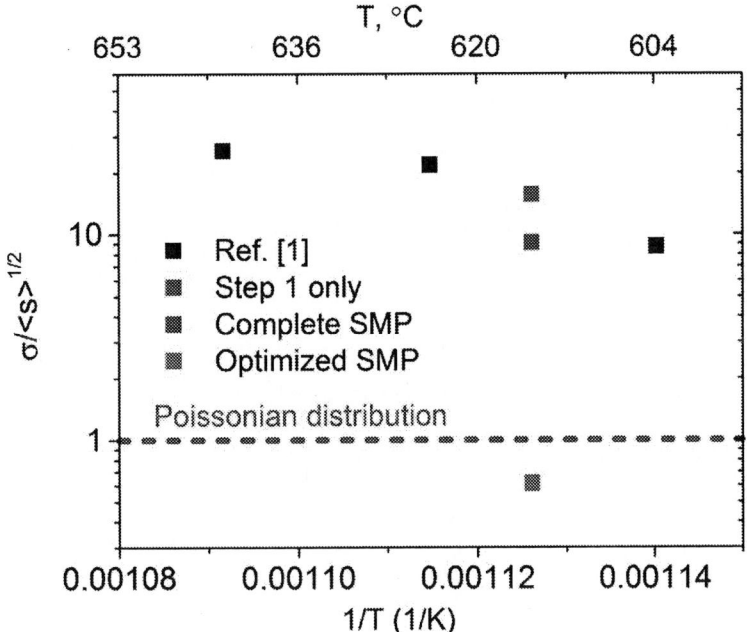

Figure 2. Standard deviations of the GaAs NW lengths of this work and Ref. [1], normalized to the Poissonian value $\sqrt{\langle s \rangle}$. Dashed line corresponds to Poisson distribution with $\sigma/\sqrt{\langle s \rangle} = 1$..

In conclusion, we have developed a model that quantitatively explains the differences in the shapes and widths of the LDs of self-catalyzed GaAs NWs grown on $SiO_x/Si(111)$ without pre-deposition of Ga and using different substrate preparation procedures. One of the samples shows sub-Poissonian narrowing of the LD. Therefore, the optimized SMP may be further used for obtaining highly-uniform ensembles of GaAs NWs and probably extended to other III-V NWs.

Acknowledgments

YB, VGD, NVS, AAK and ASS gratefully acknowledge the financial support received from the Ministry of Education and Science of the Russian Federation under grant 14.587.21.0040 (project ID RFMEFI5871X0040). TT and ED thank Joachim Wagner for the technical maintenance of the molecular beam epitaxy laboratory and gratefully acknowledge the support by the Structural Characterization Facilities at Ion Beam Center.

References
[1] Matteini F, Dubrovskii V G, Rüffer D, Tütüncüoğlu G, Fontana Y and Morral A F 2015 *Nanotechnology* **26** 105603

[2] Colombo C, Spirkoska D, Frimmer M, Abstreiter G, Fontcuberta i Morral A 2008 *Phys. Rev. B* **77** 155326

[3] Dubrovskii V G, Xu T, Díaz Álvarez A, Plissard S R, Caroff P, Glas F, Grandidier B 2015 *Nano Lett.* **15** 5580

[4] Cirlin G E, Bouravleuv A D, Soshnikov I P, Samsonenko Yu B, Dubrovskii V G, Arakcheeva E M, Tanklevskaya E M, Werner P 2010 *Nanoscale Res. Lett.* **5**, 360

[5] Tersoff J 2015 *Nano Lett.* **15** 6609

[6] Koivusalo E S, Hakkarainen T V, Guina M, Dubrovskii V G 2017 *Nano Lett.* **17** 5350

[7] Tauchnitz T, Nurmamytov T, Hubner R, Engler M, Facsko S, Schneider H, Helm M, Dimakis E 2017 *Cryst. Growth Des.* **17** 5276

[8] Dubrovskii V G 2009 *J. Chem. Phys.* **131** 164514

[9] Glas F, Ramdani M R, Patriarche G, Harmand J C 2013 *Phys. Rev. B* **88** 195304

[10] Dubrovskii V G, Sibirev N V 2017 *J. Phys. D: Appl. Phys.* **50** 254004

[11] Glas F, Dubrovskii V G 2017 *Phys. Rev. Materials* **1**, 036003

[12] Dubrovskii V G, Nazarenko M V 2010 *J. Chem. Phys.* **132**, 114507.

[13] Glas F, Harmand J C, Patriarche G 2010 *Phys. Rev. Lett.* **104** 135501

[14] Glas F 2014 *Phys. Rev. B* **90** 125406

[15] Dubrovskii V G 2013 *Phys. Rev. B* **87** 195426

Genetic algorithm application for solving X-ray diffraction inverse problem

D V Sivkov[1,2]**, V I Punegov**[1]

[1]IPM of Komi SC UrB RAS, Syktyvkar, 167982, Russia
[2]Immanuel Kant Baltic Federal University, Kaliningrad, 236041, Russia

Abstract. For solving inverse problem of scattering of X-ray with transversely restricted wavefront from the ideal crystal within the dynamical theory of X-ray diffraction taking into account the effect of the diffractometer's instrumental function the Genetic Algorithm in the form of Differential Evolution method was used. The calculations were performed.

Introduction

Today, the actual tasks are to increase quality control and improve the production of elements of nanoelectronic devices. The most effective method for analyzing the structure of this objects is X-ray diffraction. This method is sensitive to the fine details of the nanometre range. In addition, this method is non-destructive and rapid. The information on the structure of the object under study is obtained in the form of the scattering intensity distribution map in the reciprocal space, and to get the required parameters the inverse problem need to be solved. In addition to developing a theory describing the interaction of X-ray with matter, an effective algorithm for minimizing the residual functional is needed. The feature of this problem is in the large number of local minima in the parameter space and large number of parameters. Therefore, traditional iterative optimization methods are not well suited due to the considerable time required to find a solution. Evolutionary algorithms, including genetic ones [1], showed their high efficiency for searching solutions in the space of a large number of parameters in a wide range, particularly in problems of X-ray scattering [2, 3].

An experiment on X-ray scattering is simulated and an approach to the solution of the inverse problem is proposed.

Diffraction theory

In the X-ray scattering experiment, the total intensity obtained at the output of the detector is written as

$$I_{sim}(q_x, q_z) = K \cdot \overline{I_h}(q_x, q_z) + I_{bg}, \tag{1}$$

where I_{bg} – background intensity, K – intensity scaling parameter. The expression for the scattering intensity at the input of the detector, taking into account the influence of the instrumental function of the monochromator and the analyzer, has the form [1]

$$\overline{I_h}(q_x, q_z) = \frac{\int_{-\infty}^{+\infty} \int_{-\infty}^{+\infty} dq'_x dq'_z R^M(q'_x, q'_z) R^A(q'_x, q'_z) I_h(q_x - q'_x, q_z - q'_z)}{\int_{-\infty}^{+\infty} \int_{-\infty}^{+\infty} dq'_x dq'_z R^M(q'_x, q'_z) R^A(q'_x, q'_z)}, \tag{2}$$

Content from this work may be used under the terms of the Creative Commons Attribution 3.0 licence. Any further distribution of this work must maintain attribution to the author(s) and the title of the work, journal citation and DOI.

Published under licence by IOP Publishing Ltd

where R^M, R^A – monochromator and analyzer reflection coefficients, I_h – beam scattering intensity. It can be represented as two terms, reflecting the nature of the scattering, connected through the Debye-Waller factor f

$$I_h(q_x, q_z) = (1 - f^2)I_h^d(q_x, q_z) + f^2 I_h^c(q_x, q_z), \tag{3}$$

I_h^c– the coherent part, describing the ideal crystal scattering. It contains information on the thicknesses and periods of the layers in the crystal structure; I_h^d – the diffuse part. It contains information on the parameters of structural defects, such as quantum dots, pores, dislocations of different types (their size, shape, orientation, distribution in the sample, etc.).

The scattering intensity calculations were performed within the dynamical theory of X-ray diffraction expanded to the special case of transversely restricted wavefronts of the incident and reflected waves taking into account the effect of the diffractometer's instrumental function [4]. In that case the intensity of the diffracted X-ray wave (in the Bragg geometry) near the reciprocal lattice point **h** depends on the illuminated area of the crystal top surface $l_x^{(in)}$ and crystal thickness l_z as [4]

$$I_h^c(q_x, q_z) = \left| a_h \frac{\exp(i\bar{\xi}l_z) - 1}{\bar{Q}} \mathrm{sinc}\left(\frac{q_x l_x^{(in)}}{2}\right) \right|^2, \tag{4}$$

where $\quad \bar{\psi} = 2a_0 - q_x\cot\theta_B - q_z, \qquad \bar{\xi} = \sqrt{\bar{\psi}^2 - 4a_h a_{\bar{h}}}, \qquad \xi_{1,2} = (-\bar{\psi} \pm \bar{\xi})/2,$

$\bar{Q} = \xi_1\exp(i\bar{\xi}l_z) - \xi_2, \qquad a_0 = \pi\chi_0/(\lambda\sin\theta_B), \qquad a_{h,\bar{h}} = C\pi\chi_{h,\bar{h}}/(\lambda\sin\theta_B).$

Here λ – X-ray wavelength, χ_g – Fourier component of X-ray polarizability (determined by **h**), C – polarization factor.

To model the experiment, Poisson noise function in the form of

$$n(q_x, q_z) = \sqrt{\frac{I_R(q_x, q_z)}{6I_{bg}}} \ \mathrm{rand}[(-1,1)], \tag{5}$$

where $\mathrm{rand}[(-1,1)]$ – random value produced by pseudorandom number generator within the [-1,1] diapason, was also added to the expression (1) as a multiplier [3].

Residual functional
For the minimization procedure in the paper we considered the following types of residual functional

$$\rho_{abs}(q_x^i, q_z^i) = \frac{1}{N_q}\sum_{i=1}^{N_q}\left|I_{sim}(q_x^i, q_z^i; \mathbf{x}) - I_{exp}(q_x^i, q_z^i)\right|, \tag{6a}$$

$$\rho_{sqr}(q_x^i, q_z^i) = \frac{1}{N_q}\sum_{i=1}^{N_q}\frac{\left[I_{sim}(q_x^i, q_z^i; \mathbf{x}) - I_{exp}(q_x^i, q_z^i)\right]^2}{I_{sim}(q_x^i, q_z^i; \mathbf{x})}, \tag{6b}$$

$$\rho_{abs}^{log}(q_x^i, q_z^i) = \frac{1}{N_q}\sum_{i=1}^{N_q}\left|\log I_{sim}(q_x^i, q_z^i; \mathbf{x}) - \log I_{exp}(q_x^i, q_z^i)\right|, \tag{6c}$$

$$\rho_{sqr}^{log}(q_x^i, q_z^i) = \frac{1}{N_q}\sum_{i=1}^{N_q}\left[\log I_{sim}(q_x^i, q_z^i; \mathbf{x}) - \log I_{exp}(q_x^i, q_z^i)\right]^2, \tag{6d}$$

where $\mathbf{x}$ – vector of parameters in the parameter space, I_{sim} – diffracted intensity theoretical data, I_{exp} – experimental data.

The comparison between theoretical and experimental data in absolute value (6a) is effective for analysis at high values intensity, as well as for small-angle scattering in undeformed crystals. That is, in the cases when the data on the curve have the same order.

The root-mean-square deviation (6b) should be effective in cases similar to (6a). In addition, if it is used, the accuracy of the solution is slightly affected by the statistical noise and such approach is effective for analyzing highly noisy signals.

The logarithms comparison (6c) is effective for data analysis at low intensity values or when the peaks of the investigated curve differ by an order of magnitude, especially by several orders. This is true for large scattering angles, when the scattering intensity contains information on the smallest (of the order of the incident wavelength) structural features and the spectrum contains peaks of higher orders. It can also be used effectively for small-angle scattering, but when the diffuse scattering channel is considered, which also contains information on small deformations.

In addition, by analogy with (6b), the residual functional in the form of the root-mean-square deviation of the logarithms (6d) was added.

Differential Evolution algorithm

Differential Evolution (DE) algorithm is described in detail in [5]. For calculation DE-strategy in the form "rand/1/exp" was used. That means that population vector to be mutated was chosen randomly, one difference vector was used and trial vector was generated using exponential crossover.

For our task in obtaining mutant vectors, the absolute values of the corresponding expression were taken in order to exclude the search for a solution among the parameter vectors with negative parameters.

Results

In the work, the experiment of X-ray scattering on an ideal 100 μm thickness Si crystal near a lattice site (111) was simulated. The width of the slit before the sample (the wave's transverse width) $w = l_x^{(in)} \sin(\theta_B) = 100 \mu m$, the background intensity value $I_{bg} = 1$, intensity scaling parameter $K = 10^6$. The wavelength of the incident beam $\lambda = 1.54$ Å. The statistical noise was added (as in a real experiment) according to the formula (5). Calculations were performed using C++. Program was based on the algorithm template from the *Differential Evolution homepage* [6].

Structure parameters using for modeling of the scattering experiment, their search ranges and DE algorithm parameters using for the inverse problem solving and are shown in the table 1. DE algorithm parameters were chosen as the best for searching. DE algorithm parameters F and C were chosen according to the lowest goodness of fit (residual functional) ρ_{abs}^{log} [3]. In table 1 "$\mathbf{x}_0$ = random" means that target vector chosen randomly from the population.

Structure parameters	Search ranges	DE algorithm parameters
$L_z = 100$ μm	(10-200) μm	$F = 0.1$
$w = 100$ μm	(10-200) μm	$C = 0.4$
$I_{bg} = 1$	(0-10)	$\mathbf{x}_0$ = random
$K = 10^6$	$(0.1-10) \times 10^6$	$N = 40$

Table 1. Structure parameters, their search ranges and DE algorithm parameters, using for search.

The results of solving the inverse problem of X-ray diffraction within the framework of the proposed approach are presented in Figure 1. For every residual functional form (6a-6c) the inverse problem was solved. On figure 2 the dependences of the corresponding residual functional values on the number of DE algorithm cycles are presented. For curves having the same color, the same pseudo-random number generator seed was chosen. In table 2 the average values of the relative deviations from the unknown structural parameters specified in the numerical experiment are presented.

It is clearly seen that the best solution (most precise) was found for the ρ_{sqr} residual functional (figure 2b) given by expression (6b). With the residual functional in the form (6a) the less accurate result was obtained (figure 2a). It should be mentioned that the use of the residual functional ρ_{sqr}^{log} in the form (6d) allows solving the problem faster but with the worst accuracy. That makes it possible to use it for the initial narrowing of the parameter range and the subsequent search using the solution (6b). The same applies to the (6c), which is not far behind in speed (6d).

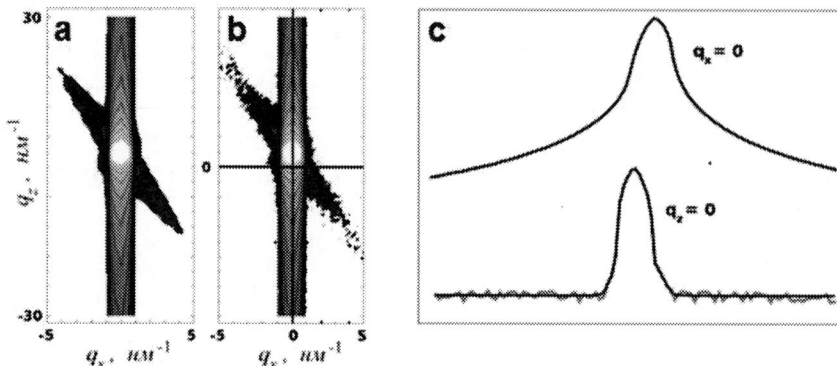

Figure 1. (a, b, c). (a) X-ray scattering intensity distribution map in reciprocal space - the result of the inverse problem solve; (b) X-ray scattering intensity distribution map in reciprocal space - experiment simulation; (c) scattering curves for $q_x = 0$ and $q_x = 0$: black - minimization of the residual functional, gray - simulation of the experiment with Poisson noise considering.

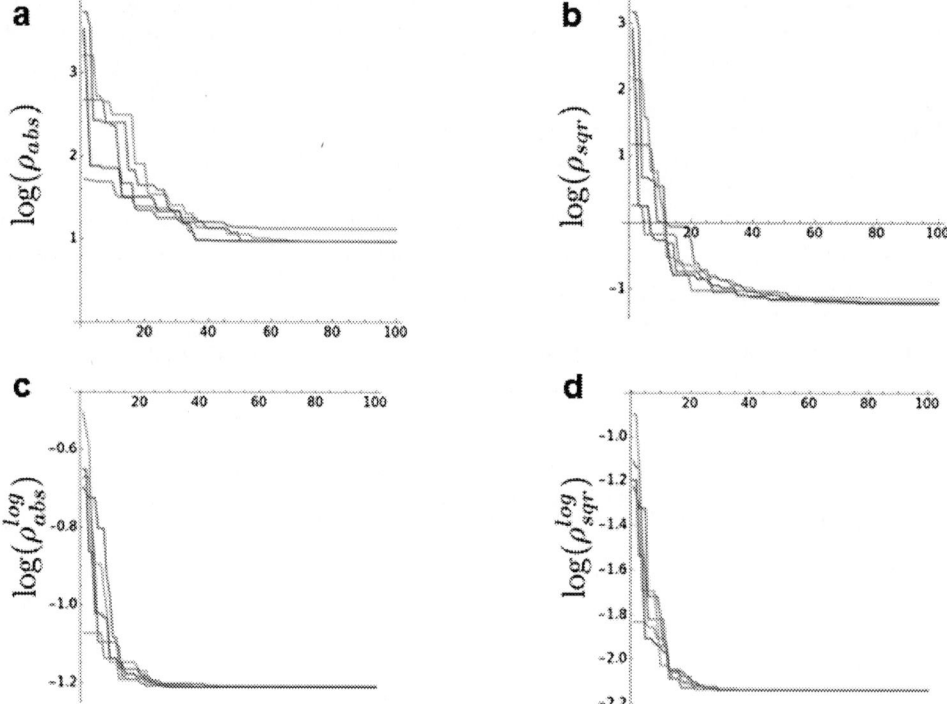

Figure 2. (a, b, c, d). Dependences of residual functional on the number of algorithm cycles in log scale. Residual functional in the form of (a) absolute value, (b) root-mean-square value, (c) absolute logarithmic value, (d) root-mean-square logarithmic value. Curves having the same color have the same seed for pseudo-random number generator.

	ρ_{abs}	ρ_{sqr}	ρ_{abs}^{log}	ρ_{sqr}^{log}
$\overline{\Delta L_z / L_z}$	0.1	0.01	0.07	0.008
$\overline{\Delta w / w}$	0.03	0.06	0.16	0.1
$\overline{\Delta I_{bg} / I_{bg}}$	1.6e-4	2.6e-4	9.2e-4	20.0e-4
$\overline{\Delta K / K}$	3.1e-2	3.0e-2	1.6e-2	5.3e-2

Table 2. The average values of the relative deviations obtained after solving diffraction inverse problem with using DE algorithm.

Conclusion

The approach for solving the inverse problem in the analysis of model X-ray scattering reciprocal space maps is developed. The proposed approach based on using DE algorithm for analysis the results of numerical simulation and experimental data of X-ray diffraction is useful and effective for nondestructive diagnostics of nanostructured media.

Acknowledgments

This research was supported by the Russian Academic Excellence Project at the Immanuel Kant Baltic Federal University, Program of UB RAS No 18-10-2-23, grant RFBR 17-02-00090, grant RFBR and Komi Republic 16-43-110350 p-a.

References

[1] Mitchell M 1996 *An introduction to genetic algorithms* (London: MIT Press) p 158

[2] Wormington M, Pannaccione C, Matney K M, Bowen D K 1999 *Phil. Trans. R. Soc. Lond. A.* **357** 2827

[3] Hannon A F, Sunday D F, Windover D, Kline J 2016 *J. Micro/Nanolith. MEMS MOEMS* **15(3)** 034001

[4] Punegov V I , Pavlov K M, Karpov A V and Faleev N N 2017 *J. Appl. Cryst.* **50** 1256

[5] Storn R, Price K 1997 *Journal of Global Optimization* **11** 341

[6] http://www1.icsi.berkeley.edu/~storn/code.html

Stress evolution in AlN layers grown on c-Al₂O₃ by plasma-assisted molecular beam epitaxy at metal-rich conditions

O A Koshelev, D V Nechaev, S V Ivanov, and V N Jmerik

Ioffe Institute, 26 Polytekhnicheskaya Str., St. Petersburg 194021, Russia

Abstract. Stress generation and relaxation in AlN nucleation and buffer layers grown on c-Al₂O₃ substrates by plasma-assisted molecular beam epitaxy at low (LT, T_S=780°C) and high (HT, T_S=850°C) substrate temperatures were studied. Oscillation behavior of the wafer curvature (stress×thickness) was revealed during pulsed deposition of AlN in metal modulated epitaxy and migration enhanced epitaxy modes under the metal-rich conditions, which corresponded to Al bilayer formation and consumption. Minimal average stress ~ +0.05 GPa in AlN buffer layers was observed at LT growth, which increased up to +0.85 GPa at HT growth. In addition, the LT AlN layers demonstrated lower densities of screw and edge threading dislocations equal to $7.8 \cdot 10^8$ and $5.1 \cdot 10^9$ cm⁻², respectively, while in the case of HT AlN layers those densities were significantly higher.

1. Introduction

To manufacture III-Nitride-based photonic and electronic devices the c-sapphire substrates are most commonly used. Such parameters as low cost, transparency, chemical and thermal stability make them most suitable for mass-production of III-N devices. However, the relatively high lattice mismatch of c-Al₂O₃ and AlN (~13%) can lead to an enormous threading dislocation (TD) density (~10^{10}cm⁻²) even in the several-microns-thick AlN buffer layers [1]. This worsens the output parameters of the devices and several techniques have been actively developing to overcome this problem. Elaboration of initial nucleation (NL) and buffer (BL) AlN layers on c-Al₂O₃ with typical thicknesses of several tens of nanometers and few microns, respectively, plays a key role in reducing the TDs in the device heterostructures.

The growth of AlN NLs on c-Al₂O₃ substrates has been studied in detail by several groups which employ high-temperature gas-phase technologies, such as metal-organic chemical vapor deposition (MOCVD). They described complex processes of generation and relaxation of internal stresses in these heterostructures, which relate to both crystal mismatch and coalescence of AlN grains during growth. For instance, Radhavan et al. have studied the temperature dependent evolution of both compressive and tensile stresses in the AlN/c-Al₂O₃ heterostructures during their growth at substrate temperatures T_S=600-1100°C [2]. They have found that coalescence of grains in NLs can lead to significant tensile stresses in the growing AlN layers, which may cause cracking of the AlN buffer layers at large enough thicknesses. An important role of grain boundaries in the origination of TDs was also c.

In contrast, mechanisms of stress evolution during low-temperature plasma-assisted molecular beam epitaxy (PA MBE) of AlN layers on c-Al₂O₃ has not been studied enough yet. Recently we have reported a decrease in the TD density in AlN NLs grown on c-Al₂O₃ by migration enhanced epitaxy (MEE) at a relatively high T_S~780°C as compared to a two-stage growth of AlN NLs, starting at low T_S~550°C [3]. In this paper, we elucidate the evolution of stresses in AlN NLs and following AlN BLs

Content from this work may be used under the terms of the Creative Commons Attribution 3.0 licence. Any further distribution of this work must maintain attribution to the author(s) and the title of the work, journal citation and DOI.

Published under licence by IOP Publishing Ltd

during their PA MBE deposition by using pulsed techniques of MEE and metal-modulated epitaxy (MME), respectively, at the metal-rich conditions and different T_S=780 and 850°C. The structural perfection of resulting hererostructures is also analyzed.

2. Experiment

AlN layers were grown by PA MBE (Riber Compact 21T) on c-Al$_2$O$_3$ substrates annealed at T_S=850°C and nitridized for 10 min at T_S=780°C. Figure 1 shows a schematic design of the studied heterostructures and the sequences of Al and plasma-activated nitrogen (N$_2$*) fluxes during MEE and MME growth modes at fixed T_S=780 or 850°C (noted as LT and HT, respectively), as has been reported in details by us previously [3,4]. To study the evolution of the AlN surface morphology during initial growth stage, the series of AlN-NLs with a thickness varied from 7 to 65nm was also grown at LT. The AlN BLs with a thickness of about 300 nm were grown atop by using a continuous (standard) growth mode at the same fluxes and T_S. The N$_2$* flux was kept constant at 0.45 ML/s, while the Al flux was varied from 0.5 to 0.64 ML/s for LT and HT AlN growth conditions, respectively.

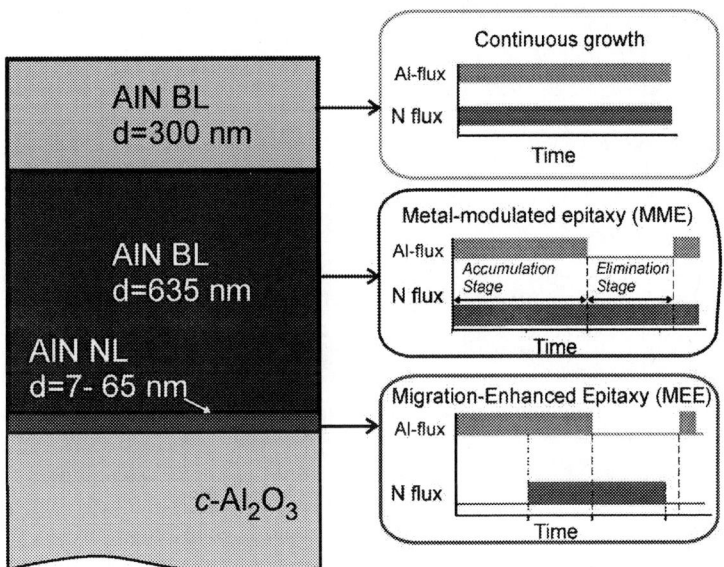

Figure 1. The design of grown structures and flux diagrams of three main growth modes.

A home-made multi-beam optical stress sensor (MOSS) based on a 10 mW laser emitting at a wavelength of 532 nm and a CCD-camera with recording frequency of 15 Hz was used to evaluate *in situ* the incremental stresses during growth of AlN NLs and BLs. The surface morphology of the AlN layers was monitored by the reflection high-energy electron diffraction (RHEED) and studied *ex situ* by atomic-force microscopy (AFM). The structural quality of the AlN layers was estimated by x-ray diffraction (XRD) analysis through measuring the full width at half maximum (FWHM) of ω-scans of symmetric AlN (0002) and skew-symmetric AlN (10-15) reflections.

3. Results and discussion

Prior to the growth of LT- and HT- AlN NLs, the RHEED images from the nitrided substrates exhibited a relatively blurred diffraction pattern, that remained unchanged during the initial growth of the NLs, but then it became brighter. Moreover, in the case LT-AlN NL the streaky or two-dimensional (2D) pattern shown in Figure 2a gradually appeared at a thickness of more than 10 nm. Previously, we studied the AlN NL grown by MEE at T_S~750°C and found that these films with a thickness of 65nm had a flat grain morphology [3]. Figure 2b illustrates the evolution of the morphology of such AlN NLs with thickness varied in the 7-65 nm range. Their AFM images show a monotonous decrease of the grain density and some increase in the surface morphology roughness up to a thickness of 40 nm, followed then by smoothing the layers. This is accompanied by the lateral 2D growth of the initial NL grains with a typical diameter expanding from several tens to hundreds of nanometers.

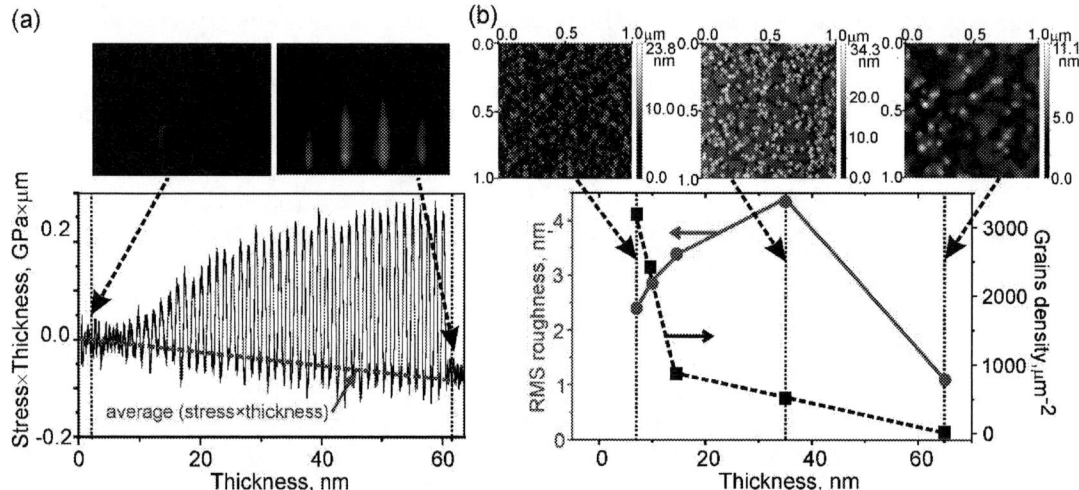

Figure 2. (a) Evolution of the stress×thickness vs thickness during the growth of 65-nm-thick LT-AlN NL by MEE. (b) The LT-AlN NL RMS roughness and grain density versus thickness. The dotted lines indicate the thicknesses of AlN NLs for which the RHEED (a) or AFM (b) images are related by arrows from the top of the figures.

Figure 2a demonstrates also the oscillatory change in the curvature (stress×thickness), with a period number corresponding to the number of MEE cycles. These oscillations begin at a layer thickness of ~10 nm and continue with the increasing amplitude up to a thickness of 40 nm, where the oscillation amplitude almost saturates. It should be noted that growth at the Al-stage of MEE results in higher values of the substrate curvature with a positive sign (higher tensile stress), while the curvature drops abruptly at the beginning of the N-stage of MEE. Despite the oscillations of the incremental curvature, the average curvature in the LT-AlN NL exhibits a negative slope (the red dashed line in Figure 2a) corresponding to a compressive stress ~-1 GPa.

In principle, there are two main sources for the stress generation in the AlN NL grown on c-Al_2O_3 substrate. In addition to the incomplete relaxation of the crystal mismatch between AlN and the substrate, leading to induce of a compressive stress, it should be taken into account the generation of the tensile stress due to grain coalescence in growing films in accordance with a model of Nix and Clemens [6]. Figure 2a indicates that the former mechanism dominates in our LT-AlN NLs despite the rather intense coalescence inside. In contrast, AlN/c-Al_2O_3 heterostructures grown by high-temperature MOCVD exhibit mainly the tensile stress related to the grain coalescence, as has been reported earlier [2].

The study of relation between metallic phase on the AlN surface and MOSS data was continued during the growth of AlN BLs by pulsed MME technique at different growth temperatures. Figures 3a and b shows oscillations of the substrate curvature for LT- and HT- AlN BLs, respectively. Both figures demonstrate the oscillations with the higher values of curvature during the stage of Al-excess accumulation on the surface of the AlN film growing by MME. This corresponds to the above observation in MEE of the introduction of incremental positive (tensile) stress after occurrence of metallic Al on the surface of growing AlN layer.

SPBOPEN 2018

IOP Publishing

IOP Conf. Series: Journal of Physics: Conf. Series **1124** (2018) 022041   doi:10.1088/1742-6596/1124/2/022041

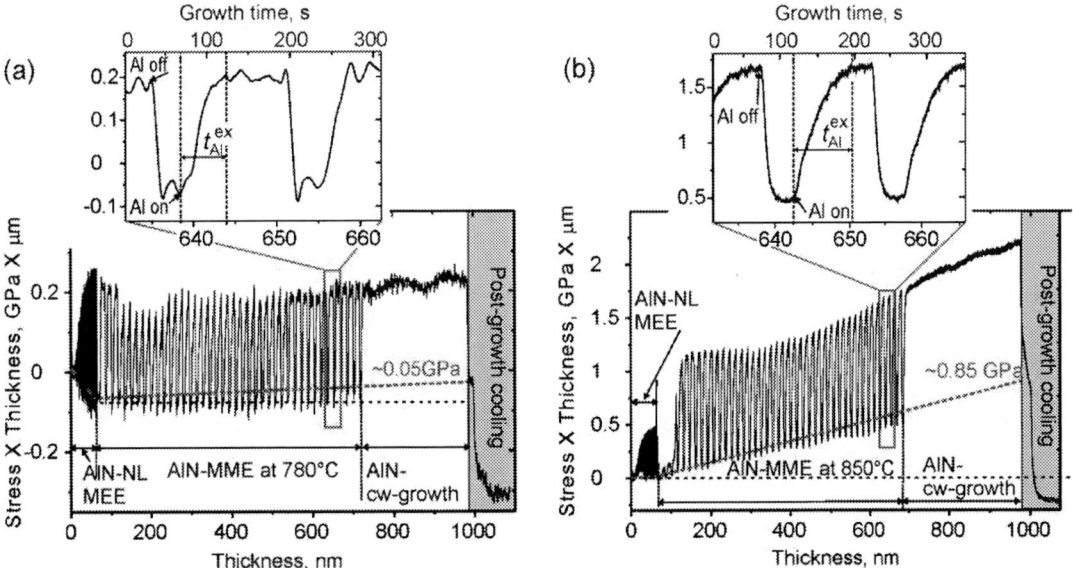

Figure 3. The stress×thickness versus film thickness during growth of AlN NL and BL on c-Al$_2$O$_3$ at LT (a) and HT (b) growth conditions. Insets show the magnified periods of MME of AlN BLs.

During the accumulation stage of the AlN BL growth by LT- and HT-MME the MOSS-data demonstrate a transition from initial positive (tensile) to zero incremental stress for a characteristic time t_{Al}^{ex}~55s and ~80s, respectively (see insets in Fig. 3a and b). Then, the films grow with a constant curvature till the beginning of the stage of Al consumption under the nitrogen flux exposure, when the curvature exhibits a compressive incremental stress for a few seconds during incorporation of the Al-excess MLs. Then, a stress-free growth proceeds until the start of a new MME period. The average stresses in that case are calculated based on the minimum stress×thickness values of MOSS-data during the Al-excess elimination stage. The nominal amount of Al-excess (D_{Al}^{ex}) accumulated on the surface of AlN films for the time t_{Al}^{ex} during the accumulation stage of MME at the Me-rich conditions can be estimated from the flux balance equation as

$$D_{Al}^{ex} = \left(F_{Al} - F_{N2*} - F_{Al}^{D}\right) \times t_{Al}^{ex},$$

where F_{Al} and F_{N2*} are the incident Al and N$_2$* fluxes (ML/s), F_{Al}^{D} is a desorbing Al flux which, at the Al-rich conditions used, corresponds to the equilibrium Al pressure over the Al melt [8]. Using the experimental data for the first two terms and reference data for the Al desorbing flux of 0.02 and 0.17ML/s at 780 and 850°C [9], respectively, gives the equal nominal Al-excess amounts at both temperatures as large as ~1.7 ML. This value corresponds to the general view on growth of wurtzite III-N films at the metal-rich conditions resulting in formation of group-III atoms bilayer on their surfaces [8,10]. Further growth of the III-N binaries occurs at this constant coverage of the surface by metal atoms, while the excess metal is accumulated in metallic droplets with gradually increasing dimensions during the accumulation stage and then either incorporated into the AlN BL or evaporated during the Al elimination stage.

It is worth noting that the generation of elastic stress in thin films by the extremely thin metallic adlayers has been reported in the past, e.g. Floro et al. [5] observed generation of compressive stress in GeSi films at the occurrence of Ge-segregated layers. At the moment, we cannot explain the reason of generation of the opposite (tensile) stress in our AlN layers and this issue will be studied elsewhere.

One should also note that LT-AlN BL shows the almost relaxed growth with a negligible average stress of ~0.05 GPa (Figure 3a), while in the case of HT-AlN BL the significant average tensile stress

195

of 0.85 GPa (Figure 3b) is generated. As a possible explanation of this difference, one can notice that Figure 3b demonstrates the slightly tensile average stress in the HT AlN-NL, which may be caused by a faster relaxation of crystal mismatch at higher T_S and, consequently, a greater contribution of coalescence in the stress generation process. The observation of the MOSS-data oscillations with an increasing amplitude and gradual transition from spotty to streaky RHEED pattern (not shown) during the initial stage of the HT-AlN BL growth can be also related to the coalescence continuing in this layer. In contrast, the LT-AlN NLs generate the average compressive stress (Fig. 3a), and the amplitude of the MOSS oscillations during LT-AlN BL does not increase.

Thus, growth of AlN NL and BL at a moderate temperatures $T_S \sim 780°C$ allows one to achieve the almost stress-free growth up to a thickness of 1 μm, that is a big challenge for AlN films grown by high-temperature MOCVD and PA MBE at the relatively high T_S of 850°C. Advantage of the moderate growth temperature in PA MBE is also confirmed by XRD data evaluating the screw and edge TD densities in LT(HT) AlN layers as $7.8 \cdot 10^8 (2.3 \cdot 10^9)$ and $5.1 \cdot 10^9 (3.2 \cdot 10^{10})$ cm^{-2}, respectively. The exact mechanisms of stress evolution in AlN films grown by PA MBE need in further studies.

4. Conclusion

In conclusion, the oscillation behavior of the wafer curvature (stress×thickness) data measured by MOSS has been found and analyzed during growth of both AlN NLs and BLs by pulsed MEE and MME modes in PA MBE at the metal-rich conditions and different substrate temperatures. It has been explained by introduction of the incremental tensile (compressive) stress during accumulation (consumption) of an excess Al-bilayer on the AlN surface. Study of the average stresses in ~ 1-μm-thick AlN films (NL & BL) grown at moderate and high temperatures revealed the significantly lower average tensile stress (~ 0.05 GPa) in the former. It was explained preliminarily by a greater role of grain coalescence in the films grown at higher temperatures. In addition, AlN films grown at LT on c-Al_2O_3 demonstrated much lower densities of screw and edge TDs equal to $7.8 \cdot 10^8$ and $5.1 \cdot 10^9$ cm^{-2}, respectively, in comparison with the HT AlN films.

Acknowledgments
The work was supported by Russian Science Foundation (#14-22-00107).

References
[1] Zhu D, Wallis D J, Humphreys C J 2013 J. Rep. Prog. Phys. **76** 106501.
[2] Raghavan S, Redwing J M 2004 J. Appl. Phys. **96** 2995.
[3] Nechaev D V, Aseev P A, Jmerik V N, Brunkov P N, Kuznetsova Y V, Sitnikova A A, Ratnikov V V, and Ivanov S V 2013 J. Crystal Growth **378** 319.
[4] Jmerik V N, Mizerov A M, Nechaev D V, Aseev P A, Sitnikova A A, Troshkov S I, Kop'ev P S, Ivanov S V 2012 J. Cryst. Growth **354** 188.
[5] Floro J A, Chason E, Lee S R, Twesten R D, Hwang R Q and Freund L B 1997 J. Electron. Mater. **26** 969.
[6] Romanov A E, Beltz G E, Cantu P, Wu F, Keller S, DenBaars S P, Speck J S 2006 Appl. Phys. Lett. **89** 161922.
[7] Nix W D and Clemens B M 1999 J. Mater. Res. **14** 3467.
[8] Koblmueller G, Averbeck R, Geelhaar L, Riechert H, Hosler W, Pongratz P 2003 J. Appl. Phys. **93** 9591.
[9] Alcock C B, Itkin V P, Horrigan M K, 1984 Canadian Metallurgical Quartely **23** 309.
[10] Heying B, Averbeck R, Chen L F, Haus E, Riechert H, Speck J S 2000 J. Appl. Phys. **88** 1855.

Fabrication method of the patterned mask for controllable growth of low-dimensional semiconductor nanostructures

L N Dvoretckaia[1], A M Mozharov[1,2], V V Fedorov[1], A D Bolshakov[1] and I S Mukhin[1,2]

[1] St. Petersburg Academic University, Khlopina 8/3, 194021 St. Petersburg, Russia
[2] ITMO University, 197101 St. Petersburg, Russia

Abstract. One of the important problems on the way to development of highly efficient optoelectronic devices such as silicon-based light-emitting diodes, hybrid solar cells as well as semiconductor lasers based on quantum dots is to obtain nanostructures with high crystalline perfection and small size dispersion. In this paper, we studied the method of thegrowth mask fabrication using photolithography through a close-packed array of microspherical lenses to obtain monodisperse nanostructures on a silicon growth substrate for the subsequent epitaxy of ordered GaP nanowires (NWs) array with a controllable morphology. Theoretical modeling of UV wave propagation demonstrated the possibility of the light focusing under a microspherical lenses array in the photoresist. With the method use the growth masks made of silicon oxide on Si (111) substrates were fabricated. Molecular-beam epitaxy was used to synthesize GaP NW ordered arrays.

1. Intoduction

Epitaxial growth of A3B5 materials on Si substrates is the problem of current interest. Solution of this problem will allow development of highly efficient and cheap LEDs, elements for solar energy and other optoelectronic devices [1]. Today the functional A3B5 structures for optonanoelectronics can be realized with the use of InGaN material system with a buffer layer on sapphire substrates. The buffer layer in this case reaches about 70% of the entire structure volume which leads to additional costs in the device manufacturing [2, 3]. The advantages of III-V materials grown on Si involve low cost and proven production and post-processing technologies of silicon substrates. However, there are several problems associated with a large lattice mismatch between most of III-V and Si, the polarity of III-V materials and the difference in their physical properties and crystal structures. These problems lead to poor electronic characteristics of the heterostructures while difference in lattice parameters and thermal expansion coefficients between the substrate and the layers leads to the mechanical damage of epitaxial layers [4]. The method of NWs epitaxial growth is a promising method for solving these problems since the area of the NW/ Si heterointerface is rather small while their surface area is large. Therefore NWs growth on Si substrates helps to reduce the number of defects and does not require the use of a buffer layer [5]. That is why these nanostructures are believed to be very promising in such applications as light-emitting diodes or optoelectronic devices.

The most common technique for NW arrays growth is the vapor-liquid-solid (VLS) method [6]. The main drawback of this method is the synthesized structures size dispersion. The dispersion is

---

Content from this work may be used under the terms of the Creative Commons Attribution 3.0 licence. Any further distribution of this work must maintain attribution to the author(s) and the title of the work, journal citation and DOI.
Published under licence by IOP Publishing Ltd

determined by the initial size distribution of the catalyst droplets formed from a thin metal layer deposited on the growth substrate prior to the NWs growth. Another drawback of the technique is the unwanted doping of the NW with atoms of the catalyst reducing electronic properties of the nanostructures. The approach allowing to solve these problems is fabrication of an ordered nanometer-sized holes pattern with a given morphology for further growth of the nanowire arrays or quantum dots. Fabrication of a large-scale growth mask technologically is a challenging task. To this end, most commonly used technique for fabrication of the patterned substrates involves use of expensive electron-beam lithography method. The time spent on one lithographic process in the latter case is proportional to the area of the sample which makes this method inapplicable for mass production. The photolithography over micro- / nano-spherical lenses allows fast fabrication of large close-packed arrays of nanostructures and their surface density variation [7]. In this paper we present the results on development of the micro-spherical lenses enhanced photolithography method, including deposition of a chemically active layer on a silicon wafer surface and selective etching of this layer. The results on the semiconductor material growth over the fabricated pattern for verification of the selective area growth capability of the method are presented.

2. Pre-processing

The photolithography through micro- and nanosphere lenses (SiO2, polystyrene (PS)) method improves resolution of the conventional photolithography which is limited by the diffraction limit and enhances fabrication capability compare to conventional methods [7, 8]. An ordered array of microspherical lenses applied to the resist surface focuses the radiation in the resist when exposed as a result sub-micron-sized elements are formed on the substrate. This approach provides incident wave focusing by microspherical lenses into a photoresist layer and it is highly efficient for obtaining ordered micro- / nanostructure arrays. In our work this method is used for fabrication of a growth mask for selective area epitaxial growth . The mask material deposition technique depends on the choice of the mask material and could be either vacuum thermal deposition or plasma-chemical deposition. After liquid or plasma chemical etching of the obtained structure on the silicon substrate surface we obtain nanoscale pits in the mask material.

GaP was chosen as a material for selective areagrowth via molecular beam epitaxy. Si (111) wafer was used as a substrate since GaP has a low lattice mismatch with Si. Epitaxially grown GaP NWs can be obtained in the wurtzite phase providing direct-band [9] and can be used as a active elements for LEDs. On the other hand, ordered cubic GaP NWs grown on Si can function as a wide-band window and an antireflection coating of solar cells [10, 11].

It is known that growth of A3B5 group materials is suppressed on silicon oxide [12, 13, 14]. Consequently, we choose SiO_x as the material for the mask fabrication. By the method of plasma chemical deposition (PECVD) 20 nm silica layer was deposited on the substrate at a temperature of 350°C. To create holes in the fabricated layer it is necessary to form a resistive mask. We selected optimal parameters for focusing of the light with a wavelength of 365 nm by microspheres into the photoresist layer and exposition of the photoresist.

The thickness of the photoresist was determined via numerical modeling and corresponded to the optimal propagation of the incident electromagnetic plane wave into the photoresist layer as shown in figure 1 (a, b). By changing the exposure parameters (for example, the exposure dose) we are able to change the diameter of the formed nano-holes obtained after the resist development. We also note that use of different diameters microspheres allows us to vary the period of the nanostructure array formed in the photoresist.

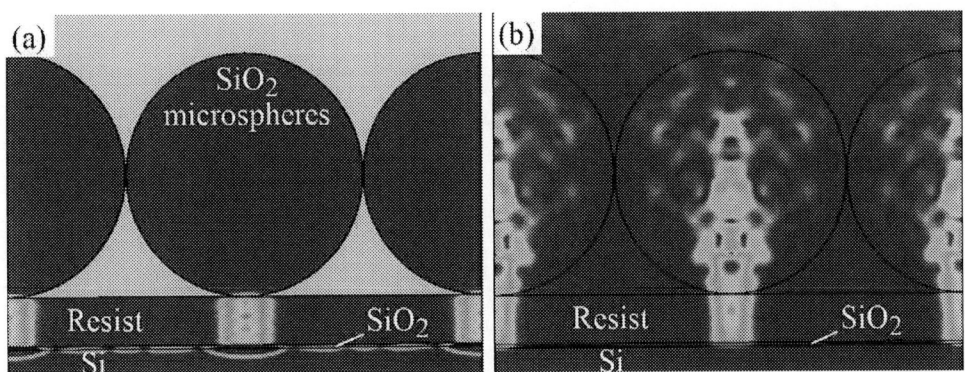

Figure 1 (a, b). The modeling results image of radiation absorption by photoresist using ordered array of SiO_2 microspheres deposited on photoresist coated Si substrate with SiO_2 layer (a), The modeling image of radiation power flow in photoresist using ordered array of SiO_2 microspheres (b).

Figure 2 (a) shows SEM image of an ordered array of SiO_2 microspherical lenses applied to the resist surface which were subsequently exposed under LED with wavelength of 365 nm. After the photolithography process etching of silicon oxide in SF_6 and removal of the residual resist the regular round holes of a fixed submicron size were formed in the SiO_2 layer as shown in figure 2 (b).

Further, the mask fabricated in SiO_2 was used as a growth pattern for epitaxy of GaP nanostructures. In molecular-beam epitaxy growth chamber degassing of the substrate occurred at 700°C measured with pyrometer and growth of GaP was carried out at a temperature of 620°C. In order to initialize the VLS mechanism 0.3 ML thick Ga layer was deposited providing formation of Ga droplets. The ratio of the equivalent partial pressures of the molecular beam (BEP) was 1/8 (GaP/P) measured with a thermal ionization vacuumeter placed on the beam path.

Figure 2 (c, d) shows SEM image of the synthesized GaP NWs. In the upper right-hand corner of figure 2 (d) the Fourier image of the obtained GaP NWs top view SEM image is shown. The hexagonal beams of Fourier image show that the nanostructures are arranged in a hexagon. It can be noted that at sprawl of GaP growth material over the mask takes place which can lead to closure of GaP nanostructures. This problem can be avoided via adjustment of the growth parameters. We also observed that the growth of GaP NW did not occur in every hole of the mask which certainly requires further optimization of the epitaxial growth parameters. Improvement of GaP NWs synthesis method will allow to fabricate ordered arrays of self-catalyzed nanostructures of a given morphology for realization of functional devices.

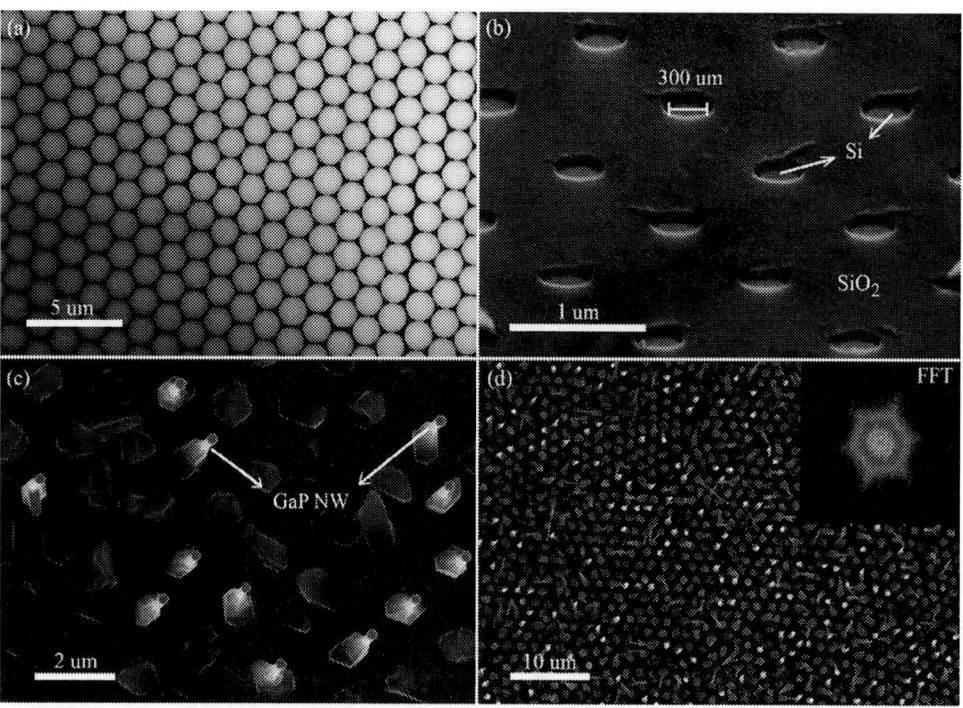

Figure 2 (a, b, c, d). Simulation SEM image of the ordered array of SiO$_2$ microspheres deposited over photoresist on Si substrate (a), nanostructures made of photoresist after development (b), ordered array of hollows in SiO$_2$ layer after lift-off (c), ordered array of epitaxial GaP nanowires and its Fourier image (FFT) showing the periodicity in nanostructure dislocations (rays of a hexahedron on inset) (d).

3. Conclusion

Combining the photolithography over micro- / nano-spherical lenses method with the technologies of plasma-chemical deposition from the gas phase materials and plasma-chemical etching enabled us to create arrays of ordered nanohollows in a SiO$_2$ layer on silicon substrates. Using numerical simulation the optimal thickness of the photoresist layer was obtained which corresponds to the focusing of the incident electromagnetic plane wave into the photoresist layer. An ordered array of nanostructures with a diameter of about 300 nm was obtained. Using molecular-beam epitaxy technique we synthesized ordered arrays of GaP NW on a silicon substrate.

Development of the substrate patterning method for synthesis of ordered semiconductor NWs with a given geometry is an actual problem due to high potential of their application as functional elements of nanophotonics and nanoelectronics.

Acknowledgements

This work was carried out with the support of the Russian Science Foundation (grant 18-72-00219), the Russian Federation President grant (MK-3632.2017.2), the Russian Foundation for Basic Research (16-32-60094mol_a_dk).

References

[1]   A Y Liu, J Peters, X Huang, D Jung, J Norman, M L Lee, et al. 2017 *Optics letters* **338-341** *42*(2)

[2]   H Li, M Khoury, B Bonef, A I Alhassan, et al. 2017 *ACS applied materials & interfaces* **9(41)**

36417-36422

[3]   C Y Shieh, Z Y Li, J Y Chang and G C Chi 2015 *Materials Chemistry and Physics* **157** 63-68

[4]   W Zhao, M Steidl, A Paszuk, S Brückner, A Dobrich 2017 *Applied Surface Science* **392** 1043-1048

[5]   A D Bolshakov, A M Mozharov et al 2018 *Beilstein Journal of Nanotechnology* **9(1)** 146-154

[6]   S Kodambaka, J Tersoff, M C Reuter and F M Ross 2006 *Physical review letters* **96(9)** 096105

[7]   Z Zhang, C Geng, Z Hao, T Wei and Q Yan 2016 *Advances in Colloid and Interface Science* **228** 105

[8]   L N Dvoretckaia, A M Mozharov and I S Mukhin 2017 *Journal of Physics: Conference Series* Vol **917** No **6** 062062

[9]   S Assali, D Kriegner, I Zardo, S Plissard et al 2014 *Nanoepitaxy: Materials and Devices VI* **9174** 917405

[10] A Mozharov, A Bolshakov, G Cirlin and I Mukhin 2015 *physica status solidi (RRL)-Rapid Research Letters* **9(9)** 507-510

[11] M Feifel, T Rachow, J Benick et al 2016 *IEEE Journal of Photovoltaics* **6(1)** 384-390

[12] K Ikejiri, T Sato, H Yoshida, K Hiruma et al 2008 *Nanotechnology* **19(26)** 265604

[13] S Hara, J Motohisa, J Noborisaka, J Takeda and T Fukui 2005 *CONFERENCE SERIES-INSTITUTE OF PHYSICS* **184** 393

[14] C Renard, N Cherkasin, A Jaffré, L Vincent et al 2013 *Applied Physics Letters* **102(19)** 191915

MBE growth of GaAs nanowires with modulated crystal structure

I V Ilkiv[1], K P Kotlyar[1], D A Kirilenko[2], S P Lebedev[2], A A Lebedev[2], P A Alekseev[2], A D Bouravleuv[1], G E Cirlin[1]

[1]St. Petersburg Academic University, St. Petersburg 194021, Russia
[2]Ioffe Physical Technical Institute RAS, St. Petersburg 194021, Russia

Abstract. In this study, we report on the Au-assisted molecular beam epitaxy of GaAs nanowires using graphene covered SiC substrates. The transmission electron microscopy characterization revealed the quasiperiodic superstructure of the nanowires.

1. Introduction

During recent years, semiconductor nanowires attract great attention due to their original optical and electronic properties. Nanometer size cross-section and high aspect ratio rise to many interesting physical properties which are not seen in bulk materials, such as the possibility to combine lattice-mismatch materials. Indeed, it is not only the possibility to grown the III-V nanowires directly on Si substrates that enables integration of optoelectronic devices with Si-based microelectronics [1], but also to make heterostructures without even the changing material. It is well known that most of III-V materials (with the exception of nitrides III-N) have a zincblende (ZB) crystal structure, whereas nanowires of the same material very often show features of polytypism, i.e. they typically exhibit random intermixing of the ZB and wurtzite (WZ) crystal structures [2]. The existence of uncontrolled intermixing has an impact on the nanowires properties and may pose challenges in future nanoelectronic devices due to electron scattering, low carrier lifetime and carrier mobility [3-5]. In turn, the ability to control the crystal structure during the nanowire growth opens up new opportunities for diverse applications in nanotechnology. For instance, an optical characterization of a crystal-phase quantum dot with demonstration of single photon emissions was reported by Akopian *et. al.* [6]. A periodic structure composed of either single twin planes or alternating WZ/ZB phases showed yield interesting superlattice effects [7-9].

In this letter, we report on the MBE growth of Au-seeded GaAs nanowires with modulated crystal structure on hybrid graphene/SiC substrates.

2. Experiment

The investigations were carried out using 6H-SiC (00001) wafers, having surface cowered with several graphene layers. In order to obtain graphene film, we used the method of the thermal decomposition of a SiC surface (detail procedure is described in ref. [10]).

The solid-source Compact 21 TM Riber MBE system equipped with additional high vacuum connected chamber for gold deposition, which also allows to perform the annealing of the samples at the temperatures up to 950°C, was used for the growth experiments. Prior to growth the samples were loaded into additional chamber, where they were degassed at temperature of 500°C. Subsequent gold deposition leading to the formation of Au catalyst particles, was performed. Then the temperature was

Content from this work may be used under the terms of the Creative Commons Attribution 3.0 licence. Any further distribution of this work must maintain attribution to the author(s) and the title of the work, journal citation and DOI.
Published under licence by IOP Publishing Ltd

decreased, the samples were transferred to the main chamber, which is equipped with standard Ga effusion cell and As cracking cell to provide As dimers. After reaching the growth temperature of 550°C, Ga and As shutters were simultaneously opened. The growth rate was 1 ML/s, the total growth time was equal to 15 minutes. Note, that the growth fluxes are calibrated as 2D equivalent growth rate units, realized by measurements of growth rate oscillations during reflection high energy electron diffraction (RHEED) experiments on GaAs (100) surfaces. Growth was terminated by switching off the Ga supply while maintaining the As supply until the substrate temperature dropped to below 300°C.

The morphology of the nanowires was investigated using Zeiss Supra 25 field-emission scanning electron microscope (SEM).

Structural characterization was performed using Jeol JEM-2100F transmission electron microscope (TEM) operated at 20 kV. Samples for TEM were obtained by depositing nanowires onto carbon film coated Cu grids, by gently rubbing the grid against the sample, in most cases breaking the nanowires off at the base.

3. Results

Typical array of GaAs nanowires formed under condition described above are shown in Figure 1. The result obtained demonstrate freestanding nanowires. Since the diameter of the seed particles can vary greatly in the method used, nanowires have a considerable variation in size. The Au catalyzed GaAs nanowires were found to have a high aspect ratio, with average length of about 4 μm and gradually tapered off cross-section size, which is decreasing from 110 to 10 nm. It is known [11], the carbon layer forming on SiC (0001) presents a highly corrugated surface that leads to a shorter diffusion length of the adatoms and restrains their mobility. This has been shown to lead also to metallic clusters of smaller diameter, and indeed Au catalyst particles with diameter of around 4 nm subsequently were found on the tips of nanowires. Thus, tapered form of nanowires could be related to diffusion limited growth mode [12]. Another salient feature that all nanowires were oriented randomly. Since nanowires typically grow perpendicularly to the substrate normal [13,14], this result suggests that here nanowires nucleate exclusively on the few-layer thick graphene regions created on facets of the SiC substrate [15].

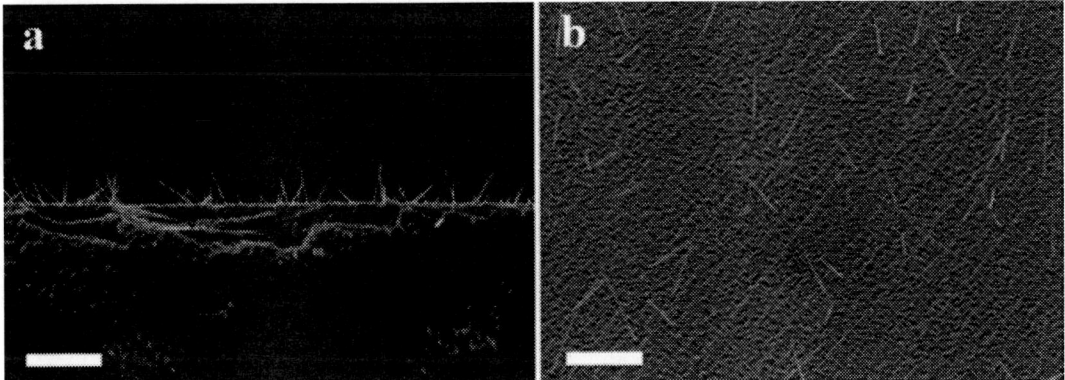

Figure 1. SEM images of GaAs nanowires: cross sectional view (a), 20 degrees tilted side view (b). Scale bars corresponds to 4 μm.

In order to locally identify the crystal formation, we perform TEM analyses. In Figure 2, we show an example of a structural characterization of GaAs nanowire. As can be seen nanowire has a very special superlattice crystal structure. Periodically appearing WZ segments (bright areas) typically exhibit a constant length of around 10 nm. It has been determined also that this structural variation is diameter-dependent, WZ phase occurring predominantly at smaller diameters, i.e. on the tip of nanowire. As a matter of fact, the numerous WZ/ZB segments stacking are probably a result of the low growth temperature of nanowires [16,17]. The diffraction pattern is a superposition of WZ and ZB diffraction patterns. Certain ZB diffraction spots appear as doublets/triplets, which corresponds to the rotation twins.

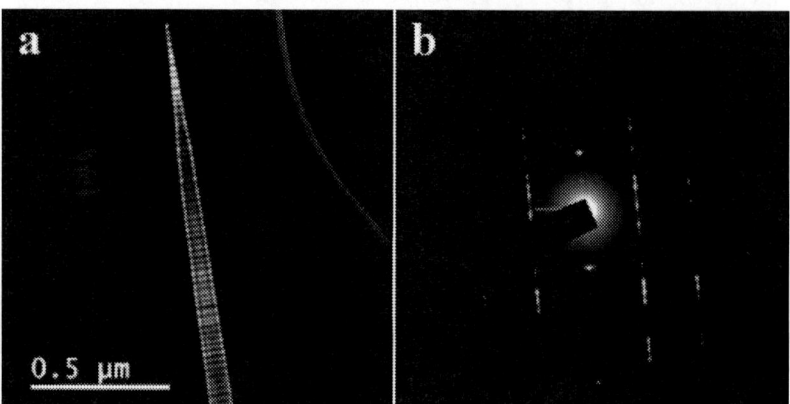

Figure 2.TEM dark-field micrograph image of GaAs nanowire (a), a global diffraction pattern (b).

4. Conclusion
In summary, the GaAs nanowire have been successfully grown directly on graphene. It was found that nanowires have superlattice ZB/WZ crystal structure. Thus, nanowires show indeed great promise for new electronic and optoelectronic devices based on the engineering of crystal phases.

Acknowledges
This work was carried out with the support of the Ministry of education and science of Russian Federation (state task, project No. 16.9791.2017/8.9)

References
[1]   Mårtensson T, Svensson P, Wacaser B, Larsson M, Seifert W, Deppert K, Gustafsson A, Wallenberg R, Samuelson L 2004 *Nano letters* **10** 1987-1990
[2]   Glas F, Harmand J C, Patriarche G 2007 *Phys. Rev. Lett.* **14** 146101
[3]   Woo R L, Xiao R, Kobayashi Y, Gao L, Goel N, Hudait M K, Mallouk T E, Hicks R F 2008 *Nano Lett.* **12** 4664-4669
[4]   Perera S, Fickenscher M A, Jackson H E, Smith L M, Yarrison-Rice J M, Joyce H J, Gao Q, Tan H H, Jagadish C, Zhang X, Zou J 2008 *Appl. Phys. Lett.* **5** 053110
[5]   Parkinson P, Joyce H J, Gao Q, Tan H H, Zhang X, Zou J, Jagadish C, Herz L M, Johnston M B 2009 *Nano Lett.* **9** 3349–3353
[6]   Akopian N, Patriarche G, Liu L, Harmand J C, Zwiller V 2010 *Nano Lett.* **4** 1198-1201
[7]   Dick K , Thelander C, Samuelson L, Caroff P 2010 *Nano letters* 9 3494-3499
[8]   Caroff P, Bolinsson J, Johansson J 2011 *IEEE* IEEE *J. Quantum Electron* **4** 829-846
[9]   Tsuzuki H, Cesar D, Rebello de Sousa Dias M, Castelano L, Lopez-Richard V, Rino J, Marques G 2011 *ACS nano* 7 5519-5525
[10]   Davydov V, Usachov D, Lebedev S, Smirnov A, Levitskii V, Eliseyev I, Lebedev A 2017 *Semicond.* **8** 1072-1080
[11]   Kang J, Ronen Y, Cohen Y, Convertino D, Rossi A, Coletti C, Heun S, Sobra L, Kacman P, Shtrikman H 2016 *Semicond. Sci. Technol.* **31** 115005
[12]   Rued1a-Fonseca P, Orrù M, Bellet-Amalric E, Robin E, Den Hertog M, Genuist Y, Andre R, Tatarenko S, Cibert J 2016 *Jour. Appl. Phys.* **16** 164303
[13]   Munshi A, Dheeraj D, Fauske V, Kim D, van Helvoort A, Fimland B, Weman H 2012 *Nano Lett.* **9** 4570-4576

[14] Tateno K, Takagi D, Zhang G, Gotoh H, Hibino H, Sogawa T 2012 *MRS Online Proc. Lib. Arch.* 45-50

[15] Fernández-Garrido S, Ramsteiner M, Gao G, Galves L, Sharma B, Corfdir P, Calabrese G, Schiave Z, Pfuller C, Trampert A, Lopes J N J, Brandt O, Geelhaar L 2017 *Nano Lett.* **9** 5213-5221

[16] Krogstrup P, Popovitz-Biro R, Johnson E, Madsen M , Nygård J, Shtrikman H 2010 *Nano Lett.* **11** 4475-4482

[17] Joyce H, Wong-Leung J, Gao Q, Tan H, Jagadish C 2010 *Nano Lett.* **3** 908-915

Nucleation of CdSe thin films: the kinetic model

A A Koryakin[1,3,4], S A Kukushkin[1-3] and A V Redkov[1,3]

[1]Institute of Problems of Mechanical Engineering RAS, St. Petersburg 199178, Russia
[2]Peter the Great St. Petersburg Polytechnic University, St. Petersburg 195251, Russia
[3]ITMO University, St. Petersburg 197101, Russia
[4]St. Petersburg Academic University, Khlopina 8/3, St.Petersburg 194021, Russia

Abstract. The growth of CdSe thin films by the method of thermal evaporation is studied. The growth model is developed and kinetic nucleation theory is applied to describe main dependencies of CdSe formation. Two mechanisms of growth are considered: 2D-growth (nucleation of disk-shaped nanoislands) and 3D-growth (nucleation of hemispherical nanoislands). The impact of growth conditions such as the substrate and evaporator temperatures and partial pressures on the nucleation mechanism is discussed. The conditions at which each mechanism prevails are estimated.

1. Introduction

In present time a lot of attention is paid to the growth technology of bulk semiconductor crystals and thin films. One of the major research areas are compound semiconductors of group A^{II}-B^{VI}, such as cadmium sulfide (CdS), cadmium telluride (CdTe), cadmium selenide (CdSe) and many others. The main interest in these materials is based on their optoelectronic properties and possible applications in solar cell industry, radiation detectors, photoconductors [1-3]. CdSe is one of prominent semiconductors due to its high photosensitive properties and it is widely used in production of thin film transistors, detectors of gamma rays and PEC cells [4]. It is usually having n-type of conductivity and direct bandgap of 1.74 eV, which is suitable for visible light conversion. Due to great importance of this material in industry, it is very topical to study principles and mechanisms of formation of CdSe crystals, thin films and quantum dots. There are many techniques to grow CdSe, such as vacuum evaporation, pyrolysis, hot wall deposition, cathodic electrodeposition and many others [5-7]. The vacuum evaporation is quite simple and doesn't require high material costs and thus will be considered in this paper. The other important question is the substrate, on which the film will be grown. In [6,8,9] authors have shown the advantages of use of SiC/Si structures, grown by the method of substitution of atoms [10], as substrates for growth of A^{II}-B^{VI} semiconductors. The main advantages are: accessibility of Si substrates, cheapness of formation of SiC/Si by the method of atom substitution, chemical stability of SiC film against precursors used in production of A^{II}-B^{VI} thin films, and a feature of method [10], namely the presence of pore system under the SiC film, which allows one to overcome stresses arising due to mismatch of crystalline lattices of CdSe, SiC and Si, and difference in their coefficients of thermal expansion.

Content from this work may be used under the terms of the Creative Commons Attribution 3.0 licence. Any further distribution of this work must maintain attribution to the author(s) and the title of the work, journal citation and DOI.
Published under licence by IOP Publishing Ltd

Despite the numerous accumulated experimental data, as far as we know, there is no quantitative theory of formation of CdSe films by the method of vacuum evaporation. This paper is devoted to the study of CdSe formation mechanisms using nucleation theory, which would allow estimating of various parameters of the resulting films depending on growth conditions: temperatures of the evaporator and the substrate. The paper is the continuation of series of articles on the growth of A^{II}-B^{VI} semiconductors [6,8,9] on SiC/Si substrates.

2. Growth of CdSe thin film via vacuum evaporation

The process of growth of CdSe films by the method of evaporation and condensation in vacuum [6] was extensively studied by Kalinkin et. al [7] and is very similar to the growth of CdTe, described in [8,9]. During the growth CdSe powder is being heated up to the temperature T_g in evaporation zone of the reactor. T_g must be high enough to ensure sublimation of CdSe and its dissociation into molecules of Cd and Se_2 due to chemical reaction:

$$2CdSe(s) = 2Cd(g) + Se_2(g)$$

In current consideration we neglect formation of single Se atoms in vapor phase, since this process is not very probable. After sublimation Cd and Se_2 are being transferred to the substrate zone, which is heated up to temperature T_s, which is lower than T_g. Since the pressure of vapor is equal to saturated one at temperature T_g, a flux of components towards the surface exceeds the flux of components evaporating from the surface of the substrate. This causes accumulation of the material at the surface of the substrate and therefore growth of the film. There are several mechanisms of crystal growth such as growth by Burton, Cabrera and Frank mechanisms (moving of individual steps and growth spirals [11]), nucleation and growth of 2D disk-shaped nanoislands of monoatomic height, and nucleation of 3D-nanoislands having a shape of hemisphere or faceted surface (see Figure 1). Since at typical growth conditions difference in T_g and T_s is quite big to ensure high growth rates of the film, we assume that the growth mechanisms are mainly 2D- and 3D nucleation. Growth via BCF mechanism is typically observed when supersaturation is small (<1-5%) and thus we will not consider it in current paper. It is quite important to understand which mechanism plays a dominant role: 2D or 3D-nucleation. During the growth in 3D-regime usually one can observe higher density of defects, grain boundaries, dislocations, and in some cases it allows growing different 3D-structures like quantum dots or nanowires. Nucleation in 2D-regime allows one to grow more homogeneous and smooth films, although at a slower growth rates.

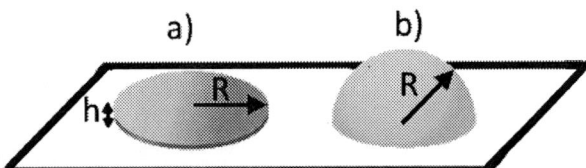

Figure 1. Nucleation of adatoms into disk-shaped particles in 2D-regime (a) and into hemispherical particles in 3D-regime (b).

Below we will estimate conditions at which 2D- and 3D-nucleation take place during the growth of CdSe film [6]. We consider three main parameters which may affect nucleation: the adsorption energies of Cd and Se atoms (in this paper we suppose that Se_2 molecules are being completely dissociated into atoms on the surface of the substrate and diffuse across it, however one should note that another mechanism may take place at elevated temperatures, namely re-evaporation of Se_2 molecules and its transfer through gas phase across the substrate towards the growing particles [12], this mechanism will be studied in next papers); diffusion coefficients and, as a consequence, the mobility of Cd and Se adatoms, and supersaturation, which is determined by difference in the fluxes to/from the surface (temperatures T_g and T_s).

3. Nucleation of CdSe film: model, parameters and results

To estimate which mechanism, 2D- or 3D-nucleation plays a key role one should calculate growth rate independently via both mechanisms at given conditions and compare them. In this section we describe the model, necessary parameters and calculation procedure.

We assume, that both types of adatoms (Cd and Se) diffuse independently and nucleate into nanoislands of CdSe according to:

$$Cd(ad) + Se(ad) = CdSe(s),$$

Nucleation rate I can be calculated using equation [9]:

$$I = N_0 W^+(n_c)\, Z \exp(-\Delta F(n_c)/kT) \qquad (2)$$

where $N_0 \sim 1/l_0^2$, n is the number of CdSe molecules in the nanoisland, n_c is the number of CdSe molecules in critical nuclei, Z is Zeldovich factor, $W^+(n_c)$ is the diffusion coefficient of the island in the space of sizes [9],

$$W^+(n_c) = - \left. \frac{kT\, dn/dt}{\partial \Delta F/\partial n} \right|_{n \to n_c}$$

$\Delta F(i)$ is the free energy of nuclei formation. For ΔF in case of 2D nucleation (disk-shaped nanoisland) one can write:

$$\Delta F(i)_{2D} = \alpha_{2D}\, n^{1/2} - \Delta\mu\, n,$$

and for the case of 3D nucleation (hemispherical nanoisland):

$$\Delta F(i)_{3D} = \alpha_{3D}\, n^{2/3} - \Delta\mu\, n,$$

where $\Delta\mu = kT \ln(C_{Cd} C_{Se}/K^{eq}) = kT \ln(\xi + 1)$, C_i are the concentrations of Cd and Se adatoms on the substrate surface; $K^{eq} = C_{Cd}^{eq} C_{Se}^{eq}$ is the equilibrium constant of the chemical reaction $Cd + Se \rightarrow$ CdSe, C_i^{eq} is the equilibrium concentration of the i-th element, $\xi = C_{Cd} C_{Te}/K^{eq} - 1$ is the supersaturation of 2D gas of adatoms, $\alpha_{2D} = 2\pi^{1/2}\gamma(h\Omega)^{1/2}$, $\alpha_{3D} = (2\pi)^{1/3}(3\Omega)^{2/3}\gamma$, γ is the effective surface energy of the island grains, Ω is the volume of the CdSe molecule, h is the height of the island (equals to height of monolayer in our case). Using the results presented in [9] and the same chain of transformations one can write equations for nucleation rates in 2D case and in 3D case.

$$I_{2D} = \frac{N_0^2 D_{CdSe}(\Delta\mu)^{3/2}}{\gamma(\hbar\Omega kT)^{1/2}} \exp\left(-\frac{\alpha_{2D}^2}{4kT\Delta\mu}\right), \tag{3}$$

$$I_{3D} = \frac{N_0^2 D_{CdSe}\Delta\mu^2}{2^{3/2}\Omega(kT\gamma^3)^{1/2}} \exp\left(-\frac{4\alpha_{3D}^3}{27kT\Delta\mu^2}\right), \tag{4}$$

where [12]:

$$D_{CdSe} = \frac{D_{Cd}D_{Se}C_{Cd}^{eq}C_{Se}^{eq}}{D_{Se}C_{Se}^{eq}\ln(\lambda_{Cd}/r_c) + D_{Cd}C_{Cd}^{eq}\ln(\lambda_{Se}/r_c)}, \tag{5}$$

λ_i is diffusion length of i-th adatom, r_c – is the critical radius of the nuclei.

In order to calculate growth rates numerically one have to determine firstly all abovementioned parameters, e.g. Cd and Se diffusion coefficients and lifetimes. These quantities can be determined the following expressions [9]:

$$\tau_i = \tau_{0i} \exp\left(\frac{E_{a,i}}{kT}\right) \tag{6}$$

$$D_i = \frac{l_i^2 \nu_i}{z} \exp\left(-\frac{E_{d,i}}{kT}\right) \tag{7}$$

where $E_{a,i}$, $E_{d,i}$ are the activation energies of adsorption and diffusion of the i-th element, respectively, τ_{0i} is inversely proportional to the frequency of vibrations of adatom ν_i which equals to $\sim 10^{13} s^{-1}$ [9], l_i is the diffusion jump distance, which is for simplicity equals to lattice parameter of CdSe in our calculations ($l_i = 0.608$ nm). z is the number of possible neighbor sites to which adatom can jump.

The activation energies of adsorption and diffusion of Cd and Se were chosen according to [13], $E_{a,Cd} = 0.3$ eV, $E_{a,Se} = 2.07$ eV, $E_{d,Cd} = 0.21$ eV, $E_{d,Se} = 0.17$ eV. Using this values and equations (1) and (2) one can calculate dependence of lifetimes, diffusion coefficients and diffusion lengths ($\lambda_i^2 = D_i\tau_i$) on temperature. The results are presented in Figure 2.

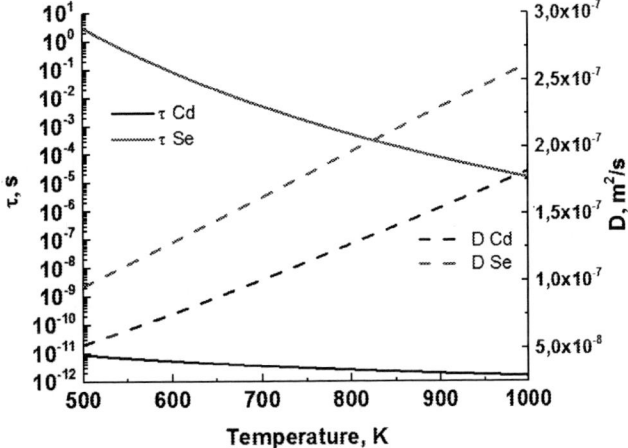

Figure 2. The dependence of Cd and Se lifetimes and diffusion coefficients on the substrate temperature.

The second step is to determine equilibrium concentrations of Cd and Se adatoms at the surface, fluxes of atoms from vapor phase J_{Cd} and J_{Se} to the substrate, and to find supersaturation ξ depending on temperature of the substrate T_s and the evaporator T_g. To do so we can use the data on the saturated vapor pressure of Cd and Se_2 over the surface of congruently sublimating crystal of CdSe [14], which can be described by the formula $\log P_{Cd}(atm) = -11138.7 \, / \, T + 6.91$. We also assume that $P_{Cd} = 2P_{Se2}$. Using the data, one can calculate fluxes of Cd and Se to the substrate: $J_{Cd} = P_{Cd}/(2\pi m_{Cd}kT)^{1/2}$ and $J_{Se} = 2P_{Se_2}/\left(2\pi m_{Se_2}kT\right)^{1/2}$, correspondently, where m_{Cd} and m_{Se_2} are the masses of Cd and Se_2, correspondently. After the flux densities are calculated, it is easy to estimate the adatom concentrations using the formula $C_i = n_i/N_0 = J_i\tau_i/N_0$, where n_i is the surface density of the i-th adatom [9]. Surface energy or CdSe γ equals to $0.246 \, J \, m^{-2}$ [13].

Thus we have a full set of data to perform calculation of nucleation rates via different mechanisms: 2D and 3D (equations 3,4). We performed calculation of I_{2D} and I_{3D} at different temperatures of the evaporator and the substrate. To determine, which mechanism plays the key role, we build the ratio of I_{2D}/I_{3D}, which is shown in Figure 3. Dashed white line denotes conditions at which $I_{2D} = I_{3D}$.

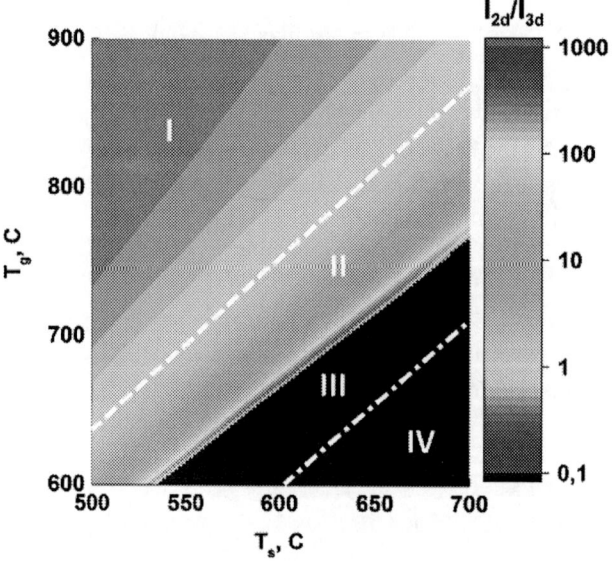

Figure 3. The dependence of the ratio I_{2D}/I_{3D} on the temperatures of the evaporator and the substrate. Dashed line marks region at which ratio I_{2D}/I_{3D} equals to 1. Dash-dotted line marks equality of temperatures T_g and T_s.

There are 4 regions in the Figure 3. The region denoted as I corresponds to growth mainly via 3D nucleation of hemispherical CdSe nanoislands ($I_{2D} \ll I_{3D}$). This regime ensures higher growth rates in more nonequilibrium conditions, which may lead to formation of various defects like boundary grains, dislocations etc. In addition, it is possible in this regime, that other three-dimensional structures, such as filamentary crystals or nanoislands of more complex shape will also be formed. The regime denoted as II corresponds to growth mainly via 2D nucleation of monolayer-height disk-shaped nanoislands of CdSe ($I_{2D} \gg I_{3D}$). This regime can be characterized as layer-by-layer growth in more equilibrium conditions, and in this regime smoother films with less number of defects will grow. The region denoted as III corresponds to extremely small nucleation rates via both mechanisms and it seems that growth should occur via different growth mechanisms (eg. Burton-Cabrera-Frank). The region denoted as IV corresponds to sublimation of the crystal, since substrate temperature T_s is higher than T_g.

4. Conclusions

The growth of CdSe via the method of evaporation and condensation in vacuum is considered. The model of growth is proposed based on nucleation theory of multicomponent systems. Estimates of the parameters of the model at different temperatures of the evaporator and the substrate are made (supersaturation, equilibrium concentrations, Cd and Se adatoms lifetimes, diffusion coefficients and lengths). Calculations are performed for the nucleation rate of CdSe nanoislands of different shapes (hemispherical and disk-shaped) and it is shown that there are two regimes at which nucleation of certain type of nanoislands prevails. The calculations presented open up new possibilities in controlling of growth rate and quality of CdSe thins films and nanostructures by the method of thermal evaporation and condensation in vacuum. The method proposed can be used also to study growth of other types of crystals and thin films.

Acknowledgements

This work was supported by the Russian Science Foundation (№ 14-12-01102).

References

[1] Bang J H, Kamat P V 2009. *ACS Nano* **3** 1467
[2] Roth M. 1989 *Nuclear Instruments and Methods in Physics Research Section A: Accelerators, Spectrometers, Detectors and Associated Equipment* **283** 291
[3] Bube R H 1955 *Proceedings of the IRE* **43** 1836
[4] Miller D J, Haneman D 1981 *Solar Energy Materials* **4** 223
[5] Meteleva Y V, Radychev N A, Novikov G F 2007 *Inorganic Materials* **43** 455
[6] Antipov V V, Kukushkin S A, Osipov A V, Rubets V P 2018 *Physics of the Solid State* **60** 504
[7] Kalinkin I P, Aleskovskii V B, Simashkevich A V, *Epitaxial Films of II–VI Compounds* (Leningrad State University, Leningrad, 1978) [in Russian].
[8] Antipov V V, Kukushkin S A, Osipov A V 2017 *Physics of the Solid State* **59** 399
[9] Koryakin A A, Kukushkin S A, Redkov A V 2017 *Materials Physics & Mechanics* **32** 262
[10] Kukushkin S A, Osipov A V 2013 *Journal of Applied Physics* **113** 024909
[11] Burton W K, Cabrera N, Frank F C 1951 *Phil. Trans. Roy. Soc.* **243** 299
[12] Kukushkin S A, Slyozov V V *Disperse Systems on the Surface of Solids (Evolution Approach): Mechanisms of Thin Films Formation* (Saint-Petersbrug, Nauka, 1996) [in Russian].
[13] Rempel J Y, Trout B L, Bawendi M G, Jensen K F 2005 *The Journal of Physical Chemistry B* **109** 19320
[14] Wösten W J 1961 *The Journal of Physical Chemistry* **65** 1949

Energy transfer in upconversion nanoparticles – phthalocyanine hybrid complexes

D A Gvozdev, E P Lukashev, V Z Paschenko

Department of Biophysics, Faculty of Biology, M.V. Lomonosov Moscow State University, 119992, Moscow, Russia

Abstract. In this work, we have studied the interaction between upconversion nanoparticles, and aluminum octacarboxyphthalocyanine in water solutions. It was shown that the self-assembled hybrid complexes are stable in water and NaCl solutions. The efficiency of nonradiative energy transfer from nanoparticles to the aluminum phthalocyanines increases with the number of phthalocyanine molecules in solution, but phthalocyanine sensitized fluorescence decreases due to phthalocyanine dimerization process. Also, singlet oxygen was generated by the phthalocyanine in the hybrid complex under infrared laser irradiation. The detected effects are of interest from the point of view of the directional search of components for a hybrid, highly efficient photosensitizers.

Introduction

Recently, lanthanide-doped upconversion nanoparticles (UCNPs) have attracted considerable attention in biological applications [1]. A feature of these particles is that they luminesce in the visible spectrum region when excited in the near-infrared (NIR) region. Utilization of NIR excitation light not only allows for deeper light penetration and reduced photodamage effects, but also offers lower autofluorescence, reduced light scattering, and phototoxicity. UCNP are promising light labels for the biodetection of various types of analytes and fluorescence bioimaging.

In addition, nanoparticles can also potentially serve as a donor of energy to other molecules, for example, photosensitizers (Ps). Photosensitizers used in photodynamic therapy absorbed light in the region of 600-750 nm, which is the lower boundary of the so-called "window of transmission" of the biological tissues. Next, Ps goes to triplet excited state and can produce reactive oxygen species when interacting with molecular oxygen. Upconversion nanoparticles have one of the luminescence band in the area of 660 nm. Hence, the excitation of UCNPs using an NIR light renders it possible to activate a photosensitizer through a Forster resonance energy transfer (FRET) mechanism, provided that the emission of UCNPs could well match with the absorption of the photosensitizer, resulting in the generation of $^1O_2$ under NIR irradiation without recourse to new fluorophores, that absorbing NIR light by themselves.

Previously, it has been demonstrated that energy transfer occurs between UCNP and different fluorophores [2, 3]. In this work, we studied the transfer of energy from the upconversion nanoparticle to the photosensitizer molecules, that forming a hybrid complex in an aqueous solution based on electrostatic interactions. We used $NaYF_4(Er/Tm,Yb)/NaYF_4$ core-shell structure UCNPs with an absorption maximum at 980 nm (Mesolight, USA), which luminescence spectra are shown in figure 1, a. The particles had a polymer PEG shell with amine groups, creating a positive charge on the surface of the particle. We used aluminum octacarboxyphthalocyanine (NIOPIK, Russia) as photosensitizer and acceptor of the energy (Pc(-8)). We choose Ps with a maximal charge for strongest electrostatic

Content from this work may be used under the terms of the Creative Commons Attribution 3.0 licence. Any further distribution of this work must maintain attribution to the author(s) and the title of the work, journal citation and DOI.

Published under licence by IOP Publishing Ltd

interaction; also, we prefer Al and not Zn due to the higher probability of ZnPc(-8) to aggregate in aqueous solutions.

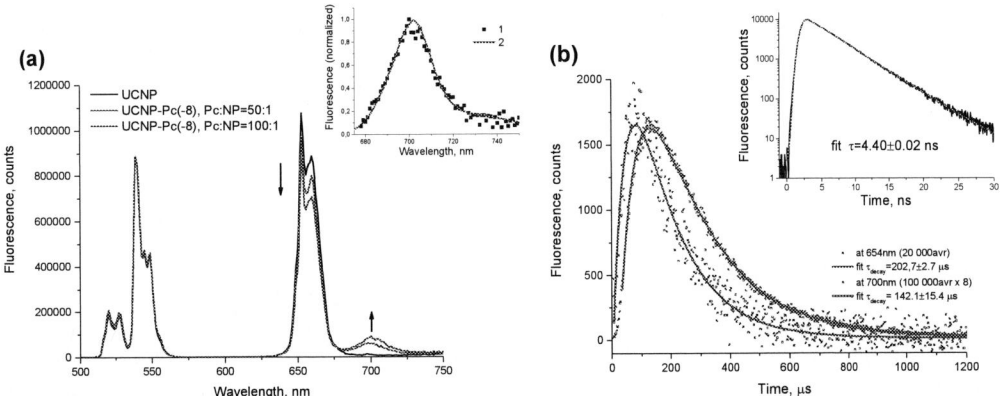

Figure 1(a, b). (a) UCNP luminescence spectrum in water (control) and with Pc(-8) of different concentrations. After addition of Pc(-8) in solution UCNP luminescence in red band (with a maximum in 654 nm) decreases, and sensitized Pc fluorescence near 700 nm appears. Excitation by 980 nm diode laser in continuous mode. Inset: Pc(-8) fluorescence spectrum as a result of energy transfer from UCNP - (1), under 980 nm laser irradiation) and (2), under 390 nm excitation of UCNP-Pc(-8) hybrid complex. **(b)** UCNP at 654 nm (red) and Pc(-8) at 700 nm (blue) luminescence decay curves under 980 nm diode laser excitation in pulse mode. Inset – Pc(-8) fluorescence decay curve under 390 nm excitation.

Experimental

Absorption spectra were recorded on USB2000 spectrometer with a DT-MINI-2-GS deuterium tungsten halogen light source (Ocean Optics, USA), in standard 10 mm quartz cuvette.

Fluorescence spectra and lifetimes were measured at room temperature using fluorolog-3 fluorimeter (Horiba-Jobin-Yvon, France) with kinetic console Fluoro-Hub (Horiba Scientific). For registration of fluorescence spectra led laser with a wavelength of 980 nm (1.6 W, 12 VDC, Laserlands, China), operating in stationary mode was used as a light source. The 200 µl sample was placed in a 5x5 mm quartz cell and was constantly stirred throughout the experiment. To increase the energy of the exciting light and the intensity of the irradiation short throw lens ware mounted between the laser and the cuvette with the sample, and the cuvette was placed in a special holder with two reflecting spherical mirrors, which allowed to increase sensitivity almost three times.

Measurement of the UCNP luminescence lifetimes was performed at a wavelength of 540 nm and 654 nm, and the Pc(-8) fluorescence lifetime at a wavelength of 700 nm. In this case, the laser radiation was modulated by an external TTL signal (5V) with a frequency of 50 Hz. Duration of laser n-shaped light pulse was determined by the time of the incoming TTL signal and we set 15 µs. The second TTL trigger pulse (5V, 1mks) was used to start the scan read of the emitted photons and it followed through 10 µs after the beginning of the laser flash. Detection of luminescence lifetimes was carried out by using DataStation v2.6 (Horiba Scientific) in mode Multi Channel Scaling (MCS). The time scale was set to 1 ms (1000 channels with step 1 µs) when UCNP emission at the wavelength of 540 nm had been recorded, and the accumulation was carried out for 10000 single pulses. In the case of luminescence lifetime registration at the wavelength of 655 nm (UCNP) or 700 nm (sensitized Pc(-8) fluorescence), the timeline was 2 ms (2000 channels in increments of 1 µs), and accumulation was carried out during 20000 and 100000 pulses, respectively.

To measure the Pc(-8) fluorescence lifetime on nanosecond time scale Fuoro-Hub was transferred to Time Correlated Single Photon Counting (TCSPC). In this case, the source of pulsed excitation light was NanoLED-390 (390 nm, ~ 1 ns, 11 pJ, Horiba Scientific), operating at a frequency of 1 MHz. Analysis of kinetic curves was carried out using the program DAC - 6, which included in the software of the fluorimeter.

For the detection of singlet oxygen generated by the photosensitizer in the complex of the hybrid complex with the nanoparticle, we used the method proposed in [4]. Briefly, imidazole was used as singlet oxygen scavenger, and we can detect decreasing of an optical density of p-nitrosodimethylaniline (RNO) upon its interaction with imidazole derivatives. As a control, a solution of Pc(-8), RNO and imidazole without UCNP was used. Irradiation was carried out by 980 nm laser in continuous mode. Concentrations of RNO and imidazole were the same as recommended in [4]. The experiment was performed in a thermostated cell Qpod 2e (Quantum Northwest, United States) at 30°C under magnetic stirring.

Results and discussion

It was shown that the hybrid complex Pc-UCNP self-assembled in aqueous solution probably due to electrostatic interactions. After adding phthalocyanine to the water solution of nanoparticles, we observe a quenching of UCNP luminescence intensity and enhance of Pc sensitized fluorescence with excitation by infrared laser with a wavelength of 980 nm (fig. 1, a). Simultaneously, the lifetime of the UCNP luminescence is also reduced by ~20% (from 84 to 64 microseconds). UCNP luminescence decay curve was monoexponential in a control without Pc(-8) and in hybrid complex, which indicates that a) UCNP sample is highly monodispersed and b) all nanoparticles in the solution containing Pc(-8) are in complex with phthalocyanine molecules. Also, one should note (fig. 2) that after interaction of UCNP with Pc(-8) only UCNP luminescence red band at 654 nm but not a green band at 540 nm was quenched as the result of energy transfer by Forster mechanism.

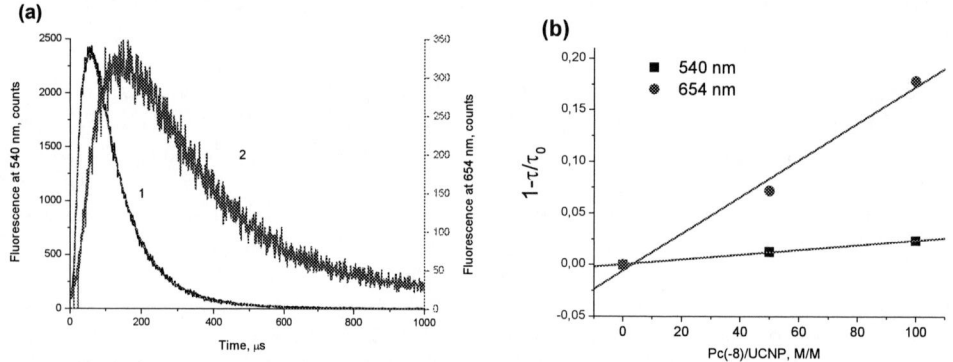

Figure 2(a, b). (a) UCNP luminescence decay curves in the maximum of green (1, at 540 nm) and red (2, at 654 nm) bands in water solution. **(b)** Changes of UCNP luminescence lifetimes τ in green and red bands upon addition of Pc(-8); τ_0 is in control without Pc(-8).

According to FRET theory, an efficiency of energy transfer from donor to acceptor molecule depends on integral overlap S between fluorescence spectrum of the donor and absorption spectrum of the acceptor:

$$S = \int_0^\infty F_d(\lambda)\varepsilon_a(\lambda)\lambda^4 d\lambda \qquad (1)$$

where $F_d(\lambda)$ is the normalized fluorescence spectrum of the donor, $\varepsilon_a(\lambda)$ denotes the absorption spectrum of the acceptor, λ is the light wavelength. The rate constant k of FRET is calculated by the relation

$$k = (R_0/R)^6/\tau_D \qquad (2)$$

where R – the distance between energy donor and acceptor molecules, and R_0 is Forster radius, which determined as

214

$$R_o = \sqrt[6]{8,8 \times 10^{-25}(\chi^2 n^{-4} \varphi_d S)} \tag{3}$$

where φ_d is the quantum yield of the donor in absence of the acceptor, n is the refractive index of the surrounding medium, χ^2 denotes the orientation factor between the transition dipole moments of the donor and acceptor (usually $\chi^2 = 2/3$ in solutions due to random orientation of molecules). The value of the Forster radius was calculated to be about 25 Å. This value is much smaller than the upconversion particle size (about 30 nm). Therefore, the energy transfer to the molecules of phthalocyanine involves primarily those emitting centers in UCNP that are closer to the surface of the particle. However, in the case of quenched emitting centers near the surface of the nanoparticle and unquenched emitting centers closer to the particle core, two time components would be observed in the kinetics of UCNP luminescence decay. Analysis of the kinetics of nanoparticles luminescence decay in the presence of Pc(-8) showed that the curve is monoexponential. Thus, the emitting centers in the core of the nanoparticle also participate in the interaction with the phthalocyanine, apparently transferring energy to the peripheral centers.

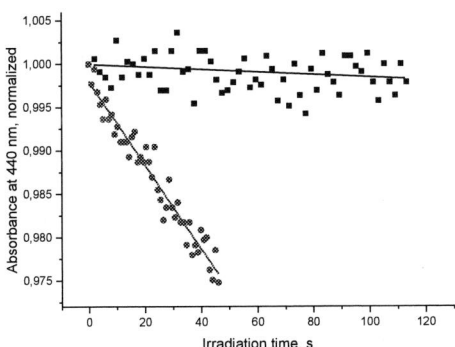

Figure 3. Time irradiation (980 nm, 1,5 W) dependence of RNO degradation in Pc(-8) solution (black points) and in presence of Pc(-8)-UCNP hybrid complexes (red points).

Of great importance is the number of Pc molecules per nanoparticle. It is known that phthalocyanine dimers are not fluoresced and cannot generate singlet oxygen, so formation of Pc dimmers is highly unfavorable for PDT application. However, after adding an excess of Pc(-8) to UCNP solution, we observed the dimerization of molecules of the phthalocyanine, which leads to the quenching of sensitized Pc fluorescence. Moreover, dimerization of the Pc leads to a change of the shape of the absorption spectrum of Pc monomeric form, which causes the quenching of the nanoparticles luminescence due to the increase of the overlap integral of the donor fluorescence spectrum and an absorption spectrum of the acceptor molecule. The detergent (LDAO or Na-cholate, 0.5%) prevents the formation of the Pc aggregates, however, the interaction between the Pc and UCNP in the complex is also impaired. Also, we found that presence of the NaCl up to 2 mM cannot affect interaction of Pc and UCNP, as well as a Pc dimerization process. Surprisingly, UCNPs tend to precipitate in PBS buffer solution, perhaps, due to phosphate anions impact.

Further, we performed an experiment to determine the rate of singlet oxygen generation by phthalocyanine in complex with an upconversion nanoparticle (Pc:UCNP was 100:1) when irradiated with NIR light. We have recorded RNO absorption spectra and normalized the optical density value to its value before sample irradiation. The results are presented in fig. 3. In the absence of UCNP, we did not observe RNO bleaching because the Pc(-8) does not absorb light in the NIR region. In the absence of imidazole, RNO bleaching does not occur, which means that RNO does not degrade in direct contact with the excited state of phthalocyanine. Also, RNO does not bleach in a system that contains all the necessary components but in the absence of light irradiation. Thus, the bleaching of the RNO occurs due to the transfer of energy of absorbed NIR photon from the UCNP to the Pc(-8) and the resulting generation of $^1O_2$.

Conclusions

We demonstrated the energy transfer from upconversion $NaYF_4(Er/Tm,Yb)/NaYF_4$ nanoparticle to aluminum octacarboxyphthalocyanine molecules in water solution, that leads to singlet oxygen generation by Pc in complex with UCNP under NIR irradiation. UCNP luminescence quenched and Pc sensitized fluorescence appears upon complexation, that reveals Forster mechanism of energy transfer. We estimate the distance between energy donors of UCNP and Pc(-8) molecules in 2.5 nm, that means close contact of components in the hybrid complex.

The main problem of such hybrid complex formation was to found the optimal ratio of [Pc]/[UCNP] because of phthalocyanine dimerization on the polymer shell of nanoparticles was occurred. The detergent (LDAO, Na-cholate) or buffer (PBS) molecules, which were used to prevent Pc aggregation, unfortunately, worsen the interaction between UCNP and Pc in the complex. We also found that the high ionic strength of the solution was not affected in any way by the degree of Pc aggregation in the presence of nanoparticles.

Finally, we can say that such photosensitizer-upconversion nanoparticle water-soluble conjugates as transducers of low energy light to toxic oxygen species may have several advantages for photodynamic therapy treatment of cancer cells.

Acknowledgments

The work was supported by the Russian Foundation for Basic Research (project №15-29-01167)

References

[1] Chen G, Qiu H, Prasad P N and Chen X 2014 *Chem. Rev.* **114** 5161–5214
[2] Zhu K, Liu G, Hu J, and Liu S 2017 *Biomacromolecules* **18 (8)** 2571–2582
[3] Bednarkiewicz A, Nyk M, Samoc M, and Strek W 2010 *J. Phys. Chem. C* **114** 17535–17541
[4] Kraljic I, Moshni S E 1978 *Photochemistry and Photobiology* **28** 577-581

Solid-state Modeling of Human Tracheobronchial Tree for 23 Generations of Airways

V V Makevnina[1]

[1]Saint-Petersburg State University, Saint-Petersburg 199034, Russia

Abstract. A three-dimensional solid-state model of the human tracheobronchial tree for 23 generations of airways is constructed in this work, using a specialized software package Solid Works. It is shown that with a significant increase in computational resources, the model obtained can be applied for computational experiments.

1. Introduction

To study the aspects of flow and transport of particles in lungs, and to increase effectiveness of medicinal aerosol preparations, it is necessary to understand the mechanisms of the air flow behavior and the transfer of aerosols in the airways. To determine the spatial geometry of the air flow region, and to study the displacement of air particles or aerosols in the lung, three-dimensional solid-state models of human tracheobronchial tree (TBT) are used. They provide three-dimensional visualization and can serve as a basis for modeling in computational fluid dynamics. Human TBT is a complex asymmetric structure of gradually branching airways. The trachea is divided into two main bronchi entering the right and left lungs (Figure 1), which, when entering the lung, are called "stem" (first order). Further, in the lungs, the main bronchi continue to divide and branch into smaller bronchi, and the latter divide into bronchioles. As branching from the trachea to the periphery, the diameter and length of subsequent generations of airways decrease, and the total area of their cross section increases in the same direction due to an increase in the number of the latter in more distant generations [1, 2, 3, 4, 5].

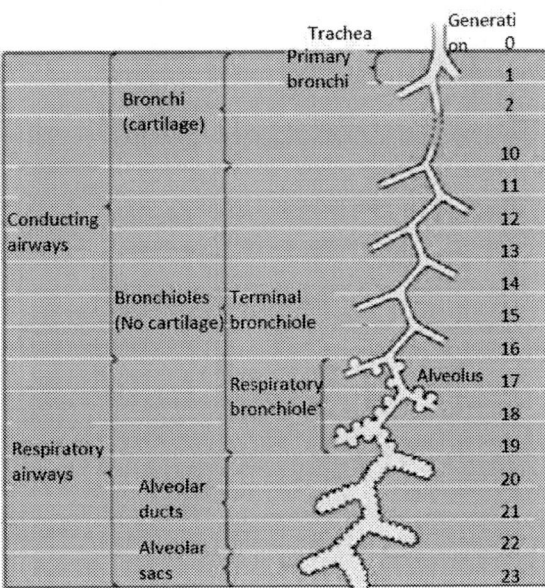

Figure 1. Branching generation of the airways [1]

Content from this work may be used under the terms of the Creative Commons Attribution 3.0 licence. Any further distribution of this work must maintain attribution to the author(s) and the title of the work, journal citation and DOI.

Published under licence by IOP Publishing Ltd

According to the TBT morphometry proposed by Weibel [2], there are 23 generations of airways, along the airway of the human lungs, the first 16 of which are known as conducting and refer to an anatomical dead space. Bronchioles of the 17th and 19th generation belong to the transition zone, in which there is a slight gas exchange. Bronchioles of the 20-23 generations belong to the respiratory zone and form the parenchyma of the lungs [2, 3]. In the 23rd generation there are 8388608 pairs of airways [2], so even the construction of a solid-state model, not to mention numerical calculations, requires huge computing resources that far exceed the capabilities available on powerful personal computers. Therefore, at the moment only simplified partial models of the TBT are used and there are no complete TBT models for all 23 generations.

The purpose of this paper was to investigate the possibility of constructing a solid-state model of human TBT for all 23 generations of airways. To achieve that the task was set: to try to build the model to the extent that it is possible.

2. Methods

Currently, the intensively developing direction of studying the biomechanics of the respiratory system is computer simulation. Thanks to modern integrated systems of geometric modeling and analysis, such as SolidWorks and ANSYS, you can create not only virtual images of models, but also explore them with the help of modern tools of engineering computer analysis. In this paper, construction of a three-dimensional model of TBT was carried out in the automated design system Solid Works, intended for solid-state parametric modeling on personal computers. Application of the specialized software package ANSYS is also allowable, but on the computer that was available to the author the system did not allow to save the model in the same (as much as possible) volume as in the SolidWorks software package, so for the final version of the simulation was chosen the latter package.

3. Results

According to Weibel's morphometric model [2], TBT can be represented as a symmetrical system of branching tubes with regular dichotomy and fixed length with respect to the diameter of the airways (Figure 2). The model proposed by him was developed on the basis of data from a study of prepared healthy lungs and bronchograms of living people, analyzed using statistical methods [2, 3].

For ease modeling, the following assumptions were made [2]:
• lungs are a set of generations of airways (channels) located parallel to each other. Each "parent" airway gives rise to two "child" branches;
• airways branch out according to the type of regular dichotomy, that is, two conjugated elements have the same dimensions and branch off from their "parent" at equal angles;
• the pressure in the respiratory structure is evenly distributed along the length of each individual air unit;
• airway elements are in the form of a hollow cylinder, two ends of which are cut to the wedge shape, in order to allow stitching with other elements during modeling. At the proximal end, it is conjugated to the "parent" airway, and at the peripheral end to the "daughter" branches.

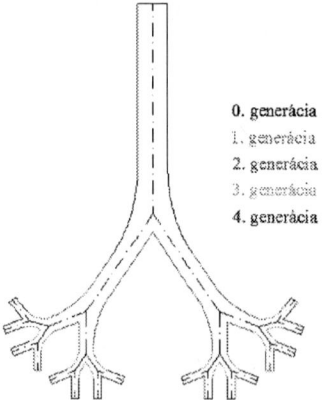

Figure 2. The symmetrical model of Weibel's TBT for the first four generations [2, 3]

The maximum possible solid-state model of human TBT for 23 generations of airways (the trachea is taken as zero generation) with the regular dichotomy (Figure 3) was constructed using a specialized software package SolidWorks. To simulate the geometric image, we used the quantitative characteristics given in Weibel's paper [2] (Table 1). The angle of bifurcation is the angle at which the "daughter" airways, formed by their axes with the "parent" axis, diverge - in this work was assumed equal to 65 ° [4].

Table 1. Dimensions of human airways with regular dichotomy [2]

Generation z	Number per generation n(z)	Diameter d(z) cm	Length l(z) cm	Total crossection S(z) cm^3	Total volume V(z) cm^3
0 (трахея)	1	1.8	12.0	2.54	30.50
1	2	1.22	4.76	2.33	11.25
2	4	0.83	1.90	2.13	3.97
3	8	0.56	0.76	2.00	1.52
4	16	0.45	1.27	2.48	3.46
5	32	0.35	1.07	3.11	3.30
6	64	0.28	0.90	3.96	3.53
7	128	0.23	0.76	5.10	3.85
8	256	0.186	0.64	6.95	4.45
9	512	0.154	0.54	9.56	5.17
10	1024	0.130	0.46	13.4	6.21
11	2048	0.109	0.39	19.6	7.56
12	4096	0.095	0.33	28.8	9.82
13	8192	0.082	0.27	44.5	12.45
14	16384	0.074	0.23	69.4	16.40
15	32768	0.066	0.20	113.0	21.70
16	65536	0.060	0.165	180.0	29.70
17	131072	0.054	0.141	300.0	41.80
18	262144	0.050	0.117	534.0	61.10
19	524288	0.047	0.099	944.0	93.20
20	1048576	0.045	0.083	1600.0	139.50
21	2097152	0.043	0.070	3220.0	224.30
22	4194304	0.041	0.059	5880.0	350.00
23	8388608	0.041	0.050	11800.0	591.00

Figure 3a shows the solid-state model of human TBT with regular dichotomy. First, a model was constructed for 10 generations of airways. Then one of the ends of the sixth generation was selected from each side (right and left). For these ends, starting from the seventh, two branches on each side (right and left), construction was completed in full up to the 23rd generation of the airway (Figure 3, b, c, d show enlarged fragments).

According to Weibel [2], the diameter of the lumen of the airway gradually decreases from 12.2 mm (main bronchus diameter) to 0.41 mm (diameter corresponding to the elements of the 23rd generation of the airways); the length decreases from 47.6 mm to 0.50 mm, respectively. For the 10th generation, there are 1024 pairs of airways, for the 23rd - 8388608 pairs, respectively [2]. In the constructed model, 263136 pairs of generations were obtained in aggregate.

It should be noted that this is the maximum that was possible to obtain. The model was built on a fairly powerful computer with good characteristics, which has, in particular, 16 [actually working] processors and 32 GB of RAM. An ordinary computer, which has, for example, 8 GB of RAM and 4 processors, opens the model within 24 hours. However, after the opening, it is no possible to perform any manipulation. So, for example, if you try to increase, decrease, rotate the model (or try to use any other functions of the program), computer hangs for about six hours. Therefore, to build a solid-state model of human TBT for all 23 generations of airways, an even more powerful computer is required, and the author unfortunately does not have it. Despite the difficulties, the author nevertheless succeeded in demonstrating that in the case of solving a technical problem, the simulation of human TBT for all 23 generations of airways has already become possible. And, as computers become more powerful, the use of numerical methods for flow modeling in all 23 generations of airways also becomes possible.

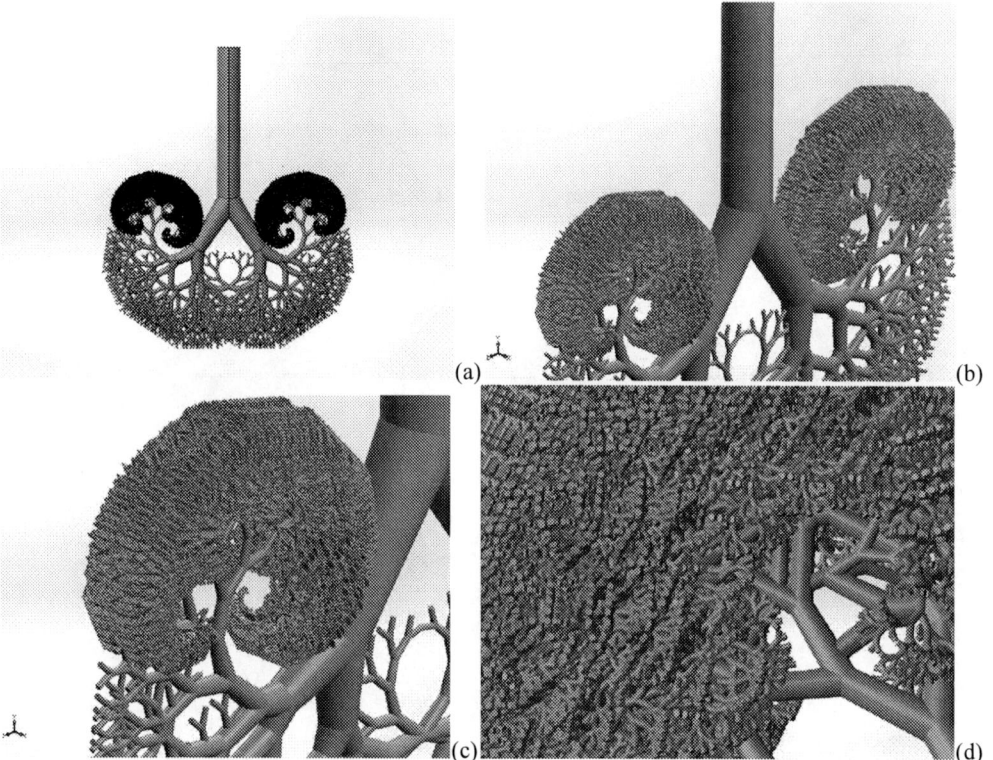

Figure 3 (a, b, c, d). Solid-state model of TBT: general view (a) and enlarged fragments of the lower elements of the airways (b, c, d)

4. Conclusions
Thus, the paper shows that with significant increase in computational resources, the construction of a solid-state model of human TBT for all 23 generations of airways is possible. In the presence of a more powerful computer, the solid-state model obtained opens up the potential for using in computational experiments.

Acknowledgements
The work is dedicated to the 50th anniversary of the chief anesthesiologist-resuscitator of the Ministry of Defense of the Russian Federation, the colonel of the medical service, Professor A.V. Shchegolev, to whom the author expresses his enormous gratitude.

References
[1] Gihad Ibrahim. CFD models of the bronchial airways with dynamic boundaries. Department of Engineering University of Leicester, Leicester, England, 2014, p. 116
[2] Ewald R.Weibel, M.D. Morphometry of the Human Lung. New York, Academic Press Inc., 1963, p. 151 with 109 figures.
[3] Michaela Chovancova and Jakub Elcner. The pressure gradient in the human respiratory tract. EPJ Web of Conferences, EFM13, 2014, p. 02047-1 – 02047-6
[4] Pastukhov A.D. X-ray Anatomy of the Trachea and Main Bronchi in Normal and with Chest Deformities. Dissertation. FSBEI of Higher Education. «Academician E.A. Wagner PSMU» of the Ministry of Health of the Russian Federation. Perm, 2017, 205 p.
[5] Lok-Tin Choi. Simulation of Fluid Dynamics and Particle Transport in Realistic Human Airways. Dissertation. RMIT University. Bundoora, Victoria, Australia, 2007, p. 91

Peculiarities of noncontact cardiac signal registration

V A Simon, V A Gerasimov, D K Kostrin, L M Selivanov and A A Uhov

Department of electronic instruments and devices, Saint Petersburg Electrotechnical University "LETI", 197376, Saint Petersburg, Russia

Abstract. The goal of actual research was to test the possibility of electrocardiogram registration through the layer of dielectric material and compare the received signal with the electrocardiogram recordings obtained through resistive contact between skin and electrodes. Several prototypes of the devices for electric cardiac signal registration were designed, assembled and tested. The comparison among recorded electrocardiograms is performed. Differences between cardiac signal recordings obtained by means of the ohmic contact and through capacitive link over dielectric environment are demonstrated.

1. Introduction

One of the relevant problems in biomedical and electronic engineering is electric cardiac signal (ECS) registration through dielectric environment such as clothing. The motivation of research in this field is to increase the number of patients coming through electrophysiological observation per unit of time thanks to decreasing the amount of work for electrodes installation: the classic contact electrocardiogram (ECG) registration devices require the maintenance of stable low-resistance contact, which leads to the necessity of conductive gel application, requires the using of suction cap electrodes (with the alternative in flat electrodes for single use) and long wires. These circumstances restrict the area of use of contact ECG systems to the clinical monitoring. Application of the contact ECG devices in the camp conditions, in smart clothing and wearable accessories is quite limited. Despite that, the minimal bill of materials in electronic circuit of the simplest contact ECG device includes just one instrumentation amplifier, several operational amplifiers and sufficient number of passive components. Also the contact ECG systems were successfully approbated during decades of manufacturing and operation for biomedical purposes.

There are many lab prototypes and very few industrially produced noncontact ECG registration devices which can record the ECS through dielectric layer between skin and sensor surface. The ideal noncontact ECG registration device shouldn't have any resistive contact with the patient's body: all electric conductors of the device should be isolated with at least one dielectric layer.

2. Problem definition

When developing an ECG registration system, an engineer has to deal with signals in the millivolt scale [1]. The frequency band of the ECS is limited to several hundred Hertz, in practice all useful information is enclosed in a band from 0.05 Hz to 100 Hz. Overlapping with the frequency of mains power line, ECS is sensitive to the corresponding 50/60 Hz interference, thus special attention is paid to the methods of mains hum reduction in the field of ECG registration. In addition to mains hum, there are several factors that complicate the development of any ECG monitoring system:

Content from this work may be used under the terms of the Creative Commons Attribution 3.0 licence. Any further distribution of this work must maintain attribution to the author(s) and the title of the work, journal citation and DOI.

Published under licence by IOP Publishing Ltd

- noise generated by the electrode-skin contact;
- patient's movements affecting the resistance of electrode-skin contact;
- electromagnetic interference from various electronic equipment;
- electric potentials generated by muscle activity – electromyographic (EMG) artefacts.

The experiments show that the last of the factors is the most significant and hard-to-remove: the cardiac muscle contractions and the activity of musculoskeletal system are phenomena of the same nature. In this regard, one of the possible methods to eliminate the influence of EMG artefacts on the quality of the resulting ECG is a simple rejection of ECS data at the time of high muscle activity. An accelerometer can be used to identify the peak EMG activity moments.

Analysis of papers [2-4], devoted to the development of systems for noncontact ECG registration, leads to the following conclusion: the noncontact technique deals with extremely high impedance of the signal source. The signal is coming through the skin-electrode capacitance of several pF to the front-end amplifier. Thus, the noncontact ECS measurement requires the direct current (DC) biasing of the signal inputs using a circuit with ultra-high ohmic resistance. The front-end amplifier should have ultra-low input leakage currents.

In ideal conditions that ensure the minimum impact of electromagnetic interferences, in particular mains hum, the noncontact ECS registration system is fairly feasible [5][6]. Herewith the distance to the electrodes may be quite long (at least centimeters), or the electrodes can be separated from the surface of the skin with several layers of cotton fabric. However the stability of noncontact ECG monitoring technique in the real world filled with electromagnetic disturbances is under discussion.

3. Results and Discussion

Several prototypes for contact and noncontact ECG registration were assembled on the printed circuit boards (PCB). All circuitry and electrodes are located on the same PCB for better mains hum elimination. First prototype (figure 1) has tinned electrodes without any covering; second and third devices have a layer of solder mask on the electrodes. Each prototype has three electrodes: two sensors and one so called driven right leg (DRL) electrode. The difference of potentials between sensors is amplified and recorded by means of a sound card inside a laptop. The DRL electrode is used for common-mode suppression. The assembled devices are supplied through USB port of the laptop. The laptop is powered by internal lithium-ion battery and disconnected from the mains power line during ECG registration. The PCB is placed on the left side of the thorax during the experiment. The appearance of the prototypes and obtained ECS recordings are shown below.

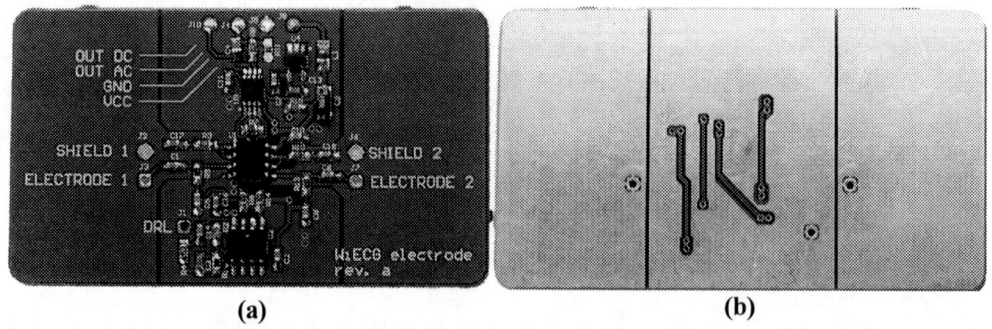

Figure 1. Top layer **(a)** and bottom layer **(b)** of the first PCB with uncovered electrodes.

The top layer of the first prototype's PCB has active shielding over sensing electrodes. The bottom layer contains several isolated areas: sensing electrodes on the left side and on the right side, DRL electrode at the centre. The linear dimensions of the PCB are 64.2 mm (length) by 36.7 mm (width). ECG recorded by first prototype through direct contact between skin and electrodes is shown in figure 2.

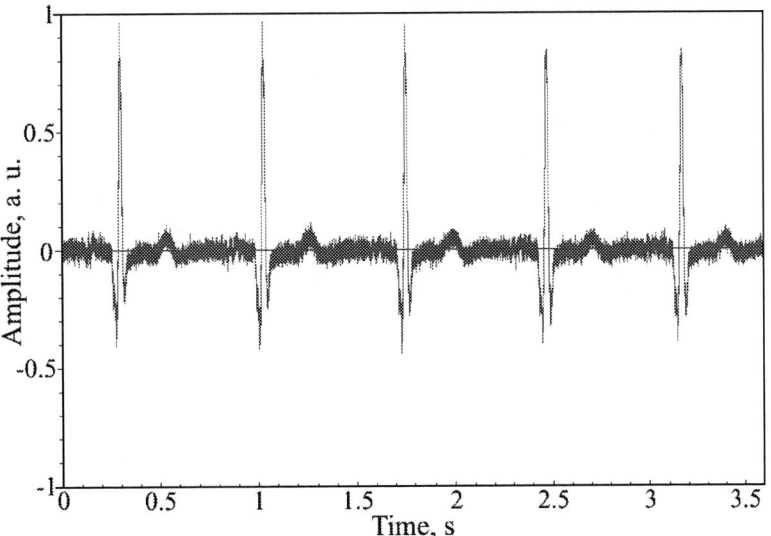

Figure 2. ECG recorded through direct contact with the skin. The DRL driver gain is -10. No filtration applied.

ECG obtained through the layer of cotton textile is presented in figure 3. The DRL electrode has direct contact to the body through right hand touch. Since the direct skin-sensor contact is absent, the signal to noise ratio is considerably decreasing.

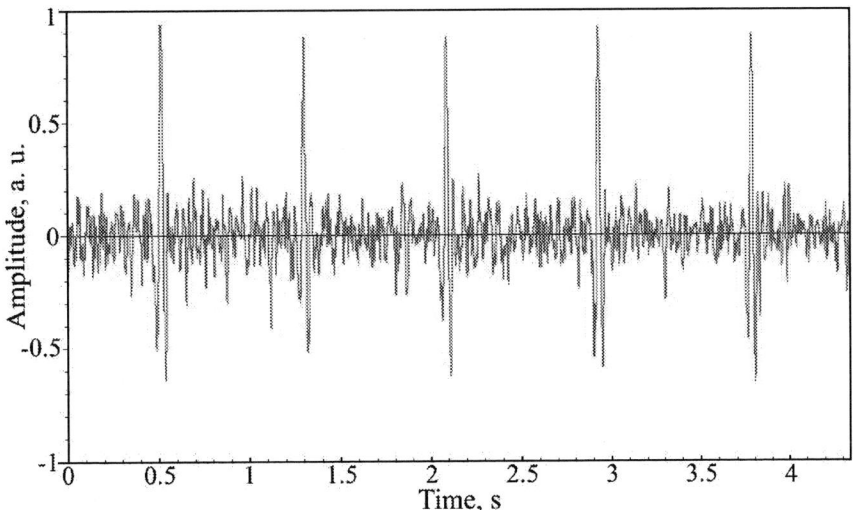

Figure 3. ECG recorded through the layer of cotton fabric. The DRL driver gain is -200. The 50 Hz notch and 100 Hz lowpass digital filters are applied.

The second prototype's PCB has solder mask covering on the electrodes. Long accordion-like tracks connected to the reference voltage source (half-supply voltage) are added to the sensors to create the DC biasing through leakage currents. The appearance of the PCB is shown in figure 4. The recorded ECGs are presented in figures 5, 6, 7. The 50 Hz notch and 100 Hz lowpass digital filters are applied to the received ECS.

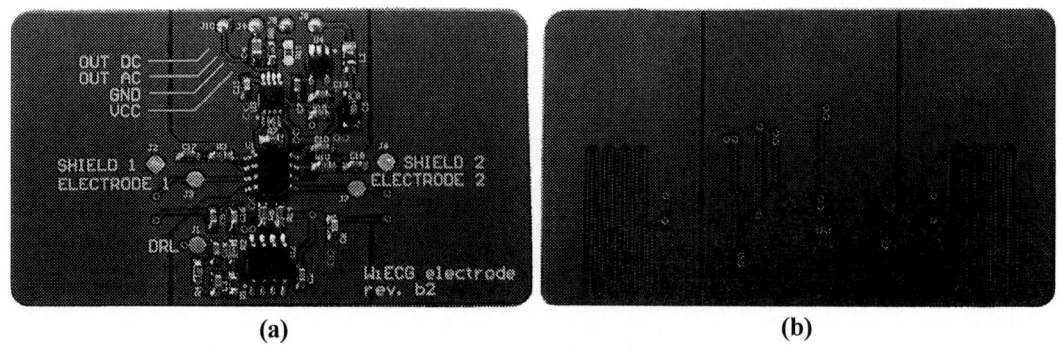

Figure 4. Top layer **(a)** and bottom layer **(b)** of the second PCB with solder mask covered electrodes.

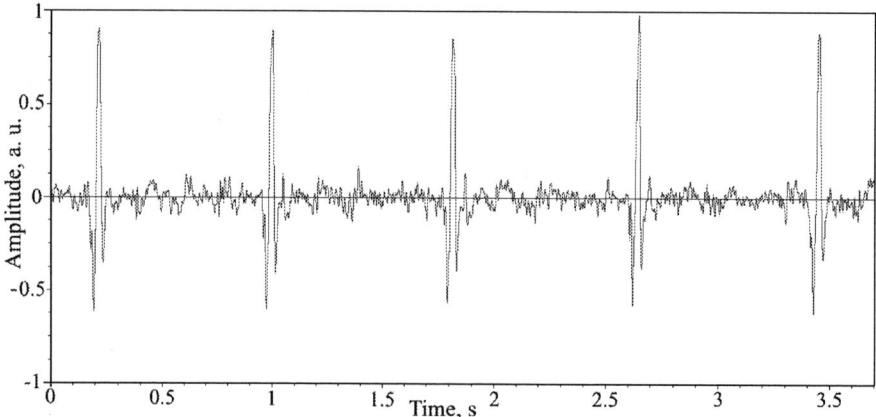

Figure 5. ECG recorded through the solder mask. The DRL electrode has direct contact to the body through right hand touch. The DRL driver gain is -10.

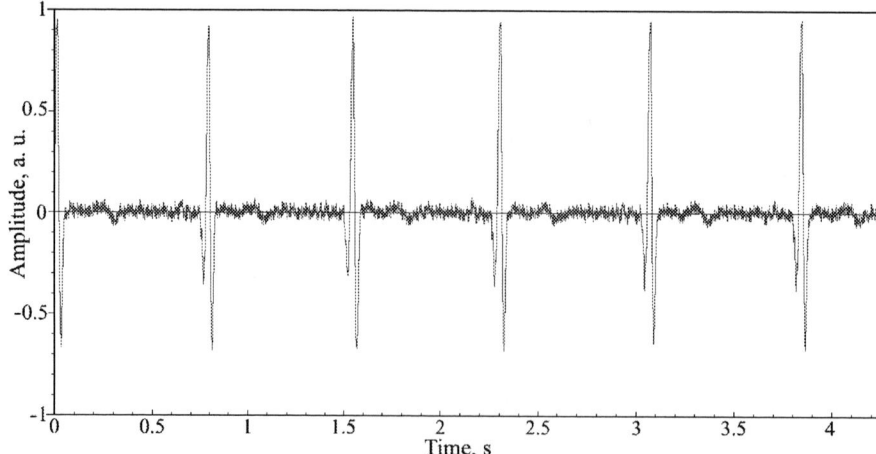

Figure 6. ECG recorded through the solder mask. All electrodes are capacitively coupled to the body through the solder mask. The DRL driver gain is -1000.

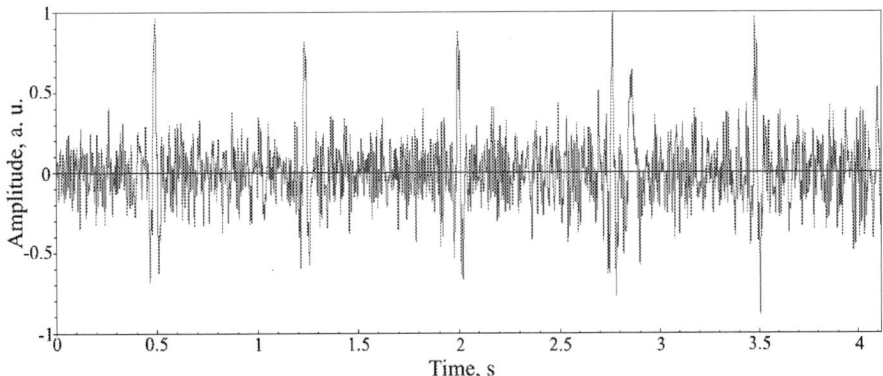

Figure 7. ECG recorded through the layer of cotton fabric and the solder mask. The DRL electrode has direct contact to the body through right hand touch. The DRL driver gain is -1000.

In the third prototype the DRL circuitry and accordion-like tracks are unified (figure 8). The DRL signal consists of the DC reference voltage plus the inverted and amplified common-mode. Obtained ECGs are shown in figures 9, 10. The hum notch and 100 Hz lowpass digital filters are applied.

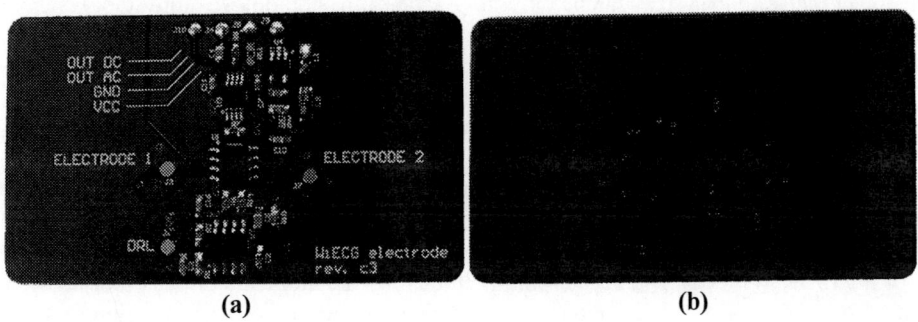

Figure 8. Top layer **(a)** and bottom layer **(b)** of the third PCB with solder mask covered electrodes.

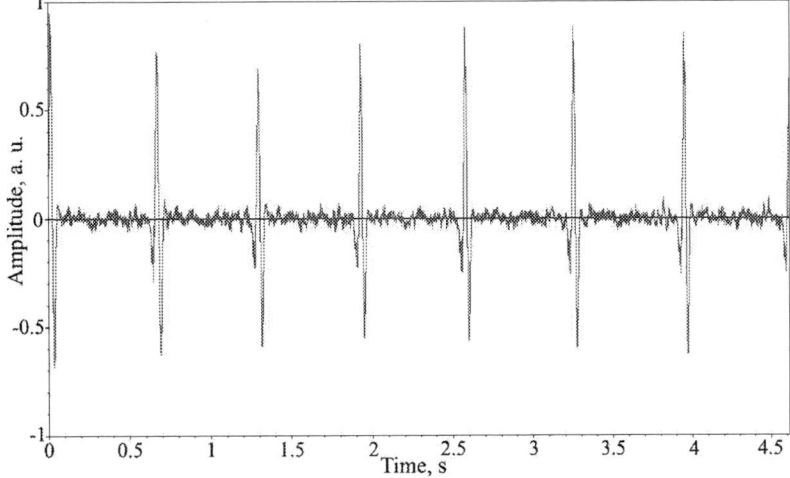

Figure 9. ECG recorded through the solder mask. All electrodes are capacitively coupled to the body. The DRL driver gain is -1000.

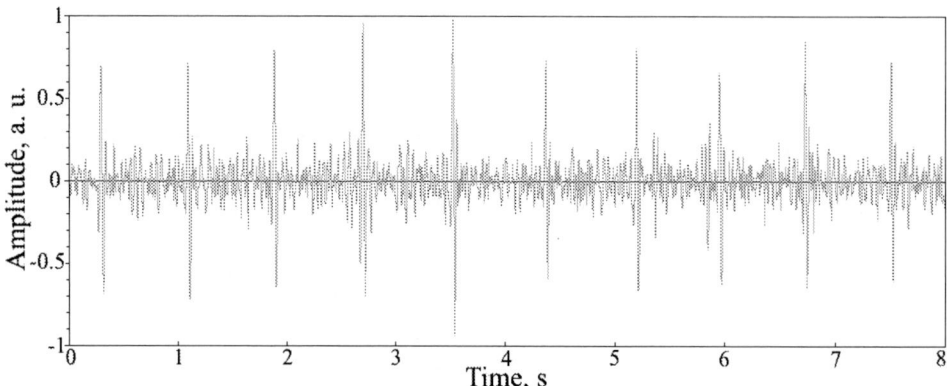

Figure 10. ECG recorded through the layer of cotton fabric and the solder mask. The DRL electrode has direct contact to the body through right hand touch. The DRL driver gain is -100.

4. Conclusion

Presented signal recordings are promising for further research in direction of increasing the signal to noise ratio to improve the quality of noncontact ECS registration technique. Obtained ECGs prove that the noncontact registration system can be put into practice and should have quite stable operation. This system will utilize generally available electronics and hence it will be sold by low cost. The reliability of the noncontact ECG technique depends on several conditions:

- the surface of each electrode should be at least 500 mm^2;
- the DRL electrode is directly connected to any limb of the patient or capacitively coupled to the body through an electrode with large surface of several thousand mm^2. This electrode can be manufactured as a part of clothing or furniture items.

One of the hindrances for noncontact ECS registration is the patient movements, especially for continuous ECG monitoring outside the clinic. As it stated above, this problem can be resolved with the use of microelectromechanical accelerometer indicating the moments of peak motion activity. The spoiled ECG data recorded in these moments can be simply put out of consideration. Also the mains hum should be deeply rejected by means of hardware notch filter [7] or by using 24-bit analog to digital converter [8][9] (32-bit ideally) and adaptive filtration algorithms.

References

[1] Gacek A and Pedrycz W 2012 *ECG Signal Processing, Classification and Interpretation: A Comprehensive Framework of Computational Intelligence* (London: Springer-Verlag)

[2] Sullivan T J, Deiss S R and Cauwenberghs G 2007 *Proceedings of the IEEE Biomedical Circuits and Systems Conference (Montreal)* pp 154–7

[3] Chi Y M and Cauwenberghs G 2009 *Proc. 31st Annual Int. Conf. of the IEEE Engineering in Medicine and Biology Society (Minneapolis)* pp 4218–21

[4] Chi Y M and Cauwenberghs G 2010 *Proc. Int. Conf. on Body Sensor Networks (Singapore)* pp 297–301

[5] Prance R J, Debray A, Clark T D, Prance H, Nock M, Harland C J and Clippingdale A J 2000 *Measurement Science and Technology* **11** 291–7

[6] Harland C J, Clark T D and Prance R J 2002 *Measurement Science and Technology* **13** 163–9

[7] Gerasimov V A, Kostrin D K, Selivanov L M, Simon V A and Uhov A A 2017 *Fundamental problems of radioengineering and device construction* **17** (4) pp 1068–70

[8] Uhov A A, Gerasimov V A, Selivanov L M, Kostrin D K and Simon V A 2016 *Proc. IEEE NW Russia Young Researchers in Electrical and Electronic Engineering Conf.* pp 797–9

[9] Simon V A, Gerasimov V A, Selivanov L M, Kostrin D K and Uhov A A 2017 *Journal of Physics: Conference Series* **929** 012016

Compact nuclear magnetic spectrometer for non-destructive condition testing of biological objects

N S Myazin[1], V V Davydov[1, 2]

[1]Higher School of applied physics and space technologies, Peter the Great Saint Petersburg Polytechnic University, Saint Petersburg 195251, Russia
[2]All-Russian Research Institute of Phytopathology, Moscow Region 143050, Russia

Abstract. Paper discusses the express control technique based on the phenomenon of nuclear magnetic resonance (NMR). For a small-sized NMR spectrometer, a new magnetic system and a signal registration circuit have been developed, which makes it possible to detect the NMR signal at different frequencies from different nuclei in biological media, thus registering the spectrum of this media. In addition, this design also allows measuring the longitudinal and transverse relaxation times of the medium. With the help of the proposed technique, studies of a media have been carried out; the results of these studies are presented.

1. Introduction

Currently, nuclear magnetic resonance is one of the most accurate and reliable methods for various studies of condensed matter, as well as the measurement of various physical quantities (for example, magnetic field induction, etc. [1–9]). A large number of different models of NMR spectrometers and relaxometers have been developed to study the media. The greatest difficulties arise during the express control of the state of condensed media at sampling site. Such studies are most in demand to control the medium condition before carrying out different physical and chemical experiments (in particular after long their storage, container changing or transportation). Also this is suitable during ecological monitoring of difficult to access places of water objects and their coastal zones, during quality control of production, etc. [8–13].

Unlike other methods, studies using NMR do not cause changes in the physical structure and chemical composition of the condensed medium. The only condition for their implementation is the presence of nuclei with magnetic moment in the studied medium. Almost all existing environments have these nuclei, but signals from most of these nuclei can be registered only in strong homogenous magnetic fields with induction of more than 10 T. These fields can only be created in stationary superconducting solenoids that require a specialized cooling system. In accordance with the international standards for express control devices, their total weight with batteries should not exceed 10 kg. Therefore, we have developed small-sized designs of NMR-relaxometers [11, 13–16], in which the NMR signal was recorded in a weak magnetic field using a modulation technique. The signal was recorded at the resonance frequency of protons. To determine the state of the condensed medium, we used a technique based on measuring the time of longitudinal T_1 and transverse T_2 relaxation. The measured values T_1 and T_2 were compared with the standard values, and then the presence of impurities in the test medium was determined using the proposed mathematical model.

However, experience with the operation of a small-sized NMR relaxometer has shown that during environmental monitoring this may not be enough to make a decision at the measurement site for the further use of the investigated medium. The conducted experiments showed that the aqueous media, depending on the natural conditions, might contain chemical elements (for example, fluorine, potassium,

---

Content from this work may be used under the terms of the Creative Commons Attribution 3.0 licence. Any further distribution of this work must maintain attribution to the author(s) and the title of the work, journal citation and DOI.

Published under licence by IOP Publishing Ltd

etc.) that are harmless in small quantities for living organisms. Their effect on the shape of the NMR signal is difficult to predict: these elements cause significant changes in T_1 and T_2. For this reason, after measuring T_1 and T_2, it may be concluded that the researched medium represents a potential hazard. Thus, it is possible to make wrong further decisions, which leads to additional spending on researching samples of this medium.

One of the solutions to this problem is to register the NMR signal on the nuclei of the elements mentioned earlier. Registration of NMR signal on fluorine nuclei is not very difficult (since fluorine has a sensitivity comparable in order of magnitude with protons and their resonant frequencies are close enough), but other nuclei have a lower sensitivity to NMR method than protons and a lower gyromagnetic ratio. In a compact magnetic system, it is extremely difficult to change magnetic field induction during media express control. Therefore, it became necessary to develop a small-size NMR spectrometer, in which the frequency of the NMR signal detection f_{nmr} varies widely. This makes it possible to record the NMR signals at different resonant frequencies f_{nmr} from nuclei mentioned above. The implementation of this device will solve the problems discussed above in most cases, which arise when investigating media with a compact NMR relaxometer in the express mode.

2. Design of a compact NMR spectrometer

Figure 1 shows the structural diagram of a compact NMR spectrometer with the new construction elements developed by us.

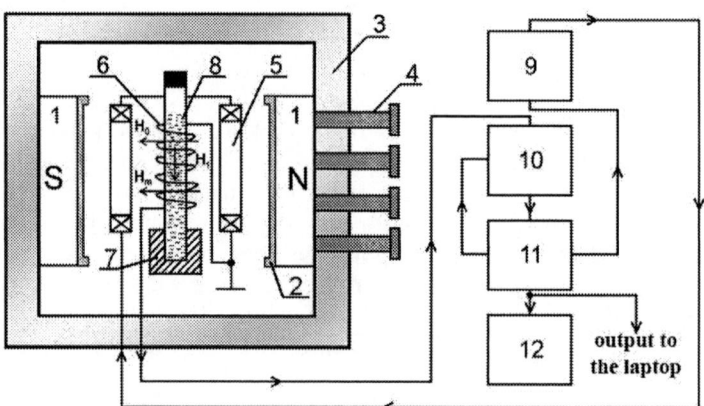

Figure 1 Structural scheme of compact NMR spectrometer: 1 — permanent magnet; 2 — inserts; 3 — neutral for the placement and alignment of the magnets; 4 — adjusting screws; 5 — modulation coil; 6 — NMR signal registration coil; 7 — locking device for the container with the researched medium; 8 — container with the researched medium; 9 — magnetic field modulation generator; 10 — registration scheme including RF autodyne generator; 11 — processing and control unit; 12 — oscilloscope

The small-sized magnetic system was made of the material NdFeB in the form of a disk 1 with a large residual induction. This allowed at the diameter of the magnet 1 poles $d_m = 92$ mm and distance between them $d_z = 16$ mm to provide in the area of the registration coil 6 the inhomogeneity $0.5 \cdot 10^{-3}$ cm^{-1} at the induction equal to $B_0 = 0.132$ T. The magnetic field modulation frequency can vary from 1 to 200 Hz. We have experimentally established that the optimal value is $f_m = 50$, since it provides a high value of the signal noise ratio and at the same time allows to investigate a large number of media [1]. The weight of the new developed magnetic system together with the coils of modulation 5 and registration 6, and the locking device of the container 7 appeared to be equal to about 3.2 kg.

To register NMR signals in weak magnetic field from nuclei with low sensitivity to the NMR method, a new scheme of a weak oscillation generator (an autodyne detector) was developed [1, 5, 8, 10]. This scheme is assembled on the basis of an amplifying cascade with drain detection of the NMR signal and its subsequent amplification using modern planar field-effect transistors. Such construction of the circuit allows to create a minimum level of oscillations in the receiver-transmission loop of the autodyne detector, in which the sample with the investigated medium is placed, to obtain the greatest sensitivity (the ratio of the generation amplitude in the loop to the change in its Q) for NMR signal registering.

In the new design of the compact NMR spectrometer, the scheme of AFC on a resonance is based on the STM32 microcontroller (ARM Cortex M3, STM32F100RBT6B). In addition, on its basis, we have created an accumulation scheme and the new auto-tuning circuits to the maximum S/N ratio for the generation level of the autodyne detector (field H_1), the modulation frequency f_m and amplitude H_m of the B_0 field.

Figure 2 shows, as an example, the registered NMR signals (the change in the voltage U_S at the output of the autodyne detector 10 from the time t) from the aqueous solution of sodium hydroxide NaOH at the resonance frequency of the sodium nuclei f_{Na} = 1488591 Hz.

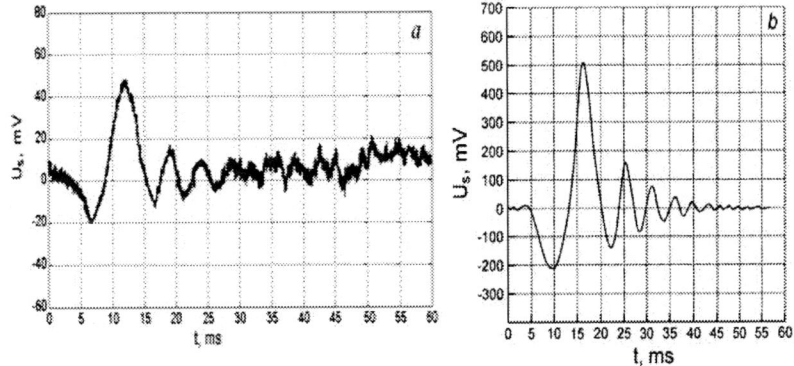

Figure 2 The registered NMR signal from sodium hydroxide at T = 291.3 K: a) without accumulation; b) output of accumulation scheme

In the earlier developed design of a compact NMR relaxometer [8, 11, 13, 14] researchment of this media was possible only by using the NMR signal from protons registered at the resonant frequency f_p. Analysis of the presented on figure 2 NMR signals shows that ratio S/N > 1.3 while registering signals on the sodium nuclei resonant frequency f_{Na}. This makes it possible to perform AFC f_{nmr} on the resonance of sodium nuclei. However, measuring of the relaxation constants T_1 and T_2 with an error less than 1.0% (which allows us to uniquely determine the medium state [5–8, 10, 12]) impossible without using accumulation scheme since the ratio S/N < 3.0 (figure 2.a). In addition, there are noises at the peaks of the NMR signal. The subsequent accumulation of the NMR signal allows us to obtain a ratio S/N> 10.0 (figure 2.b), ensuring the measurement of T_1 and T_2 with the required accuracy.

Nevertheless, when investigating media for the presence of nuclei with low sensitivity to the NMR method, a situation may occur when the S/N of the detected signal is less than 1.3. In this case, the operation of the auto-tuning circuit will be impossible, and the signal cannot be registered. Therefore, in order to increase the S/N ratio in the new developed compact magnetic system, we proposed a new technical solution to the problem posed to increase the volume of researched medium V_R and reduce the value of ΔB. At the same time, an optimum between the values of B_0, ΔB and V_R was achieved, which makes it possible to obtain S/N > 1.3 of the detected signal from different nuclei of the investigated medium with low sensitivity to the NMR method. In addition, this allows measuring the number of peaks ("wiggles") in the NMR signal of at least 5 (figure 2.b). This makes it possible to measure T_2 with an error not greater than 1.0% [5, 8–10, 13]. Reduction in the degree of inhomogeneity of the magnetic field was achieved by placing inserts 2 (figure 1) at the poles of the magnets in the form of steps (shims) of soft magnetic material (ARMCO iron). The inserts that we have developed are disks with a diameter of 92 mm, a thickness of 8 mm. From the edge of the disc, along its diameter, two strips were made in the form of steps with a recess to the center (the width of each step is 3 mm and a height is 2 mm). As a result, inside of each of the inserts 2 (figure 1) was formed a free space along the axis of action of the field B_0 in the form of hollows (of cylindrical shape with a depth of 4 mm and a diameter of 80 mm), where we have put carcasses with modulation coils. Therefore, to accommodate the manufactured inserts, the distance d_z between the poles of the magnets 1 was increased by only 10 mm compared to the original ($d_z = 16$ mm), the new value $d_z = 26$ mm. In the newly developed design of the magnetic system, the possibility of mechanical alignment of the mutual arrangement of the magnet poles, previously used in [8, 11, 13, 14], was also retained.

In addition, in the developed new design of a compact NMR spectrometer we can measure amplitude change U_S of the detected NMR signal by changing the frequency of NMR signal detection (autodyne detector 10) at a constant value of the magnetic field induction B_0. This allows us to register the absorption spectrum of the investigated medium using nuclei with magnetic moments. Figure 3–5 shows, as an example, the dependence of the amplitude U_S of the detected NMR signal on the frequency f_{nmr} at $B_0 = 0.132$ T and at temperature T = 293.3 K.

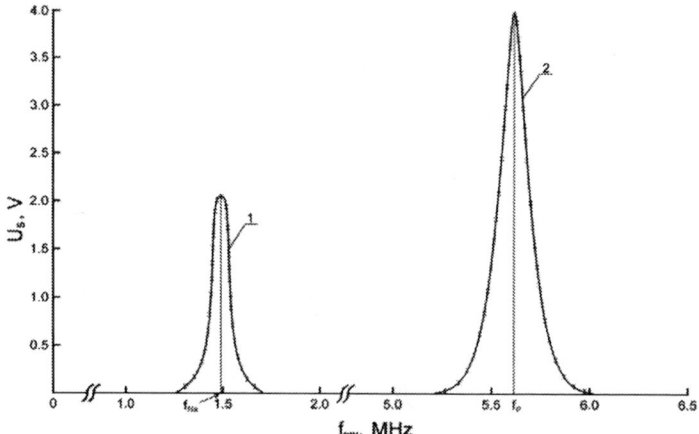

Figure 3 The absorption spectrum of an aqueous solution of sodium hydroxide. Graph 1 corresponds to the detected NMR signal from sodium nuclei, graph 2 — from protons

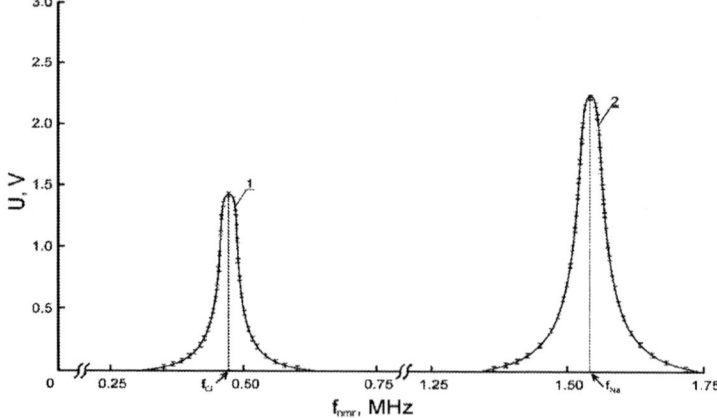

Figure 4 The absorption spectrum of sodium chloride. Graph 1 corresponds to the NMR absorption signal on chlorine nuclei, graph 2 — from sodium

IOP Conf. Series: Journal of Physics: Conf. Series **1124** (2018) 031004   doi:10.1088/1742-6596/1124/3/031004

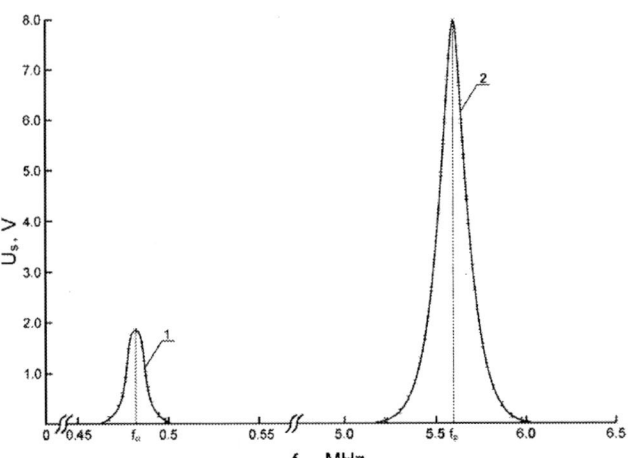

Figure 5 The absorption spectrum of sodium chloride. Graph 1 corresponds to the NMR absorption signal on chlorine nuclei, graph 2 — from protons

Using the dependence of U_S on f_{nmr} (absorption spectrum) obtained in figures 3–5 it is possible to determine which nuclei with magnetic moments are present in the investigated medium, their relative concentrations, acidity of the mediums (pH), etc. Similar dependences are obtained in stationary high-resolution NMR spectrometers while studying various mediums.

3. Conclusion

The obtained experimental results have shown that the small-size NMR spectrometer developed by us makes it possible to carry out express control over the state of a much larger number of media than the previously considered instruments in [8, 11, 13, 14].

For the first time, the possibility of registering NMR signals in a weak field from a large number of nuclei with magnetic moments in the investigated medium (except for carbon, nitrogen and sulfur nuclei) was realized for a frequency tuning f_{nmr} with its automatic tuning to resonance. This allows us to consider this device as a small-sized NMR spectrometer, since it differs from the NMR relaxometers considered in [8, 11, 13, 16].

In the future, our researches will be aimed at increasing the sensitivity of the autodyne detector circuit. This is necessary to solve the problem of registering the NMR signal from the nuclei of chemical elements that are not very sensitive to the NMR method (for example, calcium, etc.) but cause significant changes in medium's T_1 and T_2 even at their small concentrations [13, 15-19].

References

[1]   Davydov V V, Myazin N S and Velichko E N 2017 *Technical Physics Letters* **43** 607–610
[2]   Davydov V V, Dudkin V I, Petrov A A and Myazin N S 2016 *Technical Physics Letters* **42** 692–696
[3]   Davydov V V, Dudkin V I and Karseev A Y 2015 *Instruments and Experimental Techniques* **58** 787–793
[4]   Neronov Yu I and Seregin N N 2012 *Journal of Experimental and Theoretical Physics* **115** 777-781
[5]   Arkhipov V V 2012 *Instruments and Experimental Techniques* **55** 692–695
[6]   Kashaev R S and Gazizov E G 2010 *Journal of Applied Spectroscopy* 77 321–328
[7]   Mussil V V, Pilipenko V V, Lemeshevskaya E T and Keremzhanov K D 2011 *Instruments and Experimental Techniques* **54** 397–399

[8]   Karseev A Yu, Cheremiskina A V, Davydov V V and Velichko E N 2014 *J. of Physics: Conference Series* **541(1)** 012006
[9]   Alexandrov A S, Archipov R V, Ivanov A A, Gnezdilov O I, Gafurov M R and Skirda V D 2014 *Applied Magnetic Resonance* **45** 1275–1287
[10] Filippov A V, Artamonova M R, Rudakova M F, Gimatdinov R G and Skirda V D 2012 *Magnetic Resonance in Chemistry* **50** 114–119
[11] Karseev A Yu, Vologdin V A and Davydov V V 2015 *Journal of Physics: Conference Series* **643(1)** 012108
[12] Zaporozhets O V, Shkurdoda V F, Peregudov O N and Zaporozhets V K 2010 *Instruments and Experimental Techniques* **53** 718–722
[13] Davydov V V, Velichko E N, Dudkin V I and Karseev A Yu 2015 *Instruments and Experimental   Techniques* **58** 234–238
[14] Davydov V V, Dudkin V I and Karseev A Yu 2014 *Measurement Techniques* **57** 912
[15] Davydov V V and Myazin N S 2017 *Measurement Techniques* **60** 183–189
[16] Myazin N S, Logunov S E, Davydov V V, Rud' V Yu, Grebenikova N M and Yushkova V V 2017 *Journal of Physics: Conference Series* **929(1)** 012064
[17] Grebenikova N M, Smirnov K J, Artemiev V V, Davydov V V and Kruzhalov S V 2018 *Journal of Physics: Conference Series* **1038(1)** 012089
[18] Myazin N S, Davydov V V, Yushkova V V and Rud" V Yu 2018 *Journal of Physics: Conference Series* **1038(1)** 012088
[19] Davydov V V, Kruzhalov S V, Grebenikova N M and Smirnov K J 2018 *Measurement Techniques* **61** 365–372

SPBOPEN 2018   IOP Publishing

Influence of electrostatic interactions on cell-penetrating peptide-small interfering RNA complex formation and intracellular delivery efficiency

A Svirina[1], I Terterov[1,2], V Klimenko[1,2], S Shmakov[1], N Knyazev[1,2,3],
A Emelyanov[1,4], V Vysochinskaya[5], A Bogdanov[2,4,6]

[1]Nanobiotech Lab, St. Petersburg Academic University, St. Petersburg 194021, Russia
[2]Saint Petersburg Clinical Scientific and Practical Center of Specialized Types of Medical Care (Oncological), St, Petersburg 197758, Russia
[3]Institute of Cytology RAS, St. Petersburg 194064, Russia
[4]Saint Petersburg Nuclear Physics Institute, National Research Center «Kurchatov Institute», Lenoblast 188300, Russia
[5]Research Institute of Influenza, St. Petersburg 197376, Russia
[6]Bionics Lab, ITMO University, St. Petersburg 197101, Russia

Abstract. Cell-penetrating peptides (CPP) are short positively charged biopolymers that can translocate through lipid membranes. Due to their unique properties CPPs are promising agents for intracellular drug delivery. In this work we studied potency of intracellular delivery of small interfering RNA (siRNA) by means of two CPPs, namely primary amphipathic MPG-ΔNLS peptide and secondary amphipathic EB1 peptide. Optimal concentration conditions and peptide-to-siRNA ratios have been found for stable peptide-siRNA complex formation and delivery efficiency has been shown.

1. Introduction

Cell-penetrating peptides are able to cross lipid bilayer, enter into the cell and so they are considered as agents for intracellular drug delivery [1]. CPPs can carry as cargo a variety of different therapeutic molecules: small compounds, nucleotides, siRNA or proteins [1]. One common feature of cell-penetrating peptides is their positive net charge (+4 – +9) [2]. This property allows them to effectively bind cargo molecules with the opposite charge, resulting in nanocomplexes capable for intracellular delivery.

Small interfering RNAs (siRNAs) can inhibit expression of a specific genes, thus providing

Content from this work may be used under the terms of the Creative Commons Attribution 3.0 licence. Any further distribution of this work must maintain attribution to the author(s) and the title of the work, journal citation and DOI.
Published under licence by IOP Publishing Ltd

potential for new target therapeutics [3]. The main obstacle for the clinical use of siRNA is that it cannot cross cell membranes, due to its hight negative net charge, and thus auxiliary agents for the intracellular delivery are needed [4].

In this work we compare the ability for of two peptides for siRNA intracellular delivery. The frst peptide is cell-penetrating peptide EB1 [5], a Penetratin peptide derivative. It has net charge +8 and is secondary amphipathic, that is it shows amphipathic properties when acquiring an alpha-helical secondary structure on a membrane. The second is MPG peptide variant (with ΔNLS amino acid mutaion) [6], it has net charge +4 and is primary amphipathic, i.e. it amphipathicity is enclosed in the primary structure of its amino acid sequence.

We have found optimal conditions for cell penetrating peptide-siRNA complex formation, characterized these complexes and showed their ability for intracellular delivery.

2. Materials and methods

2.1 Peptides and siRNA

Two peptides had been compared: MPG-ΔNLS (ac-GALFLGFLGAAGSTMGAWSQPKSKRKV-cya) and EB1 (LIRLWSHLIHIWFQNRRLKWKKK-NH$_2$). For the delivery tests siRNA Cy3 tagged duplex directed against the BCR-ABL was used (sense 5`-GCAGAGUUCAAAAGCCCUUdTdT, antisense 5`-AAGGGCUUUUGAACUCUGCdTdT) [7].

2.2 Peptide-siRNA complex characterisation

The peptide-siRNA complex has been prepared by mixing equal volumes of peptide and siRNA water solutions, and subsequently incubation for 10 minutes at 37C°. Final siRNA concentration was 1µM in all cases, while amounts of peptide were chosen to get desired peptide-to-siRNA ratios. Size (hydrodynamic diameter) and zeta potential of the formed complexes were measured with ZetaSizer Nano ZS (Malvern, USA). siRNA binding into complexes was accessed with gel-electrophoresis on 20% polyacrylamide gel at 150V for 1 h in TBE (Tris-borate/EDTA) buffer containing ethidium bromide. Pictures were taken in Gel Doc™ XR+ Gel Documentation System (Bio-Rad, USA).

2.3 Intracellular delivery assay

For intracellular delivery tests we used CT26 cell line (*Mus Musculus* colon carcinoma, ATCC, CRL-2638). Cells were plated at a density 5×10^4 cells in 400 µL RPMI 1640 medium (HyClone, US) containing 10% FBS (HyClone, US), 100 µM of streptomycin, and 100 U/mL of penicillin, in 24 well plate. 100 µl of prepared complex solutions were added to 400 µl of cells, resulting in total siRNA concentration of 200 nM in each well. For positive control HiPerFect liposomal transfection reagent (Qiagen, USA) was used according to the manufacturer protocol with same amount of siRNA. The intracellular delivery of siRNA tagged with Cy3 dye has been controlled after 24 hours using confocal microscopy Axio Observer Z1 (Zeiss, Germany), the cells have been washed before imaging. The transfection efficiency was examined using flow cytometry with Epics XL (Beckman Coulter, USA).

3. Results and discussion

Total net charge in physiological pH differs twice for the investigated peptides being +8 for EB1 peptide and +4 for MPG peptide. To ensure comparable influence of electrostatic interactions we studied complexes with the same peptide-to-siRNA charge ratios for both peptides. For complex formation peptide-to-siRNA molar ratios were 2.5/1, 5/1, 10/1 and 20/1 for EB1 peptide, while for MPG peptide we used values of 5/1, 10/1, 20/1 and 40/1.

Size properties of obtained complexes differs for two investigated peptides (see Fig. 1a). As is seen for EB1 complex sizes are around 100 nm, while it seems that MPG peptide-siRNA complex tend to aggregate into large micron-sized structures.

Zeta potential values of the resulting complexes demonstrate same trend for both peptides with the increase of peptide-to-siRNA ratio. However, it is seen from Fig. 1b that for the same peptide-to-siRNA charge ratios zeta-potential is always lower for MPG-siRNA complex comparing with EB1-siRNA complex, and this difference is more pronounced for larger peptide-siRNA ratios.

From gel-shift experiments (Fig. 2) it is seen that siRNA is completely bind into complexes for the same charge peptide-to-siRNA ratios. Absence of band on electrophoresis represents the total capture of siRNA in complex, so we conclude that there is no free siRNA in complex solution for 10 and 20 EB1 peptide-to-siRNA ratios and 20 and 40 MPG peptide-to-siRNA ratios. This correlates with low or positive zeta potential of corresponding complexes (see Fig. 1b).

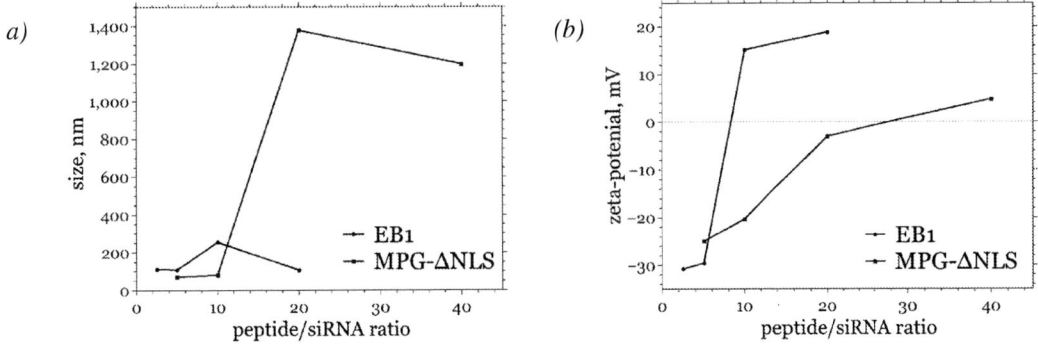

Figure 1(a, b). (a) Mean peptide-siRNA complex size (hydrodynamic diameter) measured with dynamic ligth scattering for different peptide-to-siRNA ratios for MPG and EB1 peptides; **(b)** Zeta-potential value of peptide-siRNA complex for different peptide-to-siRNA ratios for MPG and EB1 peptides.

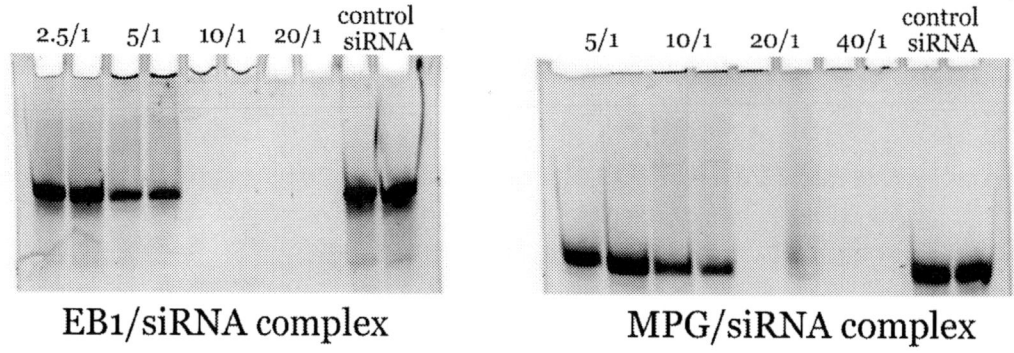

Figure 2. Electrophoresis of peptide-siRNA complexes in different peptide-to-siRNA ratios showing siRNA capture in complexes.

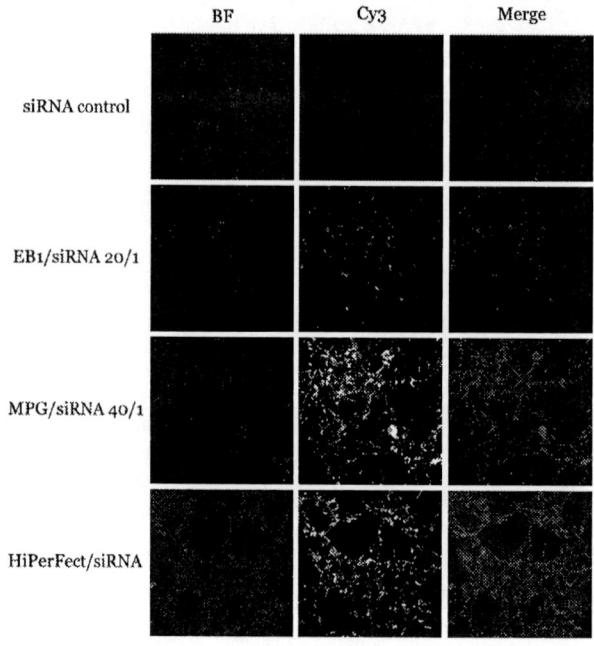

Figure 3. Confocal microscopy image of transfected cells, 24 hours after addition of transfection agents with siRNA. Three columns correspond to brightfield microscopy, Cy3 fluorescence and both channels merged. The first line corresponds to cells with naked siRNA in water solution without transfection agents, the second and the third line corresponds to siRNA transfection by peptide-siRNA complexes in given peptide-to-siRNA ratios for EB1 and for MPG peptides, the fourth line corresponds to siRNA transfection by commercially available transfection reagent HiPerFect.

For EB1 and MPG peptides we identified optimal 20/1 and 40/1 ratios respectively, as for this ratios all siRNA is captured into complexes and they have positive values of zeta potential, which is important for intracellular delivery [2]. Results of the delivery efficiency of this complexes compared with HiPerFect transfection reagent (HPF) are shown on Fig. 3 and Fig. 4.

Transfection efficiency for the investigated peptide complexes was almost the same as for liposomal delivery agent HiPerFect, as it is seen from Fig. 4b. Interestingly, while the size of complex is about 100 nm for EB1 peptide, it was up to 1.5 μm for MPG peptide as it is seen from Fig. 1a. Though it is known that micron-sized complexes are less effective for intracellular delivery than nanometer-sized ones [8], large MPG-siRNA complex showed the ability for intracellular siRNA delivery (see Fig. 3 and Fig. 4 (a, b)). A possible explanation is that such a large peptide-siRNA complexes come into contact with cells leading to increase of peptide and siRNA concentration near the cell surface, thus giving a way to peptide collective behaviour-induced translocation.

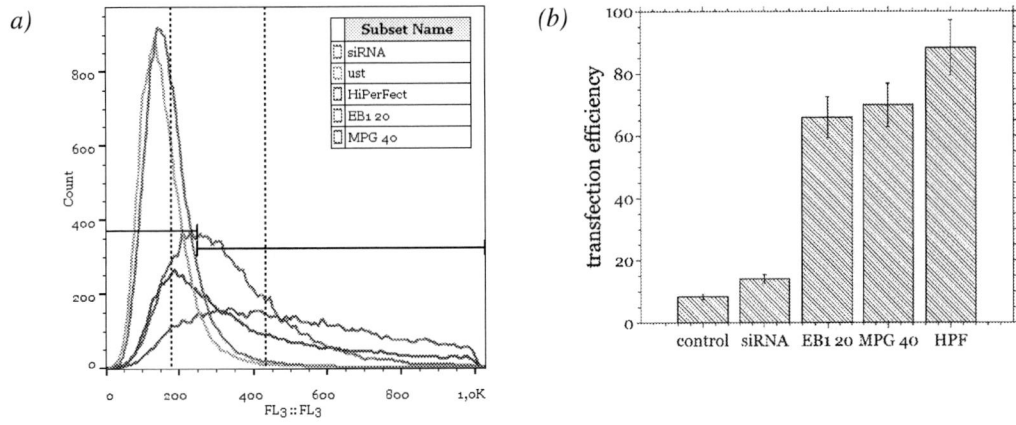

Figure 4(a, b). (a) Histogram of fluorescence distribution of cells measured by flow cytometry; **(b)** Transfection efficiency calculated from flow cytometry measurements. The first cloumn corresponds to control cells, second – shows the transfection of naked siRNA, third and fourth columns represent the transfection siRNA complexed with peptide in peptide-to-siRNA ratios 20-1 and 40-1 for EB1 and MPG respectively, fifth column – siRNA transfection by HiPerFect.

4. Conclusions

Our findings demonstrate that charge peptide-to-siRNA ratio in complex formation is more important than molar ratio. Charge ratio determines siRNA capturing into complex and correlates with its zeta potential – one of the main physico-chemical property that influence internalization into the cell. Also, we found that efficiency of intracellular delivery of siRNA is high enough for complexes with both peptides. Interestingly, micron-size structure of MPG-siRNA complex does not degrade delivery. MPG carry twice less charge than EB1, and thus complex with MPG contains twice as much peptide as complex with EB1, however delivery efficiency is comparable in both cases. This indicates that charge ratio may have a greater effect on peptide delivery properties than total amount of cell-penetrating peptide in the probe.

Acknowledgments

Anna Svirina was partially supported by FASIE.

References

[1] Milletti F. *Drug discovery today* 2012 **1;17**(15-16):850-60.

[2] Bechara C, Sagan S. 2013 *FEBS letters* **587.12** 1693-1702.

[3] Zuckerman J, Davis M. 2015 *Nature reviews Drug discovery* **14.12**: 843-56.

[4] Medarova Z, Pham W, Farrar C, Petkova V, Moore A. 2007 *Nature medicine* **13**(3):372.

[5] Lundberg P, El-Andaloussi S, Sütlü T, Johansson H, Langel Ü. 2007 *The FASEB Journal* **1;21**(11):2664-71.

[6] Simeoni F, Morris M, Heitz F, Divita G. 2003 *Nucleic acids research* **1;31**(11):2717-24.

[7] Scherr, M., Battmer, K., Winkler, T., Heidenreich, O., Ganser, A., Eder, M. 2003 *Blood* **101**(4), 1566-1569.

[8] Crombez L, Morris M, Dufort S, Aldrian-Herrada G, Nguyen Q, Mc Master G, Coll J, Heitz F, Divita G. 2009 *Nucleic acids research* **29;37**(14):4559-69.

SPBOPEN 2018

Nanoparticle tracking analysis of extracellular vesicles reveals two populations of exosomes

AV Selenina[1], VA Kulichkova[1], AN Tomilin[1,2,3], AS Tsimokha[1]

[1] Laboratory of the Molecular Biology of Stem Cells Institute of Cytology RAS, Saint-Petersburg 194064, Russia; [2] Institute of Translational Biomedicine, Saint-Petersburg State University, Saint-Petersburg 199034, Russia; [3] Institute of Nanobiotechnologies, Peter the Great Saint-Petersburg Polytechnic University, Saint-Petersburg 195251, Russia

Abstract. Exosomes are extracellular vesicles of 30 to100 nm in diameter, which are secreted by various cells of mammals. It is known that exosomes represent a specific way of cell-to-cell communication. Mechanisms of secretion and uptake of these vesicles are not clear. To address this problem, we fused the transmembrane protein CD63, a known exosomal marker, at its N-termini to the EGFP protein. We have shown that these proteins exit cells as a part of exosomes, substantiated by the fluorescent nanoparticle tracking analysis. The analysis has also shown that there are two prevailing exosome populations in general population of vesicles.

Introduction

Extracellular vesicles in biological liquids contain exosomes, small nanovesicles 30–100 nm in size, formed from multivesicular endosomes, larger microvesicles, and apoptotic bodies, 100- to 1000-nm particles budding or shedding from membrane surface [1]. Exosomes were first described in 1983, but only recently it was demonstrated that extracellular vesicles are involved in key physiological processes such as intercellular interactions, cell adhesion, migration, invasion, angiogenesis and growth of tumour cells [3]. It is important to note that these vesicles are not just "garbage bags", as previously thought, but play a significant role in intercellular communications, serving as vehicles for transferring between cells various cellular constituents, such as proteins, lipids, and nucleic acids. Mechanisms of secretion and uptake of these vesicles are not entirely clear, mainly because of the difficulties with accurate identification of these vesicles due to their variable size and lack of in vivo markers [2, 3, 4]. Immunofluorescence microscopy has been extensively used to study exosome traffic [5]. The main disadvantage of this method, besides its limited sensitivity and nonspecific fluorescence, is the need for cell fixation. Nowadays, there are more convenient ways to visualize the localization and movement of proteins and associated organelles directly in living cells, which relies on fusing of studied proteins with fluorescent proteins [6, 7]. Fluorescent proteins are members of a structurally homologous class of proteins that share the unique property of being self-sufficient to form a visible wavelength chromophore from a sequence of 3 amino acids within their own polypeptide sequence. It is common research practice to introduce a gene (or a gene chimera) encoding a fluorescent protein into living cells and subsequently visualize the location and dynamics of the proteins using fluorescence microscopy [6, 7]. The protein CD63, also known as Lamp3, belongs to the family of tetraspanins and is primarily associated with multivesicular endosomes/bodies and the exosome membranes. In present work we established a HeLa cell line, stably expressing the transmembrane protein CD63 fused at the N-terminus to the enhanced green fluorescent protein (EGFP) to define exosomes size and efficiency of sequential centrifugation for their isolation.

Figures and results

We set out to create a tool that would allow visualization and measuring GFP-labeled exosomes using nanoparticles tracking analysis (NTA). To this end, the exosome marker CD63 was fused at its N-terminus to the enhanced GFP (EGFP) protein. The recombinant protein EGFP-CD63 in cell extracts

Content from this work may be used under the terms of the Creative Commons Attribution 3.0 licence. Any further distribution of this work must maintain attribution to the author(s) and the title of the work, journal citation and DOI.

Published under licence by IOP Publishing Ltd

was next analyzed by SDS-PAGE and Western blotting. Positive signal was absent in total cell extracts of wild-type HeLa cells but readily detected in the extract of EGFP-CD63 HeLa cells (Fig.1 a, b). Intracellular localization of EGFP-CD63 protein was determined by confocal fluorescent microscopy of living cells. The EGFP signal was detected in the cytoplasm, specifically, in a structure adjacent to the nucleus, resembling the endoplasmic reticulum (Fig.1 c). This localization is consistent with previously published data [7].

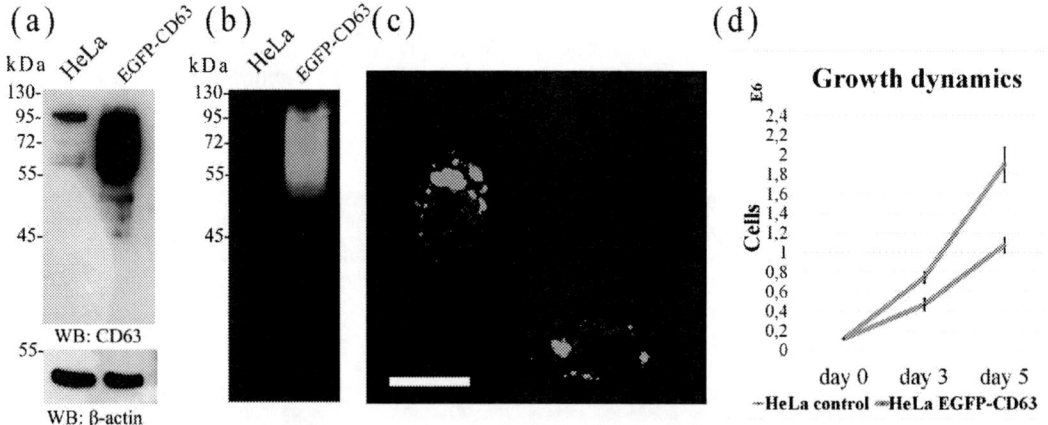

Figure 1. Expression of ectopic EGFP-CD63 and distribution of EGFP-tagged multivesicular bodies in living cells. Whole cell extracts (15 µg) of wild-type and stably expressing EGFP-CD63 HeLa cells were separated by SDS-PAGE and analyzed by Western blotting using antibodies against CD63 **(a)** and using the ChemiDoc ™ Touch System equipped with the Alexa488 filter **(b)**. Antibodies to β-actin were used for input control. **(c)** HeLa cells stably expressing EGFP-CD63 were analyzed by confocal microscopy (green); Hoechst 33528 was used to stain nuclei (blue). Observation with a 60×/1.3NA oil immersion objective of confocal microscope Olimpus Fluoview FV3000; scale - 20µm. **(d)** Growth dynamics of wild-type HeLa cells (blue line) and stably expressing EGFP-CD63 HeLa cells (green line) within 5-day period.

To assess whether transgenic EGFP-CD63 protein affects cellular proliferation, we have analyzed the dynamics of cell growth for 5 days. The cell growth curves of the wild-type and stably expressing EGFP-CD63 protein HeLa cells are shown in Fig.1 d. The stably expressing EGFP-CD63 HeLa cells showed about two-fold lower cell proliferation rate, compared to wild-type HeLa cells, consistent with previously shown involvement of tetraspanins into various cellular processes, including proliferation, via protein-protein interaction [8].

To prove that the recombinant protein EGFP-CD63 have been integrated into exosome membrane and released by cells as a part of extracellular vesicles, we performed sequential centrifugations of the cell-conditioned medium including filtration through the 0.22-µm filters (Fig.2 a). We next analyzed the obtained fraction of exosome-like vesicles (ELVs) released by EGFP-CD63 HeLa cells for particles size distribution and quantities by nanoparticles tracking analysis (NTA). This technology utilizes the properties of both light scattering and Brownian motion in order to obtain size distribution and concentration measurement of particles in liquid suspension [9]. We performed nanoparticle tracking analysis to confirm the size of the whole population of extracellular vesicles released by EGFP-CD63 HeLa cells in scatter mode (Fig.2 b) To investigate, which of the obtained fractions contains recombinant protein EGFP-CD63, equal amounts of fractions were subjected to SDS-PAGE and analyzed by Western blotting, using antibodies to protein GFP. The fluorescent protein signal was absent from a vesicle-free culture supernatant, but readily detected in precipitated by ultracentrifugation vesicle-containing fraction ELVs (Fig.2 c). These results suggest that EGFP-CD63 in the conditioned medium is not represented by a free protein, but is associated with extracellular vesicles.

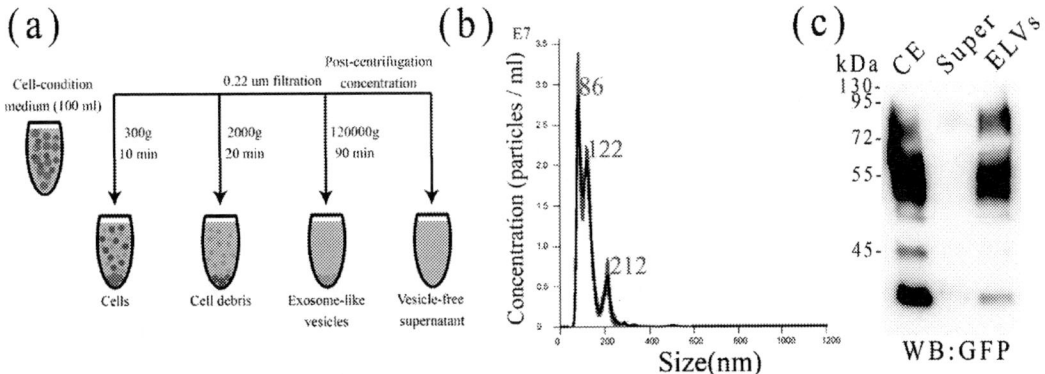

Figure 2. Protein EGFP-CD63 was released by cells as a part of extracellular vesicles. (a) A scheme of the isolation of extracellular vesicles. Two main steps included the isolation of exosome-like vesicles (ELVs) by sequential centrifugation and a vesicle-free culture supernatant by subsequently concentrating proteins. **(b)** Nanoparticle tracking analysis with the analyzer NanoSight NS300; the average mean size in scatter was 126.4 ± 5.0 nm in the whole population of ELVs released by EGFP-CD63 HeLa cells. **(c)** Samples of whole cell extract (15 μg, CE), ELVs and a vesicle-free culture supernatant (Super) were subjected to SDS-PAGE and analyzed by Western blotting using antibodies against GFP protein. Samples of ELVs and a vesicle-free supernatant were normalized onto the number of cells (40×10^6) used to condition the medium

To further verify the integration of EGFP-CD63 into exosome membrane, ELVs pellet were dissolved in PBS and loaded onto a sucrose gradient. After density gradient ultracentrifugation all collected fractions were analysed by Western blot analysis with anti-GFP antibodies (Fig.3 a). The GFP immunoreactivity was detected exclusively within fractions whose density was previously shown for exosomes [4]. Using nanoparticle tracking analysis in fluorescent mode we found EGFP-carrying vesicles in two populations of exosomes (Fig.3 b). Thus, our results confirm the efficacy of the exosome isolation approach. Despite the overlap of the sizes of microvesicles and exosomes, our method of vesicle purification allows purification of two prevailing exosome populations with ~85 and 135 nm size from general population of vesicles.

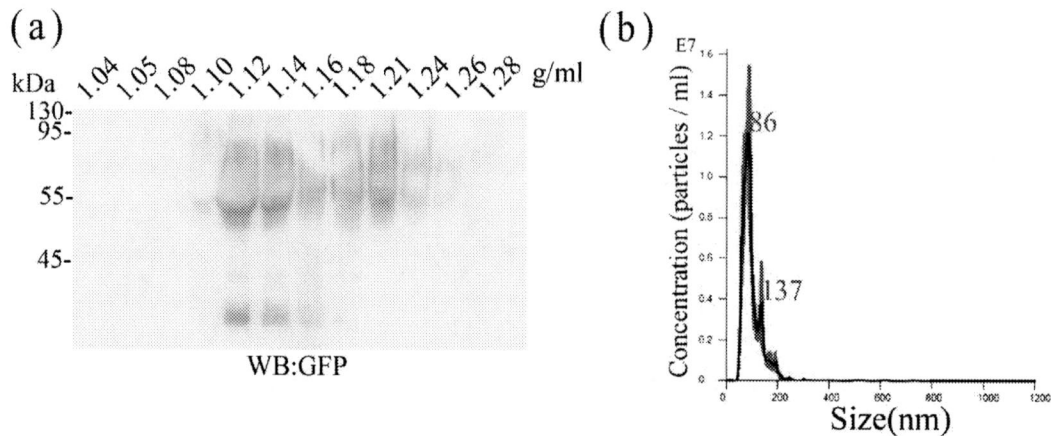

Figure 3. Quantification of EGFP-labeling exosomes revealed two populations of exosomes in size 86 and 137 nm. (a) Sucrose density gradient of ELVs pellet from conditioned medium of 60×10^6 cells after sequential centrifugation. (b) Nanoparticle tracking analysis of ELVs obtained from EGFP-CD63 HeLa cells; the average mean size of fluorescent vesicles was 101.1 ± 3.8 nm.

Acknowledgments

This work was supported by the Russian Foundation for Basic Research (#16-04-01667). Nanoparticle tracking analysis was supported by the Russian Science Foundation (# 16-14-10343).

References

[1] Théry, C., Zitvogel, L., Amigorena, S. 2002 *Nat. Rev. Immunol.* **2**, 569-579
[2] Zomer, A., Vendrig, T., Hopmans, E.S., van Eijndhoven, M., Middeldorp, J.M., Pegtel, D.M. 2010. *Com. & Integ. Biol.* **3**, 447-450.
[3] Lakhal S., Wood M. J. A. 2011 *Bioessays.* **33(10)**, 737-741.
[4] Valadi H, Ekström K 2007 *Nature cell biology* **9**, 654-659.
[5] Koumangoye, R.B., Sakwe, A.M., Goodwin, J.S., Patel, T., Ochieng, J. 2011. *PLoS ONE* **6**, e24234.
[6] Kulichkova, V. A., Artamonova, T. O., Zaykova, ... & Tsimokha, A. S. 2015 *Mol biotech*, **57(1)**, 36-44.
[7] Kulichkova, V. A., Selenina, A. V., Tomilin, A. N., & Tsimokha, A. S. 2018 *Cell & Tissue Bio*, **12(2)**, 146-152
[8] Boucheix C, Rubinstein E 2001 *Cell. & Mol. Life Sciences CMLS* **58**, 1189-1205
[9] Dragovic R, Collett G, Hole P 2015 *Methods* **87**, 64-74.

SPBOPEN 2018

IOP Publishing

IOP Conf. Series: Journal of Physics: Conf. Series **1124** (2018) 031007   doi:10.1088/1742-6596/1124/3/031007

Diagnostic potential of the oral fluid for the observation people with multiple dental caries by means of FTIR.

P V Seredin[1], D L Goloshchapov[1], Y A Plotnikova[2], Y A Ippolitov[2], J Vongsvivut[3]

[1]Department of Solid State Physics and Nanostructures, Voronezh State University, Voronezh, Russia
[2]Department of Pediatric Dentistry with Orthodontia, Voronezh State Medical University, Voronezh, Russia
[3]Australian Synchrotron (Synchrotron Light Source Australia Pty LTD), 800 Blackburn Rd Clayton, VIC 3168, Australia

Abstract. The FTIR-spectra of the oral fluid, as well as the calculated mineral-organic, carbon-phosphate, Amide II/Amide I and protein/thiocyanate ratios were compared between subjects with and without multiple caries. The complex analysis of the experimental IR-data showed that the organic-mineral balance in the oral fluid of those with multiple caries shifted towards a reduction in the mineral complexes, accompanied by an increase in the organic component. The most indicative changes in the composition of the oral fluid of those with multiple caries occurred in relation to the number of $-N=C=S$ groups associated with the presence of thiocyanate observed in the IR-spectrum at 2150–1950 cm^{-1}, which increased two-fold. The complex data analysis presented has the potential for application as both tissue markers and as a diagnostic approach for the estimation of cariogenesis in mixed saliva samples.

1. Introduction

IR-spectroscopy has been successfully applied for the detection of tissue marker pathologies [1,2]. With the development of spectroscopic express-methods of human saliva analysis, screening of diseases at the molecular level at any early stage is possible [3]. By comparing the changes in the molecular composition of saliva obtained by FTIR at the different stages of pathology in the oral cavity (caries), it is possible to obtain novel data concerning the course of this disease [2]. This information can help not only to specify the mechanisms responsible for caries development, but also to reveal their relationships with the processes of de-mineralisation/mineralisation of the hard dental tissues, as well as to specify saliva proteomics of caries development to elucidate potentially-significant tissue markers. The complex analysis related to the quantitative and qualitative data on the changes in the molecular composition of the oral fluid by FTIR, as presented in this report, has the potential to increase the accuracy of the detection of future carious processes.

2. Materials and methods

Twenty humans participated in this study: 10 men and 10 women (between 22 and 28 years of age). All participants did not take any medicines or drugs, were non-smokers and did not drink spirits. The first group of participants (5 men and 5 women) was physically healthy with caries-free teeth and without gum disorders. The second group (5 men and 5 women) were conditionally healthy but regularly snacking on easily digestible carbohydrates between meals. On examination, each participant

Content from this work may be used under the terms of the Creative Commons Attribution 3.0 licence. Any further distribution of this work must maintain attribution to the author(s) and the title of the work, journal citation and DOI.
Published under licence by IOP Publishing Ltd

in this group had teeth with lesion focuses related with primary and secondary caries at the stage corresponding to the 3rd degree according to the ICDAS scale [4] (Figure 1a,b). Participants abstained from food and did not drink for at least 2 hours before sampling of their oral fluid.

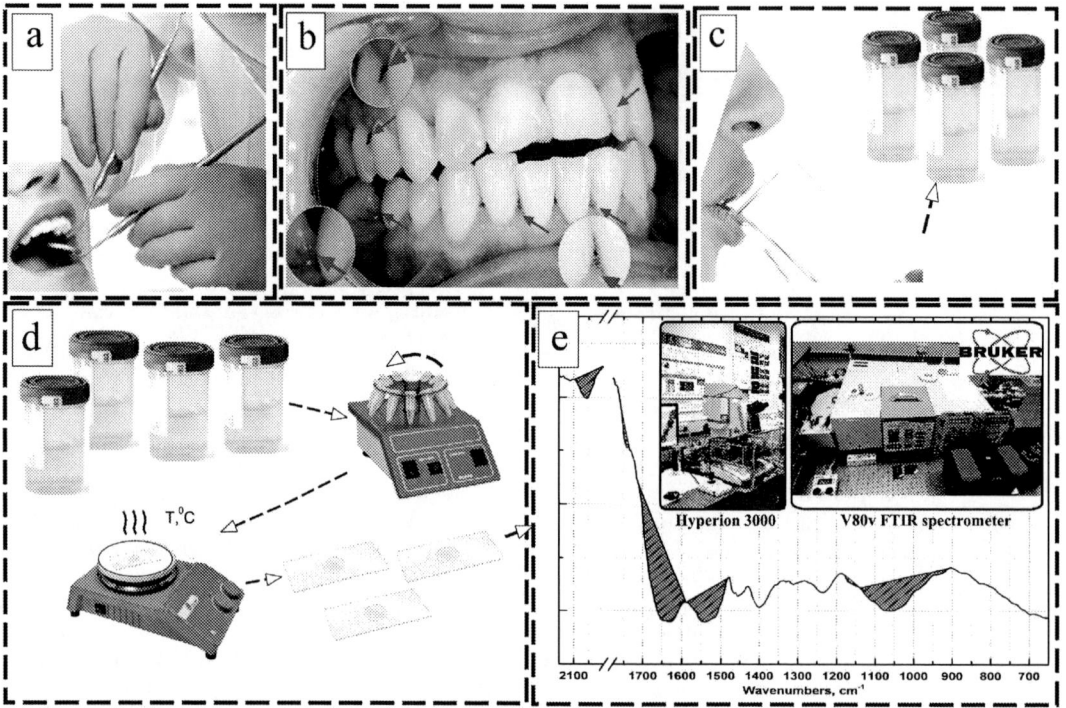

Figure 1(a, b, c, d, e). (a) - preventive examination, **(b)** - diagnosis of multiple caries, **(c)** – oral fluid collection, **(d)** – sample preparation, **(e)** typical IR-spectra of oral fluid of group with multiple caries.

Non-stimulated mixed saliva was sampled during daylight, to minimise circadian rhythm, 5 minutes after the preliminary rinsing of the oral cavity with pure water (Figure 1c). The saliva was placed into 15 ml sterile test tubes for subsequent laboratory investigation according to the standard technique [3] as shown in the insert **c** in Figure 1. After sampling, the test tubes were cooled down to 4°C, then centrifuged before drying in the oven at 36°C to remove excess moisture (Figure 1d).Analysis of the mixed saliva samples was performed with the Vertex-70 spectrometer (Bruker, Germany) using an attachment for attenuated total reflection provided with a diamond prism according to the technique described in [2,5]. In addition, samples were subjected to Infrared Microspectroscopy (IRM) beamline at the Australian Synchrotron, Victoria, Australia using a Hyperion 3000 IR microscope (Bruker, Germany) and high-pressure diamond cell for the analysis of microsamples. IR-spectra were recorded within the range of 4000–500 cm^{-1}. In order to provide the quantitative estimations with the use of FTIR data and to find the difference in the molecular composition of the oral fluid between the group of healthy patients and the group of patients with multiple caries, an approach was applied that was certified in a number of our previous works [2,6]. Spectral data processing (background subtraction, correction for the atmosphere effect, averaging of the spectra and data integration) and analysis were performed using the professional software suite OPUS (version 7.5).

3. Results

The use of FTIR for the analysis qualitatively demonstrated that the molecular composition of the mixed saliva was characterised by a specific set of vibration modes in the IR-spectra, in agreement with published data on biological fluids [1–3,7–9]. The analysis of the vibrations intensity in the IR-spectrum of the oral fluid obtained from the patients with multiple caries (Figure 2a) showed that it is considerably more intensive than for the patients from the first group. Attention should be focused to six ranges (Figure 2b): 2150–1950 cm^{-1}, 1765 – 1725 cm^{-1}, 1720-1480 cm^{-1}, 1185 – 1140 cm^{-1}, 1150-900 cm^{-1} and 870-700 cm^{-1}.

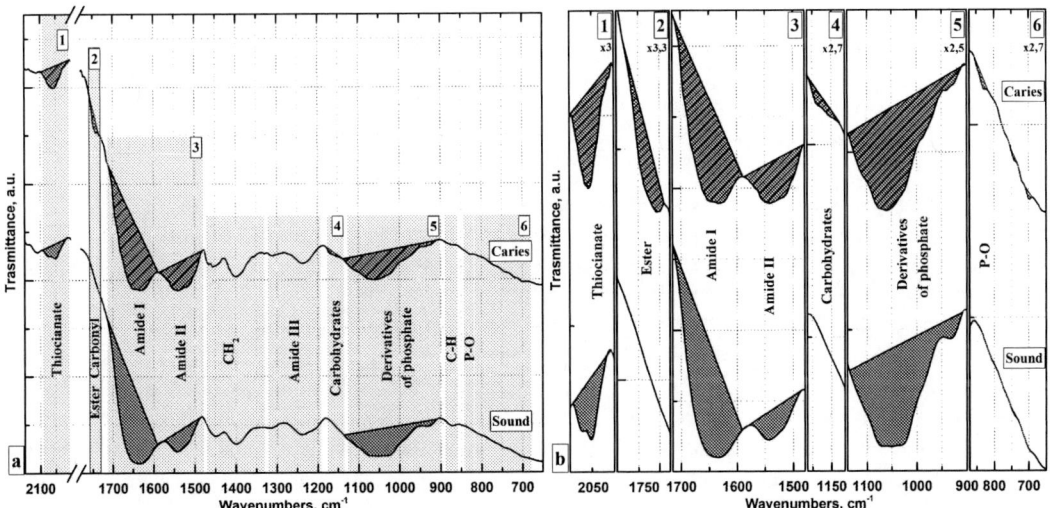

Figure 2(a, b). (a) the full range FTIR-spectra of saliva sound and multiple caries:, (b) 1,2 - sound enamel, 3 -sound dentin, 4,5 - carious enamel.

The most indicative changes in the composition of the oral fluid of those with multiple caries occurred in relation to the number of −N=C=S groups associated with the presence of thiocyanate [3] observed in the IR-spectrum at 2150–1950 cm^{-1} (Figure 1b, area 1). Vibrations in the IR-spectra in the three ranges: 1765–1725 cm^{-1} - the vibration of >C=O and is associated with the carboxylic group of ester (ester carbonyl); 1185–1140 cm^{-1} is related to carbohydrates present in the oral fluid and 870–700 cm^{-1} - C-H, P-O bonds of phosphodiesters, esters, as well as other lipids and carbohydrates of saliva were observed only in the samples of those with multiple caries. For all the samples the mathematical estimation of the changes in the molecular composition of saliva can be given on the basis of the calculations and analysis of different ratios between organic and mineral components in the oral fluid sample [2,6].

4. Discussion

The detected features in the IR spectra of the oral fluid of group members with multiple caries confirmed the known data about the change of the molecular composition of pathological processes in the oral cavity.

The significant changes in the region 2150–1950 cm^{-1} are attributed to the number of −N=C=S groups associated with the presence of thiocyanate in saliva (Figure 2b, inserts 1). The level of thiocyanates in saliva having an anti-bacterial effect on the bacterial vital products can be enhanced during pathological processes in the human organism [3]. The ratio of protein/thiocyanate, that was calculated from the ratio of the integral intensities of the amide bands (Amide I and Amide II) in the range of 1700-1500 cm^{-1} to the integral intensity of −N=C=S vibration bands, arranged at 2150 – 1950 cm^{-1}, associated with thiocyanate demonstrated in fact, a two-fold decrease. It follows from the fact that molecular composition of the mixed saliva taken from the volunteers of the second group the share of chemical bonds inherent to thiocyanate increased in relation to that of proteins.

The vibrations modes that arise only in IR spectra of people with multiple caries of the mode are also related to pathological processes (Figure 2b, area 2,4,6). The IR-band in the range of 1765 – 1725 cm^{-1} (Figure 2b, area 2), according to the data of [1,10,11] corresponds to the vibration of >C=O and is associated with the carboxylic group of ester (ester carbonyl). The presence of esters in a hard dental tissue of humans suffering from dental caries was shown in a number of works [1,11,12]. Their authors noted the fact that esters are more often observed in the carious tissue than in the intact one [13]. Keeping it in mind the experimental proof for the possibility to use this kind of information for the analysis of the oral fluid is yet absent. Also, analysis of literature showed that vibration band in the range of 1185 - 1140 cm^{-1} is related to carbohydrates that are present in the oral fluid. Carbohydrates are involved in the composition of saliva mucins covering and lubricating mucous tunic surface in the oral cavity [14], prevent adherence of anaerobic bacteria and their colonization, protect tissues from their physical damage.

Increasing the proportion between organic components in the composition of the oral fluid of people with multiple cavities alters the ratio of Amide II / Amide I. The ratio of integral intensity for the band of Amid II (CN stretching, NH bending vibrations) in the range of 1590 - 1505 cm^{-1} to the integral intensity of the band of Amid I (C=O stretching) in the interval of 1723-1590 cm^{-1} two times as higher for people with multiple caries. This fact is confirmed by calculations of the mineral-organic ratio (the ratio of the integral intensity of the phosphate bands in the IR spectrum (spectral ranges of 1078-900 cm^{-1}), to the integral intensity of vibration band 1700 – 1590 cm^{-1} associated with Amide I) and carbon-phosphate ratio - the relation of the integral intensity for vibration bands of C=O and CH$_2$/CH$_3$ bonds localized in the range of 1430 - 1360 cm^{-1} to the integral intensity of phosphate bands in the IR-spectrum within the region of 1078-900 cm^{-1}. In all the cases it can be easily seen that in the case of multiple caries in the group of patients a decrease of mineral-organic ratio can be observed meaning a reduce of the share of mineral groups and complexes in saliva composition and/or increase of organic component share in case of the presence of cariogenic bacteria in the mixed saliva [15,16].

Vibrations with the frequencies of 870 cm^{-1}, 827 cm^{-1}, 772 cm^{-1}, 740 cm^{-1} and 700 cm^{-1} in the IR-spectrum of the second group of the patients participating in the examination (patients with multiple caries) are characterized by a high relative intensity as compared with similar bands in the IR-spectrum of participants from the first group. From the analysis of literature data it is clear that these vibration modes are concerned with C-H, P-O bonds of phosphodiesters, esters and as with other lipids as with carbohydrates of saliva [17]. The reason for an increase of intensity in this group of vibrations in the spectrum of the second group of the examination's participants is an increase of lipids concentration and esters of saliva related to the same group of the components due to the caries development as it was shown in [18,19]. It should be noted that the content of these substances and the correlated molecular complexes relative to the content of proteins in saliva is quite low. Therefore, the change in intensity of vibrations in the IR-spectra within the range of 870 - 700 cm^{-1} as in regard to the changes in the range of vibrations related to proteins is also insignificant.

5. Conclusion
All features of the IR-spectra of the oral fluid suggest that the organic-mineral balance in the oral fluid of subjects with multiple caries is shifted towards a reduction in the content of the mineral groups and complexes as well as an increase in the organic component. The ratio of Amide II/Amide I was integral to these changes in molecular composition, increasing two times as high for the group with multiple caries compared with those without. By means of the calculation it was proved that the ratio of protein/thiocyanate was, in fact, decreased two-fold, indicating that the chemical bonds inherent to thiocyanate increased in relation to the share of proteins in mixed saliva from participants with multiple caries. The complex data analysis in this work shows the potential of FTIR for application as a diagnostic approach for the estimation of cariogenesis in mixed saliva samples.

Acknowledgments
This work was supported by the grant of Russian Science Foundation, grant number 16-15-00003.

References

[1] Fujii S, Sato S, Fukuda K, Okinaga T, Ariyoshi W, Usui M, Nakashima K, Nishihara T and Takenaka S 2016 Diagnosis of Periodontal Disease from Saliva Samples Using Fourier Transform Infrared Microscopy Coupled with Partial Least Squares Discriminant Analysis *Anal. Sci. Int. J. Jpn. Soc. Anal. Chem.* **32** 225–31

[2] Seredin P, Goloshchapov D, Kashkarov V, Ippolitov Y and Bambery K 2016 The investigations of changes in mineral–organic and carbon–phosphate ratios in the mixed saliva by synchrotron infrared spectroscopy *Results Phys.* **6** 315–21

[3] Mikkonen J J W, Raittila J, Rieppo L, Lappalainen R, Kullaa A M and Myllymaa S 2016 Fourier Transform Infrared Spectroscopy and Photoacoustic Spectroscopy for Saliva Analysis *Appl. Spectrosc.* **70** 1502–10

[4] Ismail A I, Sohn W, Tellez M, Amaya A, Sen A, Hasson H and Pitts N B 2007 The International Caries Detection and Assessment System (ICDAS): an integrated system for measuring dental caries *Community Dent. Oral Epidemiol.* **35** 170–8

[5] Goloshchapov D L, Kashkarov V M, Seredin P V, Ippolitov Y A, Plotnikova Y A and Bambery K 2016 The study of efficiency of endogenous and exogenous preventive methods of tooth enamel remineralisation by FTIR microscopy using synchrotron radiation *J. Phys. Conf. Ser.* **741** 012054

[6] Seredin P V, Goloshchapov D L, Ippolitov Y A and Kalivradzhiyan E S 2018 Does dentifrice provide the necessary saturation of ions in oral fluids to favour remineralisation? *Russ. Open Med. J.* **7** e0106

[7] Shaw R A and Mantsch H H 2006 Infrared Spectroscopy in Clinical and Diagnostic Analysis *Encyclopedia of Analytical Chemistry* (John Wiley & Sons, Ltd)

[8] Badea I, Crisan M, Fetea F and Socaciu C 2014 Characterization of resting versus stimulated saliva fingerprints using Middle-Infrared Spectroscopy assisted by Principal Component Analysis *Romanian Biotechnol. Lett.* **19** 9817–27

[9] Júnior C, Cesar P, Strixino J F, Raniero L, Júnior C, Cesar P, Strixino J F and Raniero L 2015 Analysis of saliva by Fourier transform infrared spectroscopy for diagnosis of physiological stress in athletes *Res. Biomed. Eng.* **31** 116–24

[10] Silverstein R M, Bassler G C and Morrill T C 1991 *Spectrometric Identification of Organic Compounds* (Wiley)

[11] Scherdin-Almhöjd U 2017 *Identification of esters in carious dentine Staining and chemo-mechanical excavation*

[12] Ulrica S. Almhöjd, Jörgen G. Norén, Anna Arvidsson, Åke Nilsson and Peter Lingström 2014 Analysis of Carious Dentine using FTIR and ToF-SIMS *Oral Health Dent. Manag.* **13** 735–44

[13] Larmas M 1972 A chromatographic and histochemical study of nonspecific esterases in human carious dentine *Arch. Oral Biol.* **17** 1121–32

[14] Baughan L W, Robertello F J, Sarrett D C, Denny P A and Denny P C 2000 Salivary mucin as related to oral Streptococcus mutans in elderly people *Oral Microbiol. Immunol.* **15** 10–4

[15] Gao X, Jiang S, Koh D and Hsu C-Y S 2016 Salivary biomarkers for dental caries *Periodontol. 2000* **70** 128–41

[16] Guo L and Shi W 2013 Salivary Biomarkers for Caries Risk Assessment *J. Calif. Dent. Assoc.* **41** 107–18

[17] Gregor Cevc, Theresa M Allen and Saul L Neidleman 1993 *Phospholipids Handbook* (CRC Press)

[18] Tomita Y, Miyake N and Yamanaka S 2008 Lipids in human parotid saliva with regard to caries experience *J. Oleo Sci.* **57** 115–21

[19] Belstrøm D, Jersie-Christensen R R, Lyon D, Damgaard C, Jensen L J, Holmstrup P and Olsen J V 2016 Metaproteomics of saliva identifies human protein markers specific for individuals with periodontitis and dental caries compared to orally healthy controls *PeerJ* **4** e2433

SPBOPEN 2018 IOP Publishing

Possible applications of bacterial cellulose in the manufacture of electrical insulating paper

N Zhuravleva[1], A Reznik[1], D Kiesewetter[1], A Stolpner[1], A Khripunov[2]

[1]Institute of Power Engineering and Transportation, Peter the Great Saint Petersburg Polytechnic University, Saint Petersburg 195251, Russia

[2]Institute of macromolecular compounds of Russian Academy of Science, Saint Petersburg, 199004, Russia

Abstract. The specimens of the electrical insulating cellulose paper made of bacterial cellulose or with the structure of which has been modified by the bacterial cellulose, were studied. It is shown that such insulating paper has a greater breakdown voltage compared to standard paper and better mechanical strength. The advantage of studied paper samples after accelerated thermal aging is most clearly seen. It is also found that the paper made of bacterial cellulose or paper with modified structure by bacterial cellulose has a greater resistance to the destructive effects of insulating liquids.

1. Introduction

Electrical insulation paper (EIP) is a widely demanded dielectric which raw materials base is deliberately and self-reproducible. In addition to the well-known use of cellulose in the production of insulating paper, the application of nanofibers of cellulose expands the possibilities of using this type of dielectric. For example, according to experts, such paper can serve as a flexible basis for electronic circuits and be used in various high-tech devices [1]. The use of nano-gel film (NGF) cellulose Glucoacetobacter xylinus (NGF CGX) [2] – the so-called bacterial cellulose (BC) (the average diameter of the crystalline fibrils of wich is 5-6 nm), opens up new opportunities for the use of cellulose in technology, in particular, creates the basis for the design of cellulose dielectrics of a new generation. The biopolymer is synthesized by bacteria (Gluconacetobacter, Rhizobium, Agrobacterium and Sarcina [3]) and accumulates in the form of NGF in different culture media. A positive feature of this biopolymer fiber is the absence of its composition of lignin, resins, fats, metal ions and other by-products. Structural features and unique properties of this material are in demand in technology. In particular, the use of bacterial cellulose in the production of electrical insulation paper can significantly improve the performance properties of paper-impregnated insulation of power ransformers.

2. The objects of study and technique of the measurements

The samples of electrical insulating paper (EIP) were made of nano-gel film of bacterial cellulose. Sample No 1 was manufactured by drying the NGF in the open air at room temperature and normal humidity. The sample № 2 was obtained from the suspension of bacterial cellulose obtained by grinding of the gel film in the industrial blender (JTC OmniBlend I, model TM-767), followed by casting the suspension on the surface of the glass. For comparison, mechanical and electrical properties of the samples of EIP paper modified by the microcrystalline cellulose (MCC) made of

Content from this work may be used under the terms of the Creative Commons Attribution 3.0 licence. Any further distribution of this work must maintain attribution to the author(s) and the title of the work, journal citation and DOI.
Published under licence by IOP Publishing Ltd

bacterial cellulose [4], the source of which was the NGF of the biopolymer (EIP – 97%, MCC BC – 3%) were investigated also. The measurement of the limit of mechanical strength of the paper was determined by using the testing machine ES series model ESM301/ESM301L "MARK-10". The monitoring of the formation of sludge particles in the dielectric fluids under thermal aging was carried out with the microscope and TV camera. Calculation of the parameters of contamination of the oil with sludge was fulfilled using the special computer program. The paper structure was investigated using the scanning electron microscope (SEM) "Supra 55VP-25-78" (Fig. 1, 2). The fiber sizes and the distance between nodes are obtained by computer processing of the obtained images by SEM.

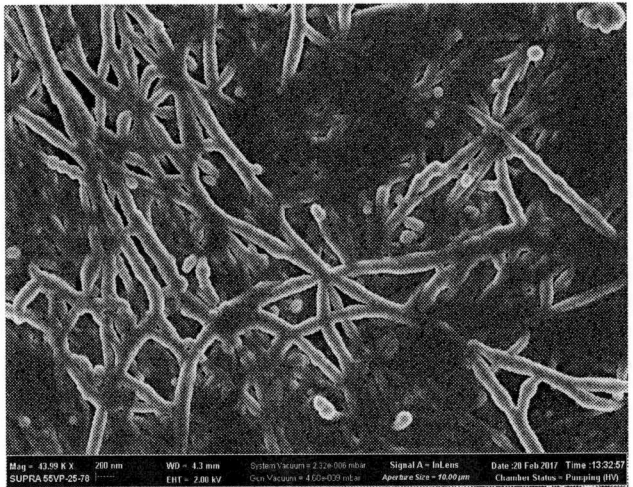 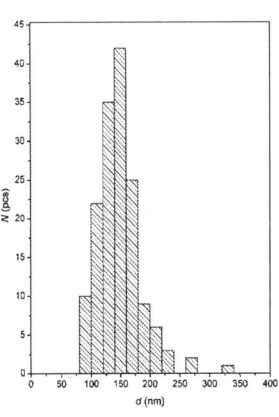

Figure 1 (a, b). (a) The structure of the BC paper (sample No 2), **(b)** The distribution of the thickness of the fibers of the sample of the bacterial cellulose paper (sample No 2)

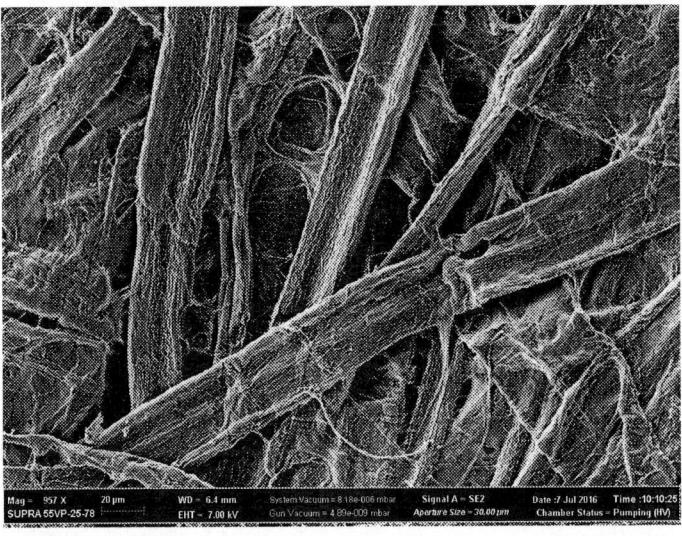 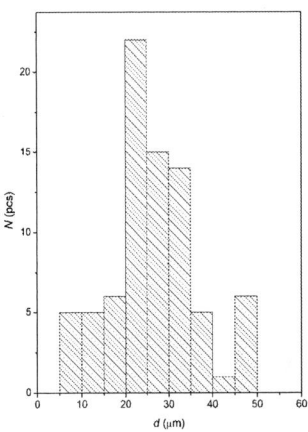

Figure 2 (a, b). (a) The structure of the of traditional electrical insulating paper, **(b)** The distribution of the thickness of the fibers of the sample

3. Obtained results

The average thickness d of the colloidal formations between the nodes of the cellulose mesh in the BC castings (No 2) was 150 nm with the standard deviation of 40 nm (Fig. 1), unlike the thickness of the fibers of coniferous pulp samples of traditional electrical insulating paper (EIC), which d was 28 μm (Fig. 2). The average distance between the nodes h of the cellulose mesh in the castings of BC was 130±70 nm.

The study of the mechanical properties of samples of cellulose dielectrics showed that the greatest value of the maximum mechanical tensile strength (MMTS) both in the initial state and after thermal aging had electrical insulating paper made of bacterial cellulose (Fig. 3 a). The results of measurements of MMTS as a function on time of thermal aging of the samples are shown in the Fig. 3. From the received data (Fig. 3) it follows that the MMTS values changes for the sample made of bacterial cellulose was negligible during the aging process, for the sample from EIC +10% BC - decreased by 6.5% after 120 h of aging, and for the sample EIC 100% - decreased by 28%. The average value of maximum mechanical tensile strength of the paper with the addition of BC after 120 hours of artificial aging was 1.5 times higher than that of standard EIP what is in good accordance with the results of works [4-8].

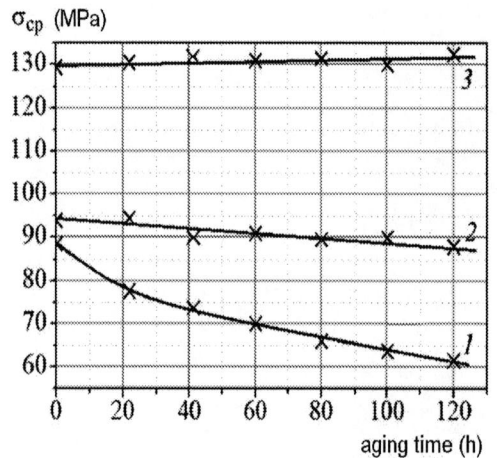 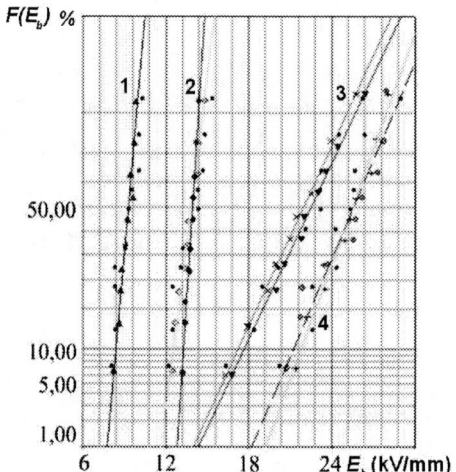

Figure 3 (a, b). (a) Dependence of average value of maximum mechanical tensile strength on thermal aging time for electrical insulating paper: 1 – standard EIP, 2 – 90% EIC+10% CGX, 3 – 100% CGX, **(b)** The distribution functions of short-term electric strength E_b of cellulose dielectrics in the initial state (before thermal aging): 1 – standard EIP, 2 – 90% EIC+10% CGX, 3 – 100% BC, 4 – NGF CGX.

The results of the study of the short-term electrical strength of four EIP samples are given in Fig 3 (b). Breakdown electric field strength (E_b) of these EIP samples (Fig. 3 b) was (kV/mm): 1 – 9.1, 2 – 13.7 and 3 – 21.7 4 – 24.9. Breakdown electric field strength of samples of the paper with BC mentioned above was 58 kV/mm for the sample No 1 and 58...69 kV/mm for the sample No 2 that is approximately 2.5 times more in comparison with the standard EIP.

On the basis of the fulfilled measurements of the geometric dimensions of fibers EIC and the structural elements (SE) of the BC, we can declare that as a result of thermo-oxidative processes occurring during thermal aging of plant cellulose the thickness and length of the fibers is reduced, and for bacterial cellulose length of the SE decrease but thickness of the SE increases. Hypothetically, such a partial disruption of the BC structure can serve to hardening of the material due to formation additional links between its fibers. The fact of the fiber structure change is confirmed by the results of the measurement of the cross-correlation function of the scattered coherent radiation by EIP [9-10] when the wavelength (λ) changes. It can be assumed that a decrease in the thickness and length of the fibers of the standard electrical insulating paper cause to an increase in the depth of penetration of optical radiation, respectively – to a narrowing of the spectral width ($\Delta\lambda$) of the cross-correlation function [9].

Research of electrophysical properties of cellulose dielectrics was carried out also for industrially made electrical insulating paper of the Russian and foreign production. It is shown that a dielectrics made of NGF CGX has significantly greater mechanical and electrical strength than an industrially manufactured insulating paper. It was also found that addition of the biopolymer increases the durability and service life of the insulating paper against to the destructive action of the liquid dielectrics, in particular, the electrical insulating oil of GK and synthetic ester MIDEL7131.

The obtained samples of the paper of BC and with BC have greater resistance to the destructive influence of heated electrical insulating liquids than the standard paper. The use of BC can also improve some parameters of conductive cable paper. However, the use of traditional paper making technology based on cellulose raw materials is complicated by the problem of splitting (disintegrating of the structure) of nano-gel film, which is very difficult due to the high structuring and low thickness of bacterial cellulose tapes, as can be seen from Fig. 1. A comparative analysis of the composition of paper samples on fiber confirmed the complexity of the qualitative splitting and grinding of raw materials of the biopolymer. However, our research on the development of technology for obtaining dielectrics from bacterial cellulose allows us to hope for the elimination of this problem.

The results of the performed studies confirmed the possibility of a significant improvement in the performance of electrical insulating paper based on bacterial cellulose or by modifying the structure of EIP by bacterial cellulose.

References

[1] Masaya Nogi, Shinichiro Iwamoto, Antonio Norio Nakagaito, Hiroyuki Yano *Advanced Materials* 2009 **21** (16), 1595–1598

[2] Baklagena Y G, Lukasheva N V, Khripunov A K, Klechkovskaya V V, Apkharova N A, Romanov D P, Tolmachev D A *Vysokomolekulyarnye soedinenia Seres A* 2010 **52**(4) 615

[3] Bielecki S, Krystynowicz A, Turkiewicz M, Kalinowska H *J Biopolymers Online* 2005 **5**

[4] Kiesewetter D V, Zhuravleva N M, Reznik A S, Tukacheva A V, Smirnova E G, Khripunov A K *Proc. 10th Electric Power Quality and Supply Reliability Conference, Tallinn* 2016 193

[5] Zhuravleva N M, Reznik A S, Kiesewetter D V, Tashlanov D O *Proc. 2017 IEEE North West Russia Young Researchers in Electrical and Electronic Engineering Conference, Saint Petersburg* 2017 1220

[6] Zhuravleva N M, Reznik A S, Kiesewetter D V, Tukacheva A V, Smirnova E G *Proc. 57th International Scientific Conference on Power and Electrical Engineering of Riga technical University, Riga* 2016 1

[7] Zhuravleva N M, Reznik A S, Kiesewetter D V, Tukacheva A V, Smirnova E G *Proc. ELEKTRO-2016, High Tatras, Slovac Rep.,* 2016 649

[8] Zhuravleva N M, Kiesewetter D V, Reznik A S, Smirnova E G, Khripunov A K *St. Petersburg polytechnic university journal engineering science and technology* 2018 **24**(1) 75

[9] Kiesewetter D, Malyugin V, Reznik A, Yudin A, Zhuravleva N *Journal of Physics: Conf. Series* 2017 **917** 042020

[10] Kiesewetter D V, Malyugin V I, Reznik A S, Yudin A V, Zhuravleva N M Proc. *XXVI International Scientific Conference Electronics - ET2017, Sozopol* 2017 1

SPBOPEN 2018 IOP Publishing

Magnetic nanoparticles for medical application with a coating deposited with various methods

V N Zorin[1], E B Naumisheva[2,3], V N Postnov[2,3], N V Evreinova[2,4], K G Gareev[1], D V Korolev[2,5]

[1]St. Petersburg Electrotechnical University, St. Petersburg 197376, Russia
[2]Almazov National Medical Research Centre, St. Petersburg 197341, Russia
[3]St. Petersburg State University, St. Petersburg 199034, Russia
[4]St. Petersburg State Technological Institute (technical university), St. Petersburg 190013, Russia
[5]ITMO University, St. Petersburg 197101, Russia

Abstract. The work is devoted to the evaluation of the effectiveness of the methods of modifying magnetic nanoparticles for the development of targeted drug delivery systems. Both chemisorption by aminated particles and adsorption by unaminated particles are considered. Various methods for the determination of albumin and the study of the kinetics of dissolution of the albumin coat are given.

1. Introduction

Nanoparticles of magnetite (Fe_3O_4) and maghemite (γ-Fe_2O_3) are most often used in biomedical research. These particles should be non-toxic and non-immunogenic, and their size should allow penetration through the capillaries into tissues and organs without the formation of thromboses for successfully *in vivo* application. In practice, most often used magnetic nanoparticles coated with organic compounds – fatty acids, dextran, starch, polyvinyl alcohol, polyethylene glycol, various proteins [1]. These coatings should provide not only electrostatic stabilization of nanoparticles, the safety of their use, but also the possibility of conjugation of medications and compounds that increase the targeting of delivery to the lesion sites. The promising material for the creation of such coatings is albumin. Albumin does not have the toxicity of many polymeric carriers, largely binds highly lipophilic compounds and provides transport of the sorbed drug substance to the cell, which in particular allows the use of the anti-cancer drug paclitaxel without the highly toxic solvent Cremophor EL [2]. Moreover, albumin carriers with a 10 nm size bind to the cell non-specifically, providing intracellular delivery of the preparation. However, the main advantage of albumin is the ability for active transport of drugs inside the endothelial cell. This is due to the fact that a specific substrate for albumin binding, the gp60 receptor, is expressed on the cell surface. Interaction with this receptor stimulates the phagocytosis of the albumin-gp60 complex by the cell. Subsequent exocytosis through the basolateral part of the cell causes the active delivery of an albumin-bound drug into the tumor interstitium [3].

To date, various methods have been developed for creating protein coatings on the surface of iron oxide nanoparticles. In [4], in order to obtain a protein coating on Fe_3O_4-γ Fe_2O_3 nanoparticles at the stage of the formation of magnetic particles, trypsin was used in an aqueous solution close to neutral pH at a temperature of $4°$ C. The authors of patent JP635019 used to create a binding layer containing

Content from this work may be used under the terms of the Creative Commons Attribution 3.0 licence. Any further distribution of this work must maintain attribution to the author(s) and the title of the work, journal citation and DOI.
Published under licence by IOP Publishing Ltd

organosilicon compounds on the surface of particles, the functional groups of which were used for carrying out the polymerization reaction and immobilization of biological components, on magnetic particles with polymer coatings containing immobilized biological components, including proteins. A method for the immobilization of trypsin on the surface of magnetite is known [5] including the treatment of the surface of magnetite with 3-aminopropyltriethoxysilane, the modification of the resulting derivative with glutaraldehyde, and subsequent binding of trypsin to the carrier. In work [6], to fix the adsorption layer of albumin macromolecules on magnetite nanoparticles, carbodiimide was used as the cross-linking agent. In the patent [7], a method for creating a stable protein coating based on the property of proteins to undergo cross-linking by the action of free radicals, which are generated on the surface of nanoparticles under the action of an initiator – hydrogen peroxide, was proposed.

It is obvious that the evaluation of the efficiency of using the above methods of modifying magnetic nanoparticles (MNPs) for the development of targeted drug delivery systems and the development of new methods is an urgent task, and this work is aimed at solving it.

2. Materials and methods
2.1. Synthesis of MNPs

The synthesis of the MNPs was carried out as follows [8]. To a solution containing a mixture of ferric sulfate and ferrous sulfate in a molar ratio of 2:1 and a volume of 700 ml, with constant stirring a mixture of 25% ammonium hydroxide solution and 1% ammonium acetate solution was added at a rate of 4 ml/min. Thus, the ratio of iron and ammonium acetate was 2:1:0.1. The synthesis was carried out until a saturated black color was fixed and a pH value of 8–9 was established. The next day, the resulting colloidal product was separated with centrifugation and washed four times with distilled water. To prepare the dry sample, the obtained MNPs were filtered and subjected to freeze-drying at a temperature of $-50\,^\circ$C and a pressure of 3 Pa for 48 hours.

2.2. Amination of MNPs

The modification of the MNPs surface was carried out according to the following procedure. In a 50 ml round-bottomed flask, 2 grams of dry MNPs and 25 ml of a 5% solution of 3-aminopropyltriethoxysilane were placed in pre-dried benzene. The reaction mixture was refluxed for two hours at 80° C using a thermostated cell connected to a LT-105a liquid thermostat (LOIP, Russia). The excess reagent was removed with repeated washing with dry chloroform using magnetic separation. At the final stage of washing, ethyl alcohol was used. During the modification and washing, the reaction mixture was vigorously stirred by means of a magnetic stirrer.

The total content of amino groups in the samples was determined as follows. 1 ml of 0.1N hydrochloric acid was added to the aminated MNPs and set to neutralize the amino groups for 15 minutes, periodically shaking. The suspension was then centrifuged for 5 minutes at 3000 min^{-1}. The supernatant was titrated with alkali (NaOH) with an acid-base indicator of methylorange 0.1. The final total content of amino groups was calculated from the amount of alkali.

The total content of amino groups does not give an idea of how much of the active substance can be chemisorbed on the modified surface. Therefore, the number of available amino groups by capacitance over the fluorescent dye indocyanine green (ICG) was determined. To this, 1 ml of a fluorescent dye solution of 1 mg/ml and 1 ml of distilled water were added to a suspension containing 50 mg of pre-aminated nanoparticles in 2 ml of water. Sorption was carried out in 15-ml polypropylene tubes on a LS-220 shaker (LOIP, Russia) at a stirring rate of 300 min^{-1} for 2 hours. The solution was then centrifuged for 5 minutes at 3000 min^{-1} and washed 5 times with distilled water and centrifuged.

Upon termination of the sorption of the ICG, its amount was determined. To this, 10 ml of a 0.1 N sodium hydroxide solution was added to the washed precipitate and stripped for 15 minutes. The solution was then centrifuged and 1 ml of supernatant was taken. The resulting solution was analyzed for the fluorescent dye content with the spectrophotometric method using the method [9] with a Unico 2802S spectrophotometer (Unico Sys, USA) at a wavelength of 700 nm.

The total amount of amino groups was 0.81 mmol/g, the amount of available amino groups was 0.048 mmol/g. The amount of available amino groups is about an order of magnitude lower than their total content in the sample. This can be explained by the fact that access of relatively large molecules to most amino groups in the shell is difficult. This fact was already considered by the authors in [10].

2.3. Formation of albumin coating on the surface of MNPs
The formation of the albumin coating on the MNPs surface was carried out with several methods:
- adsorption immobilization of albumin on the MNPs surface;
- chemisorption of albumin on the surface of aminated MNPs;
- cross-linking of the coating with glutaraldehyde.

Adsorption immobilization and chemisorption of albumin were carried out with a mixing device LS-220 at a frequency of 300 min^{-1} for 2 hours. To do this, 50 mg of MNPs was placed in a 15 ml polypropylene tube and 2 ml of a 20% albumin solution was added. Human albumin was used (Waxter AG, Vienna, Austria). After sorption, the resulting preparation was washed five times with magnetic separation.

Crosslinking of the coating with glutaraldehyde was carried out as follows. A sample of 50 mg of aminated MNPs was treated with a solution that was prepared by diluting 25 μl of 50% glutaraldehyde (Sigma Aldrich, Germany) solution with 10 ml of water followed by the addition of 1.3 ml of albumin (20% solution) for 1 hour with stirring, then repeatedly washed from the excess of reagents with water with magnetic separation.

2.4. Determination of the amount of albumin in aqueous solution
The method of determination of albumin with the aid of bromocresol green dye [11] was used in the work. It was believed that in the sample there were no substances interfering with the analysis [12]. Serum albumin, being a weak acid, binds in citrate buffer with bromocresol green to form a complex of gray-blue color. The photometric measurement is based on the presence of an absorption peak with a maximum at a wavelength of 585 nm by the molecules of the colored complex. All other components of the solution have weak absorption at this wavelength. Thus, the increase in the optical density of the reaction mixture is directly proportional to the content of albumin in the analyzed sample [13].

The preparation for determining the amount of albumin was prepared by adding to a citrate buffer of 500 ml volume and pH = 4.2 500 μl of Tween 80 and 90 mg of bromocresol green dye. To 5 ml of the preparation different amounts of albumin were added and spectrophotometric studies were performed at the 10th minute at a wavelength of 585 nm against a blank sample.

Similarly, a calibration was performed to determine the albumin in a Krebs-Henseleit solution [14]. For this, 2 ml of the preparation and 2 ml of the Krebs-Henseleit solution were taken.

2.5. Determination of the amount of albumin in the coating with thermogravimetry
The study of the amount of albumin in the coating was carried out with thermogravimetry by heating to 900 °C in the air at a rate of 10 °C/min on the SetaramSetsys Evolution (Setaram Instrumentation, France). It was assumed that the entire organic shell consists of albumin and burns in the air.

2.6. Investigation of the physical properties of samples
Microphotographs of the MNPs were obtained with a transmission electron microscope (TEM) with JEM-1400 STEM (JEOL, Japan). The static magnetic properties of the MNPs were studied using a vibrating sample magnetometer Lake Shore 7410 (Lake Shore Cryotronics Inc., USA) in air at standard temperature.

3. Results and Discussion
As can be seen from the obtained TEM images (Fig. 1), samples with albumin sorbed on the MNPs and chemisorbed on the aminated MNPs are individual coated particles. In this case, the coating is thicker in samples with albumin sorbed on unaminated MNPs. The sample with albumin cross-linked with glutaraldehyde is a cluster of which several MNPs are combined by a common coating. The average size of such clusters is 150 nm, while the size of individual nanoparticles is within 10–20 nm.

3.1. Determination of the amount of albumin adsorbed and chemisorbed on MNPs

An analysis of the amount of albumin adsorbed on the MNPs surface and chemically adsorbed on the surface of aminated MNPs was performed on the residual protein content. After immobilization, the initial solution was analyzed for the presence of albumin, which for adsorption was 7.7 mg, and for chemisorption was 7.3 mg for 50 mg of MNPs.

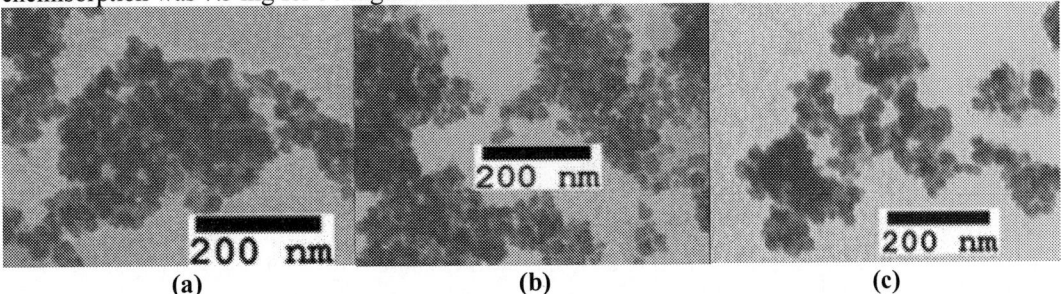

Figure 1(a,b,c). TEM images of MNPs coated with albumin with various methods: **(a)** sorption by MNPs; **(b)** chemisorption by aminated MNPs; **(c)** – crosslinking with glutaraldehyde.

3.2. Determination of the amount of albumin in the cross-linked shell by thermogravimetry

The thermogram of heating the MNPs sample with albumin crosslinked with the glutaraldehyde method is shown in Figure 2. Denaturation of the protein crosslinked on the nanoparticle surface usually occurs at temperatures below 100° C [15]. This stage coincides with the stage of water loss by the sample. Both processes are endothermic and therefore not separable. With a further increase in temperature, a two-stage burnout of albumin denaturing products is observed. In the approximation that the sample is dry, that all organic compounds burn out, it is possible to estimate the amount of albumin cross-linked on the MNPs surface, which is 16.7%. This figure correlates with the amount of adsorbed and chemisorbed protein (about 15%).

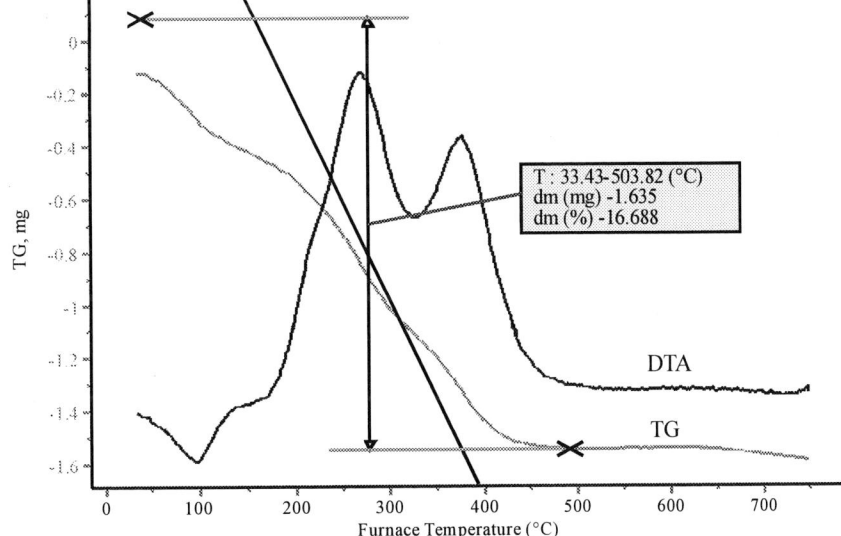

Figure 2. Thermogram of MNPs sample with albumin, cross-linked glutaraldehyde method.

3.3. Investigation of kinetics of dissolution of albumin coating in Krebs-Henseleit buffer

Samples were washed from the excess of the active substance and lyophilized. A sample of 50 mg was placed in a conical tube and poured 2 ml with a Krebs-Henseleit buffer solution, which was similar to the salt composition of the blood. Dissolution of the coating was carried out on a LOIP LS-110 agitator at a speed of 300 min^{-1}. At certain time intervals (0.5, 1, 2, 3, 5, 5 hours), the solutions were centrifuged and analyzed for albumin content as described above. However, according to the data obtained, it was not possible to establish the release of albumin. This suggests that albumin, in any method of immobilization, forms a strong coating that is insoluble in Krebs-Henseleit buffer solution.

Thus, obtained nanoparticles can be used in the development of magnetically controlled delivery products.

Acknowledgments

The study was carried out with the financial support of the Russian Foundation for Basic Research in the framework of scientific projects No. 17-00-00275 (17-00-00272) and No. 16-32-60010.

The work is based on the equipment of the Resource Center of St. Petersburg State University "Innovative Technologies of Composite Nanomaterials".

References

[1] Piotrovsky L B 2013 Essays on nanomedicine (St. Petersburg: Ed. The European House) p 204
[2] Desai N, Trieu V, Yao Z 2012 *Clin. Cancer Res.* **14** 1317
[3] Alyautdin R N, Romanov B K, Lepakhin V K, Bunyatyan N D, Merkulov V A 2014 *Safety and risk of pharmacotherapy* **2** 10
[4] Nishimura K, Hasegawa M, Ogura Y, Nishi T, Kataoka K, Handa H, Abe M 2002 *J. Appl. Phys.* **91** 8555
[5] Bendikene V G, Razyunas A A, Yuodka B A 1989 *The method of obtaining immobilized trypsin* Patent SU 1518373 ,
[6] Peng Z G, Hidajat K, Uddin M S 2004 *J.Coll. Interf.Sci.* **271** 277
[7] Rozenfeld M A, Bychkova A V, Sorokina O N, Kovarskii A L, Leonova V B, Lomakin S M, Makarov G G 2013 *Method of obtaining protein coatings on the surface of solid bodies containing ions of metals of variable valency* Patent RU 2484178
[8] Toropova Y G, Golovkin A S, Gareev K G, Malashicheva A B, Gorshkov A N, Afonin M V, Korolev D V, Galagudza M M 2017 *Int. J. Nanomedicine* **12** 593
[9] Gareev K G, Babikova K Y, Postnov V N, Naumisheva E B, Korolev D V 2017 *J.Phys. Conf. Series* **917** 042008
[10] Galagudza M, Korolev D, Sonin D 2010 *J.Manufact. Techn. Manag.* **8** 930
[11] Doumas B T, Watson W A, Biggs H G 1971 *Clin. .Chim. Acta* **31** 87
[12] Young D S 2000 Effects of drugs on clinical laboratory tests (5th ed. AACC Press)
[13] Tietz N W 1995 ed. Clinical Guide to Laboratory Tests (3rd ed. Philadelphia, Pa: WB Saunders) p. 22
[14] Krebs H A, Henseleit K, Hoppe-Seylers Z 1932 Physiol. Chem. **210** 33
[15] Kulikova G A, Ryabinina I V, Guseynov S S, Parfenyuk E V 2010 Thermochim. Acta **503-504** 65

A studying of subphase temperature and dissolved ascorbic acid concentration influence on the process of Langmuir monolayer formation

M M Qassime[1,2], V A Goryacheva[1], A J Al-Alwani[1,3],
T N Lugovitskaya[1], A B Shipovskaya[1], E G Glukhovskoy[1]

[1]Department of Nano- and Biomedical Technologies, Saratov State University,
Saratov, 410012, Russia
[2]Ministry of Science and Technology, Baghdad, Iraq
[3]Babylon University, Babylon, Iraq

e-mail: muhannadmq77@gmail.com

Abstract. Some of the prosperities of ascorbic acid (AscA) as a subphase, temperature and concentration were investigated in the Langmuir technique. We studied the forms of ascorbic acid (*L*- & *D*-AscA) actions as a subphase on monolayer formation in different concentration with different temperature in the Langmuir technique as a model of cell membranes under fixed pressure. In this study we have been noticed that there are different effects of ascorbic acid (*L*- & *D*-AscA) forms as a subphase, temperature and concentration. These results have many advantages include the possible fine control over the composition, circumstances and packing of the membrane being mimicked.

1. Introduction

Vitamin C or ascorbic acid (AscA) is a white crystalline solid soluble in water. AscA has important role in biochemical function required for the biosynthesis of collagen, L-carnitine, certain neurotransmitters. It is also involved in protein metabolism and also has important functions in physiological antioxidant mechanism [1-3] through regenerate other antioxidants within the body, including alpha-tocopherol (Vitamin-E) [4]. When we talk of AscA, we refer to the group of ascorbic acid analogs that can be both synthetic and natural molecules [5-7]. It is primarily comes in two forms: *L*-AscA and *D*-AscA (Figure 1).

The *L*-AscA, which can come in both natural (found in fruits and vegetables) and synthetic forms (found in most other supplements). Between the natural and synthetic varieties of *L*-AscA there are no known differences in how they affect our bodies. *D*-AscA, meanwhile, does not exist in nature and though chemically identical to its counterparts, is molecularly different. This molecular difference that makes *D*-AscA impossible to be synthesized by humane body and unusable in a vitamin supplements [8].

Content from this work may be used under the terms of the Creative Commons Attribution 3.0 licence. Any further distribution of this work must maintain attribution to the author(s) and the title of the work, journal citation and DOI.

Published under licence by IOP Publishing Ltd

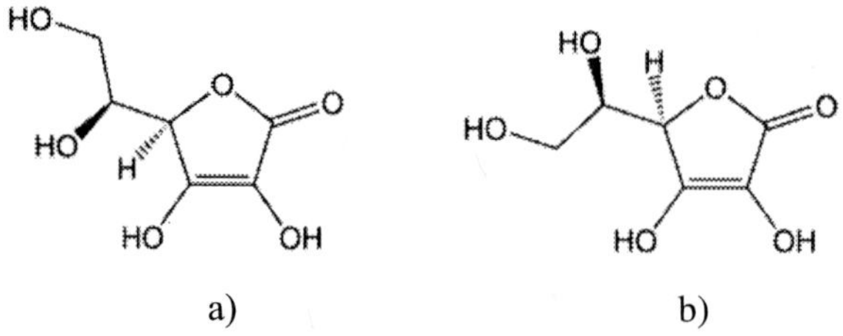

Figure 1. Ascorbic acid isomers **(a)** *L*-AscA and **(b)** *D*-AscA

AscA is actively transported into the body via two sodium-dependent vitamin C transporters SVCT1 and SVCT2 [9, 10]. Both of these transporters show significantly more affinity for the *L*- versus *D*-AscA [9, 11], and this selectivity likely explains earlier observations of significantly lower tissue accumulation and anti-scorbutic activity of *D*-AscA in guinea pigs [12, 13]. Although *D*-AscA acid is a commonly added food preservative [14], administration of *D*- and *L*-AscA together does not affect the bioavailability of the latter in humans [15].

Biological membranes contain lipid bilayers as their basic structural unit. Lipid bilayers are sheet-like assemblies of thousands of amphiphilic lipid molecules held together by hydrophobic interactions between their acyl chains. Such bilayers form the boundaries between intracellular cytoplasm and the cell's outside environment, as well as between the interior of many of the cell's organelles and their cytoplasm.

Lipid monolayers can be formed on the surface of water/buffer using a Langmuir film balance. Langmuir technique is one of the most promising techniques for preparing lipid monolayers, as it enables the precise control of the monolayer thickness, homogeneous deposition of the monolayer over large areas and the possibility to make multilayer structures with varying layer composition. In the Langmuir model system, various parameters such as *lipid composition, subphase, and temperature* can be chosen to imitate biological conditions In addition, lipid monolayers are very well-defined, stable, homogeneous bi-dimensional system with planar geometry [16, 17].

In Langmuir technique the solvent properties of water are associated with the attraction between its electrical dipole and the charge of the solute [18]. Most of the monolayer forming substances used by the pioneers of the method, and still mostly used today, is composed of two parts. One that is water loving and if alone, dissolves in water. Second part has the opposite property. Such molecules composed of a hydrophilic and hydrophobic part are called amphiphiles or surface active agents (in short surfactants). The most common prototypes of surfactant are the long chain fatty acids. An example of such a molecules to float and conveys to them the ability to form a monolayer.

The subphase is defined as the substance on which the monolayer is going to be formed [19]. The surface tension is analogous to the vapour pressure, constant at equilibrium at a solid-gas interface but changing with temperature. But, unlike the vapour pressure which increases with increasing temperature, the surface tension decreases when the temperature increases. Surface pressure is defined as the difference between the surface tension of pure subphase (for example water) and the same subphase covered with molecules [20]. The injection of few microgram of surfactant at the air-water interface will at first get the molecules to spread all over the available surface area. Applying an external force to those floating surfactant molecules will affect their positions within the trough and ultimately, if compression is sufficient, create a solid film [21]. The pressure-area isotherms are used more frequently and provide information about structure, area, interactions between lipid monolayers and various membrane-lytic nano-objects, phase transitions, compressibility and hysteresis of the monolayers.

2. Materials and methods

An arachidic acid (ArA) 99 % and chloroform 99.5 % were received from Sigma Aldrich. The ArA was dissolved in chloroform to concentration of 0.1 M is used as a work solution. The 70 µl of work solution was sprayed first on the water subphase with resistivity 18.2 MΩ×cm. second on L- and D-AscA (Meligen corp., RF and Khimreaktiv corp., RF, respectively) as subphase in concentration 10^{-2} and 10^{-3} M. After 5 minutes the monolayer was compressed by moveable barriers at the constant rate of area decreasing of 10 cm^2/min and this done in different temperatures 25, 37 and 41 °C.

3. Results and discussion

Figure 2, we can see the compression isotherms of Langmuir monolayers (L- & D-AscA) formation in concentration 10^{-3} and 10^{-2} M and temperatures were 25, 37 and 41 °C. The results show us a little different in the gas phase monolayer, liquid phase and collapses formation with different concentration of L- and D-AscA in temperature 25 °C. But, in the temperatures 37 and 41 °C we have different in the L- and D-AscA form, collapses, liquid and gas phase monolayer formation in different concentrations, this significant differences due to the effect of temperature on the surface tension of subphase will be decreased and this we can see clear in high concentration.

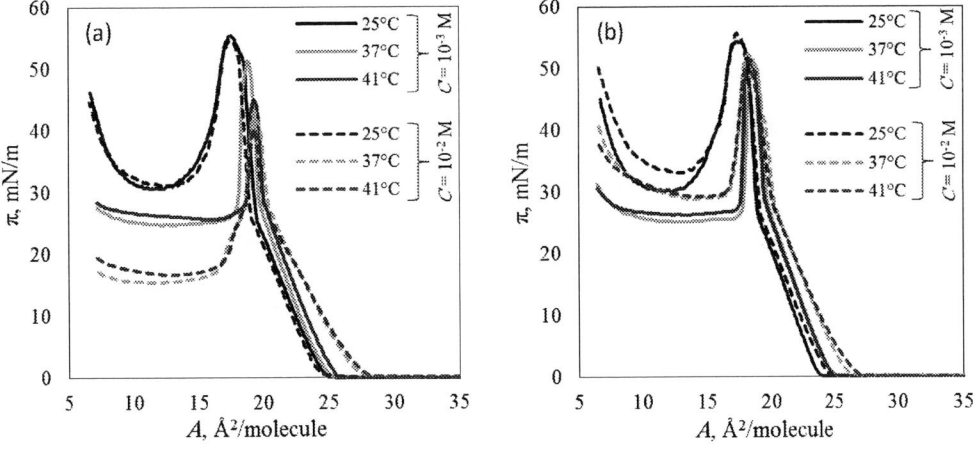

Figure 2. Compression isotherms of L-AscA **(a)** and D-AscA **(b)**.

Lipids with shorter chains are less stiff and less viscous because they are more susceptible to changes in kinetic energy due to their smaller molecular size and they have less surface area to undergo stabilizing van der Waals interactions with neighboring hydrophobic chains. While lipid chains with carbon-carbon double bonds (unsaturated) are more fluid than lipids that are saturated with hydrogen and thus have only single bonds. On the molecular level, unsaturated double bonds make it harder for the lipids to pack together by putting kinks into the otherwise straightened hydrocarbon chain, these results have been approved by other studies in the same field [22, 23].

4. Conclusion

Ascorbic acid molecular form has effect on the formation of Langmuir monolayer at high temperatures. The high concentration of ascorbic acid in subphase has a significant effect on the formation of Langmuir monolayer. There is a significant effect of temperature on the isotherm monolayer formation. No significant effect of low temperature on the formation of Langmuir monolayer, but high temperature has a significant effect with high concentration of ascorbic acid in subphase on the formation of Langmuir monolayer.

For next experimental research:

It will be interesting to study effect of high concentration of AscA 10^{-1} M on pressure-area isotherm. Of particular interest are studies the specifics of the interaction ascorbic acid molecules with a complex composition monolayer. How will be the membrane structure and their properties change in case of including vitamin E molecules.

Other questions type about the effect of temperature and concentration on monolayer phospholipid with protein and cholesterol to explain cell membrane fluidity.

Acknowledgments

This work was supported by grants from the RFBR № 17-07-00407-a and RFBR mol_nr № 17-32-50137.

References

[1] Li Y, Schellhorn H E 2007 *J Nutr* **137** 2171
[2] Carr AC, Frei B 1999 *Am J Clin Nutr* **69** 1086
[3] Frei B, England L, Ames BN 1989 *Proc Natl Acad Sci U S A* **86** 6377
[4] Jacob RA, Sotoudeh G. Vitamin C function and status in chronic disease 2002 *Nutr Clin Care* **5** 66
[5] Huh W K, Lee B H, Kim S T, Kim Y R, Rhie G E, Baek Y W, et al 1998 *Mol. Microbiol* **30** 895
[6] Raić-Mlić S, Svedruzić D, Gazivoda T, Marunović A, Hergold-Brundić A, Nagl A, Balzarini J, De Clercq E, Mintas M 2000 *J. Med. Chem* **43** 4806
[7] Yamamoto I, Tai A, Fujinami Y, Sasaki K, Okazaki S 2002 *J. Med. Chem* **45** 462
[8] Carocho M, C.F.R. Ferreira I 2013 *Food and Chemical Toxicology* **51** 15
[9] Tsukaguchi H, Tokui T, Mackenzie B, Berger U V, Chen X Z, Wang Y, Brubaker R F, Hediger M A 1999 *Nature* **399** 70
[10] Savini I, Rossi A, Pierro C, Avigliano L, Catani M V 2008 *SVCT1 and SVCT2: Amino Acids* **34** 347
[11] Rumsey S C, Welch R W, Garraffo H M, Ge P, Lu S F, Crossman A T, Kirk K L, Levine M 1999 *J. Biol. Chem* **274** 23215
[12] Goldman H M, Gould B S, Munro H N 1981 *Am. J. Clin. Nutr* **34** 24
[13] Hughes R E, Hurley R J 1969 *Br. J. Nutr* **23** 211
[14] Levine M 1996 *Am. J. Clin. Nutr* **64** 381
[15] Sauberlich H E, Tamura T, Craig C B, Freeberg L E, Liu T 1996 *Am. J. Clin. Nutr* **64** 336
[16] Brezesinski G, Mohwald H 2003 *Adv Colloid Interface Sci* **100-102** 563
[17] Marsh D 1996 *Biophys J* **70** 2248
[18] Shaw D J 1992 *Colloid and Surface Chemistry*, 4th Ed., (Oxford, Butterworth-Heineman)
[19] Martin P, Szablewski M 1995 *Operating Manual* 4th Ed, (Grunfeld, NIMA Technology ltd. Coventry)

[20] Adamson A W, Gast AP 1997 *Physical Chemistry of Surfaces*, 6th Ed., (New York, John Wiley & Sons)
[21] Atkins P 1998 *Physical Chemistry*, 6th Ed., (New York, W.H. Freeman & Co)
[22] Lee D C, Chapman D 1987 *Symp Soc Exp Biol* **41** 35
[23] Peter J Q 1988 *Symposia of the society for experimental biology* **42** 311

SPBOPEN 2018

IOP Publishing

IOP Conf. Series: Journal of Physics: Conf. Series **1124** (2018) 031011   doi:10.1088/1742-6596/1124/3/031011

Investigation optical properties and functional surface characteristics of nanoparticles based on porous silicon for applications in biomedicine

A O Belorus[1], A I Pastukhov[1], A S Lenshin[2], P V Seredin[2], Yu M Spivak[1] V A Moshnikov[1]

[1]St. Petersburg State Electrotechnical University «LETI», Saint Petersburg, Russian Federation
[2]Voronezh State University, Voronezh, Russian Federation

Abstract. In this paper optical characteristics of porous silicon layers and nanoparticles are investigated. Samples were obtained by electrochemical anodic etching of n-type conductivity monocrystalline silicon wafers. It is shown that the surface morphology porous silicon samples are depending on etching technological parameters.

1. Introduction

Nowadays porous silicon (por-Si) is one of the perspective material in field of drug delivery system (DDS). Nanocontainers based on porous silicon have attractive properties as stability, ability to high loading, high biocompatibility and the prolonged release possibility of one, two or more drugs with different physico-chemical properties [1-3]. However, to enable loading and retention of useful drugs in the porous nanoparticles we are needing to know information about internal structure, morphology and surface chemistry of the delivery drug system. Determination of nanoparticles size and form is also an important task for DDS applications in vivo. Since it is expected that these characteristics will affect bioavailability, biodistribution of nanoparticles and their interaction with the cells of the body tissues [3-6].

Thus, it is necessary to investigate changes in composition, morphology and optical properties of porous silicon nanoparticles, depending on the method of their preparation for further application in medicine and pharmacology fields [7-13].

2. Experimental

In this work, porous silicon layers were obtained by electrochemical anodic etching of n-type conductivity monocrystalline silicon wafers with [100] crystal orientation. As electrolytes are used solutions based on hydrofluoric acid with added alcohol and dimethylformamide. Por-Si powders were obtained from meso, macro and nanoporous silicon samples by using ultrasonic bath (5 min, 100 W, 35 kHz). The transmission electron microscopy powders' results are shown in Figure 1, 2.

Content from this work may be used under the terms of the Creative Commons Attribution 3.0 licence. Any further distribution of this work must maintain attribution to the author(s) and the title of the work, journal citation and DOI.
Published under licence by IOP Publishing Ltd

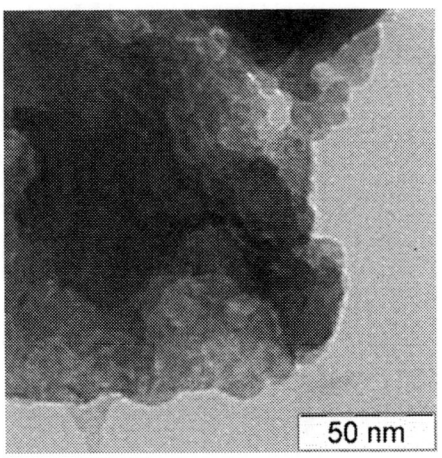 

a   b

Figure 1. TEM image of the surface of a) "mesoporous" -serial number 1 b) "macroporous" – serial number 2 silicon powder.

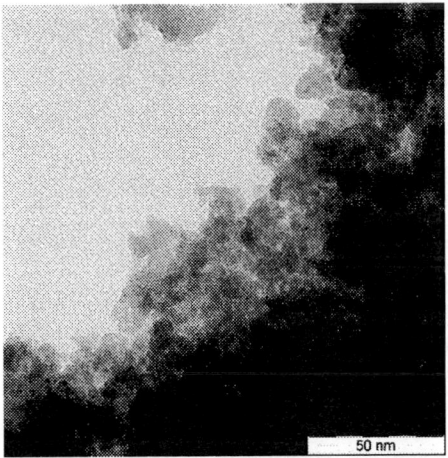

Figure 2. TEM image of the surface of "nanoporous" – serial number 3 silicon powder.

The conditions for obtaining porous silicon samples are presented in Table 1. As can be seen from Table 1 the technological parameters varied during the electrochemical process were the current density and the anodizing time, the composition of electrolyte.

Table 1. Technological parameters of electrochemical etching.

Serial number of samples	Resistivity, $\Omega \cdot cm$	Electrolyte composition	Anodizing time, min	Anodization current density, mA/cm^2
1	0.2	$HF:C_3H_8O:H_2O_2$	10	25
2	0.2	$HF:C_3H_8O:H_2O:$ $(CH_3)_2NC(O)H$	10	25
3	1	$HF:C_3H_8O:H_2O_2$	10	70

SPBOPEN 2018   IOP Publishing

Porous silicon layer surface composition was analyzed using infrared Fourier-transform spectrometer Vertex 70 (Bruker) with console for spectroscopy of disturbed total internal reflection (measurement range 4000-550 cm^{-1}, resolution 2 cm^{-1}). Analysis estimated depth of this method at wavenumbers up to 2000 cm^{-1} is limited to 1.5 μm, and in the range 2000-4000 cm^{-1} does not exceed 10 μm.

3. Results and discussion

Porous silicon powders, obtained under various electrochemical etching technological conditions, were investigated using Fourier-transform spectroscopy method. IR transmission spectra of nano-, meso- and macroporous (macro-) porous silicon nanoparticle powders is represented at Figure 3.

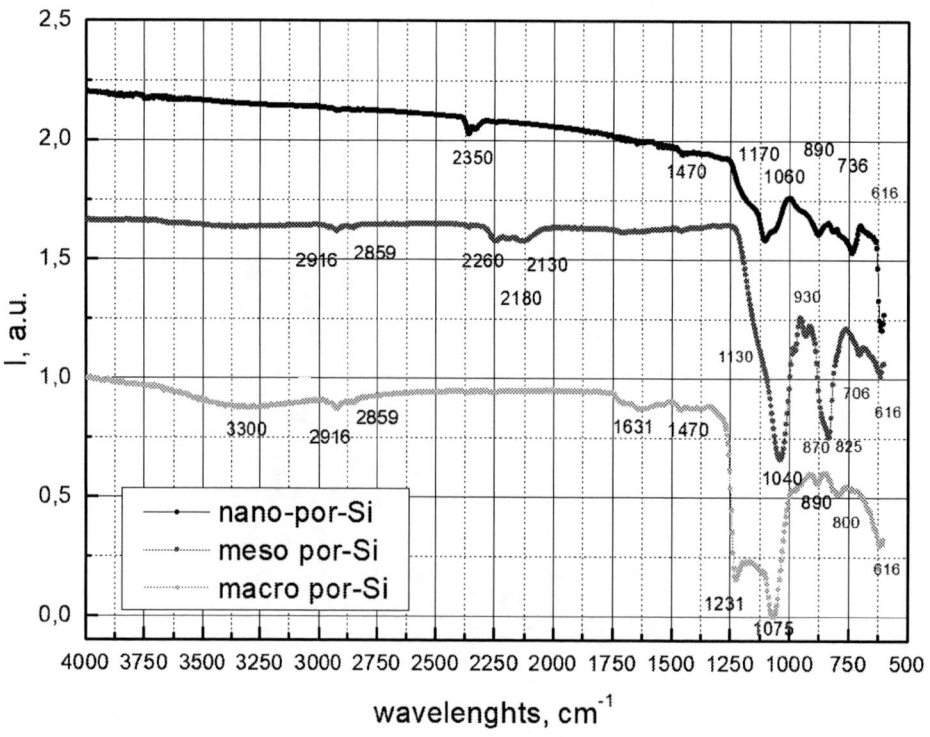

Figure 3. IR-transmission spectra of porous silicon nanoparticle samples for various series: a) series 3 nano por-Si b) series 1 meso por-Si c) series 2 macro por-Si

As seen from the spectra, all samples are characterized by sufficient number of Si-Si bonds valence symmetric vibrations (frequency 616 cm^{-1}), SiO valence antisymmetric vibrations in O-SiO and C in C-SiO (frequencies 1060, 1130, 1075 cm^{-1} for nano-, meso-, macro por-Si, respectively). Since the porous silicon nanoparticles surface is highly developed (powders specific surface area reaches 700 m^2/g), and there are hydrogen-containing components (HF, C$_3$H$_8$O) during the process of production and storage, all samples have adsorbed hydrogen, which forms bonds with silicon atoms in various forms (frequencies 800 cm^{-1} SiH$_2$ twisting, 870 and 890 cm^{-1} – scissor).

A noticeable presence of antisymmetric deformation bonds CH$_3$ vibrations at frequencies 1470 cm^{-1} for samples of series 2 and 3, as well as the valence antisymmetric CH bonds vibrations in CH, CH$_2$ at frequencies 2859, 2916 cm^{-1} for macro- and mesoporous silicon samples only, allows to make a conclusion about the dispersion storage medium influence (isopropyl alcohol) on the CH_x complexes formation on the nanoparticles surface of all samples.

4. Summary
In this work, three series of porous silicon samples were obtained by electrochemical anodic etching. As electrolytes are used solutions based on hydrofluoric acid with added alcohol and dimethylformamide. Porous silicon powders were obtained from meso, macro and nanoporous silicon samples by using crushing in ultrasonic bath.

Electrolyte with N,N-Dimethylformamide addition (samples of series 1 and 2 comparison) allows to form larger porous silicon nanoparticles due to the fluoride ions decreased concentration in the solution. An increase in anodization current density leads to a decrease in the size of the porous silicon nanoparticles to the nanometer scale (sample of series 3).

From the analysis of infrared Fourier-transform spectroscopy data, it can be concluded that the samples surface of all series is passivated by oxygen and hydrogen molecules, which is explained by the presence of water and isopropyl alcohol in the electrolyte composition.

Acknowledgments
The work was supported by a grant from the President of the Russian Federation (MK-4865.2016.2), projects 0021109 competition UMNIK 15-12 and within the framework of the project "Increase of the publication activity of scientific and scientific-pedagogical workers" of program to improve competitiveness among the world's leading scientific- educational centres. The investigation was accomplished on account of Russian Science Foundation (project N 17-79-20239).

References
[1] Song Yu, Li Yi, Xu Q, Liu Z 2017 Mesoporous silica nanoparticles for stimuli-responsive controlled drug delivery: advances, challenges, and outlook *International Journal of Nanomedicine* (№) 12 pp 97-110

[2] Santos H A 2012 Porous-based biomaterials for tissue engineering and delivery applications *Biomatter* (№) 4 (21) pp 237–238

[3] Santos H A, Hirvonen J 2012 Nanostructured porous silicon materials: Potential candidates for improving drug delivery *Nanomedicine* (№) 9 vol 7 pp 1281–1284

[4] Langer R 1998 Drug delivery and targeting *Nature* (№) 6679 vol 392 pp 5–10.

[5] Liu Q, Zhang J X, Sun W 2012 Delivering hydrophilic and hydrophobic chemotherapeutics simultaneously by magnetic mesoporous silica nanoparticles to inhibit cancer cells *Int. J. Nanomedicine* (№) 10(2147) vol 7 pp 999–1013

[6] Spivak Yu M, Belorus A O, Somov P A, Bespalova K A, Moshnikov V A, Tulenin S S 2015 Porous Silicon Nanoparticles for Targeted Drug Delivery: Structure and Morphology *Journal of Physics: Conference Series* vol 543 (1)

[7] Uhanov Ju I 1977 Opticheskie svojstva poluprovodnikov *Nauka* 366 p.

[8] Ledoux G, Guillois O, Porterat D, and Reynaud C 2000 Photoluminescence properties of silicon nanocrystals as a function of their size *Physical review B* (№) 23 vol 62 pp 15943–15951.

[9] Belorus A O, Maraeva E V, Spivak Y M, Moshnikov V A 2015 The study of porous silicon powders by capillary condensation *Journal of Physics: Conference Series* vol 586 (1)

[10] Spivak Yu M, Myakin S V, Moshnikov V A, Panov M F, Belorus A O, Bobkov A A 2016 Surface functionality features of porous silicon prepared and treated in different conditions *Journal of Nanomaterials* vol 2016

[11] Belorus A O, Bespalova K, Spivak Yu M 2016 Morphology and internal structure of porous silicon powders in dependence on the conditions of post-processing *Proceedings of the 2016 IEEE*

North West Russia Section Young Researchers in Electrical and Electronic Engineering Conference, EIConRusNW 2016 pp 23-28

[12] Pastukhov A I, Belorus A O, Bukina Y V, Spivak Y M, Moshnikov V A 2017 Influence of technology conditions on the surface energy of porous silicon using the method of contact angle *Proceedings of the 2017 IEEE Russia Section Young Researchers in Electrical and Electronic Engineering Conference, EIConRus 2017* pp 1183-1185

[13] Belorus A O, Bukina Y V, Pastukhov A I, Stebko D S, Spivak Yu M, Moshnikov V A 2017 Investigation of porous silicon obtained under different conditions by the contact angle method *Journal of Physics: Conference Series* vol 929

Fluorescent method of bacterial contamination control on meat surface

M V Mekhrengin[1], I K Meshkovskii[1], L A Kaftyreva[2], V I Guryev[1] and D A Pogorelaya[1]

[1]Department of Light-Guided Photonics, ITMO University, St. Petersburg 197101, Russia
[2]Intestinal Infections Laboratory, St. Petersburg Pasteur Institute, St. Petersburg 197101, Russia

Abstract. This work presents the fluorescent method of meat bacterial contamination control based on the influence of E. coli metabolites on the fluorescence spectrum of meat surface. The experiment revealed that proposed method allows detecting bacterial contamination with concentrations above 10^6 CFU/cm^2 on a beef surface.

1. Introduction

Food safety and quality control are the most disturbing problems of our time. Consumption of the contaminated food causes diseases associated with infected products, which are growing from year to year [1]. Thereby quality assurance is the fundamental task for preventing foodborne diseases. Conventional techniques are:

1. the culture and colony counting methods based on bacteria counting [2],
2. immunology-based methods based on antigen-antibody interactions [3],
3. polymerase chain reaction (PCR) method [4].

Although these well-established techniques are highly sensitive to low bacterial concentrations they are very time consuming (from a few tens of hours to few days) [5]. Moreover, most of the methods require a special sample preparation and/or specific reagents application [6] that increases detection time and makes it necessary to accomplish all the measurements by the specialists in the lab conditions.

Other methods should be determined to reduce time consumption and requirements for measurement conditions. Most suitable one is presumed optical taking into account its rapid response and potentially high sensitivity. In this work optical methods of diagnostics will be considered.

2. Brief overview of methods

There are four widely spread optical methods for diagnosis food contamination. All of these methods operate in real-time but vary in sensitivity, complexity of design of sensors and cost of implementation. Further, there is a brief overview of them.

- *Surface Plasmon Resonance (SPR) methods*, which are extremely sensitive (allow to measure small concentrations of bacteria less than 10 CFU/cm^2 [7]) but complex in the design to the same extent.
- *Interferometric methods* that are potentially quite sensitive but need to prepare testing sample transparent to the light. Such a requirement automatically makes this method time consuming [8].

Content from this work may be used under the terms of the Creative Commons Attribution 3.0 licence. Any further distribution of this work must maintain attribution to the author(s) and the title of the work, journal citation and DOI.
Published under licence by IOP Publishing Ltd

- *Fiber-optic methods* that can be implemented in simple design using photodetector, light source and sensing fiber. The significant drawback is invasiveness of probing since fiber must be placed into the sample [9] that greatly increases the risk of contamination of the sample by fiber.
- *Spectral methods* are Raman, Fourier-transform infrared spectroscopy and fluorescence. Fourier-transform infrared spectroscopy is insensitive to low concentrations of bacteria. According to the literature, the best result is the detection of concentrations of $\sim 10^3$–10^4 CFU/cm^2 [10]. Raman spectroscopy allows high sensitivity but requires a precisely aligned optical setup [11]. That imposes strong restrictions on the use of both methods in production environment.

Fluorescent methods stand out among spectral methods because in prospect they may be sensitive to low concentrations of bacteria and to different bacteria species but quite easy to implement. In fact, methods have instantaneous measurement time. The current state of the art technologies allow you to refuse to use fluorescence tags for labeling samples because the electronics market offers spectrum analyzers with high sensitivity [12, 13] and powerful narrow-band light emitters. Thus, nowadays fluorescent methods are most appropriate methods to control food contamination among optical spectral methods, which can be created in portable and low cost implementation.

3. The fluorescent method of the food contamination control

Bacterial contamination control based on the fluorescence method which uses the fact that infected and uninfected biological surfaces have different spectra of the fluorescence in the visible range of electromagnetic radiation. Differences in spectra are caused by the presence or absence of metabolites of bacteria E.coli on that surface. Moreover, the shape of the fluorescence spectrum, which is excited by the powerful narrow-band light source, changes with the increasing of the amount of E.coli metabolic products that was confirmed while making experiments. Thereby comparison the radiation intensities at some narrow bands of the fluorescence spectrum gives an opportunity to detect the presence of pathogenic or opportunistic bacteria (Salmonella typhimurium and Escherichia coli) on the surface of the sample.

The measurements are conducted in conditions of an uneven biological surface, a varying distance from surface to the optical probe and the changing angle of incidence of radiation. Therefore, the intensity of fluorescence fluctuates from one measurement to another and so criterion for controlling contamination is to be independent on intensity. In this article task of unambiguous determination of bacterial contamination of the test sample is solved on the basis of the proposed criterion of estimating the bacterial contamination by the ratio of particular spectrum marker wavelengths.

4. Results and discussions

Figure 1 shows the fluorescence imaging device with fiber-optic bundle, which was designed for the bacterial contamination control on a meat surface. The current device uses 450 nm 1.6 W laser diode (PL TB450B, OSRAM Opto Semiconductors GmbH, Germany) as the excitation light source, the multipurpose tunable spectrometer (ASP-150T, Avesta Ltd., Russia), 7x1 fiber bundle (art photonics GmbH, Germany). All measurements were conducted using this device.

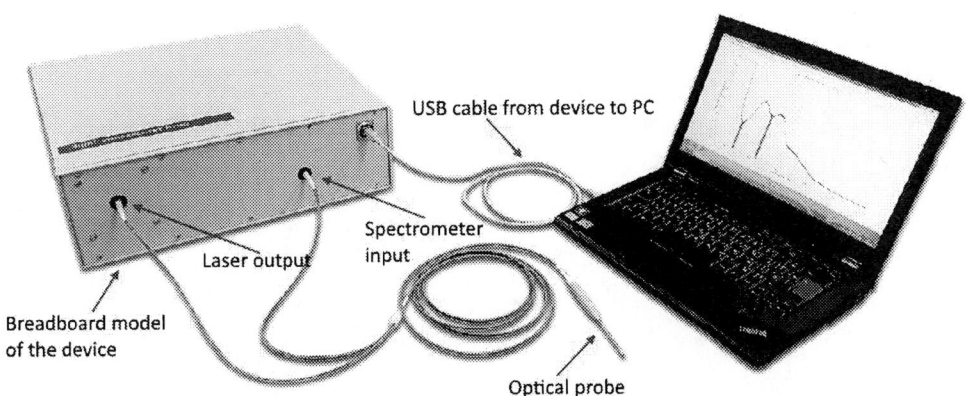

Figure 1. The fluorescence imaging device with fiber-optic bundle.

Bacteria of the *Escherichia coli* (E. coli) species are the most common pathogenic bacteria that occur on food products, particularly on beef. In this regard, experiments with bacterial culture of E. coli ATCC 29522 were conducted.

All measurements were performed in research zone of Intestinal Infections Laboratory (St. Petersburg Pasteur Institute, Russia) without thermostabilization, at room temperature (25 ± 5 °C). Test samples of beef were artificially contaminated with E. coli and put into dishes.

Figure 2 depicts the fluorescence of the contaminated sample. A peak in emitted spectrum is showed on the figure and it is growing over time. This peak characterizes the found fluorescence that is radiated by the contaminated samples. Raise is caused by the increase in the number of metabolites of the bacteria during time. Fluorescence is within visible band of the electromagnetic spectrum and has an emission peak at wavelength $\lambda = 535nm$. The microbiological measurements found concentration of bacteria to be 10^6 CFU/cm^2 is on the sample after 5 hours of the experiment.

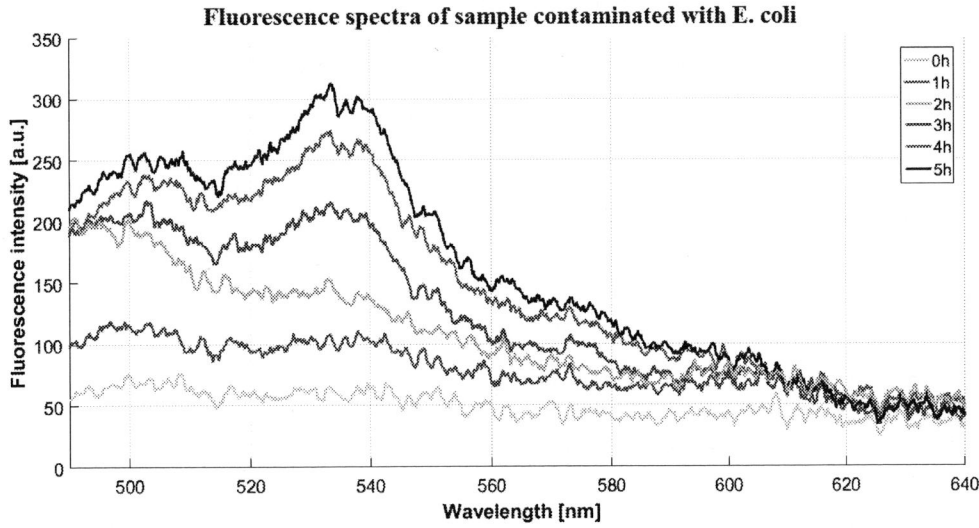

Figure 2. The impact of the bacteria on the beef fluorescent spectra of sample artificially contaminated with E. coli.

Figure 3 shows the fluorescence intensity of uninfected meat sample during the experiment. It may be noticed that there is no change in the shape of the spectrum of uncontaminated meat. This gives reason to distinguish between infected and uninfected meat.

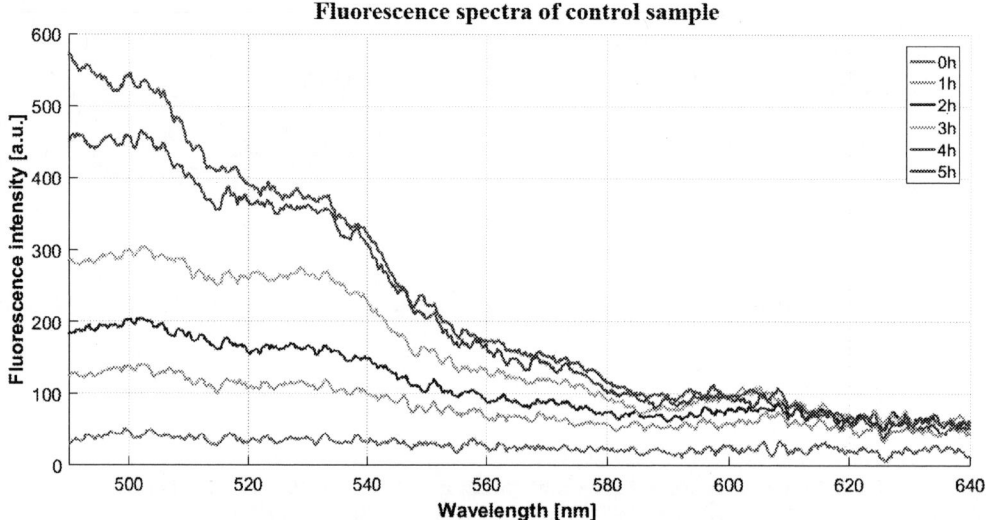

Figure 3. Fluorescent spectra of uninfected sample.

Figure 4 shows the normalized fluorescence spectra of infected and uninfected samples after 5 hours. Normalization was carried out by the intensity of fluorescence at a wavelength $\lambda_1 = 500$ nm. Such a wavelength was chosen because it is convenient to operate two marker wavelengths: $\lambda_1 = 500$ nm and $\lambda_2 = 535$ nm which comparison of intensities (I_{λ_1} to I_{λ_2}) determines the shape of the spectrum and indicates the contamination.

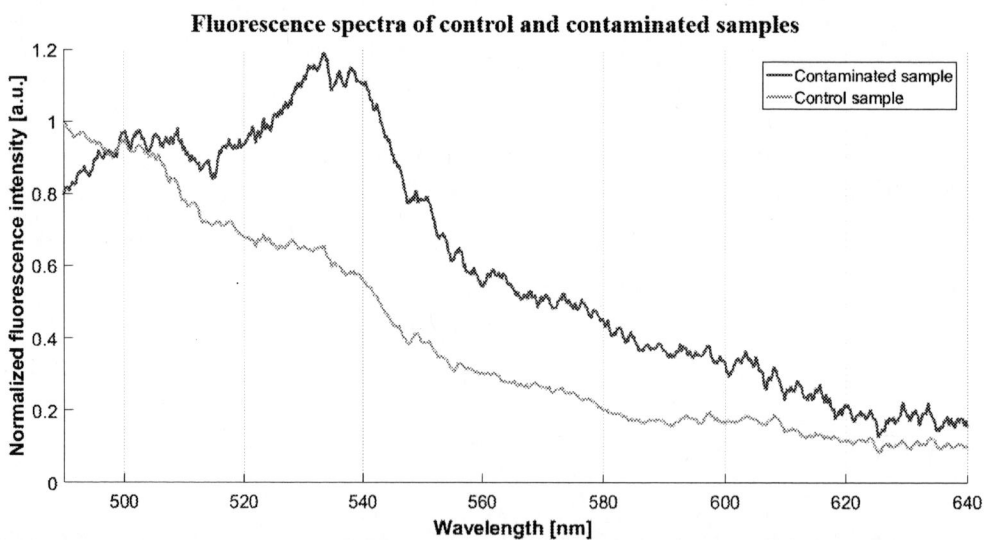

Figure 4. The impact of the bacteria on the normalized beef fluorescent spectrum: green – uninfected sample; red – sample artificially infected with E. coli bacteria (10^6 CFU/cm^2).

Thus, the essence of the proposed method is as follows. The laser irradiates the sample at wavelength 450 nm and the fluorescent spectra contain the peak with central wavelength 535 nm. The proposed criterion to decide whether the sample is infected or not is in comparison of $I_{\lambda 1}$ to $I_{\lambda 2}$. When $I_{\lambda 1} > I_{\lambda 2}$ there is given a negative assessment (the sample's surface is not contaminated), and if $I_{\lambda 1} < I_{\lambda 2}$ there is given a positive assessment (the sample's surface is contaminated). Figure 4 clearly illustrates that point. That ratio may be used as an obvious marker of bacterial contamination of meat surface.

5. Conclusion

As a result of the research it was investigated that infected with E. coli sample may be confidently detected 5 hours after the contamination by the fluorescence spectra of its irradiated surface. Proposed method allows detection of the bacterial infection by bacterial metabolites. Concentration of the bacteria is above 10^6 CFU/cm^2 after 5 hours.

The advantage of this method is that the proposed criterion of the bacterial contamination on the meat surface is not dependent on intensity of fluorescence. In addition, the device has the simple structure and measurements are carried out contactless giving instant assessment of the bacterial contamination.

Although the proposed method is able to detect high concentrations of bacteria, it may serve the meat industry as a first stage of quality assurance. Applying fluorescent labels with dyes or time-resolved spectroscopy may be perspective to improve the sensitivity of the method.

References

[1]   Authority, E. F. S. "The European Union summary report on trends and sources of zoonoses, zoonotic agents and food-borne outbreaks in 2013." EFSa Journal 13.1 (2015).

[2]   Ayçiçek H. et al., "Assessment of the bacterial contamination on hands of hospital food handlers," *Food control*, vol 15, no. 4, pp 253-9, 2004

[3]   Chen C. S., Durst R. A. "Simultaneous detection of Escherichia coli O157: H7, Salmonella spp. and Listeria monocytogenes with an array-based immunosorbent assay using universal protein G-liposomal nanovesicles," *Talanta*, vol 69, no. 1, pp 232-8, 2006

[4]   Batt C A 2007 Science **316** 1579–80

[5]   Lazcka, Olivier, F. Javier Del Campo, and F. Xavier Munoz. "Pathogen detection: a perspective of traditional methods and biosensors," *Biosensors and bioelectronics* 22.7 (2007): pp 1205-17.

[6]   Lu X. et al., "Application of mid-infrared and Raman spectroscopy to the study of bacteria," *Food and Bioprocess Technology*, vol 4. no. 6, pp 919-935, 2011

[7]   A.D. Taylor, Q. Yu, S. Chen, J. Homola, S. Jiang, Sens. Actuators B Chem. **107** (2005) 202.

[8]   A. Mathesz et al., "Integrated optical biosensor for rapid detection of bacteria," *Optofluidics, Microfluidics Nanofluidics*, vol. 2, no. 1, pp 15–21, 2015

[9]   Quero G. et al. "Long period fiber grating working in reflection mode as valuable biosensing platform for the detection of drug resistant bacteria," *Sensors and Actuators B: Chemical*, vol. 230, pp 510-20, 2016

[10]   Fan X. et al., "Sensitive optical biosensors for unlabeled targets: A review," A*nalytica Chimica Acta, vol.* 620, no. 1, pp 8-26, 2008

[11]   Félix-Rivera H., Hernández-Rivera S. P., "Raman Spectroscopy Techniques for the Detection of Biological Samples in Suspensions and as Aerosol Particles: A Review," *Sensing and Imaging: An International Journal*, vol. 13, no. 1, pp 1-25, 2012

[12]   Bridgeman J. et al., "Portable LED fluorescence instrumentation for the rapid assessment of potable water quality," *Science of the Total Environment*, vol. 524, pp 338-46, 2015

[13]   Sohn M. et al., "Fluorescence spectroscopy for rapid detection and classification of bacterial pathogens," *Applied spectroscopy*, vol. 63, no. 11, pp 1251-55, 2009

SPBOPEN 2018

Study of electric properties of self-assembled films of albumin during their dehydration

M A Baranov[1], A P Alekseenko[1], E N Velichko[1]

[1]Higher school of applied physics and space technologies, Peter the Great St. Petersburg Polytechnic University, St. Petersburg 195251, Russia

Abstract. In this paper conductivity of biomolecular films are considered. Experimental results on study of electric response of albumin solution under the influence of external magnetic field and without the field are presented. The influence of magnetic field on electrical characteristics of protein films is discussed.

1. Introduction

Development of microelectronics is concerned with search of new principles, substances and materials, study of new scientific technical ideas. Significant research trends are as follows: realization of methods of diagnostics of processes of self-organization of molecular systems, study of the structure and properties of new materials, self-organization processes, special of organic materials of biomolecular electronics. Using of organic components, in particular biomacromolecules, is a basis of the new promising technology – molecular electronics and its component – bionanoelectronics [1]. There are a number of publications on creation of biomolecular computers [2, 3].

Integration of modern electronic devices with biological and biomolecular elements is an important issue. Protein films can be considered as one of the basic elements of new electronics. Study of the processes of formation of biomolecular films is a separate task. Various structures, which determine the properties of the film, can be formed under the influence of various factors during the dehydration of the solution. This process can be controlled by these parameters.

It should be noted, that elements of any electronics are impacted by various electromagnetic fields, both natural and artificial origin [4]. Therefore, effects of electromagnetic field on biological systems, proteins, and biological solutions should be studied [5, 6].

Magnetic properties of proteins are insufficiently studied [7, 8].Thus, the aim of this work is research of the electric response of the protein albumin solution under the influence of external magnetic field and without it.

2. Methods of investigation

In this work electric properties, in particular, conductivity of albumin solution during it's dehydration and dry albumin films were studied.

Experiments were done on water solutions of albumin during their dehydration. Water solution of albumin was prepared from concentrated protein albumin solution with pH = 9.3 and distilled water with pH = 7.0. in the ratio of 1:1. A pH shift of studied solution to an acidic side to pH = 4,8 was provided with vinegar acid CH_3COOH 99%. This value of pH corresponds to isoelectric point of albumin.

The second part of this work is devoted to studies of electric response of the albumin solution under the influence of an external magnetic field.

Based on the literature analysis [6, 7], it was decided to use fields comparable with the Earth's magnetic fields (the magnetic field equals about 0.34 Oe).

Content from this work may be used under the terms of the Creative Commons Attribution 3.0 licence. Any further distribution of this work must maintain attribution to the author(s) and the title of the work, journal citation and DOI.
Published under licence by IOP Publishing Ltd

SPBOPEN 2018   IOP Publishing

In our experiments, DC magnetic fields were created by a inductive coil around the sample, and a generator supplying a voltage to the coil. The coil 4 mm in thickness and 40 turns in number, was made of copper wire of 0.1 mm cross section. The experimental setup is presented in Fig. 1.

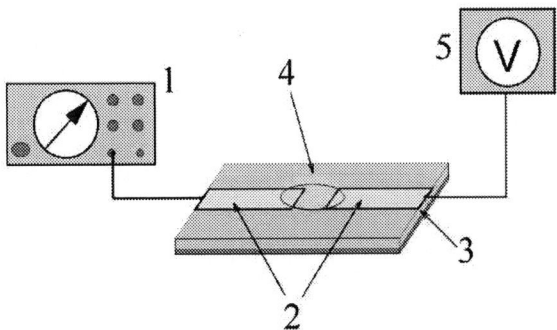

Figure 1. Experimental setup: 1 — generator; 2 — electrodes; 3 — dielectric substrate; 4 — protein solution; 5 — voltmeter.

The investigated solution (4) in volume of 5 µl was placed on the dielectric substrate (3) in the interelectrode gap (2). The signal in volume 5 V from generator (1) with various frequencies was applied to the interelectrode gap. Fig. 2 shows electric scheme of experimental setup.

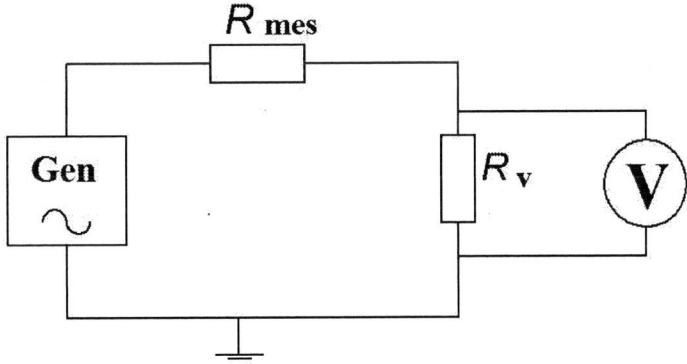

Figure 2. Electric scheme of experimental setup: Gen – generator; R_{mes} – resistance of samples; R_v – input resistance of voltmeter, V – voltmeter.

The voltage (V_{out}) at the interelectrode gap was measured by voltmeter (5) as a function of time and frequency. In accordance with the equation (1) the output voltage is proportional to sum of samples resistance and voltmeter input resistance.

$$V_{out} \sim \frac{R_v}{R_v + R_{mes}} \tag{1}$$

It is known that conductivity of an object is in reverse proportion to resistance and, thus, proportional to the output voltage in our measurements. So the described experimental setup allowed us to compare and analyze the conductivity of the samples.

3. Experimental Results

Experimental results on studies of conductivity of albumin solution during dehydration were obtained. Distilled water was measured as a control sample. Fig. 3 – 5 illustrate the transfer characteristic of distilled water, water albumin solution pH = 4.8 and pH = 9.3.

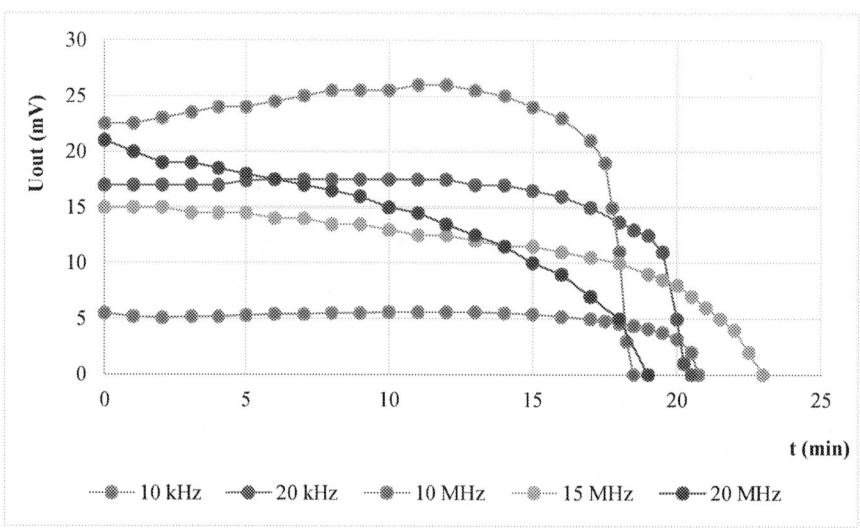

Figure 3. Dependence of voltage on time in distilled water.

Also a study of conductivity of albumin was performed on pure albumin samples (pH = 9.3) and on native albumin samples (pH = 4.8).

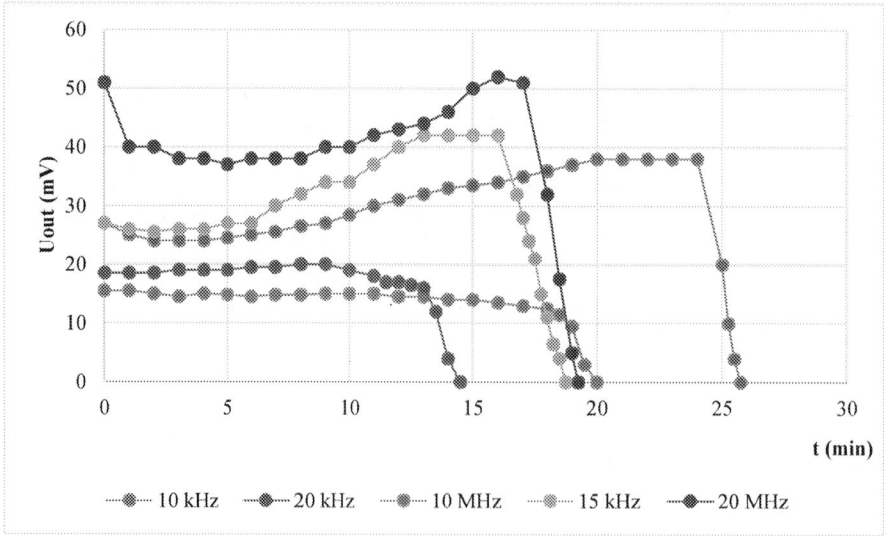

Figure 4. Dependence of voltage on time in pure albumin solution with pH = 9.3.

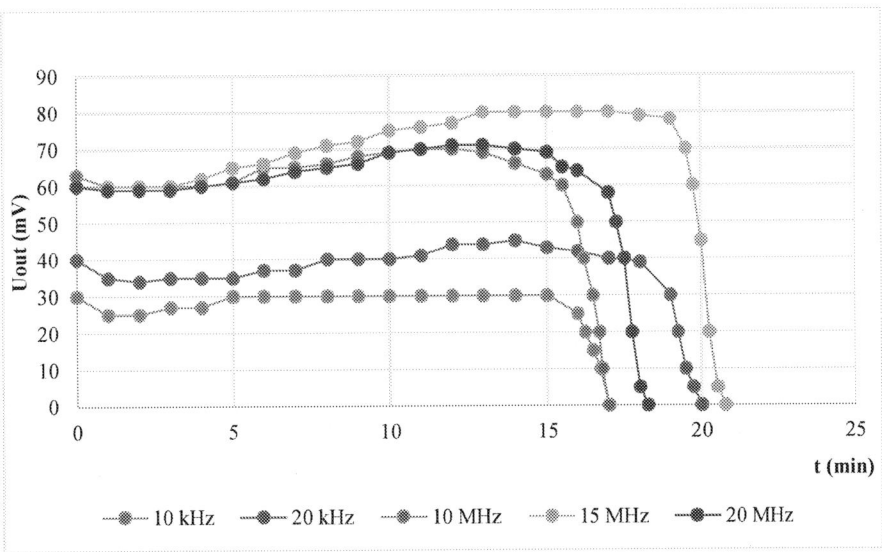

Figure 5. Dependence of voltage on time in water albumin solution with pH = 4.8.

Fig. 3 – 5 show, that the voltage on protein solution exceeds the voltage on distilled water at about 3 times. However, the voltage measured on samples of albumin in isoelectric point (pH = 4.8) is higher than voltage on pure albumin (pH = 9.3). This effect may be caused by differences in electric resistanses of the samples of albumin at different acidity.

Conductivity of solutions under the influence of external magnetic filed has been studied. Experimental results are presented in Figures 6, 7.

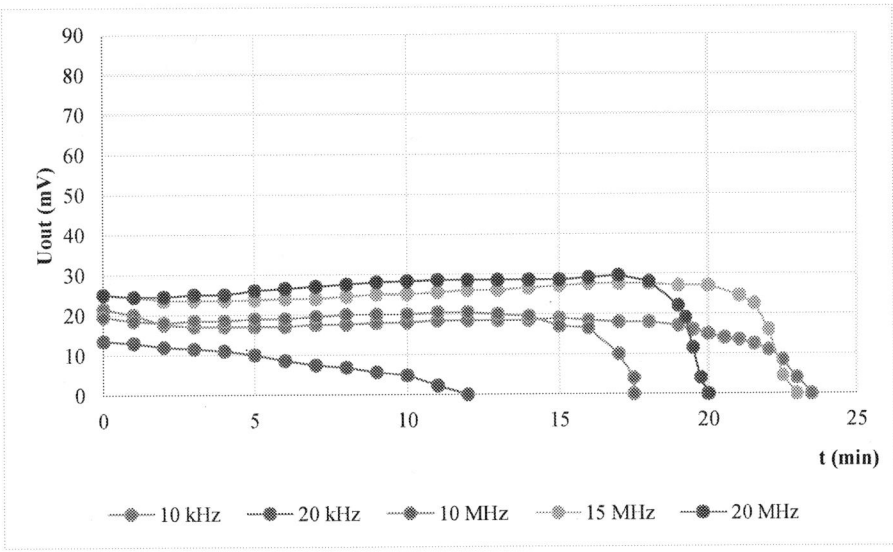

Figure 6. Dependence of voltage on time in water albumin solution with pH = 4.8 at external magnetic field 3 Oe, co-directional with the drying front of the sample.

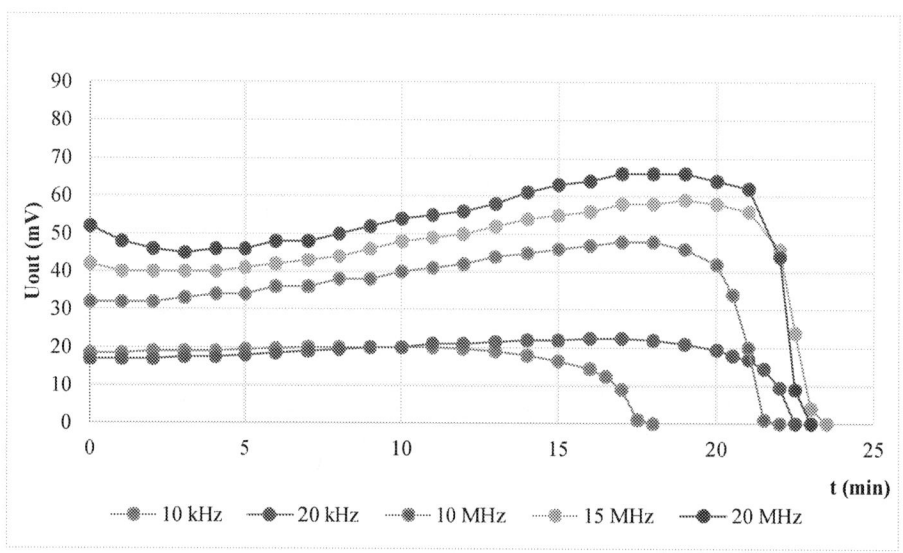

Figure 7. Dependence of signal voltage on time in the drop of water albumin solution with pH = 4.8 at external magnetic field 3 Oe, anti-directional with sample drying front.

Fig. 6, 7 show, that external magnetic field affects on electrical response of protein solution. The voltage in the interelectrode gap for albumin solution (pH = 4.8) in external magnetic field which was applied in direction "up" exceeds the voltage for albumin solution in external magnetic field which was applied in direction "down". This effect may be caused by conformation of molecules. In our opinion, a change of biomolecules conformation occures under an external magnetic field. But a character of this effect has not been established. This is task for further research.

4. Conclusion

In our studies, the experimental results on conductivity of albumin protein films and solutions in external magnetic fields and without it, were obtained. It was shown that the amplitude of output signal is higher in protein solution in comparison with distilled water. We revealed that such electrical properties as conductivity of the protein film is significantly influenced by the value of intensity of magnetic field. And amplitude of the output signal depends on the direction of the applied magnetic field. It was shown, that biological systems are sensitive to weak external influences and can be controlled by them. The experimental results obtained in this work allowed us to conclude that solutions of proteins have nonlinear magnetic properties.

References

[1] Noy A 2011 *Bionanoelecrtonics Adv. Mater*. **23** 807

[2] Farfel J, Stefanovic D 2006 *International Workshop on DNA-Based Computers* **3892** 38–54

[3] Sakowski S. et al. 2017 *Genetics and molecular biology* **40** 4 860-870

[4] Baranov M A, et al 2017 *PIERS-Spring IEEE* pp 42–44

[5] Howard M, Kruse K 2014 *J. Cell. Sci.* **127** 11

[6] Misteli T 2001 J. *Cell. Sci.* **155** 2 181

[7] Qin S, Yin H et al. 2015 *Nature Materials* 13

[8] Matsumoto Y, Chen R, Anikeeva P, Jasanoff A 2015 *Nature Communications* **8721** 10

Controlled modification of hyaluronic acid for photoinduced reactions in tissue engineering.

A.V. Sochilina[1,2], A.G. Savelyev[2,3], P.A. Demina[1,2,4], N.V. Ierusalimsky[6], D.A. Khochenkov[2,5], R.A. Akasov[1,3], N.V. Sholina[2,3,5], E.V. Khaydukov[2,3], A.N. Generalova[1,2]

[1] Shemyakin-Ovchinnikov Institute of bioorganic chemistry RAS, Moscow 117997, Russia

[2] Federal Scientific Research Centre "Crystallography and Photonics" RAS, Moscow 119333, Russia

[3] Sechenov First Moscow State Medical University, Moscow 119991, Russia

[4] Lomonosov Moscow State University of Fine Chemical Technologies, Russian Technological University, Moscow 119571, Russia

[5] Federal State Budgetary Institution "N.N. Blokhin National Medical Research Center of Oncology" of the Ministry of Health of the Russian Federation, Moscow 115478, Russia

[6] Lomonosov Moscow State University, Moscow 119991, Russia

Abstract

Implementation of up-to-date 3D printing methods for tissue engineering applications requires synthesis of novel materials possessing both biocompatibility and processability properties. Since the hyaluronic acid is an endogenous material it represents an attractive candidate for scaffold formation. In this work, chemical modification of hyaluronic acid with glycidyl methacrylate (HAGM) was used to produce derivatives that are capable of photoinduced cross-linking. Aiming to control the degree of substitution in hyaluronic acid by glycidyl methacrylate we developed a new quantitative method based on oxidation of double bonds in HAGM by permanganate ions. HAGM hydrogel scaffolds have been produced and examined as *in vitro* and *in vivo*.

1. Introduction

Tissue engineering is a branch of regenerative medicine serving to create artificial implants that can functionally replace damaged organs and tissues instead of using donor organs. Creation of an implant requires fabrication of 3-dimensional (3D) scaffolds that serve as a structural support for patient cells. Despite the rapid development of this approach, a comprehensive material for scaffolds that could meet all the requirements of native tissue has not been developed yet.

Materials (commonly polymers) for scaffold fabrication can be classified as natural, semisynthetic (derivatives of natural polymers) and synthetic. Alongside with wide variety and good reproducibility the most essential advantage of synthetic polymers is vast plasticity of physical chemical properties (e.g. poly(lactic acid)). Despite such benefits, most of synthetic polymers are often completely unsuitable for tissue engineering due to lack of biocompatibility and biodegradability that natural polymers have (e.g., collagen) [1]. Therefore, in order to combine the desired properties of both synthetic and natural polymers the development of new semisynthetic materials is needed.

Hyaluronic acid (HA), polysaccharide of extracellular matrix, appears to be promising material for scaffolding due to its endogenous origin [2]. However, it is impossible to produce hydrogel scaffolds out of non-modified HA via photocrosslinking. In this work we demonstrate the synthesis of HA derivatives, carrying double bonds, capable to crosslinking by flavin mononucleotide as a photosensitizer. The new rapid quantitative method has been implemented to control degree of

---

Content from this work may be used under the terms of the Creative Commons Attribution 3.0 licence. Any further distribution of this work must maintain attribution to the author(s) and the title of the work, journal citation and DOI.

Published under licence by IOP Publishing Ltd

substitution (DS) in HA after modification. The synthesized hyaluronic acid glycidyl methacrylate (HAGM) has been used to produce hydrogel structures via photocrosslinking under irradiation at 365 nm wavelength. *In vitro* and *in vivo* tests of the produced hydrogels confirmed their biocompatibility.

2. Materials and methods

2.1. Materials

Following materials were purchased from Sigma-Aldrich (USA) and used without further purification: sodium hyaluronate with average molecular mass 1 MDa and 100 kDa, glycidyl methacrylate (GMA), triethylamine (TEA), tetraethylammonium bromide (TEAB), acetone, dimethylformamide (DMF), phosphate buffered saline, pH 7.0 (PBS), potassium permanganate ($KMnO_4$), poly(ethylene glycol) diacrylate (PEG-DA) with average molecular weight 575 kDa, flavin mononucleotide (FMN).

2.2. Methods

2.2.1. Modification of HA

First, 0.5 g of HA was dissolved in 100 ml of PBS under vigorous stirring. Then, DMF was added with volume ratio PBS:DMF = 100:33. GMA was added in the range of concentrations to achieve mass ratio (HA:GMA) 1:6, 1:10, 1:15 and vigorously stirred. Then, 3 ml of TEA (catalyst) was added resulting to pH increase up to 12. Reaction was carried out for 7 days under stirring at room temperature. In modified method TEA was replaced with TEAB. Before introduction into the reaction mixture, 500 or 1000 mg of TEAB were dissolved in 5 ml of water and added to the mixture before GMA.

2.2.2. Detection of double bonds by $KMnO_4$ in GMA and HAGM

GMA was dissolved in deionized water at concentration 0.5 wt.%. After addition of GMA solution in the range of volumes from 10 to 70 µl to 0,025 wt.% $KMnO_4$ (standard) solution, the absorption spectra were recorded. The increase of added GMA volume was accompanied with color changing from violet to brownish yellow. Quantity of GMA and, correspondingly, double bonds needed to fully reduce permanganate ions was counted. This data will be used further to evaluate the degree of substitution (DS) of HAGM.

HAGM was dissolved in deionized water at concentration of 0,5 wt.% and then added to standard water solution of $KMnO_4$ (0,025 wt.%) until purple color completely turned into brownish yellow. To confirm complete disappearance of permanganate ions the solution was spectrophotometrically analyzed. DS of HAGM was evaluated as ratio of double bonds quantity (acquired previously from quantity of GMA needed to fully reduce 1 ml of standard $KMnO_4$ solution) to HAGM disaccharide quantity of samples.

All the absorption spectra were recorded at UV-Vis Spectrophotometer Evolution 201 (Thermo Fisher Scientific, USA).

2.2.3. Gelation of hydrogels

Photocurable compositions (PCCs) were produced by combining HAGM and PEG-DA in PBS and FMN as a photosensitizer. The final concentrations in the PCCs were as follows: 20 wt.% of HAGM, 5 wt.% of PEG-DA and 0.0004 wt.% of FMN. All the compositions were sonicated for 3 hours and left overnight under dark conditions up to the formation of uniform consistence.

Gelation of hydrogels due to activation of the crosslinking process in the produced PCCs was performed by UV diode "Vega" (Medlaz-Neva, Russia) emitting at 365 nm wavelength corresponding to the absorption peak of FMN. The intensity of irradiation was 20 mW/cm^2. 3D printing of hydrogel scaffolds was conducted according to the method described in [3], but irradiation was at 365 nm wavelength.

2.2.4. Cell culture

Immortalized human fibroblasts BJ-5ta were grown in RPMI-1640 medium at 37 °C in 5% CO_2-humidified atmosphere. RPMI-1640 medium was supplemented with 10 % FBS, 2 mM L-glutamine, 100 µg/ml streptomycin, and 100 U/ml penicillin. When reaching 80-90 % confluence, cells were detached by Trypsin-EDTA solution, subcultivation ratio was from 1:3 to 1:6.

2.2.5. In vitro study

To demonstrate cell growth on hydrogel surface in vitro hydrogel samples were placed into 24-well plates and seeded with BJ-5ta fibroblasts (20,000 cells per each sample). The cell growth was observed by light microscopy within 7 days.

2.2.6. Animal experiments

All studies were conducted in an AAALAC accredited facility in compliance with the PHS Guidelines for the Care and Use of Animals in Research and were approved by the institutional Animal Care and Use Committee. Six week old BDF1 mice (n = 6) were purchased at vivarium of N.N. Blokhin Russian Cancer Research Center. The mice were housed with free access to food and water. Pre-implantation procedures were carried out under strict sterile conditions. Scaffolds were immersed in a 0.9 % sodium chloride solution at 37 °C for at least 30 min. Prior to implantation, animals were anesthetized using a single intraperitoneal injection of Zoletil (20 mg/ml, 50 µl per animal, Virbac SA, Carros, France) and Rometar 2% (20 mg/ml, 10 µl per animal, Spofa, Czech Republic). Subcutaneous pockets (side length 15 mm) were formed on the back of mice, and implanted scaffolds were placed on the back muscle fascia. Then the pockets were closed with non-absorbable monofilament polypropylene suture (Prolene 4.0, Ethicon). After the implantation, the mice were randomly divided into three groups: 1 week (n = 2), 2 week (n = 2), and 4 week (n = 2). At the end of experiment animals were sacrificed by cervical dislocation. Immediately after the sacrifice, scaffolds were collected and fixed in 4 % paraformaldehyde for 2 h. Then the samples were dehydrated, embedded in paraffin, and cut into 3 µm thick sections. The alternate sections were stained with Hematoxylin and Eosin according to the manufacturer protocol. Three images of each animal were randomly selected to evaluate the tissue response to the implanted scaffold. ALEICA DM4000 B LED microscope, equipped with a LEICA DFC7000 T digital camera and LAS V4.8 software (Leica Microsystems, Switzerland), was used for the sample imaging and examination.

3. Results and discussions

Conjugation of GMA with HA was performed to acquire HA derivatives capable of crosslinking. Conjugation of GMA can be carried out via reaction mechanisms of reversible transesterification or irreversible epoxy ring opening with carboxyl or hydroxyl moieties of HA (Figure 1). According to the work [4] products of latter mechanism are more stable and tend to accumulate during longer periods of time substituting products of former mechanism. On the basis of these considerations HAGM was produced as described in work [4].

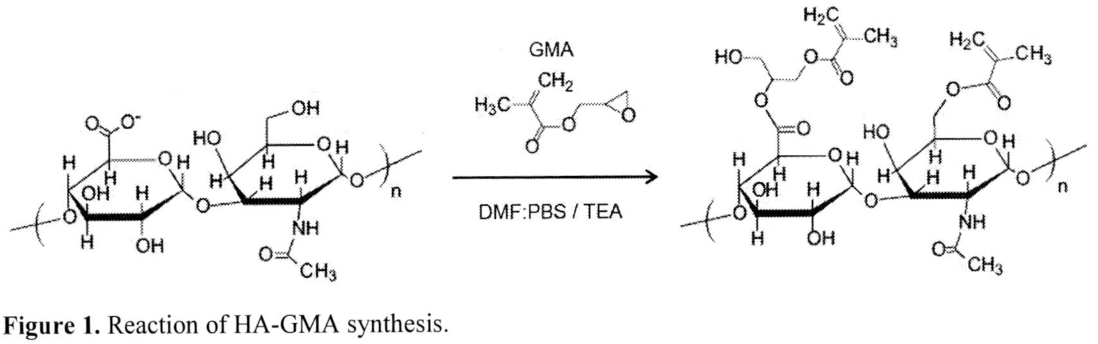

Figure 1. Reaction of HA-GMA synthesis.

The original method [4] was changed in order to investigate correlation between DS and reaction conditions such as catalyst type, HA:GMA ratio and reaction time. The ability of DS regulation enables the control of crosslinking degree and, correspondingly, mechanical and biodegradability properties of scaffolds. Thus, quantitative analysis of double bonds in HA derivatives was needed. In this work we used $KMnO_4$ and absorption spectra measurements to easy and rapidly quantify DS in HAGM instead of using NMR.

Permanganate ions oxidize double bonds in HAGM, changing the solution color from purple to brownish yellow (Figure 2 (a)). Total disappearance of peaks at 525 nm [5] confirms a complete consumption of permanganate ions in the course of reaction (Figure 2 (b)).

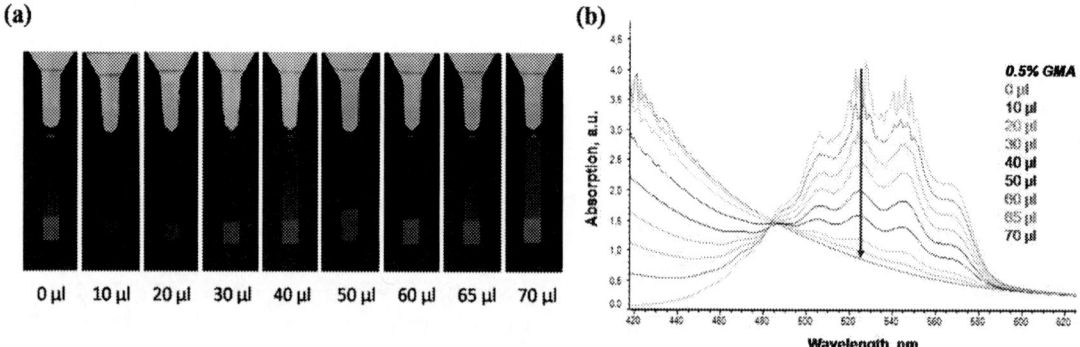

Figure 2(a,b). (a) Color change of $KMnO_4$ standard solution with different amount of added 0,5% GMA solution. **(b)** Decrease of $KMnO_4$ peaks under GMA solution addition, spectrophotometric data.

The concentration of the consumed $KMnO_4$ corresponds to the concentration of double bonds. After that quantity of different HAGM samples needed to reduce 1 ml of standard $KMnO_4$ solution were determined and their DS was counted. Results are given in Table 1.

Table 1. Dependence of DS on reaction conditions.

Changing parameter	HA:GMA, m/m			HA:catalyst, m/m			Time of reaction, days					
	1:6	1:10	1:15	HA:TEA 1:4,5	HA:TEAB 1:1	HA:TEAB 1:2	1	2	3	4	5	6
DS, %	10	13	22	22	41	45	17	21	28	40	38	38

Augmentation of GMA concentration and substitution of TEA by TEAB result in higher DS. Further increase of TEAB feebly affects HA-GMA DS. Experiment with reaction time indicated that DS achieved maximum of 40 % on 4-th day and then remained the same with insignificant changes.

The synthesized HAGM has been utilized to produce hydrogel films. For this aim the initial PCC was placed between two microscope slides. The distance between glasses was adjusted using silicon spacers to 150 μm. The exposure was performed for 12 minutes, then the sample was rotated and irradiated again from another side. Afterwards the microscope slides were gently separated and the produced crosslinked hydrogel film was placed into PBS for 7 days. Aiming to demonstrate the printability of hydrogels we produced hydrogel scaffold using 3D extrusion printing as described in [3], but irradiation wavelength was changed to 365 nm. The scaffold produced with 2 mm period of grating is presented in Figure 3.

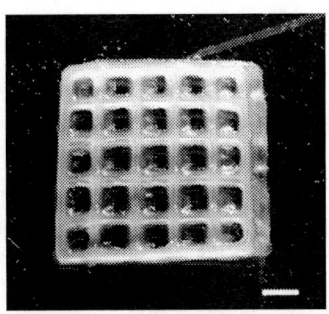

Figure 3. HAGM hydrogel scaffold produced by 3D extrusion printing. Scale bar 2 mm.

Recently, we used extract-based MTT assay to demonstrate that HAGM-based hydrogel was not toxic to cells *in vitro* [3]. In the current research, cell attachment to the hydrogel scaffolds *in vitro* as well as hydrogel biocompatibility *in vivo* were studied. For *in vitro* experiments, immortalized human fibroblasts BJ-5ta were used as model cells since fibroblasts play the leading role in tissue repairing. It was demonstrated that fibroblasts were able to grow on the hydrogel surface, confirming high *in vitro* biocompatibility of the HAGM-based hydrogels (Figure 3). It should be noted, that fibroblasts preferred to attach to the rough regions of hydrogels that usually occurred at the edge of the samples.

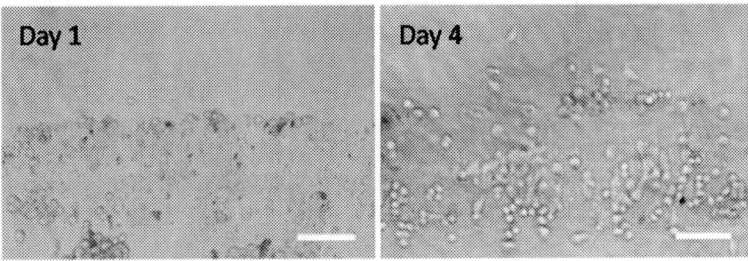

Figure 4. Human fibroblasts BJ-5ta growth on the hydrogel surface, 1 and 4 days of incubation. Scale bar 100 μm.

For *in vivo* experiments, scaffolds were implanted into mice (back muscle fascia) for 1, 2, and 4 weeks. Then the scaffolds within the surrounding tissue were collected, and the scaffold-induced inflammation, neovascularization, and fibrosis were evaluated. Thus, moderate infiltration of lymphocytes, granulocytes, and macrophages was found in a week after implantation, which indicated a weak inflammation process. Additionally, a minimal capillarity was noted, and the scaffold closely adhered to the tissue. It should be also noted that the scaffold thickness increased compared to the initial one (before implantation), which may be explained with hydrogel swelling due to enzymatic hydrolysis [6].

By the 2nd week the inflammation decreased, and granulation tissue was formed at the scaffold site (Figure 5 (a)). The scaffold fragments separated from the main mass were subjected to macrophagal resorption. Basophilic color in H&E staining indicated the presence of HAGM in the scaffold composition. The scaffold thickness was restored to the initial one. Finally, by the 4th week, separated fragments disappeared, and a very thin mature connective tissue capsule formed around the scaffold. No tendency to scarring was found. The scaffold thickness did not change, and the newly formed capillaries and fibroblasts did not penetrate the scaffold (Figure 5 (b)).

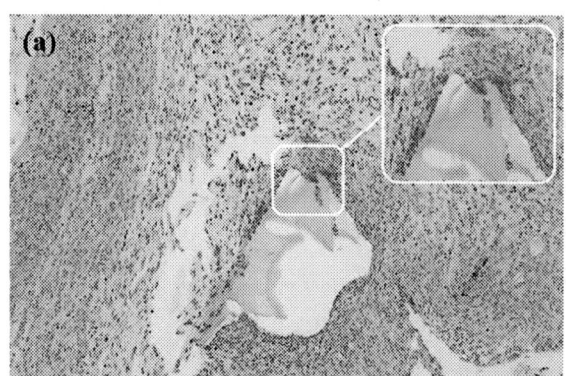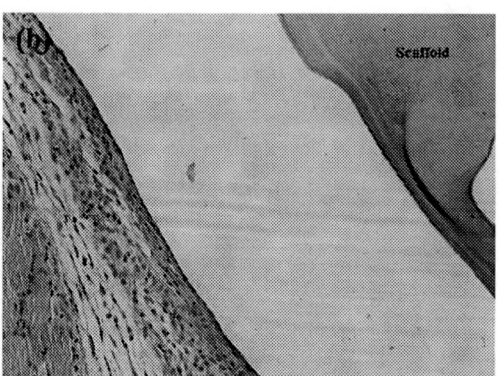

Figure 5(a,b). (a) Histological overview of a H&E stained section from mouse skin, 2 weeks after scaffold subcutaneous implantation (x100). Granulation tissue formation, scaffold fragments are surrounded with thin connective tissue capsule. An insert: macrophagal resorption of the implanted material. Large macrophages and giant cells are shown. Edema accompanied with lympho-macrophagal and weak neutrophil reaction is observed. **(b)** Histological overview of a H&E stained section from mouse skin, 4 week after scaffold subcutaneous implantation (x200). Scaffold preserves the initial shape. Formation of a relatively thin mature connective tissue capsule around the scaffold is shown.

4. Conclusion

In present study we developed approach to the controlled modification of sodium hyaluronate with GMA. We demonstrated easy and rapid quantitative analysis of DS in HAGM using $KMnO_4$. Modified HA can participate in radical photoinduced crosslinking and possesses a great potential for hydrogel scaffold production. According to *in vitro* data, *in vivo* results demonstrated high HAGM-based hydrogel biocompatibility without wound healing disorders, hypertrophic scars formation, or long-term inflammatory reactions. Furthermore, controlled introduction of double bonds can enable regulation of swelling and biodegradation rate of final hydrogel product already at the stage of preparation of PCCs for 3D printing.

5. Acknowledgements

The reported study was funded by RSCF according to the research project № 17-19-01416.

References

[1] Naahidi S, Jafari M, Logan M, Wang Y, Yuan Y, Bae H, Dixon B, Chen P 2017 *Biotechnol. Adv.* **35** 530
[2] Savelyev A G, Bardakova K N, Khaydukov E V, Generalova A N, Popov V K, Chichkov B N, Semchishen V A 2017 *J. Photochem. Photobiol., A* **341** 108
[3] Savelyev A G, Sochilina A V, Akasov R A, Mironov A V, Semchishen V A, Generalova A N, Khaydukov E V, Popov V K 2018 *Sovrem. Tehnol. v Med.* **10** 88
[4] Bencherif S A, Srinivasan A, Horkay F, Hollinger J O, Matyjaszewski K, Washburn N R 2008 *Biomaterials* **29** 1739
[5] Hallinan N 2012 U.S. Patent No. 8,293,534
[6] Park K 1988 *Biomaterials* **8** 435

Microchip device with dry-stored reagents for Loop mediated isothermal amplification

A Tupik[1], G Rudnitskaya[1], A Bulyanitsa[1,2], T Lukashenko[1], D Varlamov[3], A Evstrapov[1,4]

[1] Lab of information and measurement biosensor and chemosensor microsystems, Institute for Analytical Instrumentation RAS, St. Petersburg 198095, Russia
[2] Department of Higher Mathematics, Peter the Great St. Petersburg Polytechnic University, St. Petersburg 195251, Russia
[3] All-Russia Research Institute of Agricultural Biotechnology, Moscow 127550, Russia
[4] Department of nanophotonics and metamaterials, ITMO University, St. Petersburg 197101, Russia

Abstract. Storage of key reagents for amplification in a microfluidic device is decisive for point-of-care diagnostics. A simple approach of drying the reagents without special stabilizers by dehydration in a glass microchip is described. For loop mediated isothermal amplification of nucleic acids, a microchip device with two reactors was approbated. It was shown that the microchip device is suitable for detecting nucleic acids template of the Sacbrood virus. The experimental results obtained in microchip reactors with dry-stored primers were comparable to the results with non-dried primers. The shelf life of dehydrated primers without declined its reactivity turned out less than three months.

1. Introduction

In recent years, there has been a growing interest in developing integrated, portable microfluidic devices for nucleic acid-based diagnostic for many applications, including clinical diagnosis, food quality control, and environmental monitoring [1, 2]. The advantages of microfluidic devices include: 1) small reaction volumes that reduce sample and reagents consumption; 2) small thermal mass what ensures rapid heat transfer with low thermal inertia; 3) portability with low power consumption; 4) the possibility to combine a sample preparation, purification and extraction of nucleic acids, its amplification and detection on a microfluidic platform. If the microfluidic device contains dry or liquid reagents, simple pipetting operations are required to start the processes. This prevents contamination of samples and facilitates raped diagnostic by minimally trained personnel for point-of-care applications especially under conditions of resource constraints.

Polymerase chain reaction (PCR) is a widely used amplification technique for molecular diagnostics, which enabling to amplification of target nucleic acids templates containing just a few molecules to detectable levels. Typically, the reaction mixture is prepared immediately before the PCR by mixing all the components, some of which are stored at low temperature (-20°C). Long-term storage, as well as repeated freezing and thawing can lead to loss of reactivity of the reagents. To preserve the reactivity of PCR reagents and increase their shelf life at temperatures above 0°C, freeze-drying is used. Addition of special stabilizers allows maintaining the properties of the dried PCR mixture for more than a year when stored in a wide temperature range [3]. Successful on-chip PCR amplifications

---

Content from this work may be used under the terms of the Creative Commons Attribution 3.0 licence. Any further distribution of this work must maintain attribution to the author(s) and the title of the work, journal citation and DOI.

Published under licence by IOP Publishing Ltd

with dried reagents stored were reported [4-6]. However, the PCR requires a special temperature mode with rapid temperature transitions and precise temperature control for efficient and selective amplification of nucleic acids. Isothermal amplification techniques, such as loop mediated isothermal amplification (LAMP), may resolve the challenges associated with the accuracy of the amplification temperature mode without compromising analytical sensitivity and specificity. LAMP has a moderate incubation temperature leading to simplified heating and low power consumption, and allows direct genetic amplification from bacterial cells due to tolerance to substances that typically inhibit PCR, in addition, the LAMP result can be detected either visually or by simple detectors [10, 11]. The amplification results obtained with the storage of the LAMP reagents were also presented [7-9]. These advantages make it especially interesting at developing of microfluidic devices for point-of-care applications.

The point-of-care diagnostics include tests at home, at the bedside or in the field when resources are limited. It is relevant not only for diagnosing of human diseases, but also animals and insects. The Sacbrood virus (SBV) is deadly to broods of honey bees. A vector of the virus is adult bee, for which the disease can be asymptomatic, while a single diseased larva can infect up to 3000 healthy larvae [12]. Therefore portable detection methods are necessary to control and eradicate SBV in the field.

A one-step reverse transcription loop-mediated isothermal amplification (RT-LAMP) assay was developed for the rapid identification of SBV [13]. Isothermal amplification was carried out in a simple water bath for 30 min at 65°C, and a positive amplification reaction was visible to the naked eye due to addition of intercalating fluorescence dye (SYBR Green I). The authors concluded that color observation method (using fluorescence dye) was ten times more sensitive than the white turbidity observation.

The purpose of our research was to choose simple techniques for implementation a prototype microfluidic chip for SBV detecting with dehydrated reagents inside. We believe that developing of miniature device containing dry-stored reagents for LAMP would simplify even more this assay for application in the field diagnostics. We used the same primer system [13] and LAMP reaction mixture with an intercalating fluorescence dye to amplify the SBV target templates. Primers were chosen as reagents for storage because of they relate to the key reagents for amplification and determinate the specificity and final yield of the reaction. Dry-stored primers for different target templates in consequently different microfluidic channels allow a multiplex LAMP assay, for example, for rapid analysis of several genes [8]. At the same time, primers are much less sensitive to drying than other main components of the reaction such as polymerase enzymes.

An overview of current methods for integration various classes of reagents inside microfluidic devices are considered in [14]. There are two main approaches: 1) the immobilization of reagents on carriers (beads, hydrogels and paraffin capsules) followed by an integration of these carriers onto a microfluidic device and 2) the direct pattering of reagents onto surfaces in microstructures. The latter one includes contact deposition methods (micro and nanocontact printing, high-resolution patterns using Dip-pen nanolithography and FluidFM) and non-contact deposition methods (pipetting robots, sprayers and piezoelectric inkjet printers). Their advantages are reproducibility, high accuracy of dosing of small volumes down to femtoliters, and the possibility of industrial application with high throughput of deposition and low dead volumes. However, most of these techniques require complex technology equipments to treatment of surface at the intermediate stage of microchip manufacturing, when access to the microstructures is free. In this case at the next stage of device manufacturing it will be necessary to protect the received layers of reagents from affecting of the applied technology of bonding for hermetization of microstructures. A method of drying reagents by dehydration in microstructures was chosen as the most simple and affordable method of introducing reagents into microchip reactors. This enabled the integration of the reagents in ready-made microchip devices.

2. Experimental technique
Microfluidic chip consisted of two glass plates that were bonding together with photo-curing adhesive. There were two reaction cells with a depth of 100 μm connected to four wells for liquid input and

output on the microchip (Fig.1a). These reaction cells (reactors) were fabricated by photolithography methods and acid etching. The total volume of each reactor was 2 μl, and, in addition, the same volume was required for the input/output system.

Four microliters of primers solution (F3/B3 0.3 μM; FIP/BIP 1.2 μM; LF/LB 0.6 μM) was injected into one of the reactors via the inlet well; the other reactor was used as positive control of the amplification process. Since the glass surface was hydrophilic, the solution filled the entire reactor evenly and then evaporated completely, leaving the primers in the microchannel. The microfluidic chips with dried reagents were stored at room temperature for 24 hours, 1 week and 3 months without refrigeration and with inlet and outlet open. Sample preparation with extraction and reverse transcription of SBV nucleic acids was carried out on the base of the All-Russia Research Institute of Agricultural Biotechnology. Reagents for LAMP with intercalating Eva Green dye and SD polymerase were provided by SYNTOL (Moscow, Russia). Nine microliters of reagent solution (with or without primers) and one microliter of DNA-template solution were premixed. As the target template the cDNA fragment of SBV was used with concentrations 10^4 and 10^5 copies/μl.

Four microliters of the mixture were injected into a reaction cell. To avoid evaporation during heating, all microchip wells were sealed with mineral oil. Then the microfluidic chip was heated by a thermocycler in an isothermal temperature mode for about 45 minutes. The final result of DNA amplification was determined based on the fluorescence induced by intercalating dye. Prototype model of the scanner device (IAI RAS, Russia) was used to get result images. In parallel with the experiments on microfluidic chip, control tests were carried out in polypropylene tubes with 10 μl mixture by the analyser ANK–32 (IAI RAS, Russia). There were tubes with negative control (without DNA template), positive control (solution contained all reaction components, as in the control reactor on a microchip), control of primers lack (solution without primers, as in the reactor with dry-stored primers). Also, tubes with dehydrated primers were used, which had been prepared and stored under the same conditions as the microchip with dried reagents.

3. Result and discussion

Since any intercalating fluorescence dye is non-specific and can have a fluorescent signal from non-specific DNA targets, it was important to evaluate the level of the background fluorescent signal under our conditions. Images obtained after isothermal amplification without target templates (negative control) in one microchip cell and with the addition of 104 copies of the target DNA (positive control) in another cell are presented in Fig.1b.

The heterogeneity of the fluorescence signals within a cell area requires more careful analysis. Simulation of informative signals from the images of positive and negative control allows us to draw the following conclusions.

1. If informative signals is considered as random quantities, then the distribution of these random variables has the following general integral characteristics: the asymmetry coefficient A is small (absolute value is not more than 0.1), the excess coefficient ε for these distributions has the value 2.1±0.3. These parameters are approximate to the integral characteristics of a uniform distribution. However, the probability for hypothesis of a uniform distribution with parameters found from a sample of 10 measurements estimated based on a χ^2 criterion (Pearson-criterion) as less than 1%. Even the refinement of the parameters increases this probability to 4.5%, which also does not allow it to be accepted.

2. Such a pattern of distributions is more adequately modeled either by a mixture of normal and uniform distributions, or by a sum of 2-3 normal distributions with spaced mathematical expectations and comparable weights.

In particular, the informative signal with positive control (see Fig.1b) is well approximated by the sum of three Gaussian distributions with equal weights and variances. In this case, the mathematical expectations are shifted relative to each other by approximately 2.5 standard deviations. The value of the selective correlation coefficient between the experimental and model distributions estimated at 12 points exceeds 0.91, which indicates the adequacy of such an approximation.

This distribution law corresponds to the following scheme for the formation of an information signal: an optical signal is measured from a part of the region with a transfer function of the Gaussian type, and then similar informative signals are added from the other two parts. In this case, the field of view overlaps slightly.

The parameters considered earlier - the coefficients of asymmetry and excess - refer to the normalized characteristics and do not allow estimating the quantitative coincidence of the signals.

Eventually, a comparative estimate can be given for two parameters: amplitude (maximum of the signal after compensation of the baseline) – 6.7 and 8.9 relative units; area of the peak (area under the curve) - 23 and 32 relative units. The level of fluorescence signal obtained in positive control is more than the same in negative control. Thus, our glass microchip is suitable for detecting of Sacbrood virus by loop mediated isothermal amplification method with intercalating dye.

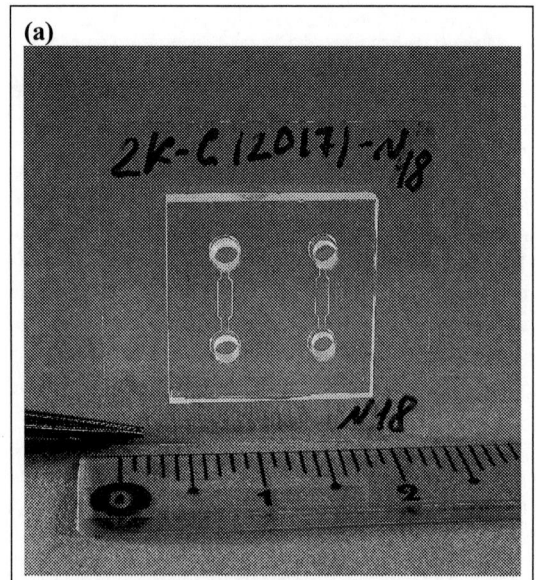

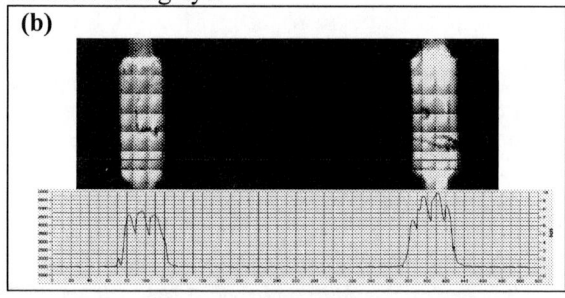

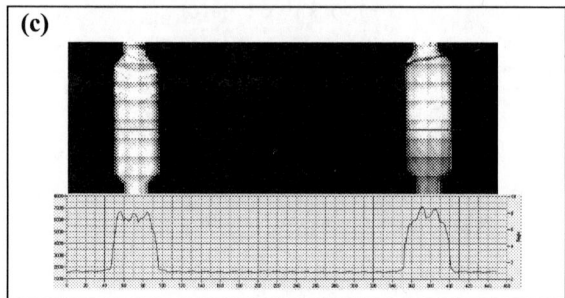

Figure 1 (a,b,c,). (a) Photo of a glass microchip with two reaction cells (reactors): the primers for Sacbrood virus were dehydrated in one cell, another cell was used as a positive control of the amplification process; **(b)** Fluorescence images of the microchip cells after isothermal amplification and the graphic dependence of the signal intensity on the cells without (left) and with (right) the addition of Sacbrood virus DNA-templates; **(c)** Fluorescence image of the microchip cells after isothermal amplification and the graphic dependence of the signal intensity both on the cell with the dehydrated primers (right) and on the control cell (left) with the same initial amount of DNA-template.

The results obtained on microchip with dry reagents stored for 24 hours are presented in Fig. 1c. In this case concentration of 10^5 copies/µl of the target DNA-template was used. The amplitudes of signals resulted were 11.6 and 10.7 relative units, consequently. Areas of the peak (area under the curve) were 46.2 and 42.5 relative units. There was a good quantitative coincidence of the estimates. Consequently, the primers stored in one of the microchip cells were successfully hydrated by injected mixture and the LAMP reaction began to progress without deterioration the efficiency.

Further experimental results showed that the activity of dehydrated primers was persistent after storage for a week, but decreased after three months of storage both in the microchip reactor and in polypropylene tubes. To determine the more accurate lifetime of dry primers research should be continued.

4. Conclusion

A simple approach to the integration of primers into the glass microchip device by dehydration technique was applied, and the results obtained after loop mediated isothermal amplification in a reactor with dehydrated primers were comparable to the results with non-dried primers. The storage life of dehydrated primers turned out less than three months. It is attractive that all the technologies used were simple and commonly available.

Acknowledgements

The work was carried out and financially supported within the framework of the State contract (no. 007-00229-18-00).

References

[1] Su W, Gao X, Jiang L, Qin J 2015 *J. Chromatogr A* **1377** 13
[2] Ahrberg C D, Manz A, Chung B G 2016 *Lab Chip* **16** 3866
[3] Shaikhaev G 2005 RU. Patent No 2259401
[4] Brivio M, Li Y, Ahlford A, Kjeldsen B G J Reimers J L, Bu M, Syvanen A C, Bang D D, Wolff A 2007 *Miniaturized Systems for Chemistry and Life Sciences* (Paris: μTAS) p. 59
[5] Suvorova A O, Mashyanov P N, Ashyna Y V, Slyadnev M N, Ganeev A A 2015 *Analytics and Control* **19** 331
[6] Kim J, Byun D, Mauk M G, Bau H H 2009 *Lab Chip* **9** 606
[7] Chen D, Mauk M, Qiu X, Liu C, Kim J, Ramprasad S, Ongagna S, Abrams W R, Malamud D, Corstjens P, Bau H H 2010 *Biomed. Microdevices* **12** 705
[8] Fang X, Chen H, Yu S, Jiang X, Kong J 2011 *Anal. Chem.* **83** 690
[9] Song J, Liu C, Mauk M G, Peng J, Schoenfeld T, Bau H H 2018 *Anal. Chem.* **90** 1209
[10] Ahmad F, Hashsham S A 2012 *Anal. Chim. Acta.* **733** 1
[11] Notomi T, Mori Y, Tomita N, Kanda H 2015 *J. Microbiology* **53** 1
[12] Grobov O F, Smirnova A M and Popov E T 1987 *Diseases and pests of honey bees* (Moscow: Agropromizdat) p 335
[13] Yang J L, Yang R, Shen K F, Peng X W, Xiong T, Liu Z H 2012 *J. Virology* **9** 47
[14] Hitzblecka M, Delamarche E 2013 *Chem. Soc. Rev.* **42** 8494

Investigation of the color effect of the test object on the pupil response

D Bobrova, N Bikberdina, M Boronenko

Yugra State University, Khanty-Mansiysk 628012, Russia

Abstract. In modern tracking systems, passive observation is used. Based on the video files identify the person. Mimics determine the elementary emotions. Terrorists are taught to hide emotions. We need a method for recognizing a person's stressed state, based on an independent reaction of the body. An active method for recognizing the reaction of a person to an external stimulus is proposed. The first results of an investigation of the eye reaction to the test object are presented. When developing test objects for emotion recognition systems, it is important not to introduce distortion. It is necessary that the test object affects only the human emotions and does not perform other actions that cause the pupil changes. The article presents the results of studies of the effect of the color of the test object on the size of the pupil.

1. Introduction

As is known, the effectiveness of using biometrics for identification systems is extremely high, since it allows precise identification of the identity, and the error probability is not more than 0.1% [1-2]. Over the past 10 years, the number of different biometric parameters used for identification has increased significantly. The main methods of recognition include three biometric methods that are currently used in practice:

- recognition by fingerprint (refers to contact methods, does not report anything about the emotional state);
- Face image recognition (contactless method that allows to recognize the emotional state, however, emotions that have learned to hide in different ways);
- recognition by the iris of the eye (contactless method, does not say anything about the emotional state, but allows to identify a person);

More advanced video surveillance systems should not only perform automatic scanning of people's faces with subsequent identification in the database of the Ministry of Internal Affairs, but also analyze the emotional state of a person. Security systems, by certain features, should identify those who are about to commit a crime. In the development of such systems, researchers are mainly focused on the use of biometric imitations. Already there are certain approaches, theoretical developments, have already learned to identify the main emotions. For the technology of image recognition and processing, the search algorithm is used to recognize human behavior and emotions [19-22]. Using the results of measuring the speed and amplitude of the pupil's reaction to light, it can be established whether a person used drugs or alcohol. It is known that the biochemical reaction of the pupil to the emotions tested is unambiguous. To use this fact, you need to move from passive biometric security systems to active ones. Active systems are based on the evaluation of the pupil's response to the test object. The test object can be sound or visual. The result of recognition should not depend on such factors as lighting, background, camera angle, number, gender and race, as well as the age of people in the frame. The test object itself should not cause any other reaction of the pupil, except for the reaction

---

Content from this work may be used under the terms of the Creative Commons Attribution 3.0 licence. Any further distribution of this work must maintain attribution to the author(s) and the title of the work, journal citation and DOI.

Published under licence by IOP Publishing Ltd

caused by the emotions. The article presents the first results of the study of the effect of the color of test objects on the pupil response.

2. Experimental method

The active method is based on the use of an unambiguous involuntary biochemical reaction of the human body to heard (seen) information. The main provisions on which the developed method of recognition of a person's psychophysical state is based:

- unconscious unambiguous biochemical reaction of the human body to the heard (seen) information
- the pupil's response to the incoming information at the time t_1, considered with respect to the time instant t_0, is interpreted unambiguously, if the pupil illumination is the same in both cases; spatial and temporal resolution of modern digital video cameras allows you to shoot video with the required resolution;
- test objects should create the same illumination of the eyes;
- do not make changes in the size of the pupil in any way other than emotional.

To study the elementary reaction of the eye to the color of the test object, a sequence of slides was used. Each slide had its own color. The duration of the slide observation is 10 s. The reaction of the pupil was recorded on the video camera. The analysis of the received video frames was carried out in the freely distributed ImageJ program. In Figure 1, a sequence of typical video frames containing information on the reaction of the eye to the color of test objects. Test objects used neutral, not causing obvious emotions.

Video-shooting of the eyes was carried out at distances in which the change in illumination from test objects is not significant. Test objects caused one of two emotions: affection or disgust. When disgust was observed narrowing of the pupil. With a positive emotion, the pupil was dilated. Large bursts of the relative value of the black level correspond to the instant of closing the eyes. Figures 1 (a, b) show a typical dependence of the response of pupils of people on test objects in daylight and artificial lighting.

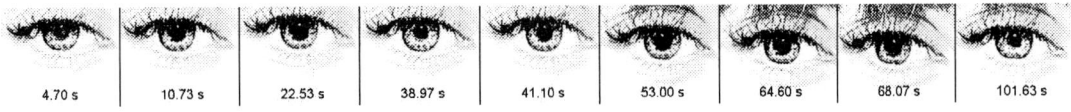

| 4.70 s | 10.73 s | 22.53 s | 38.97 s | 41.10 s | 53.00 s | 64.60 s | 68.07 s | 101.63 s |

Figure 1. The pupil 's reaction to the color of the test object

Figure 2 shows the results of examining the pupil's response to the same test objects (distances of 75 and 100 cm. It should be noted that in daylight, the same test objects were used for the experiments. The relationship between the intensity of the physical stimulus and the mental response (sensitivity) is expressed in the following formula (Zabrodin Yu, Lebedev A, 1977):

$$\frac{dS}{S} = a\frac{dx_s}{x_s^z}.$$

where S - external impact, x_s - sensation, dS - change in external influence, dx_s - change in sensation; z- expresses the level of awareness of the subject about the problems of the experiment and varies from 0 to 1. As can be seen from the graphs, the response to the test objects is present.

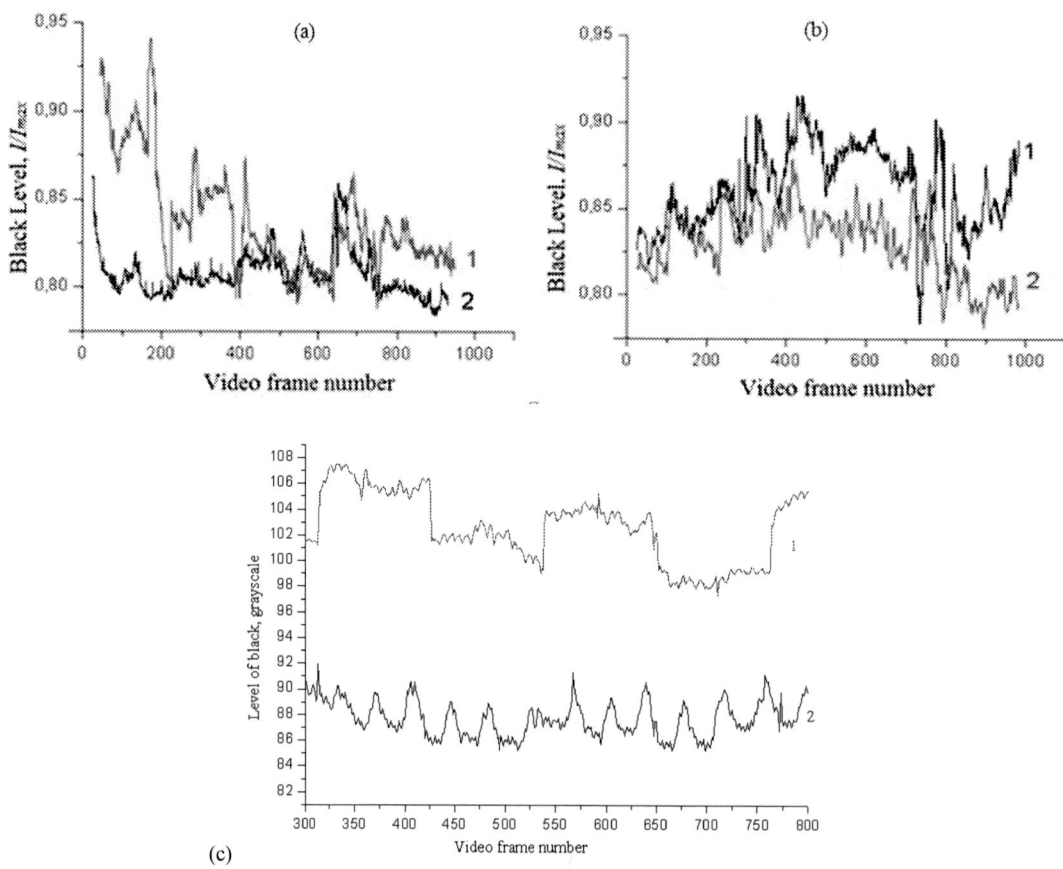

Figure 2(a, b). (a) Artificial lighting: 1-gray eyes; 2 black eyes; **(b)** Natural lighting: 1-gray eyes; 2-black eyes;. **(c)** The pupil's reaction to the test object located at (1) -75 cm and (2) -100 cm from the eye.

Thus, in order to exclude the introduction of a methodical error in the definition of the psychophysical state of a person, it is necessary to place test objects at a distance at which the introduction of distortions is negligible. The intensity of pupil illumination Ip (Trol), was introduced as a measure of retinal illumination in the daytime. The quantity I is the product of the illumination of L by the surface area of the pupil S:

$$I_p = LS.$$

Thus, by increasing the distance from the eye to the test object, you can eliminate the introduction of additional distortions. The black eyes seemed to be the most difficult to analyze. It turned out on the contrary, the graphs of the dependencies of the pupil's response to the test object are less noisy. It is necessary to check whether the "favorite" color affects the size of the pupil.

3. Conclusion

New biometric technologies using 3D-pulse and vibroimage (VibroImage) make it possible to receive more than 10 million informative samples per second about a person's psychophysiological state. Such systems can be used to build biometric systems of the third generation, simultaneously for 10 seconds carrying out identification and testing of the user's thoughts. A good complement to such systems may be the analysis of pupillary reactions in the video image.

In the process of research it was established:

1. It is possible to select a distance to the test objects at which the distortion of the pupil's reaction is small.
2. Correlation of emotions and pupil sizes is traced.

The presented results confirm the possibility of creating active security systems.

Acknowledgments

The reported research was funded by Russian Foundation for Basic Research and the government of the region of the Russian Federation, grant № 18-47-860018.

References

[1] Vekilov T 2017 *Automation in industry* **2** 45

[2] Varlataya S, Rudnykh N, Luzhin V 2016 *Young scientist* **7** 49

[3] Shleymovich M 2015 *Scientific-methodical and informational* **1** 13

[4] Bekmurzin M, Zakharov V, Zacek O 2014 Bulletin of the Moscow University of the Ministry of Internal Affairs of Russia **10** 78

[5] Volkov A, Semenova I 2012 *Actual problems of aviation and cosmonautics* **1** 8

Artificial neural network prediction of lysozyme solubility for protein crystallization

A S Sokolovskii[1], N A Besedina[2], Yu Berdnikov[1] and V G Dubrovskii[1]

[1] Department of Modern Functional Materials, ITMO University, 197101 St. Petersburg, Russia
[2] Nanobiotech Laboratory, St. Petersburg Academic University, 194021 St. Petersburg, Russia

Abstract. In current work, we use the method of artificial neural networks to predict the variation of the solubility of lysozyme protein in solution as the complex function of four crystallization parameters: temperature and pH of the bulk solution, the concentration of precipitant and the concentration of buffer solution. The artificial neural network predicts the salting-in effect at small concentrations of sodium chloride. The predicted dependence of the solubility on the concentration of sodium acetate buffer shows the minimum around 0.1M at different temperatures of the bulk solution.

1. Introduction

Understanding of interaction mechanisms and functions of macromolecules in living cells requires the knowledge of the spatial structure of the considered molecules. This information can be accessed by X-ray analysis of crystallized protein structures [1]. Within this approach, the formation of protein crystals is the limiting step in the process of establishing the tertiary structure of the protein.

The counter-diffusion technique is considered as one of the most progressive methods for protein crystallization. Within this technique the results of a single experiment are equivalent to data acquired in a large set of crystallization experiments within standard methods like sitting drop, batch crystallization, etc [2].

In the counter-diffusion process nucleation and growth of protein crystals are controlled by supersaturation of the protein solution, which is defined as the ratio of protein volume concentration to protein solubility. We can calculate the concentrations of protein molecules by modeling their diffusion in capillary and diffusion of precipitant molecules [3]. However, the solubility of protein molecules cannot be accessed directly due to its complex non-linear dependences on the temperature and pH of the bulk solution, the concentration of buffer and precipitant concentration. Within this work we use the approach of artificial neural networks to evaluate the solubility of protein molecules.

2. Method

In this paper, we use the Artificial Neural Network (ANN) with three hidden layers, with 30, 20, 10 hidden neurons in each hidden layer. The four inputs of the ANN are temperature of the bulk solution,

Content from this work may be used under the terms of the Creative Commons Attribution 3.0 licence. Any further distribution of this work must maintain attribution to the author(s) and the title of the work, journal citation and DOI.
Published under licence by IOP Publishing Ltd

sodium chloride concentration, sodium acetate buffer concentration and pH-level of the bulk solution. The only output gives the solubility of lysozyme protein molecules. Figure 1 summarizes the simplified topology of ANN. The scheme of the ANN with three hidden layers obtained from MATLAB software is shown in figure 2.

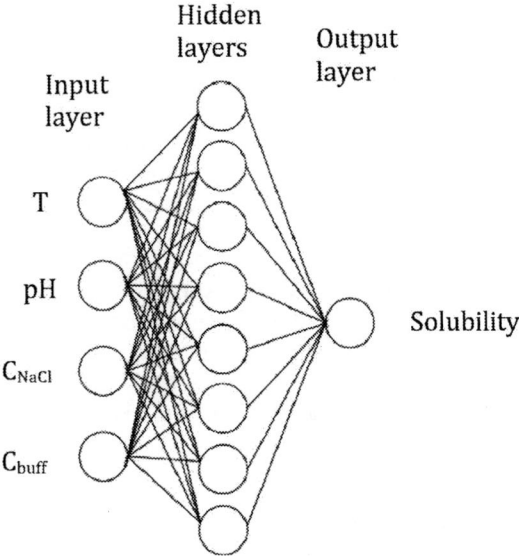

Figure 1. ANN architecture topology. The ANN use four inputs: temperature and pH of the bulk solution, sodium chloride concentration, sodium acetate buffer concentration. The only output is the solubility of lysozyme protein molecules.

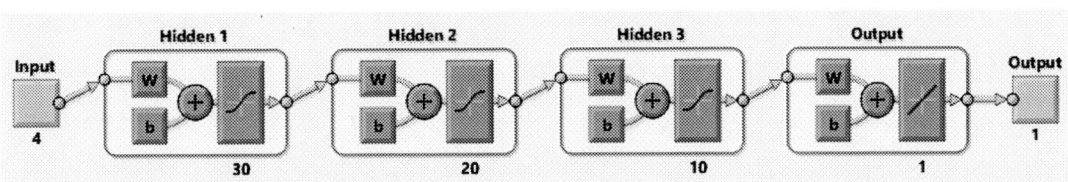

Figure 2. The scheme of the trained neural network. The ANN contains three hidden layers with 30, 20 and 10 neurons. "w" and "b" denotes the weight and the bias matrixes of each hidden layer, correspondingly.

We use the backpropagation (BP) ANN, with Levenberg-Marquardt (LM), Bayesian regularization (BR) and Quasi-Newton Broyden-Fletcher-Goldfarb-Shanno (BFGS) training algorithms. The performance of ANN is evaluated by mean-squared error (MSE). Modeling of ANN was carried out with Neural Network Toolbox of MATLAB system. We choose and compare three different training algorithms to optimize the evaluation times and value of MSE. The details of the ANN performance are shown in Table 1. BR algorithm provides the lowest MSE value in comparison to other algorithms but requires the longest computation time of 17835 seconds. LM algorithm requires 12% less computation time, but MSE value for LM algorithm is five times greater than for BR algorithm. BFGS algorithm provides the shortest computation times, which is about ten times less than for LM and BR algorithms but the MSE value of 0.0268 is ten times higher than for BR algorithm.

Table 1. Efficiency comparison of training algorithms.

Name of algorithm	Epochs	Time of evolution, sec	MSE
LM	50000	15670	0.0098
BR	50000	17835	0.002
BFGS	50000	1575	0.0268

The training of the ANN was based on the solubility data published in Refs. [4–7]. All the data correspond to crystallization conditions of lysozyme crystals with tetragonal syngony within the temperature range from 1.6 to 30.7 °C; pH of the solution varies from 4.0 to 5.4; concentration of sodium acetate buffer changes from 0.01M to 0.5M; concentration of sodium chloride remains from 0% to 7% w/v. We use 5193 values of solubility in the set of crystallization conditions for ANN training.

3. Results

The MSE after training during 50000 epochs was estimated to 0.002 for Bayesian regularization algorithm. The obtained MSE value is 2.5 times less than in ref. [8] which can be explained by a larger set of the training data. We also predict the dependence of the solubility on the concentration of sodium acetate buffer, which was not discussed previously. The set of data allow us to analyze the solubility at small sodium chloride concentrations and observe the salting-in effect as shown in figure 3.

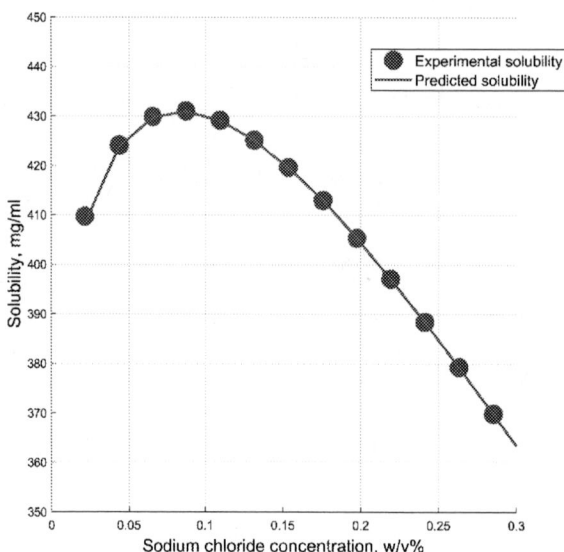

Figure 3. The solubility of lysozyme at 18 °C, at pH = 4.5, as a function of NaCl concentration. The solubility increases at small concentrations of sodium chloride which demonstrate the presence of the salting-in effect

According to ref. [4] the solubility sensitive to the amount of sodium acetate buffer with the minimum near the concentrations of the buffer of 0.1M. The predicted dependence reproduces this minimum for different temperatures and NaCl concentrations. Figure 4 (a,b) shows the obtained predicted dependences of solubility on the concentration of sodium acetate buffer.

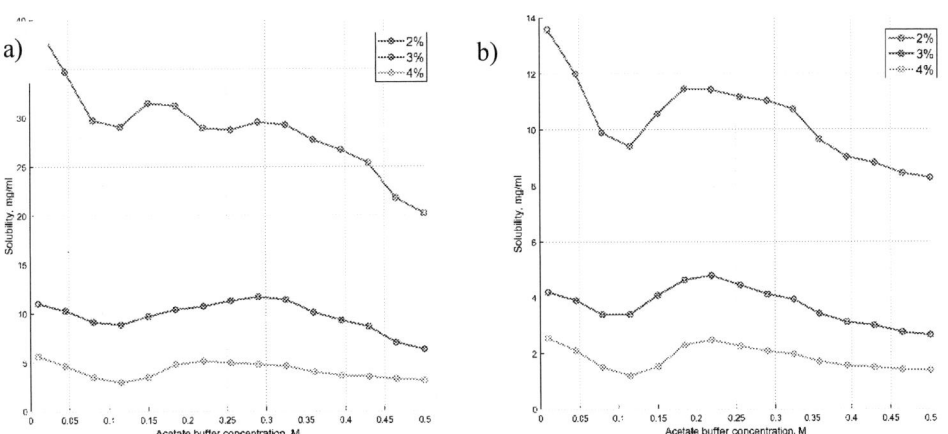

Figure 4. Solubility of lysozyme at pH = 4.0, as a function of sodium acetate buffer concentration, for different 2%, 3% and 4% of NaCl concentration: a) at 10 °C; b) at 20 °C

4. Conclusion

We demonstrate that the approach of artificial neural networks can provide an approximation for the solubility as a function of the temperature of the bulk solution, the concentration of buffer and precipitant concentration. In the predicted dependence the salting-in effect is observed at small sodium chloride concentrations. The dependence of the solubility on the concentration of sodium acetate buffer shows the minimum around 0.1M at temperatures of the bulk solution of 10 and 20°C and NaCl concentration of 2%, 3% and 4%. In future, the developed approach of artificial neural networks can be used for prediction of solubility for different protein molecules.

Acknowledgments

The authors gratefully acknowledge the financial support received from the Ministry of Education and Science of the Russian Federation under grant 14.587.21.0040 (project ID RFMEFI5871X0040).

References

[1] Franklin R E and Gosling R G 1953 *Nature* **172** 156
[2] García-Ruiz J M 2003 *Methods Enzymol.* **368** 130–154.
[3] Sokolovskiy A S, Trushin Yu V, Lubov M N, Eliseev I E, Yudenko A N and Dubina M V 2016 *J. Phys.: Conf. Ser.* **741** 012053
[4] Forsythe E L and Pusey M L *J. Cryst. Growth* 1996 **168** 112–17
[5] Forsythe E L, Judge R A and Pusey M L *J Chem Eng Data* 1999 **44** 637–40
[6] Cacioppo E and Pusey M L *J. Cryst. Growth* 1991 **114** 286–92
[7] Otalora F and García-Ruiz J M *J. Cryst. Growth* 1997 **182** 141–54
[8] Zhang X, Zhang S and He X *J. Cryst. Growth* 2004 **264** 409–16

Effective application of independent component analysis to the task of gradual EMG control

M O Shamshin[1], V A Makarov[1,2], S A Lobov[1]

[1] Lobachevsky State University of Nizhny Novgorod, Gagarin Ave. 23, 603950 Nizhny Novgorod, Russia

[2] Instituto de Matemática Interdisciplinar, Applied Mathematics Dept., Universidad Complutense de Madrid, Avda Complutense s/n, 28040 Madrid, Spain

Abstract. In this work, the possibility of a reliable application of the independent component analysis to the problem of EMG interfaces is studied. The developed algorithm takes advantage of the experimental training procedure and enables a computationally efficient identification of gradual finger movements in real time by using inexpensive EMG sensors with a low sampling rate. Experimental data show that the algorithm improves the separation of motor units by 55 % as compared to raw EMG signals.

1. Introduction

Nowadays electromyographic (EMG) interfaces find applications in many areas of our life including the game industry, bionic prostheses, exoskeleton devices, and electric wheelchairs among others. Reliable EMG control in these applications requires an accurate automatic analysis of EMG data providing high-fidelity recognition of gestures made by the user. The quality of classification of EMG patterns is primarily defined by the features extracted from the signals, rather than by the classification algorithm itself [1-3]. Therefore, in the literature an extensive attention has been paid to the development and analysis of different methods of feature extraction from EMG signals. Among the most popular approaches we can mention the fitting of autoregressive models [4,5], Fourier transform [6], and wavelet analysis [7].

Recently, the wavelet analysis has received a great attention due to its flexibility and applicability to non-stationary signals [8]. Using the wavelet transform one can isolate the motor unit action potentials (MUAPs) from EMG signals and hence identify the motor units participating in a given muscle contraction. This, in turn, enables classification of EMG patterns in a human-machine interface (HMI). Besides HMI applications, the analysis of individual MUAPs is applied in medical diagnostic of muscle injuries [9]. However, the wavelet approach requires high-quality EMG recordings. Since the duration of MUAP is about 5 ms, estimation of their shapes requires the sampling rate of at least 5 kHz. Moreover, preamplifiers with high SNR and high-resolution analog-to-digital converters (ADC) are also necessary, which leads to additional requirements to the analogue part of the EMG recording devices. In addition, the dimension of the data flow increases due to multisensor recordings. These factors demand a relatively high processor power from the on-board computers for processing EMG data in real time, a condition that is hardly met by inexpensive standard equipment widely used in practice.

In this work we explore the independent component analysis (ICA) as a computationally efficient method for the analysis of MUAPs generated by muscle contraction. Instead of the analysis of signal waveforms, the proposed method searches for statistical relationships (beyond linear correlations) among multiple EMG signals recorded simultaneously. It enables separating recorded signals into MUAPs activity by a simple computationally inexpensive linear transformation. Moreover, the synchrony in online reading of MUAPs reduces significantly the requirements to the sampling rate, which can be relaxed to 1 kHz or even less. This makes the approach applicable to practically all commercially available devices.

Content from this work may be used under the terms of the Creative Commons Attribution 3.0 licence. Any further distribution of this work must maintain attribution to the author(s) and the title of the work, journal citation and DOI.
Published under licence by IOP Publishing Ltd

2. Materials and methods

The experiments were carried out by using a myographic Thalmic Myo armband bracelet with the voltage resolution of 25 µV and sampling rate set to 200 Hz. For the sake of simplicity, in this work we aimed at identification of two types of finger movements: bending of ring and pinky fingers. The bracelet was put on the subject's forearm and located in such a way that two sensors were located over muscle flexor digitorum superficialis that participates in movements of these fingers (Fig. 1). The subject was gradually bending his fingers one by one. Note that most people cannot bend their pinky finger separately from the ring finger, which produces additional experimental complexity.

The collected data have been processed by an ICA-like algorithm (see below) and then finger gestures have been classified. To quantify the classification fidelity we evaluated the root mean square (RMS) signals both from raw EMG data and from the processed data over 100 ms windows:

$$R(t) = \sqrt{\frac{1}{N}\sum_{\tau=0}^{N-1} y(t-\tau)}, \tag{1}$$

where y is the detrended original signal (either raw EMG or processed EMG) and $N = 20$ is the number of samples in a window. The RMS signals were then additionally smoothed by a low-pass second order Butterworth filter with a cutoff frequency of 10 Hz. To measure the quality of the gradual measurement of finger bending we used Pearson correlation coefficients evaluated over RMSs of raw and processed EMG data.

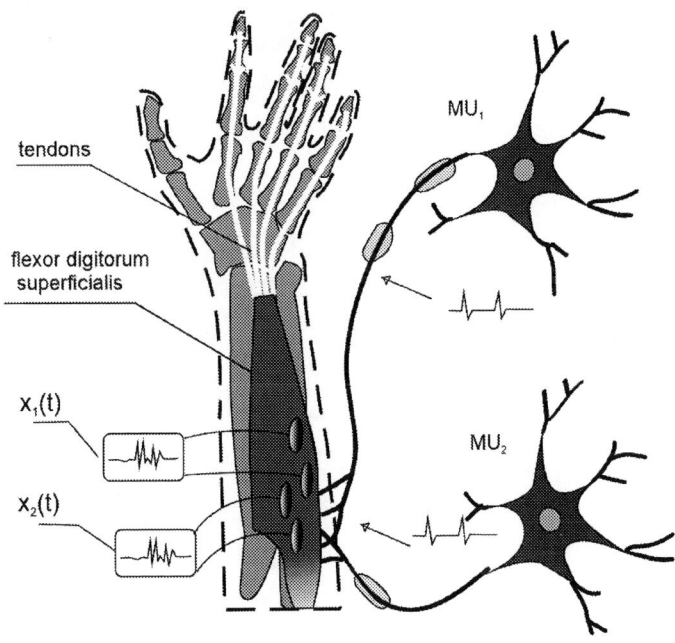

Figure 1. Sketch of the model of the recording of EMG activity from flexor digitorum superficialis muscle contributed by two motor units responsible for bending of different fingers.

3. Results

It is convenient to represent the raw EMG data recorded from two sensors $x_1(t)$ and $x_2(t)$ (Fig. 1) in the matrix form:

$$X = \begin{pmatrix} x_1(t) \\ x_2(t) \end{pmatrix}, \tag{2}$$

where $t \in \{1,2,\dots,T\}$ is the discrete sampling time. Figure 2a shows a representative example of the data obtained while a subject performed the two types of movements: bending his ring and pinky fingers one by one. We observe that such movements produce two characteristic clouds of points aligned along two inclined axes on the (x_1, x_2)-plane. Such a situation is typical for application of ICA.

Earlier ICA was applied to EMG signals for source separation [3,10]. We then can introduce the MUAPs space as a matrix of two unknown vectors $F = (f_1(t), f_2(t))^T$ describing activation of two motor units driving the ring and pinky fingers, respectively (Fig. 1). Then, the recorded data can be represented as a mixture of MUAPs with some weighting matrix A:

$$X = AF. \tag{3}$$

ICA estimates the mixing matrix and MUAPs from the EMG data, by assuming that MUAPs are statistically independent variables (i.e., their join probability can be factorized). There exist a number of ICA algorithms: Infomax, KernelICA, JADE, FastICA, etc. These algorithms employ different numerically friendly criteria of statistical independence and optimize them through iterative searching algorithms.

In general, all algorithms proceed in two steps: 1) Data whitening and 2) Rotation of the whitened data. The whitening can be achieved by applying a linear transformation $Y = WX$, where $W = D^{-1/2}V$ and D and V are the eigenvalues and eigenvectors matrices of the data covariance matrix, respectively. On the second step we rotate the data by an angle ϕ: $\tilde{F}(\phi) = R(\phi)Y$, where R is the standard rotation matrix. The angle ϕ is selected by minimizing Gaussianity of the projection. Thus obtained $\tilde{F}(\phi_{min})$ approximates the original MUAPs up to a constant. Figure 2b shows the ICA estimated MUAPs for the EMG data set represented in Fig. 2a. We note that each separate movement of one finger is represented along one single axis.

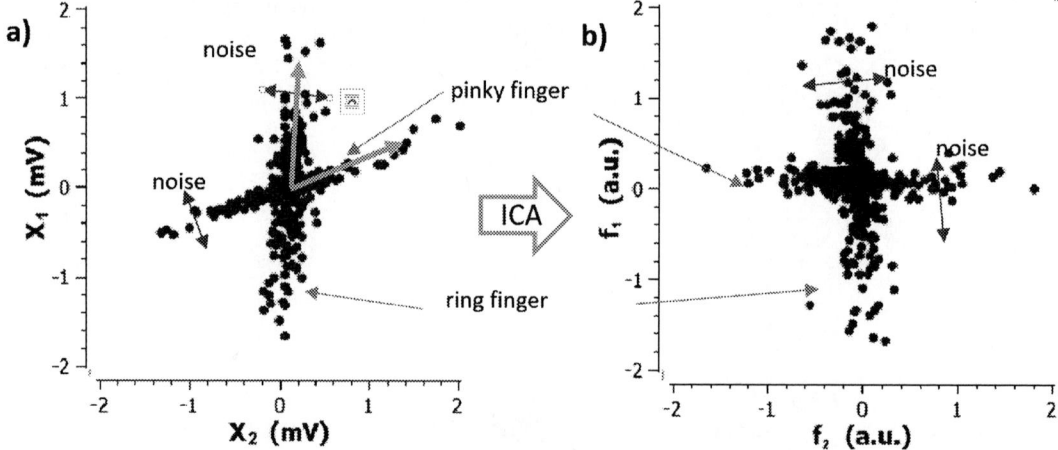

Figure 2. Example of extraction of MUAPs from an EMG recording. a) Experimental data collected from two electrodes while a subject was bending his ring and pinky fingers one by one (see also Fig. 1). Orange arrows correspond to vectors b_1, b_2 found be linear regression. b) ICA estimated MUAPs. A standard algorithm and the simplified strategy based on vectors b_1, b_2 provide the same result.

Although ICA can deal well with the blind source separation problem, we, however, aim at simplifying the procedure by designing an adequate experimental protocol and straightforward data analysis technique. We assume that the movement of each finger is caused by individual MUAPs activating flexor digitorum superficialis muscle (Fig. 1). Then, the muscle contraction produces an electric potential recorded by electrodes. The amplitude of an EMG signal generated by individual MUAP depends on the distance from the sensor to the innervated zone [12]. Then, following the derivation by Naik and colleagues [10] and taking into account different locations of the sensors we can write the data model:

$$x_i = \sum_{m,n} a_{i,m} f_m(t - \tau_{m,n}) \ , \ i \in \{1,2\}, \tag{4}$$

where $f_m(t)$ represents the m-th MUAP, $a_{i,m}$ is its amplitude at the i-th electrode, and $\tau_{m,n}$ is the n-th time instant of occurrence of the m-th MUAP.

Figure 2a suggests that only MUAPs of two main types participate in the finger movements. In an experiment, we now can ask the subject to move only one finger. Then, MUAPs of only one form will prevail in the EMG and hence we can reduce (4) to:

$$x_1 = a_{1,1} \sum_n f_1(t - \tau_{1,n}) + n_1(t), \ x_2 = a_{2,1} \sum_n f_1(t - \tau_{1,n}) + n_2(t), \tag{5}$$

where $n_{1,2}(t)$ are the signals due to residual MUAPs, interpreted as noise. Then, assuming that the noisy terms are small enough, such signals will generate a cloud in the (x_1, x_2)-plane similar to one of those shown in Fig. 2a. Applying the standard least square fit in this case we will get the straight line:

$$x_1 = k_1 x_2 \ , \tag{6}$$

where k_1 is the slope. Next, the subject will move another finger, and then we will record the activity generated by MUAPs of another type (second type of movements). Similarly we get:

$$x_1 = k_2 x_2. \tag{7}$$

From (5) we have the ratios $k_1 = \frac{a_{1,1}}{a_{2,1}}$ and $k_2 = \frac{a_{1,2}}{a_{2,2}}$. Therefore, we can construct two basis vectors $b_1 = (1, k_1)^T$ and $b_2 = (1, k_2)^T$ and use them to construct mixing matrix $\tilde{A} = [b_1, b_2]$. Finally, to unmix the data and obtain MUAPs we apply:

$$\tilde{F} = \tilde{A}^{-1} X. \tag{8}$$

Such an experimentally assisted procedure is much more robust than a blind ICA on mixed data.

We tested the algorithm in 5 male subjects. The collected data had been processed and then we evaluated Pearson correlations among RMS signals (see methods). The Pearson correlation obtained with raw data was 0.80 ± 0.08. After application of the algorithm it decreased significantly to 0.25 ± 0.14 ($p < 0.00001$, t-test). Thus, application of the linear transformation found by our algorithm to raw EMG data improves the separation of motor units by 55 %.

4. Conclusion
In this work we have studied the applicability of ICA to EMG data collected from an inexpensive armband bracelet. For online processing of EMG data we proposed and tested a computationally efficient ICA-like algorithm. It also works at relatively low sampling rates, which is important for designing wearable low-power devices. To develop the algorithm we introduced a model of EMG signals contributed by only one motor unit at a time. Such a situation can be easily achieved experimentally during the algorithm tuning. We have shown that the mathematical model agrees with the recorded EMG data, which enables recognition of dexterous motor activity. The algorithm helps reduce the Pearson correlation of RMS signals by 55%. This enables effective identification of the activity of motor units and identification of finger movements.

Acknowledgments
This study was supported by the governmental assignment of the Ministry of Education and Science of the Russian Federation (grant No. 8.2487.2017/ПЧ).

References

[1] Hargrove L, Englehart K, and Hudgins B 2008 Biomed. Signal Process. Control **3** 175-180

[2] Englehart K and Hudgins B 2003 IEEE Trans. Biomed. Eng. **50** 848-854

[3] Lobov S, Krilova N, Kastalsky I, Kazantsev V, Makarov V A 2018 Sensors **18** 1122

[4] Huang H P and Chen C Y 1999 IEEE Intl. Conf. On Robotics and Automation **1** 2392-2397

[5] Naik G R, Nguyen H T 2015 IEEE Journal of Biomedical and Health Informatics **19(2)** 478–485

[6] Jiang Y, Sakoda S, Hoshigawa S, Ye H, Yabuki Y, Nakamura T, Yokoi H 2014 International Conference on Robotics and Biomimetics **1** 1368–1373

[7] Lucas M F, Gaufriau A, Pascual S, Doncarli C and Farina D 2008 Biomedical Signal Processing and Control **2** 169–174

[8] Hramov A E, Koronovskii A A, Makarov VA 2015 *Wavelets in Neuroscience.* Springer.

[9] Merlo A, Farina D, Merletti R 2003 Biomedical Engineering **3** 316–323

[10] Naik G R, Kumar D K and Palaniswami M 2014 Expert Syst. **31** 91-99

[11] Naik G R, Kumar D K 2011 Comput. Methods Biomech. Biomed. Eng. vol. 14 p. 1105-1111

[12] De Luca C J 1993 Journal of Applied Biomechanics **13(2)** 135–163.

SPBOPEN 2018   IOP Publishing

Use of electrophoretic light scattering for investigation the parameters of macromolecules

E A Savchenko[1], E K Nepomnyashchaya[1], E N Velichko[1], A N Skvortsov[2]

[1]Higher school of applied physics and space technologies, Institute of Physics, Nanotechnology and Telecommunications, Peter the Great Saint-Petersburg Polytechnic University, Saint-Petersburg, 195251, Russia

[2]Department of Biophysics, Institute of Physics, Nanotechnology and Telecommunications, Peter the Great Saint-Petersburg Polytechnic University, Saint-Petersburg, 195251, Russia

Abstract. We present a new method for investigation of the electrophoretic mobility and radius of macromolecules based on electrophoresis in the regime of total internal reflection. The theory of dynamic light scattering and its application for the analysis of data obtained by the technique of capillary electrophoresis in the regime of total internal reflection is considered. Experimental setup for simultaneously separation and registration of macromolecules under study was developed. The results of studies of suspensions with latex microspheres are discussed.

1. Introduction

Macromolecules including proteins, nucleic acids, enzymes and ligands mediate biomolecular interactions which could be controlled by measurements of size and electrophoretic mobility [1]. The electrophoretic mobility is related to zeta potential. The degree of the mutual interaction of particles, as well as their interaction with the liquid medium, is determined by zeta potential. Such parameter of colloid dispersions as stability depends on the value of this electrokinetic potential [2]. Colloids with high electrokinteic potential are electrically stabilized, while those with low electrokinetic potential are prone to coagulation because of the electrokintec interactions between the particles [2].

Investigation of light scattering parameters for the analysis, identification, and characterization of different macromolecules are widely used in various scientific areas [1]. The advantages of light scattering, such as, minimal sample preparation, rapid and simultaneous analysis, ease of quantitation and identification, make it an attractive technique for the analysis of biological analyte [3].

A variety of methods are used to characterize electrostatic properties of macromolecules. Capillary electrophoresis was developed into a strong analytical technique for separation and calculates electrophoretic mobility. Using this technique high efficiency can be obtained with short analysis times. One of the advantages of using capillary electrophoresis is that it can be coupled to a range of different techniques of detection signal [4]. In our work the theory of dynamic light scattering and capillary electrophoresis in the regime of total internal reflection is considered. The method which based on this theory is called electrophoretic light scattering.

2. Techniques

Electrophoretic light scattering (ELS) is a technique for measuring the electrophoretic mobility of

Content from this work may be used under the terms of the Creative Commons Attribution 3.0 licence. Any further distribution of this work must maintain attribution to the author(s) and the title of the work, journal citation and DOI.

Published under licence by IOP Publishing Ltd

particles in multicomponent liquid systems with a recording of correlation characteristics of the light scattered by moving particles [5]. The ELS technique is based on the physical phenomenon known as electrophoresis, i.e., the motion of colloidal particles in an electric field. As the particles migrates, it begins to separate into its constituent components due to differences in their electrophoretic mobility [6]. This mobility is converted into zeta potential to enable comparison of molecular solution under different conditions. In the case of proteins, the measurement of protein mobility allows the calculation of protein charge, which in turn is related to factors such as activity and kinetics of chemical reaction [3].

There are two basic types of the electrophoresis cell that is used in modern electrophoretic light scattering measurements: the capillary and the parallel plate electrodes [7]. In this work the capillary type cell was used. This form of cell has the advantage that the electrodes can be large in size so reducing the current density at the surface and ensures that any gas bubbles produced [7].

3. Experimental setup

Separation and studies of physical parameters of the objects by the electrophoretic light scattering technique involved the regime of total internal reflection with the aim of increasing its sensitivity and reducing the volume of the sample and capillary. In order to carry out our studies, we developed an experimental setup presented in figure 1.

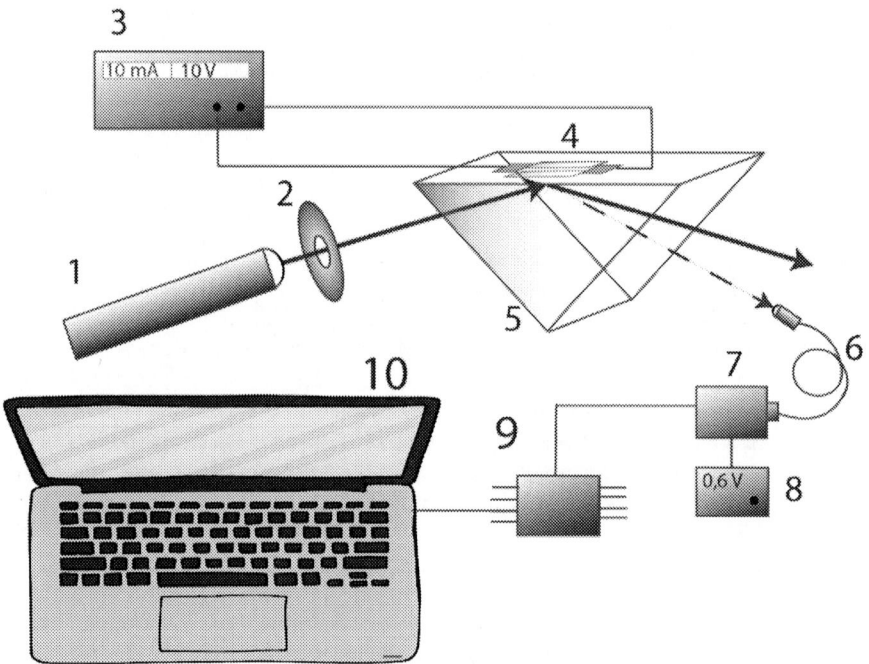

Figure 1. Experimental setup: 1 — laser, 2 — diaphragms, 3 — power supply, 4 — electrodes and sample in capillary, 5 — total internal reflection prism, 6 — optical fiber, 7 — photomultiplier, 8 — power supply of photomultiplier, 9 — ACD converter, 9 — computer.

A focused light from a semiconductor laser with wavelength $\lambda = 655$ nm and the power 2.5 mW was directed on the side face of the total internal reflection prism. The diaphragm was used to narrow the laser beam. The radiation incident at the angles greater than the critical angle was fully reflected from the internal sizes of the prism and returned back. The part of radiation penetrated into the object and scattered from the particles. Scattering light was collected by optical fiber and registered by photomultiplier which can be located in different angles. Then the signal was digitized by ADC converter and recorded by the computer. The obtained data was processed in MS Excel and calculated autocorrelation functions were analyzed. The temperature was between 18 and 20°C

4. Samples

The suspension with latex microspheres with radii of 1 μm and 320 nm were used as a test sample. The scheme of electrodes and sample in capillary is shown in figure 2. A sample was located in the capillary cell, which was placed on the top side of prism. The capillary cell consists of the two glass plates 2 cm in size with the gap between them of 90 μm and two electrodes 2*4 mm in size.

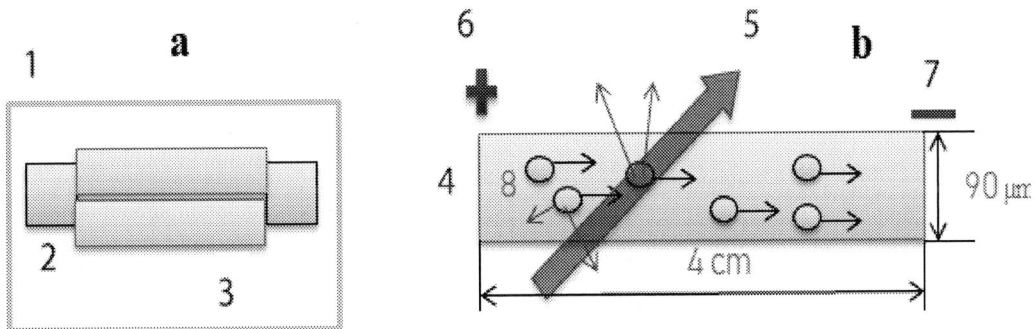

Figure 2(a, b). The scheme of area of investigation **(a)** capillary cell; **(b)** scattering volume: 1 — the top side of the prism, 2 — electrodes, 3 — glass plates, 4 — gap between glass plates, 5 — laser beam, 6 — anode, 7 — cathode, 8 — particles.

5. Results

The electrophoretic mobility in the case of a monodisperse solution can be calculated according to the relation (1)

$$\Delta t = \frac{2\pi}{\mu E K \cos \varphi/2},\qquad(1)$$

where Δt is the oscillation period of the autocorrelation function, the value of which can be found from the obtained experimental dependences, K is scattering vector, φ is the angle at which scattered radiation is recorded. The electrophoretic mobility of microspheres with a radius of 1 μm and 320 nm are $1.42 \cdot 10^{-5}$ cm^2/s·V and $2.59 \cdot 10^{-5}$ cm^2/s·V.

The autocorrelation functions for the measurements of mixtures of latex microspheres with radii of 1 μm and 320 nm without external electric field and in electric field with strength E = 1.25 V/cm, 2.5 V/cm and 5 V/cm are presented in figure 3.

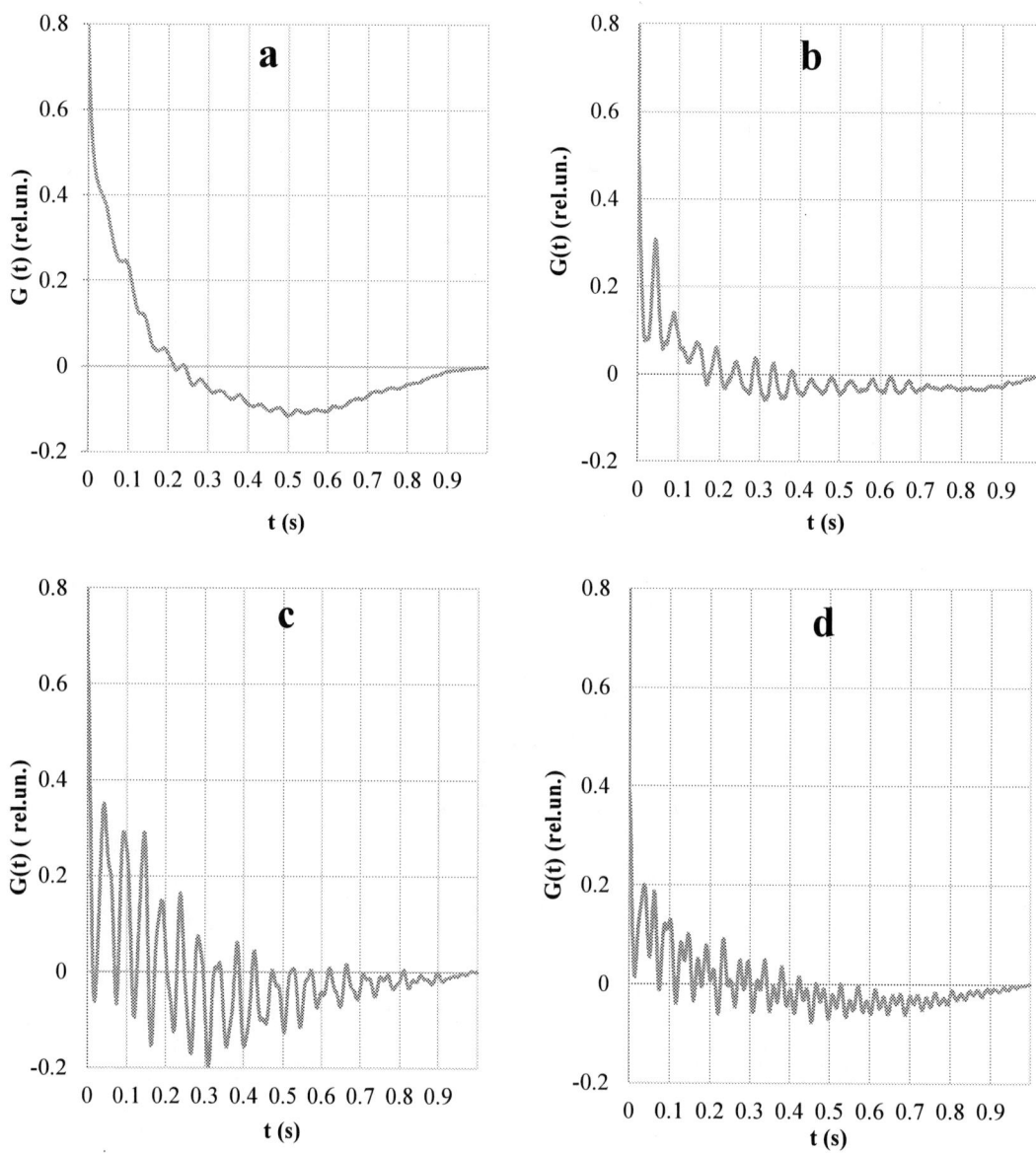

Figure 3(a - d). Autocorrelation functions of solution of latex microspheres with a radius of 1 μm and 310 nm in electric field (a) 0 V/cm; (b) 1.25 V/cm; (c) 2.5 /cm; (d) 5 V/cm.

To obtain the separation of molecules in the solution the power spectrum of the photocurrent was calculated. The power spectrum of the photocurrent from suspensions of latex microspheres with radii of 1 μm and 310 nm in electric field 5 V/cm are presented in figure 4.

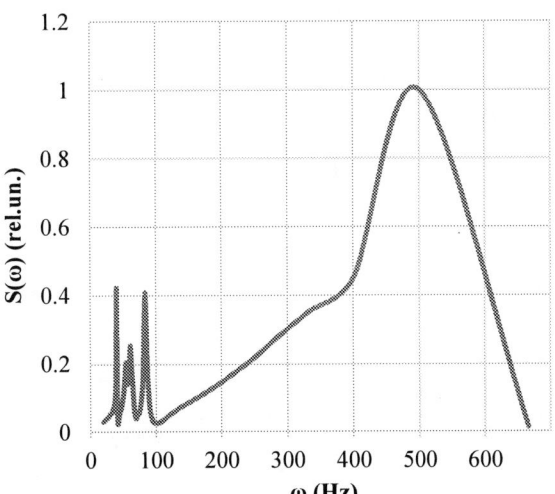

Figure 4. The power spectrum of the photocurrent from suspensions of latex microspheres with radii of 1 μm and 310 nm in electric field 5 V/cm.

According to the obtained power spectrum of photocurrent two components responsible for different sizes of particles in the polydisperse mixture are shown. The first component in the range from 0 to 100 Hz is responsible for the microspheres of 1 μm, the second in the range from 100 to 650 Hz is responsible for 320 nm. It is possible to obtain separation in the polydisperse system using this system. By analyzing calculated autocorrelation function (figure 3) we can obtain the distribution of electrophoretic mobilities in mixture of microspheres. For this purpose special processing program based on the Tikhonov regularization method is need to be used as a basis for the algorithm for the solution of the inverse problem of light scattering.

6. Conclusions

The first experimental results confirmed possibilities of the capillary ELS technique in the regime of total internal reflection to determine parameters of macromolecules (size, zeta-potential) and separate them by electrophoretic mobility in multicomponent liquid systems. The application of total internal reflection regime with small capillary allowed us to increase the sensitivity in comparison with standard methods of determine parameters of macromolecules and their separation.

References

[1] Tuchin V V 2007 *Tissue Optics: Light Scattering Methods and Instruments for Medical Diagnosis* (SPIE Tutorial Texts in Optical Engineering TT38, Bellingham, WA, USA) PM 166 p 882
[2] Terechenko S A Shalaev P V, Masloboev Yu.P., Dolgushin S A, Dechabo V A, Yudin I K 2017 *J Biom.Engeen.* **50** 333
[3] Ball S. 2013 Biopharm International **26** (5) 20
[4]Chetwynd A J, Guggenheim E J, Briffa S M, Thorn J A, Lynch I, Valsami-Jones E 2018 *Nanomaterials* **8**(2) 99
[5] Medrano M., Perez A.T., Lobry L., Peters F. 2009 Langmuir **25** (20) 12034
[6] Lewis. A., Harris. N. 2017 Multidisciplinary Digital Publishing Institute Proceedings **1** (4) 278
[7] Roberts G.C.K.2013 *Encyclopedia of Biophysics* (European Biophysical Societies' Association) p 2884

Fluorescent detection and analysis of single molecules

O B Kuznetsova[1], E A Savchenko[1], E T Aksenov[1]

[1]Higher school of applied physics and space technologies, Institute of Physics, Nanotechnology and Telecommunications, Peter the Great Saint-Petersburg Polytechnic University, Saint-Petersburg, 195251, Russia

Abstract. Total internal reflection fluorescence microscopy for detection and analysis of single molecules is considered. The experimental setup for the detection of single molecules by total internal reflection fluorescence microscopy was developed. Water solution of Rhodamine 6G was investigated.

1. Introduction

There are several methods of single molecule detection. Nowadays, researchers are mostly interested in the fluorescence microscopy method as the non-invasive technique for molecule detection [1]. In comparison with classical light microscopy, fluorescence microscopy has a number of advantages. For example, possibility to determine the distance between molecules, sufficient for interaction between them [2]. In addition, the use of sets of fluorophores has allowed researchers to identify the microscopic cellular components and specific cells among substances which are not fluorescent.

One of the most common types of fluorescence microscopy is the so-called Total internal reflection fluorescence (TIRF) microscopy. TIRF microscopy produces a thin excitation light field (the evanescent field) that nominally decays exponentially. Selective excitement of fluorophores near the sample interface is seen by this field. This method is widely used in biomedicine: researching exocytosis and endocytosis of neurotransmitters in the synapse, researching the dynamics of proteins interactions on the cell membrane [3]. The method we consider allows us to obtain images of single molecules. Thus we will be able to obtain information about the local nanometer environment of molecules. This information will allow us to study the structure and dynamics of single molecules at microscopic level. One of the advantages of TIRFM in comparison with other methods is its ability to obtain an optical layer of better quality at large depths of thick samples. In addition, the scattering of both stimulating and fluorescent light influences the signal reception in this method not so much as in confocal microscopy. The TIRF method is less toxic and has a high signal-to-noise ratio. TIRFM is also more capable of collecting data on high-throughput interactions within and between single molecules given low background noise. TIRFM tends to be more reliable in terms of single-molecular tracking, which becomes important when considering the drifting effects of microscope equipment.

2. Techniques

The considered method is based on the phenomenon of full internal reflection, which is described by the Snellius law [5]. According to which certain conditions must be fulfilled:

- the angle of incidence of radiation should exceed the value of the critical angle (1)
- the refractive index of the medium (n_1) from which the light falls to the surface shall be greater than the refractive index of the environment (n_2).

$$\theta_c = \sin^{-1}\frac{n_2}{n_1} \tag{1}$$

In our case, the angle value is $\theta_c = \arcsin(1,14) = 48,7°$.

$$d(\theta) = \frac{\lambda/n_1}{4\pi(sin^2\theta - (n_2/n_1))^{-1/2}} \tag{2}$$

The depth of penetration of electromagnetic radiation into a medium with a smaller refractive index for our object of investigation $d(\theta) = 219$ nm (2).

3. Experimental setup

The scheme of the experimental setup is shown in Fig.1. The light flux passes through the prism of total reflection, is reflected at the interface and passes through the opposite face of the prism, while a part of the electromagnetic radiation penetrates into a less dense medium, thereby exciting fluorescence in the object under investigation. The reflected light of the incident beam generates an electromagnetic field in a medium with a lower index of refraction.

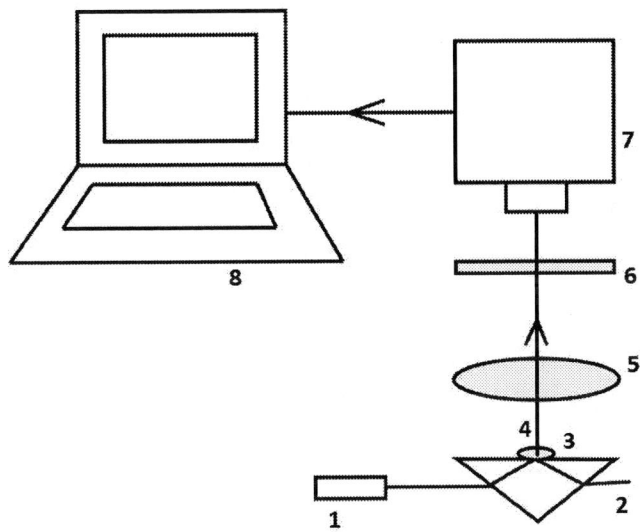

Figure 1. Experimental setup. 1 — laser, 2 — prism of total internal reflection, 3 — sample, 4 — emission, 5 — microscope lens, 6 — filter, 7 — CCD camera, 8 — computer.

Fluorophores located closest to the surface of the glass are selectively excited when interacting with this field, and the secondary fluorescence from these radiators (fluorophores) is collected by the optics of the microscope. The principle of this process is shown in Fig.2.

Then the signal passes through a filter installed in a microscope, and is detected by a CCD camera. In this work, a camera with a low noise level (62 dB), high efficiency and high dynamic (50 dB) and spectral (380 – 650 nm) ranges was chosen. The received signal is then displayed on the computer as an image in Altami Studio.

The light that hits the matrix elements is converted into an electric charge, a charge picture is formed, which is proportional to the illumination in each cell. The matrix can accumulate charges for a certain period of time. The total charge accumulated in the cell is equal to the product of charges for the exposure time. To obtain a color image, the light beam passes through a set of special filters of green, blue and red color. According to the received pictures, we can, for example, monitor the trajectories of rhodamine molecules.

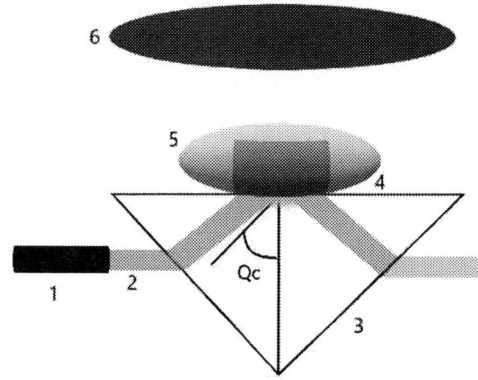

Figure 2. Total internal reflection fluorescence. 1 — laser, 2 — emission, 3 — prism, 4 — evanescent field, 5 —sample, 6 — objective.

4. Samples
As a test sample, a water solution of rhodamine 6G with fluorescence excitation length within 526 nm was used. In accordance with this, a laser with a wavelength of 532 nm was selected. Refractive indices of rhodamine ($n2=1,332$) and glass are ($n_1=1,518$) [3,4].
Rhodamine dyes are used in biotechnology in such methods as fluorescence microscopy, flow cytometry and enzyme immunoassay. The dye has surprisingly high photostability, high quantum yield of fluorescence (0.95 [6]), and its generation is close to the absorption maximum (approximately 530 nm). The dye forming range is from 570 to 660 nm with a maximum at 590 nm.

5. Results and discussion
The first results of the experiment confirmed the potential of the chosen method in the detection of rhodamine molecules 6G. The application of total internal reflection fluorescence microscopy on prism allowed us to increase the sensitivity in comparison with standard methods.

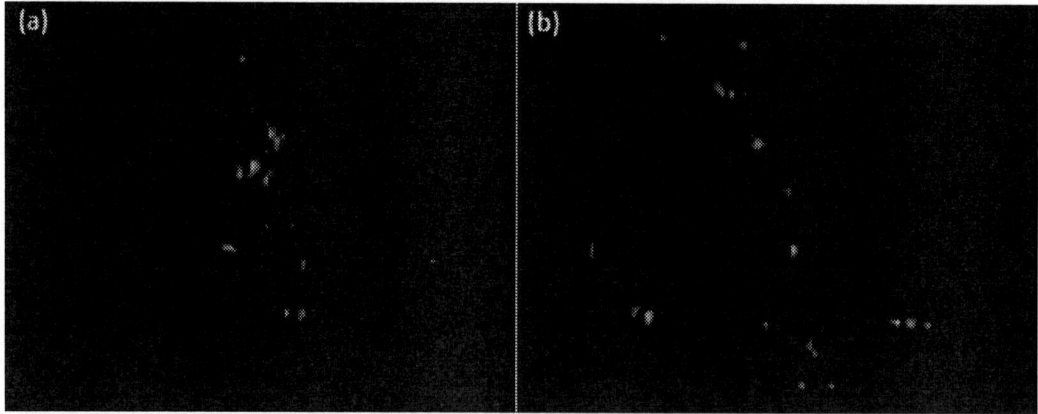

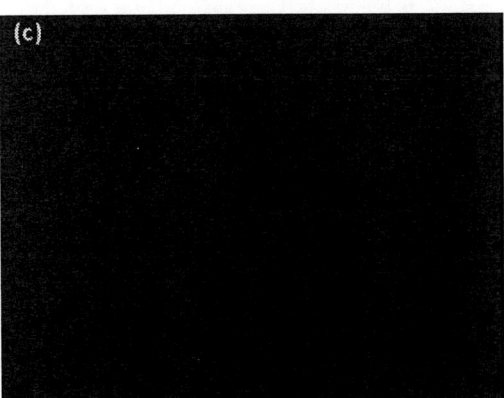

Figure 3 (a, b, c). (a, b) Fluorescence of molecules of rhodamine 6G in an water solution at the different sections $C=2*10^{-6}$mg/L; **(c)** Water solution of rhodamine in concentration $C=5*10^{-6}$mg/L.

In the Fig.3 fluorescent molecules in different concentration are presented. We can determine location of the fluorescent centers in studied samples on the basis of these pictures.

The research of single molecules by the fluorescence method in the mode of total internal reflection will help to understand how the biotic processes in the cell proceed. It will also be possible to introduce single molecules as probes, which will allow us to study the structure and dynamics of materials in which molecules will be introduced at such a microscopic level [8].

Nowadays, TIRF microscopy has a solid theoretical basis, and the practical application of this technique has been greatly simplified due to recent technological advances. Not surprisingly, it is increasingly being used in biomolecular research and research in cell biology. The configuration of TIRF systems based on direct or inverted microscopes is relatively simple when using a laser light source, but can also be performed using a conventional arc lamp, provided that the light passing through the central part of the lens is blocked. Now, complete modular microscopic systems assembled for TIRF microscopy in combination with other optical techniques are now available, and some manufacturers produce high numerical aperture lenses designed specifically for applications that use internal reflection. The TIRF method is compatible with various lighting modes, including light-field, darkfield, phase-contrast, differential interference contrast and traditional epifluorescence. A special advantage of systems with illumination through the lens is their compatibility with various mechanisms of manipulation with biomolecules, such as atomic force microscopy. Probably, the tendency of combining TIRF microscopy with other additional techniques will continue. Also, TIRFM has potential advantages in any application where visualization of small structures or single molecules is required in samples with a large number of fluorophores located outside the optical plane of interest, such as molecules in solutions with Brownian motion, vesicles undergoing endocytosis or exocytosis, or transfer of individual proteins in cells [6,7].

References
[1] Shera E B, Seitzinger N K, Davis L M, Keller R A, Soper S A 1990 *Chemical Physics Letters* **174** 553-557
[2] Truskey G A, Burmeister J S, Grapa E, Reichert W M 1992 *Journal of cell science* **103** 491-499
[3] Okhonin Bindhu C V, Harilal S S, Nampoor V P N, Vallabhan C P G 1999 *Modern Physics Letters B* (India: Cochin University of Science and Technology) 563–576
[4] Savchenko E A, Nepomnyashchaya E K, Dyubo D B, Velichko E N, Tsybin O Y 2017 *Journal of Physics: Conference Series* **917** 042010
[5] Sedyuk I, Zakkai N, Zakkai J 2010 *Methods in molecular biophysics: Structure. Function. Dinamics.* (Moscow) p 701
[6] Schneckenburger H 2005 *Current opinion in biotechnology* **16** 13-18

[7] Stelmashchuk O, Tarakanchikova Y, Seryogina E, Piavchenko G, Zherebtsov E, Dunaev A 2018. *Saratov Fall Meeting 2017: Optical Technologies in Biophysics and Medicine XIX* Proc. Of SPIE p 1071619

[8] Stelmashchuk O, Zherebtsov E, Zherebtsova A, Kuznetsova E, Vinokurov A, Dunaev A, 2017 *Laser Physics Letters* **14** 065603

SPBOPEN 2018   IOP Publishing

Non-invasive research of biological objects by the method of laser polarimetry

M V Putintseva[1], E T Aksenov[1], C C Korikov[1], E N Velichko[1]

[1]Institute of Physics, Nanotechnology and Telecommunications, Peter the Great St. Petersburg Polytechnic University, Saint Petersburg 195251, Russia

Abstract. Recent studies have displayed that the biological tissue, especially skin, alters the polarization state of the incident light. Using this property will enable the study of abnormalities and diseases that change not only the light intensity but also its polarization state. This paper briefly considers spatial speckle-correlometry and polarimetry method for measuring changes of polarization state of the light scattered from a biological tissue. This technique provides the possibility to have the most comprehensive information on the optical and polarization properties of the skin sector containing scar structures and other abnormalities and diseases. Practical application of the developed approach is shown in the research.

1. Introduction

Recently there have been increased rather disputable questions related to the optical methods for diagnosing diseases caused by alterations in morphological characteristics in biological tissues including the epidermis. The evolution of proper diagnostic techniques for monitoring skin abnormalities is currently one of the most challenging areas of research. Nearly all examination and diagnostic methods for measuring the progress of skin disease are based on visualisation [1]. Abnormal changes in the skin layers structure cause optical properties alteration of healthy and diseased tissue. In general skin shows absorption and scattering properties when it is exposed to light. The majority of the existing methods are based on measuring intensity of backscattered and transmitted light. Studies have shown that tissue affects the polarization state of the incident light. The main polarization-altering agents are scattering particles such as cell nuclei, mitochondria and collagen fibers which demonstrate birefringent effect [2]. Thus, the progress of diseases could be determined and monitored by measuring the polarization state changes. That particular property is the basis for proposed method. Non-invasive and highly sensitive three dimensional speckle-correlometry and polarimetry technique has certain perspectives. Such approach helps to get the information on properties of large scatterers in human epithelium, provide histological data about human tissues without causing damagers of the cutaneous structure [3,4]. Apart from that this method can be performed in-vivo. Moreover, our optical spatial speckle-correlometry technique may be efficiently applied in various areas, for example in experimental mechanics for obtaining the information about macro displacements, anelastic deformations, rotations, velocity of moving objects, vibration characteristics, surface finish quality, structural defects, etc. [5].

2. Theory

Correlation and statistical analysis of the speckle intensity fluctuations of scattered fields provide the additional information about the inner and morphological structure of the scattering object. The possible approach for speckle modulated scattered fields is the usage of cross-correlation functions of

Content from this work may be used under the terms of the Creative Commons Attribution 3.0 licence. Any further distribution of this work must maintain attribution to the author(s) and the title of the work, journal citation and DOI.

Published under licence by IOP Publishing Ltd

the speckle intensity fluctuations with different polarization states. This provides demonstration of the developed fine polarization structure [6]. The speckle size can be determined from calculations of the normalized autocovariance function of the speckle intensity pattern obtained in the observation plane, $R_I(\Delta x, \Delta y)$ corresponds to the normalized autocorrelation function of the intensity. It has a zero base and its width provides the "average width" of a speckle. Obviously, $R_I(\Delta x, \Delta y)$ is calculated from the intensity distribution of the measured speckle, I, as described in [7]

$$R_I = \frac{FT^{-1}\left[\left|FT\left[I(x,y)\right]\right|^2\right] - \left\langle I(x,y)\right\rangle^2}{\left\langle I(x,y)^2\right\rangle - \left\langle I(x,y)\right\rangle^2}, \tag{1}$$

where FT is the Fourier Transform, $< >$ is a spatial average, $R_I(\Delta x, 0)$ and $R_I(0, \Delta y)$ are the horizontal and the vertical profiles of $R_I(\Delta x, \Delta y)$, respectively.

3. Experimental setup
The experimental setup for studying the diffuse back-reflectance of polarized light is depicted in Fig. 1.

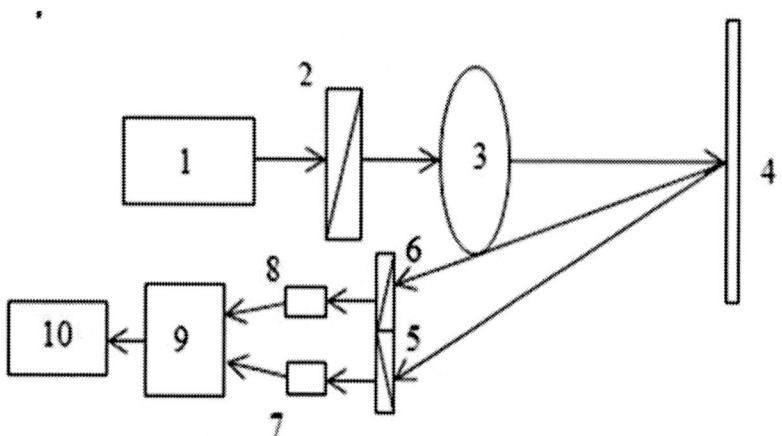

Figure 1. Scheme of the experimental setup. 1 — laser diode, 2 — polariser, 3 — lens, 4 — tissue under study, 5, 6 — polarizer, 7, 8 – photomultiplier with a pinhole, 9 – oscilloscope, 10 — computer.

In this research model experiments were performed on the leaves of trees which are at various stages of withering. In some places, there were damages on the samples under study — a small area of the leaf. During scanning with polarized light of the examined objects alterations in the recorded signals when the beam passed through the affected regions were observed. To each polarimetric state of the incident field corresponds a unique variation of intensity during the scan [8]. During this scan, the projection state that is the closest to the polarimetric state of the incident field will give rise to maximum intensity at the output of the analyser. On the contrary, there will be the minimum output when the incident state of illumination is projected on the more diametrically opposed state [9, 10]. The experiment was carried out in a darken room. A laser diode emitted a beam with $\lambda = 650$ nm through a linear polarizer oriented parallel to the sample scattering plane. After, went through a lens, it gets on the sample under study. Further the backscattered light passes through a second linear polarizers 5 and 6 oriented parallel and perpendicular to the scattering plane. The data recorder consisted of photodetectors 7, 8 and also an oscilloscope 9 and the personal computer for further processing 10.

4. Results of the research and their discussion

Our experimental study was performed in vivo and as a result the time series of speckle intensity fluctuation induced by the sample scanning were obtained. The intensity fluctuations are caused by the sample scanning across the light beam. Figure 2 shows the typical forms of the intensity cross-correlation function for different stages of the leaves withering.

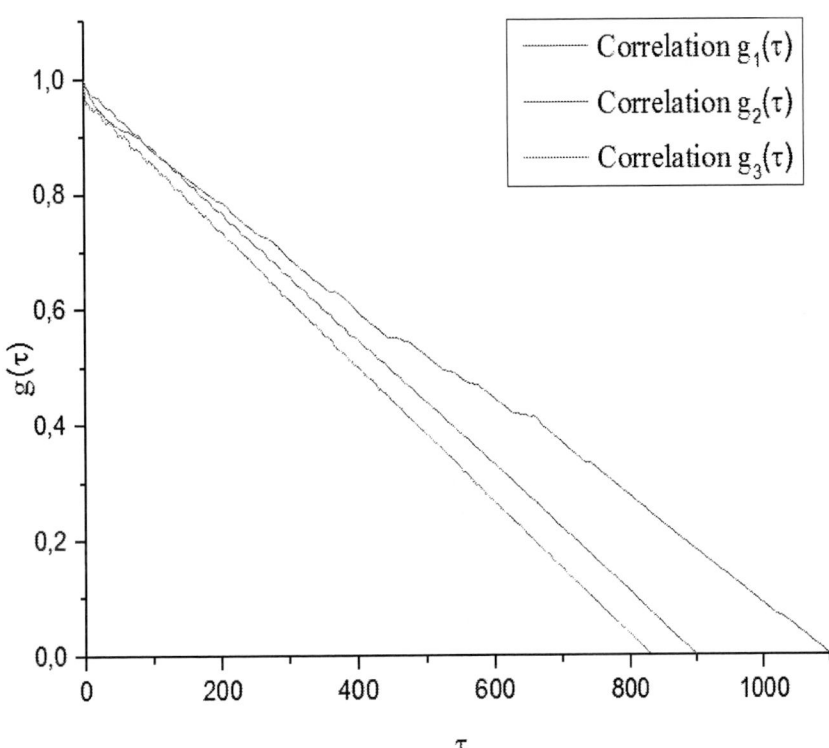

Figure 2. Normalized autocorrelation function of registered speckle intensity fluctuations of the studied leaves samples which are at various stages of withering.

Here g_1 is a correlation function for the green leaf, g_2 – yellow, g_3 – red one. Different correlation characteristics of speckle intensity fluctuations of samples that were investigated showed a high sensitivity to discussed changes in tissue structure [11]. This approach has a number of assets over the other diagnostic methods, by reason of rather high speed-of-response, accuracy and multifunctionality [12]. Thus, it can be applied to an internal and external research of structure of the human tissue in vivo. We are currently working on the analysis of the area of the skin that contains scar structures and other abnormalities and diseases. Eventually, charts of the polarization characteristics distribution of the skin will be received.

References

[1]Litvinova K S, Rafailov I E, Dunaev A V, Sokolovski S G, Rafailov E U 2017 *Progress in Quantum Electronics* **56** 1–14

[2] Wang X, Wang LV, Sun C, Yang C 2003 *Journal of Biomedical Optics* **8** 608-617

[3] Tuchin V V 1997 *Proc. SPIE* **2981** 120–159

[4] Dremin V, Zherebtsov E, Sidorov V, Krupatkin A, Makovik I et al. 2017 *J. of Biomed. Opt.* **22** 085003

[5] Vladimirov A P 2015 *Diagn.,, Res. and Mech. of mat. and str.* **6** 27–57

[6] Zimnyakov D A, Tuchin V V, Larin K V, Mishin A A 1997 *Proc. SPIE* **2981** 172–180

[7] Piederrière Y, Cariou J, Guern Y, Le Jeune B, Le Brun G, Lotrian J 2004 *Opt. Express* **12** 176–188

[8] Ghassemi P, Miranbaygi M H 2009 *J. of Zhejiang Un. Sc. B* **10** 602–608

[9] Ghabbach A, Zerrad M, Soriano G, Amra C 2014 *Opt. Express* **22** 12

[10] Goodman J W 2007 *Speckle Phenomena in Optics* (Roberts & Company Pub.) p 48–50

[11] Zimnyakov D A, Tuchin V V, and Utz S R 1994 *Opt. Spectrosc.* **76** 747–753

[12] Viktin I A, Laszlo R D, and Whyman C L 2002 *Opt. Express* **10** 222–229

In vitro model of CNS neuronal pathway recovery using microfluidic chips

O Antipova*[1], Y Pigareva[1], V Kolpakov[1], A Gladkov[1,2], A Bukatin[3], I Mukhina[1,2], V Kazantsev[1], A Pimashkin[1]

[1]Neuroengineering Laboratory, Translational technologies Center, Lobachevsky State University of Nizhny Novgorod, Nizhny Novgorod, 603022, Russia

[2]Central Research Laboratory, Cell Technology Department, Federal State Budgetary Educational Institution of Higher Education «Privolzhsky Research Medical University» of the Ministry of Health of the Russian Federation 603081, Russia

[3]Saint Petersburg Academic University Nanotechnology Research and Education Centre of the RAS, Saint Petersburg 194021, Russia

Abstract. Development of biodegradable implants using new methods in biotechnology and neural engineering is one of the most perspective approaches in rehabilitation of central nervous system after injury and in neurodegenerative diseases. In this study we propose an experimental model of brain injury by growing two weakly coupled neuronal networks in three-chamber microfluidic chip. We modeled a functional recovery by plating a new dissociated cells in the place of weak synaptic connectivity. We showed that a weak synaptic unidirectional connections between two cultures of neuronal cells can be enhanced by integrating a new cell population into the growth site of axons. The proposed microfluidic chip design may be used to create a new type of scaffold which recovers realistic heterogeneous architecture of neuronal connectivity for rehabilitation of the brain injury.

1. Introduction

One of the promising methods of neurotransplantation in CNS injury recovery is a scaffold development and implantation. The scaffold serves as a basic structure made of biodegradable materials which carries induced pluripotent cells (IPSc) or stem cells. Such structure composed of biodegradable material (polymers, hydrogels, hyaluronic acid) provides integration of the cells in the brain with defined cellular density and after few days it dissolves leaving only the cells in the area of trauma [1-4]. In a perspective such structures will be implanted into the injury area of the brain and then the restore the lost functions [5].

The fundamental problem functional integration of various types of the cells in such technology consists of two tasks.

First, the new cells should differentiate and form functional connectivity with the mature neural tissue. Such problem can be studied on a model of the neural tissue by growing clusters of dissociated neuronal cultures in microfluidic chips [6,7]. Several groups demonstrated that functional bi-directional connection form between primary neurons and stem cell-derived neurons [8,9] via chemical synapses and between human iPS cells using in vitro co-culture device [10]. Also, in our previous study [11] we showed that progenitor neurons can be co-cultured with mature neural network and form functional synaptic connectivity.

The second problem is that the structure of the scaffolds today is usually is homogeneous and can provide only homogeneous network of integrated cells, which can adversely affect the restoration of injuries in the brain. This problem can be solved studying different geometric shapes of the structure that shapes axons growth and network development [12]. In this study we used microfluidic devices with asymmetric microchannels design, which allow to isolate the cells and provide axon growth in

---

Content from this work may be used under the terms of the Creative Commons Attribution 3.0 licence. Any further distribution of this work must maintain attribution to the author(s) and the title of the work, journal citation and DOI.

Published under licence by IOP Publishing Ltd

desired direction to form heterogeneous network. We propose and test a model of unidirectional functional connectivity recovery by plating a neurons into the chip with three chambers with asymmetric design.

2. Materials and methods.

2.1. Device Fabrication
Microfluidic chips were fabricated using silicone polymer PDMS (polydimethylsiloxane). Standard two-layer lithography was used for mould fabrication [12]. Liquid PDMS was poured onto the structured mold in a thin layer (3-5 mm) and put in dry-air sterilizer for 4 hours at 70 °C. Then the PDMS chips were carefully cut from mold and inserted in dry-air sterilizer for 12 hours at 100 °C. Then the chip was mounted on the sterile glass (figure 1(a)) or on the surface of a microelectrode array (MEA).

The chips consisted of three chambers: Source (1), Implant (2) and Target (3) which were connected by 2 sets of 8 asymmetric microchannels (figure 1(b)). The length of each microchannel was equal to 600 μm and consisted of three sections to provide unidirectional axon growth.

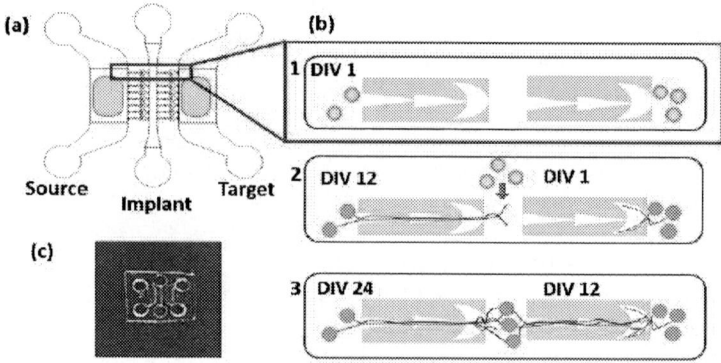

Figure 1(a, b, c). Microfluidic chip with 3 compartments (chambers) to model CNS injury as uncoupled networks and restoring connectivity by integrating new cells. (a) Chip structure and dimensions;(b) Stages of the experiment: 1. growing two uncoupled cultures; 2. integration of the new cells by plating the culture in Implant chamber; 3. Confirmation of connectivity formation; (c) Photo of the chip.

2.2. Cell Culture
Hippocampal neuronal cells were dissociated from embryonic mice (E18) and plated in the chambers of PDMS chips with initial density approximately 7,500 cells/mm2. First, the cells were plated in the Source and the Target chambers of the chip, the Implant chamber remained empty. The neurite outgrowth through the microchannels was monitored during 20 days using automated microscope system (Cell IQ, ChipMan Technologies, Finland).

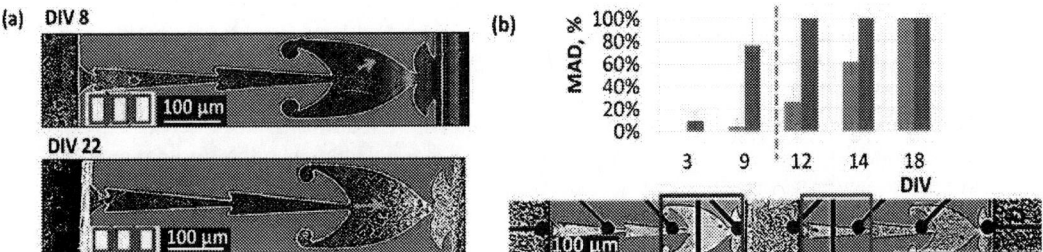

Figure 2 (a,b). Axon and dendrite growth before and after integration of new cells in Implant chamber. (a) The axon grew through the microchannels on 8 DIV and 22 DIV (12 days after plating the cells into the Implant chamber). The neurities grew dominantly in one direction (green arrow) while the microchannel structures prevented opposite growth (red arrow). (b) Estimation of connectivity formation (see Methods) in Source with Implant (red) and Implant with Target (blue).

2.3. Extracellular Recording and Stimulation

Bioelectrical activity of neuronal cultures was studied using microelectrode arrays with 60 electrodes (Multichannel Systems, Germany). We manually aligned the microfluidic chips with MEA so that 24 electrodes were placed in the microchannels between Source and Implant chambers (3 electrodes in each of 8 microchannels), 24 electrodes – in the microchannels between Implant and Target chambers, 8 electrodes - in Implant chamber. The recording of the bioelectrical activity was performed using a multichannel USB-MEA120-Inv-2-BC-System (Multichannel systems, Germany) at a sample rate of 20 kHz. The experiments were performed from 3 to 24 DIV after culture plating. Analysis of the recorded bioelectrical activity was performed with custom made scripts in Matlab [6]. The responses to electrical stimulation were recorded on 23 DIV. We used a low-frequency stimulation consisted of biphasic voltage pulses ±800 mV, 260 µs per phase, positive first, intervals between stimuli were 3 s. Series of 60 stimuli were applied through one electrode in the microchannel between Source and Implant chambers or between Implant and Target chambers.

3. Results and conclusion.

3.1. Connectivity between neuronal cultures

First, we developed a microfluidic chip which consisted of three chambers connected by asymmetric microchannels. Each microchannel had three sections with "traps" that did not allow the axons to grow in the direction from Implant to Source chamber and from Target to Implant chamber (figure 2). Hippocampal neurons were plated in the Source and the Target chambers. Within 48 h after plating the neurons started to grow neurites into the microchannels from the Source and reached the Implant chamber at least after 5 days. Neurites from the Target chamber grew in the traps of the microchannels and did not reach the middle Implant chamber. On 12 DIV the cells were plated in the Implant chamber and started to grow in the microchannels within 2 days forming connection with the axons from the Source and the dendrites from the Target .

We estimated a coupling of the cultures with neurites through the microchannels before and after plating the cells in the Implant chamber on 3, 9, 12, 14, and 18 DIV (n=10 chips) (figure 2 (b)). The percentage of the filled channels (MAD) between Source and Implant was high (76%) before planting, and reached 100% on the 1st day after new cells were plated into the Implant chamber. The first sections of microchannels between Implant and Target were partially filled (26.20% ± 12.67%) by axons from the Source. On 2 DIV after cells integration the MAD between Implant and Target was equal to 61.30% ±10.75% and reached 100% on 6 DIV.

3.2. Bioelectrical spiking activity

We recorded the spiking activity from the MEAs on 11, 17 and 24 DIV (n=3 cultures). On 11 DIV we observed spikes and sparse bursts from all microchannels (figure 3 (a), middle panel). There were no spikes in Implant chamber and in the first section of microchannels between Implant and Target. On 18 DIV (6 DIV after plating in Implant chamber) the axons passed through all microchannels and the bursts propagated from the Source to the end of the microchannels between the Implant and the Target chamber (figure 3 (a), bottom panel).

3.3. Stimulus evoked activity

Next, we applied a series of electrical stimuli to confirm that the spiking activity propagated from the Source to the Target chamber after new cells integration. The activity should pass through the microchannels reinforcing new cells in the Implant chamber, but not vice versa. The stimulus applied to the axons of the Source chamber (figure 3 (b), middle) induced the spiking response which was registered at the end of the microchannels between Implant and Target chamber. In contrast. the stimulus applied to the axons in the microchannels between Implant and Target chamber didn't evoke the spikes in the axons of the Source chamber (figure 3 (b), bottom).

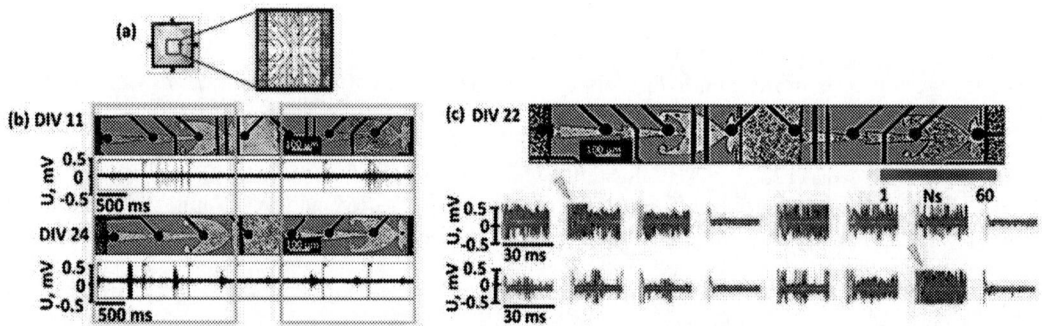

Figure 3 (a,b,c). (a) Microfluidic chip coupled with a microelectrode array; (b) Bioelectrical spiking activity on 11 DIV and 24 DIV (12 days after plating the cells in Implant chamber). (c) Stimulus evoked spiking activity. Middle - stimulus applied to the axons between Source and Implant, bottom-applied to the axons between Implant and Target. Presented signals corresponded to the electrodes of MEA on the photo (top). Each plot presented the signals from all stimuli (1-60) and color of the signal coded the stimulus number in a sequence.

3.4. Conclusion

In this work we developed a method to form a new unidirectional synaptic connectivity between two populations of cultured networks using microfluidic devices. The results show that integration of new dissociated cells between two mature networks provide a new "bridge" for spiking activity propagation. The direction of activity pathways was defined by a structure of the asymmetric microchannels of the chip. Therefore, this co-culture system can be used in study of neurorehabilitation processes using stem or IPS cells [8] and in design of new type of the scaffolds which will provide proper morphological and functional repair of the neural tissue.

Acknowledgments

This research was supported by the grant of the president of Russian Federation No. MK-6795.2018.4. Design and microfluidic chip manufacture was funded by the Ministry of Education and Science of the Russian Federation, governmental order no. 16.9790.2017/BCh

References

[1] Jin, Y., Bouyer, J., Shumsky, J. S., Haas, C., & Fischer, I. 2016 J Neuroscience. 320 69

[2] Wang, T. Y., Forsythe, J. S., Parish, C. L., & Nisbet, D. R. 2012 J Journal of biomaterials applications. 27(4) 369

[3] Akopova, T. A., Timashev, P. S., Demina, T. S., Bardakova, K. N., Minaev, N. V., Burdukovskii, V. F., Surin, N. M. 2015 J Mendeleev Communications. 25(4) 280

[4] Akovantseva, A. V. Koroleva, D. S. Asyutin, L. F. Pimenova, N. A. Konovalov, T. A. Akopova, A. B. Solov'eva, I. V. Mukhina, M. V. Vedunova, B. N. Chichkov, V. N. Bagratashvili 2016 J Applied Biochemistry and Microbiology. 52(5) 508

[5] An, Y., Tsang, K. K., & Zhang, H. 2006 J Biomedical Materials. 1(2) R38.

[6] Taylor, A. M., Rhee, S. W., Tu, C. H., Cribbs, D. H., Cotman, C. W., & Jeon, N. L. 2003 J Langmuir. 19(5) 1551

[7] Park, J. W., Vahidi, B., Taylor, A. M., Rhee, S. W., Jeon, N. L. 2006 J Nature protocols. 1(4) 2128

[8] Shimba, K., Saito, A., Takeuchi, A., Takayama, Y., Kotani, K., Jimbo, Y. 2013. Engineering in Medicine and Biology Society (EMBC), 2013 35th Annual International Conference of the IEEE 6675

[9] Takayama, Y., Moriguchi, H., Kotani, K., Suzuki, T., Mabuchi, K., Jimbo, Y.. 2012 J Biosystems. 107(1) 1

[10] Takayama, Y., Kida, Y. S. 2016 J PLoS ONE. 11(2) 1

[11] Gladkov, A., Pigareva, Y., Kolpakov, V., Malishev, E., Bukatin, A., Kazantsev, V., Pimashkin, A. 2016 J Front. Neurosci. Conference Abstract: MEA Meeting 2016 | 10th International Meeting on Substrate-Integrated Electrode Arrays

[12] Gladkov, A., Pigareva, Y., Kutyina, D., Kolpakov, V., Bukatin, A., Mukhina, I., Pimashkin, A. 2017 J Scientific Reports. 7(1) 15625

[13] Balyabin A.V., Tikhobrazova O.P., Muravyeva M.S., Klyuev E.A., Ponyatovskaya A.V., Shirokova O.M., Bardakova K.N., Minaev N.V., Koroleva A.V., Mitaeva Y.I., Mitroshina E.V., Vedunova M.V., Rochev Y.A., Chichkov B.N., Timashev P.S., Bagratashvili V.N., Mukhina I.V.2016. J Sovremennye tehnologii vmedicine 8(4) 198

SPBOPEN 2018   IOP Publishing

Investigation of Na+/K+ -ATPase role in cancer cells' functioning

S A Perkov[1], A K Emelyanov[1,4,5], S V Shmakov[1], A A Bogdanov[2-4]

[1] Nanobiotech Lab, St. Petersburg Academic University, St. Petersburg 194021, Russia
[2] Saint Petersburg clinical scientific and practical center for special types of medical care (oncology-oriented), Russia
[3] Bionics Lab, ITMO University, St. Petersburg, Russia
[4] The Petersburg Nuclear Physics Institute, Gatchina, Russia
[5] Pavlov First Saint-Petersburg State Medical University, St. Petersburg, Russia

Abstract. Na+/K+ -ATPase is an essential protein for cell functioning which has a great impact on osmotic stabilization, electrochemical potential and volume of the cell. According to the research of A Bogdanov and co-authors there is an overexpression of ATP1A1 and down-regulation of ATP1A2 in breast cancer cells [1]. Therefore, we assumed an existence of correlation between cancer cells proliferation and ATPase expression, which gives a possibility to regulate cancer cells proliferation via ATPase inhibition. In this research we have found a down-regulation of ATP1A1 in K562 cells after its processing with $AgNO_3$. We expect that our research could help in investigating Na+/K+ -ATPase role in cancer cells' functioning.

1. Introduction

One of the most important problems for a living cell is maintaining its volume and osmolarity. There are plenty of ions that are essential for cell's existence. The pressure created by the osmolarity differences between intracellular and extracellular space can easily reach 1 atm or more [2], which is too high for plasma membrane to withstand. Now it is clear that osmotic stabilization of a living animal cell is related to the active transport of Na^+ from and K^+ into the cell, which is provided by Na/K ATPase, also known as Na/K pump. Therefore, we can see a great role of Na+/K+ -ATPase in cells' existence.

Na^+/K^+ -ATPase's (NKA) main function is the creation and maintenance of electrochemical gradients for sodium and potassium ions in living cells. These gradients have a huge influence on cells' volume, osmolarity and resting potential. The minimal functional NKA consist of two associated subunits: α- and β-. The catalytic α-subunit is responsible for ATP energy conversion to transport Na^+ and K^+ ions and has ATP and cardiac glycosides binding sites. In human tissues it is presented in four isoforms (α_1, α_2, α_3, α_4). The beta subunit is responsible for delivery and insertion of α_1 subunits into the cell membrane and it presented in three isoforms (β_1, β_2, β_3).

A series of experiments has shown that cardiac glycosides actually inhibit the cellular sodium-potassium pump and due to this may have an antitumor effect [3]. From the other side, silver nitrate also inhibits the Na+/K+ -ATPase, probably acting like a potassium mimic and actually blocking the K-site of NKA [4]. Furthermore silver has antitumor effect [5]. Taking into account the fact that many cell cultures have an overexpression of NKA [1], and combining all of the above together, occurs an assumption that silver's antitumor effect is related to the NKA inhibition.

Content from this work may be used under the terms of the Creative Commons Attribution 3.0 licence. Any further distribution of this work must maintain attribution to the author(s) and the title of the work, journal citation and DOI.
Published under licence by IOP Publishing Ltd

2. Materials and methods

In present study we focused on α_1 subunit (*ATP1A1* gene) expression in K562 cells (chronic myeloid leukemia). K562, chronic myelogenous leukemia, culture was taken from the Russian vertebrates' cells collection of Cytology Institute, RAS. Cells were cultivated in RPMI-1640 medium (HyClone, USA) with the addition of 10% embryonic bovine serum (HyClone, USA) in presence 40 ug/ml of gentamicine (Sigma, USA) in temperature of 37°C in 5% CO_2 atmosphere. Then processed with $AgNO_3$ in two concentrations: $9 \cdot 10^{-5}$ (9-5) and 10^{-4} (1-4) mol/L. Control group was processed with the same amount of water. We chose this concentrations because processing cells with it causes a metabolism grow, which has shown via MTT testing. After purifying mRNA from cells (Qiagene RNeasy kit, Netherlands) and synthesis of cDNA from mRNA via reverse transcription (Thermo Scientific, RevertAid First Strand cDNA synthesis kit, USA), we used real-time PCR (Eco Real-Time PCR System, Illumina) with GAPDH as a control gene to detect an expression change of ATP1A1.

Furthermore we compared an expression of *ATP1A1* via qRT-PCR (Eco Real-Time PCR System, Illumina) with GAPDH as a control gene in two similar cell lines: 3T3b and 3T3-SV40. 3T3b is cell line of mice fibroblasts and 3T3-SV40 is the same cell line, but transformed by SV40 virus. They have highly similar structure, but different sensibility to silver (Fig. 1).

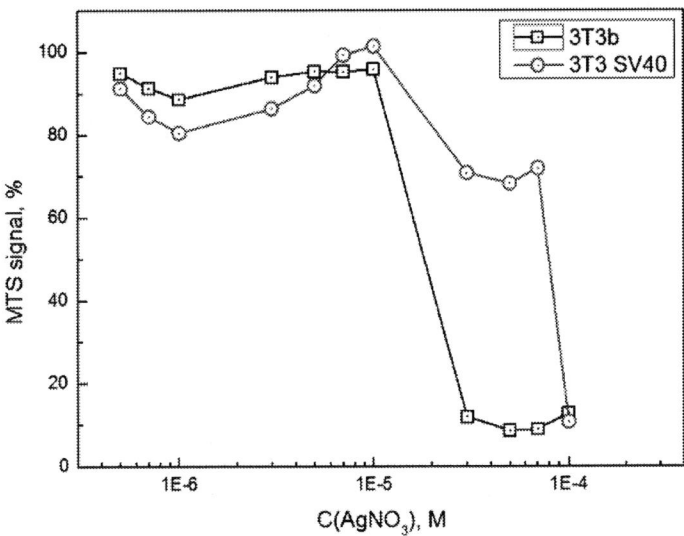

Figure 1: comparison of cells' viability after processing with silver nitrate in different concentrations.

3. Results

We have shown the downregulation of *ATP1A1* expression in investigated samples (9-6 and 1-5) after inhibition with $AgNO_3$ (Fig. 2a). This may explain an antitumor effect of silver: majority of cancer cells have an overexpression of ATP1A1 and processing cells with silver nitrate lowers the expression levels, which could be the reason of silver's antitumor effect.

An experiment with 3T3b and 3T3-SV40 cell lines has shown a difference in *ATP1A1* expression: 3T3-SV40 contain more NKA then 3T3b (Fig. 2b). This may explain why 3T3b cells more sensible to silver: silver inhibits NKA, and 3T3b have less NKA, then 3T3-SV40. This mean that in presence of silver nitrate in equal concentrations, percent of blocked NKA in 3T3b cells would be higher, than in 3T3-SV40 cells.

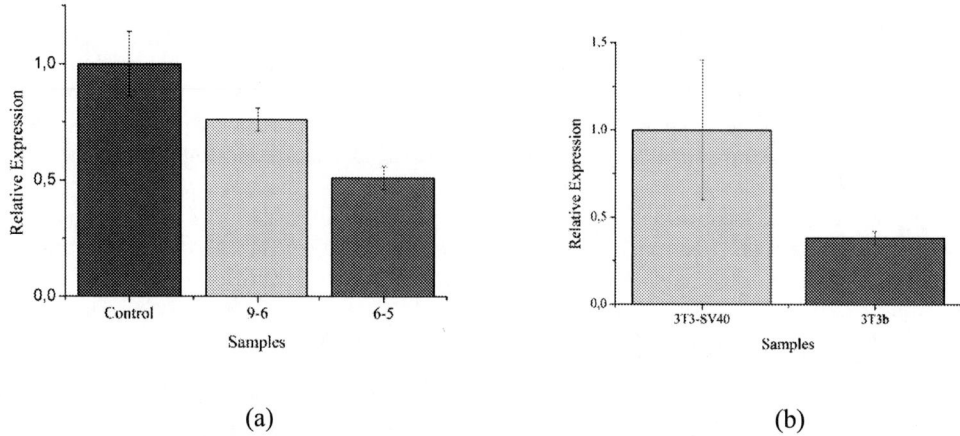

(a)   (b)

Figure 2(a,b): (a) Comparison of ATP1A1 expression in proceeded and non-proceeded samples. Each bar represents level of relative expression ATP1A1 gene to GAPDH gene; **(b)** Comparison of *ATP1A1* expression in 3T3b and 3T3-SV40 cell lines.

4. Discussion

Considering all of the above, we can assume that NKA has a great role in mechanism of antitumor activity of silver. Still it is not clear what happens after NKA inhibition and what processes lead to cancer cells' death.

References
[1] Alexey Bogdanov, Fedor Moiseenko, Michael Dubina "Abnormal expression of ATP1A1 and ATP1A2 in breast cancer", 2017.
[2] Clay M. Armstrong "The Na/K pump, Cl ion, and osmotic stabilization of cells", 2003
[3] Chen D, "Inhibition of Na+/K+-ATPase induces hybrid cell death and enhanced sensitivity to chemotherapy in human glioblastoma cells", 2014
[4] Hussain S, "Potent and reversible interaction of silver with pure Na,K-ATPase and Na,K-ATPase-liposomes", 1994
[5] Moises A Franco-Molina, "Antitumor activity of colloidal silver on MCF-7 human breast cancer cells", 2010.

SPBOPEN 2018   IOP Publishing

Accuracy increase of blood viscoelasticity measurement of the microfluidic Wheatstone bridge

N D Shipulya[1], S A Konakov[2]

[1]Department of Instrument Technology, National Research University of Information Technologies, Mechanics and Optics, St. Petersburg 197101, Russia
[2]Institute of Metallurgy, Machinery and Transport, Polytechnic University of Peter the Great, St. Petersburg 195251, Russia

Abstract. Development of methods for the analysis and control of biomaterials goes to the most important task level in the field of modern medicine. Blood viscoelasticity measurement in the microfluidic Wheatstone-bridge is one of the most promising method today. Nowadays this measurement parameter does not meet the requirements of laboratories, so the method is not used widely. We propose to improve the method by changing geometry of microfluidic Wheatstone bridge. Our method allows increase the accuracy due to statistical processing of obtained data. Designed device can be used not only to measure blood parameters, but also for other types of fluid.

1. Introduction

The microfluidic Wheatstone bridge (MWB) is used for a long time for measuring of various parameters. For example, the authors of [1] use it to measure the viscoelasticity of blood, the authors of [2] use the system, which facilitates rapid, on-demand fluid sampling in the bridge. The authors of [3] use a similar microfluidic system for electrokinetic study of miscellaneous liquid–solid interfaces. In [4], we proposed the use of a simple form of the MWB for measurement of pressure with high accuracy.

In general such the microfluidic devices use in different areas for high precision measurements. The purpose of the future device determines its material. For example, one of the most popular materials for all microfluidic devices is different polymer and epoxy resins. This choice is due to the characteristics of these materials, such as chemical resistance, strength of the material, ease of manufacture, availability and cost. Also microfluidic devices are made of silicon [5]. The article deals with the configuration of the standard MWB [1] for measuring the viscosity and elasticity of blood. The creators of this design make a measurement of the parameters, based on the visual position of the boundary between the two liquids. In this article, we describe the results of an experiment of such a structure. And also draw conclusions about the performance of the standard MWB. And we offer an improved design of the MWB, which takes into account many of the shortcomings of the already created device.

2. Description of the principle of the simple the microfluidic Wheatstone bridge for measuring of the blood viscoelasticity

To measure the viscoelasticity of blood, the authors of [1] use a system of the MWB. The system contains two input channels for blood and PBS solution. These two streams have different flow rates, different consumption and hydraulic pressures. As the fluids pass through the channels, they collide in a bridge microchannel. Different flow rates make it possible to compare the physical parameters of a liquid. The PBS solution presses on the blood with certain strength. The blood under this influence starts to move in the opposite direction and acquires an acceleration of the flow. At some point in time, the system comes to a stable equilibrium, the position of the two liquids stops relative to each other. Then the camera fixes this position. Digital imaging techniques allow to measure the area of the blood in the bridging channel. The viscoelasticity of the blood is calculated according to the size of this area. Such a system is simple but not sufficiently accurate. In this case there are measurement errors, which are caused by defects in the channels shape (imperfect geometry). In addition interphase interaction between blood and the PBS solution leads to meniscus appearance near the wall of the channel.

It is possible to carry out these measurements repeatedly, in order to increase accuracy and obtain an average result after statistical data processing. But the action increase measurement time and volume of blood and PBS solution per one experiment (Fig 1).

Content from this work may be used under the terms of the Creative Commons Attribution 3.0 licence. Any further distribution of this work must maintain attribution to the author(s) and the title of the work, journal citation and DOI.

Published under licence by IOP Publishing Ltd

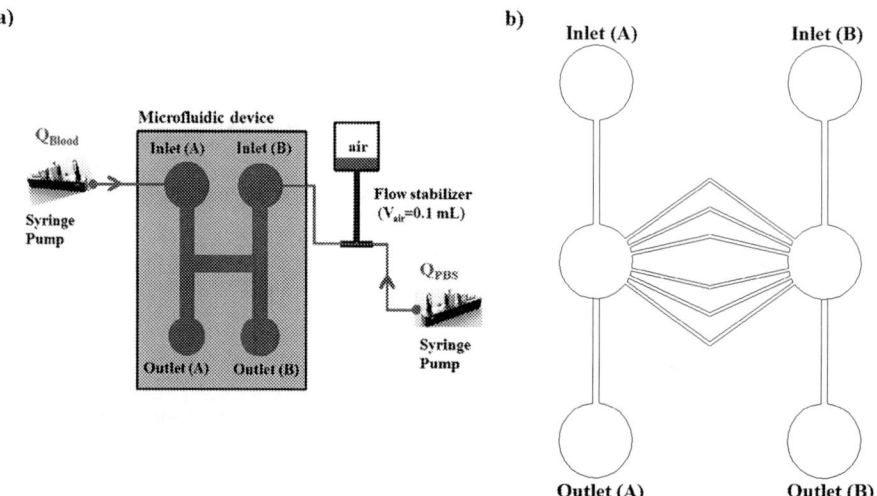

Figure 1(a, b). Scheme of the blood viscoelasticity measurement in the microfluidic Wheatstone-bridge: **(a)** measurement scheme according to article [1]; **(b)** improved scheme with additional bridge channels.

3. Technique of the experiment

The microfluidic plate was manufactured for behavior studies of measurement system, which was described in the article [4]. The plate has several structure tests; one of this is a structure of the microfluidic measurement bridge. It has four outputs, two inputs and two outputs for two different liquid (A – no color, B – black color). Width of measurement channel is 8 mm, length is 40 mm, and depth of channel is 1 mm. Appearance of this structure test and its outputs during operation is shown in the Fig 2.

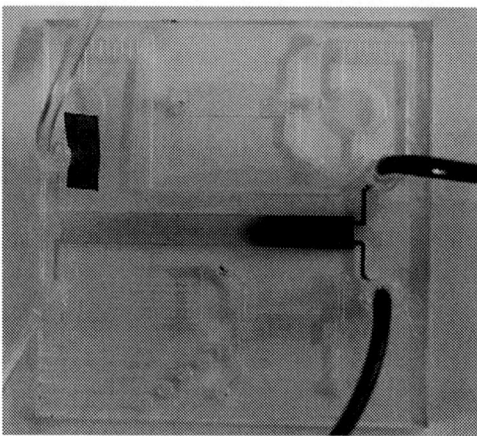

Figure 2. Appearance of this test structure and external terminals for liquids during operation. The width of the measuring channel is 8 mm; the length is 40 mm with a channel depth of 1 mm.

The essence of the experiment is as follows. Initially, the system is filled with liquid A, after which the second part is filled with a liquid B, while still pumping liquid A through the system. As a result, equilibrium is established and forms a boundary between liquids A and B. This boundary is distinguishable and this is shown in the Fig 2.

A flow occurs in the channel, when there is a pressure difference at the ends of the measuring channel. The pressure difference is due to various factors such as viscosity, flow velocity, hydraulic resistance and other characteristics of the liquids properties. The apparent boundary between liquids A and B begins to move, can be fixed visually or by another method. This motion is the output signal of the measuring system, which must be processed.

Specialized photographic equipment is used to test the system detection of the movement of liquids in the measuring channel; images are processed using computer methods. The experiment was as follows. At the beginning the system was brought into this state, when the measuring channel was completely filled with liquid B. After that increase the flow of liquid A, so that in the measuring channel the reverse movement began. The dynamics of this movement was observed in time. Fig 3 shows the image of the measuring channel at different times. They visually see the movement of the interface of the liquid, which indicates the presence of a flow in the measuring channel.

Figure 3. Image of the measuring channel at different times.

In the observed experiment, each photo was processed to obtain a normalized output signal by brightness. This became the basis of numerical analysis. The values of the output signal were averaged over the width of the entire channel. As a result, for each frame, the distribution of the normalized output signal along the length of the measuring channel over time has been obtained. These results are shown in Fig 4(a).

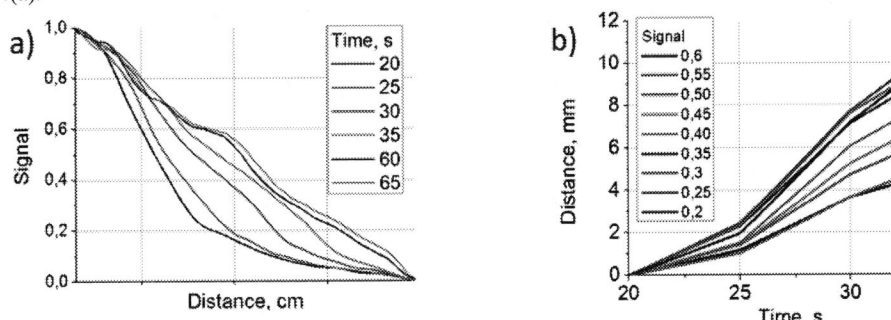

Figure 4(a, b). (a) Distribution of the normalized output signal along the length of the measuring channel over time; **(b)** The average flow rate for each output level.

The results of the measurements show that for the outermost regions a constant value of the output signal is observed (within the noise level). This corresponds to the region of pure liquids flows A and B from their inputs to the outlets in the microfluidic structure under consideration. For the central area of the measuring channel it is convenient to consider the movement of a point with a fixed level of the output signal. For levels from 0.6 to 0.2, considerable movement can be observed at a distance of 1 to 1.5 cm during the observation time 45 s. To quantify the speed, you can calculate the movement along the measuring channel of a local point, which gives an output signal of a given level. These data are shown in Figure 4 for the output signal levels from 0.2 to 0.6. If you know the movement in time, you can calculate the average flow rate for each of the levels of the output signal. These results are shown in Fig 4(b).

The presented data are very ambiguous and difficult to use to obtain information about the flow in the measuring bridge. The presence of local kinks in the graph 3 and the inequality of the rates for different levels of the output signal (in Fig 5) speak about this.

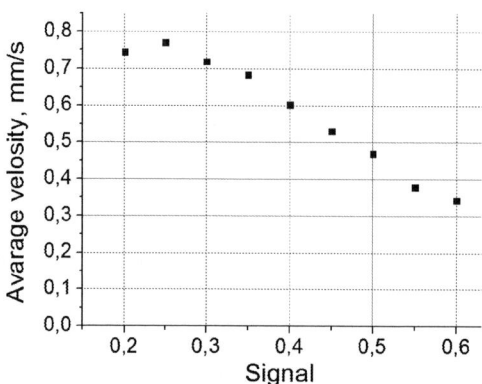

Figure 5. Indicators of different levels of the output signal.

Consider the reasons for the appearance of these difficulties. Such phenomena as the volatility of the fluxes of substances A and B during the measurement make some contribution. Minor pulsations in the liquid flow lead to pressure pulsations at the ends of the measuring channel and as a consequence this leads to an uneven flow rate in it. From this it follows that the measuring system must have a stable source of fluxes of substances. In addition, the measurement time should be less to reduce the negative impact of this factor.

However, the main source of the problems considered is the poor design of the measuring channel. The channel has a large width; therefore, a parabolic flow profile develops in it. Proof of this is shown in Figure 6.

Figure 6. Visual proof of the main source of problems of this system, the presence of a parabolic profile.

In this case, depending on the direction of flow in the measuring channel, the parabola has a convex part either in the liquid A or in the liquid B. The presence of a parabolic profile is harmful for two reasons:

1) Intensification of diffusion of substances into each other. With a parabolic profile, the total length of the interface of the two liquids is greater than at the direct boundary, and then the total diffusion flux is also greater. Secondly, the parabolic profile near the channel walls contributes to the formation of a thin boundary layer which diffuses rapidly into another liquid.

2) In the experimental procedure, averaging of the signal over the entire width of the channel is assumed. Accordingly, in the presence of a parabolic interface, such averaging will give an incorrect picture. This leads to a significant distortion in the level of the output signal and to the contradictory results shown above.

The transition from a parabolic profile to a direct one is a solution to this problem. Physically, this can be done with the appearance of slippage of the flow at the boundary. But such an effect cannot be precisely regulated; therefore its use in the measuring system is unjustified.

Another solution involves with obtaining output profiles along the measuring channel, if there is no averaging over the width. In this case a line with a width of 1 pixel will have a flat boundary between the liquids A and B. But in this case the noise in the measuring channel is significantly increased.

In this way we come to the conclusion, need to create a system which would not have a wide measuring channel, to keep the phase boundary as clear as possible, but at the same time it would be possible to carry out averaging of the measured signal to reduce the noise to an acceptable level. From this condition we arrive at the structure of the measuring bridge with a lot of narrow measuring channels. The scheme of such a microfluidic structure is shown in Figure 1.

4. Proposed solution

We propose to increase the number of measuring channels in microfluidic structure. In this way we can provide statistical data processing and keep reagents consumption on one level.

To provide a realization of this approach we propose to use an advanced form of the MWB. The MWB system has new elements in its composition. The elements are additional bridge channels in the structure of the MWB, which provides a multiple measurement.

In our new design the MWB has special expanded chambers (pressure equalizers) between which all bridge channels are situated. This is aimed at reducing the effect of the pressure difference between inlets of bridge channels. If these chambers are absent, vortices can be formed and pressure near the inlets of channels has strong nonuniformity.

The flow of two liquids in the improved structure of the MWB was simulated. Depending on the ratio of flow rate and viscosity the behavior of the interface between liquids was investigated. The simulation results show the correct operation of the improved structure, because both systems have a similar function (Fig 7).

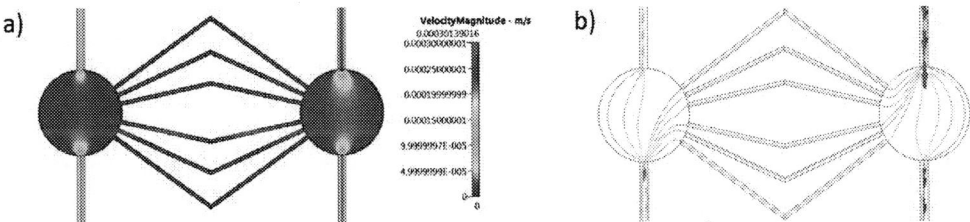

Figure 7(a, b). Velocity in the bridge channels: **(a)** total liquid flow; **(b)** current lines through the structure

5. Conclusions

This work is devoted to microfluidic structure blood viscoelasticity measurement. Considering this question, we have found and have eliminated the critical moments. In this case there are measurement errors, which are caused by defects in the channels shape (imperfect geometry). In addition interphase interaction between blood and the PBS solution leads to meniscus appearance near the wall of the channel. We proposed a method of increase in accuracy of measurement at preservation of the minimum volume of tests of blood. The method consists in change of geometry of microfluidic structure for the purpose of increase in accuracy by carrying out statistical processing of results. The advanced system is presented in Fig. 1. Such structure contains the big field of alignment of pressure; this ensures the same pressure at all the inputs to the bridge channels. The behavior of the proposed microfluidic structure by methods of mathematical modeling is studied. Its operability and possibility of application in laboratory practice is shown.

In the future we are planning to make a prototype of MWB for experimental investigation. The planned experiments will be based on the operation of real working fluids. After successful results, certified patenting is planned. The possibility of applying an improved design will be considered in a wide range of applications.

Acknowledgments
The research was partially supported by FASIE.

References
[1] Kang Y J, Lee S-J 2013 *J Biomicrofluidics* **7** 054122
[2] Tanyeri M, Ranka M, Sittipolkul N, Schroeder C M 2011 *J The Royal Society of Chemistry* **11** 4181
[3] Plecis A, Chen Y 2008 *J Analytical Chemistry* **10** 3736
[4] Shipulya N D, Konakov S A, Krzhizhanovskaya V V 2016 *J of Physics: Conference Series* **738** 012071
[5] Son I-S, Lala A, Hubbard B, Olsen T 2001 *J Sensors and Actuators A: Physical* **91** 351-356

SPBOPEN 2018

IOP Publishing

Singlet oxygen generation mechanism in the presence of excited nanoporous silicon

D M Samosvat [1], O P Chikalova-Luzina [1], V S Khromov[1], A G Zegrya[1], G G Zegrya[1]

[1]Division of Solid State Electronics, Ioffe Institute RAS, Saint Petersburg 194021, Russia

Abstract. A theoretical analysis of the mechanism of generation of singlet oxygen in the presence of photoexcited nanoporous silicon is presented. It is shown that the generation of singlet oxygen is based on nonradiative energy transfer from nanoporous silicon to an oxygen molecule by the exchange mechanism. An analytical expression is obtained and a numerical estimation of the probability of energy transfer from nanoporous silicon to an oxygen molecule is given. It is shown that its numerical value on the order of $10^3 - 10^4$ s^{-1} is in good agreement with experiment.

1. Introduction

Recent studies have shown the promise of nano-porous silicon (np-Si) for micro- and optoelectronics [1] as well as for biology and medicine [2,3]. As it was shown in a number of works [2,4,5], generation of singlet oxygen is possible in the presence of excited nanoporous silicon. This was shown both by direct measurement of luminescence [4], and indirectly, on the basis of the biological effect of singlet oxygen [5]. Therefore, clarifying the exact mechanism of this process is an important task. In [4-6], an assumption was made on the role of the exchange interaction in this process. In work [6], a mechanism is proposed for generating singlet oxygen due to the simultaneous tunneling of an electron and a hole with or without the participation of phonons. This mechanism is similar in essence to the tandem mechanism of energy transfer, considered in [7]. Nevertheless, this mechanism is not considered in this paper, because in our opinion it is less effective than the exchange interaction without charge transfer. In addition, although the appearance of a dipole mechanism of energy transfer due to violation of the spin selection rules during spin-orbit interaction in silicon and oxygen, nevertheless this effect is negligibly small due to the weakness of the spin-orbit interaction in oxygen. According to the quantum chemical calculations [8] and experiments, oxygen has a triplet form in the ground state, ie, the total spin of the oxygen molecule is S = 1. Oxygen allows for the existence of two long-lived singlet forms with energies of 0.98 eV and 1.63 eV above the ground state. The lifetime of an electron at the second excited level with an energy of 1.63 eV is of the order of 7 seconds, however, its lifetime in the first excited state with an energy of 0.98 eV is extremely large (~ 72 min for a free oxygen molecule). The last metastable form is singlet oxygen and is chemically active. It is well known that the size-quantization effect leads to an increase in the effective width of the forbidden band of the semiconductor. This leads to the possibility of resonant energy transfer from nanoporous silicon to the oxygen molecule. The purpose of this paper is a theoretical study of the process of energy transfer from excited nanoporous silicon to an oxygen molecule.

Content from this work may be used under the terms of the Creative Commons Attribution 3.0 licence. Any further distribution of this work must maintain attribution to the author(s) and the title of the work, journal citation and DOI.

Published under licence by IOP Publishing Ltd

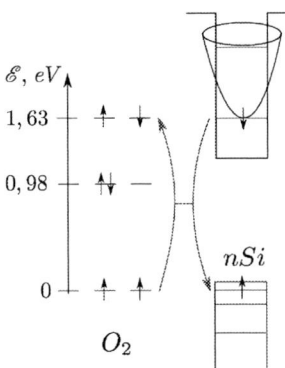

Figure 1. Molecular orbitals of the O2 molecule and the scheme of generation of singlet oxygen.

2. Theory

To calculate the probability of energy transfer in quantum mechanics, the perturbation theory method is usually used [9]. In the case of strong interaction, the density matrix method [10] is used. Nanoporous silicon can be represented by a simple quantum well model for electrons and holes. Its depth is determined by the work function of the electron from silicon. In this case, quantization exists only in one direction, in others there is free motion. The scheme of the process of energy transfer and energy levels of silicon and molecular oxygen is shown in Fig. 1,2,3. To calculate the energy transfer matrix element, it is necessary to know the wave functions of the initial and final states of the system. Since excited states are considered in the oxygen molecule, it is impossible to take them into account correctly without taking the configuration interaction into account [8]. Usually for the simplest systems, such as molecular oxygen, there is a small set of configurations to correctly estimate the energy and wave function of the electrons. Molecular orbitals (MO) can be represented as a linear combination of atomic orbitals (AO), centered on the nucleus A and nucleus B of the oxygen molecule (see Fig. 2). To compose AO, Slater orbitals are used, and it is assumed that the outer 2p electrons move in the Coulomb field with an effective charge $Z * = 4.55$. For them, $n * = 2$, therefore the general form of the wave functions of the molecular orbitals takes the following form (see also Fig. 2):

$$\pi_{\binom{g}{u}}^{\pm}(i) = C_{\binom{g}{u}}\left(\frac{1}{2\sqrt{6}}\left(\frac{Z^*}{a_0}\right)^{5/2} r_i \sqrt{\frac{3}{8\pi}} \sin\theta_i \exp(\pm i\varphi_i)\left(\exp(-\alpha r_i^A)\binom{-}{+}\exp(-\alpha r_i^B)\right)\right) \quad (1)$$

Here a_0 is the Bohr radius of the electron, $C_{g,u}$ is the normalization constant. The wave functions of the states of the oxygen molecule for the ground state $^3\Sigma_g^-$ - (triplet):

$$\Psi_{(S_z)\mu}^{3\Sigma}(\xi_1, \xi_2) = F_\mu^{\,3\Sigma}(\mathbf{r}_1, \mathbf{r}_2) \cdot \chi_{S_z}^{3\Sigma}(1,2) = \frac{1}{\sqrt{2}} \cdot \left[\pi_\mu^+(1)\pi_\mu^-(2) - \pi_\mu^-(1)\pi_\mu^+(2)\right] \cdot \chi_{S_z}^{3\Sigma}(1,2), \quad (2)$$

where $\chi_{S_z}^{3\Sigma}(1,2)$ is one of the three symmetric spin functions with total spin (-1,0,1). For the second excited state $^1\Sigma_g^+$ (singlet):

$$\Psi_\mu^{1\Sigma}(\xi_1, \xi_2) = F_\mu^{\,1\Sigma}(\mathbf{r}_1, \mathbf{r}_2) \cdot \chi_0^{1\Sigma}(1,2) = \frac{1}{\sqrt{2}} \cdot \left[\pi_\mu^+(1)\pi_\mu^-(2) + \pi_\mu^-(1)\pi_\mu^+(2)\right] \cdot \frac{1}{\sqrt{2}}[\alpha(1)\beta(2) -$$

$$\beta(1)\alpha(2)] \quad (3)$$

The effect of excited configurations can be assumed to be small. The matrix element of energy transfer is represented by the expression:

$$M_{if} = \int_{V_1 V_2 V_3} \psi_i^*(\xi_1, \xi_2, \xi_3)\hat{V}\psi_f(\xi_1, \xi_2, \xi_3)d^3r_1 d^3r_2 d^3r_3 \quad (4)$$

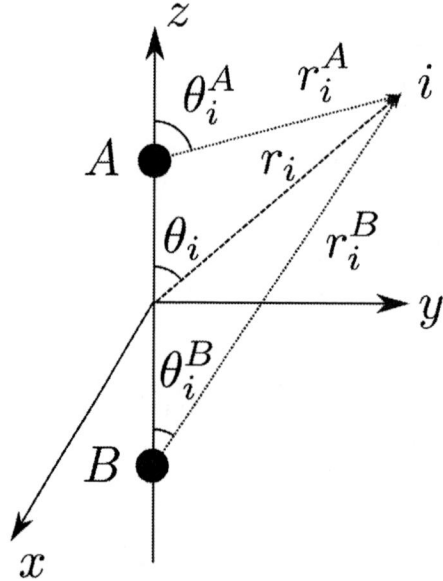

Figure 2. Oxygen molecule and coordinate notation.

Here $\xi = (r, \sigma)$ is the set of coordinate and spin variables. The interaction operator is taken to be the Coulomb interaction operator between all three electrons (an excited electron in silicon and two electrons per MO oxygen), namely $V = V_{12} + V_{23} + V_{13}$, where $V_{ij} = \frac{e^2}{\varepsilon|r_i - r_j|}$ is the Coulomb interaction operator.

The total wave function of the initial state (silicon and oxygen molecule), antisymmetric when replacing any two electrons has the form:

$$\psi_i(\xi_1, \xi_2, \xi_3) = \frac{1}{\sqrt{3}}[\Psi^c_{(+)}(\mathbf{r_1})\chi_c(1) \cdot \Psi^{^3\Sigma_g^-}_{(S_z)}(\xi_2, \xi_3) +$$

$$+\Psi^c_{(+)}(\mathbf{r_2})\chi_c(2) \cdot \Psi^{^3\Sigma_g^-}_{(S_z)}(\xi_3, \xi_1) + \Psi^c_{(+)}(\mathbf{r_3})\chi_c(3) \cdot \Psi^{^3\Sigma_g^-}_{(S_z)}(\xi_1, \xi_2)] \quad (5)$$

The antisymmetric wave function of the final state has a similar form. The substitution of the electronic $\Psi^c_{(+)}(\mathbf{r_1})$ and hole $\Psi_{h1}(\mathbf{r_1})$ wave functions in (4) yields:

$$M_{if} = 2 \cdot (M_{Cou\,l} + \frac{M'_{Cou\,l}}{2} - M_{ex}) \tag{6}$$

According to the spin selection rules, the matrix element of the Coulomb interaction without taking into account the spin-orbit interaction is zero. The nonzero matrix element of the exchange interaction is:

$$M_{ex} = \int\limits_{V_1 V_2 V_3} \overline{\Psi_{h1}}(\mathbf{r_1})\chi_h^T(1) \cdot \overline{\Psi^{1\Sigma_g^+}}(\xi_2,\xi_3) \cdot \left(\hat{V}_{31} + \hat{V}_{32} + \hat{V}_{12}\right) \cdot \Psi_{(+)}^C(\mathbf{r_3})\chi_c(3)$$

$$\cdot \Psi_{(S_z)}^{3\Sigma_g^-}(\xi_2,\xi_1)d^3\mathbf{r_1}d^3\mathbf{r_2}d^3\mathbf{r_3}$$

$$= (M_{ex(31)}(\mathbf{r}) + M_{ex(32)}(\mathbf{r}) + M_{ex(12)}(\mathbf{r})) \times \chi_h^T(1)\left(\chi_0^{1\Sigma}(2,3)\right)^T \cdot \chi_c(3)\chi_{S_z}^{3\Sigma}(2,1)$$

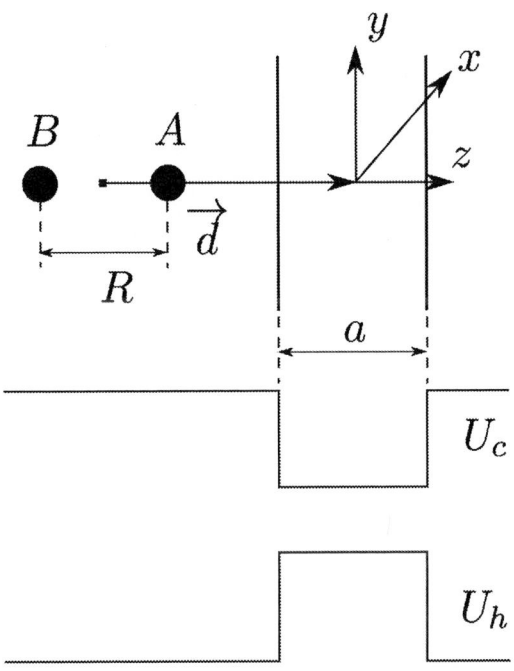

Figure 3. Geometry of the problem. The oxygen molecule next to nanoporous silicon.

A simple estimation shows that the wave function in silicon has longer "tails" than the oxygen molecule, and therefore the structure of the wave functions in silicon is not important. It is important to calculate the overlap between the tail of the silicon wave functions and the function of oxygen. In the matrix element itself, the calculation procedure is similar, taking into account the fact that the Coulomb potential is expanded in multipoles and in the final expression all contributions are summed. For simplicity, we assume that all three matrix elements are equal in order of magnitude.

We calculate the probability of energy transfer. Since in the initial and final states there is a continuous motion in the plane of the quantum well, in contrast to the case of a completely discrete spectrum, perturbation theory can be used to calculate the probability of energy transfer. To calculate the transfer rate, it is necessary to sum the probability of energy transfer across all the initial and final states. The generation rate of singlet oxygen has the following form:

$$G = \frac{2\pi}{\hbar} \int d\Gamma_i \, d\Gamma_f |M_{ex}|^2 f_c f_h \delta(E_i - E_f)$$

Since we normalize the wave functions in the box in the plane of free motion, the density of the initial states has the form (for the finite ones, similarly):

$$d\Gamma_i = \frac{L_x L_y q_c dq_c d\varphi_c}{(2\pi)^2}$$

The calculations yield that the total matrix element is proportional to $q_c q_h cos(\varphi_c - \varphi_h)$, where $q_{c(h)}$ is the wave vector of the electron(hole) in the plane of the well. This means that, in the process of scattering, firstly, there is no conservation of the quasimomentum, and secondly, the matrix element is maximal for the momenta of the electron and hole co-directed with each other. Therefore, the missing impulse should be transferred to the oxygen molecule. The speed becomes:

$$G = \frac{2\pi}{\hbar} \int q_c dq_c d\varphi_c q_h dq_h d\varphi_h \, |\widetilde{M}_{ex}|^2 f_c f_h \delta(E_i - E_f)$$

Here $\widetilde{M}_{ex}$ is the reduced matrix element equal to $\widetilde{M}_{ex} = M_{ex} L_x L_y$. Using the delta function property and integrating over the corners, this expression can be calculated:

$$G = \frac{2\pi}{\hbar} \frac{\hbar^2}{T^2} n_c n_h e^{-\frac{1}{T}\left(E_0 - \frac{\hbar^2 k_c^2}{2m_{zz}}\right)} X^3 |\langle A \rangle_0|^2 \frac{m_{xx} + m_{yy}}{24 m_{hh}^2} \tag{7}$$

Here, $\langle A \rangle_0 = \langle \widetilde{M}_{ex}/(q_c q_h) \rangle$ is the value of the reduced matrix element averaged over the directions of the hole in silicon, and $X = \frac{2m_{hh}}{\hbar^2}\left[E_0 - \frac{\hbar^2 k_c^2}{2m_{zz}}\right]$, where $E_0 = \Delta - E_g$ and Δ is the transition energy in the oxygen molecule. In this case, m_{xx} m_{yy}, m_{zz}, m_{hh} are the effective masses of charge carriers in silicon, and n_c, n_h are the two-dimensional carrier concentrations in the well.

According to expression (7), a numerical calculation was made for two-dimensional electron and hole concentrations in silicon of the order of 10^{13}sm^{-2}. In this case, the wall thickness of the porous silicon was adopted 1.2 nm. For such concentrations and such a wall thickness, in the case of an oxygen molecule adsorbed on the surface of porous silicon, $G \sim 10^4$ s^{-1}. This rate correlates well with the transfer time obtained from the experiment of 120 μs [4]. The total transfer rate rapidly decays with increasing distance between porous silicon and the oxygen molecule due to the exchange nature of the interaction of electrons and for the distance from the surface of nanoporous silicon to an oxygen molecule of 0.6 Å, the rate decreases by an order of magnitude.

The effectiveness of the transfer process depends on a number of conditions. First, the characteristic residence time of the oxygen molecule in the adsorbed state should exceed the transport time. Secondly, the lifetime of the oxygen molecule in the first excited state, as well as the lifetime of the exciton, are greater than the inverse transfer time. In this case, the electron in oxygen will pass to the second excited state, from which it relaxes to the first excited state. These conditions are realized in practice. The account of phonons and the mechanism considered in [7] are not considered here.

References

[1] Korotcenkov G 2016 *Porous silicon: from formation to application* (CRC Press, London)
[2] Kovalev D, Gross E, Kunzner N,Koch F, Timoshenko V. Yu., Fujii M. 2002 Phys. Rev. Lett. 89,137401 (2002)
[3] Moser J G 1998 *Photodynamic tumor therapy: 2nd and 3rd generation photosensitizers* (CRC Press)
[4] Gongalsky M B, Konstantinova E A, Osminkina L A et al. 2010, Semiconductors 44 89
[5] Timoshenko V Yu, Kudryavtsev A A, Osminkina L A, Vorontsov A S, Ryabchikov Yu V,Belogorokhov I, Kovalev D, Kashkarov P K 2006 JETP Letters 83 9 492
[6] Kovalev D, Fujii M 2005 Adv. Mater. 17 2531-2544
[7] Reich K V, Shklovskii B I 2016 ACS Nano 10 10267
[8] Fumi F G, Parr R G 1953 J. Chem. Phys. 21 1864
[9] Landau L D, Lifshitz E M 1965 *Quantum Mechanics (Volume 3 of A Course of Theoretical Physics)* (Pergamon Press)
[10] Agranovich V M, Galanin M D, *Electronic Excitation Energy Transfer in Condensed Matter* (North-Holland, Amsterdam, The Netherlands, 1983).

SPBOPEN 2018   IOP Publishing

IOP Conf. Series: Journal of Physics: Conf. Series **1124** (2018) 031026   doi:10.1088/1742-6596/1124/3/031026

Formation and manipulation of polyacrylamide spheroids doped with magnetic nanoparticles in microfluidic chip

D V Nozdriukhin[1,2]**, N A Filatov**[1] **and A S Bukatin**[1,3]

[1]Nanobiotech Lab, St. Petersburg Academic University, St. Petersburg 194021, Russia
[2]Peter the Great St. Petersburg Polytechnic University, St. Petersburg 195251, Russia
[3] Laboratory of Information and Measurement Biosensor and Chemosensor Microsystems, Institute for Analytical Instrumentation RAS, St. Petersburg 198095, Russia

Abstract. Nowadays isolation and sorting of biological objects plays an important role in life science and medicine. Magnetic sorting has big opportunities because of its advantages such as sufficient selectivity and bioinertity. In this work we studied formation and handling of polyacrylamide microparticles doped with magnetic nanoparticles using droplet microfluidics. Using a flow focusing droplet generator magnetic nanoparticles were encapsulated to polyacrylamide droplets which were polymerized inside the device. It was shown that these magnetic droplets could be efficiently sorted by magnetic field in a microfluidic device.

1. Introduction
Nowadays precise sorting became one of the significant objectives in biology and medicine. It is essential for separation, isolation and analysis of defined types of biological objects such as cells, vesicles, exosomes, DNA molecules, peptides and others with defined properties (productivity, size, specificity, etc.). Different sorting techniques can be divided into three trends: optical (imaging techniques and optical-detected signals like FACS [1]), field-based (magnetophoresis [2], electro- or dielectophoresis [3]) and mechanical (hydrodynamic sorting or deterministic lateral displacement method [4]). Utilization of magnetic sorting has several advantages: it is biologically non-invasive and doesn't cause harmful effects on biological objects.

Microfluidics is ideally suited for handling biological objects and new approaches have become increasingly accessible for researchers and clinicians. Magnetophoretic sorter can be implemented in microfluidic devices, which leads to reducing its size, cost and complexity. Due to small sizes of the device ordinary permanent neodymium magnets can be used to create strong magnetic fields for efficient separation.

Biological objects usually have weak magnetic properties, magnetic particles can be used to enhance magnetic sorting's efficiency. Especially, superparamagnetic nanoparticles are highly suitable for such application, because of there one-domain structure and zero coercivity, leading to lack of residual magnetization and absence of motion without any external initiation [5]. For molecular magnetic sorting these nanoparticles can be conjugated with antibodies or oligonucleotides, to which target molecules will be attached [6]. For cell sorting, magnetic particles can be affixed on the cell's membrane with a targeting ligand (extrinsic labeling), or they can get through a membrane and accumulates inside a cell (internalization via cell incapsulation) [7].

Due to small size and potential toxicity and instability of superparamagnetic nanoparticles in biological environment the common way is to encapsulate them into polymeric micro or nanoparticles. The surface of such containers can be labeled with target molecules for isolation and sorting of

Content from this work may be used under the terms of the Creative Commons Attribution 3.0 licence. Any further distribution of this work must maintain attribution to the author(s) and the title of the work, journal citation and DOI.
Published under licence by IOP Publishing Ltd

exosomes, proteins or DNA molecules, which can be attached the container's modified surface and then removed from the sample.

Recently droplet microfluidics was introduced as a high throughput technique to produce monodisperse polymeric and hydrogel particles with defined properties [8]. In this work, we studied encapsulation of magnetic nanoparticles into polyacrylamide droplets and investigated their suitability for magnetic sorting in a microfluidic device.

2. Materials and methods.

For generation of polyacrylamide droplets microfluidic chip with a flow-focusing droplet generator and twisted outlet channel (Figure 1a) was used. Its operating principle is based on the fact that the continuous phase flowing from two side channels meets the dispersed phase at the channel's intersection, where the dispersed phase get squeezed by the continuous phase and breaks up into droplets (Figure 1b). Droplet size can be precisely controlled with changing the flow rates of the liquids. For magnetic separation another microfluidic device with long strait channel and two outlets was used.

Microfluidic devices were fabricated using standard soft lithography process from PDMS Sylgard 184 (Dow Corning) [9]. The droplet formation region represents the intersection of two channels for continuous phase and one channel for dispersed phase, separated from the outlet channel by 15 μm aperture. Width of the outlet channel near the aperture is 60 μm. Depth of all inlet channels, droplet formation region and the outlet channel near the aperture is 40 μm. The twisted channel was manufactured wider and deeper (120 μm x 120 μm) than generator's channel to prevent sticking of the polymerized particles.

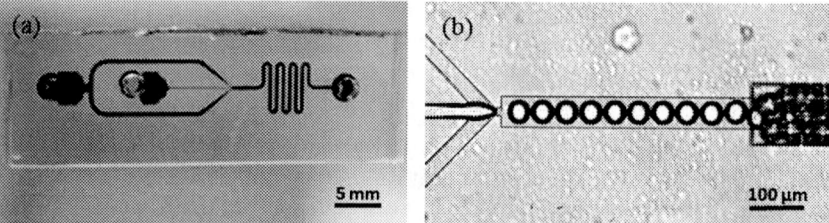

Figure 1. a) General view of a microfluidic flow focusing droplet generator with twisted outlet channel; **b)** Generation of polyacrylamide droplets in a flow-focusing droplet generator

Mineral light oil (cat. N. 330779, Sigma Aldrich) with 4% w/w ABIL EM 180 surfactant (Evonik Industries) and catalyst for polymerization – 1.5% w/v TEMED (Sigma Aldrich) were used as a continuous phase. The mix of 30% acrylamide solution (Bio-Rad), initiator for polymerization - 10% water solution of ammonium persulfate (BioRad) and 0.1% of ferrofluid (magnetite nanoparticles, 400 mg/ml aqueous solution) was used as a dispersed phase. The size distribution of magnetic iron oxide nanoparticles was measured by dynamic light scattering method using Zetasizer Nano ZS (Malvern Instruments, UK).

After the generation, droplets were re-injected to the second microfluidic device with long strait channel for magnetic sorting. To cause lateral displacement of droplets, permanent rectangular NdFeB magnet (N38 alloy) was used.

For introducing continuous and dispersed phases into the microfluidic chip the microfluidic pressure controller based on ITV001 electro-pneumatic regulators (SMC, Japan) was made. The chip was observed using optical microscope Leica DM4000 B LED (Leica Microsystems). Droplets generation and motion were recorded using a camera Pike F100B (Allied Vision Technologies).

3. Results

In our case, the magnetic force acts on every droplet [10]:

$$F = \frac{\chi_p - \chi_e}{\mu_0} V (B\nabla) B,$$

where χ_p - magnetic susceptibility of droplet with magnetic nanoparticles, χ_e - magnetic susceptibility of environment, V - volume of the droplet (m^3), μ_0 is the magnetic constant (H·m^{-1}), B - magnetic flux density (T). Also, the Stokes drag acts in the opposite direction to magnetic force:

$$F_d = 6\pi\eta r v,$$

where η is the viscosity of oil (kg·m^{-1}·s^{-1}), r – the radius of the droplet (m), v - droplet velocity in displacement direction (m·s^{-1}). Both of this forces tends to lateral displacement:

$$L \sim \frac{(\chi_p - \chi_e)V*(B\nabla)B}{6\pi\eta r \mu_0}.$$

Based on these formulas, droplet velocity and time, that droplet with magnetic particles requires for relocation from one sidewall of the channel to another under the influence of magnetic force, the length of a sorter's magnetic force zone was set to 4 mm: dimensions of magnet satisfied this parameter.

Firstly, the size distribution of magnetic nanoparticles was obtained by using dynamic light scattering method in diluted ferrofluid. Consequently, by 14 measurements the size of nanoparticles was defined in range of 35-145 nm (Figure 2a). Then, nanoparticles were packed into droplets. Due to microfluidic way of formation, size distribution of droplets is narrow. Because of stochastic loading of droplets with magnetic nanoparticles, the probability of packing one determined amount of nanoparticles can be described with Poisson statistics, where the distribution parameter

$$\lambda = \frac{quantity\ of\ nanoparticles}{quantity\ of\ droplets} \sim 5 \cdot 10^3$$

is very high , so it can be approximated by Gauss distribution with $\sigma \approx \sqrt{\lambda}$.

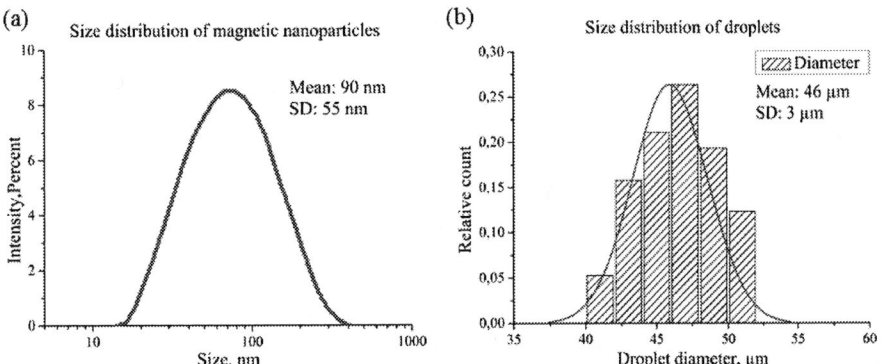

Figure 2. a) Size distribution of droplets measured by 60 droplet diameters; **b)** Size distribution of magnetic nanoparticles by intensity

Utilization of above-mentioned topology of flow-focusing droplet generator allows to produce droplets with diameters from 15 μm to 80 μm in dripping mode inside the generator channel and up to 100 μm in jetting mode, when stream of dispersed phase reaches wide outlet channel. In this work, droplets with 40-50 μm diameter were used (Figure 2b), which were generated in dripping mode. Acrylamide polymerization was initiated by TEMED diffusion from continues phase inside droplets after their formation. Twisted outlet channel of droplet generator increases the time that droplets stay in the chip, so acrylamide solution turns into gel inside it and on the way out the suspension of solid beads was collected.

The next step is to introduce magnetic gel beads into the second chip, which has a strait separation channel and two outlet channels with different hydrodynamic resistance (~1:2.5). So without any external forces, beads with magnetic particles will flow into a channel with less resistance. Droplet flow was focused to the upper wall of the chip near the entrance to the separation channel (Figure 3a) and guided to the most probable outlet (Figure 3b). In our case, velocity of droplet motion along the channel is about 200 μm/s

After that, the permanent magnet 30x5x5 mm with magnetization, perpendicular to channel's axis was used to organize a lateral displacement of droplets. As a result, droplets were guided by magnetic force into less probable channel with higher hydrodynamic resistance (Figure 3c) with speed about of 25 µm/s.

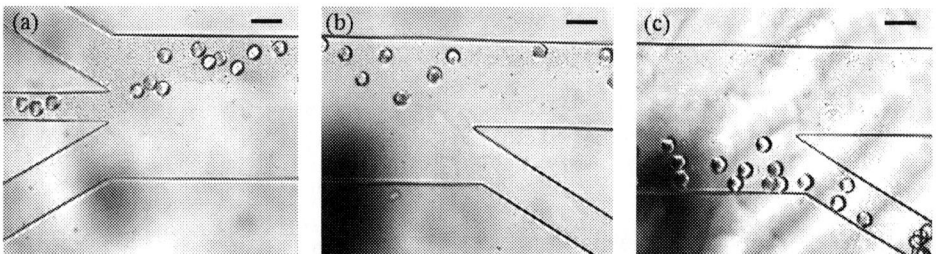

Figure 3. a) Focusing of droplet flow; **b)** Droplets motion without magnet; **c)** Droplet motion with magnet, scale bars are equal to 100 µm.

4. Conclusions
The suspension of magnetic-modified polyacrylamide beads with narrow size distribution was produced in flow-focusing microfluidic device with on-chip polymerization. The possibility of using permanent magnet to manipulate this beads was shown, what presents the suitability of polyacrylamide beads for magnetic on-chip sorting.

Acknowledgments
This work was financially supported by the Ministry of Education and Science of the Russian Federation, project no. 16.9790.2017/BCh.

References
[1] Kang N-Y, Yun S-W, Ha H-H, Park S-J & Chang Y-T 2011 *Nature Protocols* **6** 1044–1052
[2] Pamme N, Wilhelm C *Lab Chip* 2006 **6** 974-80
[3] Voldman J 2006 *Annual Review of Biomedical Engineering* **8** 425-454
[4] Shields IV C W, Reyesab C D and López G P 2015 *Lab Chip* **15** 1230-1249
[5] Taylor E N, Webster T J 2009 *Int J Nanomedicine* **4** 145–152.
[6] E J Cho, S Jung, K Lee, H J Lee, K C Nam and H-J Bae 2010 *Chem. Commun.* **46** 6557-6559
[7] H Yun, K Kim and W G Lee 2013 *Biofabrication* **5** 022001
[8] J H Kim, T Y Jeon, T M Choi, T S Shim, S-H Kim and S-M Yang 2014 *Langmuir* **30** 1473-1488
[9] Bukatin A S, Mukhin I S, Malyshev E I, Kukhtevich I V, Evstrapov A A, Dubina M V
    2016 *Technical Physics* Vol **61** No 10 pp 1566-1571
[10] Y J Sung, J Y H Kim, H I Choi, H S Kwak & S J Sim 2017 *Scientific Reports* **7** 10390

SPBOPEN 2018

IOP Publishing

IOP Conf. Series: Journal of Physics: Conf. Series **1124** (2018) 031027   doi:10.1088/1742-6596/1124/3/031027

Polylactide film deposition onto titanium surface from different solutions

A V Romashkin, N S Struchkov, Yu A Polikarpov, V A Petukhov, D D Levin, V K Nevolin

Scientific and Educational Center of Probe Microscopy and Nanotechnology, National Research University of Electronic Technology, Moscow 124498, Russia

Abstract. The dependence of the porosity, roughness and contact angle with water of polylactide films, produced by dip-coating, on the solution composition and concentration, the drawing rate, as well as the influence of the nitrogen and air atmosphere are investigated. Films with low porosity and roughness from tetrahydrofuran (THF) and dichloromethane (DM) solutions were obtained on the titanium covered glass surface. Technique of a residual solvent detection was proposed, and optimal parameters of films drying were determined.

1. Introduction

Coatings based on biodegradable polymers on titanium can significantly improve growth and attachment of cells [1], and can be used to significantly reduce the integration time of titanium implants, which are currently widely used in prosthetics, and for development of new bioactive materials for prosthetics and tissue engineering purposes. Aliphatic polyesters, in particular, polylactide (PLA), are the most used biopolymers, their biodegradation rate can be varied over a wide range by altering the degree of crystallinity [2] and the composition of copolymers based on them [3], thus it was chosen as a model polymer. Surface morphology plays an important role in the formation of coatings from biodegradable materials because it determines the effectiveness of attachment and the proliferation of cultured cells. Moreover, at the initial stages of tissue repair the formation of the submicron roughness is important [4]. In addition, functional groups contained on the surface, even in the case of its low roughness, are important factor for cell growth [5]. Thus, to investigate the adhesion and proliferation of cells, it is necessary to study the methods of formation of PLA films with controlled porosity, roughness and to investigate the influence of composition of the material and especially its near-surface layers on it. Films morphology according to theoretical and experimental studies depends on solvent concentration [6], its evaporation rate, water content in solution and in atmosphere [7]. However, there is not enough data about PLA films formation with nanoscale roughness and pore absence. Our work is devoted to solving this problem. In view of complex form of implants, it requires to use dip-coating or spray deposition methods.

2. Materials and methods

PLA polymer (4043D, NatureWorks LLC, Minnetonka, MN, USA) was dissolved in tetrahydrofuran (THF), dichloromethane (DM) and benzyl alcohol (BA) (with purity $\geq$99.8%, Komponent-Reaktiv Ltd, Moscow, Russia), the last one, known as non-solvent of PLA, was added in selected experiments [8]. Dissolution of the polymer was performed using stirring and ultrasonic treatment. The films were deposited on cover glasses, with a 15 nm layer of magnetron sputtered titanium on one side by dip-coating method at low drawing rate (~5 mm/min) under normal conditions and nitrogen

---

Content from this work may be used under the terms of the Creative Commons Attribution 3.0 licence. Any further distribution of this work must maintain attribution to the author(s) and the title of the work, journal citation and DOI.

Published under licence by IOP Publishing Ltd

atmosphere. Drying of thin films was carried out in atmosphere at temperature about 85 °C for 7−10 minutes, and for the study of thick films samples to determine the solvent evaporation parameters were additionally dried in vacuum at 55 °C for 48 hours. Drying temperatures were below 100 °C to prevent possible unwanted formation of crystallites of the polymer which can be formed even at the temperature about 90−120 °C [9], and have a substantially longer biodegradation time due to additional barriers for the diffusion of molecules and other mechanisms of biodegradation [2]. The surface morphology and thickness of films were studied by the atomic force microscopy (AFM) (Solver PRO microscope, NT-MDT Ltd, Moscow, Russia) using cantilever with resonance frequency in the range of 90−150 kHz. The presence of the residual solvent and the degree of crystallinity in the films were studied by Raman spectroscopy (Centaur U HR spectrometer, Nano Scan Technologies Ltd, Dolgoprudny, Russia). The water contact angle θ_w was measured with a deionized water droplet (volume 5 µl) produced by a micropipette. The edges of droplet were captured on camera of a horizontal optical microscope and processed in specialized software for measuring the value θ_w.

3. Results and discussions

PLA films, deposited from THF and DM solution (~4.5 wt %) by dip-coating technique, showed less porosity and pore size compared to the results of solution deposition with though higher concentration (8 wt %) reported in the literature [1]. Average pore size decreases with increase of PLA concentration (see fig. 1a and 1b), which is consistent with other studies [6].

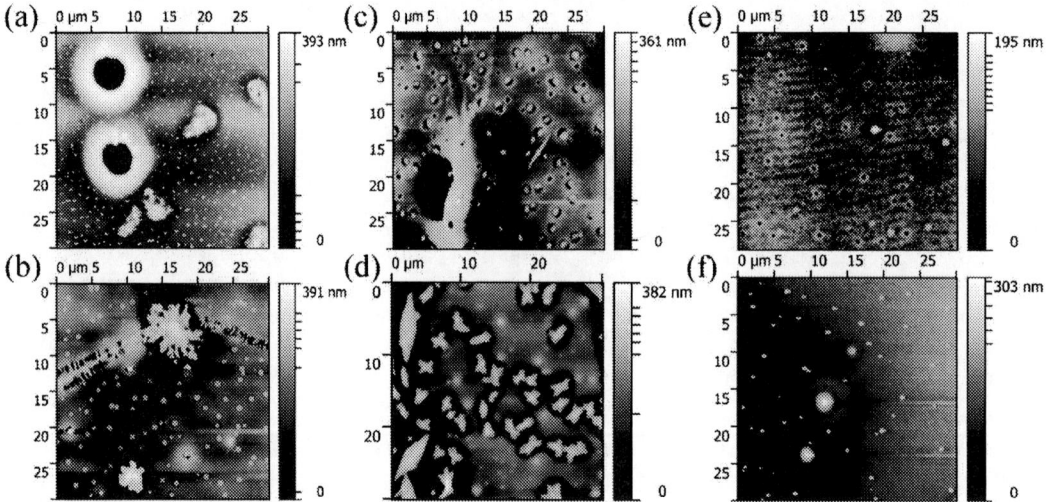

Figure 1(a, b, c, d, e, f). AFM-images of PLA films obtained by the dip-coating from solutions with different concentration and conditions: **(a)** 25 mg/ml in THF; **(b)** 45 mg/ml in THF; **(c)** 45 mg/ml in THF in a nitrogen atmosphere; **(d)** 45 mg/ml from mixture of 3.5 wt % BA and THF, **(e)** 45 mg/ml in DM, **(f)** 45 mg/ml from mixture of 1 wt % BA and DM. **(d-f)** was formed in nitrogen atmosphere.

In the case of anhydrous atmosphere, a significant reduction of microparticles number on the surface, as well as an additional decrease in the average pore size were observed (see fig. 1b with 1c). Also, the presence of residual water content in the solvent and in the film, deposited in air, has a significant effect on the pores formation. Higher evaporation rate of DM solution compared to THF as expected resulted in lower film porosity (see fig. 1c and 1e) that is correlated with literature results [7]. To prevent pores formation, we suggested to add a low volatile component (BA) to the solution. As a result, the decrease of films porosity was occurred for both THF and DM solutions (fig. 1c, 1d and 1e, 1f). The BA, as we suppose, interfered the process of phase separation of the solution into a polymer and a solvent leading to the formation of pores, by realizing the presence of a slowly evaporating component that interacts sufficiently with both the polymer and the solvent. That role can be played by a low-molecular PLA used as a component [10] of composition of film or other organic molecules corresponding to the foregoing properties. But in our work the BA was chosen as such component. However, its excessive content (3.5 %) caused the appearance of nanoparticles on the films surface (fig. 1c and 1d). The optimal concentration of BA for minimization of pores and particles number is expected to be less than 1%. Initially dip-coating were carried out in a draft chamber with airflow. The average films thickness deposited from THF solution with PLA concentration of 8, 25 and 45 mg/ml were about 10, 130 and 180 nm respectively with significant porosity for 10 nm film, it was also weakly depended on drawing rate in a range of 1−10 mm/min. Increasing of the surface roughness was observed in the case of an ultra-slow drawing rates (1 mm/min). Observed complexity of film thickness dependence on solution concentration was probably caused by predominant role of solvent evaporation in a film deposition process and pore formation mechanism [6], prevailing in the case of low concentrations and low drawing rates realized in our experiments. Our result is contrary to power law thickness dependence on drawing rate and concentrations reported in the literature for processes where wetting of the substrate, drawing rate, viscosity of the solution are critical [11].

To reduce the effect of water vapour present in the air, the formation of films was carried out in a glovebox in a nitrogen atmosphere. The average thickness of the films deposited in nitrogen atmosphere from solutions with various concentrations is shown in fig. 2. The obtained results revealed a different dependence of the thickness of the formed films on the concentration of solutions, from experimental results in the air, but it is in agreement with known models in which the viscosity of the solution has a predominant effect on the thickness of the deposited film [12], which is expressed in the fact that the experimental points are acceptably approximated by power function. In this case, in contrast to the air atmosphere, the influence of the solvent evaporation rate factor was significantly reduced due to the absence of a gas flow near the surface of the film being formed. In addition, due to the substantial difference in the mechanism of film formation in air, the pore size significantly increases in comparison with films formed in a nitrogen atmosphere. It was also shown that it is possible to proportionally increase the thickness of the film by iterative coating cycles without changing the solution and special drying.

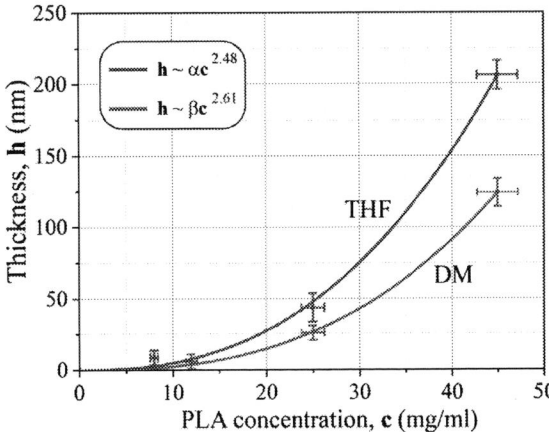

Figure 2. The results of measuring the thickness of PLA films obtained from THF and DM solutions at different concentrations at drawing rate 5 mm/min, approximated by a power function.

It was found by comparing the AFM data before and after drying that even thin films should be dried to remove the residual solvent at temperatures below 100 °C, which leads to the partial formation of open pores from microcavities formed during the deposition of films [6] and observed on AFM after drying. However, the addition of a low-volatility component to the solution can lead to its only partial removal from the film. To determine the presence of residual solvent, thicker films were formed, followed by drying in vacuum and analyzed by Raman spectroscopy. The optimum drying parameters were determined by analyzing films with 5–15 μm thickness because of the greater accuracy of determining the absence of a fluorescent background and relative intensity of PLA Raman peaks in the region of 2800–3100 cm^{-1}. It has been shown (see fig. 3) that when THF is used as the main solvent, the presence of a noticeable amount of BA in the film after drying, detected by the presence of peaks near 3060 cm^{-1} and 1000 cm^{-1} [13], is not detected, but when DM is used as the main solvent, the spectra shows the presence of BA.

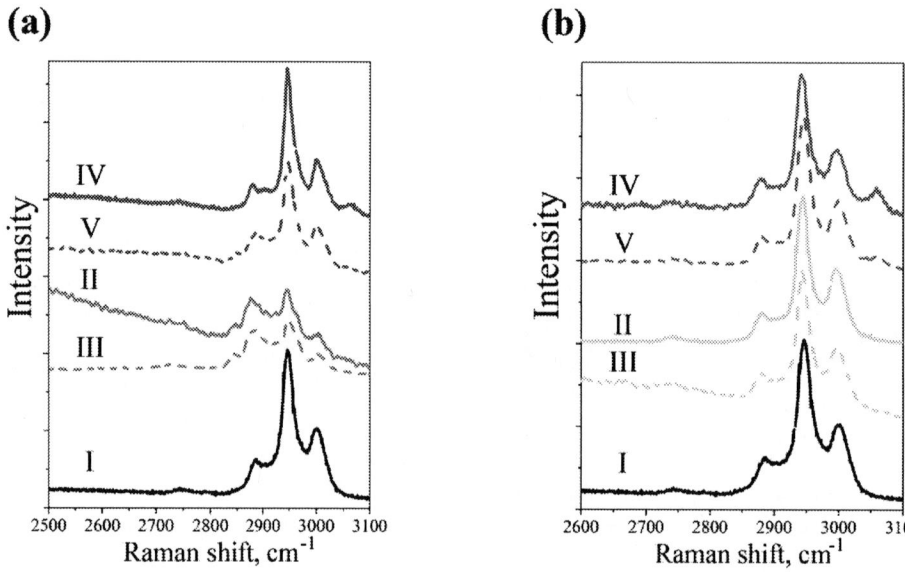

Figure 3(a, b). Raman spectra of PLA films: **(a/b)**: I – reference spectra of pure PLA; II, III – PLA from THF**(a)** / DM**(b)** solution before and after vacuum drying; IV, V – PLA from THF**(a)** / DM**(b)** and 3.5 wt % **(a)** / 1 wt % **(b)** BA solution before and after vacuum drying.

This fact indicates a significant difference in the mechanism of BA evaporation using different solvents. Presumably, DM, which is characterized by a lower boiling point and a higher vapor pressure compared to THF, evaporated from the film too quickly during film drying, apparently without forming the conditions for complete removal of the BA that remains in the film and presumably interacts with PLA replacing the evaporated DM, which in turn creates an additional deterioration in the conditions for its removal from the film upon drying, in view of its substantial interaction with PLA, the prerequisite of which are the known results on the substantial swelling of PLA in BA [8]. In addition, the reason for such behavior probably is difference in the energy of interaction between DM and BA compared with THF with BA, in combination with the difference in the interaction energies of the used molecules with PLA. The latter is evident in view of the better solubility of PLA in DM in comparison with THF [8]. Thus, based on the obtained spectral data, in the case of DM, it is impossible to effectively evaporate the BA, whose presence is detected on the spectra after prolonged vacuum drying. On the other hand, the usage of THF as the main solvent allows completely (at the level of the absence of the peak near 3060 cm^{-1} in the spectrum) to remove solvents from the film during drying.

Also, decrease in the relative intensity of 2880 cm^{-1} to 2950 cm^{-1} peak, indicates an increase in the material crystallinity [14] and microscale crystallites formation, which is undesirable for developing coatings, and shows the lowest degree crystallinity in the case of THF films and maximal in the case of DM, especially with BA (the peak of 2880 cm^{-1} is almost zero and coincides with the intensity in the region of 2900–2920 cm^{-1}), and the intermediate value in the case of THF with BA. A low degree of crystallinity can also improve the solvent removal conditions of the film, which is an additional factor for removing the BA. And additionally the presence of a significant fluorescent background indicates the presence of a residual solvent, as well as the presence of a noticeable peak near 2850 cm^{-1} (only for THF). In the case of DM, in view of the absence of preliminary drying of the films in air, were observed peaks near 700 cm^{-1} [15] and 3060 cm^{-1} that are comparable in intensity with a peak near 2880 cm^{-1}, but there was no fluorescent background, characteristic as we assume for a small number of residual solvent molecules interacting with the polymer.

These results of Raman spectroscopy correlate well with the measurement data of water contact angle and film surface (θ_w), which revealed a significant effect of the solvent composition on the value of θ_w (see fig. 4).

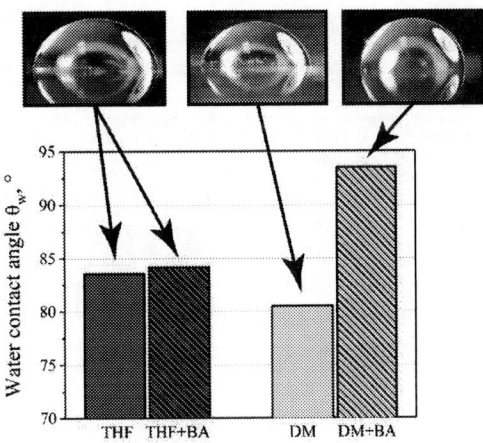

Figure 4. Results of water contact angle measurements.

The addition of BA to the PLA solution in DM led to a significant increase of the water contact angle of film from 80° to 93°, that in combination with obtained spectroscopy data from thick films, indicates that the probable cause of the observed effect is the presence of residual BA molecules near the surface of thin PLA film, the presence of which presumably forms a smaller number of polar groups on the film surface. Possibility of participation of such a mechanism in this case agrees with the data known from the literature on the effect of functional groups formed on the surface by plasma treatment on the wetting angle, as well as on the relaxation of their influence during annealing of films above the glass transition temperature [5]. The absence of this kind of change in the wetting angle in the case of films obtained from THF solution also correlates with the Raman data demonstrating the best results of the BA releasing from the formed films, and also indicates that there is no need for long drying of the films thinner than 200 nm to remove the residual solvent. The observed slight difference in the wetting angle of films obtained from pure DM and THF (see fig. 4) is apparently caused by a greater porosity in the case of THF, that is consistent with the known Cassie-Baxter model, which describes an increase in the wetting angle in view of the presence of pores that create microregions where liquid-gas boundaries are formed [16], which are not specific for a smooth surface or a surface with sparsely lying nanoparticles. And known Wenzel model for surfaces with nanoparticles or roughness, which are found on the films deposited from BA-containing solution, does not explain increasing or non-changing water contact angle on them. However, the detected surface hydrophobicity is a negative factor for cells adhesion and growth, which proves the necessity of careful selection of solution composition to ensure controllable porosity and at the same time suitability for cell culturing.

4. Conclusion

The comparison of the results of the formation of films in different atmospheres and solutions with different concentrations of THF and DM revealed significant differences in the mechanism of formation, as well as in the thickness and degree of porosity of the films, and films formed in the nitrogen atmosphere have fewer defects and smaller size of pores. It was found that the sample with a low volatile BA, added to the solution, differs by low porosity and a slightly higher degree of crystallinity at the same amount of residue solvents compared to pure THF sample so it is possible to completely remove it from the film after drying in the case of using THF as the main solvent. Maximum degree of crystallinity was detected in films deposited from DM solution, in which BA can not be fully evaporated by drying, that additionally appears also at a significant increase in the wetting angle with water. Thus, in the absence of longtime vacuum drying, main peaks of residual solvents in thick film was observed on the Raman spectra, which makes such process and films unacceptable for using as a biodegradable coating. A technique for estimating the presence of a residual solvent in a PLA film was proposed, based on observing the characteristic peaks of the solvents themselves in the case of a large amount of them in film, as well as on detecting the fluorescent background of the films, even in the case of a small amount of a residual solvent. Optimal vacuum drying time for films with thickness more than 500 nm was determined as more than 48 hours, and the drying of the films to remove the residual solvent at thicknesses less than 200 nm can be carried out at a temperature of up to 100 °C for 10 minutes. Despite the possibility of a significant change in the porosity of the film with the addition of BA, it is necessary to completely remove the residual solvent, which can also be detected by the absence of an increase in hydrophobicity with respect to the samples with a completely removed solvent. Thus, careful selection of the solution composition and quality control are required to form films suitable for cell cultivating and for use as coatings for implant.

Acknowledgments

This work was supported by the Ministry of Education and Science of the Russian Federation, agreement № 14.575.21.0125 (unique ID RFMEFI57517X0125).

References

[1] Abdal-hay A, Hwang M G, Lim J K 2012 *Journal of sol-gel science and technology* **64** 756
[2] Cai H, Dave V, Gross R A, and McCarthy S P 1996 *J. Polym. Sci. B Polym. Phys.* **34** 2701
[3] Middleton J C, Tipton A J 2000 *Biomaterials* **21** 2335
[4] Vandrovcová M, Bacakova L 2011 *Physiological Research* **60** 403
[5] Slepička P, Trostová, S, Slepičková Kasálková N, Kolská Z, Sajdl P, Švorčík V 2012 *Plasma Processes and Polymers* **9** 197
[6] Wu Y, Clark R L 2007 *Journal of colloid and interface science* **310** 529
[7] Dayal P, Liu J, Kumar S, Kyu T 2007 *Macromolecules* **40** 7689
[8] Sato S, Gondo D, Wada T, Kanehashi S, Nagai K 2013 *J. Appl. Polym. Sci.* **129** 1607
[9] Garlotta D *Journal of Polymers and the Environment* 2001 **9** 63
[10] Sun G, Chan C M *Colloid and Polymer Science* 2013 **291** 1495
[11] Fang H W, Li K Y, Su T L, Yang T C K, Chang J S, Lin P L, Chang W C 2008 *Materials Letters* **62** 3739
[12] Yang T I, Su T L, Lin P L, Tseng I H, Chang C H, Fang H W 2012 *J. Appl. Polym. Sci.* **124** 2333
[13] Prystupa D A, Anderson A, Torrie B H 1994 *Journal of Raman spectroscopy* **25** 175
[14] Qin D, Kean R T 1998 *Applied Spectroscopy* **52** 488
[15] Ramirez-Cedeno M L, Ortiz-Rivera W, Pacheco-Londono L C, Hernandez-Rivera S P 2010 *IEEE Sensors Journal* **10** 693
[16] Nosonovsky M, Bhushan B 2008 *Advanced Functional Materials* **18** 843

Comparison of step and flow-focusing emulsification methods for water-in-oil monodisperse drops in microfluidic chips

N A Filatov[1], D V Nozdriukhin[1,2], A A Evstrapov[1,3,4], A S Bukatin[1,3]

[1]Nanobiotech Lab, St. Petersburg Academic University, St. Petersburg 194021, Russia
[2]Peter the Great St. Petersburg Polytechnic University, St. Petersburg 195251, Russia
[3]Laboratory of Information and Measurement Biosensor and Chemosensor Microsystems, Institute for Analytical Instrumentation RAS, St. Petersburg 198095, Russia
[4]Department of Nano-Photonics and Metamaterials ITMO University, St. Petersburg 197101, Russia

Abstract. Emulsions are widely used in various disciplines such as food, chemistry, and pharmaceuticals, due to their efficient encapsulation properties. Recently, the emulsion was emerged as a unique tool for fabrication of multi-complex microparticles, which used as drug delivery agents, as effective manipulators for sorting proteins, DNA, and etc. The high productivity of stable monodisperse emulsions formation is of great importance for these aims. Here, the step and the flow-focusing methods for water-in-oil drops formation in microfluidic chips were compared as a tool for generating well-defined and monodispersed drops using mineral oil as a continuous phase. As a result, drops diameter compares with aperture (nozzle) dimensions for the flow-focusing generator and their diameter is much higher for the step design. Monodispersion and productivity for the step-chip strongly depend on phase densities difference. Flow-focusing method is more suitable for fabrication of monodispersed microparticles and step method for rough technologies such as investigating in food industry or cosmetics.

1. Introduction

Emulsions have long been utilized in various fields, such as food industry, cosmetics, pharmaceuticals, etc. [1,2]. Recently, the emulsions were emerged as a unique tool for fabrication of multi-complex microparticles, which used as drug delivery agents, effective manipulators for sorting proteins, DNA, and etc. [3]. To extend the opportunities of emulsions for microparticles generation, the high productivity of stable monodisperse emulsions generation is required. High monodispersity of emulsions can increase signal-to-noise ratio and improve predictability and standardization of assays [4]. There are classical bulk methods, such as colloid mills, mixers and sonicators, which give high-throughput emulsion generation but have lack control over droplet size and monodispersity.

Recently microfluidics technologies were introduced for droplet formation. These technologies are known as droplet microfluidics. Such emulsifier systems are usually based on T-junctions, co-flow or flow focusing generators, which enable to precisely control droplets size and achieve qualitative monodispersity. But these generators are significantly slower than the traditional emulsification techniques [5]. To overcome low productivity of microfluidic emulsifier systems scaling-up can be used [6].

Content from this work may be used under the terms of the Creative Commons Attribution 3.0 licence. Any further distribution of this work must maintain attribution to the author(s) and the title of the work, journal citation and DOI.
Published under licence by IOP Publishing Ltd

Besides, in the work of E. Stolovicki, R. Ziblat and D. A. Weitz [4] it was proposed to use parallel step emulsifier devises without shear stress and efficient nozzle clearance during droplet formation for high-throughput droplet generation. Their emulsifier worked well for a wide range of drops, sizing from 30 to 1000 µm (from 14 pl to 520 nl) with coefficient of variation (CV) ≤5.2% at production rates of 0.03 and 10 L per hour and achieved by 400 and 120 parallelized nozzles respectively. They used the high purity water (Milli-Q, USA) as a disperse phase and the fluorocarbon oil HFE 7500 (Novec Engineered Fluid, 3M, USA) as a continuous phase. The construction of the step-emulsification chip which allows self-cleaning of nozzles (step place) should provide both high-throughput and well-definite droplet sizes [4]. The microfluidic step-emulsifier operates with two immiscible liquids. The first liquid (dispersed phase) flow through nozzles of the device into a deep and wide reservoir (tank) with the second liquid (continuous phase). Under certain conditions, the confined stream of the disperse phase breaks into small monodisperse droplets at the step [7]. Efficient removal of the drops from the nozzle outlet is imperative to achieve high production rates while maintaining low CV. The self-cleaning method of steps is based on droplet buoyancy in continuous phase. Drops do not stay near the step exit, because they either fall under the action of gravity or float up [4].

In this work, we compared the step and the flow-focusing methods for water-in-oil drops formation in microfluidic devices. A mineral oil was used as a continuous phase due to its low cost compared with fluorocarbon oil. The flow focusing design was used due to its operating principle which is based on the fact that the continuous phase flowing through two side channels meets the dispersed phase at a channels' intersection, where the dispersed phase is squeezed by the continuous phase and breaks up into droplets. It allows avoiding a contact of forming droplets with channels' walls and helps to prevent potential negative effects on sample's components of the dispersed phase [8].

2. Materials and methods

The PDMS microchips were fabricated by standard soft lithography method [9]. In case of the flow-focusing design, to cover a PDMS replica, an oxygen plasma bonding with a glass slide was used. To cover a PDMS replica for the step emulsifier, an oxygen plasma bonding with a PDMS film was used. To create hydrophobization coating on microchannels walls, commercially available anti-rain agent Turtle Wax (the USA) was used. The mineral oil (330779 light, Sigma-Aldrich) with the 3.5% surfactant Abil EM 180 was used as a continuous medium. The oil density was 0.84 g/cm^3. The surfactant is needed to prevent the droplets coalescence and provide droplets stability. Deionized water was used as a disperse phase. In the flow-focusing droplet generator microchannels height was 40 µm, output channel width was 200 µm and aperture width was 15 µm (figure 1a). The step microchip had 256 nozzles with dimensions 45x15 μm^2 (figure 1d).

In case of the flow focusing device, two syringe pumps PHD 2000 (Harvard Apparatus) with 100 µl and 500 µl Hamilton–Microliter Series Gastight syringes for dispersed and continuous phases respectively were used. In contrast, to making an emulsion in the step generator, only one dispersed phase was pumped by the one syringe pump PHD 2000. During droplet generation in the step microchip, the dispersed phase flows into the opened reservoir with volume ~10 ml of the continuous medium (figure 1d). The tank was made like a sandwich by an oxygen plasma bonding of two glass slides 5x10 cm^2 and 5 mm PDMS layer between slides.

The water-in-oil drops formation in the flow-focusing chips was studied by optical microscope Leica DM4000 B LED (Leica Microsystems). Formation of droplets was recorded on video by the LEICA camera. MatLab script was used to recognize drops on video frames and obtain the dependences of droplet diameter and productivity for the flow-focusing method. In case of the step emulsifier, the digital USB microscope Prima Expert was used. To analyze drop diameter, also the MatLab script was used. The productivity for the step method was estimated as the water flow rate divided by the average diameter.

SPBOPEN 2018   IOP Publishing
IOP Conf. Series: Journal of Physics: Conf. Series **1124** (2018) 031028   doi:10.1088/1742-6596/1124/3/031028

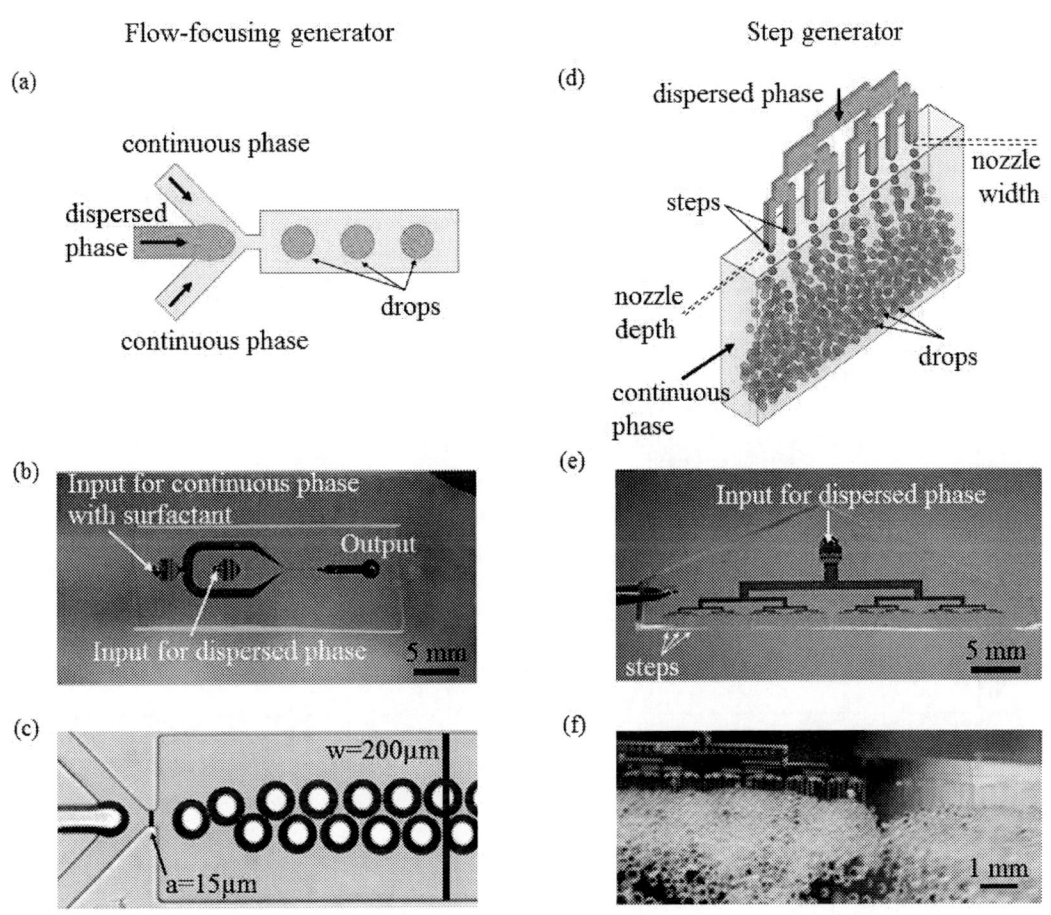

Figure 1. (a-c) Flow-focusing microfluidic droplet generator with the aperture a=15μm: **(a)** work scheme; **(b)** chip image; **(c)** camera image of real-time formation. **(d-f)** Step generator with 15 μm nozzle width: **(c)** work scheme; **(d)** chip image; **(f)** camera image of real-time formation.

3. Results and discussion

The drop formation regime (jetting, dripping) depends on capillary number, which describes a ratio of the forces of surface tension and viscous [10]. The dripping mode was in both types of droplet generators.

The flow-focusing microfluidic droplet generator with the aperture 15 μm allowed to form small volume droplets in the range of diameters from 20 to 70 μm (figure 2a). These diameters correspond to volumes in the range from 4 to 180 pl. The data indicates that droplet diameter depends only on the ratio between disperse and continuous phases flow rates, not by their values. An emulsion is monodisperse at any stable regime, the CV is ≤1%. To estimate the CV, probability as function of the drop diameter data was approximated by a Gaussian (figure 2b). Moreover, the droplet production rate was analyzed. The productivity is strongly depends on both flow rate values and their ratio. The single flow-focusing generator provided the productivity up to 200 Hz.

In case of the step emulsifier, droplet diameter (figure 2c) was in the range from 350 to 750 μm (from 20 to 220 nl) and had up to 22% CV, so it was a quite polydisperse emulsion. To find the CV, also a probability as function of the drop diameter data was approximated by a Gaussian (figure 2d). The maximum productivity was up to 1800 Hz. Note that the 1800 Hz production corresponds to 256

345

nozzles of the step microchip, consequently, on average a single nozzle have only about 7 Hz. It is according to formation of the large drop volume (from 20 to 220 nl). When droplets are generated in the step emulsifier their detachment from the nozzle occurred under the action of gravity due to differences of phase densities. So, a velocity of drops removing from the nozzle outlet is strongly depended on phases densities. The greater the difference between densities, the faster a drop disengaging from the nozzle. In the work [4], where the step method was investigated, the fluorocarbon oil [4] was used, which density is 1.6 g/cm^3, water droplets floated upward and the difference was 0.6 g/cm^3. In our case of mineral oil its density is 0.84 g/cm^3, so water drops fell down and the density difference was only 0.16 g/cm^3. Perhaps it is the reason for forming drops from 30 to 1000 μm with the CV ≤5.2% and the single nozzle productivity from 1500 to 36 Hz, respectively, in case of fluorocarbon oil [4].

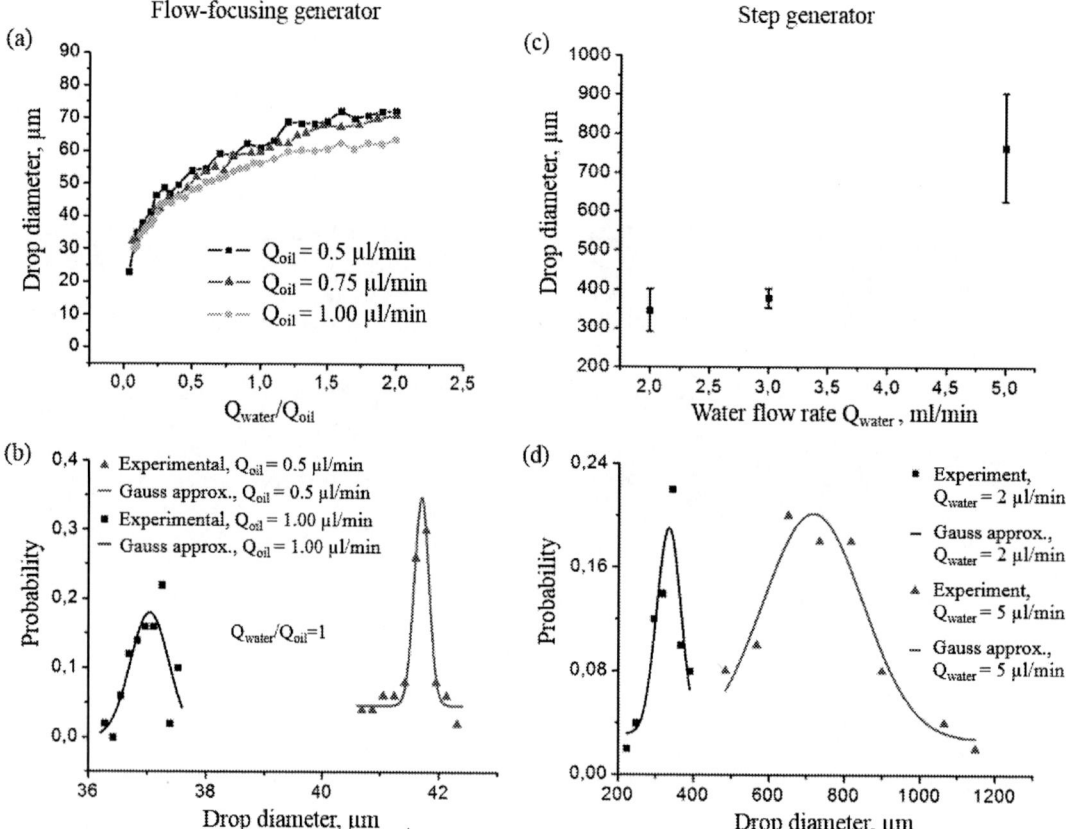

Figure 2. Droplet diameter as a function of ratio between water and oil flow rate **(a)** and probability as function of drop diameter **(b)** for the flow-focusing microfluidic droplet generator with the aperture a=15μm. The average (black squares) and standard deviation (error bars) of drop diameter from water flow rate **(c)** and probability as function of drop diameter **(d)** for the step generator with 15 μm nozzle width.

4. Conclusions

To sum up, the comparison of step and flow-focusing emulsification methods for water-in-oil monodisperse drops in microfluidic chips was done using the mineral oil as a continuous phase. The flow-focusing emulsifier have such advantages: monodispersity (≤1% CV), low consumption of continuous phase, getting small drops to 0.5 pl. Also, the single flow-focusing generator gives up to

200 Hz production rate. In contrast, the step emulsifier is convenient method for simple generating of large volumes of emulsion with drops in the range from 20 to 220 nl. The step generator with 256 nozzles give up to 1800 Hz production (or 7 Hz per single nozzle), but lack monodispersity ($\leq$22% CV). Moreover, the production of step method depends on phases densities that makes the system selective. So if the fluorocarbon oil is used as a continuous phase, it is possible to form drops from 14 pl to 520 nl with the CV $\leq$5.2% at the single nozzle production from 1500 to 36 Hz, respectively.

Note that drop diameter compared with aperture (nozzle) dimensions for the flow-focusing generator and is much higher in case of the step design for the mineral oil as a continuous phase.

The flow-focusing generator can be used for single analysis of cells and molecules, for droplet PCR, for the microparticles creation because of monodispersity. To solve drops mass production problem, scaling up and using several parallel flow-focusing generator is required [6]. It is more reliable and gives stable monodisperse emulsions. Moreover, to form drops in the flow-focusing generator, the big volume like 10ml step tank is not needed. The step microchip is more suited for rough technologies such as investigating in food industry or cosmetics.

Acknowledgments
This work was financially supported by the grant from the President of Russian Federation (project MK-2131.2017.4).

References
[1] Nozdriukhin D V, Belousov K I, Filatov N A, Bukatin A S 2017 *IOP Conf. Series: Journal of Physics: Conf. Series* **917** 042015
[2] Kim S-H, Weitz D A 2011 *Angew. Chem.* **123** 8890-8893
[3] Choi A, Seo K D, Kim D W, Kim B C and Kim D C 2017 *Lab Chip* **17** 591-613
[4] Stolovicki E, Ziblat R and Weitz D A 2018 *Lab Chip* **18** 132-138
[5] Ofner A, Moore D G, Rühs P A, Schwendimann P, Eggersdorfer M, Amstad E, Weitz D A, Studart A R 2016 *Macromol. Chem. Phys.* **218** 1600472
[6] Mulligan M K, Rothstein J P 2012 *Microfluidics and Nanofluidics* **13** 65-73
[7] Charkaborty I, Ricouvier J, Yazhgur P, Tabeling P, Leshansky A M 2017 *Lab Chip* **17** 3609-3620
[8] Filatov N A, Belousov K I, Bukatin A S, Kukhtevich I V and Evstrapov A A 2016 *J. Phys.: Conf. Ser.* **741** 012052
[9] Bukatin A S, Mukhin I S, Malyshev E I, Kukhtevich I V, Evstrapov A A and Dubina M V 2016 *Tech. Phys.* **61** 1566-71
[10] Cubaud T and Mason T G 2008 *Phys. Fluids* **20** 053302

SPBOPEN 2018   IOP Publishing

Spectroscopic investigation of vegetable pectin - fluorescent protein EGFP interaction in tumor cells *in vitro*

V V Danilov[1], V V Klimenko[2], S V Shmakov[2], A I Khrebtov[3], A S Kulagina[4]

[1]Department of Physics, Emperor Alexander I St Petersburg State Transport University, St. Petersburg 190031, Russia
[2]Nanobiotech Lab, St. Petersburg Academic University, St Petersburg 194021, Russia
[3]Epitaxial Nanotech Lab, St. Petersburg Academic University, St Petersburg 194021, Russia
[4] Lab of Epitaxial Nanostructures, ITMO University, St Petersburg 197101, Russia

Abstract Using spectroscopic methods, solutions of zosterin and CT26 cell culture were investigated. Fluorescent spectroscopy revealed a negative dynamics of growth of a malignant culture containing a green fluorescent protein in the presence of zosterin. It was observed two-component luminescence kinetics of EGFP protein.

1. Introduction

Pectin substances (pectin polysaccharides, pectins) are the most complex and interesting in terms of structural organization and functional activity in present. Despite significant advances in the field of structural studies of polysaccharides, the complexity of construction and the irregular nature of the carbohydrate chains of pectin macromolecules do not allow us to consider their structure established. Therefore, at the present time, there is no generally accepted model for the structure of pectin macromolecules. Our work is devoted to the study of a drug known under the commercial name "Zosterin-Ultra". The raw material for its production is the perennial aquatic plant Zostera marina, which grows exclusively in the pure water areas of the Russian Federation. Marine pectin (zosterin) has the formula $C_6H_8O_6$ and is 90-95% represented by a mixture of polygalacturonic and polyglucuronic acids [1, 2]. The properties of this pectin proved to be promising, some of them have already found application in medicine and biology. Currently, Aquamir CJSC, based on its own original technology, produces BADs under the commercial names "Zosterin-Ultra 60%" and "Zosterin-Ultra 30%", which are positioned as natural remedies with qualities of sorbents and immunomodulators.

The most of pectin studies are carried out by IR spectroscopy, chromatography [3]. The main aim to a more detailed study of the properties of zosterin is to obtain information on the driving forces of protein transformation when they interact with pectin. In the literature there is no full information about spectral and optical characteristics of this substance and its solutions. In this paper, we want to pay attention to the possibility of a negative effect of zosterin on the morphology of malignant cell cultures besides.

Content from this work may be used under the terms of the Creative Commons Attribution 3.0 licence. Any further distribution of this work must maintain attribution to the author(s) and the title of the work, journal citation and DOI.
Published under licence by IOP Publishing Ltd

2. Methods

2.1 Chemical substances

In the work the commercial preparation "Zosterin-Ultra 60%", which is a brown powder, was investigated. The 1 g powder was dissolved in 50 ml phosphate buffered saline (PBS, 150 mM NaCl, HyClone, USA) at the 60° C temperature using a magnetic stirrer for 10 min. The concentration of stock solution of Zosterin was 20 mg/ml. The fraction of undissolved particles and aggregates were removed by centrifugation of the solutions at 300xg rate regime for 5 minutes. The supernatant was isolated in individual 15 ml tubs and used for further in vitro experiments.

2.2 Cell cultures

A CT26 murine colon carcinoma cell line (ATCC CRL-2638) stably expressing EGFP (CT26-EGFP) was obtained by lentiviral transduction in the Shemyakin-Ovchinnikov Institute of Bioorganic Chemistry RAS headed by Dr. Sergey A. Lukyanov. The cells were cultured in RPMI-1640 medium (HyClone, USA) supplemented with 10% fetal bovine serum (FBS; HyClone, USA), 10 units/mL penicillin and 10 μg/mL streptomycin in a humidified incubator with 5% CO_2 at 37° C. For experimental study the CT26-EGFP cells were seeded in 6-well plates at a density $5 \cdot 10^5$ cells/well and incubated in RPMI medium 10% FBS. 24 hours after seeding 100 mkl zosterin solution was added into cellular medium and pipetted. The concentration of zosterin in medium was 2 mg/ml. 24 h after incubation the fluorescence signal of CT26-EGFP in control and experimental groups were analyzed.

The green fluorescent protein (GFP) is a protein composed of 238 amino acid residues (26.9 kDa) that exhibits bright green fluorescence when exposed to light in the blue to ultraviolet range. EGFP variant contains chromophore mutations that make the protein 35 times brighter than wild-type GFP. The fluorescence quantum yield (QY) of EGFP is 0.60. EGFP have a single emission peak of 509 nm by 488 nm excitation.

2.3 Fluorescence imaging

The fluorescence signal from CT26-EGFP cells was obtained using confocal microscope Carl Zeiss Axio Observer Z1 (Zeiss, Germany). The fluorescence of EGFP was measured using Yokogawa CSU-X1 spinning disk, objective 20x/0.5, the 488 nm continuous wave (CW) laser Sapphire LP (Coherent Inc., USA), exposure time is 2000 ms. The fluorescence signals were taken from CT26-EGFP cells as pixel brightness value and analyzed using the software AxioVision Rel. 4.9 (Zeiss, Germany). The results of fluorescence were reported as the mean values of fluorescence intensity with standard deviation.

2.4 Measurement of fluorescence kinetics

The luminescence kinetics was measured with MicroTime 100 (PicoQuant) laser scanning fluorescence microscope. To excite the luminescence, a pulsed diode laser was used (λ_{ex} = 405 nm, τ_{pulse} = 80 ps), to register the fluorescence a photomultiplier (185-800 nm) was used also.

3. Results and discussions

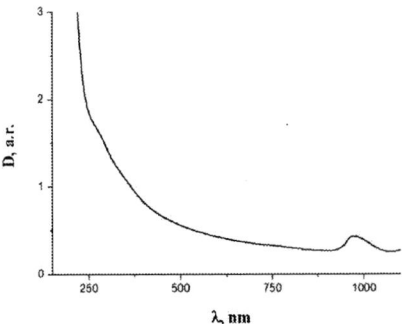

Figure 1 The absorption spectrum of zosterin aqueous solution in the visible and UV regions

The spectrum of zosterin on Figure 1 is weakly characteristic and smoothly decreases to the IR region. Aggregation plays a key role in gelling, as well as in self-assembly of proteins into supramolecular systems. Therefore, the ability to directly monitor protein aggregation is extremely important when studying the properties of protein-containing systems. Apparently, the transmission of zosterin in the visible range is due to scattering on the micellar structures.

The method of fluorescence spectroscopy is one of the most common for studying the physicochemical properties of biological systems and, in particular, the structure of proteins. This method allows one to monitor changes in the microenvironment of the protein's own fluorophores or the introduced fluorescent label.

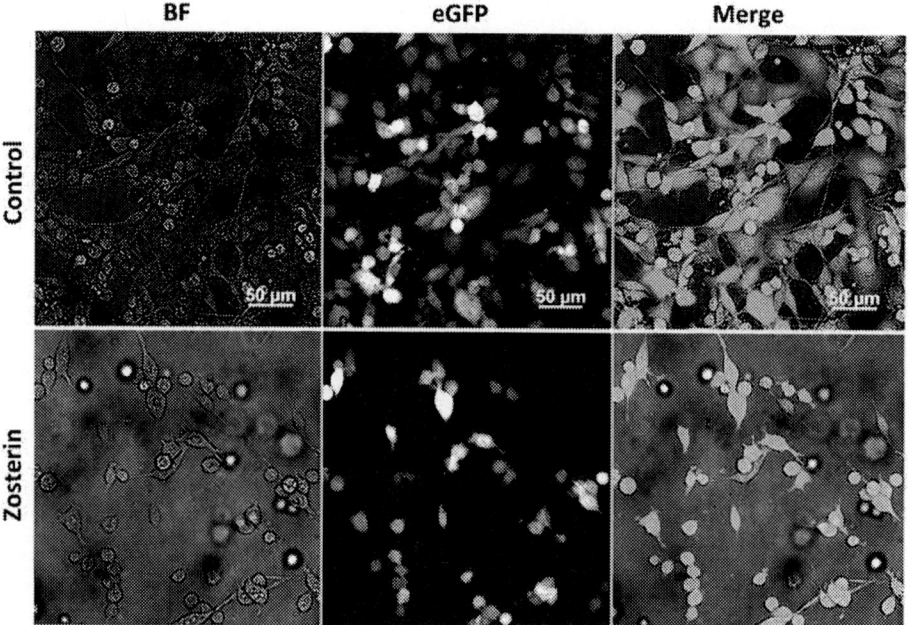

Figure 2. Confocal fluorescence imaging of CT26-EGFP. Top panel is control CT26-EGFP cells and the bottom panel CT26-EGFP cells incubated with zosterin solution 24 h. BF is bright field imaging, EGFP means fluorescence imaging of EGFP, Merge is combined image of BF and EGFP.

In the experimental sample, the cytotoxic effect of Zosterin on tumor cells CT26-EGFP is observed. The survival of cells is reduced in comparison with the control. There is a change in the morphology of the cells and their adhesion to the surface decreases. The presence of Zosterin in the experimental sample changes the viscosity of the culture medium. There is no quenching of the fluorescence signal of EGFP in cells compared to the control. There is no change in the distribution of the fluorescence signal across the cell. There is a slight increase in the average fluorescence signal EGFP from cells in the experimental sample and this may be due to the cytotoxic effect of the Zosterin preparation. Mutations of amino acid residues in the immediate vicinity of the chromophore can significantly alter the fluorescent properties of the protein [5]. On Figure 2 the luminescence kinetics of cell cultures and zosterin is shown.

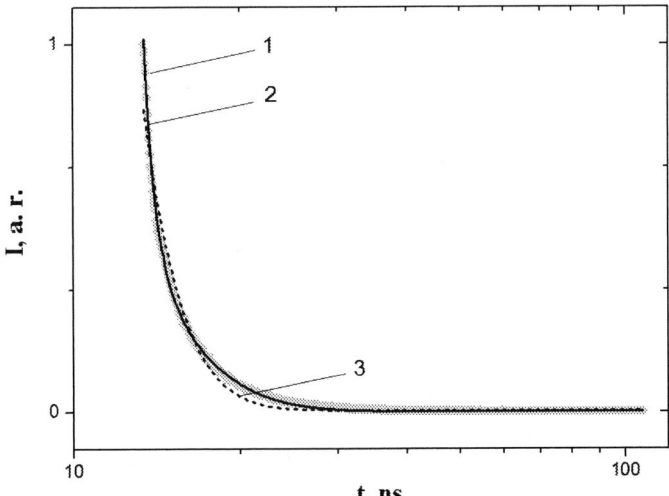

Figure 3 Luminescence kinetics decay of the cell culture containing the EGFP protein (experimental points) is 1; the approximation by two-exponential dependence (line) is 2; the approximation by exponential dependence with one decay time (dashed line) is 3.

It is known that in a green fluorescent protein, the 203 amino acid residue (Thr203 or tryptophan) is located near the chromophore and is potentially capable to influence on the spectral characteristics [6]. Fluorescent properties of tryptophan are extremely sensitive to changes in its microenvironment, and mainly, polarity. The fluorescence duration of tryptophan in water is 3 ns, and in the protein it varies from 2 to 4.6 ns [7]. Therefore, in the measured luminescence kinetics of the EGFP protein being studied, the presence of two decay life-times 3.7 ns and 0.6 ns may be due to the presence of different forms of the protein. Complexation with low molecular weight ligands and macromolecules, denaturation, aggregation, and other processes cause different connections with the environment and affect fluorescence duration of tryptophan, including. The effect of zosterin on the luminescence kinetics of cell cultures containing the EGFP protein is the subject of further study.

Conclusions
The absorption spectra of "Zosterin-Ultra 60%" aqueous solution were obtained. Fluorescent spectroscopy revealed a negative dynamics of growth of CT26 malignant culture containing a green fluorescent protein in the presence of zosterin. The luminescence decay kinetics of the EGFP protein has two components, what indicates about two forms of the protein in cells in solution.

References
[1] Loenko Yu K, Artjukov A A, Kozlovskaya E P, Miroshnichenko V A, Elyakov G B 1997 *Zosterin* (Vladivostok: Dal'nauka) p 212
[2] Suhoveeva M V, Podkorytova A V 2006 *Commercial seaweed and grasses of the seas of the Far East: biology, distribution, stocks, processing technology* (Vladivostok: TINRO-center) p 243
[3] Laskin J, Lifshitz H 2012 *Principles of mass spectrometry in application to biomolecules* (Moscow: The Technosphere) p 608
[4] Ovodov Yu S, Golovchenko V V, Guenter E A, Popov S V 2009 *Pectin substances of plants of the European North of Russia* (Ekaterinburg: UrB RAS) p 105
[5] Niwa H, Inouye S, Hirano T, Matsuno T, Kojima S, Kubota M, Ohashi M, Tsuji F I 1996 Proc Natl Acad Sci USA **93** 13617
[6] Ormö M, Cubitt A B, Kallio K, Gross L A, Tsien R Y, Remington S J 1996 Science **273** 1392
[7] Konev S V,Volotovsky I D, 1979 *Photobiology* (Minsk: BSU named after V. I. Lenin) p 384

SPBOPEN 2018

Modulation *p*-doping as the way to attain multi-state lasing in short-cavity InAs/InGaAs quantum dot lasers

V V Korenev[1,2], A V Savelyev[1], V G Dubrovskii[3], S Breuer[4], M V Maximov[1,2], A E Zhukov[1,2]

[1]St. Petersburg Academic University RAS, St. Petersburg 194021, Russia
[2]Peter the Great St. Petersburg Polytechnic University, St. Petersburg 195251, Russia
[3]ITMO University, St. Petersburg 197101, Russia
[4]Technische Universität Darmstadt, Darmstadt 64289, Germany

E-mail: *korenev@spbau.ru*

Abstract. The influence of modulation *p*-doping on the lasing threshold current densities of InAs/InGaAs quantum dot (QD) lasers is studied at different levels of acceptor concentrations. It is found that for both undoped and *p*-doped samples, there exists a critical cavity length at which threshold current densities of the lasing via ground- and excited states of QDs become equal so that multi-state lasing can be observed only in samples having longer cavities. It is shown that the usage of modulation *p*-doping is an efficient way to overcome this limitation and to obtain multi-state lasing even in case of sufficiently short samples.

1. Introduction

For many practical applications including ultrafast data transmission and optical coherence tomography, there is a need for stable optical sources emitting near 1.2-1.3 μm corresponding to the transparency window of Si-based waveguides [1 – 2]. Long-wavelength InAs/InGaAs quantum dot (QD) lasers emitting via the QD ground-state (GS) optical transition allow not only to overlap the required spectral range, but also have small threshold currents and high temperature stability [3]. Even if the lasing starts at QD GS around 1.3 μm, further increase in injection usually results in appearance of an additional short wavelength lasing line near at 1.2 μm associated with the first excited-state (ES) optical transition of QDs, the multi-state lasing, i.e. simultaneous lasing via GS and ES of QDs, takes place [4 – 8].

The multi-state lasing can be useful in view of achieving broader lasing spectra [1] and a larger number of optical channels from a single laser source [9]. However, the GS lasing band is quenched with the increase in injection current shortly after the onset of multi-state lasing [7, 10 – 11]. One of the ways to overcome this effect is the usage of *p*-doping that can be an efficient solution particularly in case of the short samples [10]. At the same time, it is also known that *p*-doping may influence not only the output power corresponding to the GS of QDs but also the lasing threshold current [10 – 12]. This result allows one to expect that the *p*-doping can influence a minimal laser cavity length at which multi-state lasing, i.e. simultaneous lasing via GS and ES of QDs, still can be seen. Studying of this question constitutes the key goal of this work.

Content from this work may be used under the terms of the Creative Commons Attribution 3.0 licence. Any further distribution of this work must maintain attribution to the author(s) and the title of the work, journal citation and DOI.
Published under licence by IOP Publishing Ltd

2. Experimental details

Three series of InAs/InGaAs QD lasers having different levels of p-doping of 0, 3 and 5×10^{17} cm^{-3} were grown using molecular beam epitaxy (MBE). The active region of each sample was comprised of 10 layers of InAs/InGaAs QDs separated by 35 nm-thick GaAs spacers. The p-doping was implemented by using of carbon atoms embedded into the central areas of barrier layers. The laser diodes were fabricated to have various lengths and the same stripe width of 50 μm. The experimental dependences were derived from the light-current curves and emission spectra measured at different injection currents.

3. Experimental results and their interpretation

To study the influence of p-doping level on the critical length of laser cavity at which multi-state lasing still can be seen, the dependence of threshold current densities of undoped and p-doped samples on laser cavity length was first measured – see Fig. 1 (a, b, c).

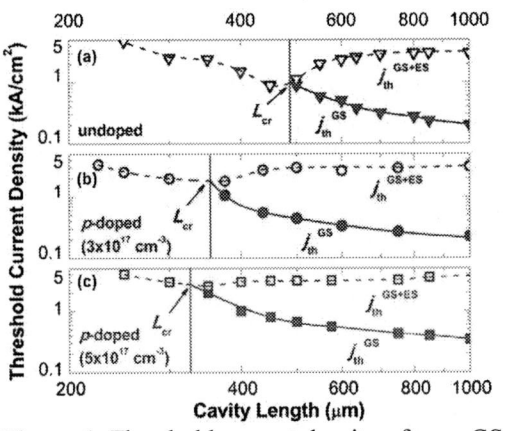 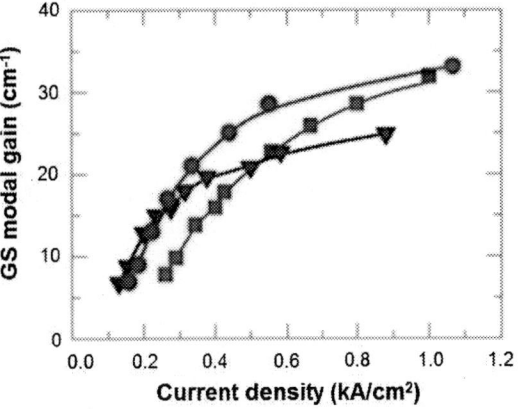

Figure 1. Threshold current density of pure GS-lasing (filled symbols, solid lines) and ES-lasing (empty symbols, dashed lines) vs cavity length. The dopant concentration is 0 (a), 3 (b) and 5 (c) $\times$ 10^{17} cm^{-3}. Vertical lines show the minimum cavity lengths (L_{cr}) at which GS-lasing component can be seen.

Figure 2. Experimental dependence of the modal gain corresponding to QD GS on the injection current density. Triangles, circles, and squares correspond to The dopant concentrations of 0, 3 and 5×10^{17} cm^{-3}.

As it can be seen in Fig. 1, the decrease in laser cavity length results in the increase in threshold current density of the onset of the lasing via QD GS ($j_{th,GS}$), while threshold current of multi-state lasing ($j_{th,GS+ES}$) decreases. The injection current density interval corresponding to the multi-state lasing, i.e. the interval between $j_{th,GS+ES}$ and $j_{th,GS}$, tends to shrink as cavity length increases. Therefore, there is a critical cavity length (L_{cr}) at which threshold current of GS-lasing and of multi-state lasing become equal – see the vertical lines in Fig. 1. The characteristic feature of L_{cr} is that there is no lasing corresponding to QD GS for the samples shorter than L_{cr}, where pure ES-lasing is only observed. Such a behavior remains true for p-doped samples as well – compare Fig. 1a, 1b and 1c. This poses a natural limit on the minimum length of laser cavity at which practically useful GS-lasing can be achieved.

However, the usage of modulation p-doping occurs to be a beneficial method to further increase the range of cavity lengths towards shorter cavities, at which simultaneous lasing via QD GS and ES still can be seen. Indeed, in accord with the experimental results shown in Fig. 1, the increase in the dopant concentration from 0 to 3×10^{17} cm^{-3} and to 5×10^{17} cm^{-3} results in the decrease of L_{cr} from 0.49 mm to 0.35 mm and to 0.33 mm respectively. The reason behind is the following: for the same cavity length, the population inversion of QD GS in the case of p-doped samples is higher than in the case of undoped samples because of the higher occupancy of the hole energy levels due to the extra acceptor atoms [7, 10]. This allows holding the ground-state lasing condition even for cavity lengths shorter than L_{cr} of the undoped sample. Besides the higher is the concentration of the acceptor atoms, the smaller is L_{cr}. At the same time, one may also note that the values of L_{cr} of the two p-doped QD lasers are close to each other as compared to L_{cr} of the undoped laser. In other words, we observe a "saturation" of the p-doping effect when doping becomes higher than 6 acceptors per QD.

It should be noted that the increase in p-doping level also contributes to the increase in the internal loss (α_{in}) in laser cavity. Indeed, the increase in α_{in} may be approximated from the experimental dependence of the reciprocal differential quantum efficiency (η_D^{-1}) on laser cavity length (L) using a well-known relationship $\eta_D \propto \alpha_{out}/(\alpha_{in} + \alpha_{out})$ as described in [10], where the output loss $\alpha_{out}= (1/L)\cdot\ln(1/R)$ and R is a reflectivity of as-cleaved mirrors. For the samples under the study $R=0.3$ and α_{in} values corresponding to the p-doping level increasing from 0 to 3 and to 5×10^{17} cm^{-3} are 0.8, 1.0 and 1.8 cm^{-1}. This trend can be explained by the enhancement of the free carrier absorption in laser cavity with the increase in p-dopant concentration.

At the same time, as p-doping level increases, the absolute value of the GS gain also tends to increase mitigating the abovementioned growth in internal loss that can be also seen from the experimental dependence of the GS modal gain on the injection current density – see Fig. 2 at sufficiently high injection current densities. For instance, GS gain of the laser structure with dopant concentration of 3×10^{17} cm^{-3} is higher than in case of the undoped sample as injection current density exceeds 0.3 kA/cm^2. It is also possible to estimate maximum optical loss at which the GS gain can still balance the ES gain. Using the values of L_{cr} derived from the Fig. 1a – 1c, these values are 25.4, 35.4 and 38.3 cm^{-1} for the dopant concentrations of 0, 3 and 5×10^{17} cm^{-3} respectively. This illustrates the abovementioned enhancement of the GS gain hat mitigates the increase in internal loss due to the p-doping.

4. Conclusion

In conclusion, the influence of p-doping level on the threshold current densities of ground- and multi-state lasing in InAs/InGaAs quantum dot (QD) lasers was studied in details. It was shown that for each sample, there exists a critical cavity length and multi-state lasing can be observed only for cavities longer than the critical one. This is due to the increased output loss in sufficiently short samples. At the same time, the usage of p-doping occurs to be a beneficial way to overcome this limit: It is shown that the higher is the p-doping level, the smaller is the critical length and, thus, the shorter can be the samples in which multi-state lasing can be observed. This effect can be qualitatively explained as interplay between the contribution of the p-doping into the total loss and GS gain. Although p-doping contributes to the increase in the internal loss, the latter is mitigated by the contribution of p-doping to the GS gain. It should be noted that at sufficiently high levels of p-doping, its effect on cavity critical length becomes limited and saturates at the dopant concentration exceeding 10 acceptors per QD.

Acknowledgements

This work is supported by the Russian Foundation for Basic Research (project #18-502-12081) and by the Ministry of Higher Education and Science of the Russian Federation (project № 3.9787.2017/8.9).

References

[1] Zhukov A E, Kovsh A R 2007 *Quantum Electronics* **38** 409

[2] Lam C F 2007 Passive Optical Networks: Principles and Practice San–Diego (Elsevier) 158

[3] Savelyev A V, Novikov I I, Maximov M V, Shernyakov Yu M, Zhukov A E 2009 Semicond. **43** 1597

[4] Markus A, Fiore A 2004 Phys. Stat. Sol. (a) **201** 338

[5] Kim Y J, Joshi Y K, Fedorov A G 2010, J. Appl. Phys. **107** 073104

[6] Sugawara M, Hatori N, Ebe H, Ishida M, Arakawa Y, Akiyama K, Otsubo T, Nakata Y 2005 J. Appl. Phys. **97** 043523

[7] Korenev V V, Savelyev A V, Zhukov A E, Omelchenko A V, Maximov M V 2013 Appl. Phys. Lett. **102** 112101

[8] Korenev V V, Savelyev A V, Zhukov A E, Omelchenko A V, Maximov M V 2013 Semiconductors **47** 1397

[9] Wojcik G L, Yin D, Kovsh A R, Gubenko A E, Krestnikov I L, Mikhrin S S, Livshits D A, Fattal D A, Fiorentino M, Beausoleil R G 2009 Proc. SPIE **7230** 72300M

[10] Korenev V V, Savelyev A V, Maximov M V, Zubov F I, Shernyakov Yu M, Kulagina M M, Zhukov A E 2017 Appl. Phys. Lett. **111** 132103

[11] Smowton P M, Sandall I C, Mowbray D J, Liu H Y, Hopkinson M 2007 IEEE J. Sel. Top. Quantum Electron. **13** 1261

[12] Korenev V V, Savelyev A V, Maximov M V, Zubov F I, Shernyakov Yu M, Zhukov A E 2017 J. Phys. Conf. Ser. **917** 052001

Injection microdisk lasers based on multilayers of InGaAs/GaAs quantum well-dot structures

N V Kryzhanovskaya[1,2], E I Moiseev[1], K P Kotlyar[1], Yu A Guseva[3],
Yu M Shernyakov[3], S A Scherbak[1,2], M V Maximov[1], A E Zhukov[1,2]

[1] St Petersburg Academic University, St Petersburg, Russia
[2] Peter the Great St Petersburg Polytechnic University, St. Petersburg, Russia
[3] Ioffe Institute, St Petersburg, Russia

Abstract. We report results of characterization of injection microdisk lasers based on active layers employing 2, 5 and 10 layers of InGaAs quantum well-dot structures. The microlasers operate in continuous wave regime at room temperature without external cooling. The minimal microdisk diameter 10 μm is obtained for 5 layers of InGaAs quantum well-dot structures. Lasing wavelength is around 1.1...1.15 μm, minimal threshold current is 1.6 mA (threshold current density 900 A/cm^2).

1. Introduction

Now active research work is conducted on microlasers for optical data transmission. As example of compact optical sources of laser emission can serve vertical cavity surface emission lasers [1]. Use of the plasmons between metal and semiconductor in the resonators covered with metal allows reducing significantly the physical size of nanolasers: the lateral sizes can be less than one micron [2]. In nanolasers with photon crystal resonator the energy for transmitting one bit of 13 fJ is achieved [3]. Recent results demonstrate that microdisk (MD) cavity based on whispering gallery mode (WMG) can be used to produce ultra small and low threshold current lasers [e.g. 4, 5]. Using novel kind of active region based on 5 layers of InGaAs quantum well-dots we demonstrated a high output power of 18 mW, differential efficiency of about 31 % and peak electrical-to-optical power conversion efficiency of 15 % in a 31 μm in diameter microdisk laser [6]. The InGaAs quantum well-dots provide the advantages of the conventional Stranski-Krastanow quantum dot active region (carrier localization in QDs, suppression of the sidewall recombination, low thresholds etc). At the same time the quantum well-dots provide rather narrow gain spectra with high modal gain (as high as 10 cm^{-1}). A possible solution to increase further the modal gain (and thus differential gain) in quantum dot based lasers is to use multiple layers of active region. In this work we discuss the characteristics of injection MD lasers with diameters varied from 10 to 30 μm containing on 2, 5 and 10 layers of InGaAs quantum well-dot structures.

2. Experiment details

The laser structures were grown by in a low pressure MOVPE reactor using hydrogen as a carrier gas in the temperature range of 500-700°C. Trimethylgallium (TMGa), trimethylindium (TMIn),

---

Content from this work may be used under the terms of the Creative Commons Attribution 3.0 licence. Any further distribution of this work must maintain attribution to the author(s) and the title of the work, journal citation and DOI.
Published under licence by IOP Publishing Ltd

trimethylaluminum (TMAl) and diethylzinc (DEZn) metalorganic compounds were used as Ga, Al, In and Zn atom precursors. Arsine (AsH_3) and silane (SiH_4) hydrides were used as As and Si atom precursors respectively. The structures consist of multiple layers of InGaAs quantum well-dots embedded in GaAs waveguide core whose thickness (0.8 μm) was adjusted to support the only fundamental mode in the vertical (growth) direction. The InGaAs quantum well-dots (QWDs) represent a dense array of indium-rich islands formed inside In-depleted residual quantum well [7]. The laser structure is also consisted of p- and n-type doped $Al_{0.34}Ga_{0.66}As$ cladding layers and a 0.2 μm-thick p-type doped GaAs top contact layer.

First, the heterostructures were processed into broad-area lasers with as-cleaved facets and ridges of 100 μm in width and different length. No facet coating was applied. The lasers were mounted p-side up and wire bonded on a copper block to enable testing.

MD lasers with diameters D varied from 10 to 30 μm were formed by photolithography and dry etching. AgMn/NiAu and AuGe/Ni/Au metallization was used for formation of ohmic contacts to the $p+$ GaAs cap layer and the underlying n-doped GaAs layer, respectively. Needle probes were used for electrical connections. A piezoelectrically adjustable Olympus LMPlan IR objective x10 was used to collect in-plane emitted light from a microlaser. The emission was detected with a Horiba FHR 1000 monochromator and a Horiba Symphony InGaAs CCD array.

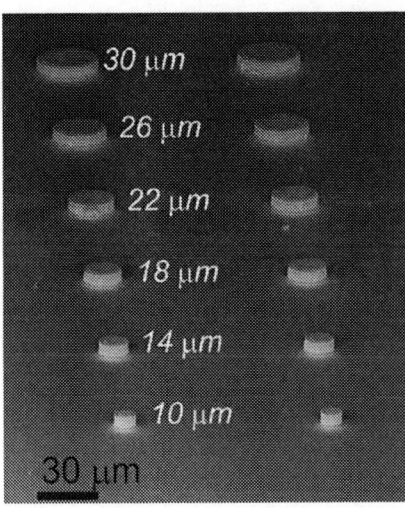

Figure 1. SEM image of injection MD lasers with 10-30 μm outer diameter.

3. Results

To evaluate the material quality first we studies characteristic of broad-area lasers. Figure 2 shows the threshold current (I_{th}) dependence versus number of QWDs layer obtained in 4 mm length lasers. The CW lasing is obtained even in laser with single layer of InGaAs QWDs with I_{th}~96 A/cm^2. It means that gain provided by single active layer is enough to overcome the loss. Increase of the number of QWDs layer results in increase of the QWDs saturated gain (G_{SAT}), transparency current (I_{tr}) and thus threshold current I_{th}~370 A/cm^2. The light emission spectra measured above the threshold current reveal the lasing wavelength in spectral range 1.08-1.2 μm at the ground state of the QWDs (Figure 2, right axis). The differential quantum efficiency of the lasers is ~80% and inner loss are estimated as low as 1.5 cm^{-1} (for 5 layer structure). The modal gain does not increase proportionally with the number of the QWDs layers (or total QWDs density) due to the different overlap of the active region with optical mode in the center and at the edges of waveguide layer as one can see from the schematics of the refractive index and TE fundamental mode profile drawn for structure with 5 active layers.

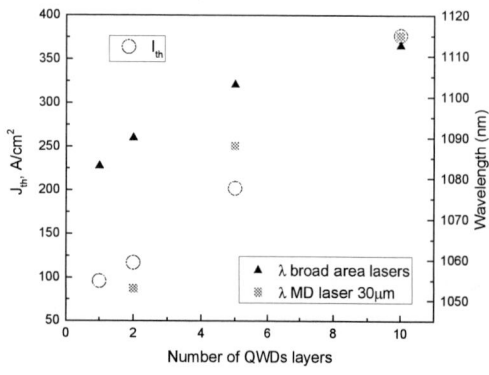

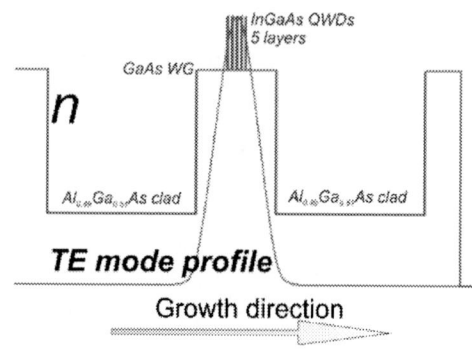

Figure 2. Threshold current of broad-area lasers (circles) and lasing wavelength (triangles) as a function of QWDs layers number.

Figure 3. Schematics of the refractive index and fundamental TE mode profile

MD lasers demonstrated room temperature lasing in cw regime via WMG. Figure 4 demonstrates electroluminescence spectra obtained for 30 μm in diameter devices with 2, 5 and 10 QWDs layers. As it was shown in our previous work [8] the narrow gain spectrum of the QWDs results in single-mode emission in optically pumped microlasers up to 9μm diameter. Here we observe single-mode emission with side mode suppression ratio more than 40 dB for 30 μm MD lasers regardless the quantity of the active layers. The emission wavelength exhibits blue-shift with decrease of the QWDs layer number (Figure 2). Compare to the broad-area lasers the observed blue-shift is larger due to the higher loss in 30 μm MD lasers. Evidently transitions via excited states of QWDs start to play significant role in the lasing for devices with 2 QWDs layers. The threshold current density dependence on MD laser diameters obtained for structures with 2, 5 and 10 QWDs layers are shown at Figure 5. The lowest threshold current densities for all diameters (10-30 μm) are obtained for microlasers with 5 QWDs layers. Though decrease of the QWDs layers number shall result in reduced transparency current we observe in opposite slight increase of the threshold current in device of the largest diameter (30 μm) with 2 QWDs layers. This observation correlates with lasing wavelength blue shift caused by filling of the QWDs excited states. 2 QWDs lasers operate down to 18 μm diameters. The devices of the smaller diameter do not come to lasing. Increase of the number of QWDs layers to 10 results in the lasing at longest wavelength (1115 nm) and increase of the threshold current due to higher absorption loss in active region. The threshold current keeps at nearly the same level (~2 kA/cm^2) for diameters 30, 26, 22 μm and then rapidly increases to ~4.8 kA/cm^2 for 18 μm in diameter device.

To conclude, lasing parameters of broad-area lasers and microdisk laser with 2, 5 and 10 InGaAs QWDs layers were studied under cw, RT conditions. It was found that the optimal number of QWDs layers is 5 for MD lasers in proposed geometry of vertical resonator formed by epitaxial layers. The gain provided by ground state of 2 QWDs layers is not enough to overcome loss in 10-30 μm MD diameter range and wavelength shift towards excited state is observed. The use of larger number of active layers helps to achieve larger emission wavelength in spectral range suitable for silicon photonics ($\lambda > 1.1$ μm). At the same time vertical waveguide structure should be optimized to realize efficient overlap of fundamental optical mode in resonator with all layer of active region. Another way to decrease threshold current in MD lasers is to use ring-type metal p-contact to reduce useless pumping of the inner MD region, since WGMs are supported at the microlaser's boundary.

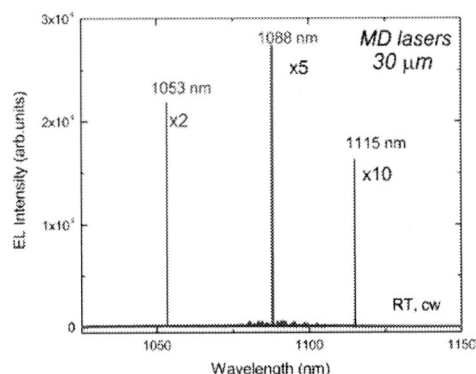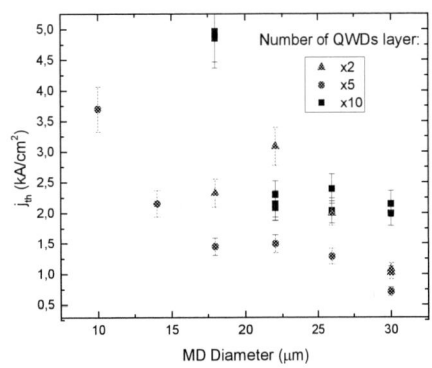

Figure 4. Electroluminescence spectra obtained in 30 μm in diameter devices with 2, 5 and 10 QWDs layers at room temperature.

Figure 5. Dependence of threshold current density on MD laser diameters for structures with 2, 5 and 10 QWDs layers

Acknowledgments
The work is supported by Russian Science Foundation (project № 18-12-00287).

References

[1] S.A.Blokhin, N.A. Maleev, M.A. Bobrov, A. G. Kuzmenkovm, A. V. Sakharov, V. M. Ustinov, High-Speed Semiconductor Vertical-Cavity Surface-Emitting Lasers for Optical Data-Transmission Systems (Review), Tech. Phys. Lett. (2018) 44: 1.

[2] K. Ding, M. T. Hill, Z. C. Liu, L. J. Yin, P. J. van Veldhoven, and C. Z. Ning, Record performance of electrical injection sub-wavelength metallic-cavity semiconductor lasers at room temperature // Optics Express. 2013. Vol. 21. 4728-4733.

[3] S.Matsuo, A.Shinya, T.Kakitsuka, K.Nozaki, T. Segawa, T.Sato, Y. Kawaguchi, M.Notomi High-speed ultracompact buried heterostructure photonic-crystal laser with 13 fJ of energy consumed per bit transmitted» // Nature Photonics. 2010. Vol. 4. 648 – 654

[4] M.V.Maximov, N.V. Kryzhanovskaya, A.M.Nadtochiy, E.I. Moiseev, I.I. Shostak, A.A. Bogdanov, Z.F.Sadrieva, A.E.Zhukov, A.A. Lipovskii, D.V. Karpov, J. Laukkanen, J. Tommila, Ultrasmall microdisk and microring lasers based on InAs/InGaAs/GaAs quantum dots, Nanoscale Research Letters 9 (2014) 657

[5] Slusher R.E., Levi A.F.J., Mohideen U., McCall S.L., Pearton S.J., Logan R.A. Threshold characteristics of semiconductor microdisk lasers // Applied Physics Letters. 1993. Vol. 63. 1310.

[6] E.Moiseev, N.Kryzhanovskaya, M.Maximov, F.Zubov, A.Nadtochiy, M.Kuilagina, N.Kalyuzhnyy, S. Mintairov, A.Zhukov Highly efficient injection microdisk lasers based on quantum well-dots, Optics letters, accepted

[7] A M Nadtochiy, M V Maximov, S A Mintairov, N A Kalyuzhnyy, S Rouvimov, A E Zhukov InGaAs/GaAs hybrid quantum well-dot nanostructures: Impact of substrate orientation and recombination mechanisms, J. Phys.: Conf. Ser., 917, 032001 (2017).

[8] N.V. Kryzhanovskaya, M.V.Maximov, A.E. Zhukov, A.M. Nadtochiy, E.I. Moiseev, I.I. Shostak, M.M. Kulagina, K.A. Vashanova, Y.M. Zadiranov, S.I. Troshkov, V.V. Nevedomsky, S.A. Ruvimov, A.A. Lipovskii, N.A. Kalyuzhnyy, S.A. Mintairov, Single-Mode Emission from 4-9-μm Microdisk Lasers with Dense Array of InGaAs Quantum Dots, IEEE J. Lightwave Technol 33(1), 171-175 (2015).

Optical properties of In$_{0.8}$Ga$_{0.2}$As quantum dots in GaAs photovoltaic convertors

R A Salii[1], V V Evstropov[1], S A Mintairov[1], M A Mintairov[1], M Z Shvarts[1] and N A Kalyuzhnyy[1]

[1] Ioffe Institute, St.-Petersburg, Russia, Polytechnicheskaya str. 26

E-mail: r.saliy@mail.ioffe.ru

Abstract. In the work, GaAs photovoltaic convertor structures with embedded In$_{0.8}$Ga$_{0.2}$As arrays of quantum dots (QDs) have been obtained by metalorganic vapor-phase epitaxy. Spectral characteristic of internal quantum yield showed two spectral peaks at 930 and 985 nm of photosensitivity beyond the GaAs absorption edge. A theoretical calculation of the band transitions in the nanoheterostructure showed that the peak at 930 nm corresponds to the electron–heavy-hole band transition in the wetting layer of QDs, while the long-wavelength peak at 985 nm corresponds to InGaAs QDs. Contributions of both QDs and wetting layer to the total photocurrent of the devices with the different number of QDs have been calculated.

1. Introduction

Photovoltaic converters (PVCs) based on AIIIBV semiconductor heterostructures are ones of the most effective solutions among renewable energy sources. However, today, in this field of photovoltaics, there are still a number of unresolved problems related to various fundamental loss mechanisms that prevent the efficiency of such PVCs from approaching the theoretical limit [1]. At the present time, one of the most successful approaches allowing to significantly reduce the fundamental losses are multijunction solar cells [2], in particular, based on the lattice-matched InGaP/GaAs/Ge material system. In such PVCs, the efficiency increases due to absorbing radiation of different parts of the solar spectrum by different subcells of a multijunction structure.

However, multijunction PVCs still have several unsolved issues such as current mismatch between their subcells due to low photogenerated current of the middle GaAs junction. It reduces total current of the whole structure and limits the maximum achievable efficiency of a PVC. One of the ways to solve the problem is the embedding the quantum-size objects in GaAs to create a subbandgap, what allows absorbing radiation of an additional part of the solar spectrum increasing the photogenerated current of the GaAs subcell. Also, in the last two decades, the models describing one-junction based PVC with an intermediate band (IB) is actively developing. Theoretically, IB PVCs allow absorbing radiation of a wide solar spectrum due to multiphoton absorption processes in semiconductor structures based on a single p-n junction [4]. For practical realization of both multijunction PVCs with current balance and IB PVC, the arrays of InAs quantum dots (QDs) embedded in GaAs [5] can be proposed.

The aim of this work is to obtain GaAs PVCs with embedded In$_{0.8}$Ga$_{0.2}$As QD arrays, which have better relaxation parameters in the GaAs matrix compared with classic InAs QDs. The electro-optical properties of In$_{0.8}$Ga$_{0.2}$As QDs are investigated, and their contribution to GaAs PVC photocurrent is determined.

Content from this work may be used under the terms of the Creative Commons Attribution 3.0 licence. Any further distribution of this work must maintain attribution to the author(s) and the title of the work, journal citation and DOI.

Published under licence by IOP Publishing Ltd

2. Experimental procedure

All structures were grown by metalorganic vapor-phase epitaxy (MOVPE) on a R&D installation with a horizontal type reactor at low pressure (100mbar). A series of structures of single-junction GaAs PVCs differed by the number of embedded QD layers in a wide range from 1 to 20 layers. The schematic structure of experimental samples is shown in Fig.1. The growth carried out according to the following technological process: after growing the $Al_{0.3}Ga_{0.7}As$ rear potential barrier, the n-GaAs base and a part of the GaAs i-region were grown at 700°C, and then the reactor was cooled down to QD material deposition temperature of 520°C. Then, the $In_{0.2}Ga_{0.8}As$ material was deposited for nucleation of coherent islands in the Stranski-Krastanov mode [6] with forming@ a wetting layer (WL). After that, the QDs were covered with GaAs cap layer at the same temperature in order to protect them from degradation during the subsequent reactor heating to the GaAs spacer layer growth temperature (600°C). The GaAs spacer thickness was 35 nm. The process repeated according to the number of embedded QD layers, and then the rest of the i-GaAs and p-GaAs emitter, $Al_{0.8}Ga_{0.2}As$ "window" and n+-GaAs contact layer were grown at the temperature of 700°C. The amount of the $In_{0.8}Ga_{0.2}As$ material for QDs formation was chosen according to the previous study, which was devoted to determination of the maximum photoluminescence (PL) intensity of $In_{0.8}Ga_{0.2}As$/GaAs heterostructures and was 2 ML (~6Å) [7]. Also a reference structure without QD layers (reference GaAs PVC) was created.

PVCs based on grown structures were created by expedited procedure due to forming metallic front and rear contacts and subsequent etching the p+-GaAs contact layer from the photoactive surface.

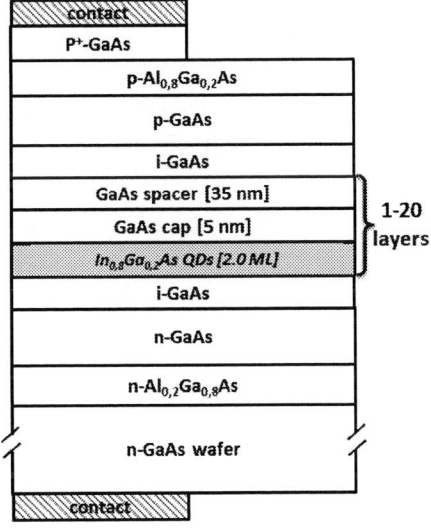

Figure 1. GaAs PVC experimental structure

A multifunctional installation has been used for measuring the internal quantum efficiency (IQE) of the PVCs. The measuring hardware included an unblocked ultra-violet halogen light source, a grating monochromator with 2 nm/mm dispersion within 300–1200 nm wavelength scanning range, and an optical chopper of 90 Hz and high sensitivity lock-in electronics. The IQE measurement was controlled by a laptop. The lock-in technique allows weakling output signals at the QD sensitivity range to be detected confidently.

3. Results and discussion

Fig.2a shows the spectral characteristics of IQE of a PVC with QD arrays in comparison with GaAs PVC reference sample. Embedding the $In_{0.8}Ga_{0.2}As$ QD layer into aGaAs PVC does not lead to the IQE drop in the GaAs absorption range up to 20 InGaAs QD layers. This indicates an improvement in the quality of the active region with the InGaAs QDs due to better relaxation of the medium in comparison with InAs QDs grown by MOVPE [8, 9].

In the long-wavelength part of the spectrum beyond the GaAs absorption edge (in the range of 880 – 1100 nm) a sharp peak at 930 nm and a long-wavelength peak in the region of ~985 nm have been observed (Fig. 2 b).

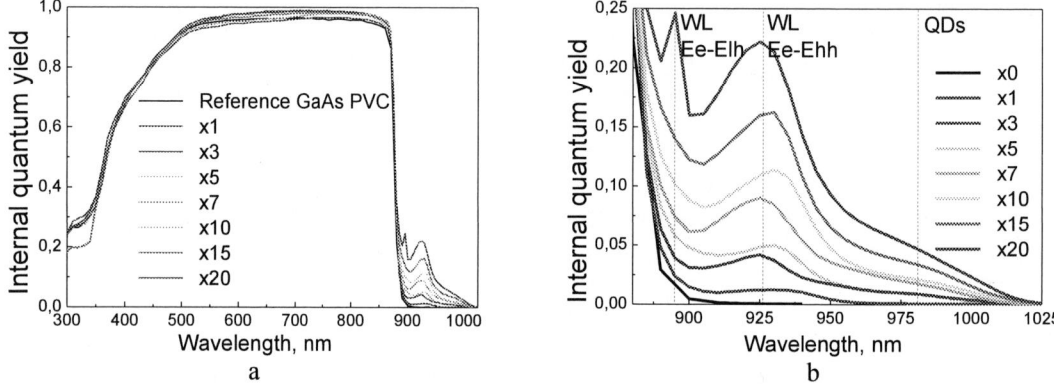

a   b

Figure 2. Spectral characteristics of the internal quantum yield of a GaAs PVC with embedded $In_{0.8}Ga_{0.2}As$ QD arrays, a – whole spectra, b – scaled part of QDs contribution.

To identify spectral peaks, a model of QW was suggested for describing WL. Theoretical calculation of energy transitions for InGaAs/GaAs QW based on model-solid theory [10] was carried out. According to the model calculation, the position of the spectral peak at 930nm (1.33eV) corresponds to the electron–heavy-hole (Ee-Ehh) transition in the $In_{0.8}Ga_{0.2}As$ wetting layer (Fig.2b "WL Ee-Ehh"). The upper estimate of the WL thickness is 5.2Å (Fig.3). For such a thickness, the energy of electron–light-hole (Ee-Elh) transition is 1.39eV (892 nm) (Fig.2b "WL Ee-Elh"), so it is not observed on the spectral characteristic due to overlapping by the blurred GaAs absorption edge [11] in the range of 870 – 900 nm. Thus, the spectral peak at ~985 nm obviously corresponds to the contribution of QDs (Fig. 2 b "QDs").

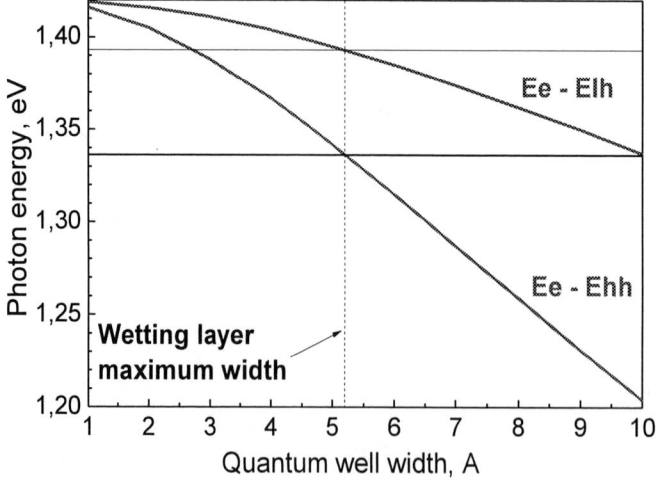

Figure 3. Theoretical calculation of energy transitions in a InGaAs/GaAs QW.

Fig.4 shows a linear increase of the maximum internal quantum yield (IQY) values for both WL peak (Ee-Ehh transition) and QD peak with increasing the quantity of QD arrays. The obtained data correlate with both current contribution of QDs and the total current contribution of QDs and WL versus a number of embedded QD arrays for AM0 and AM1.5 solar spectra after normalizing to one layer (Fig.5). These contributions were calculated separately according to the data obtained in analyzing the spectral characteristics and classification of the spectral peaks described above. The total maximum WL and QD contribution to the photocurrent of a PVC with 20 QD layers in the active region was 1.2 mA/cm^2 and 1.03 mA/cm^2 for the AM0 and AM1.5 spectra, respectively. The contribution of QDs remains constant (~0.006 mA/cm^2) with increasing the number of embedded QD layers in the PVC structure (Fig.5, blue and magenta curves). Total contribution of WL and QDs normalized to one QD layer slightly decreased from 0.06mA/cm^2 (structure with 1 and 3 QD layers) to 0.05 mA/ cm^2 (structures with 5 - 20 QD layers) for the AM0 spectra, and from 0.05 mA/cm^2 (structure with 1 and 3 QD layers) to 0.04 mA/cm^2 (structures with 5 - 20 QD layers) for the AM1.5 spectra (Fig.5, black and red curves). Thus, the total contribution of QD and WL decreases with increasing the number of QD layers up to 5 and then practically does not change with the QD layer increase up to 20.

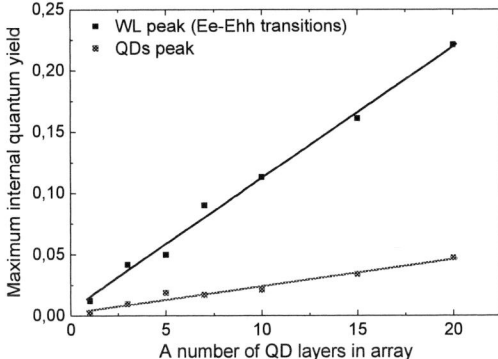

Figure 4. Maximum values of IQY for WL peak at 930 nm (black line) and QDs peak at 965 nm (red lone).

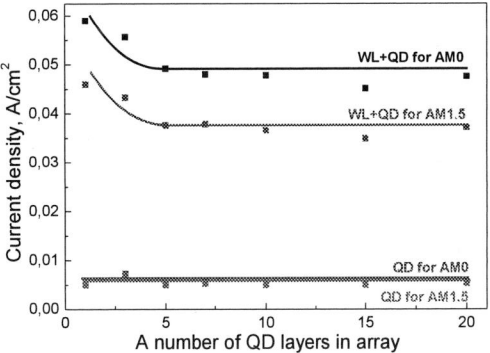

Figure 5. Total photocurrent contribution per one layer for WL and QDs (black and red curves for AM0 and AM1.5 spectra, respectively) and QDs photocurrent contribution (blue and magenta lines for AM0 and AM1.5 spectra, respectively).

This indicates that the quality of the QD array remains practically unchanged, what correlates with the data on the quality of the GaAs p-n junction with embedded QD arrays discussed above.

Conclusions

In the work, an experimental analysis of In$_{0.8}$Ga$_{0.2}$As QD arrays formation in the GaAs matrix by the MOVPE in the Stranski-Krastanov mode was carried out. The internal quantum yield of the reference GaAs PVC, as well as the PVCs with embedded 1, 3, 5, 7, 10, 15 and 20 layers of InGaAs QDs was measured. It has been shown that the p-n junction quality is practically the same even for 20 QD layers embedded in the PVC structure. It has been demonstrated that the main contribution to the photogenerated current of a PVC is done by the WL of the QDs, as shown through classification of the spectral peaks of IQY by theoretical calculations of the energy transitions by the InGaAs/GaAs quantum well model. It is established that, with increasing the number of embedded QD layers, their contribution to the photocurrent became constant after staking more than 5 QD layers up to 20 QD layers in an array. The total maximum WL and QD contribution to the photocurrent of a PVC with 20 QD layers in the active region was 1.20 mA/cm^2 and 1.03 mA/cm^2 for the AM0 and AM1.5 spectra, respectively. The developed MOVPE for growing $In_{0.8}Ga_{0.2}As$ QD arrays is promising for both IB SCs and multijunction SCs with improved current balance.

Acknowledgements
The work has been supported by the Russian Foundation for Basic Research (grant No. 18-08-01281). The authors wish to thank N. Kh. Timoshina for the SC spectral dependence measurements.

References
[1] Andreev V M, Grilikhes V A and Rumyantsev V D 1997 *Photovoltanc Conversion of Concentrated Sunlight* (John Willey & Sons Ltd)
[2] Nishioka K, Takamoto T, Agui T, Kaneiwa M, Uraoka Y And Fuyuki T 2004 *Jpn. J. Appl. Phys.* **43** pp 882–889
[3] Sasaki K, Agui T, Nakaido K, Takahashi N, Onitsuka R, and Takamoto T 2013 *AIP Conference Proceedings* **1556** pp 22–25
[4] Luque A and Marti A 1997 *Phys. Rev. Lett.* **78** pp 5014–5017
[5] Martí A, Cuadra L and Luque A 2002 Quasi-drift diffusion model for the quantum dot intermediate band solar cell transactions on electron devices **49** pp 1632–1639
[6] Stranski I N and von Krastanov L 1939 (Akad. Wiss. Lit. Mainz Abh. Math. Naturwiss. Kl.)
[7] Salii R A et al. 2017 *Proc. Int. Conf. Physica.SPb/2017* (Saint-Petersburg) http://physica.spb.ru/data/uploads/physica2017thesises.html
[8] Bailey C G, Forbes D V, Polly S J, Bittner Z S, Dai Y, Mackos C, Raffaelle R P and Hubbard S M 2012 *Journal of Photovoltaics* **2** pp 269 – 275
[9] Kalyuzhnyy N A, Mintairov S A, Salii R A, Nadtochiy A M, Payusov A S, Brunkov P N, Nevedomsky V N, Shvarts M Z, Martí A, Andreev V M and Luque A 2016 *Progress in Photovoltaics* **24** pp 1261–1271
[10] Van de Walle C G 1989 *Phys. Rev.* **39** pp 1871–1883
[11] Moss T S, Burrell G J and Ellis B 1973 *Semiconductor Opto-Electronics* (Butterworth & Co.)

SPBOPEN 2018

IOP Publishing

IOP Conf. Series: Journal of Physics: Conf. Series **1124** (2018) 041004   doi:10.1088/1742-6596/1124/4/041004

Features of direct digital synthesis applications for microwave excitation signal formation in quantum frequency standard on the atoms of cesium

A A Petrov[1,2,3], V V Davydov[2], D V Zalyotov[1], V E Shabanov[1], D V Shapovalov[4]

[1]Russian Institute of Radionavigation and Time, St. Petersburg, 192012, Russia
[2]Higher School of applied physics and space technologies, Peter the Great Saint Petersburg Polytechnic University, St. Petersburg, 195251, Russia
[3]Institute for Analytical Instrumentation Russian Academy of Sciences, St. Petersburg, 198095, Russia
[4]«Information satellite systems» named after academician M. F. Reshetnev», Zheleznogorsk, Russia

Abstract. Features of direct digital synthesis applications in quantum frequency standard on the atoms of cesium are considered. A new design of a frequency synthesizer based on direct digital synthesis method and magnetic field stabilization system are presented. Experimental research of frequency synthesizer showed improvement parameters of a microwave excitation signal, such as step frequency tuning, time frequency tuning, range of generating frequencies and spectral characteristics. Magnetic field stabilization system eliminates one of the most important perturbing factors affecting on long-term frequency stability. Experimental research of metrological characteristic of quantum frequency standard showed improvement long-term frequency stability by 15%.

1. Introduction

The accurate measurement of time and frequency is vital to the success of many fields of science and technology [1-8]. For examples success in atomic physics (atom-photon interactions, atomic collisions, and atomic interactions with static and dynamic electromagnetic fields), geodesy, radio-astronomy (very long baseline interferometry) and pulsar astronomy depends on very high stable output signal of frequency standards and uniform timescales [5-12]. The same is valid for the operation of satellite-based navigation systems, metrological services and telecommunication systems. [1, 2, 4, 5, 12-14]

Quantum frequency standards on the atoms of cesium (also named cesium atomic clock) also are used as a clock generators in the communications equipment and in a data transmission devices, applied in the satellite navigation systems GLONASS and GPS as a clock generators and in the various metrological services [12-14]. Also these standards perform a role of the reference signals with high precision and stability in radio equipment. [1, 14-17]

Considering the high importance of precision atomic clocks in science and technology and a vast area of their application, modernization of existing and development new quantum frequency standards are the urgent tasks [1, 14-17].

The process of frequency standards modernization includes various directions: reducing energy consumption, reducing weight and dimensions, improving metrological characteristics. And in present

Content from this work may be used under the terms of the Creative Commons Attribution 3.0 licence. Any further distribution of this work must maintain attribution to the author(s) and the title of the work, journal citation and DOI.
Published under licence by IOP Publishing Ltd

work some directions related to direct digital synthesis applications of cesium atomic clock modernization for improvement its metrological characteristics are considered.

2. Principles of cesium atomic clock operation

Work of a cesium atomic clock is based on the principle of adjustment a highly stable voltage-controlled quartz crystal oscillator (VCXO) to quantum frequency transition of atoms of caesium-133 [1, 5, 12, 16]. Figure 1 shows a block diagram of a cesium atomic clock.

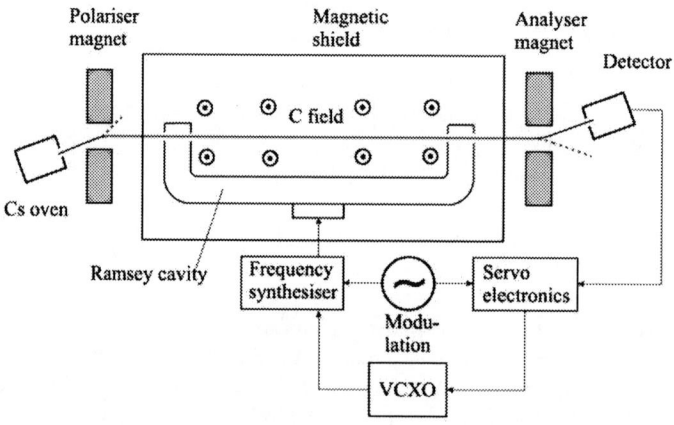

Figure 1 Block diagram of a cesium atomic clock.

Output signal frequency 5 MHz of the VCXO is supplied to the frequency synthesizer. Frequency synthesizer consists of frequency converter, mixer signals and multiplier signals. In the frequency converter input signal frequency 5 MHz is converted to the signal frequency 12,631772 MHz and supplied to the input of mixer signals. In the multiplier signals input signal frequency 5 MHz is multiplied to the frequency 270 MHz and then to frequency 9180 MHz. This signal frequency 9180 MHz is also supplied to the input of mixer signals. As a result, the output signal of the frequency synthesizer is the signal of ultrahigh frequency 9192,631772 MHz. This signal is supplied in the Ramsey cavity.

In cesium atomic clock with the help of magnet polarizer the atoms are prepared such that they are either in the |F=4, mF=0> or in |F=3, mF=0> state. Afterwards the atoms interact with an electromagnetic field that induces transitions into the former unoccupied level.

A uniform magnetic field is used to separate energetically the otherwise degenerate magnetic sub-levels in order to allow the excitation of the clock transition |F = 3, mF = 0> → |F = 4, mF = 0> isolated from the other transitions. By convention such a field is referred to as the C-field as it is applied between the fields of the polarizer and the analyzer.

The magnitude of the C-field is chosen as a compromise between two conflicting requirements. First, it has to be large enough to separate the otherwise overlapping resonances. Second, the C-field shifts the resonance frequency quadratically which has to be corrected. In the scheme of a commercial Cs clock the C-field is often generated by a coil with windings around the Ramsey resonator. Owing to the dependence of the frequency of the clock transition from the magnetic field, efficient magnetic shielding has to be provided in order to attenuate the ambient magnetic field and the magnitude of the associated fluctuations.

The atoms in the former unoccupied state are detected and allow one to determine the frequency of the interrogating field where the transition probability has a maximum. The observed transition frequency is corrected for all known frequency offsets that would shift the transition frequency from the unperturbed transition and is used to produce a standard frequency or pulse per second every 9192631772 cycles [1, 5, 12, 16].

Scanning the frequency ν of the atomic resonance leads to a detector current like the one shown on the figure 2. The signal shows the Ramsey resonance structure on a broader, so-called, Rabi pedestal.

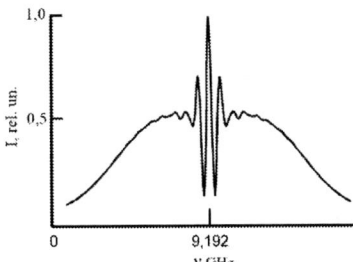

Figure 2 Ramsey resonance structure on the Rabi pedestal.

The central resonance with the frequency transition ν_0 is used to stabilize the frequency of the crystal oscillator to the atomic transition frequency. To this end, the output synthesizer's frequency is modulated. The signal from the detector is detected, integrated and used for changing the voltage of the VCXO proportional to the frequency. From this suitable output frequencies are derived, such as 5MHz or a 1 PPS signal.

3. New design of frequency synthesizer

Frequency synthesizer is one of the main blocks of quantum frequency standard. Frequency synthesizer takes a part of generating the microwave signal at ~9.2 GHz (used to interrogate the ^{133}Cs atoms hyperfine resonance transition) from the 5 MHz quartz oscillator frequency [1, 16, 17].

The main characteristic of the frequency synthesizer is ability to impact on the characteristic of frequency stability of the quantum frequency standard output signal. Frequency instability introduced by the synthesizer is determined by the lateral discrete spectrum components of the signal that occurs in dividing, multiplying, mixing frequency signals, the accuracy of the generated frequency, and the impact on the signal of natural and technical noise.

In order to provide the best possible frequency stability, it is crucial that the microwave signal which interrogates the ^{133}Cs atoms be as "clean" as possible; that is, free of unwanted sidebands and spurious signals which can cause Bloch-Siegert frequency shifts.

Experimental study showed that the present method of generating the frequency synthesizer output signal needs to increase the accuracy. The large resolution of step frequency is necessary. New scheme of the frequency synthesizer is designed by using method of direct digital synthesis (DDS - Direct Digital Synthesis). This method allows to generate the output signal of the synthesizer with accuracy about 10^{-5} Hz. Step frequency tuning ΔF_{out} calculated by the formula below:

$$\Delta F_{out} = \frac{F_{clk}}{2^N},$$ (1)

where F_{clk} is the clock frequency, N is the capacity of accumulator.

In our scheme the clock frequency is equal F_{clk}=15 MHz, the capacity of accumulator is equal N=40. Step frequency tuning is equal $\Delta F_{out} = 1,36 \cdot 10^{-5}$ Hz.

When we calculate step frequency tuning relatively resonant frequency of the microwave transitions of cesium atoms, using parity (1), we get relative step frequency tuning $\Delta F_{out\,rel}$:

$$\Delta F_{out\,rel} = \frac{\Delta F_{out}}{9.192\;GHz} = \frac{1.36^{-5}\;Hz}{9.192\;GHz} = 1.48 * 10^{-15} \tag{2}$$

The application of direct digital synthesis gave the possibility of obtaining the generated frequencies in a wide range (~1 MHz), in contrast to previous schemes, where this feature was absent. This feature gets a possible to develop a magnetic field stabilization system. Range of generated output frequencies may be calculated by the formula below:

$$F_{out} = \frac{M*F_{clk}}{2^N}, \tag{3}$$

where M is the frequency code in decimal, F_{clk} is the clock frequency, N is the capacity of accumulator.

To meet the requirements for spectral purity of output signal 10 bit DAC was used. It is possible to obtain the suppression of lateral amplitude components in the spectrum of the output signal is not worse than - 90 dB.

In figure 3, as an example, oscillograms measured in the band of 6 kHz of the output signal of a previously used design (a) and a new (b) of the frequency synthesizer are presented.

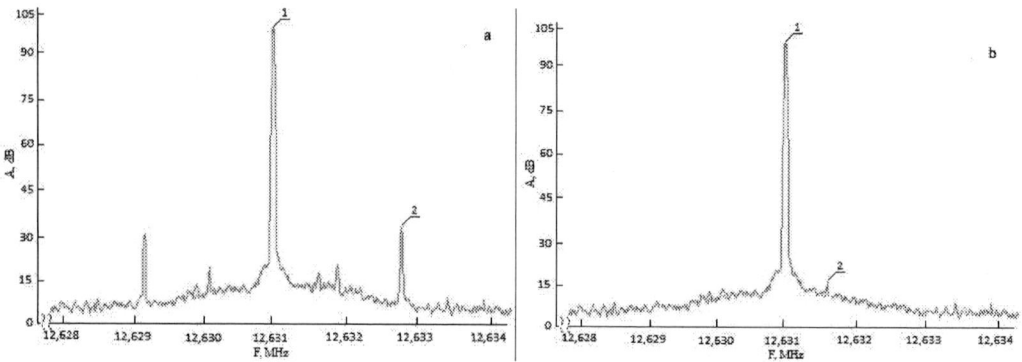

Figure 3 Suppression of the lateral components in the band of 6 kHz.

The experimental results show that the suppression of lateral components in the spectrum of microwave-excitation signal in the band of 6 kHz is improved on 24 dB.

With decrease of lateral components more fine-tuning on the center of the resonance line is occur. This leads to a more accurate determination of the value of the nominal output frequency of frequency standard.

One of the differences between the design of the frequency synthesizer developed by using DDS from the previously used is the absence of a quartz filter in it. Quartz filter with good frequency selectivity has a high temperature dependence of the output frequency. In the new design of frequency synthesizer quartz filter is absence. At the same time, the level of suppression of combinational components due to the use of the new method remained at the same level, and the temperature-dependent elements became less. This allows to improve the temperature coefficient of frequency (TCF) of the quantum frequency standard.

An experimental study of quantum frequency standard with new design of frequency synthesizer based on DDS showed an improvement in the TCF by to 2.4 times compared to the previously used designs of frequency synthesizer.

In addition new design of the frequency synthesizer allows eliminating one of the most important perturbing factors affecting on long-term frequency stability.

4. Magnetic field stabilization system

The stable isotope Cs-133 has a two hyperfine states F = 4 and F = 3 which are split in the magnetic field into 16 components. In accordance with the selection rules seven transitions between the components of hyperfine sublevels are possible [1, 16, 17]. These are represented in figure 4.

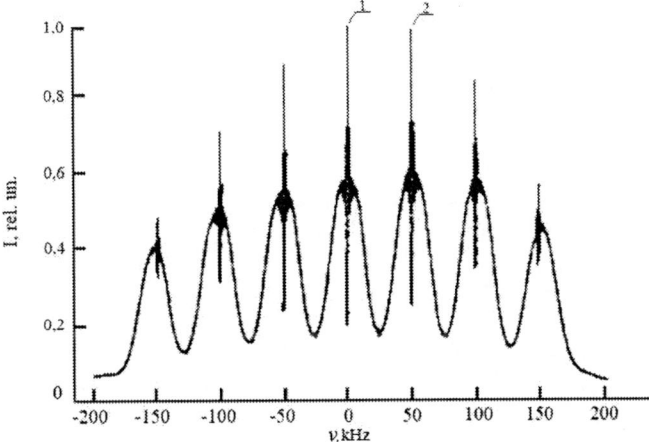

Figure 4 Microwave resonances in the Ramsey cavity.

The central resonance $|F = 3, m_F = 0\rangle \rightarrow |F = 4, m_F = 0\rangle$ (marked on the figure 4 as a «1») due to the Zeeman effect expose a quadratic frequency shift. With the help of formula (1) we can calculate a frequency shift.

$$\Delta f_{B^2} \approx 4.2745 * 10^{-2} \text{ Hz} * \left(\frac{6*10^{-6}\text{T}}{\text{uT}}\right)^2 = 1{,}5388 \text{ Hz} \qquad (4)$$

For a typical value of the C field near 8 μT the frequency shift is 2.7 Hz corresponding to a relative frequency shift of $3 \cdot 10^{-10}$.

The accuracy of the output signal quantum frequency standard is dependent on the shift of the central resonance. It should be noted that not only the central resonance is exposed the frequency shift, but also all six transitions $(3, m_F) \leftrightarrow (4, m_F)$, which $\Delta m_F = 0$. To express these changes as a function of magnetic induction B and atomic constants use the equation Bright Rabi:

$$E(F, m_F) = -\frac{h\nu}{2(2I+1)} - g_I\mu_B Bm_F + \varepsilon\frac{h\nu}{2}\left(1 + \frac{4m_F}{2I+1}x + x^2\right)^{\frac{1}{2}}, \qquad (5)$$

where $E(F, m_F)$ is the change energy of atoms in the ground state; I is the quantum number of nuclear spin; g_I is the factor Lande for electron; μ_B is the Bohr magneton; m_F is the magnetic quantum number; $x = \frac{(g_j+g_I)\mu_B B}{h\nu}$.

This formula can be used for calculation the frequency shift of any transition between two hyperfine sublevels, depending on the magnetic field. Revealing this expression, we find that the first order member is proportional to the magnetic field B. For cesium beam primary frequency standards we must consider the quadratic order member of this expression.

In theory, the frequency shifts can be taken into account in the calculation of the functional dependence on magnetic field values and the atomic constants using equation Bright-Rabi. But in practice, any changes of the magnetic field shift the resonance frequency. And values of these frequency shifts cannot be accounted for in advance.

Thanks to the development of a new frequency synthesizer the range of generated output frequencies has been expanded. It allowed detuning output synthesizer's frequency to the neighboring resonance frequency of spectral line that makes it possible to adjust the C-field in quantum frequency standard.

Now in cesium atomic clock the magnetic field is maintained by the active stabilization system. For this purpose the neighboring transition $|F =3, m_F=1> \leftrightarrow |F =4, m_F=1>$ (marked on the figure 4 as a «2») is used. The method of C-field adjustment is similar to the method of frequency adjustment to the main maximum. For this purpose, the average value of the sampling frequency v_i is changed from the value of the v_{Cs} to $v_{Cs} + \Delta v$, where v_{Cs} is the frequency of the main transition of the cesium atom, Δv - difference between the transitions for a preset value of the magnetic field. Then the value of magnetic field is adjusted in a such way that the frequency of transition $|F =3, m_F=1> \leftrightarrow |F =4, m_F=1>$ match the preset frequency. This adjustment set up automatically several times per minute. The value of the field is automatically maintained at a predetermined level.

Alternately closing the ring-locked loop at the central and the neighboring transition we adjust the frequency of the VCXO to the frequency of the central atomic transition, and support the constant value of the magnetic field inside the Ramsey cavity.

In this case effects associated with any changes in the magnetic field (for example, long-term drift of the current source, temperature dependence, effect of external magnetic field, etc.) are excluded.

5. Conclusions

Application the DDS in quantum frequency standard on the atoms of cesium allowed to improve the output characteristics of frequency synthesizer and also to develop magnetic field stabilization system. The use of new design of the frequency synthesizer and system for stabilizing magnetic field in quantum frequency standards construction allow to improve frequency stability of quantum frequency standard.

Experimental research of the metrological characteristics of the modernizing quantum frequency standard on the atoms of cesium - 133 showed improvement in the TCF by to 2.4 times and long-term frequency stability on 15 %.

References

[1] Riehle F 2004 *Frequency standards. Basics and Applications* (New Jersy: Wiley-VCH)
[2] Oduan K and Gino B 2002 *Chronometry and basics of GPS.* 400
[3] Semenov V V, Nikiforov N F, Yermak S V and Davydov V V 1991 *Soviet journal of Communications Technology and Electronics* **36** 59 – 63
[4] Glazov A I., Grigor'ev V V, Kravtsov V E, Mityurev A K, Svetlichnyi A B, Savkin K B and Tikhomirov S V 2008 *Measurement Techniques* **51** 1064 – 1070
[5] Kolmogorov O V, Shchipunov A N, Prokhorov D V, Donchenko S S, Buev S G, Malimon A N, Balaev R I and Fedorova D M 2017 *Measurement Techniques* **60** 901 – 905
[6] Davydov V V, Kruzhalov S V, Grebenikova N M and Smirnov K J 2018 *Measurement Techniques* **61** 365–372
[7] Petrov A A and Davydov V V 2015 *Lecture Notes in Computer Science (including subseries Lecture Notes in Artificial Intelligence and Lecture Notes in Bioinformatics)* **9247** 739-744

[8] Davydov V V, Ermak S N, Karseev A U, Nepomnyashchay H K, Petrov A A and Velichko E N 2014 *Lecture Notes in Computer Science (including subseries Lecture Notes in Artificial Intelligence and Lecture Notes in Bioinformatics)* **8638 LNCS** 694-702

[9] Davydov V V, Sharova N V, Fedorova E N, Gilshteyn E P, Malanin K Y, Fedotov I V, Vologdin V A and Karseev A Yu 2015 *Lecture Notes in Computer Science (including subseries Lecture Notes in Artificial Intelligence and Lecture Notes in Bioinformatics)* **9247** 712-721

[10] Ermolaev A N, Krishpents G P, Davydov V V and Vysoczkiy M G 2016 *Journal of Physics: Conference Series* **741(1)** 012171

[11] Logunov S E, Koshkin A Yu, Davydov V V and Petrov A A 2016 *Journal of Physics: Conference Series* **741(1)** 012092

[12] Petrov A A, Davydov V V, Myazin N S and Kaganovskiy V E 2017 *Lecture Notes in Computer Science (including subseries Lecture Notes in Artificial Intelligence and Lecture Notes in Bioinformatics)* **10531 LNCS** 561-568

[13] Petrov A A, Vologdin V A, Davydov V V and Zalyotov D V 2015 *Journal of Physics: Conference Series* **643(1)** 012087

[14] Balaev R I, Malimon A N, Fedorova D M, Kurchanov A F and Troyan V I 2017 *Measurement Techniques* **60** 806 – 812

[15] Pakhomov A A 2017 *Journal of Communications Technology and Electronics* **52** 1114 – 1118

[16] Petrov A A and Davydov V V 2017 *Journal of Communications Technology and Electronics* **62** 289 – 293

[17] Petrov A A, Grebenikova N M, Lukashev N A, Davydov V V, Ivanova N V, Rodygina N S and Moroz A V 2018 *Journal of Physics: Conference Series* 1038 (1) 012032

Effect of carrier localization on performance of coupled large optical cavity diode lasers

A A Serin[1,2], A S Payusov[2], Yu M Shernyakov[2], N A Kalyuzhnyy[2], S A Mintairov[2], N Yu Gordeev[2], F I Zubov[1,3], M V Maximov[1,3] and A E Zhukov[1,3]

[1]St Petersburg Academic University, 194021, St Petersburg, Russia
[2]Ioffe Institute, 194021, St Petersburg, Russia
[3] ITMO University, 197101 St. Petersburg, Russia

Abstract. Results are presented on a comparative study of InGaAs/GaAs/AlGaAs coupled large optical cavity (CLOC) laser heterostructures emitting at either 1.03 or 0.98 μm. The smaller energy of carrier localization in the active region of the latter structure leads to stronger carrier leakage out of the quantum well that in its turn results in a higher internal optical loss, higher threshold current density and more pronounced temperature sensitivity. In continuous wave regime, the maximal output power of 8 W in the 0.98-μm lasers is found to be limited by the thermal rollover, whereas in the longer wavelength laser it is 10.8 W being limited by the catastrophic optical mirror damage.

1. Introduction

A broad optical waveguide design [1] is widely exploited in diode lasers because an extension of the vertical mode profile is favorable for narrowing the laser beam divergence and decreasing optical loss caused by free carrier absorption in highly doped claddings [2, 3]. Additionally, the absolute level of output optical power at the catastrophic optical mirror damage increases as the mode size expands [4]. Still another advantage of the broad waveguide is that the mode size is almost completely controlled by the waveguide thickness rather than the refraction indexes and therefore temperature- and injection-induced variations of the indexes do not affect the mode [5]. However in broad waveguide lasers extended distance between the laser active region and the *p*-contact/heasink interface may enhance electrical and thermal resistances. Moreover, as soon as the waveguide thickness exceeds a certain cutoff value, higher-order transverse modes appear in the waveguide. This may cause high-order mode or multimode lasing, which deteriorates the vertical far-field pattern and the laser beam quality. For a simple rectangular index profile, the cutoff thickness is given as

$$d_{\text{cutoff}} = \frac{\lambda}{2\sqrt{n_{\text{WG}}^2 - n_{\text{clad}}^2}}, \tag{1}$$

where λ is the wavelength in free space, n_{WG} and n_{clad} are indices of refraction of the waveguiding and cladding layers, respectively. For near-infrared (Al)GaAs laser diodes, Eq. 1 gives d_{cutoff} of only 0.4-0.5 μm highlighting the importance of methods of higher-order mode suppression.
Although the concept of the broad waveguide laser dates back to the 90s of the last century, novel laser heterostructures, such as asymmetric location of the active region in combination with symmetric [6] or asymmetric [7] cladding layers, super-large optical cavity waveguides [8], ultra-broad waveguides

Content from this work may be used under the terms of the Creative Commons Attribution 3.0 licence. Any further distribution of this work must maintain attribution to the author(s) and the title of the work, journal citation and DOI.
Published under licence by IOP Publishing Ltd

with photonic band crystal structures [9], and extreme, double asymmetric large optical cavity structures [10], have been still emerging in attempts to find an acceptable tradeoff between optical loss, resistance, beam divergence, and other laser parameters. Recently, we have proposed a Coupled Large Optical Cavity (CLOC) laser heterostructure in which the fundamental mode lasing is stabilized owing to a resonant interaction of the higher-order mode of one (active) waveguide with the only mode of another (passive) waveguide [11]. Moreover, the CLOC laser heterostructure gives an opportunity to achieve low internal loss and thermal resistance, as was proven for InGaAs/GaAs/AlGaAs diode lasers emitting at 1.03 μm [12]. The shorter emission wavelength inevitably leads to poor localization of charge carriers in the active region and, therefore, can deteriorate the performance of a laser with a broad waveguide. In the present work, we study CLOC lasers emitting at 0.98 μm and compare their characteristics with the 1.03- μm counterpart.

2. Experimental

Two laser heterostructures were grown by metal-organic vapor phase epitaxy. In both cases, an InGaAs double quantum well (DQW) was used with the target wavelength at room temperature either 0.98 or 1.03 μm. A sequence of epitaxial layers is listed in Table 1 in order from the $n+$ GaAs substrate to the surface. The active waveguide where the DQW is located has a total thickness of 1.35 μm. Lasing on its first excited mode is eliminated by means of the mode leakage into the n-type dope GaAs passive waveguide, whereas the second excited mode lasing is suppressed by shifting the DQW towards the mode node closer to the upper cladding. Since the penetration of the fundamental mode into the upper cladding is negligible, it is grown 0.5-μm thick that is 2-2.5 times thinner than typically used. In combination with 0.15-μm thick cap layer suitable for non-alloyed ohmic metallization, this allows us to locate the active region as shallow as 920 nm from the epi-side surface. Owing to a close spacing between the active region and laser heatsink, effective heat removal is achieved under continuous wave operation that is confirmed by a low value of specific thermal resistance of 6×10^{-3} K/W×cm^2 [13].

Table 1. Layer sequences of the laser heterostructures.

Material	Designation	Doping, cm^{-3}	Thickness, nm	
			0.98μm laser	1.03μm laser
n - GaAs (Si)	Buffer	-2E+18	~ 400	~ 400
n -Al$_{0.25}$Ga$_{0.75}$As (Si)	Lower cladding	-2E+18	1200	1200
n - GaAs (Si)	Passive waveguide	-2E+18	570	550
n -Al$_{0.25}$Ga$_{0.75}$As (Si)	Optical barrier	-5E+17	250	250
GaAs	Active waveguide	Undoped	1040	1030
In$_{0.15}$Ga$_{0.85}$As or In$_{0.25}$Ga$_{0.75}$As	Active region	Undoped	~8	~8.5
GaAs	Spacer	Undoped	40	40
In$_{0.15}$Ga$_{0.85}$As or In$_{0.25}$Ga$_{0.75}$As	Active region	Undoped	~8	~8.5
GaAs	Active waveguide	Undoped	260	270
p^+-Al$_{0.25}$Ga$_{0.75}$As (Zn)	Upper cladding	+3E+17	150	150
p^+-Al$_{0.25}$Ga$_{0.75}$As (Zn)	Upper cladding	+2E+18	350	350
p^{++}-GaAs (Zn)	Cap	+4E+19	150	150

The wafers were processed into broad-area edge-emitting lasers with 100-μm-wide stripes. No facet coating was applied. Diodes of various lengths were soldered onto a copper heatsink with indium solder and tested in pulsed or continuous wave regimes.

3. Results and discussion

For both laser structures, far-field pattern measurements were done. Stable single-spatial-mode emission with the divergence of 32-33° (full width at half maximum) and a Gaussian-like profile, which closely matches the theoretically calculated pattern was revealed regardless of cavity length, injection current (up to 6A) and temperature (up to 80°C).

For laser diodes of different lengths, threshold current density J_{th} was evaluated and lasing spectra were measured from which lasing wavelength λ was extracted. The results are summarized in figure 1(a). In the 4 mm long lasers J_{th} of 230 and 275 A/cm² was found in the 1.03 and 0.98-µm structures, respectively. When cavity length L decreases, J_{th} gradually grows and reaches 480 and 750 A/cm² at $L \sim 1$ mm. Higher threshold current density in the shorter-wavelength lasers can be attributed to a higher carrier concentration in the GaAs waveguide, where they recombine radiatively or non-radiatively. In [14], a practical criterion was experimentally determined for the bandgap difference (>7 $k_B T$) sufficient to suppress carrier escape from the active region into the surrounding material. It should be emphasized that the above criterion is satisfied for the longer-wavelength laser heterostructures under study but not for the shorter-wavelength one.

As expected, the lasing peak position in long ($L > 1$ mm) diodes corresponds fairly well to the ground-state optical transition of the respective DQW. However, when the cavity length shortens below 1 mm, lasing peak is blue shifted by about 50 nm. This is a signature of the excited-state optical transition lasing. The reason for such a behavior is that the modal gain, which is achieved on the ground-state optical transition, becomes insufficient to balance optical loss. The fact that the excited-state lasing in two different laser heterostructures takes place at the same cavity length can be interpreted in terms of the similar gain-current density dependences of the 0.98-µm and 1.03-µm DQW active regions. In particular, the maximal gain corresponding to ground state transition is the same for 0.98-µm and 1.03-µm DQW active regions.

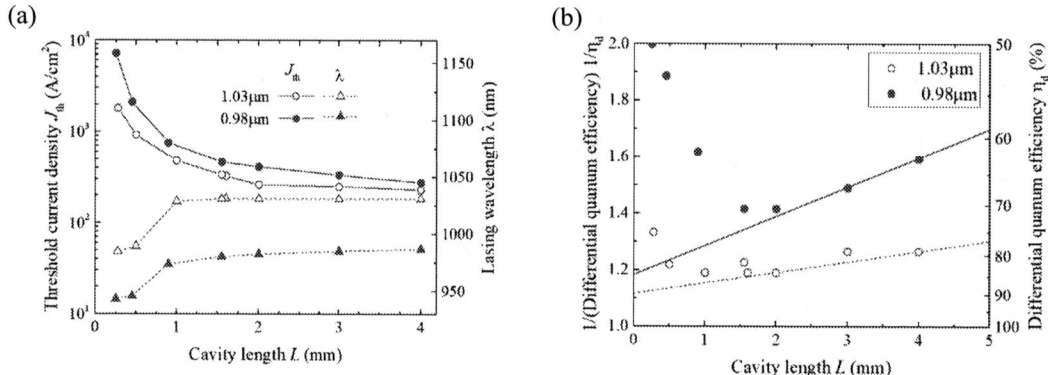

Figure 1. (a) Threshold current density (circles) and lasing wavelength (triangles); (b) differential quantum efficiency (right axis) and its inverse (left axis); lines are fits with Eq. (3). All quantities are shown as functions of cavity length. Solid and open symbols correspond to 0.98-µm and 1.03-µm laser structures, respectively.

Figure 1(b) shows differential quantum efficiency η_d evaluated from the near-threshold parts of light-current curves of various diode lasers operating under pulsed excitation. Results are presented as a function of the cavity length. The maximal value of η_d is 84% and 71% in the 1.03-µm and 0.98-µm laser heterostructures, respectively. The differential quantum efficiency of a diode laser is given by a well-known formula

$$\eta_d = \eta_i \frac{\alpha_m}{\alpha_m + \alpha_i}, \qquad (2),$$

where η_i is the internal differential quantum efficiency, α_i – internal optical loss, $\alpha_m = \ln(R^{-1})\, L^{-1}$ – output (mirror) loss, R – the mirror reflectivity, which is approximately 0.3 for as-cleaved facets. It is generally assumed that both η_i and α_i are independent of cavity length. The subsequent linear relation between $(\eta_d)^{-1}$ and L is thus frequently used to extract the experimental values of η_i and α_i:

$$\eta_{\mathrm{d}}^{-1} = \eta_{\mathrm{i}}^{-1}\left(1 + \frac{\alpha_{\mathrm{i}}}{\ln(R^{-1})}L\right) \tag{3}.$$

It is seen that $(\eta_{\mathrm{d}})^{-1}$ does linearly proportional L if relatively long lasers are only taken into consideration. Fitting the experimental data with Eq. (3) gives $\alpha_{\mathrm{i}} = 1$ cm^{-1} and $\eta_{\mathrm{i}} = 85\%$ for the 0.98-μm and $\alpha_{\mathrm{i}} = 0.4$ cm^{-1} and $\eta_{\mathrm{i}} = 89\%$ for the 1.03-μm laser structures, respectively. Eq. (3) predicts that shorter cavities should provide a higher external differential efficiency. However, this is obviously not the case as evident from figure 1(b). Higher Fermi energy, which is required to provide a higher optical gain in shorter diodes, also promotes the free carrier absorption in the waveguide. We calculated the effective internal loss for all cavity lengths from the experimentally measured η_{d} using Eq, (2) under the assumption that η_{i} remains the same as in longer samples [15]. Results are presented in figure 2(a). They can be satisfactorily fitted by the following empirical equation:

$$\alpha_{\mathrm{i}} = \alpha_0 + k_\alpha \alpha_{\mathrm{m}}{}^{n} \tag{4},$$

where n was found to be 2 for both laser structures, α_0 which has the meaning of internal optical loss in infinitely long cavities is 0.4 and 1.0 cm^{-1}, and k_α which somehow takes into account specificity of the laser structures is 0.004 and 0.013 cm the 1.03-μm and 0.98-μm laser heterostructures, respectively. Significantly higher value of the k_α parameter in the latter case reflects a stronger effect of the waveguide population there and explains a faster decrease of the external differential efficiency in shorter resonators.

We also estimated the modal gain G in both structures as a sum of the internal loss and the mirror loss (figure 2b) and then fitted G using a logarithmic relation with the current density:

$$G \equiv \alpha_{\mathrm{i}} + \alpha_{\mathrm{m}} = G_0 \ln(J//J_0) \tag{5},$$

where G_0 is the gain parameter, J_0 – the transparency current density. Two regions, which correspond to the ground-state and the excited-state lasing at lower and higher injection levels, respectively, are clearly observed in the G-J dependences. In the ground-state lasing regime, both structures are characterized with the same $G_0 = 13$ cm^{-1}, whereas J_0 is slightly higher in the shorter-wavelength laser structure (205 vs 175 A/cm^2). Both parameters are 2.5-3 times larger in the excited-state lasing regime.

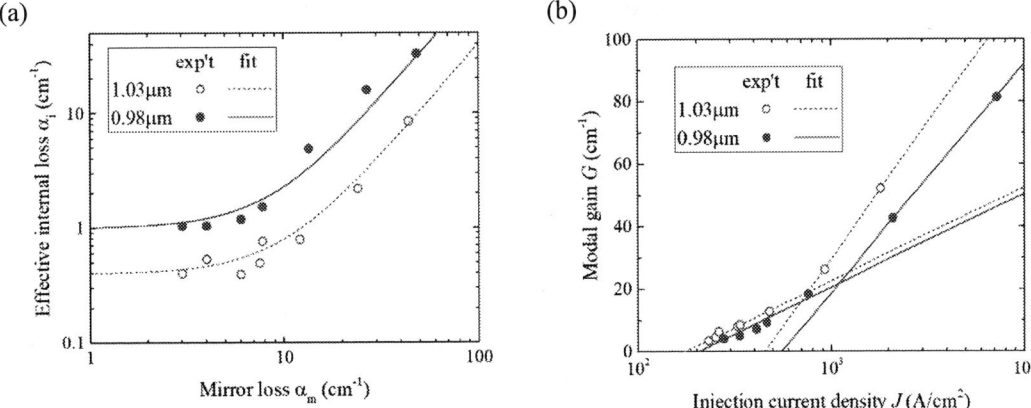

Figure 2. Effective internal loss (a) and modal gain (b). All quantities are shown as functions of injection current density. Solid and open symbols correspond to 0.98-μm and 1.03-μm laser structures, respectively. Lines are fits with Eq. (4) (figure 2a) and Eq. (5) (figure 2b).

Figure 3(a) shows a temperature induced increase of the threshold current density measured in the pulsed regime to avoid self-heating effects. The characteristic temperatures T_0 and T_1 which are introduced via the following exponential relation

$$J_{th}(T) = J_{th}(20^0C)\exp((T - 20^0C)/T_0), \eta_d(T) = \eta_d(20^0C)\exp(-(T - 20^0C)/T_1) \qquad (6),$$

were estimated to be $T_0 = 61$ K, $T_1 = 130$ K in the 0.98-μm structure and $T_0 = 110$ K, $T_1 = 250$ K in the 1.03-μm one. Worse temperature stability found for the shorter-wavelength laser heterostructure again reflects stronger carrier escape out of the quantum well in that case.

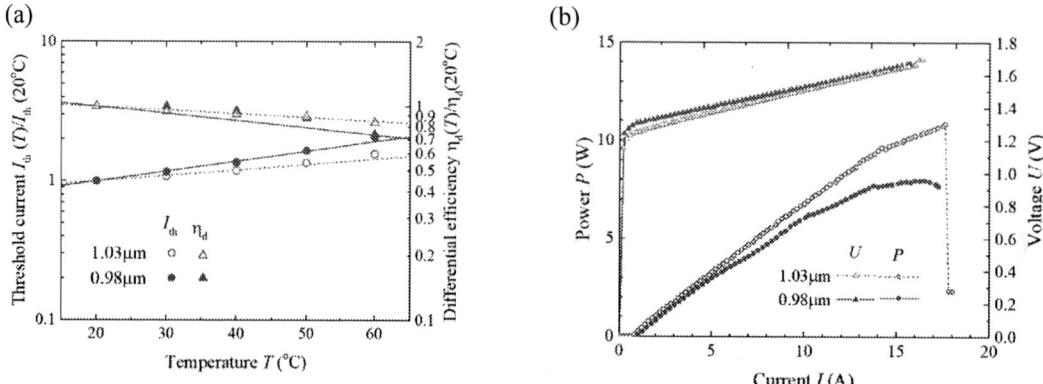

Figure 3. (a) Threshold current density (circles) and slope efficiency (triangles) against temperature and their fits with Eq. (6); (b) output power (circles) and voltage (triangles) vs injection current. Solid and open symbols correspond to 0.98-μm and 1.03-μm laser structures, respectively.

Finally, we measured voltage-current and light-current characteristics under continuous wave excitation. Results are shown in figure 3(b) for 3-mm long laser diodes. Both structures demonstrate low series resistance of 22-24 mΩ. The diode turn-on voltage was extracted to be 1.29 and 1.24 V in the 0.98-μm and 1.03-μm laser structures, respectively. In both cases, the turn-on voltage value correlates perfectly well with the optical transition energy of a given laser structure. The output power of the 1.03-μm laser remains a linear function of injection current up to approximately 13-14 A whereas in the 0.98-μm structure a non-linearity of the light-current curve becomes visible already at 5-6 A. Because we expect similar thermal resistances in both lasers, the difference in power characteristics can be explained by a higher temperature sensitivity of the longer-wavelength structure. As a result, its output power is limited by the thermal rollover. Nevertheless, the maximal power level reaches the level of 8 W. In the longer-wavelength laser, the thermal rollover scenario of power limitation is not realized and the maximal power of 10.8 W is determined by the catastrophic optical mirror damage.

4. Conclusion
In conclusion, two sets of diode lasers having different wavelengths of emission and the same layer structure with shallow (less than 1 μm) location of the active region with respect to the diode surface were fabricated and studied. The coupled large optical cavity design used for the laser waveguides allows lasing via the fundamental vertical mode, reducing the internal loss and lowering series resistance. We found several clear pieces of evidence of insufficient carrier localization in the quantum well of the shorter-wavelength ($\lambda = 0.98$ μm) laser heterostructure as compared to its longer-wavelength ($\lambda = 1.03$ μm) counterpart. Although the difference in wavelengths looks insignificant, it transforms into more than $2k_BT$ difference in optical transition energies and, thus, in localization energies with respect to the surrounding material.

Acknowledgments
This work was supported by the Ministry of Education and Science of the Russian Federation in the framework of the Federal Target Program "Research and development on priority directions of

scientific-technological complex of Russia for 2014-2020", the code 2017-14-579-0057, agreement №14.578.21.0254 of September 26, 2017, unique ID RFMEFI57817X0254.

References
[1] Garbuzov D Z, Abeles J H, Morris N A, *et al* 1996 *Proc. SPIE* **2682** 20
[2] Al-Muhanna A, Mawst L J, Botez D, *et al* 1998 *Appl. Phys. Lett.* **73** 1182
[3] Livshits D A, Kochnev I V, Lantratov V M, *et al* 2000 *Electron. Lett.* **36** 1848
[4] Botez D 1999 *Appl. Phys. Lett.* **74** 3102
[5] Gordeev N Yu, Maximov M V, Zhukov A E 2017 *Las. Phys.* **27**, 086201
[6] Pikhtin N A, Slipchenko S O, Sokolova Z N *et al* 2004 *Electron. Lett.* **40** 1413
[7] Lee J J, Mawst L J, Botez D 2002 *IEEE Photon. Tech. Lett.* **14** 1046
[8] Pietrzak A, Crump P, Wenzel H, *et al* 2011 *IEEE J. Select. Topics Quantum Electron.* **17** 1715
[9] Miah M J, Kettler T, Posilovic K, *et al* 2014 *Appl. Phys. Lett.* **105** 151105
[10] Crump P, Erbert G, Wenzel H, *et al* 2013 *IEEE J. Select. Topics Quantum Electron.* **19** 1501211
[11] Gordeev N Yu, Payusov A S, Shernyakov Yu M, *et al* 2015 *Opt. Lett.* **40** 2150
[12] Zhukov A E, Gordeev N Yu, Shernyakov Yu M, *et al* 2018 *Tech. Phys. Lett.* **44** 46
[13] Zhukov A E, Gordeev N Yu, Payusov A S, *et al* 2018 26th Int. Symp. "Nanostructures: Physics and Technology", Minsk, Belarus, June 18–24
[14] Frevert C, Crump P, Bugge F, *et al* 2016 *Semicond. Sci. Technol.* **31** 025003
[15] Zhukov A E, Kovsh A R, Ustinov V M, Alferov Zh I 2003 Las. Phys. **13** 319

Development of technological installation for obtaining functional thin film elements used in solar elements of space equipment

T O Zinchenko, V I Kondrashin, E A Pecherskaya, P E Golubkov, K O Nikolaev, F A Abdullin

Polytechnic Institute, Penza State University, Penza 440026, Russia

Abstract. The power supply system directly affects the basic characteristics of the spacecraft. The main and, in fact, the only source of energy in space is a solar battery. The main disadvantage of solar batteries is low protection from solar radiation. A solution of this problem is proposed by introducing H and BN into the transparent conductive coating of the solar cell. The influence of the concentration of embedded H and BN on the protective characteristics of transparent conducting coatings based on tin dioxide obtained by the spray pyrolysis method is considered in the article.

1. Introduction

The system of the spacecraft power supply (power supply system, PSS) - the spacecraft system providing other systems power, is one of the most important systems, in many ways it determines the geometry of space vehicles, design, mass, and the period of active existence. The failure of the power supply system leads to the failure of the entire apparatus. Nowadays, a solar battery in space is the only source of generating the required amount of electricity. In this case, solar cells have such disadvantages as high cost and in the case of using solar cells in near - Earth orbit, the radiation effect on the photocell production material is affected. Due to this negative influence, the structure of solar cells changes, which leads to a decrease in the electricity generation [1].

In this regard, the urgent task is the development of an inexpensive, easy-to-manage technological installation for obtaining radiation-protected functional thin-film coatings for solar cells.

2. Experiment

The optical and electrical properties of the transparent conductive coating depend on the production technology, the choice of which is usually associated with taking into account the optimum functioning of the coating for a particular use while minimizing production costs. The metal oxide coatings application can be performed by physical or chemical methods, as well as by their combinations. Traditional methods, which include thermal vacuum deposition, magnetron sputtering, pulse- laser deposition, chemical vapor deposition, are complicated by the vacuum use, expensive process equipment and are not easy for industrial implementation. Spray-pyrolysis is the most suitable for these requirements method, because it has the following advantages: simplicity; low cost; the possibility of varying the properties of ATO by changing their application modes; large surface of the area coverage; mass production potential [2].

Content from this work may be used under the terms of the Creative Commons Attribution 3.0 licence. Any further distribution of this work must maintain attribution to the author(s) and the title of the work, journal citation and DOI.

Published under licence by IOP Publishing Ltd

The process of obtaining ATO samples based on tin dioxide in the framework of spray pyrolysis technology consisted of several stages: preparation of glass substrates surface; preparation of precursors solutions; deposition of pure and antimony doped tin dioxide films on the substrate surface. To produce a transparent conductive oxide, tin dioxide doped with antimony was used, because this material is the most suitable for obtaining this type of coating. SnO_2 is sufficiently resistant to atmospheric conditions, chemically inert and can withstand high temperature, having high conductivity and transparency [3].

Table 1 shows the values of the technological parameters for which the ATO samples based on tin dioxide were obtained, and in Figure 1 some photographs of them are presented.

Table 1 Technological parameters of the ATO production process

Parameter	Meaning
material of substrates	soda-calcium-silicate glass
precursor 1	$SnCl_4 \cdot 5H_2O$
precursor number 2	$SbCl_3$
solvent	ethanol
carrier- gas	compressed air
C_M (precursor 1 concentration)	0,25 mol/l
N_{prec2} (precursor 2 concentration)	0,1 0,05 mol. % 0,025
V_{rast} (volume of solutions)	5–20 ml
l (distance between spray and substrate)	300 mm
p (inlet air pressure in the nebulizer)	2 bar
T_S (heater temperature)	450 °C

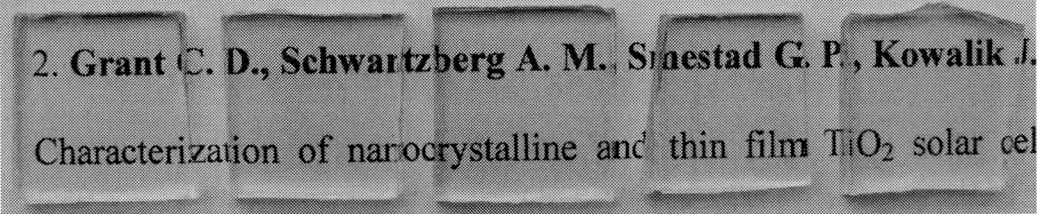

Figure 1. Photograph of samples coated with $SnO_2 +$ Sb.

In Figure 1, you can see that the samples are transparent, but you can only quantify the transparency by using spectrophotometry.

After the samples synthesis, electrophysical parameters and properties of the coatings were measured [4]. To protect against radiation, the hydrogen and boron nitride introduction was carried out on film boundaries into the near-surface region, because boron-containing fillers as thermal neutron absorbers are very well suited, and in particular such as boron nitride of hexagonal modification (BN). Bohr absorbs neutrons very efficiently: the thermal neutron absorption cross section is 3837 barns, while the neutron absorption cross section by the majority of elements is on the order of 1 barn units. To reduce the mass of the solar cell and reduce the efficiency of the secondary particles production process in them, it is necessary to use coatings consisting of elements with small values of the nuclear charge Z. Therefore, the possibility of using hydrogen-containing materials is actively investigated. As a result, it was decided to use a composite material based on BN + H.

Table 2 presents the technological parameters of the H + BN implantation process.

Table 2 Technological parameters of the H + BN in the ATO introduction process

Sample number	% of embedded H + BN
1	0
2	5
3	10
4	15
5	20

Figure 2. Photograph of samples coated with SnO2 + Sb with the H + BN introduction.

It will increase the solar cells protection from the radiation effects as these elements reduce the proton flux and the number of generated neutrons in the photocell. It is clearly seen in Fig. 3

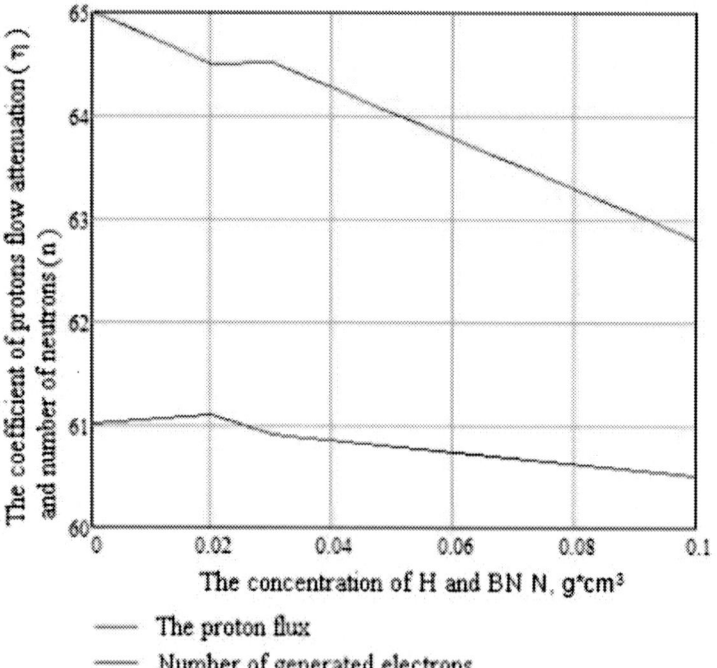

—— The proton flux
—— Number of generated electrons

Figure 3. Dependence of the coefficient of the proton flux attenuation and the generated neutrons number on the H and BN concentration.

3. Conclusion

In the interaction of heavy GCL nuclei with light elements, the primary nuclei decay on fragments with a small mileage in the screen material effectively proceeds, as a result the radiation fluxes behind the screen are greatly weakened, and their energy spectra become more "soft". Thus, the introduction of light elements in the composition of the film material increases its efficiency, which is confirmed by the results of the experiment.

References

[1] Guschin V. N. 2003 Basics of the spacecrafts device (Moscow: Mechanical engineering).

[2] T.O. Zinchenko, E.A. Pecherskaya, V.I. Kondrashin, A.S. Kozlyakov, Y.V. Shepeleva 2017 18th Int. Conf. of Young Specialists on Micro/Nanotechnologies and Electron Devices (Russia, Novosibirsk) pp. 320-324.

[3] T.O. Zinchenko, V.I. Kondrashin, E.A. Pecherskaya, A.S. Kozlyakov, K.O. Nikolaev, Y.V. Shepeleva 2017 *IOP Conf. Series: Materials Science and Engineering* **225**,. DOI:10.1088/1757-899X/225/1/012255.

[4] Raksha, S.V., Kondrashin, V.I., Pecherskaya, E.A., Nikolaev, K.O. 2015 *Journal of Nano- and Electronic Physics* vol **7** (4) 04062.

Spectral Approach to Li–ion Batteries Degradation Rate Estimation

E A Yusupova[1], V G Malyshkin[2], V V Davydov[1]

[1]Peter the Great Saint–Petersburg Polytechnic University, 195251, Russia

[2]A.F. Ioffe Physical Technical Institute, Saint Petersburg 194021, Russia

Abstract. Li–ion batteries are critically important component of modern solar power plant systems with over 30% contribution to the total cost. The uncertainty of Li-ion battery degradation rate is a very important risk source of a solar system. The estimation of degradation rate spectrum is an important practical problem. A two stage approach to degradation spectrum estimation is proposed: on the first stage a probability density is constructed, and on the second stage the signal is averaged with the probability density from the first stage. The problem is reduced to matrix spectrum analysis that does not require a L^2 norm to minimize. The technique application is demonstrated on two–stage degradation model and commercial Li-ion battery degradation data.

1. Introduction

Li-ion batteries are critically important component of modern solar power plant systems, they typically contribute to over 30% to the total cost. The uncertainty in Li-ion degradation rate is the key economic risk in solar power systems[1, 2]: when a station is built future Li-ion degradation rate has a very high uncertainty, what lead to an uncertainty in the battery lifetime, this makes degradation rate distribution study an important research topic. Because manufacturers–provided data is often incomplete or incorrect, the estimation of future degradation rate is of great practical interest for solar power stations.

The degradation of Li–ion batteries is an active field, studied both in terms of applications and theoretical modeling. The term "degradation" can be defined differently. In can be understood as:

- Battery capacity fading, this reduce energy characteristics.

- Internal resistance increase, this reduce power characteristics.

- Increased self–discharge.

- Increased risk of catastrophic failure or even battery fire.

For solar power systems battery capacity fading is the most important degradation aspect (the situation is different for electric vehicles and avionic systems). Battery capacity as a function of cycle number $C(N)$ is typically the most important characteristic in practice. The definition of "cycle" can be rather complex, it includes charge/discharge rate, temperature, humidity and other conditions. There are exist several standards[3–5] where the measurement conditions are specified in details. Among all the specified parameters these two are the most important:

- Current. It is measured in battery capacity C. $1C$ means charge/discharge the battery in 1 hour; $2C$ — in 30 minutes. For Li–ion typical testing current is $0.5C$ or $1C$. Batteries, capable of higher current, are seldom. The degradation rate depends strongly on the current.

Content from this work may be used under the terms of the Creative Commons Attribution 3.0 licence. Any further distribution of this work must maintain attribution to the author(s) and the title of the work, journal citation and DOI.

Published under licence by IOP Publishing Ltd

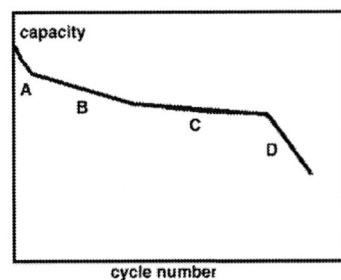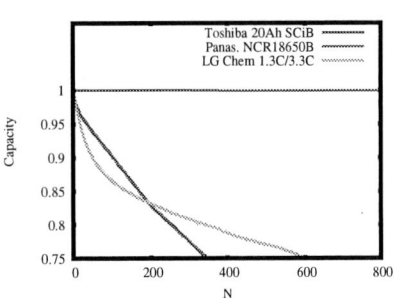

Figure 1. Left (the figure is from [6]): typical 4 stages $C(N)$ degradation. Right: $C(N)$ for three commercially available Li-ion batteries[7–9] according to manufacturer data.

- Temperature. The degradation rate is highly dependent on the temperature. Most often used temperatures are the $20°$ and $45°$. Few manufacturers provide temperature–dependent degradation data. In special applications exploitation temperature range can be critically important characteristic.

Typical $C(N)$ dependence represent two– or four– stages [6] dependence. In in Fig.1 the $C(N)$ for [6] model and for three commercially available batteries is presented. In table I the degradation rate (the number of cycles to 0.8 of the initial capacity) is presented according to manufacturers data. One can clearly see degradation rate increase with current and temperature increase.

2. Degradation Rate Spectrum

The dC/dN is the degradation rate. It can vary greatly and is typically obtained by some interpolation technique. A common approach in signal analysis is to represent a signal (in this paper the signal $f = dC/dN$) as a linear superposition of some basis functions (such as Fourier or wavelets[13]). To obtain such a superposition the difference between a signal and interpolating expression is minimized as a L^2 norm (least squares) and obtained components are interpreted as projection amplitudes(e.g. Fourier amplitudes), this consists in choosing $Q_k(x)$ $k = 0..n - 1$ basis and measure $\langle g(x) \rangle = \int g(x)\omega(x)dx$ (e.g. Fourier basis, orthogonal polynomials, etc.), then for a signal

Table I. Li-ion modules degradation characteristics at different charge/discharge current and the temperature

Manufacturer	Model	Cycles to 0.8	Wh/L	Wh/kg	W/L	W/kg
A123[10]	ANR26650	6800: (1.2C/2C , 23°)			5800	2600
		1300: (1.2C/2C , 45°)				
		670:(1.2C/8C , 45°)				
LG Chem[9]	18650HG2	340: (1.3C/3.3C)	680	240		
		160: (1.3C/6.6C)				
Samsung[11]	ICR18650-22F	1660: (0.8C/1C, 23°)	490	190		
		660: (0.8C/1C, 45°)				
		330: (0.7C/1C, 60°)				
EnerDel[12]	CE175-360 (17.5 Ah)	6400: (0.5C/0.5C, 0°)		147		1256
		2600: (0.5C/0.5C, 30°)				
		1200: (0.5C/0.5C, 45°)				

383

$f(x)$ the least squares answers is obtained by minimizing the norm $\langle (f(x) - \sum_{i=0}^{n-1} \beta_i Q_i(x))^2 \rangle \to \min$ to obtain a linear superposition of f:

$$f \approx \sum_{i=0}^{n-1} \beta_i Q_i(x) \tag{1}$$

with the answer $\beta_i = \sum_{j=0}^{n-1} \left(G^{-1}\right)_{ij} \langle fQ_j \rangle$, here $G_{ij} = \langle Q_i Q_j \rangle$ is Gram matrix. If, for example, $Q_i(x)$ is Fourier basis, then $\langle fQ_j \rangle$ are plain Fourier components, $\beta_i = \langle fQ_j \rangle$, the $\langle Q_i Q_j \rangle$ is unit matrix (Fourier basis is orthogonal). In general case any (1) type of answer requires to calculate the vector $\langle fQ_j \rangle$ and the matrix $\langle Q_i Q_j \rangle$ (this matrix can be always chosen to be unit matrix by applying e.g. Gram–Schmidt orthogonalization). Such an approach (direct interpolation of f) has a number of limitation. For example the second moment $\langle f^2 \rangle$ may be infinite, the errors may be non–Gaussian, etc. In such situation L^2 norm minimization make no sense (few outliers can make the (1) result invalid). Similar situation take place for the signals with spikes[14], when the standard deviation does not exist.

Consider an alternative. Instead of (1) linear superposition of contributions to f, introduce $\psi(x) = \sum_{j=0}^{n-1} \alpha_j Q_j(x)$ as a linear superposition of the basis, and consider the signal weighted with the $\psi^2(x)\omega(x)dx$ weight:

$$f_\psi = \frac{\langle \psi^2(x)f \rangle}{\langle \psi^2(x) \rangle} = \frac{\sum\limits_{i,j=0}^{n-1} \alpha_i \langle fQ_i Q_j \rangle \alpha_j}{\sum\limits_{i,j=0}^{n-1} \alpha_i \langle Q_i Q_j \rangle \alpha_j} \tag{2}$$

Now, instead of linear superposition to f, we consider linear superposition (defined by α_i) to $\psi(x)$, and the square of $\psi(x)$ provide averaging weight. Any vector α_i now corresponds to f_ψ, which is a **ratio of two quadratic forms** on α_i. The approach is now **two stage**: 1). For a given α_i build the weight $\psi^2(x)\omega(x)dx$ 2).Average the signal with this weight to obtain f_ψ (a ratio of two quadratic forms). The proposed (2) approach is to obtain from the data **two matrices**: $\langle fQ_i Q_j \rangle$ and $\langle Q_i Q_j \rangle$, then, for a given vector α_i, to obtain (2), the signal averaged with $\psi^2(x)\omega(x)dx$ weight. The key difference is that while the β_i (1) define contributions to the observable f, the α_i define contributions to $\psi(x)$ (the square of $\psi(x)$ define averaging weight to obtain f_ψ), without any L^2 norm involved[15]. With two matrices obtained from the data the generalized eigenvalues problem:

$$\sum_{k=0}^{n-1} \langle fQ_j Q_k \rangle \alpha_k^{[i]} = \lambda^{[i]} \sum_{k=0}^{n-1} \langle Q_j Q_k \rangle \alpha_k^{[i]} \tag{3}$$

has a unique solution: **the unique** basis in which both $\langle fQ_i Q_j \rangle$ and $\langle Q_i Q_j \rangle$ matrices are **simultaneously** diagonal. The most important is the interpretation of (3) spectrum. The situation is similar to the one in random matrix theory[16]: to interpret the distribution of (3) eigenvalues $\lambda^{[i]}$ as a distribution of f.

3. Model Data

For Li–ion battery capacity C, which fade with charge–discharge cycle number $x = N$, the degradation per cycle typically has several stages similar to those in Fig. 1. The $C(N)$ is a relaxation type of process, and we are going to use the $\langle \frac{dC}{dN} Q_j Q_k \rangle$ and $\langle Q_j Q_k \rangle$ matrices in (3) to solve generalized eigenvalues problem (3) as $\sum_{k=0}^{n-1} \langle \frac{dC}{dN} Q_j Q_k \rangle \alpha_k^{[i]} = \lambda^{[i]} \sum_{k=0}^{n-1} \langle Q_j Q_k \rangle \alpha_k^{[i]}$ then

SPBOPEN 2018   IOP Publishing

IOP Conf. Series: Journal of Physics: Conf. Series **1124** (2018) 041007   doi:10.1088/1742-6596/1124/4/041007

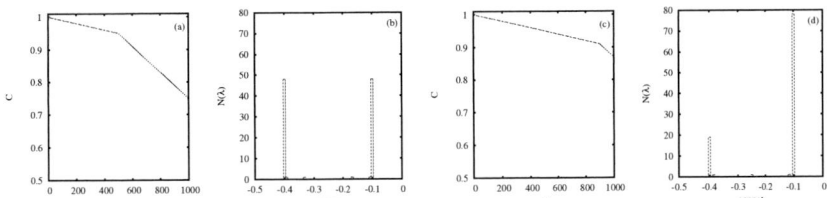

Figure 2. Two stage degradation Li-Ion model with the slope on first and second stages -0.0001 and -0.0004 respectively. The stages length is 500:500 for (a) and 900:100 for (c). The (b) and (d) are the distributions of λ from (3) corresponding to (a) and (c), calculated with $n = 100$ in polynomial basis.

interpret the distribution of eigenvalues $\lambda^{[i]}$ as the distribution of degradation rates dC/dN. In Fig. 2 a model of two–stage degradation rate is presented with different stage lengths. Eigenvalues problem solution give $\lambda^{[i]}$ eigenvalues distribution. The eigenvalues take two values (equal to exact model values) their number is related to the stage length. Changing stages length change only eigenvalues number, not their values.

4. Experimental Data

Consider degradation rate for LG Chem battery[9] shown in Fig. 1. Given the $C(N)$ the $\left\langle \frac{dC}{dN} Q_j Q_k \right\rangle$ and $\langle Q_j Q_k \rangle$ matrices are calculated and put to (3). The data is obtained from manufacturer chart by discretization, this limit maximal dimension n in Fig. 3 because of discretization errors. The two–stages degradation is clearly observed. A very important feature of treating $\lambda^{[i]}$ as dC/dN distribution is that all $\lambda^{[i]}$ are finite even when standard definition is infinite. A "proxy for an average" can be obtained as $\frac{1}{n} \sum_{i=0}^{n-1} \lambda^{[i]}$. An important feature of the approach is that the worst/best degradation rate can be directly obtained from the spectrum as minimal/maximal λ, and used for the estimation of catastrophic degradation risk.

5. Conclusion

Obtaining from the data not a vector $\left\langle \frac{dC}{dN} Q_i \right\rangle$, but a matrix $\left\langle \frac{dC}{dN} Q_i Q_j \right\rangle$, allows to estimate degradation rate distribution directly, without any apriori assumptions (such as stages number) made by

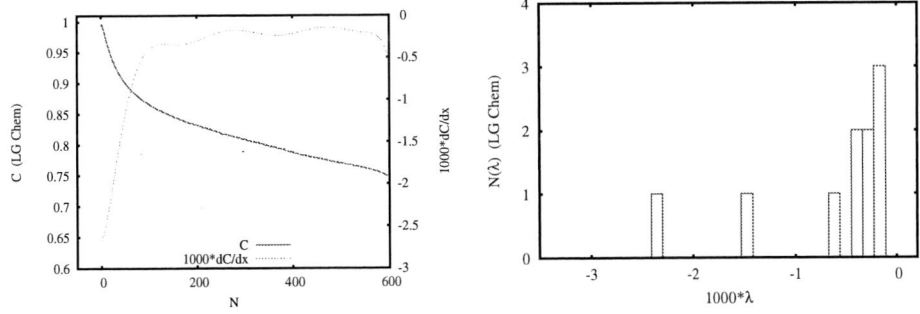

Figure 3. Left: The $C(N)$ and calculated dC/dN for LG Chem battery[9]. Right: The distribution of $\lambda = dC/dN$ calculated using $\omega(x) = 1$ and $n = 10$.

a human. Mathematically the problem is reduced to matrix generalized eigenvalues problem, not to L^2 norm minimization as in Fourier or Laplace.

A novel approach to obtain degradation rate spectrum is developed. Mathematically the problem is reduced not to typically considered interpolation problem, but to matrix spectrum analysis. In this paper the technique is demonstrated for the simplest problem: to obtain dC/dN distribution. However, same technique can be extended to correlate different attributes with each other, even if they do not have the second moment. We can, for example, correlate spikes in dC/dN with the spikes in discharge current. For this task one need to solve the (3) for two signals of interest, then project eigenfunctions on each other. The simplest case $n = 2$ correspond to introduced in ([17]) probability correlation concept $\widetilde{\rho}(f, g)$, that shows how the *probability of low/high f* correlates with the *probability of low/high g*; this is different from regular correlation $\rho(f, g)$ (values correlation), that shows how *low/high value of f* correlates to *low/high value of g*. In contrast with values correlation, probability correlation can be applied to the signals with spikes, what make the answer especially advantageous in situations, where the fact of spike carry the most important information, and the magnitude of spikes is less important.

References

[1] Bobyl A.V., Zabrodskii A.G., Malyshkin V.G., Novikova O.V., Terukov E.I., and Agafonov D.V., Generalized Radon–Nikodym Approach to Direct Estimation of Degradation Rate Distribution. (Деградация Li-ion накопителей энергии. Применение обобщенного подхода Радона–Никодима к оценке распределения скоростей деградации.), *Izversiya RAN. Energetika. (Известия РАН. Энергетика)* (2018) , 46–58.

[2] Bobyl A.V., Zabrodskii A.G., Kompan M.E., Malyshkin V.G., Novikova O.V., Terukova E.E., and Agafonov D.V., Generalized Radon–Nikodym Spectral Approach. Application to Relaxation Dynamics Study., *ArXiv e-prints* (2016) , https://arxiv.org/abs/1611.07386, arXiv:1611.07386 [math.NA].

[3] Mikolajczak C., Kahn M., White K., and Long R.T., *Lithium-ion batteries hazard and use assessment* (Springer Science & Business Media, 2012).

[4] Castillo E.C., in *Advances in Battery Technologies for Electric Vehicles* (Elsevier, 2015) 469–494.

[5] IEC 61960–3:2017 (2017), Secondary cells and batteries containing alkaline or other non-acid electrolytes - Secondary lithium cells and batteries for portable applications.

[6] Spotnitz R., Simulation of capacity fade in lithium-ion batteries, *Journal of Power Sources* (2003) , **113**, 72–80.

[7] Toshiba. The 2.9Ah SCiB.

[8] Panasonic NCR18650B. Specification.

[9] LG Chem. 18650HG2 datasheet. http://www.nkon.nl/sk/k/hg2.pdf.

[10] A123. Nanophosphate High Power Lithium Ion Cell ANR26650.

[11] Samsung SDI. Introduction of ICR18650-22F.

[12] EnerDel. CE175-360 and CE160-365 cells.

[13] Bozhokin S., Continuous wavelet transform and exactly solvable model of nonstationary signals, *Technical Physics* (2012) , **57**, 900–906.

[14] Bobyl A., Davydov V., Zabrodskii A., Kostik N., Malyshkin V., Novikova O., Urishov D., and Yusupova E., The Spectral approach to timeserie bursts analysis (Спектральный подход к анализу всплесков временной последовательности), *ISSN 0131-5226. Теоретический и научно-практический журнал. ИАЭП.* (2018) , doi:10.24411/0131-5226-2018-10010.

[15] Malyshkin V.G., Norm-Free Radon-Nikodym Approach to Machine Learning, *ArXiv e-prints* (2015) , http://arxiv.org/abs/1512.03219, arXiv:1512.03219 [cs.LG].

[16] Guhr T., Müller-Groeling A., and Weidenmüller H.A., Random-matrix theories in quantum physics: common concepts, *Physics Reports* (1998) , **299**, 189–425.

[17] Malyshkin V.G., Market Dynamics. On A Muse Of Cash Flow And Liquidity Deficit, *ArXiv e-prints* (2017) , arXiv:1709.06759 [q-fin.TR].

$CH_3NH_3PbI_3$ crystal growth, structure and composition

V E Anikeeva[1,2] , O I Semenova[2] , O E Tereshchenko[1,2]
[1]Department of Physics, Novosibirsk State University, Novosibirsk 630090, Russia
[2]Rzhanov Institute of Semiconductor Physics SB RAS, Novosibirsk 630090, Russia

Abstract. For the past five years methyl-ammonium lead trihalide perovskite solar cells have shown remarkable progress in power conversion efficiencies due to their unique properties. Single $CH_3NH_3PbI_3$ crystals were grown on the base of concentrated aqueous solution condensation of the precursor. The grown crystal structure and stoichiometry were studied by RHEED, X-ray and XPS. The perovskite crystal photocurrent spectra were measured in the temperature range of 85-330 K.

1. Introduction
In the future, the renewable energy source, including the photovoltaic (PV), one will have become a major energy resource by about 2050. To make it a competitive to fossil fuels, it is necessary for PV cells to reduce the total cost by developing new materials, device structures and processes. For the past five years, methyl-ammonium lead trihalide (perovskite - $CH_3NH_3PbI_3$) has aroused significant interest due to its unique properties, low cost and facile deposition techniques. Solution-processed perovskite solar cells reached high PCE (power conversion efficiency) values from 3.2% in 2009 [1] to the National Renewable Energy Laboratory certified value of 22.1% in 2017 [2]. Such a PCE growth has actually taken decades for other photovoltaic solar cells. Such high PCE values are attributed to the optimal optical band gap (1.5 eV), high absorption coefficient, weakly bound excitons, that easily dissociate into free carriers, and extraordinarily electron-hole diffusion length [3]. Now $CH_3NH_3PbI_3$ has been suggested as a novel low-cost solution-processed material for high efficiency hybrid solar cells. Another striking attribute of the perovskite is its low non-radiative recombination rates, compared to other thin-film poly-crystalline semiconductors. These physical properties made perovskite attractive for various research fields ranging from lasing [4, 5], light-emitting devices [6] and photodetection [7, 8] to tandem solar cell applications [9–11].
In this work we study the chemical composition and structure of methyl-ammonium lead trihalide single crystals with the aim to create an air-stable material for a further detailed study of its optical properties.

2. Experiment and results
Two precursors were prepared for the synthesis of powder $CH_3NH_3PbI_3$: CH_3NH_3I and PbI_2. The synthesized perovskite powder was dissolved at 65°C in hydro-iodic acid stabilized with 1.5% hypophosphorous acid to produce a supersaturated solution [12]. A small $CH_3NH_3PbI_3$ crystal was put in the solution as a fuse. The solution temperature was decreased within several days. The rate of this decreasing was 0.1°C an hour. The photo of the grown crystals is shown in Figure 1.

Content from this work may be used under the terms of the Creative Commons Attribution 3.0 licence. Any further distribution of this work must maintain attribution to the author(s) and the title of the work, journal citation and DOI.
Published under licence by IOP Publishing Ltd

Figure 1. The grown perovskite crystals.

The structural analysis of perovskite crystals was carried out in the Siberian Synchrotron and Terahertz Radiation Centre. The phase transition from the tetragonal to the cubic phase was detected at about 56 °C, shown in Figure 2 (a). The $CH_3NH_3PbI_3$ crystal stoichiometry was confirmed by X-ray photoelectron spectroscopy (XPS). All the main elements (carbon, nitrogen, lead and iodine) are represented on the XPS spectra (Figure 2 (b)). Because of the absence of chemically shifted components in the spectra of the lines of constituent elements inherent in oxide forms, it can be argued that oxygen is part of the adsorbed water molecules on the perovskite surface.

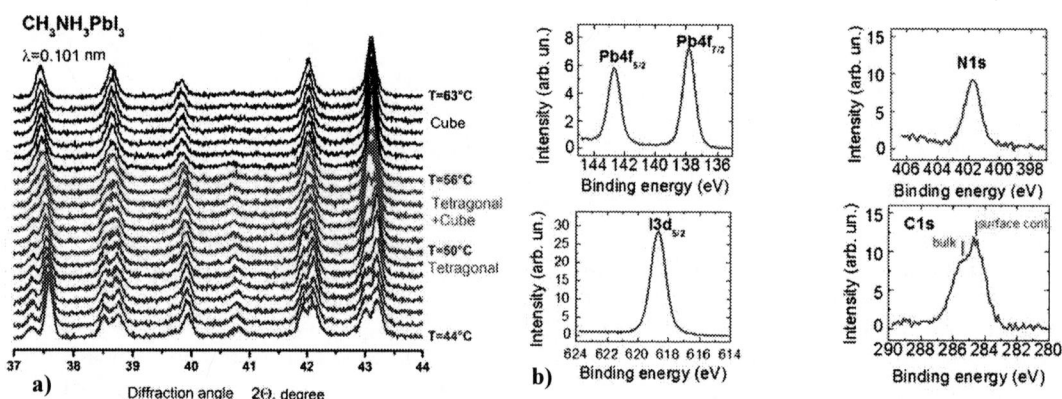

Figure 2 (a, b). (a) X-ray powder diffraction pattern for a single $CH_3NH_3PbI_3$ crystal sample; (b) XPS spectra of the Pb 4f, N1s, I3d and C1s peaks for the single $CH_3NH_3PbI_3$ crystal.

The diffraction pattern from the perovskite surface measured by the Reflection high-energy electron diffraction (RHEED) method is shown in Figure 3 (a). The observed diffraction pattern demonstrates the $CH_3NH_3PbI_3$ surface crystalinity. By the X-ray diffraction (Figure 3, b) it was established that the grown single crystals have the space group I4/mcm and the lattice constants a = 8.8776(7) Å, b = 8.8776(7) Å, c = 12.6702(8) Å, V = 998.56(17) Å^3. The obtained crystallographic parameters confirm the formation of tetragonal $CH_3NH_3PbI_3$, which well agrees to the CCDC data reported in [13].

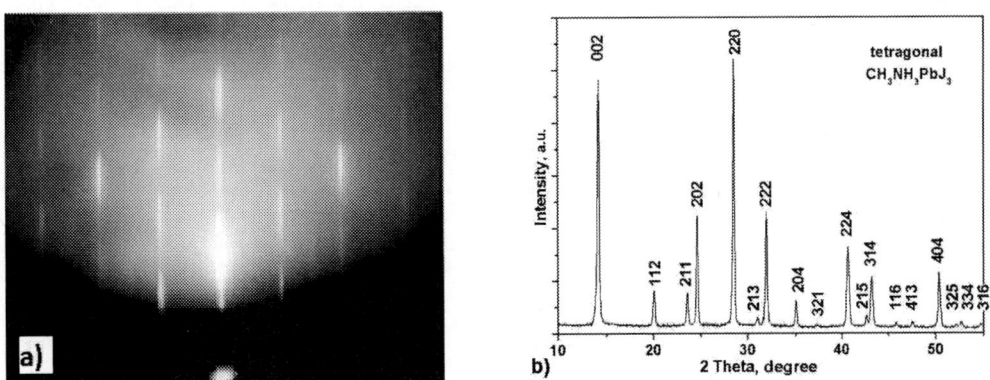

Figure 3 (a, b). (a) A RHEED pattern obtained from the electron diffraction from a clean perovskite surface; **(b)** XRD pattern of the $CH_3NH_3PbI_3$ crystals.

Besides the structural transformation, we studied the temperature dependent photoconductivity of perovskite crystals. The temperature-regulated liquid nitrogen cryostat was used to measure the temperature dependence of the photocurrent spectra. The modulated light, passing through the monochromator, was incident on a crystal placed in the cryostat, schematically shown in Figure 4.

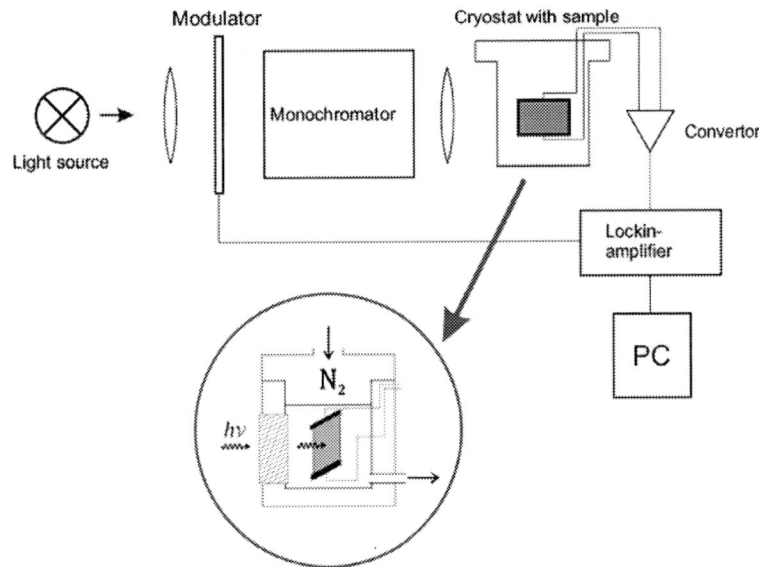

Figure 4. Experimental setup of the photocurrent spectra measurements in the 85-360 K temperature range.

The temperature dependence of the perovskite photoconductivity was studied in the range of 85-340 K. The spectral measurements of the photocurrent showed that the known structural phase transitions in the perovskite $CH_3NH_3PbI_3$ are well manifested in the photoconductivity temperature dependence. It is shown in Fig. 5 that, at temperatures below 130 K, the perovskite, being in the orthorhombic phase, has an absorption threshold of about 1.64 eV, and several excitons are observed in the photocurrent spectrum. Such pronounced exciton peaks are observed for the first time in a

$CH_3NH_3PbI_3$ crystal, which may indicate a structural perfection of the crystals. In the temperature range 85-140 K, the changes in the position of the absorption threshold did not exceed 5 meV. With a further increase in temperature in the interval 140-160 K, the second peak (threshold) appeared, shifted to the red region, which completely dominated at 160 K. In the temperature range 140-160 K, the crystal changed from the phase with an orthorhombic structure to the tetragonal phase [14], while the shift of the absorption threshold in the photocurrent spectrum was more than 110 meV. In the phase transition region the photocurrent magnitude is one or two orders of magnitude smaller than in the regions with a certain structure, which is probably due to the crystal structure disordering of in the transition region. The transition to the tetragonal phase was also accompanied by the appearance of the second threshold and its shift to the red region.

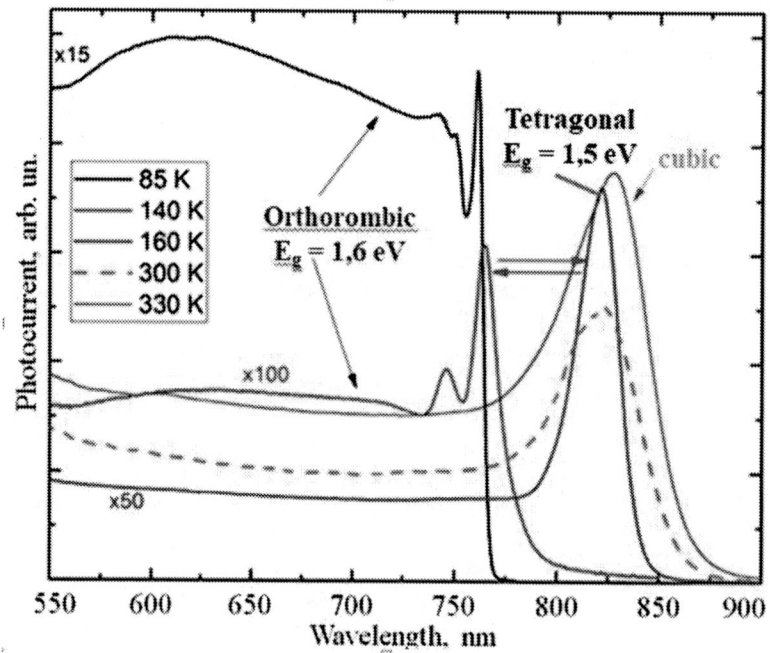

Figure 5. Photocurrent spectra of $CH_3NH_3PbI_3$ measured at various temperatures.

3. Conclusion

We have developed the technique for growing $MAPbI_3$ crystals from an aqueous HI solution (57%) stabilized with hypophosphorous acid H_3PO_2 (2%), which stoichiometry and structure were determined by XPS, RHEED, and X-ray diffraction. Our approach enables us to create single crystals with their volumes exceeding 1cubic centimetre. The grown single crystals are air-stable. This makes it possible to reliably study the structure and optical properties of this semiconductor material. The chemical composition and structure of methyl-ammonium lead trihalide single crystals were studied in details. The abrupt change in the photocurrent spectra near the 160 K region is most likely due to the phase structural transition.

Acknowledgments

The authors are grateful to A.N. Shmakov for his diffraction measurements and V.A.Golyashov for his optical measurements. This work was supported by the Federal Agency for Scientific Organizations [grant No AAAA-A17-117042110141-5].

References

[1] Kojima A, Teshima K, Shirai Y, Miyasaka T. 2009 *J. Am. Chem. Soc.* **131** 6050

[2] Yang W S, et al. 2015 *J. Science* **348** 1234

[3] Stranks S D, et al. 2013 *J. Science* **342** 341

[4] Xing, G C, et al. 2014 *J. Nature Mater.* **13** 476

[5] Saliba M, et al. 2016 *J. Adv. Mater.* **28** 923

[6] Tan Z K, et al. 2014 *J. Nature Nanotech.* **9** 687

[7] Dong R, et al. 2015 *J. Adv. Mater.* **27** 1912

[8] Domanski K, et al. 2015 *J. Adv. Functional Mater.* **25** 6936

[9] Albrecht S, et al. 2016 *J. Science* **9** 81

[10] Baena J P C, et al. 2015 *J. Science* **8** 2928

[11] Bailie C D, et al. 2015 *J. Science* **8** 956

[12] Yudanova E S, et all. 2017 J. of Str. Chemistry. **58** 1567.

[13] Allen F H. 2002 *J. Acta Crystallogr. B58* **3-1** 380

[14] Takeo O. 2015 *Book. IntechOpen* **ch. 3** 77

SPBOPEN 2018

IOP Publishing

Isolation and optoelectronic characterization of Si solar cells microstructure defects

A Gajdos, P Skarvada, R Macku, N Papez, L Skvarenina, D Sobola

Department of Physics, Brno University of Technology, Brno 61600, Czech Republic

Abstract. This research article presents results of silicon solar cell defects optoelectronic characterization based on several experimental methods. These microstructure defects have their origin mainly in the production process, but also can be caused by mechanical stress. However, some defect related spots emit light when the cell is reverse biased. Therefore, electroluminescence (EL) method is used for macroscopic localization and scanning near-field optical microscopy (SNOM) combined with photomultiplier tube in order to scan topography of defective area in microscale. Moreover, elemental analysis of the defects related spots provided by energy-dispersive X-ray spectroscopy (EDX) is presented as well. Besides that, focused ion beam (FIB) was used to isolate the defective spots by 2 μm wide and 2 μm deep barrier. Isolation pattern around the defect is avoiding leakage current flow through it. Since leakage current does not flow through defect, solar cell parameters in reverse conditions are improved.

1. Introduction
Silicon solar cells are still commercially most used photovoltaic devices [1,2]. Even though silicon solar cell production technologies are well known and enhanced for many years, some defects in fabrication process may appear [3,4]. Solar cell uses the entire wafer sheet area for conversion of solar energy to electric one, for this reason defect-free wafers are necessary. However, the presence of defect with various types of breakdown mechanism can be determined by electrical measurement on fully fabricated cell. Nonetheless, not all inhomogeneities are shunts. There are many types of inhomogeneities such as inclusions, Schotky type shunts, cracks, crystal defects, etc [5]. To locate defect related spot, the fact that various defects emit a light in visible range whereas the sample is in reverse biased condition is used [6]. Methods for defects characterization can be also used for different types of solar cells [7].

2. Methods

2.1. Electrical measurement set-up
Measurements of electrical *I-V* characteristics is performed by a source meter Keithley 2420 in thermally insulated box, which also provides basic shielding. Solar cell sample is placed between two electrodes with isolation layer in the middle to avoid electrical shunt. Thermal stability is realized by Peltier's module cooled by water circuit, module is controlled by a source meter Keithley 2510-AT. Setup is controlled by PC via GPIB-USB interface and it allows fully automated measurement. This basic measurement had to be done for a detection of imperfection as well as an estimation of a suitable voltage bias which essential for next presented methods [8].

Content from this work may be used under the terms of the Creative Commons Attribution 3.0 licence. Any further distribution of this work must maintain attribution to the author(s) and the title of the work, journal citation and DOI.
Published under licence by IOP Publishing Ltd

2.2. Macroscale localization set-up

Experimental set-up for macroscale imperfection localization is based on a CCD camera equipped by cooled 3.2MPx Si-chip which sensing radiation in a spectral range from 300 nm to 1100 nm (fig. 1a). Surface of solar cell is captured through macro lens with focal length 105 mm from minimum focusing distance 41 cm. Measurement is done in dark place, because the radiation from sample has very low intensity and it could not be visible in the daylight. Consequently, measurement is depended on voltage bias, because imperfection radiate from threshold level in reverse biased condition (determined by *I-V* measurement). A voltage bias is set by power supply Agilent E3631A.

2.3. Microscale localization set-up

Precise localization of imperfections is performed by SNOM combined with photomultiplier tube as well in reverse biased conditions (fig. 1b). Principle of this method is scanning surface of defective area by scanning probe and simultaneously detect the emitted radiation from imperfection by photomultiplier tube [9]. Nevertheless, radiation is glowing to all directions, thus scanning probe is placed between emitting spot and tube. While probe is scanning, amount of detected light by tube is affected. Emitted radiation is measured at each step of probe trajectory and as a result, probe forms a shadow map of defective area. Final topography with "shadow maps" is presented at fig. 3.

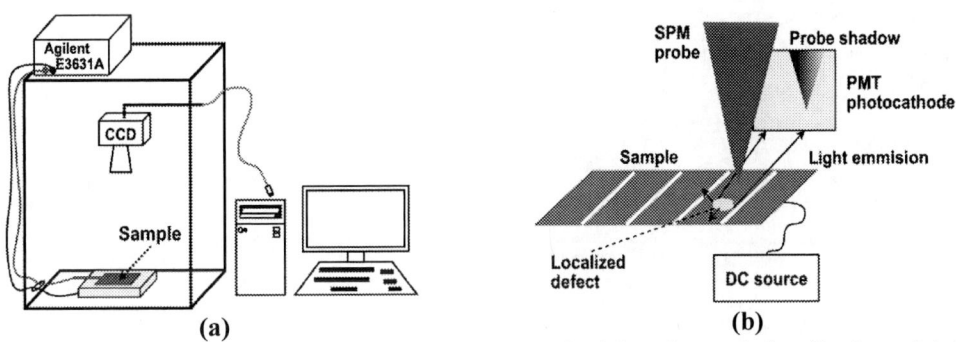

Figure 1(a,b). (a) Macroscale localization set-up; **(b)** Principle Microscale localization of defects using SNOM.

2.4. Defect-isolation method

To isolate microscopic imperfections dual-beam system (FIB-SEM) Tescan Lyra3 is used [10]. Localization of defective is possible, because of the know topography provided by SNOM. Localization only by SEM would be very difficult, since there are many microscopic inhomogeneities without effect on electrical properties of the solar cell. Even back-scattered electron (BSE) detector does not provide satisfying material contrast to enable microscopic localization of defective area. Isolation process is performed by ions of gallium that mill the potential barrier around the imperfection to avoid the leakage current flow through it. Isolation barrier is in order of tens micrometers.

3. Results and discussion

Mentioned characterization and isolation methods have been done on multiple samples, but in this article only result from one sample are presented. For investigation purpose the monocrystalline silicon solar cell is cut into small pieces (approximately 10 x 10 mm^2) containing only few imperfections. Presented sample contains numerous imperfections, that provides parasitic current pathways which caused leakage current through the solar cell. A significant leakage current could be observed above the threshold reverse voltage bias $U_r > 6$ V from the red *I-V* characteristics before isolation process shown in fig. 2b. Every measurement was performed several times for a stability verification and time independence of the obtained results.

A sensitive CCD camera were used for raw localization process of imperfections on cell sample. Dominant radiation was observed from spot marked as "A" in fig. 2a when reverse bias reached voltage 6 V. However, silicon does not produce visible radiation by common recombination process, visible radiation can be observed during avalanche or Zener breakdown [9]. Topography and "shadow map" of radiation spot "A" performed by SNOM with photomultiplier tube in reverse bias higher than 6 V is shown in fig. 3a (top – shadow map; middle – topography: bottom – shadow map overlapped on topography). Combination of these two images provides location of radiation spot "A". Imperfection related to radiation spot "A" founded by SEM (shown in Figure 4a) is common type of defect in monocrystalline silicon devices called pit defect [11]. The breakdown mechanism of this defect type is determined as avalanche type. This defect was successfully isolated by 2 μm wide circular barrier to depth of 2 μm with outer radius R = 14 μm and inner radius r = 16 μm. As a result, formed barrier prevents current flow directed to defect. Repeated I-V measurement (blue, fig. 2b) after defect "A" isolation shows that reverse current above breakdown voltage significantly decreased. Addition of parallel shunt resistance from fitted data for defect "A" is R_{defA} = 970 Ω.

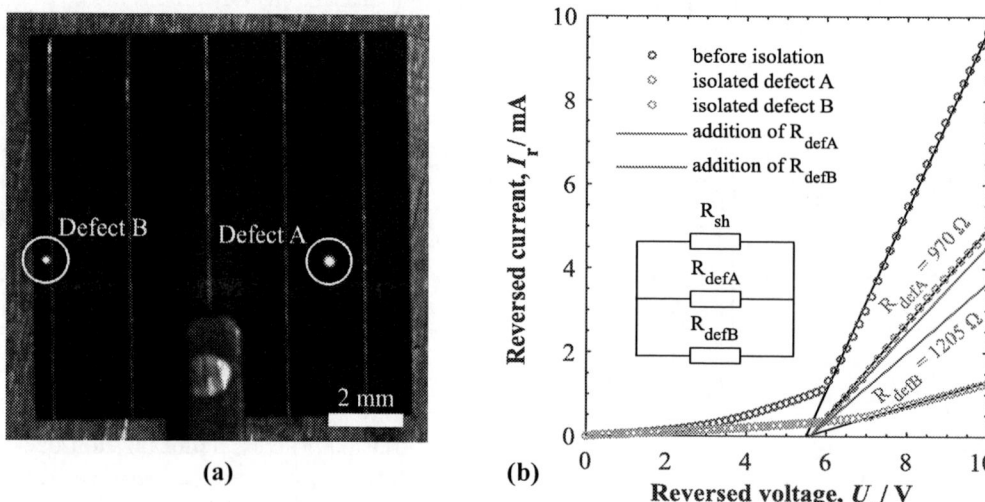

Figure 2(a,b). (a) Electroluminescence image overlapped on solar cell sample image before defects isolation (U_r = 6 V, t = 100s, T = 295 K); **(b)** Current voltage curves "before" and "after" defects isolation. Shunt resistance of each defect is also presented. T = 297 K.

Even if reverse current after breakdown significantly decreased after defect "A" isolation, repeated measurement by CCD camera shows that radiation from spot "A" decreased, but defect "B" is now dominant when bias is still U_r = 6 V. The value of the breakdown voltage coincides with threshold value for radiation of spot "B". Topography of radiation spot "B" with "shadow map" is presented in fig. 3b. Corresponding SEM micrograph of defective area to topography is in fig. 4b and it is visible that radiation spot "B" is also pit type with avalanche breakdown mechanism. Defect "B" is approximately same size as defect "A".

Defect "B" was isolated by annulus with outer radius R = 18 μm and inner radius r = 20 μm milled to a depth of 2 μm in surface and successfully form a barrier for current flow directed to defect. Radiation intensity from defect "B" rapidly decreased after isolation. Decrease of reverse current after threshold voltage is almost same in case of defect "A" isolation. Addition of parallel shunt resistance is from fitted data for defect "B" is R_{defB} = 1205 Ω.

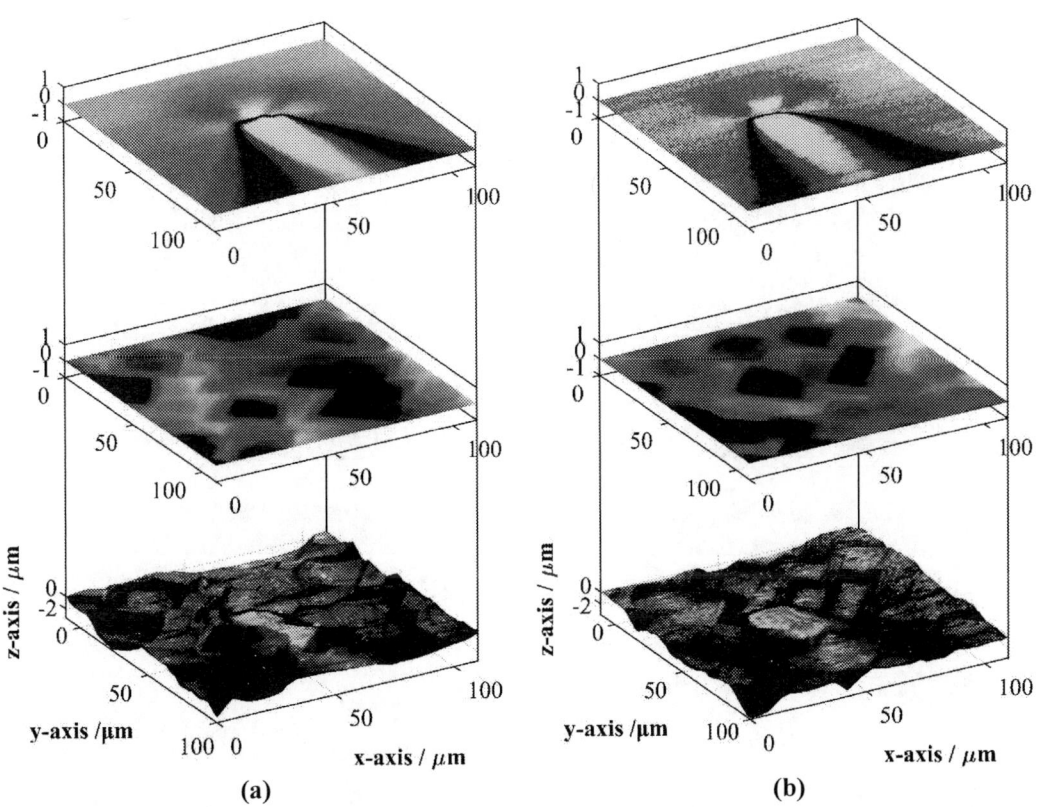

Figure 3(a,b). Topography of defective area combined with "shadow map" for **(a)** defect A ($U_r = 6.1$ V); **(b)** defect B ($U_r = 6.1$ V).

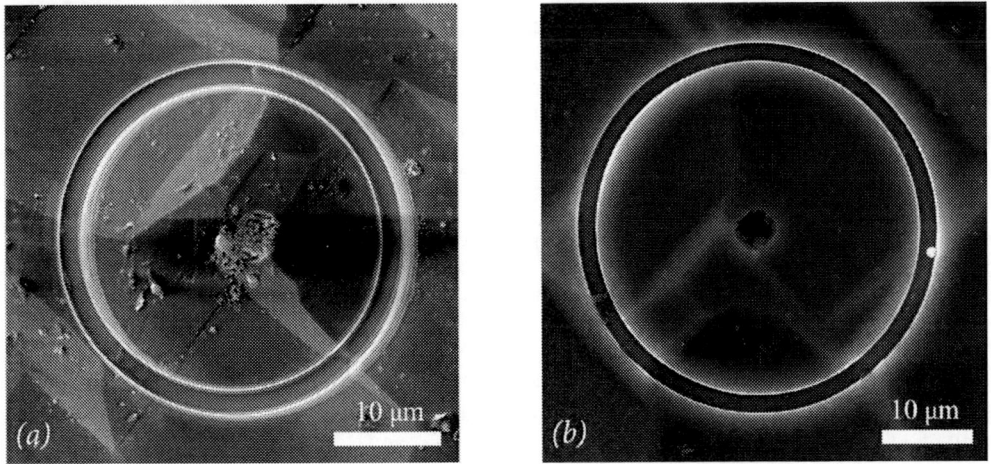

Figure 4(a,b). SEM micrograph of isolated **(a)** defect A ($U_{HV} = 5$ kV, detector SE, tilt 0°); **(b)** defect B ($U_{HV} = 20$ kV, detector SE, tilt 0°).

Last set of figures 5(a-d) shows EDX analysis of defect "A". This analysis proves that any metallic inclusions of are not involved in nature of defect. EDX analysis is not present for defect "B" because it can be observed similar results as in case of defect "A".

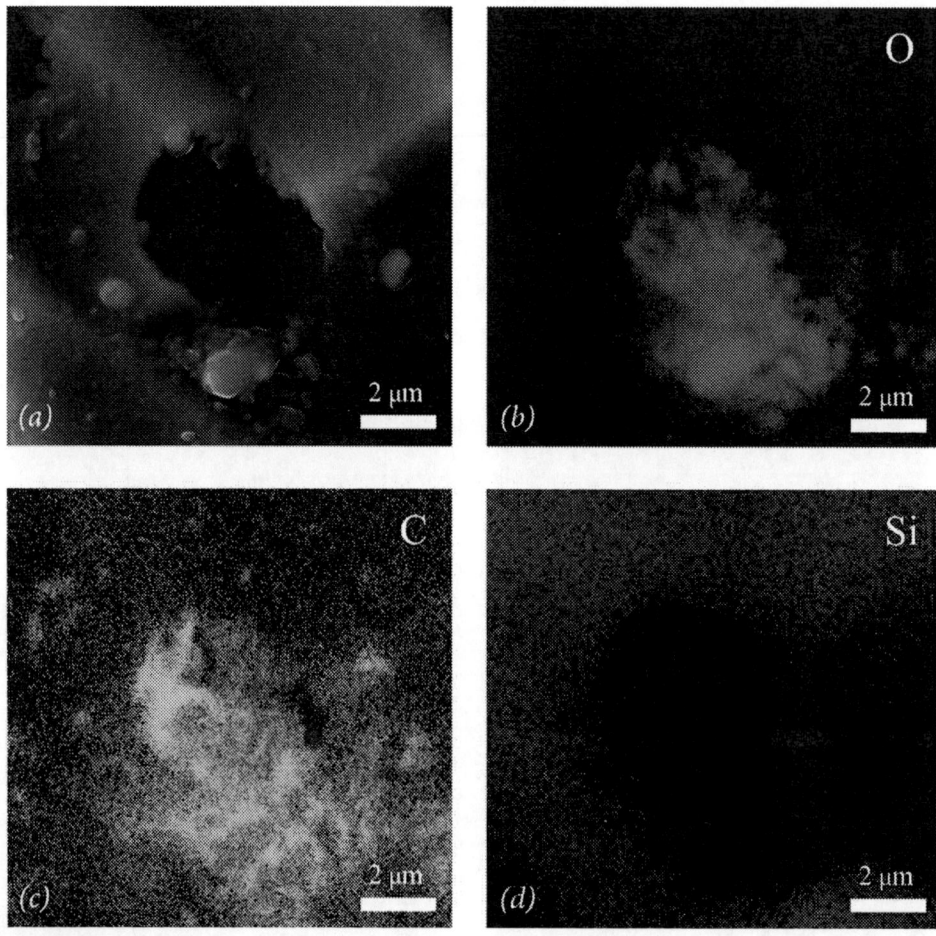

Figure 5(a-d). (a) SEM micrograph of defect A (U_{HV} = 20 kV, detector SE, tilt 0°); **(b-d)** EDX map scanning spectra for the defect A

4. Conclusion

The methods in this paper present the measurements for a detection and localization of the defects or inhomogeneities in the silicon solar cells. These methods could be applied as well on the different types of solar cells. Result from two localized defects on one sample are presented. Breakdown voltage in *I-V* characteristics strongly correlates with radiation threshold voltage obtained by electroluminescence. Both of defects are isolated by experimental method using gallium ion milling. The process of FIB isolation has been successfully repeated on multiple solar cells samples. Electrical properties in reverse biased conditions of investigated sample has been improved. Parallel shunt resistance significantly decreased with each defect isolation. EDX analysis show that inhomogeneities around defects are not metallic inclusions.

Acknowledgments

This work was supported by the Internal Grant Agency of Brno University of Technology, grant No. FEKT-S-17-4626. This support was gratefully acknowledged.

References

[1] Green M A, Hishikawa Y, Dunlop E D, Levi D H, Hohl-Ebinger J and Ho-Baillie A W Y 2018 *Prog Photovolt Res Appl* **26** 3–12

[2] ISE F 2018 Photovoltaics Report 45

[3] Riepe S, Reis I E, Kwapil W, Falkenberg M A, Schön J, Behnken H, Bauer J, Kreßner-Kiel D, Seifert W and Koch W *Phys. Status Solidi* (c) *8* 733–8

[4] Saga T 2010 *NPG Asia Materials* **2** 96–102

[5] Lausch D, Petter K, Bakowskie R, Czekalla C, Lenzner J, von Wenckstern H and Grundmann M 2010 *Appl. Phys. Lett.* **97** 073506

[6] Newman R 1955 *Physical Review* **100** 700

[7] Skvarenina L and Macků R 2017 *Solid State Phenomena* **258** 473–6

[8] Ibrahim A 2011 *International Journal of Renewable Energy Research* **1** 60–5

[9] Škarvada P, Tománek P, Koktavý P, Macků R, Šicner J, Vondra M, Dallaeva D, Smith S and Grmela L 2014 *Applied Surface Science* **312** 50–6

[10] Gajdoš A, Škvarenina L, Škarvada P and Macků R 2017 *Photonics, Devices, and Systems VII* vol 10603 p 1060316

[11] Bishop J W 1989 *Solar Cells* **26** 335–49

SPBOPEN 2018

Investigation of the effect of chemical pre-treatment on uniformity of the silicon wafer texturing for manufacturing a solar cell

D Kudryashov[1], A Gudovskikh[1,2], A Rodin[3], N Pinaev[3]

[1] St. Petersburg Academic University, St. Petersburg 194021, Russia
[2] St. Petersburg Electrotechnical University "LETI", St. Petersburg 197376, Russia
[3] Lyceum "Physical-Technical High School" (PTHS), St. Petersburg 194021, Russia

Abstract. The role of chemical pre-treatment on silicon surface quality and texturing uniformity was investigated. It was shown that for better uniformity of texturing the initial silicon surface of as-cut slurry sawed wafer should be treated to clean up from organic and metal contaminations followed by the organic solvent traces removing and saw damaged silicon layer etching. Additional measurements of photoluminescence decay confirmed high quality and uniformity of textured wafers passivated by a-Si:H.

1. Introduction

Solar cells attract a lot of attention, especially in recent years, since when governments have been trying to solve the problem of CO_2 emission. At the same time, solar cells are starting to be used at remote regions, where it is cost non-competitive to lay power lines. Although the highest efficiency value for a solar cell (46%) was reached for a multijunction III-V solar cell [1], nowadays the most common is a silicon-based solar cell due to its production costs to efficiency ratio. An efficiency of 26.7% is already reached for a silicon heterojunction solar cell with interdigitated back contacts [2]. When transferring a technology to mass production, problems arise, such as non-uniformity of solar cells characteristics for large size wafers. Texturing is one of the critical stages for silicon based solar cells fabrication. It consists of damaged layer removing and micro pyramids forming on silicon surface, which helps to increase light trapping and thus to improve solar cell efficiency. For texturing generally use as-cut Si wafers produced by standard slurry sawing (figure 1a). Its surface is characterized by dentate and fractured areas with cavities. Although the recipes for pyramids formation using heated KOH solution with isopropanol (IPA) are well known [3], pre-treatment procedures for initial surface preparation has not been described well enough. The quality of initial silicon surface strongly affects on texturing uniformity and, as a result, on solar cells characteristics. After the slurry sawing, silicon surface could content a lot of such contaminations as metal traces, slurry residuals and organic traces (figure 1b). Moreover, damaged layer of about 5 μm depth from each side is formed under the silicon surface after mechanical sawing [3]. Prior the texturing, all these contaminations and the damaged layer is to be removed to obtain uniform textured surface. In this paper, our study on the effect of surface preparation on texturing uniformity of silicon wafer is presented.

Content from this work may be used under the terms of the Creative Commons Attribution 3.0 licence. Any further distribution of this work must maintain attribution to the author(s) and the title of the work, journal citation and DOI.
Published under licence by IOP Publishing Ltd

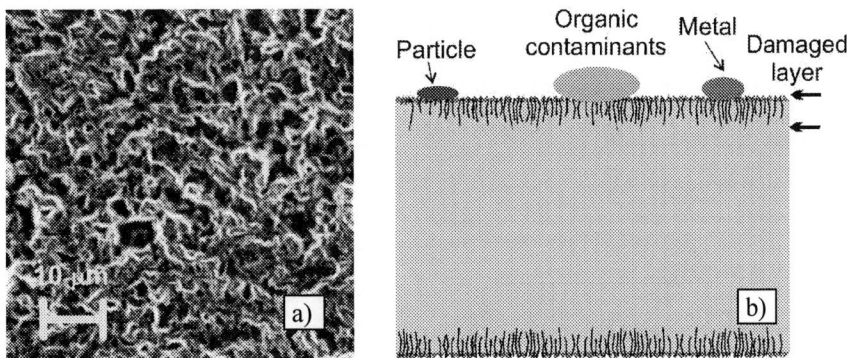

Figure 1. Optical image of as-cut silicon wafer after slurry sawing (a) and a schematic of possible contaminations on its surface (b).

2. Experimental

10x10 cm^2 Cz as-cut n-Si(100) 1-7 Ohm·cm wafer with a thickness of 210 μm was used. It was divided on four 4×4 cm^2 samples. Each sample was treated with its own procedure as shown in table 1. Subsequent texturing process for all samples was carried out by immersing the samples into 4% KOH+10% IPA solution at 85 °C for 60 minutes.

Table 1. Pre-treatment procedures used in the investigation

Sample's number	Pre-treatment
1	Without pre-treatment
2	20% KOH at 80 °C for 5 min
3	CCl$_4$ → IPA → 20% KOH at 80 °C for 5 min
4	CCl$_4$ → IPA → boiled HNO$_3$ for 10 min → 20% KOH at 80 °C for 5 min

For primary surface characterization, optical microscope Zeiss and optical CIS scanner were used. Surface scanning was provided with the same bright and contrast settings for all samples. To reveal the existence of a damaged silicon layer, after texturing a 100 nm of a-Si:H the layer was deposited on each silicon surface by PECVD with Oxford Plasmalab equipment and then photoluminescence (PL) decay measurements were carried out. PL is known as a powerful tool for characterization of interface properties in silicon based solar cells [4,5]. Moreover, PL decay in the silicon much closely reflects the kinetic of minority carriers and could be used for local lifetime values estimations. Two-dimensions PL decay time measurements were performed with the experimental setup shown in figure 2.

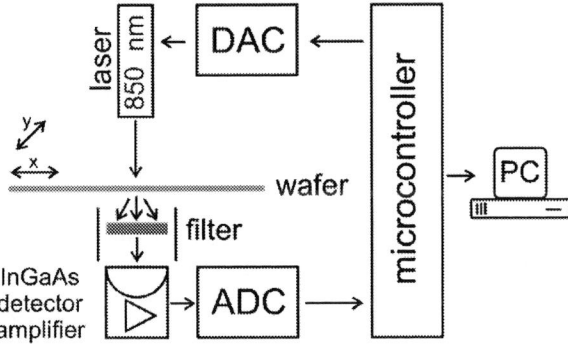

Figure 2. Schematic of experimental setup used to perform PL decay measurements.

3. Results and discussion

Optical images of silicon surfaces for samples 1-4 after texturing are shown in figure 3. On the surface of sample 1, which was textured without pre-treatment (figure 3a), there exist some areas with local non-uniformities formed predominantly due to effects from slurry sawing. Removing of 10 μm of the damaged layer in 20% KOH prior texturing has reduced the number of non-uniformities (figure 3b). Concentrated heated KOH solutions without IPA lead to silicon fast etching with surface smothering (figure 4a).

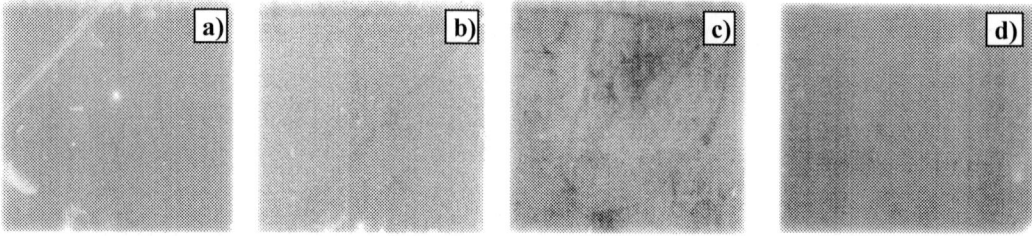

Figure 3. Plan-view optical images of Si samples 4×4 cm^2 after texturing with different pre-treatment: (a) – sample №1, (b) sample №2, (c) – sample №3, (d) – sample №4.

Including of $CCl_4 \rightarrow$ IPA cleaning stages for removing organic contamination has led to increase in non-uniformities on silicon surface after texturing (figure 3c). It might be due to the difficulties with organic solvents removing from rough surface of as-cut silicon wafer (figure 1a). Traces of organics affect on local silicon etch rate and thus lead to inhomogeneities after the texturing (figure 4b), where small-sized pyramids are forming along the big ones. Such deviation in the pyramid sizes leads to different light reflectance and visual non-uniformity. Boiling in concentrated HNO_3 completely removes any traces of organic solvents and in the majority of metal residual on the silicon surface that results in almost total disappearance of any non-uniformity (figure 3d). SEM image of textured silicon surface for sample №4 is shown in figure 4c.

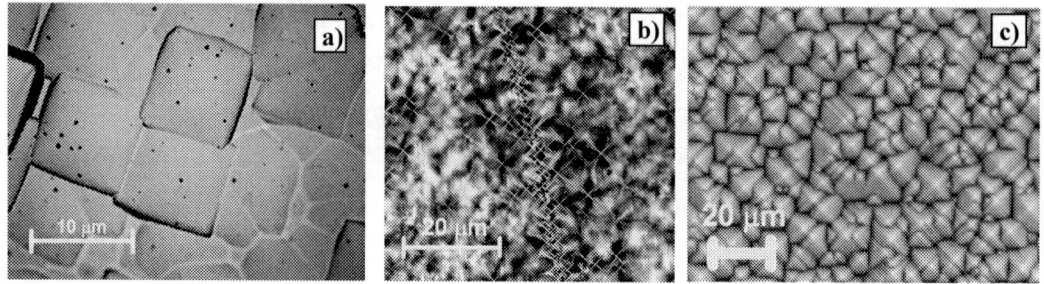

Figure 4. Optical image of silicon surface after planarization in 20% KOH (a), optical image of saw damaged area after texturing (b), SEM image of textured silicon surface for sample №4 (c).

Optical and SEM images rather badly characterize near-surface layer structure of silicon wafer, where defects from slurry sawing could remain. Local defects act there as a recombination centers for minority charge carriers reducing solar cells efficiency. To visualize the defects method of PL decay measurements was used. PL decay time is directly related to minority carrier lifetime in silicon and could be used for local lifetime values estimations. Figure 5 shows the results for lifetime measurements in the silicon samples pre-treated using different conditions.

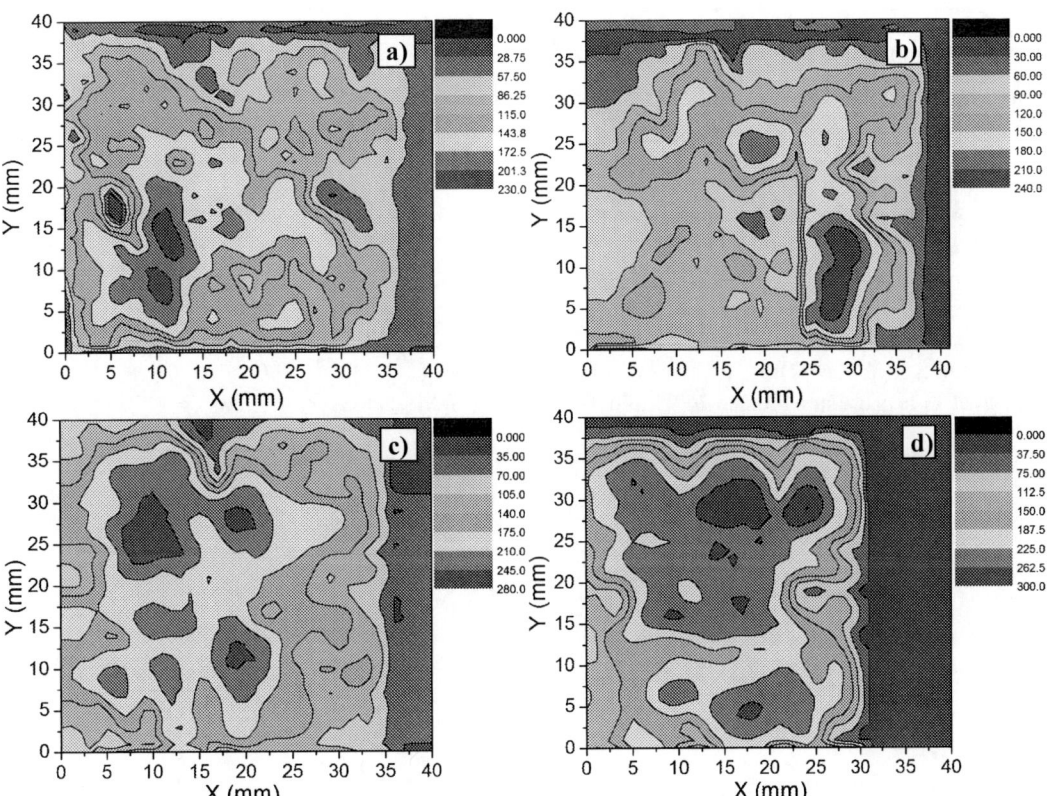

Figure 5. Surface distribution of PL decay time for sample №1 (a), №2 (b), №3 (c) and №4 (d) after texturing and a-Si:H passivation.

Sample №1 after texturing has shown the lowest carrier lifetime values and the highest non-uniformity of PL decay surface distribution (figure 5a). Removing of 10 μm-thick silicon layer and subsequent texturing led to increasing of PL decay uniformity however its absolute value has not change much (figure 5b). Additional stage of surface cleaning with organic solvents prior the planarization and texturing has resulted in increasing of PL decay values and total surface uniformity (figure 5c). The highest values of PL decay and respectively silicon surface quality were obtained for sample №4 (figure 5d) where initial as-cut silicon was preliminary cleaned with CCl_4 and IPA solvents followed by organic and metal traces removing in boiling HNO_3, then planarized with 20% KOH and finally textured in 4% KOH+IPA solution.

4. Conclusions
According to the results obtained, chemical pre-treatment strongly affects on silicon surface quality and uniformity of its texturing. Prior the texturing, as-cut slurry sawing silicon wafer should be cleaned in organic solvents to remove traces of organic contaminations, followed by carefully cleaning with boiling HNO_3 for dissolving of metal contaminations and decomposing of residual organic compounds, and finally smoothed with 20% KOH heated solution. Implementation of all listed stages gives a well-prepared silicon surface for subsequent uniform texturing in 4% KOH + IPA solution. Measurements of effective lifetime confirmed the absence of damaged layer under the silicon surface.

Acknowledgment

This work was carried out with the support of grant of government of the Russian Federation (16.2593.2017/8.9)

References

[1] Green M A, Hishikawa Y, Dunlop ED, Levi DH, Hohl-Ebinger J, Ho-Baillie AWY 2018 *Prog Photovolt Res Appl.* **26:3** 3

[2] Yoshikawa K, Kawasaki H, Yoshida W, Irie T, Konishi K, Nakano K, Uto T, Adachi D, Kanematsu M, Uzu H, Yamamoto K 2017 *Nat Energy* **2(5)** 17032

[3] Sopori B et al. 2013 IEEE 39th Photovoltaic Specialists Conference (Tampa) p 0945

[4] Tardon S, Rösch M, Brüggemann M R, Unold T, Bauer G H 2004 *J Non-Cryst Solids* **338/340** 444

[5] Bauer G H, Tardon S, Rösch M, Unold T 2004 *Phys Status Solidi (c)* **1** 1308

The optical method for condition control of flowing medium

N M Grebenikova[1], K J Smirnov[2,3], V V Davydov[1] and V.Yu. Rud'[4]

[1]Higher School of applied physics and space technologies, Peter the Great Saint Petersburg Polytechnic University, Saint Petersburg, 195251, Russia
[2] OJSC "NRI "Electron", Saint Petersburg, 194223, Russia
[3]The Bonch-Bruevich Saint Petersburg State University of Telecommunications, Saint Petersburg 193232, Russia
[4]All Russian Research Institute of Phytopathology, Moscow Region 143050, Russia

Abstract. In the article a new design of the optical part of a refractometer for monitoring the state of liquid medium is considered. The method for medium state control at the light-shadow boundary is substantiated. The experimental results of different liquid media are presented.

1. Introduction

The experimental and theoretical researches of different liquid media stream are one of the urgent problems of applied physics [1-7]. The non-contact methods are considered to be the most effective for liquid stream researches [3, 6-11]. Unlike other methods refractometric method makes it possible to carry out these researches without changing the physical structure and chemical composition of the liquid medium [9, 12-15].

Today the refractometry method, as well as refractometric instruments and systems are widely represented in the production market and are used throughout the pulp and paper industry, chemical industry, medicine, food industry, and also in laboratories for applied research.

Using refractometric methods create devices for monitoring liquid media - refractometers. Refractometers can be of different types as a model of the optical part of the device. This depends on the optical phenomenon, which is the basis of the device's operation. For example, the phenomenon of interference or total internal reflection.

Refractometers based on the work of which the phenomenon of total internal reflection lies must be in contact with the optical part with the liquid being studied. Therefore, they are installed in the vessels or on the pipelines in which the investigated liquid is located.

The advantage of this type of instrument is the measurement in real time in the liquid flow with the solution being investigated.

On the foreign market, a wide range of optical devices is available, allowing to monitoring the state of liquid fluids. But for the domestic consumer they have a number of shortcomings. For example, they need to be adapted to each specific production. And the universality of instruments of this type is accompanied by their increased dimensions. In addition, such devices are of high cost, which makes them less accessible.

Content from this work may be used under the terms of the Creative Commons Attribution 3.0 licence. Any further distribution of this work must maintain attribution to the author(s) and the title of the work, journal citation and DOI.

Published under licence by IOP Publishing Ltd

2. The optical part of the device and the results of experiments

Usually the main elements of the refractometer are the probe, which is immersed in the liquid under investigation (it contains the optical part of the device), and the data acquisition and processing unit (electronic unit, figure 1).

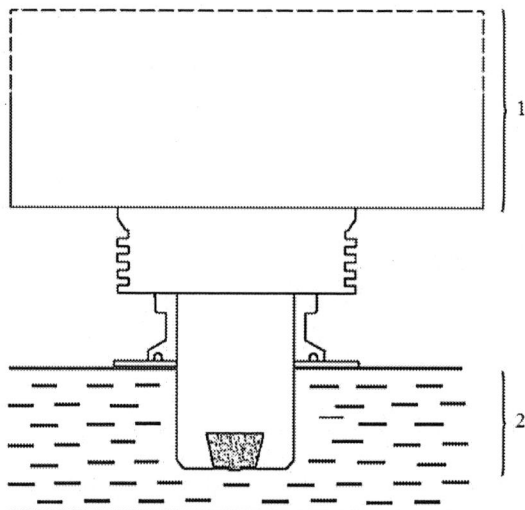

Figure 1 Structural diagram of the refractometer: 1 - the electronic unit, 2 - the optical part of the instrument.

As a rule, in liquids there are gas bubbles, insoluble components, etc. This creates additional difficulties in the measurements [2, 9, 14-17].

The results obtained earlier have shown that in case the researched liquid media contains large insoluble compounds (for example, juice with pulp, medical suspensions, biological solutions, etc.), it is most expedient to control their state in the liquid flow by registering the position of the light-shadow boundary on the photodiode ruler.

Experiments with the standard design of the refractometer [17-20] have shown that there are a number of significant limitations in determining the state of the liquid medium from the shift of the light-shadow boundary. The limitations are related to the vignetting of the laser beam and a decrease in the contrast of the light-shadow boundary. There were many problems with sealing the optical part of the refractometer when measuring in high-pressure flows in a pipeline, etc.

For solving these problems we changed the optical part design of the refractometer (figure 2).

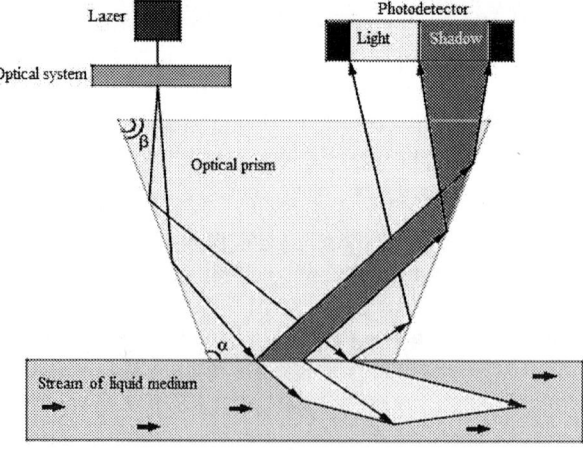

Figure 2 The Structural diagram of the optical part of the refractometer.

For this part the new the new prism design of leukosapphire prism design was calculated and manufactured (figure 3).

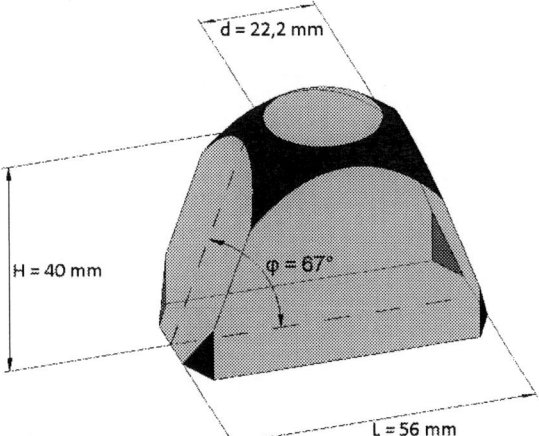

Figure 3 Geometric shape and dimensions of the prism.

Research has shown that new design of the prism allowed us to use laser radiation with a plane angle of the radiation pattern of about 22.6^0 for measurements. The position of the semiconductor laser is moved along the base of the prism within 12 mm. Because the center of the laser beam must fall on the top face of the prism, which borders with the flow at a critical angle α_c:

$$\alpha_c = \arcsin (n_m/n_p) \tag{1}$$

where n_p is the refractive index of the material from the prism is made.

In figure 4 is shown, as an example, the intensity of laser radiation recorded by a photodiode line [21] at different concentrations of potassium N_k in the liquid aqueous solution of potassium nitrate.

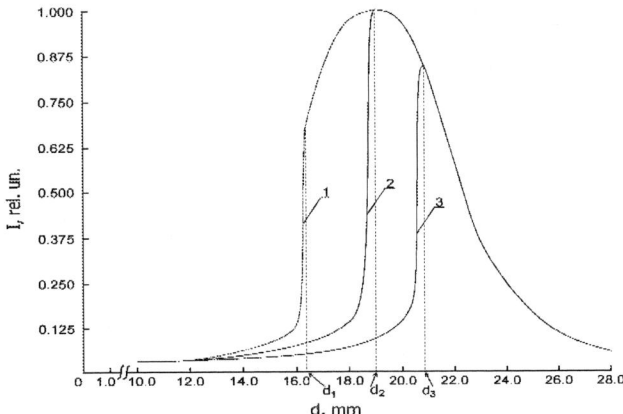

Figure 4. Distribution of the intensity I along the length d of the photodiode line. Graphics 1, 2, 3 corresponds to N_k в %: 30; 44; 52.

The results show a decrease in the contrast ratio of the light-shadow boundary as it moves along the length of the photodiode ruler (d_1 and d_3) to the first graduation (d_2) when the value of N_k changes. This makes it possible to adjust the position of the laser relative to the base of the prism. Because center of the laser radiation beam must fall on the upper face of the prism at an angle α_c. The geometry of the new prism made it possible to apply a sealing gasket of conical shape in the optical probe of the refractometer. The gasket makes the influence of the vignetting effect of the laser beam on the faces of the prism not significant as compared to the prism of the trapezoidal geometry with an annular gasket previously used in the refractometer. The conical gasket also provides greater sealing reliability because sealing of the probe is very important for fast fluid flows of high pressure in the pipeline.

For example, on the concentration of N or T. Figure 5, as an example of the refractometer operation, studies of flowing aqueous solutions of various media at a temperature T = 293.1 K.

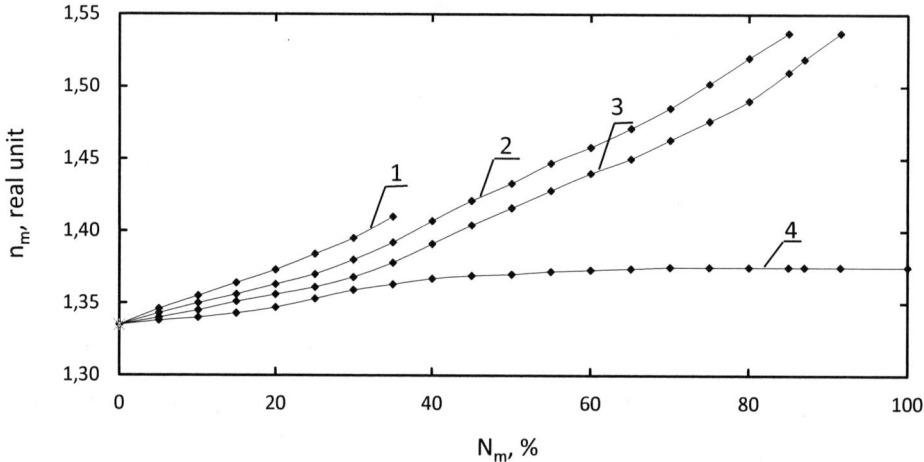

Figure 5. Dependence of the refractive index of aqueous solutions of some substances on concentration: 1-gelatin, 2 - sodium glutamate, 3 - sucrose, 4 - ethyl alcohol.

3. Conclusion

Modernization of the optical design of the refractometer has made it possible to improve the accuracy of monitoring the state of the liquid medium, expand the temperature range T of the liquid medium at which this control can be carried out. This is important during the continuous process. Expanding the T-range allows increasing the n_m measurement range for monitoring the state of transparent media. The new device design allows carrying out measurements of n_m from 1.2246 to 1.6120 with an error of 0.1% in the liquid medium. This result makes it possible to expand the use of the refractometer especially in the medical industry: the control of the state of phenylhydrazine (n_m = 1.6105) in the manufacture of medicines (eg, antipyrine, amidopyrine).

4. References

[1] Rozanov N N 2012 *Optics and Spectroscopy* **113** 613
[2] Popovac M and Hanjalic K 2007 *Flow, Turbulence and Combustion* 78 177-186
[3] Davydov V V, Dudkin V I and Vologdin V A 2015 *Journal of Applied Spectroscopy* **82** 936
[4] Davydov V V, Dudkin V I and Karseev A Yu 2014 *Optical Memory & Neural Networks (Information Optics)* **23** 170
[5] Davydov V V, Dudkin V I and Karseev A Yu 2014 *Optical Memory & Neural Networks (Information Optics)* **23** 259
[6] Pavlov I N, Rinkevicius B S and Tolkachev A V 2011 *Measurement Techniques* **53** 1130
[7] Vologdin V A, Davydov V V and Velichko E N 2016 *Journal of Physics: Conference Series* **741** 012095
[8] Davydov V V, Dudkin V I, Velichko E N and Karseev A Yu 2015 *Instruments and Experimental Techniques* **58** 234
[9] Shur V L, Naidenov A S and Lukin A Ya 2006 *Measurement Techniques* **49** 815
[10] Belov N P, Lapshov S N, Maiorov E.E, Sherstobitova A S and Yaskov A D 2012 *Journal of Applied Spectroscopy* **79** 514

[11]   Karseev A Yu, Vologdin V A and Davydov VV 2015 *Journal of Physics: Conference Series* **643** 012108

[12]   Davydov V V, Dudkin V I and Karseev A Yu 2014 *Technical Physics Letters* **40** 895

[13]   Akmarov K A, Belov N P, Smirnov Yu Yu, Sherstobitova A S, Scherbakova Yu E and Yaskov A D 2013 *Scientific and Technical Journal of Information Technologies, Mechanics and Optics* **87** 39

[14]   Fekete Z A, Hoffmannz E A, Kortvelyesi T and Penke B 2007 *Molecular Physics* **105** 19

[15]   Davydov, V.V., Kruzhalov, S.V., Grebenikova, N.M. and Smirnov, K.J 2018 *Measurement Techniques* **61** 365

[16]   Grebenikova, N.M., Smirnov, K.J., Artemiev, V.V., Davydov, V.V. and Kruzhalov, S.V. 2018 *Journal of Physics: Conference Series* **1038(1)** 012089

[17]   Tsierkezos N G and Molinou I E 1998 *Journal Chem. Eng. Data* **43** 989

[18]   Zhou Y, Li S, Zhai Q, Jiang Y and Hu M 2010 *Journal Chem. Eng. Data* **55** 1289

[19]   Belov N P, Prokopenko V T, Sherstobitova A S and Yaskov A D 2012 *Scientific and Technical Journal of Information Technologies, Mechanics and Optics* **78** 1

[20]   Myazin, N.S.Email Author, Logunov, S.E., Davydov, V.V., Rud, V.Yu., Grebenikova, N.M. and Yushkova, V.V. 2017 *Journal of Physics: Conference Series* **929(1)**,012064

[21]   Smirnov, K.J., Medzakovskiy, V.I., Davydov, V.V., Vysoczky, M.G. and Glagolev, S.F. 2017 *Journal of Physics: Conference Series* **917(6)**,062019

Study of p-type contact topography influence on characteristics of microdisk and microring lasers

A S Novikov[1], E I Moiseev[1], N V Kryzhanovskaya[1,2], B.I. Afinogenov[3],
Y A Guseva[1,4], K Kotlyar[1], A A Lipovskii[1,2], A G Nasibulin[3], M V Maximov[1,4],
V G Tikhomirov[5], A E Zhukov[1,2]

[1] St Petersburg Academic University, St Petersburg, Russia
[2] Peter the Great St Petersburg Polytechnic University, St Petersburg, Russia
[3] Skolkovo Institute of Science and Technology, Nobel str. 3, 143026 Moscow, Russia
[4] Ioffe Institute, St Petersburg, Russia
[5] Saint-Petersburg State Electrotechnical University "LETI", St. Petersburg, Russia

Abstract. The results are presented for injection microlasers with InAs/InGaAs self-organized quantum dot active region grown on a GaAs substrate. Lasing wavelength is around 1.26...1.28 μm. It is demonstrated that the use of ring-shaped p-contact for microdisk quantum dot lasers provides lowering the threshold current as compared to that of conventional microdisk lasers with round-shaped contact topology. Both spectral characteristics and thermal resistance are preserved, whereas the series electrical resistance slightly increases.

1. Introduction

Microdisk/microring lasers are considered as a part of future on-chip optical interconnect systems and planar optoelectronic circuits [1]. Owing to the total internal reflection of light traveling at a periphery of a circular resonator, this type of lasers sustains the so-called Whispering Gallery Modes (WGMs). WGMs provides low threshold current and inter-mode distance (free spectral range) sufficient for realization of singlemode or, at least, quasi-singlemode lasing [2, 3]. However, nonradiative surface recombination at microresonator sidewalls can be an issue as the resonator size is reduced down to tens of microns. Self-organized quantum dots (QDs) provide a lateral carrier transport suppression as compared to two-dimensional quantum well structures. This property allows one reducing a size of resonator or use ring-type of resonator without significant threshold increasing. Another advantage of InAs/InGaAs/GaAs QDs is their ability of emitting at wavelengths around 1.3 μm [4] that corresponds to a transparency window of silicon- or silicon-germanium-based planar waveguides as well as standard optical fibers.

Since the WGM field intensity is localized near the resonator edges, the central part of the active region interacts with the optical mode inefficiently and, hence, its pumping is useless. It has been demonstrated in optically pumped QD microring lasers [5] that an increase of the inner diameter (i.e. the diameter of the etched hole in the center) initially results in reduction of the threshold pump power. On the other hand, low thermal resistance seems to be important for CW operation of injection microlasers at room and elevated temperatures. In this respect, microdisk lasers may have certain advantage over microring counterparts owing to larger area of thermal contact between the substrate and the microcavity itself. A reasonable tradeoff can probably be achieved by combining a disk-shaped microresonator, which provides good thermal contact, with ring-shaped top electrical contact.

In this work we study quantum dot microdisk lasers emitting at 1.26...1.28 μm in CW regime at room temperature. We use ring-type top contact for reducing the threshold current and compare characteristics with those of microdisk lasers with traditional (round-shaped) top contact.

Content from this work may be used under the terms of the Creative Commons Attribution 3.0 licence. Any further distribution of this work must maintain attribution to the author(s) and the title of the work, journal citation and DOI.
Published under licence by IOP Publishing Ltd

2. Experiment

A laser structure was grown by molecular beam epitaxy on an $n+$ GaAs(100) substrate with $n+$ doped GaAs:Si epitaxial buffer. The laser active region comprises ten layers of InAs/In$_{0.15}$Ga$_{0.85}$As QDs separated from each other with 35-nm thick GaAs spacers. The active region was deposited in the middle of a 0.44-μm thick GaAs waveguiding layer confined with n-doped (bottom) and p-doped (top) Al$_{0.25}$Ga$_{0.75}$As claddings. The structure was terminated with a 0.2-μm thick $p++$ GaAs cap layer. In general, the layered structure corresponds to a conventional separate confinement heterostructure (SCH). Depending on post-growth processing, it can be used for fabrication of either edge-emitting stripe laser diodes, or (as it was done in the present work) for disk/ring microlasers.

Microresonators under study were formed by means of deep chemical plasma etching. The etch depth was about 7 μm, i.e. the etch process was done though both cladding layers with the active region and the waveguide in between, and was terminated somewhere in the buffer layer. AgMn/NiAu (AuGe/Ni/Au) metallization was used to form ohmic contacts to $p++$ cap layer ($n+$ substrate, respectively). The outer diameter D of microdisk resonators ranges from 9 to 28 μm. Micrograph is shown in Figure 1(a). In the majority of microdisk lasers, the top p-contact has a round shape with the contact diameter D_C being approximately 3 μm smaller than D (Figure 1(b)). The top contact of some microdisk lases was ring-shaped (Figure 1(c)) with the inner contact diameter d_C varied from 0 to $0.6D$. In still other devices, a ring-shaped top contact was accompanied by a micro-ring resonator shape as it was achieved by means of etching of the inner hole in the resonator center. Microlasers were mounted on a coper heatsink and tested at room temperature under CW excitation without external cooling. Electroluminescence signal was collected with a piezoelectrically adjustable ×10 Olympus LMPlan IR objective. A Horiba FHR 1000 monochromator in combination with a Horiba Symphony InGaAs array was used for spectral detection. The spectral resolution was about 30 pm.

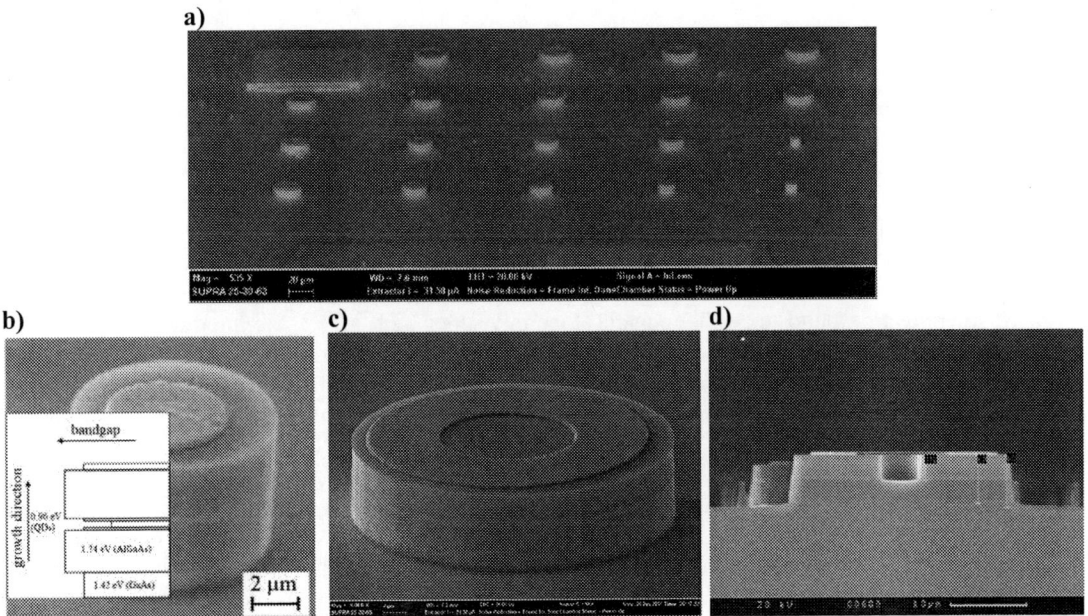

Figure 1. Scanning electron microscope images of an array of microdisk and microring lasers with different type of contact and different diameters (a); microdisk with disk-shaped contact (D = 9 μm and D_C =6 μm) (Inset: Schematic band diagram of the laser heterostructure) (b): microdisk with ring-shaped contact (D = 24μm, D_C =21 μm, d_C =7 μm) (c); microring laser (D = 28 μm, d = 6 μm, D_C = 25 μm, d_C =12 μm)).

3. Results and discussion

First we studied electrical characteristics. We found that the diode voltage U as a function of injection current I for all microlasers studied can be fitted well using equation: $U \approx I\,R_S + U_0$, where R_S is the series resistance, U_0 the diode turn-on voltage. Figure 2(a) summarizes R_S and U_0 for microlasers with disk contract. The turn-on voltage slightly increases from 1.09 to 1.228 Volts as the disk diameter gets smaller (D changes from 28 down to 9 μm). Because U_0 is directly associated with the quasi-Fermi levels separation at lasing threshold, such a behavior reflects the fact that optical loss becomes higher in smaller microresonators. The series resistance as a function of microlaser size can be satisfactorily described by two terms: $R_S \approx \rho_S /(\pi D^2/4) + r_S / (2D)$, of which the former corresponds to current flow through the etched mesa with specific resistance $\rho_S = 5.5\times10^{-5}$ $\Omega\times cm^2$, whereas the latter describes current spread in substrate [6] with the coefficient $r_S = 4\times10^{-2}$ $\Omega\times cm$. For simplicity we suggested that the mesa series resistance scales inversely with the disk (not electrical contact) area. However, the situation can be more complicated because of incomplete current spread over the mesa. The fact that the current spread in the mesa is suppressed can be illustrated by comparison of I-V curves of several microdisk lasers with ring-shaped top contact with various inner contact diameter d_C (Figure 2(b)). It is seen that while the turn-on voltage remains unchanged, the series resistance increases with decreasing area of the ring-shaped electrical contact. It is explicitly demonstrated in Figure 2(c) where the resistance is shown as a function of the inner contact diameter d_C.

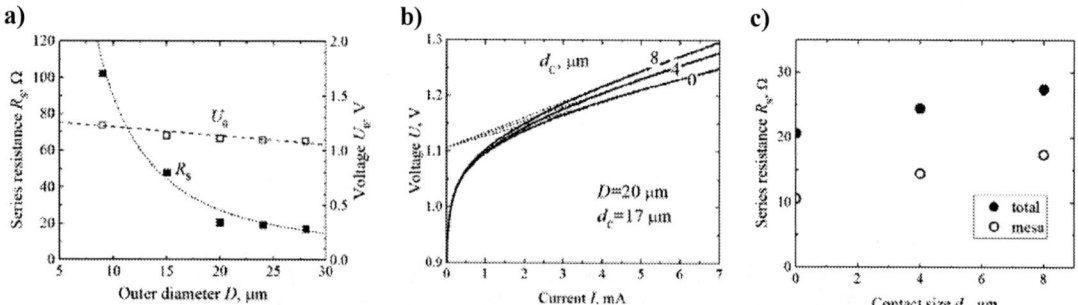

Figure 2. Turn-on voltage (open symbols) and series resistance (solid symbols) in microlasers with round-shaped top contact against disk diameters. Dotted line is the fit curve (see the text), dashed line is guide for eyes (a); I-V curves and their linear fits for disk microlasers with ring-shaped contact (b); series resistance of microlasers of Figure (b) as a function of inner contact diameter: solid symbols – total series resistance, open symbols – mesa-related term only.

Meanwhile we found that the contact topology does not affect the mode structure of the microresonator. In other words, WGM intermode distance is preserved (it is illustrated in Figure 3(a)), however their relative intensities can vary. We also did not found any noticeable impact of ring-shaped contact or even ring-shaped resonator on thermal resistance of microlasers under study. To confirm this statement, Figure 3(b) shows spectral positions of the dominant (lasing) WGMs as a function of injection current for microdisk and microring laser resonators of the same outer diameter. The red-shift of the microlaser mode is caused by the active region self-heating under CW operation. It is seen that the slope $d\lambda / dI$ is practically same (about 0.106-0.108 nm/mA for the presented size of the microresonator). This finding indicates that the main path of heat dissipation in microlasers of this sort is heat removal from the active region through the bottom cladding layer. This path is obviously independent of the top contact geometry and/or inner hole etched via the top cladding only (see Figure 1(d)).

Finally, we compared threshold characteristics of microlasers with different geometry. Figure 4(a) reveals how the threshold current I_{th} reduces with decreasing the outer diameter D of microlasers with conventional (round-shaped) top contact. For the larger microlasers ($D = 24$ and 28 μm) the threshold current density (i.e. $I_{th} / (\pi D^2/4)$) is approximately 1 kA/cm². It does not remain constant for smaller microdisks being 3 times higher for the smallest microdisk lasers ($D = 9$ μm). This finding correlates with previously mentioned increase of the optical loss in smaller resonator. This conclusion is further confirmed by behavior of the lasing wavelength. It is seen in Figure 4(a) that the wavelength of lasing WGM is gradually blue-shifted but still remains within the ground-state optical transition of QDs

(> 1.25 μm). However, for the smallest microdisk resonator lasing proceeds via the first excited-state QD optical transition (~1.2 μm).

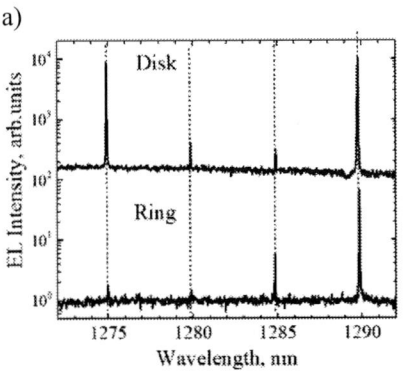 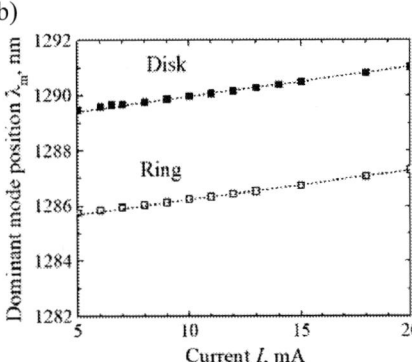

Figure 3. Emission spectra (a) and wavelength of the dominant mode against injection current (b) of disk and ring (d = 6 μm) microlasers of the same outer diameter D =28 μm. In Figure (a), the ring spectrum is shifted by 4 nm for clarity. In Figure (b), dotted lines are linear fits.

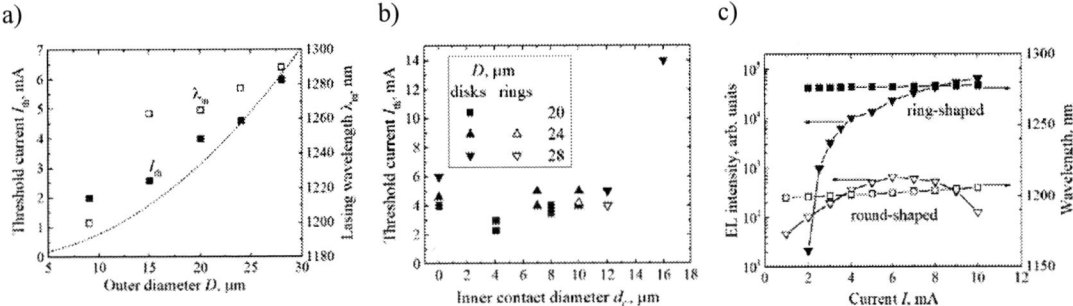

Figure 4. Threshold current (solid symbols) and lasing wavelength (open symbols) against outer diameter of microdisks with round-shaped top contact; dotted line – approximation with constant current density of 1 kA/cm² (a); threshold current against inner contact diameter for microdisk (solid symbols) and microrings (open symbols) of different size: d_c =0 corresponds to round-shaped contact (b); lasing mode intensity (triangles) and wavelength (squares) against current for microlasers with ring-shaped (D = 20 μm, d_c = 4 μm) and round-shaped (D = 9 μm) microdisk lasers (c).

Important finding is that the threshold current can be reduced by applying ring-shaped top contact rather than by bare scaling the outer diameter. It is shown in Figure 4(b) that I_{th} drops (up to 1.5 times) as the inner contact diameter increases initially. Owing to this effect, it is possible to achieve the threshold current as low as 2.3 mA (D = 20 μm, d_c = 4 μm). It is comparable with that of the smallest microlaser with round-shape top contact (D = 9 μm) where I_{th} of 2 mA was measured. At the same time, the former microlaser has two advantages over the latter one (Figure 4(c)). First, lasing still proceeds on the ground-state transition of QDs with sufficiently long emission wavelength (~1.28 μm) and, second, the intensity of the lasing mode does not saturates up to at 10 mA (whereas it start to drop already at 6 mA in the D = 9 μm round-shaped contact microlaser).

4. Conclusions

Quantum dot based injection microdisk and microring lasers operating at room temperature in CW regime were demonstrated. Although it is expected that the resonators of these types are characterized by very low optical loss, we found evidences (such as growth of the threshold current density and of turn-on diode voltage) that the loss does increase with reducing the microresonator outer diameter. As a result of this, lasing wavelength switches from the ground-state to the first excited-state optical transition of QDs as soon as the microdisk diameter reaches 9 μm. It was found that the top contact topology does not affect the WGM structure as well as the thermal resistance of the microresonator provided that the outer diameter is kept unchanged. At the same time, the electrical series resistance increases as the top contact surface area is reduced. The most important finding is that the microdisk lasers with ring-shaped top contact or microring lasers are capable of lowering the threshold current as compared to the microdisk lasers with round-shaped contact of the same outer diameter. The most noticeable effect was found in the microdisk laser having the following geometrical parameters: $D = 20$ μm, $D_C = 17$ μm, $d_C = 4$ μm. In this device, the threshold current of about 2 mA was achieved without a transition to the excited-state lasing. These properties make microlasers with ring-shaped top contact very promising for applications in future optical interconnections on a chip.

Acknowledgments

The work is supported by the Skolkovo Foundation (grant agreement for Russian educational and scientific organisation no.6 dd. 30.12.2015)" and Skolkovo Institute of Science and Technology (General agreement no. 3663-MRA dd. 25.12.2017).

References

[1] Liang D, Fiorentino M, Srinivasan S, Bowers J E and Beausoleil R G 2011 IEEE J. Selected Topics Quantum Electron. **17** 1528.

[2] McCall S L, Levi A F J, Slusher R E, Pearton S J and Logan R A 1992 Appl. Phys. Lett. **60** 289.

[3] G. C. Righini et al Whispering gallery mode microresonators: Fundamentals and applications.

[4] Zhukov A E, Kovsh A R, Maleev N A, Mikhrin S S, Ustinov V M, Tsatsul'nikov A F, Maximov M V, Volovik B V, Bedarev D A, Shernyakov Yu M, Kop'ev P S, Alferov Zh I, Ledentsov N N and Bimberg D 1999 Appl. Phys. Lett. **75** 1926.

[5] Kryzhanovskaya N V, Zhukov A E, Nadtochii A M, Maximov M V, Moiseev E I, Kulagina M M, Savelyev A V, Arakcheeva E M, Lipovskii A A, Zubov F I, Kapsalis A, Mesaritakis C, Syvridis D, Mintairov A, Livshits D 2013 Semiconductors **47** 1387.

[6] Electrical Contacts: Principles and Applications, 2nd Edition (ed. by P. G. Slade), CRC Press, 2014 (NY, USA).

SPBOPEN 2018

IOP Publishing

The ways to improve the energy conversion efficiency in erbium-doped Gd$_2$O$_3$ nanoparticles

Yu A Kuznetsova[1], A F Zatsepin[1], M A Mashkovtsev[1], V N Rychkov[1], S P Vyatkina[1]

[1]Institute of Physics and Technology, Ural Federal University, Ekaterinburg, Russia

Abstract. The basic requirements for the crystal lattice and defectiveness of Gd$_2$O$_3$ matrix as well as for the concentration of Er^{3+} dopants to achieve the enhanced parameters of energy conversion in Gd$_2$O$_3$:Er nanoparticles are summarized. The obtained data allow to optimize and improve the functional characteristics of Gd$_2$O$_3$:Er -based down-conversion layers applying in solar cells.

1. Introduction

Gadolinium oxide doped with rare-earth ions are of interest as a new type of material for energy conversion devices, in particular, for solar cells [1-3]. Improvement of functional characteristics of convertors based on rare-earths requires the optimization of synthesis technologies which determine the atomic and electronic structure of materials and, ultimately, their optical properties. In present work, we summarize the basic requirements for the crystal structure type and defectiveness of Gd$_2$O$_3$ matrix as well as the concentration of dopant ions to achieve the enhanced energy conversion efficiency.

2. Samples and Methods

The pure and erbium-doped Gd$_2$O$_3$ nanoparticles with a cubic crystal lattice were synthesized by chemical precipitation from water-alcohol solutions of gadolinium and erbium nitrates with the obtaining of layered rare-earth hydroxides as intermediate products. The method of self-propagating high-temperature combustion with glycine and ammonium nitrate as fuels was used for obtaining the monoclinic modification of Gd$_2$O$_3$ nanoparticles. XRD, SEM and Raman spectroscopy data confirmed that the chosen technologies of synthesis provide the obtaining of stable low-dimensional monophasic erbium-doped gadolinium oxide with an average particle size of 50 nm.

Optical properties of cubic and monoclinic nano-Gd$_2$O$_3$:Er polymorphs were investigated in a 8-300 K temperature range by using a McPherson VuVAS spectrometer and Perkin Elmer LS 55 spectrophotometer.

3. Results

3.1 Requirements for Gd$_2$O$_3$ matrix

Experimental and theoretical study of gadolinium oxide electronic structure by X-ray photoelectron spectroscopy (XPS) and density functional theory (DFT) calculations revealed intrinsic point defects caused by the technological features of synthesis for nanoparticles with cubic and monoclinic crystal structures [4]. In cubic polymorph, there is a violation of oxygen coordination of lattice cations (oxygen vacancies). Photoluminescence optical spectroscopy data combined with the thermoluminescence measurements indicated three types of oxygen-deficient centers with different

---

Content from this work may be used under the terms of the Creative Commons Attribution 3.0 licence. Any further distribution of this work must maintain attribution to the author(s) and the title of the work, journal citation and DOI.

Published under licence by IOP Publishing Ltd

charge states: F^{2+}, F^+ and F – centers [5]. Non-elementary luminescence in $2.0 - 3.5$ eV spectral range observed for cubic phase of nano-Gd_2O_3 is formed by radiative transitions in different types of oxygen vacancies. For the monoclinic nano-Gd_2O_3 phase no optical activity was detected even at helium temperature. The reason for this, as XPS and DFT data indicate, is presence of impurity hydroxide ions that have emerged from the precursors used in the synthesis. OH^- – groups are external quenchers of luminescence due to an increase in the probability of non-radiative deactivation of excited states of emission centers [4]. Thereby, the monoclinic polymorph is not suitable for energy conversion purposes and we focus on Gd_2O_3 nanoparticles with a cubic structure.

Figure 1 shows the photoluminescence (PL) and PL excitation spectra for cubic nano-Gd_2O_3 doped with erbium ions. The main feature is the presence of bands corresponding to the transitions in Gd^{3+} ions, that is not typical for regular cations of host lattice. Cations in regular lattice positions form extended states of conduction band, so there are no any local states of Gd^{3+} ions in the ideal Gd_2O_3 lattice. An existence of Gd^{3+} local electronic levels in the optical transparence area of Gd_2O_3 host-matrix indicates that these ions have distorted energy structure. It is reasonable explained by influence of oxygen defects in the nearest environment of Gd^{3+} ions, which are responsible for the change in the cation energy structure and the appearance of additional Gd^{3+} electronic states in the Gd_2O_3 band gap region. By this way, the Gd^{3+} ions become optically active and can be excited in the UV spectral region with the following energy transfer to the erbium dopants. This provides an additional channel for energy conversion and one of the main advantages of Gd_2O_3:Er system, because there is no need to introduce into the matrix the ions acting as donors of excitation, since this role is played by host lattice cations – Gd^{3+} ions. Down-conversion in Gd^{3+} – Er^{3+} pair is especially promising for solar energy applications. It can provide a reduction in thermalization losses that occur in a silicon solar cell when absorbing photons with an energy exceeding the Si band gap. In other words, additional quanta with UV energy will participate in the solar energy conversion [1-3].

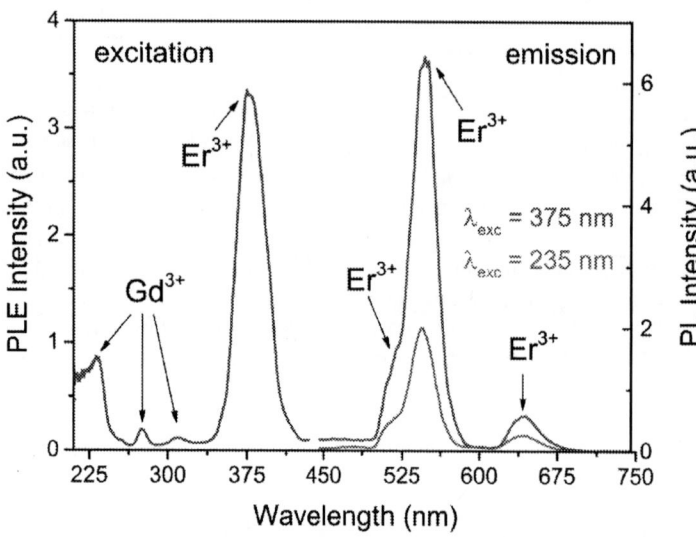

Figure 1. Photoluminescence (red and green lines, right scale) and excitation (blue line, left scale) spectra for Gd_2O_3:Er nanoparticles at room temperature. Arrows show the optical transitions in Gd^{3+} and Er^{3+} ions.

3.2 Requirements for Er^{3+} concentration
After determining the matrix characteristics, the next task is to find the optimal concentration of dopant. On the one hand, with an increase in the Er^{3+} concentration, the efficiency and the rate of $Gd^{3+} \rightarrow Er^{3+}$ energy transfer also increase due to a shortening of donor-acceptor distance. However, at a high amount of acceptor the non-radiative energy migration along the Er^{3+} ions can occur (concentration quenching), which leads to a decrease in quantum efficiency [6]. Thus, it is necessary to determine such Er^{3+} concentration, that will provide a high overall efficiency of energy conversion.

We have considered this problem by analyzing the temperature dependences of luminescence for Gd_2O_3 nanoparticles with different Er^{3+} concentration (from 0.25 to 8%).

Monitoring changes in the luminescence properties of Gd_2O_3:Er nanoparticles in temperature range of 8-300 K we found that quenching curves for Er^{3+} emission don't obey the classic Mott law and the low-temperature plateau is absent in the $I(T)$ dependences. It means the activation energy for emission quenching for Er^{3+} centers in Gd_2O_3 nanoparticles takes a dispersed rather than a discrete value. The dispersion of optical centers over the energy barrier for luminescence quenching was represented by a Gaussian with different parameters for different Er^{3+} concentrations. A detailed analysis of quenching curves is described in our work [7]. The largest values of width and maximum of the distribution were found for Gd_2O_3:Er (1%) nanoparticles. This means that for a given activator concentration, one can expect the least non-radiative losses. It is confirmed also by the phenomenon of giant phonon softening found by independent measurements of the reflection spectra for nano-Gd_2O_3:Er, where namely for Gd_2O_3:Er (1%) sample the smallest frequency of lattice vibrations was detected [8]. We performed a quantitative estimation of quantum efficiency of luminescence for all samples of the concentration series Gd_2O_3:Er (0.25-8%), following the equation that accounts the dispersion of energy barrier for luminescence quenching:

$$\eta = \frac{1}{1 + \int\limits_0^\infty \exp(-E_a/kT)g(E_a)dE_a} \, , \tag{1}$$

where η is the quantum efficiency, E_a is the activation energy for emission quenching, $g(E_a)$ is the Gaussian distribution of optical centers over the E_a and k is the Boltzmann constant. The obtained quantum efficiency values as a function of Er^{3+} concentration are shown in figure 2. Indeed, for Gd_2O_3:Er (1%) nanoparticles there is enhanced quantum efficiency of erbium luminescence under UV radiation conversion due to the minimization of thermal losses. The basic requirements for Gd_2O_3 matrix and Er^{3+} dopant for achieving the improved characteristics of energy conversion as well as the main optical parameters of Gd_2O_3:Er system are summarized in table 1.

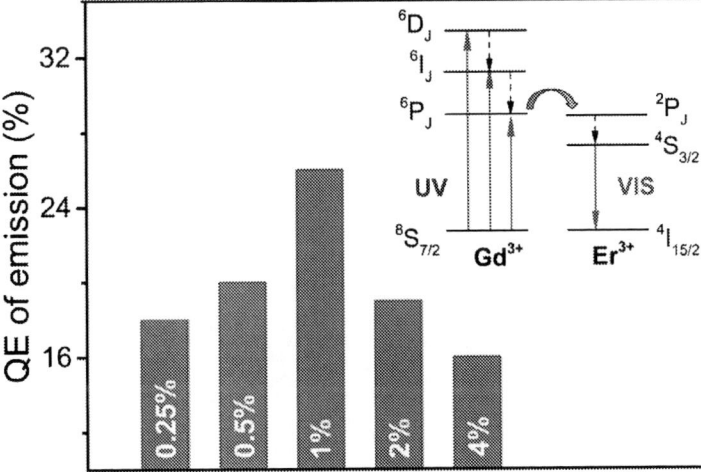

Figure 2. Quantum efficiency (QE) of erbium visible emission under UV radiation conversion due to the $Gd^{3+} \rightarrow Er^{3+}$ energy transfer in Gd_2O_3 nanoparticles doped with 0.25-4% of Er^{3+} ions.

Table 1. Structural and optical parameters of Gd_2O_3:Er nanoparticles for achieving the improved efficiency of UV-visible conversion.

Fundamental characteristics		Tunable parameters	
Crystal structure	cubic Ia-3 [4]	Er^{3+} concentration	1% [7]
Lattice parameter	10.81 Å [2]	$Gd^{3+} \to Er^{3+}$ energy transfer efficiency	50% [2]
Direct energy gap	5.38 eV [8]	$Gd^{3+} \to Er^{3+}$ energy transfer rate	29 µs [7]
Indirect energy gap	5.15 eV [8]	Quantum efficiency of emission	26% [7]
Phonon frequency	242 cm^{-1} [8]	Decay time of emission	86 µs [7]

4. Conclusion

Summarizing the data about atomic structure, electronic states and optical properties of Gd_2O_3:Er nanoparticles we established the basic requirements for matrix and activator to achieve improved characteristics of UV-visible energy conversion. The cubic structure of Gd_2O_3 host lattice is more preferable than the monoclinic phase due to the presence of optically active defective Gd^{3+} cations acting as excitation donors and providing an additional channel for UV radiation conversion. Monoclinic structure of Gd_2O_3 nanoparticles is not suitable for application as a luminescent material because of impurity hydroxide ions that arise at the synthesis stage and completely quench the emission.

The optimal concentration of Er^{3+} activator among the Gd_2O_3:Er (0.25-8%) series is 1%. Nanoparticles with such dopant amount have a larger barrier for luminescence quenching and, as a result, the highest quantum efficiency. These recommendations are the basis for further searching the ways to tunable and improve the characteristics of energy conversion devices on the basis of Gd_2O_3:Er system.

Acknowledgments

The work has been funded by the Ministry of Education and Science of the Russian Federation (Government task №3.1485.2017/4.6).

References

[1] Zatsepin A F, Kuznetsova Yu. A. 2015 *J. Phys. Conf. Ser.* **643** 012057
[2] Kuznetsova Yu A, Zatsepin A F, Pustovarov V A et all 2017 *J. Phys. Conf. Ser.* **917** 052015
[3] Trofimova E S, Pustovarov V A, Zatsepin A F 2017 *AIP Conf. Proc.* **1886** 020024
[4] Zatsepin D A, Boukhvalov D W, Zatsepin A F et all 2018 *Appl. Surf. Sci.* **436** 697-707
[5] Zatsepin A F, Kuznetsova Yu A, Spallino L et all 2016 *Energy Proc.* **102** 144-151
[6] Bünzli J-C G, Eliseeva S B 2010 *Basics of Lanthanide Photophysics* (Springer, Berlin)
[7] Zatsepin A F, Kuznetsova Yu A 2018 *Appl. Mater. Today* **12** 34-42
[8] Zatsepin A F, Kuznetsova Yu A et all 2018 *Iop Conf. Ser.: Mat. Sci. Eng.* **292** 012047

Theoretical model for spatial separation of dominating recombination region in a-Si:H/c-Si structures

A.D. Maslov, E.V. Bezuglaya

Department of Micro- and nanoelectronics, Ryazan State Radio Engineering University, Ryazan 390005, Russia

Abstract. In this paper we describe a theoretical model for spatial quantitative separation and determination of dominating recombination region in structures with amorphous/crystalline silicon heterojunction. This model is based on analysis of ideality factor, experimentally obtained from open circuit voltage dependence of generation rate; characteristic energy, corresponding to the maximum possible quasi Fermi level separation, experimentally obtained from temperature dependence of open circuit voltage. Compared to the existing approaches in this work we consider recombination in the bulk of amorphous silicon.

1. Introduction

There is a model [1] for mathematical separation of total recombination rate $R_{total} = \sum R_i$, described by SRH statistics. The obtained values correspond to the different spatial regions of photoactive layers in solar cells. Grover [1] spatially separated investigated structure of CIGS and a-Si/c-Si solar cells and marked 3 regions: interface, depletion region and quasi neutral part of the base (Figure 1). This approach allows to define dominating recombination region and calculate contribution of every region on total recombination rate. In this case discontinuity equation under open circuit conditions:

$$G = R_i + R_b + R_d, \tag{1}$$

where G – generation rate,- R_i – recombination rate at the interface $R_i = R_0^I \beta^2$, R_b – recombination rate in the quasi-neutral part of the base $R_b = R_0^b \beta^2$, and R_d – recombination rate in the depletion region $R_d = R_0^d \beta$.
Grover denoted each recombination rate as a multiplication of light dependent β and light independent values R_0^I, R_0^d, R_0^b.

$$G_{avg} = [R_0^I + R_o^b]\beta^2 + R_o^d \beta. \tag{2}$$

Here β is defined from the difference of quasi Fermi levels that equals open circuit voltage V_{oc}:

$$V_{oc} = \frac{kT}{q} \ln \frac{n_e p_h}{n_i^2} = \frac{kT}{q} \ln \beta^2, \tag{3}$$

with n_e and p_h – total concentration of electrons and holes respectively, n_i – intrinsic charge carrier concentration, k – Boltzmann constant, T – temperature, q – elementary charge.
This approach suggests to define diode ideality factor n from experimental $G(V_{oc})$ measurements and links n with β and R_0^I, R_0^d, R_0^b.

Content from this work may be used under the terms of the Creative Commons Attribution 3.0 licence. Any further distribution of this work must maintain attribution to the author(s) and the title of the work, journal citation and DOI.
Published under licence by IOP Publishing Ltd

$$n = \left[\frac{kT}{q}\frac{d(\ln G_{avg})}{dV_{oc}}\right]^{-1}. \tag{4}$$

Solving quadratic equation (2) about β and considering (3) and (4):

$$n(G) = \frac{k_2 G}{\sqrt{1+k_2 G}\left[\sqrt{1+k_2 G}-1\right]}, \tag{5}$$

with

$$k_1 = \frac{R_0^a + R_o^d}{2*(R_0^l + R_o^b)}, k_2 = \frac{4*(R_0^l + R_o^b)}{(R_0^a + R_o^d)^2}. \tag{6}$$

Considering the fact that ideality factor provides information about dominating recombination region discontinuity equation could be simplified:

$$G = (R_i^0 + R_b^0)\beta^2, \text{ or } G = R_d^0\beta. \tag{7}$$

However, despite the fact that this model developed for heterojunction amorphous/crystalline silicon structures, recombination in the bulk of amorphous silicon is not considered there. Moreover recombination in a-Si could be described on the basis of classical SRH approach and amphoteric defect model [2]. It should be considered for describing recombination in amorphous silicon.

In this paper we extended Grover's model and analytically considered recombination in the bulk of amorphous silicon layer.

2. Model

We spatially separate a-Si/c-Si structure and mark four recombination regions: interface R_i, depletion region R_d, quasi neutral part of the base R_b and the bulk of amorphous silicon R_a (Figure 1).

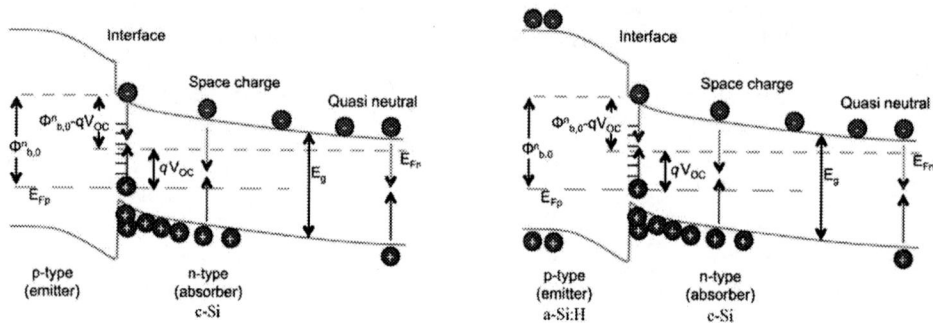

Figure 1 – Spatial separation of recombination regions in a-Si:H/c-Si cells by the Grover model (left) and proposed model (right)

According to [1] we denote recombination rate in each region as a multiplication of light dependent and light independent values. We denote recombination rate in amorphous silicon R_a similarly, that provides the extended equation of generation-recombination balance (8):

$$G = R_i + R_b + R_d + R_a. \tag{8}$$

Following Steingrube [2] and Olibet [3] recombination rate in the bulk of amorphous silicon could be defined considering classical SRH statistics and amphoteric defect model:

$$R_{SRH} = \left(\frac{1}{\sigma_p p_f} + \frac{1}{\sigma_n n_f}\right)^{-1} v_{th} N_t, \tag{9}$$

$$R_{DB} = \frac{n_f \sigma_n^0 + p_f \sigma_p^0}{\frac{p_f \sigma_p^0}{n_f \sigma_n^+} + 1 + \frac{n_f \sigma_n^0}{p_f \sigma_p^-}} v_{th} N_{DB}. \tag{10}$$

These equations consider concentration of traps N_t or dangling bonds N_{DB}, thermal velocity v_{th}, concentration of free electrons n_f or holes p_f, capture cross section of holes σ_p and electrons σ_n with positive σ_n^+, neutral σ_n^0, σ_p^0 and negative σ_n^-, σ_p^- charge for amphoteric states.

However in the case of n-type, for instance, these equations could be simplified [3]:

$$R_{DB} = p_f \sigma_p^- v_{th} N_{DB}, \quad R_{SRH} = p_f \sigma_p v_{th} N_t. \tag{11}$$

In this case there is no functional difference between SRH and amphoteric defect model. Considering that recombination in amorphous silicon corresponds to the case of non-ideal diode $n \geq 2$ [4], we denote $R_a = R_a^0 \beta$.

Replacing recombination rate by multiplication:

$$G = (R_i^0 + R_b^0)\beta^2 + (R_d^0 + R_a^0)\beta, \tag{12}$$

where R_i^0, R_b^0, R_d^0, R_a^0 light independent values.

The proposed model is based on analysis of ideality factor, experimentally obtained from open circuit voltage dependence of generation rate (Figure 2). The obtained value allows to estimate dominating recombination in the quasi neutral base or at the interface ($n \sim 1$); and in the depletion region or in the bulk of amorphous silicon ($n \sim 2$). In the case of $n \gg 2$, we assume that charge carrier recombine predominantly in a-Si [4]. Therefore, defined ideality factor provides simplified equation of generation-recombination balance (13), considering only dominating regions.

$$G = (R_i^0 + R_b^0)\beta^2, \text{ or } G = (R_d^0 + R_a^0)\beta. \tag{13}$$

To calculate the prevailing value in the sums $R_i^0 + R_b^0$ and $R_d^0 + R_a^0$, we, considering Grover [1], suggest to measure temperature dependence of open circuit voltage (Figure 2).

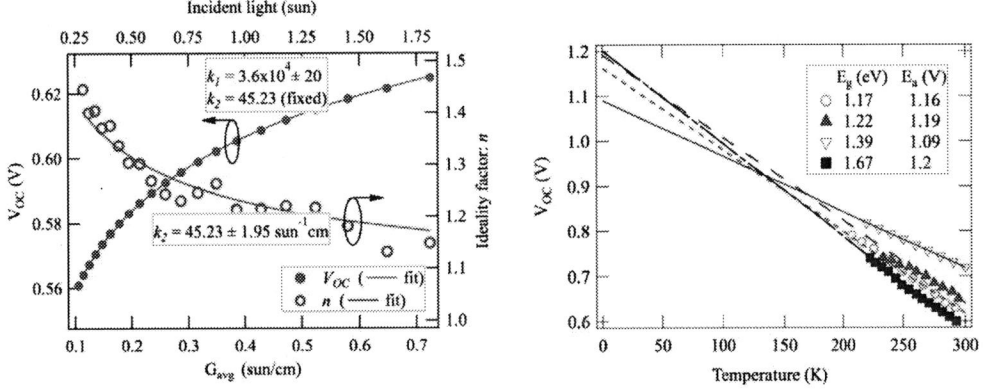

Figure 2 – Dependence of open circuit voltage from generation rate and temperature dependence of open circuit voltage with extrapolation of the obtained data to estimate maximum possible quasi Fermi level separation [1]

Extrapolation of this dependence allows to define a characteristic energy, corresponding to the maximum possible quasi Fermi level separation, which equals:

1. $E_a = E_g(c - Si)$ c-Si band gap, in the case of dominating recombination in quasi neutral part of crystalline silicon or

$E_a = \varphi_{b,0}^n(c - Si)$ interface potential, in the case of dominating recombination at the interface, $n \sim 1$;

2. $E_a = E_g(a - Si:H)$ a-Si mobility gap, in the case of dominating recombination in the bulk of amorphous silicon or

$E_a = E_g(c - Si)$ c-Si band gap, in the case of dominating recombination in the depletion region, $n \sim 2$.

3. Conclusion

The proposed model allows to separate quantitatively recombination regions and to calculate maximum value of recombination rate in the corresponding region, which significantly determines total recombination rate in heterojunction structures based on a-Si/c-Si. The described theory, extended by experimental measurements, provides these tasks. It this case the approach consists of the following steps:

- experimental determination of ideality factor from $G(V_{oc})$ dependence;
- analysis of pair of dominating regions according to ideality factor equals 1 or 2 (≥ 2);
- measuring temperature dependence of open circuit voltage and extrapolation of the obtained data to define characteristic energy;
- considering value of the characteristic energy estimation of dominating region quantitatively.

Despite that other recombination rates except the dominating value are neglected and solution is based on the simplified discontinuity equation, the physical and mathematical aspects of the model could be extended for the case of total generation-recombination balance $G = (R_i^0 + R_b^0)\beta^2 + (R_d^0 + R_a^0)\beta$, that will provide all of the recombination rate values.

References

[1] Grover S, Li J V, Young D L, Stradins P, Branz H M 2013 *Appl. Phys. Lett.* 103 093502
[2] Steingrube S, Brendel R, Altermatt P P 2012 *Phys. Status Solidi A* 209 2 390-400
[3] Olibet S, Vallat-Sauvain E, Ballif C 2007 *Phys. Rewiev B* 76 035326
[4] Kroon M A, Swaaij R A C M M 2001 *Journal of Applied Physics* 90 2 994-1000

Surface morphology after reactive ion etching of silicon and gallium arsenide based solar cells

N Papež[1], D Sobola[1], A Gajdoš[1], Ľ Škvarenina[1], R Macků[1], M Eliáš[2], A Nebojsa[2], R Motúz[3]

[1]Department of Physics, Brno University of Technology, Technická 2848/8, 616 00 Brno, Czech Republic

[2]CEITEC Nano Research Infrastructure, Brno University of Technology, Purkyňova 123, 612 00 Brno, Czech Republic

[3]Department of Theoretical and Experimental Electrical Engineering, Brno University of Technology, Technická 3082/12, 616 00 Brno, Czech Republic

Abstract. This work aims to study the surface morphology of solar cells before and after reactive ion etching (RIE) at two different pressures. Two types of solar cell based on GaAs and polycrystalline Si were processed and compared. Energy Dispersive X-ray spectroscopy (EDS) was used for the composition analysis of the samples. Raman spectroscopy showed a structural fingerprint of materials before and after processing. Atomic Force Microscope (AFM) demonstrated dimensional topography with high resolution. Optical spectrometer detected changing of reflectance the samples. Experimentally, it has been confirmed that GaAs solar cells have a very high endurance to ion bombardment in comparison to Si cells.

1. Introduction

It is known that silicon is the most used material for the fabrication of solar cells. Due to this it is also widely available with an affordable purchase price. On the other hand, GaAs based cells are mainly used in difficult conditions such as space, military or air. We focused on the morphology and structure of the surface of these two types of solar cells to demonstrate their real endurance and compare the differences between them. For surface processing, we chose ion bombardment. It has been shown that in some cases the loss of energy is caused by the reflection of light from the surface, and surface treatment or an antireflective film is required to improve light absorption efficiency [1, 2].

2. Methods

The samples of GaAs and polycrystalline Si solar cells were etched by chemically reactive plasma. This was done using a PlasmaPro 100 Cobra ICP Etch system, where activated reactive gases were made of sulphur hexafluoride (SF_6), chlorine (Cl_2) and oxygen (O_2). Etch conditions strongly depend on gas flow, pressure and RF power. Therefore, we chose two treatments under different conditions. In our case RF power was 200 watts for both processing the same, as well as the gas flow was not changed. However, the difference was in the pressure setting – for the first processing we chose 40 mTorr and for the second processing 10 mTorr. It is also important to note that these two treatments did not follow each other, but was done separately for each new specimen.

Content from this work may be used under the terms of the Creative Commons Attribution 3.0 licence. Any further distribution of this work must maintain attribution to the author(s) and the title of the work, journal citation and DOI.

Published under licence by IOP Publishing Ltd

3. Experimental results

After etching of silicon at a pressure of 40 mTorr in Figure 1b, it is noticeably visible that the surface differences are up to 2 μm greater than the surface in Figure 1a which was not processed [3]. These changes indicate a significant increase of S_a parameter, up to 576.5 nm, in Table 1. This parameter expresses the absolute value of the difference in height of each point relative to the arithmetic mean of the surface. There is also a much lower occurrence of features than in Figure 1a.

After etching at a pressure of 10 mTorr, Figure 1c shows much sharper features than before etching in Figure 1a. Sharpness is also indicated by coefficient of kurtosis of −0.386 in Table 1. In comparison between the two figures we can see the relatively similar roughness of the surface, which is also confirmed by the value of S_a in Table 1.

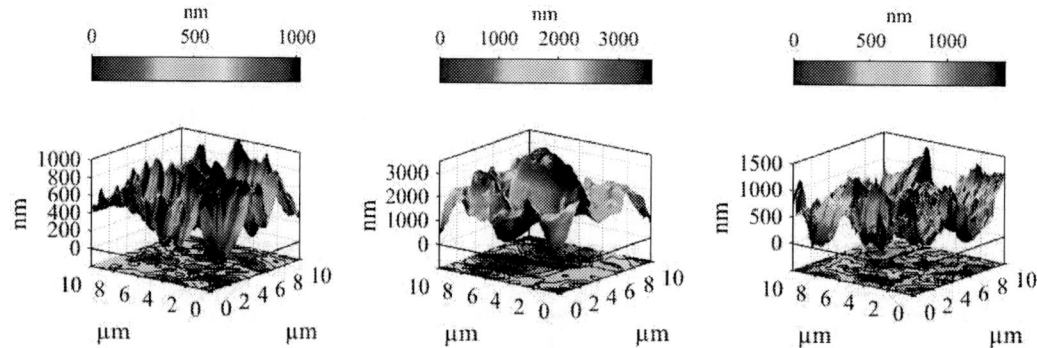

Figure 1(a, b, c). Measurement by AFM; (a) Si solar cell before processing; (b) Si solar cell after RIE processing with pressure of 40 mTorr; (c) Si solar cell after RIE processing with pressure of 10 mTorr.

Differences in morphology of etching GaAs based solar cell compared to the Si based are noticeable from the Figure 2. In particular, there are minimal peak height changes which, after etching at 40 mTorr (Figure 2b) and 10 mTorr (Figure 2c), have a difference of only 5 nm. For silicon based solar cells, the peak height change occurred within 2 μm, as written above.

With decreasing etching pressure, an increasing number of features (Figure 2b and 2c) and higher coefficient of kurtosis can be observed. It can be argued that, at pressure of 10 mTorr, the sharper edges of the features begin to form on both the silicon sample (Figure 1c) and the gallium arsenide sample (Figure 2c). However, the structural changes occurred on the Si cell surface in order of micrometers, but for the GaAs solar cell, it was in nanometer units. Thus, in GaAs sample, much higher reflectivity can be expected in this case. Reflectivity is directly related to surface roughness.

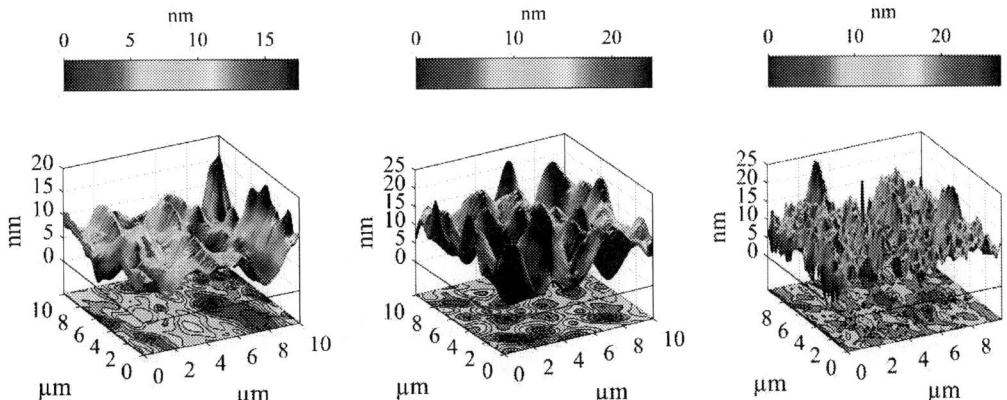

Figure 2(a, b, c). Measurement by AFM; (a) GaAs solar cell before processing; (b) GaAs solar cell after RIE processing with pressure of 40 mTorr; (c) GaAs solar cell after RIE processing with pressure of 10 mTorr.

Compared to silicon in gallium arsenide cells, minor differences were observed with AFM. The heights remained constant at similar sizes around 20 nm [4], while in the second processing with lower pressure, a larger number of features began to appear. Features achieved the highest heights during etching pressure of 10 mTorr.

Table 1. Calculated S parameters obtained by AFM.

	Average Roughness S_a [nm]		Root Mean Square S_q [nm]		Surface skewness S_{sk} [−]		Coefficient of kurtosis S_{ka} [−]	
	Si	GaAs	Si	GaAs	Si	GaAs	Si	GaAs
No etching	152.1	2.133	186.6	2.716	−0.1	0,133	−0.487	0.085
RIE, 40 mTorr	576.5	3.327	741.7	4.111	0.077	0.056	−0.102	−0.28
RIE, 10 mTorr	186.6	2.356	228.9	3.057	0.139	0.207	−0.386	0.976

Significant changes in the composition of material can be observed by using of electron dispersion spectroscopy on the cell surface in Figure 3. After etching at pressure of 10 mTorr there was a noticeable influence of SF_6 reactive gas, which is characterized by a fluorine peak in the spectrum. In contrast, at a higher pressure of 40 mTorr, there was a decrease of the fluorine peak. However, the intensity of silicon at the 1.75 keV band almost doubled against non-etched sample. This can be also attributed to the loss of silver that is used in silicon as a doped material.

In the case of GaAs sample, a much lower difference in the material composition compared to Si sample can be observed. In particular, the changes at different etch pressure in Figure 4. Here the composition of the material was almost the same. However, the decrease in intensity is noticeable before and after processing.

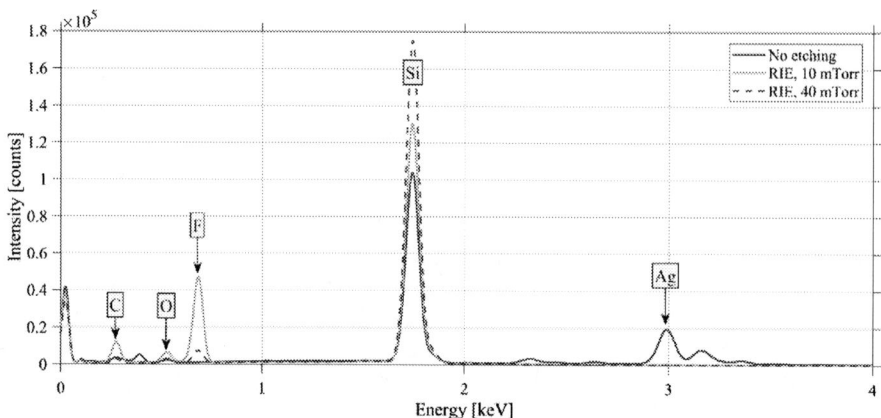

Figure 3. Material composition of Si based solar cell measured by EDS.

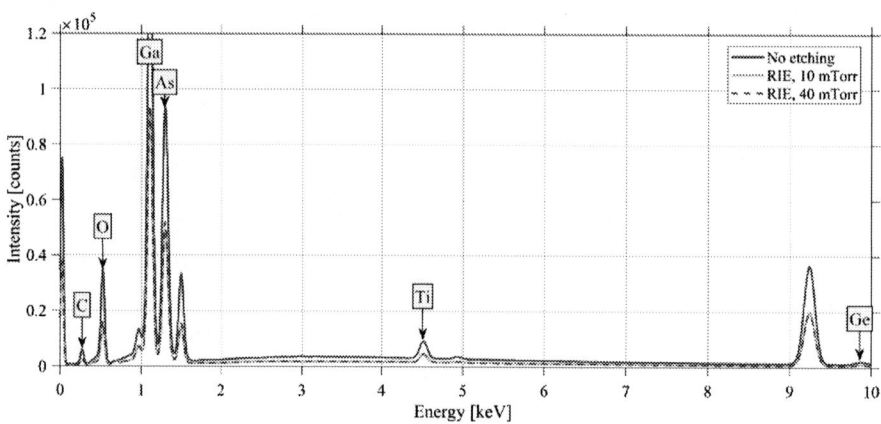

Figure 4. Material composition of GaAs based solar cell measured by EDS.

Using of Raman spectroscopy, the change in intensity for both types of cell was observed on the samples. However, each material has a different behavior within the Raman spectra. Figure 5a shows a typical silicon spectrum with several silicone modes. The decrease in spectrum intensity is evident. The largest drop was observed in the case of etching at pressure of 10 mTorr. A number of factors may lead to this effect. One of them may for example be a different roughness of the surface.

The opposite phenomenon can be seen in GaAs spectrum (Figure 5b). Band shape and intensity are strongly influenced by etching. However, unlike silicon, the intensity of the spectrum has increased after etching. There is also a larger peak with full width at half maximum (FWHM). On a longitudinal mode of GaAs of about 270 cm^{-1}, a peak slope can be observed after etching. This also indicates mechanical stresses due to structural damage. Longitudinal mode of AlAs with a value of about 395 cm^{-1} has a much smaller height after etching. Which could be expected because the GaAs sample was doped with aluminum on its surface.

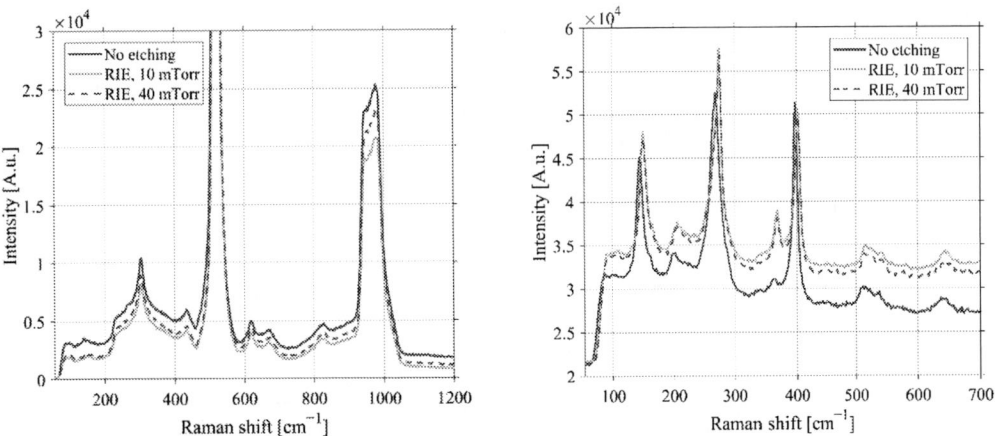

Figure 5(a, b). Measurement using Raman spectroscopy; (a) Si solar cell before and after processing; (b) GaAs solar cell before and after processing.

Measured reflectance of both types of solar cells confirms the intensity drop in both etching pressure of 40 and 10 mTorr. Interestingly, the Si based cell had a higher decline in reflectivity using etching at 40 mTorr (Figure 6a), but in the GaAs-based cell there was higher drop in etching at 10 mTorr (Figure 6b). This confirms the different behavior of the morphology of these materials. It is also noticeable from the figures that the GaAs solar cell has a much greater reflectivity than Si cell which also contributes to a much higher roughness degree as mentioned in Table 1.

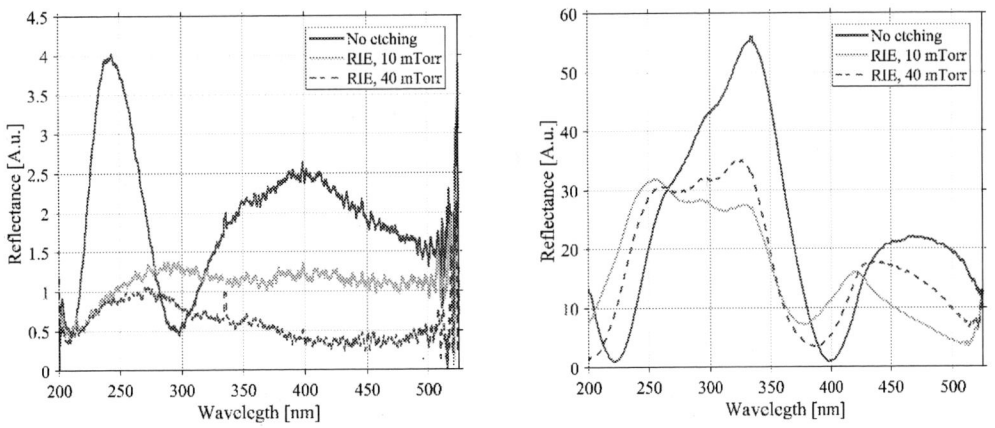

Figure 6(a, b). Measurement of reflectance in the ultraviolet region of the spectrum; (a) Si solar cell before and after processing; (b) GaAs solar cell before and after processing.

4. Conclusion

Using various experimental measurements, it has been verified that silicon based solar cells are more susceptible to surface modification than GaAs based solar cells processed by RIE. These differences are particularly noticeable in Table 1, where data were obtained using AFM, where the average values for silicon vary within hundreds of nanometers. An increasing coefficient of kurtosis was also observed for both types of cells with 10 mTorr pressure.

It was found that the material composition significantly differs in silicon after etching analysing by EDS. Compared to that, in GaAs sample was found only decrease in intensity. Any other changes between etching pressures of 10 and 40 mTorr for this cell were not observed.

Raman spectroscopy revealed the different behaviour of both types of cells. With the decreasing etching pressure on the Si specimen, the intensity of the Raman peaks has also decreased. In contrast to GaAs specimens with lower etching pressure, the total spectral intensity increased.

Optical spectrometer confirmed a difference for material modification and reflectance changing for both Si and GaAs samples. The reflectivity dropped after etching in all cases. This was already assumed according to the findings from the measured roughness degree using AFM. However, much higher reflectivity was measured on the GaAs sample surface. This is due to the fact that the features of GaAs surface rise to about 20 nm, on the other hand, the Si surface features differs by up to 3 μm which is a very significant difference.

Acknowledgments

Research described in the paper was financially supported by the Ministry of Education, Youth and Sports of the Czech Republic under the project CEITEC 2020 (LQ1601), by the National Sustainability Program under grant LO1401 and by Internal Grant Agency of Brno University of Technology, grant No. FEKT-S-17-4626. Part of the work was carried out with the support of CEITEC Nano Research Infrastructure (MEYS CR, 2016–2019). For the research, infrastructure of the SIX Center was used.

References

[1] Liu S, Niu X, Shan W, et al 2014 *Solar Energy Materials and Solar Cells* 127
[2] Papež N, Škvarenina Ľ, Tofel P, Sobola D 2017 *Proc. SPIE* 1060313
[3] Ţălu Ş, Papež N, Sobola D, et al 2017 *Journal of Materials Science: Materials in Electronics* **28** 20
[4] Sobola D, Ţălu Ş, Sadovský P, et al 2017 *Advances in Electrical and Electronic Engineering* **3** 15

Reducing of thermal resistance of edge-emitting lasers based on coupled waveguides

A A Serin[1], A S Payusov[1], Yu M Shernyakov[1], N A Kalyuzhnyy[1], S A Mintairov[1], M V Maximov[2], N Yu Gordeev[1] and A E Zhukov[2]

[1]Ioffe Institute, 194021, St Petersburg, Russia
[2]St Petersburg Academic University, 194021, St Petersburg, Russia

E-mail: spbgate21@gmail.com

Abstract. We present a study on minimization of thermal resistance of edge-emitting diode lasers by placing the active region closer to the laser surface. We demonstrate that using coupled large optical cavity (CLOC) approach one can reduce the thickness of the p-cladding down to 0.5 μm due to the strong mode localization in a broad active waveguide (1.35 μm) coupled to a passive optical cavity. Also, the advanced laser design allowed us to put the active region closer to the cladding in comparison with conventional broad waveguide lasers. As a result, a record-low thermal resistance of 2 K/W for 3 mm device directly mounted on a copper heatsink in combination with a low optical internal loss of 0.4 cm^{-1} where demonstrated.

1. Introduction

Thermal resistance (R_{th}) of edge-emitting lasers plays a crucial role in its performance under high injection levels in a continuous wave (CW) regime. As the injection level grows, the active medium heats up and consequently diminishes the maximal output power. This effect is known as a thermal rollover [1]. Lower thermal resistance would provide better heat dissipation from the active medium resulting in higher CW optical power.

In diode lasers mounted epi-side down, a substantial fraction of the thermal resistance is associated with the layers located between the active medium and the epi-side surface covered usually with a thin metal film (metal p-contact). Hence, an evident approach for lowering R_{th} is to put the active region as close as possible to the epi-side surface. In conventional separate confinement heterostructures, intermediate layers between the active region and the surface have a typical total thickness of around 2 μm and include a highly doped cap layer, top cladding layer, and top part of a waveguide (WG) layer. Thinning of those layers requires a proper laser design, otherwise it may lead to p-metal diffusion, optical mode leakage, and increased fast-axis divergence angle, respectively. Longer laser cavities are favorable for effective heat dissipation since thermal resistance is inversely proportional to the cavity length [2]. However, this approach requires reducing the internal optical loss to compensate reduced mirror loss [3]. In this work, we present a study of InGaAs/GaAs/AlGaAs lasers with an advanced waveguide design intended for lowering both thermal resistance and internal loss.

Content from this work may be used under the terms of the Creative Commons Attribution 3.0 licence. Any further distribution of this work must maintain attribution to the author(s) and the title of the work, journal citation and DOI.
Published under licence by IOP Publishing Ltd

2. Laser wafer design

Our waveguide design follows a novel concept of transverse mode engineering that is based on Coupled Large Optical Cavity (CLOC) approach allowing effective suppressing of high-order mode lasing [4]. Fundamental modes propagating in multimode waveguides would have better localization in the waveguide core resulting in a lower optical loss. Other advantages of such waveguides are the following: firstly, thick claddings are not required since the fundamental mode penetration into them is negligible; secondly, the active region can be easily shifted toward a cladding without fears of high-order mode lasing. Both things favor reducing the wafer thermal resistance. Following the basic idea, we have designed a laser structure (figure 1) comprising a GaAs active waveguide (thickness 1.35 μm) separated from 550 nm GaAs passive waveguide with 250 nm $Al_{0.25}Ga_{0.75}As$ layer, and two p- and n-doped $Al_{0.25}Ga_{0.75}As$ claddings. The active region based on two InGaAs quantum wells (wavelength about 1 μm) is strongly shifted toward the p-cladding which is as thin as 500 nm. As a result, the active region is located at the distance of only 920 nm from the epi-side surface. Transverse single-mode lasing is ensured by placing the active region in the minimum of the second mode and eliminating the first mode by using the CLOC technique.

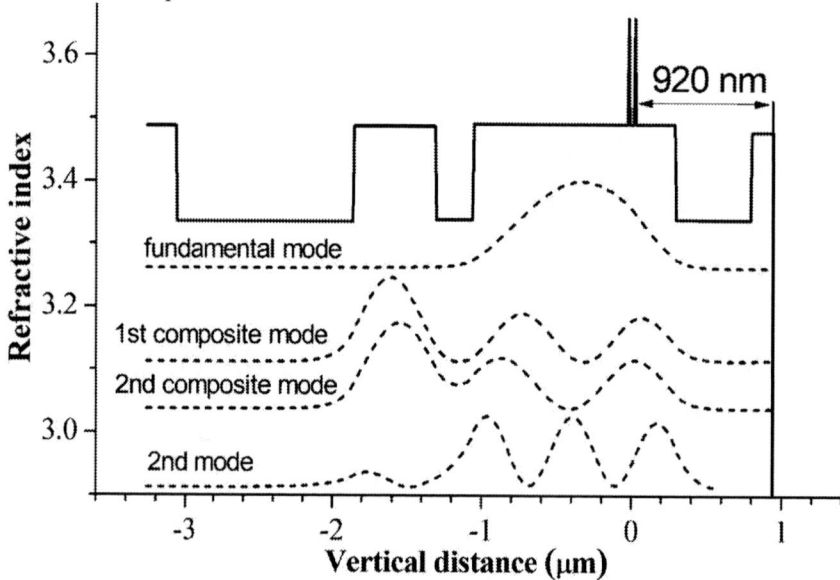

Figure 1. Scheme of the laser heterostructure: refractive index profiles (left axis) and simulated intensity profiles of the fundamental, second-order and two composite modes.

3. Experiment details and results

The laser wafer was grown with metal-organic vapor phase epitaxy on an n-GaAs (100) substrate and processed into broad-area stripe lasers with 100-μm-wide uncoated apertures. Devices with various lengths were mounted onto a copper heat-sink with indium solder and tested both in pulsed and CW regimes. At all excitation levels, the lasers showed stable single-mode emission with the divergence of 33° FWHM (full width at half maximum) fully corresponding to the simulated one. As expected, the lasers showed a low internal optical loss of 0.4 cm^{-1} being evaluated from the dependence of the reciprocal differential efficiency on the cavity length.

To calculated the thermal resistance R_{th}, we have made use of a modified technique based on the temperature dependence of the laser wavelength [5]. In the experiment, a thermal sensor providing a driving signal for a Peltier thermoelectric cooler (TEC) was placed at the heat sink 1.5 mm off the laser chip. Instead of commonly measured laser wavelength maximum, which in the case of broad-area multi-mode lasers may give an unrealistic value of R_{th}, we have investigated true spontaneous emission (TSE) spectra collected with a 200 μm optical fiber through a top-contact window (figure 2a). A long-wavelength tail of the TSE spectrum follows the active region temperature (figure 2b). Meanwhile, the lasing peak demonstrates blue shift (reflecting the fact that the quasi-Fermi levels are not fully pinned at the threshold), which can be wrongly interpreted as the laser diode cooling.

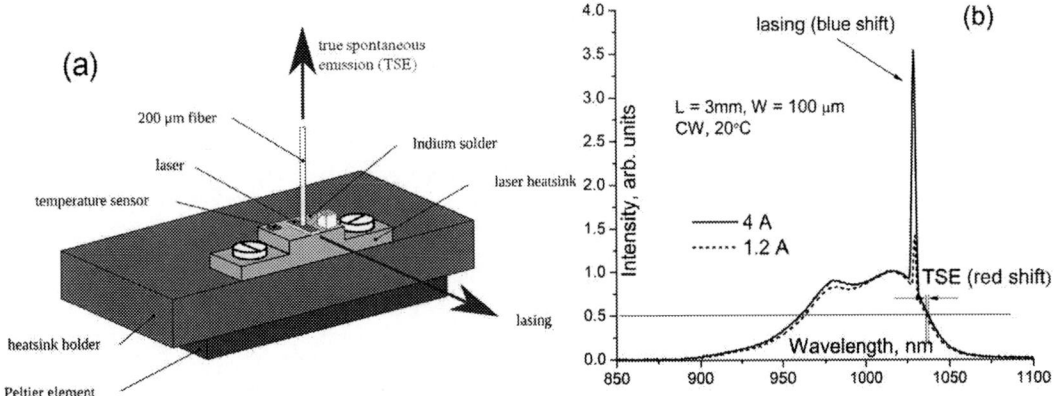

Figure 2. Scheme of the laser mounting used for thermal resistance measurements (a). Increased dissipated power at the current of 4 A causes a blue shift of the lasing wavelength and red shift of the TSE wavelength (b).

Firstly, we have measured a temperature-induced red shift of the TSE wavelength of the 3 mm diode laser under pulsed excitation varying the heatsink and correspondingly the laser temperature (figure 3a). The calculated red shift of 0.37 nm/K is quite typical for GaAs/AlGaAs materials in a given wavelength range. Then, we have gauged the dependence of the TSE spectrum shift on the dissipated power (i.e. the input electrical power after deduction of the optical power) under CW pumping (figure 3b). It was found to be of 0.76 nm per 1 W. Taking these two values we have calculated the thermal resistance as low as 2 K/W, which corresponds to the specific thermal resistance as low as 6 K/W×mm. A characteristic temperature T_0 of the threshold current was found to be ~110 K which is rather typical for lasers based on InGaAs/GaAs quantum wells.

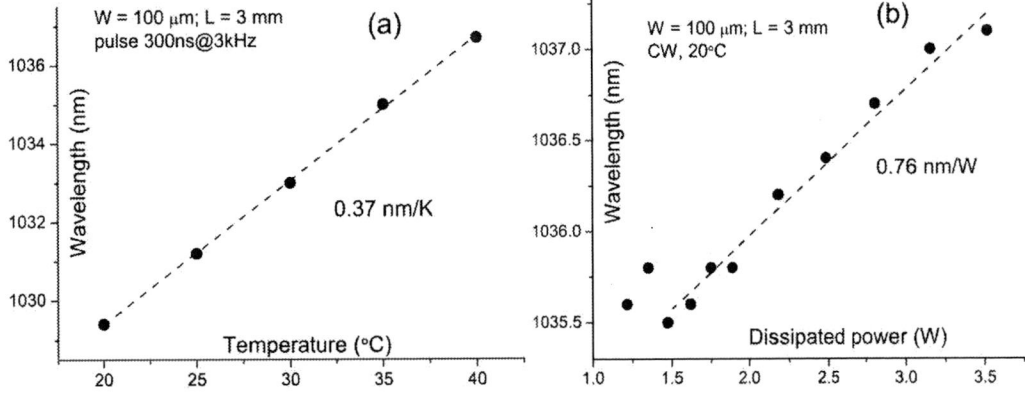

Figure 3. Temperature-induced (a) and dissipated-power-induced (b) shifts of the TSE wavelength.

Both low thermal resistance and low optical loss allowed the lasers to exceed the optical power of 10 W under CW pumping.

4. Conclusion

In conclusion, we have demonstrated the possibility of reducing significantly the depth of the active region location in diode lasers intended for CW high-power operation. Taking advantage of the CLOC concept allow locating the quantum wells on the depth as shallow as 920 nm. As a result, 3 mm long devices demonstrate the thermal resistance of 2 K/W, which is, to the best of our knowledge, the lowest for diode lasers directly mounted on copper heat-sinks. Another advantage of the laser structure is extremely low internal optical loss (0.4 cm^{-1}), which allows us to use long-cavity diodes with the minimal impact on the slope efficiency.

Acknowledgments

The work is supported by the Russian Science Foundation project #14-42-00006.

References

[1]   Crump P, Erbert G, Wenzel H, Frevert C, Schultz C M, Hasler K-H, Staske R, Sumpf B, Maassdorf A, Bugge F, Knigge S and Trankle G 2013 *IEEE J. Sel. Top. Quantum Electron.* **19** 1501211

[2]   Amann M 1987 *Appl. Phys. Lett.* **50** 4–6

[3]   Pietrzak A, Hülsewede R, Zorn M, Hirsekorn O, Sebastian J, Meusel J, Hennig P, Crump P, Wenzel H, Knigge S, Maaßdorf A, Bugge F and Erbert G 2013 *Technologies for Optical Countermeasures X; and High-Power Lasers 2013: Technology and Systems* vol 8898, ed D H Titterton, M A Richardson, R J Grasso, H Ackermann and W L Bohn p 889807

[4]   Gordeev N Y, Maximov M V and Zhukov A E 2017 *Laser Phys.* **27** 086201

[5]   Behringer M 2007 High-power diode laser technology and characteristics *Springer Ser. Opt. Sci.* **128** p 36

SPBOPEN 2018

IOP Publishing

IOP Conf. Series: Journal of Physics: Conf. Series **1124** (2018) 041017   doi:10.1088/1742-6596/1124/4/041017

All-oxide heterojunction solar cells formed by magnetron sputtering

D Kudryashov[1], A Gudovskikh[1,2], A Monastyrenko[1]

[1] St. Petersburg Academic University, St. Petersburg 194021, Russia
[2] St. Petersburg Electrotechnical University "LETI", St. Petersburg 197376, Russia

Abstract. All-oxide solar cells based on ZnO/Cu_2O heterostructure were grown by the magnetron sputtering deposition at elevated temperatures. An effect of ZnO and Cu_2O films thickness on the solar cells characteristics was demonstrated. The open circuit voltage of 315 mV and the density of a short-circuit current of 0.5 mA/cm^2 under AM1.5 illumination were measured for optimized ZnO/Cu_2O based solar cell.

1. Introduction

Heterojunction solar cells (SC) have many advantages over the traditional homojunction devices. The highest efficiencies for silicon SC had been reached due to using a-Si:H/c-Si heterojunctions [1]. The choice of materials for the next generation solar cell is to meet such requirements as high chemical stability, low toxicity and scalability on the industrial scale. Copper (I) oxide is an interesting material in terms of photovoltaic semiconductors because of its availability, non-toxicity, and a low cost of its production. Cu_2O has a band-gap value of 2.2 eV and a high optical absorption coefficient [2]. Thin films of copper (I) oxide free of external doping are characterized by p-type conductivity [3]. Contacting with zinc oxide, it is possible to form a ZnO/Cu_2O heterojunction and make a solar cell with theoretically estimated efficiency of 20% [4]. By the moment, the best efficiency of 8.1% has been reached for $ZnGeO/Cu_2O$ based solar cells [5], but extremely high temperatures (above 1000 °C) during the growth of such solar cells require its production cost to grow dramatically. Another method used for oxides growth is electrochemical deposition with subsequent annealing. This approach is a simple and a low-cost method but hardly a reproducible one. Solar cells produced with electrochemical deposition technique show an efficiency of only 0.25% [6]. The magnetron sputtering is a low-cost method with a precise thickness control. Thin films formed by the magnetron sputtering are usually more defective than the materials grown at high temperatures. But with the magnetron sputtering it is possible to precisely control the properties of oxides and heterointerface quality [7,8]. By the moment, ZnO/Cu_2O based solar cells fabricated with the magnetron sputtering deposition have reached an efficiency of 0.24% [9]. Such low values can be attributed to a low quality of oxide layers and high concentration of surface states. Also a post treatment such as annealing could lead to uncontrollable diffusion and thus to degradation of solar cells characteristics. On the other hand, the potential of the magnetron sputtering has not been exhausted. For example, it is possible to form a ZnO/Cu_2O heterostructure at one process without opening in air, which would have been decreased the probability of surface oxidation. Also the magnetron sputtering deposition is suited well for flexible substrates, and therefore it can potentially decrease the cost of solar cells. Using of sintered ZnO and metal Cu targets can potentially reduce the amount of materials involved in solar cell manufacturing process. In this paper, experimental results for ZnO/Cu_2O heterojunction solar cells formed by the magnetron sputtering are presented

Content from this work may be used under the terms of the Creative Commons Attribution 3.0 licence. Any further distribution of this work must maintain attribution to the author(s) and the title of the work, journal citation and DOI.

Published under licence by IOP Publishing Ltd

2. Experimental

All films for the solar cells were grown by rf-magnetron sputtering using Boc Edwards Auto 500 technique with a modified substrate holder at a temperature of 450 °C (figure 1). Due to the holder construction it is possible to heat only a small region up to 4×4 cm^2. Cu_2O films with a thickness of 1-5 microns were deposited on preliminary oxidized Si(100) substrates covered by Cu, Ti or Au films at fixed oxygen (99.999 %) and argon (99.999 %) partial pressures and at rf-power of 50-300 W. Copper (99.99%) and ZnO (99,9%) from LTS Chemicals were used as the magnetron targets.

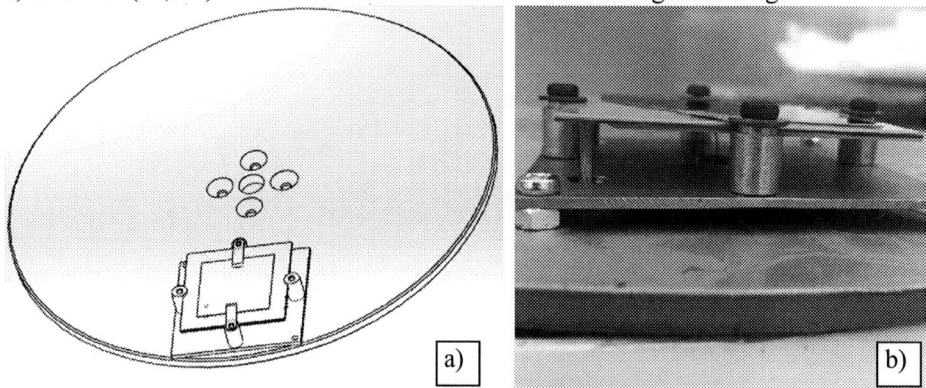

a) b)

Figure 1. Schematic of a modified substrate holder for magnetron sputtering (a) and a picture of a substrate mounted on the heater (b).

ZnO films were deposited at rf-power of 25-150 W. 100 nm of transparent conductive oxide ITO was deposited at a power of 50 W to form top electrical contact. After the growth all samples were annealed at a temperature of 100 °C for 30 min. SEM images were analyzed using a scanning electron microscope (SEM) Zeiss SUPRA 25. Current-voltage curves were measured using Keithley 2400 source meter under AM1.5 solar spectrum at 25 °C. Spectra of external quantum efficiency were recorded using equipment based on Solar Laser Systems M266 monochromator.

3. Results and discussion

Growth temperature of 450 °C was used, and this has made it possible to grow of copper (I) oxide with a microcrystalline structure [10]. The first results on the formation of a solar cell based on the Cu_2O/ZnO heterostructure with copper and titanium back contacts has showed that when it is illuminated by the AM1.5 spectrum on the current-voltage characteristics the reverse direction of the IV curve is observed, compared with that what should be obtained theoretically (based on the bandgap diagram). The samples has shown an open-circuit voltage of about 60-80 mV (figure 2). In the same configuration of deposited ZnO and Cu_2O layers, but with Cu_2O growth at room temperature, the polarity of the solar cell was the direct one, which indicates the fact that at a the low temperature growth the copper oxide is so defective that photocurrent is generated mainly in the ZnO layer. Replacement of the copper back contact on the gold at the high temperature Cu_2O growth led to a change in the form of IV curve to a forward one (figure 3) indicating that the photogeneration is attributed to Cu_2O layer. All subsequent solar cell structures were grown with Au back contacts.

SPBOPEN 2018   IOP Publishing

IOP Conf. Series: Journal of Physics: Conf. Series **1124** (2018) 041017   doi:10.1088/1742-6596/1124/4/041017

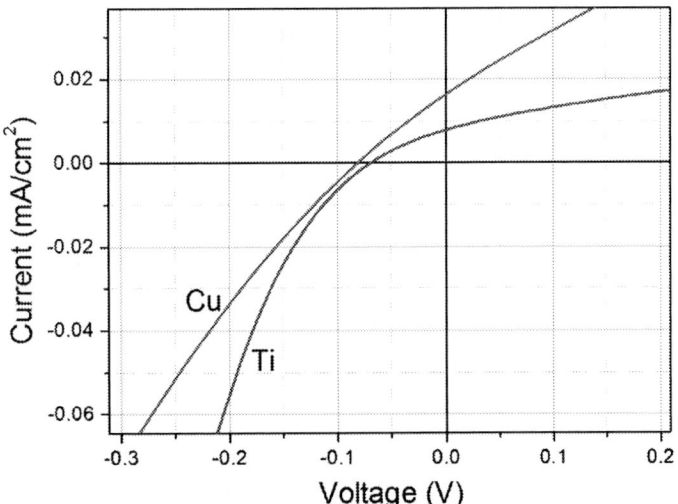

Figure 2. I-V curves under AM1.5 illumination for ZnO(20 nm)/Cu$_2$O(3000 nm) solar cells with titanium and copper back contact.

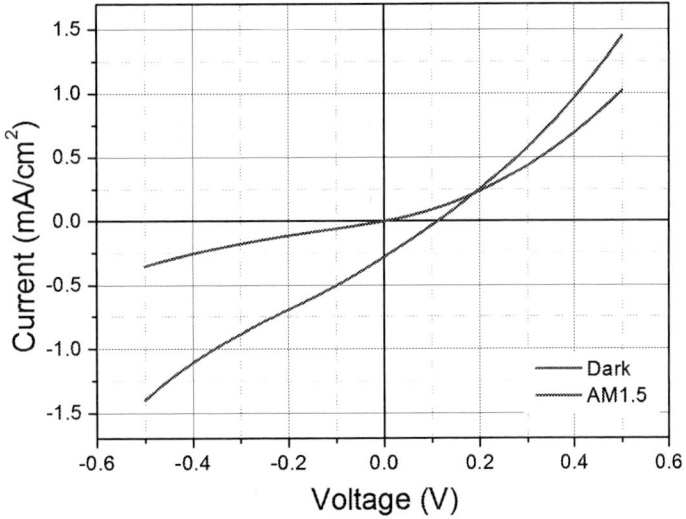

Figure 3. I-V curves in the dark and under AM1.5 illumination for ZnO(20 nm)/Cu$_2$O(3000 nm) solar cells with gold back contact grown in the same regime as in figure 2.

Figure 4 shows current-voltage curves for ZnO/Cu$_2$O solar cells with 20 nm ZnO layers grown at different sputtering power and the constant Cu$_2$O base layer thickness of 3 μm. Despite that, our previous experimental data have shown that ZnO films grown at a power of 150 W have better photoluminescence response [11], for the ZnO/Cu$_2$O solar cell the highest values of open circuit voltage (V$_{oc}$) we obtained using ZnO films grown at lower sputtering powers. On the one hand, it could be resulted to copper (I) oxide surface oxidation to CuO due to the presence in great quantity of oxygen radicals in plasma at startup of ZnO deposition at higher sputtering power. On the other hand, due to the high kinetic energy of ZnO particles, a part of them could chemically react with Cu$_2$O resulting in heterointerface degradation.

433

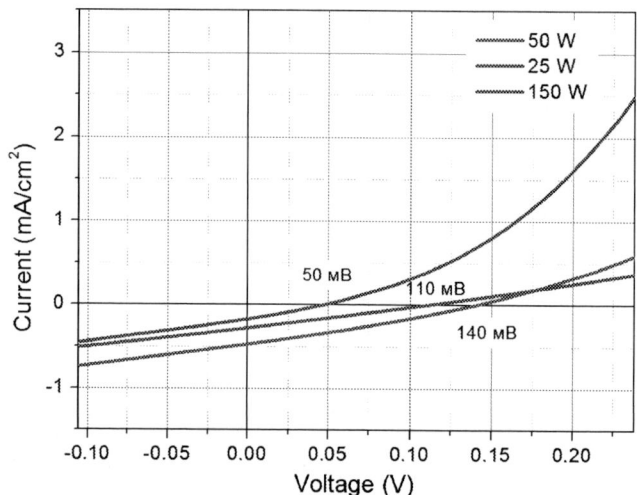

Figure 4. I-V curves under AM1.5 illumination for ZnO(20 nm)/Cu$_2$O(3000 nm) solar cells with ZnO layers grown at different sputtering power.

Cu$_2$O thickness has a non-linearly effect on ZnO/Cu$_2$O solar cells characteristics (figure 5a). The highest V$_{oc}$ value (0.315 mV) was measured for the solar cell with a Cu$_2$O thickness of 3 μm. Further increase of base layer thickness results in decrease of V$_{oc}$ and short-circuit current (I$_{sc}$). The resistivity of the grown Cu$_2$O films is quite large due to the absence of intentional doping and thus when copper (I) oxide is getting thicker the effect of built-in electrical field reduces.

Figure 5b shows the external quantum efficiency (EQE) curve for ZnO/Cu$_2$O solar cell with ZnO thickness of 20 nm and Cu$_2$O thickness of 3 μm deposited by magnetron sputtering. The long-wavelength region of the EQE spectra is quite low due to a high recombination rate in the bulk and on the interfaces.

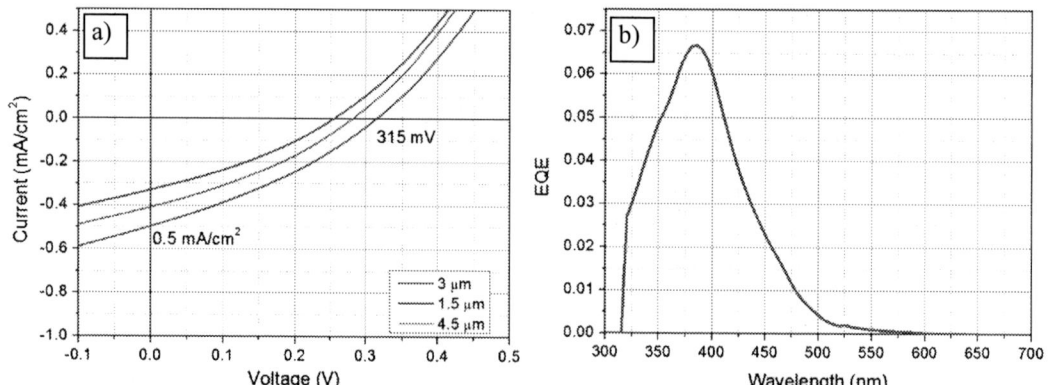

Figure 5. I-V curves under AM1.5 illumination for ZnO/Cu$_2$O solar cells with different Cu$_2$O layer thickness (a) and an EQE spectra for solar cell with a Cu$_2$O thickness of 3 μm (b).

4. Conclusions

According to the results obtained, the choice of material of back contact has a strong effect on the ZnO/Cu$_2$O based solar cell characteristics. To reach high V$_{oc}$ values, it is required to use a metal having such a high work function as gold or platinum has. Elevating the temperatures (up to 450 °C) allows to grow ZnO/Cu$_2$O based solar cell with open circuit voltage of 315 mV and a short-circuit current of 0.5 mA/cm^2 under AM1.5. Obtained data for EQE measurements points to high recombination rate in the Cu$_2$O bulk and its interfaces. To improve the copper (I) oxide characteristics, further research is to be carried out.

Acknowledgment
This work was carried out with the support of the grant of government of the Russian Federation (16.2593.2017/8.9)

References
[1] Schulze T F, Korte L, Conrad E, Schmidt M, Rech B 2010 *J. Appl. Phys.* **107** 023711
[2] Rühle S, Anderson A, Barad H-N, Kupfer B, Bouhadana Y, Rosh-Hodesh E, Zaban A 2012 *J. Phys. Chem. Lett.* **3** 3755
[3] Lee Y S, Winkler M T, Siah S C, Brandt R, Buonassisi T 2011 *Appl. Phys. Lett.* **98.19** 192115
[4] Minami T, Miyata T, Nishi Y 2014 *Solar Energy* **105** 206
[5] Minami T, Nishi Y, Miyata T 2016 *Applied Physics Express* **9** 052301
[6] Fujimoto K, Oku T, Akiyama T, Suzuki A 2013 *J. Phys.: Conf. Series* **443** 012024
[7] Sosnin D V, Kudryashov D A, Gudovskikh A S, Zelentsov K S 2015 *Tech. Phys. Lett.* **41(8)** 804
[8] Kudryashov D, Babichev A, Nikitina E, Gudovskikh A, Kladko P 2015 *J. Phys.: Conf. Ser.* **643** 012013
[9] Noda S, Shima H, Akinaga H 2013 *J. Phys.: Conf. Series* **443** 012027
[10] Agekyan V F, Borisov E V, Gudovskikh A S, Kudryashov D A, Monastyrenko A O,Serov A Yu, Filosofov N G 2018 *Semiconductors* **52** 383
[11] Kudryashov D, Babichev A, Nikitina E, Gudovskikh A, Kladko P 2015 *J. Phys.: Conf. Ser.* **643** 012013

Influence of Au surface properties on photon emission from localized metal-insulator-metal tunnel contact

V A Shkoldin[1], D V Permyakov[2], M V Zhukov[2,3], A A Vasiliev[1], T A Mamaeva[5], A O Golubok[2,3], A V Uskov[2,4], A D Bolshakov[1], A K Samusev[2], I S Mukhin[1,2]

[1] St. Petersburg Academic University, St. Petersburg 194021, Russia
[2] ITMO University, St. Petersburg 197101, Russia
[3] Institute for Analytical Instrumentation RAS, St. Petersburg 198095, Russia
[4] P. N. Lebedev Physical Institute, Russian Academy of Sciences, 119991 Moscow, Russia.
[5] Peter the Great St. Petersburg Polytechnic University, St-Petersburg 195251, Russia

Abstract. Inelastic electron tunneling in a tunnel junction may be used as an electrical nanosource of surface plasmon polaritons and photons. In this work, we investigate emission from tunnel contact between the tungsten tip with Au coating and a thin Au film on glass. The experiment has shown that intensity of this emission dramatically depends on Au surface properties.

1. Introduction

The Information technology development nowadays necessitates search for new ways for increase of performance and decrease of power consumption of the computing devices. The next stage of the information technologies evolution is associated with the transition from electronic to photonic components. This technological step can be realized using photons or surface plasmons polaritons (SPP) for communication logical bits. The optical signal propagates along the waveguide faster than the electrical signal through the metal wires. It therefore may speed up performance of the information technologies devices. In addition, use of photonic components will lead to increase of the devices efficiency and, consequently, to decrease of their heating.

Nonetheless, one of the bottlenecks of the forthcoming technologies integration is the large size difference between the existing components. The real implementation of photonic devices requires development of local nanosized photonic sources. Semiconductor lasers with Fabry-Perot or microring resonators are not an appropriate option to be implemented on a chip, due to its large size equal or greater than the emitted wavelength.

An alternative way of light generation is based on tunnel junction use. A process of light emission in the metal-insulator-metal (MIM) contact was discovered by Lambe and McCarthy in 1976 [6]. The gap size in the tunnel junction is equal to a few angstroms. Consequently, such MIM contact may be used for development of the subwavelength light source, though quantum yield of such device is not large (10^{-6}-10^{-4} photon-per-electron). Despite potential of this approach, device based on this principle was not developed, yet.

The scanning tunneling microscope (STM) is a useful tool for studying the tunnel junction effects. The MIM contact can be obtained between the metal tip and the metal surface. The STM allows to control different parameters of the tunnel junction. Quantum yield in this case can be tailored with change of electrical properties, for example. Besides electrical properties of the tunnel junction [1], efficiency of photon emission depends on the surface characteristics of the metal. This work has shown that intensity of this emission dramatically depends on Au surface properties.

2. Samples and experimental setup

The experiments were carried out with 140 µm thick glass substrates covered with 30 nm thick gold films deposited via the thermal evaporation. The experimental setup used in this work consists of a commercially available STM head operating in air (AIST-NT TriOS) coupled with an inverted optical microscope (Fig.1). The emitted light was collected through the glass substrate using oil immersion objective (100x, NA=1,4) and focused onto the single-photon counter (IDQ ID 120) based on avalanche photodiode.

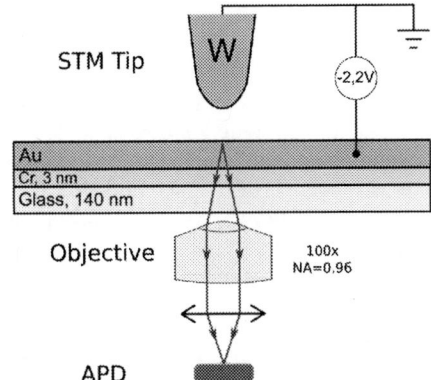

Figure 1. Schematic of the experiment. The scanning tunneling microscope (STM) mounted on inverted optical microscope.

Gold-air-gold tunnel junction was studied in this experiment. The STM tips were electro-chemically etched from a tungsten wire (150 μm diameter) in KOH solution and coated with gold. The tip radius was about 100 nm according to the SEM measurements. The second metal contact was the metal film on the glass with 150 μm thickness. The cover glasses were coated by thermal evaporation with a thin chromium film (3-6 nm) and then coated with gold (from 16 nm to 47 nm). The evaporation settings were subject to change from one sample to another. As a result, the films had different surface properties. Optical transparency of the samples was about 15%.

3. Experiment

Light emission may be observed in a tunnel junction with applied bias voltage (V_b). According to the law of conservation of energy, the tunneling electrons may undergo (elastic) horizontal transition into empty states at the metal surface. Also, the tunneling electrons may lose energy in the barrier region (Fig.2).

The energy dissipation can lead to a photon emission with energy equal eV in accordance with single-electron photon emission processes. The lowest observed photon energy value equals to 0,6 eV [7] and depends on materials of the contact. Emission spectrum corresponding to the gold-air-gold contact has maximum at 700-750 nm (~1,6-1,7 eV) [2,3].

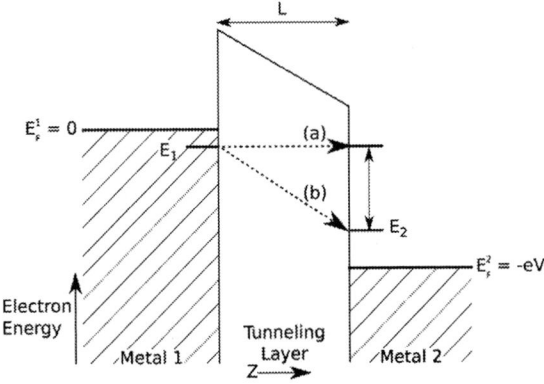

Figure 2. Schematic band diagram for tunneling between two metals (Au coated tip and Au film). Elastic processes (a) corresponding to ordinary tunneling. Inelastic processes (b) give rise to loss of energy with excitation in the barrier region.

Visible photon emission was generated by applying 1,5-3V of bias voltage. The experiment was conducted in the air atmosphere leading to unstable work of the tunnel contact. Thin water layer (<1 nm) covers the STM tip and sample surface under typical laboratory conditions. This leads to a water bridge formation between the tip and sample [8]. Also, normally STM system measures only tunnel current between the electrodes, while in the latter case the sum of tunnel current and an ionic or Faradaic current of electrochemical nature appears [4]. The standard theoretical electrochemical potential for water electrolysis is 1,23 V, which is smaller, than the applied bias voltage in our experiment [5].

It has been found experimentally that the unstable tunnel contact has a larger quantum yield, than the stable one. In the latter regime only insignificant emission can appear [1]. Feedback gain was set for generation in the feedback system. Sample was moved in the upper direction by a piezo stage until the contract was attained, and it was immediately moved down with the current detection. Frequency of the obtained oscillations was approximately equal to 52 Hz.

The investigated Au films with different thickness have been deposited on glass substrate by thermal evaporation technique. In addition, properties of these films were analyzed using AFM. Measurement of the emission intensity was carried out with the same tip and settings of the feedback system in each experiment. The current setpoint was 165 nA and the applied surface bias was 2,2 V.

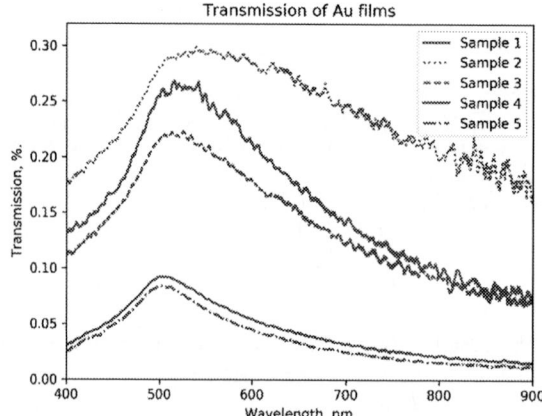

Figure 3. Transmission spectra of the samples with different thickness Au films.

The emitted light was collected in the objective after propagation through Au-Cr films and glass (see Figure 1). Therefore, optical transmission spectra were measured for each sample (Figure 3), and measured intensity was normalized to transmission coefficient at a wavelength of 740 nm. This wavelength was chosen, due to maximum intensity of the MIM contact emission near this value [9].

Properties of the films were analyzed using AFM. Typical AFM surface scan is shown in Figure 4. The surface of the films is characterized with the grain size. Table 1 shows characteristic of investigated samples and measured parameters of their surfaces. Thickness of films influence on sample transmission quality. As we can see, the intensity of luminescence proportional to the grain size. Also, the intensity inversely proportional to mean grain height.

Further investigation of the optical radiation features from the tunnel junction will be a great step for future electro-optical chips development. These chips will be basic functional elements of the optical logic devices.

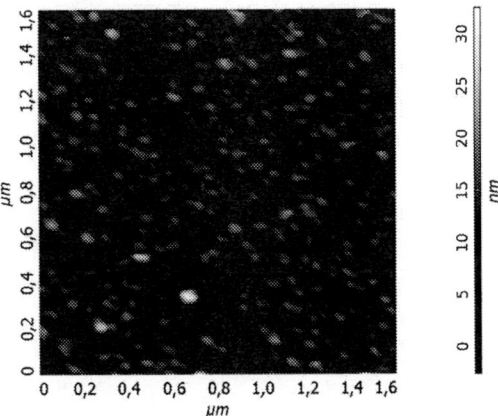

Figure 4. AFM surface image of the gold film on Sample 3.

N	Cr, nm	Au, nm	Mean height, nm	Grain size, nm	Roughness Average, nm	Normalized intensity
1	2,7	26,4	2,8	84,5	0,5	100,0%
2	4,6	16,3	4,0	60	3,3	3,8%
3	4,0	26,7	4,0	40	2,6	2,7%
4	6,1	43,1	14	40	4,2	0,4%
5	5,6	47,2	14	32	2	0,2%

Acknowledgements
This work was carried out with the support of the Russian Science Foundation (Grant 17-19- 01532).

References
[1] Rogez B, Cao S, Dujardin G, Comtet G, Le Moal E, Mayne A, Boer-Duchemin E 2016 *J. Nanotechnology* **27** 465201

[2] Wang T, Boer-Duchemin E, Zhang Y, Comtet G, Dujardin G 2011 *J. Nanotechnology* **22** 175201

[3] Hoffmann G, Aizpurua J, Apell P, Berndt R 2001 *J. Surface science* **482** 1159-1162

[4] Song M, Jang J, Bae S, Lee C 2002 *J. Langmuir* **18** 2780-2784

[5] Senftle F, Grant J, Senftle F 2010 *J. Electrochimica Acta* **55** 5148-5153

[6] Lambe J, McCarthy S 1976 *J. Physical Review Letters* **37** 923

[7] Boyle M, Mitra J, Dawson P 2009 *J. Applied Physics Letters* **94** 233118

[8] Gómez-Monivas S, Sáenz J, Calleja M, García R 2003 *J. Physical review letters* **91** 056101

[9] Parzefall M, Bharadwaj P, Jain, Novotny L 2017 *Quantum Plasmonics* (Switzerland, Springer International Publishing) p 211

SPBOPEN 2018

IOP Publishing

Diagnostics of inhomogeneities of parameters of light-emitting heterostructures by the photoelectric method

I V Frolov[1], O A Radaev[1,2], V A Sergeev[1,2], S V Vasin[1,2]

[1] Ulyanovsk Branch of Kotel'nikov Institute of Radio-Engineering and Electronics of Russian Academy of Sciences, Ulyanovsk, 432071, Russia
[2] Ulyanovsk State Technical University, Ulyanovsk, Russia

Abstract. The automated measuring complex for diagnostics of spatial inhomogeneity of parameters of light emitting heterostructures by photoelectric method with local photoexcitation is described. Results of the estimation of the homogeneity of the photocurrent value distribution on the area of crystals of blue and green commercial InGaN/GaN light-emitting diodes are presented. It is shown that near the crystal boundaries the level of inhomogeneity is higher than at the center. It is revealed that inhomogeneity of photovoltaic parameters of heterostructures of green LEDs is greater than that of blue LEDs.

1. Introduction

Despite the improvement of the technologies for manufacturing light-emitting InGaN/GaN heterostructures, problems associated with the degradation of the characteristics of semiconductor devices based on them are still unresolved due to the formation of defects in the active region of the heterostructure [1]. One of the reasons leading to an increase the degree of defectiveness of the heterostructure is clusterization (the formation of local regions with a high concentration) of indium atoms in the InGaN solid solution during the growth of the structure. Due to the fluctuation of the composition of the solid solution is formed inhomogeneous current distribution over the area of the structure, as well as the formation of sites of local overheating, which leads to acceleration of the formation of defects in the heterostructure under the influence of the injection current during operation [2]. The spatial inhomogeneity of the solid solution composition is manifested in the nonuniform distribution of electro-optical and thermophysical parameters over the area of the crystal. Methods of scanning near-field electroluminescence, infrared microscopy [3], photoluminescence mapping are used for its diagnosis [4].

In this article the automated measuring complex for the diagnosis of spatial heterogeneity of the parameters of light-emitting heterostructures by photoelectric method with local photoexcitation and the results of the investigation of inhomogeneities of photoelectric parameters of light-emitting heterostructures of commercial green and blue LEDs are presented.

2. Measuring complex

A measuring complex was developed to diagnose the lateral homogeneity of light-emitting heterostructures by photovoltaic method at local photoexcitation in static and dynamic modes [5]. The complex makes it possible to scan the working surface of a heterostructure with a step up to 10 μm when the object is irradiated with an optical spot of laser radiation in a stationary mode and in a mode of pulse or harmonic modulation of the photoexcitation intensity.

Content from this work may be used under the terms of the Creative Commons Attribution 3.0 licence. Any further distribution of this work must maintain attribution to the author(s) and the title of the work, journal citation and DOI.
Published under licence by IOP Publishing Ltd

The measuring complex consists of software and hardware parts. The structural diagram of the hardware of the complex is shown in Fig. 1. The test sample is fixed to the XYZ-positioner, which allows automatically moving the investigated sample in the XY plane in the range of 10×10 mm with steps of 10 μm using the stepping motors controlled by the microcontroller. Sony SLD3232VF laser diode emitting at a wavelength of 405 nm is used for photoexcitation of the object. The laser diode radiation is focused on the surface of the sample by the lens and the objective. The beam diameter of the laser radiation does not exceed 10 μm. The laser diode is connected to a stabilized current source when measuring the photocurrent profile on the crystal surface in a stationary mode. To register a constant photocurrent is converted by a transimpedance amplifier to voltage use a digital multimeter Rigol DM3058. For registration DC photocurrent transformed to the voltage of the transimpedance amplifier uses the digital multi-meter Rigol DM3058.

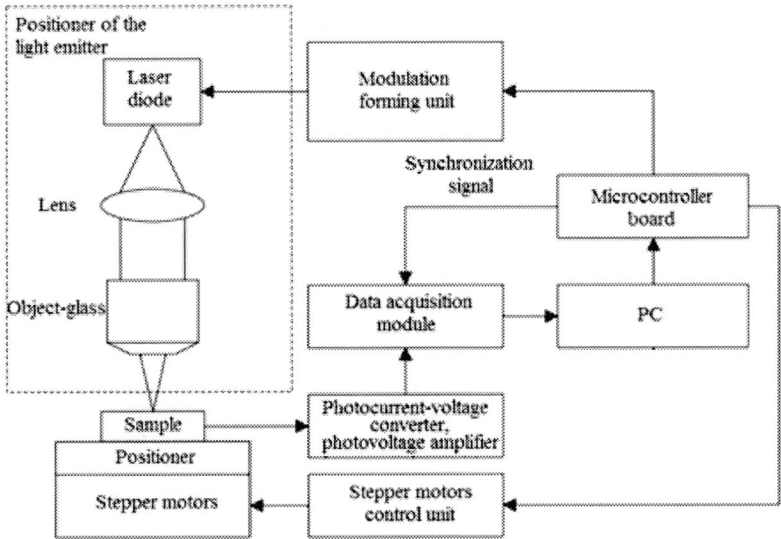

Figure 1. The block diagram of the automated measuring complex

For measuring the dynamic parameters of the structure's photoresponse (the boundary modulation frequency of the photocurrent in the harmonic mode of the photoexcitation and the rise and fall of the photocurrent pulse front in the photoexcitation pulse mode) the laser radiation intensity is modulated by a software-controlled RIGOL DG1022 generator. Harmonic modulation is performed in the frequency range from 1 kHz to 20 MHz in the mode of continuous frequency tuning with a logarithmic step during the set time of swiping. In pulse modulation mode, the pulse duration is set in the range of 100 μs...1 ms with a duty cycle 2...10 depending on the frequency properties of the investigated object. For registration of a variable signal of photocurrent the transimpedance amplifier converter constructed on the OPA656 operational amplifier (OA) is used. The La-n1USB data acquisition module performs analog-to-digital conversion of the voltage coming from the output of the transimpedance amplifier and transmits the data to the memory of the computer. On retrieved data the computer program builds a three-dimensional distribution of dynamic parameters of the photoelectric response on a surface of a crystal of the investigated sample.

The software part of the complex is developed in LabVIEW graphical programming environment.

3. Experimental results
With the help of the developed complex the inhomogeneities of photoelectric parameters of heterostructures of commercial light-emitting diodes of two types are investigated: XREBLU-L1-0000-00K01 of blue glow and XREGRN-L1-0000-00P01 of green glow. LEDs of these types are made on the

basis of EZ1000 crystals 980x980 um, which have an InGaN / GaN heterostructure. Assessment of heterogeneity of distribution of the photocurrent on the surface of the crystal was performed based on the calculation of the relative standard deviation σ/I_{Ph_mean} by calculating the standard deviation σ and the average values of $I_{Ph\ mean}$ the photocurrent. The crystal surface was divided into seven segments (Fig. 2).

On Fig. 2 the profile of distribution of the photocurrent value over the crystal surface of a blue LED is presented. It can be seen from the figure that the average level of the photocurrent at the center of the crystal is higher than near the crystal boundaries. In this case, near the crystal boundaries, the degree of inhomogeneity of the photocurrent distribution is higher than in the center. According to the estimates presented in Table 1, the inhomogeneity of the distribution of the photocurrent along the crystal surface of green LEDs is 1.5 ... 1.7 times higher than that of blue LEDs.

Table 1. Photocurrent distribution parameters

The segment number of the crystal	XREBLU-L1-0000-00K01			XREGRN-L1-0000-00P01		
	I_{Ph_mean}, μA	σ, μA	σ/I_{Ph_mean}	I_{Ph_mean}, μA	σ, μA	σ/I_{Ph_mean}
1	89.5	2.525	0.028	34.2	2.251	0.066
2	89.2	1.884	0.021	33.8	1.683	0.050
3	92.3	1.645	0.018	35.6	1.514	0.043
4	92.6	1.348	0.015	36.9	1.151	0.031
5	93.7	1.668	0.018	37.7	1.246	0.033
6	91.5	2.380	0.026	37.1	1.141	0.031
7	89.9	2.807	0.031	35.2	1.883	0.053

To estimate the dynamic parameters of the photoelectric response of the heterostructure, the boundary frequency of the modulation of the photocurrent f_{3dB} was measured with harmonic modulation of the photoexcitation intensity. In Fig. 3 the profile of the distribution of the values of the boundary frequency of modulation of the photocurrent along the surface of the crystal of the blue LED is shown.

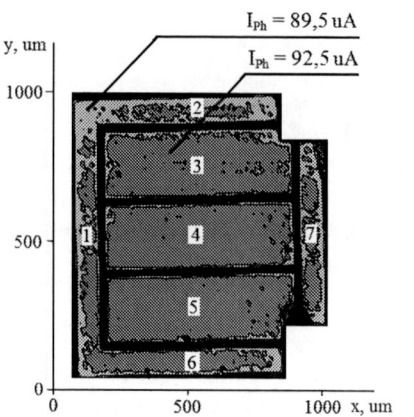 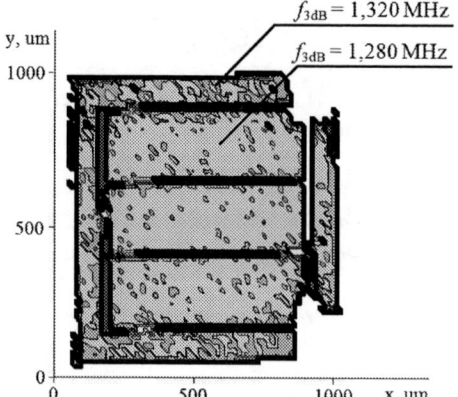

Figure 2. Distribution profile of the photocurrent value over the crystal surface LED XREBLU-L1-0000-00K01

Figure 3. Profile of the distribution of the boundary frequency of modulation of the photocurrent along the surface of the LED XREBLU-L1-0000-00K01

It is seen from the figure that near the crystal boundaries the average value of the boundary frequency of photocurrent modulation (f_{3dB} = 1.320 MHz) is approximately 3% larger than at the center of the crystal (f_{3dB} = 1.280 MHz). This means that the average lifetime of charge carriers at the center of the crystal is smaller than near the boundaries, which indicates large nonradiative losses near the crystal boundaries [3].

Table 2 presents the results of measuring the average value of the photocurrent, the standard deviation and relative standard deviation of the distribution of the photocurrent of the two samples of green and blue LEDs of 5 pieces each. From the data of the table it follows that the average value of σ/I_{Ph_mean} parameter used to estimate the inhomogeneity of the photocurrent is 1.3 times higher for green LEDs than for blue ones. The obtained results agree with the generally accepted statement that the inhomogeneity of the composition of a solid solution of heterostructures with a high content of indium emitting in the green region of the spectrum is higher than the inhomogeneity of the composition of heterostructures emitting in the blue region of the spectrum [7].

Table 2. Photocurrent parameters of the two samples of green and blue LEDs

	XREBLU-L1-0000-00K01					XREGRN-L1-0000-00P01				
	№1	№2	№3	№4	№5	№1	№2	№3	№4	№5
I_{Ph_mean}, µA	69.6	59.2	69.0	63.2	69.7	29,9	32.2	40.8	31.3	30,7
σ, µA	1.51	1.77	1.34	1.73	1.84	1,3	0.87	1.1	0.98	0,583
σ/I_{Ph_mean}	0,022	0,030	0,019	0,027	0,026	0,044	0,027	0,027	0,031	0,029

4. Conclusions

1. The automated measuring complex for registration of photoelectric response of LEDs heterostructures at their local static and dynamic photoexcitation is developed.

2. The results of the experimental testing of the complex on commercial InGaN/GaN LEDs in the static and dynamic mode of photoexcitation confirm the presence of inhomogeneities in the distribution of the photoelectric response on the surface of the LED crystal.

3. The inhomogeneity of the photovoltaic parameters of green LEDs heterostructures is greater than that of blue LEDs heterostructures.

4. The obtained results assessment of heterogeneity of distribution of the photocurrent and the boundary frequency of modulation of the photocurrent on the square of the crystal may indicate that the degree of defects in the heterostructure near the borders of the crystal is higher than in the center of the crystal.

Acknowledgments

The reported study was funded by RFBR and Government of Ulyanovsk Region according to research project № 16-47-732159.

References

[1] Frolov I V, Radaev O A, Sergeev V A *Journal of Physics: Conference Series* 2017 **917** 052017.

[2] Leungetal K K *Journal Of Applied Physics* 2010 **107** 073103.

[3] Fischer P, Christen J, Zacharias M et al. *Japanese Journal of Applied Physics, Part 1: Regular Papers and Short Notes and Review Papers* 2000 **39** no. 4 B 2414.

[4] Gelzinyte et al. *Journal Of Applied Physics* 2015 **117** 023111.

[5] Sergeev V A, Vasin S V, Radaev O A, Frolov I V 2018 *Moscow Workshop on Electronic and Networking Technologies(MWENT)* (National Research University Higher School of Economics) DOI:10.1109/MWENT.2018.8337210

[6] Schubert E F 2006 *Light Emitting Diodes* (Cambridge University Press).

[7] Jeong H et al. *Scientific reports* 2015 5:9373.

SPBOPEN 2018

IOP Publishing

Influence of coating layers on characteristics of microdisk lasers with InAs/InGaAs quantum dots active region

I Y Agafonov[1], N V Kryzhanovskaya[1,2], E I Moiseev[1], A S Dragunova[1], M V Fetisova[1], K P Kotlyar[1], S A Scherbak[1,2], I V Reduto[1,2], Yu A Guseva[3], A A Lipovskii[1,2], M V Maximov[1,3], A E Zhukov[1]

[1] St Petersburg Academic University, St Petersburg, Russia
[2] Peter the Great St Petersburg Polytechnic University, St. Petersburg, Russia
[3] Ioffe Institute, St Petersburg, Russia

Abstract. Quantum-dot microdisk lasers coated with different dielectric layers transparent in spectral diapason of laser radiation were studied. We observe that the coating influences the mode spectra of the microlasers, reduces the resonator Q factor and improves their thermal resistance.

1. Introduction

In recent years, semiconductor microlasers based on whispering gallery mode (WMG) resonators have attracted increased attention. Due to a low optical loss WMG resonators demonstrate high quality factors (Q). Moreover, WMG modes are very sensitive to the refractive index discontinuity at the resonator surfaces and in regions outside the resonator where the mode evanescent tails penetrate into [1, 2]. Microlasers are of great interest as ultra-sensitive sensors due to their narrow linewidth capable of reporting slight changes in resonance that would not be resolved by passive resonators [3]. For the onset of lasing, besides a significant increase of the signal-to-noise ratio, also a reduction in bandwidth of the lasing modes is expected, thus improving the detection limit of the sensors [4]. WGMs are based on effect of total internal reflection. From the point of view of classical optics for total internal reflection the critical angle θ_c can be defined from a ratio $\theta_c = \arcsin(n_2/n_1)$, where n_1 and n_2 are refraction indices of the resonator and surrounding material. Thus, the surrounding material influences total internal reflection, and thereby resonant spectrum and quality factor of the resonator. Various microdisk sensors are based on detecting of these changes. Presently lasing in different types of semiconductor microdisks [7-9] as well as influence of surrounding material on WGM mode frequencies in various passive microspherical resonators (sensors) [5,6] has been reported. However the impact of surrounding material on lasing characteristics of GaAs–based microdisk resonators hasn't been studied yet. These studies are very important for efficient light outcoupling from the lasers, realization of highly sensitive microdisk detectors and for proper choose of coating layers for planarization, passivation, etc. In this work we study spectral and lasing characteristics of microdisk lasers with InAs/InGaAs quantum dots active region coated with different dielectric layers transparent in spectral diapason of laser radiation (SU-8 photoresist and TiO_2) and compare the results with characteristics of the micordisks without coating (air environment).

Content from this work may be used under the terms of the Creative Commons Attribution 3.0 licence. Any further distribution of this work must maintain attribution to the author(s) and the title of the work, journal citation and DOI.
Published under licence by IOP Publishing Ltd

2. Experiment details

An epitaxial structure used for fabrication of the microdisk lasers was grown by molecular beam epitaxy on a GaAs substrate. The epitaxial structure for optically pumped microlasers contains an active region which represents 5 layers of InAs/InGaAs quantum dots (QDs) inserted into a 0.35-μm-thick GaAs waveguiding layer. The waveguide is cladded with 400-nm-thick $Al_{0.98}Ga_{0.02}As$ layer from the substrate side. Optically pumped microdisk resonators were fabricated using photolithography and Ar+ ion beam etching. The diameter of the microdisks was varied from 5.4 to 10.4 μm. The $Al_{0.98}Ga_{0.02}As$ bottom cladding layer was selectively oxidized to be transformed into an AlGaO oxide. Spectral position of quantum dot ground-state transition was located around 1.28 μm at room temperature.

An epitaxial structure for electrically pumped microlasers consists ten layers of InAs/$In_{0.15}Ga_{0.85}As$ QDs deposited in the middle of a 0.44 μm thick GaAs waveguiding layer confined with $Al_{0.25}Ga_{0.75}As$ claddings. Microdisk resonators were formed by means of chemical plasma etching to have a diameter of 30 μm. Etch depth was about 7 μm. AgMn/NiAu (AuGe/Ni/Au) metallization was used to form ohmic contacts to $p+$ GaAs cap layer ($n+$ substrate, respectively).

Spin coating was used to cover the microdisks with epoxy-based photoresist SU8 (n_{SU-8} =1.56 @ 1.3 μm). Atomic layer deposition was utilized to cover the disks with TiO_2 layer with different thickness 100, 150, 200 and 250 nm (n_{TiO2}=2.46 @ 1.3 μm).

The microdisk lasers were studied by confocal optical spectroscopy (Integra Spectra, NT MDT) at room temperature under optical pumping with YAG: Nd laser (λ =532nm) using Olympus x100 microobjective. Needle probes were exploited for electrical connections. A piezoelectrically adjustable Olympus LMPlan IR objective x10 was used to collect in-plane emitted light from a microlaser. The emission was detected with a Horiba FHR 1000 monochromator and a Horiba Symphony InGaAs CCD array.

3. Results

Microdisks were first studied in the air environment (n_{air}=1). All the lasers demonstrate lasing at room temperature. Spectra of the microlasers obtained above the threshold contain broad spontaneous emission of InAs/InGaAs quantum dots and sharp lines corresponding to the WGM resonances.

To inverstigate the influence of surrounding material the 6 μm in diameter microdisk lasers were covered with SU8 dielectric layers and studied under optical pumping (Figure 1). Use of external dielectric layers leads to change of mode spectrum, at the same time spectral position of the lasing line remains in the spectral range of the ground transition of the InAs/InGaAs QDs. A threshold values were determined from the dependence of integrated intensity of the dominant line on the pump power as 350 μW for initial microdisks and 400 μW for microdisks in SU8 layer. The quality factor of both resonators was about 30000. Thus, one can conclude that surrounding microlasers with SU-8 dielectric layers doesn't lead to deterioration in laser performance.

SPBOPEN 2018   IOP Publishing

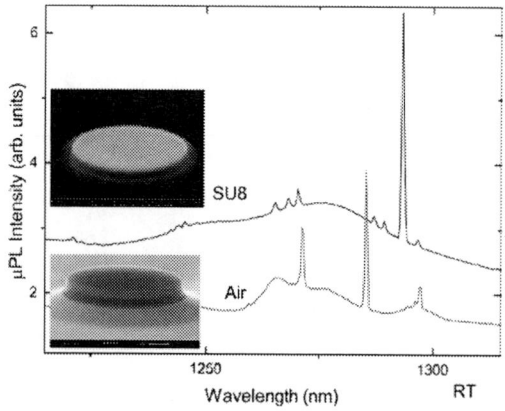

Figure 1. μPL spectra of the 6μm in diameter optically pumped microdisk lasers without coating (Air) and covered with SU8 layer obtained above threshold at room temperature. The spectra are vertically shifted for clarity.

Figure 2. EL spectra of the 30μm electrically pumped microdisk lasers covered with SU8 layer obtained above threshold at room temperature. The inset: spectral position of the lasing line vs pump power.

In the electrically pumped microlasers with planar electric contacts formed on SU8 dielectric layer room temperature lasing has been also achieved (Figure 2). The threshold current did not change compare to the initial microlaser. Lasing wavelength shifts to longer wave as bias current increases because of the laser self-heating effect. Inset in Figure 2 displays the lasing wavelength as a function of electric power dissipated inside the device. Slope of this dependence, which is 0.048 nm/mW can be used to calculate a thermal resistance of the device. Taking into account a temperature induced shift of a WGM line (0.075 nm/°C) one can calculate a thermal resistance to be 0.64 °C/mW. The obtained value is much smaller than in initial microlaser 0.9 °C/mW and becames comparable to thermal resistance of the best vertical cavity surface emitting lasers on GaAs substrates [10-11]. Thus, covering the microdisk with a transparent dielectric layer of SU8 reduces microlaser thermal resistance due to more efficient heat dissipation through the microlaser's sidewalls.

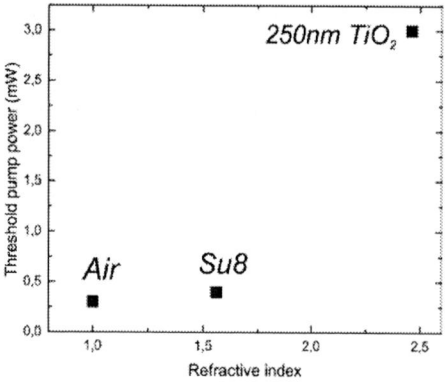

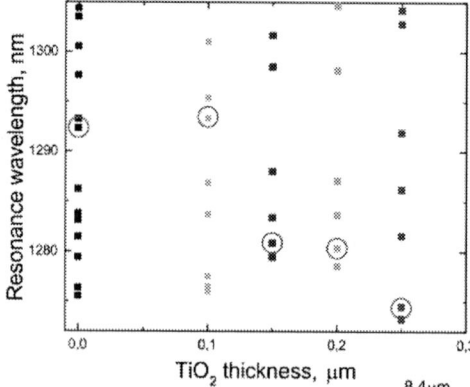

Figure 3. Threshold pump power of 6.4 μm microlaser vs refractive index of the ambient environment.

Figure 4. Spectral position of the resonances observed at PL spectra of 8.4 μm in diameter microlaser in air and covered with TiO2 layers of different thickness. The spectral position of the lasing wavelength is denoted by circles.

Next we studied spectral and threshold characteristics of the optically pumped microlasers covered with TiO_2 layer with different thickness (92 nm, 152 nm, 199 nm and 254 nm). Diameter of the studied microlasers was varied in the range from 5.4 to 10.5 μm. Figure 3 compares the threshold pump power obtained for 6.4 μm microlaser surrounded by air (the highest optical confinement), by 1 μm-thick SU8 and by 250 nm TiO_2 coating layer. We observe that the deposition of theTiO_2 dielectric layer leads to ten-times increase of the threshold power up to 3 mW. Increase of the refractive index of the environment (i.e. decrease of the refractive index discontinuity at the surface of the resonator) results also in worsening the Q-factor ($\lambda/\Delta\lambda$). Q-factor for the air-coated microdisk exceeds 30 000, and it drops down to 22000 in case of TiO_2 coating layer.

Deposition of TiO_2 layer also results in thinning of the resonances observed at photoluminescence (PL) spectra of microlasers (Figure 4). One can observe dramatic change in spectral positions of different WGM and their quantity caused by increase of refractive index of the surrounding material (layer thickness). We observe an increase in side mode suppression ratio that results and quasi-singlemode lasing at TiO_2 layer thickness more than 100 nm (Figure 5). Increase of the TiO_2 layer thickness also results in blue- shift of the lasing wavelength that can be explained by increase of radiation loss (circles on Figure 4). However room temperature lasing is still achieved even with 250 nm TiO_2 coating layer with the lasing wavelength within the quantum dots ground state optical transition.

The relative threshold power (P_{th}/P_{th}^0, where P_{th}^0 is threshold power of the initial microlaser) dependences on the TiO_2 layer thickness for 5.4, 6.4 and 10.4 µm in diameter microlasers are demonstrated in Figure 5. In case of the smallest microlaser (5.4 µm) covering with only 100 nm of TiO_2 results in 3-fold increase of threshold power. For thicker TiO_2 layers the room temperature lasing does not occur. In case of the microlaser with 6.4 µm diameter lasing is observed up to 200 nm of TiO_2. In the largest microlaser (10.4 µm) the obtained pump power does not exceed 1.5 P_{th}^0 even for 200 nm of TiO_2.

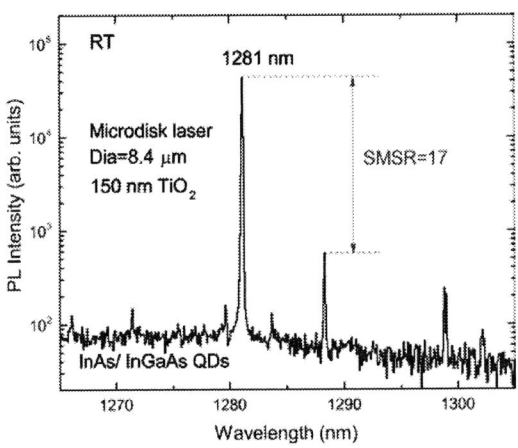

Figure 5. µPL spectra of the 8.4µm in diameter optically pumped microdisk lasers covered with 150 nm TiO_2 layer obtained above threshold (P=1.5 P_{th}) at room temperature.

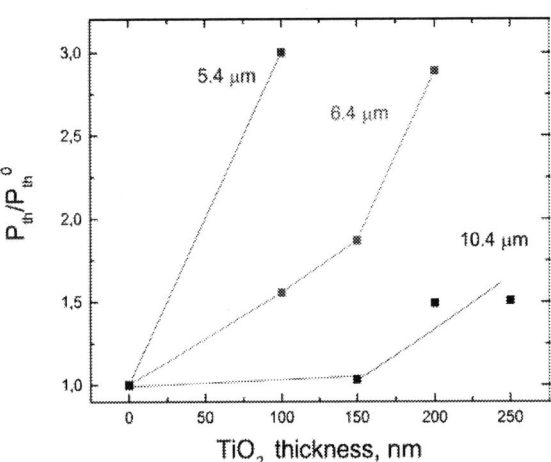

Figure 6. The relative threshold power (P_{th}/P_{th}^0, where P_{th}^0 is threshold power of the initial microlaser) dependences on the TiO_2 layer thickness for microlasers with 5.4, 6.4 and 10.4 µm obtained at room temperature.

To conclude, we observe that the SU8 coating helps to reduce microlaser thermal resistance. SU8 coating can be successfully used for top contact pads formation without deterioration of the threshold properties. Increase of coating refractive index to 2.56 (TiO_2) results in drastic spectral and threshold changes of microlasers even for 100 nm thick layer. The covering of the largest microlaser (10.4 µm in diameter) with the TiO_2 layer helps to achieve quasi-singlemode lasing while the threshold level is preserved. In other words, the coating layers with high-index can be used to improve lasing characteristics of the device. However, the covering of the smallest microlaser (5.4 µm in diameter) with the TiO_2 layer demonstrates its high sensitivity to surrounding. This means that lasers of such diameters can be successfully used for sensing purposes.

Acknowledgments
The work is supported by Russian Science Foundation (project № 18-12-00287).

References

[1] M. L. Gorodetsky, "Optical microresonators with gigantic quality factor," Fizmatlit, Moscow, 2011.

[2] E.I. Moiseev, N.V. Kryzhanovskaya, Yu.S. Polubavkina, M.V. Maximov, M.M. Kulagina, Yu.M. Zadiranov, A.A. Lipovskii, I.S. Mukhin, A.M. Mozharov, F.E. Komissarenko, Z.F. Sadrieva, A.E. Krasnok, A.A. Bogdanov, A.V. Lavrinenko, and A.E. Zhukov, ACS Photonics, 4 (2), pp 275–281 (2017).

[3] L. He, S.K. Ozdemir, J. Zhu, W. Kim, and L. Yang, Nature Nanotechnol. 6, 428–432 (2011).

[4] A. Francois and M. Himmelhaus, Appl. Phys. Lett. 94, 031101 (2009).

[5] Cho S. Y., Jokerst N. M. A polymer microdisk photonic sensor integrated onto silicon//IEEE photonics technology letters. – 2006. – T. 18. – No. 20. – Page 2096-2098.

[6] Roelkens G. et al. III-V-on-silicon photonic devices for optical communication and sensing//Photonics. – Multidisciplinary Digital Publishing Institute, 2015. – T. 2. – No. 3. – Page 969-1004.

[7] Z. Zhang, L. Yang, V. Liu, T. Hong, K. Vahala, and A. Scherer, Appl. Phys. Lett. 90, 111119 (2007).

[8] Q. Song, H. Cao, S. T. Ho, and G. S. Solomon, Appl. Phys. Lett. 94, 061109 (2009).

[9] A. C. Tamboli, E. D. Haberer, R. Sharma, K. H. Lee, S. Nakamura, and E. L. Hu, Nature Photon. 1, 61–64 (2007).

[10] A. N. Al-Omari, G. P. Carey, S. Hallstein, J. P. Watson, G. Dang, K. L. Lear, Low thermal resistance high-speed top-emitting 980-nm VCSELs//IEEE Photonics Technology Letters. 2006. Vol. 18, 1225 – 1227.

[11] A. Demir, G. Zhao, D.G. Deppe. Lithographic lasers with low thermal resistance//Electronics Letters. 2010. Vol.46, 1147-1149.

Influence of hydrogen plasma passivation on electrical and spectral characteristics of GaN nanowires / Si solar cells

K U Shugurov[1], A M Mozharov[1], V V Fedorov[1], A D Bolshakov[1], G A Sapunov[1] and I S Mukhin[1,2]

[1] Saint-Petersburg Academic University, 8/3 Khlopina st., St. Petersburg 194021, Russian Federation

[2] ITMO University, 49 Kronverksky pr., St. Petersburg 197101, Russian Federation

Abstract. Volt-ampere and spectral characteristics of the series of GaN nanowires/Si solar cells differing in interface layer preparation are obtained. Influence of hydrogen plasma passivation on electrical and spectral curves is analyzed. It is demonstrated that the interface passivation significantly affects the characteristics of the studied cells.

1. Introduction

Today, search and implementation of new ways to increase the photoelectric conversion efficiency of the modern solar cells is the research area of interest. At the same time researchers attempt not only to provide high solar cell (SC) efficiency, but also to achieve as low production cost of the SCs as possible. One of the most promising approach for development of photoconverters today is semiconductor heterostructural SCs.

One of the main features that affects the heterojunction SC characteristics is the presence of a heterointerface [1] containing a certain density of the surface states. Minority carriers crossing the interface can be trapped by these states, resulting in significant reduction of the photoelectric conversion efficiency. Recombination processes at the surface states in the case of indirect-band based SCs, for example silicon or germanium, are practically the major limiting factor affecting SCs efficiency. Thus, interface passivation can play an important role in SCs efficiency improvement [2].

It should also be noted that $A^{III}B^{V}$ compounds have several advantages over the other compounds, and the use of group III nitrides is dictated with a number of their important properties including high crystal quality and growth in the absence of catalyst [3].

The work is devoted to the investigation of hydrogen plasma passivation influence of the GaN / Si heterointerface on SCs electrical and spectral characteristics.

2. Experimental details

In our previous work [4] the characteristics of the SCs based on n-GaN nanowires (NWs) on p-Si substrate were obtained. The study has shown that the interface preparation method significantly influences the samples spectral and electrical curves. Poor characteristics of the obtained SCs can be concerned with high surface recombination level at the interface [5]. Traditionally, to decrease the effects of this phenomenon, passivation of the surface is used to reduce the density of dangling bonds at the heterointerface.

Content from this work may be used under the terms of the Creative Commons Attribution 3.0 licence. Any further distribution of this work must maintain attribution to the author(s) and the title of the work, journal citation and DOI.

Published under licence by IOP Publishing Ltd

A series of samples differing from each other with the interface layer preparation technique was investigated. Here we demonstrate characteristics of the most efficient samples. The first sample had no special buffer layer during the growth (native-grown SiN_x sample with the NWs grown on bare Si surface), while another one had (buffer SiN_x, this sampe was intentionally nitridated prior to GaN deposition).

Typical view of the GaN NWs grown by using molecular beam epitaxy is presented in Fig. 1. The presence of an emitter layer in the form of n-GaN NWs array provides good antireflection properties without the use of expensive multilayer coatings [6].

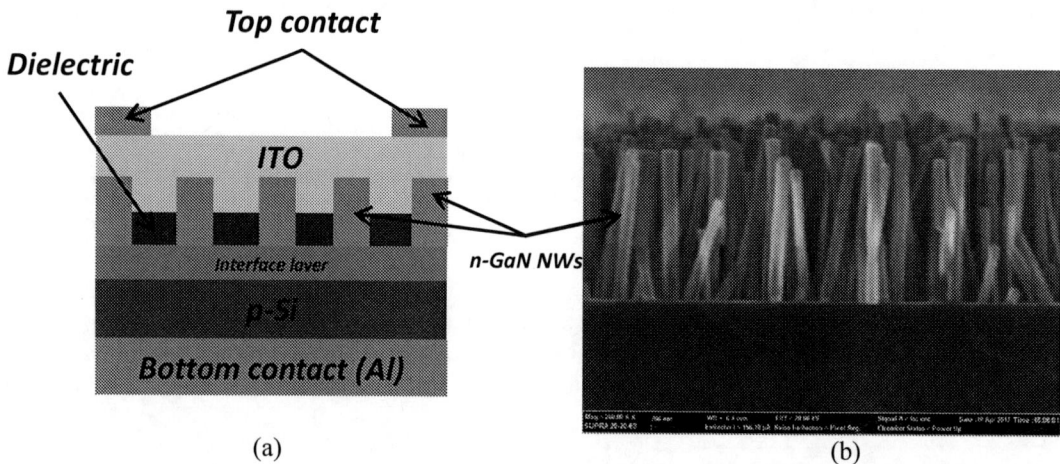

(a)   (b)

Figure 1(a, b). (a) schematics of the SC; **(b)** SEM image of the synthesized NWs

The front surface of the grown sample was treated with hydrogen plasma in PECVD machine. Hydrogen was chosen to passivate the substrate surface because of its small atom, allowing efficient penetration between crystal lattice atoms.

At the next technological step a back Al contact was fabricated on each of the synthesized GaN/Si heterostructures by thermal vacuum deposition and further thermal annealing. Using various methods of post-growth treatment, a dielectric insulating layer was deposited over the emitter layer, while tips of the NWs were left uncovered and then a conductive transparent indium-tin oxide (ITO) coating was applied via magnetron sputtering. On the basis of these heterostructures solar photoconverters with a photoactive region in the form of a circular meze 2.5 mm in diameter were fabricated.

Current-voltage characteristics (I-V characteristics) were measured using a Keithley 2400 SourceMeter multimeter and thermo-stabilized stage under illumination conditions of the AM 1.5G solar spectrum.

Spectral dependence of external quantum efficiency (EQE) was measured in the 350 − 1200 nm wavelength range using M266 (Laser Solar) monochromator and calibrated silicon photodiode with known spectral characteristics. The temperature during electric and spectral measurements was maintained at 25°C.

3. Results and discussion

The samples I-V curves measured before and after the passivation are shown in Fig. 2 demonstrating that hetero-interface passivation has significantly affected the curves. It can be also noticed that there is a big gain of short-circuit current (up to 26mA/cm^2) and double open-circuit voltage increase (up to 0.3V). The curve inclination at the reverse voltages before passivation (Fig. 2a) indicates the low shunt resistance presence, however, it is insufficient after the passivation. The dominance of hetero-interface recombination is the most likely reason the obtained I-V curve form (Fig. 2a) [7]. The practically linear shape of the forward part of I-V characteristic (Fig. 2b) indicates the presence of parasitic series resistance. Hydrogen passivation is seen to have no effect on the series resistance.

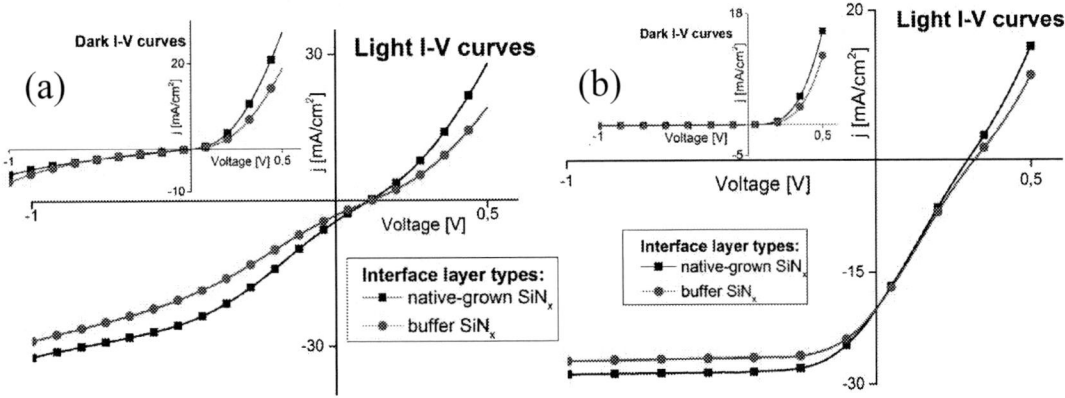

Figure 2(a, b). (a) Light and dark I-V characteristics before hydrogen passivation; **(b)** after hydrogen passivation

Fig. 3 demonstrates the samples spectral dependencies of external quantum efficiency (EQE) before and after passivation. The passivation influence can be easily noticed.

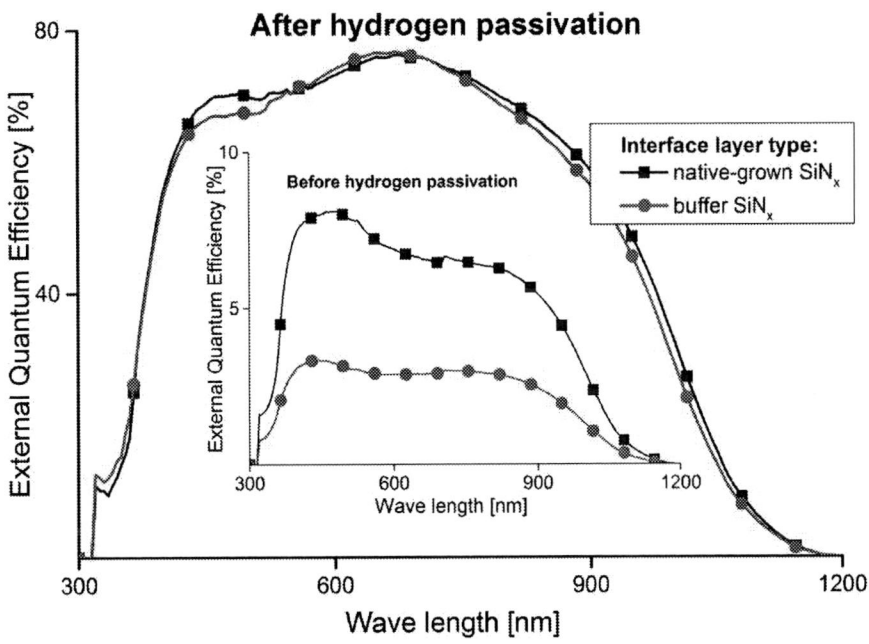

Figure 3. Experimental spectral characteristics before and after the passivation.

The obtained curves demonstrate drop of the EQE in the infrared part of spectrum. However, this effect can be observed both before and after the passivation. So we can conclude that this phenomenon can be concerned not with the heterointerface. Low EQE values in longwave range are dictated by recombination in the SC base and therefore low minority carriers lifetime in p-Si [8, 9]. This effect can be explained with bottom Al contact annealing at high temperature which might have increased the number of defects at the Si/Al interface. Thus IR-photons could be trapped by these defects more effectively.

4. Summary
In the course of the work, the characteristics of a series of n-GaN NWs / p-Si heterostructures synthesized at various technological parameters, before and after passivation, were compared. To determine the electrical and spectral characteristics of these structures, ohmic contacts were fabricated by thermal vacuum deposition of metals, magnetron sputtering and thermal annealing methods. I-V curves of the samples were obtained using the AM1.5G solar spectrum imitator. Spectral dependencies of external quantum efficiency (EQE) were obtained in the 300 – 1200 nm range. It is concluded that hydrogen passivation reduces current losses at the interface.

Acknowledgments
This work was carried out with the support of the Russian Federation President grants (MK-6492.2018.2, MK-3632.2017.2 and SP-2324.2018.1), the Russian Foundation for Basic Research (16-32-60094 mol_a_dk, 163200560 mol_a and 18-32-00899 mol_a), the leading universities of the Russian Federation (grant 074U01), grant of government of the Russian Federation (3.9796.2017/8.9 and 16.2593.2017/4.6).

References
[1] Gudovskikh A S, Kalyuzhnyy N A, Lantratov V M, Mintairov S A, Shvarts M Z and Andreev V M 2009 *Semiconductors* **43** 475-80

[2] Solovan M N, Brus V V and Maryanchuk P D 2014 *Semiconductors* **48** 1540-42.

[3] Ivantsov V A, Sukhoveev V A, Nikolaev V I, Nikitina I P and Dmitriev V A 1997 *Physics of the Solid State* **39** 858-60

[4] Shugurov K U, Mozharov A M, Fedorov V V, Bolshakov A D, Sapunov G A and Mukhin I S 2018 *Journal of Physics: Conference Series* **993** 012034

[5] Kosyachenko L A 2006 *Semiconductors* **40** 730-46.

[6] Mozharov A M, Bolshakov A D, Cirlin G E and Mukhin I S 2015 *Phys. Status Solidi RRL* **9** 507-10

[7] Brus V V 2012 *Solar Energy* **86** 786-91

[8] Sharma B L and Purohit R K 1974 *Pergamon Press* Semiconductor heterojunctions

[9] Brus V V, Ilashchuk M I, Kovalyuk Z D, Maryanchuk P D and Ulyanytsky K S 2011 *Semiconductor Science and Technology* **26** 125006

Mixed halide perovskite light emitting solar cell

D Gets[1], A Ishteev[1,3], T Liashenko[1], D Saranin[2], S Makarov[1] and A Zakhidov[1,3]

[1]ITMO University, 49 Kronverksky Pr., St. Petersburg, 197101, Russia
[2]National University of Science and Technology, NUST "MISIS", 2 Leninsky Pr., Moscow, 119049, Russia
[3]The University of Texas at Dallas, 800 W Campbell Rd, Richardson, TX 75080, USA

Abstract. We demonstrate that the halide perovskite planar solar cells with the architecture of ITO/PEDOT:PSS/Perovskite/PCBM/LiF/Al show a switchable dual operation of descent photovoltaic and quite bright electroluminescence in visible range. In our experiments, the active layer is made of a mixed halide perovskite ($MAPbBr_2I$) and the device is properly cycled upon light and bias exposure. We argue that this curious effect of switchable double functionality between solar cell and light-emitting device in one architecture is caused by photoinduced segregation in the perovskite. It is shown that the bright red electroluminescence at low voltage of ~ 2 (3) eV appears only after cycling the device in PV regime. On the other hand, electroluminescence operation also effects the following PV mode. This effect is caused by redistribution of photoactivated ions I-/Br- and their vacancies during photoexcitation in PV regime.

1. Introduction

Solar cells (SCs) and light-emitting diodes (LEDs) may share similar structural architecture designs, but each device specifically configured energetically to provide one specific function most effectively. In SCs the positions of highest occupied molecular orbital (HOMO) and lowest unoccupied molecular orbitals (LUMO) of perovskite (PS) and transport layers are selected for an efficient harvesting of photogenerated electrons and holes from the PS photoactive layer to the electron transport layer (ETL) and hole transport layer (HTL) towards the contacts. At the same time, in organic or perovskite LEDs HOMO and LUMO of transport layers are selected for the efficient injection of electrons and holes into the perovskite (PS) emission layer. This difference in the device designs does not allow to create a dual functional device (SC-LED) based on conventional materials owing to the mismatch of HOMO and LUMO levels for organic LED and SC will experience additional potential barriers in the reciprocal working regime. In order to create efficient optoelectronic device with the dual functionality, one has to adjust the energy band structure of the optoelectronic device via manipulation the height and the width of the potential barriers.

In turn, halide perovskites have emerged as promising materials for optoelectronic devices development owing to their high absorption, low exciton binding energy and solution processed synthesis technology [1, 2]. These advantages allow for the cost-effective production of highly efficient solar cells (SCs) and light emitting diodes (LEDs). For the last 5 years, photovoltaic parameters of perovskite SCs enhanced and even reached efficiency values of well-established solar cell material like silicon SCs [1]. Since the perovskites possess direct band gap, they also can be used for light generation. Nowadays parameters of perovskite LEDs are among the best and these LEDs capable of concurring with LEDs based on metal-organic complexes and different conjugated polymers [2, 3]. Solution processed synthesis of perovskites allows to gradually change the band gap value in the range of 1.5 eV to 2.3 eV by the quantitative change of its halides concentration $MAPbBr_xI_{3-x}$ ($0<x<3$) perovskite.

In turn, development of the dual functional devices are already underway, and certain progress has been achieved for the creation of such devices based on well-known material such as silicon [4] and recently emerged perovskite materials [5]. Although efficiency of obtained devices was relatively low: $MAPbBr_3$ perovskite based SC demonstrated ~1% PCE and 0.12% EQE_{EL}, silicon based SC demonstrated 2.1% efficiency, but it illuminated in IR region (~ 1100nm), $MAPbI_3$ perovskite based SC demonstrated ~12.8%

Content from this work may be used under the terms of the Creative Commons Attribution 3.0 licence. Any further distribution of this work must maintain attribution to the author(s) and the title of the work, journal citation and DOI.

Published under licence by IOP Publishing Ltd

PCE and 0.04% EQE_{EL} (~ 750nm) [6]. Therefore, one of the most attractive opportunities to realize the dual functional device is based on mixed halide perovskites $MAPbBr_xI_{3-x}$ (0<x<3) which will irradiate in visible range and have relatively high PCE and EQE_{EL} efficiencies. However, mixed halide perovskite exhibit one crucial property – segregation [7], which degrades device characteristics. Here we demonstrate a possibility to create such "dual functional" device and show that segregation can aid for switching of the device working regimes by p-i-n structure formation inside of the perovskite layer [8].

2. Method

Perovskite based planar p-i-n devices were fabricated in the ambient atmosphere. Design of perovskite light emitting solar cells had the following sequence of layers ITO/PEDOT:PSS/Perovskite/PCBM/LiF/Al (figure 1a) where band gap diagram is shown in the figure 1b. Device functional layers were subsequently deposited onto ITO covered substrates. Glass substrates with ITO pixels were cleaned in ultrasound bath in DI water, DMF, toluene, acetone and IPA consequently. The method of obtaining layers for these devices includes two techniques. PEDOT:PSS, perovskite and PCBM were deposited by a spin coating technique. Whereas LiF and Al were deposited by a vacuum deposition technique.

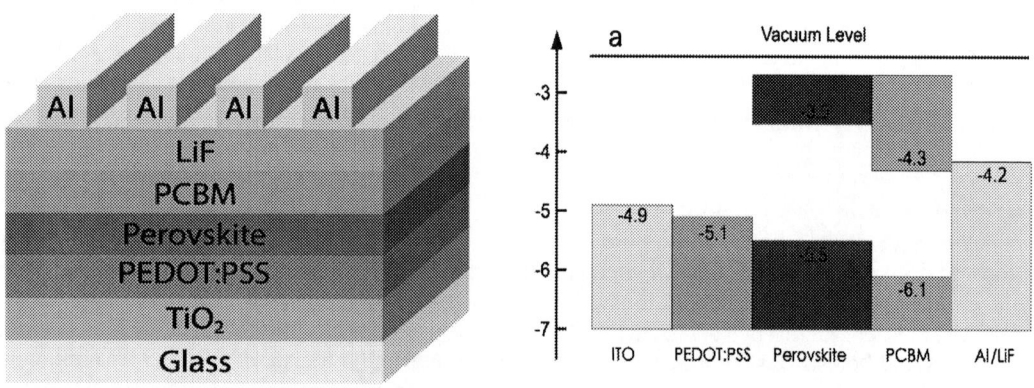

Figure 1. (a) Dual function device structure and (b) corresponding band diagram

Water dispersion of PEDOT: PSS 4083 was used as a hole transport material (HTL). It was annealed on a heating plate for 600 s at 120 ℃ in ambient atmosphere. Photoactive layer based on $CH_3NH_3PbBr_2I$ was prepared by the consequent dissolving MAI and $PbBr_2$ in DMF:DMSO (7:3) respectively. The solutions were steered overnight until dissolved at room temperature. Obtained perovskite solution was deposited by the single step solvent engineering technique [9] on top of HTL in 2 spin-cycles. Diethyl ether (DE) as antisolvent was slowly dripped on the rotating substrate in 10 s past the jumping from 1000 to 3000 rpm. Perovskite films were annealed on the heating plate in dry atmosphere for 600s at 100C. Fullerene derivative material $PC_{60}BM$ was used as electron transport layer (ETL). $PC_{60}BM$ was dissolved in chlorobenzene (CB) (25 mg/ml) and filtered through PTFE 0,45 syringe filter. Filtered solution was deposited onto perovskite layer by the spin coating. Before aluminium cathode deposition set of samples was divided into 2 groups. The aluminium cathode was deposited onto $PC_{60}BM$ layer in the first group and the second group had 3nm LiF sublayer on $PC_{60}BM$ layer. The thickness of the ETL, perovskite and HTL was measured by stylus profilometry. Acquired thickness of HTL equal 30nm, $PC_{60}BM$ − 50nm, perovskite − 200 nm. The thickness of Al cathode was in situ controlled by Inficon deposition controller.

3. Results

We investigated influence of different working regimes on the performance of reciprocal working regime. The device was switched between PV regime and LED regime. PV regime characterized by measurement of J-V characteristics under 1 sun (AM1.5 G). LED regime characterized by application of external forward voltage bias much higher than V_{OC}. The device demonstrated relatively good PV parameters (figure 2). In both regimes segregation will occur [10, 11] but rates of segregation will be different since regimes utilize different charge concentrations. In case of PV-regime it will manifest significantly stronger in comparison with a LED-regime, since segregation rates under 1 sun illumination usually considerably higher than in case of charge injection with equivalent density [11].

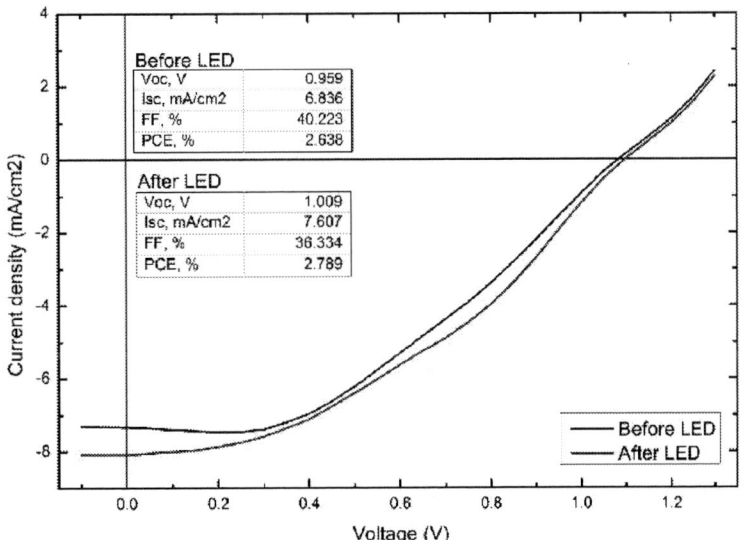

Figure 2. J-V characteristics of the device under 1 sun radiation before (black) and after (red) LED-regime

To transfer device to LED regime p-i-n structure should be created otherwise device will show weak EL and high voltage/current and eventually burn out. To ensure that p-i-n structure was formed in perovskite layer device was soaked under 1 sun illumination and at 2 Volts (figure 3a). Measured PV characteristics device was transferred into LED-regime (figure 2). Exposure device to ~2.0 V resulted in electroluminescence (LED regime) ignition (figure 3b).

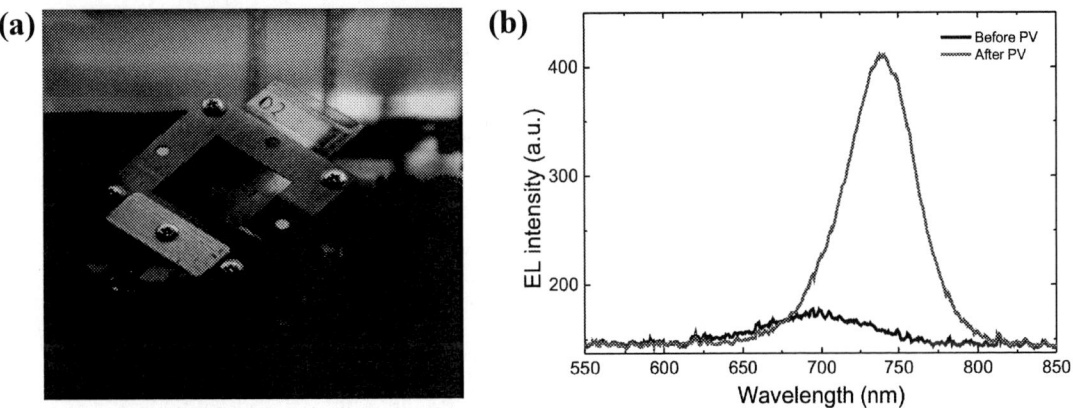

Figure 3. (a) Electroluminescence of the devices under 2V (b) corresponding intensity of the electroluminescence

The electroluminescence under 2.0 V demonstrated optical power ~ 130 μW. After the removal of external bias and device relaxation, the device was switched into the PV regime. Following PV measurements also demonstrated good PV parameters (figure 2). Cycling PV and LED regime has demonstrated the ability of the device to work in two reciprocal regimes. It should be noted, that non-soaked device demonstrated very weak EL at high voltages ~ 4V.

This, in turn, argues in favor of lowering potential barriers in the device band structure due to ions migration towards contacts (figure 4). Application of external voltage results in migration of ions and their vacancies generated by white light illumination towards interface layers of PS/ETL and HTL/PS. Amount of available ions and vacancies is greater than in case of the non-soaked device. Accumulation of ions and their vacancies at interface layers results in doping of interface regions of photoactive layer with following band bending and p-i-n structure formation like in case of light emitting electrochemical cells (LEECs). In LEECs light generation starts after application of high external voltage. This voltage is necessary to move ions towards contacts in polymer [12]. When p-i-n structure starts to form and potential barriers become lower, charge carriers can easier get into emission layer through potential barriers.

Segregation aids for the formation of the p-i-n structure in the device. Since it manifests in creation of photoinduced Br- and I-ions that tend to form enriched regions. Carriers injected into perovskite layer recombine in I-rich regions. Moreover, segregation is a temporal effect. Removal of external illumination or bias leads to backward redistribution of photoinduced ions and their vacancies.

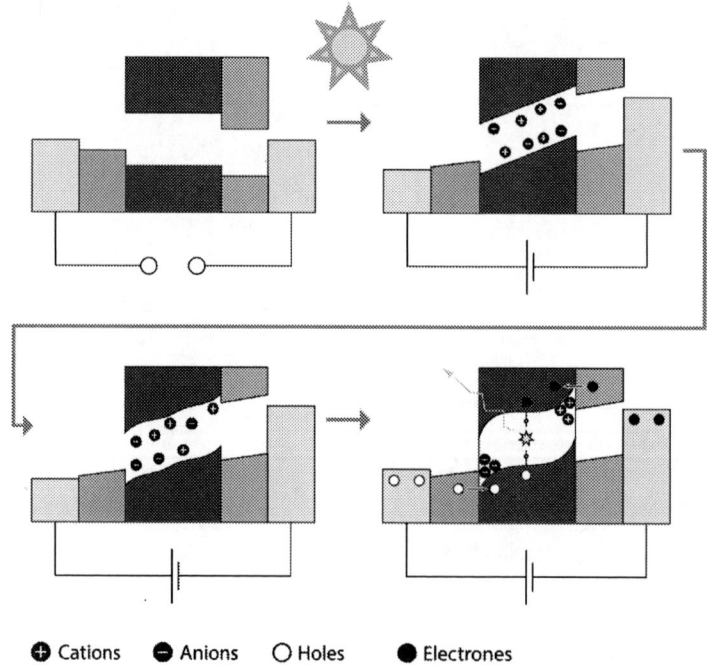

⊕ Cations ⬤ Anions ◯ Holes ● Electrones

Figure 4. An example of segregation in of device structure used in present work

Further accumulation of ions and their vacancies near interfaces leads to the formation of pronounced p-i-n structure, where bent potential barriers become lower and therefore electrons and holes can easily get into perovskite layer and radiatively recombine there. Segregation and p-i-n junction formation are temporal, the removal of external bias will lead to diffusion of ions and their vacancies back to the bulk of perovskite layer (figure 4).

4. Conclusion

The work demonstrates the ability of the mixed halide perovskite optoelectronic device to work in two different regimes. In this case, segregation of halogen ions upon light irradiation, which usually negatively affects PV device performance, helps to the dual functionality by the ions movement towards the interfaces between the perovskite photoactive layer and transport layers. As a result, p-i-n structure formed within mixed halide perovskite, which allows electrons and holes to be injected into perovskite layer and radiatively recombine there in I-rich areas. The removal of external bias leads to the reduction of ions/vacancies concentration at interfaces and causes their redistribution back to the mixed halide perovskite layer, resulting in the Br- and I-rich regions dissociation.

References

[1] Martin A. Green, Anita Ho-Baillie and Henry J. Snaith 2014 Nature Photonics **8** 506-514

[2] Brandon R. Sutherland and Edward H. Sargen 2016 Nature Photonics **10** 295-302

[3] Zheng-Guang Wu, Yi-Ming Jing, Guang-Zhao Lu, Jie Zhou, and You-Xuan Zheng, Liang Zhou, Yi Wang, and Yi Pan 2016 Scientific reports **6** 38478

[4] D.M. Zhigunov, A.S. and Il'in, and P.A. Forsh, V.N. Verbitskii, E.I. Terukov, and P.K. Kashkarov 2017 Technical Physics Letters **43** 496-498

[5] Hak-Beom Kim, Yung Jin Yoon, Jaeki Jeong, Jungwoo Heo, Hyungsu Jang, Jung Hwa Seo, Bright Walker, and Jin Young Kim 2017 Energy & Environmental Science **10** 1950-1957

[6] L. Gil-Escrig, G. Longo, A. Pertegás, C. Roldán-Carmona, A. Soriano, M. Sessolo, and H. J Bolink 2015 Chemical Communications **51** 569-571

[7] E. T. Hoke, D. J. Slotcavage, E. R. Dohner, A. R. Bowring, H. I. Karunadasa, and M. D. McGehee 2015 Chemical Science **6** 613-617

[8] Y. Deng, Z. Xiao, and J. Huang 2015 Advanced Energy Materials **5** 1500721

[9] J.-H. Im, H.-S. Kim and N.-G. Park 2014 Apl Materials **2** 081510

[10] A. J. Barker, A. Sadhanala, F. Deschler, M. Gandini, S. P. Senanayak, P. M. Pearce, E. Mosconi, A. Pearson, Y. Wu, A. R. Srimath Kandada et al. 2017 ACS Energy Letters **2** 1416â˜AS,–424

[11] I. L. Braly, R. J. Stoddard, A. Rajagopal, A. R. Uhl, J. K. Katahara, A. K.-Y. Jen and H. W. Hillhouse 2017 ACS Energy Letters **2** 1841–1847

[12] S. B. Meier, D. Tordera, A. Pertegas, C. Roldan-Carmona, E. Ortí and H. J. Bolink 2014 Materials Today **17** 217–223

SPBOPEN 2018

IOP Publishing

Non-radiative coupled donor-acceptor pair recombination in nanostructures based on nitrides and related phenomena

N A Talnishnikh[1], E I Shabunina[2], N M Shmidt[2], A E Chernyakov[1], D S Arteev[2], A A Zybin[3]

[1]Submicron Heterostructures for Microelectronics Research and Engineering Center RAS, St Petersburg 194021, Russia
[2]Ioffe Physical Technical Institute, St Petersburg 194021, Russia
[3]JSC «Svetlana-Electronpribor», 27 Engels av., St Petersburg, 194156, Russia

Abstract. It has been clarified that several peculiarities of I-V characteristics at U < 1.5 V and low frequency noise dependences in InGaN/GaN nanostructures for green and blue spectral range as well as in Schottky barriers in AlGaN/GaN HEMT nanostructures are evoked by the non-radiative coupled deep donor-acceptor pair recombination. The strong contribution of the non-radiative donor-acceptor recombination in the degradation of the nanostructures properties has been confirmed experimentally.

Introduction

Nanostructures such as InGaN/GaN and AlGaN/GaN with different Al and Ga concentration are promising materials for the realization of HEMTs, LEDs and photo-detectors in a wide spectral range. The potential of these materials is still partially implemented [1] due to the lack of understanding of the non-radiative recombination. The peculiarities of the process in nitride-based devices are still under discussion [2]. The vast majority of articles consider the non-radiative recombination in the framework of statistics Shokley-Read-Hall [3], while peculiarities of carrier transport related to the nanostructured fractal organisation of nitrides [2] are usually ignored due to the fact that they occur at current and voltage values significantly lower than operational ones. In various works of recent years, it has been shown that radiative deep donor-acceptor pair (DAP) in GaN and nitride-based nanostructures takes place in the regions related to complicated fractal structure in nitrides [4, 5]. According to the theoretical estimations for other materials carried out in [6], DAPs might be a source of not only a radiative DAP recombination, but also non-radiative DAP one. Experimental results presented in this paper allow us to assume the existence of the DAP recombination and related phenomena in light-emitting green and blue InGaN/GaN nanostructures and Schottky barriers of AlGaN/GaN HEMT nanostructures.

Experimental

The HEMT topology with an Ni-Au barrier was fabricated in JSC «Svetlana-Electronpribor» on standard AlGaN/GaN HEMT nanostructures grown by the metal-organic chemical vapor deposition in Ioffe Institute on a (0001) sapphire substrates [7]. Commercial LED nanostructures with the external quantum efficiency η in the range of 40 – 50 % for blue (450 – 460 nm) and 15 – 30 % for green (520 – 530 nm) LEDs were studied. The current-voltage (I-V) characteristics in the region of $10^{-13} \div 1$ A

Content from this work may be used under the terms of the Creative Commons Attribution 3.0 licence. Any further distribution of this work must maintain attribution to the author(s) and the title of the work, journal citation and DOI.
Published under licence by IOP Publishing Ltd

were measured. The power spectral noise density dependences on current were investigated to clarify the presence of the current crowding effect in all nanostructures. The nanostructures properties was studied after several stages of accelerated ageing test under the injection current of 65 A/cm² at 110 °C.

Results and discussion

The common properties of all investigated nanostructures are the divergence of the forward branch of I-V characteristics obtained experimentally from the Shokley-Read-Hall (SRH) model at U < 1.5 V, the values of ideality factor n > 2 related to the presence of the parallel non-linear conductive shunt at a depletion space (figure 1, a). The increase of such shape in I-V characteristics is observed with the growth in nanostructural disorder (ND) (figure 1, b). The growth in ND disorder is due to the conversion from 2D to 3D growth mode. ND disorder is quantitatively characterized by a multifractal parameter – the degree of disorder (Δ_p) [2, 8]. Such a correlation is typical for both blue and green LEDs, but is more evident in blue LED structures as shown earlier in [2].

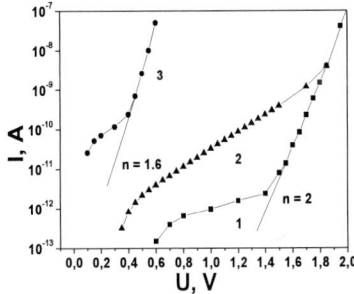 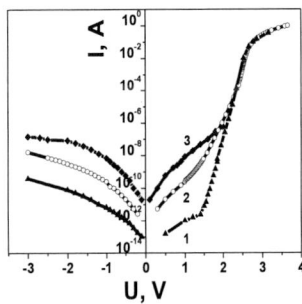

Figure 1 (a, b). (a) Forward I-V characteristics of: 1 – blue InGaN/GaN LED nanostructures, 2 – green InGaN/GaN LED nanostructures, 3 – Shottky barrier of HEMT nanostructures; **(b)** I-V characteristics of blue LEDs with different Δ_p values: 1 – 0.330, 2 – 0.348, 3 – 0.355.

Beside these peculiarities contradicting the SRH statistics, there are also current crowding effect (figure 2, a) identified by the deviation of voltage fluctuation dependences on current S_v (I) from the theoretical shape of ~ 1/I (dash line) corresponding to the homogeneous current distribution and the absence of generation-recombination noise at the current noise density spectra for all investigated samples (figure 2, b). The results obtained allow us to assume the existence of more complicated centers in the nanostructures, related to the fractal nature of nitrides along with the SRH ones. It should be noted that the I-V characteristics with n > 2 at low bias were observed in various materials [6]. It was clarified that such shape of I-V characteristics according to the model in [6] is due to the presence of DAPs with a high local concentration at the depletion space in case if the distance between deep donor and deep acceptor centers is less than 3 nm and the activation energy of centers E_A and E_D around middle gap with $E_A - E_D$ near 100 meV has an almost symmetric distribution. These conditions are implemented in nanostructures based on nitrides: GaN barriers are separated by a 3 nm InGaN layer with random compositional fluctuations and the deep centers with such E_A and E_D are present in nitrides [9]. Moreover, V-defects contain poorly bound atoms of Ga and In whose concentration is remarkably higher than that of SRH centers. In the local regions with irregular alloy composition, a concentration of excess In and Ga atoms can exceed a concentration of uniformly distributed local SRH centers by a several orders of magnitude.

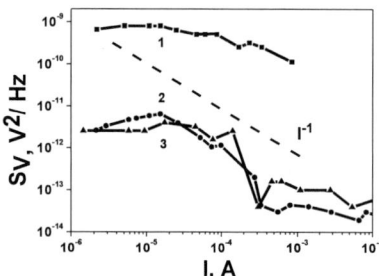 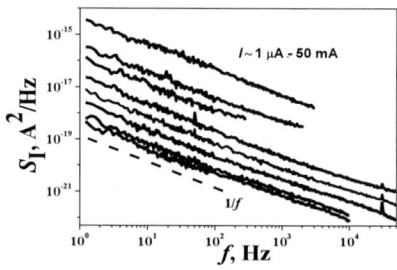

Figure 2 (a, b). (a) The voltage fluctuation dependences on current in: 1 – Shottky barrier of HEMT nanostructures; 2 – blue InGaN/GaN LED nanostructures; 3 – green InGaN/GaN LED nanostructures. **(b)** Typical low-frequency current noise spectra in the investigated samples.

The theory of defects developed for other materials shows that the donor-acceptor coupled centers are excited (metastable). According to [10] they are the source for the development of the phonon-assisted tunneling and multi phonon recombination, local heating, the defect displacement, Frenkel pair generation or recombination, defects diffusion, defects annealing under the injection current. All these phenomena are observed in LED nanostructures during the ageing test. The obtained results being in agreement with those published in many papers [11] suggest that several mechanisms of the degradation processes take place in InGaN/GaN LED structures. The common mechanism for all LEDs is the current crowding effect intensifying during aging which results in local overheating identified by the methods of low frequency noise measurements [12] and infrared microscopy. The control of I-V characteristics after different stages of aging reveals the complicated process of defects generation under injection current [13]. The alternation of defects generation in two local channels at early stage of aging is observed. One of them is the defect generation in local InGaN regions with non-equilibrium alloy composition and another one is defect generation in extended defect system. The start of defects generation happening in one of the channels only can be identified by peculiarities of I-V characteristics. A primary increase in values of reverse current and conductivity of the paths relates to the defects generation occurring in the extended defect system, which is more common in LED structures with $\Delta_p > 0.345$ [12, 13]. A growth in forward current values at U < 1.5 V, which is not accompanied by any noticeable modification in reverse current, is related to the defects generation in local InGaN regions with non-equilibrium alloy composition. At early aging stage, this modification can either reduce or increase external quantum efficiency (EQE) values by 5 – 10 %. At this stage, a great variety of changes in I-V characteristics and EQE values for LEDs (especially blue LEDs) of the same lot is observed. Moreover, an increase in forward current values only at V < 2 V during an aging test shorter than 100 h enables us to identify LEDs with an unpredictable failure [14]. This affect was not observed in green LEDs. However, uniform behavior at late stage of degradation (Figure 3. a), when the defect generation occurs in both channels simultaneously, leading to a growth in conductivity of the paths, and reverse and forward current values at U < 1.5 V only, is common for all LEDs under the study. The uniform behavior is related to a non-radiative DAP and results in a more than 20 % decrease in EQE values. It should be noted that a similar behavior, including that in nitride-based lasers, was observed in many papers [2, 15]. At the final stage of degradation, the redistribution of In between local regions in InGaN alloy, which is reflected in wave-length redistribution, is observed (Figure 3, b). The work [16] provides experimental data obtained by TEM in a structure after degradation, revealing In migration along dislocations and adjacent MQW areas.

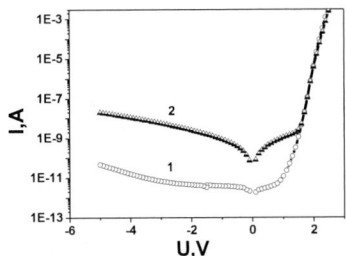 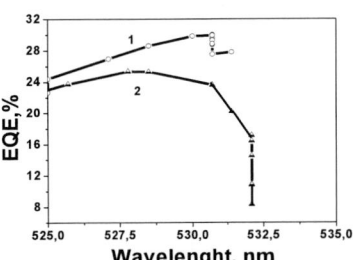

Figure 3 (a, b). (a) The evolution of I-V characteristics after an ageing test leading to a decrease in the EQE values from 30 % to 22 %: 1 – before, 2 – after; **(b)** The change of the EQE and the peak value of wavelength after ageing test in green LEDs: 1 – before, 2 – after.

Conclusion

The obtained results allow us to assume the participation of the coupled deep donor-acceptor pairs in non-radiative recombination. The local concentration of such metastable centers can be higher that of SRH ones in V-defects and in the regions with irregular alloy composition. Their presence is identified by the peculiarities of I-V characteristics at U < 1.5 V and by a non-linear development of EQE degradation.

Acknowledgments

The work is partially supported by the Russian Foundation for Basic Research (grant № 17-08-01000 A). The measurements were carried out at the Center of Multi-User Facilities "Element Base of Microwave Photonics and Nanoelectronics: Technology, Diagnostics, Metrology".

References

[1] De Santi C et al 2016 *Journal of Applied Physics* 119 094501
[2] Petrov V.N. et al 2016 *Semiconductors* 50 1173
[3] Joachim Piprek et al 2015 *Applied Physics Letters* 106 101101
[4] M.Matus et al 2017 *Journal of Applied Physics* 121 065104
[5] A.V.Andrianov et al 2002 *Semiconductors* 36 641-646
[6] Steingrube S et al 2011 *Journal of Applied Physics* 110 014515
[7] Shalygin V A et al 2011 *Journal of Applied Physics* 109 073108
[8] A.I.Besyulkin et al 2005 *Physica Status Solidi C* 2 837
[9] In-Hwan Lee et al 2017 *Appl.Phys.Lett.* 111 062103
[10] Yassievich I.N., 1994 *Semicond.Sci. Technology*, 9 1433
[11] Yung K. C. 2011 *J. of Appl.Phys.* 109 094509 .
[12] Natalia Shmidt et al 2013 *Physica Status Solidi C* 10 332
[13] Natalia Shmidt et al 2015 *Physica Status Solidi C* 12 349
[14] A.E.Chernyakov et al 2012 *Micrioelectronics Reliability* 52 2180
[15] P.Y. Wen et al 2015 *Journal of Physics D: Applied Physics* 48 415101
[16] Leung K.K. *et al.* 2010 *J.of Appl.Phys.* 107 073103.

SPBOPEN 2018

IOP Publishing

IOP Conf. Series: Journal of Physics: Conf. Series **1124** (2018) 041024   doi:10.1088/1742-6596/1124/4/041024

Formation of nanoscale structures on the surface of gallium arsenide by local anodic oxidation and plasma chemical etching

V S Klimin[1], M S Solodovnik[1], S A Lisitsyn[1], A A Rezvan[1], S V Balakirev[1]

[1] Southern Federal University, Department of Nanotechnology and Microsystems, Taganrog 347922, Russia

Abstract. This work is devoted to the analysis of problems of the present methods of surface treatment and the preparation of structures with nanoscale. The urgency of the work is caused by the fact that it uses the method of obtaining structures without the standard operations of optical lithography. The main methods of forming nanoscale structures in the work is the formation of an oxide layer on the surface of a semiconductor by the method of local anodic oxidation performed on a probe microscope. After the formation of oxide structures, plasma chemical etching was carried out in combined chloride plasma using inductively coupled and capacitive plasma discharges. As a result, nanoscale structures with a height of 3.5 to 70 nm and a width of 60 to 180 nm were obtained. Structures with a high aspect ratio were obtained.

1. Introduction

The increased interest in optoelectronic devices accelerated the development of advanced technologies for processing semiconductor structures based on GaAs. Mastering the technology of selective etching of GaAs as compared to AlGaAs was the key process for the formation of GaAs-based electronic devices, for example, such as heterojunction bipolar transistors (HBTs) and high-mobility electron transistors (HEMTs) [1-10].

The most common method of forming the relief on the surface of GaAs-based structures is the method of reactive ion etching. This method is based on the use of high ion energy, which leads to the destruction of bonds in molecules and the desorption of sputtered etch products from the surface. This method is very effective, however, due to high ion energies, this effect can damage the structure, which leads to a deterioration in the electrical and optical characteristics of the finished devices [11-18]. With a decrease in ion energy or an increase in the chemical activity in the plasma, the growth rate slows down considerably or the aspect ratio of the profile worsens, which considerably limits the application of these structures. In this regard, the actual task is to search for modes of nanoscale topography formation, which combine high-quality etching characteristics, such as (roughness of etched surface, high aspect ratio) with low damage caused by plasma treatment. One of their promising solutions to this problem is the use of a combined plasma discharge of reactive ion etching and inductively coupled plasma. Combining the two types of discharges significantly improves the etching characteristics, this is due to the increased plasma density, which is 2-4 orders of magnitude higher, which improves the efficiency of III-V bond rupture, spraying rate and desorption of etch products formed on the surface [19-22]. The main advantages of this system are the control of ion energy and their density, separately from the capacitive discharge, which leads to the least damage to the surface of the structure and the fact that the DC bias is controlled separately from the plasma

Content from this work may be used under the terms of the Creative Commons Attribution 3.0 licence. Any further distribution of this work must maintain attribution to the author(s) and the title of the work, journal citation and DOI.

Published under licence by IOP Publishing Ltd

source, which provides greater control over the etching parameters. These advantages make it possible to provide the necessary uniformity of plasma, but high density at low cost of operation, which is a very important factor in the production of modern micro- and nanoelectronic devices [22-28].

The aim of the work was to study the modes of processing and profiling structures based on gallium arsenide by a combination of local anodic oxidation methods and the method of plasma chemical etching taking into account the formation voltage of the oxide masking layer and the etching time.

2. Description of the method

In this paper, a combination of local anodic oxidation and plasma-chemical etching was used to treat near-surface structures based on gallium arsenide. Local anodic oxidation was carried out by means of scanning probe microscopy using a silicon probe. Plasma-chemical treatment was carried out in a discharge by combined plasma using two types of capacitive and inductively coupled discharge. This technology allows you to combine the advantages of two types of categories. The high concentration of electrons and ions provides a high etching rate, and the capacitive discharge allows the ions to be guided along the field and directed perpendicular to the substrate to be treated, allows to reduce energy costs for production, and also to increase the yield of suitable products, which leads to a reduction in the cost of devices based on this technology. Eliminating the operations associated with liquid lithography, the relative toxicity of the processes of nanoscale profiling of the surface of semiconductor structures decreases. Obtained oxide nanostructures were used as masking layers during plasma chemical etching in STE ICPe68. The reaction gas was BCl_3, which has a number of advantages over other chlorine-containing gases in the processes of plasma chemical treatment of structures based on gallium arsenide. The pressure of the gas atmosphere of the gases in the reactor was 1 Pa. The flow rate of the buffer gas-transport, which makes it possible to increase the controllability of the process by reducing the ionization potential

Modification of the surface was carried out with the help of atomic force microscopy on the surface of chemically purified substrates a thin layer of oxide was formed. Local anodic oxidation was carried out under the following conditions: probe travel speed 1.5 μm/sec, humidity 90%, Set point 0.5 nA. The formation voltage of nanostructures varied from 6 to 10 V. After formation of oxide structures on the surface, structures were formed by plasma chemical etching with parameters, the pressure in the reactor was 2 Pa, gas flows $N_{Ar} = 100$ cm^3/min, $N_{BCl3} = 10$ cm^3/min, power The source of the capacitive plasma was $W_{RIE} = 35$ V, the bias voltage $U_{bias} = 95$ V. The power of the inductive-coupled plasma source $W_{ICP} = 400$ V, the etching time varied from 0.5 to 1 minutes.

Surface control was carried out using NTegra probe laboratory. AFM images were processed with the help of a specialized software package "Image Analysis", and with the help of this software, surface profiles were obtained on which the height of the obtained nanostructures was determined.

3. Results and discussion

Figure 1 shows the results of experimental studies. The height of the structures obtained depends on the thickness of the oxide layer on the surface formed initially. The thickness of the resulting oxide layer depends on many parameters. Figure 1 shows the dependence of the thickness of such a layer on the formation stress at local anodic oxidation.

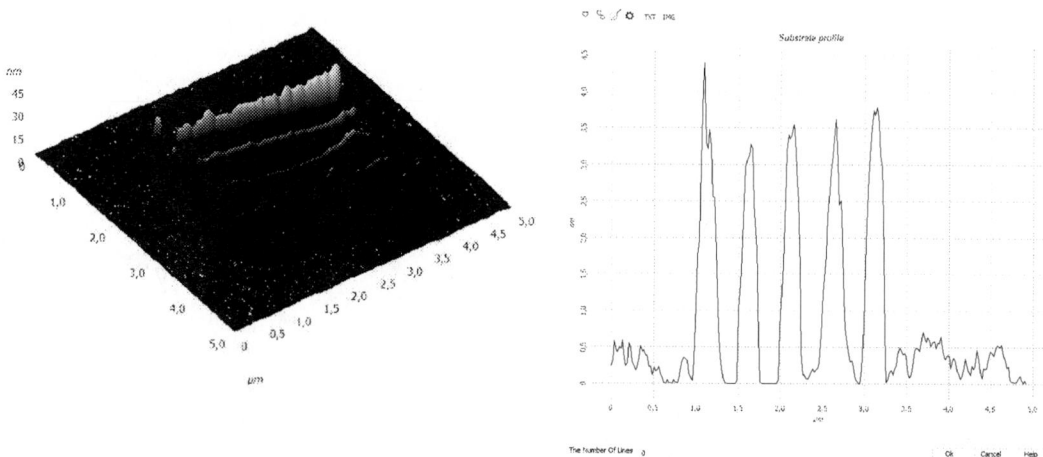

Figure 1. AFM - image and surface profile of formed structures on the surface of gallium arsenide by the method of local anodic oxidation and plasma chemical etching.

It can be concluded from the obtained AFM images that the oxide layer on the surface of the GaAs substrate is also removed by the plasma chemical method. Figure 2 shows the dependence of the height of the formed structures on the formation voltage of the oxide layer and on the time of plasma chemical etching in chloride plasma.

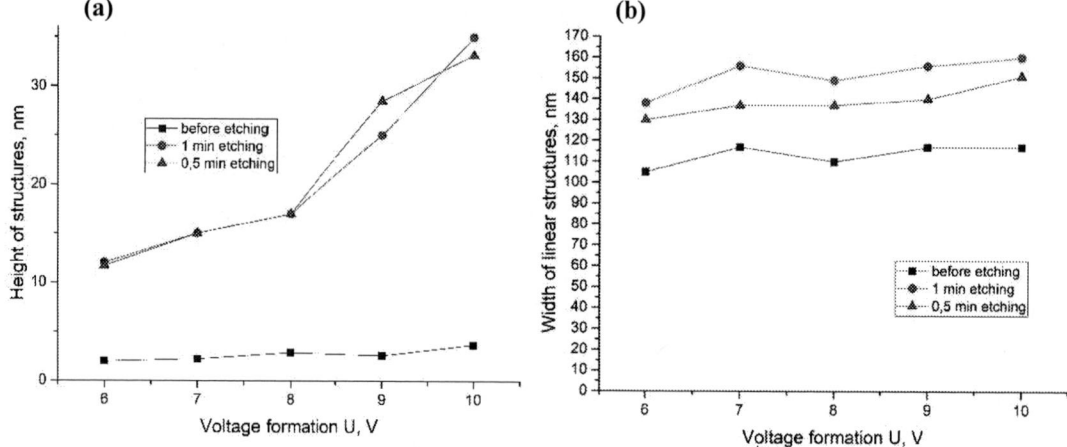

Figure 2(a, b). (a) Dependence of the height of the resulting structure on the formation voltage of the oxide layer and the time of plasma chemical etching; **(b)** Dependence of the width of the resulting structure on the formation voltage of the oxide layer and the time of plasma chemical etching.

It can be seen from the dependences that the height of the obtained structures depends on the thickness of the oxide obtained and the plasma-chemical etching regimes. Initially, oxide structures on the surface had the same thickness. The width of the formed structure from the etching time and very little depends on the formation stress of the oxide structure.

4. Conclusion

In the work, studies were carried out on the influence of the formation stress under local anodic oxidation with subsequent plasma-chemical etching on the parameters of structures based on GaAs. As a result, nanoscale structures with a height of 3.5 to 70 nm and a width of 60 to 180 nm were obtained. Structures with a high aspect ratio were obtained. Taking into account the carried out researches, it is possible to select the optimum modes of formation of structures that will allow, most accurately control the thickness of the etched layer, the roughness of the etched surface and the verticality of the resulting structure. Experimental studies have been carried out on the influence of the modes of forming a masking coating for subsequent plasma chemical etching, using the method of local anodic oxidation. It is shown that the surface after removal of the oxide has a low roughness, suitable for the formation of nanoscale structures based on gallium arsenide. Also, the results of experimental studies of the dependence of the height of structures on the formation voltage of the masking oxide coating for different times of plasma chemical treatment were obtained. These dependences showed that there is a direct dependence of the height of the structures on the thickness of the oxide layer on the surface of the gallium arsenide substrate and on the formation stress of such a layer, respectively. The etching time also influences the height of the structures, however, with the greatest etching time, the roughness of the etched surface also increased.

Based on the results of the experimental studies, it is possible to select optimal regimes for the formation of nanoscale structures that will allow controlling the height of the structures, the aspect ratio and the roughness of the treated surface.

Acknowledgments

This work was supported by Grant of the President of the Russian Federation No. MK-2629.2017.8. The results were obtained using the equipment of the Research and Education Center and Center for Collective Use "Nanotechnologies" of Southern Federal University.

References

[1] Liu Y, Lehman V, Wang L. 2015 *Com. Net.* **83** 85-99
[2] Chekurov N, Grigoras K, Peltonen A, Franssila S, Tittonen I 2009 *Nanotechnology.* **20** 5
[3] Il'Ina M V, Blinov Y F, Il'In O I, Klimin V S, Ageev O A 2016 *Proc. of SPIE* **10224** 102240U-1
[4] Nishinaga T and Shen X Q 1994 *Appl. Surf. Sci.* **82-83** 141
[5] Klimin V S, Il'Ina M V, Il'In O I, Rudyk N N, Ageev O A 2017 *J. Phys.: Conf. Ser.* **917** 092023
[6] Higuchi Y, Uemura M, Masui Y, Kitada T, Shimomura S, Hiyamizu S 2003 *J. Cryst. Growth* **251** 80
[7] Ageev O A, Solodovnik M S, Balakirev S V, Mikhaylin I A 2016 *J. Phys.: Conf. Ser.* **681** 012036
[8] Schmidt M, Johari Z, Ismail R, Mizuta H, Chong H 2012 Microelectron. Eng. 98 313
[9] Tseng A 2004 J. Micromech. Microeng. 14 15
[10] Ageev O A, Solodovnik M S, Balakirev S V, Eremenko M M, Mikhaylin I A 2017 *J. Cryst. Growth* **457** 46-51
[11] Han J, Lee H, Min B, Lee S 2010 Microelectron. Eng. 87 1
[12] Tominov R V, Zamburg E G, Khakhulin D, Klimin V S, Smirnov V A, Chu Y H, Ageev, O.A. 2017 *J. Phys.: Conf. Ser.* **917** 032023
[13] Orloff J, Swanson L W, Utlaut M 2003 Springer New York
[14] Klimin V S, Tominov R V, Eskov A V, Krasnoborodko S Y, Ageev O A 2017 *J. Phys.: Conf. Ser.* **917** 092005
[15] Nagase M, Nakamatsu K, Matsui S, Namatsu H 2005 Japanese J. of Appl. Phys. 44(7) 5409
[16] Ageev O A, Solodovnik M S, Balakirev S V, Mikhaylin I A 2016 *Technical Physics* **61** (7) 971–977
[17] Taillon J A, Pellegrinelli C, Huang Y-L, Wachsman E D, Salamanca-Riba L G 2018 *Ultramicroscopy* **184** Part A 24-38

[18] Rudyk N N, Il'In O I, Il'Ina M V, Fedotov A A, Klimin V S, Ageev O A 2017 *J. Phys.: Conf. Ser.* **917** 082008

[19] Tominov R V, Bespoludin V V, Klimin V S, Smirnov V A, Ageev O A 2017 *IOP Conf. Ser.: Mater. Sci. Eng.* **256** 012023

[20] Schaffer M, Schaffer B, Ramasse Q 2012 Ultramicroscopy 114 62

[21] Avilov V I, Ageev O A, Smirnov V A, Solodovnik M S, Tsukanova O G 2015 *Nanotech. in Russia* **10** (3-4) 214-219

[22] Salvati E, Brandt L R , Papadaki C, Zhang H, Mousavi S M, Wermeille D, Korsunsky A M, 2018 *Mater. Lett.* **213** 346–349

[23] Korsunsky A M , Salvati E , Lunt A G J , Sui T , Mughal M Z , Daniel R, Keckes J, Bemporad E, Sebastiani M 2018 *J. Matdes* **145** 55–64

[24] Ageev O A, Klimin V S, Solodovnik M S, Eskov A V, Krasnoborodko S Y 2016 *J. Phys.: Conf. Ser.* **741** 012178

[25] Sabouri A, Anthony C J, Bowen J, Vishnyakov V, Prewett P D ,2014 *Microelectron. Eng.* **121** 24-26

[26] Ostadi H, Jiang K, Prewett P D 2009 *Microelectron. Eng.* **86** (4–6) 1021-1024

[27] Klimin V S, Solodovnik M S, Smirnov V A, Eskov A V, Tominov R V, Ageev O A 2016 *Proc. of SPIE* **10224** 102241Z-1

[28] Wang Y-c, Xie D-g, Ning X-h, Shan Z-w 2015 *Appl. Phys. Lett.* **106** 081905

Quantum autonomous magnetic field sensor

S E Logunov[1], A Yu Koshkin[1], V V Davydov[1,2]

[1]Peter the Great Saint-Petersburg Polytechnic University, Saint Petersburg 195251, Russia

[2]Department of Ecology, All-Russian Research Institute of Phytopathology, 143050, Moscow Region, Odintsovo district, B.Vyazyomy, Russia

Abstract. A new design of a quantum sensor based on a ferrofluid cell for recording variations in the magnetic field is considered. The possibility of determining the position of the magnetic object in the zone of placement of a quantum sensor by changing the structure of the diffraction pattern of laser radiation is established. The results of experimental studies are presented.

1. Introduction

One of the tasks of applied physics is the development of devices for detecting various magnetic objects by measured magnetic field parameters [1-4]. These objects can be located both in a stationary and mobile state. Moreover, the measurements of the magnetic field parameters must be carried out with a high degree of reliability in different conditions: geographic, seismological, etc. Also in a complex magnetic environment, in the presence of a large number of interference of various kinds [3, 5, 7]. The among all types of developed magnetometers and sensors for monitoring the parameters of the magnetic field the quantum magnetometers are have the greatest versatility for carrying out measurements and the best indicators for measuring accuracy and sensitivity [1-8].

At present, the marine mobile magnetic objects detection is actual task. For the detection of mobile offshore facilities, a large number of different instruments have been developed. For example, devices that measure the variation in pressure fluctuation in the field of a non-uniform sea wave. According to this change, the presence of a mobile marine object in a given area is determined [8]. Fiber-optic antennas, radar stations and acoustic sensors are also used to detect mobile objects [9]. All of them have advantages and disadvantages. Of particular interest are devices that can operate in an off-line mode. But, such devices are subject to rather stringent requirements in various areas (the complexity of their detection, high noise immunity, continuous operation of at least 12 months on an autonomous power source, etc.). In addition, the device must have a high degree of reliability in terms of identifying the position of the object itself in a given area (for example, when solving security tasks or eliminating unauthorized intrusion). None of the devices in operation meet the above requirements. Therefore, the search for possible solutions to the task is extremely urgent and in demand.

2. Quantum sensor and measurement technique.

The paper considers one of the possible solutions to the task at hand - the creation of a quantum sensor based on a ferrofluid cell made of quartz glass. Previous studies have shown that in the case of placing a ferrofluid cell in a magnetic field, nanoparticles of a ferromagnetic liquid are located on magnetic field lines, forming transparent and non-transparent zones [11, 12]. As a magnetic fluid, it is most expedient to use an aqueous solution of single-domain hematite nanoparticles with a concentration concentration

---

Content from this work may be used under the terms of the Creative Commons Attribution 3.0 licence. Any further distribution of this work must maintain attribution to the author(s) and the title of the work, journal citation and DOI.

Published under licence by IOP Publishing Ltd

of 0.054 with a surfactant (surfactant) tetramethylammonium hydroxide. For laser radiation with $\lambda =$ 613 nm, passing through a ferromagnetic liquid placed in a cell, this configuration of nanoparticles is similar to a diffraction grating. The period of this lattice is determined by the distance between the force lines of the magnetic field [11, 12]. The use of these cells makes it possible to investigate the structure of the force lines of various magnetic systems [11-14]. If the ferrofluid cell is located in a weak uniform magnetic field (for example, with an induction of $B_0 = 0.206$ mT and an inhomogeneity of 10^{-5} cm^{-1}), which is created by special inductors. Then the diffraction pattern from the laser radiation transmitted or reflected from the cell is detected with a photodetectors [13]. The resulting picture relative to the central maximum is symmetric [11, 12].

In fig. 1, as an example, the scheme for the propagation of laser radiation through a ferrofluid cell located in a magnetic field B is shown. The structure of the magnetic field lines in this case is controlled by the diffraction pattern from the laser radiation transmitted through the ferrofluid cell [11, 12]. At the boundaries d: air-glass media, glass-magnetic fluid, magnetic liquid-glass and glass-air - refraction of laser radiation is taken into account. It is established that the intensity of doubly reflected laser radiation from the boundaries of media sections (glass-magnetic fluid and glass-air) to the formation of a diffraction image in the transmitted radiation has no significant effect. In the experiment, the photoreceiver (screen 3) is placed at a distance L from the side face of the ferrofluid cell. In its plane, a diffraction image is recorded. The position of each maximum on the screen relative to its center (point 0) will be determined by the diffraction order k and depends on the period of the diffraction grating d_r formed. In this case, the transparent faces of the cell are located perpendicular to the laser radiation incident on them and the lateral planes of the inductor. In addition, the transparent faces of the cell are parallel to the force lines of the magnetic field that are created by the coils of inductance.

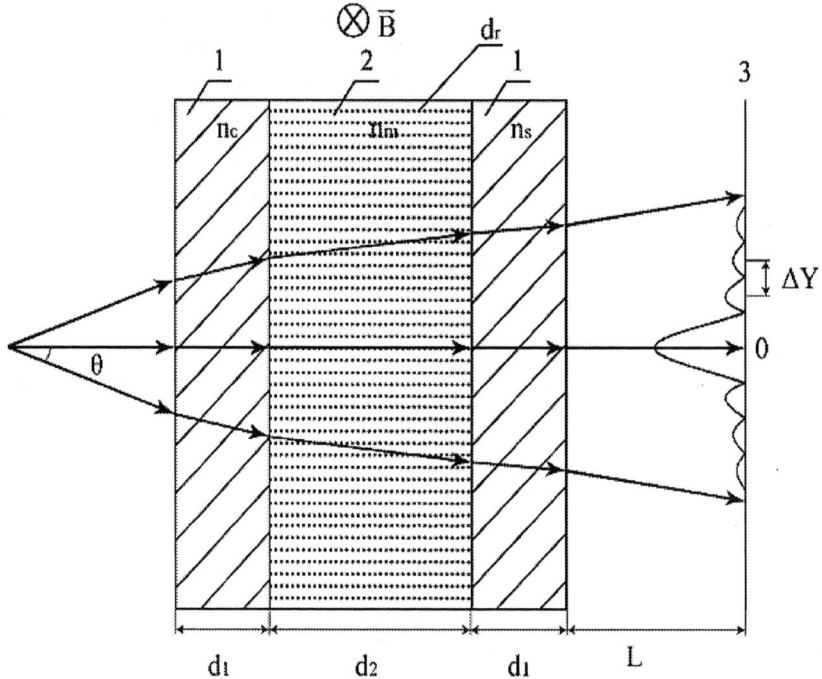

Figure 1. Scheme of propagation of laser radiation rays in a ferrofluid cell when a diffraction pattern is recorded in transmitted light: 1 - walls of a ferrofluid cell; 2 - magnetic fluid layer in the direction perpendicular to the magnetic field with the thickness equal to d_r; 3 - the screen.

In the case of a permanent magnet with induction B_m (imitation of the motion of a magnetic object), the position of the central maximum shifts next to the inductance coils, the shape of the maxima changes to a diffraction pattern, and its symmetry is violated. In fig. 2 as an example, the type of diffraction pattern recorded by the camera located in the zone where the screen 3 was located (fig. 1) is presented. In fig. 2a corresponds to the absence of a magnetic object near the inductance coils in which the ferrofluid cell is located. The magnetic field in the ferrofluid cell is homogeneous. In fig. 2.b - corresponds to one of the moments of motion next to the inductance coils of the magnetic object. Diffraction patterns in fig. 2 are presented after computer processing.

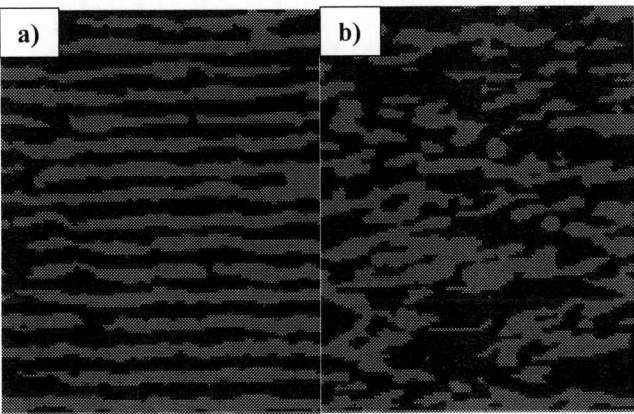

Figure 2 (a, b). The diffraction pattern of the laser radiation in the case of the magnetic fluid placing: (a) in a uniform magnetic field; (b) in an inhomogeneous magnetic field.

The result obtained in fig. 2 shows that the magnetic field created by the mobile object made a change in the structure of the force lines of the magnetic field of the inductor, in which the ferrofluid cell is located. The inhomogeneity of the magnetic field has changed significantly, the symmetry of the diffraction pattern has disrupted. The conducted experiments and the results of their analysis showed that the information presented in this form is not very convenient for the operation of electronic systems. A quantum sensor will be placed on an autonomous object, for example, an underwater buoy. Therefore, the term in the diffraction pattern is chosen and along it (along the width or height of the diffraction cell) a distribution of the intensity I, the recorded laser radiation, is constructed. In fig. 3, as an example, the distribution data in the diffraction pattern for the laser radiation transmitted through the ferrofluid cell is presented. In one case, the magnetic object was absent in the zone of placement of the inductor with a ferrofluid cell. In the other case (fig. 3b) he made a move. The intensity distribution I corresponds to a diffraction pattern fixed at a certain time t. For example, when the distortions in the structure of the force lines in the inductor reached a maximum value.

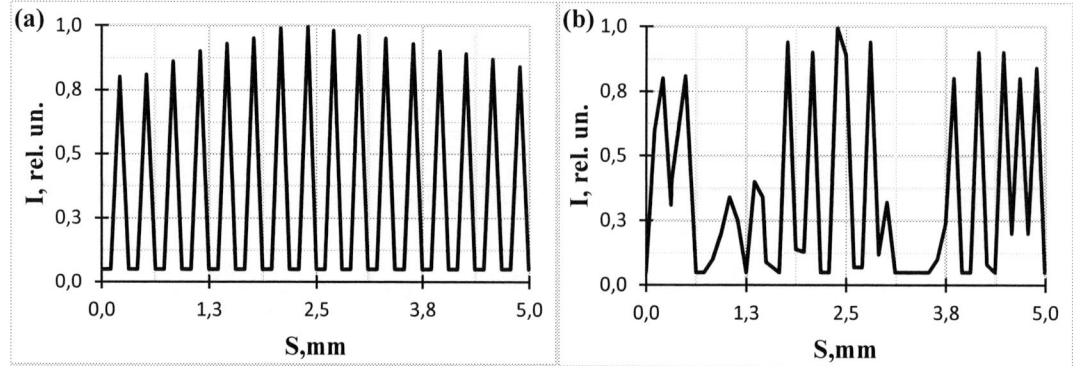

Figure 3. The dependence of the intensity I on the distance between the force lines of the magnetic field: a) in a homogeneous field B_0; b) the magnetic field B_m from the mobile object is additionally present in the inductor.

The conducted studies have shown that the change in the position and amplitude of the maxima in the recorded diffraction pattern (fig. 2b) depends on the trajectory and speed of the magnetic object relative to the position of the ferrofluid cell in the solenoid. In addition, a number of features related to the geometric arrangement of the plane in which a diffraction image was recorded in a quantum sensor, relative to the direction of the magnetic field lines of the mobile object, and also the size and type of nanoparticles in the ferromagnetic liquid from which the cell was made was established. In addition, during the experiments it was established that electromagnetic radiation when scanning the radiation pattern of various types of radar stations, which are currently used to solve various problems in water areas [10]. Do not cause distortions in the structure of the force lines in the ferrofluid cell that would correspond to the finding near it a magnetic object. This shows the high degree of noise immunity of the quantum sensor developed by us.

3. Conclusion

The obtained results show that the method developed by us with the use of a quantum sensor on a ferrofluid cell allows us to determine the presence of a magnetic object in the zone of its location as a result of changes in the structure of the recorded diffraction pattern of laser radiation. For the reliability of determining the presence of a mobile magnetic object in a given zone, it is necessary to install three quantum sensors that detect the change in the magnetic field lines in three planes. This completely eliminates the measurement error. The developed quantum sensor and the electronics necessary for its operation have a high compactness. Low power consumption of the quantum sensor and electronics allows them to operate continuously in an autonomous mode from the batteries (depending on the battery life) to three years. These characteristics, as well as a high degree of protection against interference, fully meet the requirements for autonomous systems that were considered earlier.

References

[1] Zaiko A I, Vorob'ev A V, Ivanova G A and Shakirova G R 2016 *Measurement Techniques* **59** 532
[2] Vershovskii A K, Dmitriev S P and Pazgalev A S 2013 *Technical Physics Journal* **83** 879
[3] Davydov V V, Dudkin V I and Karseev A Yu 2015 *Instrument and Experimental Techniques* **58** 787
[4] Davydov V V, Velichko E N, Dudkin V I and Karseev A Yu 2014 *Measurement Techniques* **57** 684
[5] Davydov V V, Velichko E N, Dudkin V I and Karseev A Yu 2015 *Measurement Techniques* **58** 556

[6] Myazin N S, Logunov S E, Davydov V V, Rud' V Yu, Grebenikova N M and Yushkova V V 2017 *Journal of Physics: Conference Series* **929(1)** 012064

[7] Davydov V V, Dudkin V I and Karseev A Yu 2015 *Technical Physics* **60** 456

[8] Aruev N N and Neronov Y I 2012 *Technical Physics* **57** 1579

[9] Kadykov I F 2016 *Measurement Techniques* **59** 193

[10] Davydov R V, Saveliev I V, Lenets V A, Tarasenko M Yu, Davydov V V and Rud' V Yu 2017 *Lecture Notes in Computer Science (including subseries Lecture Notes in Artificial Intelligence and Lecture Notes in Bioinformatics)* **10531 LNCS** 177

[11] Logunov S E, Koshkin A Yu, DavydovV V and Petrov A A 2016 *Journal of Physics: Conference Series* **741(1)** 012092

[12] Logunov S E, Davydov V V, Vysoczky M G, Koshkin A Yu and Rud' V Yu 2017 *Journal of Physics: Conference Series* **917(5)** 052028

[13] Myazin N S, Smirnov K J, Davydov V V and Logunov S E 2017 *Journal of Physics: Conference Series* **929(1)** 012080

[14] Davydov V V, Dudkin V I and Karseev A Yu 2014 *Measurement Techniques* **57** 912

Stabilization TiO₂ anatase by F-ion doping for solar panel producing

S.N. Fedorov[1], R.V. Kurtenkov[1], V.V. Vasiliev[1]
[1]Department of Metallurgy, St. Petersburg Mining University, St. Petersburg 199106, Russia

Abstract. The influence of modifying F-ion doping at various technological mode of the dopant on multi-state TiO₂ is studied in details. It is so called Anatase to Rutile Transformation (ART) shown that the effect of F-doping has one the best stabilization function at these days. Basing on this, an optimal design process of F-ion modifying and an optimal doping level are discussed in details. The anatase-rutile phases were determined by X-ray method scanning with drive axis Theta-2Theta, continuous scan range 10-80 and scan speed 2 deg/min.

Introduction

Titanium dioxide has been attracting significant research interest in the field of renewable energy. Owing to the particular energetic levels of its conduction and valence bands, titanium dioxide is able to function as a semiconductor photocatalyst and facilitate reactions involved in applications including hydrogen production through water splitting (solar hydrogen), sunlight-driven water purification, self cleaning coatings and self sterilizing coatings [1-4].

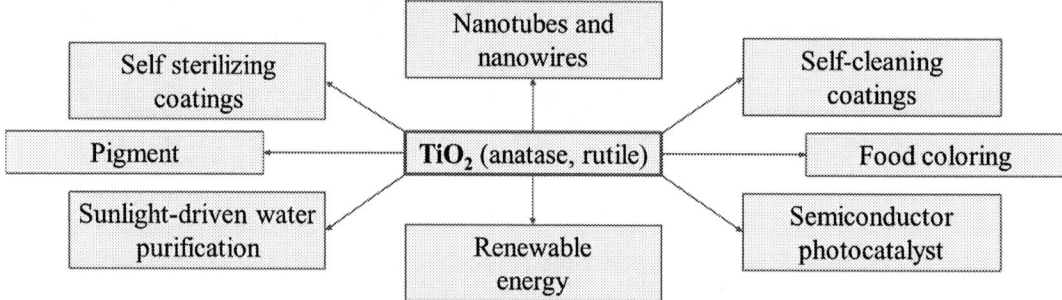

Figure 1. Various of using titanium dioxide

The possibility of using nonmetal dopants, such as carbon, nitrogen, and sulfur, may be more appropriate for the extension of photocatalytic activity into the visible-light region because the related impurity states are near the valence band edge. From the available experimental evidences and our theoretical results, authors [5-10] conclude that both bandgap narrowing and the overlap between the O 2p state and dopant-introduced states strongly affect the photoactivities of doped TiO₂ materials. Authors [5, 11] propose that carbon is the most promising candidate for second-generation photocatalytic materials based on TiO₂ but in comparison to F-ion as a doping agent there are not full TiO₂ stabilization data to conclude that other dopant are most perspective [12]. According to the work [E.S. Gorlanov, V.Yu. Bazhin, S.N. Fedorov "Low-temperature phase formation in a Ti-B-C-O system" in Tsvetnye metally], we determined that one of the intermediate products is titanium dioxide in the form of anatase doped by F-ion [12].

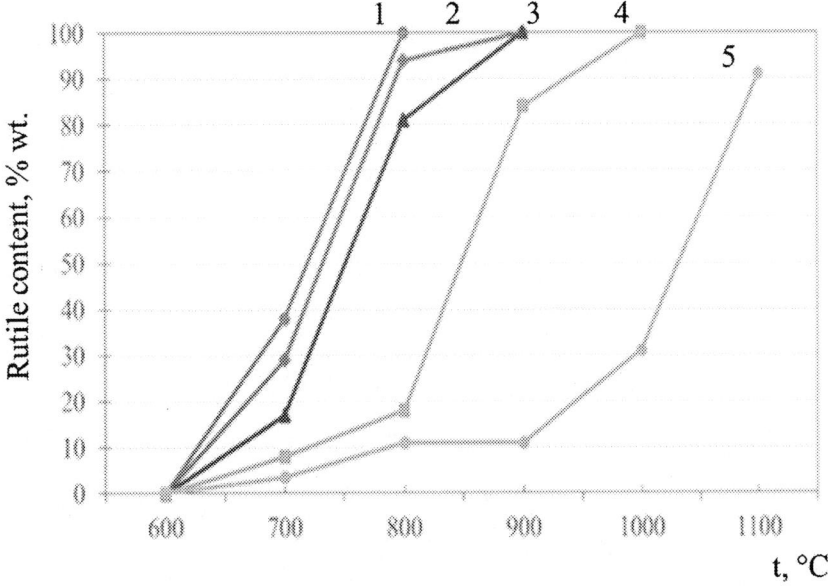

Figure 2. Dependence of the transformation from anatase to rutile at different temperatures and atmosphere (1 – argon TiO_2 doped, 2 – vacuum TiO_2 original, 3 - air TiO_2 original, 4 – vacuum TiO_2 doped, 5 – air TiO_2 doped)

More over in this field there are not insufficient data on the thermodynamics of the doping process by fluorine ion. In this work we have tried to fill these shortcomings and to study the necessary data of the process.

Experiment and results
Raw materials preparation included hydrolyse of $TiCl_4$ for producing H_2TiO_3. According to reaction:

$$TiCl_4 + NH_4OH + 3H_2O = H_2TiO_3 + NH_4OH(g) + 4HCl(g) \qquad (1)$$

The hydrolyse operation is necessary in order to obtain pure metatitanic acid with active particle surface. Otherwise the reaction (2) will not work by using sol-gel method:

$$H_2TiO_3 + 2HF = TiOF_2 + 2H_2O(g) \qquad (2)$$

Here is the doping process reaction. The doping allows get stabile anatase TiO_2 structure till temperature more than 1000 °C (Figure 3).

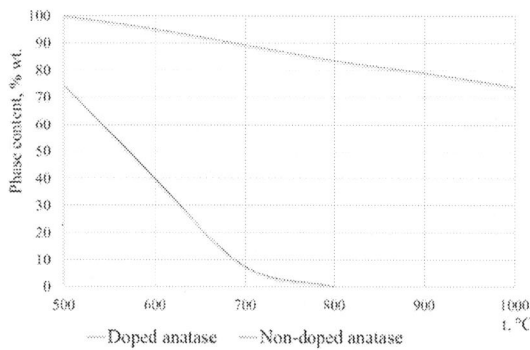 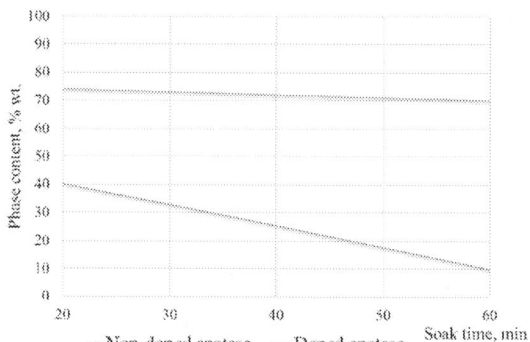

Figure 3. Dependence of the transformation from anatase to rutile at different temperatures

Figure 4. Dependence of the transformation from anatase to rutile with increasing soak time

The original TiO_2 has 74.89 % anatase after heating at 500 °C while doped TiO_2 only starts to transform to rutile at 550 °C and keeps anatase phase till temperature more than 1000 °C that confirms is X-ray analyze method scanning with drive axis Theta-2Theta, continuous scan range 10-80 and scan speed 2 deg/min, radiation of CuKα (Fig. 5). Also with increasing of soak time at given temperatures the samples have a different rate of transformation to rutile (Fig. 4). The samples of non-doped (original) and doped anatase by F-ion were soaked at 600 °C and 1000 °C respectively. Doped anatase has significantly slower transformed to the rutile modification at higher temperature.

Table 1. Anatase and rutile content at different temperatures

	t, °C	500	600	700	800	900	1000
Non-doped for 20 min	Anatase, % wt.	74.89	40.11	7.54	0	0	0
	Rutile, % wt.	25.11	59.89	92.46	100	100	100
Doped for 20 min	Anatase, % wt.	100	95.23	89.41	83.65	79.12	74.07
	Rutile, % wt.	0	4.77	10.59	16.35	20.88	25.93

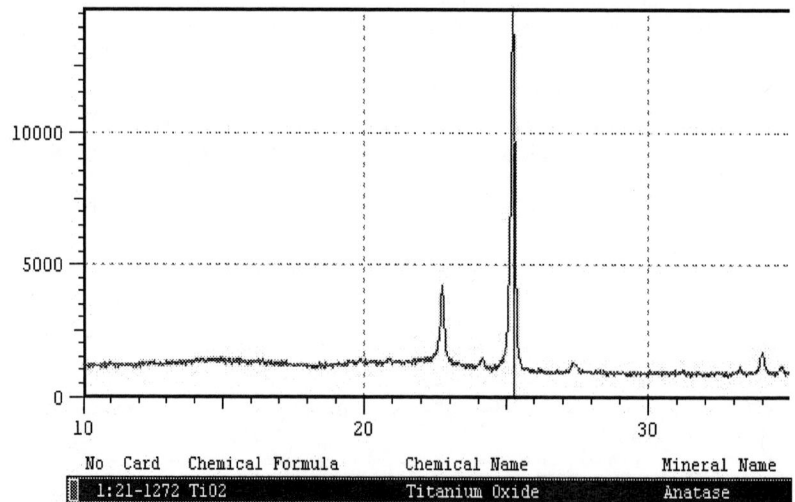

Figure 5. X-ray analyse of TiO_2 sample after 1000 °C heating

The current investigation demonstrates effective F-ion doping of TiO_2 to stabilize anatase phase that can be used in semiconductor photocatalyst and facilitate reactions involved in applications including hydrogen production through water splitting (solar hydrogen) or for solar panel producing as a renewable energy.

Acknowledgments
The work is supported in different parts by the state project № 11.4098.2017 from January, 1st, 2017.

References
[1] D.A.H. Hanaor, M.H.N. Assadi, S. Li, A. Yu, and C.C. Sorrell, Computational Mechanics 50, (2) 185-194 (2012)

[2] A.G. Syrkov, L.A. Yachmenova, E.V. Remzova. Interrelation of reactivity and hydrophobicity of surface-modified filler metals with tribological characteristics of lubricants based on them. Journal of Mining Institute. № 206. 245-248 (2013)

[3] A.G. Syrkov. Synergetic changes of tribochemical properties in heterogeneous systems containing surface-modified metals. Journal of Mining Institute. № 216. 122-130 (2015)

[4] Dorian A. H. Hanaor, Charles C. Sorrell. Review of the anatase to rutile phase transformation. J Mater Sci (2011) 46:855–874 (2010)

[5] H. Wang and J. P. Lewis. Second-generation photocatalytic materials: anion-doped TiO_2. Journal of Physics: Condensed Matter 421–434 (2006)

[6] A.A. Tyutrin, N.V. Nemchinova. The computerization of the silicon oxygen refining. European journal of natural history, № 3. 54-55 (2013)

[7] S. Ito, S.cYoshida, T. Watanabe. Preparation of Colloidal Anatase TiO_2 Secondary Submicroparticles by Hydrothermal Sol-Gel Method. Chem. Lett., № 1. 70-71 (2000)

[8] US5811192 B01J35/00. Titanium dioxide film having photocatalytic activity and substrate having the same. Takahama Koichi, Kishimoto Hirotsugu, Nakgawa Takaharu, Deki Shigenito, Hashimoto Noboru (1998)

[9] Kuz'min M.P. et al. Changing the properties of indium tin oxide by introducing aluminum cations. Electrochemistry Communications. V. 67. 35–38 (2016)

[10] Yuanzhi Li, Doo-Sun Hwang, Nam Hee Lee, Sun-Jae Kim. Synthesis and characterization of carbon-doped titania as an artificial solar light sensitive photocatalyst. Chemical Physics Letters 404 (2005) 25–29

[11] Kuz'min M.P. et. al. Preparation of aluminum–carbon nanotubes composite material by hot pressing. Metallurgist. Vol. 61. 815–821 (2018)

[12] E.S. Gorlanov, V.Yu. Bazhin, S.N. Fedorov Low-temperature phase formation in a Ti-B-C-O system. Tsvetnye metally, № 8. 76-82 (2017)

Alpha-factor in multimode lasers

I K Boikov[1], A V Savelyev[2]

[1] Department of Applied Mathematics and Physics, St. Petersburg Academic University of the Russian Academy of Sciences, St. Petersburg 194021, Russia
[2] Nanophotonics Lab, St. Petersburg Academic University of the Russian Academy of Sciences, St. Petersburg 194021, Russia

E-mail: ik.boikov@gmail.com

Abstract. A theory of the spectral linewidth for a multi-mode laser is presented and used to explain experiment results. It turned out that linewidth is much greater than predicted by Schawlow-Townes theory. Henry attributed this enhancement to variation of real part of refractive index with carrier density. Spontaneous emissions cause changes to imaginary part of refractive index, which is coupled to changes in a real counterpart. This approach is expanded to cover multi-mode lasers and interactions between modes.

1. Introduction

The main forte of a laser is the ability to generate monochromatic light. However, in reality, spectrum's linewidth has a minimum. Several factors contribute to it, and spontaneous emission is one of them. Schallow and Townes in [1] showed that it indeed causes broadening due to randomness of phase of an emitted photon. However, experiments showed that semiconductor laser has much broader linewidth than predicted (by ~50 times).

Then Henry has built a theory around carrier density alteration caused by emissions in [2]. The main idea is that a single emission changes imaginary part of a refractive index profile. This change is coupled to a change in a real part via Kramers-Kronig relation. The ratio between them he called the alpha-factor. It turned out that linewidth broadening is proportional to the square of alpha-factor.

However Henry's theory was built for single-mode lasers. At present numerous applications of multimode lasers (such as comb lasers) are being developed. For example, high-precision spectroscopy [3], wavelength-division multiplexing [4] and sub-pm distance measurements [5]. Yet excessive linewidth broadening can be the key factor in their success. Thus we aim for expanding Henry's theory to lasers with multiple modes.

2. Theory

2.1. Phase deviation for a single emission

There are two ways through which spontaneous recombination can change phase: via immediate phase change due to radiated photon's phase randomness ($\Delta\phi'_k$) and via alpha-factor ($\Delta\phi''_k$).

Consider β_k as k-th mode's complex amplitude, for which $|\beta_k|^2 = I_k$ equals amount of photons in k-th the mode. Then, if there was i-th emission in k-th mode we find from geometry

$$\Delta\phi'_k = I_k^{-1/2} \sin\theta_i , \tag{1}$$

Content from this work may be used under the terms of the Creative Commons Attribution 3.0 licence. Any further distribution of this work must maintain attribution to the author(s) and the title of the work, journal citation and DOI.
Published under licence by IOP Publishing Ltd

where θ_i is emitted photon's phase.

To find $\Delta\phi''$ assume ω_k and k_k - k-th mode's frequency and wavenumber. For simplicity it's aligned with z axis. Then, the mode's electric field is written as:

$$\boldsymbol{E_k} \sim \boldsymbol{\beta_k} \exp(i\omega_k t - ik_k z) + \boldsymbol{\beta_k^*} \exp(-i\omega_k t + ik_k z) \qquad (2)$$

Wave equation for it:

$$\frac{\partial^2 \boldsymbol{E_k}}{\partial z^2} = \frac{1}{c^2}\frac{\partial^2}{\partial t^2}(\epsilon_k \boldsymbol{E_k}) \qquad (3)$$

In order to take into account ϵ_k time dependency due to dispersion and time dependence of $\boldsymbol{\beta_k}$, we write:

$$\epsilon_k \boldsymbol{E_k} \sim \left(\epsilon_k \boldsymbol{\beta_k} + i\frac{\partial \epsilon_k}{\partial \omega_k}\frac{\partial \boldsymbol{\beta_k}}{\partial t}\right) \exp\left(i\omega_k t - ik_k z\right) + c.c. \qquad (4)$$

Going back to (3) and assuming that high order derivatives are negligibly small:

$$\frac{2i\omega_k}{c^2}\left(\epsilon_k + \frac{\omega_k}{2}\frac{\partial \epsilon_k}{\partial \omega_k}\right)\frac{\partial \boldsymbol{\beta_k}}{\partial t} = \left(\frac{\omega_k^2}{c^2}\epsilon_k - k_k^2\right)\boldsymbol{\beta_k} \qquad (5)$$

A permittivity is related to a refractive index via $\epsilon_k = (n_k' - in_k'')^2$. Then the left part of the equation can be rewritten with a group velocity v_g:

$$\epsilon_k + \frac{\omega_k}{2}\frac{\partial \epsilon_k}{\partial \omega_k} = n_k'\left(n_k' + \omega_k\frac{\partial \epsilon_k}{\partial \omega_k}\right) = \frac{c}{v_g}n_k' \qquad (6)$$

The imaginary part of n_k can be expressed with g_k and α_k - k-th mode's material gain and media's attenuation respectively:

$$g_k - \alpha_k = -\frac{2\omega_k}{c}n_k'' \qquad (7)$$

If k-th mode radiates, $n_k'' = 0$ in equilibrium and so $k_k = n_k'\omega_k/c$.

Spontaneous emissions cause variations in density of charge carriers. Let's consider a emission in m-th mode which caused change in n_m''. Due to homogeneous broadening a "bump" in $n''(\omega)$ around ω_m will appear:

$$n''(\omega) \to n''(\omega) + F(\omega - \omega_m)\Delta n_m''$$

We assume $F(\Delta\omega) = \left(1 + (\Delta\omega/\Gamma)^2\right)^{-1}$ where Γ is a broadening parameter [6]. Due to Kramers-Kronig relation k-th mode's permittivity will change (see Fig.1)

$$\begin{aligned}\epsilon_k &= (n_k + \Delta n_{km}' - i\Delta n_{km}'')^2 \\ &\approx n_k^2 - 2i\Delta n_m''\left(F_{km} + i\alpha_{km}\right)\end{aligned} \qquad (8)$$

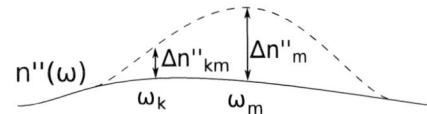

Figure 1. $n''(\omega)$ change upon emission.

Here we substituted $F_{km} = \Delta n_{km}''/\Delta n_m''$ which corresponds to homogeneous broadening, and $\alpha_{km} = \Delta n_{km}'/\Delta n_m''$ - inter-mode alpha-factor. With (6) and (8) $\partial\boldsymbol{\beta_k}/\partial t$ can be factored out from (5):

$$\frac{\partial \boldsymbol{\beta_k}}{\partial t} = -\frac{\omega_k}{c}v_g\Delta n_m''\left(\Gamma_{km} + i\alpha_{km}\right)\boldsymbol{\beta_k} \qquad (9)$$

Because m-th mode radiates, $n''_m = 0$, upon emission we can write $\Delta n''_m = -\dfrac{c}{2\omega_m}(g_m - \alpha_m)$. Also it's known that $(g_m - \alpha_m)v_g = G_m - \gamma_m$, where G_m and γ_m are m-th mode's modal gain and modal attenuation respectively:

$$\frac{\partial \boldsymbol{\beta_k}}{\partial t} = \frac{\omega_k}{\omega_m}\frac{G_m - \gamma_m}{2}\left(\Gamma_{km} + i\alpha_{km}\right)\boldsymbol{\beta_k} \tag{10}$$

For the clearer picture we will convert $\boldsymbol{\beta_k}$ to intensity and phase:

$$I_k = \boldsymbol{\beta_k}\boldsymbol{\beta_k^*} \Rightarrow \dot{I}_k = \frac{\omega_k}{\omega_m}\left(G_m - \gamma_m\right)\Gamma_{km}I_k \tag{11}$$

$$\phi_k = \frac{1}{2i}\ln\left(\frac{\boldsymbol{\beta_k}}{\boldsymbol{\beta_k^*}}\right) \Rightarrow \dot{\phi}_k = \frac{1}{2i}\frac{\dot{\boldsymbol{\beta}}_k\boldsymbol{\beta_k^*} - \boldsymbol{\beta}\dot{\boldsymbol{\beta}}_k^*}{I} = \frac{1}{2}\frac{\omega_k}{\omega_m}(G_m - \gamma_m)\alpha_{km} \tag{12}$$

Since we don't know how $G_m - \gamma_m$ changes over time, we can bind it to intensity, which behaviour is known:

$$\dot{I}_m = (G_m - \gamma_m)I_m \Rightarrow G_m - \gamma_m = \frac{\dot{I}_m}{I_m} \tag{13}$$

Assume that i-th emission into m-th mode happened at $t = 0$. Then the intensity will increase by 1: $I_m \to I_m + 1$. After certain period of time I_m will return to its equilibrium value. This way with (12) and (13) we can find total phase change:

$$\Delta\phi''_k = \int\limits_0^\infty \dot{\phi}_k dt = \frac{1}{2}\frac{\omega_k}{\omega_m}\alpha_{km}\int\limits_0^\infty (G_m - \gamma_m)dt = \frac{1}{2}\frac{\omega_k}{\omega_m}\alpha_{km}\ln\left(\frac{I_m}{I_m + 1}\right) \tag{14}$$

For radiating modes $I_m \gg 1$, also for simplicity $\omega_k/\omega_m \approx 1$. This way (14) may be simplified:

$$\Delta\phi''_k \approx -\frac{\alpha_{km}}{2I_m} \tag{15}$$

2.2. Linewidth broadening

Together with out-of-phase component $\Delta\phi'_k$ (which non-zero only for $k = m$) we have found a total phase change for a single emission:

$$\Delta\phi_k = -\frac{\alpha_{km}}{2I_m} + \frac{1}{I_m^{1/2}}\left[\delta_{km}\sin\left(\theta_i\right) - \alpha_{km}\cos\left(\theta_i\right)\right] \tag{16}$$

Assume $m \neq k$:

$$\Delta\phi_k = -\frac{\alpha_{km}}{2I_m} - \frac{\alpha_{km}}{I_m^{1/2}}\cos\left(\theta_i\right) \tag{17}$$

First summand is negligibly small for $I_m \gg 1$. Then if R_m is a spontaneous emission rate into m-th mode, after time t average square of k-th mode deviation due to m-th mode will be:

$$\langle\Delta\phi_k^2\rangle = \frac{\alpha_{km}^2}{2I_m}R_m t \tag{18}$$

Since emission is a random process, mode's phase will do Brownian motion [2] and thus mode's power spectrum will be Lorentzian with FWHM Δf_k

$$\Delta f_k = \frac{\alpha_{km}^2}{4\pi I_m}R_m \tag{19}$$

For $k = m$ from [2]:

$$\Delta f_k = \frac{1 + \alpha_{kk}^2}{4\pi I_k} R_k$$

Assuming that contribution from each emission is independent. Then in order to get full FWHM for mode, all modes' contributions should be considered:

$$\Delta f_k = \frac{1}{4\pi I_k} R_k + \sum_m \frac{\alpha_{km}^2}{4\pi I_m} R_m \tag{20}$$

2.3. Alpha-factor

Here $\mathrm{P}\int \ldots$ stands for principal value integral. In our $n(\omega)$ notation Kramers-Kronig relation states that

$$n'(\omega) = 1 - \frac{2}{\pi} \mathrm{P} \int_0^\infty \frac{\omega' n''(\omega')}{\omega'^2 - \omega^2} d\omega' \tag{21}$$

Using the property of refractive index $n''(-x) = -n''(x)$, we change integration domain:

$$n'(\omega) = 1 - \frac{1}{\pi} \mathrm{P} \int_{-\infty}^\infty \frac{\omega' n''(\omega')}{\omega'^2 - \omega^2} d\omega' \tag{22}$$

As shown earlier, spontaneous emission into m-th mode will cause change $n''(\omega) \to n''(\omega) \pm F(\omega \mp \omega_m)\Delta n''_m$. Particularly, for k-th mode $n'(\omega_k) \to n'(\omega_k) + \Delta n'(\omega_k)$.

$$\Delta n'_k = \Delta n'(\omega_k) = -\frac{\Delta n''_m}{\pi} \mathrm{P} \int_{-\infty}^\infty \frac{\omega' \left[F(\omega' - \omega_m) - F(\omega' + \omega_m) \right]}{\omega'^2 - \omega_k^2} d\omega' \tag{23}$$

This integral can be computed with methods of complex analysis:

$$\alpha_{km} = \frac{\Delta n'_k}{\Delta n''_m} = -\frac{\xi}{1 + \xi^2}, \text{ where } \xi = \frac{\omega_m - \omega_k}{\Gamma} \tag{24}$$

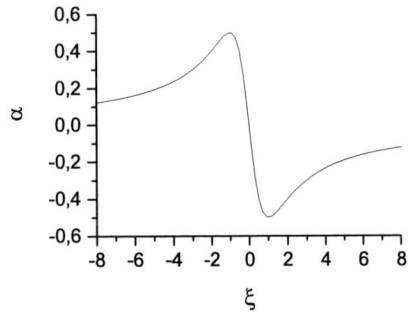

Figure 2. Alpha factor dependency on frequency difference between modes

3. Examples

In this section we estimate linewidth broadening due to alpha-factor for a QD laser with maximal modal gain 273 cm^{-1}, $\Gamma = 4$ meV and 0.2 eV frequency distance between modes. For clarity we approximate comb laser spectra with a stack of modes of equal intensity which is usually the case.

As seen on Fig.3, as number of active modes increases, overall linewidith broadening reduces. This can be explained by the fact that the strongest sources of noise are on sides of a spectra as the intensity here is smaller. And further away these sources are the less noise there will be.

On Fig.4 as $\Gamma/\Delta\omega$ increases, broadening increases as well due to expanding of $n''(\omega)$ "bump" (see Fig.1). However, as it reaches hundreds, opposite effect takes place - ξ becomes too small for neighbouring modes (see Fig.2) and alpha-factors never reach significant values. However, such Γ can't be reached in practice.

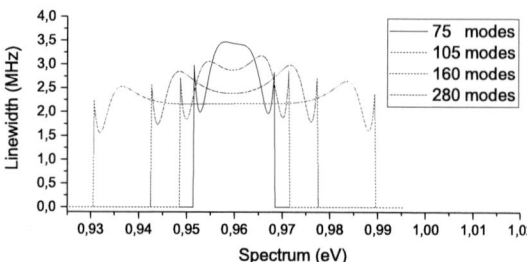

Figure 3. Number of radiating modes impact on linewidth broadening.

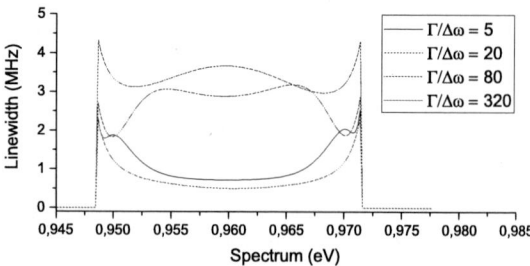

Figure 4. Homogeneous broadening impact on linewidth broadening.

4. Results

The method of calculating linewidth broadening of semiconductor lasers caused by spontaneous emissions was derived. This method requires usage of inter-mode alpha factor, which represents impact of spontanous emission into one mode on refractive index of neighbouring modes. It turns out that inter-mode alpha-factor is zero for mode interaction with itself and reaches maximum for modes, whose frequencies difference equals to homogeneous broadening parameter.

We have shown that in order to reduce linewidth broadening one should increase a number of radiating modes, reduce homogeneous linewidth broadening and a reach higher intensity for each mode.

Acknowledgments
This work was supported by RFBR-DFG, project no. 18-502-12081.

References
[1] Schawlow A L and Townes H, 1958 Infrared and optical masers *Phys. Rev.* **112 (6)** 1940
[2] Henry C H 1982 Theory of the linewidth of semiconductor lasers *IEEE J. Quantum Electron.* **18 (2)** 259
[3] Mandon J et al. 2009 Fourier transform spectroscopy with a frequency comb *Nature Photon.* **3** 99
[4] Marin-Palomo P et al. Microresonator-based solitons for massively parallel coherent optical communications 2017 *Nature* **546** 274
[5] Schibli T R et al. Displacement metrology with sub-pm resolution in air based on a fs-comb wavelength synthesizer 2006 *Opt. Express* **14 (13)** 5984
[6] Savelyev A V, Korenev V V, Maximov M V and Zhukov A E 2015 Spatial hole burning and spectral stability of a quantum-dot laser *Semiconductors* **49** 1499

Heterointerfaces in the bottom tunnel part of GaInP/GaAs/Ge solar cells

M A Mintairov[1], V V Evstropov[1], S A Mintairov[1], M Z Shvarts[1], N A Kalyuzhnyy[1]

[1] Ioffe Institute, St.-Petersburg, Russia, Polytechnicheskaya str. 26

E-mail: mamint@mail.ioffe.ru

Abstract. The tunnel (intergenerator) part between the middle (GaAs) and bottom (Ge) subcell of a GaInP/GaAs/Ge triple-junction solar cell was investigated. It is shown that the previously observed features of the multi-junction solar cells IV-curves, namely the inflection of the characteristics, arises from heterointerfaces in the region between the GaAs p-n junction and the bottom tunnel p^{++}-n^{++} junction. A theoretical study was carried out, and various ways of optimizing such heterointerfeces are proposed.

1. Introduction

At present time, triple-junction solar cells (TJ SCs) are actively analyzed, developed and improved. It is worth emphasizing that this activity is greatly hampered by the many-link topological structure of such SCs. This structure is based on three basic parts: *GaInP*, *GaAs* and *Ge* photovoltaic homo p-n junctions. In addition to the basic parts, the TJ SC structure also has a set of so-called connecting elements [1]. This set includes: resistive contacts, substrate, intergenerator layers and others.

This work continues previous ones [2, 3] aimed in studying similar parts of the solar cells structure. First, it is a part located between *GaInP* and *GaAs* p-n junctions (photovoltaic generators); it includes: a tunnel *GaAs* homo *p++-n++* junction, framed by two separation barriers - front and back ones. Such a part was called a top tunnel part [2, 3], otherwise the top intergenerator part. Secondly, it is a similar section located between the *GaAs* and *Ge* p-n junctions. It includes the tunnel hetero *p++(Al₀₄Ga₀.₆As)-n++(GaAs)* junction, framed by several barriers including the front and back ones. Such part is called bottom intergenerator part or bottom tunnel part.

As previously established [2, 3], the properties of the intergenerator parts affect the characteristics and parameters of a multi-junction (MJ) SC, both positively [4, 5] and negatively [1-3,6]. In particular, it was established that the top tunnel part under certain conditions can counteract the basic photovoltaic *GaInP*, *GaAs*, and *Ge* p-n junctions. This is due to the presence of an opposite-connected tunnel junction.

In the paper, the IV-curves of the bottom tunnel part of the *GaInP/GaAs/Ge* SC has been obtained and analyzed. It was determined experimentally that this part does not generate counteracting photo-emf. Also it contains heterobarriers responsible for inflection appearance on the SC IV-curve. Such effect was previously observed in a *GaInP/GaAs/Ge* SC [1,7], but location of heterobarier was not determined.

Content from this work may be used under the terms of the Creative Commons Attribution 3.0 licence. Any further distribution of this work must maintain attribution to the author(s) and the title of the work, journal citation and DOI.

Published under licence by IOP Publishing Ltd

2. Object of study and energy diagram of the bottom tunnel (intergenerator) part

To study the bottom tunnel part, a double-junction *GaAs/Ge* SC has been used. The sample has been grown by metal-organic vapor phase epitaxy. The intergenerator part had a similar structure as in triple-junction SC used in [1] and included layers listed in Fig. 1.

.

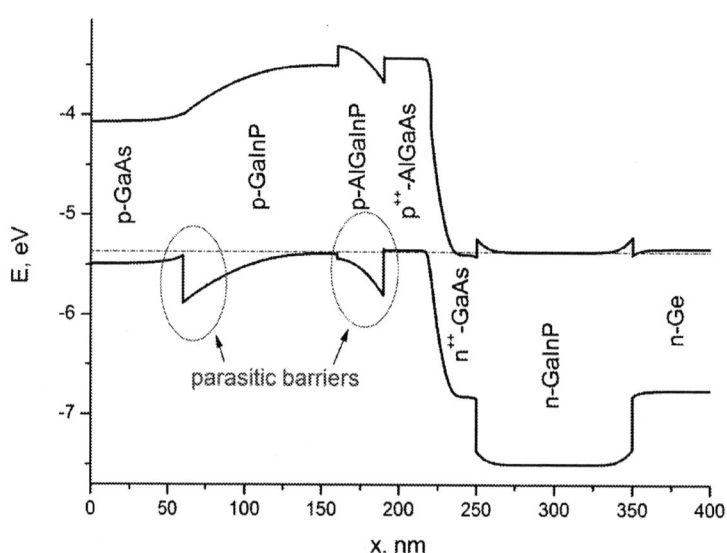

Figure 1. Bottom tunnel (intergenerator) part band diagram of a *GaInP/GaAs/Ge* SC.

Fig. 1 shows the bottom intergenerator part energy diagram. The emphasized attention in Figure 1 is paid to isotype heterointerfaces creating energy barriers. According to our interpretation, they are responsible for the inflection appearance on the light IV-characteristic (Fig. 2b). A similar inflection was already observed earlier for the *GaInP/GaAs/Ge* SC [1]. In that work, an IV-curve of the heterointerfaces responsible for the inflection was obtained, but the heterointerface location was not established.

3. Light IV-curves of the bottom tunnel (intergenerator) part

Figure 2a shows the experimental V_{oc}-J_{sc} dependence for the sample under study. As is known, such a dependence after a voltage shift by a certain value $V_{a,oc}$ [8, 9] coincides with the resistiveless dark IV-curve. In its turn, a resistiveless light IV-curve is produced from the dark one by current shift to the photogenerated current value J_g and by using V_a correction [8, 9]. As shown in [8], this correction depends on the imbalance of the photogenerated currents (at full balance it becomes zero) and usually does not exceed 0.05 V. The obtained tunnel part IV-curves was determined in the voltage range from 0.1 to 0.5 V. The magnitude of these voltages is significantly exceeds V_a, therefore, to simplify the calculations, it was not taken into account. As can be seen (Fig. 2a), the V_{oc}-J_{sc} dependence, and, consequently the dark IV-curve, is well described by a two-exponential model:

$$J_{sc} = J_{02}\left(\frac{V_{OC}}{2kT} - 1\right) + J_{03}\left(\frac{V_{OC}}{3kT} - 1\right) \tag{1}$$

where k – Boltzmann constant, T – absolute temperature. Approximation by formula (1) of the V_{oc}-J_{sc} dependence allows to determining the parameters J_{02}=5.5·10^{-13} A/cm^2 J_{03}=5.5·10^{-9} A/cm^2. These parameters have been used to obtain a resistiveless light IV-curves (dashed lines in Fig 2b).

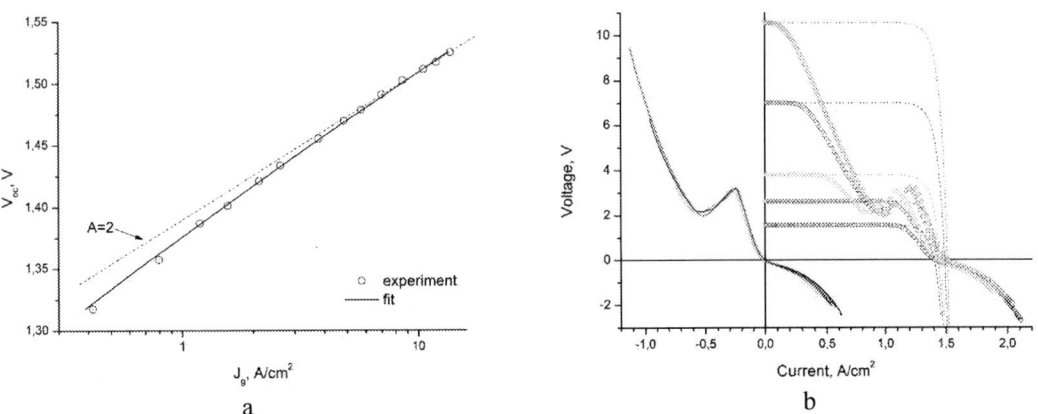

Figure 2. (a) Experimental (symbols) and theoretical (solid line) J_gV_{oc}-curve of the generator part of a *GaAs/Ge* SC. (b) Extraction of the tunnel part IV-curves (solid lines) by voltage subtracting of experimental IV-curves (symbols) from generating ones (dash lines).

As can be seen (Fig. 2b), the obtained IV-curves do not coincide with the experimental ones. The voltaic subtraction of the restiveless IV-curves from the experimental ones results in a set of IV-characteristics of the intergenerator (tunnel) part for different J_g (Fig. 2b on the left). One can see that the tunnel part IV-curve does not depend on J_g, what implies that the bottom tunnel part is not photoactive in contrast to the top one [2]. From the shape of the obtained IV-curves (Fig. 3b on the left), it is seen that they contain the inflection of the curve observed earlier for the *GaInP/GaAs/Ge* SC [1]. Such an inflection is associated with the presence of a heterointerface in the inter-generator part. The heterointerfaces IV-curve has been extracted from the tunnel part IV-curve and compared with the previously observed characteristic.

4. IV-curve of heterointerfaces in the bottom tunnel (intergenerator) part
From the shape of the tunnel part IV-curves (Fig. 2a on the left), it is clear that they contain a voltaic sum of two characteristics: one – IV-curve of tunnel p^{++}-n^{++} junction, the another one - heterointerface IV-curve. For heterointerface IV-curve extraction, it has been suggested that in the region below the tunnel diode peak current, the IV-curve is determined by the voltaic sum of the linear part of the tunnel diode, series resistance, and the heterointerface. It is seen that, in the current range from 0 to the peak current the IV-curve is linear. Consequently, the sum of the tunnel diode and the resistive component gives a linear IV-curve. Such a linear characteristic, as is well known, is described by the $I·R$ function, where R is the lumped linear equivalent of the series resistance. Thus, in order to obtain the heterointerface IV-curve, it is sufficient to determine the value of the lumped resistive equivalent R and subtract from the tunnel part IV-curve the linear function $I·R$. As a result, the obtained IV-curve in the current region below the tunnel junction peak current will be determined only by the heterointerface.

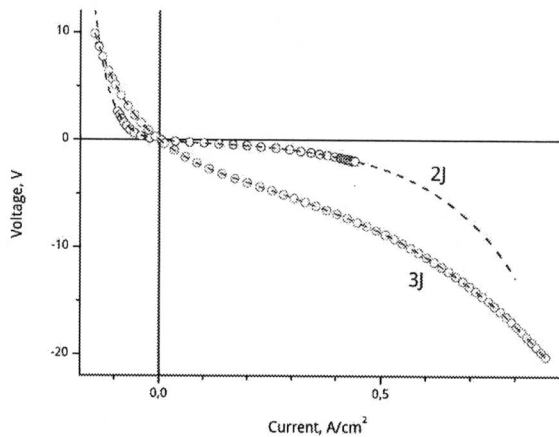

Figure 3. Heterointerfaces IV-curves extacted from full IV-curves of double-junction *GaAs/Ge* (blue) and triple-junction *GaInP/GaAs/Ge* (green) solar cells. Symbols - extracted from experimental IV-curves, dashed lines – approximation by eq.(2). Approximation parameters are given in Table 1.

The above described procedure has been carried out. The lumped resistive equivalent R has been determined by the method described in [8, 9], R=0.4 *Ohm·cm²*. The extraction result is shown in Fig. 3. The figure also contains the IV-curve of the heterointerface extracted from the full IV-curve of a triple-junction *GaInP/GaAs/Ge* SC[1]. Note that both IV-curves are well approximated by formula (2) proposed in [1] to describe the heterointerfaces characteristics.

$$J = J_0 \left(\exp\left(-\frac{V}{E_1} \right) - \exp\left(\frac{V}{E_2} \right) \right)$$

(2)

The result of the approximation is shown in Fig. 3 (dashed lines). The obtained parameters are given in Table 1.

Table 1. Approximation parameters for heterointerfaces IV-curves of *GaInP/GaAs/Ge* and *GaAs/Ge* SC

Solar cell	E_1, V	E_2, V	J_0, A/cm²
GaInP/GaAs/Ge	0.1	0.44	2.75
GaAs/Ge	0.035	0.19	0.19

Investigation of the triple-junction structures has shown that samples, which do not contain a p-*GaAs/p-GaInP*, and *p-AlGaInP/p⁺⁺-AlGaAs* heterointerfaces (Fig 1 circled in red) in the bottom tunnel part do not have an inflection on its IV-curves. Thus, such heterointerfaces are responsible for forming an inflection on the full IV-curves of MJ SCs. From the energy diagram (Fig. 1), it can be seen that these heterointerfaces can be optimized either by reducing the thickness or by increasing the doping level of a *p-GaInP* and *p-AlGaInP* layers or by replacing them with other materials. Under these conditions, the IV-curves of MJ SCs do not contain an inflection. For example, in [9, 10], instead of the *p-GaInP* and *p-AlGaInP* layers, the *AlGaAs* barrier was used, and the presented IV curves did not contain a region with an inflection.

Conclusions

The connecting parts of multi-junction solar cells sometimes significantly affect the perfomance of such devises. Therefore, the definition of the main properties of these parts is important for the purpose   of   further   increasing   the   number   of   sub-cells. The bottom intergenerator tunnel part (between *GaAs* and *Ge* photovoltaic p-n junctions) has been experimentally analyzed. For this, the method of extracting the IV-curve of the tunnel part previously developed for top tunnel part (between *GaInP* and *GaAs* p-n junctions) IV-curve extraction, has been used. It has been found that, in contrast to the top, the bottom tunnel part has no counteracting photovoltaic effect.

A correlation was also established between the presence of an inflection on the IV-curve and the presence of heterobarriers on the bottom part energy band diagram. A comparison of IV-curves having a similar inflection in double-junction (*GaAs/Ge*) and triple-junction (*GaInP/GaAs/Ge*) solar cells has been performed. The shape of the heterointerfaces IV-curve significantly affects the shape of the full IV-characteristic and is approximated by a two-exponential function. Optimizing such heterointerfaces (increasing the doping level, reducing the thickness of the layers or replacing them with other materials) make it possible to eliminate the nonlinearity of the IV-curves. The obtained properties of the tunnel inergenerator part should be taken into account in the development of multi-junction solar cells.

Acknowledgements

The work has been supported by the Russian Foundation for Basic Research (grant No. 17-08-01056).

References

[1]   M.A.Mintairov, V.V.Evstropov, N. A. Kalyuzhnyy, S. A. Mintairov, N.Kh.Timoshina, M.Z.Shvarts, AIP Conf. Proc., Aix-les-Bains (France) , 2015, v. 1679, p. 050007

[2]   M.A. Mintairov, V.V.Evstropov, S. A. Mintairov, M.Z.Shvarts, .A. Kozhukhovskaia, N. A. Kalyuzhnyy, Proc. of 33 PVSEC, Amsterdam (Netherlands) 2017, pp. 1275 – 1277

[3]   M.A. Mintairov, V.V.Evstropov, S. A. Mintairov, M.Z.Shvarts, .A. Kozhukhovskaia, N. A. Kalyuzhnyy, JOP Conf. Series, Saint Petersburg (Russia), v. 917, (2017), p 052034

[4]   M.Z.Shvarts, V.M.Emelyanov, M.A.Mintairov, V.V.Evstropov, N.Kh.Timoshina, AIP Conf. Proc., Aix-les-Bains (France) , 2015, v. 1679, p. 120003

[5]   M.Z. Shvarts, V.M. Emelyanov, V.V. Evstropov, M.A. Mintairov, E.D. Filimonov, S.A. Kozhukhovskaia, AIP Conf. Proc., Aix-les-Bains (France) , 2015, v. 1679, p. 120003

[6]   M.A. Mintairov, V.V. Evstropov, N.A. Kalyuzhnyy, S.A. Mintairov, M.Z.Shvarts, N.Kh.Timoshina, S.A. Salii, V.M. Lantratov, Semiconductors, v.48, 5, 2014, pp: 653-658

[7]   R. Hoheisel and A.W. Bett, IEEE Journal of Photovoltaics, vol.2, No.3, 2012

[8]   M.A. Mintairov, V.V. Evstropov, N.A. Kalyuzhnyy, S.A. Mintairov, V.D. Rumyantsev, N.K. Timoshina, M.Z. Shvarts, V.M. Lantratov, Semiconductors, v.46, 8, 2012, pp: 1051-1058

[9]   M. A. Mintairov, N. A. Kalyuzhnyy, V. V. Evstropov, V. M. Lantratov, S. A. Mintairov, M. Z. Shvarts, V. M. Andreev, A. Luque, IEEE Journal of Photovoltaics, v.5, 4, 2015, pp: 1229 - 1236

[10]   M.A. Mintairov, V.V. Evstropov, S.A. Mintairov, R.A. Salii, M.Z. Shvarts, N.K. Timoshina, N.A. Kalyuzhnyy, Proc. of 31 PVSEC, Hamburg (Germany) 2015 pp. 1470-1473

SPBOPEN 2018   IOP Publishing

Growth and optical characterization of 7.5 μm quantum-cascade laser heterostructures grown by MBE

A V Babichev[1], A G Gladyshev[2], E S Kolodeznyi[1], A S Kurochkin[1],
G S Sokolovskii[1,3], V E Bougrov[1], L Ya Karachinsky[1,2,3], I I Novikov[1,2,3],
V V Dudelev[3], V N Nevedomsky[3], S O Slipchenko[3], A V Lutetskiy[3],
A N Sofronov[4], D A Firsov[4], L E Vorobjev[4], N A Pikhtin[3], A Yu Egorov[1]

[1]ITMO University, Saint Petersburg 197101, Russia
[2]Connector Optics LLC, St. Petersburg 194292, Russia
[3]Ioffe Institute, St. Petersburg 194021, Russia
[4]Peter the Great St. Petersburg Polytechnic University, St. Petersburg 195251, Russia

Corresponding author: andrey.babichev@connector-optics.com

Abstract. We present the results on growth, characterization and optical properties of quantum-cascade laser heterostructures grown by molecular-beam epitaxy. The double-phonon resonance design for forming of active region together with thick InP-based top cladding was used to increase the quantum-cascade laser performance. The results of electroluminescence studies of 4-cleaved samples are presented. The room temperature lasing at 8.0 um with threshold current density about 1.98 kA/cm^2 was achieved.

1. Introduction

Quantum-cascade lasers (QCL) emitting in the spectral range of 7-8 μm can be effectively used for medical applications and remote gas analysis. In particular, lasers with a wavelength of 7.3-7.9 μm are using for the detection of SO_2, CH_4, H_2S, C_2H_2, N_2O, and TNT. SO_2 is one of the most common impurities present in the medical gas supply lines in hospitals. Due to the high toxicity of SO_2, the special attention is paid to the development of compact systems for bedside monitoring of this oxide that can be based on QCLs [1]. In turn, using the QCLs is possible for effective remote monitoring of the CH_4 leaks in gas pipelines and chemical plants. Moreover, the presence of CH_4 in human exhale can indicate a number of diseases of the gastrointestinal tract [2].

2. Samples and experimental setup

The QCL heterostructures were grown by Connector Optics LLC using a production molecular-beam epitaxy system Riber 49 [3]. The Riber 49 is equipped with a solid-state source of arsenic and ABI 1000 cells for forming of gallium and indium fluxes. The InP substrate with orientation (001) was doped with sulfur to a level of 1×10^{17} cm^{-3}. The lower confinement layer was formed by thick $In_{0.53}Ga_{0.47}As$ layer (0.5 μm thick). An active region including 50 quantum cascades consisted of unstrained $In_{0.53}Ga_{0.47}As$ quantum wells and $Al_{0.48}In_{0.52}As$ barrier layers. A double-phonon resonance design was used to fabricate the cascades [4]. The top cladding layers were formed by InP and $In_{0.53}Ga_{0.47}As$ layers with a doping level of 1×10^{17} cm^{-3} and thicknesses of 3.9 μm and 0.1 μm, respectively. The contact $In_{0.53}Ga_{0.47}As$ consisted of heavy doped layer (doping level is about 1×10^{19} cm^{-3}) and had thickness of 20 nm.

Content from this work may be used under the terms of the Creative Commons Attribution 3.0 licence. Any further distribution of this work must maintain attribution to the author(s) and the title of the work, journal citation and DOI.
Published under licence by IOP Publishing Ltd

Recently, we have shown that multi-cascade heterostructures with thin top cladding allows to reach room-temperature lasing [5]. Here in, we proposed the modified structure, in relation with Ref. 4, with thick top cladding based on binary solid alloy (InP). It is well known that InP solid alloy has better thermal conductivity that allows to increase the efficiency of heat removal from multi-cascade active region. QCL heterostructures were characterized by XRD and TEM.

The optical quality of QCL heterostructure was verified by fabrication of four-cleaved laser samples, based on well-known approach presented in ref. 6,7. The top metallization was formed by deposition of Ti/Pt/Au. The bottom metallization based on AuTe/Au was used after the lapping of substrate down to 150 μm. Laser chips with typical size of 0.5×0.5 mm^2 were mounted on copper heatsinks.

The samples were measured in a cryostat with ZnSe optical window with 70% optical transmission in the desired spectral range. The bias was applied to the structure in the form of $0.05 - 0.4$ μs pulses at a frequency of 1-5 kHz with gate circuit based on the power MOSFET. Sample in the cryostat was physically connected to the circuit with low-impedance micro-strip line. Current through the sample was controlled as the voltage on the series-connected 1 Ohm resistor. Waveforms of voltage and current through the sample were recorded with digital oscilloscope.

The spectral measurements were performed with the Bruker Vertex 80v Fourier spectrometer operating in a step-scan mode with a liquid nitrogen cooled HgCdTe photodetector. Sample in the cryostat was placed in the focus of the spectrometer input port. The photodetector signal was measured with a boxcar averager in the case of spontaneous emission studies, and directly recorded with the spectrometer ADC in the case of stimulated emission (experimental setup schematics is presented in [8]). Spectral resolution was 8 cm^{-1} (0.98 meV) and 0.2 cm^{-1} (0.025 meV) for spontaneous and stimulated emission measurements, respectively. The total dynamic range of the experimental setup is more than 4 orders of magnitude.

3. Experimental Results

TEM cross-section is presented on Figure 1. Obtained results indicate the high structural quality and precision of the applied growth technique. Fluctuations of composition and thickness were less than 1% for studied heterostructure.

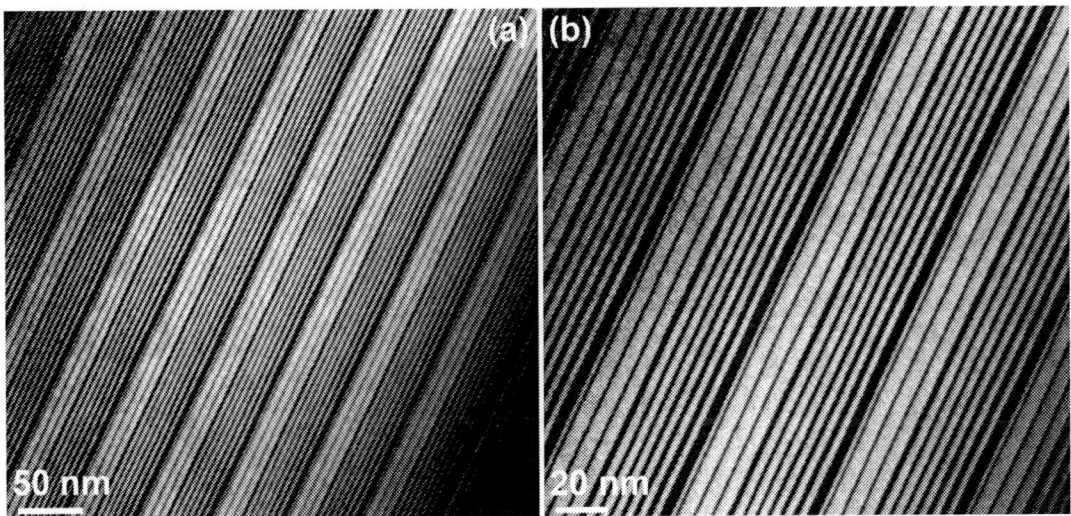

Figure 1. The dark-field TEM image of the one cascade of QCL heterostructure (cross section (1-10)). (a) low magnification. (b) high magnification.

The typical L-I-V dependences are presented in Figure 2a. Threshold voltages of 15 V and 18 V are measured at 80 and 300 K, respectively. These values are in coincidence with the previous published results for the same active region QCL [9]. The temperature dependence of threshold current density, j_{th}, is presented on Figure 1b (inset) and can be approximated by expression (1) [10]:

$$j_{th}(T) = j_0 \exp\left(\frac{T}{T_0}\right) \qquad (1);$$

where T_0 is the characteristic temperature and $j_0 = j_{th}(0)$. The determined values of T_0 and j_0 are 146 K and 0.26 kA/cm^2, respectively. The threshold current density measured at 300 K is about 1.98 kA/cm^2. This value is higher than values that have been published for long ridge lasers based on the same design of active region [11,12].

The below-threshold intersubband electroluminescence spectrum at 80 K is shown on Figure 2b. The full-width at half maximum (FWHM) of the spectrum is 13 meV that in a good agreement with previous published results [10].

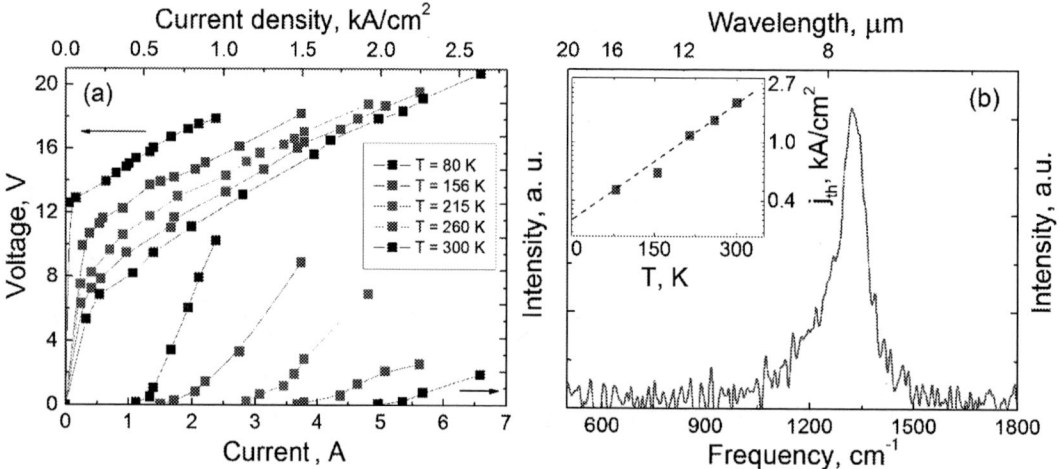

Figure 2. (a) The typical L-I-V curves of 7.5 μm quantum-cascade laser. Pulse duration – 50 ns, frequency – 5 kHz; (b) Intersubband electroluminescence spectra measured at T = 80 K. Pulse duration – 400 ns, frequency – 5 kHz, current = 0.92 A. Inset depicts j_{th}(T) dependence (logarithmic scale).

The lasing spectra measured at different temperatures are shown on Figure 3a (low pumping current case) and Figure 3b (high pumping current case). Increase the temperature results in wavelength red shift. The emission wavelength shifts from 7.5 μm up to 8.0 μm at low pumping current and from 7.7 μm to 8.0 μm at high pumping current. Small red shift of wavelength measured at 80 K with increase of current from 1.3 A to 2.9 A can be related with heating of sample. The typical FWHM of lasing spectra is 1.5 – 2 nm (0.03-0.04 meV).

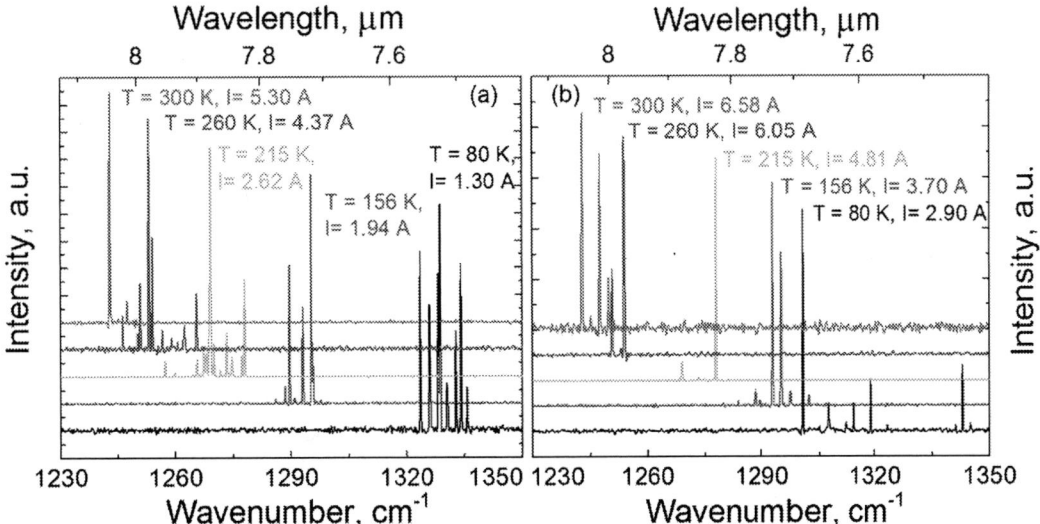

Figure 3. (a) Lasing spectra at different temperatures. Pulse duration – 50 ns, frequency – 5 kHz. Intensities are normalized; (b) Lasing spectra at different temperatures. Pulse duration – 50 ns, frequency – 5 kHz. Intensities are normalized.

Conclusions

QCL heterostructures was grown by MBE. The high structural quality of grown heterostructures were proved by X-ray diffraction and transmission electron microscopy studies (fluctuations of composition and thickness < 1%). The heterostructures were used for manufacturing of 4-cleaved lasers sized 0.5×0.5 mm^2. Laser samples mounted on copper heatsinks have shown room temperature lasing at ~ 8 μm. The threshold current density measured at 300 K was about 1.98 kA/cm^2. The estimated values of T_0 and j_0 were 146 K and 0.26 kA/cm^2, respectively.

Acknowledgments

This work was supported by the Ministry of Education and Science of the Russian Federation in the framework of the Federal target program "Research and development in priority areas of Russian science and technology complex for 2014 – 2020", code 2016-14-579-0009, agreement No. 14.578.21.0204, unique ID: RFMEFI57816X0204.

References

[1] Zimmermann H, Wiese M, Fiorani L and Ragnoni A 2017 J. Sens. Sens. Syst. **6** 155
[2] Kunkel D, Basseri R J, Makhani M D, Chong K, Chang C and Pimentel M 2011 Dig. Dis. Sci. **56** 1612
[3] Babichev A V et al. 2016 Semiconductors **50** 1299–3
[4] Babichev A V et al. 2017 Tech. Phys. Lett. **43** 666
[5] Babichev A V, Gladyshev A G, Kurochkin A S, Kolodeznyi E S, Sokolovskii G S, Bougrov V E, Karachinsky L Ya, Novikov I I, Bousseksou A G and Egorov A Yu 2018 Semiconductors **52** 1082–5
[6] Maulini R, Beck M, Faist J and Gini E 2004. Appl. Phys. Lett. **84** 1659–1
[7] Egorov A Y et al. 2015 Semiconductors **49** 1527-30
[8] Gusev G A et al. 2017 J. Phys.: Conf. Ser. **917** 052019
[9] Kurochkin A S, Babichev A V, Denisov D V, Karachinsky L Ya, Novikov I I, Sofronov A N, Firsov D A, Vorobjev L E, Bousseksou A and Egorov A Yu 2018 J. Phys.: Conf. Ser. **993** 012031

[10] Sirtori C, Faist J, Capasso F, Sivco D L, Hutchinson A L, Chu S N G and Cho A Y 1996 Appl. Phys. Lett. **68** 1745–7
[11] Yu J S, Slivken S and Razeghi M 2010 Semicond. Sci. Technol. **25** 125015
[12] Bidaux Y, Terazzi R, Bismuto A, Gresch T, Blaser S, Muller A and Faist J 2015 J. Appl. Phys. **118** 093101

SPBOPEN 2018

Modeling of the internal quantum yield for solar cell with quantum dots.

A. Panchak, N Kalyuzhnyy, S. Mintairov and M. Shvarts
Photovoltaic Lab Ioffe Institute, Saint Petersburg 194021, Russia

Abstract. By modeling solar cells with quantum dots based on A3B5 materials it is necessary to take into account the large number of localized quantum levels arising in the forbidden band of the host material. In this paper Empiric k-p Hamiltonian model have been used to describe InAs/GaAs structure. For the model two shape of quantum dot: box and cylinder have been considered.

Introduction

Quantum dots (QDs) being implant into the space charge region of a solar cell (SC) can form localized energy levels within the forbidden band of the host material. At irradiation of a wide spectrum light these levels can serve as a stepping stone to absorb additional low-energy photons and increase photocurrent. There is no decrease in voltage here, since the energy levels of the quantum dot are localized and the charge carriers from there cannot take a part in the current flow. Thus increasing the current generation without decreasing the voltage can increase the efficiency of SC. Such a concept is well known as intermediate band solar cell (IBSC) [1].

Implanting QDs was realized in SC prototypes based on GaAs/InAs structures [2, 3]. Strictly speaking such material set do not reproduce IBSC concept fully. InAs QD produces in bandgap of GaAs host material not very deep energy levels. Charge carriers could go out of localized states to conduction band or valence band of host material with only phonon absorption. So there is mainly photon and phonon transition but not IBSC two-photon transition [4].

Usually the shape of QD in the lateral direction is greater than along the direction of structure growth. Also both lateral sizes are comparable [5]. Thus the shape of the QD can be described as a parallelepiped or a cylinder.

The amount and position of the energy levels within the bandgap of the host material depend on QDs material and size. In the GaAs/InAs system quantum dot of a size about 10 nm produce more than 150 energy levels in the forbidden band. Simulation of light absorption in a QD requires the calculation of transitions probability between all the levels. Since GaAs and InAs are materials of a zinc blende structure type, for accurate calculation it is necessary to use an 8-band semiconductor model. In this case modeling of the internal quantum efficiency for a single dot without the use of a supercomputer is not advisable, because an ordinary laptop takes about a month for this task. Another practical way to calculate the absorption is the Empiric k•p Hamiltonian (EKPH) use [6]. This method could give an approximate result with accuracy comparable to 8-band model.

Modeling of the internal quantum efficiency

EKPH method is a four band k•p method for modeling crystals of zincblende type. It involves conduction band (*cb*) states and three sub-bands of valence band states: the heavy holes (*hh*) the light holes (*lh*) and the split off states (*so*).

Content from this work may be used under the terms of the Creative Commons Attribution 3.0 licence. Any further distribution of this work must maintain attribution to the author(s) and the title of the work, journal citation and DOI.
Published under licence by IOP Publishing Ltd

In k•p methods the one-electron Hamiltonian is developed in matrix form in a basis

$$\left|0, v, \mathbf{k}\right\rangle = u_{v,0}(\mathbf{r})e^{i\mathbf{k}\cdot\mathbf{r}} / \sqrt{\Omega} \qquad (1)$$

where $u_{v,0}(\mathbf{r})$ are the Bloch functions in the Γ point of the Brillouin zone, v labels the *cb*, *hh*, *lh* and *so* bands and $\mathbf{k}$ is an arbitrary wavevector. In the EKPH modeling starts by using the simplest Hamiltonian H0 without spin orbit coupling (see it for example in Datta [7]) but then the dispersion functions implied by this Hamiltonian are changed into the empiric dispersion functions characterized by their energies at the Γ points and the measured effective masses for each sub-band. However the eigenvectors remain unchanged so leaving unchanged the diagonalization matrix of this Hamiltonian (which made with these eigenvectors).

In this paper InAs QD in GaAs host material has been modeled. The size of QD was 16 nm in lateral and 6 nm vertical dimensions. It should be noted the QDs varied in size. Here the median size with 10% variation is taken. If the QDs are characterized by a constant offset potential extended among the whole QD and there is assume the QDs to have the shape of a box or cylinder, then the calculation of the energy levels is done in few seconds. But the energy spectra for a different QD shape are different. Both cases are shown on figure 1.

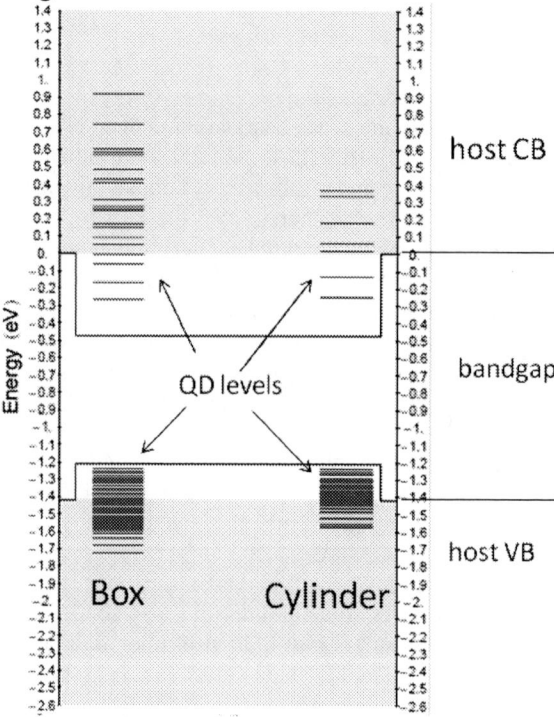

Figure 1. Energy spectra for a different QD shapes (box of cylinder). QD material is InAs, host material is GaAs. Both modeled QDs have lateral size of 16 nm and height of 6 nm.

The difference between box and cylindrical models appears because of wavefunctions parts. If the QD has a shape of box, wavefunction could be considered as a product of periodic trigonometric functions (Sin or Cos), inside and exponents outside the quantum dot. This representation could be done by using separation of variables [7], which goes usually by rectangular coordinates.

If the shape of a quantum dot is taken as cylinder, the use of cylindrical coordinates is more reasonable. The wave functions here are the product of a periodic function and Bessel function of the first kind inside the quantum dot and the exponent and Bessel function of the second kind outside the quantum dot [8].

The calculation of the internal quantum yield (IQY) of a QD SC was done for both shapes. The difference between energy spectrums of box and cylindrical shape didn't lead to big difference in calculated IQY curves. Such similar result is possible because of energy levels density in valence band. Indeed number of QD levels in cylindrical model is fewer. But for subband absorption only transition of energy less than host bandgap are actual. Furthermore the density of energy levels in valence band is high due to the high effective mass of the *hh*. So there are many transition of very similar energy in the host sub-bandgap region (0.9-1.4 eV).

The accuracy estimate was carried out relatively to real and well characterized QD SC prototype [3]. Calculated curves are given in figure 2 together with the measured IQY in the prototype (called SB in reference 3).

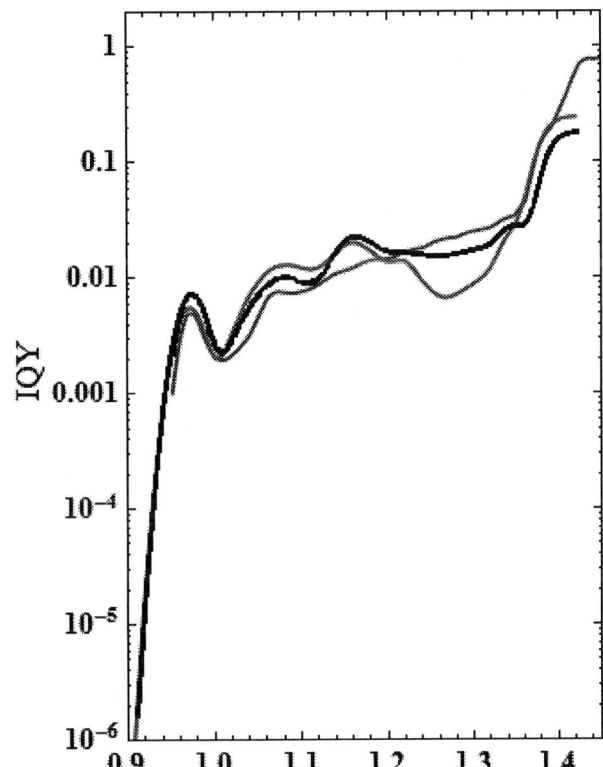

Figure 2. Internal Quantum Yield vs. photon energy. Blue line is drawn with the measured data in the QD SC prototype; red line is drawn with the data calculated for box-shaped QD, Green line is drawn with the data calculated for cylindrical-shape QD.

For the calculation of the IQY in the host sub-bandgap region (0.9-1.4 eV) numerous transitions have to be calculated. In photon absorptions there are around 200 confined hh states (depending on the size of the QD) and some few lh states that act as initial states; for final states there are the some few dozens of cb states, of which only few ones are within the host semiconductor bandgap and constitute the energy levels. The rest are within the conduction band forming the so called virtual bound states. There are more than 1000 transitions contributing to the sub-bandgap absorption. Once the absorption coefficients are calculated the calculation of the IQY is straightforward in about 2 hours with a simple laptop.

In the area of 0.9 eV there is quite good matching, but there is mismatch near the edge of host material absorption. This mismatch is the result of QD size unification. In modelled structure only QDs of a size 16x16x6 nm^3 ±5% were considered. The variation of the QDs size supposed to be a Gaussian function. In real structure the distribution of the QD size could be different. For example in [2] there is shown a bimodal distribution of the QDs. For a better matching calculated and practical data each QD should be taken into account. This requires an AFM and TEM measurements for each single structure in a whole volume and modelling of each single QD.

Conclusion

In this paper IQY of the InAs/GaAs solar cell was modelled. Two shapes of QDs: box and cylinder were considered. 4-band EKPH method was used as the theoretical background of calculation. For the box model there is more QD levels in the host material bandgap. Due to the high effective mass of the *hh* in InAs QD levels are closely packed in the VB area. In CB area box QD produce more levels in the host bandgap than cylinder QD. Nevertheless final calculated curve in both cases goes very close. The reason for this is the smooth variation of transition energy in sub-bandgap region. The final calculated result was obtained in about two hour for both cases. Both simplification of QD shape give close results in terms of accuracy to the measured data. Thus the box approximation preferable to use in multithreading computer architecture due to higher amount of transition. The cylinder approximation can give faster result in a single core high performance computer.

References

[1] A.Luque, A.Marti, C. Stanley 2012 *Nature Photonics* **6** 146
[2] N.A. Kalyuzhnyy, S.A. Mintairov, R.A. Salii, A.M. Nadtochiy, A.S. Payusov, P.N. Brunkov, V.N. Nevedomsky, M.Z. Shvarts, A. Marti, V.M. Andreev, A. Luque 2016 *Progress in Photovoltaics* **24** 1261.
[3] E. Antolín, A. Marti, C. D. Farmer, P. G. Linares, E. Hernández, A. M. Sánchez, T. Ben, S. I. Molina, C. R. Stanley, and A. Luque., 2010 *J. Appl. Phys.* **108** 064513.
[4] Mikhail Mintairov, Valery Evstropov, Maxim Shvarts, Sergey Mintairov, Roman Salii and Nikolay Kalyuzhnyy 2016 *AIP Conference Proceedings* 1748
[5] R.A. Salii, S.A. Mintairov, P.N. Brunkov,A.M. Nadtochiy, A.S. Pauysov, N.A. Kalyuzhnyy, 2015, *Semiconductors* **49** 1111
[6] A. Luque, A. Mellor, E. Antolin, P. G. Linares, I. Ramiro, I. Tobias, and A. Marti 2012 *Solar Energy Materials and Solar Cells* **103**, 171.
[7] S. Datta 1989 *Quantum Phenomena* (Addison Wesley).
[8] R Nedzinskas, V Karpus, B Čechavičius, J Kavaliauskas and G Valušis 2015 *Phys. Scr.* **90**, 065801

Effect of temperature on dry etching of III-V structures

I A Morozov [1] A S Gudovskikh [1,2], D A Kudryashov[1], K P Kotlyar[1], K Y Shubina[1]

[1]Saint-Petersburg Academic University RAS, 194021, St.-Petersburg, Russia
[2]Saint-Petersburg Electrotechnical University "LETI", 197376, St.-Petersburg, Russia

Abstract. For the dry etching of complex compounds III-V a mixture of Cl_2 / BCl_3 gases is used as the main etching technique. This chemistry for deep etching requires the use of hard resist such as metals or SiO_2. Unfortunately, these resists suggest the use of acids to remove them that can damage the structures. In this paper, we try to use standard lithographic resists.

1. Introduction
Plasma etching is a powerful method of etching III-V compounds [1–5]. The main etching method is etching in chlorine plasma. Recently for the deep etching method of silicon the etching at cryogenic temperatures has been began to use. In this paper, we investigated the effect of cryogenic temperatures on the etching of GaAs / AlGaAs. It was assumed that low temperatures would make it possible to obtain smooth and vertical etching walls.

2. Experimental details
Masks were prepared by photolithography using photoresist AZ MIR 701 with micron size colomns and plates. Etching was performed with ICP etching mode ICP with Cl_2/BCl_3 chemistry. Etching temperature was varied from -140°C to 20°C. Pressure and ICP power were in the range of 2-7 mTorr and 800-1500W, respectively. Bias voltage was measured in the range of 50-150V. Etching time was the same for all the samples and set equal to one minute. Structural properties were studied by SEM Carl Zeiss Supra.

3. Results
In this paper, different etched samples were obtained. Temperatures of etched substrate were -140°C; -80°C and 20°C was used as a reference. The resist was not deleted after the etching and the samples were transferred to SEM "As Is". SEM images are shown in fig 1-4. Sample 1(fig.1) was etched at -140°C and bias voltage of -100V. The etching rate was very low, about 200 nm per minute. The bottom of the etched surface was obtained smooth with low defects density. The selectivity of etching of the resist and the substrate turned out to be low approximately 1: 1. The effect turned out to be the opposite of what was originally intended.

Content from this work may be used under the terms of the Creative Commons Attribution 3.0 licence. Any further distribution of this work must maintain attribution to the author(s) and the title of the work, journal citation and DOI.
Published under licence by IOP Publishing Ltd

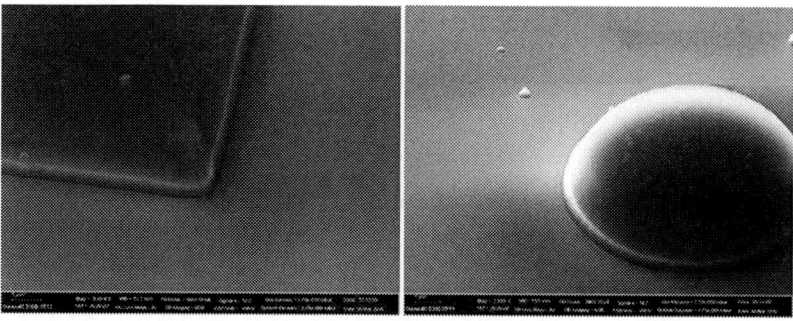

Figure 1. SEM image of etched profile of -140 °C and and bias voltage -100V.

Sample 2(fig.2) was etched at -80°C and bias voltage of -100 V. The etching rate being equal to about 1000 nm per minute was higher compared to the previous process. The obtained surface of the bottom was rough with high defects density. The selectivity of etching of the resist and the substrate turned out to be low.

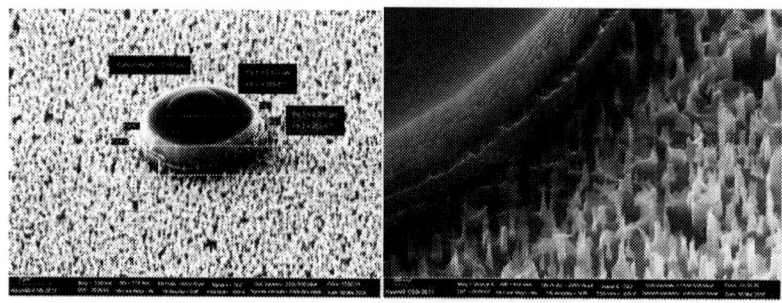

Figure 2. SEM image of etched profile of -80 °C and and bias voltage -100V.

After etching of the second sample we obtained a "black" surface and supposed that the bias voltage could be too high. Therefore the next process was performed with bias voltage of -60V. Sample 3 (fig.3) was etched at -80°C and bias voltage of -60 V. The etching rate was low and equal about 600 nm per minute. The bottom of the etching was obtained with rough surface and high defects density but the morphology of the surface was different compared to the sample 2. The selectivity of etching of the resist and the substrate turned out to be low.

Sample 4 (fig.4) was etched at 20 °C and bias voltage of -100 V. The etching rate was high, about 2000 nm per minute. The smooth surface of the bottom of the etching was obtained with low defects density. The selectivity of etching of the resist and the substrate turned out to be well enough and equal about 1:10. However the edge of the resist mask was etched faster compared to the middle.

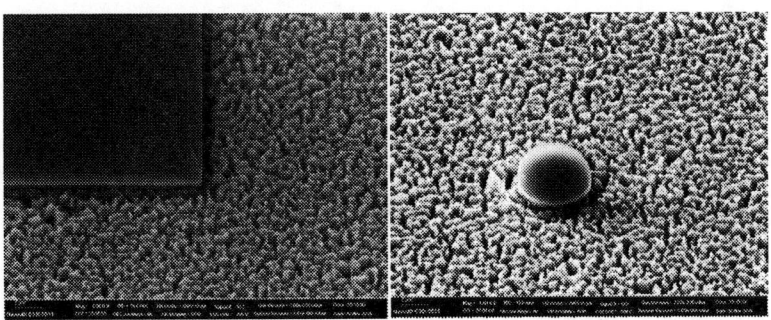

Figure 3. SEM image of etched profile of -80 °C and and bias voltage -60V.

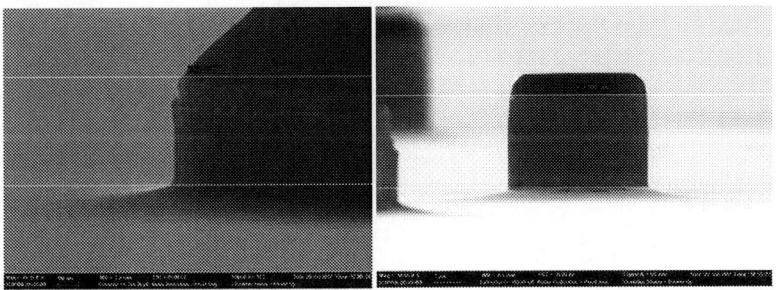

Figure 4 SEM image of etched profile of 20 °C and and bias voltage -100V.

Comparison of the etching rates as a function of temperature is shown in the figure 5. As can be seen, the etching rate decreases linearly with decreasing temperature.

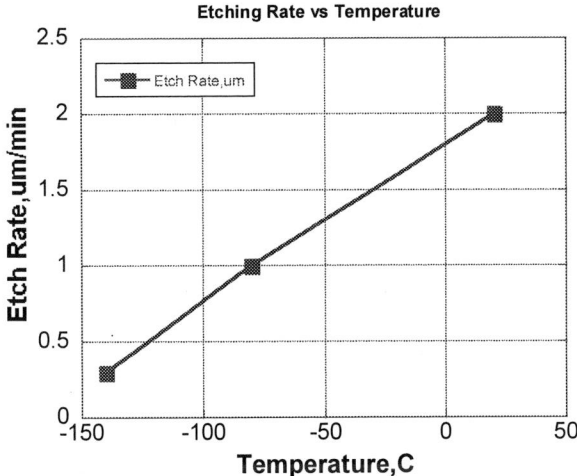

Figure 5 Etching rate for different etching temperatures .

It has been shown that temperature decrease affects dramatically the etching rate of III-V compounds. Resist etching rate demonstrate the same behavior with temperature. It is also shown that as the temperature decreases uniformity also decreases significantly. Low temperatures needs to be deeply investigated

Acknowlegements This work was supported by the Russian Scientific Foundation under Grant No.17-19-01482.

References

[1]   Suzuki K, Okudaira S, Sakudo N and Kanomata I 1977 Microwave Plasma Etching *Jpn. J. Appl. Phys.* **16** 1979

[2]   Chapman B and Vossen J L 1981 *Glow Discharge Processes: Sputtering and Plasma Etching Phys. Today* **34** 62–62

[3]   Gregus J A, Vernon M F, Gottscho R A, Scheller G R, Hobson W S, Opila R L and Yoon E 1993 Low-temperature plasma etching of GaAs, AlGaAs, and AlAs *Plasma Chem. Plasma Process.* **13**

[4]   Lee J W, Hong J, Lambers E S, Abernathy C R, Pearton S J, Hobson W S and Ren F 1997 Plasma etching of III-V semiconductors in BCl3 chemistries .1. GaAs and related compounds *Plasma Chem. Plasma Process.* **17** 155–67

[5]   PEARTON S J, HOBSON W S, ABERNATHY C R, REN F, FULLOWAN T R, KATZ A and PERLEY A P 1993 DRY-ETCHING CHARACTERISTICS OF III-V SEMICONDUCTORS IN MICROWAVE BCL3 DISCHARGES *Plasma Chem. Plasma Process.* **13** 311–32

[6]   Dussart R, Tillocher T, Lefaucheux P and Boufnichel M 2014 Plasma cryogenic etching of silicon: from the early days to today's advanced technologies *J. Phys. D. Appl. Phys.* **47** 123001

SPBOPEN 2018

IOP Publishing

IOP Conf. Series: Journal of Physics: Conf. Series **1124** (2018) 041032   doi:10.1088/1742-6596/1124/4/041032

Chromaticity coordinates temperature dependence for blue laser diodes for solid state laser lighting

A Aluyev[1], Yu Akhmerov[1], N Kourova[1, 2], V Logachev[1, 3], M Mezhenny[1], A Chelny[1], A Savchuk[1, 2], V Murashev[1, 2], O Rabinovich[2], S Didenko[2]

[1] Optron JSC, Moscow 105187, Russia
[2] National university of science and technology "MISiS" Moscow, 119049, Russian Federation
[3] Lomonosov Moscow State University, Moscow, 119991, Russian Federation

Abstract. Colour characteristic temperature dependence of such as X- and Y-coordinates (CIE 1931) for high power blue laser diode was investigated. It was shown, that thermal stability of colour characteristic of high power blue laser diode allows to create solid state lighting.

1. Introduction

Every year the lighting market based on "white" LEDs is growing dynamically. This is due to several reasons: high lifetime, compactness, usage simplicity, high efficiency, low power consumption [1]. However, despite their advantages, blue LEDs that are used as optical radiation sources for exciting the phosphor luminescence, based on the yttrium aluminum garnet doped with cerium (YAG:Ce^{3+}), have several fundamental drawbacks: internal quantum efficiency drop at high current densities, that with the total internal reflection effect, reduce the external quantum efficiency[2]. Laser diodes (LD), as sources of optical radiation, do not have such disadvantages as described above, as compared with LEDs [3, 4].

Previously, it was shown that the color characteristics of solid state laser lighting systems can correspond to the color characteristics of an absolutely black body [5]. From the point of view of laser lighting systems operation, the color characteristics stability as a function of temperature is a great practical interest.

The goal of current work was to investigate the color characteristics temperature dependence of the blue high-power laser diodes versus temperature.

2. Experiment

In this investigation Nichia NDB7A75 LD was used as laser light source. LD was fixed on copper heat sink which in turn was mounted on thermal-electro cooler (TEC). LD spectral characteristics were measured in temperature range 15-30 °C by S-150HR spectrometer. This spectrometer has a very high resolving power (up to ± 0.01 nm).

3. Result and Discussion

In Fig.1a the dependence of the laser diodes spectral characteristics at a fixed temperature vs the pump current are shown. In Fig. 1b the dependence of emission spectral characteristics at fixed optical

---

Content from this work may be used under the terms of the Creative Commons Attribution 3.0 licence. Any further distribution of this work must maintain attribution to the author(s) and the title of the work, journal citation and DOI.
Published under licence by IOP Publishing Ltd

radiation power versus temperature are shown. Then, calculations, based on these spectral characteristics have been provided.

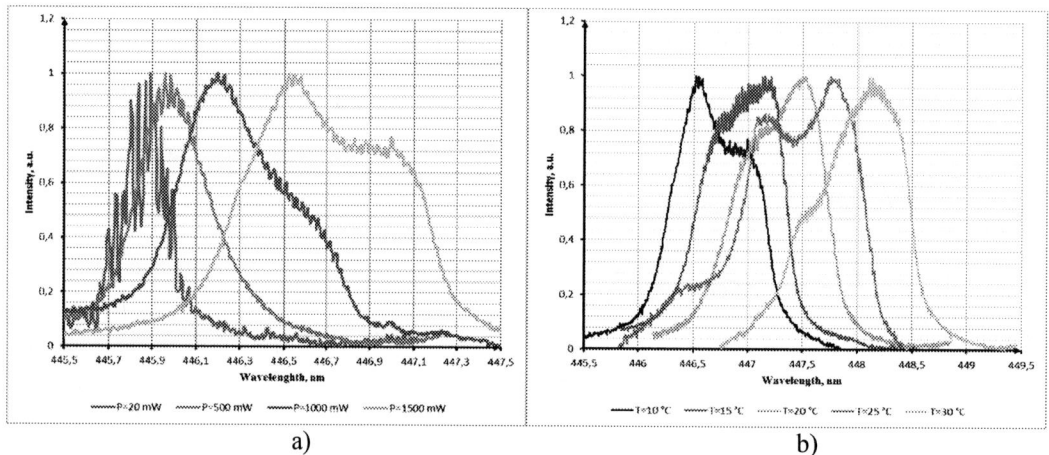

Figure 1. Spectral characteristics LD : a) at a fixed temperature (20°C) vs the forward current; b) fixed optical radiation power (1500mW) vs temperature

In the Fig. 2 X- and Y- chromaticity coordinate (CIE 1931) dependence versus output power at fixed temperature are shown.

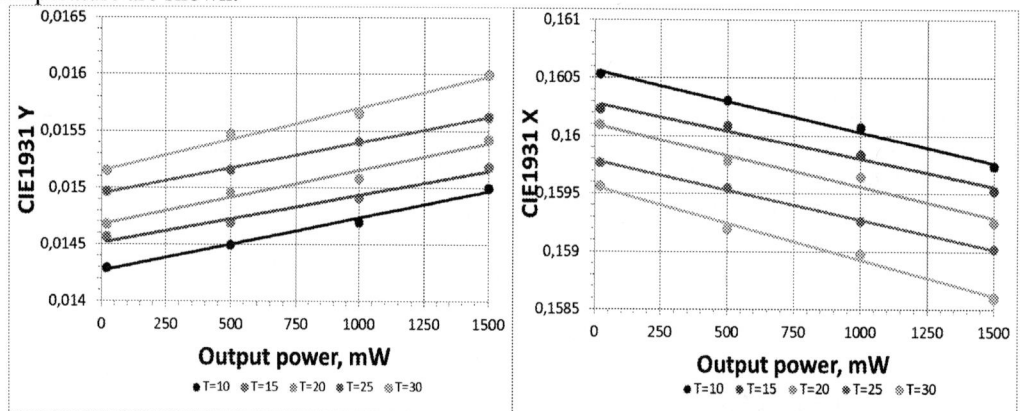

Figure 2. CIE 1931 X- and Y- chromaticity coordinate dependence versus output power at fixed temperature

It can be seen that increasing temperature in range 10-30 °C led to a decrease X- and increase Y-coordinates. For curve corresponding to 10 °C, 15 °C and 30 °C angle of inclination changes insignificantly. For curve corresponding to 20 °C and 25 °C slight deviation of inclination angle was observed. This can be explained by a LD spectrum fluctuation due to indium fluctuation in quantum wells [6]. The difference between the X-coordinate values at 10 °C and 30 °C does not exceed 0.0015, the difference between the Y-coordinate values at 10 °C and 30 °C does not exceed 0.002. The values for the color coordinates change versus the temperature and the output power of the LD are in the range of the tolerance values for the color coordinates determination (less than 0.01) for the light sources in accordance with CIE1931. Based on this, it can be concluded that the temperature change has a slight effect on the solid state laser lighting color parameters.

Further, we investigated the wavelength deviation dependence at the maximum of the LD emission spectrum (λ_{max}) and the LD dominant wavelength (λ_d) versus the temperature at different output optical power. The results are shown in Fig. 3.

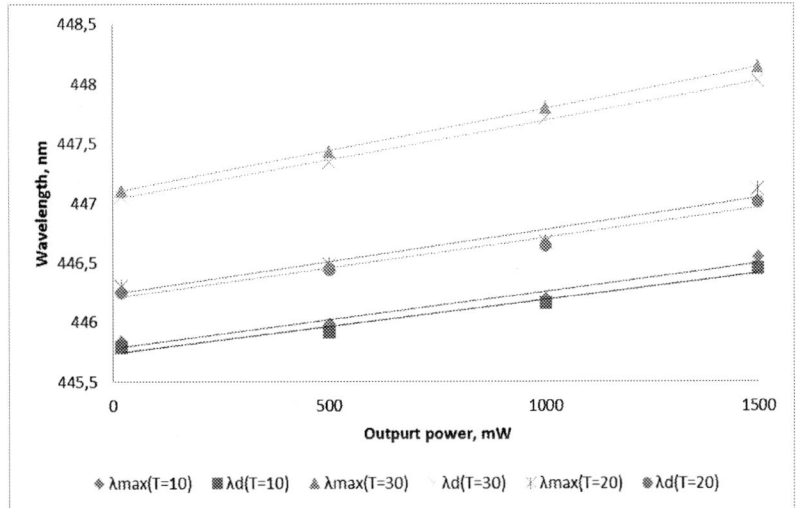

Figure 3. Maximum (λ_{max}) and dominant (λ_d) wavelength dependence versus output power LD

It can be seen that difference between λ_{max} and λ_d increases with increasing LD output power. It can be explained by spectrum line width rise. Difference between λ_{max} and λ_d in output power range 20 – 1500 mW does not exceed 1 nm at difference temperatures. Figure 4 shows the area in the CIE1931 color space for LD at 1000 mW and 1500 mW.

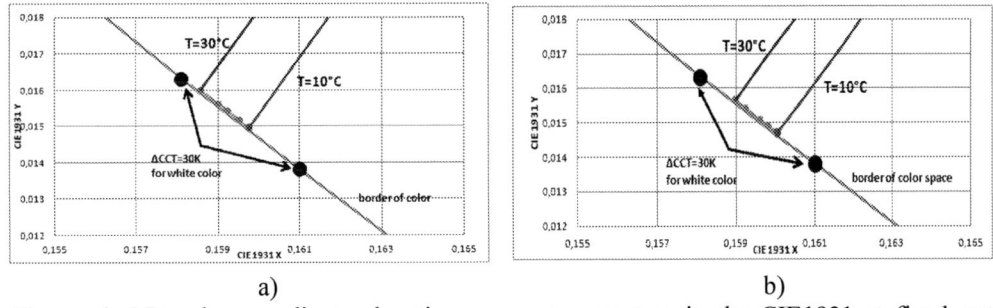

a)   b)

Figure 4. LD color coordinates location versus temperature in the CIE1931 at fixed output power: a) 1000 mW; b) 1500 mW

The red and blue lines connect the LD color coordinates and the phosphor based on YAG:Ce^{3+}. The color temperature value for this source is CCT = 4500 K, the black dots on the border of the color space indicate the area that corresponds to the deviation from the CCT by 30 K. It can be seen in the figure that the LD color coordinates lie entirely in this region, which indicates a high stability of the color temperature.

4. Conclusion

LD color coordinates dependences at different output optical power values versus temperature are investigated. It is shown that at the temperature change, the LD color coordinates change insignificantly, the deviations of the X- and Y- color coordinates according to CIE1931 do not exceed 0.01. The dependence of the dominant wavelength and the wavelength at the maximum of the laser radiation spectrum at the temperature versus different values of the output optical power was investigated. It is shown that the dominant wavelength and the wavelength at the maximum of the emission spectrum vary slightly with temperature. It is detected that the difference in values between λ_{max} and λ_d is negligible, therefore, only λ_{max} can be used to estimate the color parameters of laser radiation systems. Changing the color coordinates at different temperatures leads to a slight deviation of the CCT, not more than 30 K. It can be concluded that high power blue LDs provide stable colorimetric characteristics such as X- and Y- coordinates (CIE 1931) and CCT. Blue LD has a promising future as a light source for solid state lighting creation.

References

[1] Jy Bhardwaj, John M. Cesaratto, Isaac H. Wildeson, Henry Choy, Ashish Tandon, 2017 *Phys. Status Solidi A*, **2** 1600826
[2] Cantore M, Pfaff N, Farrell R M, Speck James S, Nakamura S, Den Baars S P 2016 *Optics Express* **24(2)** A215
[3] Daniel Feezell, Shuji Nakamura 2018 Comptes Rendus Physique **19(1–2)** 1–84
[4] Chelny A 2017 *Semiconductors lasers and system* 83
[5] Akhmerov Y L 2017 *Semiconductors lasers and system* 88
[6] Chichibu S, Nakamura S 1996 *Appl. Phys. Lett.* **69** 4188

SPBOPEN 2018

Optical coupling upset in multijunction solar cells with built-in Bragg reflectors

S A Levina, E D Filimonov, V M Emelyanov and M Z Shvarts

Ioffe Institute, St. Petersburg, 194021, Russia

e-mail: levina@mail.ioffe.ru

Abstract. In the present work, an opportunity to block the negative influence of luminescent coupling between subcells in multijunction solar cells with help of built-in Bragg reflectors is investigated. Temperature modes, at which the blocking of optical interaction in the GaAs-Ge pair of subcells is possible, have been determined.

1. Introduction

Distributed Bragg reflectors (BRs) have found wide application in devices of modern optoelectronics [1, 2], providing control over the transmission, recycling and optical absorption of radiation in semiconductor structures. BRs are successively applied in solar cells to raise their performance and radiation resistance owing to increase the efficiency of generation and separation of charge carriers in the photoactive layers with repeated reflection (recycling) of radiation [3, 4].

In multijunction solar cells (MJ SCs) based on semiconductors of high crystallographic perfection with the dominance of radiative processes over non-radiative ones, the radiation recycling processes are usually considered in a wider context, not limited to areas of separate subcells. In such MJ SCs, recombination (luminescent) radiation is cycled not only within layers of a "native" subcell but also penetrates in adjacent photoactive regions forming processes of optical coupling between subcells [5]. Clearly, in the modes of current matching between subcells, excessive charge carriers having no way to escape into the external circuit recombine radiatively and form light flux with radiation wavelength within the range of the narrowband p-n junction sensitivity. With some probability such luminescent photons will be absorbed in the adjacent narrowband subcell with the generation of an additional photocurrent. The investigations show that, in some cases, the optical coupling effect can play a positive role of self-balancing, transferring a MJ SC into the current matching mode for subcells [6]. However, in determining rated photovoltaic characteristics of MJ SC, the negative contribution of such luminescent interaction among subcells complicates substantially the experimental methodology and techniques, and makes interpretation of results difficult [7-9].

The pronounced influence of the optics coupling is manifested in studying the spectral dependencies of photoresponse of narrowband subcells. So, in the narrowband p-n junction spectral sensitivity range, an uncontrollable drop of sensitivity is observed, and beyond it, on the contrary, an anomalous photoresponse is registered [10].

Different methods for eliminating the mentioned "negative" effect of the optical coupling on the measurement results have been proposed [5-9]. Most of them are based on simplified models, which are varied depending on a type or peculiarities of solar cells. For this reason, they cannot be considered universal ones. Another radical way to decrease the effectiveness of optical coupling or, practically, its complete elimination is high-energy (about several MeV) electron irradiation of samples [11]. Such an approach results in changing a dominant recombination mechanism for non-

Content from this work may be used under the terms of the Creative Commons Attribution 3.0 licence. Any further distribution of this work must maintain attribution to the author(s) and the title of the work, journal citation and DOI.

Published under licence by IOP Publishing Ltd

radiative one, what leads to optical coupling suppression. It is obvious that all considered examples have their own limitations and disadvantages.

In the present work, mechanisms of selective optical blocking for luminescent radiation propagating between MJ SC subcells owing to using built-in BRs were investigated.

2. Experimental results

Let us consider the mechanisms of sunlight propagation and absorption in the monolithic multilayer structure on the example of the GaAs-BR-Ge stack. External radiation, spectrally covering the GaAs sensitivity range, initiates processes of charge carrier (electron-hole pair) generation in the wideband subcell, which, then, recombine radiatively with producing secondary photons in the range of 860-870nm. In the structure, such a secondary radiation will be either reflected by a Bragg mirror with initiating the recycling mechanism in the "native" GaAs subcell or penetrates the Ge subcell determining optical coupling between GaAs and Ge subcells.

If a BR completely reflects the luminescent flux in the direction to the Ge subcell, optical interaction will be perfectly blocked, and the values being registered for the Ge subcell photoresponse spectral dependence will have the true value without a necessity for any correction in correspondence with [10].

In case if the mirror reflection spectrum overlaps the wavelength range of recombination radiation only partially, the photon flux Φ_1 or Φ_2 (fig.1) will penetrate beyond a BR with generating an additional photocurrent at absorption in Ge subcell. In circumstances like this, optical coupling between subcells can be upset (blocked), if a shift of the BR reflection spectrum and luminescent radiation with respect to each other is initiated by heating/cooling the structure.

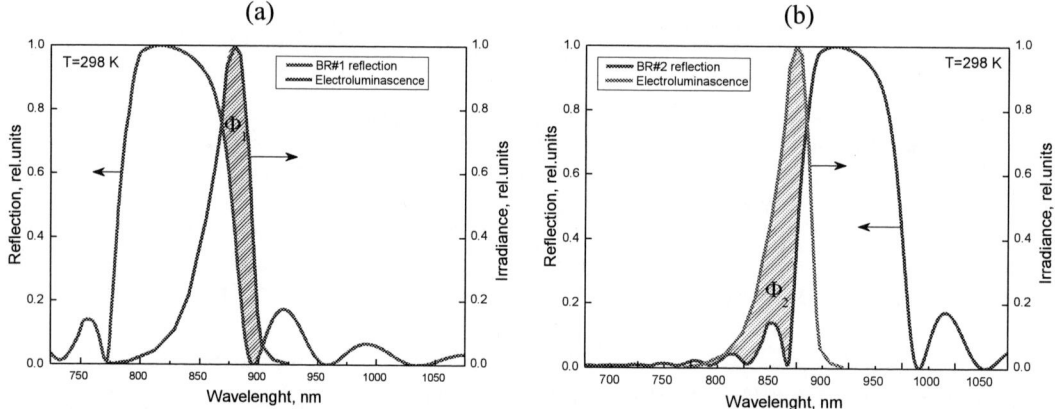

Figure1. Spectra of electroluminescent radiation of the GaAs subcell (green and orange lines) and of reflection of built-in Bragg mirrors (green and blue lines) of two samples: (a) – BR#1; (b) – BR#2. All spectral dependences was obtained at 298K. Fluxes $\Phi1$ and $\Phi2$ (filled area) correspond to radiation passing into the bottom Ge subcell and forming optical coupling between two junctions.

In the work, temperature conditions essential for blocking optical coupling, if such an effect is observed at standard conditions (298K), in two different MJ SCs with the BR (structures BR#1, BR#2), have been determined. It is obvious that, to initiate mechanisms of blocking the optical coupling, the BR#1 structure should be cooled and BR#2 one be heated promoting a shift of the luminescence emission peak towards to the short- or long-wavelength spectral range, respectively.

It has been established during investigation that shift of the electroluminescent peak of GaAs subcell (fig.2a) with temperature is approximately 0.4nm/K (fig.2c, blue). As the peak shifts toward longer wavelengths with temperature gain, the half-width of the electroluminescence spectrum increases with a coefficient of the order of 0.1nm/K (fig.2d, blue line). Thus, it is more likely that a complete upset of optical coupling could be observed in ordinary SC with BR at cooling (low temperature mode of operation) rather than heating.

As for the BR reflection spectra (fig.2b), with a temperature variation, its half-width remains practically unchanged (fig.2d, green), as well as the shift of the resonance reflectance is practically negligible and are only 0.08nm/K (fig.2c, green). Figure 2b shows theoretical and experimental temperature dependences of the BR reflection spectra. The discrepancies between of experimental data and model results (fig. 2b on the left) are due to a part of the reflected radiation was absorbed in the upper (relative to the BR) photoactive layers of the structure. With increasing temperature, the reflection range of the Bragg mirror becomes more and more "covered" by GaAs absorption curve, which is most pronounced on the left side of fig.2b. Consequently, at some temperature (393K) the ability to observe/distinguish the reflection spectrum of the BR is completely lost. In such a case the change in the spectral half-width and the shift of the absorption edge of BR resonance reflectance with temperature can be obtained only through theoretical consideration.

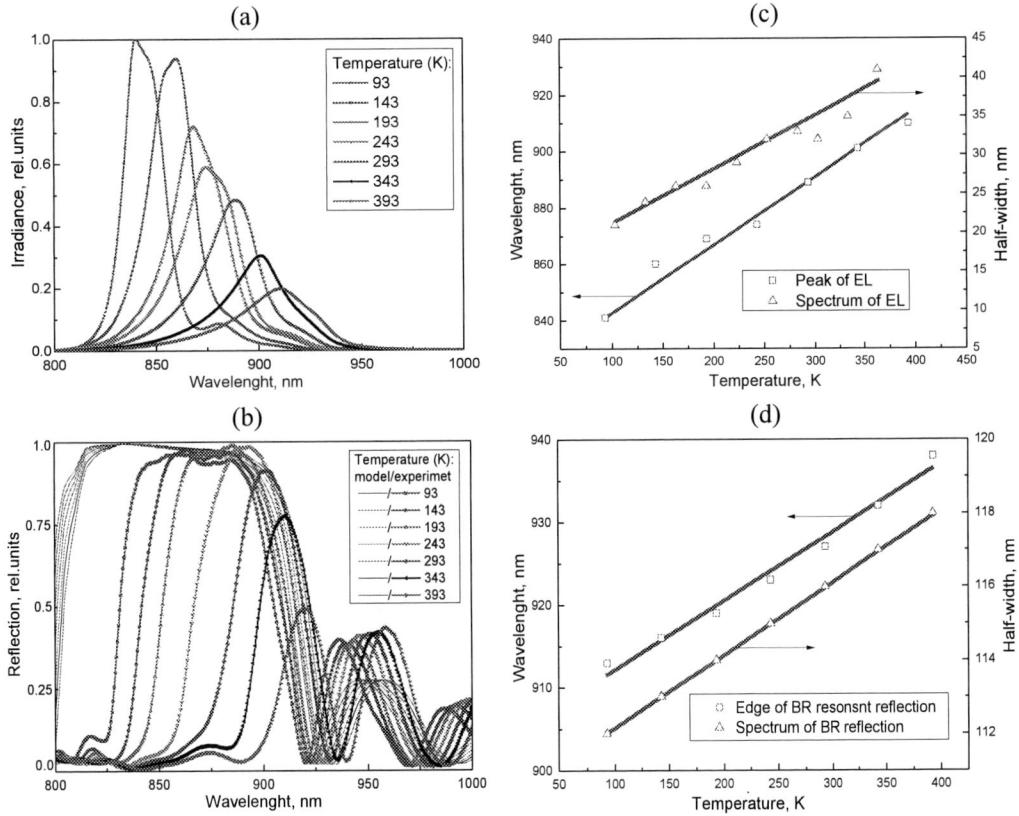

Figure2. Temperature dependencies of: (a) – of GaAs electroluminescence spectra; (b) – of BR reflection spectra; (c) – the GaAs electroluminescence peak position (blue) and half-widths of GaAs electroluminescence spectrum (red); (d) – position of the edge of BR resonance reflection (violet) and half-widths of BR reflection spectrum (green).

Hence, the temperature-controlled transition of the SC into the blocking mode of optical coupling is possible if:
- the luminescence peak is apart from the right edge of BR resonance reflectance less than 70-80nm (fig.1a), that is enough to provide EL light to be reflected backwards to GaAs subcell at cooling.

- the luminescence peak is located at the distance less than 30-40nm from the left edge of BR resonance reflectance to get overlapping the spectra in heating.

3. Results

In this article, possibilities to block optical interaction between p-n junction in multijunction solar cells with use of built-in Bragg reflectors were investigated. Temperature modes, at which can be achieved violation of optical interaction in the GaAs-Ge pair of subcells, have been determined. It was found that if discrepancy between spectra of GaAs electroluminescence and BR reflection is less then 80nm (the luminescence peak is apart the right edge of the BR resonance reflection) or 40nm (left edge) cooling/heating of the sample is able to overcome optical coupling.

Acknowledge

The reported study was funded by RFBR according to the research project № 18-32-00561.

References

[1]   Tsuda S, Knox W H, Cundiff S T, Jan W Y, and Cunningham J.E 1996 Mode-locking ultrafast solid-state lasers with saturable *Bragg reflectors IEEE J. of Selected Topics in Quantum Electron* **2** 454-464

[2]   Shen J L, Jung T, Murthy S, Chau T, Wu M C, Lo Y H, Chua C L and Zhu Z H 1999 Mode locking of external-cavity semiconductor lasers with saturable Bragg reflectors *J. Opt. Soc. Am.* **16** 1064-1067

[3]   García I, Geisz J, Steiner M, Olson J, Friedman D and Kurtz S 2012 Design of Semiconductor-based Back Reflectors for High Voc Monolithic Multijunction Solar Cells *in Proceedings of the 38th IEEE Photovoltaic Specialists Conference* **1** 002042–002047

[4]   Shvarts M, Chosta O, Kochnev I, Lantratov V, Andreev V 2001 Radiation resistant AlGaAs/GaAs concentrator solar cells with internal Bragg reflector *Solar Energy Materials and Solar Cells* **68/1** 105-122

[5]   Meusel M, Baur C, Letay G, Bett A W, Warta W, Fernandez E. 2003 Spectral response measurements of monolithic GaInP/Ga(In)As/Ge triple-junction solar cell: measurement artifacts and their explanation *Progress in Photovoltaics: Research and Applications* **11** 499–514

[6]   Brown A S, Green M A. 2002 Radiative coupling as a means to reduce spectral mismatch in monolithic tandem solar cell stacks—theoretical considerations *Proceedings of the 29th IEEE Photovoltaic Specialists Conference, New Orleans* **1** 868–871

[7]   Pravettoni M, Galleano R, Virtuani A, Dunlop E D 2011 Spectral response measurement of double-junction thin-film photovoltaic devices: the impact of shunt resistance and bias voltage *Measurement Science and Technology* **22** 1–10

[8]   Barrigón E, Espinet-González P, Contreras Y and Rey-Stolle I 2015 Implications of low breakdown voltage of component subcells on external quantum efficiency measurements of multijunction solar cells *Prog. Photovolt., Res. Appl.* **23** 1597–607

[9]   Steiner A and Geisz J F 2012 Non-linear luminescent coupling in series-connected multijunction solar cells *Appl. Phys. Lett.* **100** 251106-1–251106-5

[10]   Shvarts M, Emelyanov V, Evstropov V, Mintairov M, Filimonov E and Kozhukhovskaia S 2016 Overcoming of Luminescent Coupling Effect in Experimental Search for True Quantum Efficiency Value in MJ SCs *AIP Conf. Proc.* **1** 1766 060005

[11]   Baur C, Hermle M, Dimroth F and Bett A W 2007 Effects of optical coupling in III-V multilayer systems *Appl. Phys. Lett.* **90** 192109

Quantum efficiency measurement of subcells in multi-junction solar cells based on III-V/Si

A I Baranov, A S Gudovskikh, D A Kudryashov, I A Morozov, A M Mozharov, K S Zelentsov

St Petersburg Academic University of RAS, 194021 St Petersburg, Russia.

Abstract. The work is devoted to development of the post-growth technology for direct study of the top subcell based on p-i-n junction with the GaPAs i-layer in the double-junction solar cell grown on silicon wafer. It allowed to more precisely measure quantum efficiency (QE) of the top subcell without additional IR-illumination for saturation of the bottom subcell in wafer. In result, measured QE of mesa-structure shows more reliable results without artefacts in spectra.

1. Introduction

Nowadays, growth of III-V compounds on silicon wafers is the attractive task in the world. The development of such technology will allow to grow multi-junction solar cells (MJSC) with high efficiency on low cost silicon wafers. Gallium phosphide (GaP) is one of these semiconductors for growth on Si wafer since their lattice-mismatch is only 0.37% [1]. Furthermore, silicon can be effectively passivated by GaP so it opens attractive perspectives for the pseudomorphic epitaxial growth of GaP-based layers. The incorporation of 0.43 % atomic fraction of nitrogen into the GaP lattice leads to direct band transition, and a bandgap energy of GaPN decreases drastically with growth of nitrogen content in small interval of few percent [2,3]. Moreover, an addition of arsenic allows to vary the bandgap energy in a range of 1.5 to 2.1 eV while remaining lattice-matched to Si or GaP wafers [4]. According to simulation 1.1 eV and 1.7 eV are the most optimal values of the bandgap energy for the bottom and the top subcells in double-junction SC, respectively [5]. The first one corresponds to silicon E_g=1.12 eV [1], and the second one can be grown as layers of GaPAsN with arsenic content of 20-30%. In this case the theoretical limit of the InGaPAsN/Si system achieves efficiency of 37.4% for AM1.5G [4], which is higher than in SC based on GaInP/GaAs system. In this work, we grew the double-junction SC with active layer of i-GaP$_{0.70}$As$_{0.30}$ on Si wafer as initial step of growth 1.7 eV-GaPAsN layer. Further, we develop the post-growth processing for measurement its photoelectric properties.

2. Sample fabrication

The double-junction SC was grown by molecular-beam epitaxy using Veeco GEN III setup. The bottom subcell was fabricated by pre-growth phosphorous flux exposition of p-type (p=1.0×10^{16}cm^{-3}) silicon wafer for 10 minutes at 500 °C to form silicon p-n homojunction. In the initial step of growth, buffer layers of n-GaP(As) were formed by migration-enhanced epitaxy method described in previous work [6]. Then, the tunnel junction and the n-i-p structure with active GaPAs i-layer were subsequently grown. The schematic view of as-grown structure is presented in Figure 1a.

Further, the post-growth technology was applied to fabricate mesa-structures where both metal contacts are formed directly to n-GaPAs and p-GaP(As) layers of the top subcell. Initially, alloys of Au/Ge were deposited on the front layer of n-GaPAs in BOC Edwards Auto500 equipment for vacuum evaporation (Figure 1b). The used mask allows one to form circular contacts in the form of points for future

---

Content from this work may be used under the terms of the Creative Commons Attribution 3.0 licence. Any further distribution of this work must maintain attribution to the author(s) and the title of the work, journal citation and DOI.

Published under licence by IOP Publishing Ltd

capacitance measurements and the contact grid for photoelectrical measurements described in this study (mesa-structure and solar cell in Figure 1f, respectively). The key task of technology is the precise controlled wet etching of P-rich III-V compounds down to p-GaP(As). As a result, optimal conditions were found for their etching in a wet solution of $KIO_3:HCl:H_2O$, and the resist protecting Au/Ge top contact did not degrade. The etching rate was accurately controlled by SEM measurements: structures before and after etching are presented in Figure 2a and Figure 2b, respectively. According to images, we stopped exactly in p-GaP(As) (Figure 1c) as supposed. In next step, alloys of Au/Zn were deposited on the structure to form metal contact to p-GaP(As). Therefore, we fabricated both contacts to adjacent layers of i-GaPAs in the top subcell (Figure 1d). The photograph of fabricated sample is shown in Figure 1f. Furthermore, contacts for transfer line measurements (TLM) were also formed to investigate their ohmic behaviour during the deposition of metals to the top and to the bottom contact: they are located in the first and third rows, respectively, in the photograph of the sample. The ohmic behaviour of Au/Ge and Au/Zn alloys was obtained after rapid thermal annealing at 360-400 °C using the Jipiec JetFirst 100 equipment. Also, an ohmic contact was fabricated to the p-Si wafer by an indium deposition (Figure 1e).

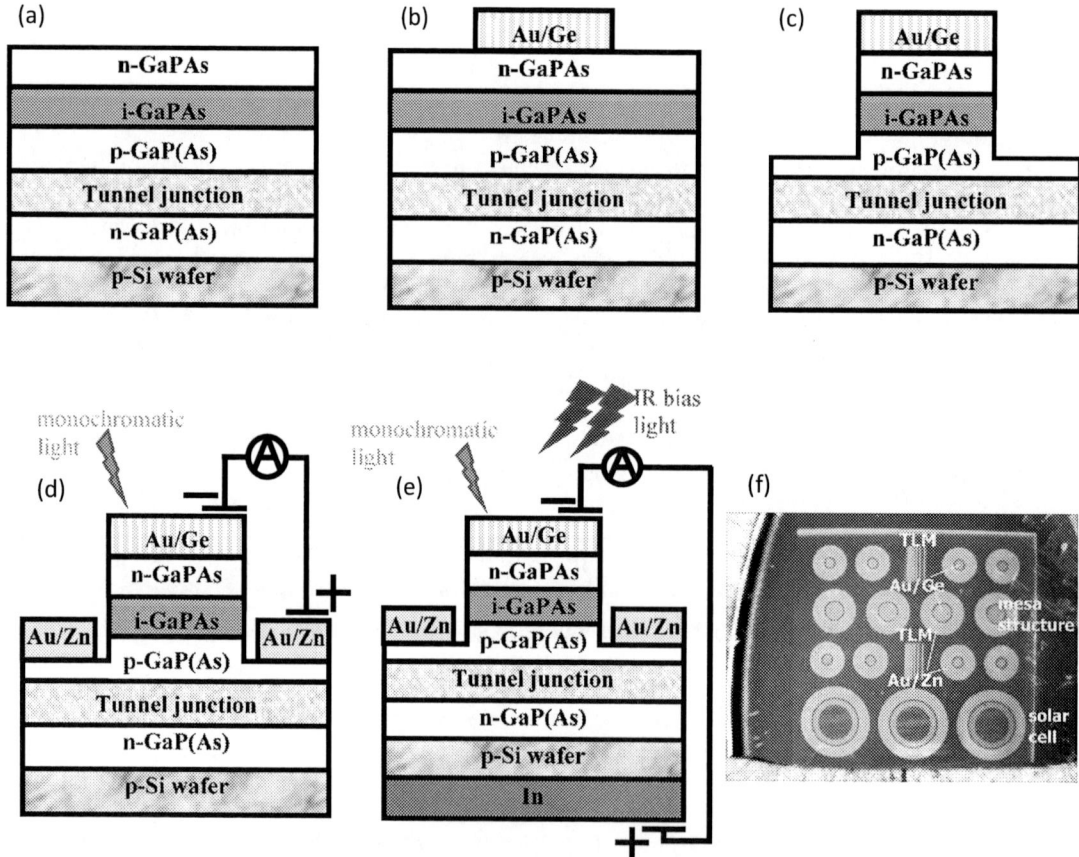

Figure 1 (a, b, c, d, e, f). Schematic view of initial **(a)**, after Au/Ge deposition **(b)** and after wet etching **(c)** structure. Schematic view of quantum efficiency measurements of prepared sample without **(d)** and with **(e)** infra-red bias light. The photograph of sample after the post-growth technology **(f)**.

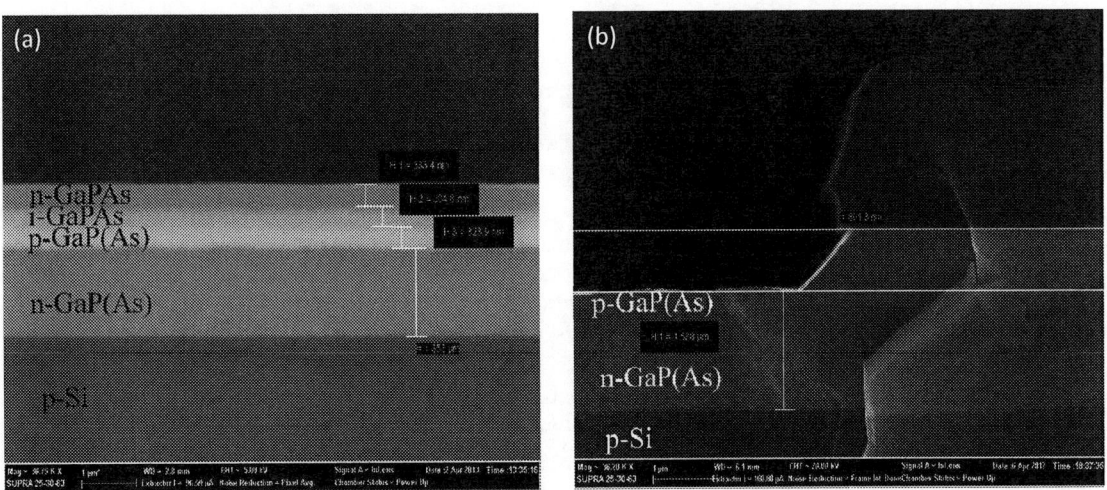

Figure 2 (a, b). SEM-image of the initial structure **(a)** and after etching **(b)**.

3. Measurements and results.

Further, the external quantum efficiency (EQE) was measured for the samples. Conventionally, an additional illumination must be used for separation of QE measurements from different subcells in MJSC [7]. In our case, the infra-red illumination of high flux was used for saturation of silicon junction in the bottom subcell to measure QE of the top subcell with the i-GaPAs layer: the current between Au/Ge and indium contacts was measured (Figure 1e). However, there are several sources of errors, which could affect the accuracy of the QE measurements using this technique. In particular, leakage dark current of the measured junction could lead to detection of excessive current due to responses from the other subcell. One of the way to reduce this effect is to apply external bias voltage, but in this case there is ambiguity of voltage distribution between junctions. Therefore, the developed post-growth technology allowed us to directly measure photoelectric properties of the top subcell: the current between Au/Ge and Au/Zn contacts was measured (Figure 1d). The latter should be more reliable since it does not require any additional external conditions. External QE was measured at room temperature in wavelength range of 300..1200 nm by both methods (Figure 3).

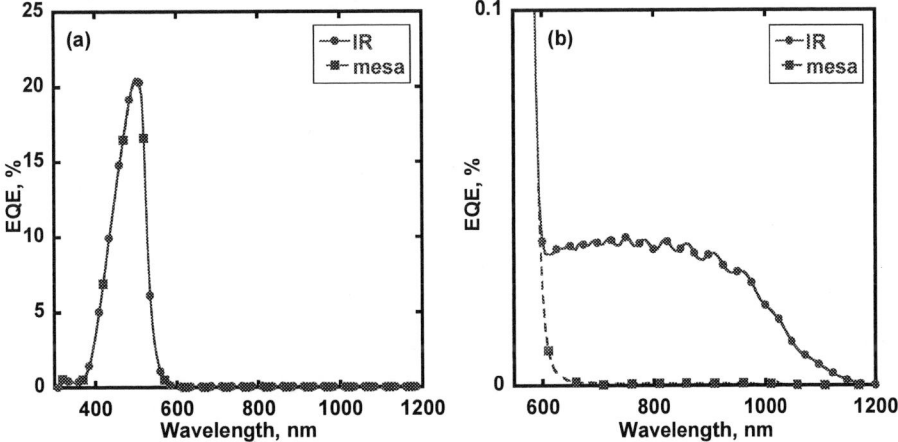

Figure 3 (a, b). The external quantum efficiency in the top subcell with the active layer of i-GaPAs **(a)** and zoomed long wavelength region **(b)**.

First of all, the top subcell exhibits a photovoltaic behaviour but value of the EQE is very low due to the low thickness of i-GaPAs (400 nm) and a possible existence of deep defects in the layers. According to curves, the shape of the QE is the same for both methods in range of 400..600 nm. However, QE under infra-red illumination is not equal to zero at wavelength higher 600 nm where a photon energy is lower than the bandgap energy of the i-GaPAs layer (Figure 3b). The observed shoulder could be associated with the response from the bottom subcell since it is similar to the shape of QE for silicon junction. Thus, classical method did not allow us to fully exclude its influence. Possible reason is shunt currents in the structure.

On the other hand, the mesa-structure has not such problems: the value of the EQE is less than the accuracy of the equipment so it is equal to zero as supposed for the top subcell in this region of wavelength (Figure 3b). Consequently, the post-growth technology allowed us to obtain more reliable result for grown double-junction SC. Moreover, the fabricated mesa-structure can be explored by capacitance method to obtain information about properties of the top subcell since it allows one to exclude influence of the capacitance of the bottom subcell. Its possibilities will be studied in the future experiments.

Acknowledgments

This work was carried out with the support of grant of government of the Russian Federation (16.2593.2017/8.9).

References

[1]   Adachi S 2005 *Properties of Group-IV, III-V and II-VI Semiconductors* (Chichester, UK: John Wiley & Sons, Ltd)

[2]   Skierbiszewski C, Perlin P, Wisniewski P, Knap W, Suski T, Walukiewicz W, Shan W, Yu K M, Ager J W, Haller E E, Geisz J F and Olson J M 2000 *Appl. Phys. Lett.* **76** 2409–11

[3]   Buyanova I A, Rudko G Y, Chen W M, Xin H P and Tu C W 2002 *Appl. Phys. Lett.* **80** 1740–2

[4]   Geisz J F and Friedman D J 2002 *Semicond. Sci. Technol.* **17** 769–77

[5]   Kurtz S R, Faine P and Olson J M 1990 *J. Appl. Phys.* **68** 1890–5

[6]   Sobolev M S, Lazarenko A A, Nikitina E V., Pirogov E V., Gudovskikh A S and Egorov A Y 2015 *Semiconductors* **49** 559–62

[7]   Burdick J and Glatfelter T 1986 *Sol. Cells* **18** 301–14

Temperature annealing effect on ITO film

K P Kotlyar[1], D A Kudryashov[1], I P Soshnikov[1,2,3], A V Uvarov[1], R R Reznik[4]

[1]St. Petersburg Academic University of the RAS, Khlopina 8/3, St. Petersburg 194021, Russia

[2]Ioffe Physical-Technical Institute of the RAS, Politekhnicheskaya 26, St. Petersburg 194021, Russia

[3]Institute for Analytical Instrumentation of the RAS, Rizhsky 26, St. Petersburg 198095, Russia

[4]ITMO University, Kronverkskiy pr. 49, St. Petersburg 197101, Russia

Abstract. In the work we investigate processing and optoelectronic properties of ITO films coated on glass and Si substrates using RF magnetron deposition .Temperature annealing modifies the properties of the ITO film that affects the final characteristics of devices where they are used. It was discovered that a great change of characteristics is observed by annealing under the temperature of 200°C: the increase of the transmission up to 90% and decrease of film resistance from 160 to 40 Ohm·$\square^{-1}$. The ρ_c is $4\cdot10^{-3}\Omega\cdot cm^2$ for Au and $2\cdot10^{-3}\Omega\cdot cm^2$ for Ag contacts.

1. Introduction

Thin films of indium tin oxide (ITO) are a unique material that combines good electrical conductivity and optical transparency [1-7]. ITO is used as a transparent ohmic contact and current spreading layer in different types of solar cells, LEDs, to make transparent conductive coatings for displays, etc. The ITO also has good chemical stability [5]. In development of devices with using ITO films it is critical to consider the initial properties of the films and their compatibility with the device, as well as the method and conditions of their depositing [1, 2]. All these parameters influence the final characteristics of the device; it is especially important for devices with increased temperature sensitivity (organic LEDs, solar cells based on nanowires, etc.). One of the ways of processing devices is deposition of conducting ITO film without the heating of the sample and modification of the properties of films received by low-temperature heating. This work is dedicated to the research of optoelectronic properties of ITO films under thermal annealing.

2. Experiment

ITO thin films were deposited on glass (2×2 cm^2) and Si(111) substrates by RF magnetron sputtering (BOC Edwards Auto 500 RF) at a power of 50 W and argon pressure of 1.3 mTorr [1]. 3 inch target with a composition of 90% In_2O_3-10%SnO_2 was used. It is deposited without heating the samples. Distance between target and substrate was 100 mm. The sputtering chamber was evacuated to less than $5\cdot10^{-6}$ mbar prior to deposition. All the ITO samples were with film thickness about 100 nm. Film thickness was measured with a profiler AMBiOS XP-1. The surface morphology is studied using scanning electron microscopy (SEM). Same glasses were heated on the hot plate at 100 and 200°C for 1 hour under atmospheric pressure. The resistivity of ITO films was measured with a TLM probe method. Au and Ag contacts were formed by photolithography methods. The optical transmittance of

Content from this work may be used under the terms of the Creative Commons Attribution 3.0 licence. Any further distribution of this work must maintain attribution to the author(s) and the title of the work, journal citation and DOI.

Published under licence by IOP Publishing Ltd

the films was measured using spectrophotometer based on Solar Laser Systems M266 monochromator. Then, the ITO contact was formed and investigated for semiconductor devices: InGaN/GaN LEDs and piezonanogenerators.

3. Results and discussion
3.1. Electrical properties
At an annealing temperature below 200°C, the resistance of ITO layer did not change. As a result of annealing at 200°C, the resistance decreased from 160 to 38 Ohm·$\square^{-1}$. This result correlates with the data for the annealing of films obtained by other deposition methods [5]. We formed two types contacts: samples with Ag or Au contact for ITO. Resistance of Ag/ITO turned out to be less than Au/ITO ($4 \cdot 10^{-3}$ vs. $2 \cdot 10^{-3}$ Ohm·cm^2) and slightly depended on the annealing temperature of the ITO film. This may be due to the higher relative affinity of work function from silver and ITO than to gold (4.62 eV for ITO [3], 4.7 eV for Ag and 4.8 eV for Au). Also note that the measured film resistance for samples with Ag contacts was lower and equalized for both types of metallization after annealing. The increase in annealing temperature may have led to oxygen-deficient films and changing crystyline structure of films [2, 5].

3.2. Optical properties
As can be seen from figure 1 for samples that are not subjected to heat treatment and annealed at 100°C, the optoelectronic properties remained practically unchanged. As a result of annealing at 200°C, the transmission coefficient of the film increases, especially in the short-wave region (350-480 nm) important for blue LEDs. A study of the morphology of the ITO film deposited on Si before and after annealing by the SEM method did not reveal any differences. The film has a weakly expressed texture and microcrystalline structure. In [5-7], a similar change in the optoelectronic properties is explained by a change in the structure of the films. Firstly, the number of oxygen vacancies is changed and consequently changed light absorption and free-electrons. Secondly, the film crystalline structure changes (from amorphous to crystalline). Thus, in order to improve the conductivity and the transmission of the ITO film, it must be annealed at higher than 200°C. Also, based on the results presented in the paper, one can choose the preferred type of metallization.

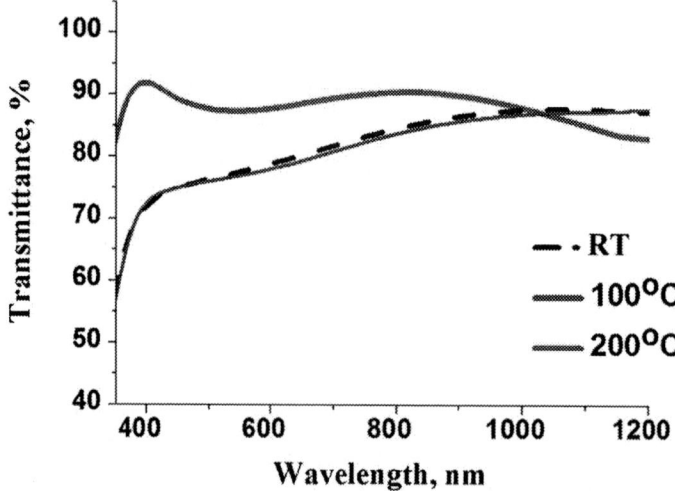

Figure 1. Dependence of the transmission coefficient of the ITO film on the annealing temperature.

3.3. Formation of ITO contact for semiconductor devices
Figure 2 shows the LED design (a) based on planar technology and the transition to three-dimensional nanostructures - nanorods or nanowires (NWs) [8]. When investigating the optoelectronic properties

of such structures, it is necessary to form contacts. One of the ways - the formation of the top contact - coating ITO. The space between NWs is filled with a polymer. Note that such a design is suitable not only for the formation of vertically radiating structures but also for photovoltaic and vertical integrated nanogenerators. The presence of a complex structure implies the setting annealing modes. Therefore, annealing at 200°C is suitable for us. Figure 3 shows SEM images in the cross-section geometry of 100 nm ITO on Si and an example of the top contact of the ITO for the array of GaAs NWs. In the image, you can see the tops of the NWs covered by the ITO sticking out of the ITO layer. This technology is tested and works well for creating test prototypes of various devices.

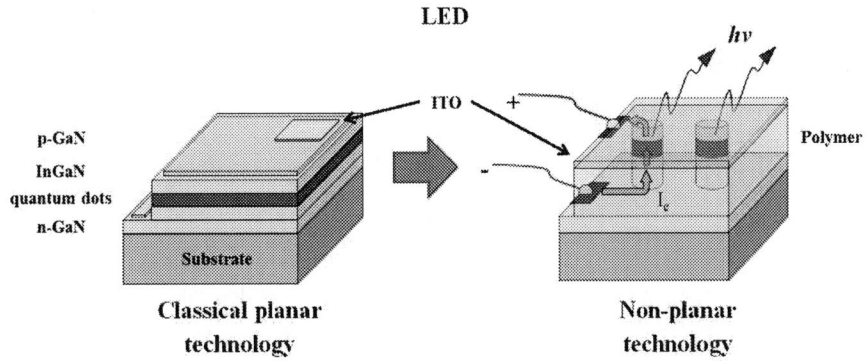

Figure 2. Designs of LED.

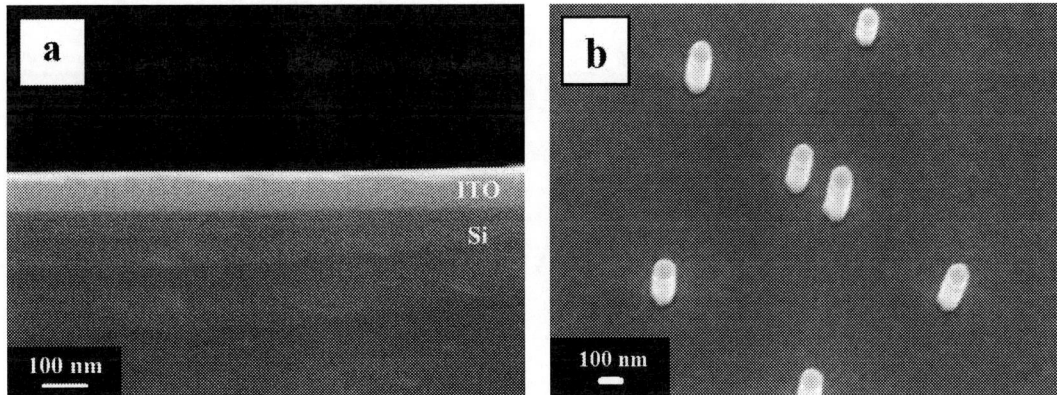

Figure 3(a, b). The cross section **(a)** view of the ITO film; The isometric view **(b)** of the top ITO contact for GaAs nanowire array.

4. Conclusion
The change of characteristics is observed by annealing under the temperature of 200°C: the increase of the transmission up to 90% and decrease of film resistance from 160 to 40 Ohm·$\square^{-1}$. The ρ_c is $4 \cdot 10^{-3} \Omega \cdot cm^2$ for Au and $2 \cdot 10^{-3} \Omega \cdot cm^2$ for Ag contacts. This result can be used for create nanostructure devices with increased temperature sensitivity

Acknowledgments
The study of ITO properties were financially supported by the Russian Foundation for Basic Research (project 18-07-01364 A) and FASIE. The nanowire samples were grown under the support of the Ministry of education and science of Russian Federation (state task, project No. 16.9791.2017/8.9).

References

[1] Kudryashov D, Gudovskikh A, Zelentsov K 2013 *Solid State Phenomena* **200** 10

[2] Yakovlev S P, Soshnikov I P, Korablev V V, Poloskin D S, Yagovkina M A 2004 *J. Surf. Invest.: X-Ray, Synchrotron Neutron Tech.* **4** 57

[3] Sihm K H, Paek M C, Lee B T, Kim C, Kang J Y, 2001 *Appl. Phys. A* **72** 471

[4] Lin Y C, Chang S J, Su Y K, Tsai T Y, Chang C S, Shei S C, Kuo C W, Chen S C 2003 *Solid-State Electron.* **47** 849

[5] Reza Fallah H, Ghasemi M, Hassanzadeh A, Steki H 2007 *Mater. Res. Bull.* **42** 487

[6] Koseoglu H, Turkoglu F, Kurt M, Yaman M D, Akca G K, Gulnur A, Ozyuzer L 2015 *Vacuum*

[7] Reza Fallah H, Ghasemi M, Javad Vahid M 2010 *Renewable Energy* **35** 1527

[8] Tsai Y.-L., Lai K.-Y., Lee M.-J., Liao Y.-K., Ooi B. S., Kuo H.-C., He J.-H. 2016 Prog Quantum Electron. **49** 1

Silicon doping of GaP layers grown by time-modulated PECVD

A I Baranov[1,2], I A Morozov[1], A V Uvarov[1], G E Yakovlev[3], J-P Kleider[2]

[1]St Petersburg Academic University of RAS, 194021 St Petersburg, Russia.
[2]GeePs (Group of electrical engineering – Paris), UMR CNRS 8507, CentraleSupélec, Univ. Paris-Sud, Université Paris-Saclay, Sorbonne Universités, UPMC Univ Paris 06, 91192 Gif-sur-Yvette CEDEX, France.
[3]St Petersburg Electrotechnical University "LETI", 197376 St Petersburg, Russia.

Abstract. We study the doping in GaP layers grown on n-type silicon wafer by time-modulated plasma-enhanced chemical vapour deposition with additional flow of silane. Classical and electrochemical (ECV) capacitance-voltage methods were performed on GaP/Si heterostructures and they demonstrate high electron concentration in the GaP layer and similar profiles. In addition, glow-discharge optical emission spectroscopy revealed high silicon content in GaP, which should be responsible for the detected high n-type doping.

1. Introduction

Nowadays, many efforts are devoted to the fabrication of various optoelectronic devices based on III-V compounds on cheap silicon wafers, for example, multi-junction (MJ) solar cells (SCs). Gallium phosphide (GaP) has only 0.37% lattice mismatch with Si and exhibits good passivation properties [1] so it is one of the most perspective semiconductors for epitaxial growth on Si wafer. Also, the efficiency of single-junction SCs based on the GaP/Si heterojunction should be by 1.1% higher than in a standard silicon diffused homojunction, because the open circuit voltage, V_{OC}, will be higher by 49 mV and could reach up to 0.7 V according to theoretical estimations [2]. This is why a lot of work is being devoted to the development of dislocation-free growth of GaP on silicon wafers [3–7]. However, the vapor-phase and molecular beam epitaxy (VPE and MBE, respectively) methods that are currently used to produce GaP on silicon require high temperatures of 500-800°C leading to unsatisfactory quality of the heterointerface, layers deformation due to differences in thermal expansion coefficients, and deterioration of the bulk properties due to creation of a large number of threading dislocations. In addition, catastrophic drop of minority charge carriers lifetime was observed directly in the silicon wafer due to the annealing procedure of the wafer in the VPE chamber [8]. Also, high temperatures could lead to an inter-diffusion of the III, V and IV elements, during the epitaxial growth process. Note that in this case they are doping impurities for each other. Besides the above mentioned problems, there are economic ones due to strong requirements of ultra-high vacuum in the chamber, ultra-pure source of atoms, more complicated maintenance of equipment, etc. Recently, the novel method of time-modulated plasma-enhanced chemical vapor deposition (PECVD) was proposed and successfully applied to the growth of GaP layers on silicon [9]. It is based on the alternative deposition of Ga and P atoms at low temperature (<400°C). In this work, we present an initial characterization of doping in grown GaP layers by three independent measurement techniques: classical capacitance-voltage (C-V) measurements, electrochemical C-V (ECV) profiling and glow-discharge optical emission spectroscopy (GDOES).

Content from this work may be used under the terms of the Creative Commons Attribution 3.0 licence. Any further distribution of this work must maintain attribution to the author(s) and the title of the work, journal citation and DOI.

Published under licence by IOP Publishing Ltd

2. Experiments and methods

Thin films of GaP with thickness of 50-75 nm were grown at 380 °C on n-type (phosphorus-doped, 2-7 $\Omega \cdot cm$) (100) silicon wafers by time-modulated PECVD methods using an Oxford PlasmaLab System 100 PECVD (13.56 MHz) setup. In this time-modulated method phosphine (PH_3) and trimethylgallium (TMG) were alternatively changed with continuous plasma discharge due to constant hydrogen (H_2) flow during the growth and purge steps. Also, additional steps of silane (SiH_4) flow were introduced as a source of silicon for desirable n-type doping of GaP. The total flow and pressure were kept constant and equal to 100 sccm and 350 mTorr, respectively. Schematic views of the used characterization methods are presented in Figure 1.

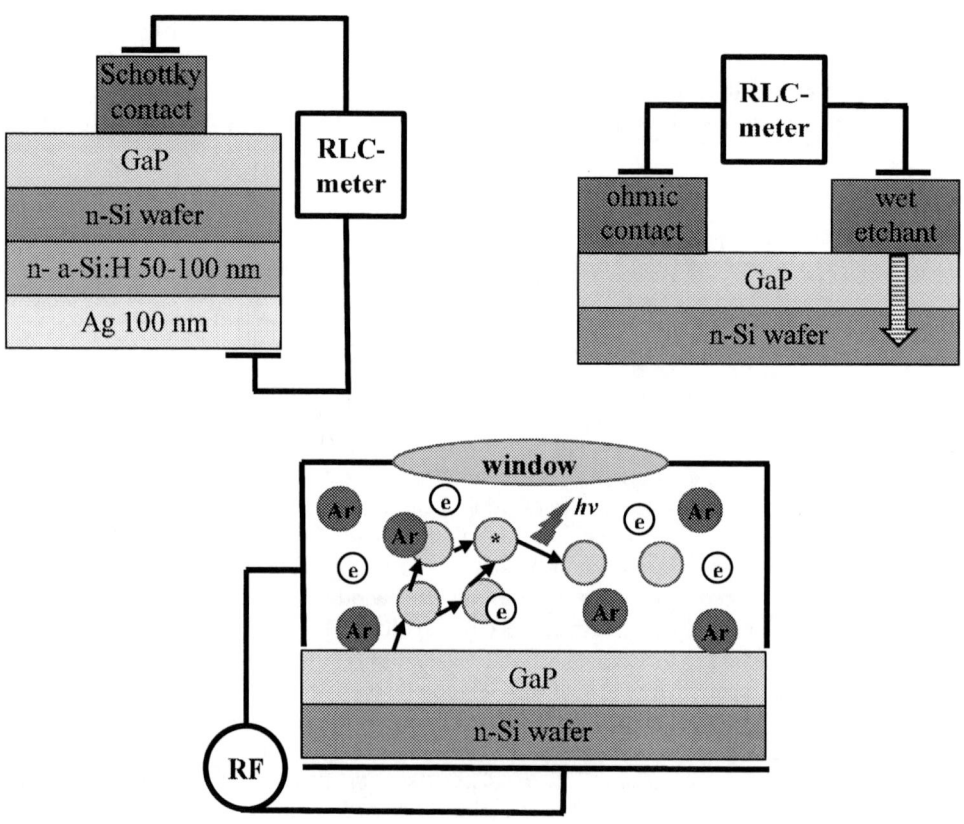

Figure 1 (a, b, c). Schematic view of classical **(a)** and electrochemical **(b)** C-V, and GDOES **(c)** techniques.

Firstly, for classical capacitance measurements we fabricated a Schottky barrier on the front side of the GaP layer from vacuum evaporated gold in a BOC Edwards Auto500 equipment. Then, an ohmic contact to the bottom side of the c-Si wafer was formed by PECVD of n-type (phosphine doped) hydrogenated amorphous silicon, followed by silver deposition and further thermal annealing during 20 min at 170 °C in the air atmosphere. We measured the dependence of the device capacitance on applied reverse bias voltage using an Agilent E4980A RLC-meter [10] (Figure 1a). The obtained C-V profiling is a widely used technique to characterize semiconductor materials and structures with highly reproducible results. However, in some cases these classical C-V measurements do not allow one to obtain high resolution deep in the investigated structure [11]. For this reason, it is sometimes preferable to use a modified C-

V technique, namely the ECV profiling. Here an electrolyte is used as a rectifying contact, which could be used in a classical C-V (depletion) mode, and also for controlled dissolution of the semiconductor that helps to overcome the above mentioned limitations and obtain the full information about the concentration distribution [12]. The schematic view of the ECV technique is presented in Figure 1b. ECV measurements were performed at room temperature using an ECVPro profiler (Nanometrics). The 0.2M NaOH solution was chosen as the electrolyte [13], and the etching current was maintained at a level of 0.5 mA/cm^2. During ECV profiling, the sample was etched gradually with a 1 nm step. Here, also the Agilent E4980A RLC-meter was used [14].

Finally, GaP/n-Si heterojunctions were explored by GDOES [15]. The schematic view of the method is presented in Figure 1c. The sample is placed in a chamber under argon RF-plasma. Argon ions are used to progressively sputter the sample and the removed atoms enter the plasma where they are excited. Then, they relax and emit photons with characteristic wavelengths for each atom, which are registered by the spectrometer. As a result, the profile of atomic content can be obtained.

3. Results

The C-V dependence (not shown here) was measured for Schottky barriers on GaP/n-Si at 1 MHz, and the free charge carrier concentration profile (N_{CV}-W) was estimated (Figure 2a). Firstly, high electron concentration is observed near the surface, which further quickly decreases with depth. Then, it becomes almost constant with a value of 1×10^{15} cm^{-3} that exactly corresponds to the doping concentration of the silicon wafer (2-7 $\Omega \cdot$cm). The estimated high concentration at low depth is corresponding to the free carrier concentration in the GaP layer rather than in silicon since we showed the absence of high phosphorous diffusion into a p-Si wafer during the same growth process of GaP in our previous work [16]. However the absolute value of the depth obtained from C-V measurements is larger than the total thickness of the GaP layer. This discrepancy could be explained by an imperfect bottom contact to the wafer that leads to a shift of the capacitance and thus of the estimated probed depth. However, the GaP layer could be fully depleted and we clearly observe the part of the GaP/n-Si heterojunction profile lying in the wafer.

The results of the ECV profiling technique is shown on Figure 2b. The global shape of the N_{CV}-W curve is similar to that of the classical C-V of Figure 2a, however there are two main differences. The first one is the shift of the depth scale to lower values by 30-50 nm, which can partially confirm our suggestion of imperfection of bottom contact to n-Si for the sample with Schottky barrier explored by the classical C-V. The second one is the peak on the N_{CV}-W curve observed at a value of W=60 nm corresponding to the thickness of GaP and position of the GaP/Si interface [17]. Furthermore, we observe high electron concentration in the GaP layer ($>1 \times 10^{17}$ cm^{-3}) as we assumed. This demonstrates that the used growth process by the time-modulated PECVD method allows one to obtain desirable highly doped n-GaP layers. The ECV method shows a more complete quantitative picture of the concentration profiling in the GaP/n-Si heterojunction compared to the classical C-V measurements in our structures with very thin n-GaP layer.

The results of GDOES profiling for Ga, P and Si atoms are shown in Figure 2c. The GaP layer was removed very rapidly (2 seconds) due to its low thickness, and for larger etching time the silicon response has a constant value indicating that the Si wafer is probed. Despite this fact, we could definitely conclude the existence of a significant silicon concentration in GaP. It should be larger than 10^{18} cm^{-3} but the precise value will be estimated after complete calibration of the used equipment in future experiments. The GDOES spectra demonstrate the possibility of high silicon incorporation in the GaP layers during the described growth process, and it is likely to be the origin of the high electron concentration estimated above from C-V methods.

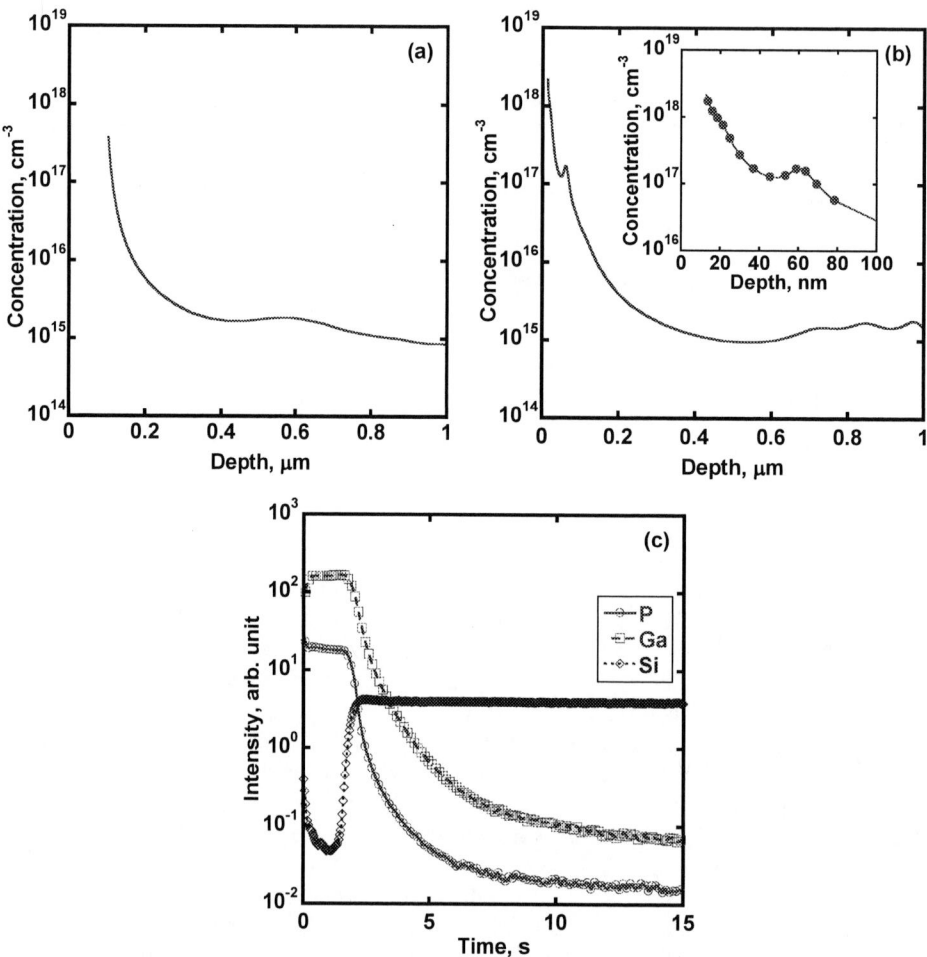

Figure 2 (a, b, c). N_{CV}-W profiling obtained by classical **(a)** and electrochemical **(b)** C-V techniques (the zoomed peak is in the inset), and GDOES spectra of **(c)** of GaP/Si structure.

4. Conclusion

We explored GaP layers grown on n-Si wafers with additional flow of silane by time-modulated PECVD method. The GaP/n-Si heterojunction was explored by classical and electrochemical C-V techniques that showed similar qualitative dependence of free electron concentration on thickness of the space charge region, however the ECV method showed more reliable values for the probed thickness. Both techniques demonstrate that the GaP layer has high electron concentration. In addition, the GDOES technique showed that a high silicon content was incorporated in the GaP layer which is thus likely to be the cause of the high electron concentration detected by the capacitance techniques.

Acknowledgments

This work was carried out in the frame of the PACSiFIC project under financial support of the Russian Foundation for Basic Research (project #16-58-150006) and CNRS (PRC no. 1062). The authors thank Patrick Chapon and Sofia Gaiaschi from HORIBA Scientific for the GDOES measurements.

References

[1] Adachi S 2005 *Properties of Group-IV, III-V and II-VI Semiconductors* (Chichester, UK: John Wiley & Sons, Ltd)

[2] Wagner H, Ohrdes T, Dastgheib-Shirazi A, Puthen-Veettil B, König D and Altermatt P P 2014 *J. Appl. Phys.* **115** 44508

[3] Döscher H, Borkenhagen B, Lilienkamp G, Daum W and Hannappel T 2011 *Surf. Sci.* **605** L38–41

[4] Yamane K, Kawai T, Furukawa Y, Okada H and Wakahara A 2010 *J. Cryst. Growth* **312** 2179–84

[5] Volz K, Beyer A, Witte W, Ohlmann J, Nmeth I, Kunert B and Stolz W 2011 *J. Cryst. Growth* **315** 37–47

[6] Lin A C, Fejer M M and Harris J S 2013 *J. Cryst. Growth* **363** 258–63

[7] Sobolev M S, Lazarenko A A, Nikitina E V., Pirogov E V., Gudovskikh A S and Egorov A Y 2015 *Semiconductors* **49** 559–62

[8] Varache R, Darnon M, Descazeaux M, Martin M, Baron T and Muñoz D 2015 *Energy Procedia* **77** 493–9

[9] Gudovskikh A S, Morozov I A, Uvarov A V, Kudryashov D A, Nikitina E V, Bukatin A S, Nevedomskiy V N and Kleider J 2018 *J. Vac. Sci. Technol. A Vacuum, Surfaces, Film.* **36** 21302

[10] Sze S M and Ng K K 2006 Frontmatter *Physics of Semiconductor Devices* (Hoboken, NJ, USA: John Wiley & Sons, Inc.)

[11] Yakovlev G E, Dorokhin M V., Zubkov V I, Dudin A L, Zdoroveyshchev A V., Malysheva E I and Danilov Y A 2018 *Semiconductors* **52** 873

[12] Yakovlev G E, Frolov D S, Zubkova A V., Levina E E, Zubkov V I, Solomonov A V., Sterlyadkin O K and Sorokin S A 2016 *Semiconductors* **50** 320–5

[13] Lee H J, Kim J H and Lee C H 2013 *ECS J. Solid State Sci. Technol.* **2** R100–4

[14] Frolov D S and Zubkov V I 2016 *Semicond. Sci. Technol.* **31** 125013

[15] Grimm W 1968 *Spectrochim. Acta Part B At. Spectrosc.* **23** 443–54

[16] Gudovskikh A S, Uvarov A V., Morozov I A, Baranov A I, Kudryashov D A, Nikitina E V., Bukatin A A, Zelentsov K S, Mukhin I S, Levtchenko A, Le Gall S and Kleider J-P 2018 *J. Renew. Sustain. Energy* **10** 21001

[17] Kroemer H, Chien W, Harris J S and Edwall D D 1980 *Appl. Phys. Lett.* **36** 295–7

1.06 µm wavelength photodetectors with metamorphic buffer layers grown on GaAs substrates

I V Samartsev, S M Nekorkin, B N Zvonkov, N V Dikareva, A V Zdoroveyshchev, A V Rykov, N V Baidus

Research Institute of Physics and Technology, Lobachevsky State University, Nizhni Novgorod, 603950, Russia

woterbox@mail.com

Abstract. The result of investigation of InGaAs photodetectors grown on GaAs substrate with metamorphic InGaP buffer is shown in this paper. It has been shown experimentally that the use of the gradient InGaP metamorphic layer in an InGaAs photodiode structure improves the spectral characteristics and crystalline quality of the structure and leads to reduction of dark current by an order of magnitude compared to photodiodes fabricated with metamorphic InGaAs layers.

1. Introduction

The InGaAs compound has become the most suitable material for creation of active region for photodetectors at 1.06 µm wavelength during the last 20 years [1, 2]. $In_xGa_{1-x}As$ solid solution with indium content x = 53% is matched to a lattice constant with InP substrates and has bandgap of 0.74 eV. The device can operate at range of wavelengths from 0.9 to 1.7 µm. However, InGaAs/InP heterostructures have a significant disadvantage: InP substrates are expensive and fragile. An alternative heterosystem for the creation of photodetectors operating at the near IR range can be InGaAs structures grown on GaAs substrates [3].

The growth of InGaAs structures on GaAs substrates has specific features. When we increase the content of indium in the $In_xGa_{1-x}As$ solid solution the bandgap of the InGaAs decreases and the of lattices mismatch between InGaAs and GaAs increases. This leads to an increase of elastic stresses in the structure. In the first stages of growth an isomorphous growth of the layer is observed. If a critical thickness is exceeded, a part of the stresses relax with the formation of misfit dislocations [4]. An increase in the number of defects can negatively affect the characteristics of photoelectronic devices. It is proposed to use the perspective method of growth of heterostructures with the use of metamorphic layers to solve this problem.

The structure for a photodetector at 0.94-1.06 µm wavelengths with an InGaAs active region grown on a GaAs substrate with an InGaP buffer layer, which consisted of tens of thin layers with constant composition, was presented earlier [5]. In the case of illumination through a substrate, spectral photosensitivity of such photodetectors is determined from the long-wave side by the width of the bandgap of the solid solution, and from the short-wavelength side by the bandgap of the substrate. The results of investigations of InGaAs photodetectors with metamorphic buffer layers InGaP and InGaAs are presented in this paper. A metamorphic buffer is supposed to reduce the concentration of defects in the active region and, thereby, reduce the dark current of the photodiodes. We also compare the

Content from this work may be used under the terms of the Creative Commons Attribution 3.0 licence. Any further distribution of this work must maintain attribution to the author(s) and the title of the work, journal citation and DOI.
Published under licence by IOP Publishing Ltd

surface roughness of photosensitive structures with the metamorphic layers InGaP, InGaAs, and the electrophysical characteristics of photodiodes fabricated on their basis.

2. Experimental samples

Photosensitive structures were obtained by the method of MOCVD at atmospheric pressure in a horizontal quartz reactor. The structures were grown as follows. A GaAs buffer layer (100 nm) was grown on the GaAs substrate, then a InGaAs metamorphic layer (100 nm) or InGaP (1000 nm) was grown with a stepwise change in composition, then the InGaAs active region (450 nm) was grown (Table 1). The temperature of substrate was 600°C and growth rate of the metamorphic layers was 0.55 nm/sec. The III/V ratio was 8 for the InGaAs metamorphic layer, and 21 for the InGaP metamorphic layer.

Au/Pd/Ti ohmic contacts were deposited on the side of the structure by electron-beam evaporation in a vacuum, then 1 mm mesas were formed by photolithography and chemical etching. Sn ohmic contacts on the substrate side were applied by the method of spark-ignition. The surface of the samples was processed to reactive ion etching in an oxygen atmosphere and to subsequent thermal oxidation in order to minimize the surface dark current.

Table 1. The design of structures.

	Layer	Thickness, nm
	n+-GaAs (substrate)	
1	n-GaAs	150
2	n-InGaAs or n-InGaP	100 or 1000
3	n-InGaAs	150
4	i-InGaAs	450
5	p-InGaAs	300

3. Results and discussion

The morphology of surfaces of produced photosensitive structures with InGaAs and InGaP metamorphic buffer layers was investigated (figures 1 and 2). Surface roughness measurements were carried out using Taylor&Hobson Talysurf CCI 2000 interference microscope. The values of the average surface roughness for structures with InGaAs and InGaP metamorphic buffer layers were 5 nm and 40 nm, respectively. The relief of the structures' surfaces has lines in various directions, which can indicate the formation of dislocations in the top layer. Lines in mutually perpendicular directions are observed on surface of the structure with the InGaP buffer layer.

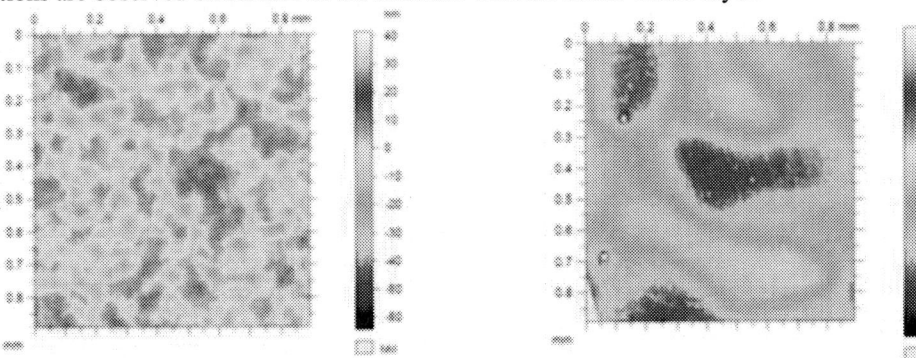

Figure 1. The relief of the structure surface with the InGaAs metamorphic buffer layer

Figure 2. The relief of the structure surface with the InGaP metamorphic buffer layer.

Photo-emf was measured for produced structures at a wavelength of 1 μm. Photo-emf values for structures with InGaAs and InGaP buffer layers were 450 mV and 650 mV respectively.

The spectral characteristics of photocurrent were measured at room temperature. The halogen lamp was used as a source of radiation in the experiment. The modulated lamp light was focused by the collecting lens on the input slit of a monochromator. After passing the monochromator, the light shone on a photodetector sample.

Photocurrent spectra of structures with InGaAs and InGaP metamorphic buffer layers are in the range of 0.9 - 1.05 μm and the maximum photosensitivity is at wavelengths of 1.05 μm and 1.03 μm, respectively (figures 3 and 4). The spectra of sample based on the structure with the InGaAs metamorphic layer have a close to triangular shape, which can be caused by light absorption in the InGaAs metamorphic layer. Samples with the gradient InGaP metamorphic buffer layer have the form more approximate to rectangular as InGaP layer doesn't absorb radiation in the given range of wavelengths.

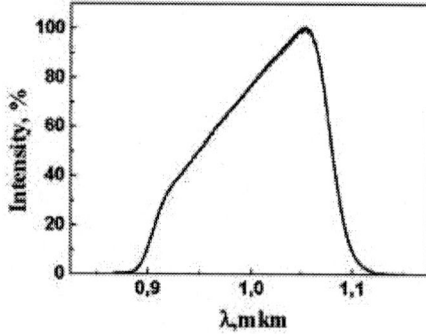 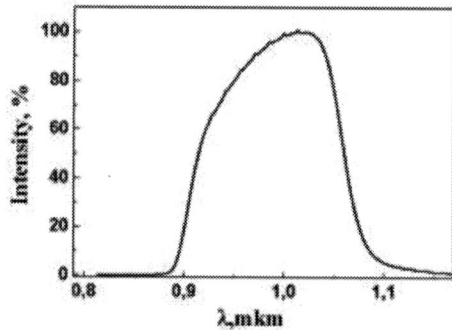

Figure 3. The spectral dependence of the photocurrent of photodiodes fabricated on the basis of heterostructures with the InGaAs metamorphic buffer layer.

Figure 4. The spectral dependence of the photocurrent of photodiodes fabricated on the basis of heterostructures with the InGaP metamorphic buffer layer.

Current–voltage characteristic were measured using Keathley 2440 automated system. Graphs of dark currents of photodiodes with InGaP and InGaAs buffer layers are shown on figure 3. Dark current of photodiodes with the InGaP metamorphic layer is 10 times lower than reverse current of photodiodes made of heterostructures with the InGaAs metamorphic layers. Perhaps this is due to the higher crystalline quality of the active region grown on the InGaP metamorphic layer. Values of dark current of photodiodes made from structures with the InGaP metamorphic buffer layer were about 50 nA at voltage of 3 V.

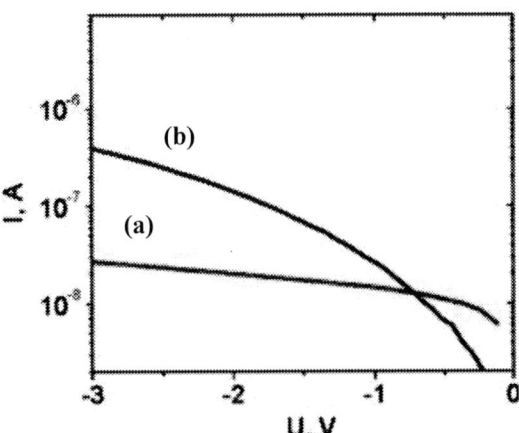

Figure 5. Dark current of photodiodes with metamorphic buffer layers InGaP (a) and InGaAs (b).

4. Conclusion

It was shown experimentally that 1.06 μm wavelength photodiode structures grown by the MOCVD on a GaAs substrate with the gradient InGaP metamorphic buffer layer with have higher surface roughness value than photodiodes of similar design with the InGaAs metamorphic buffer layer. Despite this, studies of dark currents of photodiodes showed that dark current in structures with the InGaP metamorphic layer by an order of magnitude lower than in structures with the InGaAs metamorphic layer. Thus, further structural studies of the photosensitive region are necessary to explain the effect of decreasing dark current in photodiodes with a gradient InGaP buffer layer. At the same time, worse quality of the surface can be associated with dislocations formed during growth of the top layer

We established that photocurrent of structures with the InGaP metamorphic buffer layer is characterized by rapid increase with wavelength, since InGaP transmits radiation in the range of 0.91 ÷ 1.05 μm.

It has been experimentally shown that the use of the InGaP metamorphic layer with a gradient composition is preferable to the InGaAs metamorphic layers for creation of photodiodes with the InGaAs active region for operation at 1 μm.

Acknowledgments

This study was performed within state contract no. 16.7443.2017/BCh of the Ministry of Education and Science of the Russian Federation.

References

[1]   A.M. Filachev, I.I. Taubkin, M.A. Trishenkov 2010 *Current state and main directions of development of solid-state photoelectronics* (Moscow: Fizmatkniga)

[2]   A.M. Filachev, I.I. Taubkin, M.A. Trishenkov 2011 *Solid-state photoelectronics. Photodiodes.* (Moscow: Fizmatkniga)

[3]   L.M. Kanskaya, A.Yu. Kulikov 1995 *Technical Physics Letters* **21** 21

[4]   J.W. Matthews, A.E. Blakeslee 1974 *Journal of Crystal Growth* **27** 118

[5]   B.N.Zvonkov, S.M. Nekorkin, M.V. Karzanova, N.V. Dikareva 2014 *Proceedings of the conference of Yu.E. Sedakov Research Institute* (Nizhni Novgorod: Yu.E. Sedakov Research Institute) p.102

Sunlight simulation in investigating characteristics of small-size sunlight concentrators and multijunction solar cells

E D Filimonov, S A Levina and M Z Shvarts

Ioffe Institute, St. Petersburg 194021, Russia
e-mail: efilimonov@mail.ioffe.ru

Abstract. Presented are technical solutions on constructing optical systems for experimental simulating the sunlight parameters. Designs of two types of solar simulators are proposed. In the first one, the spectrum and the radiation angular divergence are simulated on the basis of a continuous light source, what corresponds to the principle "necessity and sufficiency" of simulated parameters of the light flux in investigating optical and power characteristics of small-size sunlight concentrators (Fresnel lenses). In the second type simulator, based on a pulsed radiation source, the emphasis has been placed on high-quality simulation of irradiance and spectrum for satisfying requirements in recording current-voltage characteristics of multijunction solar cells and in determining their efficiency.

1. Introduction

Development of high-efficient photovoltaics based on the principle of concentrated solar radiation conversion has designated a number of requirements to the simulation of the sunlight parameters under laboratory conditions. So besides exact simulation of irradiance and spectral composition of incident radiation, what is necessary also in investigating of non-concentrator solar cells, the necessity to ensure the radiation divergence of 32 angular minutes in the light flux arises. However, simultaneous simulation of three main parameters "irradiance-spectrum-divergence" on solar simulators (SSs) appears to be rather complicated technical task requiring substantial material expenses. For this reason, depending on the purpose, on application regimes and on the type of measurement tasks being solved, specialized SSs are developed and created.

By virtue of wide application of lens concentrating optics in PV modules with inherent to it chromatic aberration and of multijunction solar cells (MJ SCs) sensitive to spectral characteristics of incident concentrated radiation, it obvious that it is necessary to determine mandatory groups of sunlight parameters for simulation:
- "spectrum + angular divergence" in investigating optical and power characteristics of refracting sunlight concentrators (Fresnel lenses);
- "irradiance + spectrum" in estimating PV parameters of MJ SCs.

Accordingly, in the work proposed are solutions for simulating sunlight parameters under laboratory conditions, which are necessary and sufficient for solving tasks on investigation of optical and power characteristics of lens concentrators and MJ SC I-V characteristics.

Content from this work may be used under the terms of the Creative Commons Attribution 3.0 licence. Any further distribution of this work must maintain attribution to the author(s) and the title of the work, journal citation and DOI.
Published under licence by IOP Publishing Ltd

2. Sunlight simulator for investigating Fresnel lens characteristics

Fresnel lenses are rather complicated objects for experimental estimation of their concentrating capability and optical efficiency.

At sunlight concentrating, occurs spatial (due to radiation angular divergence in the primary light flux) and spectral (due to chromatic aberrations) irradiation redistribution in the focal plane. Thus, in each point of the focal spot, the radiation spectral composition differs from the solar one. The direct influence and the immediate dependence of the concentration ratio, focal spot size and of the FL optical efficiency (capability to focus radiation in a small-size spot) from parameters of radiation incident on the lens surface are obvious. In total, the mentioned above peculiarities of the object under study and of the experimental infrastructure change the radiation power distribution character in the focal plane "blurring" the focal spot.

In simulating the pair of parameters "spectrum + angular divergence" on the SS, one can ensure conditions for correct determination of optical and power characteristics of Fresnel lenses, and obtain true information on both the irradiance distribution profile within the focal spot and the lens efficiency.

The "second", but no less significant, level of requirements includes such parameters of the light flux, as uniformity of irradiance on the tested object surface and long-term stability at the period of the optical and power characteristic recording. Success in producing uniform irradiance will depend substantially on the choice of the initial radiation source and of an optical scheme for forming the light flux. Temporal stability is easily ensured in using standard continuous Xe bulbs and a stabilized power supply source together with irradiance feedback.

To realize successively procedures for recording spatial and spectral power distribution in the FL focal plane, the SLS has been developed, scheme of which is depicted in Fig.1. A high pressure Xe discharge bulb playing a role of the initial light source is placed in the ellipsoidal reflector focus. The choice of the source is determined by a small size of the plasma area within a discharge area, by high luminous efficiency and by the proximity of the radiation spectral composition to the solar spectrum. The light flux is focused on the input of an integrating sphere playing a role of a homogenizer of the initial light flux and, at the same time, of a spatially uniform emitter. Such a constructive layout solution ensures formation of a light flux with heterogeneity of irradiance distribution in the plane of location of the tested lens within 3% (class B in correspondence with [1] by the "nonuniformity" criterion).

Tuning of the light flux spectral composition is provided by optical filters installed above the integrating sphere input port. The estimates of the "color" balance show a possibility to satisfy the "class A" requirements on the "radiation spectral composition" parameter in the range of 400-1400nm in correspondence with [1]. The light flux is focused towards the tested object by a collimating mirror, and the radiation angular divergence (about 32 min. of arc) is created owing to the matching between sphere output port diameter with collimator focal length.

Optical axis tracking along the path "sphere-collimator-object" and then towards the detector registering the irradiance distribution profile is ensured by a laser sight. A PV converter with precise (50-500µm) aperture is used as a profiling detector. This allows receiving the irradiance spatial distribution in the concentrator focus with a high resolution. By virtue of a low level of irradiance on the tested sample and, hence, in the focal spot, registration of the current signal from the PV detector is carried out by lock-in technique, and the obtained results are reduced to spectral irradiance standard conditions ($1000W/m^2$) by normalizing in correspondence with the initial indicators of the light flux incident on a Fresnel lens.

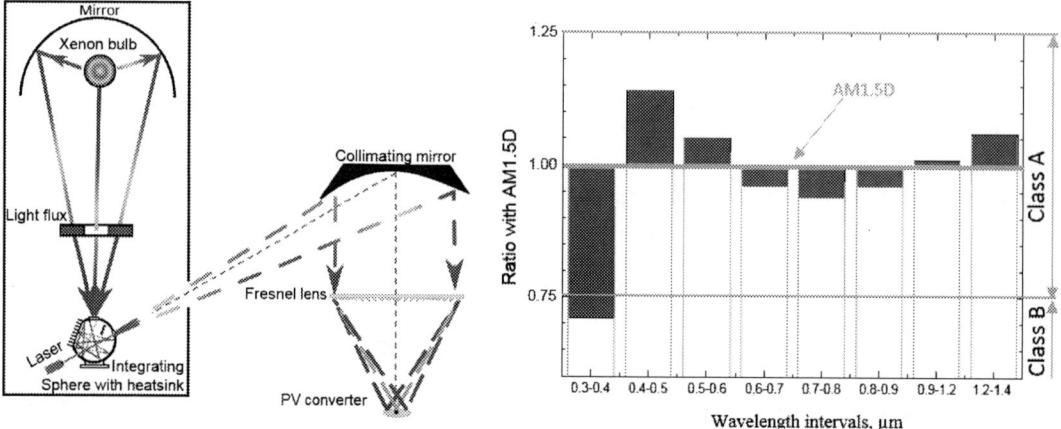

Figure. 1 Optical scheme of OPC measuring installation.

Figure. 2 Compliance of spectral irradiance produced by SS with Class A requirements (Spectral match to AM1.5D spectrum is in 100-nm wide spectral ranges). Spectral measurement results taken with UV/VIS/IR1 Multiple CAS140CT array spectroradiometer from Instrument systems GmbH.

3. Sunlight simulator for investigating MJ SCs

For investigating characteristics of MJ SCs converting the concentrated sunlight, SSs are necessary, in which the main parameter to be simulated is the spectral irradiance conserved for a wide range of variation of the total irradiance (up to 5000 concentration ratio or 500W/cm²). At the same time, it is required to ensure light flux stability within the "flat" part of the pulse, during which I-V characteristic is recorded (not less than 1msec according to [2]). Strict requirements for the correspondence of the radiation angular divergence of 32 angular minutes are not listed. However, in case of difficulties in eliminating spectral deviations in definite wavelength bands, one can use the M-factor method for re-counting and correcting in correspondence with [3].

To register MJ SC I-V characteristics in the mode of concentrated irradiation, a pulsed SS having an option of adjustment of spectral composition has been chosen (Fig.3). Two closely located U-shape Xe bulbs operating in a mode of short single light pulses are used as the initial emitter. Thanks to the use of double-bulb source and a white reflector located behind them, a powerful light flux can be formed, which, in passing through an optical window with a correcting interference light-filter, simulates quite well the sunlight spectral composition. A close mutual location of the bulbs allows realizing additionally an option to tune the spectral irradiance using the principle of radiation reabsorption by one bulb in the gas discharge plasma of the second one, and vise a versa. In increasing voltage on the bulbs power supply, variation of the efficiency of such absorption takes place: the "blue/red" ratio in the radiation spectrum shifts towards increasing a portion of "blue" light at partial smoothing out the xenon peaks. Thus, precise realignment of the "blue/red" ratio in the light flux is ensured. As a summary, spectral mismatch in the light flux with respect to standard terrestrial conditions (AM1.5D) in the range of 0.4 – 1.2 µm is presented in Fig.4. It is seen that, in increasing voltage on the bulbs, improvement of "quality" of the radiation spectral composition to the "class A" takes place (at voltages on the bulbs of 700-800V) [1].

As the distance between the cell and the illuminator decreases (in moving the latter along vertical direction), the irradiance level on SC increases. Control of the radiation concentration is performed by a single-junction GaAs photo-detector with a known dependence of the photocurrent on illumination determined by the procedure described in [4, 5]. The monitor photo-detector was located in the vicinity of the tested MJ SC to ensure similar irradiation conditions. An electronic system for controlling the SS ensures recording the MJ SC I-V curve during the "flat" part of the light pulse.

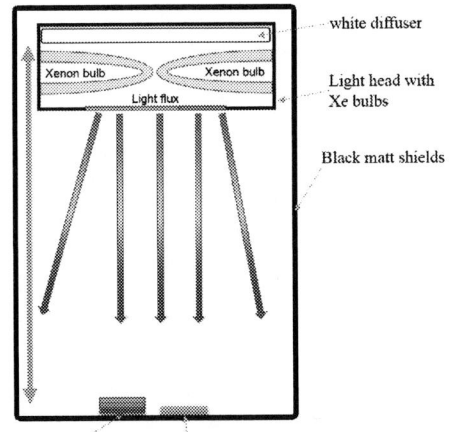

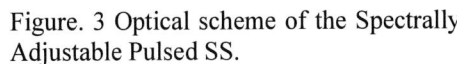

Figure. 3 Optical scheme of the Spectrally Adjustable Pulsed SS.

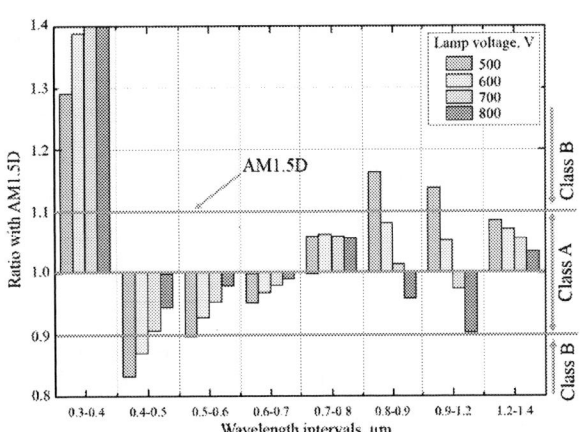

Figure. 4 Spectral irradiance produced by Spectrally Adjustable Pulsed SS with respect to Class A requirements (Spectral match to AM1.5D spectrum is in 100-nm wide spectral ranges).

4. Conclusions

The Solar Simulators presented in the work allow to simulate sunlight parameters required for solving specific research tasks. To investigate spatial and spectral power distribution in the focal plane of sunlight concentrators, one of the developed simulator reproduces two necessary parameters – angular divergence and spectral composition. In this case, the requirements for irradiance uniformity on the surface of the investigated concentrator and long-term stability of incoming light during the full cycle of the optical-power characteristic registration remain to be high. The second simulator is focused on investigating MJ CS I-V characteristics. Such a device having in basis a pulsed radiation source allows simulating the spectral distribution in the light flux with tuning in a wide range of values the total irradiance (up to 500X) in the plane of the tested sample location.

Acknowledgments

The reported study was funded by RFBR according to the research project № 18-38-00697

References

[1] International Standard. Photovoltaic devices - Part 9: Solar simulator performance requirements IEC 60904-9:2007.

[2] International Standard. Photovoltaic devices - Part 1: Measurement of photovoltaic current-voltage characteristics IEC 60904-1:2006.

[3] Carl R. Osterwald and Gerald Siefer, Chapter 10 "CPV Multijunction Solar Cell Characterization" in Handbook of Concentrator Photovoltaic Technology edited by C. Algora, I. Rey-Stolle (John Wiley & Sons, Ltd, 2016), pp. 589–614.

[4] M.Z.Shvarts, V.M. Emelyanov, N.Kh.Timoshina, V.M.Lantratov "Nonlinearity Effects in III-V Multi-Junction Solar Cells", Proceedings of the 34th IEEE Photovoltaic Specialists Conference, Philadelphia, PA, June 7-12, 2009, pp. 1412-1417.

[5] M.Z.Shvarts, E.D.Filimonov, S.A.Kozhukhovskaia, M.A.Mintairov, N.Kh.Timoshina, V.M.Andreev "Current Mismatch Violation in Concentrator Multijunction Solar Cells", AIP Conf. Proceedings, v.1881, 2017, pp. 040006.

A novel approach to characterization of bottom sub-cell in multijunction solar cell using photoluminescence.

K S Zelentsov[1], A S Gudovskikh[1,2], D A Kudryashov[1], N V Kryzhanovskaya[1], N A Kalyuzhnyy[3], S A Mintairov[3]

[1] St. Petersburg Academic University, 194021, St. Petersburg, Russia
[2] St. Petersburg Electrotechnical University "LETI", 197376, St. Petersburg, Russia
[3] Ioffe Institute, 194021, St. Petersburg, Russia

e-mail: veorg00@gmail.com

Abstract. Multijunction solar cells grown on group IV substrates using III-V compounds are very promising due to their high efficiency (over 40%). Further development of these structures is mainly focused on III-V layers while bottom sub-cell based on p-n junction in the substrate is disregarded. However, high temperatures required for III-V epitaxy as well as diffusion over III-V/IV interface may affect minority carrier lifetime in the substrate. GaAs/GaInP/(AlAs)/Ge heterostructures were grown by MOCVD. Thermal annealing at the conditions of triple-junction solar cells was performed and photoluminescence spectra for these structures were investigated.

1. Introduction

Group IV substrates are widely used for photovoltaic applications because of their excellent electrical and mechanical properties. Multijunction III-V solar cells grown on Ge substrates are of the great interest due to their supreme efficiency that has already exceeded 46% [1]. The development of these structures is mainly focused on the optimization of III-V layers parameters while germanium sub-cell is disregarded. However, Ge sub-cell contributes up to 10% of total solar cell efficiency [2] but the properties of III-V interface are not studied enough. The p-n junction in bottom sub-cell is conventionally formed by the diffusion during epitaxial growth of III-V layers, thus the parameters of the sub-cell are dependent on the temperature and growth time. III-V epitaxy and especially top sub-cell based on GaInP requires high temperatures so the parameters of III-V/Ge interface alter during subsequent layers growth. Accordingly, direct characterization becomes complex because of the number of layers grown. We have previously shown that diffusion of group III atoms during epitaxial growth of III-V layers leads to the formation of undesirable potential barrier for charge carriers at the III-V/IV heterointerface [3]. Thus, charge carrier lifetime which depends on impurity level can be also affected by diffusion processes. In this paper the influence of III-V epitaxy on the minority carrier lifetime in Ge substrates is discussed.

2. Samples

Structure that represents bottom sub-cell (figure 1a) of multijunction solar cell was fabricated using Aixtron 200 MOCVD equipment. P-type Ga-doped ($N_A = (2\text{-}4)\cdot 10^{17}$ cm^{-3}) germanium substrates <100> with 6° offcut toward <111> plane were used to ensure APD-free growth of epitaxial layers. Epitaxy of germanium can lead to a high level of background doping, thus 100 nm n-GaInP layer was grown on the substrate. The p-n junction was formed by the diffusion of P atoms into the substrate during epitaxial growth. The structure was capped with 300 nm n$^+$-GaAs contact layer.

In order to suppress the simultaneous diffusion of group III atoms over III-V/IV interface another structure with thin AlAs layer was grown (figure 2b). AlAs is reported to be an efficient diffusion barrier [4]. The p-n junction in Ge is formed by As diffusion during AlAs nucleation process. High value of

Content from this work may be used under the terms of the Creative Commons Attribution 3.0 licence. Any further distribution of this work must maintain attribution to the author(s) and the title of the work, journal citation and DOI.
Published under licence by IOP Publishing Ltd

conduction band offset (ΔE_C = 0.3-0.5 eV) [5] at the AlAs/Ge interface may affect charge carrier transport. AlAs layer was heavily doped with Te ($N_D > 10^{19}$ cm^{-3}) to ensure tunnelling through the barrier. This structure was also capped with 300 nm n$^+$-GaAs contact layer.

(a)   (b)

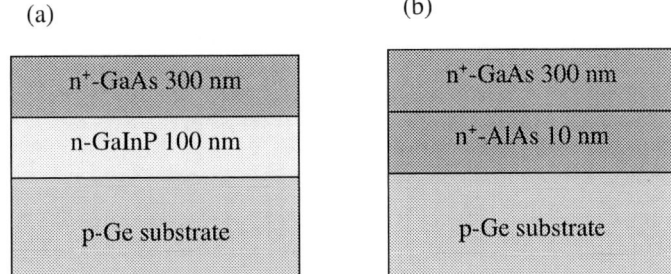

Figure 1. Schematic representation of conventional Ge sub-cells (a) and the sub-cell with AlAs diffusion barrier (b).

Both structures with and without AlAs layer exhibit close values of V_{OC} = 0.2V, however GaAs/AlAs/Ge demonstrate lower value of I_{SC} (figure 2). The diffusion coefficient of As in Ge is higher than that of P [6] which results in thicker emitter and thus higher recombination rate of photogenerated electron-hole pairs.

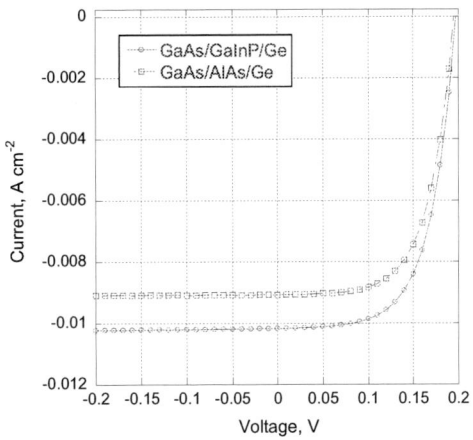

Figure 2. I-V curves for Ge sub-cell and the sub-cell with AlAs diffusion barrier.

3. Experiment

For the study of solar cells, non-destructive research methods are preferable, since it is possible to control parameters of interest of the same structure during epitaxial growth. Minority carrier lifetime in Ge substrate can be estimated by different measurement techniques such as QSSPC [7], however the results obtained by it are not reliable at high doping levels ($>10^{17}$) of the substrates. Also, the top side of the substrate near III-V/IV heterointerface is of the most interest so the photoluminescence technique was used to investigate grown structures.

To verify the applicability of selected method a PL spectrum (figure 3a) was obtained for clean Ge substrate using 914nm laser diode. The substrate was preliminary passivated using HF to reduce recombination losses and increase PL intensity. PL peak was observed with a maximum around 1600 nm. This behaviour is related to InGaAs photodetector which has an absorption edge at 1800 nm. To demonstrate the possibility of obtaining a signal from a substrate in structures with grown III-V layers, thick 2000 nm layer of GaAs was formed on Ge substrate and PL spectrum was measured (figure 3b). Germanium peak was detected although it has lower amplitude. The signal in range of 1000-1400 nm was observed, which is related to defects in GaAs.

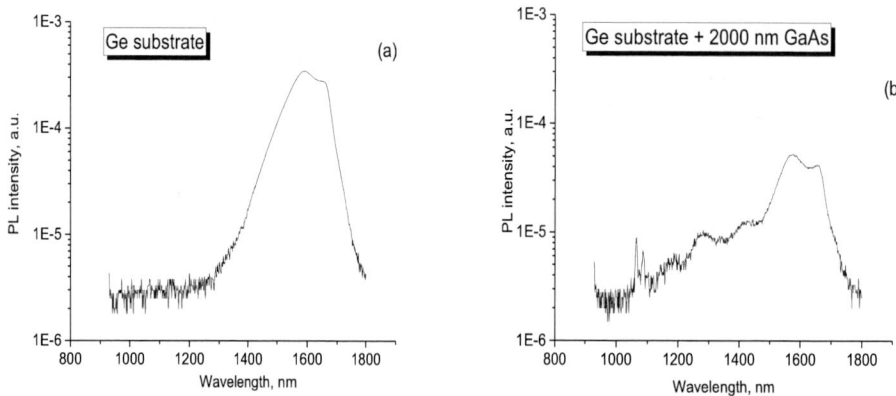

Figure 3. Photoluminescence spectra of Ge substrate (a) and GaAs/Ge structure (b).

4. Results

Photoluminescence spectra measured for germanium sub-cells described in section 2 demonstrate the similar behavior with a slightly higher intensity for GaAs/AlAs/Ge structure (figure 4a) which can be explained by a smaller total thickness of III-V layers. Thus, no difference between the effect of P and As diffusion on minority carrier lifetime in the substrate was observed. However, high temperatures required for epitaxial growth of subsequent sub-cells of multijunction solar cells affect the diffusion over III-V/IV interface. To investigate the effect of triple-junction GaInP/GaAs/Ge solar cell growth in MOCVD process on substrate lifetime test structures were annealed under corresponding conditions (T = 500..700°C, t > 1 hour). No significant change in PL intensity for conventional Ge sub-cell after thermal annealing was observed. However, the structure with AlAs barrier layer demonstrated PL intensity from Ge substrate and therefore a charge carrier lifetime drastically reduced. Presumably, this is caused by dopant dispersion from AlAs layer into the substrate.

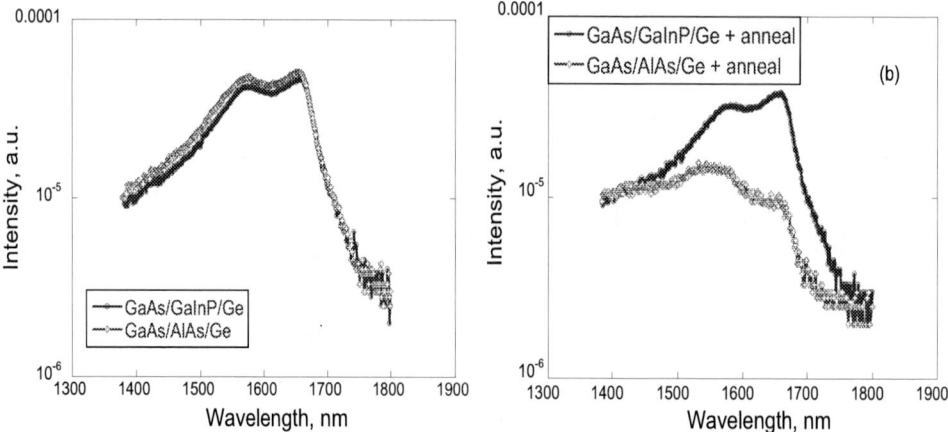

Figure 4. Photoluminescence spectra for Ge sub-cells before (a) and after (b) thermal annealing.

5. Conclusion

A novel approach to characterization of bottom sub-cell in multijunction (GaInP/GaAs/Ge) solar cells was proposed. The ability to detect signal from germanium substrate through epitaxial III-V layers was demonstrated. The results on minority carrier lifetime obtained using photoluminescence technique were discussed.

6. Acknowledgments

This work was supported in part by the Russian Foundation for Basic Research under grant number 17-08-00474.

References

[1] M.A. Green *et al*, *Prog. Photovolt. Res. Appl.* 23, 805, 2015
[2] N.A Kalyuzhnyy *et al*, *Semiconductors* 44(11) : 1520-1528, 2010
[3] A.S. Gudovskikh *et al*, *J. Phys. D: Appl. Phys. 45 495305*, 2012
[4] C. K. Chia *et al*, *Appl. Phys. Lett.* 92 141905, 2008
[5] A. Mujica *et al*, *Phys. review. B, Condensed matter* 11, 1992
[6] F.A. Trumbore, *The Bell System Technical Journal*, vol. 39 issue 1, pp 205-233, 1960
[7] R.A. Sinton, *Proc of the 25th IEEE photovoltaics specialists conference*, pp 457-460, 1996

SPBOPEN 2018   IOP Publishing

Numerical simulation of the carbon nanotubes transport layer influence on performance of GaAs solar cell

S A Raudik[1], A M Mozharov[1], D M Mitin[2], A D Bolshakov[1], P M Rajanna[3], A G Nasibulin[3] and I S Mukhin[1,4]

[1]St.Petersburg Academic University, Khlopina 8/3, 194021 St. Petersburg, Russia
[2]Ioffe Institute, 194021 St. Petersburg, Russia
[3]Skolkovo Institute of Science and Technology, 121205 Moscow, Russia
[4]ITMO University, 197101 St. Petersburg, Russia

E-mail: raudik_sa@mail.ru

Abstract. In this paper we present the results on simulation of the photovoltaic properties of the conventional one junction GaAs solar cell modified with a transparent transport layer. The use of the transport layer allows reduction of the top layer lateral resistance leading to rise of the cell efficiency. In the modeling, we considered three different materials of the transport layer namely, ITO, AZO and carbon nanotubes. The optimum values of the transport layer thickness and distance between the metallic grid bars corresponding to the highest theoretical efficiency are obtained. With light concentration, 17.3%, 10.5 % and 15.1% efficiency was reached for SC with CNT, ITO and AZO transport layers respectively.

1. Introduction

High cost of the solar cells (SCs) based on expensive substrates such as Ge and GaAs synthesized via MOCVD or MBE dictates necessity of the light concentration system use in order to reduce area of the photoactive element. In addition, 10-100 times increase of the luminous flux allows rise of the solar cell efficiency due to increase of the cell nominal voltage [1]. The light concentration requires optimization of the carriers collection in the top grid contact and improved heat dissipation.

Lateral spreading of the carries in the SC takes place mainly in the upper emitter layer and wide band gap window layer. Correspondingly, resistances of the latter layers play the main role in the total resistance of the cell, limitation of fill factor and SC current. Conventional method for improvement of the ohmic losses in SC upper layers is fabrication of the dense metal contact grid on the SC surface allowing carriers collection from these layers. However, the metal grid is not optically transparent limiting the radiation absorption. An alternative approach is fabrication of additional transparent layer over the wide band gap window providing high electrical conductivity such as graphene [2]. Use of this layer allows to increase distance between the grid bars and consequently to improve transparency of the contact layer.

In this report, we study theoretically the effects of different material transport layers used in conventional GaAs cell on the SC photovoltaic properties.

Content from this work may be used under the terms of the Creative Commons Attribution 3.0 licence. Any further distribution of this work must maintain attribution to the author(s) and the title of the work, journal citation and DOI.
Published under licence by IOP Publishing Ltd

2. Calculation

In our simulations we considered AlGaAs/GaAs heterostructure SC, synthesized on 300 µm thick GaAs substrate. The calculations were carried out in Silvaco TCAD software package. Material parameters (carriers life time and mobility) were approximated according the concentration dependent. Shockley-Read-Hall model [3] [4] and Lombardi constant voltage and temperature (CVT) model [5] respectively. We also considered radiative and Auger recombination [6].

We performed the calculation with a different top transparent conductive layers. Considered materials for the top conductive layer were carbon nanotubes (CNT), indium tin oxide (ITO) and aluminum zinc oxide (AZO). ITO and AZO was approximated as a semiconductor with known parabolic carriers dispersion law. Linear dispersion law typical for graphene was taken for CNT. Carbon nanotubes were considered a three-dimensional semiconductor with dummy parameters. The parameter values of the materials are shown in Table 1. In order to calculate the light dissipation in the considered layer we took into account refractive index and light absorption coefficient of the materials. The data was taken from [7], [8] and [9].

Table 1: Model parameters of the transport layer materials

	CNT	ITO	AZO
Electron affinity	4.2 eV	4.7 eV	4.3 eV
Bandgap	10 meV	4.0 eV	3.37 eV
Electron density of states	8.75×10^{15} cm^{-3}	2.5×10^{19} cm^{-3}	2.5×10^{19} cm^{-3}
Hole density of states	8.75×10^{15} cm^{-3}	8.87×10^{18} cm^{-3}	8.87×10^{18} cm^{-3}
Base doping level	6.4×10^{17} cm^{-3}	4.5×10^{19} cm^{-3}	2.0×10^{19} cm^{-3}
Doping type	p-type	n-type	n-type
Permittivity	15.0	11.1	11.1

Increase of the transport layer thickness leads to decrease of the SC resistance but, on the other hand, leads to rise of the unwanted light absorption in the transport layer. Due to the latter facts, an optimal thickness of the transport layer exists. The relation between transparency and the sheet resistance for the considered materials is presented in Figure 1.

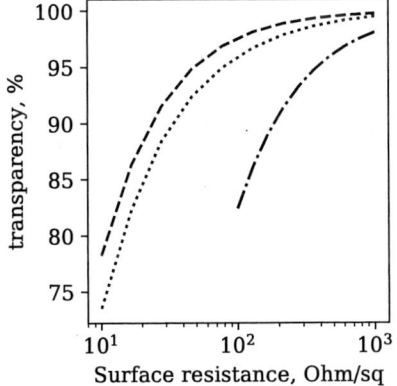

Figure 1: Relation between transparency and materials sheet resistance

We considered the SCs without antireflection coating since the studied effects relates to conduction properties of the transport layers. Consideration of the anti-reflection coating is a simple numerical problem that was out of the investigation scope.

Due to n-type conductivity of ITO and AZO and p-type conductivity of the CNTs we modeled SCs with n-type window for ITO or AZO and p-type window with CNTs layer.

In the first part of the modeling we optimized thickness, doping level and composition of the one-junction SC layers without the transport layer. Our simulations showed that SCs with n-doped emitter and without it have similar efficiencies. We assume that this effect relates to the absence of lateral current. The design of the optimized structure is presented in tables 2-3.

Table 2: Structure parameters for CS with CNT.

Layer		Material	Thickness	Doping type	Doping level
4th	Transport layer	CNT			
3rd	Window	$Al_{0.5}Ga_{0.5}As$	20 nm	p-type	5×10^{18}
2nd	Base	GaAs	2.6 μ	n-type	2×10^{16}
1st	BSF	$Al_{0.3}Ga_{0.7}As$	20 nm	n-type	5×10^{18}
	Substrate	GaAs	300 μ	n-type	1×10^{18}

Table 3: Structure parameters for CS with ITO and AZO.

Layer		Material	Thickness	Doping type	Doping level
4th	Transport layer	ITO or AZO			
3rd	Window	$Al_{0.5}Ga_{0.5}As$	20 nm	n-type	5×10^{18}
2nd	Base	GaAs	2.0 μ	p-type	6×10^{16}
1st	BSF	$Al_{0.3}Ga_{0.7}As$	20 nm	p-type	5×10^{18}
	Substrate	GaAs	300 μ	p-type	1×10^{18}

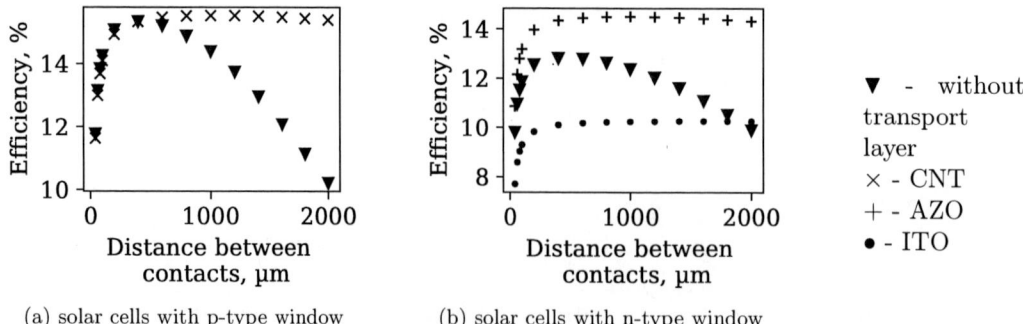

(a) solar cells with p-type window   (b) solar cells with n-type window

Figure 2: Efficiency depending on distance between metal contacts in case of different transport layers

After the design optimization, we performed simulations of the SC with the different transport layers and without it varying the distance between the contact bars. The results of the calculations are shown in Figure 2. For the SC without the transport layer, an optimum distance of about 500nm for both n- and p-type window SCs corresponding to the maximum efficiency exists. The latter phenomenon relates to influence of two opposite effects: increase of ohmic losses with the distance and corresponding drop of the light reflection from the contacts.

In the SCs with the transport layer, ohmic losses in the upper layers are insufficient so efficiency does not drop with the distance between the bars. In the further simulations of the SCs with the transport layer, the distance between the contacts was set to 1300 µm corresponding to the SC efficiency saturation.

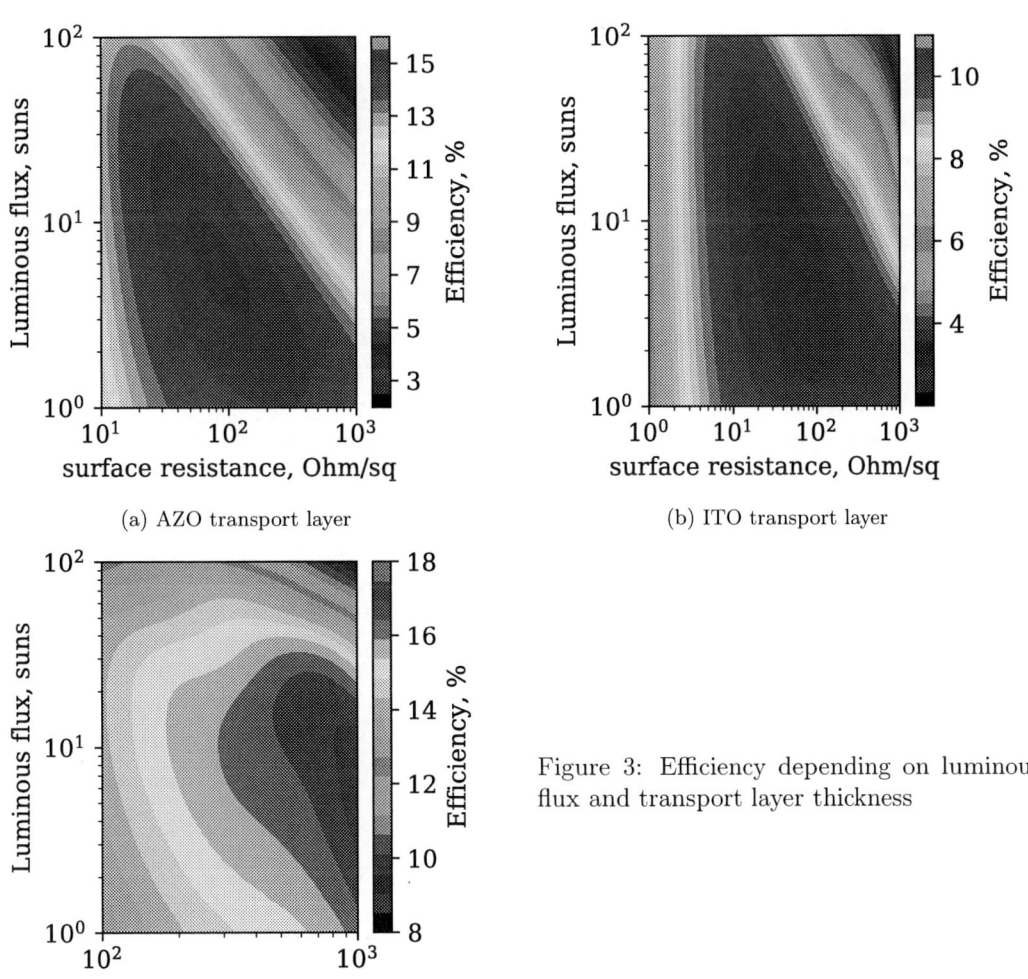

(a) AZO transport layer

(b) ITO transport layer

(c) CNT transport layer

Figure 3: Efficiency depending on luminous flux and transport layer thickness

In the final part of our investigation, we simulated the efficiency of the SCs with different transport layers under the concentrated light. The numerical modeling results of the SC efficiency dependence on the light concentration and sheet resistance of the transport layer are shown in Figure 3. Due to monotonic relation between the sheet resistance and the layer thickness an optimized value of the latter can be obtained. Solar cells with n-type window (ITO and AZO) demonstrate worse efficiency due to lower hole mobility in the p-type substrate. Worth noting, that in case of ITO transport layer, the major factor limiting the efficiency was high electron affinity of the ITO and consequently existence of the potential barrier between the transport layer and window.

3. Summary

In this work, we carried out the numerical simulations of the conventional GaAs SC enhanced with the additional transport layer placed at the top surface of the SC. The optimal thickness of different materials conductive layer, as well as distance between the metal grid bars have been calculated. Without the light concentration, efficiency grows from 12.8 % to 14.5% when AZO is added and from 15.3% to 15.5% when CNTs layer is added. The maximum efficiency for the SC with ITO transport layer was 10.3%. With light concentration, 17.3%, 10.5 % and 15.1% efficiency was reached for SC with CNT, ITO and AZO transport layers respectively. Use of the transport layers allows increase of the distance between the contact bars and their width. Consequently, this approach is assumed very promising for the factory application due to possibility of the screen printing use instead of the lithography methods.

Acknowledgments

The work was supported by the Skolkovo Foundation (grant agreement for Russian educational and scientific organisation no.7 dd. 19.12.2017) and Skolkovo Institute of Science and Technology (General agreement no. 3663-MRA dd. 25.12.2017)

References

[1] MacMillan H F et al. 1988 *IEEE Photovoltaic Specialists Conf. (Las Vegas)* 462–8
[2] Marinho B et al 2012 *Powder Technology* **221** 351–58
[3] Shockley W and Read W T 1952 *Phys. Rev.* **87** 835–42.
[4] Hall R N 1952 *Phys. Rev.* **87** 387.
[5] Lombardi C et al 1988 *IEEE Trans. on CAD* **7** 1164–71
[6] Selberherr S 1984 *Analysis and Simulation of Semiconductor Devices* (New York: Springer-Verlag Wien)
[7] G Xiao G, Tao Y, Lu J and Zhang Z 2010 *Proc. 3rd International Nanoelectronics Conference (INEC)* p 208–9
[8] A. F. Knig et al 2014 *ACS Nano* **8(6)**, 6182-92
[9] Treharne R E et al 2011 *J. Phys.: Conf. Series* **286** 012038

SPBOPEN 2018

IOP Publishing

IOP Conf. Series: Journal of Physics: Conf. Series **1124** (2018) 041041   doi:10.1088/1742-6596/1124/4/041041

Investigating the output characteristics of Ring Laser in Laser Dynamic Goniometer

N.A. Eno[1], P.A. Pavlov[1], J.T. Bagshaw[1]

[1]Laser measurement and Navigation System Dept., Saint Petersburg Electrotechnical University "LETI", Saint Petersburg 197376, Russia.

Abstract: Lately, it becomes a pressing need for increasing the accuracy in measurements and calibration of optical polygon using laser dynamic goniometer, therefore the need to understudy the output characteristics of Ring laser which is used as the reference scale during this measurement of refractive indices, calibration of optical polygon and in building of laser dynamic goniometer using high precision goniometer along with the optical angle encoder. This paper covers algorithms, procedures of operation, and results of the investigations, as well as the calculations of coefficient characterizing the null shift of ring laser output, the scale factor, coefficient characterizing the nonlinearity of ring laser output and graph of dependency. This article also shows that, the algorithm for compensating of the ring laser generalized zero shifts increases the accuracy of the angular measurements and makes it possible to determine the coefficients of the ring laser output characteristic.

1.Introduction

RL is He-Ne laser of GL-1 (He-Ne laser plasma tube of type '*GL*-111' made in the former USSR. Helium Neon Gas Laser with 12V DC Power Supply) type in a ring form, despite its name, it's often in square form or triangular cavity, two light waves of the same wavelength traveling in opposite direction to each other is generated in RL and interference pattern is observed that is used to determine rotational speed; RL is use as a reference scale in the building of LDGs due to its high rate of accuracy and speed of operations. LDG is a device used for calibration of angles, measurement of Refractive indices [2] and so on, in general, our final objective is creating possible methods of increasing accuracy of LDG, as it has been seen that the use of gyros reduces the time of measurement and increases accuracy, reliability and measurement re-productivity [3-6]. Systematic errors of RL make up the systematic errors encountered during operation in LDG, hence a strong need to investigate RL output characteristics. Using high precision DG made it possible for this understudy [1-7]. Algorithms and equation were created following procedures based on the already created algorithms. processing of measurements results using a special software like Mathlab, originlab etc., in order to obtain the respective values of the three component of the output characteristics of RL which are: coefficient characterizing the null shift of the RL output characteristic, scale factor, coefficient characterizing the nonlinearity of the RL output and plotting a dependency graph of N2π (RL pulse) against T (time of RL complete revolution)

2. Methods

Applying a D.C voltage to the motor of the DG enable the spindle to rotate, note that the speed of rotation depends largely on the voltage. The spindle of a dynamic goniometer rotates in bearings with optical angle encoder and RL attached to it. Spindle is brought into rotation by a drive consisting of a motor and a motor control system. output signals from the optical ngle encoder and RL are fed to the electronics unit that performs pre-processing of data and their transfer to the computer. The data obtained from the computer are post processed to generate two columns of data consisting of time recorded by the counter and the RL pulses. These two columns is use to calculate rotation of angles. The investigation took place within the angular velocity of rotation range of 360degs clockwise and counterclockwise, it was carried out with the period of 25 revolutions. The output signal from optical encoder passed through the counting frequency divider and produced the RL output signal phase reading pulse (*Ni*) and time moment reading pulse (*ti*). Readable pulses were produced at the interval of 1 deg. The measurements were taken within the period of not less than 10 rotations of RL. Nonlinear output characteristics decrease with increase of the speed of revolution.

Content from this work may be used under the terms of the Creative Commons Attribution 3.0 licence. Any further distribution of this work must maintain attribution to the author(s) and the title of the work, journal citation and DOI.

Published under licence by IOP Publishing Ltd

3. Algorithm and Equation

From the RL pulling zone or lock in zone, the frequency of the output RL signal is proportional to the input as written below:

$$v(t) = K_1 \Omega + K_0 + K_{-1}/\Omega \cdot E(t) \tag{1}$$

where K_0 is the coefficient characterizing the null shift of the RL output characteristic,

k_1 is the RL scale number, K_{-1} is the coefficient characterizing the nonlinearity of the RL output characteristic, Ω and $\Omega^*{}_E$ are RL rotation velocity in the laboratory system of coordinates and the vertical component of the Earth rotation velocity respectively.

On measuring the angular output signal of RL periods which integrated within time interval formed by output pulses of optical encoder. Integrating (1) gives the number of period Ni in the time interval from 0........ ti

The angle φ_i which RL rotates can be written as:

$$\varphi_i = 2\pi N_i / N_{2\pi} \tag{2}$$

Where $\quad N_i = \dfrac{1}{2\pi} \int_0^t v(t)\partial t; \quad$ and $\quad N_{2\pi} = \dfrac{1}{2\pi} \int_0^t v(t)\partial t$

N_i and $N_{2\pi}$ are the numbers of periods in the RL output signal occurring in a time interval defined by the angle and during a complete rotation respectively;

$v(t)$ is the output signal frequency.

t = time of measurement of angle φ_i

2π is the number of RL output signal periods within the complete revolution.

$$\int_O^T V dt = \int_O^T K_1 \Omega dt + K_O \int_O^T dt + \int_O^T \frac{K_{-1}}{\Omega} dt \tag{3}$$

From equation (2)

$$\varphi_i = \varphi_i + \frac{k_0 + k_1 \Omega^*{}_E}{K_1} X(\frac{\omega t - \omega i}{\omega t \omega i}) \varphi_i + \frac{k_{-1}}{k_1}(Ji - Jt\frac{\varphi_i}{2\pi}) \tag{4}$$

But $\quad Ji = \int_O^{ti} \dfrac{dt}{\omega(t) + \Omega^*{}_E} \tag{5}$

$$\phi_i = \omega i.ti$$

Where ti is ϕ_i measurement time

Component proportionality to the vertical component of the Earth rotation velocity is the major contribution

$$\Delta\phi_i = \overset{-}{\phi}i - \phi i = \frac{k_0 + k_1 \Omega^*{}_E}{K_1} X(\frac{\omega t - \omega i}{\omega t \omega i})\phi_i + \frac{k_{-1}}{k_1}(Ji - Jt\frac{\phi_i}{2\pi}) \tag{6}$$

As $\quad k_1\Omega^*{}_E \geq K_0, \dfrac{K_{-1}}{\Omega 0}$

It implies that the major contribution to the measurement error is the vertical component of the earth rotation

since the latitude of the site is well known, the systematic error from the earth rotation can be reduced logically

$$\phi_i^* = 2\pi \frac{Ni \pm N_{2\pi}\Omega^* E. \, ti/_{2\pi}}{N_{2\pi}(1 \pm \Omega^* E \, T/_{2\pi})} \tag{7}$$

Measurement equation with compensation of the earth component and it's called phase–time angle measurement technique algorithm with compensation of the vertical component of the earth's rotational speed.

The "plus" and "minus" sign depends on the rotational direction.

"Plus" is the clockwise direction of the RL, while "minus" is the anticlockwise direction.

After compensation of the earth rotation, this ellinminate the instability of the RL roatational speed

$$F = (\frac{1}{2\pi})(Ko + K_1\Omega * E) \tag{8}$$

where F =general zero shift, describing the corresponding general null shift of the output characteristics of RL

3.1 Equation of measurement with compensation of generalized zero shift

By the (F) we mean the sum of the terms characterizing the vertical component of the earth's rotation speed and the actual zero shift of the output characteristic of the RL

$$F' = \frac{1}{2\pi}(K_0 + K_1\Omega_E^*) \tag{9}$$

We introduce a new measurement equation

$$\widetilde{\varphi}_i^{**} = 2\pi \frac{N_i \pm F't_i}{N_{2\pi} \pm F'T}, \tag{10}$$

Where the sign "plus" or "minus" is determined by the direction of rotation RL

For definiteness, in what follows we shall take the minus sign and consider terms of the first order of smallness with respect to $\Delta\varphi_i / \varphi_i$

Then the expression (1.32) takes the form:

$$\tilde{\phi}_i^{**} = \tilde{\phi}_i(1 - \frac{2\pi}{N_{2\pi}}F'\frac{\varpi_T - \varpi_i}{\varpi_i\varpi_T}) \cong \tilde{\phi}_i(1 - \frac{2\pi}{K_1}F'\frac{\varpi_T - \varpi_i}{\varpi_i\varpi_T}) \tag{11}$$

Substituting the expression $\widetilde{\varphi}_i$ in (1.33) [7], we obtain:

$$\Delta\varphi_i^{**} = \widetilde{\varphi}_i^{**} - \varphi_i = \frac{1}{\omega_0'}\frac{K_{-1}}{K_1}\left[\varphi_i\frac{(\varpi_T - \varpi_i)}{\varpi_i\varpi_T} + \frac{1}{\omega_0'}(\delta J_i - \frac{\varphi_i}{2\pi}\delta J_T\right] \tag{12}$$

Thus, the error in measuring the angle using the measurement equation with compensation of the generalized zero shift is determined by the instability of the rotational speed and the value of the nonlinearity coefficient in the output characteristic of the RL. From comparison of expressions (6) and (10) it follows that the use of an algorithm with compensation of generalized zero shifts reduces the error of angular measurements by a ring laser.

4. Results

Due to the uncertainty of the generalized zero shift, before practical use of the measurement equation (11), it is necessary to determine experimentally the value of the generalized zero shift (GNS), which, in general, is not a constant value. The generalized zero shift (GNS) is found from the dependence of the number of periods of the output signal of the RL on the angle 2π on the turn-over time T. Indeed, from expression (12) under the assumption that $\omega(t) \gg \Omega_E^*, \omega(t) \cong const$ follows that

$$N_{2\pi} = K_1 + F'T + K_{-1}'T^2,$$

Where $K'_{-1} = \dfrac{1}{(2\pi)^2} K_{-1}$, $F' = F/2\pi$

To find the value of GNS, it is necessary, using the experimental data, to construct the dependence of $N_{2\pi} = f(T)$, from which the value of GNS is determined by regression analysis. It is known [7] that when using the method of least squares, the error in finding the regression coefficients decreases with increasing data array involved in processing. Therefore, it is expedient to use the so-called "sliding" values of $N_{i2\pi}$, which are obtained at 2π, but with a shift less than 2π:

$$N_{i2\pi} = N_{i+k} - N_i$$

Where k is the number of uniform shifts within the turnover.

One of the ways of realizations of GNS measurements with slightly accelerated rotation is shown in Figure 4(a) below, Approximation of the dependence presented in Fig. 4(a) with the use of the method of least squares allowed us to determine $K_1 = 987922.935 \pm 0,001$; $F' = -10.033 \pm 0.001 s^{-1}$; $K_{-1}' = 0.01759 \pm 0.0001 s^{-2}$.

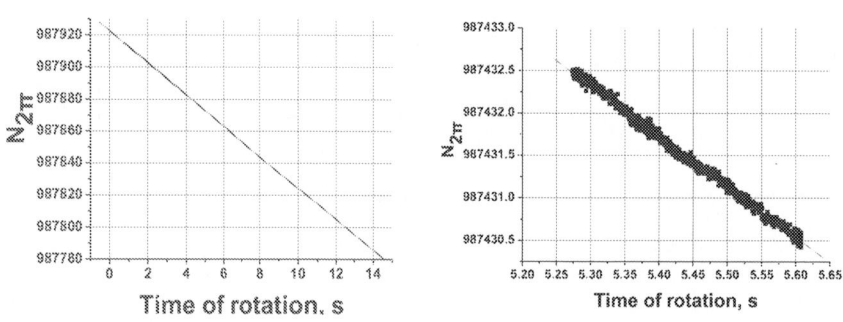

Figure 4.1 (a, b). (a) slightly accelerated rotation   (b) un-stabilized rotation speed

Figure 4.1 (a) Dependence of the number of periods of RL on the angle of 2π from the period of revolution with slightly accelerated rotation.

This example shows that in the regime with uniformly accelerated rotation of the RL, the GNS can be determined with high accuracy. However, its implementation requires an additional experiment, which is not always possible with the practical use of the laser goniometric system (LGS).

4.1 Determination of the generalized zero shift without additional experiment

In real LGS, there is always instability in the rotational speed of the RL, which makes it possible to construct the dependence $N_{i2\pi} = f(T)$ without resorting to an additional experiment. Figure 4(b) shows the characteristic dependence of $N_{i2\pi} = f(T)$ for the LGS without stabilizing the rotational speed of the RL, from which the GNS was found.

Figure 4(b) Dependence of the number of periods of RL on the angle of 2π from the period of revolution with un-stabilized rotation speed.

From Figure 4(b), the range of change in the revolution period is not as great as in the case of accelerated rotation, so the nonlinearity of the output characteristic is not manifested. In this case, linear approximation was use to find the GNS. The result of the calculations is - F = -5.94 +/- 0.01s^{-1}.

From the results of the definition of GNS, it follows that; the error in measuring the GNS is smaller, the larger the change in the period of the RL revolution, i.e. the greater the instability of the rotation speed.

Conclusions

The phase-time measurement method, consisting of simultaneous measurement of the phase of the output signal of the RL and its measurement time, allow the use of measurement equations that reduce the error caused by the instability of the RL rotational speed. The highest measurement accuracy was obtained by using the equation with compensation of the generalized zero shift. The considered method of finding a GNS is based on the use of the experimental dependence $N_{2\pi}(T)$, from which the regression analysis using the dependence $N_{2\pi} = K_1 + F'T + K'_{-1}T^2$ determines the value of the GNS. The error in the result of the measurements of the GNS decreases with the use of "sliding" values of $N_{i2\pi}$ and an increase in the range of the change in the RL. Values of K_0, k_1 were calculated, from our results it can be stated nonlinearity output of RL (K_{-1}) could not be calculated due to non-detection, was smaller the noise. Average RL velocity Ω as 13arc/seconds means that we had a stable zero shifts for RL output, after knowing the accurate latitude of the site, and from the graph we can see the dependency of the output characteristics as the scale factor increases with time(s).

It is of utmost importance to understudy the output characteristics of RL from time to time taking into account the vertical component of the earth rotation.

References

[1] Watanabe,T., Fujimoto H., Nakayama K., Masuda T., Kajitani M., Automatic high precision calibration system for angle encoder, Lasers in Metrology and Art Conservation, Munich, Proc. SPIE 4401, 2001, 267–274
[2] Filatov Yu., Loukianov D., Probst R. Dynamic Angle Measurement by Means of a Ring Laser // Metrologia, 1997, №34. p. 343-351.
[3] Burnashev M., Filatov Yu, Kirianov K., Loukianov D., Mesentcev A, Pavlov P. «Precision angle measurement in a diffractional spectrometer by means of a ring laser». Measurement Science and Technology, 1998 v.9, N7, pp.1067-1071
[4] Ivashchenko E. M. and Pavlov, P. A. "A method of eliminating the influence of magnetic field in a dynamic laser goniometer linear and angular measurements," Meas. Tech. 55(10), 2013, 1141–1147
[5] Peshekhonov, V.G., Gyroscopic navigation systems: Current status and prospects, Gyroscopy and Navigation, 2011, vol. 2, issue 3, p.111
[6] Filatov Y.V, et al. «Dynamic ring laser goniometer» in «Optical Gyros and their Applications», NATO RTO Agardograph 339, pp.12-1–12-30, May 1999.
[7] Pavlov, P.A. Analysis of measurement algorithms with a laser dynamic goniometer-Measuring technique, 2008, No. 1, P.17

Investigation of the electrical and optical properties of nickel oxide films produced by RF magnetron sputtering method

A Aglikov[1,2], A Mozharov[1], D Kudryashov[1], D Sosnin[2], A Vasilev[1], A Bolshakov[1], S Makarov[2] and I Mukhin[1,2]

[1]Renewable Energy Laboratory, St. Petersburg Academic University, St. Petersburg, 194021, Russia

[2]The Metamaterial Laboratory, ITMO University, St. Petersburg, 199034, Russia

Abstract. Nickel oxide (NiO) thin films were deposited by an RF magnetron sputtering process in different atmospheres: one with a mixture of nitrogen and argon, and another with oxygen and argon. The structural, optical and electrical properties of NiO films were investigated using the spectroscopy, atomic-force microscopy and resistivity measurements. The dependencies of the film properties on atmosphere composition were studied. Optimization of the NiO thin film properties was carried out for further fabrication of effective hole transport layers for perovskite solar cells.

1. Introduction

The main challenges in developing alternative energy sources are decrease of the solar cells (SC) production costs and increase of their efficiency and reliability. The state of the art low cost solar cells base on the materials like organic compounds and perovskites combined in bulk heterojunction structure. This type of structure is formed by electron and hole transport layers (HTL & ETL) as well as a light absorbing layer there between. Perovskites excels other materials at low-cost production, good electric and optical properties. The laboratory samples of the perovskite-based SCs with efficiencies up to 20% have already been obtained [1].

To improve the SC efficiency, it isn't enough to form good active layer like perovskite but it is important to choice right transport ones to provide good carrier separation resulting in high electrical current and voltage. One of the most suitable HTL material to fabricate the perovskite SCs is nickel oxide [2].

Nickel oxide (NiO) is a promising binary semiconductor due to its optical, electrical and magnetic properties as well as excellent chemical stability. It has been used as a functional layer material for electrochromic displays and magnetic devices [3]. Furthermore, NiO is a p-type semiconductor having wide band gap energy around $3.6 \div 4$ eV. NiO thin films can be fabricated by various physical and chemical techniques such a spray pyrolysis, sol-gel method, atomic-layer deposition, molecular beam epitaxy, magnetron sputtering and so on. Among these techniques reactive sputtering is the convenient method being the most widely used. Several studies have been carried out devoted to investigation of the film properties dependence on sputtering parameters. Reference data and previous studies have shown that good electric and optical properties can be obtained in rf-sputtered NiO thin films in a pure oxygen atmosphere [4].

Content from this work may be used under the terms of the Creative Commons Attribution 3.0 licence. Any further distribution of this work must maintain attribution to the author(s) and the title of the work, journal citation and DOI.
Published under licence by IOP Publishing Ltd

This work is focused on sputtering NiO films in the mixed atmosphere of oxygen with argon as well as nitrogen with argon in various ratios. Influencing the growth atmospheric composition on NiO film electrical and optical properties has been investigated.

2. Experimental methods

Nickel oxide thin films (NiO) were deposited by an rf-sputtering system from NiO target of 99.00% purity onto 1 mm thick glass substrates with the BOC Edwards Auto 500. The deposition was carried out in various atmosphere compositions. The distance between the target and the substrate was approximately 100 mm. The chamber was pumped to a pressure below 1×10^{-6} mbar before deposition. The deposition was performed at a gas pressure of 1.7×10^{-3} mbar for 1 hour with 100 W power at room temperature. The mixtures had the following composition: 5%, 15% and 30% of nitrogen or oxygen in argon residue as well as a pure argon atmosphere.

The thickness of NiO films were measured by profilometer Ambios XP-1. The surface morphology and roughness were studied with atomic force microscopy (AFM) using a Bruker Bioscope Catalyst microscope in tapping mode. Additionally, resistivities of all films were obtained through conventional volt-ampere characteristics measurements.

3. Results and discussion

Fig. 1 shows the growth rate of NiO films at different atmosphere mixtures during growth. Hereinafter all the graphs to the left from zero (corresponding to pure argon atmosphere) correspond to the films grown in a mixture of nitrogen and argon. In the right part of the graphs the data corresponding to oxygen and argon mixture is shown. Fig. 2 shows the variation of NiO films surface root mean square (RMS) roughness measured by AFM. Maximum growth rate was detected in a pure argon atmosphere, most probably due to strong activation of the atoms mobility on the substrate surface in argon plasma. The presence of oxygen or nitrogen in the mixture reduces the atoms mobility. However, further increase of oxygen or nitrogen concentration in the mixture enhances the growth rate. The maximum root-mean-square roughness was obtained with the growth of films in a mixture of argon (95%) and nitrogen (5%) and amounted to 2.53 nm.

Fig. 3 shows the results of the films resistivity measurements via volt-ampere characteristics. As well as being measured, a trivial formula with the shape and dimensions of the thin film (width, height & length) was used for calculation resistivity from VAC according to formula (1). Thin films grown in N_2 and Ar atmosphere have a higher resistivity in comparison with "oxygen" films. To be using the NiO thin films as hole transport layer resistivity of films should be low. This measurement have been carried out for find the best growth atmosphere since lowest resistivity of films.

$$R = \rho \cdot l/S \qquad (1)$$

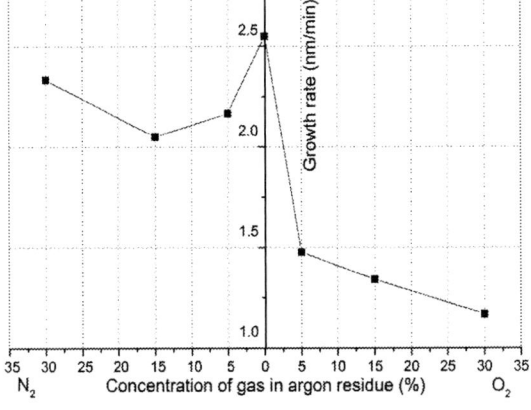
Figure 1. Growth rate vs. atmosphere mixture

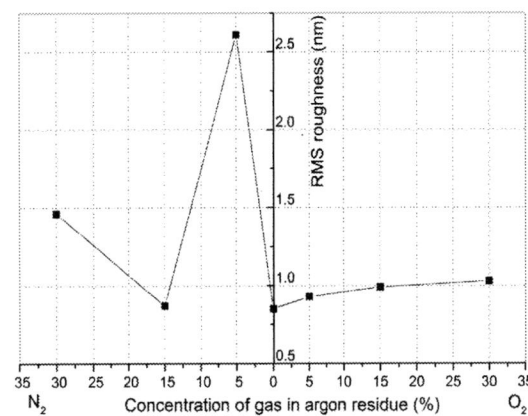
Figure 2. RMS roughness vs. atmosphere mixture

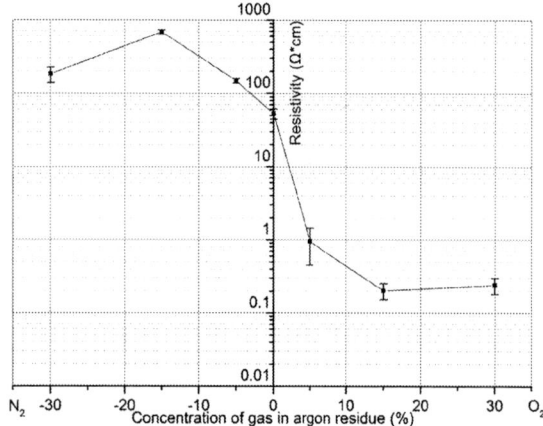

Figure 3. Resistivity vs. atmosphere mixture

To find dispersion of extinction coefficient, the reflectance (R) and transmittance (T) of the thin films were measured in a visible part of spectrum. Next, absorption (α) and extinction (k) coefficients were calculated using formulas (2) and (3). Fig. 4a and Fig. 4b show spectral dependences of extinction coefficient on mixture composition for nitrogen with argon, and oxygen with argon respectively. This measurement have been carried out for find the most transmittance films.

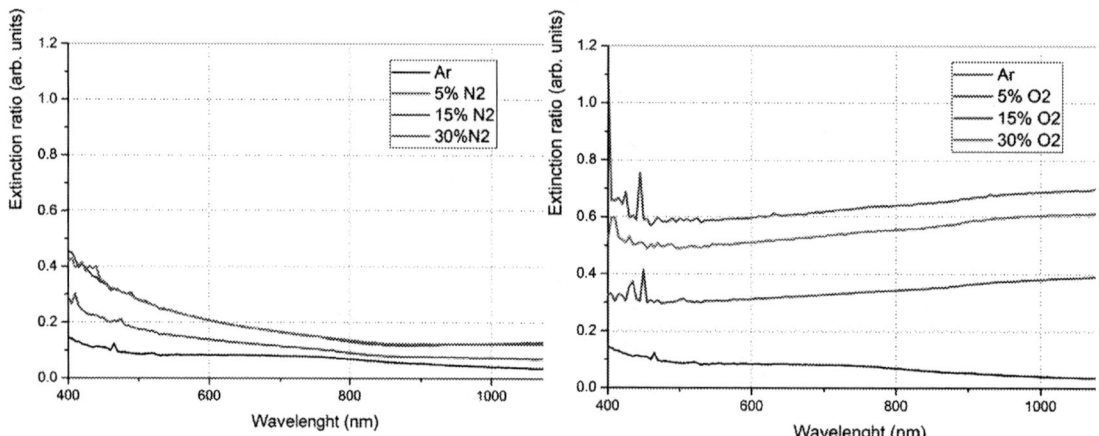

Figure 4(a, b). Spectral dependences of extinction coefficient

$$T = (1 - R) \cdot exp(-\alpha \cdot d) \tag{2}$$

$$k = \alpha \cdot \lambda / 4\pi \tag{3}$$

$$MEC = \int \left(k \cdot S_{sun} \cdot S_{perv}\right) d\lambda \bigg/ \int \left(S_{sun} \cdot S_{perv}\right) d\lambda \tag{4}$$

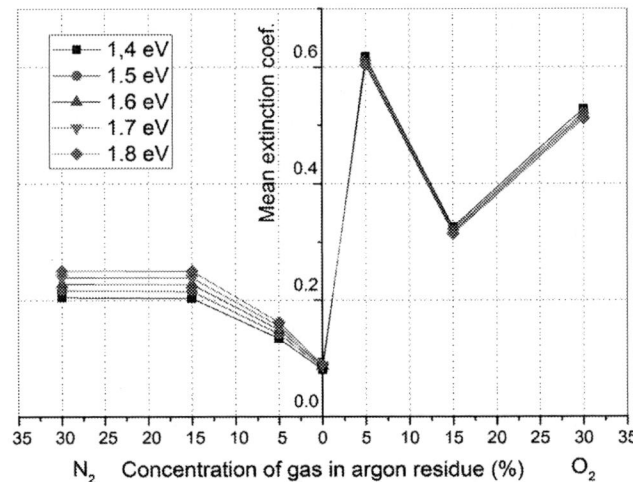

Figure 5. Mean extinction coefficient vs. atmosphere mixture

The mean extinction coefficient (MEC) being integral characteristic, was calculated for correct comparison of the optical quality of the thin films. Fig. 5 shows MEC as a function of atmospheric mixture composition. To calculate the mean extinction coefficient, the convolution of the solar spectrum S_{sun}, the spectral sensitivity of perovskite S_{per} (as a conventional direct-band semiconductor with a band gap 1.4-1.8 eV), and the spectral dependence of the extinction coefficient k were calculated. Next, the convolution between solar spectra with perovskite spectral response was found, and their ratio was calculated according to formula (4). MEC shows the losses of energy in HTL. Coefficient was found to evaluate the quality of films in a solar cell, taking into account the solar spectrum and the perovskite absorption spectrum.

4. Conclusions

The following conclusions can be drawn from the results of the study:

(1) The deposition rate was maximum, and the root-mean-square roughness was minimum during the growth in a pure argon atmosphere. The deposition rate was higher during the growth in a nitrogen-argon atmosphere compare to an oxygen-argon atmosphere;

(2) Films grown in the oxygen-argon atmosphere have the lowest resistivity. The resistivity decreases with increase of the oxygen in the atmosphere mixture;

(3) "Argon" films have the best transparency and efficiency. Films grown in the atmosphere with low nitrogen in argon residue have acceptable characteristics for using as hole transport layer in perovskite solar cells.

In the work have been investigated the electrical and optical properties of sputtering nickel oxide thin films in the mixed atmosphere of oxygen with argon as well as nitrogen with argon in various ratios. It was found that best films as HTL in perovskite SC growing in pure argon atmosphere. Mixtures of argon with nitrogen and argon with oxygen are not suitable for growing, since the optical transmittance of films obtained in these atmospheres is lower than "pure argon" films. Resistivity of "pure argon" films take average values, but it's not critical property for such use.

Acknowledgments

Research has been supported by the RFBR grants No. 16-32-60094; No. 16-3200560; No. 17-03-00621; No. 18-32-00899. Work was carried out with the support of the Russian Federation President grants MK-6492.2018.2 and MK-3632.2017.2 as well as Grant for the Leading Universities of the Russian Federation No. 074U01 & Grant of Government of the Russian Federation No. 3.9796.2017/8.9; No. 16.2593.2017/4.6 and No. 16.2483.2017/4.6.

References

[1] Kojima A, Teshima K, Shirai Y, Miyasaka T. *Organometal halide perovskites as visible-light sensitizers for photovoltaic cells*. J. Am. Chem. Soc. 131, 6050–6051 (2009)

[2] Dong-Ho Kim et. al. *Sputter deposited p-type nickel oxide thin films as an anode buffer layer in organic solar cells,* Proceedings of 37th Photovoltaic Specialists Conference (PVSC IEEE 2011)

[3] E. Aytan, B. Debnath, F. Kargar et. al. *Spin-phonon coupling in antiferromagnetic nickel oxide* Appl. Phys. Lett. 111, 252402 (2017)

[4] M. Guziewicz et al. *Optimized hydrogen sensing properties of PLD-grown nanocomposite NiO:Au and NiO:Pd thin films at ppb-concentration levels Opt. Appl.*, 41, 431440 (2011)

Lateral mode engineering in diode lasers based on coupled ridges

A S Payusov[1], A A Serin[1], Yu M Shernyakov[1], D A Rybalko[1,2], M M Kulagina[1], M V Maximov[2], N Yu Gordeev[1]

[1]Ioffe Institute, 26 Polytechnicheskaya, St Petersburg 194021, Russia
[2]St. Petersburg Academic University, 8/3 Khlopina, St Petersburg 194021, Russia

Abstract. We present an experimental study on the edge-emitting lasers based on coupled ridges. The main idea of these structures is to ensure fundamental mode lasing in broadened multi-mode ridges by means of using a high-order mode filtering based on the resonant optical tunneling into a nearby passive stripe. For our experiments, we used a conventional InAs/InGaAs quantum dot (QD) laser wafer ($\lambda \sim 1.28 \mu m$) and placed a 3 μm passive dielectric-covered ridge at the distance of 4 μm from 10 μm main active ridge. The devices demonstrated stable far-field patterns with suppressed first-order mode lasing and without any deterioration of the main laser parameters. However, side lobes in the far-field patterns indicated second-order mode traces attributed to the current spread from the main stripe, which increase the effective stripe width. We assume that optimization of the laser hafer and etching technique may lead to a pure lateral single-mode lasing in the coupled ridge devices.

1. Introduction

Today spatial single-mode edge-emitting lasers with increased optical power are essential for various applications requiring high quality output beam. Often the maximal optical power is limited by the active region volume, which depends on the laser ridge width. Therefore, increasing optical power in conventional ridge-waveguide laser diodes requires the waveguide broadening in the lateral direction, which, in its turn, usually leads to multi-mode lasing and poor beam quality. A number of approaches have been developed to improve the beam quality of lasers with broadened ridge waveguides. Among them are tapered waveguides [1] and external cavities [2]. However, each approach has some drawbacks. For example, lateral far-fields of tapered lasers are quite unstable against current and temperature changes, external cavities are sensitive to adjustment and usually require additional optical components besides diffraction gratings. Hence, approaches allowing expanding ridge waveguide up to 10-20 μm ensuring fundamental mode lasing are still desired.

Recently a new approach of transverse mode engineering has been proposed. This approach utilizes coupled large optical cavity (CLOC) structures for effective suppression of high-order transverse

---

Content from this work may be used under the terms of the Creative Commons Attribution 3.0 licence. Any further distribution of this work must maintain attribution to the author(s) and the title of the work, journal citation and DOI.

Published under licence by IOP Publishing Ltd

modes in edge-emitting lasers with broadened waveguides [3]. The waveguide of the CLOC laser consists of a single-mode narrow passive waveguide optically coupled to a broadened active multi-mode waveguide. A high-order vertical mode of the broad active waveguide is suppressed due to the resonant tunneling into a coupled single mode passive waveguide. The idea can be modified for increasing the width of the lateral single-mode waveguide. However, practical implementation of the proposed design is not so trivial since the difference in the effective refractive index for the lateral modes is much smaller than for the vertical ones. In this paper, we present an experimental study on the edge-emitting lasers based on two laterally coupled ridges.

2. Experiment details and results

In order to determine optimal parameters for coupled ridge waveguide i.e. the distance between stripes and the etch depth, we carried on etch tests and numerical simulations with the FIMMWAVE mode solver. The aim of the etch tests was to find out the ridge sidewall profiles and their dependence on the ridge width. Then we allowed for the real ridge profiles in our simulations. Figure 1a shows the simulated intensity profile for the first-order mode of the broadened 10 μm laser stripe. When we put the single-mode ridge at a distance sufficient for optical coupling, the first-order mode of the broad ridge and the fundamental mode of the narrow ridge form two composite modes (figure 1b). One can see that the intensity redistributes between the two ridges. Thus, a 2D optical confinement factor for the high-order mode is reduced in favor to the fundamental mode of the broad ridge. Electrical isolation of the additional passive ridge introduces an extra optical loss to the high-order mode. As a result, the width of single-mode stripe could be increased approximately by a factor of two (e.g. from 4 μm to 10 μm as shown in fig. 1).

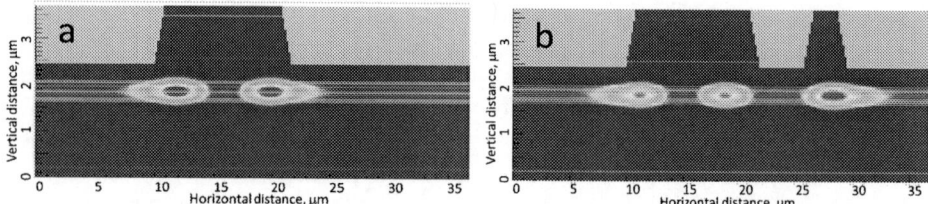

Figure 1 Simulated intensity profiles for the first-order mode of the 10 μm single ridge (a) and the composite mode of the coupled ridges (b). Vertical and horizontal scales are different.

The laser wafer under study was grown by molecular beam epitaxy. Ten layers of InAs quantum dots capped with InGaAs and separated by 35 nm GaAs were sandwiched between 1.5 μm $Al_{0.35}Ga_{0.65}As$ claddings. The wafer was processed into shallow mesa ridge-waveguide lasers using standard photolithography. The stripes were formed with the reactive-ion etching through the p-contact and p-cladding layers. Two laser types were processed: reference 10 μm single-ridge samples and coupled-ridge lasers having the 10 μm active ridge and 3.5 μm passive ridge separated with a 4 μm-width trench (figure 2). Additional dielectric layer was deposited onto the passive ridge. All studied devices were mounted on copper heatsinks using indium solder in order to minimize overheating in continuous wave (cw) regime and improve current spread.

We have not found any significant differences in the basic parameters of both single-ridge and coupled-ridge devices. The 2 mm long lasers showed the threshold current density of 100 A/cm^2 and the lasing wavelength of 1.26 μm in cw regime, which corresponds to the lasing via the QD ground state. Plotting reciprocal differential quantum efficiency versus cavity length yielded the internal quantum efficiency of 77% and the internal loss of 2.8 cm^{-1}.

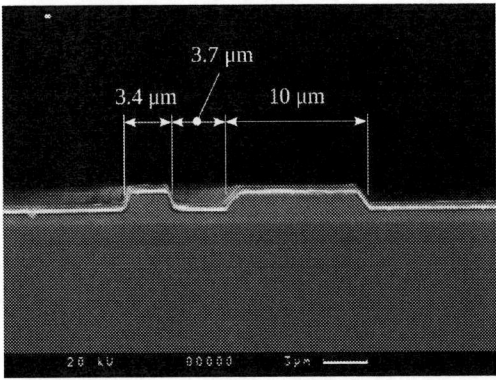

Figure 2. SEM image of the facet of the double ridge waveguide

In order to confirm the elimination of the first-order mode lasing in the coupled-ridge devices we studied far-field patterns of both reference and coupled-ridge devices. The results are shown in figure 3. Single 10 μm-width stripes demonstrate pronounced lateral multimode lasing (figure 3a) while the similar stripes possessing adjoined passive waveguide show one dominant lobe in the lateral far-field patterns (figure 3b). Its divergence of 4.5 deg. FWHM corresponds well to the simulated value for fundamental mode of the 10 μm-width waveguide. However, side lobes in the far-field pattern of the coupled-ridge device indicate traces of second-order mode lasing. We explain this effect by the current spreading from the main stripe into the additional stripe through the unetched highly doped p-cladding between the ridges. This current spreading increases effective stripe width and provides conditions for the second-order mode lasing. We assume that optimization of the laser heterostructure and the etching technique may lead to a pure lateral single-mode lasing in the coupled--ridge devices. In fact, two passive stripes can be located side by side with the active stripe, so further increasing of the laser stripe is feasible.

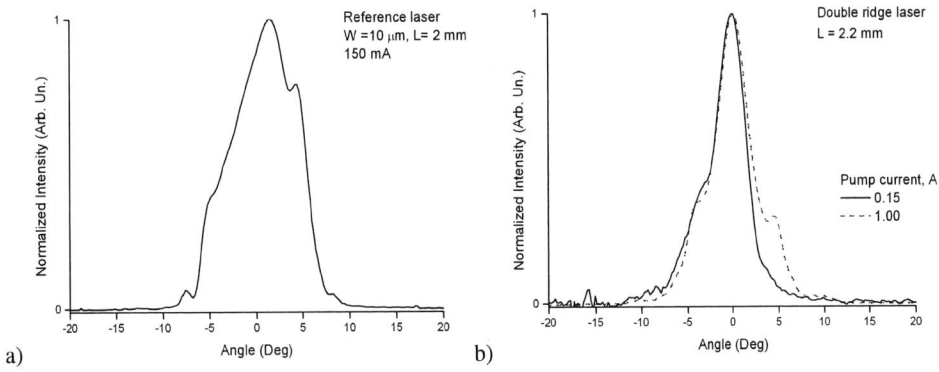

Figure 3. Far-field patterns of the single ridge (a) and coupled-ridge (b) lasers measured in pulsed regime. The length of the devices were 2 and 2.2 mm respectively

3. Conclusion

In conclusion, using a novel approach for lateral mode engineering we were able to modify far-field patterns of the conventional InAs/InGaAs QD lasers with coupled ridges without noticeable deterioration of the device performance. We assume that optimization of the laser heterostructure and etching technique may lead to a pure lateral single-mode lasing in the coupled ridge devices having higher optical power.

Acknowledgments

This work was supported by the Russian Science Foundation (project No 17-72-10060).

References

[1]   Sumpf B *et al* 2008 *Proc. SPIE* 68760M
[2]   Glebov L 2017 *Proc. SPIE* **1012319** 1012319
[3]   Gordeev N Yu *et al* 2015 *Opt. Lett.* **40** 2150

Au-SiO₂-Si hybrid plasmonic waveguide micro-ring resonator sensor

M A Butt[1], S A Degtyarev[1,2], S N Khonina[1,2]

[1]Samara National Research University, Moskovkoye Shosse 34, Samara 443086, Russia
[2]Image Processing Systems Institute of the RAS – Branch of the "Crystallography and Photonics" Federal Research and Development Center of the RAS (FRDC RAS), 151, Molodogvardeiskaya, Samara, 443001, Russia

Abstract. A refractive index sensor based on straight silicon waveguide and a hybrid plasmonic micro-ring resonator of radius 1 µm is presented. The hybrid waveguide is made up of a Metal-Insulator-Silicon structure, where the electric field is significantly enhanced in the thin insulator layer. The micro-ring design is optimized where silicon micro-ring waveguide width and a total height of the structure is 400 nm and 220 nm, respectively. The transmission spectrum and electric field distribution of this sensor structure are simulated using Finite Element Method (FEM). Despite the small size of hybrid micro-ring resonator and sensing medium, the sensitivity is relatively high as compared to conventional dielectric micro-ring resonator sensors. The resonance shift of $\Delta\lambda = 27$ nm is obtained for $\Delta n = 0.15$ and yields the best sensitivity of 180 nm/RIU in the near infrared range.

1. Introduction

Optical ring resonators have attracted a lot of attention as one of the most promising biological sensors [1-3]. It measures the target molecules through evaluating the deviations in light behaviour, which is caused by an interaction between electromagnetic (EM) wave and biological molecules such as proteins, bacteria or DNA samples [4-6]. The change in the light behaviour appears due to the interaction between an evanescent field of the resonating light inside the ring and bio-particles that exist in the sensing area. The presence of bio-particles in the medium modifies the effective refractive index of the surrounded medium, which results in deviation of resonance conditions of the resonator.

Recently, a new type of plasmonic waveguides known as hybrid plasmonic waveguides has developed a significant interest due to their ability to provide both subwavelength confinement and long propagation lengths [7,8]. These waveguides are capable of sub-wavelength confinement in two dimensions with low propagation loss. The hybrid mode is highly confined to sizes $\ll 100$ times than the area of a diffraction-limited spot while maintaining propagation distances exceeding those of surface plasmon polaritons [7]. These waveguides resemble the Insulator-Metal (IM) waveguide structure, except that a very thin layer of low-index material separates the metal layer and a higher index dielectric layer. Theoretical studies show that this thin layer can support a low loss compact mode whose propagation length strongly depends on the layer thickness.

Hybrid plasmonic waveguides incorporate the merits of the dielectric and plasmonic waveguides [7-10]. The excitation of surface electron oscillation in the hybrid waveguide concentrates the photons to stay close to the metal surface as a result, the radiation loss is minimized. These waveguides offer a high optical confinement beyond the diffraction limit whereas the propagation loss is relatively low.

Content from this work may be used under the terms of the Creative Commons Attribution 3.0 licence. Any further distribution of this work must maintain attribution to the author(s) and the title of the work, journal citation and DOI.

Published under licence by IOP Publishing Ltd

Silicon photonics has become very attractive because of its fabrication compatibility with the standard CMOS microelectronics technology [11-15]. Therefore, it is interesting to develop a silicon-based hybrid plasmonic waveguide with basic fabrication processes.

In this work, we propose a micro-ring resonator based on hybrid plasmonic waveguide for refractive index sensing. The high sensitivity and miniaturized hybrid plasmonic micro-ring resonator sensors are suitable for lab-on-chip sensing applications.

2. Sensor design and simulation parameters

In our sensor design, the micro-ring resonator is composed of Au-SiO$_2$-Si as shown in figure 1. Gold (Au) is preferred as the metal layer because of its biocompatibility and resistant to oxidization as compared to silver (Ag). The hybrid waveguide micro-ring radius is R=1 μm (from the outer edge). The ring waveguide has a width of 400 nm and a total height of the ring is 220 nm. As the ring is made up of three layers (Si, SiO$_2$ and Au), therefore each layer is optimized carefully because the total performance of the sensor is dependent on the optimized dimensions. Hence, these parameters should be handled separately else the overall performance of the sensor will be affected. The Si ring waveguide is side-coupled with a single mode Si strip waveguide (Width= 400 nm, Height= 220 nm) for resonance excitation at the TM-polarized light. The gap between straight waveguide and micro-ring is fixed at 100 nm.

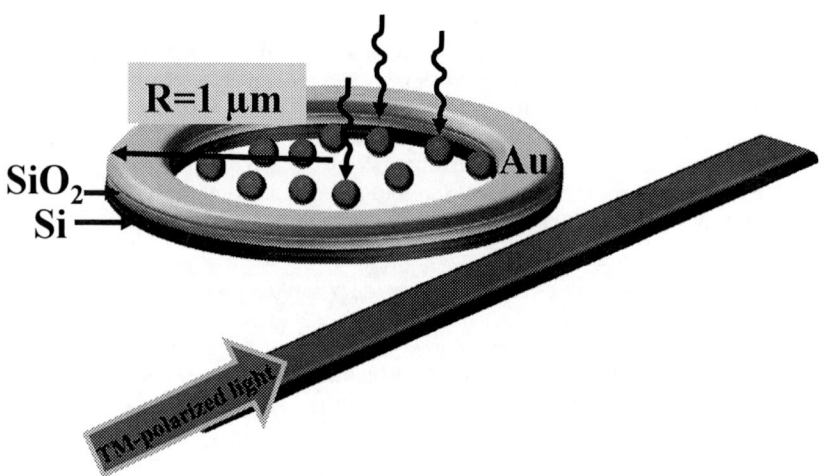

Figure 1. Hybrid plasmonic micro-ring waveguide structure.

In COMSOL simulations, the subdomains in the waveguide cross section were divided into triangular mesh elements. The grid size is set to $\lambda/5$ for the air medium and $\lambda/15$ for the waveguides geometries in order to obtain precise simulation results within the computational resources. Moreover, scattering boundary conditions were also applied at the outer edges of the FEM simulation window to estimate an open geometry. Si, SiO$_2$ and Au layer height plays a significant role in the coupling strength. Therefore, we optimized the height of each layer in order to obtain the maximum resonance excitation in the SiO$_2$ layer. The electric field in the SiO$_2$ layer is significantly enhanced by the excitation of surface plasmon wave and the discontinuity of the electric field across the Si strip wall. Due to the large lap over between the waveguide mode and the cladding medium, a small refractive index change (Δn) can bring in a large resonance wavelength shift ($\Delta\lambda$), compared to regular micro-ring sensors.

At first, we emphasize on the optimization of Si and Au layer height and maintain SiO$_2$ layer height at 30 nm. From figure 2, it can be seen that maximum resonance in the micro-ring is obtained at λ_{res}=1293.5 nm TM-polarized light when Si and Au layers have a height of 128 nm and 62 nm, respectively.

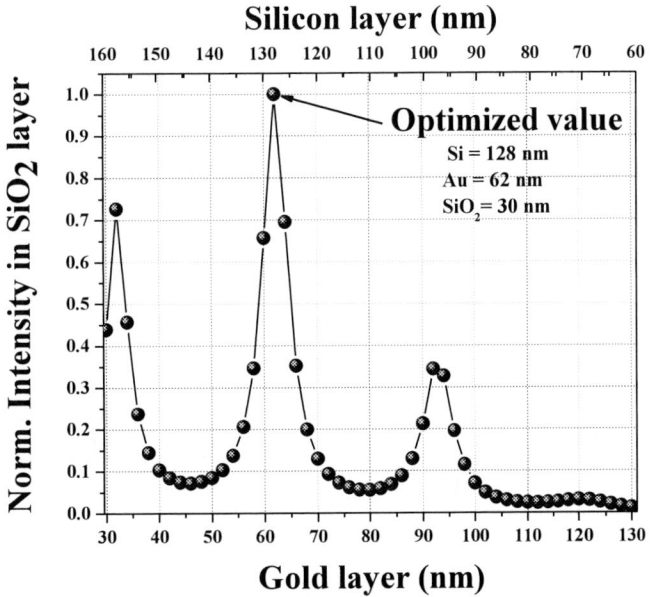

Figure 2. Optimization of silicon and gold layer thickness

Moreover, the resonance efficiency strongly depends on the SiO_2 layer height. For the optimization of the SiO_2 layer height, Si and Au are maintained at 128 nm and 62 nm, respectively. We simulated a design at λ_{res}=1293.5 nm by varying the layer thickness of SiO_2 from 10 nm- 60 nm with a step size of 2 nm. The maximum resonance is attained when the SiO_2 layer thickness is 30 nm as shown in figure 3. The insets of figure 3 show the normalized intensity distribution (plane cut at H_{SiO2} /2) in the ring at a resonance wavelength.

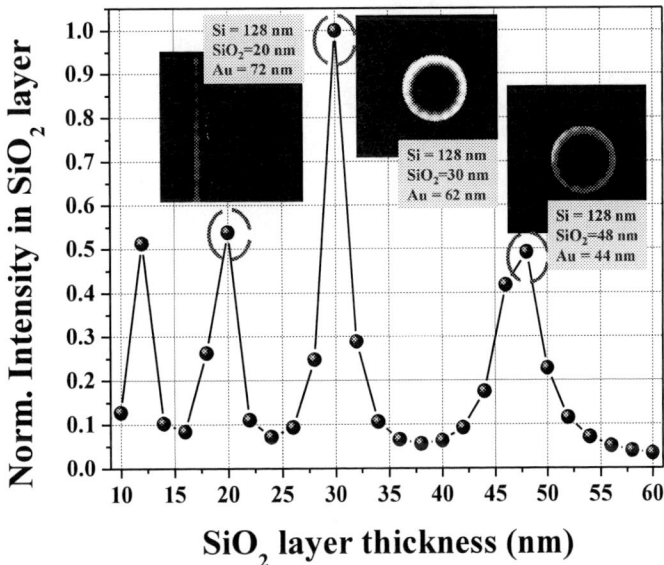

Figure 3. Optimization of SiO_2 layer thickness

3. Sensor application

The hybrid mode resonance and refractive index sensitivity were obtained by filling the surrounding medium with dielectrics having various refractive index values. The normalized intensity in the hybrid plasmonic waveguide for the wavelength range of 1260-1350 nm was measured at n=1.0, 1.05, 1.1 and 1.15. The parametric sweep function is used to calculate the electric field in the ring with a wavelength increment of 0.5. From figure 4, it can be seen that the resonance peak shifts towards the larger wavelength with increasing refractive index. For n=1.0, the resonance peak was obtained at 1293.5 nm which shifts at 1302 nm for $\Delta n = 0.05$.

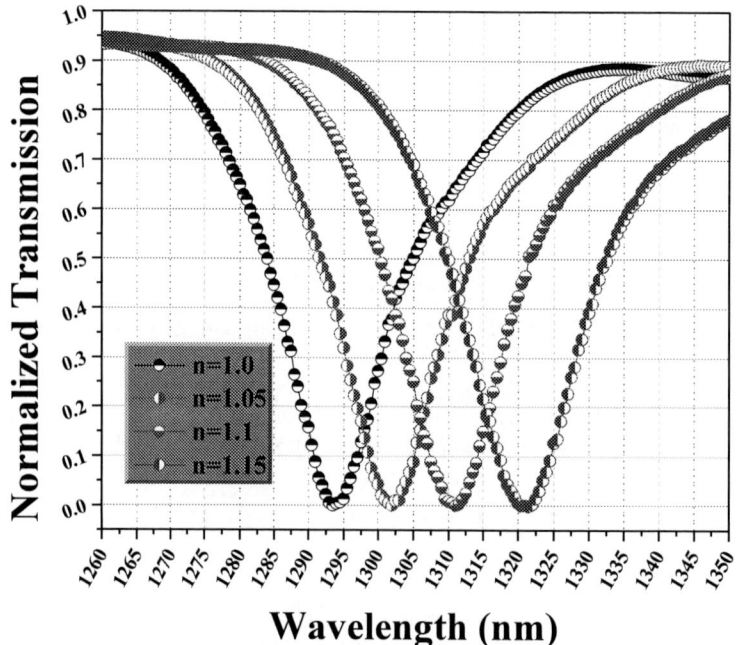

Figure 4. Spectral response of a hybrid plasmonic micro-ring resonator for various refractive indices medium.

Sensitivity is one of the most significant figures of merit to consider. There are two factors which contribute to the total sensitivity of the sensor: the WG sensitivity and the device sensitivity. The device sensitivity is defined as the ratio of the change in the resonance wavelength ($\Delta\lambda$) to the change of the refractive index (Δn). A sensitivity of our designed sensor is 180 nm/RIU which is greater than the value obtained by dielectric micro-ring of radius 100 μm [16]. The sensitivity of hybrid plasmonic micro-ring resonator is calculated in figure 5. Moreover, the sensitivity of an optical sensor is enhanced by improving the sensor design.

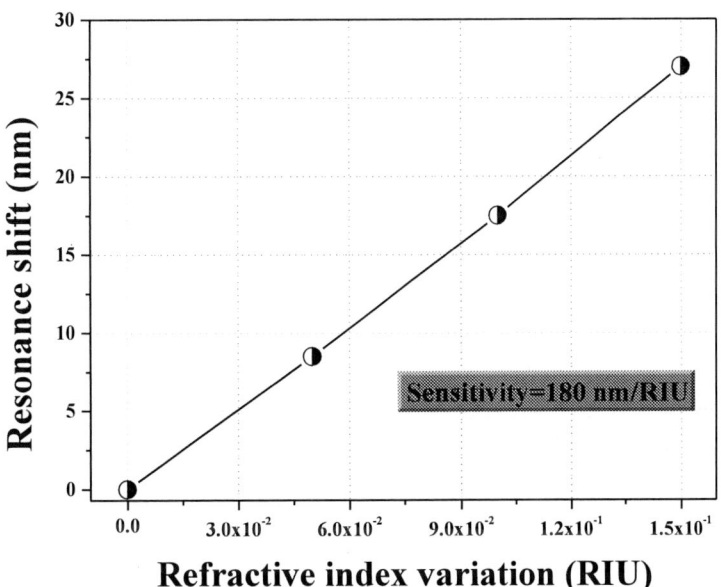

Figure 5. Sensitivity of a hybrid plasmonic waveguide micro-ring resonator

The sensing performance of a sensor is efficiently assessed by focusing on other parameters such as the figure of merit (FOM), detection limit (DL) and quality factor (Q-factor). FOM is the ratio between sensitivity and full width at half maximum (S/FWHM). The calculated FOM of our designed RI sensor is ~8.57 which is quite large for the 1 μm ring radius. Considerably large FOM shows higher sensitivity and better resolution of refractive index sensor.

The DL is a critical parameter to evaluate the performance of a sensor. It is estimated by the resonance Q-factor as well as the detector system noise level. DL is the smallest refractive index change that can be detected by the system and can be calculated by DL=f/FOM. Where f is the fraction of the resonance linewidth that can be resolved. From the resolution analysis in [17], we can approximate this value as 1/400 to evaluate the property of our ring resonator. The detection limit of the sensor is around 2.917 x10^{-4} RIU.

Integrated resonators with high Q-factors are highly desirable for a wide range of applications such as narrow-bandwidth filters, high-performance lasers, high-efficiency non-linear optic devices and high sensitivity sensors. The Q-factor is defined as λ_{res}/FWHM, where λ_{res} is the resonance wavelength and FWHM is the full width at half maximum of the resonance peak. The Q-factor of our designed hybrid micro-ring resonator is ~61.6; it is relatively larger than that of ring resonators composed merely of dielectric/plasmonic WGs with same cavity size. The Q-factor can be improved by increasing the radius of the ring in order to reduce the bending loss of the ring.

4. Conclusion

We proposed a refractive index sensor based on hybrid plasmonic waveguide micro-ring resonator of radius 1 μm. The hybrid-mode in the micro-ring resonator waveguide is obtained by the mutual coupling between the photonic mode in the silicon waveguide and the plasmonic mode at the metal surface. Au, SiO$_2$ and Si ring waveguide heights are carefully optimized to obtain the maximum coupling of photonic mode and the plasmonic mode which results in the high electric field in SiO$_2$ layer. Our designed sensor shows the best sensitivity of 180 nm/RIU. Due to its miniaturized size and sensitivity, this sensor can be used in future applications in label-free lab-on-chip biochemical diagnoses.

Acknowledgments

This work is performed with the financial support of the Ministry of Education of the Russian Federation in the framework of the government-assigned task No. 3.5319.2017/8.9.

References

[1] Lin N, Jiang L, Wang S, Xiao H, Lu Y, Tsai H L 2011 *Appl. Opt.* **50** 3615.

[2] Zhu H, White I M, Suter J D, Zourob M, Fan X 2007 *Anal. Chem.* **79** 930.

[3] Xu Z, Luo Y, Liu D, Shum P P, Sun Q 2017 *Scientific Reports* **7** 1.

[4] Butt M A, Kozlova E S, Khonina S N, Skidanov R V 2016 CEUR Workshop Proceedings **1638** 16, DOI: 10.18287/1613-0073-2016-1638-16-23.

[5] Chao C Y, Fung W, Guo L J 2006 *IEEE J. Sel. Top. Quantum Electron.* 12 134.

[6] Vos K D, Bartolozzi I, Schacht E, Bienstman P, Baets R 2007 *Opt. Express.* 15 7610.

[7] Oulton R F, Sorger V J, Genov D, Pile D, Zhang X 2008 Nat. Photon. **2** 496.

[8] Butt M A, Khonina S N, Kazanskiy N L 2018 J. Mod. Opt. **65** 1135.

[9] Dia D, He S 2009 *Opt. Exp.* **17** 16646.

[10] Wu M, Han Z, Van V 2010 *Opt. Exp.* **18** 11728.

[11] Butt M A, Degtyarev S A, Kazanskiy N L 2017 J. Mod. Opt. **64** 1892.

[12] Butt M A, Khonina S N, Kazanskiy N L 2018 J. Mod. Opt. **65** 174.

[13] Thomson D, Zilkie A, Bowers J E, Komljenovic T, Reed G T, Vivien L 2016 *J. Opt.* 18 1

[14] Gavela A F, Garcia D G, Ramirez J C, Lechuga L M 2016 *Sensors* 16 1

[15] Butt M A, Khonina S N, Kazanskiy N L 2018 Optik. https://doi.org/10.1016/j.ijleo.2018.04.134.

[16] Guider R, Gandolfi D, Chalyan T, Pasquardini L, Samusenko A, Pederzolli C, Pucker G, Pavesi L 2015 *Sensing and Bio-Sensing Research* **6** 99.

[17] Claes T, Molera J G, de Vos K, Schacht E, Baets R, Bienstman P 2009 *IEEE Photon. J* 1 197.

Magnetron sputtered MoS₂: optical and structural analysis

A.I. Belikov, Kyaw Zin Phyo

Department of electronic technologies in mechanical engineering, Bauman Moscow State Technical University (BMSTU), Moscow 105005, Russia

Abstract. Transition Metal Dichalcogenide (TMDC) such as MoS_2 and WS_2, due to unique electronic and optical properties, have sizable band gap, allowing applications such as transistors, photo detectors and other electronic devices. So, fabrication and analysis of these films are important for new generation devices. The substrate temperature strongly effects on the sputtered MoS_2 films structure and were weakly influence on optical properties of MoS_2 films. The band gap (Eg) depends on thickness of the MoS_2 films.

1. Introduction

In recent years, various kinds of two-dimensional (2D) materials have received considerable attention as a promising material group for variety electronic applications. Among these materials, MoS_2 has attracted much attention, because of its remarkable physical properties and, a stable atomically thin structure has promoted novel applications [1]. MoS_2 is a member of the group transition metal dichalcogenides (TMDs) and crystallise with a layer-type structure which is connected in the intermolecular by Van-der-Waals forces. The crystal structure of MoS_2 is described as a stacking the sulphur and molybdenum atoms in layers are hexagonally packed with three layers of S-Mo-S. Within the layer each Mo atom are held together by strong covalent bond with six sulphur atoms, while between the layers the bonding is much weaker.

MoS_2 has an indirect band gap of 1.3eV, which increases as a function of decreasing film thickness. In monolayer MoS_2 (thickness≈0.65 nm), the band gap becomes direct with a width about 1.8eV [2]. MoS_2 has been extensively studied for their semiconducting properties as a channel material in conventional field-effect transistors, as well as phototransistors and other optoelectronic devices and has also shown promise in, sensors device for biomedicine. Group of Radisavljevic B et. al 2011 [3] obtained a significant improvement in the device performance with the use of a HfO_2 under gate dielectric and demonstrated a large mobility up to about 200 cm²/Vs along with an on/off ratio of 10^8. Qing Zhang et. al 2014 [4] shown molybdenum disulfide (MoS_2) is one of the most typical TMDCs and direct band gap of 1.8 eV in monolayer and layer dependence in band structure. This property of MoS_2 is inspiring, which will largely to compensate the weakness of gapless graphene, thus making it possible for MoS_2 material to be used in the next generation switching and optoelectronic devices. Therefore, recently many works have been studied to develop high-quality and large scale continuous MoS_2 thin films.

The synthesis of MoS_2 thin films of better quality is highly required to increase the performance of the electronic devices. Different methods such as ALD, CVD and magnetron sputtering are usually adopted to deposit the MoS_2 layer on the silicon substrate. Among them, radio-frequency magnetron sputtering (RFMS) has several advantages such as low deposition temperature, ease to handle and the capability of depositing thin films of better quality [5].The parameters of RFMS, such as power, working gas pressure, substrate temperature and substrate to target distance,

Content from this work may be used under the terms of the Creative Commons Attribution 3.0 licence. Any further distribution of this work must maintain attribution to the author(s) and the title of the work, journal citation and DOI.

Published under licence by IOP Publishing Ltd

have a strong influence on the structure and the properties of the deposited thin films. In traditional physics, it is known as the influence of substrate temperature on the structures and phases of films. As these results, the electrical, optical and electronic properties can change too.

There are different deposition techniques to prepare thin films in which the deposition temperature is one of the main parameters that should be controlled to get high quality films. In that article [6] analysed the effect of the substrate temperature on the structural, morphological, electrical and optical properties of thin films deposited by different techniques.

In this study, the effect of substrate temperature on the structural features that can be measured as a roughness and optical properties of the sputtered MoS_2 thin films has been studied.

2. Substrate Temperature Effect on Structural and Optical properties

The substrate temperature effect on the structural properties of the films including orientations of crystals growth and roughness surface, and the electrical properties including the density of charge carriers, and the optical properties including absorption coefficient, will change.

2.1. Structural properties

A large number of experimental results showed that the crystals growth orientations and preferential orientation of MoS_2 thin films are sensitive to the substrate temperature and found about the changes from amorphous to polycrystalline structure with substrate temperature increasing. Nozhenkov M V [7] researched and found that MoS_2 films on the Al_2O_3 substrate prepared at substrate temperatures less than 110°C are quasi-amorphous and around 200°C are textured polycrystalline structures. Miika Mattinen et al. [8] demonstrated MoS_2 films are deposited on a variety of substrates, which reveal notable differences in growth rate, surface morphology, and crystallinity and also explored how the deposition temperature affects film crystallinity. As noted earlier, self-limiting growth seemed to occur between 250°C and 325°C. The crystallinity of the deposited films seemed to improve with increasing deposition temperature.

V. Weiss et al [9] were prepared MoS_2 thin films by reactive magnetron sputtering from a molybdenum target and textured structure (001) orientation occurs at high substrate temperatures. Zhan et al. [10] used lower temperatures of up to 750°C to grow MoS_2 on the SiO_2 substrate. The grain sizes were found to be 10-30 nm, and the devices fabricated with this grown MoS_2 displayed p-type behaviour with field-effect mobilities between 0.004 and 0.04 cm^2/Vs.

2.2. Optical properties

For optoelectronic device applications of MoS_2 to be fully understood, it is necessary to know the refractive index (n) and the absorption coefficient (α). The change in the absorption coefficient will change the reflectance of the films. Its complex refractive index in visible range is important, because many of its novel properties are closely related to this wavelength range. Beal et al. [11] measured the complex refractive index of bulk MoS_2 in 1979. Spectrophotometer is a powerful technique to measure the optical properties of thin films.

R Bichsel et.al [12] have been investigated as influence of process conditions on the electrical and optical properties of RF magnetron sputtered MoS_2 films and has been determined that the optical properties were weakly influenced by substrate temperature. Another report [13] of Mo films on soda lime glass substrates deposited by magnetron sputtering have been investigated the effect of sputtering parameters on the film properties and found coefficient of reflectance were weakly influenced by substrate temperature and demonstrated that the working pressure has a significant affect. In 2007, Blake et al. [14] successfully visualized graphene under an optical microscope by utilizing the contrast of graphene on a SiO_2/Si substrate. Given the refractive index of graphene, the contrast can further be calculated based on the Fresnel law.

We have analyzed the pronounced difference between the obtained complex reflectance of MoS_2 film at the different substrate temperatures. Furthermore, we have calculated bandgap of MoS_2 as a function incident light wavelength for using the obtained number of reflectance.

3. Experimental

Molybdenum disulphide films were prepared by magnetron sputtering MoS_2 (98% purity) target on silicon substrates at different substrate temperature (at 200 °C, 250 °C and 300 °C) and different deposition time (2 minutes and 60 minutes). Before the deposition, substrates were chemically cleaned, consisting of two stages: stage one – the substrates were cleaned in an ultrasonic bath in sequence: in alkaline solution and in ethanol for 2 minutes; the stage two is the treatment of the substrate surface by the flow of argon ions from an independent ion source in a vacuum, immediately before sputtering of the target material. The distance between the target and substrate was maintained at 50 mm for all the depositions. The different deposition parameters that are used for growing MoS_2 films on various parameters are listed in Table-1.

Table 1. Variation process conditions of magnetron sputtered MoS_2 films.

Sample code	Substrate temperature, °C	Deposition time, min
A	200	60
B	250	60
C	300	60
D	300	2
E	200	3
F	300	3

4. Results and discussion

Surface morphology and roughness of MoS_2 thin film grown on Si were examined through AFM. The three-dimensional AFM images (30×30μm) of samples are shown in Fig.1.

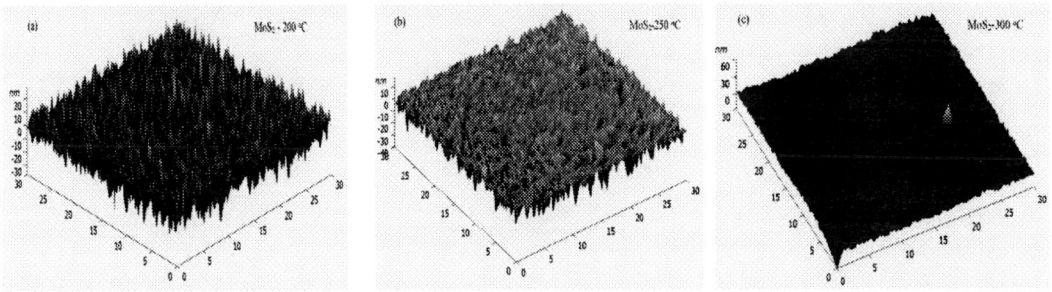

Figure 1(a, b, c). AFM images 3D (30×30μm) of MoS_2 films deposited using various substrate temperatures: (a) experimental conditions of sample A: Ra = 4.85 nm; (b) sample B: Ra = 3.68 nm caption; (c) sample C: Ra = 2.79 nm.

So, the measured average surface roughness of these samples A, B, C was 4.85 nm, 3.68 nm and 2.79 nm. This verified the surface roughness of MoS_2 film is smoothly with the increase of substrate temperature and the grain size increases at a higher of substrate temperature.

Samples A, B and C are deposited 60 minutes for thick coating about 500 nm and samples D, E and F are deposited during 2 and 3 minutes for thin films MoS_2 on Si with a thickness 15 nm and 20 nm respectively. Fig.2 shows 3D AFM image and profile of sample E that used for films thickness determinate.

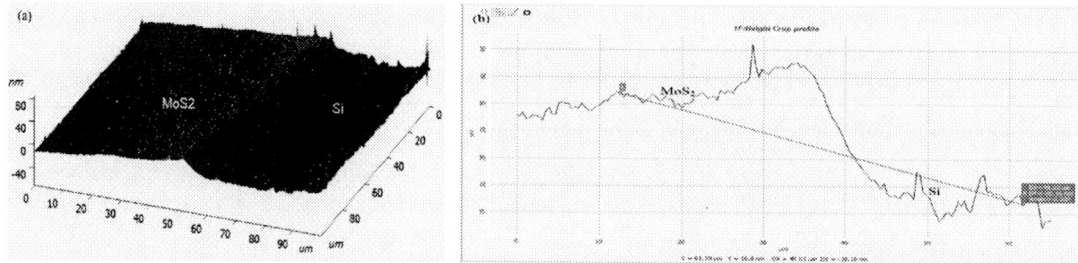

Figure 2(a, b). (a) 3D AFM image for film edge; (b) film edge profile for sample E (20nm).

For example, in Fig. 2 show the height of the step thin film MoS_2 on Si substrate formed by layering the growth planes with a clear outline of the island boundaries gives characteristic values of their thickness on the order of 20 nm.

We measured reflectance five samples MoS_2 thin films on Si substrate with the differing in substrate temperature and thickness. Fig. 3 represents the variation of reflectance of all samples with respect to wavelength from 378 to 800 nm.

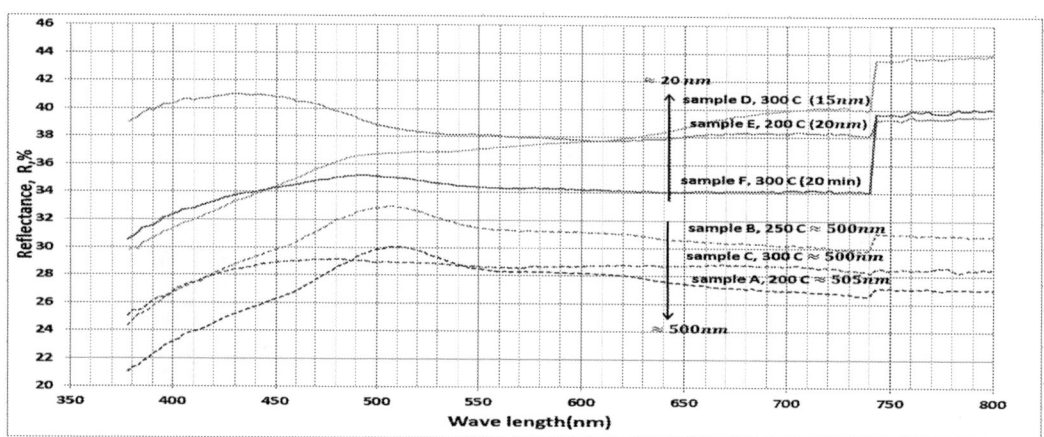

Figure 3. The schematic of spectra reflectance measurement for MoS_2 films on Si-substrates sputtered at different substrate temperatures and different thicknesses.

Figure 3 shows that the thinner films MoS_2 ($\approx$20nm) deposited at lower deposition time (samples D, E and F) demonstrate better reflectivity than thick films MoS_2 ($\approx$500 nm) deposited at higher deposition time (samples A, B and C). Figure 3 shows that the MoS_2 films deposited 60 min at different substrate temperature 200 °C, 250 °C and 300 °C is nearly equally 30% reflectance.

The absorption coefficient has been determined using the equation from the Kubelka-Munk (K-M) model [15] to solve the relation of reflectance and absorption coefficients of the sample. Band gap energies of the MoS_2 samples (Fig. 4) were estimated from their diffuse reflectance spectra utilizing Kubelka-Munk treatment:

$$\frac{\alpha}{s} = \frac{(1-R)^2}{2R}$$

(1)

Where α is the absorption coefficient of the sample. R is reflectance. The scattering coefficient, S is, in fact, dominated by particle size and refractive index of the sample. It is not a strong function of the wavelength or the absorption coefficient, so the K-M model considers it a constant. From Fig. 3, the band gap energies of films can be determined using absorption coefficient α and then to solve the following equation:

$$(\alpha h v)^{1/2} \approx (h v - E g) \quad \text{(for indirect band gap)} \qquad (2)$$

Where Eg is the optical band gap and hv is the photon energy. So we can take the variation of the optical absorption coefficient, $(\alpha h v)^{1/2}$ to zero, then will be allowed us to determine the energy gap. If $(\alpha h v)^{1/2}$ equal to zero in the following equation 2, will be equally the photon energy and band gap of films MoS_2. The spectral dependence of absorption coefficient sputtered MoS_2 film of sample C at thickness 510 nm is shown in figure 4.

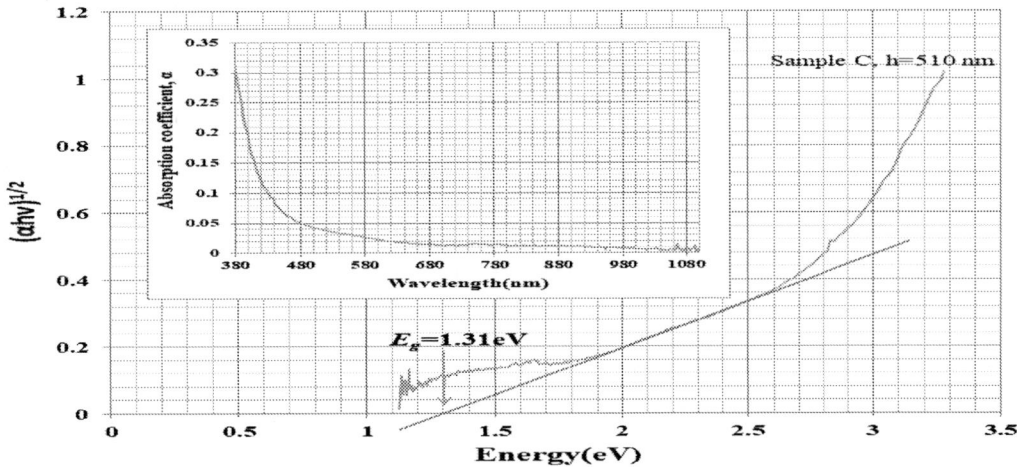

Figure 4. The graphs of absorption coefficient on wavelength (inset) and shows $(\alpha h v)^{1/2}$ vs the photon energy (Kubelka-Munk) plots of the sample C (thickness 510 nm).

Figure 4 presents the Kubelka-Munk plots for sample used to determine their band gap energy associated with an indirect transition. By extrapolating the different plots to zero absorption, the obtained energy gap is 1.31 eV for MoS_2 of sample C on Si substrate.

Using a Kubelka-Munk plot we obtained an indirect band gap for all samples MoS_2 on Si substrate, produced at different substrate temperatures and various deposition times. It can be observed that the indirect band gap increases gradually with the decrease of thickness. Therefore, the estimated indirect band gap values (1.38 eV) for sample D was very closely to the reported indirect band gap value of MoS_2 [16]. The effect of substrate temperature were weakly influenced on optical properties of MoS_2 films and the band gap (Eg) depends on thickness of the MoS_2 films. Consolidated experimental results included in table 2.

Table 2. Variation of energy band gap for samples prepared with different substrate temperatures and deposition times.

Sample code	Substrate temperature, °C	Deposition time, min	Thickness, nm	Roughness, nm	Bandgap, eV
A	200	60	505	4.85	1.29
B	250	60	500	3.68	1.30
C	300	60	510	2.79	1.31
D	300	2	15	0.465	1.38
E	200	3	20	1.201	1.37
F	300	3	20	0.974	1.37

As it can be seen from the Table 2, the optical band gap energy of samples A, B and C are 1.3eV $\pm$ 0.01eV that the same value as it for bulk MoS_2 and samples E and F are 1.37eV with the different substrate temperatures at the same deposition time. It can be seen band gap energy of samples increase with the decreasing thin films thicknesses. In can also be seen roughness of MoS_2 depend on thickness of thin films.

5. Conclusions

Molybdenum disulfide (MoS_2) films were grown on silicon substrate using a direct current (dc) magnetron sputtering system. MoS_2 films were deposited at different substrate temperatures and different deposition times by keeping other experimental parameters such as working pressure, target to substrate distance. Received samples were studied and compared for various characteristics using different techniques AFM and spectrophotometry. AFM results reveal that lower roughness can be observed on the surface of the deposited MoS_2 films for higher substrate temperatures. On the other hand, the band gap energy were weakly changed at the different substrate temperature. The band gap values of the MoS_2 films show a clear dependence on the MoS_2 film thickness, indicating that thinner MoS_2 films have a larger Eg.

References

[1] Xiao Li, Hongwei Zhu 2015 Journal of Materiomics **1** 33-44.

[2] Mak K F, Lee C, Hone J, Shan J and Heinz T F 2010 Phys. Rev. Lett. **105** 136805

[3] Radisavljevic B, Radenovic A, Brivio J, Giacometti V, Kis A 2011 Nature Nanotechnology **6** 147

[4] Ganatra R, Zhang Q 2014 ACS Nano **8** 4074 -99

[5] R Bichsel, F Levy 1986 J. Phys. D: Appl. Phys. **19** 1809-1819

[6] Ikhmayies Shadia J 2012 Modern Aspects of Bulk Crystal and Thin Film Preparation p 338

[7] Nozhenkov M V 2013 Friction & Lubrication in Machines and Mechanisms **6** 2

[8] Miika Mattinen, Timo Hatanpaa, Tiina Sarnet, Kenichiro Mizohata, Kristoffer Meinander, Peter J. King, Leonid Khriachtchev, Jyrki Räisänen, Mikko Ritala and Markku Leskela 2017 Adv. Mater. Interfaces **4** 1700123

[9] Weiss V, Bohne W, Röhrich J, Strub E, Bloeck U, Sieber I and Ellmer K 2004 Journal of Applied Physics **95** 7665

[10] Y Zhan, Z Liu, S Najmaei, P M Ajayan, J Lou 2012 Small **8** 966–971

[11] A R Beal, H P Hughes 1979 J. Phys. C: Solid State Phys **12** 881

[12] R Bichsel, F Levy 1986 J. Phys. D: Appl. Phys **19** 1809-181

[13] Peng-cheng Huang, Chia-ho Huang, Mao-yong Lin, Chia-ying Chou, Chun-yao Hsu, Chin-guo Kuo 2013 International Journal of Photoenergy **1** 8

[14] P Blakea, E W Hill, A H Castro Neto, K S Novoselov, D Jiang, R Yang, T J. Booth, A. K. Geim 2007 Appl. Phys. Lett. **91** 063124

[15] Kubelka and Munk Zeit 1931 Für Tekn. Physik **12** p593

[16] K. F. Mak, C. Lee, J. Hone, J. Shan, T. F. Heinz 2010 Phys. Rev. Lett. **105** 136805

SPBOPEN 2018

IOP Publishing

IOP Conf. Series: Journal of Physics: Conf. Series **1124** (2018) 051003   doi:10.1088/1742-6596/1124/5/051003

Effect of defocusing and apodization on the resolution of incoherent optical system beyond the Rayleigh limit

Andra Naresh Kumar Reddy[1,3], M.A. Butt[1], S.N. Khonina[1,2]

[1]Samara National Research University, Moskovkoye Shosse 34, Samara 443086, Russia
[2]Image Processing Systems Institute of the RAS – Branch of the "Crystallography and Photonics" Federal Scientific Research Center of the Russian Academy of Sciences (FSRC-RAS), 151, Molodogvardeiskaya, Samara, 443001, Russia
[3]School of Engineering, Anurag Group of Institutions, Venkatapur, Ghatekesar, Medchal District, Hyderabad, 500088, Telangana, India.

Abstract. We have investigated the resolution of the two overlapping point sources produced by the defocused incoherent system. By employing the quadratic amplitude filter in the pupil plane, the resultant intensity distributions produced by the incoherent optical system have been modified into a required component, while two-point sources are separated by the distance less than that of the diffraction limit and a comparison between them has been made by analyzing the position of the central dip in the curves. The results of the study indicate that, the apodization mask produces a higher spatial frequency under the Rayleigh limit, which means that the resolution limit is improved.

1. Introduction

The Rayleigh criterion [1] has been made use in assessing the performance of the optical systems, but they can be quite misleading since the diffraction field characteristics are not revealed. It has best seen by calculating the resultant intensity distributions from two-point objects [2]. The process of apodization, deliberately modifies the response of the optical system in the resolution of one – and two-point objects [3-6] [9-10] and this technique is also applied for improving the resolution of two line objects [7]. The current study focuses on modifying the aperture into the suitable form [7-8]. In the present study, we report the variation of the resolving power of the apodized optical system, illuminated by incoherent light. In addition to this we investigated the combine effect of defocus aberration and apodization on the resolution of two point sources separated by the small distances less than that of the Rayleigh limit. The resolution of the resultant intensity distribution in the Gaussian focal plane ($\phi = 0$) and defocused planes ($\phi \neq 0$) is investigated for the Airy and apodized cases. Here it is observed that the influence of defocus is pronounced in resolving the two incoherent point sources for $\phi = 2\pi$ (maximum defocused image plane). Principally, this study plays a significant role in situations where the resolution of the two-point objects have to be taken essentially in the spectroscopy and astronomical observations for detecting the spectral images of double-lines, points and far-field objects.

2. Formulation

For a two-dimensional incoherent imaging system, the incident plane wave propagating in a direction perpendicular to the diffracting screen and results, the spatial intensity distribution of two overlapping point sources in the focal region of the optical system [2] [9]:

$$I(Z) = |X(Z-B)|^2 + \alpha|X(Z+B)|^2 + 2\sqrt{\alpha}\gamma(Z_0)|X(Z-B)||X(Z+B)| \qquad (1)$$

Where $2B = Z_0$ is the separation between two mutually incoherent point sources, α is the intensity ratio of point sources ($\alpha \rightarrow 1$, for equal intensity two point objects), $\gamma(Z_0)$ is the real part of the complex degree of the coherence of the illumination ($\gamma \rightarrow 0$, defines the wave incoherency), Z is the

Content from this work may be used under the terms of the Creative Commons Attribution 3.0 licence. Any further distribution of this work must maintain attribution to the author(s) and the title of the work, journal citation and DOI.

Published under licence by IOP Publishing Ltd

dimensionless diffraction parameter, X(Z+B) and X(Z–B) are the amplitude impulse response of the incoherent imaging system corresponding to the two point sources. The two points are located at the distance of $Z_0/2$ on either side of the optical axis. The amplitude response of the pupil function is given by [8-9]:

$$X(Z \pm B) = 2 \int_0^1 T \, J_0\{(Z \pm B)\rho\} \rho \, d\rho \qquad (2)$$

Where J_0 is Bessel function of first kind and zero order; ρ is the radial pupil coordinate; T is the pupil function which combines incident wave, aberration ($\Phi(\rho)$) and the quadratic amplitude apodization function (ρ^2). However, here we regard rather a paraxial-optical system, so numerical aperture (NA) is small (for free space $\ll 1$).

$$\Phi(\rho) = \exp\left\{-i\phi\frac{\rho^2}{2}\right\}; \phi, \text{represents the degree of defect-of-focus} \qquad (3)$$

$$X(Z \pm B) = 2 \int_0^1 \rho^2 \, \exp\left\{-i\phi\frac{\rho^2}{2}\right\} \, J_0\{(Z \pm B)\rho\} \rho \, d\rho \qquad (4)$$

3. MTF calculation and Intensity Distributions

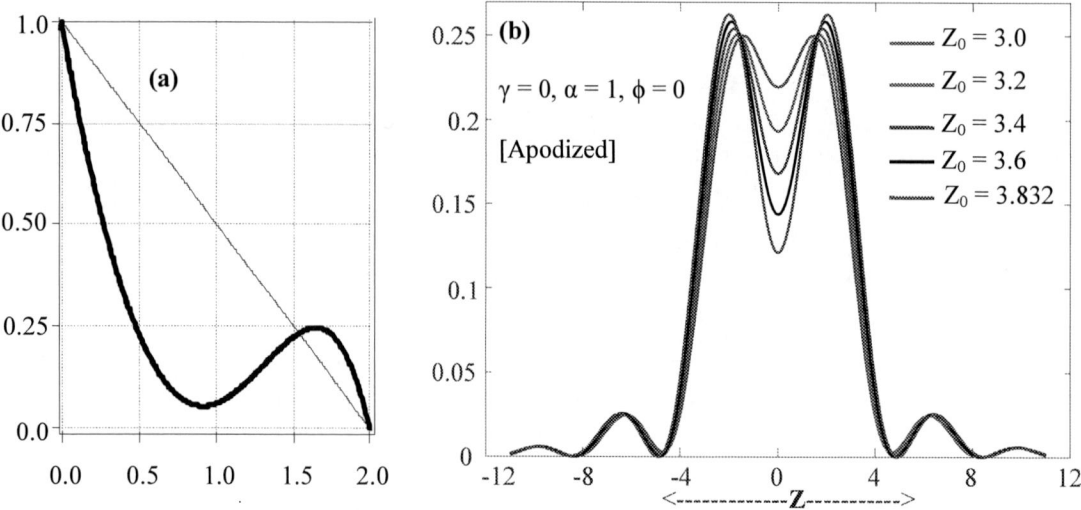

Figure 1 (a, b). (a) MTF for a conventional system (thin line) and the aberration-free apodized system (thick line) **(b)** Intensity distributions of two incoherent point sources are produced by an apodized optical system for various limits of separation (Z_0)

In this paper, we investigated the insertion of an amplitude filter at the aperture stop of the incoherent imaging system in order to address the impact of apodization function.Calculating intensity distributions from two point sources that are incoherent with respect to each other are used in the assessment of the optical systems. Equation (1), the equation for resultant-intensity distributions from two incoherent point sources have been evaluated for the dimensionless parameter Z_0. However, the difference between both Airy and apodized optical systems seems important to reveal information concerning the characteristics of the composite image intensity distribution. In Figure 1(a), we present the simulated Modulation Transfer Function (MTF) of the conventional optical system and the apodized one using the amplitude quadratic filter. While it is clear that the introduction of the quadratic amplitude apodization filter provides higher spatial frequency under the Rayleigh limit ($3.832 = 1.22\lambda/D_{\text{Aperture diameter}}$), which means that the resolution limit is improved. Figure1(b) illustrates a set of curves from two mutually incoherent point sources for various distance separations (Z_0) less than that of the Rayleigh limit. In the presence of an apodization pupil-mask, the position of the two principal maxima in the intensity distribution move outwards, the overlapping point sources are essentially well resolved with a clear presence of a dip in the resultant spatial intensity distributions. It is also observed that for certain Z_0 values (equal to 3.4, 3.6 and 3.832), the optical system provides super-two-point-resolution phenomena.

3.1 Defocus

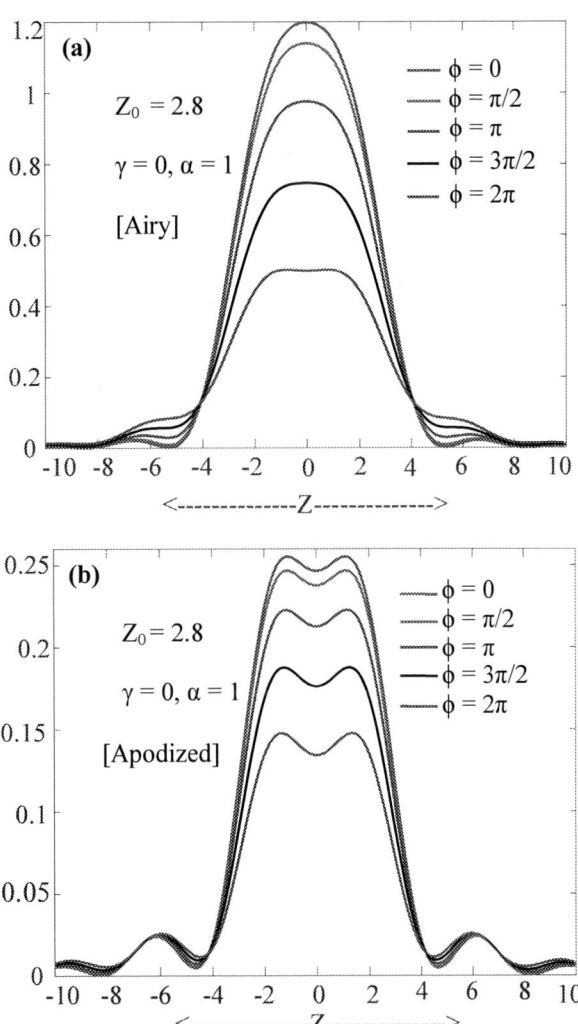

Figure 2 (a, b). **(a)** Influence of defocus in resolving the two incoherent point sources beyond the Rayleigh limit (Z_0= 2.8) in the case of the Airy pupil **(b)** Resolution is increasing in the apodized system for all defocused planes ($\phi = \pi/2$, π, $3\pi/2$ and 2π)

The defocus aberration has the functional form $\Phi(\rho)$. The coefficient ϕ is the defocusing parameter and also defined as the amount of aberration, whose non-zero value specifies the out of focus plane. Defocus is the widely experienced aberration for a telescopic objectives in which the pupil function forms the image not at the location of Gaussian image point. Due to defocus aberration, the principal maxima longitudinally shifts away from the diffraction focus, to either side, results the interior energy of the principal maxima decreases and spreads out to the sidelobes region [7-9]. Here $\phi = 0$, $\pi/2$, π, $3\pi/2$ and 2π are the different image planes investigated.

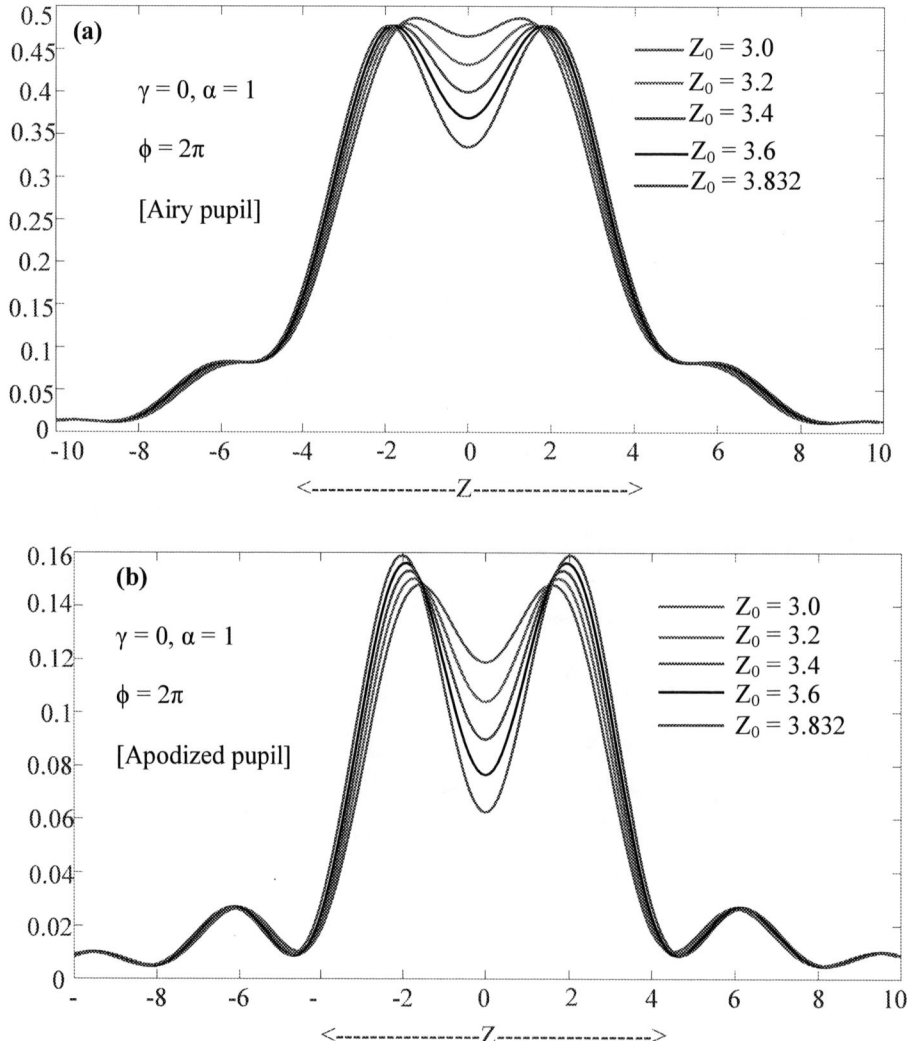

Figure 3 (a, b). Composite Image Intensity distribution of two mutually incoherent point sources separated by various distance values Z_0 for the maximum defocused plane ($\phi = 2\pi$) **(a)** Airy pupil **(b)** Apodized pupil

In Figure 2(a), the value $Z_0 = 2.8$, which is a separation very less than that of the Rayleigh limit. The points are not resolved for the values of ϕ varying from 0 to 2π in steps of $\pi/2$, the curves are essentially with no indication of a dip in intensity. For the unapodized optical system, as the defocusing effect (ϕ) increases, the curves are becoming flatter at the peak position. This effect indicates that the position of the two maxima in the resultant intensity distribution has a tendency to shift away from each other in the maximum defocused plane ($\phi = 2\pi$). However, when the apodization is applied across the pupil, the two submaxima in the resultant intensity distribution move outward for all values of ϕ producing a definite measured separation. As is evidenced by the presence of dips, therefore, the points are said to be just resolved, which has shown in Figure 2(b). Figure 3 (a) shows the profile of the intensity distribution for various distance separations (Z_0) less than that of the Rayleigh limit, when the image plane is fully defocused ($\phi = 2\pi$). The defocusing aberration is found to be effective in improving the resolution, i.e. the points are said to be resolved, but the interior energy in their principal maxima is longitudinally shifted. For the apodized system (Figure 3 (b)), in the presence of the maximum defocusing effect ($\phi = 2\pi$), the limiting separation is increasing with the values of Z_0. As 'Z_0' value increases, a distinct dip in the resulting irradiance curve approaches to the lower intensity maximum value. It is observed that by employing the quadratic amplitude apodization, the position of the two maxima in the resultant intensity plot move outwards, the size and shape of the points in the resultant intensity distribution has been modified into a required component. It concludes that the maximum defocused plane ($\phi = 2\pi$) is found to be effective in resolving the two equal intensity point sources separated in the case of apodized aperture, i.e. the resolution below the Rayleigh limit is found to be excellent for the apodized image intensity distribution in the maximum defocused plane ($\phi = 2\pi$).

4. Conclusion

By employing the amplitude apodization mask, the maxima in the intensity distribution shift away from each other producing a large measured separation. As is evidenced by an indication of dips in the resultant intensity distributions, and therefore, the two-point objects are well resolved. It is emphasized that in the presence of quadratic amplitude apodization in the pupil plane the characteristics of the maximum defocused image plane ($\phi = 2\pi$) are more effective than corresponding lower defocused planes in resolving the two-point sources. The current work concludes that the composite image intensity distribution in the defcoused plane can achieve the superresolution effect with the quadratic amplitude apodization. This study is very useful to solve the problem of two-point resolution below the diffraction limit in the presence of optical aberrations.

Acknowledgments. This work was financially supported by the Ministry of Education and Science of Russian Federation, the Russian Foundation for Basic Research (grants 16-07-00825, 16-29-11698) and by the Federal Agency of Scientific Organizations (Agreement No 007-GZ/Ch3363/26).

References

[1] Lord Rayleigh, *Collected Papers* (Cambridge U. P., London.,1902), Vol 3, p 84.
[2] Asakura T, Ueno T 1974 *Nouv. Rev. Optique* **5 (6)** 349.
[3] Keshavulu M, Komala R, Reddy ANK, Goud SL 2012 *Acta Physica Polonica A.* **122(1)** 90.
[4] Khonina SN 2012 *Computer Optics.* **36(3)** 357.
[5] Reddy ANK, Khonina SN, Sagar D 2017 *Computer Optics.* **41(4)** 484.
[6] Reddy ANK, Sagar D 2013 *Opt. Pura. y Apl.* **46 (3)** 215.
[7] Karuna Sagar D, Bhikshamiah G, Keshavulu M, Lacha Goud S 2006 *J. Mod. Opt.* **53(14)** 2011.
[8] Sagar D, Keshavulu M, Sayanna R, Goud SL 2003 *Opt. Commun.* **217(1-6)** 59.
[9] Reddy ANK, Sagar D, Khonina SN 2017 *Optics and Spectroscopy* **123 (6)** 940.
[10] Reddy ANK, Verma P, Khonina SN, Hashemi M, Martinez-Corral M 2017 *IEEE Conf. Proc.* pp. 1008-1012, C. No: CFP17CIT-ART, doi: 10.1109/ICIT.2017.7915499.

SPBOPEN 2018

IOP Publishing

Increasing the resolution of the aberrated optical system based on quadratic amplitude apodization

P A Khorin[1], S A Fomchenkov[1,2]

[1]Department of Engineering Cybernetics, Samara National Research University, Samara 443086, Russia

[2]Micro- and Nanotechnologies Laboratory, Image Processing Systems Institute of the RAS – Branch of the Federal Scientific Research Centre "Crystallography and Photonics" of RAS, Samara 443001, Russia

Abstract. Optical imaging systems have a resolution limit due to the diffraction nature of light. Moreover, point spread function is distorted and blurred in the presence of both the wavefront aberrations, and the optical system. One way to overcome the diffraction limit is the amplitude or phase apodization of the optical system. However, as a rule, apodization allows not only reducing the size of the light spot, but also leads to the appearance of side lobes which degrade depicting properties. Thus, it is necessary to observe a compromise between reducing the size of the light spot and the side lobe level, which requires appropriate studies. In this paper we simulate the formation of spread functions from two closely spaced point light sources. We carried out a study of the change in the resolving power of a focusing system with its amplitude apodization by a quadratic function in the absence and with various aberrations.

1. Introduction

Typically, optical imaging systems, such as optical communication systems, microscopes, as well as human vision systems, suffer from resolution limitations. This is related to the nature of light diffraction, as well as on the presence of aberrations both in the wavefront and in the optical system [1-7]. The main causes of wavefront aberrations are: nonideality of the forms of the optical systems elements, errors in the alignment of the system, etc. A generally accepted representation of the wave front is the basis of the Zernike polynomials [8-10]. Earlier, for direct measurement of wavefront aberration coefficients, multichannel DOEs [11-13] were approved, which are coordinated with a set of Zernike polynomials [14-17]. The coefficients of the wavefront expansion in the Zernike polynomials allow us to determine the magnitude of the deviation from the ideal front and the types of aberrations that are present in the distortion. We note that the Zernike polynomials are a convenient analytic approximation of the eigenfunctions of the correlation operator [18], known as Karhunen-Loève functions [19-20].

One way to overcome the diffraction limit is the amplitude or phase apodization of the optical system [21-27]. However, as a rule, apodization allows not only to reduce the light spot size, but also leads to the appearance of side lobes [28-31], which degrade depicting properties. Thus, it is necessary to observe a compromise between reducing the size of the light spot and the side lobes' level, which requires appropriate studies.

In particular, when asymmetric apodization [21-23] is introduced into the pupil plane, it is possible to eliminate low-frequency side lobes in the distribution of the incident field. This leads to the solution of

Content from this work may be used under the terms of the Creative Commons Attribution 3.0 licence. Any further distribution of this work must maintain attribution to the author(s) and the title of the work, journal citation and DOI.
Published under licence by IOP Publishing Ltd

a practical problem known as two-point resolution [32], when two closely located point light sources are observed in the presence of geometric aberrations.

2. Theoretical foundations

In this paper we consider the formation of the point spread functions (PSF) from two closely spaced point light sources. A study was also made in the change of the system's resolution with its amplitude apodization by the quadratic function in the absence and in the presence of various aberrations.

The wavefront is usually described as follows:

$$W(r,\varphi) = \exp\left[2\pi i \psi(r,\varphi)\right],$$ (1)

where ψ is the phase of the wavefront. We define the phase as follows: $\psi(x) = \arg[\cos(\alpha x)]$ for modeling tasks, known as resolution of two points, where α is a variable parameter.

$$\psi(x) = \begin{cases} 0, \overset{\circ}{\psi}(x) > 0 \\ \pi, \overset{\circ}{\psi}(x) < 0 \end{cases}, \overset{\circ}{\psi}(x) = \cos(\alpha x)$$ (2)

To construct the pictures of the spread function we used the simple optical system Fourier-correlator. The simulation results are shown in figure 1.

Figure 1 (a, b, c). The amplitude and phase (α=30, x8): **(a)** wavefront phase ψ, **(b)** wavefront W, **(c)** Fourier plane

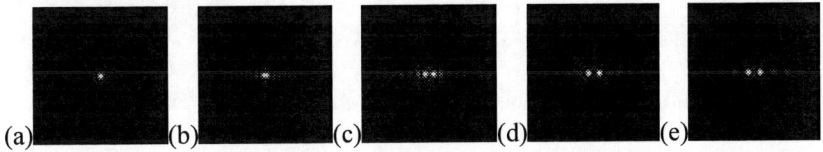

Figure 2 (a, b, c, d, e). The amplitude in Fourier plane (x8), α: **(a)** 1, **(b)** 2, **(c)** 3, **(d)** 4, **(e)** 5

From the series of figures (Fig. 2 (a) - (e)), which are presented above, it is evident that the larger the variable parameter α, i.e. the more diffraction orders are encoded in the phase, the further apart from each other are the maximum peaks of the point spread function.

3. Numerical modeling

We perform a number of experiments and find the limiting parameter α, according to the Rayleigh criterion. To distinguish between the two points can be considered limiting parameter $\alpha = 2.5$ (Fig. 3), because Airy spot radius of the PSF and the distance between the maximum peaks in the Fourier plane are approximately equal. Consequently, it can be concluded that the two points considered in the Fourier plane will be distinguished by the Rayleigh criterion for $\alpha \geq 2.5$.

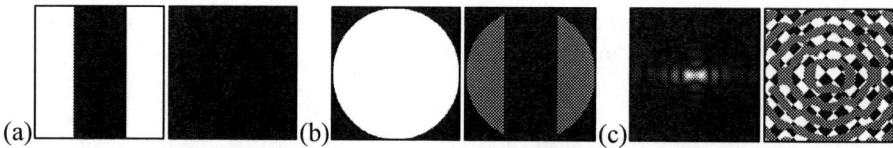

Figure 3 (a, b, c). The amplitude and phase (α=2.5, x16): **(a)** wavefront phase ψ, **(b)** wavefront W, **(c)** Fourier plane

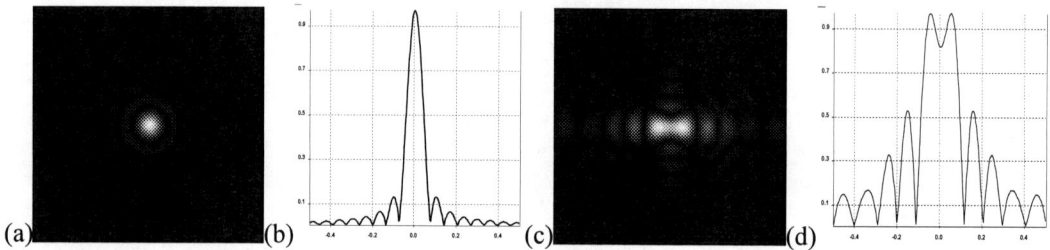

Figure 4 (a, b, c, d). (a)-(b) The amplitude of the point spread functions for optical system from a point light source and **(c)-(d)** two closely spaced point light sources; **(b)** and **(d)** – cross-section graphs

To overcome the diffraction limit, we used the quadratic amplitude apodization of the optical system. Apodization of this kind will give an effect similar to what we get in the presence of spherical aberration (Z_{20} - in terms of Zernike, Fig. 5).

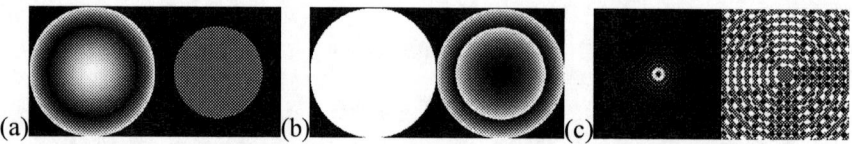

Figure 5 (a, b, c). The amplitude and phase of Zernike function $Z_{20} = r^2 \exp[im\varphi]$: **(a)** wavefront phase ψ, **(b)** wavefront W, **(c)** PSF

Let's perform numerical simulation, where we submit a test image - the cross (Fig. 6 (a)) to the input to the Fourier optical scheme, and then add additional spherical aberration to the lens surface (Fig. 6 (b)). We will show how the image will change.

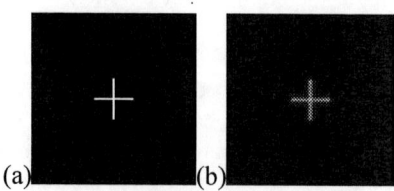

Figure 6 (a, b). The test image - cross: **(a)** ideal, **(b)** distorted by spherical aberration

However, distortion of the wave front of the proposed species will bring a positive effect when solving the problem of distinguishing two points. We expect an increase in the distance between the maximum peaks of the PSF in the Fourier plane. Let us perform a numerical experiment in which the phase of the wavefront will be supplemented by a quadratic function, like a spherical aberration (Fig. 7). Let there be given a field with a phase that is coded as follows: $\psi(x) = r^2 \arg[\cos(\alpha x)]$.

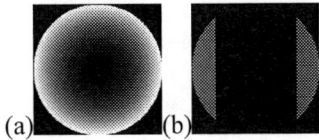

Figure 7 (a, b,). (a) The amplitude and **(b)** phase of the field ψ

4. Results and discussion

As expected, there is an improvement in image quality, from the point of view of overcoming the diffraction limit. From Fig. 8, it can be seen that the distance between the central peaks has increased to a value at which the points are separable according to the Rayleigh criterion.

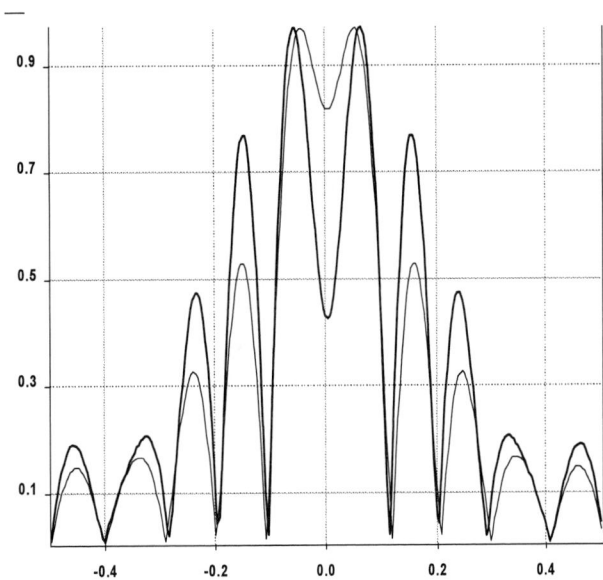

Figure 8. (simple) The amplitude of the point spread functions for optical system without apodization and **(bold)** with apodization – cross-section graphs

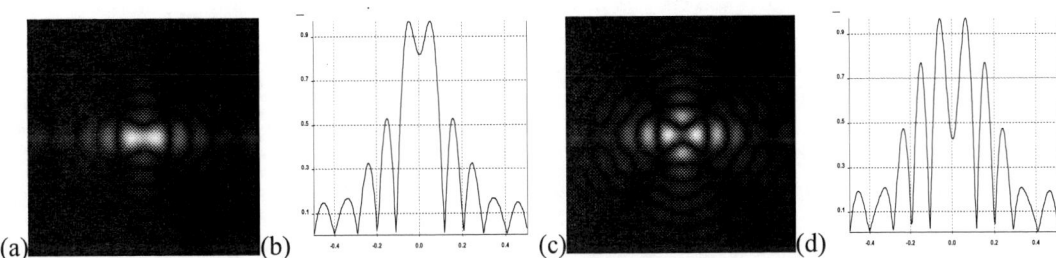

Figure 9 (a, b, c, d). The amplitude of the point spread functions for optical system **(a)-(b)** without apodization and **(c)-(d)** with apodization; **(b)** and **(d)** – cross-section graphs

Similarly, we simulated situations where the wavefront was distorted by various aberrations. In each of the cases (the first four orders of aberration in terms of Zernike) we observe a merger of two peaks in the PSF. Moreover, the next diffraction order is clearly visible (fig.10).

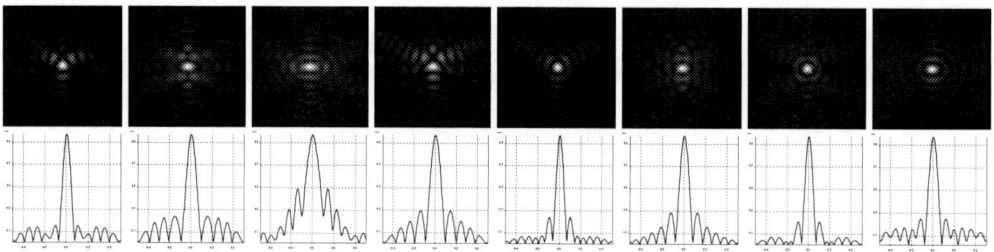

Figure 10. The amplitude in Fourier plane: Z_{1-1}, Z_{2-2}, Z_{20}, Z_{3-3}, Z_{3-1}, Z_{4-4}, Z_{4-2}, Z_{40} (top, left to right), (bottom, left to right) cross-section graphs

5. Conclusion

To overcome the diffraction limit, we used the quadratic amplitude apodization of the optical system. An increase in the distance between the maximum peaks of the PSF was expected. Numerical modeling confirmed the hypothesis of the possibility of distinguishing closely spaced points. In this paper, a study was conducted changing the resolution of the focusing system, adding considered apodization. In a number of experiments, it was found that, in the case of an ideal lens, indistinguishable points can be identified, using the Rayleigh criterion, by modifying the phase features of the wave front. As for the consideration of a nonideal optical system, we simulated situations when the wave front is distorted by some specific aberration. The first four orders of wave aberrations in terms of Zernike were considered. It was found that there is a central peak in the Fourier plane in each of the above cases. If you treat it as a high-frequency noise, the next diffraction order (maximum) is informative in terms of distinguishing between the two points. Thus, it is shown that the considered quadratic amplitude apodization improves resolution of an optical system with small aberrations.Они всегда присутствуют в таких точных приборах, как наземные астрономические телескопы, системы оптической коммуникации, микроскопы и измерительные приборы, используемые в офтальмологии.

Acknowledgments

This work was financially supported by Russian Foundation for Basic Research (grants Nos. 15-29-03823, 16-41-630542, 16-47-630546_r_a, 18-37-00056_mol_a) and by the Federal Agency of Scientific Organizations (agreement No. 007-GZ/C3363/26)

References

[1] Welford W T 1986 *Aberrations of optical systems* Adam Hilger Press (Bristol and Philadelphia)
[2] Charman W.N. 1991 *Optom. Vis. Sci.,* **68**574-583
[3] Hardy JW *Oxford, New York,* 1998.
[4] D. A. Atchison, *Clin. Exp. Optom.* 2009; **92**171–172
[5] Klebanov I M, Karsakov A V, Khonina S N, Davydov A N, Polyakov K A 2017 *Computer Optics* **41** 30-36
[6]Khorin P A, KhoninaS N, Karsakov A V, Branchevskiy S L 2016 *Computer Optics*, **40**810-817
[7]Soifer VA, Korotkova O, Khonina SN, Shchepakina EA 2016 *Computer Optics***40**605-624
[8] R. J. Noll, 1976 *J. Opt. Soc. Am.* **66**207–211
[9] J. Y. Wang and D. E. Silva, 1980 *Appl. Opt.* **19**1510–1518
[10] C. J. R. Sheppard, 2015 *Journal of the Optical Society of America A,*. **32**928-933
[11] Kotlyar, V.V., Khonina S.N., Soifer V.A. 1998 *Journal of Modern Optics*, **45**1495-1506

[12] Khonina S.N., Kotlyar V.V., Soifer V.A., Jefimovs K., Turunen J. 2004, *Journal of Modern Optics*, **51**761–773

[13] Khonina S.N., Savelyev D.A., Kazanskiy N.L., 2015 *Optics Express,* **23**17845-17859

[14] Ha, Y. Zhao D., Wang Y., Kotlyar V.V., Khonina S.N., Soifer V.A. 1998 *Proc. SPIE* **3557**191-197

[15] Khonina S.N., Kotlyar V.V., Wang Y., Zhao D., P 1998 *Computer Optics*, **18**52-56

[16] Porfirev A.P., Khonina S.N. 2016 *Proc. SPIE* **9807**E-9p

[17] Degtyarev S. A., Porfirev A. P., Khonina S. N. 2017*, Proc. SPIE,* **103370Q**-8pp

[18] G. M. Dai, 1995 *J. Opt. Soc. Am. A* **12**2182–2193

[19] Soifer V.A., Golub M.A., Khonina S.N 1993 *Pattern Recognition and Image Analysis*, **3**289-295

[20] Soifer V.A., Khonina S.N. 1994*, Pattern Recognition and Image Analysis*, **4**137-148

[21]Jacquinot P, Roizen-Dossier B 1964*Progress in Optics***3** 29-32

[22]Cheng L, Siu GG 1991 *Measurement and Technology***2** 198-202

[23] Roddier F. Cambridge, U.K., New York, NY: 1999 *Cambridge University Press.*

[24] E.W. Justh, P.S. Krishnaprasad, M.A. Vorontsov, 2004 *Automatica* **40**1129 – 1141

[25] Khonina SN, Ustinov A V 2015*Pattern Recognition and Image Analysis***25** 626-631

[26]Reddy A N K, Sagar D K, Khonina S N 2017 *Computer Optics***41**484-488

[27] Reddy A N K, Sagar D K, Khonina S N2017 *Optics and Spectroscopy***123** 940–949

[28]Siu GG, Cheng L, Chiu DS 1994 *J. Phys. D: AppliedPhysics***27** 459-463

[29]Khonina S N, Volotovsky S G 2011*Computer Optics***35** 438-451

[30]Khonina SN, Pelevina E A2011 *Optical Memory and Neural Networks***20** 155-167

[31] Khonina SN, Ustinov AV, Pelevina E A 2011*J Opt***13**095702

[32]Barakat R1962 *JOSA***52** 276-283

SPBOPEN 2018   IOP Publishing

Optical properties and energy band parameters of luminescent CaMoO₄:Bi ceramics

R A Parulin[1], I V Timoshenko[1], Yu A Kuznetsova[1], A F Zatsepin[1], E S Buyanova[2], Z A Mikhaylovskaya[2], M S I Koubisy[1]

[1]Institute of Physics and Technology, Ural Federal University, Yekaterinburg 620002, Russia

[2]Institute of Natural Sciences and Mathematics, Ural Federal University, Yekaterinburg 620026, Russia

Abstract. We studied the role of intrinsic defects of matrix and Bi dopant in the formation of optical properties and energy structure of CaMoO₄:Bi ceramic. Non-elementary luminescence was detected in a pure CaMoO₄ matrix due to radiative transitions in intrinsic vacancy-type defects, which are associated with non-stoichiometry in calcium. The experiment showed that Bi ions act as quenchers of luminescence.

1. Introduction
Scheelite-like oxides of the ABO₄ type are promising research objects due to the relative ease of doping and, as a result, the variety of chemical composition and functional characteristics of the compounds obtained. Scheelite-like compounds are perspective materials for applications in lasers, optoelectronics, photonics, etc [1].

2. Samples and methods

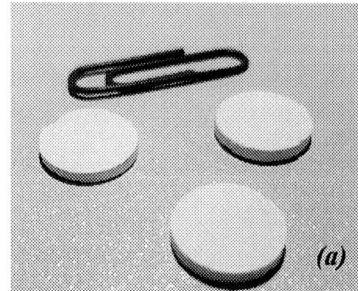 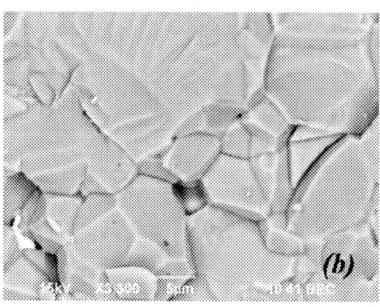

Figure 1 (a, b). (a) Samples of CaMO₄:Bi; **(b)** SEM image of Ca₀.₇Bi₀.₂MoO₄ ceramics.

Synthesis of CaMoO₄:Bi was made in terms of the standard ceramics models of for Bi₂O₃, MoO₃, CaCO₃ stoichiometric compounds grinded in the agate mortar box with the use of alcohol as homogenizer and annealed in 500-700°C temperature range (fig. 1, a). Samples were certificated by using x-ray fluorescence analyzer (Bruker Advance D8 diffractometer, VÅNTEC detector, CuKα radiation, Ni-filter, θ/θ geometry). Elementary cell parameters were calculated within structureless way in Fullprof suite.

Content from this work may be used under the terms of the Creative Commons Attribution 3.0 licence. Any further distribution of this work must maintain attribution to the author(s) and the title of the work, journal citation and DOI.
Published under licence by IOP Publishing Ltd

Microstructure analysis of sintered samples was made with the use of raster electron microscope JEOL JSM 6390LA that includes an energy dispersive microanalyzer JED 2300 (fig. 1, b).

Reflection spectra of $CaMO_4$:Bi were obtained on a Perkin Elmer Lambda 35 spectrometer with a spectral step of 1 nm and a slit of 2 nm. Using the Kubelka-Munk model provides the conversion of the initial experimental data to the absorption spectra. In accordance with Kubelka-Munk equation:

$$F(hv) = \frac{(1-R)^2}{2R} \qquad (1)$$

where $F(hv)$ is the Kubelka-Munk function that is directly proportional to the absorption coefficient, R – spectral reflection coefficient. At the same time, the Tauc power-law expression was used for determining of energy gap [2, 3]:

$$\alpha(hv) \cdot hv = A \cdot (hv - E_g)^n \qquad (2)$$

where A is a constant, E_g is the energy gap, n is the exponent that determines the type of interband transitions (1/2; 3/2; 2 and 3 for direct and indirect allowed and forbidden transitions respectively). Luminescent properties were investigated on a Perkin Elmer LS 55 spectrometer. The emission was registered in the 1.8-3.5 eV spectral range after the excitation by photons with energies of 4-6 eV (this region is most effective for the exciton generation).

3. Results and discussion

3.1 Reflection spectra

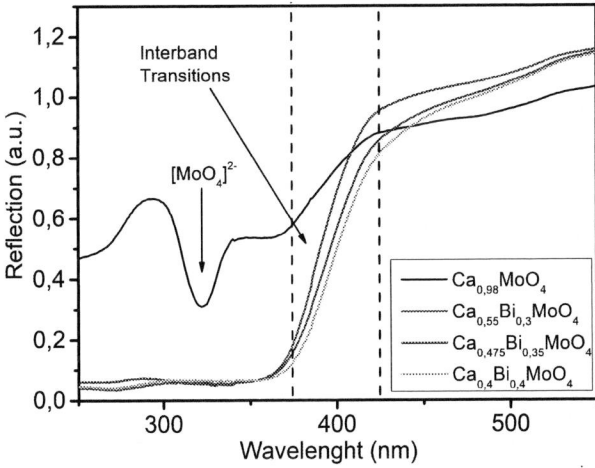

Figure 2. Reflection spectra for $Ca_{1-3x}Bi_{2x}\Phi_xMoO_4$. The dotted lines show the region that was used for obtaining the absorption spectra. Arrow shows a MoO_4^{2-} anion complex for undoped sample.

Reflection spectra (fig. 2) showed that $CaMoO_4$:Bi samples do not contain absorption band in area of 320 nm in contrast with undoped $Ca_{0.98}MoO_4$, which can be associated with the fact that Bi is changing energetic parameters.

3.2 Interband transitions

Results of Kubelka-Munk and Tauc equations were used to understand the dependence between energy gap for direct and indirect electron transitions and concentration of Bi. The spectral dependences of the Kubelka-Munk function in the coordinates for direct and indirect transitions are shown using the example of one of the samples (figure 3, a, b). The approximation of linear range by straight line up to the crossing with the abscissa axis allows determining the energy gap and also energy of phonons for indirect transitions. Bismuth adding affects energy structure of $CaMoO_4$ which is more expressed for indirect transitions than for direct transitions at lower concentration of dopant (table 1, figure 4).

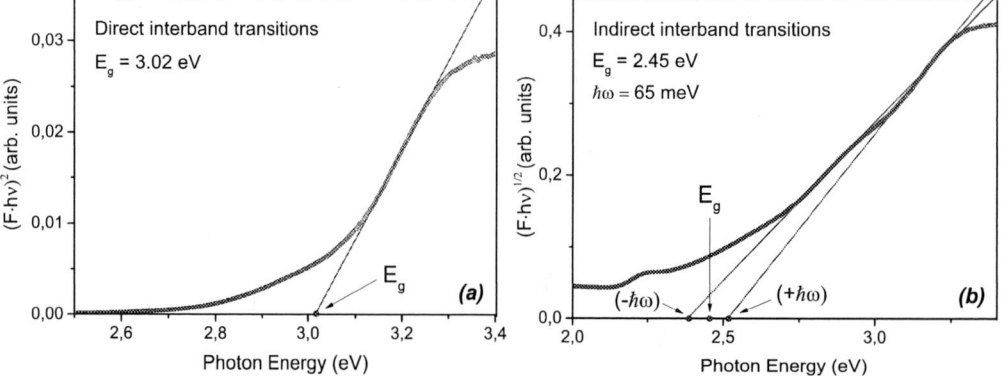

Figure 3 (a,b). Scheme of determination of energy band for direct **(a)** and indirect **(b)** transitions using Tauc equation.

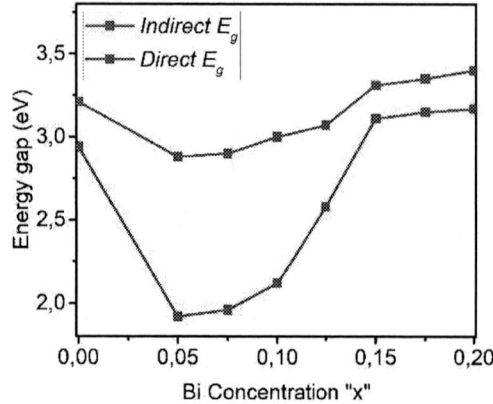

Figure 4. Dependence between energy gap for direct and indirect transitions and dopant concentration.

3.3 Excitation and luminescence spectra

For undoped sample at 4 eV excitation the non-elementary luminescence was observed and can be described as superposition of three gauss bands with maximums at 2.18, 2.45, 2.75 eV (figure 5, a). As showed in [4] non-elementary luminescence can be associated with recombination of self-trapped excitons on anion complex, which are firstly in singlet state and thermally transit to triplet state with lower energy. Singlet energy level is involved to excitation process, while triplet levels are involved to emission due to effective non-radiative relaxation from singlet states to triplet states (figure 5, b).

4. Conclusion

In present work we determining, that adding Bi affects (energy parameters) energy structure of $CaMoO_4$ by changing it is energy gap. At the same time, for indirect transitions this effect is more expressed at lower concentration of dopant, than for direct transitions. Also, this leads to quench of autolocalized excitons luminescence, that was observed for undoped sample.

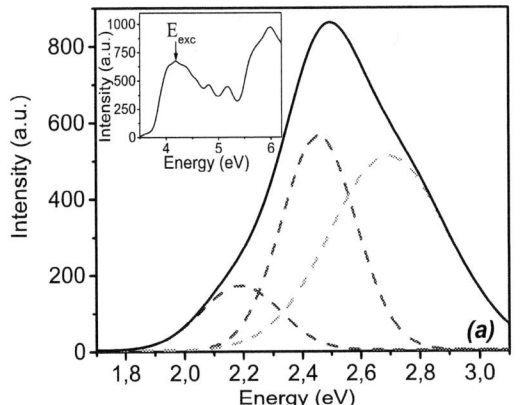 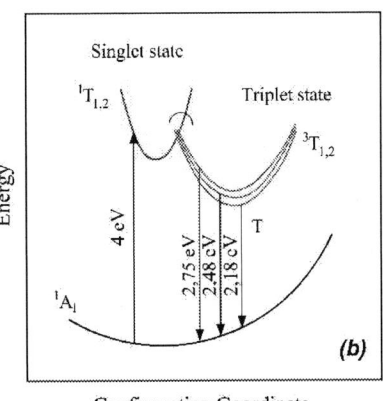

Figure 5 (a, b). (a) Emission and excitation spectra of $Ca_{0,98}MO_4$ with non-elementary luminescence; **(b)** Scheme of radiative transitions in $CaMoO_4$.

Table 1. The energy gap for direct and indirect transitions and phonon energy of materials of $Ca_{1-3x}Bi_{2x}\Phi_xMoO_4$, obtained from analysis of reflection spectra.

Composition	Indirect		Direct
	Energy gap, eV	Phonon energy, meV	Energy gap, eV
$Ca_{0.98}MoO_4$	2.94	-	3.21
$Ca_{0.80}Bi_{0.10}MoO_4$	1.92	78	2.88
$Ca_{0.775}Bi_{0.15}MoO_4$	1.96	85	2.90
$Ca_{0.70}Bi_{0.20}MoO_4$	2.12	63	3.00
$Ca_{0.625}Bi_{0.25}MoO_4$	2.58	73	3.07
$Ca_{0,55}Bi_{0,3}MoO_4$	3.11	65	3.31
$Ca_{0,475}Bi_{0,35}MoO_4$	3.15	-	3.35
$Ca_{0,4}Bi_{0,4}MoO_4$	3.17	-	3.4

Acknowledgments

The work has been funded by the Ministry of Education and Science of the Russian Federation (Government task №3.1485.2017/4.6).

References

[1] Mikhailik V B, et all 2005 *J. Appl. Phys.* **97** 083523
[2] Kuznetsova Yu A, Zatsepin A F 2017 *J. Phys.: Conf. Ser.* **917** 062001
[3] Kubelka P, Munk F 1931 *J. Tech. Phys.* **12** 593
[4] Mikhailik V B, Kraus H, Itoh M, Iri D, Uchida M 2005 *J. Phys.: Cond. Mat.* **17** 7209

SPBOPEN 2018   IOP Publishing

Polarization plane rotation for higher order modes in twisted optical fibers with discrete rotationally symmetric core

C N Alexeyev, M C Alexeyeva, B P Lapin, D V Vikulin, M A Yavorsky

V.I. Vernadsky Crimean Federal University, Simferopol 295007, Russia

Abstract. We have studied arising of polarization plane rotation for higher-order modes that propagate in weakly twisted optical fibers with core, which possesses discrete rotational symmetry. We have demonstrated that the spin-orbit interaction causes emergence of circular birefringence for hybrid modes of this system. We have obtained analytical expression for the coefficient of optical activity related to propagation of such higher-order modes.

1. Introduction

Polarization plane rotation of light refers to basic notions of classical optics [1]. Generally, this phenomenon is associated with the presence of a circular birefringence in media. It is revealed in materials, in which left- and right circularly polarized light has differing propagation velocities, or refractive indices n_+ and n_-, respectively. In such media the polarization plane rotates through an angle γ that turns to be dependent on the propagation distance z and the wavelength λ: $\gamma = \pi(n_+ - n_-)z / \lambda$. Such material's ability to rotate polarization plane is connected with the chirality of molecules in optically active media.

Recent progress in fabrication of nano-engineered metamaterials enabled creation of optically active media by introducing structural chirality into a locally non-chiral medium [2]. One of the most interesting artificially created chiral media, which manifest an optical activity, are twisted solid-core photonic-crystal fibers (PCFs) [3,4]. The presence of the optical activity in such fibers is usually connected with the existence of the form or material anisotropy in the initially untwisted fibers. These anisotropies in the untwisted fiber exhibit themselves in the existence of two directions, along which electrical field of linearly polarized fundamental modes is directed. Rotation of these directions causes emergence of circular birefringence [5]. But the optical activity of the twisted PCF studied in [3,4] cannot be explained by the rotation of two singled out directions, because a choice of the unique pair of them is unrealizable due to a multifold symmetry of the fiber's core. A possible way to explain rotation of the polarization plane of the fundamental mode is to allow for coupling of this mode to higher order modes with orbital numbers *l* and known as optical vortices (OVs) [6]. Due to the same symmetry type OVs are sensitive to the rotation of the cross-section of the twisted PCF thus imparting such sensitivity to the coupled modes with *l*=0. It was analytically demonstrated [7] for the simplest model of twisted PCF presented by a multihelical fiber (MHF) [8] and the significant role of spin-orbit interaction on arising of optical activity was pointed out. Typical values of $n_+ - n_-$ for Gaussian beams in such systems proved to be very small (of order 10^{-9} refractive index units), so that to facilitate observation of the phenomenon it was suggested to study this effect near resonances where fundamental modes are converted in the twisted fiber lattice into OVs [9]. It was established that the

---

Content from this work may be used under the terms of the Creative Commons Attribution 3.0 licence. Any further distribution of this work must maintain attribution to the author(s) and the title of the work, journal citation and DOI.

Published under licence by IOP Publishing Ltd

SPBOPEN 2018

constant of optical activity near resonances can greatly exceed the one for MHFs not supporting such type of conversion.

A natural question emerges: "Is it there exists a reciprocal influence of coupling between zero-order and higher-order modes on the polarization plane rotation for the higher order modes?" Therefore, the aim of this thesis is to study the structure of higher-order modes of those MHFs, which couple the fundamental modes through form-induced perturbation to OVs, and demonstrate the presence of optical activity for such modes.

2. Polarization plane rotation for higher order modes

The refractive index distribution in the MHF (Figure 1) can be theoretically modelled in a number of ways, but we prefer to use one of its simplest forms [10]:

$$n^2(r,\varphi) \approx \tilde{n}^2 - n_{co}^2 \Phi(r)\cos l(\varphi - qz) \equiv \tilde{n}^2 - v^2, \qquad (1)$$

where $\tilde{n}^2 = n_{co}^2 (1 - 2\Delta f(r))$, $\Delta \ll 1$ is the height of the refractive index profile function f, n_{co} is the core's refractive index, $\Phi(r) = 2\Delta\delta r f'_r$, $\delta \ll 1$ is the dimensionless parameter of the cross-section's deformation, l is the order of rotational symmetry of the MHF, $q = 2\pi / H$ and H is the pitch of the lattice. Here we use cylindrical-polar coordinates (r,φ,z).

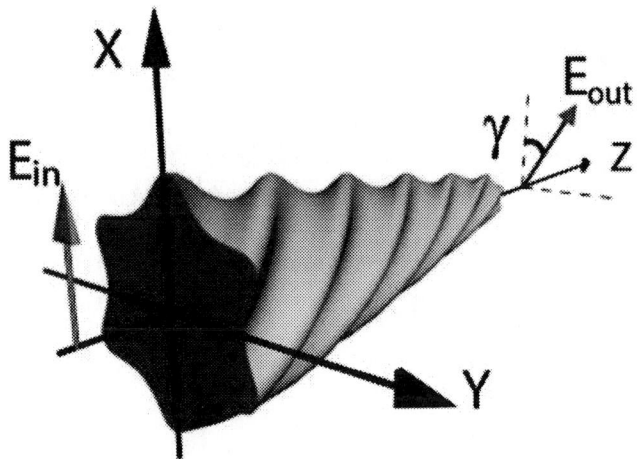

Figure 1. The model of an MHF (the order of rotational symmetry $l = 6$) and the geometry of the problem.

The standard way to study propagation of light in MHFs with the refractive index (1) is to solve the vector wave equation [11] for the case of relatively large twist period [6]:

$$(\vec{\nabla}^2 + k^2 n^2)\mathbf{E}_t = -\vec{\nabla}_t \left(\mathbf{E}_t \cdot \vec{\nabla}_t \ln n^2 \right), \qquad (2)$$

where k is the wave number, $\vec{\nabla}^2$ is the Laplacian, $\vec{\nabla}_t = (\partial / \partial x, \partial / \partial y)$ and $\mathbf{E}_t$ is the transverse component of the electric field.

As is evident, the refractive index (1) is not translational invariant, and, therefore, the vector wave equation (2) lacks invariance with respect to the longitudinal coordinate z. To restore it one has to follow the scheme suggested in [7]. This scheme also allows one to account for a vector nature of electromagnetic field and to write analytical expressions for modes of the MHF with the refractive index (1). After a little algebra one can obtain expressions for oppositely polarized higher-order modes $\xi_{1,4}$ and $\xi_{3,6}$ in the laboratory frame:

$$\xi_{1,4} \propto \sin[l(\varphi - qz)]\begin{pmatrix}1\\ \pm i\end{pmatrix}\exp\left[i\left(\tilde{\beta}_l + \varepsilon_\pm\right)z\right],$$

$$\xi_{3,6} \propto \cos[l(\varphi - qz)]\begin{pmatrix}1\\ \pm i\end{pmatrix}\exp\left[i\left(\tilde{\beta}_l - \frac{Q^2}{2\varsigma\tilde{\beta}_0} - \varepsilon_\pm\right)z\right] \tag{3}$$

where we have omitted insignificant radial factors, $\varepsilon_\pm = \varsigma l^2\left(2q\tilde{\beta}_0 \pm \alpha\right)^2 / 2Q^2$, $\varsigma = \tilde{\beta}_l - \tilde{\beta}_0$, $\tilde{\beta}_{0,l}$ are the scalar propagation constants, α is the constant of the spin-orbit interaction and Q characterizes interaction between fundamental mode and OVs.

To demonstrate polarization plane rotation for higher order modes let us consider excitation of the input end of the fiber by an x-polarized field with azimuthal dependence of $\sin l\varphi$. Then using of the expressions for $\xi_{1,4}$ allows us to write down the coefficient of optical activity A

$$A_l \equiv \pi(n_+ - n_-)/\lambda = \frac{(\beta_1 - q) - (\beta_4 + q)}{2} \tag{4}$$

in the explicit form

$$A_l \approx 2\alpha q\tilde{\beta}_0\varsigma l^2 Q^{-2}. \tag{5}$$

Here we assume, that β_l is the propagation constant of the corresponding mode from (3).

As is seen from (5), the coefficient of optical activity for higher-order mode is proportional to the twist rate q, as is observed for the one for $l=0$ modes. An important distinction from the case of $l=0$ modes consists in the dependence of the coefficient A_l on the square of azimuthal number l. Direct comparison of A_l with A_{FM}, where A_{FM} is the coefficient of the optical activity for the fundamental modes, gives that A_l exceeds A_{FM} at least by six orders of magnitude at $l=3$ for the fiber parameters given under figure 2.

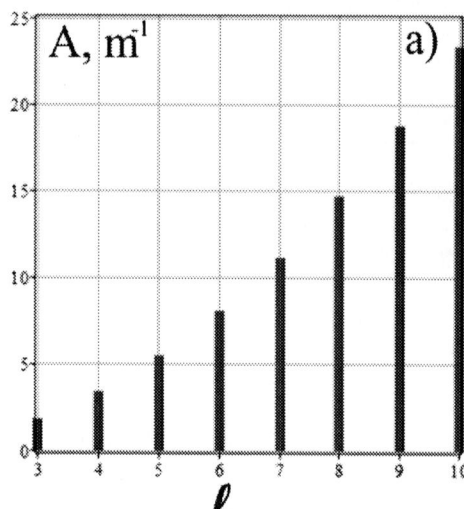 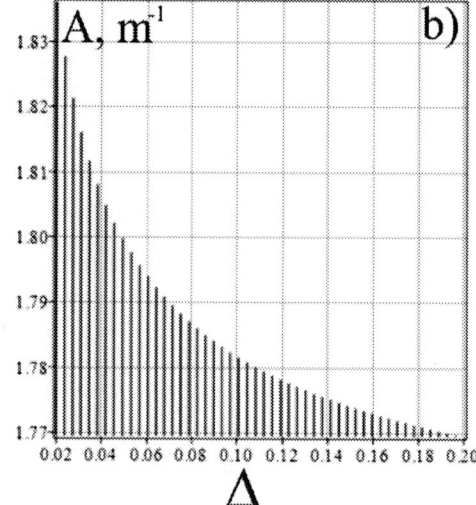

Figure 2. The optical activity A for: a) different l at $\Delta = 0.02$; b) different Δ at $\ell = 3$. The fiber parameters: $n_{co} = 1.5$, the core's radius $r_0 = 10\lambda_0$, $\lambda_0 = 632.8 nm$, $\delta = 0.05$, $q = 5m^{-1}$.

3. Conclusion

In conclusion, we have studied the polarization plane rotation for the higher-order modes in the fiber with multihelical distribution of the refractive index. We have shown that the coefficient of optical activity for such modes is proportional to the twist rate q of the fiber and to the square of azimuthal number l. A direct comparison of coefficients of optical activity for $l = 0$ modes and higher-order modes shows that polarization plane rotation for higher-order modes is much more intensive than the one for $l = 0$ modes.

Acknowledgments

Authors acknowledge partial financial support of grant №ВГ01/2017 of V.I. Vernadsky Crimean Federal University and grant RFBR № 17-42-92014.

References

[1] Hecht E 1987 *Optics, 2nd ed.* (Massachusetts: Addison-Wesley)
[2] Lakhtakia A, Messier R 2005 *Sculptured thin films: nanoengineered morphology and optics* (Bellingham: SPIE Press)
[3] Xi X M, Weiss T, Wong G K L, Biancalana F, Barnett S M, Padgett M J, P. Russell S J 2013 *Phys. Rev. Lett.* **110**, 143903
[4] Russell S J, Beravat R, Wong G K L 2017 *Phil. Trans. R. Soc. A* **375**, 20150440
[5] Barlow A J, Ramskovhansen J J, Payne D N 1981 Appl. Opt. 20, 2962
[6] Alexeyev C N, Volyar A V, Yavorsky M A 2007 *Fiber optical vortices, in Lasers, Optics and Electro-Optics Research Trends* (New York: Nova Publishers)
[7] Alexeyev C N, Lapin B P, Milione G, Yavorsky M A 2015, *Phys. Rev. A* **92**, 033809
[8] Alexeyev C N 2012 *Appl. Opt.* **51**, 6125.
[9] Alexeyev C N, Lapin B P, Yavorsky M A 2016 *Opt. Lett.* **41**, 962
[10] Alexeyev C N, Volyar A V, Yavorsky M A, 2007 *J. Opt. A: Pure Appl. Opt.* **9** 537
[11] Snyder A W, Love J D 1985 *Optical Waveguide Theory* (London, New York: Chapman and Hall)

SPBOPEN 2018   IOP Publishing

Room temperature lasing from microdisk laser in aqueous medium

M V Fetisova[1,2], N V Kryzhanovskaya[1,2], I V Reduto[1,2,3], E I Moiseev[1],
S A Blokhin[4], K P Kotlyar[1], S A Scherbak[1,2], A A Lipovskii[1,2], A A Kornev[1],
A S Bukatin[1], M V Maximov[1], A E Zhukov[1]

[1] St Petersburg Academic University, St Petersburg, Russia

[2] Peter the Great St Petersburg Polytechnic University, St. Petersburg, Russia

[3] Institute of Photonics, University of Eastern Finland, Joensuu 80101, Finland

[4] Ioffe Institute, St Petersburg, Russia

Abstract. Lasing of optically pumped semiconductor microdisks immersed in aqueous medium is demonstrated for the first time. Microlasers containing quantum dot active region were placed into the transparent polydimethylsiloxane chamber filled with distilled water at room temperature. The spectral and threshold characteristics of the lasers are compared in both air and aqueous environments. We suppose that such high-Q microlasers can be used as highly sensitive biosensors.

1. Introduction

Optical whispering-gallery-mode (WGM) microresonators have unique properties, such as small mode volumes, extremely low total optical loss, which leads to ultrahigh quality factors and low lasing thresholds. WGM resonators are also prospective for biomolecular detection due to high sensitivity of the WGMs to surrounding environmental conditions. In past decades, significant results were obtained in this field of application with passive microspherical WGM resonators [1]. Biochemical samples typically require aqueous solution, and the resonator should be placed into a cuvette with water or in a microfluidic chip [2]. The dimensions of such microspherical based biosensor are typically large (several cm^3). Recently, low-threshold lasing in optically pumped semiconductor III-V microdisk lasers with very small diameter (down to 1 µm) has been demonstrated [3]. Operation at room and elevated temperatures of injection microdisk lasers were also reported in [4]. Such WGM microlasers coupled with a photodetector could be an ideal platform to realize small-sized sensors of single nano-sized particles or molecules. In this work we study lasing characteristics of microdisk lasers in air and in water environment. Room temperature lasing in optically pumped semiconductor microdisks in aqueous medium is demonstrated for the first time. The obtained results open wide prospects for use of the microlasers as building block of the high-sensitive biosensors.

2. Experiment details

Epitaxial structures used for fabrication of the microdisk lasers were grown by molecular beam epitaxy on n-doped GaAs substrate. An active region represents 5 layers of InAs/In$_x$Ga$_{1-x}$As quantum dots inserted into a 0.35-µm-thick GaAs waveguiding layer cladded with 400-nm-thick Al$_{0.98}$Ga$_{0.02}$As

Content from this work may be used under the terms of the Creative Commons Attribution 3.0 licence. Any further distribution of this work must maintain attribution to the author(s) and the title of the work, journal citation and DOI.
Published under licence by IOP Publishing Ltd

layer from the substrate side. The spectral position of quantum dot ground-state transition was located around 1.28 µm at room temperature. Microdisk resonators were fabricated using photolithography and Ar$^+$ ion beam etching. Outer diameter of the microdisks was 7 µm. The $Al_{0.98}Ga_{0.02}As$ bottom cladding layer was selectively oxidized to be transformed into an AlGaO layer providing effective optical confinement from the bottom side [5].

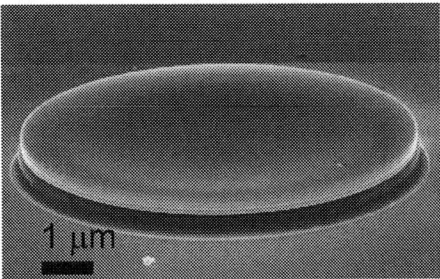

Figure 1. SEM image of a microdisk laser with 7 µm outer diameter.

Microdisks were tested in air and after that placed into a chamber made of transparent polydimethylsiloxane (PDMS) using Sylgard 184 silicone elastomer. Then the chamber was filled with distilled water (figure 2) and lasers were tested in aqueous medium. The microdisk lasers were studied at room temperature under optical pumping with cw-operating YAG: Nd laser (λ=532nm). The signal was collected by microobjective x100 and detected by ANDOR iDus multi-channel detector and confocal optical spectroscopy setup (Integra Spectra, NT MDT).

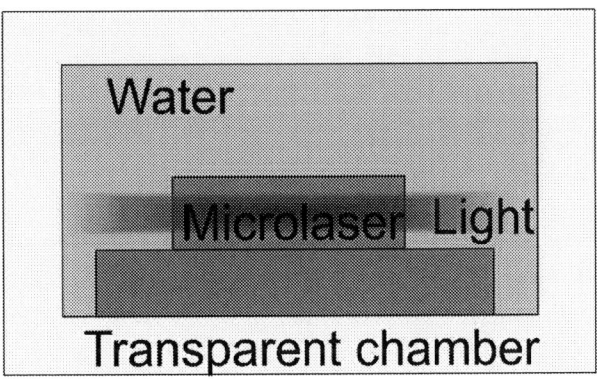

Figure 2. The schematics of the PDMS chamber construction with a sample and distilled water.

3. Results

First, microdisk lasers were characterized in air surroundings. The room temperature spectra of initial microdisks contain a series of sharp lines corresponding to WGMs of different radial and azimuthal order (Figure 3). One WGM (λ ~1297 nm), which spectral position is the closest to the quantum dots ground-state transition peak, demonstrates lasing behavior. The increase of pump power results in distinct knee at the light in - light out dependence for this WGM at threshold pump power $P_{th} \sim 0.3$mW (Figure 4). The full width at half maximum value of the line (Δλ~30pm) near the threshold gives the resonator quality factor Q ($\lambda/\Delta\lambda$) as high as 3×10^4. Then the microlasers were inserted into the PDMS chamber without water filling and tested to estimate the influence of the chamber's walls (light absorption and reflection) on optical pump power and signal collection. Spectral positions of the WGM lines remain almost the same as in the initial microlasers (Figure 3). Due to the certain limitations in device processing, the diameters of microdisks under study were slightly varied from sample to sample. According to SEM data, the actual diameter of microdisks with nominal diameter of 7 μm may vary from 6.9 μm to 7.1 μm. Thus emission wavelengths of the same WGM also may vary from disk to disk in the range of several nanometers.

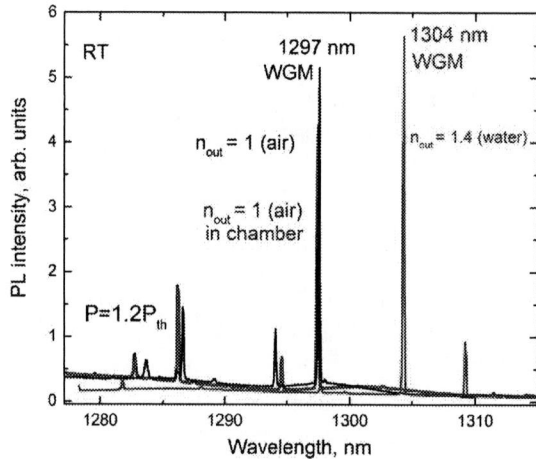

Figure 3. The room temperature μPL spectra of the microdisk lasers in the air and in the water at $P \sim 1.2\,P_{th}$.

Placing the microlasers into the empty PDMS chamber results in increase of the threshold pump power to $P_{th} \sim 0.7$ mW (as measured on the chamber input window) due to the loss of the pumping laser power in PDMS (Figure 4). When the PDMS chamber was filled with deionized water we observed dramatic change in the microdisk mode spectrum. We attribute this modification to the change of the refractive index of the surrounding material (1.33 for water at 1.3 μm). The wavelength of lasing WGM is now shifted by +7 nm to 1304 nm (Figure 3) and the threshold pump power increases to 1.5 mW (Figure 4). The observed two-times increase of threshold pump power cannot be caused just by light absorption in the water since the water absorption coefficient at $\lambda \sim 0.53$ μm is low ($\alpha \sim 3 \cdot 10^{-4}$ cm^{-1}). The observed increase of the threshold pump power can be partially explained by worsening of the resonator quality factor. Indeed, the WGM linewidth becomes three times broader after water filling. Nevertheless, the resonator quality factor remains sufficiently high ($Q \sim 10\ 000$) for sensing purposes.

Spectral and threshold characteristics of the microlasers recovered to their initial state when the water evaporated from the PDMS chamber.

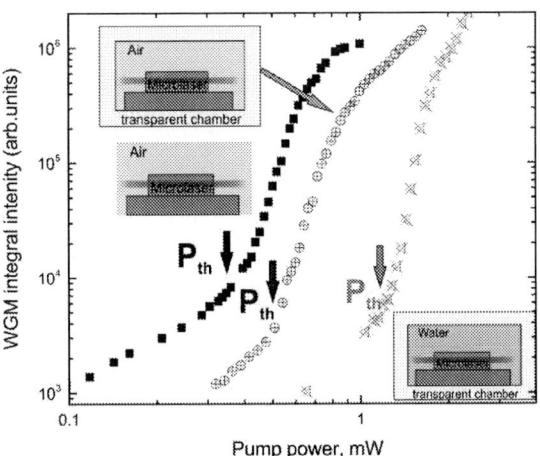

Figure 4. The dependence of the dominant WGM intensity on the pump power.

It is worth mentioning that the water ambient also results in the improvement of the thermal conduction compare to the air ambient. The thermal resistance of laser was estimated from the dependence of the spectral position of the lasing WGM line on the pump power. Considering the temperature induced shift of a WGM line 0.075 nm/°C, the thermal resistance R_{th} of 10 and 5 °C/mW was estimated for the 7 μm microdisk laser operating in the air or in the water, respectively.

To conclude, characteristics of the semiconductor microdisk lasers operating at room temperature under optical pumping in aqueous medium are studied for the first time. The obtained results (threshold, quality factor) demonstrate the possibility to use microdisk lasers for compact sensors requiring aqueous solution.

Acknowledgments
The work is supported by FRBR (18-02-00895), Program of fundamental studies of the Presidium of RAS, Ministry of Higher Education and Science of the Russian Federation (project № 3.9787.2017/8.9, № 16.9790.2017/BCh).

References
[1] Bog, U. et al. On-chip microlasers for biomolecular detection via highly localized deposition of a multifunctional phospholipid ink. Lab Chip 13, 2701–2707 (2013).
[2] G.C. Righini, et al..Whispering gallery mode microresonators: fundamentals and applications, Riv. Nuovo Cimento Soc. Ital. Fis, 7, 34, (2011).
[3] N.V. Kryzhanovskaya, A.E. Zhukov, M.V. Maximov, E.I. Moiseev, I.I. Shostak, A. M. Nadtochiy, Yu.V. Kudashova, A.A. Lipovskii, M.M. Kulagina, S.I. Troshkov «Room temperature lasing in 1-μm microdisk quantum dot lasers» IEEE Journal of Selected Topics in Quantum Electronics, 21(6), 1-5 (2015).
[4] N. V. Kryzhanovskaya, E. I. Moiseev, Yu. V. Kudashova, F. I. Zubov, M. M. Kulagina, S. I. Troshkov, Yu. M. Zadiranov, D. A. Livshits, M. V. Maximov , and A. E. Zhukov "High-temperature continuous wave operation (up to 100°C) of InAs/InGaAs quantum dot electrically injected microdisk lasers ", Proc. SPIE 9767, Novel In-Plane Semiconductor Lasers XV, 97670J (March 7, 2016).
[5] Kryzhanovskaya N. V., Blokhin S. A., Maximov M. V., Nadtochy A. M., Zhukov A. E., Fedorova K. V., Ledentsov N. N., Ustinov V. M., Il'inskaya N. D., Bimberg D., "Effect of AlGaAs-(AlGa)(x)Oy Pedestal Parameters on Characteristics of a Microdisk Laser with Active Region Based on InAs/InGaAs Quantum Dots", Semiconductors, 45 , 962-965 (2011).

SPBOPEN 2018   IOP Publishing

Self-assembly of lines of microscopic photonic crystals

M S Ashurov[1], A A Ezhov[2], T A Kazakova[3], S O Klimonsky[1]

[1]Faculty of Materials Science, Lomonosov Moscow State University, Moscow, 119991 Russia
[2]Faculty of Physics, Lomonosov Moscow State University, Moscow, 119991 Russia
[3]Faculty of Biology, Lomonosov Moscow State University, Moscow, 119991 Russia

Abstract. Lines of identical photonic crystals (PhCs) were obtained by the method of intermittent motion of the meniscus during evaporation of a colloidal solution. Periodically repeating ridges consisting of close-packed SiO_2 microspheres were formed parallel to the contact line. The height of the ridges strongly depend on the solution concentration. Local transmission measurements were performed to prove that the ridges formed at high concentrations are high enough to show the properties of PhCs. The period of the structures was about 140-300 µm. Both the period and width of the ridges depend on the solution concentration and the evaporation temperature. The obtained lines of PhCs can probably be used as masks for lithography, diffraction optical elements and microscopic chemical sensors.

1. Introduction

The aim of the work is a self-assembly of lines of microscopic photonic crystals by the intermittent deposition of colloidal SiO_2 microspheres. Although complex patterned structures can be formed using top-down approaches such as prepatterning of substrates and soft lithography, methods of self-assembly are strongly desired. Such methods aimed at the deposition of discrete narrow stripes of colloidal microspheres were previously developed in [1-4]: in [1-3] it was the method based on concavely curved meniscus and in [4] – the method based on intermittent, "stick–slip" meniscus motion during vertical deposition of microspheres. Nevertheless, no one reported on the photonic crystal properties of obtained stripes. The main reason was connected with their insufficient thickness. Here we demonstrate, however, that formation of ridges with high enough height to be PhCs is possible with a slow meniscus motion and a sufficiently high SiO_2 concentration.

2. Experimental

SiO_2 colloidal microspheres with a diameter of 200-300 nm were synthesized by the method presented in [5]. The method of fabrication of lines of identical PhCs was based on intermittent, "stick–slip" motion of the meniscus during evaporation of a colloidal solution [4, 6]. We used sufficiently concentrated ethanol-based suspensions of SiO_2 microspheres (up to 1 g/l) and low temperatures of evaporation (31-34 °C). Glass coverslips (Menzel-Gläser) thoroughly washed in ethanol and acetone were vertically immersed into the suspension for the deposition of the microspheres.

The fabricated structures were investigated by optical microscopy (Nikon ECLIPSE E600 P04), scanning electron microscopy (LEO SUPRA 50VP) and local transmission spectroscopy. The last was performed using confocal laser scanning microscope (CLSM) Olympus FluoView FV1000 based on IX81 inverted microscope equipped with scanning unit with a spectral type detection system. External Schott ACE halogen light source was used with the internal IR interference filter removed to expand the

Content from this work may be used under the terms of the Creative Commons Attribution 3.0 licence. Any further distribution of this work must maintain attribution to the author(s) and the title of the work, journal citation and DOI.
Published under licence by IOP Publishing Ltd

spectrum. Measurements were performed using objectives 10x (NA 0.40) and 20x (NA 0.75) and the spatial resolution of at least of 3 microns was regulated by the size of the confocal aperture. NA of microscope condenser was varied from 0.7 down to less than 0.2. Spectra are recorded as CLSM raster and post-processed with CLSM and other software. Transmission of the samples was normalized by the transmission of the glass coverslip.

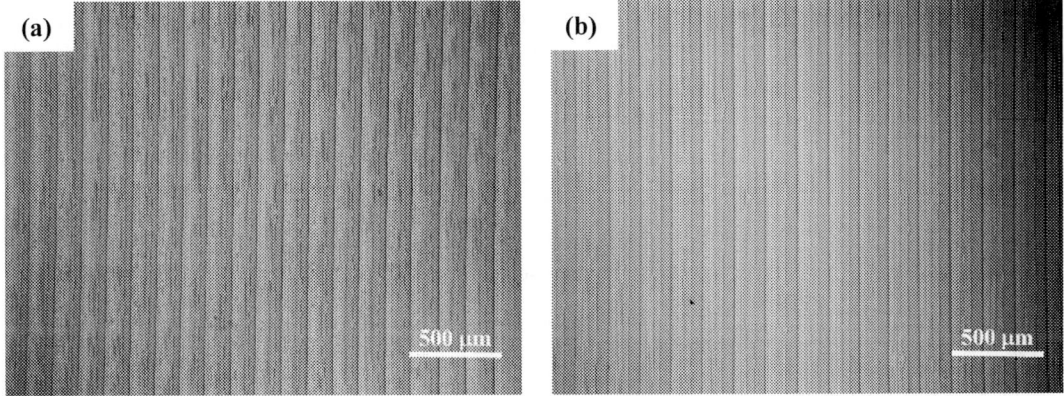

Figure 1(a, b). Microphotographs of the structures prepared at different evaporation temperatures using suspensions with different SiO_2 concentrations: **(a)** 31 °C, 0.9 g/l; **(b)** 34 °C, 0.45 g/l.

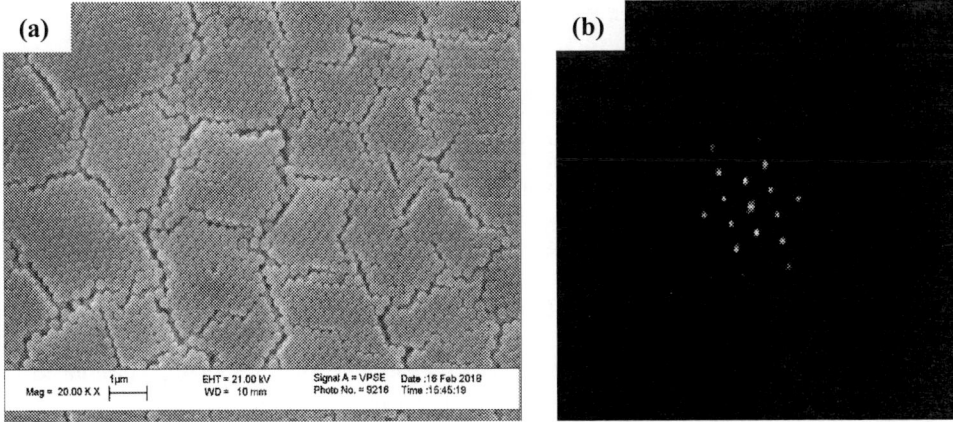

Figure 2(a, b). **(a)** Scanning electron microscopy image of the top of one of the ridges shown in Fig. 1(a); **(b)** The Fourier transform of the image (a).

3. Results

The structures consisting of parallel ridges of deposited SiO_2 microspheres repeating with a period of 140–300 µm and empty gaps between them have been prepared. Their main peculiarities previously reported in [4] are illustrated by two examples shown in Figure 1: the period of the structures increases with increasing temperature and the height of ridges decreases with decreasing a microsphere concentration. The height is proportional to the number of microsphere layers differ from each other by their colour. The highest ridges were obtained at the lowest temperature (31 °C) and highest concentrations (up to 1 g/l). The width of the ridge grows with concentration. As a result, at concentrations above 1 g/l, the ridges begin to touch each other, forming a continuous film, especially at the end of the deposition process, when the volume of the suspension is reduced due to evaporation. The microspheres within each ridge form a close-packed array, as shown in Fig. 2(a). The Fourier transform of the image (Fig. 2(b)), consisting of well localized spots, indicates a high degree of hexagonal ordering of the microspheres on the surface.

Each ridge shown in Fig. 1(a) has 8-9 layers of close-packed microspheres, as seen in Fig. 3(a). Their height is sufficient to show the properties of PhCs. In order to test them we carried out local transmission measurements for one of the ridges (Fig. 3(b)). The approximate location of the area from which the local spectrum was recorded is shown in Fig. 3(a) with a white rectangle. Its position was chosen on the top of the ridge. The drop in the transmittance spectrum related with the photonic band gap is clearly seen at 516 nm (Fig. 3(b)).

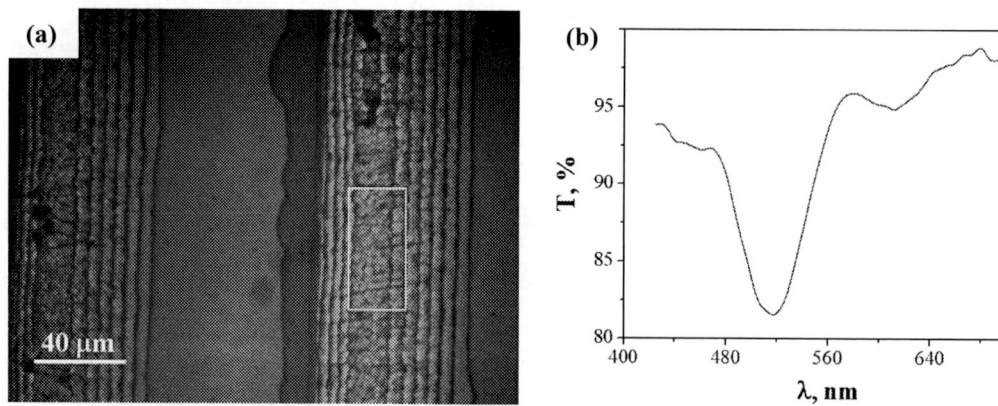

Figure 3(a, b). (a) The high resolution microphotograph of two of the ridges shown in Fig. 1(a); **(b)** The local transmittance spectrum for the right ridge.

4. Conclusions

Driven by the concept of lab on a chip, we develop lines of microscopic PhCs by a self-assembly approach. Our method, based on intermittent motion of suspension meniscus, is more simple and non-destructive compared to cutting methods with the help of machine tools. Periodically repeating ridges consisting of close-packed SiO_2 microspheres are formed parallel to the contact line. The height of the ridges strongly depends on the solution concentration. The ridges formed at high concentrations are high enough to show the properties of PhCs. The drop in the transmittance, specific for PhCs, is clearly observed in local spectra. The obtained lines of PhCs can probably be used, for example, as masks for lithography, diffraction optical elements [4] and microscopic chemical sensors.

Acknowledgments

The work was supported by the Russian Foundation for Basic Research (grant number 16-03-00853). A.A.E. acknowledge support by Russian Ministry of Education and Science (Grant No. 14.W03.31.0008).

References

[1] Watanabe S, Inukai K, Mizuta Sh and Miyahara M T 2009 *Langmuir* **25** 7287
[2] Mino Y, Watanabe S and Miyahara M T 2011 *Langmuir* **27** 5290
[3] Mino Y, Watanabe S and Miyahara M T 2012 *ACS Appl. Mater. Interfaces* **4** 3184
[4] Ashurov M S, Kazakova T A, Stepanov A L, Klimonsky S O 2016 *Appl. Phys. A* **122** 1054
[5] Klimonsky S O, Bakhia T, Knotko A V, Lukashin A V 2014 *Dokl. Chem.* **457** 115
[6] Teh L K, Tan N K, Wong C C, Li S 2005 *Appl. Phys. A* **81** 1399

SPBOPEN 2018

IOP Publishing

IOP Conf. Series: Journal of Physics: Conf. Series **1124** (2018) 051009   doi:10.1088/1742-6596/1124/5/051009

Broadband nanoantenna

D Poletaev[1], B Sokolenko[2], V Voytitsky[3], A Nudga[1]

[1]Department of radiophysics and electronics V.I. Vernadsky crimean federal university, Simferopol 295007, Russia

[2]Department of general physics V.I. Vernadsky crimean federal university, Simferopol 295007, Russia

[3]Department of mathematical analysis V.I. Vernadsky crimean federal university, Simferopol 295007, Russia

Abstract. In this paper a broadband nanoantenna was proposed. Its equivalent circuit and a model were constructed. The analysis of the nanoantenna's effectiveness was carried out. As a result, a conclusion was made about the possibility of applying this nanoantenna to photovoltaics.

1. Introduction

Nanoantennas are the wide class of instruments that are used in various fields of modern science and technology. These devices are able to directly convert light into electrical energy [1]. The advantage of such devices is the simplicity of manufacturing, which reduces the cost of final devices. However, the range of wavelengths in which a nanoantenna operates effectively is rather narrow [2]. To extend the range of nanoantennas wavelengths, the sets of nanoantenas were used [3]. Each of them is tuned to its specific wavelength. The disadvantage of such system is a large surface area occupied by a single element. In scientific works [4 – 5], fractal nanoantenas, and tie-bow nanoantennas are described. However, they can not operate over a wide range of wavelengths. It is advisable to propose a construction of a nanoantenna that capable efficiently operate over a wide range of wavelengths. The aim of the work is a theoretical analysis of the construction of the proposed broadband nanoantenna.

2. Theoretical part and simulation results

2.1. Theoretical data

The proposed nanoantenna consists of two connected monopole antennas formed by square prisms of a conducting material (figure 1). The height of the first prism is d_1, the second – is d_2. The sides of the square of the base of the prisms are the same and equal to b. The Pointing vector of the incident electromagnetic wave directed perpendicular to this structure. Each square prism of this nanoantenna has its own capacitance and inductance. Their values depend on the geometric and electrophysical parameters (relative permittivity, conductivity and relative permeability) of the elements. The capacitance and inductance of each square prism form an oscillatory circuit with its own resonant wavelength. The degree of connection of these resonant circuits varies depending on the wavelength, because of the presence of the skin layer in real conductors. The equivalent circuitry of this nanoantenna is shown in figure 2.

A solitary square prism made of a conductive material has an inductance and capacitance. These values are calculated with sufficient accuracy, for example in [6]. The magnitude of the resistances R_1

Content from this work may be used under the terms of the Creative Commons Attribution 3.0 licence. Any further distribution of this work must maintain attribution to the author(s) and the title of the work, journal citation and DOI.

Published under licence by IOP Publishing Ltd

and R_2 for the first and second prisms, respectively, are nonlinearly dependent on the length of the electromagnetic wave, due to the presence of the skin effect [7]. The impedance of such a nano-antenna can be written as:

$$Z = \left[\frac{1}{R_1 + j2\pi cL_1/\lambda} + \frac{1}{R_1 - j\lambda/(2\pi cC_1)} + \frac{1}{R_2 + j2\pi cL_2/\lambda} + \frac{1}{R_2 - j\lambda/(2\pi cC_2)} \right]^{-1} \quad (1)$$

where j – is the imaginary unit; c – is the speed of light in a vacuum; λ – is the length of the electromagnetic wave.

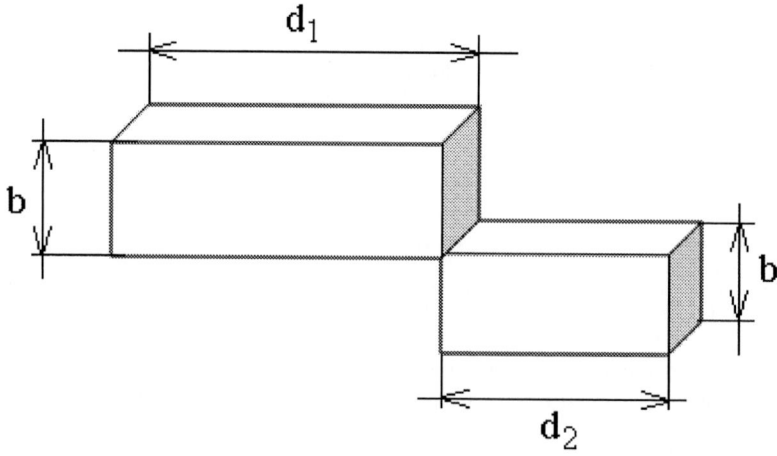

Figure 1. Structure of the nanoantenna.

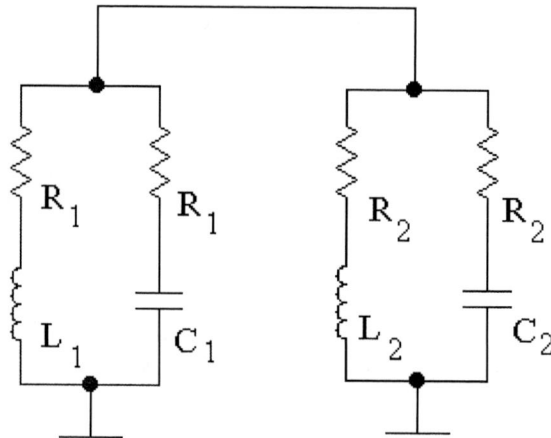

Figure 2. Equivalent circuit of a nanoantenna.

2.2. Results and discussion

Antenna operates efficiently in a range of wavelengths where its impedance is purely active [7]. The graph of the dependence of the imaginary part of the impedance of a nanoantenna (1) with optimized dimensions $d_1 = d_2 = 35$ nm, $b = 48$ nm for a conductive material – copper from the wavelength is shown in figure 3. In this graph other curve is shown for comparison the dependence of the imaginary part of a single monopole antenna formed by a square prism of conductive material with height $d_1 = 35$ nm with the side of the base square $b = 48$ nm.

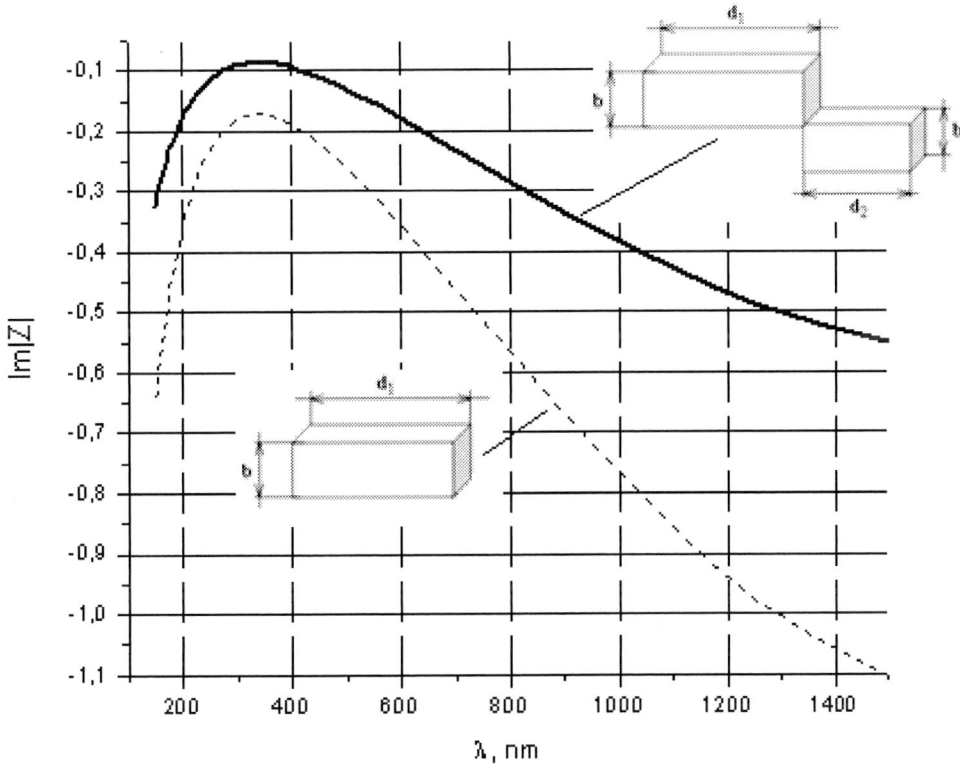

Figure 3. Dependence of the imaginary part of the nanoantenna's impedance from the wavelength.

As can be seen from the graph (figure 3) in the solar radiation wavelength range [1], the proposed nanoantenna has more than 2 times less reactive resistance than the monopole antenna. The area occupied by the proposed nanoantenna compared with monopoly antenna increases only twice. However, the effectiveness of the proposed nano antenna is much greater than the efficiency of the monopole antenna, due to a wider wavelength range. The minimum of the reactance (figure 3) is in the near ultraviolet region [1]. This makes it possible to compensate the absorption of the ultraviolet part of the spectrum in the atmosphere. The proposed nanoantenna can be used in photovoltaics.

In the future, it is planned to built numerical model of the antenna array of such nanoantenas. It is advisable to use for numerical model boundary element method.

The described nanoantenna can be made by method of deep ultraviolet lithography with double exposure. In the future, it is planned to carry out experimental studies of the antenna array of such nanoantenas.

A patent for this construction of nanoantenna had been applied. A positive decision regarding the patent in the Russian Federation was received.

Acknowledgments

This work was partially supported by the V.I. Vernadsky Crimean federal university development program for 2015 – 2024 within the framework of grant support for young scientists.

References

[1] Spinelli P et al. 2012 *J. Opt.* **14**
[2] Olmon R L, RaschkeM B 2012 *Nanotechnology* **23**
[3] Zhengtong L et al. 2008 *Metamaterials* **2**
[4] Curto A G et al. 2010 *Science* **329** 930
[5] Huang J et al 2009 *Nano Lett.* **9** 1897
[6] Balanis C A 1982 *Antenna theory: analysis and design* (New York: Harper & Row)
[7] Schelkunoff S A, Friis H T 1952 *Antennas: theory and practice* (New York: Wiley)

SPBOPEN 2018 IOP Publishing

Novel highly efficient blue-emitting branched oligoarylsilanes

M S Skorotetcky[1], O V Borshchev[1], N M Surin[1], S A Ponomarenko[1,2]

[1]Enikolopov Institute of Synthetic Polymeric Materials of the Russian Academy of Sciences 117393 Moscow, Russia

[2]Lomonosov Moscow State University, Chemistry Department 119992 Moscow, Russia

Abstract. In this work synthesis and study of novel nanostructured organosilicon luminophores (NOLs) based on *p*-terphenyl and POPOP (1,4-*bis*(5-phenyloxazol-2-*yl*) benzene) moieties with unique optical properties is reported. NOLs are branched molecules composed from a donor-acceptor-donor framework with different optical bandgap luminescence moieties connected *via* silicon atoms. As a result effective intramolecular Förster resonance energy transfer can be observed. The main advantages of such materials in comparison with conventional organic luminophores are high molar absorption coefficient, excellent photoluminescence quantum yield, fast luminescence decay time, good processability and low toxicity. Using NOLs in plastic scintillators, widely utilized for radiation detection and in elementary particles discoveries, led to significant improving their efficiency, which combines both high light output and fast decay time. In the best samples the scintillation light yield was 20% higher in comparison with the standard commercially available plastic scintillator UPS-89.

1. Introduction

Over the last years branched and dendritic luminescent molecules have been studied as promising functional materials for various organic optoelectronic devices: organic light-emitting diodes, photovoltaic cells and wavelength shifters. Such molecules exhibit so-called "molecular antennas" effect of the light absorbing by their functional groups, transfer the excitation energy nonradiatively from the periphery to the core and emit light from the one [1].

Previously, we introduced a new class of highly efficient organic luminophores – Nanostructured Organosilicon Luminophores (NOLs), which are branched or dendritic molecules with efficient Förster resonance energy transfer (FRET) between them. [2]. One of the promising applications of this class of compounds is their usage as active components in plastic scintillators (PS) – composite materials which are able to emit photons (scintillation) when excited by ionizing radiation. The classical PS is a solution of two types of organic luminophores (activator and wavelength sifter) in a polymer matrix - usually polystyrene or polyvinyl toluene. In this case, the transfer of the electron excitation energy occurs mainly radiatively from the polymer matrix sequentially to the activator, and then to the spectrum shifter, which leads to decrease in the efficiency of the final device [3]. Due to the unique structure of NOLs it is possible to combine activator and wavelength shifter in one molecule in such a way that effective non-radiative energy transfer can occur between them and thereby eliminate losses in light re-emission [4]. This requires the usage of organic luminophores with good ability to absorb the light in the ultraviolet range corresponding to the spectrum of the radioluminescence of the polymer matrix and effective luminescence. Nowadays, *p*-terphenyl as activator and 1,4-*bis*(5-phenyloxazol-2-*yl*) benzene (**POPOP**) as wavelength sifter are widely used [5] in the industry. Therefore, the development of NOLs based on

---

Content from this work may be used under the terms of the Creative Commons Attribution 3.0 licence. Any further distribution of this work must maintain attribution to the author(s) and the title of the work, journal citation and DOI.

Published under licence by IOP Publishing Ltd

p-terphenyl and **POPOP** as new functional materials for organic photonics is an important scientific and technical task, the solution of which is the aim of this work.

2. Results and discussion

For this purpose we applied *p*-terphenyl and **POPOP** luminescence moieties in accordance with the principles underlying the creation of high-performance nanostructured organosilicon luminophores (NOLs). These phosphors are well known for their highly efficient emission in UV and blue spectral regions, respectively, and high Stokes shift values [6]. We successfully synthesized and isolated in an individual state novel fluorescent molecule **NOL I-B** with *p*-terphenyl units on the periphery and fluorescent POPOP fragments in the center [7]. It is symmetrical molecule composed by a donor-acceptor-donor framework with different optical bandgap luminescence moieties connected by silicon atom (Fig. 1). As the result, an effective Förster inductive-resonance transfer of the electronic excitation energy can be observed.

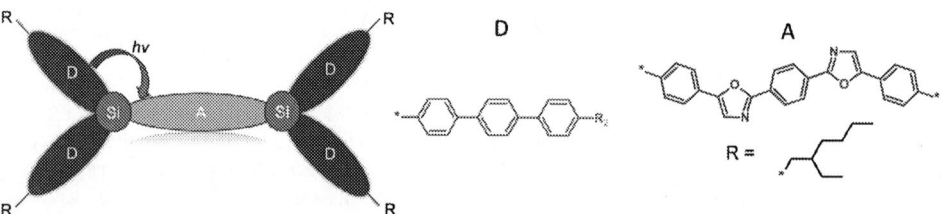

Figure 1. Schematic representation of synthesized NOL consisting of inner acceptor (A) and the peripheral donor (D) parts.

The photoluminescence study revealed that **NOL I-B** possesses high luminescence quantum yield up to 96% with the emission maximum at 425 nm and efficient intramolecular energy transfer up to 99%. Employment of several absorbing fragments makes it possible to achieve high molar extinction coefficients up to $1.8*10^5$ M^{-1}cm^{-1}. At the same time good solubility in common organic solvents was observed. One more interesting feature of this compound is the short photoluminescence decay time (0.95 ns) comparable to the acceptor luminophore itself (1.19 ns).

A scintillation efficiency study was carried out for **NOL I-B** in thin polystyrene films obtained in several steps. Initially, a mixture of polystyrene and **NOL I-B** of a different concentration, was manufactured in the extruder ("MicroCompounder DACA"). Due to its good solubility both in most organic solvents and in polystyrene, it is possible to obtain materials with a high loading of this luminophore. For comparison, *p*-terphenyl dissolves in polystyrene only in an amount of several percent [8]. After that, the polymer composites was melted and processed in a hydraulic press ("Specac"), which made it possible to obtain films of the desired thickness (420 µm). On figure 2 there is comparison of the absorption/luminescence spectra of the NOL I-B and polystyrene luminescence which consists of the luminescence of ordinary phenyl fragments and its excimers. Their good overlap is observed which is necessary for the creation of an effective PS.

A study of scintillation efficiency was carried out using the α-particle (^{241}Am) source for various concentrations of **NOL I-B** (Figure 3). As a standard, a commercially available plastic scintillator based on *p*-terphenyl and **POPOP** (UPS-89, Amcrys-H, Ukraine) with scintillation light yield 1000 photon per MeV was used (red curve). At low concentrations of **NOL I-B**, a low efficiency is observed. This is apparently connected with insufficient number of donor fragments which absorb only a part of the energy from the polystyrene matrix. With increasing concentration, the value of the light yield increases and leaves on the plateau which is connected with the attainment of the maximum value of reabsorption. In the best samples the scintillation efficiency was 20% higher in comparison with the standard.

SPBOPEN 2018

IOP Publishing

IOP Conf. Series: Journal of Physics: Conf. Series **1124** (2018) 051010   doi:10.1088/1742-6596/1124/5/051010

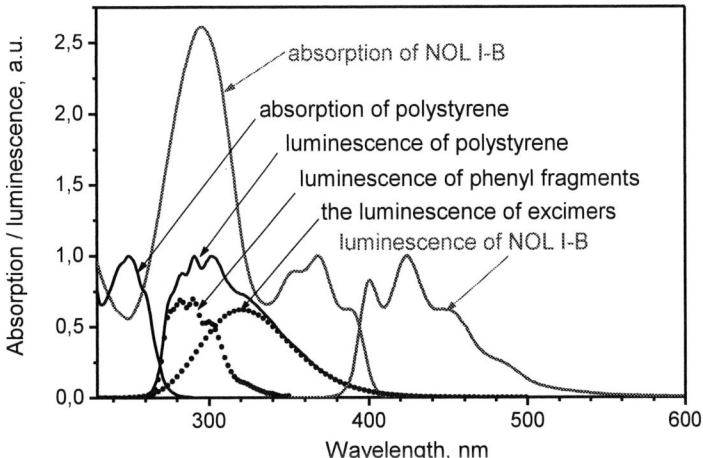

Figure 2. Absorption and luminescence spectra of **NOL I-B** (red and blue curves) and polystyrene (black curves).

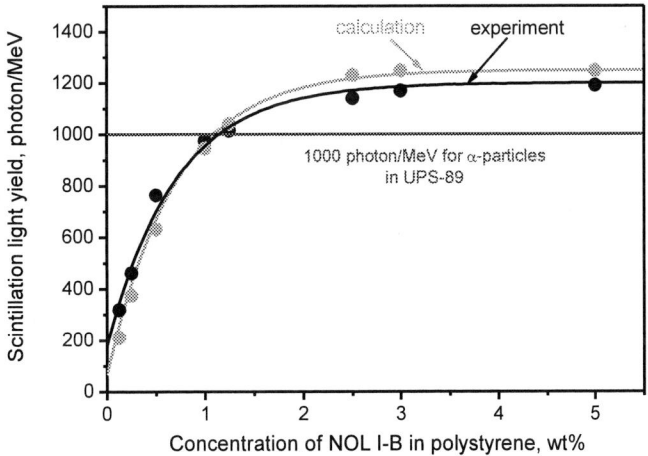

Figure 3. The dependence of the scintillation light yield on the phosphor concentration of **NOL I-B** (black curve). The green curve is the theoretical simulation of the experiment. Red line is the maximum efficiency of a standard plastic scintillator for α-particles (UPS-89, Amcrys-H, Ukraine).

The theoretical calculation of the scintillation light yield of scintillators based on NOL I-B was performed according to the scheme given below. Firstly, we calculated the parameter A:

$$A = \{1 - \exp(-[L_1] \times \varepsilon_{L_1} \times \lambda_\alpha\} \times K_R \times K_{EET} \times Q_L \qquad (1)$$

595

where, L_1 - concentration of the primary phosphor (activator); ε_{L_1} - average value of the molar extinction coefficient of activator in the spectral range corresponding to the luminescence of the polymer matrix of the scintillator; λ_α - mean free path of an α-particle with an energy of 1 MeV in a scintillator; K_R - self-luminescence reabsorption coefficient of activator or secondary phosphor (L_2) in the case of using of wavelength sifter; K_{EET} - value of the efficiency of energy transfer from the activator to the wavelength sifter; Q_L - quantum yield.

For the **NOL I-B** K_{EET} is 1, and for the standard scintillator 0.7-0.75 [3]. The value $\{1-\exp(-[L_1]\times\varepsilon_{L_1}\times\lambda_\alpha\}$ characterizes the efficiency of collecting photons emitted by matrix (polystyrene). Then, based on the known light output value of the standard scintillator, the expected light yield of the test composition is:

$$N_{ph}^x = 1000 \times \frac{A_x}{A_{s\,tand}} \qquad (2)$$

A good agreement between calculation and experiment confirms the correctness of the principle combining activator and wavelength sifter of conventional scintillator into one molecule. The gain in the scintillation light yield is achieved due to the higher efficiency of energy transfer by the non-radiative mechanism in comparison with the radiative transfer from *p*-terphenyl to **POPOP** after the equality of the values of $\{1-\exp(-[L_1]\times\varepsilon_{L_1}\times\lambda_\alpha\}$ for **NOL I-B** and standard scintillator. The further growth of the scintillation light yield is limited by the reabsorption as a result of concentration increase.

Thus, in this work an efficient method for the synthesis of novel NOL containing four *p*-terphenyl donor and **POPOP** acceptor chromophoric fragments separated by silicon atoms is developed. The systematical study of their properties showed their high potential for application as functional materials in particles photodetectors and UV-Vis wavelength shifters for efficient down-conversion of light [9].

Acknowledgments
This work was financially supported by Federal Agency of Scientific Organizations and Ministry of Science and Education of the Russian Federation (grants NSh-5698.2018.3 and MK-5235.2018.3).

References
[1] Ziessel R, Harriman A. 2011 *Chem. Comm.* **47** 611
[2] Ponomarenko S A, Surin N M, Borshchev O V, Luponosov Y N, Akimov D Y, Alexandrov I S, Burenkov A A, Kovalenko A G, Stekhanov V N, Kleymyuk E A, Gritsenko O T, Cherkaev G V, Kechek'yan A S, Serenko O A, Muzafarov A M. 2014 *Sci. Rep.* **4** 6549
[3] Rozman I M, Kilin S F. 1959 *Successes of physical sciences* **3** 459
[4] Surin N M , Ponomarenko S A, Borshchev O V, Luponosov Y N, Muzafarov A M. 2010 *Patent №* RU **2380726**
[5] Grinev B V, Senchishin V G. 2003 *Plastic scintillators* (akta)
[6] Skorotetcky M S, Borshchev O V, Surin N M, Odarchenko Y, Pisarev S A, Peregudova S M, Törnroos K W, Chernyshov D, Ivanov D A, Ponomarenko S A. 2017 *Dyes Pigm.* **141** 128
[7] Ponomarenko S A, Surin N M, Borshchev O V, Skorotetcky M S, Muzafarov A M. 2015 *Proc. of SPIE* **9545** 954509
[8] Yemam H A, Mahl A, Tinkham J S, Koubek J T, Greife Uwe, Sellinger A. 2017 *Chem. Eur. J.* **23** 8921
[9] Ponomarenko S A, Borshchev O V, Surin N M, Skorotetcky M S, Kleymyuk E A, Starikova T Yu, Tereshenko A S. 2017 *Proc. SPIE* **103440N**

AUTHOR INDEX

Abdullin, F A 378
Abdullin, Kh. A. 77
Afinogenov, B. I. 408
Agafonov, I Y 444
Ageev, O A 1, 149, 158, 162, 964, 1087, 1141
Aglikov, A. 542
Akasov, R. A. 277
Akhmerov, Yu 499
Akimov, A. 761
Aksenov, E T 306, 311
Al-Alwani, A J 91, 257, 1018, 1136
Alekhina, O V 1096
Alekseenko, A P 272
Alekseev, A A 984
Alekseev, P A 202, 1038, 1121
Alekseev, P 1083
Alekseev, V 956, 972
Alexeev, A N 1087
Alexeyev, C N 578
Alexeyeva, M C 578
Aleynik, A S 852
Almaev, A V 895
Alonso, E T 905
Aluyev, A 499
Alyabyeva, N I 87
An, P 765, 769, 784, 788
Anikeeva, V E 387
Anokhin, A 659
Anoshkin, I 602
Antipova, O 315
Antonov, I N 1106
Antonov, V I 1151
Arsenov, P V 1171
Artamonov, D V 1043
Arteev, D S 458
Artyukhov, I 968
Arustamyan, D A 915
Arzamastsev, S. 956
Ashurov, M S 586
Avdeev, S P 118, 158
Averkin, S 1156
Avilov, V I 87
Avilova, M M 1183
Babich, E S 642, 678
Babichev, A V 486
Babkin, A V 1110, 1115
Babkina, Anastasiia N 856
Bagayev, S N 630

Bagshaw, J. T. 537
Baidus, N V 171, 520
Bakarov, A K 848
Bakulin, E 998
Balakirev, S V 1, 81, 112, 462, 1087
Baldycheva, A 905
Balezin, M 813
Baranov, A I 507, 515
Baranov, M A 272
Baryshev, A M 874
Beginin, E N 900
Begletsova, N 1018, 1026
Bektemirova, M 870
Belanova, A 659
Belenkov, E A 6, 43, 49
Belenkov, M E 43
Belikov, A. I. 557
Belorus, A O 262
Belov, A I 30
Belyakov, M 813
Benemanskaya, G V 722
Benimetskiy, F A 1121
Berdnikov, Y 181
Berdnikov, Yu 292
Berezovskaya, T N 980
Bert, N A 122, 127
Berzhansky, V 838, 844, 870
Besedina, N A 292
Bezpaly, A D 651
Bezuglaya, E. V. 417
Bikberdina, N 288
Blokhin, S A 582
Bobrova, D 288
Bogatov, N. M. 96, 915
Bogdanov, A 233, 320, 808, 813, 817
Bogomazova, N 1126
Boikov, I K 476
Bokova, A 838
Bolshakov, A 154, 197, 436, 449, 532, 542, 976
Bolshedvorskii, S 761
Bondarev, A D 176
Borodin, B R 1121
Boronenko, M 288
Boronin, A I 107
Borshchev, O V 593
Bougrov, V 486, 1013
Bouravleuv, A D 202, 980
Breuer, S 352

Bukatin, A 315, 333, 343, 582
Bulyanitsa, A 283
Burakov, A E.............................1096, 1110, 1115
Burakova, I V...............................1096, 1110, 1115
Burtsev, A A...948
Butt, M A 551, 563
Buyanova, E S 574
Bykova, E M..87
Carapezzi, S 1165
Chebotarev, S N 68, 915
Chekhonin, I A..630
Chekhonin, M A...630
Chelny, A 499
Cherkasov, K V...942
Chernikov, E V 895
Chernov, V M..43
Chernova, R K 1026
Chernyakov, A E 458
Chernyshev, A..659
Chikalova-Luzina, O P 328
Chipouline, A 597
Chu, Y H 1146
Chumakov, A 1018, 1026, 1136
Cirlin, G................................. 20, 202, 1083
Clemente, I............................... 149, 1156
Craciun, M F..905
Danilina, E. M. 96
Danilov, V V..348
Danilov, Yu 865
Danilushkin, A V..948
Danishevskaya, Y 870
Davydov, R V 1151, 1160
Davydov, V V 63, 227, 365, 382, 403, 467
Davydova, T A..895
Degtyarev, S A...551
Deligeorgis, G...1070
Demchenko, P..602, 905
Dementev, P A...722
Demidov, A A................................928, 952
Demidov, E...860
Demina, P. A. 277
Denisenko, M. V. 673
Denisov, S A..171
Derevyannikova, E A......................................107
Dermeneva, M. 1013
Didenko, S...499
Dikareva, N V..520
Dimakis, E 181
Divochiy, A V 689, 699
Dolzhikova, A...956
Dorokhin, M 171, 860, 865
Dostanko, A P 149

Dragunova, A S 444
Drakin, A Yu...928, 952
Dubrovskii, V G 142, 181, 292, 352
Dudelev, V V 486
Dukelskii, K V 827
Dunaeva, T. 972
Dunaevskiy, M 1038, 1083, 1106, 1121
Dvoreckaia, L N 154
Dvoretckaia, L N 197, 976
Dzyubanenko, S 998
Efimov, A A 1132, 1171
Efimov, Yu P 743
Egorkin, V I 1074
Egorov, A Yu 486
Egorov, I 1049
Egorov, V S 630
Egorova, M N..984
Elezov, M. S. 611
Eliáš, M 421
Eliseev, S A 743
Elizarov, V 718
Emelyanov, A 233, 320
Emelyanov, V M 503
Enns, Y. B. 878
Enns, Y. 1008
Eno, N. A. 537
Eremenko, M M 1, 81, 112
Ermak, S V 747
Ermakov, A 1018
Eseev, M K 1078
Evreinova, N V... 252
Evseenkov, A S 937
Evstrapov, A A..343
Evstrapov, A 283
Evstropiev, S K 827
Evstropov, V V 360, 481
Evstropyev, K S 827
Evtikhiev, V P 621
Ezhov, A A 586
Faradzhev, S P 994
Faradzheva, M P 994
Fatkhutdinova, L I 747
Fedorov, G E 797
Fedorov, G F 779
Fedorov, S. N. 472
Fedorov, V V 39, 154, 197, 449
Fedoseev, M 35
Fedotov, A A 133
Ferrari, S 756, 765, 769, 784, 788
Fetisova, M V 444, 582
Filatov, D O..1106
Filatov, N A...333, 343

Filimonov, E D 503, 524
Filippov, S V 1165
Firsov, D A 486
Fomchenkov, S A 568, 616, 710, 739
Fomin, A 53, 968, 1049, 1053, 1057
Fomin, E V 176
Fomina, M 35, 53, 968
Frizyuk, K S 647
Frolov, I V 440
Gabitov, I 597
Gajdoš, A 392, 421
Galkovsky, T 1126
Galunin, E V 1115
Gangrskaia, E 642, 823
Gareev, K G 252
Garmash, V I 1074
Gavrilov, S A 1102
Gavrus, I V 68
Gayduchenko, I A 779, 797
Generalova, A. N. 277
Gerasimov, V A 221
Gets, D 453
Girshova, E I 1193
Gladkov, A 315
Gladyshev, A G 486
Gladyshev, S. A. 817
Glagolev, S F 63
Glechik, D 844
Glukhovskoy, E 91, 257, 1018, 1026, 1136
Golikov, A 664, 765, 784, 788
Goloshchapov, D 72, 243
Golosov, D A 149
Goltsman, G 611, 664, 756, 761, 765, 769, 779, 784, 788,
797
Golubkov, P E 378, 1043
Golubok, A O 436, 976
Gomon, D 602
Goncharova, L M 68
Gorbyak, V V 727
Gordeev, N Yu 372, 427, 547
Gorokh, G 57, 1126
Gorshkov, O N 1106
Goryacheva, I Yu 1061
Goryacheva, V A 257
Goshev, A A 1078
Grebenchukov, A N 793, 905
Grebenikova, N M 403
Greshnyakov, V A 6
Gridin, G E 616
Grigoryev, M N 1179
Grishkanich, A 718
Gromov, D G 1102
Gubaidullin, A R 1188
Gubaydullin, A R 736, 803, 1199
Gudkov, V A 1074
Gudovskikh, A 39, 398, 431, 495, 507, 528
Guina, M 1205
Guryev, V I 267
Gusev, E Yu 118, 158
Guseva, Y A 408
Guseva, Yu A 356, 444
Gvozdev, D A 212
Helm, M 181
Hennrich, F 779
Ierusalimsky, N. V. 277
Il'In, O I 133
Il'Ina, M V 919
Ilkiv, I V 202
Inyushov, A 705
Iorsh, I. V. 808
Ippolitov, Y A 243
Ishteev, A 453
Isoaho, R 1205
Ivanov, K A 803, 1188, 1193, 1199
Ivanov, S V 192
Ivanov, V V 1132, 1171
Ivonin, M 149
Javadzade, J 761
Jean, T 142
Jityaeva, J Y 158
Jmerik, V N 192
Kaasik, V 668, 823
Kadykova, Yu 956
Kaftyreva, L A 267
Kahl, O 756
Kaidashev, E 659
Kalganova, S 956, 968, 972
Kaliteevski, M A 736, 803, 1188, 1193, 1199
Kalyuzhnyy, N 137, 360, 372, 427, 481, 491, 528
Kamalov, R V 15
Kapitanova, P V 813, 817
Kapitonov, A N 984
Kapitonov, Yu V 743
Kapustin, S N 1078
Karachinsky, L Ya 486
Karakchieva, O 832
Karavainikov, A 838
Kardash, T Yu 107
Karenkikh, O G 1087
Karzanov, V 860
Katin, K. P. 1070
Kaydashev, V 659
Kazakin, A. N. 878
Kazakov, V 636

Kazakova, T A 586
Kazantsev, V 315
Khaydukov, E. V. 277
Khochenkov, D. A. 277
Khodzitsky, M K 793, 905
Khodzitsky, M. 602
Khonina, S N 551, 563, 616
Khorin, P A 568
Khrebtov, A I 20, 348
Khripunov, A 248
Khromov, V S 328
Kiesewetter, D 248
Kinev, N V 874
Kirilenko, A 154
Kirilenko, D A 122, 202
Kivshar, Y. S. 817
Kleider, J-P 515
Kleimanov, R. 1008
Klenov, N. V. 673
Klimenko, V 233, 348
Klimin, V S 81, 149, 162, 462, 960, 964, 1141
Klimonsky, S O. 586
Klunnikova, Y V 11
Knyaginin, D A 952
Knyazev, N 233
Kolesnikov, I E 684
Kolesnikova, A S 91, 1136
Kolodeznyi, E S 486
Kolomiytsev, A S 87
Kolosko, A G 1165
Kolpakov, V 315
Kolykhalova, E D 1193
Komarevtsev, I. 1008
Komrakova, S 761
Konakov, S. 323, 998
Kondrashin, V I. 378
Konkov, O I 994
Konopel'Ko, L 718
Konshin, A A 919
Kopyev, V V 933
Korenev, V V 352
Koriazhkina, M N 1106
Korikov, C C 311, 731
Korneev, A 664, 756, 761, 765, 769, 784, 788
Kornev, A A 582
Korolev, D S 30, 1106
Korolev, D V 252
Korotkov, A 998
Korsakov, A S 752
Koryakin, A A 166, 181, 206
Koshelets, V P. 874
Koshelev, K. L. 808, 817

Koshelev, O A 192
Koshkin, A Yu 467
Koshuro, V 35, 1057, 1066
Kostrin, D K 221
Kotliar, K P 176
Kotlyar, K P 202, 356, 444, 495, 511, 582
Kotlyar, K. 408
Kots, I N 960, 1141
Koubisy, M S I 574
Kourova, N. 499
Kovalskiy, V A 171
Kovalyuk, V 664, 756, 761, 765, 769, 784, 788
Kozlowski, A V 1003
Krivenkov, R Yu 625
Kryzhanovskaya, N V ... 20, 356, 408, 444, 528, 582, 695, 1205
Kudrin, A 860
Kudryashov, D 398, 431, 495, 507, 511, 528, 542
Kukushkin, S A 24, 166, 206
Kulagina, A S 348, 827
Kulagina, M 547, 1205
Kulichkova, V A 239
Küppers, F 597
Kupriyanov, A N 910
Kurnosov, D A 1110, 1115
Kurochkin, A S 486
Kurtenkov, R. V. 472
Kuzhelev, M V 133
Kuznetsov, Y 860, 865
Kuznetsova, I A 884
Kuznetsova, O B 306
Kuznetsova, Yu A 413, 574
Lapin, B P 578
Lavrentyev, V 972
Lazarenko, A A 100
Lazarenko, P I 1102
Lebedev, A A 202, 1121
Lebedev, E A 1102
Lebedev, S P 202, 1121
Lenshin, A 72, 262
Lesnikov, V 860
Levin, D D 337
Levina, S A 503, 524
Levitskii, I V 621
Lewes-Malandrakis, G 756
Liashenko, T. 453
Lihachev, A I 923
Limonov, M. F. 817
Lioubtchenko, D. 602
Lipovskii, A 408, 444, 582, 642, 668, 678, 695, 823
Lisitsyn, S A 462
Lobino, M 664
Loboda, V 998

Lobov, S A 296
Logachev, V 499
Loganchuk, S M 915
Logunov, S E 467
Lozovenko, A A 57
Lozovskii, V N..68
Lubenchenko, A V 689
Lugovitskaya, T N 257
Lugovskoy, N 844
Lukashchuk, A 597
Lukashenko, T 283
Lukashev, E P 212
Lutetskiy, A V 486
Lyanguzov, N 659
Lyashko, S 838
Macku, R 392, 421
Makarenko, D. P. 77
Makarov, S...............................453, 542, 607
Makarov, V A 296
Makeev, M O 942
Makevnina, V V 217
Maksimova, N K 895
Malekizandi, M 597
Malyar, I V 1003
Malysheva, E 865
Malyshkin, V G 382
Malyukov, S P 11
Mamaeva, T A..436
Mamonkina, A N 747
Manshina, A A 684
Markin, A V 1061
Markina, N E 1061
Martemyanov, N A..15
Martynov, A V 1043
Mashkovtsev, M A..413
Maslov, A. D. 417
Maslov, M. M. 1070
Masyukov, M S 793
Matyushkin, Y E 797
Maximov, M V.....352, 356, 372, 408, 427, 444, 547, 582, 1205
Mekhov, I B 630
Mekhrengin, M V 267
Melezhik, A V 1096, 1115
Meshkov, S A 942
Meshkovskii, I K 267
Mezentsov, S. 1057
Mezentsov, S 1175
Mezhenny, M 499
Miakonkikh, A 149, 1156
Mikhailova, T 838, 1179
Mikhaylin, I A....................................1, 81, 112
Mikhaylov, A N....................................30, 1106

Mikhaylov, A O 1038
Mikhaylovskaya, Z A 574
Mikheev, G M 625
Mikheev, K G 625
Mingaleva, A E 714
Minkov, K N 1171
Mintairov, M A..360, 481
Mintairov, S A 137, 360, 372, 427, 481, 528
Mintairov, S. 491
Mitin, D M 532
Mitrofanov, M I 621
Mizerov, A..72, 980
Mizerov, M N 621
Mkrtchyan, E S 1096, 1115
Mogileva, T N 625
Mohamed, A A A 915
Moiseenko, I M 655
Moiseev, E I 356, 408, 444, 582, 1205
Mokhov, D V 980
Monastyrenko, A 431
Morozov, I A 495, 507, 515
Morozov, K M 736, 803, 1199
Morozov, M Y 655
Morozov, P V 689
Morozov, S V 20
Moshnikov, V A 262
Moskaletz, O 636
Moskotin, M V 779, 797
Motúz, R 421
Mozharov, A.............154, 197, 449, 507, 532, 542, 976
Mukhin, I.................154, 197, 449, 532, 542, 976
Mukhin, S 436
Mukhina, I 315
Murashev, V 499
Murashkevich, A 1126
Muravijova, D. 1013
Myasoedova, T N 1179
Myazin, N S 227
Mylnikov, I L 1032
Mynbaeva, M. 1013
Nadtochiy, A M 137, 1205
Naidenko, N A 960
Nashchekin, A V 176, 923
Nasibulin, A G 408, 532
Naumisheva, E B 252
Nebel, C 756
Nebojsa, A 421
Nechaev, D V 192
Nedviga, A 870
Nekipelov, S V 714
Nekorkin, S M 520
Nemtseva, S Yu 1102

Nenasheva, E A 813
Nepomnyashchaya, E K 301, 731
Neskorniuk, V 597
Neskoromnaya, E A 1096, 1110
Nevedomskiy, V N 127, 137
Nevedomsky, V N 486
Nevolin, V K 337
Nezhentsev, A V 1074
Niemi, T 1205
Nikiforov, K 103
Nikitina, E 72, 100, 980, 1188
Nikolaev, D N 621
Nikolaev, K O 378
Nikolaev, V G 630
Nikolaeva, A A 647
Nikolskaya, A A 30
Nikonorov, N V 827, 856
Novikov, A S 408
Novikov, I I 486
Novopashin, S A 107
Nozdriukhin, D V 333, 343
Nudga, A 589
Nunn, N 625
Obraztsova, E D 797
Oleinik, V L 933
Olekhno, N A 647
Osipov, A V 24
Ovchinnikov, G 597
Pakharukov, Yu V 990
Palkanov, P 1053
Panchak, A 491
Panchenko, I V 1146
Pankin, D V 684
Papež, N 392, 421
Papshev, V 53
Parafin, A 865
Parulin, R A 574
Paschenko, V Z 212
Pashchenko, A. S. 96
Pastukhov, A I 262
Pavlov, D A 30, 1106
Pavlov, P. A. 537
Pavlov, S I 176
Payusov, A S 372, 427, 547
Pecherskaya, E A 378, 1043
Pelipenko, M I 1146
Perin, A 651, 705, 774
Perkov, S A 320
Permyakov, D V 436
Pernice, W 756, 765, 769, 784, 788
Petrenyov, I A 15
Petrov, A A 365

Petrov, M I 647
Petrov, S I 1087
Petrov, V. V. 77, 1183
Petrova, J 890
Petrova, O V 714
Petukhov, V A 337
Phyo, Kyaw Z. 557
Pigareva, Y 315
Pikhtin, N A 486
Pimashkin, A 315
Pinaev, N 398
Pirogov, E V 100
Pligovka, A 1126
Plotnikova, Y A 243
Podoskin, A A 1038
Pogorelaya, D A 267, 852
Pogosov, A G 848
Pokhabov, D A 848
Poletaev, D 589
Polikarpov, Yu A 337
Polischuk, O V 655
Poltavtsev, S V 743
Polyakova, V V 1141
Ponomarenko, S A 593
Ponomaryova, N S 928
Popov, E O 1165
Popov, V V 655
Porfirev, A P 616, 739
Poshivalova, E 1092
Postnov, V N 252
Pozharov, M V 1136
Pozina, G 1188
Poznyak, A A 57
Prikhodko, A V 994
Prokhodtsov, A 788
Prokopov, A 838, 844
Proskuryakov, V 1175
Protas, N V 1132, 1171
Prudaev, I A 933
Prudkovskiy, V. S. 1070
Puech, P. 1070
Punegov, V I 187
Pustozerov, A V 774
Putintseva, M V 311
Pyatishev, E. 878, 1008
Pyatkov, F 779
Qassime, M M 257
Rabinovich, O 499
Radaev, O A 440
Rajanna, P M 532
Raskhodchikov, A V 695
Raskhodchikov, D V 668

Raskhodchikov, D 823
Raudik, S A 532
Reddy, Andra N. K. 563
Redkov, A V 24, 206
Reduto, I 444, 582, 642, 823
Reshetov, I V 668
Reznik, A 248
Reznik, R 20, 511, 1083
Rezvan, A A 162, 462, 960, 964, 1141
Rodin, A 398
Rodionov, I 968, 1049, 1053, 1057, 1092, 1175
Romanov, N R 699
Romashkin, A V 337
Rud', V. Y. 403, 1160, 1160
Rudakov, K I 874
Rudenko, K 149, 1156
Rudnitskaya, G 283
Rudyk, N N 133
Russo, S 905
Ruzankina, J 718
Ryabova, M A 1106
Ryazanov, R M 1102
Rybalka, S B 928, 952
Rybalko, D A 547
Rybin, M G 797
Rybin, M. V. 817
Rychkov, V N 413
Rykov, A V 171, 520
Ryzhuk, R V 1087
Rzhevskiy, S S 684
Sadreev, A F 813
Sadrieva, Z 808, 813, 817
Safargaliev, R F 990
Sakhadzhi, G. V. 948
Salii, R A 137, 360
Salimgareev, D D 752
Samartsev, I V 520
Samosvat, D M 328
Samsonenko, Yu B 20
Samusev, A K 436
Samusev, K. B...817
Sannikov, D A 1205
Santer, S 1018
Sapunov, G A 154, 449
Saranin, D 453
Satanin, A. M. 673
Savchenko, E A 301, 306
Savchuk, A 499
Savelyev, A V 352, 476
Savelyev, A. G. 277
Savelyev, D A 710
Savenko, O V 884

Sayenko, A V 11
Scherbak, S A 356, 444, 582, 678, 695
Scherbatenko, M. L. 611
Schneider, H 181
Sedykh, S V 928
Selenina, A V 239
Selifonova, E I 1026
Selivanov, L M 221
Semenova, O I 387
Semyonov, V V 747
Seredin, P 72, 243, 262
Sergeev, V A 440
Serin, A A 372, 427, 547
Sevastyanov, E Yu 895
Sevostyanov, V P 91, 1026, 1136
Shabanov, V E 365
Shabiev, F K 990
Shabunina, E I 458
Shaman, Yu P 1102
Shamshin, M O 296
Shandarov, V M 651
Shandyba, N A 1146
Shaposhnikov, A 838
Shapovalov, D V 365
Sharaevskaya, A Yu 900
Sharov, V 1083
Shchelkunov, A 1049
Shenderova, O A 625
Shengurov, V G 171
Shenina, M E 1106
Shepeleva, Y V 1043
Shernyakov, Yu M 356, 372, 427, 547
Shevyrin, A A 848
Shinkarenko, O 1018, 1136
Shipovskaya, A B 257
Shipulya, N D 323
Shishkina, O A 928
Shklyaev, A A 848
Shkoldin, V A 436
Shmakov, S 233, 320, 348
Shmidt, N M 458
Shmygalev, A S 752
Sholina, N. V. 277
Shomysov, N N 714
Shostka, N 832
Shubina, K Y 495, 980
Shugurov, K U 449
Shumikhin, K. V. 948
Shumilin, A 53, 968
Shurkaeva, I V 689
Shvarts, M. 360, 481, 491, 503, 524
Sibirev, N V 181

Sidorov, A I	727	Svirina, A	233
Simon, V A	221	Svirko, Yu	823
Sitnikova, A A	122, 154	Sych, D. V.	611
Sivkov, D V	187, 714	Sychev, S. K.	808
Sivkov, V N	714	Tagantsev, D K	668
Skandakov, R N	714	Talnishnikh, N A	458
Skaptsov, A	53	Tauchnitz, T.	181
Skarvada, P	392	Terekhov, V I	752
Skibina, Y S	1061	Tereshchenko, O E	387
Skorotetcky, M S	593	Terterov, I	233
Škvarenina, L	392, 421	Terukov, E I	1160
Skvortsov, A N	301	Tetelbaum, D I	30
Slavinskaya, E M	107	Tikhomirov, V G	408, 937
Slipchenko, S O	486, 1038	Timoshenko, I V	574
Smagulova, S A	984	Timoshnev, S N	722
Smirnov, E	664	Tingaev, M I	49
Smirnov, K J	63, 403	Titova, N	779
Smirnov, K V	689, 699	Tkachev, A G	1110
Smirnov, V A	1146	Tolstunov, M I	11
Smirnova, T E	895, 933	Tomilin, A N	239
Smolovik, M A	852	Tomilin, S	838
Snigirev, L A	122	Tominov, R V	87, 1146
Sobola, D	392, 421	Tomskaya, A E	984
Sobolev, Dmitrii I	856	Touel, L	915
Sobolev, M S	100	Trigorly, S	972
Sochilina, A. V.	277	Trushnikov, I	705
Sofronov, A N	486	Tsimokha, A S	239
Sokolenko, B	589, 832	Tsvetkova, O Yu	91, 1136
Sokolovskii, A S	181, 292	Tsymbalov, A	890
Sokolovskii, G S	486	Tupik, A	283
Sokura, L A	127	Tushavin, G V	63
Solodovnik, M S	1, 81, 112, 462, 1087	Tutova, M S	1022
Solomin, E G	919	Uhov, A A	221
Solovev, I A	743	Uskov, A V	436, 976
Soshenko, V	761	Uvarov, A V	39, 511, 515
Soshnikov, I P	176, 511	Uvarov, I V	910
Sosnin, D	542	Vaganov, M	636
Spivak, Yu M	262	Vakhtomin, Yu B	689, 699
Stadnichenko, A I	107	Vakulov, Z	149, 1141
Starkov, A S	1032	Varlamov, D	283
Starkov, I A	1032	Varnavskaya, A V	68
Starnikova, A. P.	77	Vasilev, A	542
Stetsyura, S V	1003	Vasileva, A A	684
Stolpner, A	248	Vasiliev, A A	436
Strigalev, V E	852	Vasiliev, V. V.	472
Stroganov, B V	743	Vasin, S V	440
Struchkov, N S	337	Vasinkina, E	956, 972
Stupin, D D	923	Ved, M	171, 865
Sun, Y	607	Velichko, E N	272, 301, 311
Surin, N M	593	Vergeles, P S	171
Surnin, K A	176	Vetter, A	756
Sushkov, A A	30	Vikulin, D V	578

Vlasov, A A	852
Vokhmintsev, A S	15
Volik, D P	149
Volkova, N A	827
Volodin, A	659
Vongsvivut, J	243
Vorobjev, L E	486
Vorobyov, V	761
Voronov, B	664, 779
Voyko, A	53, 968
Voytitsky, V	589
Vozianova, A V	793
Voznyuk, G V	621
Vyatkina, S P	413
Vysochinskaya, V	233
Weinstein, I A	15
Yagafarov, T	1205
Yagovkina, M	1013
Yakimov, E B	171
Yakimova, R.	1070
Yakovenko, N A	68
Yakovlev, G E	515
Yashina, N Yu	91
Yavorsky, M A	578
Yugay, A	998
Yulmetova, O S	1022
Yulmetova, R F	1022
Yushkanov, A A	884
Yusupova, E A	382
Zabalueva, Z A	731
Zaikovskii, A V	107
Zaitsev, A D	905
Zakharevich, A	53, 1061
Zakhidov, A	453
Zakhlebayeva, A	1126
Zalyotov, D V	365
Zamburg, E G	149, 1146
Zatsepin, A F	413, 574
Zavadskiy, S M	149
Zdoroveyshchev, A	171, 520, 865
Zegrya, A G	328
Zegrya, G G	328
Zelentsov, K S	507, 528
Zemlyakov, V E	1074
Zemlyanoy, V K	1132
Zhdanov, E Yu	848
Zhevlakov, A	718
Zhilkin, B P	752
Zhukov, A E	20, 352, 356, 372, 408, 427, 444, 582, 695, 1205
Zhukov, M V	436
Zhuravleva, N	248
Zinchenko, T O	378, 1043

Zolotov, P	664, 689, 699
Zolotukhin, D	72
Zolotukhin, P.	659
Zorin, V N	252
Zubkova, E	765, 769, 784, 788
Zubov, F I	372
Zuev, D A	607
Zvonkov, B N	520
Zybin, A A	458

Institute of Physics
Dirac House, Temple Back
Bristol BS1 6BE UK

ISSN: 1742-6588
ISBN 978-1-5108-8099-3

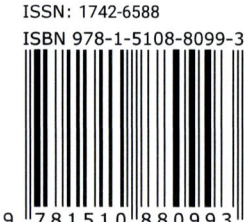

5th International School and Conference "Saint Petersburg OPEN" 2018

Optoelectronics, Photonics, Engineering and Nanostructures

Journal of Physics: Conference Series Volume 1124

St. Petersburg, Russia
2-5 April 2018

Part 2 of 2

5th International School and Conference "Saint Petersburg OPEN" 2018

Optoelectronics, Photonics, Engineering and Nanostructures

Journal of Physics: Conference Series Volume 1124

St. Petersburg, Russia
2-5 April 2018

Part 2 of 2

ISBN: 978-1-5108-8099-3
ISSN: 1742-6588

Printed from e-media with permission by:

Curran Associates, Inc.
57 Morehouse Lane
Red Hook, NY 12571

Some format issues inherent in the e-media version may also appear in this print version.

This work is licensed under a Creative Commons Attribution 3.0 International Licence.
Licence details: http://creativecommons.org/licenses/by/3.0/.

No changes have been made to the content of these proceedings. There may be changes to pagination and minor adjustments for aesthetics.

Printed with permission by Curran Associates, Inc. (2026)

For permission requests, please contact the Institute of Physics
at the address below.

Institute of Physics
Dirac House, Temple Back
Bristol BS1 6BE UK

Phone: 44 1 17 929 7481
Fax: 44 1 17 920 0979

techtracking@iop.org

Additional copies of this publication are available from:

Curran Associates, Inc.
57 Morehouse Lane
Red Hook, NY 12571 USA
Phone: 845-758-0400
Fax: 845-758-2633
Email: curran@proceedings.com
Web: www.proceedings.com

TABLE OF CONTENTS

VOLUME 1

Preface

Peer Review Statement

CRYSTAL GROWTH AND STRUCTURAL PROPERTIES OF SEMICONDUCTOR MATERIALS AND NANOSTRUCTURES

Analytical–Monte Carlo Model of the Growth of in Nanostructures During Droplet Epitaxy on the Triangle-Patterned GaAs Substrates 1
S V Balakirev, M S Solodovnik, I A Mikhaylin, M M Eremenko, O A Ageev

Investigation on Structural Transitions of Graphenes into Diamond Polymorphs at High Pressure 6
V A Greshnyakov, E A Belenkov

Investigation of Titanium Oxide Film on Sapphire Substrate for Gas Sensor 11
Y V Klunnikova, S P Malyukov, A V Sayenko, M I Tolstunov

Nanostructural Features of Anodic Zirconia Synthesized using Different Temperature Modes 15
I A Petrenyov, R V Kamalov, A S Vokhmintsev, N A Martemyanov, I A Weinstein

Structural Properties of Multilayer Heterostructure for Quantum-Cascade Lasers Grown by MBE Growth 20
R R Reznik, N V Kryzhanovskaya, A E Zhukov, A I Khrebtov, Yu B Samsonenko, S V Morozov, G E Cirlin

Spiral Growth of a Crystal Due to Chemical Reaction 24
A V Redkov, S A Kukushkin, A V Osipov

Formation of Hexagonal Silicon Regions in Silicon 30
A A Nikolskaya, D S Korolev, A N Mikhaylov, A I Belov, A A Sushkov, D A Pavlov, D I Tetelbaum

Microstructure and Composition of Zirconium Coatings Obtained by Electro-Spark Alloying 35
V Koshuro, M Fomina, M Fedoseev

Investigation of Deposition Conditions on the Structural Properties of μc-Si:H 39
A V Uvarov, A S Gudovskikh, V V Fedorov

Structure of Fluorographene and Its Polymorphous Varieties 43
M E Belenkov, V M Chernov, E A Belenkov

Carbon Materials Formed by Polymerization of C_{20} and C_{24} Fullerites 49
M I Tingaev, E A Belenkov

Metal Oxide (Ta-TaOx)-Coatings Obtained by Magnetron Sputtering and Heat Treatment with High-Frequency Currents 53
M Fomina, A Voyko, A Shumilin, V Papshev, A Zakharevich, A Skaptsov, A Fomin

Synthesis of Bismuth Nanowires for Thermoelectric Applications 57
A A Lozovenko, A A Poznyak, G G Gorokh

High Speed Near-Infrared Range Sensor Based on InP/InGaAs Heterostructures 63
K J Smirnov, V V Davydov, S F Glagolev, G V Tushavin

Features of a Zone Thermal Crystallization Semiconductor Thin Films Grown from a Discrete
Liquid Source 68
*S N Chebotarev, A V Varnavskaya, I V Gavrus, V N Lozovskii, L M Goncharova, N A
Yakovenko*

Comprehensive Study of the Optical and Physical Properties of InGaN Nanocolumns Grown on
Si(111) and por-Si/Si(111) Substrates by Plasma-Assisted Molecular Beam Epitaxy 72
D Zolotukhin, P Seredin, A Lenshin, D Goloshchapov, A Mizerov, E Nikitina

Features of the Mechanism of Gas Sensitivity of the Zinc Oxide Nanorods Arrays to Carbon
Monoxide 77
V. V. Petrov, A. P. Starnikova, Kh. A. Abdullin, D. P. Makarenko

Droplet Epitaxy of In/AlGaAs Nanostructures on the As-Stabilized Surface 81
S V Balakirev, M M Eremenko, I A Mikhaylin, V S Klimin, M S Solodovnik

Investigation of the Electrode Material Influence on the Titanium Oxide Nanosize Structures
Memristor Effect 87
V I Avilov, A S Kolomiytsev, R V Tominov, N I Alyabyeva, E M Bykova

Properties of Semiconductor GaAs Nanoparticles, Synthesized by Combination of Mechanical
Milling Methods and Chemical Etching 91
*N Yu Yashina, A J K Al-Alwani, O Yu Tsvetkova, A S Kolesnikova, E G Glukhovskoy, V P
Sevostyanov*

Research of Influence Al on Luminescence and Dark Current-Voltage Characteristics of InAs/GaAs
Heterostructures.. 96
A. S. Pashchenko, E. M. Danilina, N. M. Bogatov

Doping of GaP Layers Grown by Molecular-Beam Epitaxy on Silicon Substrates 100
A A Lazarenko, M S Sobolev, E V Pirogov, E V Nikitina

Multiscale Simulation of Surface Characteristics of Field Emitter Tip 103
K Nikiforov

Plasma-Arc Sputtering for Synthesis of Pt/CeO$_2$ and Pd/CeO$_2$ Nanosystems 107
*E A Derevyannikova, T Yu Kardash, A I Stadnichenko, E M Slavinskaya, A V Zaikovskii, S A
Novopashin, A I Boronin*

Study of the Geometrical Parameters of in Nanostructures During Droplet Epitaxy on the As-
Stabilized GaAs(001) Surface 112
S V Balakirev, M M Eremenko, I A Mikhaylin, M S Solodovnik

Preparation of Flat Surfaces on Silicon Carbide Substrate using Electron Beam Processing 118
E Yu Gusev, S P Avdeev

Transformation of the Defect Structure of InGaAs and InAlAs Metamorphic Buffer Layers
Depending on Indium Concentration.. 122
L A Snigirev, A A Sitnikova, D A Kirilenko, N A Bert

The Features of Moiré Pattern on Electron-Microscope Images of Free-Standing Quantum Dots Containing Dislocations 127
  L A Sokura, V N Nevedomskiy, N A Bert

Development of Local Catalytic Centers Positioning Technology for Carbon Nanotubes Growth 133
  O I Il'In, N N Rudyk, M V Kuzhelev, A A Fedotov

Investigation of InAs/InGaP Nano-Heterostructures Grown by MOVPE for Intermediate Band Solar Cells 137
  N A Kalyuzhnyy, S A Mintairov, A M Nadtochiy, V N Nevedomskiy, R A Salii

Suppression of Miscibility Gaps in Vapor-Liquid-Solid InGaAs and InGaN Nanowires 142
  T Jean, V G Dubrovskii

Size Effects in LiNbO3 Thin Films Fabricated by Pulsed Laser Deposition 149
  Z Vakulov, M Ivonin, E G Zamburg, V S Klimin, D P Volik, D A Golosov, S M Zavadskiy, A P Dostanko, A V Miakonkikh, I E Clemente, K V Rudenko, O A Ageev

GaN Nanostructures Grown on Si(111) by PA-MBE via Droplet Epitaxy: SEM and HRTEM Study 154
  G A Sapunov, V V Fedorov, A D Bolshakov, A M Mozharov, L N Dvoreckaia, A Kirilenko, A A Sitnikova, I S Mukhin

Effect of Substrate Temperature on the Properties of Plasma Deposited Silicon Oxide Thin Films 158
  E Yu Gusev, J Y Jityaeva, S P Avdeev, O A Ageev

Study of the Effect of Carbon-Containing Gas Pressure on the Geometric Parameters of an Array of Carbon Nanostructures 162
  V S Klimin, A A Rezvan, O A Ageev

Influence of Elastic Stresses on the Vapor-Solid-Solid Growth Mechanism of Au-Catalyzed GaAs Nanowires 166
  A A Koryakin, S A Kukushkin

Structural Investigation of Light-Emitting A3B5 Structures Grown on Ge/Si(100) Substrate 171
  A V Rykov, M V Dorokhin, P S Vergeles, V A Kovalskiy, E B Yakimov, M V Ved', N V Baidus, A V Zdoroveyshchev, V G Shengurov, S A Denisov

The Effect of AlGaAs/GaAs Laser Heterostructure Ionic Surface Cleaning on the Structure and Properties of AlN Films Deposited by the Method of Reactive Ion-Plasma Sputtering 176
  E V Fomin, K A Surnin, A D Bondarev, I P Soshnikov, K P Kotliar, S I Pavlov, A V Nashchekin

Theoretical Analysis of the Length Distributions of Ga-Catalyzed GaAs Nanowires 181
  Y Berdnikov, V G Dubrovskii, N V Sibirev, A A Koryakin, A. S. Sokolovskii, T Tauchnitz, H Schneider, M Helm, E Dimakis

Genetic Algorithm Application for Solving X-Ray Diffraction Inverse Problem 187
  D V Sivkov, V I Punegov

Stress Evolution in AlN Layers Grown on c-Al2O3 by Plasma-Assisted Molecular Beam Epitaxy at Metal-Rich Conditions 192
  O A Koshelev, D V Nechaev, S V Ivanov, V N Jmerik

Fabrication Method of the Patterned Mask for Controllable Growth of Low-Dimensional Semiconductor Nanostructures 197
  L N Dvoretckaia, A M Mozharov, V V Fedorov, A D Bolshakov, I S Mukhin

MBE Growth of GaAs Nanowires with Modulated Crystal Structure 202
I V Ilkiv, K P Kotlyar, D A Kirilenko, S P Lebedev, A A Lebedev, P A Alekseev, A D Bouravleuv, G E Cirlin

Nucleation of CdSe Thin Films: The Kinetic Model 206
A A Koryakin, S A Kukushkin, A V Redkov

NANOBIOTECHNOLOGY, BIOPHYSICS AND BIOPHOTONICS

Energy Transfer in Upconversion Nanoparticles – Phthalocyanine Hybrid Complexes 212
D A Gvozdev, E P Lukashev, V Z Paschenko

Solid-State Modeling of Human Tracheobronchial Tree for 23 Generations of Airways 217
V V Makevnina

Peculiarities of Noncontact Cardiac Signal Registration 221
V A Simon, V A Gerasimov, D K Kostrin, L M Selivanov, A A Uhov

Compact Nuclear Magnetic Spectrometer for Non-Destructive Condition Testing of Biological
Objects 227
N S Myazin, V V Davydov

Influence of Electrostatic Interactions on Cell-Penetrating Peptide-Small Interfering RNA Complex
Formation and Intracellular Delivery Efficiency 233
A Svirina, I Terterov, V Klimenko, S Shmakov, N Knyazev, A Emelyanov, V Vysochinskaya, A Bogdanov

Nanoparticle Tracking Analysis of Extracellular Vesicles Reveals Two Populations of Exosomes 239
A V Selenina, V A Kulichkova, A N Tomilin, A S Tsimokha

Diagnostic Potential of the Oral Fluid for the Observation People with Multiple Dental Caries by
Means of FTIR 243
P V Seredin, D L Goloshchapov, Y A Plotnikova, Y A Ippolitov, J Vongsvivut

Possible Applications of Bacterial Cellulose in the Manufacture of Electrical Insulating Paper 248
N Zhuravleva, A Reznik, D Kiesewetter, A Stolpner, A Khripunov

Magnetic Nanoparticles for Medical Application with a Coating Deposited with Various Methods 252
V N Zorin, E B Naumisheva, V N Postnov, N V Evreinova, K G Gareev, D V Korolev

A Studying of Subphase Temperature and Dissolved Ascorbic Acid Concentration Influence on the
Process of Langmuir Monolayer Formation 257
M M Qassime, V A Goryacheva, A J Al-Alwani, T N Lugovitskaya, A B Shipovskaya, E G Glukhovskoy

Investigation Optical Properties and Functional Surface Characteristics of Nanoparticles Based on
Porous Silicon for Applications in Biomedicine 262
A O Belorus, A I Pastukhov, A S Lenshin, P V Seredin, Yu M Spivak, V A Moshnikov

Fluorescent Method of Bacterial Contamination Control on Meat Surface 267
M V Mekhrengin, I K Meshkovskii, L A Kaftyreva, V I Guryev, D A Pogorelaya

Study of Electric Properties of Self-Assembled Films of Albumin During Their Dehydration 272
M A Baranov, A P Alekseenko, E N Velichko

Controlled Modification of Hyaluronic Acid for Photoinduced Reactions in Tissue Engineering................... 277
A. V. Sochilina, A. G. Savelyev, P. A. Demina, N. V. Ierusalimsky, D. A. Khochenkov, R. A. Akasov, N. V. Sholina, E. V. Khaydukov, A. N. Generalova

Microchip Device with Dry-Stored Reagents for Loop Mediated Isothermal Amplification........................... 283
A Tupik, G Rudnitskaya, A Bulyanitsa, T Lukashenko, D Varlamov, A Evstrapov

Investigation of the Color Effect of the Test Object on the Pupil Response... 288
D Bobrova, N Bikberdina, M Boronenko

Artificial Neural Network Prediction of Lysozyme Solubility for Protein Crystallization............................... 292
A S Sokolovskii, N A Besedina, Yu Berdnikov, V G Dubrovskii

Effective Application of Independent Component Analysis to the Task of Gradual EMG Control 296
M O Shamshin, V A Makarov, S A Lobov

Use of Electrophoretic Light Scattering for Investigation the Parameters of Macromolecules........................ 301
E A Savchenko, E K Nepomnyashchaya, E N Velichko, A N Skvortsov

Fluorescent Detection and Analysis of Single Molecules... 306
O B Kuznetsova, E A Savchenko, E T Aksenov

Non-Invasive Research of Biological Objects by the Method of Laser Polarimetry..311
M V Putintseva, E T Aksenov, C C Korikov, E N Velichko

In Vitro Model of CNS Neuronal Pathway Recovery using Microfluidic Chips.. 315
O Antipova, Y Pigareva, V Kolpakov, A Gladkov, A Bukatin, I Mukhina, V Kazantsev, A Pimashkin

Investigation of Na+/K+-ATPase Role in Cancer Cells' Functioning 320
S A Perkov, A K Emelyanov, S V Shmakov, A A Bogdanov

Accuracy Increase of Blood Viscoelasticity Measurement of the Microfluidic Wheatstone Bridge................ 323
N D Shipulya, S A Konakov

Singlet Oxygen Generation Mechanism in the Presence of Excited Nanoporous Silicon................................ 328
D M Samosvat, O P Chikalova-Luzina, V S Khromov, A G Zegrya, G G Zegrya

Formation and Manipulation of Polyacrylamide Spheroids Doped with Magnetic Nanoparticles in
Microfluidic Chip 333
D V Nozdriukhin, N A Filatov, A S Bukatin

Polylactide Film Deposition onto Titanium Surface from Different Solutions 337
A V Romashkin, N S Struchkov, Yu A Polikarpov, V A Petukhov, D D Levin, V K Nevolin

Comparison of Step and Flow-Focusing Emulsification Methods for Water-In-Oil Monodisperse
Drops in Microfluidic Chips.. 343
N A Filatov, D V Nozdriukhin, A A Evstrapov, A S Bukatin

Spectroscopic Investigation of Vegetable Pectin - Fluorescent Protein EGFP Interaction in Tumor
Cells *in Vitro* 348
V V Danilov, V V Klimenko, S V Shmakov, A I Khrebtov, A S Kulagina

LASERS, SOLAR CELLS AND OTHER OPTOELECTRONIC DEVICES

Modulation P-Doping as the Way to Attain Multi-State Lasing in Short-Cavity InAs/InGaAs Quantum Dot Lasers 352
    V V Korenev, A V Savelyev, V G Dubrovskii, S Breuer, M V Maximov, A E Zhukov

Injection Microdisk Lasers Based on Multilayers of InGaAs/GaAs Quantum Well-Dot Structures 356
    N V Kryzhanovskaya, E I Moiseev, K P Kotlyar, Yu A Guseva, Yu M Shernyakov, S A Scherbak, M V Maximov, A E Zhukov

Optical Properties of In$_{0.8}$Ga$_{0.2}$As Quantum Dots in GaAs Photovoltaic Convertors 360
    R A Salii, V V Evstropov, S A Mintairov, M A Mintairov, M Z Shvarts, N A Kalyuzhnyy

Features of Direct Digital Synthesis Applications for Microwave Excitation Signal Formation in Quantum Frequency Standard on the Atoms of Cesium 365
    A A Petrov, V V Davydov, D V Zalyotov, V E Shabanov, D V Shapovalov

Effect of Carrier Localization on Performance of Coupled Large Optical Cavity Diode Lasers 372
    A A Serin, A S Payusov, Yu M Shernyakov, N A Kalyuzhnyy, S A Mintairov, N Yu Gordeev, F I Zubov, M V Maximov, A E Zhukov

Development of Technological Installation for Obtaining Functional Thin Film Elements Used in Solar Elements of Space Equipment 378
    T O Zinchenko, V I Kondrashin, E A Pecherskaya, P E Golubkov, K O Nikolaev, F A Abdullin

Spectral Approach to Li–ion Batteries Degradation Rate Estimation 382
    E A Yusupova, V G Malyshkin, V V Davydov

CH$_3$NH$_3$PbI$_3$ Crystal Growth, Structure and Composition 387
    V E Anikeeva, O I Semenova, O E Tereshchenko

Isolation and Optoelectronic Characterization of Si Solar Cells Microstructure Defects 392
    A Gajdos, P Skarvada, R Macku, N Papez, L Skvarenina, D Sobola

Investigation of the Effect of Chemical Pre-Treatment on Uniformity of the Silicon Wafer Texturing for Manufacturing a Solar Cell 398
    D Kudryashov, A Gudovskikh, A Rodin, N Pinaev

The Optical Method for Condition Control of Flowing Medium 403
    N M Grebenikova, K J Smirnov, V V Davydov, V. Y. Rud'

Study of P-Type Contact Topography Influence on Characteristics of Microdisk and Microring Lasers 408
    A S Novikov, E I Moiseev, N V Kryzhanovskaya, B. I. Afinogenov, Y A Guseva, K Kotlyar, A A Lipovskii, A G Nasibulin, M V Maximov, V G Tikhomirov, A E Zhukov

The Ways to Improve the Energy Conversion Efficiency in Erbium-Doped Gd$_2$O$_3$ Nanoparticles 413
    Yu A Kuznetsova, A F Zatsepin, M A Mashkovtsev, V N Rychkov, S P Vyatkina

Theoretical Model for Spatial Separation of Dominating Recombination Region in a-Si:H/c-Si Structures 417
    A. D. Maslov, E. V. Bezuglaya

Surface Morphology After Reactive Ion Etching of Silicon and Gallium Arsenide Based Solar Cells 421
    N Papež, D Sobola, A Gajdoš, L Škvarenina, R Macku, M Eliáš, A Nebojsa, R Motúz

Reducing of Thermal Resistance of Edge-Emitting Lasers Based on Coupled Waveguides............................ 427
A A Serin, A S Payusov, Yu M Shernyakov, N A Kalyuzhnyy, S A Mintairov, M V Maximov, N Yu Gordeev, A E Zhukov

All-Oxide Heterojunction Solar Cells Formed by Magnetron Sputtering 431
D Kudryashov, A Gudovskikh, A Monastyrenko

Influence of Au Surface Properties on Photon Emission from Localized Metal-Insulator-Metal Tunnel Contact 436
V A Shkoldin, D V Permyakov, M V Zhukov, A A Vasiliev, T A Mamaeva, A O Golubok, A V Uskov, A D Bolshakov, A K Samusev, S Mukhin

Diagnostics of Inhomogeneities of Parameters of Light-Emitting Heterostructures by the Photoelectric Method.. 440
I V Frolov, O A Radaev, V A Sergeev, S V Vasin

Influence of Coating Layers on Characteristics of Microdisk Lasers with InAs/InGaAs Quantum Dots Active Region.. 444
I Y Agafonov, N V Kryzhanovskaya, E I Moiseev, A S Dragunova, M V Fetisova, K P Kotlyar, S A Scherbak, I V Reduto, Yu A Guseva, A A Lipovskii, M V Maximov, A E Zhukov

Influence of Hydrogen Plasma Passivation on Electrical and Spectral Characteristics of GaN Nanowires / Si Solar Cells.. 449
K U Shugurov, A M Mozharov, V V Fedorov, A D Bolshakov, G A Sapunov, I S Mukhin

Mixed Halide Perovskite Light Emitting Solar Cell.. 453
D Gets, A Ishteev, T Liashenko, D Saranin, S Makarov, A Zakhidov

Non-Radiative Coupled Donor-Acceptor Pair Recombination in Nanostructures Based on Nitrides and Related Phenomena 458
N A Talnishnikh, E I Shabunina, N M Shmidt, A E Chernyakov, D S Arteev, A A Zybin

Formation of Nanoscale Structures on the Surface of Gallium Arsenide by Local Anodic Oxidation and Plasma Chemical Etching 462
V S Klimin, M S Solodovnik, S A Lisitsyn, A A Rezvan, S V Balakirev

Quantum Autonomous Magnetic Field Sensor.. 467
S E Logunov, A Yu Koshkin, V V Davydov

Stabilization TiO$_2$ Anatase by F-Ion Doping for Solar Panel Producing... 472
S. N. Fedorov, R. V. Kurtenkov, V. V. Vasiliev

Alpha-Factor in Multimode Lasers.. 476
I K Boikov, A V Savelyev

Heterointerfaces in the Bottom Tunnel Part of GaInP/GaAs/Ge Solar Cells.. 481
M A Mintairov, V V Evstropov, S A Mintairov, M Z Shvarts, N A Kalyuzhnyy

Growth and Optical Characterization of 7.5 μm Quantum-Cascade Laser Heterostructures Grown by MBE 486
A V Babichev, A G Gladyshev, E S Kolodeznyi, A S Kurochkin, G S Sokolovskii, V E Bougrov, L Ya Karachinsky, I I Novikov, V V Dudelev, V N Nevedomsky, S O Slipchenko, A V Lutetskiy, A N Sofronov, D A Firsov, L E Vorobjev, N A Pikhtin, A Yu Egorov

Modeling of the Internal Quantum Yield for Solar Cell with Quantum Dots.. 491
A. Panchak, N Kalyuzhnyy, S. Mintairov, M. Shvarts

Effect of Temperature on Dry Etching of III-V Structures 495
I A Morozov, A S Gudovskikh, D A Kudryashov, K P Kotlyar, K Y Shubina

Chromaticity Coordinates Temperature Dependence for Blue Laser Diodes for Solid State Laser
Lighting 499
A Aluyev, Yu Akhmerov, N Kourova, V Logachev, M Mezhenny, A Chelny, A Savchuk, V
Murashev, O Rabinovich, S Didenko

Optical Coupling Upset in Multijunction Solar Cells with Built-In Bragg Reflectors 503
S A Levina, E D Filimonov, V M Emelyanov, M Z Shvarts

Quantum Efficiency Measurement of Subcells in Multi-Junction Solar Cells Based on III-V/Si ... 507
A I Baranov, A S Gudovskikh, D A Kudryashov, I A Morozov, A M Mozharov, K S Zelentsov

Temperature Annealing Effect on ITO Film 511
K P Kotlyar, D A Kudryashov, I P Soshnikov, A V Uvarov, R R Reznik

Silicon Doping of GaP Layers Grown by Time-Modulated PECVD 515
A I Baranov, I A Morozov, A V Uvarov, G E Yakovlev, J-P Kleider

1.06 µm Wavelength Photodetectors with Metamorphic Buffer Layers Grown on GaAs Substrates 520
I V Samartsev, S M Nekorkin, B N Zvonkov, N V Dikareva, A V Zdoroveyshchev, A V Rykov, N
V Baidus

Sunlight Simulation in Investigating Characteristics of Small-Size Sunlight Concentrators and
Multijunction Solar Cells 524
E D Filimonov, S A Levina, M Z Shvarts

A Novel Approach to Characterization of Bottom Sub-Cell in Multijunction Solar Cell using
Photoluminescence. 528
K S Zelentsov, A S Gudovskikh, D A Kudryashov, N V Kryzhanovskaya, N A Kalyuzhnyy, S A
Mintairov

Numerical Simulation of the Carbon Nanotubes Transport Layer Influence on Performance of
GaAs Solar Cell 532
S A Raudik, A M Mozharov, D M Mitin, A D Bolshakov, P M Rajanna, A G Nasibulin, I S
Mukhin

Investigating the Output Characteristics of Ring Laser in Laser Dynamic Goniometer 537
N. A. Eno, P. A. Pavlov, J. T. Bagshaw

Investigation of the Electrical and Optical Properties of Nickel Oxide Films Produced by RF
Magnetron Sputtering Method... 542
A Aglikov, A Mozharov, D Kudryashov, D Sosnin, A Vasilev, A Bolshakov, S Makarov, I
Mukhin

Lateral Mode Engineering in Diode Lasers Based on Coupled Ridges 547
A S Payusov, A A Serin, Yu M Shernyakov, D A Rybalko, M M Kulagina, M V Maximov, N Yu
Gordeev

NANOPHOTONICS, SPECTROSCOPY, MICROCAVITIES, OPTICS, PLASMONICS

Au-SiO$_2$-Si Hybrid Plasmonic Waveguide Micro-Ring Resonator Sensor 551
M A Butt, S A Degtyarev, S N Khonina

Magnetron Sputtered MoS_2: Optical and Structural Analysis 557
A. I. Belikov, Kyaw Z. Phyo

Effect of Defocusing and Apodization on the Resolution of Incoherent Optical System Beyond the Rayleigh Limit 563
Andra N. K. Reddy, M. A. Butt, S. N. Khonina

Increasing the Resolution of the Aberrated Optical System Based on Quadratic Amplitude Apodization 568
P A Khorin, S A Fomchenkov

Optical Properties and Energy Band Parameters of Luminescent $CaMoO_4$:Bi Ceramics 574
R A Parulin, I V Timoshenko, Yu A Kuznetsova, A F Zatsepin, E S Buyanova, Z A Mikhaylovskaya, M S I Koubisy

Polarization Plane Rotation for Higher Order Modes in Twisted Optical Fibers with Discrete Rotationally Symmetric Core 578
C N Alexeyev, M C Alexeyeva, B P Lapin, D V Vikulin, M A Yavorsky

Room Temperature Lasing from Microdisk Laser in Aqueous Medium 582
M V Fetisova, N V Kryzhanovskaya, I V Reduto, E I Moiseev, S A Blokhin, K P Kotlyar, S A Scherbak, A A Lipovskii, A A Kornev, A S Bukatin, M V Maximov, A E Zhukov

Self-Assembly of Lines of Microscopic Photonic Crystals 586
M S Ashurov, A A Ezhov, T A Kazakova, S O Klimonsky

Broadband Nanoantenna 589
D Poletaev, B Sokolenko, V Voytitsky, A Nudga

Novel Highly Efficient Blue-Emitting Branched Oligoarylsilanes 593
M S Skorotetcky, O V Borshchev, N M Surin, S A Ponomarenko

VOLUME 2

Nonlinear Pulses in Dispersion-Managed Fiber-Optic Systems in Presence of High Losses 597
V Neskorniuk, A Lukashchuk, I Gabitov, A Chipouline, M Malekizandi, G Ovchinnikov, F Küppers

Study of Optical Pumping Influence on Carbon Nanotubes Permittivity in THz Frequency Range 602
P Demchenko, D Gomon, I Anoshkin, D Lioubtchenko, M Khodzitsky

Numerical Study of Optical Properties of Sphere-Gap-Cone Hybrid Nanoantenna 607
Y Sun, S. V. Makarov, D A Zuev

Active and Passive Phase Stabilization for the All-Fiber Michelson Interferometer 611
M. S. Elezov, M. L. Scherbatenko, D. V. Sych, G. N. Goltsman

Binary Diffractive Optics for 3D-Demultiplexing of OAM Beams 616
A P Porfirev, S A Fomchenkov, G E Gridin, S N Khonina

The Study of Photoluminescence Properties of AlGaAs/GaAs Heterostructure After Ga+ Focused Ion Beam Etching 621
G V Voznyuk, I V Levitskii, M I Mitrofanov, M N Mizerov, D N Nikolaev, V P Evtikhiev

Optical Limiting in Nanodiamond Suspension: Shortening of the Laser Pulses 625
R Yu Krivenkov, K G Mikheev, T N Mogileva, N Nunn, O A Shenderova, G M Mikheev

New Ultra High-Speed All-Optical Coherent D-Trigger 630
  S N Bagayev, V S Egorov, V G Nikolaev, I B Mekhov, I A Chekhonin, M A Chekhonin

Comparison of Optical Spectral Devices in the Framework of System Approach 636
  V Kazakov, O Moskaletz, M Vaganov

Four-Step Fabrication of SERS-Active Microfluidic Channels 642
  E S Babich, E S Gangrskaia, I V Reduto, A A Lipovskii

Second Harmonic Generation and Spontaneous Parametric Down-Conversion in Mie
Nanoresonators 647
  M I Petrov, A A Nikolaeva, K S Frizyuk, N A Olekhno

Spatially Modulated Channel Waveguide Elements Optically Written in Photorefractive Lithium
Niobate 651
  A D Bezpaly, A S Perin, V M Shandarov

Plasmonic Amplification of Terahertz Radiation in a Double-Layer Graphene Nanoribbon Array 655
  I M Moiseenko, O V Polischuk, M Y Morozov, V V Popov

SERS Induced by Two Coupled Monolayers of Au Plasmonic Nanoparticles... 659
  *V Kaydashev, P Zolotukhin, A Belanova, A Anokhin, A Volodin, A Chernyshev, N Lyanguzov, E
  Kaidashev*

Superconducting Nanowire Single-Photon Detector on Lithium Niobate.. 664
  E Smirnov, A Golikov, P Zolotov, V Kovalyuk, M Lobino, B Voronov, A Korneev, G Goltsman

Study of Charge Relaxation in Poled Silicate Glasses.. 668
  D V Raskhodchikov, I V Reshetov, D K Tagantsev, A A Lipovskii, V P Kaasik

Initialization of the Bell States of Two Qubits by Unipolar Pulses 673
  M. V. Denisenko, N. V. Klenov, A. M. Satanin

Second Harmonic Generation by Metal Core - Dielectric Shell Spherical Nanoparticles: Spatial vs.
Plasmon Resonances 678
  S A Scherbak, A A Lipovskii, E S Babich

Laser-Deposited Hybrid Au-Ag@C Nanoparticles as Efficient SERS & Adsorption Material........................ 684
  A A Vasileva, D V Pankin, I E Kolesnikov, S S Rzhevskiy, A A Manshina

Influence of Sputtering Parameters on the Main Characteristics of Ultra-Thin Vanadium Nitride
Films.. 689
  *P I Zolotov, A V Divochiy, Yu B Vakhtomin, A V Lubenchenko, P V Morozov, I V Shurkaeva, K
  V Smirnov*

Dielectric Surrounding Decimates Eigenmodes of Microdisk Optical Resonators 695
  A V Raskhodchikov, S A Scherbak, N V Kryzhanovskaya, A E Zhukov, A A Lipovskii

Electron Diffusivity Measurements of VN Superconducting Single-Photon Detectors 699
  N R Romanov, P I Zolotov, Yu B Vakhtomin, A V Divochiy, K V Smirnov

Phase Diffraction Patterns Optically Induced by Laser Bessel-Like Beams in Photorefractive
Lithium Niobate with a Doped Surface Layer.. 705
  I Trushnikov, A Inyushov, A Perin

Focusing of High-Order Vortex Laser Beams by Binary Axicon with Different Numerical Aperture 710
  D A Savelyev, S A Fomchenkov

NEXAFS C1s-Spectra Apparatus Distortion Study 714
A E Mingaleva, O V Petrova, D V Sivkov, N N Shomysov, S V Nekipelov, R N Skandakov, V N Sivkov

Raman Lidar with for Geoecological Monitoring 718
J Ruzankina, V Elizarov, L Konopel'Ko, A Zhevlakov, A Grishkanich

Soft X-Ray Photoemission Study of Ba Adsorption on the Ceramic Multiferroic BiFeO$_3$ 722
P A Dementev, G V Benemanskaya, S N Timoshnev

UV-Induced Refractive Index Changes Due to Silver Molecular Clusters in Photo-Thermo-Refractive Glass 727
V V Gorbyak, A I Sidorov

Solution of the Multiple Light Scattering Problem in Turbid Suspensions via Cross-Correlation Processing 731
Z A Zabalueva, E K Nepomnyashchaya, C C Korikov

Quality Factor Enhancement in Gold-Based Tamm Plasmon Cavity by Sub-Wavelength Structuration 736
A R Gubaydullin, K M Morozov, M A Kaliteevski

Development of Diffractive Optical Elements with Low Surface Roughness by Direct Laser Writing 739
S A Fomchenkov, A P Porfirev

Separation of Inhomogeneous and Homogeneous Broadening Manifestations in InGaAs/GaAs Quantum Wells by Time-Resolved Four-Wave Mixing 743
I A Solovev, Yu V Kapitonov, B V Stroganov, Yu P Efimov, S A Eliseev, S V Poltavtsev

Magneto-Resonance Methods of the Relaxation Rate Measuring for the Proton-Containing Flowing Fluids Composition Studying 747
L I Fatkhutdinova, A N Mamonkina, S V Ermak, V V Semyonov

Transmission of Thermal Imaging by using Infrared Bundle Based on Silver Halide Solid Solution 752
A S Shmygalev, D D Salimgareev, A S Korsakov, B P Zhilkin, V I Terekhov

On-Chip Single-Photon Spectrometer for Visible and Infrared Wavelength Range 756
V Kovalyuk, O Kahl, S Ferrari, A Vetter, G Lewes-Malandrakis, C Nebel, A Korneev, G Goltsman, W Pernice

On-Chip Controlled Placement of Nanodiamonds with a Nitrogen-Vacancy Color Centers (NV) 761
S Komrakova, J Javadzade, V Vorobyov, S Bolshedvorskii, V Soshenko, A Akimov, V Kovalyuk, A Korneev, G Goltsman

Experimental Optimisation of O-Ring Resonator Q-Factor for On-Chip Spontaneous Four Wave Mixing 765
P An, V Kovalyuk, A Golikov, E Zubkova, S Ferrari, A Korneev, W Pernice, G Goltsman

Optimization of Contra-Directional Coupler Based on Silicon Nitride Bragg Rib Waveguide 769
E Zubkova, P An, V Kovalyuk, A Korneev, S Ferrari, W Pernice, G Goltsman

Investigation of Influence Incoherent Background Illumination on the Nonlinear Response of a Lithium Niobate Crystal Sample at Low Light Intensity 774
A V Pustozerov, A S Perin

Bolometric Effect for Detection of sub-THz Radiation with Devices Based on Carbon Nanotubes............... 779
M V Moskotin, I A Gayduchenko, G N Goltsman, N Titova, B M Voronov, G F Fedorov, F Pyatkov, F Hennrich

Silicon Nitride Nanophotonic Circuit for On-Chip Spontaneous Four-Wave Mixing..................................... 784
A Golikov, V Kovalyuk, P An, E Zubkova, S Ferrari, W Pernice, A Korneev, G Goltsman

Optimization of On-Chip Photonic Delay Lines for Telecom Wavelengths... 788
A Prokhodtsov, P An, V Kovalyuk, E Zubkova, A Golikov, A Korneev, S Ferrari, W Pernice, G Goltsman

Graphene-Based Tunability of Chiral Metasurface in Terahertz Frequency Range 793
M S Masyukov, A V Vozianova, A N Grebenchukov, M K Khodzitsky

Graphene-Layer and Graphene-Nanoribbon FETs as THz Detectors.. 797
Y E Matyushkin, I A Gayduchenko, M V Moskotin, G N Goltsman, G E Fedorov, M G Rybin, E D Obraztsova

Purcell Effect in GaN-Based Waveguiding Structures 803
K M Morozov, K A Ivanov, A R Gubaydullin, M A Kaliteevski

Strong Coupling Between Excitons in Transition Metal Dichalcogenides and Optical Bound States in the Continuum 808
S. K. Sychev, K. L. Koshelev, Z. F. Sadrieva, A. A. Bogdanov, I. V. Iorsh

Experimental Observation of Symmetry Protected Bound State in the Radiation Continuum in the Periodic Array of Ceramic Disks... 813
A A Bogdanov, M Balezin, P V Kapitanova, Z Sadrieva, M Belyakov, E A Nenasheva, A F Sadreev

High-Q States and Strong Mode Coupling in High-Index Dielectric Resonators. 817
S. A. Gladyshev, A. A. Bogdanov, P. V. Kapitanova, M. V. Rybin, K. L. Koshelev, Z. F. Sadrieva, K. B. Samusev, Y. S. Kivshar, M. F. Limonov

RIE for Structuring E-Field Processed Glasses 823
I Reduto, D Raskhodchikov, E Gangrskaia, V Kaasik, Yu Svirko, A Lipovskii

Photodecomposition of Organic/Inorganic Composite Materials Based on Polyvinylpyrrolidone 827
A S Kulagina, S K Evstropiev, K V Dukelskii, N A Volkova, K S Evstropyev, N V Nikonorov

Generation of the Trapping Light Structures Based on Vector Fields 832
N Shostka, O Karakchieva, B V Sokolenko

Tamm Magnetophotonic Structures with Bi-Substituted Iron Garnet Layers at Oblique Incidence 838
T Mikhailova, S Tomilin, S Lyashko, A Shaposhnikov, A Prokopov, A Karavainikov, A Bokova, V Berzhansky

Epitaxial Films of Garnet Ferrite with Anisotropy "easy Plane" for Magneto-Optical Eddy Current Flaw Detection 844
N Lugovskoy, V Berzhansky, D Glechik, A Prokopov

SPINTRONICS, ELECTRO- AND MAGNETOOPTICS

Electrically Controlled Spin Polarization in Suspended GaAs Quantum Point Contacts............................... 848
D A Pokhabov, A G Pogosov, E Yu Zhdanov, A A Shevyrin, A K Bakarov, A A Shklyaev

The Study of Mechanical Resonances of the Phase Electro-Optic Modulator Based on LiNbO$_3$ for Noise Reduction of Fiber-Optic Gyroscope 852
  M A Smolovik, D A Pogorelaya, A A Vlasov, A S Aleynik, V E Strigalev

Magneto-Optical Properties of Metaphosphate and Borate Glasses 856
  Dmitrii I Sobolev, Anastasiia N Babkina, Nikolay V Nikonorov

Investigation of the Magnetic Properties of Manganese Silicide Grown on i-GaAs Substrate by Pulsed Laser Deposition 860
  Y Kuznetsov, M Dorokhin, A Kudrin, V Lesnikov, E Demidov, V Karzanov

On the Mechanism of Spin-Polarized Injection in (Ga,Mn)As/n+GaAs/InGaAs Zener Tunnel Diode 865
  M Ved, M Dorokhin, E Malysheva, A Zdoroveyshchev, Yu Danilov, A Parafin, Yu Kuznetsov

Morphology of Garnet Films for Thermo-Magnetic Recording 870
  V Berzhansky, Y Danishevskaya, A Nedviga, M Bektemirova

ELECTRIC, MAGNETIC AND MICROWAVE DEVICES

A 0.3-0.7 THz Flux-Flow Oscillator Integrated with the Slot Antenna and Elliptical Lens 874
  N V Kinev, K I Rudakov, A M Baryshev, V P Koshelets

The Development of the Bistable Micromechanical Actuator for Optical Relay 878
  Y. B. Enns, E. N. Pyatishev, A. N. Kazakin

Calculation of High-Frequency Conductivity and Hall Constant of a Thin Conductive Layer in the View of Equal Specularity Coefficients of Its Surfaces 884
  O V Savenko, I A Kuznetsova, A A Yushkanov

Electrical and Photoelectric Characteristics of Gallium Oxide Films Obtained by RF Magnetron Sputtering 890
  A Tsymbalov, J Petrova

Effect of Metal Modifiers on the Characteristics of Resistive Hydrogen Sensors Based on Thin Films of Tin Dioxide 895
  A V Almaev, N K Maksimova, E Yu Sevastyanov, E V Chernikov, T A Davydova, T E Smirnova

Discrete Diffraction in Network of Magnonic Crystals 900
  E N Beginin, A Yu Sharaevskaya

The Study of Optical Properties of Graphene Intercalated with Ferric Chloride for Application in Terahertz Photonics 905
  A D Zaitsev, A N Grebenchukov, P S Demchenko, E T Alonso, M F Craciun, S Russo, A Baldycheva, M K Khodzitsky

Lifetime Testing of a MEMS Switch with Pt-Pt Contact 910
  I V Uvarov, A N Kupriyanov

Simulation of Electrical Conductivity of Silicon Diodes with Bismuth Implanted-Ion Profiles 915
  S M Loganchuk, S N Chebotarev, D A Arustamyan, A A A Mohamed, L Touel, N M Bogatov

Flexoelectrical Nanogenerator Design using Aligned Carbon Nanotubes 919
  M V Il'Ina, A A Konshin, E G Solomin

How to Take Fractional-Order Derivative Experimentally? 923
  D D Stupin, A I Lihachev, A V Nashchekin

Ti/4H-SiC Schottky Diode Breakdown Voltage with Different Thickness of 4H-SiC Epitaxial Layer............ 928
    S V Sedykh, S B Rybalka, A Yu Drakin, A A Demidov, N S Ponomaryova, O A Shishkina

The Influence of the Deep Level Type on a Switching Time Delay of GaAs Avalanche S-Diodes 933
    V V Kopyev, T E Smirnova, V L Oleinik, I A Prudaev

Accelerated Degradation HEMT Based on AlGaN / SiC 937
    A S Evseenkov, V G Tikhomirov

Software Complex for Calculating the Initial Section of the Current-Voltage Characteristics of a
Resonant-Tunneling Diode with the Possibility of Computer Statistical Experiment...................................... 942
    K V Cherkasov, S A Meshkov, M O Makeev

Development of the Sheet Electron Beam Focusing System Based on Thermionic and Field
Emission Cathodes 948
    A V Danilushkin, A A Burtsev, K. V. Shumikhin, G. V. Sakhadzhi

Increase of Accuracy of Capacitance Parameters Measurements of Power Semiconductor Modules
on Base IGBT and FRD.. 952
    D A Knyaginin, A Yu Drakin, S B Rybalka, A A Demidov

Influence of Microwave Electromagnetic Field on the Structure of Polymers.. 956
    E Vasinkina, S Kalganova, V Alekseev, Yu Kadykova, S Arzamastsev, A Dolzhikova

Investigation of the Influence of Parameters of Nanoscale Profiling of the Surface of GaAs
Structures by a Combination of Local Anodic Oxidation and Plasma Chemical Etching Methods 960
    V S Klimin, A A Rezvan, I N Kots, N A Naidenko

Research of using Plasma Methods for Formation Field Emitters Based on Carbon Nanoscale
Structures.. 964
    V S Klimin, A A Rezvan, O A Ageev

Numerical Simulation of Induction Heating of a Carburizing Container with a Titanium Sample 968
    A Voyko, M Fomina, A Shumilin, I Rodionov, S Kalganova, I Artyukhov, A Fomin

Measurements in Microwave Electrotechnology 972
    S Kalganova, E Vasinkina, V Alekseev, V Lavrentyev, S Trigorly, T Dunaeva

Effect of Metallic Nanoantennas on the Efficiency of the Surface Plasmon-Polariton Generation via
Excitation of Electromagnetic Waves in a Tunnel Junction.. 976
    L N Dvoretckaia, A M Mozharov, A V Uskov, A D Bolshakov, A O Golubok, I S Mukhin

OTHER ASPECTS OF NANOTECHNOLOGY

Metal-Assisted Photoenhanced Wet Chemical Etching of GaN Epitaxial Layers.. 980
    K Yu Shubina, D V Mokhov, T N Berezovskaya, A M Mizerov, E V Nikitina, A D Bouravleuv

Investigation of Luminescence Quantum Yields of Carbon Dots Synthesized from Ethylene Glycol,
Citric Acid and Berries 984
    M N Egorova, A E Tomskaya, A N Kapitonov, S A Smagulova, A A Alekseev

Formation of a Microheterophase State from Planar Nanoparticles of Graphene at the Oil-Water
Interface.. 990
    Yu V Pakharukov, F K Shabiev, R F Safargaliev

The Nanosecond Studies of Granular Carbon Nanostructures Based on High Temperature Superconductors 994
M P Faradzheva, A V Prikhodko, O I Konkov, S P Faradzhev

Thermoelectric Peltier Micromodules Processed by Thin-Film Technology 998
E Bakulin, S Dzyubanenko, S Konakov, A Korotkov, V Loboda, A Yugay

The Influence of Photo-Stimulated Adsorption of Polyelectrolyte Molecules on Electro-Physical Characteristics of Structures Based on Single Crystal Silicon Substrates 1003
A V Kozlowski, S V Stetsyura, I V Malyar

Electromechanical Bending Microactuator as Optical Shutter 1008
R. Kleimanov, Y. Enns, E. Pyatishev, I. Komarevtsev

Application of XRD Methods for the Pilot Studies of New Functional Materials for Photonics 1013
M. Dermeneva, D. Muravijova, M. Mynbaeva, V. Bougrov, M. Yagovkina

The Formation of Arachidic Acid Langmuir Monolayers on the NiCl$_2$ Solution 1018
A Chumakov, Ammar J Al-Alwani, A Ermakov, O Shinkarenko, N Begletsova, E Glukhovskoy, S Santer

Chemical Analysis of Thin-Film's Colour Generation During Surface Laser Oxidation of TiN-Coating 1022
O S Yulmetova, M S Tutova, R F Yulmetova

Studies of the Formation of Copper Nanoparticles Monolayers on the Water Subphase 1026
N N Begletsova, E I Selifonova, R K Chernova, A S Chumakov, V P Sevostyanov, E G Glukhovskoy

Multicaloric Effect in Barium Titanate Nanotube 1032
I A Starkov, I L Mylnikov, A S Starkov

Scanning Probe Microscopy of AlGaAs/GaAs Diode After Partial Electrical Breakdown 1038
A O Mikhaylov, P A Alekseev, A A Podoskin, S O Slipchenko, M S Dunaevskiy

Methods of Applying the Reliability Theory for the Analysis of Micro-Arc Oxidation Process 1043
P E Golubkov, E A Pecherskaya, Y V Shepeleva, A V Martynov, T O Zinchenko, D V Artamonov

Structure of "Chromium Steel-Base – Ti-Coating" and Its Production by the Contact Welding 1049
I Egorov, A Shchelkunov, A Fomin, I Rodionov

Structural Transformations on the Surface of 1.3343 Tool Steel and 12Cr18Ni10Ti Stainless Steel After Induction Heat Treatment and Quenching 1053
P Palkanov, A Fomin, I Rodionov

Formation of Wear-Resistant Oxide-Carbide Coatings on Titanium by Electrospark Alloying 1057
S Mezentsov, V Koshuro, A Fomin, I Rodionov

Modified Multicapillary Glass Structure for SERS Application 1061
N E Markina, A V Markin, A M Zakharevich, Y S Skibina, I Yu Goryacheva

Carbide Coatings Obtained by Electro-Spark Alloying and Finishing 1066
V Koshuro

High Robustness of Epitaxial 4H-SiC Graphene to Oxidation Processes 1070
V. S. Prudkovskiy, K. P. Katin, M. M. Maslov, P. Puech, R. Yakimova, G. Deligeorgis

Ohmic Contacts to N-Type 4H- and 6H-SiC 1074
  V I Egorkin, A V Nezhentsev, V E Zemlyakov, V A Gudkov, V I Garmash

Investigation of the Metal Surfaces Destruction Due to Electrochemical Corrosion and Cavitation,
Methods of Protection with the Use of Polymer Composite Coatings Based on CNT 1078
  A A Goshev, M K Eseev, S N Kapustin

Composition-Dependent Conductivity of $In_xGa_{1-x}As$ Nanowires 1083
  V Sharov, P Alekseev, M Dunaevskiy, R Reznik, G Cirlin

MBE Formation of Self-Catalyzed GaAs Nanowires using ZnO Nanosized Films 1087
  M S Solodovnik, O G Karenkikh, S V Balakirev, S I Petrov, R V Ryzhuk, A N Alexeev, O A Ageev

Cathodic Transformation of Bactericidal Silver-Containing Bioceramic Coatings for Implants 1092
  E Poshivalova, I Rodionov

Synthesis and Properties of a Polyamine-Cumulene/Carbon Nanotubes for Removing Harmful
Substances from Aqueous Solutions 1096
  E A Neskoromnaya, A V Melezhik, O V Alekhina, I V Burakova, E S Mkrtchyan, A E Burakov

Nano-Sized Al-Ni Energetic Powder Material for Heat Release Element of Thermoelectric Device 1102
  S Yu Nemtseva, E A Lebedev, Yu P Shaman, P I Lazarenko, R M Ryazanov, S A Gavrilov, D G Gromov

Investigation of Local Charge Accumulation in Yttria Stabilized Zirconia Films with Au
Nanoparticles by Scanning Kelvin Probe Microscopy 1106
  D O Filatov, O N Gorshkov, A N Mikhaylov, D S Korolev, M N Koriazhkina, M A Ryabova, I N Antonov, M E Shenina, D A Pavlov, M S Dunaevskiy

Adsorption of the Methylene Blue Dye on Carbon Nanocomposites Under Dynamic Conditions: A
Kinetic Study 1110
  I V Burakova, A V Babkin, E A Neskoromnaya, A E Burakov, D A Kurnosov, A G Tkachev

Kinetics of the Adsorption of Synthetic Dyes on a Polyhydroquinone/Graphene Carbon
Nanocomposite 1115
  A V Babkin, A V Melezhik, D A Kurnosov, E S Mkrtchyan, I V Burakova, A E Burakov, E V Galunin

Kelvin Probe Microscopy of $MoSe_2$ Monolayers on Graphene 1121
  B R Borodin, M S Dunaevskiy, F A Benimetskiy, S P Lebedev, A A Lebedev, P A Alekseev

Photosensitive Sulphide Heterostructures Obtained by using Successive Ionic Layer Adsorption
and Reaction on Planar and Profiled Substrates 1126
  N Bogomazova, G Gorokh, A Zakhlebayeva, A Pligovka, A Murashkevich, T Galkovsky

Computer Simulation of Aerosol Nanoparticles Focusing and Deposition Process for Functional
Microstructure Fabrication 1132
  N V Protas, A A Efimov, V K Zemlyanoy, V V Ivanov

Study of the Possibility of using Arachidic Acid as a Monocrystalline Substrate for the Formation
of Monolayers of Aromatic Hydrocarbons 1136
  O A Shinkarenko, A S Chumakov, M V Pozharov, A S Kolesnikova, A J K Al-Alwani, O Yu Tsvetkova, V P Sevostyanov, E G Glukhovskoy

Masking Coating Formation by the Focused Ion Beams Method for Plasma Chemical Treatment 1141
  I N Kots, V S Klimin, V V Polyakova, A A Rezvan, Z E Vakulov, O A Ageev

Size Effect on Memristive Properties of Nanocrystalline ZnO Film for Resistive Synaptic Devices 1146
  N A Shandyba, I V Panchenko, R V Tominov, V A Smirnov, M I Pelipenko, E G Zamburg, Y H
  Chu

Computer Simulation of Femtosecond Pulsed Laser Ablation of Aluminium and Copper 1151
  R V Davydov, V I Antonov

Nanometer-Scale Oxidation of Silicon Surface by ICP Plasma 1156
  I Clemente, A Miakonkikh, S Averkin, K Rudenko

Polarization Photosensitivity of n-P-CdSiAs$_2$ Photodiode 1160
  R V Davydov, V Yu Rud', Yu V Rud', E I Terukov

Spatial Point Pattern Analysis of the Local Current Distribution on the Surface of Multi-Tip Field
Emitters 1165
  S V Filippov, S Carapezzi, E O Popov, A G Kolosko

Investigation of Sintering of Silver Lines on a Heated Plastic Substrate in the Dry Aerosol Jet
Printing 1171
  A A Efimov, K N Minkov, P V Arsenov, N V Protas, V V Ivanov

Structure and Hardness of the Zirconium Surface After Laser Modification 1175
  V Proskuryakov, S Mezentzov, I Rodionov

The Electrochemical Deposition of Silicon - Carbon Thin Films from Organic Solution 1179
  M N Grigoryev, T N Myasoedova, T S Mikhailova

Investigation of Interaction of Molecules of Inorganic Gases with Surface of Copper-Containing
Polyacrylonitrile 1183
  M M Avilova, V V Petrov

Optical Studies of InAs/GaAs Monolayer Bragg Superlattices...1188
  K A Ivanov, A R Gubaidullin, G Pozina, E V Nikitina, M A Kaliteevski

Multiple Frequency Bloch Oscillations in Natural Superlattices..1193
  K A Ivanov, E I Girshova, E D Kolykhalova, M A Kaliteevski

Enhancement of Spontaneous Emission Probability in a Disordered Media..1199
  K M Morozov, K A Ivanov, A R Gubaydullin, M A Kaliteevski

Room Temperature Lasing in Injection Microdisks with InGaAsN/GaAs Quantum Well Active
Region 1205
  E I Moiseev, M V Maximov, A M Nadtochiy, N V Kryzhanovskaya, D A Sannikov, T Yagafarov,
  M Kulagina, T Niemi, R Isoaho, M Guina, A E Zhukov

Author Index

SPBOPEN 2018

IOP Publishing

IOP Conf. Series: Journal of Physics: Conf. Series **1124** (2018) 051011   doi:10.1088/1742-6596/1124/5/051011

Nonlinear Pulses in Dispersion-Managed Fiber-Optic Systems in Presence of High Losses

V Neskorniuk[1]**, A Lukashchuk**[2]**, I Gabitov**[3,1]**, A Chipouline**[1,4]**, M Malekizandi**[4]**, G Ovchinnikov**[1]**, F Küppers**[4]

[1]Skolkovo Institute of Science and Technology, Skolkovo Innovation Center, Moscow 143026, Russia

[2]École Polytechnique Fédérale de Lausanne, CH-1015 Lausanne, Switzerland

[3]Department of Mathematics, The University of Arizona, 617 N. Santa Rita Avenue, Tucson, Arizona 85721, USA

[4]Institut für Mikrowellentechnik und Photonik, Technische Universität Darmstadt, Merckstrasse 25, Darmstadt 64283, Germany

Abstract. A dispersion managed fiber-optic system can support dynamically stable pulses - DM-solitons and two branches of stable pairs of coupled DM-solitons – lower and upper bisolitons. Using semi-analytical model, we have found the dependency of the shape of these pulses on the level of losses and other parameters of the system and have verified the validity of our findings both numerically and experimentally. The obtained data can be used to define whether a dispersion-managed fiber-optic system can be considered as the one with constant dispersion.

1. Introduction

Dispersion management is widely used in fiber-optic systems to suppress dispersive broadening of the pulses. In a dispersion-managed (DM) system optical elements with high absolute values of normal and anomalous group velocity dispersion (GVD) are combined in a periodically repeating dispersion map to minimize the average value of GVD in the system. Propagation of an electromagnetic pulse through this system is well described by Nonlinear Schrödinger Equation (NLSE).

For NLSE applied to describe the DM optical fiber system, three groups of nonlinear dynamically stable solutions have been found both numerically and experimentally. There are: dispersion-managed solitons (DM-solitons) [1], two branches of stable two-soliton anti-phase compounds - lower and upper bisolitons [2,3], and three-solitons - bound compositions of three DM-solitons [4,5]. The shape of these pulses constantly evolves during propagation over the system regaining its original shape only at the end of each dispersion map period.

This work focuses only on DM-solitons, lower and upper bisolitons. Despite stable propagation of DM-solitons has been experimentally found in lossy systems rather long time ago [1], the existence of bisolitons has been experimentally confirmed only in the experiments deliberately excluding losses [3,4]. Similarly, the dependence of the shape of DM-soliton, lower and upper bisoliton on the parameters of a DM system has been described numerically only for the case of lossless systems [2,6]. Nevertheless, losses of optical power in a typical fiber-optic communication line are substantial. Although the losses are compensated by periodically installed optical amplifiers, propagation of DM-solitons and bisolitons is strongly affected by these losses. Therefore, for possible practical applications, it is essential to take

Content from this work may be used under the terms of the Creative Commons Attribution 3.0 licence. Any further distribution of this work must maintain attribution to the author(s) and the title of the work, journal citation and DOI.

Published under licence by IOP Publishing Ltd

into account the effect of losses on the shape of DM-solitons and bisolitons in the DM fiber-optic systems. In particular, bisolitons can be used to increase the capacity of the existing DM long-haul fiber-optic communication lines. We can introduce lower and upper bisolitons into communication alphabet as two extra letters (correspondingly, '2' and '3') in addition to the already employed DM-soliton ('1') and empty slot ('0') [2]. Moreover, bisolitons were suggested for usage in the optical retiming scheme [7].

2. Method

To study dynamically stable pulses in lossy DM systems we solved numerically the Gabitov-Turitsyn equation [8] describing the slow dynamics of pulses in a DM nonlinear system, which is governed solely by an averaged dispersion and nonlinearity. DM-solitons, lower and upper bisolitons are solitary solutions of this equation. To find them, we adapted the existing computer algorithm [6] to account for losses.

The obtained results were verified both numerically and experimentally. To check them numerically, we have examined the propagation of the obtained pulses by solving NLSE, and found that the obtained pulses are indeed dynamically stable solutions of NLSE - they returned to their original shapes having passed any number of dispersion map periods.

To check experimentally the predictions of our computer algorithm we compared the numerically obtained solutions with ones that emerge in a particular real-world DM fiber system. The experimentally investigated line had been chosen as close as possible to the real terrestrial systems. The modelled line had the length of about 3300 km, the period length 47.3 km, and losses over the period 15.2 dB. We have shown that the relationship between the peak power and the full-width at half maximum (FWHM) of the DM-solitons generated in our experiment agrees with the one predicted by our algorithm (Fig. 1).

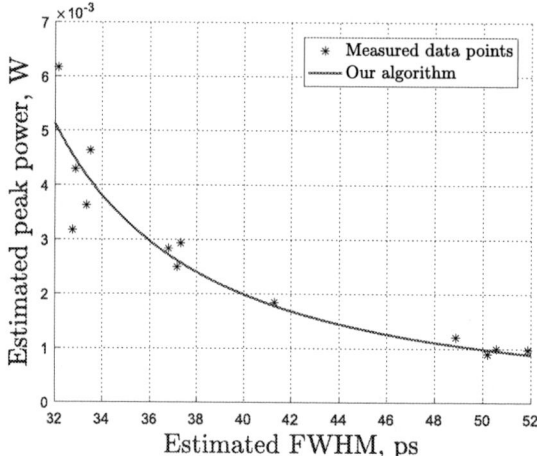

Figure 1. Comparison between the relation of estimated peak power and FWHM for experimentally measured DM-solitons and the numerically predicted ones.

3. Numerical results

Dependence of the shape of DM-solitons (lower and upper bisolitons) on parameters of a DM fiber-optic system (excepting losses) manifests itself through the dependence on the single dimensionless combination - reduced dispersion $\bar{d}_0$[2]:

$$\bar{d}_0 = \frac{2\langle\beta_2\rangle}{\gamma P L \langle\beta_2^{TF}\rangle}, \tag{1}$$

where $\langle \beta_2 \rangle$ is the averaged GVD of the DM system, $\langle \beta_2^{TF} \rangle$ is the averaged GVD of fibers in the system with anomalous dispersion ($\beta_2 < 0$) in the DM system, γ is the effective nonlinear coefficient, P is the characteristic pulse peak power, and L is the period length of the system.

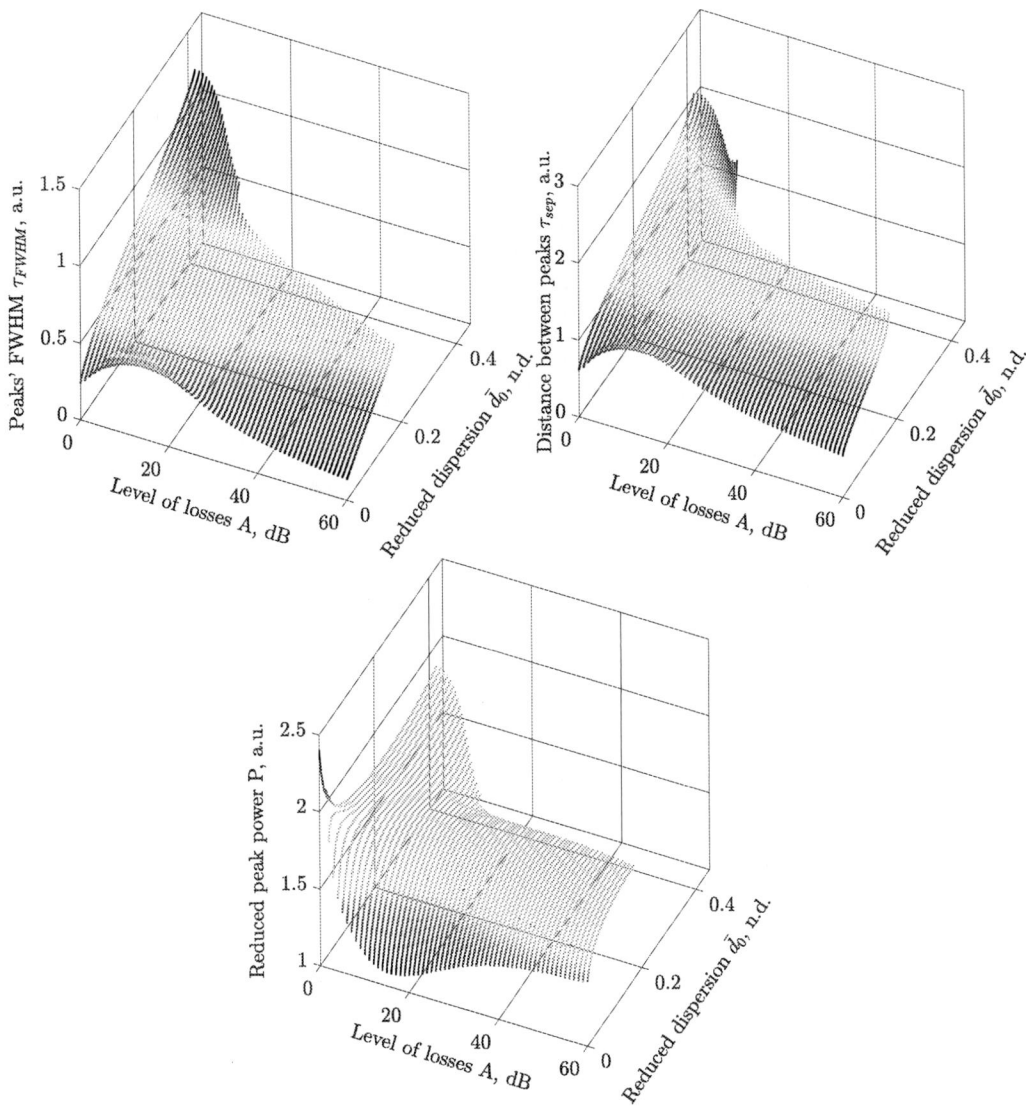

Figure 2. Dependence of the parameters describing the shape of lower bisoliton on the level of the losses on the period of DM fiber-optic system and on other parameters of the system expressed through the reduced dispersion $\bar{d}_0$. As the parameters we measured reduced peak power P and FWHM of each of the pulses forming the lower bisoliton along with the distance between the centers of mass of each of the pulses τ_{sep}.

To make conclusions about the dependence of this link on losses we studied numerically the dependence of characteristic parameters of the lower and upper bisolitons on both reduced dispersion $\bar{d}_0$ and level of losses. We examined reduced peak power P and FWHM of each of the pulses forming the bisolitons, temporal separation between the pulses τ_{sep}, and found that dependence of these parameters on reduced dispersion $\bar{d}_0$ is affected by the level of losses (Fig. 2).

$$P = \max_{t}(|u(t)|^2)/\aleph, \quad \aleph = \frac{1-e^{-\alpha}}{\alpha}, \quad \alpha = A\frac{\ln 10}{10}; \quad \tau_{sep} = \frac{\int_0^{\infty} dt\, t|u(t)|^2}{\int_0^{\infty} dt\, |u(t)|^2} - \frac{\int_{-\infty}^0 dt\, t|u(t)|^2}{\int_{-\infty}^0 dt\, |u(t)|^2}, \quad (2)$$

where $u(t)$ is the wavepacket of the pulse and t is the retarded time frame moving with the group velocity of the bisoliton, so that $t = 0$ point is always situated at the point of bisoliton antysimmetry ($u(t) = -u(-t)$).

Moreover, we have found that the position of the upper limit on the reduced dispersion $\bar{d}_0$ for which lower and upper bisolitons can stably propagate, referred to as the bifurcartion point $\bar{d}_0^{bif}$ [2], is very sensitive to the level of losses (Fig. 3). Since bisolitons are not dynamically stable solutions of NLSE with constant GVD, we suggest usage of bifurcation point $\bar{d}_0^{bif}$ as a limit on DM system parameters for which the real GVD distribution cannot be approximated by the average GVD.

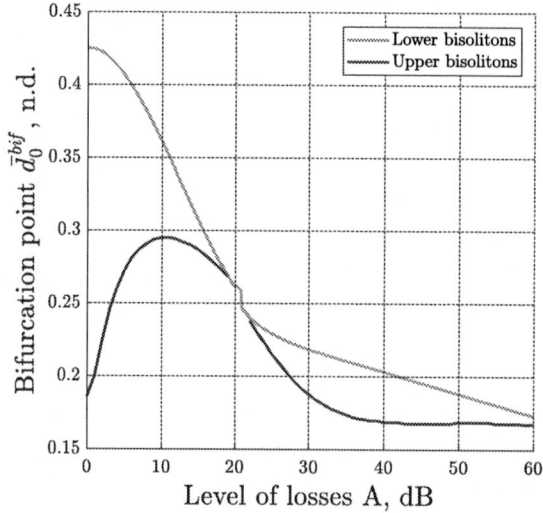

Figure 3. Dependence of the position of bifurcation point $\bar{d}_0^{bif}$ (green line) and the existence region of upper bisolitons (between green and blue lines) on the level of losses A on the period of a DM fiber-optic system. Bifurcation point $\bar{d}_0^{bif}$ is defined in [2]. For the levels of losses for which the blue line is drawn upper bisolitons exist for reduced dispersion $\bar{d}_0$ values between the ones given by the green and the blue lines. For the levels of losses for which blue line is not depicted we didn't find upper bisolitons. Lower bisolitons for a given level of losses exist for the reduced dispersion values $\bar{d}_0 \in \left[0, \bar{d}_0^{bif}\right]$

Acknowledgements

The authors are highly grateful to Skolkovo Institute of Science and Technology for funding and to Institut für Mikrowellentechnik und Photonik of Technische Universität Darmstadt for the provided experimental equipment.

References

[1] Turitsyn S. K., Bale B. G., Fedoruk M. P., 2012 *Phys. Rep.* **521** 135-203.

[2] Gabitov I., Indik R., Mollenauer L., Shkarayev M., Stepanov M., Lushnikov P.M. 2007 *Opt. Lett.* **32** 605–607

[3] Stratmann M., Pagel T., Mitschke F. 2005 *Phys. Rev. Lett.* **95** 143902.

[4] Rohrmann P., Hause A., Mitschke F. 2013 *Phys. Rev. A* **87** 043834.

[5] Alamoudi S. M., Al Khawaja U., Baizakov B. B. 2014 *Phys. Rev. A* **89** 053817.

[6] Shkarayev M., Stepanov M. G 2009 *Physica D* **238** 840-845.

[7] Johnson S., Pau S., Küppers F. 2011 *J. Lightwave Technol.*, **23** 3493-3499.

[8] Gabitov I. R., Turitsyn S. K. 1996 *Opt. Lett.* **21** 327-329.

SPBOPEN 2018   IOP Publishing

Study of optical pumping influence on carbon nanotubes permittivity in THz frequency range

P Demchenko[1], D Gomon[1], I Anoshkin[1,2], D Lioubtchenko[1,2], M Khodzitsky[1]

[1] THz Biomedicine Laboratory, Department of Photonics and Optical Information Technologies, ITMO University, St. Petersburg, 197101, Russian Federation
[2] Microwave/THz microsystem Laboratory, KTH – Royal Institute of Technology, Stockholm SE-100 44, Sweden

Abstract. Equivalent complex permittivity of carbon nanotubes (CNT) was measured with/without light illumination at the frequency range of 0.2-1 THz. It was shown that we can tune the dispersion of the CNT complex conductivity during varying of optical pumping (wavelength of 980 nm). These results mean that CNT is perspective candidate for development of THz tunable attenuators and phase shifters.

Introduction

Nowdays terahertz (THz) radiation is widely used in such areas, as THz scanning for security applications[1], THz tomography[2], THz imaging for analysis of historic paintings and manuscripts[3], THz sensing and imaging in manufacturing quality control[4], THz biomedicine[5-7] and THz Communication Systems[8]. At the moment, tunable components are being developed for use in terahertz spectroscopy devices, and therefore materials that can effectively change their properties in the THz frequency range at various influences (optical, electrical, mechanical, etc.) are investigated[9]. In this work influence of optical pumping (wavelength of 980 nm) on CNT complex permittivity is studied.

Experimental samples

The carbon nanotubes samples were synthesized by aerosol chemical vapor deposition method[10].

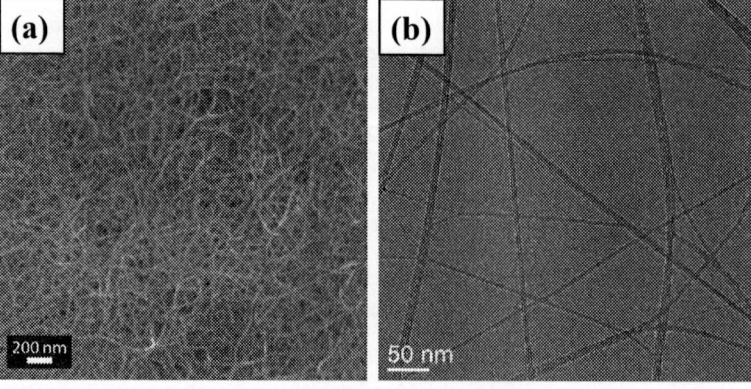

Figure 1(a, b). Typical **(a)** SEM and **(b)** TEM images of the CNTs.

Content from this work may be used under the terms of the Creative Commons Attribution 3.0 licence. Any further distribution of this work must maintain attribution to the author(s) and the title of the work, journal citation and DOI.
Published under licence by IOP Publishing Ltd

Figure 1 shows a typical high-resolution scanning electron microscope (SEM) JEOL JSM-7500FA image of the CNT network on glass substrate and spherical aberration corrected transmission electron microscope (TEM) JEOL JEM-2200FS image of the CNTs with diameters of 1.3–2.0 nm. SEM was realized at 2 kV (Figure 1 (a)) and 2.5 mm working distance and TEM was realized at 80 kV (Figure 1 (b)).

The single walled carbon nanotubes (SWCNTs) diameters were 1-1,3 nm. SWCNT layers were deposited by manual dry transfer from the nitrocellulose filter onto polytetrafluoroethylene (PTFE) substrates. The samples were simultaneously illuminated with an infrared laser in continuous wave mode (CW) at a wavelength of 980 nm with optical power of 1.1 W. The illumination area was approximately 1,6 cm^2.

Experimental setup

A 3D-image of the THz TDS setup is shown in Figure 2 [11, 12]. The femtosecond IR laser parameters are the following: a central wavelength of 1040 nm; a full width at half maximum (FWHM) of 5 nm; an average output power of 1.1 W; and a pulse width of 120 fs. The femtosecond infrared laser beam is split in two beams: probe and pump with a ratio of 10% to 90%, respectively. The optical delay line controls optical path of the pump beam. The beam is modulated by a chopper (Ch) at 667 Hz. The generator of the THz radiation (InAs crystal) takes place in a magnetic system with a field of 2 T. After passing through an IR radiation filter, the THz beam incidents on the sample. The sample may be simultaneously illuminated by an infrared laser (IR laser). The parameters of the illuminating laser are the following: continuous wave mode (CW) at a wavelength of 980 nm; a varying optical power from 1 mW to 8 W; and a sample area illumination of 0.94 cm^2.

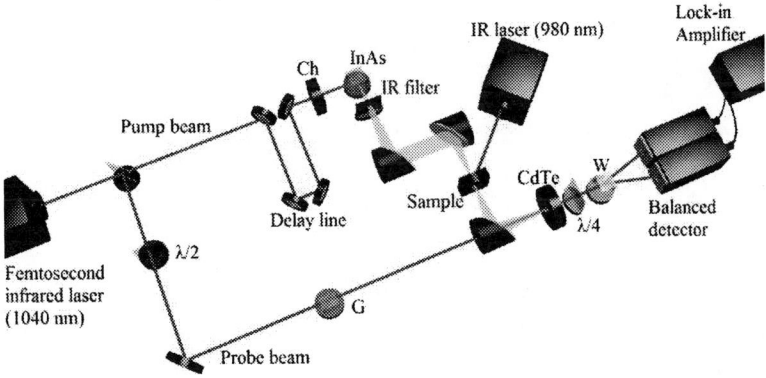

Figure 2. Schematic of the THz TDS setup with optical illumination. Ch – optical chopper, InAs – THz radiation generator (InAs crystal in a magnetic field of 2 T), G – Glan prism, CdTe – nonlinear crystal, $\lambda/2$ and $\lambda/4$ – half-wave and quarter-wave plates, W – Wollaston prism.

The THz radiation induces birefringence in the electro-optical crystal (CdTe) for the infrared probe beam. Probe IR beam is split by Wollaston prism (W) into two beams with orthogonal polarization, which are detected by balanced photodiodes (balanced detector). The Fourier transform of the temporal waveform $E(t)$ gives the spectral distribution of the THz pulse in the frequency domain. The THz field in the frequency domain is a complex value, containing the amplitude and the phase information which may be used to calculate all complex optical properties of a sample.

Results

THz TDS waveforms of irradiated and non irradiated samples were measured by terahertz time-domain spectroscopy (THz TDS). Based on the standard thin-film approximation [13] we can extract the dispersion of complex sheet conductivity of the CNT from the measured transmission spectra by application of the relation:

$$\hat{\sigma}_{CNT}(f) = \frac{1}{Z_0}(n_{Sub}+1)(\frac{\hat{E}_0(f)}{\hat{E}(f)}-1),$$

where Z_0 =377 Ω is impedance of free space, f is a frequency, n_{Sub} is a complex refractive index of substrate, $\hat{E}(f)$ and $\hat{E}_0(f)$ are complex frequency-domain electric fields of the THz wave transmitted through the CNT on the PTFE substrate, and, as a reference, through the bare substrate, respectively. Equivalent complex permittivity of CNT may be derived as [14]:

$$\hat{\varepsilon}_{CNT}(f) = 1 + \frac{i\hat{\sigma}_{CNT}(f)}{2\pi f \varepsilon_0 d_{CNT}},$$

where $\varepsilon_0 = 8.85 \cdot 10^{-12}$ F/m is the permittivity of free space, d_{CNT} is a thickness of CNT layer ($d_{CNT} = 55$ nm).

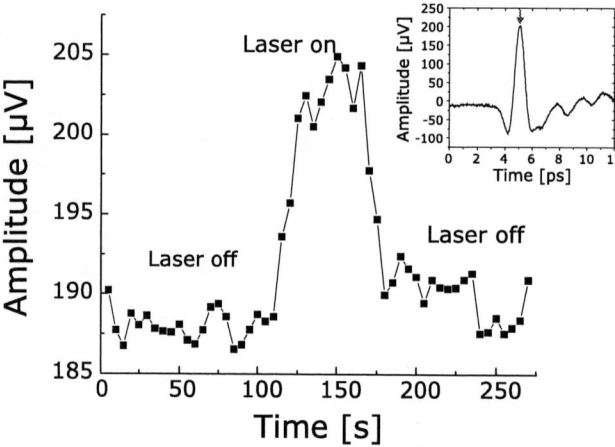

Figure 3. Amplitude variation of the THz TDS waveform in the on and off illumination states. Inset: THz pulse in the time domain.

During illumination of 0.7 W/cm^2 dynamics of the THz waveform changed in the amplitude of the radiation. For this purpose, the delay line was set to the maximum of the waveform amplitude (~5.3 ps) (inset in Figure 3). To increase the accuracy of the amplitude measurement, the time constant of the lock-in-amplifier was increased up to 1 second.

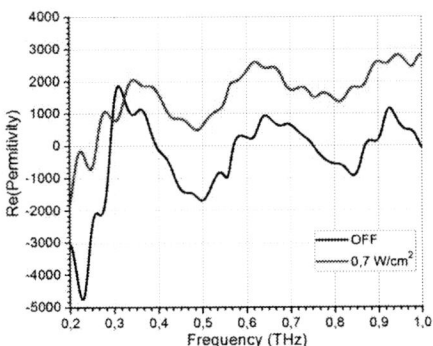 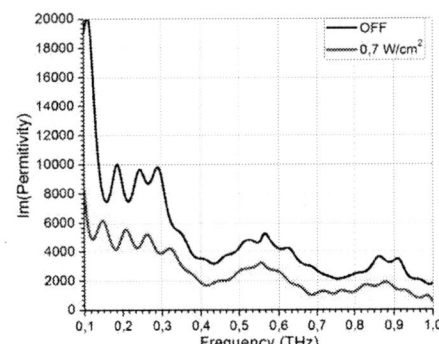

Figure 4(a, b). (a) Real and **(b)** imaginary parts of the CNT sample's relative permittivity, extracted from THz TDS measurements with/without illumination power levels.

The real part of complex permittivity is responsible for waveform time delay. Figure 4(a) shows small difference between the real part of relative permittivity of the illuminated and not illuminated samples. That why waveform time delay between illuminated and not illuminated samples is insignificant. The imaginary part of complex permittivity is responsible for waveform amplitude. Figure 4(b) shows that imaginary part of relative permittivity of the illuminated sample is lower that not illuminated one. Therefore this leads to increase of waveform amplitude at optical pumping.

Conclusions
Opportunity for the optically control of the CNT permittivity in the THz frequency range was demonstrated. These results mean that CNT is perspective candidate for element of THz attenuators and phase shifters.

Acknowledgments
This work was financially supported by Government of Russian Federation (Grant 08-08).

References
[1] I. Hosako, N. Sekine, M. Patrashin, S. Saito, K. Fukunaga, Y. Kasai, P. Baron, T. Seta, J. Mendrok, S. Ochiai, and H. Yasuda, "At the dawn of a new era in terahertz technology," *Proc. IEEE*, vol. 95, no. 8, pp. 1611–1623, 2007.
[2] Wallace, V. P., Taday, P. F., Fitzgerald, A. J., Woodward, R. M., Cluff, J., Pye, R. J., & Arnone, D. D. "Terahertz pulsed imaging and spectroscopy for biomedical and pharmaceutical applications," *Faraday Discussions*, 126, 255-263, 2004.
[3] Fukunaga, Kaori, et al. "Terahertz imaging for analysis of historic paintings and manuscripts," Infrared, Millimeter and Terahertz Waves, 2008. IRMMW-THz 2008. 33rd International Conference on. IEEE, 2008.
[4] Zeitler, J. Axel, and Yao-Chun Shen. "Industrial applications of terahertz imaging," Terahertz spectroscopy and imaging. Springer Berlin Heidelberg, pp. 451-489, 2012.
[5] Borovkova M., Serebriakova M., Fedorov V., Sedykh E., Vaks V., Lichutin A., Salnikova A., Khodzitsky M. "Investigation of terahertz radiation influence on rat glial cells" Biomedical Optics Express, Vol. 8, No. 1, pp. 273-280, 2017.
[6] M. Borovkova, M. Khodzitsky, P. Demchenko, O. Cherkasova, A. Popov, I. Meglinski "Terahertz time-domain spectroscopy for non-invasive assessment of water content in biological samples" Biomedical Optics Express, Vol. 9, No. 5, pp. 2266-2276, 2018.

[7] Gusev S, Demchenko P, Cherkasova O "Influence of glucose concentration on blood optical properties in THz frequency range" *Chinese Optics*, 11(2): 182-189, 2018.

[8] Koenig, S., et al. "Wireless sub-THz communication system with high data rate," *Nature Photonics* 7(12): 977, 2013.

[9] Grebenchukov A, Zaitsev A, Khodzitsky M, "Optically controlled narrowband terahertz switcher based on graphene" *Chinese Optics*, 11(2): 166-173, 2018.

[10] I. V. Anoshkin, A. G. Nasibulin, Y. Tian, B. Liu, H. Jiang, and E. I. Kauppinen, "Hybrid carbon source for single-walled carbon nanotube synthesis by aerosol cvd method," *Carbon*, vol. 78, pp. 130–136, 2014.

[11] V.Y. Soboleva, D.A. Gomon, E.A. Sedykh, V.K. Balya, and M.K. Khodzitskii, "Development of narrow bandpass filters based on cross cavities for the terahertz frequency range," *Journal of Optical Technology*, vol. 84, no. 8, pp. 521-524, 2017.

[12] D.A. Gomon, E.A. Sedykh, M.K. Khodzitsky, K.I. Zaitsev, I.T. Monroy, S. Rodriguez, A.V. Vozianova, "Influence of the Geometric Parameters of the Electrical Ring Resonator Metasurface on the Performance of Metamaterial Absorbers for Terahertz Applications," *Chinese Optics*, vol. 11, no. 1, pp. 47-59, 2018.

[13] G. Jnawali, Y. Rao, H. Yan, and T.F. Heinz, "Observation of a transient decrease in terahertz conductivity of single-layer graphene induced by ultrafast optical excitation," *Nano letters*, vol. 13, no. 2, pp. 524-530, 2013.

[14] C. How Gan, "Analysis of surface plasmon excitation at terahertz frequencies with highly doped graphene sheets via attenuated total reflection," *Applied Physics Letters*, vol. 101, no. 11, pp. 111609-4, sep 2012.

SPBOPEN 2018   IOP Publishing

Numerical study of optical properties of sphere-gap-cone hybrid nanoantenna

Y Sun*, S.V. Makarov, D A Zuev

Department of Nanophotonics and Metamaterials, ITMO University, Saint-Petersburg, Russia

Abstract. Hybrid nanostructures exhibit extraordinary optical properties achieved by the combination of metal and dielectric components in a single nanostructure. Here we numerically study optical properties of metal-dielectric (hybrid) sphere-gap-cone nanostructure. Numerical simulation related to scattering cross section demonstrates that this nanoantenna possesses strong magnetic and electric response in the visible range. In addition, looking into resonant wavelengths indicate this nanostructure is capable of unidirectional scattering. We also reveal electric field enhancement in the gap between dielectric and metal components of the nanoantenna.

1. Introduction

The metal nanoparticles are proven to be efficient systems for light controlling considering the ability to modify the near-field distribution [1]. On the other hand, low-loss high-index dielectric nanoparticles with inherent magnetic and electric Mie resonances offer a great opportunity for light controlling via designing their scattering properties [2]. Recently, experiments have demonstrated the opportunity to simultaneously obtain high localization of the near field and magnetic response in hybrid metal/dielectric nanostructures [3-4]. So far plenty of hybrid nanostructures such as core-shell, adjacent nanoparticles are investigated [5-7].

In this paper, we numerically study novel type (sphere-gap-cone) of hybrid nanostructure made from plasmonic (gold) and dielectric (crystalline silicon) components. The investigation of optical properties reveals that this nanostructure provides distinct resonant properties, opportunity for unidirectional scattering as well as enhancement of electrical field in the gap between nanoparticles composing nanostructure.

2. Results and Discussion

A schematic of the sphere-gap-cone hybrid nanostructure is presented in Fig. 1a. As can be seen from the figure, the gold spherical nanoparticle is located on the truncated crystalline silicon cone nanoparticle with a 10 nm gap between them. The bottom base's diameter as well as the height of the cone is 190 nm. The diameter of the cone upper base is half of the bottom base, i.e. in the proposed model of 95 nm. By applying Computer Simulation Technology (CST) calculation method, we have numerically studied scattering properties of the hybrid nanostructure in the dark-field geometry under oblique s-polarized incidence at 68 degrees normal to the symmetry axis. Fig.1b indicates the simulated scattering cross section of the hybrid sphere-gap-cone, in which points A and B are the resonances, respectively. By looking into the field distribution, we can verify that resonance A

Content from this work may be used under the terms of the Creative Commons Attribution 3.0 licence. Any further distribution of this work must maintain attribution to the author(s) and the title of the work, journal citation and DOI.

Published under licence by IOP Publishing Ltd

indicates strong magnetic resonance, while resonance B demonstrates electric dipole resonance. Consequently, the hybrid nanostructure support strong magnetic and electric response in the visible range.

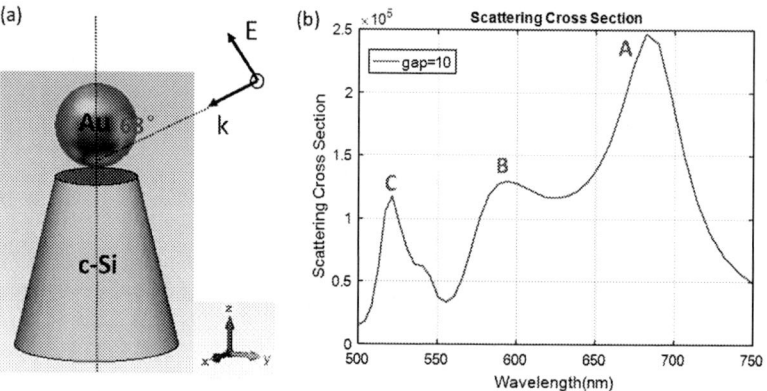

Fig.1 (a) Schematic of hybrid sphere-gap-cone nanostructure with s-polarized incidence; (b) scattering cross section as a function of wavalength

Besides the scattering cross section in the whole range, we also demonstrate that this nanoantenna can be engineered for unidirectional scattering realisation. It is well known from Mie theory, that unidirectional scattering occurs when electric dipole resonance and magnetic resonance have comparable intensities. As can be learned from the scattering properties and field distribution, resonance A is the magnetic dipole resonance with high intensity, while resonance B is the electric dipole resonance with low intensity. Thus we are supposed to achieve unidirectional scattering somewhere between resonance A and resonance B. Fig. 2 depicts the polar and 3D scattering at the resonant wavelength of 682 nm (point A). As can be seen from the figure, this resonance also possesses high directivity in one direction, which makes this nanostructure a good candidate for unidirectional scattering.

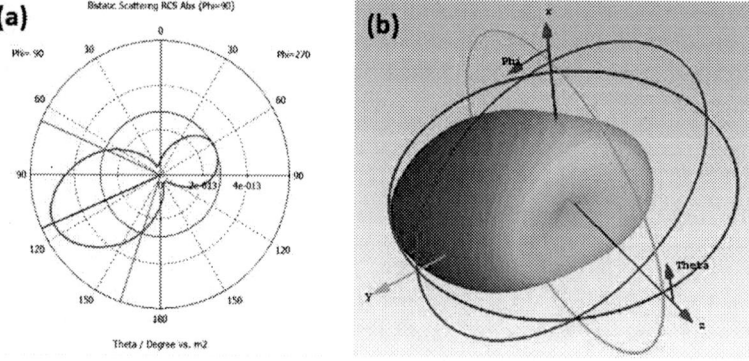

Fig. 2 Far-field results at wavelength 682 nm (a) polar; (b) 3D pattern

In order to look into the electric field enhancement, the system is illuminated with a P-polarized electromagnetic plane wave with 68 degrees oblique incidence. Fig. 3 shows the electric field enhancement with the variation of wavelength in the visible range. This system provides high value of the E-filed enhancement almost approaching 14 in this excitation conditions. Besides, simulations are also carried out to study the scattering cross section in the same excitation. Different from S-polarized excitation, we can only observe the strong magnetic resonance A (682 nm) since the electric vector is perpendicular to the incident surface. However, in this case, light can be strongly constrained within the subwavelength gap, as shown in the set of figure 3.

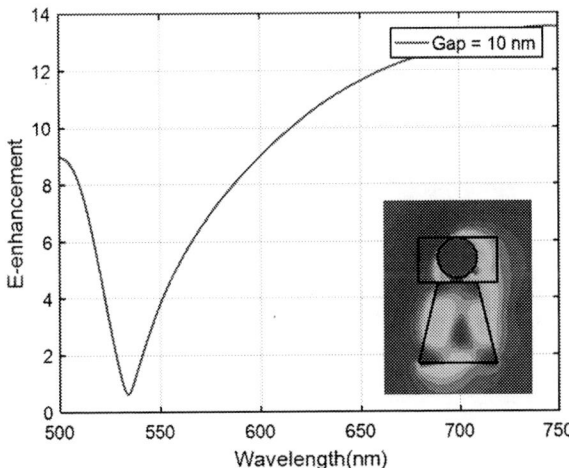

Fig. 3 Electric filed enhancement in the gap as a function of wavelength. Inset: distribution of electric field at resonant wavelength (682 nm). The outside rectangular border of the nanosphere is demonstrated for well establishing the mesh cells, and thus achievement of relatively reliable calculating results

3. Conclusion

In summary, we have investigated numerically optical properties of sphere-gap-cone hybrid nanostructure. We demonstrate, that this nanostructure provides optical properties inherent to plasmonic and dielectric components (resonant scattering, directivity in one direction), but also enhancement of electrical field in the gap between nanostructure components. We believe, the carried out research paves the way for further investigation of optical properties of such nanostructure as well as its experimental realisation.

Acknowledgments

This work was supported by the Ministry of Education and Science of Russian Federation (Project 2.2267.2017/4.6 and 16.8939.2017/8.9) as well as Russian Foundation for Basic Research (Projects 16-37-60101 and 17-02-00538). Y Sun acknowledges the support from the China Scholarship Council (201706160135). D. Zuev acknowledges the scholarship of the President of the Russian Federation.

References

[1] Giannini V, Fernández-Domínguez A I, Heck S C, and Maier S A 2011 *Chemical reviews* **111** 3888
[2] Kuznetsov A I, Miroshnichenko A E, Fu Y H, Zhang J, and Luk'yanchuk B 2012 *Scientific reports* **2** 492
[3] Nakamura S, Senoh M, Nagahama S, Iwase N, Yamada T, Matsushita T, Kiyoku H and Sugimoto Y 1996 *Japan. J. Appl. Phys.* **35** L74

[4] Miroshnichenko A E, Luk'yanchuk B, Maier S A, and Kivshar Y S 2011 *Acs Nano* **6** 837

[5] Shibanuma T, Grinblat G, Albella P, and Maier S A 2017 *Nano letters* 17 2647

[6] Timpu F, Hendricks N R, Petrov M, Ni S, Renaut C, Wolf H, Isa L, Kivshar Y and Grange R 2017 *Nano letters* **17** 5381

[7] Sun S, Li M, Du Q, Png C E and Bai P. Metal–Dielectric 2017 *The Journal of Physical Chemistry C* **121** 12871

SPBOPEN 2018   IOP Publishing

Active and passive phase stabilization for the all-fiber Michelson interferometer

M.S. Elezov[1], M.L.Scherbatenko[1], D.V. Sych[1,2], G.N. Goltsman[1]

[1]Department of Physics, Moscow State University of Education, 29 Malaya
Pirogovskaya St, Moscow, 119435, Russia
[2]P. N. Lebedev Physical Institute, Russian Academy of Sciences, Leninskiy Prospekt
53, Moscow, Russia

Abstract. We put forward two methods for phase stabilization in the all-fiber Michelson
interferometer. To perform passive phase stabilization, we use a heat bath for all fibers and
electro-optical components, and put the interferometer in a hermetic case. To perform active
phase stabilization, we monitor output power of the interferometer and develop an electronic
feedback control. The phase stabilization methods enable stable interference pattern for several
minutes, and can be helpful for the development of the optimal quantum receiver for coherent
signals.

1. Introduction

Interferometry is a very useful technique of combining waves and observing the result of their
interference in order to diagnose the parameters of interest that influence the interference pattern. This
technique has numerous applications in physics, engineering, and applied sciences. For example,
interferometry is recognized to be a crucial prerequisite in the field of ultimately sensitive detection of
optical quantum signals. The signals can have very broad origin: from the extremely weak fluctuations
caused by gravitational waves, to the long-haul classical and quantum telecommunication. The later
case is mostly employed in optical fibers, hence the practical need to deal with the optical fiber
interferometry.

One of the shortcoming of optical fibers is their high sensitivity to the external influences, such as
mechanical stress, bends, or airflows. In order to build a practical device utilizing optical fiber
interferometer, we need to reduce the interference instability and increase the stable operation time. A
particular application that we have in mind is the development of the all-fiber optimal quantum
receiver for weak coherent signals. The problem of optimal signal detection dates back to the late 60's,
when quantum-mechanical properties of the signal were recognized as an unavoidable bottleneck for
the ultimate measurement precision limit [1].

Traditionally, heterodyne or homodyne receivers are used to measure the phase and the amplitude of
the signal without *a priori* knowledge of the signal parameters. Though in the simplest case, a receiver
should be able to discriminate just two different signal states, e.g. coherent states with a fixed
amplitude and binary modulated phase. In 1967 Helstrom found the theoretical limit for the maximum
quantum signal detection precision, or the minimum signal detection error [1]. This limit is called the
Helstrom bound, and a receiver able to attain this limit is called the optimum quantum receiver. The
optimum quantum receiver has a higher signal detection precision compared to the heterodyne or
homodyne receivers. Unfortunately, Helstrom did not propose a concrete realization of the optimum
quantum receiver.

Content from this work may be used under the terms of the Creative Commons Attribution 3.0 licence. Any further distribution
of this work must maintain attribution to the author(s) and the title of the work, journal citation and DOI.
Published under licence by IOP Publishing Ltd

In 1973, Kennedy proposed a quantum receiver [2], which is based on the optical displacement of the input signal state and a single-photon detection. The Kennedy receiver exceeds the precision of the homodyne and heterodyne receivers in the limit of strong signals, but does not approach the Helstrom bound. Soon after the Kennedy's proposal, Dolinar improved the scheme and shown the first proposal of an optimum quantum receiver [3], though his proposal is difficult to implement in reality. A more realistic approach to build the optimal quantum receiver was proposed only recently [4], which we plan to realize in the future. The Kennedy receiver is based essentially on the same principle (optical displacement of the signal and photon counting), but allows an easier implementation. Several proof-of-principle demonstrations of the Kennedy receiver and its slightly modified versions were already shown with help of the free-space interferometry [5,6]. However, the all-fiber realization of the Kennedy receiver has not yet been demonstrated.

We develop a Kennedy receiver based on the all-fiber Michelson interferometer. On the way to achieve the goal, we came across with some difficulties. The main difficulty of developing all-fiber interferometer is instability of the interference due to varying local temperature of a fiber, vibrations, polarization distortion, to name a few. To auto-compensate the polarization instability, we utilize the Michelson-Faraday interferometric scheme. Thus the phase instability remains the main challenge which we solve here. In this work, we show several technical solutions for passive and active phase stabilization of the Michelson interferometer.

2. Experimental setup and passive phase stabilization

We use a fiber-based DFB laser with 1548.7770 nm wavelength. The width of the spectrum is 2 MHz. The drift of central wavelength is negligible, due to the current and temperature stabilizations of the laser. The polarization is linear. Radiation is guided through the polarization controller PC1 on a polarization-dependent circulator C1, which is installed in a hermetic Box #1 (see Fig 1). The circulator plays the role of an optical isolator capable of transmitting radiation only in one direction from the laser to the interferometer. However, because of the reflections from the ends of the circulator, undesirable interference occurs, which leads to instability of the optical power at the input of the Michelson interferometer. On ports 1 and 2 of the circulator, two short SM-fibers were additionally added to stabilize the "parasitic" interference. At the input of the interferometer, a polarization controller PC2 is placed to adjust the polarization and improve the contrast. The Michelson interferometer consists of a 90/10 splitter BS1, two phase modulators PM1 and PM2, and two Faraday mirrors FM1 and FM2.

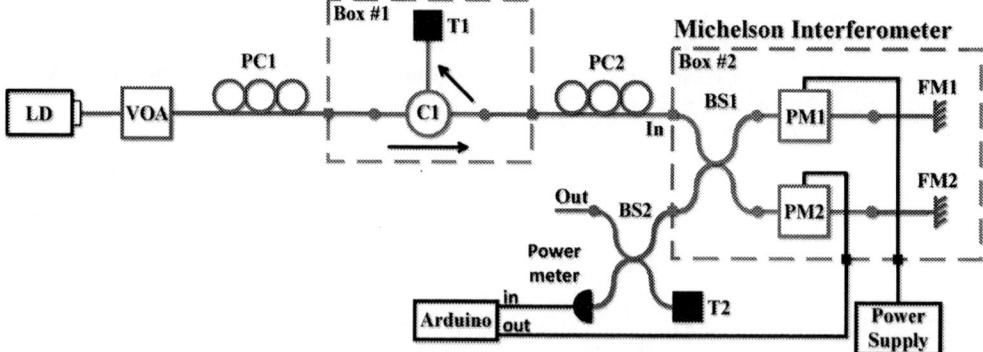

Figure 1. Experimental setup. LD - Laser Diode at 1550 nm; PC1 and PC2 are the polarization controllers FPC031; C1 is a circulator; T1 and T2 are fiber light traps (terminators); BS1 and BS2 - beam splitters 50/50 and 90/10, respectively; PM1 and PM2 - phase modulators; FM1 and FM2 - Faraday's mirrors; Power meter Ophir VEGA with power head PD300-IRG-V1. Red line is fiber SMF-28e; Blue line - PM fiber; Black line - electrical wire.

We use single-mode (SM) fibers in the Michelson interferometer. The behavior and analysis of multi-mode fiber interferometers are complex, because the multimode nature of the radiation creates more complex interference phenomena, which are very difficult to take into account and control. The characteristics of interferometers based on multi-mode fiber are much worse than ones based on SM fibers. Though even in the case of SM fibers, there is a number of problems associated with polarization phenomena in fibers. The polarization state of radiation in SM fibers can change rapidly and unpredictably due to external influences. Mismatch and fluctuations of the polarization state cause interference contrast fluctuations. Therefore, it is better to use polarization-maintaining (PM) fibers. In the case of employment SM fibers, we have to use reliable mechanical fixation of the fibers, which leads to a more stable polarization of the propagating radiation.

The main task of adjustment of the Michelson interferometer is phase difference stabilization between two arms. The delay phase φ on a waveguide length l is $\varphi = 2\pi \frac{nl}{\lambda}$, where λ is radiation wavelength, n is average index of refraction of the fiber core. The temperature instability of individual optical components is the main effect that influences the phase delay of the optical fiber due to changing refractive index of quartz and geometric parameters of the fiber (length and core diameter). The change of phase can be represented in the form $\delta\varphi = \frac{2\pi}{\lambda}\left(n\frac{\partial l}{\partial T} + l\frac{\partial n}{\partial T}\right)\delta T$. The first item is the contribution of geometric stretching. The second item is the contribution of changing index of refraction. The main contribution to the phase changing is the fluctuation of the index of refraction. The phase sensitivity of the two-arm interferometer is $K = \frac{\delta\varphi_2 - \delta\varphi_1}{\delta T} = \frac{2\pi}{\lambda}\frac{\partial n}{\partial T}\Delta l$. Where Δl is difference length of two arms. To reduce phase sensitivity, we should use arms of the interferometer with equal lengths ($\Delta l \approx 0$). The arms of the interferometer should be in the same external physical conditions, so that the phase sensitivities of fiber arms are the same. But, in reality, it is difficult to achieve equality of the arms. Thus the contribution of temperature fluctuations can be significantly reduced, but does not disappear.

An additional influence on the interference stability is the mechanical stress on the fibers and on the joints of the interferometer connectors: air convection, acoustic waves, pressure, mechanical vibrations transmitted from the table, etc. To ensure the same physical conditions for the arms of the interferometer, the optical elements of the Michelson interferometer are fixed by means of a sticky aluminum tape to a massive brass plate, which has the role of a thermal bath. The plate is placed in a hermetic metal Box #2. The electrical contacts of the phase modulators are connected to the case. For additional shielding against thermal radiation, the Box #2 is wrapped in aluminum foil and in a dense black textile and placed on four bolts fastened to the optical table. The bolts act as anti-vibration mounts that reduce mechanical vibrations coming from the optical table. The polarization controller PC2 is placed at the input of the interferometer for adjusting polarization.

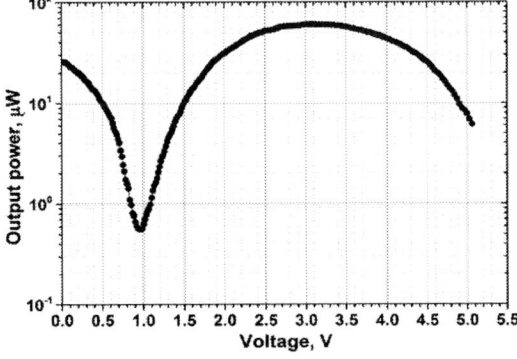

Figure 2. Output optical power from the Michelson interferometer as a function of voltage on the phase modulator PM2.

We supplied the phase modulator PM2 by voltage from 0 to 5.1 V for the interference contrast measurement (see Fig. 2). The minimum output power is 0.57 uW, and the maximum output power is 60.3 uW, which correspond to the phase difference of 0° and 180° between the arms of the interferometer, respectively. In this case we achieve the interference contrast $C = \frac{P_{max}}{P_{min}} = 106$.

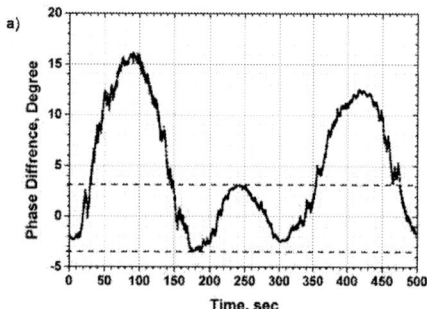 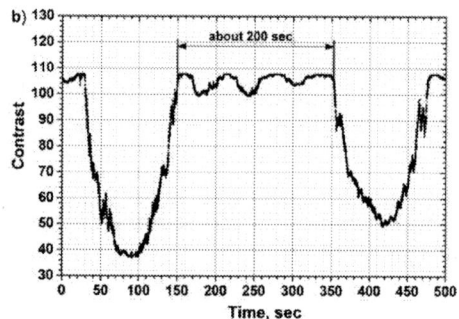

Figure 3(a,b). Phase **(a)** and contrast **(b)** instability of the Michelson interferometer.

3. Active phase stabilization

After performing the passive phase stabilization, we obtain the slow-varying drift of the interference pattern (see Fig. 3a). To achieve high interference contrast (more than 100), we have to manually adjust the phase difference by the phase modulator PM1. Then the task of the interference stabilization can be formulated not as "What should be done to make the phase difference between the arms of the interferometer constant?", but rather "What should be done to change the phase difference as slowly as possible?" We obtain the stabilization of interference contrast by all the above methods of passive phase stabilization on a timescale of a few minutes (see Fig. 3b). This turns out to be enough for employing the active phase stabilization system. We use a feedback control in the Michelson interferometer for active phase stabilization. To do this, the output of the interferometer through the BS2 beam splitter is applied to the Ophir VEGA power meter with the PD300-IRG-V1 detector head. From the power meter, an analog signal carrying the voltage amplitude, which directly proportional to the measured optical power, is fed to the hardware platform "Arduino Uno". After the signal processing, the platform supplies the needed voltage to the phase modulator. The maximal phase adjustment rate is about 70 ms. The rate is limited by the detector head rate. After the active phase stabilization, we can maintain the required phase difference between two arms of the Michelson interferometer for much longer time far exceeding the data acquisition time of the quantum receiver.

4. Conclusions

We developed passive and active methods for the phase stabilization in an all-fiber Michelson interferometer that operates at the conventional telecom C-band (1550 nm). The methods extend the stable operation time for sufficiently long period and can be used for the development of the all-fiber optimal receiver for coherent signals [4].

5. Acknowledgement

The research has been carried out with the support of the Russian Science Foundation (project No. 17-72-30036).

6. References

[1] C.W. Helstrom, Inf. Control 10, 254 (1967).
[2] R. Kennedy, MIT Res. Lab. Electron. Quart. Prog. Rep. 108, 219 (1973).

[3] S. Dolinar, MIT Res. Lab. Electron. Quart. Prog. Rep. 111, 115 (1973).
[4] D. Sych and G. Leuchs, "Practical receiver for optimal discrimination of binary coherent signals", Phys. Rev. Lett., 117, 200501 (2016).
[5] C. Wittmann, et al, Phys. Rev. Lett. 104, 100505 (2010).
[6] C. Wittmann, et al, Phys. Rev. A 81, 062338 (2010).

SPBOPEN 2018   IOP Publishing

Binary diffractive optics for 3D-demultiplexing of OAM beams

A P Porfirev[1,2], S A Fomchenkov[1,2], G E Gridin[1], S N Khonina[1,2]

[1]Samara National Research University, Samara 443086, Russia
[2]Image Processing Systems Institute of RAS - Branch of the FSRC "Crystallography and Photonics" RAS, Samara 443001, Russia

Abstract. We propose to use a combination of spiral Fresnel zone plates (SFZPs) and binary diffractive gratings in order to realize 3D-demultiplexing of vortex beams – laser beams having an orbital angular momentum (OAM). It is known that binary and blazed fork-shaped diffractive gratings are widely used for demultiplexing of OAM beams and their analysis. In contrast to the fork-shaped gratings, in the case of SFZPs the diffraction orders differ in their focus positions along the propagation axis. A combination of these two approaches allows us to perform the so-called 3D-demultiplexing of OAM beams.

1. Introduction

Multiplexing and demultiplexing of laser beams having an orbital angular momentum (OAM) is of great interest in the field of optical communications [1] and optical manipulation [2]. Currently, numerous techniques are used to perform these operations such as conventional diffractive optical gratings, Dammann gratings [3] etc. However, these elements only allow the multiplexing or demultiplexing of OAM beams solely in the single transverse plane. Indeed, the method of multiplexing of OAM beams proposed by Zhu et al. is limited by the fact that generated multiple beams have only equal topological charges [4].

In this paper we use a combination of such well-known elements like spiral Fresnel zone plates (SFZPs) and binary diffractive gratings in order to realize 3D-demultiplexing of OAM beams with different topological charges. In contrast to the fork-shaped gratings, in the case of SFZPs the diffraction orders differ in their focus positions along the propagation axis. A combination of a SFZP and conventional binary diffractive gratings allows us to perform the so-called 3D-demultiplexing of OAM beams. It can be used for design of compact OAM- beam detection systems for optical communications or dynamic switching systems for application of active sorting of microparticles.

2. Modelling

2.1. Description of the proposed technique

The well-known binary fork-shaped gratings allows one to generate the OAM-optical vortex beams with the opposite topological charges in conjugate diffraction orders (see Fig. 1a). Such an element irradiated with an OV beam having the specific topological charge l produces OV beams with topological charges defined as `$l + pn$ (n is the diffraction order number, p is an integer showing the number of the internal "teeth" of the fork-shaped grating), as well as an additional zero-order OV beam with a topological charge equal to l [6-9]. For example, if a vortex beam with a topological charge of +1 passes through a fork-shaped grating with $p = 1$, then in the -1 diffraction order, the helical phase on the target beam is

Content from this work may be used under the terms of the Creative Commons Attribution 3.0 licence. Any further distribution of this work must maintain attribution to the author(s) and the title of the work, journal citation and DOI.
Published under licence by IOP Publishing Ltd

removed, this beam eventually evolves into a Gaussian-like beam. The Gaussian-like beam can be isolated from the other OAM beams, which still have helical phase fronts, using a spatial mode filter at the focal point of a lens to couple the power only of the Gaussian mode due to the mode-matching constraints[10].

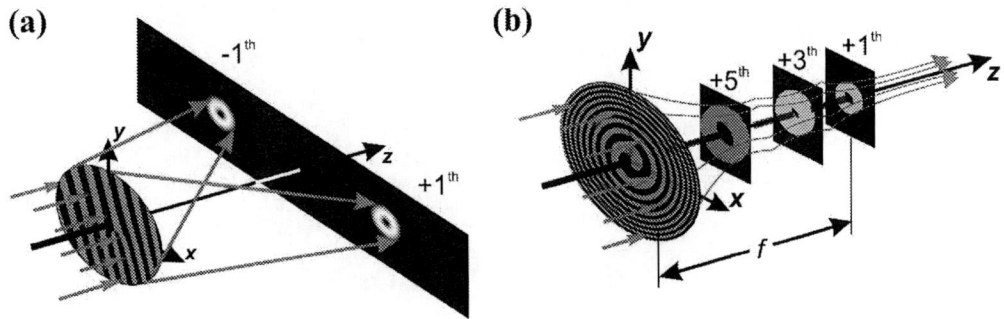

Figure 1. OAM beams generated with **(a)** a fork-shaped phase grating and **(b)** a SFZP.

A SFZP has the transmission function in the following form:

$$\tau(r,\varphi)=exp\left\{i\,arg\left[\,cos\left(\alpha r^2+p\varphi\right)\right]\right\}$$ (1)

where (r, φ) are the polar coordinates in the hologram plane, $\alpha = \pi/\lambda f$ gives the radial scale, f is the first-order focal length of the SFZP, which corresponds to the incident radiation wavelength λ, and p is the topological charge of the generated OV beam. As mentioned above, in sharp contrast to the fork-shaped hologram, a SFZP generates an OV beam on its optical axis (see Fig. 1b), with the generated diffraction orders pattern located along the propagation axis at various focal lengths [11]. The detection of the OAM beams in this case is the same as in the case of fork-shaped holograms, the initial vortex beam with a topological charge of l transforms to the beam with a topological charge defined as $l + pn$ at the nth diffraction order.

The combination of these two elements in one diffractive optical element (DOE) allows the generation of optical vortex beams with different topological charges in different diffraction orders, also permitting focusing and selection at different transverse planes located along the propagation axis, that is, the 3D-demultiplexing of OAM beams.

2.2. Modelling methods

The Fresnel transformation to perform the numerical simulation of the propagation of the OAM beam with defined topological charges after passing through a SFZP took the following form:

$$E(\xi,\eta,z)=\frac{1}{i\lambda z}exp(ikz)\iint\limits_{\Omega}\tau(x,y)exp\left\{i\frac{k}{2z}\left[(\xi-x)^2+(\eta-y)^2\right]\right\}dxdy,$$ (2)

where (x, y) are the Cartesian coordinates in the plane of a DOE ($x = r\cos\varphi$, $y = r\sin\varphi$), (ξ, η) are the Cartesian coordinates at a distance z from the plane of the DOE, ω_0 is the waist beam radius, $k = 2\pi/\lambda$ is the wavenumber, and R is the radius of the DOE.

2.3. Results

Figure 2 shows the numerically obtained longitudinal intensity profiles of OAM beams with different topological charges passed through a SFZP. For different topological charges, an intensity maximum on the optical axis appears in different locations corresponding to the different diffraction orders, while OAM beams with a non-zero topological charge are formed at the other diffraction orders.

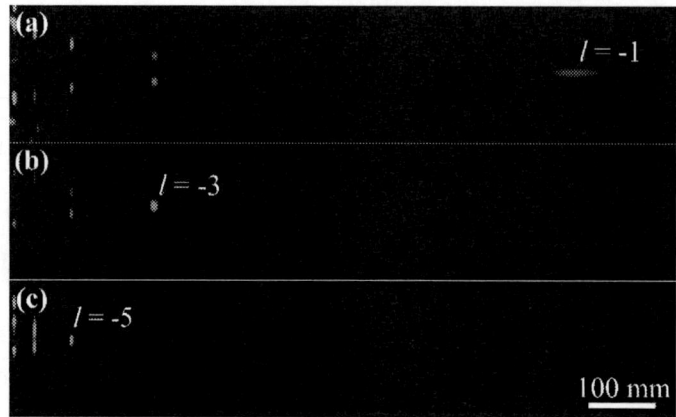

Figure 2. Numerically obtained longitudinal intensity profiles of OAM beams with different topological charges l: **(a)** -1, **(b)** -3 and **(c)** -5, passed through a SFZP with $p = 1$.

3. Experiments

3.1. Experimental details

The optical setup used for the experimental investigation of the demultiplexing of OAM beams performed by SFZPs is shown in Fig. 3. We used radiation from a solid-state laser with a wavelength of 532 nm. An output laser beam collimated and spatially filtered with the help of a system consisting of the microobjective MO_1, a pinhole P and a lens L_1 was modulated using the first spatial light modulator (SLM) (SLM_1, HOLOEYE) implementing the phase transmission function of a vortex beam generator in the form of the blazed fork-shaped grating. The designed SFZP implemented using the second SLM made it possible to perform demultiplexing of the generated optical vortex beams. The CMOS video-camera (TOUPCAM) was used to record the formed intensity distributions at different distances z. Insets in Fig. 3 show the interferogram patterns obtained from the interference of the generated OAM beams and a Gaussian beam with a spherical wavefront. The spiral-shaped interference patterns clearly demonstrate the presence of the helical wavefront inherent to the OAM beams: in this case, the number of the spirals corresponds to the topological charge of the generated OAM beam.

3.2. Results

Figure 4 shows the longitudinal profiles of the intensity distributions shaped by a microobjective MO_3 for optical vortices with topological charges of $l = -1$, -3, and -5 respectively. The presented longitudinal profiles are cross-sections of the laser beam along the vertical axis. It is evident, that for the different optical vortex beams, the bright spots appear in different diffraction orders. In addition, the brightness of the spots is proportional to the diffraction order analogous to the conventional fork-shaped gratings.

SPBOPEN 2018

IOP Publishing

IOP Conf. Series: Journal of Physics: Conf. Series **1124** (2018) 051015   doi:10.1088/1742-6596/1124/5/051015

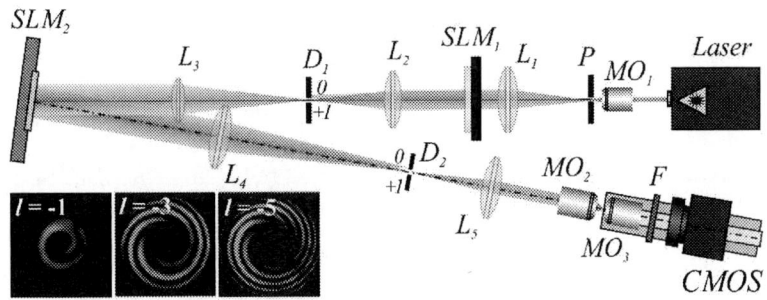

Figure 3. Experimental optical setup: SLM_1 and SLM_2 are two spatial light modulators (HOLOEYE), MO_1, MO_2 and MO_3 are the microobjectives, L_1, L_2, L_3, L_4 and L_5 are the convex lenses, D_1 and D_2 are two diaphragms, P is a pinhole, F is a neutral density filter, $CMOS$ - is a videocamera (TOUPCAM). The insets show the spiral-shaped interference patterns of the generated OAM beams with different topological charges l.

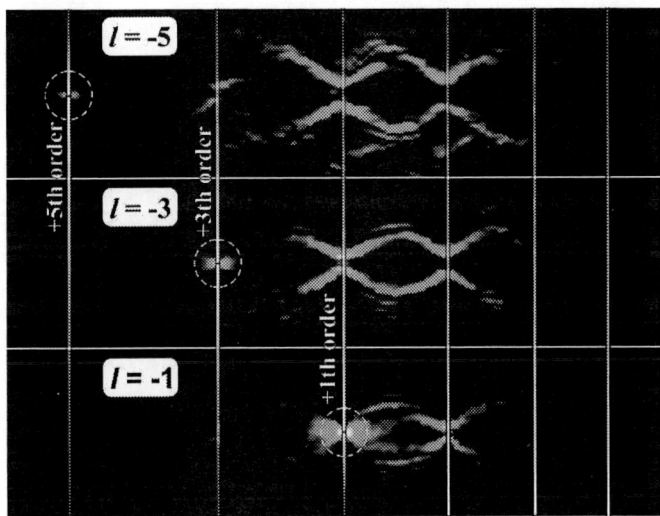

Figure 4. The longitudinal profiles are composed of data from cross-sections of the laser beam along the vertical axis for the cases of the optical vortices with topological charges of l = -1, -3, and -5 shaped with the help of a single SFZP.

4. Acknowledgments
This work was supported by Russian Federation Presidential grant for support of the leading scientific schools (NSh-6307.2018.8) in part of design of the utilized optical elements, and by Russian Federation Presidential grants for support of young candidates of sciences (MK-2390.2017.2) and Russian Foundation for Basic Research (grants No. 17-42-630008, 16-47-630677, 16-29-11744) in part of experimental results.

References

[1] Wang J, Yang J-Y, Fazal I M, Ahmed N, Yan Y, Huang H, Ren Y, Yue Y, Doilnar S, Tur M, Willner A E 2012 *Nat. Photon.* **6** 488

[2] Grier D G 2003 *Nature* **424** 810

[3] Lei T, Zhang M, Li Y, Jia P, Liu G N, Xu X, Li Z, Min C, Lin J, Yu C, Niu H, Yuan X 2015 *Light Sci. Appl.* **4** e257

[4] Zhu L, Sun M, Zhu M, Chen J, Gao X, Ma W, Zhang D 2014 *Opt. Express* **22** 21354

[5] Bazhenov V Y, Vasnetsov M V, Soskin, M S 1990 *Pis. Zh. Eksp. Teor. Fiz.*. **52** 1037

[6] Padgett M, Courtial J, Allen L, Franke-Arnold S, Barnett S M 2002 *J. Mod. Opt.* **49** 777

[7] Moreno I, Davis J A, Pascoguin B L, Mitry M J, Cottrell D M 2009 *Opt. Lett.* **34** 2927

[8] Bekshaev A Y, Orlinska O V 2010 *Opt. Commun.* **283** 1244

[9] Stoyanov L, Topuzoski S, Stefanov I, Janicijevic L, Dreischuh A 2015 *Opt. Commun.* **350** 301

[10] Willner A E, Huang H, Yan Y, Ren Y, Ahmed N, Xie G, Bao C, Li L, Cao Y, Zhao Z, Wang J, Lavery M P J, Tur M, Ramachandran S, Molisch A F, Ashrafi N, Ashrafi S 2015 *Adv. Opt. Photon.* **7** 66-106

[11] Soskin M S, Polyanskii P V, Arkhelyuk O O 2004 *New J. Phys.* **6** 196

The study of photoluminescence properties of AlGaAs/GaAs heterostructure after Ga+ focused ion beam etching

G V Voznyuk[1], I V Levitskii[2], M I Mitrofanov[2], M N Mizerov[3], D N Nikolaev[2], V P Evtikhiev[2]

[1]Department of Light Technologies and Optoelectronics, ITMO University, Saint Petersburg 197101, Russia
[2]Laboratory of Semiconductor Luminescence and Injection Emitters, Ioffe Institute RAS, Saint Petersburg 194021, Russia
[3]SHM R&E Ctr RAS, Saint Petersburg, 194021, Russia

Abstract. One of the promising tools for monolithic photonic integrated circuits (PICs) fabrication is focused ion beam (FIB) lithography. It's well-known that FIB etching process induces radiation defects formation and consequently leads to losses in internal quantum efficiency of luminescence. We demonstrate the possibility of restoring luminescence properties of etched AlGaAs/GaAs heterostructure by means of annealing. Achieved results give an opportunity for fabrication of active PICs elements by FIB.

1. Introduction

Photonic Integrated Circuit (PIC) is a complex integrated circuit, which combine a plenty of optical devices in a single photonic chip [1]. Integrated photonics components can be divided in two groups: active - lasers, modulators, active filters, and passive - low-loss waveguides, passive filters, splitters and couplers [2,3].

There are two basic methods of photonic elements integration - hybrid and monolithic. Hybrid coupling allows integration of silicon photonic circuits with microlasers based on direct-band A3B5 semiconductors [4,5]. It is possible to realize monolithic PIC design, completely based on A3B5 heterostructures.

Focused ion beam (FIB) is a potential tool for prototyping microcircuit devices. FIB is a direct nanolithography technique based on the physical impact on the material by high-energy ions focused in nanoscopic scale. The technology is capable of realising elements with size about ~ 10 nanometers and about several nanometeres gap between them, etching at different depths (three-dimensional lithography), avoiding differences in the etching rate depending on the geometry of the lithographic pattern being formed [6,7]. FIB can be easily integrated into other technological processes with UHV requirements. The main downside of FIB is radiation defects formation during etching process. High-energy ions induce radiation defects that lead to decrease of heterostructure internal quantum efficiency of the luminescence. This problem is well-researched for the ion implantation process with typical energies 100-300 keV [8]. However, the influence of radiation defects, produced by 30 keV FIB, on active PICs based on A3B5 semiconductors has been investigated insufficiently.

Content from this work may be used under the terms of the Creative Commons Attribution 3.0 licence. Any further distribution of this work must maintain attribution to the author(s) and the title of the work, journal citation and DOI.

Published under licence by IOP Publishing Ltd

2. Experiment

To study the effect of FIB etching on photoluminescence (PL) properties we prepared test double

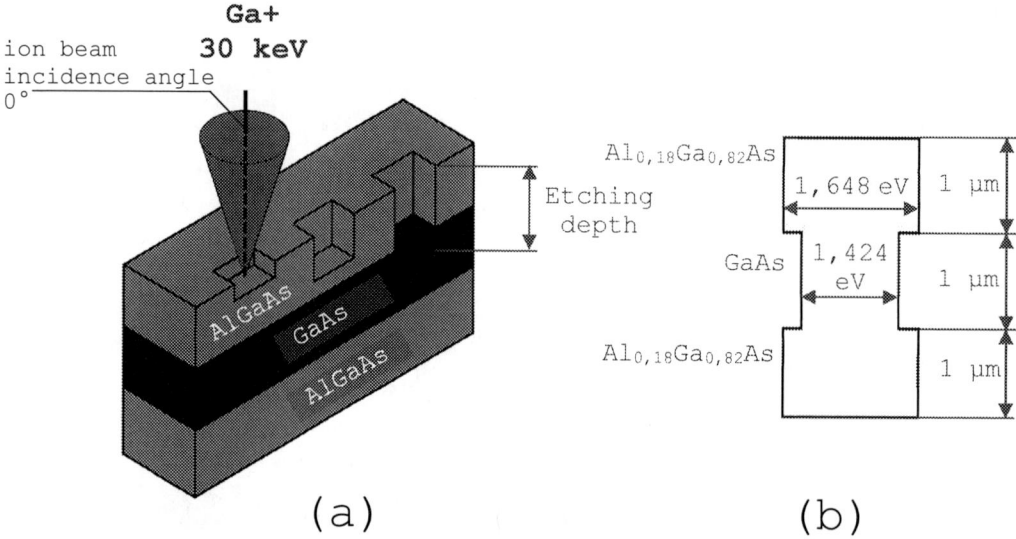

Figure 1. Scheme of the sample (a) and bandgap diagram of the etched sample (b).

heterostructure (AlGaAs/GaAs) consisting of a 1000 nm thick GaAs layer enclosed between two 1000 nm thick $Al_{0.18}Ga_{0.82}As$ emitter layers (figure 1). On sample surface 7 squares 50 x 50 μm with depths of 1, 150, 180, 330, 850, 1200 and 1500 nm were etched using Ga+ ions at 30 keV ion energy. The effect of FIB etching on the heterostructure was studied by photoluminescence. The pump source was a solid-state laser with a wavelength of 671 nm, which was focused into 10 μm diameter spot.

3. Results and discussions

All measured spectra (example on inset figure 2) contain interband luminescence of GaAs and 808 nm peak of the laser diode pumping of the solid state laser. Luminescence of $Al_{0.18}Ga_{0.82}As$ layer was not observed. In all PL measurements the shape of the spectrum from the etched region fully fits the shape of the non-etched region. The change in relative integral intensity after FIB etching normalized on the PL intensity of non-etched surface is presented on figure 2 (squares). The measurements show a strong decrease in the PL signal starting from the first square. Deeper etching leads to the quenching of PL signal and for the samples with depth of 850 nm and deeper PL signal was not observed. Such kind of behavior could appear due to the accumulation of radiation defects during etching. The nonequilibrium carriers generated by laser in the emitter layer recombine nonradiatively that leads to the suppression of photoluminescence signal.

To restore the PL properties of etched samples, the heterostructure was annealed in vacuum condition at 300 ° C during 20 minutes. Photoluminescence from the first three squares (etching depths 1, 150 and 180 nm) was restored almost completely to the initial level. At the 4th and the 5th etched squares (depths of 330 and 850 nm), a strong drop in signal intensity was observed. PL signal after

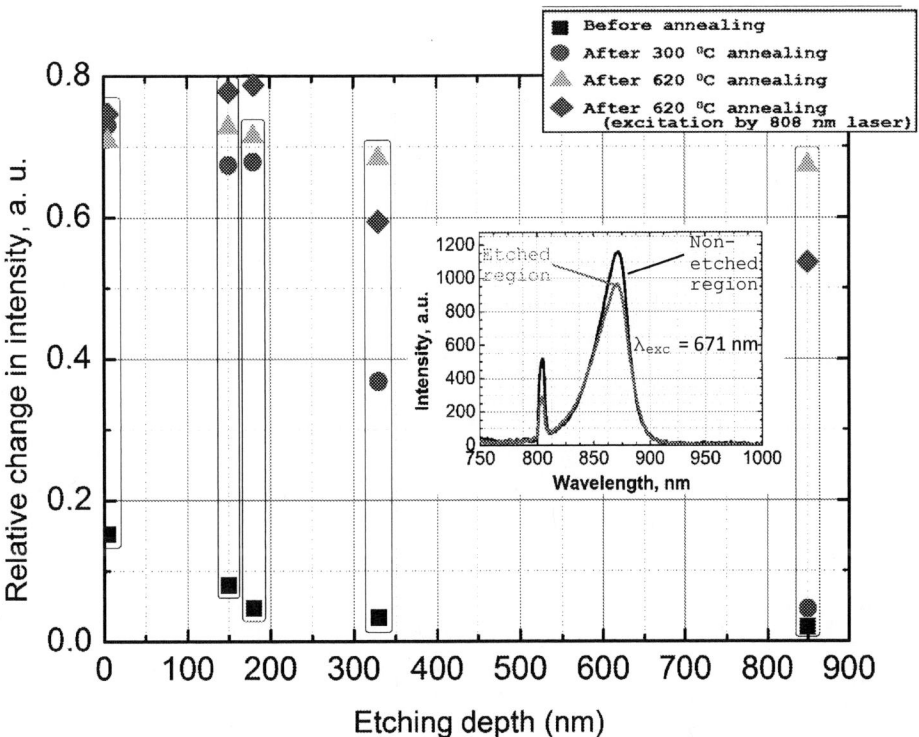

Figure 2. The evolution of luminescence intensity for five different etching depths (squares), 300 ° C annealing during 20 minutes (circles) and after 620 ° C annealing in an arsenic atmosphere during 20 minutes (triangles). *Inset* demonstrates the PL spectra of etched (850 nm depth) and non-etched surface.

first annealing is presented on figure 2 (circles). At the next stage, the structure was annealed in an arsenic atmosphere at the temperature close to the non-congruent decomposition (620 ° C) during 20 minutes. The annealing led to the recovery of the PL intensity up to 75-80% of luminescence of the non-etched surface from the 1th to the 5th etched square (triangles fig. 2). PL from the 6th and the 7th etched squares (etching depth of 1200 nm and 1500 nm) was not observed.

To analyse the experimental data we used SRIM software package for calculation of the distribution of vacancies and implanted atoms in the AlGaAs layer (fig.3). According to these calculations, the penetration depth of all kind of radiation defects does not exceed 70 nm.

The result of calculations does not correlate with our PL measurements. One of the possible reasons is that SRIM calculations do not include local heat and gradual damage accumulation. These factors make lattice more loosely coupled and atoms are easier to dislodge, so the damage may be underestimated. We suppose that a high local temperature during etching process and a high concentration of nonequilibrium radiation defects produce conditions for the diffusion and accumulation of point defects closer to the heterointerface $Al_{0.18}Ga_{0.82}As$ emitter-GaAs layer. As previously mentioned the PL intensity of the annealed samples does not exceed 75-80%. To understand these limits we conducted PL measurements with excitation by 808 nm laser (rhombus on fig. 2) in order to generate nonequilibrium carriers only in GaAs layer. The comparison of the PL signal of etched and nonetched surface shows the same decrease like in case with 671 nm excitation. Therefore, the limitation of the PL signal occurs from the surface damage.

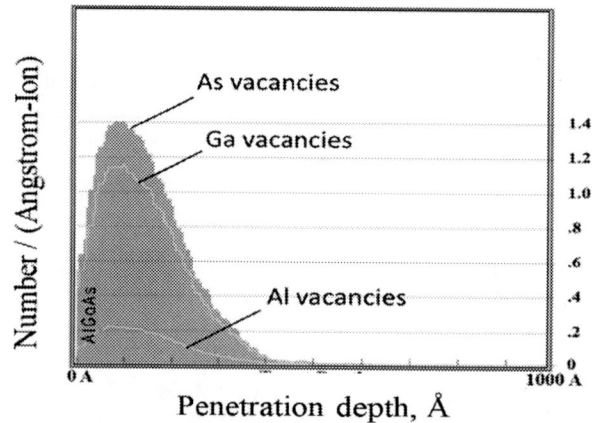

Figure 3. Distribution profiles of Al, Ga, As vacancies calculated by SRIM with the following parameters: Ga+ ion source, accelerating voltage 30 keV and ion beam incidence angle 0°.

Conclusion

A possibility of recovering of the luminescent properties of the AlGaAs/GaAs heterostructure etched down to the 150 nm distance from the heterointerface was demonstrated. Our experimental findings confirm the prospects of using FIB etching technique for making active PICs elements.

References

[1] Nagarajan R, Joyner C H, Schneider RP et al. 2005 Ieee Journal Of Selected Topics In Quantum Electronics **11** 50
[2] Ye WN, Xiong YL, JOURNAL OF MODERN OPTICS **60** 1299
[3] Lawniczuk K, Augustin LM, Grote N et al ADVANCED OPTICAL TECHNOLOGIES **4** 157
[4] Roelkens G, Liu L, Liang D et al. 2010 Laser & Photonics reviews **4** 751
[5] Reithmaier JP, Benyoucef M 2016 ECS Transactions **72** 171
[6] Michael GS, Brett BL, Kyle M, Jason DF, Philip DR 2017 J. Vac. Sci. Technol. **35**
[7] Kannegulla A, Cheng LJ 2016 Nanotechnology **27**
[8] Steckl A J, Chen P, Choo A G et al., 1992 MRS Online Proceedings Library (OPL) **281** 319

SPBOPEN 2018

IOP Publishing

Optical limiting in nanodiamond suspension: shortening of the laser pulses

R Yu Krivenkov[1], K G Mikheev[1], T N Mogileva[1], N Nunn[2], O A Shenderova[2] G M Mikheev[1]

[1]Institute of Mechanics, Udmurt Federal Research Center of the UB RAS, Izhevsk, Russia 426067

[2]Adámas Nanotechnologies, Inc., 8100 Brownleigh Dr, Suite 120, Raleigh, NC 27617, USA

Abstract. Optical limiting (OL) in an aqueous suspension of nanodiamonds (NDs) with an average nanoparticle size of 20 nm is studied and the results are presented. NDs are obtained by grinding the micron-sized diamond powders manufactured by static high-pressure, high-temperature (HPHT) synthesis. The experiments are performed by the z-scanning technique with a closed aperture. The temporal parameters of nanosecond laser pulses at 1064 nm passing through the suspension in the OL mode are measured. It is found that the closer the cuvette with suspension to the laser beam waist, the shorter both the trailing edge of the laser pulses passing through the suspension and the laser pulse duration. The smooth control capability of the pulse duration in the range from 22 to 15 ns is demonstrated.

1. Introduction

Optical limiting is a phenomenon in which the power of a laser beam passing through an optical medium is nonlinearly attenuated. The physical mechanisms leading to OL can be two-photon (multiphoton) absorption, free-carrier absorption, inverse saturable absorption, nonlinear light scattering, nonlinear refraction, etc. [1, 2]. In the nanosecond range of laser pulses, OL was observed in suspensions of carbon-black [3], carbon nanotubes [4–7], onion-like carbon [8, 9], graphene [10, 11], and detonation nanodiamonds (DNDs) [12–14]. Recently, we have carried out studies of the nonlinear optical properties of aqueous suspensions of single-digit DNDs and have found a saturable absorption that transforms into OL at high laser influence [15]. However, the studies of OL in HPHT NDs suspensions with larger primary crystallites sizes have not been previously carried out. It is known that OL is accompanied by a change of the temporal shape of the transmitted laser pulses. The OL in the carbon nanotube (CNT) suspension due to nonlinear scattering of light by vapor bubbles leads to the shortening of the laser pulse duration [16]. The shortening of the pulse duration is due to the preferential scattering of the trailing edge of the laser pulse. The OL due to two-photon absorption leads to strong absorption of laser power at the top of the laser pulse. As a result, the duration of the laser pulse passing through the nonlinear medium increases. Therefore, it is important to study the temporal shape of laser pulses to reveal the mechanism of the OL in the aqueous suspension of HPHT NDs.

In this paper, we demonstrated that by scanning a cell with HPHT ND aqueous suspension along the axis of a focused laser beam, it is possible to smoothly control the duration of nanosecond laser pulses at 1064 nm. The shortening of the duration of light pulses occurs due to the shortening of the trailing edge of laser pulses passing through the suspension.

Content from this work may be used under the terms of the Creative Commons Attribution 3.0 licence. Any further distribution of this work must maintain attribution to the author(s) and the title of the work, journal citation and DOI.

Published under licence by IOP Publishing Ltd

2. Experiments

NDs purchased from Van Moppes, Ltd were produced by grinding the micron-sized diamond powder manufactured by static high-pressure, high-temperature synthesis in hydraulic presses. To ensure the formation of a stable HPHT ND suspension in water the sample was chemically treated in acidic environment, resulting in zeta potential of -45 mV. Obtained ND suspension had been stable for a long time (more than 3 years). Examination of the nanoparticles on an electron microscope confirmed that the most of the particles were of 20 nm in diameter.

To characterize the HPHT ND particles by Raman spectroscopy, the appropriate aqueous HPHT suspension was used. The sample was deposited onto the glass substrate and dried at room temperature. Raman spectra were measured with Horiba Jobin Yvon HR 800 Raman spectrometer at 632.8 nm excitation. In order to prevent NDs from graphitization and blackening the input laser intensity was set below 10 kW/cm^2 [17–19]. From Figure 1, one can see that Raman spectra of HPHT ND clusters have a clear nanodiamond band at approximately 1330 cm^{-1}. This value exactly coincides with the Raman shift for nanodiamond films synthesized by chemical vapor deposition (CVD) [20], but is 2 cm^{-1} less than for bulk crystalline diamond [21].

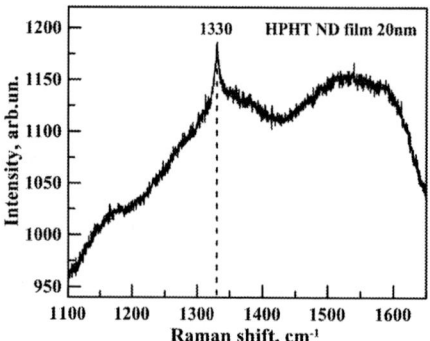

Figure 1. Raman spectrum of the HPHT NDs obtained upon excitation by light at a wavelength of 632.8 nm.

Figure 2 shows the optical density spectrum of the 1 wt % aqueous HPHT ND suspension placed in an optical cuvette with a thickness of 1 mm. The absorption spectrum was measured with a Perkin ELMER

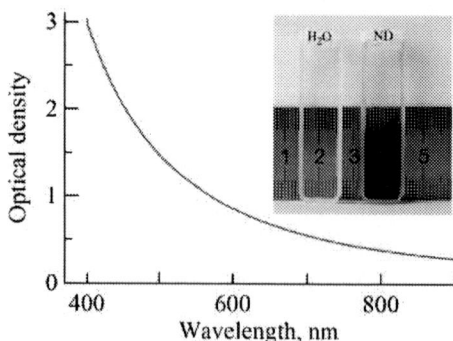

Figure 2. The optical density of the 1 wt % aqueous HPHT ND suspension in the 1 mm path length optical cuvette and photograph of the 1 mm cuvettes with the (right) suspension and (left) distilled water (inset).

LAMBDA 650 double-beam UV/Vis spectrophotometer relative to the same cuvette filled with distilled water. It is seen that in the range of 400-900 nm the optical density of the suspension decreases monotonically with wavelength increase. Such a spectrum of optical density in the visible and near infrared ranges is characteristic for suspensions of carbon nanotubes, onion-like carbon and DNDs [8, 9, 13, 16, 22, 23]. Photographs of the 1 mm cuvettes with the suspension and distilled water are shown in Figure 2 (see inset). Linear transmittance of the 1 mm thick 1 wt % HPHT ND suspension at 1064 nm laser wavelength was 62.9 %. Experiments were carried out using a z-scan technique with a close aperture z-scan system [24] according to the optical scheme shown in Figure 3. In the experiments we used passive Q-switch YAG:Nd^{3+} - laser, which generates TEM$_{00}$ mode single-frequency radiation [18]. The τ_0 pulse duration of the laser, operated at $\lambda = 1064$ nm, was 22.1 ns.

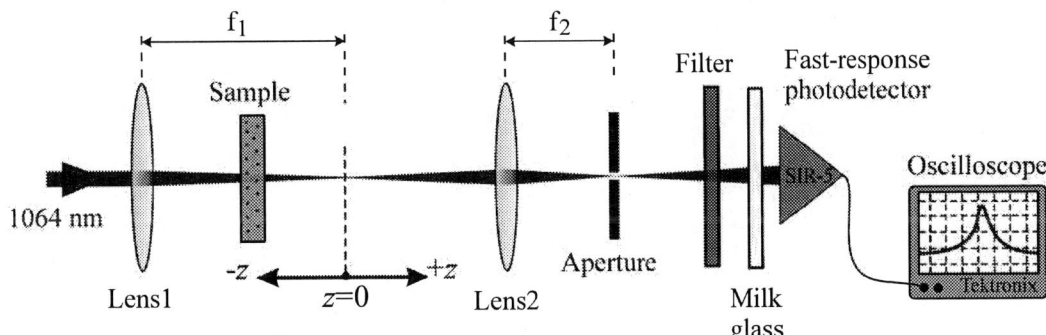

Figure 3. Schematics of the experimental setup.

The laser beam was focused by a lens (Lens 1) with a focal length of 150 mm. The diameter of the laser beam $2w_0$ in the beam waist ($z = 0$) was 114 μm. For the Gaussian beam $w^2(z) = w_0^2[1+z^2/z_0^2]$, where $z_0 = \pi w_0^2/\lambda$ is true. The 1 mm cell with the suspension was located on a one-coordinate motor stage. An aperture was used to select the scattered radiation from the transmitted laser beam. It was located in the focal plane of the output lens (Lens 2). The focal length of the output lens was 100 mm. At the position of the cuvette far from the waist of the focused beam, the laser radiation passed through the diaphragm completely. The laser pulses passed through the aperture and then, after attenuation by neutral filters, passed through the milk glass and entered the fast speed SIR-5 photodetector (ThroLab) with a rise time less than 60 ps. The electrical pulses generated at the output of the photodetector were recorded by a broadband TDS7704B digital oscilloscope (TEKTRONIX). The amplitude and the temporal shape of output laser pulses were studied as a function of the cuvette position z/z_0 along the focused laser beam axis. The duration of the input and output laser pulses τ_{hw} was defined with respect to the ratio 0.5 of its maximum. The temporal evolution of the laser pulses was characterized by measurement of the rise time τ_{rise} and fall time τ_{fall}, which were defined with respect to the 0.1 and 0.9 of the pulse amplitude.

3. Results and discussion

Figures 4a shows oscillograms (normalized to the maximum value) of the output laser pulses at various values of z/z_0, where $z_0 = \pi w_0^2/\lambda$, $z_0 = 9.6$ mm (z_0 is Rayleigh length). One can see that the less the z/z_0, the smaller the transmitted laser pulse amplitude (Figure 4b). This indicates that OL occurs. One can also observe that the less the z/z_0, the shorter the output laser pulse duration (Figures 4c). As a result, the duration of the laser pulse decreases from 22 to 15 ns. The pulse duration can be approximated by the formula $\tau_{out} = \tau_0\left[1 - a\exp(-(z/z_0)^2/2b^2)\right]$, where $a = 0.35$, $b = 0.84$. It can also be observed that the decrease of the amplitude is accompanied by the decrease of the pulse duration (Figures 4b, c). Figure 4d shows the fall time of the output laser pulse as a function of z/z_0.

From this figure, as well as from the oscillograms presented in Figure 4a, it follows that the shortening of the transmitted pulses duration occurs due to the output laser pulse trailing edge being cut off. This means that the OL in HPHT ND suspensions does not arise due to the two-photon absorption, but due to the cumulative effect, leading to a nonlinear light scattering. As shown in Figure 4b, the dependence of the transmitted laser pulse on z/z_0 is not symmetric with respect to the point $z/z_0 = 0$. This means that nonlinear refraction also occurs along with nonlinear scattering in the suspension.

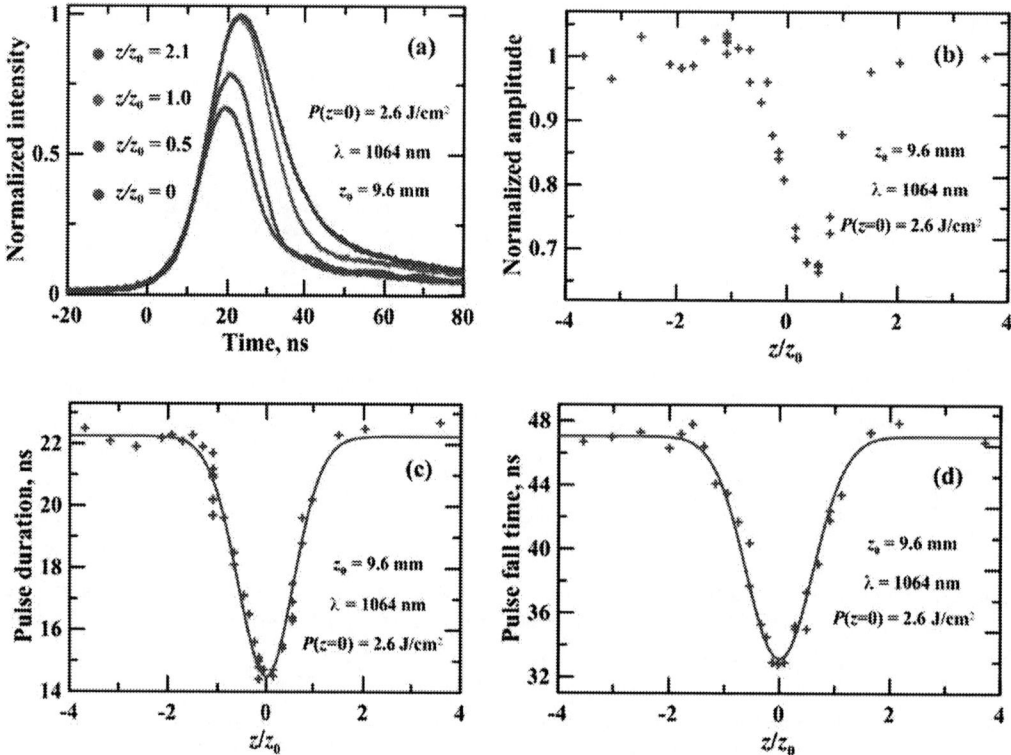

Figure 4. (a) Oscillograms of the output laser pulse at fixed input pulse energy for different z/z_0; the transmitted laser pulse **(b)** amplitude, **(c)** duration and **(d)** fall time as a function of z/z_0.

Thus the obtained results show that the OL in the HPHT ND suspension is due to nonlinear scattering and nonlinear refraction of light.

Acknowledgments

This work was supported by the RFBR (Grant no. 16-42-180147).

References

[1] Tutt L W, Boggess T F 1993 *Prog. Quant. Electr.* **17** 299
[2] Dini D, Calvete M J F, Hanack M 2016 *Chem. Rev.* **116** 13043
[3] Mansour K, Soileau M J, Van Stryland E W 1992 *J. Opt. Soc. Am. B* **9** 1100
[4] Vivien L, Anglaret E, Riehl D, Hache F, Bacou F, Andrieux M, Lafonta F, Journet C, Goze C, Brunet M, Bernier P 2000 *Opt. Commun* **174** 271
[5] Izard N, Billaud P, Riehl D, Anglaret E 2005 *Opt. Lett.* **30** 1509
[6] Wang J, Chen Y, Blau W J 2009 *J. Mater. Chem.* **19** 7425

[7]   Wang J, Früchtl D, Sun Z, Coleman J N, Blau W J 2010 *J. Phys. Chem. C* **114** 6148

[8]   Koudoumas E, Kokkinaki O, Konstantaki M, Couris S, Korovin S, Detkov P, Kuznetsov V, Pimenov S, Pustovoi V 2002 *Chem. Phys. Lett.* **357** 336

[9]   Mikheev G M, Kuznetsov V L, Bulatov D L, Mogileva T N, Moseenkov S I, Ishchenko A V 2009 *Quant. Electron.* **39** 342

[10]   Lim G – K, Chen Z - L, Clark J, Goh R G S, Ng W – H, Tan H - W, Friend R H, Ho P K H, Chua L – L 2011 *Nature Photonics* **5** 554

[11]   Chen Y, Bai T, Dong N, Fan F, Zhang S, Zhuang X, Sun J, Zhang B, Zhang X, Wang J, Blau W J 2016 *Prog. Mater. Sci.* **84** 118

[12]   Mikheev G M, Puzyr' A P, Vanyukov V V, Purtov K V, Mogileva T N, Bondar' V S 2010 *Tech. Phys. Lett.* **36** 358

[13]   Vanyukov V V., Mikheev G M, Mogileva T N, Puzyr A P, Bondar V S, Svirko Y P 2014 *Opt. Mater.* **37** 2990

[14]   Pichot V, Muller O, Seve A, Yvon A, Merlat L, Spitzer D 2017 *Sci. Rep.* **7** 14086

[15]   Mikheev G M, Krivenkov R Y, Mogileva T N, Mikheev K G, Nunn N, Shenderova O A 2017 *J. Phys. Chem. C* **121** 8630

[16]   Mikheev G M, Mogileva T N, Okotrub A V, Bulatov D L, Vanyukov V V 2010 *Quant. Electron.* **40** 45

[17]   Osswald S, Behler K, Gogotsi Y. 2008 *J. Appl. Phys.* **104** 074308

[18]   Mikheev G M, Mikheev K G, Mogileva T N, Puzyr A P, Bondar V S 2014 *Quant. Electron.* **44** 1

[19]   Mikheev K G, Krivenkov R Yu, Mogileva T N, Puzyr A P, Bondar V S, Bulatov D L, Mikheev G M 2017 *J. Nanophotonics* **11** 032502

[20]   Zolotukhin A A, Ismagilov R R, Dolganov M A, Obraztsov A N 2012 *Journal of Nanoelectronics and Optoelectronics* **7** 22

[21]   Prawer S, Nemanich R J 2004 *Phil. Trans. R. Soc. Lond. A* **362** 2537

[22]   Vivien L, Lancon P, Riehl D, Hache F, Anglaret E 2002 *Carbon* **40** 1789

[23]   Vanyukov V V, Mikheev G M, Mogileva T N, Puzyr A P, Bondar V S, Svirko Y P 2015 *Applied Optics* **54** 3290

[24]   Mikheev G M, Krivenkov R Y, Mikheev K G, Okotrub A V, Mogileva T N 2016 *Quant. Electron.* **46** 719

SPBOPEN 2018   IOP Publishing

New ultra high-speed all-optical coherent D-trigger

S N Bagayev[1], V S Egorov[2], V G Nikolaev[3], I B Mekhov[2,4,5], I A Chekhonin[2], M A Chekhonin[2]

[1]Institute of Laser Physics, Novosibirsk, Russia
[2]St.Petersburg State University, St. Petersburg, Russia
[3]ITMO University, St. Petersburg, Russia
[4]University of Oxford, UK
[5]University Paris-Saclay, France

E-mail: chekhonin@mail.ru

Abstract. We study the interaction of two counterpropagating unipolar video pulses of electromagnetic radiation in a dense resonant two-level medium. The pulse durations are less than one oscillation period of an atomic transition. We show that a *polariton cluster* (i.e. the compact long-living strongly coupled state of electromagnetic field and matter polarisation) is created, when two unipolar video pulses collide in a resonant medium of the frequency ω_o (the pulses correspond to self-induced transparency solitons of the same amplitudes and opposite polarities).
We studied for the first time multiple recording and erasing of a polariton cluster in a thin layer of a resonant medium (quantum dots) placed on the mirror surface. We showed that dynamics of the medium population difference $N(x,t)$ is analogous to the operation of a D-trigger of the pulse rate 60 000 GHz and higher. We found such a method of the polariton cluster recording and erasing that excludes the accumulation of erasing errors. Therefore, the total duration of the optical D-trigger operation time can strongly exceed the phase relaxation time T_2.

1. Introduction

Recently, the generation, coherent propagation, and interaction in dense resonant media between ultrashort (few-cycle, single-cycle, sub-cycle) pulses of the electromagnetic field attract strong interest. Modern methods of generating unipolar video pulses are reviewed in [1,2]. Effects of coherent interaction between counterpropagating pulses are considered in a significantly smaller number of works, e.g. [3-5].

Here we theoretically study the processes of formation and dynamics of *polariton clusters*, which are created during collisions between unipolar pulses of the amplitudes of opposite signs (solitons of the self-induced transparency).

The correct description of propagation and interaction of ultrashort (single-cycle, sub-cycle) solitons is possible only by eliminating the approximations imposed by the slowly varying envelope (SVEA) and rotating-wave approximations (RWA) both in time and spatial coordinate. The theory of propagation of such solitons in dense resonant media is well established, cf. for example [6-12].

Dynamics of a quantum two-level particle with the dipole moment d and transition frequency ω_o is described by the Bloch equations [6,7] for the pseudospin vector projections $s = (s_1, s_2, s_3)$:

Content from this work may be used under the terms of the Creative Commons Attribution 3.0 licence. Any further distribution of this work must maintain attribution to the author(s) and the title of the work, journal citation and DOI.
Published under licence by IOP Publishing Ltd

$$\dot{s}_1 = -\omega_0 s_2 - \frac{1}{T_2} s_1 \tag{1}$$

$$\dot{s}_2 = \omega_0 s_1 + 2\frac{d}{\hbar} E(t,z) s_3 - \frac{1}{T_2} s_2 \tag{2}$$

$$\dot{s}_3 = -2\frac{d}{\hbar} E(t,z) s_2 - \frac{1}{T_1}(s_3 + 1) \tag{3}$$

$$P(t,z) = N_0 \cdot d \cdot s_1 . \tag{4}$$

Here $E(t,z)$ is the electric filed amplitude of a pulse propagating along the coordinate z, T_1 и T_2 are the longitudinal and transversal relaxation times, $P(t,z)$ is the medium polarization, N_0 is the resonant particle density. The physical meaning of the pseudospin projection s_3 is the normalized population difference of a two-level medium: $s_3(z,t) = N(z,t)/N_0$.

Dynamics of the electromagnetic field propagation is described by Maxwell's equations:

$$\frac{1}{c}\frac{\partial H_y}{\partial t} = -\frac{\partial E_x}{\partial z} \tag{5}$$

$$\frac{1}{c}\frac{\partial E_x}{\partial t} = -\frac{\partial H_y}{\partial z} - \frac{4\pi}{c}\dot{P} \tag{6}$$

or by the wave equation

$$\frac{\partial^2 E}{\partial t^2} - c^2 \frac{\partial^2 E}{\partial z^2} = -4\pi\ddot{P} . \tag{7}$$

Here E_x and H_y are the projections of electric and magnetic fields of the pulse on the transverse axes x and y, c is the speed of light in vacuum.

In the absence of relaxation ($T_1 = T_2 = \infty$), the Maxwell-Bloch equations have analytical single-soliton solutions of the self-induced transparency theory [7-12]:

$$E = \pm \frac{\hbar}{dt_p} sech\left(\frac{t - zV^{-1} - t_0}{t_p}\right) \tag{8}$$

$$V = \frac{c}{\sqrt{1 + \dfrac{2\alpha\tau_p^2}{1 + \tau_p^2}}} \tag{9}$$

$$\alpha = \frac{4\pi}{\hbar} \cdot \frac{N_0 d^2}{\omega_0} = \frac{1}{t_c \omega_0} . \tag{10}$$

Here V is the soliton propagation velocity along the coordinate z, t_p is the soliton duration, $\tau_p = \omega_0 t_p$.

2. Formation of a polariton cluster during the collision of two solitons

In this work we have studied the process of inelastic collision between two self-induced transparency solitons (8) in a resonant medium. We theoretically considered a problem of two identical counterpropagating unipolar solitons of either the same or different polarities. We solved numerically the Maxwell-Bloch equations (1)-(3) and (5)-(6) without using the slow amplitude approximations (SVEA, RWA) both in time and spatial coordinate. We took into account the finite relaxation times of a medium T_1 и T_2. The details of the simulation methods are described in [4].

In Figure 1, we show dynamics of the normalized population difference $s_3 = N(z,t)/N_o$ during the collision of two unipolar solitons. One can see that the long-living polariton cluster is formed during the inelastic collision between the solitons of the opposite polarity (right panel) near the collision point z_0 ($Kz_0 = 15$). The cluster has a double spatial structure. By varying the soliton duration t_p, we have shown that the quantity $\langle N(z)\rangle$ averaged over the cluster existence area $Kz = [14...16]$ has a well-defined maximum at the soliton duration $\omega_o t_p = 0.75$.

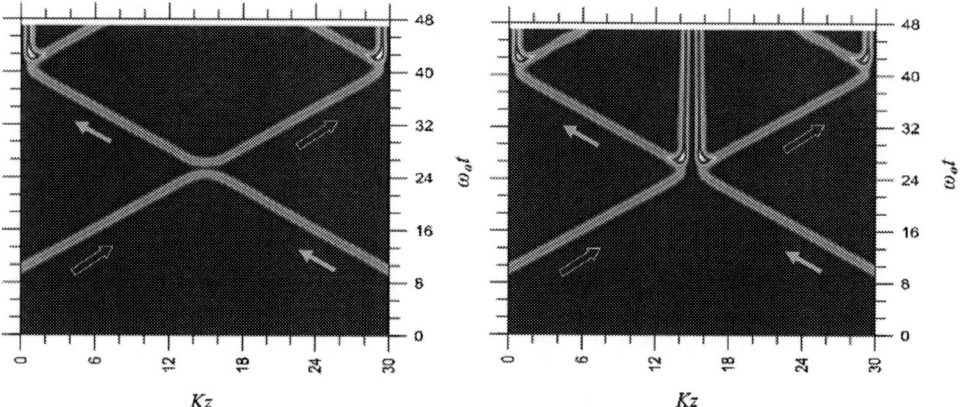

Figure 1. Dynamics of the normalized population difference $s_3 = N(z,t)/N_o$ during the collision of two unipolar solitons. Arrows show the directions of pulse propagation. Left panel: collision between the pulses of the same polarity; right panel: collision of the pulses of the opposite polarity.
$K = 2\pi/\lambda$ is the wave vector. $\omega_a t_p = 1$, $T_1 = 10$ psec, $T_2 = 10$ psec,
$d = 5D$, $N_o = 1.77 \cdot 10^{20}$ cm^{-3}, $\omega_o = 1 \cdot 10^{15}$ rad/sec, $2dE_o/\hbar\omega_o = 2$.

Such features have a simple physical explanation. Analysing the temporal behaviour of the sum fields of two solitons $E(t)$ in various spatial points, one sees that there exist two special points with the coordinates z' and z'' (cf. Figures 2). In these points, the field $E(t)$ has a form of bipolar single-cycle pulse, which can resonantly excite the medium at the atomic transition frequency ω_o. The condition of resonant excitation corresponds to the optimal soliton duration $\omega_o t_p = 0.75$.

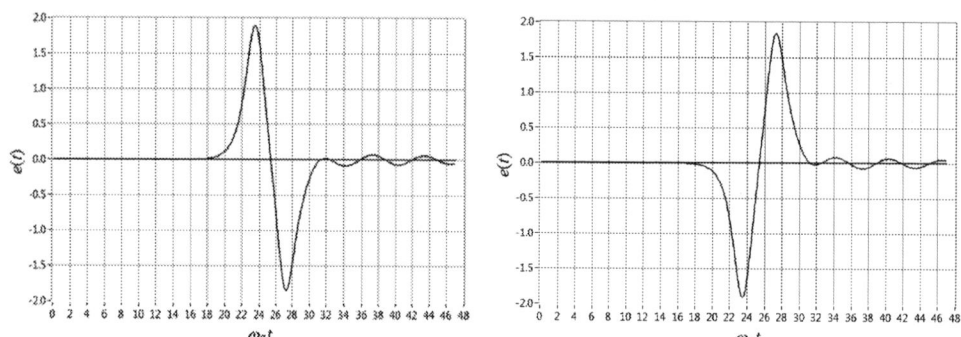

Figure 2. Temporal behaviour of the field $e(t)$ in two points of the polariton cluster z' и z'' near the point of the pulse collision z_o ($Kz_o = 15$). Left panel: $Kz' = Kz_o - 1.8$, right panel: $Kz'' = Kz_o + 1.8$.
$e(t)$ is the dimensionless field amplitude: $e(t) = 2dE(t)/\hbar\omega_o$.

Figure 3 shows the spatial distribution of the medium polarization s_1 and population difference s_3 immediately after the collision between the solitons of opposite polarities. One sees that s_3 has a double structure (cf. also Figure 1), while s_1 oscillates with the amplitude close to the maximal one ($s_1 \approx 1$). Nevertheless, the radiation of cluster in the far field is negligibly small. The reason is that the spatial Fourier spectrum of the medium polarization $P(t,z) = N_o \cdot d \cdot s_1$ is concentrated at the wave vectors $Q \ggg 2\pi/\lambda$, which do not satisfy the phase matching conditions. Therefore, the cluster field $E(t)$ exists in the near field only. An analogous mechanism of the forbidden radiation for long-living polariton clusters, which are formed during the collision between the few-cycle self-induced transparency solitons, has been discussed in [4].

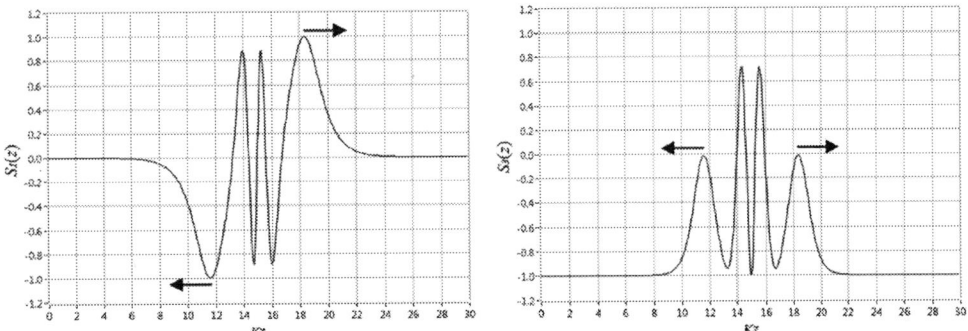

Figure 3. Spatial distribution of the normalized medium polarization s_1 (left panel) and normalized population difference s_3 (right panel) in a polariton cluster at the time moment $\omega_o t = 28.85$ immediately after the pulse collision. The collision happens at $\omega_o t = 25.38$, $Kz_0 = 15$. The arrows show the propagation directions.

3. Recording and erasing polariton clusters. D-trigger.

In this part of the work, we studied the propagation of a periodic sequence of counterpropagating unipolar pulses through a dense resonant medium. The repeated action of two pulses of opposite polarities on the polariton cluster leads to its complete destruction (cf. Figure 4). The next pair of pulses reproduces the polariton cluster again. Such a process can be repeated many times, and dynamics of the population difference $N(t)$ is analogous to the action of an electronic D-trigger.

Such recording and erasing processes are coherent, because they happen at time intervals $\Delta t \lll T_2$. Therefore, the pulse repetition period T should satisfy the condition $T = (2n + 1)T_0/2$, where $T_0 = 2\pi/\omega_o$ is the oscillation period of the two-level atomic system, $n = 0, 1, 2 \ldots$ is an integer.

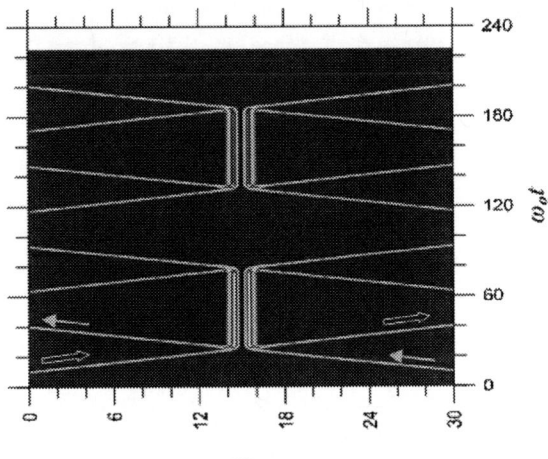

Figure 4. Dynamics of the normalized population difference $s_3 = N(z,t)/N_o$ during the recording and erasing polariton cluster by unipolar pulses of the opposite polarities. The repetition period T of recording and erasing pulses is 8.5 of the atomic oscillation period ($\omega_o T = 53.5$). $\omega_o t_p = 0.8$, $T_1 = 10$ psec, $T_2 = 10$ psec, $d = 5D$, $N_o = 1.77 \cdot 10^{19}$ cm^{-3}, $\omega_o = 1 \cdot 10^{15}$ rad/sec, $2dE_o/\hbar\omega_o = 2.5$.

From a practical point of view, a convenient method to obtain two counterpropagating unipolar pulses of opposite polarities is based on the pulse reflection from a mirror (cf. Figure 5).

Below we consider a case, where a thin layer ($\approx \lambda/2$, 300 nm) of a resonant medium is placed on the mirror surface, while the counterpropagating pulse of the duration t_p equal to 1/8 of the atomic transition oscillation period ($t_p = 0.24$ fsec) is created as a result of the reflection from the mirror (cf. Figure 5). The light-induced stationary polariton cluster is situated at the distance of $\approx \lambda/10$ (60 nm) from the mirror. Within the cluster, the medium polarisation oscillates with the maximal amplitude, nevertheless its radiation in the far field is forbidden.

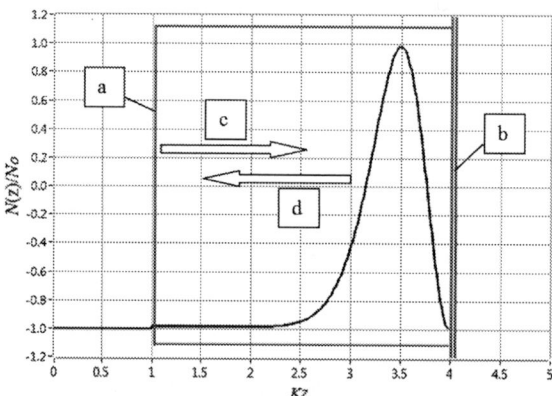

Figure 5. Spatial distribution of the normalized population difference $N(z)/N_o$ in polariton cluster. a – region occupied by resonant medium, b – mirror, c – incident pulse, d - reflected pulse. $K = 2\pi/\lambda$ is the wave vector, $\lambda = 600$ nm.

In Figure 6, we show the result of multiple recording and erasing the polariton cluster by a periodic sequence of unipolar pulses incident on the mirror. It turns out that, if one uses the pulse sequence with polarity inversion (+1, +1, -1, -1, +1, +1, -1, -1, ...), then the cluster erasing error approaches zero. Moreover, the time Δt of the stable operation of the D-trigger increases multiple times: $\Delta t >>> T_2$.

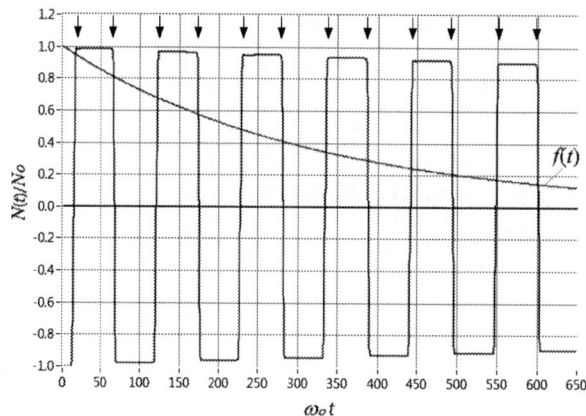

Figure 6. Dependence $s_3 = N(t)/N_o$ for the cluster point $Kz = 3.5$. The arrows show the moments of pulse actions. Pulse rate is 60 000 GHz (60 THz). The function $f(t) = \exp[-(t/T_2)]$. Pulse duration is $\tau_p = 0.24$ fsec ($\omega_o\tau_p = 0.75$), $T_1 = 10$ psec, $T_2 = 100$ fsec, $d = 25D$, $N_o = 2.22 \cdot 10^{18}$ cm^{-3}, $\omega_o = 3.142 \cdot 10^{15}$ rad/sec, $2dE_o/\hbar\omega_o = 2.67$.

To confirm this, we performed calculations for very small relaxation times: $T_1 = 10$ psec, $T_2 = 100$ fsec.

Figure 6 shows the influence of the finite relaxation time T_2 on the decrease of the value of $|N(t)/N_o|$ in the regime ON/OFF of the D-trigger. The degradation of the cluster coherence happens at the moments of the cluster switch, i.e. during the time interval $\approx \tau_p$ and it is absent in the interval T between the pulses. At the chosen pulse repetition frequency, the decrease rate of $|N(t)/N_o|$ is 18 times smaller than the decrease rate of the function $f(t) = \exp[-(t/T_2)]$.

Varying the amplitude reflection coefficient R_a has shown that the D-trigger operates stably at the values $R_a = 0.8 \ldots 1.0$.

4. Conclusion

In conclusion, we have systematically studied an effect of creation of *polariton cluster* (i.e. a compact long-living strongly coupled state of the electromagnetic field and medium polarization). We performed numerical solution of the problem of inelastic collision between two counterpropagating solitons of the self-induced transparency effect.

We physically explained the fact, that for creating a polariton cluster, the best choice is the use of a pair of solitons with the field amplitudes of opposite signs. We have shown that the reason of the long life-time of the cluster is due to the forbiddance of its radiation in the far field.

We studied the processes of multiple recording and erasing of the polariton cluster, and showed that dynamics of the population difference of the medium is analogous to the operation of an electronic D-trigger of high pulse rate 60 000 GHz and higher.

We studied the processes of recording and erasing of a polariton cluster in a thin layer of a resonant medium (quantum dots) near a mirror, and showed that the cluster size is of the order of $\approx \lambda/10$. This makes it a very perspective basic element for all-optical signal processing.

We found a method of cluster recording and erasure, which excludes the erasure error accumulation. Therefore the total time of the optical D-trigger operation can multiply exceed the phase relaxation time T_2.

Acknowledgements

This work was supported by Russian Science Foundation (project № 17-19-01097).
I.B.M. acknowledges additional support by the Russian Foundation for Basic Research (project № 18-02-01095).

References

[1] Frantzeskakis D J, Leblond H, Mihalache D 2014 Rom. J. Phys. **59** (7-8) 767-784
[2] Arkhipov R M, Pakhomov A V, Arkhipov M V, Babushkin I, Tolmachev Yu A, Rosanov N N 2017 JETP Lett. **105** (6) 408-418
[3] Rosanov N N, Semenov V E, Vyssotina N V 2007 Laser Physics **17** (11) 1311–1316
[4] Bagaev S N, Egorov V S, Nikolaev V G, Chekhonin I A, Chekhonin M A 2015 Russian Journal of Physical Chemistry B **9** (4) 582-586
[5] Novitsky D V 2012 Phys. Rev. A **85** (4) 043813-1 - 043813-7
[6] Allen L, Eberly J H, *Optical resonance and two-level atoms* (Wiley, New York, 1975)
[7] Bullough R K, Caudrey P J, Eilbek J C, Gibbon J D 1974 Opto-electronics **6** 121-140
[8] Eilbeck J C, Gibbon J D, Caudrey P J, Bullough R K 1973 J. Phys. A **6** 1337-1347
[9] Bullough R K, Jack P M, Kitchenside P W, Saunders R 1979 Phys. Scripta **20** 364-381
[10] Maimistov A I, Caputo J G Optics and Spectroscopy 2003 **94** (2) 245–250
[11] Kalosha V P, Herrmann J Phys. Rev. Lett. 1999 **83** 544-547
[12] Bullough R K, Ahmad F 1971 Phys. Rev. Lett. **27** (6) 330-333

SPBOPEN 2018   IOP Publishing

Comparison of optical spectral devices in the framework of system approach

V Kazakov, O Moskaletz, M Vaganov

St. Petersburg State University of Aerospace Instrumentation, St. Petersburg 190000, Russia

Abstract. In the framework of solving the problem of comparison and evaluation problem of the spectral devices performance and quality two methods of optical spectra measurement are considered: by using diffraction grating spectral device and multichannel optical spectrometer. Comparison has been performed based on the matrix representation of the measuring results of the optical radiation energy spectrum by each device. Complex and power spectrum spread functions are obtained. Parameters of spectral devices which are defined its spectral resolution are established.

1. Introduction

The problem of comparing and evaluating the performance and quality of spectral devices, formulated in [1], has a multicriterial character. Development of existing traditional methods and the emerging new methods and technical devices of spectroscopy, for example [2], stimulates its further development.

Two types of spectral devices are compared, which has different principles of working: a diffraction spectral device based on a transmission grating and a multichannel optical spectrometer [2], which carries out spectral decomposition based on the resonance phenomenon, i.e. the spectral decomposition is implemented by the principle of the narrow-band optical filtration in n parallel channels.

Comparison of devices of different classes is possible if there is a common description of the spectra and the unity approach to the estimation of its main characteristics. This problem is solved on the basis of the developed system approach [3] to the description of the working of the above mentioned spectral devices, which makes it possible to establish the input-output connection of the spectral device, which is the most important problem in the theory of spectral measurements [4].

The system approach is based on the methods of the signals theory and the linear systems theory where the connection between an input and output of a device is given as a linear integral operator, the kernel of which is spread function, which is exhaustive characteristic of linear system. In the case of an optical spectral device the spread function is as a device's reaction to monochromatic radiation. It allows determining its most important characteristic – its spectral resolution, as well as the errors in spectral measurements. The result of this approach is the matrix representation of the spread functions of spectral devices, which is the basis for their comparison and evaluation of their effectiveness and quality.

2. Diffraction grating spectral device

The optical scheme of the diffraction grating spectral device is shown in Figure 1.

Content from this work may be used under the terms of the Creative Commons Attribution 3.0 licence. Any further distribution of this work must maintain attribution to the author(s) and the title of the work, journal citation and DOI.
Published under licence by IOP Publishing Ltd

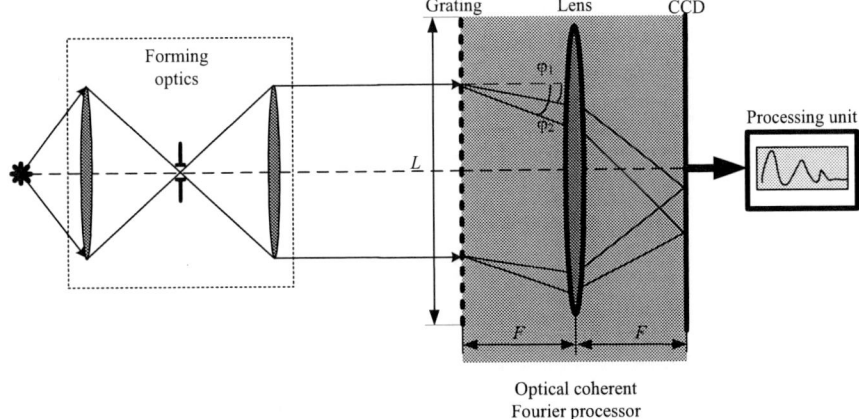

Figure 1. The optical scheme of the diffraction grating spectral device.

In contrast to the well-known methodology for describing the action of spectral devices, for example [5], in this work the obtaining of the energy spectrum spread function of a diffraction grating spectral device is based on the principles of wave optics, the fundamental positions of the theory of linear systems [6] and signal theory [7,8] and methods of radiooptics [9] and bases on a consistent consideration of the passage of optical radiation through the whole spectral device from its input aperture to the result of photodetection.

The diffraction grating performs spatial modulation of the incident wave in accordance with its transparency function $T(\xi)$, represented in the form of an expansion in the Fourier series at its aperture L [10]:

$$T(\xi) = \sum_{n=-\infty}^{\infty} \frac{\sin(n\Omega_g \frac{\tau_g}{2})}{n\Omega_g} \cdot \frac{\sin[(N+\frac{1}{2})n\Omega_g T_g]}{L \cdot \sin\left(n\Omega_g \frac{T_g}{2}\right)} \exp(in\Omega_g \xi), \qquad (1)$$

where L – grating aperture, $\Omega_g = 2\pi / T_g$, T_g – diffraction grating period, N – the number of transparency element in the diffraction grating.

Further transformation of the field by the optical system of the device can be described by using optical coherent Fourier processor and thus obtain an input-output connection of the device for the complex apparatus spectrum in +1 diffraction order:

$$S_a(\omega_x, t) = M \int_{\Delta\Omega} S(\omega') \exp(i\omega't) \frac{\sin[(\omega(x)-\omega')\frac{T_a}{2}]}{(\omega(x)-\omega')\frac{T_a}{2}} d\omega', \qquad (2)$$

where $\Delta\Omega$ – analyzed frequency band, $S(\omega')$ – complex spectrum of analyzed radiation, M – proportional coefficient, $T_a = Lx / 2c_0 F$, c_0 – velocity of light.

The relationship between the spatial coordinate x and the spectral frequency ω is given by the expression:

$$\omega(x) = 2\pi c_0 F / T_g x, \qquad (3)$$

where F – focal length of the lens.

The processing of the complex apparatus spectrum is carried out using a linear CCD, where each element performs narrow-band filtering, determined by the length of the element along the axis of spatial frequencies $\Delta\omega_n$, equal to:

$$\Delta\omega_n = \frac{(2\pi c_0 F)2\Delta x_n}{T_g x_n^2},$$ (4)

or

$$\Delta x_n = \frac{T_g x_n^2(\Delta\omega_n(x_n))}{4\pi c_0 F}.$$ (5)

According to the general theory of photodetection, the equivalent photodetection scheme by one element of the linear CCD can be described by the structural scheme, which is shown in Figure 2.

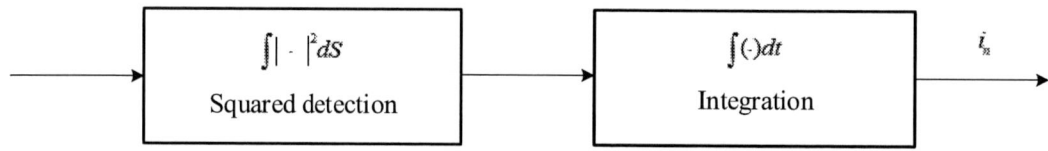

Figure 2. The equivalent scheme of the photodetector

The photocurrent i_f is given by:

$$i_f = \gamma q_e \frac{p}{\hbar\omega'},$$ (6)

here γ denotes the quantum efficiency; q_a is the charge of the electron; $\hbar$ is Planck's constant; p corresponds to the power of an optical radiation; ω' is the angular frequency of an optical radiation fallen on the photo-detector.

The power of an optical radiation p fallen on the photo-detector is defined as

$$p = \iint_{S_f} \mathbf{P}\, \mathbf{ds},$$ (7)

where S_f is area of the sensitive surface of the photodetector; $\mathbf{ds}=\mathbf{n}ds$, $\mathbf{n}$ is a single normal to the sensitive surface of the photodetector; $\mathbf{P}$ is Poynting vector which is given by,

$$\mathbf{P} = \mathbf{E}\times\mathbf{H} = |\mathbf{E}\times\mathbf{H}|\mathbf{s} = \sqrt{\varepsilon/\mu}\cdot|\mathbf{E}|^2\,\mathbf{s} = \sqrt{\mu/\varepsilon}\,|\mathbf{H}|^2\,\mathbf{s},$$ (8)

here $\mathbf{E}$ denotes the electric field vector; $\mathbf{H}$ is magnetic field vector, $\mathbf{s}$ means a unit vector; ε is the permittivity of the material; μ is the permeability of the material.

Substitution instead of the values of the vector E of the complex apparatus spectrum, as well as the spatial and temporal integration operations, according to the algorithm shown in Fig. 2, allows us to represent the result of the photodetection in the form:

$$\bar{i}_n = B \int_{\omega_n-\Delta\omega_n}^{\omega_n+\Delta\omega_n} \frac{\sin^2[(\omega(x)-\omega')\frac{T_a}{2}]}{\left[(\omega(x)-\omega')\frac{T_a}{2}\right]^2} G(\omega')d\omega' = B \int_{\omega_n-\Delta\omega_n}^{\omega_n+\Delta\omega_n} \operatorname{sinc}^2(\cdot)\cdot G(\omega')d\omega'.$$ (9)

The obtained expression for the photocurrent in each element of the linear CCD allows us to write a mathematical expression that describes the input-output connection of the spectral device in the matrix form for the energy spectrum when the spectrum is detected on all elements of the linear CCD:

$$\|G(\omega_n)\| = B \int_{\omega_n - \Delta\omega_n}^{\omega_n + \Delta\omega_n} diag\{A_n(\omega,\omega')\} \cdot \|G(\omega')\| d\omega' , \qquad (10)$$

where $G(\omega')$ – the energy spectrum of analyzed signal; $diag\{A_n(\omega,\omega')\}$ – energy spectrum spread function of a spectral device of the optical range in a matrix form; $A_n(\omega,\omega') = \text{sinc}^2(\cdot)$.

Thus, the results of the reading of the spectrometric information are given in the form of the reference values of the energy spectrum averaged over the size of the sensitive element of the linear CCD and the integration time T_R.

3. Multichannel optical spectrometer
The optical scheme of the multichannel optical spectrometer is shown in Figure 3.

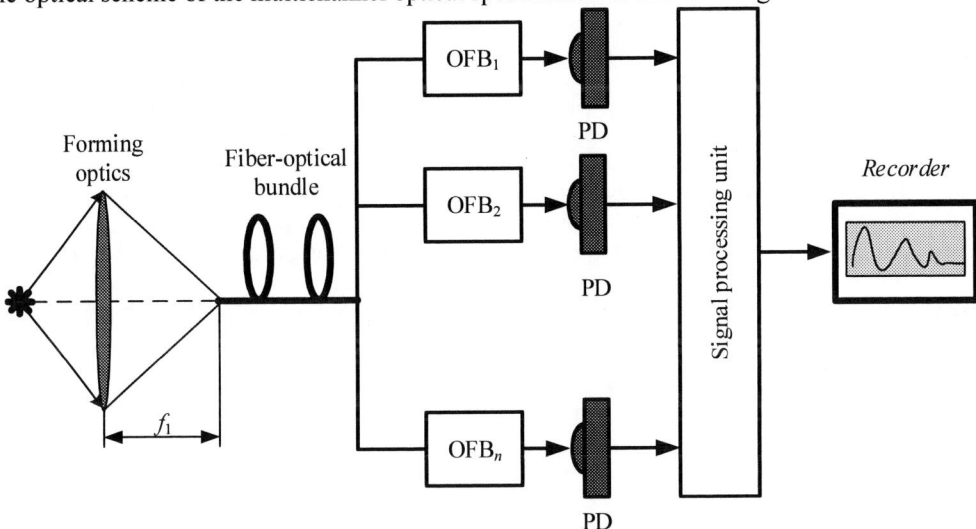

OFB – optical filtration block; PD - photodiode

Figure 3. The optical scheme of the multichannel optical spectrometer

The operation of the multichannel optical spectrometer with transmitting analyzed optical signals by the optical fiber is described as follows: A forming optics transmits optical radiation from surroundings to the common input of the fiber-optical bundle which is situated in its focal length. The fiber-optical bundle is used for transmitting radiation on at the given distance from source of an optical radiation. The optical radiation passed through the fiber-optical bundle is transmitted to each channel of spectrometer for the further spectral decomposition. Each channel contains the narrow-band optical filter, which has been set on the certain wavelength. Receiving of spectrometric information in a multichannel optical spectrometer is carried out by a photodetector set in each channel. After spectral decomposition and photodetecting analyzed signal is processed by the signal processing unit. The received spectroscopic information about analyzed optical radiation is displayed on the recorder. The results of receiving the spectrometric information in a multichannel optical spectrometer are presented in matrix form. These are the sampling values of the energy spectrum averaged in the frequency band of each narrowband interference filter.

Taking into account the specifics of the considered device the process of receiving of the energy spectrum estimation of the optical signal in one channel can be represented by a functional scheme shown in figure 4 [11].

The complex spectrum of analyzed signal is proportional to the intensity of the electrical components of the optical radiation, so the mathematical form of the sequence of operations presented in the figure 4 has the form:

$$G_k(\omega) = \int_{-\frac{T_R}{2}}^{\frac{T_R}{2}} i_k(t)dt = P_k(\omega_k)\int_{-\frac{T_R}{2}}^{\frac{T_R}{2}} |S_{ak}(\omega,t)|^2 dt, \tag{11}$$

where $i_k(t)$ is current of the photodetector for the channel k; $P_k(\omega_k)$ is coefficient of the spectral sensitivity of the photodetector; T_R is integration time; $t_0 = -\frac{T_R}{2}$.

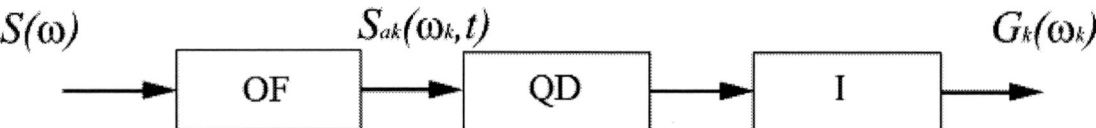

OF – an optical filter, QD – quadratic detector, I – integrator, $S(\omega)$ – the input signal for channel k of spectrometer, $S_{ak}(\omega_k,t)$ – the complex spectrum from the output of optical filter k, $G_k(\omega_k)$ – the energy spectrum of an output optical signal in channel k of spectrometer

Figure 4. Functional scheme of receiving of the energy spectrum estimation

In paper [11] it is shown that the expression describing receiving of an energy spectrum of an optical signal in the k channel of the spectrometer is defined as:

$$G_k(\omega) = \int_{-\Delta\omega_k}^{\Delta\omega_k} W_{kk}(\omega_k,\omega') \cdot G(\omega')d\omega' . \tag{12}$$

Here $G(\omega')$ represents the energy spectrum of analyzed signal, $G(\omega')=|S(\omega')|^2$; $2\Delta\omega_k$ designates the bandwidth of optical filter, $W_{kk}(\omega_k,\omega')$ denotes a "partial" the energy spread function of channel k of the multichannel optical spectrometer.

Then the receiving of an energy spectrum of an optical signal by a multichannel spectrometer can be represented in matrix form:

$$\|G_{ak}(\omega)\| = \int_{-\infty}^{\infty} diag\{W_{kk}(\omega_k,\omega')\} \cdot \|G(\omega')\|d\omega' , \tag{13}$$

where $diag\{W_{kk}(\omega_k,\omega')\}$ is energy spectrum spread function of a multichannel optical spectrometer in a matrix form.

Expression (13) describes the energy spectrum in the form of its reference values and establishes a connection between the real energy distribution by the frequencies (energy spectrum) and the spectral energy distribution by the frequencies obtained experimentally using a multichannel spectrometer.

4. Conclusion

The comparison of the spectral instruments of a different class: the diffraction grating spectral device and a multichannel optical spectrometer based on their presentation in the form of multichannel linear systems has been considered in this paper.

Despite the different principle of operation, the result of spectrum measurement by each device was obtained in a unified matrix form. Thus, the requirement to generality of the spectra description and the unity of approach to the estimation of the main characteristics of the spectral devices was accomplished. Received the energy spread functions of spectral devices in matrix form allow to compare spectral devices, in particular, on criterion of their resolution capability.

The frequency filtration in the diffraction grating spectral device is performed by each element of the CCD line. Thus, the resolution is determined by the parameters of the device: the period of the grating, the focal length of the lens and the geometric dimensions of the sensitive element of the CCD line.

The resolution of a multichannel optical spectrometer is determined only by the bandwidth of the optical filter set in each channel.

It should also be noted that the matrix representation of the measurement results in the form sampling values of the energy spectrum by each device is the basis for the application of the sampling theorem in the frequency domain and allows establishing a continuous function describing the energy spectrum.

Acknowledgments

This work was supported by RFBR (Russian Foundation for Basic Research): projects № 17-07-00826, 17-07-00554, 18-07-00636, 16-07-00549 and grant of President of Russian Federation MK-5894.2018.8

References

[1] Kiselev B.A., Malykhin A.V. 1982 Proc. scientific. articles: Modern trends in the technique of spectroscopy (Novosibirsk) pp. 125-152.

[2] Moskaletz O.D., Vaganov M.A., Kulakov S.V., Preslenev L.N., Arkhipov I.N., Kotlikov E.N. and Prokashev V.N. 2009 Patent of Russian Federation № 86734.

[3] M. A. Vaganov, V. I. Kazakov, O. D. Moskaletz. 2017 *J. Aut.n and Rem. Control.* vol. 78, Iss. 6, p.1144

[4] Gorelik G. S. 1959 *Oscillations and waves. Introduction to acoustics, radiophysics and optics.* (Moscow: Publishing house of Physico-mathematical literature) p 656.

[5] Tarasov K I 1977 *Spectral devices* (The Publishing House of the Mechanical engineering. Leningrad) p. 388

[6] Zadeh L and Desoer C 1994 *Linear System Theory*, (Springer; Corrected edition) p 509

[7] Franks L E 1969 *Signal theory* (Englewood Cliffs, N.J., Prentice-Hall) p 317

[8] Gonorovskiy I S 1963 *Radio Circuits and Signals* (M.: Sovetskoe Radio) p 608

[9] Stroke G W 1966 *An introduction to coherent optics and holography* (Academic press, New-Jork – London) p 347

[10] Kazakov V.I., Moskaletz D.O., Moskaletz O.D. 2016 Proc. SPIE (SPIE, Bellingham) vol. 9889, p 988924

[11] Vaganov M.A., Moskaletz O.D. 2011 Inf.-upr. Sis. vol. 5, p 231.

SPBOPEN 2018   IOP Publishing

Four-step fabrication of SERS-active microfluidic channels

E S Babich[1], E S Gangrskaia[1], I V Reduto[1,2,3] and A A Lipovskii[1,3]

[1] Peter the Great St. Petersburg Polytechnic University, St. Petersburg 195251, Russia
[2] University of Eastern Finland, Joensuu 80101, Finland
[3] St. Petersburg Academic University, St. Petersburg 194021, Russia

Abstract. We present a technique to fabricate SERS substrates fully integrated with microfluidic channels. The fabrication of microsized dielectric channels was based on selective wet etching of a poled/unpoled silicate glass while silver nanoisland film was grown on the bottom of the channels via out-diffusion technique. Different combinations of fabrication steps were studied to optimize the channels' depth, formation of the metal nanoislands and the roughness of the channels' bottom.

1. Introduction

Surface enhanced Raman scattering (SERS) is a powerful analytical tool for label-free biological and chemical sensing [1, 2]. Because of a high sensitivity, this phenomenon is widely applied, in particular, in biomedicine, pharmacy analysis, food safety inspection and environment monitoring. SERS-active substrates can provide the enhancement of Raman signal as high as $10^8 - 10^{11}$ [3]. The prospective ones are large-scale ensembles of self-assembled metal nanoparticles (NPs) [4, 5]. Optically excited surface plasmon resonance in metal NPs increases local electric field of incident light-wave in their vicinity. In its turn, this E-field enhances Raman scattering from a sample medium adsorbed on the NPs ensembles' surface.

Microfluidic chip is another perspective analytical tool widely used for chemical/biological sensing [6]. Devices based on microfluidics are able to sustain continues sampling of microliters of an analyte and solution mixing. The integration of SERS-active substrates with the microfluidic platform opens a way to in situ analysis of biological systems and chemical compounds [7]. The combination of the high speed of the analysis and minimal sample consumption is perfectly fulfilling the needs of aforementioned applications. Several implementations of the on-chip detection system were suggested by now. Among them are synthesized SERS-active colloids which are injected simultaneously with an analyte and mixed in glassy microfluidic channels [8] or lithographically made plasmonic structures (nanodisks [9], nanopillars [10] etc.) exploited in polymeric microfluidics. However, the controllable in situ fabrication of SERS-active substrates within microchannels is in demand [11, 12]. Here we present a new four-step technique for the in situ fabrication of glass microfluidic channels with SERS-active silver nanoisland film formed on the bottom of the channels.

2. Experimental

The channels were fabricated by chemical etching of a "Menzel" commercial soda-lime glass thermally poled with a profiled anodic electrode. The glassy-carbon electrode was patterned as a set of

Content from this work may be used under the terms of the Creative Commons Attribution 3.0 licence. Any further distribution of this work must maintain attribution to the author(s) and the title of the work, journal citation and DOI.
Published under licence by IOP Publishing Ltd

650 nm deep periodic rectangle hollows separated by 200 μm gaps. The studied channels were 10, 30, 60 and 100 μm in width with the fixed length, 2000 μm. In the poling, while DC voltage of 300 V was applied at 300°C for 30 min to the 1 mm thick glass slide with pressed anodic and cathodic electrodes, positive ions of the glass were drifting from the glass anodic surface towards the depth. Finally, a depleted with the ions (poled) glass layer was formed, which thickness depended on the electric field strength. Since electric field in the glass region under the electrode hollows is weaker than in the regions of the electrode-glass contact, the poled layer in the regions under the hollows was thinner [13]. Etching of the samples in a polishing etchant, $NH_4F:8H_2O$, allowed forming of grooves (channels) in these regions due to the slower etching rate of poled comparatively to unpoled glass [14] and to the thinner poled layer under the hollows. The duration of the etching varied from 10 to 40 minutes. We used lasted for 20 minutes ion-exchange processing of the samples in $AgNO_3$(5 wt.%)/$NaNO_3$(95 wt.%) melt heated up to 325°C for doping the subsurface layer of the glass with silver ions. A silver nanoisland film was self-assembled on the samples' surface via the reduction of silver ions penetrated into the samples. The reduction was performed via annealing the samples in hydrogen atmosphere at 250°C for 10 min. These steps are illustrated by figure 1a.

Three series of samples were fabricated with the change in the order of the aforementioned steps, as shown in figure 1b. We performed thermal poling, etching, ion-exchange of the sample 1 (**S1**), thermal poling, ion-exchange, etching of the sample 2 (**S2**), and ion-exchange, poling, etching of the sample 3 (**S3**). The formation of the nanoisland film via annealing was the last step, the same for all the samples. All sequences of these steps resulted in glass substrates with microsized channels with silver nanoisland film on their bottom. The dependence of the channels' depth on the etching time (10, 20, 30 and 40 min) was studied using tip profilometer measurements (Dektak 150, Veeco). The bottom relief and the nanoparticles' morphology in the channels etched for 10 min were characterized using atomic force microscope (AFM, Dimension 3100, Veeco) with RTESP tip and with scanning electron microscope (SEM, Leo 1550 Gemini, Oberkochen).

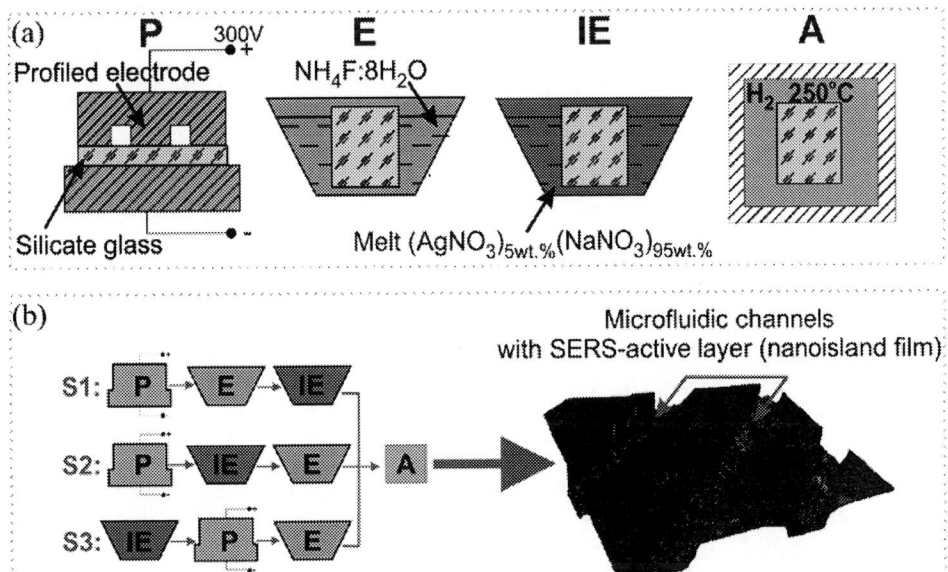

Figure 1. Schematics of the microfluidics channels fabrication: (**a**) thermal poling (**P**) with the structured electrode, wet etching (**E**) in the polishing etchant, silver-sodium ion-exchange (**IE**) and silver nanoisland film formation in hydrogen annealing (**A**). (**b**) Three routines of the samples fabrication and overall look of a substrate with microfluidic channels integrated with SERS-active layer.

3. Results and discussion

We studied the etching processing of the samples **S1** and **S3** series, assuming that the etching of the **S2** was similar to the **S1**. Both **S1** and **S2** were firstly poled, which was the crucial glass modification resulted in selective etching as it is stated above. We etched the **S1** and **S3** samples for 10, 20, 30 and 40 min and measured the depth of the channels of 10, 30, 60 and 100 μm width prior the fabrication steps followed the etching. It should be noted that both poled (between the electrode hollow) and weaker poled/unpoled (under the electrode hollow) regions of the glass are etched in the processing. Thus, the measured depth of a channel is the difference of the thicknesses of the layers etched off between and under the hollows. When the thin poled layer under the hollows is completely etched off, the thicker layer between the hollows starts working as a mask because of a slower etching rate comparatively to the unpoled glass region beneath the hollow.

The results of the etching experiments are presented in figure 2a. One can see that the channels' depth increased linearly with the etching time, while the channels in **S3** were generally deeper than in **S1**, excluding the point corresponding to 10 min etching. The channels' depth did not significantly change after the long lasting etching. The reason is that when in the etching the poled layer outside the channels was completely etched off, its masking effect disappeared and the etching rates in both areas outside and inside the channels became equal [14]. The maximal depth in **S1** was 0.9 μm (after 20 min etching) and in **S3** it was 1.3 μm (after 30 min etching).

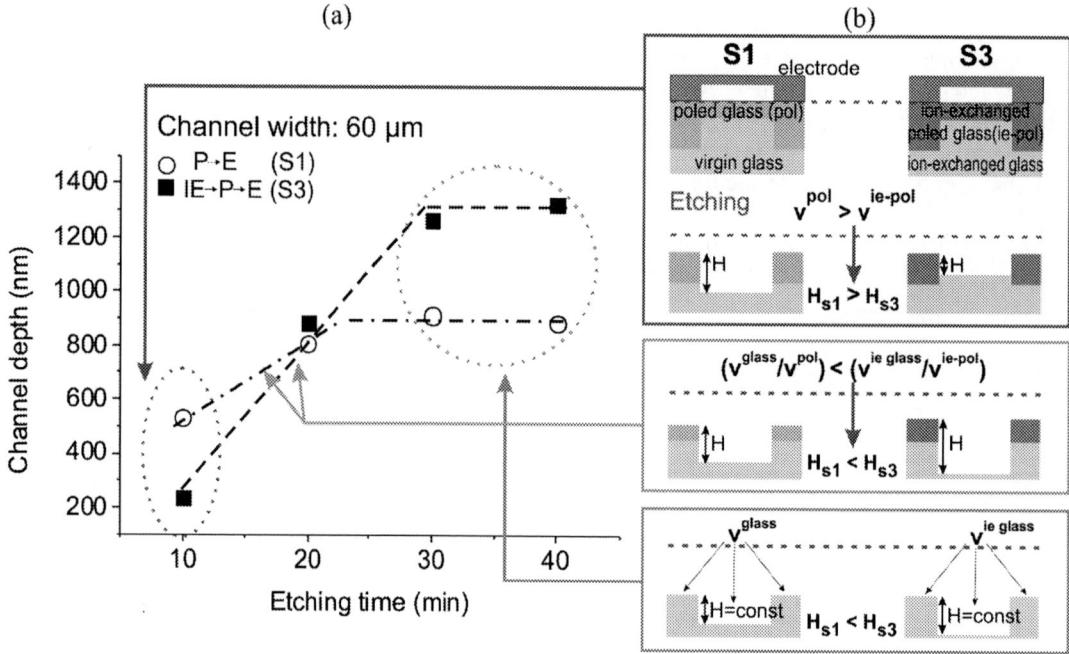

Figure 2. (a) Temporal evolution of the channels depth in the course of the chemical etching. **(b)** The schematic explaining of the depth evolution, where v^{pol}, $v^{ie\text{-}pol}$, v^{glass}, $v^{ie\ glass}$ are the etching rates of poled, ion-exchanged and then poled, virgin and ion-exchanged glass, correspondingly.

We assume that the poled layers formed in the ion-exchanged glass (**S3**) are more etching-resistive comparatively to ones formed in the virgin glass (**S1**). This could be due to a lower mobility of silver ions in **S3** compared to alkali ones in **S1** and corresponding differences in the poling-induced restructuring of the poled regions of the ion-exchanged and virgin glasses. These differences could result in a longer time necessary to etch off a thin poled layer corresponding to the electrode hollow and, respectively, in the slower increase in the channel depth at the initial stage of the **S3** etching, up to

10 min. This consideration is illustrated in figure 2b. In longer etching, when the poled layer under the electrode hollow is completely etched off, lower etching rate of the poled ion-exchanged layer of the **S3** out of the channel provides longer time of the removal of this layer, which corresponds to the longer masking effect, which is responsible for the selective etching of the channel bottom. This resulted in the deeper channel in **S3**. With the following increase of the etching duration, the complete removal of the poled layer out of the channels takes place both in **S1** and **S3**. It is worth to note that here we do not account for possible difference in the etching rates of the virgin and ion-exchanged glasses. Besides, we did not find differences in the depth of channels of different widths in all the samples.

The AFM and SEM characterization of the **S1-S3** samples was performed for 10 min etching only, corresponding images of the nanoisland films integrated with the 60 μm channels in **S1-S3** samples are shown in figure 3. Silver nanoisland films on the bottom of the channels were observed in all the samples (see insets in figure 2a-c). Notably, the film in **S3** treated as follows: ion-exchange – poling – etching – annealing, essentially differed from ones in **S1** and **S2**. The fraction of area occupied by silver nanoparticles (fill factor) was 37% for **S3** while only 18% and 17% for **S1** and **S2**, respectively. The overall nanoislands' shape was a truncated sphere however the average size, D, and polydispersity, σ, of the nanoislands were the highest for the **S3** as well, 25 nm and 44%, correspondingly. The average sizes of the formed nanoislands in the **S1** and **S2** were 16 nm and 18 nm, while their polydispersity was 34%. The high density and NPs size of the film in the **S3** (ion-exchanged prior the poling) indicates a higher silver content compared to **S2** and **S3** which had been firstly poled. The reason is a lack of alkali ions in the subsurface layer of the poled glass under the electrode hollow, which inhibits the ion-exchange processing. The nanoparticles did not grow outside the channels in all the samples (see figure 3d) because of the essential thickness of alkali/silver ions depleted layer formed under the strong electric field in the regions of the electrode-glass contact.

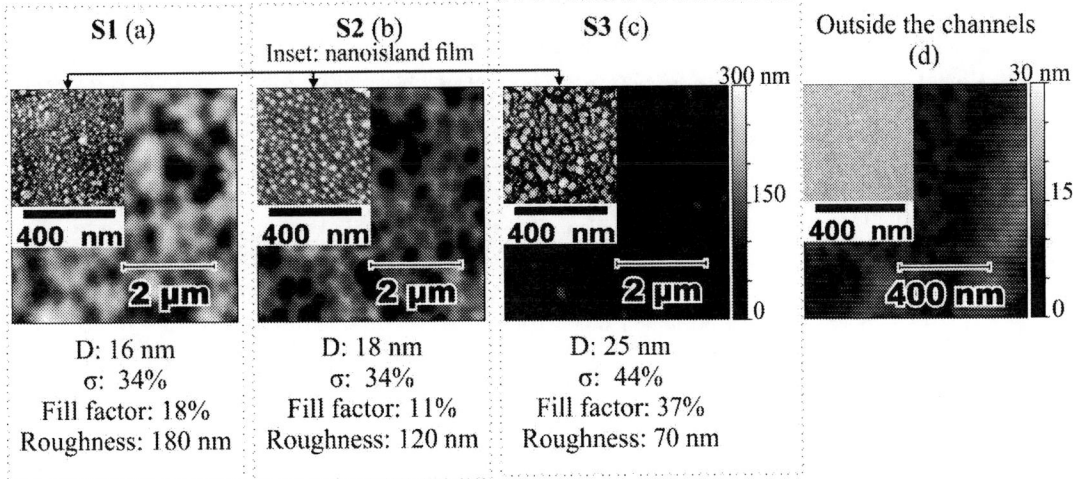

Figure 3. (a-c) SEM and AFM images of the 60 μm channels' bottom in the samples **S1-S3** etched for 10 min; **(d)** SEM and AFM images of the area outside the channel. Average size D, polydispersity σ, fill factor and roughness are denoted near corresponding images of the samples.

One can see in figure 3 that the relief of glass on the bottom of the channels differs in differently fabricated channels. The average roughness of the channels bottom was 180 nm in the **S1**, 120 nm in the **S2** and 70 nm in the **S3**. Finally, processing order of the sample **S3** has provided both formation of denser silver nanoisland film and smoother bottom of the etched channel.

4. Conclusions

We believe that ion-exchange – poling – etching – annealing order of glass slides processing is optimal for the formation of silver nanoparticles integrated into microfluidic channels. This four-step fabrication resulted in the formation of the silver nanoisland film with the highest fill factor, 37%, on the bottom of the microfluidic channel while the bottom was the smoothest, 70 nm roughness. Essentially, the nanoparticles did not grow outside the channels. We also found a difference in the depth of the channels formed in the differently treated samples. The maximal depth, 1.3 µm, has been registered after 30 min etching in the glass sample, which was ion-exchanged prior the thermal poling. The longer etching resulted in the stationary depth of the channels because of the complete removal of the poled layer of the glass outside the channels.

Acknowledgment

The reported study was supported by Russian Science Foundation grant #16-12-10044. AFM characterization were performed using equipment owned by the Federal Joint Research Center "Material science and characterization in advanced technology".

References

[1]   Culha M 2013 *Appl. Spectr.* **67** 355
[2]   Liu Y, Zhou H, Hu Z, Yu G, Yang D and Zhao J 2017 *Biosens. Bioelectron.* **94** 131
[3]   Camden J P, Dieringer J A, Wang Y, Masiello D J, Marks L D, Schatz G C and Van Duyne R P 2008 *J. Am. Chem. Soc.* **130** 12616
[4]   Zhurikhina V V, Brunkov P N, Melehin V G, Kaplas T, Svirko Yu, Rutckaia V V and Lipovskii A A 2012 *Nanoscale Res. Lett.* **7** 676
[5]   Zhong L B, Yin J, Zheng Y M, Liu Q, Cheng X X and Luo F H 2014 *Anal. Chem.* **86** 6262
[6]   Watanabe T, Sassa F, Yoshizumi Y and Suzuki H 2017 *Electron. Commun. Jpn.* **100** 25-32
[7]   Tycova A, Prikryl J and Foret F 2017 *Electrophoresis* **38** 1977
[8]   Wang C and Yu C 2015 *Nanotechnology* **26** 092001
[9]   Zhao Y, Zhang Y, Huang J, Zhang Z, Chen, X and Zhang W 2015 *J. Mater. Chem. A* **3** 6408
[10] Li M, Zhao F, Zeng J, Qi J, Lu J and Shih W C 2014 *J. Biomed. Opt.* **19** 111611
[11] Parisi J, Dong Q and Lei Y 2015 *RSC Adv.* **5** 14081–89
[12] Xie Y, Yang S, Mao Z, Li P, Zhao C, Cohick Z, Huang P H and Huang T J 2014 *ACS Nano* **8** 12175–84
[13] Redkov A V, Melehin V G, Statcenko V V and Lipovskii A A 2015 *J. Non-Cryst. Solids* **409** 166–9
[14] Reduto I, Kamenskii A, Redkov A and Lipovskii A 2017 *J. Electrochem. Soc.* **164** E385

SPBOPEN 2018   IOP Publishing

Second harmonic generation and spontaneous parametric down-conversion in Mie nanoresonators

M I Petrov[1,2], A A Nikolaeva[1], K S Frizyuk[1] and N A Olekhno[1]

[1]Department of Nanophotonics and Metamaterials, ITMO University, 199034 St. Petersburg, Russia
[2]Department of Physics of Condensed Matter, St Petersburg Academic University, 194021 St. Petersburg, Russia

E-mail: m.petrov@metalab.ifmo.ru

Abstract. The recent active research in the field of nonlinear Mie nanoresonators has led to significant increase of harmonic generation efficiency from subwavelength structures. The progress in the second order nonlinear processes became evident after switching to nanoresonators made of materials that possess a bulk nonlinearity such as semiconductor and perovskite-type materials. Despite of that, the theory of second-harmonic generation from spherical Mie nanoparticles with a bulk nonlinearity has not been developed yet. In this work, we present a theoretical approach based on Mie theory for describing the second order nonlinear processes in all-dielectric nanoresonators. We focus on the second harmonic generation (SHG) process as well as on the inverse one, the process of spontaneous parametric down-conversion (SPDC). We demonstrate the SHG-SPDC correspondence in terms of Mie resonances and identify the selection rules governing these second order nonlinear processes. We show how one can control the spatial correlations of photons in SPDC from a single Mie-nanoparticle by properly choosing the parameters of the system, in particular, allowing for Kerker-type unidirectional emission of photons.

1. Introduction

The progress of all-dielectric photonics has led to the development of a new class of nonlinear optical devices [1]. This became possible not in the last turn due to a variety of dielectric and semiconductor materials being used, which include those with a non-zero bulk second order dielectric susceptibility tensor $\hat{\chi}^{(2)}$. This has resulted in the drastic enhancement of SHG from Mie-resonant nanostructures reported in the past two years. The experiments have immediately shown that SHG can be enhanced in dielectric structures with an intrinsic bulk nonlinear susceptibility up to the scale of $10^{-5} - 10^{-4}$ [2, 3, 4, 5, 6] comparing to $10^{-9} - 10^{-8}$ typical for plasmonics [7]. The observed values of the efficiency became achievable due to the high field of Mie modes concentrated inside the nanoresonators, allowing an overlap with bulk sources. However, to control the efficiency of SHG, it is vitally important to identify the mode coupling mechanism in a nonlinear process: eigenmodes of a resonator at the fundamental frequency couple to the modes at the second harmonic (SH) taking into account the specific form of the $\hat{\chi}^{(2)}$ tensor. The symmetry of both fundamental modes and SH modes combined with the crystalline symmetry either allows or forbids the coupling channels.

Within the prospective of enhanced SHG efficiency in a single Mie resonant structure, it becomes accessible to observe the inverse process referred to as the spontaneous parametric

Content from this work may be used under the terms of the Creative Commons Attribution 3.0 licence. Any further distribution of this work must maintain attribution to the author(s) and the title of the work, journal citation and DOI.
Published under licence by IOP Publishing Ltd

SPBOPEN 2018   IOP Publishing

IOP Conf. Series: Journal of Physics: Conf. Series **1124** (2018) 051021   doi:10.1088/1742-6596/1124/5/051021

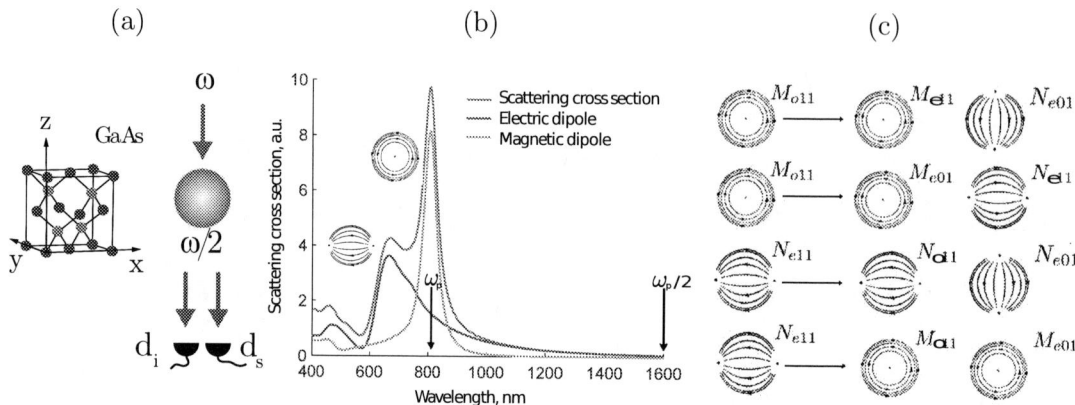

Figure 1. (a): SPDC by a single nanoparticle made from a material with a bulk nonlinearity (e.g., GaAs) in the collinear geometry. (b): The scattering cross-section of a GaAs nanoparticle with the diameter of 220 nm. (c): Allowed dipole channels of the SPDC decay and SHG in a GaAS nanosphere.

down-conversion (SPDC) of photons. This effect is well-known for its applications to the entangled photon pairs generation in bulk nonlinear crystals. The recent studies were focused on observing SPDC in one-dimensional on-chip microsystems [8] as well as on the photon pair generation from quantum dots [9], and now the first studies of the SPDC process in a single dielectric nanostructure have been reported [10].

In this work, we theoretically consider the problem of SHG-SPDC correspondence from the Mie theory point of view. We identify the selection rules, which show the mode coupling during the nonlinear emission of photons.

2. Results

Second harmonic generation. We describe the SHG process classically relying on the weak coupling approach, also known as the non-depleted pump approximation. The intensity of the second harmonic generation as of a coherent process is determined by the cross density of states [11]:

$$I_{2\omega} = \frac{(2\omega)^3 \mu\mu_0}{2} \left[\int \int dV' dV'' \mathbf{P}^*(\mathbf{r}', 2\omega, \omega) \mathrm{Im}\left(\hat{\mathbf{G}}(\mathbf{r}', \mathbf{r}'', 2\omega)\right) \mathbf{P}(\mathbf{r}, 2\omega, \omega) \right], \qquad (1)$$

where $\hat{\mathbf{G}}(\mathbf{r}', \mathbf{r}'', 2\omega)$ is the dyadic Green's function, $P_i(\mathbf{r}, 2\omega, \omega) = \chi_{ijk}^{(2)} E_j(\mathbf{r}, \omega) E_k(\mathbf{r}, \omega)$ is the induced polarization vector at the doubled frequency, $E_i(\mathbf{r}, \omega)$ is the component of the pump electric field inside the nanoparticle, and $\chi_{ijk}^{(2)}$ is the second order nonlinear susceptibility tensor. From (1) one can derive that the intensity of SHG is given by the coefficients $D_{\mathbf{q},\mathbf{q}'\to\mathbf{q}''}$, defining the matrix elements of the transitions:

$$D_{\mathbf{q},\mathbf{q}'\to\mathbf{q}''} = \int_V \chi_{\alpha\beta\gamma}^{(2)} \mathbf{W}_{\mathbf{q}}^{\alpha}(\omega, \mathbf{r_0}) \mathbf{W}_{\mathbf{q}'}^{\beta}(\omega, \mathbf{r_0}) \mathbf{W}_{\mathbf{q}''}^{\gamma}(2\omega, \mathbf{r_0}) d^3 r_0. \qquad (2)$$

Here, the vector $\mathbf{q} = (n, m, p, t)$ defines the particular resonant mode through four quantum numbers: t defines whether the mode is magnetic (M - harmonic) or electric (N - harmonic), $p =$ odd/even defines the parity of the mode, n and m define the multipolar order and the

648

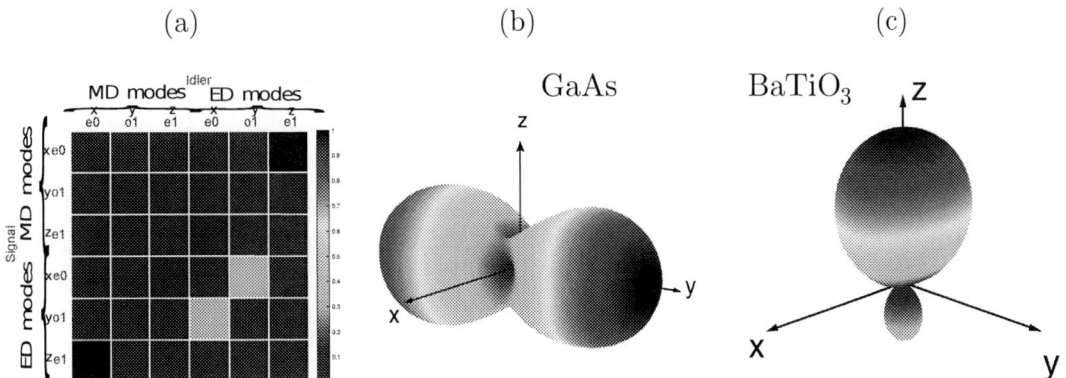

Figure 2. (a): The excitation of different resonant modes during the SPDC decay in a GaAs nanoparticle. (b,c): The directionality diagram of collinear two-photon detection rate $|T(\mathbf{r}_i, \mathbf{r}_i)|^2$ for GaAs (b) and BaTiO$_3$ (c) nanoparticles.

azimuthal number of the mode. The vectors $\mathbf{q}$ and $\mathbf{q}'$ describe the mode at the fundamental wavelength, while $\mathbf{q}''$ corresponds to the SH state. Here, the summation over different components α, β, γ is assumed. The intensity of the second harmonic is defined by these coefficients summed over all alowed transitions $\mathbf{q}, \mathbf{q}' \to \mathbf{q}''$: $I_{2\omega} \sim \sum_{\mathbf{q}, \mathbf{q}' \mathbf{q}''} \left| D_{\mathbf{q}, \mathbf{q}' \to \mathbf{q}''} \right|^2$.

Spontaneous parametric down-conversion. The degenerate SPDC process considers the decay of a photon at the frequency ω into two photons, the signal and the idler, both having the frequency $\omega/2$, see figure 1a. Its consideration requires a quantum treatment that has been done in a general way in Ref.[12] and involves the two-photon amplitude

$$T_{is}(\mathbf{r}_i, \omega/2, \mathbf{d}_i; \mathbf{r}_s, \omega/2, \mathbf{d}_s) = \int_V \left\langle d_i^* | \hat{\mathbf{G}}(\mathbf{r}_i, \mathbf{r}_0, \omega/2) \hat{\Gamma}(\mathbf{r}_0, \omega) \hat{\mathbf{G}}(\mathbf{r}_0, \mathbf{r}_s, \omega/2) | d_s^* \right\rangle d^3 r_0, \qquad (3)$$

where the integration is performed over the volume of a nonlinear nanoparticle, $\mathbf{d_i}$, $\mathbf{d_s}$ are dipole moments of the idler and signal detectors, and $\hat{\Gamma}$ is the generation matrix given by the expression $\hat{\Gamma}_{\alpha\beta}(\mathbf{r}_0, \omega) = \chi^{(2)}_{\alpha\beta\gamma} E_\gamma(\mathbf{r}_0, \omega)$. The magnitude of $|T_{is}|^2$ defines the rate of simultaneous detection of two photons at two given coordinates. The analysis of this expression shows that the detection probability is defined by the same coefficients $|T_{is}|^2 \sim \left| D_{\mathbf{q}, \mathbf{q}' \leftarrow \mathbf{q}''} \right|^2$, which represent the efficiency of a mode $\mathbf{q}''$ to decay into modes $\mathbf{q}$ and $\mathbf{q}'$. In the following, we consider a *collinear detection* when the signal and idler photons are detected at the same position $\mathbf{r}_i = \mathbf{r}_s$.

Considering a resonant pump at the magnetic dipole resonance (see figure 1b), one excites dominantly magnetic and electric dipoles. They can decay into two-photon states via different dipole channels shown in figure 1c. These are allowed transitions for SPDC and reciprocally for SHG. The correlations between different dipole modes can also be seen from figure 2a, where the simultaneous generation of down-converted modes is shown taking into account all modes excited at the pump frequency. The obtained mode coupling results in specific spatial correlations in the two-photon amplitude. The angular diagrams of the collinear two-photon detection rate are shown in figure 2b for GaAs and in figure 2c for BaTiO$_3$. One can see that in the case of GaAs, the simultaneous detection of two photons is most probable in the direction of y-axis. It is worth noting that in the case of a BaTiO$_3$ nanoparticle that has a perovskite type crystalline structure [3], one can observe a unidirectional emission of correlated photons in the backward direction. This can be considered as a Kerker-type nonlinear generation of correlated photons.

3. Conclusion

We have built a model of the second order nonlinear process in resonant Mie nanoparticles and identified the mechanisms of mode coupling and allowed resonant transitions. We applied this model to the SPDC process and found correlations between different resonant modes simultaneously generated during the decay. Our simulations show that one should expect high directionality of photons emitted during the SPDC process in subwavelength nanostructrues. In particular cases, a unidirectional emission is possible. We believe that our results can be of high interest for experimentalists working in the field of nonlinear nanophotonics.

Acknowledgments

We are grateful to A.N. Poddubny and A.A. Sukhorukov for fruitful discussions. The work is supported by the Russian Foundation for Basic Research (proj. №18-32-01052 and 18-02-01206). MIP and NAO acknowledge the support of the Foundation for the Advancement of Theoretical Physics and Mathematics "BASIS".

References

[1] Kruk S and Kivshar Y 2017 *ACS Photonics* **4** 2638–49
[2] Camacho-Morales R et al 2016 *Nano Lett.* **16** 7191–7
[3] Timpu F, Sergeyev A, Hendricks N and Grange R. 2017 *ACS Photonics* **4** 76–84
[4] Makarov S V et al 2017 *Nano Lett.* **17** 3047–53
[5] Carletti L, Locatelli A, Neshev D and De Angelis C 2016 *ACS Photonics* **3** 1500–7
[6] Liu S, Keeler G A, Reno J L, Sinclair M B and Brener I 2016 *Adv. Opt. Mater.* **4** 1457–62
[7] Butet J, Brevet P F and Martin O J F 2015 *ACS Nano* **9** 10545–62
[8] Solntsev A S, Sukhorukov A A 2017 *Reviews in Physics* **2** 19–31
[9] Orieux A, Versteegh M A M, Jons K D and Ducci S 2017 *Rep. Prog. Phys.* **80** 076001
[10] Solntsev A S et al 2016 *2016 Photonics and Fiber Technology Congress* paper NT3A.4
[11] Caze A, Pierrat R and Carminati R 2013 *Phys. Rev. Lett.* **110** 63903
[12] Poddubny A N, Iorsh I V and Sukhorukov A A 2016 *Phys. Rev. Lett.* **117** 123901

Spatially Modulated Channel Waveguide Elements Optically Written in Photorefractive Lithium Niobate

A D Bezpaly, A S Perin, V M Shandarov

Department of Quantum Electronics, Tomsk State University of Control Systems and Radioelectronics, Tomsk 634050, Russia

Abstract. Optical inducing of the channel waveguide elements with spatial modulated parameters in lithium niobate sample with Cu-doped layer is experimentally demonstrated. The channel waveguides have been formed via the point-by-point exposure at light wavelengths of 532 nm and 450 nm. It is shown that the distance between exposing light spot centers affects to the homogeneity and width of channel waveguides.

1. Introduction

The improvement of quantum and integrated optical devices is directly related to the development of photonics, nonlinear optics and laser technologies. Therefore, the studies of such aspects as influence of laser radiation on matter, methods of formation and design of light control elements is important today [1–3]. The light fluxes can be controlled by means of channel waveguides and diffraction gratings based on photorefractive materials [4–6]. One of the most promising materials in this area is lithium niobate crystal ($LiNbO_3$) which demonstrates the strong photorefractive nonlinearity especially at its doping with some impurities like iron (Fe), copper (Cu), manganese (Mn) and others [7, 8]. The photonic elements may be realized due the photorefractive properties of material using optical inducing which makes it possible to set topology and control it during the formation process of such structures.

The main aim of this work is experimental studies of different channel waveguide elements with spatial modulation of their parameters, optically induced in lithium niobate sample with Cu-doped surface layer.

2. Experimental setups and conditions

Channel waveguides were formed by step-by-step exposure of laser radiation to the doped surface of lithium niobate as it is shown in Figure 1. The radiation sources were solid-state YAG:Nd^{3+} laser with λ = 532 nm and semiconductor laser with λ = 450 nm. The dimensions of the sample used are $30 \times 3 \times 15$ mm^3 along X, Y, and Z axes. The sample has been thermally doped with Cu ions from a film deposited onto the wafer surface perpendicular to the crystal Y surface by vacuum sputtering. The doped layer thickness makes up about 100 μm. The light beam is focused onto the doped crystal surface using spherical lens. In different experiments the time of formation was 1-10 seconds. The step between exposing light spot centers is varied from 20 to 60 μm. It allows us to induce channel waveguides and their systems with different spatial parameters, change waveguide width and longitudinal homogeneity of its borders.

Content from this work may be used under the terms of the Creative Commons Attribution 3.0 licence. Any further distribution of this work must maintain attribution to the author(s) and the title of the work, journal citation and DOI.

Published under licence by IOP Publishing Ltd

SPBOPEN 2018   IOP Publishing
IOP Conf. Series: Journal of Physics: Conf. Series **1124** (2018) 051022   doi:10.1088/1742-6596/1124/5/051022

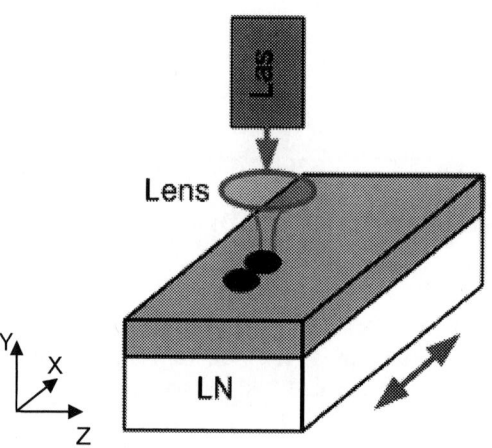

Figure 1. Experimental setup for optical inducing of waveguide channels within lithium niobate surface layer: LN is lithium niobate sample.

Figure 2 shows the schematic image of the formed structures studies by optical probing method using the radiation of He-Ne laser ($\lambda = 633$ nm) with extraordinary light polarization. To probe the exposed areas we collimate light radiation and direct it in the Y axis direction of the sample. The diameter of collimated laser beam is ~1 mm. Then, the light image from exit facet of lithium niobate sample is transmitted to laser beam analyzer by focusing lens. So from the obtained light patterns we study optically induced structures with different spatial modulated parameters in the sample surface doped layer.

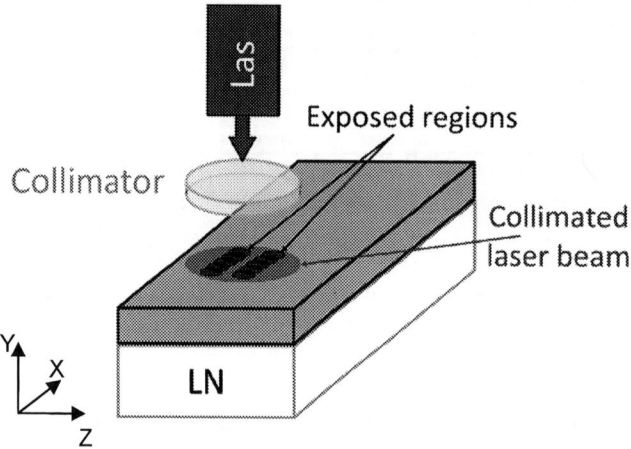

Figure 2. Experimental setup for optical probing of induced waveguide structures within lithium niobate surface layer: LN is lithium niobate sample.

3. Experimental results and discussions
Some experimental results are illustrated by light field patterns in Figure 3. Light field pattern (Figure 3 a) shows the result of induced element consisting of three longitudinally homogeneous exposed

652

stripes. Two light areas enclosed between these dark stripes are waveguide regions. Such dark stripe determining the border of waveguide consists of 30 exposed points with a distance between their centers 25 μm. In the case illustrated in Figure 3 b, the induced element also consists of two waveguides, but one border of the lower waveguide is longitudinally inhomogeneous (number 3 in Figure 2 b). It consists of 16 discrete exposed points with a distance between their centers 50 μm. In both cases shown in Figure 2 a and 2 b, the widths of induced waveguide regions is ~20 μm.

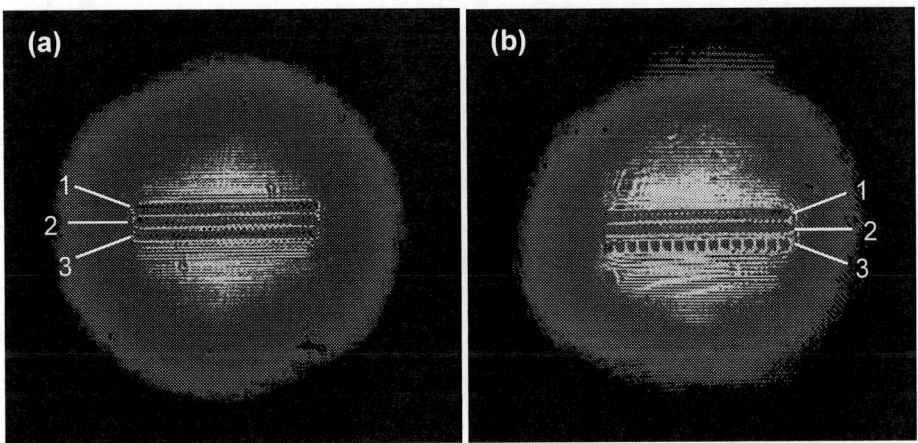

Figure 3(a, b). Light images at optical probing of the induced channel waveguide elements. **(a)** Homogeneous in longitudinal direction; **(b)** With different longitudinal homogeneity (different distances between centers of exposed points).

To check the waveguide properties of optically induced elements, we excite light in waveguides along their directions. Light beam of He-Ne laser (λ=633 nm) is focused onto the entrance surface of waveguide channel using spherical lens with focal length of 4 centimeters. The light images at the exit surface of the waveguide is illustrated in Figure 4a and Figure 5a. These images shows that the light is localized in the induced waveguides enclosed between dark regions. The dark areas in figure 4a and 5a correspond to the exposed stripes shown in Figure 3a and 3b.

To study the light propagation in the formed elements with different spatial homogeneities we use the intensity distribution profiles illustrated in Figure 4b and 5b. Two intensity maxima shown in Figure 4b and 5b correspond to the localized beams from Figure 4a and 5a. Figure 4b shows that the light beams, propagated in the induced waveguide element with homogeneous borders, have the same intensity level.

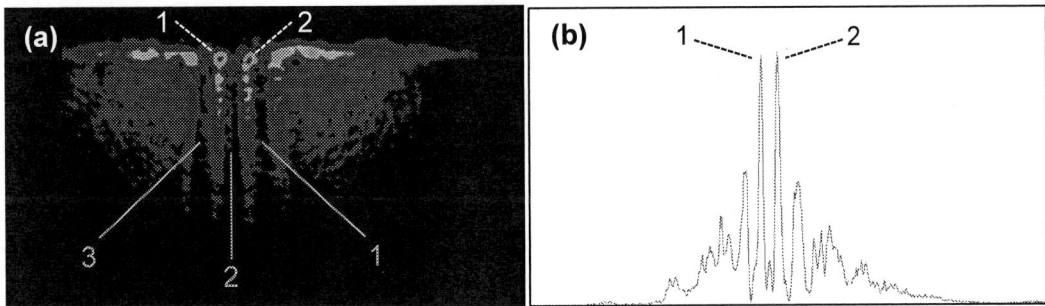

Figure 4(a, b). The light patterns at the output facet of channel waveguide elements with longitudinal homogeneity (a) and its intensity profile (b).

However, the intensity level of light beam, propagated in the waveguide with longitudinal inhomogeneity, is lower than the intensity level of propagating light in the longitudinally homogeneous waveguide (Figure 5b). Perhaps this is due to the fact that the longitudinally homogeneity influence to the propagation loss level of the light in such structures.

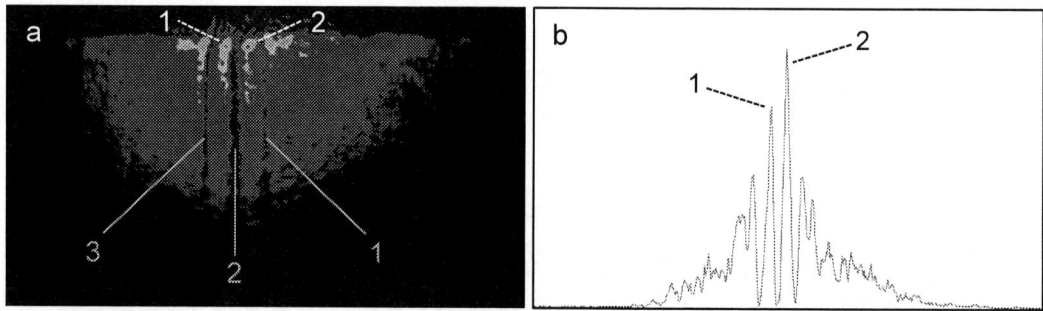

Figure 5(a, b). The light patterns at the output facet of channel waveguide elements with different longitudinal homogeneity (a) and its intensity profile (b).

In the near future we are going to study the influence of longitudinal homogeneity to light energy transition of the narrow focused laser beam from one waveguide to other when one of them is excited. Such method will make it possible to research the interconnection between induced channel waveguides with different spatial homogeneities.

4. Conclusion
In conclusion, we have demonstrated that channel waveguide elements may be optically induced in doped surface layers of lithium niobate. Waveguide configuration is not limited by simple straight elements, the waveguide channels may be longitudinally modulated with their parameters or curvature. Such optically controlled and reconfigured elements are perspective components of modern photonic devices.

Acknowledgments
This study was carried out with the financial support of Ministry of Education and Science of Russia (the project on request 3.1110.2017/PCh).

References
[1] Yamada H and Chu T, Ishida S, Arakawa Y 2006 *IEEE JSTQE* **12** 1371
[2] Rabus D G, Bian Z and Shakouri A 2007 *IEEE JSTQE.* **13** 1349
[3] Bazzan M and Sada C 2015 *Appl. Phys. Rev.* **2** 040603-1
[4] Kroesen S, Horn W, Imbrock J and Denz C 20 *Opt. Express* **22** 23339
[5] Vittadello L, Zaltron A, Argiolas N, Bazzan M, Rossetto N and Signorini R 2016 *J. Phys. D: Appl. Phys.* **49** 125103
[6] Perin A, Shandarov V and Ryabchenok V 2016 Physics of Wave Phenomena **24** 7
[7] Kip D 1998 *Appl. Phys. B.* **67** 131
[8] Krätzig E 1990 *Ferroelectrics* **104** 257

Plasmonic amplification of terahertz radiation in a double-layer graphene nanoribbon array

I M Moiseenko[1,2], O V Polischuk[1], M Y Morozov[1], V V Popov[1,2]

[1]Kotelnikov Institute of Radio Engineering and Electronics (Saratov Branch), Russian Academy of Sciences, Saratov 410019, Russia
[2] Saratov State University, Saratov 410012, Russia

Abstract. The plasmonic absorption/amplification spectrum of terahertz radiation in a double-layer graphene nanoribbon array is theoretically studied. It is shown that this graphene structure exhibits strong plasmon response and giant amplification in a broad terahertz-frequency range.

1. Introduction
Graphene exhibits strong plasmonic response at terahertz (THz) frequencies due to both high density and small "relativistic" effective mass of the charge carriers [1, 2]. Stimulated emission of near-infrared and THz photons from population inverted graphene was recently observed [3]. As compared with the stimulated emission of the electromagnetic modes (photons), the stimulated emission of plasmons by the interband transitions in the population inverted graphene exhibits a much higher gain due to a small group velocity of the plasmons in graphene and strong confinement of the plasmon field in the vicinity the graphene layer [4].

2. Theoretical model
In this paper, we consider the amplification of a THz wave by the stimulated generation of resonant plasmons in a patterned population inverted graphene structure. The response of graphene is described by the complex-valued dynamic conductivity in the local approximation [4]

$$\sigma_{w,d(0)}(\omega) = \sigma_{\text{intra}}(\omega) + \sigma_{\text{inter}}(\omega), \tag{1}$$

where

$$\sigma_{\text{intra}}(\omega) = \frac{e^2 8k_B T\tau}{4\pi\hbar^2(1-i\omega\tau)}\ln\left[1+\exp\left(\frac{\mathcal{E}_F}{k_B T}\right)\right]$$

and

$$\sigma_{\text{inter}}(\omega) = \frac{e^2}{4\hbar}\left[\tanh\left(\frac{\hbar\omega - 2\mathcal{E}_F}{4k_B T}\right) - \frac{4\hbar\omega}{i\pi}\int_0^\infty \frac{G(\mathcal{E},\mathcal{E}_F) - G(\hbar\omega/2,\mathcal{E}_F)}{(\hbar\omega)^2 - 4\mathcal{E}^2}d\mathcal{E}\right].$$

Here ω is the frequency of the incoming electromagnetic wave, e is the elementary charge, $\hbar$ is the reduced Planck constant, k_B is the Boltzmann constant, $\mathcal{E}_F$ is the quasi-Fermi energy, and

$$G(\mathcal{E},\mathcal{E}') = \sinh(\mathcal{E}/k_B T)/\left[\cosh(\mathcal{E}/k_B T) + \cosh(\mathcal{E}'/k_B T)\right].$$

Content from this work may be used under the terms of the Creative Commons Attribution 3.0 licence. Any further distribution of this work must maintain attribution to the author(s) and the title of the work, journal citation and DOI.
Published under licence by IOP Publishing Ltd

The conductivity term $\sigma_{\text{intra}}(\omega)$ in equation (1) describes a Drude-like response involving the intraband processes with the phenomenological electron and hole scattering time τ. The conductivity term $\sigma_{\text{inter}}(\omega)$ describes the interband transitions. The real part of the graphene conductivity (1) is responsible for the absorption/amplification regimes depending on its sign (see figure 1.)

Consider two graphene nanoribbon arrays separated by a thin dielectric barrier slab (see inset on figure 2(a)). External THz wave is incident normally to the plane of the structure along y-direction with the polarization of the electric field $E_x^{(0)}$ across the graphene strips. The optical problem of interaction between THz wave and the double-layer graphene structure has been solved by using a self-consistent electromagnetic approach and integral equations method. The system of two coupled integral equations for the surface current density $j_x^{(w)}(x,y)$ in the graphene nanoribbon at $y=d$ and $y=0$ planes are

$$j_x^{(w)}(x,d) - \sigma_{w,d}(\omega)Z^{(1)}E_x^{(0)}\delta_{m0} = \int_0^w j_x^{(w)}(x',d)G_m^{(1)}(x,x',d)dx' + \int_0^w j_x^{(w)}(x',0)\,G_m^{(2)}(x,x',d)dx',$$

$$j_x^{(w)}(x,0) - \sigma_{w,0}(\omega)Z^{(2)}E_x^{(0)}\delta_{m0} = \int_0^w j_x^{(w)}(x',d)G_m^{(1)}(x,x',0)dx' + \int_0^w j_x^{(w)}(x',0)G_m^{(2)}(x,x',0)dx',$$

(2)

where

$$G_m^{1(2)}(x,x',d) = \sigma_{w,d}(\omega)\frac{1}{L}\sum_m Z_{m,d}^{1(2)}\exp\left(iq_m(x-x')\right),$$

$$G_m^{1(2)}(x,x',0) = \sigma_{w,0}(\omega)\frac{1}{L}\sum_m Z_{m,0}^{1(2)}\exp\left(iq_m(x-x')\right)$$

are the kernels of the integral equations, $q_m = 2\pi m / L$ ($m = \pm1, \pm2, \pm3...$) is the longitudinal wave number of the mth spatial harmonic and $Z_{m,d}^{1(2)}, Z_{m,0}^{1(2)}$ are the explicit algebraic coefficients, calculated from the Maxwell equations:

$$Z_{m,d}^{(1)} = -\frac{\alpha_m^{(a)}\alpha_m^{(b)}\alpha_m^{(s)}\varepsilon_b}{\omega\varepsilon_0\chi_m}\left(1+\exp(2i\alpha_m^{(b)}d)\right) + \frac{\alpha_m^{(a)}\alpha_m^{(b)}\alpha_m^{(b)}\varepsilon_s}{\omega\varepsilon_0\chi_m}\left(1-\exp(2i\alpha_m^{(b)}d)\right),$$

$$Z_{m,d}^{(2)} = -\frac{\alpha_m^{(a)}\alpha_m^{(b)}\alpha_m^{(s)}\varepsilon_b}{\omega\varepsilon_0\chi_m}2\exp(i\alpha_m^{(b)}d),$$

$$Z_{m,0}^{(1)} = Z_{m,d}^{(2)},$$

$$Z_{m,0}^{(2)} = -\frac{\alpha_m^{(b)}\alpha_m^{(b)}\alpha_m^{(s)}\varepsilon_a}{\varepsilon\omega_0\chi_m}\left(1-\exp(2i\alpha_m^{(b)}d)\right) - \frac{\alpha_m^{(a)}\alpha_m^{(b)}\alpha_m^{(s)}\varepsilon_b}{\omega\varepsilon_0\chi_m}\left(1+\exp(-2i\alpha_m^{(b)}d)\right),$$

$$Z^{(1)} = Z_{m,d}^{(1)}\frac{\varepsilon_0\varepsilon_a\omega}{\alpha_m^{(a)}}\delta_{m0}2\exp(-i\alpha_m^{(a)}d)\Big|_{m=0},$$

$$Z^{(2)} = Z_{m,0}^{(1)}\frac{\varepsilon_0\varepsilon_a\omega}{\alpha_m^{(a)}}\delta_{m0}2\exp(-i\alpha_m^{(a)}d)\Big|_{m=0},$$

where

$$\chi_m = \left(\alpha_m^{(b)}\varepsilon_a + \alpha_m^{(a)}\varepsilon_b\right)\left(\alpha_m^{(s)}\varepsilon_b - \alpha_m^{(b)}\varepsilon_s\right) + \exp(2i\alpha_m^{(b)}d)\left(\alpha_m^{(b)}\varepsilon_a - \alpha_m^{(a)}\varepsilon_b\right)\left(\alpha_m^{(s)}\varepsilon_b + \alpha_m^{(b)}\varepsilon_s\right).$$

Here $\alpha_m^{(j)} = \pm\sqrt{k_0^2\varepsilon_j - q_m^2}$ ($j = a, b, s$) is the transverse wave number of the mth spatial harmonic in the jth medium, where a, b, s corresponds to ambient, barrier and substrate medium respectively.

The system (2) of the Fredholm integral equations of the second kind was solved numerically by the Galerkin method with the expansion of two desired functions (x component of the density of the surface electric current $j_x^{(w)}(x,d)$ and $j_x^{(w)}(x,0)$ in $y=d$ and $y=0$ planes) in terms of the Legendre orthogonal polynomials. In the results, the induced electric and magnetic fields at the any point of the structure can be found.

The wavelength of the resonance plasmon mode excited in graphene by the incident THz wave is determined by the period L in the graphene nanoribbon array. Since the plasmon wavelength in graphene is much (by two–three orders of magnitude) shorter than the length of the electromagnetic wave at the same frequency [1, 2], it is necessary to fulfill the condition $L << 2\pi / k_0$ for excitation of the plasmon resonances in the structure. In this case, only the zero-order Fourier harmonics E_{x0} correspond to the traveling waves emitted into the ambient medium and the substrate, while all higher-order Fourier harmonics describe evanescent fields decreasing at $y \to \pm\infty$. Hence, the reflectance R and transmittance T of the structure under investigation in the far-field region can be calculated as $R = |E_{x0}(d)|^2 / |E^{(0)}|^2$ and $T = |E_{x0}(d)|^2 \sqrt{\varepsilon_s} / |E^{(0)}|^2$, respectively. The absorption/amplification coefficient is calculated as $A = P_{abs} / P_0$, where P_0 is the THz power incident on the structure period (per unit of structure width);

$$P_{abs} = \frac{1}{2}\int_0^L \mathrm{Re}\left[\sigma_{w,d}(\omega)\right]|E_x(x,d)|^2 dx + \frac{1}{2}\int_0^L \mathrm{Re}\left[\sigma_{w,0}(\omega)\right]|E_x(x,0)|^2 dx,$$

is the THz power absorbed at the structure period (per unit of the structure width). The law of energy conservation $R + T + A = 1$ is fulfilled both in the absorption regime ($A > 0$) and in the amplification regime ($A < 0$).

The calculations are performed for realistic parameters of the structure under investigation (mentioned in the caption of figure 2 (a, b)) at room temperature. The values of the complex-valued dynamic surface graphene conductivity are assumed to be equal for different graphene nanoribbon arrays, $\sigma_{w,d}(\omega) = \sigma_{w,0}(\omega)$.

3. Results and discussion
Figure 1 shows the calculated value of the real part of the graphene dynamic surface conductivity given by equation (1) as a function of the quasi-Fermi energy and frequency for $\tau = 1\,\mathrm{ps}$ at room temperature.

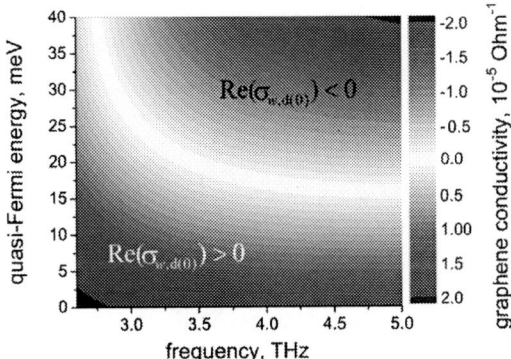

Figure 1. Contour map of the real part of graphene conductivity as a function of the quasi-Fermi energy and frequency.

For sufficiently strong pumping (large absolute value of the quasi-Fermi energy $\mathcal{E}_F$), the interband transitions in graphene prevail over Drude losses (a real part of its conductivity becomes negative at a broad range of THz frequencies, $\mathrm{Re}(\sigma_{w,d(0)}) < 0$, red color region on the contour map of the figure 1), that corresponds to the energy gain (negative absorbance) in graphene. Figure 2 shows the calculated spectrum for first plasmon resonance as a function of the quasi-Fermi energy and frequency at room temperature.

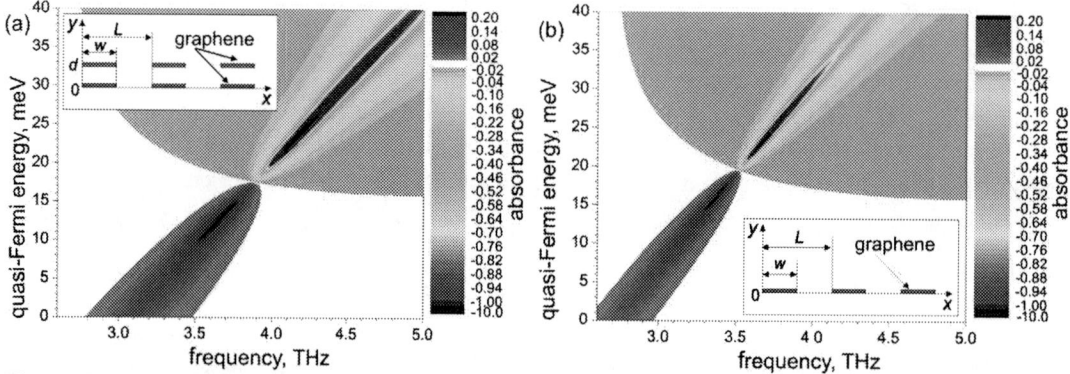

Figure 2(a, b). (a) Absorbance spectrum for the considered graphene structure with period L=500 nm, strip width w=0.25 nm, barrier layer thick d=50 nm, dielectric constant of the substrate and barrier layer is 11.7. The electron scattering time in graphene is 1 ps for temperature 300K. Schematic view of the double-layer graphene nanoribbon array and the coordinate system is shown in the inset. **(b)** The same for single array of graphene nanoribbons.

It is shown that such THz amplifier has much broader dynamic range as compared to that based on a single planar array of graphene nanoribbons (figure 2(b)) and larger amplifications at the same resonant frequency are achieved at lower quasi-Fermi energy.

Acknowledgments
Financial support was provided by the Russian Foundation for Basic Research (Project No. 16-02-00814).

References
[1] Chen J, Badioli M, Alonso-González P, Thongrattanasiri S, Huth F, Hillenbrand R, Koppens F 2012 *Nature* **77** 487
[2] Fei Z, Rodin A S, Andreev G O, Bao W, McLeod A S, Wagner M, Zhang L M, Zhao Z, Thiemens M, Dominguez G, Fogler M M, Castro Neto A H, Lau C N, Keilmann F, Basov D N 2012 *Nature* **82** 487
[3] Boubanga-Tombet S, Chan S, Watanabe T, Satou A, Ryzhii V and Otsuji T 2012 *Phys. Rev. B* **85** 035443
[4] Dubinov A A, Aleshkin V Y, Mitin V, Otsuji T and Ryzhii V 2011 *J. Phys.: Condens. Matter* **23** 145302

SPBOPEN 2018   IOP Publishing

SERS induced by two coupled monolayers of Au plasmonic nanoparticles

V Kaydashev[1], P Zolotukhin[2], A Belanova[2], A Anokhin[1], A Volodin[5], A Chernyshev[3], N Lyanguzov[4] and E Kaidashev[1]

[1]Laboratory of Nanomaterials, Southern Federal University, 200/1 Stachki Ave., Rostov-on-Don, 344090, Russia

[2]Academy of Biology, Southern Federal University, 194/1 Stachki Ave., Rostov-on-Don, 344090, Russia

[3]Institute of Physical and Organic Chemistry, Southern Federal University, 194/2 Stachki Ave., Rostov-on-Don, 344090, Russia

[4]Physics Faculty, Southern Federal University, 5 Zorge St., Rostov-on-Don, 344090 Russia

[5]Laboratory of Solid State Physics and Magnetism, KU Leuven, Celestijnenlaan 200D, Leuven, 3001, Belgium

Abstract. Highly efficient SERS substrates were prepared PLD and chemical methods by combining two quasi-homogeneous ~1 cm^2 monolayers of 5-8 nm Au plasmonic coupled nanoparticles separated by chitosan based polymer film spacer thinner than 10 nm. Plasmonic modes of individual Au particles in each monolayer are coupled and form a 2D array of "hot-spots". Moreover, upon combining the monolayers in stack the plasmonic modes of neighbouring nanoparticles in different layers show also strong coupling. The variation of optical transmission of the nanosystem and its efficiency to enhance Raman spectra of Methylene Blue molecules are studied when polymer spacer thickness is altered.

1. Introduction

Strong coupling of plasmonic modes in metallic nanoparticles separated by sub-10nm gaps results in efficient focusing of the incident electromagnetic energy in the near field. Several plasmonic coupled nanosystems namely self-similar nanoparticles [1,2], bowtie antennas [3], self-assembled noble metal nanoparticle chains [4], coupled nanorods [5] or multi-layer stacks made of monolayers of isolated nanoparticles [6] were proposed as highly efficient systems to detect single chemical and biological molecules by using SERS and fluorescence enhancement [2, 6, 7, 15].

Gaint SERS with estimated enhancement factor (EF) as high as 10^6-10^9 was reported for a number of plasmonic and plasmonic coupled nanosystems [3, 8-11]. However the estimation of that factor often needs special remarks and direct comparison of the SERS efficiency of different systems is often erroneous. SERS enhancement factor given as $EF=(I_{SERS}/I_{ref})/(C_{SERS}/C_{ref})$, where $I_{SERS,ref}$ denote integrated intensities of some Raman line and $C_{SERS,ref}$ are concentrations of analyte adsorbed on the surface of SERS substrate, depends on both the ability of plasmonic system to enhance the local electric field and on the effective surface area of the system. Often these parameters are difficult to measure directly. A minimal quantity of analyte which can be detected by SERS when its deposited on the substrate is more practical parameter for comparison of SERS efficiency. In this case only the

Content from this work may be used under the terms of the Creative Commons Attribution 3.0 licence. Any further distribution of this work must maintain attribution to the author(s) and the title of the work, journal citation and DOI.

Published under licence by IOP Publishing Ltd

measurements done using the same protocol can be directly compared or the amount of deposited molecules should be precisely controlled. Anyway the SERS based detection of the following analytes traces is claimed. Ag/ZnO nanotrees with vast surface area allow detect R6G molecules at 10^{-9} M upon sample washing, the corresponding EF was estimated to be 10^6 [8]. Recently, SERS substrate made of three-dimensionally distributed hot spots showed superior detection limit of 10^{-12} M with EF=10^9 [9] and AAO-based Au porous structure showed detection limit of 10^{-10} M with EF=2.3×10^7 for analyte dried on the surface [10]. The detection limit of 3D systems is normally goes down to ~10^{-9} M when the analyte is removed from the surface by washing upon the substrate incubation in solution [11], however is still much better compared to 2D not coupled plasmonic systems. SERS substrate made of 2D array of plasmonic coupled objects having much smaller active surface area compared to 3D systems however show also high performance because of higher local electric fields. In particular, it was shown that one can detect analyte traces at 10^{-8} M with estimated EF value of 10^7 using an array of bowtie antennas [3]. Note, that these values were obtained when analyte of corresponding concentration was dried on the substrate and the detection limit is likely 10-100 times worse upon substrate washing.

A preparation of well reproducible, highly efficient and cheap SERS substrates based on homogeneous array of "hot-spots" with active area of ~1 cm^2 is urgent challenge for many fields of modern bio-sensing. Strong localization of the incident electromagnetic energy in a "hot spot" requires a precise control of the gaps between adjacent plasmonic nanoparticles. Direct e-beam lithography fabrication provides superior precision in design of plasmonic coupled nanosystems, however, it is a time consuming and expensive approach [3, 11]. Several self-assembling based methods were proposed recently as a cost-effective compromise to e-beam lithography, namely, the use of molecular ligands to produce nanogaps [2], a template assisted assembling of nanoparticle chains [4], controlled percolation of Au nanopillars in anodized aluminum matrix [10] and organizing monolayers of isolated particles to multi-layer stacks [6].

Recently we developed PLD technique to design ~1 cm^2 quasi-homogeneous monolayers of Au and Ag 5-8 nm plasmonic coupled nanoparticles separated by several nm gaps [12]. To further boost the efficiency of our SERS substrates we combined two monolayers of such densely positioned particles to bi-layer using several-nm-thick chitosan-based polymer spacer. The thickness of a polymer film is varied to find an optimal SERS response of Methylene Blue dye molecules chemisorbed on a nanostructure.

2. Experimental

The monolayers of plasmonic coupled nanoparticles were prepared by PLD technique [12]. Briefly, the vacuum chamber was evacuated to 2×10^{-4} mbar and Ar gas flow was introduced to maintain a pressure of 70 Pa. Au particles were deposited at room temperature. A monolayer of ~5-8 nm Au nanoparticles separated by gaps smaller than their sizes is obtained for 500 laser shots. Two identical monolayers of such nanoparticles were organized in bi-layer stack by using several-nm-thick chitosan film. A monolayer of nanoparticles deposited on SiO$_2$ substrate was submerged in the polymerization mixture in dust-free conditions at controlled temperature of 25 °C. After 18 hours of incubation, the sample was washed and dried in dust-free ambient. FInally, the second monolayer of Au nanoparticles was deposited. The extent of plasmonic coupling between nanoparticles in different monolayers was systematically varied by alteration of the polymer spacer thickness.

The morphology of Au nanoparticles prepared at similar conditions on Si substrate was characterized by using FE-SEM Zeiss SUPRA 25. A typical SEM image of a monolayer of plasmonic coupled nanoparticles on Si substrate similar to those used for preparation of bi-layer stacks is shown in Fig.1a. Atomic force microscopy (AFM) measurements of the nanoparticle bi-layer thickness were performed by using a Bruker Dimension 3100D scanning probe microscope in soft tapping mode. The polymer film thickness was determined as following. The bi-layer structure was gently scratched with a "blade" and AFM imaging was then performed across a region including one of the trough borders to establish the depth of the trough, which is considered to be a particle bi-layer film thickness. An example of the AFM scratch profiling of the trough is shown in the inset in Fig. 2b.

From the depth profile measurements performed by averaging of 128 cross sections over 1 μm² region the bi-layer thickness was measured. The thickness of a polymer film was estimated knowing the nanoparticle monolayer height, which is ~2-3 nm. This method is applicable because the polymeric film is less stiff than the substrate and the silicon tip. To verify the blade had not damaged the surface of the silicon substrate, roughness of the trough bottom and the non-coated silicon substrate were compared. No noticeable difference in roughness was observed. The film thickness is uniform across the whole film. Optical transmittance was studied by using UV/Vis Spectrophotometer Agilent Technologies HP-8453. SERS response was characterized by using Renishaw inVia Reflex Raman spectrometer with spectral resolution better than 1 cm⁻¹. The samples were excited by light of HeNe laser at 632.8 nm with power density less than ~5.4×10^5 W/cm² using ×50 long focal distance objective (NA=0.5). Raman scattering was collected in the backward direction. The following measurement protocol was applied for the SERS measurements. Upon the short incubation in water solution of MB the sample was well washed in bi-distilled water and dried. Thus, only chemisorbed molecules could remain on the surface. The used excitation wavelength is ~57 nm blue shifted from the wavelength of the best overlap of absorption and fluorescence spectra, thus the light absorption enhancement induced by the particles is more efficient than the fluorescence enhancement. In opposite, the fluorescence is heavily quenched by the system when the dye molecules are chemisorbed directly onto a metal or only slightly enhanced because the distance between any nanoparticle and phosphor molecule on the surface does not exceed 5 nm, which is not optimal for fluorescence enhancement. In contrast to fluorescence a Raman spectrum is enhanced for any metal-to-molecule separations in the range of ~0-10 nm. The sketch of a SERS study is illustrated in Fig.1b.

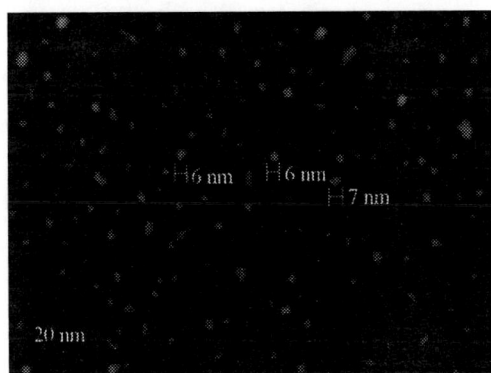 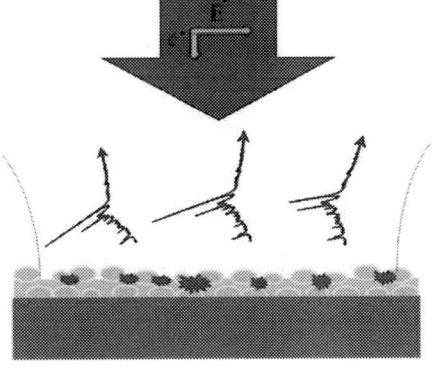

Figure 1(a, b). (a) SEM image of a monolayer of plasmonic coupled nanoparticles on Si substrate; **(b)** Schematic of SERS induced by bi-layer stack of two monolayers of plasmonic coupled nanoparticles.

3. Results and discussion

A monolayer of Au particles reveals a broad absorption band centered near 730 nm and a blue/UV shoulder in the range of ~250-450 nm (Fig 1a), which are the superposition of "bonding" and "anti-bonding" modes of many nanoparticle pairs correspondingly, i.e., the products of the individual nanoparticles hybridized plasmons [13]. When the monolayers are organized in stack the "bonding" mode is blue shifted to 704, 678 and 672 nm and its intensity is increased upon the hybridization of plasmons of particles in different monolayers when spacer thickness becomes larger than ~3 nm evidencing that strong plasmonic coupling occurs (Fig.2a). Similar effect of the optical absorption increase was recently reported for nearly touching Au nonopillars in strong plasmonic coupling regime

[14]. SERS response alteration caused by the variation of the polymer film thickness in the range of 2-10 nm was studied by using Methylene Blue (MB) dye molecules chemisorbed from water solution. The SERS study was carried out for 10^{-5} M/l solutions of MB, i.e. for which the analyte concentration was ~2 orders of magnitude higher than a detection limit for the studied system. At this concentration the substrate shows an excellent SERS response with high signal-to-noise ratio which allows to readily follow the evolution of SERS response caused by systems geometry change. Note, that chitosan based films normally show a sponge-like structure. Thus, the target molecules could supposedly also penetrate inside the spacer layer and be positioned in between the two monolayers of metal nanoparticles. If so, than the alteration of polymer film thickness could influence both the near electric field localization efficiency and the population of target molecules inside the porous film, which both may enhance the Raman signal. The SERS intensity reveal local maximum when chitosan layer thickness is ~ 5 nm which is ~ 10.3 and 4 times larger compared to samples with ~6 nm and ~4 nm thick spacers (Fig.2b). The thicker spacer does not result in larger SERS signal. Thus, the SERS efficiency passes through the local maximum and the optical absorption is shifted and increased when the geometry of a system corresponds to strong coupling regime.

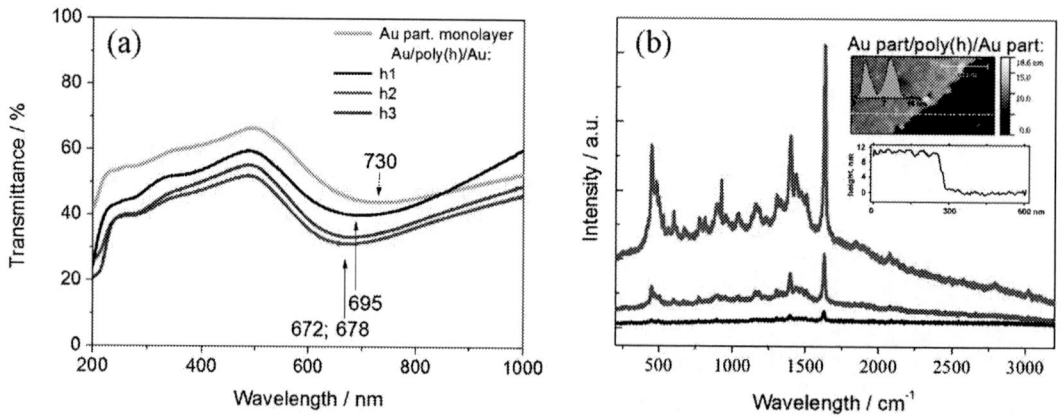

Figure 2(a, b). (a) Optical transmittance, **(b)** SERS of chemisorbed MB dye molecules for bi-layers with 4 nm (black), 5 nm (red) and 6 nm (blue) spacers. A typical AFM scan used to evaluate the polymer film thickness is shown in the inset.

Acknowledgments
Study was funded by RFBR according to the research project №18-02-00151A "Study of photothermal effect in plasmonic coupled nanosystems" and by Russian Education and Science Ministry project No.16.5405.2017/8.9

References
[1] Li K, Stockman M I, Bergman D J 2003, *Phys. Rev. Lett.*, **91** 227402
[2] Lim D K, Jeon K S, Kim H M, Nam J M, and Suh Y D 2010 *Nat. Mater.* **9** 60
[3] Zhang J, Irannejad M, Cui B. 2015 *Plasmonics* **10** 831
[4] Hanske C, Tebbe M, Kuttner C, Bieber V, Tsukruk V V, Chanana M, König T A F, and Fery A 2014 *Nano Lett.* **14** 6863
[5] Aizpurua J, Bryant G W, Richter L J, García de Abajo F J 2005 *Phys. Rev. B* **71** 235420
[6] Feng A L, You M L, Tian L, Singamaneni S, Liu M, Duan Z, Lu T J, Xu F, and Lin M 2015 *Sci. Reports* **5** 7779
[7] Law W C, Yong K T, Baev A, Hu R, Prasad P N 2009, *Opt. Express*, **17** 19041
[8] Cheng C, Yan B, Wong S M, Li X, Zhou W, Yu T, Shen Z, Yu H, Fan H J 2010, *Appl. Mater. Interfaces*, **2** 1824

[9] Wu L, Wang W, Zhang W, Su H, Liu Q, Gu J, Deng T, Zhang D 2018, *NPG Asia Materials*, **10** e462

[10] Sui C, Wang K, Wang S, Ren J, Bai X, Bai J 2016, *Nanoscale*, **8**, 5920

[11] Lim L K, Ng B K, Fu C Y, L Y M Tobing, Zhang D H 2017 *Nanotechnology*, **28** 235302

[12] Kaydashev V E, Lyanguzov N, Zhilin D, Tsaturyan A, Raspopova E A, Kaidashev E M 2016 *J. Phys: Conference Series,* **741** 012145

[13] Nordlander P, Oubre C, Prodan E, Li K, Stockman M I 2004, Nano Lett. **4** 899

[14] Liu L, Zhang Q, Lu Y, Du W, Li B, Cui Y, Yuan C, Zhan P, Ge H, Wang Z, Chen Y 2017 *AIP Adv.*, **7**, 065205

[15] Zhang T, Gao N, Li S, Lang M J, Xu Q H 2015, *J. Phys. Chem. Lett.* **6** 2043

Superconducting nanowire single-photon detector on lithium niobate

E Smirnov[1], A Golikov[1], P Zolotov[2,7], V Kovalyuk[1,3], M Lobino[4,5], B Voronov[1], A Korneev[1,6,7], G Goltsman[1,2,7]

[1]Department of Physics, Moscow State Pedagogical University, 119992, Russia
[2]LLC Superconducting nanotechnology (Scontel), 119021, Russia
[3]Zavoisky Physical-Technical Institute of the Russian Academy of Sciences, 420029, Russia
[4]Centre for Quantum Dynamics, Griffith University, Brisbane, 4111, Australia
[5]Queensland Micro- and Nanotechnology Centre, Griffith University, Brisbane, 4111, Australia
[6]Moscow Institute of Physics and Technology (State University), 141700, Russia
[7]National Research University Higher School of Economics, Moscow 101000, Russia

Abstract. We demonstrate superconducting niobium nitride nanowires folded on top of lithium niobate substrate. We report of 6% system detection efficiency at 20 s^{-1} dark count rate at telecommunication wavelength (1550 nm). Our results shown great potential for the use of NbN nanowires in the field of linear and nonlinear integrated quantum photonics.

1. Introduction

It was shown that for optical quantum computations it is necessary to have a photon source, a single-photon detector and only linear optical elements (phase shifters and splitters) [1]. The most promising method for implementing optical calculation is the use of quantum-photonic integrated circuits (QPICs) that combine all the necessary components on a single chip [2]. To date, there are several main platforms used for the QPICs: silicon, silicon nitride, gallium arsenide, diamond, silicon carbide and lithium niobate (LN). Each platform has its advantages and disadvantages for quantum information protocols. therefore, work on QPICs is going in parallel [3]. Unlike other materials, LN has a strong $\chi^{(2)}$ nonlinearity, birefringence, high electro-optic effect and can be used not only for linear optical elements, but also for high efficient on-chip single-photon sources, active phase shifters and modulators. Here we show our recent results on the fabrication of one of the key QPIC element – superconducting nanowire single-photon detector (SNSPD or SSPD) [4] on a lithium LN substrate.

2. Device fabrication and experimental results

For SNSPD fabrication we used commercially available Z-cut LN substrate (from Gooch and Housego). After cleaning the surface of the 12x12 mm^2 LN substrate we deposit an ultra-thin niobium nitride (NbN) film with a nominal thickness of 7 nm $\pm$ 0.5 nm by a reactive magnetron sputtering in Ar and N$_2$ gases atmosphere. We reached a maximum critical temperature T_c = 11.3 K for the films deposited at a substrate temperature T_S = 800°C with partial pressures of Ar and N$_2$ of 6×10^{-3} and 2.5×10^{-4} mbar, respectively (Figure 1a). The sheet resistance of the deposited NbN film measured at room temperature was 471 Ω/sq. We used photolithography and standard lift-off technique for preparing Au-contact pads and alignment marks as well as e-beam lithography and reactive ion etching (RIE) in SF$_6$ for meander type NbN nanowire formation.

Dependency of the system detection efficiency on the bias current, measured at a telecommunication wavelength λ=1550 nm is shown in Figure1b. We found efficiency at the level of 6 % at 20 s^{-1} dark count rate (DCR). Unlike amorphous superconducting detectors, such as recently demonstrated WSi on lithium niobate [5], NbN detectors have a higher operating temperature, which significantly reduces the cryogenic equipment cost, as well as a higher bias current, which improves jitter due to better signal-to-noise ratio.

Content from this work may be used under the terms of the Creative Commons Attribution 3.0 licence. Any further distribution of this work must maintain attribution to the author(s) and the title of the work, journal citation and DOI.
Published under licence by IOP Publishing Ltd

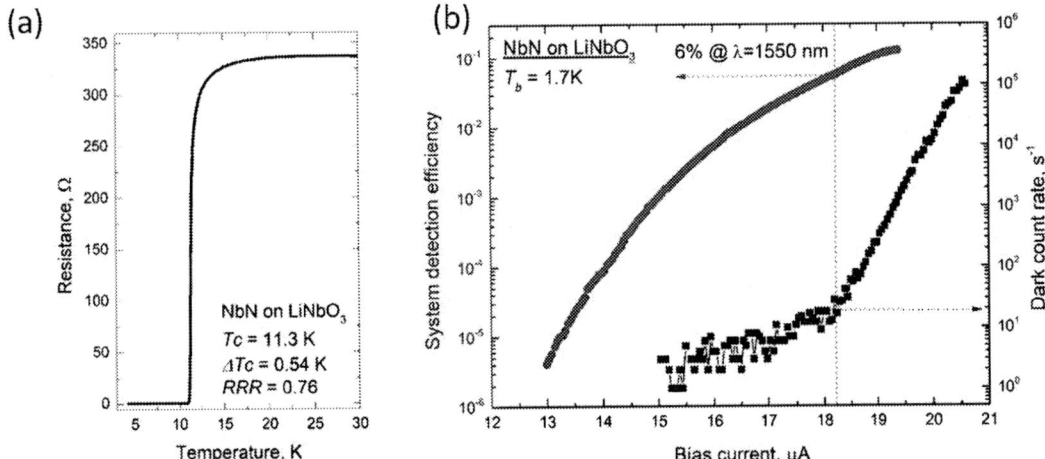

Figure 1 (a-b). (a) Measured dependence of the resistance on temperature for NbN film on LN substrate and **(b)** system detection efficiency (blue circles) and dark count rate (black squares) of SNSPD vs bias current.

Additionally to the successful implementation of classical SNSPDs for the fabrication of integrated devices on a LN substrate, it is necessary to realize waveguide integrated detectors, which are usually performed as U- or W-shaped NbN nanowire on other photonic platforms [3]. Waveguide integrated superconducting single photon detectors (WSSPDs) allow to detect radiation, which spreads within waveguide absorbing of evanescent waveguide field. For the properly designed WSSPDs, combining high absorption and internal detection efficiency it is possible to achieve close to 100% on-chip detection efficiency.

There is a serious obstacle associated with the sequence of the fabrication steps in WSSPD technology. If you first fabricated the integrated detectors, then the ultrathin NbN film degrades during the manufacture of waveguides. In contrast, if first to make waveguides, then by heating the substrate up to 800°C during the NbN film deposition, waveguides are destroyed. For this reason, NbN deposition temperature should be decreased and high quality superconducting films should be produced at room temperature or at a low heating.

We investigated the surface roughness of LN by atomic force microscopy, which is a very important characteristic for thin NbN superconducting films (4 - 10 nm). In Figure 2 shown an AFM scan of the LN surface showing a roughness $R_a = 0.15$ nm, suitable for NbN film deposition.

Index number	Sheet resistance, Ω/sq
1	386
2	473
3	602
4	285

Table 1. List of sheet resistance values of NbN thin films.

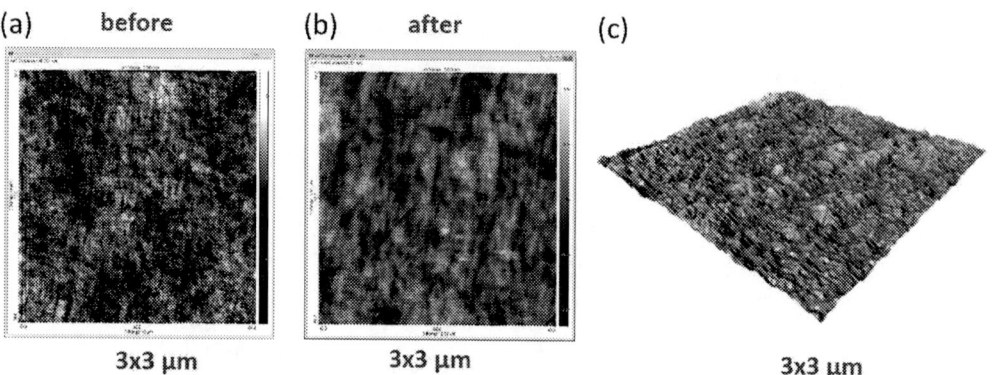

Figure 2 (a -b). AFM scan (3×3 μm) of the LN surface **(a)** before and **(b)** after NbN thin film deposition; **(c)** 3D plot of the LN surface after NbN thin film deposition.

We obtained a critical temperature of 9.4 K for a film grown at room temperature with the sheet resistance equal to 285 Ω/sq shown great potential for the use of NbN nanowires in the field of linear and nonlinear integrated quantum photonics.

In order to determine the most suitable films prepared at ambient temperature for SSPD pattering, we deposited four types of films with different sheet resistance (Table.1). Using e-beam lithography and negative resist mask (ma-N 2403), we patterned detectors, varying their sizes to determinate the most suitable geometry for detection of radiation within waveguide (Figure 2(b-d)).

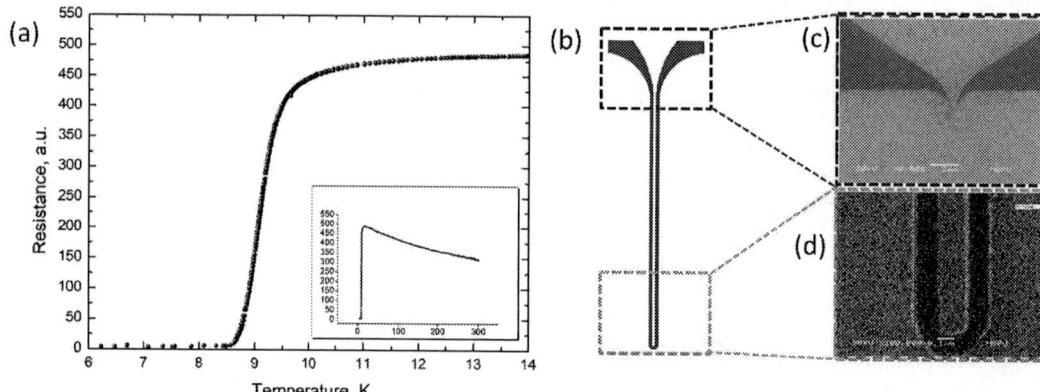

Figure 3 (a - d). (a) Measured dependence of the resistance on temperature for NbN film on LN substrate. Film grown at room temperature with the sheet resistance equal to 285 Ω/sq; **(b)** Schematic view of the U-shaped NbN nanowire pattern; **(c)** Zoom-in image of the NbN nanowire part for contact pads connection, obtained by a scanning electron microscope (SEM); **(d)** Zoom-in image of the nanowire turn obtained by SEM.

3. Conclusion

We demonstrated an SNSPD fabricated on lithium niobate substrate. We shown the system detection efficiency of 6 % at 20 s^{-1} dark count rate. According to our data, this is the highest value for SSPDs on LN that has been demonstrated to date. Direct comparison with the data presented in [6] shows significant increase of the detection efficiency of the NbN detectors. We associate this result with the better quality of the superconductor film grown at a higher substrate temperature. The second part of our work is formation of SSPD on waveguide. For this we changed parameters of NbN deposition process, including the ambient substrate temperature. Our results shown great potential for the use of NbN nanowires in the field of linear and nonlinear integrated quantum photonics on LN substrates.

Acknowledgments

G. Goltsman and V. Kovalyuk acknowledge support by Ministry of Education and Science of the Russian Federation №14.583.21.0065 (RFMEFI58317X0065).

References
[1] Knill E, Laflamme R, Milburn G J, 2001 *Nature* **409** 6816
[2] Aspuru-GuzikA, Walther P, 2012 *Nat. Physics* **8** 4, 2012
[3] Bogdanov S, Shalaginov M Y, Boltaseeva A, and Shalaev V M 2016 *Opt. Mater. Express***7** 1
[4] Gol'tsman G N, Okunev O, Chulkova G, Lipatov A, Semenov A, Smirnov K, Voronov B, Dzardanov A, Williams C, Sobolewski R 2001 *Appl. Phys. Letters***796**
[5] Hopker J P, Bartnick M, Meyer-Scott E, Thiele F, Stephan K, Montaut N, Santandrea M, Herrmann H, Lengeling S, Ricken R, Quiring V, Meier T, Lita A, Verma V, Gerrits T, Nam S W, Silberhorn C, and Bartley T J B 2018 *arXiv:1708.06232v1.*
[6] Tanner M G, Alvarez L S E, Jiang W, Warburton R J, Barber Z H, Hadfield R H 2012 *Nanotechnology***23** 50

SPBOPEN 2018

IOP Publishing

Study of charge relaxation in poled silicate glasses

D V Raskhodchikov[1,2], I V Reshetov[2], D K Tagantsev[2], A A Lipovskii[1,2],V P Kaasik[2]

[1]Department of Physics and Technology of Nanostructures, St. Petersburg Academic University RAS, St. Petersburg 194021, Russia
[2] Institute of Physics, Nanotechnology and Telecommunications, Peter the Great St. Petersburg Polytechnic University, St. Petersburg 195251, Russia

Abstract. We have characterized relaxation peaks observed in the thermally stimulated relaxation current (TSDC) of thermally-poled soda-lime glass in 250-780°C temperature range. We relate the low-temperature (below glass transition) peak to chemical potential driven relaxation that is ionic diffusion. This peak tends to shift towards higher temperatures and to disappear in stronger-poled glasses. Two registered high-temperature (above glass transition, Tg) peaks are supposedly related to electric potential driven processes. Studies of the compositional profiles of poled glasses after secondary thermal processings allowed us concluding that the lower temperature peak found above Tg is supposedly related to the migration of univalent ions (sodium and potassium), while the higher-temperature one is because of the migration of bivalent calcium ions.

1. Introduction

Investigations dedicated to the study of the processes taking place in thermal poling of glasses are of great interest due to the possibility to use thermal poling for fabricating micro- and nanostructures for photonics, microfluidics, integrated optics and others. This is because the poling results in breaking central symmetry of initially isotropic glasses and in their structural and compositional modification. However today there is no full understanding of the mechanisms of these processes.

It is known that glass polarization followed by the glass sample cooling under applied voltage results in the formation of "frozen" spatial charge under anodic surface of the poled glass [1] because of the redistribution of positive ions, the main charge carriers in glasses. At room temperature, uniform redistribution (in other terms, charge relaxation) of the frozen spatial charge is impossible due to kinetic restrictions (low mobility of the ions at room temperature), so that poled glasses stay in a thermodynamically non-equilibrium state. However, at elevated temperatures the mobility of the ions increases and, therefore, at certain increased temperatures, charge relaxation can take place. This results in the movement of poled glassy system towards thermodynamic equilibrium via frozen spatial charge redistribution that behaves in the appearance of depolarization currents.

Temperature dependences of depolarization currents of poled glasses recorded at a constant heating rate are called TSDC spectra. The TSDC spectra of poled glasses provide useful information about the mechanisms of glass polarization and relaxation [2]. In particular, recently, on the base of TSDC spectra studies it has been found that the most part of the frozen in poling spatial charge proves to relax at temperatures above glass transition temperatures [3]. This has explained the problem related to the difference in the total charge passed through the glass sample in the polarization and depolarization processes [4]. In the present paper, we report our study on revealing the mechanisms of charge relaxation processes taking place in poled multicomponent silicate glasses.

Content from this work may be used under the terms of the Creative Commons Attribution 3.0 licence. Any further distribution of this work must maintain attribution to the author(s) and the title of the work, journal citation and DOI.
Published under licence by IOP Publishing Ltd

2. Experimental

A soda-lime glass slides "Menzel" purchased from Agar Scientific was used in this study. The glass composition is presented in Table 1.

Table 1. Composition of Menzel glass in molar % of oxides [5].

SiO$_2$	Al$_2$O3	Na$_2$O	K$_2$O	MgO	CaO	others
72.2	1.2	14.3	1.2	4.3	6.4	0.33

The samples were slides 1 mm in thickness. Glass polarization was performed at 300 C for 30 min under DC electric voltage of 500 V. After the poling, the samples were cooled down to room temperature under the voltage applied and then voltage was off. The schematic of the poling experiment is shown in Figure 1a. We used stainless steel electrodes pressed to the glass surface with evaporated 5 nm chromium sublayer covered with 50 nm gold film on both surfaces of the slides. Step-like increasing of the poling voltage was used to prevent electric breakdown of the glass. Time dependences of the electrical current through the glass samples were recorded, which were used to calculate the electric charge passed through the glass slides in the course of the poling.

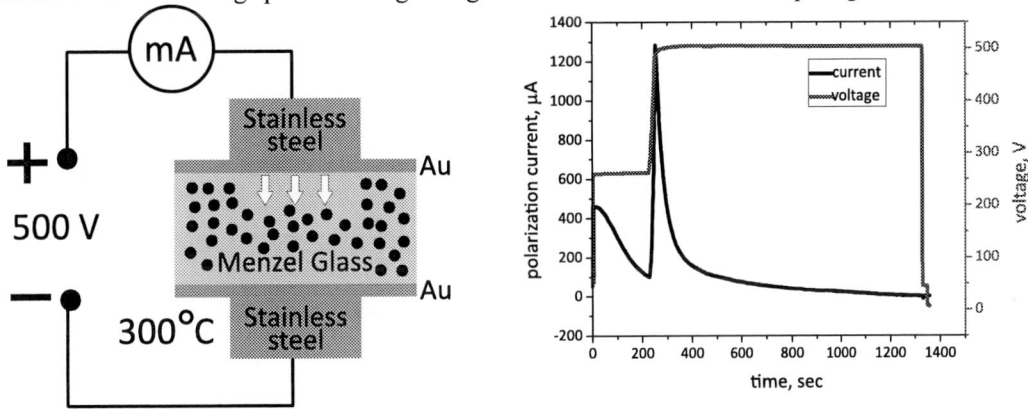

Figure 1. Schematic of the glass polarization process (left) and a typical temporal dependence of the polarization current (right).

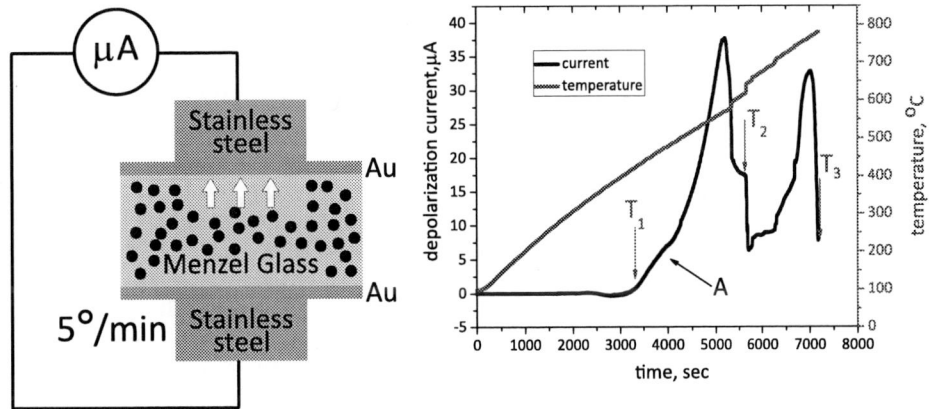

Figure 2. Schematic of the TSDC measurements (left) and the TSDC spectrum of the poled Menzel glass slide which poling current temporal dependence is presented in Figure 1 (right).

The TSDC spectra of the poled samples were recorded at heating rate 5°C/min. The schematic of the TSDC measurements is shown in Figure 2. Above the glass transition temperature, which is ~530°C, for the glass sample poled up to the almost complete disappearance of the polarization current, two depolarization current peaks were observed in the TSDC spectrum of poled glass that evidenced two different ions (with different mobilities) taking part in the charge relaxation processes.

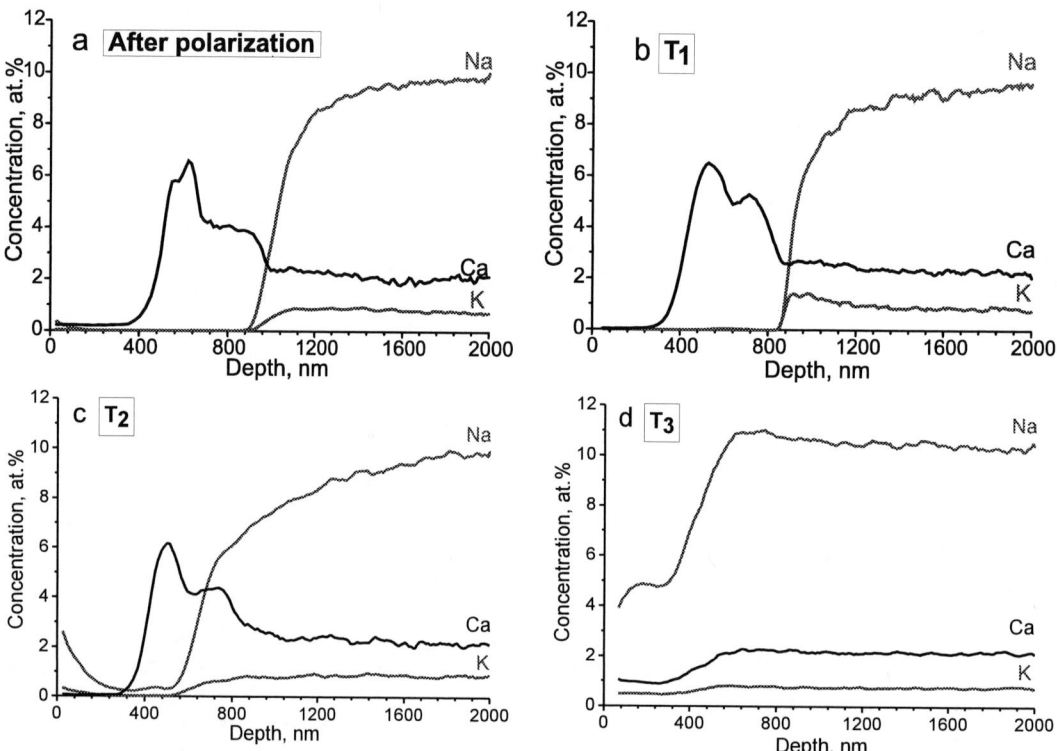

Figure3. Concentration profiles of the main charge carriers (Na, K and Ca) in the reference sample after the polarization (a) and in the samples which depolarization was interrupted at the temperatures T_1 (b), T_2 (c) and T_3 (d) as shown in Figure 2.

The TSDC spectra were taken from three samples poled at the same temperature, time and voltage, but their depolarization processes were interrupted at different temperatures T_1, T_2 and T_3 indicated in Figure 2. Concentration profiles of alkalis and alkaline-earth elements in these samples were measured with the secondary ion mass spectrometry (SIMS) – station CAMECA IMS7f. These profiles are presented in Figure 3.

The existence of a shoulder A at the left slope of the lower-temperature TSDC peak shown in Figure 1 allowed us supposing the existence of one more relaxation peak below the glass transition temperature. To reveal this we performed a set of glass poling experiments with varying density of the electric charge passed through the samples in the poling procedure. The results of the TSDC characterization of these samples in the temperature range 250-480°C are presented in Figure 4. Here one can see clear relaxation peak in the TSDC spectrum of the sample with the minimal charge density, and this peak shifts towards higher temperatures and degrades with increasing the density of the passed charge.

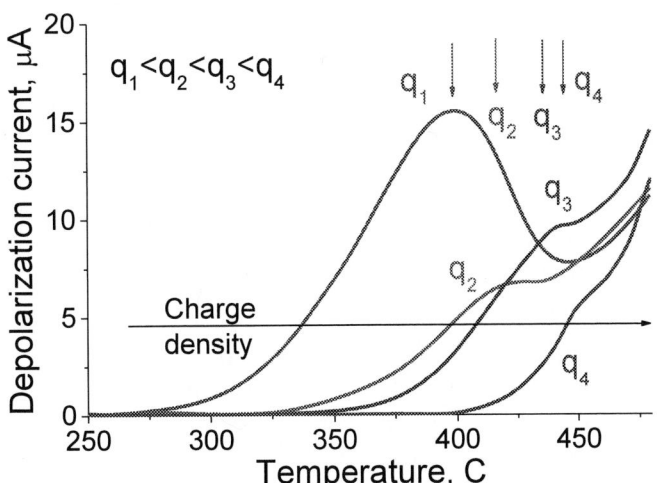

Figure 4. TSDC spectra of glasses poled with different passed charge densities, $q_1 < q_2 < q_3 < q_4$.

3. Discussion

The comparison of the SIMS-measured concentration profiles of the poled soda-lime glass samples, depolarization process of which was interrupted at different temperatures, leads to conclusion that the two TSDC peaks observed above the glass transition temperature should be associated with the relaxation (migration) of different types of charge carriers, namely: univalent alkali ions and bivalent alkali-earth ions. While the absence of essential changes in alkali and alkali-earth elements was registered at T_1 (below glass transition) temperature, the redistribution of the alkalis was seen at T_2, alkali-earth ions being at about the same positions. At T_3 temperature, both alkalis and alkali-earth ions shifted towards the anodic surface of the sample. It is worth to note that even at the highest, ~780°C, temperature the glass did not return in the initial state, and the concentrations of the glass modifiers stayed essentially below ones in the depth of the sample. This is probably related to the restructuration of the glass in poling, which be haves in the increase of bridging oxygen atoms and, respectively, the glass connectivity, followed by the formation of molecular oxygen in the poled glass [6].

The consideration of the relaxation peak (shoulder A on the left slope of the lower-temperature TSDC peak in Figure 2) observed below the glass transition temperature (q_1 in Figure 4) allows concluding that the TSDC peak position shifts towards higher temperatures and, in general, decreases in amplitude with increasing the charge passed through the samples in poling. Formally, this means that more severe poling condition degrades the low-temperature relaxation process. Since "frozen" spatial electric charge and, respectively, frozen electric field increases with the poling duration at initial stages of the poling [7], and, the same, with passed charge, one can conclude that drift mechanism of the relaxation starts prevailing when the "frozen" charge growth. Taking into account the temperature position of the q1 peak and the start of the peak at 250°C we can relate it to the process driven by the gradient of chemical potential of ions, that is, sodium ions diffusion with the activation energy of ~0.8 eV [8], rather than the drift of these ions under the "frozen" electric field. The degradation and the shift of this peak to higher temperatures evidences decreasing in the number (or cross-section) of the channels related to diffusion transport (that is, gradient of chemical potential) with simultaneous increasing of the temperature at which the peak takes place. The latter indicates increasing the activation energy of this kind of relaxation, which could be because of the restructuring and compositional changes. Besides, the same behavior of TSDC spectra was observed in glasses differing in sodium-silicon relation [9,10]. In general, the behavior of the peak under discussion corresponds to the replacement of channels of diffusion-driven relaxation below the glass transition with electric drift driven relaxation channels above the glass transition.

4. Conclusion

Finally, we have proposed an explanation of the TSDC relaxation peaks observed in poled soda-lime glass below and above the glass transition temperature. Below the glass transition the diffusion-related TSDC peak exists, which degrades with the poling strength increase being replaced by two drift-related peaks. Lower temperature peak occurring at ~600 $^{\circ}$C we associate with the drift of alkalis while the higher temperature peak at ~750 $^{\circ}$C – with alkaline-earth drift.

Acknowledgements

This work was supported Russian Science Foundation, grant #16-12-10044. The SIMS studies were performed using the equipment of the Joint Research Centre «Material science and characterization in advanced technology» (Ioffe Institute, St. Petersburg, Russia).

References

[1] Xu W, Arentoft J, Wong D and Fleming S 1999 *IEEE Photon. Techn. Lett.* **11** 10 1265-67
[2] Obata A, Nakamura S and Yamashita K 2004 *Biomaterials* **25** 21 5163-69
[3] Lipovskii A, Morozova A and Tagantsev D 2016 *J. Phys. Chem.* C **120** 40 23129
[4] Mariappan C, Roling B 2008 *Solid State Ionics* **179** 671-8
[5] http://www.agarscientific.com/microscope-slides.html
[6] Redkov A V, Melehin V G and Lipovskii A A 2015 *J. Phys. Chem.* C **119** 30 17298–307
[7] Petrov M I, Lepen'kin Ya A and Lipovskii A A 2012 *J. Appl. Phys.* **112** 4 043101
[8] Zhurikhina V V, Petrov M I, Sokolov K S and Shustova O V 2010 *Techn. Phys.* **55** 10 1447–52
[9] Hong Chi-Ming and Day D E 1981 *J. Am. Ceram. Soc.* **64** 2 61-8
[10] Nascimento M, Nascimento E, Pontuschka W, et al 2006 *Cerâmica* **52** 321 22-30

SPBOPEN 2018

IOP Publishing

Initialization of the Bell states of two qubits by unipolar pulses

M. V. Denisenko[1, *], N.V. Klenov [2-4] and A. M. Satanin[1]

[1] Lobachevsky State University of Nizhny Novgorod, Nizhny Novgorod 603950, Russia
[2] Faculty of Physics, Lomonosov Moscow State University, Moscow 119991, Russia
[3] Moscow Technological University (MIREA), 119454, Moscow, Russia
[4] Moscow Technical University of Communications and Informatics, Moscow 111024, Russia

*Corresponding author: mar.denisenko@gmail.com

Abstract. Simulation of the entangled (Bell) states generation of two qubits by using of unipolar picoseconds pulses was performed. As an example, a system of two coupled superconducting flux qubits interacting with fluxons that integrated with a Josephson transmission line has been considered. The influence of the pulse shape and quantum noise on the accuracy of the Bell states initialization and the way to control of nonlocal entangled states is discussed.

1. Introduction
Recently the superconducting circuit quantum electrodynamics is an actively developing field, in which significant progress has been achieved in the manipulation of quantum bits (qubits). One of the interesting directions in this area is the use of new generation of energy-efficient logic family for fast read-out and control of quantum registers [1-8].

In this paper, we are going to study the simplest two qubits register whose transition frequencies are located in the micro- or millimeter ranges, and the times of longitudinal and transverse relaxation are microseconds [1]. It is well known that in this field of quantum logical manipulation usually the Rabi-technique is involved. We propose here a new approach for the implementation of quantum logic which is based on the control of the qubit system by unipolar sub-nanosecond solitary-like pulses of the rectangular shape. Such kind of magnetic field pulses may be generated by fluxons in transmission lines [2-4]. The analysis was carried out by numerical solution of the master equation for the density matrix operator and the populations of qubit levels were calculated. It was shown that the optimal way to control a system of two qubits can be achieved when two control pulses with proper delay may be applied. We found the parameters of the control pulses for the initialization of the entangled state with fidelity of 99%. Also, the initialization problem for the Bell states is analyzed and the fidelity 98% for these states is found.

2. The model of the system and the basic equations
Hamiltonian for two coupled flux qubits can be represented as

$$H(t) = -\frac{1}{2}\left(\varepsilon_1(t)\sigma_z^{(1)} + \Delta_1\sigma_x^{(1)}\right)\otimes I^{(2)} - \frac{1}{2}I^{(1)}\otimes\left(\varepsilon_2(t)\sigma_z^{(2)} + \Delta_2\sigma_x^{(2)}\right) - \frac{1}{2}J\sigma_z^{(1)}\otimes\sigma_z^{(2)}, \qquad (1)$$

Content from this work may be used under the terms of the Creative Commons Attribution 3.0 licence. Any further distribution of this work must maintain attribution to the author(s) and the title of the work, journal citation and DOI.

Published under licence by IOP Publishing Ltd

where Δ_i is tunnel splitting in the i-th qubit ($i = 1,2$), J is the interaction constant, $\varepsilon_i(t)$ are the form for the pulses; $\sigma_x^{(i)} = \begin{pmatrix} 0 & 1 \\ 1 & 0 \end{pmatrix}$ and $\sigma_z^{(i)} = \begin{pmatrix} 1 & 0 \\ 0 & -1 \end{pmatrix}$ are Pauli matrixes, and $I^{(i)}$ is unit matrix of the i-th qubit, the symbol "$\otimes$" is used to indicate the Kronecker product.

We considered two coupled qubits with slightly different tunnel splitting: $\Delta_2 = \Delta_1 + \delta\Delta$, where $\delta\Delta \ll \Delta_i$. The two coupled qubits gives us a four-level system ($j = 1,2,3,4$). The eigenvalues E_j and eigenstates $|\varphi_j\rangle$ in the stationary case ($\varepsilon_i(t) = 0$) can be determined from the stationary Schrödinger equation: $H(\varepsilon_i(t) = 0)|\varphi_j\rangle = E_j|\varphi_j\rangle$. We assume that at the initial time $t = 0$ the qubits system is initialized in the ground state ($j = 1$) with energy $E_1 = -\frac{1}{2}\sqrt{J^2 + (2\Delta_1 + \delta\Delta)^2}$ and eigenvector $\langle\varphi_1| = \left(\sqrt{2(k^2 + (2\Delta_1 + \delta\Delta)^2)}\right)^{-1}(k, 2\Delta_1 + \delta\Delta, 2\Delta_1 + \delta\Delta, k)$, where $k = J + \sqrt{J^2 + (2\Delta_1 + \delta\Delta)^2}$. The changes in the state of the system of coupled qubits occur due to interaction with unipolar pulses $\varepsilon_i(t)$. Control pulses can act at different instants of time and have different parameters (amplitudes and durations) because of different magnetic coupling between a fluxon and a qubit and different shapes of fluxons in tunable transmission lines. For our calculations, we used square pulses with smoothed fronts of the following form:

$$\varepsilon_i(t) = A_i \begin{cases} (t - t_{in,i})/t_0, & t_{in,i} \leq t < t_{in,i} + t_0, \\ 1, & t_{in,i} + t_0 \leq t \leq t_{off,i} - t_0, \\ (t_{off,i} - t)/t_0, & t_{off,i} - t_0 < t \leq t_{off,i}, \end{cases} \qquad (2)$$

We found the parameters of the control pulses for the initialization of the entangled state with fidelity of 99%. Also, the initialization problem for Bell states is analyzed and the fidelity 98% for these states is found.where A_i is an amplitude, $t_{in,i}, t_{off,i}$ are the turn-on and turn-off times for the pulse, which determine the duration $\tau_i = t_{off,i} - t_{in,i}$, and t_0 is time smoothed fronts ($t_0 \ll \tau_i$).These parameters can be controlled in an experiment, thus controlling the evolution of the quantum system (by analogy with a single qubit [11]).

In conditions close to real experiments, there are many channels of decoherence that affect the measurement results of the coupled qubits. Following of general approach [9], we describe the relaxation in a system of two coupled qubits, assuming that each of them interacts with an infinite bosonic bath. The master equation of two qubits, which is averaged over the reservoir variables, has the form ($\hbar = 1$):

$$i\frac{\partial\rho}{\partial t} = [H(t), \rho] + \frac{\gamma_1}{2}\left(\sigma_z^{(1)} \otimes I^{(2)}\rho\sigma_z^{(1)} \otimes I^{(2)} - \rho\right) + \frac{\gamma_2}{2}\left(I^{(1)} \otimes \sigma_z^{(2)}\rho I^{(1)} \otimes \sigma_z^{(2)} - \rho\right) \qquad (3)$$

where ρ is the density matrix of two coupled qubits, γ_i is the relaxation rate of the i-th qubit, $H(t)$ is the Hamiltonian in the form Eq. (1).

We will be interested in the dynamics of the diagonal elements ρ_{jj} of the density matrix, which correspond to the level populations W_j. We introduce the concept of concurrence for a two-qubit system [10]. The concurrence is an entanglement monotone defined for a mixed state of two qubits as:

$$C(\rho) = \max\{0, \lambda_1 - \lambda_2 - \lambda_3 - \lambda_4\}. \tag{4}$$

In Eq. (4) λ_j is the non-negative eigenvalue of the Hermitian matrix $R = \sqrt{\sqrt{\rho}\tilde{\rho}\sqrt{\rho}}$ in decreasing order. Here, $\tilde{\rho} = (\sigma_y \otimes \sigma_y)\rho^*(\sigma_y \otimes \sigma_y)$, and ρ^* are the complex conjugate of ρ, σ_y is Pauli matrix.

3. Results and discussion

For definiteness we will assume that the duration of the first pulse $\varepsilon_1(t)$ is fixed value τ_1 . The second unipolar pulse $\varepsilon_2(t)$ has a delay δT relative to the first one ($t_{in,2} = t_{in,1} + \delta T$) and the pulse duration τ_2 may be changed. The numerical analysis of Eq. (4) showed that there are ranges of amplitudes and durations of unipolar action, when the system is initialized in nonlocal entangled states (with the population of the 2nd and the 3d levels) with a fidelity of 98%. The characteristic dynamics of excitation of entangled states (the level populations) $W_2(t), W_3(t)$ is shown in Fig. 1 (a) and (b). The time range of the action of the pulses is shown in Fig. 1 by vertical arrows. Note that the ranges for the adjustable parameters of the controllable pulse (A_2, τ_2) are spaced, which makes it possible to realize an "isolated" initialization in each of the states with large fidelity. We note that the effect of noise affects the process at times $\tau_i \sim \dfrac{1}{\gamma_i}$.

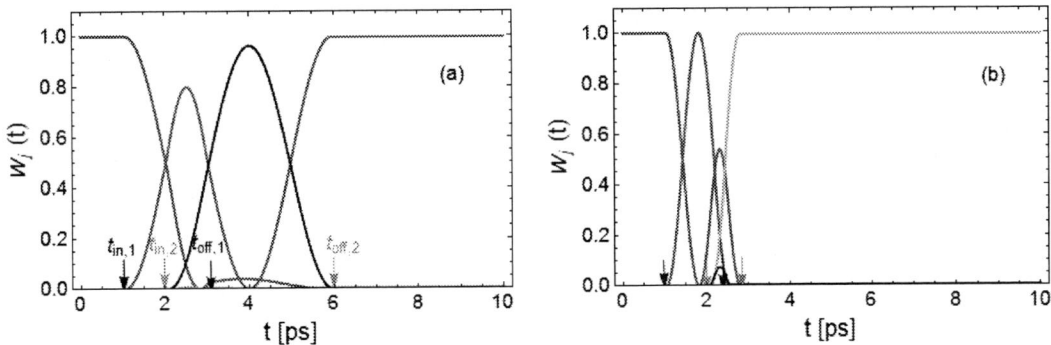

Figure 1. The time-dependant dynamics of the level populations $W_j(t)$ of two coupled qubits when the parameters were chosen as: $A_{1,2} = 1.58$ GHz and $t_{off,2} = 6$ ps (a); $A_{1,2} = 4.1$ GHz and $t_{off,2} = 2.75$ ps.(b). The blue curve characterizes the behavior of $W_1(t)$, the red curve – $W_2(t)$, the green – $W_3(t)$, the black – $W_4(t)$. Vertical arrows show the range of turn-on and off for unipolar pulses. The parameters of the pulses and qubits are as follows: $\Delta_1 = 0.01$ GHz, $J = 0.01$ GHz, $\delta\Delta = 0.02$ GHz, $\gamma_{1,2} = 0.0001$ GHz, $t_0 = 0.1$ ps, $t_{in,1} = 1$ ps, $\delta T = 1$ ps, $t_{off,1} = 3$ ps (a) and $t_{off,1} = 2.5$ ps.

It is interesting to study the initialization of Bell's states, which are very important for the implementation of two-qubit quantum logic. The Bell states are four specific maximally entangled quantum states for two qubits. In this case, the intermediate states of the qubit system have the same the level populations $W_2(t) = W_3(t) \sim 0.5$. The degree of entanglement (the concurrence) of the individual qubits states, which can be described by the quantity Eq. (4), is maximal: $C(\rho) \rightarrow 1$ [10].

In Fig. 2 it is shown that it is possible to select parameters for the Bell states generation. At the same time, the degree of entanglement reaches its maximum value.

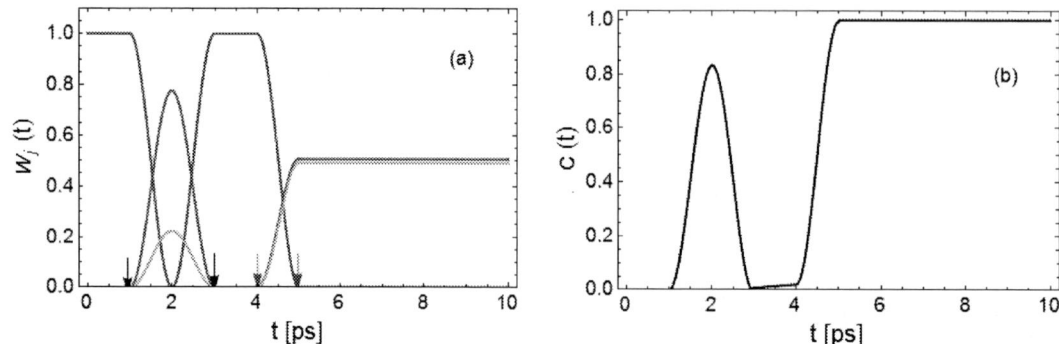

Figure 2. Evolution of the level populations of two coupled qubits in the recording of Bell states (a)and the dynamic of the concurrence (b). The colors of the curves on (a) are identical to Fig. 1. The parameters of the pulses and qubits are as follows: $\Delta_1 = 0.01$ GHz, $J = 0.065$ GHz, $\delta\Delta = 0.02$ GHz, $\gamma_{1,2} = 0.0001$ GHz, $A_{1,2} = 3.15$ GHz, $t_0 = 0.1$ ps, $t_{in,1} = 1$ ps, $\delta T = 3$ ps, $t_{off,1} = 3$ ps and $t_{off,2} = 5$ ps.

4. Conclusion

In this paper, we showed that a system of two coupled superconducting flux qubits can be controlled by means of unipolar sub-nanosecond pulses. It is shown that by varying the parameters of such impacts (amplitude, duration), one can realize initialization in nonlocal intermediate as well as Bell states. These ranges of operating parameters were obtained numerically, based on the solution of the master equation. The effects of quantum noise on these manipulations were studied and it was found that decoherence affects only when the duration of the signal is increased (i.e., at times $1/\gamma_i \sim 0.1$ μs).

The degree of entanglement (the concurrence) was estimated and it was demonstrated that for Bell states it can be realized with accuracy of up to 98%, and in nonlocal states with an accuracy of 99%.

Acknowledgments

This work was supported by the Ministry of Education of the Russian Federation (contract No. 3.3026.2017), by the RFBR (grants No. 18-07-01206 and 16-07-01012), and by the RNF No. 18-72-00158.

References

[1] M. H. Devoret and R. J. Schoelkopf, *Science* **339**, 1169-1174 (2013).
[2] D. V. Averin, K. Rabenstein, and V. K. Semenov, *Phys. Rev.* B **73**, 094504 (2006).
[3] A. Fedorov, A. Shnirman, G. Schoen, A. Kidiyarova-Shevchenko, *Phys. Rev.* B **75**, 224504 (2007).
[4] A. Herr, A. Fedorov, A. Shnirman, E. Ilichev, and G. Schon, *Supercond. Sci. Technol.* **20**, S450 – S454 (2007).
[5] R. McDermott, M.G. Vavilov, B.L.T .Plourde, F.K. Wilhelm, P.J. Liebermann, O.A. Mukhanov, and T.A. Ohki, *Quantum science and technology* **3**(2), 024004 (2018).
[6] I. I. Soloviev, N. V. Klenov, A. L. Pankratov, E. Il'ichev, and L. S. Kuzmin, *Phys. Rev.* E **87**, 060901(R) (2013).
[7] R. McDermott, M. G. Vavilov, *Phys. Rev.* Applied 2, 014007 (2014).
[8] I. I. Soloviev, N. V. Klenov, S. V. Bakurskiy, A. L. Pankratov, L. S. Kuzmin, *Appl. Phys. Lett.* **105**, 202602 (2014).
[9] M.O. Scully, M.S. Zubairy. *Quantum Optics* (Cambridge University Press, 1997).
[10] W. K. Wootters, *Phys. Rev. Lett.* **80**, 2245 (1998).

[11] N. V. Klenov, A. V. Kuznetsov, I. I. Soloviev, S. V. Bakurskiy, M. V. Denisenko, and A. M. Satanin, *Low Temperature Physics* **43**(7), 789–798 (2017).

SPBOPEN 2018   IOP Publishing

IOP Conf. Series: Journal of Physics: Conf. Series **1124** (2018) 051028   doi:10.1088/1742-6596/1124/5/051028

Second harmonic generation by metal core - dielectric shell spherical nanoparticles: spatial vs. plasmon resonances

S A Scherbak[1,2], A A Lipovskii[1,2] and E S Babich[2]

1 Department of Physics and Technology of Nanoheterostructures, St. Petersburg Academic University RAS, St. Petersburg 194021, Russia
2 Institute of Physics, Nanotechnology and Telecommunications, Peter the Great St. Petersburg Polytechnic University, St. Petersburg 195251, Russia

E-mail: sergeygtn@yandex.ru

Abstract. An analytical study of the second harmonic generation by metal core - dielectric shell spherical nanoparticles was performed. Dependences of the second harmonic intensity on the dielectric shell thickness were calculated. We demonstrated that the thin coating allows tuning surface plasmon resonance wavelength, whereas in presence of the thick shell spatial Mie resonances occur. Estimated efficiency of these resonances surpasses the impact of the plasmon enhancement in the second harmonic generation. Possibility to combine plasmon (metal) and spatial (dielectric) resonances to achieve even more advantage in the second harmonic generation was shown.

1. Introduction

The second harmonic generation (SHG) by nanoparticles (NPs) is under extensive study due to its applicability in sensing [1]. Metal NPs in which the nonlinear response can be greatly enhanced by the surface plasmon resonance (SPR) phenomenon [2, 3] are of particular interest. Their plasmonic properties are very sensitive to the NPs composition/structure, environment, size and shape [4-7] that allows spectral positioning of the SPR. Particularly, a thin dielectric covering of a metal NP offers a possibility to tune the SPR wavelength that was multiply demonstrated [8, 9]. At the same time, a thick coating does not affect NPs plasmonic properties more than the thin one, but supports spatial Mie resonances. Using these effects, significant advantage in the SHG can be achieved. In this study, we theoretically analyze the influence of the dielectric cover on the efficiency of the SHG by spherical metal NPs focusing on the potential of combining the plasmon (metal) resonances with the spatial (dielectric) ones. To follow general features of the process we limit our consideration by a model with a Drude-metal core and a non-dispersive isotropic dielectric shell.

2. Theory

Solution of the problem of nonlinear scattering by NPs is based on the nonlinear Mie theory [10-12] and includes three steps: 1) solution of the classical Mie problem to calculate fundamental electromagnetic fields, 2) definition of nonlinear sources using these fields, and 3) calculation of

---

Content from this work may be used under the terms of the Creative Commons Attribution 3.0 licence. Any further distribution of this work must maintain attribution to the author(s) and the title of the work, journal citation and DOI.

Published under licence by IOP Publishing Ltd

the second harmonic (SH) fields. A similar problem for the inverted structure (dielectric core - metal shell) has recently been considered [10]. We partly follow their methodology and all the details can be found in Ref. 10, despite that solution cannot satisfy the present problem since different fields expansions should be used.

We consider a spherical metal particle (core) of radius a coated with a spherical layer (shell) of outer radius b and thickness $h = b - a$ illuminated by the X-polarized and Z-directed plane light-wave of frequency ω (wavelength λ). Using the solution of the linear Mie problem for core-shell spherical NPs (expansions of the electromagnetic fields via vector spherical harmonics and expressions for the expansion coefficients can be found elsewhere [13, 14]) we define spatial distribution of nonlinear sources. Only surface second order susceptibility, χ_s, is considered because the materials are centrosymmetric. We also presume that the core-shell interface is the only nonlinear surface and that the tensor component $\chi^s_{\perp\perp\perp}$ dominates. In this case, the surface second order polarization is [10]:

$$P^s_\perp(2\omega) = \chi^s_{\perp\perp\perp} E_\perp(\omega) E_\perp(\omega), \tag{1}$$

where $E_\perp(\omega)$ is the normal-to-the-surface component of the fundamental (at frequency ω) electric field at the core-shell interface. The expression for the nonlinear polarization given by equation (1) needs to be expanded via certain components of the vector spherical harmonics. Here we omit this technical step, details of which can be found elsewhere [10, 11].

Next, we derive the SH electromagnetic fields generated by the calculated nonlinear sources. The following boundary conditions for transversal components of the SH electric, $\mathbf{E}(2\omega)$, and magnetic, $\mathbf{H}(2\omega)$, fields at the core-shell interface are used:

$$\mathbf{E}^{\theta,\phi}_{shell}(2\omega) - \mathbf{E}^{\theta,\phi}_{core}(2\omega) = -\frac{4\pi}{\varepsilon_{shell}(2\omega)}\nabla_{\theta,\phi}P^s_\perp(2\omega), \tag{2}$$

$$\mathbf{H}^{\theta,\phi}_{shell}(2\omega) - \mathbf{H}^{\theta,\phi}_{core}(2\omega) = \frac{4\pi}{c}2i\omega[\hat{\mathbf{e}}_\mathbf{r} \times \mathbf{P}^s_\perp]^{\theta,\phi}, \tag{3}$$

where r, θ and ϕ denote the orts of the spherical coordinate system, $\hat{\mathbf{e}}_\mathbf{r}$ is a corresponding unit basis vector, $\varepsilon_{shell}(2\omega)$ is the dielectric permittivity of the shell. Boundary conditions at the outer interface are trivial. Using these equations, we determine the SH fields distribution throughout the space. Finally, we present the derived equation to calculate the total SH intensity:

$$W_{2\omega} = \frac{c}{8a^2K_{out}^2}\sum_{n=1}^{\infty}\left[\frac{2n(n+1)}{2n+1}|C_{0n}a_n^{2\omega}|^2 + \frac{2(n-1)n^2(n+1)^2(n+2)}{2n+1}|C_{2n}a_n^{2\omega}|^2\right], \tag{4}$$

where

$$a_n^{2\omega} = \frac{\psi_n(m_1x)}{\psi'_n(m_1x)}\frac{im_1[\psi'_n(m_2x) - A_n\chi'_n(m_2x)]}{m_2\xi'_n(y)[\psi_n(m_2y) - A_n\chi_n(m_2y)] - \xi_n(y)[\psi'_n(m_2y) - A_n\chi'_n(m_2y)]}\frac{4\pi}{\varepsilon_{shell}}\frac{x}{a}, \tag{5}$$

$$C_{mn} = -\frac{1}{2}\left(\int_{-1}^{1}P_n^m(t)P_n^m(t)dt\right)^{-1}\chi^s_{\perp\perp\perp}$$

$$\cdot \sum_{n_1,n_2=1}^{\infty}n_1(n_1+1)n_2(n_2+1)E_{n_1}E_{n_2}d_{n_1}^\omega d_{n_2}^\omega\frac{j_{n_1}(m_1x)j_{n_2}(m_1x)}{(m_1x)^2}\int_{-1}^{1}P_n^m(t)P_{n_1}^1(t)P_{n_2}^1(t)dt, \tag{6}$$

$$d_n^\omega = \frac{1}{\psi'_n(m_1x)}\frac{im_1[\psi'_n(m_2x) - A_n\chi'_n(m_2x)]}{m_2\xi'_n(y)[\psi_n(m_2y) - A_n\chi_n(m_2y)] - \xi_n(y)[\psi'_n(m_2y) - A_n\chi'_n(m_2y)]}, \tag{7}$$

$$A_n = \frac{m_2\psi'_n(m_1x)\psi_n(m_2x) - m_1\psi_n(m_1x)\psi'_n(m_2x)}{m_2\psi'_n(m_1x)\xi_n(m_2x) - m_1\psi_n(m_1x)\xi'_n(m_2x)}, \tag{8}$$

$x = \frac{2\pi n_{out} a}{\lambda}$, $y = \frac{2\pi n_{out} b}{\lambda}$ are dimensionless size parameters; $m_1 = \frac{n_{core}}{n_{out}}$, $m_2 = \frac{n_{shell}}{n_{out}}$ are relative refractive indices; ψ_n , ξ_n, χ_n are, respectively, Riccati-Bessel spherical functions of the first kind, Riccati-Bessel spherical functions of the second kind and Riccati-Hankel spherical functions of the first kind (sometimes called Riccati-Bessel spherical functions of the third kind); K_{out} is the SH wavenumber in the surrounding medium; j_n - a spherical Bessel function of the first kind; P_n^m - associated Legendre polynomial; $E_n = E_0 i^n \frac{2n+1}{n(n+1)}$, E_0 is the incident field magnitude further taken as unit; n_1, n_2, n are multipole orders of two fundamental and one SH modes, respectively. Note that all the frequency-dependent factors in equation (5) should be taken at frequency 2ω.

3. Results

We calculate the spectra of the SHG by a spherical NP of Drude metal covered with a differently thick non-dispersive dielectric shell with permittivity ε_{shell}=4.0 using the analytical solution given by equation (4). The plasma wavelength of the metal was chosen λ_p=200 nm and relaxation rate γ=3 · 10^{13} sec^{-1} to consider the behaviour of SPRs in the visible range.

3.1. Thin shell case

We fix the radius of the NP, 5 nm, while the thickness of the shell is varied up to 20 nm. Dependences of the SH intensity on the fundamental wavelength are presented in figure 1a.

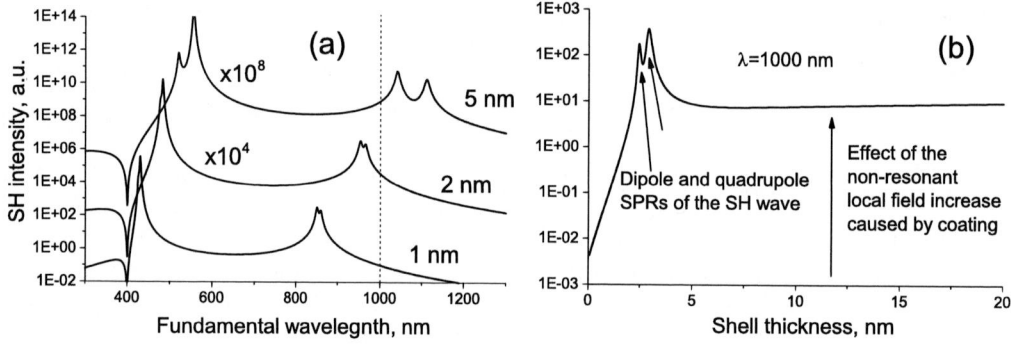

Figure 1. a) SH intensity spectra of 5 nm in radius spherical NP of Drude-metal covered with a differently thick dielectric shell (ε_{shell}=4.0 ; the shell thickness is labelled near each spectrum; the curves are separated along y-axis with the offset of 10^4. b) SH intensity, fundamental wavelength λ=1000 nm, vs. shell thickness.

These spectra demonstrate four resonance peaks, which origin in a non-coated NP was discussed elsewhere [12]. Briefly, short-wavelength peaks are quadrupole and dipole SPRs of the fundamental wave, whereas long-wavelength ones are resonances of the SH wave. The quadrupole peaks appear in the nonlinear spectra of the small NP because the pure dipole mechanism of the nonlinear interaction (two dipole fundamental modes beget a dipole SH mode) is forbidden by the spherical symmetry [11, 12] and the quadrupole impact is essential. The SPRs red-shift with the shell thickness increasing that, albeit being a well-known phenomenon, offers a way to tune SPR wavelength - see figure 1b where the dependence of the SH intensity on the shell thickness at the fundamental wavelength λ=1000 nm is shown. The peaks marked with the arrows in figure 1b correspond to the shell thicknesses at which the fundamental wavelength coincides with doubled dipole and quadrupole SPR wavelengths. Essentially, in figure 1b it can be seen that the SH signal for thicker coatings stays three orders higher than one from the non-coated NP in spite of detuning the resonance. This originates not only from the proximity of the resonance but also from the monotonic increase of both resonant and non-resonant local fields when the particle is embedded in an optically denser medium. This effect arises from a plain fact that the local electric field of a spherical inclusion grows with increasing permittivity of the outer medium in accordance with formula:

$$E_{loc} \sim \frac{3\varepsilon_{out}}{\varepsilon_{Me} + 2\varepsilon_{out}} E_0, \tag{9}$$

where ε_{Me} and ε_{out} are permittivities of the inclusion and the outer medium, respectively. ε_{out} in the numerator of equation (9) acts as a general scale factor for both resonant and non-resonant fields. In the case of a core-shell NP, the shell can be treated as an effective outer medium, which permittivity monotonically grows up to ε_{shell} with coating thickness. Note that this effect does not depend on the SPR position and, therefore, it is of general nature. Moreover, in a more lossy metal, like gold, it dominates the resonant enhancement. We discussed this effect in details in our previous paper [15].

3.2. Thick shell case

Above we considered thin, up to a few particles radii, shells. Such shells affect NPs plasmonic properties while the electromagnetic fields are not fully localized in the coating layer and the NP feels the surrounding depending on the shell thickness. A different kind of influence takes place in a thick, about an order of optical wavelengths, shells because of spatial Mie resonances. In figure 2 we present the dependence of the SH intensity on shell thickness at fundamental wavelength λ=1000 nm of a covered (ε_{shell}=4.0) spherical, 5 nm in radius, NP of Drude-metal (λ_p=200 nm, γ=3 · 10^{13}sec^{-1}). In all the further calculations the metal core is the same. Essentially, figure 2 represents an extension of figure 1b.

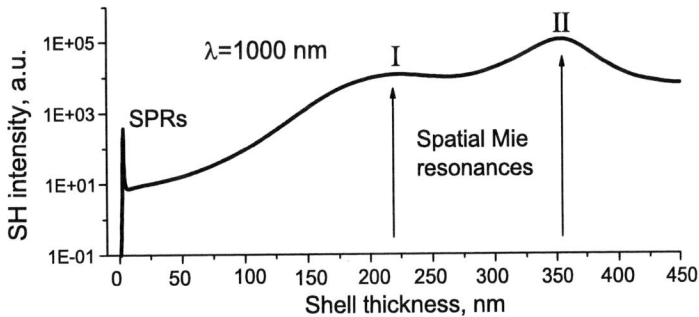

Figure 2. SH intensity vs. shell thickness, fundamental wavelength λ=1000 nm.

Two thick-shell peaks in figure 2, marked with I and II, are the spatial Mie resonances. Peak I corresponds to the spatial resonance of the SH wave, whereas peak II corresponds to the spatial resonances of both SH and fundamental waves (the SH wave resonates in the next order). Important to note that the efficiency of these Mie resonances surpasses the plasmonic enhancement (the thin-shell peaks) by 1.5-2 orders. This follows from the well-known general dependence of the SHG intensity: $W_{2\omega} \sim |E_{2\omega}|^2 |E_\omega|^4$ [3]. The SPR enhancement relates only to one factor (the SH field $E_{2\omega}$ in this case) if the NP is not specifically tailored [16], whereas spatial resonance II affects both fields at once; basically, the 6th power beats the 2nd. Thus, thicker coating is more efficient for the SHG regardless the SPR position, and about 5-6 orders total increase compared to non-covered NPs can be provided.

The shell which thickness is above a few NPs radii barely affects NPs plasmonic properties and the SPR position in particular. This offers an appealing possibility to combine the plasmonic (metal) enhancement with the spatial (dielectric) one by specific choice of the excitation wavelength. In figure 3 we demonstrate dependences of the SH intensity on the shell thickness at fundamental wavelengths λ=603 nm (dipole SPR wavelength of the NP covered with the thick shell with ε_{shell}=4.0) and λ=1206 nm (doubled dipole SPR wavelength).

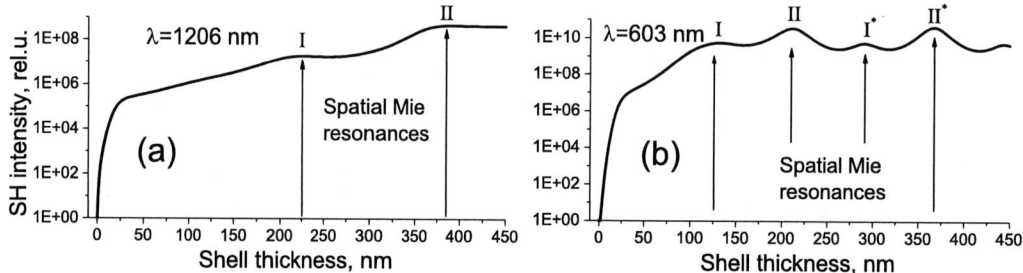

Figure 3. The SH intensity vs. shell thickness, fundamental wavelengths λ=1206 nm (a) and λ=603 nm (b). The dependences are normalized by the SHG by the non-covered NP.

The marks, I and II, of the peaks refer to ones introduced in figure 2 (marks I* and II* in figure 3b indicates the next orders of the corresponding resonances). Note that in figure 3 we normalize the dependences by the SH signal from the non-covered NP. In the case of combining the resonances the advantage in the SHG is more significant: almost 8 orders when the SH wave is SPR-enhanced and 10 orders when the fundamental wave is SPR-enhanced. However, one should note that adjusting the fundamental wavelength to the SPR results in the SH wave being ultraviolet (UV), and many dielectric materials typically have substantial losses in the UV region. We consider this effect introducing an imaginary part to the dielectric permittivity of the shell at the SH frequency: $\varepsilon_{shell}(2\omega)$=4.0+3.0i, the dielectric permittivity at fundamental frequency being the same: $\varepsilon_{shell}(\omega)$=4.0. In this case, the dependence of the SH intensity on the shell thickness at fundamental wavelengths λ=603 nm is presented in figure 4.

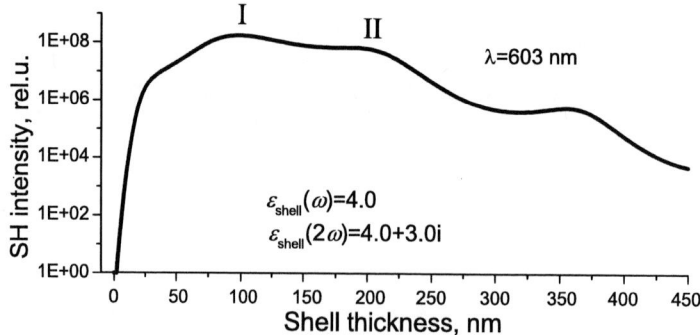

Figure 4. The SH intensity vs. shell thickness, fundamental wavelength λ=603 nm. The losses in the shell at the SH frequency are introduced: $\varepsilon_{shell}(2\omega)$=4.0+3.0i; $\varepsilon_{shell}(\omega)$=4.0.

The SH intensity drastically drops for thicker coatings if the absorption in the shell is taken into account. Therefore, despite the fact that the SPR enhancement of the fundamental wave provides more intense SHG, the UV losses of the shell can supress this effect.

4. Conclusion

We theoretically analyzed the influence of the dielectric cover on the efficiency of the SH scattering by spherical metal NPs. The possibility to adjust SPR to a certain wavelength by covering a NP with a thin shell was demonstrated. A few orders non-resonant growth of the SHG resulted from the increase of the NPs local fields under the thin coating was shown. We considered thicker, up to an order of optical wavelengths, shells supporting spatial Mie resonances, which efficiency surpasses the SPR impact in the SHG. Moreover, since the thicker shell does not affect the NPs SPR position, the SPR can be easily combined with the spatial Mie resonances. Advantage in the SHG in this case relatively to non-coated NPs was evaluated: up to 8 orders when the SH wave is SPR-enhanced and about 10 orders when the fundamental wave is SPR-enhanced. However, we demonstrated that in the latter case the SHG efficiency can be supressed by UV losses which are typical for many dielectrics.

Acknowledgments

This study was funded by RFBR according to the research project 18-32-00097

References

[1] Eisenthal K B 2006 Second harmonic spectroscopy of aqueous nano- and microparticle interfaces *Chem. Rev.* **106** 146277
[2] Kauranen M and Zayats A 2012 Nonlinear plasmonics *Nat. Photonics* **6** 73748
[3] Shahbazyan T V and Stockman M I 2013 *Plasmonics: Theory and Applications* (Dordrecht: Springer Netherlands)
[4] Heisler F, Babich E, Scherbak S, Chervinskii S, Hasan M, Samusev A and Lipovskii A 2015 Resonant Optical Properties of Single Out-Diffused Silver Nanoislands *J. Phys. Chem.* C **119** 266927
[5] Wang Z L and Cowley J M 1987 Size and shape dependence of the surface plasmon frequencies for supported metal particle systems *Ultramicroscopy* **23** 97107
[6] Kelly K, Coronado E, Zhao L and Schatz G 2003 The optical properties of metal nanoparticles: The influence of size, shape, and dielectric environment *J. Phys. Chem.* B **107** 66877
[7] Noguez C 2007 Surface Plasmons on Metal Nanoparticles: The Influence of Shape and Physical Environment *J. Phys. Chem.* C **111** 380619
[8] Okamoto T, Yamaguchi I and Kobayashi T 2000 Local plasmon sensor with gold colloid monolayers deposited upon glass substrates *Opt. Lett.* **25** 3724
[9] Scherbak S, Kapralov N, Reduto I, Chervinskii S, Svirko O and Lipovskii A 2017 Tuning Plasmonic Properties of Truncated Gold Nanospheres by Coating *Plasmonics* **12** 190310
[10] Butet J, Russier-Antoine I, Jonin C, Lascoux N, Benichou E and Brevet P-F 2012 Nonlinear Mie theory for the second harmonic generation in metallic nanoshells *J. Opt. Soc. Am.* B **29** 2213-21
[11] Gonella G and Dai H L 2011 Determination of adsorption geometry on spherical particles from nonlinear Mie theory analysis of surface second harmonic generation *Phys. Rev.* B **84** 15
[12] Dadap J I, Shan J and Heinz T F 2004 Theory of optical second-harmonic generation from a sphere of centrosymmetric material: small-particle limit *J. Opt. Soc. Am.* B **21** 132847
[13] Bohren C F and Huffman D R 1983 *Absorption and Scattering of Light by Small Particles* (New York: Willey)
[14] Aden A L and Kerker M 1951 Scattering of electromagnetic waves from two concentric spheres *J. Appl. Phys.* **22** 12426
[15] Chervinskii S, Koskinen K, Scherbak S, Kauranen M and Lipovskii A 2018 Nonresonant Local Fields Enhance Second-Harmonic Generation from Metal Nanoislands with Dielectric Cover *Phys. Rev. Lett.* **120** 113902
[16] Thyagarajan K, Rivier S, Lovera A and Martin O J F 2012 Enhanced second-harmonic generation from double resonant plasmonic antennae *Opt. Express* **20** 12860

SPBOPEN 2018

Laser-deposited hybrid Au-Ag@C nanoparticles as efficient SERS & adsorption material

A A Vasileva[1,2], D V Pankin[2], I E Kolesnikov[2], S S Rzhevskiy[1] and A A Manshina[1]

[1]Institute of Chemistry, Saint Petersburg State University, Universitetskii pr. 26, Saint Petersburg 198504 Russia
[2]Center for Optical and Laser Materials Research, Saint Petersburg State University, Uljanovskaya 5, Saint Petersburg 198504 Russia

Abstract. In this work, we present an approach to the synthesis of hybrid multifunctional nanomaterials with pronounced functional properties (SERS, adsorption properties). One-step laser-induced deposition method was used to synthesize hybrid Au-Ag@C nanoparticles. Physicochemical and functional properties of obtained nanostructures were studied using various experimental techniques. Adsorption properties were studied via Raman spectroscopy using organic chromophore – rhodamine 6G.

1. Introduction

Carbon/metal hybrids are regarded as a new family of multifunctional nanomaterials, which are highly promising for multifarious applications thanks to enhanced characteristics resulting from synergetic effect of components, mutual influence of their electronic structures, their physical and chemical interaction. The main impact of nanometal phase of such hybrid materials is typically governed by localized surface-plasmon resonance to give surface-enhanced Raman scattering (SERS) [1-3]. Silver nanoparticles show the largest effect in these phenomena but gold is also considered as a prospective component due to chemical stability and compatibility with various target materials (especially biological), that stimulates growing attention to bimetallic Au-Ag nano-aggregates and associated expansion of the application areas [4,5]. The carbon phase functionality is determined by its electronic structure related to both allotropic form and morphology. These observations clearly point to various ways of bimetallic (in particular Au-Ag)/carbon hybrid nanomaterials applications, which can cover a wide range of sensing technologies.

The method used for obtaining of multiphase nanomaterials is laser-induced self-organization processes. Unlike other known techniques for synthesizing multiphase nanomaterials [6,7], this approach is a one-step process based on the effect of laser irradiation at the interface between the solid substrate and a solution, resulting in the formation of hybrid nanostructures (metallic particles embedded in a carbonaceous matrix) of well-defined morphology and structure. The laser irradiation influences to molecules of organometallic complex dissolved in a chosen solvent, and as a result, the organometallic molecules destruct with formation of hybrid nanoparticles. In dependence on experiment geometry the nanoparticles may be deposited onto different types of substrates: cover glass, ITO films, anodic aluminium oxide nanoporous membranes, silicon, etc [8,9]. The morphology

Content from this work may be used under the terms of the Creative Commons Attribution 3.0 licence. Any further distribution of this work must maintain attribution to the author(s) and the title of the work, journal citation and DOI.
Published under licence by IOP Publishing Ltd

of obtained nanostructures mainly depends on choice of solvent as well as time of laser deposition procedure. The composition of nanostructures is determined by type of used organometallic complex. The metal loading of nanostructures depends on the solubility of complex in a chosen solvent and deposition time.

2. Experimental part

The experimental work included two main steps: synthesis of hybrid nanoparticles and investigation of their functional properties. Both of them are discussed below.

2.1. Laser-induced deposition of hybrid Au-Ag@C nanoparticles

The solution of the organometallic complex $[Au_{12}Ag_{12}(C_2Ph)_{18}Cl_3(PPh_2(C_6H_4)_3PPh_2)](PF_6)_3$ in dichlorethane was used for the laser-induced deposition procedure. The methodology of the organometallic complex synthesis is presented in [10]. The concentration of organometallic complex was 4 mg/ml that is close to the solubility limit. To enchance the dissolving process the ultrasonic bathing of the solution for 5 min was utilized. The undissolved species were removed with centrifugation of the solution for 5 min at 10000 rpm. Then the cuvette with the organometallic solution was covered with substrate and illuminated with laser beam. Figure 1 shows the laser-induced deposition procedure. As a light source He-Cd laser (CW, $\lambda = 325$ nm and P = 15 mW) was chosen as the laser wavelength coincides with the organomenallic complex absorption band [11]. The laser-deposition time was varied from 5 to 15 min.

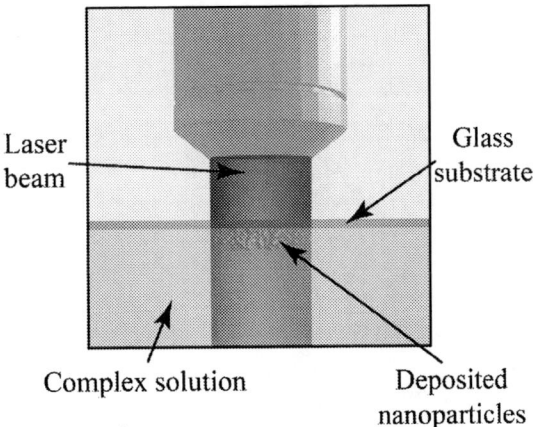

Figure 1. Scheme of experimental procedure for thelaser-induced deposition process

2.2. Characterization of NPs

The morphology and composition of obtained nanostructures were investigated by scanning electron micrographs (SEM) and energy-dispersive X-ray spectroscopy (EDX) on Zeiss Merlin Scanning Electron Microscope with Field Emission cathode, GEMINI II electron-optics column, oil-free vacuum system. As the synthesized samples could contain carbon, the SEM images were obtained with charge compensation to avoid the preliminary carbon spattering of the studied sample.

To study the availability of the deposited NPs for the SERS analysis the solution of rhodamine 6G was chosen as a model substance. The concentration of rhodamine 6G in acetone was 10^{-6} M. The SERS spectra presented in this paper were obtained with confocal express Raman spectrometer Senterra (Bruker) with 50 μm iris aperture set and spectral resolution about 3 cm^{-1}. The Stokes part of Raman spectra was recorded in the region 80-4500 cm^{-1}. Solid-state laser with 532 nm wavelength and 0.2 mW power was used for excitation of Raman scattering. The adsorption properties were also studied by Raman spectroscopy. The substrate with hybrid nanoparticles was immersed in rhodamine

6G solution for defined time. Then the Raman spectrum was measured. After that, the immersion process and Raman spectrum measurement were repeated for several times.

3. Results and discussion

3.1. Characterization of hybrid Au-Ag@C nanostructures
The irradiation of the interface of the substrate and organometallic solution results in formation of the agglomerated nanoparticles that form rather porous layer (Figure 2a). It is important to note that the deposition time increase effects the number of the deposited nanoparticles and the morphology of the layer on the whole. The most porous structure is formed as a result of 10 min irradiation, further increase of the irradiation time results in formation of dencier packed nanoparticles.

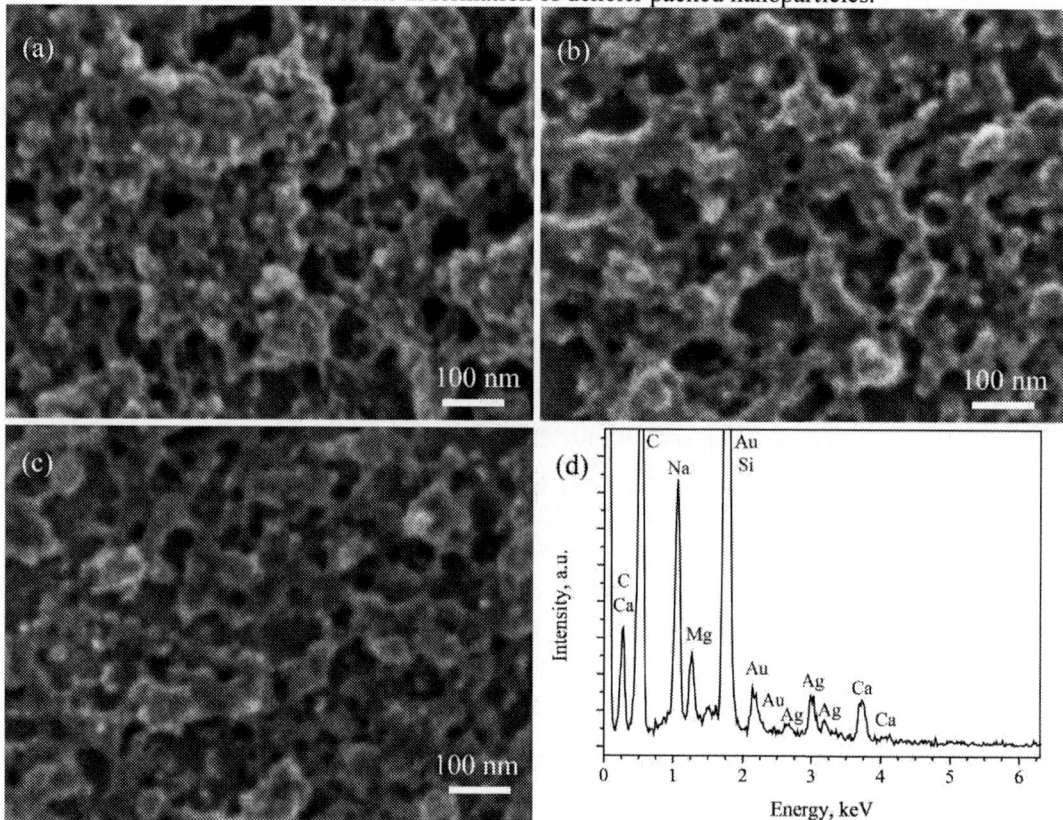

Figure 2. Scanning electron micrographs of hybrid Au-Ag@C nanostructures deposited during (a) 5 min, (b) 10 min, (c) 15 min, (d) typical EDX spectra from area with nanoparticles.

The composition of metal clusters was checked with EDX. The results are presented in Figure 2d. It shows pronounced peaks of C, Au and Ag that originate from the deposited nanoparticles, and in a good agreement with the composition of the organometallic precursor while Ca, Na, Mg and Si signals originate from the glassy substrate. The absorption spectrum of the deposited nanoparticles demonstrate single pronounced peak at 540 nm testifying to the bimetal Au-Ag nanoinclusions in the structure of the deposited nanoparticles, weak absorption band at 470 nm can indicate also presence of some amount of silver nanoinclusions (Figure 3).

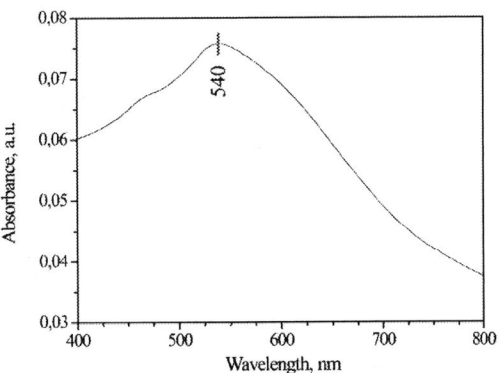

Figure 3. Absorption spectrum of deposited nanostructures

3.2. SERS and Adsorption properties of the deposited nanostructures

The chosen for SERS measurements laser wavelength is quite closed to the position of surface plasmon resonance of obtained nanostructures. The measurements were carried out for different adsorption times. "0 min" means that substrate with nanostructures was put out of Petri dish immediately after put in. The SERS spectra for different times of adsorption are presented in Figure 4a.

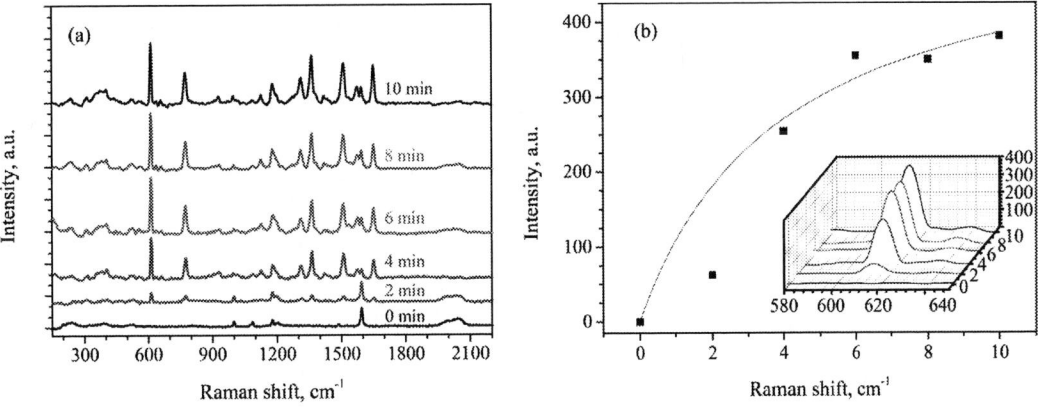

Figure 4. (a) SERS of rhodamine 6G with different adsorption times; (b) Isotherm of adsorption on hybrid Au-Ag@C nanostructures.

To determine the dynamics of adsorption effect, the dependence of SERS signal at 610 cm^{-1} (related to C-C-C ring in-plane bending mode [12]) on the adsorption time was examined. This peak has several advantages for quantitative estimations. Among them it has relatively high intensity, comparing with other ones of rhodamine 6G and there are absent of other peaks nearby which can influence on the 610 cm^{-1} peak intensity. The intensity of SERS signal shows proportional behaviour with adsorption time that seems to be proportional of adsorbed molecules amount. In assumption that in case of such low concentrations (10^{-6} M) adsorbs monolayer of analyte and there is no interaction between adsorbed molecules, the obtained data was approximated with Langmuir isotherm (Figure 4b).

$$I = \Gamma_\infty \frac{kT}{1+kT} \qquad (1)$$

where I is intensity of 610 cm^{-1} peak, T is time (in minutes in our case), k is adsorption velocity, Γ_{∞} is adsorption limit value. For the fitted curve presented at figure 4b, the obtained parameters are: Γ_{∞}=537, k = 0.254 min^{-1}.

From presented data, we can observe that saturation of SERS-active substrate is achieved after 6 min. Because of uniform distribution of metal clusters in carbonaceous matrix takes place the synergetic effect of adsorption and SERS.

4. Conclusion

In conclusion, we have demonstrated that laser-induced deposition method allows depositing of hybrid Au-Ag@C nanoparticles with controlled morphology possessing both adsorption and SERS properties. Optimal time for synthesis of hybrid Au-Ag@C nanoparticles with high-developed surface area was determined to be 10 min. Isotherm of low concentrated rhodamine 6G adsorption on hybrid Au-Ag@C nanostructures was successfully fitted with Langmuir model. Combination of metal and carbon phases is promising for decrease of the detection limit of various chemical substances.

Acknowledgements

This work was supported by the RFBR grant # 17-03-01284. The experimental investigations were performed at Research Park of Saint Petersburg State University at "Centre for Optical and Laser materials research" and "Interdisciplinary Resource Centre for Nanotechnology".

References

[1]   Fateixa S, Helena N and Tito T 2015 Physical Chemistry Chemical Physics 17(33) 21046.
[2]   Wang C, Tang F, Wang X and Li L 2015 ACS applied materials & interfaces 7(24) 13653.
[3]   Kolesnikov I E, Ivanov D A, Kireev A A, Mamonova D V, Golyeva E V, Mikhailov M D and Manshina A A 2018 Journal of Solid State Chemistry 258 835.
[4]   Wu H, Wang P, He H and Jin Y 2012 Nano Research 5(2) 135.
[5]   Kumari M M, Jacob J and Philip D 2015 Spectrochimica Acta Part A: Molecular and Biomolecular Spectroscopy 137 185.
[6]   Sahay R, Reddy V J and Ramakrishna S 2014 Journal of Mechanical and Materials Engineering 9(1) 25.
[7]   Povolotskaya A V, Povolotskiy A V and Manshina A A 2015 Russian Chemical Reviews 84(6) 579.
[8]   Bashouti M Y, Povolotckaia A V, Povolotskiy A V, Tunik S P, Christiansen S H, Leuchs G and Manshina A A 2016 RSC Advances 6(79) 75681.
[9]   Bashouti M Y, Manshina A, Povolotckaia A, Povolotskiy A, Kireev A, Petrov Y and Christiansen S 2015 Lab on a Chip 15(7) 1742.
[10] Koshevoy I O, Karttunen A J, Tunik S P, Haukka M, Selivanov S I, Melnikov A S, Serdobintsev P Yu and Pakkanen T A 2009 Organometallics, 28(5), 1369.
[11] Manshina A A, Grachova E V, Povolotskiy A V, Povolotckaia A V, Petrov Y V, Koshevoy I O, Makarova A A, Vyalikh D M and Tunik S P 2015 Scientific reports 5 12027.
[12] Hildebrandt P and Stockburger M 1984 The Journal of Physical Chemistry 88(24) 5935.

Influence of sputtering parameters on the main characteristics of ultra-thin vanadium nitride films

P I Zolotov[1, 2, 3,] A V Divochiy[2], Yu B Vakhtomin[2, 3], A V Lubenchenko[4], P V Morozov[2], I V Shurkaeva[4], K V Smirnov[1, 2, 3]

[1]National Research University Higher School of Economics, Moscow 101000, Russia
[2]LLC «Superconducting Nanotechnology» (SCONTEL), Moscow 119021, Russia
[3]Moscow State Pedagogical University, Moscow 119991, Russia
[4]National Research University Moscow Power Engineering Institute, Moscow 111250, Russia

Abstract. We researched the relation between deposition and ultra-thin VN films parameters. To conduct the experimental study we varied substrate temperature, Ar and N_2 partial pressures and deposition rate. The study allowed us to obtain the films with close to the bulk values transition temperatures and implement such samples in order to fabricate superconducting single-photon detectors.

1. Introduction

Over the last years dramatic influx of interest was received by the field of superconducting devices. They already found its place in many important research areas such as radio astronomy and quantum processing experiments [1, 2]. In all of these devices the key role is given to a thin (<100 nm) superconducting films which are fabricated in different incarnations. Main parameters of the films such as superconductor energy gap (Δ), resistivity (ρ) and thickness (d) form a device specifications and by that its implementation area. Many decades niobium nitride (NbN) was the material of choice for such applications because of its relatively high transition temperature (T_c), which has been demonstrated on a wide range of substrate materials and deposition techniques. This material happens to have a wide range of each parameter [3]. Although it was discovered recently that only specific range of the NbN film parameters leads to the high performance of superconducting devices, in particular – superconducting single-photon detectors (SSPD) [4]. It was also found that the characterization of the NbN superconducting properties could be sophisticated due to the presence of few upper and under layers of the film [5], and because of that, most of the material properties calculations, which are thickness-dependent, are either unreliable or complex to perform. To test the versatility of these recent discoveries, we researched a new material – vanadium nitride. The material shows no strong correlation between the deposition and superconducting parameters, and has a relatively high T_c with respect to popular amorphous material such as WSi [6, 7]. In this work we show that VN has a potential as SSPD material as well as its peculiar properties which need further understanding.

Content from this work may be used under the terms of the Creative Commons Attribution 3.0 licence. Any further distribution of this work must maintain attribution to the author(s) and the title of the work, journal citation and DOI.
Published under licence by IOP Publishing Ltd

2. Production and experimental study

The properties of superconducting vanadium nitride films were studied earlier [6, 8]. Although, this material has never been used for superconducting devices fabrication, such as hot-electron bolometers (HEB) and SSPDs, before our investigation. Therefore, there is no information related to research of ultra-thin VN films and its features. Our study was focused on the films with thicknesses <15 nm, which we used for SSPD fabrication [9]. And the films 15-40 nm thick were used to define main characteristics of the films avoiding its suppression by the thickness [10].

As well as many other materials, we deposited our films using reactive magnetron sputtering. Deposition was performed in AJA International Inc. Orion series system using vanadium (99.9 %) 2" target that was sputtered in Ar and N_2 atmosphere. The deposition was performed onto commonly-used Si, Si/SiO_2 and Si/Si_3N_4 substrates with the typical background pressure of $<8 \cdot 10^{-8}$ Torr. In order to study the dependencies of the main thin films characteristics on deposition parameters we changed total pressure (P_{tot}), N_2 concentration, substrate temperature (T_{sub}) and deposition rate (by discharge current). These parameters are considered to influence the stoichiometry of the films and are precisely controlled in case of NbN deposition in order to obtain films with desired parameters [3, 11]. Besides standard for thin and ultra-thin films sheet resistance (R_s) dependency of T_c, new tendencies, which are related to study of the disordered films, also require obtaining the T_c dependency on residual resistivity ratio (RRR=R_{300K}/R_{20K}) and ρ ($\rho = R_s /d$) [4, 12]. These dependencies for the VN ultra-thin films are presented on figure 1. To obtain this data, the variation of P_{tot} (in the range 0.5 – 9 mTorr), T_{sub} (400-800°C) and N_2 concentration (11-28 %) in the gas mixture was performed, as well as the variation of the deposition time of the films. As it follows from the graphs, these changes did not have strong influence on the films parameters. However, we should note that the highest values of T_c were obtained at higher deposition rates (higher discharge currents) of ~3 nm/min. Obtained maximal T_c of 8.3 K for ~34 nm thick film is very similar to the one obtained with ≥0.5 μm films in works [6, 8]. The major difference of the VN films RRR with respect to the NbN films with thicknesses of <10 nm could be noticed on figure 1b. Even for the films with suppressed values of the transition temperature, the RRR remained above 1.

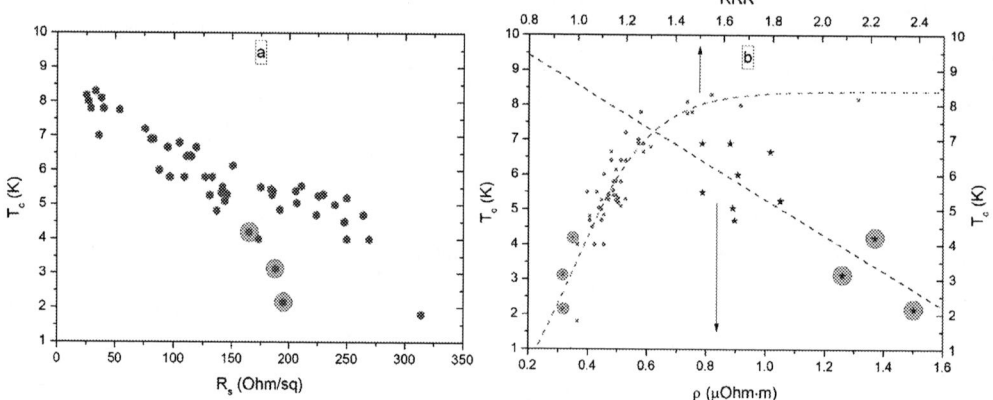

Figure 1. Transition temperature dependencies on the main VN films parameters: **(a)** sheet resistance; **(b)** resistivity and residual resistivity ratio. Green circles highlight the films, which were deposited in pure N_2 atmosphere

To research the extreme way of changing the films characteristics, we excluded the Ar from the gas mixture and performed the series of films depositions in N_2 atmosphere. This experiment resulted in films (highlighted on fig 1 by the green circles) that differ from the majority of the data points by its parameters. In overall these tests showed that thin VN films have only slight change of its parameters even when the deposition process is drastically changed. We believe that such behavior could be related to the formation of the single-phase material among all of the deposition parameters. This contrasts with the well-known phase mixture that exists in thin NbN films deposited with non-optimal parameters or substrates which do not ensure agreement between the lattices [11]. This result could benefit to the VN films over other polycrystalline materials such as NbN (NbTiN), TaN and TiN.

As it was pointed out in many works, in case of ultra-thin superconducting films there are upper and under layers of the superconducting stratum [4, 5]. Our results on the resistivity measurements were realized with the data obtained by X-ray Photoelectron Spectroscopy (XPS), which allows the definition of depth profile of the films [5, 13] (excluding the cases when the film thickness is more than 13 nm). The novelty and reliability of the method bring major refinements in the main superconducting parameters calculations and therefore is well appreciated, especially in case of the new thin-film materials. In our experiments it allowed to utilize the thickness data more precise than the standard AFM-based measurements. For example, in 6 nm deposited film only 2.4-3 nanometers remained supposedly superconductive (see Table 1) and therefore only this thickness should be considered in ρ calculations.

XPS study of the films was performed with the help of the electron-ion spectroscopy module based on Nanofab 25 (NT-MDT) platform. In the analysis chamber was an ultrahigh oil-free vacuum about 0.75 nTorr. All spectra were recorded with use of Mg anode of the X-ray source without a monochromator. The spectra were recorded with an electrostatic hemispherical energy analyzer SPECS Phoibos 225. All survey spectra scans were recorded at a pass energy of 80 eV. The detailed scans of distinctive lines were in most cases recorded as wide as needed just to encompass the peak(s) of interest and were obtained with a pass energy of 20 eV.

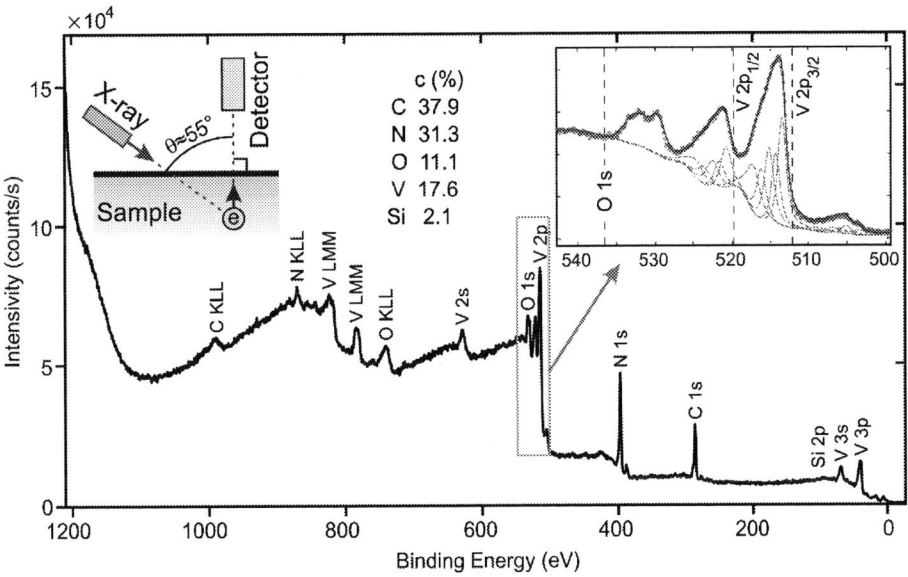

Figure 2. XPS survey spectrum of a VN film. Enlarged image in the upper right corner: decomposition of lines into partial peaks; image in the upper left corner: principle scheme of the experiment; in the upper middle relative atomic concentrations are presented.

Non-destructive chemical and phase depth profiling of the nano-scaled films in this investigation was carried out on the base of method [13], that includes a new approaches for background subtraction of multiple inelastically scattered photoelectrons considering depth inhomogeneity of electron inelastic scattering; a new method of photoelectron line decomposition into component peaks considering physical nature of different decomposition parameters; joint solution of the background subtraction and photoelectron line decomposition problems; control of line decomposition accuracy with the help of a suggested performance criterion; calculation of layer thicknesses for a multilayer target using a simple formula. In total, this method enables to determine depth profiles with sub-monolayer accuracy using XPS data.

Figure 2 displays experimental XPS data as calculated using the mentioned method. The line N1s is decomposed into three peaks corresponding to different phases of vanadium nitride: VN, VN_x and VNO_x (not shown). The line Si2p is decomposed into three peaks: Si, SiO_2 and VN_y-SiO_x (not shown).

Table 1 shows computed results for the thicknesses following the formula from work [13], which accounts for partial intensities obtained by decomposition of the lines N1s, C1s and V2p.

Table 1. Chemical and phase depth profiling of a VN film on Si substrate.

	d, nm	Formula
Σ (deposited)	6.04	
8	0.58	hydrocarbons
7	0.75	V_2O_5
6	0.42	VO_2
5	0.6	VNO_x
4	0.56	VN_x
3	2.45	VN
2	1.26	VN_y-SiO_x
1	2.67	SiO_2
Substrate (0)	Inf	Si

The data above shows that there is an interface between the vanadium nitride layer (#3) and the substrate (#0, #1), as between the vanadium nitride and the oxide layers (#6, #7). These transition layers contain different phases (layers #2, #4, #5). Such result is obtained for the first time.

In order to study the possibility of obtained VN films implementation in superconducting nanoelectronics, we fabricated SSPDs out of the films with different T_c and R_s. We used the same fabrication route as for our NbN SSPDs [4, 14] with one slight modification – instead of PMMA A3 electron-beam resist we used ZEP 520 A7 because of its higher resistance to plasma-chemical etching process in SF_6, which proved to be less selective to VN with respect to the NbN films. The geometries of the devices varied in nanostripe width and length. The widths were 100-130 nm, and the active area was located on 3x3, 7x7, 11x11 or 15x15 μm^2 areas with fill factor ~0.6.

Device characterization consisted of critical current (I_c) measurements of the batches that showed single-photon response, from which we then calculated critical current density (J_c). With respect to the XPS data, the average value of J_c was ~$1 \cdot 10^{10}$ A/m^2. This value was uncharacteristically increasing along with the nanostripe length. This phenomenon was explained by the latching problem of the SSPD, that was observed on NbN detectors before [15]. The latching disappeared at certain kinetic inductance (L_{kin}) of the nanostripe of ~ 1.5 µH, which was extracted from detector pulse shape using oscilloscope (figure 3a). To make further analysis of the VN specific kinetic inductance we studied the dependency of L_{kin} on the detectors normal-state resistance, which showed its universality in work [16]. It appeared that in average, VN devices have 2.3 times larger kinetic inductance with respect to NbN-based devices – figure 2b). Perhaps this fact could be related to the difference in the magnetic field penetration depth of these materials. Indeed, this parameter could decrease the maximal counting rates of the SSPD [15], but in contrast could benefit in superconducting imagers that were presented recently [17].

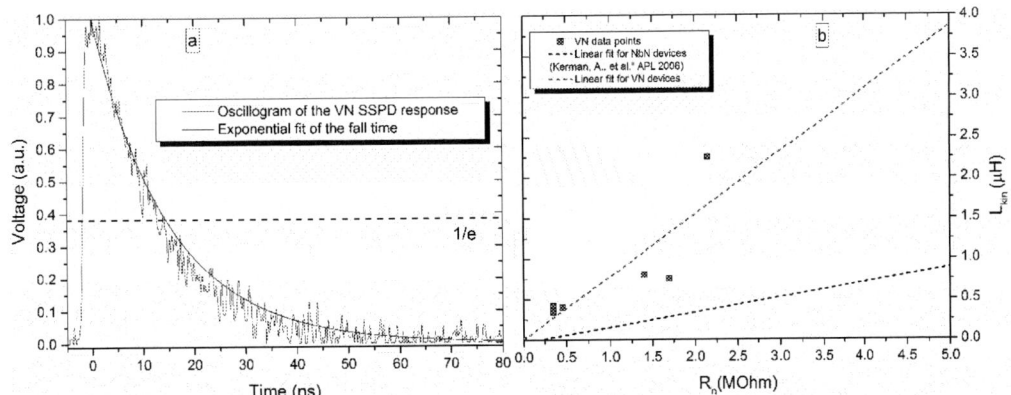

Figure 3. (a) The single-photon response of VN SSPD and its exponential decay. Dotted line marks the amplitude decrease down to 1/e. (b) SSPD resistance dependency of kinetic inductance. The kinetic inductance for VN devices was calculated indirectly with the method presented in work [16] and was mapped with the linear fit of the data for NbN devices from the same work. Each VN data point corresponds to the approximated value within the batch (~15 devices with the same geometry).

3. Conclusion

We have demonstrated main superconducting vanadium nitride films (<50 nm) parameters. The T_c dependencies on the main characteristics of the films, such as R_s, ρ and RRR were obtained. Our investigation showed that characteristics of VN films remain stable along with the variation of most of the sputtering parameters. In addition to that we applied novel method of films thickness and compound characterization. Obtained VN films were implemented in superconducting single-photon detectors and its peculiar properties were discovered.

Acknowledgments

The work is supported by the Russian Science Foundation (RSF) Project No. 18-12-00364.

References

[1] Huebers, H.W., Semenov, A., Schubert, J., Gol'tsman, G.N., Voronov, B.M., Gershenzon, E.M., Krabbe, A. and Roeser, H.P., 2000, June. NbN hot-electron bolometer as THz mixer for SOFIA. In Airborne Telescope Systems (Vol. 4014, pp. 195-203). International Society for Optics and Photonics.

[2] Hadfield, R.H., 2009. Single-photon detectors for optical quantum information applications. Nature photonics, 3(12), p.696.

[3] Chockalingam, S.P., Chand, M., Jesudasan, J., Tripathi, V. and Raychaudhuri, P., 2008. Superconducting properties and Hall effect of epitaxial NbN thin films. Physical Review B, 77(21), p.214503.

[4] Smirnov, K., Divochiy, A., Vakhtomin, Y., Morozov, P., Zolotov, P., Antipov, A. and Seleznev, V., 2018. NbN single-photon detectors with saturated dependence of quantum efficiency. Superconductor Science and Technology, 31(3), p.035011.

[5] Lubenchenko, A.V., Batrakov, A.A., Krause, S., Pavolotsky, A.B., Shurkaeva, I.V., Ivanov, D.A. and Lubenchenko, O.I., 2017, November. XPS Depth Profiling of Air-Oxidized Nanofilms of NbN on GaN Buffer-Layers. In Journal of Physics: Conference Series (Vol. 917, No. 9, p. 092001). IOP Publishing.

[6] Gray, K.E., Kampwirth, R.T., DW Capone, I.I., Vaglio, R. and Zasadzinski, J., 1988. Superconducting properties of VN x sputtered films including spin fluctuations and radiation damage of stoichiometric VN. Physical Review B, 38(4), p.2333.

[7] Seleznev, V.A., Divochiy, A.V., Vakhtomin, Y.B., Morozov, P.V., Zolotov, P.I., Vasil'ev, D.D., Moiseev, K.M., Malevannaya, E.I. and Smirnov, K.V., 2016, August. Superconducting detector of IR single-photons based on thin WSi films. In Journal of Physics: Conference Series (Vol. 737, No. 1, p. 012032). IOP Publishing.

[8] Zasadzinski, J., Vaglio, R., Rubino, G., Gray, K.E. and Russo, M., 1985. Properties of superconducting vanadium nitride sputtered films. Physical Review B, 32(5), p.2929.

[9] Zolotov, P., Divochiy, A., Vakhtomin, Y., Seleznev, V., Morozov, P. and Smirnov, K., 2018. Superconducting Single-photon Detectors Made of Ultra-thin VN Films. KnE Energy & Physics, 3(3), pp.83-89.

[10] Kang, L., Jin, B.B., Liu, X.Y., Jia, X.Q., Chen, J., Ji, Z.M., Xu, W.W., Wu, P.H., Mi, S.B., Pimenov, A. and Wu, Y.J., 2011. Suppression of superconductivity in epitaxial NbN ultrathin films. Journal of Applied Physics, 109(3), p.033908.

[11] Marsili, F., 2009. Single-photon and photon-number-resolving detectors based on superconducting nanowires.

[12] Burdastyh, M.V., Postolova, S.V., Baturina, T.I., Proslier, T., Vinokur, V.M. and Mironov, A.Y., 2017. Superconductor–Insulator Transition in NbTiN Films. JETP Letters, 106(11), pp.749-753.

[13] Lubenchenko A. V. et al. 2017 Applied Surface Science. 427. 711–721.

[14] Zolotov, P.I., Divochiy, A.V., Vakhtomin, Y.B., Morozov, P.V., Seleznev, V.A. and Smirnov, K.V., 2017, November. Development of high-effective superconducting single-photon detectors aimed for mid-IR spectrum range. In Journal of Physics: Conference Series (Vol. 917, No. 6, p. 062037). IOP Publishing.

[15] Yang, J.K., Kerman, A.J., Dauler, E.A., Anant, V., Rosfjord, K.M. and Berggren, K.K., 2007. Modeling the electrical and thermal response of superconducting nanowire single-photon detectors. IEEE transactions on applied superconductivity, 17(2), pp.581-585.

[16] Kerman, A.J., Dauler, E.A., Keicher, W.E., Yang, J.K., Berggren, K.K., Gol'Tsman, G. and Voronov, B., 2006. Kinetic-inductance-limited reset time of superconducting nanowire photon counters. Applied physics letters, 88(11), p.111116.

[17] Zhao, Q.Y., Zhu, D., Calandri, N., Dane, A.E., McCaughan, A.N., Bellei, F., Wang, H.Z., Santavicca, D.F. and Berggren, K.K., 2017. Single-photon imager based on a superconducting nanowire delay line. Nature Photonics, 11(4), p.247.

SPBOPEN 2018   IOP Publishing
IOP Conf. Series: Journal of Physics: Conf. Series **1124** (2018) 051031   doi:10.1088/1742-6596/1124/5/051031

Dielectric surrounding decimates eigenmodes of microdisk optical resonators

A V Raskhodchikov[1,2], S A Scherbak[1,2], N V Kryzhanovskaya[1,2], A E Zhukov[1,2] and A A Lipovskii[1,2]

[1] St. Petersburg Academic University RAS, St. Petersburg 194021, Russia
[2] Peter the Great St. Petersburg Polytechnic University, St. Petersburg 195251, Russia

Abstract. We performed a numerical study of optical whispering gallery modes in microdisk resonators modified via their embedding in a homogeneous dielectric surrounding or covering with a thin dielectric layer. Mode spectra and electromagnetic field distributions were calculated through the solution of the Helmholtz equation using COMSOL Multiphysics environment. It is shown that the modification results in the decimation of the resonator modes.

1. Introduction

Semiconductor microlasers using resonators with whispering gallery modes (WGMs) are of a high interest for intra-chip optical communications, sensing and quantum informatics [1]. WGMs usually have high quality factor, up to 10^{11} [2], and, therefore, narrow spectral peaks. Numerous researches are aimed at planar structures, first of all microdisks (MD) [3-5], for they are very promising in sense of integration with semiconductor-based electronics. Typically, emission spectra of MD lasers of several microns in diameter contain a rich set of the resonator modes of a high (40-50) azimuthal order and up to 4th or 5th radial order. This restricts information capacity of optical channels where the MD lasers can be used. Nowadays, various methods to select lasing modes or even to achieve single-mode emission are being developed, e.g. etching a groove on a resonators surface [6], piercing a hole through it [7], or applying periodical roughness on sidewalls [8]. However, destructive nature of these techniques limits their generality. The other promising way is to outcouple the resonator with external cavity, waveguide or antenna [9-11], though it typically requires a very fine tuning. In the present study, we consider a simple way of the radial modes decimation via covering the MD resonators with dielectric layers or their embedding in dielectric media.

2. Approach

To analyze WGMs of the MD resonators we performed their numerical simulation in COMSOL Multiphysics environment. The calculations were made using the finite element method. We placed a randomly oriented dipole antenna inside the resonator as an excitation source. The similar approach is widely used in time-domain modeling [12,13]. To analyze mode spectra and mode field distributions we solved frequency-dependent Helmholtz equation

$$\nabla^2 \mathbf{E} + k^2 \mathbf{E} = \mathbf{F}(\mathbf{r}, \omega)$$

for each desired frequency and calculated the energy in the resonator, W, as the integral of the electromagnetic energy density over the resonator volume. When the excitation is close to any eigenfrequency of the resonator, the energy increases greatly. Thus, the energy spectrum coincides with the MD mode spectrum, which being constructed with this approach contains only real, physical

---

Content from this work may be used under the terms of the Creative Commons Attribution 3.0 licence. Any further distribution of this work must maintain attribution to the author(s) and the title of the work, journal citation and DOI.
Published under licence by IOP Publishing Ltd

modes unlike direct numerical solution of the eigenvalue problem [14]. Importantly, this is less demanding for computational power, whereas eigenvalue problem is solved by a memory-intensive direct solver. Besides, we artificially introduced losses in the resonator via setting the imaginary part of its material index, k=5e-5. This provided physically meaningful mode spectral width and allowed us selecting a reasonable wavelength calculations grid, ~0.1 nm, which gave us the possibility to detect all the modes.

We considered 2D geometry (a cylinder of the infinite width), since the phenomena under discussion are qualitatively the same as in 3D, whereas computational burden decreases greatly. Because of the planar geometry of the problem, the exciting dipole locates in the plane. Therefore, **E**-field is in-plane and **H**-field is out-of-plane. Thus, all the modes calculated are of TM kind. Note that the consideration of TE-modes does not add any additional results.

3. Embedding MDs in a homogeneous dielectric medium

We simulated mode spectra of a 6 μm in diameter GaAs MD resonator in the air surrounding. This MD possesses large number of eigenmodes that should be reduced. The examples of |**H**|-field distribution of typical calculated modes up to 3rd radial order are presented in figure 1. The notation of modes is TM(x, y) where x is a radial number, y is an azimuthal number.

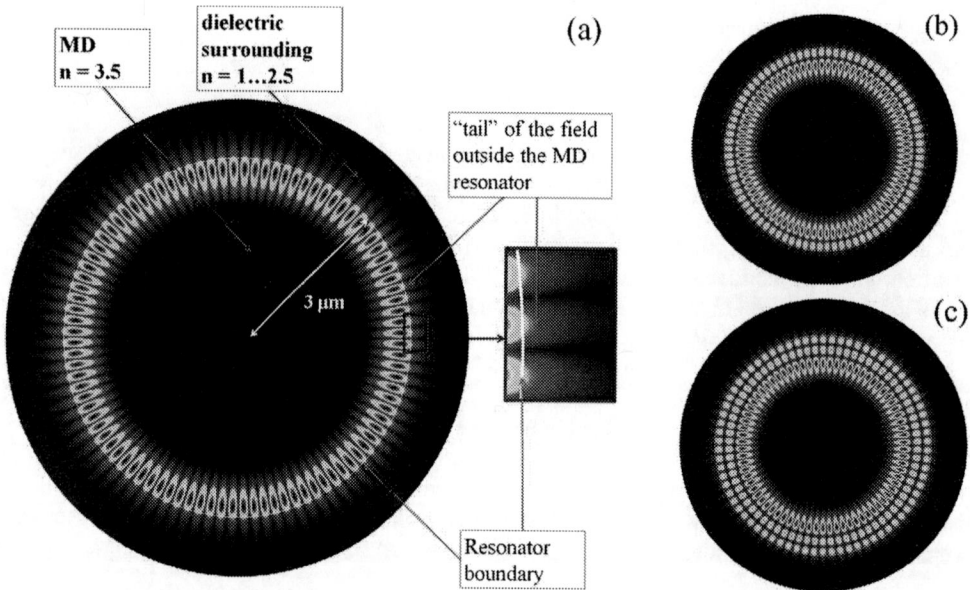

Figure 1. Spatial distribution of |H|-field of the first radial mode TM(1,48) (a), the second radial mode TM(2,42) (b) and the third radial mode TM(3, 38) (c) in the 6 μm in diameter microdisk resonator.

Next, we varied refractive index of the outer medium, n_{out}, from 1 to 2.5 (index of the MD is 3.5) and simulated mode spectra of the MD resonator embedded in this medium. The calculated spectra of the electromagnetic energy inside the resonator are presented in figure 2a. The latter demonstrates that the number of supported modes significantly reduces with increasing index of the outer medium: the modes of higher radial orders vanish. This is seen in figure 2b where the dependence of the quality factor of the second and the third radial order modes on n_{out} are presented. The dependences have a representative threshold behaviour: Q-factor drastically drops and the modes vanish at a certain threshold value of n_{out}. For higher radial modes the threshold is lower, and these modes are the first ones to disappear. Note that the calculated quality factors are just estimations since additional losses were introduced artificially. Nevertheless, this barely affects the revealed tendency. The Q-factor was calculated as $\lambda_c/\Delta\lambda$, where λ_c is the resonant frequency and $\Delta\lambda$ is the resonance width that is the full width at half maximum.

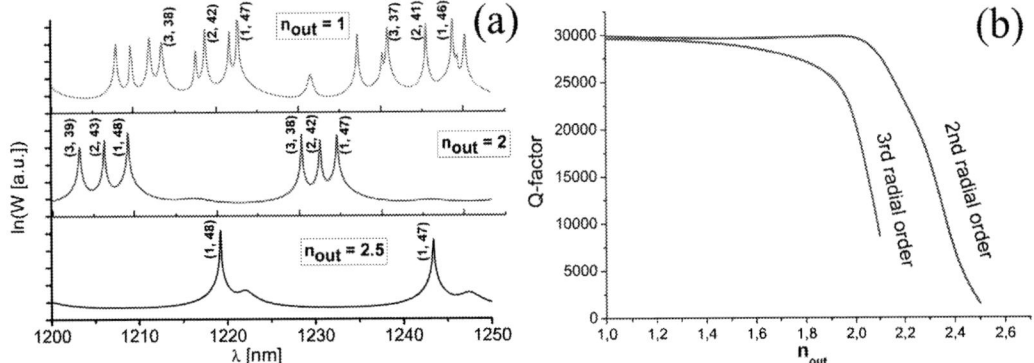

Figure 2. Mode spectra of the MD with different indices n_{out} of the outer medium, the mode numbers of the lowest radial modes are indicated near corresponding maxima (a); quality factor of the second and the third radial order modes vs. n_{out} (b).

We believe that the origin of this phenomenon is a break of the total internal reflection (TIR) conditions at the MD-surrounding interface. Indeed, the disturbance of TIR directly results in a mode degradation that corresponds to the observed threshold behaviour. Moreover, higher radial modes, which azimuthal number is typically lower and, therefore, an angle of the internal reflection is higher, should be first to disappear. This reasoning relates only to the geometrical optics interpretation of the problem, but, albeit very rough, helps to understand the phenomenon under discussion.

4. Covering MDs with a thin dielectric layer

The same effect takes place when instead of the embedding the MD in a homogeneous medium the resonator is covered with a dielectric layer of a finite thickness. This is schematically shown in figure 3a. In the figure 3b we show the evolution of the energy spectrum of the MD resonator with the thickness of covering layer increase from 50 to 250 nm, the index of the cover being 2.5.

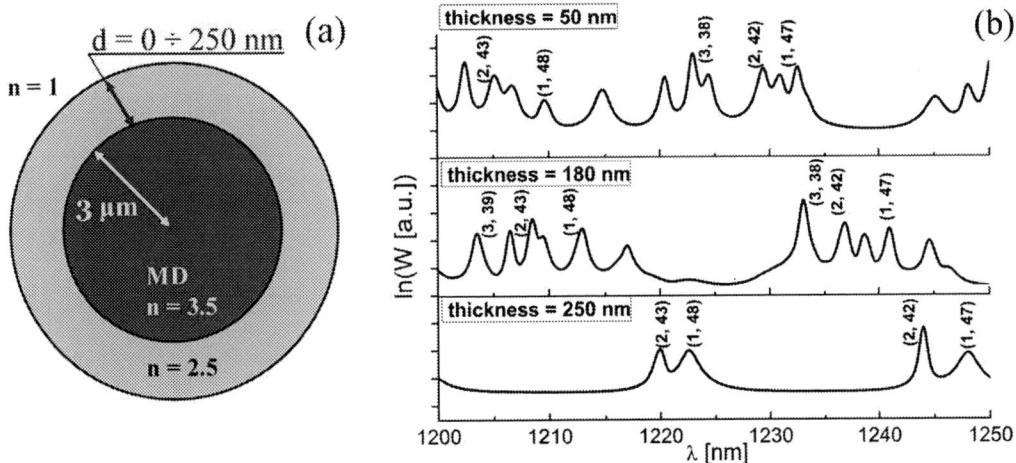

Figure 3. Schematics of the MD with a dielectric layer (a) and the mode spectra of the 6-μm MD (n = 3.5) covered with 50 nm (upper), 150 nm (middle) and 250 nm (lower) thick dielectric layer (n = 2.5) (b). Mode numbers of the lowest radial modes are indicated near corresponding maxima.

Qualitatively, the process is very similar to the one demonstrated in figure 2. Modes of higher radial orders start to vanish when a certain thickness is reached. For example in the figure 3, only the first and the second order modes remain in the resonator with 250 nm thick dielectric film. Note that the effect of the cover saturates with the film thickness since electromagnetic fields of WGMs are strongly localized in the MD. After some point, further thickness increase hardly affects the mode spectrum. Generally, the influence of a dielectric film can be interpreted as the embedding of the MD in an effective medium the index of which growth up to the index of the film material while the film thickness increases. Therefore, the reasonings regarding break of the TIR condition are also valid here.

5. Conclusion

We numerically simulated the microdisk resonators using 2D model and considered the transformation of their mode spectra under the influence of embedding the MD in dielectric media or covering it with dielectric layers. Both approaches provide vanishing higher-order radial modes. This is because of the threshold-like drop of the quality-factor of the higher radial modes of the resonator with increasing index of a surrounding medium and the thickness of a deposited dielectric film. The origin of this phenomenon is a break of the total internal reflection conditions at the MD-surrounding interface. Thus, a simple technique to decimate mode spectrum of microdisk resonators is proposed.

Acknowledgements

This study was supported by Russian Foundation for Basic Research (project #16-29-03111) and Russian Ministry of Education and Science (3.9787.2017/8.9).

References

[1] Kryzhanovskaya N V, Maximov M V, Zhukov A E 2014 *Quantum Electron.* **44** (3) 189
[2] Savchenkov A A, Matsko A B, Ilchenko V S and Maleki L 2007 Optical resonators with ten million finesse *Opt. Express* **15** 6768
[3] McCall S L, Levi A F J, Slusher R E, Pearton S J and Logan R A 1992 Whispering-gallery mode microdisk lasers *Appl. Phys. Lett.* **60** 289–91
[4] Gayral B, Gérard J M, Lemaître A, Dupuis C, Manin L and Pelouard J L 1999 High-Q wet-etched GaAs microdisks containing InAs quantum boxes *Appl. Phys. Lett.* **75** 1908–10
[5] Ide T, Baba T, Tatebayashi J, et al Y 2005 Room temperature continuous wave lasing in InAs quantum-dot microdisks with air cladding *Opt. Express* **13** 1615–20
[6] Bogdanov A A, Mukhin I S, Kryzhanovskaya N V. et al 2015 Mode selection in InAs quantum dot microdisk lasers using focused ion beam technique *Opt. Lett.* **40** 4022
[7] Zhen-Nan Tian, Feng Yu, Yan-Hao Yu, Jun-Jie Xu, Qi-Dai Chen, and Hong-Bo Sun *Optics Letters* **42** (8) 1572 (2017)
[8] Boriskina S V, Benson T M, Sewell P and Nosich A I 2004 *J. Opt. Soc. Am. B-Optical Phys.* **21** 1792–6
[9] Moiseev E I, Kryzhanovskaya N, Polubavkina Y S et al 2017 Light Outcoupling from Quantum Dot-Based Microdisk Laser via Plasmonic Nanoantenna *ACS Photonics* **4** 275–81
[10] Gotzinger S, Benson O and Sandoghdar V 2001 Towards controlled coupling between a high-Q whispering-gallery mode and a single nanoparticle *Appl. Phys. B Lasers Opt.* **73** 825–8
[11] Yariv A. 2002 Critical coupling and its control in optical waveguide-ring resonator systems *IEEE Photonics Technol. Lett.* **14** 2001–3
[12] Mintairov A M, Chu Y, He Y, Blokhin S et al 2008 *Phys. Rev. B - Condens. Matter Mater. Phys.* **77** 1–7
[13] Xu Y, Lee R K and Yariv A 2000 *Phys. Rev. A* **61** 33808
[14] https://www.comsol.com/model/download/344161/models.woptics.fabry_perot.pdf

Electron diffusivity measurements of VN superconducting single-photon detectors

N R Romanov[1,3], P I Zolotov[1,2,3], Yu B Vakhtomin[1,3], A V Divochiy[3] and K V Smirnov[1,2,3]

[1]Moscow State Pedagogical University, Moscow 119991, Russia
[2]National Research University Higher School of Economics, Moscow 101000, Russia
[3]LLC Superconducting Nanotechnology (SCONTEL), Moscow 119021, Russia

Abstract. The research of ultrathin vanadium nitride (VN) films as a promising candidate for superconducting single-photon detectors (SSPD) is presented. The electron diffusivity measurements are performed for such devices. Devices that were fabricated out from 9.9 nm films had diffusivity coefficient of 0.41 cm²/s and from 5.4 nm – 0.54 cm²/s. Obtained values are similar to other typical SSPD materials. The diffusivity that increases along with decreasing of the film thickness is expected to allow fabrication of the devices with improved characteristics. Fabricated VN SSPDs showed prominent single-photon response in the range 0.9-1.55 μm.

1. Introduction

Since 2001 when Gol'tsman *et al.* [1] demonstrated the first superconducting single-photon detector, its performance has been improved greatly [2, 3, 4]. This SSPD improvement led to the wide spread of the devices and now they are involved in many applications [5, 6]. Partly this improvement was related to the appearance of new materials. When the first devices were fabricated using niobium nitride films, the modern ones could involve new for this sphere materials such as NbTiN, WSi, MoSi [7, 8, 9]. On the one hand, exploration of new materials allows to improve SSPD technology. For example, modern devices have detection efficiency of ~90% in wide spectrum range [3, 10], GHz counting rates [11], temporal resolution around tens picoseconds [2]. On the other hand, researches in this field allow further understanding of physical operation principles for such devices. Indeed, despite the fact that SSPDs are one of the most popular single-photon detection technologies, there is no comprehensive theory that explains its functioning principles [12]. Therefore, the search of the new SSPD materials remains an actual problem, which involves many scientific groups. With this work we begin a presentation of results of VN SSPDs fabrication and further research of its characteristics. Here we present the electron diffusivity (*D*) measurements of the VN devices. This parameter determines many features of SSPDs and therefore it remains one of the defining material parameters that shows the prospects of the superconducting materials for the implementation in the SSPDs. Specifically, as shown in [13], diffusivity plays a significant role in formation and subsequent dynamics of resistive region when a photon is absorbed by nanostripe. Hence, it defines the efficiency of the SSPD and its operation wavelengths range. It can be noted that diffusivity also has its contribution to the depairing current [14], which influence on the performance of the detectors.

Content from this work may be used under the terms of the Creative Commons Attribution 3.0 licence. Any further distribution of this work must maintain attribution to the author(s) and the title of the work, journal citation and DOI.
Published under licence by IOP Publishing Ltd

2. Fabrication of VN structures, experimental methods and setup

2.1 VN structures and methods

The films for our work were deposited over silicon substrates with additional 250 nm thick thermally grown SiO_2 layer. The deposition was performed in AJA International Inc. sputtering system with typical background pressure ~$5 \cdot 10^{-8}$ Torr.

In order to study the effect of the deposition rate of VN films on its characteristics and in particular on the electron diffusion coefficient, we conducted several processes with different deposition rates of the films. The deposition rate was changed by variation of the discharge current and was controlled by atomic force microscopy measurements of the thickness of the deposited films. The thicknesses (h) of obtained films were set by sputtering time. We should note that along with decrease of the deposition rate, the transition temperature (T_c) decreased as well for the films with the same thickness. For the diffusivity measurements we selected two films 5.4 and 9.9 nm thick that had the same transition temperature, but were deposited with different rate (see table 1).

Table 1. VN films' parameters

Film #	h, nm	Deposition rate, nm/min	$R_s(300\ K)$, Ohm/□	T_c, K
1	9.9	2.2	111	5.5
2	5.4	3	202	5.5

To carry out measurements VN structures with standard topology for the SSPD [3] were made. The superconducting stripe was manufactured by electron-beam lithography in the form of meander had width of ~110 nm and filled up square area of 15×15 μm² with filling factor of ~0.6. It should be noted that the T_c of SSPDs has significantly decreased compared to the values of the uniform films and amounted to 2.9 and 4.65 K for detectors based on 5.4 and 9.9 nm films respectively. This decrease could be related to the equal oxidization of the films, which resulted in more apparent T_c reduction in thinner film. It also could be caused by the local constrictions of the films on the scale comparable to the nanostripe width of the devices. These hypotheses are confirmed by the differences in the calculated and real values of the detectors' resistances, which are in better correspondence for the devices made of thicker film. The single-photon detection performance of such VN devices at different wavelengths could be found elsewhere [15].

The diffusivity was determined directly in SSPDs using the standard method [16] – measurements of dependence of resistance (R) on temperature in magnetic field and further estimations of the D as:

$$D = -\frac{4k_B}{\pi e}\left[\frac{dB}{dT_c}\right]^{-1} \tag{1},$$

where k_B is Boltzmann constant, e is the elementary charge and dB/dT_c is derivative of magnetic field with respect to superconducting transition temperature.

2.2 Experimental setup

For the measurements of R(T) dependences in magnetic field the SSPDs were placed into special holder of cryogenic insert. The insert was put into double-walled cryogenic dipstick. The dipstick was placed into standard liquid helium storage Dewar. Liquid helium entered the inner volume of the dipstick through capillary. This construction allowed reaching minimal temperature of 1.6 K by pumping of helium vapors out from inner volume of dipstick. In addition, the detector holder had a heater for increasing temperature above 1.6 K and temperature sensors, which could operate in magnetic field. To create magnetic field up to 4 T a superconducting magnet coil was used. The samples were oriented normally to lines of magnetic field. We managed values of the magnetic field B

by regulating of superconducting magnet coil current. We biased a sample with direct current which was much smaller than critical current of the device (about 0.1 µA at 1.6 K and at B=0 T). Such low value of bias current did not shift critical temperature of sample. We measured voltage values from the device by Lakeshore Temperature Monitor 218 and then we were calculating the R. The scheme of the experimental setup is presented at figure 1.

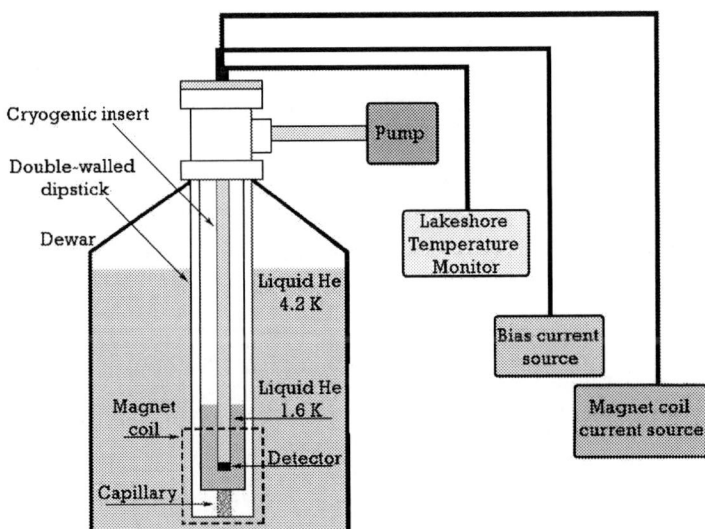

Figure 1. Scheme of the experimental setup for diffusivity measurements.

3. Experimental results

We carried out series of R(T) measurements for samples with two different thicknesses at various values of magnetic field. Normalized to 1 at R_{20K} dependences of R(T) of the SSPD sample with thickness of 5.4 nm is presented on figure 2a. As it was expected superconducting transition temperature, which was determined as $0.5 \cdot R_{20K}$, reduced along with increasing of magnetic field. The similar dependences of R(T) in magnetic field were measured for VN SSPD with thickness of 9.9 nm. According to presented R(T) measurements we plotted curves showing the variation of the T_c with respect to the magnetic field (see figure 2b). Calculated from these dependences values of dB/dT_c and the D (according to equation 1) for two samples are shown in table 2. Diffusivity values were 0.54 cm^2/s for 5.4 nm VN SSPD and 0.41 cm^2/s for 9.9 nm VN device

First of all, we note that the specified diffusion coefficient values have shown that the D of ultrathin superconducting VN SSPDs are close to values for other materials used for SSPD fabrication. It confirms that the VN films are potentially applicable for creation of SSPDs. It is noteworthy that the SSPD based on thicker superconducting VN film had lower diffusivity coefficient. Typically, as the thickness of the film increases the electron diffusion coefficient increases as well. That is usually associated with decreased influence of quantum corrections to the conductivity for thicker films. Let us note that the obtained R(T) dependences in the magnetic field demonstrate qualitative behavior typical for films in which quantum corrections to conductivity have a noticeable effect. For both detectors increasing of the magnetic field did not led to the R(T) curves shifting by parallel transfer towards decreasing of the transition temperature, but showed a significant increase in the width of the superconducting transition associated with the displacement of its lower temperature boundary. However, equal values of the transition temperatures of the unstructured films may indicate that the degree of disorder of the film with thickness of 9.9 nm is higher in comparison with the 5.4 nm thick film.

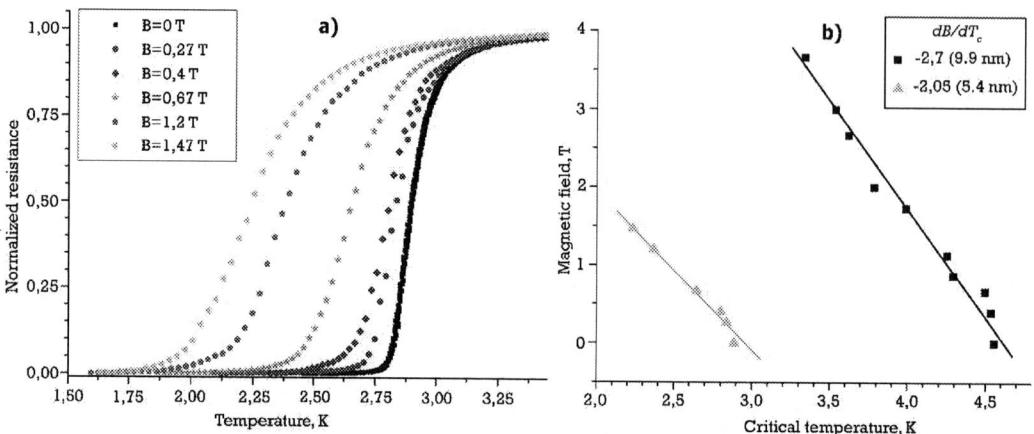

Figure 2. a) Dependences of R vs T for 5.4 nm sample. Critical temperature T_c was determined as temperature where $R=0.5 \cdot R_{20K}$; **b)** Dependences of B vs T_c for two samples with thicknesses of 5.4 nm and 9.9 nm.

Table 2. The values of dB/dT_c and the D for used samples

Detector #	h, nm	dB/dT_c	D, cm²/s
1	9.9	-2.7	0.41
2	5.4	-2.05	0.54

Doubtless further researches of diffusivity variation at different thicknesses of VN films are necessary. At the same time present study demonstrated that: a) diffusivity of VN SSPD detectors are close to values for other materials used for SSPD fabrication; b) by changing the deposition rate of VN films it is possible to obtain thicker VN films with reduced diffusivity. Likewise, for thicker superconducting films (and/or films with a lower value of R_s) the radiation absorption coefficient increases. Moreover, with respect to the results presented in work [13], probability of appearance of voltage pulse in the superconducting stripe must increase after an absorbance of a photon with reduction of diffusivity. Authors [13] considered a fact based on both reduction of possibility of hot electron thermalization by diffusivity at initial stage of hot spot formation as well as decreasing of time of inelastic electron-electron interaction τ_{e-e}. It reduces both superconductor`s volume in which energy of an absorbed photon is distributed and electrons thermalization time through their interaction.

In conclusion we want to note that two samples presented in this work had single-photon responses in the spectrum range 0.9-1.55 µm. Therefore, we experimentally demonstrated suitability of the VN films for the implementation as SSPD material. More detailed research of the influence of sputtering parameters on the VN films, quantum efficiency, typical jitter values and counting rates of VN SSPDs will be presented by the authors elsewhere soon.

4. Conclusion

We investigated the electron diffusivity for SSPDs based on vanadium nitride ultra-thin superconducting films. Obtained values of 0.41 cm²/s for devices made out of 9.9 nm film and 0.54 cm²/s for 5.4 nm film device show similar results with the other popular SSPD materials. Atypical decrease of the diffusivity along with increase of the VN film thickness we associate with the change of the film deposition rate, which influence on the superconductor's properties. Presented results allows us to consider vanadium nitride as a promising candidate for superconducting devices fabrication.

Acknowledgements

The work is supported by the Russian Science Foundation (RSF) Project No. 18-12-00364.

References

[1] Gol'tsman, G.N., Okunev, O., Chulkova, G., Lipatov, A., Semenov, A., Smirnov, K., Voronov, B., Dzardanov, A., Williams, C. and Sobolewski, R., 2001. Picosecond superconducting single-photon optical detector. Applied physics letters, 79(6), pp.705-707.

[2] Shcheslavskiy, V., Morozov, P., Divochiy, A., Vakhtomin, Y., Smirnov, K. and Becker, W., 2016. Ultrafast time measurements by time-correlated single photon counting coupled with superconducting single photon detector. Review of scientific instruments, 87(5), p.053117.

[3] Smirnov, K., Divochiy, A., Vakhtomin, Y., Morozov, P., Zolotov, P., Antipov, A. and Seleznev, V., 2018. NbN single-photon detectors with saturated dependence of quantum efficiency. Superconductor Science and Technology, 31(3), p.035011.

[4] Zolotov, P., Divochiy, A., Vakhtomin, Y., Moshkova, M., Morozov, P., Seleznev, V. and Smirnov, K., 2018, February. Photon-number-resolving SSPDs with system detection efficiency over 50% at telecom range. In AIP Conference Proceedings (Vol. 1936, No. 1, p. 020019). AIP Publishing.

[5] Zhao, Q., Xia, L., Wan, C., Hu, J., Jia, T., Gu, M., Zhang, L., Kang, L., Chen, J., Zhang, X. and Wu, P., 2015. Long-haul and high-resolution optical time domain reflectometry using superconducting nanowire single-photon detectors. Scientific reports, 5, p.10441.

[6] Grein, M.E., Willis, M., Kerman, A., Dauler, E., Romkey, B., Rosenberg, D., Yoon, J., Molnar, R., Robinson, B.S., Murphy, D. and Boroson, D.M., 2014, June. A fiber-coupled photon-counting optical receiver based on NbN superconducting nanowires for the Lunar Laser Communication Demonstration. In CLEO: Science and Innovations (pp. SM4J-5). Optical Society of America.

[7] Dorenbos, S.N., Reiger, E.M., Perinetti, U., Zwiller, V., Zijlstra, T. and Klapwijk, T.M., 2008. Low noise superconducting single photon detectors on silicon. Applied Physics Letters, 93(13), p.131101.

[8] Seleznev, V.A., Divochiy, A.V., Vakhtomin, Y.B., Morozov, P.V., Zolotov, P.I., Vasil'ev, D.D., Moiseev, K.M., Malevannaya, E.I. and Smirnov, K.V., 2016, August. Superconducting detector of IR single-photons based on thin WSi films. In Journal of Physics: Conference Series (Vol. 737, No. 1, p. 012032). IOP Publishing.

[9] Korneeva, Y.P., Mikhailov, M.Y., Pershin, Y.P., Manova, N.N., Divochiy, A.V., Vakhtomin, Y.B., Korneev, A.A., Smirnov, K.V., Sivakov, A.G., Devizenko, A.Y. and Goltsman, G.N., 2014. Superconducting single-photon detector made of MoSi film. Superconductor Science and Technology, 27(9), p.095012.

[10] Marsili, F., Bellei, F., Najafi, F., Dane, A.E., Dauler, E.A., Molnar, R.J. and Berggren, K.K., 2012. Efficient single photon detection from 500 nm to 5 μm wavelength. Nano letters, 12(9), pp.4799-4804.

[11] Esmaeil Zadeh, I., Los, J.W., Gourgues, R.B., Steinmetz, V., Bulgarini, G., Dobrovolskiy, S.M., Zwiller, V. and Dorenbos, S.N., 2017. Single-photon detectors combining high efficiency, high detection rates, and ultra-high timing resolution. APL Photonics, 2(11), p.111301.

[12] Renema, J.J., Gaudio, R., Wang, Q., Zhou, Z., Gaggero, A., Mattioli, F., Leoni, R., Sahin, D., De Dood, M.J.A., Fiore, A. and Van Exter, M.P., 2014. Experimental test of theories of the detection mechanism in a nanowire superconducting single photon detector. Physical review letters, 112(11), p.117604.

[13] Vodolazov, D.Y., 2016. Theory of single photon detection by 'dirty' current-carrying superconducting strip based on the kinetic equation approach. arXiv preprint arXiv:1611.06060.

[14] Semenov, A.D., Gol'tsman, G.N. and Korneev, A.A., 2001. Quantum detection by current carrying superconducting film. Physica C: Superconductivity, 351(4), pp.349-356.

[15] Zolotov, P., Divochiy, A., Vakhtomin, Y., Seleznev, V., Morozov, P. and Smirnov, K., 2018. Superconducting Single-photon Detectors Made of Ultra-thin VN Films. KnE Energy & Physics, 3(3), pp.83-89.

[16] Gershenzon, E.M., Gershenzon, M.E., Gol'tsman, G.N., Lyul'kin, A.M., Semenov, A.D. and Sergeev, A.V., 1990. Electron-phonon interaction in ultrathin Nb films. Sov. Phys. JETP, 70(3), pp.505-511.

Phase diffraction patterns optically induced by laser Bessel-like beams in photorefractive lithium niobate with a doped surface layer

I Trushnikov, A Inyushov, A Perin

Tomsk State University of Control System and Radioelectronics, Tomsk, 634050, Russia

Abstract. The evolution of the characteristics of one-dimensional phase diffraction structures during their optical induction by bessel-like monochromatic beams in photorefractive samples of lithium niobate is studied experimentally. Both, one-dimensional (1D) and two-dimensional (2D) Bessel-like beams with different topology of 2D beam cross-sections are formed from Gaussian laser beams using the amplitude masks with rectangular and annular apertures. Also, the property of reconstructing the formed bessel-like laser beams after they have passed through an obstacle has been experimentally verified. These almost diffraction-free light fields with wavelengths of 457 and 532 nm can change the refractive indices of photorefractive lithium niobate samples and form within them the nonlinear photonic diffraction structures.

1. Introduction

In the most cases, any laser source generates Gaussian-shaped light beam, which may be tightly focused with beam waist region in longitudinal direction depending on the light wavelength and minimal transverse waist size. However, some applications require not conventional shapes of light beams, which demonstrate some properties distinct of Gaussian laser beams. These are diffraction-free beams including Bessel-like ones [1], Airy beams [2] and some other non-diffracting shape-preserving light beams [3]. The Bessel-like beams are close to theoretical diffraction-free fields which are not limited in the transverse directions. The real Bessel beams cannot exist because of the infinite optical power they should carry. However, there are some configurations that may form Bessel-like beams in the bounded space area.

The usual ways to form two-dimensional almost diffraction-free light fields exploit so called axicon lenses, annular apertures or optical fiber elements [1]. However, in some cases not only two-dimensional light fields are required but also one-dimensional ones. Strictly speaking they are quasi-one-dimensional Bessel-like beams because in the second transverse dimension these fields display the Gaussian-like profiles if they are formed by laser beams.

The main aim of this study is formation of one-dimensional and two-dimensional Bessel-like fields with different shapes of transverse light patterns using diffraction grating-like amplitude transparencies. The obtained longitudinally homogeneous light patterns generate photonic structures, e.g. waveguide or diffraction systems in photorefractive lithium niobate (LiNbO$_3$) samples [4-6]. We use for this purpose of amplitude masks including their couples rotated with respect to each other at some angles. Every mask contains the metal screen with two rectangular slits in it. Amplitude masks with an annular aperture were also used.

Content from this work may be used under the terms of the Creative Commons Attribution 3.0 licence. Any further distribution of this work must maintain attribution to the author(s) and the title of the work, journal citation and DOI.

Published under licence by IOP Publishing Ltd

2. Experimental conditions and experimental results

The solid-state YAG:Nd^{3+} laser with light wavelength of λ=532 nm and semiconductor laser (λ=457 nm) are used as CW light sources in experiments. The near to parallel laser beam illuminates the amplitude mask that is located in the focal plane of a lens (cylindrical or spherical, depending on one-dimensional or two-dimensional light field is formed).The longitudinally uniform interference pattern appears after this lens which may be used to generate the photonic phase structures in the photosensitive material [6]. In the scheme with amplitude mask two light beams produced by the slits in a screen interfere in an area after lens (Figure 1). To form two-dimensional Bessel-like field, we use two amplitude masks with required angle between directions of their slits. For the formation photonic structures we use the photorefractive samples LiNbO$_3$:Fe and LiNbO$_3$:Cu. To test properties of the phase photonic structure generated within the sample, we use laser radiation with λ=532 nm and λ=457 nm. The near field and far field diffraction patterns are studied with a CCD camera at this stage. The formed patterns were also studied with the help of a microscope.

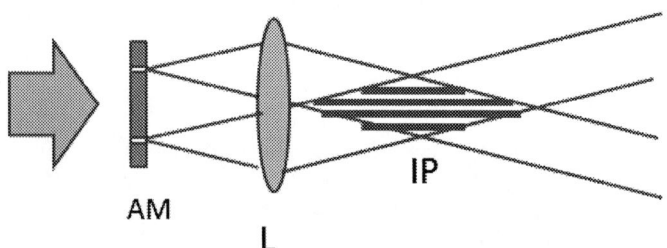

Figure 1. Experimental setup for formation of Bessel-like beams with amplitude masks. AM – amplitude mask, L – lens, IP – interference pattern

2.1. Bessel-like light beam formation with amplitude masks

As the particular results, images in Figure 2 show 1D (λ=457 nm) and 2D light patterns (λ=532 nm) near the back focal plane of a cylindrical lens with focal length 100 mm (a), and spherical lenses (b - c) with focal length of 180 mm at different dimensions of slits and different slit orientation with respect to each other. To form 1D beam, we use the metal mask with two parallel rectangular slits. The slit cross-section measures 0.1×1.5 mm^2 with a distance between slit centers of 0.6 mm. Similar masks are used to form 2D Bessel-like beams with slit width 0.2 mm and their length 2 mm. The distances between their centers range from 0.5 to 1 mm.

To obtain interference patterns with longitudinal displacement, amplitude masks with an annular aperture with a slit width of 0.05 mm and a diameter of 0.5 mm It is illustrated by images in Figure 3 with dependences of intensity cross-sections of Gaussian beam and 2D interference field on position of CCD camera along the light propagation direction measured in centimeters. Figure 4 demonstrates corresponding dependences of Gaussian beam diameter at half maximum intensity level and the width of interference field central maximum for the case when they are comparable in dimensions. It is clearly seen that variation of transverse dimensions of this Bessel-like beam is much less when compared with that for the usual Gaussian beam.

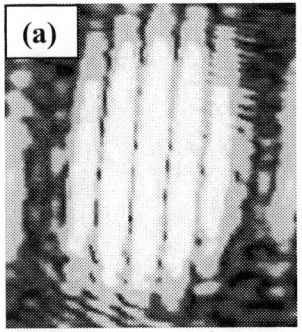 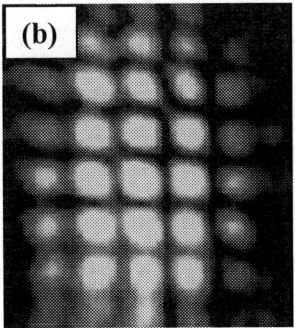 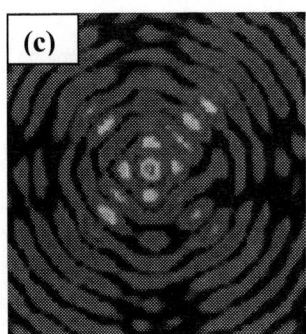

Figure 2. Light cross-sections of 1D and 2D Bessel-like fields for horizontal widths of central maxima: a – 90 μm; b – 75 μm (90° angle between slit directions); c – 70 μm (annular slit).

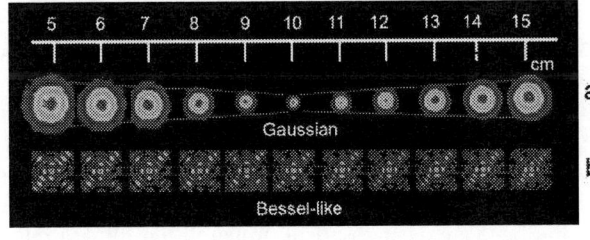

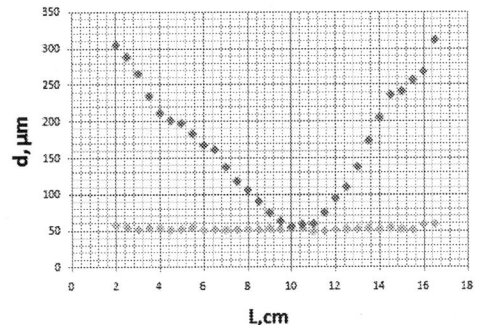

Figure 3. Intensity cross-sections of Gaussian beam (a) and 2D interference field (b) at different positions along light propagation direction.

Figure 4. Gaussian beam diameter (blue) and the central maximum diameter of interference pattern (green) along light propagation direction.

2.2. Formation of photonic structures by Bessel-like beams in photorefractive lithium niobate

Bessel-like 1D and 2D beams are used to generate phase diffraction patterns in photorefractive samples of copper-doped plate LiNbO₃:Cu (0.05 wt%). The near-surface region is doped with Cu ions by thermal diffusion at a temperature of 900 ° C from a film deposited on the surface of the sample by sputtering in a vacuum. The thickness of the doped layer is about 100 μm. The dimensions of this plate are $10 \times 6 \times 3$ mm³ along X, Y, and Z axes. The presence of copper does not lead to a significant increase in the dark conductivity of LiNbO₃, which ensures the long-term storage of optically induced elements in such a material. The light polarization at phase structure induction corresponds to the crystal extraordinary or ordinary waves. At the structure readout we use the extraordinarily polarized light waves of YAG or semiconductor lasers. The light pattern in Figure 5a illustrates near-field diffraction of light (λ=532 nm) on the 1D few-element phase grating with spatial period 180 μm induced within the LiNbO₃:Cu plate. The optical power and exposure time at this grating creation are 1 mW and 3 minutes. Image in figure 5b shows cross-section of 2D interference pattern obtained with two amplitude masks 90° rotated with respect to each other in the same crystal plate. Figure 5c shows the generated two-dimensional diffraction structure on the output plane of the crystal with an increase of 80x. It should be noted that exposure time to induce this 2D structure makes up 30 minutes at the same light power as it was used at 1D grating formation.

SPBOPEN 2018   IOP Publishing

IOP Conf. Series: Journal of Physics: Conf. Series **1124** (2018) 051033   doi:10.1088/1742-6596/1124/5/051033

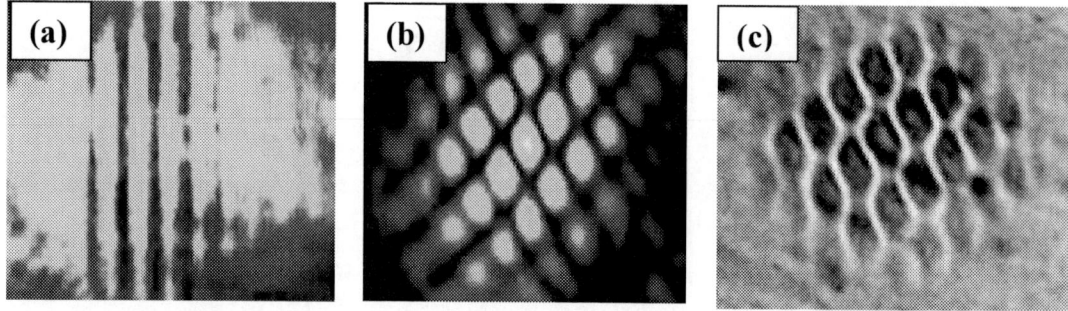

Figure 5. Light diffraction patterns at the output facet of LiNbO$_3$:Cu plate (a), 2d Bessel-like field (b), formed diffraction pattern by light Bessel-like field two-dimensional (perpendicularly superimposed masks are located at an angle of 45 ° to the optical axis) (c)

Images in Figure 6 demonstrate the change of 1D diffraction grating profile in time during its generation in photorefractive plate LiNbO$_3$:Cu. They are obtained with amplitude mask scheme. The spatial period of phase grating in this case is 100 µm. At the initial stage of the grating formation (less than 10 seconds) the near field diffraction image and its intensity profile are practically the same as in the light field. However, at the exposure time increase (t=80 s) the arising of maximal intensity within side lobes of near field diffraction image as well as in intensity profile of that image are observed. Indeed, the photorefractive optical nonlinearity is saturable and that results in possible creation of phase diffraction and waveguide elements with required profiles in photorefractive lithium niobate. Image in Figure 7 demonstrate light picture in the far zone during optical probe on the formed phase structure with a spatial period of 190 µm.

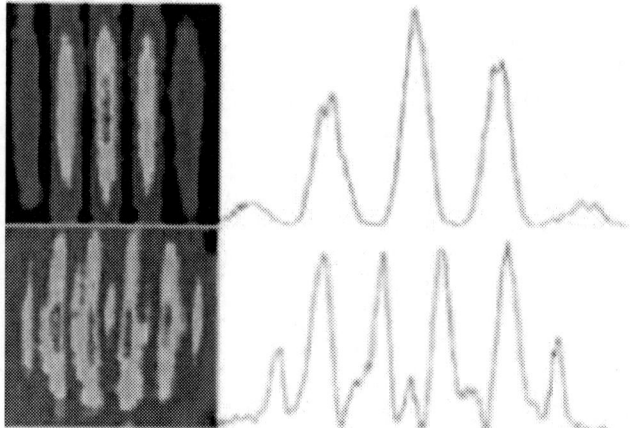

Figure 6. Near field diffraction images at the exit surface of LiNbO$_3$:Cu plate (left) and their intensity profiles (right) for exposure times less than 10 s (upper line) and 80 s (lower line).

The time evolution of the grating diffraction efficiency (the ratio of the intensities of the diffraction maxima of the first and zero orders) for light with a wavelength of 532 nm (Figure 8) made it possible to estimate the magnitude of the change in the refractive index of the material in the lattice region. At a measured value of the diffraction efficiency at an exposure time of about 100 s, the change in the refractive index is 0.000103.

708

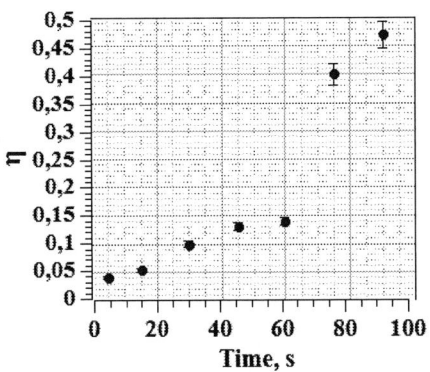

Figure 7. Light picture in the far zone during optical probe of the phase structure in the crystal (the spatial period of the phase structure is 190 μm)

Figure 8. The graph of the dependence of the diffraction efficiency on the exposure time

3. Conclusion

In conclusion, our experimental results confirm the possibility creation of few-element phase diffraction structures and more complicated photonic waveguide circuits with required profiles within photosensitive materials like photorefractive crystals by light fields with Bessel-like shapes. Variation of parameters of optical schemes gives the additional degree of freedom to create Bessel-like light beams with needed characteristics. Experimentally demonstrated the change in the refractive index profile of one-dimensional phase diffraction structures during their optical induction in crystal photorefractive lithium niobate samples by one-dimensional Bessel-like beams.

Acknowledgments

This study was carried out with the financial support of Ministry of Education and Science of Russia (within the task N 3.1110.2017/PCh of the project part).

References

[1]   Duocastella M and Arnold C B 2012 Laser Photonics Rev. **6** 607
[2]   Siviloglou G A and Christodoulides D N 2007 Optics Letters **32** 979
[3]   Bandres M A and Rodrıguez-Lara B M 2013 New Journal of Physics **15** 013054
[4]   Zhu X, Schülzgen A, Li L, and Peyghambarian N 2009 Applied Physics Letters **94** 01102
[5]   Kip D 1998 Appl. Phys. B **67** 131
[6]   Perin A S, Shandarov V M, Chen F 2011 Physics of wave phenomena **19** 251

SPBOPEN 2018   IOP Publishing

Focusing of high-order vortex laser beams by binary axicon with different numerical aperture

D A Savelyev [1,2], S A Fomchenkov [1,2]

[1]Department of Engineering Cybernetics, Samara National Research University, Samara 443086, Russia
[2]Laboratory of Laser Measurements, Image Processing Systems Institute - Branch of the Federal Scientific Research Centre "Crystallography and Photonics" of Russian Academy of Sciences, Samara 443001, Russia

Abstract. The spatial distribution of the focused high-order vortex Gauss-Laguerre beams is investigated subject to the change of the numerical aperture of the diffraction axicon and order of the vortex phase in the 3D model. Modeling of diffraction is numerically investigated by the finite difference time domain (FDTD) method using high-performance computing.

1. Introduction

It is known that the phase singularity can be used to amplify the longitudinal component of the electric vector of laser radiation [1-8] in the case of homogeneous polarization, which is produced by most modern lasers. In particular, vortex phase functions were considered in [1, 2, 5, 7, 8], and in [3, 4, 6] - superposition of vortex phase functions, both co-axial [3, 4] and spatially separated [6].
The presence of a powerful longitudinal component in the focus region makes it possible to improve the optical resolution and is used for optical manipulation, electron acceleration, material processing, microscopy and other applications [9-13].
Note that the energy of the longitudinal component becomes significant only at high numerical aperture [4, 14]. This requirement can be reduced by using high-order laser beams [15-18]. Also, it was shown that the use of the axicon makes it possible to strengthen the focusing of laser beams in comparison with the lens theoretically [19, 20] and experimentally [21, 22].]. Applications of a diffractive axicon with high numerical aperture provide generation of longitudinally polarized laser needles [23-26]. In [27, 28], a comparison was made of the effect of a diffraction axicon and a conical taper for laser beams focusing. Especial attention was paid to near-field diffraction by micro-axicons [29-33].
In this paper, we investigate the focusing of vortex beams using diffractive axicons with numerical aperture from 0.5 to 0.95. For the numerical simulation of diffraction we use the FDTD-method with high-performance computing [34]. Calculations were made on the computational cluster with power of 850 GFlops.

2. Investigation of vortex laser beam focusing

Simulation parameters: the wavelength $\lambda = 0.532$ microns, the size of the computational domain x, y, z $\in [-3.8\lambda; 3.8\lambda]$. The thickness of the absorbing layer PML $\sim 1.3\lambda$, the sampling step of space $- \lambda/21$,

---

Content from this work may be used under the terms of the Creative Commons Attribution 3.0 licence. Any further distribution of this work must maintain attribution to the author(s) and the title of the work, journal citation and DOI.
Published under licence by IOP Publishing Ltd

the sampling step of time $- \lambda/(42c)$, where c is the velocity of light. The Gauss-Laguerre beams were considered as the input laser radiation with circular polarization and order of the vortex phase m from 1 to 3. The vortex direction was opposite to the direction of circular polarization. The refractive index of the axicon and the substrate is n = 1.5. The numerical aperture (NA) of the focusing binary axicon was 0.95 and 0.5. The results of numerical simulation in the xz plane are shown in figure 1.

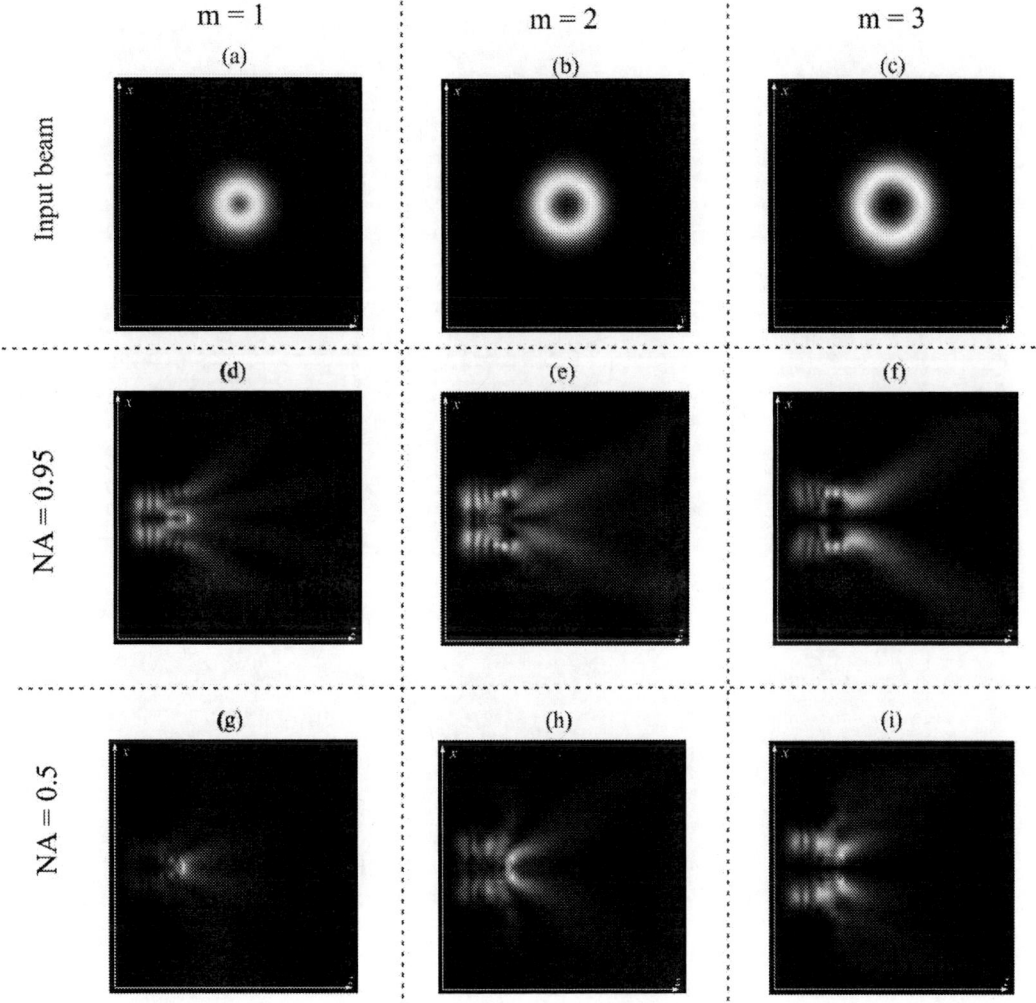

Figure 1. The propagation of a high-order vortex Gauss-Laguerre beams through the axicons with a different NA, total intensity: input beam (plane xy) – (a), (b), (c); the longitudinal cross section (xz), NA = 0.95 – (d), (e), (f); the longitudinal cross section (xz), NA = 0.5 – (g), (h), (i).

The results of numerical simulation in the xz plane showed that for a numerical aperture NA = 0.95, two cones are formed for the case m = 1 at a small angle and angle corresponding to the numerical aperture of the axicon (a "conical" vortex). For the same numerical aperture at m = 2, an extended light segment is formed on the optical axis: its formation begins upon the completion of the propagation of the outer cone. For the case when m = 3, only one light cone is formed. For the case of a numerical aperture equal to 0.5, an increase in the order of m leads to defocusing.

Consider the cross-section for the axicons considered earlier. In figure 2 shows the simulation results at a distance of λ from the axicon. It should be noted that an increase in m for the case NA =

0.5 leads to a steady increase in the diameter of the central dip. Comparable values of the central dip for both axicons were obtained for the case m = 3.

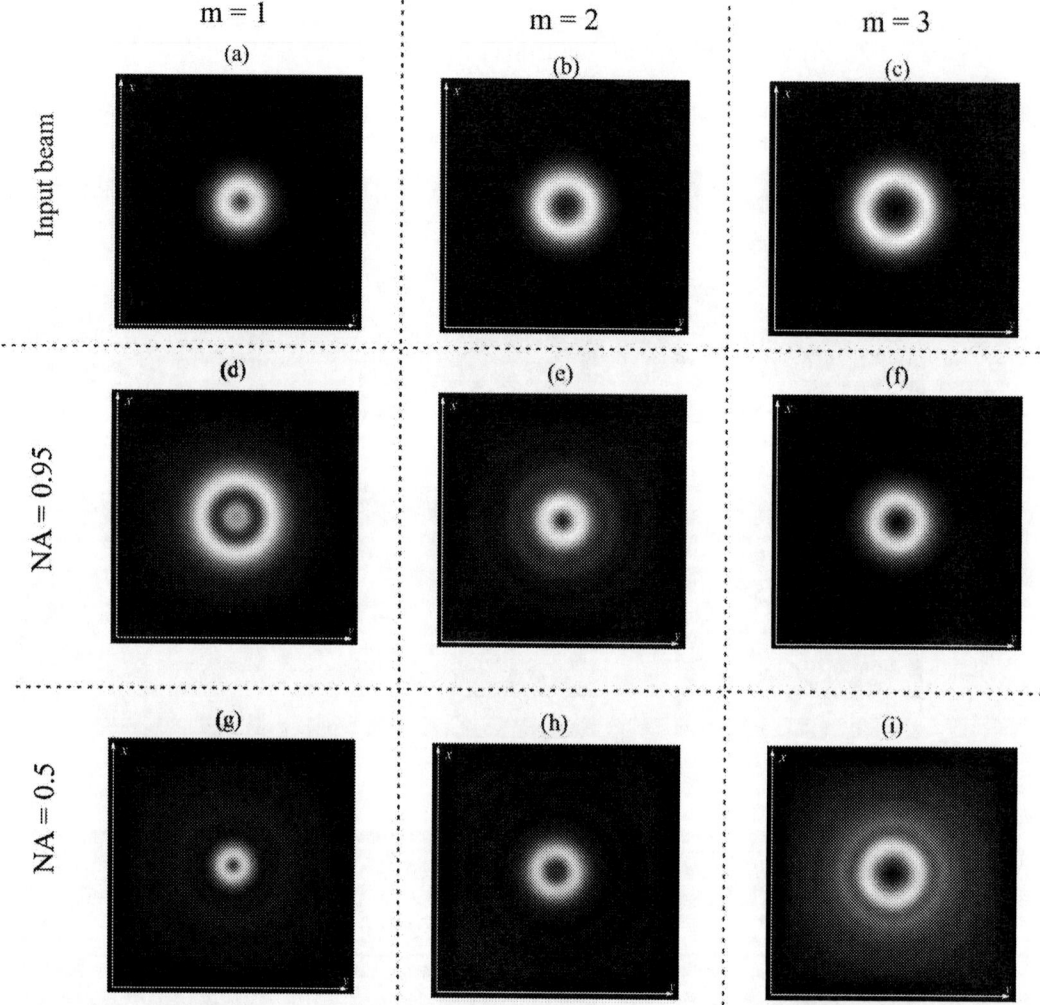

Figure 2. The propagation of a high-order vortex Gauss-Laguerre beams through the axicons with a different NA, total intensity, cross section at a distance of λ from axicon, plane xy: input beam – (a), (b), (c); NA = 0.95 – (d), (e), (f); NA = 0.5 – (g), (h), (i).

Thus, numerically, using the FDTD method, a comparative study of the diffraction of a vortex Gauss-Laguerre beams with circular polarization on diffraction axioms with NA = 0.95 and NA = 0.5 was performed.

3. Conclusion

In this paper we show the effect of the diffraction of vortex laser beams when changing the order of the vortex phase from 1 to 3 and numerical aperture of the focusing binary axicon in the 3D model. To numerically simulate the diffraction of the laser radiation under consideration the finite difference method in the time domain using high-performance computing are used.

It is shown that for a numerical aperture NA = 0.95, two cones are formed for the case m = 1 at a small angle and angle corresponding to the numerical aperture of the axicon (a "conical" vortex). For the same numerical aperture at m = 2, an extended light segment is formed on the optical axis: its formation begins upon the completion of the propagation of the outer cone.

Acknowledgments

This work was financially supported by the Russian Foundation of Basic Research (RFBR) (grants Nos. 16-07-00825, 16-29-11698), by Federal Agency of Scientific Organizations (agreement No. 007-GZ/C3363/26), by grant of the President of the Russian Federation for the scientific school NSh-6307.2018.8 and by the Ministry of Science and Higher Education of Russian Federation.

References

[1]   Ganic D, Gan X, Gu M 2003 *Optics Express* **11** 2747
[2]   Zhang Z, Pu J, Wang X 2008 *Optical Engineering* **47** 068001
[3]   Khonina S N, Volotovsky S G 2010 *Journal of the Optical Society of America A* **27** 2188
[4]   Khonina S N, Golub I 2011 *Optics Letters* **36** 352
[5]   Khonina S N, Kazanskiy N L, Volotovsky S G 2011 *Journal of Modern Optics* **58** 748
[6]   Chen Z, Pu J, Zhao D 2011 *Physics Letters A* **375** 2958
[7]   Huang K, Shi P, Cao G, Li K, Zhang X, Li Y 2011 *Optics Letters* **36** 888
[8]   Khonina S N 2013 *Optical Engineering* **52** 91711
[9]   Novotny L, Beversluis M, Youngworth K, Brown T 2001 *Phys. Rev. Lett.* **86** 5251
[10]   Hayazawa N, Saito Y, Kawata S 2004 *Appl. Phys. Lett.* **85** 6239
[11]   Yew E, Sheppard C 2007 *Opt. Commun.* **275** 453
[12]   Zhan Q 2009 *Advances in Optics and Photonics* **1** 1
[13]   Khonina S N, Golub I 2012 *Journal of the Optical Society of America A* **29** 2242
[14]   Lerman G M, Levy V 2008 *Opt. Express* **16** 4567
[15]   Kozawa Y, Sato S 2007 *Journal of the Optical Society of America A* **24** 1793
[16]   Khonina S N, Alferov S V, Karpeev S V 2013 *Optics Letters* **38** 3223
[17]   Savelyev D A, Khonina S N 2015 *Computer Optics* **39** 654
[18]   Savelyev D A, Khonina S N and Golub I 2016. AIP Conference Proceedings vol. 1724 (AIP Publishing) p 020021.
[19]   Khonina S N, Savelyev D A 2013 *Journal of Experimental and Theoretical Physics* **117** 623
[20]   Khonina S N, Degtyarev S A 2016 *Journal of Optical Technology* **83** 197
[21]   Khonina S N, Karpeev S V, Alferov S V, Savelyev D A 2013 *Computer Optics* **37** 76
[22]   Khonina S N, Karpeev S V, Alferov S V, Savelyev D A, Laukkanen J, Turunen J 2013 *J. Opt.* **15** 085704
[23]   Dehez H, April A, Piché M 2012 *Opt. Express* **20** 14891
[24]   Wang H, Shi L, Lukyanchuk B, Sheppard C, Chong C T 2008 *Nat. Photonics* **2** 501
[25]   Khonina S N 2010 *Computer Optics* **34** 461
[26]   Khonina S N, Degtyarev S A 2015 *Journal of Russian Laser Research* **36** 151
[27]   Kotlyar V, Stafeev S 2009 *Computer Optics* **33** 52
[28]   Khonina S, Degtyarev S, Savelyev D, Ustinov A 2017 *Optics Express* **25** 19052
[29]   Babadjanyan A J, Margaryan N L, Nerkararyan K V 2000 *J. Appl. Phys.* **87** 3785
[30]   Yu Y J, Noh H, Hong M H, Noh H R, Arakawa Y, Jhe W 2006 *Opt. Commun.* **267** 264
[31]   Grosjean T, Fahys A, Suarez M, Charraut D, Salut R, Courjon D 2008 *J. Microsc.* **229** 354
[32]   Berweger S, Atkin J M, Olmon R L, Raschke M B 2012 *J. Phys. Chem. Lett.* **3** 945
[33]   Gorodetski Y, Drezet A, Genet C, Ebbesen T W 2013 *Physical review letters* **110** 203906
[34]   Savelyev D A, Khonina S N 2014 *Journal of Physics: Conference Series* **490** 012213

SPBOPEN 2018   IOP Publishing

NEXAFS C1s-spectra apparatus distortion study

A E Mingaleva[1,2], O V Petrova[1,2], D V Sivkov[1,3], N N Shomysov[1], S V Nekipelov[1,2], R N Skandakov[1], V N Sivkov[1]

[1]Komi Science Centre of Urals Branch, RAS, 167982, Syktyvkar, Russia
[2]Pitirim Sorokin Syktyvkar State University, 167001, Syktyvkar, Russia
[3]Immanuel Kant Baltic Federal University, 236041, Kaliningrad, Russia

Abstract. The results of the comparison of TEY and transmission methods of X-ray absorption spectra measurements as well as "thickness effect" modeling in the linear absorption coefficient spectral dependences of the C_{60}-films in the NEXAFS C1s-spectra are represented. The calculations were performed with using linear absorption coefficient spectra obtained in TEY mode as the true (undistorted) data. The modelling results are in good agreement with the experiment. All NEXAFS C1s-spectra was carried out at the BESSY II using synchrotron radiation from the Russian-German dipole beamline.

Introduction

Absorption spectroscopy with soft X-ray (25-1500 eV) analyzes the near edge X-ray absorption fine structure (NEXAFS) for identification of condensed mater atomic and chemical composition and electron structure. NEXAFS is formed by electron excitation from core to unoccupied states of mater structural unit (atom, molecule, cluster or quasimolecule) and investigate mater in nanosize scale.

Traditionally, the X-ray absorption coefficient $\mu(E)$ is determined in transmission mode using the sample of known thickness d by measuring the intensities of incident I_0 and transmitted I through a sample monochromatic X-ray beam:

$$\mu(E) = \frac{1}{d} ln\left(\frac{I_0}{I}\right), \tag{1}$$

where the intensities I_0 and I are measured with correct evaluation of background radiations. This method requires the sample in the form of very thin film because of the radiation's short absorption length. More over it is well known that the intensities and form of X-ray absorption $\mu(E)$ thin structure is distorted together with sample thickness d growth [1-3]. Such spectra distortions results from spectrometer apparatus function are known in literature as "thickness effect" [1, 4]. "Thickness effect" distortion express in intensity reduction and broadening of sharp resonances within contrasts structure and unification of smaller peaks.

Due to apparatus function $\varphi(E-E')$ which is normalized per unit the experimentally measured I_0, I and μ would differ from their true values I_0', I' and μ' [2,3]:

$$\mu(E) = -\frac{1}{d} \cdot ln\left[k \int e^{-\mu'(E') \cdot d} \cdot \varphi(E - E')dE'\right], \tag{2}$$

where k - normalization constant. Due to thin lines absence in soft X-ray emission spectra the form of apparatus function in this region as well as $\mu(E)$ true value can't be identified. Therefore, in modeling, this function is chosen as the dispersion.

At present time the absorption coefficient measurements are commonly carried out by registration of total electron yield (TEY) from the investigated sample. NEXAFS realized in TEY mode probe only thin near-surface layer up to 50 $\mathring{A}$ [5]. This fact allows neglecting the "thickness

Content from this work may be used under the terms of the Creative Commons Attribution 3.0 licence. Any further distribution of this work must maintain attribution to the author(s) and the title of the work, journal citation and DOI.
Published under licence by IOP Publishing Ltd

effect" distortion influence for NEXAFS-spectra measured in TEY mode which is proportional to linear absorption coefficient and incident radiation intensity [6].

In current research a comparison between NEXAFS $C1s$-spectra measured as in transmission and in TEY mode and correct evaluation of background radiation is represented on the example of nanostructured carbon material – fullerene C_{60}. Moreover the modelling of NEXAFS $C1s$-spectra "thickness effect" distortion for thin C_{60}-layers using as the true value σ' in equitation (2) the TEY-spectra was made. This modelling allow to estimate the optimal C_{60}-layers thickness for better correlation between transmission and TEY spectra.

The choice of polycrystalline fullerite C_{60} as a test compound for current resource is limited by following factors: (i) the fullerite C_{60} has one absorption edge and narrow multi-peaks contrasts fine structure and (ii) it can be prepared as thin film for transmission method and fine crystalline powder for TEY method.

Experimental and simulation details

All NEXAFS $C1s$-spectra of fullerene C_{60} measured in TEY and transmission mode was carried out at the Berliner Elektronenspeicherring fur Synchrotronstrahlung II (BESSY II) using radiation from the Russian-German dipole beamline (RGBL).

For suppression and estimation of background radiation in incident and transmitted through thin sample film synchrotron radiation and measured TEY-signal the specific absorption 220 nm Ti-film filter was used [7,8]. In TEY-experiment, Ti-filter was fixed on gold mesh with a small cell and placed between exit slit and analytical chamber. In transmission the Ti-filter was used as a film for C_{60}-layers deposition. The photon energy was calibrated by first narrow peak (285.38 eV) in NEXAFS $C1s$-spectra of HOPG [9].

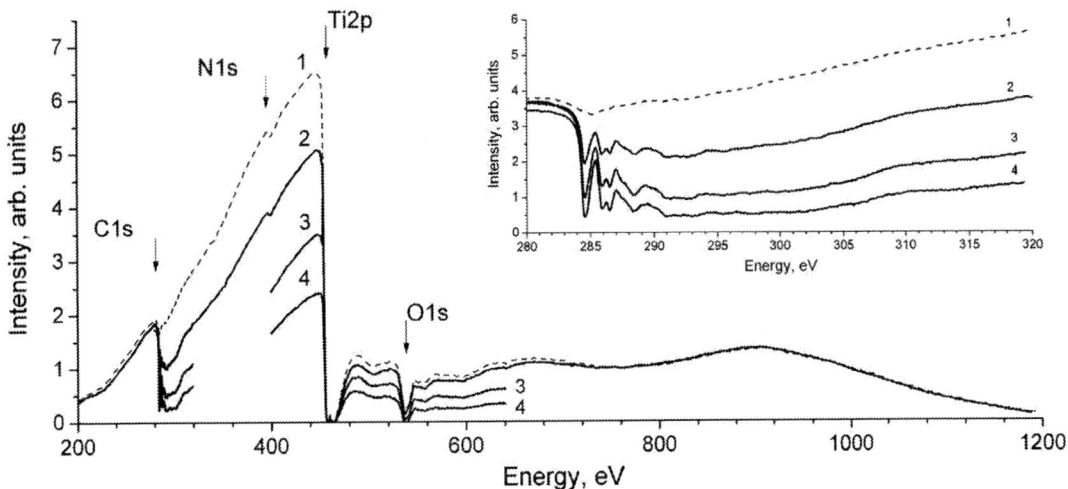

Figure 1. Transmitted intensities of Ti-film with thickness d=220 nm (curve 1) and deposited on it C_{60}-layers with thickness d=60 nm (curve 2), d=145 nm (curve 3), d=260 nm (curve 4).

For transmission experiments the samples were prepared in the form of 60, 145 and 260 nm C_{60}-layers deposited by vacuum thermal evaporation to 220 nm free Ti-films and then fixed on Cu-holder with round slits, which linear sizes considerably exceeding the diameter of the radiation beam on the sample. The linear absorption coefficient was determined by expression (1). For measurement of incident radiation the 220 nm free Ti-films was used. Incident and transmitted radiation intensities were measured by TEY-signal registration from clear Au-photocathode. In Fig.1 the spectral dependencies of monochromatic (i.e. without background) intensities of transmitted throw 220 nm free

Ti-films (I_0) and throw 60, 145 and 260 *nm* C_{60}-layers deposited on *Ti*-films in wide energy range and in *C1s*-adge (the inset) are presented.

The samples for the investigation by TEY method were prepared by fullerite C_{60} pressing on the surface of *Cu*-holder. The incident photon flux was measured using the clean *Au*-photocathode. The incident monochromatic intensity in arbitrary units was obtained by means of the division the TEY monochromatic signal of *Au* plate by the *Au* atomic cross section [8].

The modelling of NEXAFS *C1s*-spectra "thickness effect" was carried out by the formula (2) using the TEY-spectra as true (undistorted) value and a dispersion apparatus function:

$$\varphi\left(E - E^{'}\right) = A_0 \left[\frac{1}{(E - E^{'})^2 + \left(\frac{\Delta E}{2}\right)^2} \right], \tag{3}$$

where ΔE – spectrometer energy resolution, which was chosen to be $\Delta E = 0,15 eV$, $A_0 = \Delta E/2\pi$ – normalization factor.

Result and discussion

Figure 2 include the results of the comparison study of $\mu(E)$ spectral dependences of C_{60} in NEXAFS *C1s*-spectra, measured by transmission methods for C_{60}-film thickness $d = 260$ *nm* and in TEY mode. In Figure 3 partial *C1s* $\mu(E)$ spectral dependencies in arbitrary units and after normalization to unity at 315 *eV* are shown. The figure demonstrates well correlation of TEY and transmitted spectra in numbers of peaks and its energy position within discrete area before the *C1s*-edge (a-d peaks). However in this area the clear differences in the relative peak intensities are observed. In the transmission spectra, the structure located above the absorption edge (e-k) is low-contrast and has a significantly lower intensity in comparison with similar structure in TEY-spectra.

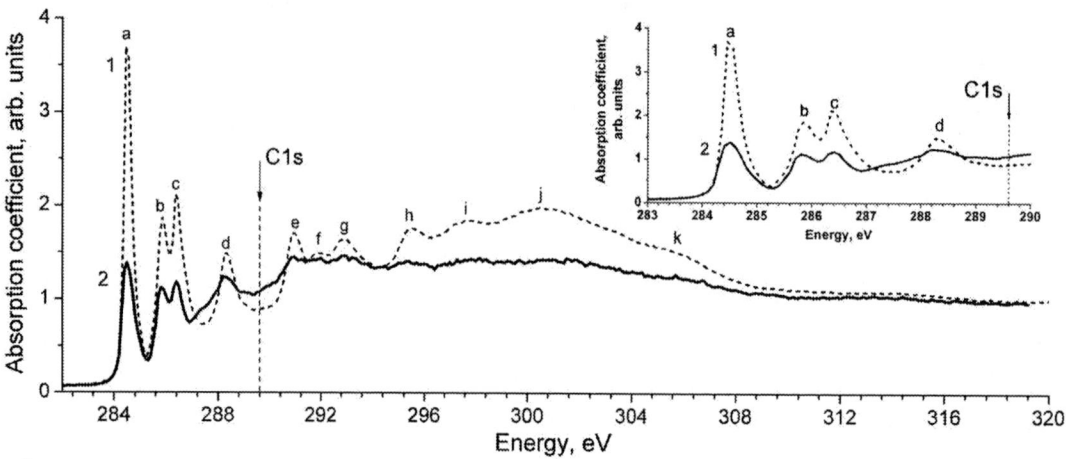

Figure 2. Linear absorption coefficient spectral dependences of C_{60} in NEXAFS *C1s*-spectra, measured by TEY (curve 1) and transmission methods for C_{60}-film thickness $d = 260$ *nm* (curve 2). The *C1s*-edge is indicated by the arrow.

The results of "thickness effect" modeling for series of sample thickness $d = 60$ *nm* (curve 2), $d = 145$ *nm* (curve 3) and $d = 260$ *nm* (curve 4) are shown in figure 3. Figure 3 demonstrates the strong influence of the thickness effect in the spectrum of fullerite in the NEXAFS *C1s*-absorption edge, which is expressed in the decreasing of the area of narrow peaks and its intensity in the maximum with an increase of the sample thickness. The developed modeling program allows evaluating the optimal sample thickness for studying it by the photoabsorption method. This program can be adapted for the study of thickness effect in other samples spectra. Calculation results are in good agreement with NEXAFS *C1s*-absorption spectra of fullerite C_{60} obtained by the transmission method.

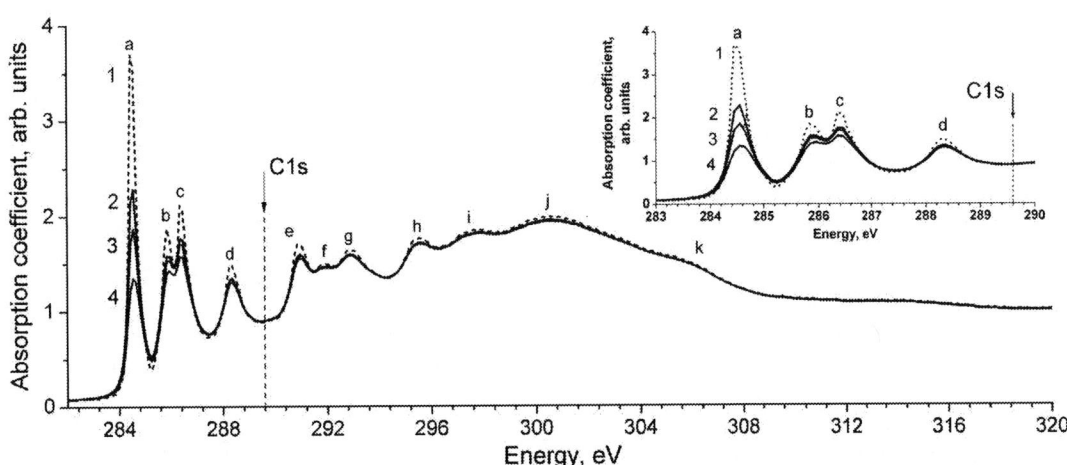

Figure 3. Linear absorption coefficient spectral dependences of C_{60} in NEXAFS *C1s*-spectra, measured by TEY method (true spectrum) (curve 1) and calculated for different C_{60}-film thickness d=60 *nm* (curve 2), d=145 *nm* (curve 3) and d=260 *nm* (curve 4). The *C1s*-edge is indicated by the arrow.

Conclusion

The current study allows us to conclude that transmission method cannot be used for correct X-ray linear absorption coefficient $\mu(E)$ measurements especially for the NEXAFS included narrow selective peaks and contrast structure, since apparatus distortions arise even at sample thickness of several tens *nm*. The modelling results indicate that due to "thickness effect" the narrow peaks are reduced in intensity and broadening while the continuum peaks undistorted. The modelling show the good agreement between simulating and experimental spectra measured in transmission mode for C_{60}-film thickness d=260 *nm*.

Acknowledgments

This work was supported by Program of UB RAS №18-10-2-23; grants RFBR and Komi Republic 16-42-110610 p-a, 16-43-110350 p-a; the Bilateral Program of the Russian-German Laboratory at BESSY II and the Russian Academic Excellence Project at the Immanuel Kant Baltic Federal University.

References

[1] Parrat L G, Hempstead C F, Jossem E L 1957 *Phys. Rev.* **105** 1228
[2] Sivkov V N, Vinogradov A S, Nekipelov S V, Sivkov D V, Vyalikh D V, Molodtsov S L 2006 *Opt. Spectrosc.* **101** 724
[3] Sivkov V N, Vinogradov A S 2002 *Opt. Spectrosc.* **93** 395
[4] Lukirskii A P, Zimkina T M 1963 *Izv. Akad. Nauk SSSR* **27** 324
[5] Stöhr J 1992 *NEXAFS Spectroscopy* (Berlin: Springer Verlag) p 403
[6] Gudat W, Kunz C 1972 *Phys. Rev.* **29** 169
[7] Kummer K, Sivkov V N, Vyalikh D V, Maslyuk V V, Bluher A, Nekipelov S V, Bredow T, Mertig I, Mertig M, Molodtsov S L 2009 *Phys. Rev.* B **80** 155433
[8] Sivkov V N, Ob''edkov A M, Petrova O V, Nekipelov S V, Kremlev K V, Kaverin B S, Semenov N M, Gusev S A 2015 *Physics of the Solid State* **57** 197
[9] Batson P E 1993 *Phys. Rev.* **48** 2608

SPBOPEN 2018

Raman lidar with for geoecological monitoring

J Ruzankina[1,3]**, V Elizarov**[2]**, L Konopel'ko**[1]**, A Zhevlakov**[2]**, A Grishkanich**[3]

[1]Department of Ecology and Technosphere Safety, ITMO University, St. Petersburg, 197101, Russia
[2]Research center of laser physics, ITMO University, St.Petersburg, 197101, Russia
[3]Department of Quantum Electronics and Opto-Electronic Devices, St.Petersburg State Electrotechnical University, St.Petersburg, 197376, Russia

Abstract. Application of CCD array photodetector technology is shown to provide record synchronously hundreds of spectral intervals, increase essentially the relative aperture and intensity of echo-signals in the input optical path as well as to reduce the mass – dimension parameters of hyperspectral Raman LIDAR as a whole for geoecological monitoring by aerial and underwater unmanned vehicles.

1. Introduction

The detection of methane gas is extremely important for health and safety reasons and its monitoring is required in many areas, such as water-treatment plants, the oil and gas industry, landfill sites, and commercial or domestic environments. There is increasing need to quantify methane being lost in production processes, so analytical detection methods will become important if public policies on climate change and greenhouse gas emissions are tightened to combat global warming.

As a consequence of the increasing amounts of methane in the atmosphere posing an environmental and health hazard, many researchers have focused their attention upon quantifying the amount of methane in a sample of air. There is a plethora of analytical techniques available to detect methane, ranging from infrared (IR) spectroscopy to electrochemistry, through to gas chromatography (GC); each technique carries its own advantages and disadvantages.

The authors of article propose to use hyperspectral anti-Stokes Raman spectrometry for geoecological monitoring. Raman LIDAR allows to register simultaneously a wide composition of chemical substances under sensing at one lasing wavelength. LIDAR s with ultraspectral polycromator are used in geoecology for studying the dangerous gases of technogenic origin that appear in atmosphere under disposing of wastes, trash, poisonous and explosive substances, and for searching the growth of narcotic plants. CARS technique is able to measure the concentration on level of 3-10 molecules and determine hydrocarbons under presence of impurities in real atmosphere and hydrosphere [1,2].

2. Experiments and results

A few ultraspectral Raman LIDARs detecting an iodine radionuclides, heavy hydrocarbons as well as methane leaks on the gas pipelines are elaborated and tested [3,4]. The sensing is carried out at wavelength of 261.7 nm by 7ns - pulses with energy of 10 mJ and 200 Hz repetition rate of Nd:ILF laser. The installed optical head provides the combined scanning of space by laser beam linearly through the movable mirror as well as in spiral trajectory using deflector. Polycromator have ensured spectral resolution $\Delta v = 0.6$ cm-1 and selectivity $v/\Delta v \gg 1000$. The swing speed of mirror in angle of $\pm 10o$ at

Content from this work may be used under the terms of the Creative Commons Attribution 3.0 licence. Any further distribution of this work must maintain attribution to the author(s) and the title of the work, journal citation and DOI.
Published under licence by IOP Publishing Ltd

frequency of 20 Hz allows to track the items space on inclined courses. Deflector scans in narrow angle field (±5o) with 15' sensitivity during 1 ms.

Background noises on receiving canal are eliminated practically to naught due to the detection of echo-signals at the anti-Stoks frequencies.

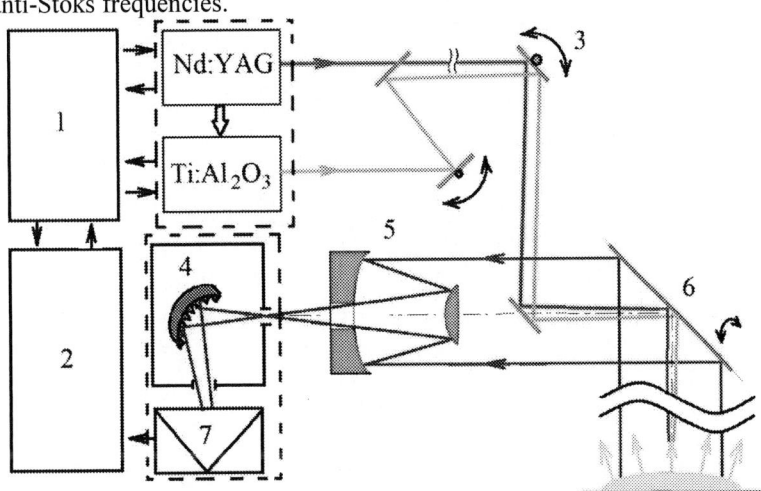

Figure 1. Functional diagram KARS lidar

Unfortunately application of conventional photodetectors (PMT, photodiodes or streak tubes) limits the input lens relative aperture (f/4÷f/3) and number of recorded spectra in receiving channels of Raman LIDAR. Photodetectors should be located at a distance to each other in polychromator. Therefore a small aperture is disadvantage of Raman LIDAR. The authors propose the use of highly sensitive matrices. Modern progress in the technology of CCD array photodetector with high sensitivity in wide spectral range provides to record synchronously hundreds of spectral intervals, increase substantially the intensity of echo-signals in the input optical path (f/2÷f/1.5) and to reduce the mass and dimensions of LIDAR system as a whole. Fig. 2 presents some frequency shifts of the molecules-indicators for hydrocarbon gases.

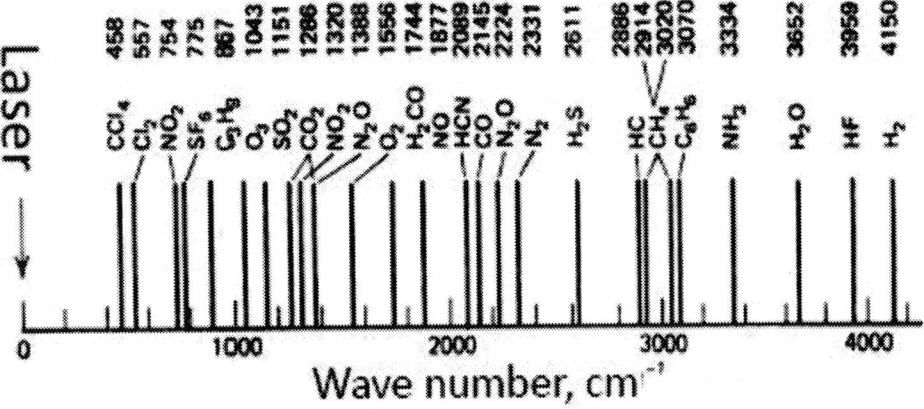

Figure 2. Raman shifts for hydrocarbon gases

For example, the model of S10140-1109-01Back-thinned type CCD area (2048 × 122; 2048 × 506 pixels (table 1)) ensures a high UV sensi-tivity and stable characteristics un-der UV light irradiation (Fig.3) [5].

Figure 3. CCD area sensor

Table 1. The matrix parameters

Image size	24.576 x 6.072 mm
Pixel size	12 x 12 μm
Pixel pitch	12 μm
Number of effective pixels	2048 x 506 pixels
Package	Metal
Line rate (typ.)	40 line/s
Line rate (max.)	60 line/s
Spectral response range	200 to 1100 nm
Dark current (typ.)	30 e-/pixel/s
Readout noise (typ.)	4 e-r ms

Back-thinned type CCD area image sensors deliver high quantum efficiency (90% or more at the peak wavelength) in spectral range up to VUV region, and nave great stability for UV region (Fig.4). Moreover, these also feature low noise and are therefore ideal for low-light level detection.

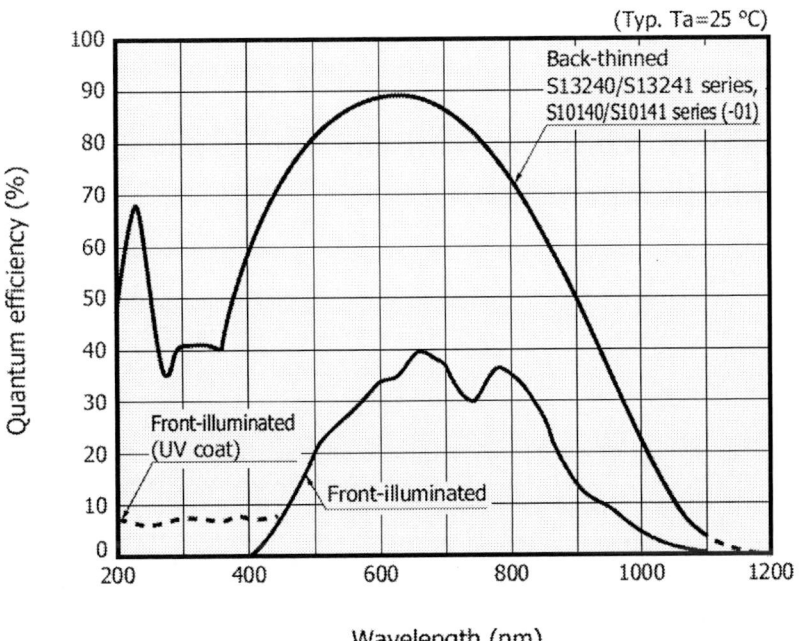

Figure 4. Spectral response

The authors develop a hyperspectral Raman LIDAR for geoecological monitoring for both unmanned and unmanned underwater vehicle. A polychromator with $\lambda/\Delta\lambda \gg 1000$ is used to ensure the registration of a wide range of pollutant substances in the lidar receiving channel. The application of the matrix receiver enables the use of more than 100 spectral channels. Information about the intensity of each spectral channel is extracted as a TTL signal from a group of pixels of the matrix with a frequency of at least 10 MHz by means of a multichannel deserting head C10150 or C10151. The detection head converts the spectrum image into an analogy video signal, which is transmitted via the C7557-01 controller to the PC via a USB interface. The resulting sequence of frames is stored on the PC and period is decrypted by specialized software.

3. Conclusion

Studies have shown the prospects of Raman spectroscopy and, in particular, coherent anti-Stokes Raman spectroscopy as a highly sensitive method of remote detection and recognition of a wide class of radionuclides contaminants in the environment. The matrix is used as a photodetector in Raman LIDAR. So it will allow us to achieve extremely high sensitivity and selectivity with a possibility of registration the spectrum of more than 100 indicator substances on-line.

It should be noted that the results given in the article are preliminary. Therefore, the authors propose to continue their research to develop the technology of hyperspectral anti-Stokes Raman spectrometry.

References

[1] Zhevlakov A, Bespalov V, Elizarov V, Grishkanich A, Kascheev S, Makarov E, Bogoslovsky S, Il'inskiy A 2014 *Proc. SPIE* 92451U

[2] Petrov N, Bespalov V, Makarov E, Zhevlakov A, Soldatov 2014 *Laser Optics, International Conference*

[3] Alimov S, Danilov O, Zhevlakov A, Kascheev S, Kosachiov S, Mak A, Petrov S, Ustyugov V, 2009 *J. Opt.Technol.* 76 (4)

[4] Kascheev S, Elizarov V, Grishkanich A, Bespalov V, Vasil'ev S, Zhevlakov A 2014 *Proc. SPIE* 92741K

SPBOPEN 2018   IOP Publishing

Soft X-ray photoemission study of Ba adsorption on the ceramic multiferroic BiFeO$_3$

P A Dementev[1], G V Benemanskaya[1], S N Timoshnev[2]

[1]Laboratory of Surface Optics, Ioffe Institute, Saint Petersburg 194021, Russia
[2]Laboratory of Nanoelectronics, St. Petersburg Academic University, St. Petersburg 194021, Russia

Abstract. Electronic structure of the ceramic multiferroic BiFeO$_3$ and ultrathin Ba/BiFeO$_3$ interface was studied *in situ* using photoelectron spectroscopy with energies in the range of 120–900 eV. The photoemission from the Bi 4*f*, O 1*s*, Fe 2*p*, and Ba 4*d* core levels was studied. An effect of Ba adsorption is found to induce a significant change in all spectra that is originated from the strong interaction with the charge transfer between Fe, O, Bi surface atoms and Ba adatoms. The recharge between $Fe^{3+} \leftrightarrow Fe^{2+}$ ions in surface region induced by Ba adsorption is found that provide enhancing the ferroelectric polarization gain.

1. Introduction

Multiferroics are the arresting group of materials with combining ferroelectric and magnetic properties. Multiferroics produce a lot of attention due to their potential application in spintronic and data storage devices. Multiferroic BiFeO$_3$ (BFO) is a suitable candidate for attaining ferroelectric and antiferromagnetic domain coupling owing to its high Curie temperature of ~1100 K and high Neel temperature of ~643 K [1, 2]. BFO has attracted much attention on account of simultaneously exhibiting magnetic and ferroelectric ordering. BiFeO$_3$ is characterized by a relatively simple crystal structure as the rhombohedrally distorted perovskite. Large ferroelectric polarization was predicted and observed in the BFO thin film in recent publications [3-6]. Electronic properties of BFO studied by X-ray photoelectron spectroscopy have been discussed in a series of experimental works [7-11]. The fine structure of the Fe $2p_{3/2}$ core level was reveal to be on account of both the Fe^{2+} and Fe^{3+} ions [4, 9, 11]. It was found that ferroelectric polarization tends to grow with increasing Fe^{2+} ions [4, 11]. Most of the works focused on study of the BFO core levels, and no works are devoted to the valence band spectra at low excitation energies, when surface and near-surface layers can be studied.

Applications of multiferroic materials in nanoscale devices necessitate studies of electronic properties of both the surface and interface. Studies of these systems provide insight into fundamental mechanisms of coupling between the lattice, spin, and electronic degrees of freedom and resulting order parameters in the bulk and at the interfaces. However, metal/BFO interfaces are still not investigated.

Recently, photoelectron spectroscopy (PES) experiments for the clean BFO have been carried out [1-4]. Most of BFO works focused on core level spectra, and no study is devoted to the valence band. This work present first PES study of Ba/BFO interface.

This paper reports PES studies of evolution in the electronic structure of the Ba/BiFeO interface relevant to charge transfer and to exist of the Fe^{2+} ions connected with the ferroelectric polarization. The data obtained through change in the valence band, Fe 2*p*, Bi 4*f*, O 1*s*, Ba 4*d* core levels spectra show some mechanisms of interface formation and significant interaction between adatoms (Ba) and

Content from this work may be used under the terms of the Creative Commons Attribution 3.0 licence. Any further distribution of this work must maintain attribution to the author(s) and the title of the work, journal citation and DOI.
Published under licence by IOP Publishing Ltd

surface atoms (Bi, Fe, O). We demonstrate that a mechanism of the Ba/BiFeO interface formation is originated from the excited charge transfer in surface region that leads to effect of recharge of $Fe^{3+} \leftrightarrow Fe^{2+}$ ions and increase in amount of the Fe^{2+} ions.

2. Experimental

Photoemission studies were carried out at BESSY II (Helmholtz Zentrum, Berlin) via synchrotron radiation with photon energies in the range of 120–900 eV. Experiments were performed *in situ* in an ultrahigh vacuum of 5×10^{-10} Torr. Ceramic polycrystalline sample of BFO was obtained from initial fine-dispersed powder by sintering processing at a temperature of ~ 1200 K. The BFO sample was preliminary heated *in situ* at a temperature of ~1000 K. Photoelectrons in a cone oriented along the normal to the surface were detected. The normal photoemission spectra from the valence band (VB) and from Bi 4f, O 1s, Fe 2p, Ba 4d core levels were recorded. The total energy resolution was 50 meV.

Step-by-step deposition of Ba submonolayer coverage up to 2 monolayers (ML) was performed onto the BFO surface. To ascertain the Ba coverage, the amount of 1 ML of Ba atoms can be analyzed from the intensity saturation of the Ba 4d core level peak. This made possible to determine the Ba dosage deposited onto the sample to better than 25%. The Ba overlayer is presented in monolayer units because of Ba sticking coefficient is equal to one at least up to 1 ML. Note that 1 ML is defined as one complete layer of Ba atoms and equal to ~ 6.5×10^{14} atoms/cm^2. The normal photoemission spectra from the Bi 4f, O 1s, Fe 2p, Ba 4d core levels were recorded.

3. Results and discussion

Overview of the normal photoemission spectrum for the clean BiFeO$_3$ ceramic sample at excitation energy of *hv* = 400 eV is presented in Figure 1a, curve 1. The spectrum clearly shows the principal Bi 4f peak and denotes a lack of foreign contaminations. As can been seen, the photoemission in the valence band region (VB) has too little intensity and could be measured at lower excitation energies (see Figure 1b).

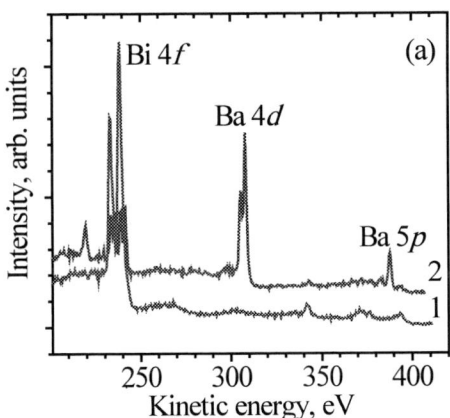 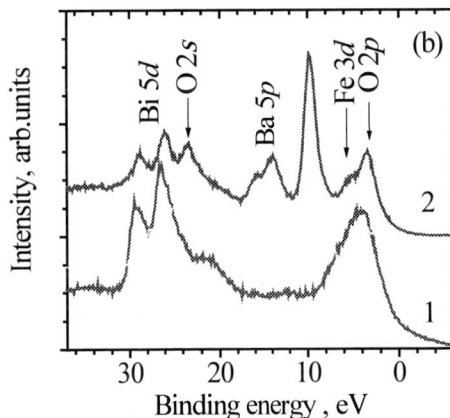

Figure 1(a, b). (a) Normal photoemission spectrum for the clean BiFeO$_3$ (curve 1) ceramic sample and after 0.9 ML of Ba adsorption (curve 2). Excitation energy *hv* = 400 eV; **(b)** Normal photoemission spectra in the valence band region of the BiFeO$_3$ ceramic sample at different Ba coverages: 1 – clean sample, 2 – 0.9 ML of Ba. Excitation energy *hv* = 120 eV.

The normal photoemission spectrum of the ceramic BFO sample in the VB region and Bi 5d core level is shown in Fig. 1b, curve 1. The excitation energy is 120 eV. The position of the valence band maximum $E_{VBM} = 0$ eV is determined by extrapolating a linear fit to the leading edge of the VB photoemission. A broad band between binding energies of 0 and 10 eV below the E_{VBM} is obtained.

Calculations show that the width of the VB is about 8 eV [4, 8]. It is slightly less than obtained experimentally. Two small features located at binding energies of ~4 eV and ~6 eV can be related to the photoemission from the O $2p$ and Fe $3d$ core levels [5, 8, 12]. The broad feature between 20 and 30 eV is believed to consist of two separate structures, of which at least one is due to Bi $5d$ core level. It is evident that the Bi $5d$ are composed of two components, the lower-binding energy ones Bi $5d_{3/2}$ at 26 eV and the higher-binding energy ones Bi $5d_{5/2}$ at 29 eV. An unresolved shoulder at about 22 eV can be attributed to the O–related state, namely, O $2s$ as shown in [13].

Figure 2a presents photoemission spectra for the Bi $4f_{5/2}$ and Bi $4f_{7/2}$ core levels doublet. Shape of Bi $4f$ spectrum for the clean BFO (curve 1) coincides well with the data known from literature, for example [2]. The Ba adsorption on BFO surface is found to cause splitting the double structured spectrum of Bi $4f$ into two double structured ones, Figure 2a, curve 2. The initial Bi $4f$ core level peaks are shifted slightly to higher binding energy. The appearance of extra fine structure demonstrates the obvious chemical shift of the Bi $4f$ core level peaks corresponding to the photoemission from Bi atoms in the interface area.

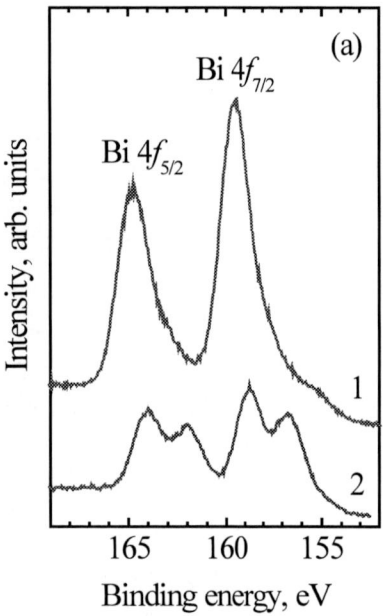 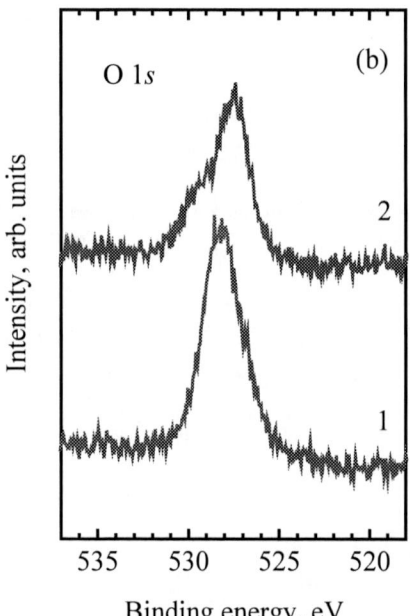

Figure 2(a, b). (a) Normal photoemission spectra of Bi $4f$ core level: 1 – clean BFO, 2 – Ba/BFO interface at Ba coverage 0.5 ML. Excitation energy $hv = 400$ eV; **(b)** Normal photoemission spectra of O $1s$ core level: 1 – clean BFO, 2 – Ba/BFO interface at Ba coverage 0.9 ML. Excitation energy $hv = 900$ eV.

A pronounced change induced by Ba adsorption is found to occur in the O $1s$ core level spectrum, as illustrated in Figure 2b. For the clean BFO surface the O $1s$ core level spectrum becomes a single peak at binding energy of 527 eV, Figure 2b, curve 1. The result characterizes the oxygen-metal bond in a single chemical state and the BFO surface without contaminations. Upon Ba adsorption, additional structure is found to appear on the high-binding-energy side of the O $1s$ peak at binding energy of about 530 eV, Figure 2b, curve 2. The emerge of the additional peak may be assigned to Ba adsorption providing interaction of oxygen surface atoms with Ba adatoms that leads to the appreciable chemical shift [9, 14, 15]. The energy shift of ~1 eV for the initial key component of the O $1s$ peak to the lower binding energy indicates that the leading contribution to the chemical shift is the charge transfer, so the character of oxygen covalency is increased [9]. Besides, it should be considered that after Ba deposition oxygen atoms interact with modified Fe ions.

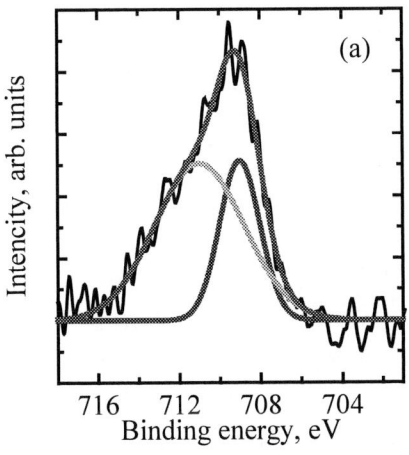 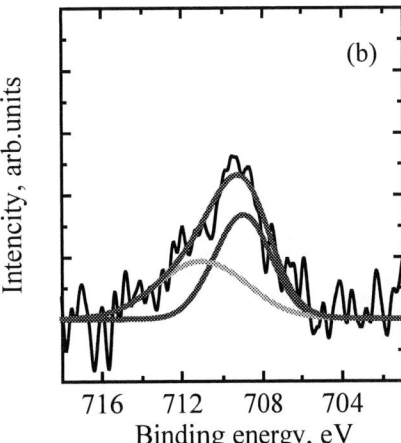

Figure 3(a, b). Decomposition of normal photoemission spectra of Fe $2p_{3/2}$ core level for clean ceramic BFO **(a)** and for Ba/BiFeO$_3$ interface at Ba coverage of 0.9 ML **(b)**. Thin red lines represent experimental data and thick black lines represent fitting results with fitting components. Excitation energy $hv = 850$ eV.

The magnetic and ferroelectric properties of BFO are due to the Fe ions octahedrically surrounded with the oxygen ions, and due to not only Fe^{3+} ions but especially also due to Fe^{2+} ions [4, 7, 8]. So, studies of the photoemission from the Fe $2p$ core level can be used diagnostically to reliably identify the presence both the Fe^{3+} and Fe^{2+} ions. Figure 3 represents photoemission from the Fe $2p_{3/2}$ core level for the clean BFO surface. The broad band Fe $2p_{3/2}$ was decoupled into two subbands using Gaussian function. According to the fitting result, the atomic ratio of Fe^{2+} and Fe^{3+} ions corresponds to $Fe^{2+}/Fe^{3+} \approx 1$ in clean BFO. The Ba adsorption on BFO surface is found to cause significant change in the Fe $2p_{3/2}$ spectrum (Figure 3). As can been seen, the atomic ratio is increased up to the value of $Fe^{2+}/Fe^{3+} \approx 1.5$ for Ba/BFO interface at Ba coverage of 0.9 ML. The observed effect of Ba adsorption signifies the improving ferroelectric polarization. The Ba ultrathin layer on the BFO surface is found to produce the strong interaction between Ba adatoms and Fe ions in interface area that leads to increase the ratio $Fe^{2+}/Fe^{3+} \approx 1.5$. The Fe $2p_{3/2}$ core level evolution is obviously caused by the recharge between $Fe^{3+} \leftrightarrow Fe^{2+}$ ions and may be one of the reasons for enhanced ferroelectric and magnetic properties of BFO due to increasing in the amount of the Fe^{2+} ions.

4. Conclusions

To summarize, the effect of Ba adsorption on the ceramic BFO surface was studied by PES using different excitation energy 120-900 eV. The Ba adsorption is found to modify the valence band spectrum as well as the Fe $2p$, O $1s$, Bi $4f$ core level spectra. It is found that the atomic ratio of the Fe^{2+}/Fe^{3+} ions increases from ~1 up to ~1.5 in the process of the Ba/BFO interface formation. The data point out that the BFO ferroelectric polarization could be enhanced due to increase in the amount of the Fe^{2+} ions under Ba adsorption. The appearance of additional shoulder on the high-binding-energy side of the O $1s$ photoemission peak is found that cause strong interaction of surface oxygen atoms with Ba adatoms. The negative energy shift of the key O $1s$ photoemission component indicates increasing the covalence of oxygen atoms. The Ba adsorption is found to induce splitting of the double structured spectrum of Bi $4f$ core level into two double structured ones accompanied by the chemical shift. It could be attributed to the charge transfer from the Ba adsorbed atoms to the Bi surface atoms with increasing both the Bi-ionicity and electron density in the interface area. The data point out that the complex process of the charge transfer between Ba-Fe, Ba-Bi, Ba-O, Bi-O induced by Ba adsorption occurs in the condition of recharge between $Fe^{3+} \leftrightarrow Fe^{2+}$ ions in the interface area that provide enhancing the ferroelectric polarization gain.

Acknowledgments
We thank the Russian-German Beamline, Synchrotron BESSY II, Helmholtz Zentrum, Berlin. The authors thank M. N. Lapushkin for help in the preparation of the experiment.

References
[1] Martin L W, Crane S P, Chu Y-H, Holcomb M B, Gajek M, Huijben M, Yang C-H, Balke N, Ramesh R 2008 J. Phys.: Condens. Matter **20** 434220
[2] Baettig P, Ederer C, Spaldin N A 2005 Phys. Rev. B **72** 214105
[3] Li J, Wang J, Wuttig M, Ramesh R, Wang N, Ruette B, Pyatakov A P, Zvezdin A K, Viehland D 2004 Appl. Phys. Lett. **84** 5261
[4] Gao F, Cai C, Wang Y, Dong S, Qiu X Y, Yuan G L, Liu Z G 2006 J. Appl. Phys. **99** 094105
[5] Neaton J B, Ederer C, Waghmare U V, Spaldin N A, Rabe K M 2005 Phys. Rev. B **71** 014113
[6] Gareeva Z V, Zvezdin A K 2009 Phys. Stat. Solidi RRL **3** 79
[7] Higuchi T, Liu Y-S, Yao P, Glans P-A, Guo J, Chang C, Wu Z, Sakamoto W, Itoh N, Shimura T, Yogo T 2008 Phys. Rev. B **78** 085106
[8] Mandal S, Ghosh C K, Sarkar D, Maiti U N, Chattopadhyay K K 2010 Sol. Stat. Sci. **12** 1803
[9] Kozakov A T, Kochur A G, Googlev K A, Nikolsky A V, Raevski I P, Smotrakov V S, Yeremkin V V 2011 J. Electron Spectrosc. Relat. Phenom. **184** 16
[10] Infante I C, Juraszek J, Fusil S, Dupe B, Gemeiner P, Dieguez O, Pailloux F, Jouen S, Jacquet E 2011 Phys. Rev. Lett. **107** 237601
[11] Yu B, Li M, Wang J, Hu Z, Liu X, Zhu Y, Zhao X 2012 Thin Solid Films **520** 4089
[12] Li S, Morasch J, Klein A, Chirila C, Pintilie L, Jia L, Ellmer K, Naderer M, Reichmann K, Groting M, Albe K 2013 Phys. Rev. B **88** 045428
[13] Jeon Y, Liang G, Chen J, Croft M, Ruckman M W, Marzo D D, Hegde M S 1990 Phys. Rev. B **41** 4066
[14] Bocquet A E, Fujimori A, Mizokawa T, Saitoh T, Namatame H, Suga S, Kimizuka N, Takeda Y, Takano M 1992 Phys. Rev. B **45** 1561
[15] Lindberg P A P, Shen Z-X, Wells B O, Dessau D S, Mitzi D B, Lindau I, Spicer W E., Kapitulnik A 1989 Phys. Rev. B **39** 2890

SPBOPEN 2018

UV-induced refractive index changes due to silver molecular clusters in photo-thermo-refractive glass

V V Gorbyak, A I Sidorov

Department of Optical Informational Technologies and Materials, ITMO University, St. Petersburg 199034, Russia

Abstract. It is shown experimentally that in photo-thermo-refractive glasses the transformation of charged silver subnanosized molecular clusters to neutral state by UV irradiation results in the increase of glass refractive index. Quantum chemistry calculations suggest that such refractive index change is mainly dominated by silver neutral cluster polarizability. In addition to discussion of proposed mechanism, the examples of possible applications are also presented.

1. Introduction

Local modulation of the refractive index of glasses and other transparent optical materials is used for recording of holograms, optical waveguides and other elements of integrated optics. In glasses without sensitizers local modulation of refractive index can be obtained by focused IR femtosecond laser [1]. On the other hand, photo-thermo-refractive (PTR) glasses, doped with silver ions, are an attractive photosensitive material for recording commercially available Bragg gratings and for developing new polyfunctional materials for photonics and integrated optics. Here we experimentally demonstrate that the local refractive index change (RIC) in PTR glasses during low-intensity continuous UV irradiation is sufficient for optical devices fabrication (optical waveguides, holograms, etc.).

2. Experimental and results

The glasses, composed of $Na_2O–ZnO–Al_2O_3–SiO_2–NaF–NaCl$, and doped with Ag_2O (0.12 mol%), CeO_2 (0.07 mol%) as photosensitizer, and Sb_2O_3 (0.04 mol%) as reductant were used for investigations. As is shown in [2], this type of glasses initially contains silver ions and charged silver molecular clusters (SMCs), which have weak visible luminescence. To direct writing of waveguide, we photosensitized the glass sample by irradiation by UV mercury lamp through the mask. For the measurements of RIC in the other group of samples the volume holograms were recorded using continuous wave 325 nm He–Cd laser (Kimmon IK3501R-G). Laser power density during hologram recording was 0.75 mW/cm², irradiation dose was 6 J/cm². After UV exposure, coinciding with Ce^{3+} ions absorption, and thermal treatment at 350 °C during 1h, the charged SMCs are transformed into neutral ones possessing intense light emission in the visible and the positive RIC can reach $0.76 \cdot 10^{-4}$.

Content from this work may be used under the terms of the Creative Commons Attribution 3.0 licence. Any further distribution of this work must maintain attribution to the author(s) and the title of the work, journal citation and DOI.

Published under licence by IOP Publishing Ltd

SPBOPEN 2018   IOP Publishing
IOP Conf. Series: Journal of Physics: Conf. Series **1124** (2018) 051038   doi:10.1088/1742-6596/1124/5/051038

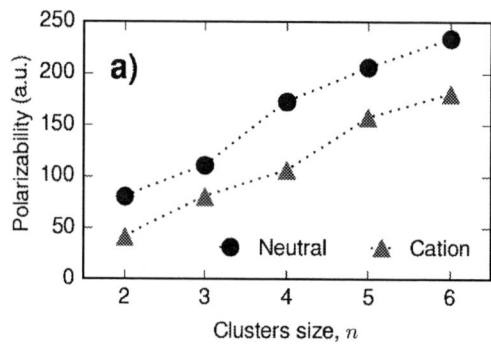

Figure 1. Calculated static polarizabilities of neutral and positively charged SMCs.

In order to gain insight into the possible source of the RIC, the static electronic dipole polarizabilities of neutral and charged SMCs were calculated (Fig. 1) using time-dependent density functional theory. All calculations were carried out with Amsterdam Density Functional and Dalton quantum chemistry programs. Employing the well-known Lorentz-Lorenz equation, we estimated the relative change of the molar refractivity between exposed and unexposed areas. According to our results, the higher polarizability values for neutral SMCs mainly determine the RIC.

As an example, we demonstrate the effect of self-focusing of Gaussian continuous UV laser beam in the glass bulk. The sample was irradiated by UV mercury lamp to demonstrate lack of self-focusing the semiconductor laser beam ($\lambda = 405$ nm). It shows on the figure 2-right.

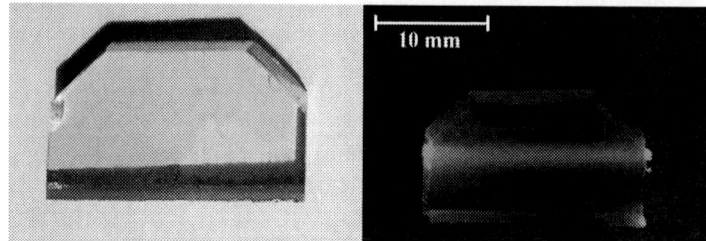

Figure 2. Photo of the explored sample. Left – virgin glass; right – luminescence of PTR glass excited by semiconductor laser with parallel beam and $\lambda = 405$ nm before He-Cd laser irradiation. Excitation beam is entering through the left edge

Then, we were irradiating the sample by continuous UV single-mode He-Cd laser ($\lambda = 325$ nm) with parallel beam. After irradiation the luminescence cone is formed in the glass bulk (Fig. 3). The wavelength of laser radiation matches the absorption band of Ce^{3+} ions. In the process of the irradiation Ce^{3+} ions give away the electrons and the Sb^{5+}, Ag^+, Ag_n^+ trap them. The charged SMCs are transformed to neutral state. The polarizability of neutral SMCs is higher than of the charged ones. In this way in the area with neutral SMCs the refractive index is increased. By this the extended positive lens is formed.

728

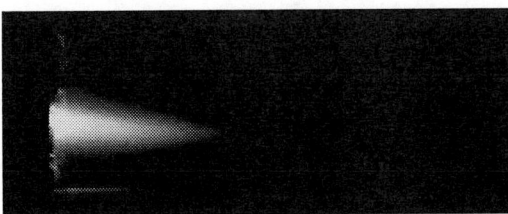

Figure 3. Luminescence of PTR glass excited by semiconductor laser with parallel beam and $\lambda = 405$ nm after He-Cd laser irradiation. Excitation beam is entering through the left edge

After that the sample was heat treated at $T = 350$ °C (less than T_g) during for 1 hour. On the figure 4 you can see that the area of the luminescence is increased as compared with the luminescence area in the Fig. 3. Rise of the luminescence area and increase of the luminescence intensity take place because the complex $(Sb^{5+})^-$ give away electrons during the thermal treatment and the Ag^+ and Ag_n^+ trap them. Fig. 5 demonstrates the difference between luminescence intensity before and after laser irradiation and thermal treatment.

Figure 4. Luminescence of PTR glass excited by semiconductor laser with parallel beam and $\lambda = 405$ nm after He-Cd laser irradiation and thermal treatment at $T = 350$ °C during 1 h. Excitation beam is entering through the left edge

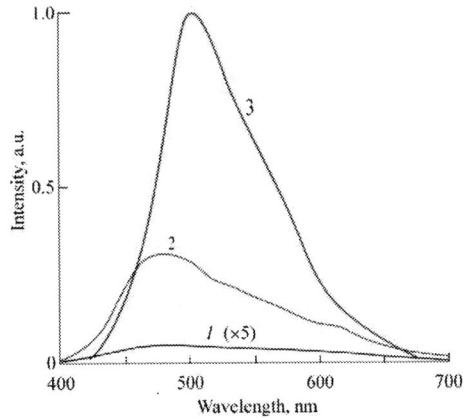

Figure 5. Luminescence spectra of the sample. 1 – before He-Cd laser irradiation; 2 – after He-Cd laser irradiation; 3 – after He-Cd laser irradiation and thermal treatment less than T_g. Excitation wavelength is 360 nm

Fig. 6 shows the dependence of luminescence area diameter changing via the distance in glass bulk.

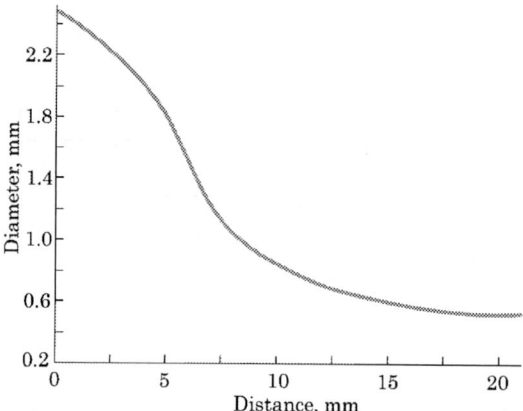

Figure 6. Diameter of luminescent area via distance after He-Cd laser irradiation and thermal treatment at T = 350 °C during 1 h.

3. Conclusion

It is shown that UV exposure without any chemical processing or high-temperature treatment can be used to fabricate optical waveguides, holograms, etc in the glasses under study. Photoinduced RIC is mainly caused by photochemical charge-exchange reactions, which reduce charged SMCs into neutral species. Besides general applications in the integrated optics, we believe that considered relatively new technique has potential to be applied for creating variety of photonics devices based on highly emissive SMCs.

Acknowledgments

This work was supported by the Ministry of Education and Science of Russian Federation (Project 16.1651.2017/4.6).

References

[1] Streltsov A M, Borelli N F 2002 *J. Opt. Soc. Am. B* **19** 2496
[2] Dubrovin V D, Ignatiev A I, Nikonorov N V, Sidorov A I, Shakhverdov T A and Agafonova D S 2014 *Opt. Mat.* **36** 753

Solution of the multiple light scattering problem in turbid suspensions via cross-correlation processing

Z A Zabalueva[1], E K Nepomnyashchaya[1], C C Korikov[1]

[1]Institute of Physics, Nanotechnology and Telecommunications, Peter the Great St.Petersburg Polytechnic University, St.Petersburg195251, Russia

Abstract. The original scheme of the cross-correlation technique for assessing the size of colloid particles in suspensions of high turbidity without any dilution is proposed. In the designed experimental scheme the illumination with two in-phase coherent laser beams are used. The scattered radiation signalsare registered by the square-law photodetectors and processed without any special correlators on a computer. It allowed us to simplify the experimental setup. Parameters of the setup were established by calibrating experiments. Cross-correlation functions were obtained for monodisperse suspensions of latex microspheres, and their sizes were assessed.

1. Introduction

The dimensions of particles suspended in a fluid can be determined by such methods as electron beam microscopy, sedimentation analysis, static or dynamic light scattering, cross-correlation method, etc.

Electron beam microscopy is a direct method for determination of the dimensions of nanoparticles. In an electron beam microscope instead of light rays, directed electron rays are used, and the role of optical lenses is played by electromagnetic lenses, which control the movement of electrons by means of a magnetic field. Electron beam microscopy makes it possible to quickly make photomicrographs of particles from different angles [1], but it is inconvenient that the sample requires a long pre-treatment, such as drying and deposition on the substrate.

In sedimentation analysis, particles in a fluid are settled by gravity or centrifugal force. By the rate of mass increase of the sediment, a kinetic curve is constructed — the dependence of the mass of the settled matter on the deposition time.The sedimentation rate determines the size of the microparticles. The advantage of the method of sedimentation analysis for the study of disperse systems is its high accuracy in the study of highly diluted samples, since the phenomena of coagulation arising from the co-precipitation of particles of various sizes are completely excluded. But this method is not applicable in the case of studies of concentrated and turbid media. Experiments take a long time, and when graphically processing the results, the probability of errors is high.

Nowadays the most common methods for determining the size of colloidal particles are optical methods. This is explained not only by the convenience and quickness of their performing, but also by the accuracy of the obtained results [2]. These methods include the method of static light scattering, the method of dynamic light scattering, and the cross-correlation spectroscopy. The drawbacks of these methods include the fact that the sample must be liquid, so only solutions, suspensions and emulsions can be investigated. The choice of solvent agent for a solid can be extremely difficult because of the possibility of chemical and physical interaction between the dispersed phase and the dispersion medium. Also, these methods are poor in that they give accurate results only for particles which shape is close to spherical.

Content from this work may be used under the terms of the Creative Commons Attribution 3.0 licence. Any further distribution of this work must maintain attribution to the author(s) and the title of the work, journal citation and DOI.
Published under licence by IOP Publishing Ltd

Static light scattering, that is also called laser diffraction, can be used to determine the radii of particles with sizes from few nanometers to millimeter, suspended in a dispersive medium. In the experiments on static light scattering, the angular dependences of the time-averaged intensity of the scattered light are analyzed. Calculations are derived from the Mie's scattering theory describing scattering by homogeneous spheres. This method has a number of advantages, such as short duration and simplicity of measurements, the absence of the need for careful preparation of the sample, a large range of particle diameters which dimensions can be investigated without adjusting the equipment. Static light scattering also does not require user calibration with standard samples and gives a high measurement accuracy. A very small amount of sample is needed to carry out the experiments. It should be noted that it may be difficult to detect low concentrations of small scatterers at large concentrations of large particles. Also this method is not completely correct when the highly turbid suspensions are studied.

Among the optical methods the method of dynamic light scattering and the cross-correlation method should be distinguished [3]. The first one, based on the analysis of the time-depended autocorrelation function of scattered light fluctuations, is successfully used to assess the size and distribution of particle sizes in monodisperse and polydisperse systems. Though under conditions of strong turbidity of a colloidal sample classical dynamic light scattering becomes inapplicable [4]. This limitation can be overcome with the cross-correlation method. Dynamic light scattering and cross-correlation have the same disadvantages as static light scattering. The difference lies in the way in which the experimental data are processed and in the fewer photodetectors necessary for the measuring apparatus [5]. In this paper, we suggest an original scheme of the cross-correlation method, which makes it possible to determine the sizes and diffusion coefficients of nanoparticles in highly turbid suspensions. In our work we consider the task of assessing sizes of dispersed particles in turbid monodisperse suspensions, so we decided to use the cross-correlation method. It, unlike most standard methods, is used for liquid samples and allows us to study turbid suspensions without diluting them.

2. The theory of cross-correlation spectroscopy

In a dispersion medium, the particles are in continuous chaotic motion. Consequently, when a laser beam is incident on such a disperse system, the light that is scattered randomly, changes its amplitude, frequency, polarization, and phase. Singly scattered light carries information about the parameters of the particle on which it was scattered. A method of dynamic light scattering is used to investigate dilute suspensions. The cuvette with the sample under investigation is illuminated by a laser beam and the scattered light is recorded under a certain angle or angels sequentially. The signal received from the square-law photodetector corresponds with the fluctuations in the intensity of the scattered light. For this signal, a temporal autocorrelation function is calculated, in the form of which it is possible to estimate the dimensions of the scatterers. However the method of dynamic light scattering cannot be used for turbid samples, since the photodetector will detect mainly multiply scattered light, which is a significant obstacle. To study turbid suspensions using dynamic light scattering, samples have to be diluted to sufficiently low concentrations [6].

The cross-correlation method solves this problem. With its help, it is possible to study turbid suspensions without diluting them [7]. In the method of cross-correlation two experiments on dynamic light scattering are simultaneously carried out. Two equal laser beams are focused on the same scattering volume inside the cuvette [8]. From the particles of the disperse system, scattering occurs, after which the scattered light is detected at points in the far field by two square-law photodetectors. The idea is that single scattering can be detected from a tightly focused incident beam, whereas multiple scattering tends to arise from a larger fuzzy sort of halo around the incident beam [9]. Based on two signals received from photodetectors, a time-dependent cross-correlation function is calculated. Contributions to the cross-correlation function are created only by singly scattered light. This is the suppression of multiple scattering, which makes it possible to study turbid suspensions without diluting them.

For the normalized cross-correlation function one can use the following approximation [10]. It comes from the theory of Relay scattering on particles in liquids

$$g_{12}(\tau) = A \exp[-2K_1\tau + K_2\tau^2]. \tag{1}$$

Here A, K_1 and K_2 are fitting coefficients. K_1 is called the first cumulant and uniquely is related to the diffusion coefficient D by the ratio [11]

$$K_1 = Dq^2, \tag{2}$$

where q is the magnitude of the scattering vector given by

$$q = \frac{4\pi n}{\lambda} \sin\frac{\theta}{2}. \tag{3}$$

Here θ is the scattering angle, n is the refractive index, and λ is the wavelength of laser radiation. Using the calculated diffusion coefficients from the Stokes-Einstein relation [12] for spherical particles, the hydrodynamic radius r can be determined

$$D = \frac{KT}{6\pi\eta r}, \tag{4}$$

where K is Boltzmann's constant, T is the absolute temperature, and η is the dynamic viscosity of the solvent.

3. Experimental setup

In Fig. 1 a diagram of the developed experimental setup is shown. The beam splitter separates the laser radiation into two equal in-phase beams, which are focused by the lens into the same scattering volume. Photodetectors are located in the plane of the laser radiation at a very small angle to each other symmetrically at an angle of 90° to the direction of the laser beam. They detect the fluctuations in the scattering intensity caused by the Brownian diffusion of colloidal particles through the scattering volume. Scattering light intensity dependence on time is digitized by a digital oscilloscope and processed by a computer. In this scheme we avoided the use of correlator due to specific computer processing [12].

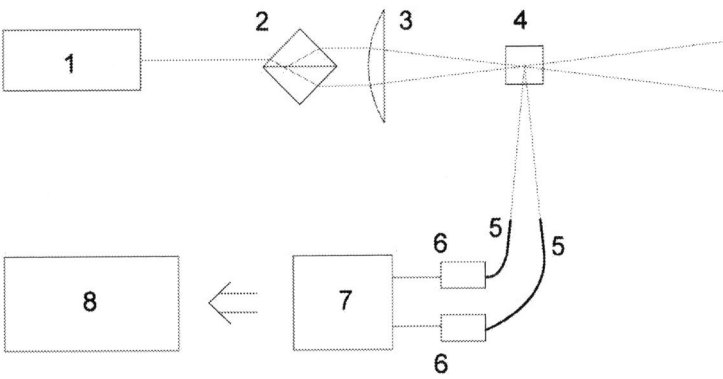

Figure 1. Experimental setup. 1 — laser, 2 — beam splitter, 3 — converging lens, 4 — cuvette with the studied suspension, 5 — multimode optical fiber, 6 — square-law photodetectors, 7 — digital oscilloscope, 8 — computer.

Monodisperse suspensions of latex microspheres with different diameters were chosen as model objects. The cross-correlation functions of the scattering signals were calculated for various parameters of the experimental scheme: the focal spot dimension, angle between the probing beams, the duration of the recorded signals, the digitizing frequency and the position of the photodetectors. The accuracy of the determination of the dimensions of monodisperse particles was studied for various concentrations and sizes of microspheres and was compared with the classical method of dynamic light scattering [13]. As a result of our investigation the following parameters of the experimental setup were established: the angle between the beams focused on the cuvette, the angle between the detected beams, the distance of the points of detection of the scattered light, the distance from the center of the cuvette to the detection plane, and the size of the scattering volume. In the case where the photodetectors are located as shown in Fig. 1, the cross-correlation function tends to decrease with increasing delay time τ. This means that it can be approximated by a curve described by expression (1). Also, our experience shows that in order to obtain the expected form of cross-correlation function, the recording time of the signals must not be less than a certain value. For particles of diameter on the order of a micron, it is sufficient that the duration of the recorded signal is half a second.

Cross-correlation functions are shown in Fig. 2. They are calculated for monodisperse suspensions. It turned out that the concentration of scatterers has almost no effect on the shape of the cross-correlation function, and for particles of different sizes, the rates of decrease of cross-correlation functions was different. The sizes of disperse particles were estimated from the shape of cross-correlation functions. The fitting coefficients for the approximation were selected by the iterative method using MATLAB.

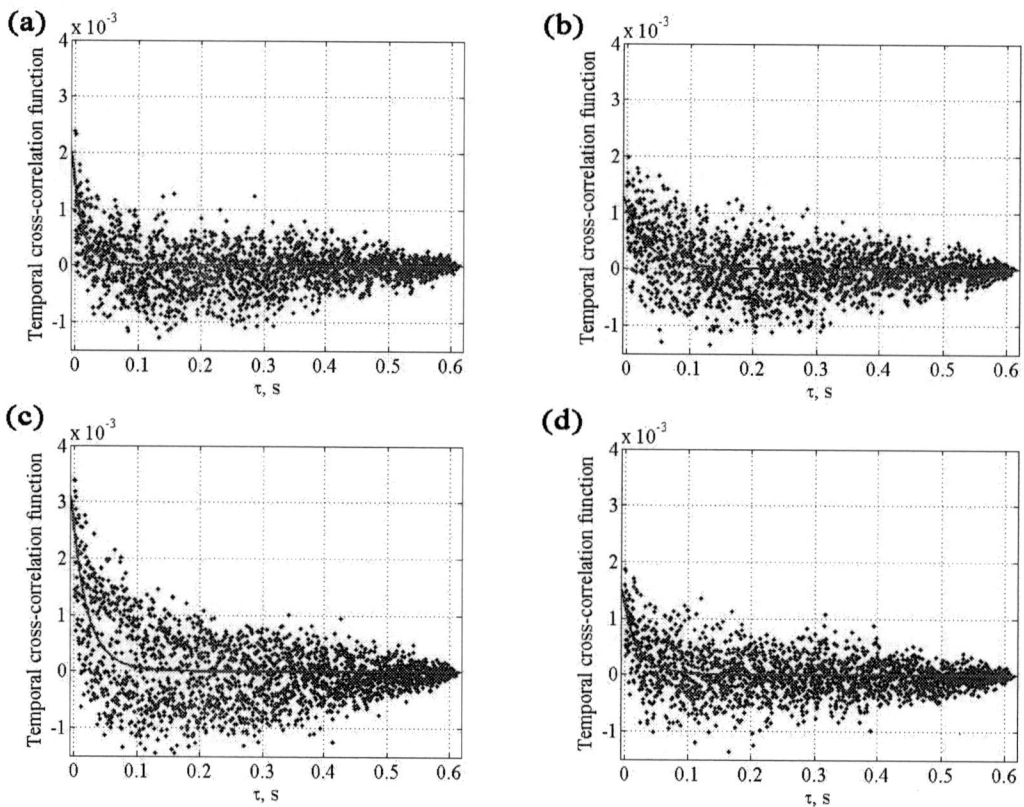

Figure 2(a, b, c, d). Cross-correlation functions for water suspensions of monodisperse microspheres with different diameters and concentrations. **(a)** Diameter 0.32 µm, concentration 0.5%; **(b)** diameter 0.32 µm, concentration 1.5%; **(c)** diameter 0.97 µm, concentration 0.05%; **(d)** diameter 0.97 µm, concentration 0.25%. The dots represent the experimental cross-correlation functions. Smooth curves are the approximations obtained by the formula (1).

4. Results and conclusion

We developed a scheme of an experimental setup for the investigation of turbid suspensions using the cross-correlation method and assembled a measuring stand. Carried out some calibration experiments with the use of suspensions of latex microspheres, which allowed us to select the optimal parameters of the cross-correlation setup.

Now we are developing a technique for processing experimental data to optimize the cross-correlation method and to improve the accuracy of the microparticle parameters estimation especially for polydisperse suspensions. Then we plan to extend the method of cross-correlation to the study of not only the dimensions of the microspheres, but also the velocities of the particles and their shapes in the case of nonspherical scatterers. Cross-correlation is a promising method for studying various turbid liquid samples, combining the advantages of optical methods and wider applicability limits than most other methods.

Acknowledgments

This work is supported by State Assignment in science activity for universities (project № 3.5469.2017).

References

[1] Spivak G V, Saparin G V, Bykov M V 1970 Sov. Phys. Usp. **12** 6
[2] Xu K 2015 Particuology **18** 11-21
[3] Burdíková J, Mravec F, Wasserbauer J, Pekař M 2017 Coll. and Pol. Sc. **295** 1 67–74
[4] Scheffold F 2014 SPIE Digital Library **9232** 923203
[5] Zuccolotto-Bernez AB, Braham NB, Haro-Pérez C, Rojas-Ochoa L F 2016 Ap. Opt. **55** 31 8806-8812
[6] Photocor Mini User Manual. Photocor. Moscow. 2017. 83 p.
[7] Haro-Pérez C, Ojeda-Mendoza G, Rojas-Ochoa 2011 J. of Chem. Phys. **134** 24 244902
[8] Urban C, Schurtenberger P 1998 J. Colloid Interface Sci. **207** 1
[9] Meyer W V, Cannell D S, Smart A E, Taylor T W, Tin P 1997 Appl. Opt. **36** 30
[10] Mazza M G, Giovambattista N, Stanley H E, Starr F W 2007 Phys. Rev. E **76** 021203
[11] Stetefeld J, McKenna S A, Patel T R 2016 Biophys Rev **8** 409–427
[12] Nepomnyashchaya E, Velichko E, Aksenov E. 2016 J. of Phys. Conf. Ser. **769** 1 012025
[13] Nepomnyashchaya E, Zabalueva Z, Velichko E, Aksenov E 2017 EPJ Web Conf. **161** 02017

SPBOPEN 2018   IOP Publishing

Quality factor enhancement in gold-based Tamm plasmon cavity by sub-wavelength structuration

A R Gubaydullin[1,2], K M Morozov[1,2*], M A Kaliteevski[1,2]

[1]St. Petersburg Academic University, St. Petersburg, 194021 Russia
[2]ITMO University, St. Petersburg, 197101 Russia

Abstract. We have demonstrated theoretically using a finite elements method that losses in Tamm plasmon structures can be reduced by using a sub-wavelength structuration of the metal layer. The structures consist of a GaAs/$Al_{0.95}Ga_{0.05}As$ Bragg reflector covered with a sub-wavelength golden grating. The results demonstrate that the quality factor of the Tamm plasmon mode with grating increases substantially, with respect to the similar structure without a grating. Also, we have demonstrated that resonance frequency of Tamm plasmon mode can be tuned by varying the filling factor parameter.

1. Introduction

During the past decade, there are growing interest of researchers to nanostructures where Tamm plasmon modes can be realised. Tamm plasmon is an electromagnetic state localized at the interface between the metal and the specially designed Bragg mirror was predicted recently [1]. Tamm plasmon provides new possibilities for utilizing metallic parts in modern optoelectronic devices. Tamm plasmon modes are forming in both TE and TM polarizations and demonstrate lower losses than conventional plasmon modes. Recently, polarization-controlled confined Tamm plasmon lasers [2] and control of spontaneous emission rate of quantum dots placed in Tamm structure [3] have been demonstrated.

Despite progress in study of Tamm plasmons and many ways of potential applications of them in modern devices, the main obstacles of improving performance of Tamm plasmon modes in lasing optoelectronic devices are losses and heating of metallic parts due to absorption as in the others plasmonic structures [4]. A number of experimental studies on plasmonic devices aimed at the reduction of losses in metal by structuring the metal layer. One of the approaches to reduce losses is a formation of a subwavelength grating on the metallic layer.

This paper is aimed at theoretical analysis of the possibility to tune the Tamm mode resonance position and quality factor by covering the distributed Bragg reflector (DBR) with a periodical golden sub-wavelength grating.

Content from this work may be used under the terms of the Creative Commons Attribution 3.0 licence. Any further distribution of this work must maintain attribution to the author(s) and the title of the work, journal citation and DOI.
Published under licence by IOP Publishing Ltd

2. Results and discussion

The structure under study is Tamm plasmon structure with the sub-wavelength structuration of the covering golden layer (figure 1). Conventional Tamm Plasmon structure is based on the distributed Bragg reflector (DBR). In our case DBR consists 30 pairs GaAs/Al$_{0.95}$Ga$_{0.05}$As quarter-wavelength layers with high refractive index GaAs layer on the top (to realize electromagnetic field localization on the edge between metal and DBR). Scheme of the structure is shown in figure 1 (a). To reduce absorption in the metallic layer the sub-wavelength grating on the top of DBR could be formed [5]. Here we consider parameters and materials generally used in fabrication of sub-wavelength gratings by electron beam lithography– polymer resist material PMMA (polymethyl methacrylate). We investigated a series of Tamm plasmon structures with gold-PMMA gratings with different ratio of filling factors. The grating could be defined by two parameters: a period (in our case it's 250 nm) and a filling factor (ff), i.e. relation between thickness of a golden part of a grating period to a PMMA part. For instance, when the filling factor is 50% it means that half of the period (125 nm) is covered by the golden layer and another half is covered by the PMMA layer with gold on the top (figure 1 (a)). Therefore, we are analyzing how the presence of the grating with different filling factors influence on the properties of Tamm plasmon mode: the resonance frequency position and the quality factor.

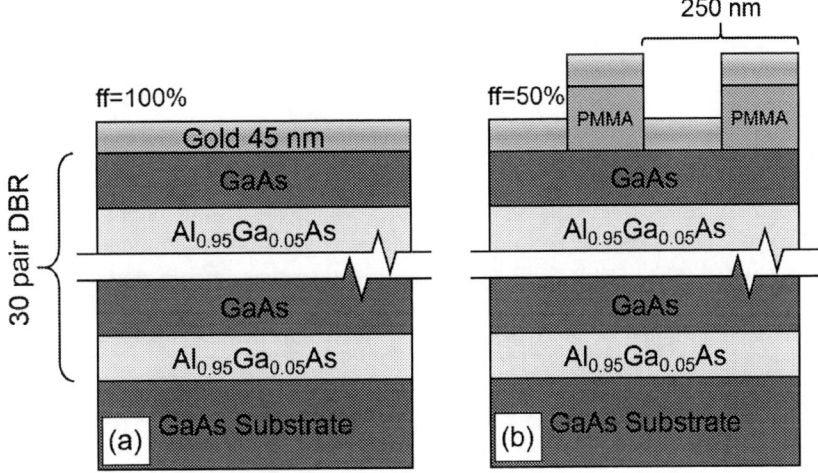

Figure 1(a, b). (color online) **(a)** Scheme of Tamm plasmon based structure without structuration (ff=100%) **(b)** Scheme of Tamm plasmon based structure with structuration (ff=50%) of covering layer.

Two-dimensional electromagnetic simulations were carried out using a finite element method (FEM) in COMSOL Multiphysics modelling environment. The thickness of the golden layer is 45 nm, and the thickness of the PMMA layer is 90 nm. Real and imaginary parts of a gold refractive index was obtained by fitting the experimental data of Johnson and Christy [6]. Reflection spectra were calculated for each structure with different filling factor in case of TE polarized wave and normal incidence. From obtained spectra we estimate the frequency position of the Tamm resonance and the quality factor by fitting the Tamm plasmon resonance by using the Lorentz function.

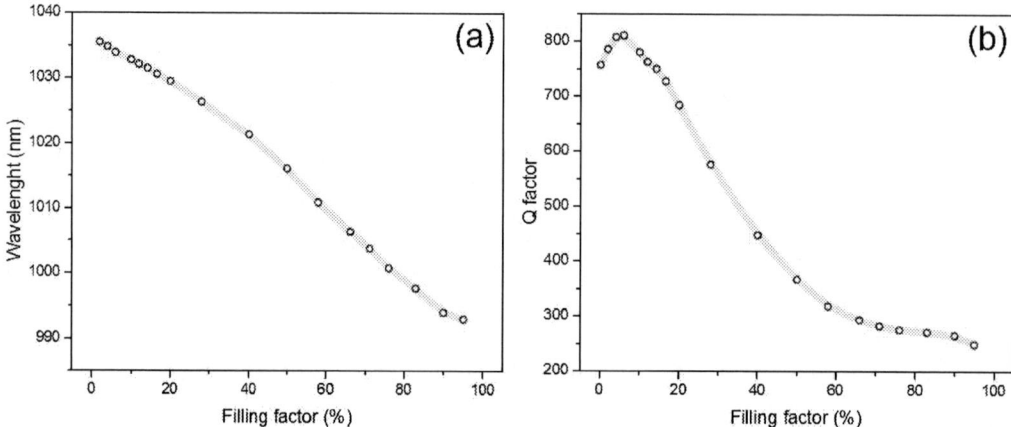

Figure 2(a, b). (color online) **(a)** Dependence of Tamm Plasmon resonance frequency on filling factor. **(b)** Dependence of investigated Tamm plasmon cavity quality factor on filling factor.

Results of simulations are demonstrated in figure 2. Figure 2 (a) demonstrates how the Tamm plasmon resonance wavelength depends on the filling factor. We observe the maximum resonance shifting around 42 nm, when filling factor is 0%. Figure 2 (b) shows the quality factor dependence on the filling factor. The quality factor increases substantially while decreasing the filling factor (which also means a decrease of the gold/GaAs contact area) due to reduced absorption in the metallic layer. Dependence has maximum value of the Q factor =811 at the filling factor value = 6%. Relatively high quality factor values (more than 650) could be achieved in structures with filling factors less than 20%. In structures with filling factors more than 60% influence of sub-wavelength structuration is insignificant due to the dominant losses in metal.

3. Conclusions

We provided theoretical analysis of the sub-wavelength structuration influence on the properties of Tamm plasmon structure formed by GaAs/Al $_{0.95}$Ga$_{0.05}$As DBR with the gold metal coating layer on the top. Several FEM electromagnetic simulations were carried out with different parameters of the golden layer structuration. We demonstrated that the resonance frequency of Tamm plasmon mode can be tuned by changing the filling factor parameter. Realisation of the sub-wavelength structuration in Tamm plasmon structure can significantly increase the quality factor of Tamm plasmon mode.

Acknowledgments

The work has been supported by the RFBR grant No 18-32-00800/18.

References

[1] Kaliteevski M, Iorsh I, Brand S, Abram R A, Chamberlain J M, Kavokin A V and Shelykh I A 2007 *Phys. Rev. B* **76**(16)
[2] Symonds C, Lheureux G, Hugonin J P, Greffet J J, Laverdant J, Brucoli G, Lemaitre A, Senellart P and Bellessa J 2013 *Nano Lett.* **13**(7) 3179
[3] Gubaydullin AR et al. 2017 *Sci. Rep.* **7**(1)
[4] Khurgin J B 2015 *Nature Nanotech.* **10**(1) 2
[5] Gubaydullin A R et al 2017 *Appl. Phys. Lett.* **111**(26) 26110
[6] Johnson P B, Christy R W 1972 *Phys. Rev. B* **6**(12) 4370

Development of diffractive optical elements with low surface roughness by direct laser writing

S A Fomchenkov[1,2], A P Porfirev[1,2]

[1]Department of Engineering Cybernetics, Samara National Research University, Samara 443086, Russia

[2]Micro- and Nanotechnologies Laboratory, Image Processing Systems Institute of the RAS – Branch of the Federal Scientific Research Centre "Crystallography and Photonics" of RAS, Samara 443001, Russia

Abstract. A novel method of forming the phase diffractive optical elements (DOEs) by direct laser writing in thin films of aluminum was presented. The quality of the writing process in aluminum films were investigated depending on the parameters of magnetron sputtering process. This method of phase diffraction optical elements forming substantially reduces the time, fabrication steps, costs of production and significantly improves the quality of the elements in comparison with traditional methods.

1. Introduction

Diffractive optical elements (DOE) allow many functions such as multiplication and beam formation, optical signal distribution through processing channels, wave front formation [1,2]. One of the traditional methods of the phase diffractive optical elements fabrication is a method based on microstructure forming inside the substrate [3,4]. But the most problem part of this method is low fabrication accuracy because of plasma etching involving in this process.

Typically, this process begins with the deposition of thin metal film such as chromium on the quartz substrate, which acts as a hard mask. Moreover, the patterns are formed on the hard mask with either by photolithography or by laser writing system. Further, the desired patterns are obtained by etching the substrate with the help of plasma-chemical etching. For example, a quartz substrate is physically etched in a gas (SF_6) through a chromium oxide (Cr_2O_3) mask. After the mask is removed by chemical etching in a special solution. All steps of this method shown in figure 1.

2. Fabrication

The main idea of presented method is removing of low controlling plasma etching process which is responsible for the substrate's top surface roughness and consequently reduces its efficiency. Furthermore, it will allow us to reduce the number of fabrication steps and consequently reduce the time and costs of production.

In details, this process consists of the following steps:

1) Cleaning of a fused silica substrate.

2) Chromium thin film sputtering on the quartz substrate.

3) Hardmask's direct laser writing in the chromium thin film by circular laser writing station. The chromium thin film exposed by focused laser radiation oxidizes into Cr_2O_3.

4) Unexposed chromium is removed using a special solution.

Content from this work may be used under the terms of the Creative Commons Attribution 3.0 licence. Any further distribution of this work must maintain attribution to the author(s) and the title of the work, journal citation and DOI.

Published under licence by IOP Publishing Ltd

5) Transfer of the DOE's profile to the silica substrate achieved using a plasma etching system. This step is one of the most non-controlled.

6) The depth of etching is controlled by the profilometer.

7) Removal of the hard mask in a solution. Washing in distilled water. Blowing with compressed air and again the depth of etching is controlling.

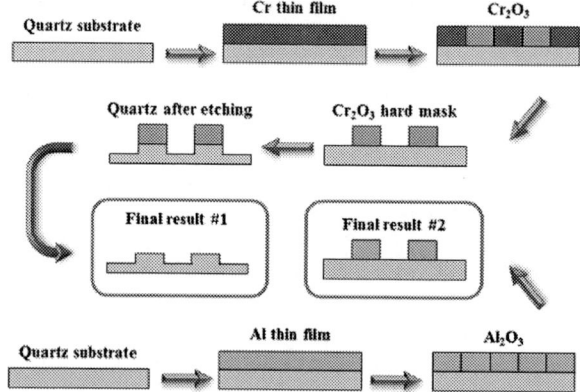

Figure 1. General scheme of the phase DOE manufacturing steps: typical method and presented method.

The main idea of presented method is removing of a low controlling plasma etching process which is responsible for the negative effects. External view of plasma etching system "Caroline PE15" shown on figure 2.

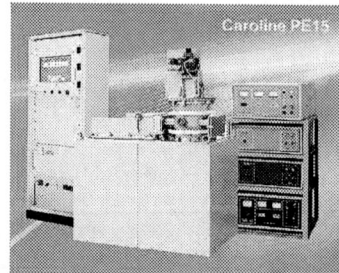

Figure 2. Plasma etching sestem using for manufacture by typical method.

Aluminum thin films were deposited by magnetron sputtering system "Caroline D12A". Then direct laser writing process was conducted by circular laser writing system "CLWS-200S". When aluminum thin film was exposed by focused laser radiation, it oxidized into aluminum oxide.

Unexposed aluminum metal was removed with the help of solution. Laser writing parameters were chosen in such a way that the effect was similar to the annealing process at temperature 700 °C [5,6]. The obtained Al_2O_3 film was optically transparent and allows visible radiations. In the visible wavelength spectrum, Al_2O_3 has a refractive index of 1.8 which is higher than the refractive index of quartz 1.5 that provides a significant refractive index contrast. As a result, this fact reduces the requirement of high aspect ratio of the diffractive element structures in the Al_2O_3 film as compared to the structures in quartz [7]. The general scheme of the presented technological manufacturing process is shown also in figure 1.

Total view of element step by step after each manufacture process is shown in figure 3.

Figure 3. Total view of element after manufacture step: cleaning, Al deposition, laser writing and development, final result.

3. Advantages of the presented method

The quartz etching is the most non-controlled part, however the etching process provides high level of roughness. Surface of substrate, which is not covered by chromium oxide mask, interacts directly with plasma inside a vacuum chamber. This kind of etching is not homogeneous, so therefore roughness of open surfaces increasing linear with time and power of etching [8,9].

So therefore developed method also better in providing of high quality elements, with smooth surface. To confirm these conclusions, measurements of elements manufactured in different ways were made by profiler. Results of measurements made by profiler are shown on figure 4.

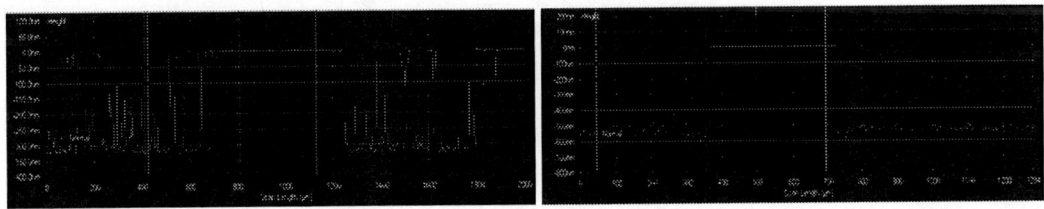

Figure 4. Profile of single step (left – etched element, right – direct write elemenent).

4. Conclusion

We presented a method in order to reduce fabrication time and significantly improve the quality of the elements. This allows obtaining greater efficiency in comparison with similar structures, made on quartz, using plasma etching systems.

Acknowledgments

This work was financially supported by Russian Federation Presidential grants for support of young candidates of sciences (MK-2390.2017.2) and Russian Foundation for Basic Research grant No. 17-42-630008, 18-07-01122, 18-07-01380, 16-47-630677 in part of design and experimental investigation of multilayer diffractive optical elements and by Russian Federation Presidential grant for support of the leading scientific schools (NSh-6307.2018.8) in part of manufacturing of diffractive optical elements.

References

[1] Bykov D A, Dokolovich L L, Diffraction of an optical beam on a Bragg grating with a defect layer, 2014 Computer Optics 38(4), 590–597.

[2] Rubinsztein-Dunlop H, Forbes A, Roadmap on structured light, 2017 J. Opt. 19(1), 1–55.

[3] Fomchenkov S A, Butt M A, Podlipnov V V, Poletaev S D, Skidanov R V and Kazanskiy N L, E-beam lithography exposure conditions for the fabrication of RGB filter based on metal/dielectric subwavelength grating, 2016 J. of Physics: Conf. Ser. 741(1), 012150.

[4] Butt M A, Fomchenkov S A, Ullah A, Habib M, Ali R Z, Modelling of multilayer dielectric filters based on TiO_2/SiO_2 and TiO_2/MgF_2 for fluorescence microscopy imaging, 2016 Computer Optics 40(5), 674–678.

[5] Butt M A, Fomchenkov S A, Thermal effect on the optical and morphological properties of TiO_2 thin films obtained by annealing a Ti metal layer, 2017 Journal of the Korean Physical Society 70(2), 169–172.

[6] Butt M A, Fomchenkov S A, Khonina S N, Dielectric-Metal-Dielectric (D-M-D) infrared (IR) heat reflectors, 2017 J. of Physics: Conf. Ser. 917(6), 062007.

[7] Fomchenkov S A, Butt M A, Fabrication of amplitude-phase type diffractive optical elements in aluminium films, 2017 J. of Physics: Conf. Ser. 917(6), 062026.

[8] Juneja S, Poletayev S D, Fomchenkov S A, Khonina S N, Skidanov R V, Kazanskiy N L, Reactive ion etching of indium-tin oxide films by CCl_4-based Inductivity Coupled Plasma, 2016 J. of Physics: Conf. Ser. 741(1), 012105.

[9] Glyanko M S, Volkov A V, Fomchenkov S A, Assessment of surface roughness of substrates subjected to plasma-chemical etching, 2014 J. of Physics: Conf. Ser. 541(1), 012100.

SPBOPEN 2018

IOP Publishing

IOP Conf. Series: Journal of Physics: Conf. Series **1124** (2018) 051042   doi:10.1088/1742-6596/1124/5/051042

Separation of inhomogeneous and homogeneous broadening manifestations in InGaAs/GaAs quantum wells by time-resolved four-wave mixing

I A Solovev[1], Yu V Kapitonov[2], B V Stroganov[2], Yu P Efimov[2], S A Eliseev[2] and S V Poltavtsev [1,3]

1 Spin Optics Laboratory, Saint Petersburg State University, 198504 Saint Petersburg, Russia
2 Saint Petersburg State University, 198504 Saint Petersburg, Russia
3 Experimentelle Physik 2, Technische Universität Dortmund, D-44221 Dortmund, Germany

Abstract. Coherent dynamics of excitons in single InGaAs/GaAs quantum well was investigated experimentally. In order to carry out this study a three-pulse setup capable of measuring time-resolved four-wave mixing was build. Measured signal consists of two distinguishable components of similar amplitude corresponding to the fast decaying free polarization decay and the long-lived primary photon echo, which proves comparability of homogenous and inhomogeneous broadenings of the exciton resonance.

1. Introduction

Coherent optical control of localized states in semiconductor nanostructures has promising prospective in optical information processing [1, 2]. Moreover, experiments involving coherent control conveniently provide information on fundamental properties of optical excitations in media under study. The time-resolved four-wave mixing (FWM) could be used in order to explore coherent optical dynamics. This powerful method is able to separate inhomogeneous and homogeneous broadening manifestations [3], to distinguish polarization interference from the quantum beats [4], and to determine important coherent properties such as irreversible (T_2) and reversible (T_2^*) dephasing times.

In recent years many new systems were studied using FWM-based techniques including halide perovskites [5], transition metal dichalcogenide monolayers [6] InGaAs/GaAs quantum dots embedded in Tamm-plasmon microcavity [7]. The one thing in common in all these experiments is a prevalence of the inhomogeneous broadening of studied resonances over the homogeneous linewidth measured at low temperatures.

On the other hand, there is a special interest in semiconductor nano-heterostructures where inhomogeneous broadening becomes comparable with homogeneous linewidth or even less. Fabrication of such structures requires the usage of the molecular beam epitaxy growth method. Recently we have shown successful growth of InGaAs/GaAs single quantum wells with inhomogeneous linewidth smaller than the homogeneous one [8]. Comparable inhomogeneous and

Content from this work may be used under the terms of the Creative Commons Attribution 3.0 licence. Any further distribution of this work must maintain attribution to the author(s) and the title of the work, journal citation and DOI.

Published under licence by IOP Publishing Ltd

homogeneous linewidths could give rise to the phenomena such as coexistence of the free polarization decay (FPD) and the photon echo (PE) in the same experiment [4].

In this work we present a time-resolved study of four-wave mixing from InGaAs/GaAs single quantum wells.

2. Sample and experimental setup

The sample P551 under study was grown by the molecular beam epitaxy [8]. It contains single 3 nm thick $In_{0.03}Ga_{0.97}As$/GaAs QW. The sample was rotated during the growth to achieve better uniformity across its surface. Basic characterization of the sample was done by the reflection spectroscopy in the Brewster geometry [8].

The experimental setup (see figure 1) is designed to detect the time-resolved degenerate FWM signal in reflection geometry. Optical excitation was performed by few-picosecond laser pulses. This regime is optimal for spectral separation of the coherent dynamics of individual quasiparticle resonance. Laser spectrum is monitored by spectrometer with CCD-detector. In our experiments laser was tuned to the heavy-hole exciton resonance. Laser beam is split into three paths denoted in the figure 1 as 1st, 2nd and reference (Ref) beams. 2nd and Ref pulses pass through optical delay lines. 1st and 2nd pulses are focused by a concave mirror to the same spot on the surface of the sample cooled down to T=1.4 K in a closed-cycle helium cryostat. Coherent response from the sample (Signal in figure 1) is comprised of several components – reflected 1st and 2nd pulses with wavevectors k_1 and k_2 respectively, and FWM response with wavevector $(2k_2−k_1)$. After being collimated by the same spherical concave mirror these components can be separated by shifting transversally the retroreflector installed after the mirror. Selected component is mixed with Ref pulse on a non-polarizing beamsplitter. The pair of mixed beams exiting beamsplitter is focused on the balanced photodetector.

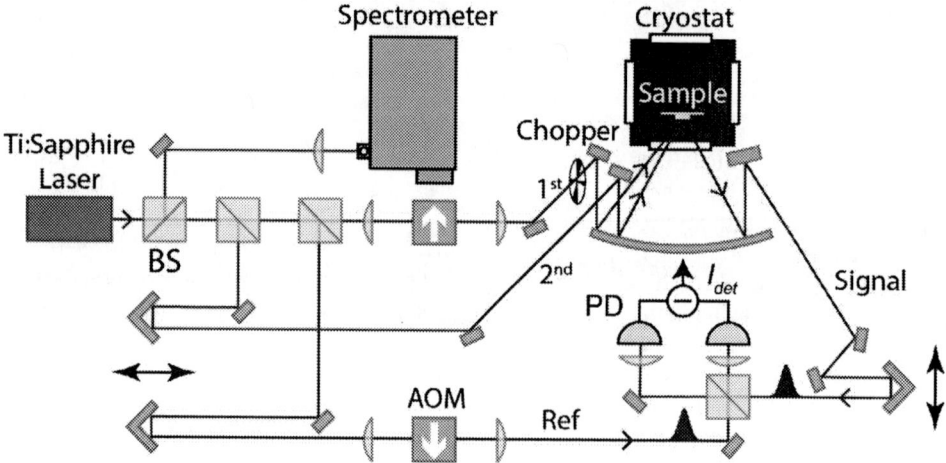

Figure 1. The scheme of the experimental setup.

The optical frequencies of 1st and Ref beams are shifted by two optical shifters based on travelling wave acousto-optic modulators with independently selected radio frequencies. Two beam frequencies are shifted in opposite directions so that interference signal detected by the balanced photodetector is modulated on the differential frequency. Additional signal-to-noise improvement is achieved by a slow modulation of the 1st pulse by mechanical chopper. Such a dual modulation provides high-sensitive background-free detection of the FWM amplitude.

SPBOPEN 2018

IOP Publishing

IOP Conf. Series: Journal of Physics: Conf. Series **1124** (2018) 051042   doi:10.1088/1742-6596/1124/5/051042

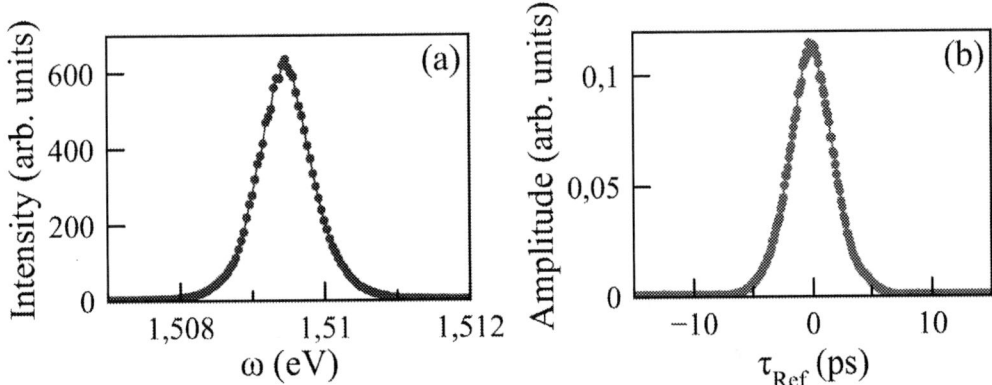

Figure 2. The spectrum (a) and auto-correlation function (b) of pumping picosecond pulses.

Figure 2 illustrates spectral- and time-domain profiles of the laser pulse. The spectral width is about 900 μeV and temporal duration extracted from auto-correlation function is approximately 3.1 ps. Time-bandwidth product corresponds to the nearly transformed-limited pulse.

3. Results and discussion

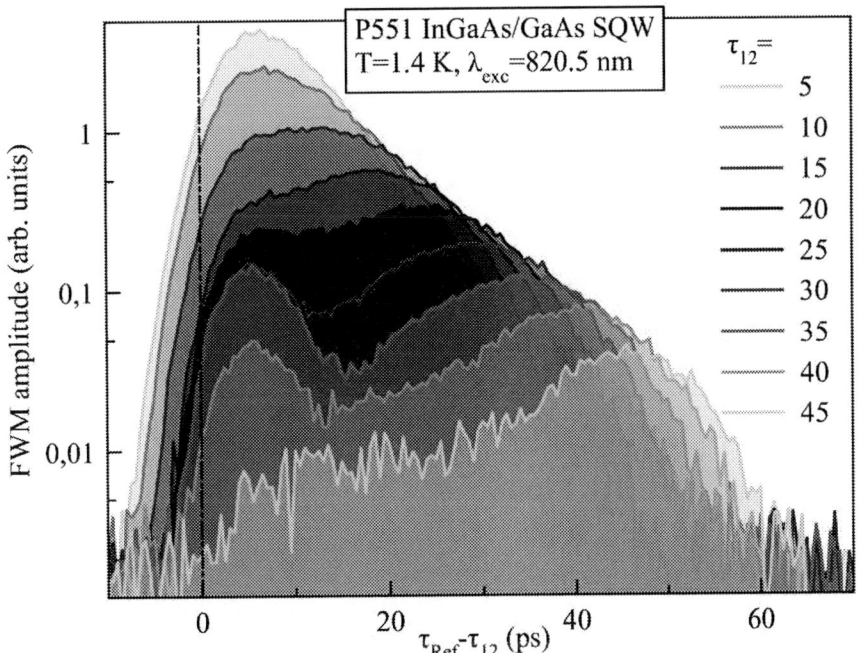

Figure 3. Temporal profile of FWM signal as a function of delay between pumping pulses.

The set of temporal profiles of FWM signal from excitons detected at a number of delays between the 1st and 2nd pulses τ_{12} is illustrated on the figure 3 with delays indicated in picoseconds (note the logarithmic Y axis). At delay $\tau_{12} \geq 20$ ps the FWM signal is comprised of two distinguishable features.

745

The first one is prompt with respect to the second pulse and corresponds to the free polarization decay (FPD). The second one is delayed and has maximum at $\tau_{Ref}=2\tau_{12}$ corresponding to the two-pulse primary PE. Similar temporal behavior was detected in GaAs/AlGaAs multiple QW [3]. Temporal width of PE is about 10 ps revealing relatively small value of inhomogeneity which generally agrees with previous study of the same sample by another technique [8].

4. Conclusion

Experimental study of the coherent optical dynamics of excitons in InGaAs/GaAs quantum well was performed. Time-resolved four-wave mixing demonstrates coexistence of the free-polarization decay and the photon echo signals. We explain this behavior as clear manifestation of comparability of inhomogeneous and homogeneous broadenings of the exciton resonance. This fact indicates high optical quality of the quantum well sample.

Acknowledgements

The reported study was carried out on the equipment of SPBU Resource Center "Nanophotonics" and supported by Russian Foundation for Basic Research (RFBR) projects 16-29-03115 ofi_m and 18-32-00684 mol_a. We acknowledge St. Petersburg State University for the research grants 11.34.2.2012. The authors appreciate financial support by Russian Foundation of Basic Research (RFBR) through the International Collaborative Research Centre TRR-160 (Contract No. 15-52-12016 NNIO_a). Authors thank Valentin Davydov for constructive discussions and reviewing the text of this work.

References

[1] Langer L et al 2012 *Phys. Rev. Lett.* **109** 157403
[2] Langer L et al 2014 *Nature Photonics* **8** 851–57
[3] Webb M D, Cundiff S T, Steel D G 1991 *Phys. Rev. Lett.* **66** 934-37
[4] Koch M et al 1992 *Phys. Rev. Lett.* **69** 3631
[5] March S A et al 2016 *Scientific Reports* **6** 39139
[6] Jakubczyk T et al 2016 *Nano Lett.* **16** (9) 5333–39
[7] Salewski M et al 2017 *Phys. Rev.* B **95** 035312
[8] Poltavtsev S V et al 2014 *Solid State Communications* **199** 47–51

SPBOPEN 2018

IOP Publishing

IOP Conf. Series: Journal of Physics: Conf. Series **1124** (2018) 051043   doi:10.1088/1742-6596/1124/5/051043

Magneto-resonance methods of the relaxation rate measuring for the proton-containing flowing fluids composition studying

L I Fatkhutdinova[1], A N Mamonkina[1], S V Ermak[1], V V Semyonov[1]
[1]Peter the Great St. Petersburg Polytechnic University, 195251, St. Petersburg, Russia

Abstract. In the present work, the existing pulsed nuclear magnetic resonance methods for determining the longitudinal relaxation time in a condensed medium are analyzed. There is considered an alternative method of the nuclear magnetic momentums relaxation times determining in the current proton-containing liquid by the functional dependence of the detected signal intensity from the amplitude of the magnetic field modulation in the magnet-analyzer.

1. Introduction

Determination of the relaxation constants of proton-containing media is one of the key problems of applied nuclear magnetic resonance spectroscopy, the solution of which allows obtaining information on the internal structure of the investigated working substance used in various branches of science and technology. A typical example of such use is magneto-resonance tomography, where the contrast of the object elements image under investigation essentially depends on the accuracy of measurements of the longitudinal and transverse relaxation times. In the practice of magnetic resonance in a condensed medium, to measure these characteristics, the pulse methods for detecting of the free precession and the spin echo signals are used [1]. Three radio pulses π, $\pi/2$, π (or series π, $\pi/2$, π, π, $\pi/2$, π,...) are used to measure the longitudinal relaxation time T_1. The first π-pulse inverts the magnetization vector, then it is set to a new value according to the law

$$-M(t) = M_0 \left[1 - 2exp\left(-\frac{t}{T_1}\right)\right]. \tag{1}$$

The second radio pulse $\pi/2$ induces a free precession signal after a time of τ, the amplitude of which is proportional to the value of $M(t)$. In this case, the spin-lattice relaxation time is determined from expression (1) by fixing the free precession signal for different time intervals τ.

The method of measuring the longitudinal relaxation time measuring also applies to the pulse methods of measuring [2], according to which the ratio of the signals corresponding to two fast consecutives (in comparison with the longitudinal relaxation rate) passing, divided by a time interval τ, is measured. The signal observed during the first passing is proportional to the transverse macroscopic magnetization, which at the optimum transmission speed and the corresponding intensity of the radio-frequency field is $M_0 = \chi H_0$, where χ – is the static nuclear susceptibility, H_0 – is a constant magnetic field. The signal of the second rapid passing through the resonance in the opposite direction is proportional to the longitudinal component of the macroscopic magnetization $-M_z = -M_z \left[1 - 2exp\left(\frac{\tau}{T_1}\right)\right]$. For a symmetric passing of a constant magnetic field,

---

Content from this work may be used under the terms of the Creative Commons Attribution 3.0 licence. Any further distribution of this work must maintain attribution to the author(s) and the title of the work, journal citation and DOI.

Published under licence by IOP Publishing Ltd

the nuclear magnetic resonance signal corresponds to the stationary value of the nuclear magnetization M, defined by expression

$$M = M_0 \frac{1-exp\left(-\frac{\tau}{T_1}\right)}{1-exp\left(-\frac{\tau}{T_1}\right)}.\tag{2}$$

Making two measurements of the nuclear magnetic resonance signal amplitude with different values of τ, we can thus determine the longitudinal relaxation time T_1.

The methods of determining the longitudinal relaxation time considered above are fundamentally ineffective in the nuclear magnetic resonance with the flowing proton-containing sample, because of the discrepancy between the experimental parameters and the required measurement accuracy.

The latter is due to the fact that the formulas (1) and (2) are valid for a flowing sample only if the time τ is much shorter, than the time of finding of the flowing fluid in the analyzer sensor θ, which is directly determined by the sensor size and the fluid flow rate (otherwise , at $\theta \ll \tau$, a sample change occurs in the analyzer's sensor, which leads to a loss of information on the dynamics of macroscopic nuclear magnetization during the first passage through the resonance). As an example, let us take the case of an analyzer sensor volume of 1 cm^3 at a flow rate for a flowing sample of 50 cm^3/s, (corresponding to the maximum signal to noise ratio of nuclear magnetic resonance in a flowing medium). In this case, the time $\theta = 0.02$ s, that is, the time τ, must be at least several times less than this value, for example, 0.005 s. It is easy to calculate from formula (2) that the change in time τ, twice from 0.005 s to 0.01 s, corresponds to a relative change in the amplitude of the signals in only 0.2%, which is substantially lower than the error limits for measuring the amplitude of the nuclear magnetic resonance signal in the flowing sample. Therefore, to measure its longitudinal relaxation time, the method of preliminary polarization of the flowing liquid in a magnetizer is used and the magnetic resonance signal is processed, the value of which is determined by both the relaxation constants and the parameters of the experimental scheme.

In figure 1 is shown variant of such scheme, where the digits indicate: 1 – pipeline with the flowing liquid, 2 – magnet-polarizer, 3 - storage volume, 4 – magnet- analyzer, 5 – receiving coil, 6 – modulation coil, 7 – autodyne detector, 8 – sound generator. When the working fluid flows through the pipeline 1 in the storage volume 3, the flowing sample proton polarization takes place, which is detected by the magnetic resonance signal with the autodyne detector 7 and the standard magnetic field scanning technique in the magnet-analyzer 4.

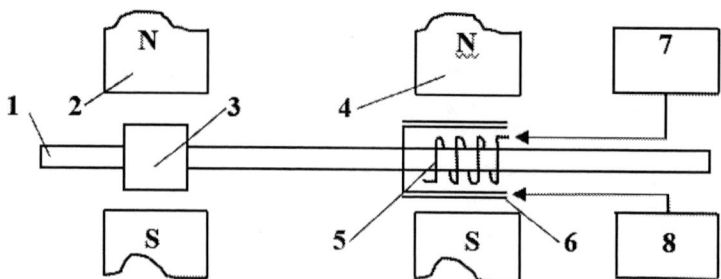

Figure 1. Experimental setup.

In this case, the longitudinal relaxation time T_l is determined by the following empirical formula [3]

$$T_1 = \frac{V_P}{q \ln\left(\frac{V_P + V_T}{V_T}\right)}\tag{3}$$

where V_P – is the storage volume, V_T – is the volume of the connecting pipeline between the storage volume and the receiving coil location area, q – is the flow rate near its values, corresponding to the maximum intensity of the NMR-signal in the autodyne detector.

Expression (3) makes it possible to estimate only the order of the T_1 -magnitude, since the functional dependence of the detected signal intensity on the flow rate near its optimal value is very shallow. For example, for the parameters of the experimental setup, $V_P = 95$ cm^3, $V_T = 4$ cm^3, and variations in the optimal value of q within the range of 20 to 30 cm^3/s, the longitudinal relaxation time T_1 of water protons in accordance with expression (3) varies from 1 to 1,5 seconds.

2. Investigated method of T_1 measuring

In this paper, we proposed and tested the method of the relaxation times determining by comparing the NMR-signals from the standard flowing medium with the known value of T_1 and the flowing test sample according to the functional dependence of the detected signal amplitude by expression (1). To uniquely determine of the test sample longitudinal relaxation time deviation from the standard sample, it is necessary to take into account the dynamics of the behavior of the nuclear magnetic resonance signal when the amplitude of the radio-frequency field produced by the receiving coil of the autodyne detector is varied. Such dynamics was manifested in the experiment when the amplitude of the magnetic field modulation in the magnet-analyzer changed. In this case, under conditions of a slow passing of the magnetic field near the resonant value the saturation factor α equal to $(\gamma H_1)^2 T_1 T_2$ should be affected on the intensity of the detected signal, where γ – is the gyromagnetic ratio of the working substance atoms, H_1 – is the amplitude of the radio-frequency field in the receiver coil of the autodyne detector, T_1 and T_2 – are the longitudinal and transverse relaxation [4].

Obviously, the parameter α should be inversely proportional to the modulation amplitude $\beta = 1/\alpha$. For this case, under the assumption of the Lorentz form of the absorption line, it is easy to approximate the observed signal intensity dependence I in the function of the parameter β by the following formula

$$I = \frac{6.7\beta^2}{(1+\beta^2)^3}, \qquad (4)$$

where the coefficient 6,7 in the numerator is chosen from the normalization condition for the maximum signal intensity per unit. In accordance with this expression, the change in the test sample is equivalent to the change in the parameter α, the value of which can be varied by changing the amplitude of the modulation of the magnetic field H_1. Comparing the dependence (4) for two samples - the reference (with the known relaxation times T_1 and T_2) and the investigated sample at a constant value of H_1, it is not difficult to determine the required relaxation constants. The reliability of this method can be demonstrated by the example for one flowing sample with fixed constants T_1 and T_2, but with different values of H_1.

The figure 2 (a) shows a typical oscillogram of the observed signal on water protons from the main line, where the constant T$_2$ is determined from the time of the observed relaxation beats decrease.

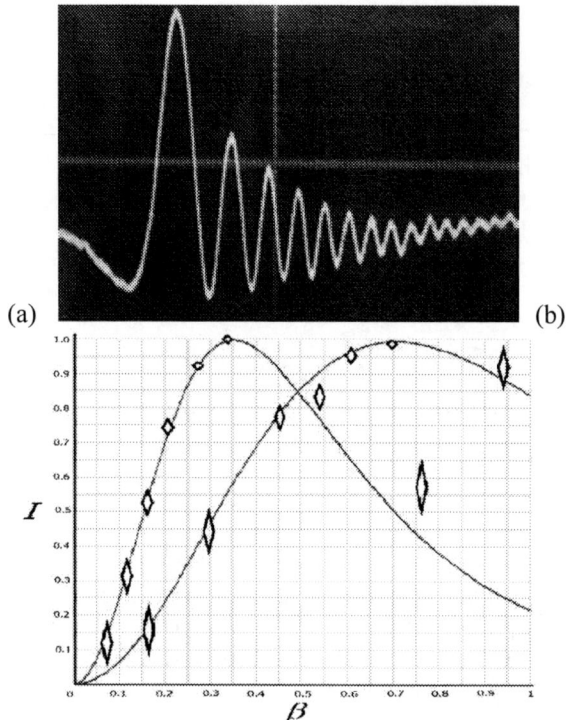

Figure 2 (a, b). (a) Oscillogram of the observed signal. (b) Experimental (points) and calculated (continuous line) dependences of the observed signal.

The extremely low modulation frequencies ($\sim$ 7 Hz) of the working magnetic field for two values of the radio-frequency field amplitude are used in the experiment – H_1 and $2H_1$. The amplitude of the observed signal was fixed (pick-to-pick) and the square of this amplitude was plotted as a function of the parameter β.

The corresponding experimental and calculated dependences of the observed signal are presented in figure 2 (b), which demonstrate their close correspondence near the β values to the left of the values $\beta = 0.35$ и $\beta = 0.7$. The observed spread of experimental and calculated dependences to the right of the values $\beta = 0.35$ and $\beta = 0.7$ is definitely explained by noncompliance with the conditions of slow passage of the magnetic field through the resonance.

The transverse relaxation time T_2 was measured from the decay of the beat amplitude and was of the order of 1 ms for all test samples, which was caused by the influence of the magnetic field gradient in the magnet-analyzer. The longitudinal relaxation time of the standard sample was taken $T_1 = 0.65$ s, determined from expression (4) and with respect to which the values of T_1 of the test samples were measured. As the tested samples, alcohol and degassed water were chosen for which the measured relaxation times were found equal to 1.7 s and 3.8 s, respectively, which is in good agreement with the literature data.

3. Conclusion

The proposed modulation technique of the longitudinal relaxation time measuring can be required for the development of the proton-containing liquids small-sized relaxometers without attracting of the expensive high-resolution NMR equipment.

References

[1] Chizhik V I 2000 Nuclear Magnetic Relaxation (St. Petersburg: St. Petersburg State University Press) p 388

[2] Abragam A 1963 Nuclear Magnetizm (Moscow) p 551

[3] Zhernovoy A I 1964 Nuclear Magnetic Resonance in a Flowing Fluid (Moscow) p 215

[4] Leshe A 1963 Nuclear Induction (Moscow) p 684

SPBOPEN 2018   IOP Publishing

Transmission of thermal imaging by using infrared bundle based on silver halide solid solution

A S Shmygalev[1], D D Salimgareev[1], A S Korsakov[1], B P Zhilkin[2], V I Terekhov[3]

[1]Institute of Chemical Engineering, UrFU, Yekaterinburg 620002, Russia
[2]Ural Power Engineering Institute, UrFU, Yekaterinburg 620012, Russia
[3]Kutateladze Institute of Thermal Physics SB RAS, Novosibirsk 630090, Russia

Abstract. In this study, infrared bundle consisting of seven single fibres was manufactured. Experiences on transmission of the heated object thermal image was performed. The experimental data show that there is a fundamental possibility of thermal image transmission through the infrared bundle.

1. Introduction

Remote temperature detection of the heated objects by means crystalline infrared fibers based on silver halide solid solution [1] is a novel direction that opens up broad application prospects. It is primarily related with the properties of infrared fibers [2], in particular, with their wide spectral transmission from 2.5 to 25.0 μm, which corresponds to the temperature range from -200 to $+1000$ °C, according to Wien displacement law. Obtaining temperature distribution on a heated body surface using infrared bundle in real-time represent interest from both fundamental and practical point of view. Today, similar technology is implemented in endoscopes with fiber optics used in medicine for visual inspection of the internal cavities of the human body, biopsy, surgical and therapeutic influences on biological tissues by laser radiation, washing the cavity and filling it with air or liquid, injection of medicinal solutions, removal of tumors and foreign bodies, etc. [3]. Obtaining a thermal (infrared) image has several noticeable advantages, in contrast an endoscopic image in the visible range of the spectrum.

Thus, obtaining data on the temperature distribution in medicine will localize and assess the degree of any inflammatory process, including cancer, which will further contribute to the correct choice of treatment methods and simplify the procedure of exposure to the seat of infection. Using the infrared fiber optic cable in defectoscopy of power equipment and its components due to the temperature difference, it is possible to detect the presence of various kinds of damage and timely perform maintenance, thereby reducing the failure risk of the entire device. In addition, this equipment can be used for quality control of welds, measurement and analysis of thermal parameters inside pipelines and many other applications.

2. Experimental data

Infrared bundle consisting of seven crystalline fibres with $AgCl_{0.75}Br_{0.25}$ composition (length 150 mm, diameter 1.12 mm) were made for thermal imaging survey. Single optical fibres were packed by hexagonal type with compliance the straightness of their structure (Figure 1). The type of stacking we used provides stability of fibres in the bundle and has a small idle area, which 2.5 times less than in

---

Content from this work may be used under the terms of the Creative Commons Attribution 3.0 licence. Any further distribution of this work must maintain attribution to the author(s) and the title of the work, journal citation and DOI.
Published under licence by IOP Publishing Ltd

square type stacking [4]. The final trimming of the bundle on two sides by a glassy cutter was carried out to create a plane-parallel surface of the thermal transfer image.

(a)   **(b)**

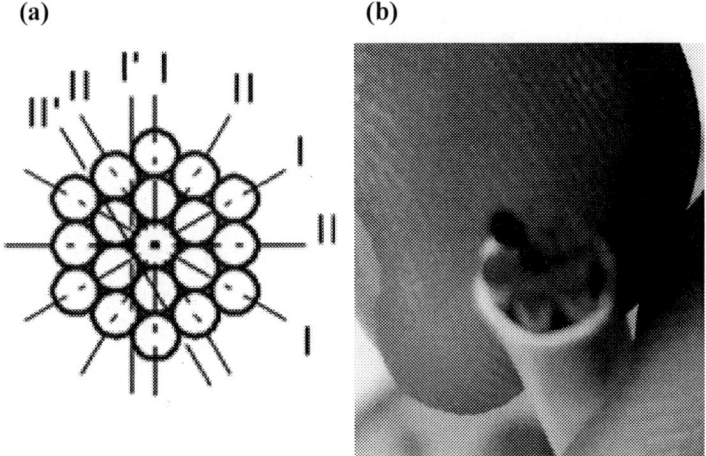

Figure 1(a, b). (a) Hexagonal type packaging of optical fibers **(b)** Imaging of the infrared bundle receiving end.

The experimental setup for investigation of the thermal image transfer possibility through infrared bundle was constructed. The infrared bundle *(3)* in a protective fluoropolymer shell was fixed in a holder and placed vertically. The thermal imager *(1)* NEC 7102WV was placed above the top end of the bundle. Along the perimeter of the bundle upper end to eliminate the influence of ambient light, there was a protective screen *(2)*. Under the lower end of the infrared bundle at a fixed distance x = 1 mm was Peltier's element *(4)*, which worked as a heat source. The change in the value of the voltage supplied to the element by means of a DC *(5)* source allowed to change the surface temperature. Preliminary thermal imaging survey (without details 2 and 3) showed that the temperature distribution on the upper surface of the Peltier's element at different degrees of heating was uniform (surface temperature variation of less than 0.5 K).

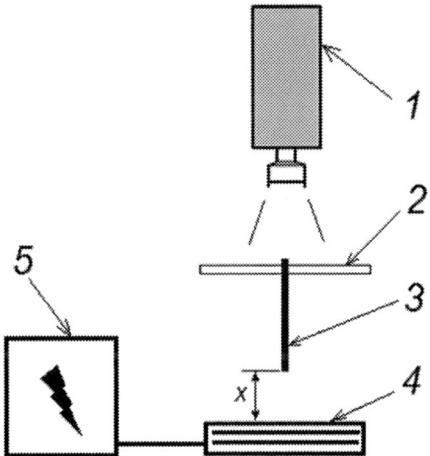

Figure 2. Scheme of the experimental setup.

The procedure of the experiment was the following. As soon as the temperature of the upper surface of the Peltier's element (hereinafter the sample temperature) reached a constant value, it was placed under the lower end of the infrared bundle. At the same time, the thermal imager lens was focused on the upper end of the bundle. Experiments showed that immediately after the appearance of the heated surface under the lower end of the infrared bundle, the thermal imager recorded an instantaneous change in the thermal image on the upper end.

As the observation object we used a copper wire with diameter three times smaller than the diameter of the single fiber (approximately 300 μm) which was wound on surface of the Peltier's element horizontally and vertically. The Peltier's element was the source of thermal radiation. The heating of the cooper wire was made at the expense of heat conductivity that is due to the thermal contact resistance did not allow the object to reach the temperature of the source during the observation time. The temperature of the thermal radiation source was equal to 80 °C in all experiments. That was done to maintain equable and constant heating for a long time, which was a primary condition for the experiment. The obtained thermal imaging (Figure 3) clearly show the temperature difference of the object and source. This indicates a good degree of visualization.

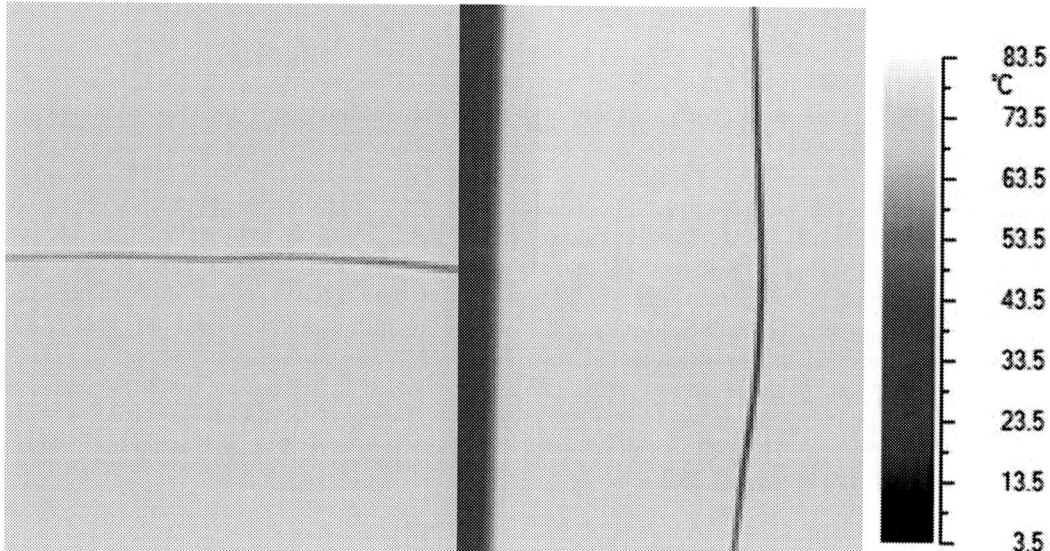

Figure 3. Thermal imaging of the cooper wire.

The infrared bundle was placed perpendicularly relative to the surface of the thermal radiation source and it moved left-right and up-down with the vertical and horizontal location of the wire respectively. Testo-882 thermal imager in real time recorded the temperature changes of the single fibers. The temperature of the cooper wire was significantly different from the Peltier's element temperature, which allowed to easily capture its temperature field through the infrared bundle. The thermal images of the infrared bundle receiving end with the vertical location of the object are shown in figure 4a, with the horizontal location - in figure 4b.

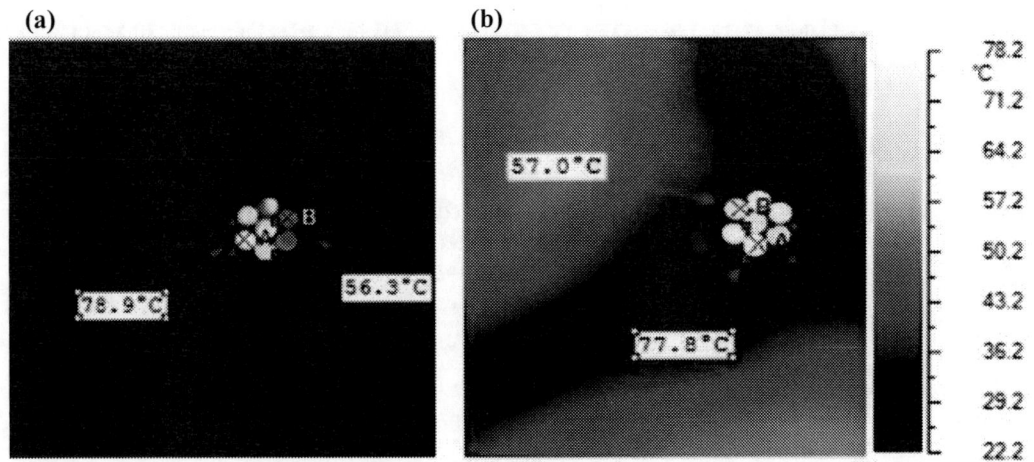

Figure 4(a, b). (a) Thermal imaging of the infrared bundle end with vertical location of the object **(b)** Thermal imaging of the infrared bundle end with horizontal location of the object.

3. Conclusion

Thus, the obtained data show that there is a fundamental possibility of thermal image transmission through the infrared bundle consisting from single fibres based on silver halide solid solutions. The results of the experiments reveal that such a small bundle is not capable of transmitting clear thermal picture of the object, and can only record the temperature likewise a single fiber. A solution to this problem is to increase resolution of infrared bundle due to increase in the number of single fibres in the structure, as well as reducing their diameter. Therefore, the manufacture of fiber bundle, and further investigation of thermal images transmission requires additional scientific and engineering research.

Acknowledgments

The research has been supported by the grants of President of the Russian Federation SP-2455.2018.1.

References

[1] Shmygalev A., Zhilkin B., Korsakov A., Nizovtsev M., Sterlyagov A., Terekhov V. 2016 *Tech. Phys. Lett.* **42(9)** 883
[2] Lewi T., Katzir A. 2012 *Opt. Lett.* **42** 2733
[3] Khatskevich T N 2002 *Endoscopes* (Novosibirsk: SGGA) p 196
[4] Marukovich E I 2011 Fiber-optic copying in casting and metallurgy (Minsk: Belarusian science) p 330

SPBOPEN 2018   IOP Publishing

IOP Conf. Series: Journal of Physics: Conf. Series **1124** (2018) 051045   doi:10.1088/1742-6596/1124/5/051045

On-chip single-photon spectrometer for visible and infrared wavelength range

V Kovalyuk[1,2,3]**, O Kahl**[2,3]**, S Ferrari**[2,3]**, A Vetter**[3,4]**, G Lewes-Malandrakis**[4]**, C Nebel**[5]**, A Korneev**[6,7]**, G Goltsman**[6,7]**, W Pernice**[2]

[1]Department of Physics, Moscow State Pedagogical University, 119992, Russia
[2]Institute of Physics, University of Münster, 48149, Germany
[3]Institute of Nanotechnology, Karlsruhe Institute of Technology, 76344, Germany
[4]Institute of Theoretical Solid State Physics, Karlsruhe Institute of Technology, 76131, Germany
[5]Fraunhofer Institute for Applied Solid State Physics, 79108, Germany
[6]Moscow Institute of Physics and Technology (State University), 141700, Russia
[7]National Research University Higher School of Economics, Moscow 101000, Russia

Abstract. Here we show our latest progress in the field of a single-photon spectrometer for the visible and infrared wavelengths ranges implementation. We consider three different on-chip approaches: a coherent spectrometer with a low power of the heterodyne, a coherent spectrometer with a high power of the heterodyne, and an eight-channel single-photon spectrometer for direct detection. Along with high efficiency, spectrometers show high detection efficiency and temporal resolution through the use of waveguide integrated superconducting nanowire single-photon detectors

1. Introduction

The detection of photons by superconducting nanowire single-photon detectors (SNSPDs or SSPDs) [1] is an inherently binary mechanism, revealing either absence or presence photons, but losing their spectral information. For some state-of-the-art microscopy techniques such as fluorescence resonance energy transfer (FRET) or fluorescence lifetime imaging (FLIM) wavelength discrimination is essential and required spectral demultiplexing prior to detection [2, 3]. Here we show our latest progress on the field of a single-photon spectrometer of the visible and infrared wavelengths range implementation. We use an approach adopted from quantum photonic integration to realize a compact and scalable on-chip single-photon spectrometer with waveguide integrated single-photon detectors (WSSPDs) [4] and implemented three nanophotonic configurations that allow us to produce a wide bandwidth heterodyne detection with ultra-high resolution greater then 10^9 [5,6] as well as the parallel detection on eight wavelength channels, but with reduced wavelength resolution up to 10^3 [7].

2. Device design and fabrication

The multilayer wafers consisting of silicon nitride (450 nm) and silicon dioxide (2.6 μm) thin films on top of a silicon carrier wafer were used as starting materials. On top of the wafers we deposited the ultra-thin NbN superconducting film with a nominal thickness of 4 nm by a reactive magnetron sputtering in Argon and Nitrogen atmosphere. We reached a maximum critical temperature T_c = 9.5 K for films deposited at a substrate temperature T_S = 800 °C with partial pressures of Argon and Nitrogen of 6×10^{-3} and 2.5×10^{-4} mbar, respectively. The sheet resistance of the deposited NbN films measured at room temperature was 620 Ohms/sq.

We fabricated different types of on-chip spectrometers using several steps of e-beam lithography with dry etching and thin-film deposition. In the first step, Au-contact pads and alignment marks were formed using PMMA resist and standard lift-off technique. In the second step, U-shaped NbN nanowires were made with use of HSQ resist by reactive ion etching (RIE) in CF_4. Finally, negative electron resist maN-2403 and etching in atmosphere of CHF_3 were employed for waveguide formation. A schematic

Content from this work may be used under the terms of the Creative Commons Attribution 3.0 licence. Any further distribution of this work must maintain attribution to the author(s) and the title of the work, journal citation and DOI.
Published under licence by IOP Publishing Ltd

view of fabricated nanophotonic circuits for quantum and spectral photonic applications is shown in Fig.1.

The integrated devices include a waveguide and a detector parts. For the first two types of spectrometers, the waveguide part (in blue) includes two focusing grating couplers (FGCs) for 1550-nm light coupling, 50:50 Y-splitter for calibration of optical power as well as U-shaped superconducting NbN nanowire (80 nm width) as a single-photon detector (Fig. 1a,b). For a spectrometer of the third type (Fig. 1c), the waveguide part includes arrayed waveguide grating (AWG) [8], adapted to operate close to the central wavelength of 1550 nm, and 740 nm for IR and visible spectrometer implementations. The eight-channel of the spectrometer are terminated by WSSPDs connected to RF contact pads (highlighted in yellow). The available optical FGCs at the input of the circuit serve to calibrate the efficiency of the optical coupling, and at the output ones to separate the losses in AWG and detectors (dotted color arrows).

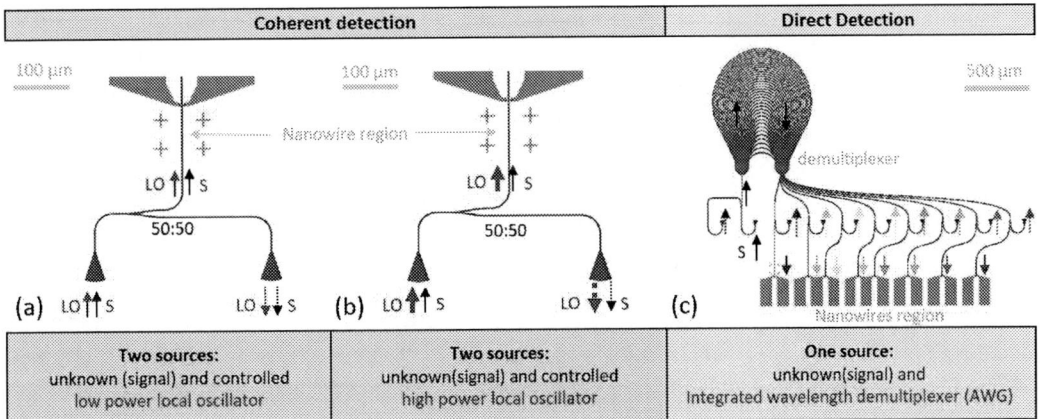

Figure 1 (a-c). Schematic view of nanophotonic circuits for quantum and spectral photonic applications. The waveguide is highlighted in blue, RF contact pads and lithographic markers (crosses) are shown in yellow; S is signal laser; LO is local oscillator; dotted color arrows show the reference outputs for a calibration procedure. **(a)** Scheme of a nanophotonic device for coherent detection with low power LO. **(b)** Scheme of nanophotonic device for coherent detection with high power local oscillator. **(c)** Scheme for nanophotonic device for eight-channel spectrometer with arrayed waveguide grating as an optical demultiplexer and WSSPDs as detectors.

3. Operation principles

In all types of on-chip spectrometers presented here, the active non-linear element providing light detection are WSSPDs integrated with the silicon nitride optical waveguides [9]. However, the principle of operation of superconducting nanowires in these devices is different.

For the operation of the first type of spectrometer (Fig.1a) we routed to WSSPD two weak signals with slightly different frequencies: well-controlled local oscillator (LO) and unknown signal (S), whose spectrum needs to be measured. In this case, WSSPDs operates below the superconducting transition temperature and registers the total electromagnetic field of two sources (beating). Since the probability of photocounts in this mode is proportional to the optical power, the frequency of the "clicks" follows the beating of the total electromagnetic field. Knowing the frequency and phase of the LO and measuring the beating frequency (intermediate frequency), one can determine the frequency and phase of the unknown signal [5].

For the operation of the second type of device (Fig.1b), the LO power is significantly increased by heating the electronic WSSPD sub-system to the superconducting transition temperature. In this case, WSSPD operates as a hot-electron bolometer (HEB). If such a bolometer is biased by a direct current, then the voltage follows the change in the electron sub-system temperature, which is proportional to the square of the amplitude of the beating [6]. The frequency and phase of the unknown signal is found by analyzing the voltage at the bolometer.

In the third type of spectrometer (Fig.1c), the unknown signal is divided by the wavelength by AWG connected to the eight WSSPDs. By measuring the photocounts in each of the channels, one can recover the original spectrum. To calibrate the device we used additional FGCs, connected with output waveguides through the integrated 50:50 Y-splitter. The light direction is indicated by dotted color arrows (Fig.1c).

4. Experimental results

4.1. On-chip efficiency and timing jitter

At first, using a tunable, fiber-coupled lasers and a calibrated photon flux, restricted by attenuators, we measured the efficiencies of devices. The nanophotonic devices placed on a motorized stage (AttoCube Systems) in a cryostat at 1.6 K temperature with 12-channel optical and 8-channel electrical access. The on-chip detection efficiency was measured as the ratio of WSSPD count rate with the exception of the dark count rate to the photon flux propagating in the waveguide. The on-chip detection efficiency is determined by the product of the internal detection efficiency (IDE) and the absorption coefficient. While the absorption of the NbN waveguide's evanescent mode by nanowire due to its length can be made close to unity, the internal detection efficiency is limited by the constrictions that arise in the fabrication of nanowires. For this reason, even for devices with the same geometry, the measured detection efficiency at the same wavelength can differ due to changing IDE. The measured on-chip detection efficiency for on-chip IR and visible spectrometer were found at the level of 19% and 21%, including internal losses of the AWG device and detectors. All detection efficiencies are given for dark count rate less than 10 s^{-1}. At the second stage, using a femtosecond laser sources we determined the timing jitter performance. The timing jitter values of the devices were measured as below 50 ps in both wavelength regimes. The timing jitter was limited by the room temperature readout scheme and independent of the wavelength range.

For the coherent spectrometer shown in Fig. 1a, the on-chip efficiency was defined as the minimal detectable signal at an intermediate frequency. By decreasing the signal at a fixed LO power below a pico-watt level, we were able to detect a signal at an intermediate frequency close to the quantum limit Φph / RBW = 1.6-2.3hv. Here Φph is the photon flux reaching the detector, and RBW is the resolution bandwidth of the spectrum analyzer, used for analyzing the voltage trace from WSSPD [5].

For the coherent spectrometer shown in Fig. 1b, we measured the conversion efficiency as the ratio of the power at the intermediate frequency to the power reaching the bolometer. We found the conversion efficiency of such a spectrometer at the level of -22 dB, as well as the conversion bandwidth up to 3.5 GHz [6].

After measuring the on-chip efficiency, the spectral resolution of the spectrometers was found.

4.2. Wavelength resolution

For coherent detection, the spectral resolution was limited by the stability of the LO. We obtained a line width at an intermediate frequency of Δf = 4 MHz, whereas in the homodyne scheme, when the role of the signal and the local oscillator was performed by the same laser source but shifted in frequency with the help of an acousto-optic modulator, we reached the signal width at an intermediate frequency equal to Δf = 1 kHz. Thus, the spectral resolution of the first two devices ($f/\Delta f$) was 10^6 and 10^9 for heterodyne and homodyne reception, respectively.

For the spectrometer shown in Fig. 1c, the resolution is determined by the configuration of the waveguide circuit [7, 8]. We designed the AWG with the spectral range of 16 nm and 45 nm and the channel resolution 2.2 nm and 6.4 nm for the IR and visible spectrometer, respectively. In this case, an increase in the resolution of the spectrometer leads to an increase in its dimensions, and, consequently, to an increase in optical losses.

4.3. Practical implementation

We investigated nanodiamond cluster created by drop-casting a colloidal solution of nanocrystals onto a microscope slide using the AWG visible range single-photon spectrometer. We analyze the emitted fluorescence spectrum using the eight-channel on-chip single photon spectrometer upon continuous-wave excitation at 532 nm. Taking advantage of the SNSPDs high timing resolution and fast response time, in a complementary experiment we replaced the exciting 532 nm continuous wave laser with a passively mode-locked laser, which produces pulses of 32 ps duration (FWHM) at 440 nm wavelength. In combination with the scanning confocal microscope setup, a lifetime map of the diamond nanocluster was obtained alongside the spectral information [7].

5. Conclusion

We implemented and studied three different single-photon spectrometers. The coherent spectrometers shown in Fig. 1a,b, allow to achieve a huge spectral resolution, which is not available for the direct-detection spectrometers of Fig 1c.making them suitable for applications where high spectral resolution is required, such as spectrometry of gas lines. In this case, an exact and smooth tuning of the local oscillator is necessary, which might leads to an increased acquisition time and post processing. The advantage of the first type of spectrometer over the second is the quantum limited sensitivity, as well as the low required power of the LO (below the picowatt level). While the second type of spectrometer has a wide conversion band reaching 3.5 GHz, but not capable of counting individual photons events. In addition, along with spectral information, signal phase information can be available for analysis

The third kind of spectrometers presented here, based on direct detection can obtain both spectral and temporal information over several channels simultaneously. On-chip spectrometer with AWG arrays can be used similar to CCD arrays in conventional spectrometers. A similar type of integrated spectrometers can be realized adopting a wide variety of devices for light demultiplexing, such as, photonic crystals, planar diffraction gratings, etc. [10, 11]. These integrated circuits enable live monitoring of numerous fluorophores over a wide spectral range. Highly precise infrared and thermal imaging becomes possible owing to the huge spectral range of the SNSPDs. In addition demonstrated on-chip single-photon spectrometers can be used in high-speed quantum optical implementations, which require reliable detection of individual photons on tailored wavelengths. Further work will be related to the practical demonstration of the capabilities of the first two types of spectrometers for working with real single-photon sources (quantum dots, nanotubes, etc.) as well as combining all of the demonstrated approaches in a single device.

Acknowledgments

V. Kovalyuk acknowledges support by the Russian Science Foundation (project No. 16-19-10633; design, NbN film deposition and testing). A. Korneev acknowledges support of the Russian Science Foundation (project No. 17-72-30036; nanophotonic circuits modeling) and W. Pernice acknowledges support by the DFG grants PE 1832/1-1 & PE 1832/1-2 and the Helmholtz society through grant HIRG-0005 (fabrication of nanophotonic circuits and testing).

References

[1] Gol'tsman G N, Okunev O, Chulkova G, Lipatov A, Semenov A, Smirnov K, Voronov B, Dzardanov A, Williams C, Sobolewski R 2001 *Appl. Phys. Lett.* **79** 6 705–707
[2] Okabe K, Inada N, Gota C, Harada Y, Funatsu T, Uchiyama S 2012 *Nat. Commun.* **3**, 705
[3] Borst J W, Visser A J W G 2010 *Meas. Sci. Technol.* **21** 102002
[4] Sprengers J P, Gaggero A, Sahin D, Jahanmirinejad S, Frucci G, Mattioli F, Leoni R, Beetz J, Lermer M, Kamp M, Hofling S, Sanjines R, Fiore A 2011 *Appl. Phys. Lett.* **99** 181110
[5] Kovalyuk V, Ferrari S, Kahl O, Semenov A, Shcherbatenko M, Lobanov Y, Ozhegov R, Korneev A, Kaurova N, Voronov B, Pernice W, Gol'tsman G 2017 *Sci. Rep.* **7**, 4812

[6]   Kovalyuk V, Ferrari S, Kahl O, Semenov A, Lobanov Y, Shcherbatenko M, Korneev A, Pernice W, Gol'tsman G 2017 *J. Phys. Conf. Ser.* **7** 917 62042 1-6
[7]   Kahl O, Ferrari S, Kovalyuk V, Vetter A, Lewes-Malandrakis G, Nebel C, Korneev A, Goltsman G, Pernice W 2017 *Optica* **4** 5 557–562
[8]   Smit M K, Member A, Van Dam C 1996 *IEEE J. Sel. Top. Quantum Electron.* **2** 2 236–250
[9]   Kahl O, Ferrari S, Kovalyuk V, Goltsman G, Korneev A, Pernice W H P 2015 *Sci. Rep.* **5**, 10941
[10]   Kyotoku B, Chen L, Lipson M 2010 *Opt. Express* **18** 1 102–107
[11]   Momeni B, Hosseini E S, Askari M, Soltan M, Adibi A 2009 *Opt. Commun.* **282** 15 3168–3171

SPBOPEN 2018

IOP Publishing

On-chip controlled placement of nanodiamonds with a nitrogen-vacancy color centers (NV)

S Komrakova[1], J Javadzade[4,7], V Vorobyov[3], S Bolshedvorskii[4], V Soshenko[3], A Akimov[3,7,6], V Kovalyuk[1,2], A Korneev[1,2,4], G Goltsman[1,2,5]

[1] Department of Physics, Moscow State Pedagogical University, 119992, Russia

[2] Zavoisky Physical-Technical Institute of the Russian Academy of Sciences, 420029, Russia

[3] P.N. Lebedev Physical Institute, 53 Leninskij Prospekt, Moscow, 119991 Russia

[4] Moscow Institute of Physics and Technology (State University), 141700, Russia

[5] National Research University Higher School of Economics, Moscow 101000, Russia

[6] Texas A & M University, 4242 TAMU, College Station, TX 77843, USA

[7] Russian Quantum Center, 100 Novaya St., Skolkovo, Moscow, 143025, Russia

Abstract. Here we studied the fabrication technique of a kilopixel array of nanodiamonds with a nitrogen-vacancy color centers (NV) on top of the chip and measured the second-order correlation function deep, clearly demonstrated the presence of single-photon sources. The controlled position of nanodiamonds, determined from the measurement of second-order correlation fiction, was realize, as well as the yield of optimized technique equals 12.5% is shown.

1. Introduction

The theoretical and experimental studies of quantum mechanics have moved to the use of various physical systems for use in metrology, quantum cryptography, quantum communication, and teleportation, quantum computer science, and others. For the practical implementation of quantum technology, single photon sources are of interest, since photons are ideal carriers of information. In addition, such sources are promising for integration with quantum photonic integrated circuits (QPICs) [1] for the implementation of fully functional nanophotonic devices. QPICs combine several advantages over table-top components and devices, including small size, high performance, no need of alignment, high temperature stability as well as the possibility of integration with electronic circuits on the same platform. Along with single-photon detectors [2] and integrated photonic components [3], a fast and effective single-photon source (SPS) for the implementation of such circuits is required [4]. A promising candidate for an integrated SPS is a nanodiamond with a nitrogen-vacancy (NV) color center.

To improve the performance of NV-centers as a single-photon source and gaining the ability to scale, one should be able to place them in a controlled manner. Here we experimentally study method to deterministically position nanodiamond basing on lift-off process.

2. Device design and fabrication.

To create an array of nanodiamonds, we used a glass substrate and a reactive magnetron sputtering method, to deposit a 60 nm TiN layer on top. The choice of titanium nitride was determined because of its good adhesion to nanodiamonds as well as the possibility of creating a nano-antenna from it in the future work.

Content from this work may be used under the terms of the Creative Commons Attribution 3.0 licence. Any further distribution of this work must maintain attribution to the author(s) and the title of the work, journal citation and DOI.

Published under licence by IOP Publishing Ltd

For controlled placement of nanodiamonds we applied a layer of polymethyl methacrylate 100 nm thick and baked at 150 °C for 2 minutes on a hot plate. Using the electron-beam lithography system, we created an array of 33 by 33 holes with a diameter of 170 nm. For a successful placement in the holes in the resist, previously, we varied the diameter of the holes and thickness of the resist. The ratio of the thickness of the resist to the height of the holes should have been no less than 1 to 1.5.

If the resist is too thick, for example with a ratio of 1 to 1, after it removing nanodiamonds removes also. If the resist is too thin, then it will not be able to remove the unnecessary diamonds. For the resist development, we pierced the chip in a mash of isopropanol and water (8: 1) for 20 seconds.

At the next step, we put a drop of nanodiamonds solution and incubated it for 30 minutes. To avoid unnecessary particles, we first removed a drop of the solution with a pipette and sent a stream of isopropanol on the sample. Finally, for removing polymethyl methacrylate (PMMA), we sent a stream of acetone heated at 55 °C on the chip, washed the chip in isopropanol and drying in a stream of N_2. The main steps of the fabrication process are shown in Figure 1.

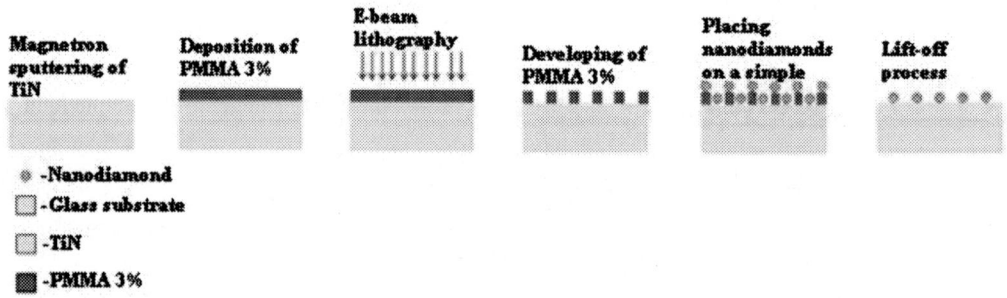

Figure 1 Schematic view of the fabrication process, including (from left to right): magnetron sputtering of TiN film, spin coating of PMMA 3%, e-beam lithography, developing of PMMA, placing nanodiamonds on the chip as well as a lift-off process.

We had a problem of spalling nanodiamonds in large agglomerates, which prevented the decomposition of holes in 1-3 particles. In order to avoid this, we crushed the solution with nanodiamonds in an ultrasonic bath for 3 hours at frequency of sound oscillations of 40 kHz, every hour we took out a test tube and shaken it for five minutes in a centrifuge.

The optical micrograph of one of the fabricated array of nanodiamonds is shown in Figure 2(a).

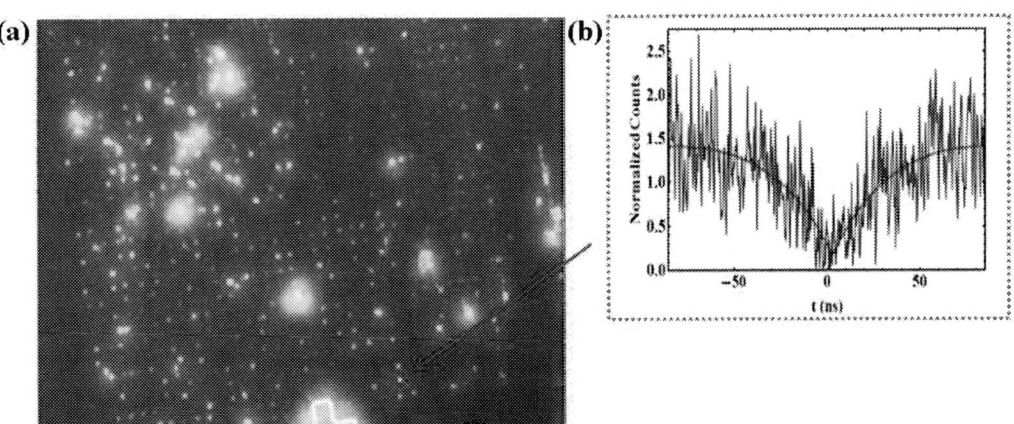

Figure 2(a,b). (a) The optical image of nanodiamond array; **(b)** Second-order correlation function $g^{(2)}(\tau)$ demonstrates the strong deep and represented the nanodiamond with single NV center on top of the chip.

3. Experimental setup and results

The presence of nanodiamonds with NV centers was detected using a confocal microscope technique [5]. The excitation of an NV center for measurement of the second-order correlation functions $g^{(2)}(\tau)$ was performed by 532 nm continuous wave laser (Coherent; Compass 315M–100). Beam scanning across the sample was realized using Galvano mirrors (Cambridge Technology; 6215H). The fluorescent radiation from NV centers was collected with a wide aperture (NA = 1.49) oil immersion lens (Nikon; CFI Apo TIRF 100X Oil) and separated from the excitation radiation by the dichroic mirror (Semrock; LPD01–633RU-25), notch filter (Semrock; NF03–532E25) and longpass filter (Semrock; LP02–633RU-25). A single-photon avalanche photodiode, SPAD (PerkinElmer; SPCM–AQRH–14–FC) was used as a detector of the NV center single-photon emission [5]. The installation works were made according to the following algorithm. The laser radiation is focused on the substrate surface in a spot about half the wavelength in diameter with a nanodiamond and excites an optical transition in the NV center. The radiation from the NV center is collected by means of an objective and an intermediate layer of immersion oil. The NV center radiation is separated from the reflected green laser light by a filter and detected by a Brown-Twiss interferometer, which includes a 50:50 beam splitter with two channels and avalanche photodiodes.

Digital signal processing and calculation of the autocorrelation function $g^{(2)}(\tau)$ are performed using a digital oscilloscope and software. Second-order correlation functions $g^{(2)}(\tau)$ was obtained for different points of the nanodiamond array. The measured dependence of $g^{(2)}(\tau)$ for one of the nanodiamond, demonstrated the presence of single NV center, is shown in Figure 2(b). Approximately in every eighth point of the array was a single-photon source, demonstrating the achieved yield of the studied method.

Further work will be devoted to the integration of NV centers with nanoantennas from TiN film. In Figure 3 shown a schematic image of such the array, as well as a single nanoantenna with specified dimensions. Titanium nitride is highlighted in blue, as well as alignment crosses are shown in yellow.

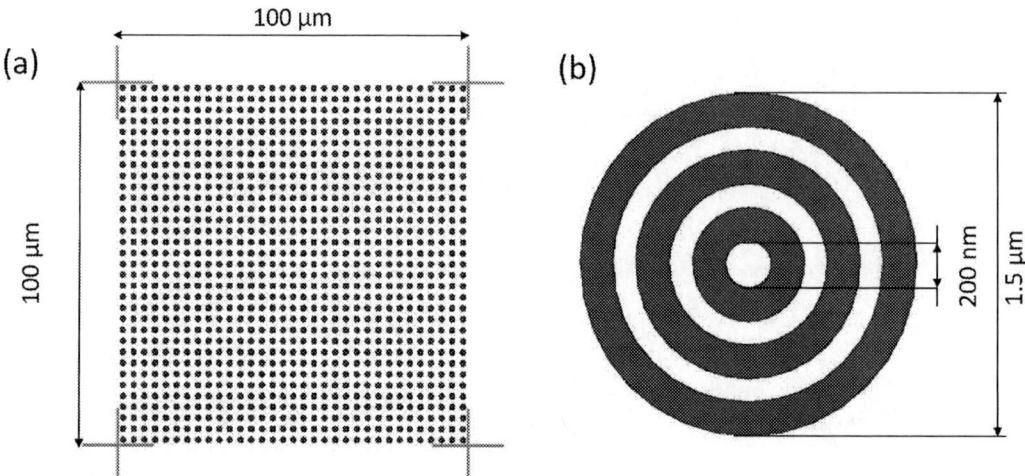

Figure 3 (a, b). (a) Schematic view of the kilopixel array of TiN annular nanoantennas; **(b)** Schematic view of the single annular nanoantenna for NV color center in a nanodiamond placed in the center.

4. Conclusion

In conclusion, we fabricated on-chip array of nanodiamonds using e-beam lithography and liftoff technique, measured the second order correlation function for a large number of single-photon emitters and achieved the yield of 12.5%. The further work will be concentrated on increasing the yield of this method as well as on increasing the speed of spontaneous emission by approaching NVcenters with hyperbolic metamaterials (HMM) and increasing the efficiency of collecting radiation from the NV-center, by placing the nanodiamond at the center of the annular nanoantenna [4-6].

5. Acknowledgments

A. Korneev, V. Kovalyuk and G. Goltsman acknowledge support by Ministry of Education and Science of the Russian Federation № 14.586.21.0063 (RFMEFI58618X0063).

References

[1]   Aspuru-Guzik A and Walther P 2012 *Nat. Phys.* **8** 285–91

[2]   Pernice W H P, Schuck C, Minaeva O, Li M, Goltsman G N, Sergienko A V, Tang H X 2012 *Nat. Commun* **3** 1325

[3]   Silveerstone J, Bonneau D, O'Brien J L, Thompson M G 2016 *IEEE J. Sel. Top. Quantum Electron.* **22** 390

[4]   Bogdanov S, Shalaginov M Y, Boltasseva A, and Shalaev V M 2016 *Opt. Mater. Express* **7** 111
[5]   Shalaginov M Y, Vorobyov V V, Liu J, Ferrera M, Akimov A V, Alexei Lagutchev A, Smolyaninov A N, Klimov V V, Irudayaraj J, Kildishev A V, Boltasseva A, Shalaev V M 2015 *Laser Photonics Review* **9** 1

[6]   Galfsky T, Krishnamoorthy H N S, Newman W, Narimanov E E, Jacob Z, Menon V M 2015 *Optica* **2** 1

[7]   Bermu'dez-Uren E, Gonzales-Ballestero C, Geiselmann, Marty R, Radko I P, Holmgaard T, Alavedyan Y, Moreno E, Garci'a-Vidal F J, Bozhevolnyi S I, Quidant R 2015 *Nature communications* **6** 1

SPBOPEN 2018

IOP Publishing

IOP Conf. Series: Journal of Physics: Conf. Series **1124** (2018) 051047   doi:10.1088/1742-6596/1124/5/051047

Experimental optimisation of O-ring resonator Q-factor for on-chip spontaneous four wave mixing

P An[1,2], V Kovalyuk[1,2], A Golikov[1], E Zubkova[1,2], S Ferrari[3,4], A Korneev[1,5], W Pernice[3,4], G Goltsman[1,2,6]

[1]Department of Physics, Moscow State Pedagogical University, 119992, Russia
[2]Zavoisky Physical-Technical Institute of the Russian Academy of Sciences, 420029, Russia
[3]Institute of Physics, University of Münster, 48149, Germany
[4]CeNTech - Center for Nanotechnology, University of Münster, 48149, Germany
[5]Moscow Institute of Physics and Technology (State University), 141700, Russia
[6]National Research University Higher School of Economics, Moscow 101000, Russia

Abstract. In this paper we experimentally studied the influence of geometrical parameters of the planar O-ring resonators on its Q-factor and losses. We systematically changed the gap between the bus waveguide and the ring, as well as the width of the ring. We found the highest $Q = 5 \times 10^5$ for gap 2.0 µm and the ring width 2 µm. This work is important for further on-chip SFWM applications since the generation rate of the biphoton field strongly depends on the quality factor as Q^3

1. Introduction
Planar optical ring resonators (ORRs) are very powerful devices for different applications, including optical filtering [1], biology [2], photon pair generation [3] and generation of high-dimensional entangled quantum states [4] . One of the most important parameters of the ORR are waveguide losses and quality factor (Q-factor). In this work we experimentally study dependence of ORR Q-factor and internal losses on two main geometrical parameters: gap (G) between single mode bus waveguide and O-ring waveguide width (W_r).

2. Device designing and fabrication
We fabricated a 2D array of devices with two sweep parameters: $G = 0.1 \div 2.0$ µm and $W_r = 1.2 \div 2.0$ µm and designed them using a self-made Python programming script for positive e-beam lithography processing. The ttypical element of the array consisted of two focusing grating couplers (FGCs) for input and output light awaveguide bus with fixed width of 1µm width and a O-ring with fixed radius of 50µm, located at a variable distance (gap) from the bus waveguide as depicted inFigure 1(a). The total number of devices inside the array was 200. We used commercial available Si wafers with 450 nm Si_3N_4 surface as an optical layer, and SiO_2 layer with 2600 nm as an optical buffer between Si and Si_3N_4 .

Fabrication process included one step e-beam lithography with current of electron beam of 25 pA and 30 kV accelerate voltage and included a proximity effect correction using NanoMaker software. A positive e-beam resist ZEP 520A was spin coated at a speed of 4000 rpm to achieve a nominal thickness of 400 nm. We developed ZEP 520A in O-Xylene for 50 sec at a temperature of 25°C after the exposure. Dark field optical microcopy was used for controlling of the development procedure. Reactive ion etching process (RIE) was used for waveguide finalizing in argon and CHF_3 gas mixture. We etched only the half of height of the optical layer (rib waveguides) due to better geometry parameters reproducibility. Finally, ZEP 520A was removed by oxygen plasma cleaning.

Content from this work may be used under the terms of the Creative Commons Attribution 3.0 licence. Any further distribution of this work must maintain attribution to the author(s) and the title of the work, journal citation and DOI.
Published under licence by IOP Publishing Ltd

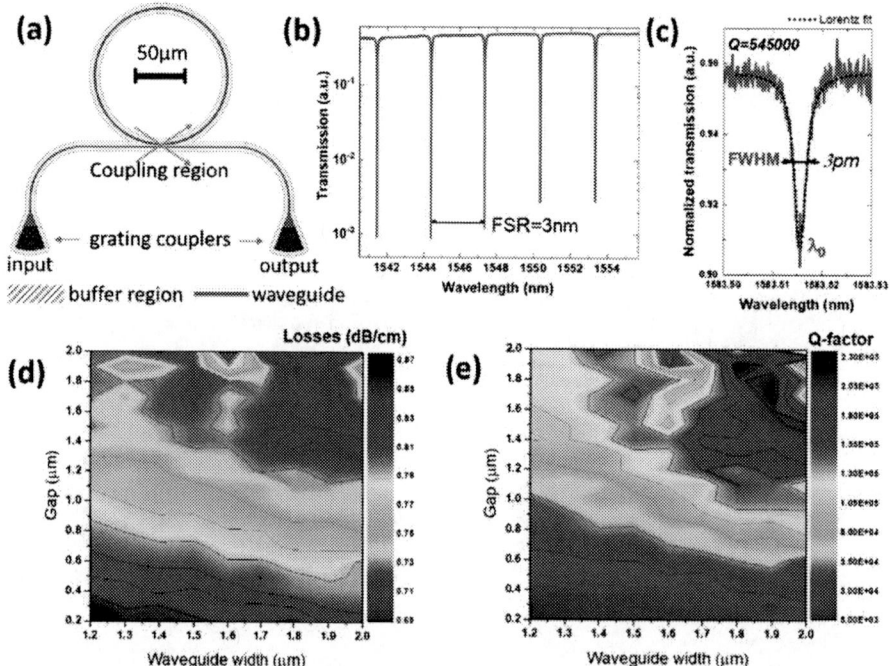

Figure 1(a-e). (a) Design of the planar optical resonator (ORR); **(b)** Transmission spectrum of ORR; **(c)** Single resonance peak demonstrated Q-factor over 5×10^5; **(d)** Color contour map of the optical losses vs gap and ring waveguide width; **(e)** Color contour map of the Q-factor vs gap and ring waveguide width

3. Experimental setup

Our experimental setup consisted with tunable laser source (New Focus TLB-6600) for tune the light in the range of 1510÷1620 nm, polarization controller for adjusting polarization, 3D mechanical stage with piezo motors for precision geometrical aligning of FGCs and light from fibers array [5], as well as a fast photodetector for optical power measurementsAdditionally to the main ORRs, we measured test devices without rings, fabricated in the same technology process. We subtract the spectrum of FGC from ORRs spectra, obtaining the normalized O-ring transmission spectra.

4. Methods and results

Typical optical transmission of one of the ORR is shown in Figure 1(b). We found, that the measured free spectral range (FSR) has a good agreement with the theoretical equation:

$$FSR = \Delta\lambda = \frac{\lambda^2}{n_g L}, \tag{1}$$

where $\Delta\lambda$ is the wavelength difference between two resonances, n_g is group refractive index and $L = 2\pi R$ is length of the ring ($R = 63.73$ μm and $n_g \approx 2$).

The Q-factor of the O-ring resonator is proportional to numbers of oscillations of the field before the stored optical power deplets to 1/e on respect to the initial power: $P(x) = P_0 e^{-\alpha x}$.

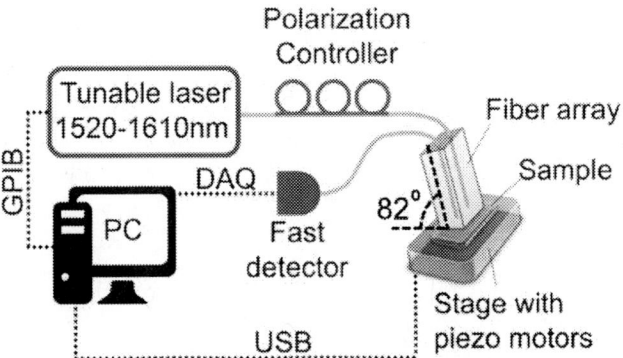

Figure 2. Experimental setup for the measurements of the transmission spectra of ORRs. Light from the tunable laser source is aligned with the ORRs using a fiber array and 3D stage with piezo motors. Light, passing thought structures, was detected by a fast photodetector. The photodetectros signal was recorded using a NI DAQ system. A polarization controller serves to match the the direction of polarization of the incident wave and the FGC.

The number of oscillations, are determinated by the losses α in the ring. The formal expression for the intrinsic Q-factor is:

$$Q_{int} = \frac{2\pi n_g}{\alpha \lambda}. \tag{2}$$

Q_{int} cannot be directly measured from the transmission spectra, but can be extrapolated by determining the loaded quality factor Q_{loaded}.

In order to extrapolate Q_{int} we first determined the loaded quality factor of the resonance peak directly from transmission spectra:

$$Q_{loaded} = \frac{\lambda_0}{FWHM}, \tag{3}$$

where FWHM is the full width half maximum of a resonant peak and λ_0 is the resonance wavelength (Figure 1(c)). We chose the λ_0 near 1550 nm for all of the structures and resonance Lorentz fitting for the best extracting λ_0 and peak FWHM.

In a second step, we calculated the intrinsic Q-factor of the ring [6]:

$$Q_{int} = \frac{2Q_{loaded}}{1+\sqrt{T_0}}, \tag{4}$$

where T_0 is the fraction of transmitted optical power measured by the photodetector at the resonant wavelength λ_0.

Finally, combining equation (1), (2) and (4) we calculated optical losses α in the ring:

$$\alpha = \frac{2\pi n_g}{Q_{int}\lambda_0} = \frac{\pi \cdot \lambda_0}{FSR \cdot L}\left[FWHM(1+\sqrt{T_0})\right]. \tag{5}$$

In Figure 1(d,e) are shown a color contour map of Q-factor and losses dependencies on G and W_r. The highest Q-factor corresponds the region with the lowest losses pointing to the close to the critical coupling regime [7]. we found devices with higher quality factor value of $Q = 5.45 \cdot 10^5$, at the resonance wavelength of 1583.5 nm.

5. Conclusions

In conclusion, we studied dependence of ORR Q-factor vs gap between single mode bus waveguide and O-ring with different waveguide width (W_r). We found that lowest losses (~ 0.6 dB/cm) and highest Q-factor (over $2.3 \cdot 10^5$) at wavelength 1550 nm has the ring with $G = 1.8 \mu m$ and $W_r = 1.9 \mu m$. This results is important for further on-chip SFWM applications since the generation rate of the biphoton field strongly depends on the quality factor as Q^3.

Acknowledgments

P. An, V. Kovalyuk, E. Zubkova and G. Goltsman acknowledge support by the Russian Science Foundation (project No. 16-12-00045; device design and testing). A. Korneev acknowledges support of the Russian Science Foundation (project No. 17-72-30036; nanophotonic circuits modeling) and W. Pernice acknowledges support by the DFG grants PE 1832/1-1 & PE 1832/1-2 and the Helmholtz society through grant HIRG-0005 (fabrication of nanophotonic circuits).

References

[1]   Rabus, D. G. et.al. 2002. *IEEE Journal of Selected Topics in Quantum Electronics* **8** 6 1405–1411
[2]   Juan-Colás, J. et.al. 2016. *Nature Communications* **7**
[3]   E. Engin et.al. 2013 *Opt. Express* **21** 27826-27834
[4]   Kues, M. et.al. 2017. *Nature* **546** 7660 622–626
[5]   Taillaert D. et. al. 2016 *Japanese Journal of Applied Physics* **45** 8-A 6071–6077
[6]   P. Barclay et.al 2005 *Opt. Express* **13** 801-820
[7]   Yariv, A. 2002. *IEEE Photonics Technology Letters* **14** 4 483–485

SPBOPEN 2018   IOP Publishing

Optimization of contra-directional coupler based on silicon nitride Bragg rib waveguide

E Zubkova[1,2], P An[1,2], V Kovalyuk[1,2], A Korneev[1,3], S Ferrari[4,5], W Pernice[4,5], G Goltsman[1,2,6]

[1]Department of Physics, Moscow State Pedagogical University, 119992, Russia
[2]Zavoisky Physical-Technical Institute of the Russian Academy of Sciences, 420029, Russia
[3]Moscow Institute of Physics and Technology (State University), 141700, Russia
[4]Institute of Physics, University of Münster, 48149, Germany
[5]CeNTech - Center for Nanotechnology, University of Münster, 48149, Germany
[6]National Research University Higher School of Economics, Moscow 101000, Russia

Abstract. We report on the development and fabrication of a contra-directional coupler based on the Bragg waveguide on Si_3N_4 platform. Transmitted and reflected by the contra-directional coupler spectra were measured. The reflected spectra exactly matches the one notched by the main channel of the coupler. Losses are about 3dB, coupling to the directing branch of the coupler is practically lossless. FWHM of the transmitted (reflected) spectra is 3.46 nm.

1. Introduction

Quantum photonic integrated circuits (QPICs) enables the implementation of complex architectures on a single chip [1] allowing to overcome the high losses and large footprint of free-space optical systems. The research interest is growing towards the integration of single-photon sources, single-photon detectors and passive elements, such as filters, splitters and resonant cavities. The development of low-loss passive components capable of spectrally select and distribute the optical modes propagating into a photonic waveguide would enable efficient filtering of pump light in case of optically excited single photon sources and for multi-channel (i.e. parallel) processing of quantum information.

Bragg's grating has been proposed as key-element for a large range of applications, such as: lasers [2], wavelength-division multiplexing (WDM) filters [3] and optical sensors [4].Here we present the realization of nanophotonic contra-directional coupler (CDC) based on the Bragg waveguide [5] capable of spectrally split the propagating signal with low insertion losses.

2. Device design and fabrication

For the devices fabrication a multi-layered substrate, with thicknesses of silicon (Si), silicon oxide (SiO_2), and silicon nitride (Si_3N_4) of 450 μm, 2.6 μm, and 450 nm respectively, has been used. The rib waveguide has been realized in the Si_3N_4 layer. Si_3N_4 was chosen because it has low absorption in the visible and IR wavelengths, a high refractive index ($n \approx 2$) and also demonstrates good mechanical properties.

With use of the COMSOL Multiphysics, numerical simulation of the Bragg waveguide properties has been carried out, including the analysis of the electric field intensity distribution along the waveguide CDC for both TE_1 and TE_2 modes (Figure 1(b)). Effective mode indices (n_1 and n_2) and the reflection wavelength have been evaluated using the phase-matching condition [6] (Figure 1(a)):

$$\lambda_C = 2\Lambda n_{av} = \Lambda(n_1 + n_2) \tag{1}$$

Content from this work may be used under the terms of the Creative Commons Attribution 3.0 licence. Any further distribution of this work must maintain attribution to the author(s) and the title of the work, journal citation and DOI.
Published under licence by IOP Publishing Ltd

$$\lambda_1 = 2\Lambda n_1 \qquad\qquad (2)$$

$$\lambda_2 = 2\Lambda n_2 \qquad\qquad (3)$$

where λ is the central wavelength of the reflected spectrum, Λ is the Bragg grating period, and n_1, n_2, n_{av} are the effective refractive index and average refractive index, respectively. For the simulation, the width of the waveguide (W_{in}, W_{out}) and their separation G have been varied. The structure has been optimized for operation in the telecommunication wavelength range, around 1550 nm. Calculation of the parameters of the Bragg grating, such as the Bragg grating period (Λ), filling factor ($ff = a/\Lambda$), height of teeth (h), and study of the waveguide structure transmittance depending on these parameters, are given elsewhere [7].

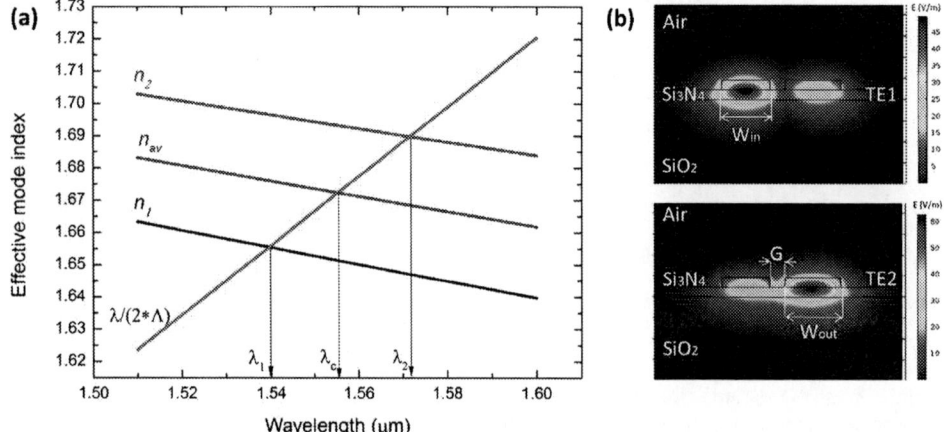

Figure 1 (a, b,). **(a)** Calculated effective mode indices with the condition of phase matching and certain reflected wavelengths; **(b)** The calculated intensity distribution of the electric field along the CDC waveguides for the first (TE$_1$) and second (TE$_2$) modes.

The CDC consists of two adjacent anti-phase Bragg waveguides [5] One of these waveguides, referred to as the main waveguide (W_{in}), serves as a carrier, supporting propagation of the signal in a wide spectral range, while the other waveguide (W_{out}) is used for directing the notched part of the spectrum, which is reflected back and propagates in the backward direction ,.relative to the original direction of the wave(Figure 2(b)). Focusing grating couplers (FGC) have been used for the single mode fiber – to– waveguide and waveguide – to – single mode fiber coupling of light (Figure 2(c, d)). The main waveguide has an input and output couplers allowing for the transmission spectrum control. The coupled waveguide is equipped with the output coupler only allowing for the reflected wavelength measurement and control. The integrated device had three optical ports (an input, an output with a transmitted signal, and an output with a reflected signal) and two unconnected Bragg waveguides (Figure 2(a)).

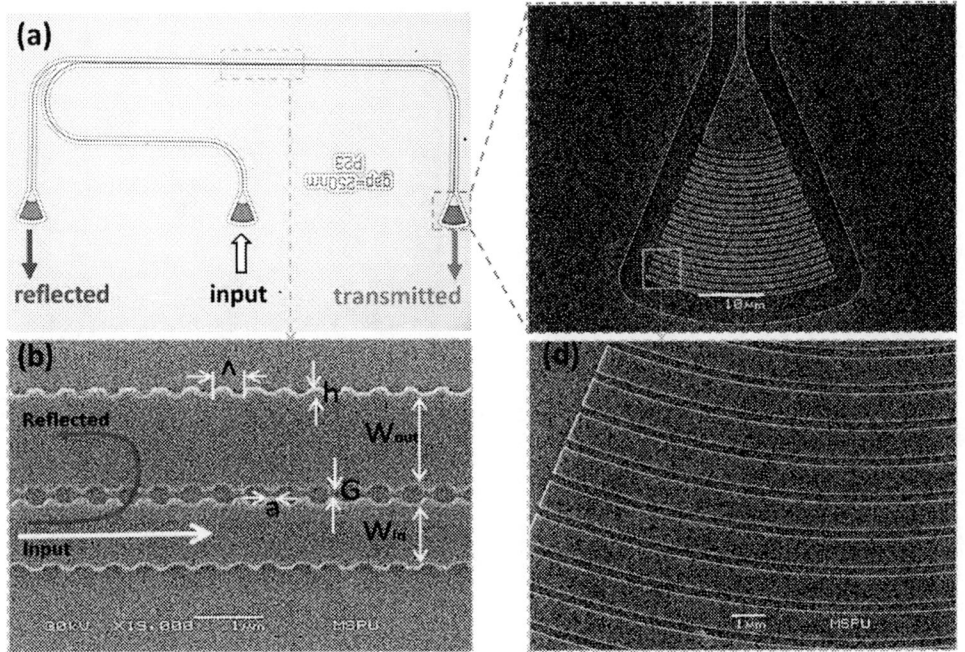

Figure 2 (a-d). (a) Optical micrograph of a fabricated device; **(b)** SEM image of CDC section based on the Bragg waveguide (Λ is a period of the Bragg waveguide, W_{in} is a width of the main waveguide, W_{out} is a width of the branch with reflected signal, h is a height of the teeth, a is a width of a tooth, G is a spacing between the teeth of the two waveguides); **(c)** SEM image of the FGC; **(d)** Zoom-in of the FGC (SEM).

The structures were fabricated by means of electron-beam lithography using a positive resist ZEP 520A, providing good contrast during the lithography. NanoMaker software [8] was used for proximity effect correction and precise adjustment of the e-beam exposure time. Removing of the exposed material was done by reactive-ion-etching (RIE) in CHF_3-Ar mixture. The resist residuals were cleaned out with help of the oxygen plasma.

3. Experimental setup and obtained data

In order to characterize properties of the fabricated devices, transmitted and reflected spectra were measured (Figure 3(b)). A piezo-actuated positioning system has been used to optimize the alignment between the FGC and the fiber array thus maximizing the transmitted signal. A tunable narrow line wide-band laser was used as a light source which was carried through a polarization controller and then delivered to the device using one channel of fiber array (FA) placed above on-chip focusing grating coupler (FGC) [9] (Figure 3(a)). The grating couplers are located at a distance of 250 μm from each other, the same distance is given between the individual optical fibers in the array. A wide spectrum of radiation from a tunable laser is supplied into the main waveguide of the structure through an input coupler. Spreading along the Bragg waveguide, the narrow spectral range determined by the parameters of the Bragg grating (Λ, ff, h), is reflected back and passes into the diverging waveguide. As a result, a transmission spectrum is eliminated from the coupler output, and the reflection spectrum is measured at the diverging coupler output (Figure 3(b)). As was found, the reflected spectra coincide with the notched spectra in the main waveguide. A PC was used to control the laser and power detector which was used to measure the transmitted and reflected power.

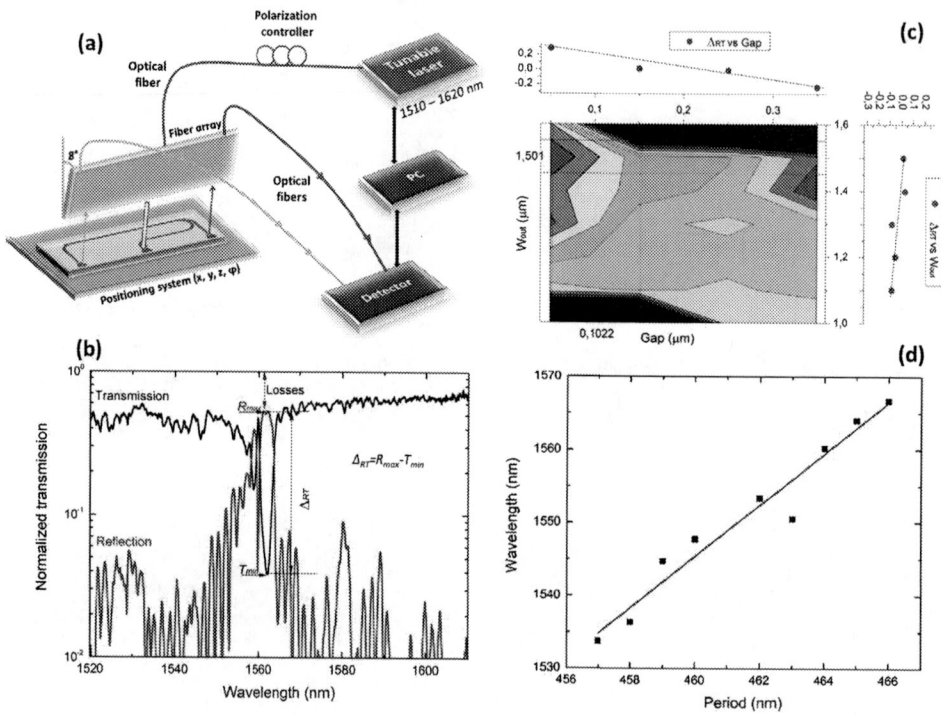

Figure 3 (a-d). (a) Schematic view of the experimental setup; **(b)** Normalized transmitted (black line) and reflected spectra (red line); **(c)** Dependence of the notched wavelength on the period of the Bragg waveguide (dots are the experimental points, line represents the linear approximation); **(d)** Dependence of the notched wavelength on the period of the Bragg waveguide (dots are the experimental points, line represents the linear approximation).

Device insertion losses, extracted via transmission normalization of on-chip reference circuit with two FGCs and waveguide between them, were about -3 dB. As depicted in Figure 2(b), FWHM of the transmitted (reflected) spectra is 3.46 nm. Coupling to the diverging branch of the coupler is practically lossless. By optimizing the width of the coupled waveguide (W_{out}) and the spacing between the waveguides (G) aiming at the maximization of the reflected power into the coupled waveguide.

We estimated the difference between the normalized values of the maximum reflected power (R_{max}) and the minimum transmission power (T_{min}) for the reflected spectral component ($\Delta_{RT} = R_{max} - T_{min}$) (Figure 3(c)). We found as optimal parameters for the structure: $W_{in} = 1\ \mu m$, $W_{out} = 1,5\ \mu m$, $Gap = 50\ nm$, corresponding to the maximum transition of the reflected spectral component from the main to the outgoing waveguide.

In the colorplot if Figure 3(c) is reported the transmitted signal at different Wout sidth and gap. The red region indicates the maximum transmission region, whilethe green and blue areas correspond to increasingly lossy reagions.

By varying the Bragg grating period at a fixed filling factor $ff = 0.5$ and teeth height $h = 100\ nm$, we have measured dependence of the notched wavelength on the period (Figure 2(d)).

4. Conclusion

It was demonstrated that optimized contra-directional coupler based on the Si$_3$N$_4$ waveguide can be successfully used for notching aspecified wavelength range and directing it into another waveguide with very low losses. Further work will be devoted to the improvement of the spectral characteristics of the contra-directional coupler, development of the multi-channel demultiplexer as well as its integration with the waveguide integrated superconducting nanowire single-photon detectors (WSSPDs) on silicon nitride platform [10].

Acknowledgments
E. Zubkova, P. An acknowledge support by the Russian Science Foundation (project No. 16-19-10633; design, and testing). A. Korneev acknowledges support of the Russian Science Foundation (project No. 17-72-30036; nanophotonic circuits modeling) and W. Pernice acknowledges support by the DFG grants PE 1832/1-1 & PE 1832/1-2 and the Helmholtz society through grant HIRG-0005 (fabrication of nanophotonic circuits and testing). We also are thankful to Yu. Lobanov (MSPU) for helpful discussion and proofreading of the English version of the text.

References
[1] Silverstone J W, Bonneau D, O'Brien J, Thompson M G 2016 Silicon Quantum Photonics *IEEE J. Sel. Top. Quantum Electron.* 22 1–13
[2] Fang A W, Lively E, Kuo Y-H, Liang D, and Bowers J E 2008 *Opt. Express* **16** 4413–4419
[3] Wang X, Shi W, Vafaei R, Jaeger N A F and Chrostowski L 2011 *IEEE Photon. Technol. Lett.* **23** 290–292
[4] Chrostowski L, Grist S, Flueckiger J, Shi W, Wang X, Ouellet E, Yun H, Webb M, Nie B, Liang Z, Cheung K C, Ratner A S S D M and Jaeger N A F 2012 *Proceedings of SPIE* 8236 823620
[5] Shi W. *et al.* 2013 *Opt. Express* **21** 6733–6738
[6] Shi W, Wang X, Lin C, Yun H, Liu Y, Baehr-Jones T, Chrostowski L 2013 *Optics Express* **21** 3 3633.
[7] Zubkova E, An P, Kovalyuk V, Korneev A, Ferrari S, Pernice W, Goltsman G 2017 *IOP Conf. Series: Journal of Physics: Conf. Series* **917** 062042
[8] http://www.nanomaker.com
[9] Van Laere F, Claes T, Schrauwen J, Scheerlinck S, Bogaerts W, Taillaert D, O'Faolain L, Van Thourhout D, Baets R 2007 *Optics Express* **19** 21–24 1919–1921
[10] Kahl O, Ferrari S, Kovalyuk V, Goltsman G N, Korneev A, Pernice W H P 2015 *Scientific Reports* **5** 10941

SPBOPEN 2018   IOP Publishing

IOP Conf. Series: Journal of Physics: Conf. Series **1124** (2018) 051049   doi:10.1088/1742-6596/1124/5/051049

Investigation of influence incoherent background illumination on the nonlinear response of a lithium niobate crystal sample at low light intensity

A V Pustozerov, A S Perin

Department of Microwave and Quantum Radio Engineering, Tomsk State University of Control Systems and Radioelectronics, Tomsk, 634050, Russia

Abstract. The total compensation of nonlinear diffraction of coherent laser beams of He-Ne laser (wavelengths 633 nm) with diameters on full width of half maximum near to 15 - 20 μm due of assistance of incoherent background have been experimentally demonstrated. Incoherent background had a shorter wavelengths (455 – 465) nm and much lower intensity compared with coherent signal beam with wavelengths 633 nm. Obtained the dependences of the refractive index change in the $LiNbO_3$:Fe crystal under the conditions of such incoherent background.

1. Introduction

Effects of self-action may occur in nonlinear optical media. Depending on the response of the optical medium can be self-focusing and self-defocusing of light beams are possible besides in some conditions it consequences diffraction-less propagation of these beams accompanied with regimes of spatial solitons [1, 2]. The one of most unique materials among photorefractive crystals is lithium niobate ($LiNbO_3$) doped with some impurities, for example with iron (Fe), copper (Cu), manganese (Mn), cerium (Ce) and their combinations [3] The base response of the lithium niobate crystal is self-defocusing. To observe self-focusing of light beams instead their self-defocusing within $LiNbO_3$ crystal samples, the externally applied electric field, the thermo-optic and pyroelectric effects have been used [3 – 6]. For attain that we use one of alternative method, it's a use photovoltaic properties of this material by combinations of light fields with different wavelengths [7, 8].

The main aim of this work is experimental study of optical nonlinearity sign cross-over from self-defocusing to self-focusing in conditions of incoherent background within $LiNbO_3$:Fe sample.

2. Experimental conditions and results

Two distinct experimental approaches are used in our study. In our first experiment, we check value of refractive index change in $LiNbO_3$:Fe sample under influence of incoherent light produced by LED`s with central wavelengths of 455 nm, 470 nm, 525 nm. Along with this, light sources with different power and spatial coherence are used. This change is due to the photorefractive properties of Fe-doped $LiNbO_3$. The experimental setup exploits the interference between red beams reflected from the entrance and exit surfaces of $LiNbO_3$:Fe sample (Figure 1). Optical power of red beam used ranges from 10 to 20 μW at beam diameter near to 1 mm to exclude photorefractive self-action of this beam within the crystal. Red light propagates within the sample (LN) under small angle to crystal X axis. Interference patterns formed by partially reflected from the slightly nonparallel entrance and exit

Content from this work may be used under the terms of the Creative Commons Attribution 3.0 licence. Any further distribution of this work must maintain attribution to the author(s) and the title of the work, journal citation and DOI.
Published under licence by IOP Publishing Ltd

surfaces of the sample. A phase of the light beam within the sample can vary in this scheme due to incoherent background influence or air temperature variation within the room.

LED with average λ=455 nm its light power obtains 0,5 W but angular divergence (FWHM) makes up 120° In experiment used sample of $LiNbO_3$:Fe (0,005 wt%) with dimensions of $5\times10\times10$ mm^3 along X, Y, and Z axes. Light beam of He-Ne laser with linear polarization is incident onto the sample entrance surface (YZ plane) with angle ~10° to X axis in XZ plane of the crystal. Red beam polarization corresponds to the extraordinary wave of the crystal. Incoherent light is introduced into the sample through the back YZ surface of the sample. For introduce the background light into the crystal, we use hollow metal tubes with mirror-like internal surface. Observation of light pictures was made by means of the analyser of light beams.

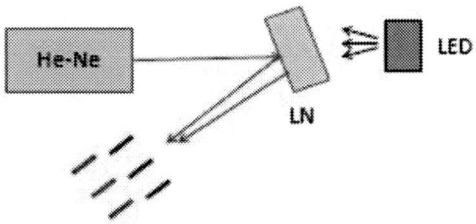

Figure 1. A scheme of experiments on studies of incoherent background influence to the refractive index of $LiNbO_3$:Fe; He-Ne – helium-neon laser; LN – crystal sample; LED – light-emitting diode.

Some particular results on a change of interference pattern caused by reflections of He-Ne laser beam from the entrance and exit surfaces of $LiNbO_3$:Fe sample under its illumination with incoherent light are shown in Figure 2. The difference of these patterns is practically the same for all LED's used with only quantitative distinction of interference maxima shift velocity. Images in Figure 2 shows the fastest shift of interference maxima for using LED with 455 nm and optical power of 500 mW and with optical power of 633nm He-Ne laser is ~12 μW. Due to the sample exposure for time near to 6 minutes we observe the spatial shift of interference maxima, within the whole light pattern, corresponding to phase change of red beam on π in the sample. It is confirmed by intensity profiles of interference patterns at the start of incoherent illumination (t=0) and after 6 minutes of exposure (c and d of Figure 2).

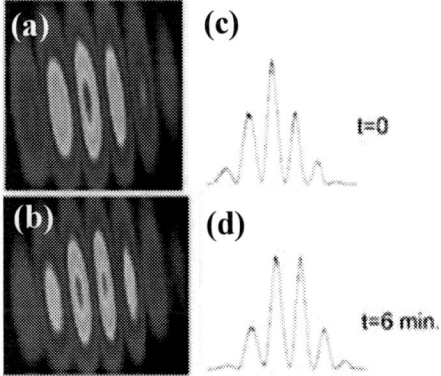

Figure 2(a, b, c, d). Images of interference patterns of He-Ne laser beams reflected from entrance and exit surfaces of $LiNbO_3$:Fe sample for the background light wavelength 455 nm at initial moment **(a)** and at exposure time 6 minutes **(b)** and intensity profiles of these patterns along Z direction: **(c)** t=0; **(d)** t=6 minutes.

For LED`s with a lower light intensity similar shift requires ~30 minutes of exposure. The results observed are explained by the arising of photovoltaic field E_{pv} within $LiNbO_3$:Fe sample at its illumination by incoherent light with almost uniform intensity distribution over the sample surface. This electric field changes the material refractive indices by means of linear electro-optic (Pockels) effect on the value $\Delta n = -0.5n^3rE_{pv}$, where n and r are refractive index and electro-optic coefficient corresponding to light polarization and propagation direction. Accordingly, the phase shift of He-Ne laser beam within the sample changes on the value $k\Delta nL$, where k is the wave number of light in free space and L is the light propagation distance.

Figure 3 shown dependence of the change of the refractive index of the crystal from the time of exposure of incoherent background illumination LED with λ=455nm and light power of 0.5W. Experimentally observed shift of interference maxima gives value $\Delta n_e \approx (0.7 - 1.2) \cdot 10^{-4}$ for extraordinarily polarized light beam in $LiNbO_3$ crystal.

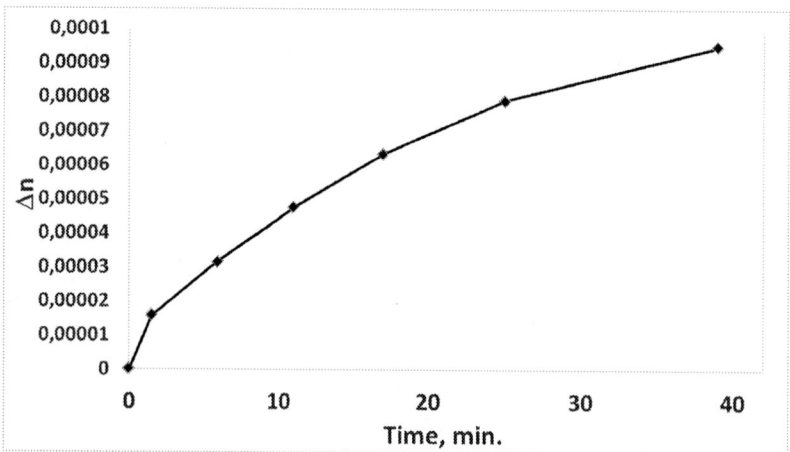

Figure 3. Graph of change in the refractive index of the photorefractive crystal of lithium niobate from time with incoherent background illumination LED with λ=455 nm and light power 0,5 W.

In our second experiment, we are possibility of red beam self-focusing using incoherent background (Figure 4). As incoherent background we use laser diode with wavelength 450 nm. The laser diode beam has elliptical shape with dimensions ~4×10 mm^2 in its cross-section at the distance 60 centimeters from diode lens. The light beam of He-Ne laser with λ=633 nm is focused by 8× and 4× microscope objectives (SL) onto the input YZ surface of crystal sample. Light spot diameter (FWHM) makes up ~15-20 µm on this surface.

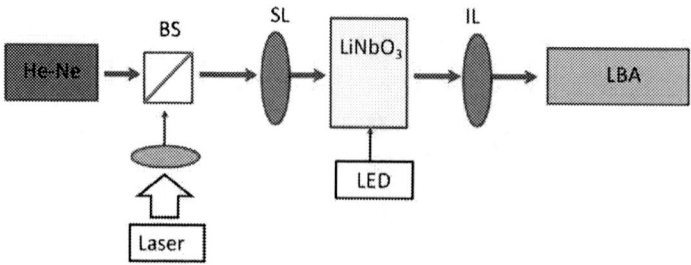

Figure 4. A scheme of experimental setup for investigation of red beam nonlinear diffraction in $LiNbO_3$:Fe: He-Ne, helium-neon laser; Laser, laser diode; BS, beam splitter; LED, light-emitting diode; SL, spherical lens; IL, imaging lens; LBA, CCD camera.

The light beam propagates practically along crystal X axis and increases its diameter to 40 – 60 μm at the sample exit surface due to natural diffraction. Light patterns from entrance or exit surfaces of sample are imaged by lens (IL) on laser beam analyzer (LBA). It is inserted into the sample along direction parallel to red beam propagation. Temporal dependences of red light patterns on the sample exit surface YZ are studied using LBA.

For the compensation of red beam natural (linear) diffraction requires increasing of background intensity which can be obtained with proper shaping of LED light field. Using laser diode with light wavelength 450 nm we also obtain the total compensation of nonlinear light diffraction caused by photovoltaic current produced within the crystal by narrow red beam and only very partial compensation of red beam linear diffraction. Background light is introduced into crystal sample along direction parallel to the signal (λ=633 nm) beam. Red beam power ~100 μW and its diameter is ~20 μm in that case. Power density of background light makes up about ~1 W/cm^2 and for red signal beam it is much higher again. Light images in Figure 5 show temporal evolution of light patterns for red beam at exit surface of the sample LiNbO$_3$:Fe without background and with background. It is seen that incoherent background with shorter wavelength than signal light beam, totally compensates the nonlinear diffraction of this signal beam at conditions described. And it is clear from analysis of these results that proper intensity of incoherent background will allow compensation of both, as linear as nonlinear beam diffraction that means stronger self-focusing of red beam.

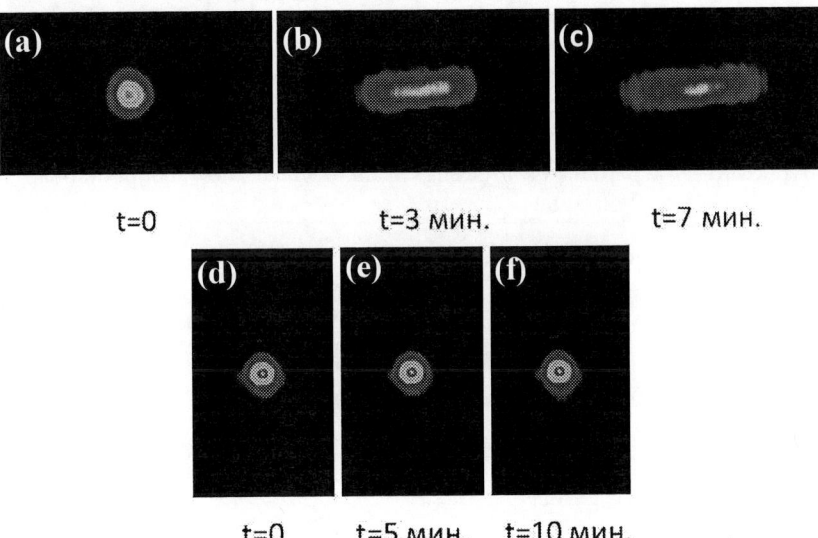

Figure 5(a, b, c, d, e, f). Images of light spots of red beam at exit surface of the sample at exposure beginning (a) and after some times when incoherent background is absent (b, c) and after start in conditions with incoherent background (d, e and f). Input FWHM is ~15 μm, red beam power is 20 μW, intensity of background with λ=450 nm is more than order less than average intensity of red beam.

The graph shows the dependence of the width of the red beam of a He-Ne laser on time in conditions with background illumination and without it (Figure 6).

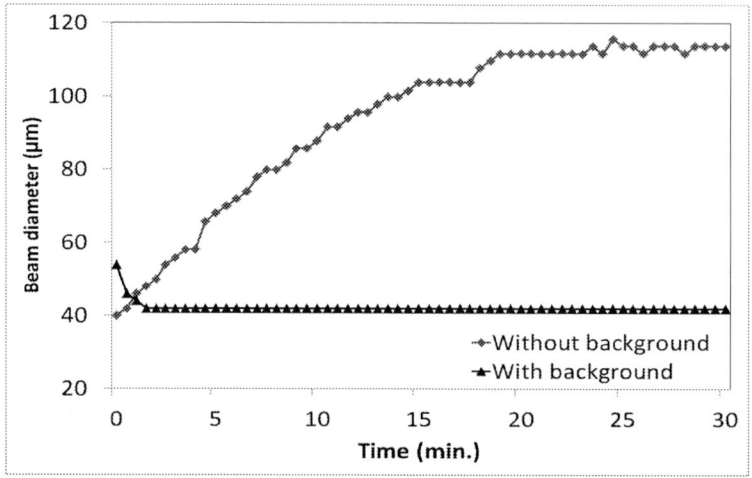

Figure 6. Evolution of beam diameter (λ=633 nm) for light propagation with no background ($\Diamond$) and with background (Δ); λ_{bg}=450 nm; d=20 μm.

3. Conclusion

In conclusion, we have demonstrated that influence of spatially incoherent background with a more short wavelength on the narrow coherent signal beam with longer wavelength when it distributed in a LiNbO$_3$:Fe sample provides total compensation of nonlinear beam diffraction caused by photorefractive effect and partial self-focusing. It may create new configurations of all-optical photonic elements based on this crystal.

Acknowledgments

This study was carried out with the financial support of Ministry of Education and Science of Russia (within the task N 3.1110.2017/PCh of the project part and the task N 3.8898.2017/BCh of the basic part).

References

[1] Tiemann M, Halfmann T and Tschudi T 2009 Optics Communications **282** 17
[2] Stegeman G 1999 Science **286** 5444
[3] Kip D 1998 Applied Physics **67** 131
[4] Kip D, Krätzig E, Shandarov V and Moretti P 1998 Optics Letters **23** 343
[5] Safioui J, Devaux F and Chauvet M 2009 Optics Express **17** 205
[6] Perin A, Shandarov V and Ryabchenok V 2016 Physics of Wave Phenomena **24** 7
[7] Anastassiou C, Shih M, Mitchell M, Chen Z and Segev M 1998 Optics Letters **23** 12
[8] She W, Xu C, Guo B and Lee W 2006 Journal of the Optical Society of America B **23** 10

Bolometric effect for detection of sub-THz radiation with devices based on carbon nanotubes

M V Moskotin[1], I A Gayduchenko[1], G N Goltsman[1], N Titova[1], B M Voronov[1], G F Fedorov[1], F Pyatkov[2,3] and F Hennrich[2]

[1]Moscow State University of Education, Moscow, 119991, Russia
[2]Institute of Nanotechnology, Karlsruhe Institute of Technology, Karlsruhe 76021, Germany
[3]Department of Materials and Earth Sciences, Technische Universität Darmstadt, Darmstadt 64287, Germany

Abstract. In this work we investigate the response on THz radiation of a FET device based on an individual carbon nanotube conductance channel. It was already shown, that the response of such devices can be either of diode rectification origin or of thermoelectric effect origin or of their combination. In this work we demonstrate that at 77K and 8K temperatures strong bolometric effect also makes a significant contribution to the response.

1. Introduction

Carbon nanotubes (CNTs) are a promising optoelectronic material with unique electronic, optical properties [1]. It is well established that CNT make a particularly promising material in the terahertz region of the electromagnetic spectrum (defined as 0.1−10 THz) which is one of the least developed regimes lying in the gap between efficient manipulation with electronics and photonics. Development of the optoelectronic devices in THz region is important for a variety of potential applications, ranging from medical diagnostics to security [2]. Recently the concepts of several THz optoelectronic devices based on carbon nanotube (CNT) arrays were proposed, evaluated, and some of them realized: detectors, modulators and sources [3,4,5,6]. While rectification of the electromagnetic radiation gives good results at room temperatures [6] the traditional bolometric response of CNT based detectors may result in very high performance of such devices at cryogenic temperatures [3,7]. Bolometric response requires firstly that the sensing element has resistance that strongly depends on the temperature and secondly, temperature of the sensing element is significantly elevated as it is exposed to radiation.

In this work we explore this opportunity using devices based on quasimetallic nanotubes (qmCNT). Our approach is to use devices in which the interface between the nanotube and the electrode is not perfect so that thermal activation is required for transmission of charge carriers through the interface.

We investigate two types of devices: Type I devices are based on individual qmCNT with one contact having non-ohmic properties due to imperfect CNT-metal interface. Type II devices are made with multiple qmCNTs deposited over predefined metal pads using dielectrophoresis technique. In the latter case transmission through the CNT-metal interface is hampered due to mechanical deformation of the nanotubes in the contact area.

In both cases relatively good coupling to the radiation can be achieved since the channel itself has metal type conductivity.

Content from this work may be used under the terms of the Creative Commons Attribution 3.0 licence. Any further distribution of this work must maintain attribution to the author(s) and the title of the work, journal citation and DOI.
Published under licence by IOP Publishing Ltd

2. Experimental setup and device characterization

2.1. Device fabrication

Our experimental devices are made in a configuration of field-effect transistors in which the conduction channel is formed by carbon nanotubes. In this work two types (fig 1) of devices were tested, which differ in fabrication methodology of the conductance channel and the contacts.

In the first case (fig 1a) carbon nanotubes are grown on a silicon substrate with a dielectric layer of 500nm silicon oxide by CVD technique described in [5]. The conductance channel is continued by a double-purpose spiral antenna, which is used both to couple THz radiation and to conduct direct current. The elements of the antenna are made of different metals: the source electrode is made of nickel, whereas the drain electrode is made of gold. This asymmetry is achieved with the technique of electron-beam evaporation at different angles. The silicon substrate of the device is used as the gate.

In the second case (fig 1b) sorted metallic carbon nanotubes are dielectrophoretically deposited directly across the predefined golden electrodes as described in [8,9]. In both devices unnecessary CNTs are etched in oxygen plasma.

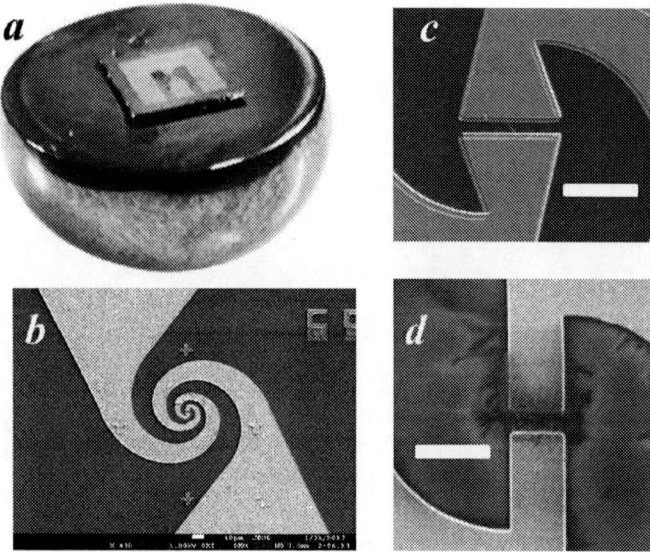

Figure 1. Device configurations a) Chip with a device on a silicon lens. b) SEM image of the log-periodic spiral antenna. c) SEM image of an individual nanotube with asymmetric metallization. Scale bar is 5 μm. d) SEM image of nanotubes deposited across the electrodes using dielectrophoresis. Scale bar is 5 μm.

2.2. Experimental setup

Our experimental device chip was placed in the centre of the flat surface of a silicon lens, which was located inside a cryostat. The device was irradiated through a high-density polyethylene window. To measure current-voltage and transfer characteristics we used a DAQ system with a self-made module to apply voltage on the source electrode and to measure the current in the channel. The gate voltage was applied with Keithley 2400 source-meter. A series of devices with similar design had been tested.

The terahertz radiation was provided by three backward wave oscillators (BWO). The first BWO was used as a source of sub-THz radiation with a frequency of 129 GHz. The maximum intensity of the radiation at the cryostat window was 400 μW/cm2 and was adjusted by an attenuator. The second BWO was used for 250-450 GHz frequency range, and the third BWO was used for 600GHz frequency. These BWO's maximum power was found to be equal ~50 μW.

2.3. Device characterization

First, we characterize electrical properties of our FET devices by measuring their transfer characteristics. Fig. 2 shows typical G(Vg) curves of devices type I and II obtained at a constant bias voltage of 10meV at three different temperatures: 300K, 77K and 8K. We note that in case of type I devices conductance has a pronounced minimum as a function of gate voltage with minimal conductance rapidly decreasing as the temperature is going down. Conductance of device type II slightly decreases as gate voltage is increased indicating p-doping of the conductance channel. As the temperature is decreased conductance in the whole range of gate voltages is going down. In case of both devices dependence of minimal conductance on temperature is very strong. In case of device type I $G_{MIN}(T)$ is consistent with thermally activated transport with a bandgap of about 20meV (see Fig. 2c). In case of device II Arrhenius plot of $G_{MIN}(T)$ is not linear that may be explained with tunnelling through the barrier at the contact [10] Such a barrier should appear when a metallic CNT is placed across two predefined electrodes due to its inevitable bending close to the CNT-metal contact.

We show for comparison in the Fig. 2d temperature dependence of conductance of a tunnelling barrier calculated using Landauer-Buttiker approximation.

$$I \sim \int_{-\infty}^{+\infty} T(E) \cdot (f(E + V_{SD}) - f(E)) dE$$

where T(E) is transmission coefficient through a barrier calculated in a quasi-classical approximation and $f(E)$ is the Fermi distribution function. Transport in such a system is governed by thermal activation across the barrier along with tunnelling that becomes more important at low temperatures.

We thus conclude that in both cases conductance of our devices is most sensitive to temperature at low temperatures. In both case this is due thermally activated transport.

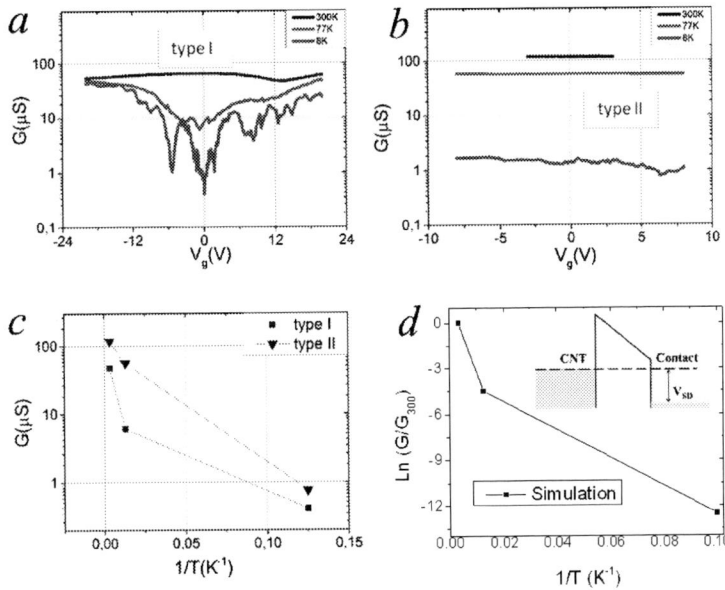

Figure 2. Characterization of THz detectors. (a), (b) conductance as a function of gate voltages for devices type I and type II respectively.
(c) Arrhenius plots of temperature dependence of minimal conductance of devices type I and type II.
(d) Temperature dependence of conductance normalized to its value at 300K for a single contact barrier simulated using Landauer-Buttiker approximation.

3. Results and Discussion

Next, we study the IV curves of type I devices with and without radiation. Figure 3 (a) and (b) show the IV curves of type I device measured at a gate voltage corresponding to conductance minimum. Figure 3 (c) shows IV curve of type II device at zero gate voltage. We note that at room temperature effect of radiation on the slope of the IV curve is very week. We also note that at temperatures approaching 10K the IV curves become strongly non-linear. At the same time radiation with relatively small power of few microwatts significantly increases the IV slope making it almost linear. Our interpretation of this effect is that radiation results in strong elevation of electron temperature and thus reduces the device resistance facilitating activation above the corresponding barrier. Importance of cooling the device down to cryogenic temperatures follows from the general expression for the bolometric responsivity (ratio of registered voltage ΔU to the radiation power P) of a device [3]:

$$\frac{\Delta U}{P} \sim \frac{dT}{dP} \cdot \frac{dR}{dT}$$

First term in the product increases upon cooling down due to reduced heat capacitance and thermal conductance of the electronic subsystem. As for the second term we note that in case of thermally activated transport with $R \sim \exp\left(-\Delta/k_B T\right)$ with Δ being effective energy and k_B – Boltzmann constant we get

$$\frac{dR}{dT} \sim -\frac{\Delta}{k_B T^2} R$$

which strongly increases upon reducing the temperature.

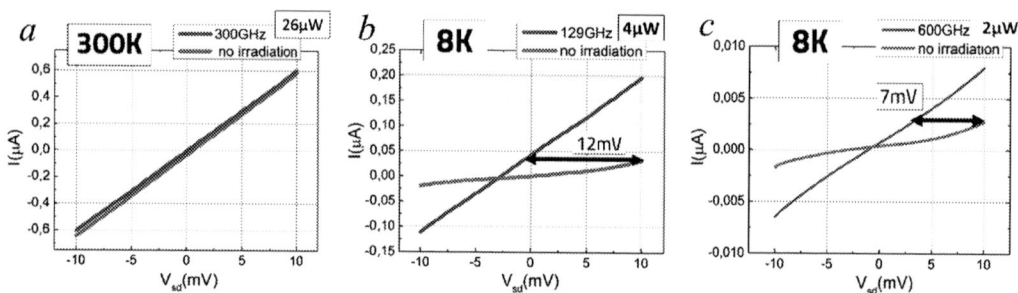

Figure 3. IV-curves measured in two conditions: with and without radiation for a) device type I at 300K; b) device type I at 8K; c) device type II at 8K. The response voltage is shown with the arrows.

We now evaluate the responsivity of our devices. As seen from Fig. 3 (b) and (c) maintaining constant current through the device (30nA in case of device I and 3 nA in case of device II) results in a change of the bias voltage by a value depending on the radiation power. The corresponding responsivity is about 3 to 4 kV/W. Estimating the noise spectral density using Nyquist formula we get the noise equivalent power (NEP) of about 2pW/√Hz for Device I and 7pW/√Hz for Device II. These values refer to the so called external responsivity that assumes perfect coupling of the device to the radiation. Following the methodology described in [7] we estimate the power scattered in the device to be about 4 orders below the power incident onto the antenna and find the responsivity of our devices to be about $\sim 10^7$ V/W resulting in the NEP values below 1fW/√Hz which compares to the best commercially available THz radiation detectors [11]. We note that also type II device shows slightly worse characteristics its fabrication route is more straightforward and can be adjusted for mass production. It also can be further improved via using better impedance match with the antenna used.

4. Conclusion

At low temperatures a strong modification of the IV-curves under the radiation indicates a strong bolometric effect making a significant contribution to the response origin.

The estimated responsivity of our devices is ~10^7 V/W. We thus have shown that efficient THz bolometers can be fabricated using CNT based devices with imperfect CNT-metal interfaces. Such devices operate as bolometers at cryogenic temperatures. Their performance is compatible with that of the state-of the-art HEB devices and is much better than results obtained with CNT based bolometers in this frequency range before.

5. Acknowledgments

This work was supported by the Ministry of Education and Science of Russian Federation (contract №14.583.21.0069, RFMEFI58317X0069). SEM imaging was performed using equipment of MIPT Shared Facilities Center. We thank Prof. M. Kappes for preparation of mCNT-suspension (Type II). We also thank Prof. R. Krupke for dielectrophoretical deposition of mCNTs on our devices.

References
[1] P. Avouris, M. Freitag, and V. Perebeinos, *Nat. Photonics 2*, **341–350** (2008);
[2] D. Saeedkia,Handbook of Terahertz Technology for Imaging, Sensing and Communications, 1st edition, Woodhead Publishing, 2013;
[3] K. Fu, R. Zannoni, C. Chan, S. H. Adams, J. Nicholson, E. Polizzi, and K. S. Yngvesson, *Appl. Phys. Lett.* **92**, 033105 (2008).;
[4] V. Ryzhii, T. Otsuji, M. Ryzhii, V. G. Leiman, G. Fedorov, G. N. Goltzman, I. A. Gayduchenko, N. Titova, D. Coquillat, D. But, W. Knap, V. Mitin, and M. S. Shur, *J. Appl. Phys.* **120** (2016) 044501-1-13;
[5] I. Gayduchenko, A. Kardakova, G. Fedorov, B. Voronov, M. Finkel, D. Jimenez, S. Morozov, M. Presniakov, G. Goltsman, *J. Appl. Phys.* **118** (2015) 194303;
[6] G.Fedorov, I.Gayduchenko, N.Titova, A.Gazaliev, M.Moskotin, N.Kaurova, B. Voronov, G.Goltsman, Carbon Nanotube Based Schottky Diodes as Uncooled Terahertz Radiation Detectors, *Physica Status Solidi (b)* **255** (1);
[7] Abdel El Fatimy, Rachael L. Myers-Ward, Anthony K. Boyd, Kevin M. Daniels, D. Kurt Gaskill& Paola Barbara 2016 *Nature Nanotechnology* **volume 11**, pages 335–338;
[8] R. Krupke, F. Hennrich, H. B. Weber, M. M. Kappes, H. V. Löhneysen, *Nano Lett.* **2003, 3**, 1019 – 1023;
[9] A. Vijayaraghavan, S. Blatt, D. Weissenberger, M. Oron-Carl, F. Hennrich, D. Gerthsen, H. Hahn, R. Krupke, *Nano Lett.* **2007, 7**, 1556 – 1560;
[10] J. Appenzeller, M. Radosavljevic´, J. Knoch, and Ph. Avouris. Tunnelling Versus Thermionic Emission in One-Dimensional Semiconductors, *Phys Rev Lett*, **92**, 048301 (2004);
[11] Seliverstov, S., Maslennikov, S., Ryabchun, S., Finkel, M., Klapwijk, T. M., Kaurova, N., ... & Goltsman, G. (2015). Fast and sensitive terahertz direct detector based on superconducting antenna-coupled hot electron bolometer. *IEEE Transactions on Applied Superconductivity*, **25(3)**, 1-4.

SPBOPEN 2018   IOP Publishing

Silicon nitride nanophotonic circuit for on-chip spontaneous four-wave mixing

A Golikov[1], V Kovalyuk[1,2], P An[1,2], E Zubkova[1,2], S Ferrari[3,4], W Pernice[3,4], A Korneev[1,5], G Goltsman[1,2,6]

[1]Department of Physics, Moscow State Pedagogical University, 119992, Russia
[2]Zavoisky Physical-Technical Institute of the Russian Academy of Sciences, 420029, Russia
[3]Institute of Physics, University of Münster, 48149, Germany
[4]CeNTech - Center for Nanotechnology, University of Münster, 48149, Germany
[5]Moscow Institute of Physics and Technology (State University), 141700, Russia
[6]National Research University Higher School of Economics, Moscow 101000, Russia

Abstract. Here we present an integrated nanophotonic circuit for on-chip spontaneous four-wave mixing. The fabricated device includes an O-ring resonator, a Bragg noch-filter as well as a nine-channel arrayed waveguide gratings (AWG) operated in the C-band wavelength range (1550 nm). The measured optical losses of the device (-6.8 dB) as well as a high Q-factor ($> 1.2\times10^5$) shows a good potential for realizing the spontaneous four-wave mixing on the silicon nitride chip.

1. Introduction

Quantum photonics integrated circuits (QPICs) is the one of the most promising approach for realization of long term quantum computation processing [1]. QPICs combine several advantages over table-top schemes: small size, high performance, and no need of alignment, high temperature stability as well as the possibility of integration with electronic circuits on the same platform. Quantum processing applications would require to fully integrating single-photon sources, logical elements and single-photon detectors onto the same platform. Because of its large optical bandwidth, thermal and mechanical properties, silicon nitride (Si3N4 has) been demonstrated to be a promising platform [2]. In this paper we present the design and characterization of a nanophotonic circuitry which can be used for single photons generation and on-chip spectral demultiplexing.

2. Device design and fabrication

Block diagram of the nanophotonic devices for nonlinear pair generation and filtering is shown in Figure 1. Such circuit consists of three main elements: Bragg waveguide [4, 5] as a notch wavelength filter, O-ring resonator [6] as a biphoton field source and the arrayed waveguide grating (AWG) as an optical demultiplexer [7]. Biphoton field is generated by optical pumping the integrated O-ring resonator. The pump light is filtered out by the Bragg waveguide and photons are distributed to different waveguide channels for a further correlated detection. The Figure 4(e) shows the expected transmission spectra of the nanophotonic device and its individual components. The dashed line indicates the transmission spectrum of the O-ring resonator, the solid black line shows the Bragg pumping filter, and the individual channels of the AWG are shown as colored semi-ovals. The channels labels correspond to the one shown in Figure 4(a). For the correct device operations, all the

---

Content from this work may be used under the terms of the Creative Commons Attribution 3.0 licence. Any further distribution of this work must maintain attribution to the author(s) and the title of the work, journal citation and DOI.
Published under licence by IOP Publishing Ltd

spectral channels of the AWG, as well as the pumping filtration, should be aligned with O-ring resonances.

The main fabrication steps are shown in Figure 2. The integrated circuit is realized on commercially available silicon wafers Si (350–400 μm) with a thermal silicon oxide of SiO_2 (2600 nm) and silicon nitride Si_3N_4 (450 nm) grown on top. A spin coated high resolution ZEP 520A

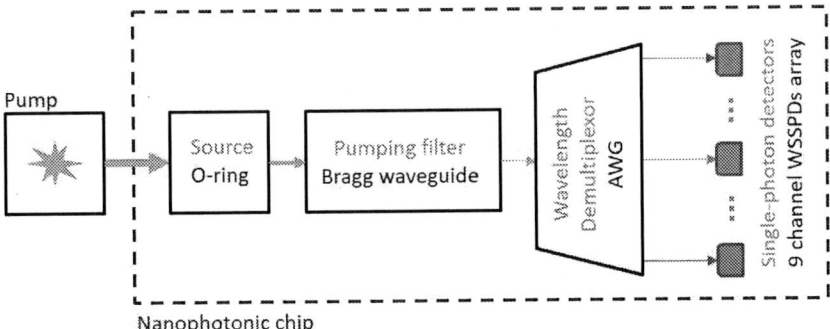

Figure 1. Block diagram of a nanophotonic circuit utilizing nonlinear photon-pair sources includes. The blue color shows schematically waveguide integrated superconducting single-photon detectors, the realization of which is planned in the future work.

resist was used for providing e-beam lithography. The chip exposure of a waveguide pattern was made by e-beam lithographer system Cable 9000C. For waveguide formation the method of plasma-chemical etching of Si_3N_4 layer in the atmosphere of Ar and CHF_3 was used. By cleaning the residual resist in the oxygen plasma, nanophotonic chip was finalized.

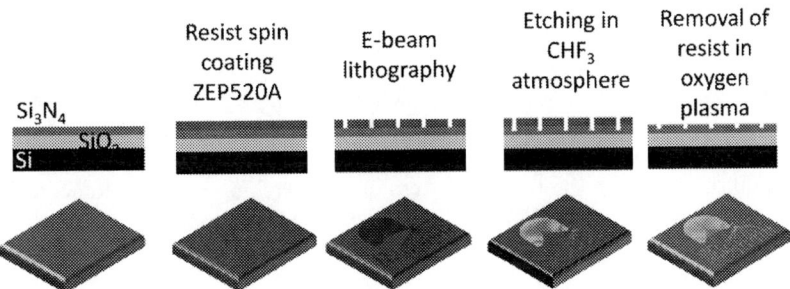

Figure 2. Technological route of manufacturing the device

Fabricated nanophotonic circuit is shown in Figure 4(a) and includes the SEM images of the main integrated components: Bragg waveguide (Figure 4 (b)), O-ring resonator (Figure 4 (c)), the star coupler of the arrayed waveguide grating (Figure 4 (d)).

3. Experimental setup and results
At this stage, the transmission spectrum of the circuit and found optical losses is tested.

The experimental setup is shown in Figure 3. We used a tunable laser source (TLB-6600) to generate light in a wavelength range of 1510-1620 nm, a polarization controller (Thorlabs FPC032) to adjust the polarization of the fiber mode which is coupled onto the input of focusing grating coupler (FGC). Using an optical fiber array, connected via FGCs we collected light that passed through all of the components, including the O-ring resonator and the Bragg waveguide (output 1) as well as the O-ring resonator, the Bragg waveguide and the AWG (outputs 2-10). The light was detected using a infrared photodiode Hamamatsu G9801.

The transmission spectrum of the nanophotonic circuit is shown in Figure 4(f). We analyzed the data and extracted the coupling losses (-7.8 dB), the filtering losses (-2.7 dB) and also the losses for a wavelength demultiplexer (-6.8 dB) separately.

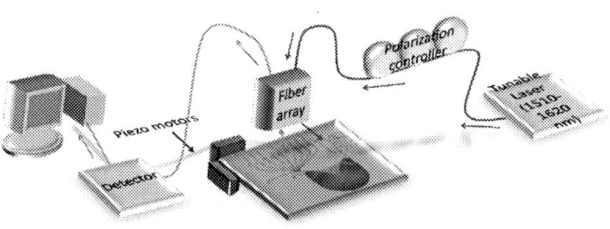

Figure 3. The scheme of the experimental setup including the tunable laser source, polarization controller, fiber array, room-temperature detectors as well as nanophotonic chip placed on a piezo stage.

Although the transmission spectrum is qualitatively similar to the one that was incorporated in the design of the nanophotonic device, the position of the spectral channels, as well as their spectral alignment, is still not sufficient to provide the generation of the biphoton field and its detection. This is due to the lack of optimized technology for the fabrication of nanophotonic devices. In the future, the work will be aimed at optimizing the fabrication technology by the increasing a resolution and improving the star coupler to minimize cross talk as well as combining with superconducting nanowire single-photon detectors.

4. Conclusion
In conclusion, we proposed nanophotonic architecture for on-chip single-photon emission through spontaneous four-wave mixing process. The measured optical losses of the device (-6.8 dB) as well as a high Q-factor ($> 1.2 \times 10^5$) is demonstrated. Further work will be based on improving the optical losses of the circuit by optimize the reflow process and the etching process to achieve spontaneous four-wave mixing as well as integration with on-chip superconducting nanowire single-photon detectors (SNSPDs) [8].

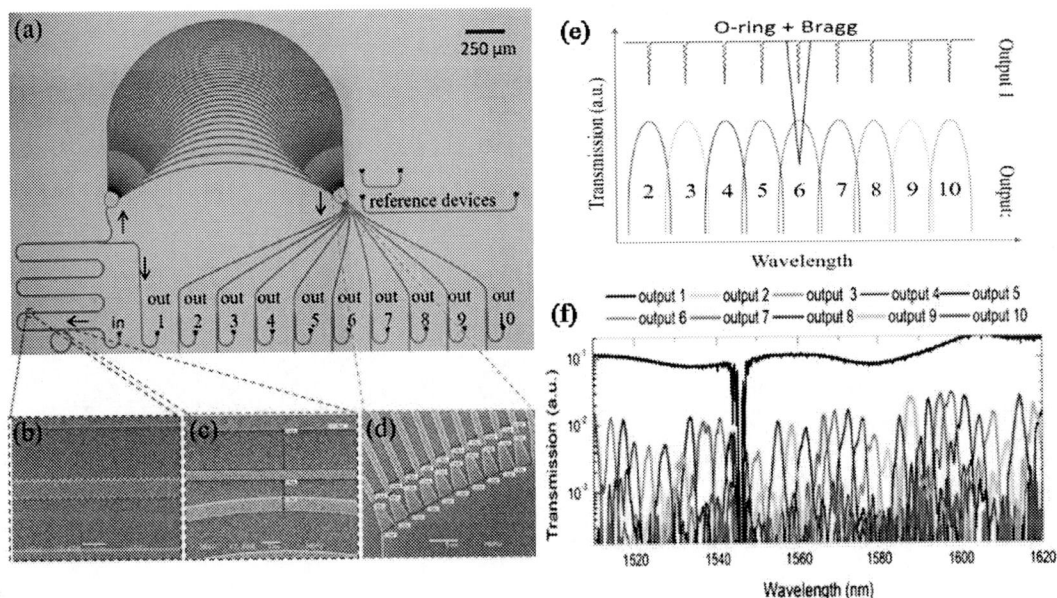

Figure 4 (a-f). (a) Optical image of the fabricated nanophotonic device; **(b)** A SEM image of the Bragg waveguide; **(c)** A SEM image of the gap between the waveguide and the O-ring; **(d)** A SEM image of the star coupler in AWG; **(e)** A schematic representation of the transmitted spectrum via device; **(f)** Measured transmitted spectrum via device.

Acknowledgments

A. Golikov, P. An and V. Kovalyuk acknowledge support by the Russian Science Foundation (project No. 16-19-10633; design, NbN film deposition and testing). A. Korneev acknowledges support of the Russian Science Foundation (project No. 17-72-30036; nanophotonic circuits modeling) and W. Pernice acknowledges support by the DFG grants PE 1832/1-1 & PE 1832/1-2 and the Helmholtz society through grant HIRG-0005 (fabrication of nanophotonic circuits and testing).

References

[1]   Aspuru-Guzik A and Walther P 2012 *Nat. Phys.* **8** 285–91
[2]   Ramelow S, Farsi A, Clemmen S, Orquiza D, Luke K, Lipson M, and Gaeta A L 2015 *ArXiv:1508.04358v1*
[3]   Silverstone J, Bonneau D, O'Brien J and Thompson M 2016 *IEEE J. Sel. Top. Quantum Electron.* **22** 6 390–402.
[4]   Spencer D , Davenport M, Srinivasan S, Khurgin J, Morton P and Bowers J 2015 *Opt. Express* **23** 30329
[5]   Zubkova E, An P, Kovalyuk V, Korneev A, Ferrari S, PerniceW, GoltsmanG, Integrated Bragg waveguides as an efficient optical notch filter on silicon nitride platform
[6]   Bogaerts W, de Heyn P, van Vaerenbergh T, de Vos K, Kumar Selvaraja S, Claes T, Dumon P, Bienstman P, van Thourhout D and Baets R 2012 *Laser Phot. Rev.* **6** 47–73
[7]   Smit M, Member A and Van Dam C 1996 *IEEE J. Sel. Top. Quantum Electron.* **2** 236–50
[8]   Kahl O, Ferrari S, Kovalyuk V, Vetter A, Lewes-Malandrakis G, Nebel C, Korneev A, Goltsman G, and Pernice W 2017 *Optica,* **4** 5 557–562

SPBOPEN 2018   IOP Publishing

Optimization of on-chip photonic delay lines for telecom wavelengths

A Prokhodtsov[1,2], P An[2,3], V Kovalyuk[2,3], E Zubkova[2,3], A Golikov[2,4], A Korneev[2,4], S Ferrari[5,6], W Pernice[5,6], G Goltsman[1,2,3]

[1]National Research University Higher School of Economics, Moscow 101000, Russia
[2]Department of Physics, Moscow State Pedagogical University, 119992, Russia
[3]Zavoisky Physical-Technical Institute of the Russian Academy of Sciences, 420029, Russia
[4]Moscow Institute of Physics and Technology (State University), 141700, Russia
[5]Institute of Physics, University of Münster, 48149, Germany
[6]CeNTech - Center for Nanotechnology, University of Münster, 48149, Germany

Abstract. In this work, we experimentally studied optical delay lines on silicon nitride platform for telecomm wavelength (1550 nm). We modeled the group delay time and fabricated spiral optical delay lines with different waveguide widths and radii as well as measured their transmission. For the half etched rib waveguides we achieved the losses in the range of 3 dB/cm

1. Introduction

Integrated optical delay lines (DL) can be successfully used for time delay in microwave photonics, highly stable microwave generators, in the processing of optical signals as well as in the construction of quantum-photonic integrated circuits (QPICs) [1, 2]. Optical true time delays (OTTD) can also be used for broadband phased array antenna applications to reduce beam distortion due to frequency variation and electromagnetic interference [3]. Different fully-integrated OTTD have been demonstrated [4-6], where the maximum time delay was limited to the picosecond-few nanoseconds range because of high propagation losses. To reduce the overall footprint, usually the delay lines fabricated in the form of a long waveguide are twisted into a spiral. Low-loss (0.08 dB/m) spiral DL have been already demonstrated on silicon chip, and the possibility of realizing 250m long delay line have been discussed [1]. Here we propose as waveguide material low pressure chemical vapor deposited (LPCVD) silicon nitride (Si_3N_4), which is a promising platform for nanophotonic circuits and combines good mechanical properties, low optical absorption in the infrared (IR) and visible wavelength ranges as well as possibility for creation of single-photon sources based on four-wave mixing [7].

2. Device design and modeling

In a planar technology we designed a DL as an Archimedean spiral. The radius of consecutive half-arcs is gradually increased from the center thus forming a spiral. In this realization, the delay line begins with two half-circles, routing in opposite directions with increasing radius, starting from a minimum value r. Figures 1(a-c) show the single delay line consisting of the spiral path a strait segment and two focusing grating couplers for input and output light. The schematic view of 2D array of delay lines with different waveguide widths $W = 0.9 - 1.3$ μm (X-axis) and full lengths $l = 1.5 - 40$ mm (Y-axis) is shown in Figure 1(d).

Content from this work may be used under the terms of the Creative Commons Attribution 3.0 licence. Any further distribution of this work must maintain attribution to the author(s) and the title of the work, journal citation and DOI.
Published under licence by IOP Publishing Ltd

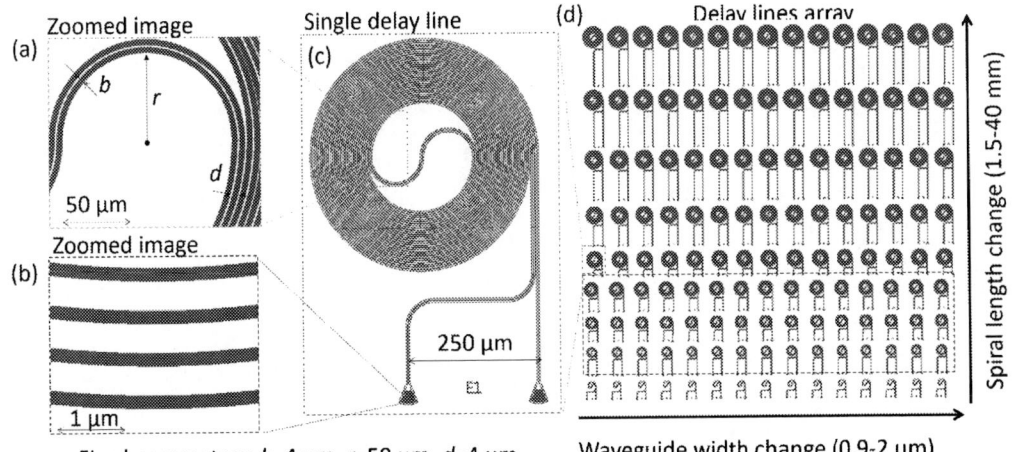

Fixed parameters: b=4 μm, r=50 μm, d=4 μm — Waveguide width change (0,9-2 μm)

Figure 1. (a-d). Schematic view of a mask layout for e-beam lithography; **(a)** Zoom-in of a central half-circle; **(b)** Zoom-in of a focusing grating coupler; **(c)** Mask layout of a single spiral delay line; **(d)** 2D array of delay lines with different parameters: waveguide width (*X*-axis) and length (*Y*-axis).

The total length of the spiral lines (L_s) was calculated by summing the lengths of the individual clockwise (l_{cw}) and anticlockwise waveguides (l_{aw}):

$$L_s = l_{cw} + l_{aw}. \tag{1}$$

Individual semicircles for each turn form complete circles, adding its length to the first ones placed in the center:

$$L_S = 2\pi r + 2\pi \sum_{k=1}^{n} R_k, \tag{2}$$

where r is the radius of the smallest arc, k is the coefficient of increase in radius ($k = 1, 2...n$), R_k are the radii of half-arcs. The increment of R_k depends on the separation between the waveguides (d) and a waveguide width (w):

$$R_{k+1} = R_k + \Delta R = R_k + d + w. \tag{3}$$

In most of the cases, there is a need for designing a delay line with a specified and accurate full length. Equation (2) shows that the L_s grows discretely with the number of arcs. In general, the full length can be controlled by the different ways, for example, by the waveguide output angle or by rotating the entire device [8]. In this work, we used the addition straight segments and arcs with a total length L_{SS} calculated as difference between full length L_f and length of the spiral part (L_S):

$$L_{ss} = L_f - L_s. \tag{4}$$

Since the full length increases with the number of half-arcs, we need stop the loop iteration when the length approaches the desired length, while residula length is added to straight segments.

In the first step, we use a numerical calculation of the effective mode index (n_{eff}) in COMSOL Multiphysics. In Figure 3(b) is shown the calculated fundamental quasi-TE mode for a silicon nitride waveguide with a width of $w = 1$ μm and a half-etched waveguide height of 450 nm plotted for 1550 nm wavelength. Using a dispersion formula for silicon nitride refractive index we calculated $n_{eff}(w)$ for slightly higher and slightly lower wavelength than 1550 nm ($d\lambda_0$) and extracted group refractive index (n_g) using the well-known formula

SPBOPEN 2018   IOP Publishing
IOP Conf. Series: Journal of Physics: Conf. Series **1124** (2018) 051052   doi:10.1088/1742-6596/1124/5/051052

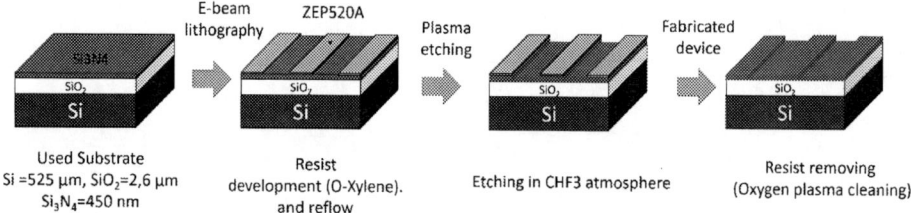

Figure 2. Schematic view of device fabrication rout.

$$n_g = n_{eff} - \lambda_0 \left(\frac{dn_{eff}}{d\lambda_0} \right), \tag{5}$$

After that, we extracted the group delay found the group delay

$$\tau_g = \frac{L_f}{(c/n_g)}, \tag{6}$$

where c is the speed of light.

3. Device fabrication

For the devices fabrication, a multi-layered substrate wase used. The thicknesses of silicon (Si), silicon oxide (SiO_2), and silicon nitride (Si_3N_4) were chosen as 450 µm, 2.6 µm, and 450 nm, respectively. The rib waveguides were formed from Si_3N_4 layer by means of one e-beam lithography step with a high-contrast positive resist ZEP 520A. The exposed structures have been developed in O-Xylene for 50 sec and isopropanol has been used as a stopper. Removal of the material below the exposed resist was done by reactive-ion-etching (RIE) in CHF_3 and Ar mixture. The residual resist was cleaned out with oxygen plasma, thus finalizing the fabrication process.

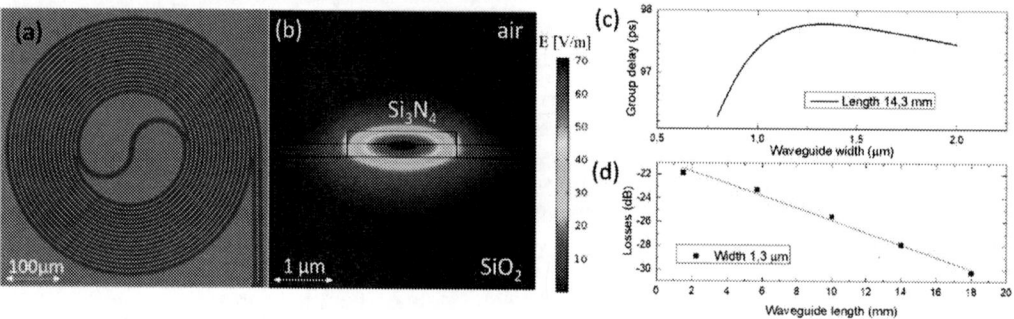

Figure 3. (a-d). (a) Micrograph of the spiral part of DL; **(b)** The simulation of the quasi TE optical mode, propagated inside the waveguide; **(c)** Calculated group delay on the waveguide width for DL with the length of 14.3 mm; **(d)** Losses in spiral waveguide vs waveguide length at fixed waveguide width of 1.3 mm and a minimum radius of $r = 50$ µm.

4. Experimental setup and results

In the second step, we characterized the optical transmission of integrated DLs in the spectral range of 1510 - 1620 nm. We use the experimental setup shown in Figure 4. Light from tunable laser source

790

(New Focus, TLB 6600) was coupled into the chip using an optical fiber array. Incoming light is adjusted with polarization controller and collected again at the output fiber by a low-noise photodetector and acquired by a PC.

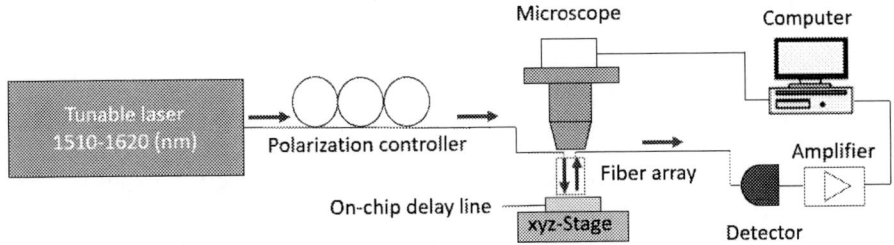

Figure 4. Schematic view of the experimental setup for DLs transmission measurements.

An optical microscope with a USB CMOS camera served to pre-align the device with an fiber array, while a more accurate alignment occurred by adjusting the stage position to maximize the transmitted signal. The dependence of optical losses on the DL length with a minimum radius $r = 50$ µm is shown in Figure 2(d). Although such a device has a small footprint, due to large losses ≈ 3dB/cm, the use in integrated optics is strongly limited and reducing of losses needed.

To reduce the optical losses, we systematically changed both the minimum spiral radius and the width of the waveguide. We found, when the width of the waveguide increases, increases the group delay (due to the increase in the effective refractive index) but the losses are reduced (by removing the maximum of the mode from the edges of the waveguide where the greatest scattering of light occurs).

In the third step, we measured the dependence of the optical losses on minimum spiral radius (Figure 5). Using a minimum radius below 40 µm leads to significant optical losses, whereas at large radii they become negligibly small and practically no difference with the losses onto a straight waveguides of the same length can be observed.

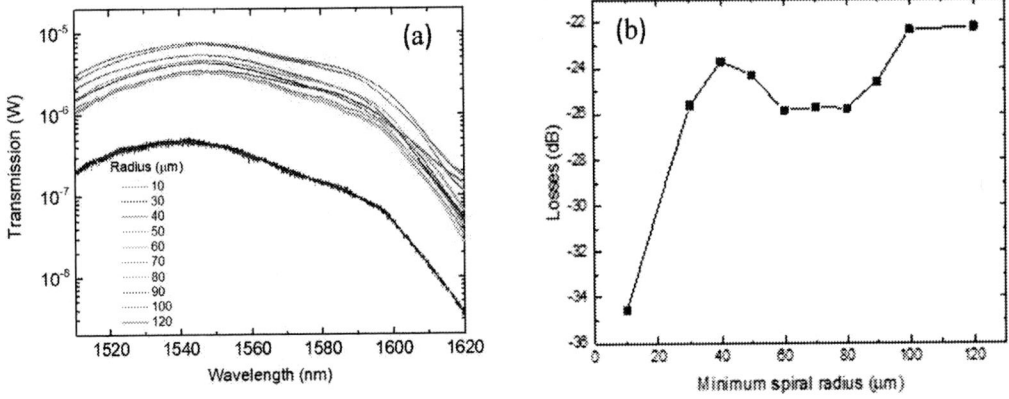

Figure 5. (a, b). (a) The measured transmission spectra of DLs together with focusing grating couplers; **(b)** Optical losses in spiral waveguide vs minimum spiral radius in a range of 20-120 µm at a fixed waveguide length of 10 mm and waveguide width of 1.4 µm.

5. Conclusion

We fabricated delay lines with different parameters of the length and width of the waveguide and studied their optical transmission and time delay performance. We found the spiral losses in a range 3 dB/cm as well as simulated the dependence of group delay vs waveguide width in order of 100 ps. Further work will be devoted to the fabrication of delay lines for integrated optical switchers using in spontaneous four-wave-mixing [9], as well as their integration with other integrated optical elements including sources, logic elements and single-photon detectors [10].

Acknowledgments

P. An, V. Kovalyuk, E. Zubkova and G. Goltsman acknowledge support by the Russian Science Foundation (project No. 16-12-00045; device design and testing). A. Prokhodtsov and A. Korneev acknowledge support of the Russian Science Foundation (project No. 17-72-30036; nanophotonic circuits modeling) and W. Pernice acknowledges support by the DFG grants PE 1832/1-1 & PE 1832/1-2 and the Helmholtz society through grant HIRG-0005 (fabrication of nanophotonic circuits).

References

[1]   Lee H, Chen. T, Li J, Painter O, and Vahala K J 2012 *Nat. Commun* **3** 867
[2]   Silveerstone J, Bonneau D, O'Brien J L, Thompson M G 2016 *IEEE J. Sel. Top. Quantum Electron.* **22** 390
[3]   Frigyes I, Seeds A J 1995 *IEEE Transactions on Microwave Theory and Techniques* **43** 2378
[4]   Wang X, Howley B, Chen M, Basile P, and Chen R 2006 *Proc, Integr, Photon. Res. Appl. Nanophoton.* 1
[5]   Sumida D S, Wang S, and Pepper D M 2005 *Proc. CLEO* 734
[6]   Zhuang L, Roeloffzen C G H, Hiedeman R G, Borreman A, Meijerink A, and Etten van W 2007 *IEEE Photonics Technology Let* **19** 1130
[7]   Bogdanov S, Shalaginov M Y, Boltasseva A, and Shalaev V M 2016 *Opt. Mater. Express* **7** 111
[8]   Stopinski S, Malinowski M, Piramidowicz R, Kleijn E, Smit M K, Leijtens X J M 2013 *IEEE Photonics J.* **5** 5
[9]   Ramelow S, Farsi A, Clemmen S, Orquiza D, Luke K, Lipson M, and Gaeta A L 2015 *ArXiv:1508.04358v1*
[10] Pernice W H P, Schuck C, Minaeva O, Li M, Goltsman G N, Sergienko A V, Tang H X 2012 *Nat. Commun* **3** 1325

Graphene-based tunability of chiral metasurface in terahertz frequency range

M S Masyukov[1], A V Vozianova[1], A N Grebenchukov[1], M K Khodzitsky[1]

[1]Terahertz Biomedicine Laboratory, ITMO University, St. Petersburg 197101, Russia

Abstract. This paper is devoted to bi-layer chiral metasurface with graphene half-petals. Numerical simulation in CST Microwave Studio confirms that changing of chemical potential of graphene leads to changing in transmission coefficients for this metasurface. Due to chirality, these coefficients are different for right-handed and left-handed circularly polarized wave. Thus, such metasurface can be used as tunable polarization-converter.

Introduction

Recent decades THz frequency range has become very popular in scientific society due to its unique properties and applications in space exploration, tomography [1] and biomedicine [2-4], etc. Despite to this fact, a deficit of passive components still exists for terahertz frequency range. The solution of the problem could be found in development of metamaterials with different functionalities. Metamaterials or artificial effective media are consisted of an array of unit cells and show different effects that cannot be found in nature, for example, negative refraction index. Such phenomenon had been predicted theoretically by V. Veselago in 1967 [5], although it was confirmed experimentally many years later by D. Smith and others in 1999 [6]. This pioneer work initiated a big increase in research on metamaterials. Metamaterials have different applications, such as tunable reflectors, switchers, filters, and perfect absorbers [7-9] and especially they are used in polarization components. Chiral planar metamaterials, or chiral metasurfaces, show negative refractive index, circular dichroism, etc. These unique properties allow applying chiral metamaterials in polarization optics. In this work we propose a tunable polarization converter – a bi-layer chiral metasurface that is composed of conjugated gammadion resonators with graphene inclusions.

The metasurface under the study

The unit cell of the investigated metasurface [10-11] is shown in Fig. 1. The geometrical parameters are following: the side of the unit cell a=600 μm, a width of the planar gammadion petal w=10 μm (its inner radius R_{min}=140 μm), the silicone substrate thickness is 45 μm and permittivity ε=11.34. The back side resonator is made entirely of perfect electric conductor (PEC), while the front resonator half-part has been replaced by graphene.

Content from this work may be used under the terms of the Creative Commons Attribution 3.0 licence. Any further distribution of this work must maintain attribution to the author(s) and the title of the work, journal citation and DOI.

Published under licence by IOP Publishing Ltd

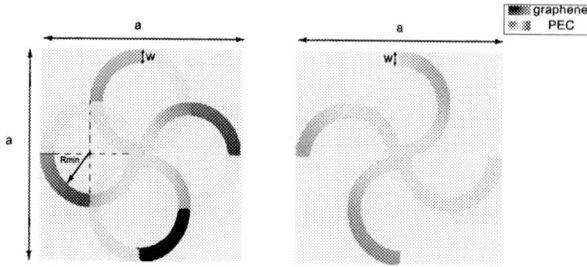

Figure 1. The schematic representation of the original metasurface unit cell, the views of the front (a) and back (b) layers: yellow color corresponds to PEC, grey one corresponds to graphene and blue one is silicone.

Two variations of the metasurface were studied. In the first case the center of the top resonator was made of PEC and the edges of the gammadion petals were made of graphene (as shown in Fig. 1-a). In other case the metallic and graphene parts were reversed.

The numerical simulation and polarizing properties calculation approach

The transmission simulations for linearly polarized waves were performed in frequency domain using CST Microwave Studio based on Finite Elements Method. For each design the unit cell was translated along x and y axes directions. A scheme of the numerical simulation can be found in Fig. 3.

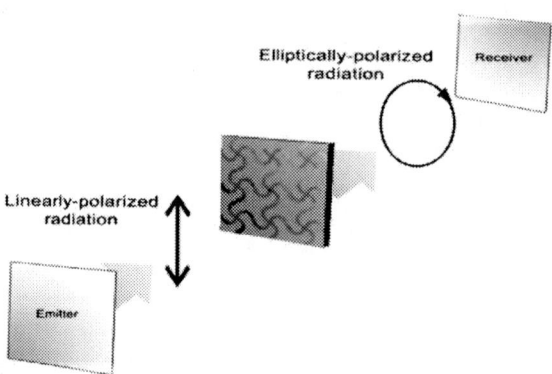

Figure 2. The scheme of the virtual experiment. After the passing through the metasurface, the polarization state of the terahertz waves was changed from the linear to the elliptical one.

To calculate the polarization properties of the metasurface one needs to evaluate the transmission spectra for circularly polarized waves. These spectra can be simply calculated from the simulated co- and cross-polarization transmission coefficients T_{xx} and T_{xy} respectively by using Jones calculus approach [12] for the structures with fourfold (C4) rotational symmetry:

$$\begin{pmatrix} T_{++} & T_{+-} \\ T_{-+} & T_{--} \end{pmatrix} = \begin{pmatrix} T_{xx} + iT_{xy} & 0 \\ 0 & T_{xx} - iT_{xy} \end{pmatrix}, \tag{1}$$

where T_{++} refers to the transmission amplitudes for right-handed circularly polarized waves, T_{--} to the left-handed ones respectively. Having the transmission amplitudes for circularly polarized waves calculated, the ellipticity angle η was found using the next formula:

$$\eta = \frac{1}{2}\arcsin\left(\frac{|T_{++}|^2 - |T_{--}|^2}{|T_{++}|^2 + |T_{--}|^2}\right). \tag{2}$$

When the ellipticity angle reaches $\eta = +45$ degrees, it corresponds to the right-handed circular polarization of the transmitted wave, while $\eta = -45$ degrees corresponds to the left-handed circular polarization of the transmitted wave. The intermediate values of the ellipticity refers to the elliptically polarized waves, and when $\eta = 0$ the polarization is linear.

Results

The ellipticity angle spectra were calculated for six values of graphene chemical potential for both metasurface types in the frequency range of $0.1 - 0.2$ THz.

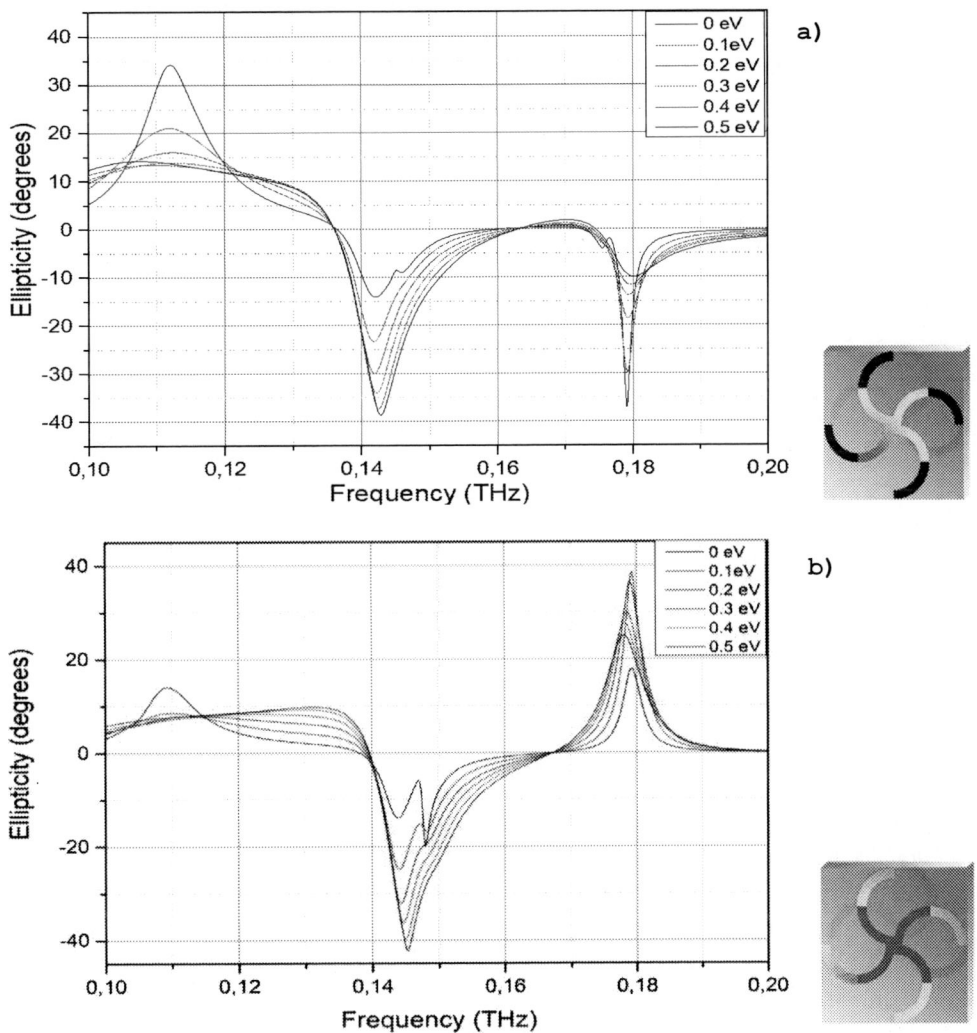

Figure 3. The ellipticity angle spectra for two types of the metasurface: a) original metasurface; b) metasurface with inverted materials of the top side resonator

The results shown in Fig. 2 represent the strong dependence of polarizing properties on chemical potential of graphene. The polarization state of the transmitted wave can be changed by varying of the chemical potential of graphene from 0 to 0.5 eV. As we can see, this dependence is almost the same for both structures around the frequency of 0.145 THz, but quite different for the resonance at 0.178 THz: for the original metasurface design maximal value of the ellipticity angle is $|\eta_{max}|$=38 degrees, for the second design $|\eta_{max}|$=43 degrees. This effect is caused by replacing parts of graphene by PEC on the front side resonator. Due to the fact that the material of the back side resonator was not changed, we can suppose that the ellipticity extreme at 0.145 was caused by the resonances from this resonator. The second extreme in this frequency range may depend on the resonance from the hybrid front resonator.

Conclusions

In summary, the influence of graphene includings on chiral metasurface polarizing properties has been studied. It was found that the ellipticity angle depends on the Fermi level of the graphene includings. The noticeable changes of ellipticity provide a possible wide usage of the metasurface as a tunable polarization converter in many applications, for example, terahertz polarimetry, terahertz time-domain spectroscopy, etc.

Acknowledgments

This work was supported by Government of Russian Federation (Grant 08–08) and MTT-s Undergraduate Scholarship.

References

[1] Woolard D L, Loerop W R and Shur M 2003 *Terahertz sensing technology: emerging scientific applications & novel device concepts* vol. 2. World scientific
[2] Borovkova M, Serebriakova M, Fedorov V, Sedykh E, Vaks V, Lichutin A, Salnikova A, Khodzitsky M 2017 *Biomedical Optics Express* **8**(1) 273
[3] Borovkova M, Khodzitsky M, Demchenko P, Cherkasova O, Popov A, Meglinski I 2018 *Biomedical Optics Express* **9**(5) 2266
[4] Gusev S I, Demchenko P S, Cherkasova O P, Fedorov V I, Khodzitsky M K 2018 *Chinese Optics* **11**(2) 182
[5] Veselago V G 1968 *Soviet physics uspekhi* **10**(4) 509
[6] Smith D R et.al. 2000 *Physical review letters* **84**(18) 4184
[7] Grebenchukov A, Zaitsev A, Khodzitsky M 2018 Chinese Optics **11**(2) 166
[8] Soboleva V Y, Gomon D A, Sedykh E A, Balya V K, Khodzitskii M K 2017 *Journal of Optical Technology* **84**(8) 521
[9] Gomon D A, Sedykh E A, Rodriguez S, Monroy I T, Zaitsev K I, Vozianova A V, Khodzitsky M K 2018 *Chinese Optics* **11**(1) 47
[10] Korolenko S Y, Grebenchukov A N, Masyukov M S, Vozianova A V, Khodzitsky M K 2016 *Journal of Physics: Conference Series* **735**(1) 012089
[11] Masyukov M S, Vozianova A, Grebenchukov A, Khodzitsky M K 2017 *Proceedings of SPIE – The International Society for Optical Engineering* **10343** 1034338
[12] Menzel C, Rockstuhl C, Lederer F 2010 *Physical Review A* **82**(5) 053811

SPBOPEN 2018

IOP Conf. Series: Journal of Physics: Conf. Series **1124** (2018) 051054

IOP Publishing

doi:10.1088/1742-6596/1124/5/051054

Graphene-layer and graphene-nanoribbon FETs as THz detectors

Y E Matyushkin[1][*], I A Gayduchenko[1], M V Moskotin[1], G N Goltsman[1], G E Fedorov[1], M G Rybin[2], E D Obraztsova[2]

[1]Moscow State University of Education, Moscow, 119991, Russia

[2]Prokhorov General Physics Institute, Moscow, 119991, Russia

Abstract. We report on detection of sub-THz radiation (129-430 GHz) using graphene based asymmetric field-effect transistor (FET) structures with different channel geometry: monolayer graphene, graphene nanoribbons. In all devices types we observed the similar trends of response on sub-THz radiation. The response fell with increasing frequency at room temperature, but increased with increasing frequency at 77 K. Our calculations show that the change in the trend of the frequency dependence at 77 K is associated with the appearance of plasma waves in the graphene channel. Unusual properties of p-n junctions in graphene are highlighted using devices of special geometry.

Introduction

In recent decades there has been an active growth of interest in devices capable of working in the terahertz region of the electromagnetic wave spectrum. This is due, on the one hand, to the development of experimental equipment, and, on the other hand, to a wide variety of possible applications for terahertz devices (medicine, security and communication systems, space exploration, etc.) [1]. A separate important area is the development of high-efficiency terahertz detectors. Carbon nanomaterials like graphene and carbon nanotubes, thanks to its unique optoelectronic properties [2,3], are excellent candidates for the role of a sensitive part of such detectors [4-10]. In this paper, we experimentally compare the response of the detectors with asymmetric contacts (one is the Schottky contact and one – the Ohmic contact) based on monolayer graphene (MLG) and graphene nanoribbons (GNR) to sub-THz radiation. Detection of the radiation in such devices takes place through rectification: due to the asymmetry of the current-voltage (I-V) curve mean current under a harmonically changing non-zero voltage. The latter can be attributed to the unusual properties of the Schottky barrier on graphene-vanadium interface. To study these unusual properties of the barrier, we've made devices with asymmetric contacts based on graphene angled-nanoribbons (GANR: graphene nanoribbons which are at an angle to the metal contacts).

Device fabrication and characterization

The graphene, acting as the conducting channel of a field-effect transistor (FET), was put on top of an oxidized silicon wafer. This silicon substrate was a 480 μm thick silicon wafer covered with a 500 nm thick thermally grown SiO_2 layer. The doped silicon (with the room temperature resistivity of 10 Ω·cm) formes a gate electrode transparent for the sub-THz and THz radiation. Graphene was synthesized using home-made cold-wall CVD-reactor [11]. The growth procedure was adjusted to

---

Content from this work may be used under the terms of the Creative Commons Attribution 3.0 licence. Any further distribution of this work must maintain attribution to the author(s) and the title of the work, journal citation and DOI.

Published under licence by IOP Publishing Ltd

always give single layer graphene. We made two types of detectors with different channel geometry were further defined using e-beam lithography and etching in oxygen plasma (Figure 1(a,b)) [6].

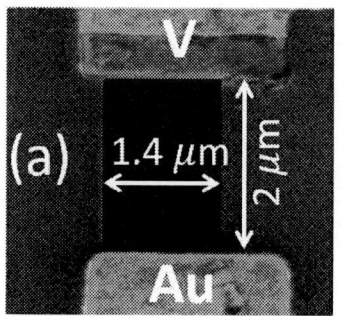 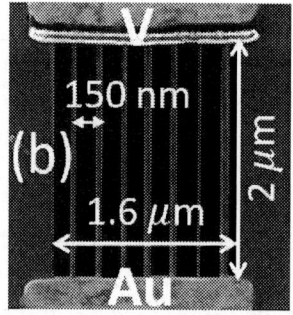 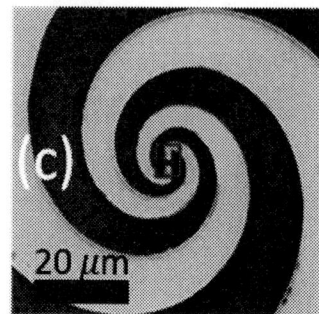

Figure 1(a,b,c). SEM-images of the devices channels made of: **(a)** Graphene monolayer; **(b)** Graphene nanoribbons; **(c)** Logarithmic spiral antenna SEM-image.

The structures of the first type use graphene channel with a width of 1.4 mcm. In the GNR-FETs, graphene was partitioned to form an array of the parallel ribbons connecting the source and drain perpendicular to the electrodes. Width of the ribbons is about 100 nm. The devices (Figure 1(a,b)) are coupled to the radiation with a logarithmic spiral antenna which also serves for DC contacts (Figure 2). The channel of GANR devices is formed by array with the parallel GNRs connecting the source and drain at 45^0 to the electrodes. The source and drain electrodes are made of metals with different work function: vanadium and gold.

The source and drain contacts are connected to a spiral antenna (see Figure 1(a,b)) to ensure a sufficiently broad band device response. We have chosen the log-spiral antenna defined in polar coordinates as $R = R_0 \cdot e^{\varphi/\beta}$ with following parameters: the inner radius of the spiral R_0 equals 5.5 µm, outer radius $R_{max} = 68$ µm the parameter determining the rate of spiral $\beta = 3.2$. Further details of the antenna parameters and its characteristics are presented in Ref. [12]. Importantly we have extensive experience in the manufacture and operation of such antennas (Figure 1(c)).

Experimental setup and results

The fabricated device chip was fixed on flat surface of silicon lens for better radiation coupling (Figure 2). Then lens with chip was put inside optical cryostat. The THz radiation was generated by two backward wave oscillators (BWO) in the frequency range 129-450 GHz. The total losses in the silicon lens and the cryostat optical window do not exceed 5–6 dB.

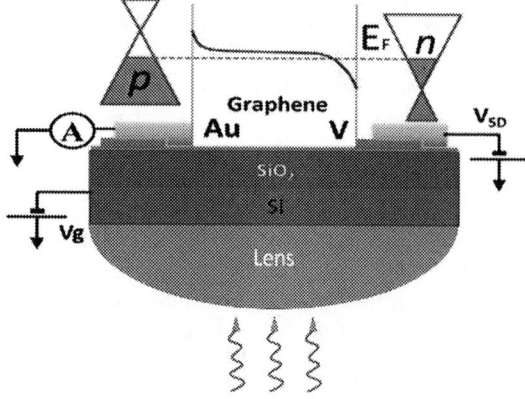

Figure 2. Experimental setup.

As the device is exposed to the radiation the IV curves shifts so that non-zero voltage corresponds to zero current. Figure 3 illustrates the response of the devices to sub-THz radiation at frequency 129 GHz. Comparing the absolute values of the non-zero voltage at zero current for MLG and GNR, it is clearly seen that for GNR this effect is much weaker.

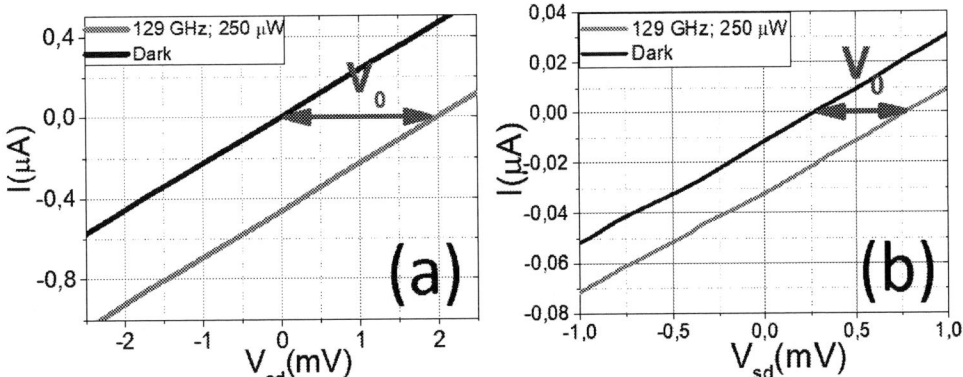

Figure 3(a,b). IV curves of the devices measured with and without sub-THz radiation. **(a)** – monolayer graphene: $V_0 = 2\ mV$; **(b)** – graphene nanoribbon: $V_0 = 0.75\ mV$.

The MLG-FET detector exhibit the room-temperature responsivity from R = 14 V/W at f = 129 GHz to R = 3 V/W at f = 450 GHz (Figure 4). The GNR-FET detector exhibits markedly lower responsivity at this frequency range at 300 K.

The most important feature observed in our experiments is the qualitative change of the frequency dependence of the device responsivity upon a decrease of the temperature from 300 K to 77 K (Figure 4(a)). As the temperature is lowered, the character of the dependence of the responsivity on frequency changes from a decreasing to an increasing one.

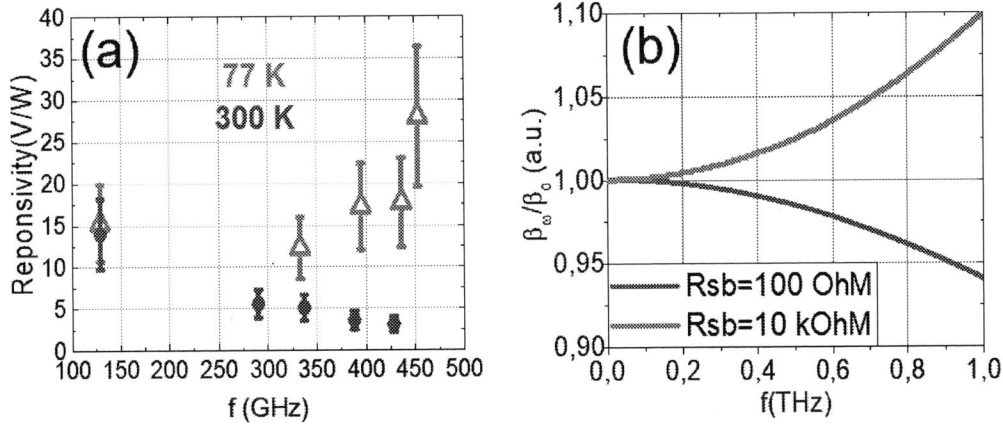

Figure 4(a,b). Responsivity of the graphene monolayer devices as a function of the radiation frequency. **(a)** – experimental results; **(b)** - theoretical calculation with equation (1).

Discussion

In our previous work [6] we have shown that the elastic scattering rate in the CVD-grown graphene used in our experiments does not change in the temperature range between 77K and 300K. We thus can rule out the simple plasmon resonance model of detector response discussed in [13]. Further development of this model accounting for the Schottky barriers capacitance brings to the following formula for the frequency dependence of detector response

$$\frac{\beta_0}{\beta_\omega} = \left| cos\left(\frac{\pi\sqrt{(\omega+i\nu)\omega}}{2\Omega}\right) - ib\frac{(\omega+i\nu)}{\nu}(1-i\omega\tau_s)\right|^{-2} \tag{1}$$

In Eq. (1):

$$\Omega = \frac{\pi^{5/4}v_w}{L}\left[\frac{e^2}{k_g v_w \hbar}W_g\sqrt{n}\right]^{\frac{1}{2}} \tag{2}$$

Ω is the plasma frequency characteristic for the gated channel [14], n is the electron density in the quasi-neutral section of the channel with the length close to the net channel length L, W_g is the gate layer thickness, $v_w \cong 10^8\ cm/s$ is the characteristic electron velocity in graphene, k is the effective dielectric constant (depending on the dielectric constants above and beneath the graphene (or for the graphene nanoribbon), $\gamma = \tau^{-1} + \xi(2\pi/\lambda)^2$ [15] is the plasma oscillations decay rate, τ^{-1} is the frequency of electron collisions with impurities, phonons, and edges (in graphene), ξ is the electron velocity, $\lambda = 4L$ is the plasma wavelength, β_0 low frequency responsivity, $\tau_s = r_s \cdot C_s$ and $\cdot C_s$ are the Schottky junction recharging time (through the Schottky junction) and capacitance, respectively, and $b = L/\sigma_0 r_s = r_{2DES}/r_s$, $r_s = (dJ_s/dV)^{-1}$ is the Schottky junction differential resistance, where $J_s(V)$ is the junction current-voltage characteristics, and r_{2DES} is the dc resistance of the quasi-neutral channel section. The AC conductivity of the 2DES channel is equal to $\sigma_\omega = \sigma_0\tau^{-1}/(\tau^{-1} - i\omega)$, where σ_0 is the DC conductivity.

Based on the temperature evolution of the DC transport, we argue that the only temperature-dependent parameter in the above equations is the Schottky junction recharging time, which is proportional to the barrier resistance. Figure 4(b), show the frequency evolution of the parameter β_0/β_ω for two values of τ_s. We see that the responsivity is a rising function of frequency for large enough values of τ_s and decreasing otherwise. Other parameters input into the calculations are provided in the figure caption.
In order to explain why a Schottky barrier is formed at the graphene/metal interface we note that the transport of carriers through the p-n junction in graphene depends on the angle between the normal to the junction and the carrier momentum. For normally incident electrons, there is no energy barrier and they shunt the nonlinear transport. At the same time, the current of non-normally incident electrons is a non-linear function of the applied voltage [16,17].
This scenario is confirmed in our case with much smaller signal obtained in the case of the GNR devices. In the case of the nanoribbons, a smaller channel width should lead to a more collimated motion of the charge carriers, so that the fraction of normally incident carriers is larger, causing smaller response value.

To verify these conclusions, we fabricated GANR devices (Figure 5(a)) In order to fabricate these devices we used the same fabrication route as in the case of GNR FETs. The only difference is that the metal edge is not normal to the ribbons. The designed angle is 45 degrees. Such a design ensures that the overall current in the ribbons is directed at an angle of 45 degrees to the p-n junction at the vanadium interface. Following the previous discussion, we argue that in this case larger fraction of carriers encounter an energy barrier at the GANR/V interface and their impact on the device conductance is larger than that in the case of plane graphene. In such case stronger temperature dependence is expected. This prediction is confirmed by the transport measurements reported in the Figure 5(b,c), where conductance of GNR and GANR as a function of gate voltage is shown. We see a

30% drop of conductance in case of GANR as temperature is decreased from 300 to 77 K compared to ~ 15% drop in case of GNR.

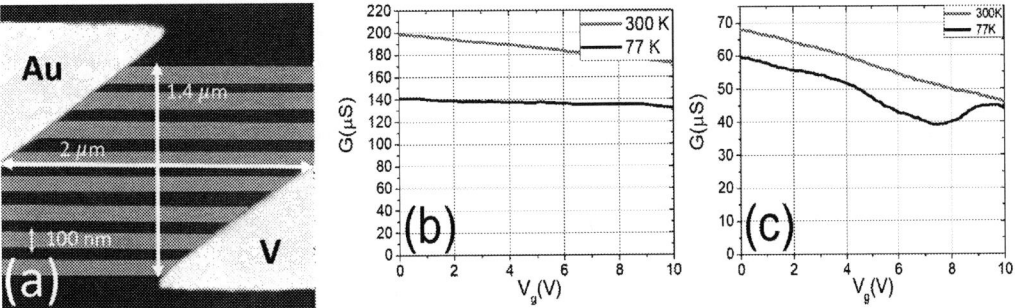

Figure 5(a,b,c). (a) SEM-image of graphene-angled nanoribbons; **(b)** Transfer characteristic of the GANR at 300 K and 77 K; **(c)** Transfer characteristic of the GNR at 300 K and 77 K.

A stronger drop in conductivity at 77K in GANR-devices (30%) than in GNR devices (15%) indirectly suggests the hypothesis that the main detection mechanism is rectification at the Schottky barrier. Investigation of THz radiation detection in GANR asymmetric structures will be subject of our further studies.

Conclusion

Difference between the transport characteristics of the GNR and GANR devices confirms that the properties of the Schottky barrier for the charge carriers in graphene at a p-n junction depends on the angle between the carrier drift velocity and the junction. We have shown that plasma waves can affect the frequency dependence of the response of a graphene lateral Schottky diode even in the graphene devices with rather modest electron mobilities far from the first plasmon resonance, which should be observed at 4.9 THz for our channel geometry and carrier concentration. A strong enhancement of the response is observed at moderately low temperatures when the frequency is increased towards the plasma resonance frequency.

Acknowledgments
This work was supported by the Ministry of Education and Science of Russian Federation (contract №14.583.21.0069, RFMEFI58317X0069). SEM imaging was performed using equipment of MIPT Shared Facilities Center.

References

[1] Masayoshi Tonouchi, 2007 *Nat. Photonics* **1** 97–105
[2] F. Bonaccorso, Z. Sun, T. Hasan, A. C. Ferrari 2010 *Nat. Photonics* **4** 611
[3] A.N. Grigorenko, M. Polini, K.S. Novoselov 2012 *Nat. Photonics* **6** 749
[4] Vicarelli L, Vitiello M S, Coquillat D, Lombardo A, Ferrari A C, Knap W, Polini M, Pellegrini V and Tredicucci A 2012 *Nat. Mat.* **11** 865
[5] Xinghan Cai, Andrei B. Sushkov, Ryan J. Suess, Mohammad M. Jadidi, Gregory S. Jenkins, Luke O. Nyakiti, Rachael L. Myers-Ward, Shanshan Li, Jun Yan, D. Kurt Gaskill, Thomas E. Murphy, H. Dennis Drew & Michael S. Fuhrer 2014 *Nature Nanotechnology* **9** 814
[6] I A Gayduchenko *et al* 2018 *Nanotechnology* **29** 245204
[7] D. A. Bandurin, I. Gayduchenko, Y. Cao, M. Moskotin, A. Principi, I. V. Grigorieva, G. Goltsman, G. Fedorov, and D. Svintsov 2018 *Appl. Phys. Lett.* **112** 141101

[8] Gayduchenko I, Kardakova A, Fedorov G, Voronov B, Finkel M, Jimenez D, Morozov S, Presniakov M, Goltsman G 2015 *J. Appl. Phys.* **118** 194303

[9] Fedorov G, Gayduchenko I, Titova N, Gazaliev A, Moskotin M, Kaurova N, Voronov B and Goltsman G 2018 *Phys. Status Solidi B* **255** 1700227

[10] Xiaowei He *et all* 2014 *Nano Lett.* **14** (7) 3953–3958

[11] M. Rybin, A. Pereyaslavtsev, T. Vasilieva, V. Myasnikov, I. Sokolov, A. Pavlova, E.A. Obraztsova, A. Khomich, V. Ralchenko, E.D. Obraztsova 2016 *Carbon* **96** 196

[12] Alexander Shurakov, Sergey Seliverstov, Natalia Kaurova, Matvey Finkel, Boris Voronov, and Gregory Goltsman 2012 *Ieee transactions on terahertz science and technology* **2**(4) 400-405

[13] Ryzhii V and Shur M S 2006 *Jpn. J. Appl. Phys.* **45** 42-45

[14] Tomadin A Polini M 2013 *Phys. Rev.* B **88** 205426

[15] Dyakonov M and Shur M 1993 *Phys. Rev. Lett.* **71** 2465

[16] M.I. Katsnelson, K.S. Novoselov, A.K. Geim 2006 *Nature Physics* **2** 620

[17] Sutar S, Comfort E S, Liu J, Taniguchi T, Watanabe K, and Lee J U 2012 *Nano Lett.* **12** 4460–4464

Purcell effect in GaN-based waveguiding structures

K M Morozov[1,2]*, K A Ivanov[2], A R Gubaydullin[1,2], M A Kaliteevski[1,2,3]

[1]St. Petersburg Academic University, St. Petersburg, 194021 Russia
[2]ITMO University, St. Petersburg, 197101 Russia
[3]Ioffe Institute, St. Petersburg, 194021 Russia

Abstract. We provide an analysis of Purcell coefficient dependence on frequency, wavevector and emitter position in nitride-based waveguide structures. Was shown that spontaneous emission of emitter placed in one-dimensional slab waveguide structure could lead to a modification of spontaneous emission rate in the both considered structures. Results for symmetric and asymmetric nitride-based waveguides in TE polarization case was demonstrated.

1. Introduction

Using nitride material system allows to utilize some important properties, like a wide transparency range [1], and suitability for harsh environments in optoelectronic devices. Various optical components have already been demonstrated based on the nitride material system like directional couplers [2], all-optical modulators, waveguide gratings [3,4]. In the nitride materials system highlights Gallium Nitride (GaN). GaN is perspective, high refractive index material to using in integrated optics (fabricate low loss waveguides in ultraviolet region and other components to optical circuits systems). In the last decade improvement in technological methods allow to fabricate nanoscale waveguide structures with high quality of interfaces with different geometries, like slab, ridge and rib on different kind of substrates [5,6].

Modification of spontaneous emission rate of emitters, coupled with different types of resonant modes has attracted attention of researchers due to perspectives of fabrication of high efficiency compact light sources, single-photon generators etc. The rate of spontaneous emission can be estimated by a number of various methods. Here, for the analysis of Purcell effect in layered waveguide structures, was used approach, called S-quantization method. S-quantization formalism allows to calculate the spontaneous emission rate in arbitrary layered dielectric structure for states inside and outside the light cone [7]. These rigorous and self-consistent method is based on analysis of scattering matrix eigenvalues and don't require to solve integral-differential equations and apply perturbation theory methods.

The aim of that paper is to calculate the modification of the spontaneous emission rate of known GaN core-based waveguide structures by S-quantization method and to get analysis of the results with compare to analytics.

Content from this work may be used under the terms of the Creative Commons Attribution 3.0 licence. Any further distribution of this work must maintain attribution to the author(s) and the title of the work, journal citation and DOI.
Published under licence by IOP Publishing Ltd

2. Results and discussion

Was considered the well-known slab waveguide model. As model structures was chosen slab GaN waveguide layer (n_{GaN}=2.33) of thickness 400 nm in two different cladding semi-infinite materials: in two semi-infinite AlN claddings(n_{AlN}=2.075) and semi-infinite sapphire substrate layer ($n_{Al_2O_3}$=1.76) with air on the top. Schemes of two waveguide structures are pictured on figure 1.

As known, both cladding and core materials of first structure (GaN and AlN) has birefringence phenomenon and describes by ordinary and extraordinary refractive indexes. In our paper we limit consideration only on transfer electric (TE) waveguide modes that describes only with ordinary refractive index for simplicity.

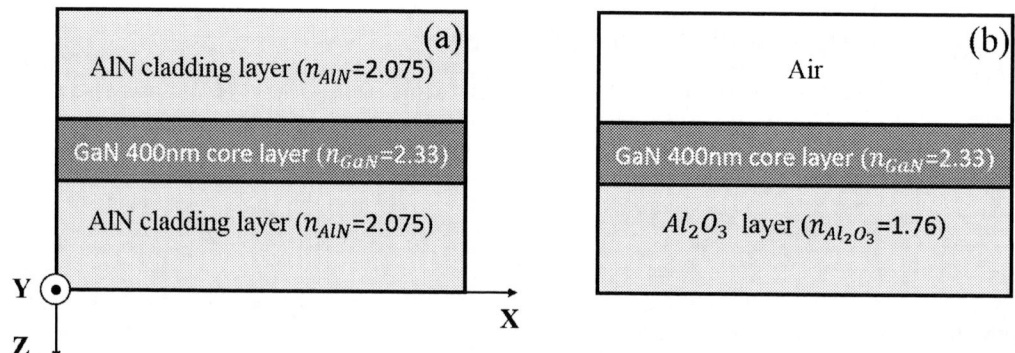

Figure 1. (color online) Schemes of considered waveguide structure. (a) Slab GaN waveguide structure in AlN claddings. (b) Asymmetric GaN waveguide layer in air and sapphire claddings.

S-quantization method uses transfer and scattering matrix formalism in the traveling-wave basis and in case of TE-polarized field, E_y component can be written as:

$$E_y(x,z) = E(z)exp(iK_x x). \tag{1}$$

Scattering matrix $\hat{S}$ interrelate incident and runaway waves at boundary of quantization box as:

$$\begin{pmatrix} E_{K_x}^+(L) \\ E_{K_x}^-(0) \end{pmatrix} = \hat{S} \begin{pmatrix} E_{K_x}^+(0) \\ E_{K_x}^-(L) \end{pmatrix}. \tag{2}$$

Field in S-quantization formalism quantizes by equating $\hat{S}$ matrix eigenvalues to unity:

$$\beta^{(1,2)} = 1. \tag{3}$$

In waveguide regime, when the wavevector component K_x is outside the light cone, equation (3) becomes:

$$\left(1 \pm i\sqrt{M_{12}M_{21}}\right) = 0, \tag{4}$$

where M_{12}, M_{21} - the components of transfer matrix through the quantization box and "+" sign corresponds to symmetric eigenvector, and sign "-" corresponds to antisymmetric eigenvector. Equations (4) are easily covered to known form of dispersion equations for even and odd waveguide modes. Mode Purcell factor could get from Fermi Golden rule and reduces to relation between squared dipole matrix elements, in waveguide case to homogeneous media case respectively.

SPBOPEN 2018   IOP Publishing
IOP Conf. Series: Journal of Physics: Conf. Series **1124** (2018) 051055   doi:10.1088/1742-6596/1124/5/051055

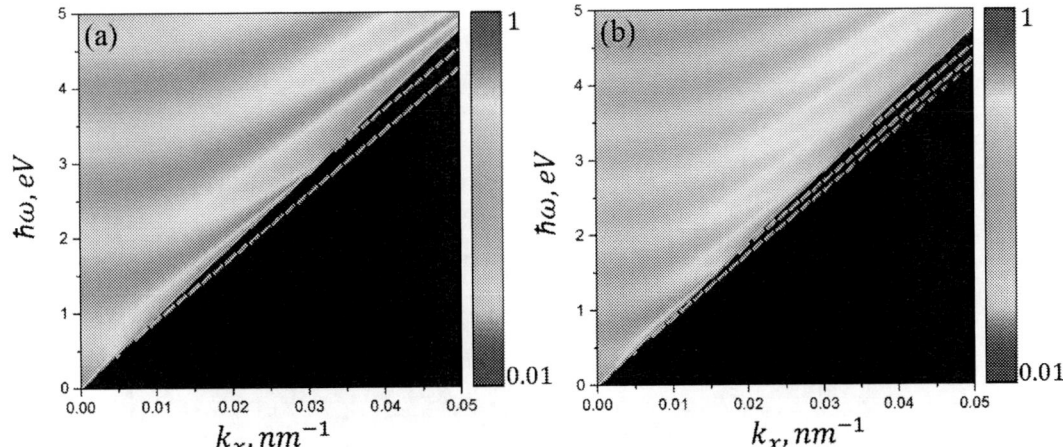

Figure 2. (color online) Dependence of the mode Purcell coefficient for the dipole oriented along the y axis and (a) placed at the centre of a 400-nm-thick GaN slab waveguide, limited by semi-infinite AlN layers, and (b) moved by 150 nm with respect to the centre, on the emission frequency and wave-vector component kx parallel to the layer interfaces (case of TE polarization).

The patterns of the modal Purcell factor in case of TE polarization are shown on figure 2. Figure 2 (a) shows distribution of mode Purcell coefficient on energy and wave vector component kx when the dipole is placed at the centre of GaN layer and oriented along Oy axis. It can be seen, that in area with Purcell factor much more than zero (inside the light cone) are interleaving of local maximums and minimums of Purcell factor. The maximums are corresponded to Fabry-Perot modes of the slab. On the edge of light cone (between light cones of core and cladding materials), shows two branches of considered even waveguide TE modes. Figure 2 (b) shows the case of dipole emitter placed at 150 nm from the structure centre. In comparison with (a) there are more waveguide modes: both even and odd. Because in this case, emitter is placed in area where odd modes electric field nonzero. The maximum value of mode Purcell factor in current structure is close to 1 and corresponded to fundamental mode (TE$_0$) and dipole position in the centre of waveguide (where placed maximum magnitude of electric field too).

It is known, that difference between core and cladding refractive indexes defines the confinement of the field in waveguide. To compare values of mode Purcell factors was considered the situation of known waveguide system GaN core layer in sapphire and air claddings. Figure 3 shown the pattern of modal Purcell factor dependence on wave vector component kx and energy, when dipole is placed at centre of waveguide layer. So as the second structure is the asymmetric waveguide, there are three light cone curves, that corresponding to each layer. Three even waveguide modes are shown in area 3. Areas 1 and 2 are areas within light cone for air and Al_2O_3 layers, where as in previous case, figured interleaving of Fabry-Perot modes. In both considered structures, maximum value of mode Purcell factor, that corresponds to waveguide modes, does not exceed to value of 2. It can be explained by parameters of structure – there are not huge difference between core and cladding refractive indexes, as analytical equation for maximum mode Purcell factor from [7] says: $F_p \approx 1.3 n_{core}^2 (n_{core}^2 - n_{cladding}^2)$.

SPBOPEN 2018   IOP Publishing

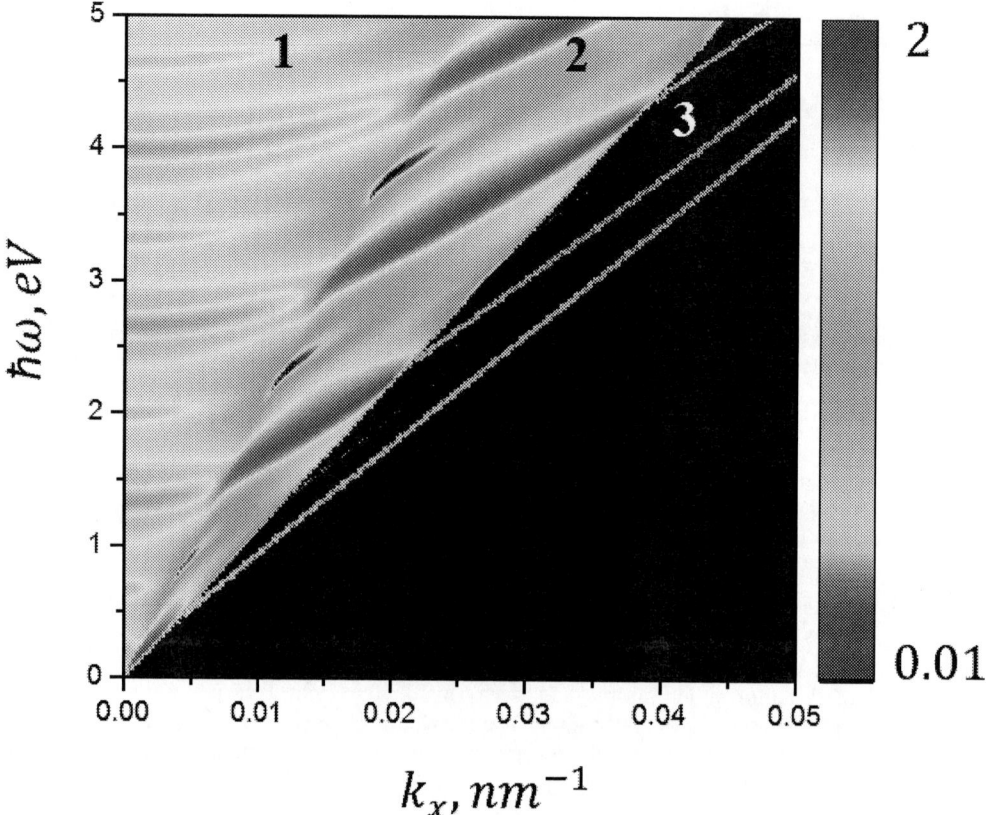

Figure 3. (color online) Dependence of the mode Purcell coefficient for the dipole oriented along the y axis and placed at the centre of a 400-nm-thick GaN slab waveguide, limited by semi-infinite Al_2O_3 and air layers on the emission frequency and wave-vector component kx parallel to the layer interfaces (case of TE polarization).

3. Conclusions

Calculation of spontaneous emission rate for emitter placed in two different slab GaN-based waveguide structures using S-quantization formalism was carried out. Cases of symmetric and asymmetric waveguide structures were considered. Maximum calculated value of the mode Purcell factor in both considered structures doesn't exceed value of 2, that could be explained by low refractive indexes difference between core and cladding layers.

Acknowledgments

This work has been supported by Russian Science Foundation (Project no. 16-12-10503) and by the grant of Minobrnauka № 16.9789.2017/BCh.

References

[1] Chowdhury A, Ng H M, Bhardwaj M, Weimann N G 2003 *Appl. Phys. Lett.* **83**, 1077
[2] Zhang Y, McKnight L, Engin E, Watson I M, Cryan M J, Gu E, Thompson M G, Calvez S, O'Brien J L, Dawson M D 2011 *Appl. Phys. Lett.* **99**, 161119
[3] Hui R, Wan Y, Li J, Jin S, Lin J, Jiang H 2005 *IEEE J.Quantum Electron.* **41**, 100
[4] Gromovyi M et al. 2014 Journal of the European Optical Society - Rapid publications, Europe, v. 9
[5] Westreich O, Katz M, Paltiel Y, Ternyak O, Sicron N. 2015 *Phys. Status Solidi A*, **212**, 1043-1048
[6] Chen H, Fu H, Huang X et al. 2017 *Opt Express* **25**(25):31758-31773
[7] Ivanov KA, Gubaydullin A R, Morozov K M, Sasin M E, Kaliteevski M A 2017 *Opt. and Spectrosc.* **122**.5: 864-872

SPBOPEN 2018

IOP Publishing

IOP Conf. Series: Journal of Physics: Conf. Series **1124** (2018) 051056   doi:10.1088/1742-6596/1124/5/051056

Strong coupling between excitons in transition metal dichalcogenides and optical bound states in the continuum

S. K. Sychev[1,2], K. L. Koshelev[2,4], Z. F. Sadrieva[2], A. A. Bogdanov[2,3], I. V. Iorsh[2]

[1]St. Petersburg Academic University RAS, Saint Petersburg, Russia
[2] International Research Centre for Nanophotonics and Metamaterials, ITMO University, Saint Petersburg, Russia
[3]Ioffe Institute, Saint Petersburg, Russia
[4] Nonlinear Physics Centre, Australian National University, Canberra, Australia

E-mail: [1]sychevstanislav@gmail.com

Abstract. Being motivated by recent achievements in the rapidly developing fields of optical bound states in the continuum (BICs) and excitons in monolayers of transition metal dichalcogenides, we analyze strong coupling between BICs in Ta_2O_5 periodic photonic structures and excitons in WSe_2 monolayers. We demonstrate that giant radiative lifetime of BICs allows to engineer the exciton-polariton lifetime enhancing it three orders of magnitude compared to a bare exciton. We show that maximal lifetime of hybrid light-matter state can be achieved at any point of **k**-space by shaping the geometry of the photonic structure.

1. Introduction

Monolayers of transition metal dichalcogenides (TMDCs) are a certain class of post-graphene two-dimensional materials, attracting vast research interest in recent years. Excitons in TMDC monolayers shows variety of unique features, such as large binding energy [1], leading to strong excitonic response at room temperature, and significant oscillator strength, providing substantial exciton-photon interaction. These properties allow the observation of the so-called strong coupling regime, leading to the emergence of the new quasiparticles, exciton-polaritons [2] even at room temperatures [3].

Strong coupling of TMDC excitons with light has been observed in the structures resembling the conventional microcavities, where the monolayer has been sandwiched between two Bragg mirrors. However, since fabrication of hight quality TMDC monolayers is based on mechanical exfoliation techniques, with are not compatible with the standard epitaxial techniques used for the Bragg mirror fabrication, the realisation of structures supporting exciton-polaritons is quite technologically demanding.

In this work we suggest using photonic crystal slab, supporting optical bound states in the continuum (BICs), to achieve strong exciton-photon coupling. BICs demonstrate exceptional properties [4], in particular, giant radiative lifetime, limited only be surface roughness, and could be excited by plane wave, what favourably distinguishes them from waveguide modes.

Content from this work may be used under the terms of the Creative Commons Attribution 3.0 licence. Any further distribution of this work must maintain attribution to the author(s) and the title of the work, journal citation and DOI.
Published under licence by IOP Publishing Ltd

2. Results

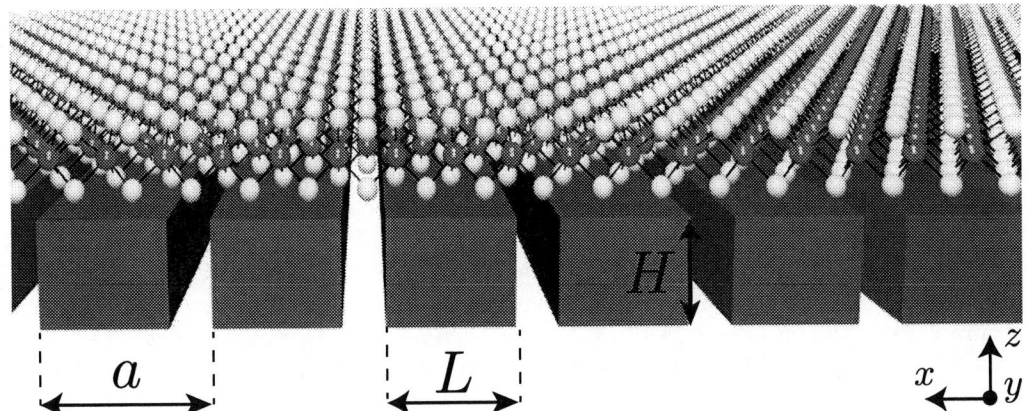

Figure 1. (a) Sketch of a studied structure. The refractive index of Ta$_2$O$_5$ bars $\varepsilon = 2.1$. The TMDC made of WSe$_2$ is laid on top of the PhC slab. The calculations performed for the PhC with period $a = 1.03H$ and bars width $a = 0.9L$

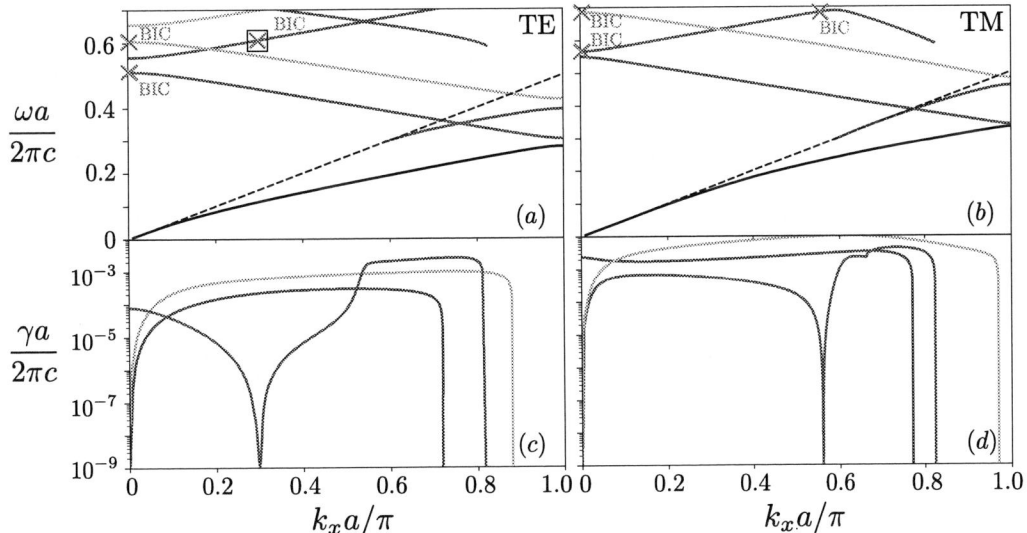

Figure 2. Eigenmode spectrum of air-suspended Ta$_2$O$_5$ one-dimensional grating. Band structure for (a) TE-polarized and (b) TM-polarized modes, respectively. Dimensionless inverse radiation lifetime for (c) TE-polarized and (d) TM-polarized modes, respectively. BICs are marked by orange crosses.

We use the guided-mode expansion [5] method to obtain exciton photonic crystal slab (PCS) eigenmodes. Photonic crystal slab consists of Ta$_2$O$_5$ rectangular bars with height H and width L being spaced equidistant with period a. We put refractive index of Ta$_2$O$_5$ equal to 2.1 which is appropriate for the red band of the visible spectrum range [7]. The calculation are performed for PCS with $a = 1.03H$, $L = 0.9a$. The spectrum of eigenfrequencies and inverse radiation lifetimes of the PCS for in-plane wavevectors along the x direction of the first Brillouin zone is shown in Fig. 2(a,c) for TE-polarized, Fig. 2(b,d) for TM-polarized modes, respectively. Dispersion curves under the light line $\omega = ck_x$ describe pure guided modes with zero diffraction losses while photonic states above the light line are leaky provided their radiation lifetime is finite. The BICs represent unusual leaky modes with $\gamma = 0$ and can be formed both at the center of the Brillouin zone (at-Γ BIC) and at specific points between the zone edge and center (off-Γ BIC).

In order to tune the off-Γ BIC frequency to a resonance with exciton energy $E_{exc} = 1.74$ eV [6] PCS hight equal to $H = 418$ nm was taken. Fig. 3(a) demonstrates strong coupling between the exciton and the off-Γ BIC which manifests itself as an avoided resonance crossing with Rabi splitting of order of 3 meV. Exciton-polariton inverse lifetime are shown in Fig. 3(b) in comparison with the inverse lifetimes of bare excitonic and photonic modes. As it can be seen, for specific values of k_x the lifetime of polariton modes can exceed the bare exciton lifetime by almost three orders of magnitude and reaches 0.66 ns. The most important, Fig. 3(b) shows the maximal lifetime can be realized not at the center or the edge of the Brillouin zone,but at the pointof phase space,where the group velocity of the mode is finite.

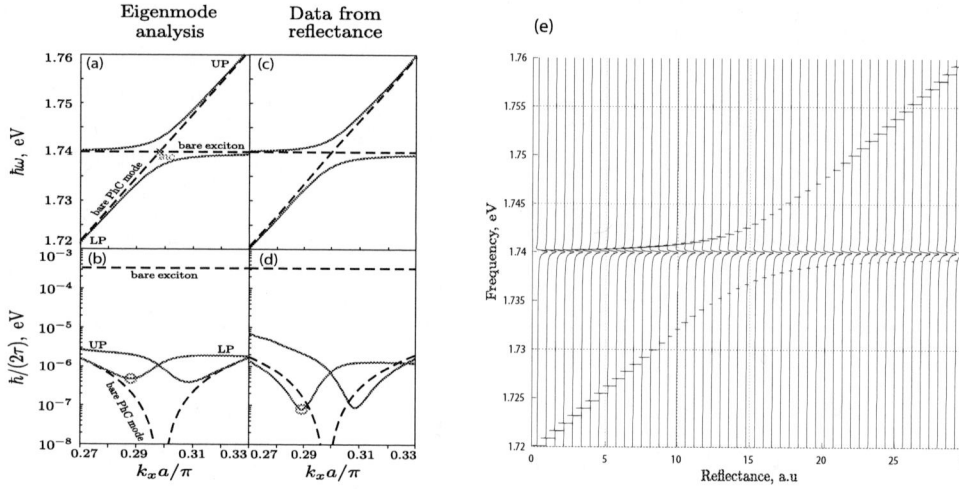

Figure 3. Dispersion and inverse lifetime of exciton-polaritons at the conditions of strong coupling between the TE-polarized photonic mode supporting an off-Γ BIC and in-plane polarized exciton (a,b) and the fitting of Fano lineshape of reflection spectrum (c,d). (e) Reflectance spectrum of WSe$_2$ placed on top of photonic crystal slab. Each spectrum is shifted in horizontal dirction by 0.6

The reflectance spectrum, which could be experimentally observed using angle-resolved reflection spectroscopy, was obtained via using Fourier modal method [8] is shown in Fig. 3(e). We fit the reflectance spectrum to Fano lineshape using Levenberg-Marquardt algorithm to obtain real and imaginary parts of eigenfrequency shown in Fig. 3(c,d). We see that the GME method and the reflectance spectrum calculations show good agreement providing the same value of Rabi splitting and the 3-fold enhancement of LP lifetime with respect to a bare exciton. However, the maximal values of the damping rate extracted by the fitting method are one order

SPBOPEN 2018
IOP Publishing

IOP Conf. Series: Journal of Physics: Conf. Series **1124** (2018) 051056   doi:10.1088/1742-6596/1124/5/051056

smaller than those for the eigenmode analysis. The difference is the result of approximations used for the GME calculations.

Finally, we calculate the dependence of lower-polariton lifetime on in-plane wavevector $\boldsymbol{k} = k_x, k_y$ in the two-dimensional Brillouin zone by means of the eigenmode analysis. One can see that lifetime value exhibits maximum at $k_x \pm 0.29\pi/a$, $k_y = 0$ and decreases smoothly in the vicinity of these points.

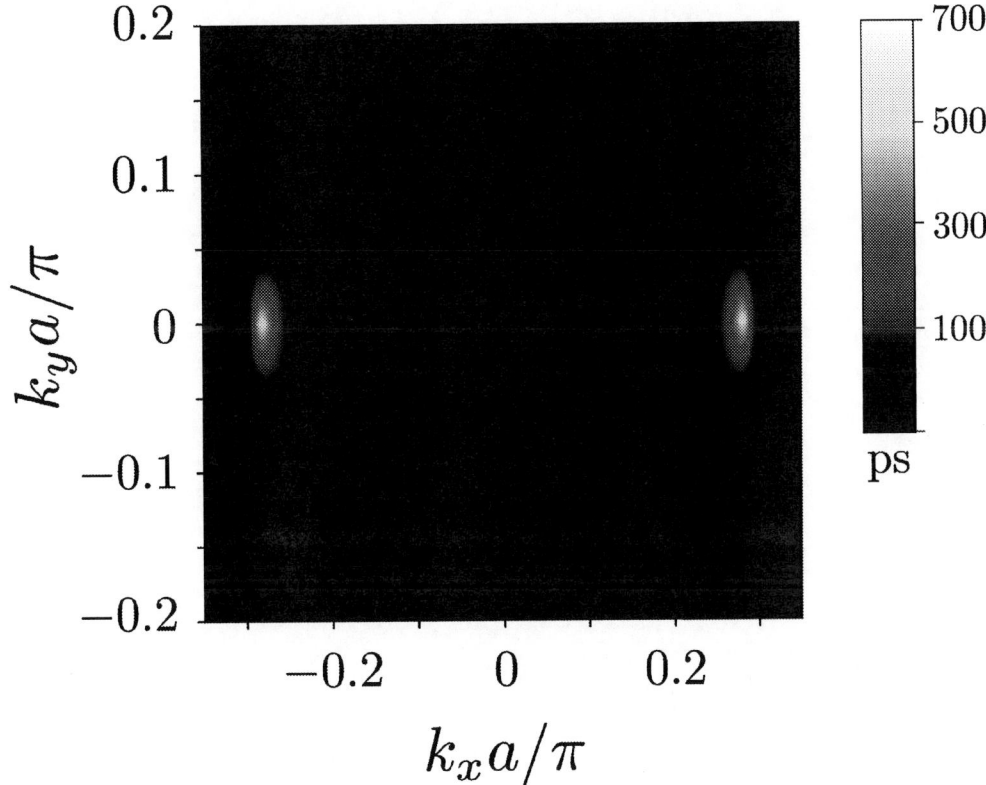

Figure 4. Map of dependence of lifetime of the lower polariton mode on in-plane wavevector.

3. Conclusion

We have proposed an experimentally feasible scheme to achieve strong coupled exciton-photon system in a two-dimensional nanostructure comprising a TMDC monolayer and a periodic photonic nanostructure. Importantly, this scheme does not require the growth of Bragg mirrors, which substantially simplifies the fabrication. Moreover, we have shown that it also allows the polariton condensation at the finite momenta, which opens possibilities for the non-resonant excitation of moving polariton condensates

Acknowledgements

This work was supported by the Russian Foundation for Basic Research (16-37-60064, 17-02-01234), the Ministry of Education and Science of the Russian Federation (3.1668.2017/4.6), the President of Rus- sian Federation (MK-403.2018.2).

References

[1] G. Wang, A. Chernikov et al, arXiv:1707.05863
[2] Kavokin, Alexey V., et al. Microcavities. Vol. 21. Oxford University Press, 2017.
[3] Sun, Zheng, et al. Nature Photonics 11.8 (2017): 491.
[4] Hsu, C. W., et al, Nature Reviews Materials 1, 16048 (2016)
[5] Andreani, L. C., et al, Physical Review B, 73(23), 235114 (2006).
[6] Wang, G. et al., Nature communications 6, 10110 (2015).
[7] Gao, L. et al., Optics express 20.14, 15734-15751(2012).
[8] Li, L. JOSA A, 13(9), 1870-1876 (1996).

SPBOPEN 2018   IOP Publishing

IOP Conf. Series: Journal of Physics: Conf. Series **1124** (2018) 051057   doi:10.1088/1742-6596/1124/5/051057

Experimental observation of symmetry protected bound state in the radiation continuum in the periodic array of ceramic disks

A A Bogdanov[1,2]**, M Balezin**[1]**, P V Kapitanova**[1]**, Z Sadrieva**[1]**, M Belyakov**[1,3]**, E A Nenasheva**[4]**, and A F Sadreev**[5]

[1] Department of Photonics and Metameteials, ITMO University, St.Petersburg 197101, Russia
[2] Ioffe Institute, St. Petersburg 194021, Russia
[3] St.Petersburg Academic University, Khlopina 8/3, St.-Petersburg, 194021, Russia
[4] Giricond Research Institute, Ceramics Co., Ltd., Saint Petersburg 194223, Russia
[5] Kirensky Institute of Physics, Federal Research Center KSC SB RAS, 660036 Krasnoyarsk, Russia

E-mail: `mikhail.beliakov@metalab.ifmo.ru`

Abstract. We observe experimentally symmetry protected bound states in the continuum (BICs) in the stack of ceramic disks. We analyze transformation of the resonant state into BIC with increasing of the number of disks theoretically and experimentally. In the experiment we measure the transmission spectra through the array of ceramic disks at GHz frequencies. We observe a quadratic growth of the Q-factor of quasi-BIC with the number of disks due to material losses. We cross-check the observation of BIC in the transmission spectra by measurement of the field profiles which are a good agreement with the results of the numerical simulations.

1. Introduction
It is well known that a dielectric rod or slab supports waveguide modes formed under the condition of total internal reflection from the waveguide boundaries. The wavenumber of the waveguide modes lies under the light line of the surrounding space making them orthogonal to the radiation continuum[1]. The spectrum of the leaky modes lies above the light line. However recently it was acknowledged that introduction of periodic modulation of the refractive index along the rod's axis discretisize the radiation continuum, that could result in enormous supress of leakage. With a certain set of parameters the leakage is completely blocked and the resonant state is localized, i.e. decoupled from the radiation continuum. Such localized solutions are known as bound states in the continuum (BICs) [2, 3]. Our goal was to reseach possibility of recreating resonant state in the finite rod array with simular features.

2. Transmission spectra and scattering cross section measurements
Resonant state can be exited by simple wave and Q-factor can be obtained by measurment of total cross section. But this method have one major flaw.

Content from this work may be used under the terms of the Creative Commons Attribution 3.0 licence. Any further distribution of this work must maintain attribution to the author(s) and the title of the work, journal citation and DOI.
Published under licence by IOP Publishing Ltd

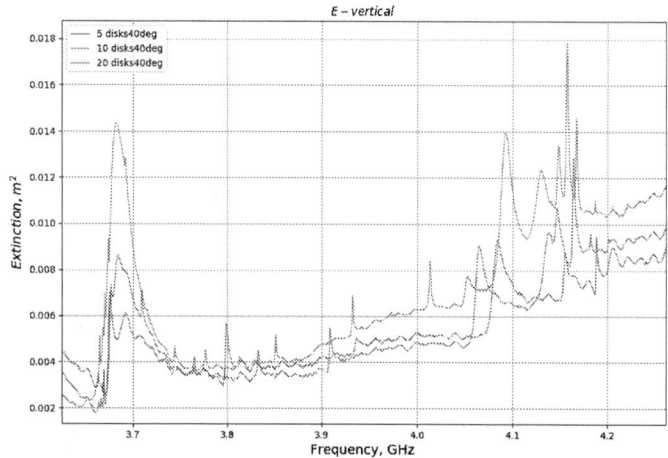

Figure 1. Measured extinction for 5, 10 an 20 disks chain

Fig. 3 shows total cross section for 5, 10 an 20 disks chain. It is nearly impossible to distinguish one peak and measure Q-factor. As shown on Fig. 4, modes with $m = 0$ and $m = 2$ become mixed with each other, when we increase amount of disks. This problem solved by usage of loop antennas, because it can exite only modes with $m = 0$

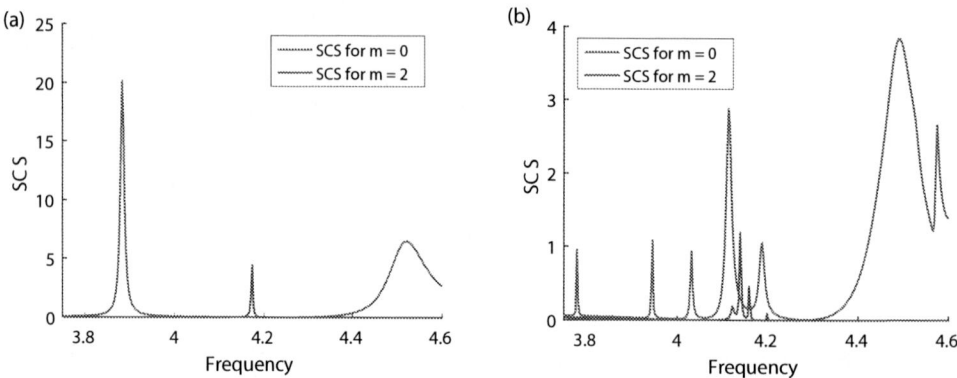

Figure 2. (a) Theoretical SCS for one disk (b) Theoretical SCS for ten disks

The fabricated prototype is shown in Fig. 3. The disks are fabricated from BaO-TiO$_2$ microwave ceramic with the permittivity of ε=40 and $\tan(\delta)$=0.0001 (measured at 1 MHz). The radius and the height of the disks are D=10.2 mm and h=10.1 mm, respectively.

SPBOPEN 2018   IOP Publishing

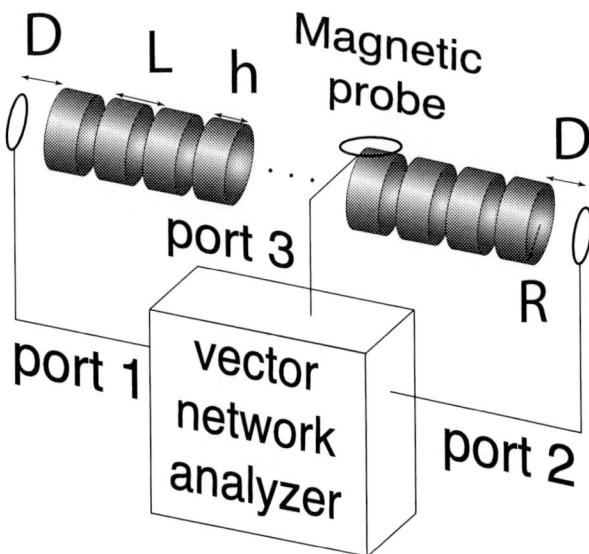

Figure 3. Experimental setup for transmission spectra and field profile measurment

We characterize the resonances of the chain by measurement its transmission with two loop antennas connected to the vector network analyzer. The results of the measurements are shown in Fig. 4. One can see that the spectrum consists of two bands. The dashed line divides the spectrum on waveguide modes and resonant states in the continuum. Fig. 4 (b) shows the the transmission in the region of leaky resonances. The last peak in the series corresponds to the quasi-BIC.

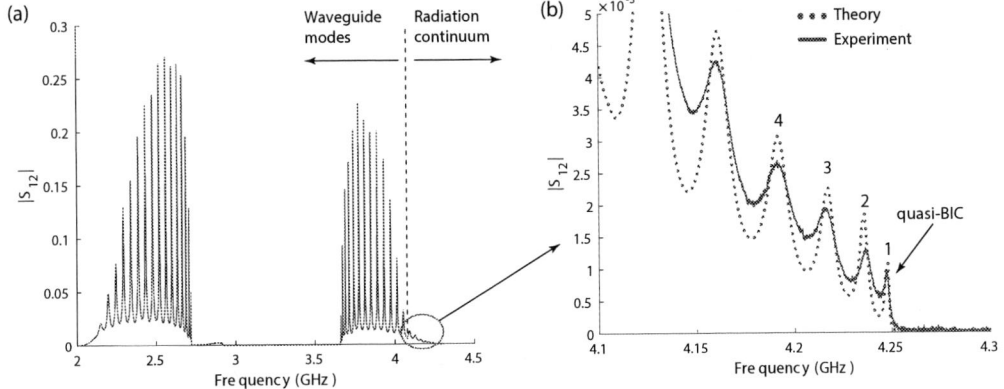

Figure 4. (a) The transmission spectra of the array of 20 disks between two loop antennas as sketched in Fig. 1. (b) The transmission spectra in the radiation continuum.

3. Results

The dependence of its Q-factor on the number of disks is shown in Fig. 5. (mode 1). In the absence of losses the Q-factor growth quadratically with N. However, in the experiment the Q-factor saturates because of the material losses in ceramics. Therefore, at low N the total losses are mainly determined by radiation leakage but at N of several tens absorption makes the main contribution.

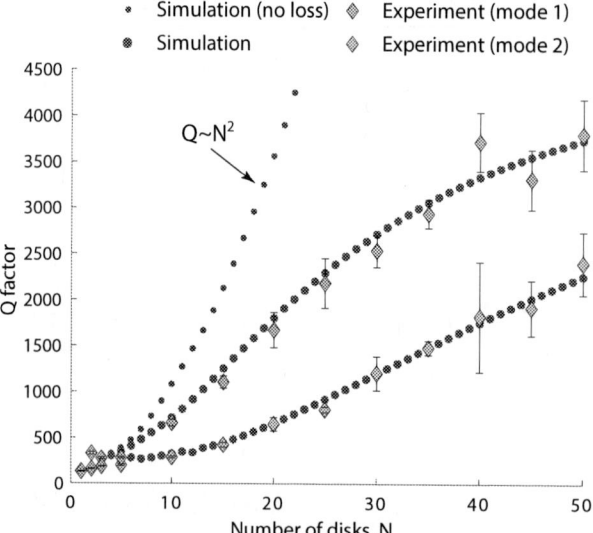

Figure 5. Comparison of experimental results and theoretical simulation for first 2 modes. Theoretical simulation made in COMSOL.

4. Conclusion
We experimentally analyzed the transformation of a resonant state into BIC with increase the number of the disks in one dimensional chain of scatterers for the first time. The obtained results are very important for practical implementation of BIC in radiofrequency and photonic devices.

Acknowledgments: This work was supported by the Russian Foundation for Basic Research (16-37-60064, 17-02-01234), the Ministry of Education and Science of the Russian Federation (3.1668.2017/4.6), the President of Russian Federation (MK-403.2018.2).

References
[1] J. D. Jackson, *Classical Electrodynamics* (John Wiley and Sons, Inc., New York, 1962).
[2] Z F Sadrieva and A A Bogdanov, Bound state in the continuum in the one-dimensional photonic crystal slab, J. Phys.: Conf. Ser. **741** 012122 (2016).
[3] E.N. Bulgakov and A.F. Sadreev, "Bloch bound states in the radiation continuum in a periodic array of dielectric rods", Phys. Rev. A**90**, 053801 (2014).

SPBOPEN 2018   IOP Publishing

High-Q states and Strong mode coupling in high-index dielectric resonators.

S. A. Gladyshev[1], A. A. Bogdanov[2,3], P. V. Kapitanova[2], M.V. Rybin[2,3], K. L. Koshelev[2], Z. F. Sadrieva[2], K.B. Samusev[2,3], Y.S. Kivshar[2] and M.F. Limonov[2,3]

[1]St. Petersburg National Research Academic University of the RAS, 194021 Saint Petersburg, Russia
[2]International Research Centre for Nanophotonics and Metamaterials, ITMO University, St. Petersburg, Russia
[3]Ioffe Institute, St. Petersburg 194021, Russia

E-mail: `sergey.gladishev@metalab.ifmo.ru`

Abstract. We study strong coupling between eigenmodes of a single subwavelength high-index dielectric resonator and analyse the mode transformation and Fano resonances by varying resonator's aspect ratio. We demonstrate that the strong mode coupling is associated with the physics of bound states in the continuum when the radiative losses are almost suppressed due to the Friedrich-Wintgen scenario of destructive interference. We confirm our theoretical findings with microwave experiments by using a high-index cylindrical resonator with tunable aspect ratio.

Introduction

Optical resonators form the basis of lasers, optical sensors, switches and amplifiers. The most important characteristic of the resonator is Q factor, which characterize amplification degree of the electromagnetic field in the resonator. To date the record Q factor of optical resonators today is about 10^{11} [1]. It is achieved in spherical resonators with the whispering gallery mode. However, large sizes do not allow them to be installed into optical integrated circuits. A more attractive way to confine light is to use destructive interference in the regime of the strong mode coupling. This mechanism is related to the physics of bound states in the continuum (BICs) [2]. The BIC-inspired mechanism of light localization provides make possible realization of high-Q states in photonic crystal cavities and slabs [3, 4], coupled waveguide arrays [5, 6, 7], dielectric gratings [8], core-shell spherical particles [9], and dielectric resonators [10, 11]. Using this mechanism, we have shown that even a subwavlength dielectric resonator could demonstrate high Q factors.

Results

We consider a subwavelength dielectric cylindrical resonator with permittivity $\varepsilon_1 = 80$, radius r, and length l placed in vacuum ($\varepsilon_2 = 1$), as shown in Fig. 1a, and analyze its spectrum by calculating the maps of the scattering cross-section (C_{sca}), depending on the aspect ratio r/l. The spectra are calculated by using the CST Microwave Studio software and T-matrix

Content from this work may be used under the terms of the Creative Commons Attribution 3.0 licence. Any further distribution of this work must maintain attribution to the author(s) and the title of the work, journal citation and DOI.
Published under licence by IOP Publishing Ltd

SPBOPEN 2018   IOP Publishing

IOP Conf. Series: Journal of Physics: Conf. Series **1124** (2018) 051058   doi:10.1088/1742-6596/1124/5/051058

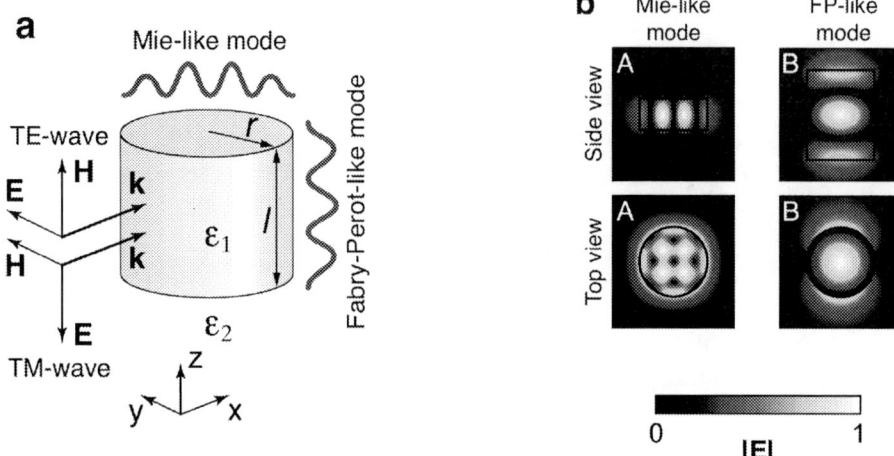

Figure 1. (a) TE- and TM-polarized waves incident on a dielectric cylindrical resonator with permittivity $\varepsilon_1 = 80$, radius r, and length l placed in vacuum ($\varepsilon_2 = 1$). (b) Distribution of the electric field amplitude $|\mathbf{E}|$ for the Fabry-Perot-like mode $TM_{1,1,1}$ (point A) and Mie-like mode $TE_{1,1,0}$ (point B).

computations [12, 13]. The electric field of the incident wave is assumed to be perpendicular to the axis of the cylinder (see Fig. 1a). To compare C_{sca} for cylinders with different aspect ratios, we normalize C_{sca} by the projected cross-section of the resonator, $S = 2rl$. The maps of the normalized C_{sca} calculated for cylinders with different aspect ratio r/l excited by TM and TE-polarized wave are shown in Figs. 2a and 2b, respectively. We denote the modes of a cylindrical resonator as $TE_{n,k,p}$ and $TM_{n,k,p}$, where n, k, p are the indices denoting the azimuthal, radial, and axial wavenumbers, respectively. Generally speaking, distinguishing between $TE_{n,k,p}$ and $TM_{n,k,p}$ modes for a cylinder of a finite length is justified only for $n = 0$. For other cases, the polarisation is hybrid [14]. In the case of arbitrary n, k, p the mode polarization is mixed. Thus, under the terms TE or TM we further imply the dominant polarization of the modes.

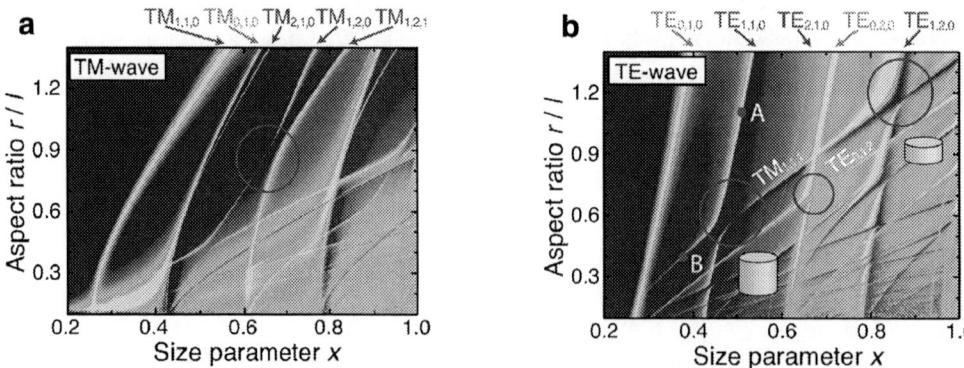

Figure 2. Dependencies of the total scattering cross-section (SCS) of the cylinder σ normalized to the projected cross-section $S = 2rl$ on the aspect ratio of the cylinder and size parameter $x = r\omega/c = 2\pi r/\lambda$ for TM and TE-polarized incident wave, respectively.

The low-frequency spectrum of the dielectric cylinder under consideration consists of three types of modes. The modes with the axial index p = 0 and azimuthal index n = 0, 1 demonstrate a small frequency shift with changing changing r/l. They are formed mainly due to reflection from a side wall of the cylinder, and they could be associated with the Mie resonances of an infinite cylinder (see Figs. 1a and 3a). The modes with the indices $p > 0$ and n = 0,1 demonstrate a strong shift to higher frequencies with increasing aspect ratio r/l. They are formed mainly due to reflection from the faces of the cylinder, and they could be associated with the Fabry-Perot modes (see Figs. 1a and 3a). The modes with the azimuthal index n = 2, 3, ... are formed due to the wave incident on the side wall of the cylinder at the angles bigger than the total internal reflection angle, which is about 6.4 degrees for $\varepsilon_1 = 80$. Therefore, they are close in nature to the whispering gallery modes (see Fig. 3a) and their high Q factor is explained by total internal reflection but not by destructive interference as we have for quasi-BIC. Properties of WGMs are well-studied (see, e.g., Refs. 15, 16, and 17) and further we focus on the Mie-like ($TE_{1,1,0}$) and Fabry-Perot-like ($TE_{1,1,1}$) modes. Their electric field distributions are shown in Fig. 1b.

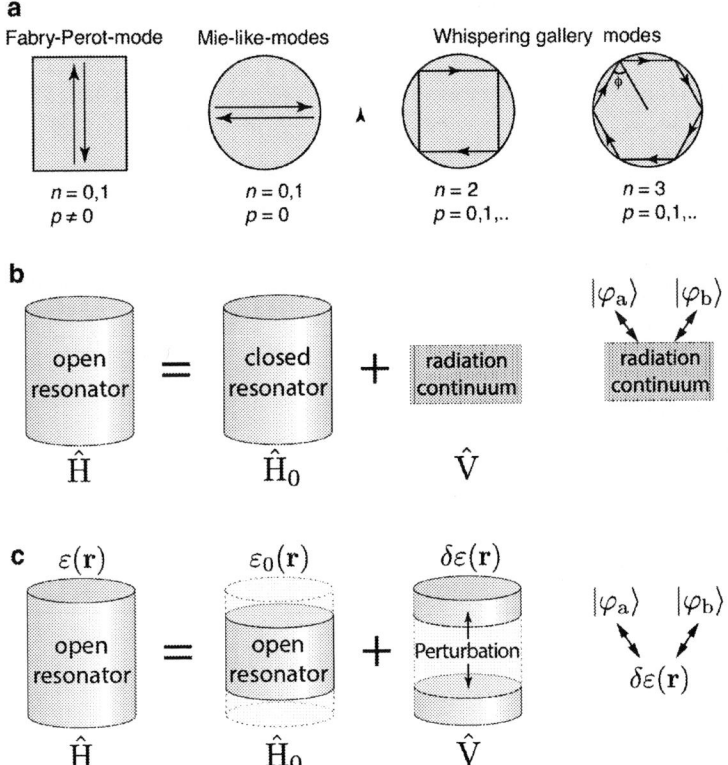

Figure 3. Modes of a dielectric resonator and models of their coupling. (a) Classification of eigenmodes of a dielectric resonator. (b) FriedrichWintgen approach describing an open cylindrical resonator as a closed resonator and a radiation continuum. Eigenmodes of the resonator interact via the radiation continuum. (c) Non-Hermitian approach describing an open cylindrical resonator by a complex spectrum of eigenfrequencies. Eigenmodes of the resonator interact via perturbation $\delta\varepsilon(r)$ responsible for change of resonator aspect ratio.

In quantum mechanics, in the simplest case, the system with light-matter interaction is described by a sum of Hamiltonian without interaction $\hat{H}_0$ and an interaction potential $\hat{V}$ (see, e.g., Ref. 18). The diagonal components of $\hat{V}$ are responsible for energy shift and the off-diagonal components are responsible for the coupling. The interaction results in a mixing of the light and matter states and in appearance of an avoided resonance crossing the characteristic feature of the strong coupling regime [19].

In electromagnetism, due the fact that a resonator is an open system, description of the interaction between the modes becomes more complicated. There are two main approaches describing the interaction between the modes in open system. The first one considers an open system (dielectric cylindrical resonator in our case) as a closed system with non-radiating modes $|\phi_a\rangle$ and $|\phi_b\rangle$ interacting with a continuum of the radiation modes outside of the resonator in accord with the Friedrich-Wintgen mechanism [22] (Fig. 3b). The difficulty of this method is to correctly define the basis of the non-radiating modes and their coupling constants with the radiation continuum. In the second approach, the resonator is primordially considered as an open non-Hermitian system, characterised by a complex eigenfrequency spectrum. In this approach, a small change of resonator shape could be described as a perturbation $\delta\varepsilon(\boldsymbol{r})$ playing a role of the interaction potential $\hat{V}$ between modes $|\phi_a\rangle$ and $|\phi_b\rangle$. In our case, a perturbation $\delta\varepsilon(\boldsymbol{r})$ is responsible for change of the aspect ratio of the cylindrical resonator (Fig. 3c). This method is well-developed for quantum mechanics and electrodynamics [20, 21]. It allows to find spectrum, eigenmodes, and interaction constants straightforwardly from the Maxwells equations.

For the cylindrical resonator, the strong coupling between the Mie-like and Fabry-Perot-like modes is clearly manifested in the map of the SCS as avoided resonance crossing points (Figs. 2a and 2b). The most pronounced regions of the avoided resonance crossing are marked by red ellipses in Fig. 2b. More detailed analysis shows that in the vicinity of the avoided resonance crossing, the Q factor of one the coupled mode becomes very high that corresponds to the appearance of a quasi-BIC. The dramatical increase of the Q factor is a result of destructive interference between the modes with similar radiation patterns in far field.

Experimental results

Finally, we perform the experimental study to demonstrate the existence of the avoided crossing regime between the $TE_{1,1,0}$ and $TM_{1,1,1}$ resonances in the microwave frequency range. In the experiment, the plastic cylindrical vessel filled with water is placed in the middle between two antennas. The aspect ratio of the cylindrical resonator is defined by the amount of water. The photo of the experimental setup is shown in Fig. 4a. The resonator is excited by TE polarized electromagnetic wave incident perpendicular to the cylinder axis z (see Fig. 4a). The measured dependence of the SCS of the cylindrical resonator depending on its aspect ratio is shown in Fig. 4b. The results of the numerical simulations taking into account the losses in water are shown in Fig. 4c. One can see that the experimental positions of the resonances are in a good agreement with the real part of eigenfrequencies (marked by white circles) calculated using the resonant state expansion method. In spite of losses in water, which broaden the resonances, the avoided crossing regime between the $TE_{1,1,0}$ and $TM_{1,1,1}$ modes and suppression of SCS clearly manifest themselves for the aspect ratio in the range of $0.5 < r/l < 0.6$. Discrepancies between the measured and calculated maps of SCS could be explained by not perfect plane wave radiated by a horn antenna and parasitic scattering from the auxiliary equipment (holder of the resonator and plastic cylindrical vessel).

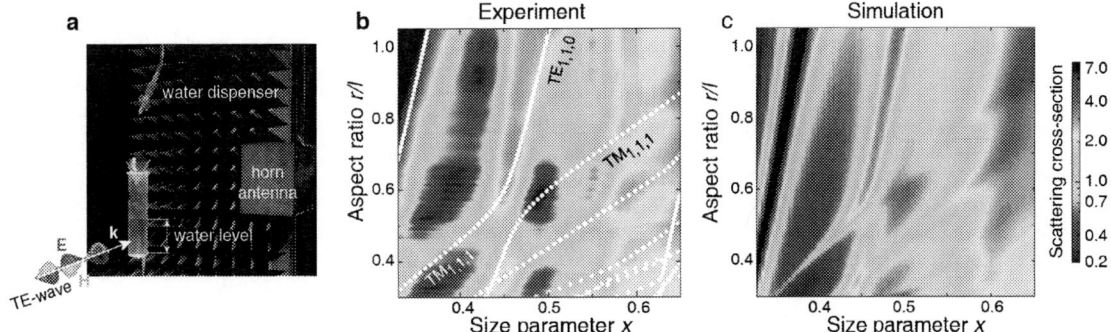

Figure 4. (a) Experimental setup for the measurement of SCS spectra of the cylindrical resonator filled with water depending on its aspect ratio r/l and size parameter x. (b) Measured SCS map demonstrating the avoided crossing regime between $TE_{1,1,0}$ and $TM_{1,1,1}$ resonances. The circles are the real part of eigenfrequencies obtained from the resonant state expansion method for a dielectric cylinder with the permittivity $\varepsilon_1 = 80$ embedded in air ($\varepsilon_2 = 1$). (c) Calculated SCS map of the cylindrical resonator filled with water depending on the size parameter x and aspect ratio r/l.

Conclusion

We have demonstrated that a subwavelength homogeneous dielectric resonator can support strongly interacting modes. We have confirmed our theoretical results in microwave experiment by using a cylindrical resonator filled with water. Our results open new horizons for active and passive optical nanodevices including efficient biosensors, low threshold nanolasers, perfect filters, waveguides, and nanoantennas.

Acknowledgement

This work was supported by the Russian Foundation for Basic Research (16-37-60064, 17-02-01234), the Ministry of Education and Science of the Russian Federation (3.1668.2017/4.6), the President of Russian Federation (MK-403.2018.2).

References

[1] Savchenkov, Anatoliy A., et al. "Optical resonators with ten million finesse." Optics Express 15.11 (2007): 6768-6773.

[2] Von Neuman, J. and Wigner, E. Uber merkwurdige diskrete Eigenwerte. Uber das Verhalten von Eigenwerten bei adiabatischen Prozessen. Phys. Zeitschrift 30, 467470 (1929).

[3] Bulgakov, E. N. Sadreev, A. F. Bound states in the continuum in photonic waveguides inspired by defects. Phys. Rev. B 78, 75105 (2008).

[4] Hsu, C. W., Zhen, B., Lee, J., Chua, S.-L., Johnson, S. G., Joannopoulos, J. D. and Soljacic, M. Observation of trapped light within the radiation continuum. Nature 499, 188191 (2013).

[5] Plotnik, Y., Peleg, O., Dreisow, F., Heinrich, M., Nolte, S., Szameit, A. Segev, M. Experimental Observation of Optical Bound States in the Continuum. Phys. Rev. Lett. 107, 183901 (2011).

[6] Molina, M. I., Miroshnichenko, A. E. Kivshar, Y. S. Surface bound states in the continuum. Phys. Rev. Lett. 108, 70401 (2012).

[7] Corrielli, G., Della Valle, G., Crespi, A., Osellame, R. Longhi, S. Observation of surface states with algebraic localization. Phys. Rev. Lett. 111, 220403 (2013).

[8] Marinica, D. C., Borisov, a. G. Shabanov, S. V. Bound States in the Continuum in Photonics. Phys. Rev. Lett. 100, 183902 (2008).

[9] Monticone, F. Al, A. Embedded Photonic Eigenvalues in 3D Nanostructures. Phys. Rev. Lett. 112, 213903 (2014).

[10] Lepetit, T. Kant, B. Controlling multipolar radiation with symmetries for electromagnetic bound states in the continuum. Phys. Rev. B 90, 241103 (2014).

[11] Lepetit, T., Akmansoy, E., Ganne, J.-P. Lourtioz, J.-M. Resonance continuum coupling in high-permittivity dielectric metamaterials. Phys. Rev. B 82, 195307 (2010).

[12] Mishchenko, M. I. Light scattering by sizeshape distributions of randomly oriented axially symmetric particles of a size comparable to a wavelength. Applied optics 32, 46524666 (1993).

[13] Mishchenko, M. I., Travis, L. D. T-matrix computations of light scattering by large spheroidal particles. Optics communications 109, 1621 (1994).

[14] Jackson, J. D. Classical electrodynamics (John Wiley and Sons, 2007).

[15] Matsko, A. , Ilchenko, V. Optical resonators with whispering-gallery modes-part I: basics. IEEE J. Sel. Top. Quantum Electron. 12, 314 (2006).

[16] Oraevsky, A. N. Whispering-gallery waves. Quant. Electron. 32, 377400 (2002).

[17] Ilchenko, V. S. and Matsko, A. B. Optical resonators with whispering-gallery modes-part ii: applications. IEEE J. Sel. Top. Quantum Electron. 12, 1532 (2006).

[18] Landau, L. D. and Lifshitz, E. M. Quantum Mechanics : Non-Relativistic Theory (Pergamon, Oxford, 1989), third edn.

[19] Scully, M. and Zubairy, M. Quantum Optics (Cambridge University Press, Cambridge, UK, 1997).

[20] Ching, E. S. C. et al. Quasinormal-mode expansion for waves in open systems. Rev. Mod. Phys. 70, 15451554 (1998).

[21] Muljarov, E. A., Langbein, W. and Zimmermann, R. Brillouin-Wigner perturbation theory in open electromagnetic systems. EPL (Europhysics Lett.) 92, 50010 (2011).

[22] Friedrich, H., and D. Wintgen. "Interfering resonances and bound states in the continuum." Physical Review A 32.6 (1985): 3231.

RIE for structuring E-field processed glasses

I Reduto[1,2,3], D Raskhodchikov[2,3], E Gangrskaia[3], V Kaasik[3], Yu Svirko[1], A Lipovskii[2,3]

[1] Institute of Photonics, University of Eastern Finland, Joensuu 80101, Finland
[2] St. Petersburg Academic University RAS, St. Petersburg 194021, Russia
[3] Peter the Great St. Petersburg Polytechnic University, St. Petersburg 195251, Russia

Abstract. We present studies of reactive ion etching of glasses modified by electric field applied to the glasses at elevated temperature. Voltage dependence of the thickness of the modified glass layers is revealed. Secondary ion mass spectrometry data are used to establish the relation between the ion etching rate of the electric field modified glasses and their local composition resulted from the modification. Using structured anodic electrodes allows formation of nanoscale relief structures on the glass surface because of the volume relaxation of the modified regions. This is thermal electric field imprinting procedure. It is shown that the reactive ion etching allows effective deepening the glass surface relief formed in the course of the imprinting. The reactive ion etching allows keeping a high lateral resolution of the surface relief formed in the imprinting. The formation of gratings with micron-scale periodicity is demonstrated.

Introduction

Glasses are widely used for the formation of devices for photonics and sensing, like phase diffraction gratings and microfluidic channels. Many of these applications require formation of a given relief pattern on the surface of a glass plate. Apart from conventionally used techniques for the patterning, like laser micromachining, moulding and lithography followed by an etching, a novel one-step thermal-electric-field imprinting (TEFI) technique has recently been developed [1]. TEFI is based on the modification of a glass at elevated (but below glass transition) temperature by shaped electric field (E-field) generated using a structured anodic electrode. Occurring structural/compositional changes of the glass result in the formation surface relief reproducing the electrode pattern and define patterned chemical stability, respectively. Importantly, the surface relief of the E-field modified glass, which results from the volume relaxation (shrinkage) of the structurally/compositionally modified layer [3], presents height-scaled stamp of the anodic electrode. In conjunction with acidic [3] or alkalic [4] chemical etching TEFI allows fabricating relief structures as high as several microns, the pattern of the etched structure being direct or reversed profile to the electrode relief, respectively. Applicability of plasma-chemical etching to TEFI-processed glasses for deepening their surface relief has recently been demonstrated [5]. Here we present our studies of the reactive ion etching (RIE) of similarly modified glasses.

Thermal Electric Field Imprinting

In the experiments we used one millimeter thick "Menzel" soda-lime glass microscope slides (chemical composition is presented in the Table 1). Surface-structured glassy carbon plate pressed to the glass surface or deposited Cr-Au films were used as the anodic electrodes. The TEFI was carried out at 300°C under 300-1000 V voltage. In the TEFI, the processing stopped when the current through the samples dropped down to 10% of initial (maximal) current and the slides were cooled under the voltage applied. Processed slides were characterized with secondary ion mass spectrometry (IMS-7f, Cameca, France), stylus (Dektak 150, Veeco Instruments, USA) and optical (Wyko NT9300, Veeco Instruments, USA) profilometry, atomic force microscopy (Dimension 3100, Veeco, USA) using RTESP tips and subjected to 12HCF$_3$:38Ar gas mixture RIE (Plasmalab 80, Oxford Plasma Technology, UK). Schematic of the slides processing is presented in Figure 1.

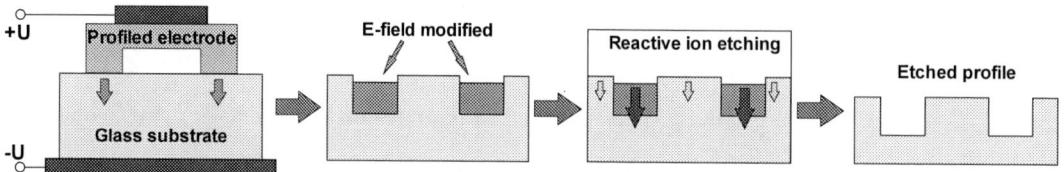

Figure 1. Schematics of thermal electric-field modification and reactive ion etching.

Table 1. Chemical composition of the Menzel glass microscope slides [6].

Oxide	SiO$_2$	Na$_2$O	K$_2$O	CaO	MgO	Al$_2$O$_3$	Fe$_2$O$_3$	SO$_3$
wt%	72.20	14.30	1.20	6.40	4.30	1.20	0.03	0.30

Secondary-ion mass spectrometry

Data obtained using secondary-ion mass spectrometry (SIMS) showed essential changes in the glass composition (relatively to deeper region of the virgin glass) up to one micron depth, which resulted from the TEFI modification. Importantly, the SIMS data showed strong depth dependence of the concentrations of all constituents of the glass slides including silicon. Since the current depths were calculated basing on the final depth of the crater formed in the slide by ions in the SIMS and the duration of the measurements, registered depth dependence of silicon concentration data allowed us presuming that the rate of ion etching in the course of the SIMS measurements was depth-dependent. This is because silicon, which is glass network former in silicate glasses, is not capable of drifting under electric field. This allowed us calculating the SIMS etching rate assuming silicon concentration has to be constant. (Fig. 2a). Using this rate we nonlinearly changed the depth scale of the SIMS data and accounted for this in the concentrations data in Figure 2b. Comparing Figures 2a and 2b we established that the absence of movable (Na$^+$, K$^+$, Ca^{2+}) ions, which makes this initially multicomponent glass to be closer to fused silica, essentially influences the etching rate. This stimulated us to perform RIE processing of the glass slides subjected to the TEFI with 12HCF$_3$:38Ar gas mixture conventionally used for RIE of silica [7].

Reactive Ion Etching

We characterized the surface profile of the TEFI-processed glass slides in the vicinity of the anodic electrode edge before and after the RIE. Measured depth of the steps between TEFI modified and virgin areas of the slides after the RIE and the measured rate of the virgin glass RIE, 6 nm/min, allowed us deducing the thickness of the TEFI modified glass regions, which were 180, 250, 330 and 370 nm for the TEFI processing at 300, 500, 700 and 1000 V, respectively. We supposed the modified glass region completely etched off when the etching rates of both modified and virgin glasses became equal. Note that the actual relief of the slides was deeper because of the etched depths summation with the TEFI-formed relief (80, 130, 150 and 400 nm for the same voltages). Thus, maximal obtained relief steps were 240, 350, 460 and 730 nm, respectively – see Figure 2c, here the decrease in the curves slope corresponds to the etching out of the modified glass region.

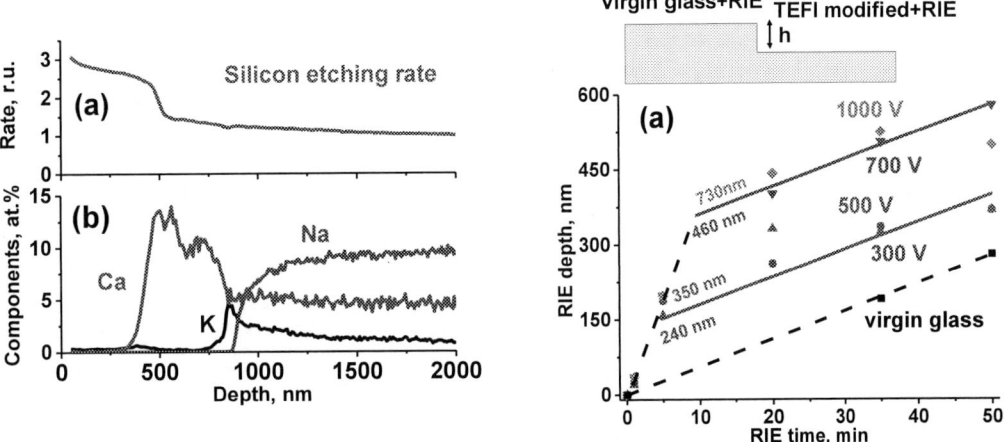

Figure 2. SIMS measured relative etching rate **(a)** and depth dependence of movable ions concentration in the glass **(b)** after 500 V TEFI processing with pressed glassy carbon electrode. Reactive ion etching of the TEFI glasses processed with deposited chromium electrode **(c)**. Inset: schematic of the RIE depth measurements.

Surface structures imprinting

Using the TEFI combined with the RIE we successfully fabricated micron-scale lateral resolution periodic relief structures on the glass surface, which can behave as diffraction gratings (Figure 3). The temporal evolution of the gratings profiles under RIE were characterized with optical profiler (Figure 3a). After TEFI we found surface profiles depth being of 30-40 nm (Figure 3b). Maximal profile depth obtained with the RIE was over 100 nm (Figure 3c). Important to note the existence of "whiskers" at the positions corresponding to the anodic electrode edges after the first 4 minutes of the etching and their complete absence after 10 min RIE. This "whiskers" correspond to stronger electric field treatment in the vicinity of the electrode edges [3,8]. The depth of this "whiskers" is ~110 nm that is also the maximal depth of the etched structure. This means that the etching of stronger modified glass regions is faster, however, finally the glass relief precisely repeats the electrode profile. This is strong distinction in comparison with the wet acidic etching which leaves the "whiskers" untouched [3].

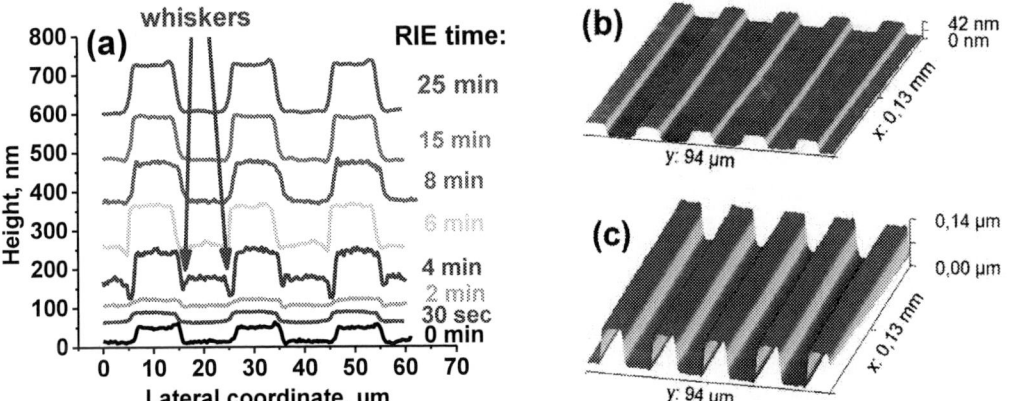

Figure 3. Reactive ion etching of the TEFI glass. Surface profile of RIE TEFI-processed 20 μm in period grating **(a)** and 3D plot of optical profiler data after the TEFI **(b)** and after the RIE **(c)**. The glass was modified at 300°C under 300 V voltage; the TEFI with pressed glassy carbon electrode.

Figure 4a presents temporal evolution of the RIE-formed gratings. The peculiarities during first 1-2 minutes of the RIE are because of the existence of ~20 nm damaged subsurface layer in the initial glass, which is generally typical for glasses [9]. Degradation of the surface relief at the initial stages of the RIE is also shown in Figure 3a. The thickness of the damaged layer according to our SIMS data is about 20-30 nm, which corresponds to the graph in Fig. 4a. The maximal relief depth was achieved after 10 minutes of the RIE. Importantly, the RIE surface profile was not degrade even after 50 minutes of the treatment. Figure 4b,c illustrates the AFM image and surface profile of fabricated 2 μm in period grating.

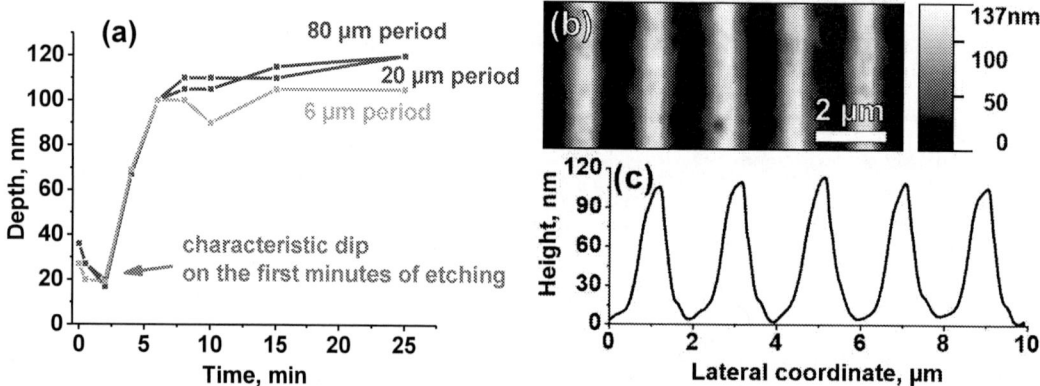

Figure 4. Depth of the gratings during the RIE process **(a)**; AFM image **(b)** and surface profile **(c)** of 2 μm in period grating of glass RIE TEFI-modified at 300°C under 300 V voltages. TEFI with pressed glassy carbon electrode.

Conclusion

Finally we have characterized the thickness and chemical composition changes in ~1 μm thick region of the TEFI-modified soda-lime glass. Varying the TEFI voltage allows one changing the thickness of the chemically and structurally modified region. Different durability of the virgin and modified glass regions allows using reactive ion etching for the surface profiling of the TEFI-processed glasses. The gratings with the period down to 2 microns were successfully formed. Contrary to wet etching, in the RIE the glass relief precisely repeats the profile of used in the TEFI anodic electrode.

Acknowledgments

This study was supported the Ministry of Education and Science of the Russian Federation (project #16.1233.2014/K). AFM characterization were performed using equipment owned by the Federal Joint Research Center "Material science and characterization in advanced technology" (Ioffe Institute, St. Petersburg, Russia).

References

[1] Lipovskii A A, Melehin V G, Petrov M I, Svirko Yu P 2011 *Lithography: Principles, Processes and Materials* (Ed Hennessy T C, New York: Nova Science Publishers) p 149
[2] Lipovskii A A, Kuittinen M, Karvinen P, et al 2008 *Nanotechnology* **19**(41) 415304
[3] Reduto I, Kamenskii A, Redkov A, Lipovskii A 2017 *J. Electrochem. Soc.* **164** 13
[4] Ikutame N, Kawaguchi K, Ikeda H, et al 2013 *J. Appl. Phys.* **114** 4
[5] Alexandrov S E, Lipovskii A, Osipov A A, et al 2017 *Appl. Phys. Lett.* **111** (11) 111604
[6] http://www.agarscientific.com/microscope-slides.html
[7] Bazylenko M V, Gross M 1996 *J. Vac. Sci. Technol., A* **14** (6) pp. 2994-3003
[8] Zhan Z, Li W, Yu L, Wang L, Sun D 2017 *Materials* **10** 158
[9] Maslov V 1983 *Strength of Materials* **15**(8) 1108

Photodecomposition of organic/inorganic composite materials based on polyvinylpyrrolidone

A S Kulagina[1,2], S K Evstropiev[3], K V Dukelskii[3,4],
N A Volkova[3], K S Evstropyev[3], N V Nikonorov[3]

[1] Epitaxial Nanotech Lab, St. Petersburg Academic University, 194021, Russia
[2] Lab of Epitaxial Nanostructures, ITMO University, St Petersburg 197101, Russia
[3] Department of Optoinformatics Tech and Materials, ITMO University, St Petersburg 197101, Russia
[4] The Bonch-Bruevich St Petersburg State University of Telecommunications, St Petersburg 191186, Russia

Abstract. The photodecomposition under UV light and powerful laser (λ = 532 nm) irradiations of organic/inorganic solutions and composite coatings containing organic diazo dye has been studied. The presence of zinc nitrate significantly changes the behavior of spectral properties evolution at UV irradiation and significantly accelerates the photolysis of organic dye. It was found that prepared composite $CSB/Zn(NO_3)_2/PVP$ and CSB/PVP coatings demonstrate the high transparency and non-linear optical properties in visible spectral range and can be used as limiters of powerful laser irradiation.

1. Introduction

The organic/inorganic composite materials based on polyvinylpyrrolidone (PVP, Fig.1) are promising for different optical applications. The aim of this work is the study of the photodecomposition under UV light and under powerful laser (λ = 532 nm) irradiations of organic/inorganic solutions and composite coatings based on PVP and containing organic diazo dye. The presence of organic dye Chicago Sky Blue (CSB) in the solutions and $CSB/Zn(NO_3)_2/PVP$ composite coatings allows to estimate photocatalytic properties of these materials. The structure and spectral properties of this dye was described in [1] in detail.

It's known that the addition of aromatic hydrocarbons into the materials structure can enhance non-linear optical properties of obtained composites [2]. The composite based on PVP demonstrate high non-linear optical properties and transparency in visible spectral range [2,3]. Moreover, the composite coatings may be deposited on the surface of solid material by simple technique (spraying; dipping process). Therefore, PVP-based transparent composite coatings could be considered as promising material for thin temporal non-linear optical limiters.

2. Materials and Methods

The aqueous solutions of $Zn(NO_3)_2$ (0.11 M), organic diazo dye Chicago Sky Blue (CSB; Sigma-Aldrich, Fig. 2) (8.7 10^{-6} M) and the solution of PVP (M_w=1300000) in propanol-2 (1.1·10^{-5} M) were used as raw materials. The composite $CSB/Zn(NO_3)_2/PVP$ solutions were prepared by mixing these initial solutions.

Content from this work may be used under the terms of the Creative Commons Attribution 3.0 licence. Any further distribution of this work must maintain attribution to the author(s) and the title of the work, journal citation and DOI.
Published under licence by IOP Publishing Ltd

The composite CSB/Zn(NO$_3$)$_2$/PVP coatings were prepared on the glasses by the dipping method. The specimens were first put into the solution, then were withdrawn from it and were dried under 50°C for 24 hours. The absorption spectra were measured with Shimadzu UV-3600 spectrophotometer. The UV irradiation was carried out using high pressure Hg lamp (DRT-250 (Russia)). The power density of light irradiation incident on the experimental sample was 0.25 W/cm^2.

The non-linear optical measurements were performed using z-scan technique [1]. The source was a frequency doubled Nd:YAG laser with pulse parameters: λ_{ex}=532 nm, τ_{pulse}=5 ns, E_{pulse}= 0.5-2 10^4 µJ. Beam with Gaussian profile was focused by the lens with F=15 cm. There are two photometers (OPHIR) was used for energy detecting; the output energy was measured in two regimes "open aperture" and "closed aperture". 2 mm cell was in case of sols. The optical scheme consisted of laser source, beam splitter, filters, expander, lens, automated shift with sample holder, detectors.

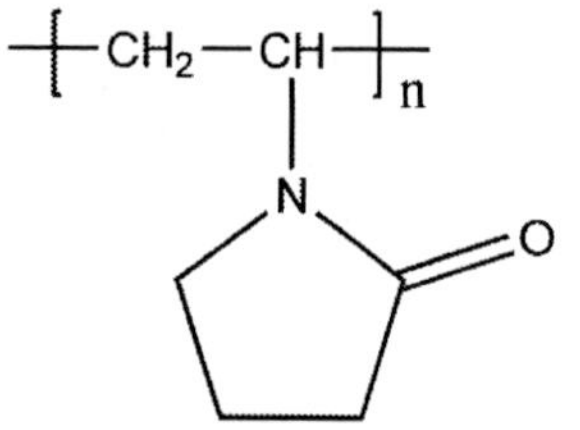

Figure 1. The structure of the PVP

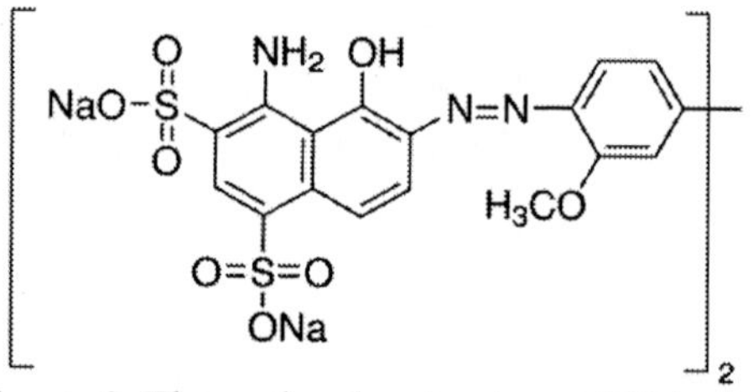

Figure 2. The molecular structure of Chicago Sky Blue dye

3. Experimental results and discussions

3.1. Spectral properties

The absorption spectrum of composite coatings containing CSB dye is shown in Fig.1. The intensive absorption band which is characteristic for this dye is observed in red part of visible spectral range. Fig.1 shows that two maximums are observed in the spectrum of the composite coating. This phenomenon is related to the intermolecular interaction and the formation of the dimers and trimers of CSB molecules [1]. At low dye concentration CSB molecules in solutions and organic polymer are in monomer form, but at the concentration more than ~ 10^{-6} M they interact with each other and form the dimers and trimers of dye molecules [1]. These molecular aggregates have the spectral properties slightly differs from monomers and maximum of their absorption band is located at longer wavelengths.

UV irradiation decomposes diazo dye CSB both in aqueous solutions and composite coatings. Fig.3 demonstrates the changes of CSB/Zn(NO$_3$)$_2$/PVP composite coating during UV irradiation. The intensity of CSB absorption band significantly is reduced after UV irradiation. Also, the shape of absorption spectra changes significantly demonstrating photodecomposition of dimers and trimers of CSB molecules.

The kinetics of CSB photodecomposition in the mixed solutions is shown in Fig.4. .It was found that CSB photolysis proceeds significantly faster in the pure CSB solution (Fig.4, curve 1) and in the CSB solution with metal nitrate addition (Fig.4, curve 2) than in coating (Fig.4, curve 3). Almost full decomposition of CSB dye in the solution containing zinc nitrate is observed after UV irradiation during 15 min (Fig.4, curve 2). In the composite solution without zinc nitrate the photodecomposition proceeds very slowly. Full decomposition CSB in composite coating isn't observed after UV irradiation during 60 min.

SPBOPEN 2018

IOP Publishing

IOP Conf. Series: Journal of Physics: Conf. Series **1124** (2018) 051060   doi:10.1088/1742-6596/1124/5/051060

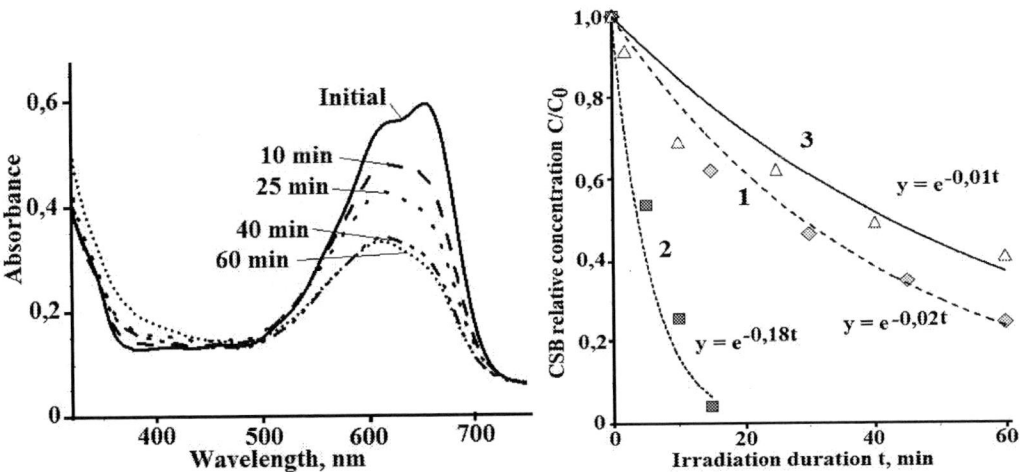

Figure 3. Photolysis of CSB in composite coatings   **Figure 4.** Kinetics of CSB photolysis

It was found that the presence of zinc nitrate in the coating composition hasn't influence on the spectral properties of CSB and on the kinetics of dye decomposition.

3.2. Nonlinear transmittance at 532 nm

Synthesized $CSB/Zn(NO_3)_2/PVP$ and CSB/PVP composite coatings and sols were investigated in two modes of laser action. The sample was placed in the focal plane of the positive lens (F=+15 cm). In the first case, the sample was fixed, and input energy was changed by means of calibrated neutral filters. In the second case, vice versa, input energy was fixed and the sample was moved along the optical axis, passing successively before the focal plane positions and after (the z-scan method). The threshold of sensitivity, when nonlinear effects appeared, was about 0.2 GW/cm^2 for all samples (Fig. 5).

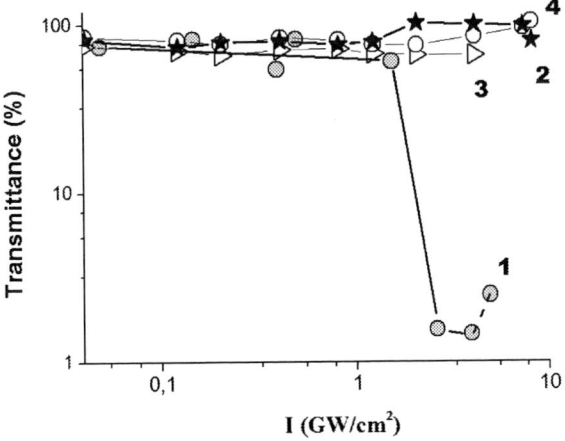

Figure 5. Experimental dots correspond to limiting measurements in the single pulse mode: *1* is CSB/PVP coating; *2* is $CSB/Zn(NO_3)_2/PVP$ coating; *3* is CSB/PVP sol; *4* is $CSB/Zn(NO_3)_2/PVP$ sol (lines are drawn for clarity). Beam parameters were λ=532 nm, τ=5 ns.

Investigations of samples resistance showed, that starting from 2 GW/cm^2, the optical limiting (nonlinear transmittance reducing) in the case of the CBS/PVP coating and the bleaching effect in the

829

case of the $CSB/Zn(NO_3)_2/PVP$ coating occurred irreversibly (Fig. 5). The maximum attenuation rate for the coating without nitrates reached 50 times at the intensity $I=2$ GW/cm^2, and its evaporation occurred at intensities of about 5 GW/cm^2. A comparison of the coatings of the two compositions in the frequency regime of 10 Hz and at an intensity of 5 GW/cm^2 showed that the ratio of the absorbed energies to the breakdown point was 2.5, and CSB/PVP coating is harder (not shown). Consequently, the presence of nitrates negatively affects the resistance to laser radiation, which, as noted above, is associated with the acceleration of the degradation of dye molecules under exposure. Unlike coatings, solutions of similar compositions were only bleaching at the high energies (Fig.5, 3 and 4 measurements), therefore their use as nonlinear optical filters [5] is difficult. We note, that using of synthesized CSB/PVP coatings as laser radiation limiters of nanosecond duration is possible in case of high intensities about 2-5 GW/cm^2, including for several seconds at a repetition rate of 10 Hz. Investigations by z-scan method have shown differences between sols and coatings, between presence of nitrates and absence (Fig. 6).

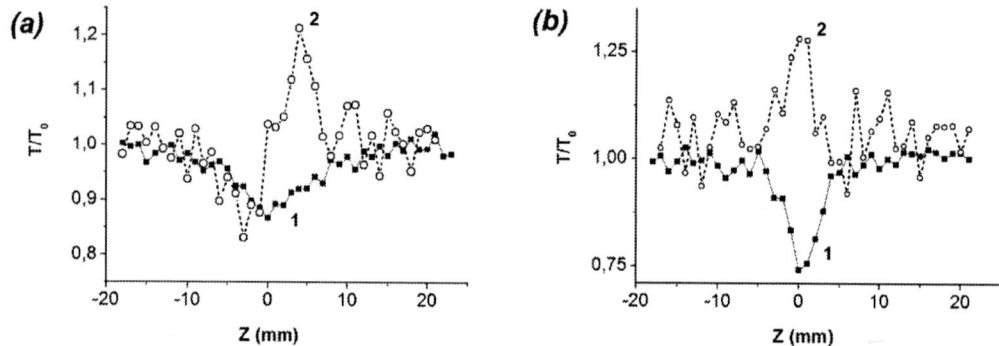

Figure 6. Z-scan measurements of sols: *(a)* is CSB/PVP; *(b)* is $CSB/Zn(NO_3)_2/PVP$. *(1)* is open aperture regime of z-scan; *(2)* is close aperture regime of z-scan (normalized to open). $I_{\text{in focus, Z=0}} = 0.2$ GW/cm^2. T/T_0 is normalized transmittance; Z is distance from focal plane to the sample (the movement was from lens to detector in each case).

As can be seen from the spectra on Fig. 3, the synthesized samples absorb radiation with 532 nm wavelength weakly. Consequently, transmission decrease with increasing power density on Fig. 6 at open aperture regime is due to nonlinear scattering primarily. During z-scan measurements, it was observed strong scattering from samples. We found, that nonlinear properties of coatings and sols are similar in many ways. Principled difference appeared in the regime of open aperture. In $CSB/Zn(NO_3)_2/PVP$ sol the "valley" is more narrow and dip than in CSB/PVP sol. In case of coatings we registered weak bleaching at nitrate presence, in case of sols it was absorbance (Fig.6 (a), (b)). It is explained by the small number of scaterring centers in a thin coating layer and additional saturable absorption of dye molecules at nitrates present. Curves 2 on Fig. 6 indicate, that important mechanism changing transmittance trend is nonlinear refraction also. The change of valley to peak in closed aperture regime corresponds to positive refraction. The shape of the transmittance curve is asymmetric for both the coatings (not shown) and the sols, what requires more detailed consideration of mechanisms. In the case of coatings, the z-scan curves measured with a closed aperture are almost identical; they indicated a positive character of refraction and are asymmetric with respect to the focus. Measurements of sols allow us to make a more qualitative assessment of the differences in the mechanisms of nonlinear transmission for the presence of nitrates and absence. From fig. 6, it can be seen that in a liquid medium, nitrates negatively affect the formation of a positive lens at the focus of exposure. Thus, the presence of nitrates reduces ability to optical limiting. Calculation of the coefficients of nonlinear scattering and refraction is an important task for these coatings and will be carried out in future studies.

4. Conclusions

The presence of zinc nitrate significantly changes the behavior of spectral properties evolution at UV irradiation and significantly accelerates the photolysis of organic dye. It was found that prepared composite CSB/PVP coatings demonstrate the high transparency and non-linear optical properties at the 532 nm wavelength and can be used as limiters of nanosecond laser pulses at intensities of about 2-5 GW/cm^2, in including within a few seconds with frequency action (10 Hz).

Acknowledgments

Some experiments were carried out with the support of the project 16.9791.2017/8.9 of Ministry Education and Science of Russia. The authors are grateful to I M Belousova and S I Vavilov State Optical Institute for technical support of the experiments on nonlinear transmittance measurements.

References

[1] Abbott L C, Batchelor S N, Oakes J, Lindsay Smith J R, Moore J N 2004 *J. Phys. Chem. B* **108** 13735.

[2] Ganeev R A, Ryasnyanskii A I, Kodirov M K, Kamalov Sh R, Li V A, Tugushev R I, Usmanov T 2002 *Techn. Phys.* **47** 991.

[3] Kulagina A S, Evstropiev S K, KhrebtovA I 2017 *J Phys: Conference Series* **917** 062044.

[4] Bagrov I A, Danilov V V, Evstrop'ev S K, Kiselev V M, Kislyakov I M, Panfutova A S, KhrebtovA I 2015 *Techn. Phys. Lett.* **41** 65.

[5] Danilov V V, Panfutova A S, Shilov V B, Belousova I M, Ermolaeva G M, Khrebtov A I, Videnichev D A 2015 *Russian Journal of Physical Chemistry B* **9** 561

SPBOPEN 2018

IOP Publishing

Generation of the trapping light structures based on vector fields

N Shostka[1], O Karakchieva[1], B V Sokolenko[2]
[1] Scientific research department, V.I. Vernadsky Crimean Federal University, Simferopol 295007, Russia
[2] Institute of Physics and Technology, V.I. Vernadsky Crimean Federal University, Simferopol 295007, Russia

Abstract. The paper discusses the main mechanisms for the formation of arrays of singular beams with non-uniform polarization for capture and particle control. The article considers a new system of vector beams with an ideal dark area, suitable for trapping and manipulating light-absorbing particles in the air. In this case, the trapping properties in the light structure can be controlled. For this purpose, the mechanism of the formation of light structures based on a single-axis crystal scheme was considered in detail. The obtained results provide a new way for creating and controlling the properties of beams suitable for trapping and will expand the possibilities of micromanipulation of absorbing particles. Besides the possibility of generating multidimensional lattices of polaritons are considered.

1. Introduction

Researches pay great attention to trapping and manipulation of particles with optical tweezers because of broad areas of their practical applications [1-9]. Usually optical tweezers manipulate microscopic objects in planar geometry, and the particles are limited by a thin layer determined by the focal region of the beam, but to create an optical potential well and achieve a fully three-dimensional capture, a so-called «optical bottle» was proposed [1]. The term «optical bottle beam» describes a beam with a finite axial region of low intensity surrounded in all dimensions by light [1-4]. Once in such a light construction, the absorbing particle is held in its axial region and under certain conditions cannot penetrate through the light walls [10]. Any shift of a particle leads to a heating of that part of it that is farther from the centre of the beam. The resulting pressure difference on the cold part of the particle close to the centre of the beam and the hot peripheral one returns the particle to a position of stable equilibrium inside the optical "bottle". A number of different techniques has been proposed to generate bottle beams such as mechanical angular scanning of a laser beam [2] and the use of optical diffractive elements such as holograms and phase plates to form a desired hollow axial structure [3].

Most of realizations of optical tweezers and bottle beams employ coherent light, however it was shown, that the light with partial spatial and temporal coherence could be used for realization bottle beams. For example, it was shown that focusing of the spatially partially incoherent light beam by an axicon leads to the formation of a bottle beam [4, 5]. A great number of researches in this area was developed for trapping particles in liquids. However stable trapping of absorbing particles in air has been achieved only recently in a new trap created by two counter-propagating optical vortex beams [6]. Generated after uniaxial crystal vector beams could create intensity distribution as in bottle beam [7-9, 11].

Content from this work may be used under the terms of the Creative Commons Attribution 3.0 licence. Any further distribution of this work must maintain attribution to the author(s) and the title of the work, journal citation and DOI.
Published under licence by IOP Publishing Ltd

In this paper, we consider the formation of 3D trap arrays when a Gaussian beam array and arrays of singular optical beams pass along the crystal axis, when the axis of each beam in the array will be inclined at a small angle relative to the propagation axis and we show method of generation periodical incoherent and coherent bottle beams created by using periodical diffractive screen, and controlled shaping of bottle beams with uniaxial crystal.

From another hand, polariton models allow us to describe the dynamics of the generation and propagation of linear and nonlinear waves in different media over a wide frequency range. Depending on the parameters of the initial polariton wave, the form of the system of equations changes. In the case of linear polarization, a single spatial soliton (one polariton flow) or a cnoidal wave (several streams) forms depending on the beam width: for a narrow beam with a width on the order of the wavelength, one stream is formed, and for a broad beam with a thickness of several tens of wavelengths, multiple threads [12,13]. It was considered the process of propagation of an array of polariton waves in dependence on the polarization of the incoming wave. As a result, a significant increase in signal power was obtained, which makes it possible to use such arrays in various communication devices.

2. Bottle beam structures

Consider the model of formation bottle beams by uniaxial crystal. The main advantage of the method based on using inhomogeneous medium is the possibility of highly efficient generation of a number of practically important optical beams, whose power is limited only by the threshold for the destruction of glass elements. Also, an anisotropic medium makes it possible to form not only a grid of phase and polarization singularities in the beam, but also to control their shape and mutual position [14].

The problem of generating necessary beam structures in this method reduces to the following scheme. Primarily will describe the properties of the beam array consisting of the N paraxial local gaussian beams. The array is simultaneously a complex beam that propagates along the crystal axis and a superposition of individual beams propagating at an angle to the optical axis [15]. The axis of the single beam is shifted by the distance r_0 relative to the center of the array and is tilted at the small angle to the z-axis of the array. Without analysis, it is impossible to say whether the integrity of the optical bottle will be preserved when the beam is tilted, and also what pattern will be observed when the beam angle in the array is increased. In connection with this, in Ref. [16] it was shown experimentally and analytically the effect of focusing of a Gaussian beam propagated under small angle α with respect to the optical axis of a uniaxial crystal on the generation of a bottle beam. At $\alpha = 0°$ two foci that correspond to ordinary and extraordinary parts of a beam form a closed 3D structure of a bottle beam. Starting from the value of $\alpha = \pm 2°$ the closed 3D symmetric structure of a bottle beam breaks down.

We impose the following conditions on the initial array: each beam in the array should propagate at an angle to the crystal axis not exceeding two degrees, and the beams must be at the necessary distance from each other so that interference effects do not arise between them.

As it was shown in [17], the slope relatively to the propagation axis z is displayed for small angles as $y \rightarrow y' + i\alpha z_0$, where α – the angle between the optical axis of the crystal and the propagation direction (axis) of the beam in the array, $z_0 = \dfrac{k\omega_0^2}{2}$, k – wave number.

To generate arrays of bottle beams we use the following experimental setup: initial Gaussian beam from He-Ne laser with a wavelength of $\lambda = 632.8$ nm is transformed into array by diaphragm D with N pinholes. With the lens system the formed array is focused into c-cut uniaxial crystal of $LiNbO_3$ with $n_o = 2.286$, $n_o = 2.203$. Due to the birefringence in the anisotropic medium of the crystal each Gaussian beam splits into ordinary and extraordinary beams with different radii of curvature. The light after crystal is focused by the lens with f = 7 cm. The foci points locate at the different positions Z_o and Z_e, spaced by a distance $\delta = Z_e - Z_o$. Intensity patterns are presented on Figure 1 (a–d).

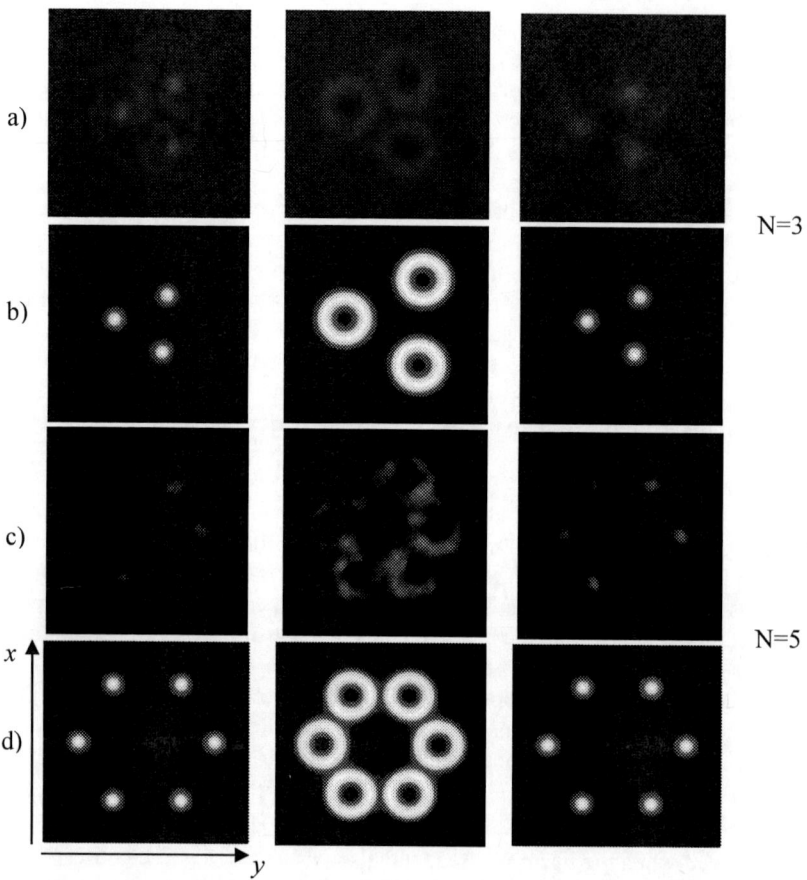

Figure 1 (a–d). Experimental **(a, c)** intensity distributions and theoretical calculation **(b, d)** in the formed array with N=3 and N=5 singular beams respectively

3. Vortex beam structures

Polarization of light can also affect the quality of capture, so the ability to control the polarization states of optical beams gives an additional degree of freedom in manipulating the trapped particles in space, which will allow more controlled change in the positions of trapped particles in the array.

Thus, at the present time, singular beams with a spatial variation of the polarization over the beam cross section are of particular interest. The simplest examples of such beams are radially and azimuthally polarized beams. Recently, a number of studies devoted to the generation of azimuthally and radially polarized beams have been presented in connection with the potential for wide application in microscopy, laser lithography, data storage, and capture devices. As it was shown, a single circularly polarized optical vortex with opposite signs of topological charge and polarization propagating along the optical axis of the crystal is a superposition of radially and azimuthally polarized beams [3]. In this work we form arrays with spatial variation of polarization similar to the previous case with one difference - at the entrance to the crystal an array of optical vortices is generated on the hologram.

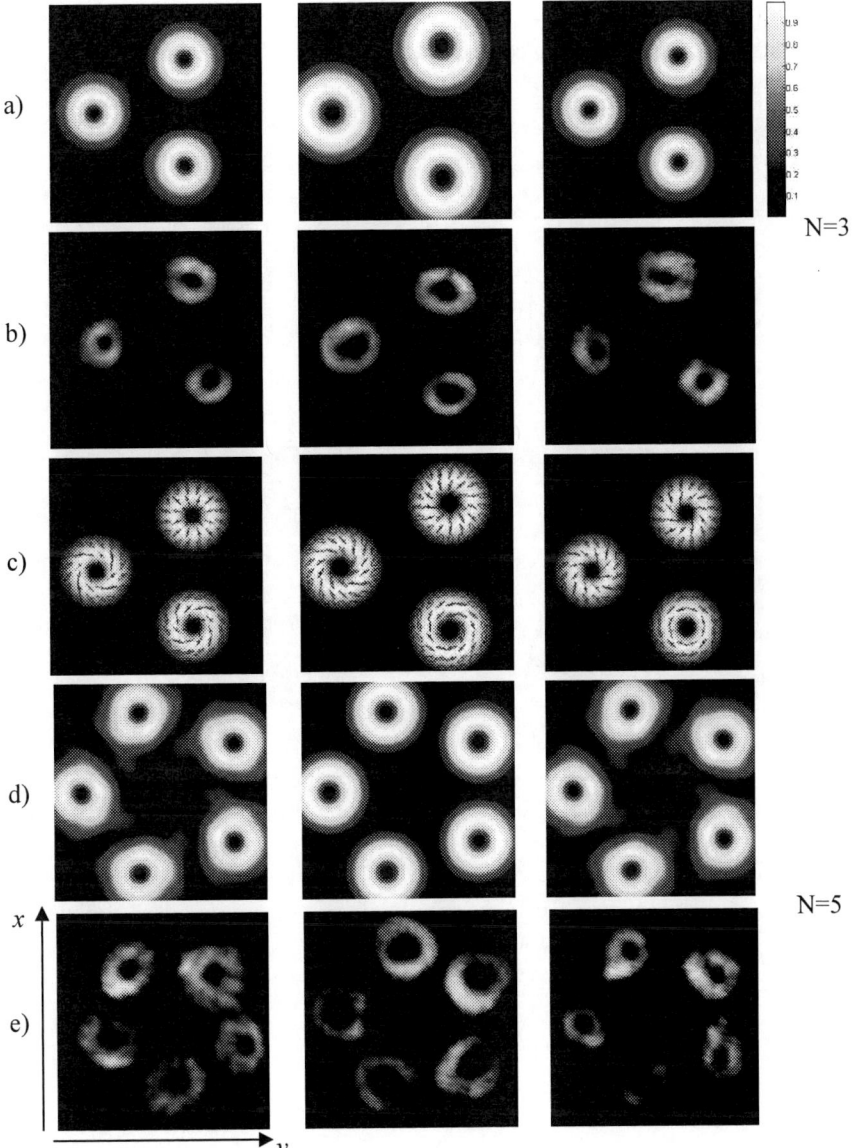

Figure 2 (a–e). Theoretical calculation of intensity distribution **(a, d)** and polarization **(c)**; **(b)** corresponds to experimental intensity patterns in the waist region of a vector beams array (N=3 and N=5 respectively) formed after uniaxial crystal

Analysis of patterns of polarization distribution shows the following:

– with an increase of the number of beams in the array the effect of spherical aberration increases;
– when focusing an array with N>3 beams, the beams begin to overlap, destructive interference occurs between them, resulting in the polarization distribution patterns lose pure radial and azimuthal distributions.

4. Structured polariton flows

In a dielectric medium with cubic nonlinearity electromagnetic wave with a frequency lying in spectral gap can generate a plane polariton wave, which as a result of the stability transforms into a spatial soliton. In such medium the polariton flux representing a wave with a plane wave front with linear polarization stratifies (splits into plane flows) as a result of self-focusing.

Polariton flow with a plane wave front $\dfrac{d^2 e}{d\bar{\eta}^2} + \alpha_1 e + \alpha_3 e^3 = 0$, where $\bar{\eta} = (x + iy)$, α_1 and α_3 – medium coefficients.

We obtain the general solution for the vector polariton wave front $e(\bar{\zeta}, \bar{\eta})$ depending on the right and left spirality coordinates $\left(\bar{\bar{\xi}}, \bar{\bar{\eta}} \right)$, where $\bar{\bar{\xi}} = \sqrt{|\alpha_1|}\bar{\xi}$, $\bar{\bar{\eta}} = \sqrt{|\alpha_1|}\bar{\eta}$,

$$e\left(\bar{\bar{\xi}}, \bar{\bar{\eta}} \right) = i \sqrt{\frac{|\alpha_1|}{|\alpha_3|}} \tanh\left\{ C_1 + C_2 \bar{\bar{\xi}} + \frac{i}{2}\left[C_2(i+1) - \left(i - 1 - i2C_2^2\right)^{1/2} \right]\bar{\bar{\eta}} \right\}.$$

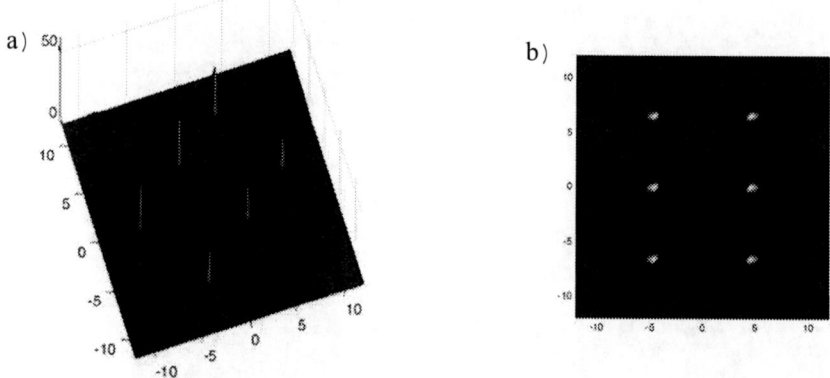

Figure 3 (a,b). Number of polariton flows of polariton wave with initial circular polarization

The constants C_1 and C_2 are determined by the boundary conditions at $\bar{\bar{\xi}} = 0$ and $\bar{\eta} = 0$ for the polariton wave. For example, if the value of envelope at longitudinal axis z is $e(0,0) = \mathrm{Re}\left[i\sqrt{|\alpha_1|/|\alpha_3|}\tanh(C_1) \right] = 0$, it allows to determine the constant $C_1 = 0$ and $C_2 = 1$ and

$$e\left(\bar{\bar{\xi}}, \bar{\bar{\eta}} \right) = i \sqrt{\frac{|\alpha_1|}{|\alpha_3|}} \tanh\left\{ \bar{\bar{\xi}} + \frac{1}{2}\left[i - 1 + \sqrt{i+1} \right]\bar{\bar{\eta}} \right\}.$$

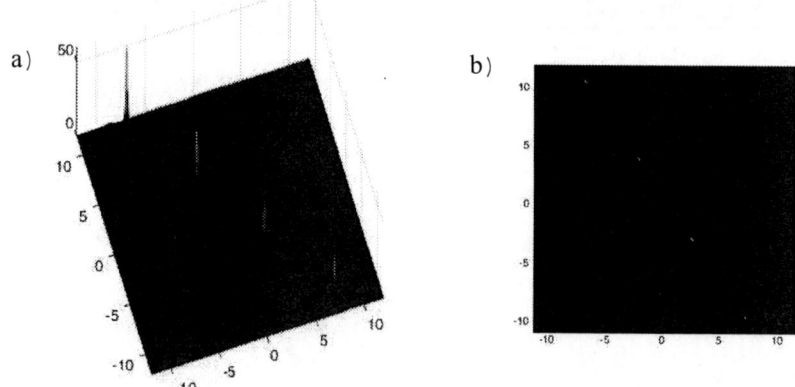

Figure 4 (a,b). Array of polariton flows at the scalar polariton wave with elliptical polarization

The weak intensity waves with the frequency lying in a spectrum gap don't propagate through the linear medium. On the contrary the waves of new spectrum branches appear in the gap as the nonlinear periodic cnoidal waves or spatial solitons, depending on the sign of nonlinear susceptibility and value of wave perturbation. These nonlinear waves propagate through the medium. The given effects are similar to the nonlinear (self-induced) transparency for power wave or pulse with carrier frequency close to resonance transition of the medium atoms. This effect can be used for design and creation of the nonlinear filter that transforms the harmonic wave to nonlinear cnoidal wave or spatial soliton.

Acknowledgments

This work was supported by the Russian Foundation for Basic Research and the government of the region of the Russian Federation grant № 17-42-92020 and partially supported by the V.I. Vernadsky Crimean Federal University Development Program for 2015 – 2024.

References

[1] Arlt J and Padgett M 2000 *Opt. Lett.* **25 (4)** 191

[2] Alpman C, Esseling M, Rose P, Denz C 2012 *Applied Physics Letters* **100 (11)** 111101

[3] Fadeyeva T, Shvedov V, Shostka N, Alexeyev C and Volyar A 2010 *Optics Letters* **35 (22)** 3787-3789

[4] Karpeev S, 2016 *Computer optics* **40 (4)** 583

[5] Ahluwalia B, Cheong W, Yuan C, Zhang L, Tao S, Bu, Wang H 2006 *Optics Letters* **31** 987.

[6] Shvedov V, Davoyan A, Hnatovsky C, Engheta N and Krolikowski W 2014 *Nature Photonics* **8** 846

[7] Shvedov V, Hnatovsky C, Shostka N and Krolikowski W 2011 *J. Opt. Soc. Am. B* **30** 1

[8] Gong L, Liu W, Zhao Q, Ren Y, Qiu X, Zhong M, Li Y 2016 *Scientific Reports* **6** 29001

[9] Vickers J, Burch M, Vyas R, Singh S 2008 *J. Opt. Soc. Am. A* **25 (3)** 823

[10] Desyatnikov Anton S, Shvedov V G, Rode Andrei V, Krolikowski Wieslaw and Kivshar Yuri S 2009 Optics Express **17** (10) 8201 – 8211

[11] Shostka N, Ivanov M, Shostka V 2016 *Technical Physics Letters* **42 (9)** 944

[12] Dzedolik I, Karakchieva O 2013 *Applied Optics* **52 (13)** 3073

[13] Dzedolik I, Karakchieva O 2013 *J. Opt. Soc. Am. B.* **30 (4)** 843

[14] Shvedov V G, Hnatovsky C, Shostka N, Krolikowski Wieslaw 2013 *JOSA B* **30** 1 – 6

[15] Izdebskaya Ya, Fadeyeva T, Shvedov V and Volyar A 2006 *Opt. Lett.* **31**(17) 2523 – 2525

[16] Ivanov M O and Shostka N V 2016 *J. Opt.* **18** 075603

[17] Fadeeva T A, Shvedov V G, Izdebskaya Ya V, Volyar A V, Desyatnikov A S, Kivshar Yu S, Krolikowski Wieslaw 2010 *Optics Express* **18** 10848

Tamm magnetophotonic structures with Bi-substituted iron garnet layers at oblique incidence

T Mikhailova, S Tomilin, S Lyashko, A Shaposhnikov, A Prokopov, A Karavainikov, A Bokova and V Berzhansky

Institute of Physics and Technology, V. I. Vernadsky Crimean Federal University, Simferopol 295007, Russia

Abstract. Optical and magnetooptical spectra of three magnetophotonic crystals coated with Au layer were investigated theoretically at oblique incidence of light. The properties of proposed structures were studied experimentally at normal incidence. The two magnetophotonic crystals consisting of a nonmagnetic dielectric photonic crystal and top layers of Bi-substituted iron garnet (Bi:IG) $-SiO_2-Au$ show a blue shift of spectra and Tamm plasmon-polaritons resonances with increasing angle of incidence without spectral splitting of resonances for TE and TM incident light. The coupling between cavity mode and Tamm plasmon-polaritons in the spectra of microcavity with Bi:IG and the top SiO_2-Au layer gives spectral splitting of resonances for different polarizations of incident light.

1. Introduction

The present work has focused on the investigation of magnetophotonic structures supporting optical surface states, so-called Tamm plasmon-polaritons. Tamm plasmon-polaritons arise in the magnetophotonic crystals interfaced with a uniform medium with the negative dielectric constant, such as a metallic layer of gold or silver [1-3]. The structures are similar to the resonant cavity. The electromagnetic field are concentrated near the interface between magnetophotonic crystal and metallic layer. This state reveals itself as a narrow peak in the transmittance spectra and is responsible for the enhancement of the magnetooptical effects for the corresponding wavelength.

2. Experimental technique and measured optical and magnetooptical spectra

For consideration at oblique incidence the three Tamm magnetophotonic structures were proposed

No. 1 – GGG / $[TiO_2/SiO_2]^7$ / M1 / SiO_2 / Au,

No. 2 – GGG / $[TiO_2/SiO_2]^7$ / M1 / M2 / SiO_2 / Au,

No. 3 – GGG / $[TiO_2/SiO_2]^4$ / M1 / M3 / $[SiO_2/TiO_2]^4$ / SiO_2 / Au,

where GGG is substrate of gadolinium gallium garnet with crystallographic orientation of (111). M1, M2 and M3 are Bi:IG of compositions of $Bi_{1.0}Y_{0.5}Gd_{1.5}Fe_{4.2}Al_{0.8}O_{12}$, $Bi_{1.5}Gd_{1.5}Fe_{4.5}Al_{0.5}O_{12}$ and $Bi_{2.8}Y_{0.2}Fe_5O_{12}$, respectively. All three magnetophotonic crystals were modelled, synthesized and investigated at normal incidence. Dielectric Bragg mirrors were synthesized by electron beam evaporation on hot (400 °C) substrate. The thickness of layers was optically controlled during deposition. Iron garnet layers were fabricated by reactive ion beam sputtering of corresponding ceramic targets in argon-oxygen mixture and crystallized in the annealing process at the air. Au thin films were deposited by thermal evaporation in vacuum. The description of synthesis techniques in detailed can be found in [4,5]. Garnet and Au thicknesses were determined by MII-4 microinterferometer. Investigation of transmittance was carried out using an automated

---

Content from this work may be used under the terms of the Creative Commons Attribution 3.0 licence. Any further distribution of this work must maintain attribution to the author(s) and the title of the work, journal citation and DOI.

Published under licence by IOP Publishing Ltd

spectrophotometer KFK-3. Measurements of Faraday rotation were performed using hand-made computer-control spectropolarimeter by compensation method in saturation fields. The Figure 1 shows considered magnetophotonic crystals. The Figures 2 and 3 shows measured and fitted transmittance and Faraday rotation spectra of structure configurations. The following thicknesses were used in calculations for structures:

No. 1 – h_{TiO2} = 73 nm and h_{SiO2} = 115 nm for Bragg mirror layers, h_{M1} = 108 nm and h_{bSiO2} = 140 nm for M1 and buffer SiO$_2$ layers, h_{Au} = 28.3 nm for top Au layer;

No. 2 – h_{TiO2} = 73 nm and h_{SiO2} = 115 nm for Bragg mirror layers, h_{M1} = 55 nm, h_{M2} = 178 nm and h_{bSiO2} = 80 nm for M1, M2 and buffer SiO$_2$ layers, h_{Au} = 28.3 nm for top Au layer;

No. 3 – h_{TiO2} = 79 nm and h_{SiO2} = 117 nm for Bragg mirror layers, h_{M1} = 66 nm, h_{M3} = 166 nm and h_{bSiO2} = 185 nm (or h_{bSiO2} = 220 nm) for M1, M3 and buffer SiO$_2$ layers; h_{Au} = 40 nm for top Au layer. We considered two configurations of structure No. 3 with different h_{bSiO2}.

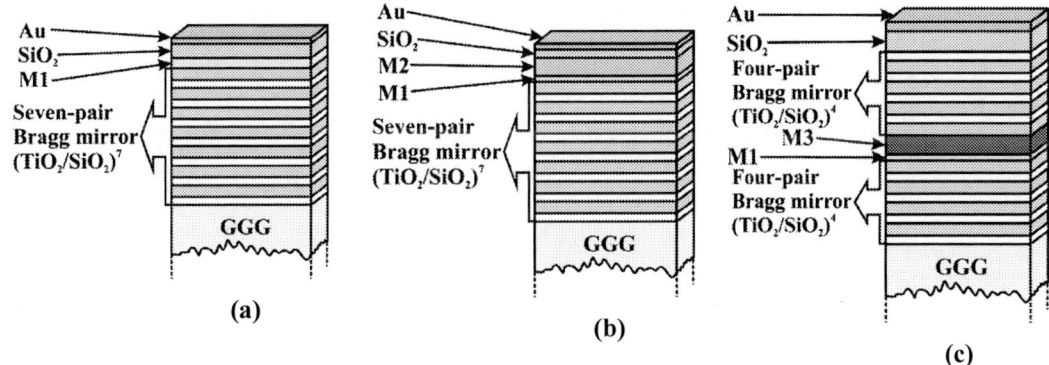

Figure 1(a, b, c). Schematic diagrams of synthesized and considered configurations of Tamm magnetophotonic structures No. 1 **(a)**, No. 2 **(b)** and No. 3 **(c)**.

A one resonance on Tamm optical state is formed inside photonic band gab at the wavelength of 666 nm and 648 nm for the structures No. 1 and No. 2, correspondingly. At the spectra of transmittance and Faraday rotation angle of structure No. 3, the hybrid state of Tamm and microcavity modes manifests itself in the form of two resonances that are anticrossing and cannot be combined at the same wavelength with the change of buffer SiO$_2$ thickness. The value of the mode repulsion is 20 nm for h_{bSiO2} = 185 nm – in this configuration resonances are closest to each other. For this configuration with the strongest hybridization, the microcavity mode has a 1.5-fold smaller Faraday rotation than the Faraday rotation of configuration with a low degree of hybridization (for example, h_{bSiO2} = 220 nm). Since the double-layer magnetic film of the structures No. 2 and No. 3 have compositions of iron garnets with a significant substitution of bismuth, they have a large values of Faraday rotation angle than the crystal No. 1.

3. Modelling

The 4×4 transfer matrix method were used to consider the optical and magnetooptical spectra of proposed structures at oblique incidence of light [6]. The realistic components of permittivity tensors of layers were taken into account. We changed the angle of incidence of light in the range from 0 to 60° and the polarization of the incident wave (TM and TE) in the calculations. The calculations were performed for structures without a substrate. The incident end medium and the exit end medium are the same. The medium has a refractive index 1. Figures 4 and 5 show the simulated spectra of the structures as a function of the angle of incidence. The transmission of structures No. 1 and No. 2 with an increase in the angle of incidence changes in a similar way – blue shift of resonance (spectrum) occurs, the intensity of TM mode increases and the intensity of TE mode decreases (Figure 4).

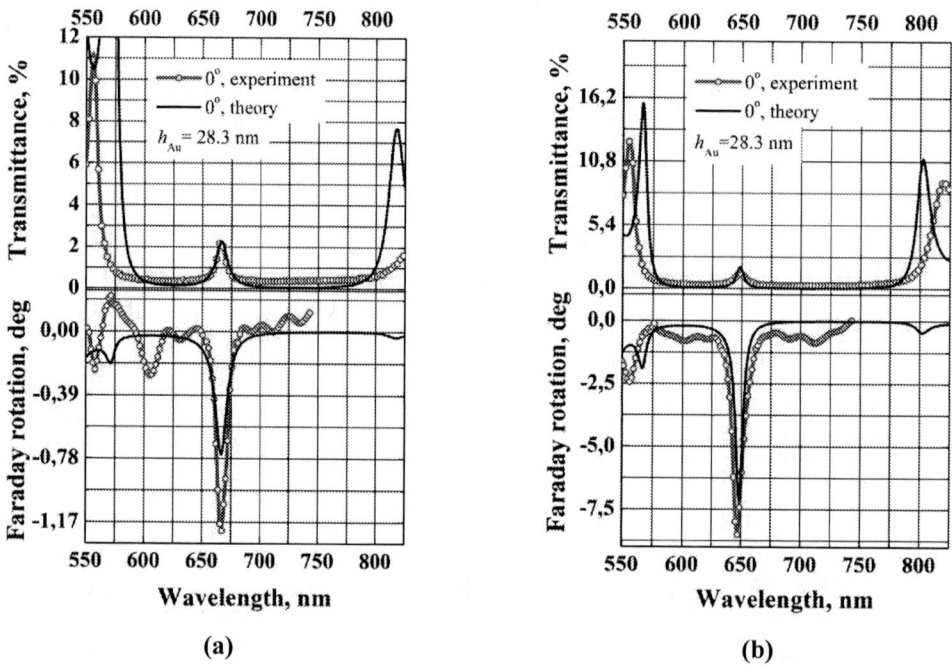

Figure 2(a, b). Measured (red symbols and lines) and fitted (black lines) transmittance and Faraday rotation spectra of structure No. 1 **(a)** and No. 2 **(b)** for TM-polarized light at normal incidence.

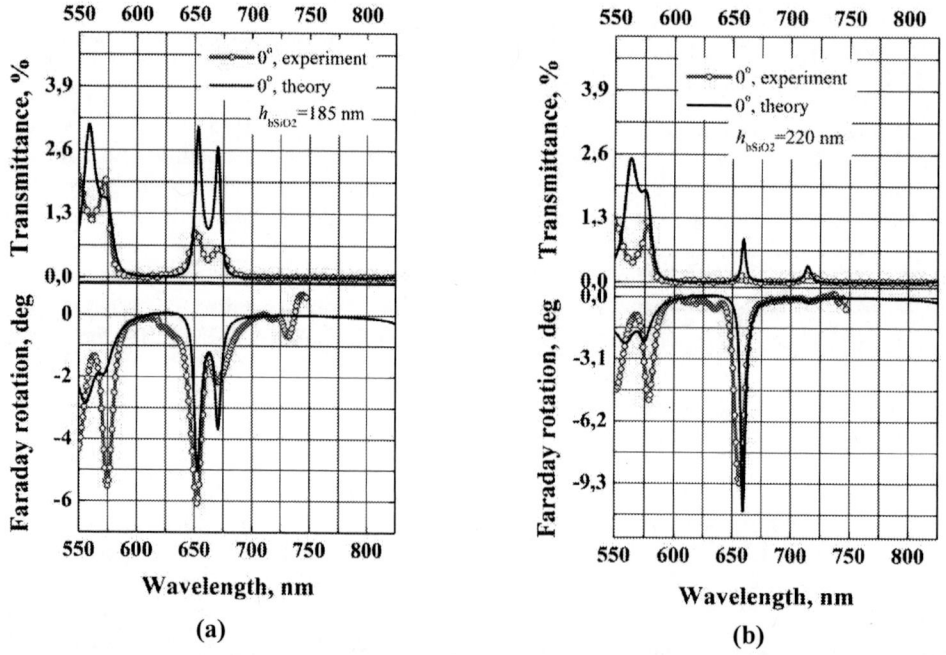

Figure 3(a, b). Measured (red symbols and lines) and fitted (black lines) transmittance and Faraday rotation spectra of structure No. 3 with thicknesses of buffer SiO_2 layer h_{bSiO2} of 185 nm **(a)** and 220 nm **(b)** for TM-polarized light at normal incidence. The value of transmittance are reduced sixfold for calculated spectra.

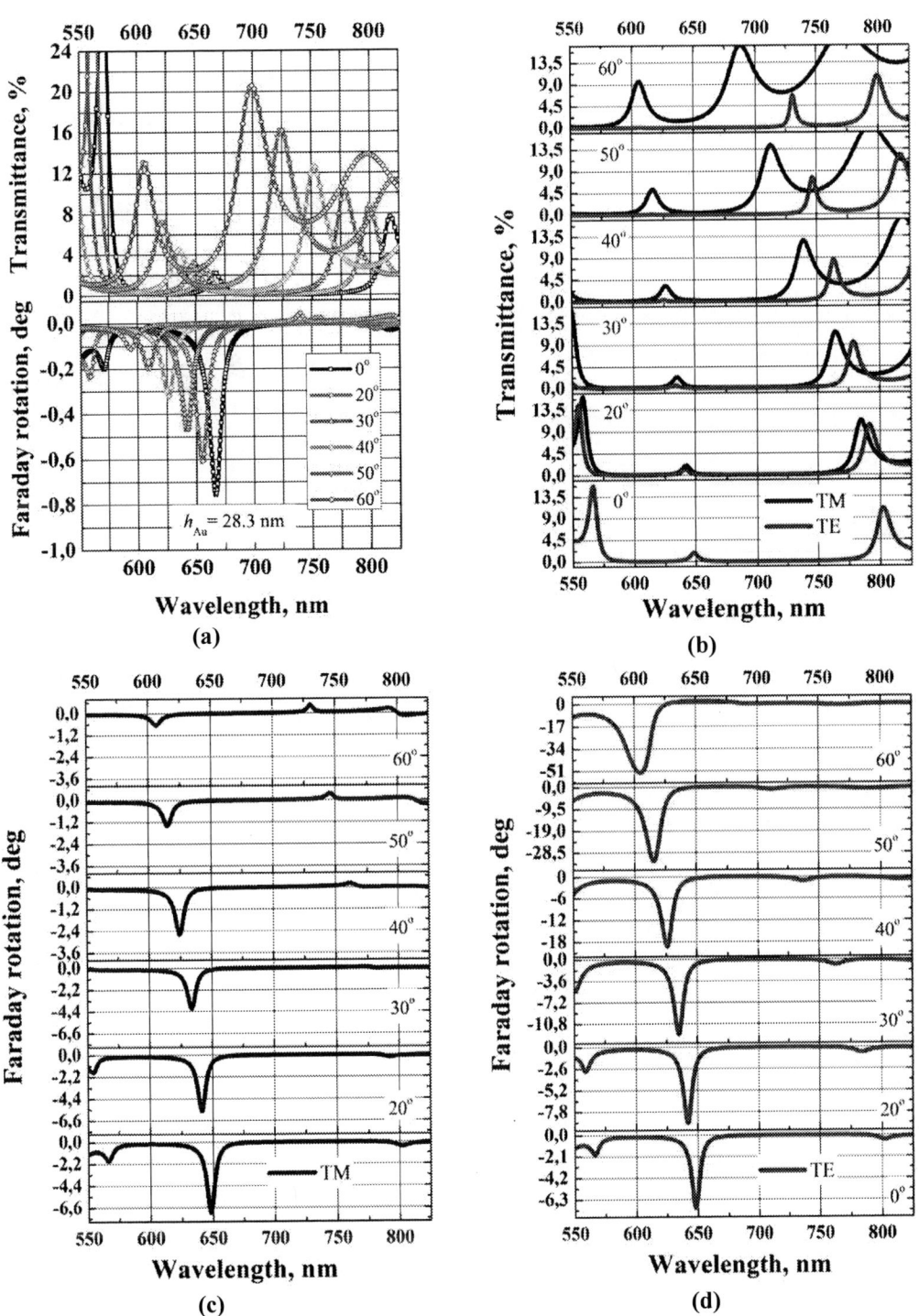

Figure 4(a, b, c, d). Calculated spectra as a function of the angle of incidence (0°, 20°, 30°, 40°, 50°, 60°) : transmittance and Faraday rotation spectra of structure No. 1 for TM incident light **(a)**; transmittance **(b)** and Faraday rotation **(c, d)** spectra of structure No. 2 for TM (black line) and TE (purple line) incident light.

SPBOPEN 2018   IOP Publishing
IOP Conf. Series: Journal of Physics: Conf. Series **1124** (2018) 051062   doi:10.1088/1742-6596/1124/5/051062

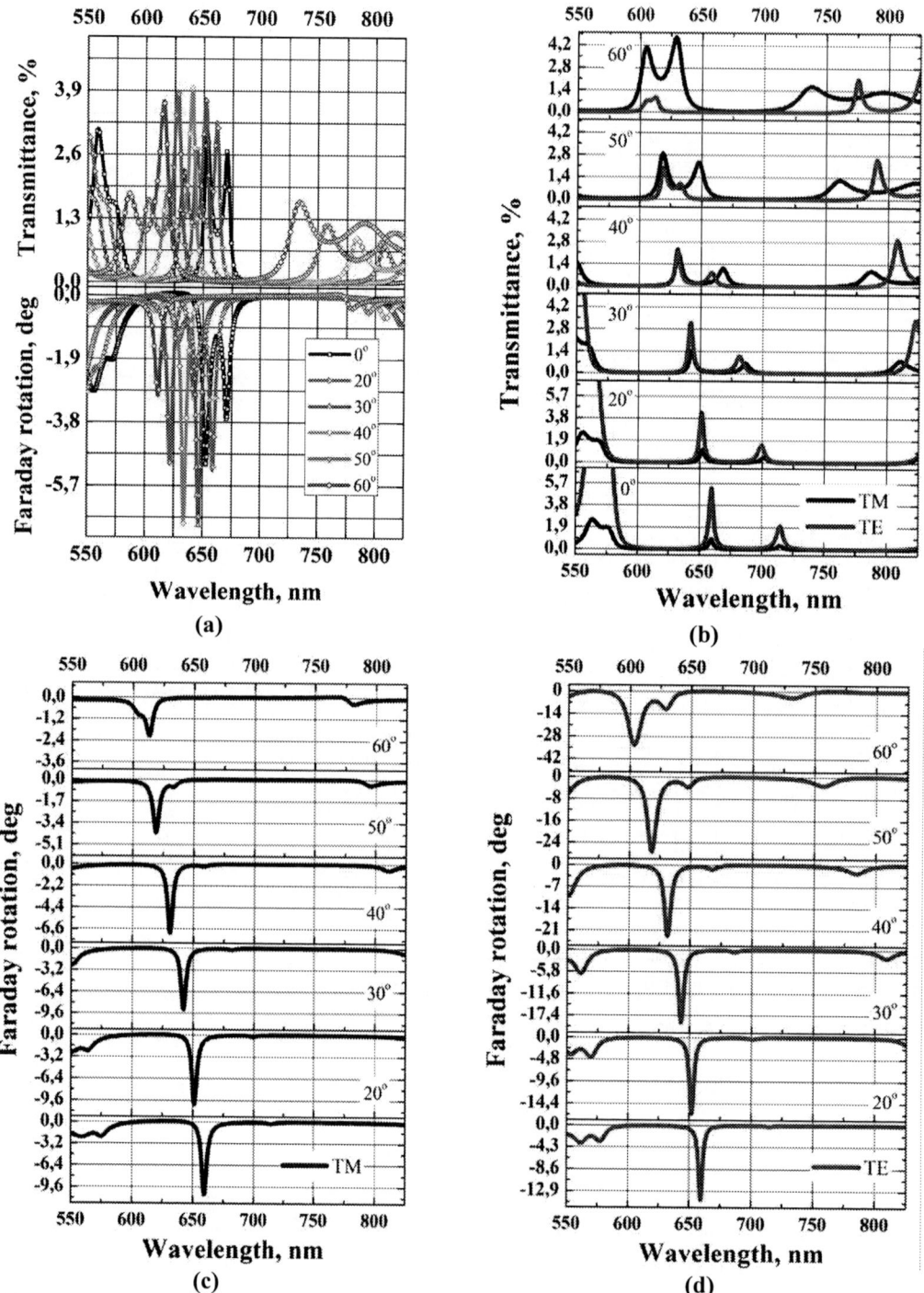

Figure 5(a, b, c, d). Calculated spectra as a function of the angle of incidence (0°, 20°, 30°, 40°, 50°, 60°): transmittance and Faraday rotation spectra of structure No. 3 with h_{bSiO2} = 185 nm for TM incident light **(a)**; transmittance **(b)** and Faraday rotation **(c, d)** spectra of structure No. 3 with h_{bSiO2} = 220 nm for TM (black line) and TE (purple line) incident light. The value of transmittance are reduced sixfold.

The behavior of spectra of Faraday rotation angle is similarly for two structures. For structure No. 1 and TM incident light transmittance increases from 2.1 to 13%, and Faraday rotation decreases from -0.75 to -0.11 deg (Figure 4, a). For structure No. 2 and TM incident light transmission increases from 1.9 to 9 %, and Faraday rotation decreases from -7.06 to -0.73 deg (Figure 4, b, c). For TE mode transmittance decreases to 0.3 %, Faraday rotation increases to -51.5 deg (Figure 4, b, d). The shift for TE and TM modes is the same and amounts to the structure No. 1 61 nm and to the structure No. 2 43 nm. The peak positions for TM and TE modes are the same for each angle of incidence.

Completely different behavior with spectral splitting of resonances for TM and TE incident light is observed for the structure No. 3 (Figure 5). The spectral shift and amplitude of resonances of hybrid state shows the dependence on the configuration of structure (in our case on the buffer SiO_2 thickness). We showed the results only for TM incident wave for $h_{bSiO2} = 185$ nm. In the transmission spectra, Tamm peak is shifted by 85 nm from 670 nm to 585 nm, microcavity peak is shifted by 36 nm from 653 nm to 617 nm (Figure 5, a). Positions of resonances for transmission spectrum and the spectrum of Faraday rotation are differ. As the angle of incidence increases, the spectral splitting for TE and TM modes increases. As a consequence, the difference between the wavelengths of transmission and Faraday rotation peaks increases. This behavior is well illustrated by data for configuration with $h_{bSiO2} = 220$ nm. Let us consider the behavior of spectra for TM incident light. Transmission peaks of Tamm and microcavity modes shift respectively by 110 nm (from 714 to 604 nm) and 30 nm (from 659 to 629 nm). Faraday rotation peaks of Tamm and microcavity modes shift respectively by 109 nm (from 714 to 605 nm) and 46 nm (from 659 to 613 nm). For TE incident light the shift values and transmission peaks of Tamm and microcavity modes are respectively 109 nm (from 714 to 605 nm) and 46 nm (from 659 to 613 nm). The shift values of Faraday rotation peaks of Tamm and microcavity modes are respectively 84 nm (from 714 to 630 nm) and 56 nm (from 659 to 603 nm). So, the coupled state with the strongest hybridization (convergence) of Tamm and microcavity modes is most exposed to spectral splitting.

4. Conclusions

The obtained results can be used to design of magnetooptical devices, promised to further application in optical information process, photonics, spintronics and sensors.

Acknowledgments

The authors acknowledge support by the RF Ministry of Education and Science in the framework of the state task (project no. 3.7126.2017/8.9).

References

[1] Inoue M, Baryshev A V, Goto T, Baek S M, Mito S, Takagi H, and Lim P B 2013 Magnetophotonic Crystals: Experimental Realization and Applications Magnetophotonics eds. M. Inoue et al. (Springer-Verlag Berlin Heidelberg) chapter 7 pp 163–190
[2] Vinogradov A P, Dorofeenko A V, Erokhin S G, Inoue M, Lisyansky A A, Merzlikin A M and Granovsky A B 2006 *Phys. Rev. B* **74** 045128
[3] Kaliteevski M, Iorsh I, Brand S, Abram R A, Chamberlain J M, Kavokin A V, and Shelykh I A 2007 *Phys. Rev. B* **76** 165415
[4] Berzhansky V N, Shaposhnikov A N, Prokopov A R, Karavainikov A V, Mikhailova T V, Lukienko I N, Kharchenko Yu N, Golub V O, Salyuk O Yu, Belotelov V I 2016 *Journal of Experimental and Theoretical Physics* **123** 744
[5] Tomilin S V, Yanovsky A S 2013 *J. of Nano- and Electronic Physics* **5** 03014
[6] Yin C P, Wang T B and Wang H Z 2012 Eur. Phys. J. B **85** 104

SPBOPEN 2018

IOP Publishing

IOP Conf. Series: Journal of Physics: Conf. Series **1124** (2018) 051063   doi:10.1088/1742-6596/1124/5/051063

Epitaxial films of garnet ferrite with anisotropy "easy plane" for magneto-optical eddy current flaw detection

N Lugovskoy, V Berzhansky, D Glechik, A Prokopov

Institute of Physics and Technology, V. I. Vernadsky Crimean Federal University, Simferopol 295007, Russia

Abstract. The possibilities of magnetooptical sensors based on ferrite garnet films with various types of magnetic anisotropy are studied in the visualization of eddy current images of linear defects in the test object from aluminum alloys. Potentially high efficiency of sensors based on an easy-plane anisotropy films is shown.

1. Introduction

In the modern world with new technologies, the diagnosis and detection of all the smaller defects of various designs become more important. In some areas where structures are under heavy loads, different pressures and high speeds, it is important not to allow the use of an industrial design, even with micron damage. To determine such defects, it is best to use the eddy current (EC) method [1-2]. An important factor in the development of this method can be considered the creation of new composite materials, their implementation in many important areas of industry and the fact that most of these materials are conductive and can therefore be subjected to EC flaw detection method. But this method has a number of disadvantages. The possibility of applying magneto-optical (MO) sensors to the eddy current method will eliminate some of them and will allow adding to the EC method the observation of defects in real time, visualization of the shape and topology of the defect, the implementation of faster defectometry, which will shorten the total time of investigations.

2. Installation and objects of research

The method of magnetooptical eddy current (MOEC) flaw detection is based on the reaction of the magnetic structure of a thin-film sensor to the magnetic fields of eddy currents. The registration of these changes is carried out using the Faraday Effect. As indicators of the magnetic field, transparent epitaxial films of ferrites-garnets (FGEF) with a given type of magnetic anisotropy are used.

The sensitivity of this method and the contrast of the resulting image essentially depend on the film parameters. These parameters include the type of magnetic anisotropy of the film ("easy axis" or "easy plane"), the period of the domain structure (DS), the saturation field H_s, and the value of the Faraday specific rotation. The frequency f and the amplitude H_i of the alternating magnetic field exciting the eddy currents, as well as the value of the bias field H_- also affect the sensitivity of the MOEC method [3-5]. The purpose of this paper is to compare the possibilities of using films with the "easy plane" (EP) and "easy axis" (EA) anisotropy on the example of MOEC visualization of linear defects in non-magnetic material.

Films, used as sensors, with the general composition $(Bi,Re)_3(Fe,Me)_5O_{12}$, where Re - Y, Lu, Pr, Me - Al, Ga, Sc were grown by liquid phase epitaxy (LPE) on paramagnetic GGG (gadolinium gallium garnet) substrates. The films had the following characteristics: EA films - thickness h = 7,2 μm, period DS 2ω = 17 μm, saturation field H_s = 77 Oe; EP films: thickness h = 12 μm, saturation field H_s = 235

Content from this work may be used under the terms of the Creative Commons Attribution 3.0 licence. Any further distribution of this work must maintain attribution to the author(s) and the title of the work, journal citation and DOI.
Published under licence by IOP Publishing Ltd

SPBOPEN 2018   IOP Publishing

IOP Conf. Series: Journal of Physics: Conf. Series **1124** (2018) 051063   doi:10.1088/1742-6596/1124/5/051063

Oe. For the study test samples of linear defects of the "through-slot" type in aluminum with the width of 40 μm were taken. The study was conducted under the same conditions in the frequency range **f** from 8 to 80 kHz. The intensity of the alternating field of the inductor $\mathbf{H_i}$ could vary from 2 to 400 Oe, and the bias field **H_** from 0 to 40 Oe.

3. Experiment and results
Figures 1-2 show MOEC comparative images obtained using EA and EP films at 15 and 30 kHz. The results of analysis of these images for epitaxial films with anisotropy of the easy axis are presented below.

- At low intensity of the alternating field $\mathbf{H_i}$ linear defects are visualized by light bands (Fig. 1a, d). With an increase of the field **H~**, the intensity of the bright bands increases. For fields larger than some critical value $\mathbf{H_{\sim cr}}$ = 14 Oe at **f** = 15 kHz, the defect is displayed in a dark band, and the image acquires a clear binary structure (Fig. 1b, c, e, f).

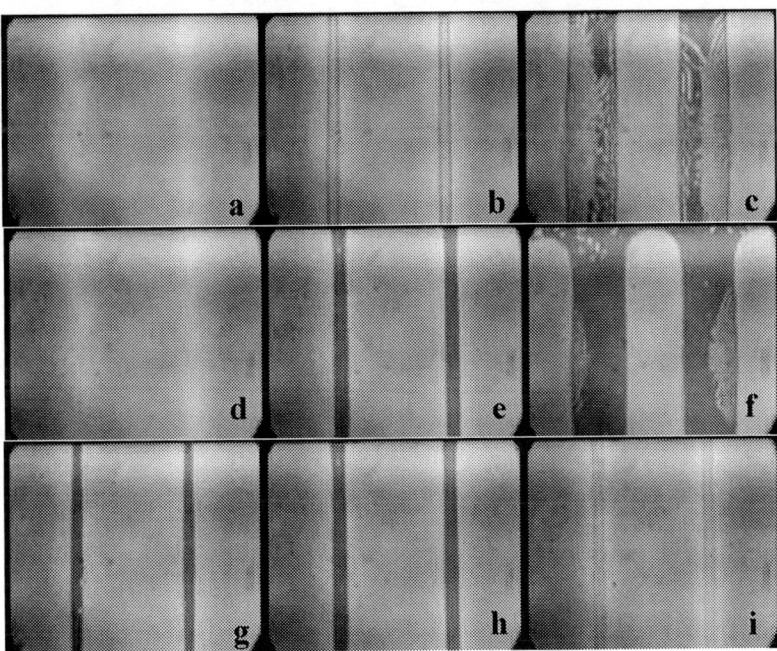

Figure 1. Effect of the alternating field **H~** (a, b, c), the bias field **H_**(a-c and d-f), frequency f (g, h, i) on the detection efficiency of a 40 μm slit in Al sample by means of the EA film MO sensor: **f** = 30 kHz, **H~** = 12 (a, d), 16 (b, e), 32 Oe (c, f), **H_** = 2 (a, b, c) and 6 Oe (d, e, f); **f** = 15 (g), 30 (h), 60 (i) kHz , **H~** = 16 Oe, **H_** = 6 Oe.

- An increase in the amplitude of the excitation field **H~** leads to an increase in the width of the binary image of the MO. Thus, when $\mathbf{H_i}$ increases by a factor of 2, the width of the magneto-optical image increases by a factor of 6 (Fig. 1e, f). This effect indicates a potential possibility of detecting defects much smaller than 40 μm.
- When the frequency increases, fragments of the labyrinthine domain structure appear in the dark band, complicating the binary image and reducing the clarity of the magneto-optical image (Fig. 1 g - i).
- Increasing the bias field **H_** leads to partial or complete suppression of the structure of the labyrinth domain and restoration of a clear binary image (Fig. 1b, e; 1c, f).

845

The results of visualization for films with anisotropy of the easy axis:
- At any amplitudes of the alternating field linear defects are displayed by light bands. An increase in the amplitude of the alternating field leads to an increase in the contrast (brightness) and an increase in the bandwidth (Fig. 2 a-d).
- Increasing the frequency of the alternating field leads to an increase in contrast, increasing the sensitivity of the epitaxial sensor film (Fig. 2 e-g).
- Increasing the bias field does not affect the magneto-optical image of defects.

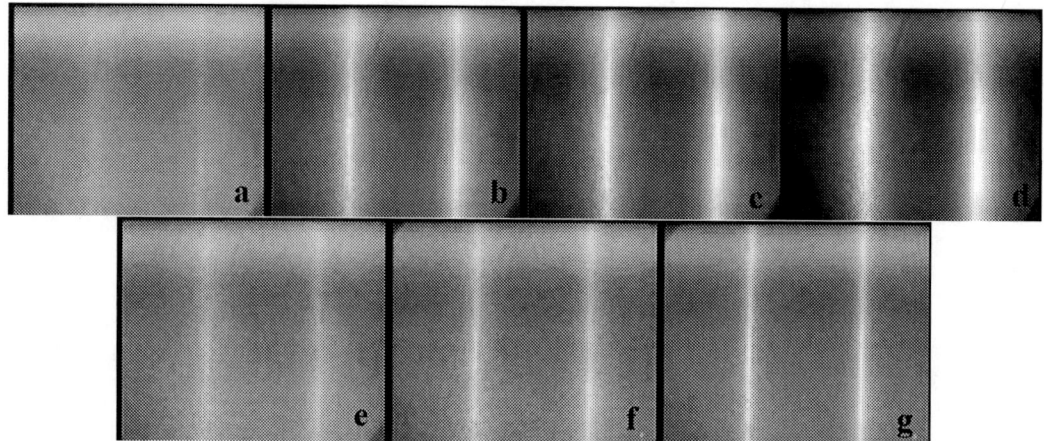

Figure 2. Effect of the alternating field $H_{\sim}$ (a-d) and frequency f (e-g) on the detection efficiency of a 40 μm slit in Al sample by means of the EP film MO sensor. Top row: f = 30 kHz, $H_{\sim}$ = 12 (a), 16 (b), 20 (c), 32 Oe (d); bottom row: f = 15 (e), 30 (f), 60 kHz (g) in the fields $H_{\sim}$ = 8 Oe, $H_{\_}$ = 2 Oe.

All these results indicate that both films with the easy axis and easy plane anisotropy can be used as sensors in magneto-optical eddy current flaw detection. Each of the types of anisotropy has its own peculiarities, advantages and disadvantages, which will enable them to solve various problems. For example, EA films show complex patterns with a domain structure and EP films are a simple binary picture, and therefore more optimal for transferring the image to a digital signal for further automation of the flaw detection process.

4. Calculations and modeling
The modeling and distribution of eddy currents and magnetic fields induced by them in the vicinity of a linear defect was carried out by the finite element method in the CAD system Comsol Multiphysics. Objects of modeling: the aluminum sample (plate) measuring 100 mm by 100 mm, 0.15 mm thick with a 40 μm defect (slit) in the center; inductor with diameter of 50 mm, a height of 25 mm, a wire diameter of $3 \cdot 10^{-6}$ m^2, the number of turns is 80, the conductivity of the wire is $6 \cdot 10^7$ sm/m, the current strength is up to 1 A.

Figure 3 shows the model distribution of the normal component of the vortex magnetic field in the vicinity of the defect for different frequencies of eddy currents excited by the inductor. It is seen that as the frequency increases, the maximum of B_z-component of the EC field increases, and the width of the distribution decreases. For comparison with the experimental data, it was necessary to present in the form of a digital dependence the brightness of the MO of the defect image recorded by the easy-plane MO sensor as a bright light line.

For processing and analysis, obtained experimental magneto-optical images, an image analysis program written in the Matlab package was compiled. It is based on the translation of a graphic file into a special matrix, which contains the brightness value of all pixels of the MO image. Further with this matrix it was possible to perform various mathematical operations. In particular, the brightest fragment of the image (corresponding to the defect) and the darkest (background) are selected to determine the image MO contrast, and the matrices of the selected regions are constructed, then the average value is found for these matrices. Contrast is defined as the ratio of these mean values.

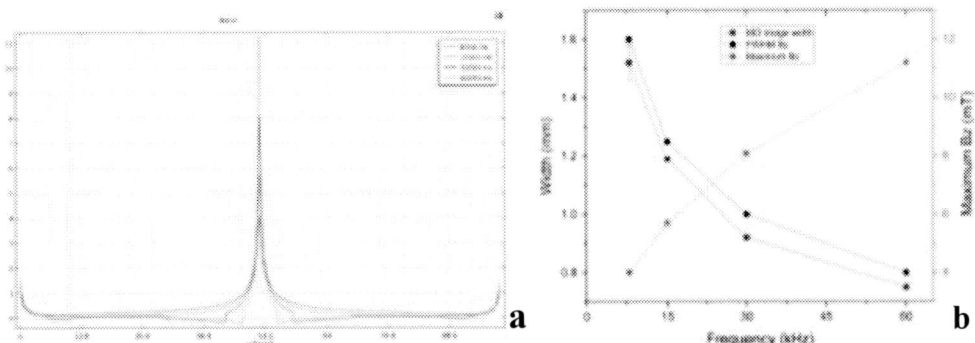

Figure 3. The frequency dependences of the B_z–component of the EC field distribution near the defect (theory) **(a)**, the maximum and the width of the B_z (FWHM) distribution (theory) and the width of the EP film MO image (experiment) **(b)**.

The electrodynamic simulation of the EC magnetic fields in the sample under study adequately describes the change in the brightness of the MO of the image of the sensor obtained by the EP film with increasing amplitude and frequency of the alternating field. For example, Fig. 3b shows the dependence of the half-width of the experimental MO image in the frequency range from 8 to 60 kHz and theoretically predicted values of this parameter found by mathematical processing. We can see that they are in good agreement. In accordance with theoretical calculations, an increase in the amplitude of the maximum with a frequency (Fig. 3b) also leads to an increase in the contrast of the magneto-optical image in the experiment (Fig. 2).

5. Conclusion
The type of magnetic anisotropy of FGEF determines the character of the MO image. The reason for this phenomenon lies in the various processes that occur when magnetization of such films is reversed. Thin magnetic films with an easy-plane anisotropy make it possible to obtain contrasting MOEC images of defects at a lower excitation field intensity, they do not require the imposition of large displacement fields and they do not cause visual interference when operating at high frequencies. Eddy current MO sensors, created based on these films, will have a sufficiently high sensitivity and low power consumption in the conduct of flaw detection of structures made of conductive materials.

Acknowledgments
The authors acknowledge support by the RF Ministry of Education and Science in the framework of the state task (project no. 3.7126.2017/8.9).

References
[1] García-Martín J, Gómez-Gil J, Vázquez-Sánchez E 2011 Sensors 11(3) 2525
[2] Janousek L, Capova K, Yusa N Miya K 2008 IEEE Trans. Magn. 44 1618
[3] Vishnevskii V, Berzhansky V, Lugovskoy N, Prokopov A, Pankov F 2015 Solid State Phenomena 230 273
[4] Berzhansky V, Filippov D, Lugovskoy N 2016 Phys. Procedia 82C 27
[5] Grechishkin R, Kustov M, Ilyashenko S, Gasanov S, Dumas-Bouchiat F and Dempsey N 2016 J. Appl. Phys. 120 174502

Electrically controlled spin polarization in suspended GaAs quantum point contacts

D A Pokhabov[1,2], A G Pogosov[1,2], E Yu Zhdanov[1,2], A A Shevyrin[1,2], A K Bakarov[1,2] and A A Shklyaev[1,2]

[1] Institute of Semiconductor Physics, SB RAS, Novosibirsk, Russia
[2] Novosibirsk State University, Novosibirsk, Russia

Abstract. We report on the observation of the lateral electric spin polarization effect in a suspended GaAs-based quantum point contact (QPC) separated from the substrate. The effect manifests itself in the experiment as the appearance of an additional half-integer plateaus at $0.5 \times 2e^2/h$ when the asymmetric voltage is applied to the side gates in zero magnetic field. The appearance of the plateaus has been attributed to the spin degeneracy lifting caused by the spin-orbit coupling associated with the lateral electric field in the asymmetrically biased QPC. We have experimentally demonstrated that, despite the relatively small g-factor in GaAs, the observation of the spin polarization in the GaAs-based QPCs becomes possible after the suspension due to the enhancement of the electron-electron interaction and the effect of the electric field guiding. These features are caused by a partial confinement of the electric field lines within a suspended semiconductor layer with a high dielectric constant.

1. Introduction

An opportunity to manipulate the electron spin by purely electric means without ferromagnetic materials and external magnetic field seems attractive for future spintronic devices [1, 2]. It has been shown earlier that electron spin in a quantum point contact (QPC) can be controlled via lateral spin-orbit coupling (LSOC) effect caused by lateral electric field applied to QPC by means of side gates [3, 4]. This effect has been steadily observed in QPCs fabricated from InAs-based materials, where the intrinsic g-factor is high. Previously, it has been concluded that the electron-electron interaction plays a crucial role in the observation of the effect. Namely, calculations show that in the absence of the electron-electron interaction, the effect is not observed [5]. On the other hand, if the electron-electron interaction is strong enough, this effect should be observable even in low g-factor materials, such as for example GaAs. However, in QPCs, fabricated from GaAs, there has been no clear experimental evidence of the effect. It should be noted that, since the spin coherence length in GaAs is much higher than that in InAs, it seems more interesting for practical applications to control the spin polarization in GaAs.

The lateral spin-orbit coupling [3], associated with the electric field, is described by the term in the Hamiltonian

$$\hat{H}_{\mathrm{LSOC}} = \gamma \boldsymbol{\sigma} \cdot \left[\mathbf{k} \times \mathbf{E} \right],$$

where γ is the lateral spin-orbit interaction constant, $\boldsymbol{\sigma}$ is Pauli vector, $\mathbf{k}$ is electron quasimomentum, and $\mathbf{E}$ is electric field in the QPC channel.

Content from this work may be used under the terms of the Creative Commons Attribution 3.0 licence. Any further distribution of this work must maintain attribution to the author(s) and the title of the work, journal citation and DOI.
Published under licence by IOP Publishing Ltd

In previous studies, we have shown that one possible way to enhance the electron-electron interaction is to suspend a nanostructure [6-8], i.e. to detach it from a substrate. Electron transport in such suspended nanostructures has features originating from additional mechanical degrees of freedom [9], poor heat coupling to the bulk [9], and the electron-electron interaction [6-8] being enhanced due to electric field confinement in a high-dielectric membrane. Many types of the semiconductor suspended nanostructures, including quantum dots [6, 7], ring electron interferometers [11], Hall bars [10], antidot lattices [8, 12, 13] have been implemented.

Suspended QPCs have been studied earlier [14-18]. The sensitivity of the QPC conductance to mechanical deformations has been successfully used to detect mechanical vibrations of resonators [17, 18]. In Ref. [14], integer quantization in suspended QPC was demonstrated, and it was also shown that the suspension leads to an increase in the spacing between the one-dimensional subbands. Earlier, we have found that the suspension of the QPCs results in the appearance of an additional anomalous $0.7 \times 2e^2/h$ conductance plateau, which also could be explained by the enhancement of the electron-electron interaction after suspension [15].

In the present paper, it is experimentally shown that QPCs fabricated from GaAs with low intrinsic spin-orbit coupling and being detached from substrate (suspended), demonstrate the LSOC effect and, thus, a pure electric spin manipulation.

2. Experimental details

The experimental samples were fabricated on the basis of GaAs/AlGaAs heterostructures with a two-dimensional electron gas (2DEG), grown by means of molecular-beam epitaxy. First, the 400 nm-thick $Al_{0.8}Ga_{0.2}As$ sacrificial layer were grown on the substrate. Then the 166 nm-thick GaAs/AlAs heterostructure was grown above the sacrificial layer. The heterostructure contains the 2DEG in the 13 nm-thick GaAs layer, which forms a symmetric square quantum well for electrons, and resides at a depth of 90 nm beneath the surface (see Figure 1). The electron mobility and the 2DEG density are 2×10^6 cm^2/Vs and $(5-7) \times 10^{11}$ cm^{-2} at 4.2 K, respectively. The lateral form of the samples was defined using electron-beam lithography followed by reactive ion etching. The samples were suspended, i.e. detached from the substrate, by means of selective wet etching of the sacrificial layer from under the created nanostructures in 1:100 HF water solution. QPCs represented adiabatic 2DEG constrictions with a lithographic width of 800-900 nm (see Figure 2). The QPCs were equipped with two side gates separated from the channel by 100-150 nm wide trenches. The measurements were carried out using the lock-in technique in the linear response regime with the alternating voltage of the magnitude 30 μV and the frequency 70 Hz at the temperature 4.2 K. Conductance G ($= I_{AC}/V_{AC}$) was measured as a function of DC voltages V_{G1} and V_{G2} applied to the gates, and the source-drain voltage V_{SD}. To create spin polarization in the channel, an asymmetric voltage was applied between the gates $\Delta V_G = V_{G1} - V_{G2}$. We have checked that leakage currents to both side gates are absent by means of dc measurements in the range from −10 V to +10 V.

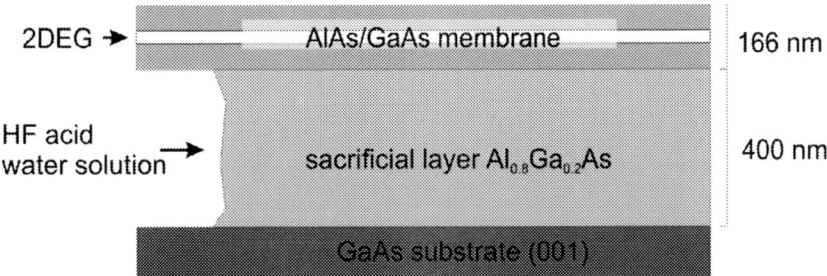

Figure 1 Schematic representations of the heterostructure with a sacrificial layer.

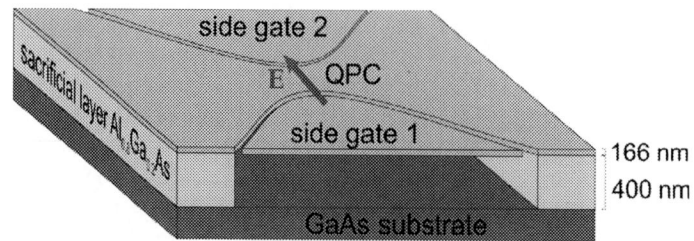

Figure 2 Schematic representations of a suspended QPC with applied lateral electric field.

3. Results and discussion

At zero bias between the side gates ($\Delta V_G = 0$) the conductance G of the QPC as a function of $V_G = V_{G1} = V_{G2}$ demonstrates conventional integer quantization [19, 20] in the units of $2e^2/h$, while non-integer plateaus are not observed. An anomalous plateau $0.5 \times 2e^2/h$ emerges when an asymmetric voltage $\Delta V_G \neq 0$ is applied to the side gates of the suspended QPC (see Figure 3 (b)). In non-suspended QPCs (the samples before the suspension) an asymmetric lateral bias $\Delta V_G \neq 0$ does not lead to an appearance of the 0.5-plateau (see Figure 3 (a)).

The appearance of the 0.5-plateau can be attributed to the spin degeneracy lifting caused by the spin-orbit coupling associated with the lateral electric field in the asymmetrically biased QPC. We believe that observation of spin polarization after the QPC suspension becomes possible due to the following two features of the suspended QPC compared to their non-suspended analogues: (i) enhanced electron-electron interaction, and (ii) amplified value of the lateral electric field in the QPC at given ΔV_G. Both of these features can be attributed to the separation of the QPC channel from the substrate, leading to a confinement of the electric field lines inside a suspended semiconductor layer with a high dielectric constant $\varepsilon_{GaAs} \approx 13$ (this can be compared with magnetic field lines confined and guided in a core material with high magnetic permeability). The enhancement of the electron-electron interaction can be also explained by the removal of a part of the polarizing medium from under the QPC channel, that attenuates the electron-electron coupling.

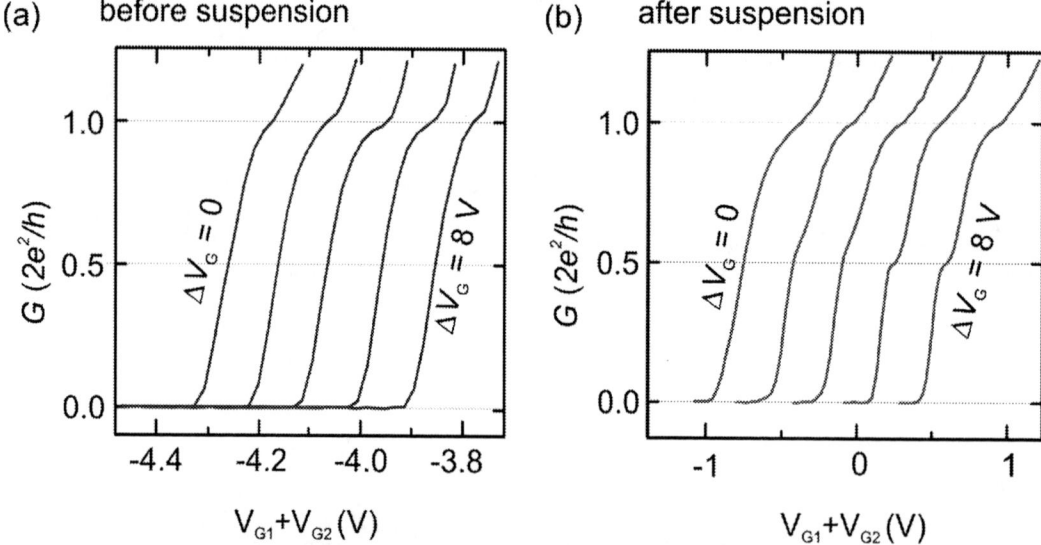

Figure 3 Conductance of the QPC before (a) and after (b) suspension. Different curves correspond to different ΔV_G ranging from 0 to 8 V in 2 V steps. Curves corresponding to different values of ΔV_G are shifted in the horizontal direction.

4. Conclusions

In conclusion, we have experimentally observed $0.5 \times 2e^2/h$ plateaus in QPCs based on low g-factor GaAs material in zero magnetic field. The 0.5-plateaus are observed only with an asymmetric bias $\Delta V_G \neq 0$, leading to a lateral electric field in the QPC channel. We associate the appearance of this feature with the spin polarization caused by LSOC-effect. The half-integer plateaus are observed only in the suspended QPCs, originating from the spin polarization that can be attributed to the enhancement of the electron-electron interaction in suspended QPCs.

Acknowledgements

The work is supported by Russian Foundation for Basic Research (RFBR) (Grant No 16-02-00579 — electron transport measurements), Russian Science Foundation (Grant No 18-72-10058 — experimental samples fabrication) and State Programme (Grant No 0306-2016-0015 — epitaxial growth of structures).

References

[1] Datta S and Das B 1990 *Appl. Phys. Lett.* **56** 665
[2] Chuang P, Ho S C, LSmith L W, Sfigakis F, Pepper M, Chen C H, Fan J C, Griffiths J P, Farrer I, Beere H E, Jones G A C, Ritchie D A and Chen T M 2015 *Nature Nanotechnology* **10** 35
[3] Debray P, Rahman S M S, Wan J, Newrock R S, Cahay M, Ngo A T, Ulloa S E, Herbert S T, Muhammad M 2009 *Nature Nanotechnology* **4** 759
[4] Karlsson H, Yakimenko I I and Berggren K-F 2018 *J. Phys.: Condens. Matter* **30** 215302
[5] Wan J, Cahay M, Debray P, and Newrock R S 2009 *Phys. Rev. B* **80** 155440
[6] Pogosov A G, Budantsev M V, Lavrov R A, Plotnikov A E, Bakarov A K, Toropov A I and Portal J C 2006 *JETP Lett.* **83** 122
[7] Pogosov A G, Budantsev M V, Shevyrin A A, Plotnikov A E, Bakarov A K and Toropov A I 2008 *JETP Lett.* **87** 150
[8] Zhdanov E Yu, Pogosov A G, Budantsev M V, Pokhabov D A and Bakarov A K 2017 *Semiconductors* **51** 8
[9] Shevyrin A A, Pogosov A G, Bakarov A K and Shklyaev A A 2016 *Phys. Rev. Lett.* **117** 017702
[10] Pogosov A G, Budantsev M V, Zhdanov E Yu, Pokhabov D A, Bakarov A K and Toropov A I 2012 *Appl. Phys. Lett.* **100** 181902
[11] Pokhabov D A, Pogosov A G, Shevyrin A A, Zhdanov E Yu, Bakarov A K, Shklyaev A A, Ishutkin S V, Stepanenko M V and Shesterikov E V 2018 *J. Phys.: Conf. Ser.* **964** 012008
[12] Zhdanov E Yu, Pogosov A G, Budantsev M V and Pokhabov D A 2013 *AIP Conf. Proc.* **1566** 211
[13] Zhdanov E Yu, Pogosov A G, Budantsev M V, Pokhabov D A, Bakarov A K and Toropov A I 2015 *J. Phys.: Conf. Ser.* **643** 012079
[14] Rössler C, Herz M, Bichler M and Ludwig S 2010 *Solid State Commun.* **150** 861
[15] Pokhabov D A, Pogosov A G, Zhdanov E Yu, Shevyrin A A, Bakarov A K and Shklyaev A A 2018 *Appl. Phys. Lett.* **112** 082102
[16] Shevyrin A A, Pogosov A G, Budantsev M V, Bakarov A K, Toropov A I, Ishutkin S V and Shesterikov E V 2014 *Appl. Phys. Lett.* **104** 203102
[17] Cleland N A, Aldridge J S, Driscoll D C and Gossard A C 2002 *Appl. Phys. Lett.* **81** 1699
[18] Okazaki Y, Mahboob I, Onomitsu K, Sasaki S and Yamaguchi H 2013 *Appl. Phys. Lett.* **103**, 192105
[19] van Wees B J, van Houten H, Beenakker C W J, Williamson J G, Kouwenhoven L P, van der Marel D and Foxon C T 1988 *Phys. Rev. Lett.* **60** 848
[20] Wharam D A, Thornton T J, Newbury R, Pepper M, Ahmed H, Frost J E F, Hasko D G, Peacock D C, Ritchie D A and Jones G A C 1988 *J. Phys. C: Solid State Phys.* **21** L209

SPBOPEN 2018   IOP Publishing

The study of mechanical resonances of the phase electro-optic modulator based on LiNbO$_3$ for noise reduction of fiber-optic gyroscope

M A Smolovik, D A Pogorelaya, A A Vlasov, A S Aleynik and V E Strigalev

Department of Light-Guided Photonics, ITMO University, St. Petersburg 197101, Russia

Abstract. We report about the study of Ti diffused LiNbO$_3$ x-cut electro-optic modulator (EOM). This paper describes a study of mechanical resonances induced by applied voltage in EOM and their influence on the EOM functioning. The method for reduction of mechanical resonances influence is presented. Described method can be used for noise reduction of fiber-optic interferometric sensors and fiber-optic gyroscope in particular.

1. Introduction

EOMs are widely used in fiber-optic communication systems as well as in fiber-optic sensors [1, 2]. Most of these EOMs are based on a lithium niobate (LN, LiNbO$_3$) crystal because of its unique properties [3, 4]. The most sensitive fiber-optic sensors are interferometric ones. An additional phase modulation is applied in these devices by using EOMs. It allows to increase sensitivity, stability and dynamic range of sensors [2, 5]. However EOMs have some disadvantages what affects the accuracy of fiber-optic sensors. A LN crystal possesses both electro-optic (Pockels) and inverse piezoelectric effects. Second one can negatively impact to stability of EOM [4]. It causes resonant mechanical oscillations in LiNbO$_3$ and, consequently, leads to spurious phase shift of optical radiation, which contains information about measured value. Therefore a suppression of resonant mechanical oscillations in LiNbO$_3$ is very important in order to use an EOM in the precision sensors.

   The present paper is dedicated to the EOM based on LN crystal (x-cut) implemented by Ti diffusion technology [6] and it is considered as a part of a closed-loop fiber-optic gyroscope (FOG) for navigation grade applications. Such type of FOG was described in [7].

2. Experiment
The study was implemented in three stages.

2.1. The electrical method of the EOM study
When modulating voltage is applied to electrodes of EOM the electrical impedance of the EOM is changed because of piezoelectric effect. Thus, on the first stage the study of mechanical resonances of the EOM has been implemented by means of the measurement of electrical impedance as follows. An electrical sinusoidal signal with different frequencies is applied to the electrodes through high frequency connector and the reflected signal is estimated (S$_{11}$-parameter) [8]. If the frequency of the electric signal matches the frequency of natural oscillations of the EOM then energy of this signal is partially consumed for mechanical oscillations and, consequently, an amplitude of reflected signal is decreased. These measurements are implemented for two frequency ranges: 1 – 600 kHz and 300 kHz

Content from this work may be used under the terms of the Creative Commons Attribution 3.0 licence. Any further distribution of this work must maintain attribution to the author(s) and the title of the work, journal citation and DOI.

Published under licence by IOP Publishing Ltd

– 10 MHz. The S_{11}-parameter for low-frequency range was measured by set-up based on the developed electronic circuit, precision digital oscilloscope (Rohde & Schwarz, RTO1024) and arbitrary waveform generator (RIGOL DG4102). The same characteristic for high-frequency range was measured by vector network analyzer "OBZOR TR1300/1". As a result we've received graphs which demonstrate a presence of resonant frequencies of the EOM sample.

After that we've fixed the EOM on the substrate by various materials in order to reduce or eliminate resonances in the EOM. Epoxy adhesive, UV curing optical adhesive, urethane adhesive, liquid sodium silicate have been used as fixing materials. The experiment results for the liquid sodium silicate are shown in figure 1.

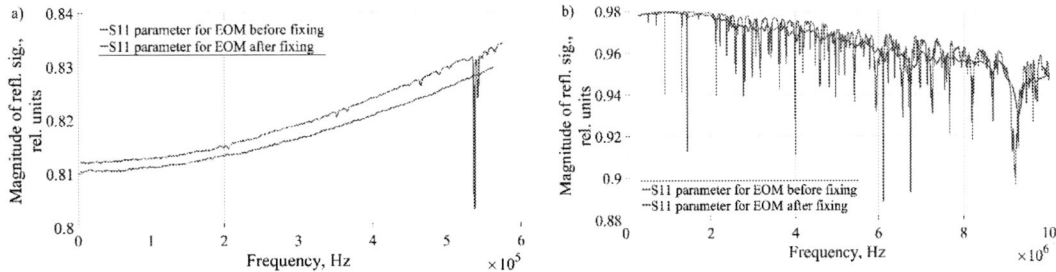

Figure 1 (a, b). S_{11} parameter before and after fixing the EOM by liquid sodium silicate: **(a)** 1 – 600 kHz frequency range; **(b)** 300 kHz – 10 MHz frequency range.

2.2. The optical method of the EOM study

The EOM is part of the high precision fiber-optic gyroscope (FOG). Therefore, on the second stage of the research an influence of the resonant oscillations of the EOM on output signal of FOG is estimated in this paper as well. Such FOG is an interferometric sensor based on Sagnac effect. An operating principle of it is well known [5, 7]. The basic circuit of the EOM as a part of a FOG is presented in figure 2.

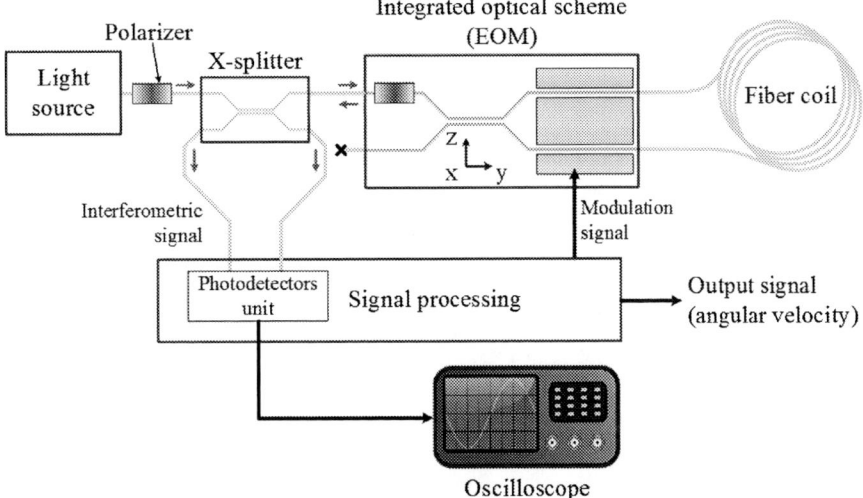

Figure 2. The measurement circuit of the influence of mechanical resonances on interferometric signal in the FOG scheme.

As shown in figure 2 the EOM was included in FOG circuit and special modulating signal was applied to the EOM electrodes. The form of this modulating signal is meander, which peak-to-peak amplitude contributes the resulting phase shift is equal to π radians. The high precision digital oscilloscope "Rohde & Schwarz RTO1024" registered interferometric output signal with sample rate value of 5 GSa/s. The photodetector signal, which was registered by oscilloscope, is shown in figure 3, where τ corresponds to eigenfrequency (or proper frequency) of the fiber coil and depends on coil length [5]. In our case, τ is equal to 7.3 microseconds. The τ_{samp} is time when analog-to-digital converter (ADC) operates, i.e. ADC acquisition implements during the τ_{samp}. Signal ripple during the "τ_{samp}" can enter distortions to the ADC signal, so it is important to reduce mechanical resonances of the EOM. The spectrum of the signal during "τ_{samp}" is shown in figure 3 as well.

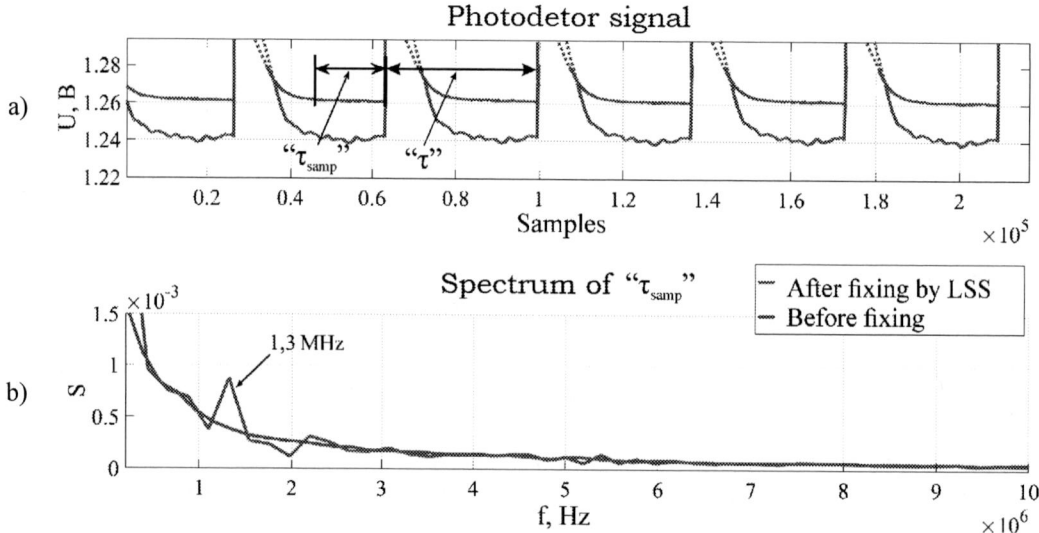

Figure 3 (a, b). The measurement of the interferometric FOG signal before and after fixing of the EOM by liquid sodium silicate (LSS): (a) photodetector voltage vs number of sample; (b) the spectrum of "τ_{samp}" signal.

Experimental results show presence of intensity fluctuation of the FOG interferometric signal before fixing the EOM (figure 3, red curve). This fluctuation frequency is about 1.3 MHz that agrees with previous experiment. Suppression of resonant oscillations by means of the fixing by liquid sodium silicate significantly reduces such fluctuations of the interferometric signal.

2.3. Noise reduction of the FOG output signal
On the third stage we've measured the FOG output signal (angular velocity, see figure 2) before and after fixing of the EOM to confirm the obtained results. FOG remained stationary during the experiment, thus the measured angular rate was constant and equal to the projection of the Earth's rate on the axis of fiber-optic coil. The signal noise level has been estimated by computing the standard deviation of FOG signal with and without fixing of the EOM. The obtained output data were averaged over 1 second. The output FOG signals for the both cases are presented in figure 4.

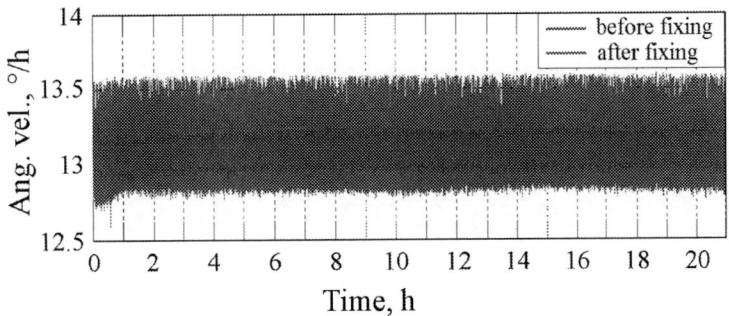

Figure 4. The output FOG signal before and after fixing of the EOM by liquid sodium silicate.

Standard deviation of the obtained signals is 0.17 °/h before fixing (figure 4, red curve) and 0.046 °/h after fixing (figure 4, blue curve). These measurements show reduction of standard deviation of the FOG data after fixing the EOM. In particular, standard deviation of the output signal can be reduced by means of such fixing by 3.7 times.

3. Conclusion

Influence of natural resonant oscillations on operation of the electro-optic modulator based on lithium niobate was experimentally revealed. Such behaviour of EOM is explained by piezoelectric effect which result in change of waveguide and polarization parameters of EOM and result in change of electro-optic coefficients of the crystal as well. The fixing of EOM on the substrate by various materials was implemented to reduce the noise of a fiber-optic gyroscope. Thus, we've demonstrated efficiency of this method for noise reduction of fiber-optic sensors. In particular, fixing of the EOM on the substrate by liquid sodium silicate has led to decrease of standard deviation of the output FOG signal by 3.7 times.

Acknowledgments

This research was carried out at ITMO University and was supported by the Ministry of Education and Science of the Russian Federation (Project No.03.G25.31.0245).

References
[1] Agrawal G P 2010 Fiber-Optic Communication Systems (New York: John Wiley & Sons, Inc.) p 626
[2] Shizhuo Yin, Paul B. Ruffin, Francis T.S. Yu 2008 (New York: CRC Press) p 477
[3] Wooten Ed L, Kissa K M et al. 2000 IEEE J. Sel. Top. Quantum Electron. 6(1) 69-82
[4] Murphy E and Chen A 2012 Broadband optical modulators (Boca Raton: CRC Press) p 568
[5] Lefevre H C 2014 Fiber Optic Gyroscope (London: Artech House) p 405
[6] Bazzan M, Sada C 2015 Appl. Phys. Rev. 2(4) 040603
[7] Aleinik A S et al. 2016 Gyroscopy and Navigation 7(3) 214-222
[8] Hewlett Packard 1967 S-parameter Techniques for Faster, More Accurate Network Design (Application note 95-1) p 12

SPBOPEN 2018   IOP Publishing

IOP Conf. Series: Journal of Physics: Conf. Series **1124** (2018) 061003   doi:10.1088/1742-6596/1124/6/061003

Magneto-optical properties of metaphosphate and borate glasses

Dmitrii I Sobolev, Anastasiia N Babkina, Nikolay V Nikonorov

Department of Optical Information Technologies and Materials, ITMO University, Saint Petersburg 197101, Russia

Abstract. The paper shows results of Faraday Effect study of potassium-alumina-borate and metaphosphate glass under uniform longitudinal magnetic field. The Verde constant of glass under study is obtained based on the experimental data for several wavelengths. Magneto-optical properties of two types of glass: $K_2O-Al_2O_3-B_2O_3$ glass doped with $MnFe_2O_4$ nanocrystals and $P_2O_2-Tb_2O_3-Ga_2O_3$ glass are compared with industrial glass. Application prospects of borate and metaphosphate glass in fields of isolation of laser radiation, current and magnetic field sensing is shown.

1. Introduction

Faraday Effect, which has a number of significant practical applications, is the rotation of the plane of polarization of a light beam passing through a transparent medium in a magnetic field. Magneto-optical materials (crystals, glasses, films) are promising for use in modern high-tech devices such as current sensors [1-3], optical isolators [4-5] in semiconductor devices when measuring the effective mass of charge carriers, for example, in homogeneity degree studies of semiconductor wafers, aimed at rejecting defective samples. The optical isolators greatly benefits from high Verdet constant materials. In these devices the incident plane polarized light is passed through a Faraday rotating material in a magnetic field so as to produce a 45° rotation of the polarization plane. It is then transmitted through polarization analyzer oriented at 45° with respect to the first polarizer. Back reflected light is blocked by this system which is important for sensitive laser systems where back reflected radiation is detrimental to the laser source as well as in fiber laser schemes. The fiber laser systems, which have become very powerful during last decades, also require fiber optical elements including optical isolators.

Ferromagnetic and paramagnetic glasses are good candidates for all these applications. Glass with high rare-earth elements concentration (up to 20 mol %) possesses large magneto-optical rotation with high transparency in the visible region [6]. The main restriction of its application is almost linear spectral dependence of the Verdet constant which becomes insignificant in long wavelength region. Use of glass with ferrite particles removes this restriction.

The present paper demonstrates the results of the Faraday Effect study in optical active glass medium, comprising of the Verde constant measurements, optical system adjustment, experimentation and physical processes simulation. As a result of the work, an experimental setup was assembled, the influence of the magnetic field on the radiation propagating in the glass was recorded, the angle of rotation of the plane of light polarization was measured, and the Verde constant for samples of synthesized glass was determined.

Content from this work may be used under the terms of the Creative Commons Attribution 3.0 licence. Any further distribution of this work must maintain attribution to the author(s) and the title of the work, journal citation and DOI.

Published under licence by IOP Publishing Ltd

2. Materials and methods

In the present work two types of glass: the 25 K_2O-20 Al_2O_3-55 B_2O_3 glass system doped with 3% Fe_2O_3 and 2% MnO (mol %) and the P_2O_2-x Tb_2O_3-(25-x) Ga_2O_3 (where x varied from 5 to 25) (mol %) glass system were investigated. Both types of glass were synthesized by melt-quenching technique with stirring the melt by the platinum stirrer. Different crucibles were used to synthesis phosphate glass with zero Ga_2O_3 concentration: silicate (F-237) and platinum (F-246). After synthesis the fabricated samples of each composition were cut to several pieces, thereafter borate glass samples were subjected to additional treatments to nucleate manganese ferrite $MnFe_2O_4$ nanocrystals (NCs) dispersed in matrix.

Standard experimental setup was used for carrying magneto-optical measurements. Several semiconductor coherent sources of radiation (λ = 405 nm, 532 nm, 632.8 nm, 808 nm, 980 nm) were used. Coil of more than 2 000 turns of copper wire provided 70 T magnetic field.

There were two modes of the experimental setup operation: without current (mode 1) and with current (mode 2). The intensity of light transmitted through two polarizers, according to the Malus law, depended on the relative angle between the transmission axes of the polarizers (α).

$$I = \frac{1}{2}I_0 \cos^2 \alpha \qquad (1)$$

This formula was valid when there was no current in the coil (mode 1). In regime 2, the external magnetic field introduced an additional shift in the polarization of the radiation $\Delta\alpha$.

$$I = \frac{1}{2}I_0 \cos^2(\alpha + \Delta\alpha) \qquad (2)$$

Using the experimental setup I dependence on the relative angle between the polarizer axes in two modes: with the current turned off and with the current turned on – was revealed.

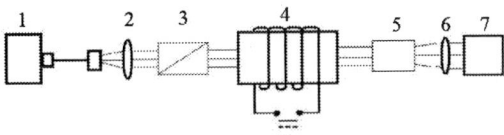

Figure 1. Schematic diagram of experimental setup. 1 - laser, 2 - lens, 3 - polarizer, 4 - magnetic coil, 5 - analyzer, 6- collimator, 7 - photodetector

Using the accepted values, two curves were constructed and approximated by function (3) for mode 1 and fucntion (4) for mode 2.

$$\psi = const \cdot \cos^2(\alpha + k_1) \qquad (3)$$

$$\psi = const \cdot \cos^2(\alpha + k_2) \qquad (4)$$

The distance between the curves along the X- axis (the axis of angles) was $\Delta\alpha$, i.e. the angle at which the plane of polarization rotated by the magnetic field, $\Delta\alpha$ was found as the difference between the coefficients k1 and k2. Thereby the value of the Verde constant could be obtained. Generally the Verde constant is measured in rad / (T $\cdot$ m). In this work, a current of 1.5 A was used. Using the mathematical package MathCad, based on the experimental data and equations above, the value of the rotation angle of the plane of polarization was calculated and averaged. The Verde constant was calculated as follows:

$$V = \frac{\theta}{L \cdot B},\qquad(5)$$

where L – optical path of the light beam inside the glass sample, B - magnetic field, θ- rotation angle.

3. Results

Table 1 and Figure 2 present the results of the Verde constant measurements at various wavelengths for potassium-alumina-borate, metaphosphate glasses and industrial glass "MOS-15", which is a silicate glass doped with terbium ions. It can be seen that as the terbium concentration increases, the metaphosphate glass Verdet constant values approach those for industrial glass. The borate glass with $MnFe_2O_4$ nanocrystals has high absorption in the ultraviolet and visible regions being transparent in the near IR therefore the Verdet constant at 980 nm is presented only. However the advantage is the big Verdet constant value obtained for this type of glass.

In contrast, metaphosphate glasses activated by terbium have high transmittance in the ultraviolet and visible ranges. It is worth noting that in metaphosphate environment terbium ions do not possess luminescence thus not distorting data from detector. This opens the prospect of using these materials in a wide spectral range from ultraviolet to infrared, to create fiber-optic current and magnetic field sensors, as well as Faraday isolators.

Table 1. Mean Verdet constant (rad/T m) at various wavelengths for 5%, 10%, 15%, 20%, 25% Tb metaphosphate glass, borate glass with $MnFe_2O_4$ NCs (3% Fe_2O_3 and 2% MnO) and MOS-15 at room temperature.

Samples	5% Tb_2O_3 F-263	10% Tb_2O_3 F-265)	15% Tb_2O_3 F-264	20% Tb_2O_3 F-262	25% Tb_2O_3 F-237	25% Tb_2O_3 F-246	MOS-15	3% Fe_2O_3 2% MnO
Thickness, mm	38.5	27	37	25.5	25	24	42	10
Verdet constant at 405 nm	-51.81	-113.86	-172.11	-319.79	-324.13	-192.92	-343.59	-
Verdet constant at 535 nm	-29.14	-68.34	-95.69	-381.33	-162.17	-118.43	-201.84	-
Verdet constant at 632.8 nm	-12.95	-55.4	-67.4	-117.3	-119.6	-37.1	-	-
Verdet constant at 808 nm	-12.3	-27.7	-40.4	-46.9	-59.8	-43	-74.8	-
Verdet constant at 980 nm	-9.4	-24.9	-38	-46	-59.8	-70	-74.2	+228.57

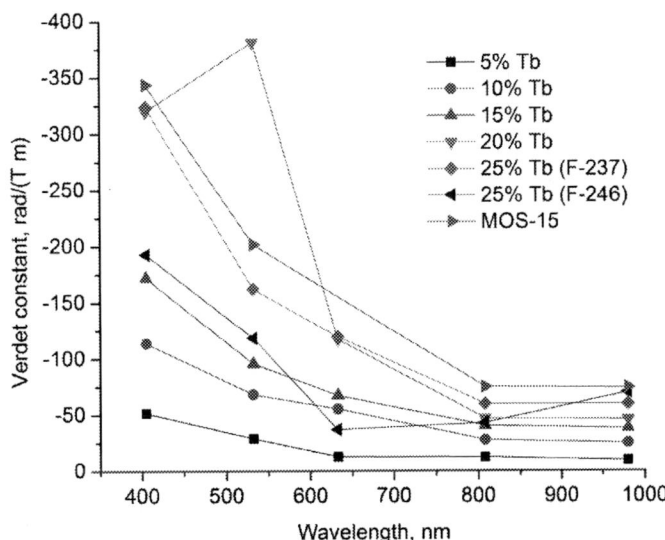

Figure 2. Spectral dependence of the Verdet constant for different types of terbium doped glass at room temperature (connecting lines are provided for convenience)

4. Conclusions

The Faraday Effect in potassium-alumina-borate doped with iron and manganese ions, and terbium doped metaphosphate glass was studied, and the Verde constant value for these materials was measured. As a result of work, an experimental setup has been assembled; the effect of a longitudinal magnetic field on the radiation propagation in glass was recorded. The experimental data was in good agreement with the approximation curves. Big values of the Verdet constant for the high-terbium metaphosphate glass in blue region and $MnFe_2O_4$-doped borate glass in near IR region were obtained. This opens the prospect of using these materials in a wide spectral range from ultraviolet to infrared, to create fiber-optic current and magnetic field sensors, as well as Faraday isolators.

References
[1]   Shen Y, Lu Y, Liu Z, Yu X, Zhang G, Yu W 2015 *Journal of Magnetism and Magnetic Materials* **389** 180
[2]   Chen Q, Ma Q, Wang H, Chen Q 2015 *Journal of Non-Crystalline Solids* **429** 13
[3]   Chen Q, Wang H, Perero S, Wang Q, Chen Q 2015 *Journal of Magnetism and Magnetic Materials* **408** 43
[4]   Starobor A, Zheleznov D, Palashov O, Savinkov V, Sigaev V 2016 Optics Communications **358** 176
[5]   Chen Z, Yang L, Wang X, Wang J, Hang Y 2015 *Materials Letters* **161** 93
[6]   Ballato J, Snitzer E 1995 Applied Optics 34 6848

SPBOPEN 2018   IOP Publishing

IOP Conf. Series: Journal of Physics: Conf. Series **1124** (2018) 061004   doi:10.1088/1742-6596/1124/6/061004

Investigation of the magnetic properties of manganese silicide grown on i-GaAs substrate by pulsed laser deposition

Y Kuznetsov[1], M Dorokhin[1], A Kudrin[1], V Lesnikov[1], E Demidov[2], V Karzanov[2]

1 Physical Technical Research Institute of Lobachevsky State University of Nizhny Novgorod, Nizhny Novgorod, 603950 Russia
2 Lobachevsky State University of Nizhny Novgorod, 603950, Gagarin ave.23, Nizhniy Novgorod, Russia

Abstract. In this paper we have investigated the properties of the MnSi ferromagnetic layer grown on a semi-insulating GaAs (100) substrate by pulsed laser deposition method. Basing on the analysis of the temperature and magnetic-field dependencies of the Hall voltage and magnetization, as well as on analysis of the electron paramagnetic resonance spectra, it was shown that the manganese silicide film with Mn content of about 45 % is ferromagnetic with a Curie temperature of ~ 370 K.

1. Introduction

Spintronics is one of the main trends for the development of microelectronics. The important task of the spintronics technology is the fabrication of ferromagnetic materials with Curie point close to room temperature or above. Silicon - the main element of all semiconductor electronics, is typically a non-magnetic semiconductor. A number of papers have shown [1]-[3] that the introduction of manganese atoms into the silicon with high enough concentration films allows one to confer the magnetic properties to the material. In the present work we report on the investigation of thin manganese silicide films with near 45x55 % Mn and Si composition.

2. Method of experiment

The magnetic properties of the structure were investigated by analyzing the magnetic field dependence of the Hall voltage, the Nernst-Ettingshausen effect [4], the EPR spectrum [2] and the magnetic field dependence of the magnetization recorded by the variable field gradient method [5].
Figure 1 shows a schematic diagram of the mounting of a sample for recording the magnetic field dependences of Hall voltage and Nernst-Ettingshausen voltage. The Hall voltage was measured in the Wan-der-Pauw geometry with the sample temperature kept uniform along the sample.

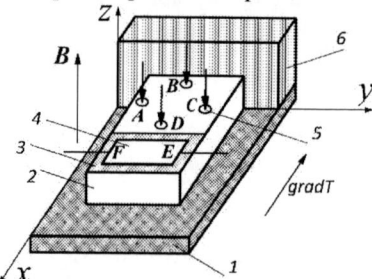

Figure 1. Schematic picture of mounting the sample when recording the magnetic field dependences of the Hall voltage and Nernst-Ettingshausen; 1 - holder, 2 - sample, 3 - thermal grease, 4 - heating element (resistor), 5 - contacts, 6 - cooling radiator.

For the thermoelectric and thermomagnetic measurements one of the sample edges was kept under an increased temperature. For that reason an electric current was supplied on the resistor (FE) thus the Joule

Content from this work may be used under the terms of the Creative Commons Attribution 3.0 licence. Any further distribution of this work must maintain attribution to the author(s) and the title of the work, journal citation and DOI.
Published under licence by IOP Publishing Ltd

heat raised the temperature of the corresponding edge. Another edge of the structure was kept in thermal contact with the radiator. The temperature gradient created this way causes a number of effects: a thermal current, a Seebeck voltage generation and the Nernst-Ettingshausen (NE) effect under the applied magnetic field. The latter phenomenon consists in the generation of a transverse potential difference under the applied both the temperature gradient at the edges of the sample and the external magnetic field normal to sample surface (Fig.1). The analysis of Nernst-Ettingshausen voltage, in conjunction with the Hall voltage, gives information on the scattering behavior of free carriers.

According to [4], the Hall and Nernst-Ettingshausen voltages are equal to:

$$U_{NE} = QB\Delta T, \tag{1}$$
$$U_H = R_H BI, \tag{2}$$

where Q – Nernst-Ettingshausen constant, R_H – is the Hall constant, B –external magnetic field, I – electric current passing through the sample (in the Hall measurements), and ΔT – temperature difference between the edges of the structure under study (for NE measurements).

The Hall and Nernst-Ettingshausen constants depend on the true mobility of free charge carriers:

$$Q = \left(\frac{1}{2} - r\right) A_r \frac{k}{e} \mu, \tag{3}$$

$$R_H = -\frac{A_r \cdot \mu}{\sigma}, \tag{4}$$

where k – Boltzmann constant, e – elementary charge, μ – true mobility of the free charge carriers, and σ – electric conductivity, r – is a Hall factor and A_r – is the scattering factor.

The Hall factor and the scattering factor are related by:

$$A_r = \frac{3\sqrt{\pi}}{4} \frac{\Gamma(\frac{3}{2} + 2r)}{\Gamma^2(2 + r)}, \tag{5}$$

where $\Gamma(x)$ – is gamma function.

Solving the obtained system with respect to mobility and the scattering factor, we obtain the expression:

$$r = \frac{1}{2} - \frac{QeR_{sl}}{R_H k} \rightarrow \mu = \mu(r). \tag{6}$$

From expression (6), one can determine the true mobility. Knowing the scattering factor one can also determine the contribution of each scattering center (impurity or phonons).

In ferromagnetic semiconductors, the scattering of the charge carrier depends on its spin orientation, therefore, an additional term appears in expressions (1) and (2), leading to distortion of the magnetic field dependences of the Hall and Nernst-Ettingshausen voltages:

$$U_H(B) = R_H BI + R_s M(B)I, \tag{7}$$
$$U_{NE}(B) = Q_0 B\Delta T + Q_M M(B)\Delta T, \tag{8}$$

where R_H and Q_M – are the constants of the anomalous Hall effect and anomalous Nersent-Ettinshausen effect, respectively.

By analyzing the characteristic form of these magnetic field dependences, it is possible to determine the magnetic state of the structure under study.

3. Experimental

Figure 2 illustrates the magnetic-field dependences of the Hall voltage recorded for the investigated sample at $T = 300K$, $T = 77K$ and $T = 370K$.

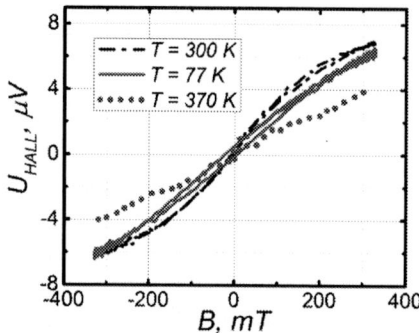

Figure 2. The magnetic field dependence of the Hall voltage recorded at 300K (dashed line) and 77K (solid line) and 370K (dot).

The 77 K and 300 K dependences are non-linear evidencing on the presence of an anomalous Hall component in the Hall constant. A clear hysteresis loop can be seen at the $U_{Hall}(B)$ dependence measured at 77 K. With the temperature increase the coercive field decreases and thus no hysteresis can be detected at 300 K. At a temperature of $T = 370\ K$ the $U_{Hall}(B)$ dependence can be approximated by a linear function. This temperature can be estimated as the Curie point of investigated MnSi film.

To further confirm the presence of ferromagnetic properties, the magnetic field dependence of the magnetization of the structure was recorded by the alternating force gradient magnetometer [5]. Figure 3 illustrates the dependences of the magnetization of the investigated structure on the magnitude of the external magnetic field with the B applied perpendicular to the film surface (in-plane geometry) and parallel to the film surface (out-of-plane).

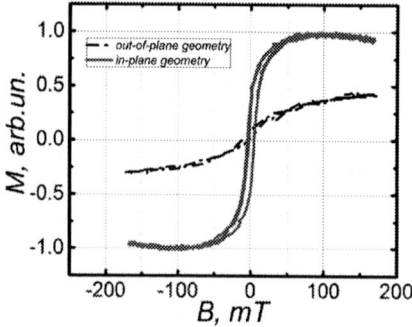

Figure 3. The magnetic field dependence of the magnetization of the structure under study with the external field is oriented in-plane (solid line) and out-of-plane (dashed line).

The presence of a hysteresis loop within the in-plane measurements indicates the presence of ferromagnetic ordering in the structure under study. In the out-of-plane measurement no hysteresis loop was detected which is in agreement with the anomalous Hall effect measurements (Figure 2). This fact is believed to be due to the easy magnetization axis of the MnSi lies mostly in the plane of the film. Figures 4a and 4b illustrate the magnetic field dependences of the Hall and Nernst-Ettingshausen voltages recorded at an average sample temperature $T_{av} = 196.75$ K and $T_{av} = 390$ K, respectively. The average temperature was calculated by the formula: $T_{av} = (T_{hot} + T_{cold})/2$.

SPBOPEN 2018

IOP Publishing

IOP Conf. Series: Journal of Physics: Conf. Series **1124** (2018) 061004   doi:10.1088/1742-6596/1124/6/061004

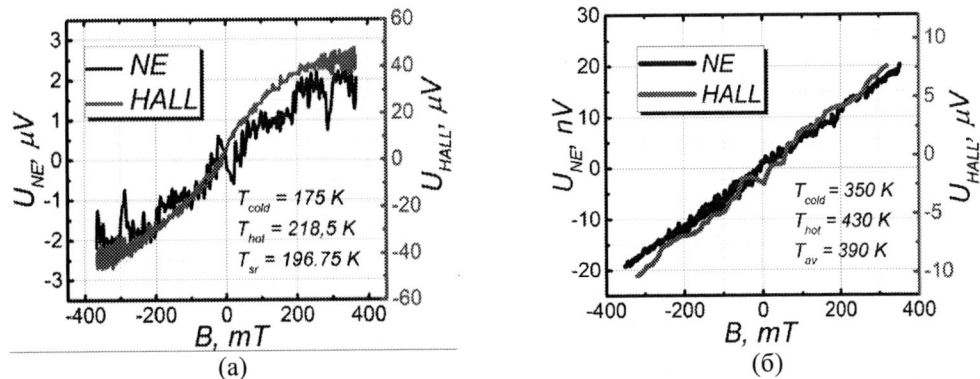

Figure 4. Magnetic field dependences of the Hall voltage and Nernst-Ettingshausen, registered with: (a) - T_{av} – 196.75 K, (b) - T_{av} = 390 K.

From the graph illustrated at Figure 4a it can be seen that the magnetic-field dependences of the Hall and Nernst-Ettingshausen voltages have a clearly nonlinear form, whereas at 390 K a linear dependence is clearly visible (see Figure 4b). The change of the magnetic field dependence of the Nernst-Ettingshausen voltage, along with the change of the magnetic-field dependence of the Hall voltage, mentioned above, additionally attests to the presence of the "ferromagnet-paramagnet" phase transition point in the structure under study. From the analysis of the recorded magnetic field dependences of the Hall voltage and the Nernst-Ettingshausen voltage over a wide temperature range (20 ± 430) K, the temperature dependences of the true mobility of free charge carriers (Figure 5a), the scattering factor and Hall factor (Figure 5b) were calculated. In this case the R_H and Q_0 coefficients below the Curie point were derived by analyzing the linear part of $U_H(B)$ and $U_{NE}(B)$ dependences respectively. The temperature dependence of the mobility is given in the Arrhenius plot.

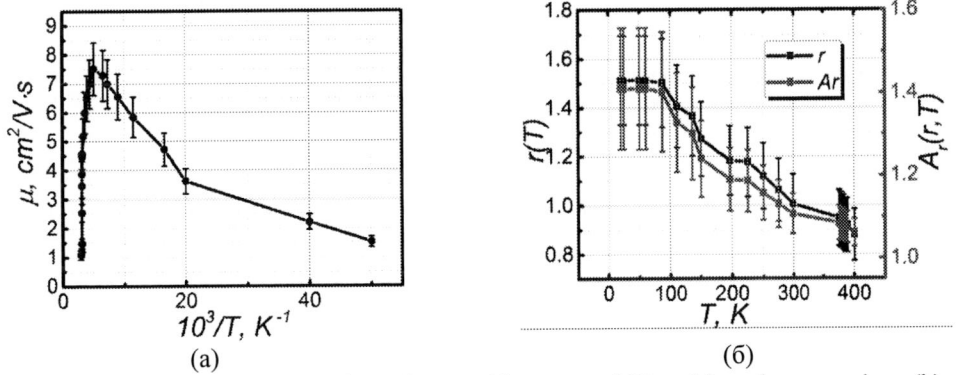

Figure 5. Calculated temperature dependences: (a) - true mobility of free charge carriers, (b) - scattering factor and Hall-factor.

From the graphs illustrated in Figures 5a and 5b it is seen that up to 70 K the carrier scattering was due to impurity ions only - the mobility increases with increasing temperature, the scattering factor is about 1.5. After passing through this temperature point, the contribution of scattering by acoustic phonons increases, which leads to a decrease in the scattering factor and a decrease in the mobility of the charge. The last consequence leads to the fact that the values of the Hall voltage signal and the Nernst-Ettingshausen voltage decrease, which can be seen on the graphs above.

The "ferromagnet-paramagnet" phase transition was also studied by analyzing the spectra of electron paramagnetic resonance of EPR recorded in the temperature range of 300-500 K. Figure 6a illustrates the EPR spectrum of the sample under investigation at 300 K, and Figure 6b shows the temperature

863

dependence of the magnitude of the peak amplitude on the spectrum, which is normalized to the maximum value in the investigated range (by the peak amplitude at room temperature).

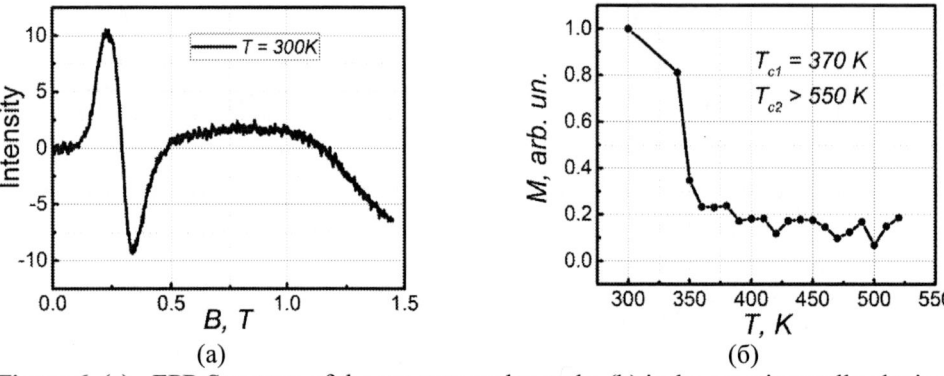

Figure 6. (a) - EPR Spectrum of the structure under study, (b) is the experimentally obtained temperature dependence of the peak value on the EPR spectrum of the sample under study.

One can see the rapid drop of the magnetization at ~ 350 K, however above 350 K the value of M does not drop down to zero and is preserved even at the highest temperature used. The latter is believed to be due to the presence of at least two magnetic phases in the film. The Curie temperature of the first phase, as shown by other studies, is about 370K, the Curie point of the other phase is at a temperature above 525K. We note that the presence of the second magnetic phase was not detected neither via magnetotransport measurements, nor via anomalous Nernst-Ettingshausen effect measurements. This can lead us to conclusion that the ratio of the second magnetic phase with respect to the first one is relatively low thus it does not have a significant influence on the electronic transport.

In conclusion

In this work, the magnetic properties of manganese silicide film were studied. It was shown that the presence of a large amount of manganese in silicon can lead to a ferromagnetic ordering of a structure with a Curie temperature of T = 370 K, which opens prospects for the instrumental use of this material.

Acknowledgments

This study was supported by the Russian Science Foundation (project 17-79-20173).

References

[1]. Demidov E, Podolsky V, Lesnikov V, Pavlova E, Bobrov A, Karzanov V, Malehonova N, Tronov A, High-temperature diamond-like Si-based ferromagnet with self-organized superlattice distribution of Mn impurity, JETP letters, 2013, volume 9, 706-709.

[2]. Schroeter D, Steinki N, Scarioni A, Schumacher H, Süllow S, Menzel D, Electronic transport properties of nanostructured MnSi-films, Physica B (2017), 98.

[3]. Jansen R, Silicon spintronics, Nature mat. 11 (2012) 400-408.

[4]. Tsidilkovskii I, Thermomagnetic phenomena in semiconductors (in Russia), Moscow, (1960), 365.

[5]. Flanders P, An alternating-gradient magnetometer, J. Appl. Phys. 63 (1988) 3940-3945.

On the mechanism of spin-polarized injection in (Ga,Mn)As/n+GaAs/InGaAs Zener tunnel diode

M Ved[1], M Dorokhin[1], E Malysheva[1], A Zdoroveyshchev[1], Yu Danilov[1], A Parafin[2] and Yu Kuznetsov[1]

[1]Physical Technical Research Institute of Lobachevsky State University of Nizhny Novgorod, Nizhny Novgorod, 603950 Russia

[2]InstituteofAppliedPhysics, Russian Academy of Sciences, Nizhny Novgorod, 603950 Russia

Abstract. In the present work we have investigated the spin light-emitting diode structures based on InGaAs/GaAs quantum well and a (Ga,Mn)As dilute magnetic semiconductor as an injecting layer. The luminescent properties and the magnetic-field dependence of the circular polarization degree of the initial and annealed samples were measured. It was experimentally shown that the temperature dependence of circular polarization degree below the Curie point (~ 110 K) is determined by the thermally activated tunnelling of bound electrons from energetically split spin-up and spin-down subbands of magnetized (Ga,Mn)As. The approximation parameters allowed us to derive the values of the spin energy splitting in (Ga,Mn)As subband and the spin depolarization during the transfer from (Ga,Mn)As to the device active region.

1. Introduction

Diluted magnetic semiconductors (DMS) that combine semiconductor and ferromagnetic properties are promising elements of spintronic devices, since they are characterized by a high degree of carriers spin polarization (which is the main property), and the compatibility with the semiconductor technology [1]. The (Ga,Mn)As solid solution is considered as one of the most promising DMS because of its good compatibility with GaAs, which is the basic material of optoelectronic devices. The disadvantage of DMS, and in particular of (Ga,Mn)As, is the relatively low Curie temperature. The main method for obtaining (Ga,Mn)As layers with increased Curie temperature (~110K) is the epitaxial growth (Ga,Mn)As/GaAs layers by the low-temperature molecular-beam epitaxy technique [2]. An alternative method for obtaining DMS is pulsed laser deposition (PLD). Earlier it was shown that the layers obtained by this method are ferromagnetic with a Curie temperature not exceeding 40K. This is due to the fact that only a small fraction of the Mn atoms are embedded in the Ga substitutional position in GaAs lattice [3]. However it was recently demonstrated that the annealing of (Ga,Mn)As-based structures by a short-wave pulsed laser promotes the redistribution of Mn in the (Ga,Mn)As layer and, as a result, an increase of the Curie temperature [4].

The present work is devoted to the study of spin light-emitting diodes including (Ga,Mn)As layers. The spin LEDs were fabricated by the combined technique of a metal-organic vapor phase epitaxy (MOVPE) and pulsed lased deposition. The diode structure was based on the InGaAs/GaAs quantum

---

Content from this work may be used under the terms of the Creative Commons Attribution 3.0 licence. Any further distribution of this work must maintain attribution to the author(s) and the title of the work, journal citation and DOI.

Published under licence by IOP Publishing Ltd

well (QW) and the (Ga,Mn)As layer which was used to inject the spin polarized holes into QW. We study the modification of SLED properties after the pulsed laser annealing of the structure.

2. Experimental technique

The as-grown samples were formed by combined epitaxial method. This method is described in detail in [2]. The following layers were sequentially grown by metal organic vapour-phase epitaxy at atmospheric pressure and a temperature of 600°C on p-GaAs (100) substrates: p-GaAs buffer layer; $In_xGa_{1-x}As$ QW (concentration $p \sim 8 \cdot 10^{17} cm^{-3}$, width d_{QW}= 10 nm, In content $x \sim 0.12$); i-GaAs, n - GaAs layer with a gradient doping, heavily doped n^{++} layer ($n \sim 7 \cdot 10^{18}$ cm^{-3}) and δ <Si> -layer (total electron concentration in n^{++} GaAs and δ <Si> was 10^{19} cm^{-3}). Highly doped layer thickness was 30 nm. Then, GaMnAs layer was grown by the pulsed laser deposition of Mn and GaAs targets at a temperature of 290°C in the same reactor.

In order to modify the (Ga,Mn)As structure and to increase the Curie temperature the pulsed laser annealing of the samples surface was used [3]. The annealing experiments were performed using the excimer laser LPX-200 on KrF (wavelength 248 nm, pulse duration ~ 30 ns, energy density in the pulse 290 mJ/cm²). It was earlier demonstrated that the laser annealing leads to an increase of the concentration of electrically active Mn in diluted magnetic semiconductor layers, and, as a consequence, to an increase of the Curie temperature [3]. In our work, we have studied both the as-grown and annealed structures in order to characterize the influence of annealing on the electroluminescence properties.

To form the diode structure an Au Ohmic contact was deposited onto the surface of the samples and the sparking of In foil was used to create the Ohmic contact to the base. The electroluminescent radiation of the samples was detected in the reverse bias mode (negative voltage onto top (Ga,Mn)As contact with respect to the base). The formed structures are shown in Figure 1.

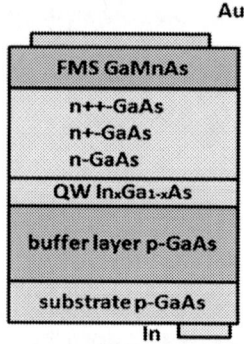

Figure 1. Scheme of the studying structures.

When spin light-emitting diodes are introduced into an external magnetic field, their emission becomes partially circularly polarized. Studies of circularly polarized electroluminescence were carried out using standard measurement techniques with a quarter wave plate and a polarizer [1]. The degree of circular polarization of EL (P_{EL}) was calculated by the formula:

$$P_{EL} = \frac{I(\sigma^+)-I(\sigma^-)}{I(\sigma^+)+I(\sigma^-)} \cdot 100\% , \qquad (1)$$

where P_{EL} — is the degree of circular polarization of electroluminescence; $I(\sigma^+)$, $I(\sigma^-)$ are the intensities of the left and right circularly polarized EL components respectively.

3. Results and discussion

Measurements of magnetic field dependences of P_{EL} for as-grown sample and for annealed sample at different temperatures were performed. It was shown, that for the as-grown structure, the maximum operating temperature of the spin light-emitting diode (i.e., the temperature at which non-linear magnetic field dependence of a polarization degree is conserved) is 25K. The detailed measurements of as-grown samples were performed in [4].

After the pulsed laser annealing of the structures, the diode operating temperature has increased to ~ 110 K, due to an increase of the Curie temperature of the (Ga,Mn)As layer. The magnetic field dependences of structure after laser annealing is shown in Figure 2.

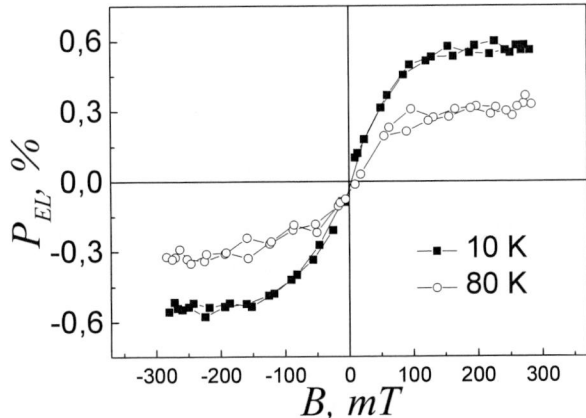

Figure 2. Magnetic field dependence of P_{EL} for annealed sample. Diode current – 10mA.

The increase of the Curie temperature is presumably due to the redistribution of Mn in the layer of a dilute magnetic semiconductor: most of Mn becomes embedded into the lattice sites. In addition, the luminescence intensity, as well as the magnitude of the degree of circular polarization after laser exposure, remained almost without changes. It means that the laser impact affected basically the ferromagnetic injector, while maintaining the parameters of the light-emitting diode.

The comparison of the temperature dependence of the degree of circular polarization for the as-grown and annealed structures is shown at Figure 3.

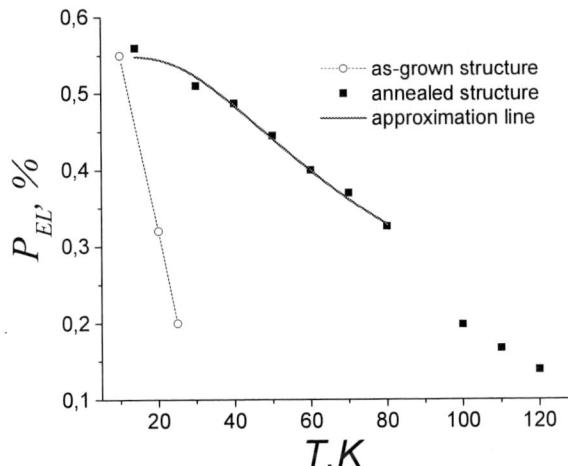

Figure 3. Temperature dependence of the degree of circular polarization for as-grown and annealed structures (I = 10mA, B = 300mT) and a curve approximating dependence for annealed structure by the formula 2 (red line).

Let us discuss the mechanism of spin injection from (Ga,Mn)As into semiconductor structure. The band diagram of studied structure is shown in Figure 4.

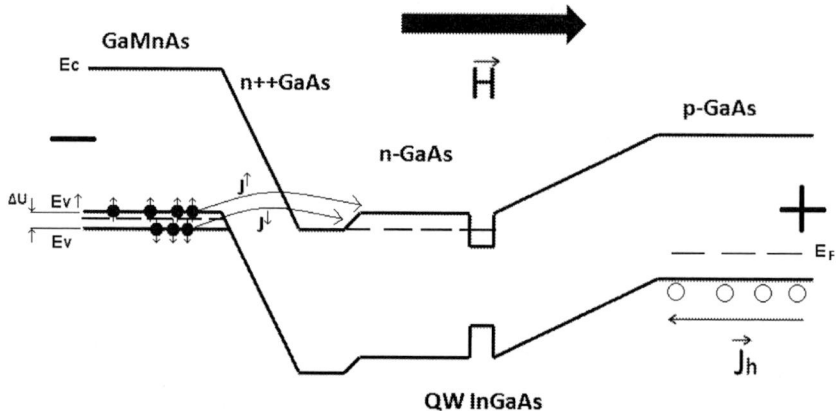

Figure 4. The scheme of the spin injection process in the studied samples.

The rapid drop of the polarization degree near the Curie point can obviously be attributed with the decrease of spin polarization in (Ga,Mn)As because of the demagnetization of the sample. At the temperature, which is far below the Curie point the relatively small decrease of P_{EL} with increasing T was obtained. We believe that in the low-temperature range (10-80 K) the magnetization value of (Ga,Mn)As does not change significantly and the detected P_{EL} decrease is due to the decrease of the current spin polarization. The latter can be defined as

$$\eta = \frac{J^{\uparrow} - J^{\downarrow}}{J^{\uparrow} + J^{\downarrow}} \cdot 100\%, \qquad (2)$$

where $J^{\uparrow}(J^{\downarrow})$ - is a spin-up and spin-down current respectively. These currents are determined by the probability of tunnelling and thermal activation through a potential barrier for charge carriers with different spins (shown schematically at Fig.4). When the structure is introduced into an external magnetic field (to a saturation of magnetization), energy levels of charge carriers with parallel and antiparallel spins in the DMS layer split and, as a result, the probability of tunnelling for carriers with the majority spin increases. Then the spin polarized current and the EL polarization degree are determined only by the thermal energy of the charge carriers.

To evaluate the spin polarization degree in low temperature range the following formula was used:

$$P_{EL}(T) = P_0 \frac{1 - exp\left(\frac{-q\Delta U}{kT}\right)}{1 + exp\left(\frac{-q\Delta U}{kT}\right)} \cdot 100\%, \qquad (3)$$

where P_0 – the carriers spin polarization in magnetized (Ga,Mn)As divided by the depolarization factor [5]; ΔU – the difference of energy levels of charge carriers with parallel and antiparallel spins in the DMS layer. The approximation of the experimental $P_{EL}(T)$ dependence by the formula (3) with the P_0 and ΔU being used as the adjustable parameters is given at Fig.3 (solid line). One can see that there is a good agreement between the experimental data and the approximation curve taking into account the experimental error.

The energy splitting value used for approximation was $\Delta U = 9.55$ meV. This value correlates well with the values of the energy splitting between spin-up and spin-down bands of magnetized (Ga,Mn)As in [6] – 5-11 meV and in [7] – 9-12 meV. This, in turn, is in good agreement with the spin injection picture plotted at Fig.4 since within the thermally activated tunnelling process the spin polarized current should indeed be dependent on the potential barrier difference for spin-up and spin-down carriers.

The P_0 value derived from the approximation was 0.56 %. Within the known value of spin polarization in (Ga,Mn)As (calculated in [8] from the Andreev Spectroscopy measurements;

$P \sim 57\ \%$) one can estimate the depolarization factor as high as 100. This value corresponds to the decrease of electron spin polarization due to the spin scattering during the injection through a highly doped n+GaAs and p-GaAs layers.

Thus the energy difference ΔU, which was revealed from the approximation of experimental curve is in good agreement with the values obtained from completely different experiments [6,7]. We believe that such agreement is the evidence of the validity of the theory used.

It should be noted that the approximation by formula (3) is only applicable at temperatures far below the Curie temperature, because of when approaching the Curie point, the (Ga,Mn)As demagnetization becomes the most significant factor influencing the polarization degree.

In conclusion we have investigated the temperature dependence of circular polarization degree of spin light-emitting diodes based on p-GaAs/InGaAs/n-GaAs/n+GaAs Zener tunneling diode structure with the (Ga,Mn)As layer deposited on the top n+GaAs. The laser annealing of the top (Ga,Mn)As have led to the increase of the SLED operating temperature. The analysis of P_{EL} temperature dependence has revealed the mechanism of thermally activated tunneling of bound electrons from energetically split spin-up and spin-down subbands of magnetized (Ga,Mn)As.

Acknowledgments
Authors would like to acknowledge Dr. B.N. Zvonkov for the fabrication of investigated samples. This study was supported by the Russian Foundation for Basic Research (projects 18-37-00358 (growth of samples by metal organic vapour-phase epitaxy and pulsed laser deposition), 16-07-01102), Ministry of Education and Science of Russian Federation (project 8.175.2017/PP) and the scholarship of the president of the Russian Federation (SP-2450.2018.5).

References
[1] Holub H, Bhattacharya P 2007 *J. Phys. D: Appl. Phys* **40** R179
[2] Dietl T, Ohno Y 2014 *Rev. Mod. Phys.* **86** 187.
[3] Vikhrova O, Danilov Yu, Zvonkov B, et.al. 2017 *Physics of the Solid State* V.**59**(11) 2150.
[4] Malysheva E, Dorokhin M, Ved' M, et.al. 2015 *Semiconductors* V.**49**(11) 1448.
[5] S.H. Liang, T.T. Zhang, P. Barate, et.al. 2014 *Phys.Rev.B* **90** 085310.
[6] Dietl T 2001 *Phys. Rev. B* **63** 195205.
[7] Linnarsson M, Janzen E, Monemar B, et.al. 1997 *Phys.Rev.B* **55** 6938.
[8] S. Piano, R. Grein, C.J. Mellor, et.al. 2011 *Phys.Rev.B* **83** 081305(R).

SPBOPEN 2018   IOP Publishing

Morphology of garnet films for thermo-magnetic recording

V Berzhansky[1], Y Danishevskaya[1,2], A Nedviga[1], M Bektemirova[1]

[1]Institute of Physics and Technology, V. I. Vernadsky Crimean Federal University, Simferopol 295007, Russia
[2]Crimean University of Culture, Art and Tourism, Simferopol 295017, Russia

Abstract. The work deals with the investigation of sensitivity of epitaxial garnet films for thermo-magnetic recording. The spatial resolution of replicas of inhomogeneous fields was associated with the morphological features of the films. Surface morphology of the intense epitaxial films was studied by optical methods and atomic force microscopy. It was shown that the changes in the block structure with an increase in the mismatch between the film-substrate constants occur.

1. Introduction

Uniaxial anisotropy high-coercive garnet films (GF) are an optimal media for thermo-magnetic recording by laser beams or contact printing [1]. For example, contact printing magneto-optic (MO) films were used [2] for criminalistics examinations of magnetic recordings. For these purposes the films had a low Curie temperature ($T_C < 100^0 C$). The effective technological way to increase coercivity H_c is to make strains by increasing mismatch $\Delta a = (a_f - a_s)$ between crystalline lattices of film a_f and substrate a_s. Stresses are proportional to the mismatch and networks of misfit dislocations are formed usually to relief them [3].

2. High-Coercive Garnet Films Synthesis

The films of composition $(Bi,Lu,Sm,Ca)_3(Fe,Ga,Al,Sc,Zr)_5O_{12}$ were investigated. The films was synthesized by liquid-phase epitaxy on substrates $Gd_3Ga_5O_{12}$ with crystallographic orientation (111) and thickness 0.6 mm. The crystal lattice parameter of all applied substrates was standard ($a_s = 12.383$ Å). The thickness of the initial epitaxial layer was $h = 4\text{-}7$ μm. According to the results of the X-ray diffraction analysis, the absolute mismatch of the crystal constants was within the limits of $\Delta a = a_f - a_s = 0.040 - 0.113$ Å corresponded to the relative mismatch of $f = 0.3\text{-}0.9\%$. Samples after epitaxy were polished by an Al_2O_3 abrasive of fraction 1/0. Some of them were polished to get a wedge.

3. Magnetic and Morphology Properties of the Films

All samples have a block structure that forms a kind of mosaic with an element of the order of 1-5 μm in size (Fig. 1). The stress distribution is obtained in images of a polarization-optical microscope with a phase-contrast objective. Stresses are seen on the cross section of the film in reflected light (differential interference-contrast microscopy).

Content from this work may be used under the terms of the Creative Commons Attribution 3.0 licence. Any further distribution of this work must maintain attribution to the author(s) and the title of the work, journal citation and DOI.
Published under licence by IOP Publishing Ltd

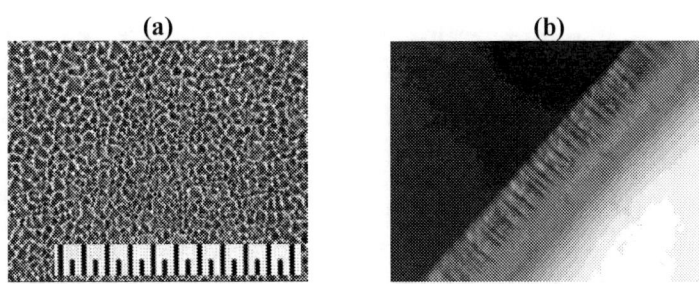

Figure 1 (a, b). Domain structure and stress in the film. Phase contrast microscopy **(a).** The cross section of the film. Differential interference-contrast microscopy **(b).**

The dimensions of the blocks have a dependence on the value Δa (Fig. 2, Fig. 3). We assume that such a block structure affects the sensitivity and spatial resolution of the magneto-optical converter.

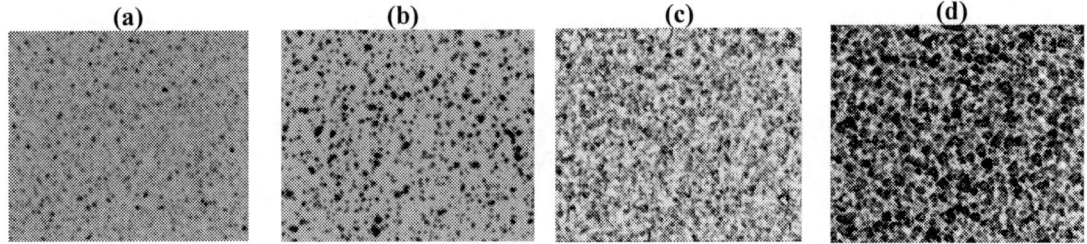

Figure 2 (a, b, c, d). Surfaces of samples with Δa : 0.074 Å **(a),** 0.089 Å **(b),** 0.099 Å **(c)** and 0.102 Å **(d).** Optical Microscopy.

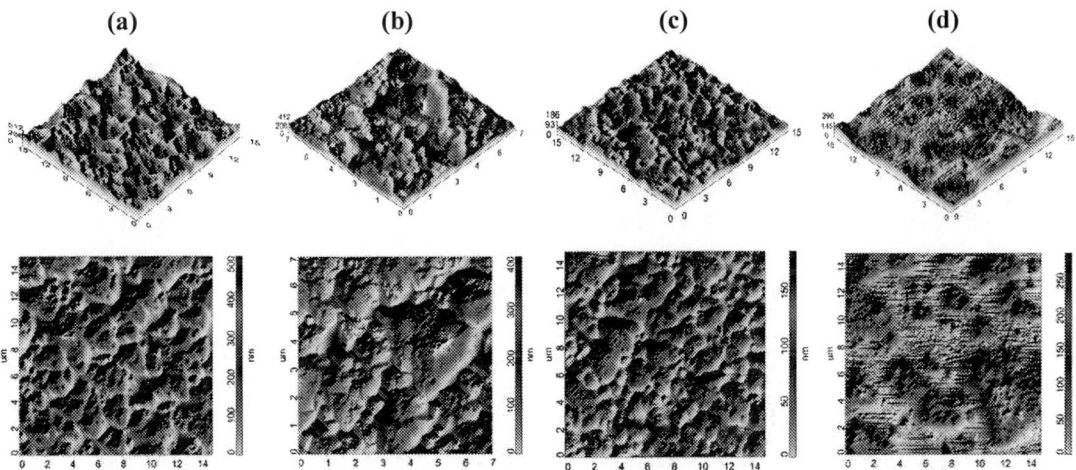

Figure 3 (a, b, c, d). Surfaces of samples with Δa : 0.074 Å **(a),** 0.089 Å **(b),** 0.099 Å **(c)** and 0.102 Å **(d).** Atomic Force Microscopy.

The existing of two types of domain structure (DS) (i.e. hysteresis) is a standard property of strained GF [4,5]. Magnetooptical images of two types DS are obtained after demagnetizing of the same film's area by variable magnetic field (stable DS) and heating up to $T > T_C$ (meta-stable one).

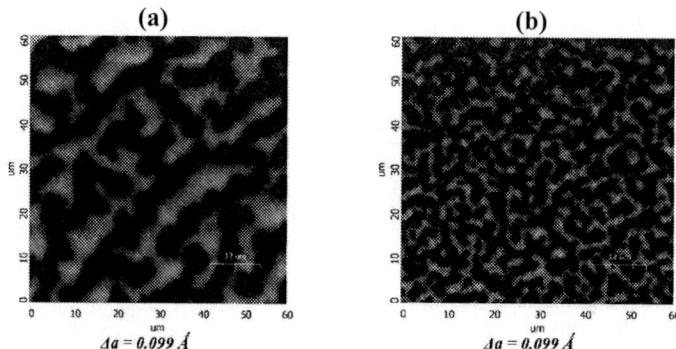

Figure 4 (a, b). Magnetooptical images of two types DS. Stable DS **(a)**, meta-stable DS **(b)**.

Images of the films were registered (Fig. 5) with the help of polarization microscopy in crossed polarizers. These images are showed the distribution of internal stresses in films with different $\varDelta a$ and the domain boundaries in the form of black curve. It is seen that the contrast of stresses image in the films is increased with increasing Δa. Domain boundaries are torn at places of strong stress contrast.

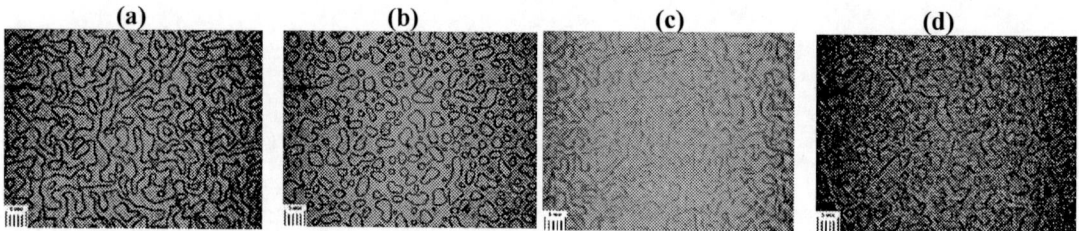

Figure 5 (a, b, c, d). Domain walls and stress in the films after demagnetization by heating up to $T > T_C$ (samples with $\varDelta a$: 0.074 Å **(a)**, 0.089 Å **(b)**, 0.099 Å **(c)** and 0.102 Å **(d)**). Polarization microscope (polarizers crossed).

It is possible to use these films for contact printing due to the existence of a metastable domain structure. The film and magnetic recording media are pressed and heated to the Curie point of the film. Then the system cools down. As a result, we have a copy of the signal on the ferrite-garnet film.

To investigate the sensitivity of epitaxial garnet films for thermo-magnetic recording, records of signals with spatial periods from 150 to 1.9 μm were used. The amplitude of the residual magnetization is varied within more than 60 dB. Thus, a sequence was formed on the carrier consisting of at least 11 differently magnetized sections – wave packets with free (demagnetized) gaps.

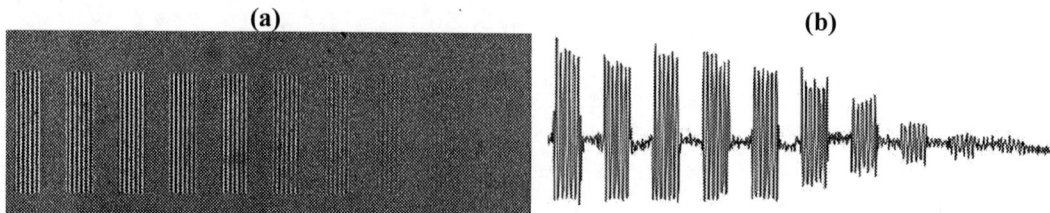

Figure 6 (a, b). Thermomagnetic replicas **(a)** and intensity profiles for recording period of 15 μm **(b)**. Amplitude of the field increases from right to left.

Copies of both "complete" and "incomplete" replicas are actually observed. It is advisable to analyze the ratio of the number of the first to the number of the second (or vice versa). If this ratio is the lower, then the sensitivity of the sample to the field at a given spatial frequency is lower.

The sensitivity of the film was calculated by counting the number of N_r packets reproduced by the film in the form of complete replicas, as well as the total number of observed N_v packets. As a result, the signal with a period of 5.95 μm is best copied. Signals with periods of 1.98 μm, 2.06 μm, 3.6 μm are copied well with films with $\Delta a = 0.098$ Å.

The spatial resolution of thermal replicas of magnetic records is limited by the dimensions of the segments of the block structure of the films. As a result, the sensitivity of the magneto-optical converter is lost in replicas of signals with periods 1-2 μm. Areas of concentration of dislocations are nonmagnetic regions. Replica of the magnetic field of the information carrier takes the form of a curved or discontinuous line.

4. Conclusions

It has been established that all samples have a block structure that forms a kind of mosaic with an element of the order of 1-5 μm in size. The dimensions of the blocks have a dependence on the value Δa. We assume that such a block structure affects the sensitivity and spatial resolution of the magneto-optical converter. Therefore, samples with $\Delta a > 0,099$ Å with large segments are not appropriate for replication purposes.

Acknowledgments

The authors from V. I. Vernadsky Crimean Federal University acknowledge support by the RF Ministry of Education and Science in the framework of the state task (project no. 3.7126.2017/8.9).

References

[1] Inoue F, Itoh A, Kawanishi K 1980 *Japan. J. Appl. Phys.* **19** 2105

[2] Vishnevskii V, Nesteruk A, Nedviga A, Dubinko S and Prokopov A 2007 *Sens. Lett.* **5** 29

[3] Nesteruk A G, Vishnevskii V G, Dubinko S V, Nedviga A S, Mikhailova H V 2005 Abstracts of International Conference ICFM'2005 (Partenit) p 159

[4] Mikherskii R, Dubinko S, Vishnevskii V, Nedviga A and Prokopov A 2000 *Ukrainian J. Phys.* **45** 368

[5] Danishevskaya Y V, Krikun A S, Nedviga A S, Mikhailova T V and Berzhansky V N 2017 *J. Phys. : Conf. Ser.* **917** 072004

SPBOPEN 2018   IOP Publishing

A 0.3-0.7 THz flux-flow oscillator integrated with the slot antenna and elliptical lens

N V Kinev[1], K I Rudakov[1,2,3], A M Baryshev[3] and V P Koshelets[1]

[1]Kotel'nikov Institute of Radio Engineering and Electronics of RAS, 125009 Moscow, Russia
[2]Moscow Institute of Physics and Technology, 141701 Dolgoprudny, Russia
[3]University of Groningen, 9712 CP Groningen, Netherlands

Abstract. We present a new implementation for a terahertz flux-flow oscillator based on a long Josephson junction, in this concept the oscillator is integrated with the lens antenna providing the THz emission in the open space. The slot planar antenna is used for radiation and matched to the oscillator (by the input) and to the elliptical silicon lens with a width of 10 mm (by output). Three designs of antenna coupled with the oscillator by a microstrip line are developed and calculated, the obtained operating ranges are $250 - 410$ GHz, $330 - 570$ GHz and $420 - 700$ GHz. The impedance and beam pattern of antenna designs are calculated.

1. Introduction
The lack of wideband sources in terahertz (THz) frequency range nowdays is one of the significant problems of radio physics, astronomy and spectroscopy. A flux-flow oscillator (FFO) based on a long Josephson junction with length $l \gg \lambda_J$ is promising source in THz region. Such oscillators based on tunnel structures Nb/AlOx/Nb or Nb/AlN/NbN have the operating range from 200 to 750 GHz which is about 100% of central frequency. It was successfully implemented as a local oscillator in heterodyne receiver [1-2] for pumping the SIS mixer, in such layout the FFO and mixer are located on a single chip. The output power of such oscillator is not so large – from about 1 µW up to a several µW, which is enough for some application. There were no reports of utilization FFO as an external source, however such oscillator could be useful for many applications due to its ultra-wideband frequency range, simplicity of fabrication, relatively low cost and enough power for different tasks – heterodyne receivers, spectroscopy.

2. Concept of the THz oscillator
The principle of operation of FFO is discussed in classical papers (e.g. [3-4] and others), its modern characteristics as part of the superconducting integrated receiver of 450-650 GHz range are presented in [2]. The operating range of the specimen with dimensions 400×16 µm^2 based on the Nb/AlN/NbN trilayer is from 200 GHz up to 750 GHz with the spectral linewidth of about 1 MHz. The upper border of operating range is limited by a half of superconductor energy gap Δ and can potentially reach 1 THz. The frequency f of oscillation is defined by the Josephson equation $hf = 2eV_{DC}$, where V_{DC} is the DC voltage of the junction. The phase-lock loop (PLL) is used for the phase locking of the emission, the PLL system narrows the actual spectral line down to about 40 kHz and collect up to 97% of the power in the peak (Spectral ratio).

The principal idea of this work is the integration of such oscillator with a transmitting slot antenna on a single chip (fig. 1a,b) placed on the back surface of the silicon elliptical lens (fig. 1c). The substrate

Content from this work may be used under the terms of the Creative Commons Attribution 3.0 licence. Any further distribution of this work must maintain attribution to the author(s) and the title of the work, journal citation and DOI.
Published under licence by IOP Publishing Ltd

of the chip is also made of silicon with dielectric constant ε = 11.7 to avoid the beam refractions on the chip-lens border. Thus, the main task is the coupling of the oscillator having low output impedance (less than 1 Ω) to the lens antenna having high impedance (tens of Ω) and forming a beam pattern required for applications. A similar type of slot antenna as shown in fig. 1a was used in [5] as the receiving antenna coupled to a SIS detector in the receiver for 100 GHz, 246 GHz and 500 GHz. This type of antenna was chosen in the present work for the first effort to couple FFO with the antenna. Such antenna has the much less bandwidth than the oscillator, hence the band 300 − 700 GHz is to be realized with three antenna designs with central frequencies of 350 GHz, 450 GHz and 600 GHz. The designs differ from each other in geometry − the length and the width of slots and connections between them, and also in the layout of the microstrip line including the impedance transformer; the microstrip line is the feeder for the antenna. Another type of slot antenna shown in fig. 1b is also proposed with central frequency of 450 GHz, such antenna is more narrowband than the previous one according to calculations. A similar antenna was used in [6] also as the receiving antenna for SIS receiver with the central frequency of 500 GHz, the SIS-mixer having impedance of 4 Ω was connected to microstrip line.

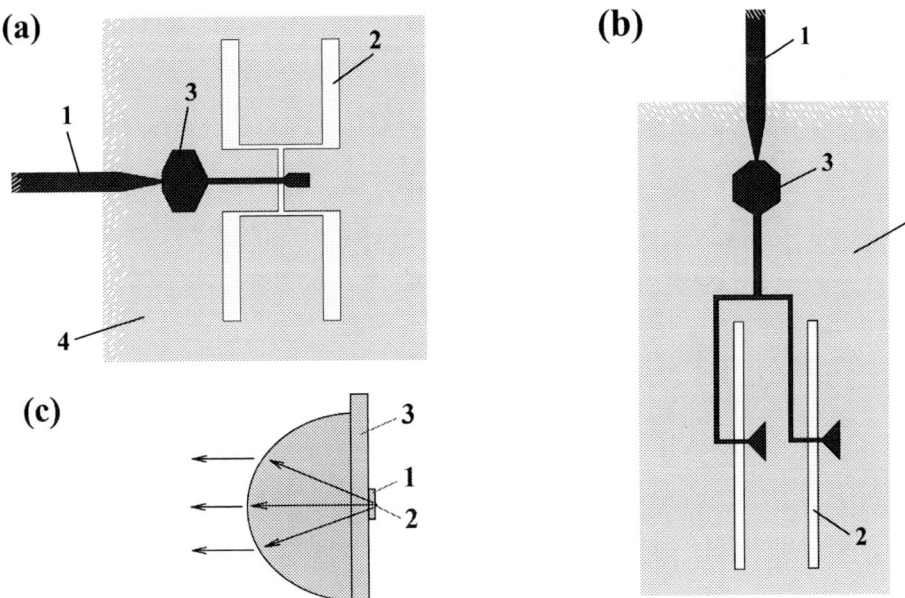

Figure 1. **(a,b)** layout of the planar structure of THz oscillator (1) based on a long SIS junction coupled to slot antenna (2) by microstrip line (3), metallization layer (4) of antenna is also the bottom electrode of the microstrip and the SIS junction; **(c)** scheme of the chip (1) with oscillator and antenna shown on **(a)** or **(b)** placed at the far focus (2) of the silicon lens (3). The scheme (c) is not to scale.

All the slot antennas presented will be fabricated from superconducting Nb thin film with the thickness of 200 nm. The elliptical converging lens with the width of 10 mm will be used, the same lens was used for superconducting receiver of 450 − 650 GHz in the TELIS instrument [2]. It was shown in the paper [5] that such lens matched to the antenna in its focus can be used in the wide frequency band 0.1 − 1 THz without changing the lens dimensions. As the FFO based on Nb/AlOx/Nb or Nb/AlN/NbN requires cooling for superconducting state, the chip with the lens are placed in the liquid helium cryogenic system with the temperature 4.2 K, in this setup the antenna and all the microstrip lines made of Nb with critical temperature of about 9 K are also in the superconducting state.

3. Numerical simulations

The calculations of power coupling of oscillator with the antenna are made using the specialized software for microwave 3D modeling – CST Studio. For taking into account the superconducting state of the antenna and mictrostrip lines London penetration depth for thin Nb films $\lambda_L = 85$ nm is used. All the results presented below are made for slot antenna of the first type shown in fig. 1a. The results of the numerical simulations of the coupling for three designs developed for the central frequencies 350 GHz, 450 GHz and 600 GHz are presented in fig. 2, the radiated part in the open space is shown. More than 70% of oscillator output power is radiated in the ranges of $250 - 410$ GHz, $330 - 570$ GHz and $420 - 700$ GHz for three designs; the dependence is rather flat around the central frequencies. The microstrip line is designed for the matching of the low output impedance of the FFO which was set equal to 0.5 Ω in the calculation model with the high input impedance of the slot antenna. In the fig. 3 the impedances of the antennas of 450 GHz and 600 GHz in the point of connection to microstrip line are presented. For both designs the frequency dependence of the impedance is rather flat, without any jumps in the operating region: full impedance is changing from 30 Ω to 40 Ω for 450 GHz design and from 26 Ω to 40 Ω for 600 GHz design.

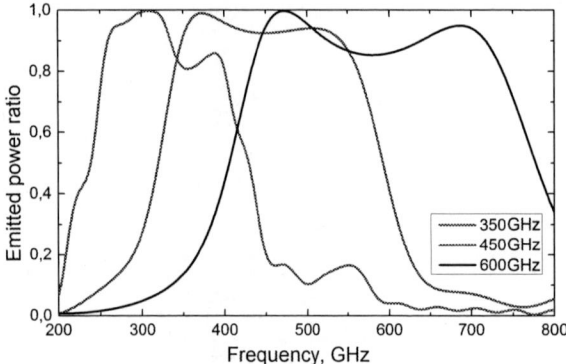
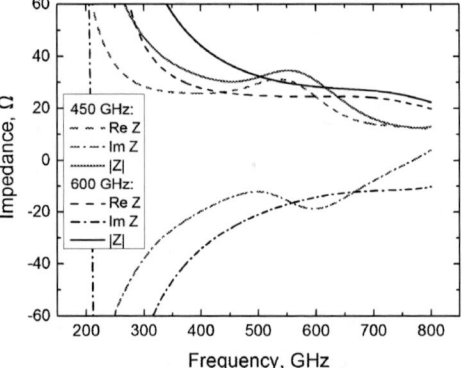

Figure 2. Emitted power ratio into solid angle 4π by antennas of three different designs

Figure 3. The impedances of the slot antenna of designs for 450 GHz and 600 GHz.

In the fig. 4a the calculated beam patterns at the fixed frequency for 350 GHz, 450 GHz and 600 GHz antenna designs are shown, the main power is concentrated in the center lobe. In the fig.4b the set of beam patterns for the 450 GHz antenna design at the frequencies from 350 GHz up to 650 GHz with the step of 50 GHz are shown. This set at the different frequencies demonstrates the weak dependence on the operating frequency. Beam pattern numerical simulations are made not taking into account the elliptical lens which greatly narrows and gains the central lobe, and has no influence on the dependence in fig.2 [5]. The overall operating frequency range for all the proposed designs of slot antenna is from 250 GHz to 700 GHz.

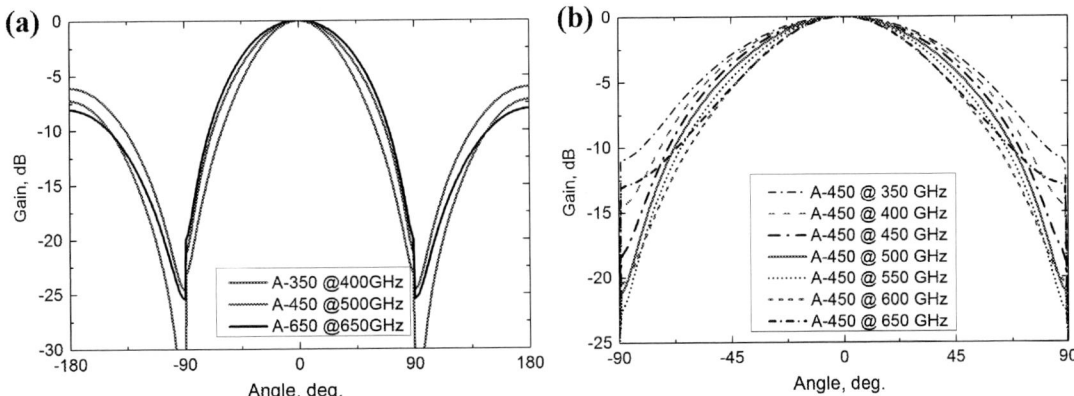

Figure 4. (a) beam patterns of three antenna designs at the fixed frequency in the region of operation; (b) set of beam patterns for 450 GHz antenna design at different frequencies.

4. Conclusion

The superconducting THz oscillator based on the long Josephson junction with unidirectional flow of the fluxons (flux-flow oscillator or FFO) is an encouraging solution of the THz source for the tasks where wideband frequency tuning is required and the high power is not necessary. The applications of such oscillator are the heterodyne receiving and gas spectroscopy, for instance, in the radio astronomy, Earth atmosphere monitoring, medicine devices and security systems, the control of the technological processes in manufacture, telecommunications etc. In this paper we propose the idea and implementation for the external THz source based on FFO integrated with the transmitting slot antenna integrated on a single chip with the oscillator. The lens is used forming the narrow beam pattern of the lens antenna. The first results are presented – the numerical simulations of different antenna designs developed for $250 - 410$ GHz, $330 - 570$ GHz and $420 - 700$ GHz regions, that overall covers the 0.25 - 0.7 THz range. The next step of this work is fabricating of the FFO-antenna samples and their experimental research. Another important aspect of this elaboration is the possibility of the frequency and phase locking of the oscillator, for this task a more complex design should be developed, in which the part of the FFO output power branches to the harmonic mixer coupled to the feedback (frequency and phase locking) loop.

Acknowledgments

This work is supported by Russian Science Foundation (project № 17-79-20343).

References

[1] Koshelets V P, Shitov S V 2000 *Supercond. Sci. Technol.* **13** R53

[2] Lange G, Boersma D, Dercksen J, Dmitriev P, Ermakov A B, Filippenko L V, Golstein H, Hoogeveen R, Jong L, Khudchenko A V, Kinev N V, Kiselev O S, Kuik B, Lange A, Rantwijk J, Sobolev A S, Torgashin M Y, Vries E, Yagoubov P A, Koshelets V P 2010 *Supercond. Sci. Technol.* **23** 045016

[3] Nagatsuma T, Enpuku K, Iri F, Yoshida K 1983 *J. Appl. Phys.* **54** 3302

[4] Zhang Y 1991 *Theoretical and experimental studies of the flux-flow type Josephson oscillator* (Gothenburg: Chalmers University of Technology)

[5] Filipovic D F, Gearhart S S, Rebeiz G M 1993 *IEEE Trans. Microw. Theory Techn.* **41** 10

[6] Zmuidzinas J 1992 *IEEE Trans. Microw. Theory Techn.* **40** 1797

SPBOPEN 2018

IOP Publishing

IOP Conf. Series: Journal of Physics: Conf. Series **1124** (2018) 071002   doi:10.1088/1742-6596/1124/7/071002

The development of the bistable micromechanical actuator for optical relay

Y.B. Enns, E. N. Pyatishev, A. N. Kazakin

Peter the Great Saint-Petersburg Polytechnic University, St. Petersburg, Russia

Abstract. In this work is presented the results of the development of the bistable micromechanical actuator for optical MEMS switch, whereby the mentioned actuator is fabricated on the bulk technique of the microsystems engineering. The influence of local profile stiffening on the dynamic behavior of the arch-shaped suspension was experimentally and theoretically research. This paper is described theoretical and experimental results, obtained during numerical simulation and measuring of the optical switch amplitude-frequency dependence for AC and DC control voltage. Using dynamic methods gives an option of reducing control voltage level.

1. Introduction

In recent decade, microelectromechanical systems (MEMS) became widely used as switching equipment components, specifically, electrical and optical relays and switches. The differential feature of such components consists in the absence of optoelectronic conversion of the switching signal, low input power, miniature overall size, low weight and fabrication cost. The switching of the optical signal is performed by the electromechanical system, which enables micromirror motion in the optical channel. The key component of such systems is flexible beam suspension. The vast majority of flexible suspensions of micromechanical structure are linear elements, wherein linearity provides easy energy conversion. More interest is generated in nonlinear stiffening elements. The reason of the interest in nonlinear suspensions is the opportunity to build multistable systems. Special attention is given to arch-shaped suspensions, which profile is determined by the shape of the mechanical stability loss. Nonlinearity of such suspensions is conditioned by the spur increase of the axial load under lateral beam load [1]. Such structure has two mechanical power minimums, in other words, has two stable states. The presence of two power minimums is determined by stability loss. However, there is bifurcation irregularity in such systems, in other words, significant difference between potential well depths of the first and the second stable states [2]. The method of the local stiffening of arms of the arch-shaped suspension is used to increase the state stability [3]. To ensure effective switching performance between stable states, dynamic methods with AC and DC control voltage are used. The resonance frequency of the control voltage has amplitude-dependent characteristic.

2. Development of design and fabrication method

The bistable actuator of the optical relay is the microelectromechanical system (Fig. 1), designed to perform the optical switching through optical fiber channels. The signal switching process is realized by stopping or passing the optical signal by the micromirror displacement (Fig. 2). The micromirror displaces by 38 μm, which allows arranging the optical signal transmission. The distance between the

Content from this work may be used under the terms of the Creative Commons Attribution 3.0 licence. Any further distribution of this work must maintain attribution to the author(s) and the title of the work, journal citation and DOI.

Published under licence by IOP Publishing Ltd

optical fibers is 40 μm. In order to reduce the losses the collimating microlens is formed at the optical fiber end using wet etching method [4], herewith, the optical losses amounts 0.7 dB.

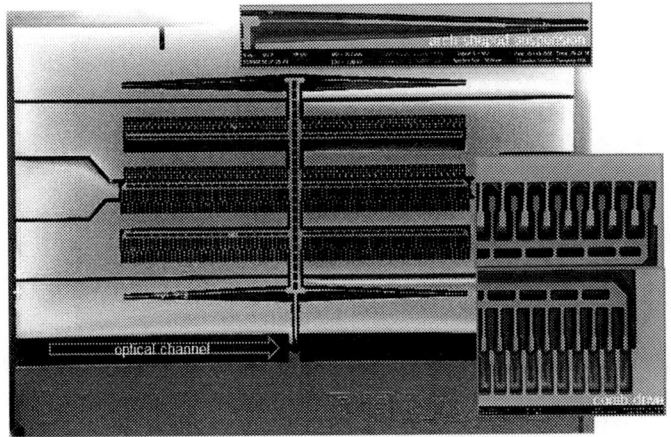

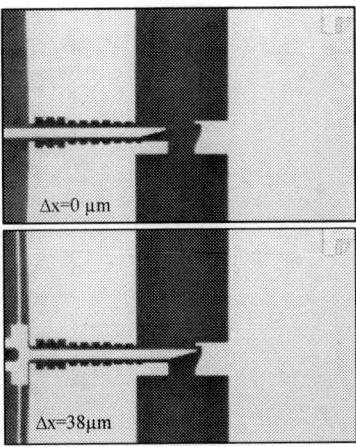

Figure 1. SEM image of the prototype of the bistable microelectromechanical actuator of the optical relay.

Figure 2. Image of the micromirror: open and closed positions

The micromechanical system of the actuator is fabricated on the bulk technique of the microsystems engineering using deep reactive-ion etching (DRIE). Furthermore, DRIE provides one-cycle formation of the micromechanical system as well as the channel for optical fiber, which significantly simplifies the positioning of the optical system. As a structural layer the single-crystal silicon is used, fixed on the glass substrate using the anode merging method. The structural layer thickness is chosen based on the optical fiber diameter (125 μm) and equals 90 μm. The mechanical structure is released by the etching of the glass substrate, herewith, forming the cavity for the deep mounting of the optical fiber.

The main element of the bistable micromechanical actuator is the nonlinear flexible suspension (arched beam), where the availability of two stable states is conditioned by the mechanical stability loss. The heavy increase of the axial load on the beam under the lateral load leads to the formation of two mechanical energy minimums. The lateral load is forming at the central section of the arched beam by joining the comb actuator. To eliminate non-symmetrical (rotational) buckling mode the flexible suspension is made of two arched beams attached in parallel. For said arched flexible elements the small depth of the second potential hole and asymmetrical load values of bifurcation points are common [2]. In order to increase the mechanical stability of the second stable state, method of the local stiffness increase of the arched beam was applied, the beam profile has irregular longwise thickness (Fig. 1). This method allows significantly increasing the ratio between the upper and the lower levels of the mechanical energy near the second stable state (Fig. 3). Herewith, the absolute value of the upper mechanical energy level depends less on the element thickness [3].

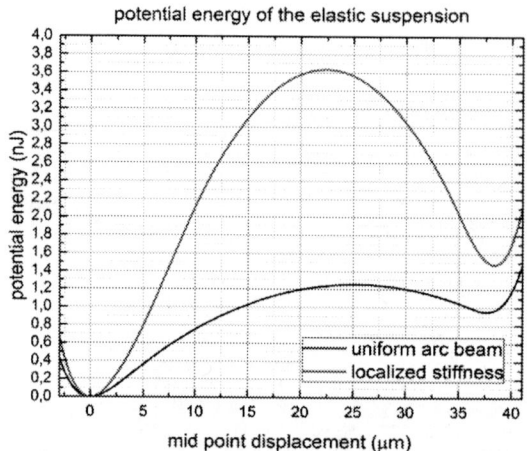

Figure 3. The dependence of the potential energy from the actuator displacements.

The electrode system of the actuator comprises two opposite systems of comb actuators and has nonlinear behavior. The profile of the comb electrodes is also formed having irregular thickness, which provides the gap changing between electrodes surfaces when the actuator moves. The changeable gap allows reducing the load level after abrupt switching, which reduces the probability of the rebound motion when the voltage rapidly releases, and simplifies the electrical control system. Furthermore, the changeable gap provides the better penetration of the etching components and DRIE reaction wastes ejection, which simplifies the etching of narrow cavities at the small deviation of the specific capacity of the actuator [5].

3. Experimental studies

3.1. Studies of the static behavior
To perform the experimental study on the bistable actuator the prototypes were fabricated. To measure the displacements optical and electrical methods were used involving capacity-voltage converter, which allows recording changes of the capacity of the comb structure of the actuator when moving. Herewith, the DC voltage was applied to one of the arrays of the comb actuator, and the capacity was measured at the opposite one. The actuator displacement and action forces of the flexible suspension are shown in Figs. 4–5. The experimental results show effective conformance with the analytical dependence and computational model of the flexible arched suspensions [3].

Fig. 4 depicts the dependence of the force for the arched beam with regular profile thickness; the non-symmetry with respect to the zero of the displacement force is clearly seen. The use of the irregular profile beam allows increasing the symmetry of the characteristic points of the displacement and stability of the second stable state (Fig. 5).

In the process of this research the switching rate of the actuator was defined: 1.5 ms to block the optical channel (Up) and 2 ms to open the optical channel (Down). Herewith, the local increase of the stiffness of the flexible suspension brings switching time closer as well: 1.25 ms to block the optical channel (Up) and 1.3 ms to open the optical channel (Down). The transient processes are shown in the Figs. 4–5 as well.

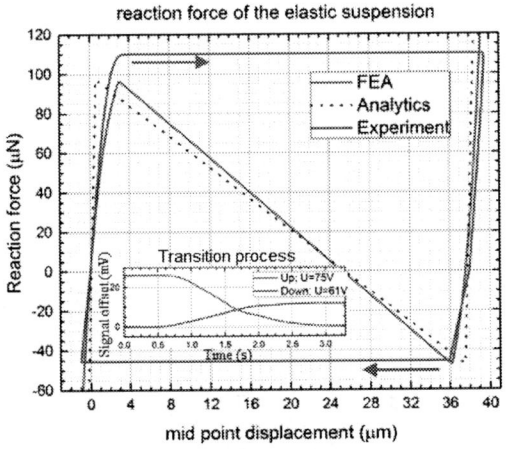
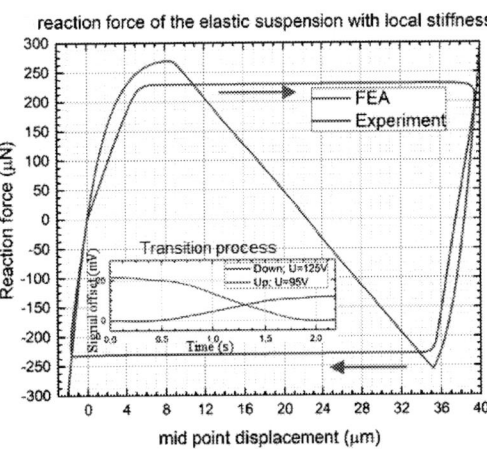

Figure 4. The dependence between the force and displacement, and the transient process of the arched beam with the regular profile thickness.

Figure 5. The dependence between the force and displacement, and the transient process of the arched beam with the irregular profile thickness.

3.2. Studies of the dynamic behavior

The method of the local increase of the stiffness of the arched suspension leads to the increase of the actuator switching load and control voltage. Dynamic switching methods can be used to decrease the control voltage [6]. Herewith, to control the actuator we use resonant signal step-up, excited by the AC voltage with addition of the DC bias voltage. The system of the arched flexible suspension has quadratic and cubic nonlinearities, in other words amplitude dependent resonance frequency [7].

To define the frequency characteristic the prototypes of the optical relay actuator were tested under variable load. The oscillation excitation performed by the AC voltage without DC bias allows achieving the separation of the mechanical oscillations from electrical actuating signal. The frequency of the electrical actuating signal equals to the half of the self-resonant frequency of the actuator.

To provide the quality oscillations the resonator was placed in the vacuum chamber. The quality was defined based on the bandwidth amplitude-frequency response of the resonator under different pressure and small actuating voltage (from 1–0.05 V) (Fig. 6). The actuating voltage value was chosen based on the limits of the linear response range of the resonator. At the high amplitude of the AC voltage (5–10 V) the frequency dependence has clearly nonlinear dependence (softing) (Fig. 7). The characteristics available from experiments correspond to the theoretical dependencies.

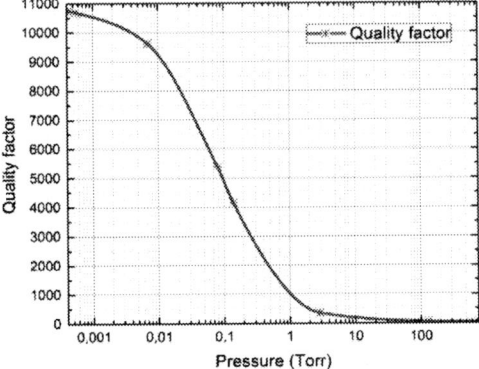
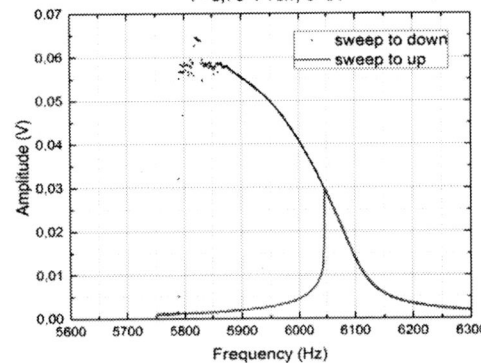

Figure 6. Quality – pressure dependence of the resonator oscillations.

Figure 7. The frequency response of the bistable actuator with the arched suspension $U_{ac} = 5$ V

The amplitude and phase characteristic of the actuator of the optical relay under different load at decreasing frequency is shown in Fig. 8. Herewith, the failure of high-quality oscillations occurs at the phase change much lesser than 90°, and doesn't depend on the oscillation amplitude. It can be explained by the increase of the oscillation sensitivity to the frequency slew rate of the AC signal.

When measuring the frequency response of the resonator with AC actuating voltage with DC bias, a disturbance on the measuring electrodes can be observed from the electrical actuation signal. The obtained amplitude-frequency characteristics are shown in Fig. 9. Herewith, the resonance frequency bias is observed within 1.35 kHz range at the bias voltage $U_{dc} = 45$ V.

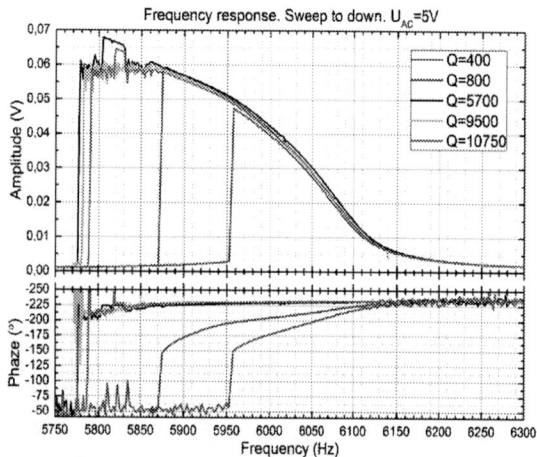

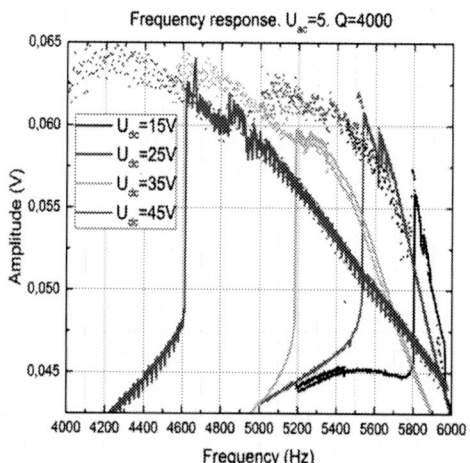

Figure 8. The amplitude and phase characteristic of the actuator of the optical relay under different load, and voltage $U_{ac} = 5$ V; Sweep down by frequency

Figure 9. The amplitude and phase characteristic of the actuator of the optical relay at voltage $U_{ac} = 5$ V and different bias voltages.

By using the obtained frequency dependences we can provide significant reduce of the control voltage when exciting the oscillations by the AC voltage with the DC bias. The nonlinear behavior of the resonator oscillations allows exciting the oscillations with the frequency, which significantly differs from the self-resonant frequency within the linear response range. The AC voltage frequency is defined by the frequency at which the jump of the amplitude frequency dependence occurs for the correspondent bias voltage. When the bias voltage is absent, the forced oscillations don't lead to the significant increase of the oscillations of the micromirror, as well as the bias voltage doesn't lead to its switching. After the abrupt switching to the second stable state, the resonance oscillations decay due to the resonance frequency change.

In the process of the testing the micromechanical resonator was excited under atmospheric pressure by the voltage at the frequency of 4.6 kHz and amplitude of $U_{ac} = 10$ V. The quality of the mechanical resonator in this case amounts 40. The supplying of the bias voltage of 50 V leads to the abrupt switching and displacement of the micromirror to the second stable state. In order to decrease the pressure the resonance system was placed in the vacuum. As it is seen in the Fig. 8, the quality increase for the nonlinear oscillations doesn't lead to the significant increase of the amplitude, as it is for the linear systems. At the quality factor of 8000 the switching between the stable states is provided at 35 V of the DC bias and 5 V of the AC voltage.

4. Conclusion

In the process of this research, the dynamic characteristics of the bistable electromechanical optical switch were studied. Experimental and numerical results demonstrate that the system of the arch-shaped suspension has quadratic and cubic nonlinearity. Herewith, the initial profile of the arch-shaped beams and external and internal axial loads are overriding factors, which determine the behavior and nonlinear features of the arch-shaped beams. The use of the dynamic methods to perform switching makes it possible to reduce control voltage level.

References

[1]   Zhang Y., Wang Y., Li Z., Huang Y., Li D., 2007, *J. Microelectromech. Syst.* **16** 684

[2]   Qiu J., Lang J. H., Slocum A. H., 2004 *J. Microelectromech. Syst.* **13** 353

[3]   Enns Y. B., Pyatishev E N, Glukhovskoy A., 2017, *J. Phys.: Conf. Ser.* **917** 082013

[4]   Ma N., Ashok P. C., Stevenson D. J., Gunn-Moore F. J., and Dholakia1 K., 2010, *Biomed Opt Express,* **1(2),** *694*

[5]   Pyatishev E. N., Enns Ya. B., Kazakin A. N., Kleimanov R. V., Korshunov A. V., 2017, *Conf. 24th ICINS, St. Petersburg, 383*

[6]   Casals-Terre J., Fargas-Marques A., Shkel A. M., 2008, *J. Microelectromech. Syst.,* **17(5),** 1082

[7]   Ramini A. H.; Hennawi Q. M.; Younis M. I., 2016, *J. Microelectromech. Syst.,* **25(3),** 570

SPBOPEN 2018

IOP Publishing

IOP Conf. Series: Journal of Physics: Conf. Series **1124** (2018) 071003   doi:10.1088/1742-6596/1124/7/071003

Calculation of high-frequency conductivity and Hall constant of a thin conductive layer in the view of equal specularity coefficients of its surfaces

O V Savenko[1], I A Kuznetsova[2], A A Yushkanov[3]

[1,2]Department of Microelectronics and General Physics, P G Demidov Yaroslavl State University, Yaroslavl 150003, Russia
[3]Department of Theoretical Physics, Moscow State Regional University, Moscow 105005, Russia

Abstract. The kinetic task about high-frequency conductivity and Hall constant of a thin conductive layer placed in the transverse stationary magnetic field and longitudinal alternative electric field is solved. The ratio between the layer thickness and the mean free path of charge carriers is supposed to be arbitrary. The skin effect isn't taken into account. We consider the diffuse-specular charge carrier reflection mechanism from layer boundaries. Specularity coefficients of the upper and lower layer surfaces are presumed to be equal. The limited cases of the degenerated and non-degenerated electron gas are examined. The dependences of conductivity and Hall constant on the electric field frequency, magnetic field induction, layer thickness and specularity coefficient of layer surfaces are studied.

1. Introduction
Electrical, optical and galvanomagnetic properties of conductive objects with the characteristic linear dimension that is comparable to and less than charge carrier mean free path differ essentially from the properties of macroscopic samples. At room temperature the values of charge carrier mean free path λ for typical semiconductors range between 50 and 1000 nm, and de Broglie wavelength λ_B is proportional to 10 nm [1, 2]. For metals with high conductivity $\lambda = 10 \div 100$ nm, and de Broglie wavelength is proportional to the interatomic distance $\lambda_B \approx 0.3$ nm [1, 2]. Present technologies permit to produce materials with the linear dimension of the order to a few dozen nanometers. Therefore, the situation, when we must take into consideration the classical size effects and may neglect quantum effects for describing of sample electric properties, is realizable in practice.

The first known work devoted to calculation of static conductivity of a thin metal film placed in the perpendicular magnetic field in the view of the diffuse and diffuse-specular electron surface reflection was published by Sondheimer in 1950 [3] and republished later in 2001 [3]. In next works [4, 5] the Sondheimer task was complicated to the cases of arbitrary magnetic field direction [4], non-spherical Fermi surface and different specularity coefficients of layer surfaces [5].

The theoretical investigations devoted to these subjects are continued in present time. In the works [6, 7] the tasks about high-frequency conductivity of a thin metal film in the view of different surface specularity coefficients for the cases of a homogeneous [6] and nonhomogeneous [7] electric field was

Content from this work may be used under the terms of the Creative Commons Attribution 3.0 licence. Any further distribution of this work must maintain attribution to the author(s) and the title of the work, journal citation and DOI.
Published under licence by IOP Publishing Ltd

considered. In the works [8, 9] the theoretical models of high-frequency conductivity and Hall constant of a thin metal film in the view of diffuse [8] and diffuse-specular [9] electron surface scattering mechanisms for equal film surface specularity coefficients was built.

2. Task statement

We consider a thin conductive layer, the material of which is metal or n- or p-type semiconductor. The layer is placed in the transverse constant magnetic field with induction **B**. The alternative electric voltage with the frequency ω is applied to the ends of the layer. The electric and magnetic fields are supposed to be homogeneous. The skin effect is neglected (we assume, that $a < \delta$, where a is the layer thickness, δ is the skin layer depth). Quantum effects aren't taken into account, because $a >> \lambda_B$, where λ_B is de Broglie wavelength of charge carriers.

The time-periodic electric field

$$\mathbf{E} = \mathbf{E}_0 \exp(-\mathrm{i}\omega t) \tag{1}$$

induces the deviation of charge carrier distribution function f_1 from Fermi distribution function f_0:

$$f(z, \mathbf{v}, t) = f_0(\varepsilon) + f_1(z, \mathbf{v}, t) = f_0(\varepsilon) + f_1(z, \mathbf{v}) \exp(-\mathrm{i}\omega t), \tag{2}$$

where $\varepsilon = m\upsilon^2 / 2$ is the electron (hole) kinetic energy in the case of spherically-symmetric energy band, $\mathbf{v}$ and m are the electron (hole) velocity and effective mass respectively.

The distribution function f_1 is determined by solving of the kinetic Boltzmann equation in the relaxation time τ approximation and linear approximation at the external field:

$$\frac{\partial f_1}{\partial z} + \frac{\nu}{\upsilon_z} f_1 + \frac{eB}{m\upsilon_z}\left(\upsilon_y \frac{\partial f_1}{\partial \upsilon_x} - \upsilon_x \frac{\partial f_1}{\partial \upsilon_y}\right) + \frac{e}{m\upsilon_z}\left(E_x \frac{\partial f_0}{\partial \upsilon_x} + E_y \frac{\partial f_0}{\partial \upsilon_y}\right) = 0. \tag{3}$$

Here $\nu = \tau^{-1} - \mathrm{i}\omega$ is the complex scattering frequency.

As the boundary condition to the equation (3) we use the model of charge carrier diffuse-specular reflection from layer surfaces. The specularity coefficients of upper and lower layer boundaries are supposed to be equal.

$$\begin{cases} f_1(\upsilon_z, 0) = q \cdot f_1(-\upsilon_z, 0); \\ f_1(-\upsilon_z, a) = q \cdot f_1(\upsilon_z, a). \end{cases} \tag{4}$$

If we know the non-equilibrium distribution function, we can calculate the current density, conductivity and Hall constant:

$$\mathbf{j} = 2e\left(\frac{m}{h}\right)^3 \int \mathbf{v} f_1 d^3\upsilon, \tag{5}$$

$$\sigma = \frac{j_x}{E_x}, \qquad A_H = \frac{E_y}{Bj_x}. \tag{6}$$

By conducting the mathematical calculation series, we obtain the following expressions of the conductivity and Hall constant:

$$\sigma(x_0, y_0, \beta_0, q, U_\mu) = \sigma_0 \Sigma(x_0, y_0, \beta_0, q, U_\mu); \tag{7}$$

$$A_H(x_0, y_0, \beta_0, q, U_\mu) = A_{H,0} R_H(x_0, y_0, \beta_0, q, U_\mu); \tag{8}$$

$$\Sigma(x_0, y_0, \beta_0, q, U_\mu) = x_0 \frac{(z_0 - b_1)^2 + (\beta_0 - b_2)^2}{(z_0 - b_1)(z_0^2 - \beta_0^2) - 2z_0\beta_0(b_2 - \beta_0)}; \quad (9)$$

$$R_H(x_0, y_0, \beta_0, q, U_\mu) = \frac{1}{\beta_0} \cdot \frac{(z_0^2 - \beta_0^2)(b_2 - \beta_0) + 2z_0\beta_0(z_0 - b_1)}{(z_0 - b_1)^2 + (\beta_0 - b_2)^2}; \quad (10)$$

$$b_{1,2} = \frac{1-q}{I_0\tilde{\upsilon}_1} \int\limits_0^\infty \int\limits_1^\infty \left(\frac{1}{\gamma^3} - \frac{1}{\gamma^5}\right) \frac{U^2 \exp(U - U_\mu)}{\left(1 + \exp(U - U_\mu)\right)^2} \cdot \frac{A_{1,2}}{A_0} d\gamma dU; \quad (11)$$

$$A_0 = 1 - 2q \exp(-p_1) \cos p_2 + q^2 \exp(-2p_1);$$

$$A_1 = 1 - (1+q) \exp(-p_1) \cos p_2 + q \exp(-2p_1);$$

$$A_2 = (1-q) \exp(-p_1) \sin p_2;$$

$$p_1 = z_0\gamma\tilde{\upsilon}_1/\sqrt{U}; \quad p_2 = \beta_0\gamma\tilde{\upsilon}_1/\sqrt{U}; \quad I_0 = \int\limits_0^\infty \frac{\sqrt{U}dU}{1 + \exp(U - U_\mu)}; \quad \gamma = 1/\cos\theta;$$

where $\sigma_0 = ne^2\tau/m$ is the static conductivity, $A_{H,0} = 1/(en)$ is Hall constant in the classic case.

We introduce dimensionless parameters:

$$z_0 = \frac{\upsilon a}{\upsilon_1} = \frac{a}{\upsilon_1\tau} - i\frac{a\omega}{\upsilon_1} = x_0 - iy_0; \quad \beta_0 = \frac{eaB}{m\upsilon_1}; \quad U = \frac{m\upsilon^2}{2k_0T}; \quad U_\mu = \frac{\mu}{k_0T}; \quad \tilde{\upsilon}_1 = \sqrt{\frac{m\upsilon_1^2}{2k_0T}}. \quad (12)$$

The parameters x_0, y_0, z_0 and β_0 are dimensionlessed to characteristic charge carrier velocity introduces by the following view:

$$n\upsilon_1^2 = \frac{5}{3}\int \upsilon^2 f_0 \frac{2d^3(m\upsilon)}{h^3}. \quad (13)$$

In the case of degenerated electron gas $\upsilon_1 \to \upsilon_F$ where υ_F is Fermi velocity. In the case of non-degenerated Fermi gas $\upsilon_1 \to \upsilon_T = \sqrt{5k_0T/m}$. Here υ_T is proportional to the averaged thermal velocity of charge carriers.

3. Limited cases

3.1. The case of degenerated electron gas

Consider the case of degenerated electron gas, i.e. $\exp(U_\mu) \gg 1$. The expressions of the dimensionless conductivity and Hall constant have the view $(9) - (10)$ with the following designations:

$$b_{1,2} = \frac{3}{2}(1-q)\int\limits_1^\infty \left(\frac{1}{\gamma^3} - \frac{1}{\gamma^5}\right) \frac{A_{1,2}}{A_0} d\gamma, \quad p_1 = z_0\gamma, \quad p_2 = \beta_0\gamma. \quad (14)$$

3.2. The case of non-degenerated electron gas

Consider the case of non-degenerated electron gas, i.e. when the condition $\exp(U_\mu) \ll 1$ is satisfied. The expressions of the dimensionless conductivity and Hall constant are determined by the formulae $(9) - (10)$ with the following designations:

$$b_{1,2} = \sqrt{\frac{8}{5\pi}}(1-q)\int_0^\infty w\exp(-w^2)\frac{A_{1,2}}{A_0}dw , \quad p_1 = \frac{z_0}{w}\sqrt{\frac{5}{2}}, \quad p_2 = \frac{\beta_0}{w}\sqrt{\frac{5}{2}}, \quad w = \frac{\sqrt{U}}{\gamma}. \quad (15)$$

4. Results analysis

In figure 1 the dependences of layer dimensionless conductivity module (a) and argument (b) on the dimensionless electric field frequency are represented. The solid and dashed curves are built for the cases of a metal and semiconductor layer respectively. With frequency increasing the conductivity module decreases. This is due to the fact that charge carriers haven't time to respond to external electric field oscillations and behave as the complexity of charge carriers not contributing to layer conductivity. The conductivity argument grows and seeks to $\pi/2$, i.e. the conductivity becomes a purely imaginary value at the high-frequency region.

In figure 2 the dependences of layer dimensionless conductivity module (a) and argument (b) on the dimensionless magnetic field induction are imaged. The solid and dashed curves are built for the cases of a metal and semiconductor layer respectively. In this figure we observe oscillations damped with magnetic field induction and layer surface specularity coefficient increasing. The oscillations of conductivity dependences on the magnetic field induction are less pronounced for the case of non-degenerated electron gas than those for the case of degenerated electron gas due to the charge carrier thermal velocity dispersion.

In figures 3 and 4 the dependences of layer dimensionless Hall constant module (a) and argument (b) on the dimensionless electric field frequency (figure 3) and magnetic field induction (figure 4) are built. The solid and dashed curves are built for the cases of a metal and semiconductor layer respectively. We observe the oscillations which are analogue to figure 2 and damped with electric field frequency and magnetic field induction increasing. When the electric field frequency is equal to the magnetic field induction we observe the resonant-like phenomenon for the Hall constant argument.

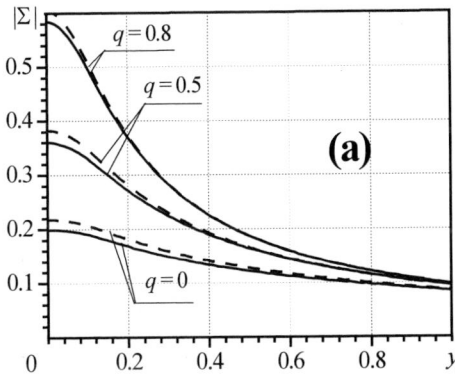 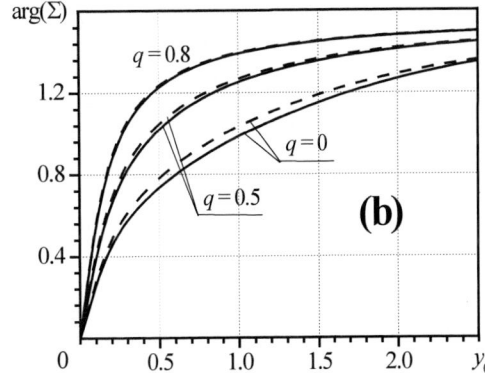

Figure 1(a, b). The dependences of dimensionless conductivity Σ module **(a)** and argument **(b)** for a metal (solid curves) and semiconductor (dashed curves) layer on the dimensionless electric field frequency y_0 at the values of the dimensionless magnetic field induction β_0 and layer thickness x_0 equal to 0.1.

SPBOPEN 2018

IOP Publishing

IOP Conf. Series: Journal of Physics: Conf. Series **1124** (2018) 071003   doi:10.1088/1742-6596/1124/7/071003

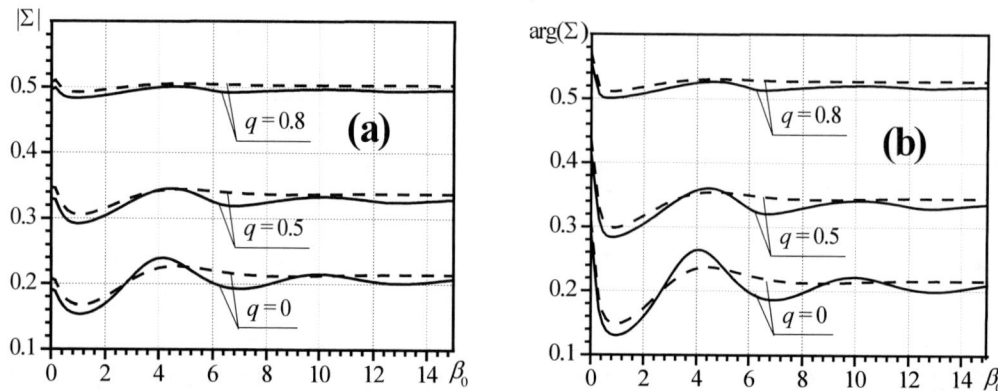

Figure 2(a, b). The dependences of dimensionless conductivity Σ module **(a)** and argument **(b)** for a metal (solid curves) and semiconductor (dashed curves) layer on the dimensionless magnetic field induction β_0 at the values of the dimensionless electric field frequency y_0 and layer thickness x_0 equal to 0.1.

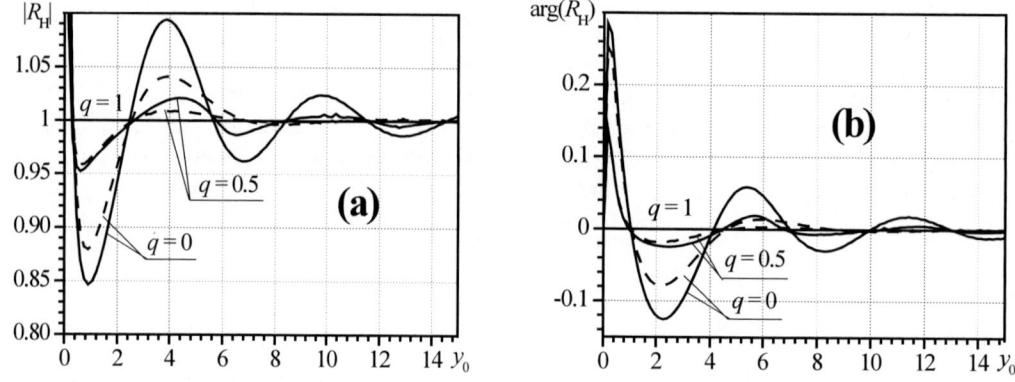

Figure 3(a, b). The dependences of dimensionless Hall constant R_H module **(a)** and argument **(b)** for a metal (solid curves) and semiconductor (dashed curves) layer on the dimensionless electric field frequency y_0 at the values of the dimensionless magnetic field induction β_0 and layer thickness x_0 equal to 0.1.

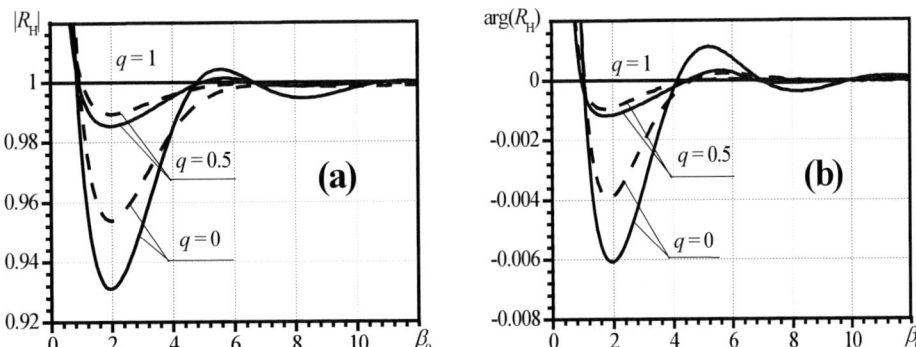

Figure 4(a, b). The dependences of dimensionless Hall constant R_H module **(a)** and argument **(b)** for a metal (solid curves) and semiconductor (dashed curves) layer on the dimensionless magnetic field induction β_0 at the values of the dimensionless electric field frequency y_0 and layer thickness x_0 equal to 0.1.

5. Conclusions

The theoretical model of high-frequency conductivity and Hall constant of a thin conductive layer for the case of equal layer surface specularity coefficients is built. The behavior of conductivity and Hall constant dependences on the non-dimensional parameters: electric field frequency, magnetic field induction, layer thickness and layer surface specularity coefficient, is analyzed. It is detected the oscillations of the dependences of conductivity on the magnetic field induction, also those of the dependences of Hall constant on the electric field frequency and magnetic field induction. The conductivity and Hall constant oscillations for the case of a semiconductor layer are less pronounced than those for the case of a metal layer due to charge carrier thermal velocity dispersion. It is shown that Hall constant argument sharply grows when the dimensionless electric field frequency is equal to the dimensionless magnetic field induction.

References

[1] Anselm A I 1982 *Introduction to Semiconductor Theory* (Englewood Cliffs, NJ: Prentice Hall) p 616
[2] Lifshits I M, Azbel M Ya and Kaganov M I 1973 *Electron Theory of Metals* (New York: Plenum) p 415
[3] Sondheimer E H 2001 *Adv. in Phys.* 2001 **50** 499
[4] Gurevich V L 1959 *Sov. Phys. JETP* **35** 464
[5] Grishin A M, Lutsishin P P, Ostroukhov Yu S and Panchenko O A 1979 *Sov. Phys. JETP* **49** 673
[6] Utkin A I, Zavitaev E V and Yushkanov A A 2016 *J Surf Inv* **10** 962
[7] Utkin A I and Yushkanov A A 2016 *Russian Microelectronics* **45** 357
[8] Kuznetsova I A, Savenko O V and Yushkanov A A 2017 *Tech Phys* **62** 1766
[9] Kuznetsova I A, Savenko O V and Yushkanov A A 2017 *J Surf Inv* **11** 1159

SPBOPEN 2018   IOP Publishing

Electrical and photoelectric characteristics of gallium oxide films obtained by RF magnetron sputtering

A Tsymbalov[1], J Petrova[1,2]

[1]Department of Semiconductor Electronics, Tomsk State University, Tomsk 634050, Russia

[2]Department of Semiconductor Electronics, Tomsk State University, Tomsk 634050, Russia

Abstract. In order to investigate dependence of conductivity on the temperature and photoreponse of β-Ga_2O_3 thin films, films were grown by a radio frequency magnetron sputtering on sapphire substrates. The conductivity of films without annealing depends weakly on the temperature to 560 °C and increase exponentially at T >560 °C. After annealing at 900 °C growth region shifts to 350 °C. Films without annealing are solar-blind and have a photoresponse to illumination with a wavelength of 222 nm.

1. Introduction.

At present, attention is focused on gallium oxide. There are five phases in which Ga_2O_3 can exist: α-, β-, γ-, δ-, ε-phases [1]. Of all five phases, two are distinguished: α- and β-phases. They are the most stable. The oxide has a transparency of about 90% over the entire visible range, to the ultraviolet region and wide bandgap about 4,9eV [2]. The conductivity of gallium oxide depends on many factors: the method of preparation, the temperature of the substrate, the partial pressure of oxygen in the chamber, the ambient temperature, the doping, the chemical composition of the substrate, the further treatment of Ga_2O_3, etc. Wide band gap materials offer the possibility of fabricating high power density electronic devices, deep-ultraviolet transparent electrode, short wavelength optical emitter and deep-ultraviolet transparent photodetector [3].

2. Experiment details.

This report discusses the results of investigations of gallium oxide films obtained by radio frequency magnetron sputtering. Films 150-200 nm in thickness were deposited by magnetron sputtering of a Ga_2O_3 target (99.9999%) onto non-heated sapphire substrates in an AUTO-500 unit (manufactured by Edwards) in an Ar / O_2 gas mixture. The oxygen concentration in the mixture was maintained at (56.1 $\pm$ 0.5) vol.%. The distance between the target and the substrate was 70 mm. The pressure in the chamber during the deposition was $7 \cdot 10^{-3}$ mbar.

After deposition of gallium oxide, the substrate with the film was divided into two parts. One part was left without treatment. The second part was annealed in argon for 30 minutes at a temperature of 900 ° C. The structure and phase composition of the gallium oxide films were determined by X-ray diffraction analysis (RDA) using a Lab-X XRD 6000 Shimadzu X-ray diffractometer. Analysis of the surface of the sputtered films was carried out using an atomic force microscope "Solver HV".

Content from this work may be used under the terms of the Creative Commons Attribution 3.0 licence. Any further distribution of this work must maintain attribution to the author(s) and the title of the work, journal citation and DOI.
Published under licence by IOP Publishing Ltd

The conductivity was recorded in a special chamber. Samples were placed in holders and placed in a chamber whose relative humidity was maintained at 32%. The temperature was increased from 50° C to 590° C.

Photoelectric characteristics were taken using Keithley 2611. At first measurements were carried out without applying radiation to the samples, then the samples were isolated from the light and subsequently irradiated at a wavelength of 400 nm and 222 nm using a light-emitting diode and a special ultraviolet lamp, respectively.

3. Results and discussion.

According to X-ray diffraction analysis, films obtained under the conditions described above and not subjected to annealing at high temperature turn out to be polycrystalline, contain crystallites of α- and β-phases of gallium oxide (Fig. 1a). The α-phase disappears after annealing in argon at 900 ° C for 30 minutes, and the β-phase crystallites remain with different crystallographic planes (Fig. 1b).

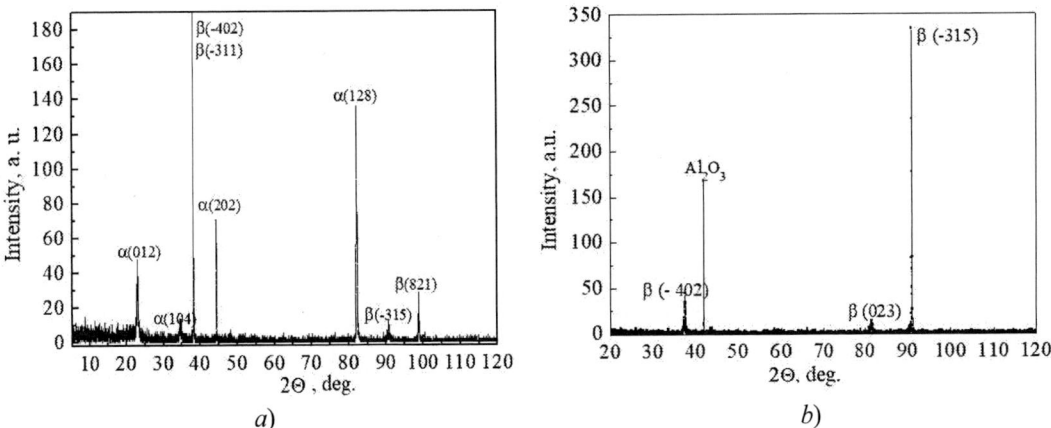

Figure 1. Spectra of X-ray diffraction of Ga_2O_3 films before annealing in Ar (a) and after 30 minutes of annealing in argon at a temperature of 900 ° C (b).

The polycrystalline structure of gallium oxide films not subjected to any treatment is confirmed by AFM data (Fig. 2). The average size of the crystallites is 40 nm (Fig. 2c).

SPBOPEN 2018   IOP Publishing
IOP Conf. Series: Journal of Physics: Conf. Series **1124** (2018) 071004   doi:10.1088/1742-6596/1124/7/071004

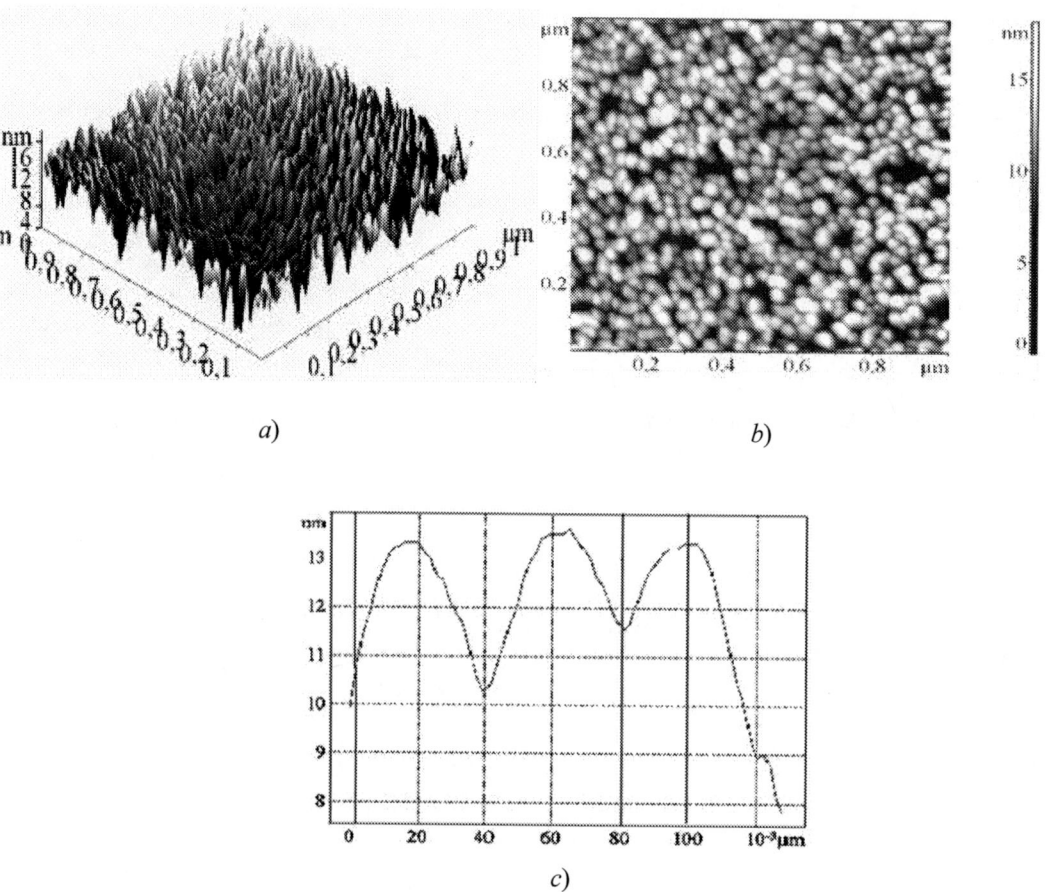

a)   *b)*

c)

Figure 2. The surface of the oxide film not subjected to high-temperature annealing: *a)* – image of the surface of the film; *b)* – the volumetric dimensions of the film; *c)* – the crystallite size.

The oxide film after annealing in argon at 900 contains crystallites whose average size is 90 – 100 nm (Fig. 3).

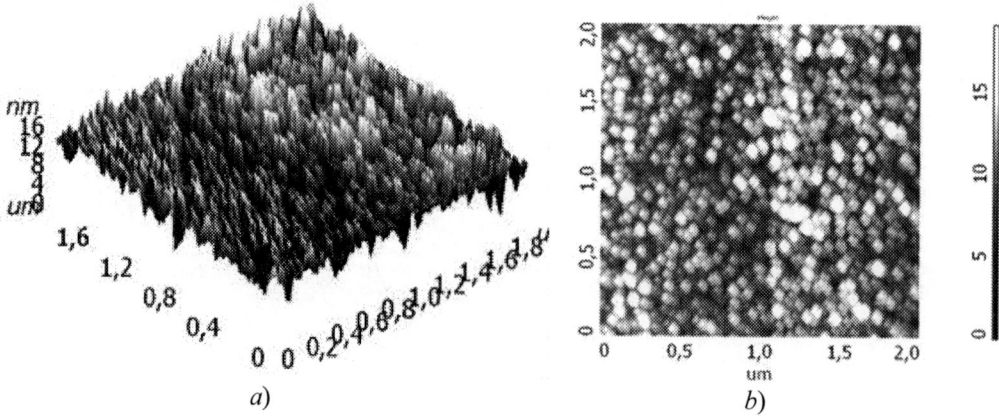

a)   *b)*

SPBOPEN 2018   IOP Publishing

IOP Conf. Series: Journal of Physics: Conf. Series **1124** (2018) 071004   doi:10.1088/1742-6596/1124/7/071004

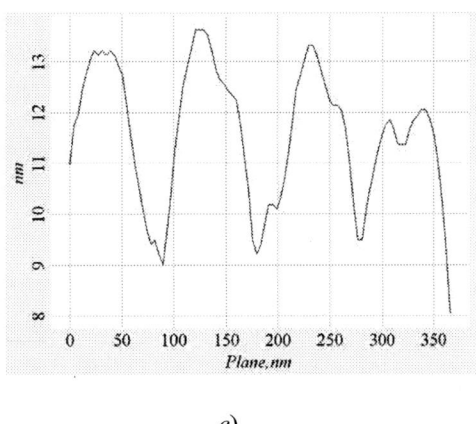

c)

Figure 3. The surface of the oxide film after annealing at 900 ° C: *a)* – image of the surface of the film; *b)* – the volumetric dimensions of the film; *c)* – the crystallite size.

Measurements of electrical and photoelectric characteristics were carried out on samples of a planar design . Films obtained on sapphire substrates and not subjected to thermal annealing had resistances of the order of $10^9 – 10^{10}$ Om. In the interval 50 – 500 ° C, the conductivity of the samples (G) depends weakly on the temperature T and increases exponentially with a further increase in temperature (Fig. 4*a*). The activation energy of conduction growth in the high-temperature region is 0.7 – 1.0 eV.

After annealing the gallium oxide films in argon for 30 minutes at 900 ° C, the region of sharp increase in conductivity at the temperature dependence of G (1 / T) shifts to lower temperatures and begins at T ≈ 350 ° C (Fig. 4*b*). The activation energy of the increase in G with increasing temperature for most of the samples is 0.3 – 0.5 eV. The curve of the dependence of G on 1 / T shows a maximum in the range 470-520 ° C.

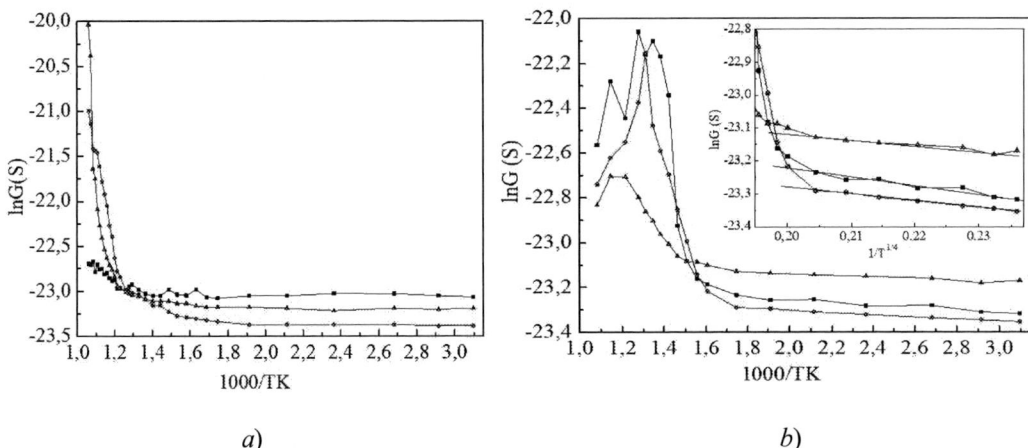

a)   *b)*

Figure 4. Temperature dependence of the conductivity of gallium oxide films on Al_2O_3 without (*a*) and after annealing (*b*) at 900 ° C; on the inset: the temperature dependence of the conductivity in the coordinates ln(G) from $1 / T^{1/4}$.

In the course it was found that samples without annealing are solar-blind and have no photoresponse on wavelength 222 nm (Fig. 5a).

Ga_2O_3 remains solar-blind after annealing. Lighting with $\lambda = 400$ nm does not lead to a change in the VAC, whereas radiation with $\lambda = 222$ nm causes a noticeable increase in current, which depends on the voltage on the sample (Fig. 5b). The absence of a photoresponse when exposed to light with $\lambda = 400$ nm is most likely due to the fact that the energy quantum of this radiation is $h\nu = 3.1$ eV less than the width of band gap of gallium oxide $Eg = 4.8 - 4.9$ eV.

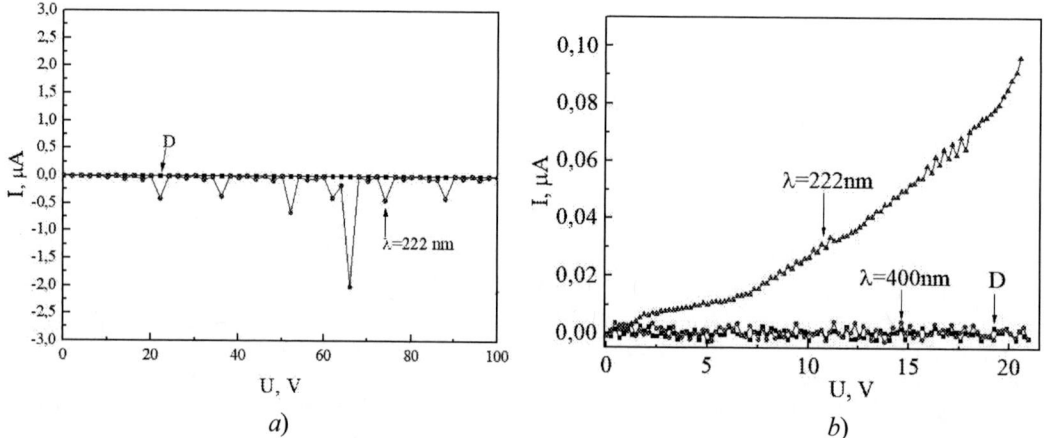

Figure 5. The volt-ampere characteristics of gallium oxide before annealing in Ar and after annealing; without (a) and during exposure to radiation (b). The wavelength is shown in the figure; the dark current is denoted by the letter D.

4. Conclusions

Annealing at 900 ° C causes a change in the surface of the film: the size of the crystallites increases twofold in comparison with the film without annealing.

The structures obtained on a dielectric substrate turn out to be solar-blind in the visible wavelength range and are sensitive to the effects of radiation in the UV-range.

After annealing Ga_2O_3 in Ar at 900 °C growth region of conductivity shifts to 350 °C.

References

[1] A.A. Dakhel Solid State Science 20 (2013) 54-58

[2] S.I. Stepanov, V.I. Nikolaev, V.E. Bougrov and A.E. Romanov Rev. Adv. Master. Sci. 44 (2016) 63-86

[3] Fabi Zhang1, Katsuhiko Saito1, Tooru Tanaka1, Mitsuhiro Nishio, Qixin Guo J Mater Sci: Mater Electron (2015) 26:9624–9629

SPBOPEN 2018

IOP Publishing

Effect of metal modifiers on the characteristics of resistive hydrogen sensors based on thin films of tin dioxide

A V Almaev[1], N K Maksimova[2], E Yu Sevastyanov[1,2], E V Chernikov[2], T A Davydova[2], T E Smirnova[1]

[1]Department of Semiconductor Electronics, Tomsk State University, Tomsk 634050, Russia
[2]Laboratory of Physics of Semiconductor Devices, Kuznetsov Siberian Physical Technical Institute, Tomsk State University, Tomsk 634050, Russia

E-mail: almaev_alex@mail.ru

Abstract. This work presents the results of investigation of the effect of the complex modifiers of Ag + Y introduced into bulk of SnO_2 thin films on the properties of hydrogen sensors and the stability of the devices at long-term test. Two types of the films with different deposited on the surface dispersed catalysts Pt/Pd/SnO_2:Sb, Ag, Y and Ag/SnO_2:Sb, Ag, Y were studied. It is shown that additives of Ag, Y in the presence Pt/Pd on the surface provide the maximum values of the response to hydrogen at the temperature 670 K. In the case of the deposited catalytic Ag the response to hydrogen is considerably lower and the temperature of the maximum response > 713 K. A common feature of two types of the films is the high stability of their properties under the periodical influence of hydrogen in long-term tests.

1. Introduction

It is known that the gas-sensitive characteristics of the sensors based on metaloxide semiconductors can be controlled by introducing metal modifiers into the bulk and depositing catalytic layers on the surface. In our works [1-3] it was shown that the presence of a two layer Pt/Pd catalysts on the tin dioxide surface promotes an increase in the density of chemisorbed oxygen, the sensors are characterized by a high adsorption response to low concentrations (10-10^3 ppm) of reduction gases (CO, H_2). Further studies [3] have shown that in the process of long-term tests of sensors at periodic influence of hydrogen, the decrease in the conductivity in clean air and the increase in the response to H_2 sensors are observed. The most significant changes occur in the first month of testing, and then the sensor's parameters are stabilized. Interestingly, the long-term exposure of other reducing gases, such as hydrogen sulphide and acetone [4], carbon monoxide [5], leads to drop in the sensitivity of the sensors. We suggested [3] that in addition to the interaction with chemisorbed oxygen atomic hydrogen interacts with lattice oxygen at the dissociative adsorption of H_2 molecules. As a result, during testing the number of excess tin atoms and hence the density of the sites of oxygen adsorption increase, there is a growth of the density of chemisorbed oxygen on the surface of tin dioxide and of the response to H_2.

In this work the purpose of the investigation is the establishment of the effect the metal modifiers of Ag and Y introduced into the bulk of SnO_2 thin films on the electrical and gas-sensitive properties of hydrogen sensors and the stability of sensors at long-term test. Two types of the films with different deposited on the surface dispersed catalysts Pt/Pd/SnO_2:Sb, Ag, Y (series 1) and Ag/SnO_2:Sb, Ag, Y

Content from this work may be used under the terms of the Creative Commons Attribution 3.0 licence. Any further distribution of this work must maintain attribution to the author(s) and the title of the work, journal citation and DOI.

Published under licence by IOP Publishing Ltd

(series 2) were studied. It should be noted that such materials as hydrogen sensors investigated for the first time.

2. Experiment

The sensitive material of sensor obtained by magnetron sputtering of tin – antimony alloy target (0.5 at.% of Sb) at the direct current. The antimony acts as shallow donor impurity and it's presence in the bulk of the films leads to the decrease of the working resistance of the sensitive semiconductor layer. Ultra dispersed layers of metals (palladium, platinum and silver) were deposited on the surface of tin dioxide by magnetron sputtering. The fabrication process of sensors is described in detail in [1,3].

The additive content in the bulk of the tin dioxide films was estimated by the ratio of the areas of the sputtered target S_{Sn} and metal pieces S_m (m=Ag, Y). Based on special studies, the optimum ratios S_m/S_{Sn} allowing directional effect on the sensor properties were determined as $S_{Ag}/S_{Sn} = 3 \times 10^{-3}$ and $S_Y/S_{Sn} = 3 \times 10^{-3}$. Manufacturing technology of the sensors includes a stabilizing annealing in air at 723 K for 24 h. In order to establish the equilibrium condition between the surface of sensors and gas mixture in the measuring camera before each measurement the samples were exposed to in the heating during 30 minutes at 673 K.

The principle of operation of gas sensors and methods of measurements of parameters of the devices are described in detail in works [1,2]. The resistance R_0 (conductance G_0) of films in pure air and similar parameters R_1 (G_1) at exposure to hydrogen were measured as functions of the operating temperature T and the concentration of hydrogen n in air. The ratio G_1/G_0 was taken as the adsorption response. The measurements were carried out at the same relative humidity RH level of the gas mixture $RH = 30 \div 35$ %.

3. Results and discussions

At first, microstructure of the film's surface of series 1 and 2 was studied with atomic-force microscopy. The thickness of films are about $120 \div 150$ nm. There are two types of microcrystals with characteristic sizes $d_1 = 12 \div 14$ nm and $d_2 = 32 \div 35$ nm in the films and these microcrystals can form large agglomerates with characteristic size $d_3 = 150 \div 185$ nm.

The results in the present work are compared with the characteristics of the sensors Pt/Pd/SnO$_2$:Sb (series 3) and Au/SnO$_2$:Sb, Au (series 4), which do not contain additives of silver and yttrium in the bulk and which have been studied by us earlier [1] (table 1). The introducing of silver and yttrium in the bulk of thin films of tin dioxide (series 1) leads to rise of resistance R_0 of the sensitive layer in compare with sensors of series 3. Magnitude of R_0 for films without deposited Pt/Pd dispersed layers on the surface (series 2, 4) is significantly lower.

Table 1. Parameters of hydrogen sensor in the pure air and at the exposure of 100 ppm hydrogen.

Type of sensor	R_0 (T=300 K), MΩ	ΔE_1, eV	ΔE_2, eV	Tmax, K	G_1/G_0
Pt/Pd/SnO$_2$:Sb, Ag, Y (series 1)	26.4	0.23	0.69	670	30
Ag/SnO$_2$:Sb, Ag, Y (series 2)	0.23	0.22	0.47	750	6
Pt/Pd/SnO$_2$:Sb (series 3)	4.14	0.11	0.17	670	25.7
Au/SnO$_2$:Sb, Au (series 4)	0.25	0.03	0.08	780	5

The temperature dependence of the sensor resistance R_0 in pure air is defined by three independent values: the electron concentration and mobility in the film bulk and the negative charge density on the surface [1]. During heating from room temperature to $T = 470 \div 500$ K, the resistance of all samples decreases mostly due to the ionization of shallow and deep centres in the film bulk, and the dependencies lnR_0 on 1000/T can be approximated by Arrhenius curves (Figure 1).

Arrhenius curves contain two linear portions from which the activation energies ΔE_1 and ΔE_2 were determined. The value of activation energy depends on the type of modifiers and the type of surface catalysts (table 1). It can be assumed that the introduction Ag and Y in the bulk of tin dioxide contributes to the formation of deep centers, characterized by the increased values of the activation energy in the samples from series 1, 2, and value of ΔE_1 coincides.

The increase of the resistance of samples in the region of $T > 470$ K is caused by an increase in the negative surface charge due to water molecule desorption from the surface and the transition of chemisorbed oxygen from the molecular form O^{2-} to the atomic form O^-. As a result, the space charge region widens and the film resistance accordingly increases. The effects of a change in the surface charge come to an end at high temperatures (~ 700 K), and then R_0 decreases again. The type of modifiers in the bulk and the type of surface catalysts has a significant effect on the gas-sensitive properties of the sensors. The dependencies of the adsorption response on the operating temperature for the sensors of series 1 and 2 are shaped as curves with a maximum (Figure 2). The temperatures T_{max} at which sensors have the maximal response at exposure to 100 ppm of hydrogen and the values of G_1/G_0 at this concentration of gas are presented in table 1.

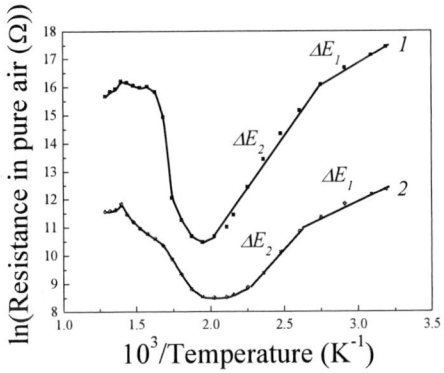

Figure 1. Temperature dependencies of the resistance of sensitive layer in pure air for sensors of series: 1 - 1, 2 - 2

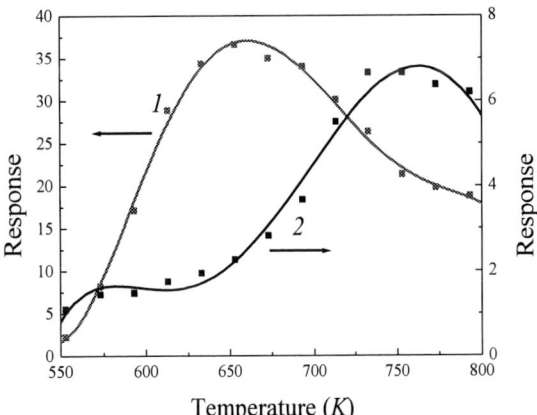

Figure 2. Temperature dependencies of the response at exposure to 100 ppm hydrogen for sensors of series: 1 - 1, 2 – 2.

Based on these studies it was found that the $T_{max} = 670$ K for the sensors of series 1 and series 3. T_{max} is in the range from 713 K to 750 K for the sensors of series 2 and series 4. At the all investigated operation temperatures of the sensors the responses of series 1 and series 3 are higher than the responses of the devices of the series 2 and series 4. The introducing of silver and yttrium in the bulk of thin films with deposited Pt/Pd doesn't lead to changes of the T_{max} and the sensor response increases.

The characteristics of thin films of tin dioxide with deposited Ag dispersed layer on the surface (series 2) are similar to the characteristics of the thin films of Au/SnO_2:Sb, Au (series 4). These samples exhibit the low resistance and the low response at exposure to hydrogen. Value of T_{max} is shifted at area of the higher operation temperature.

Figure 3 shows the concentration dependencies of responses to the hydrogen at $T = 673$ K for the films of series 1 and series 2. The response of sensor of series 1 rises as superlinear low and the response of sensors of series 2 increases as sublinear low with increasing of hydrogen concentration. The response of sensors of series 1 is larger than the response of the sensors of series 2 in all studied hydrogen concentration. The superlinear growth with increase of the hydrogen concentration was observed for sensors based on $Pt/Pd/SnO_2$:Sb (series 3), and the sublinear growth is observed for sensors based on Au/SnO_2:Sb, Au (series 4) [1]. In the work [6] it was shown that characteristics of these samples are described by means of the over – barrier model of conductivity. It is assumed that the over – barrier model of conductivity is applicable to describe of the characteristics of the thin films of tin dioxide with modifiers of silver and yttrium.

Of most interest are the results of a study of the characteristics of the sensors of series 1 and series 2 depending on the duration of the test (table 2). Sensors of series 1 and series 2 demonstrate high temporal stability. The response on the hydrogen effect is virtually unchanged with increasing operating time of the sensor. It is important that during the 8 day daily the sensors was subjected to repeated (5 cycles) the impact of 1000 ppm hydrogen. Each cycle consisted of heating the sensors at the operating temperature T_{max} for 30 minutes, filing into the chamber of hydrogen, the exposure to hydrogen for 25 minutes and pumping through the camera of clean air for 4 minutes. It can be seen that all the parameters has changed slightly. Therefore Ag and Y additives during tin dioxide deposition significantly improves the stability of characteristics of the hydrogen sensors.

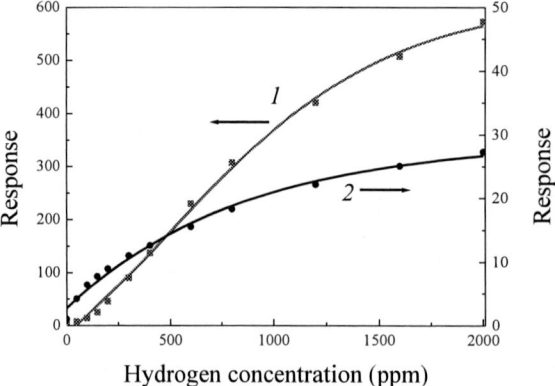

Figure 3. The concentration dependencies of responses to hydrogen for sensors of series: $1 - 1, 2 - 2$.

Table 2. Changes of sensor's characteristics at long-term test.

Duration of test	Pt/Pd/SnO$_2$:Sb, Ag, Y (series 1)			Ag/SnO$_2$:Sb, Ag, Y (series 2)		
	$G_0 \times 10^7$, S	G_1, mS	G_1/G_0	$G_0 \times 10^6$, S	G_1, mS	G_1/G_0
1	0.80	0.046	579.5	6.05	0.076	12.6
29	1.30	0.074	559.8	5.07	0.104	20.6
55	1.42	0.078	545.8	4.52	0.087	19.3
85	1.32	0.075	571.2	4.78	0.087	18.1

A processes causing instability of sensors at long-term test are described in the papers [4,5]. However the deciding factor determining the instability of the hydrogen sensors is the process of reduction of tin from oxide by hydrogen at high operating temperatures. A detailed discussion on the effect of modifiers of silver and yttrium on the stability of the sensor's characteristics is difficult. Analysis of absorption spectra produced by UV-visible spectroscopy shows that after annealing of films silver in the structure of tin dioxide is presented in the form of a metal clusters Ag^0 with size 10 nm. To explain the role of Ag it is possible to use the data of x-ray photoemission spectroscopy (RFS) presented in paper [7]. The decrease of the bonding energy of Sn3d and O1s in SnO$_2$:Ag films by $0.5 \div 0.7$ eV compared to pure SnO$_2$ takes place. This process can facilitate the interaction of atomic hydrogen with lattice oxygen. By means of photoelectronic spectroscopy (XPS) [8] it was found that the bond length Y–O in SnO$_2$ does not correspond to the oxide Y$_2$O$_3$, i.e. the yttrium oxide is absent in the structure of tin dioxide, the ions Y^{3+} are introduced into the lattice of SnO$_2$. By comparison of values of the bonding energy of Y–O and Sn–O, we can assume that in thin films of tin dioxide after annealing the yttrium segregates on the surface of the nanocrystals of SnO$_2$ and forms a strong bond with lattice oxygen. In such condition increase the density of superstoichiometric tin atoms and consequently centers for adsorption of oxygen take place. In case of samples of series 1 and series 2 there are metal silver and yttrium ions Y^{3+} in the film of tin dioxide. Due to the presence of metallic silver ions yttrium Y^{3+} more actively form bond with lattice oxygen. As a result, hydrogen interacts only with chemisorbed oxygen on the surface of tin dioxide nanocrystals, recovery of tin dioxide to tin does not occur.

4. Conclusion

The introduction of modifiers Ag, Y in the bulk of tin dioxide thin films in the presence of Pt/Pd on the surface provides the maximum values of the response to hydrogen at $T_{max} = 670$ K. In the case of the deposited catalytic Ag the response to hydrogen is considerably lower and $T_{max} > 713$ K. A common feature of the films is the high stability of their properties under the periodical influence of hydrogen in long-term tests. In literature data on the stability properties of the sensors after long-term use are lacking. The complex modifiers of silver and yttrium interact with lattice atoms of tin and oxygen contribute to the emergence of deep centers in the semiconductor, and these additives prevent the process of reduction of tin dioxide at hydrogen adsorption and ensure the stability of the parameters of the sensors in operation.

Acknowledgments

This work was performed as part of the State Task of the Ministry of Education and Science of the Russian Federation (No 3.2068.2017/4.6).

References

[1] Sevastyanov E Y, Maksimova N K, Chernikov E V, Novikov V A, Rudov F V, Sergeychenko N V 2012 *Semiconductors* **46** 801

[2] Gaman V I, Almaev A V, Maksimova N K, Sergeychenko N V 2016 *Key Eng. Mat.* **683** 353

[3] Gaman V I, Almaev A V, Maksimova N K 2015 Proc. Int. Siberian Conf. on Control and Communications (Omsk) (IEEE conference publications), pp 1-4

[4] Lee Ingun, Choi Seon-Jin, Park Kwang-Min, Lee Sun Sook, Choi Sungho, Kim Il-Doo, Park C O 2014 *Sensor Actuat. B* **197** 300

[5] Vahdatifar Sahar, Khodadadi Abbas Ali, Mortazavi Yadollah 2014 *Sensor Actuat. B* **191** 421

[6] Almaev A V, Gaman V I 2017 *Russ. Phys. J.* **60** 1081

[7] Matsushima S, Teraoka Y, Miura N, Yamazoe N 1988 *Jpn. J. Appl. Phys.* **27** 1798

[8] Cheng L, Ma S Y, Li X B, Luo J, Li W Q, Li F M, Mao Y Z, Wang T T, Li Y F 2014 *Sensor Actuat. B* **200** 181

SPBOPEN 2018

IOP Publishing

Discrete diffraction in network of magnonic crystals

E N Beginin[1], A Yu Sharaevskaya[1,2]

[1]Laboratory "Metamaterials", Saratov State University, Saratov 410012, Russia

[2]Kotel'nikov Institute of Radioengineering and Electronics, Moscow 125009

Abstract. We report on theoretically study of spin waves propagating through the system of side – coupled one-dimensional magnonic crystals. Using transfer matrix method (TMM) and implement for magnetic system acquainted coupled mode approach we obtained transmission coefficients in magnonic system. We considered the discrete diffraction of spin waves in lattice of periodic structures. Results of calculating introduce complicated behaviour of spin wave package in Bragg bandgap area. The main advantage of our theoretical results is possibility to create selective devices of proposed 2D magnonic network.

1. Introduction

Concepts to use spin waves for data processing on the basis of ferromagnetic periodically structured films have created a revival interestt [1 – 7, 12]. Investigations of magnetostatic spin waves (MSW) in periodically micro - and nanostructured media which are called magnonic crystals (MCs) are getting more attention in area of magnonics. MCs are microwave analogy of photonic and phononic crystals. In addition, some other important properties of MSW propagation in MCs are interesting and important for microwave signal processing, e.g. presence of magnonic band gaps, which are extremely sensitive to various imperfections of a periodic lattice of a MC [1 – 3]. Recent works of studying coupled ferromagnetic films and coupled MCs show possibility of control spin wave spectra and band gaps in such coupled MCs [10, 11]. In this case, we present results of obtaining characteristics of propagating magnetostatics spin waves in system of laterally coupled magnonic network.

Content from this work may be used under the terms of the Creative Commons Attribution 3.0 licence. Any further distribution of this work must maintain attribution to the author(s) and the title of the work, journal citation and DOI.
Published under licence by IOP Publishing Ltd

2. Model

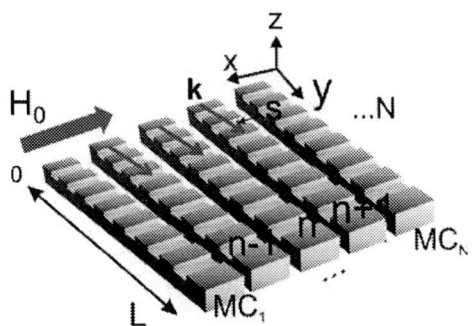

Figure 1. Schematic model of coupled magnonic network.

Let us consider scheme of coupled array of magnonic periodic waveguides (figure 1). The investigated structure consists of 41 1D side – coupled magnonic crystals of yttrium iron garnet ($Y_3Fe_5O_{12}$) material, where L is length of one magnonic element. The uniform static magnetic field was applied in the plane of the waveguide along the x - direction for the excitation of the magnetostatic surface spin waves (MSSWs). Plane theoretical analysis is based on the set of coupled equations relayed on the coupled mode approximation [8]:

$$
\ldots
$$

$$
i\frac{dA_n(\omega; y)}{dy} + k(\omega)A_n - K(\omega)(A_{n+1} + A_{n-1}) = 0
$$

$$
i\frac{dB_n(\omega; y)}{dy} - k(\omega)A_n + K(\omega)(B_{n+1} + B_{n-1}) = 0 \tag{1}
$$

$$
\ldots
$$

Where solutions of (1) are presented by

$$
B_n(y) = \sum_{m=-\infty}^{\infty} i^{(n-m)} J_{n-m}(2Ky) e^{-iky} B_m(0) \tag{2}
$$

$$
A_n(y) = \sum_{m=-\infty}^{\infty} i^{(n-m)} J_{n-m}(-2Ky) e^{iky} A_m(0)
$$

where J_{n-m} is the [n]th order Bessel function, $A(0, y)$, $B(0, y)$ – amplitudes of spin waves. In what follows we assume identical single mode waveguides and weak coupling between them and write coupling equation between k_x and k_y like $k_y(k_x, \omega) = k(\omega) - 2K(\omega)\cos(k_x s)$. Thus, we arrive at the dispersion relation [8], where $k(\omega)$ is dispersion relation for magnetostatics surface spin wave [9]. Phenomenological coupling coefficient between MC is presented by

$$K = \frac{\pi}{2L}Z \tag{3}$$

Also, it can be shown, that the amplitudes of waves in the $y = 0$ and $y = L$ cross sections are related by

$$
\begin{vmatrix} \cdots \\ B_{n-1}(y) \\ A_n(y) \\ B_n(y) \\ A_{n+1}(y) \\ \cdots \end{vmatrix} =
\begin{vmatrix}
\cdots & \cdots & \cdots & \cdots & \cdots & \cdots \\
\cdots & i^0 J_0(2Ky)e^{-iky} & 0 & i^{(-1)}J_{-1}(2Ky)e^{-iky} & 0 & \cdots \\
\cdots & 0 & i^0 J_0(-2Ky)e^{iky} & 0 & i^{(-1)}J_{-1}(-2Ky)e^{iky} & \cdots \\
\cdots & i^{(+1)}J_{+1}(2Ky)e^{-iky} & 0 & i^0 J_0(2Ky)e^{-iky} & 0 & \cdots \\
\cdots & 0 & i^{(+1)}J_1(-2Ky)e^{iky} & 0 & i^0 J_0(-2Ky)e^{iky} & \cdots \\
\cdots & \cdots & \cdots & \cdots & \cdots & \cdots
\end{vmatrix}
\begin{vmatrix} \cdots \\ B_{n-1}(0) \\ A_n(0) \\ B_n(0) \\ A_{n+1}(0) \\ \cdots \end{vmatrix} \tag{4}
$$

Matrices of boundaries are [11, 12] presenting as

$$
\ddot{G}_s =
\begin{bmatrix}
.. & .. & .. & .. & .. & .. \\
.. & \rho & \eta & 0 & 0 & .. \\
.. & \eta & \rho & 0 & 0 & .. \\
.. & 0 & 0 & \rho & \eta & .. \\
.. & 0 & 0 & \eta & \rho & .. \\
.. & .. & .. & .. & .. & ..
\end{bmatrix} \tag{5}
$$

Where $\rho = (1 - sB)^{-1}, \eta = sB(1 - sB)^{-1}$, s= +1, -1, $B = d_1 - d_2 / d_1 + d_2$ (thickness for crests is d_1 and for grooves is d_2 of MC). Matrix of one period is $\ddot{M}_D = \ddot{M}_1 \ddot{G}_+ \ddot{M}_2 \ddot{G}_-$ (where M_1 and M_2 are matrices of different thicknesses). Thus, we constructed transfer matrix using relations (1, 2, 3 ,4 ,5) for all N periods of system [11, 12]. The following step is obtaining frequency dependences for transmission coefficients for coupled MCs media taking into account the dispersion relation in y and x axis and coupling (3) between waveguides.

3. Results and discussion

Unique features of discrete refraction and diffraction in ferromagnetic array are based on band structure of the allowed modes. For pulsed beams exhibiting a finite spectrum, both in the spatial and temporal domain, consist of an ensemble of such Bloch modes, and the band structure completely determines the beam dynamics [8]. Hence, results for transmission coefficients in the structure of coupled MCs are presented on figure 2 and figure 4. Period of structures is $D = 200$ μm, thickness of crests is $d_1 = 15$ μm, thickness of grooves is $d_2 = 13$ μm. Number of periods is 40, length of crests and groove is $L_1 = L_2 = 100$ μm. Saturation magnetization is $M_0 = 1750/(4\pi)$ G, applied magnetic field is $H_0 = 300$ Oe.

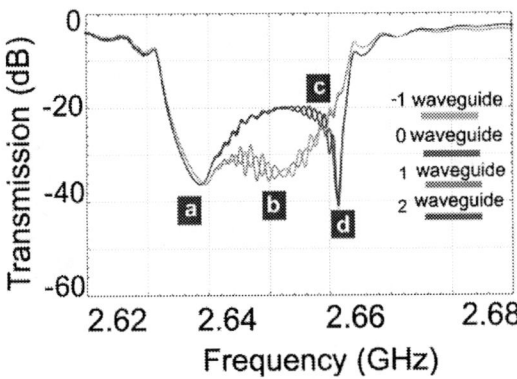

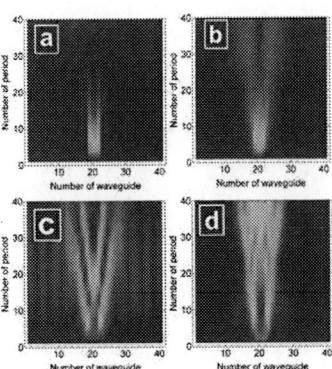

Figure 2. Transmission characteristics toward band gap frequencies for input excitation of 4 ports with Gaussian input beam for system of N (N = 41) magnonic crystal with parameter of coupling Z = 1.5.

Figure 3 (a,b,c,d). Diffraction patterns of N (N = 41) magnonic crystals with parameter of coupling Z = 1.5 at frequencies **(a)** 2.63 GHz **(b)** 2.65 GHz **(c)** 2.66 GHz **(d)** 2.67 GHz.

Figures (2, 3) demonstrate the results of discrete diffraction of MSSW with weak couple $Z = 1.5$. We generate four middle ports with Gaussian input signal for effective propagation of signal. In transmission characteristics we showed, that pumping is remarked from ports (-1, 1) to other both ports (0, 2). We notice complicated quality of signal behaviour from waveguides in band gap (f ≈ 2.65) area. Distribution of amplitude of magnetostatic wave (diffraction pattern) at different frequencies is shown on figure 3. Thus, from the graph above we can see that for $f = 2.63$ GHz and $f = 2.65$ GHz (figures (3(a, b)) signal is damping across length of structure. It can be explained by geometrical variation and features of wave propagation toward band gap. Further, it has conclusively been shown that propagating signal splits into four local wave packages with direction at centrally to length and along the edges (figure (2(c)). In the range at close of band gap signal implodes near both local central of wave packages (figure (2(d)). For the purpose increasing of Z (Z = 2.5) transmission characteristics illustrate that coupling leads to complex behaviour within Bragg bandgap (figures 4). Diffraction pattern without bandgap demonstrates beam spread across length (figures 5(a)). Nextly, at frequency = 2.64 GHz (figure 5(b)) pattern provides damping of signal. At $f = 2.66$ GHz and $f = 2.67$ GHz we obtain diffraction patterns for four (figure 5(c)) and two (figure 5(d)) maximums of amplitudes in the centers of spin wave beams. However, array modification from a continuous to a discrete translational invariance will lead to so-called dynamical localization, which can provide essentially diffraction-free propagation [8]. Such interesting behavior opens new selective opportunities for magnonic application.

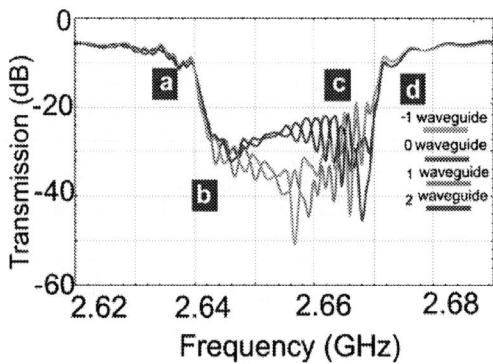

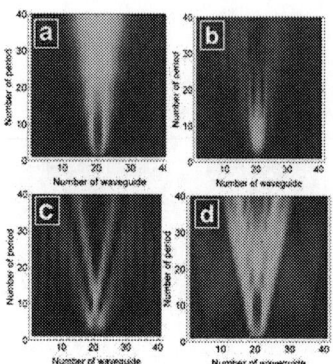

Figure 4. Transmission characteristics toward band gap frequencies for input excitation of 4 ports with Gaussian input beam for system of N (N = 41) magnonic crystal with parameter of coupling Z = 2.5.

Figure 5 (a,b,c,d). Diffraction patterns of N (N = 41) magnonic crystals with parameter of coupling Z = 2.5 at frequencies **(a)** 2.63 GHz **(b)** 2.64 GHz **(c)** 2.67 GHz **(d)** 2.68 GHz.

4. Summary

In conclusion, we present the results of theoretically investigations of magnonic 2D network using transfer matrix method and coupled mode approach. We considered coupling lattice of 41 1D magnonic crystals with Gaussian input beam for four waveguides and studied some features of formation discrete diffraction for spin – wave – packet. We noticed interesting properties and effects by changing coupling coefficient in area of magnonic bandgaps. Interest to such complicated structure is based on possibility to applied of narrow switching elements. Moreover, control of spin wave spectra can be realized in two dimensions with planar array of periodic structures. It should be pointed, that the results can be used for creating controlled devices for planar magnonic.

Acknowledgments

The support from RFBR (18-37-00373) and grant from Russian Science Foundation (Project # 16-19-10283) is acknowledged.

References

[1]   Nikitov S A, Tailhades Ph, Tsai C S 2001 *JMMM* **236** 320-330
[2]   Nikitov S A et al. 2015 *Phys. Usp.* **58** 1099
[3]   Kruglyak V V, Demokritov S O, and Grundler D 2010, *J. Phys. D: Appl. Phys.* **43** 264001
[4]   Chumak A V , Serga A A and Hillebrands B J. 2017 *Phys. D: Appl. Phys.* **50** 244001
[5]   Beginin E N, Sadovnikov A V, Sharaevskaya A Yu, Stognij A I, and Nikitov S A 2018 *Appl. Phys. Lett.* **112** 122404
[6]   Chumak A V, Vasyuchka V V, Serga A A, and Hillebrands B 2015 *Nat Phys* **11** 453–461
[7]   Lenket B. et al. 2011 *Phys. Rep.* **507** 107
[8]   Lederer F, Stegeman G I et al. 2008 *Physics Reports* **463**
[9]   O'Keeffe T W, Patterson R W 1978 *J. Appl. Phys.* **49** 4886 – 4895
[10]   Morozova M A, Sharaevskaya A Yu, Sadovnikov A V, Grishin S V, Romanenko D V, Beginin E N, Sharaevskii Yu P, and Nikitov S A 2016 *Journal of Applied Physics* **120** 223901
[11]   Sharaevskaya A Yu, Beginin E N, Sharaevskii Yu P 2017 *IEEE Transactions on Magnetics* **99** 1-1
[12]   Chumak A V , Serga A A, Wolff S, Hillebrands B, and Kostylev M P *J. Appl. Phys* 2009 **105** 083906

SPBOPEN 2018   IOP Publishing

The study of optical properties of graphene intercalated with ferric chloride for application in terahertz photonics

A D Zaitsev[1], A N Grebenchukov[1], P S Demchenko[1], E T Alonso[2], M F Craciun[2], S Russo[2], A Baldycheva[2], M K Khodzitsky[1]

[1] Terahertz Biomedicine Laboratory, Department of Photonics and Optical Information Technologies, ITMO University, St. Petersburg 197101, Russia

[2] Opto-Electronics Systems Laboratory, University of Exeter, Exeter EX4 4SB, UK

Abstract. The investigation and development of new functional two-dimensional materials is one of the most important challenges for terahertz (THz) photonics and optoelectronics. These materials allow to dynamically manipulate the properties of the THz radiation. Graphene and graphene-based materials are the promising candidates for this task as they are efficient and fast-acting in the THz frequency range. In this work we have experimentally studied the properties of novel material based on few-layered graphene intercalated with ferric chloride (FeCl$_3$-FLG) in the THz frequency range. In particular, the influence of infrared optical pumping intensity (using 980 nm continuous-wave (CW) laser) on the spectral properties of FeCl$_3$-FLG was investigated. The experimental results have shown the efficiency of the suggested method of radiation characteristics control.

1. Introduction

Terahertz (THz) waves are defined as the electromagnetic radiation in the frequency from 0.1 to 10 THz. This spectral domain has low frequency crystalline lattice vibrations and other intermolecular vibrations in many chemical and biological materials, including explosives, drugs, and other biomolecules. Therefore this radiation has become very popular in the wireless communication [1-3] and biomedicine [4-6]. Since graphene invention, two-dimensional (2D) materials have attracted increasing interest in research community both from the point of view of fundamental physics studies, also from the point of view of application in the field of THz science and technology. The vast majority of the works are devoted to the first 2D material, graphene [7], because there are very few materials that can effectively interact with the THz radiation. However, graphene has some disadvantages, such as absence of a band gap and low efficiency of control by optical pumping. To overcome these disadvantages, other 2D graphene-based modifications were proposed in recent researches [8, 9]. In this work, we perform measurements of optical properties of FeCl$_3$-FLG using THz time-domain spectroscopy (THz-TDS) system under an external infrared (980 nm) continuous-wave excitation.

2. Sample preparation and the THz conductivity measurements

For the samples under investigation, the CVD (chemical vapor deposition) graphene grown on Ni substrate was used. Then, the Ni substrate underneath of few-layer graphene film was etched. After that, graphene film was transferred to the glass substrate. Graphene on the glass substrate was placed in the furnace tube with FeCl$_3$ powder and was intercalated at 300 °C. Basically, molecules of FeCl$_3$

Content from this work may be used under the terms of the Creative Commons Attribution 3.0 licence. Any further distribution of this work must maintain attribution to the author(s) and the title of the work, journal citation and DOI.

Published under licence by IOP Publishing Ltd

were inserted between atomic layers of graphene. Then, the intercalated graphene was transferred from the glass to PET and quartz substrates.

The measurements of transmission spectra of FeCl$_3$-FLG on PET and quartz substrates for the different pumping intensities were performed using typical THz-TDS system. The scheme of this setup is shown in Fig. 1. The 980 nm CW laser was used as the infrared optical pumping source. To obtain complex sheet conductivity dispersion it is appropriate to use an effective medium model and a thin-film approximation [10]:

$$\frac{\hat{E}_{sam}(\omega)}{\hat{E}_{sub}(\omega)} = \frac{\hat{n}+1}{\hat{n}+1+Z_0\,\hat{\sigma}(\omega)},$$

(1)

where $\hat{E}_{sam}(\omega)$ and $\hat{E}_{sub}(\omega)$ are the complex amplitudes of the signals transmitted through the sample on a substrate (full structure) and bare substrate correspondingly, $\hat{n}$ is substrate refractive index, $Z_0 = 377$ Ohm is the free space impedance. The measured waveforms for the graphene on PET and quartz substrates for the different optical pumping intensities are shown in Fig. 2 (there are also shown the waveforms for the corresponding bare substrates). The photos of the used samples are depicted in Fig. 3.

The complex conductivity dispersions were extracted from the experimental data using an Equation 1. The dependencies of the real and imaginary parts of the complex sheet conductivity of the samples on the infrared continuous-wave 980 nm pumping intensity are shown in Fig. 4 (PET substrate) and Fig. 5 (quartz substrate) for the different frequencies of the THz radiation.

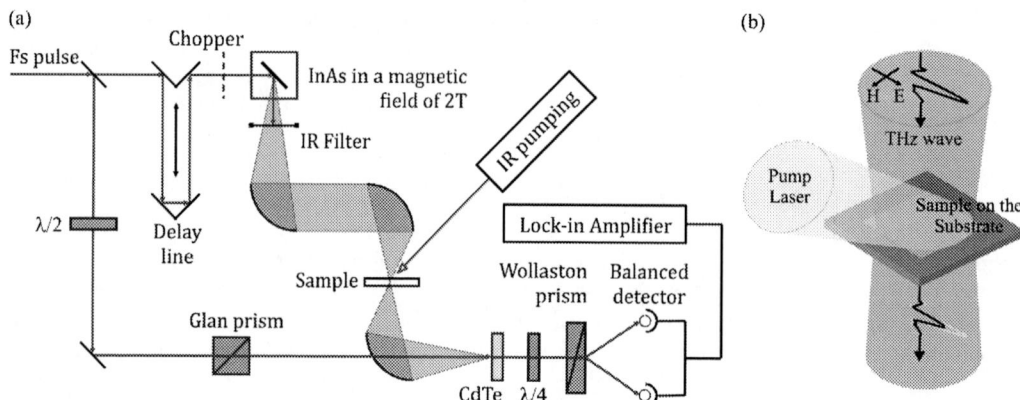

Figure 1. (a) Scheme of the experimental setup; **(b)** Sample optical pumping scheme (the parameters of the THz wave transmitting through the structure are controlled by optical pumping of the sample).

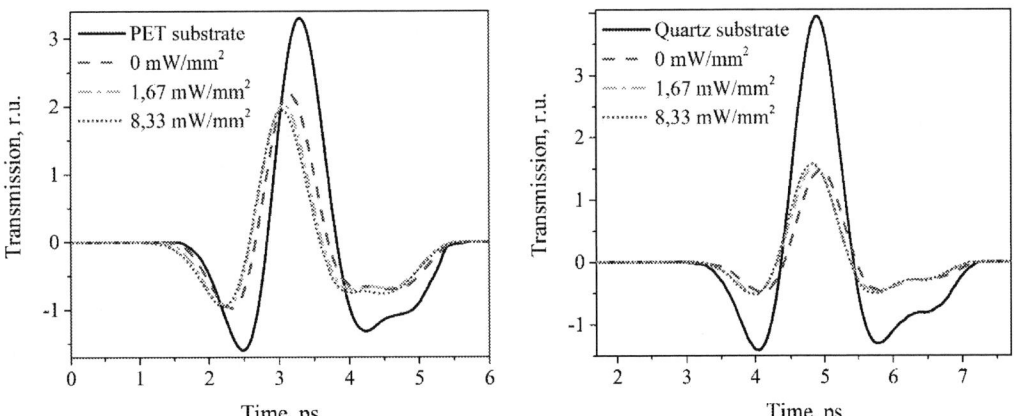

Figure 2. Measured THz waveforms of FeCl₃-FLG on PET (left graph) and quartz (right graph) substrates under infrared optical pumping of different intensities.

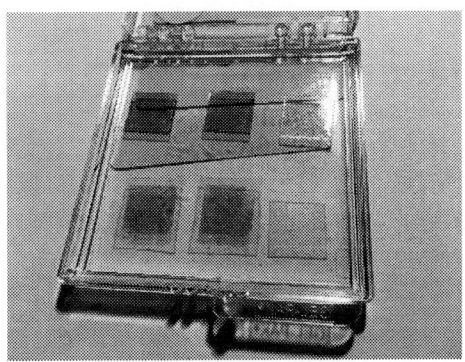

Figure 3. The samples of FeCl₃-FLG on PET (bottom – two left samples) and quartz (top – two left samples) substrates. The corresponding bare substrates are placed at right side.

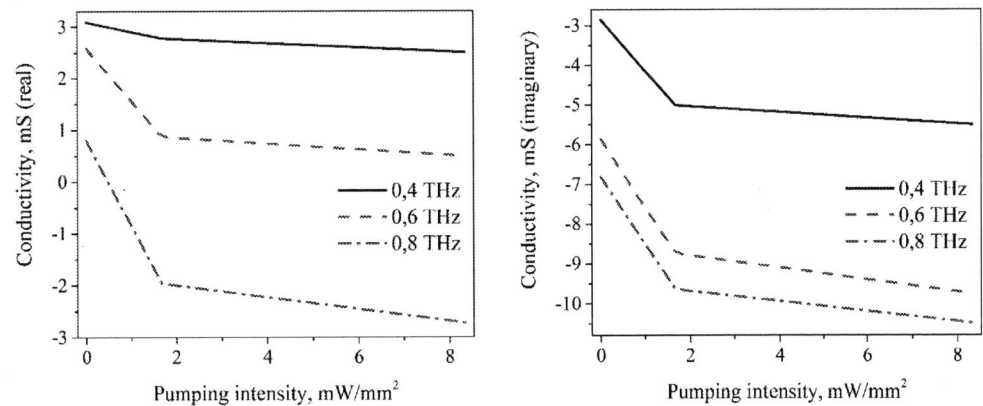

Figure 4. The dependencies of the real (left graph) and imaginary (right graph) parts of complex sheet conductivity of FeCl₃-FLG on the infrared pumping intensity (three-point curves) for the different frequencies of the THz radiation (PET substrate).

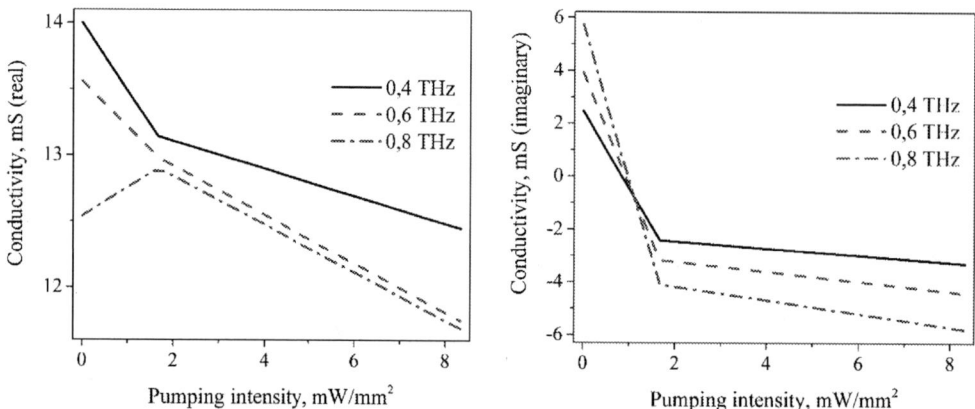

Figure 5. The dependencies of the real (left graph) and imaginary (right graph) parts of complex sheet conductivity of $FeCl_3$-FLG on the infrared pumping intensity (three-point curves) for the different frequencies of the THz radiation (quartz substrate).

It can be clearly seen that conductivity behaviour of $FeCl_3$-FLG under optical pumping depends on the substrate type. The significant changes in real and imaginary parts of complex sheet conductivity are observed under infrared optical pumping. There is also a saturation of changes in complex conductivity under relatively low optical pumping. The tuning of the complex conductivity depends on the frequency of the THz radiation: the greatest changes are achieved at higher frequencies (0,8 THz, in case of this work).

For $FeCl_3$-FLG on PET substrate, the changes mostly in real part of sheet conductivity are observed. For the pumping intensities in range from 0 to 8.33 mW/mm², the change in real part of complex sheet conductivity is equal to 3.52 mS at 0.8 THz and 0.58 mS at 0.4 THz. There are also significant changes in imaginary part of complex conductivity, which are equal to 3.67 mS at 0.8 THz and 2.65 mS at 0.4 THz. It means that $FeCl_3$-FLG on PET substrate can be used for the efficient amplitude modulation of the THz radiation at frequencies around 0.8 THz, and phase modulation of the THz radiation at different frequencies in range 0.4-0.8 THz. The losses in this structure are low.

For $FeCl_3$-FLG on quartz substrate, the changes mostly in imaginary part of sheet conductivity are observed. For the pumping intensities in range from 0 to 8.33 mW/mm², the change in real part of complex sheet conductivity is equal to 1.55 mS at 0.4 THz and 0.85 mS at 0.8 THz. The losses in this structure are high, because the real part of complex sheet conductivity has a values of about 0.12-0.14 mS. There are significant changes in imaginary part of complex conductivity, which are equal to 5.76 mS at 0.4 THz and 11.52 mS at 0.8 THz. The $FeCl_3$-FLG on quartz substrate can be used for the efficient phase modulation of the THz radiation at different frequencies in range 0.4-0.8 THz.

The change of parameters of the structures at 1.67 mW/mm² pumping intensity is from 77% to 85% relative to the full change (at 8.33 mW/mm² pumping intensity).

These structures can be used in high-speed communication systems, spectroscopy, contactless diagnostics, visualization systems, and in medicine (including tomography of the surface layers of the body), as the active components that modulate the amplitude and phase of the THz radiation.

3. Conclusions

In summary, we have experimentally demonstrated the efficient optical properties tunability of $FeCl_3$-FLG on different substrates by relatively low-intensity infrared optical pumping. It was shown that tunability character depends on the substrate type. The $FeCl_3$-FLG on PET substrate can be used for the efficient amplitude modulation of the THz radiation at frequencies around 0.8 THz, and phase modulation of the THz radiation at different frequencies in range 0.4-0.8 THz. The FeCl₃-FLG on quartz substrate can be used for the more efficient phase modulation of the THz radiation at different frequencies in range 0.4-0.8 THz in comparison to the FeCl₃-FLG on PET substrate. The proposed material can be used in high-speed, efficient and tunable THz photonic and optoelectronic systems.

Acknowledgments
This work was supported by Government of Russian Federation (grant 08-08).

References

[1] Borovkova M, Serebriakova M, Fedorov V, Sedykh E, Vaks V, Lichutin A, Salnikova A, Khodzitsky M 2017 *Biomedical Optics Express* **8**(1) 273
[2] Borovkova M, Khodzitsky M, Demchenko P, Cherkasova O, Popov A, Meglinski I 2018 *Biomedical Optics Express* **9**(5) 2266
[3] Gusev S I, Demchenko P S, Cherkasova O P, Fedorov V I, Khodzitsky M K 2018 *Chinese Optics* **11**(2) 182
[4] Grebenchukov A, Zaitsev A, Khodzitsky M 2018 *Chinese Optics* **11**(2) 166
[5] Soboleva V Y, Gomon D A, Sedykh E A, Balya V K, Khodzitskii M K 2017 *Journal of Optical Technology* **84**(8) 521
[6] Gomon D A, Sedykh E A, Rodriguez S, Monroy I T, Zaitsev K I, Vozianova A V, Khodzitsky M K 2018 *Chinese Optics* **11**(1) 47
[7] Perreault F, Fonseca de Faria A, Elimelech M 2015 *Chem. Soc. Rev.* **44** 5861
[8] Junpeng L, Hongwei L 2018 *Optics Communications* **406** 24
[9] Fan Z 2017 *Scientific reports* **7**(1) 14828
[10] Zhou Q L, Shi Y L, Jin B, Zhang C L 2008 *Appl. Phys. Lett.* **93** 102103

Lifetime testing of a MEMS switch with Pt-Pt contact

I V Uvarov, A N Kupriyanov

Laboratory of Micro- and Nanosystem Technology, Yaroslavl Branch of the Institute of Physics and Technology RAS, Yaroslavl 150007, Russia

Abstract. This paper presents the lifetime testing results of an electrostatically actuated microelectromechanical systems (MEMS) switch with the resistive contact and the active contact breaking mechanism. Moveable electrode of the switch is an aluminium beam with platinum contact bumps located on its bottom surface, which comes in contact with platinum thin-film electrodes. The switch operates in a cold DC mode in a standard laboratory environment. Testing is performed at three levels of the input current: 0.05, 0.5 and 5 mA. The dependence of the resistance in the "on" state on the number of actuation cycles is measured. Resistance of the "fresh" samples is in the range from 150 to 350 Ω. Instability of the resistance is observed during cycling, that is probably related to contamination of the contacts. Lifetime of the switch is limited by the sharp increase of the on-resistance up to 100 MΩ and varies from 2×10^3 to 5×10^4 cycles depending on the switch design and the input current.

1. Introduction

Resistive contact MEMS switches are used in radio frequency and microwave systems for signal routing and control [1]. Attenuators [2], phase shifters [3], filters [4] and other devices based on the MEMS switches are presented. Among various actuation mechanisms, electrostatic actuation is the most popular. In comparison with switches based on pin-diodes and field-effect transistors, electrostatically actuated MEMS switches have low power consumption, good RF performance and high radiation resistance. Compared to conventional electromechanical relays, they are significantly smaller and have shorter switching time. MEMS switches are also used in integrated circuits and are considered as an alternative to semiconductor switches [5]. The simplest logic elements [6] and memory cells [7] are demonstrated. Furthermore, MEMS switches can be fabricated using conventional microtechnology, which allows them to be integrated into CMOS circuits [8].

However, incorporation of MEMS switches into commercial products is limited by their relatively low reliability [9]. The most common reason of failure is an increase of the contact resistance during cyclic operation, which occurs due to contamination, oxidation, and mechanical damage of contacting surfaces [10]. At present, a lot of research is carried out to increase the reliability of contacts. Proper selection of the contact material plays an important role. One of the most widely used materials is gold [11-13]. Its chemical inertness and high conductivity provides insensitivity to contamination and low contact resistance. However, low hardness and low melting point of Au makes a switch susceptible to stiction and unable to transfer high-power signals [14, 15]. Harder platinum group metals such as Ru, Rh, Ir and Pt are considered as an alternative to Au [16, 17]. They allow to overcome the shortcomings of the gold contacts and to provide acceptable contact resistance and high reliability simultaneously. In this work the lifetime of an electrostatically actuated MEMS switch with Pt-Pt contact is investigated.

Content from this work may be used under the terms of the Creative Commons Attribution 3.0 licence. Any further distribution of this work must maintain attribution to the author(s) and the title of the work, journal citation and DOI.
Published under licence by IOP Publishing Ltd

2. Design of the switch and experimental setup

SEM image of the switch is shown on figure 1. The device is based on a 2-μm-thick aluminum beam (source) attached to the torsion springs. The beam has a length of 100 μm and a width of 8, 16, 24 or 32 μm depending on the design. Gate and drain electrodes are made of 100 nm thick Pt film and placed under the each arm of the beam, so the switch has two symmetric output channels. The gap between the beam and the gate electrodes is 1.5 μm. Platinum contact bumps of 0.5 μm in height are located on the bottom side of the beam (one bump per arm, figure 2). Thus, Pt-Pt contact was obtained. The switches were fabricated by surface micromachining on the oxidized silicon substrates. Details of the design and the fabrication process can be found in our previous works [18, 19].

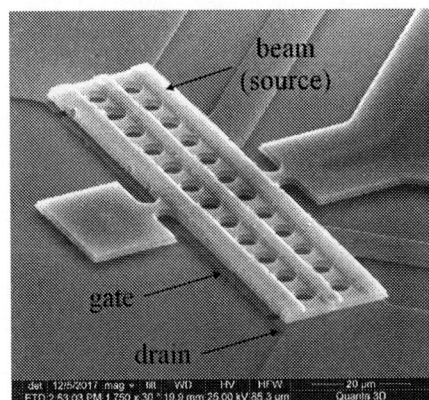

Figure 1. SEM image of the switch.

Figure 2. Contact bump located on the bottom surface of the beam.

The switch operates as follows. Initially the beam has the horizontal position. When the driving voltage is applied to one of the gates, the beam tilts towards the electrode under the electrostatic force and comes in contact with the corresponding drain. Thus, the switch goes to the "on" state. When the voltage is removed, the beam returns to the initial state under the elastic force of the springs. In case of stiction, i.e. when the elastic force is insufficient to overcome the forces acting between the bump and the drain, the switch retains the "on" state. In order to detach the beam from the drain, the voltage is applied to the opposite gate. The beam tilts to the opposite direction and comes in contact with another drain. Therefore, the design of the switch provides the active contact breaking mechanism, which protects the device against stiction.

Testing of the switches was performed in a standard laboratory environment, the samples were unpackaged. Measurement equipment was connected to the switch as shown on figure 3. Driving voltage (V_{G1}, V_{G2}) was periodically applied to the gates from the DC power supply Agilent E3647A in such a way the channels were actuated alternately with the frequency of 2.5 Hz. The minimal gate-to-source voltage needed to close the switch was measured previously and the range of 29-46 V was obtained [19]. The amplitude of pulses was chosen to be 60 V in order to ensure the actuation. The switches operated in a cold DC mode. Input voltage (V_S) was applied to the source from the analog output module National Instruments PXI-6711. It was turned on 40 ms after closing and turned off 60 ms before opening the switch in order to avoid hot operation and excessive wear of the contacts. All the samples were tested at $V_S = 5$ V. Output voltage (V_{D1}, V_{D2}) was registered at the drain electrodes by the oscilloscope PicoScope 5442B and the multifunction input/output module NI PXI-6143. Typical gate and drain signals observed during the test are shown on figure 4. The current I_D flowing through the switch was adjusted by the load resistors R_1 and R_2. It was measured at the one of the channels (I_{D1}) by the multimeter Keysight 34461A. The equipment was controlled by LabView software. On-resistance R_{ON} of the both channels was calculated at each actuation cycle from the resistive divider circuit. The experiments were performed at three levels of the current: 0.05, 0.5 and 5 mA.

911

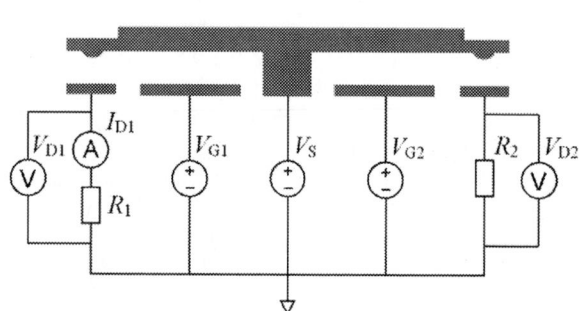

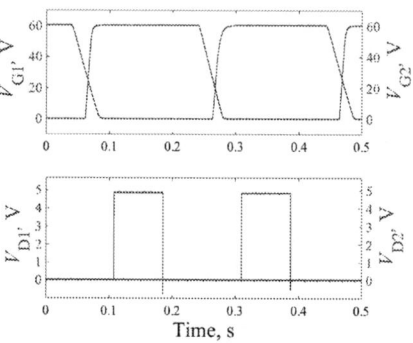

Figure 3. Connection of the measurement equipment to the switch.

Figure 4. Gate and drain signals of the switch during the test.

3. Results and discussion

Typical dependence of the switch resistance in the "on" state on the number of actuation cycles is shown in figure 5. R_{ON} was changing between 100 and 2000 Ω during the test. Such instability was observed for all the samples. After several thousands of cycles R_{ON} sharply increased up to 100 MΩ that was considered as a failure of the switch. The on-resistance of the "fresh" devices was in the range of 150÷350 Ω. R_{ON} was approximately two times lower and more stable at $I_D = 5$ mA than at 0.05 and 0.5 mA (figure 6a). This fact was probably connected with the calculation method, whose accuracy depends on the relation between the load resistance and resistance of the switch. Another probable reason is that at high current level the asperities of the contacting surfaces are melted easier, thus increasing the effective contact area which results in lowered contact resistance.

It is worth noting that R_{ON} was rather high for ohmic MEMS switch, typical value should be less than 5 Ω [1]. Measured resistance consisted of two main parts. One part is the sheet resistance of the thin metal film of the drain electrode, the beam and the signal lines. This resistance cannot be eliminated from the measurement because of practical limitations on the geometry and placement of the contact pads. High resistance of the switch was partly caused by the small thickness of the drain electrodes and their connecting lines, which have the resistance of about 100 Ω. The other part is the contact resistance caused by the current flowing through a small contact area (it is also called constriction resistance). It includes a resistance coming from the contamination thin film between the metal contacts. It is known that Pt-group metals are susceptible to contamination and frictional polymerization [10, 15-17] that can lead to unstable and increased contact resistance during the switching cycles. Probably, this component was responsible for the instability of R_{ON}. As a result, both parts contributed sufficiently to the total on-resistance. Further we plan to minimize the sheet resistance by increasing the thickness of the signal lines. This will reduce the total resistance of the switch and allow to investigate the contact resistance more precisely. It is also necessary to inspect the contacting surfaces for wear and contamination in order to identify the reason of the on-resistance instability.

Contact resistance of the switch is usually estimated using the following equation [20]:

$$R_C = \frac{\rho}{2a},$$ (1)

where ρ is the resistivity of the contact material (15.5 $\mu\Omega$ cm for Pt [14]), a is the radius of the contact spot. For plastic deformation of contacting surfaces a is given by [20]

$$a = \sqrt{\frac{F_C}{\pi H}},$$ (2)

where F_C is the contact force and H is the Meyer indentation hardness of the contact material (5.1 GPa for Pt [14, 15]). Combining equations (1) and (2) it can be seen that R_C is inverse proportional to $F_C^{1/2}$.

Contact force is determined by the electrostatic force acting between the beam and the drain electrode in the "on" state and is proportional to the beam width. Therefore, switches with the widest beam should have the lowest on-resistance. Nevertheless, there was no clear dependence of the R_{ON} on the beam width (figure 6a). Careful estimation of the contact force and precise measurement of the contact resistance is needed to explain this result.

Lifetime of the switches was limited by the sharp increase of the on-resistance and varied from 2×10^3 to 5×10^4 cycles. The samples with the narrowest beam ($w = 8$ μm) failed with the least number of cycles at all current levels (figure 6b). Probably, it was due to the lowest contact force that was insufficient to break the continuously growing contamination film. Switches with $w = 24$ μm typically showed the longest lifecycle. At $I_D = 5$ mA the lifecycle was slightly lower than at 0.05 and 0.5 mA, that was connected with the intensified wear of the contacting surfaces. In general, the obtained data corresponded to the endurance of the switches with the Pt-Pt contact available in the literature [15, 17]. Although the switches were able to withstand a relatively small number of actuation cycles, the stiction was not observed even at the current of 5 mA, which corresponded to the switching power of 25 mW. In the future we plan to increase the current and evaluate the capabilities of switching high-power signals.

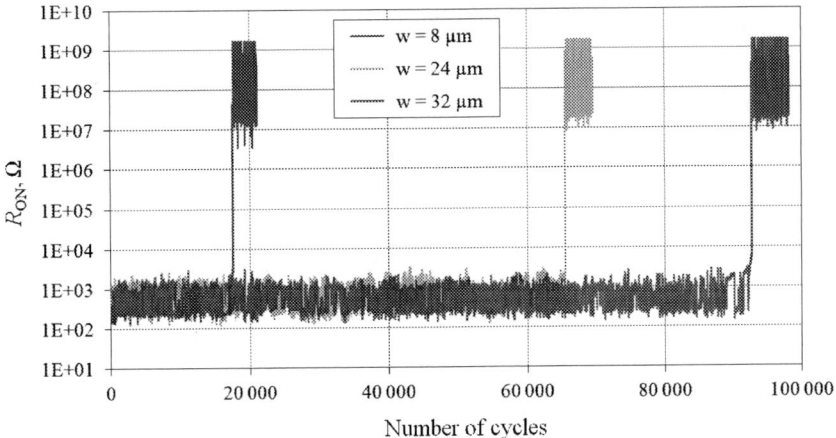

Figure 5. Dependence of the on-resistance on the number of actuation cycles for three samples having different width of the beam w. Measurements are performed at $I_D = 0.5$ mA.

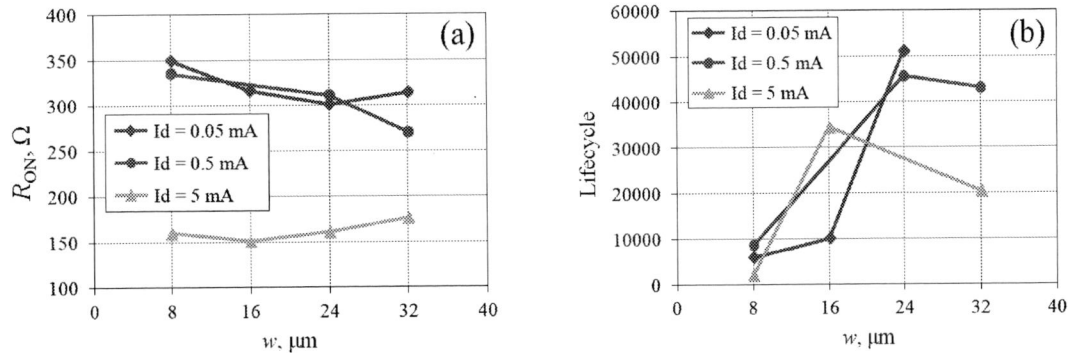

Figure 6. Dependence of the initial on-resistance (a) and the lifecycle (b) on the beam width.

4. Conclusions

Electrostatically actuated MEMS switch with Pt-Pt contact was tested for reliability. The samples operated in a standard laboratory environment under cold switching conditions. Testing was performed at the input current from 0.05 to 5 mA. The dependence of the on-resistance on the number of switching cycles was measured. The initial contact resistance was from 150 to 350 Ω and did not depend on the switch design. The resistance was unstable during the test, varying from 100 to 2000 Ω. The possible reason of instability was the contamination of contacting surfaces that is typical for platinum group metals. After several thousands of cycles the resistance increased sharply up to 100 MΩ that was considered as a failure of the device. The lifetime of the switches was from 2×10^3 to 5×10^4 cycles. Switches with the narrowest beam had the shortest lifecycle, no longer than 1×10^4, probably because of the lowest contact force. The main reason of failure was the increase of the resistance during cycling. The stiction was not observed even at the highest current. Further, we plan to increase the input current and test the switch for transmission of high-power signals.

Acknowledgments

This work was supported by RFBR research project No. 16-37-60065 mol_a_dk and performed using the equipment of Facilities Sharing Centre "Diagnostics of Micro- and Nanostructures". Ilia Uvarov thanks Denis Pukhov and Leonid Mazaletskiy from YB IPT RAS for the SEM measurements.

References

[1]   Rebeiz G M 2003 *RF MEMS: Theory, Design, and Technology* (Hoboken, New Jersey: John Wiley & Sons, Inc.)
[2]   Guo X, Gong Z, Zhong Q, Liang X and Liu Z 2016 *J. Micromech. Microeng.* **26** 074002
[3]   Sharma A K, Gautam A K, Farinelli P, Dutta A and Singh S G 2015 *J. Micromech. Microeng.* **25** 035014
[4]   Sekar V, Armendariz M and Entesari K 2011 *IEEE Trans. Microw. Theory Techn.* **59** 866-76
[5]   Pott V, Kam H, Nathanael R, Jeon J, Alon E and Liu T-J K 2010 *Proc. IEEE* **98** 2076
[6]   Chakraborty S and Bhattacharyya T K 2010 *J. Micromech. Microeng.* **20** 105026
[7]   Chua G L, Singh P, Soon B W, Liang Y S, Jayaraman K G, Kim T T-H and Singh N 2014 *Appl. Phys. Lett.* **105** 113503
[8]   Dai C-L and Chen J-H 2006 *Microsyst. Technol.* **12** 1143-51
[9]   Rebeiz G M, Patel C D, Han S K, Ko C-H and Ho K M J 2013 *IEEE Microw. Mag.* **14** 57-67
[10]   Toler B F, Coutu R A and McBride J W 2013 *J. Micromech. Microeng.* **23** 103001
[11]   Song Y-H, Choi D-H, Yang H-H and Yoon J-B 2011 *J. Microelectromech. Syst.* **20** 204-11
[12]   Chow L L W, Schrader S A and Kurabayashi K 2006 *Appl. Phys. Lett.* **89** 133501
[13]   Majumder S, McGruer N E, Adams G G, Zavracky P M, Morrison R H and Krim J 2003 *Sensors and Actuators A* **93** 19-26
[14]   Kwon H, Park J-H, Lee H-C, Choi D-J, Park Y-H, Nam H-J and Joo Y-C 2008 *Jpn. J. Appl. Phys.* **47** 6558-62
[15]   Song Y-H, Kim M-W, Lee J O, Ko S-D and Yoon J-B 2013 *J. Microelectromech. Syst.* **22** 846-54
[16]   Czaplewski D A, Nordquist C D, Dyck C W, Patrizi G A, Kraus G M and Cowan W D 2012 *J. Micromech. Microeng.* **22** 105005
[17]   Chen L, Lee H, Guo Z J, McGruer N E, Gilbert K W, Mall S, Leedy K D and Adams G G 2007 *J. Appl. Phys.* **102** 074910
[18]   Uvarov I V, Naumov V V, Koroleva O M, Vaganova E I and Amirov I I 2016 *Proc. SPIE* **10224** 102241A
[19]   Uvarov I V, Naumov V V, Kupriyanov A N, Koroleva O M, Vaganova E I, Amirov I I 2017 *J. Phys.: Conf. Ser.* **917** 082001
[20]   Holm R 1967 *Electric Contacts: Theory and Application* (New York: Springer)

SPBOPEN 2018 IOP Publishing

Simulation of electrical conductivity of silicon diodes with bismuth implanted-ion profiles

S M Loganchuk[1], S N Chebotarev[1], D A Arustamyan[1], A A A Mohamed[1], L Touel[1], N M Bogatov[2]

[1]Department of Physics and Electronics, Platov South-Russian State Polytechnic University (NPI), Novocherkassk 346421, Russia

[1]Department of Optoelectronics, Kuban State University, Krasnodar, 350040, Russia

Abstract. Numerical simulation of electrical conductivity of silicon diodes doped with bismuth during ion implantation was carried out. The input parameters were experimental data on the distribution profiles of bismuth implanted with energy from 40 to 360 keV. It was also taken into account that the concentration of bismuth affects the mobility of charge carriers. The effect of ion energy, doping level and temperature on the current-voltage characteristics of a silicon diode structure with a Gaussian distribution profile of bismuth was theoretically studied. Also, the saturation current and the non-ideality factor of the diode structure were calculated.

1. Introduction

Silicon diodes remain the basic components of modern electronics [1]. Improving the characteristics of diodes occurs both due to the expansion of technological solutions, and by modeling their properties [2]. There are many different technological variations in the formation of p-n junctions, one of which is ion doping [3,4]. Note that for the formation of pn-junctions the most suitable method is ion doping, which provides the necessary level of process control. The main advantage of this method is that it is possible to obtain concentrations of implanted impurities greater than the limit of their equilibrium solubility in the substrate. The subsequent temperature treatment allows to get rid of radiation defects. Therefore, studies of substrates implanted with metal impurities become important for the development of the electronic component base. An intensive study of silicon doped with bismuth during ion implantation is observed. Recently, a number of interesting experimental results have been obtained that contain information on the distribution profiles of bismuth as a function of the energy and density of the ion beam flux. The aim of this paper is to simulate the electrical conductivity of silicon doped with bismuth, as well as to study the effect of temperature and ion energy on the current-voltage characteristics of diode structures.

2. Results and discussion

Diode structures shown on Fig.1 have a thickness of 1.5 μm with transverse dimensions of 2.5x2.5 μm were used for modeling. In all of the experiments un-doped single-crystal silicon wafers with a resistivity of 20 Ω-cm was used. The profiles of the distribution of bismuth in silicon during ion implantation, taken from [5], are shown on Fig. 2. The ion energy varied from 40 to 360 keV. The distribution of

Content from this work may be used under the terms of the Creative Commons Attribution 3.0 licence. Any further distribution of this work must maintain attribution to the author(s) and the title of the work, journal citation and DOI.

Published under licence by IOP Publishing Ltd

bismuth depends essentially on the energy of the implanted bismuth ions. At the minimum energy, the maximum concentration of the dopant was at a depth of 2.5 μm from the surface of the structure. Implantation of bismuth ions with maximum energy shifted a maximum to a depth of 12.5 μm. The simulation was carried out by numerical methods by solving the Poisson equation and the standard diffusion-drift model [6]. Profiles were set using the Gaussian function, which was built on the basis of experimental data. Figure 1 clearly reflects the distribution of the electron density in the diode structure in a steady state. In the lower part of the figure, a dark red band of increased electron concentration is clearly visible. In the calculations it was taken into account that the mobility of the charge carriers depends on the level of doping [7].

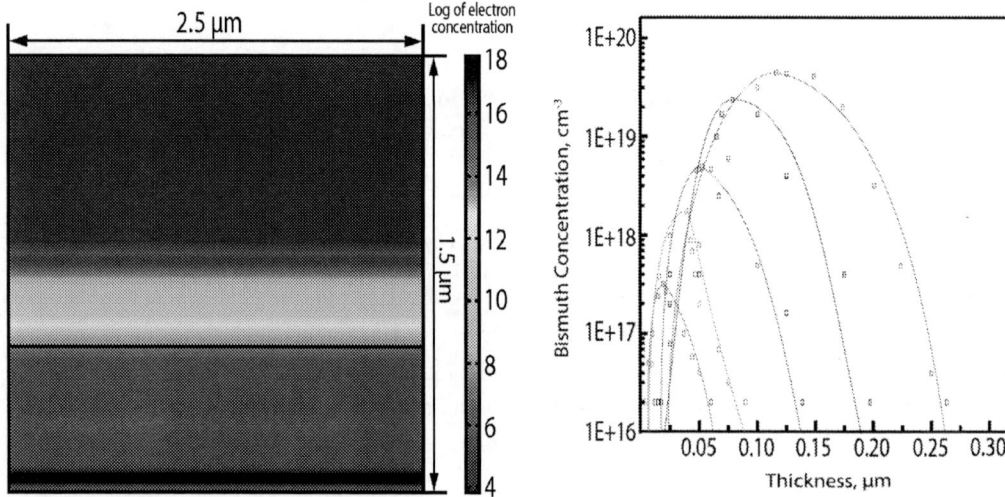

Figure 1 Diode structure   **Figure 2** The profiles of the distribution of bismuth in silicon

The simulation was carried out by numerical methods by solving the Poisson equation and the standard diffusion-drift model [6]. For simulation of dependence of electrical current density from voltage bias Poisson equation has been solved with drift-diffusion. The system of the basic equation for simulation model is as follows:

$$\frac{\partial}{\partial x}\left(\varepsilon_r \frac{\partial \varphi}{\partial x}\right) + \frac{\partial}{\partial y}\left(\varepsilon \frac{\partial \varphi}{\partial y}\right) = -q\left(p - n + N_D^+ - N_A^-\right), \qquad (1)$$

$$J_n = q\mu_n n \mathrm{E}_n + q D_n \nabla n, \qquad (2)$$

$$J_p = q\mu_p p \mathrm{E}_p + q D_p \nabla p, \qquad (3)$$

Where x – length of diode, y – width of diode, ε_r – the permittivity of Si(Bi), φ – the electrostatic potential, ε – vacuum permittivity, q – the elementary charge, p and n – the concentration of electrons and holes, N_D^+ – the donor impurity concentration, N_A^- – the ionized acceptors concentration, J_n and J_p – the electrons' and holes' electric current density, μ_n and μ_p – the electron and the hole mobilities, E_n and E_p – the effective driving electrical field of electrons and holes, $D_n = \dfrac{kT}{q}\mu_n$ and $D_p = \dfrac{kT}{q}\mu_p$ – the electron and hole diffusivities according to Einstein's relationship, k – the Boltzmann's constant, T – absolute temperature.

Based on this model, direct branches of volt-ampere characteristics for various temperatures and mobilities of charge carriers were calculated, as well as the dependence of the current density on temperature.

Figure 3 shows a series of current-voltage characteristics of an Si(Bi) diode obtained at different temperatures from 250 to 450 K, formed by implantation of bismuth with energy 150 keV. The behavior of the curves fits into the classical theory of pn-junctions. A characteristic feature of this series is a substantial increase in the current density for doping with energies above 300 keV. The insert in the upper left corner reflects the temperature dependence of the current density (for a forward voltage of 1 V) calculated for the doping profiles formed at different ion energies.

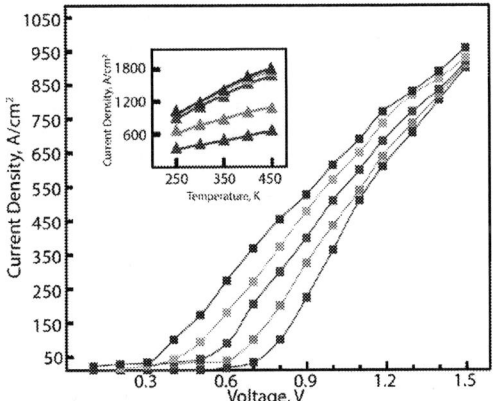

Figure 3 Dependence of current density from voltage for different temperatures

Figure 4 Dependence of current density from voltage for different electron mobilities

Figure 4 shows the calculated current-voltage characteristics for a diode doped with bismuth in the entire energy range. In these calculations, the temperature was assumed to be 300 K and the electron mobility varied from 540 to 40 cm²/(V·s).

An inverse branch of the current-voltage characteristic was also constructed for various mobilities at a temperature of 300 K. The reverse branch of the current-voltage characteristic was used to calculate the saturation current, which was necessary for calculating the non-ideality factor. To calculate the non-ideality coefficient, the following equation was used:

$$J = J_0 \left(\exp\left(\frac{qV}{nkT} \right) - 1 \right), \tag{4}$$

where J_0 is saturation current density, q is elementary charge, V is voltage, n is ideality factor, k is Boltzmann's constant, T is absolute temperature.

From the equation (4), a formula was obtained for the nonideality coefficient:

$$n = \frac{qV}{\ln\left(\frac{J + J_0}{J_0} \right) kT}, \tag{5}$$

The results of calculations of saturation currents and ideality coefficients for different ion energies and electron mobilities are presented in Table 1. As can be seen from Table 1, the average value of the non-ideality coefficient for a given diode structure is 8.3976.

Table 1. Saturation current and non-ideality coefficient

Ion energy, keV	Electron Mobility, cm^2/(V·s)	Saturation Current Density, 10^{-8} A/cm^2	Ideality factor
40	540	7.0977	9.071
80	280	7.1000	8.188
150	190	7.1024	8.199
250	80	7.1188	8.233
360	40	7.2111	8.297

3. Conclusions

A numerical model is developed and the influence of the energy of implanted bismuth ions and the temperature of the current-voltage characteristics of silicon diodes is calculated. It is shown that a significant increase in the current density is observed at implantation energies of more than 300 keV. In the calculations it was taken into account that the mobility of charge carriers depended on the concentration of the dopant and varied from 540 to 40 cm^2/(V·s). The temperature dependence of current-voltage characteristics for diodes obtained during ion implantation of bismuth with energies in the range from 40 to 360 keV was also investigated. The saturation current and the nonideality coefficient of the Si(Bi) diode structure were also calculated.

References

[1] Aryaee Panah M E, Han L., Norrman K., Pryds N., Nadtochiy A., Zhukov A E., Lavrinenko A V., Semenova E S. 2017 *Opt. Mat. Exp.* **7** 2260

[2] Chebotarev S N, Pashchenko A S, Lunin L S, Zhivotova E N, Erimeev G A, Lunina M L 2017 *Beilstein J. Nanotech.* **8** 12

[3] Chebotarev S N, Pashchenko A S, Lunin L S, Irkha V A 2016 *Nanotech. in Russia.* **11** 435

[4] Yatsenko A N, Chebotarev S N, Lozovskii V N, Mohamed A A, Erimeev G A, Goncharova L M, Varnavskaya A A 2017 *J. Phys: Conf. Ser.* **917** 032008

[5] Liang J H 1999 *Nucl. Instr. and Meth. in Phys. Research*, **153** 436

[6] Lozovskii V N, Chebotarev S N, Irkha V A, Valov G V 2010 *Tech. Phys. Lett.* **36** 737

[7] Ferreira da Silva A. 1996 *J. Appl. Phys.* **79** 3453

SPBOPEN 2018   IOP Publishing

Flexoelectrical nanogenerator design using aligned carbon nanotubes

M V Il'ina[1], A A Konshin[1] and E G Solomin[2]

[1]Southern Federal University, Institute of Nanotechnologies, Electronics and Electronic Equipment Engineering, Taganrog 347922, Russia
[2]AT&S India Pvt Ltd., Mysore, Karnataka 571302, India

Abstract. The experimental studies of the current and surface potential generated during deformation and vibration of carbon nanotubes (CNTs) were carried out. It is shown that an individual CNT generate a current of -24 nA by pressing force of 0.155 nN. It is also established that a CNT array generates a surface potential up to 0.32 V when it vibrates with a frequency of 10 kHz. The obtained results show the ability of CNTs to transform external mechanical influences, including minor mechanical vibrations of the environment, into electric current and potential. These results can be used for the design of high-efficiency flexoelectrical nanogenerator based on aligned carbon nanotubes.

1. Introduction

The rapid development of nanotechnology leads to gradual size reduction and energy consumption of electronics devices, which opens the possibility of using the energy of the environment as sources to power devices [1]. One of the promising devices capable of converting mechanical energy into electrical energy is a flexoelectric nanogenerator that uses not only piezoelectric, but also flexoelectric properties of nanostructures [2, 3]. Currently the search for nanostructures that can be used for creation of high-efficiency nanogenerators is performed. In this case, not only nanostructures based on traditional piezoelectric materials, but also nanostructures based on materials that do not exhibit volumetric piezoelectric properties, come under consideration [3]. Recent work shows that the flexoelectric effect in nanostructures is comparable to piezoelectric, and exceeds it several times in some cases [4]. This fact makes it possible to substantially increase the output voltage of nanogenerators due to the simultaneous influence of flexo- and piezoelectric [4–7]. Through high strength and elasticity values, piezoelectric and flexoelectric properties the carbon nanotubes (CNTs) are given great attention [8, 9]. First works in this field appeared only in 2016 [10]. Thus, the development of a flexoelectrical nanogenerator based on CNTs is an important and urgent task of modern nanoelectronics and requires further research.

The aim is to study the dependence of current and surface potential on a CNT deformation for flexoelectrical nanogenerator design using aligned carbon nanotubes.

2. Experimental studies

As the experimental sample we used vertically aligned CNTs array grown by plasma-enhanced chemical vapor deposition using NANOFAB NTC-9 (NT-MDT, Russia) [11, 12]. The studies of the geometrical parameters of the CNTs array were carried out by using a Nova NanoLab 600 scanning

---

Content from this work may be used under the terms of the Creative Commons Attribution 3.0 licence. Any further distribution of this work must maintain attribution to the author(s) and the title of the work, journal citation and DOI.
Published under licence by IOP Publishing Ltd

electron microscope (SEM) (FEI, the Netherlands). Analysis of the SEM image showed that diameter of a CNT at 34±3 nm, length at 370±40 nm and density of CNTs in an array at 47 μm^{-2} (Figure 1a). The structural analysis of the CNT array by a Tecnai Osiris transmission electron microscope (FEI, Netherlands) showed that the experimental sample consists of multi-walled carbon nanotubes having bamboo-like structure defects (Figure 1b).

 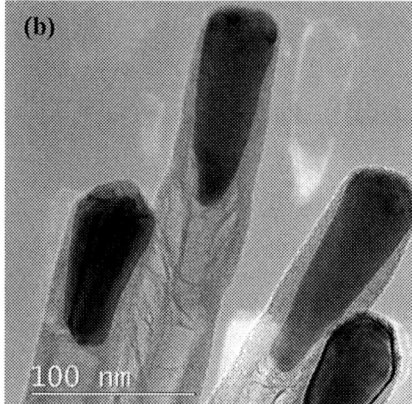

Figure 1 (a, b). (a) SEM image and **(b)** TEM image of the experimental sample of the vertically aligned CNTs array.

The study of the current generated during deformation of CNT was carried out by atomic force microscopy (AFM) in the force spectroscopy mode with parallel detection of the current using the Ntegra probe nanolaboratory (NT-MDT, Russia) (Figure 2a). Commercial cantilever with a platinum coating NSG11/Pt was used as the AFM probe. The pressing force the AFM probe to the CNT top was 155 nN. The study of the surface potential generated during mechanical vibrations of CNTs was performed by oscilloscope Wavepro 7100A (LeCroy, USA) (Figure 3a). The mechanical vibrations of CNTs were generated by the piezoscanner of the Ntegra probe nanolaboratory. A frequency of the mechanical vibrations was set up from 100 Hz to 5 kHz.

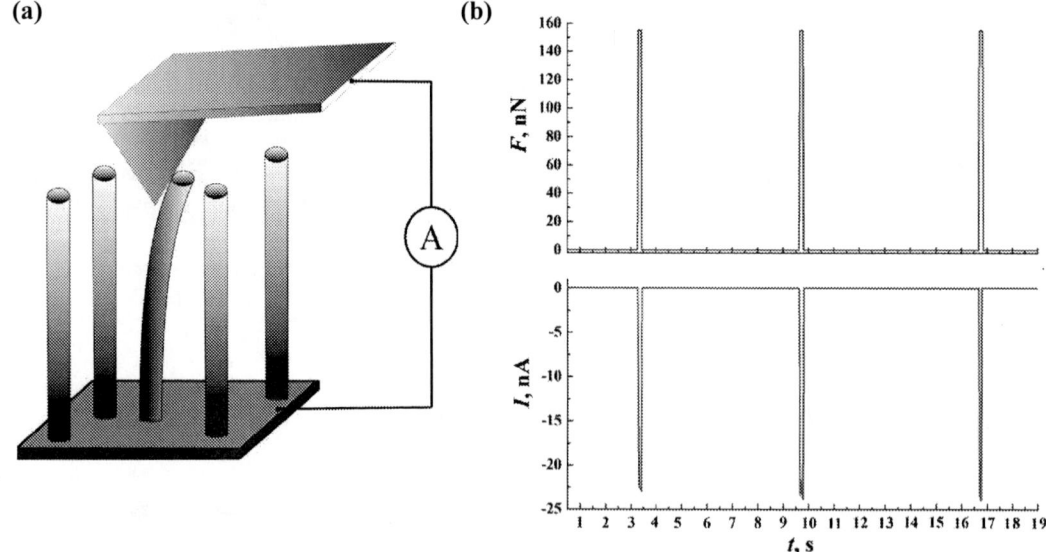

Figure 2 (a, b). (a) The schematic representation of the measuring process; **(b)** Time dependent current of CNT when applying pressing force pulses.

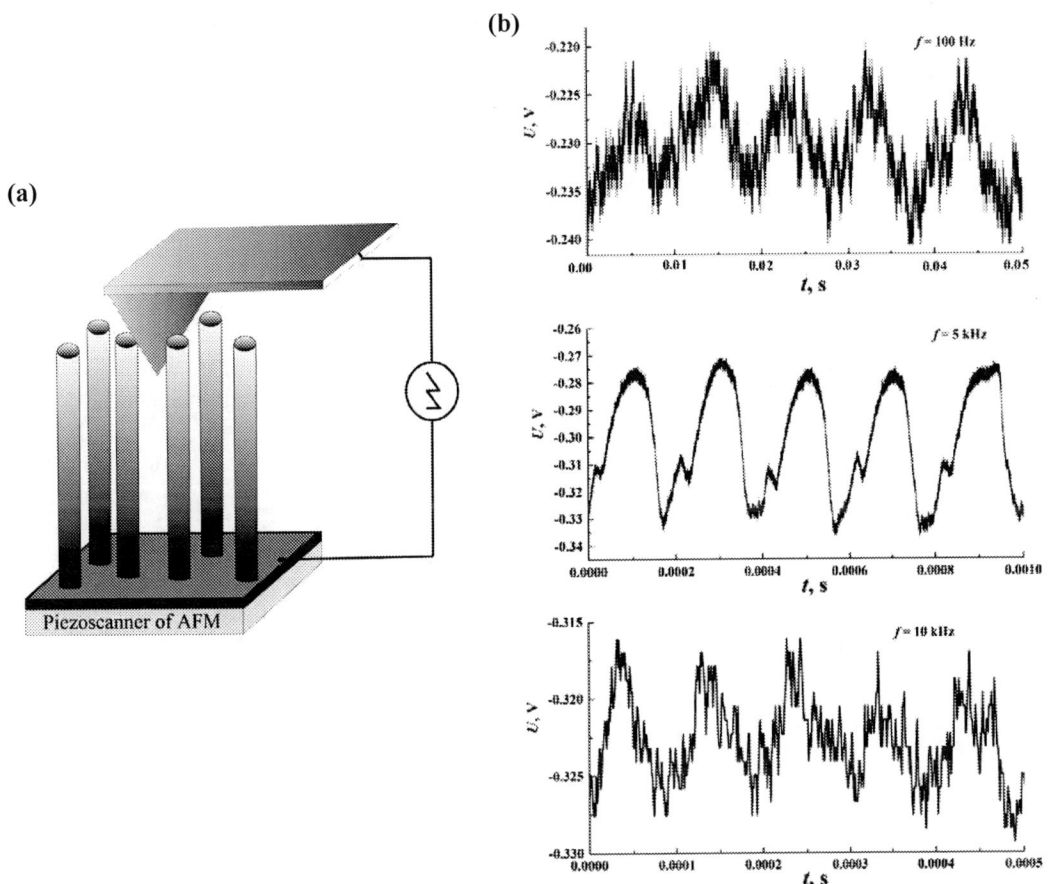

Figure 3 (a, b). (a) The schematic representation of the measuring process; **(b)** Time dependent potential of CNTs generated during mechanical vibration.

The results of experimental studies of the current generated during deformation of an individual CNT and the surface potential generated during mechanical vibrations of CNTs are shown in Figure 2b and Figure 3b, respectively.

3. Results

The results of the experimental studies show that the strained CNT generates a current up to -24 nA (Figure 2b). Thus, at the initial state of the CNT current did not flow. The value of the current varied from 0 to -24 nA when the pressing force the AFM probe to the CNT top increased from 0 to 155 nN (Figure 2b), respectively. The value of the current decreased back to zero with the subsequent removal of the force. This effect is reproducible and is not related to the properties of the substrate and the measuring system [13]. In addition, it is necessary to take into account that the value of the generated current also depends on the resistivity of CNTs [14].

The generation of CNT current is related to anomalous flexo- and piezoelectric properties of carbon nanotubes having non-uniform strain occurring in the process of AFM force spectroscopy [13, 15, 16]. The non-uniform strain of CNT leads to an asymmetric redistribution of the electron density along the nanotube axis and the formation of a nonzero total electric moment in the nanotube [13]. In addition, the presence of the nickel catalytic centers on CNT tops and bamboo-like structure defects can increase the value of the total electric moment of the strained CNT [13].

The analysis of the obtained by mechanical vibration of the CNT results confirmed the generation of surface potential of the CNT array. The potential value varied from -0.22 to -0.32 V with increasing a vibration frequency from 100 Hz to 10 kHz (Figure 3b). In addition, there was a clear dependence of the amplitude of the generated potential on the vibration frequency (Figure 3b). The maximum amplitude of the generated potential (~ 0.05 V) was observed at a frequency of 5 kHz. Obtained results allow us to suggest the sensitivity of CNT to external environment vibrations of the frequency from 100 Hz to 10 kHz.

Thus, carbon nanotubes are capable of generating an electric current and a potential during their deformation and vibration, which are associated with the manifestation of the flexo- and piezoelectric effects.

4. Conclusion

In summary the ability of carbon nanotubes to transform external mechanical influences, including minor mechanical vibrations of the environment, into electric current and potential has been experimentally proven. It is shown that an individual carbon nanotube generates a current of -24 nA at a pressing force of 155 nN. The CNT array generates a surface potential up to 0.32 V when it vibrates with a frequency of 10 kHz.

The obtained results open a number of possibilities for further fundamental and applied research of aligned carbon nanotubes for design of high-efficiency flexoelectrical nanogenerators.

References

[1] Hu Y and Wang Z L 2014 *Nano Energy* **14**, 3–14
[2] Yudin P V and Tagantsev A K 2013 *Nanotechnology* **24** (43), 432001
[3] Zhang J, Wang C and Bowen C 2014 *Nanoscale* **6** (22), 13314–13327
[4] Wang K F and Wang B L 2017 *Int. J. Eng. Sci.* **116,** 88–103
[5] Jiang X, Huang W, Zhang S 2013 *Nano Energy* **2** (6), 1079–1092
[6] Deng Q, Kammoun M, Erturk A, and Sharma P 2014 *Int. J. Solids Struct.* **51** (18), 3218–3225.
[7] Liang X, Hu S and Shen S 2017 *Smart Mater. Struct.* **26** (3), 35050
[8] Kundalwal S I, Meguid S A and Weng G J 2017. *Carbon* **117**, 462–472
[9] Kvashnin A G, Sorokin P B and Yakobson B I 2015. *J. Phys. Chem. Lett.* **6** (14), 2740–2744.
[10] Wang H, Shi M, Zhu K, Su Z, Cheng X, Song Y, Chen X, Liao Zh, Zhang M and Zhang H 2016 *Nanoscale* **8** (43), 18489–18494
[11] Klimin V S, Il'ina M V, Il'in O I, Rudyk N N and Ageev O A 2017 *IOP Conf. Series: Journal of Physics: Conf. Series* **917**, 092023
[12] Il'in O I, Il'ina M V, Rudyk N N, Fedotov A A and Ageev O A 2018 *Nanosystems: Phys. Chem. Math.* **9** (1), 92–94
[13] Il'ina M V, Il'in O I, Blinov Y F, Konshin A A, Konoplev B G and Ageev O A 2018 *Materials* **11**, 638
[14] Ageev O A, Il'in O I, Rubashkina M V, Smirnov V A, Tsukanova O G and Fedotov A A 2015 *Tech. Phys.* **60** (7), 1044–1050
[15] Il'ina M V, Il'in O I, Blinov Yu F, Smirnov V A, Kolomiytsev A S, Fedotov A A, Konoplev B G and Ageev O A 2017 *Carbon* **123**, 514–524
[16] Ilina M V, Blinov Yu F, Ilin O I, Rudyk N N and Ageev O A 2017 *IOP Conf. Ser.: Mater. Sci. Eng.* **256**, 012024

SPBOPEN 2018

IOP Publishing

How to take fractional-order derivative experimentally?

D D Stupin[1], A I Lihachev[1,2], and A V Nashchekin[1,2]

[1]St. Petersburg Academic University, Khlopina 8/3, 194021 St. Petersburg, Russia
[2]Ioffe Institute, Russian Academy of sciences, 194021 St. Petersburg, Russia

E-mail: Stu87@ya.ru, Stupin@spbau.ru

Abstract. In this study we demonstrate a simple and compact implementation of the fractional-order differentiator. At the heart of the proposed approach lies the "universal power law" of the capacitance dispersion of the interface between nanoporous TiN micro-electrode and NaCl 0.9% solution. Using of this phenomenon we take experimentally 0.68- and 0.77-order derivatives of various functions.

1. Introduction

The great interest in nowadays material science is connected with unusual properties of the micro- and nanoscale structures. Such properties as quantum-size effects [1, 2], inverse population in heterojunctions [1], negative differential resistance [3], negative refraction [4], photons band gap in photonic crystals [5, 6] and many others are hot topics in modern electronics and photonics. In this paper we use unique behavior of the electrical signal on the interface between the nanoporous micro-electrode and electrolyte for creating analogue device, which realize fractional-order derivation (FOD) [7, 8, 9] – a progressive tool for nowadays materials modeling, signal processing, automatic control, and bioelectrical measurements [10].

2. Working principle

The interface between nanoporous metal electrode and electrolyte in general could be considered as constant phase element (CPE) [11, 12, 13, 14], which admittance Y is described by "universal power law" of capacitance dispersion [12, 15]. Namely, the current response J through CPE depend on excitation voltage (EV) V as

$$J(t) \doteqdot J(\omega) = YV(\omega) = Q(i\omega)^\alpha V(\omega) \doteqdot Q\frac{d^\alpha}{dt^\alpha}V(t), \tag{1}$$

where ω is frequency, α – nonideality parameter, Q is multiplier with $\Omega^{-1}\text{Hz}^{-\alpha}$ dimension, $\doteqdot$ means Fourier transformation. Thus the current through CPE element is proportional to FOD of the excitation voltage. This phenomenon is used in this study for analogue fractional differentiator creating.

3. Materials and methods

As electrochemical cell we have used multielectrode array MEA 200/30 ITO (Multichannel Systems, Germany), which was filled with 1 ml 0.9% NaCl solution (Biolot, Russia).

Content from this work may be used under the terms of the Creative Commons Attribution 3.0 licence. Any further distribution of this work must maintain attribution to the author(s) and the title of the work, journal citation and DOI.
Published under licence by IOP Publishing Ltd

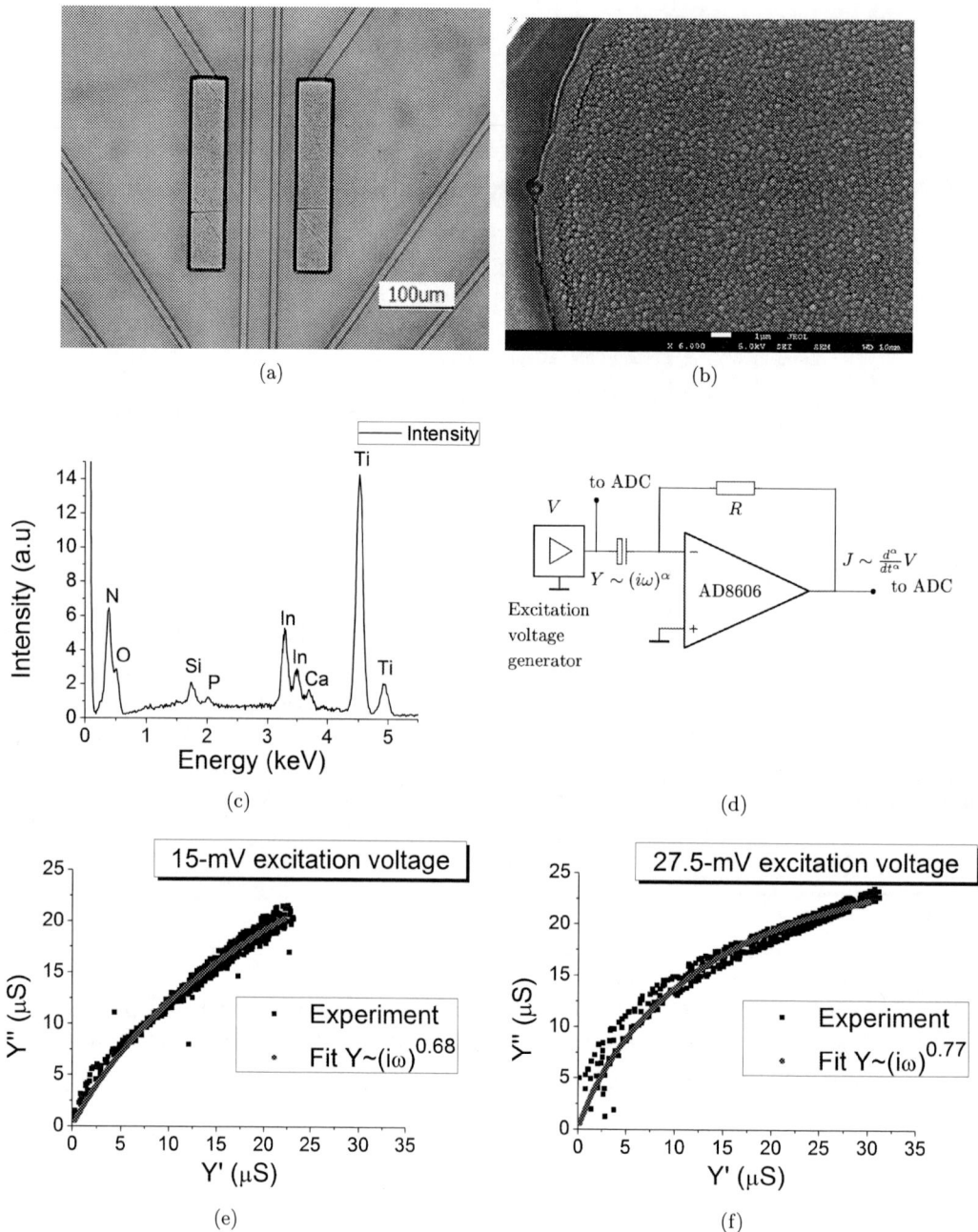

Figure 1. Micro-electrodes characterization and differentiator's properties. (a) Electrodes micro-photography; (b) electrodes morphology; (c) electrodes elemental composition obtained by X-ray microanalysis; (d) differentiator's electronic scheme; (e) and (f) – admittance spectra for different EV amplitudes.

Optical photograph Fig. 1(a) of the MEA 200/30 ITO electrodes was made on IMN100 microscope (Vistec Semiconductors Systems GmbH, Germany). Electrodes surface characterization and X-ray microanalysis were performed using scanning electron microscope JSM 7001F (JEOL, Japan) [Fig. 1(b,c)].

The proposed differentiator scheme is based on operational amplifier AD8606 (Analog Devices, USA) connected into current-to-voltage converter circuit [Fig. 1(d)]. The output voltage of this scheme is equal to $-JR$, where $R = 100$ kΩ is feed-back resistor, J is current through electrode/electrolyte interface with admittance Y. Current J shown in last section is proportional to fractional-order derivative of EV (Eq. 1) that's why scheme presented on Fig. 1(d) realizes fraction-derivation.

For producing EV was used AKIP-3413/3 generator (AKIP, Russia), which has build-in wave-forms library. The EV and output voltage of the differentiator were recorded by analogue-to-digital converter (ADC) L-Card E20-10 (L-Card, Russia, 4 channels). Hardware setup and the MATLAB code NELM which were described in ref. [16] were used for admittance measurement and for it's complex non-linear least square fitting [17] by Eq. 1. The measurement frequency range was 2 Hz $\div$ 1 kHz.

4. Results and discussion

The data presented on Fig. 1(a-c) indicate that electrodes of multi-electrode array have developed surface and mainly consist of TiN, oxygen and indium. Thus they are a classical systems where "universal power law" of the capacitance dispersion should be observed [11, 14] that confirmed by presented on Fig. 1(e,f) admittance measurement results.

Obtained by proposed differentiator fraction-order derivatives are shown on Fig. 2. These results indicate FOD qualitatively corresponds to Heaviside's FOD definition [7]. It should be noticed, that FOD of the step function is not identically equal to zero and FOD of the exponent is not equal to exponent. The FOD of the parabolic function is asymmetric, because the excitation voltage is equal to zero at negative time. It is worth to emphasize that FOD of the cardiac-shape function does not dramatically different from the original cardiac shape function.

SPBOPEN 2018   IOP Publishing
IOP Conf. Series: Journal of Physics: Conf. Series **1124** (2018) 071011   doi:10.1088/1742-6596/1124/7/071011

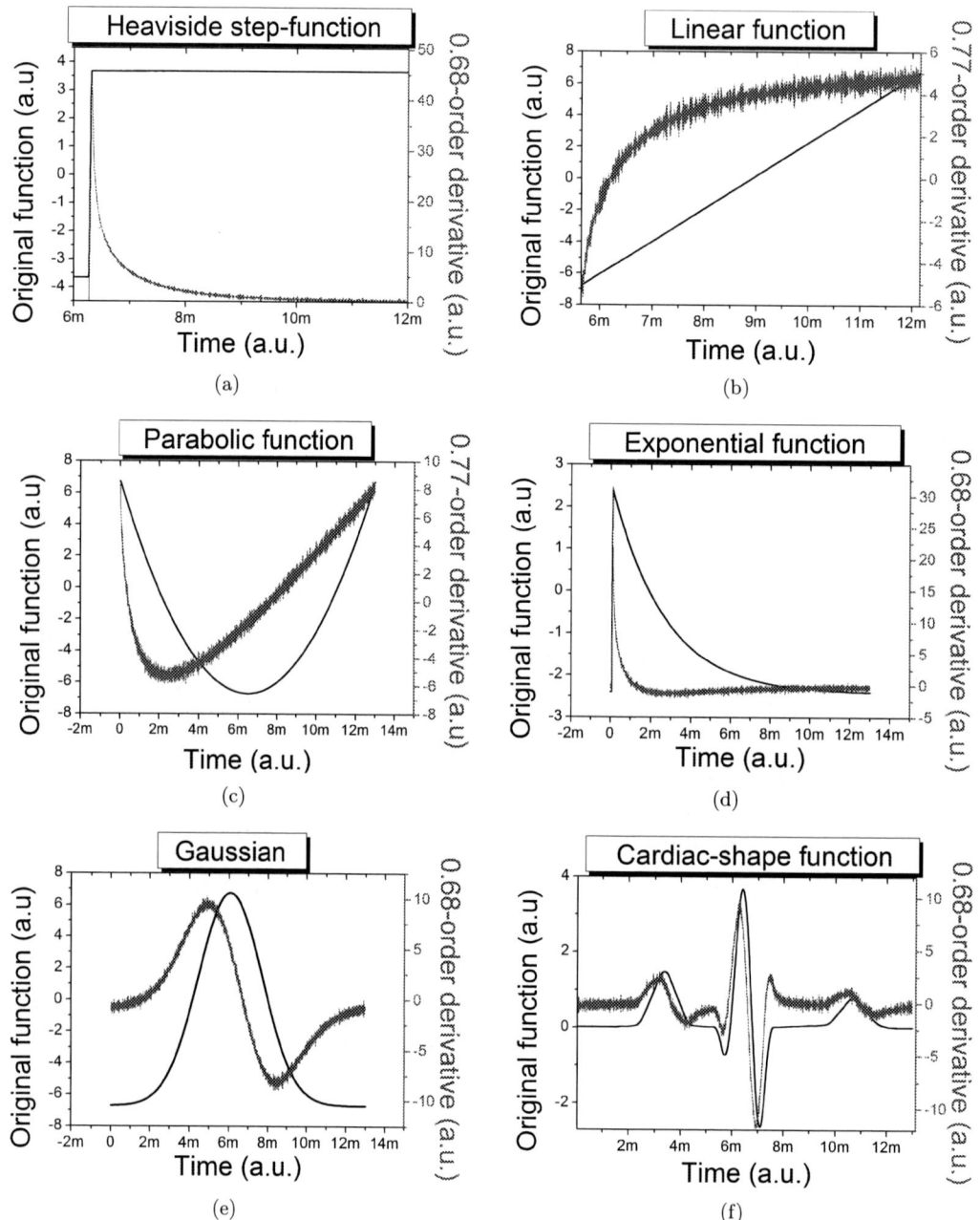

Figure 2. Experimentally obtained fractional orders derivatives (red lines, J) for various functions (black lines, V). The order 0.68 is for 15-mV EV, order 0.77 is for 27.5-mV EV.

5. Conclusion

Here we have proposed a simple implementation of the fractional-order differentiator and experimentally confirmed by using it unusual properties of the FOD, namely non-zero value of the FOD of constant, non-exponential shape of the FOD of the exponent, and non-linear shape of FOD of parabolic function. Developed device is low cost and could be assembled in micro-scale that is important for portable applications, *e.g.* biosensors. The FOD obtained by proposed differentiator is found to be in perfect qualitatively agreement with Heavisides FOD definition. Finally, the observed little distortion of the cardiac-shape function after fraction-derivation opens a direct way of the current-domain measurements usage in cardiology, that could be simply realized in comparison with voltage-domain.

We believe, that results of our study will be useful in the area of the modern audio signal processing, as sample for fractional calculus, in the bioelectrical experiments and biosensorics.

Acknowledgments

The authors acknowledges SPbOPEN Organizing Committee for holding wonderful conference, and Sergei V. Koniakhin, Nikolay A. Verlov, Anton S. Bukatin, and Michael V. Dubina for multifaceted assistance and support. SEM characterization were performed using equipment owned by the Federal Joint Research Center "Material science and characterization in advanced technology" with financial support by Ministry of Education and Science of the Russian Federation (id RFMEFI62117X0018). This study was funded by Ministry of Education and Science of Russian Federation, governmental order 16.9790.2017/BCh.

References

[1] Alferov Z I 2001 *Rev. Mod. Phys.* **73**(3) 767–782
[2] Ivchenko E L and Pikus G 2012 *Superlattices and other heterostructures: symmetry and optical phenomena* vol 110 (Springer Science & Business Media)
[3] Léonard F m c and Tersoff J 2000 *Phys. Rev. Lett.* **85**(22) 4767–4770
[4] Veselago V and Narimanov E 2006 *Nature materials* **5** 759
[5] Yablonovitch E, Gmitter T and Leung K 1991 *Physical review letters* **67** 2295
[6] Joannopoulos J D, Johnson S G, Winn J N and Meade R D 2011 *Photonic crystals: molding the flow of light* (Princeton university press)
[7] Kenneth S Miller B R 1993 *An introduction to the fractional calculus and fractional differential equations* 1st ed (Wiley)
[8] Charef A 2006 *IEE Proceedings-Control Theory and Applications* **153** 714–720
[9] Sheng H, Sun H, Coopmans C, Chen Y and Bohannan G 2011 *The European Physical Journal Special Topics* **193** 93–104
[10] Krishna B 2011 *Signal Processing* **91** 386–426
[11] Kerner Z and Pajkossy T 2000 *Electrochimica Acta* **46** 207–211
[12] De Levie R 1965 *Electrochimica Acta* **10** 113–130
[13] Singh M B and Kant R 2013 *Journal of Electroanalytical Chemistry* **704** 197–207
[14] Pajkossy T and Nyikos L 1990 *Physical Review B* **42** 709
[15] Martin M and Lasia A 2011 *Electrochimica Acta* **56** 8058–8068
[16] Stupin D D, Koniakhin S V, Verlov N A and Dubina M V 2017 *Phys. Rev. Applied* **7**(5) 054024
[17] Macdonald J R and Garber J 1977 *Journal of the Electrochemical Society* **124** 1022–1030

SPBOPEN 2018   IOP Publishing

Ti/4H-SiC Schottky diode breakdown voltage with different thickness of 4H-SiC epitaxial layer

S V Sedykh, S B Rybalka, A Yu Drakin, A A Demidov, N S Ponomaryova, O A Shishkina

Bryansk State Technical University, Bryansk, Boulevard 50 let Oktyabrya 7, Russia

Abstract. Breakdown voltage for Ti/4H-SiC type Schottky diode with six guard rings have been calculated theoretically and by mean of numerical simulations. It is shown that the breakdown voltage can be increase at the minimum on 100 V in case when thickness of the n-type 4H-SiC epitaxial layer increase from 18 up to 22 μm. It is established that the breakdown voltage value for Ti/4H-SiC type Schottky diode with guard rings calculated by mean simulation in ATLAS program and theoretically have good approximation. Thus, above approach gives the possibility for projection of diode structure with different 4H-SiC epitaxial layer thickness with higher breakdown voltage value.

1. Introduction

It is known that the silicon carbide belongs to a class of materials commonly referred to as wide-bandgap semiconductors and now silicon carbide represents an excellent candidate for high-temperature power electronic device applications because of its high breakdown voltage, low series resistance and stability under high temperature conditions [1]. Silicon carbide Schottky diodes are of special interest since these unipolar devices avoid reverse recovery effects of bipolar devices, thereby offering higher frequency operation. In particular, SiC Schottky diodes for power electronics in future must be produced by domestic company the ZAO «GRUPPA KREMNY EL» (Bryansk). It is obviously that for development of component base on the base of SiC studying and optimization of such important device as Schottky diode it is necessary.

Earlier, in our previous paper have been studied 4H-SiC type Schottky diodes with Ni and Ti Schottky anode contacts without guard rings [2-5] and 4H-SiC type MOS transistors [6].

Therefore in present paper the main goal is investigation of thickness of the epitaxial layer (4H-SiC) effect on breakdown voltage 4H-SiC Schottky diode with Ti Schottky anode contact with guard rings for increasing of breakdown voltage value.

2. Materials and methods

Figure 1 shows the schematic silicone carbide Schottky diode structure for calculation. Thus, for calculation and numerical simulation were chosen the following the Schottky diode parameters: the concentration of donors (nitrogen) in the substrate equals $N^+ = 10^{18}$ cm^{-3}, in the n-type epitaxial layer (nitrogen) equals $N^- = 3 \times 10^{15}$ cm^{-3}, in the guard rings (boron, depth of guard about 2 μm) regions $N_{p+} = 10^{18}$ cm^{-3}, anode material is Ti (titanium), the thickness d of the epitaxial layer (4H-SiC) was chosen equals 18 μm, 20 μm and 22 μm, the radius of the structure equals r=140 μm. For simulation

---

Content from this work may be used under the terms of the Creative Commons Attribution 3.0 licence. Any further distribution of this work must maintain attribution to the author(s) and the title of the work, journal citation and DOI.
Published under licence by IOP Publishing Ltd

model of current-voltage characteristics has been solved electrostatic Poisson's equation in cylindrical coordinates together with continuity equations for electrons and holes in ATLAS program.

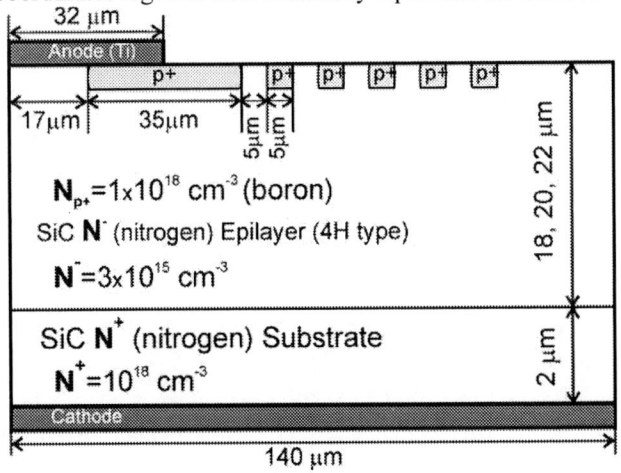

Figure 1. Schematic silicone carbide Schottky diode structure in cylindrical coordinates for calculation.

3. Results and discussion

The breakdown voltage in classical theory depends on critical field, epilayer doping and thickness, edge termination etc [7]. For instance, in general case the breakdown voltage V_{BV} of strongly asymmetric junction can be calculated by following approximation formula [1,7,8]:

$$V_{BV} = \frac{E_c W}{2}, \tag{1}$$

where E_c – critical electric field, W – the space-charge region thickness. In the limit case we can believed that the space-charge region thickness W equals the thickness d of the 4H-SiC epitaxial layer (18 μm, 20 μm and 22 μm in our case, respectively).

On the other hand, for a plane-parallel p–n junction the critical breakdown field strength (E_c) can be determined from the condition for equality to unity of the ionization integral:

$$\int_0^d \alpha_n \exp\left[-\int_x^d (\alpha_n - \alpha_p) dx'\right] dx = 1, \tag{2}$$

where α_n and α_p are the ionization coefficients for electrons and holes. In 4H-SiC, the ionization coefficients depend exponentially on the reciprocal field:

$$\alpha_n = \alpha_{n0} \exp\left(-\frac{E_n}{E}\right), \tag{3}$$

$$\alpha_p = \alpha_{p0} \exp\left(-\frac{E_p}{E}\right), \tag{4}$$

where α_{n0}=1.76×10^8 cm^{-1}, α_{p0}=3.41×10^8 cm^{-1}, E_n=3.3×10^7 V/cm, E_p=2.5×10^7 V/cm [9,10]. Then, substitute (3) and (4) into equation (1) obtain the following integral equation in E_c:

$$\int_0^d \alpha_{n0} \exp\left(-\cfrac{E_n}{E_c - \cfrac{q}{\varepsilon_0 \varepsilon_r} N^- x}\right) \exp\int_0^d \alpha_{n0} \exp\left(-\cfrac{E_n}{E_c - \cfrac{q}{\varepsilon_0 \varepsilon_r} N^- x}\right) -$$

$$\left. -\alpha_{p0} \exp\left(-\cfrac{E_p}{E_c - \cfrac{q}{\varepsilon_0 \varepsilon_r} N^- x}\right) dx'\right] dx = 1, \tag{5}$$

where $\varepsilon_0 = 8.85 \times 10^{-12}$ F/m – the dielectric constant, $\varepsilon_r = 9.7$ – the dielectric relative permeability of 4H-SiC, $N^- = 3 \times 10^{15}$ cm^{-3} – the concentration of donors in the 4H-SiC epitaxial layer, $d=W$ – the thickness of the 4H-SiC epitaxial layer, q – the elementary charge. Further, after solving numerically Eq. (5) we can obtain the critical breakdown field strength E_c and then calculate the breakdown voltage V_{BV} value using the Eq. (1). The critical breakdown field strength E_c and breakdown voltage V_{BV} value at concentration of donors in the 4H-SiC epitaxial layer $N^- = 3 \times 10^{15}$ cm^{-3} and various thicknesses d of the 4H-SiC epitaxial layer are generalized in Table 1.

Table 1. Breakdown voltage V_{BV} value calculated by Eq. (1) for various thickness d of the 4H-SiC epitaxial layer.

d, thickness of the 4H-SiC epitaxial layer (μm)	N^-, concentration of donors in the 4H-SiC epitaxial layer (cm^{-3})	E_c, critical breakdown field calculated by Eq. (5) (V/m)	V_{BV}, breakdown voltage calculated by Eq. (1) (V)
18	3×10^{15}	2.377220×10^8	2.139×10^3
20	3×10^{15}	2.377290×10^8	2.377×10^3
22	3×10^{15}	2.377217×10^8	2.615×10^3

Afterward, with aim to compare theoretically calculated breakdown voltage has been carried out numerical simulation. For simulation model of reverse current-voltage (*I-V*) characteristics has been used physical analytical model in ATLAS program where has been solved electrostatic Poisson's equation in cylindrical coordinates together with drift–diffusion and continuity equations [11,12]. Above-mentioned numerical model was described in detail in previous works [2-5], but in our case the incomplete impact ionization has been taking into account. Further, for simulation of reverse current-voltage characteristics on structure of Schottky diode from Figure 1 were chosen temperature 300 K and thickness of the 4H-SiC epitaxial layer – 18 μm, 20 μm and 22 μm. Simulation results of reverse current-voltage characteristics for Ti/4H-SiC Schottky diodes with different thickness of the epitaxial layer in ATLAS program are presented in Figure 2.

As can be seen from Figure 2 the breakdown voltage value V_{ATLAS} corresponds to 2.332 kV – for thickness of the 4H-SiC epitaxial layer equals 18 μm, 2.380 kV – for 20 μm, 2.412 kV – for 22 μm. In addition, in ATLAS program log also there is possibility find the critical breakdown field E_c value in case when ionization integral value exceeds unity that corresponds to diode breakdown voltage V_{II} condition.

In Table 2 are generalized the breakdown voltage V_{ATLAS} calculated in ATLAS from Figure 1, breakdown voltage V_{II} when ionization integral value in ATLAS exceeds unity and breakdown voltage V_{BV} calculated theoretically by Eq. (1). As can be seen from Table 1, theoretical value of breakdown voltage V_{BV} calculated by Eq. (1) for Ti/4H-SiC Schottky diode is in good agreement with breakdown voltage data V_{ATLAS} by simulation in ATLAS (see Figure 2) and breakdown voltage V_{II} value from ATLAS program log.

IOP Conf. Series: Journal of Physics: Conf. Series **1124** (2018) 071012   doi:10.1088/1742-6596/1124/7/071012

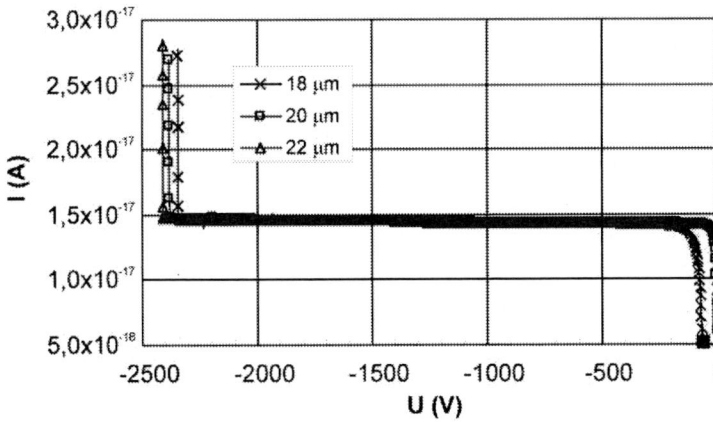

Figure 2. Reverse *I-V* characteristics of Ti/4H-SiC Schottky diode calculated in ATLAS for various thickness of the epitaxial layer (4H-SiC) 18 μm, 20 μm and 22 μm.

Thus, it is established that the breakdown voltage value for Ti/4H-SiC type Schottky diode with guard rings can be calculated in ATLAS program with good approximation.

Table 2. Breakdown voltage value calculated theoretically and in ATLAS program for various thickness *d* of the 4H-SiC epitaxial layer.

d, thickness of the 4H-SiC epitaxial layer (μm)	V_{ATLAS}, breakdown voltage calculated in ATLAS from Figure 1 (V)	V_{II}, breakdown voltage when ionization integral value in ATLAS exceeds unity (V)	V_{BV}, breakdown voltage calculated theoretically by Eq. (1) (V)
18	2.332×10^3	2.302×10^3	2.139×10^3
20	2.380×10^3	2.354×10^3	2.377×10^3
22	2.412×10^3	2.404×10^3	2.615×10^3

Because of this as follows from Table 1 in case when thickness of the 4H-SiC epitaxial layer increase on 4 μm (from 18 up to 22 μm) it lead to increase of the breakdown voltage at the minimum on 100 V for Ti/4H-SiC type Schottky diode with six guard rings.

4. Conclusions

Thus, investigation of thickness of the epitaxial layer (4H-SiC) effect on breakdown voltage 4H-SiC Schottky diode with Ti Schottky anode contacts with guard rings for increasing of breakdown voltage value has been carried out. It is shown that breakdown voltage Ti/4H-SiC Schottky diode can be calculated theoretically and by mean of numerical simulation in ATLAS program. For Ti/4H-SiC type Schottky diode with six guard rings it is established that in case when thickness of the 4H-SiC epitaxial layer increase from 18 up to 22 μm it lead to increase of breakdown voltage at the minimum on ~100 V in accordance with ATLAS model numerical calculation and theoretical calculation.

Therefore, it is suggested that matches of numerical model results with theoretically obtained results indicate that the ATLAS simulation model is correct and can be used for calculation of such type SiC Schottky diode current-voltage characteristics and the optimal design of diode structure determination.

Acknowledgements

Authors would like to thank Dr. Ivanov P.A. (Ioffe Physicotechnical Institute, Russian Academy of Sciences, St. Petersburg) for help in carrying out of simulation in used program. This work has been supported by the Russian Ministry of Education and Science (task No. 8.1729.2017/4.6).

References

[1] Kimoto T, Cooper J A 2014 *Fundamentals of Silicon Carbide Technology. Growth, Characteriztion, Devices, and Applications* (New York: Wiley–IEEE Press.)

[2] Panchenko P V, Rybalka S B, Malakhanov A A, Krayushkina E Yu, Rad'kov A V 2016 *Proc. SPIE "International Conference on Micro- and Nano-Electronics"* **10224** 102240Y-1

[3] Panchenko P V, Rybalka S B, Malakhanov A A, Demidov A A 2017 *Book of Abstract 4th International School and Conference on Optoelectronics, Photonics, Engineering and Nanostructures "Saint-Petersburg OPEN 2017"* (Saint-Petersburg: St. Petersburg Academic University) p. 546

[4] Panchenko P V, Rybalka S B, Malakhanov A A, Demidov A A, Krayushkina E Yu, Shishkina O A 2017 *J. Phys.: Conf. Series.* **917** 082010

[5] Rybalka S B, Krayushkina E Yu, Demidov A A, Drakin A Yu, Zotin V F 2017 *Belgorod State University Scientific Bulletin: Mathematics & Physics* **48** (20) 93

[6] Ivanov P A, Potapov A S, Rybalka S B, Malakhanov A A 2017 *Journal of Radio Electronics* **6** 1

[7] Baliga B 2008 *Fundamentals of Power Semiconductor Devices* (Berlin: Springer Science–Business Media)

[8] Sze S M, Ng K K 2007 *Physics of Semiconductor Devices* (New Jersey: John Wiley & Sons Int.)

[9] Hatakeyama T, Watanabe T, Shinohe T, Kojima K, Arai K, Sano N 2004 *Appl. Phys. Lett.* **85** (8) 1380

[10] Hatakeyama T 2009 *Phys. Stat. Sol. (a)* **206** (10) 2284

[11] Bakowski M, Gustafsson U, Lindefelt U 1997 *Phys. Stat. Sol. (a)* **162** 421

[12] Chandan Kumar Sarkar 2013 *Technology Computer Aided Design: Simulation for VLSI MOSFET* (Boca Raton: CRC Press)

SPBOPEN 2018

IOP Publishing

The influence of the deep level type on a switching time delay of GaAs avalanche S-diodes

V V Kopyev, T E Smirnova, V L Oleinik, I A Prudaev

Functional Electronics Laboratory, Tomsk State University, 36 Lenin Avenue, Tomsk, 634050, Tomsk, Russia

E-mail: viktor.kopev@gmail.com

Abstract. In this work, the influence of the deep level type on a switching time delay of GaAs avalanche S-diodes is investigated. It is shown that the dependence of time delay on voltage can be approximated by an exponential function. The 1.5 times increase of voltage leads to significant decrease of time delay (up to 1000 times). The avalanche S-diodes doped with Fe and Cr/Fe impurities exhibit better stability than diodes doped with Cr. The switching time instability (jitter) for the S-diodes based on GaAs doped Fe or Cr/Fe was 14 times less than for the S-diodes based on GaAs doped with Cr at the same switching conditions.

1. Introduction

Transition metals, such as manganese, chromium, iron, and copper create deep acceptor levels in GaAs energy band gap. Iron and chromium impurities are used for production of semi-insulating material with resistivity of 10^5 and 10^9 Ω·cm, respectively. It makes possible to produce π-v-n structures, with hole- and electron- conductivity layers of high resistance. Semiconductor structures based on GaAs with deep levels are used for production of avalanche S-diodes [1,2]. The reverse current-voltage characteristics (I-V-curves) of the S-diodes have a region with negative differential resistance (NDR). For the first time, this device was described in [3], where GaAs structures doped with iron were used. The authors of different works explained the S-type I-V-curve using the mechanism of recharge of deep levels [1,2,4]. Recent results allow to associate the NDR with the generation of collapsing Gunn domains in the avalanche regime [5,6].

Traditionally, in solid-state pulsed power technology, the field-effect transistors and insulated-gate bipolar transistors are used. As a rule, the switching times of this devices are relatively large (more than 5-10 ns). For the switching in subnanosecond range, another pulsed power devices are used: drift step recovery diodes [7], or avalanche devices, such as sharper diodes [8] and high-gain photoconductive switches [9,10]. The S-diode is an avalanche device which have typical switching times of 0.05-2 ns, and the threshold voltages of 50-2000 V. This makes possible to use it in pulse power suppliers for high power semiconductor lasers [11]. Another promising applications are pulse power suppliers for ultrawideband radars and lidars or electrooptical modulators with Pockels cells.

It is well known, that all avalanche devices suffer from poor switching stability which is caused by random nature of switching. According to statistics, to increase the stability (to decrease the deviation of the time delay) the time delay of switching should be decreased. In this study we investigate the influence of deep level impurities on a time delay and stability (jitter) of switching of the avalanche S-diodes.

Content from this work may be used under the terms of the Creative Commons Attribution 3.0 licence. Any further distribution of this work must maintain attribution to the author(s) and the title of the work, journal citation and DOI.
Published under licence by IOP Publishing Ltd

2. Experimental conditions

The GaAs (100) wafers were grown by vertical gradient freeze (VGF) method. The three types of structures were made from this material. The first structure (GaAs:Cr) was made in the process of diffusion of chromium into the n-GaAs wafer. The second structure (GaAs:Fe) were made by diffusion of iron into n-GaAs, and the third one (GaAs:Cr, Fe) was made by diffusion of chromium and iron into n-GaAs. Initial concentration of donor impurity in GaAs was $5 \cdot 10^{16}$ см$^{-3}$. The avalanche S-diodes were made by the planar mesa-technology. The diffusion was performed using the techniques which are described in [12, 13]. The thickness of the π-layer (doped with iron) was $d_\pi \approx 36$ μm (the average resistivity $\rho_\pi \approx 2 \cdot 10^5$ Ω·cm), the thickness of the v-layer (doped with chromium) $d_v \approx 51$ μm ($\rho_v \approx 2 \cdot 10^9$ Ω·cm).

After process of diffusion, substrates were cut to chips of square form with 800 μm × 800 μm size. Each chip was placed into microwave metal/ceramic package. Contacts were made from the Pb (6%)-Sn (91%)-Ag (3%) alloy. The contacts were formed at $T = 580°C$ in hydrogen atmosphere. The schematic view of the S-diode structure is shown in the Figure 1.

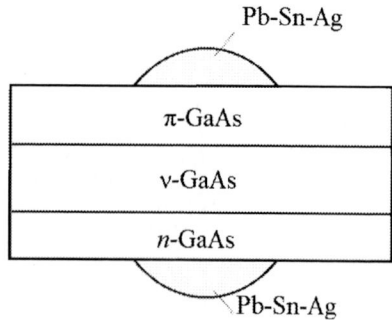

Figure 1. Schematic view of the S-diode structure.

The time delay characteristics were measured in sharper circuit with a load of 51 Ω (figure 2). The rising edge of a triggering voltage pulse was 5 ns. Pulse frequency was 10 Hz. In the measurements, the LeCroy WaveSurfer 104 Xs oscilloscope (1 GHz bandwidth) was used. We measured the time delay of S-diodes as a function of applied pulse voltage. A time delay was considered as a time between a 0.5-level rising edge of a triggering pulse and a 0.5-level rising edge of the pulse after S-diode switch-on.

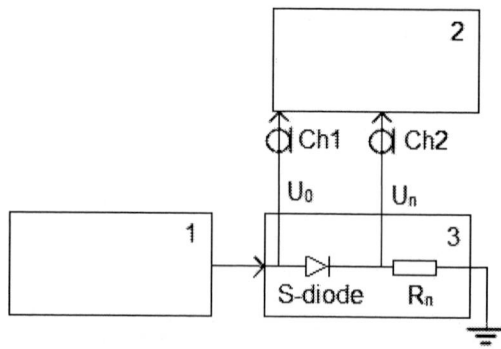

Figure 2. Schematic view of the measurement setup: 1 – triggering generator; 2 – S-diode holder and load resistance $R_n = 51$ Ω; 3 - (Ch1, Ch2) first and second channels of the oscilloscope, U_0 and U_n – amplitudes of voltage for triggering pulse and pulse at the load, respectively.

3. Experimental results

Figure 3(a) shows the time delay (t_d) dependence on a triggering pulse voltage. These curves are well approximated by equation $\lg t_d = A - B \cdot U^{-1}$ (were A and B are the constants) These data indicate that time delay decrease when voltage grows up. It should be noted, when voltage increase in 1.5 times, the time delay changes from 30 μs to 50 ns. It is known from the literature that for S-diodes based on π-ν-n structures, the most probable mechanism of time delay dependence on voltage is the process of recharging of deep centers during the microplasma avalanche breakdown in π–ν junction [1]. The slopes of the curves depicted in figure 3 are different for different types of structures. As illustrated in figure 3(a), the slope of the curve for GaAs:Fe structure less than for structure with chromium impurities (GaAs:Cr and GaAs:Cr, Fe).

Figure 3(b) shows jitter dependence on voltage (U^{-1}). Here the jitter was measured for 600 switching events and it represented the total range of measured time instability. Thus, jitter equals to $6 \cdot \sigma$ (σ - standard deviation of time delay). As we can see, jitter decreases when the pulse voltage increases. For 100 ns time delay the S-diodes based on GaAs:Fe and GaAs:Cr, Fe structures have 1.4 ns and 2 ns jitter, respectively. The diodes based on GaAs:Cr structures have 20 ns jitter at the same time delay.

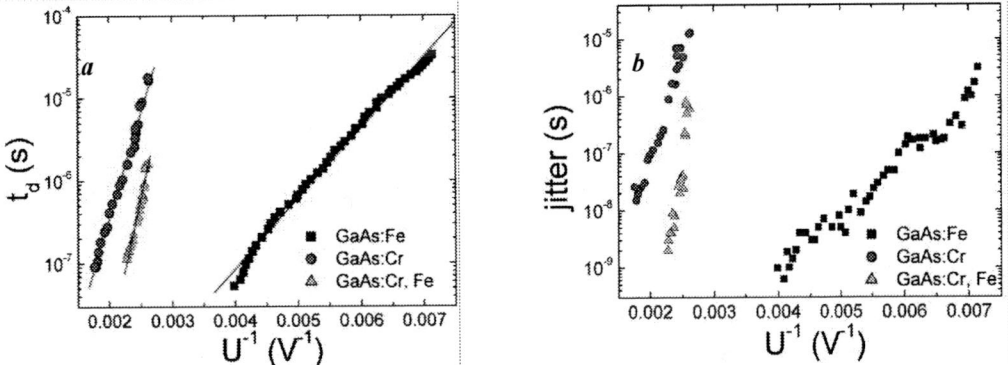

Figure 3(a, b). (a) Dependences of the time delay on voltage; **(b)** Dependences of the jitter on voltage

Figure 4 shows dependence of jitter to time delay ratio (j/d ratio) on triggering pulse voltage. The ratio characterizes relative switching stability and it can be used as a critical parameter for comparative analysis of different types of S-diodes. Obviously, to increase the stability, the time delay of switching should be decreased and the j/t ratio should be a constant. However, in experiment the j/d ratio decreases at low voltages and saturates. This fact can be explained by a change of switching delay mechanism: under saturation at high voltages we have single mechanism with constant j/d ratio.

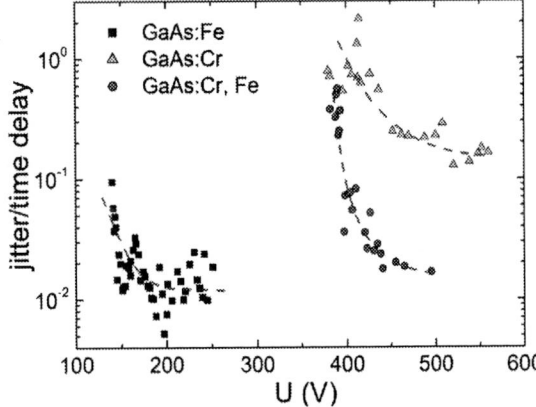

Figure 4. Dependences of j/d ratio on triggering pulse voltage.

935

For application it is important to provide a lower j/d ratio. The experiment shows that switching instability of S-diodes doped with iron is less by 14 times in comparison with diodes doped with Cr. Assuming a constant j/t ratio we can estimate the jitter for different time delays. For example, at time delays of 1-10 ns the jitter should be 100-1000 ps for S-diodes doped with Cr (about 10 % of time delay). This values correspond to 17-170 ps standard deviation which are very close to rms jitter of GaAs high-gain photoconductive switches (14-96 ps [14-17]). In this regard, S-diode structures doped with iron (GaAs:Fe and GaAs:Fe, Cr) looks more promising because of lower jitter (about 1 % of time delay).

4. Conclusion
The experimental results of study of deep level type influence on switching time delay and jitter for S-diodes were presented. It was found that the switching time delay and jitter depend on triggering pulse voltage. As a critical parameter for comparative analysis of switching stability the jitter to time delay ratio (j/d ratio) was proposed. In experiment, the j/d ratio decreasing and saturation were found under triggering pulse voltage increase. Avalanche S-diodes doped with Fe and Cr/Fe impurities exhibit much better stability than diodes with Cr. Thus, S-diode structures doped with iron are more promising because of lower jitter which riches about 1% of time delay. Our future experimental work will be aimed at the study of j/d ratio for a shorter time delays. Also, the physical mechanisms responsible for switching delay and stability of S-diodes doped with different deep levels will be investigated in our further work.

Acknowledgments
This work was supported by Ministry of Education and Science of Russia (Project No 11.2247.2017).

References
[1] Khludkov S S 1983 Soviet Physics Journal **26** 928
[2] Prudaev I A, Khludkov S S, Skakunov M S and Tolbanov O P 2010 Instruments and Experimental Techniques **4** 530
[3] Yamashita S, Hosokawa Y, Anbe T, and Nakano T 1970 Proc. of the IEEE **58** 1279
[4] Prudaev I A and Khludkov S S 2009 Russ. Phys. J. **52** 163
[5] Prudaev I A, Verkcholetov M G, Koroleva A D, Tolbanov O P 2018 Tech. Phys. Lett. **44**
[6] Vainshtein S N, Yuferev V S, and Kostamovaara J T 2005 IEEE Trans. Electron Devices **52** 2760
[7] Ivanov P A, Kon'kov O I, Samsonova T P, Potapov A S, Grekhov I V 2016 Semiconductros **49** 1511
[8] Grekhov I V, Lyublinskiy A G, Yusupova S A 2017 Tech. Phys. **62** 812
[9] Williamson S, Albrecht G F, and Mourou G 1982 Rev. Sci. Instrum. **53** 867
[10] Hu L, Su J, Ding Z, Hao Q, Yuan X 2014 J. Appl. Phys. **115** 094503.
[11] Kopyev V V, Prudaev I A, Romanov I S 2014 J. Phys.: Conf. Ser **541** 012055
[12] Prudaev I A and Khludkov S S 2008 Russ. Phys. J. **51** 1157
[13] Ardyshev M V, Prudaev I A, Tolbanov O P and Khludkov S S 2008 Inorganic Materials **4** 918
[14] Zutavern F J, Loubriel G M, McLaughlin D L, O'Malley M W, Helgeson W D, and Denison G J 1993 Proc. SPIE **1873** 50
[15] Gaudet J A, Skipper M C, Abdalla M D, Ahern S M, Romero S P, Mar A, Zutavem F J, Loubriel G M, O'Malley M W, and Helgeson W D 2000 Proc. SPIE **4031** 121
[16] Shi W, Zhang L, Gui H, Hou L, Xu M, and Qu G 2013 Appl. Phys. Lett. **102** 154106
[17] Gui H, Shi W, Ma C, Fan L, Zhang L, Zhang S, and Xu Y 2015 IEEE Photon. Tech. Lett. **27** 2001

Accelerated degradation HEMT based on AlGaN / SiC

A S Evseenkov[1], V G Tikhomirov[1]

[1]Saint-Petersburg Electrotechnical University "LETI", Prof. Popova 5, St. Petersburg 197376, Russia

Abstract. Created HEMT based on the structures of the AlGaN/SiC. Was studied all the basic characteristics of the transistors and determine their operating parameters, including power, Gm, delay time, current saturation, etc. A study was made of the degradation of the characteristics of transistors under stress conditions of operation. Modes of express testing were found that allow to estimate the reliability of HEMT. Based on the results of the field distribution simulation was carried out near the gate for a more detailed description of the degradation mechanisms.

1. Introduction

Currently heteroepitaxial structures of group III nitrides are actively replacing traditional $A^{III}B^{V}$ materials on the basis of gallium arsenide in the optical and microwave electronics [1-3]. Large values of electron concentration in the channel in combination with high breakdown fields make it possible to provide a microwave power density in GaN-based field effect transistors 5-10 times larger than in GaAs-based devices. In the transition to a new group of materials, the use of silicon nitride as a dielectric becomes particularly topical. There are a number of problems arising from the operation of HEMT based on nitrides in microwave modes. In connection with the increasing current densities, the issue of heating the device and its degradation resistance is acute [4]. The temperature of the working area of the device can exceed 150 degrees Celsius. One of the most important problems is the presence in the transistor structures of defects of various types, which substantially reduce both the speed and power of the device. This is especially evident with prolonged operation of the device. Possible degradation can cause both changes in the instrument parameters beyond the specification limits, as well as a complete failure of the device.

2. The samples

For the experiment, several AlGaN/SiC wafers were made, on which HEMT with a gate width of 720 μm and a length of 0.5 μm are placed. SiC was used as a substrate, on which the epitaxial transistor structures, shown in figure 1, were grown. The main functional layers are: barrier layer $Al_{0.25}Ga_{0.77}N$, sub-buff layer of AlN, GaN channel layer and the top layer $Al_{0.35}Ga_{0.65}N$. As the dielectric silicon nitride was used. The wafer was divided into individual transistors from which for further experiments control samples were selected on the following parameters: the saturation current - 310 to 360 mA and a cut-off voltage - not less than 5 V, the leakage current at 60 V – no more than 800 microamps.

Content from this work may be used under the terms of the Creative Commons Attribution 3.0 licence. Any further distribution of this work must maintain attribution to the author(s) and the title of the work, journal citation and DOI.

Published under licence by IOP Publishing Ltd

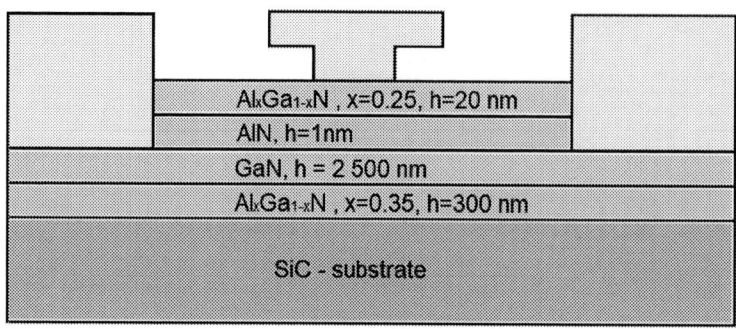

Figure 1. Scheme of epitaxial structure

3. Experiment

Created HEMTs passed extensive testing, including a study of the basic characteristics and identification of operating parameters. In particular, their power parameters, delay times, saturation currents, etc. were determined. The test was conducted for continuous operation of transistors. The voltage at the source-drain was 0 Volts, while the voltage from -10 to -50 V was applied to the gate-drain. It should be noted that the voltage did not go to 0 V between the readings at the measurement points, but kept on a constant bar for 1 minute.

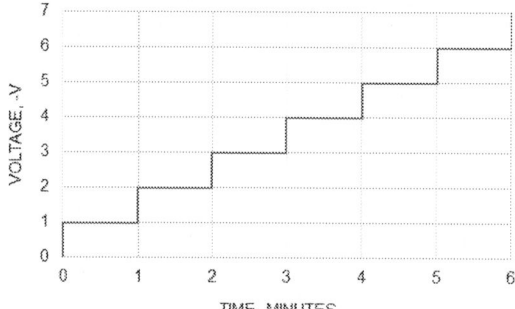

Figure 2. Voltage form of Step-stress

The volt-ampere characteristic for comparison with the degradation tests was investigated and without a slight delay between the measurements. After a cycle of staggered degradation, there was a sharp increase in reverse currents through the gate at a critical voltage equal to 32 V (Figure 3). The experiment included a series of samples of 10 transistors, most of which showed similar dependencies. On average, the sample changed its characteristics as follows: reduction of saturation currents from 360 to 345 mA, leakage currents increased from 40 to 160 µA, the drainage losses increased from 17 to 21%.

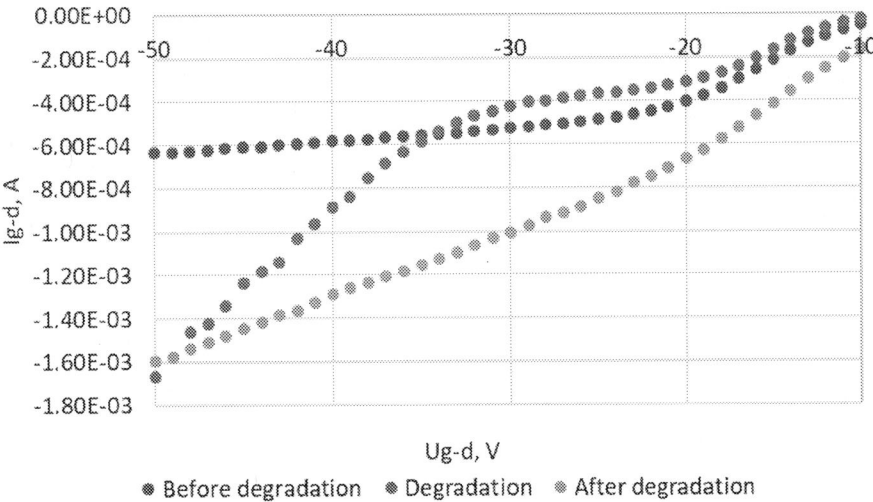

● Before degradation ● Degradation ● After degradation

Figure 3. Volt-ampere characteristic Gate-drain transistor

A sharp increase in the leakage current under the indicated experimental conditions was observed by groups of researchers from leading scientific laboratories and described in [7,8]. To date, one of the common explanations is that when a high electric field intensity occurs under the gate due to the inverse piezoelectric effect, crystallographic defects appear in the material of the barrier layer, which serve as trap centers for the electrons of the shutter material and allow them to enter the channel layer, creating a leakage current through the shutter. The described mechanism is clearly shown in Fig.5.

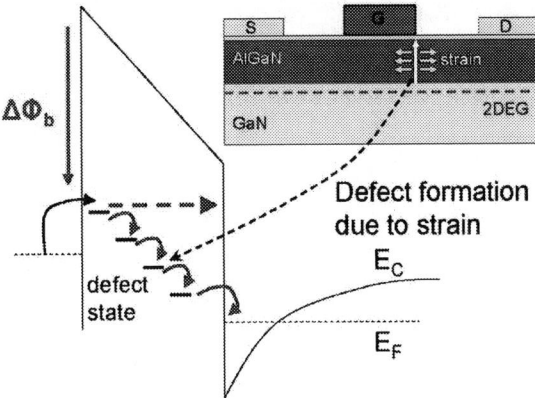

Figure 4. Scheme of the mechanism of electrons entering the channel layer and the occurrence of leakage current through the gate

From the consideration of the leakage current paths in this image, it can be seen that the key initial factor triggering this mechanism is the distribution of the electric field under the gate. However, in known published works, there is no detailed analysis of the electric field in the gate region at a depth of the order of a dozen nanometers. Despite the seeming obviousness of the problem, a reliable calculation of the distribution of the electric field under the gate with reverse bias has a number of complex features. For example, layers of different semiconductor and dielectric materials under the gate have the dimensions of one and tens of nanometers. In this case, at a depth of 20-30 nm there is a heteroboundary directly below which is a 2D electron gas (2DEG) of very high concentration. The layers have essentially different properties and the exact calculation of the depletion region under the gate, taking into account the whole set of their characteristics and the influence of the interface between the layers, is a very difficult task. In turn, the distribution of the electric field strength determines the force and the area of mechanical action on the barrier layer due to the inverse piezoelectric effect and, as a consequence, the probable paths of leakage current flow. It can be, as vertical breakdown, and lateral, or some intermediate, directed towards the nearest electrode with zero potential. With a strictly symmetrical arrangement of the gate between the source and the drain, it is obvious that the field will also have a symmetrical character and the path of the breakdown current will be determined by the inhomogeneities at the edge of the shutter, fig.5.

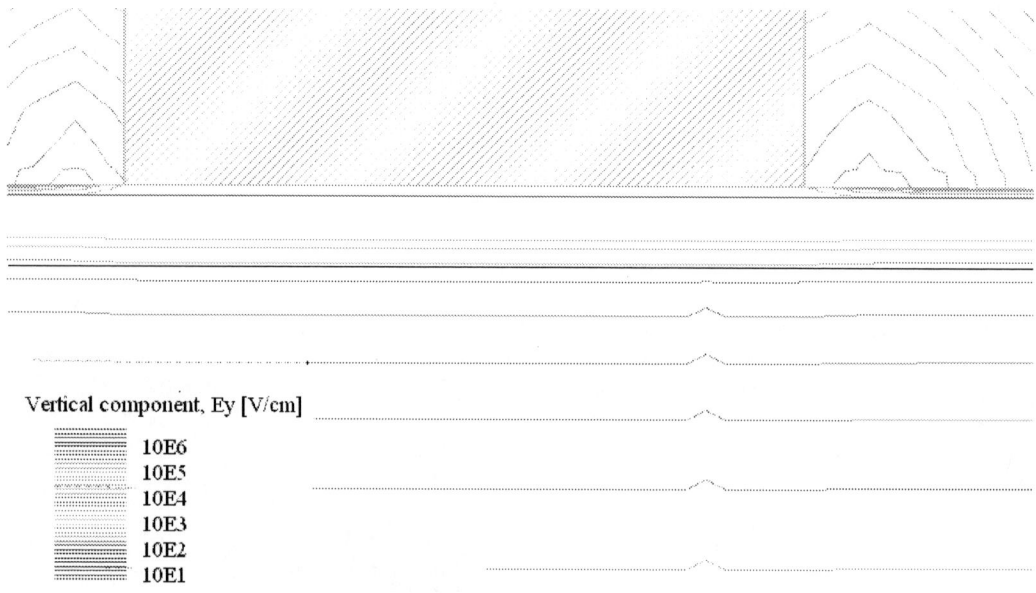

Figure 5. The result of the numerical calculation of the electric field of the gate region of one of the basic variants of the design of the transistor.

To confirm the observed effect, numerical modeling was carried out using a developed package of subprograms that worked together with known numerical modeling systems and described by the authors in previous publications [5].

Figure 6 shows the result of such a numerical calculation of the electric field of the gate region of one of the basic variants of the design of the transistor. Analyzing the results of the calculation, we can conclude that the main gradient of the field is located almost entirely in the barrier layer, and the electric field strength reaches under the gate and near the channel several megavolts per centimeter.

Thus, it becomes clear that one of the main mechanisms of degradation is the inverse piezoelectric effect [6-8], which leads to irreversible damage to the structure of the transistor when the field strength is reached above the threshold, which ultimately can lead to the formation of defects on the edge of the shutter on side drain [9]. One of the other factors that degrade performance is the hot charge carriers, which pass into the gate region as a result of the Paul-Fraenkel mechanisms [10]. The current-voltage gate-drain characteristic of the transistor after a degradation is shown in Figure 6.

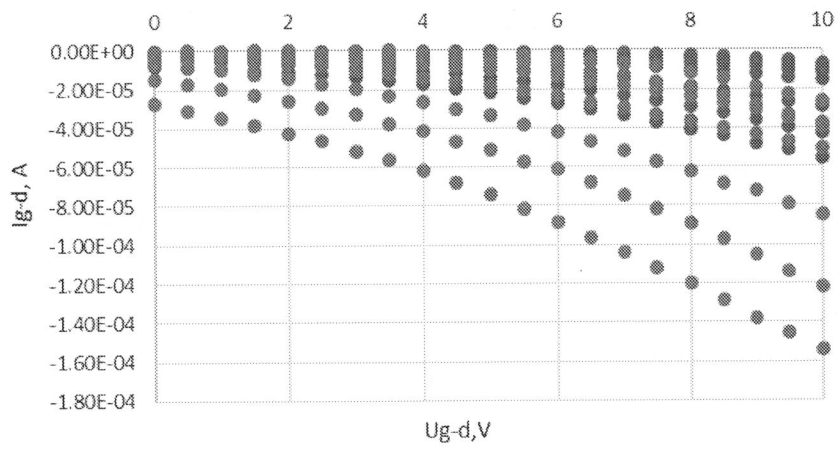

Figure 6. Volt Ampere characteristic gate-drain transistor after a different number of cycles of degradation

4. Conclusion

Thus, during work, transistors with high electron mobility based on AlGaN / SiC structures were created. The gate width was 720 μm, the length was 0.3 μm. All the basic characteristics of transistors were studied and their operating parameters were determined: steepness, source lags, saturation currents, etc. To evaluate the reliability of transistors, a stepwise express degradation test was developed, which is especially important for this class of devices. Intensive degradation began in the samples, beginning with a critical voltage equal to 31 volts. The sample changed its characteristics as follows: reduction of saturation currents from 360 to 345 mA, leakage currents increased from 60 to 160 μA, lags increased from 17 to 21%. With the subsequent repetition of degradation cycles, the degradation process continued until the transistor failed.

Список литературы

[1]   Evseenkov A S, Lamkin I A, Tarasov S A, Solomonov A V 2015 Journal of Physics: Conference Series T 643 № 1 C 012033

[2]   Evseenkov A S, Tarasov S A, Lamkin I A, Solomonov AV, Kurin S Y 2015 Proceedings Of The 2015 IEEE North West Russia Section Young Researchers In Electrical And Electronic Engineering Conference, Elconrusnw 27-29

[3]   Zubkov V I, Evseenkov A S, Orlova T A, Zubkova A V 2015 Russian Physics Journal 2015 T 58 № 8 C 1172-1180

[4]   Faraclas E, Anwar A AlGaN/GaN 2006 Solid-StateElectron 2006 50(6):1051–6

[5]   Tikhomirov V, Zemlyakov V, Volkov V, ParnesYa, Vyuginov V, Lundin W, Sakharov A, Zavarin E, Tsatsulnikov A, Cherkashin N, Mizerov M, and Ustinov V 2016 Semiconductors Vol. 50 N 2 pp 244–248

[6]   V. Tilak et al. In: Proc. IEEE 27th Int. Symp. on Compound Semiconductors (2000) p. 357.

[7]   J. Joh and J. A. del Alamo, IEDM Tech. Dig., Dec. 2006, pp. 415–418.

[8]   J. Joh and J. A. del Alamo, IEEE Electron Device Lett., vol. 29, no. 4, pp. 287–289, apr. 2008.

[9]   J. L. Jimenez, U. Chowdhury, M. Y. Kao, A. Balistreri, C. Lee, P. Saunier, P. C. Chao, W. W. Hu, K. Chu, A. Immorlica, J. del Alamo, J. Joh, and M. Shur, Proc. ROCS,May 2006, pp. 78–81.

[10]   Y. Ling, M. Jing-Jing et al., Field. CHIN. PHYS. LETT. Vol. 27, No. 2 (2010) 027102

SPBOPEN 2018　　　　　　　　　　　　　　　　　　　　IOP Publishing

Software complex for calculating the initial section of the current-voltage characteristics of a resonant-tunneling diode with the possibility of computer statistical experiment

K V Cherkasov[1], S A Meshkov[1], M O Makeev[1]

[1]Bauman Moscow State Technical University, Moscow 105005, Russia

Abstract. The object of research is a resonance-tunneling diode (RTD) based on multilayer AlGaAs heterostructures. A software complex with high-speed algorithm modeling the RTD current-voltage (I-V) characteristic's initial section with the possibility of carrying out a computer statistical experiment for studying the effect of technological errors of the diode design parameters on its I-V characteristic is submitted; the results of such study are shown

1. Introduction

One of the ways of improving the indices of nonlinear ~~radio~~ radio-signal converters is to use nanoelectronic devices, such as resonant-tunneling diodes (RTDs) based on AlGaAs/GaAs heterostuctures [1-12]. Today's microwave electronic products (including the solid-state electronics) markets' growth speed has caused the microwave electronics production volumes to move to the large-scale (and in some segments – the mass) area. This trend is observed for Russian and abroad markets alike. This task requires a software tool for technological errors simulation and carrying out a statistical analysis of device parameters. The objective is to develop a software tool for RTD current-voltage (I-V) characteristic's modelling with the possibility of computer statistical experiment.

2. RTD I-V characteristic's high-speed simulation module

The Tsu-Esaki formula is used as RTD I-V characteristic calculation mathematical model [13]. Diode's resonant-tunneling structure's (RTS) tunneling transparency is calculated by transfer matrix method [14].

The input arguments are RTS construction's parameters: layers' thickness in monolayers (ML) for two spacers, two barriers and one well areas, Al doping percent for barrier areas and spacers potential height. The bottom profile of the conduction band for different bias voltages is calculated according to the entered RTS construction parameters (Fig.1a).

The next step is to calculate RTS tunnel transparency using transparency matrix method (Fig. 1b). Tunnel transparency is calculated for each bias voltage value (after the input the maximum bias voltage value Umax is used to initialize a vector [0; Umax] with determined discretization step).

Then the RTS I-V characteristic is calculated using Tsu-Esaki formula. The influence of ohmic contacts' resistance and mesa dimensions is also taken into account in the RTD I-V characteristic calculation.

Content from this work may be used under the terms of the Creative Commons Attribution 3.0 licence. Any further distribution of this work must maintain attribution to the author(s) and the title of the work, journal citation and DOI.
Published under licence by IOP Publishing Ltd

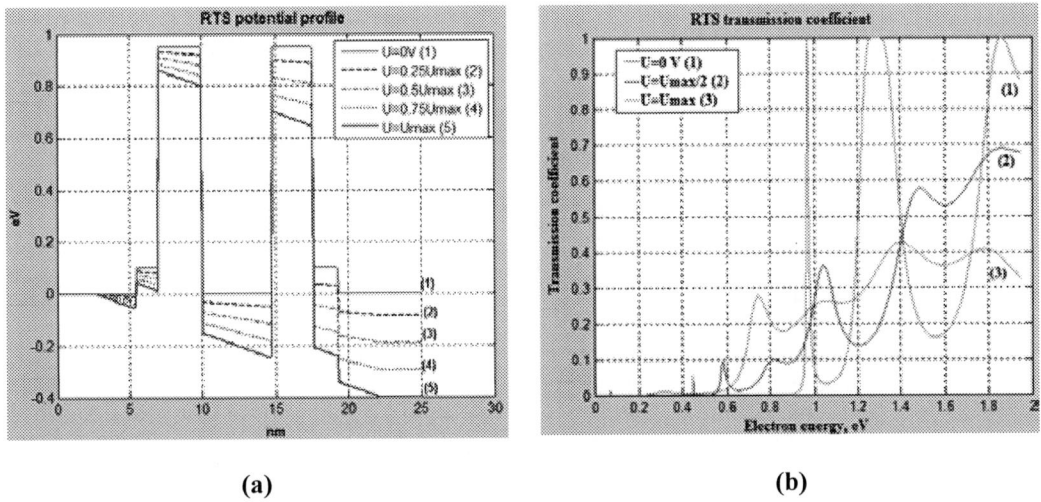

(a)   (b)

Figure 1. RTS parameters simulated: **(a)** potential profile for different bias voltages; **(b)** tunnel transparency curves.

3. Comparison between RTD I-V characteristic simulation results and experimental data

To estimate the simulation accuracy, a comparison between RTD I-V characteristic simulation results and the experimental data obtained by measuring a batch of 27 diodes was carried out (Fig.2, diode RTS parameters is shown in Table 1).

Table 1. Diode RTS parameters

Layer	Chemical composition	Conductance	Thickness, Å
Spacer	GaAs	i	21
Barrier	AlAs	i	29
Well	GaAs	i	49
Barrier	AlAs	i	29
Spacer	GaAs	i	21

I-V characteristics measurements of RTDs were performed using microprobe bench, which consists of a microprobe device, power supply Agilent E3641A and personal computer. This bench provided I-V characteristics measuring of the diodes in voltage range from 0 to 35 V (accuracy $\Delta U = \pm 1$ mV) and current range from 0 to 0,8 A (accuracy $\Delta I = \pm 10$ μA).

Dispersion of I-V characteristic lies within the measurement error range. Therefore, an averaged I-V characteristic value is used for all following comparisons. Maximum current difference between the calculated and experimental I-V characteristic in the 0...0.4 V range is 3.49%.

The performance benchmark of the developed software complex (named RTSVAC) and its analogues (WinGreen [15], Nanohub [16] and dif2RTD [17]) displayed that the calculating of the I-V characteristics using RTSVAC takes 20 seconds, while it takes 90-120 minutes in dif2RTD for the same task, 25-35 seconds in WinGreen, and 35-40 seconds in Nanohub. It is evident that the developed algorithm has 100 times higher performance in comparison with dif2RTD. Also, it is not inferior to WinGreen and Nanohub in the calculation speed.

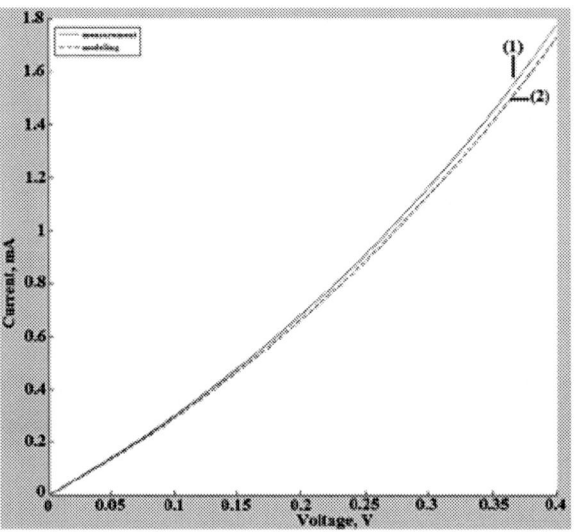

Figure 2. Measured (curve 1) and simulated (curve 2) RTD I-V characteristics

The improved performance is obtained by implementing the described algorithm in Matlab environment, which is highly optimized for working with vector and matrix data, and by some algorithmic optimizations, related with memory allocation and general simulation flow, e.g. by moving duplicated source code in independent subroutines.

The achieved acceleration of the algorithm allows us to proceed to the statistical analysis of RTD I-V characteristic.

4. Computer statistical experiment module
During the computer statistical experiment module development the following technological errors groups were identified as affecting the RTD I-V characteristic: diode RTS parameters (thickness and chemical composition of the layers), ohmic contacts' resistances and mesa area errors. Basing on the RTD I-V characteristic initial section calculation algorithm, a module of computer statistical experiment was developed.

Listed parameters are considered stochastic and are represented in pairs «nominal value – standard deviation». All parameters are considered as continuous random variables and are varied randomly by Gaussian distribution law. Mesa dimensions are correlated with each other with +1 correlation coefficient. This fact is taken into account when modeling rectangle mesa area errors.

RTD I-V characteristic's simulation with random parameters is carried out according to algorithm described in point 2 on every iteration loop. Each iteration's simulation results are saved in different files.

5. The study of RTD technological errors' influence on diode I-V characteristic's variance
The main RTD technological errors influence on RTD I-V characteristic was studied by using the developed software package. The studied errors are diode's RTS technological errors, ohmic contacts resistance errors, and mesa area errors. The study was based on the samples of 100 RTD I-V characteristics obtained by computer statistical experiment. The influence of RTS technological errors and combined influence of ohmic contacts' resistance and mesa area errors on RTD I-V characteristic were studied separately. Parameters' maximum deviations values are given in Table 2.

Table 2. Maximum deviations values of RTD parameters

Group	Parameter	Deviations
RTS	Layers thickness	+/- 0.5 ML
RTS	Al percentage in barriers	+/- 1 %
Ohmic contacts	Ohmic contacts resistance	+/- 0.29 Ohm (nominal value 1.32 Ohm)
Mesa	Mesa dimensions	+/- 3 μm (nominal value 30x30 μm)

Two batches were modeled (Fig.3a, b). Only RTS parameters errors were taken into account in the first one, while ohmic contacts errors and mesa area errors were considered zeros. In the second batch RTS parameters errors were considered zeros while combined influence of ohmic contacts' errors and mesa area errors was studied.

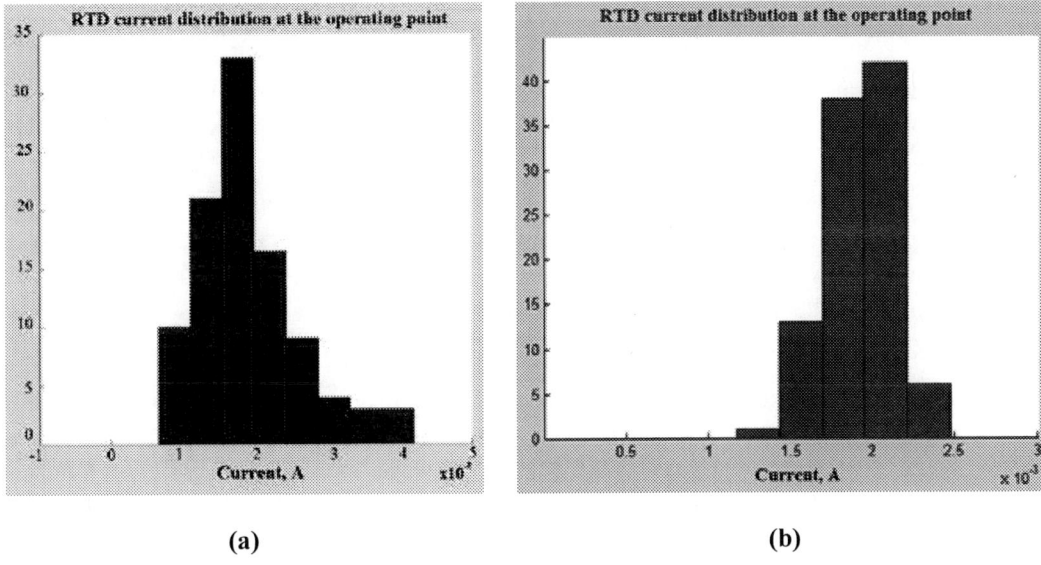

(a)   (b)

Figure 3. Distributions of the current at the RTD operating point: **(a)** RTS technological errors influence; **(b)** combined influence of ohmic contacts resistances and mesa area errors.

Comparison of obtained RTD current distributions revealed that the maximum contribution to the I-V characteristics variance is made by the diode RTS technological errors.

6. Results adequacy assessment

To evaluate the adequacy of the current distributions obtained in point 5, the most realistic case was modeled – a combined influence of all factors listed in Table 2 (Fig.4).

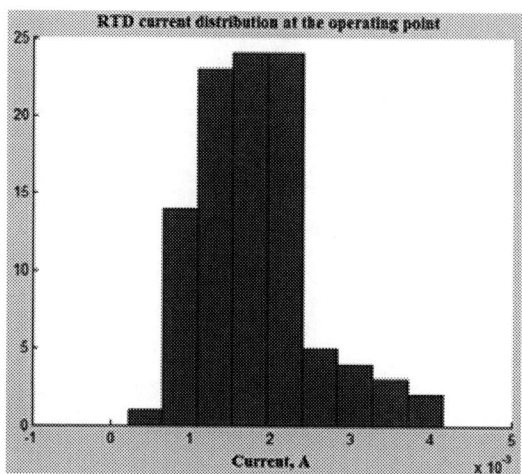

Figure 4 Distributions of the current at the RTD operating point under combined influence of RTS technological errors, ohmic contacts' resistance and mesa area errors.

I-V characteristics of 30 diodes batch were measured experimentally. Both modeled and measured current distributions parameters are listed in Table 3.

Table 3. Distributions parameters of current at the RTD operating point.

	Expected value, mA	Variance, mA^2	Standard deviation, mA
Modeling result	1.97	0.5064	0.7116
Experimental data	1.86	0.3098	0.5565

Mean values' and variances equality hypotheses verification were tested by Student and Fisher criteria respectively at the 0.05 significance level. It was determined that the tested hypotheses do not confront with the experimental data. Hence, the RTD statistical model can be considered adequate.

7. Conclusion

An RTD I-V characteristic's modelling software package allowing carrying out computer statistical experiment was developed. The developed tool allows to study technological errors influence on I-V characteristic's variance. RTD I-V characteristic modelling algorithm provides high modeling accuracy and the speed sufficient for computer statistical experiment implementation based on it.

The study of various error groups' influence on RTD I-V characteristic's variance using the developed software revealed that the maximum contribution to the RTD batch's I-V characteristics variance is made by diode RTS technological errors.

Comparison between experimental and simulated current distributions statistical parameters proved that simulated data is adequate to experiment results.

Acknowledgments

The research work was supported by Ministry of Education and Science of the Russian Federation under state task № 16.1663.2017/4.6.

References

[1] Ivanov Yu A, Meshkov S A, Fedorenko I A, Fedorkova N V and Shashurin V D 2010 *J. Commun. Technol. Electron.* **55** 921

[2] Ivanov Yu A, Meshkov S A, Fedorenko I A, Fedorkova N V and Shashurin V D 2011 *Microwave & Telecommunication Technology (CriMiCo), 2011 21th Int. Crimean Conf.* p 181

[3] Ivanov Yu A, Gudkov A G, Agasieva S V, Meshkov S A, Sinyakin V Yu and Makeev M O 2014 *Microwave & Telecommunication Technology (CriMiCo), 2014 24th Int. Crimean Conf.* p 1063

[4] Sinyakin V Yu, Makeev M O and Meshkov S A 2016 *J. Phys.: Conf. Ser.* **741** 012160

[5] Makeev M O, Meshkov S A, Sinyakin V Yu and Razoumny Yu N 2017 *Adv. in Astronautical Sci.* **161** 475

[6] Kanaya H, Shibayama H, Suzuki S and Asada M 2012 *Appl. Phys. Express* **5** 124101

[7] Maekawa T, Kanaya H, Suzuki S and Asada M 2016 *Appl. Phys. Express* **9** 024101

[8] Wang J, Al-Khalidi A, Alharbi K, Ofiare A, Zhou H, Wasige E and Figueiredo J 2017 Proc. European Microwave Conf. (London) vol.1 (New York: Curran Associates, Inc.) p 341

[9] Mizuta H and Tanoue T *High-speed and functional applications of resonant tunnelling diodes. In The Physics and Applications of Resonant Tunnelling Diodes* 2006 (New York: Cambridge university press) p 133

[10] Nagatsuma T, Fujita M, Kaku A, Tsuji D, Nakai S, Tsuruda K and Mukai T 2014 Proc. Int. Conf. Telecomm. and Rem. Sens. (Luxembourg) vol. 1 (Bulgaria: SciTePress) p 41

[11] Srivastava A 2015 *Eur. J. of Adv. in Eng. and Techn.* **2** 54

[12] Diebold S, Tsuruda K, Kim J-Y, Mukai T, Fujita M and Nagatsuma T 2016 Proc. SPIE 9856, Terahertz Physics, Devices, and Systems X: Advanced Applications in Industry and Defense (Baltimore) vol. 9856 (Washington: SPIE) 98560U

[13] Esaki L and Tsu R 1970 *IBM J. of Res. and Dev.* **14** 1 61

[14] Pérez-Álvarez R and Garcia-Molliner F 2004 *Transfer Matrix, Green Function and Related Techniques: Tools for the Study of Multilayer Heterostructures* (Castelló de la Plana: Publicacions de la Universitat Jaume I) 285 p

[15] Hochschule RheinMain https://www.hs-rm.de/en/rheinmain-university/people/indlekofer-klaus-michael/research-and-development/wingreen/

[16] Nanohub: largest nanotechnology online resource https://nanohub.org/resources/rtd/

[17] Makeev M O, Litvak Yu N, Ivanov Yu A, Meshkov S A and Migal D E 2012 dif2RTD: certificate of state registration of a computer program № 2012661001

Development of the sheet electron beam focusing system based on thermionic and field emission cathodes

A V Danilushkin[1,2], A A Burtsev[1,2], K.V. Shumikhin[1], G.V. Sakhadzhi[1]

[1] Joint Stock Company "NPP "Almaz", Saratov, 410033, Russia
[2] Yuri Gagarin Saratov State Technical University, Saratov, 410054, Russia

Abstract. Synthesis method has been developed for modeling the electron-focusing systems formed compressed sheet electron beam with a field emission cathode. The results of modeling by synthesis of a convergent sheet electron beam with a cross section of $0.15 \cdot 0.8$ mm^2 and a current of 100 mA with magnetic shielding of the thermionic and field emission cathodes are presented. The estimated electric field on the cathode is $0.2 \cdot 10^5$ V/cm at the anode voltage of 20 kV. Experimental investigations of the current-voltage characteristic of an electron gun with impregnated flat cathode in a pulsed mode are conducted, where the collector current is 100 mA.

1. Introduction

Compact amplifiers of average power at the frequencies 0.2–0.3 THz can be realized on the basis of miniaturized devices of vacuum microwave electronics, such as traveling wave tube (TWT), traveling wave klystron. Many studies discuss the prospects of creating a THz TWT with sheet or multiple electron beams. One of the main problems on the development of vacuum devices at the terahertz range is the necessity to employ thin electron beams with a high current density due to the decrease in the transverse dimensions of the tunnels in the slow-wave systems. In most cases, a current density of 500 A/cm^2 is required in the interaction space, which is difficult to achieve for contemporary cathodes. Therefore, electron-optical systems (EOS) are promising, in which compression of a sheet beam is used. EOS with compression of the electron beam and magnetically shielded cathode makes it possible to obtain sufficiently large current densities in the beam with a lower current load on the cathode and with a lower value of the magnetic field. EOS with the high compression of sheet or cylindrical beams and the full magnetic shielded cathode has a great potential of using them in mm and THz linear vacuum electron devices where the current density achieves tens and hundreds of amperes per square centimeter, and the transverse dimensions of the beam are tenths, hundredths of a millimeter [1-3]. Also the promising EOSs are the systems that use the phenomenon of a field emission. In this paper, we present the results of design, simulation of EOS with compression of sheet electron beams.

2. Modeling

Usually analysis programs are used for simulation of a converging sheet beam EOS. The beam is focused by a magnetic field, increasing several times the Brillouin field. But the use of the analysis method for the modeling of EOS is associated with repeated calculations on seeking for the geometry of EOS and the distribution of the magnetic field. Although this method is the most stringent, but its drawback is non-operability. An alternative method of analysis is the synthesis. The synthesis permits to calculate these systems operatively to obtain the geometric dimensions of the gun's electrodes, following the function describing the boundary of a beam, and distribution of the magnetic field. Profile of the anode and the focusing electrode of the gun are calculated employing the equation of the external problem of synthesis and translation of curvilinear coordinates into Cartesian.

The equations of internal and external problems of synthesis in curvilinear coordinates associated with the shape of the beam in the paraxial approximation for the model of an infinitely wide sheet beam have the form [4]:

Content from this work may be used under the terms of the Creative Commons Attribution 3.0 licence. Any further distribution of this work must maintain attribution to the author(s) and the title of the work, journal citation and DOI.

Published under licence by IOP Publishing Ltd

$$\left(u'\varphi\right)' + 2\varphi''u + ih^2\,\frac{\varphi}{\varphi_0} = \frac{i}{\sqrt{u}}$$

$$V(x,q_2) = u + \frac{\mu^2 i\varphi}{2\sqrt{u}}(2q_2 - 1) - \frac{\mu^2 q_2^2}{2}\left[\varphi^2 u'' + \varphi\varphi'u'\right]$$

where $\varphi(x)$ - function describing the boundary of a beam, x-longitudinal, q_2-transverse curvilinear coordinates; $u(x)$, $h(x)$ -normalized potential and magnetic field, a beam width of s, and the thickness of $d=2\varphi_0\Phi_0$, $i=0.095 p_\mu/\mu\mu_1$, $\mu=\Phi_0/l$, $\mu_1=s/l$, p_μ- micro-perveance, Φ_0, l–the normalizing value of the transverse and longitudinal dimensions of the focusing system. The distribution of the magnetic field in a magnetically shielded gun in the region of potential growth to $u=1$, is taken in the following form:

$$h = \sin^2(x - x_m),$$

where x_m– the coordinate of the beginning of the increase of the magnetic field in the gun. The Fig.1 a shows the simulation of EOS with a sheet beam of cross-section $0.15\cdot0.8\ \text{mm}^2$ with a linear convergence 14 and a current of 0.1 A.

The difference between the guns with the field emission cathode and with a thermionic cathode is the presence of a potential in an electron gun near the cathode. Modeling of such systems can be performed by synthesis using the model with a smooth surface of the cathode and potential gradient on its surface that can reach a value, which is necessary for a field emission by the enhancement of an electric field on tips. The distribution of a potential in the region near the cathode in electron gun is defined by the equation:

$$u'(x) = \sqrt{\left[2i\sqrt{u}(1 - \varphi_c'x) + (u_c')^2(1 - \varphi_c'x)^2\right] + 2{,}7i\varphi_c'x\sqrt{xu_c'}}$$

The potential is defined from the polynomial:

$$u(x) = \sum_0^5 a_n(x - x_1)^n$$

In the formulas φ_c' is the angular coefficient of the boundary trajectory of the beam. The start of electron trajectories is solved from the plane with an initial condition $\varphi(x_1) = 1 + \varphi_c'x_1$. The coefficients a_n provide the stitching of the potential and its first two derivatives. The equation of the internal synthesis problem is solved numerically with some values of φ_c', up to the crossover plane, where $\varphi_c'(x_{cr})=0$, and the normalized half-thickness of the beam φ_0 in the crossover is determined. Distribution of the gun's magnetic field along the normalized coordinate x defined from formula $B(x)=h(x)B_{br}$, where: $B_{br} = 1.04\cdot10^{-3}\sqrt{p_\mu U_0/sd}$ (T, V, mm).

The Fig.1b shows the simulation of EOS with a field emission cathode forming a sheet beam of $0.15\cdot0.8\ \text{mm}^2$ with a linear convergence 14 and a current of 0.1 A. The obtained value of the electric field at the cathode's plane determines the required field enhancement factor of real field emission cathode for providing the required current density according to the Fowler-Nordheim formula. It is assumed that the extracting grid in a field emission gun can occupy the same position as the equipotential in diode gun under approximately equal voltage. The electric field on the cold cathode is $0.2\cdot10^5$ V/cm at anode voltage 20 kV.

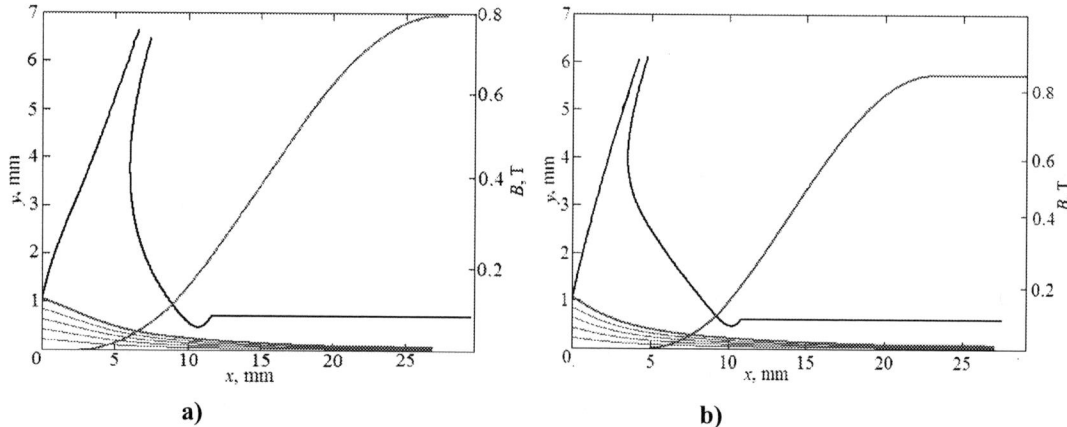

a)   b)

Figure 1. (a) The configuration of EOS with the thermionic cathode, transporting a sheet beam at voltage 20 kV, dimension of a beam $0.15 \cdot 0.8$ mm^2; **(b)** The configuration of EOS with a cold cathode.

An impregnated flat cathode with dimension of $2.1 \cdot 0.8$ mm^2 and focusing grid for experimental investigations of the EOS of a sheet beam formation were made by means of an electric discharge machining (Fig. 2a).

During the investigations of the I-V characteristic the specimen of the electron gun consisted of the cathode-heating assembly, the focusing grid, the anode and the collector with a diaphragm of 0.01 mm placed in a vacuum demountable layout. The anode potential varied from 0 to 10000 V. The experimental investigations of the I-V of the electron gun were conducted in the pulsed mode (filling factor of 0.05%, duration of 10 μs), a current of 100 mA on collector at the cathode temperature of 1200°C was obtained, and the measured beam thickness in the crossover of the gun was 70 μm.

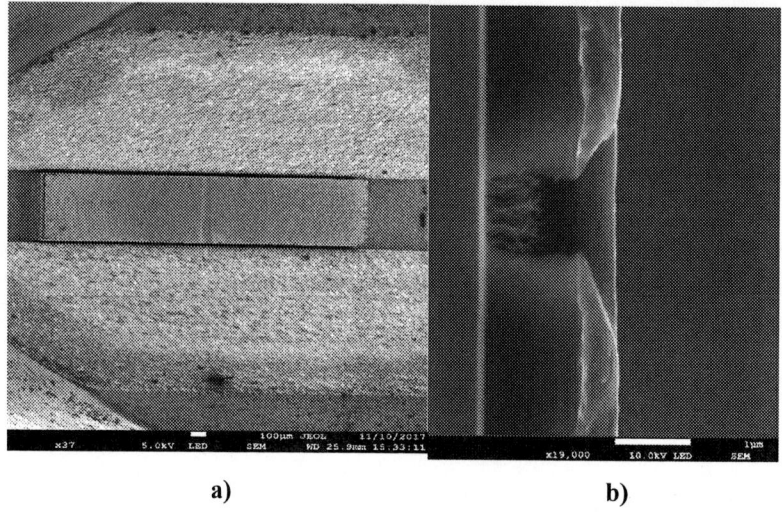

a)   b)

Figure 2 (a, b). (a) SEM-image of a cathode-grid assembly of the electron gun; **(b)** SEM-image of a cell of the field emission cathode with CNTs [5]

Fig. 2b shows a cell of a field emission triode microstructure based on verticalized CNT bundles obtained by PE-CVD locally at the center of a 200 nm-thick control gate deposited on 1.3-μm-thick SiO$_2$ [5]. A cathode–gate structure was studied in the diode mode at various gaps; the current of a single cell in matrix reached 1 μA at the voltage of 118 V and the 1 μm cathode-anode gap.

3 Conclusion

In summary, we have realized synthesis method to find electrodes configurations, the distribution of the magnetic field, providing conformed input of the compression sheet beam with preset parameters to a regular magnetic field. Synthesis method has been developed for modeling the electron-focusing systems formed compressed sheet electron beam with a field emission cathode. The experimental investigations of the I-V of the electron gun with impregnated flat cathode with dimension of $2.1 \cdot 0.8$ mm^2 are conducted in a pulsed mode (filling factor of 0.05%, duration of 10 µs), a current of 100 mA on the collector at the cathode temperature of 1200^{0}C has been obtained.

References

[1] Burtsev, A. A., Grigor'ev, Yu. A., Navrotsky, I. A., Rogovin, V. I., Sakhadzhi, G. V. and Shumikhin, K. V. 2016 Tech. Phys. Lett. 42 543 DOI: https://doi.org/10.1134/S1063785016050229

[2] Burtsev, A. A., Pavlov, A. A., Kitsyuk, E. P., Grigor'ev, Yu. A., Danilushkin, A. V., and Shumikhin, K. V. 2017 Tech. Phys. Lett. 43 542 DOI: https://doi.org/10.1134/S1063785017060062

[3] Yuan Zheng; Diana Gamzina; Branko Popovic; Neville C. Luhmann IEEE Trans. El. Dev. Volume: 63, Issue: 11, 2016 P. 4466 DOI: 10.1109/TED.2016.2606322

[4] P.V. Nevsky / Obzori po electronnoi tekhnike. Seriya 1 – Electronika SVCH. - 1989. - №15. - 48 p.

[5] A. A. Burtsev; N. A. Bushuev; Yu. A. Grigoriev; A. V. Danilushkin; A. A. Pavlov; E. P. Kitsyuk IVEC, 2017 DOI: 10.1109/IVEC.2017.8289540

SPBOPEN 2018

IOP Publishing

Increase of accuracy of capacitance parameters measurements of power semiconductor modules on base IGBT and FRD

D A Knyaginin, A Yu Drakin, S B Rybalka, A A Demidov

Bryansk State Technical University, Bryansk, 50 let Oktyabrya 7, Russia

Abstract. This investigation presents an overview of questions related to the increase in accuracy of measurements of capacitance parameters of power semiconductor modules when developing a device for automated measurements. Result addresses the matters of circuit engineering of primary components, technical characteristics and software of a meter designed to automate the process of testing of capacitance parameters of power modules based on IGBT (insulated gate bipolar transistors) and FRD (fast recovery diodes). Unit testing results that prove the efficiency of the proposed method of the technical implementation are studied. Software implementation is described; ways to improve software-hardware solution are proposed. The obtained results can be used for automated testing of electric parameters of power semiconductor devices.

1. Introduction

The national standard and programs for in-process and final control of electrical parameters provide for measurements of input and output capacitance of IGBT and the total capacitance of FRD in the course of manufacturing of IGBT- and FRD-based semiconductor modules. That said, the desired measurement range of capacitance is usually 50 pF–50 nF; while frequency of the current used for measurements shall be 1 MHz. The above-mentioned development environments were analyzed in the context of this research which implies the development of original open hardware and top level software. For the purpose of testing process control, MATLAB was selected as the product which has no FRD total capacitance shortcomings, as compared to peers, within the framework of this application-oriented measurement task.

2. Materials and methods

In order to eliminate problems which arise during the "remote" measurements of capacitance parameters of power modules, a method similar to [1] is proposed here. The essence of this method is clarified by the circuit diagram (Figure 1) using measurements of IGBT input (a) and output (b) capacitance as an example. The proposed method for measuring capacitance parameters of IGBT and FRD was implemented in the capacitance parameters measuring unit which is controlled by the external computer system. Also, this unit functions as a part of the equipment for automated parameter testing of the IGBT- and FRD-based power semiconductor modules. In order to use the capacitance parameters measuring unit as a part of the automated testing equipment within the local information network, an individual address for the controller can be assigned by means of microswitches. In terms of design, the capacitance meter was executed as a unified device and contains functional nodes in the form of printed modules as shown in the diagram. Serial converter AC/DC of Hengfu Corporation 25W Open Frame Switching Power Supply HF25W-SPL-12 was used as a converter of the mains AC

Content from this work may be used under the terms of the Creative Commons Attribution 3.0 licence. Any further distribution of this work must maintain attribution to the author(s) and the title of the work, journal citation and DOI.

Published under licence by IOP Publishing Ltd

voltage into the intermediate DC voltage. Full functional structure of capacitance measuring block shown on Figure 2. The authors developed and made the remaining modules. The unit provides for the measurements of capacitance from 50 pF to 50 nF in the below three sub-ranges: 50–500 pF, 500–5000 pF, and 5–50 nF. The sub-range used for the generation of the final result is selected automatically.

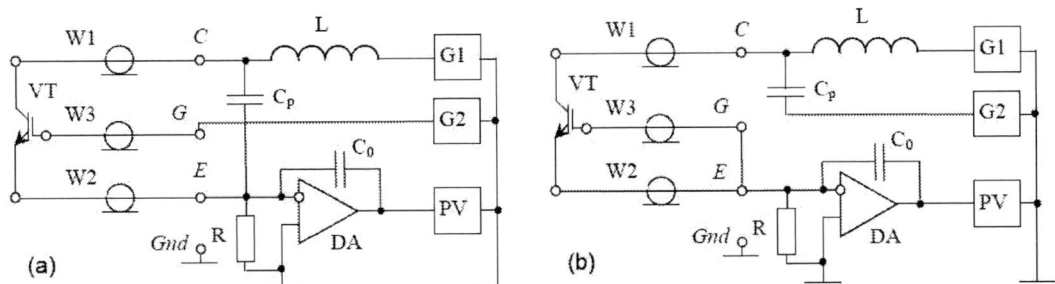

Figure 1 (a,b). a) Diagram for measuring IGBT input capacitance; b) Diagram for measuring IGBT output capacitance.

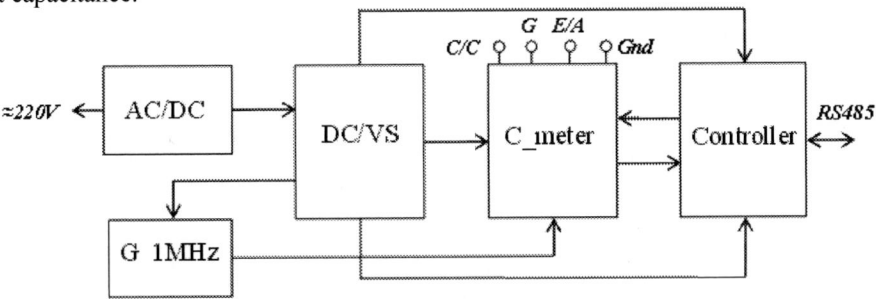

Figure 2. Functional structure of capacitance measuring block

3. Results and discussion

It is established that in the measuring converter circuit (Fig. 1 ab) the frequency characteristic of the optimum amplitude should be performed to ensure the stability of operation within measurement sub-ranges 500 – 5000 pF and 5 – 50 pF. These findings do not fully conform to the results of the device modeling in NI Multisim 12.0 environment [2, 3]. It was experimentally shown that it is possible to perform such correction by connecting a low-ohmic resistor in series with the reference capacitor C circuit resistance on the frequency of 1 MHz does not basically change, which facilitates the process of calibration of the meter. The testing result (Table 1) is similar with [3, 4] and indicatives of the efficiency of the proposed method remote measurement of the IGBT/FRD capacitance parameters in case of small values of the measured capacitance and relatively big distance between the meter and IGBT/FRD.

Table 1. The results measuring capacity of the capacitor with a nominal value of 27 pF.

Input capacity, pF		Output capacity, pF	
C/C – E/A	At a distance	G – E/A	At a distance
27,2	27,3	27,8	28,0

Figure 3 shows the results of the experiment that characterize the relative error in the measurement modes of both input and output capacitance IGBT.

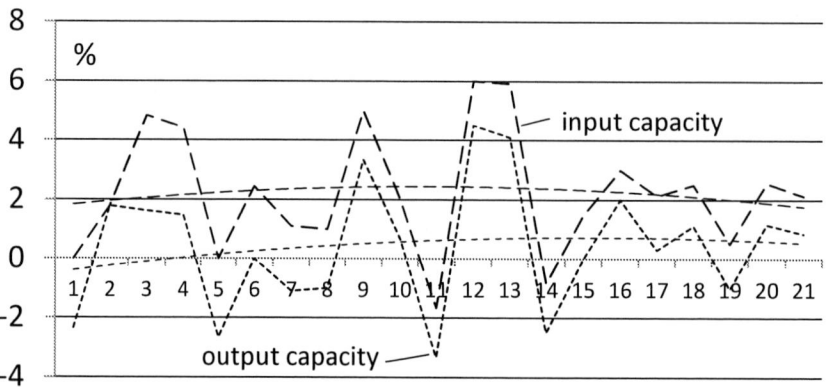

Figure 3. Relative error of measurement, %

A number of experiments were carried out with the described capacitive parameters meter including 100 measurements of FRD capacity, input and output IGBT capacity, which results shown on fig. 4, 5. The dashed line in the figures shows the measurement results of capacitance and the solid line the smoothed values using polynomial of the 3rd order. Experiments were carried out at constant ambient temperature.

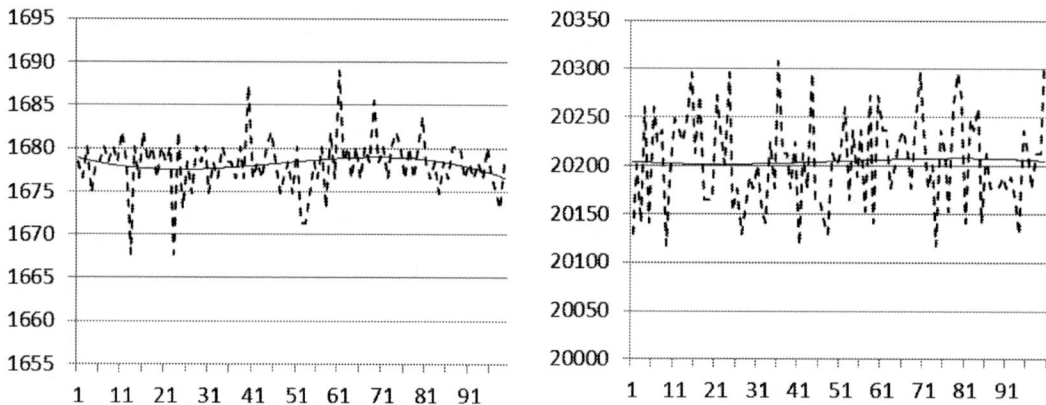

Figure 4. FRD capacity, pF   **Figure 5.** IGBT input capacity, pF

In addition, an experiment was carried out, which included 100 measurements of FRD capacity, as well as input and output IGBT capacity during heating from 25°C to 50°C. The results of measurements are shown in Fig. 6-7.

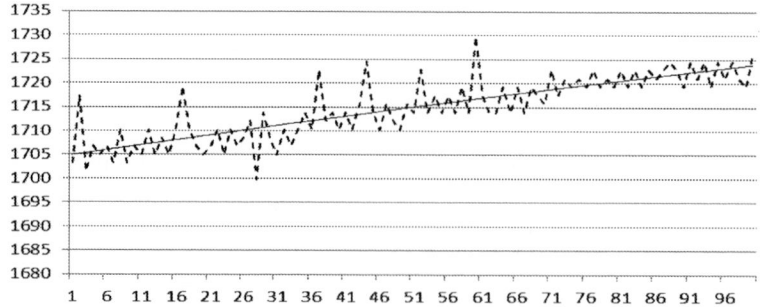

Figure 6. FRD capacity, pF during heating from 25°C to 50°C

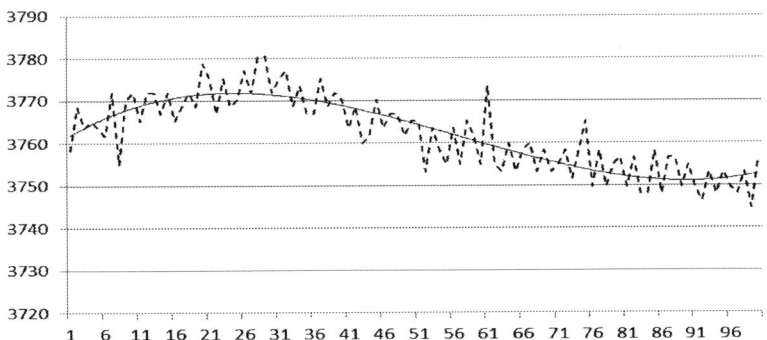

Figure 7. IGBT input capacity, pF during heating from 25°C to 50°C

4. Conclusions

The proposed method of remote measurement of capacitance parameters can be applied not only in the automation systems to test parameters of semiconductor devices but also in other spheres, for example, during the development of capacitance sensors. The results it is possible to use behavioral models based on experimental data in automated test systems because little attention is given to temperature dependent effects of capacitance parameters in the available theoretical models. The study is relevant and important because the task of automation of the testing of electrical parameters of power semiconductor modules (IGBT, FRD) at the manufacturing stage is very common today.

Acknowledgements
This work has been supported by the Russian Ministry of Education and Science (task No. 8.1729.2017/4.6).

References

[1] Levshina Ye S, Novitskiy P V 1983 *Electrical measurements of physical values (Measurement transducer)* (Leningrad: Energoatomizdat–Leningrad division publishing house)
[2] Polderman J W, Willems J C 1998 *Introduction to mathematical systems theory: a behavioral approach* (New York: Springer–Verlag)
[3] Patil N et al. 2009 *IEEE Transactions on Reliability* **58(2)** 271
[4] Palmer P R et al. 2003 *IEEE Transactions on Power Electronics* **18(5)** 1220

SPBOPEN 2018   IOP Publishing

Influence of microwave electromagnetic field on the structure of polymers

E Vasinkina[1], S Kalganova[1], V Alekseev[1] Yu Kadykova[1], S Arzamastsev[1], A Dolzhikova[1]

[1]Yuri Gagarin Saratov State Technical University, Saratov, 410054, Russia

Abstract. The results experimental studies of the microwave electromagnetic field effect on the structure and physical and mechanical properties of polymeric fibrous materials are presented. These methods and tools for studying the microwave effect on this object. The possibility for the nonthermal microwave modification of thermoplastic polymers has been proved.

Introduction

Treatment with high-frequency and ultrahigh-frequency (microwave) electromagnetic field (EMF) is used in a metallic materials processing, particularly when modifying the surface of small-sized titanium products, and polymeric materials, in particular for synthetic fibers nonthermal modification [1,2]. However, an information on this modifying effect mechanism of the microwave electromagnetic field on polymeric materials is fragmentary. In the authors opinion, this explanation of the non-thermal microwave modification nature of polymers can be found in polarization effects, their specific features at ultrahigh frequencies. Due to this effect of an external microwave electromagnetic field, without disruption of chemical bonds, conformational changes occur in the polymer macromolecules. They are related to a change in the interdomain regions molecular packing density, as a result which the crystallinity degree of the polymer changes and, as a consequence, these properties are modified. Obtaining the experimental studies results showing this influence of the nonthermal microwave treatment technological regimes on its structure and the physical and mechanical properties of polymer materials is an actual and important scientific research component.

Methodology

To conduct research on this modifying microwave effect on dielectrics, an automated microwave conveyor system was developed (Figure 1). A source of microwave energy with a 3 kW power and a 2450 MHz frequency is applied. The power supply is equipped with a rectifier and a filter, which ensures a stable continuous operation mode of the magnetron [3].

At this magnetron exit the protecting ferrite gate for the generator of microwave oven is used. The microwave power adjustment is carried out smoothly with the help of a thyristor converter and is registered by the anode current magnitude. This exact microwave power is set using a variable attenuator (Figure 2 a). At the entrance to the microwave module there are devices for measuring the incident and reflected power (Figure 2 b). Through the microwave camera passes a transport tape from the radio-transparent material located at the maximum of the electric field E of the electromagnetic wave.

Content from this work may be used under the terms of the Creative Commons Attribution 3.0 licence. Any further distribution of this work must maintain attribution to the author(s) and the title of the work, journal citation and DOI.
Published under licence by IOP Publishing Ltd

Figure 1. Specialized microwave installation for scientific research

(a) (b)

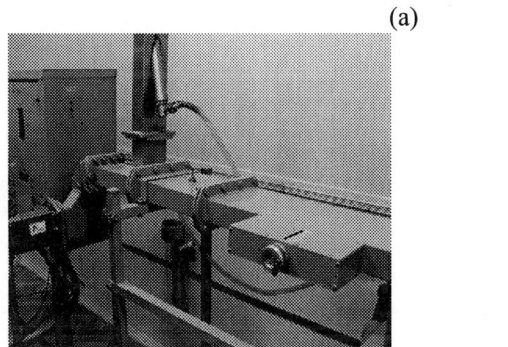

Figure 2 (a, b). (a) Variable attenuator; **(b)** Measurement of incoming and reflected microwave power.

The study objects was synthetic chemical fibers with anisotropic properties. To study the structure of polymer materials, the following methods were used:

- IR spectroscopy using the «Specord-75 IR» spectrophotometer and the «Infraium FT-801» Fourier spectrometer in the 400-4000 cm-1 region;

- the method of electronic raster microscopy using the YSM-5300 LV microscope from JEOL (Japan); gold was deposited in a plasma discharge vacuum at a temperature of $25 \pm 2 °$ C to create a conductive layer on the sample surface, the elemental composition was studied using an analytical attachment to the Microscope Link OXFORD (England) [4].

Results

We studied this non-thermal microwave modes effect on fiber polycaproamide. It has been found that unit tenacity of fiber increases by 12-15 % and thermal stability, on average, by 4.6 times (Figure 3 a). The effect is observed under the short-term microwave influence for 5-10 seconds, while the object temperature remains steady. This fact indicates non-thermal nature of microwave electromagnetic field modes influence on polymeric structure of polycaproamide fiber. It was found a slight relaxation of unit tenacity by 3-6 %, though the value of residual effect points to polycaproamide non-thermal modification.

As the microwave treatment result of the textured polycaproamide fiber, an increase in the fiber specific breaking load is observed in comparison with the original fiber not processed in microwave fields by 11-19% at 500-800 W power levels and the microwave exposure time in 20-30 s (Figure 3 b). The fiber relative elongation decreases after microwave exposure and to a greater extent with the microwave treatment time of 30-40 s [5].

(a)   (b)

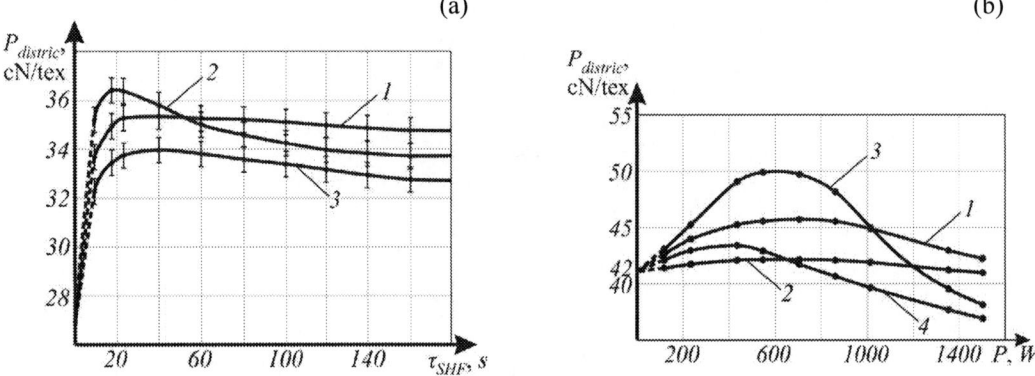

Figure 3 (a, b). (a) Specific strength of fiber; (b) Effect of microwave power on the heat resistance of fiber

We studied the use efficiency of microwave electromagnetic field non-thermal effect on polycaproamide fiber, doped with fire retarder T-2, that is due to the increase of fiber sorption capacity by 14 %. According to electron microscopy, under microwave effect fire retarder penetrates into the fiber volume and lays on the surface as 0.5-8.0 nanometers fine particles, whereas without microwave effect fire retarder mainly lays on the fiber surface as more coarse 0.20-0.30 mm particles and aggregated forming of fire retarder molecules (Figure 4) [6,7].

(a)   (b)

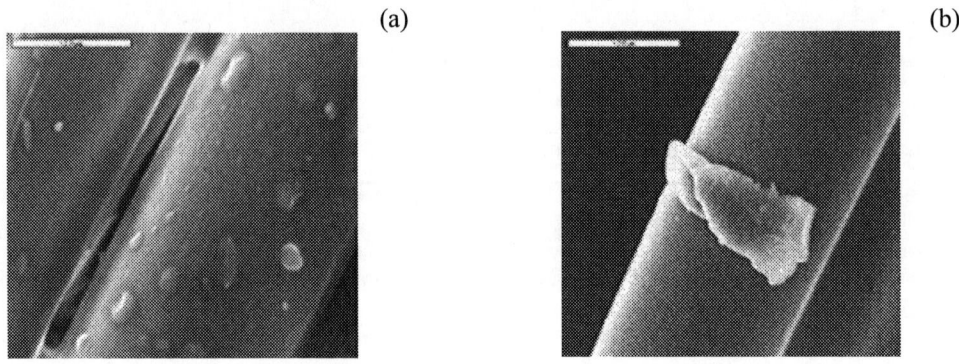

Figure 4(a, b). Fire retarder T-2 distribution on the surface of polycaproamide fiber: processed in microwave electromagnetic field (a); without microwave processing (b), (2000 x)

Conclusions

By infrared spectroscopy method it was found that non-thermal microwave effect leads to the change in the polycaproamide structure connected both with intramolecular and intermolecular hydrogen bonds and conformational changes in polymer molecule. These infrared spectroscopy results fit the electron microscopy data. We discovered the order increase in the polycaproamide structure which can cause the fiber hardening and the increase of its sorption capacity under the non-thermal microwave effect.

This experimental study proved the existence of microwave electromagnetic field non-thermal modulating effect on polymeric materials. We studied the influence of duration and microwave electromagnetic field effect power on physical and mechanical properties of objects. We obtained calculated values of electric field intensity at which best modulating effect is achieved after the processing in a microwave electromagnetic field.

References
[1]   Fomin A., Fomina M., Koshuro V., Rodionov I., Zakharevich A. and Skaptsov A. 2017 *Structure and mechanical properties of hydroxyapatite coatings produced on titanium using plasma spraying with induction preheating Ceramics International* 43(14) pp 11197-204

[2]   Kalganova S, Arkhangelskiy Yu, Lavrentyev V, Trigorly S, Artyukhov I and Stepanov S 2017 *Electrotechnology of non-thermal modification of polymeric materials in a microwave electromagnetic field* XVIII Int. UIE-Congress on Electrotechnologies for Material Processing, (Hannover) pp 333-337 (in Germany)

[3]   Arkhangelskiy Yu S, Kalganova S G and Yafarov R K 2018 *Measurement in microwave electrotechnological installations* (Saratov: JSC «Amerit») p 322 (In Russian)

[4]   Plakunova E V, Tatarintseva E A, Mostovoy A S and Panova L G 2013 Structure and properties of epoxy thermosets *Perspective materials* **3** p 57-62 (In Russian)

[5]   Tarutina L I and Pozdnyakova F O 1986 *Spectral analysis of polymers* (Leningrad: Chemistry) p 248 (In Russian)

[6]   Vasinkina E Yu, Kalganova S G, Lavrentyev V A, Trigorly S V and Alekseev V S 2018 *Phase transitions in polymers under the impact of microwave electromagnetic fields* Int. Conf. on Innovations and Perspectives Development of Mining Engineering and Electromechanics: IPDME-2018, St. Petersburg, Russia p 49 (In Russian)

[7]   Morozova M Yu and Kalganova S G 2004 Fibre Chemistry. **36** 186

SPBOPEN 2018   IOP Publishing

IOP Conf. Series: Journal of Physics: Conf. Series **1124** (2018) 071019   doi:10.1088/1742-6596/1124/7/071019

Investigation of the influence of parameters of nanoscale profiling of the surface of GaAs structures by a combination of local anodic oxidation and plasma chemical etching methods

V S Klimin[1], A A Rezvan[1], I N Kots[1], N A Naidenko[2]

[1] Southern Federal University, Department of Nanotechnology and Microsystems, Taganrog 347922, Russia
[2] Scientific Design Bureau of Computing Systems" (JSC SDB CS), Taganrog 347922, Russia

Abstract. This work is devoted to analysis of problems of present methods of surface treatment of nanostructures based on gallium arsenide. The idea of a combination of local anodic oxidation and plasma chemical etching methods was proposed to solve it. Oxide layers were used as negative masks for subsequent plasma chemical etching by the STE ICPe68 unit. BCl_3 was chosen as the chlorine containing gas, which differs from analogs in some parameters for the effect of etching of nanostructures based on gallium arsenide in the low temperature plasma. The gas mixture of reaction chamber consisted from a buffer gas $N_{Ar} = 100$ cm^3/min and a chlorine containing gas $N_{BCl_3} = 15$ cm^3/min at a pressure $P = 2$ Pa. The influence of these methods modes, which are formation voltage and etching time, on the roughness and geometric parameters, and corresponding dependences are demonstrated. Probe nanotechnology was used for surface analysis.

1. Introduction

The current pace of electronics development lead to a magnification in the density of elements arrangement on a single crystal, as well as a significant reduction in their technological dimensions [1-4]. Due to this fact, there is a toughening of requirements to precision methods of forming elements of micro- and nanoelectronics. Thus, the main problem of technology of forming devices based on gallium arsenide is to find and improve methods of surface treatment. Moreover, it must possess the resolving power and high accuracy of the process, which directly corresponds to given parameters [5-8]. In addition, surface of structures obtained by this technology should be characterized by a high degree of quality. Traditional methods of nanolithography based on liquid etching of photoresist films have significant drawbacks. Processes of this surface treatment method are limited by a number of factors, such as properties of used masks, the diffraction and inhomogeneity of liquid etching, a combination of which leads to serious deviations from the specified parameters in the formation of nanoscale structures [9-12].

The combination of local anodic oxidation with subsequent chemical etching in low temperature plasma is a promising technology for formation of a complex surface relief of nanoscale structures, their modification and profiling. The main feature of this method is possibility of high precision variation of nanoscale processing. In addition, extra operations of surface cleaning of residual reaction products and various impurities from the starting reagents are not required for this technology [13-16].

Content from this work may be used under the terms of the Creative Commons Attribution 3.0 licence. Any further distribution of this work must maintain attribution to the author(s) and the title of the work, journal citation and DOI.
Published under licence by IOP Publishing Ltd

The main objective of these experimental studies is to research parameters of processes of modification and profiling surfaces of nanoscale structures based on gallium arsenide, using a combination of local anodic oxidation methods and subsequent plasma chemical etching [17-20].

2. Description of the method

The substrate material was plates of peculiar gallium arsenide. Geometric parameters of surface were improved by standard liquid polishing. Further, substrate's surfaces were subjected to modification by the method of local anodic oxidation, and, as a result, a layer of oxide nanostructures was formed. These oxide nanoscale structures were obtained at the following formation parameters: relative humidity RH = 90%, probe travel speed V_P = 2.5 μm/sec, feedback current of the microscope I_F = 1 nA, and formation voltage varied from 7 to 10 V.

The formed oxide layers were used as negative masks for subsequent plasma chemical etching by the STE ICPe68 unit. BCl_3 was chosen as the chlorine containing gas, which differs from analogs in some parameters for the effect of etching of nanostructures based on gallium arsenide in the low temperature plasma. The gas mixture of reaction chamber consisted from a buffer gas N_{Ar} = 100 cm^3/min and a chlorine containing gas N_{BCl3} = 15 cm^3/min at a pressure P = 2 Pa. The power of source consisted from inductively coupled plasma W_{ICP} = 400 W, capacitive plasma W_{RIE} = 35 W and bias voltage U_{bias} = 10^2 V. The processing time in the plasma ranged from 0.5 to 2 minutes. Rectangular oxide structures with dimensions of 0.3x0.3μm were formed.

Surface topology control was carried out at each stage of the study by using the atomic force microscopy method of the NTegra probe laboratory. Parameters of the aspect ratio of formed structures were estimated by measuring the angle of deviation from the vertical by obtained profiles. The measurements were made in the crystallographic directions [110] and [111].

3. Results and discussion

When using this technology, it is necessary to take into account that after local anodic oxidation, an oxide layer is formed on the treated surface. Investigations of the thickness of the resulting oxide layer, as well as the surface roughness after removal of the oxide layer, have been carried out, since this region will be the surface of the finished nanostructures. The thickness of the obtained oxide layer was h_0 = 6.6 ± 0.2 nm, the height of the oxide layer above the surface of the untreated region of the GaAs surface was h_h = 2.7 ± 0.09 nm, the thickness of the oxide layer below the untreated surface was h_u = 3.9 ± 0.14 nm. In the further implementation of the technological process, it is necessary to take into account the dimensions of the oxide layers on the surface of the structures. Also, the roughness of the surface under the oxide layer is of great importance. The roughness value should have permissible values for the formation of devices of micro- and nanoelectronics, in this experiment it was S_d = 0.51 ± 0.09 nm.

Based on the results of study, AFM scans of formed structures on the GaAs surface, obtained at a formation voltage of 7 V (figure 1, a) and an etching time of 1 minute and at a formation voltage of 9 V (figure 1, d) and an etching time of 1 minute, were obtained. To determine the influence of etching time and formation voltage of anode masks on the anisotropy of process, structure profiles in crystallographic direction [110] (figure 1, b, e) were obtained and structural profiles were also obtained and studied in the crystallographic direction [111] (figure 1, c, f).

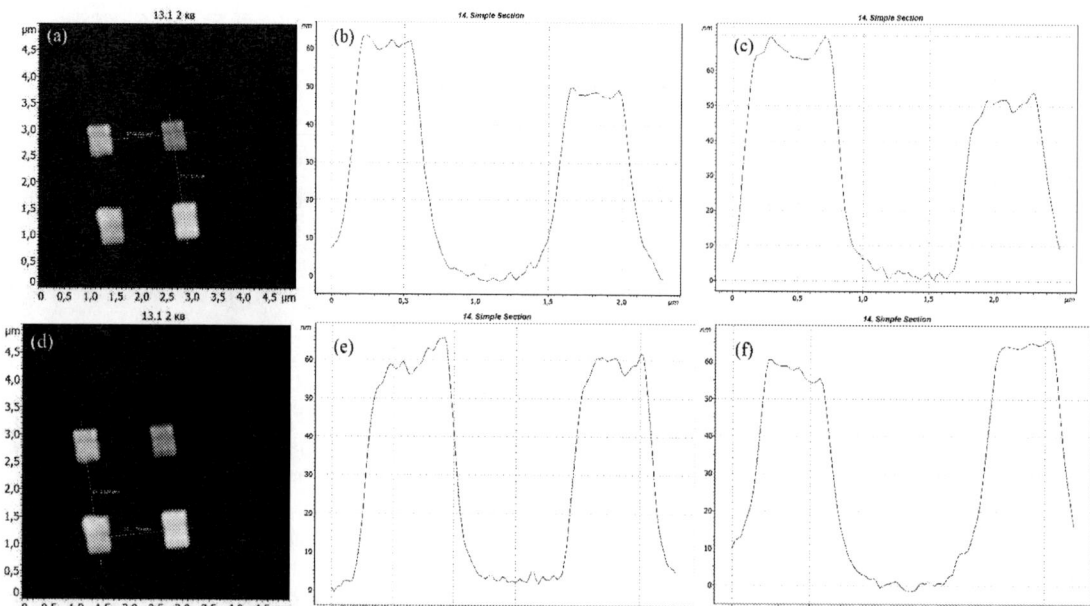

Figure 1 (a, b, c, d, e, f). (a, d) AFM-scans of volumetric structures and **(b, e)** profiles of resulting structures in the crystallographic directions [110], and **(c, f)** [111], **(b, c)** structures formed at voltage formation of masking layer 7 V and **(e, f)** 9 V.

The angle of deviation from the vertical was an internal angle of the wall of formed volumetric nanostructure. The angle was measured as a difference of $180°$ - α. Based on the results of measurements, dependencies of influences of formation voltage of the masking layers at different etching times on anisotropy of processes were obtained in the crystallographic direction [110] (figure 2, a) and the crystallographic direction [111] (figure 2, b).

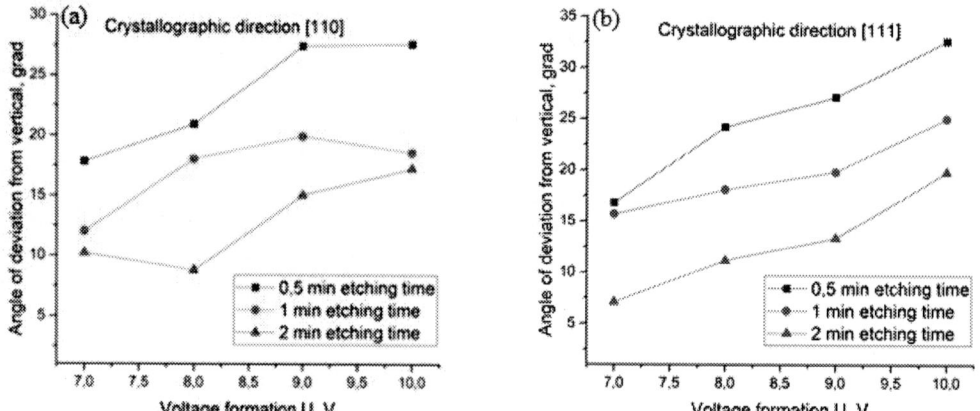

Figure 2 (a, b). Dependence of formation voltage of masking coating and different times of plasma chemical treatment on the deviation angle from the vertical: **(a)** crystallographic direction [110]; **(b)** crystallographic direction [111].

From shown dependences, it is seen that the crystallographic direction effects on the etching rates and the verticality of obtained structures. By selecting optimal regimes, it is possible to obtain structures with a high aspect ratio for use in optical and quantum electronics obtained without using liquid lithography operations.

4. Conclusion

In this experimental work, studies were made of the effect of formation regimes of oxide masking layers obtained by the local anodic oxidation method on the anisotropy of the plasma chemical etching process. It is also shown that the surface of the obtained structure, after removal of the oxide layer and plasma chemical etching, has a roughness and a relief comparable to the surface of the initial GaAs plate. In the work there were structures with a height of 30 to 60 nanometers. From the dependences obtained, it can be seen that in the crystallographic direction [110] the angle of inclination of the deviation from the vertical does not exceed 30°, and in the crystallographic direction [111] 35°. When the etching time is increased, the deviation angle from the vertical will be minimal. In this connection, it allows us to conclude that this technology can be introduced into standard processes of micro- and nanoelectronics for production modern optical and quantum devices of nanoelectronics based on GaAs structures. The results obtained in the course of experimental studies showed that using this technology to obtain nanoscale structures, it is necessary to take into account the crystallographic direction.

Acknowledgments

The results were obtained using the equipment of the Research and Education Center and Center for Collective Use "Nanotechnologies" of Southern Federal University.

References

[1] Kangawa Y, Ito T, Taguchi A, Shiraishi K, Irisawa T and Ohachi T 2002 *Appl. Surf. Sci.* **190** 517

[2] Ageev O A, Klimin V S, Solodovnik M S, Eskov A V, Krasnoborodko S Y 2016 *J. Phys.: Conf. Ser.* **741** 012178

[3] Daweritz L and Ploog K 1994 *Semicond. Sci. Tech.* **9** 123 [2] Klimin V S, Il'Ina M V, Il'In O I, Rudyk N N, Ageev O A 2017 *J. Phys.: Conf. Ser.* **917** 092023

[4] Rudyk N N, Il'In O I, Il'Ina M V, Fedotov A A, Klimin V S, Ageev O A 2017 *J. Phys.: Conf. Ser.* **917** 082008

[5] Tok E S, Neave J H, Zhang J, Joyce B A and Jones T S 1997 *Surf. Sci.* **374** 397

[6] Klimin V S, Tominov R V, Eskov A V, Krasnoborodko S Y, Ageev O A 2017 *J. Phys.: Conf. Ser.* **917** 092005

[7] Kley A, Ruggerone P and Scheffler M 1997 *Phys. Rev. Lett.* **79** 5278

[8] Tominov R V, Bespoludin V V, Klimin V S, Smirnov V A, Ageev O A 2017 *IOP Conf. Ser.: Mater. Sci. Eng.* **256** 012023

[9] Klimin V S, Solodovnik M S, Smirnov V A, Eskov A V, Tominov R V, Ageev O A 2016 *Proc. of SPIE* **10224** 102241Z-1

[10] Il'Ina M V, Blinov Y F, Il'In O I, Klimin V S, Ageev O A 2016 *Proc. of SPIE* **10224** 102240U-1

[11] Ageev O A, Solodovnik M S, Balakirev S V, Mikhaylin I A 2016 *J. Phys.: Conf. Ser.* **681** 012036

[12] Shiraishi K and Ito T 1998 *Phys. Rev.* B **57** 6301

[13] Avilov V I, Ageev O A, Smirnov V A, Solodovnik M S, Tsukanova O G 2015 *Nanotech. in Russia* **10** (3-4) 214-219

[14] Tominov R V, Zamburg E G, Khakhulin D, Klimin V S, Smirnov V A, Chu Y H, Ageev, O.A. 2017 *J. Phys.: Conf. Ser.* **917** 032023

[15] Morgan C G, Kratzer P and Scheffler M 1999 *Phys. Rev. Lett.* **82** 4886

[16] Ageev O A, Solodovnik M S, Balakirev S V, Mikhaylin I A 2016 *Technical Physics* **61** (7) 971–977

[17] Murdick D A, Wadley H N G and Zhou X W 2007 *Phys. Rev.* B **75** 125318

[18] Amrani A, Djafari Rouhani M and Mraoufel A 2011 *Appl. Nanosci.* **1** 59

[19] Ageev O A, Solodovnik M S, Balakirev S V, Eremenko M M 2016 *J. Vac. Sci. Technol.* B **34** (4) 041804

[20] Foxon C T and Joyce B A 1977 *Surf. Sci.* **64** 293

Research of using plasma methods for formation field emitters based on carbon nanoscale structures

V S Klimin[1], A A Rezvan[1] and O A Ageev[1]

[1]Department of Nanotechnology and Microsystems, Southern Federal University, Taganrog 347922, Russia

Abstract. This paper is devoted to development of technology of formation and obtaining of the design of a field emitter of electrons with an active region based on carbon nanostructures. The main idea of this technology is using of plasma methods, such as plasma chemical etching and plasma chemical vapor deposition. Using a magnetron sputtering method, sublayers V, Cr and Al nm were obtained with different thickness based on its material characteristics for prevent formation of silicates with catalytic centers material. This thing allows forming compounds, which serve as a metallic contact to field emission structure. A catalytic layer of nickel was obtained by magnetron sputtering. Based on the studies, a prototype of a field emitter with work characteristics, which are corresponding to the specified formation parameters, was obtained. Those characteristics are threshold field strength $6.7 \cdot 10^8$ V/m, field amplification factor $7.9 \cdot 10^7$ and electron work function in range from 3.1 to 5.3 mA/cm^2.

1. Introduction

In recent years, there has been an intensive development of science and technology, new discoveries in the field of materials and their formation are made every year [1-4]. More recently, allotropic modifications of carbon have been discovered, but they have already found their application in various branches of micro- and nanoelectronics [5-9]. Thus, carbon nanotubes, which have the ability to low voltage electron emission, have become a promising material for vacuum electronics and are capable of supplanting existing structures of thermionic cathodes, in terms of their emission characteristics [10-12]. However, the use of existing formation methods leads to a degradation of emission properties of active region of these structures [13-17]. Due to this fact, there is a problem of searching of formation technologies of field emitters based on carbon nanostructures, which could use plasma for correcting the ongoing processes of obtaining in each step [18-24].

2. Description of the method

In the work, studies of formation regimes of a field emitter with an active region based on carbon structures by using plasma etching and deposition technologies were made. Dielectric layer Si$_3$N$_4$, which had obtained by the method of plasma chemical vapor deposition, was used as diffusion barrier. Parameters of this process were flow rate of nitrogen N_{N2} = 60 cm^3/min and N_{SiH4} = 40 cm^3/min, inductively coupled plasma power source W_{ICP} = 600 W under pressure of working chamber, temperature of sample T = 70°C. Using the above mentioned formation regime, it is possible to achieve a rate of dielectric layer growth V_G = 11.5 nm/min.

On surfaces of experimental samples, a complex relief was formed for the growth of carbon nanostructures. An important material is sublayer material, which provides contact to future carbon nanostructures.

Content from this work may be used under the terms of the Creative Commons Attribution 3.0 licence. Any further distribution of this work must maintain attribution to the author(s) and the title of the work, journal citation and DOI.

Published under licence by IOP Publishing Ltd

Using a magnetron sputtering method, sublayers V and Cr with a thickness of 20 nm, and also Al with 500 nm were obtained. The presence of a sublayer makes it possible to prevent formation of silicates with material of catalytic centers, and also, when using ammonia in the process, forms compounds that serve as a metallic contact to the autoemission structure. A catalytic layer of nickel was obtained by magnetron sputtering. The thickness of deposited layer was varied and amounted to 10 nm. The parameters of formation were the following: temperature of sample $T = 250°C$, power of plasma source $W_S = 300$ W, pressure of chamber $P = 10$ Pa. The whole process was going on under the supply of argon in a volume of 10 cm^3/min. After that, by the plasma chemical vapor deposition method, an additional Si_3N_4 dielectric layer in thickness of 5 μm was formed on the surface of formed structure. A combination of lithography and plasma chemical etching methods allowed to form a field emitter model, that provides voltage supply and formation of upper electrode, as well as control outputs. The frame, namely walls of field emitter made by dielectric material, allow not only to protect the active region of emission from mechanical interference from the outside, but also to prevent formation of an electrostatic voltage, that prevents flow of emission current. The etching process was carried out in a fluorine containing medium. As a fluorine containing gas, SF_6 was used with average flow rate 15 cm^3/min. However, at this process power of an induced coupled plasma source was $W_{ICP} = 300$ W, pressure of unit 2.25 Pa and bias voltage $U_{bias} = 80$ V. Temperature of treatment was 27°C. Observing this formation regime, an etching rate of 400 nm/min was obtained.

The growth of carbon nanoscale structures was carried out in several stages. The structures were heated to high temperatures (750°C) in a medium, which is containing from ammonia and argon, with concentration $N_{NH3} = 15$ cm^3/min and $N_{Ar} = 70$ cm^3/min. Heating was carried out until the end of catalytic centers formation, the activation of which was made at a temperature in range 700-800°C. During activation, at these temperatures nickel is reduced to a metal. In this process, the flow rate of ammonia $N_{NH3} = 170$ cm^3/min at chamber pressure 4.5 Torr. The total heating and activation time was about 21 minutes. Subsequent control of the surface of formed structures was performed by using atomic force and scanning electron microscopy at each step of research.

3. Results and discussion

Carbon nanotubes were obtained by the method of plasma chemical vapor deposition. Acetylene was used as carbon containing gas, the concentration of which was $N_{C2H2} = 70$ cm^3/min, also the composition of working mixture included ammonia at a flow rate $N_{NH3} = 170$ cm^3/min. Temperature of this process was 750°C. The pressure of atmosphere in the reactor was 4.5 Torr. Figure 1 shows arrays of carbon nanostructures obtained on samples with Ni/Cr/Si (figure 1, a) and Ni/V/Si (figure 1, b) layers systems.

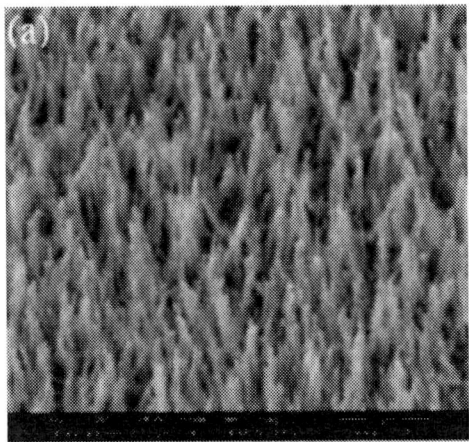 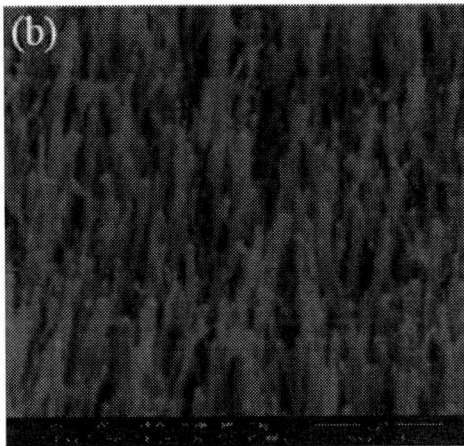

Figure 1 (a, b). SEM images of carbon nanotubes arrays obtained on structures with different sublayer material: **(a)** Ni/Cr/Si; **(b)** Ni/V/Si

Carbon nanoscale structures, formed with sublayer of chromium, were chosen for obtaining of field emitter model. This is due to the fact, that they grew mainly along the vertex mechanism and average thickness of carbon structures in the array was in range from 30 to 40 nm, which is much less than average thickness of the formed arrays with a sublayer of vanadium and aluminum. At the final stage, an upper electrode and electrical contacts were formed to field emitter model.

The obtained experimental models are characterized by a high degree of adhesion, and formed array of carbon nanostructures of active emission region has a sufficient degree of perpendicularity to substrate, which allows obtaining structures with a high aspect ratio.

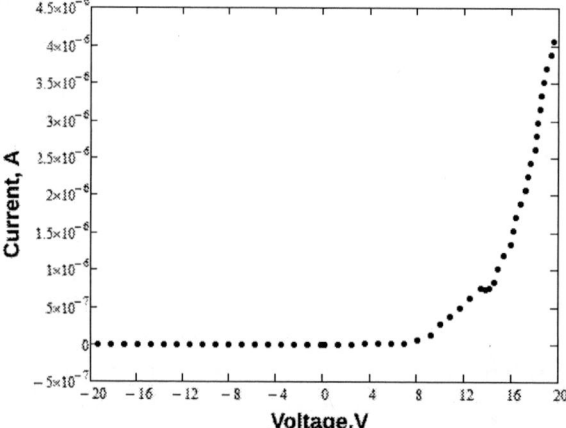

Figure 2. Current voltage characteristic of the obtained structure in the international system

Measurement of electrical characteristics was carried out by supplying a potential difference of -20 to +20 V. The obtained field emission characteristics allow us to judge the high efficiency of field emission (figure 2). The characteristics of this device are not inferior to analogs devices with field emitters made of semiconductors or graphene.

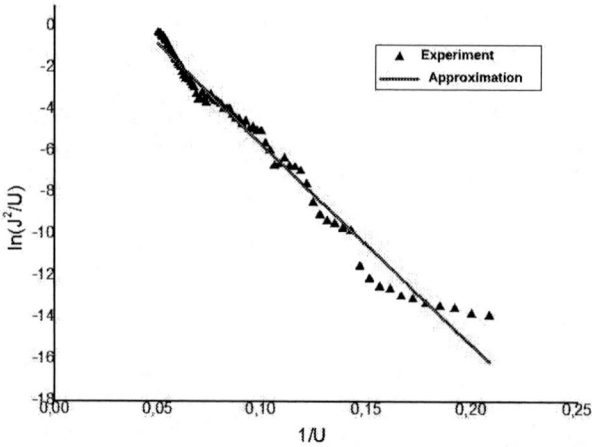

Figure 3. Current voltage characteristic of the obtained structure in the Fowler-Nordheim system

To determine the amplification factor of the field emission field, it is necessary to reconstruct the current-voltage relationship in the Fowler-Nordheim system (figure 3). Having approximated the obtained dependence, we obtain the slope, which is equal to the slope of the slope.

4. Conclusion

In this paper, field emission cell with an active element of oriented perpendicular to the substrate carbon nanostructures was formed. Characteristics of layout were investigated and it was shown, that the use of carbon nanostructures as an active element of the emission cell is promising. This technology can be used to form modern devices for vacuum microelectronics and for formation of pressure and gas sensors. In the course of this experimental work, the field emission properties of an element based on carbon nanostructures were investigated. It is shown that the field gain factor for this sample is $\beta = 7.9 \cdot 10^7$, the threshold field strength of the field emission origin is $E = 6.7 \cdot 10^8$ V/m, work function of electrons in the range from 3.1 to 5.3 mA/cm^2. Such a model of the field emission device can also be used as a vacuum gauge and has a sensitive procedure of $2.4 \cdot 10^{-8}$ A/Pa, a speed of ~4 s, with U = 20 V, and a power consumption of $1.5 \cdot 10^{-7}$ W. These data were obtained with an area of the emitting surface equal to the area of the emitting surface, $S_v = 1.09 \cdot 10^{-6}$ cm^2.

Acknowledgments

This work was supported by the Southern Federal University (grant VnGr-07/2017-02). The research carried out at Research and Educational Center of "Nanotechnologies" of Southern Federal University.

References

[1] Ageev O A, Klimin V S, Solodovnik M S, Eskov A V, Krasnoborodko S Y 2016 *J. Phys.: Conf. Ser.* **741** 012178

[2] Klimin V S, Il'Ina M V, Il'In O I, Rudyk N N, Ageev O A 2017 *J. Phys.: Conf. Ser.* **917** 092023

[3] Jeong H E, Lee S H, Kim P and Suh K Y 2008 *Colloid Surface A* **313-314** 359

[4] Rudyk N N, Il'In O I, Il'Ina M V, Fedotov A A, Klimin V S, Ageev O A 2017 *J. Phys.: Conf. Ser.* **917** 082008

[5] Qu L and Dai L 2007 *Adv. Mater.* **19** 3844

[6] Klimin V S, Tominov R V, Eskov A V, Krasnoborodko S Y, Ageev O A 2017 *J. Phys.: Conf. Ser.* **917** 092005

[7] Tominov R V, Zamburg E G, Khakhulin D, Klimin V S, Smirnov V A, Chu Y H, Ageev, O.A. 2017 *J. Phys.: Conf. Ser.* **917** 032023

[8] Majidi C, Groff R E, Autumn K, Baek S, Bush B, Gravish N, Maboudian R, Maeno Y, Schubert B, Wilkinson M and Fearing R S 2006 *Phys.Rev.Lett.* **97** 076103

[9] Klimin V S, Solodovnik M S, Smirnov V A, Eskov A V, Tominov R V, Ageev O A 2016 *Proc. of SPIE* **10224** 102241Z-1

[10] Tominov R V, Bespoludin V V, Klimin V S, Smirnov V A, Ageev O A 2017 *IOP Conf. Ser.: Mater. Sci. Eng.* **256** 012023

[11] Il'Ina M V, Blinov Y F, Il'In O I, Klimin V S, Ageev O A 2016 *Proc. of SPIE* **10224** 102240U-1

[12] Avilov V I, Ageev O A, Smirnov V A, Solodovnik M S, Tsukanova O G 2015 *Nanotech. in Russia* **10** (3-4) 214-219

[13] Rouhrig M, Thiel M, Worgull M and Houlscher H 2012 *Small* **8** 3009

[14] Autumn K, Liang Y A, Hsieh S T, Zesch W, Chan W P, Kenny Th W, Fearing R and Full R J 2000 *Nature* **405** 681

[15] Morgan C G, Kratzer P and Scheffler M 1999 *Phys. Rev. Lett.* **82** 4886

[16] Murdick D A, Wadley H N G and Zhou X W 2007 *Phys. Rev.* B **75** 125318

[17] Shiraishi K and Ito T 1998 *Phys. Rev.* B **57** 6301

[18] Amrani A, Djafari Rouhani M and Mraoufel A 2011 *Appl. Nanosci.* **1** 59

[19] Kangawa Y, Ito T, Taguchi A, Shiraishi K, Irisawa T and Ohachi T 2002 *Appl. Surf. Sci.* **190** 517

[20] Daweritz L and Ploog K 1994 *Semicond. Sci. Tech.* **9** 123

[21] Foxon C T and Joyce B A 1977 *Surf. Sci.* **64** 293

[22] Tok E S, Neave J H, Zhang J, Joyce B A and Jones T S 1997 *Surf. Sci.* **374** 397

[23] Kley A, Ruggerone P and Scheffler M 1997 *Phys. Rev. Lett.* **79** 5278

[24] Nagase M, Nakamatsu K, Matsui S, Namatsu H 2005 Japanese J. of Appl. Phys. 44(7) 5409

SPBOPEN 2018

Numerical simulation of induction heating of a carburizing container with a titanium sample

A Voyko[1], M Fomina[1], A Shumilin[1], I Rodionov[1], S Kalganova[1], I Artyukhov[1], A Fomin[1]

[1]Yuri Gagarin State Technical University of Saratov, Saratov 410054, Russia

Abstract. Induction heating of a carburizing container using induction heating was studied. The dependency of heating up to the temperature of 1660 °C was determined depending on the current values I from 0.5 to 2 kA on the inductor at the frequency of 112.5±2.5 kHz. Experimental results were compared to the data of numerical simulation of heat transfer in the carburizing container.

1. Introduction

Induction heating is widely used in the treatment of metals, e.g. in heat treatment of steel, aluminium and titanium [1-3]. Simulation of the physical processes is frequently applied when it is necessary to assess the distributed parameters in the elements of sensors, electronic components and devices [4-7].

Control of technological parameters is an important component of any technology. For example, in the production of micron-sized particles by granulation, monitoring and control of treatment parameters is important [8-10]. Tasks regarding the determination of the physical fields distribution are quite frequent and they can not be investigated by the direct methods. In this case, theoretical methods of system analysis are applied, in particular numerical methods of calculation. The initial information is the process data, e.g. power consumption, inductor current and frequency, as well as the temperature on the surface of the technological volume, i.e. the container for chemical-thermal processing (carburization).

The carburization process using a solid carburizer (carbon-containing medium) is performed in sealed chambers made of refractory metal. Products after carburization are characterized by higher hardness and changes in the chemical composition of the surface, in particular the appearance of carbides. In this paper, the effect of induction heating on a metal (titanium) container containing metallic samples and a carburizer is investigated.

2. Methodology

The technological area elements included an inductor, a dielectric chamber, a carburizing container and heat-insulating limiting parts. The current on the inductor changed from 0.5 to 2 kA on the inductor at the frequency of 112.5±2.5 kHz [11].

Numerical simulation was performed using "Elcut 6" software. In this program the geometry of the "inductor – carburizing container" system was drawn, further a finite element mesh was generated and AC magnetic field distribution and non-stationary heat transfer were modeled depending on the peculiarities of heat source and heat loss due to convection and radiation (Figure 1).

As the main elements of the "inductor – carburizing container" system were selected: an inductor (1); water (2); a quartz chamber (3); a metal container (4) with the outer diameter of 14 mm; a stopper (5); dielectric inserts (6); carbon-containing medium (7); titanium samples-disks (8) with the diameter

Content from this work may be used under the terms of the Creative Commons Attribution 3.0 licence. Any further distribution of this work must maintain attribution to the author(s) and the title of the work, journal citation and DOI.

Published under licence by IOP Publishing Ltd

of 6 mm. The following values and indices were determined in the simulation: the distribution of the current density j and the temperature $T(t)$ at different time t.

3. Results

Modeling the current density j and the temperature field T showed that the optimum current I for heating the container in the "inductor – carburizing container" system was the current $I = 1.5$ kA, since this value provided the required temperature for the carburizing process.

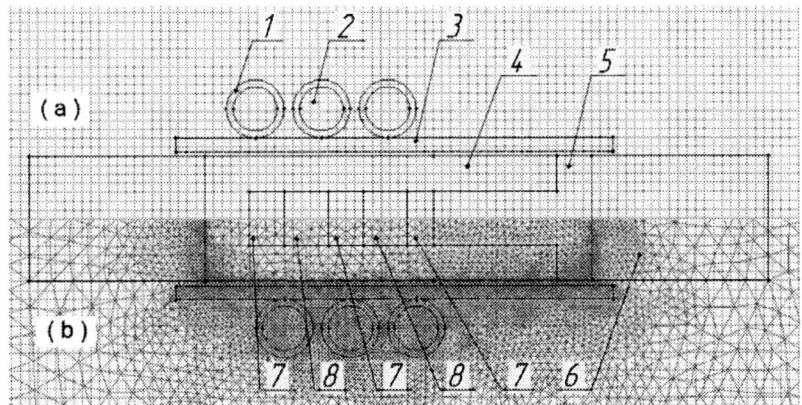

Figure 1(a,b). Geometric model of the "inductor – carburizing container" system (grid spacing 1 mm) **(a)**; finite element grid (triangular shape) **(b)**.

The obtained model of the current density distribution showed that the current density in the inductor reached a maximum $j = 7 \cdot 10^7 \text{A/m}^2$ (Figure 2).

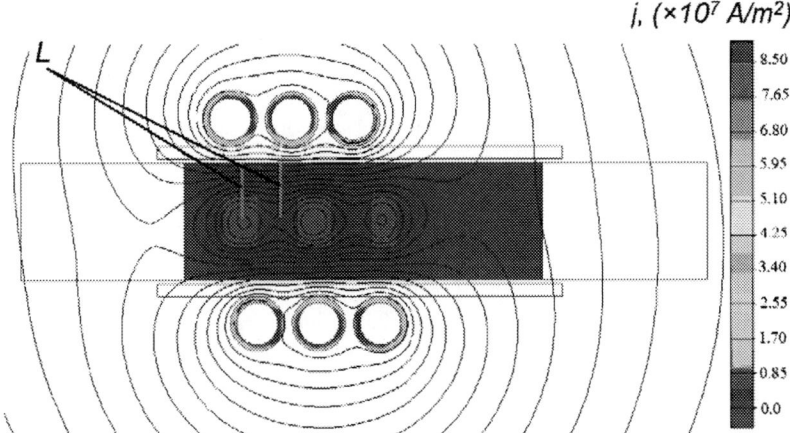

Figure 2. Distribution of the current density j in the elements of the "inductor – carburizing container" system at a current on the inductor $I = 1.5$ kA (container diameter 14 mm).

In the container, three regions with different current density were distinguished, which correspond to the titanium samples and carbon-containing medium (Figure 3). The current density in the carbon-containing medium reached a larger value $j = 0.85 \cdot 10^7 A/m^2$ (Figure 3a). Thus, in these areas more heat was emitted compared to the titanium elements of the system (Figure 3b). It is reasonable to assume that during the induction heating the core of the container will be heated to a higher temperature than its surface. The current density difference in the carbon-containing medium and titanium was clearly shown on trajectories L from the periphery to the center of the container (Figure 2,3). The obtained graphs characterize the substantial non-uniformity of the current density j in the elements of the system.

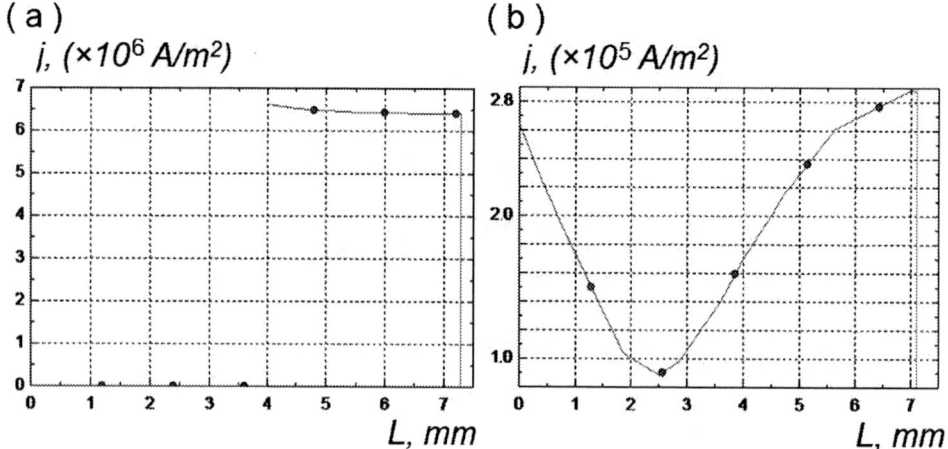

Figure 3(a,b). Graphs of the current density j as a function of the depth L in the container and in the carbon-containing medium **(a)**; in a titanium container and a sample disk **(b)**.

The current density in the container wall (section L from 0 to 4 mm) was characterized by a small value. In the region of the carbon-containing medium (section L from 4 to 7 mm), the current density was an order of magnitude higher and amounted to $j = 6.5 \cdot 10^6 A/m^2$. The current density on the container surface reached $j = (2.6-2.8) \cdot 10^5 A/m^2$ and at a depth $L = 2.6$ mm it was minimal $j = 0.8 \cdot 10^5 A/m^2$.

The results of the preliminary experiment and modeling of the temperature field showed that the temperature in the container was distributed uniformly (Figure 4).

4. Conclusions

Thus, the simulation data on induction heating at 0.6–1.5 kA corresponded to the experimental results. At a minimum inductor current of 0.6 kA, the carburization temperature was reached (about 900–950 °C), however, for a higher rate of carburization of small-sized products, it is advisable to provide a current value of the inductor of at least 1.5 kA. At the inductor current $I = 1.5-2$ kA, there was a rapid heating of the container up to the melting temperature $T \geq 1660$ °C. This processing mode can be used only for accelerated heating of the container, however, it is recommended to perform the exposure at the inductor current $I = 0.6-1.5$ kA.

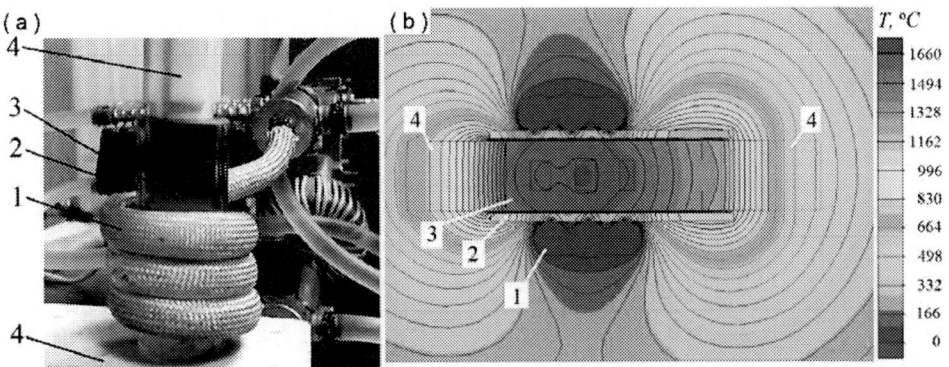

Figure 4(a, b). Arrangement of the elements of the system including an inductor (1), a dielectric chamber (2), a carburizing container (3) and heat-insulating limiting parts **(a)**; the result of the thermal field numerical simulation at $I = 1.5$ kA and $t = 300$ s **(b)**.

Acknowledgments
The research was supported by the Russian Science Foundation (project No. 18-79-10040). A. Fomin expresses his gratitude to the company "TOR" for the test access to "Elcut" software.

References
[1] Demidovich V B, Chmilenko F V and Rastvorova I I 2015 *Acta Tech. CSAV.* **60** 107
[2] Kuvaldin A B, Lepeshkin S A and Lepeshkin A R 2014 *Acta Tech. CSAV.* **59** 279
[3] Fomin A, Dorozhkin S, Fomina M, Koshuro V, Rodionov I, Zakharevich A, Petrova N and Skaptsov A 2016 *Ceram. int.* **42** 10838
[4] Aman A, Majcherek S, Schmidt M-P and Hirsch S 2014 *Procedia Eng.* **87** 124
[5] Aman A, Majcherek S, Hirsch S and Schmidt B 2015 *J. Appl. Phys.* **118** 164105
[6] Majcherek S, Aman A and Fochtmann J 2016 *J. Micromech. Microeng.* **26** 025013
[7] Aman S, Aman A and Morgner W 2013 *Compos. Sci. Technol.* **84** 58
[8] Bück A, Palis S and Tsotsas E 2015 *Powder Technol.* **270B** 575
[9] Palis S and Kienle A 2014 *J. Process Control.* **24(3)** 33
[10] Palis S and Kienle A 2013 *Ind. Eng. Chem. Res.* **52** 408
[11] Fomin A A, Fomina M A, Steinhauer A B, Petrova N V, Poshivalova E Yu and Rodionov I V 2014 Proc. 55th Int. Sci. Conf. on Power and Electrical Engineering of Riga Technical University (Riga), (Riga: IEEE) p 111

SPBOPEN 2018   IOP Publishing

Measurements in microwave electrotechnology

S Kalganova[1], E Vasinkina[1], V Alekseev[1], V Lavrentyev[1], S Trigorly[1], T Dunaeva[1]

[1]Yuri Gagarin Saratov State Technical University, Saratov, 410054, Russia

Abstract. The classification of measurements in microwave electrical engineering is given, which takes into account all measurements at different stages of creation and operation of both a microwave electrical installation and its individual elements

Introduction

Intensive and extensive research on the use of electrophysical methods of processing materials and products showed the effectiveness of using for this purpose the energy of microwave electromagnetic oscillations [1,2]. The experience of scientific research in the field of microwave electrotechnology, development and operation of microwave electrotechnological installations, shows the important role of measurements of all at the stages of the life cycle of the installation. Data on measurements and their results are given in many scientific articles [1-3]. However, all this information is fragmentary, not systematized, they do not indicate the problem of measurements in microwave electrotechnology, they do not define a complete list of measured parameters. During scientific research, during the development of microwave installations, during their testing and operation, various measurements are made of the parameters of the installation, its elements and the properties of the treated object. These measurements are specific, substantially different from similar measurements in microwave technology, in materials science, and sometimes have no analogues in related fields. This allows us to consider that a systematic approach to the measurement problem in microwave electrotechnology is very relevant.

Methodology

A variant of the classification of measurements is proposed, which takes into account the different stages of creation and operation of both the microwave electrotechnological unit and its structural elements. A distinctive feature of the developed classification is the presence in its structure of measurements at the stage of scientific research, including measurements of technological properties of the processed object.

The authors do not consider measurements within the framework of specific technologies that are of independent significance.

Results and discussion

With all the diversity of microwave electrothermal installations and the technological processes implemented in them, all these installations have the same structural scheme (Figure 1).

There can be several variants of the technical classification. One of them is classical if we add to it the studies of the physico-mechanical, physico-chemical properties of the processed object:
- measurement of current strength;
- measurement of voltage;

Content from this work may be used under the terms of the Creative Commons Attribution 3.0 licence. Any further distribution of this work must maintain attribution to the author(s) and the title of the work, journal citation and DOI.
Published under licence by IOP Publishing Ltd

- power measurement;
- frequency measurement;
- measurement of standing wave ratio;
- Measurement of Q-factor of oscillatory systems and attenuation of the lines of the trans-response;
- measurement of electrical properties of materials;
- measurement of field strength;
- research of physical-mechanical, physical-chemical and technological properties of the processed object.

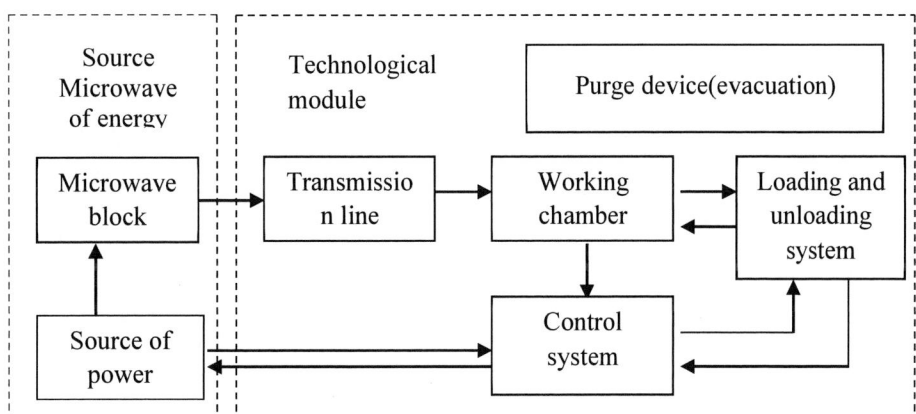

Figure 1. Structural diagram of microwave electrothermal unit

However, this classification option does not take into account that at different stages of creation and operation of both the microwave electrotechnological unit and its structural elements, different measurements have to be made. This circumstance is taken into account in the classification variant shown in Figure 2.

At the stage of developing the energy source, it is necessary to measure the current intensity I, the voltage at the electrodes U, the power P and frequency f of the magnetron P, and also the full-load characteristic P($KcmU$), f ($KcmU$) and the efficiency of the magnetron η; $KcmU$ max – he maximum permissible value of $KcmU$, at which the normal operation of the magne-tron is ensured (the parameters specified in its passport are retained); $KcmU$ - standing wave ratio to the magnetron loading tension (process of the module). For the technological module: T, U, p – temperature, moisture content and vapour pressure in the processed object; To – ambient temperature environment temperature; c, $\rho, \varepsilon', tg\delta$ - specific heat capacity, density, relative permittivity, the tangent of the dielectric loss angle of the processed object; h_e – heat transfer coefficient of evaporation; τ – time of the microwave oven of processing of an object; $K_{11}, K_{12}, \ldots, K_{33}$ – heat and mass transfer parameters included in the Lykov-Mikhailov equation (heat and mass transfer equation for capillary-porous bodies); υ- the velocity of an object in the working chamber; P_{rad} – the magnitude of the microwave radiation from the installation into the surrounding space.

At scientific researches the following parameters: ξ_1, ξ_2, ξ_3 - point coordinates in volume of the processed object; α - the attenuation coefficient of the electromagnetic wave in the transmission line with the processed object; Vmin – the minimum allowable volume of the object being processed; E – the electric field of the electromagnetic wave in the processed object; P_{Micro} - microwave energy scattered in the object; m – mass of object being processed.

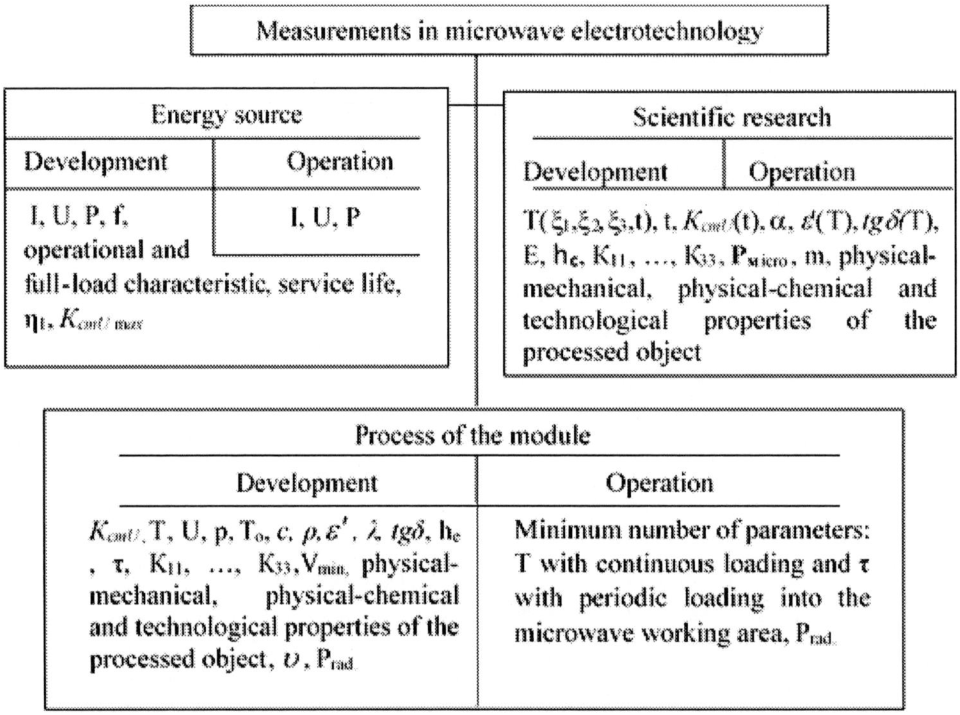

| Measurements in microwave electrotechnology | | | |

Energy source

Development	Operation
I, U, P, f, operational and full-load characteristic, service life, η_1, $K_{cont\,max}$	I, U, P

Scientific research

Development	Operation
$T(\xi_1,\xi_2,\xi_3,t)$, t, $K_{cont}(t)$, α, $\varepsilon(T)$, $tg\delta(T)$, E, h_c, K_{11}, ..., K_{33}, P_{micro}, m, physical-mechanical, physical-chemical and technological properties of the processed object	

Process of the module

Development	Operation
K_{cont}, T, U, p, T_o, c, ρ, ε', λ, $tg\delta$, h_c, τ, K_{11}, ..., K_{33}, V_{min}, physical-mechanical, physical-chemical and technological properties of the processed object, υ, P_{rad}	Minimum number of parameters: T with continuous loading and τ with periodic loading into the microwave working area, P_{rad}

Figure 2. Classification of measurements in microwave electrotechnology

As for the measurements related to the determination of physicomechanical, physico-chemical and technological properties of the processed object, they are very diverse and are determined by the aggregate state and purpose of this object. So, for example, solid bodies often measure hardness, strength, ductility, specific electrical resistance, and other specific parameters. In liquid bodies - the coefficient of kinematic viscosity, surface tension, in food - biochemical and microbiological characteristics, etc.

Other variants of classification are possible, but the scheme shown in Figure 2 makes it possible to consider quite often the most frequently occurring measurements in microwave electrotechnology. Measurements in microwave electrotechnology have their own peculiarities. First, they are very diverse: electrical, including such specific ones as microwave measurements; thermal, including mass transfer, as well as measurements related to the determination of the operational properties of the object being treated. Secondly, measurements, as a rule, have to be carried out in a relatively short time, since microwave electrotechnological processes are highly intensive. Third, for measuring in a working microwave camera, traditional measuring means (thermocouples, thermometers) are not suitable, since it affects the distribution of electromagnetic field in the chamber and, consequently, on the processing of the object. Fourth, we have measurements that have no analogues in the areas adjacent to the microwave electro-technology. For example, unlike traditional microwave technology, the coordination of the working chamber and the end load for the magnetron during the heat treatment varies in time. And fifth, to calculate the working chamber and mathematical modeling of the technological process, it is necessary to measure the dependence of the electrophysical and thermophysical parameters of the treated object on temperature and moisture content (humidity).

These features are very significant and as a result we have two consequences:

- it is not very promising to plan and carry out measurements in microwave electrotechnology, only in the field of electrical, thermal measurements or in the field of materials science or a specific technology;

- often in microwave electrotechnology, it is not possible to provide the same accuracy as measured by classical measurement methods (for example, as a measurement ε' and $tg\delta$ as a function of temperature) when measuring. However, in many cases, such high accuracy is not required. Thus, in measuring $\varepsilon'(T)$ and $tg\delta(T)$ for solving the problem of synthesis of the working chamber and mathematical modeling of the technological process, the nature of these dependences

Conclusions

Measurements in microwave electrotechnology have their own peculiarities, they are very diverse and specific. For example, electrical measurements include measurements at ultra-high frequencies, thermal measurements for mass transfer, and measurements related to determining the properties of the object being treated.

Other variants of classification are possible, but also shown in figure 2 makes it possible to consider quite often the most frequently metered in microwave electronics measurements.

The paper systematically describes the measurement in microwave electronics at both the research stage and during the development and operation of microwave electrotechnological installations.

References
[1] Vasinkina E Yu, Kalganova S G, Lavrentyev V A, Trigorly S V and Alekseev V S 2018 *Phase transitions in polymers under the impact of microwave electromagnetic fields* Int. Conf. on Innovations and Perspectives Development of Mining Engineering and Electromechanics: IPDME-2018, St. Petersburg, Russia p 49 (In Russian)
[2] Kalganova S, Arkhangelskiy Yu, Lavrentyev V, Trigorly S, Artyukhov I and Stepanov S 2017 *Electrotechnology of non-thermal modification of polymeric materials in a microwave electromagnetic field* XVIII Int. UIE-Congress on Electrotechnologies for Material Processing, (Hannover) pp 333-337 (in Germany).
[3] Didenko A N 2003 *Microwave Power Engineering: Theory and Practice* (Moscow: Nauka), p 446 (In Russian).
[4] Artyukhov I I, Fursaev M A 2000 *Magnetron generators for microwave heating plants* – (Saratov: JSC «Amerit») p 322 (In Russian)48 c.
[5] Arkhangelskiy Yu S, Kalganova S G and Yafarov R K 2018 *Measurement in microwave electrotechnological installations* (Saratov: CSTU) p 48 (In Russian)

SPBOPEN 2018   IOP Publishing

Effect of metallic nanoantennas on the efficiency of the surface plasmon-polariton generation via excitation of electromagnetic waves in a tunnel junction

L N Dvoretckaia[1], A M Mozharov[1], A V Uskov[3], A D Bolshakov[1], A O Golubok[2] and I S Mukhin[1,2]

[1] St. Petersburg Academic University, Khlopina 8/3, 194021 St. Petersburg, Russia
[2] ITMO University, 197101 St. Petersburg, Russia
[3] Lebedev Physical Institute, Moscow, Russia

E-mail: mozharov@spbau.ru

Abstract. Inelastic tunneling of electrons can be used for generation of the surface plasmon-polaritons (SPPs). In this paper, we study numerically the surface waves generation during the electrons tunneling through the barrier between tip of the scanning-tunneling microscope probe and golden substrate. To study the process the emission of the tunneling electrons was approximated with an optical dipole emission. First, we demonstrate that the efficiency of the SPP generation in terms of energy conversion cannot exceed 1 % when no antenna is present in the system. We then carry out the simulation of the SPP generation when the probe tip is in a close proximity with the golden nanoantenna in the shape of sphere or cylinder. We show that the surface plasmon-polariton generation efficiency can be increased more than two orders of magnitude with the use ofnano-antennas. Nanoantennas in the form of spherical nanoparticles are found to be more effective for the SPPs generation than nanoantennas in the shape of cylinders.

1. Introduction

Inelastic tunneling of electrons can be used for generation of the surface plasmon-polaritons (SPPs). In this paper, we study numerically the surface waves generation during the electrons tunneling through the barrier between tip of the scanning-tunneling microscope probe and golden substrate. To study the process the emission of the tunneling electrons was approximated with an optical dipole emission. First, we demonstrate that the efficiency of the SPP generation in terms of energy conversion cannot exceed 1 % when no antenna is present in the system. We then carry out the simulation of the SPP generation when the probe tip is in a close proximity with the golden nanoantenna in the shape of sphere or cylinder. We show that the surface plasmon-polariton generation efficiency can be increased more than two orders of magnitude with the use ofnano-antennas. Nanoantennas in the form of spherical nanoparticles are found to be more effective for the SPPs generation than nanoantennas in the shape of cylinders.

2. Numerical modeling

We calculated the axially symmetric problem of the STM probe tip with a metallic surface interaction in Comsol Multiphysics program. In the proposed model, the radiation in the tunnel junction was

---

Content from this work may be used under the terms of the Creative Commons Attribution 3.0 licence. Any further distribution of this work must maintain attribution to the author(s) and the title of the work, journal citation and DOI.
Published under licence by IOP Publishing Ltd

approximated with the emission of an optical dipole having vertical polarization. The presence of the STM probe and its influence on the waves propagation was not considered in the study. The location of the point dipole was fixed at a distance of 1 nm from the gold half-infinite substrate surface. The efficiency of the radiation in a tunnel junction was considered in the first part of the study. For this purpose, the ratio of all the propagating emitted electromagnetic waves energy (the sum of the plasmon and scattered waves) to the dipole source radiation power was investigated. The obtained spectral dependence of this ratio in the frequency range from 200 to 600 THz is shown in figure 1. Each point of the graph corresponds to the efficiency of the waves generation at a fixed frequency of the dipole radiation. As can be seen from the graph in figure 1 the efficiency of the dipole source energy conversion into the energy of propagating waves has a spectral maximum of 1 % at 400 THz. In this case, most of the energy goes into the structure heating.

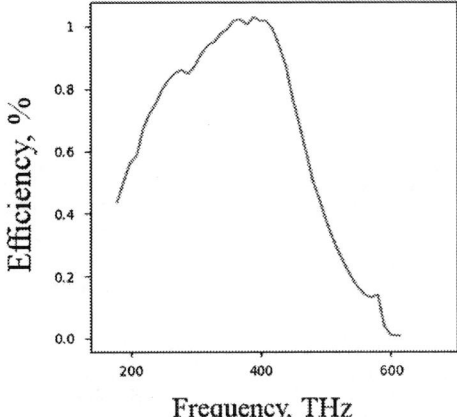

Figure 1. The ratio of all the propagating waves power to the dipole source power.

In work [8] it was shown that positioning of a gold nanoantenna in the shape of a nanorod having diameter D = 21, 25 and 31 nm and height H = 51 nm in the tunnel junction under the tungsten STM tip increases the plasmon wave generation efficiency. To study this effect in our work we simulated the propagation of electromagnetic radiation from a point dipole source when the spherical (D = 60 – 300 nm) and cylindrical (where D and H = 60 – 300 nm) nanoantenna are located on the surface of gold substrate.

To comprehensively study influence of the nanoparticle shape and dimensions on the SPP generation in the modeling we varied the geometry of the nanoparticles: in case of a gold sphere nanoantenna its radius was varied from 30 nm to 150 nm and in the case of cylindrical nanoantenna the radius and height values were modeled in the same range. Figure 2 (a) shows the spectral dependence of the ratio of scattered and plasmon wave energies sum to the dipole emission energy evaluated for the spherical gold nanoantenna in a wide dimensions range. The graph shows that the maximum efficiency can be achieved in the frequency range from 350 to 450 THz with the corresponding maximum generation efficiency of 30 %. In case of cylindrical Au nanoantenna the maximum efficiency can be obtained in the 400 to 450 THz range and takes the value of 22 % it show figure 2 (b). These maxima most likely correspond to the nanoantennas resonance with the dipole emission at small nanostructure radius of 30 – 60 nm.

Figure 3 shows the graphs of the efficiency of the dipole energy conversion into SPP energy in presence of nanoantennas. Comparing the graphs in figure 2 and figure 3 we can see that most of the source energy is directed at plasmon wave generation. It is also shown, that the spherical Au nanoantenna provides 30 % higher efficiency compare to the cylindrical Au nanoantennas.

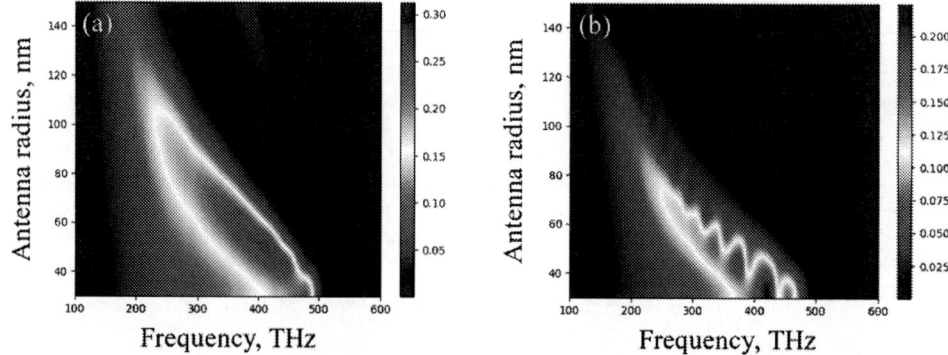

Figure 2 (a, b). The energy conversion efficiency to the sum of scattered and plasmon waves evaluated for the spherical Au nanoantenna (a) and for the cylindrical Au nanoantenna (b).

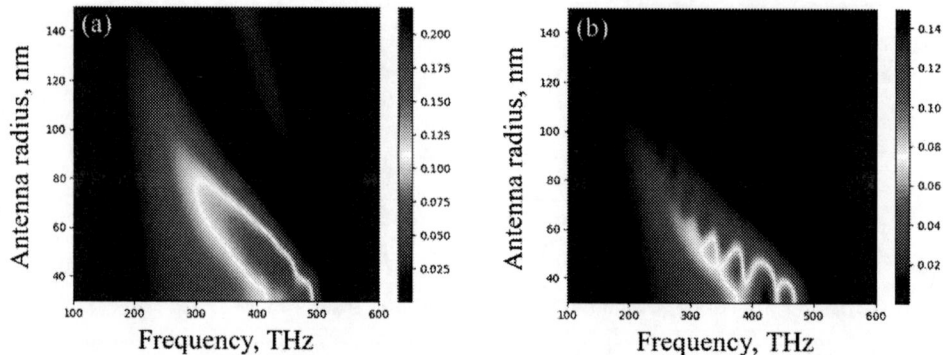

Figure 3 (a, b). Efficiency of the source energy conversion into the plasmon wave evaluated for the spherical Au nanoantenna (a) and for the cylindrical Au nanoantenna (b).

To demonstrate the effect of enhancement of the SPP generation using nanoparticles the comparison between the cases of presence and absence of the antenna was carried out. Figure 4 shows the ratio of plasmon wave energy in the presence of nanoantenna (spherical and cylindrical) to the sum of plasmon and scattered waves energy without nanoantenna. Considering the case with Au sphere the maximum enhancement was found to be 400 times and in the case of a cylindrical nanoantenna its value was 120 times. Concluding, it was shown that the introduction of a nanoantenna into the tunnel gap junction leads to increase of the excitation efficiency of the surface waves and efficiency of the SPP generation using a metallic spherical nanoantenna with radius R = 30 nm is more than 3 times higher than of a cylindrical nanoantenna.

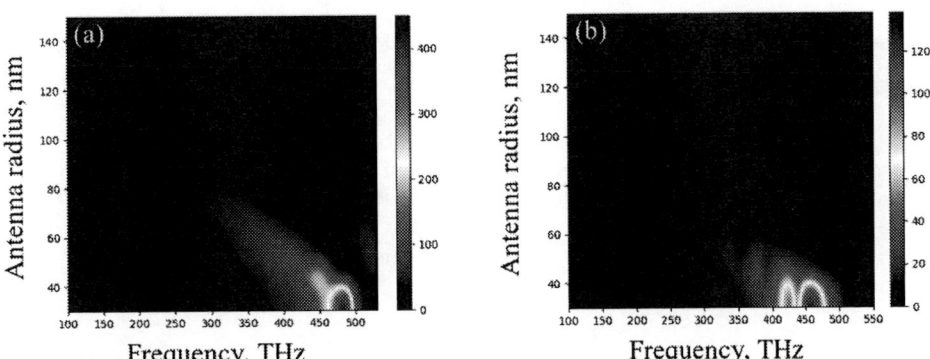

Figure 4 (a, b). The power ratio of the plasmon wave with the Au nanotenna to the power of the dipole source without the nanoantenna (a) with spherical antenna, (b) with cylindrical antenna.

3. Conclusion

To conclude, the numerical modeling of the SPP generation via excitation of the electromagnetic waves in the tunnel junction between STM probe tip and metallic substrate with and without the nanoantenna was carried out. The optimal geometric parameters of the metallic nanoantenna, corresponding to the most efficient excitation of the SPP wave propagating along the gold substrate surface were obtained: for both spherical and cylindrical antennas the optimal radius value is 30 nm at 450 – 500 THz. The comparison of the energy conversion efficiency of a dipole source to a plasmon wave demonstrates strong enhancement with the use of nanoantennas.

Acknowledgements

The authors are thankful to the Russian Science Foundation (Grant 17-19-01532) for support.

References

[1]   J Kern, R Kullock, J Prangsma, M Emmerling, M Kamp, and B Hecht 2015 *Nature Photonics* **582** 9(9)

[2]   M Parzefall, P Bharadwaj and L Novotny 2017 *Quantum Plasmonics* **211-236**

[3]   P Avouris, R Martel, T Hertel and R Sandstrom 1998 *Applied Physics A: Materials Science & Processing* **S659-S667** 66

[4]   F Tajaddodianfar, S R Moheimani, J Owen, and J N Randall 2018 *Review of Scientific Instruments* **013701** 89(1)

[5]   B Rogez, S Cao, G Dujardin, G Comtet, E Le Moal, A Mayne and E Boer-Duchemin 2016 *Nanotechnology* **465201** 27(46)

[6]   N Cazier, M Buret, A U Uskov, L Markey, J Arocas, G C Des Francs and A Bouhelier 2016 *Optics express* **3873-3884** 24(4)

[7]   Z Dong, H S Chu, D Zhu, W Du, Y A Akimov, W P Goh, T Wang, K E J Goh, C Troadec, C A Nijhuis and J K Yang 2015 *ACS Photonics* **385-391** 2(3)

[8]   F Bigourdan, J P Hugonin, F Marquier, C Sauvan and J J Greffet 2016 *Phys. Rev. Lett.* **106803** 116(10)

[9]   T Wang, E Boer-Duchemin, Y Zhang, G Comtet and G Dujardin 2011 *Nanotechnology* **175201** 22(17)

[10] D Canneson, E Le Moal, S Cao, X Quélin, H Dallaporta, G Dujardin and E Boer-Duchemin 2016 *Optics express* **26186-26200** 24(23)

SPBOPEN 2018   IOP Publishing

Metal-assisted photoenhanced wet chemical etching of GaN epitaxial layers

K Yu Shubina, D V Mokhov, T N Berezovskaya, A M Mizerov, E V Nikitina and A D Bouravleuv

Nanoelectronics Lab, St. Petersburg Academic University, St. Petersburg 194021, Russia

Abstract. In our work we successfully achieved selective removal of GaN epitaxial layer from the Si(111) substrate by photoenhanced wet chemical etching in $K_2S_2O_8$:KOH solution. Ti, Ni, Ti/Au, Cr/Au metal films were tested as etch masks. In case of thin Ti and Ni films the etch rate was quite low and metals were slightly dissolved. The use of mask containing Au film (noble metal) increased etch rate for 2-4 times. As a result, GaN was selectively removed from the substrate and highly anisotropic etch profile was formed.

1. Introduction

The wide-bandgap semiconductor material GaN has a number of unique properties, which are important for the manufacturing of modern opto- and nanoelectronics devices, and microelectromechanical systems (MEMS). However, natural substrates for GaN are currently very expensive, and silicon carbide (SiC), sapphire (c-Al_2O_3) and silicon (Si(111)) substrates are typically used for epitaxial growth of GaN. The most commercially attractive is low-cost GaN-on-Si technology.

It is well-known that crystallographic polarity is an essential feature of the wurtzite GaN, which affects structural, optical and electrical properties of the material [1]. It should be noted that GaN epitaxial films which are grown along [0001] direction (or Ga-polar films) are the most chemical resistant (figure 1).

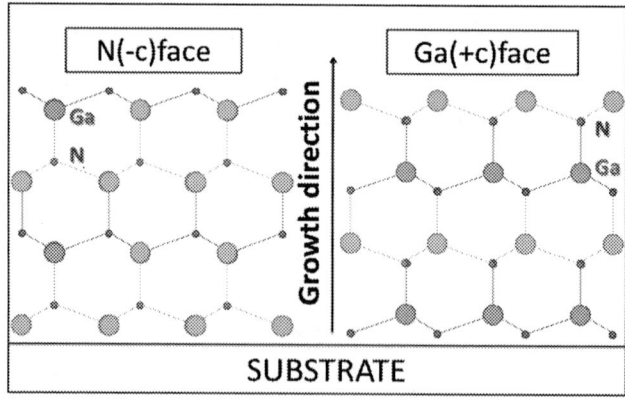

Figure 1. Schematic view of wurtzite GaN lattice exhibiting the polarity along the c-axis.

Content from this work may be used under the terms of the Creative Commons Attribution 3.0 licence. Any further distribution of this work must maintain attribution to the author(s) and the title of the work, journal citation and DOI.
Published under licence by IOP Publishing Ltd

Due to its extremely high thermal and chemical stabilily, GaN is very promising material for MEMS applications in harsh environments [2]. Thus, GaN-on-Si technology becomes more promising for MEMS industry. On the other hand, GaN durability makes postgrowth processing with traditional wet chemical etching challenging. For a long time, the etching of GaN was possible only by reactive ion etching (RIE) or inductively coupled plasma (ICP) etching [3]. However, the development of wet chemical etching techniques for nitrides goes on. Stimulation of chemical reactions with auxiliary illumination, current transmission and metals as catalysts, made possible wet chemical etching of GaN [4-5]. In this work we elaborated the technique for the complete removal of GaN epitaxial layers from silicon substrate by selective wet chemical etching.

2. Samples

The GaN/Si(111) samples were obtained by plasma assisted molecular beam epitaxy (PA-MBE) on Veeco Gen 200 MBE system equipped with RF plasma source. GaN epitaxial films 0.6-0.8 mkm thickness were grown on the high-resistance (R > 10 kOhm) silicon substrates with (111) crystallographic orientation. Preliminarily Si(111) substrates were chemically cleaned according to Shiraki method [5]. Before the start of growth process, clean Si(111) substrates were annealed in the growth chamber at 1000 °C and then nitridated in the growth chamber. Formed by nitridation on the substrate surface, thin Si_xN_y layer was used as a buffer layer. Crystallographic polarity of the samples was identified by previously developed technique (patent application №2016144845 15.11.2016), based on wet chemical etching of GaN surface by hot aqueous KOH solution. All the samples demonstrated Ga-polarity. SEM images of Ga-polar sample before and after etching in hot KOH solution are demonstrated in figure 2.

(a)   (b)

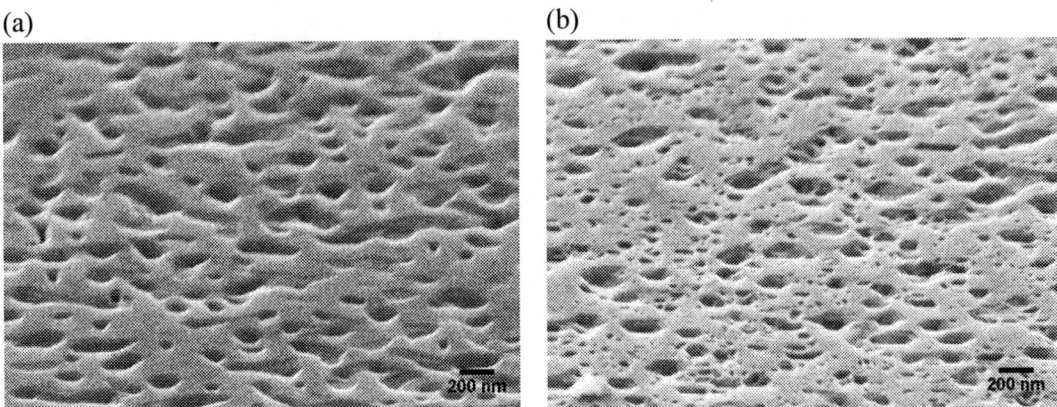

Figure 2. SEM image of Ga-polar GaN/Si_xN_y/Si(111) sample before (a) and after etching in hot aqueous KOH solution (b). Formation of hexagonal etch pits is a result of defect-selective etching of GaN layer.

3. Experimental details

As it was mentioned above, great chemical stability of Ga-polar GaN films makes traditional approaches to wet chemical etching inapplicable. At first, we tried to use photoenhanced wet chemical etching method. The etching solution contained potassium persulfate and potassium hydroxide ($K_2S_2O_8$:KOH) mixture. To induce the hole generation, necessary for the etching process, Xe lamp light source was used. The samples, coated with photoresist mask, were immersed in etching solution at room temperature. Unfortunately, the etching rate was extremely low. Moreover, we found that photoresist mask could be damaged by the etchant after 1 hour of etching. Thus, we decided to use metal mask. Samples with Ti, Ni, Ti/Au, Cr/Au mask were prepared. Metal films were formed by e-beam vacuum deposition on BOC Edwards AUTO 500 vacuum coating system. The etching procedure was similar to the one mentioned above. After etching, the surface of the samples was examined with contact profilometer Ambios XP-1 and scanning electron microscope Supra 25 Zeiss.

4. Results and discussion

Since the rate of photoenhanced wet chemical etching for GaN epitaxial layers is extremely low, the chemical resistance of the mask is very important to avoid unintentional etching. As a result of the experiments it was found that the use metal films allows to avoid fast damage of mask by etchant. However, not all masks appear to be resistant to $K_2S_2O_8$:KOH solution. In figure 3 surface of Ga-polar sample with Ti mask with 70 nm thickness on the initial stage of etching is demonstrated.

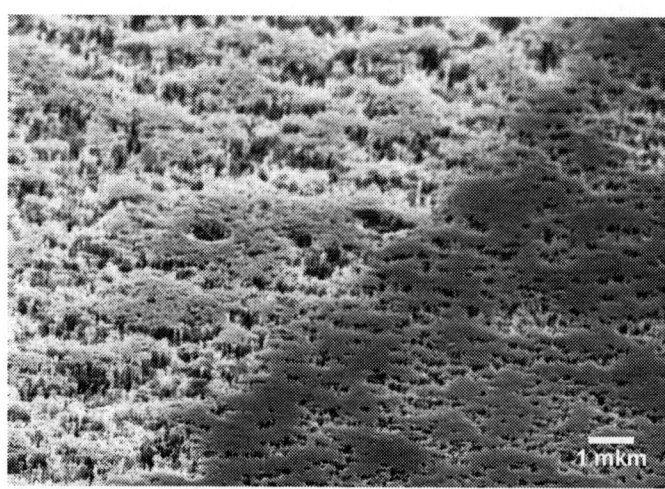

Figure 3. SEM image of the surface of GaN/Si(111) sample, partially coated with Ti mask after etching in $K_2S_2O_8$:KOH solution. Slight etching of Ti mask (right half) is observed.

It is obvious that during the etching process thin Ti film is gradually etched by $K_2S_2O_8$:KOH solution. So it turns out that the etching rate of thin Ti mask is higher than etching rate of GaN epitaxial layer. It leads to the emergence of local etching points in the area protected with Ti mask. Moreover, it reduces the etching rate of unprotected GaN area. Thus, we can assume that Ti is not a suitable mask material for the etching of GaN in $K_2S_2O_8$:KOH solution. It is worth to note that similar behavior was observed for Ni mask. However, in case of Ti mask with 1 mkm thickness, no local points of etching were found in protected area, but the etch rate of GaN was still extremely low.

Surprisingly, completely different results were obtained with the masks, containing Au film. In this case, etching rate of GaN increased in 2-4 times. Our results, obtained with noble metal (Au) are consistent with work [7]. It confirms the assumption of Bardwell et al. [8] that a noble metal working as a cathode increases the reaction rate, acting as a catalyst, on the surface of which a radical ion SO_4^{*-} forms from the persulphate ion $S_2O_8^{2-}$. These radicals are by-products of photolysis of $K_2S_2O_8$ under light with wavelength 310 nm or smaller. As a result, higher etching rate allowed to insure complete removal of GaN from the Si(111) substrate surface (see figure 4). It is worth to note that highly anisotropic etch profile was formed. In addition, no local etching points were observed in the protected area of GaN epitaxial layer. Finally, figure 4 demonstrates that the silicon substrate is not damaged. Thus, we have succesfully ensured the etch selectivity to silicon.

Some material residues are clearly visible in figure 4a. It turned difficult to remove them from the substrate surface. Later studies have shown that these were residues of Si_xN_y layer, which was formed during the nitridation of Si(111) substrate.

(a)   (b)

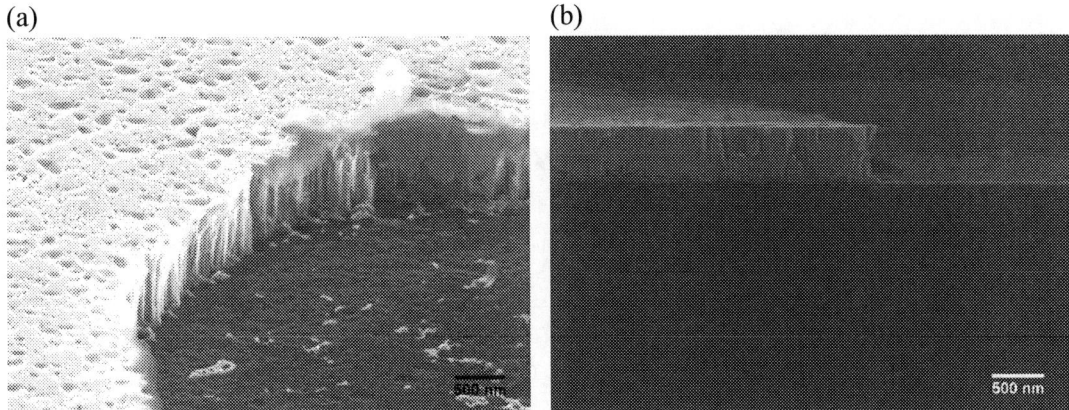

Figure 4. SEM image of Ga-polar GaN/Si$_x$N$_y$/Si(111) sample after etching in K$_2$S$_2$O$_8$:KOH. (a) Some material residues are observed, (b) highly anisotropic etch profile is formed.

5. Conclusion
In our experiments we successfully achieved selective removal of GaN epitaxial layer from the Si(111) substrate by photoenhanced wet chemical etching in K$_2$S$_2$O$_8$:KOH solution with auxiliary illumination by Xe lamp. We tried to use Ti, Ni, Ti/Au, Cr/Au metal films as etch masks. In case of Ti and Ni masks the etch rate was lower and metals were slightly dissolved and that significantly inhibited the etch process. In case of use Au film (noble metal) the etch rate was 2-4 times higher. Finally, highly anisotropic etch profile was formed. Thus, the relatively cheap method for anisotropic etching of GaN layer was elaborated. At the same time, the rate of metal-assisted photoenhanced wet chemical etching of GaN is quite low compared with RIE rate.

Acknowledgments
The work was supported by the grant of the Ministry of Education and Science of the Russian Federation № 16.9789.2017/BCh.

References
[1]   Hellman E S 1998 *MRS Internet Journal of Nitride Semiconductor Research* **3** 11
[2]   Rais-Zadeh M, Gokhale V J, Ansari A, Faucher M, Théron D, Cordier Y, and Buchaillot L 2014 *J. Microelectromech. Syst.* **23** 1252
[3]   Minsky M S, White M and Hu E L 1996 *Appl.Phys.Lett.* **68** 1531
[4]   Weyher J L, Tichelaar F D, van Dorp D H, Kelly J J and Khachapuridze A 2010 *J. Cryst. Growth* **312** 2607
[5]   Zhuang D and Edgar J H 2005 *Mater Sci Eng R Rep* **48** 1
[6]   Ishizaka A and Shiraki Y 1986 *J. Eltrochem. Soc.* **133** 666
[7]   Hwang J M, Ho K Y, Hwang Z H, Hung W H, Kei May Lau and Hwang H L 2004 *Superlattices Microstruct* **35** 45
[8]   Bardwell J A, Webb J B, Tang H, Fraser J and Moisa S 2001 *J. Appl. Phys.* **89** 4142

SPBOPEN 2018   IOP Publishing

Investigation of luminescence quantum yields of carbon dots synthesized from ethylene glycol, citric acid and berries

M N Egorova, A E Tomskaya, A N Kapitonov, S A Smagulova[1], A A Alekseev[2]
[1] Graphene nanotechnologies Lab, North-Eastern Federal University, Yakutsk, 677000
[2] North-Eastern Federal University, Yakutsk, 677000

Abstract

The paper describes the investigation of luminescencequantum yields of carbon dots (CDs) synthesized from different carbon precursors. In the synthesis, were used precursors such as ethylene glycol (EG), citric acid (CA) and juice of berries grown in the Far North.Synthesis was carried out by two methods of hydrothermal synthesis using microwave and a polytetrafluoroethylene autoclave. The photoluminescence quantum yields of these CDs were determined to befrom 0.008 to 0.46. The obtained CDs can be used to develop detectors for chemical elements and various optoelectronic applications, as well as in biomedicine and bioimage production.

Introduction

Synthesis of organic luminophores with a high quantum yield in the visible region, low-toxic, sensitive to heavy metals ions is an urgent task of applications in biomedicine and ecology. One of the most promising materials used as luminophores are carbon dots.

Carbon dots (CDs) are biocompatible and easily functionalized nanoparticles attracting considerable interest of researchers around the world as replacement of semiconductor quantum dots. This interest is related to their nanoscale, low toxicity and unique physical and chemical properties. CDs can potentially be used in such areas as bioimaging, photocatalysis, optoelectronic devices, delivery of drugs and genes, chemical sensors and biosensors. [1-3]

Carbon dots can be synthesized from a variety of environmentally friendly and inexpensive materials without the use of expensive equipment in various ways. The possibility of adjusting the photoluminescence (PL) of CDs with simple changes in the conditions of synthesis makes them promising materials in various applications, depending on the necessary photoluminescence conditions (radiation wavelength, radiation intensity, quantum yield, width of the forbidden band, etc.).

Experimental

Synthesis of carbon dots from ethylene glycol (EG-CDs) and citric acid (CA-CDs)

The synthesis of EG-CDs and CA-CDs was carried out by means of microwave synthesis. Microwave heating provides simultaneous, homogeneous and rapid heating, which leads to a uniform distribution of quantum dots dimension. [4] The technology of obtaining CDs used by us generally corresponds to that of described in [5], except that the microwave heating method is applied instead of heating on a tile.

A solution of citric acid (CA solution) was prepared by mixing 2 g of CA, 0.5 mL of a solution of PANI-graphene, 2 ml of 26% aqueous ammonia solution and 15 ml deionized (DI) water. A solution of ethylene glycol (EG solution) was prepared by mixing 10 ml of EG with 2 ml of aqueous ammonia and 2 ml of DI water. The solutions were then heated in a microwave oven for about 5 minutes at a power of 900 W. After the solutions changed color, 20 ml of DI water was added and suspensions of green CDs (G-CDs) and yellow CDs (Y-CDs) were obtained. To prepare brown CDs

Content from this work may be used under the terms of the Creative Commons Attribution 3.0 licence. Any further distribution of this work must maintain attribution to the author(s) and the title of the work, journal citation and DOI.
Published under licence by IOP Publishing Ltd

(B-CDs) the CA solution was mixed with 2 ml of H_3PO_4, then heated in microwave oven for 5 minutes (CA solution + H_3PO_4). To prepare red CDs (R-CDs) EG solution was mixed with 4 ml of H_3PO_4, then heated as well as above for 5 minutes (EG solution + H_3PO_4). To prepare orange CDs (O-CDs) 10 ml of EG were mixed with 3 ml of H_3PO_4 and heated in microwave oven for 5 minutes (EG + H_3PO_4). The obtained solutions were purified by the addition of ethyl alcohol, by centrifugation and subsequent removal of the alcohol.

Synthesis of CDs from Berry Juice

CDs were synthesized from the juice of red bilberry, bog blueberry and red currant. Juices were made by pressing berries and subsequent centrifugation. 2 ml of the obtained juice of each berry was mixed with 6 ml of 26% aqueous ammonia solution and 15 ml of DI water. The solutions was transferred to a poly (tetrafluoroethylene) autoclave and heated for 1 hour at 180 °C. The resulting mixture was then purified with a dialysis bag (MWCO 3.5 kDa) for 12 hours for further characterization.

Measurement of the quantum yield

The simplest way to determine the PL quantum yield (QY) of is the comparison method, which involves the use of a sample with a known quantum yield as a reference. Rhodamine 6G with a quantum yield of ethanol of 96% was chosen as the reference. Thus, the PL quantum yield of the test substance can be determined as follows:

$$\varphi = \varphi_R \frac{I \cdot D_R \cdot n^2}{I_R \cdot D \cdot n_R^2}$$

where I is the integral luminescence intensity; D is the optical density at the excitation wavelength of the luminescence; n is the refractive index of the solvent (to take into account the effect of the solvent on the optical properties of the phosphor). The index R refers to the parameters of the reference sample (standard) [13].

Results and discussion

In several series of experiments with EG-CDs and CA-CDs, we obtained the PL spectra shown in Fig. 1.

For sample №1, the shift of luminescence spectra maxima from the change in the excitation wavelength (300, 350, 400, 450 nm) is observed. The shift of spectra maxima is from 487 nm to 534 nm. In sample №2, the dependence of the luminescence spectra maxima shift on the excitation wavelength is also observed. The maxima shift is from 425 nm to 498 nm. Sample №3 exhibits the same properties as the previous samples. The luminescence spectra maxima in this sample shift from 431 nm to 507 nm. The most interesting result is for sample №4 that is a mixture of CA solution + H_3PO_4. In this sample, the luminescence spectra maxima do not depend on the change in the excitation wavelength. This agrees with the data given in [5].

SPBOPEN 2018   IOP Publishing

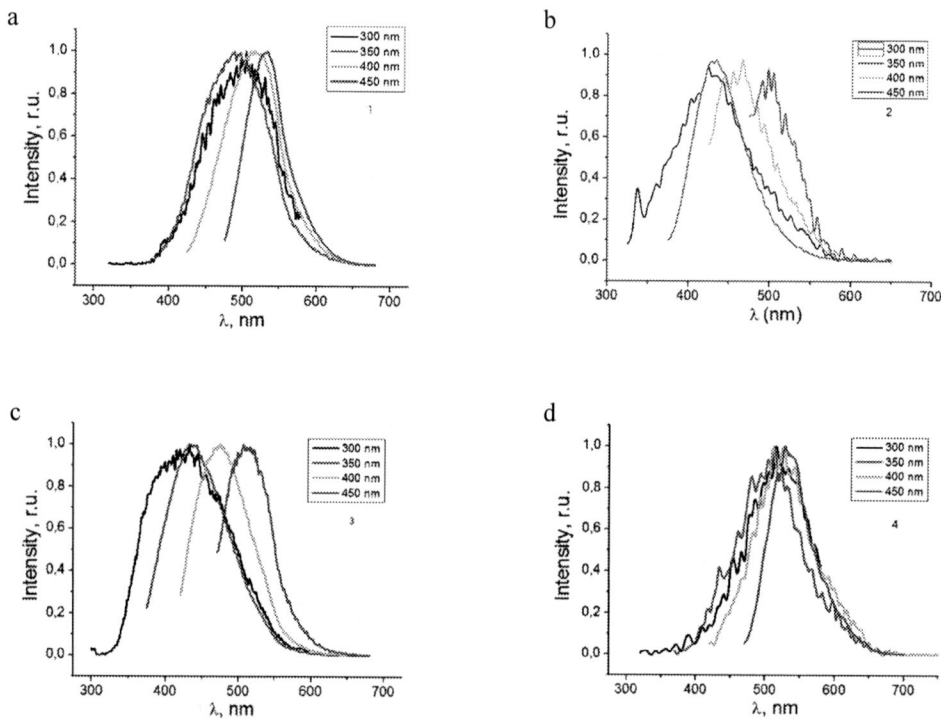

Fig. 1 (a, b, c, d) Normalized PL spectra of CDs: (a) №1 (CA+NH$_3$ + PANI-graphene+H$_2$O); (b) №2 (EG+H$_3$PO$_4$); (c) №3 (EG solution+H$_3$PO$_4$); (d) №4 (CA solution+H$_3$PO$_4$).

Figure 2 shows the absorption spectra of EG-CDs and CA-CDs. In sample №1, one peak at 206 nm is observed, which can be attributed to the p-p* transition of C = O [14] bonds. The shoulder at 336 nm refers to the n-π* -transition of C = O-bonds [15].

In sample №2, peaks at 195 nm and 220 nm are observed related to the π-π* C=C transition and the p-p* C=O transition, respectively. The shoulder in the region of 352 nm refers to the n-π* -transition of C=O-bonds [15].

Sample №3 has two shoulders about 211 and 310 nm, one peak at 270 nm in the absorption spectrum and extends to 600 nm without noticeable structures. According to [16, 17], the shoulder at 211 nm is attributed to the π-π* transition of C=C aromatic bonds, whereas the 320 nm shoulder refers to the π-π* -transition of C=O-bonds. The peak at 270 nm refers to the π-π* transition of C=C aromatic bonds. [15]

Sample №4 has a peak at 202 nm related to the n-p* transition C=O bonds [14] and one shoulder at 328 nm, referring to the n- π* -transition of C=O-bonds [15].

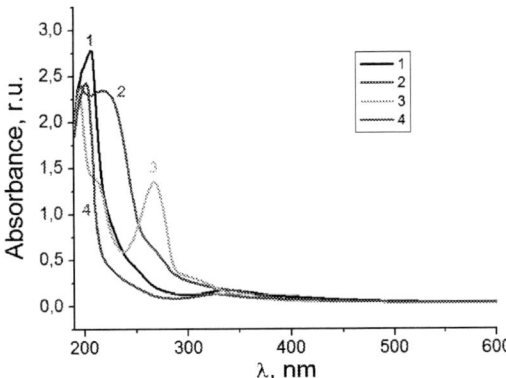

Fig. 2 Absorption spectra of CDs: 1 - №1 (CA+NH$_3$ + PANI-graphene+H$_2$O); 2 - №2 (EG+H$_3$PO$_4$); 3 - №3 (EG solution+H$_3$PO$_4$); 4 - №4 (CA solution+H$_3$PO$_4$).

Table 1 shows the results of measuring the PL QY of carbon dots. Rhodamine 6G was used as a standard. For the sample the CDs №1 (CA + NH$_3$ + PANI-graphene+H$_2$O) QY was 0.12, for the CDs № 2 (EG + H$_3$PO$_4$) - 0.15, for the CDs №3 (EG solution + H$_3$PO$_4$) - 0.46 , for CDs № 4 (CA solution + H$_3$PO$_4$) - 0.10. As can be seen from the results, the largest quantum yield is observed for sample №3 (EG solution + H$_3$PO$_4$).

The photoluminescence QY of CDs from the red bilberry juice was 0.021, of bog blueberry - 0.017, of red currant - 0.008. Low OYs for CDs from berries may be due to the small number of functional groups on the surface of CDs under these conditions.

Table 1. Quantum yields of the obtained carbon dots

Sample	Integrated emission intensity (I)	The optical density at the excitation wavelength 300 nm (D)	Refractive index of solvent (n)	Quantum yield (ф)
Rhodamine 6G (ethanol)	72.58	0.104	1.36	0.96
CDs № 1	11.25	0.122	1.33	0.12
CDs № 2	30.83	0.272	1.33	0.15
CDs № 3	113.5	0.323	1.33	0.46
CDs № 4	6.944	0.094	1.33	0.10
CDs from red bilberry	1.907	0.118	1.33	0.021
CDs from bog blueberry	2.031	0.150	1.33	0.017
CDs from redcurrant	0.967	0.150	1.33	0.008

Fig. 3 shows the PL spectra of CDs from berries when excited at different wavelengths. They look identical, which indicates a similar composition of functional groups on the CDs surface. As can be seen from the figure, all CDs samples have the property of the luminescence spectra maxima shift dependence on the excitation wavelength. For CDs from red bilberry, the luminescence spectra maxima shift was observed from 419-508 nm, for CDs from bog blueberry - from 407-511 nm, for CDs from red currant - from 396-503 nm.

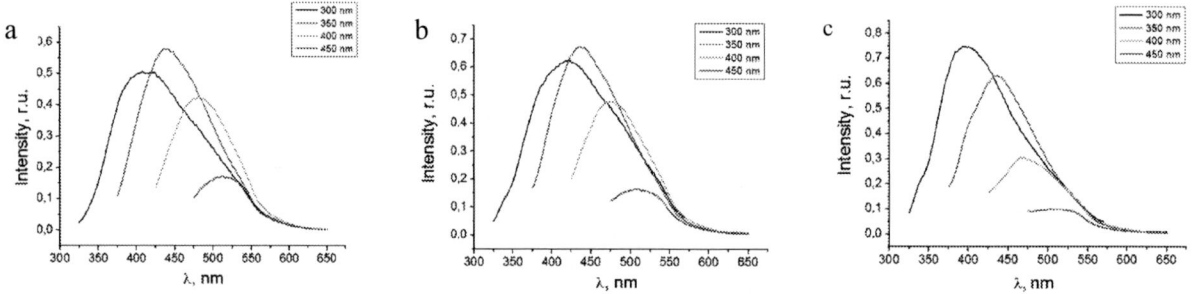

Fig. 3 (a, b, c) PL spectra of CDs from: (a) red bilberries; (b) bog blueberries: (c) red currants.

Fig. 4 shows the absorption spectra of CDs from berries. The spectra look the same and have one peak at 269 nm. This absorption band was attributed to the π-π * transition of C = C aromatic bonds [15]. The results obtained are consistent with the results shown in Fig. 3, which speak of the functional groups composition similarity on the CDs surface.

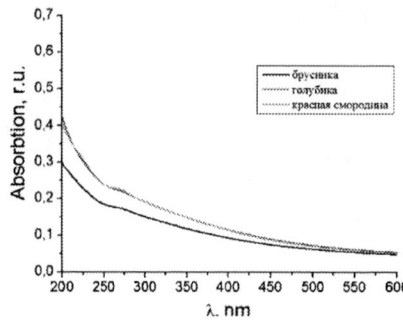

Fig. 4 Absorption spectra of UV-VIS CDs from berries.

Acknowledgments
The article was written with the support of the RFBR grant № 18-02-00449 and «Scientific and Educational Foundation for Young Scientists of Republic of Sakha (Yakutia) wthin the Project №17-2-009477» Scientific and Educational Foundation as an instrument for young scientists to develop their professional competence and science popularization 20171201005

References
[1] Zhu H, Wang X, Li Y 2009 *Chemical Communications* **34** P 5118-5120
[2] Wang C, Xu Z, Cheng H 2015 *Carbon* **82** P 87-95
[3] Wu G, Zeng F, Yu C, Wu S and Li W 2014 *J. Mater. Chem. B* **2** 8528-8537
[4] Xu J, Zeng F, Wu H, Hu C, Yu C and Wu S 2014 *Small* **10** 3750-3760
[5] Hu S, Trinchi A, Atkin P and Cole I 2015 *Angew. Chem. Int. Ed.* **54** 1 – 6
[6] Shen C, Sun Y, Wang J and Lu Y 2014 *Nanoscale* **6** 9139-9147
[7] Li H, Liu R, Lian S, Liu Y, Huang H and Kang Z 2013 *Nanoscale* **5** 3289-3297

[8] Choi H, Ko S-J, Choi Y, Joo P, Kim T, Lee B R, Jung J-W, Choi H J, Cha M, Jeong J-R, Hwang I-W, Song M H, Kim B-S and Kim J Y 2013 *Nat.Photonics***7** 732-738

[9] Liu C, Zhang P, Zhai X, Tian F, Li W, Yang J, Liu Y, Wang H, Wang W and Liu W 2012 *Biomaterials***33** 3604-3613

[10] Zheng M, Liu S, Li J, Qu D, Zhao H, Guan X, Hu X, Xie Z, Jing X and Sun Z 2014 *Adv. Mater.***26** 3554-3560

[11] Yu C, Li X, Zeng F, Zheng F and Wu S *Chem. Commun.* 2013 **49** 403-405

[12] Guo Y, Zhang L, Zhang S, Yang Y, Chen X and Zhang M *Biosens. Bioelectron.* 2015 **63** 61-71

[13] А К Вишератина, А П Литвин, А О Орлова 2016 *Введение в спектроскопию наноструктур, Лабораторный практикум* (Санкт-Петербург) с 54

[14] Maiti R, Mukherjee S, Haldar S, Bhowmick D , Ray S K Carbon 104 226–232

[15] Luo Z, Lu Y, Somers L A, Johnson A T C *J. Am. Chem. Soc.* 2009 **131 (3)** 898–899.

[16] Clark B J, Frost T, Russell M A 1993 *UV Spectroscopy: Techniques, Instrumentation, Data Handling/UV Spectrometry Group* (Chapman & Hall: London, New York)

[17] Eda G, Lin Y Y, Mattevi C, Yamaguchi H, Chen H A, Chen I S, Chen C W, Chhowalla M, *Adv. Mater.* 2010 **22** 505– 509

SPBOPEN 2018   IOP Publishing

Formation of a microheterophase state from planar nanoparticles of graphene at the oil-water interface

Yu V Pakharukov[1,2], F K Shabiev[1,2], R F Safargaliev[2]
[1]Tyumen Industrial University, Tyumen, 625000 Russia
[2]Tyumen State University, Tyumen, 625000 Russia

Abstract. It has been found that N aqueous suspension of planar graphite nanoparticles exhibits properties of displacement fluid at the oil–water interface. Experiments with the Hele–Shaw cell showed that the process of oil displacement from the interface is not accompanied by the formation of viscous "fingers" as a result of development of instability at the oil–water interface.

1. Introduction

Most of oil deposits are known to be entering the final stage of development [1]. To completely displace oil from a seam, physicochemical oil recovery methods are combined with flooding [2]. Microemulsion flooding is considered to be the most efficient technique, but the flooding agent must have a required mobility and a low surface tension in oil. In this case, the oil-water interface is stable and viscous fingers do not form. Microemulsions are, however, sensitive to aggressive media of the seam. As a result, a stable laboratory microemulsion state becomes unstable in field operations [2].

Thus, the research objective was to form, at the oil-water interface, a transition zone with a low surface tension, which would not be a microemulsion and would have a low sensitivity to deposit water temperature and hardness. To meet these requirements, a water suspension based on planar nanoparticles of graphene was chosen [3, 4].

It is known that a low surface tension at the oil–water interface is related to the formation of a structure consisting of liquid-crystalline layers of macromolecules [2]. These layers can be formed, in particular, by planar graphite nanoparticles with dimensions below 400 nm. Stability of this suspension is determined by the condition [4]

$$\Delta G_{mix} = \Delta H_{mix} - T\Delta S \leq 0 \qquad (1)$$

where ΔG_{mix} is the change in Gibbs' energy of the mixture, ΔH_{mix} is the change in enthalpy of the system, and ΔS is the change in entropy. Therefore, a solvent for suspension should possess a specific surface energy (σ) close to the energy of a monolayer of graphite nanoparticles. This requirement is satisfied by aqueous ethanol solutions [5]. By varying the concentration of ethanol, it is possible to make σ close to that of planar graphite particles.

Under real conditions, liquids frequently exhibit the phenomenon of bedding (e.g., motion of oil in water). In this case, solution of the problem of possible motion of the oil–water interface depends on the coefficient of proportionality of the bulk of the reservoir and the angle of the bed slope. The shape of the oil–water interface also depends on the ratio of viscosities of oil and water. The formation of transition regions leads to a decrease in the velocity of motion on the bed bottom and increase in that at the bed roof. The difference of the velocities of motion at the inner and outer contours depends on the permeability of bed solid. Even simple schemes of oil displacement exhibit distortion of the shape of

Content from this work may be used under the terms of the Creative Commons Attribution 3.0 licence. Any further distribution of this work must maintain attribution to the author(s) and the title of the work, journal citation and DOI.
Published under licence by IOP Publishing Ltd

the oil–water interface with the formation of water fingers. The degree of stability is determined by the coefficient of mobility defined as,

$$\lambda = \frac{K_{w0}\mu_{oil}}{K_{ow}\mu_B} \qquad (2)$$

where K_{w0} is the permeability for water in the presence of residual oil, K_{ow} is the permeability for oil in the presence of residual bound water, μ_{oil}l is the viscosity of oil, and μ_B is the viscosity of brine. The contact front is stable provided that $\lambda < 1$, which implies that μ_B must not significantly increase. Therefore, a suspension of graphite (planar) particles must have a small surface tension σ at the oil–water interface, while viscosity μ_B and density ρ should obey the empirical relation [4]

$$0.1 \geq \frac{\mu_B}{\sqrt{\sigma\rho d}} \qquad (3)$$

where d is the average pore size (or capillary diameter) in the porous structure, or

$$K_{ow}\mu B < K_{wo}\mu_{oil} \qquad (4)$$

2. Results and discussion

It was discovered that, at the oil-water interface, the water graphene suspension forms a transition multilevel microheterophase state from planar nanoparticles of graphene, oil hydrocarbon molecules.

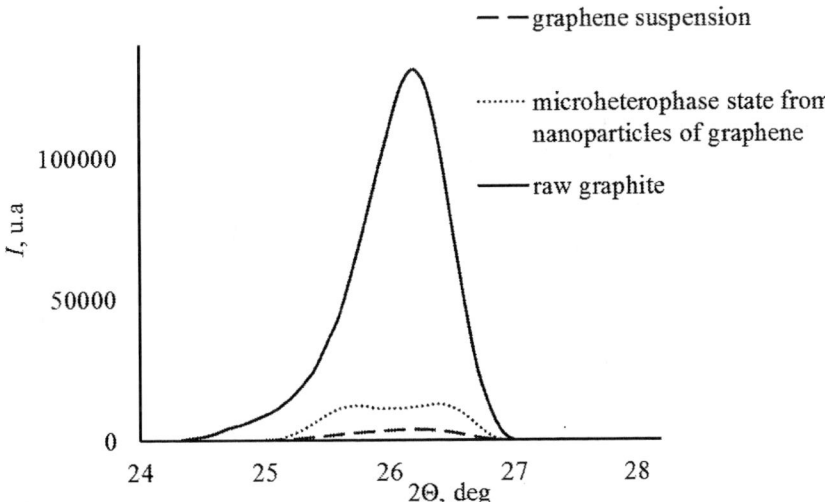

Figure 1. X-ray diffraction (peaks 002) of raw graphite and carbon material containing graphene nanoparticles

Analysis of the X-ray diffraction patterns of the carbon material obtained on a DRON-7 X-ray diffractometer (CuKα radiation) showed a significant decrease in the 002 peak in the carbon material of the graphene suspension. The decrease in the 002 peak in the graphene suspension indicates the presence of a single-layered graphene in the suspension [5]. However, in the formation of the microheterophase state, the magnitude of the peak 002 increases compared to the peak that gives the carbon material isolated from the graphene suspension. In addition to the increase in peak 002, the distance between the planes d_{002} increases from 0.337nm to 0.347nm (Figure 1.). This can be explained by the fact that in the formation of a microheterophase state, graphene particles form a multilayer structure of sheets of graphene and hydrocarbons contained in oil.

To study the behavior of the water-oil interface, the radial Hele-Shaw cell was used, the geometric parameters of which were D_0 = 2 mm, D_∞ = 120 mm, b = 0.6mm (Figure 2.) [6,7]. Displacement was carried out at a constant pressure of p = 10 kPa.

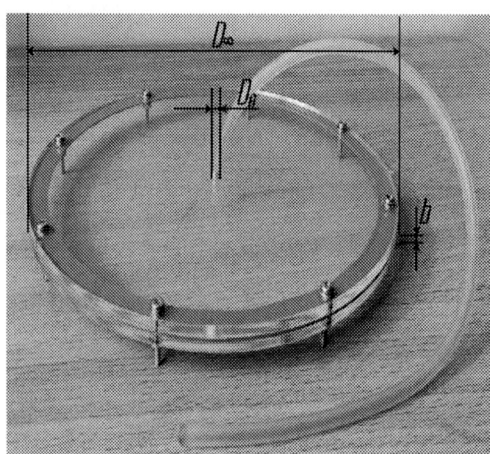

Figure2. The radial Hele-Shaw cell

As a result of synthesis of planar carbon nanostructures, a suspension with a particle size of 200÷400 nm with a low surface tension σ = 43 mN/m was formed. The electron-microscopic examination of this carbon material on a UEMV-100K transmission electron microscope using standard techniques showed the presence of perfect graphite crystals with various thicknesses (Figure. 3).

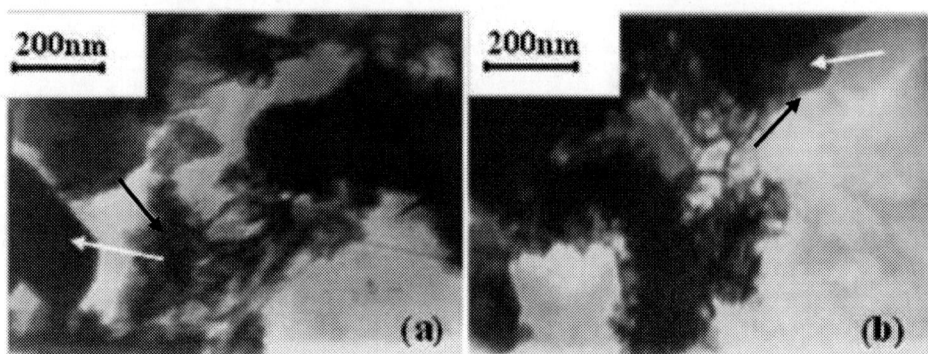

Figure. 3 Electron-microscopic images of a carbon material obtained by quenching polycrystalline graphite from 1000°C in distilled water. Arrows indicate the region containing perfect graphite crystallites (planar nanoparticles) with thicknesses (a) much below the diameter or (b) comparable with the diameter.

The addition of this suspension to an oil phase reduced the initial viscosity by 0.7%, which led to an increase in the stability of the contact front (λ < 1). When water was injected without graphene nanoparticles into the Hele-Shaw cell (with a constant pressure p = 10 kPa), viscous fingers formed due to the instability of the interface, indicating the breakthrough of water through the oil (Fig. 4, a). When graphene suspension is added to water with a concentration of 0.04 g/l, the displacement is a stable front without forming viscous fingers under the same regime with a constant pressure p = 10 kPa. In Figure 2, b, the oil-microhetero-phase-water interface is clearly visible.

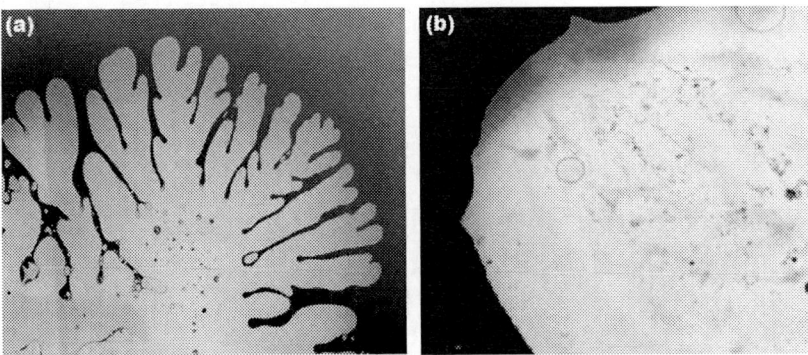

Figure 4 (a, b). Displacing oil with water in a Hele-Shaw cell **(a)** with viscous fingering; **(b)** without viscous fingering with a stable oil-graphene nanoparticles-water interface

3. Conclusion

The results of the research works prove promising the development of the technology involving the use of the graphene-based suspension to displace residual oil from the seam.

References
[1] N. M. Emanuel' and G. E. Zaikov 1987 *Chemical Methods in Oil Production Processes* (Moscow: Nauka) p 238
[2] S. E. Friberg and P. Bothorel 1987 *Microemulsions. Structure and Dynamics* (CRC, Boca Raton, FL,) p 320
[3] E. A. Belenkov and F. K. Shabiev 2015 Pis'ma Mater. **5** 459
[4] Yu.V. Pakharukov, F.K. Shabiev, R.F. Safargaliev 2018 Tech. Phys. Lett. **44**. 130
[5] Łos S., Duclaux L., Alvarez L., Hawełek Ł.,Duber S., Kempinski W. 2013 Carbon. **55**. 53
[6] L. M. Martyushev and A. I. Birzina, 2014 JETP Lett. **99**. 446
[7] L. M. Martyushev and A. I. Birzina, 2008 Tech. Phys. Lett. **34**. 213

SPBOPEN 2018

IOP Publishing

IOP Conf. Series: Journal of Physics: Conf. Series **1124** (2018) 081004   doi:10.1088/1742-6596/1124/8/081004

The nanosecond studies of granular carbon nanostructures based on high temperature superconductors

M P Faradzheva[1], A V Prikhodko[1], O I Konkov[2], Sh P Faradzhev[3]

[1]Department of Experimental Physics, SPBSTU, Saint Petersburg 194021, Russia
[2] Physicochemical Properties of Semiconductors, Ioffe Institute RAS, Saint Petersburg 194021, Russia
[3]Department of Solid State Physics, Dagestan state university, Makhachkala 367000, Russia

Abstract. The peculiarities of current percolation in two-component structures (HTSC- MWNTs) depending on the concentration of components are established. It is established that at 20% nanopowder content there is an increase in the critical temperature of the TC transition to the superconducting state. Discusses the mechanism of destruction of the percolation cluster with changing concentration of the nanopowder.

Introduction

Studies of multicomponent systems with reduced dimensionality are interest because their properties dependence of the composition, the size of the components particle, and their interaction with each other. The kinetic phenomena in such structures are well described by percolation theories [1].

The investigation of the effect of morphological factors on the capture of magnetic flux and critical currents in the superconducting structure, which is a percolation superconductor with pinning centers, is presented in [2]. The results of this studies show that the clusters of the normal phase in the superconductor increase pinning of the magnetic field and slow down the superconductivity destruction. It was established in [3,4] that when the nanoparticles $BaTiO_3$ [3] and $BaHfO_3$ [4] are added to $YBa_2Cu_3O_{7-\delta}$ (YBCO), the superconducting transition temperature increases to 103 and 107 K, respectively. The introduction of nanoparticles of other elements in YBCO, as a rule, reduces the critical temperature of the transition to the superconducting state T_C [5-7], whereas for our nanostructured materials, T_C increases.

In this paper are presented the new investigation results of the transport characteristics of HTSC materials based on $YBa_2Cu_3O_{7-\delta}$ which are modified with multi-walled carbon nanotubes (MWNTs) in the nanosecond voltage duration interval.

Experimental details

The measurements were carried out on samples consisting of a microcrystalline powder $YBa_2Cu_3O_{7-\delta}$ obtained by solid-phase sintering and a nanocrystalline powder $YBa_2Cu_3O_{7-\delta}$ obtained by the glycine-nitrate method [8, 9], MWNTs with a diameter of 20-70 nm and a length of more than 150 nm [10].

The modified structures were obtained by mixing the two powdered components in a weight ratio that

Content from this work may be used under the terms of the Creative Commons Attribution 3.0 licence. Any further distribution of this work must maintain attribution to the author(s) and the title of the work, journal citation and DOI.
Published under licence by IOP Publishing Ltd

were not sintered to avoid mutual diffusion and chemical reactions of the components. We consider a model system-a mixture of powders with artificially created Josephson weak bonds formed at natural intergranular boundaries.

In order to evaluate the transport characteristics, the temperature dependences of the resistance in the ns-interval (1-20 ns) of voltage durations up to amplitude of 1 V at a frequency of 100 Hz were investigated [11]. The nanosecond experiment was chosen because it gives is the possibility to manage the thermal overheat control percolation channel. It allows to identify the percolation characteristics in heterogeneous systems without destroying them.

Results

It is known that in the multicomponent percolation systems the transport characteristics are depended on the impurity concentration and its threshold value (N_c-critical concentration at which the percolation threshold is observed). In the investigated structures the nanodispersed component (MWNTs and YBCO nanocrystalline powder) acts as a binding substance, forming a grid of weak Josephson-type junction in microcrystalline HTSC. Being located between the microparticles of HTSC, the nanodispersed filling mixture, provides the formation of a superconducting percolation cluster in an heterogeneous medium as well as the formation of a non-superconducting percolation cluster based on MWNTs.

Actually, the performed researches of the samples at room temperature made it possible to establish that at the threshold value of the nanotube concentration N = 10%, the main percolation cluster begins to "break down" in the HTSC- MWNTs system and the conductivity is switch over the second non-superconducting component up to 20% (Figure 1a, curve 3). We can assume that starting from N = 20%, a quench to a percolation cluster on the basis of MWNTs occurs within N = 20-100%.

It should be noted that in the system HTSC - nanocrystalline powder at room temperature is observed a minimum resistance at 20%. [11]

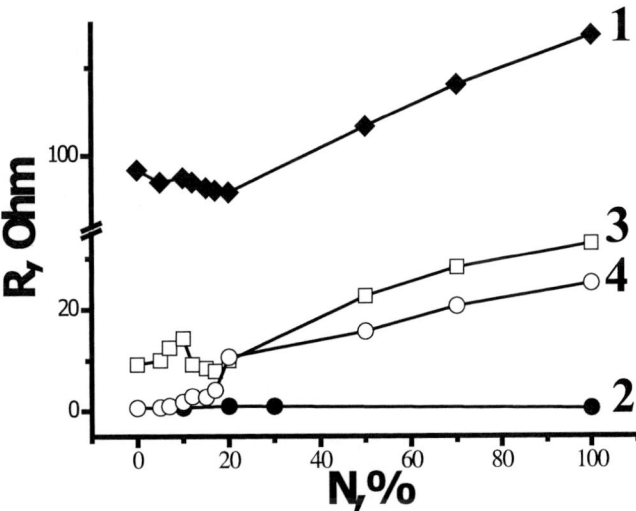

Figure 1 The resistance dependence on the concentration of nanopowder at room(1) and nitrogen(2) temperatures, MWNTs at room(3) and nitrogen(4) temperatures

The results of measurements at a nitrogen temperature showed that the state of superconductivity in the HTSC- MWNTs system changes continuously and disappears at N = 20% (Figure 1, curve 4). In this case, the system switch over to the percolation cluster based on the MWNTs, as we can observe it at room temperatures (Figure 1, curve 3). By adding of nanopowder $YBa_2Cu_3O_{7-\delta}$ to the microcrystalline HTSC medium, the "destruction" of the percolation cluster can be seen in the changing of the width (ΔT) of superconducting transition and superconducting critical temperature.

The results of the investigation of the dependence of the critical temperature (T_C, K), the width of the superconducting transition (ΔT, K), and the slope of the resistance-temperature curve (B, K^{-1}) for samples with different nanopowder content are presented in Table 1.

Table 1. Parameters of the superconducting transition (Tc, ΔT, B) for the modified $YBa_2Cu_3O_{7-\delta}$ samples with different content of nanopowder N.

N, %	0	10	20	30	100
T_c, K	94,7	96	99	95	93
ΔT, K	14	14	10	12	13
B, K^{-1}	0,064	0,064	0,09	0,07	0,071

According to percolation theory, the conductivity in inhomogeneous media depends on the impurity concentration in the initial sample and its threshold value (Nc- the critical concentration at which the percolation threshold is observed). In the structures studied by us, the nanodispersed powder acts as a binder, forming a network of weak Josephson-type bonds in the microcrystalline HTSC. The nanodispersed filler, located between the microparticles of HTSC, provides the formation of a superconducting percolation cluster in an inhomogeneous medium.

Thus, the study of transport characteristics for nanosecond voltage durations made it possible to reveal new features of the conductivity mechanism when components were changed:

– under normal conditions, when is not appear superconductivity, it is established that the switch over of mechanism conductivity in the HTSC- MWNTs system starts at 10%; by adding of nanopowder the maximum value of current percolation in the system is carried out at a concentration of 20%;

– at temperatures within the limits of the existence of T_c, it is established that by addition of a nanocrystalline powder can control the width and critical temperature of the superconducting quench. We assume that the modification of HTSC by means of carbon nanotubes will make it possible to obtain modified materials with predetermined performance characteristics in the nanosecond voltage duration interval and, accordingly, in the microwave interval.

References

[1] Kanonenko V V, Tarenkov V U, Dyachenko A I, Varukhin V N 2014 Ukraine *J. Low temperature physics*. 40 3 p.247

[2] Kuzmin Y I 2001 J.Physics of the Solid state, 43 7 p.1157

[3] Rejith P P, Vidya S, Thomas J K 2015 IOP Conf. Series: Mater. Sci. Engin. 73 p.1

[4] Rejith P P, Vidya S, Thomas J K 2012 J. Supercond. Nov. Magn. 25 p.1817

[5] Hamrita A, Ben Azzouz F, Dachraoui W, Ben Salem M 2013 J. Supercond. Nov. Magn. 26 p.879

[6] Missak S Raju P, Devendra K N, Pavan K N S., Rajasekharan T, Seshubai V. 2014 J. Supercond. Nov. Magn. 27 p. 2277

[7] Turkoz M B, Nezir S, Terzioglu C, Varilci A Yildirim G 2013 J. Mater Sci.: Mater Electron. 24. p. 896

[8] Gadzhimagomedov S K, Faradzheva M P, Tabit A F et al. 2014 J. Vestnik DGU 1 p.36

[9] Gadzhimagomedov S K, Palchaev D K et al. 2016 J. Technical physics letters, 42 1 p.9

[10] Prikhodko A V, Konkov O I, 2014 J.Physics of the Solid state, 56 7 p.1411

[11] Faradzheva M P, Prikhodko A V 2016 Proc. Int. Conf. Science week SPbSTU (Saint-Petersburg) p.298

SPBOPEN 2018   IOP Publishing
IOP Conf. Series: Journal of Physics: Conf. Series **1124** (2018) 081005   doi:10.1088/1742-6596/1124/8/081005

Thermoelectric Peltier micromodules processed by thin-film technology

E Bakulin[1], S Dzyubanenko[1], S Konakov[2], A Korotkov[3], V Loboda[3], A Yugay[1]

[1]JSC "Avangard", Saint-Petersburg, 195271, Russia
[2]JSC "Gyrooptics", Saint-Petersburg, 194044, Russia
[3]Peter the Great St. Petersburg Polytechnic University, Saint-Petersburg, 195251, Russia

Abstract. In this paper we consider the structure and manufacturing technology of thermoelectric Peltier micromodule consisting of 288 thermocouples enclosed between two silicon wafers. Method for obtaining thermoelements i.e the method of electrochemical deposition from an electrolyte solution, has been proposed. Elemental and structural analysis of the obtained semiconductor thermoelements of bismuth telluride and antimony telluride has been carried out. The obtained composition is 39.5% Bi-60.5% Te for Bi_2Te_3 and 39% Sb-61% Te for Sb_2Te_3.

1. Introduction

A large number of studies have been focused on research and development of micro capacity devices with low power and high output voltage to replace bulky batteries with a limited lifetime. Among the available power sources there has recently been much interest for thermal-to-electrical energy conversion, i.e., thermoelectric power generation, due to its attractive characteristics such as no moving parts, long lifetime and high reliability for low-power applications.

Continuous complication and miniaturization of opto-, radio- and microelectronic elements along with the increase in device performance leads to rise in operating temperature which contributes to active development of cooling systems for electronic devices [1].

Thermoelectric Peltier thin-film micromodules can be used either for power generation or for cooling microelectronic devices, as well as for effective heat removal from electronics components, since they have high refrigeration power depending on the number of thermocouples in the module with a small device size [2-4].

2. Micromodule structure

The proposed thin-film Peltier micromodule consists of 288 thermocouples (576 thermoelements) and contact layers formed between two silicon wafers. Each of the pairs consists of an element of bismuth telluride (n-type conductivity) and antimony telluride (p-type). All semiconductor thermoelements are connected sequentially. Tin is used as an intermediate layer for joining wafers. The micromodule is mounted onto a printed circuit board with an aluminum base, and a radiator is mounted on the top of the micromodule.

Figure 1 shows the device diagram (**left**) and the general view (**right**).

Content from this work may be used under the terms of the Creative Commons Attribution 3.0 licence. Any further distribution of this work must maintain attribution to the author(s) and the title of the work, journal citation and DOI.
Published under licence by IOP Publishing Ltd

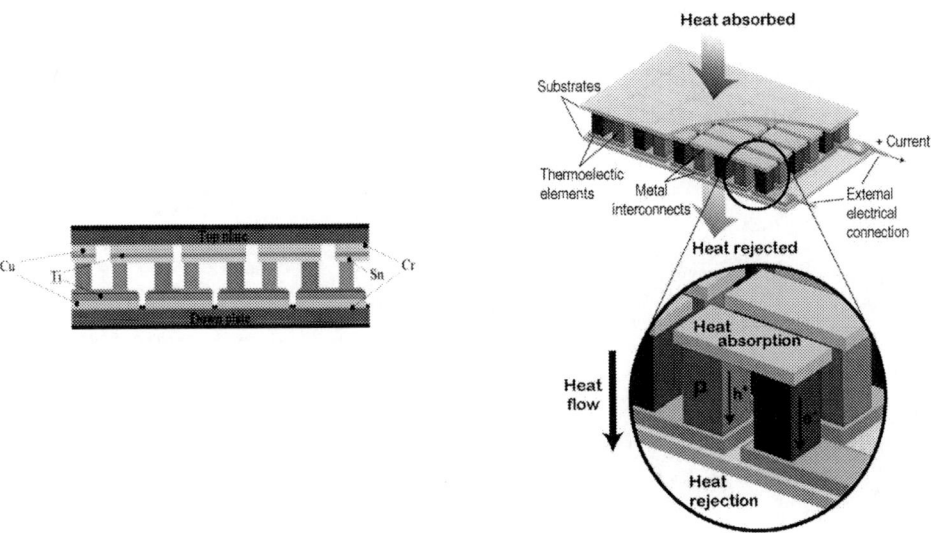

Figure 1. Thermoelectric Peltier micromodule structure [5].

3. Thermoelectric module manufacturing process

In order to make bottom substrate with thermoelements, 0.03-µm-thick Cr and 1-µm-thick Cu were sequentially sputtered on silicon wafer with dielectric oxide layer, as shown in Figure 2, a. After that, lift-off photolithography with 1-µm-thick Ti was used to form bottom electrodes, that was also used as a seed layer for further electrochemical deposition (Figure 2, b). AZ P 4620 photoresist of 28-30 µm thickness was used to form mask for thermoelectric elements deposition in developed holes (Figure 2, c). Bismuth telluride Bi_2Te_3 was deposited from nitric acid aqueous solution consisting of 1 M HNO_3, 9 mM Te, 7 mM Bi at constant current density of 5.3 mA/cm^2 until it reached 25 µm thickness (approx. 2.5 hours), as shown in Figure 2, d. After that, photoresist was removed and then applied over again to form a mask for antimony telluride deposition (Figure 2, e, f). Antimony telluride was also deposited in nitric acid aqueous solution, consisting of 1 M HNO_3, 0.1 M $C_4H_6O_6$, 7 mM Te, 12.5 mM Sb and 60 mg/l of sodium lignosulfonate at the same current density for a same deposition time (Figure 2, g). Then, mechanical polishing was done to obtain smooth and flat surface of thermoelements. Metallization etching was conducted to form bottom Ti-covered electrodes. After that, 3-µm-thick Sn bump was formed on top of each thermoelement by magnetron sputtering and lift-off photolithography for further assembly.

Top substrate was produced by using same technics of sputtering, photolithography and etching to form Ti-covered electrodes with Sn bumps on top of them. Top substrate was then flip-chip bonded to a bottom substrate with a maximum force of 5 kN for 15 minutes and cooled down under applied external pressure of 500 N.

Assembled thermomodule was then mounted on a PCB with an aluminum base using Epo-tek thermo-conductive glue. To make an electric contact, conductive wires were bonded on the electrodes of thermomodule and PCB. After that, aluminum radiator was mounted on the top of thermomodule using thermopaste to obtain good thermal contact (Figure 3).

SPBOPEN 2018   IOP Publishing

IOP Conf. Series: Journal of Physics: Conf. Series **1124** (2018) 081005   doi:10.1088/1742-6596/1124/8/081005

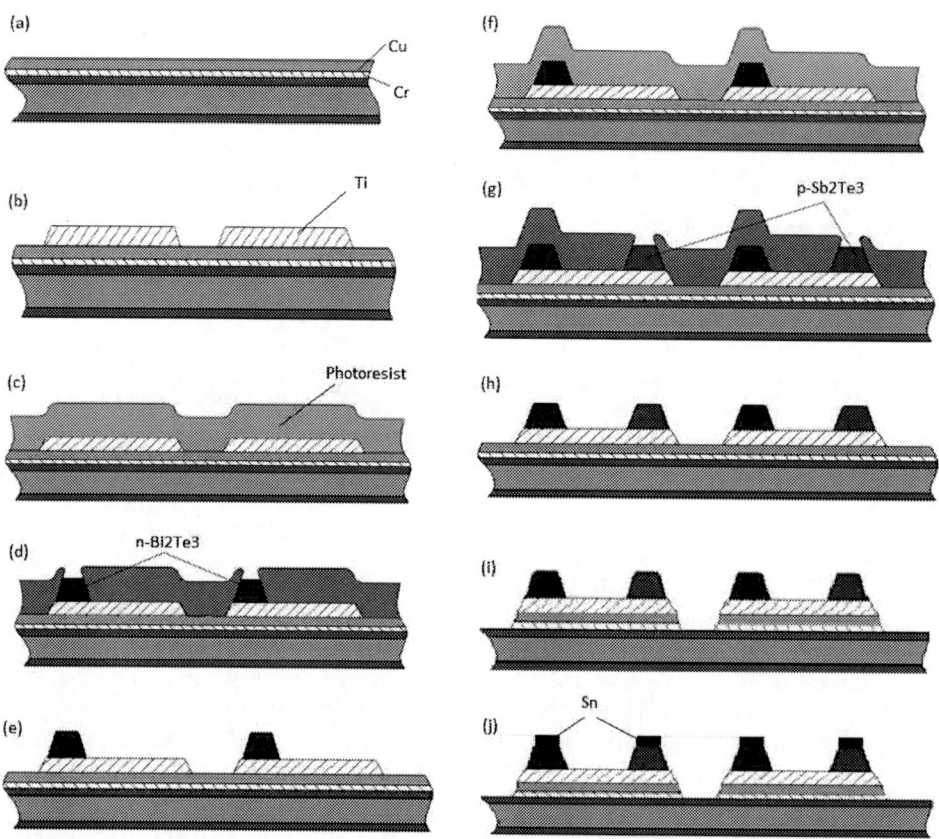

Figure 2 (a – j). Thermoelectric Peltier micromodule fabrication process: **(a)** sputtering of Cr/Cu metallization; **(b)** formation of Ti electrodes; **(c)** photoresist patterning; **(d)** electrochemical deposition of Bi_2Te_3; **(e)** photoresist removing; **(f)** photoresist patterning; **(g)** electrochemical deposition of Sb_2Te_3; **(h)** mechanical polishing and photoresist removing; **(i)** metallization etching; **(j)** Sn bumps patterning.

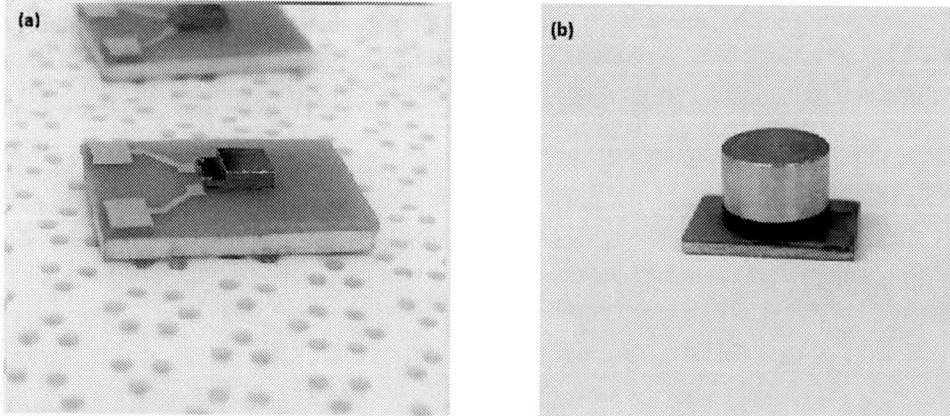

Figure 3 (a, b). Assembled thermoelectric Peltier thin-film micromodule: **(a)** mounted on a PCB without radiator; **(b)** final assembly.

1000

A key feature of the proposed technological process is the formation of semiconductor functional layers by the method of electrochemical deposition, which significantly increases the overall manufacturability of the product in comparison with bit-by-bit assembly, allows the composition of precipitated compounds to vary easily and in a wide range, changing the composition of the electrolyte, contributes to increase of thermocouple density on a unit of area, which increases the efficiency of the thermal module.

4. Results

Analysis of the images obtained with an electron microscope showed that the resulting thermoelements have a close-packed structure with crystal sizes ranging from 4 μm (bismuth telluride) to 10 μm (antimony telluride), as illustrated in Figure 4.

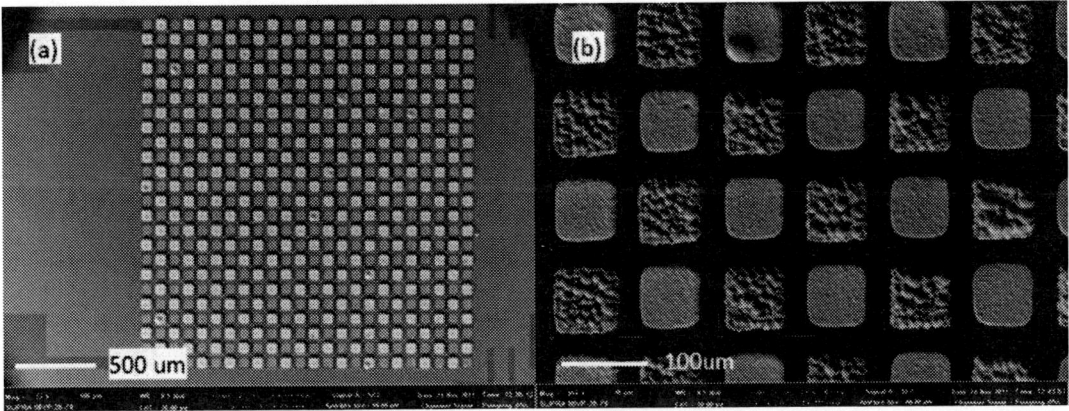

Figure 4 (a, b). Electrodeposited bismuth telluride and antimony telluride: **(a)** general view; **(b)** thermoelements prior to planarization.

Elemental composition of the compounds is close to the given stoichiometric and constitutes 39.5% Bi-60.5% Te for Bi_2Te_3 and 39% Sb-61% Te for Sb_2Te_3 (Figure 5).

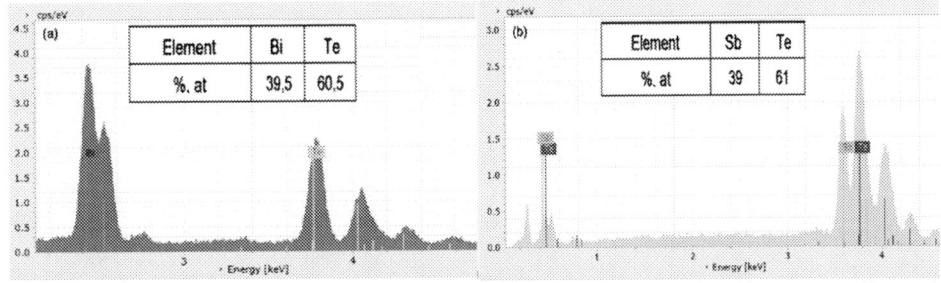

Figure 5(a, b). Ratio of components in deposited thermoelements, obtained by energy-dispersive X-ray spectroscopy: **(a)** bismuth telluride; **(b)** antimony telluride.

Testing of manufactured generator samples was carried out with a temperature differences from 25 to 100K, as it shown in Figure 6. The values of the external load ranged from 10 Ohm to10 kOhm. During the experiment, the output voltage was measured on the load; the output power and current were calculated. The voltage-current characteristics and the dependence of the output power on the current for various temperature differences of the generator sides were prepared. It has been shown that the range of power is from 3 μW to 56 μW while the temperature difference is mentioned above.

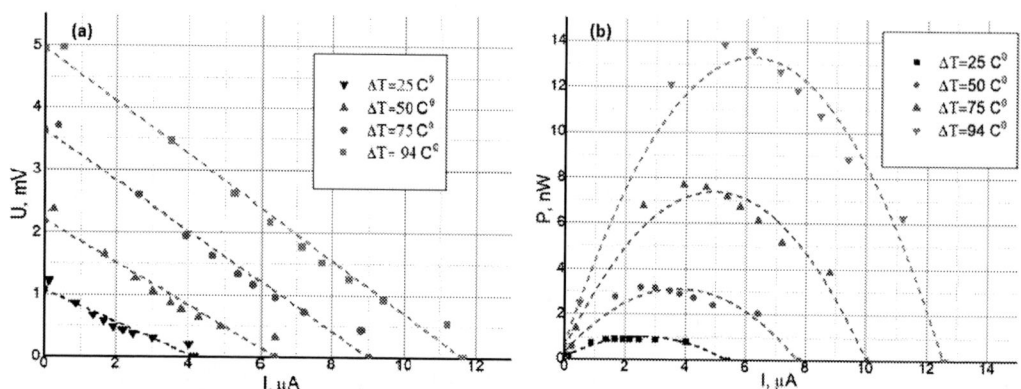

Figure 6 (a, b). Measured characteristics of thin-film Peltier micromodule: **(a)** current-voltage characteristics; **(b)** current-power characteristics.

5. Conclusion

Microminiature Peltier thermomodules, manufactured according to thin-film technology, have a number of advantages that distinguish them from analogs manufactured with the help of bit-by-bit assembly, namely: miniaturization and wide integration possibilities.

In this paper a design and a description of the manufacturing process of microminiature Peltier thermomodule using the electrochemical deposition of functional semiconductor layers were presented, the results of the surface morphology analysis and the elemental composition of the obtained compounds, measured characteristics of assembled micromodule were shown.

References

[1] Ssenoga T, Jie Z, Yuying Y, Bo L 2016 *J. Renewable and Sustainable Energy Reviews* **65** 698
[2] Kim M, Oh T 2012 *J. Materials Transactions* **12** 2160
[3] Korotkov A.S., Loboda V.V., Makarov S.B., Feldhoff A. 2017 *J. Russian Microelectronics* **2** 131–38.
[4] Volvenko S.; Dong Ge, Zavjalov S., Gruzdev A., Rashich A., Svechnikov E. 2017 *J. Progress In Electromagnetics Research Symposium - Spring (PIERS)*
[5] Snyder G, Lim J, Huang C, Fleurial J 2003 *J. Nature materials* **2** 528

SPBOPEN 2018   IOP Publishing

The influence of photo-stimulated adsorption of polyelectrolyte molecules on electro-physical characteristics of structures based on single crystal silicon substrates

A V Kozlowski[1], S V Stetsyura[1], I V Malyar[1]

[1] Department of Nano- and Biomedical Technologies, Saratov State University, Saratov 410012, Russia

Abstract. The effect of photo-stimulated adsorption of cationic (polyethylenimine, PEI) and anionic (glucose oxidase, GOx) polyelectrolyte molecules on C-V characteristics of a electrolyte/insulator/semiconductor structure was investigated. We obtained that illumination of semiconductor during the adsorption of PEI increases a shift of the flat band potential by 29% and 45% for n-Si and p-Si, respectively. Furthermore, depending on a Si conductivity type, the photo-stimulated adsorption of GOx molecules either increases (for n-Si/PEI) or decreases (for p-Si/PEI) the C-V shift, which was explained by corresponding changes of enzyme molecules on the surface, respectively. Our research is useful in fabrication of potentiometric biosensors since the growth of enzyme molecules density on the surface of the semiconductor transducer increases the threshold sensitivity of the sensor.

1. Introduction

Analysis of semiconductor capacitance characteristics (capacitance-voltage (C-V), capacitance-frequency (C-f), etc.) is a effective technique for characterization of semiconductor materials and devices. In particular, the C-V curves allow us to determine its characteristics, such as an average doping concentration, doping profiles, electrically active defect densities, carrier lifetimes, etc. In addition, this technique gives information about the interface and boundary layers. The C-V curves are based on the electron theory of the subsurface region of the space charge and differential capacitance, which also provides the opportunity for studying and analyzing of electrolyte-insulator-semiconductor (EIS) structures [1].

W. Brattein and C. Garrett showed [2] the determining role of the semiconductor (Ge) at occurrence of photo-potential and the straightening effect on the contact of Ge with the electrolyte. Since then, studies of the characteristics of the EIS-structure have been carried out quite intensively. It should be noted that contact of semiconductor with electrolyte has fundamentally different conductivity characteristics (electronic and ionic) and various aggregate states (solid and liquid). In addition, at contact metal with electrolyte, all charge carriers are localized on the metal surface. However, in semiconductors, concentration of free charge carriers is much less, which leads to the redistribution of free electrons and holes near the surface and the formation of a space-charge region (SCR). Also an important difference is that the electrolyte have only two types of charges - anions and cations, and both are present in equal concentrations. In a semiconductor, in addition to mobile charges (electrons and holes), there are also fixed charges (donors and acceptors).

Biosensors [3] and photoelectrochemical cells [4] for solar energy converters are the most promising directions for using EIS-structures. In Figure 1a it is shown the energy diagram of the EIS-structure.

Content from this work may be used under the terms of the Creative Commons Attribution 3.0 licence. Any further distribution of this work must maintain attribution to the author(s) and the title of the work, journal citation and DOI.
Published under licence by IOP Publishing Ltd

Adsorption of anions or cations onto a semiconductor surface from electrolyte solution leads to change in the band bending of semiconductor. It results in a shift of semiconductor flat band potential, which can be detected by the C-V curves.

The surface of the semiconductor transducer can be modified by various functional molecules (enzymes, DNA) to increase the selectivity, as well as by metal or carbon particles [5]. These objects can be layer-by-layer adsorbed to the surface of a semiconductor from a solution to fabricate protective or functional coatings as well as to change the properties of the substrate. When using EIS-structure as an enzyme sensor, its sensitivity and response time will depend on the concentration of enzyme molecules on the transducer surface since the enzymatic reaction rate depends on this.

Previously, we report on photo-stimulated adsorption of enzyme molecules [6] and gold nanoparticles [7] on silicon substrates modified with polyethylenimine (PEI). In addition, the efficiency of electric passivation of a single crystal Si surface by polymer coating deposited by photo-stimulated layer-by-layer adsorption were studied [8]. In this work, we studied the effect of photo-stimulated layer-by-layer adsorption of PEI and glucose oxidase (GOx) enzyme molecules on the electrical characteristics of field-effect-based capacitive EIS sensors.

2. Experimental details

The experiments were performed with single-crystal Si wafers, after chemical and mechanical polishing, of n-type ($\rho \cong 4$ Ω cm) and p-type ($\rho \cong 8$ Ω cm). Initially, the substrates were boiled in a peroxide–ammonia solution and rinsed in deionized water. This treatment leads to "reconstruction" of a native oxide layer while the silicon surface acquires negative charge in deionized water due to activation of OH-groups. Polyethylenimine (PEI) with a molecular weight of 25 kDa was used as a cationic polyelectrolyte. PEI solution was prepared in deionized water (resistivity of 18 MΩ) at a concentration of 1 mg/ml.

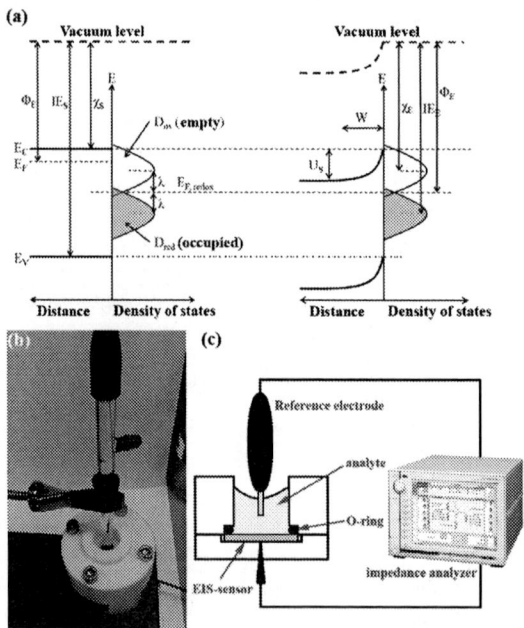

Figure 1(a, b, c). (a) - Schematic energy band diagram of the Si/SiO$_2$/electrolyte contact before (left) and after (right) equilibrium: D$_{red}$ and D$_{ox}$ - density distribution of states of the oxidation-reduction system in the electrolyte, λ - energy of reorganization, Φ, χ and IE are the work function, electron affinity, and ionization energy, respectively, of the semiconductor (index S) and electrolyte (index E); **(b)** - Electrochemical cell; **(c)** - Schematic of a capacitive EIS-sensor

The prepared Si substrates were mounted in a electrochemical cell (Figure 1b) for subsequent measurements. A 150 mM solution of NaCl was used as a electrolyte solution for subsequent electrochemical measurements. Each sample was immersed in the electrolyte solution for 30 min to equilibrate before each measurement. After 30 min, an initial C–V measurements were carried out for unmodified sample [9]. Then PEI was adsorbed from its aqueous solution in the dark or under illumination by a halogen lamp (Philips 13186 EPX/EPV). The deposition time was 10 min. Hereafter, GOx enzyme was adsorbed on the $Si/SiO_2/PEI$ structure surface from its 0.5 mg/ml aqueous solution. The electrical characteristics were measured using a probe station (Cascade Microtech) controlled by a semiconductor device analyzer (Agilent B1500A). The C-V curves were measured at frequency of 1 kHz and bias voltage of 20 mV. All measurements were performed relative to the reference electrode. An atomic-force microscope (AFM) (NTEGRA Spectra, NT-MDT Spectrum Instruments) was used to characterize the topography of the samples. Scanning was performed in a tapping mode using HA_NC/W_2C cantilevers, ETALON series.

3. Result and discussions

Figure 2(a, b) illustrates C–V curves for p-Si and n-Si substrates which were measured before and after PEI deposition. C–V curves obviously demonstrate a shift along the U axis towards negative values after PEI deposition which does not depend on the Si conductivity type. According to [9, 10], the shift of the flat-band potential in C-V measurement correlates with the number of adsorbed PEI molecules. In the case of p-Si, the photo-stimulated deposition of PEI increases a shift of the flat band potential (ΔU) by 45 % compare to PEI deposition in the dark. While for n-Si, the flat band potential shifts only by 29% compare to PEI deposition in the dark. Adsorption of the subsequent layer of anionic polyelectrolyte (GOx enzyme molecules) onto p-Si/PEI and n-Si/PEI structures leads to shift of C-V curves along the U axis towards positive values. For p-Si/PEI, the photo-stimulated deposition of GOx decreases a shift of the flat band potential by 21 % compared to GOx deposition in the dark. While for n-Si/PEI, the shift of the flat band potential increases by 124% compare to GOx deposition in the dark. This results correlates with measured AFM scans (Figure 3).

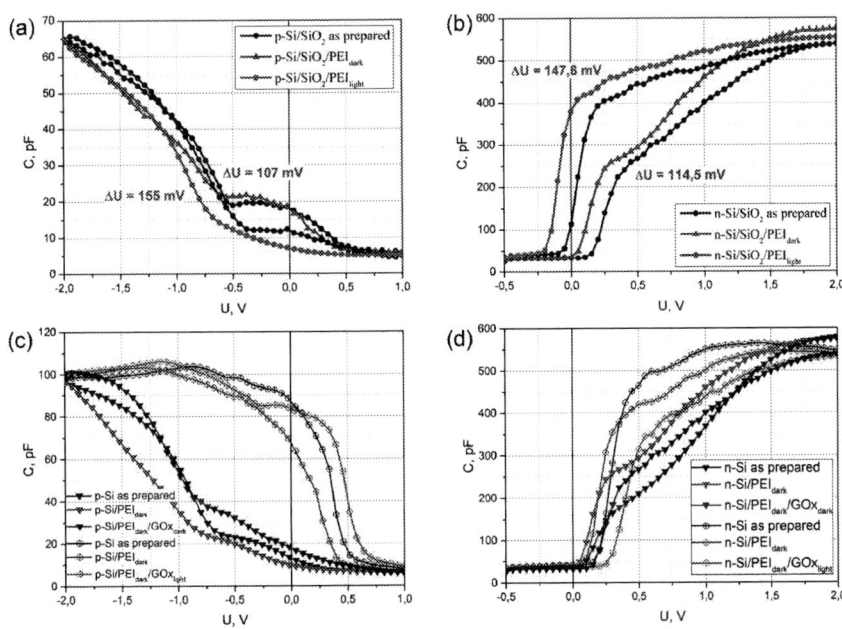

Figure 2(a, b, c, d). C–V curves at a frequency of 1 kHz for EIS structures based on p-Si (a, c) and n-Si (b, d) with PEI and GOx layers adsorbed in the dark and under illumination

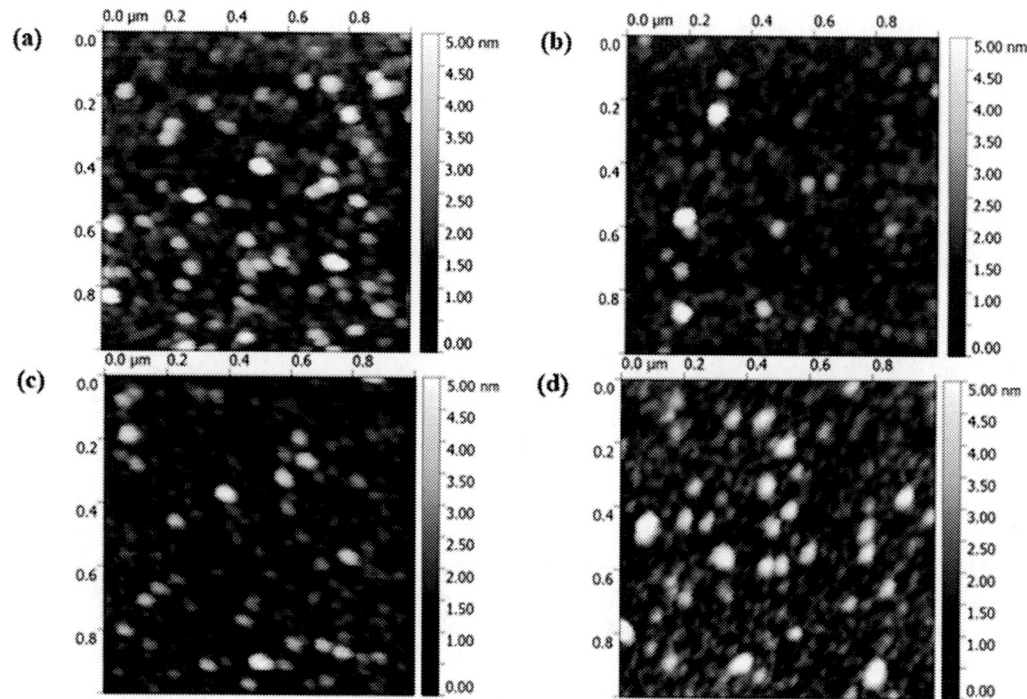

Figure 3(a, b, c, d). AFM-image of substrates after GOx adsorption onto p-Si/SiO₂/PEI (a, b) and n-Si/SiO₂/PEI (c, d) in the dark (a, c) and under illumination (b, d)

This can be explained by changes (a decrease in the first case and an increase in the second one) of adsorbed GOx molecule number as a result of charge exchange of "fast" states at the Si/SiO₂ interface and "slow" states at the SiO₂/PEI interface. It leads either to a decrease for p-Si or an increase for n-Si of electrostatic interaction between GOx molecules and the surface of the Si/SiO₂/PEI structure.

4. Conclusion

Thus, it was shown that the white-light illumination of Si substrate during adsorption of PEI molecules from the aqueous solution increases a shift of the flat band potential by 29-45 % compare to adsorption in the dark. Furthermore, depending on a silicon conductivity type (n-Si or p-Si), the amount of photo-assisted adsorbed enzyme molecules either increases (n-Si/PEI) or decreases (p-Si/PEI) under illumination. Our research is useful in fabrication of potentiometric biosensors since the growth of enzyme molecules density on the surface of the semiconductor transducer increases both the threshold sensitivity and selectivity of the sensor.

Acknowledgments

The work was supported by the Russian Foundation for Basic Research, projects no. 16-08-00524.

References

[1] Shashkin V I, Karetnikova I R, Murel A V, Nefedov I M, Shereshevskii I A 1997 *Semiconductors* **31** 789

[2] Brattain W H, Garrett C G B 1955 *Bell Syst. Tech. J.* **34** 129

[3] Poghossian A, Weil M, Cherstvy A G, Schöning M J 2013 *Anal. Bioanal. Chem.* **405** 6425

[4] Hellman A, Wang B 2017 *Inorganics* **5** 37

[5] Poghossian A, Bäcker M, Mayer D, Schöning M 2015 *Nanoscale* **7** 1023

[6] Stetsyura S, Kozlowski A 2017 *Tech. Phys. Lett.* **43** 285

[7] Malyar I, Gorin D, Santer S and Stetsyura S 2017 *Appl. Phys. Lett.* **110** 133104

[8] Stetsyura S, Kozlowski A, Malyar I 2015 *Tech. Phys. Lett.* **41** 168

[9] Poghossian A, Abouzara M, Amberger F, Mayer D, Han Y, Ingebrandt S, Offenhäusser A, Schöning M 2007 *Biosensors and Bioelectronics* **22** 2100

[10] Garyfallou G, de Smet L, Sudhölter E 2012 *Sensors and Actuators B***168** 207

SPBOPEN 2018   IOP Publishing

Electromechanical bending microactuator as optical shutter

R. Kleimanov, Y. Enns, E. Pyatishev, I. Komarevtsev

Peter the Great Saint-Petersburg Polytechnic University, St. Petersburg, Russia

Abstract. This paper presents the design, fabrication methods and characterization of a thin film electromechanical optical shutter. The overlap of the optical channel 1.4 mm high is provided by a multilayer film structure of the cantilever, which profile is determined by the residual internal stresses of the films. In the process of design analysis, the actual parameters of the strained structural layers, the temperature dependence of the curvature, the undamped natural frequencies and the damping parameters of the film cantilever were determined. Correlation of the data shows that the experimental results fit the theoretical data sufficiently closely. The developed control circuit is designed to compensate for the induced dielectric polarization and prevent the control voltage rise and the cantilever sticking effect. The design presented in this publication enables development of a simple and effective electromechanical optical shutter having interruption aperture height which exceeds 1.4mm.

1. Introduction

The development of microsystem technology leads to a stable miniaturization of sensors and actuators. Thus, with the development of broadband optical networks, micromechanical optical switches and optical switching devices become a frequent use. The vast majority of such like devices operate in-plane [1]. An optical switch represents an optical waveguide arrangement and a micromirror, driven by a micromechanical actuator, switching optical radiation. As the micromirror itself and its displacement are small it causes the search of particular approaches to the formation of a microlens collimating device. Therewith, the fabrication of such microactuators involves expensive processing procedures. At the same time, micromechanical optical switches can be fabricated using surface technology involving basic procedures of microelectronics and optical systems. Suchlike out-of-plane thin film actuators are mostly applicable to HF systems in the form of "zipping" capacitors [2]. The development of suchlike devices in the quality of optical switches requires a particular approach by reason of the necessity of achieving significant out-of-plane displacements of the switching element. Thus, an optical micromechanical shutter should guarantee the overlapping of an optical channel of more than 1-2 mm diameter. The beam diameter is determined by the size of the output section of the collimator lenses, designed to minimize the signal loss. This paper describes the design and fabrication of an out-of-plane active electromechanical element of an optical MEMS shutter of an optical channel with 1.2 mm diameter. The optical signal diameter determines the target displacement of the drive and the overlap area.

2. Development

Using of the bimetallic thin films allows fabricating structures with the high aspect ratio. The formation of such structures is based on internal stresses initiation of film layers σ_i during their deposition. The total internal stresses of the multilayer cantilever σ_s are determined by the values of the internal mechanical stresses of the layers, as well as by their thickness t_i[3]:

Content from this work may be used under the terms of the Creative Commons Attribution 3.0 licence. Any further distribution of this work must maintain attribution to the author(s) and the title of the work, journal citation and DOI.
Published under licence by IOP Publishing Ltd

$$\sigma_s = \frac{\sum_{i=1}^{n} \sigma_i t_i}{\sum_{i=1}^{n} t_i} \tag{1}$$

After the multilayer cantilever beam is released from the substrate by removing the sacrificial layer, axial internal stresses cause the occurrence of the deflection moment M_b, proportional to the total value of internal stresses and to the formation of a structure with curvature K:

$$K = \frac{M_b}{\sum_{i=1}^{n} E_i I_i} = \frac{1}{R} \tag{2}$$

Herewith, the deviation value for the cantilever is determined by the internal stresses and stiffness properties of the structural layers (Young modulus E_i) of Cr and Cu layers, as well as by the geometry of the movable element (inertia values I_i of the corresponding structural layers). Suchlike structure allows interrupting the optical signal with the required cross sectional area.

3. Design
Optical micromechanical shutter includes an optical signal transmission system and an electromechanical shutter system (Figure 1). The beam light is formed by collimating lenses and is interrupted by the microactuator. The actuator structure comprises thin film being the movable electrode, which together with the silicon substrate, forms the drive capacitor. Capacitor electrodes are isolated by the dielectric SiO_2 layer.

The cantilever structure represents a multilayer combination of structural Cr and Cu layers with different thicknesses (Figure 2). The actuator profile is determined by the top chrome layer, having high stiffness, and significant internal stresses. The bottom chrome layer is compensation layer. The aimed positioning of the cantilever neutral surface can be obtained by the layers thicknesses adjustment y_b:

$$y_b = \frac{\sum_{i=1}^{n} E_i t_i (y_i + y_{i-1})}{2 \sum_{i=1}^{n} E_i t_i}. \tag{3}$$

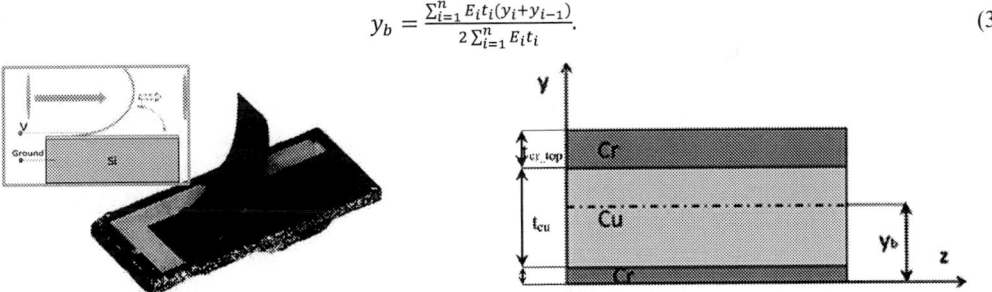

Figure 1. MEMS optical shutter and its schematic structure

Figure 2. Cantilever cross-section at arbitrary x-location

Herewith, the cantilever strain ε_b is caused by the deformation of the structural layers ε_i due to the internal stresses, and can be defined as follows [3]:

$$\varepsilon_b = \frac{\sum_{i=1}^{n} E_i t_i \eta_i \varepsilon_i}{\sum_{i=1}^{n} E_i t_i}, \tag{4}$$

where $\eta_i = 1 + \nu$ is for high strain values.

The strain forces the film cantilever to deflect. The curvature K value and its direction (sign) depend on the deflection of the neutral surface. Cantilever curvature value can be determined proceeding from the following (1)-(4) [3]:

$$K = \frac{3 \sum_{i=1}^{n} E_i t_i (y_i + y_{i-1} - 2 y_b)(\varepsilon_b - \eta_i \varepsilon_i)}{2 \sum_{i=1}^{n} E_i t_i [y_i^2 + y_i y_{i-1} + y_{i-1}^2 - 3 y_b (y_i + y_{i-1} - y_b)]} \tag{5}$$

In response to a potential difference applied between the shutter electrodes, an electrostatic load is generated, opposite in sign to the initial deflection moment. The load increase results in cantilever straightening and its fixation on the dielectric layer. To compensate for the induced polarization of the dielectric, the control circuit has been designed, which makes possible to minimize it.

4. Fabrication

The micromechanical actuator is fabricated on the silicon substrate. Thin film cantilever structure is arranged using magnetron sputter deposition of Cr/Cu/Cr layers on the aluminium sacrificial layer. Thin film cantilever is separated from the silicon substrate by the dielectric layer of thermal SiO_2. Layer thicknesses of the SiO_2 and sacrificial aluminium define the interelectrode gap, and are 0.8 μm each. The thickness of the multilayer thin film structure approximately equals 1.6 μm, where the thicknesses of the cooper layer and top chrome layer equal 1 μm and 0.5 μm, respectively. To reinforce the most mechanically loaded section of the drive, the copper layer is galvanically applied to the cantilever embedding section in the preliminary etched cavity Cr.

The occurrence of the internal stresses is caused by material structure and composition [4]. Moreover, the deposition process itself is prominent in the formation of internal stresses, which results in defect formation in the thin film as well as the initial substrate morphology. The contribution of such stresses is often greater in magnitude and may differ in sign from the temperature stresses. Herewith, the stresses caused by the difference between the thermal expansion coefficients may result in the cantilever deviation from the required shape. However, the bimorph thermal effect can be minimized by applying layers with different thicknesses [4, 5]. It is a challenging task to define the values of the internal stresses, considering all factors involved, at the same time the study of the film structure allow determining film parameters and optimizing the design.

5. Results

The actuator analysis allows determining both the mechanical properties of deposited structural layers and the internal stresses components. Elastic characteristics of structural layers may deviate from the table values due to the structural features of the layers applied by the magnetron sputter deposition. Thus, the Young modulus value, defined using the Hysitron nanoindenter, for Cu and Cr structural layers deviates from the table values and equals 60 GPa and 184 GPa, respectively. The stiffness reduction of metal films is attributed to the deposition modes, which lead to the defect formation in the films. Herewith, the results are confirmed by static measurements of the curvature profile of the film cantilever (Figure 3). The curvature radius of the multilayer cantilever under normal conditions is 1.38 mm. Furthermore, the height of the overlap area of the beam light is 2.5 mm. The internal stresses value for Cu and Cr for magnetron sputter deposition for given thicknesses of structural layers can range within 40-60 MPa and 400-600 MPa, respectively [4]. The internal stresses value in the formed multilayer cantilever can be estimated from the ratio of the theoretical values of the internal stresses of Cu and Cr for the given curvature of the cantilever (Figure 4).

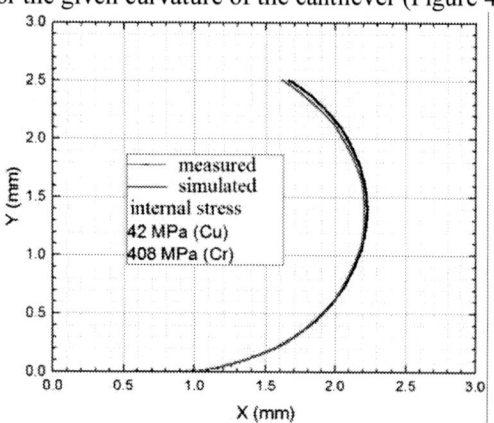

Figure 3. Profile of the film cantilever 3.5 mm in length; R=1.38 mm

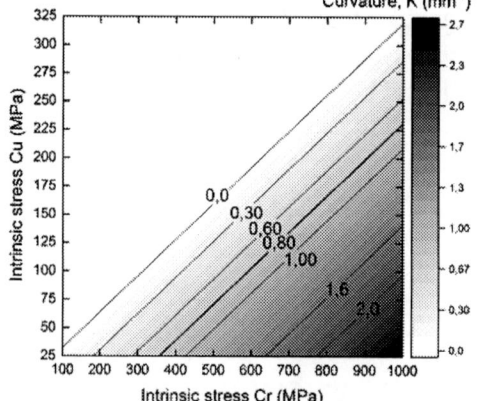

Figure 4. Curvature factor as a function of intrinsic stress ratio of chromium and cooper layers.

The estimation of the resonance frequencies, amplitude-frequency response and circuit phase response allows estimating the volumetric properties of the movable element (mass characteristics) and attenuation parameters. To provide dynamic measurements, the actuator is mounted on piezoelectric vibration drive. The oscillation excitation of the piezoelectric element is provided by alternating voltage in the frequency range from 20 to 1000 Hz. Herewith, the cantilever resonance frequencies were defined using the Laser Doppler Vibrometer (LDV) Polytec close to the free end of the cantilever. The oscillation excitation through external vibration action of small amplitude allows excluding the impact of mechanical "hardening" and electrical "softening" nonlinearities. The absence of nonlinearities in the amplitude-frequency dependence of the film cantilever allows for theoretical estimation of the resonance frequency value, ignoring its curvature [6]. Thus, in the studied frequency range, three resonance peaks were recorded (Figure 5), with frequencies within the ranges of 104-137 Hz, 400-476 Hz, 675-760 Hz, respectively (resonance frequencies scattering is associated with heterogeneous properties of films on samples, positioned in different parts of the substrate). The tests were performed at atmospheric pressure, i.e. under conditions, when the mechanical system has high energy dissipation level. The mechanical quality of the cantilever under given conditions was 19-21, 46-48 and 20-25 for the first, second and third oscillation modes, respectively. Herewith, the experimental results correlate well to the FEM model.

The relation between profile curvature and temperature fluctuation allows estimating the internal stresses arising due to thermal expansion, and undesirable deflections of the cantilever within the operating temperature range. To ensure the efficiency of the optical shutter within the required temperature range, the overlap area should not decrease below 1.4 mm. Measurements of the film cantilever profile were performed using an optical microscope with ambient temperature control within the range of 1-60 °C. The film cantilever curvature variates within 0.57-1.08 mm-1 for this temperature range (Figure 6). This corresponds to beam overlap variation of 1.86-3.35 mm, which is beyond the required deviations.

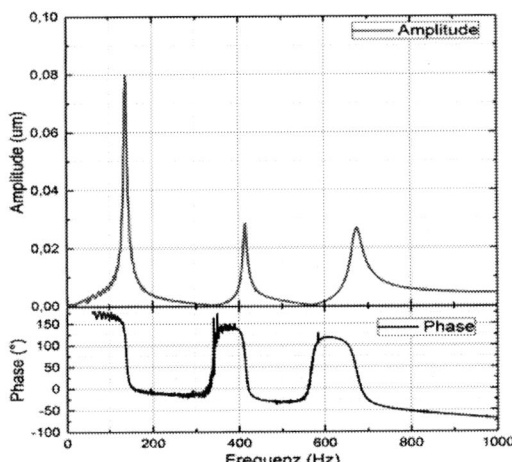
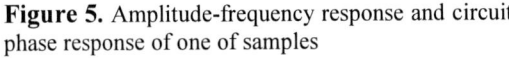
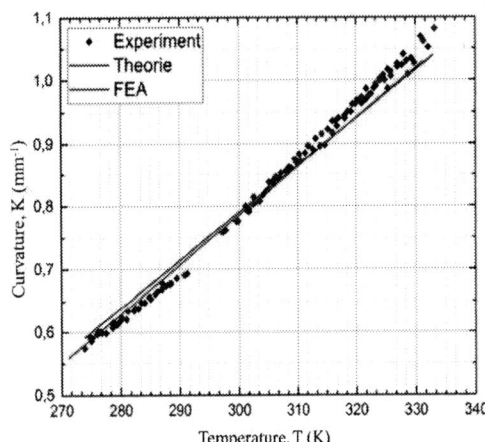

Figure 5. Amplitude-frequency response and circuit phase response of one of samples

Figure 6. Cantilever curvature as a function of temperature

The opening of the optical shutter is enabled by applying constant voltage difference to the control electrodes. The required control voltage value is in the range from 60 to 100 V. Thereat, the impact of the induced polarization of the dielectric results in a number of undesirable effects. The increase in the voltage, required for opening, is observed, during the cyclic supply of control signals, and sticking of the cantilever. To suppress the undesirable effects associated with the polarization of the separating dielectric, the circuit for supplying the control voltage as packets of alternating voltage was proposed, designed to depolarize the dielectric. The suggested circuit allows reducing the impact of the induced polarization of the dielectric, thereby, eliminating the increase in the control voltage and the sticking effect.

The experimental static and dynamic characteristics allowed defining the parameters of structural layers, which provided the validation of the theoretical model of the actuator. To validate and optimize the parameters of the device, finite element modeling was implemented.

6. Conclusions

In the process of this research the multilayer film structure of the optical shutter was created, providing to the cantilever the ability to move up to 3.35 mm apart the substrate, which allows overlapping the optical channel entirely. The actual elastic parameters of the structural layers were defined. During the dynamic tests, the natural frequencies and attenuation parameters of the film cantilever were defined. The design parameters of the multilayer structure allow overlapping the optical signal at 1.86-3.35 mm above the substrate surface, which far exceeds the required minimum 1.4 mm.

Correlation of the data shows that the experimental results fit the theoretical data sufficiently closely. The combination of tests performed under the effect of static, dynamic, thermal and acoustic loads, allow defining the mechanical constants of the obtained layers and the mechanical properties of the cantilever. The control circuit was developed to compensate for the induced dielectric polarization, and prevent the control voltage rise and the cantilever sticking effect.

References

[1]   Tsai Ch., Tsai J., 2015, *Displays* **37** 33
[2]   Pu S. H., Darbyshire D. A., Wright R. V., Kirby P. B., Rotaru M. D., Holmes A. S., and Yeatman E. M., 2016, *IEEE Electron Dev. Lett.*, **37(10)** 1340
[3]   Nikishkov G. P., 2003, *J. Appl. Phys.*, **94(8)** 5333
[4]   Misra A., Kung H., Mitchell T.E., Nastasi M., 2000, *Journal of Materials Research,* **15** 756
[5]   Dong K., Lou S., Choe H. S., Liu K., You Z., Yao J., Wu J., 2016, *App. Phys. Let.* Vol. **109**
[6]   Rua A., Cabrera R., Coy H., Merced E., Sepulveda N., 2012, *J. of App. Phys.* **(111)**

SPBOPEN 2018   IOP Publishing

IOP Conf. Series: Journal of Physics: Conf. Series **1124** (2018) 081008   doi:10.1088/1742-6596/1124/8/081008

Application of XRD methods for the pilot studies of new functional materials for photonics

M. Dermeneva[1], D. Muravijova[1,2], M. Mynbaeva[2,], V. Bougrov[1], M. Yagovkina[2]

[1] ITMO University, Saint-Petersburg 197101, Russia
[2] Ioffe Institute, Saint-Petersburg 194021, Russia

Abstract. We report on the detailed study of the structural properties of the large-area GaN slabs grown by HVPE method over ceramic support with the use of X-ray analysis with powder diffraction technique. The impact of V/III ratio on the specifics of the crystal structure of the material was studied. It was shown, that depending of growth condition either texture along (00.2) axis, i.e., along c axis in polar GaN, or along (11.0) axis, i.e., normal to so-called m-plane GaN are formed.

1. Introduction

III-nitride group materials are employed in short-wavelength light-emitting diodes, lasers and optical detectors, as well as for high temperature, high power and high frequency semiconductor devices. Despite the fact that monocrystalline III-nitride compounds remain the most hot topic, the last decade has seen progressively developed polycrystalline nitride materials, which are easier and less expensive to fabricate. Among different group-III nitrides, GaN has shown great potential in the above applications, yet very little has been reported on polycrystalline GaN so far [1-6]. We have already reported on large-area crystalline gallium nitride slabs produced with hydride vapor-phase epitaxy (HVPE)-like deposition directly on a ceramic support, without using traditional substrates made of sapphire (Al_2O_3), silicon (Si) or silicon carbide (SiC). It was found that the obtained material was n-type semiconductor with optical, thermal, and mechanical properties close to those of single-crystalline GaN produced with traditional growth techniques [7]. At the same time, preliminary studies indicated that the nucleation process for our material strongly differed from that typical of HVPE of GaN layers on conventional substrates. In particular, for our material, a liquid phase formed at the initial stage of the growth process, and this resulted in self-oriented gallium nitride growth with an ordered block structure. In this work, we report on the detailed study of the structural properties of the large-area GaN slabs with the use of X-ray analysis with powder diffraction technique. This is the basic approach in the studies of polycrystalline materials, including textured ones, where «texture» (or preferred orientation) is termed for the samples whose grains are aligned in particular orientations. The preferred orientation has a dominant effect on the intensities of diffraction peaks in the XRD patterns. In this work, we used X-ray diffraction analysis for the study of the material of large-area GaN slabs, and in particular, the impact of V/III ratio during the growth on the specifics of the crystal structure of the material.

Content from this work may be used under the terms of the Creative Commons Attribution 3.0 licence. Any further distribution of this work must maintain attribution to the author(s) and the title of the work, journal citation and DOI.
Published under licence by IOP Publishing Ltd

2. Experimental details

A set of samples with thicknesses 200–1500 μm was prepared. The samples were cut from GaN slabs which were grown under different conditions (different V/III ratio) and represented monolithic plates with a mechanically polished top surface (samples ## 1, 3 and 4). Parts of the samples adjacent to the ceramic substrate (back sides) were not polished, but chemically treated after the processing. Specimen #2 with thickness 200 μm was studied as-grown.

X-ray diffraction (XRD) analysis was performed with the use of Bruker D2 Phaser diffractometer (CuKα line was used with the wavelength 0.1540518 nm). The analysis of the diffraction patterns was carried out with DIFFRAC.EVA (Bruker Corp.) software using data from Powder Diffraction File ICCD PDF-2 release [JCPDS-International Centre for Diffraction Data (http://www.icdd.com)].

3. Results

Figure 1 shows optical images of 1500μm-thick samples. The images were obtained with an optical scanner in a transmitted-light mode. It can be seen that the samples are transparent and have a yellowish tinge. For the material grown under higher V/III ratio (shown in Fig. 1 (b)), one can note a peculiarity of the GaN growth mechanism at the stage of nucleation. In particular, nucleation islands are clearly visible at the back side of the plate in the transmitted light.

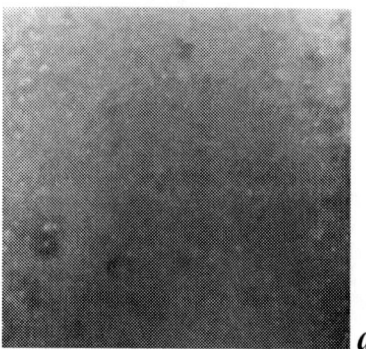 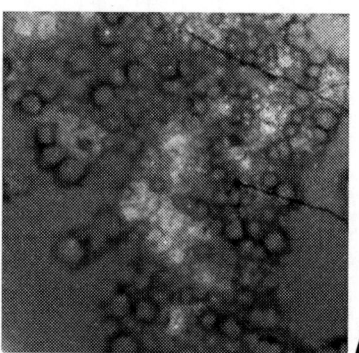

a *b*

Figure 1. Images of two samples obtained in a transmitted-light mode: *a*, sample #3; *b*, sample #4. The area shown in the images is approx. 1 cm². Higher saturation of Ga (III) was used at the initial stage of HVPE process in the case of the sample shown in image *b*, where islands at the back side of the sample are visible.

Figures 2 and 3 show diffraction patterns from the investigated samples, which were obtained for both their sides. Black curves correspond to the top surface of the samples, red curves correspond to their back sides. By blue color lines with corresponding Miller indexes a diffractogramm, corresponded to the hexagonal GaN, are presented. Peaks corresponded to additional phase are marked by "Met" for metallic gallium and "Cub" for to the cubic phase of GaN. The «β» symbol shows the position of CuKβ peaks, which correspond to (00.2) and (11.0) reflexes. These patterns correspond to diffraction from polycrystalline material. In particular, diffraction patterns from all the samples show a set of diffraction maxima with various intensities in the range of angles from 20 to 100 degrees at the 2θ scale (for clarity, a smaller angle range is shown in both figures). Table 1 shows sizes of coherent-scattering regions (CSR), which were calculated both for the average size of the crystallites and for the preferred directions depending on the type of texture.

X-ray phase analysis allowed for unambiguous identification of all the maxima in the diffraction patterns of the top surface of sample #1. Comparison of experimental data with line diagrams acquired with X-ray phase analysis showed that the position of all peaks corresponded to the table data for hexagonal GaN (PDF file #01-073-7289). It is known that the most strong (10.1) maximum for polycrystalline non-textured GaN has 2θ of 36.8 degrees. At the same time, for this sample the strongest peak, whose intensity exceeded those of other peaks by almost two orders of magnitude, was observed at

57.7 degrees. Let us note that for this maximum, there is a corresponding CuK$_\beta$ peak at the diffraction pattern. Such incompatibility of the intensity of the observed maximum with the table data can be explained by the presence of crystallites preferably oriented along (11.0) axis and their large dimension in this direction. As is shown in Table 1, for this sample the size of CSR along (11.0) axis constitutes 50 nm with averaged size of CSR calculated for all maxima being 35 nm. Diffraction pattern from the back side of sample #1 was almost identical to that from its top surface. The former also demonstrated similar, in terms of position and intensity, set of maxima, which corresponded to hexagonal GaN textured along (11.0) axis, with CSR size of ~120 nm. Weak maxima observed at 2θ of 33, 40 and 47 degrees may be related to metallic Ga (PDF file #00-005-0601) and cubic GaN (PDF file #00-052-0791).

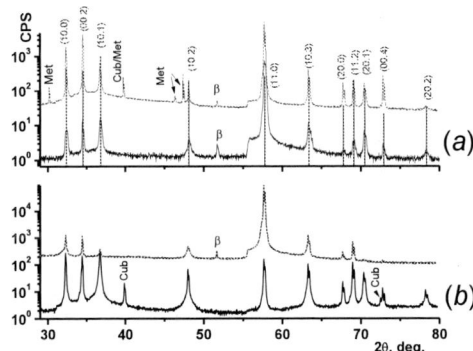

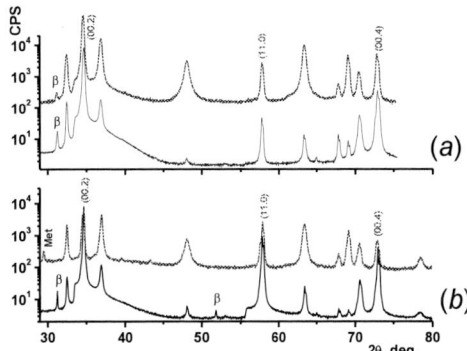

Figure 2. Diffraction patterns: (*a*), sample #1, (b), sample #2.

Figure 3. Diffraction patterns: (*a*), sample #3, (*b*), sample #4.

Table 1 Coherent-scattering regions in the studied samples

Sample #	Investigated area	Sample thickness, μm	SCR size, nm		
			The average size of the crystallites	The size of crystallites along (00.2) axis	The size of crystallites along (11.0) axis
1	top surface	1000	35	-	50
	back side		120	-	120
2	top surface	200	80	-	80
	back side		70	-	120
3	top surface	1500	45	45	54
	back side		35	36	36
4	top surface	1500	50	100	63
	back side		35	38	47

Diffraction pattern from the back side of sample #2 (which was thinner than sample #1) shows only maxima typical of hexagonal GaN with clearly expressed texture along (11.0) axis and CSR size of 70-120 nm. The pattern obtained from the top surface of this sample also agrees well with that typical of hexagonal GaN. The texture was expressed somewhat weaker, with CSR size being 80 nm. As follows from the data in figure 2(*b*), the main feature of this pattern is the presence of cubic GaN, which is expressed in extra maxima with high intensity at 2θ of 40, 85 and 97 degrees (PDF file #00-052-0791).

The results of X-ray diffraction analysis of samples ## 3 and 4 showed that these samples represented hexagonal GaN. The main difference between these samples related to the size of crystallites (see table 1). The position of the most strong peak at 2θ of 34 degrees indicates preferable orientation of the crystallites along (00.2) axis. At the same time, some texture along (11.0) axis is also present. This pattern is most clearly expressed for the top surface of sample #4, where the size of the crystallites along these two axes exceeds the average size of the crystallites: CSR along (00.2) axis is 100 nm, and along (11.0) axis, 63 nm, while the average CSR is only 50 nm. As a result of this, the strongest peaks at the diffraction pattern correspond to diffraction from (00.2) (2θ=34.56), (11.0) (2θ=57.76) and (00.4) (2θ=72.90) planes. Note that for the first two maxima at the diffraction pattern the corresponding Cu K_β peaks are also present.

Thus, X-ray diffraction and X-ray phase analysis of the studied samples showed that the material under study was in fact hexagonal GaN with a certain texture. Samples # 1and #4 grown under higher III/V ratio with thickness larger than 1 mm, had texture along (11.0) axis or a mixed-type texture. These samples also contained some metallic Ga. Samples obtained at low III/V ratio (lower Ga saturation) had the majority of grains oriented along (00.2) axis (sample #3) All the samples demonstrated increase in texture grain size towards the surface. In the thin sample #2 that contained both hexagonal and cubic GaN phase the texture was pronounced to the least degree.

4. Discussion

It was mentioned in our previous work [7] that the nucleation process of the material under investigation strongly differed from that typical of HVPE of GaN on conventional substrates; specifically, the process considered in our work involved a liquid phase formed at the initial stage of the growth. From the one hand, this favors free detachment of GaN slabs from the surface of the ceramic support, but from the other hand, it affects the specifics of self-arranged orientation in the resulting textured crystal, as follows from the results of the present study. Thus, one may suggest that self-arranged orientation of texture structure reported here is favored by Ga/N ratio at the initial stage of nucleation of GaN. In the presence of excessive gallium, the preferred orientation is (11.0); texture (00.2) is formed under smaller Ga/N ratios. This is illustrated by the images in figure 1. There, one can clearly see the difference in morphology of back sides of samples #3 and #4. Figure 1(b) shows island-like morphology of the back side of sample #4, indicating that eventually excessive Ga droplet formation occurred. In this case, a mixed (11.0) and (00.2) orientation texture is formed. In contrast to this, sample #3 with a uniform (00.2) texture has a uniform planar back side.

The above made conclusion is supported by the existing knowledge on the kinetics of Ga adsorption on various types of substrates, as well as on the influence of Ga/N flux ratio on the growth mechanism, and thus, on the crystalline quality and chemical composition of GaN [8-11]. In particular, it was shown that under certain conditions, at high temperatures, Ga may adsorb layer-by-layer at the growth front up to continuous film, which may evolve into metallic Ga droplets. Spontaneous nucleation and growth of self-oriented GaN layers or textures on top of the molten gallium may occur if the gallium is exposed to nitrogen. For the case of GaN (00.1), the optimum growth conditions are related to the formation of just a metallic Ga bilayer at the growth front. In contrast to this, GaN (11.0) growth is provided by excesses of Ga adlayer (or droplets), which shifts the growth conditions towards Ga-rich ones.

Concerning the observation of cubic phase, we should note that in our previous work we performed the analysis of defect structure in gallium nitride crystals with the use of high-resolution transmission electron microscopy (HRTEM). These studies demonstrated that the GaN material with an ordered micro-block structure of hexagonal phase contained blocks with a different degree of mis-orientation, up to 7° at the earlier stage of the growth. The predominant types of defects observed in the structure of conjugated mis-oriented blocks were multiple stacking faults (SFs) in the (00.1) plane with displacement vector (00.1) 1/3[1−100], and it could be considered as a fragment of cubic-in-hexagonal structure. Such type of defect is known to be metastable, and is formed by phase transformation provided by interfacial mismatch strain relaxation in GaN lattice.

5. Conclusion

In conclusion, X-ray diffraction analysis, which allowed for large-area scanning, enabled us to study the structure of textured GaN samples as depending on the growth conditions. The impact of V/III ratio on the specifics of the crystal structure of the material was established. In particular, it was shown that under different values of Ga saturation (Ga rich condition), evolution of GaN structure resulted either in formation of texture along (00.2) axis, i.e., along c axis in polar GaN, or along (11.0) axis, *i.e.*, normal to the so-called *m*-plane in GaN. The later observation seems to be quite important, since it is known that building GaN-based devices along the polar c-axis results in charge separation, spontaneous polarization, and degraded device performance. Building devices on the *m*-plane, a non-polar one, where there is no net polarization, radiative efficiencies should be higher, and no wavelength shift occurs, is much more preferable.

Acknowledgement

The authors are grateful to M.V. Baidakova for fruitful discussions. The XRD study has been carried out at the Joint Research Center 'Materials science and characterization in advanced technology' with financial support from the Ministry of Education and Science of the Russian Federation " (id RFMEFI62114X0007).

References

[1] Asahi H, Iwata K, Tampo H, Kuroiwa R, Hiroki M, Asami K, Nakamura S and Gonda S 1999 *J. Cryst. Growth* **201/202** 371

[2] Bour D P, Nickel N M, Van De Walle C G, Kneissl M S, Krusor B S, Mei P and Johnson N M 2000 *Appl. Phys. Lett.* **76** 2182

[3] Yamada K, Asahi H, Tampo H, Imanishi Y, Ohnishi K and Asami K 2001 *Appl. Phys. Lett.* **78** 2849

[4] Chang S H, Fang Y K, Ting S F, Lin C Y, Chen S F, Kuan H and Liang C Y 2006 *J. Electron. Mater.* **35** 1837

[5] Ariff A, Zainal N and Hassan Z 2016 *Superlattice Microstruct.* **97** 193

[6] Hsiao F M, Schnedler M, Portz V, Huang Y C, Huang B C, Shih M C, Chang C W, Tu L W, Eisele H, Dunin-Borkowski R E, Ebert Ph and Chiu Y P 2017 *J. Appl. Phys.* **121** 015701

[7] Mynbaeva M G, Pechnikov A I, Sitnikova A A, Kirilenko D A, Lavrent'ev A A, Ivanova E V and Nikolaev V I 2015 *Tech. Phys. Lett.* **41** 246

[8] Li H, Chandrasekaran H, Sunkara M K, Collazo R, Sitar Z, Stukowski M and Rajan K 2004 *MRS Symposium Proceedings* **831** E11.34

[9] Lee C D, Feenstra R M, Northrup J E, Lymperakis L and Neugebauer J 2003 *Appl. Phys. Lett.* **82** 1793

[10] Brandt O, Sun Y J, Däweritz L and Ploog K H 2004 *Phys. Rev. B* **69** 165326

[11] Lim C B, Ajay A and Monroy E 2017 *Appl. Phys. Lett.* **111** 022101

The Formation of arachidic acid Langmuir monolayers on the NiCl$_2$ solution

A Chumakov[1], Ammar J Al-Alwani[1,2], A Ermakov[3], O Shinkarenko[1], N Begletsova[1], E Glukhovskoy[1], S Santer[4]

[1]National Research Saratov State University named for N.G. Chernyshevsky, Astrakhanskaya st. 83, 410012, Saratov, Russia
[2]Babylon University, Babylon, Iraq
[3]Institute of Materials Research and Engineering, A*STAR, 2 Fusionopolis Way, Singapore 138634
[4]University of Potsdam, Institute of Physics and Astronomy Campus Golm, Karl-Liebknecht st. 24/25, Golm 14476

Abstract. The article describes the features of the formation of arachidic acid Langmuir monolayers on a water or NiCl$_2$ solution subphase surface. Monolayers structure was investigated by analyzing the compression isotherms and Brewster angle microscopy methods. Films on the solid substrate (glass plates) were investigated by atomic force microscopy method for morphology. Film structure was studied as a function of the concentration of salt in subphase and exposure time. It is shown that the structure of films obtained by using NiCl$_2$ solution as subphase is not homogeneous and we can see some grains in thin films.

1 Introduction

Today, one of the most important tasks in nanotechnology is the production of composite materials with predetermined properties. In particular, in micro and nanoelectronics, there is a problem of obtaining active and conducting layers (electrodes) in one process. Therefore, various methods aimed at obtaining composite thin films are actively developed and improved [1].One of the solutions to the above problem is the inclusion in the composition of organic films of various inorganic objects, nanoparticles [2], carbon nanotubes [3], graphene [4], etc.

The Langmuir-Blodgett method is the simplest method that allows to obtain ordered layers of nanoobjects, using the principle of self-organization of objects on the interface of phases. This method is cheap, and also does not require special conditions (vacuum, elevated temperatures, etc.) for obtaining thin-film coatings.

In our work we obtained Langmuir monolayers of arachidic acid using a different subphase. In the first stage (to obtain reference data) deionized water was used as a subphase. In the next stage (for the addition of metal ions to the subphase), the NiCl$_2$ solution with different concentration was used as a subphase. Researchers from Khomutov group [5] have previously found that metal ions from the subphase can react with surfactant molecules. And under specified conditions, this salt formation reaction leads to the formation of metallic nanoparticles under the monolayer (in the course of its compaction when the barriers are compressed) [6]. Nickel and lead salts are the most popular materials

Content from this work may be used under the terms of the Creative Commons Attribution 3.0 licence. Any further distribution of this work must maintain attribution to the author(s) and the title of the work, journal citation and DOI.
Published under licence by IOP Publishing Ltd

for the formation of such type nanoparticles. In our work, we used nickel salt as less toxic (relative to lead).

2 Materials and methods

In all experiments we use deionized water (R about 18Mom×cm) as a subphase basis. For the addition of metal ions to the subphase we prepare $NiCl_2$ (Sigma Aldrich) solution in deionized water with concentration 10^{-3} and 10^{-2} M. To change the acidity of the solution a 10^{-2} solution of NaOH was used. Solution of arachidic acid (Sigma-Aldrich) prepared in chloroform (Vecton, Russia) with concentration 10^{-3} M.

For formation and deposition of monolayers we used KSV Nima LB trough KN 1003 (KSV Nima, Finland). Exposure time (it is time between the injection of the surfactant and the beginning of the compression) was different in few series of experiments: 5, 20, 40 minutes. For measyre structure of monolayers we used Brewster angle microscopy method with Accurion Nanofilm Ultrabam microscope (field of view is 800*430 μm). Atomic force microscopy (AFM) picture taken in semicontant mode by Nanoeducator II AFM microscope.

3 Results and discussion

In the first series of experiments, we obtained monolayers compression isotherms for arachidic acid on the surface of pure deionized water and surface of $NiCl_2$ 10^{-3} solution by different exposure time (figure 1).

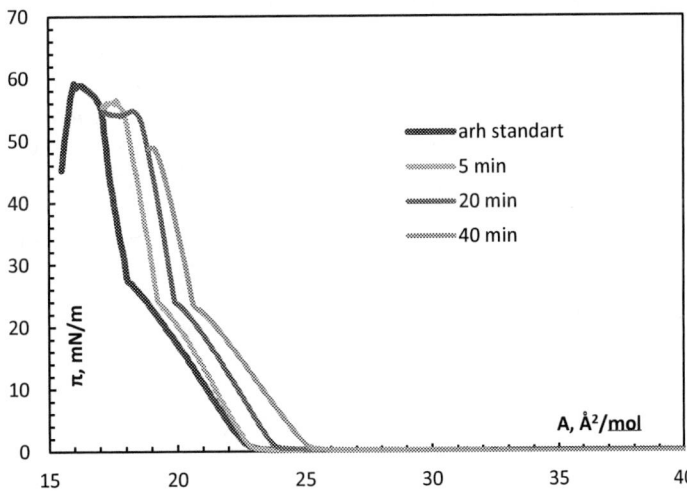

Figure 1. Compression isotherms for arachidic acid monolayers on $NiCl_2$ solution by different time of exposure monolayers (arh standart – arachidic acid compression isotherm on pure deionized water surface)

How we can see in this isotherms, they are shifted to the right. This region corresponds to a larger area per molecule of the monolayer. Since in all series of experiments the same arachic acid solution was used in the same concentration, this corresponds to an increase in the structural unit size in monolayer. In this case, this behavior of the monolayer can be explained by the course of the salt-formation reaction between surfactant and the metal ions from the subphase. In addition, monolayers also change their mechanical characteristics. In particular, the pressure of the phase transition between the phase states of the monolayer (fracture point at 27 mN/m for a monolayer of arachic acid on pure water) drops by 12%. The magnitude of the change is independent of the exposure time of the monolayer. In addition to the pressure of the phase transition, the collapse pressure of the monolayer also decrease. All these changes indicate serious structural rearrangements in the monolayer (under the conditions described above).

Therefore, in the second stage, the Brewster angle microscopy method was used to study monolayers directly in the process of formation. Figure 2 shows images of a monolayer for various composition of subphase by the same surface pressure value.

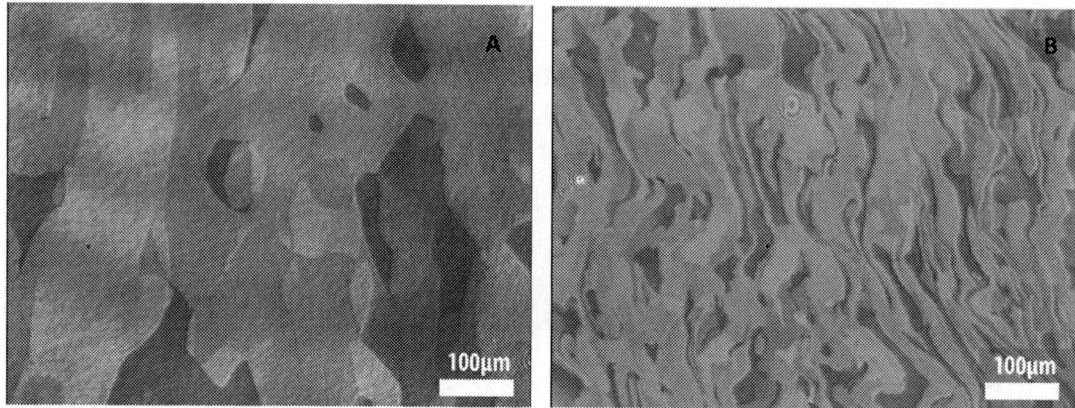

Figure 2. Brewster angle microscopy pictures of arachidic acid monolayers for different compositions of the subphase (A – on pure water subphase, B – on 10^{-3}M NiCl$_2$ solution) by same value of surface pressure (about 10 mN/m).

We can see in these images approximately the same nature of the structure of monolayers – the presence of a large number of clusters with identically oriented surfactant molecules (in this case, a different color indicates different optical properties of the sections of the monolayer, which in this case can be caused only by differences in the orientation of the molecules). However, in the presence of nickel ions in the subphase, the size of the clusters themselves is much smaller. And we did not find the dependence of their size on the exposure time of the monolayer. The effect of changing the size of clusters can be explained by the change in mechanical characteristics. Since the ratio "number of molecules at the cluster boundary / number of molecules in the cluster volume" gradually increases with decreasing size, all processes associated with lateral pressure occur at a lower value. Thus, the transmission of the impact to the cluster volume is faster (at lower values of the surface pressure of the monolayer).

To study the properties of films on solid substrates, they were transferred to glass plates. After that they were investigated by atomic force microscopy. Typical images are shown in Figure 3.

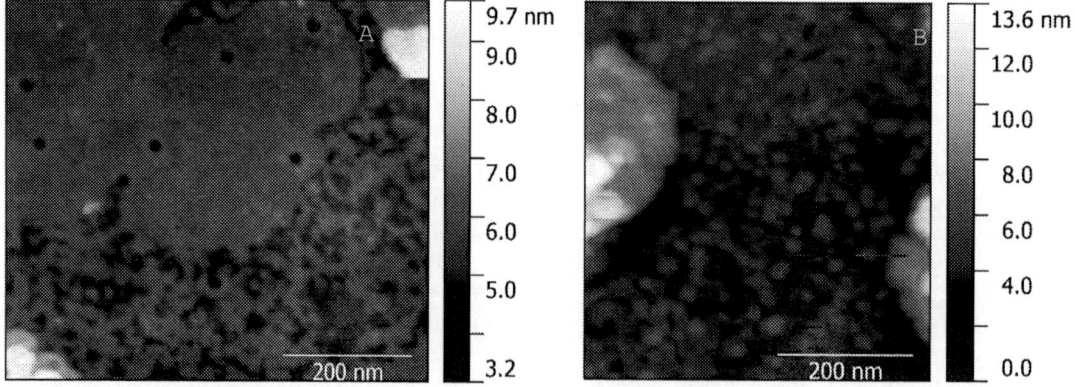

Figure 3. AFM pictures for arahidic acid film on pure water (A) and NiCl$_2$ solution (B)

The deposition of both films was carried out under the same conditions (surface pressure) by the Langmuir-Schaefer method. We applied a single monolayer. In these images we can see sections of a monolayer film. In figure 3A, such a part is located above the center of the image, and also characteristic circular discontinuities that are formed when the film dries out. The rest of the film is fairly homogeneous, and it does not have any grains or inclusions. In contrast, the film depicted in Figure 3B shows within itself certain grains or their conglomerates. In this case, their appearance is explained by the presence in the film of nanoparticles of nickel or their clusters, which give the film an inhomogeneous structure.

Conclusion

Thus, the results of the study showed that at a certain ratio between the composition of the subphase and the exposure time of the monolayer, it is possible to achieve the formation of metallic nanoparticles under the monolayer. We also established by microscopy of the Brewster angle the changes in the clustering of the monolayer upon addition of metal ions to the subphase. These changes provoke changes in the mechanical characteristics of the monolayer. The AFM method shows the structural differences in arachidic acid films obtained on pure water and $NiCl_2$ solution.

Acknowledgments

This work was supported by grant of RFBR17-32-50137

References

[1] Azad I, Manoj K, Goswami D Y, Stefanakos E 2016 *Langmuir* **32(33)** 8307
[2] Chumakov A S, Ammar J Al-Alwani, Gorbachev I A, Ermakov A V, Kletsov A A, Glukhovskoy E G, Kazak A V, Usol'tseva N V, Shtykov S N 2017 *BioNanoScience* **7(4)** 666
[3] Scholl F A, Morais P V, Gabriel R C, Schöning M J, Siqueira J R. Jr, Caseli L 2017 *ACS Appl Mater Interfaces* **9(36)** 31054
[4] Ammar J K Al-Alwani, Chumakov A S, Shinkarenko O A, Gorbachev I A, Pozharov M V, Venig S B, Glukhovskoy E G 2017 *Applied Surface Science* **424** 222
[5] Khomutov G, Bykov I, Gainutdinov R, Polyakov S, Sergeyev-Cherenkov A, Tolstikhina A 2002 *Coll. and Surf. A: Phychem. and Eng. Asp.* **198(200)** 559
[6] Khomutov G, Obydenov A, Yakovenko S, Soldatov E, Trifonov A, Khanin V, Gubin S 1999 *Materials Science and Engineering* **8(9)** 309

SPBOPEN 2018   IOP Publishing

Chemical analysis of thin-film's colour generation during surface laser oxidation of TiN-coating

O S Yulmetova[1,2], M S Tutova[1], R F Yulmetova[1]

[1] ITMO University, St. Petersburg 197101, Russian Federation
[2] Concern CSRI Elektropribor, JSC, St. Petersburg 197046, Russian Federation

Abstract. The phenomenon of metal surface colour change under laser exposure is quite unique. It's most common explanation is based on the connection between the colour and the phase-composition of the generated films. However, the phase-composition analysis is a complicated and expensive process. As an alternative option we present a calculational method, based on a chemical thermodynamic approach. According to this method colour generation can be described by chemical reactions that lead to the formation of different compaunds. And the formation of the most probable compaund have the lowest level of the Gibbs Energy, which can be calculated. The proposed method was tested on the example of laser colouring of the surface of TiN-coating in the air environment. Calculations revealed that the most probable compound is titanium dioxide. Phase analysis based on X–ray diffraction proved the results of calculations.

1. Introduction

Titanium nitride thin-film coatings have chemical inertness, thermodynamic stability, high wear resistance and hardness. Such a wide variety of properties makes it possible to use it in the microelectronic industry, during manufacture of cardiovascular implants, as an absorbent layer in a selective solar absorber, the material of the coating for cutting tools or a wear-resistant and decorative coating [1-3].

At the same time formation of different colours on titanium, titanium nitride (figure 1, a), chromium (figure 1, b) and steel (figure 1, c) surfaces after laser treatment is a unique phenomenon which is a subject of the scientific interest [4].

Figure 1. Thin film's colour generation on TiN-coating (a), chromium (b) and steel (c) surfaces.

Some authors [5] suggest that the colour of the surface depends not only on the interference in thin-film layers but also on the metal oxides' intrinsic colours. These effects are determined by the thickness, transparency, and composition of thin films, formed during laser heating in the atmosphere.

Content from this work may be used under the terms of the Creative Commons Attribution 3.0 licence. Any further distribution of this work must maintain attribution to the author(s) and the title of the work, journal citation and DOI.
Published under licence by IOP Publishing Ltd

Therefore it is important to determine the composition of the oxide films. At the same time, a compositional analysis is a complicated problem. That is why it is very important to develop various analytical and calculational methods for predicting results of the surface laser oxidation process. In the present study, the calculational method based on chemical thermodynamics was used for predicting the phase-chemical composition of colourful thin films formed by a laser oxidation of titanium nitride coating in the air.

2. Theoretical calculations

Thin film's colour generation under laser exposure can be considered as a product of chemical reactions between metal surface and components of gas environment. The chemical thermodynamic method can be used for the identification of the possible and occurring interactions and their quantitative determination. According to the thermodynamic approach the best way to assess the probability of the occurrence of any interaction is to determine the Gibbs Energy [6] of this reaction according to the formula:

$$\Delta G_T^0 = \Delta H_{298}^0 - T \cdot \Delta S_{298}^0 + \int_{298}^{T} C_p dT' - T \cdot \int_{298}^{T} 1/T' \Delta C_p dT' \tag{1},$$

where $\Delta H^o{}_T$ is the thermal effect of the process (enthalpy variation), $\Delta S^o{}_T$ is the entropy variation of the system, C_p is the thermal capacity, and T is the working temperature.

laser treatment is carried out in the atmosphere, the interaction between the titanium nitride and the most active air component – oxygen with formation different oxides (TiO, TiO_2, Ti_2O_3) is most likely to happen. To determine if the oxide formation is thermodynamically permissible is possible by calculating the Gibbs Energy $\Delta G^0{}_T$, using equation (1). Results of calculations of the Gibbs Energy $\Delta G^0{}_T$ for possible chemical reactions between TiN and O_2 are presented in Table 1.

Table 1. Results of calculations of the Gibbs Energy $\Delta G^0{}_T$, for possible chemical reactions between TiN and O_2.

	Reaction	Product	$\Delta G^0{}_T$, (kJ/mol)	
			298 K	1000 K
1	$TiN+O_2=TiO+NO$	TiO	- 376	-355
2	$TiN+1/2O_2=TiO+1/2N_2$	TiO	- 610	- 466
3	$TiN+5/4O_2=1/2Ti_2O_3+NO$	Ti_2O_3	- 603	- 574
4	$TiN+3/4O_2=1/2Ti_2O_3+1/2N_2$	Ti_2O_3	- 697	- 652
5	$TiN+3/2O_2=TiO_2+NO_2$	TiO_2	- 805	- 763
6	$TiN+O_2=TiO_2+1/2N_2$	TiO_2	- 815	- 797

Negative values of the Gibbs Energy $\Delta G^0{}_T$ mean that all reactions can be completed in the temperature range 298-1000 K. However formation of TiO_2 has the highest probability of occurrence as its Gibbs Energy is lower than that of TiO or Ti_2O_3. According to calculations based on the thermodynamic approach, formation of TiO_2 during laser oxidation of titanium nitride TiN is most likely to occur.

3. Experimental results

Laser treatment processes were studied on beryllium samples covered by TiN-coating. Samples were in the form of plates with a diameter 20 mm and a thickness of 4 mm. A thin film of titanium nitride coating was formed by cathodic arc deposition (PVD method). Nitrogen pressure was selected in the range from 0.1 to 0.2 Pa to create golden colour [7]. The samples were irradiated by exposure of a scanning beam of a fiber laser with a wavelength 1.06 μm and surface area with a diameter of 4 mm. The exposure was done under normal laboratory conditions in air. During experiments one mode was revealed that can change the colour of the titanium nitride coating from golden to blue by changing only the number of laser passes. A golden colour was obtained after the first pass, brown - after the second laser scanning of exactly the same place, purple - after the third pass, and blue - after the forth pass. In the process, the pulse width τ was 4 ns, the average power P_{aver}, and the pulse repetition frequency f were 5 W and 99 kHz, correspondingly. According to the theoretical calculations based on thermodynamic approach the most probable compound obtained after laser surface oxidation of titanium nitride coating is titanium dioxide.

XRD phase analysis was done using a Bruker D8 Discover difractometer, and it revealed that generated colourful films consist of titanium oxide TiO_2 after the first, second, third and fourth laser passes. However, every next laser pass changed the sample colour as it's shown in figure 2.

Chemical elemental analysis was made using a scanning electron microscope Mira 3 Feg Sem, and it revealed the tendency of increasing level of oxygen in the titanium nitride coating with increasing number of laser passes: from 6% of oxygen for the first pass with the golden colour to 26% of oxygen for the last pass with the blue colour (figure 2).

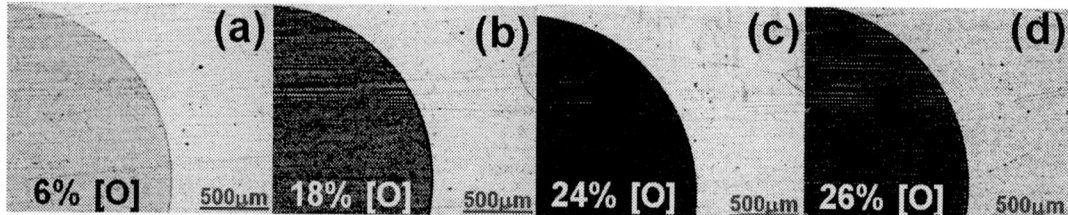

Figure 2. Optical images of TiN-coating: (a) after the first laser pass, (b) after the second laser pass, (c) after the third laser pass, (d) after the forth laser pass.

Visual observation of laser oxidation process and analysis of optical images from figure 2 reveal that the image created by laser is formed by filling of lines. Every line consists of a set of point pulses. Each pulse is a cause of a sharp point heating that initiates a reaction between the titanium nitride coating and oxygen. As a result, there are points of titanium oxide inclusions in the titanium nitride coating.

Figure 3 shows SEM micrographs of TiN-coating before and after laser treatment, obtained using a scanning electron microscope Mira 3 Feg Sem.

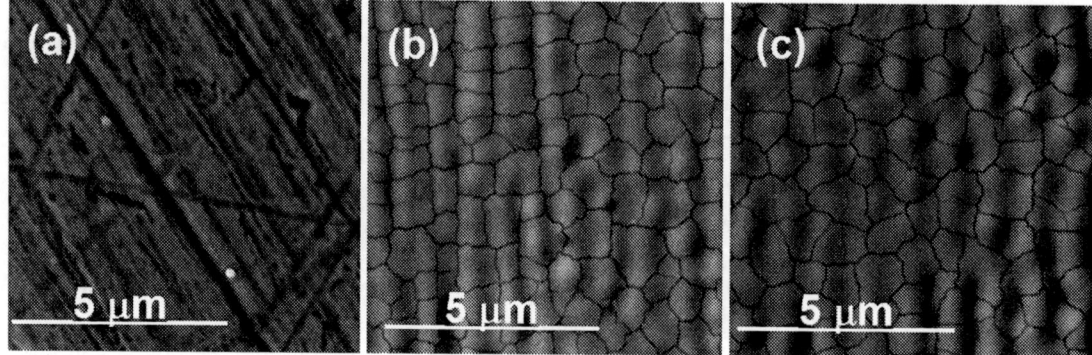

Figure 3. SEM micrographs of TiN-coating: (a) before laser treatment, (b) after the first laser pass, (c) after the forth laser pass.

According to the images, presented in figure 3 (a), a thin film of titanium nitride coating has a homogeneous structure, repeating the substrate microrelief. Figure 3 (b) shows the SEM image of the TiN-coating obtained after the first laser pass. Figure 3 (c) describes the surface structure after the forth laser pass. It can be noted that laser treatment leads to the appearance of a network structure. Formation of such structure is most likely the result of phase transformation from titanium nitride to titanium dioxide. The network structure is caused by the replacement of nitrogen with oxygen in the coating composition.

4. Discussion

Thermodynamic calculations predicted the formation of titanium oxide TiO_2 as the most probable compound after laser treatment of titanium nitride coating in the air. Phase analysis detected the appearance of titanium oxide TiO_2 after the first, the second, the third and the fourth laser pass. Other oxides and compounds were not detected meaning that the experimentally obtained laser mode creates conditions for forming only TiO_2. However, every next laser pass changes the sample colour. During experiments it was noticed that every next pass increases the size and the depth of the heated area where titanium nitride reacts with oxygen. Compositional analysis proved the tendency of increasing the level of oxygen in the TiN-coating with the increasing number of laser passes. It can be assumed that the color change is associated with the depth of oxidation process and the colour of the generated thin film is strongly depends upon its thickness and homogeneity. According to the publication [8], the color of titanium oxides varies with film thickness, since it is due to light interference phenomena taking place at the metal-oxide-air interfaces. So the thermodynamics approach can be used as a calculational method for prediction of formation of the most probable compound the amount of which can be increased by means of multipass laser processing.

Presented scientific results demonstrate that analysis of the phase and chemical transformations occurring at the surface of heated metals are generally governed by thermodynamics laws. Comparison of the calculated data with the experimental results demonstrates their good qualitative agreement. Furthermore, presented calculational method makes it possible to obtain the important detailed information about chemical reactions occurring under laser heating of a titanium nitride coating, niobium and chromium in a chemically active atmosphere.

5. Practical application

It should be noted that investigation of the phase and chemical modifications of the surface under laser exposure is important for many fields because of the variety of application areas of thin-film's color generation technology. It is successfully used as a technology image formation for providing functionality of optoelectronic read-out systems [7]. Medical implants are required to be marked for keeping them from falsification using the technology that keeps implants from surface damage during marking process [9-10]. Laser oxidation is applied as a technology intended for formation of dielectric layers [11] when manufacturing of microelectronic sensors. Laser coloration is also a very powerful tool for jewelry decoration.

Acknowledgments

The authors acknowledge the support of Centre for X-ray Diffraction Studies, St. Petersburg State University.

References

[1] Hasan Elmkhah, Faridrez Attarzadeh et al. 2018 *J. of Alloys and Compounds* **735** 422
[2] Yuchen Wang, Liqun Zhu et al. 2017 *Surf. and Coat. Technol.* **331** 90
[3] C. Mendoza, Z. Gonzalez et al. 2018 *J. of the European Ceramic Society.* **38** 495
[4] E.H. Amara, F.Haïd, A.Noukaz. 2015 *Appl. Surf. Sci.* **351** 1
[5] Z.L. Li, H.Y. Zheng, K.M. Teh et al. 2009 *Appl. Surf. Sci.* **256** 1582
[6] Gibbs J W et al. 1928 *The Collected Works of J. Willard Gibbs: In two volumes* (New York: Longmans, Green)
[7] O S Yulmetova and M A Tumanova 2017 *J. Phys.: Conf. Ser.* 917 052007
[8] Diamanti M. V., Del Curto B. and Pedeferri M. 2008 *Color Res. Appl.* **33** 221
[9] Chi-Wai Chan, Louise Carson et al. 2017 *Appl. Surf. Sci.* **404** 67
[10] Krzysztof Czyż, Jan Marczak et al. 2016 *Diamond and Related Materials* **67** 26
[11] Jaffar Moideen, Yacob Ali, Vinodh Shanmugam et al. 2018 *Solar Energy* **164** 287

Studies of the formation of copper nanoparticles monolayers on the water subphase

N N Begletsova[1,2], E I Selifonova[2], R K Chernova[2,3], A S Chumakov[1,2], V P Sevostyanov[4], E G Glukhovskoy[1,2]

[1]Department of Nano- and Biomedical Technologies, Saratov State University, Saratov, 410012, Russia

[2]Education and Research Institute of Nanostructures and Biosystems, Saratov State University, Saratov, 410012, Russia

[3]Institute of chemistry, Saratov State University, Saratov, 410012, Russia

[4]Scientific Research Institute of Technology of Organic, Inorganic Chemistry and Biotechnology, Saratov, 410005, Russia

Abstract. The article is devoted to the study of the formation of copper nanoparticles monolayers on the water subphase by analyzing the compression isotherms depending on the temperature of the water subphase and the volume of aliquot spread on its surface. The size of the copper particles stabilized by a surfactant is in the range from 21 to 33 nm. The light absorption of the copper nanoparticles solution was studied by spectrophotometric method. Using the method of dynamic light scattering the results of the copper particles size in water solution were obtained. It is shown that an increase of the temperature of the water subphase leads to a transition of copper particles and surfactant from the water surface to volume.

1. Introduction

The nanoparticles layers lined up as monolayers (MLs) are used in many applications such as conducting layers in micro- and nanoelectronics, sensitive layers in sensor systems, etc. To form such systems the liquid surface can be used as a template [1-4]. At the same time the connection of such a "liquid substrate" with nanoobjects contributes to their more efficient alignment due to their higher mobility than in the case of a solid substrate.

The Langmuir-Blodgett (LB) method is one of the most widely used methods for the formation of MLs of various substances [1-4]. The advantages of this method are that it is simple in execution because it does not require special technical conditions (creating a high vacuum, conducting an experiment using complex equipment). LB technology allows creating a unique composition of each of the layers. In

---

Content from this work may be used under the terms of the Creative Commons Attribution 3.0 licence. Any further distribution of this work must maintain attribution to the author(s) and the title of the work, journal citation and DOI.

Published under licence by IOP Publishing Ltd

turn this makes it possible to create multilayer heterostructures by sequentially transferring MLs of various compositions on solid substrates.

There are two types of MLs. The first type is Langmuir ML. It is formed from insoluble substances. The second type is Gibbs ML. It is formed by sorption on the gas-liquid interface of components partially soluble in the liquid phase [5]. In both cases, the processes of MLs formation are rather complicated and depend on many factors. pH and temperature of subphase have an important effect on the result.

Depending on the selected subphase parameters it is possible to control the structure of the ML forming a dense or porous film surface. Therefore, controlling the formation parameters as well as the transfer of metal films on solid substrates is a very important task when creating on their basis mono- and multilayer structures.

We studied the formation of MLs of copper nanoparticles (@Cu) stabilized by an anionic type surfactant sodium dodecyl sulfate (SDS) on the water subphase surface using the compression isotherm method. Changes in the states of MLs were studied depending on the temperature of the water subphase and the spread aliquot volume.

2. Material and methods

The preparation of @Cu was carried out according to the procedure [6] using the method of chemical reduction of copper ammonia in an aqueous solution of surfactant SDS. The molar ratio of salt (copper chloride) and reducing agent (hydrazine) was 1:150 at value of pH = 11.0. The obtaining of @Cu was confirmed by spectrophotometric method. The absorption spectra of the surface plasmon resonance (SPR) of the @Cu solution was registered on the spectrophotometer SHIMADZU UV-2550 in the wavelength range 190-900 nm. Determination of the copper particle size in colloidal suspension was carried out by means of dynamic light scattering (DLS) measurements using a Zetasizer Nano-ZS (Malvern Instruments Ltd, UK). To reduce the absorption intensity in studying the absorption spectra and the copper particles size the colloidal suspension was diluted with deionized water in a ratio of 1:5.

Formation and investigation of the behavior of @Cu MLs at the gas-liquid interface was carried out using the Langmuir-Blodgett bath KSV Nima LB Trough Medium KN 2002 (KSV Nima, Finland). Deionized water with a specific resistance of 18 M$\Omega \times$cm (prepared by Thermo Scientific Barnstead Smart2Pure, USA) and pH = 7 (measured by means of pH-meter pH-410, Russia) was used as a subphase. We obtained compression isotherms at aliquots volumes of the studied @Cu suspension from 100 to 1000 µl. The water subphase temperature was 22 °C at a volume of 150 ml.

We carried out a study of the compression-expansion isotherms of the formed particles MLs at the above parameters of the water subphase at the optimum aliquot volume of 400 µl.

Also the effect of the water subphase temperature (22, 29, 39 and 45 °C) on the formation @Cu MLs was studied. For these studies, the volume of the water subphase was 130 ml. Its heating was carried out using a circulating thermostat LOIP LT-105 (Russia). The temperature was controlled by means of a thermocouple placed in the LB bath. The solution of nanoparticles was deposited on the water surface. After that, we waited 5 minutes to distribute the solution evenly over the water subphase surface. Then the surface with ML was symmetrically compressed by moving barriers at a speed of 15 mm/min.

3. Results and discussion

Spectrophotometric method of measurement of the UV-visible spectrum of @Cu solution showed that in the wavelength range $\lambda = 450\text{-}750$ nm an absorption peak is observed. It corresponds to the SPR at a wavelength $\lambda_{max} = 566$ nm (Figure 1).

SPBOPEN 2018

IOP Publishing

IOP Conf. Series: Journal of Physics: Conf. Series **1124** (2018) 081011   doi:10.1088/1742-6596/1124/8/081011

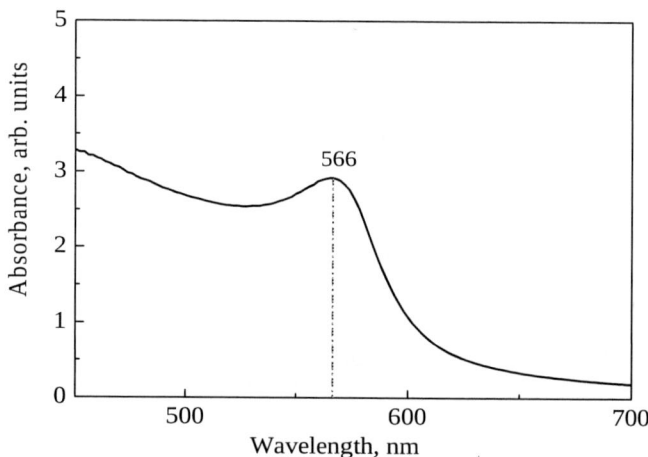

Figure 1. UV-visible spectrum of @Cu solution obtained at the day of the experiment.

According to [7] the presence of an absorption peak in the visible region corresponds to the presence of @Cu in the aqueous solution. Studies have shown that particle sizes range from 21 to 33 nm. The range of particle size dispersion was estimated as full width at half maximum (Figure 2).

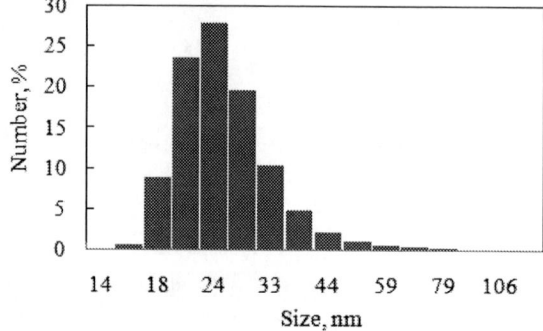

Figure 2. The histogram of particle size distribution.

In the process of studying the dependence of MLs formation on a different aliquot volume of @Cu suspension a series of compression isotherms was recorded. The aliquot volume varied in the range from 100 to 1000 μl with a step of 100 μl (Figure 3). It can be seen that with an increase in the applied amount of the substance from 100 to 1000 μl the surface pressure at the end of the compression increased from units mN/m to 30 mN/m. The area per ML in a close-packed state varied from 63 to 116 cm^2. This behavior of the obtained dependences (surface pressure from area per ML) indicates an increase in the number of particles located at the gas-liquid interface.

1028

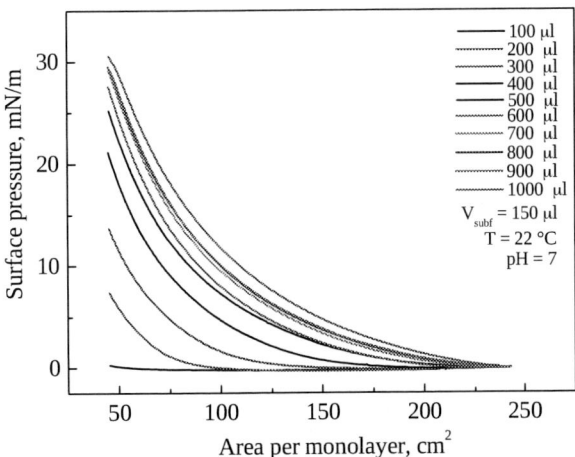

Figure 3. Compression isotherms of @Cu MLs at different aliquots volumes obtained solution spread on the water subphase.

Figure 4 shows the hysteresis obtained by successively recording compression-expansion isotherms at an aliquot volume of 400 μl. The area per copper particles ML in the process of its compression is greater than in the reverse process (expansion). This behavior of the ML may indicate the presence of interaction between the nanoparticles stabilized by the surfactant SDS. This can be the reason for the prolonged process of destruction of the formed film with an increase in the area per ML.

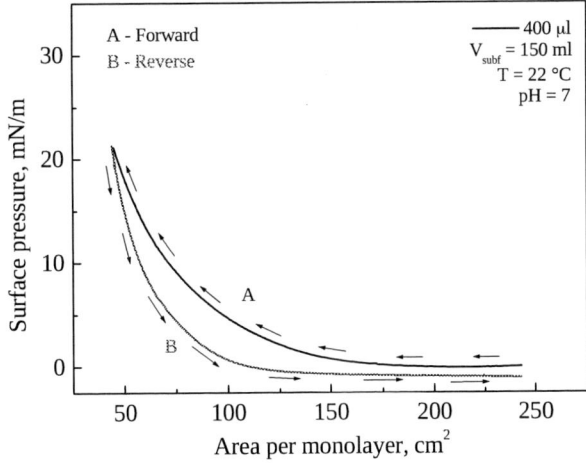

Figure 4. Compression-expansion isotherms of @Cu MLs on the water subphase.

Further studies of the formation and behavior of @Cu at the gas-liquid interface show that with an increase of the water subphase temperature from 22 to 45 °C the area per ML decreases from 100 to 60 cm^2, respectively (Figure 5a and Figure 5b). The formation of @Cu ML and SDS surfactant are determined by two competing processes. Their sorption at the gas-liquid interface, and their transition from the interface to the subphase volume (i.e. the dissolution process). It follows from the obtained isotherms that the dissolution process becomes more intense with increasing temperature.

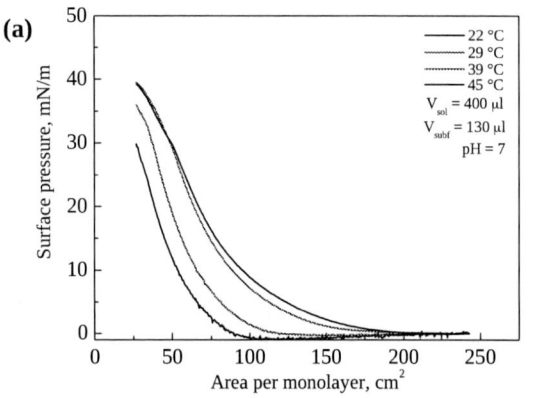 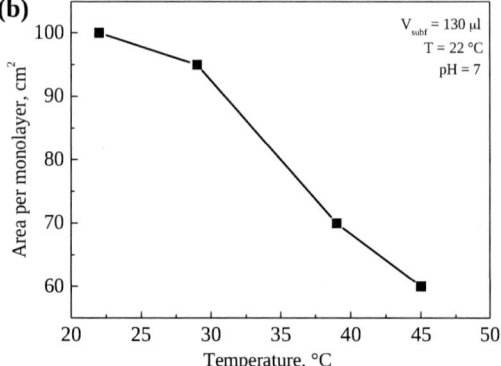

Figure 5(a, b). (a) Compression isotherms of @Cu MLs at an aliquot volume of 400 µl and subphase water temperatures: 1 – 22; 2 – 29; 3 – 39; 4 – 45 °C; **(b)** the dependence of the area per ML in the condensed state on the water subphase temperature.

4. Conclusions

The presented studies illustrate the possibility of forming on the water subphase (on the gas-liquid interface) MLs of @Cu with dimensions from 21 to 33 nm stabilized by the surfactant SDS. The absorption spectrum of the @Cu solution shows that the peak position of the SPR is at a wavelength of 566 nm. It was found that with an increase in the amount of the substance spread on the water subphase the area per ML increases. An increase of the water subphase temperature in the LB bath leads to an acceleration of the solubility of @Cu and SDS due to the dominance of this process over the sorption of particles at the gas-liquid interface. Therefore, the decrease of the particles area per ML is observed.

Acknowledgments

This work was supported by RFBR grants № 17-07-00407-a and № 17-32-50137-mol_nr.

References

[1] Li S, Zheng Y, Qi Z, Li X, Chen C 2016 *Physics Procedia* **85** 41

[2] Gorbachev I A, Goryacheva I Y, Glukhovskoy E G 2016 *BioNanoSci* **6** 153

[3] Al-Alwani A J, Chumakov A S, Albermani M S, Shinkarenko O A, Begletsova N N, Vostrikova A M, Gorbachev I A, Venig S B, Glukhovskoy E G *IOP Conf. Series: Journal of Physics: Conf. Series* vol 917 (Saint Petersburg: SPbOPEN2017) 032026

[4] Chumakov A, Al-Alwani A J, Gorbachev I, Ermakov A, Shinkarenko O, Begletsova N, Kolesnikova A, Glukhovskoy E 2017 *IOP Conf. Series: Journal of Physics: Conf. Series* vol 917 (Saint Petersburg: SPbOPEN2017) 092002

[5] Adamson A W, Gast A P 1967 *Physical Chemistry of Surfaces, Interscience* (New York)

[6] Begletsova N N, Shinkarenko O A, Chumakov A S, Al-Alwani A J K, Selifonov A A, Selifonova E I, Pozharov M V, Zakharevich A M, Chernova R K, Kolesnikova A S, Glukhovskoy E G *IOP Conf. Series: Journal of Physics: Conf. Series* vol 917 (Saint Petersburg: SPbOPEN2017) 092014

[7] Granata G, Yamaoka T, Pagnanelli F, Fuwa A 2016 *J. Nanopart. Res.* **18** 133

SPBOPEN 2018

IOP Publishing

IOP Conf. Series: Journal of Physics: Conf. Series **1124** (2018) 081012   doi:10.1088/1742-6596/1124/8/081012

Multicaloric effect in barium titanate nanotube

I A Starkov[1,2], I L Mylnikov[3], A S Starkov[2]

[1]Nanotechnology Center, St. Petersburg National Research Academic University RAS, St. Petersburg 194021, Russia
[2]National Research University of Information Technologies, Mechanics and Optics, St. Petersburg 197101, Russia
[3]Department of Physical Electronics and Technology, St. Petersburg Electrotechnical University, St. Petersburg 197376, Russia
starkov@spbau.ru

Abstract. One of the characteristic features of multiferroic materials is the dimensional effect, when the physical properties of a sample depend on its size. To describe this effect, it is necessary to take into account not only the basic physical quantities: the electric field, polarization, strain tensor, stresses, temperature, but also their spatial derivatives (gradients). In this case, both the order of differential equations describing the behaviour of the sample and the number of boundary conditions increase. As a result, a surface layer appears near the boundary, in which the above-mentioned physical quantities change abruptly. The thickness of this layer is 1-5nm. The properties of a similar layer for a barium titanate nanotube and its effect on its thermal properties are investigated. It is shown that as the nanotube size decreases, the multicaloric effect (µCE) increases. At a nanotube thickness of 5nm µCE is 3 times greater than the effect for bulk materials.

1. Introduction

Caloric effects (CEs) consist in changing the temperature or entropy of the sample when the field is applied or removed. The main caloric effects are the magnetocaloric, electrocaloric (ECE), and elastocaloric ones which correspond to the influence of a magnetic, electric, and elastic field, respectively. The magnitude of CEs is determined by the temperature dependence of the magnetic, electric, or elastic constants. The greatest values are achieved near phase transitions, where the temperature dependence of the above coefficients is especially significant. If a change in thermal properties of a sample depends on several fields, then the presence of the multicaloric effect (µCE) [1] is indicated. Interaction of fields of different nature can lead to a significant increase in magnitude of CEs (synergistic phenomena) and µCE can exceed single components. Since this interaction increases with a decrease in the dimensions of the sample, a multiple increase in µCE should be expected for nanometer-scale objects. One of the possibilities of this interaction is the flexoelectric effect (FEE). Direct FEE is the occurrence of polarization due to inhomogeneous strain. The inverse FEE is defined as the appearance of a strained bending of a thin plate when it is polarized. Experiments carried out on BaTiO3 single crystals (size of the order of tens of micrometers) confirmed the existence of both direct and inverse FEE [2].

The dependence of the flexoelectric coefficients on temperature results in the appearance of a flexocaloric effect (FCE), as first predicted in [1]. According to the calculations for the PMN ceramics with a strain gradient of $1m^{-1}$, the temperature change in the FCE depends on the ambient temperature

Content from this work may be used under the terms of the Creative Commons Attribution 3.0 licence. Any further distribution of this work must maintain attribution to the author(s) and the title of the work, journal citation and DOI.
Published under licence by IOP Publishing Ltd

and varies from 5μK at T=240K to -6μK at T=299K. Calculations based on a first principles approach confirmed the existence of FCE in the barium strontium titanate (BST) ceramics [3]. The effect is 1.5K at T = 289K under the application of a strain gradient of 1.5μm^{-1}. The highest known value of FCE is 60K and it was obtained as a result of calculations for BST ceramics. However, it is reached in a layer of thickness on the order of 1nm [4]. In this case the magnitude of the FEE is determined by enormous values of the strain gradient (10^6-10^7m^{-1}) because of the presence of a misfit strain existing at the boundaries of two crystals with different atomic lattice constants. From the results reported in literature, it follows that the FCE increases with decreasing film thickness. For instance, the magnitude of flexocaloric effect is less than 1% of the electrocaloric effect for BST film of 100nm thickness. In turn, the values of FCE and ECE coincide for 16nm films, while FCE already exceeds ECE by 3 times for thin 5nm films.

When describing any physical phenomena in nanoscale multiferroic materials, one should take into account not only the basic physical quantities: the electric and magnetic field strengths, polarization and magnetization, the strain and strain tensor, temperature, but also their spatial derivatives (gradients). A theory that takes into account the gradients of deformation in elastic bodies was developed in the mid-sixties of the last century [5]. A general theory describing the joint action of electric, magnetic, elastic, and thermal phenomena with allowance for gradient terms near phase transitions has been developed in recent years by a number of authors (see, for example, [2,6,7]). The equations of state of these materials, as a rule, are derived on the basis of variational principles from the condition of minimality of the thermodynamic potential W. The boundary conditions, according to the calculus of variations, must also be determined by the same potential.

In the presence of four fields (electric, magnetic, elastic, and thermal) there are altogether 10 gradient effects. This makes it difficult to estimate them accurately. In view of this, in order to simplify the subsequent calculations, we exclude from consideration the magnetic field and the temperature gradients. The magnetic field can be taken into account just as an electric field. For nonlinear ferroelectrics, both the gradients of the electric field strength and the polarization or electric-displacement gradients can be considered. However, in the sequel we confine ourselves to the case of only the presence of the polarization gradient in W. Besides, it is natural to assume that the thermodynamic potential, which is a scalar quantity, does not include the polarization itself, but its square. As a result, there are 3 gradient effects. The contribution to W of these effects is given by the squares of the gradients of the components of the polarization vector and the strain tensor, as well as products of the gradients of these quantities. Since the 3 gradient effects depend on the temperature, they must contribute to the entropy or temperature of the sample. This means that there are 3 caloric effects associated with gradient phenomena, which we will call gradient-caloric. As an example of the application of the new model, μCE in a barium titanate nanotube is investigated.

2. The approach

Let us describe a sample occupying the volume V bounded by the surface S. For such a purpose, we use the energy density w and introduce displacement vectors $\boldsymbol{u}$ with components u_i, (i = 1, 2, 3), and the potential φ. In the usual way, we define the electric field $E = - \varphi_{,i}$, the strain tensor $u_{ij} = (u_{i,j} + u_{j,i})/2$, and the gradient of deformations $v_{ijk} = u_{k,ij}$. Hereinafter, the subscript after the comma means differentiation with respect to the corresponding variable in the Cartesian coordinate system x_1, x_2, x_3. We believe that the energy density w, in addition to the electric field E_i, depends both on the strain tensor u_{ij} and on its gradient v_{ijk}. Moreover, it is assumed that w depends on the polarization components P_i and their derivatives $P_{i,j}$. Then the total energy W stored in the volume V has the form

$$W \equiv \int_V w(u_{ij}, v_{ijk}, E_i, P_i, P_{i,j})\, \mathrm{d}V. \tag{1}$$

The equation (1) means that the sample under consideration has both elastic and electrical properties that depend not only on the deformation and the electric field, but also on the gradients of deformation and polarization. Such a medium for brevity will be called the gradient-electroelastic. Varying (1) leads to the relation

$$\delta W = \int_V (\sigma_{ij}\delta u_{ij} + \tau_{ijk}\delta v_{ijk} + D_i\delta E_i + \frac{\partial w}{\partial P_i}\delta P_i + J_{ij}\delta P_{i,j})\, dV, \tag{2}$$

in which σ_{ij}, τ_{ijk}, D_i, J_{ij} are, respectively, the stress tensor, the higher-order strain tensor [5], the electric displacement, and the conjugate quantity

$$\sigma_{ij} = \frac{\partial w}{\partial u_{ij}}, \quad \tau_{ijk} = \frac{\partial w}{\partial v_{ijk}}, \quad D_i = \frac{\partial w}{\partial E_i}, \quad J_{ij} = \frac{\partial w}{\partial P_{i,j}}. \tag{3}$$

In (2),(3) we use the Einstein summation convention, whereby we agree to sum over repeated indices. Note that the variables σ_{ij}, τ_{ijk}, D_i, J_{ij} and u_{ij}, v_{ijk}, E_i, $P_{i,j}$ are generalized forces and coordinates conjugated to one another – the generalized forces are associated with the generalized coordinates. Using the Ostrogradskii-Gauss theorem allows one to transform the volume integrals in (2) to surface integrals

$$\delta W = \int_V \left[(\sigma_{jk,j} - \tau_{ijk,ij})\delta u_k + D_{i,i}\delta\varphi + \left(\frac{\partial w}{\partial P_i} - J_{ij,j}\right)\delta P_i \right] dV +$$

$$+ \int_S [(\sigma_{jk} - \tau_{ijk,i})n_j\delta u_k + \tau_{ijk}n_i\delta u_{k,j} + D_j n_j\delta\varphi + J_{ij}n_j\delta P_i]dS, \tag{4}$$

where n_j are the components of the normal vector to S. The extremality condition (1), according to (4), gives us the equations

$$\varsigma_{jk,j} = 0, \quad D_{i,i} = 0, \quad \frac{\partial w}{\partial P_i} = J_{ij,j}, \tag{5}$$

in which the generalized stress ζ_{jk} is defined by the equality $\zeta_{jk} = \sigma_{jk} - \tau_{ijk,i}$. To obtain the boundary conditions, we emphasize that the quantities $\delta u_{k,j}$ cannot be considered independent, since they are determined by the values δu_k on the surface S. In view of this, we represent $\delta u_{k,j}$ as

$$\delta u_{k,j} = d_j^{\parallel}\delta u_k + n_j d^{\perp}\delta u_k, \tag{6}$$

i.e. decompose the derivative into the normal and tangential components [5-6]

$$d^{\perp} \equiv n_k\frac{\partial}{\partial x_k}, \quad d_j^{\parallel} \equiv (\delta_{jk} - n_j n_k)\frac{\partial}{\partial x_k}, \tag{7}$$

where δ_{jk} is the Kronecker symbol. After substituting (6) in (4) and taking into account (5), the surface part of the total energy variation δW_{uS} containing δu_{jk} can be written out from (4) as

$$\delta W_{uS} = \int_S (T_k\delta u_k + R_k d^{\perp}\delta u_k)\, dS. \tag{8}$$

Here we use the notation

$$T_k \equiv n_i\varsigma_{ik} + n_i n_j\tau_{ijk}(d_l^{\parallel}n_l) - d_j^{\parallel}(n_i\tau_{ijk}), \quad R_k \equiv n_i n_j\tau_{ijk}. \tag{9}$$

From (8), it follows that the following 20 boundary conditions must be satisfied at the interfaces of the flexoelectrics:

$$[\varphi] = 0, \quad [u_k] = 0, \quad [P_k] = 0, \quad [d^{\perp}u_k] = 0, \quad [R_k] = 0, \quad [T_k] = 0, \quad [n_k D_k] = 0, \quad [n_k J_{ik}] = 0. \tag{10}$$

The symbol [X] denotes the jump in the quantity X when passing through the interface. The first three conditions are the standard continuity of the potential, displacements, and polarization. Continuity of the normal component of the electric displacement $n_k D_k$ is also included in the ordinary boundary conditions of electrostatics. The continuity condition of T_k is a generalization of the continuity condition for $n_i \sigma_{ij}$ in the usual theory of elasticity. New are the conditions of continuity for $d^\perp u_k$ and R_k. Thus, the electroelastic field in a flexoelectric must satisfy 5 equations (5), 20 conditions on the internal interfaces of media (10) and 10 conditions on the outer boundaries, which can consist of the specification u_k, $d^\perp u_k$, φ, P_i or T_k, R_k, n_k, D_k, $n_k J_{ik}$, or a combination of the listed conditions. In particular, in accordance with (9), the following equalities must be satisfied on the external free boundaries

$$n_k D_k = 0, \quad T_k = 0, \quad R_k = 0, \quad n_k J_{ik} = 0. \tag{11}$$

We emphasize that the above derivation of equations and boundary conditions does not depend on the form of w. It is a generalization to the case of a flexoelectric of a similar derivation for an elastic body [5].

For the thermal properties description, we assume that the energy (1) depends also on the temperature T. Then the specific entropy $s = -\partial w / \partial T$ and the total entropy $S = -\partial W / \partial T$ are determined in a standard manner. Variation of entropy occurs by analogy with the variation of energy and gives us

$$\delta S = \int_V \left[\left[\left(\frac{\partial \sigma_{jk}}{\partial T} \right)_{,j} - \left(\frac{\partial \tau_{ijk}}{\partial T} \right)_{,ij} \right] \delta u_i - \left(\frac{\partial D_i}{\partial T} \right)_{,i} \delta \varphi - \left(\frac{\partial^2 w}{\partial P_i \partial T} - \frac{\partial J_{ij,j}}{\partial T} \right) \delta P_i \right] dV -$$
$$- \int_S \left[\left(\frac{\partial \sigma_{jk}}{\partial T} - \frac{\partial \tau_{ijk,i} n_j}{\partial T} + \frac{\partial T_k}{\partial T} \right) \delta u_k + \frac{\partial R_k}{\partial T} d^\perp \delta u_k + \frac{\partial D_j n_j}{\partial T} \delta \varphi + \frac{\partial J_{ij} n_j}{\partial T} \delta P_i \right] dS. \tag{12}$$

It is important to underline that the relation (12) is derived for the first time and allows determining all the thermal characteristics of the gradient-electroelastic medium. One may conclude that there are 2 types of caloric effects – volume and surface. Surface effects, if we neglect the temperature dependence of the normal vectors, do not give a contribution to the change in entropy for free boundaries (when (11) is fulfilled). Such a contribution exists for a fixed boundary. In addition to volume and surface caloric effects, dynamic effects due to the time dependence of the variables occurring in w must also exist by analogy with the FEE [2]. To take them into account, we should add time derivatives to (1). Nonetheless, we will not do this in this study since dynamic CEs deserve consideration in a separate article.

3. Multicaloric effect in barium titanate nanotube

As a simple example of using the equations derived above, let us consider the problem of calculating the electroelastic field in a nanotube from a gradient-electroelastic material – barium titanate. We denote the internal radius by R_1, the outer radius by R_2, and the height of the nanotube by H. We use a cylindrical coordinate system r, θ, z, whose origin is located in the center of the bottom base (see figure 1(a)). The electric potential is equal to 0 at the bottom base and to the given value V at the upper base (for $z = H$). On the outer boundary (at $r = R_2$) we will consider the given mechanical pressure p. The internal pressure is set to 0. The remaining boundary conditions are considered to be free (11). Because of the axial symmetry of the problem, only the components of the displacement vectors u_r, u_z and the polarizations P_r, P_z are different from 0. The energy density in the model under consideration can be written in the form

$$w = w_{\text{LGD}} + w_{\text{elast}} + w_{\text{str}} + w_{\text{grad}}, \tag{13}$$

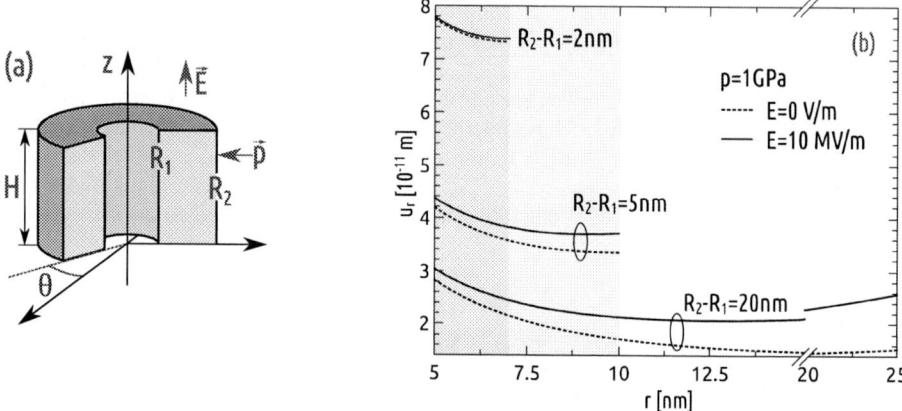

Figure 1. (a) The barium titanate nanotube under consideration. (b) The coordinate dependence of the displacement u_r for different values of nanotube thickness.

where

$$w_{\text{electr}} = a_1(P_1^2 + P_2^2 + P_3^2) + a_{11}(P_1^4 + P_2^4 + P_3^4) + a_{12}(P_1^2 P_2^2 + P_1^2 P_3^2 + P_2^2 P_3^2) +$$
$$+ a_{111}(P_1^6 + P_2^6 + P_3^6) + a_{112}[P_1^4(P_2^2 + P_3^2) + P_2^4(P_1^2 + P_3^2) + P_3^4(P_2^2 + P_1^2)] + a_{123}P_1^2 P_2^2 P_3^2, \tag{14}$$

$$w_{\text{elast}} = \frac{1}{2}c_{11}(u_{11}^2 + u_{22}^2 + \sigma_{33}^2) + c_{12}(u_{11}u_{22} + u_{11}u_{33} + u_{22}u_{33}) + \frac{1}{2}c_{44}(u_{32}^2 + u_{13}^2 + u_{12}^2), \tag{15}$$

$$w_{\text{str}} = q_{11}(u_{11}P_1^2 + u_{22}P_2^2 + u_{33}P_3^2) + q_{12}\left(u_{11}(P_2^2 + P_3^2) + u_{22}(P_1^2 + P_3^2) + u_{33}(P_2^2 + P_1^2)\right) +$$
$$+ q_{44}(u_{23}P_2 P_3 + u_{13}P_1 P_3 + u_{12}P_1 P_2). \tag{16}$$

$$w_{\text{grad}} = f_{ijkl}(u_{ij}P_{k,l} - u_{ij,l}P_k) + g_{ijkl}P_{k,l}P_{i,j} + h_{ijklmn}u_{ij,k}u_{lm,n}. \tag{17}$$

Here a are the Ginzburg-Landau coefficients, c_{ij} are the elastic constants/moduli of elasticity, q_{ij} are the coefficients of electrostriction, f_{ijkl}, g_{ijkl}, h_{ijklmn} are the gradient coefficients. Calculation results for solving the equations (5) with the above boundary conditions are presented in figures 1(b), 2.

4. Conclusion

The model outcome allows us to draw several important conclusions. Among them is the fact that the multicaloric effect in nanometer-sized samples can significantly (several times) exceed the effect in structures larger than 1mm. The latter finding can be used for creating a chip-size solid-state cooler as the obtained values of the temperature change for the multicaloric effect are sufficient for the operation of the device. The attention of the study has been directed toward the flexocaloric effect. As the nanotube size decreases, the FCE increases. Moreover, the flexocaloric effect can be either positive or negative and strongly dependent on the direction of the strain gradient. The difference can reach dozens of times. In this study we have considered only the strain gradient in the radial direction. The sign and magnitude of FCE may be different for other directions.

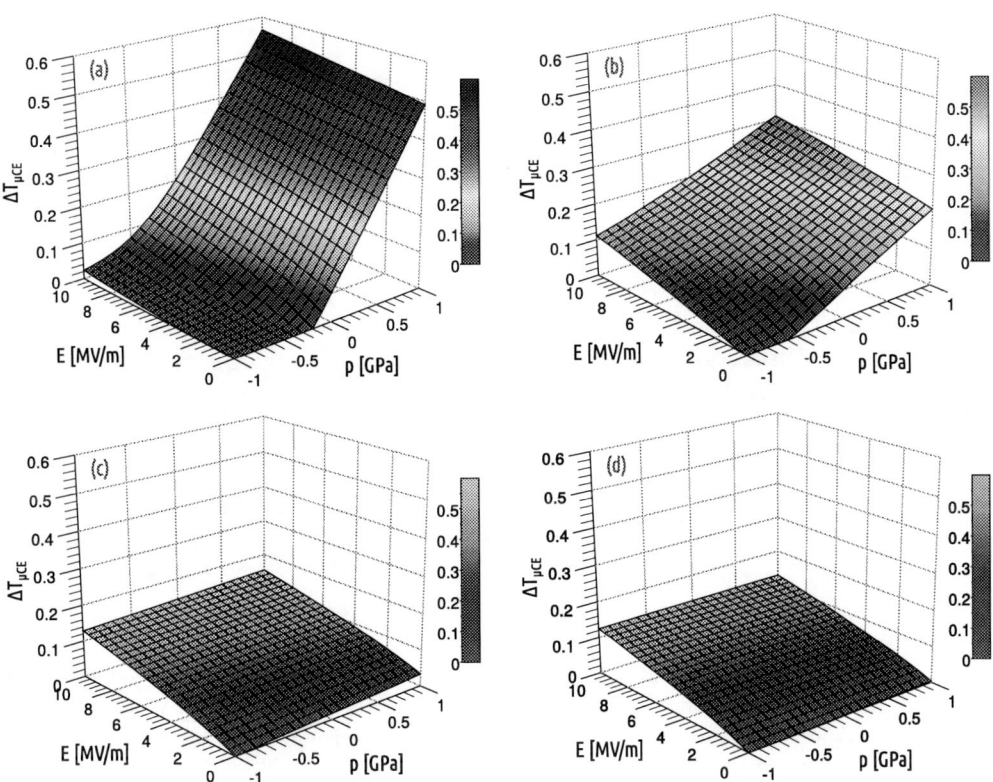

Figure 2. The temperature change in the multicaloric effect for a barium titanate nanotube **(a)** 2nm, **(b)** 5nm, **(c)** 20nm, and **(d)** 50nm thick.

Acknowledgments

The reported study was supported by RSCF, research project No.18-19-00512.

References

[1] Starkov A S, Starkov I A 2014 *J. Exp. Theor. Phys.+* **119** 258
[2] Tagantsev A K 1987 *Sov. Phys. Uspekhi* **30** 588
[3] Patel S, Chauhan A, Cuozzo J, Lisenkov S, Ponomareva I, Vaish R 2016 *Appl. Phys. Lett.* **108** 162901.
[4] Starkov A S, Starkov I A 2016 *Phys. Solid State +* **58** 1798
[5] Mindlin R D 1965 *Int. J Solids Struct.* **1** 417
[6] Majdoub M S, Sharma P, Cagin T 2008 *Phys. Rev. B* **77** 125424.5
[7] Zubko P, Catalan G, Tagantsev A K 2013 *Ann. Rev. Mater. Res.* **43** 387

SPBOPEN 2018   IOP Publishing

Scanning probe microscopy of AlGaAs/GaAs diode after partial electrical breakdown

A O Mikhaylov[1], P A Alekseev[1], A A Podoskin[1], S O Slipchenko[1] M S Dunaevskiy[1]

[1]Ioffe Institute, 194021, Saint-Petersburg, Russia

Abstract. Scanning probe microscopy allows to study surface processes in semiconductor heterostructures. This work shows the investigation of topography and surface potential distribution in AlGaAs/GaAs p-n diode after applied reverse voltage. Formation of oxide on the surface in the area of p-n junction was found due to increasing of the applied reverse voltage. Formed oxide increases a breakdown voltage by 20%.

1. Introduction.

GaAs semiconductor heterostructures are used to create electronic devices such as diodes, transistors, laser diodes, HEMT[1]. Main drawback of GaAs heterostructers is high surface state density, which leads to a pinning of the Fermi level on the surface and increasing of surface leakage current[2]. Usually, increasing of surface leakage current relate to formation of natural oxide on the surface of GaAs. Conventionally, a surface passivation is used to prevent these negative effects[3].

Recently we have found an increasing of break-down voltage effect of GaAs/AlGaAs diodes with unpassivated surface. This effect was realized by cycle increasing of break-down voltage. In this work the process of increasing the break-down voltage of GaAs/AlGaAs heterostructers with appearance of surface leakage current will be studied by using a Scanning Probe Microscope (SPM).

2. Samples and methods.

Heterostructure was created on n-GaAs substrate by MOCVD methods. The following layers were grown on the substrate: $Al0.35Ga0.65As$ (n = $1.5 * 1018$ cm-3, with thickness of 1.47 um); GaAs (p = $3-3.5 * 1016$ cm-3, with thickness of 2 um); GaAs (p = $6*1019$ cm-3, with thickness of 0.4 um). Ohmic contacts were formed on the compositions of AuGe/Ni/Au for n type and AuZn/Au for p type. The structure was cleaved on the elements with size: 400x400 um. The elements were installed on the cuprum base, using an indium solder. Figure 1. a) shows the topography of a surface of the cleavage in location of p-n junction. The difference between AlGaAs and GaAs layers on the topography is explained by the greater layer's thickness of surface oxide of AlGaAs[4]. Height difference in this location is about 1 nm.

SPM methods allows to investigate physical processes on the surface of structure. In addition to standard topography measuring, a surface potential distribution was measured. The surface potential was obtained by Gradient Kelvin-probe microscopy (GKPM)[5]. The study was carried out on the scanning probe microscope Ntegra AURA (NT – MDT company, Zelenograd, Russia). In this experiment we use silicon probes with conductive Pt/Ir coating. One part of experiment was carried out in the vacuum (P= 10-2 mbar), other part in the atmospheric conditions (P=1 bar, humidity 24%).

Content from this work may be used under the terms of the Creative Commons Attribution 3.0 licence. Any further distribution of this work must maintain attribution to the author(s) and the title of the work, journal citation and DOI.
Published under licence by IOP Publishing Ltd

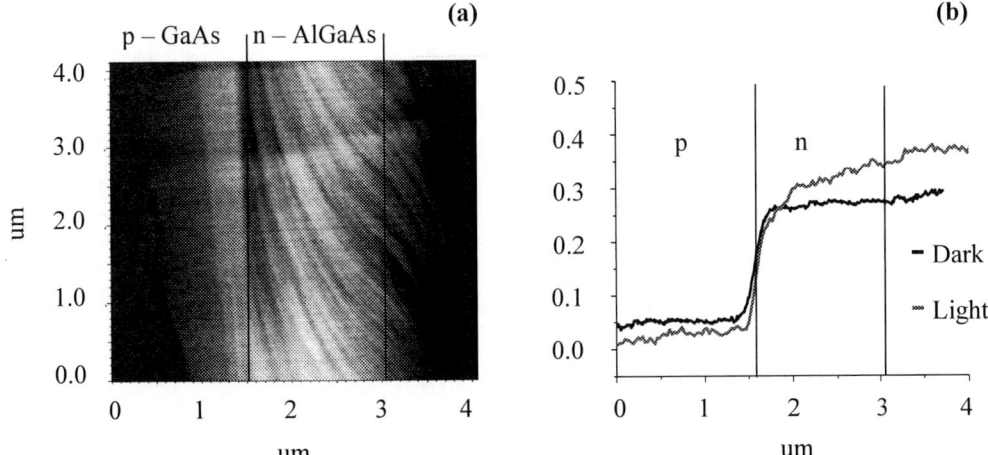

Figure 1. GaAs/AlGaAs diode structure. **a)** Topography of the chip's surface in location of p-n junction. **b)** Average profile of the surface potential distribution in the dark and in the light.

The figure 1. b shows the profiles of the surface potential distribution. In this case the surface potential is the difference between a work function of the SPM probe and the surface in the measurement point. Without light illumination the maximum contrast of the surface potential distribution is of 0.2 eV. This is explained by a pinning of the Fermi level on the surface in the middle of bandgap. Indeed, difference between the Fermi level pinning on the surface in p+ and n+ layers is about 0.2 eV in the AlxGa1-xAs alloys[6,7].

Under the light illumination, nonquilibrium charge carriers under the effect of near-surface field move to the surface and reduce the near-surface band bending. This leads to increasing the difference between Fermi level pinning of p and n layers. As the near-surface field has opposite directions in the p and n layers, the illumination increases the surface potential in n layer, and reduces in p layer[8]. From the Figure 1.a, b it follows that location of p-n junction coincides with GaAs/AlGaAs heterointerface with accuracy of 100 nm.

3.Results and discussion.

For the study of the effect of a break-down voltage increasing, it was used a 'training' procedure. The training is a process of increasing of a break-down voltage of the structure, using cyclic increasing of external reverse voltage during the measurement of an I-V curve. Figure 2 shows topography (a), surface's profiles (b) and I-V curves (c), which are corresponding to the three stages of training.

The first training was performed in vacuum conditions (Fig. 2.a, 2 picture; Fig. 2.b, c, blue lines). Using the measured I-V curves (Fig. 2.c, blue line), it's possible to find the break-down voltage, which was of 20 V. Profiles and topographies which are corresponding to first training are different from profiles and topographies of not trained structure (Fig. 2.a, 1 picture; Fig. 2.b, red line). The changes of structure topography are explained by the process of oxide formation in location of p-n junction. After the first training the height of oxide was 4 nm (Fig. 2.a, 2 picture; Fig. 2.b, blue line). The process of oxide formation will be explained below. After the second training the height of oxide was increased by 1 nm (Fig. 2.a, 3 picture; Fig. 2.b, green line). I-V curve (Fig. 2.c, green line) has changed compared to the first training (Fig. 2.c, blue line). Indeed, the break-down voltage was increased.

SPBOPEN 2018   IOP Publishing

IOP Conf. Series: Journal of Physics: Conf. Series **1124** (2018) 081013   doi:10.1088/1742-6596/1124/8/081013

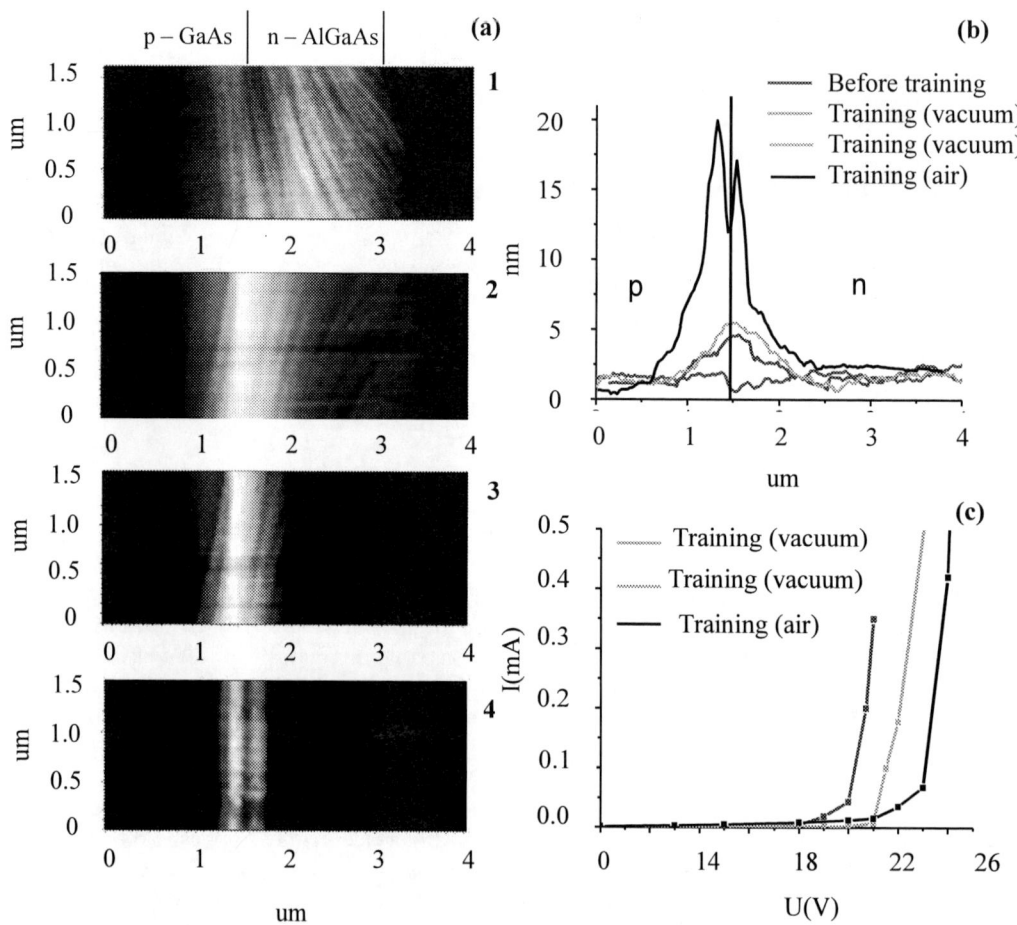

Figure 2. a) Surface topography: **1.** Before training. **2.** After the first training (vacuum). **3.** After the second training (vacuum). **4.** After third training (atmospheric conditions). **b)** Surface profiles which is corresponding to topography. **c)** I-V curves, which is corresponding each training process.

Training in the atmospheric conditions leads to the most increasing of oxide height in location of p-n junction (Fig. 2.a, 4 picture; Fig 2.b, black line). The height of the oxide was 15-20 nm. The break-down voltage was also increased (Fig. 2.c, black line).

From the figure 2 it is follows that oxide formation depends on the conditions of training and the applied reverse voltage. Break-down voltage depends on the height of oxide. For the explanation, one can assume that the formation of the oxide blocks the channel of surface leakage current.

Figure 3 schematically shows the oxidation process. This process is similar to a process of local anodic oxidation [9]. Oxide formation in location of p-n junction depends on several factors: presence of a surface absorbed water film, high electric field strength for water dissociation, and conditions for current flowing.

Conventionally, the water film presence on the surface of the semiconductors in the atmospheric ambient conditions. However, the formation of the oxide was observed also during the training in vacuum conditions. This oxidation was due to a moderate vacuum pressure (10-2 mbar). It was shown that the water film remains on the surface in these conditions [10]. The oxide formation processes are described by the following reactions:

1040

For n layer:

$$2AlGaAs + 6H_2O \rightarrow Al_2O_3 + Ga_2O_3 + 2As_2O_3 + 6H_2 \qquad (1)$$

For p layer:

$$2GaAs + 3H_2O \rightarrow Ga_2O_3 + 2As_2O_3 + 6H_2 \qquad (2)$$

In this case, the presence of strong electric field is necessary for oxide formation. Molecule of water dissociates on the hydroxyl group and hydrogen under the influence of this field. Under the reverse voltage of 20 V, the electric field in the area of p-n junction increases to the level of 105 V/cm which is leads to water dissociation[9].

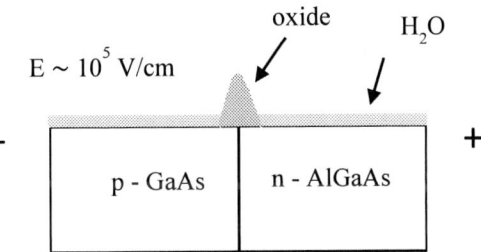

Figure 3. Model of the surface anodic oxidation in the area of p-GaAs/n-AlGaAs junction.

Increasing of the water thickness on the surface in the atmospheric conditions increases the surface leakage current. Indeed, in a voltage range of 10-18 V (Fig. 2) the current in the atmospheric conditions (black line) is bigger than current in the vacuum conditions. From the other hand, a thicker water film leads to thicker oxide formation during the training in the atmospheric conditions.

4. Conclusion.
Thus, the surface of the cleavage of the p-GaAs/n-AlGaAs diode was studied by scanning probe microscopy methods. From the topography and surface potential distributions, a location of the p-n junction was find. Applying of the reverse voltage exceeding a breakdown leads to a formation of surface oxide in the area of the junction. This oxidation is followed by an increasing of the breakdown voltage by 20%. Local anodic oxidation mechanism was proposed for the oxide formation. Due to this mechanism a surface water film dissociates in the p-n junction electric field. Products pf the dissociation oxidizes GaAs and AlGaAs. Formation of the oxide blocks a surface leakage current.

References
[1]   Vinokurov D, Zorina S, Kapitonov V, Murashova A, Nikolaev D, Stankevich A, Khomylev M, Shamakhov V, Leshko A Y, Lyutetskii A and others 2005 High-power laser diodes based on asymmetric separate-confinement heterostructures *Semiconductors* **39** 370–373
[2]   Hasegawa H, Akazawa M, Domanowska A and Adamowicz B 2010 Surface passivation of III–V semiconductors for future CMOS devices Past research, present status and key issues for future *Applied surface science* **256** 5698–5707
[3]   Geydt P, Alekseev P, Dunaevskiy M, Haggrén T, Kakko J-P, Lähderanta E and Lipsanen H 2016 Influence of surface passivation on electric properties of individual GaAs nanowires studied by current–voltage AFM measurements *Lithuanian Journal of Physics* **56**
[4]   Ankudinov A, Evtikhiev V, Tokranov V, Ulin V and Titkov A 1999 Nanorelief of an oxidized cleaved surface of a grid of alternating Ga 0.7 Al 0.3 As and GaAs heterolayers *Semiconductors* **33** 555–558

[5]   Dunaevskiy M, Alekseev P, Girard P, Lashkul A, Lahderanta E and Titkov A 2012 Analysis of the lateral resolution of electrostatic force gradient microscopy *Journal of Applied Physics* **112** 064112
[6]   Spicer W E, Lindau I, Skeath P, Su C Y and Chye P 1980 Unified Mechanism for Schottky-Barrier Formation and III-V Oxide Interface States *Phys. Rev. Lett.* **44** 420–423
[7]   Alekseev P, Dunaevskiy M, Cirlin G, Reznik R, Smirnov A, Davydov V Y and Berkovits V 2017 Unified mechanism of the surface Fermi level pinning in III-As nanowires *arXiv preprint arXiv:1710.06227*
[8]   Ankudinov A, Evtikhiev V, Ladutenko K, Rastegaeva M, Titkov A and Laiho R 2007 Kelvin probe force and surface photovoltage microscopy observation of minority holes leaked from active region of working In Ga As/ Al Ga As/ Ga As laser diode *Journal of applied physics* **101** 024504
[9]   García R, Calleja M and Pérez-Murano F 1998 Local oxidation of silicon surfaces by dynamic force microscopy: Nanofabrication and water bridge formation *Applied Physics Letters* **72** 2295–2297
[10]   Dunaevskiy M, Alekseev P, Girard P, Lahderanta E, Lashkul A and Titkov A 2011 Kelvin probe force gradient microscopy of charge dissipation in nano thin dielectric layers *Journal of Applied Physics* **110** 084304

SPBOPEN 2018   IOP Publishing

Methods of applying the reliability theory for the analysis of micro-arc oxidation process

P E Golubkov, E A Pecherskaya, Y V Shepeleva, A V Martynov, T O Zinchenko, D V Artamonov[1]

[1]Polytechnic Institute, Penza State University, Penza 440026, Russia

Abstract. To improve the handling of micro-arc oxidation (MAO) process in this paper, the analysis of factors influencing the properties and operating characteristics of MAO-coatings and their classification on the basis of methods of reliability and quality theory (cause-effect diagram and the relationship diagram) were done. The main factors influencing the MAO process were revealed and recommendations on the use of obtained results in theoretical and experimental studies were done.

1. Introduction

Micro-arc oxidation (MAO) technology exists since the 70-ies of XX century, but still here the handling of the MAO process and its automation remains the problem. Currently, this lack is eliminated by practicing the process. However, this approach is long and time-consuming, so the challenges of MAO process handling and the building of MAO automated systems are becoming increasingly real [1].

The main difficulty, with which the scientists face in solving these problems, is a large number of factors that affect properties of the MAO-coating (current density, anode and cathode currents ratio, MAO processing time, temperature and development of electrolyte, etc.) which directly affects the quality of the received product.

To eliminate these difficulties, it is advisable to systematize all factors according to the influence degree on the MAO process and identify the main ones. The similar systematization was proposed in the paper [2], but it is not sufficiently complete and does not take into account some interrelations. Reliability and quality theory methods, such as the cause-effect diagram (Ishikawa diagram) and the relationship diagram, on the other hand, can give a fairly exhaustive description of factors interrelations affecting the MAO process. All initial data to build these diagrams are set out in [3].

2. Cause-effect diagrams

Analysis of cause-effect diagram makes it possible to identify the main factors (reasons) that affect this coating property or one of its operation characteristics (quality parameters). In this work the Ishikawa diagrams were built for all quality parameters of MAO-coatings on aluminium and its alloys in silicate-alkaline electrolyte: microhardness, wear resistance, corrosion resistance, electrical strength, thermal conductivity (Figures 1 and 2) and also factors influencing each of them were identified.

Ishikawa diagram for the microhardness of the MAO-coatings is shown in Figure 1 (a). The stem charts the cause-effect relationship are shown by arrows, they are directed from cause to effect. For example, the arrow "workpiece surface finish - microhardness" indicates that the surface finish affects the microhardness, and not vice versa. If the arrow points in both directions, it means that the cause and effect have the same effect on each other. From Figure 1 we see that the microhardness of the

---

Content from this work may be used under the terms of the Creative Commons Attribution 3.0 licence. Any further distribution of this work must maintain attribution to the author(s) and the title of the work, journal citation and DOI.

Published under licence by IOP Publishing Ltd

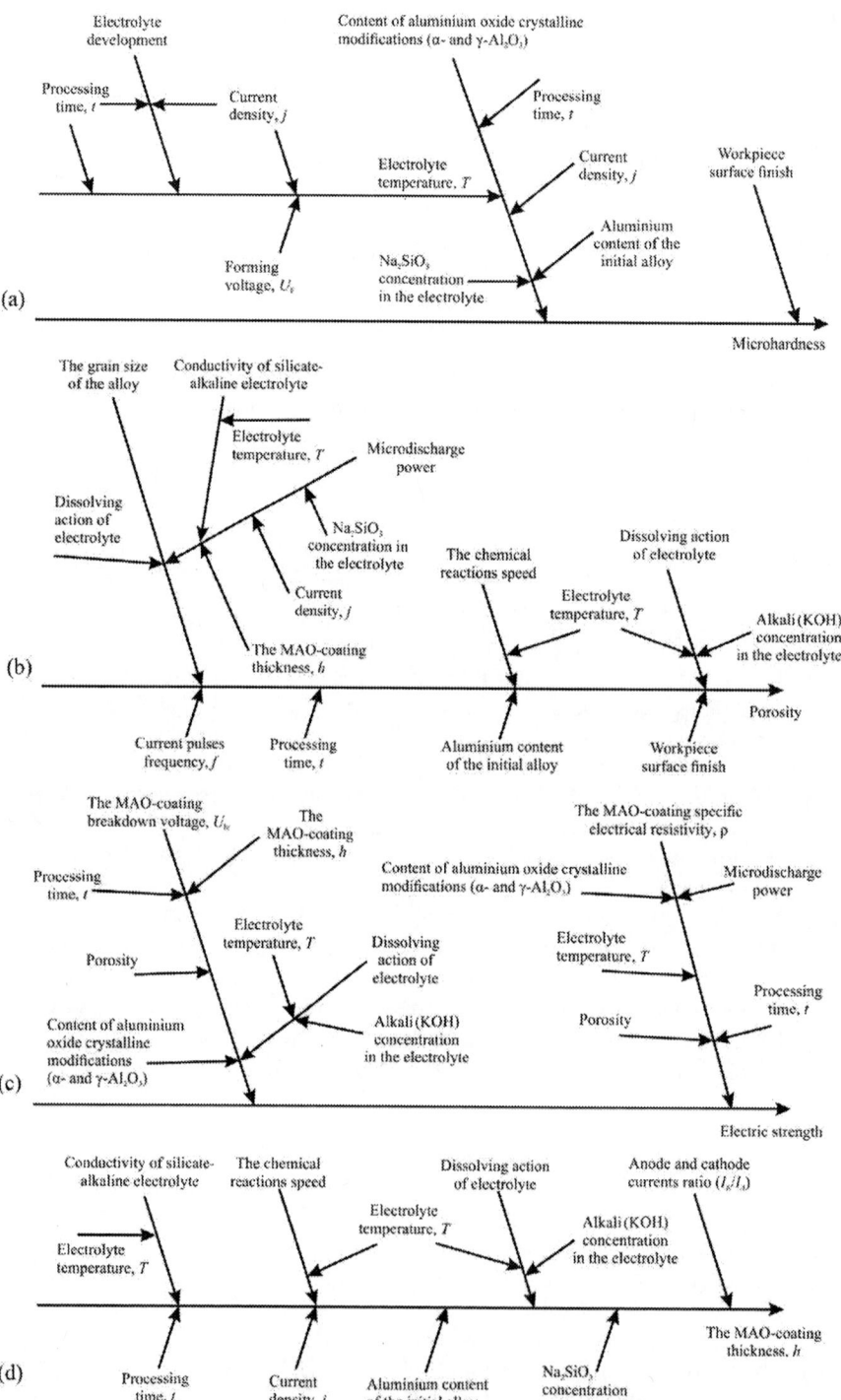

Figure 1(a, b, c, d). Cause-effect diagrams of MAO-coating quality parameters: **(a)** microhardness; **(b)** porosity; **(c)** electrical strength; **(d)** thickness.

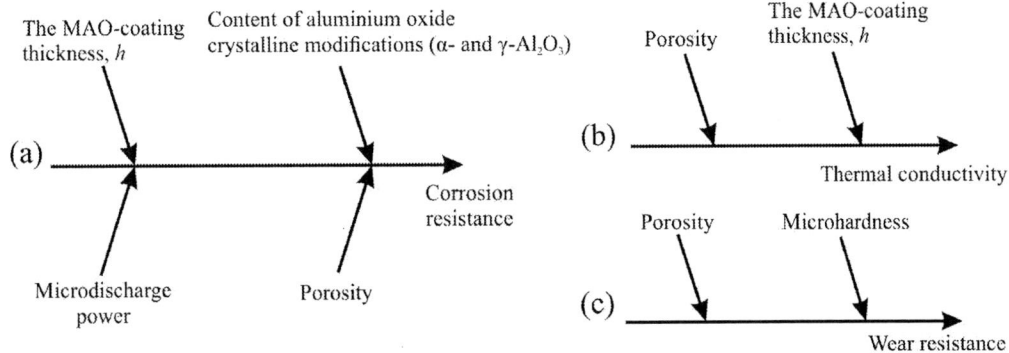

Figure 2(a, b ,c). Cause-effect diagrams of MAO-coating quality parameters: **(a)** corrosion resistance; **(b)** thermal conductivity; **(c)** wear resistance.

MAO-coating is influenced by the composition of the initial alloy and surface finish of sample, the liquid glass Na_2SiO_3 concentration in the electrolyte, current density, forming voltage, processing time and the electrolyte temperature. Factors that affect all the other properties of coatings and the quality parameters of the MAO process were identified (Table 1).

Table 1. Interrelation of MAO-coating quality parameters with the MAO process influencing factors.

Quality parameter/ coating property	Influencing factors
Microhardness	j, t, T, Al content in the initial alloy, Na_2SiO_3 concentration, surface finish of the sample
Thickness, h	$j, t, T, I_K/I_A$, Al content in the initial alloy, Na_2SiO_3 and KOH concentration
Porosity	j, t, T, f, h, Na_2SiO_3 concentration, surface finish of the sample
Electric strength	h, porosity
Corrosion resistance	h, porosity
Thermal conductivity	h, porosity
Wear resistance	microhardness, porosity

In Table 1, j is the current density, t is the treatment time, T is the electrolyte temperature, f is the current pulse frequency, I_K/I_A is the ratio of the anode and cathode currents, and h is the MAO coating thickness. If the quality parameter depends on the coating property (for example, the thermal conductivity depends on the thickness and porosity), then it automatically inherits all the causes of this property. Thus, the thermal conductivity (as well as the electrical strength and corrosion resistance) will depend on the current density j, the treatment time t and the electrolyte temperature T, the current pulse frequency f, the anode and cathode currents ratio, the electrolyte composition (KOH and Na_2SiO_3 concentrations) and the initial alloy composition (aluminium percentage in it). It should be noted that the porosity in this case also depends on the thickness, but indirectly on the microdischarges power.

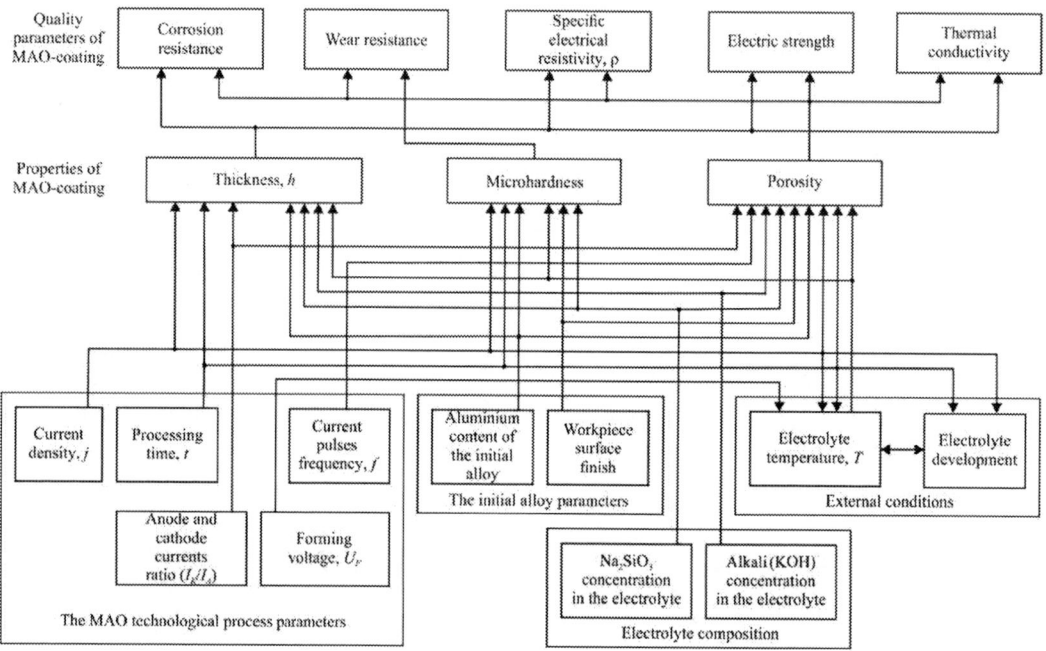

Figure 3. Relationships diagram.

3. Relationships diagram

In accordance with Table 1, a diagram of factors influencing the MAO process with the properties of the coatings obtained and their quality parameters was constructed (Figure 3). The directions of the arrows were chosen on the same principle as for the Ishikawa diagrams. The relationships diagram makes it possible to identify the main influencing factors for the entire MAO process, and not just for a particular coating property or quality parameter.

When analyzing the relationships diagram, the following was found. All quality parameters are affected only by three MAO coating properties: microhardness, thickness and porosity, and microhardness only affects wear resistance. However, as follows from the Ishikawa diagrams, the intermediate reason for the "crystalline aluminium modifications content", which mainly determines microhardness, also holds for electrical strength and corrosion resistance.

All the factors of the MAO process, which can be divided into the parameters of the technological process and the initial alloy, the electrolyte composition and the external conditions, somehow influence the MAO-coatings properties. The main parameters of the MAO technological process are the current density and processing time. They are associated with all coatings properties, and also provide external conditions (temperature and electrolyte development). Forming voltage is also one of the main factors of the MAO process, but in many cases it affects the coating properties only indirectly, through the electrolyte temperature. The current pulses frequency only affects the porosity. The anode and cathode currents ratio affects thickness and porosity, but these coating properties determine all products quality parameters (wear resistance in part) with MAO-processing.

In the case of the initial alloy parameters, the following regularities can be distinguished. First, the higher the aluminium percentage in the material workpiece, the more high-quality MAO-coating on it is formed. Thus, products made of technical AD0 aluminium (99.8 % aluminium content) are best suited for MAO-processing, followed by AMg3 (94.9 % aluminium content) and duralumin D16

(92.8 % aluminium content) [4]. This fact is explained by the one that in the aluminium alloys composition there are chemical elements that discourage the aluminium oxide formation, as a result of which the MAO-coating initial structure has a number of "weak" places, the electric breakdown occurs at lower voltage values, and the conditions for γ-Al$_2$O$_3$ – α-Al$_2$O$_3$ phase transition are realized to a lesser voltage [3]. Secondly, the workpiece surface finish affects the microhardness and porosity differently: the higher the surface finish, the microhardness is greater and the porosity is less. This should be taken into account when planning the MAO process.

A special interest is the influence of external conditions (temperature and electrolyte development) on the MAO-coating properties. As can be seen from Figure 3, the electrolyte temperature affects all properties and, consequently, all the quality parameters of the MAO-coating. However, it, as well as the electrolyte development, depends on the current density and the processing time (and also on the forming voltage), and besides, the temperature and the electrolyte development are interdependent. It can be explained as follows (Figure 1 (a)). The electrolyte development is a decrease of the ions concentration in it, due to their inclusion in the structure of the MAO-coating. With increasing current density in microdischarge, the electrolyte is more impoverished by ions, its conductivity decreases [5], and the resistance and voltage drop increases, hence the power dissipated by the electrolyte increases, which leads to its heating. On the other hand, as the temperature of the electrolyte increases, the rate of chemical reactions increases, more molecules of the electrolyte dissociate into ions, and more of them are incorporated in the coating, which further increases the electrolyte temperature. Similarly, the external conditions are affected by the time of MAO-treatment.

The effect of the electrolyte composition is also ambiguous. In general, the concentration of substances forming the electrolyte is set once before the beginning of the MAO process and does not change during processing (this is true for KOH). At the same time, the concentration of technological liquid glass Na$_2$SiO$_3$, apparently, is a function of the current density, processing time and electrolyte temperature and determines its development, since silicate ions are incorporated into the coating [3]. In addition, this parameter can indirectly, through the electrolyte temperature, affect all MAO-coatings properties.

From the point of view of the MAO process mathematical description, the quality parameters can be represented as functions of several variables, the parameters of the technological process are independent variables, while the external conditions are also functions of these parameters. The parameters responsible for the composition and the initial alloy purity, as well as the concentration of alkali (KOH) in the electrolyte, can be represented in the form of constants, since they do not change during the MAO-process. Thus, for effective control of the MAO-process it is necessary to take into account the following regularities:

$$T = F\left(j, t, U_F\right)$$
$$C_{Na_2SiO_3} = F\left(j, t, T\right)$$
$$HV = F\left(j, t, T, p_{Al}, C_{Na_2SiO_3}, S\right)$$
$$h = F\left(j, t, T, I_K/I_A, p_{Al}, C_{Na_2SiO_3}, C_{KOH}\right)$$
$$P = F\left(j, t, T, f, I_K/I_A, p_{Al}, C_{Na_2SiO_3}, C_{KOH}, S\right), \tag{1}$$
$$E = F\left(h, P\right)$$
$$CR = F\left(h, P\right)$$
$$TC = F\left(h, P\right)$$
$$WR = F\left(HV, P\right)$$

where HV - microhardness;
h and P - the MAO-coating thickness and porosity, respectively;
E - electric strength;
CR - corrosion resistance;

TC - thermal conductivity;

WR - wear resistance;

U_F - forming voltage;

I_K/I_A - anode and cathode currents ratio;

$C_{Na2SiO3}$ - liquid glass concentration in the electrolyte;

$C_{KOH} = const$ - alkali concentration in the electrolyte;

$p_{Al} = const$ - aluminium percentage in the initial alloy;

$S = const$ - the workpiece surface finish.

Considering the fact that T and $C_{Na2SiO3}$ are functions, we get:

$$
\begin{aligned}
HV &= F(j,t,U_F,p_{Al},S) \\
h &= F(j,t,U_F,I_K/I_A,p_{Al},C_{KOH}) \\
P &= F(j,t,U_F,I_K/I_A,f,p_{Al},C_{KOH},S) \\
E &= F(h,P) \\
CR &= F(h,P) \\
TC &= F(h,P) \\
WR &= F(HV,P)
\end{aligned}
\tag{2}
$$

From (2) it follows that the main parameters of the MAO process are the current density, the processing time, the forming voltage, the anode and cathode currents ratio, and the frequency of the current pulses. All other parameters either depend on the above, or are constants determined before the MAO process.

4. Conclusion

Thus, the methods of the reliability theory are presented by the informative way to search for cause-effect relationships in complex multifactor processes, such as MAO. Together with the quality control methods, they serve as the basis for the construction of the most common, primary mathematical models, as well as the hypothesis requiring experimental confirmation while studying factors affecting the quality of MAO-coatings.

References

[1] Golubkov P E, Pecherskaya E A, Karpanin O V, Shepeleva Y V, Zinchenko T O, Artamonov D V 2017 *J. of Phys.: Conf. Series* **917** 092021

[2] Mikheev A E, Trushkina T V, Girn A V, Ravodina D V 2015 *Bulletin of the Sib GAU* **16** 2

[3] Kazantsev I A and Krivenkov A O 2007 *Technology of obtaining composite materials by the micro-arc oxidation* (Penza: Information publishing center PSU) p 240

[4] Interstate Standard GOST 4784-97 2009 *Aluminium and aluminium alloys, deformable. Stamps* (Moscow: Standartinform) p 21

[5] Kuchmin I B, Nechaev G G and Solovieva N D 2013 *Chemistry and Chemical Technologies. Bulletin of the CSTU* **4** 73 57-62

SPBOPEN 2018   IOP Publishing

IOP Conf. Series: Journal of Physics: Conf. Series **1124** (2018) 081015   doi:10.1088/1742-6596/1124/8/081015

Structure of "Chromium steel-base – Ti-coating" and its production by the contact welding

I Egorov[1], A Shchelkunov[1], A Fomin[1], I Rodionov[1]

[1] Yuri Gagarin State Technical University of Saratov, Saratov 410054, Russia

Abstract. This paper describes an innovative approach for the production of a weld joint between a steel base and a titanium plate using a method of contact welding. As a result, a "Chromium steel-base – Ti-coating" structure with high strength and hardness about 450 HV was formed. The possibility of further machining, including drilling, turning and grinding for the production of rectangular plates was shown.

1. Introduction

To improve the performance characteristics of metal products, their surface is treated by various methods, including the application of functional films and coatings [1,2]. To improve the corrosion resistance and operational reliability of products that are used at high temperatures, heat resistant materials are deposited on their surface, e.g. high-alloyed chromium-nickel steels, cermet compounds (nitrides, oxides) and some pure refractory metals. Functional coatings can be applied by various gas-thermal methods (physical vapour deposition (PVD), high velocity oxygen fuel (HVOF) and plasma spraying [1], electrospark alloying and deposition [2]), but there are difficulties in combining non-homogeneous metals, in particular a steel base and a titanium coating.

One of the methods for obtaining functional metal layers is welding and surfacing. To increase the strength of the welded joint between steel and titanium, thin vanadium interlayers are used [3]. Vanadium, which is more refractory compared to titanium and iron, provides a higher strength of the welded joint. In electric arc welding, e.g. TIG, intermetallic phases with high hardness (~1112 $HV_{0.01}$) appear in the transition region of the interaction of dissimilar materials (Fe-Ti, Fe-V). The average hardness of the welding area is 275–400 HV. To improve the quality of Fe-Ti compound, eutectic Cu-V alloys are also used [4]. The formed Fe_2Ti intermetallic phase and FeTi+β-Ti mixtures have a high hardness from 10 to 23 GPa.

The problem of welding low-carbon steel and titanium is also solved in the case of friction welding [5]. In the tensile test, the fracture of welded joint samples occurs over steel, rather than in the area of titanium or welding. In the case of diffusion and laser welding, Nb [6,7] is used as the interlayer. For example, diffusion welding of AISI316L stainless steel and Ti-6Al-4V titanium alloy is conducted at a temperature of about 900±50 °C for the time not exceeding 90 min at the pressure of 30 MPa. The welded joint strength reaches a maximum value of about 500 MPa, while the microhardness of the welded joint is 100–200 HV. In the case of welding TC4 titanium alloy to stainless steel SUS301L, the strength of the joint equals 370 MPa.

Therefore, in this paper, the possible formation of "Chromium steel-base – Ti-coating" structure by the method of contact welding is discussed.

Content from this work may be used under the terms of the Creative Commons Attribution 3.0 licence. Any further distribution of this work must maintain attribution to the author(s) and the title of the work, journal citation and DOI.

Published under licence by IOP Publishing Ltd

2. Methodology

Experimental samples were made in the form of 0.5 mm thick titanium Grade 2 and 3.5 mm thick chromium steel plates (carbon – 0.9–1.0 wt.%, chromium – 17–19 wt.%). The layered "Steel – Ti" structure with a total thickness of 4–4.2 mm was obtained by contact welding (Figure 1a).

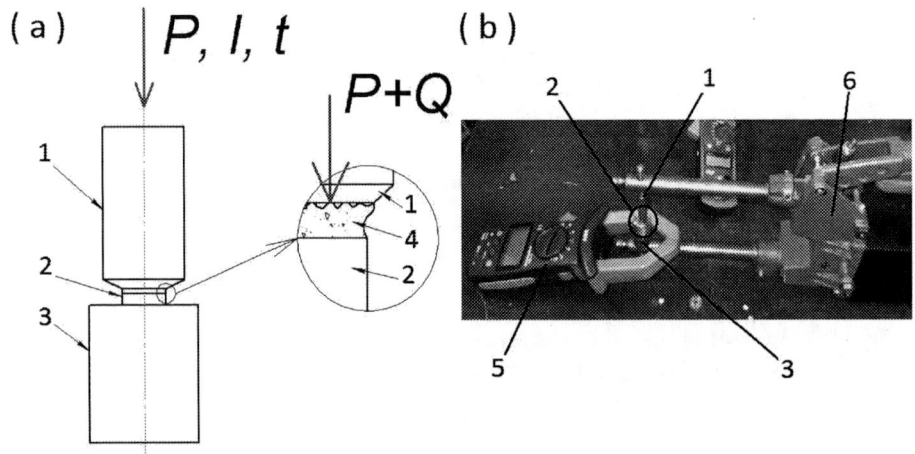

Figure 1(a, b). The scheme showing the contact welding and the arrangement of electrodes *1,3* and welded dissimilar materials *2,4* **(a)**; measuring the current provided by the contact welding device *6* using current clamps *5* **(b)**.

The working current was measured using current clamps (Figure 1b). Current strength in contact welding was 1.95 ± 0.05 kA, and the diameter of the wire electrode was not less than 5 mm. In this work, the influence of pulse duration from 250 to 1000 ms during the contact welding on the structural and mechanical characteristics was studied.

Further, the samples of the joints were processed by drilling, turning and grinding, while the total thickness of the titanium layer cut from the surface was 0.1 mm. In the course of machining, the parameters under which the obtained "Chromium steel-base – Ti-coating" system retained the required geometric parameters were determined.

The surface morphology of the samples was studied using optical microscopy (OM). To analyze the structural changes, the cross-sections of the composite "Steel – Titanium" structure were made. The Vickers method (at 50–200 gf) was used to test hardness. Measurement of the microhardness HV ensured the control of residual stresses in the welding region and the region of thermal influence.

3. Results

With a pulse duration of about 250 ms, a weld joint was formed with a low adhesive strength, which was observed during the subsequent machining. The titanium plate at the turning was separated from the steel base. When the pulse duration was increased to 500 ms, a stable attachment of titanium to the steel was observed, and during further machining the composite structure retained its integrity (Figure 2a). An increase in pulse to 750 ms also ensured a stable attachment of titanium on the steel, however, a more intense heating was observed in the area exposed to the electrode. This led to the oxidation of titanium and growth of hardness, which worsened the subsequent machining, i.e. there was an increased wear of the metalworking tool. With a pulse duration of about 1000 ms, the titanium plate melted to the full depth, which was accompanied by significant thermal deformations.

SPBOPEN 2018   IOP Publishing

IOP Conf. Series: Journal of Physics: Conf. Series **1124** (2018) 081015   doi:10.1088/1742-6596/1124/8/081015

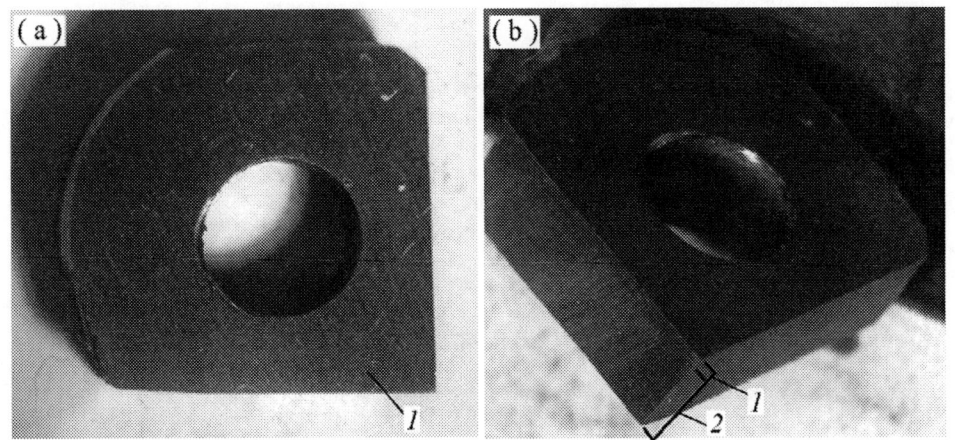

Figure 2(a, b). Top view of the sample of the weld joint after machining from the side of the titanium plate *1* **(a)**; a two-layer structure consisting of a steel base *2* and a titanium plate *1* **(b)**.

After machining, which included drilling, turning and grinding, a prismatic structure with a central hole was formed (Figure 2b). As a result, the contact welding modes were determined, in which the structure retained the required shape without collapsing under dynamic loads.

Thus, the optimal pulse duration equalled 500 ms. At the same time the hardness of the titanium plate was in the range from 290 to 320 HV. In the weld boundary, the hardness reached 450 HV, and the steel base had an average hardness of about 470–490 HV.

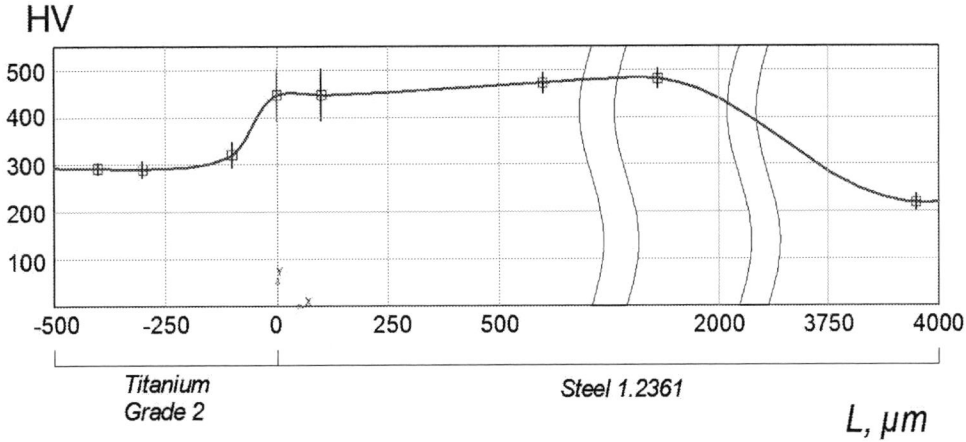

Figure 3. The microhardness HV distribution along the section in "Steel – Titanium" structure.

The initial hardness of the steel base was about 220–230 HV. Thus, under the influence of a powerful electric current pulse with an energy from 1 to 4 kJ, the following processes occurred in the welded joint: heating of the contact surfaces, local melting, phase transitions in the solid state,

1051

including quenching and average tempering. The achieved strengthening of the contact area did not prevent further machining, including drilling and reaming.

4. Conclusion

Thus, the possibility of an effective joining of chromium steel (instrumental class) with commercially pure titanium was shown. As a result, "Chromium steel-base – Ti-coating" composite structure with a high adhesion strength was obtained, which can be further modified, e.g. by gas-thermal methods, to obtain cermet layers and coatings [8].

Acknowledgments

The research was supported by the Ministry of Education and Science of the Russian Federation and German Academic Exchange Service (DAAD) in the framework of the program "Mikhail Lomonosov" (project No. 11.12785.2018/12.2). The authors express their gratitude to Mrs. Marina Fomina and Dr. Vladimir Koshuro for the discussion and analysis of the results, as well as their assistance in carrying out experimental work.

References

[1] Koshuro V A, Nechaev G G, Lyasnikova A V 2014 *Tech. Phys.* **59** 1570
[2] Fomin A A, Fomina M A, Koshuro V A, Rodionov I V, Voiko A V, Zakharevich A M, Aman A, Oseev A, Hirsch S and Majcherek S 2016 *Tech. Phys. Lett.* **42** 932
[3] Chu Q, Zhang M, Li J, Yan C and Qin Z, 2017 *J. Mater. Process. Technol.* **240** 293
[4] Chu Q, Bai R, Zhang M, Li J, Lei Z, Hu N, Bell J M and Yan C, 2017 *Mater. Charact.* **132** 330
[5] Kimura M, Iijima T, Kusaka M, Kaizu K and Fuji 2016 *J. Manuf. Process.* **24(1)** 203
[6] Song T F, Jiang X S, Shao Z Y, Fang Y J, Mo D F, Zhu D G and Zhu M H 2017 *Vacuum* **145** 68
[7] Zhang Y, Sun D Q, Gu X Y and Li H M 2016 *Mater. Lett.* **185** 152
[8] Fomin A, Dorozhkin S, Fomina M, Koshuro V, Rodionov I, Zakharevich A Petrova N and Skaptsov A 2016 *Ceram. Int.* **42** 10838

SPBOPEN 2018 IOP Publishing

Structural transformations on the surface of 1.3343 tool steel and 12Cr18Ni10Ti stainless steel after induction heat treatment and quenching

P Palkanov[1], A Fomin[1], I Rodionov[1]

[1]Yuri Gagarin State Technical University of Saratov, Saratov 410054, Russia

Abstract. The article describes the use of induction heat treatment for products made of 1.3343 tool steel (analogue R6M5) and 12Cr18Ni10Ti chromium-nickel steel. The changes in the surface morphology parameters as well as in hardness after strengthening heat treatment were studied. The hardness of tool steel after quenching (from 1150–1200 °C) reached 940–1080 HV (9.2–10.8 GPa). During induction treatment (temperature 800 °C and duration not less than 120 s) of 12Cr18Ni10Ti steel, a metal oxide coating with a hardness of at least 9.55±1.97 GPa was formed.

1. Introduction

In machinery and instrumentation, as well as medicine, stainless steels and tool materials are often used, e.g. 12Cr18Ni9Ti, 12Cr18Ni10Ti, 316L, AISI 1.2361, 1.3343, etc. They are used for the manufacture of tools, including the surgical ones, as well as various orthopedic designs [1], implants, functional sensor elements, metal or metal-ceramic tips of indentors for various purposes [2,3]. The surface of these products must have a special combination of physical, chemical and mechanical characteristics. The metal substrate ensures resistance to mechanical loads of a distributed type, however, when interacting with hard tissues (bone), large concentrated mechanical stresses appear. Under these conditions, the main characteristics of the surface layer are hardness and wear resistance [4-7]. In addition, the functional coating on metal implants should also have certain indicators of roughness, porosity, micro- and nanocrystalline structure. As a rule, modification of the surface of metal products is performed by vacuum-condensation or electrochemical methods, as well as by gas-thermal oxidation. In this paper, the study results of structural changes in the surface and hardness of steel samples subjected to induction heat treatment (IHT) are presented.

2. Methodology

The samples were disks with a diameter of 10–15 mm and length of 2–3 mm fabricated from 12Cr18Ni10Ti chromium-nickel stainless steel and 1.3343 tool steel (Figure 1a). Their surface was subjected to the texturing air-abrasive (corundum with an average fineness of 150–500 μm) treatment. The surface of 1.334 tool steel was modified by fine grinding and polishing in order to obtain the necessary roughness parameter Ra = 0.08–0.16. The resulting metal substrates of samples were also subjected to ultrasonic cleaning in an ethanol solution.

Induction heating was performed in the temperature range from 800–850 to 1200–1300 °C. The samples of carbon steel were treated with paste and since their size was small, cooling of the samples in the air with shell residues was considered to be the quenching process (Figure 1b).

Content from this work may be used under the terms of the Creative Commons Attribution 3.0 licence. Any further distribution of this work must maintain attribution to the author(s) and the title of the work, journal citation and DOI.
Published under licence by IOP Publishing Ltd

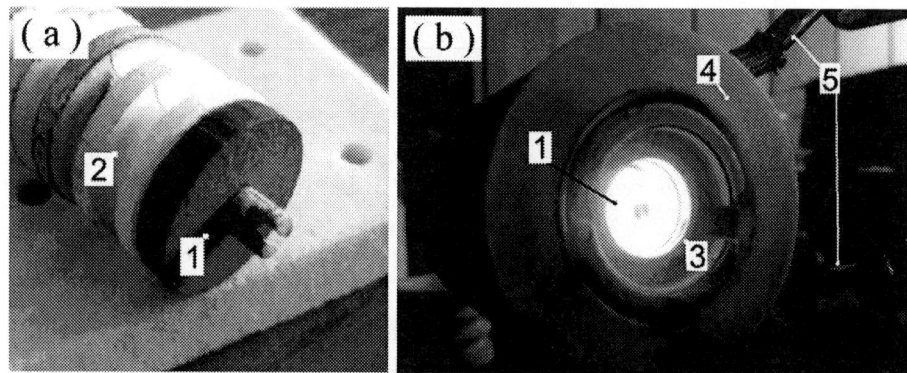

Figure 1(a, b). Samples of 1.3343 tool steel: 1 – a sample; 2 – a ceramic holder; 3 – a quartz chamber; 4 – a protective shell; 5 – a copper inductor **(a)**; as well as quenching with HFC **(b)**.

Scanning electron microscopy (SEM) with energy-dispersive X-ray analysis (EDX) was used to study the surface morphology and chemical element composition. To test hardness, the Vickers method was used.

3. Results

SEM results of the surface of stainless steel samples showed the presence of metal oxide coatings (Figure 2a). At the IHT temperature of 800 °C, grains with an average size of 180–450 nm were formed on the protrusions of the steel surface, and those of 80–150 nm were observed in the cavities (Figure 2b). Increasing the temperature to 1000–1200 °C lead to the formation of coatings with a loose structure and low adhesion.

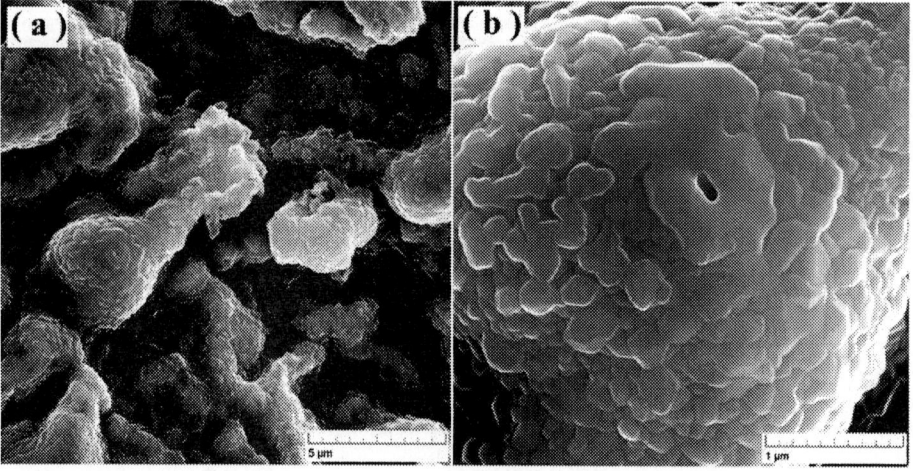

Figure 2(a, b). Morphology of the steel surface after IHT **(a)**; nanostructure of the metal oxide coating **(b)**.

The conducted chemical EDX analysis showed that in the coated samples, the main elements were Ni (42–67 at.%), Cr (29–37 at.%) and O (18–19 at.%). There were also impurities of Mn (1.2–3.4%), Fe (0.52–0.95%) and traces of Ti, Si, Al – less than 1.5 at.%.

The microstructure of tool steel had its distinctive features (Figure 3). A sample with the original structure was characterized by the presence of inclusions of refractory metal carbides, e.g. carbides of tungsten, molybdenum and chromium (Figure 3a). Their morphology was represented by round and oval formations with the size varying from 0.2–0.4 to 2–4 μm. When exposed to the temperature above 800 °C, carbides coagulated. However, when the temperature reached 1100–1200 °C, the microstructure was transformed (Figure 3b). Increased exposure promoted the formation of a carbide mesh and alloyed martensite. The chemical composition fully corresponded to that of 1.3343 tool steel (Figure 3b). The main elements were: W – 5.5–6 wt.%, Mo – 4–5 wt.%, Cr – 3.5–4 wt.%, V – 1–1.5 wt.%, C – 6.1–7.2 wt.% (elevated content typical for carbides), Fe – balance.

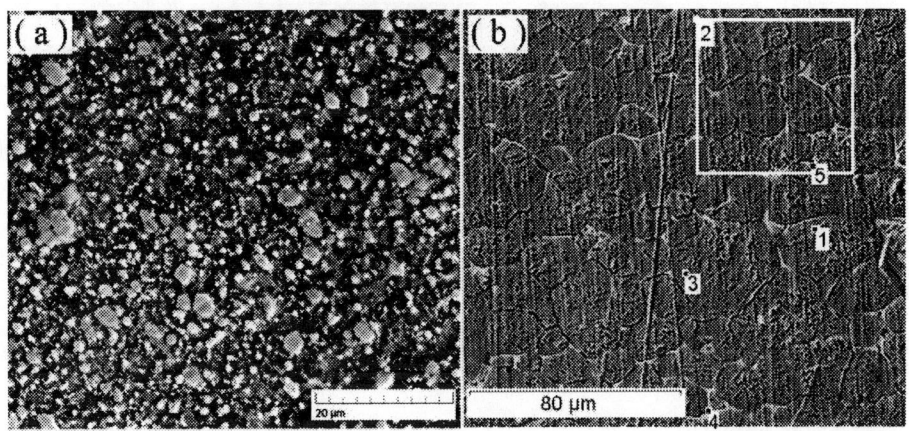

Figure 3(a, b). Microstructure of 1.3343 tool steel without heat treatment **(a)**; after high-temperature quenching with HFC (treatment temperature of 1300 °C) **(b)**.

During IHT (the temperature of 800 °C and exposure of about 90–120 s) of 12Cr18Ni10Ti steel, a metal oxide coating with a hardness of at least 9.55±1.97 GPa was formed. The hardness of 12Cr18Ni10Ti steel surface increased due to the formation of oxides of a complex composition. It is known that iron oxides are characterized by moderate hardness (5–6 levels according to the Mohs scale of hardness). Thus, high hardness values can be ensured by the oxides of alloying elements, e.g. chromium, nickel and titanium.

In the course of quenching of 1.3343 tool steel, the hardness also grew to 68–72 HRC (Figure 4). High hardness was achieved at a lower HFC quenching temperature. Thus, at T = 600 °C the steel hardness was maintained at the initial value of about 35 HRC. However, with an increase in the quenching temperature to T = 800 °C the hardness equaled 42–44 HRC. These hardness values do not show that steel could be used for a metalworking tool. Therefore, the quenching temperature must be above the phase transition, i.e. it must be at least T = 1000 °C. At this temperature the hardness reached 66–67 HRC, which is typical for high-speed tool steels of the tungsten-molybdenum group. However, due to quenching with HFC higher hardness values can be achieved, so at the temperature T = 1200 °C the hardness of steel reached 70–72 HRC. At the same time, an increase in the exposure duration at a given temperature did not contribute to an increase in hardness. Thus, it is possible to limit the duration of heat treatment with HFC to one cycle of heating and quiescent cooling.

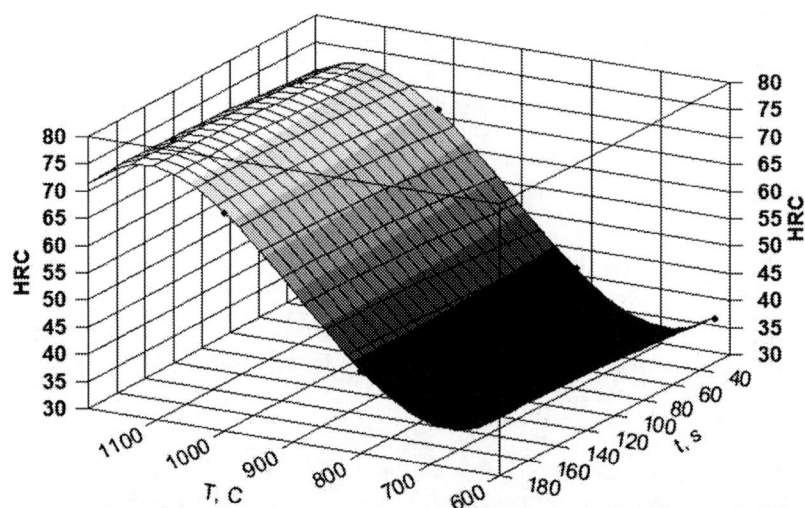

Figure 4. Dependency of hardness HRC of tool steel on the temperature T and exposure time t of the quenching with HFC.

At the high temperature induction treatment (quenching) of 1.3343 tool steel, the microstructure changed (carbides were allocated along the grain boundaries and martensite appeared). This was accompanied by an increase in hardness to 940–1080 HV (equivalent to 9.2–10.8 GPa).

4. Conclusions

Thus, the use of strengthening induction treatment of steel products ensured the formation of the required micro- and nanostructure of the surface and the near-surface layer. As a result of IHT the hardness of tool steel grew, however, to increase the hardness of stainless steel, it is necessary to provide the conditions for the formation of a metal oxide coating (gas-thermal oxidation).

Acknowledgments

The research was supported by the Ministry of Education and Science of the Russian Federation in the framework of the Program of Scientific Research in Universities (project No. 11.1943.2017/4.6). The authors express their gratitude to Mrs. Marina Fomina and Dr. Vladimir Koshuro for the discussion and analysis of the results, as well as their assistance in carrying out experimental work.

References

[1] Rodionov I V, Fomin A A, Fomina M A, Poshivalova E Yu and Zakharevich A M 2015 *Proc. SPIE Microtechnologies 2015* (Barcelona, Spain), **9519** p 951917
[2] Aman S, Aman A and Morgner W 2013 *Comp. Sci. Technol.* **84** 58
[3] Majcherek S, Aman A, Hirsch S and Schmidt B 2015 *Sensors and Actuators A.* **233** 267
[4] Koshuro V A, Nechaev G G and Lyasnikova A V 2014 *Tech. Phys.* **59** 1570
[5] Xiong J, Guo Z, Yang M, Wan W and Dong G. 2013 *Ceram. Int.* **39** 337
[6] Wang B. and Liu Z. 2016 *Int. J. Refract. Met. Hard Mater.* **55** 24
[7] Brzhozovskii B, Martynov V, Zinina E and Brovkova M. 2016 *IOP Conf. Series: Mater. Sci. Eng.* **116** 012007

SPBOPEN 2018   IOP Publishing

IOP Conf. Series: Journal of Physics: Conf. Series **1124** (2018) 081017   doi:10.1088/1742-6596/1124/8/081017

Formation of wear-resistant oxide-carbide coatings on titanium by electrospark alloying

S Mezentsov[1], V Koshuro[1], A Fomin[1], I Rodionov[1]

[1]Yuri Gagarin State Technical University of Saratov, Saratov 410054, Russia

Abstract. In this paper the peculiarities of formation of wear-resistant coatings on the surface of VT1-00 titanium by electrospark alloying are described. Sequential treatment with titanium and graphite electrodes was used, as a result of which carbide and carbon-nitride coatings were obtained. The chemical composition, surface morphology and microhardness parameters were studied. The surface after electrospark alloying with a titanium electrode had a microhardness of 1270–1300 HV (12.45–12.75 GPa), and after treatment with a graphite electrode it equalled 2040–2080 HV (20.0–20.4 GPa).

1. Introduction

High-strength metals and alloys are used for the production of different items operating under high mechanical loads. Good results are ensured due to the application of titanium alloys, however, they have some disadvantages, namely, a possible contact seizure (during friction) and low wear resistance [1,2]. In order to increase the functional characteristics of metal products, coatings with the required phase composition, structure and mechanical properties are deposited on the surface. The application of electrospark alloying (ESA) allows the formation of coatings with high mechanical properties, in particular adhesion strength and hardness [3-9]. Based on the analysis of the available works on ESA technology, it was established that in order to increase the efficiency of strengthening the surface of products, complex processing is required. In this paper, the results of the study of chemical composition, structural changes in the surface, and microhardness of titanium samples after ESA were presented.

2. Methodology

The formation of coatings by the ESA method was performed on the end surface of cylindrical samples 2–3 mm long with the diameter of 6.5–7 mm and fabricated from VT1-00 titanium alloy. The working electrodes were rods (2.0±0.1 mm in diameter and 20–25 mm long) from VT1-00 titanium alloy and MPG-6 graphite. Current density j and specific processing time t were chosen as the controlled factors of the ESA process.

Before coating deposition, titanium samples were cleaned in an aqueous ethanol solution. Scanning electron microscopy (SEM) was used to study the morphology. To study the phase composition, X-ray diffraction analysis (XRD) was performed. When measuring hardness, the Vickers method was applied (the load on the indentor equaled 100 gf).

3. Results

The chemical analysis showed that the main elements in the samples after electrospark alloying with titanium were Ti (43–48 at.%), O (31–43 at.%) and N (13–25 at.%). The subsequent ESA with graphite promoted chemical changes, namely, C (up to 72 at.%) and Ti (20–26 at.%) prevailed in the

Content from this work may be used under the terms of the Creative Commons Attribution 3.0 licence. Any further distribution of this work must maintain attribution to the author(s) and the title of the work, journal citation and DOI.

Published under licence by IOP Publishing Ltd

coating composition. The composition of the modified coatings also contained Si (up to 4 at.%), which was explained by the initial composition of the alloying electrode.

Due to XRD the spectra showing the peaks of different phases were obtained (Figure 1). In the course of ESA with a titanium electrode, on the surface of the treated product, oxide and nitride compounds of titanium were formed, which were caused by heating and interaction with the surrounding atmosphere. As a result of the subsequent exposure to a graphite electrode, titanium oxide (TiO), titanium disilicide ($TiSi_2$), titanium carbide (TiC) and silicon carbide (SiC) were found on the surface.

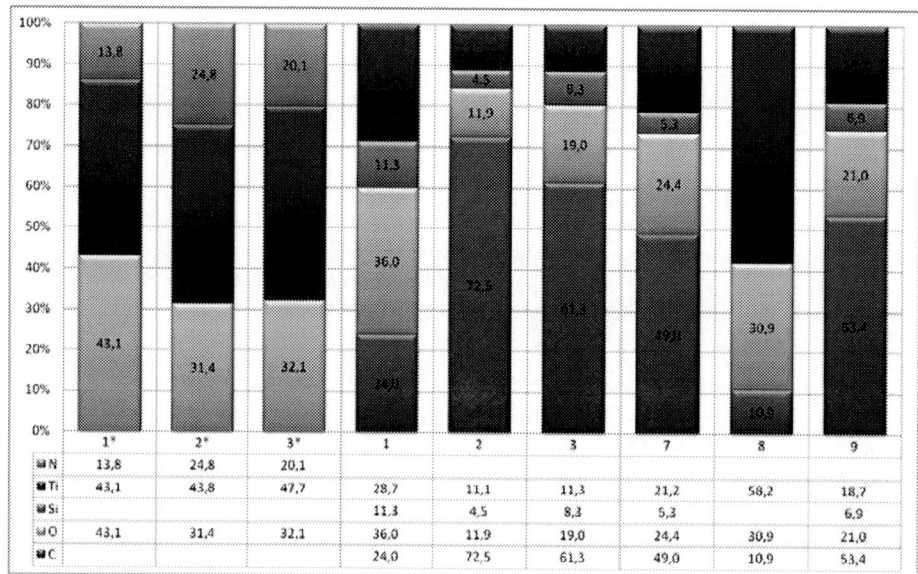

	1*	2*	3*	1	2	3	7	8	9
N	13,8	24,8	20,1						
Ti	43,1	43,8	47,7	28,7	11,1	11,3	21,2	58,2	18,7
Si				11,3	4,5	8,3	5,3		6,9
O	43,1	31,4	32,1	36,0	11,9	19,0	24,4	30,9	21,0
C				24,0	72,5	61,3	49,0	10,9	53,4

Figure 1. The diagram of distribution of chemical elements on the surface of the coatings subjected to ESA.

SEM results of titanium samples showed the presence of structurally heterogeneous surface morphology. During the treatment with a titanium electrode, the main microrelief of the coating surface was formed (Figure 2a), and subsequent treatment with graphite ensured the surface modification (Figure 2b).

The resulting coating was a composite structure. The upper layer of the coating was a film of the anode material modified by the elements of the cathode material and the interelectrode medium. These structures were located on the surface in the form of separate agglomerates, the distribution on the surface of which depended on the treatment conditions, the anode material and the exposure time. There was an area consisting of a mixture of anode and cathode materials under the agglomerates observed. Then was the layer formed due to the diffusion of the chemical elements of the anode material in the cathode material. And beneath the latter layer was the lowest one having the largest thickness, formed as a result of the pulse heat action. It was represented by a granular structure that was formed from the product material. At a bigger depth, this structure gradually changed into the original structure of titanium. Depending on the ESA modes, the significance of the first three layers can be significant or insignificant, and the role of the structure formed under the pulse heat action, is always fundamental.

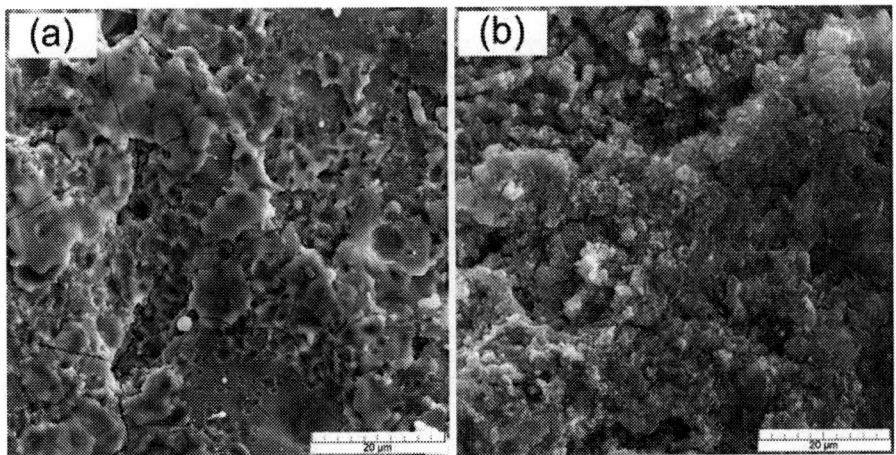

Figure 2(a, b). Surface morphology of the samples after ESA with titanium **(a)**; and graphite **(b)**.

The thickness of the deposited composite coating depended on the technological modes of ESA (Figure 3). The greatest influence was due to the current density j in the course of treatment with a titanium electrode. With subsequent exposure to a graphite electrode, there was no significant increase in the coating growth, and modification of the deposited titanium layer by the penetration of carbon was observed. Depending on the treatment mode, 8–15 to 40–70 μm thick coatings can be produced. The minimum coating thickness was formed at $j = 11$ A/mm^2 and $t = 10$ min/cm^2; the maximum thickness was ensured at $j = 18$ A/mm^2 and $t = 3.5$ min/cm^2.

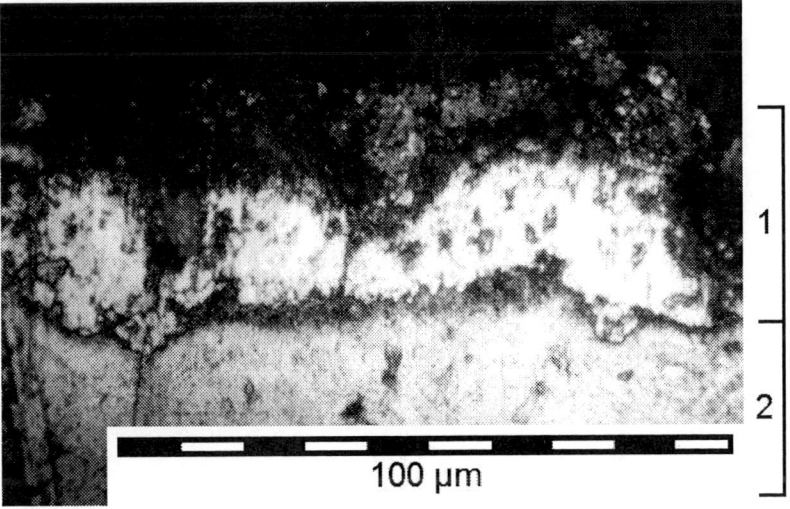

Figure 3. The microsection of the coating (1) after ESA with titanium and graphite produced on a titanium sample (2) with the greatest current density j and treatment time t.

In the course of treatment with a titanium electrode, the microhardness grew to 1270–1300 HV (12.45–12.75 GPa) at a current density of 14 A/mm^2 and a specific processing time of 5 min/cm^2. When using a graphite electrode, the microhardness reached 2040–2080 HV (20.0–20.4 GPa) at a current density of 18 A/mm^2 and specific processing time of 3.5 min/cm^2.

4. Conclusions
Thus, the sequential ESA with titanium and carbon of the surface of titanium products provided the formation of a hard coating (up to 20.0–20.4 GPa), the morphology of which had the required parameters for the effective contact interaction in rubbing elements of various types.

Acknowledgments
The research was supported by the Ministry of Education and Science of the Russian Federation in the framework of the Program of Scientific Research in Universities (project No. 11.1943.2017/4.6).

References
[1] Nikolenko S V and Syui N A 2017 *Protection of Metals and Physical Chemistry of Surfaces* **53** 889
[2] Frangini S, Masci A and Di Bartolomeo A 2002 *Surf. Coat. Technol.* **149** 279
[3] Kiryukhantsev-Korneev P V, Sheveyko A N, Shvindina N V, Levashov E A and Shtansky D V 2018 *Ceram. Int.* **44** 7637
[4] Fomin A A, Fomina M A, Koshuro V A, Rodionov I V, Voiko A V, Zakharevich A M, Aman A, Oseev A, Hirsch S and Majcherek S 2016 *Tech. Phys. Lett.* **42** 932
[5] Koshuro V, Fomin A, Fomina M, Rodionov I, Brzhozovskii B, Martynov V, Zakharevich A, Aman A, Oseev A, Majcherek S and Hirsch S 2016 *J. Phys. Conf. Ser.* **741** 012197
[6] Fomina M A, Koshuro V A, Fomin A A, Rodionov I V, Skaptsov A A, Zakharevich A M, Aman A, Oseev A, Hirsch S and Majcherek S 2016 *Proc. SPIE.* **9917** p 99171M
[7] Ivannikov A Yu, Kalita V I, Komlev D I, Radyuk A A, Bagmutov V P, Zakharov I N and Parshev S N 2016 *J. Alloys. Compd.* **655** 11
[8] Kalita V I, Komlev D I and Radyuk A A 2016 *Inorg. Mater. Appl. Res.* **7** 536
[9] Kreivaitis R, Žunda A, Kupčinskas A and Jankauskas V 2016 *Tribol. Int.* **103** 236

SPBOPEN 2018

Modified multicapillary glass structure for SERS application

N E Markina[1], A V Markin[1,2], A M Zakharevich[2], Y S Skibina[3], I Yu Goryacheva[1]

[1] Institute of Chemistry, Saratov State University, Saratov 410012, Russia
[2] Education and Research Institute of Nanostructures and Biosystems, Saratov State University, Saratov 410012, Russia
[3] SPE Nanostructured Glass Technology Ltd, Saratov, 410033, Russia

Abstract. The aim of the work is the fabrication and analytical testing of SERS substrate by modification of inner surface of multicapillary glass structure (MCGS). Silica based composite with embedded silver nanoparticles (AgNPs) was used for MCGS modification. The testing was carried out by performing solid phase extraction of the analyte (rhodamine 6G) with further its SERS registration. The results showed that modified MCGS possesses superior Raman enhancement compare to pure AgNPs and provides 12 times larger SERS signal. The growth of SERS efficiency was explained by improved sorption characteristics.

1. Introduction

Currently, investigations dedicated to surface-enhanced Raman scattering (SERS) are shifted from the fields of fundamental physics, physical chemistry, and materials science to solving practical problems. For example, application of SERS was proposed for study of corrosion [1], catalysis [2], and chemical analysis [3]. Because SERS has high sensitivity and enables identification of target analytes via analysis their vibration spectra, it is widely tested for biochemical analysis, *e.g.*, detection of cancer cells [4], therapeutical drug monitoring of antibiotics in urine [5], saliva [6], and blood [7], *etc.*

SERS-active composites with high specific surface area are of particular interest because they enable extraction of the target molecules that additionally improves effectiveness (sensitivity and selectivity) and reliability of SERS based analysis. This approach was recently realized by combination of SERS with solid phase extraction (SPE) using SERS-active sorbents based on inorganic (silica [8], calcium carbonate [9], *etc.*) and organic matrices (polysaccharides [10]). Various micro- and nanostructured glass fibres and multicapillary glass structures (MCGS) with high specific surface area were also proposed as support for fabrication of SERS platforms [11,12] or luminescent sensors [13]; however only their unique optical properties were accounted while sorption capability was not used for improvement of SERS efficiency yet.

Therefore, the aim of this study is the fabrication of SERS-active MCGS (SERS-MCGS) and estimation of its potential for SERS-based analysis. SERS-activity was achieved by modification of MCGS with silver nanoparticles (AgNPs) embedded to silica sol. Besides of strong binding of AgNPs to the MCGS, silica also plays an important role because of following advantages: (i) physical and chemical protection of AgNPs from aggregation, fusion, and oxidation during the preparation and application, (ii) transparency for probing laser light and negligible Raman cross-section, i.e. low background signal, and (iii) excellent sorption properties and facilitation of analyte adsorption (improvement of extraction efficiency).

Content from this work may be used under the terms of the Creative Commons Attribution 3.0 licence. Any further distribution of this work must maintain attribution to the author(s) and the title of the work, journal citation and DOI.
Published under licence by IOP Publishing Ltd

2. Experimental part

2.1. Materials

Silver nitrate, sodium citrate ($Na_3C_6H_5O_7 \cdot 2H_2O$), tetraethyl orthosilicate (TEOS), rhodamine 6G (R6G) were purchased from Sigma-Aldrich and have chemical grade purity. Ammonia solution (25 wt%), sulphuric acid (98 wt%), and ethanol (96 wt%) were purchased from ZAO "Vekton" (Russia). Borosilicate MCGS consisted of 4293 capillaries was kindly provided by SPC Nanostructured Glass Technology Ltd. Deionized water was used for preparation of all solutions.

2.2. Methods of characterization

SERS-active composite and MCGS before and after modification were characterized by scanning electron microscopy (SEM; Mira II LMU, Tescan, Czechia). Registration of hydrodynamic (HD) diameter and surface charge (zeta potential) of AgNPs and modifying solution were performed by dynamic light scattering (DLS) method using Zetasizer Nano (Malvern, UK). UV-visible absorbance and fluorescence measurements were performed with hybrid reader Synergy H1 (BioTek, USA).
SERS spectra were recorded using probing nanolaboratory Ntegra Spectra (NT-MDT, Russia) with integrated Raman spectrometer (Solar) using 473 nm laser and registration scheme 180°. Raman spectrometer was calibrated before SERS measurement using silicon wafer (Raman band at 520 cm^{-1}). All SERS spectra were normalized to laser power and acquisition time (counts per 1 mW and 1 sec).

2.3. Preparation of MCGS-SERS composite and SERS testing

The preparation of solution for MCGS modification includes synthesis of AgNPs and their embedding into silica sol. The synthesis of SERS-active AgNPs was performed by well known procedure based on reduction of silver nitrate with citrate ions in aqueous environment [14]. The suspension of as-prepared AgNPs was 100 times concentrated before further application using centrifugation at 10^4 g for 10 min. Solution of silica sol with embedded AgNPs was prepared by mixing of 2 ml of ethanol, 20 µl of AgNPs, and 20 µl of TEOS and further sonication of the mixture for 15 min. The silica sol with embedded AgNPs was 10 times diluted by water for DLS measurements in order diminish influence of ethanol on the DLS results.
The modification of MCGS was performed by passing of the modifying solution through MCGS with further rinsing by water and drying at 50°C. MCGS was cut before modification to the short pieces (~2 cm length) and fixed by glue in a tip for micropipette in order to facilitate operation with it.
Testing of SERS-MCGS was performed by R6G as the model analyte with the well known spectrum. The SERS analysis procedure includes passing of 25 µl of R6G solution (10^{-6} M) through SERS-MCGS with further SERS measurements. The effectiveness of SERS-MCGS was estimated by control experiment with pure AgNPs which was performed by mixing of equivalent portions (10 µl) of AgNPs, R6G, and NaCl (1M) solutions with further SERS measurements.

3. Results and discussions

3.1. SERS-MCGS: preparation and characterization

Citrate stabilized AgNPs were used as SERS-active material due to simplicity of their preparation and strong Raman enhancement. We have widely used this SERS substrate earlier for detection of pharmaceuticals in water and urine [5,9]. The successfulness of the synthesis and quality of as-prepared AgNPs were verified by DLS measurements of the size and zeta potential (Figure 1a,b) the average values of which were found as ~65 nm and –45 mV. UV-visible spectrum of AgNPs solution (Figure 1c) demonstrates the presence of pronounced plasmon band at ~400 nm which is responsible for colour of modified MCGS (Figure 2a). DLS study also showed that the size and charge of silica sol with embedded AgNPs are different from these parameters of pure AgNPs. Modification of AgNPs by silica led to growth of the size of the detected particles and improvement of the size distribution (more homogeneous) (Figure 1a). We associate changing of surface charge with high contribution of OH-groups which are poorly dissociated compare to citrate ions used for stabilization of as-prepared AgNPs (Figure 1b).

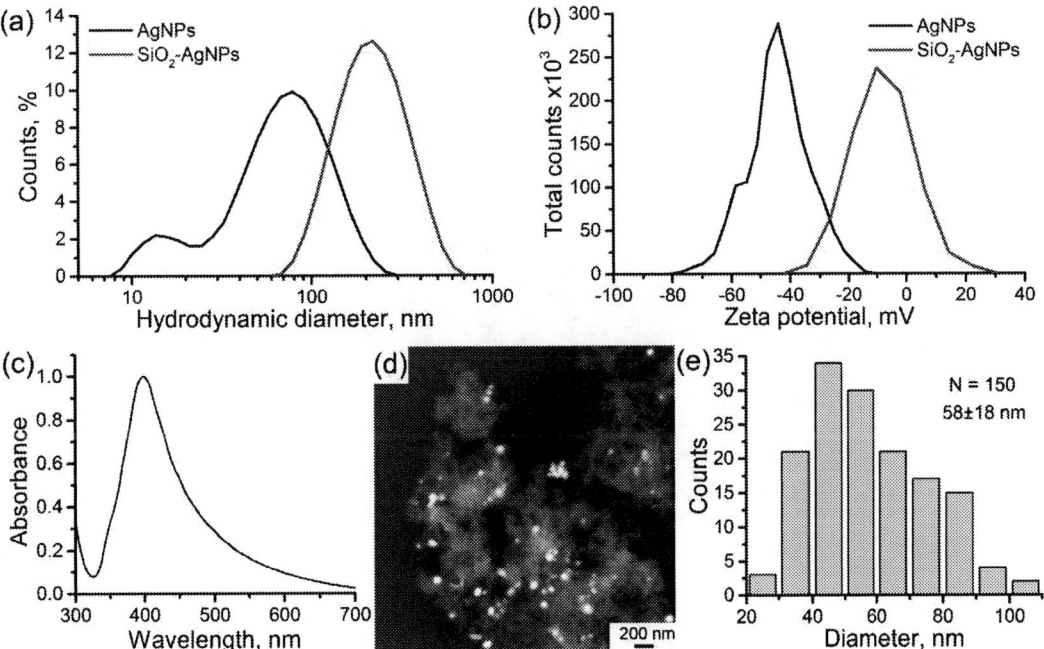

Figure 1(a-e). Size **(a)** and zeta potential **(b)** of initial citrate stabilized AgNPs and SiO_2-AgNPs composite (according to DLS measurements). **(c)** UV-visible spectrum of the suspension of as-prepared AgNPs. **(d)** SEM image of SiO_2-AgNPs composite and **(e)** size distribution of AgNPs inside the composite.

SEM studies of the dried silica sol with embedded AgNPs demonstrated that AgNPs inside silica matrix are partially aggregated (Figure 1d), while their size is in appropriate agreement with one determined by DLS (Figure 1e). However, the partial aggregation is an advantage in the case of SERS because aggregated AgNPs possess stronger Raman enhancement due to presence of "hot spots" (the space between aggregated AgNPs) [15]. SEM measurements of MCGS before and after modification (Figure 2a) demonstrate uniform distribution of AgNPs and SiO_2 inside MCGS and negligible influence of modification on inner diameter of a single capillary (~20 μm) (Figure 2b-d). Although the modification coating is not dense, the colour of modified MCGS is significantly changed (Figure 2a) due to remarkable plasmonic properties of AgNPs.

3.2. SERS detection of model analyte
The results of SERS testing demonstrate 12 times larger SERS signal for SERS-MCGS compare to pure AgNPs (Figure 2a). We explain this by better sorption properties which were improved by (i) contribution of specific surface area of MCGS and (ii) application of silica as the matrix for AgNPs. Additionally, sorption capacity was estimated by fluorescence measurements of R6G solution before and after passing through SERS-MCGS (Figure 3b). The results support our guess about improvement of sorption capacity and SERS efficiency and demonstrate almost complete extraction of R6G molecules (the sorption is ~98%).

SPBOPEN 2018

IOP Publishing

IOP Conf. Series: Journal of Physics: Conf. Series **1124** (2018) 081018   doi:10.1088/1742-6596/1124/8/081018

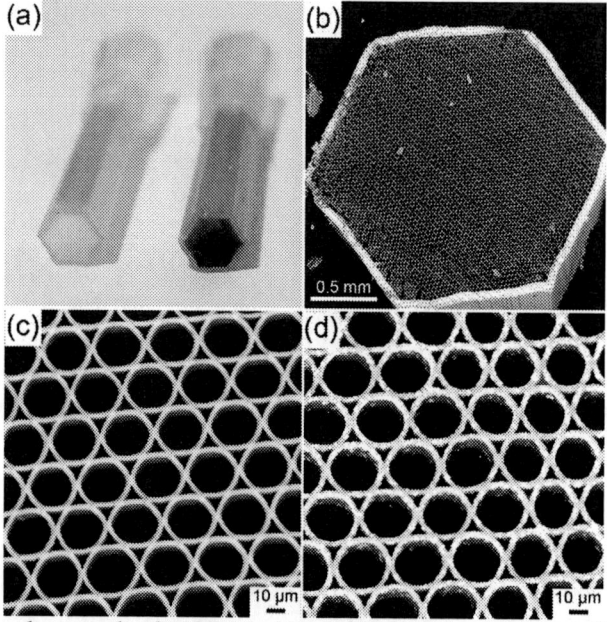

Figure 2(a-d). (a) The photograph of MCGS before (left) and after (right) modification. SEM images of initial **(b, c)** and modified **(d)** MCGS.

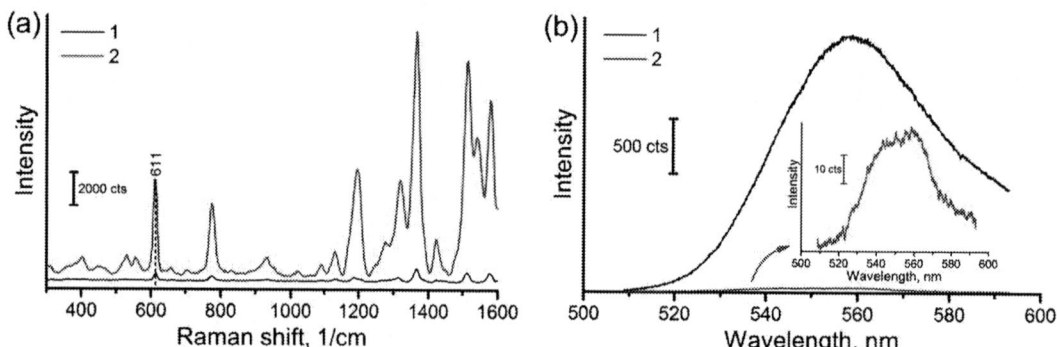

Figure 3(a, b). (a) SERS spectra of R6G (10^{-6} M) using aggregates of AgNPs (1) and modified MCGS (2); **(b)** Fluorescence spectra of R6G (10^{-6} M) before (1) and after (2) passing through modified MC.

4. Conclusions

Summing the results we can conclude that modified MCGS enables significant improvement of SERS sensitivity due to increased sorption characteristics. Moreover, SERS-MCGS can be used as SPE column which enables speed-up of SERS analysis due to direct detection of the adsorbed analytes in the column immediately after extraction. Finally, the analysis requires 25 µl of the analyte only that is comparable with sample amount required for the most advanced analytical techniques such as HPLC and LC-MS.

Acknowledgments

The work related with multicapillary systems and SERS detection was supported by Russian Science Foundation (project 14-13-00229). Markin AV also wants to thank for support the German Academic Exchange Service and the Russian Ministry of Education and Science (project 16.12789.2018/12.2).

References

[1] Cao P G, Yao J L, Zheng J W, Gu R A, Tian Z Q 2002 *Langmuir* **18** 100
[2] Li J, Liu J, Yang Y, Qin D 2015 *J. Am. Chem. Soc.* **137** 7039
[3] Fan M, Andrade G F S, Brolo A G 2011 *Anal. Chim. Acta* **693** 7
[4] Vendrell M, Maiti K K, Dhaliwal K, Chang Y T 2013 *Trends Biotechnol.* **31** 249
[5] Markina N E, Markin A V, Goryacheva I Y 2018 *Anal. Bioanal. Chem.* **410** 2221
[6] Hernández-Arteaga A, Nava J J Z, Kolosovas-Machuca E S, Velázquez-Salazar J J, Vinogradova E, José-Yacamán M, Navarro-Contreras H R 2017 *Nano Research* **10** 3662
[7] Bonifacio A, Dalla Marta S, Spizzo R, Cervo S, Steffan A, Colombatti A, Sergo V 2014 *Anal. Bioanal Chem.* **406** 2355
[8] Markina N E, Markin A V, Zakharevich A M, Gorin D A, Rusanova T Y, Goryacheva I Y 2016 *J. Nanopart. Res.* **18** 353
[9] Markina N E, Markin A V, Zakharevich A M, Goryacheva I Yu 2017 *Microchim. Acta* **184** 3937
[10] Abalde-Cela S, Auguie B, Fischlechner M, Huck W T S, Alvarez-Puebla R A, Liz-Marzan L M, Abell C 2011 *Soft Matter* **7** 1321
[11] White I M, Yazdi S H, Yu W W 2012 *Microfluid. Nanofluid.* **13** 205
[12] Markin A V, Markina N E, Goryacheva I Yu 2017 *Trends Anal. Chem.* **88** 185
[13] Pidenko S A, Burmistrova N A, Shuvalov A A, Chibrova A A, Skibina Y S, Goryacheva I Y 2018 *Anal. Chim. Acta* **1019** 14
[14] Lee P C, Meisel D 1982 *J. Phys. Chem.* **86** 3391
[15] Camden J P, Dieringer J A, Wang Y, Masiello D J, Marks L D, Schatz G C, Van Duyne R P 2008 *J. Am. Chem. Soc.* **130** 12616

SPBOPEN 2018   IOP Publishing

Carbide coatings obtained by electro-spark alloying and finishing

V Koshuro[1]

[1] Yuri Gagarin State Technical University of Saratov, Saratov 410054, Russia

Abstract. In this study it is proposed to increase the mechanical properties of HSS 1.3343 tool steel by electro-spark alloying with a hard carbide alloy. As a result, the hardness of coatings obtained by electro-spark alloying reached at least 18–22 GPa, the substrate hardness reached 3.7 GPa. After induction quenching in the air, the hardness of the coating equaled 12.0±0.1 GPa. The hardness of the steel substrate increased to 9.1±0.6 GPa. The subsequent sizing (finish grinding) ensured a reduction in roughness Ra to 0.16–0.32 and in open porosity. After grinding, the hardness of the coating and the substrate was 10.3±0.5 GPa and 8.0±0.6 GPa, respectively. The proposed solution can improve the functional qualities of various frictional surfaces used in tool-making facilities.

1. Introduction

Carbide tool materials, e.g. WC-Co, WC-TiC-Co and WC-TiC-TaC-Co hard alloys, are applied in engineering due to their high hardness and wear resistance. These materials are mainly used for the production of tool and frictional elements operating at high temperatures and pressure [1,2]. In order to increase the strength and wear resistance of various products fabricated from steel tools, metal-ceramic coatings are deposited on their surface, e.g. oxide [3-6], carbide or carbonitride [7,8] ones.

Various methods are used to deposit coatings, e.g. plasma spraying, PVD and CVD, however, they are rather complicated. It is known that the method of electro-spark alloying (ESA) is quite efficient [9,10]. Hence, the present study describes the possibility of using ESA for the production of hard WC-TiC-Co coatings on the surface of HSS 1.3343 tool steel, and the subsequent sizing that ensures the reduction in the parameters of roughness and open porosity.

2. Methodology

Samples of HSS 1.3343 tool steel were subjected to ESA with hard carbide alloy WC-TiC-Co (composition: WC – 85 wt.% , TiC – wt.6 %, Co – wt.10 %; hardness HRA 88.5, equivalent of HRC 72–73). The samples had a disc shape with a diameter of 14 mm and thickness of 2 mm. ESA was performed in a pulse mode with the following parameters: pulse duration t within 10–20 ms, operating current I from 0.8 to 2.5 A. In this work the effect of 3 steps of current ranging was studied: 0.8–1.2, 1.5–2 and 2–2.5 A. The duration of ESA was 2, 5 and 10 min.

The coated samples were subjected to induction quenching [11]. The samples were heated to 1000–1100 °C and then cooled in the air. The exposure time for quenching was 10–20 s. The coating surface after ESA and ESA followed by quenching was subjected to semifinishing and finishing in order to obtain the required roughness parameter Ra 0.16–0.32.

The surface morphology of the samples was studied using optical microscopy (MBS-10). The hardness of the resulting coatings was evaluated by microindentation using "PMT-3M" (at the load of 100 gf).

Content from this work may be used under the terms of the Creative Commons Attribution 3.0 licence. Any further distribution of this work must maintain attribution to the author(s) and the title of the work, journal citation and DOI.
Published under licence by IOP Publishing Ltd

3. Results

The surface structure of the WC-TiC-Co coatings on HSS treated at the minimum value of the current strength $I = 1.0$ A and with the minimum treatment duration $t = 2$ min was characterized by a noticeable heterogeneity (Figure 1a). Numerous areas containing open pores were observed on the surface (Figure 1b).

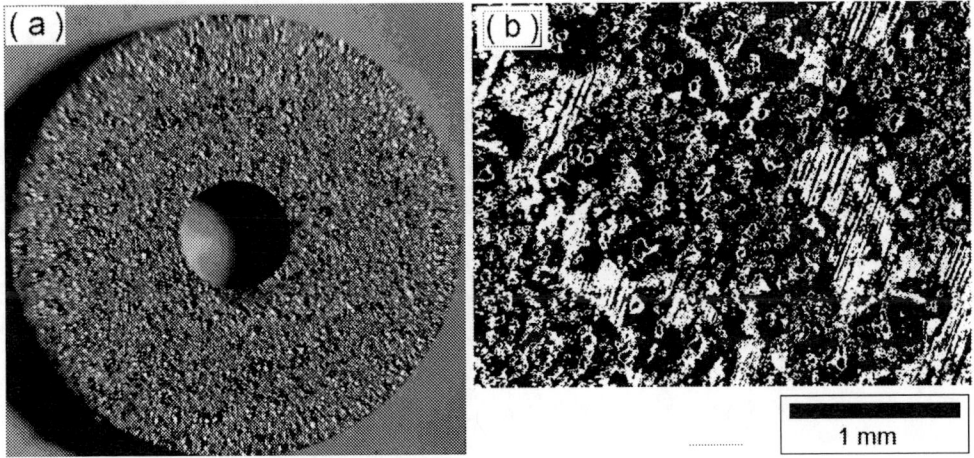

Figure 1(a, b). WC-TiC-Co coatings after ESA **(a)**; open pores after semifinishing (black and white image, the pores are shown by dark areas) **(b).**

When the current strength of electrical discharges grew to $I = 2$–2.5 A, the uniformity of the coating improved. The average size of pores on the coatings produced at the minimum treatment duration was $D = 31\pm10$ μm, and the open porosity of the coatings reached a high value $P = 65$ %. With an increase in the treatment duration to $t = 5$ min, the uniformity of the coating grew as well. However, to improve the quality of the coating, the treatment time should be increased to $t = 10$ min. As the discharge current increased to $I = 2$–2.5 A, the uniformity grew accordingly. The average size of pores increased to $D = 45\pm15$ μm, and porosity of the coatings increased as well reaching a very high value $P = 72\%$. The coating produced at the maximum treatment duration $t = 10$ min was considered the most suitable for subsequent finishing, e.g. grinding.

Induction quenching did not practically affect the parameters of the surface morphology of the carbide coatings. Visually, the coating surface, which was formed at $I = 1.0$ A and duration $t = 2$ min, became more homogeneous (Figure 2a). After heat treatment, the porosity of the coatings slightly decreased compared to the untreated coating by an average of 5% (Figure 2b). The average pore size was 53.7 ± 10 μm. The increase in current during the formation of the initial coating to $I = 2$–2.5 A led to a decrease in porosity to $50\pm5\%$. The maximum pore size also reduced to 42.3 ± 10 μm. At the same time, the minimum porosity characterized the ESA coatings at different current values and a duration of 5 min. So the surface of samples doped at the current of $I = 1.0$ A and duration of 5 min after heat treatment was characterized by a porosity of 54%. Increasing the current to $I = 1.75$ A made it possible to reduce the porosity to 45%.

After finishing the coating surface acquired a metallic luster. Defects in the form of pores became clearly visualized (Figure 3a). Apparently, this was due to the opening of the volume of closed pores. On average, the porosity of the coatings after grinding increased by 5–10%. The surface of samples subjected to ESA at $I = 1.0$ A and duration $t = 2$ min was characterized by the porosity of 56%

(Figure 3b). The average pore size after machining decreased by 5±2% and it did not depend on ESA modes.

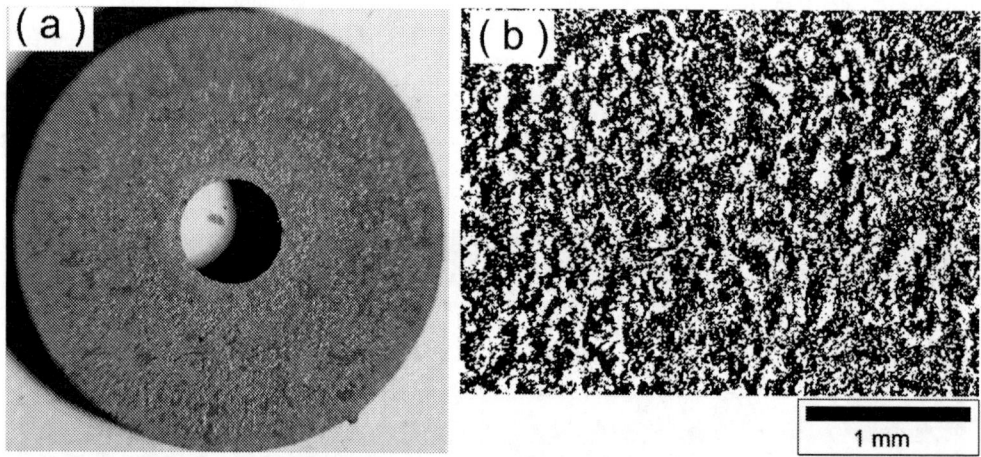

Figure 2(a, b). WC-TiC-Co coatings after ESA and IHT **(a)**; open pores (black and white image, the pores are shown by dark areas) **(b).**

The hardness of WC-TiC-Co coatings on HSS 1.3343 tool steel depended strongly on the discharge current I and duration t of ESA treatment. A minimum hardness of about 68–70 HRC (about 8–11 GPa) was observed at the maximum treatment duration $t = 10$ min. High hardness 72–73 HRC (12–13 GPa) corresponded to the minimum and medium treatment duration $t = 2$ min and $t = 5$ min. The hardness of the substrate after ESA increased from 33–34 HRC (3.2–3.3 GPa) to 37–38 HRC (about 3.6–3.7 GPa).

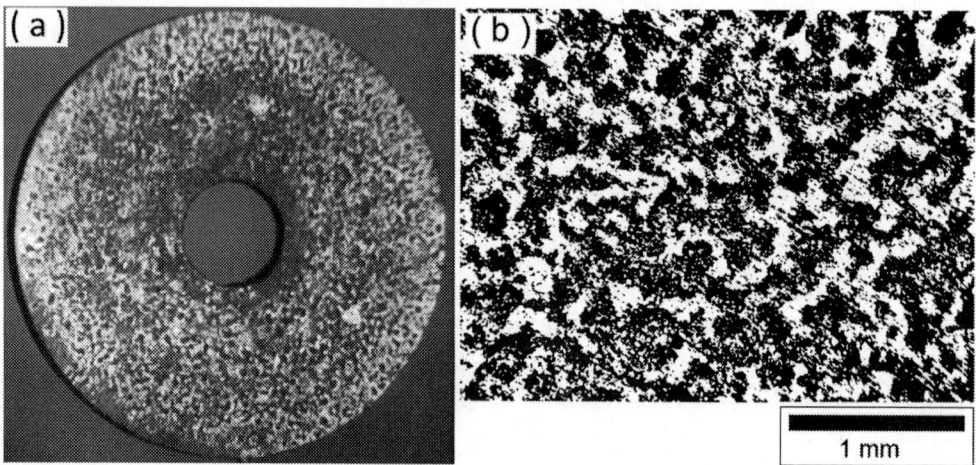

Figure 3(a, b). WC-TiC-Co coatings after ESA, IHT and finishing **(a)**; open pores after semifinishing (black and white image, the pores are shown by dark areas) **(b).**

After induction quenching, the hardness of the carbide coatings remained unchanged whereas the hardness of steel increased significantly. The hardness of 67–68 HRC (9.1–9.3 GPa) was observed in the substrates subjected to ESA at the current $I = 1.0$–1.75 A in the entire range of the treatment duration. After machining (grinding) of the surface, the hardness of the coatings decreased. The maximum hardness of 68–70 HRC (9.3–10.6 GPa) was observed in the coatings formed at low current values (I = 1.0–1.5 A) and short treatment. The hardness of steel substrates after quenching and subsequent finishing decreased to 64–66 HRC (7.8–8.5 GPa).

4. Conclusion
The surface morphology of HSS 1.3343 after ESA had high morphological heterogeneity and mechanical properties, e.g. hardness. WC-TiC-Co coatings were characterized by the pore size D = 30–45 μm. Thus, the obtained results can find application in the manufacture of metalworking tools and elements of friction pairs. The optimal combination of morphology parameters, hardness of carbide coatings (68–73 HRC) and steel (64–66 HRC) was ensured due to the ESA duration t = 5 min and operating current I = 1.0–2.5 A, as well as subsequent induction quenching and machining (finish grinding).

Acknowledgments
The research was supported by the Ministry of Education and Science of the Russian Federation and German Academic Exchange Service (DAAD) in the framework of the program "Mikhail Lomonosov" (project No. 11.12784.2018/12.2).

References
[1] Xiong J, Guo Z, Yang M, Wan W and Dong G. 2013 *Ceram. Int.* **39** 337
[2] Fomin A, Dorozhkin S, Fomina M, Koshuro V, Rodionov I, Zakharevich A, Petrova N and Skaptsov A 2016 *Ceram. Int.* **42** 10838
[3] Fomin A A, Steinhauer A B, Rodionov I V, Petrova N V, Zakharevich A M, Skaptsov A A and Gribov A N 2013 *Biomed. Eng.* **47(3)** 138
[4] Fomin A, Fomina M, Koshuro V, Rodionov I, Zakharevich A and Skaptsov A 2017 *Ceram. Int.* **43** 11197
[5] Fomin A A, Steinhauer A B, Rodionov I V, Fomina M A, Zakharevich A M, Skaptsov A A, Gribov A N and Karsakova Ya D 2014 *J. Frict. Wear.* **35(1)** 32
[6] Fomin A A, Fomina M A, Rodionov I V, Koshuro V A, Poshivalova E Yu, Shchelkunov A Yu, Skaptsov A A, Zakharevich A M and Atkin V S 2015 *Tech. Phys. Lett.* **41(9)** 909
[7] Wang B and Liu Z 2016 *Int. J. Refract. Met. Hard. Mater.* **55** 24
[8] Yang W, Xiong J, Guo Z, Du H, Yang T, Tang J and Wen B 2017 *Ceram. Int.* **43** 1911
[9] Fomin A A, Fomina M A, Rodionov I V, Koshuro V A, Poshivalova E Yu, Shchelkunov A Yu, Skaptsov A A, Zakharevich A M and Atkin V S 2016 *Tech. Phys. Lett.* **42(9)** 932
[10] Koshuro V, Fomin A, Fomina M, Rodionov I, Brzhozovskii B, Martynov V, Zakharevich A, Aman A, Oseev A, Majcherek S and Hirsch S 2016 *J. Phys. Conf. Ser.* **741(1)** 012197
[11] Fomin A A and Fomina M A 2016 Proc. 57th Int. Sci. Conf. on Power and Electrical Engineering of Riga Technical University (Riga, Latvia) (Riga: IEEE), p 1

SPBOPEN 2018   IOP Publishing

High robustness of epitaxial 4H-SiC graphene to oxidation processes

V.S. Prudkovskiy[1,2]**, K.P. Katin**[2,3]**, M.M. Maslov**[2,3]**, P. Puech**[4]**, R. Yakimova**[5]**, G. Deligeorgis**[6]

[1]Department of Physics, University of Crete, Heraklion, 71003, Greece
[2]Research Institute for the Development of Scientific and Educational Potential of Youth, Aviatorov str. 14/55, Moscow, 119620, Russia
[3]National Research Nuclear University "MEPhI", Kashirskoe sh. 31, Moscow, 115409, Russia
[4]Centre d'elaboration des Materiaux et d'etudes Structurales (CEMES), UPR-8011 CNRS, University of Toulouse, BP 94347, 31055, Toulouse, France
[5]Department of Physics, Chemistry and Biology, Linkoping University, SE-58183, Linkoping, Sweden
[6]Foundation for Research and Technology Hellas (FORTH), P.O. Box 1527, Vassilika Vuton, Heraklion, 71110, Crete, Hellas, Greece

Abstract. We present an experimental prove of high robustness of epitaxial 4H-SiC graphene to oxidation processes. During a post-fabrication cleaning procedure we noticed that epitaxial graphene is extremely stable to ozone treatment. We analyse graphene properties using both electron transport measurements and numerical calculations. We ascribe this effect to the substrate topography, which significantly affects the graphene stability under UV/ozone treatment.

1. Introduction

Great attention has been attributed to graphene as a perceptive material for new generation of electronic devices [1, 2]. Main obstacle of utilizing graphene for nanoelectronic is its significant sensitivity to contamination. Different polymers routinely used for nanofabrication leave residual traces on graphene. Majority of the research devoted to graphene-based electronics highlights the necessity of a cleaning procedure. The most efficient graphene decontamination procedure in our days is ozone cleaning [3], which is aggressive enough to clean graphene surface without destroying graphene crystal lattice for a proper chosen timescale of treatment.

2. Results

The experiment was carried out on graphene grown by sublimation process on 4H-SiC Si terminated substrate. Devices were fabricated in four-probe Van der Pauw configuration (Fig.1).

Content from this work may be used under the terms of the Creative Commons Attribution 3.0 licence. Any further distribution of this work must maintain attribution to the author(s) and the title of the work, journal citation and DOI.
Published under licence by IOP Publishing Ltd

SPBOPEN 2018   IOP Publishing
IOP Conf. Series: Journal of Physics: Conf. Series **1124** (2018) 081020   doi:10.1088/1742-6596/1124/8/081020

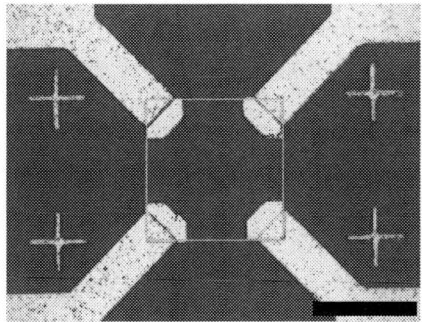

Figure 1. Optical image of a typical device, the scale bar is 50 µm. Graphene area is outlined by green line.

After fabrication all measured devices demonstrate electron doping. We associate this doping with contamination by polymers' radicals remaining on the graphene surface after fabrication process. The first two minutes of ozonation lead to a significant rise of the mobility, followed by the mobility saturation region that corresponds to 2 to 4 minutes of ozone exposure (Fig. 2). A large mobility drop after fifth minute of ozone exposure is observed. The increase of the mobility is most noticeable for the most contaminated devices. Initially low contaminated devices samples conserve their high initial mobility values and remain almost unaffected by the ozone treatment up to the fifth minute of ozonation, after which all the samples show a drop of the mobility to a range around ~250 cm^2/(V·s).

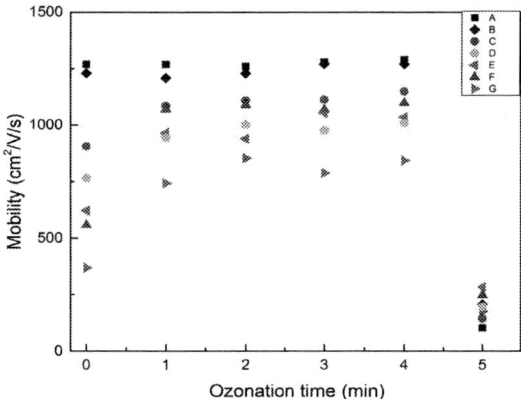

Figure 2. Graphene charge-carrier mobility as a function of ozonation time for seven devices with different doping level. Devices are named from A to G.

In order to understand this behavior we ask reader to focus on dependence of the carrier density on ozonation time (Fig. 3). The initial devices are electron doped with charge-carrier density in the range of -1.5·10^{12} to -0.5·10^{12} cm^{-2}. The carrier density of all devices after two minutes of ozonation reaches level of -4·10^{11} cm^{-2}.

1071

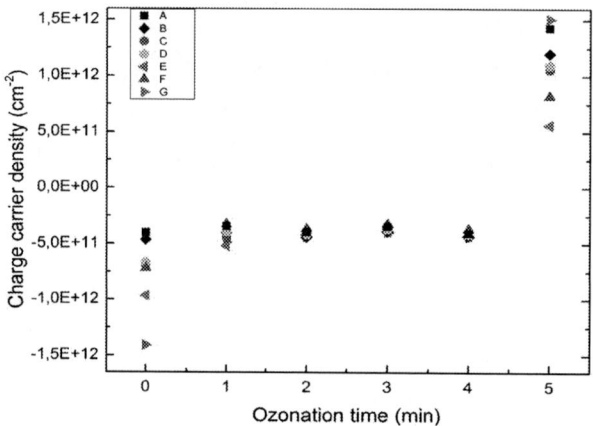

Figure 3. Graphene charge-carrier density a function of ozonation time for seven devices with different doping level.

Decrease of electron concentration coincides with the mobility rise (Fig. 2) and corresponds to the phase of residual polymer decomposition. In following, we observe a saturation of charge density at this level. After the fifth ozonation minute the density sharply rises up to $1.5 \cdot 10^{12}$ cm^{-2} along with change of charge carriers type from electrons to holes. This event coincides with the drop of the mobility (Fig. 2). The whole process of mobility and density evolution on ozonation time could be seen in Figure 4.

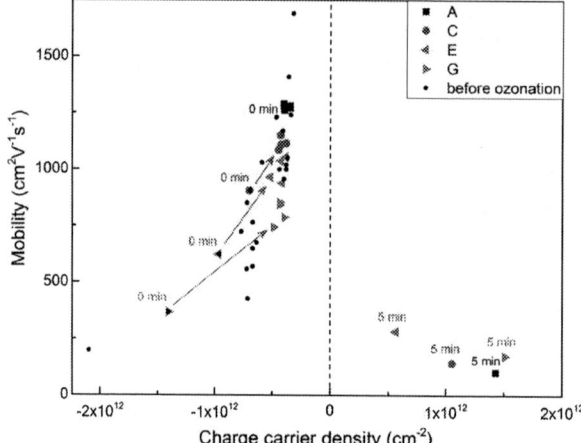

Figure 4. Graphene charge-carrier mobility μ versus density n. Black dots represent the μ(n) distribution of the all measured devices before ozonation. Colored markers correspond to devices: A (dark-red square), C (red circle), E (magenta triangle) and G (green triangle) after different ozonation time, labeled by corresponding color: "0 min" labels mark the initial states, arrows point toward the states after 1-4 min of ozonation, "5 min" labels mark the states after fifth minute of ozonation.

Thus the samples sustain up to 4 minutes of ozone cleaning procedure, which allows to increase the mobility and decrease the doping of contaminated samples. Recent experiments on exfoliated and

CVD graphene [4, 5] show that even one minute of ozonation leads to significant increase of the intensity of the D band in Raman spectra, originated from epoxy groups formation. This high SiC graphene stability is associated with the extremely flatness of epitaxial graphene contrary to exfoliated or transferred CVD graphene.

Our calculations revealed that surface roughness of graphene sheet could change the energy gain from epoxy group adsorption on a few tens of kilocalories per mole (Figure 5).

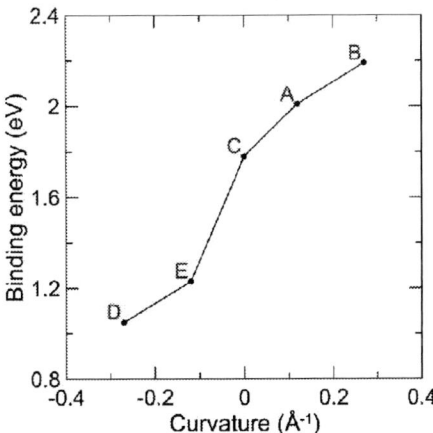

Figure 5. Dependence of the binding energy of the epoxy group on curvature of a graphene surface.

Such variations of energy are sufficient to do the two-epoxy groups adsorption more or less energetically favourable in comparison with the O2 molecule formation depending on local sheet curvature. Thus, surface roughness can significantly affect the graphene stability under UV/Ozone treatment.

3. Conclusion
We found that graphene prepared by sublimation on 4H-SiC Si terminated substrate is extremely stable to radical oxygen atoms. Our measurements reveal the absence of defects in graphene lattice after four minutes of ozonation, while exfoliated and CVD graphene cannot sustain even a minute [4, 5]. This high stability is associated with an extremely flatness of epitaxial graphene contrary to exfoliated or transferred CVD graphene. Our calculations reveal that surface roughness can change the energy gain from epoxy group adsorption by a few tenths of electron volts.

Acknowledgments
The reported study was funded by RFBR, according to the research project No. 16-32-60081 mol_a_dk.

References
[1] C. Berger, Z. Song, T. Li, X. Li, A.Y. Ogbazghi, R. Feng, Z. Dai, A.N. Marchenkov, E.H. Conrad, P.N. First, W.A. de Heer 2004 J. Phys. Chem. B **108** 19912
Berger, C.
[2] J. Baringhaus, M. Ruan, F. Edler, A. Tejeda, M. Sicot, A.Taleb-Ibrahimi, A. Li, Z. Jiang E.H. Conrad, C. Berger, C. Tegenkamp, W.A. de Heer 2014 Nature **506** 349
[3] V.S. Prudkovskiy, K.P. Katin, M.M. Maslov, P. Puech, R. Yakimova, G. Deligeorgis 2016 Carbon **109** 221
[4] Y. Mulyana, M. Uenuma, Y. Ishikawa, Y. Uraoka 2014 J. Phys. Chem. C 2014 **118** 27372
[5] M.G. Chung, D.H. Kim, H.M. Lee, T. Kim, J.H. Choi, D.K. Seo, J. Yoo, S. Hong, T.J. Kang, Y.H. Kim 2012 Sensors and Actuators B **166** 172

SPBOPEN 2018   IOP Publishing

Ohmic contacts to n-type 4H- and 6H-SiC

V I Egorkin, A V Nezhentsev, V E Zemlyakov, V A Gudkov and V I Garmash
National Research University of Electronic Technology, Moscow, 124498, Russia

Abstract. In this paper, the electrical properties of TiAl- and Ni-based ohmic contacts formed on n-type 4H- and 6H-SiC were studied. The dependence of the contact resistance on technological process was investigated. Ohmic contacts were formed by electron-beam deposition to different surfaces of silicon carbide. Short-term plasma-chemical etching of the active layer was used to improve TiAl-based contact to C-surface 4H-SiC. Using of Ni-based contacts to the Si-surface and TiAl-based to C-surface makes it possible to obtain a resistance of about 10^{-4} Ohm*cm^2. We got the process of making good ohmic contacts to 4H- and 6H-silicon carbide.

1. Introduction

Researchers make low-resistance contacts for the majority of wide-spread semiconductor materials. But for materials that have been relatively recently used for serial production of semiconductor devices, in particular for silicon carbide, the problem of obtaining low-resistance ohmic contacts, despite a considerable amount of experimental data, is still relevant. Work is under way to obtain ohmic contacts to silicon carbide with improved performance characteristics. Nickel-based metallization is the most common used for ohmic contacts manufacturing to the n- and p-layers of silicon carbide. It makes possible to obtain ohmic contact with a resistivity of about 10^{-4} Ohm*cm^2. Annealing leads to the reaction of nickel with silicon carbide, resulting in the formation of a Ni_2Si compound, which plays a decisive role in the ohmic contact formation to n-type SiC [1-3]. Ti-based metallization the second frequently encountered. The characteristic resistance is 10^{-4} Ohm*cm^2. Polytype and crystallographic surface affects the level of the contact resistance. This article shows the results of obtaining ohmic contacts to different faces and polytypes of SiC. We determined the most optimal combination of parameters for ohmic contacts to 4H- and 6H- silicon carbide.

2. Experiment

For the experiments, 6H-SiC Lelie crystals were taken with a donor concentration of $2,5*10^{18}$ cm^{-3}. We used doping with nitrogen ions (800 mkCl/cm^2), to increase the carrier concentration. Implantation was performed on the Lada-30 ion implantation unit. As a result, we obtained a concentration of donors in the near-surface region $>5*10^{19}$ cm^{-3}. We also used 4H-SiC crystals with initial carrier concentration $5*10^{19}$ cm^{-3}. We used lithographs to create a pattern of metallization. We also used a mask if the dimensions of the sample did not allow us to perform lithographic processes. The contacts were made by an electron-beam evaporation Kurt J. Lesker unit. We used two types of metallization: nickel-based and titanium-based. Contact metallization in the first case consisted of 100 nm of nickel, in the second 25 nm of titanium and 150 nm of aluminum. Contacts are four circles in one straight line with a diameter of 200 microns in steps of 1000 microns. We performed heat treatment (1000°C) after deposition. The annealing was carried out in a nitrogen atmosphere. The contact resistance was determined by the four-probe method [4]. Schematic diagram of the method is shown in Figure 1 [4].

Content from this work may be used under the terms of the Creative Commons Attribution 3.0 licence. Any further distribution of this work must maintain attribution to the author(s) and the title of the work, journal citation and DOI.
Published under licence by IOP Publishing Ltd

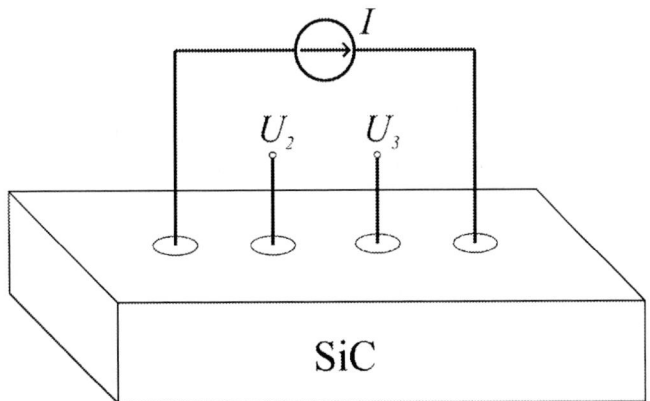

Figure 1. Schematic diagram of the four probe method.

The resistance is calculated:

$$R_{cont} = \pi a^2 \left\{ \frac{U_2}{I} - \left(\frac{U_3}{I}\right) \frac{\ln\left[(3d/2a) - 1/2\right]}{2\ln 2} \right\}$$

where a is the contact radius, d is the distance between the contact centers.
This equality holds for $a \ll d$.

In all cases, the contacts after heat treatment were ohmic, except for one. TiAl-based contact to C-surface 4H-SiC had nonlinear I-V characteristic. The manufacturing process was repeated with special surface treatment. Short-term plasma-chemical etching of the active layer was used. Plasma-etching of silicon carbide was carried out on a Corial 200L unit in an inductively coupled plasma (ICP). Etching included two stages: Surface cleaning in argon plasma (300 seconds) and etching in a fluorine-containing plasma (SF6) with the addition of oxygen (120 seconds),

3. Results
The measured resistances for different crystallographic surfaces, metallizations and polytypes of silicon carbide are shown in table 1.

Table 1. Results

Surface	Metallization (nm)	Resistance (Ohm*cm^2)	
		6H- SiC	4H-SiC
Si	Ni (100)	$2,56*10^{-4}$	$2,91*10^{-4}$
C	Ni (100)	$5,16*10^{-4}$	$4,15*10^{-3}$
Si	TiAl (25/150)	$3.71*10^{-4}$	$5,43*10^{-3}$
C	TiAl (25/150)	$2.50*10^{-4}$	$5,37*10^{-4}$

An optical microscope was used to study the surface morphology. Images are shown in figure 2 and figure 3. The unevenness of the edge is caused by the unevenness of the mask used in the deposition. The presented contacts have a good surface morphology.

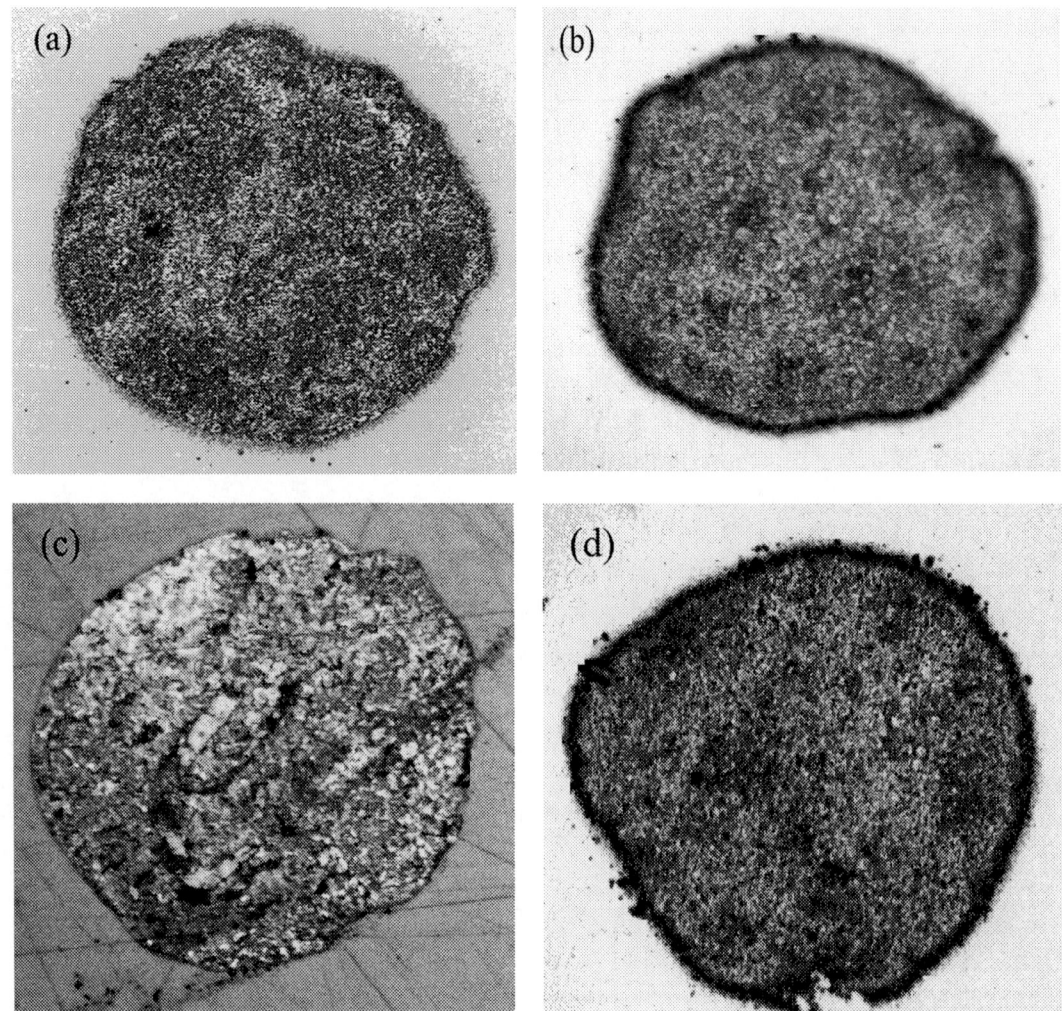

Figure 2. Surface morphology of ohmic contacts to SiC **(a, b, c, d)**. **(a)** Al/Ti, C-surface 4H-SiC; **(b)** Ni, Si-surface 4H-SiC; **(c)** Al/Ti, C-surface 6H-SiC **(d)** Ni, Si-surface 6H-SiC.

In Figures 2 it can be seen that the morphology determine by the type of metallization. The silicon carbide polytype has practically no noticeable effect. Small differences are observed for titanium-based metallization. After the annealing, it will behave differently for polytypes of silicon carbide.

4. Conclusion

We investigated the influence of crystallographic surface, metallization and polytype of silicon carbide on the ohmic contacts characteristics. It was established that it is possible to obtain low-resistance contacts to both polytypes and crystallographic surfaces. Ni-based metallization is more suitable for Si-surface. It allows to get resistance of about $3*10^{-4}$ Ohm*cm^2 for both polytypes. TiAl-based metallization is more suitable for C-surface. It allows to get resistance $2,5*10^{-4}$ Ohm*cm^2 for 6H-polytype. We used short-term plasma-chemical etching of the active layer for 4H- polytype, and it allows to get resistance $5,37*10^{-4}$ Ohm*cm^2. Resistance to 6H-SiC is slightly lower. In our opinion, this is a consequence of the difference in the band gap of the semiconductor. Also, the contacts have an acceptable surface morphology. This makes it possible to carry out lithographic processes in the future.

Acknowledgments

This work was supported by the Ministry of education and science of the Russian Federation (grantagreement 14.581.21.0021). Unique identificator RFMEFI58117X0021.

References

[1] Gao Y, Tang Y, Hoshi M, Chow T P 2000 Solid-State Electronics **44** 1875-1978
[2] Hui G, Yi-Men Z 2006 Chinese Physics **15** 2142
[3] Lee S K, Zetterling C M, Palmquist J P, Hogberg H, Jansson U 2000 Solid-State Electronics **44** 1179-1186.
[4] Shur M S 1987 GaAs devices and circuits (Minneapolis: University of Minnesota)

Investigation of the metal surfaces destruction due to electrochemical corrosion and cavitation, methods of protection with the use of polymer composite coatings based on CNT

A A Goshev, M K Eseev, S N Kapustin

Northern Arctic Federal University named after M.V. Lomonosov, Severnaya Dvina Emb. 17, 163002, Arkhangelsk, Russia

Abstract. The study is devoted to the dynamics of the appearance and development of defects at the nanoscale level for stainless steel and screw bronze of certain grades using electron, atomic force microscopy. The main mechanisms of destruction have been identified and a comparative analysis of the materials studied has been carried out. The results are presented on the possible protection of the material by applying a metal/polymer coating based on epoxy resin with different CNTs concentrations.

1. Introduction

One of the crucial problems in modern shipbuilding and ship repair is the increase of durability and the prolongation of the service life of a critical parts group (elements of screw-propellers, shafts, turbines). Cavitation and electrocorrosion are among the main causes of screws destruction of sea-going ships [1-2]. The stage of the appearance, dynamics and evolution of defects at the nanoscale is the least studied from the point of mechanism cavitation action view. Changing the time and regimes of the cavitation and electrocorrosion action on the samples, the authors observe the dynamics of the defects appearance and development. This allowed us to identify the main mechanisms of destruction and to carry out a comparative analysis of the materials studied to develop recommendations for their use in shipbuilding and ship repair, as well as to draw a conclusion about the functional options for protection.

2. Method

The essence of the new technique is that the samples of the metal alloys used for the production of screw-propellers are subjected to cavitation and electrocorrosion by a specially assembled stand (Fig. 1). As a source of cavitation effect, an ultrasonic magnetostrictive vibrating piezoelectric driver is used, placed in an electrolytic bath with a sample in water with an adjustable salt content. Varying the time of the main mechanisms destruction action – cavitation and electrochemical corrosion allowed in the laboratory to simulate the destruction of screw-propellers surface of the sea vessels. The state of the surface, the change in the relief and the roughness due to the appearance of various types defects of nanometer resolution were controlled by the methods of electron and scanning probe microscopy (ESM, SPM) according to the approach [3]. The combination of these methods of investigation made it possible to reveal temporal dynamics and mechanisms of destruction, as well as to conduct a comparative analysis of various alloys samples.

Content from this work may be used under the terms of the Creative Commons Attribution 3.0 licence. Any further distribution of this work must maintain attribution to the author(s) and the title of the work, journal citation and DOI.

Published under licence by IOP Publishing Ltd

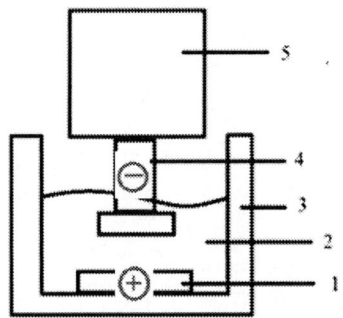

Figure 1. Researching stand. 1 – metal sample, 2 – water (salinity 0÷30‰), 3 – boxing, 4 – dispersant tip, 5 – ultrasonic disperser.

3. Cavitation and electro corrosive destruction of screw-propeller materials

On the basis of the above procedure, a study was made of the metal alloys samples (corrosion-resistant steel grade 08X18H10T (AISI 321), screw bronze BrA9Zh4N4) subjected to cavitation and electro corrosive hydrodynamic effects of varying duration. Figure 2 shows the surface roughness as a function of the ultrasound treatment time of steel and bronze samples measured by SPM microscopy.

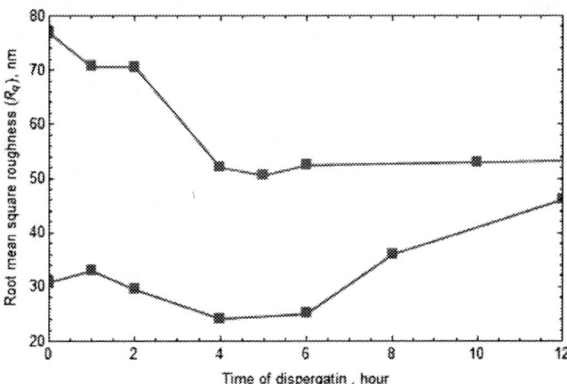

Figure 2. Dependence of the surface roughness on the time of ultrasound treatment for steel and bronze samples (red and blue colors, respectively).

The presented results allow dividing the effects of the cavitation action into two stages: surface smoothing for four hours (grinding effect) and subsequent development and evolution of defects. The destruction of steel is slow enough compared to a bronze sample. For a bronze specimen, noticeable changes in the relief and roughness occur only after 6 hours of cavitation action. Damage to the surface has local character. Only after 6 hours of ultrasound treatment the cavitation effect significantly changes the near-surface layer the appearance of deep cavities begins, after 12 hours of cavitation action, the average roughness increases one and a half times to values of the 50 nanometers order (Fig. 3). The study of cavitation fracture of steel samples revealed high anticavitational resistance, in relation to samples of screw bronze. At the initial stage, cavitation smooth the initial roughness, then further slightly increasing the average roughness to 40–50 nanometers for 15 hours of exposure (Fig. 4). The stages and character of steel destruction have distinctive from the bronze sample. The main contribution to the destruction of the surface layer is made by electrochemical corrosion. The authors conducted a study of the development defect dynamics through such destruction action. Salinity of water in the electrolytic bath is 3%, potential is 3V, the anode current is from the sample of 0.2 A. For a bronze sample, electrochemical corrosion primarily destroys the passivation layer on the surface, which is created by the presence of nickel in the alloy.

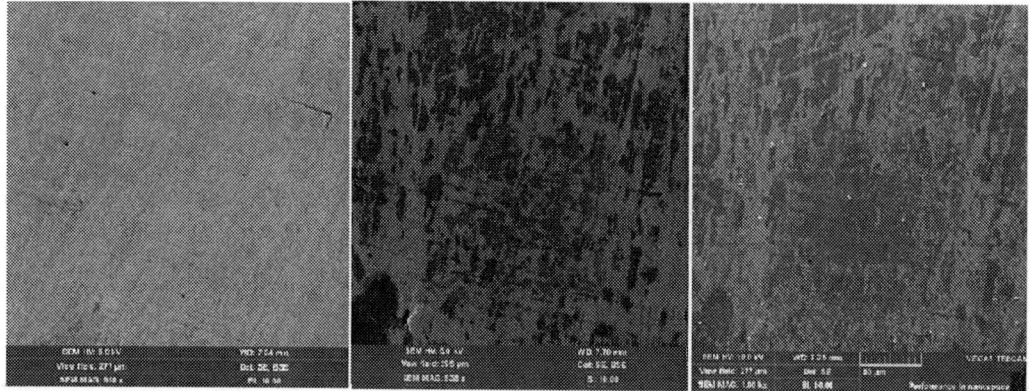

Figure 3. ESM photo of the bronze samples after the cavitation action. Exposure time is 0, 6, 10 hours respectively.

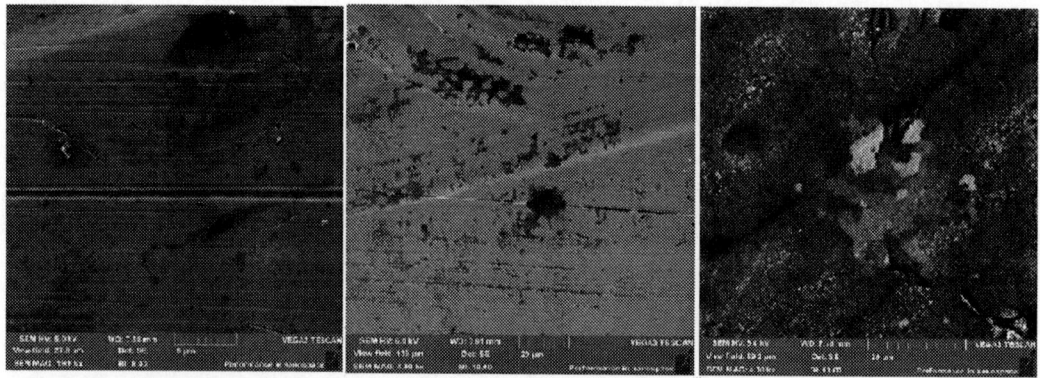

Figure 4. ESM photo of the steel samples after cavitation action. Exposure time is 0, 5, 15 hours respectively.

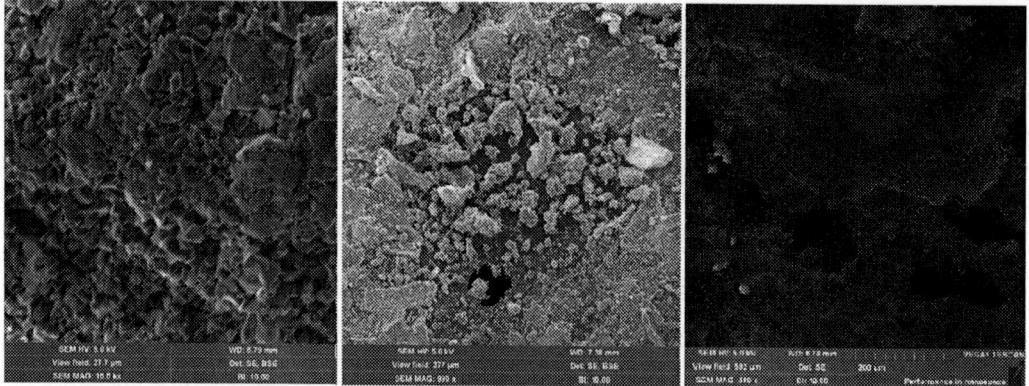

Figure 5. ESM photo of the bronze samples after electrochemical corrosion. The exposure time is 10, 30, 60 minutes, respectively.

For 30 minutes, the roughness increases twofold. During this time, a passivation layer with a thickness of the order of 50 nanometers completely disappears. Then, intercrystalline and intracrystalline corrosion develops, which leads to considerable surface degradation with an increase in the average roughness to values greater than 5 μm in a time of the order of 60 minutes (Fig. 5). With respect to electrochemical corrosion, steel samples are destroyed more intensively. At comparable values of the anode current, the destruction of the passivating layer occurs two times faster. In this case, the grain structure is more clearly exposed and intercrystalline corrosion begins, leading to an increase in the average roughness above 5 μm in a time of the order of 30 minutes (Fig. 6).

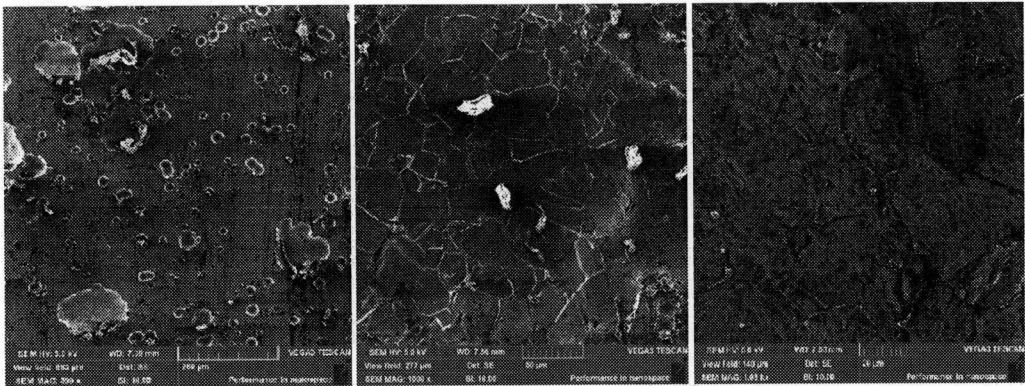

Figure 6. ESM photo of the steel samples after electrochemical corrosion. The exposure time is 10, 30, 60 minutes, respectively.

Table 1 shows the mass loss values for two samples (steel, bronze) with two action regimes (electrocorrosion and electrocorrosion + cavitation). Note that for an hour of electrochemical corrosion, both samples experience an equal mass loss of 15%. However, the combined effect of electrocorrosion and cavitation lads to different results. For the bronze sample, the loss effect is enhanced, while for the steel sample, the loss effect is reduced. This effect is attributed to the competition between the intercrystal and the intracrystal of electrocorrosion in these alloys. Steel is dominated by fracture due to intercrystalline corrosion, which is slowed by cavitation flows. In bronze, intragrain corrosion prevails, which is accelerated by the washing out of the passivation layer due to the cavitation effect.

Table 1. Mass loss Δm of the samples after an hour of exposure

Δm%	electrochemical corrosion	electrochemical corrosion + cavitation
Bronze	15%	23%
Steel	15%	7%

4. Development of a nanocomposite CNT coating for the protection of materials from destruction in hydrodynamic media

Currently, screw-steering columns are being developed in which the transfer of torque from the engine to the screw goes through the generator and electric motor system, while the screw is attached directly to the motor armature, so the screw is in the area of intensive alternating electromagnetic field induced by the windings of the electric motor. Therefore, for the development of coatings with carbon nanotube (CNT), it is necessary to study such materials for the electrophysical properties of such polymers depending on the CNT concentration. The study of concentration effects, the percolation threshold and the electrophysical properties of polymer composite materials is given in [4]. Based on the results of [4, 5] the authors propose to use as a anticorrosion coating a polymer composite material with the CNTs addition. Here the CNTs would perform the role of both the reinforcing element and the intermediate link in the metal/polymer system to increase adhesion with the surface.

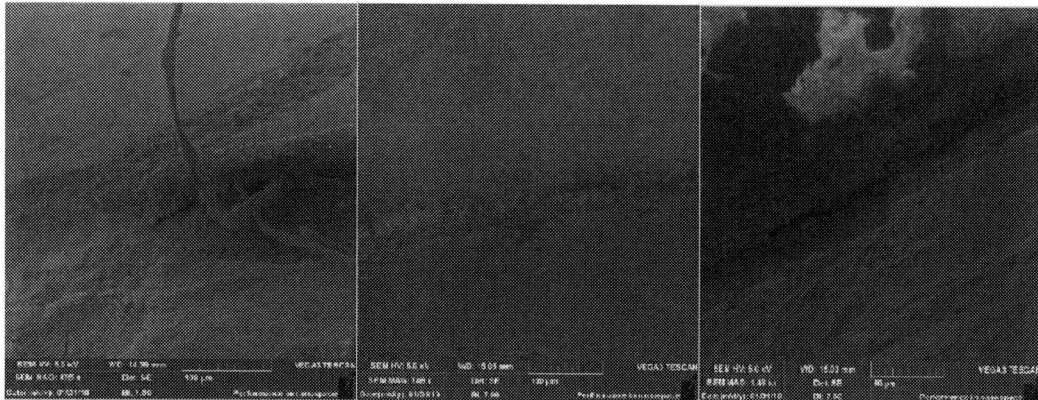

Figure 7. Boundary laye of ther metal/polymer sample after electrochemical exposure with a concentration of CNT in the polymer of 0, 1, 2%, respectively.

The results of corrosion damage study of the samples with coatings based on epoxy resin with different concentration of CNT are shown in Fig. 7. The samples were subjected to intensive electrochemical exposure in a salt bath after that the boundary of metal/polymer was investigated by the ESM method. Samples of a small (up to 1%) concentration of CNTs were found to be more resistant to the distraction. In a sample with a polymer without CNTs, large polymer fibers are torn off at the boundary, but it is more resistant to destruction than the sample with 2% of the CNT additive. In this case (beyond the percolation threshold), nanotubes obviously act as conductive defects that accelerate the destruction. We also note that the methods and regimes for applying such a coating, the exact concentration, the type of CNTs require further investigation and refinement.

5. Conclusions

We formulate main conclusions and results.

- A technique for studying the dynamics of defect evolution and development in the structure of screw-propeller metals due to cavitation and electrochemical corrosion action in a hydrodynamic medium are developed.
- Investigations of the defects evolution were carried out on the example of two materials widely used in the shipbuilding industry: corrosion-resistant steel grade 08X18H10T (AISI 321), screw bronze BrA9Zh4N4. The results of SPM and ESM of the defect dynamics and development are presented. Characteristic features of defect formation are revealed.
- The research of metal protection new methods by coating them with a composite polymer material with the addition of CNTs was proposed and started.

Acknowledgements

This work was supported by the project of the Russian Foundation of Basic Research No. 17-42-290138 "The dynamics of defects in the structure of the metal due to corrosion and cavitation action of hydrodynamic environments".

References

[1] Buravova S N, 1998 *Tech. Phys.*, **43** 1107
[2] Prikhod'ko V 2015 Met. Sci. **57**. 300
[3] Horodek P, Eseev M K et al 2015 *Nukleonika*, **60** 721
[4] Goshev A A, Eseev M K, 2017 *J. Phys.: Conf. Series* **917** 092013
[5] Eletskii A V, et al 2015 *Phys. Usp.* **58** (3) 209

SPBOPEN 2018

IOP Publishing

IOP Conf. Series: Journal of Physics: Conf. Series **1124** (2018) 081023   doi:10.1088/1742-6596/1124/8/081023

Composition-dependent conductivity of $In_xGa_{1-x}As$ nanowires

V Sharov[1], P Alekseev[1], M Dunaevskiy[1], R Reznik[2], G Cirlin[2]
[1]Laboratory of surface optics, Ioffe Institute, Saint Petersburg 194021, Russia
[2]Laboratory of epitaxial technologies, St. Petersburg Academic University, St. Petersburg 194021, Russia

Abstract. Using the methods of scanning probe microscopy, I-V characteristics of individual $In_xGa_{1-x}As$ nanowires with different In content (x) were measured. A sharp decrease in the conductivity was observed at x=0.7. It is shown that type of contact between nanowire and probe changes from ohmic at x=1 to Schottky at x=0.85. These changes were explained by the formation of surface conductive channel, induced by surface Fermi level pinning in conduction band for x>0.85.

1. Introduction

GaAs, InAs, and InGaAs nanowires each exhibit significant potential to drive new applications in electronic and optoelectronic devices due to direct bandgap, high electron mobility and compatibility with silicon-based device structures [1]. Semiconductor conductivity is usually controlled by doping. However, electronic properties of nanowires strongly depend on surface properties due to high surface/volume ratio [2] (see figure 1). In InGaAs semiconductors surface states density can reach 10^{13} cm^{-2} [3]. It is known that in InAs the Fermi level on the surface is pinned in the conduction band [4] (Figure 1, a) and in GaAs is approximately in the middle of the energy gap [5] (see Figure 1, b). Recently it was shown that in $In_{0.8}Ga_{0.2}As$ the surface position of the Fermi level coincides with the bottom of the conduction band [6] (figure 1, c).

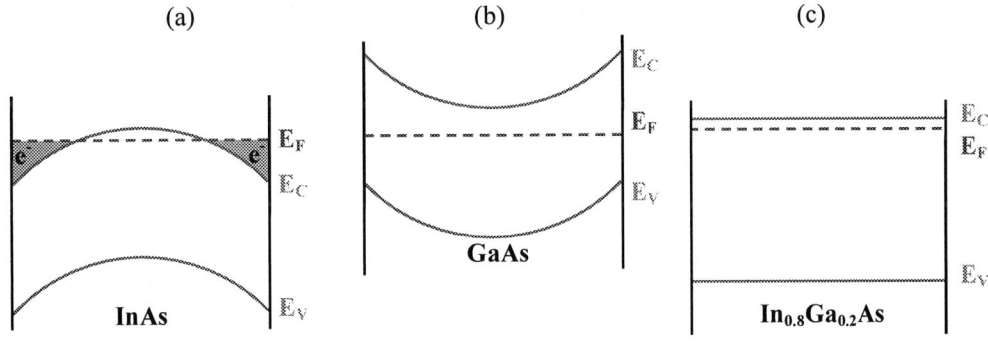

Figure 1. Band diagrams for n-type InGaAs nanowires. Formation of: **(a)** Surface conduction channel in InAs; **(b)** Depleted GaAs; **(c)** Flatband $In_{0.8}Ga_{0.2}As$.

Content from this work may be used under the terms of the Creative Commons Attribution 3.0 licence. Any further distribution of this work must maintain attribution to the author(s) and the title of the work, journal citation and DOI.
Published under licence by IOP Publishing Ltd

It is known that for $In_xGa_{1-x}As$ the pinning position is equidistant from the vacuum level for all x in metal-semiconductor interface [7]. Recently, we have shown a validity of this model for III-As nanowires surface with native oxide [8]. III-V nanowires can be grown with wurtzite crystal structure [9]. Conductivity of InAs nanowires depends on the crystal structure[10]. These effects are less studied for InGaAs nanowires. The aim of this work was to study a conductivity of the nanowires with wurtzite crystal structure using scanning probe microscopy. We measured I-V curves of single undoped oxidized $In_xGa_{1-x}As$ nanowires with different x.

2. Experiment

Scanning probe microscopy (SPM) was used for measuring of I-V curves[11]. Measurements were performed in ambient conditions on the Ntegra Aura (NT-MDT) microscope using the FM-W_2C probes with a conductive W_2C coating. Scheme of the experiment is presented in figure 2, a. For measuring of I-V curves of the NW two electrical contacts were created. One contact was connected to the surface of the substrate with nanowires array, and the second contact was created by SPM probe. $In_xGa_{1-x}As$ nanowires were grown by VLS technique using an Au as a catalyst in a molecular beam epitaxy chamber. NWs were grown on high doped n$^+$-Si substrates. A series of four samples with an In content (x) ranging from 0.7 to 1 was fabricated. Lengths of NWs were of 1-3 µm and diameters were of 20-40 nm (see figure 2, b). NWs exhibit a wurtzite crystal structure confirmed by transmission electron microscopy.

3. Results and discussion

Figure 3 shows I-V curves of $In_xGa_{1-x}As$ nanowires with different indium composition. At x=1 (pure InAs) the curve represents ohmic behavior with resistance of 50 MΩ (typical resistance of the probe), while at x=0.85–0.8 the contact switches to Schottky. At x=0.7 the resistance of NW sharply increases to 100 GΩ.

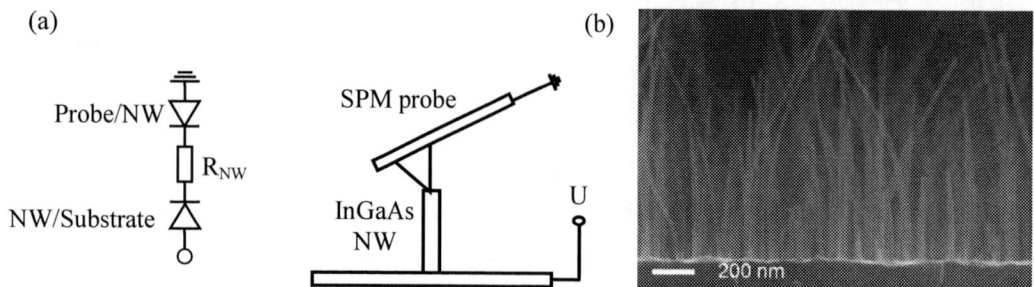

Figure 2. (a) Scheme of the experiment and equivalent electric scheme; **(b)** SEM image of the sample.

To analyse measured I-V curves it is necessary to consider equivalent electrical scheme of the measurement setup (see inset in the figure 2, a). During the measurement an electrical current pass through a barrier between the probe and NW, then through the NW itself and then through the barrier between NW and substrate. The last barrier has a zero-height due to linear shape of the I-V curve measured for InAs NW. Thus, the barrier between NW and the substrate can be neglected in the analysis [12]. Resistance of the nanowire and height of the Schottky barrier probe-NW strongly depend on the surface Fermi level pinning (see figure 1). Indeed, for the InAs NW the Schottky barrier height is zero and the resistance of the NW is low due to surface conduction channel. For the GaAs

NW the Schottky barrier height is about 0.6 eV [5] and the NW resistance is very high due to full depletion of the conductivity channel.

For the detailed explanation one should consider the $In_xGa_{1-x}As$ band diagram from [8] which is shown in figure 3, b. It is worthy to note, that the band structure of the wurtzite $In_xGa_{1-x}As$ alloys is poor studied. Recently, it was shown that conduction band offset between zinc blende and wurtzite InAs is in a range of 30-80 meV, depending on the crystal orientation of the wurtzite surface [13]. Position of the CBM for wurtzite $In_xGa_{1-x}As$ alloy is shown as grey band in figure 3, b. For x=0.7 the surface Fermi level (red line) is pinned in the band gap and the NW is fully depleted by charge carriers due to a high surface states density. In this case an electrical scheme of the experiment represents two Schottky barriers connected in opposite direction. This leads to reducing of the current down to the sensitivity of our device (30 pA). With the increasing of indium content the height of Schottky barrier is reduced with simultaneous increasing of the electrical current. For the x>0.85 the Fermi level is pinned in the conduction band that leads to the formation of surface conduction channel.

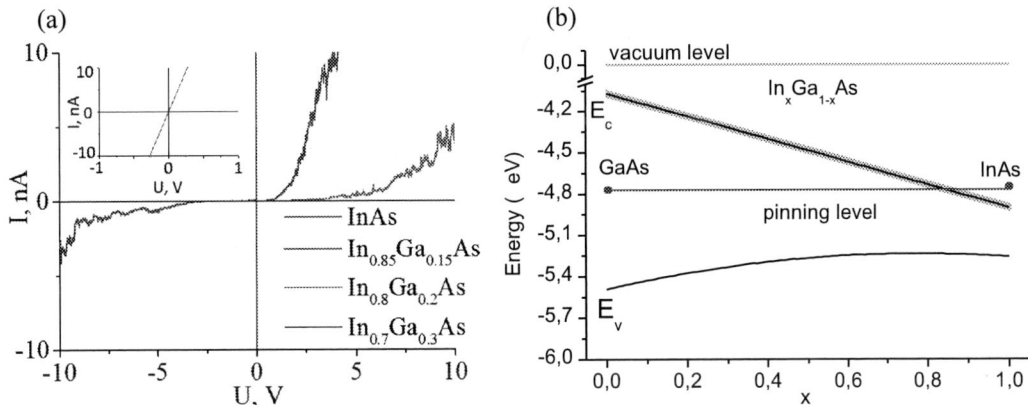

Figure 3. (a) I-V curves of InGaAs NWs; **(b)** Fermi level surface positions of InGaAs depending on indium content x.

4. Conclusion

Thus, a measurement of the I-V curves of undoped wurtzite $In_xGa_{1-x}As$ nanowires with 0.7<x<1 reveal a surface states-controlled conductivity of the nanowires. For x>0.85 an electronic surface conductive channel is formed. With a decreasing of indium content the conductivity of the nanowires is reduced by 4 orders of magnitude due to the formation of Schottky barrier to the depleted nanowire.

References

[1] Saxena D, Mokkapati S, Parkinson P, Jiang N, Gao Q, Tan H H, Jagadish C 2013 *Nature Photonics* **7** 963
[2] Alekseev P A, Dunaevskiy M S, Ulin V P, Lvova T V, Filatov D O, Nezhdanov A V, Mashin A I, Berkovits V L 2015 *Nano letters* **15** 63
[3] Hasegawa H, Akazawa M, Domanowska A, Adamowicz B 2010 *Applied surface science* **256** 5698
[4] Baier H-U, Koenders L, Mönch W 1986 *Solid state communications* **58** 327
[5] Spicer W E, Lindau I, Skeath P, Su C Y, Chye P 1980 *Phys. Rev. Lett.* **44** 420
[6] Speckbacher M, Treu J, Whittles T J, Linhart W M, Xu X, Saller K, Dhanak V R, Abstreiter G, Finley J J, Veal T D and others 2016 *Nano letters* **16** 5135
[7] Wieder H 2003 *Journal of Vacuum Science & Technology B: Microelectronics and Nanometer Structures Processing, Measurement, and Phenomena* **21** 1915

[8] Alekseev P, Dunaevskiy M, Cirlin G, Reznik R, Smirnov A, Davydov V Y, Berkovits V L 2018 *Nanotechnology* **29(31)** 314003

[9] Glas F, Harmand J-C, Patriarche G 2007 *Physical review letters* **99** 146101

[10] Dayeh S A, Susac D, Kavanagh K L, Yu E T, Wang D 2009 *Advanced Functional Materials* **19** 2102

[11] Alekseev P, Geydt P, Dunaevskiy M, Lähderanta E, Haggrén T, Kakko J-P, Lipsanen H 2017 *Applied Physics Letters* **111** 132104

[12] Geydt P, Alekseev P, Dunaevskiy M, Haggrén T, Kakko J-P, Lähderanta E, Lipsanen H 2016 *Lithuanian Journal of Physics* **56** 92

[13] Hjort M, Lehmann S, Knutsson J, Zakharov A A, Du Y A, Sakong S, Timm R, Nylund G, Lundgren E, Kratzer P and others 2014 *ACS nano* **8** 12346

MBE formation of self-catalyzed GaAs nanowires using ZnO nanosized films

M S Solodovnik[1,2], O G Karenkikh[1], S V Balakirev[1], S I Petrov[3], R V Ryzhuk[4], A N Alexeev[3], O A Ageev[1,2]

[1]Department of Nanotechnologies and Microsystems, Southern Federal University, Taganrog 347922, Russia
[2]Research and Education Center "Nanotechnologies", Southern Federal University, Taganrog 347922, Russia
[3]SemiTEq JSC, Saint-Petersburg 194156, Russia
[4]Institute of Nanoengineering in Electronics, Spintronics and Photonics, Research Nuclear University MEPhI, Moscow 115409, Russia

Abstract. We studied the influence of nanoscale ZnO films deposited onto GaAs (001) on the process of GaAs epitaxial growth. We took into consideration the most important control parameters of molecular beam epitaxy, such as substrate temperature, As_4/Ga effective flux ratio and growth rate and different thicknesses of ZnO films. We found that a ZnO film deposited on GaAs surface acts as a native GaAs oxide when the thickness of the film is being decreased, and can be removed thermally at a temperature of 620-630°C. We showed that it is possible to use nanoscale ZnO films with thickness ~5 nm in order to create horizontal GaAs nanowires grown by a self-catalytic mechanism.

1. Introduction

One-dimensional nanostructures, including semiconductor nanowires, are considered to be one of the most promising candidates for the role of active elements of device structures to be used in nanoelectronics, photonics, sensor technology and MEMS/NEMS [1–4]. Having unique structural, transport, optical and electrical characteristics, they can be used both to improve existing systems and to create new types of devices [5–7]. Self-catalytic nanowires formed by a vapor-liquid-crystal mechanism seem to be the most appropriate means to achieve the abovementioned purposes [8, 9]. The process of autocatalytic nanowire growth, as opposed to the case of heterocatalytic growth, is characterized by the fact that catalyst droplets are formed directly in the growth process, which requires the creation of specific surface conditions to initiate the formation of droplets of one of the components [10–16]. To make sure this happens, thin (a few to tens of nanometers) layers of heterogeneous material are formed on the surface, which gives rise to the liquid phase. In most cases, a thin SiO_x film acts as such a sublayer [17, 18]. However, the presence of a SiO_x sublayer on the surface and its partial destruction which takes place during the growth process may negatively affect the electrical characteristics of the nanowires obtained [19].

In the present work, our aim was to study the possibility of nanosized ZnO films use as sublayers initiating the formation of self-catalytic GaAs nanowires during molecular beam epitaxy (MBE), which is necessary to investigate the impact that ZnO films have on GaAs epitaxial growth processes.

Content from this work may be used under the terms of the Creative Commons Attribution 3.0 licence. Any further distribution of this work must maintain attribution to the author(s) and the title of the work, journal citation and DOI.
Published under licence by IOP Publishing Ltd

2. Experiment

We studied the influence of nanoscale ZnO films on GaAs epitaxial growth processes using an MBE system SemiTEq STE35 which was equipped with a solid-state sources of the elements of the group III and valved As source. We used "epi-ready" GaAs (001) wafers as substrates. After removed GaAs native oxide in the MBE growth chamber we transferred the samples to a pulsed laser deposition (PLD) system Neocera Pioneer 180 where we deposited amorphous ZnO films. To avoid possible interaction between ZnO surface and sputtered GaAs, the PLD process was carried out at a substrate temperature of 50°C. The pressure of the residual atmosphere was $1 \cdot 10^{-3}$ Pa, the laser power was 260 mJ, the laser pulse frequency was 10 Hz and ZnO (Kurt J. Lesker) was used as a target. The ZnO films had thicknesses of 5 and 20 nm at 3000 and 10 000 pulses, respectively. Then we placed the samples back into the MBE growth chamber where GaAs epitaxial growth was carried out from molecular fluxes of Ga and As$_4$. The effective deposition thickness for all the samples was 100 nm. We varied the deposition rate V from 1 to 1.5 ML/s, the effective ratio of Ga and As fluxes ($J_{As/Ga}$) ranged from 1 to 4 and the substrate temperature T during the growth ranged from 500 to 650°C. We controlled the process by recording the patterns of reflected high-energy electron diffraction (RHEED). Prior to the beginning of GaAs growth, we made a preliminary deposition of gallium (about 5-10 ML) and then opened the source of arsenic. After completed GaAs deposition, we closed the sources of Ga and As$_4$ simultaneously and cooled the sample rapidly in order to avoid any redistribution of the material on the surface. Then we studied the samples by means of scanning electron (SEM) and atomic force (AFM) microscopy using the FEI Nova Nanolab 600 and NT-MDT Ntegra systems, respectively [20–24].

3. Results and discussion

When we preheated the samples ZnO(20 nm)/GaAs(001) in a vacuum ($1 \cdot 10^{-7}$ Pa), there was no change in the RHEED pattern across the entire growth temperature range and the diffuse background we registered proved that there was ZnO on the surface of the amorphous film. Therefore, GaAs was being deposited over the ZnO layer. At the same time, at low temperatures (below 550°C) we observed the formation of a GaAs films with polycrystalline structure. For instance, at $T = 500$°C we observed the formation of an array of GaAs crystallites on the ZnO surface (Figure 1a). The crystallites were characterized by a density of $4 \cdot 10^9$ cm^{-2}, lateral dimensions of 100-250 nm and a height of 40-100 nm. An increase in the growth temperature to 580°C resulted in a fusion of the generated GaAs crystallites into a GaAs film with a rough surface, which was caused by an increase in the surface mobility of adatoms.

Unlike the previous case, when we heated ZnO(5 nm)/GaAs(001) heterostructures above 550°C, the RHEED system recorded a dot pattern with a very weak brightness appearing in the diffuse background. A similar pattern was observed in the case of GaAs native oxide film. Notably, a prolonged annealing (up to 30 min) at $T = 550$-600°C did not cause any significant change neither in the brightness or in the structure of the diffraction pattern. An increase in temperature up to 600°C led to a gradual increase in the brightness of RHEED pattern, and at $T = 620$-630°C, the pattern was similar to the one we observed during the process of GaAs native oxide thermal desorption at 580°C, i.e. there was a sharp increase in the brightness of the diffraction pattern followed by a dot RHEED pattern related to GaAs (001) surface. This can be accounted for by the thermal removal of the ZnO thin film from the surface of GaAs.

Epitaxial growth of GaAs on the surface of a 5 nm thick ZnO film within the temperature range of $T = 620$-640°C at $V = 1$ ML/s resulted into the formation of a GaAs layer with a rough surface. Reducing the temperature to 600°C led to a increase in the density of the pits from $4 \cdot 10^9$ cm^{-2} up to $2.2 \cdot 10^9$ cm^{-2} and a simultaneous decrease in their size from 300-600 nm to 100-200 nm.

Having increased the deposition rate up to 1.5 ML/s and reduced $J_{As/Ga}$ ratio to 1 within the same temperature range, we formed deep pits on the GaAs surface. The pits had a diameter of 200-300 nm, a depth of 150-200 nm and a significantly lower density of $2 \cdot 10^7$ cm^{-2} (Figure 1b).

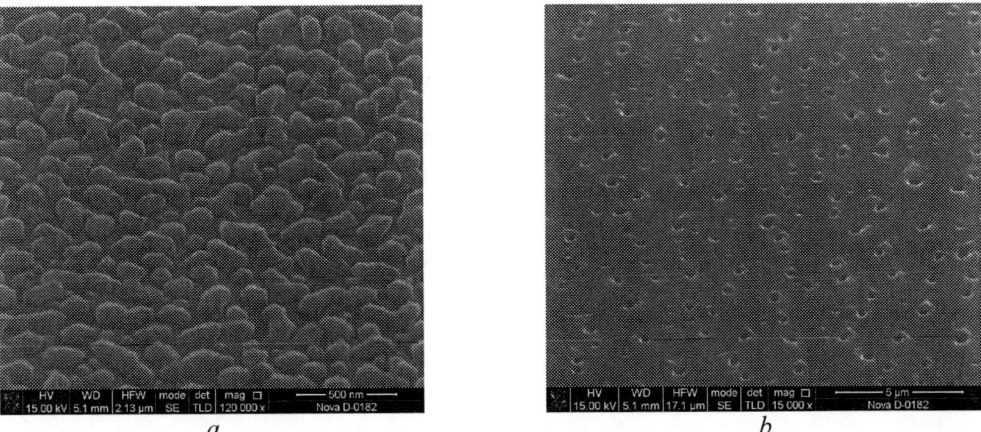

Figure 1. A SEM images of the surface of ZnO/GaAs (001) structure after the deposition of GaAs at $T = 500°C$ and 20 nm thick ZnO film (*a*) and $T = 600°C$ 5 nm thick ZnO film (*b*).

Having reduced the growth temperature further while maintaining the values of the other parameters within the temperature range of 570-590°C, we obtained horizontal non-oriented GaAs nanowires (Figure 2*a*). The density of the array of horizontal GaAs nanowires reached $1.2 \cdot 10^7$ cm^{-2}. The nanowires had a length of 0.5-1 μm, a diameter of 250-500 nm and a height (above surface level) of 120-160 nm. The AFM analysis of the structures confirmed the liquid origin of the droplets on the tops of GaAs nanowires since switching to the contact scan mode led to the destruction of spherical formations on crystal tops prominent in the phase contrast picture and the formation of pits with a depth of 60 nm on their site (Figure 2*b*). The droplet material redistributed over the surface around the nanowires according to the current AFM technique.

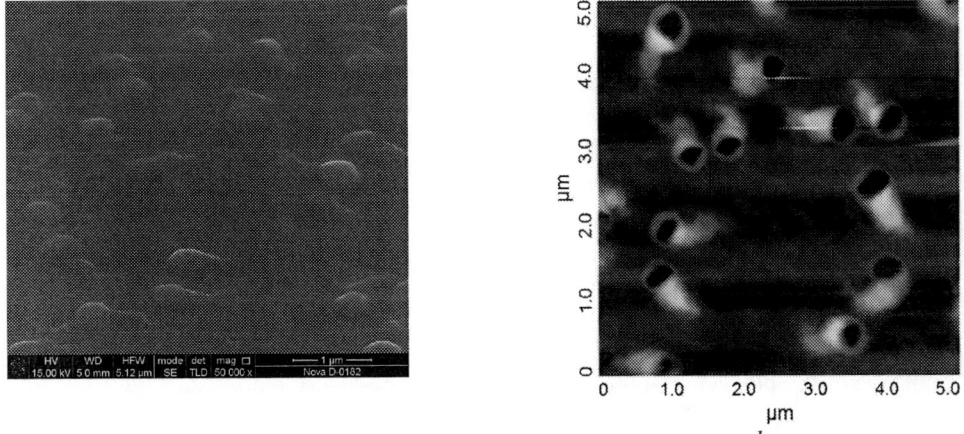

Figure 2. Surface of ZnO/GaAs (001) structure after the deposition of GaAs at $T = 580°C$: (*a*) SEM image, (*b*) AFM image in a contact mode.

The presence of liquid Ga droplets on crystals tops indicates a self-catalytic mechanism of crystal growth. Horizontal crystal growth may be predominant due to a small difference between the values of the surface energy and/or specific conditions on the surface during the formation of the catalytic centers (droplets) and the subsequent crystal growth [25]. In addition, the effective thickness of the deposition can also have a significant impact on the final morphology of the crystal obtained by self-catalytic growth [26].

Our experimental results are similar to those presented in [27, 28]. In any case, the formation of any type of GaAs surface morphology including nanowires can be explained not only by kinetic effects on the surface but also by chemical interactions between ZnO film and GaAs substrate. The presence of such an interaction helps to explain the effect of ZnO film thickness on the morphology of the growing layer of GaAs as when the film is thick, the effective area of interaction between materials decreases and, consequently, there is a decrease in the intensity of chemical reactions at the interface. This also accounts for the observed effect of thermal desorption of thinner ZnO films. Its mechanism is similar to the mechanism of GaAs native oxide removal happening during MBE [29]. Notably, the activation of chemical processes leading to ZnO film degradation, according to RHEED, starts at a temperature of 550°C.

4. Conclusion
In summary, we studied the influence of nanoscale ZnO films deposited on GaAs surface on GaAs epitaxial growth processes taking into account the technological parameters of MBE. We found that there is a significant effect of ZnO film thickness, growth temperature, effective deposition rate and flux ratio on the resulting morphology. We obtained self-catalytic GaAs horizontal nanowires and defined the conditions of their formation. We also found that in addition to kinetic factors, the chemical interaction at ZnO/GaAs interface has a considerable influence on the morphology of GaAs epitaxial layer. Notably, the activation of chemical reactions starts at a temperature of 550°C, and a 5 nm thick ZnO film gets desorbed from the surface at temperatures of 620-630°C.

Acknowledgements
The results were obtained using the equipment of the Research and Education Center and Center for Collective Use "Nanotechnologies" of Southern Federal University.

References
[1] Thelander C, Agarwal P, Brongersma S, Eymery J, Feiner L, Forchel A, Scheffler M, Riess W, Ohlsson B, Gösele U, Samuelson L 2006 *Nano Today* **10** 28-35
[2] Hayden O, Agarwal R, Lu W 2008 Nano Today 3 12–22
[3] Klimin V S, Il'Ina M V, Il'In O I, Rudyk N N, Ageev O A 2017 *J. Phys.: Conf. Ser.* **917** 092023
[4] Rudyk N N, Il'In O I, Il'Ina M V, Fedotov A A, Klimin V S, Ageev O A 2017 *J. Phys.: Conf. Ser.* **917** 082008
[5] Chen R, Tran T D, Ng K W, Ko W S, Chuang L C, Sedgwick F G, Chang-Hasnain C 2011 *Nature Photonics* **5** 170–175
[6] Huang B, Yang Y, Lin T, Yang W 2012 *Solar Energy Mat. & Solar Cells* **98** 357–362
[7] Il'Ina M V, Blinov Y F, Il'In O I, Klimin V S, Ageev O A 2016 *Proc. of SPIE* **10224** 102240U-1
[8] Johansson J, Wacaser B, Thelander K D, Seifert W 2006 *Nanotechnology* **17** 355–361
[9] Dubrovskii V G, Sibirev N V 2009 *Tech. Phys. Lett.* **35** 380–383
[10] Balakirev S V, Solodovnik M S, Ageev O A 2018 *Phys. Status Solidi B* **255** 1700360
[11] Ageev O A, Solodovnik M S, Balakirev S V, Eremenko M M 2016 *J. Vac. Sci. Technol. B* **34** 041804
[12] Ageev O A, Solodovnik M S, Balakirev S V, Mikhaylin I A 2016 *Technical Physics* **61** 971–977
[13] Ageev O A, Solodovnik M S, Balakirev S V, Mikhaylin I A, Eremenko M M 2017 *J. Cryst. Growth* **457** 46–51
[14] Balakirev S V, Solodovnik M S and Ageev O A 2017 *J. Phys. Conf. Ser.* **917** 032033
[15] Ageev O A, Solodovnik M S, Balakirev S V, Mikhaylin I A 2016 *J. Phys. Conf. Ser.* **681**

012036

[16] Balakirev S V, Solodovnik M S, Eremenko M M, Mikhaylin I A, Ageev O A 2017 *J. Phys. Conf. Ser.* **917** 032034

[17] Colombo C 2008 *Phys. Rev. B* **77** 1-5

[18] Yua S, Miaoa G, Jina Y, Zhanga L, Songa H, Jianga H, Lia Z, Lia D, Sun X 2010 *Physica E* **42** 1510-1543

[19] García Núñezn C, Braña A F, López N, García B J 2015 *J. Cryst. Growth* **430** 108–115

[20] Ageev O A, Smirnov V A, Solodovnik M S, Rukomoikin A V, Avilov V I 2012 *Semiconductors* **46** 1616–1621

[21] Klimin V S, Solodovnik M S, Smirnov V A, Eskov A V, Tominov R V, Ageev O A 2016 *Proceedings of SPIE* **10224** 102241Z

[22] Ageev O A, Klimin V S, Solodovnik M S, Eskov A V, Krasnoborodko S Y 2016 *J. Phys.: Conf. Ser.* **741** 012178

[23] Avilov V I, Ageev O A, Smirnov V A, Solodovnik M S, Tsukanova O G 2015 *Nanotech. in Russia* **10** 214–219

[24] Solodovnik M S, Balakirev S V, Eremenko M M, Mikhaylin I A, Avilov V I, Lisitsyn S A, Ageev O A 2017 *J. Phys. Conf. Ser.* **917** 032037

[25] Dubrovskii V G, Sibirev N V, Cirlin G E, Ustinov V M 2008 *Phys. Rev. E* **77** 031606

[26] Soshnikov I P, Tonkikh A A, Cirlin G E, Samsonenko Y B, Ustinov V M 2004 *Tech. Phys. Lett.* **30** 765–768

[27] Ageev O A, Solodovnik M S, Balakirev S V, Eremenko M M, Mikhaylin I A 2016 *J. Phys.: Conf. Ser.* **741** 012012

[28] Ageev O A, Konoplev B G, Rubashkina M V, Rukomoikin A V, Smirnov V A, Solodovnik M S 2013 *Nanotech. in Russia* **8** 23-28

[29] Ageev O A, Solodovnik M S, Balakirev S V, Eremenko M M 2016 *Phys. Solid State* **58** 1045–1052

Cathodic transformation of bactericidal silver-containing bioceramic coatings for implants

E Poshivalova[1], I Rodionov[1]

[1]Yuri Gagarin State Technical University of Saratov, Saratov 410054, Russia

Abstract. In the article the micro- and nanostructures of the biocomposite coating on titanium VT1-00, cathodically transformed by silver, were studied by the method of scanning electron microscopy. There is given a brief characteristic of the structural state of the coating, depending on its chemical composition. It is established that in the structure there are evenly distributed micrometric sprayed particles of a bioceramic dust adhered to the surface. The structure of individual particles is characterized by nanometric elements with an average size of 40 ± 10 nm, as well as single agglomerates of nanograins with an average size of 200 to 500 nm. The technology for obtaining high-performance coatings and the construction of an adapted electrochemical bath are proposed. Moreover, it is revealed that the concentration of the alloying silver additive in the coating is about 0.5–1.0%, which allows the increase of the percentage of engraftment of titanium implants with such bioceramic coatings.

1. Introduction

At present an actual problem of modern medical and technical products (implants) is to improve mechanical properties of the functional coatings [1-3] and to minimize their rejection. One of the reasons of the rejection is inflammation of the bone around the implant (periimplantitis), which is characterized by a gradual loss of bone tissue (resorption). An undesirable consequence of this disease is the rejection of the implanted structure. To prevent such a complication, it is advisable to use alloying additives (lanthanum, silver, copper and other elements) that impart antimicrobial and antiplatelet properties to the surface of intraosseous implants [4-8]. The most favorable condition for increasing the characteristics of biocompatibility will also be the formation of a biocompatible coating on the surface characterized by the presence of submicrometry and nano-sized elements of morphology [4,5]. The aim of the work is to determine the conditions for obtaining nanostructural characteristics of a biocomposite coating obtained by plasma spraying (PS) and following cathodic implantation of bactericidal silver-containing components.

2. Methodology

For electrochemical studies titanium plates VT1-00 with dimensions $10\times10\times2$ mm were used as samples. On their surface calcium-phosphate coatings were formed. This technology provides ultrasonic degreasing of the titanium base of medical and technical products in distilled water and alcohol, air-abrasive treatment with electrocorundum, as well as the preparation of matrix coatings, which are subsequently subjected to cathodic modification. The cathode saturation of the pores of calcium-phosphate coating with silver deposited on the titanium base was carried out from an aqueous solution of 0.4 M $AgNO_3$ in a galvanostatic regime for 10 minutes at different cathode current densities ($i_k=0.2$–0.5 mA/cm^2) and the temperature of 20 °C. Electrochemical measurements were conducted on potentiostat P-5848 complete with self-recording potentiometer KSP-4 in a standard

Content from this work may be used under the terms of the Creative Commons Attribution 3.0 licence. Any further distribution of this work must maintain attribution to the author(s) and the title of the work, journal citation and DOI.
Published under licence by IOP Publishing Ltd

glass three-electrode cell with a titanium auxiliary counter electrode and a nonaqueous silver-chloride comparison electrode. Non-current chronopotentiograms were recorded on electrodes before and after cathodic polarization.

Experimental investigations of the micro- and nanostructure of the obtained coatings were carried out using the method of scanning electron microscopy (SEM) with the possibility of conducting energy-dispersive X-ray fluorescence analysis (EDX).

3. Results

Potentiodynamic curves allow the registration of all possible processes that occur on the electrodes in the investigated interval of potentials most demonstrably. The presence of peaks and areas on i,E-curves allows the evaluation of the expected composition of products in the process of electrochemical transformations and their stability. The maximum residue limit (MRL) studies were carried out in the electrolyte of the intercalation in the potential range from -3 to +1 V with a linear sweep speed of 20 mV/s. E-t curves for Ti/HAp coatings are presented in Figure 1. Analysis of the E-t curves of Ti/HAp coatings covered with silver is quite definite. However, the value of non-current potential is affected by the value of the cathode current density. The value of the cathode process potential is monotonically shifted to a region of more negative quantity with increasing of the i_k. This is also seen in the analysis of non-current chronopotentiograms. The values of the potentials of the investigated coatings absolutely correspond to the value of the potential of pure silver.

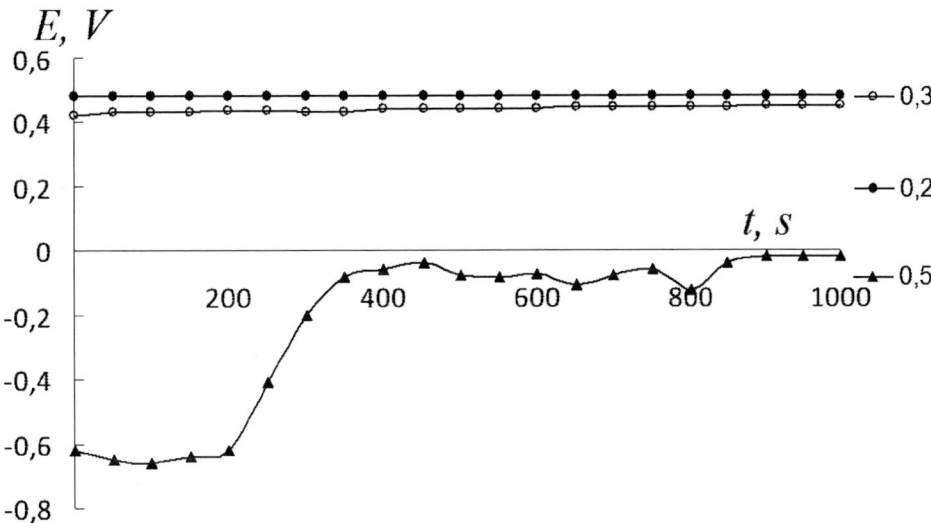

Figure 1. E-t curves of Ti/HAp coatings when they are saturated with silver.

The reconstruction of silver ions proceeds practically with a 100% current yield in the absence of nitrate ions. However, their presence reduces the current efficiency up to 80–90% due to the side processes of reconstruction of NO^{3-} to ammonia and hydroxylamine. This allows the estimation of the approximate amount of silver in the pores of the nanostructured coatings. The values of the amount of silver in the nanostructured coatings are given depending on the penetration current density (i_k). From the analysis of the conducted experiments it is established that with the increase in the current density, silver is introduced more and more stable.

SEM data show that all titanium samples with the coatings are characterized by a typical microstructure formed at the PS [4]. In the structure there are evenly distributed micro-sized sprayed particles (splats) of the bioceramic powder adhered to the surface (Figure 2a). The structure of individual splats is characterized by nano-sized elements, which are evenly distributed over the surface with an average size of 40±10 nm, as well as single agglomerates of nanograins with an average size of 200 to 500 nm (Figure 2b).

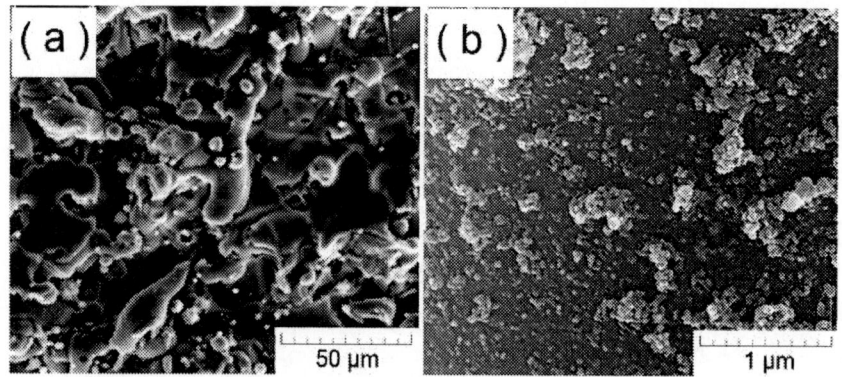

Figure 2(a, b). SEM of the coating micro- **(a)**; and nanostructure **(b)**.

EDX showed that, in addition to the main components of silver-containing PS coatings O, P and Ca with the corresponding content (40±15)%, (10±4)% and (25±15)%, there are additives of Ti, Cr, Fe and Ag. The concentration of the alloying silver additive in the coating is about 0.5–1.0%.

Based on the experimental and theoretical data obtained, we have developed a special galvanic bath construction for cathodic insertion of modifying additives (silver, lanthanum) into calcium-phosphate coatings (Figure 3).

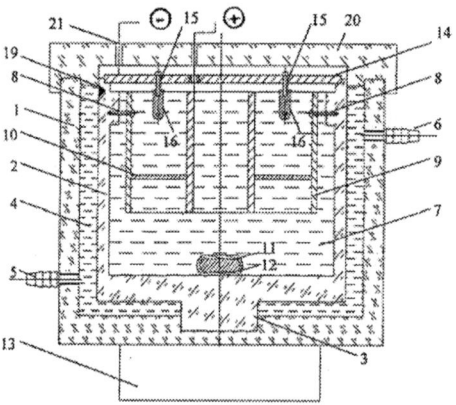

Figure 3. The construction of a cathodic introduction of lanthanum into the coatings of cylindrical titanium dental implants (explanations are in the text).

The bath consists of two cylindrical bodies of different diameters, made of polytetrafluorethylene (PTFE) – an external 1 having a deepening in the middle of the base and an inner 2 with a special protrusion 3. The protrusion 3 of the body 2 provides its fixation inside the body 1 at some distance from its inner walls, which causes the formation of a water jacket 4, through which water circulates having a thermostatically set temperature. The circulation of water through the jacket takes place through the inbuilt in an external body inlet and outlet adapter sleeves 5 and 6, which are connected by plastic hose with the thermostat. The presence of the water jacket in this bath is necessary because the temperature of the water moving in it allows the control of the temperature of the electrolyte 7 inside the body 2. In the body 2, using bolts 8 made of titanium, a double cylindrical hollow titanium anode 9 is mounted, consisting of two annular anodes of different diameter, connected with each other by titanium conductors 10. This construction feature of the anode makes it possible to ensure uniformity of the embedding process. At the bottom of the body 20 a magnet 11 is located, sealed in a glass tube 12, isolating it from the chemical action of the electrolyte. The magnet rotating by means of a magnetic stirrer 13 allows the stirring of the electrolyte solution.

As a cathode electrode, a special cover-bracket 14 was developed, into which in which screws are twisted in a certain circumference 15, on which titanium blanks of dental implants are fixed with a threaded connection 16, which are serving as working cathode elements. The screw-cap system is made of titanium.

The required level of electrolyte should correspond to a special level mark applied to the outer wall of the housing 2. The bath cover 20 is made of Teflon and has symmetrical apertures coaxial with the cathode lid 14. An additional aperture 21 in the lid of the bath is designed for the output of the electrical whisker of the cover-cathode. In addition, the complex can be composed in both potentiostatic and galvanostatic variants.

4. Conclusions

As a result of the investigation, the possibility of cathodic introduction of the bactericidal component (silver) into the matrix plasma-sprayed coating of medical and technical function (implants) was established. Additional electrochemical impact, e.g. cathodic modification, allows the increase of the percentage of engraftment of titanium implants with bioceramic coatings.

Acknowledgments

The research was supported by the Ministry of Education and Science of the Russian Federation in the framework of the Program of Scientific Research in Universities (project No. 11.1943.2017/4.6).

References

[1] Kalita V I, Komlev D I and Radyuk A A 2016 *Inorg. Mater. Appl. Res.* **7** 536
[2] Ivannikov A Yu, Kalita V I, Komlev D I, Radyuk A A, Bagmutov V P, Zakharov I N and Parshev S N 2016 *J. Alloys. Compd.* **655** 11
[3] Fomin A A, Steinhauer A B, Rodionov I V, Fomina M A, Zakharevich A M, Skaptsov A A, Gribov A N and Karsakova Ya D 2014 *J. Frict. Wear.* **35(1)** 32
[4] Fomin A, Fomina M, Koshuro V, Rodionov I, Zakharevich A and Skaptsov A 2017 *Ceram. Int.* **43** 11197
[5] Fomin A, Dorozhkin S, Fomina M, Koshuro V, Rodionov I, Zakharevich A, Petrova N and Skaptsov A 2016 *Ceram. Int.* **42** 10838
[6] Fomin A A, Steinhauer A B, Rodionov I V, Petrova N V, Zakharevich A M, Skaptsov A A and Gribov A N 2013 *Biomed. Eng.* **47(3)** 138
[7] Koshuro V, Fomina M, Fomin A and Rodionov I 2018 *Compos. Struct.* **196** 1
[8] Koshuro V, Fomin A, Fomina M, Rodionov I, Brzhozovskii B, Martynov V, Zakharevich A, Aman A, Oseev A, Majcherek S and Hirsch S 2016 *J. Phys. Conf. Ser.* **741(1)** 012197

SPBOPEN 2018   IOP Publishing

Synthesis and Properties of a Polyamine-Cumulene/Carbon Nanotubes for Removing Harmful Substances from Aqueous Solutions

E A Neskoromnaya[1], A V Melezhik[1], O V Alekhina[1], I V Burakova[1*], E S Mkrtchyan, A E Burakov[1]

[1] Department "Technology and Methods of Nanoproduct Manufacturing", Tambov State Technical University, Tambov, 392000, Russia

* iris_tamb68@mail.ru

Abstract. In the current work, a method for obtaining an effective adsorption material (nanocomposite) through polycondensation of hexamethylenetetramine in sulfuric acid in the presence of multi-walled carbon nanotubes as texture component is proposed, and the properties of this material are studied. The elemental composition of the material was determined by energy-dispersive X-ray analysis, and the morphology was evaluated by scanning electron microscopy. Its adsorption activity was estimated towards Cu(II) ions and synthetic dyes – methylene blue (MB) and methyl orange (MO). It was established that the adsorption capacity of the studied material for the dyes is 1,255 and 801 mg g^{-1} (for MB and MO, respectively), whereas for Cu(II) ions, it is 24.5 mg g^{-1}. Besides, the kinetics of the adsorption of all those species was studied, thereby making it possible to determine the expediency of using the developed nanocomposite for purification of aqueous solutions.

1. Introduction

Authoritative research works in the field of synthesis and application of new types of cabon-nanomaterial-based adsorbents prove that graphene nanostructured materials are more effective in comparison with currently used commercial adsorbents [1-3]. Graphene-based adsorbents are assumed to become an alternative to the conventional materials such as rocks, clays, hydrated aluminosilicates, activated carbons, low-cost biosorbents prepared from agricultural waste, etc [4,5]. Among the existing physical and chemical processes of removal and subsequent isolation of harmful contaminants, a significant role is played by adsorption-desorption cycles. Their practical implementation allows removal, transfer and accumulation of toxic compounds with almost 100-% efficiency [2].

The synthesis of aminocumulene, a soluble polycondensation product of hexamethylenetetramine (HMTA) in anhydrous sulfuric acid, is described by the authors in [6]. This substance presumably contains cumulated double carbon-carbon bonds and amino groups. It was established that during the HMTA polycondensation in sulfuric acid at an elevated temperature (180-200 °C), a polymeric insoluble sulfur-containing product (polyamine cumulene - PAC) is formed, and it demonstrate high adsorption properties with respect to heavy metal ions and organic dye molecules. Considering this, herein, the authors study the adsorption properties of a PAC/carbon nanotubes (CNTs) nanomaterial regarding the removal of acidic and basic organic dyes – methylene blue (MB), and methyl orange (MO), respectively, as well as heavy metal ions – Cu(II).

Content from this work may be used under the terms of the Creative Commons Attribution 3.0 licence. Any further distribution of this work must maintain attribution to the author(s) and the title of the work, journal citation and DOI.

Published under licence by IOP Publishing Ltd

2. Experimental Details

2.1. Material Synthesis

The nanocomposite was synthesized as follows. First, HMTA was slowly added to anhydrous sulfuric acid; after that, the mixture was stirred and cooled. Next, Taunit-M CNTs (NanoTechCenter Ltd., Tambov, Russia) were added to the resulting viscous solution with thorough stirring. The mixture was heated to 180 °C and held at 180-200 °C for 2 h. The product (PAC/CNTs) was washed with water until the sulfuric acid was completely removed, and dried in a drying oven at 110 °C. The CNT content in this material was 14-15 wt.%.

2.2. Kinetic Study

1,500 mg L^{-1} stock (initial) solutions were separately prepared by dissolving 1.5 g of MO and MB (both – reagent grade, Laverna Ltd., Moscow, Russia) in 1 L of distilled water and standing for several hours. To determine kinetic parameters of the MO and MB adsorption, experiments were carried out in tubes containing 0.03 g of PAC/CNTs and 30 mL of the 1,500 mg L^{-1} MO and MB solutions. To calculate kinetic parameters of the Cu(II) adsorption, tests were performed with 30 mL of a 100 mg L^{-1} Cu(NO$_3$)$_2$ 3H$_2$O solution (Laverna Ltd.) and 0.03 g of the adsorbent. All the tubes were shaken for equilibration at 100 rpm and room temperature on a Multi Bio RS-24 programmable end-over-end rotator (Biosan, Riga, Latvia). After that, the suspensions were centrifuged at 10,000 rpm for 10 min using a 5810 R device (Eppendorf, Hamburg, Germany). After the adsorption procedure, dye amounts were determined spectrophotometrically on a PE 5400V instrument (Ekros, St. Petersburg, Russia) at 400 nm (MO) and 570 nm (MB), whereas equilibrium Cu(II) adsorption concentrations were determined by atomic absorption spectrometry (AAS) on an MGA-915MD instrument (Atompribor Ltd., Saint-Petersburg, Russia). The adsorbent morphology was elucidated by scanning electron microscopy (SEM) using a Neon 40 equipment (Carl Zeiss AG, Oberkochen, Germany). The same instrument was employed for energy-dispersive X-ray elemental analysis of the material.

3. Results

3.1. Material Characterization

According to the SEM image of the PAC/CNTs nanocomposite shown in Figure 1, it can be seen that the HMTA polycondensation polymeric product is formed on the CNT surface, resulting in the spongy structure favorable for using this material as an adsorbent.

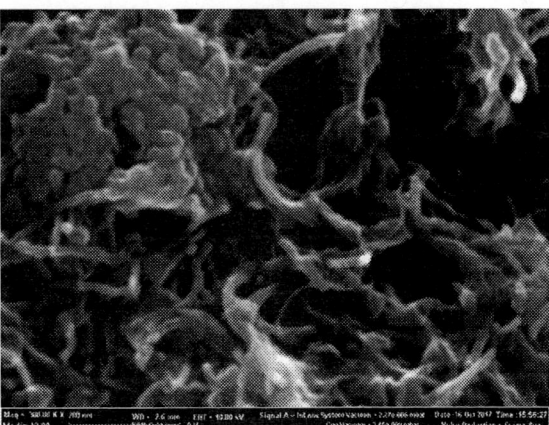

Figure 1. SEM image of the PAC/CNTs material.

The energy-dispersive X-ray analysis of the nanocomposite elemental composition gave the following results: C - 76.76 %, O - 9.54%, and S - 13.70 wt.%, corresponding to the "oxygen : sulfur" atomic ratio of 1.39:1. Sulfur is presumably present not in the form of sulfonic or sulfate groups, for which there should be 3 or 4 oxygen atoms per sulfur atom. It can be supposed that the material contains sulfide sulfur, which may have an affinity for heavy metal ions and organic matters.

3.2. Kinetic Results

Based on the results of the experimental studies, kinetic dependences of the dye and Cu(II) adsorption on the nanocomposite were constructed (Figure 2).

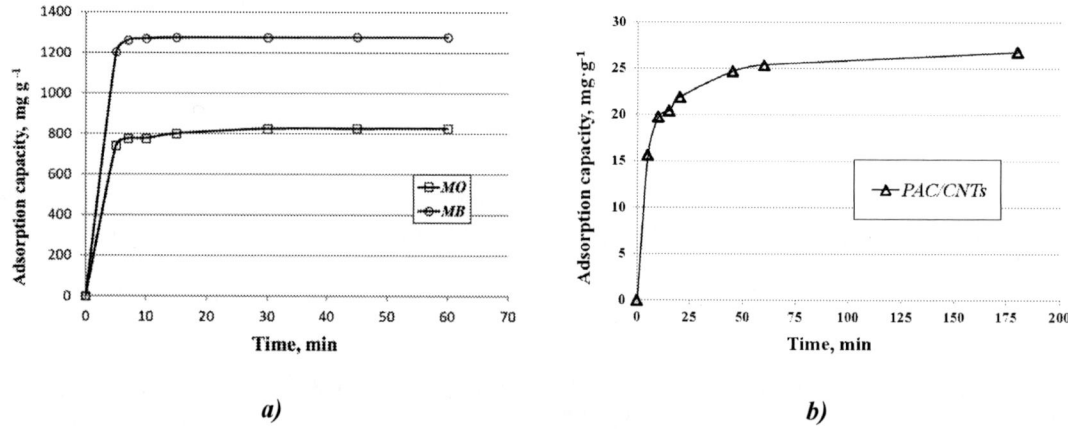

a)   *b)*

Figure 2. Kinetics dependences of the dye adsorption (*a*) and of the copper ions (*b*) on the PAC/CNTs nanocomposite.

It should be noted that the studied material exhibited a high sorption capacity for both the MB (1,255 mg g^{-1}) and the MO (801 mg g^{-1}). At the same time, equilibrium was achieved much faster (within 5 min) in comparison with the conventional materials (e.g., for activated carbons – within 30-50 min). The Cu(II) adsorption proceeded with the following parameters: capacity 24.5 mg g^{-1}, and equilibrium time 50 min.

The experimental data obtained were fitted to the diffusion, pseudo-first-order, pseudo-second-order and Elovich models (Figures 3 and 4).

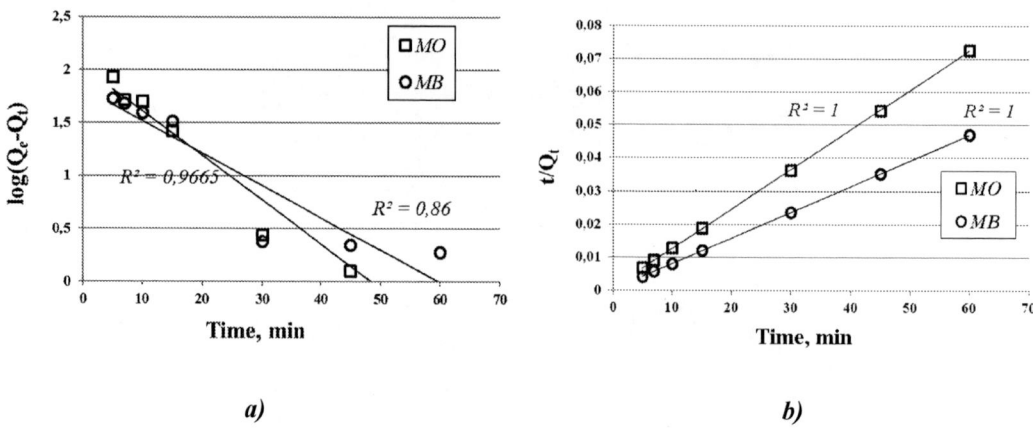

a)   *b)*

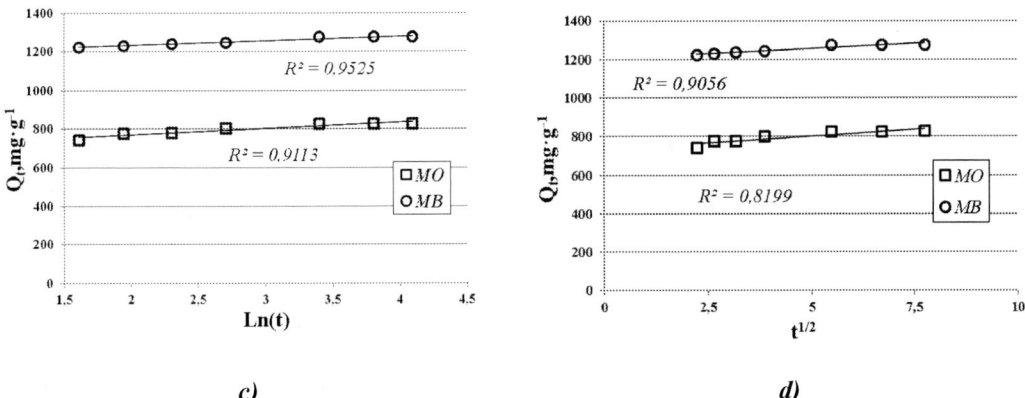

c)   *d)*

Figure 3. Kinetic models used to fit the data obtained for the dye adsorption: a) – pseudo-first-order, b) – pseudo-second-order, c) – Elovich models, and d) – intraparticle diffusion.

The adsorption interaction of the material under study with the organic dyes can be described quite well by almost all the mathematical models implemented herein. Analyzing the data using these models makes it possible to reveal a significant contribution of the chemisorption to the dye removal process (Figure 3b) (high correlation of the data with the pseudo-second-order equation), as well as chemical affinity of the adsorbent surface for the compounds being removed (data approximation using the Elovich model) (Figure 3c). Table 1 below presents the basic calculated parameters of the kinetic equations implemented.

Table 1. Kinetic parameters of the adsorption of the dyes and Cu(II) ions on the PAC/CNTs nanocomposite.

Model	Pseudo-first-order:			Pseudo-second-order:		
	$\log(Q_e - Q_t) = \log Q_e - \dfrac{k_1}{2.303}t$			$\dfrac{t}{Q_t} = \dfrac{1}{k_2 Q_e^{\,2}} + \dfrac{1}{Q_e}t$		
Adsorbate	Q_e	k_1	R^2	Q_e	k_2	R^2
MO	107.28	-0.0965	0.9665	833.33	0.0018	1
MB	66.53	-0.0702	0.86	1250	0.0032	1
Cu(II)	17.73	-0.0055	0.6737	27.47	0.00796	0.999
Model	Elovich:			Intraparticle diffusion:		
	$Q_t = \dfrac{1}{\beta}\ln(\alpha\beta) + \dfrac{1}{\beta}\ln t$			$Q_t = k_{id}\, t^{0.5} + C$		
				Step 1/Step 2		
Adsorbate	α	β	R^2	k_{id}	C	R^2
MO	$6 \cdot 10^{10}$	0.03	0.9113	13.969	733.11	0.8199
MB	$1.2 \cdot 10^{26}$	0.04	0.9525	10.322	1205.3	0.9056
Cu(II)	$1.52 \cdot 10^2$	0.324	0.9386	1.859/0.291	12.9/22.93	0.909/0.9646

* Q_e – adsorbate amount adsorbed onto the adsorbent surface at equilibrium (mg g^{-1}); Q_t - adsorbate amount adsorbed onto the adsorbent surface at time t (mg g^{-1}); k_1 - pseudo-first-order adsorption rate constant (min^{-1}); k_2

– pseudo-second-order adsorption rate constant (g mg^{-1}min^{-1}); α - initial adsorption rate constant (min^{-1} mg g^{-1}); β - degree of surface coverage and activation energy of chemisorption (g mg^{-1}); k_{id} - internal diffusion coefficient (mg g^{-1}min$^{-0.5}$); C - boundary layer thickness (mg g^{-1}).

It should also be noted that a high correlation of the experimental data with the Q_t *vs.* $t^{1/2}$ approximation (Figure 3d) indicates a contribution of the intraparticle diffusion into the porous adsorbent structure to the dye removal (Table 1).

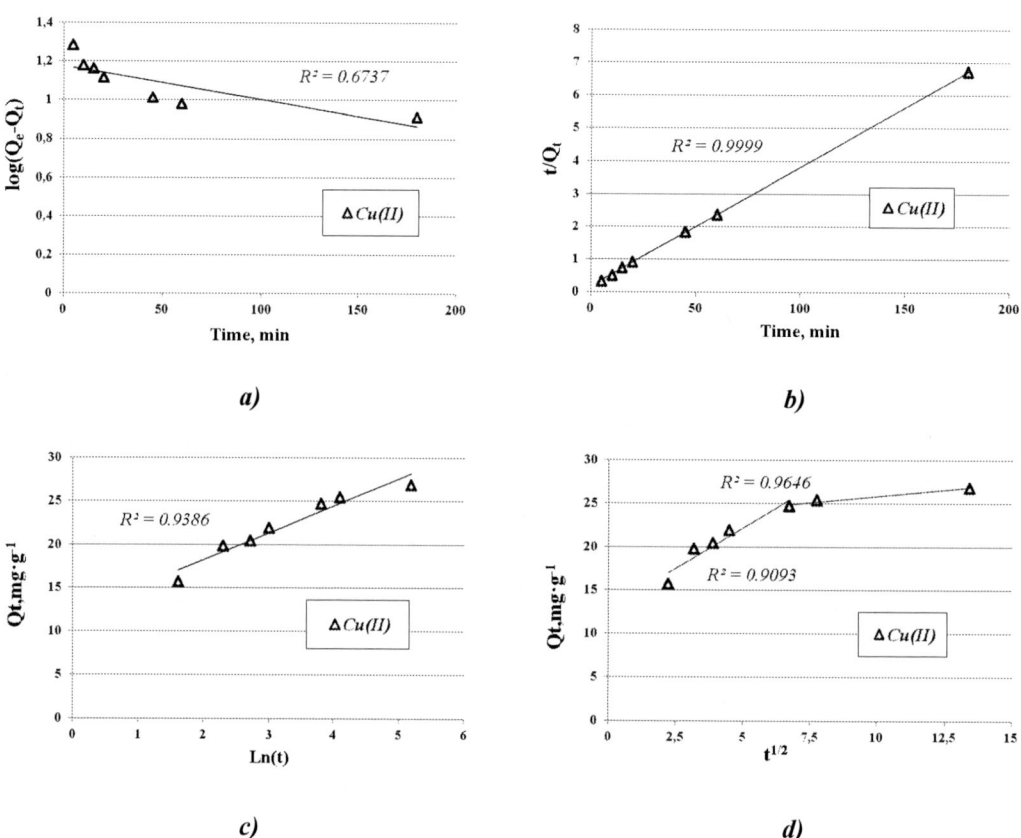

Figure 4. Kinetic models used to fit the data obtained for the Cu (II) adsorption: a) – pseudo-first-order, b) – pseudo-second-order, c) – Elovich models, and d) – intraparticle diffusion.

As for the Cu(II) adsorption, Figure 4a shows the approximation straight line constructed from the pseudo-first-order model. A sufficiently low value of the correlation coefficient (R^2=0.6737) suggests that the studied process depends little on the film diffusion of the Cu(II) ions. The pseudo-second-order model (Figure 4b, Table 1) implemented shows the best correlation with the experimental data (R^2=0.9999). The results obtained indicate a significant contribution of the chemical reaction to the Cu(II) adsorption from the aqueous solution. Heterogeneity of the adsorption sites available on the nanocomposite surface, which is due to the existence of various types of the chemical affinity for the adsorbate, can be evidenced by a good correlation of the experimental data with the Elovich model; in this case, R^2=0.9386 (Figure 4c, Table 1). Besides, the Cu(II) adsorption on the PAC/CNTs nanocomposite can successfully be described by the diffusion model, the calculated data of which are presented in Table 1. The removal process proceeds in two steps, thereby indicating the mixed-diffusion nature of the interaction. High values of the correlation coefficient (R^2=0.9093 and 0.9646) demonstrate a significant contribution of both the external and intraparticle diffusion to the Cu (II) adsorption (Table 1).

4. Conclusion

The possibility of using the adsorption material (PAC/CNTs) based on a CNT matrix modified with an organic ligand for selective removal of impurities of different chemical nature from aqueous media is considered herein. Kinetic curves describing the peculiarities of the interaction between the contaminants and the PAC/CNTs nanocomposite were constructed. The time of the organic dye adsorption was found to be 5 min at the adsorption capacity of the material being equal to 1,255 mg g^{-1} (MB) and 801 mg g^{-1}(MO), and 50 min at 24.5 mg g^{-1} (Cu(II) ions). The experimental results were processed using the known mathematical equations describing of adsorption kinetics. The analysis of the approximation straight lines elucidated the significant contribution of the chemisorption and the intraparticle diffusion of the studied contaminants into the material structure to the adsorption process (in the case of the organic compounds). The data obtained unequivocally indicate the prospect of using the material proposed as a highly effective universal adsorbent.

Acknowledgment

The research was supported by the Ministry of Education and Science of the Russian Federation under Project No. 16.1384.2017/PCh.

References

[1] Jones M G, Blonder R, Gardner G E [et al.] 2013 *Int. J. Sci. Ed.* **35** 1490-1512.
[2] Fialova D, Kremplova M, Melichar L [et al.] 2014 *Materials* **7**(3) 2242-2256.
[3] Gupta A K, Gupta M 2005 *Biomaterials* **26** 3995-4021.
[4] Pei Z, Li L, Sun L [et al.] 2013 *Carbon* **51** 156-163.
[5] Xu J, Wang L, Zhu Y 2012 *Langmuir* **28** 8418-8425.
[6] Melezhik A.V., Alekhina O.V., Gerasimova A.V., Tkachev A.G. 2017 TSTU Transactions (Vestnik TGTU) **23**(3) 461-470 (in Russian).

SPBOPEN 2018

IOP Publishing

Nano-sized Al-Ni energetic powder material for heat release element of thermoelectric device

S Yu Nemtseva[1], E A Lebedev[1], Yu P Shaman[2], P I Lazarenko[1], R M Ryazanov[2], D G Gromov[1,3], S A Gavrilov[1]

[1] Institute AMT, National Research University of Electronic Technology, Zelenograd 124498, Russia
[2] Scientific Manufacturing Complex "Technological Centre", Zelenograd 124498, Russia
[3] Institute for Bionic Technologies and Engineering, I.M. Sechenov First Moscow State Medical University, Moscow 119991, Russia

Abstract. In this paper, a method of fabrication of an Al-Ni powder energetic material with a 90-100 nm particles is proposed and experimental results of a study of its properties are presented. High values of specific energy and the rate of its release make it possible to use this material as an heat release element in thermoelectric power generation devices. It has been demonstrated experimentally that it is possible to maintain a voltage value higher than 1 V for 45 seconds as a result of combustion of a 3 gram Al-Ni sample that using a simple DC-DC converter will allow charging supercapacitors or accumulators.

1.   Introduction

In recent years scientists are focused on development of environmentally friendly energy sources capable of operating under extreme conditions such as Far North and Arctic regions. Thermoelectric generators convert heat from various sources into electrical current. They are highly reliable due to the simplicity of construction and the absence of moving parts [1]. Currently, they are used in conjunction with various sources of heat: from the heat of the human body [4], solar energy [2], ending with radioisotope fuel [3]. However, the listed heat sources are not without shortcomings - they are periodic in time, ineffective and environmentally and biologically unsafe. In this regard, the development of new sources of heat is an urgent task.

Thermites are a well-studied class of materials with a centuries-old history [5] and up to now they have been used for various surfaces joining [6]: from railroad rails to microelectronic components [7]. These energy materials are able to maintain in their volume a self-propagating exothermic reaction after primary initiation, resulting in a significant amount of heat released.

The most common thermite mixtures are the metal-oxide pairs, for example, $Al-Fe_2O_3$, $Al-CuO$ and $Al-Cu_2O$, as well as bimetallic pairs: $Al-Pt$, $Al-Ni$, $Al-Zr$. The last of these are distinguished by the absence of gas evolution in the combustion process while maintaining a relatively high specific energy density. Due to this, a thermoelectric device based on them (a schematic image of which is shown in figure 1) consisting of a source of heat generation, a thermoelectric generator and an accumulation device of generated electric energy can be created.

Content from this work may be used under the terms of the Creative Commons Attribution 3.0 licence. Any further distribution of this work must maintain attribution to the author(s) and the title of the work, journal citation and DOI.
Published under licence by IOP Publishing Ltd

Figure 1. Schematic representation of the construction of a thermoelectric device

The state-of-the-art technology allows the production of powder materials with a particle size of less than 100 nm, which makes it possible to create energy materials with improved characteristics: lower initiation energy and faster propagation velocity of the wave front [8].

In the present article, a method of fabrication of heat generation element   based on nanosize commercially available aluminum and nickel powder materials is proposed. The results of experimental studies of energy composites using DSC and high-speed video recording are presented, as well as the output characteristics of a thermoelectric generator under the influence of the investigated aluminum-nickel energy material.

2.   Experiments and results

The initial aluminum and nickel powders with particles sizes from 70 to 100 nm were obtained from LLC (APT) Advanced Powder Technologies. Energetic composite was prepared by mixing the components in an Al: Ni ratio of 50:50 at% in an ultrasonic bath for 2 hours. The hexane was added to the mixture of the initial powders as process control agent for better mixing, after that it was removed by means of a rotary evaporator. Finally mixtures were pressed without binder component in the form of a disk 25 mm in diameter with a porosity of 0.53 by means of a hydraulic press AE&T T61220M.

The study of thermal effects in compressed materials by differential scanning calorimetry was carried out using TA Instruments Q600 equipment. A 10 mg sample was placed in a ceramic (Al_2O_3) crucible and heated at 10 °C/min rate from room temperature to 800 °C in an argon flow (50 ml/min). The absolute value of the thermal effect was determined by the integration of the DSC curve over time in the region of the main exothermic peaks. The calculated value was about 792 J / g, which is lower than the theoretical value, but it may be due to the incompleteness of the chemical reaction between aluminum and nickel, and also due to the presence of an oxide layer on the surface of the aluminum particles.

Measurement of the reaction front propagation velocity after electrical initiation in pressed samples was carried out using high-speed video at a speed of 10,000 fps. The initiation of wave combustion was controlled with the help of two tungsten probes, located at a distance of 2-4 mm. For the initiation, a short-time power supply was used. The numerical value of the front velocity was defined as the ratio of the distance traveled by the front to the time during which this process occurred. During the combustion, the front of the chemical interaction was clearly visible and spread throughout the sample area at a velocity of 0.056 m / s, which is shown in Figure 2.

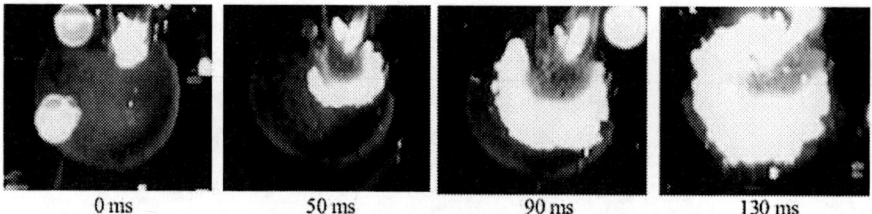

| 0 ms | 50 ms | 90 ms | 130 ms |

Figure 2. Storyboard of the reaction front propagation

To study the output characteristics of the TEG under the influence of the heat-release element based on the Al-Ni energetic powder material, a special stand was designed. The thermoelectric generator TMG-127-1.4-1.5 was installed with a cold side on an aluminum radiator. On the cold side the compacted energy material in the form of a disk 25 mm in diameter and not more than 5 mm thick was placed. The sample weight ranged from 0.3 to 3 grams. For a more uniform heat distribution and to avoid thermal shock, a copper plate with dimensions 40 * 40 * 0.6 mm was placed between the energy material and the TEG. To improve thermal contacts, thermal paste was applied to both sides of the TEG. To avoid heat loss, the energy material and TEG were covered with a heat-insulating material based on gypsum. The TEG contacts were connected to the measuring circuit with a Keithley 2700 multimeter in series or in parallel to measure the output characteristics (U_{oc}, I_{sc} and U, I without load and with load, respectively) depending on the fixed value.

Figure 3 shows the graphs of the change in the output voltage of the TEG (with 1.8 Ohms load) in the combustion process of energetic materials of different masses. In all cases, the graphs had a similar shape - rapid growth to a maximum value and a slow decline. The maximum value of the voltage and current were 2 V and 0.8 A respectively for a 3 grams sample and it was achieved in 10 seconds. The rate of increase in the magnitude of the voltage was the same in all experiments, and it is obviously determined by the thermal resistance of the TEG.

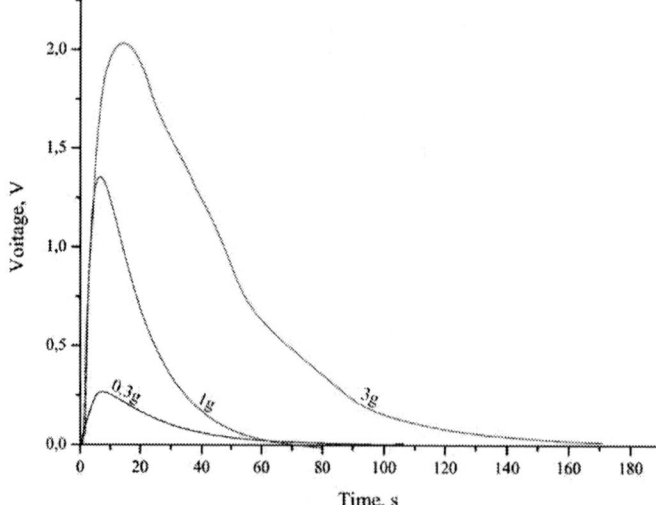

Figure 3. Graph of changes in the output voltage of TEG over time

From a practical point of view, it is important to note that the voltage above 1 V was maintained for 45 seconds, i.e. using a simple DC-DC converter it is possible to charge electrochemical capacitors.

The efficiency of the thermoelectric device was calculated by the formula 1, where Qg - is the value of the generated by TEG electric energy (the integration of the power function over time), and Qm - is the specific value of the energy released for the Al-Ni energetic material, measured with DSC. The calculated values are shown in Table 1.

$$\eta = \frac{Qg}{Qm} 100\%$$ (1)

Table 1. The calculated value of the generated energy and the efficiency coefficient for different masses of the energetic material

Weight (g)	Energy (J)	Coefficient of efficiency (%)
0.3	0.66	0.3
1	14.82	1.9
3	81.55	3.4

Dependence of the coefficient of efficiency on the mass of the element of heat generation was found. It can be explained in terms of heat loss due to dissipation in thermal insulation. The absolute value of losses is the same in all cases, but for a case with a low mass of energetic material, it will constitute a significant fraction of the total amount of heat released. It is expected that with a further increase in mass, the efficiency will go to saturation. Nevertheless, the maximum efficiency will be limited by the efficiency of the thermoelectric generator, which currently is no more than 10%.

3.   Conclusion

In the article, a thermoelectric device that converts the heat generated as a result of the combustion of a thermite-like Al-Ni energetic material was first time considered. The ability to maintain the output voltage of the TEG above 1 V for 45 seconds has been demonstrated that using a simple DC-DC converter will allow charging of batteries or supercapacitors or feeding other electronic components and devices. The value of the output voltage of 1 V was reached in only 2.5 seconds, which will allow using this type of device in extreme situations, when it is necessary to quickly provide power to the emergency notification system.

Further increase in the efficiency of the device (the maximum achieved was 3.4%) is possible due to optimization of the thermal insulation system to reduce heat losses. In addition, it is necessary to optimize the energy material in terms of the initiation energy, since The material considered in this article is characterized by a relatively high value of the initiation energy. The solution to this problem can be the use of various additives that stimulate the combustion process.

Acknowledgments
The Russian Science Foundation supported this work (project №16-19-10625)

References
[1] Yang Z, PradoGonjal J, Phillips M, Lan S, Powell A, Vaqueiro P, Gao M, Stobart R, Chen R 2017 SAE Techn. Pap. Ser 8
[2] Eswaramoorthy M, Shanmugam S 2013 En. Sour **35** 487
[3] Bennett G, Lombardo J, Hemler R, Silverman G, Whitmore C, Amos W, Johnson E, Zocher R, Hagan J, Englehart R, El-Genk M 2008 AIP Conf Proc **969** 663
[4] Lay-Ekuakille A 2009 IEEE Int Work on Med Meas and Ap. 1-4
[5] Goldschmidt H, . Vautin C 1898 J. Soc. Chem. Ind. 6 (17), 543
[6] Duckham A, Spey S J, Wang J, Reiss M E, Weihs T P, Besnoin E, Knio O M 2004 J. Appl. Phys. **96** (4) 2336
[7] Jun Z , Feng-shun W, Jian Z, Bing A, Hui L 2009 El. Pack. Techn.& High Dens. Pack. 838
[8] Hunt E M, Malcolm S, Pantoya M L, Davis F 2009 Int. J. Impact Eng. **36** (6) 842

SPBOPEN 2018   IOP Publishing

Investigation of local charge accumulation in yttria stabilized zirconia films with Au nanoparticles by Scanning Kelvin Probe Microscopy

D O Filatov[1], O N Gorshkov[1], A N Mikhaylov[1], D S Korolev[1], M N Koriazhkina[1], M A Ryabova[1], I N Antonov[1], M E Shenina[1], D A Pavlov[1], M S Dunaevskiy[2]

[1] Lobachevsky State University of Nizhny Novgorod, 23 Gagarin Ave., Nizhny Novgorod, 603950, Russia
[2] Ioffe Institute, Russian Academy of Sciences, 26 Polytechnicheskaya St., Saint-Petersburg, 194021, Russia

Abstract. The time dynamics of the spatial distribution of the potential induced by the electrons locally injected from an atomic force microscope (AFM) probe into the ultrathin (< 10 nm thick) yttria stabilized zirconia (YSZ) $ZrO_2(Y)$ films with embedded Au nanoparticles (NPs) on Si substrates was studied using Scanning Kelvin Probe Microscopy (SKPM). The SKPM images and profiles of surface potential induced by the electrons confined inside the Au NPs subject to the time elapsed after the injection have been measured and analyzed. The parameters of the charge relaxation in the YSZ:NP-Au films were determined.

1. Introduction

In recent years, the studies of the charge accumulation in the metal nanoparticles (NPs) embedded in dielectric films have attracted considerable attention due to potential applications in novel non-volatile memory devices [1]. It is expected that the use of the NP arrays instead of traditional floating gates will result in a greater reliability, smaller sizes, and lower power consumption [2].

In the present work, the relaxation of charge locally injected from the atomic-force microscope (AFM) probe into the ultrathin (< 10 nm) yttria stabilized zirconia (YSZ) $ZrO_2(Y)$ films with embedded Au NPs was studied by Scanning Kelvin Probe Microscopy (SKPM) [3, 4]. The goal of the present study is to explore experimentally the prospects of development of the non-volatile memory cells based on the metal-oxide-semiconductor field-effect transistors (MOSFETs) utilizing the YSZ:NP-Au films as the floating gates.

2. Materials and Methods

The YSZ films with single-layered Au NP arrays were formed on the n^+-Si(100) substrates covered by native oxide (SiO_2) by Alternating Magnetron Deposition. The Au films (0.5 ÷ 1 nm thick) were sandwiched between the YSZ (≈ 12% mol. Y_2O_3) layers of 2 to 8 nm in thickness and annealed in Ar for 1 hour at 450 °C. Also, the $YSZ/SiO_2/Si$ stacks with the YSZ layers thicknesses 4 ÷ 10 nm (equal to the total nanocomposite film thickness in the YSZ:NP-Au/SiO_2/Si samples) were fabricated to serve as the reference samples.

Content from this work may be used under the terms of the Creative Commons Attribution 3.0 licence. Any further distribution of this work must maintain attribution to the author(s) and the title of the work, journal citation and DOI.
Published under licence by IOP Publishing Ltd

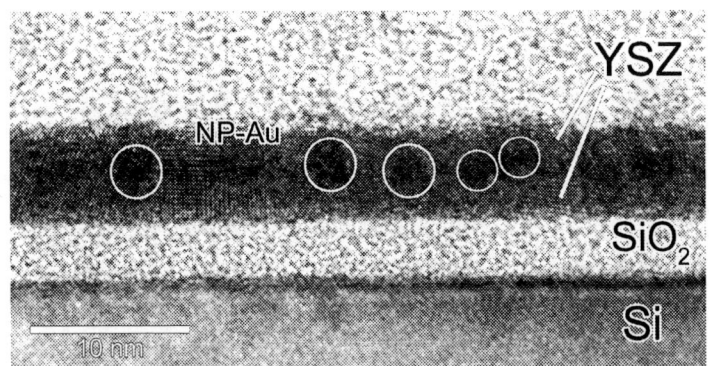

Figure 1. HR X-TEM image of the YSZ:NP-Au/SiO$_2$/Si stack.

The structure of the resulting YSZ:NP-Au/SiO$_2$/Si stacks was examined by High Resolution Cross-sectional Transmission Electron Microscopy (HR X-TEM) using Jeol® JEM-2100F transmission electron microscope at the accelerating voltage of 180 kV. More details on the sample preparation procedure and procedure of the TEM investigations can be found elsewhere [5, 6].

The SKPM measurements were carried out in ambient conditions using the NT-MDT® Solver Pro™ AFM. The charge injection in the Au NPs was performed in Contact Mode by applying a bias voltage $V_g = 1 \div 3$ V between the Pt-coated AFM probe and the substrate in the following modes:

- the pulsed mode in a single point on the sample surface (the pulse duration was ~ 1 s);
- drawing a line on the sample surface in the vector charge lithography mode.

The results of the charge injection were examined by SKPM using the two-pass technique.

3. Results and Discussion

A HR X-TEM image of an YSZ(3 nm)/Au(0.5 nm)YSZ(3 nm)/SiO$_2$/Si stack is shown in Figure 1. The X-TEM results revealed the Au films to coagulate into nearly spherical NPs. The NP diameter and the average spacing between the NPs were $2 \div 3$ nm.

Figure 2 shows a SKPM image of a charged line drawn at $V_g = -3$ V on the surface of the YSZ(2 nm)/Au(1 nm)/YSZ(2 nm)/SiO$_2$/Si stack after annealing. It should be noted that V_g was applied to the n^+-Si substrate relative to the AFM probe i. e. negative V_g corresponded to the injection of the electrons from the n^+-Si substrate into the Au NPs.

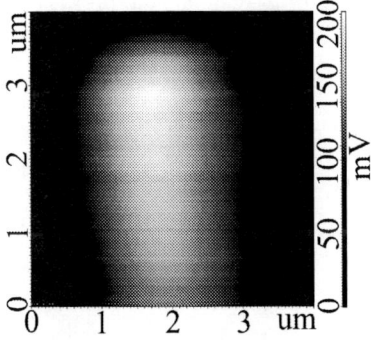

Figure 2. SKPM image of a charged line on the surface of the YSZ:NP-Au/SiO$_2$/Si film.

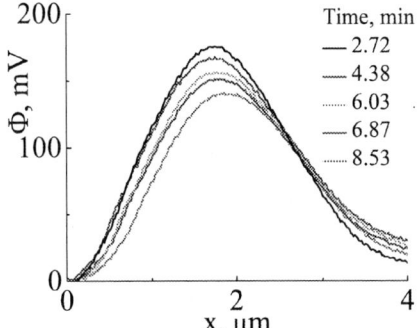

Figure 3. The SKPM profiles across the charged line shown in Figure 2.

SPBOPEN 2018   IOP Publishing
IOP Conf. Series: Journal of Physics: Conf. Series **1124** (2018) 081028   doi:10.1088/1742-6596/1124/8/081028

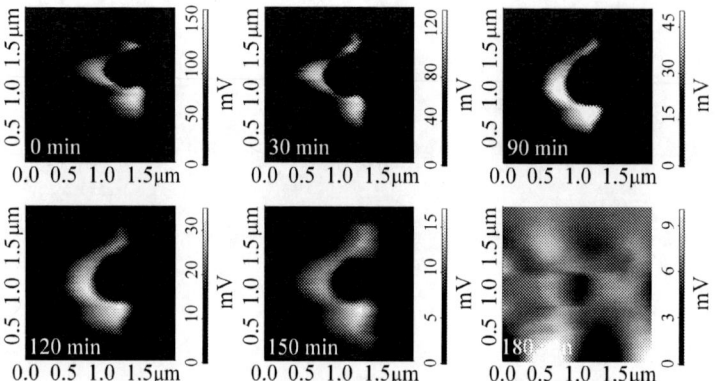

Figure 4. The SKPM images of the YSZ:NP-Au/SiO₂/Si film surface measured after different times since the point charge injection at $V_g = -3$ V.

Figure 3 shows the SKPM profiles across the charged line at various points along the line corresponding to different times elapsed since the charge injection. The decrease of the maximum values of the potential profiles Φ_m with increasing time t was related to the leakage of the injected electrons from the Au NPs into the Si substrate, most likely, via the trap-assisted tunnelling. The width of the potential profiles was almost constant that points to weak lateral charge spreading.

Figure 4 shows a series of SKPM images of an area on the YSZ:NP-Au/SiO₂/Si film surface after the point charge injection measured at different moments of time t elapsed from the charge injection. Note that the shape of the spot of increased potential is not round, and the charge spreading is not axially symmetric unlike the ones in the uniform dielectric films [4, 7] as well as in the Ge NPs formed in the SiO₂/Si films by ion implantation [8]. The angular asymmetry of the charge spreading in the YSZ:NP-Au films can be attributed to the non-uniformity of the lateral spatial distribution of the NPs, which is seen clearly in Figure 1. It seems reasonable to assume that the electrons, injected into an Au NP, tunnel to their nearest neighbor (NN) NP first. So far, the asymmetric charged spot on the YSZ:NP-Au film surface reflects the non-uniform spatial distribution of the Au NPs in the array plane. In contrary, the traps in the uniform dielectric films [7] as well as the Ge NPs in SiO₂ formed by ion implantation [8] are distributed rather uniformly in the film plane. As a consequence, the probability of the tunnel jumps of the electrons to the NN trap or Ge NP is symmetric axially in these two cases that results in the round shapes of the charged spots.

Figure 5 shows the time dependence of the maximum values of the potential profiles Φ_m measured in the charged spots presented in Figure 4.

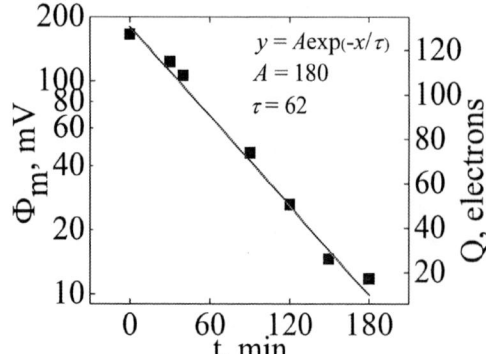

Figure 5. The time dynamics of the maximum values of the potential profiles Φ_m and of the charge Q confined in the Au NPs.

1108

The time dynamics of Φ_m obeys the exponential decay law. The charge retention time τ for this measurement series was ~ 1 hour. The highest value of τ obtained on the sample with increased YSZ film thickness (up to ≈ 10 nm) was ≈ 3 days [9].

Also, Figure 5 shows the estimates of the charge Q trapped in the Au NPs. The estimates of Q were made from the measured values of Φ_m (see Figure 5, the left ordinate axis) according to the formula [9]:

$$Q \sim \frac{2\pi\varepsilon_0(\varepsilon_d + 1)\Phi_m}{C[(z_0 + R_p + d_c)^{-2} - (2d + z_0 + R_p - d_c)^{-2}]} \frac{\partial C}{\partial z} \tag{1}$$

Here ε_0 is the vacuum permittivity, ε_d is the dielectric constant of YSZ, d is the total YSZ:NP-Au film thickness, d_c is the thickness of the cladding YSZ layer, R_p is the AFM tip curvature radius, C is the probe-to-sample capacity, z is the coordinate in the normal direction to the sample surface, and z_0 is the lifting height (the tip-sample separation) at the second pass. The dielectric screening effect of SiO_2 layer was neglected. The results presented in Figure 5 show that the surface potential variations observed in Figure 4 were induced by a countable number of electrons ($10 \div 100$) confined in the Au NPs.

It is worth noting that equation (1) is an approximate one [9], so the values of Q presented in Figure 5 should be treated as the estimates within the order of magnitude.

4. Conclusions
The results of the present study demonstrate experimentally a possibility to create the non-volatile memory cell based on the MOSFETs utilizing the Au NP arrays embedded into the YSZ-based gate insulators. However, for the practical implementation of such devices, further improvement of the MOS stacks is required in order to increase the charge retention time up to ~ 10 years (as it is typical in today's flash memory).

Acknowledgments
The authors gratefully acknowledge the financial support from Russian Megagrant Program (14.Y26.31.0021). The TEM and SKPM investigations were carried out using the shared research facilities of Research and Educational Center for Physics of Solid State Nanostructures at Lobachevsky State University of Nizhny Novgorod.

References
[1]   Lee J-S 2010 *Gold Bulletin* **43** 189
[2]   Hong S, Auciello O & Wouters D 2014 *Emerging non-volatile memories* (Berlin-Heidelberg, Springer)
[3]   Escasain E, Lopez-Elvira E, Baro A M, Colchero J & Palacios-Lidon E 2011 *Nanotechnology* **22** 375704
[4]   Dunaevskiy M S, Alekseev P A, Girard P, Lahderanta E, Lashkul A & Titkov A N 2011 *J. Appl. Phys.* **110** 084304
[5]   Gorshkov O, Antonov I, Filatov D, Shenina M, Kasatkin A, Bobrov A, Koryazhkina M, Korotaeva I & Kudryashov M 2017 *Adv. Mat. Sci. Eng.* 1759469
[6]   Gorshkov O N, Antonov I N, Filatov D O, Shenina M E, Kasatkin A P, Pavlov D A & Bobrov A I 2016 *Tech. Phys. Lett.* **42** 36
[7]   Gushchina E V, Dunaevskii M S, Alekseev P A, Durğun Özben E, Makarenko I V & Titkov A N 2014 *Tech. Phys.* **59** 1540
[8]   Dunaevskii M S, Alekseev P A, Dement'ev P A, Gushchina E V, Berkovits V L, Landeranta E & Titkov A N 2015 *Tech. Phys.* **60** 680
[9]   Koryazhkina M N, Filatov D O, Antonov I N, Ryabova M A & Dunaevskiy M S 2019 *J. Surf. Investigation: X-ray, Synchrotron, and Neutron Techn.* **13** [in press]

Adsorption of the Methylene Blue Dye on Carbon Nanocomposites Under Dynamic Conditions: A Kinetic Study

I V Burakova[1*], A V Babkin[1], E A Neskoromnaya [1], A E Burakov[1], D A Kurnosov[1], A G Tkachev[1]

[1] Department "Technology and Methods of Nanoproducts Manufacturing", Tambov State Technical University, Tambov, 392000, Russia

Abstract. The present paper describes kinetic regularities of the removal of the typical synthetic dye, methylene blue (MB), from aqueous solutions by two graphene-organic composites, polyhydroquinone (PHQ)/graphene and polyamine-cumulene (PAC)/carbon nanotubes (CNTs). To study the adsorption kinetics under dynamic conditions simulating the real industrial environment, a column method was implemented using the facility developed. The MB adsorption capacity of the studied nanocomposites was found to be very high (2,938 mg g^{-1} for the PAC/CNTs at the contact time of 75 min, and 2,610 mg g^{-1} for PHQ/graphene at 40 min). Finally, the kinetic parameters of the process were determined, and the mechanisms proceeding during the dye adsorption on the carbon nanocomposites were elucidated.

1. Introduction

Currently, in Russia, about 20 % of water samples taken from various water supply sources does not correspond to the maximum permissible concentration limits set for hazardous biological and chemical components. In fact, every second Russian citizen, in one way or another, consumes water containing a huge number of harmful components, thereby inevitably subjecting himself/herself great health risks [1].

In modern wastewater treatment processes, more and more attention is being paid to controlling the content and efficient removal of synthetic organic dyes. This is due to their negative impact on the human body - almost all of them are harmful and possess different degrees of toxicity (allergens, carcinogens, mutagens). Thus, the removal of these contaminants is a very important and urgent issue.

Adsorption removal appears to be one of the main techniques of purifying aqueous media from dyes down to practically zero levels. Wastewater treatment facilities are the ones most widely used for this purpose; their functioning is based on adsorption processes. This is due not only to the efficiency of the purification technology, but also to the simplicity of their technological implementation and equipment design. It is worth noting that high adsorption capacity of materials used to remove contaminants allows effective implementation of selective purification of aqueous and even gaseous media [2-4].

The authors of the present research have studied the kinetics of the adsorption of methylene blue (MB), representing a basic thiazine dye, on graphene nanocomposites synthesized under laboratory conditions at the Research Department "Technology and Methods of Nanoproduct Manufacturing", (Tambov State Technical University, Tambov, Russia). Nanomaterials seem to be good adsorbents due to their small size, catalytic potential, high reactivity, ease of separation, and a large number of active sites to interact with different contaminants [5-8].

Content from this work may be used under the terms of the Creative Commons Attribution 3.0 licence. Any further distribution of this work must maintain attribution to the author(s) and the title of the work, journal citation and DOI.
Published under licence by IOP Publishing Ltd

Laboratory kinetic studies are usually carried out by mixing the adsorbent with a limited volume of the solution ("batch" tests). However, under real conditions, the purification process is implemented by pumping the solution through a fixed-bed column containing the adsorbent ("column" dynamic tests). In this regard, the authors have developed a facility for conducting adsorption studies in a dynamic mode.

2. Experimental Details
2.1. A Column Facility for Studying Adsorption Under Dynamic Conditions
To determine equilibrium time in the adsorbent-solvent-sorbate system, kinetic studies were performed in a dynamic mode using a laboratory facility, the scheme and photo of which are presented in Figure 1.

The solution (100 mL) is transported from the initial solution container along two loops (1.1) and (2) through control valves. The liquid is circulated by means of a peristaltic pump. The first loop supplies the aqueous solution to adsorption cell (1) containing the porous adsorbent sample, and then, the purified liquid is returned into the initial solution container along loop (1.2).

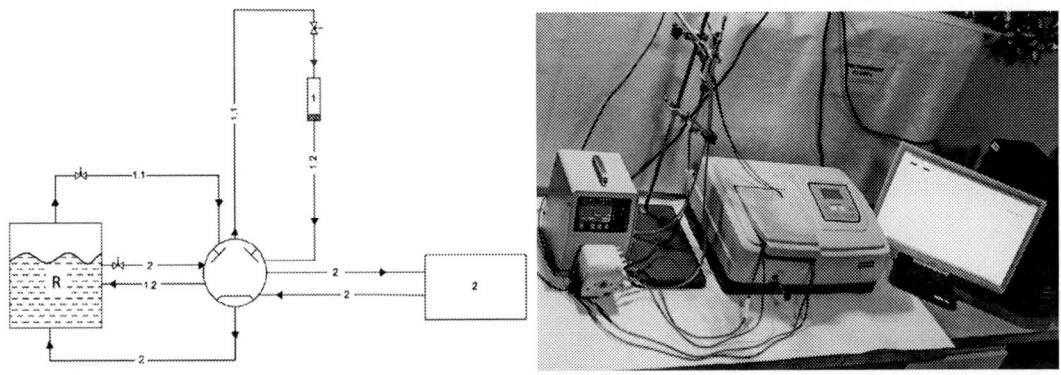

Figure 1. A laboratory facility to study the dynamic adsorption (R – initial solution container, 1-adsorption cell, 2- spectrophotometer, peristaltic pump).

Along the parallel loop, the liquid is supplied into a cuvette of spectrophotometer (2) to record optical density values at predetermined intervals, and then it is returned.

2.2. Adsorbent
Two nanocomposites – polyhydroquinone (PHQ)/graphene and polyamine-cumulene (PAC)/carbon nanotubes (CNTs) - were synthesized and used as adsorption materials. PHQ represents a molecule chemically bonded to few-layered graphene sheets, whereas PAC is a product of hexamethylenetetramine polycondensation in sulfuric acid in the presence of CNTs as texture component. The energy-dispersive X-ray analysis shows the availability of significant amounts of sulfur, presumably as sulfide groups.

2.3. Kinetic study
The initial MB solution (concentration 1,500 mg L^{-1}, and volume 100 mL) was supplied by means of the peristaltic pump into the adsorption cell containing the adsorbent (weight 0.03 g), and then was returned into the initial solution container along the reverse loop, where the concentrations were mixed. Along the parallel loop, the solution was supplied into the cuvette located in a PE 5400V spectrophotometer (Ekros, St. Petersburg, Russia) (wavelength 570 nm). The optical density of the solutions was recorded every 180 s.

3. Results and Discussion

The results of the kinetic studies under the dynamic conditions are graphically presented in Figure 2.

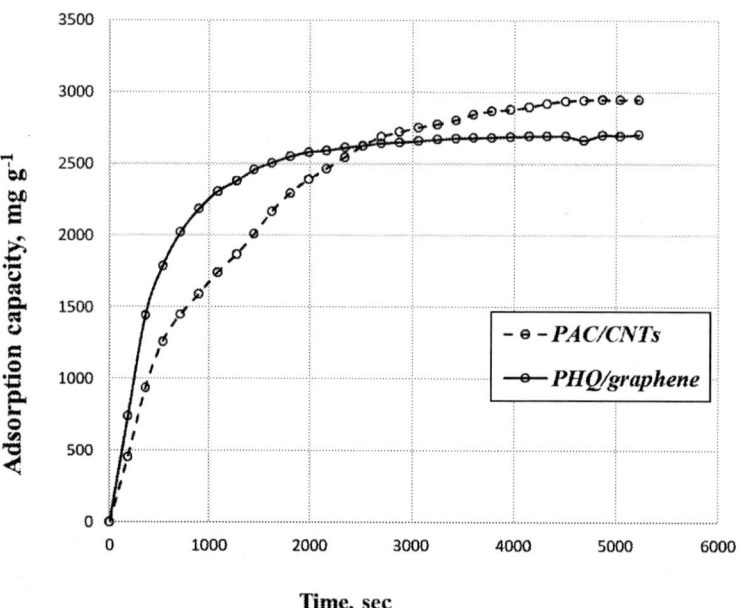

Figure 2. Kinetics of the dye adsorption on the PAC/CNTs and PHQ/graphene nanocomposites.

Due to the implementation of the adsorption process as a flow, very high adsorption capacity values were achieved for the MB removal: 2,938 mg g^{-1} at the contact time of 75 min (in the case of PAC/CNTs), and 2,610 mg g^{-1} at 40 min (in the case of PHQ/graphene). The experimental data were fitted to the following kinetic models: pseudo-first- and pseudo-second-order, Elovich, and intraparticle diffusion) (Table 1).

Figure 3 shows the model fits with the correlation coefficients, the values of which indicate the applicability of each models to describe the adsorption on the materials under study. The pseudo-second-order model, with R^2=0.9985 and 0.9972 determined for PHQ/graphene and PAC/CNTs, respectively (Figure 3b) shows a better correlation with the experimental data, in comparison with the pseudo-first-order model, with R^2=0.9210 and 0.8617 for PHQ/graphene and PAC/CNTs, respectively (Figure 3a). Based on the results obtained, a chemical interaction between the adsorbent functional groups and the dye molecules can be assumed. Theoretically, this interaction is stoichiometric, meaning that one molecule occupies only one position on the adsorbent surface.

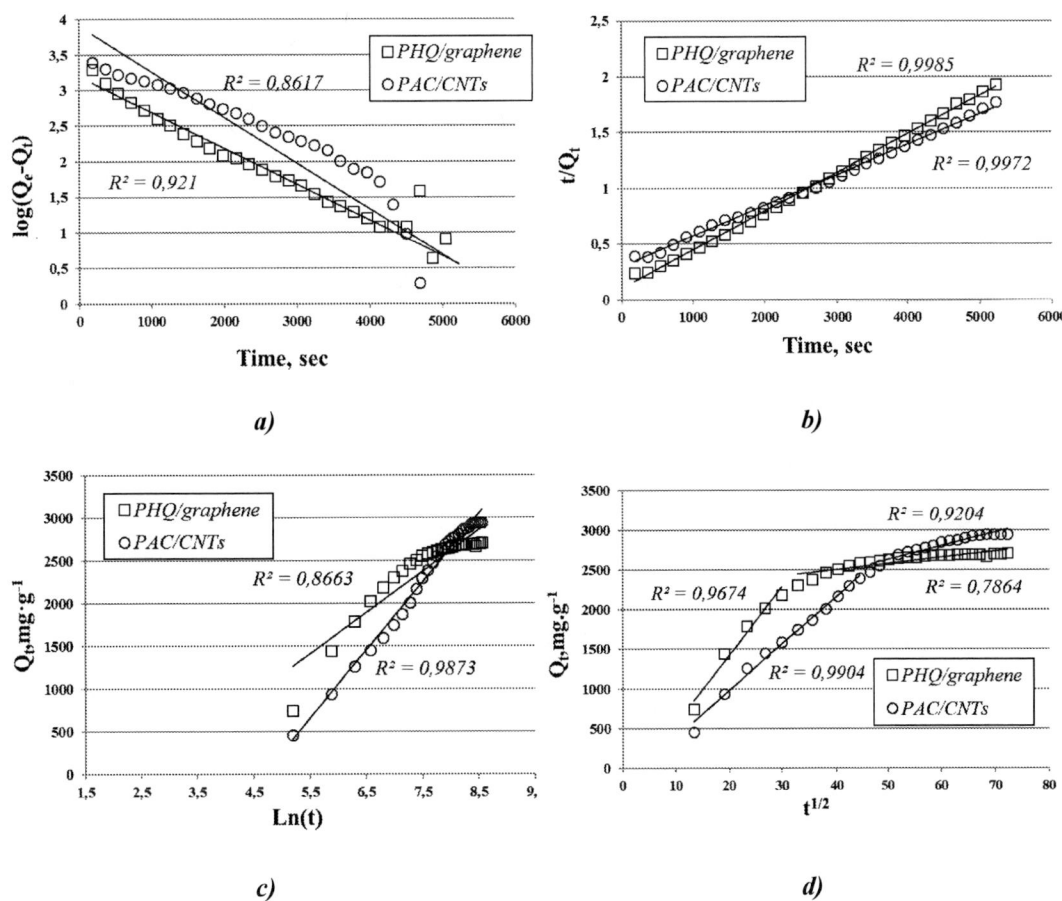

Figure 3. Experimental data fitted to the following kinetic models: a) – pseudo-first-order, b) – pseudo-second-order, c) – Elovich models, and d) – intraparticle diffusion.

The experimental data can also be described quite well by the Elovich model (Fig. 3c), which implies inhomogeneity of adsorption sites (having different energy and chemical affinity for the extracted compound) on the material surface. High correlation coefficients indicate the possibility of employing this model to describe the experiment under the conditions selected.

Table 1. Kinetic parameters of the MB adsorption.

	Pseudo-first order: $$\log(Q_e - Q_t) = \log Q_e - \frac{k_1}{2.303} t$$			Pseudo-second order: $$\frac{t}{Q_t} = \frac{1}{k_2 Q_e^2} + \frac{1}{Q_e} t$$		
	Q_e	k_1	R^2	Q_e	k_2	R^2
PHQ/ graphene	1558	-0.0012	0.921	3333.3	$9 \cdot 10^{-7}$	0.9985
PAC/ CNTs	7962	-0.0014	0.8617	3333.3	$3 \cdot 10^{-7}$	0.9972

	Elovich: $$Q_t = \frac{1}{\beta} \ln (\alpha\beta) + \frac{1}{\beta} \ln t$$			Intraparticle diffusion: $$Q_t = k_{id}\, t^{0.5} + C$$ Step 1/Step 2		
	α	β	R^2	k_{id}	C	R^2
PHQ/ graphene	35.71	0.002	0.8663	86.251/7.74	-304.4/2191	0.967/0.786
PAC/ CNTs	7.33	0.0013	0.9873	59.25/17.47	-210.7/1755	0.99/0.92

* Q_e – dye amount adsorbed onto the adsorbent surface at equilibrium (mg g^{-1}); Q_t - dye amount adsorbed onto the adsorbent surface at time t (mg g^{-1}); k_1 - pseudo-first-order adsorption rate constant (min^{-1}); k_2 – pseudo-second-order adsorption rate constant (g $mg^{-1}min^{-1}$); α - initial adsorption rate constant (min^{-1} mg g^{-1}); β - degree of surface coverage and activation energy of chemisorption (g mg^{-1}); k_{id} - intraparticle diffusion coefficient (mg $g^{-1}min^{-0.5}$); C - boundary layer thickness (mg g^{-1}).

The intraparticle diffusion model constructed from the experimental data represents a multi-linear dependence (Figure 3d). As can be seen, the adsorption removal consists of two distinct stages, which indicates its mixed-diffusion mechanism. High correlation coefficients determined for the initial region on the curve indicate a significant contribution of the external diffusion to the total process time.

From the analysis of the adsorption rate constant values (Table 1) calculated for the diffusion and chemical kinetic models, it can be found that the adsorption on PHQ/graphene proceeds several times faster than on PAC/CNTs. Besides, the chemical interaction between the dye molecules and the adsorbent functional groups is the limiting stage of the process.

4. Conclusions

The parameters of the organic dye (MB) adsorption by the composite materials based on graphene structures modified with organic products - PHQ/graphene and PAC/CNTs - are analyzed herein. To study the adsorption capacity under dynamic conditions, the authors developed a laboratory facility allowing for continuous measurement of solution optical density values using a spectrophotometer. The maximum MB adsorption capacity was experimentally found to be 2,610 (at 40 min) and 2,938 mg g^{-1} for PHQ/graphene (at 40 min) and PAC/CNTs (at 75 min), respectively. With the kinetic data approximation using the diffusion and chemical kinetic equations, it was established that the mixed-diffusion-contributed chemical interaction stage limits the process.

Acknowledgment

The research was funded by the Ministry of Education and Science of the Russian Federation (Project No. 16.1384.2017/PCh).

References

[1] Information available at http://www.water.ru/bz/likbez/regionalnye_-problemy_kachestva.shtml.
[2] Wang H J, Zhou A L, Peng F [et al.] 2007 *Mater. Sci. Eng., A* **466** 201-206.
[3] Han R, Lu Z, Zou W [et al.] 2006 *J. Hazard. Mater B* **137** 480.
[4] Mohan D, Pittman C U 2006 *J. Hazard. Mater. B* **137** 762.
[5] Wenhao Wu, Wei Chen, Daohui Lin [et al.] 2012 *Environ. Sci. & Technol.* **46** 5446-5454.
[6] Liu X Y, Ji Y S, Zhang Y H [et al.] 2007 *J. Chromatogr. A* **1165** 10-17.
[7] Kaur A, Gupta U 2009 *Mater. Chetu.* **19** 8279-8289.
[8] Yang K, Wu W, Jing Q [et al.] *Environ. Sci. Technol.* 2008, **42**(21), 7931-7936.

SPBOPEN 2018   IOP Publishing

IOP Conf. Series: Journal of Physics: Conf. Series **1124** (2018) 081030   doi:10.1088/1742-6596/1124/8/081030

Kinetics of the Adsorption of Synthetic Dyes on a Polyhydroquinone/Graphene Carbon Nanocomposite

A V Babkin[1], A V Melezhik[1], D A Kurnosov[1], E S Mkrtchyan[1], I V Burakova[1]*, A E Burakov[1], E V Galunin[1]

[1] Department "Technology and Methods of Nanoproduct Manufacturing", Tambov State Technical University, Tambov, 392000, Russia

*iris_tamb68@mail.ru

Abstract. The paper considers the possibility of applying the innovative p-benzoquinone-modified graphene-oxide-based nanocomposite material – polyhydroquinone/graphene- as an adsorbent of organic compounds of different nature on the example of synthetic dyes. The graphene content in the composite is about 30-40 %, whereas the polyhydroquinone content is 60-70 %. The kinetics of the adsorption of the dyes - methyl orange (MO) and methylene blue (MB) - from aqueous solutions was studied. The adsorption capacities towards the MB and MO were found to be 890 and 601 mg g^{-1}, respectively, for the contact time of 10 min. Kinetic dependencies were constructed, and the expected mechanisms of the adsorption on the carbon nanocomposite were described.

1. Introduction

The adsorption of organic compounds of various origins is widely studied throughout the world. The problem of complete purification of industrial effluents from water-soluble organic substances is one of the most important and, at the same time, difficult-to-solve issues. Adsorption is a universal technique that makes it possible to extract almost all impurities from the liquid phase. It is based on preferential removal of contaminant molecules under the action of a force field in adsorbent pores and due to chemically active adsorption sites [1,2].

To purify water from molecularly dissolved organic substances, activated carbons are usually used as conventional material [3]. The other conventional materials include synthetic zeolites such as NaX [4], plant-origin waste such as wheat straw, sawdust, shell husks, etc. [5], carbon nanomaterials such as carbon nanotubes (CNTs) and their functionalized forms [6-8].

The maximum degree of purification can be achieved by increasing time of the contact between the solution and the adsorbent up to 30-50 min [5]. However, under modern conditions, it is required to develop new types of adsorption materials capable of removing hard-to-remove impurities at a higher degree of adsorption and a short contact time. These requirements are met by using carbon nanomaterials such as nanotubes and graphene, both in their native or modified form. A wide range of publications are known to confirm the prospects of employing nanomaterials for the adsorption of dyes like methylene blue (MB) and methyl orange (MO).

The authors of the present work synthesized a nanocomposite material (polyhydroquinone (PHQ)/graphene) based on graphene oxide (GO) and modified with p-benzoquinone - an organic compound, the chemical formula of which is $C_6H_4O_2$. To assess the adsorption capacity of this material for organic substances, kinetic studies on the adsorption of the cationic (methylene blue - MB) and anionic (methyl orange - MO) dyes from aqueous solutions were carried out.

Content from this work may be used under the terms of the Creative Commons Attribution 3.0 licence. Any further distribution of this work must maintain attribution to the author(s) and the title of the work, journal citation and DOI.
Published under licence by IOP Publishing Ltd

SPBOPEN 2018   IOP Publishing
IOP Conf. Series: Journal of Physics: Conf. Series **1124** (2018) 081030   doi:10.1088/1742-6596/1124/8/081030

2. Experimental Details

2.1. PHQ/Graphene Nanocomposite

The adsorbent proposed herein was synthesized by polymerization of p-benzoquinone in an aqueous solution in the presence of a GO dispersion. Simultaneously, a partial reduction of GO to graphene occurs during the process, so that the product obtained probably represents PHQ molecules chemically bonded to few-layered graphene sheets (as elucidated by the X-ray diffraction data). In this material, the graphene content is about 30-40 %, the rest is PHQ. The PHQ weight content can be varied by choosing ratios of the initial reagents. Due to the presence of phenolic groups and the well-developed surface (at the expense of the texture-forming properties of graphene nanoplatelets), this material could be an effective adsorbent of organic substances and heavy metals.

2.2. Adsorbent Characterization

To analyze the crystalline structure of the PHQ/graphene nanocomposite, Raman spectra were recorded by a DXR ™ Raman microscope using Array Automation software (Thermo Fisher Scientific Inc., Waltham, MA USA).

To qualitatively determine the phase composition of the PHQ/graphene nanocomposite, a Difrey 401 X-ray diffractometer (Scientific Instruments CJSC, St. Petersburg, Russia) was used.

2.3. Kinetic Study

To carry out experiments on the organic dye removal, 30 mL of 1,500 mg L^{-1} MO and MB aqueous solutions were separately added in tubes containing the weighed amount of the adsorbent (0.03 g), then shaken on a Multi Bio RS-24 programmable rotator (Biosan, Riga, Latvia) at time intervals of 5, 15, 30, and 60 min. Next, the solutions were passed through paper filters, and optical density measurements were made on a PE 5400V spectrophotometer (Ekros, St. Petersburg, Russia) at 400 (for the MO) and 570 (for the MB) nm, followed by calculating concentration ("after-adsorption") values.

3. Results and Discussion

3.1 Characterization

Figure 1 shows the spectra with individual peaks indirectly confirming the availability of structured carbon.

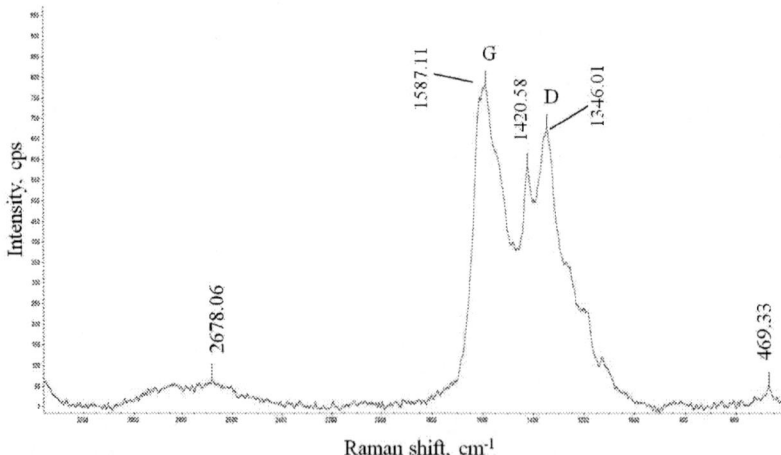

Figure 1. Raman spectra recorded for the PHQ/graphene nanocomposite.

The spectra of the synthesized nanocomposite have characteristic peaks at two frequencies: ~1,587 cm^{-1} representing the G band related to sp^2-hybridized carbon atoms possessing higher energy, and

~1,346 cm^{-1} representing the D band associated with sp^3-hybridized carbon atoms having lower energy. These peak frequencies are identical with those ones fixed for the GO. The synthesized nanocomposite is also characterized by the absence of peaks typical of deformation vibrations of the C-H bonds in the benzene ring probably related to the oxidation of polymerized quinone particles during the modification of the GO surface. High intensity of the spectral bands can be observed at a frequency of ~ 1,420 cm^{-1} typical of vibrations of the double C=C bonds in the molecular structure of the benzene ring.

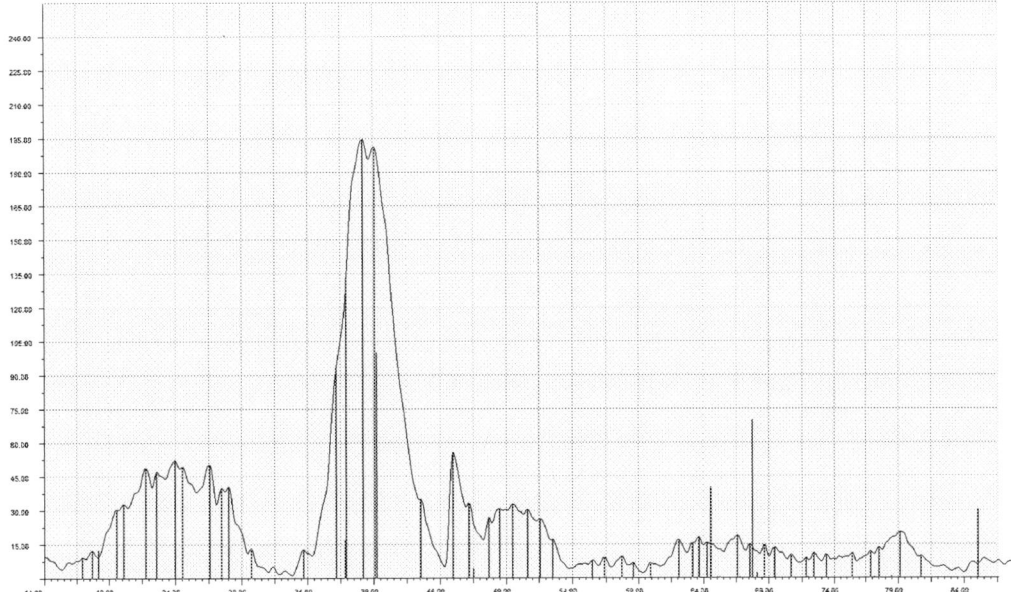

Figure 2. X-ray diffraction pattern recorded for the PHQ/graphene nanocomposite material.

The X-ray diffraction pattern presented in Figure 2 shows the availability of few-layered graphene in the studied sample (the band at 2teta = 38.0°). The bands at 2teta = 20.0-27.0° indicate the presence of organic compounds in the composite structure, which are most likely formed as a result of the chemical interaction between the PHQ and the GO surface functional groups. The unidentifiable peaks in the X-ray pattern presumably correspond to an amorphous phase formed by p-benzoquinone decomposition products.

3.2 Kinetic results
After the experiments, kinetic dependencies of the MO and MB adsorption on the PHQ/graphene nanocomposite were constructed (Figure 3). According to the data obtained, the adsorbent exhibited high adsorption activity towards the organic molecules (890 and 601 mg g^{-1} in the cases of the MB and MO, respectively). Adsorption equilibrium was reached within 10 min in both cases.

SPBOPEN 2018   IOP Publishing

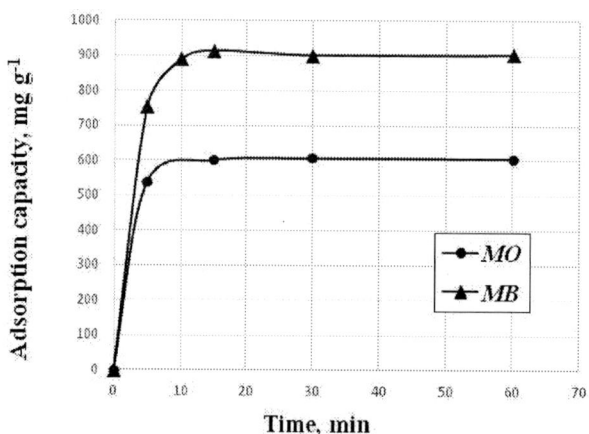

Figure 3. Kinetic of the dyes adsorption on the PHQ/graphene nanocomposite.

The adsorption proceeding in the adsorbate-adsorbent system represents a multi-stage process, which makes its description rather difficult. To calculate the kinetic parameters and elucidate possible dye adsorption mechanisms taking place on the PHQ/graphene nanocomposite, the following known models were implemented to process the experimental data: pseudo-first- and pseudo-second order, Elovich, and intraparticle diffusion.

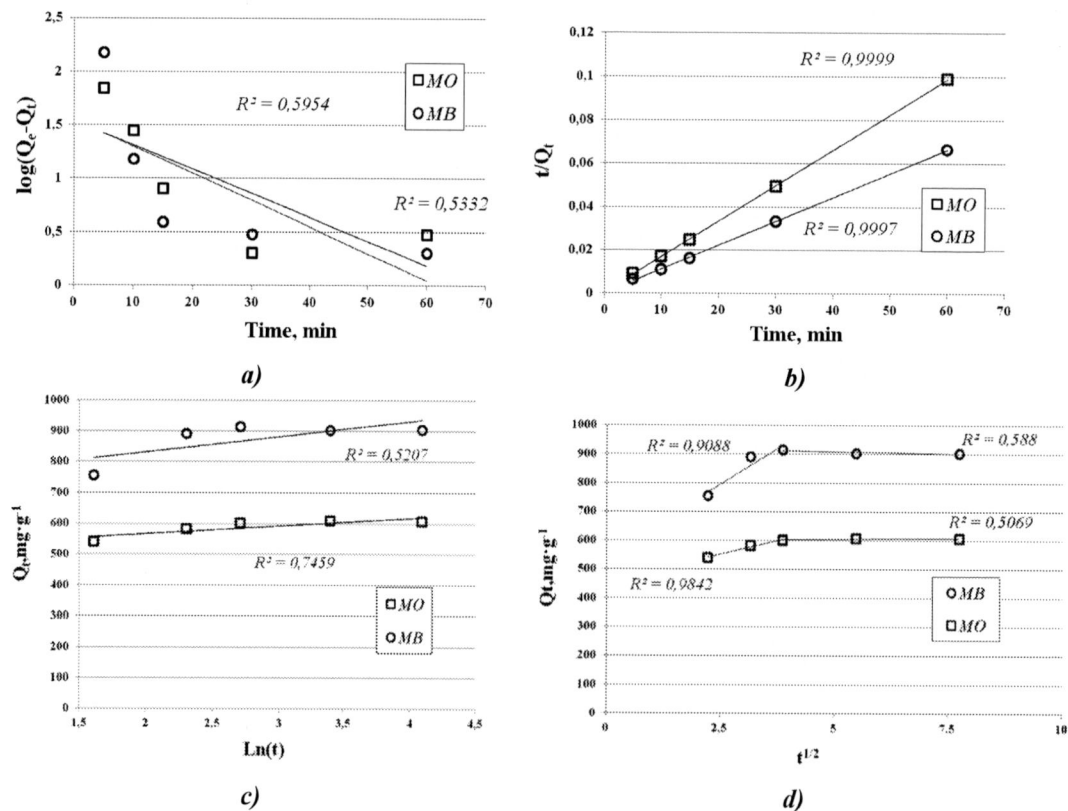

Figure 4. Kinetic models: pseudo-first-order (**a**), pseudo-second-order (**b**), Elovich (**c**), and intraparticle diffusion (**d**).

1118

Figure 4a shows the straight line of the experimental data approximation using the pseudo-first-order model. Given sufficiently low determination coefficient (R^2) values for both data sets (R^2=0.5954 – for the MB, and R^2=0.5332 – for the MO), it can be assumed that the process under study is not limited by the film diffusion of the contaminant molecules, which determines the process rate during the initial period of time. As a rule, such a model can be used when the concentration of the contaminant on the adsorbent surface is much lower than that of the adsorbent surface groups, which is possible only at the initial adsorption stage. Table 1 below presents the main calculated parameters of the kinetic equations.

Table 1. Kinetic parameters of the organic dyes adsorption onto PHQ/graphene*.

	Pseudo-first-order: $$\log(Q_e - Q_t) = \log Q_e - \frac{k_1}{2.303}t$$			Pseudo-second-order: $$\frac{t}{Q_t} = \frac{1}{k_2 Q_e^{\,2}} + \frac{1}{Q_e}t$$		
	Q_e	k_1	R^2	Q_e	k_2	R^2
MO	34.39	-0.052	0.595	625	0.0032	0.9999
MB	35.22	-0.058	0.533	909.1	0.0024	0.9997
	Elovich: $$Q_t = \frac{1}{\beta}\ln(\alpha\beta) + \frac{1}{\beta}\ln t$$			Intraparticle diffusion: $$Q_t = k_{id}\, t^{0.5} + C$$ Step 1/Step 2		
	α	β	R^2	k_{id}	C	R^2
MO	$1.16 \cdot 10^{10}$	0.039	0.7459	38.25/1.18	456.5/599	0.984/0.507
MB	$1.16 \cdot 10^{8}$	0.02	0.5207	99.57/-2.64	545.3/921.4	0.909/0.588

* Q_e – dye amount adsorbed onto the adsorbent surface at equilibrium (mg g^{-1}); Q_t – dye amount adsorbed onto the adsorbent surface at time t (mg g^{-1}); k_1- pseudo-first-order adsorption rate constant (min^{-1}); k_2 – pseudo-second-order adsorption rate constant (g mg^{-1}min^{-1}); α - initial adsorption rate constant (min^{-1} mg g^{-1}); β - degree of surface coverage and activation energy of chemisorption (g mg^{-1}); k_{id} - intraparticle diffusion coefficient (mg g^{-1}min$^{-0.5}$); C - boundary layer thickness (mg g^{-1}).

In its turn, the pseudo-second-order model is applicable to the description of the entire contaminant removal process, and indicates chemical adsorption. In the case considered herein (Figure 4b), the determination coefficient values of the approximation straight line for the MO and MB dyes are R^2 = 0.9999 and 0.9997 (Table 1), respectively, thereby confirming the chemical interaction between the dye molecules and the functional phenolic groups of the nanocomposite.

In order to evaluate the adsorption and desorption contributions, the well-known Elovich model was applied. The classical definition of the determination coefficient by the data linearization method makes it possible to evaluate the suitability of the model for using in a particular process. A low coefficient indicates the absence of heterogeneous chemisorption on the surface of the material synthesized, or it shows that this type of interaction is not limiting and does not determine the kinetics of adsorption removal (Figure 4c, Table 1).

Furthermore, to describe the kinetics of the adsorption of the organic substances on the nanocomposite under study, the diffusion model, was used herein. Fig. 4d shows the experimental data approximation in the corresponding coordinates. The trend of the dependence constructed indicates impossibility of an unambiguous interpretation of the process nature. The adsorption proceeds in a mixed diffusion mode, which confirms the stepwise variation of the approximation straight line slope.

4. Conclusion
In the present research, the innovative p-benzoquinone-modified graphene-oxide-based material – PHQ/graphene nanocomposite – is proposed as an adsorbent for removing dye molecules from

aqueous media. The analysis of the kinetic mechanisms of the dye adsorption on this nanocomposite was carried out in accordance with the commonly accepted methods allowing evaluation of the material functional characteristics. The results obtained show a significant superiority of PHQ/graphene over the existing analogs (e.g., those ones reported in [3-7]). The comprehensive study on the physical and chemical characteristics of the material made it possible to identify the crystalline structure of carbon, similar to that of graphene in type and energy, and to establish the presence of characteristic peaks related to the vibrations of constituent elements in the molecular structure of the organic modifier. The mathematical analysis of the experimental kinetic dependencies allows elucidation of the mixed diffusion mechanism of adsorption removal affected by the chemical interaction between the dye molecules and the phenolic groups of the adsorbent.

Acknowledgment

The research was funded by the Ministry of Education and Science of the Russian Federation (Project No. 16.1384.2017/PCh).

References

[1] Mishra S P 2014 *Current Science* **107**(4) 601-612.
[2] Abas S N A, Ismail M H S, Kamal M L [et al.] 2016 *Appl.Sci.J.***28**(11) 1518-1530.
[3] Belhachemi M 2011 *Appl. WaterSci.* 111–117.
[4] Shakti D, Sanghamitra B 2013 *Int. J.Sci. Environ. Technol.* **2**(4) 735-747.
[5] RafatullahM,SulaimanO, HashimR, AhmadA 2010 *J. Hazard. Mater.* **177**(1–3) 70-80.
[6] Zhao D, Zhang W, Chen Ch, Wang X 2013 *Proc. Environ. Sci.* **18** 890–895.
[7] Kang D, Yu X, Ge M, Xiao F, Xu H 2017 *Environ. Sci.* **54** 1–12.
[8] Karkeh-Abadi F, Saber-Samandari S, Saber-Samandari S 2016 *J.Hazard. Mater.* **312** 224–233.

Kelvin probe microscopy of MoSe₂ monolayers on graphene

B R Borodin[1,3], M S Dunaevskiy[1], F A Benimetskiy[2], S P Lebedev[1], A A Lebedev[1] and P A Alekseev[1]

[1] Ioffe Institute, Saint-Petersburg 194021, Russia
[2] ITMO University, Saint-Petersburg 197101 , Russia
[3] St. Petersburg Electrotechnical University "LETI", St. Petersburg 197376, Russia

Abstract. This work presents the results of an investigation of thin layers of MoSe₂ on graphene by Scanning Probe Microscopy (SPM) methods. Dependences of the surface potential and work function on the number of monolayers of the structure are presented. The dependence of the surface potential on air humidity is shown. The band structure and doping of the MoSe₂ monolayer on graphene was determined. These data can be important for detecting of the number of MoSe₂ layers and for designing nanodevices, because the surface potential has a strong influence on the operation of such devices.

1. Introduction

Properties of a transition metal dichalcogenides (TMD) are of great interest for study. These materials could be used in electronic, optoelectronic and valleytronic devices [1, 2]. The most studied TMD material is the MoS₂ [3]. The electronic properties of the TMD strongly depend on the number of layers [4] and the electronic state of the surface. For example, an adsorbed water on the surface of such devices leads to a change in the electronic properties [5]. Typically, the number of TMD layers and the effect of the dielectric environment on their electronic properties can be studied using a photoluminescence and Raman spectroscopy [6]. However, the electronic properties of van-der-Waals (vdW) materials can also be investigated by Scanning Kelvin Probe Microscopy (KPM). For example, the possibility of determining the number of layers in graphene and MoS₂ was demonstrated by the KPM method. The effect of the adsorbed water on the surface potential in MoS₂ also was studied [4]. It should be noted that TMD represents a wide class of materials with poor studied properties. Additionally, a heterostructures formed from vdW materials are of interest. The aim of this work was to study the electronic properties of the MoSe₂ layers transferred to mono- and bilayer graphene on a SiC substrate. It should be noted that MoSe₂ is of great interest for use in optoelectronic devices. For example, the possibility of creating an exciton-polariton condensate at room temperature was recently predicted for this material [7].

2. Samples and methods

The experiment was performed on the Ntegra Aura (NT-MDT) scanning probe microscope using a Si (NT-MDT) probes with a tip diameter of 30 nm in air conditions with a controlled humidity. The sample was a monolayer graphene of high quality with a small fraction (~ 10%) of two-layered islands with submicron dimensions, obtained by thermal destruction in argon of the Si-face of the 6H-SiC substrate (0001) [8]. Thin layers of MoSe₂ were obtained by a micromechanical exfoliation and then were transferred to the SiC substrate with graphene on the surface.

Content from this work may be used under the terms of the Creative Commons Attribution 3.0 licence. Any further distribution of this work must maintain attribution to the author(s) and the title of the work, journal citation and DOI.
Published under licence by IOP Publishing Ltd

The study was based on measurements by a Kelvin probe microscopy method. Figure 1 shows schemes of measurement setup (a) and probe-sample interaction (b) in Kelvin probe mode.

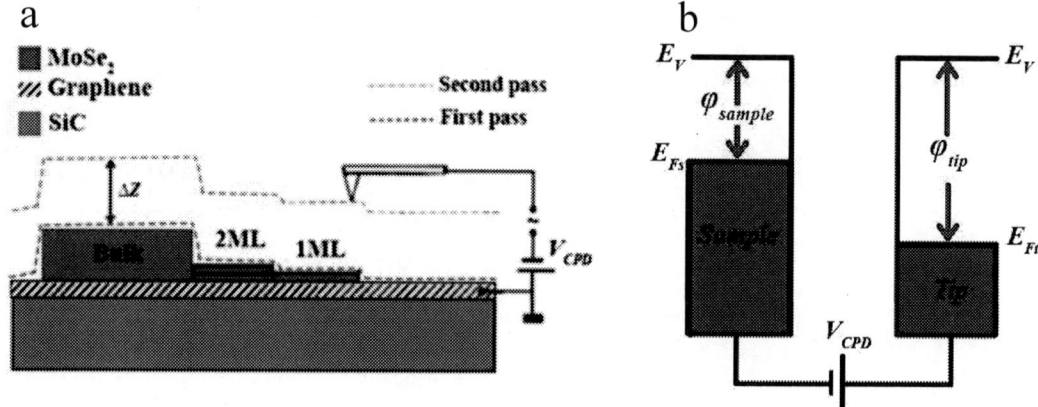

Figure 1(a, b). (a) The scheme of measurement in Kelvin probe mode; **(b)** Scheme of probe and sample interaction in Kelvin probe mode.

Figure 1 (a) shows a method of obtaining image in the Kelvin probe mode. KPM in a double-pass configuration was used. In the first pass probe in semicontact mode obtain topography. In the second pass probe repeats the relief at some distance (ΔZ) from the surface and measures a surface potential (V_{CPD}) between the probe and the surface. It is known that the surface potential it is the difference between the work function of the surface (φ_{sample}) and the probe (φ_{tip}). Figure 1 (b) shows the scheme of interaction of the probe and the surface. Thus, the interaction of the probe and the surface can be described by the equation:

$$\varphi_{sample} = \varphi_{tip} - eV_{CPD} \tag{1}$$

For the sample with a surface consisting of two materials with different surface potentials ($V_{CPD\,1}$, $V_{CPD\,2}$) it is possible to obtain a work function value ($\varphi_{sample\,1}$) if the ($\varphi_{sample\,2}$) is known, without the work function of the probe:

$$\varphi_{sample\,1} = \varphi_{sample\,2} + e(V_{CPD\,2} - V_{CPD\,1}) \tag{2}$$

Thus, the work function of the surface area can be determined by comparing its surface potential with the surface potential of the area with a known work function. The difference between surface potentials is the difference between work functions. The proposed method can be used for studied sample, because its surface has not only MoSe$_2$ areas with an unknown work function, but also the areas of mono- and bilayer graphene. Thus, work function of MoSe$_2$ was determined with respect to mono- and bilayer graphene on SiC, which work function was known as 4.55 and 4.44 eV, respectively [9].

3. Results and discussion
Figure 2 (a) shows the topography of surface obtained by atomic force microscopy in semi-contact mode. The distribution of the surface potential measured by Kelvin probe microscopy is shown in Fig. 2 (b).

IOP Conf. Series: Journal of Physics: Conf. Series **1124** (2018) 081031   doi:10.1088/1742-6596/1124/8/081031

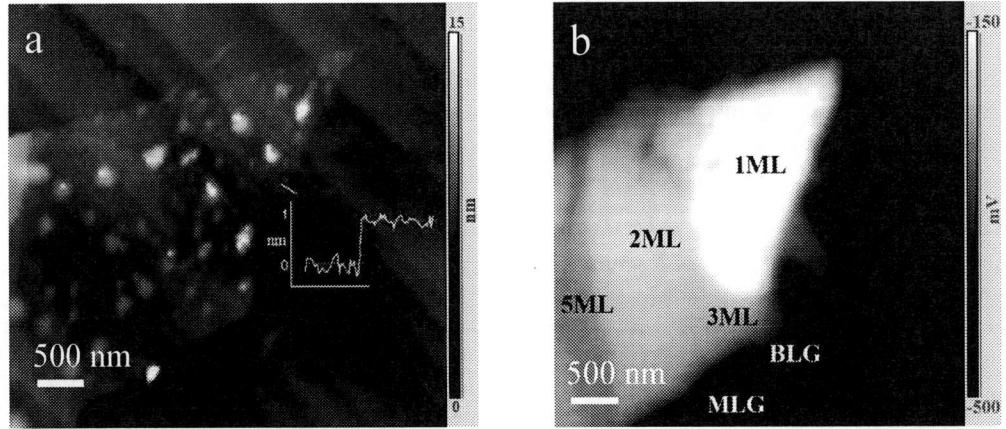

Figure 2(a, b). MoSe$_2$ layers on the graphene, **(a)** Topography image; **(b)** Surface potential distribution. MLG and BLG are the regions with monolayer and bilayer graphene, respectively. 1-5ML indicate the number of the MoSe$_2$ layers

As shown in Fig. 2, the image contains a different number of MoSe$_2$ layers. The number of the MoSe$_2$ layers were determined from the surface topography image (Fig. 2a). MoSe$_2$ monolayer thickness is of 0.7 nm. When the number of layers increases, the thickness increases by the same amount. Figure 2 (b) shows that the monolayer has the highest potential. As the number of layers increases, the surface potential decreases. Figure 2 (b) also exhibits the stripes of bilayer graphene (BLG), whose potential is higher than monolayer graphene (MLG). It can be seen that a MoSe$_2$ monolayer lies on the surface with a monolayer and a bilayer graphene. However, the distribution of the surface potential of the MoSe$_2$ monolayer is a uniform, which indicates a complete screening of the substrate potential even by a monolayer of MoSe$_2$.

Figure 3 (a) shows the dependence of the surface potential on the number of layers at different levels of air humidity. The dependence of the work function on the number of layers for a minimum air humidity (20%) is shown in Fig. 3 (b). Since the measurements were carried out by the standard Kelvin probe microscopy with amplitude modulation, an accuracy in determining the work function was of about 0.05 eV. Using a gradient Kelvin probe microscopy can increase the accuracy of such measurements [9, 10].

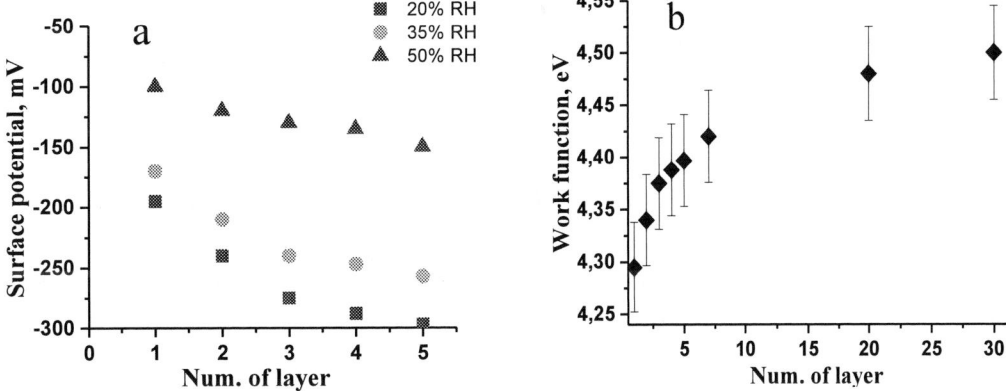

Figure 3(a, b). (a) Dependence of the surface potential on the number of layers, at different air humidity levels; **(b)** Dependence of the work function on the number of layers.

As shown in Fig. 3 (a), increase of the level of air humidity leads to a screening of the surface potential of the MoSe$_2$ layer by a liquid adsorbed layer, that agrees with the results obtained for MoS$_2$ [4]. These data show that the accuracy of determining the number of MoSe$_2$ layers depends on the experimental conditions in particular on the air humidity. Thus, when carrying out such studies it is necessary to control this parameter. The decrease in the surface potential with increasing number of layers is determined by an increase in the work function. The increase of number of MoSe$_2$ layers leads to a nonlinear increase of their work function (Fig. 3b) and can be explained by the features of the interlayer screening [4]. As well, the transition from a monoatomic layer to a bulk material leads to tune of the band structure, namely, to a narrowing of the band gap and a decrease in the energy of the valence band and conductivity band, which leads to an increase in the work function [11]. The work function of the monolayer MoSe$_2$ is obtained as of 4.3 eV. The work function is increased with the increasing of number of layers and saturated in a value of 4.5 eV that corresponds to bulk MoSe$_2$. The measured values of the work function for areas with a different number of layers were reproduced quite well. Thus, number of MoSe$_2$ layers can be determined by Kelvin probe microscopy.
Figure 4 shows the band structure of the MoSe$_2$ monolayer on graphene.

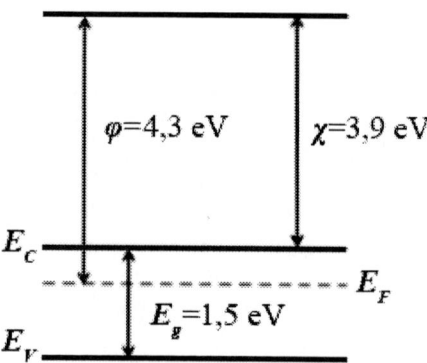

Figure 4. The band structure of the MoSe$_2$ monolayer on graphene.

Based on the data on the band gap (Eg) [11] and the electron affinity (χ) [12], a position of the Fermi level for the MoSe$_2$ monolayer on graphene was determined (3):

$$\varphi - \chi = E_c - E_F \qquad (3)$$

From the Figure 4 it is follows that a monolayer of the MoSe$_2$ on graphene is the n-doped.

4. Conclusion
To conclude, thin layers of MoSe$_2$ on graphene were studied by scanning probe microscopy. The dependences of the surface potential and work function on the number of layers and the level of air humidity were determined. The work function of the MoSe$_2$ monolayer is of 4.3 eV and nonlinearly increases with the number of layers. Measured work function are in good agreement with the data obtained by other methods. The band structure and n-type doping of the MoSe$_2$ monolayer on graphene was determined. The results of the work allow to consider the KPM as an effective method for determining the number of layers in MoSe$_2$.

References
[1] Fontana M, Deppe T, Boyd A, Rinzan M, Liu A Y, Paranjape M, Barbara P 2013 Sci. Rep. **3** 1634
[2] Zeng H, Dai J, Yao W, Xiao D, Cui X 2012 Nat. Nanotechnol. **7** 490
[3] Duan X, Wang C, Pan A, Yu R 2015 Chem. Soc. Rev. **44** 8859

[4] Feng Y, Zhang K, Li H, Wang F, Zhou B, Fang M, Wang W, Wei J, Wong H 2017 Nanotechnology **28**

[5] Schmidt H, Giustiniano F & Eda G 2015 Chem. Soc. Rev. **44** 7715–7736

[6] Tonndorf P, Schmidt R, Böttger P, Zhang X, Börner J, Liebig A, Albrecht M, Kloc C, Gordan O, Zahn D, Michaelis de Vasconcellos S, Bratschitsch R 2013 Opt. Express **21** 4908–4916

[7] Lundt N, Maryński A, Cherotchenko E, Pant A, Fan X, Tongay S, Sęk G, Kavokin A, Höfling S, Schneider C 2017 2DMater. **4** 015006

[8] Davydov V, Usachov D, Lebedev S, Smirnov A, Levitskii V, Eliseyev I, Alekseev P, Dunaevskiy M, Rybkin A, Lebedev A 2017 Semiconductors **51**, 1072–1080

[9] Panchal V, Pearce R, Yakimova R, Tzalenchuk A., Kazakova Olga 2013 Sci. Rep. **3** 2597

[10] Dunaevskiy M, Alekseev P, Girard P, Lashkul A, Lahderanta E, Titkov A 2012 Jour. Appl. Phys. **112** 064112

[11] Tongay S, Zhou J, Ataca C, Lo K, Matthews T, Li J, Grossman J, Wu J 2012 Nano Lett. **12** 5576–5580

[12] Kim D, Lim D 2016 Curr. Appl. Phys. **17** 321-325

SPBOPEN 2018   IOP Publishing

Photosensitive sulphide heterostructures obtained by using Successive Ionic Layer Adsorption and Reaction on planar and profiled substrates

N Bogomazova[1], G Gorokh[2], A Zakhlebayeva[2], A Pligovka[2], A Murashkevich[1], T Galkovsky[1]

[1]Department of Chemistry, Electrochemical Production Technology and Materials for Electronic Equipment, Belarusian State Technological University, Minsk, Belarus
[2]Nanotechnology Research Laboratory, Belarusian State University of Informatics and Radioelectronics, Minsk, Belarus

Abstract. The photosensitive heterostructures of ZnS/SnS_x were formed by Successive Ionic Layer Adsorption and Reaction method on the nanoporous anodic alumina matrixes, on tantalum oxide nanocolumns and on planar ITO/glass substrate. Investigation of morphology of ZnS/SnS_x films layered on the substrates with different surfaces has show which the layering of functional layers has dominated horizontal deposition mechanism. It has been established that the layering of sulfides on anodic alumina proceeds with a higher rate than the layering on ITO/glass. Also ZnS/SnS_x heterostructure formed on nanoporous anodic alumina has pronounced interface between layers of sulphide compounds while in the heterostructure on ITO such interface is missing. The layering of ZnS/SnS_x heterostructure on the tantalum oxide nanocolumns occurs locally. $ZnS/SnS_x/ITO/glass$ structure has the electrical resistance in the order of 140-80 Ohm. During illumination of the $ZnS/SnS_x/ITO/glass$ by white light with radiation power about 11 W in structure is formed of photoelectromotive force about 230 mV.

1. Introduction

An important trend in the development of the modern optoelectronic devices is the transition from the simple materials and homojunctions to the composite materials and heterojunctions. The formation of semiconductor heterostructures with nanoscale layers allows improving the parameters of optoelectronic devices due to effects of optical quantization, over-injection also wide-gap window, as well as allows creating devices with new functional capabilities [1]. The most promising among the chalcogenide semiconductors used in optoelectronics are sulphide materials, especially the wide-gap $A^{II}B^{VI}$ semiconductor ZnS with Eg = 3,7 eV and the narrow-gap $A^{IV}B^{VI}$ semiconductor SnS with Eg = 1,1 eV. These sulphide compounding are characterized by low values of solubility products, which allows to use chemical deposition technologies to form optically active heterostructures based on them [2, 3].

One of the most promising and technological chemical method for the formation of active layers is the SILAR (Successive ionic layer adsorption and reaction). The SILAR method is based on layer-by-layer growth of films by periodic treatment of substrates in solutions of metal salts or polyelectrolytes. Cations and anions adsorbed on the surface of the substrate form nanolayer of low-solubility compound. Monolayer application of materials with different chemical kind by SILAR method allows precise control of the thickness and composition of the forming functional layers [4]. Earlier SILAR

Content from this work may be used under the terms of the Creative Commons Attribution 3.0 licence. Any further distribution of this work must maintain attribution to the author(s) and the title of the work, journal citation and DOI.

Published under licence by IOP Publishing Ltd

method has already been successfully used in the filling of porous matrixes of anodic alumina with metal oxide compounds [5].

Profiled substrates such as porous anodic alumina and tantalum oxide nanocolumns are interesting for forming of structured systems on their base. Due to their high mechanical strength this substrates are used as matrixes for filling with different materials. The possibility to change the morphology of profiled matrices in a wide range of structural parameters allows to forming on their base of systems with defined measurements of structural elements, at that properties of those systems are determined by their measurements [6, 7].

This paper presents the results of the layering of ZnS/SnS_x heterostructures by SILAR method on the profiled anodic alumina matrix and tantalum oxide nanocolumns also on the planar indium-tin-oxide on glass substrate and investigation of morphology, electrophysical and photovoltaic properties of formed structures.

2. Experimental

2.1. Substrates for forming of ZnS/SnS_x heterostructures

Sulphide heterostructures ZnS/SnS_x were formed on the three types of substrates: 1) anodic alumina matrixes with thickness of 1 μm and pore diameters of 400 nm, 2) tantalum oxide nanocolumns with high 340 nm and diameters of columns 270 nm, 3) ITO film ($In_{1.8}Sn_{0.2}O_3$) with thickness 200 nm.

Nanoporous anodic alumina matrixes (AAM) were formed by multi-step anodizing of the 1.5 μm aluminium film magnetron-sputtered on the Si substrate in aqueous solution of 0.2 mol·dm^{-3} tartaric acid. On the first step anodizing were performed in galvanostatic mode at current 6 mA/cm^2 on thickness of Al 700 nm. After first anodizing step formed anodic alumina was selective dissolved in aqueous phosphoric-chrome solution (H_3PO_4:CrO_3:H_2O) at 80°C, which allows to structurized of aluminium surface of oxide cells imprints. On the second step anodizing were performed in potentiostatic mode at anodic potential of 213 V to form anodic alumina of thickness 1 μm.To increase the surface-to-volume ratio, AAM were subjected to pore modification in aqueous mixture of phosphoric acid and chromic oxide at 60°C during 12.5 minutes. The pore diameters were ~ 400 nm.

Tantalum oxide nanocolumns array was formed by electrochemical anodizing of two-layer Al/Ta thin-film system (1.2 μm Al and 50 nm Ta) on the Si substrate. Anodizing of the aluminum layer was performed in aqueous solution of 0.2 mol·dm^{-3} tartaric acid at voltage of 216 V. Tantalum oxide nanocolumns were formed by high-voltage reanodizing of the niobium layer via formed AAM in the mixed solution of 0.5 mol·dm^{-3} boracic acid and 0.05 mol·dm^{-3} sodium tetraborate in potentiodynamic mode at increase of potential until 450 V. Further AAM was selective dissolved in 50% phosphoric acid at 50°C.

2.2. SILAR method for layering of ZnS/SnS_x on substrates with different surfaces

ZnS/SnS_xheterostructures were deposited on the anodic alumina matrixes, array of tantalum oxide nanocolumns and ITO/glass substrate by SILAR method. The main stages of the formation of film structures are shown in Figure 1. On the first step substrates were cleaning in deionized water (T = 90−100°C) or in polar organic solvents (Figure 1, a). Next on the substrates were layered Sn-cations from solution 0.1 mol·dm^{-3} $SnCl_2$ (Figure 1, b) and S-anions from solution 0.1 mol·dm^{-3} Na_2S (Figure 1, d). After cationic and anionic layering steps the substrates were washed in deionized water at T = 20°C (Figure 1, c, e).So was formed one microlayer of SnS_x. The layering of cations and anions was repeated 20 times for each layer, as a result were formed 20 microlayers of SnS_x. After that substrates were dried at T = 50°C for 30 min and annealed at T = 350°C for 30 min (Figure 1, f). On SnS_x layers were formed 20 microlayers of ZnS by sequential layering of Zn-cations from solution 0.1 mol·dm^{-3}$ZnSO_4$·$7H_2O$ (Figure 1, g) and S-anions from solution 0.1 mol·dm^{-3} Na_2S (Figure 1, i) with washing substrates in deionized water(T = 20°C) after each layering step (Figure 1, h, j). After layering of cations and anions substrates were dried at T = 50°C for 30 min and annealed at T = 350°C for 30 min (Figure 1, k).

SPBOPEN 2018   IOP Publishing
IOP Conf. Series: Journal of Physics: Conf. Series **1124** (2018) 081032   doi:10.1088/1742-6596/1124/8/081032

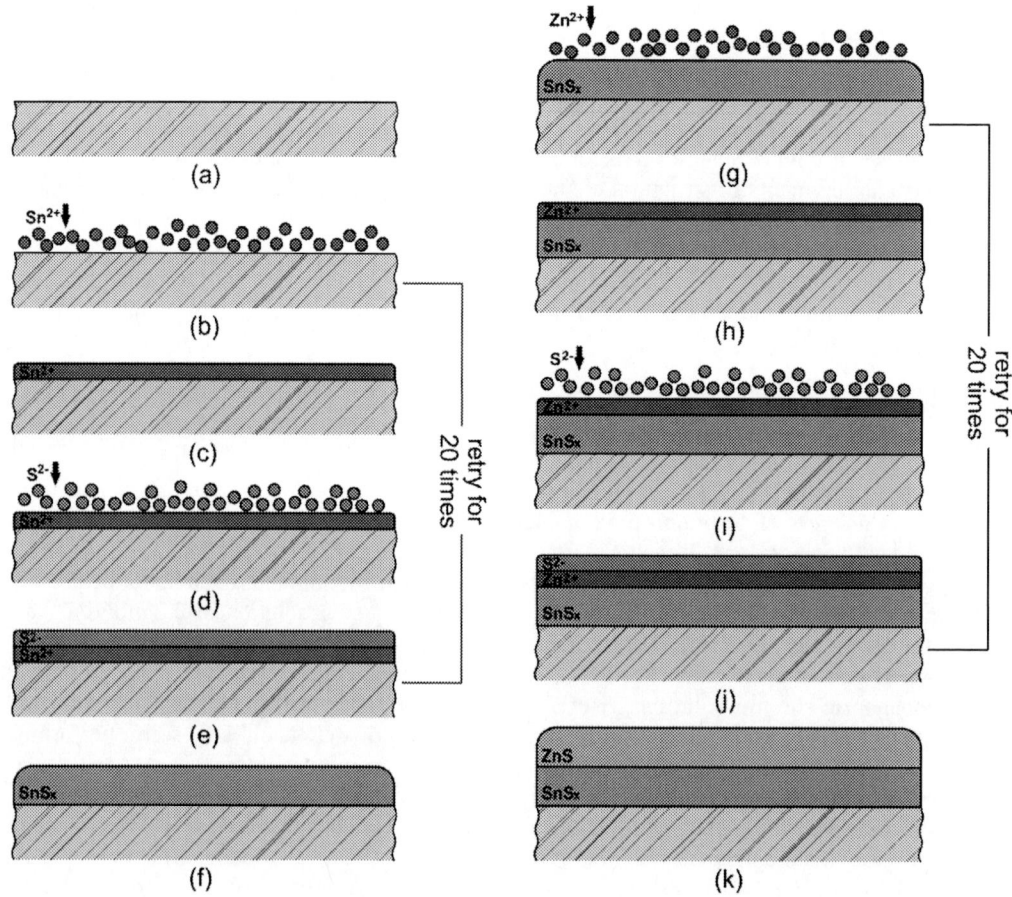

Figure 1. Schematic steps for forming of ZnS/SnS$_x$ heterostructures: **(a)** cleaning of the substrate surface; **(b)** layering of Sn^{2+} ions from solution 0.1 mol·dm^{-3}SnCl$_2$; **(c)** washing in deionized water; **(d)** layering of S^{2-} ions from solution 0.1 mol·dm^{-3}Na$_2$S; **(e)** washing in deionized water; **(f)** drying at T = 50°C, annealing at T = 350°C; **(g)** layering of Zn^{2+} ions from solution 0.1 mol·dm^{-3}ZnSO$_4$·7H$_2$O; **(h)** washing in deionized water; **(i)** layering of S^{2-} ions from solution 0.1 mol·dm^{-3}Na$_2$S ; **(j)** washing in deionized water; **(k)** drying at T = 50°C, annealing at T = 350°C.

3. Results and discussion

3.1. Investigation of morphology of ZnS/SnS$_x$ films layered on the substrates with different surfaces

In the Figure 2 presented the cross-sections of the nanoporous AAM (Figure 2, a), tantalum oxide nanocolumns (Figure 2, b) and ITO/glass substrate (Figure 2, c) with layered thin film system of ZnS/SnS$_x$. Investigation of the ionic layering of ZnS/SnS$_x$ films on the substrates with different surface morphology has shown that the layering of functional layers has dominated horizontal deposition mechanism, which leads to the formation of planar layer on the profiled matrix (Figure 2, a, b). It should be noted that formation of ZnS/SnS$_x$ on the planar ITO substrate (Figure 2, c) occurs only after the washing in polar organic solvents and ultrasonic treatment of the substrate surface.

SPBOPEN 2018   IOP Publishing

IOP Conf. Series: Journal of Physics: Conf. Series **1124** (2018) 081032   doi:10.1088/1742-6596/1124/8/081032

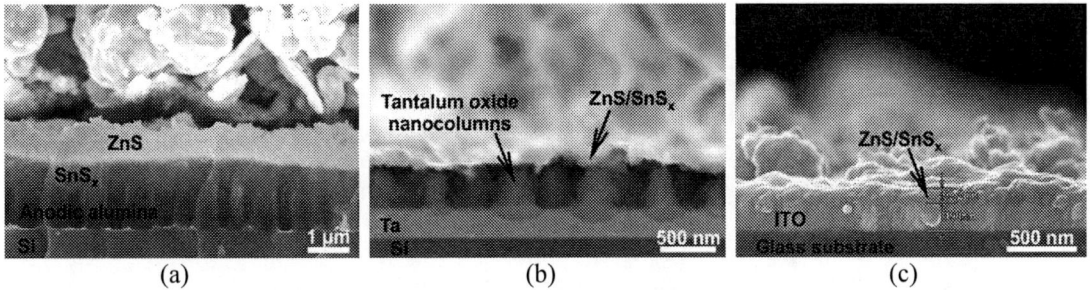

Figure 2. SEM microphotographs of the cross-section of theZnS/SnS$_x$ heterostructures on **(a)** anodic alumina matrix, **(b)** tantalum oxide nanocolumns, **(c)** ITO/glasssubstrate.

ZnS/SnS$_x$ heterostructure formed on nanoporous AAM has pronounced interface between layers of sulphide compounds (Figure 2, a). That was allows estimated thicknesses of sulphide layers, which amounted to about 600 nm for SnS$_x$ and about 870 nm for ZnS. Given that the nominal quantity of the microlayers for the each sulphide compounds was the same and amounted to 20, it can be noted that the layering of SnS$_x$ on the profiled surface of the chemically foreign Al$_2$O$_3$ proceeds with significantly less speed than the of layering of ZnS on the surface of chemically similar SnS$_x$. The calculated thicknesses of microlayers for SnS$_x$ was about 30 nm and for ZnS – about 43,5 nm. We note that in our previous experiments in separate layering of SnS$_x$ and ZnS layers on the surface of porous AAM, we obtained an analogous result about more less speed of deposition of SnS$_x$ [8].

The total thickness of the sulphide heterostructures formed on the surface of the ITO film with thickness of 200 nm was about 100 nm (Figure 2, c). So there has been significant decrease of layering speed on the planar ITO substrate in comparison with the profiled AAM substrate, which leads to decrease of the average thicknesses of microlayers of SnS$_x$ and ZnS to 2.5 nm. Such result can be accounted for removal of excess precursor solutions from surface of planar substrate, which allows realize the ionic layering in mode of monolayer film growth with precision control of its thickness.

The layering of heterostructures on the tantalum oxide nanocolumns array occurred locally. On the local areas of the substrate surface were formed ZnS/SnS$_x$ films (Figure 2, b), the formation of continued film did not occur. Probably, as in the case of planar ITO substrate, the hydrate treatment of the nanocolumns array do not allows to get surface, which is chemically usable to ion layering.

3.2. The electrophysical investigations of formed heterostructures
Investigations of the electrophysical and photosensitive properties of sulfide heterostructures were pursued on the planar ITO/glass substrates, which has the necessary transparency.

The electron transport properties of heterostructures were researched by measuring of the temperature dependences of the surface electrical resistance of sulphide films (Figure 3).

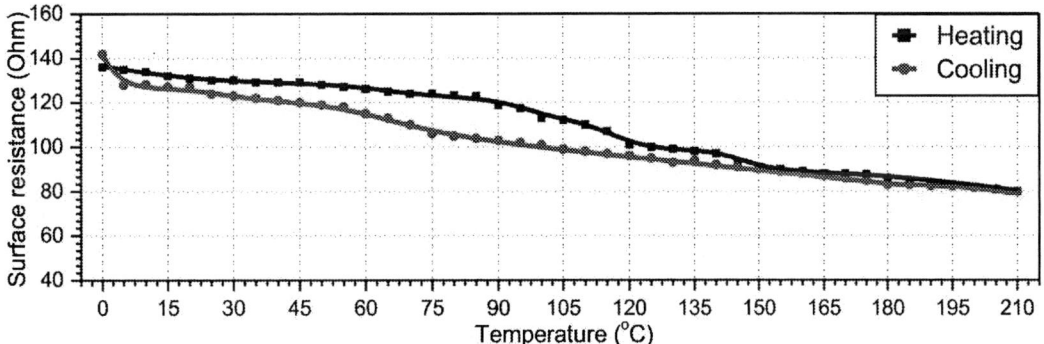

Figure 3. Temperature dependences of the surface resistance of ZnS/SnS$_x$/ITO/glass structure

The measurements were pursued in special installation, which included measuring cell, two voltmeters and current source. Heterostructures were placed in a measuring cell with clamping electrodes on the resistive heating element based on a nichrome wire. The measurements were carried out in the temperature range from 0 to 210°C. Investigations of the electrophysical properties showed that the planar structure of ZnS/SnS$_x$/ITO/glass substrate is characterized by a low electrical resistance in the order of 140-80 Ohm. Temperature dependence of the surface electrical resistance of formed structures has semiconductor character with weak hysteresis phenomena.

Photosensitive ZnS/SnS$_x$ heterostructure amounts to wide-gap window in the form of ZnS and absorptive narrow-gap layer SnS$_x$. The effect of the radiation power on the value of the generated photoelectromotive force on illumination of the heterostructure by white light was investigated. An incandescent lamp with an adjustable radiation power from 0 to 12 W was used as light source. The voltage difference, which appeared in the heterostructure during illumination by incandescent lamp, was measured. In the Figure 4, a shown schematic view of investigational heterostructure. Value of the photoelectromotive force was measured under illumination of different areas of heterostructure: ZnS/SnS$_x$/ITO structure with large (Figure 4, a, contacts a-c) and small (Figure 4, a, contacts a-b) working surface area, surfaces film ZnS (Figure 4, a, contacts b-c), ITO (Figure 4, a, contact a). Dependence of the photoelectromotive force of heterostructure on the radiation power shows in Figure 4, b

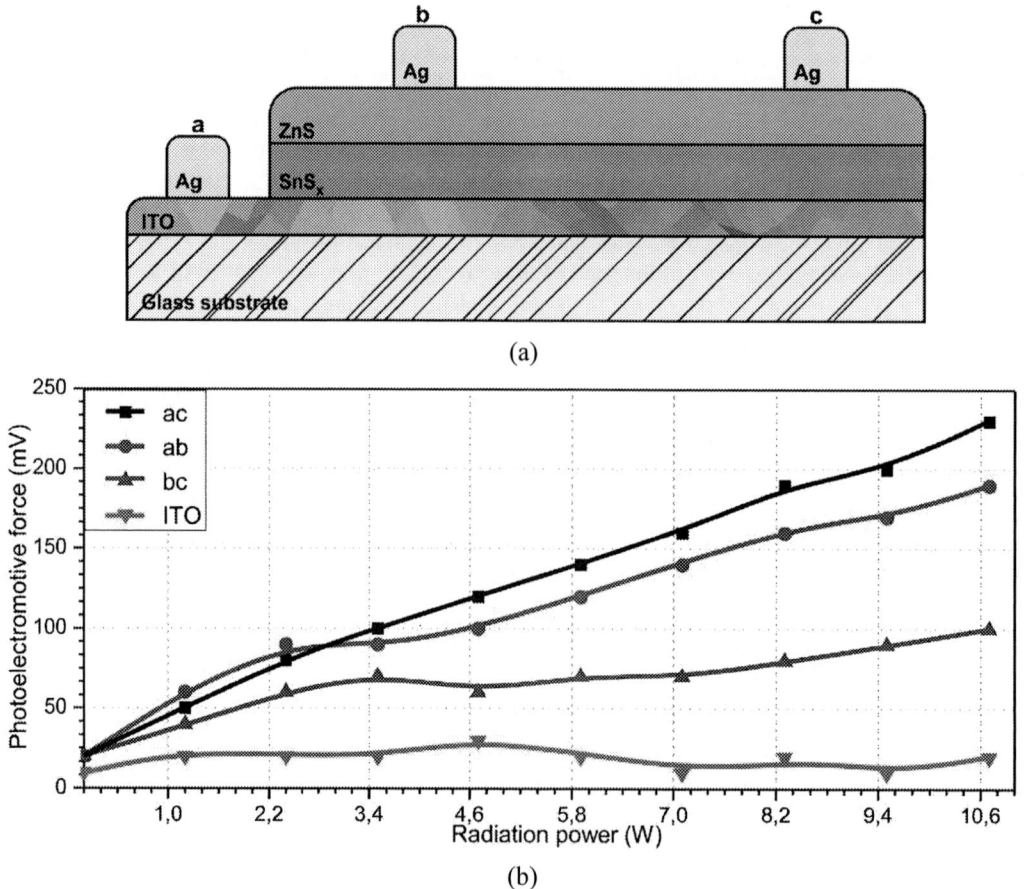

Figure 4.(a) Schematic view of ZnS/SnS$_x$/ITO/glass heterostructure with different working areas; (b) Dependence of the photoelectromotive force, generated in the ZnS/SnS$_x$ heterostructure, on the radiation power during illumination of different working areas.

As is seen from received dependences photoelectromotive force was increasing under increase of the power of the white light source. Character of photoelectromotive force increasing was close to monotonic. The most significant photoelectromotive force close to 230 mV was generated during illumination of the large working surface area of the structure (Figure 4, b, curve a-c). Value of photoelectromotive force, which was generated during illumination of the small working surface area of the structure, was about 190 mV (Figure 4, b, curve a-b). Voltage differences which were generated during illumination surfaces film ZnS (Figure 4, b, curve b-c) and ITO (Figure 4, b, curve ITO) did not exceed 100 mV (about 90 mV for ZnS and about 20 mV for ITO). This can be explained by the appearance of a volumetric photoelectromotive force which has significantly lower value than barrier-layer photoelectromotive force appearing near the potential barrier of a functional contact.

4. Conclusion

So, in this work were presented the results of the obtaining of sulphide ZnS/SnS_x heterostructures by using Successive Ionic Layer Adsorption and Reaction method on the planar ITO/glass and profiled anodic alumina also tantalum oxide nanocolumns substrates. Layering of the ZnS/SnS_x on the ITO/glass substrate was realized in monolayer films growth mode, thickness of the formed heterostructure was about 100 nm. ZnS/SnS_x heterostructure formed on nanoporous anodic alumina was characterized by pronounced interface between layers of sulphide compounds. Thicknesses of sulphide layers were about 600 nm for SnS_x and about 870 nm for ZnS. The continued ZnS/SnS_x films did not occur during the layering on the tantalum oxide nanocolumns. Forming of the heterostructures was occurred on the local areas of the substrate surface.

Investigations of the electrophysical properties of planar transparent $ZnS/SnS_x/ITO/glass$ structures shown the electrical resistance in the order of 140-80 Ohm. Photovoltaic measurements during illumination of the ZnS/SnSx/ITO/glass structure by white light with radiation power about 11 W shown the forming of photoelectromotive force about 230 mV in the functional contacts. This value is 35% higher than the result for similar structure obtained by electrochemical deposition [2]. The obtained results show the prospects of using the SILAR method for obtaining photoactive sulfide heterostructures on planar and profiled substrates. ZnS/SnS_x heterostructures can be used in low-cost thin-film solar cell.

5. Acknowledgments

The authors gratefully acknowledge the contributions of Reseach & Design Center of JSC «INTEGRAL» and Zhigulin Dmitriy, for providing images from a scanning electron microscope.

References

[1]   Tolstoy V 2006 *Uspekhi Khimii* **75(2)** 183.
[2]   Dickerson J, Boccaccini A 2005 J. *Chem. Shi. Techn.* **23(5)** 241.
[3]   Ramakrishna K 2015 *Curr. Appl. Phys.* **5** 101.
[4]   Haneefa M, Kandasamy S 2015 *Ind. J. Chem. Pharmac. Res.* **15(6)** 232.
[5]   Gorokh G, Zakhlebayeva A, Metla A et al. 2017 *J. Phys. Conf. Ser.* **917** 092011.
[6]   Mozalev A, Smith A, Borodin S 2009 *Electrochim. Act.* **54** 935.
[7]   Gorokh G, Mozalev A, Solovei D 2006 *Electrochim. Act.* **52** 1771.
[8]   Bogomazova N, Komarenko A, Galkovsky T 2017 *Modern electrochem. techn. equip.* (Minsk / Belarusian State Technological University) p 103 (in Rus.).
[9]   Lasisi A et al. *Asian J. Sci. Tech.* 2016 **7(11)** 3887.

SPBOPEN 2018   IOP Publishing

Computer simulation of aerosol nanoparticles focusing and deposition process for functional microstructure fabrication

Protas N V, Efimov A A, Zemlyanoy V K, Ivanov V V

Department of Physical and Quantum Electronics, Moscow Institute of Physics and Technology, Dolgoprudny 141700, Russia

Abstract. The possibility is shown of using computer simulation to predict the size of the structures formed by the dry aerosol jet printing method. Computer modeling consisted in determining the model parameters and solving the system of equations to determine the trajectories of Ag nanoparticles of size up to 45 nm in the process of focusing and deposition through a coaxial nozzle with an outlet diameter of about 100 μm on the substrate surface located 0.5 mm away from the nozzle. As a result of the comparative analysis, a satisfactory agreement between the model and the experiment is established.

1. Introduction

Recently, printing methods have been widely developed as means for the production of inexpensive electronic circuits with the printing resolution of up to 10 μm. These methods have the prospects for the manufacture of devices such as printed antennas [1], thin-film solar cells [2], flexible displays [3], etc. Among the existing printing methods, dry aerosol jet printing (AJP) is one of the most promising methods for manufacturing functional microstructures [4]. The main advantage of dry AJP in comparison with other printing methods is the absence of the need to prepare ink to produce a flow of nanoparticles. At the same time, the dry AJP method is a relatively new and insufficiently investigated method. Thus, there is a need to develop models of dry aerosol printing processes to predict the size of the structures to be formed and to find ways to improve this technology in order to minimize the printing resolution. Therefore, in this paper, a study is presented of using computer modeling to predict the size of the structures formed by the dry aerosol printing method.

2. CFD Model

To predict the size of the structures formed by the dry aerosol printing process using computational fluid dynamics (CFD), a 2D model of the experiment on focusing and deposition of aerosol nanoparticles was constructed. The first step in constructing the experimental model was to set the geometry of the gas channel in which the focusing and deposition of aerosol nanoparticles were carried out (Fig. 1 a). For detailed data on the linear dimensions and geometry of the gas channel, the data of the patent were used. [5] For the given geometry, the parameters of the dry aerosol printing process were specified. Those parameters were: the aerosol carrier gas flow rate Q_a, the sheath gas flow rate Q_{sh}, the outlet diameter of the nozzle D_n, the distance from the nozzle to the substrate S. In the experiment, argon was used for both the carrier gas Q_a and the sheath gas Q_{sh}. The carrier gas flow rate Q_a was set to 140 sccm. The sheath flow rate Q_{sh} was varied from 90 sccm to 190 sccm in steps of 20 sccm. The distance from the nozzle to the substrate S was set to 0.5 mm. To simplify the modeling process, and to reduce the time required for a single simulation, the entire modeling process was separated into two stages. At the first stage, the gas velocity field $\vec{u}$ in the nozzle and beyond the outlet

Content from this work may be used under the terms of the Creative Commons Attribution 3.0 licence. Any further distribution of this work must maintain attribution to the author(s) and the title of the work, journal citation and DOI.

Published under licence by IOP Publishing Ltd

was computed. The velocity field was computed on the assumption that the concentration of nanoparticles does not affect the general character of the gas flow. To find the velocity field in the nozzle and outside it by the finite element method, the following system of equations was solved:

Navier–Stokes momentum equation for compressible fluid [6], [7]:

$$\frac{\partial(\rho\vec{u})}{\partial t} + \mathrm{div}(\rho\vec{u}\otimes\vec{u}) + \vec{\nabla}p = \rho\vec{F} + \mathrm{div}\left(\mu\left[(\vec{\nabla}\otimes\vec{u}) + (\vec{\nabla}\otimes\vec{u})^{T} - \frac{2}{3}I\,\mathrm{div}\,\vec{u}\right]\right) \qquad (1)$$

Continuity equation and equation of state for perfect gas:

$$\frac{\partial\rho}{\partial t} + \mathrm{div}\rho\vec{u} = 0; \qquad p = \rho RT \qquad (2)$$

where ρ is the medium density, p is the pressure, $\vec{u}$ is the medium velocity, $\vec{F}$ is external force mass density, μ is dynamic viscosity, R is the universal gas constant, T is the temperature.

At the second stage of modeling, the nanoparticle trajectories in the stationary medium velocity field were determined. In both the simulation and the experiment, silver particles with a density ρ_p and a diameter d_p of 10.5 g/cm^3 and 45 nm, respectively, were used. In the developed model, nanoparticles were released in a combination chamber at a velocity equal to the velocity of the carrier gas. Trajectories of the nanoparticles in the nozzle and during their deposition on the substrate were determined. To compute the particle trajectories from the velocity field obtained earlier, the following forces acting on the particle were taken into account:

Drag force [8]:

$$\overrightarrow{F_{Drag}} = \frac{1}{\tau_p}m_P(\vec{u}-\vec{v}); \qquad \tau_P = \frac{\rho_P d_P^2}{18\mu} \qquad (3)$$

Saffman force [9]:

$$\overrightarrow{F_{Saff}} = -20.3d_p^2\overrightarrow{L_V}\left(\mu\rho\frac{|\vec{u}-\vec{v}|}{|\overrightarrow{L_V}|}\right)^{1/2}; \qquad \overrightarrow{L_V} = (\vec{u}-\vec{v})\times\mathrm{rot}(\vec{u}-\vec{v}) \qquad (4)$$

where m_P is the particle mass, $\vec{v}$ is the particle velocity, d_p is the particle diameter.

Thus, as a result of the second stage of modeling, the positions of the nanoparticles on the substrate are determined. The line interval in which 68.2% of the particles lie ($\pm\,\sigma$ from the average position) is a "contact spot" that determines the size of the structures formed. The dimensions of the contact spot were further compared to the width of the printed line, measured with an optical microscope in the real experiment.

3. Experimental

In order to verify the correctness of the model, we conducted a series of real experiments. We used the multi-spark discharge generator with a tube furnace to form silver nanoparticles [10]. The parameters of the m-SDG operation were set to generate nanoparticles with an average diameter of 45 nm. The carrier flow rate of argon Q_a was supplied to an internal axisymmetric channel. To the external conically converging axisymmetric channel, the sheath gas Q_{sh} was supplied. Then the primary collimation of the flow took place in the combination chamber, and the final focusing of the nanoparticle beam occurred in the nozzle with an outlet diameter D_n = 100 μm. A focused beam of particles was deposited on the moving glass substrates. The speed of the substrates v_s was 65 μm/s. Thus, the parameters of the experiment on the focusing and deposition of nanoparticles replicated the parameters of the simulation. The width of the printed lines W_p was measured, using digital microscope VHX-1000.

4. Results and discussions

Figure 1 (b) shows the computed and the experimental dependence of the printed line width W_p on the sheath flow rate Q_{sh}. Also presented are the computed trajectories of particle motion during the focusing and deposition process in the dry aerosol printing method (Figs 2a, b and c) and optical images of the lines formed from Ag nanoparticles deposited on the glass substrates (Figures 2d, e and f). Figures 1 (b) and 2 (a, b, d and e) show that, both in the model and in the experiment, with an increase in the sheath gas flow rate Q_{sh} from 90 to 150 sccm, the formed line width W_p decreases, indicating a qualitative agreement between the model and the experiment.

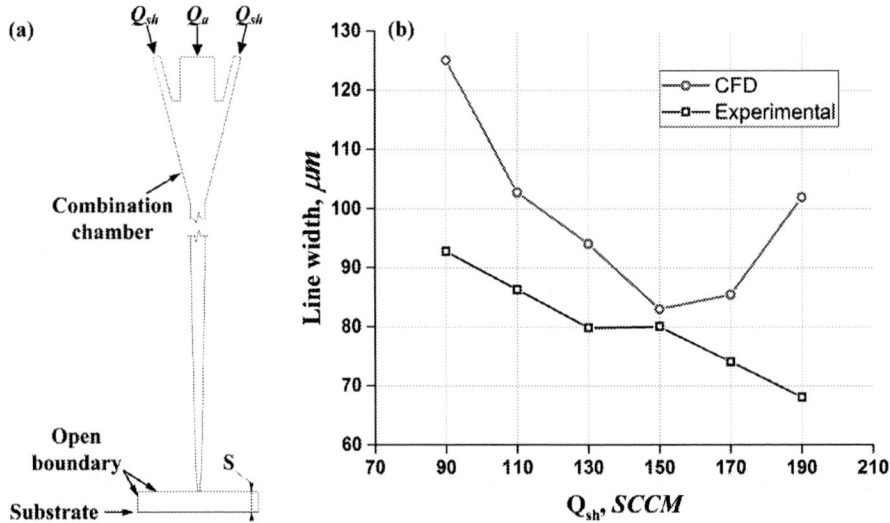

Figure 1(a, b). (a) The drawing of a part of the printing head and the nozzle, used for the simulation; **(b)** The dependence of the printed line width on the sheath flow rate Q_{sh}.

From Fig. 1 (b) and 2 (c and f) it is also seen that with the sheath gas flow rate $Q_{sh} > 150$ sccm, the results of the simulation and the real experiment differ qualitatively. This discrepancy, as well as the quantitative differences in the values of the line width of the simulation and the real experiments by several tens of percent, are probably due to the imperfections of the developed model, which does not take into account the three-dimensional nature of the nanoparticle focusing and deposition process. In particular, in the simulation, an increase in the printed line width W_p is observed with an increase in the sheath gas flow rate $Q_{sh} > 150$ sccm. At the same time, in the experiment, defocusing of the beam is not observed up to $Q_{sh} = 190$ sccm, and, consequently, the width of the formed W_p structures does not increase, but on the contrary decreases, with increasing the sheath flow rate Q_{sh}, see Fig. 1 (b) and Fig. 2 (d, e, and f). Thus, a satisfactory qualitative agreement between the results of the simulation and the experiment is established for the nanoparticle focusing and deposition process at the sheath flow rate Q_{sh} of up to 150 sccm. At the same time, for sheath flow rate $Q_{sh} > 150$ sccm, the applicability of the developed model is limited.

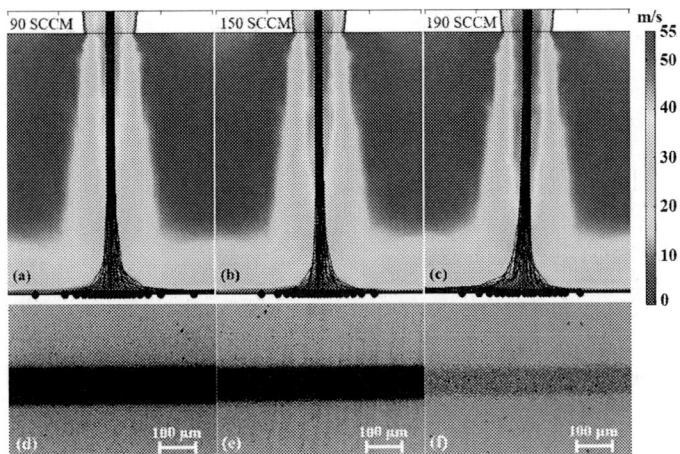

Figure 2(a) – (f). (a, b, c) Computed particle trajectories in the deposition process; **(d, e, f)** The images of the printed lines taken with a digital microscope.

5. Conclusion

A computer model has been developed of the dry aerosol printing process, which allows to determine the size of the microstructures formed from nanoparticles from the printing process parameters, specifically the sheath gas flow rate. Development of the model was based on specifying the gas passage geometry and solving the system of equations to determine the trajectories of Ag nanoparticles up to 45 nm in diameter in the process of focusing and deposition on the surface of the substrate via a coaxial nozzle with an outlet diameter of about 100 μm. From the comparison of the results of the model and the experiment, it is found that a qualitative agreement between the model and the experiment is present with the sheath flow rate of up to 150 sccm.

Acknowledgments

This work was supported by Ministry of Education and Science of the Russian Federation (project № RFMEFI57517X0160).

References

[1] Yuxiao He, Oakley C, Chahal P, Albrecht J and Papapolymerou J 2017 Aerosol Jet printed 24 GHz end-fire quasi-Yagi-Uda antenna on a 3-D printed cavity substrate (IEEE) pp 179–82

[2] Yang C, Zhou E, Miyanishi S, Hashimoto K and Tajima K 2011 Preparation of Active Layers in Polymer Solar Cells by Aerosol Jet Printing *ACS Appl. Mater. Interfaces* **3** 4053–8

[3] Han H-V, Lin H-Y, Lin C-C, Chong W-C, Li J-R, Chen K-J, Yu P, Chen T-M, Chen H-M, Lau K-M and Kuo H-C 2015 Resonant-enhanced full-color emission of quantum-dot-based micro LED display technology *Opt. Express* **23** 32504

[4] Efimov A, Potapov G, Nisan A, Urazov M and Ivanov V 2017 Application of the Spark Discharge Generator for Solvent-free Aerosol Jet Printing *Orient. J. Chem.* **33** 1047–50

[5] King, B. H. 2014 Miniature Aerosol Jet and Aerosol Jet Array U.S. Patent No. 8640975 B2.

[6] Sedov L I 1997 *Mechanics of continuous media* (Singapore: World Scientific)

[7] Landau L D and E. M. Lifshitz 2013 *Fluid Mechanics.* (Elsevier Science)

[8] Clift R, Grace J R and Weber M E 2005 *Bubbles, drops, and particles* (Mineola, N.Y.: Dover Publications)

[9] Saffman P G 1965 The lift on a small sphere in a slow shear flow *J. Fluid Mech.* **22** 385

[10] Lizunova A A, Efimov A A, Arsenov P V and Ivanov V V 2018 Influence of the sintering temperature on morphology and particle size of silver synthesized by spark discharge *IOP Conf. Ser. Mater. Sci. Eng.* **307** 012081

SPBOPEN 2018   IOP Publishing

Study of the possibility of using arachidic acid as a monocrystalline substrate for the formation of monolayers of aromatic hydrocarbons

O A Shinkarenko[1,2], A S Chumakov[1,2], M V Pozharov[3], A S Kolesnikova[2], A J K Al-Alwani[2], O Yu Tsvetkova[2], V P Sevostyanov[4], E G Glukhovskoy[1,2]

[1]Department of Nano- and Biomedical Technologies, Saratov State University, Saratov, 410012, Russia
[2]Education and Research Institute of Nanostructures and Biosystems, Saratov State University, Saratov, 410012, Russia
[3]Institute of Chemistry, Saratov State University, Saratov, 410012, Russia
[4]Scientific Research Institute of Technology of Organic, Inorganic Chemistry and Biotechnology, Saratov, 410005, Russia

Abstract. The paper discusses one of the promising approaches for the synthesis of graphene formed at the interface in the Langmuir bath. The results of the investigation of the formation of pyren monolayers (both initial elements of graphene) and a mixture of pyrene and arachidic acid are presented. It is shown how the mechanical characteristics of a monolayer change when pyrene is added to a mixture with arachidic acid: the rigidity of the monolayer increases and the compression modulus decreases. Also studied the hypothesis that arachidic acid can act as a single crystal substrate on which it is possible to form an ordered layer of pyrene.

1. Introduction

At the moment, among the most famous and studied materials is graphene, which has very attractive properties. It has an extremely high charge mobility, record conductivity, adjustable bandgap width, etc. The velocity of its electrons is about a hundred times higher than that of silicon.Already there are many directions where graphene could replace silicon. For example, a transistor based on it [1]. As its developers suggest, graphene taken as the basis, will allow working faster than its silicon analogs [2], which will significantly affect the performance of microprocessors and other devices. A large number of works are devoted to methods of obtaining graphene and they are very diverse. As the most promising, various modifications of the method of chemical vapor deposition from the gas phase [3,4], as well as methods based on the cross-linking of nanographene to obtain large-scale graphene sheets [5,6] can be distinguished.

But, despite the large number of publications in this field, there is no acceptable method for obtaining high-quality monolayers of a large area. In this regard, the development of new methods for the synthesis of graphene is interesting and relevant.

In this paper, we propose the following stage in the development of the method, which was described in [7-9]. Taking as a basis the reaction of Shole [10], a method of obtaining graphene by crosslinking molecules of polycyclic aromatic hydrocarbons (such as naphthalene, pyrene, coronene, etc.) is possible. The main factor in the implementation of this approach is the distance and orientation of the molecules that enter into the reaction.In the known methods for the production of graphene, solid

Content from this work may be used under the terms of the Creative Commons Attribution 3.0 licence. Any further distribution of this work must maintain attribution to the author(s) and the title of the work, journal citation and DOI.
Published under licence by IOP Publishing Ltd

supports are used for these purposes, which are both catalysts and surfaces that orient the nanographene molecules. But at the same time for the realization of such methods of production, high vacuum conditions are required, the use of plasma and etc. as a stimulating effect.

In this connection, the "water-air interface" was used as a "substrate" in the work. Such an approach can be realized in Langmuir baths using the Langmuir-Blodgett method. Therefore, in this paper we investigated the possibility and effect of various conditions on the formation of monolayers of aromatic hydrocarbons on the surface of the water subphase using Langmuir-Blodgett technology.

2. Materials and methods

As a starting compound for the further formation of graphene, pyrene belonging to the class of aromatic hydrocarbons was selected (fig. 1 a). It is chosen because of its large structural similarity with graphene and the availability of the required physico-chemical characteristics.

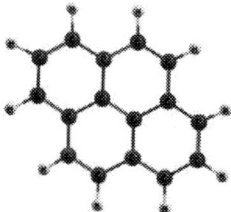 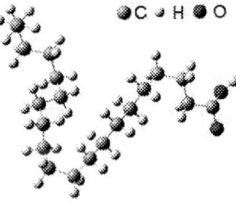

Figure 1(a, b). Molecules of **(a)** pyrene; **(b)** arachidic acid.

A study of a mixture of pyrene and arachidic acid was carried out by studying the hypothesis, where arachidic acid can act as a single-crystal substrate that promotes the formation of an ordered pyrene layer above it. The study of this mixture was carried out by analogy with [8], where the formation of naphthalene molecules over the surface of arachidic acid was studied (fig. 2).

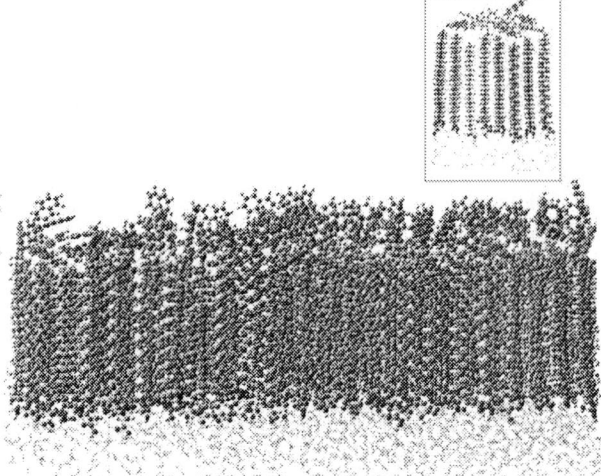

Figure 2. Ordered structure of arachidic acid layer with a layer of naphthalene molecules applied to its surface [8].

3. Results and discussion

To determine the possibility of formation of pyrene monolayers at the interface, solutions of arachic acid, pyrene, and mixtures of arachic acid and pyrene in various proportions were prepared. Then these solutions were spread on the surface of water and left for 5 minutes until the solvent completely evaporated. After that, the monolayer was compressed by two movable barriers, the compression rate was 10 mm/min. The temperature of the aqueous phase was 23 °C. The surface tension measured during compression was used to obtain compression isotherms (fig.3).

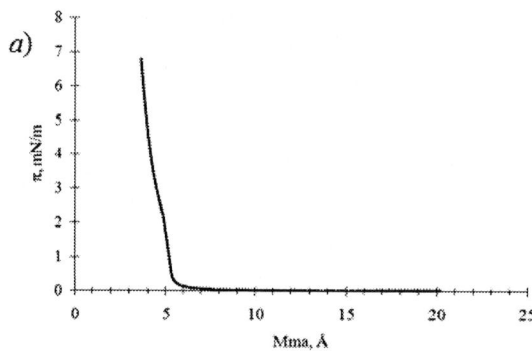

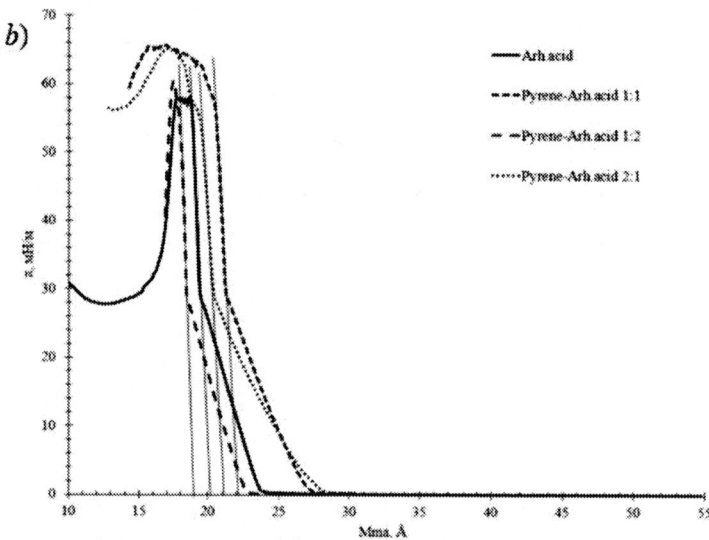

Figure 3(a, b). Compression isotherms of monolayers of **(a)** pyrene; **(b)** arachidic acid with pyrene.

Analyzing the isotreme for pyrene (fig. 3a), it can be seen that the surface pressure reaches a value about 7 mN/m, this is due to the fact that pyrene molecules do not have pronounced surface-active properties, like aromatic hydrocarbon molecules. Comparing the obtained isotherm with the isotherms of a mixture of pyrene and arachidic acid (fig. 3b), it can be seen that arachidic acid exerts a major influence on the change in the surface energy of the system and an increase in the surface pressure. It is worth noting that the areas of the monolayers compression of the mixture were recalculated to the number of molecules of only arachidic acid. The monolayer of the mixture, unlike a monolayer of pure pyrene, is able to form a close-packed solid-crystalline state on the surface of the water subphase. It can be seen that the isotherms for different mix ratios are similar to each other. The main difference consists in the displacement of the graphs and the features of some sections in the liquid-condensed and solid-crystalline state.

The isotherms were used to determine areas occupied by monolayers in «close package state» by drawing a tangent line to the point of inflexion (corresponding to pressure range of 28-55 mN/m). The crossing point between the tangent line and horizontal axis was designated as the target area.

If we assume that the change in the surface tension is due only to the presence of the amphiphilic component of the mixture (arachidic acid) on the water surface, and the presence of pyrene does not contribute to this change and is not integrated into the monolayer, then the isotherms should not shift. However, the opposite was observed experimentally: the area values for the ratio 0:1 (pure arachidic acid) 1:1, 1:2 and 2:1 were 21, 22.1, 19 and 21.2 Å, respectively. Thus, it can be assumed that some of the pyrene molecules are present in the monolayer of arachidic acid. This can also be evidenced by several transitions in the solid-crystalline state of the monolayer, which may be associated with the reorientation and incorporation of pyrene molecules in the composition of arachidic acid. This is most noticeable for the ratio 1:1.

In addition, to compare the obtained curves, the A/A_0 ratio for the solid-crystalline and liquid-condensed states was analyzed. This isotherm is shown in Fig. 4. The tangents on the sections of the liquid-condensed or solid-crystalline phases crossed the abscissa axis at one point.

The physical meaning of such a normalization is as follows. For a liquid-condensed phase, the onset of linear growth of surface pressure (or a decrease in surface energy) indicates the initiation of the interaction of surfactant molecules and a significant change in the orientation of the hydrocarbon chains of these molecules with respect to the interface. With a common starting point, it becomes possible to compare the rate of flow of such changes, or the relative compressibility of the monolayer in this state. For the solid-crystalline phase, the linear growth of the surface pressure is more closely related to the optimization of the mutual arrangement of the hydrocarbon chains of the surfactant molecules relative to each other without a significant change in the orientation of the hydrocarbon chains. In this case, it is also possible to compare the slope of the isotherms in this section or the relative compressibility of the monolayer in the solid-crystalline state.

For the presented data, the compressibility of the monolayer (K) and the compression modulus (χ) for each state of the mixtures was also calculated, the results of the calculations are presented in Table 1. These parameters were calculated by the following formulas:

$$K = -\frac{1}{A_0}\left(\frac{dA}{d\pi}\right)_{T=\text{const}}$$

and

$$\chi = \frac{1}{K}$$

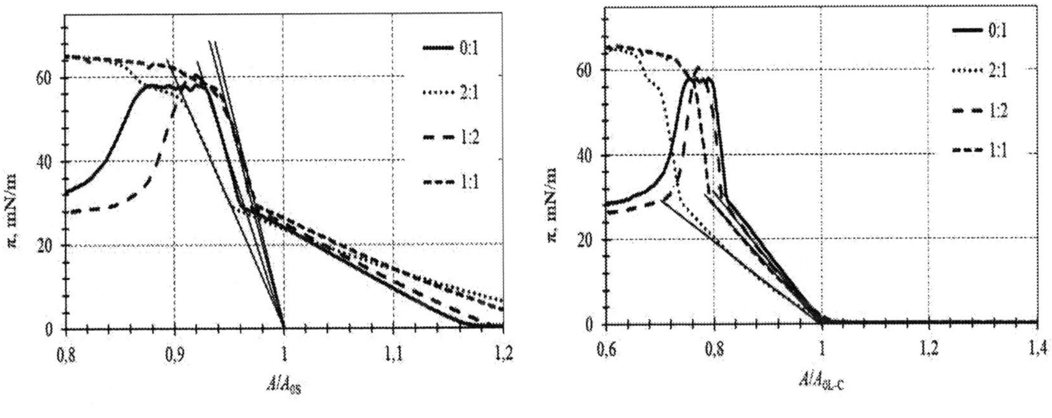

Figure 4. Compression isotherms of monolayers of arachidic acid with pyrene.

Table 1. Characteristics of monolayers of a mixture of arachidic acid and pyrene.

Ratio	Solid-crystalline states			Liquid-condensed states	
	K_S, m/mN	χ_S, mN/m	A_0, Å^2	K_{L-C}, m/mN	χ_{L-C}, mN/m
0:1	0,0009	1060	21	0,0062	160
1:2	0,0010	980	19	0,0064	157
1:1	0,0013	746	22,1	0,0079	127
2:1	0,0018	552	21,2	0,0100	100

Analyzing the mechanical characteristics of the monolayer (K and χ) for various ratios of a mixture of pyrene and arachidic acid, it can be noted that when it is added to arachidic acid, the mechanical properties of the monolayer change noticeably. In the solid-crystalline state, the rigidity of the monolayer increases, indicating an increase in compressibility from 0.009 to 0.0013 m/mN, with the compression modulus correspondingly decreasing. The rigidity of the monolayer also changes in the liquid-condensed state, which is confirmed by a change in the compressibility from 0.0062 to 0.0079 m/mN.

4. Conclusion

The study of mechanical characteristics suggests that the addition of polycyclic aromatic hydrocarbons to arachidic acid leads to an increase in the compressibility of the monolayer, and a decrease in the modulus of compression. Analysis of graphs allowed to conclude that some of the molecules of aromatic hydrocarbons are embedded in the space of hydrocarbon chains of arachidic acid and have a significant effect on the formation of the solid-crystalline state of the monolayer of the mixture. In addition, it can be assumed that some of the pyrene molecules go over the surface of the monolayer of arachidic acid.

Acknowledgments

This work was supported by grants from the Russian Foundation for Basic Research (projects №17-07-00407-a) and RFBR 17-32-50137.

References

[1] Rummeli M H, Bachmatiuk A 2010 *ACS Nano.* **4** 4206
[2] Krupka J, Strupinski W 2010 *Appl. Phys.Lett.* **96** 82
[3] Zhang J, Wang Z, Niu T 2014 *Sci Rep.* **4** 4425
[4] Zhang X, Li H, Ding F 2014 *Adv. Mater.* **26** 5479
[5] Narita A, Wang X W 2015 *Chem. Soc. Rev.* **44** 6616
[6] Feng C N, Hsieh Y C, Wu Y T 2015 *Chem. Rec.* **15** 253
[7] Kolesnikova A S, Safonov R A, Shinkarenko O A,Chumakov A S, Soldatenko E M, Glukhovskoy E G 2017 *The Journal of Surface Investigation. X-ray, Synchrotron and Neutron Techniques* **5** 33
[8] Shinkarenko O A, Safonov R A, Kolesnikova A S, Al-Alwani A J K, Pozharov M V, Glukhovskoy E G 2017 *Applied Surface Science* **424** 177
[9] Shinkarenko O A, Pozharov M V, Kolesnikova A S, ChumakovA S, Al-Alwani A J K, Tsevtkova O Yu, Glukhovskoy E G 2018 *Nanosystems: Physics, Chemistry, Mathematics* **9** 106
[10] Rempala P, Kroulı´k J 2004 *Am. Chem. Soc.* **126** 15002

SPBOPEN 2018

IOP Publishing

IOP Conf. Series: Journal of Physics: Conf. Series **1124** (2018) 081035   doi:10.1088/1742-6596/1124/8/081035

Masking coating formation by the focused ion beams method for plasma chemical treatment

I N Kots[1], V S Klimin[1,2], V V Polyakova[1], A A Rezvan[1], Z E Vakulov[1] and O A Ageev[1,2]

[1] Southern Federal University, Institute of Nanotechnologies, Electronics, and Electronic Equipment Engineering, Taganrog 347928, Russia

[2] Southern Federal University, Research and education center "Nanotechnologies", Taganrog, 347928, Russia

Abstract. The application of the focused ion beams method makes it possible to obtain modified surfaces serving as a masking layer for subsequent processing of semiconductor materials. The use of plasma chemical etching as a method for profiling modified surfaces makes it possible to obtain cleaner surfaces than for liquid etching. In this paper, experimental studies of a masking coating formation were carried out using a focused ion beams method on a scanning electron microscope, followed by a relief formation by the plasma chemical etching method. The effect of local surface treatment of silicon by the focused ion beams method on the etching rate of the treated regions by the method of plasma chemical etching was examined. Samples with the formed regions were obtained, which demonstrated masking and stimulating properties during plasma chemical etching. The structures height after plasma chemical etching varied from 10 to 266 nm.

1. Introduction

Currently, there is an active development of nanoelectronics due to the reduction in the size of functional elements, accompanied by a decrease in their energy consumption and an increase in performance. Optical lithography is the most common method for the micro- and nanoscale structures formation in connection with the simplicity of this method and its wide possibilities [1]. However, at present, this method reaches its technological limit in the possibility of obtaining nanoscale structures in the hard ultraviolet region. In this regard, exploration of new methods of nanoscale profiling, which in the future will reduce the resulting structures geometric parameters is an urgent task [2-5]. One of such methods is the methods of focused ion beams and plasma chemical etching [6-9]. The combination of these methods will solve the problems that arise when using liquid etching and obtaining masking coatings using optical lithography. The application of the focused ion beams method makes it possible to obtain modified surfaces serving as a masking layer for subsequent processing of semiconductor materials [10-21]. The use of plasma chemical etching as a method for profiling modified surfaces makes it possible to obtain purer surfaces, does not require processing after the etching process, has high resolution, has a minimal lateral undercut effect and has high productivity [22-27]. With this technique, there are two types of interaction. In the first case, the working gas ions enter into a chemical reaction with the surface being treated, forming volatile compounds at low temperatures, followed by the removal of the reaction products from the reaction chamber through a pumping system [28-33]. In the other case, the ions are accelerated by an external field and the atoms are knocked out from the surface of the sample being processed, and a reactive ion-etching regime is realized [34-38]. With the help of the focused ion beams method, it is possible to

---

Content from this work may be used under the terms of the Creative Commons Attribution 3.0 licence. Any further distribution of this work must maintain attribution to the author(s) and the title of the work, journal citation and DOI.

Published under licence by IOP Publishing Ltd

form modified regions with nanometric lateral dimensions, and by means of plasma chemical etching it is possible to form structures over modified regions with nanometer vertical dimensions. Based on the above study, a combination of these methods is an urgent task.

2. Description of the experiment and research methods

In this paper, we conducted experimental studies of the mask coating formation by the focused ion beams method on a scanning electron microscope (SEM) with an ion column Nova NanoLab 600 (FEI company, Netherlands), etching the modified regions by the method of plasma chemical treatment in a combined fluoride plasma of a capacitive and inductive discharge in a plasma chemical etching unit in an inductively coupled plasma STE ICP E68 (NTO, Russia), study of the obtained morphology on a scanning probe microscope (SPM) Ntegra Vita (NT-MDT, Russia).

For experimental studies, n-type silicon was used, with a resistivity of 0.001-0.005 ohm/cm, with an orientation of 100. The modification of the silicon substrates surface was carried out on 5x5 microns with minimal parameters of the ion beam diameter, accelerated ions energy 30 keV and with a different number of its passes, which varied from 10 up to 100.

In the next step, the samples were processed in a fluorine-containing plasma at different etching times. With constant processing parameters such as: the flux of fluorine-containing gas N_{SF6} is 15 cm^3/min, the N_{Ar} flow is 100 cm^3/min, the pressure in the chamber during the process was about 2 Pa and the fixed powers of the sources of inductively coupled and capacitive plasma.

3. Results and discussions

After modification of the silicon surface ions by gallium ions, the samples were etched in plasma. In Fig. 1 there is an image from a scanning electron microscope of the modified region after plasma chemical etching at 100 passes of ion beam exposure and 120 seconds of etching in plasma. The image shows that the modified region after plasma chemical etching has a more developed relief than the rest region. This can be explained by the modification of Ga$^+$ ions by the focused ion beam region. Ga in the fluorine-containing plasma is etched worse than silicon, so the part of the surface which contained the silicon was removed, and the part where gallium predominated remained. The samples obtained during the experiments were investigated by scanning electron microscopy and atomic force microscopy (Fig. 1).

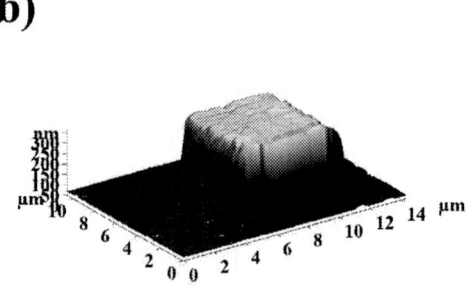

Figure 1 (a, b). a) SEM image, **b)** SPM image of the modified region at 100 passes of the ion beam and etching time in plasma 120 seconds

As a result, the rate of plasma chemical etching of regions modified by the method of focused ion beams turns out to be lower than the etching rate of the untreated region (Fig. 2).

SPBOPEN 2018   IOP Publishing
IOP Conf. Series: Journal of Physics: Conf. Series **1124** (2018) 081035   doi:10.1088/1742-6596/1124/8/081035

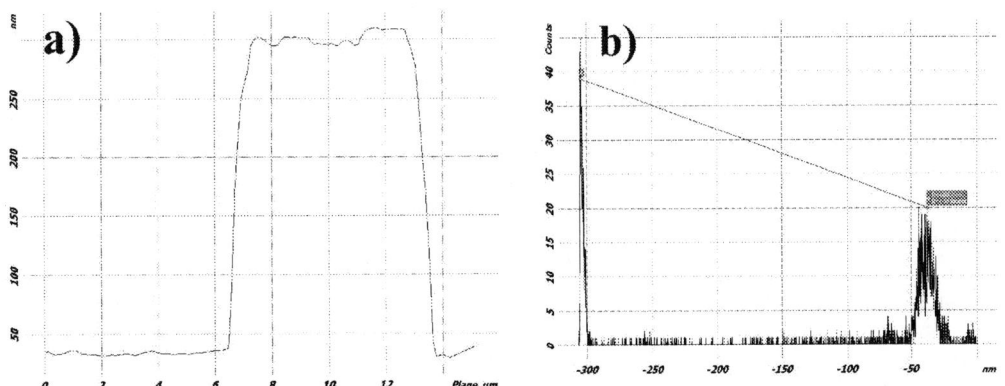

Figure 2 (a, b). a) a structure profile, **b)** a histogram from the image of an atomic force microscope
of a modified region after plasma chemical etching for 120 seconds

With an increase of the ion beam passes number from 10 to 100, at a plasma etching time of 120
seconds, the height of the obtained structures ranged from 7 to 266 nm. This is due to the greater
exposure to the sample surface and, as a consequence, to the different etching rates of the modified
regions. Also, when the plasma etching time is increased from 30 to 120 seconds with the same
number of ion beam passes, the structures height after plasma chemical etching increases from 10 to
266 nm. This is explained by the fact that, with a shorter etching time, the etching depth in the plasma
becomes smaller, due to which the difference in height between the modified and the rest of the area
decreases. Fig. 3 shows the dependence of the height of the obtained structures on the ion beam passes
number at different etching times in a plasma on a silicon substrate.

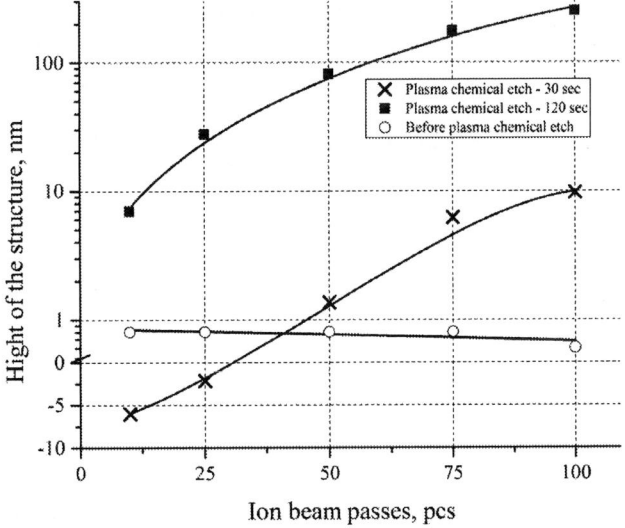

Figure 3. Dependence of height on the ion beam passes number

The graph shows that when etched for 120 seconds with an increase in the number of passes from 10
to 100, the structures height increases from 7 to 266 nm. (Figure 3). With plasma etching time for 30
seconds at 10 ion beam passes, stimulation of etching is observed, the modified regions are obtained
deeper by 6 nm than the rest of the substrate. This is due to the surface layer violation, which in the
process of plasma chemical etching is removed faster than the rest surface. However, at 100 ion beam
passes during plasma etching for 30 seconds, the modified regions exhibit a masking effect. The

1143

structures height was 10 nm. This is explained by the fact that the region of action is saturated with gallium ions, which in turn form a masking effect.

4. Conclusion

In accordance with the results obtained, it can be seen that when the region is modified by the focused ion beams method, gallium ions are introduced into the semiconductor structure. A layer with embedded gallium ions, in turn, is a masking layer for plasma chemical etching in a fluorine-containing plasma. With an increase of the ion beam passes number, the modified regions exhibit large masking properties. In the course of the work, the structure height as a passes number function and the etching time in the plasma were obtained. These dependences show that at 100 ion beam passes, with an increase in the etching time from 30 to 120 seconds, the height of the resulting structures increases from 10 to 266 nm. The number of passes also affects the height of the structures, but with a longer etching time in the plasma, the difference in height is greater. So with an etching time of 120 seconds, with an increase in the number of passes 10 to 100, the height of the structures increases from 10 to 266 nm. With a short etching time in the plasma, with a small ion beam passes number over the surface, etching is stimulated.

This technology can be used to form the structures and functional layers of quantum and optical nanoelectronics devices that require high resolution.

Acknowledgments

The study was carried out using the equipment of the Research and Education Center and the Centre of Collective Usage "Nanotechnologies" of Southern Federal University. This work was supported by the Southern Federal University (grant VnGr-07/2017-02).

References

[1] Tseng A. 2004 *J. Micromechanics Microengineering* **14** 15–34.
[2] Gusev E Y, Jityaeva J Y, Geldash A A, Ageev O A 2017 *J. Phys.: Conf. Ser.* **917** 032029
[3] Ageev O A, Klimin V S, Solodovnik M S, Eskov A V, Krasnoborodko S Y 2016 *J. Phys.: Conf. Ser.* **741** 012178
[4] Schmidt B, Oswald S, Bischoff L, 2005 *J. Electrochem.* Soc. **152** 875-879
[5] Klimin V S, Il'Ina M V, Il'In O I, Rudyk N N, Ageev O A 2017 *J. Phys.: Conf. Ser.* **917** 092023
[6] Riel B J, Hinzer K, Moisa S, Fraser J, Finnie P, Piercy P, Fafard S and Wasilewski Z R 2002 *J. Cryst. Growth* **236** 145
[7] Rudyk N N, Il'In O I, Il'Ina M V, Fedotov A A, Klimin V S, Ageev O A 2017 *J. Phys.: Conf. Ser.* **917** 082008
[8] Tominov R V, Zamburg E G, Khakhulin D, Klimin V S, Smirnov V A, Chu Y H, Ageev, O.A. 2017 *J. Phys.: Conf. Ser.* **917** 032023
[9] LaBella V P, Bullock D W, Ding Z, Emery C, Harter W G and Thibado P M 2000 *J. Vac. Sci. Tech.* A **18** 1526
[10] Tominov R V, Bespoludin V V, Klimin V S, Smirnov V A, Ageev O A 2017 *IOP Conf. Ser.: Mater. Sci. Eng.* **256** 012023
[11] Klimin V S, Solodovnik M S, Smirnov V A, Eskov A V, Tominov R V, Ageev O A 2016 *Proc. of SPIE* **10224** 102241Z-1
[12] Il'Ina M V, Blinov Y F, Il'In O I, Klimin V S, Ageev O A 2016 *Proc. of SPIE* **10224** 102240U-1
[13] Vakulov Z, Zamburg E, Golosov D A, Zavadskiy S M, Dostanko A P, Miakonkikh A V, Klemente I E, Rudenko K V, Ageev O A, 2017 *J. Phys. Conf. Ser.* **917** 032024
[14] Klimin V S, Tominov R V, Eskov A V, Krasnoborodko S Y, Ageev O A 2017 *J. Phys.: Conf. Ser.* **917** 092005
[15] Avilov V I, Ageev O A, Smirnov V A, Solodovnik M S, Tsukanova O G 2015 *Nanotech. in Russia* **10** (3-4) 214-219

[16] Ageev O A, Solodovnik M S, Balakirev S V, Eremenko M M 2016 *Phys. Solid State* **58** (5) 1045–1052

[17] Ageev O A, Kolomiytsev A S, Bykov A V, Smirnov V A, Kots I N 2015 *J. Microelectronics Reliability* **55** 2131–2134

[18] Wilhelmi O FEI Company application note 2007 URL: http://www.fei.com.

[19] Sievilä P, Chekurov N, Tittonen I 2010 *Nanotechnology* **21** 145301

[20] Vakulov Z E, Zamburg E G, Khakhulin D A, Ageev O A, 2017 *Mater. Sci. Semicond. Process.* **66** 21

[21] Yongqi Fu, Ngoi Kok Ann Bryan, Ong Nan Shing, Hung Nguyen Phu 2000 *Sensors and Actuators* **79(03)** 230-234.

[22] Ageev O A, Solodovnik M S, Balakirev S V, Mikhaylin I A 2016 *Technical Physics* **61** (7) 971–977

[23] Giannuzzi L A, Stevie F A 2004 *New York: Springer* 357

[24] Ledentsov N N *et al.* 1996 *Sol. St. Electron.* **40** 785

[25] Gusev E Y, Ageev O A, Gamaleev V A, Mikhno A S, Mironenko O O, Pronin E A 2014 *Advanced Materials Research* **893** 539-542

[26] Orloff J, Utlaut M, Swanson L 2003 *Springer US.* 36-67

[27] Tapobrata S, Dinakar K 2012 *Pan Stanford Publishing Pte Ltd.* 140-143

[28] Korsunsky A M, Guénolé J, Salvati E, Sui T, Mousavi M, Prakash A, Bitzek E, 2016 *Mater. Lett.* **185** 47–49

[29] Wang Y-c, Xie D-g, Ning X-h, Shan Z-w, 2015 *Appl. Phys. Lett.* **106** (8) 081905.

[30] Lunt A J G, Korsunsky A M, 2015 *Surf. Coat. Technol.* **283** 373–388

[31] Salvati E, Brandt L R, Papadaki C, Zhang H, Mousavi S M, Wermeille D, Korsunsky A M, 2018 *Mater. Lett.* **213** 346–349

[32] Korsunsky A M, Salvati E, Lunt A G J, Sui T, Mughal M Z, Daniel R, Keckes J, Bemporad E, Sebastiani M 2018 *J. Matdes* **145** 55–64

[33] Sabouri A, Anthony C J, Bowen J, Vishnyakov V, Prewett P D ,2014 *Microelectron. Eng.* **121** 24-26

[34] Ostadi H, Jiang K, Prewett P D 2009 *Microelectron. Eng.* **86** (4–6) 1021-1024

[35] Wang Y-c, Xie D-g, Ning X-h, Shan Z-w 2015 *Appl. Phys. Lett.* **106** 081905

[36] Taillon J A, Pellegrinelli C, Huang Y-L, Wachsman E D, Salamanca-Riba L G 2018 *Ultramicroscopy* **184** Part A 24-38

[37] Liu Y, Lehman V, Wang L. 2015 *Com. Net.* **83** 85-99

[38] Chekurov N, Grigoras K, Peltonen A, Franssila S, Tittonen I 2009 *Nanotechnology.* **20** 5

SPBOPEN 2018

IOP Publishing

Size effect on memristive properties of nanocrystalline ZnO film for resistive synaptic devices

N A Shandyba[1], I V Panchenko[1], R V Tominov[1], V A Smirnov[1], M I Pelipenko[2], E G Zamburg[3], Y H Chu[4,5,6,7]

[1]Department of Nanotechnology and Microsystems, Southern Federal University, Taganrog, 347928, Russia
[2]UAC company BERIEV, Taganrog, 347923, Russia
[3]Department of Electrical & Computer Engineering, National University of Singapore, 117582, Singapore
[4]Department of Materials Science and Engineering, National Chiao Tung University, Hsinchu, 1001, Taiwan
[5]Department of Electrophysics, National Chiao Tung University, Hsinchu, 1001, Taiwan
[6]Institute of Physics, Academia Sinica, Taipei, 11529, Taiwan
[7]Material and Chemical Research Laboratories, Industrial Technology Research Institute, Hsinchu, 31040, Taiwan

Abstract. Size effect on memristive properties of nanocrystalline ZnO film was investigated. It was shown, ZnO film thickness increase from 6.23±1.54 nm to 47.60±8.12 nm leads to high-resistance state (HRS) increase from 3.26±2.14 MΩ to 700.32±300.83 MΩ and low-resistance state (LRS) from 0.03±0.02 MΩ to 0.09±0.03 MΩ, respectively. The HRS/LRS ratio increases from 108 to 7742. The results can be useful for based on nanocrystalline ZnO films resistive synaptic devices manufacturing.

1. Introduction

A resistive switching effect in metal/oxide/metal structures is attractive for resistive synaptic devices (RSD) manufacturing [1-9]. RSD offer significant advantages over conventional computers, such as an effective processing of unstructured data, high-density storage, low voltage operation [10, 11]. RSD technology based on nanocrystalline thin oxide films has great prospects for synaptic computer systems manufacturing [12-20]. ZnO is the one of the promising oxides, which widely used in electronic element developments, sensors and microsystem technology [5, 13]. Also ZnO demonstrates effect of resistive switching and it is compatible with semiconductor technology. To fabricate ZnO based resistive synaptic elements, it is necessary to study resistive switching in ZnO films and today there are insufficiently experimental results about. Also it is important to get reliable information about morphology and thickness of ZnO films during size effect investigations. Atomic Force Microscopy (AFM) is one of the promising techniques for surface diagnostics and analysis. Reliability of AFM images defines by many factors, most of which are related to the shape and quality of the preparation of probes [21]. So, it is important to prepare probe tip, which allows getting reliable information about morphology and thickness of ZnO films.

Thus, investigation of resistive switching dependence on the thickness of nanocrystalline ZnO films is the goal of this work.

Content from this work may be used under the terms of the Creative Commons Attribution 3.0 licence. Any further distribution of this work must maintain attribution to the author(s) and the title of the work, journal citation and DOI.
Published under licence by IOP Publishing Ltd

2. Experiment details

To carry out experimental studies of the size effect on memristive properties of nanocrystalline ZnO film, five samples were prepared. Sapphire substrate Al_2O_3 with a crystallographic orientation (0001) as a wafer was used. As the bottom electrode titanium nitride (TiN) was used. TiN was deposited by pulsed laser deposition technique for the following regimes: temperature 600 °C, number of pulses: 10000, frequency: 10 Hz, argon pressure: 1 mTorr. TiN film thickness was about 30.3±5.1 nm. Nanocrystalline ZnO films were grown also by pulsed laser deposition technique. To provide electrical contact to the bottom TiN electrode, the ZnO films were deposited through a special mask pattern. So, on a template-protected surface area of the TiN films, the ZnO did not precipitate.

To investigate ZnO films morphology, advanced probe tip for AFM was fabricated using the Focused Ion Beam (FIB) local milling. The fabrication of probe tip was performed with a FEI Company DualBeam system Nova NanoLab 600, combining a Ga+ FIB and a field emission scanning electron microscope. At the first step, the probe was built by FIB-induced deposition of tungsten on a commercial Si cantilever NSG 10 with broken tip (resonant frequency: 287 kHz, force constant: 38 N/md) and then sharpening it using a focused ion beam milling. The following FIB parameters were used: the accelerating voltage of the ion beam – 28 keV; the ion beam current – 32 pA; and the dwell time of the ion beam — 0.9 µs. The chamber pressure after introducing $W(CO)6$ gas was 1×10^{-4} Pa. A bitmap of the desired probe structure was created by using Unigen 3.2 software, and then uploaded into the FIB software. Figure **1** shows secondary electron image (SEM) of the FIB-fabricated probe tip.

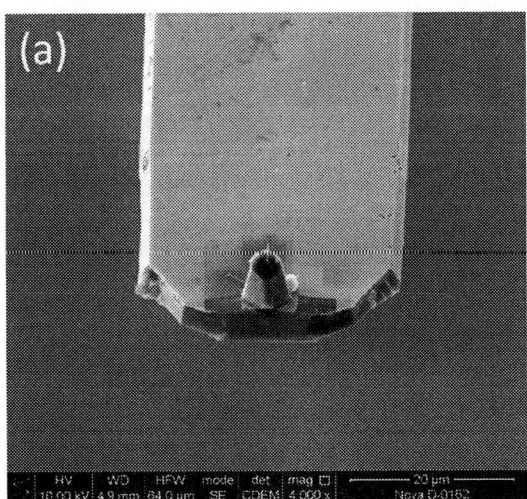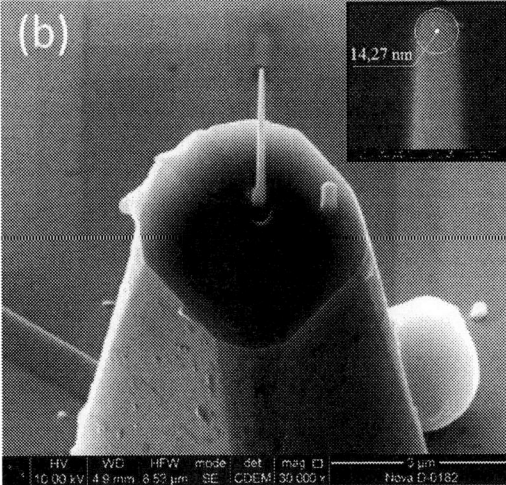

Figure 1(a, b). SEM of the FIB-fabricated AFM probe: **(a)** cantilever; **(b)** tip.

AFM images of surface of ZnO films were obtained using Probe Nanolaboratory Ntegra (NT-MDT, Russia) (figure **2**). Electrical measurements were carried out using semiconductor characterization system Keithley 4200-SCS (Keithley, USA) with W probes. During experiment, TiN film was grounded. Current-voltage (IV) characteristics of the bottom electrodes of all the samples were obtained to confirm Ohmic behavior of TiN.

Initially, ZnO films of all five samples exhibited dielectric properties. So, current-voltage characteristics measurement took place in two steps. At the first step, the electroforming of the samples was carried out at the point after which the films began to exhibit a memristor effect at this point.

The electroforming voltage of the sample was determined as follows: after the probe was landed to the surface of the ZnO film, a continuous symmetrical linear voltage signal with an initial amplitude of 1V was fed to it at a point. At the same time, the values of the currents flowing through the film were investigated. If an electrical breakdown of the film occurred, it was assumed that the given amplitude of the voltage is the electroforming voltage. If the electrical breakdown did not occur, the amplitude of the sweep increased by 1 V, and the process was repeated anew until the electroforming voltage was reached. It should be noted, that at voltages above the electroforming voltage, irreversible breakdown of the film occurred, and at voltages below the electroforming voltage, the film remains an insulator. The results of the electroforming voltage investigations of all five samples are shown in Table 1.

At the second step, the current-voltage characteristics were measured at the point (figure 3 a). For each sample, the average statistical IV characteristics at different points on the surface of the ZnO films for different number of electroforming cycles (0, 1, 2, 4, 6, 8, 10, 12) were received. IV curves were obtained depending on the thickness of ZnO films at the range from -1 to 1 V and from –5 to +5 V voltage sweep. 0.5 V was used as the read voltage. Curves analysis was implemented using Origin 8.1 software.

Table 1. Electroforming voltages of nanocrystalline ZnO films.

ZnO film thickness (nm)	6±1	20±3	40±6	45±8	47±8
Electroforming voltage (V)	3	9	14	15	16

3. Results

The analysis of SEM image (fig. **1**) shows, that a probe with a tip radius of about 14.27 nm, cone angle 1° and aspect ratio 1:30 was obtained after fabrication. Figure **2** shows experimental investigation of ZnO film morphology. It is shown that ZnO film surface has a granular structure (figure **2** a) with 0.7±0.3 μm^2 grain size and 3.2±1.1 nm grain height (figure **2** b). The ZnO film thicknesses were investigated using AFM by scanning bottom TiN/ZnO film boundary, and were in range from 6±1 to 47±8 nm (table **1**).

The number of electroforming cycles was chosen such that the maximum HRS/LRS ratio for the given sample was observed. Based on the results obtained, the dependences of the HRS and LRS on the thickness of the ZnO films (figure **3** b) were built.

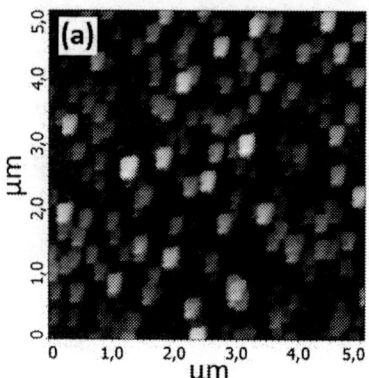 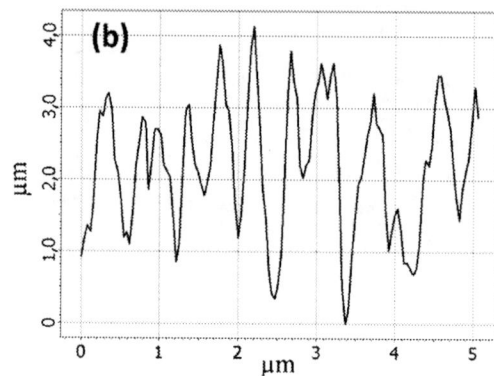

Figure 2(a, b). Investigation of nanocrystalline ZnO film surface: **(a)** AFM image of nanocrystalline ZnO film and; **(b)** average profilogram of (a).

It was shown, ZnO film thickness increase from 6.23±1.54 nm to 47.60±8.12 nm leads to high-resistance state increase from 3.26±2.14 MΩ to 700.32±300.83 MΩ and low-resistance state from 0.03±0.02 MΩ to 0.09±0.03 MΩ, respectively. The HRS/LRS ratio increases from 108 to 7742. It can be explained by increasing of the oxygen nanofilament dissolution length on metal/ZnO interface due to increasing of ZnO film thickness.

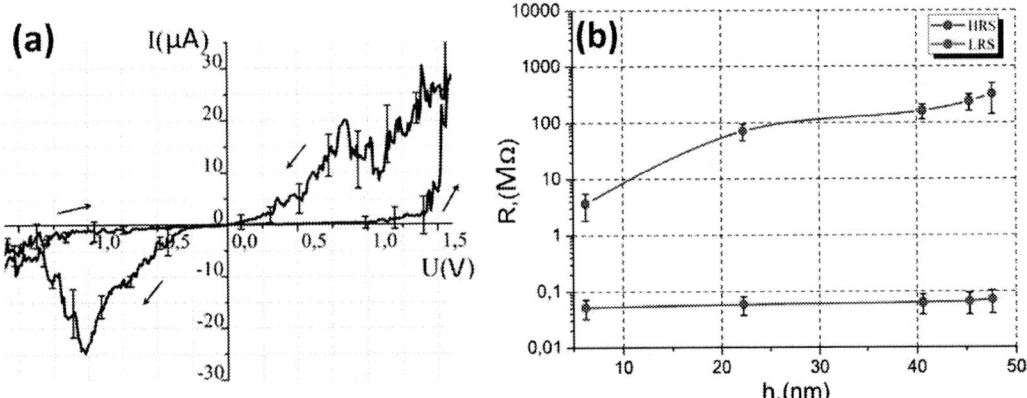

Figure 3(a, b). Size effect on memristive properties of nanocrystalline ZnO film investigation: **(a)** Average IV curve of TiN/ZnO/W structure; **(b)** HRS and LRS dependence from ZnO film thickness.

The obtained results can be used for development of technological processes of nanocrystalline ZnO film based resistive synaptic devices.

Acknowledgements
The results were obtained using the equipment of the Research and Education Center and Center for Collective Use "Nanotechnologies" of Southern Federal University.

References

[1] Dongale T D, Desai N D, Khot K V, Volos C K, Bhosale P N, Kamat R K 2018 *Journal of Nanoelectronics and Optoelectronics* **13** 68
[2] Dongale T D, Desai N D, Khot K V, Mullani N B, Pawar, P S, Tikke R S, Patil P S 2017 *Journal of Solid State Electrochemistry* **21** 2753
[3] Srivastava S, Thomas J P, Heinig N F, Leung K T 2017 *ACS applied materials & interfaces* **9** 36989
[4] Khanal G M, Acciarito S, Cardarilli G C, Chakraborty A, Nunzio L D, Fazzolari R, Susi G 2017 Electronics Letters **53** 296
[5] Khakhulin D A, Vakulov Z E, Smirnov V A, Tominov R V, Yoon J G, Ageev O A 2017 Journal *of Physics: Conference Series* **917** 092008
[6] Avilov V I, Ageev O A, Kolomiitsev A S, Konoplev B G, Smirnov V A 2014 *Semiconductors* **48** 1757
[7] Ageev O A, Alyab'eva N I, Konoplev B G, Polyakov V V, Smirnov V A 2010 Semiconductors **44** 1703
[8] Avilov V I, Ageev O A, Jityaev I L, Kolomiytsev A S, Smirnov V A 2016 Proceedings of SPIE **10224** 102240T
[9] Perkins J D, Cueto J A, Alleman J L, Warmsingh C, Keyes B M, Gedvilas L M, Parilla P A, To B, Readey D W, Ginley D S 2002 *J.Thin Solid Films* **411** 152
[10] Ohkubo I, Christen H M, Khalifah P, Sathyamurthy S, Zhai H Y, Rouleau C M, Mandrus D G, Lowndes D H 2004 *J. Appl. Surf. Sci* **223** 35
[11] Au K, Gao X S, Wang J, Bao Z Y, Liu J M, Dai J Y 2016. *J. Appl. Phys.* **114** 027019
[12] Liu Q, Long S, Wang W, Zuo Q, Zhang S, Chen J, and Liu M 2009 *J.Elect. dev. lett.* **30** 12

[13] Tominov R V, Zamburg E G, Khakhulin D A, Klimin V S, Smirnov V A, Chu Y H, Ageev O A 2017 *Journal of Physics: Conference Series* **917** 032023

[14] Ageev O A, Gusev E Yu, Jityaeva J Y, Ilina M V, Bykov A V 2016 *J. Phys.: Conf. Ser.* **741** 012001

[15] Klimin V S, Tominov R V, Eskov A V, Krasnoborodko S Y, Ageev O A 2017 *J. Phys.: Conf. Ser.* **917** 092005

[16] Tominov R V, Bespoludin V V, Klimin V S, Smirnov V A, Ageev O A 2017 *IOP Conf. Ser.: Mater. Sci. Eng.* **256** 012023

[17] Klimin V S, Solodovnik M S, Smirnov V A, Eskov A V, Tominov R V, Ageev O A 2016 *Proc. of SPIE* **10224** 102241Z

[18] Gusev E Y, Ageev O A, Gamaleev V A, Mikhno A S, Mironenko O O, Pronin E A 2014 *Advanced Materials Research* **893** 539

[19] Rudyk N N, Il'In O I, Il'Ina M V, Fedotov A A, Klimin V S, Ageev O A 2017 *J. Phys.: Conf. Ser.* **917** 082008

[20] Il'Ina M V, Blinov Y F, Il'In O I, Klimin V S, Ageev O A 2016 *Proc. of SPIE* **10224** 102240U-1

[21] Ageev O A, Kolomiytsev A S, Bykov A V, Smirnov V A, Kots I N 2015 *Microelectronics Reliability* **55** 2131

Computer simulation of femtosecond pulsed laser ablation of aluminium and copper

R V Davydov[1], V I Antonov[1]

[1]Department of Applied Mathematics and Mechanics, Peter the Great Saint Petersburg Polytechnic University, Saint Petersburg 195251, Russia

Abstract. In this paper a mathematical model for laser ablation of metals by femtosecond laser pulses is developed, based on two-temperature model with hydrodynamic equations and wide-range equation of state for metals. The simulation results for aluminium and copper are compared with experimental data for ablation depth at different laser fluence and duration of pulses. A good agreement for both metals with the results of simulation and the experiment is received

1. Introduction

Femtosecond laser material processing has been demonstrated now as an effective way for surface structural modification [1] nano/micro machining [2] and material removal (ablation) [3] of solid materials because of its minimal heat affected zone, good reproducibility, and less debris contamination. The ultra-short pulse ends before the expanding of plasma take place, leading to increased efficiency as the laser energy is absorbed at the interface of the material rather than in the generated plasma. Another advantage of femtosecond laser ablation is that the use of femtosecond laser pulses enables the deposition of energy and removal of a thin layer in time scales much faster than required for transferring the energy into the main part of the material. For metals, it happens because the finite heat conduction time, for dielectrics the removal occurs on a time scale short compared to the electron-phonon coupling [4]. Based on this, femtosecond laser ablation proceed with low collateral damage to the surrounding material, in contrast to nanosecond or picosecond laser ablation [5].

Despite a lot of research work, it is still difficult to make an accurate prediction of ablation processes, because they are significantly depending on laser parameters, target material characteristics and the surrounding medium [6-10]. For computer simulation of laser ablation, there are two main approaches are used now - molecular dynamic models and continuous thermodynamic models. In the molecular dynamic models trajectories (positions and velocities) are found by numerical solution of the equations of motion for all atoms in the system with the use of chosen interatomic interaction potential that defines the equilibrium structure and thermodynamic properties of the material [11]. The main advantage of this technique is that no further assumptions on the processes are required for it, but simulation requires a lot of computational resources, which significantly limits the size of the study area or time frame of the processes under research [12]. In thermodynamic approach a two-temperature model is widely employed for the simulation of ultra-short laser processing [13]. This continuous model describes the energy transfer inside a metal with two coupled generalized heat conduction equations for the temperatures of the electrons and the lattice. To describe the material removal processes this model can be used together with a system of hydrodynamic equations. However, when interaction between material and laser radiation is studied, an appropriate description of the thermodynamic properties of matter is

Content from this work may be used under the terms of the Creative Commons Attribution 3.0 licence. Any further distribution of this work must maintain attribution to the author(s) and the title of the work, journal citation and DOI.

Published under licence by IOP Publishing Ltd

required over a broad region of states including plasma at high pressures and temperatures and normal conditions. So the choice of an equation of state, required for solving the hydrodynamic equations, can significantly affect the results [14].

Therefore, there is still a need to develop a mathematical model with which computer simulation of laser ablation can be conducted at various parameters enough quickly and accurately. In our work we propose a mathematical model which is based on two-temperature model with developed wide-range equation of state for receiving better accuracy.

2. Mathematical model

In this work we describe the evolution of material parameters using the conservation of mass, momentum and energy of electron and ion subsystems in a two-temperature form:

$$\frac{\partial}{\partial t}\left(\frac{1}{\rho}\right) + \frac{\partial v}{\partial m} = 0 \tag{1}$$

$$\frac{\partial v}{\partial t} + \frac{\partial P}{\partial m} = 0, v = \frac{\partial x}{\partial t} \tag{2}$$

$$\frac{\partial \varepsilon_i}{\partial t} + P_i \frac{\partial v}{\partial m} = \frac{\alpha_{ei}}{\rho}(T_e - T_i) \tag{3}$$

$$\frac{\partial \varepsilon_e}{\partial t} + P_e \frac{\partial v}{\partial m} = \frac{\partial}{\partial m}\left(k\rho \frac{\partial T_e}{\partial m}\right) - \frac{\alpha_{ei}}{\rho}(T_e - T_i) + J_L \tag{4}$$

where m is the mass coordinate, $dm = pdx, m = \int_{x_0}^{x} pdx$, direction of x-axis is chosen perpendicular to the irradiated surface of the metal, v is the velocity, t is the time, ρ is the density, ρ_0 is the initial density, P_e and P_i are the pressures of electrons and ions, T_e and T_i are the temperatures of electrons and ions, ε_e and ε_i are the internal energies of electrons and ions, $P = P_e + P_i$ and $\varepsilon = \varepsilon_e + \varepsilon_i$ are the full pressure and the internal energy, α_{ei} - coefficient of electron-ion relaxation [13], J_L - energy of the absorbed laser radiation, described by formula:

$$J_L = \frac{F_{abs}}{\tau_L \delta \sqrt{\pi}\rho} \exp\left(-\frac{t^2}{\tau_L^2}\right) \exp\left(-\frac{x(m,t)-x(m_0,t)}{\delta}\right) \tag{5}$$

where F_{abs} is the laser radiation energy, τ_L is the laser pulse duration, δ is the skin depth of metal.

To solve the system of hydrodynamic equations in the two-temperature model we use wide-range semi-empirical two-temperature equation of state [15]. The solution of the system of equations is carried out by interation method with splitting into physical processes, we consider separately the hydrodynamic motion of matter upon absorption of laser radiation, electronic thermal conductivity, and energy exchange between electrons and ions. At the first step we take into account only the hydrodynamics and the absorption of the laser radiation by the substance. For approximation we use fully conservative finite-difference scheme, which was described by Samarskii [16] and modified for two-temperature situation. Oscillations arising in the calculation of solutions with discontinuities are corrected by introducing an artificial viscosity. At the second step, the process of heat transfer by electrons is considered and the one-dimensional nonlinear heat conduction equation is solved. And at the third step we solve system of ordinary differential equations for electron-ion exchange.

3. Results and discussion

To verify the proposed mathematical model, the ablation depth for metals (copper and aluminum) is calculated and compared with the experimental data for an average depth of machined grooves after using laser pulses with 170 fs duration. As we can see in Figure 1 and Figure 2 numerical results using our model are close to experimental data for both metals.

SPBOPEN 2018   IOP Publishing

IOP Conf. Series: Journal of Physics: Conf. Series **1124** (2018) 081037   doi:10.1088/1742-6596/1124/8/081037

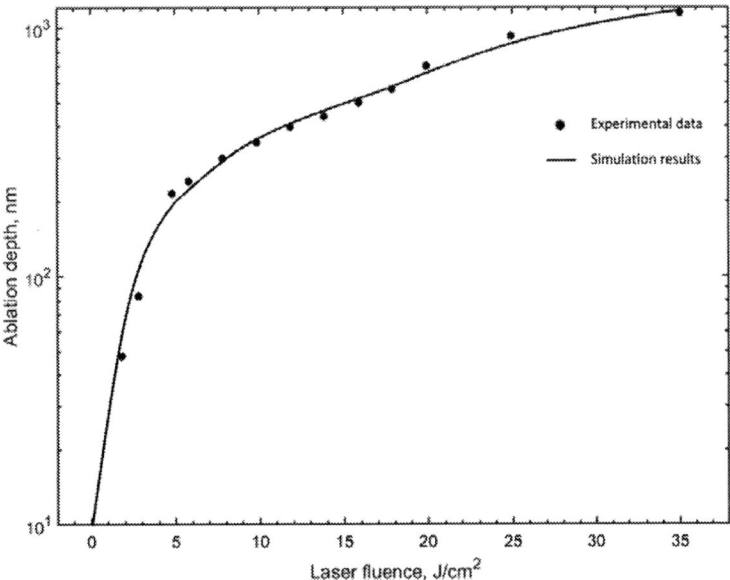

Figure 1. Experimental and numerical ablation depth as a function of the laser fluence on an aluminum target (170 fs laser pulse duration).

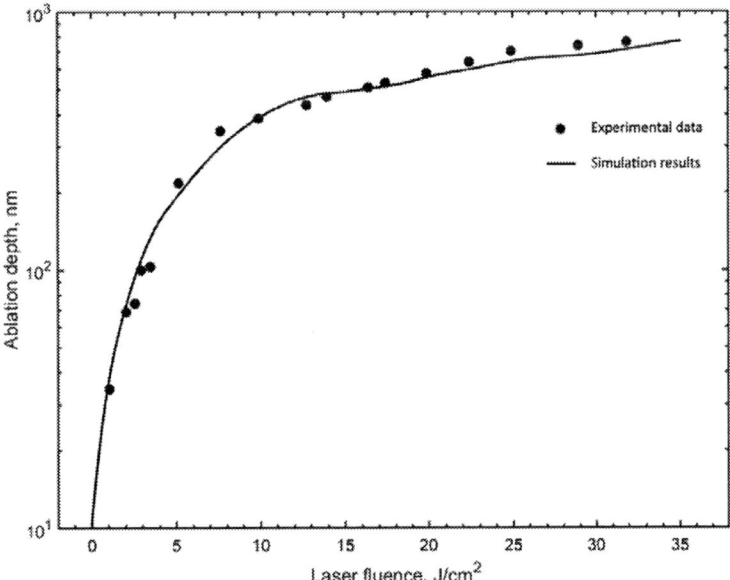

Figure 2. Experimental and numerical ablation depth as a function of the laser fluence on a copper target (170 fs laser pulse duration).

For copper we also test our model on experimental data for a wide range of laser fluence (up to 1037 J/cm^2), which are received for single laser shots and machined grooves by multiple laser pulses with 100 fs duration (Figure 3).

1153

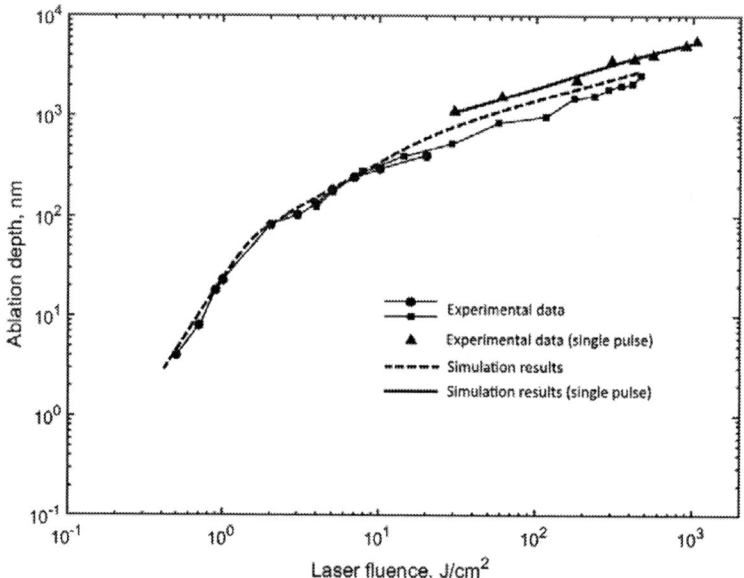

Figure 3. Experimental and numerical ablation depth as a function of the laser fluence on an copper target (100 fs laser pulse).

For single pulse numerical results well matched with experimental data, but we also receive a good agreement for other experimental data. Slight differences may be due to that the experimental ablation depth per pulse for them is estimated by dividing the measured groove depth by the number of shots. Multi-pulse irradiation of the same spot could lead to a change in the surface optical properties due to increasing roughness. Moreover, the grooves are deep and inclined groove walls could significantly change the laser beam reflection. Further investigations of these problems are suggested and the results may be implemented in this model in future.

4. Conclusion
In this paper a mathematical model for laser ablation of metals by femtosecond laser pulses is developed. A new wide-range equation of state for metals is used to increase the accuracy of calculations. The simulation results for aluminium and copper are compared with experimental data for ablation depth at different laser fluence and duration of pulses. The numerical results for both metals are in a good agreement with the experiment.

References
[1] Vorobyev A Y, Chunlei G 2015 *Journal of Applied Physics* **117(3)** 033103
[2] Chu W S, Kim C S, Lee H T, Choi J O, Park J I, Song J H, Jang K H, Ahn S H 2014 *International Journal of Precision Engineering and Manufacturing-Green Technology* **1(1)** 75
[3] Zeng H, Du X W, Singh S C, Kulinich S A, Yang S, He J, Cai W 2012 *Advanced Functional Materials* **22(7)** 1333
[4] Balling P, Schou J 2013 *Reports on Progress in Physics* **76(3)** 036502
[5] Makarov G N 2013 *Physics-Uspekhi* 2013 **56(7)** 643
[6] Tan D, Sharafudeen K N, Yue Y, Qiu J 2016 *Progress in Materials Science* **76** 154
[7] Davydov V V, Kruzhalov S V, Grebenikova N M, Smirnov K J 2018 *Measurement Techniques* **61(4)** 365-372

[8] Grebenikova N M, Smirnov K J, Artemiev V V, Davydov V V, Kruzhalov S V 2018 *J. Phys.: Conf. Ser.* **1038(1)** 012089

[7] Myazin N S, Smirnov K J, Davydov V V, Logunov S E 2017 *J. Phys.: Conf. Ser.* **929(1)** 012064

[9] Smirnov K J, Medzakovskiy V I, Davydov V V, Vysoczky M G, Glagolev S F 2017 *J. Phys.: Conf. Ser.* **917(6)** 062019

[10] Ivanov D S, Lipp V P, Veiko V P, Yakovlev E, Rethfeld B 2014 *Applied Physics A* **117(4)** 2133

[11] Wu C, Zhigilei L V 2014 *Applied Physics A* **114(1)** 11

[12] Ren Y, Cheng C W, Chen J K, Zhang Y, Tzou D Y. 2013 *International Journal of Thermal Sciences* **70** 32

[13] Cheng C W, Wang S Y, Chang K P, Chen J K 2016 *Applied Surface Science* **361** 41

[14] Davydov R V, Antonov V I 2016 *J. Phys.: Conf. Ser.* **769(1)** 012060

[15] Davydov R V, Antonov V I 2017 *J. Phys.: Conf. Ser.* **929(1)** 012040

[16] Samarskii A A, Popov Yu P 1992 *Difference schemes for gas dynamics* (Moscow: Nauka)

SPBOPEN 2018 IOP Publishing

Nanometer-scale oxidation of Silicon surface by ICP plasma

I Clemente[1,2], AMiakonkikh[1,2], S Averkin[1], K Rudenko[1,2]
[1]Institute of Physics and Technology of RAS, Moscow 117218, Russia
[2]Moscow Institute of Physics and Technology, Dolgoprudny 147001, Russia

Abstract. An experimental study of growth kinetic of silicon oxide during the initial stage of plasma enhanced oxidation is performed. Measurements were done in situ by means of spectroscopic ellipsometry with time resolution of 1 s. It was shown that oxidation of silicon in plasma does not comply with Deal-Grove model of oxidation. Measurements of concentration of oxygen in plasma and ion flux were performed to define mechanisms of oxidation. Obtained results could be applied for the development of multistage plasma processes involved in formation of nanostructures in advanced processes like atomic layer etching (ALE) and area selective atomic layer deposition.

1. Introduction

Precise and controllable modification of surface by oxygen plasma is becoming crucial step in atomic layer etching of some materials [1]. On the other hand there are promising three-step (ABC-type) area selective ALD cycles with O_2 plasma step in which oxygen plasma should remove adsorbed precursor from non-growth area [2] and should not oxidise this surface. These tasks arouse problem of measurement of oxidation kinetics with sub0nanometer precision. The growth kinetics of oxidation in different plasmas was studied elsewhere [3] for thicknesses up to 20 nm.
Our work is focused on initial stages of oxidation which allows obtaining films with thickness 0.5-1 nm within time of 2-10 sec.

2. Experiment

All experiments were carried out on (100) silicon wafers (n-type, 10-20 Ohm·cm and electron density $n=3-5 \cdot 10^{14}$ cm^{-3}), which were dipped in 5% HF solution immediately before loading to the vacuum chamber of ICP tool.
Oxidation processes were performed in plasma etching/deposition tool which detailed description is given in [4]. Plasma was excited in the reactor by RF-power introduced to the chamber through quartz window (Fig. 1). The tool is equipped with ellipsometric windows, quartz OES window and Langmuir probe port. Chamber walls were conditioned and cleaned chemically before the experiment. Plasma parameters were as follows: ICP power was 600W, pressure – 18-58 mTorr, oxygen flow – 80-150 sccm. Sample was chucked to the cooled table (18-20°C) with flow of helium under wafer to improve thermal contact. Bias voltage was not applied during the experiment.
As an *in situ* measurement tool a spectral ellipsometer Woollam M-2000X was used (246-998 nm spectral range, 73.2° incidence angle). Data acquisition takes about 0.5 s with pause between measurements of 0.5 s. Data were proceeded after end of measurement procedure using an optical model for native oxide and dependency of silicon optics from temperature which allows simultaneous measurements of oxide thickness and wafer temperature. In all experiments silicon temperature didn't exceed 100°C.

Content from this work may be used under the terms of the Creative Commons Attribution 3.0 licence. Any further distribution of this work must maintain attribution to the author(s) and the title of the work, journal citation and DOI.
Published under licence by IOP Publishing Ltd

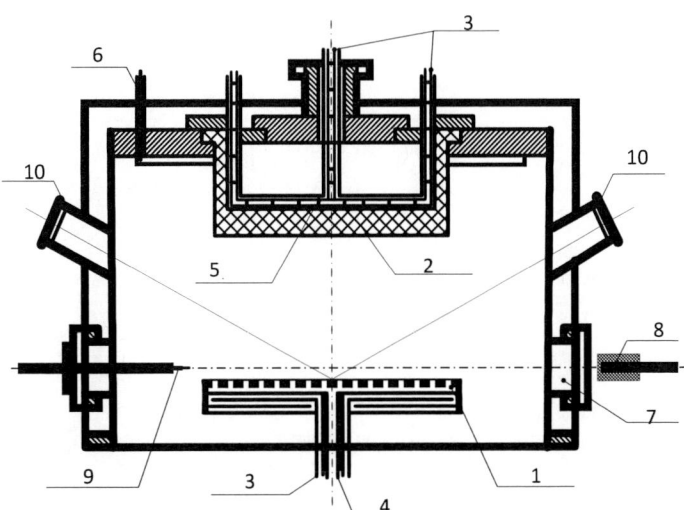

Figure 1. Plasma etching tool with ICP source for 6 inch wafers (1 – cooled chuck, 2- quartz window for RF input, 3- water cooling, 4 - He inlet, 5 - planar RF coil, 6 - gas inlet (gas shower), 7 - OES port, 8 - OES sensor, 9 - Langmuir probe, 10 – ellipsometric ports).

Concentration of oxygen ions were measured by means of Langmuir probe. Cylindrical probe with surface area of 9 mm^2 and diameter of 0.15 mm was used with ESPion system. Parameters of plasma were estimated by using OML theory. EEDF were calculated by double differentiating of IV-curve with preliminary smoothing by Savitsky-Golay method.

Concentration of oxygen radicals O* were measured by optical emission spectroscopy (OES). Spectrometer Quartz-4 with instrumental line width of 0.3 nm was used. Absolute values of concentration were measured by optical actinometry using intensity of 750.39 nm line of Ar and 777.74 nm line of O* [5]. Argon was added to the oxygen in proportion of 3% which do not lead to disturbance in plasma parameters.

3. Results

Silicon oxide thickness was measured after loading the sample and during plasma treatment. Immediately after sample loading oxide thickness measurement of oxide thickness gives value of 0.8 nm which corresponds to fast oxidation of HF dipped silicon surface in atmosphere during the sample transfer and adsorbed water from solution.

Observed growth of silicon oxide during plasma enhanced oxidation is shown on Fig. 2(a). It proves that rate is dependant on pressure and could not be described by Deal-Grove model. On the graph of oxidation rate vs film thickness (see Fig. 2(b)) one can see exponential dependence showing that diffusion of oxygen species through oxide is limiting stage. On the other hand it is obvious that the rate of oxidation is dependant also on plasma pressure. In turn the mechanism of oxidation could be related to effect of charged particles bombardment or oxygen radicals flow. Radical concentration, ion flow and ion energy are dependant on gas pressure. In range of pressures from 6 to 58 mTorr oxidation rate increases with decrease in pressure. Also, both graphs on Fig. 2 (a) and (b) show no saturation of oxidation rate with increasing in oxide thickness. But the lower pressure, the higher oxide thickness is achieved.

Results of measurements of 750.39 nm line of Ar and 777.74 nm line of O* for plasma with same parameters, except for 3% addition of Ar as actinometer, are shown on Fig 4. Ratio of spectral line intensities is equal to ratio of concentrations in case of constant electron temperatures. According to Langmuir probe measurements electron temperatures in pressure range of 2-8 mTorr is equal to 2.1-2.4 eV. Estimation of oxygen concentration vs pressure is given on Fig. 4(a). The oxygen radical concentration is increased with pressure almost linearly.

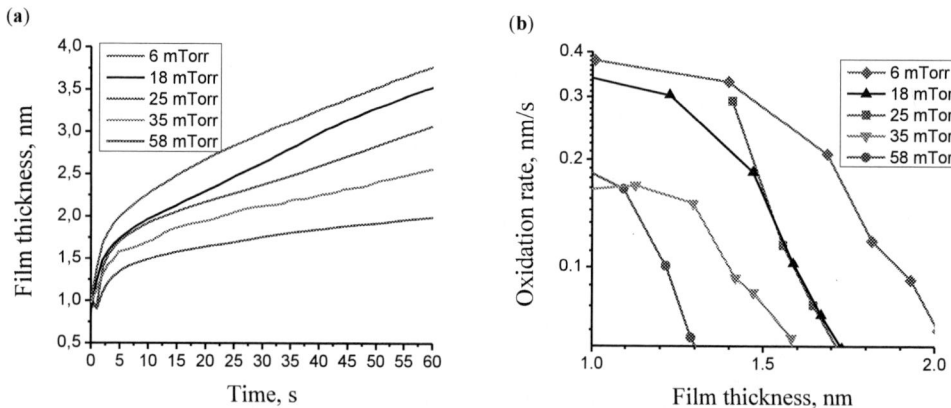

Figure 2(a, b). (a) Oxide growth depending of oxygen pressure in plasma; **(b)** dependence of oxidation rate from oxide thickness on silicon.

The behavior of oxygen ions of plasma and other plasma parameters were measured by Langmuir probe. Measurement shows negligible dependence of electron temperature and plasma potential on plasma gas pressure. That means that in that pressure range the energy distribution of particles bombardment of surface remains constant. Estimated EEDF of plasma is close to Maxwellian, that proves applicability of actinometry technique to that case.

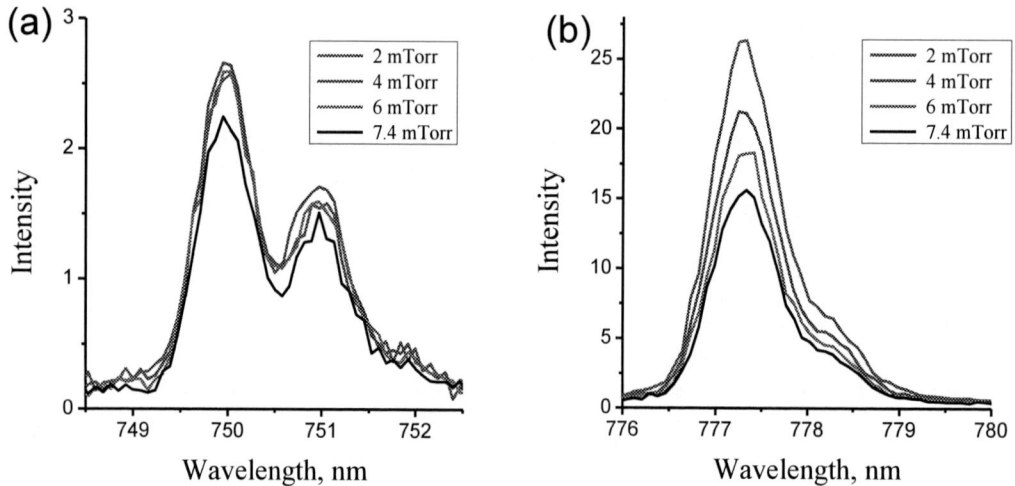

Figure 3(a, b). Emission spectrum of a O_2 plasma with addition of 3% Ar. **(a)** A 750.39 nm line of Ar;**(b)** A 777.74 nm line of O*.

The concentration of plasma (electron and positive ion concentrations) decrease with growth of pressure. Estimated flow of positive oxygen ions to the flat surface of sample vs pressure is shown on the Fig. 4(b). The graph has negative slope, which is typical for this pressure range. That allows to conclude that flow of oxygen ions from plasma, accelerated by the plasma potential (about 10 eV in that pressure range) is critical for plasma oxidation of silicon surface.

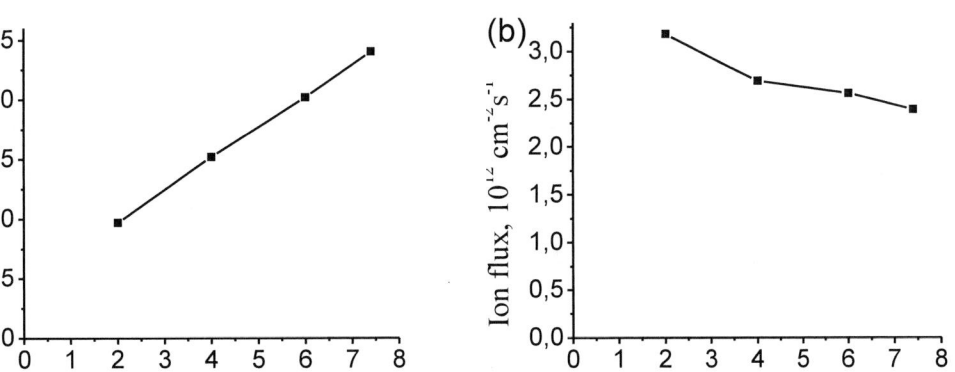

Figure 4(a, b). (a) Concentration of oxygen radicals O* by optical emission spectroscopy; **(b)** Oxygen ion flux calculated by Langmuir probe data.

4. Conclusion

By means of spectral ellipsometry rate of silicon oxidation in ICP plasma without bias was measured. It was shown that oxidation in plasma does not comply with Deal-Grove law of oxidation. Moreover oxide thickness up to 4 nm could be achieved at room temperature within 60 seconds.

The lower the pressure the more effective oxidation, and oxide thickness up to 4 nm could be obtained within 1 min time.

OES actinometry was applied to measure concentration of atomic oxygen radicals and Langmuir probe was used to estimate EEDF, electron temperature and concentration of charged particles, and flow of ions to the surface.

Obtained results could be applied for the development of multistage plasma processes involved in formation of nanostructures in advanced processes like atomic layer etching and area selective atomic layer deposition or other advanced techniques relying on controllable nanoscale surface modification.

Acknowledgments

The reported study was carried out under Program of FASO of Russia and was partially supported by RFBR, research project # 18-37-00354.

References

[1] Kanarik K J, Lill T, Hudson E A, Sriraman S, Tan S, Marks J, Vahedi V, and Gottscho R A 2015 Journal of Vacuum Science & Technology A **33** 020802

[2] Mameli A, Merkx M J M, Karasulu B, Roozeboom F, Kessels W M M, Mackus A J M 2017 ACS Nano **11** 9303

[3] Antonenko A K, Volodin V A, Efremov M D, Zazulya P S, Kamaev G N, Marin D V 2011 Optoelectronics, Instrumentation and Data Processing **47** 459

[4] Averkin S N, Valiev K A, Myakonkikh A V 2005 Tr. Fiz.-Tekhnol. Inst. Ross. Akad. Nauk **18** 121

[5] Rudenko K V, Myakonkikh A V, Orlikovsky A A 2007 Russ Microelectron **36** 179.

SPBOPEN 2018   IOP Publishing

Polarization photosensitivity of n-p-CdSiAs₂ photodiode

R V Davydov[1], V Yu Rud'[2,3,4], Yu V Rud'[4], E I Terukov[4]

[1]Department of Applied Mathematics and Mechanics, Peter the Great Saint Petersburg Polytechnic University, Saint Petersburg 195251, Russia

[2]Department of Advanced Manufacturing Technologies, Peter the Great Saint Petersburg Polytechnic University, Saint Petersburg 195251, Russia

[3]Department of Ecology, All-Russian Research Institute of Phytopathology, 143050, Moscow Region, Odintsovo district, B.Vyazyomy, Russia

[4]Department of Solid State Physics, Ioffe Physico-Technical Institute, Russian Academy of Sciences, St Petersburg 194021, Russia

Abstract. The work is devoted to the study of natural photopleochroism of n-p-CdSiAs₂ structures. In this paper the main results of polarization studies about photosensitivity of n-p-CdSiAs₂ based diode structures (electrochemical cells and fine structure), are presented. We also conclude that such structures can be used as polarimetric photodetectors for optical communications lines and photonics

1. Introduction

It is well known that intensity and polarization are basic characteristics of a light wave. A complete analysis of polarized radiation with the help of polarization intensitive photodetectors is possible only after the recorded radiation is converted with the help of an external polarization device, as a result of which the polarization characteristics of the radiation are calculated from the measured intensity. The use of polarization insensitive photodetectors for determining polarization parameters requires that the polarization element be spectrally matched with the photodetector, which gives rise to additional losses of radiation. A photodetector which does not require external polarization devices – a polarimetric photodetector – is obviously most suitable for such purposes.

The importance of such studies has increased in recent years because of the fact that the method developed for performing polarization studies of photoconductivity has proven to be very useful for studying the electronic spectrum of anisotropic semiconductors and the perfection of their structure and has opened up the possibility of creating such device with new functional possibilities. It turns out that in order to create polarization sensitive photodetectors it is no longer sufficient to establish the general characteristics of the polarization photosensitivity (PS): the effect of the nature of the positional ordering of the atoms, the behavior of specific impurities, and the characteristic parameters of semiconductors and instrumental structures must also be studied in detail.

The cadmium silicium diarsenide CdSiAs₂ (Figure 1) is the electronic analog of gallium arsenide. The paper presents a comprehensive study of the polarization photosensitivity straight forward uniaxial crystals CdSiAs₂ (E_G= 1.55 eV at 300 K). These single crystals are of interest from the point of view of creation of new optical systems in which the state of polarization of incident radiation becomes the main parameter (Figure 2). It is found that deviation of the composition of stoichiometric CdSiAs₂ toward an

Content from this work may be used under the terms of the Creative Commons Attribution 3.0 licence. Any further distribution of this work must maintain attribution to the author(s) and the title of the work, journal citation and DOI.
Published under licence by IOP Publishing Ltd

SPBOPEN 2018   IOP Publishing

IOP Conf. Series: Journal of Physics: Conf. Series **1124** (2018) 081039   doi:10.1088/1742-6596/1124/8/081039

increase in the concentration of As or Cd is accompanied by formation of additional radiative and non-radiative recombination centres, causing quenching of recombination radiation, and especially of exciton band. These single crystals can be used as the basis for sources of linearly polarized radiation and other functional devices that could not be produced with gallium arsenide crystals with isotropic crystal lattice.

Optical system is an optoelectronic product, which includes polarimetric photodiode that is designed to convert the input optical signals to electronic equipment in electrical. A typical block diagram of such a system (Figure 3) includes an optical radiation source, a modulator, a polarizer, a fiber optical line, polarimetric photodetector, an amplifier. As such, the device allows converting optical signals coming from optical linear path, electrical signals in crucial device, for example, electronic equipment reception station. Using the phenomenon of photopleochroism this allows detecting the change of the coefficient photopleochroism at the slightest change in the polarization of the incident radiation.

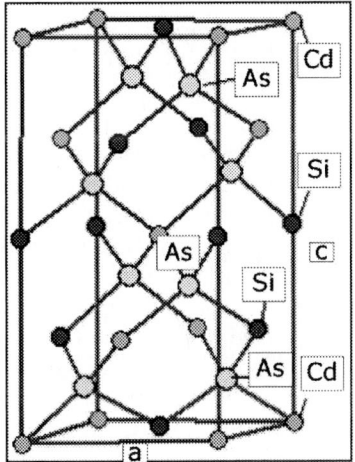

Figure 1. The crystal lattice of chalcopyrite for CdSiAs$_2$, where a-lattice constant, c is a tetragonal compression of the lattice.

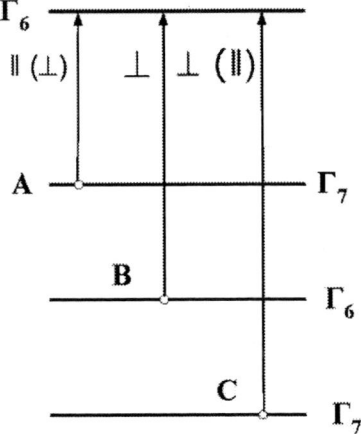

Figure 2. The diagram of the optical interband transitions with k=0 (where A-transitions 1.55eV (at 300K) and 1.85 eV (at 77K), B-transitions 1.74eV (at 300 K) and 1.85 eV (at 77K), and C-transitions 1.99 eV (at 300 K), 2.07 eV (at 77K).

1161

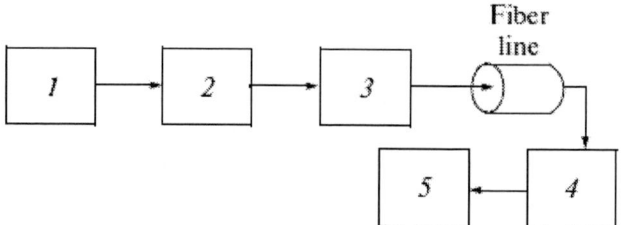

Figure 3. The block diagram of the optical system with polarimetric photodiode: (1) emitter, (2) modulator, (3) polarizer, (4) polarimetric photodiode, (5) amplifier.

2. The method

Electrochemical cells were prepared on (100) oriented uniform monocrystalline wafers of $CdSiAs_2$ doped by In. This method made it possible to reproducibly grow crystals with holes concentrations of 10^{11}-10^{23} cm^{-3} at room temperature. When compared with undoped crystals ($\rho = 10^{16}$ to 10^{17} cm^{-3}) in cell employing $CdSiAs_2$:In a higher photosensitivity could be reached as shown by the measurements. Following orientation, the plates were mechanically and chemically polished. An ohmic contact was made by chemical deposition of copper and a current lead soldered with In. The side bearing the ohmic contact was protected with a non-conducting varnish, then the plates were put into a cuvette containing an electrolyte. The measuring cell comprised a platinum counter electrode of large area facing the side of the semiconductor-electrolyte boundary on which the radiation was incident. Distilled water was usually taken as an electrolyte.

Under illumination a photo e.m.f. is generated in the H_2O-$CdSiAs_2$ structures. It is positive with respect to the electrolyte if a p-type photoelectrode is used. The sign of the photo-e.m.f. does not change with variation of the illumination intensity or incident photon energy, nor does it change from point to point on the structure. The open circuit photovoltage is logarithmic in intensity and it high intensities tends to saturate, which gives an estimate of the band bending value from 0,8 to 1,0 V for different electrodes. The maximum voltage photosensitivity of H_2O-$CdSiAs_2$:In reached 10 V/W at room temperature.

The diodes were produced by heat treatment of p-type plates with a free holes concentrarions 10^{16} cm^{-3} at room temperature. The heat treatment was carried out in evacuated quartz ampules. The method allows for conversion of the conductivity type to a specified depth. To be able to study the anisotropy of photosensitivity homojunctions were obtained on the (100) and (001) oriented plates of $CdSiAs_2$. The experimental apparatus for studying polarization photosensitivity induced falling along the normal to the receiving plane of $CdSiAs_2$ oriented plates [1].

3. Results and discussion

Under illumination, normal to the (100) surface of the photoelectrode, with linearly polarized light in the energy range of the A-transition (band-to-band) the photocurrent in found to depend on the orientation of the polarization plane of incident radiation relative to the tetragonal c-axis of the $CdSiAs_2$ crystal in electrochemical cells [2, 3]. This dependence follows to the Malus periodic law. The peak position on the polarization indicatrixs of the photocurrent exactly corresponds to the polarization **E II c**. In this respect such structures are identical to their solid state analogues. The photocurrent spectra of the H_2O-$CdSiAs_2$:In structures also give evidence of the anisotropy of the photoactive absorption. In the region of impurity absorption and interband A-transition of the photoelectrode material the photocurrent in the structure is larger at polarization **E II c** the long-wavelength exponential edge is split and the photocurrent difference at two polarizations is positive and has a peak at hw=E_A.

A typical spectral dependence of the coefficient of natural photopleochroism for such structure is given in Figure 4. As to the value and sign of natural photopleochroism, these structures are similar to their solid-state analogues. The most important feature of the photopleochroism spectra of such barrier is that the absolute maximum at hw=1.47 eV as well as the long –wavelength peak at hw=1.35 eV both fall in the range of impurity absorption of $CdSiAs_2$. Hence, doping of the photoelectrode material is always accompanied by an increased polarisational photosensitivity in the long-wavelength region whereas features of the photopleochroism spectra inside the fundamental absorption region experience nosubstantial changes compared with undoped crystals.

Also we find that in electrochemical cells based on indium doped $CdSiAs_2$ crystals a maximum azimuthal photosensitivity of $\approx 10^3$ V/WK may be reached near hw=1.55 eV at T=300K. Such structures could be used as photodetectors for LPL. [4-7].

The polarization indicatrices of a photocurrent suggest that in n-type diffusion layers the structural perfection of the original p-type plates as maintained. When illuminated along (001), the photocurrent is polarization – independent, such n-p transitions corresponding functionally to GaAs diodes. $CdSiAs_2$ diodes prepared on (001) plates exibit polarization indicatrices of photocurrent in agreement with the generalized Malius Iaw [8-10].

The natural photopleochroism spectra exhibit an extremely specific reaction as long as the exciton features are manifested in photosensitivity processes in the form of two narrow peaks in the photoconversion efficiency spectra, each predominant in a certain polarization. In the example of an n-p-$CdSiAs_2$ homodiode containing a thin film of n-type of conductivity, during illuminations from the side of the p-type conductivity substrate, a sharp oscillation of the photopleochroism arises on the photopleochroism spectrum near the energy position of the fine structure which determined from the photosensitivity spectrum before. Here the maximum of the positive photopleochroism lies near energy position belongs to the A interband transition, and its short-wavelength decay is due to the onset of B interband transitions, which are allowed predominantly in the **E II c** polarization. As it follows from Figure 5, the lowering of the temperatures causes a parallel displacement of the photopleochroism spectra and the energy position of photopleochroism oscillations in accordance with a value of dE_G/dT for $CdSiAs_2$ single crystals. We have found that the azimuthal light sensitivity of such diodes reaches 36 mA/W• deg, which indicates the prospects of their use as photoanalyzers of linearly polarized light.

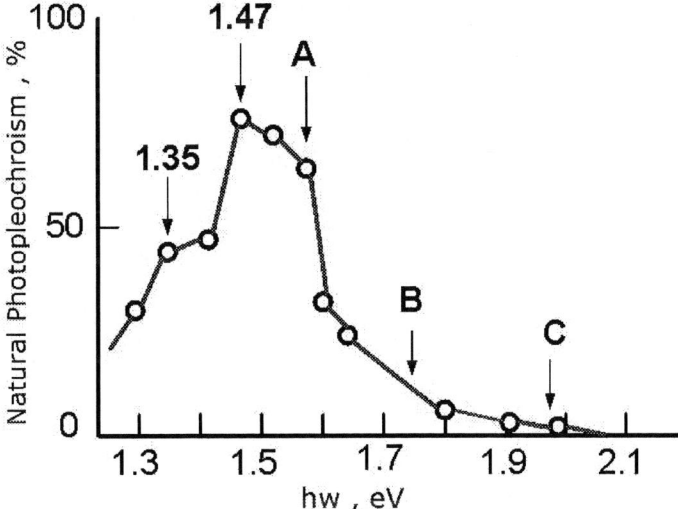

Figure 4. Spectral dependence of the coefficient of natural photopleochroism of a H_2O-$CdSiAs_2$:In photodiode at 300 K.

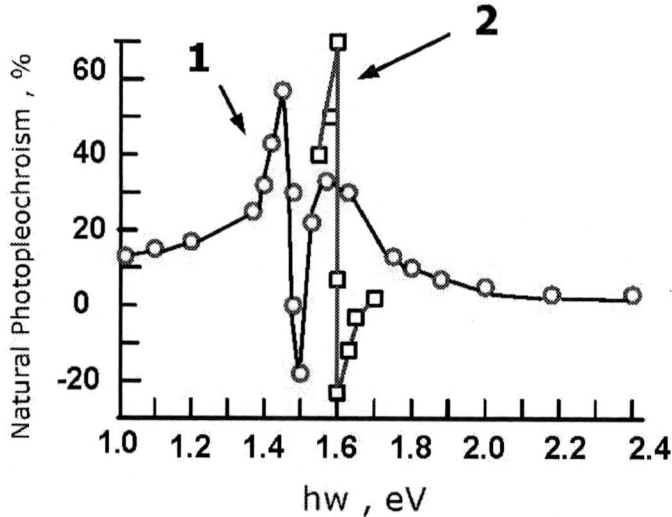

Figure 5. Natural photopleochroism spectra for n-p-CdSiAs$_2$ eliminated from the p-type substrate (curve 1 - at 300 K, curve 2 - at the T=77 K).

4. Conclusion
We conclude that such based on n-p-CdSiAs2 homodiode structures can be used as polarimetric photodetectors for optical communications lines and photonics.

References
[1] Rud' V Yu, Rud' Yu V 1993 *Jpn. J. Appl. Phys.* **32(32-3)** 672
[2] Rud' V Y, Rud' Y V, Shpunt V C, Terukov E I 2016 *Optical Memory & Neural Networks (Information Optics)* **25(1)** 40
[3] Kesamanly F P, Rud' V Yu, Rud' Yu V, Shock G V 1999 *Semiconductors* **33(3)** 483
[4] Myazin N S, Smirnov K, Davydov V V, Logunov S E 2017 *J. Phys.: Conf. Ser.* **929(1)** 012064
[5] Rud' V Y, Rud' Y V, Terukov E I 2016 *J. Opt. Technol.*, **83(5)** 275
[6] Davydov V V, Kruzhalov S V, Grebenikova N M, Smirnov K J 2018 *Measurement Techniques* **61(4)** 365-372
[7] Grebenikova N M, Smirnov K J, Artemiev V V, Davydov V V, Kruzhalov S V 2018 *J. Phys.: Conf. Ser.* **1038(1)** 012089
[8] Bairamov B Ch, Rud' V Yu, Rud' Yu V 1998 *MRS Bull* **23** 91
[9] Kesamanly F P, Rud' V Yu, Rud' Yu V 1996 *Semiconductors* **33(5)** 483
[10] Davydov R, Antonov V, Kalinin N 2015 *J. Phys.: Conf. Ser.* **643(1)** 012107

SPBOPEN 2018

IOP Publishing

Spatial point pattern analysis of the local current distribution on the surface of multi-tip field emitters

S V Filippov[1], S Carapezzi[2,3], E O Popov[1], A G Kolosko[1]

[1]Ioffe Institute, ul. Polytechnitscheskaya 26, St.-Petersburg, 194021, Russia
[2]Department of Physics and Astronomy, University of Bologna, Viale Berti Pichat 6/2, 40127 Bologna, Italy
[3]ARCES-DEI, University of Bologna, Viale Risorgimento 2, 40136 Bologna, Italy.

Abstract. A further development of the method for obtaining local characteristics of multi-tip field emitter using spatial point pattern analysis was presented. The field emission characteristics of nanocomposite emitter single-walled carbon nanotubes / polystyrene were calculated. Two sets of emission sites with different current load distribution were discovered. The application of the of spatial point pattern analysis allow to determine the spatial distribution of emission sites over the cathode surface. The results of the analysis confirm the cluster character of the emitting sites distribution on the cathode surface.

1. Introduction

The development of effective field emitters is one of the trends in the modern vacuum nanoelectronics. Now they have found application as electron sources in computer tomography, X-ray tubes, compact mass spectrometers and other devices [1-3]. The most promising emitters are microscopic structures with a distributed current load - arrays of conducting tips created using various technological methods. The main disadvantage of such structures is the thermal instability of individual emission sites on their surface, which undergo an excessive current load during the field emission and explode in an unpredictable manner, creating a vacuum discharge in the interelectrode gap. Therefore, an optimized distribution of the current load over the emission sites is one of the main technological issues for creating efficient cathodes.

The solution of this problem is to control the geometric parameters of the tips (radius, height and relative spatial arrangement) using the following techniques: scanning electron microscopy, scanning anode field emission microscopy, field emission scanning microscopy. However, all these techniques are capable of providing information on a limited portion of the cathode surface, while modern field cathodes have an area in square centimeters. In this latter case, because the analysis is required to be carried out already for the entire cathode with thousands of tips as a whole, it is preferable to use the IMLS (Integrated Measurement System with Luminescent Screen) method with Field Emission Projector (FEP).

In previous works [4-6], we reported about a method for estimating microscopic emission characteristics based on online processing of macroscopic electrical signals and corresponding field emission images obtained via IMLS. In [4] the correlation of the total brightness level of the luminescence pattern with the emission current level was shown. The approach presented in [5] made it possible to construct histograms of the brightness distribution of the emission sites and of the current load density of the surface regions. These experimental distributions were numerically compared with

---

Content from this work may be used under the terms of the Creative Commons Attribution 3.0 licence. Any further distribution of this work must maintain attribution to the author(s) and the title of the work, journal citation and DOI.
Published under licence by IOP Publishing Ltd

the ones of the simulated "ideal emitter". A histogram of the current load distribution for emission sites for a multi-tip emitter based on multiwall carbon nanotubes / polystyrene was presented in [6].

In the present paper, we investigated the current load distribution of a single wall carbon nanotubes / polystyrene (SWCNT / PS) based emitter. Additionally, we studied the spatial maps of emission sites obtained by the IMLS method by means of the spatial point pattern (SPP) analysis.

Spatial analysis is a statistical technique generally used to examine spatial entities. In case of point patterns (PPs), it is used to explore their degree of spatial clustering or ordering or randomness. It is well known that the geometry of the ensemble of individual emitters has the influence on the electron emission efficiency in large area field emitters. Then, the use of SPP analysis to study the spatial relationships between emission sites can be helpful to fully grasp their overall emission properties. SPP analysis was already used in [7] to study the spatial properties of ensembles of nanoscaled field emitters.

2. Experimental

The cathode was a metal tablet 10 mm in diameter with a SWCNT / PS nanocomposite film deposited on it. The diameter of the nanotubes was 2-4 nm, and the length was up to 10 μm. To measure field emission properties, the computerized FEP described in [4] was used. A phosphor glass screen covered with tin oxide was used as an anode. The distance between the cathode and the anode was 350 μm. The residual pressure in the projector's chamber was no more than $3 \cdot 10^{-5}$ Pa.

To obtain current-voltage characteristics (IVCs), a high-voltage power supply with a pulse frequency of 50 Hz was used. The pulses of a half-sine wave voltage of 10 ms duration were applied to the cathode. The amplitude of the pulses was kept constant for each of the selected levels of emission current.

3. Discussion.

The sample was examined at two levels of the emission currents: 320 μA and 530 μA (see figure 1a). The obtained IVCs (see figure 1b) were processed using the Elinson equation (1) [8].

$$I = A_e 1,4 \cdot 10^{-6} \varphi^{-1} \exp(10,17/\sqrt{\varphi}) F^2 \exp(-6,49 \cdot 10^9 \varphi^{3/2}/F) \qquad (1)$$

where A_e - the effective emission area, $F = \gamma U/d_{sep}$ - the electric field strength, and φ - the work function of the emitter, which here was assumed to be of 4,6 eV, d_{sep}=350 μm, γ – the effective field enhancement factor (FEF). The corresponding plots in the Fowler-Nordheim coordinates are shown in the inset figure 1 b.

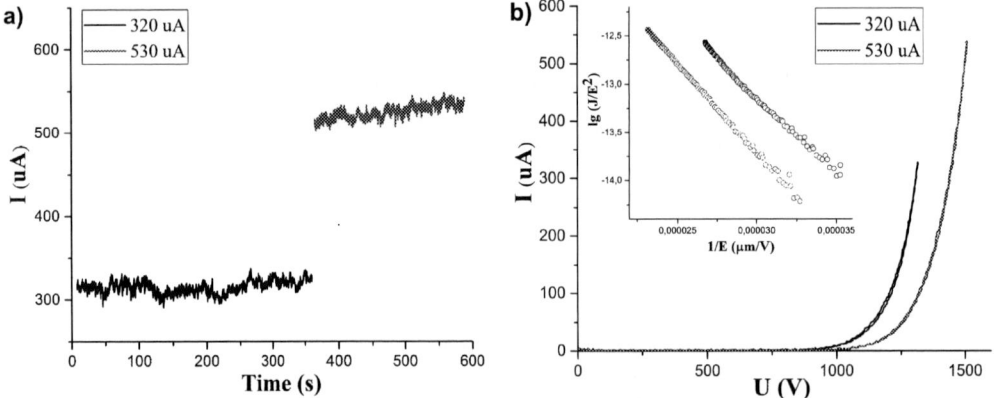

Figure 1 (a, b). Emission current levels during the experiment (a). Corresponding current-voltage curves (b) and Fowler-Nordheim plots present on the inset.

From registered IVCs basic emission parameters - the macroscopic threshold field $F_{m,th}$ and γ - were determined. At the current density I = 0,1 μA the threshold field was of 2 V/μm. The effective field enhancement factor γ, determined from the slope of the linear fit for the Fowler-Nordheim plots was of

1710 and 1490 for current levels of 320 µA and 530 µA, respectively. The decreasing of γ with the increasing of the current level is due to the inclusion of new lower emission sites.

We assumed that the brightness level of each registered emission site Y_i^{loc} on the luminescence image was proportional to the corresponding local emission current I_i at each instant of time. Thus, the current load of the sites could be calculated using the proportionality factor C, which is calculated from the total current I and the total brightness $\sum Y_i^{loc}$. To find C, the values I and $\sum Y_i^{loc}$ recorded at the beginning of the experiment was used. Using the maximum brightness values of the sites Y_i^m, the distributions of the emitting tips were constructed for their optimum brightness and the optimum current load (see figure 2 a, b):

$$Y_i^m = \max_t(Y_i^{loc}) \tag{2}$$

$$I_i = Y_i^m C = Y_i^m \left(\frac{I}{\sum Y_i^{loc}}\right)_{|t=o} \tag{3}$$

The plotted histograms represent for two current levels the emission activity of all the tips without the influence of adsorbates. Two clearly distinguished maxima on each histogram indicate the presence of at least two sets of emission sites with different field enhancement factor distribution.

On the insets of figure 2, the registered glow patterns on the luminescent phosphor screen of FEP are shown, which exhibit instantaneous brightness distribution of active emission sites over the emitter surface. The point diagrams of the optimum site distribution over the surface, corresponding to the state of the emitter "without adsorbates", are also shown.

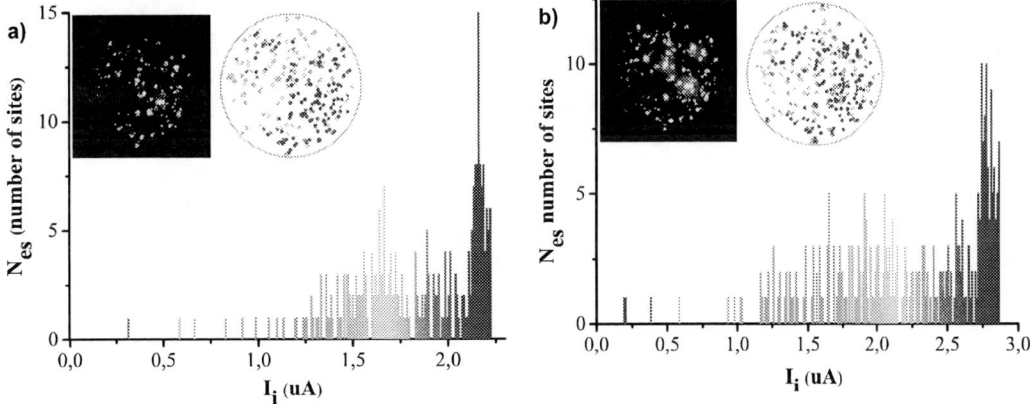

Figure 2 (a, b). Histograms of current load of emission sites for 320 µA (a) and 530 µA (b) current levels. The corresponding spatial distribution and registered brightness maxima of emission sites are shown on the insets.

Point diagrams of the optimal site distribution were subjected to SPP analysis. An analysis of the emission site arrangement was carried out using the pair-correlation function g(r) [9, 10]. The behavior of the function g(r) allows to identify the spatial structure of the emission sites: randomly located, clustered or hyper-dispersed. By definition of the pair correlation function, g(r)=1 in the case of complete spatial randomness, for which the site distribution is uniform and independent from each other. The value g(r)>1 could be indicative of the tendency of the sites to cluster. On the contrary, g(r)<1 could imply a relatively regular placement of sites.

Figure 3a, b shows the g(r) function of the PPs of the emission sites for two different current levels, 320 µA and 530 µA. In both cases, the experimental g(r) is greater than 1 in the smaller r range (r<0,028, for figure 3a and r<0,045 for figure 3b), and such departure is statistically significant (the probability that this happens in case of a random PP is less than 2%). This is indicative of clustering of the PPs in these ranges, where the order of magnitude of the cluster size is given by the upper value of r such that $g(r)>1$ and $g(r)$ lies outside the envelopes ($r=0,028$, correspondent to 280 μm, for PP of current level 320 μA, and $r=0,045$, correspondent to 450 μm, for PP of current level 530 μA). The first maximum of $g(r)$ is located at $r\approx0,01$ for the PP of current level 320 μA, and at $r\approx0,016$ for the PP of current level 530 μA. Such r values represent the most frequent distance between emission centers in each PP.

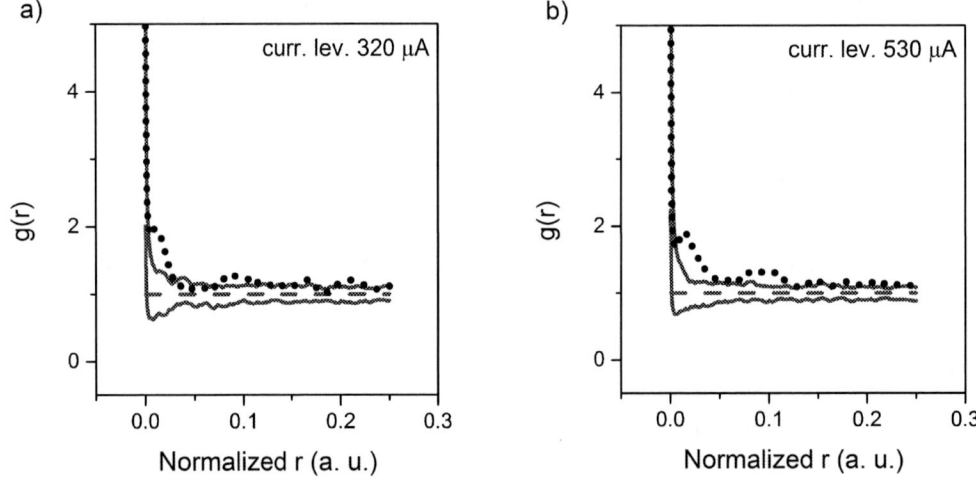

Figure 3(a, b). Plots of the computed *g*-function for all emission sites at 320 μA (a) and 530μA (b) current levels (short dotted black line). The short dashed red line represents the complete spatial randomness value of *g(r)*. The solid blue line are the upper and lower bounds of the generated simulation envelopes.

Since the experimental histograms of the current load had two clearly distinguished peaks, we divided the emission sites into two sets. For a current of 320 μA, set 1 – corresponded to emission sites with $I_i>1,83$ μA and set 2 – to emission sites with $I_i<1.83$ μA. For a current of 530 μA, the separating boundary was set at around 2.35 μA. The corresponding g(r) functions are shown in figure 4.

It is apparent that the PPs associated to these two sets possess different features, as reflected from the diversity of the behavior of their *g* functions. For both current levels, the PP of set 2 is clustered just in the smaller r range, this range being $r<0,02$ (200 μm) for current level of 320 μA, and $r<0,03$ (300 μm) for current level of 530 μA. For the PP of set 1 $g(r)>1$ in a statistical significant way for all r values, that is the clustering is present at all the distances between the emission sites. However, in the range of clustering of set 2 the associated *g* function has mostly larger values than the *g* function of set 1. That is, the clustering is more evident for set 2 on this smaller r range. Finally, the *g* function has a first maximum only for the set 1, located approximately at $r_1\approx0,014$, for which the same interpretation applies that was given for the *g*-function of the whole PP.

SPBOPEN 2018   IOP Publishing

IOP Conf. Series: Journal of Physics: Conf. Series **1124** (2018) 081040   doi:10.1088/1742-6596/1124/8/081040

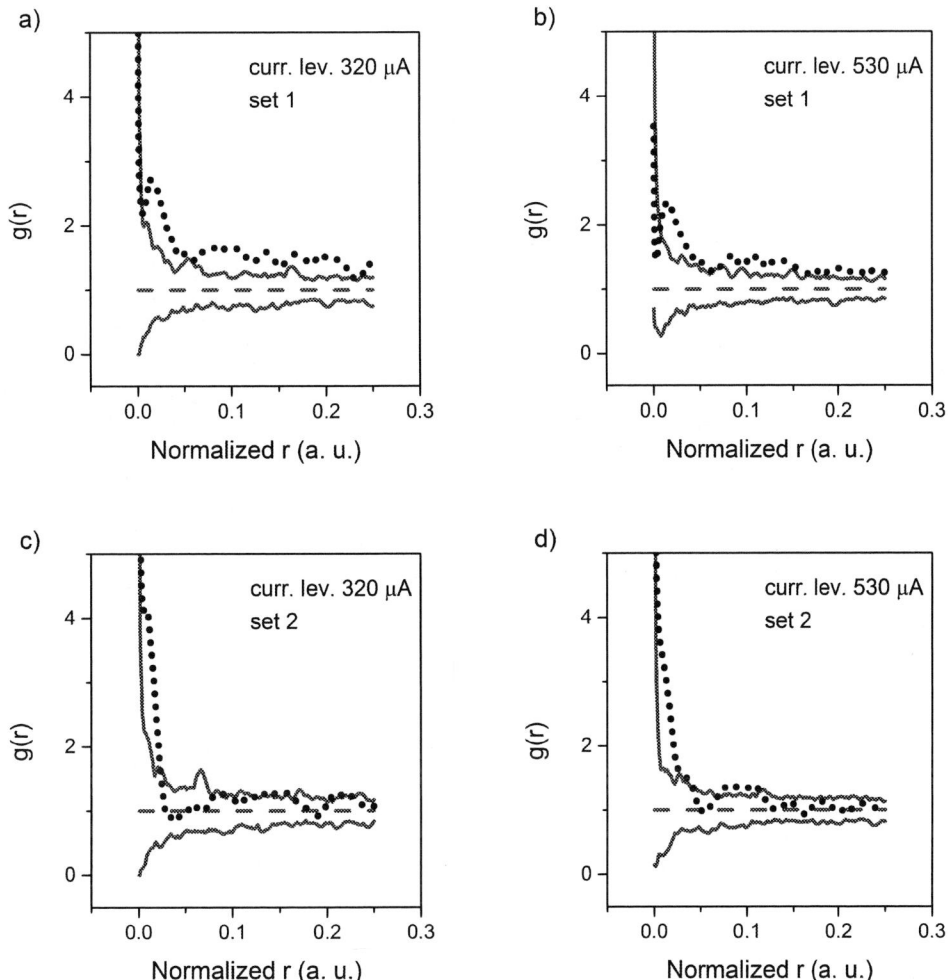

Figure 4 (a, b, c, d). Plots of the computed g-function for two sets of emission sites for 320 μA (a, c) and 530 μA (b, d) current levels (short dotted black line). The short dashed red line represents the CSR value of *g(r)*. The solid blue line are the upper and lower bounds of the generated simulation envelopes.

4. Conclusion

Emission characteristics of the SWCNT / PS nanocomposite emitter were studied. The values of the threshold field and the effective field enhancement factor for two levels of emission current were determined. SPP analysis of glow patterns obtained using FEP was carried out. The results showed a tendency of the emission sites to cluster over the cathode surface, with order of magnitude of the clusters being 280 μm for the current level of 320 μA, and slightly larger - 450 μm for the current level of 530 μA. The obtained histograms of the distribution of emission sites over the current load showed the presence of two sets of emission sites with different field enhancement factor distribution. The SPP analysis applied to these two sets showed that they possessed distinctive features. The sites with high current load (set 1) showed indication of clustering at all the distances between the emission sites. For the emission sites with low current load (set 2) the clustering was present only on smaller *r* range, the

1169

range being dependent on the current level, but in this range the degree of clustering was higher than for the sites with high current load.

References

[1] Gupta A P, Park S, Yeo S J, Jung J; Cho C, Paik S H, Park H, Cho Y C, Kim S H; Shin J H; Ahn J S, Ryu J 2017 *Materials* **10** 878

[2] Hwang J W, Mo C B, Jung H K, Ryu1 S and Hong S H 2013 *J Nanosci Nanotechnol.* **13 (11)** 7386-90

[3] Szyszka P, Grzebyk T, Krysztof M, Gorecka-Drzazga A and Dziuban J A 2017 *Technical Digest of 30th Int.Vacuum Nanoelectronics Conf.* (Regensburg: IEEE) pp 186-187

[4] Filippov S V, Popov E O, Kolosko A G 2017 *Jour. of Phys.: Conf. Ser.* **929** 012057

[5] Filippov S V, Popov E O, Kolosko A G and Vinnichek R N 2017 *Jour. of Phys.: Conf. Ser* **917** 092022

[6] Popov E O, Kolosko A G, Filippov S V and Terukov E I 2018 *J. Vac. Sci. Technol. B* **36** 02C106

[7] Carapezzi S, Castaldini A, Fabbri F, Rossi F, Negri M and Salviati G 2016 *J. Mater. Chem. C* **4** 8226

[8] Elinson M I 1974 *Cold Cathodes* (Moscow: Sov. Radio) p 336

[9] Illian J, Penttinen A, Stoyan H and Stoyan D 2008 *Statistical Analysis and modeling of Spatial Point Patterns* (Chichester: John Wiley&Sons) p 560

[10] Diggle P J 2003 *Statistical Analysis of Spatial Point Patterns* (London: Arnold) p 159

Investigation of sintering of silver lines on a heated plastic substrate in the dry aerosol jet printing

A A Efimov[1], K N Minkov[2], P V Arsenov[1], N V Protas[1], V V Ivanov[1]

[1]Department of Physical and Quantum Electronics, Moscow Institute of Physics and Technology, Dolgoprudny 141701, Russia
[2]Moscow Institute of Electronics and Mathematics, National Research University of Higher School of Economics, Moscow 123458, Russia

Abstract. The study on dry aerosol jet printing with simultaneous sintering of deposited Ag nanoparticles 20-80 nm in size on a heated plastic substrate is presented. It is established that to achieve minimal electrical resistivity of microstructures produced of ~200 μΩ·cm, the recommended temperature of the heated substrate is to be ~200 °C. It is also found that when increasing the heated substrate temperature above 225 °C, formed microstructure electrical resistivity rises abruptly due to plastic substrate deformation. The result obtained is of interest in developing a one-step technology for producing electronic components on a plastic substrate.

1. Introduction

In recent years, new technologies for the creation of electronic circuits with printing equipment are actively developing [1]. In conventional aerosol jet printing technology, functional ink is used as a source of aerosol particles [2]. The use of ink causes several problems related to its preparation, storage, removal of solvents, etc. Dry aerosol jet printing (AJP) is a recently developed additive technology for producing functional microstructures [3]. This technology does not use ink, unlike conventional aerosol jet, inkjet, and screen printing methods. In dry AJP technology, solid aerosol nanoparticles with the size of a few tens of nanometers are produced by a multi-spark discharge generator [4]. Then aerosol nanoparticles are directed into the printing head for focusing and deposition on a substrate [5]. After deposition, the structures printed are thermally sintered to improve conductive properties [6]. The sintering process may take a long time. Thus producing electronic components may require a prolonged period of time. In this study, it is proposed to combine the deposition and sintering processes to shorten the time to produce the printed lines. This approach is realized by using a hotplate and a commercial aerosol jet printer together with a custom-build multi-spark discharge generator.

2. Experimental

Silver nanoparticles in the size range from 20 to 80 nm were produced using the multi-spark discharge generator as a result of electrical erosion of silver electrodes [7]. The generator parameters were as follows: energy, repetition rate and flow rate were 3 J, 4 Hz and 2 l/min, respectively. The scheme of the experiment on the formation of conductive silver lines using dry aerosol printing on a heated plastic substrate is shown in figure 1.

Content from this work may be used under the terms of the Creative Commons Attribution 3.0 licence. Any further distribution of this work must maintain attribution to the author(s) and the title of the work, journal citation and DOI.
Published under licence by IOP Publishing Ltd

SPBOPEN 2018

IOP Publishing

IOP Conf. Series: Journal of Physics: Conf. Series **1124** (2018) 081041   doi:10.1088/1742-6596/1124/8/081041

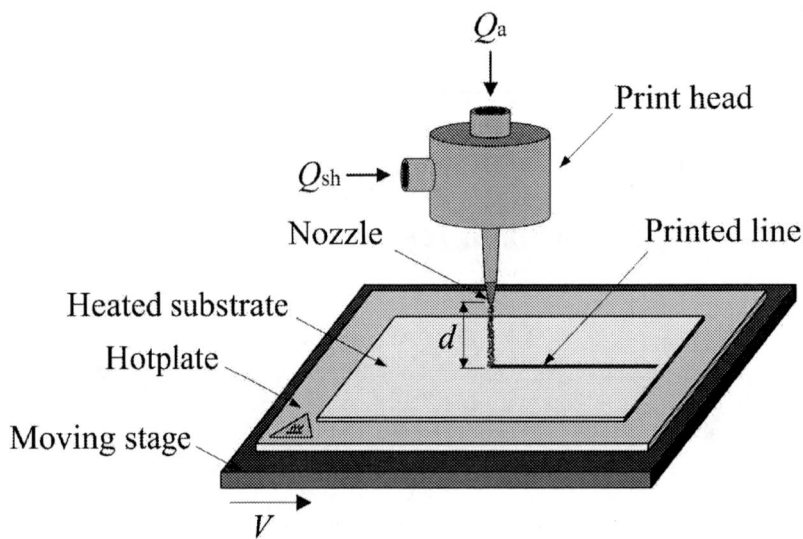

Figure 1. Experimental scheme of the formation of conductive silver lines on a heated plastic substrate, where Q_a – aerosol flow rate, Q_{sh} – sheath flow rate, d – distance from nozzle to substrate and V – substrate moving speed.

Further, the nanoparticles stream was focused into a narrow beam by means of a coaxial nozzle with an aerosol flow rate Q_a and a sheath flow rate Q_{sh} of 30 and 90 sccm, respectively. Nanoparticles beam, after focusing, was deposited on a heated plastic substrate from polyethylene terephthalate stated on a hotplate. The temperature of the hotplate was regulated in the range from 25 to 250 °C. Lines of silver nanoparticles were printed at a distance from the nozzle to the substrate S, the speed of the substrate moving V and the number of printing layers equal to: 0.5 mm, 7 mm/min and 5 layers, respectively. The electrical resistivity ρ of printed lines was calculated according to formula (1).

$$\rho = \frac{RS}{l} \qquad (1)$$

where R – resistance;
$\qquad$ S – cross-sectional area from the line 3D-profile;
$\qquad$ l – length.

The width and length of the formed lines were measured using an optical and scanning electron microscope (SEM). The line 3D-profile S was measured using an optical profilometer Leica DCM 3D. The resistance R of the printed and sintered lines was measured by the multimeter Agilent U1253B.

3. Results and discussions

As a result of Ag nanoparticles deposition on a heated plastic substrate, it is found that with the substrate temperature T_s increasing from 50 to 200 °C, there is a monotonous decrease in the printed line electrical resistivity ρ by more than 6 orders of magnitude from $6 \cdot 10^8$ to $2 \cdot 10^2$ μΩ·cm, see Figure 1a. This decrease in electrical resistivity is probably caused by an increase in the number of conducting bonds between particles, formed as a result of diffusion process at the increased temperature of the heated substrate. Increase in the number of conducting bonds between Ag particles also results in reduced porosity of the microstructure formed, as observed with a SEM, see Figure 3 (a, b, c, d).

SPBOPEN 2018   IOP Publishing

IOP Conf. Series: Journal of Physics: Conf. Series **1124** (2018) 081041   doi:10.1088/1742-6596/1124/8/081041

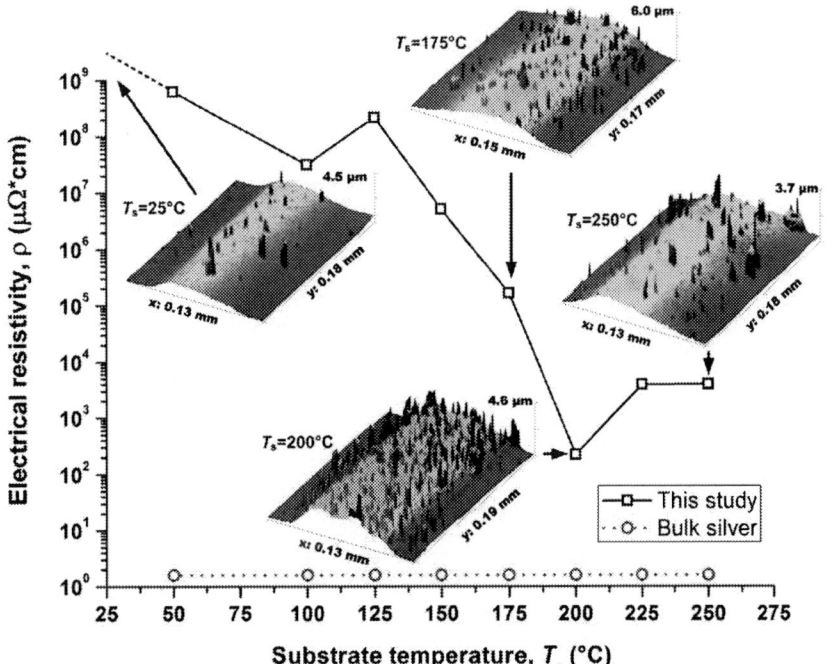

Figure 2. The dependency of the silver line resistivity on the heated plastic substrate temperature, insertions: 3D-profiles of the sintered lines.

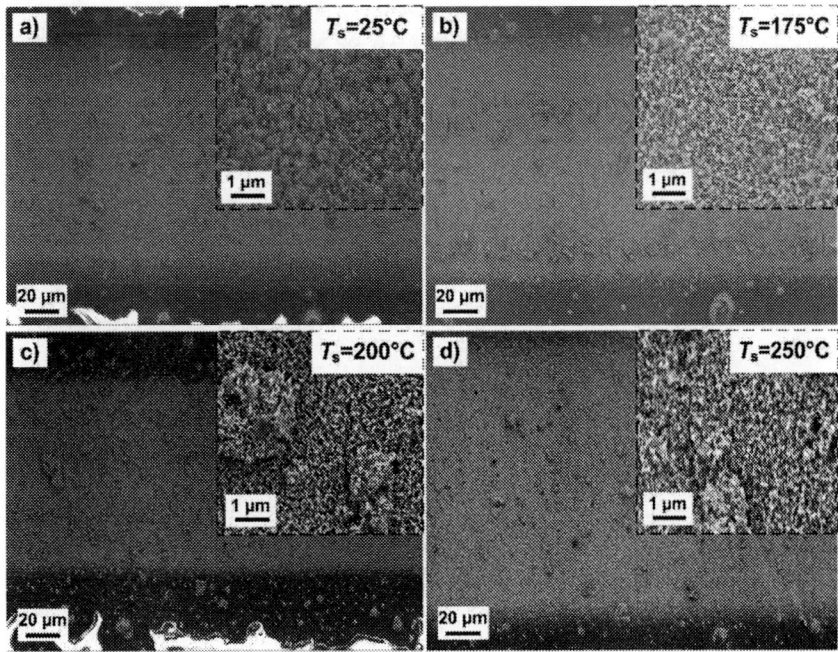

Figure 3 (a, b, c, d). SEM-images of the silver line microstructure depending on the heated substrate temperature T_s of 25°C (a), 175°C (b), 200°C (c) and 250°C (d), in the dry aerosol jet printing process.

It is also found that when the heated substrate temperature T_s is increased above 225 °C, an increase of the line resistivity to $4 \cdot 10^3$ μΩ·cm is observed, which is probably related to the process of microstructure destruction as a result of different thermal expansion coefficients for silver and the plastic substrate. Therefore, from the results of the experiment conducted it is established that the heated plastic substrate recommended temperature is to be ~200 °C to form silver lines of minimal resistivity.

4. Conclusion
A method is investigated of producing conducting microstructures of silver on a heated plastic substrate by dry aerosol jet printing method. The method allows forming conductive lines in one step without a separate post-deposition sintering operation. It is found that the heated plastic substrate recommended temperature is to be ~200 °C to provide for minimal resistivity of the microstructures produced.

Acknowledgments
This work was supported by the grant of the President of Russian Federation for young scientists MK-2302.2017.8 and carried out using the Shared Use Equipment Center for high-precision measuring in photonics (ckp.vniiofi.ru, VNIIOFI).

References
[1]   Stoukatch S, Laurent P, Dricot S, Axisa F, Seronveaux L, Vandormael D, Beeckman E, Heusdens B and Destiné J 2012 Evaluation of Aerosol Jet Printing (AJP) technology for electronic packaging and interconnect technique *2012 4th Electronic System-Integration Technology Conference* 2012 4th Electronic System-Integration Technology Conference pp 1–5
[2]   Seifert T, Baum M, Roscher F, Wiemer M and Gessner T 2015 Aerosol Jet Printing of Nano Particle Based Electrical Chip Interconnects *Mater. Today Proc.* **2** 4262–71
[3]   Efimov A, Potapov G, Nisan A, Urazov M and Ivanov V 2017 Application of the Spark Discharge Generator for Solvent-free Aerosol Jet Printing *Orient. J. Chem.* **33** 1047–50
[4]   Efimov A A, Ivanov V V, Bagazeev A V, Beketov I V, Volkov I A and Shcherbinin S V 2013 Generation of aerosol nanoparticles by the multi-spark discharge generator *Tech. Phys. Lett.* **39** 1053–6
[5]   Efimov A A, Potapov G N, Nisan A V and Ivanov V V 2017 Controlled focusing of silver nanoparticles beam to form the microstructures on substrates *Results Phys.* **7** 440–3
[6]   Wang F, Nie N, He H, Tang Z, Chen Z and Zhu W 2017 Ultrasonic-Assisted Sintering of Silver Nanoparticles for Flexible Electronics *J. Phys. Chem. C* **121** 28515–9
[7]   Lizunova A A, Efimov A A, Arsenov P V and Ivanov V V 2018 Influence of the sintering temperature on morphology and particle size of silver synthesized by spark discharge *IOP Conf. Ser. Mater. Sci. Eng.* **307** 012081

SPBOPEN 2018 IOP Publishing

Structure and hardness of the zirconium surface after laser modification

V Proskuryakov[1], S Mezentzov[1], I Rodionov[1]

[1]Yuri Gagarin State Technical University of Saratov, Saratov 410054, Russia

Abstract. The change of hardness and structure of the zirconium surface after laser modification in air has been analyzed. The dependences of the hardness parameters and the nature of the surface structure on the laser processing regimes were determined. The hardness of the hard-facing layer of zirconium was increased to 25–34 GPa.

1. Introduction

Due to its physical and chemical properties, zirconium and its alloys have found wide application in machine-building and instrument-making, as well as in the manufacture of medical and technical products. Under the influence of mechanical loads, the ductility and strength characteristics of zirconium products are reduced [1]. To improve performance, zirconium and its alloys are subjected to thermal and chemical-thermal treatment, gas and steam thermal oxidation [2].

Laser hardening of various steels, titanium, tantalum and zirconium alloys [3-10]. In the process of hardening, there is a significant change in the structure of the near-surface layer of the metal and, consequently, of the physical and mechanical properties, such as hardness and wear resistance. Laser processing of various materials is characterized by:

- Versatility - the ability to simultaneously perform thermal and chemical-thermal treatment;
- Locality - the ability to process a specific area of the surface;
- Also, in some cases, laser processing can reduce and sometimes completely eliminate a number of preliminary operations for the preparation of products, such as grinding, polishing, cleaning from process contaminants [11].

The purpose of this work was to determine the dependence of the structure and hardness of the surface layer of zirconium on the conditions of pulsed laser action.

2. Methodology

The processing of zirconium samples was performed using an automated installation for thermophysical coherent surface modification "LRS-50A". The test samples were zirconium plates of E110 grade, the size of which is $10 \times 10 \times 3$ mm. The laser action was carried out at the following voltage parameters U, experimentally selected: 350, 375, 400, 425, 450, 475 V. The duration t of the pulse action was set to 0.70, 0.85 and 1.00 ms. The laser ray was focused to a spot diameter of 1 mm. The scanning of surface was performed at a pulse recurrence frequency of radiating of 15 Hz and an overlap ratio of 0.1. The structure of the paving surface was investigated by optical microscopy using the computer program of graphic processing of microimages "Metallograph". In this case, the coverage areas of 5 mm^2 were analyzed. The microhardness was measured using a "PMT-3m" hardness tester with a Vickers indentor at an indentor load of 1.96 N (ISO 6507-1: 2005). Statistical processing of the results of the study was carried out using the "DataFit" program.

Content from this work may be used under the terms of the Creative Commons Attribution 3.0 licence. Any further distribution of this work must maintain attribution to the author(s) and the title of the work, journal citation and DOI.

Published under licence by IOP Publishing Ltd

3. Results

According to the microscopic analysis, the surface of zirconium samples as a result of pulsed laser action acquires a uniformly distributed structure at elevated values of voltage $U = 475$ V and duration $t = 0.70$ ms (Figure 1). With a pulse duration of $t = 1.00$ ms and a voltage of $U = 475$ V, a heterogeneity of the structure and a strong fusion of the laser action zone are observed.

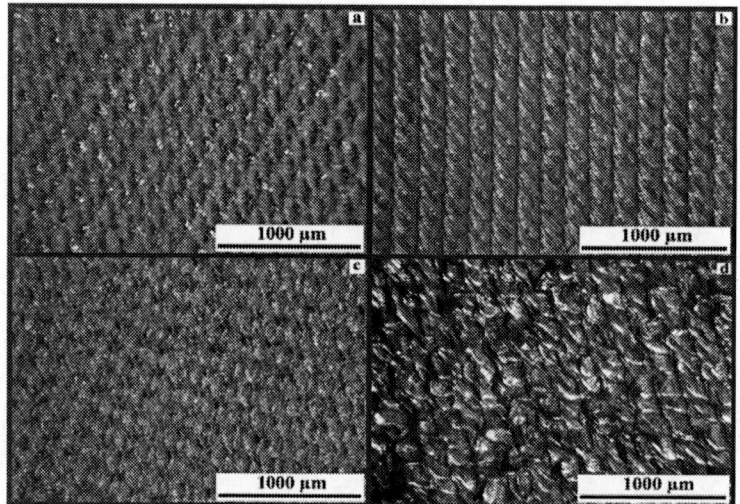

Figure 1. Morphology of the surface after pulsed laser action for a duration $t = 0.70$ ms and a voltage U: 350 V (a), 475 V (b) and duration $t = 1.00$ ms and a voltage U: 350 V (c), 475 V (d)

With a low voltage $U = 350$ V and a short duration of $t = 0.70$ ms, the sample surface was uniform, uniformly distributed. On the prepared microsections there were no microcracks (Figure 2a). With increasing voltage and duration of impulse exposure, the formation of microcracks and their growth into the interior of the metal are visualized (Figure 2b). For example, at a voltage of $U = 475$ V and a duration of $t = 1.00$ ms, the size of the cracks reaches 180 μm.

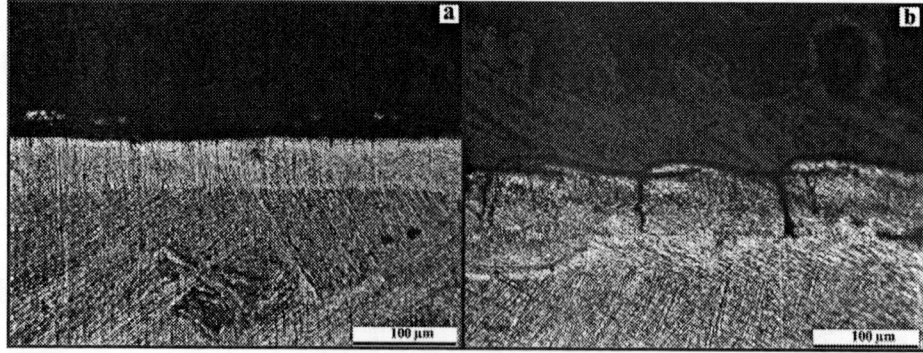

Figure 2 (a,b). Microsections of samples after laser modification: $t = 0.70$ ms and $U = 370$ V **(a)**; $t = 1.00$ ms and $U = 450$ V **(b)**.

The results of hardness measurements showed that the hardness of zirconium samples increases significantly during the laser pulsed operation. The maximum value of H = 33.9 GPa was obtained at a voltage of 450 V and a pulse duration of 1.00 ms. The hardness value of the original zirconium sample was 2.6 GPa. Based on the obtained results, an empirical model was constructed for the dependence of the hardness of the zirconium surface H on the modes of laser modification (Figure 3).

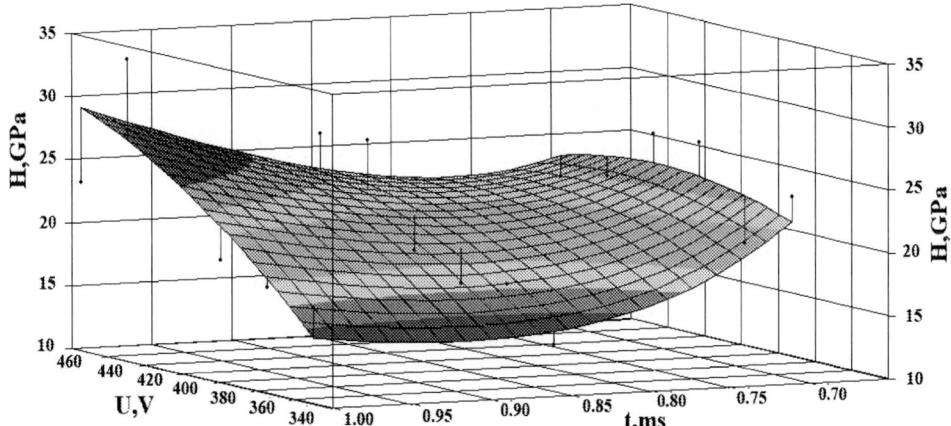

Figure 3. Dependence of the hardness of the zirconium surface on the modes of laser chemical-thermal hardening.

On this model you can see that its maximum value reaches hardness at a voltage from 420 to 460 V and a duration of 0.95–1.00 ms, and the minimum value it takes at a voltage of 340 V and duration from 0.80–1.00 ms.

The constructed empirical model is described by a regression equation:

$$H = -82.6473 + 0.689 \times x_1 - 86.244/x_2 - 4.054 \times x_1^2 + 77.44/x_2^2 - 0.2427 \times x_1/x_2$$

where x_1 – voltage (U, V); x_2 – pulse duration (t, ms).

One of the important characteristics of the coating is porosity P. Based on the measurement results, an empirical model was constructed for the dependence of the open porosity on the voltage and duration of the impulse action (Figure 4). The maximum open porosity is 68% with a duration of 0.95–1.00 ms and a voltage of 440 V, and a minimum value of 43% with a pulse duration of 0.70 ms and a voltage of 340 V.

The constructed empirical model is described by the regression equation:

$$P = 282064.5 - 3459.98 \times x_1 + 16.91 \times x_1 - 0.04 \times x_1^3 + 4.99 \times x_1^4 - 2.42 \times x_1^5 - 107.41 \times x_2 + 81.48 \times x_2^2$$

where x_1 – voltage (U, V); x_2 – pulse duration (t, ms).

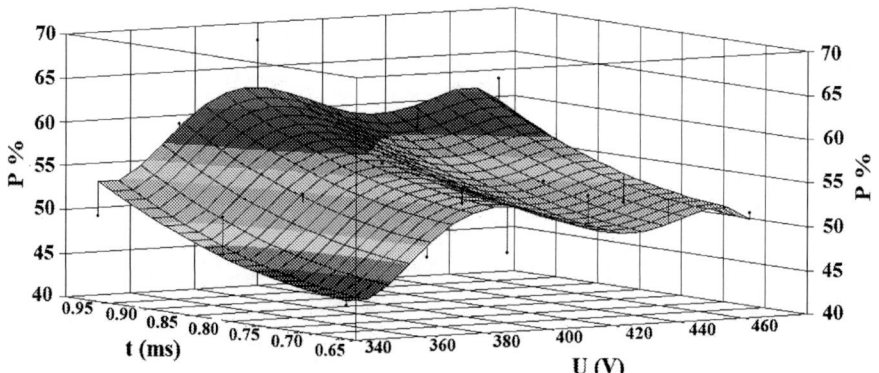

Figure 4. Dependence of open porosity on the duration of impulse action and stress.

Conclusions

It is established that under certain regimes of pulsed laser influence on the zirconium surface a highly hard coating with a homogeneous structure is formed. Laser modification of the surface of zirconium is similar to the chemical-thermal treatment by the results of hardness measurements. Rational regimes for the formation of a highly solid homogeneous structure were recommended.

Acknowledgments

The research was supported by the Ministry of Education and Science of the Russian Federation in the framework of the Program of Scientific Research in Universities (project No. 11.1943.2017/4.6).

References

[1] Chai L, Chen B, Wang S, Zhou Z and Huang W 2015 *Mater. Charact.* **110** 25
[2] Chai L, Chen B, Wang S, Guo N, Huang C, Zhou Z and Huang W 2016 *Appl. Surf. Sci.* **364** 61
[3] Koshuro V A, Fomina M A, Voyko A V, Shelkunov A A, Egorov I S, Zakharevich A M and Rodionov I V 2017 *J. Phys. Conf. Ser.* **917** 032009
[4] Fomin A, Fomina M, Koshuro V, Rodionov I, Zakharevich A and Skaptsov A 2017 *Ceram. Int.* **43** 11197
[5] Fomin A, Dorozhkin S, Fomina M, Koshuro V, Rodionov I, Zakharevich A, Petrova N and Skaptsov A 2016 *Ceram. Int.* **42** 10838
[6] Fomin A A, Steinhauer A B, Rodionov I V, Petrova N V, Zakharevich A M, Skaptsov A A and Gribov A N 2013 *Biomed. Eng.* **47(3)** 138
[7] Fomin A A, Steinhauer A B, Rodionov I V, Fomina M A, Zakharevich A M, Skaptsov A A, Gribov A N and Karsakova Ya D 2014 *J. Frict. Wear.* **35(1)** 32
[8] Fomin A A, Fomina M A, Rodionov I V, Koshuro V A, Poshivalova E Yu, Shchelkunov A Yu, Skaptsov A A, Zakharevich A M and Atkin V S 2015 *Tech. Phys. Lett.* **41(9)** 909
[9] Koshuro V A and Fomin A A 2014 Proc. 10th Int. Vacuum Electron Sources Conf. and 2nd Int. Conf. on Emission Electronics (Saint-Petersburg) (Saint-Petersburg: IEEE) p 145
[10] Koshuro V, Fomin A, Fomina M, Rodionov I, Brzhozovskii B, Martynov V, Zakharevich A, Aman A, Oseev A, Majcherek S and Hirsch S 2016 *J. Phys. Conf. Ser.* **741(1)** 012197
[11] Saeidi K, Gao X, Zhong Y and Shen Z J 2015 *Mat. Sci. Eng. A* **625** 221

SPBOPEN 2018   IOP Publishing

The electrochemical deposition of silicon - carbon thin films from organic solution

M N Grigoryev[1], T N Myasoedova[2], T S Mikhailova[2]

[1]JSC"Taganrog Scientific-Research Institute of Communication", Taganrog 347913, Russia
[2]Institute of Nanotechnologies, Microelectronics and Equipment Engineering, Southern Federal University, Taganrog 347900, Russia

Abstract. Silicon - carbon thin films were obtained on silicon substrate by using electrochemical deposition at room temperature and atmospheric pressure. Solution of electrochemical deposition was consisted of methanol or ethanol and hexamethyldisilazane $((CH_3)_3\text{-Si-NH-Si-}(CH_3)_3$, HMDSN). It was shown that a type of electrolyte is strongly effect on maximum current density and the morphology of the deposited film.

1. Introduction

The synthesis of silicon-carbon films has been widely studied due to their attractive properties, such as wear resistance, chemically inertness, and wide band gap, which have already provided optical, electronic and other applications in aggressive ambient. Silicon - carbon films are very interesting and important material for microelectronics. These films are used for gas sensors, ultracapacitors, field emission devices and other applications in aggressive ambient [1]. There are many methods for obtaining these films, such as magnetron sputtering [2], ion sputtering, chemical vapor deposition, pulsed laser deposition, electrochemical deposition from molten salt, sol – gel method [3-5]. The applications of these methods, however, have been limited, owing to the complex equipment and rigorous experimental conditions, including high vacuum and high temperature.

There is experimental evidence that most materials which can be deposited from the vapor phase can also be deposited in liquid phase using electroplating techniques and vice versa [6]. Enlightened by this conclusion, Namba [7] first attempted to grow diamond phase carbon films in the ethanol solution at a temperature less than 70 °C. From the application viewpoint, the liquid deposition techniques have many advantages such as availability for large area deposition on intricate surfaces, low deposition temperature, low consumption of energy, and simplicity of the set up, over the vapor deposition techniques.

The possibility of deposition of diamond-like carbon films by means of electrolysis of organic liquids such as methanol [8], acetonitrile [9], dimethylsulfoxide [10], and lithium acetylide in dimethylsulfoxide [11, 12], at atmospheric pressure and low temperature, has been demonstrated recently.

However, not much has been reported on the profound of electrochemical deposition of silicon-carbon films from a methanol or ethanol solution of hexamethyldisilazane. So, only Yan [13] performed a successful electrochemical deposition of silicon carbide nitride nanocomposite films from methanol solution of hexamethyldisilazane that led to the formation of the Si_3N_4 and SiC crystalline grains.

Content from this work may be used under the terms of the Creative Commons Attribution 3.0 licence. Any further distribution of this work must maintain attribution to the author(s) and the title of the work, journal citation and DOI.

Published under licence by IOP Publishing Ltd

SPBOPEN 2018   IOP Publishing

IOP Conf. Series: Journal of Physics: Conf. Series **1124** (2018) 081043   doi:10.1088/1742-6596/1124/8/081043

In this work, an electrochemical route to deposit silicon - carbon films from a methanol and ethanol solution of HMDSN at atmospheric pressure and low temperature has been reported

Methanol was also selected because its polarizability and conductivity are stronger than those of ethanol.

2. Experimental

The electrolytic deposition system was used to obtain silicon - carbon films. A schematic diagram of the system is shown in Figure 1.

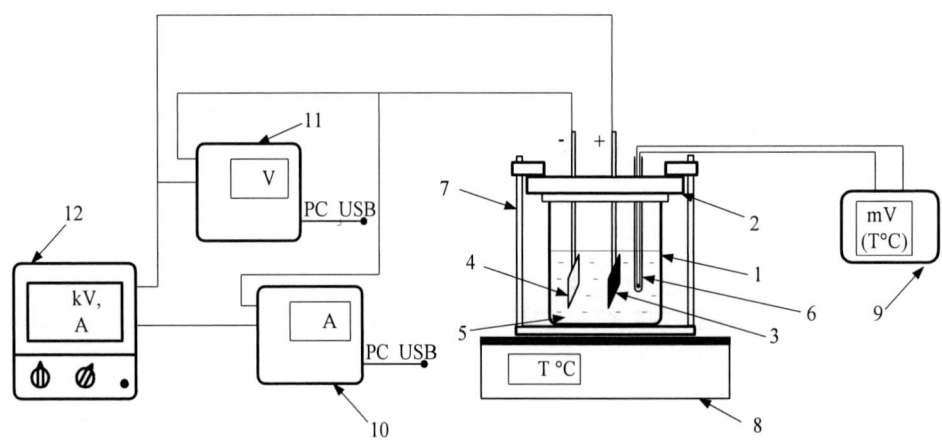

Figure 1. Schematic structure of electrolytic deposition system (1- glass cell, 2 - dielectric cover, 3 – graphite anode, 4 - cathode – substrate, 5 – solution, 6 – thermocouple, 7 – clamps, 8 - thermal table, 9 – voltmeter of thermocouple, 10 – ammeter, 11 – HV - voltmeter, 12 - power supply).

Silicon - carbon film was deposited on silicon substrate (100), with a size of 12x17x0.4 mm^3 and resistivity of 4.5 Om·cm. First, the substrate was dipped in the HF solution for a few minutes. Silicon substrate was mounted on the negative electrode and graphite was mounted on the positive electrode. The distance between the substrate and positive electrode was set to 10 mm. The deposition was done from two types of solution: 1 - a methanol solution of HMDSN; 2 – an ethanol solution of HMDSN. HMDSN was dissolved in analytically pure methanol (ethanol), with the volume ratio of HMDSN to methanol (ethanol) to be 1:9. The cleaned substrate was placed into the above solution, with an area of 10 · 15 mm^2 to be immerged therein. The films were deposited for 4 h at 60°C. The applied potential was 270 and 450V, for methanol and ethanol solutions respectively.

Figure 2 shows the curves of current dependence on the deposition time for two types of solution.

During the deposition for the nanocomposite film, we found that the current density decreased from 70 mA/cm^2 to less than 35 mA/cm^2 with increasing the deposition time for a methanol solution of HMDSN and from 42 mA/ cm^2 to 15 mA/ cm^2 for an ethanol solution of HMDSN. In both cases, a gray film was obtained on the cathode Si substrate after deposition. The thickness was measured to be about 800-900 nm using interferometer technique. It means the average deposition rate is ~212 nm/h.

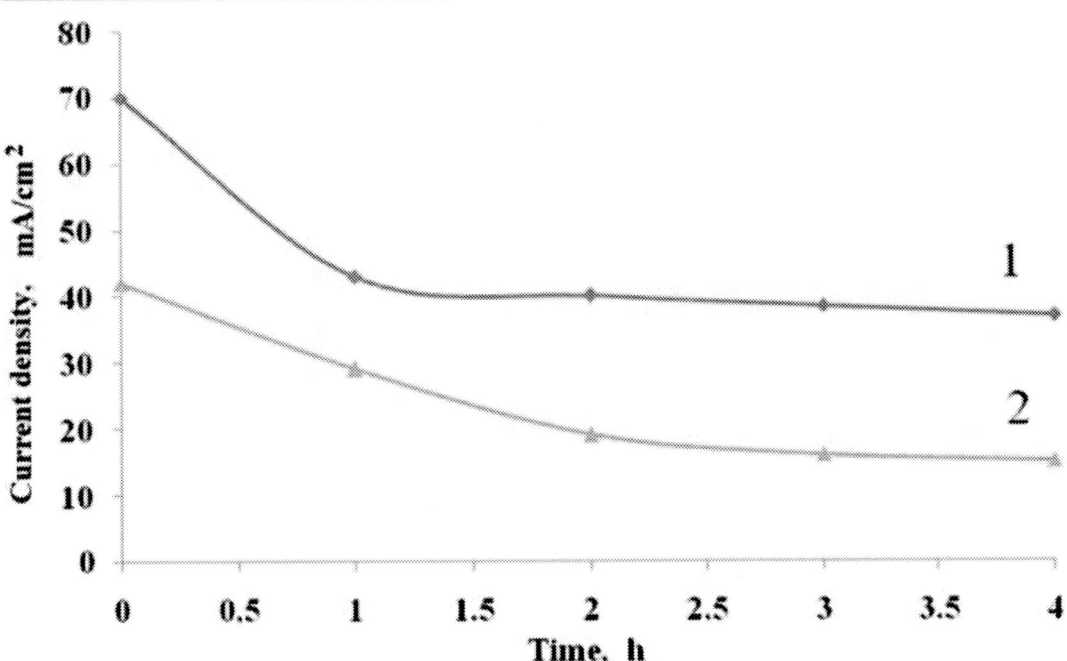

Figure 2. Current dependence on time deposition (1 - methanol - HMDSN solution; 2 - ethanol - HMDSN solution).

Most part of the current decreased within the first hour for methanol – HMDSN solution while significant current decrease ethanol – HMDSN solution is observed during the first two hours. This fact may be explained by a decrease of reaction products, increase of film thickness and different deposition rate for two types of solutions.

Namba [7] thought that the substrate current played an important role in film formation from an organic solution. The conductivity of electrolyte is indeed an important factor during electrolysis. Namba used ethanol as electrolyte, and the maximum current density was only 5 mA/cm^2, whereas in our experiment, the current density exceeded 42 mA/cm^2. Yan [13] used methanol as electrolyte, and the maximum current density was about 40 mA/cm^2, whereas in our case the current density exceeded 70 mA/cm^2. So, we see that methanol electrolyte is more promising for silicon-carbon films deposition. Methanol has active CH_3 species which concentrate near the negative electrode surface and prefer the formation of diamond-like carbon film by electrochemical reactions.

The molecule of HMDSN also contains electron-donating methyl group, which displays somewhat positive charge due to their low electro-negativity. Consequently, the molecules are liable to be absorbed on the surface of the negative electrode owing to the sorption of the electrons on the cathode surface. So the Si–C and Si–N bonds in HMDSN molecule will be dissociated to form some silicon-containing clusters like (SiC)x and (SiN)x. The clusters are deposited on the cathode substrate, and form b-Si_3N_4, a-Si_3N_4, and SiC nanoparticles finally [13].

Figure 3 shows SEM micrographs of the deposited films which were taken by SEM Zeiss Merlin compact VP-60-13. From the Fig.3a it can be seen that film deposited from ethanol solution is composed of small, compact grains. The average grain size is about 130 nm. The morphology of the film deposited from methanol solution is rather different and characterize with a rougher surface emerges with nodular-like features (Fig.3 b).

Figure 3 (a,b). SEM micrograph of the films deposited for 4h **(a)** at 270 V from ethanol solution of HMDSN; and **(b)** at 450 V from methanol solution of HMDSN.

3. Conclusions

Silicon - carbon films can be obtained from methanol-HMDSN and ethanol - HMDSN) solutions by the electrochemical deposition. This simple method may extend the field of application silicon - carbon films in future**.** This synthesis strategy may be transferred to fabricating other inorganic nanocomposite films, using appropriate organic reagent and electrolysis condition.

Acknowledgments

The work was supported in part by the Ministry of Science and Education Russia, under Contract №14.575.21.0126 (unique identifier of the contract is RFMEFI57517X0126).

References

[1] Alper J, Kim M, Vincent M, Hsia B, Radmilovic V, Carraro C, Maboudian R 2013 Journal of Power Sources **230** 302

[2] Komateu S, Hirohata Y, Fukuda S, Hino T, Yamashina T, Hata T, Kusakabe K, 1990 Thin Solid Films 917

[3] Manocha S, Ankur Darji, Manocha L M 2011 Eurasian Chem Tech Journal **13** 277

[4] Halac E B, Huck H, Oviedo C, Reinoso M E, M.A.R. de Benyacar 1999 Surface and Coatings Technology **122** 337

[5] Pusch H, Riedel C, Fasel R, Klein C 2010 Surface and Coatings Technology **205** 443

[6] Maissel L and Glang R 1970 Handbook of Thin Film Technology (New York: McGraw-Hill Book Company)

[7] Namba Y 1992 J. Vac. Sci. Technol. A **10** 3368

[8] Wang H, Shen M R, Ning Z Y, Ye C 1996 Appl. Phys. Lett. **69** 1074

[9] Guo D, Cai K, Li L T, Zhu H S, 2000 Chem. Phys. Lett. **3** 499

[10] Jiang H Q, Huang L N, Zhang Z J, Xu T, Liu W M, 2004 Chem. Lett. **33** 378

[11] Shevchenko E, Matiushenkov E, Kochubey D, Sviridov D, Kokorin A, Kulak A, 2001 Chem. Commun. **4** 317

[12] Kulak A I, Kokorin A I, Meissner D, Ralchenko V G, Vlasov I I, Kondratyuk A V, Kulak T I, 2003 Electrochem. Commun. **5** 301

[13] Yan X B 2006 Electrochemistry Communications **8** 740

Investigation of interaction of molecules of inorganic gases with surface of copper-containing polyacrylonitrile

M M Avilova[1], V V Petrov[1]

Southern Federal University, Research and Education and Centre "Microsystem technics and multisensor monitoring systems", 2, Chekhov St., Taganrog, 347922, Russia

Abstract. The possibility of adsorption of inorganic gases onto the surface of copper-containing polyacrylonitrile (PAN) films was studied by molecular modeling and quantum chemical calculations. For this, the model of a copper-containing PAN cluster was obtained by these methods, which interacted with the adsorbed gas molecule. Based on the results of the simulation, it was established that when adsorbing gas molecules, intermolecular bonds arise between them and the surface of a cluster of copper-containing PAN. A high probability of adsorption on the surface of copper-containing PAN is manifested in chlorine, hydrogen sulfide, and carbon dioxide molecules. Low probability of adsorption - for molecules of ozone and nitrogen dioxideIntroduction

1. Introduction

Thermally stabilized polyacrylonitrile (PAN) is an electrically conductive material that exhibits sensitivity to various gases at room temperature [1-3]. Based on this material, high-sensitivity gas sensors (nitrogen dioxide, hydrogen sulfide and chlorine) have been constructed [1]. Actual is the possibility to add PAN with various modifying additives, as a rule, with metal compounds [4].

The object of our study is the copper-containing PAN. The structure of the copper-containing PAN is a PAN with incorporated molecules of copper oxide, copper chloride and metallic copper [1]. Copper compounds that are part of this material have a positive effect on the growth of the film surface and lead to a decrease in the resistivity of the films [5]. Therefore, copper-containing PAN is a promising material for use in gas sensors.

In the available scientific works there is no explanation of the sensitivity of copper-containing PAN to various gases. In [6-8], we explained the gas sensitivity of PAN and cobalt-containing PAN films by mathematical modeling. The purpose of this study is to study the possibility of adsorption of inorganic gaseous pollutants to the surface of copper-containing PAN by quantum chemical calculations and molecular modeling.

2. Experimental

It was found in [5] that the structure of the copper-containing PAN consists of a PAN with molecules of copper oxide, copper chloride and metallic copper embedded in it. Therefore, for a theoretical study of the copper-containing PAN, a cluster was modeled which was formed at an IR annealing temperature of 350-500 °C. The cluster consists of two parallel arranged conjugate macromolecules of

---

Content from this work may be used under the terms of the Creative Commons Attribution 3.0 licence. Any further distribution of this work must maintain attribution to the author(s) and the title of the work, journal citation and DOI.

Published under licence by IOP Publishing Ltd

PAN pentamers with the molecules of copper oxide, copper chloride and a copper atom atom built in between them. Initially, spatial configurations of the PAN pentamer macromolecules were obtained in the Hyper Chem program. To estimate the thermodynamic parameters of the obtained structures, the software package GAUSSIAN07 was used using the 6-31 G * basic set in the framework of density functional theory (DFT) [9-11]. Semiempirical calculations of the spatial configurations of macromolecules were carried out using the B3LYP exchange-correlation functional [9, 10]. Then, a simulation of the cluster of copper-containing PAN was carried out using the meteoric calculation of the steric energy of the system in the Chemoffice 2010 software package (Chem3D subroutine). The Chem3D subroutine uses one of the methods of molecular modeling - the method of minimizing the potential energy of the system in a modified version of the force field (MM2), developed by Allinger [12-16].

In the next stage, Chemoffice subroutines Chem3, Chem, the molecules of copper and copper compounds in different versions were built between the layers of the PAN cluster. As a result of the simulation it was established that the energy minimum is characteristic for a cluster of copper-containing PAN, in which the molecules of copper compounds are located between PAN macromolecules in the sequence: a molecule of copper chloride - a molecule of copper oxide - a copper atom. The distance between the particles of the dopants and the layers of the PAN is -3.6-4.2 Å (Fig. 1).

Then the possibility of adsorption of molecules of inorganic gases (nitrogen dioxide, methane, ammonia, sulfur oxide (II), hydrogen sulfide, ozone, carbon monoxide, carbon monoxide (II), chlorine) to the surface of a cluster of copper-containing PAN was analyzed in the Chem3D program [12, 13]. For this purpose, different positions of the gas molecules relative to the cluster were set: (1) - (12) - see Fig. Position 1 assumed that the molecules of inorganic gases were built into the structure of PAN, and positions 2 - 12 showed the location of the water molecule above the surface of the cluster.

Figure 1. The configuration of a copper-containing PAN cluster.

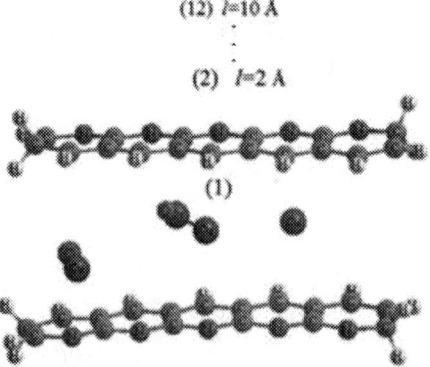

Figure 2. The arrangement of the gas molecule upon interaction with the surface of a copper-containing PAN cluster.

Further, the steric energy was calculated and the more favorable location of the molecules in question relative to the PAN cluster was found. In addition, the calculation of the energy of formation of the bond between the cluster and the gas molecule ΔE was carried out, as the difference in the energies of the system with the distance between the molecule and the cluster at the maximum distance and at the point of the energy minimum. At the same point, the distance (lmin) between the gas molecule and the cluster surface was estimated.

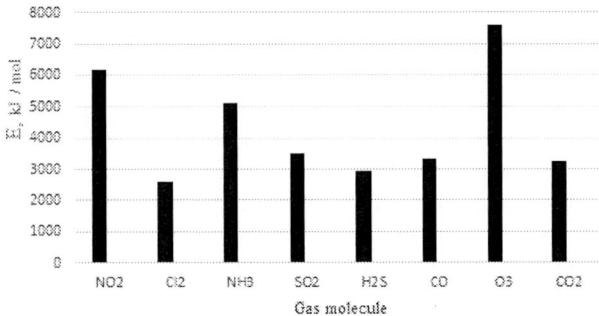

Figure 3. Dependence of the steric energy of the system "molecule of inorganic gas - cluster of copper-containing PAN" at the point of energy minimum.

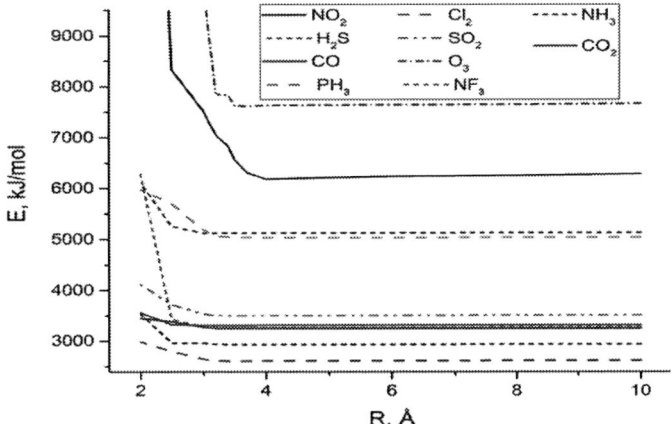

Figure 4. Dependence of the value of the steric energy of the system of a molecule of inorganic gas - a cluster of copper-containing PAN "on the distance between the molecule of inorganic gas and the surface of the cluster.

Table 1. Thermodynamic parameters of the system "copper-containing PAN cluster - gas molecule"

№	Gas molecule	E_{min} (kJ/mol)	ΔE (kJ/mol)	l_{min}.Å
1.	NO_2	6182,76	107,68	3,7
2.	Cl_2	2615,40	5,49	3,2
3.	NH_3	5120,93	7,37	2,5
4.	SO_2	3501,16	3,85	3,5
5.	H_2S	2931,58	2,68	3,5

6.	CO	3310,06	2,89	2,5
7.	O_3	7610,59	89,33	3,5
8.	CO_2	3251,94	2,81	3,2

3. Result and Discussion

Based on the results of the simulation, it was established that the cluster of copper-containing PAN is a structure of PAN macromolecules, between which the molecules of copper and copper compounds are located in the sequence: a molecule of copper chloride - a molecule of copper oxide - a copper atom. Based on the results of molecular modeling, it is determined that chemical interactions between PAN macromolecules and dopants do not occur.

Table 1 presents the thermodynamic parameters of the system "molecule of inorganic gas - cluster copper-containing PAN" results of calculations of the steric energy of the given system (E). Figure 4 shows the energy values of the system "gas molecule - a cluster of copper-containing PAN" from the distance of the gas molecule and the surface of the cluster.

According to the results presented in Table. 1 and Fig. 4 that the smallest value of steric energy in the "copper-containing PAN cluster-the gas molecule" system is characteristic for the adsorption of the chlorine molecule (Emin = 2615.40 kJ / mol). Also, low values of steric energy are inherent in hydrogen sulfide molecules (Emin = 2931.58 kJ / mol) and carbon dioxide (Emin = 3251.94 kJ / mol). High values of the steric binding energy are characteristic for ozone molecules (Emin = 7610.59 kJ / mol) and nitrogen dioxide (Emin = 6182.76 kJ / mol).

Based on the results of molecular modeling, it was established that the energetically favorable arrangement of gas molecules is above the cluster surface with a distance from the cluster lmin of 2.5-3.7 Å in the region where the lowest values of the steric energy of the system for the gases under study are observed. As the gas molecule approached the cluster for a shorter distance, the steric energy of the system grew. At the same time, the adsorption energy does not exceed 108 kJ / mol (Table 1). This means that in the system "molecule of inorganic gas-cluster copper-containing PAN" intermolecular bonds arise. Moreover, the strongest connection with the cluster is observed in molecules of ozone and nitrogen dioxide. The remaining gases, on the other hand, have weak intermolecular bonds with a cluster of copper-containing PAN.

4. Conclusion

A theoretical study of the gas sensitivity of copper-containing PAN films by quantum-chemical calculations and molecular modeling was carried out. As a result of modeling, a copper-containing PAN cluster was obtained and the possibility of adsorption of molecules of various inorganic gases to this cluster was estimated numerically.

Based on the results of the simulation, it was established that when adsorbing gas molecules, intermolecular bonds arise between them and the surface of a cluster of copper-containing PAN. A high probability of adsorption on the surface of copper-containing PAN is manifested in chlorine, hydrogen sulfide, and carbon dioxide molecules. Low probability of adsorption - for molecules of ozone and nitrogen dioxide.

References

[1] Petrov VV, Semenistaya T.V. Metal-containing polyacrylonitrile: composition, structure, properties .- Taganrog. SFedU: 2015. 169 p.

[2] Semenistaya TV, Petrov VV, Bedna TA Energy-efficient gas sensors based on nanocomposite organic semiconductors. - Taganrog: Publishing house SFU, 2013. 120 p.

[3] Semenistaya TV, Petrov VV, Ladygina AA // News of SFedU. Technical science. 2014. T.153. №4. Pp. 219-229.

[4] Semenistaya TV, Petrov VV, Kalazhokov Kh.X., et al. Investigation of the properties of nanocomposite films of cobalt-containing IR-pyrolyzed polyacrylonitrile // Electronic processing of materials. 2015. V.51. №1. C.9-18.

[5] Al-Khadrami IS, Korolev AN, Semenistaya T.V. Nazarova TN, Petrov VV Investigation of Gas Sensitive Properties of Copper-Containing Polyacrylonitrile // Izvestiya Vysshikh Uchebnykh Zavedenii. Electronics. - 2008. - No. 1. - P. 20 - 25.

[6] Avilova M.M., Petrov V.V. Investigation of gas-sensitive properties of polyacrylonitrile films modified by cobalt compounds, molecular modeling and quantum-chemical calculations // Khimicheskaya Fizika, 2017, vol. 36, No. 7, p. 90-96.

[7] Avilova MM, Petrov VV Investigation of the interaction of inorganic gases with the surface of cobalt-containing polyacrylonitrile in the presence of water molecules, Chemical Safety, 2017, vol. 1, No. 2, p. 108-116.

[8] Avilova MM, Petrov VV Investigation of the gas sensitive properties of polyacrylonitrile films modified with cobalt compounds, molecular modeling and quantum-chemical calculations, Khimicheskaya Fizika, 2018, v. 37, No. 4, p. 69-73.

[9] M.Volkenstein. Configuration statistics of polymer chains. M., L .: Publishing House of the USSR Academy of Sciences, 1959.

[10] Ignatov, SK, Quantum-chemical modeling of molecular structure, physico-chemical properties and reactivity. Part 1. Review of modern methods of electronic structure and the theory of density functional. 2007: Nizhny Novgorod, the UNN. 84 sec.

[11] Clark. T. Computer Chemistry. - Moscow: Mir, 1990.-382p.

[12] Hoenberg P., Kohn W./Inhomogeneous Electron Gas // Phys. Rev., 1964.-V. 136.-B864-B871.

[13] Parr R.G., Yang W. Density Functional Theory of Atoms and Molecules. -Oxford: Oxford University Press, 1989. - 352 p.

[14] Kittel, C. Quantum theory of solids. - Moscow: Nauka, -1967.-491 p.

[15] Dashevsky V.G. Conformational analysis of organic molecules. - Moscow: Chemistry, 1982.-347 p.

[16] Rafikov SR, Pavlova SA, Tverdokhlebova II, Methods for determination of molecular weights and polydispersity of high-molecular compounds, M., 1963.-336 p.

SPBOPEN 2018

IOP Publishing

Optical studies of InAs/GaAs monolayer Bragg superlattices

K A Ivanov[1,2], A R Gubaidullin[1], G Pozina[3], E V Nikitina[2,4], M A Kaliteevski[2]

[1]ITMO University, St. Petersburg, 197101, Russia
[2]St. Petersburg Academic University, St. Petersburg 194021, Russia
[3]Department of Physics, Chemistry and Biology (IFM), Linkoping University, Linkoping, Sweden
[4]Ioffe Institute, St. Petersburg 194021, Russia

Abstract. Periodic sequences of InAs monolayer in GaAs have been grown by molecular beam epitaxy. The properties of the structure have been modelled by transfer matrix calculation and experimentally studied by time-resolved photoluminescence. The structure obtained provides good localisation of excitons in quantum wells and opens possibilities for various optical effects. Coupling of excitons in quantum wells with electromagnetic field confined by a Bragg arrangement of layers produces a peculiar emission mode – a superradiant mode. Measurements and calculations show that for specific values of frequency of light and angle of emission, enhancement of spontaneous emission rate occurs.

1. Introduction

The topic of resonance Bragg quantum structures wells (BQWs) attracted attention spurred by a pioneering theoretical work [1]. In BQW structures with period d, if an angle of light propagation θ and its frequency ω satisfy Bragg condition

$$\hbar\omega = \frac{\pi\hbar c}{d\sqrt{n^2-\sin^2\theta}} \tag{1}$$

giant reflection coefficient has been observed [2]. Half a century ago it has been proposed, that quantum emitter can interact through electromagnetic field forming collective radiative modes, that lead to variety of fascinating effects. One of the effects is superradiance [3], a manifestation of a collective mode of tightly bound light emitters coupled by electromagnetic field. Numerous effects suggesting coupling in Bragg structures have been recently observed [2,4-6]. Recently, similar behaviour has been observed in BQW structures with ML InAs quantum well confined in GaAs matrix [7,8]. It has been suggested that such ML structures can be used in various optical devices for different applications, including proposed Bloch oscillators or any device utilizing ultra-fast optical switching [9].

In this paper we present the results of the study of a high-quality Bragg superlattice grown by MBE and containing InAs MLs in GaAs matrix. This paper also studies further the phenomenon of superradiance in ML structures with respect to more peculiar properties of this mode.

2. Growth, measurements and calculations

Experimental sample shown in fig. 1(a) was fabricated by MBE in Riber 49 chamber on rotating (100) GaAs substrate. The process was controlled by the means of high energy electron diffraction. The sample consists of 60 periods of triple InAs-monolayer QWs separated by bulk GaAs, with the thickness about 100 nm (see [7] for additional details of the fabrication process). Each triplet of QWs consists of three monolayers of InAs embedded in GaAs matrix. The spacing between adjacent MLs was chosen to be 10 nm.

Content from this work may be used under the terms of the Creative Commons Attribution 3.0 licence. Any further distribution of this work must maintain attribution to the author(s) and the title of the work, journal citation and DOI.
Published under licence by IOP Publishing Ltd

The sample was studied as to its photoluminescence spectrum and afterwards the time-resolved spectroscopy measurements were conducted. The general scheme of an experimental setup is shown in fig. 1(b).

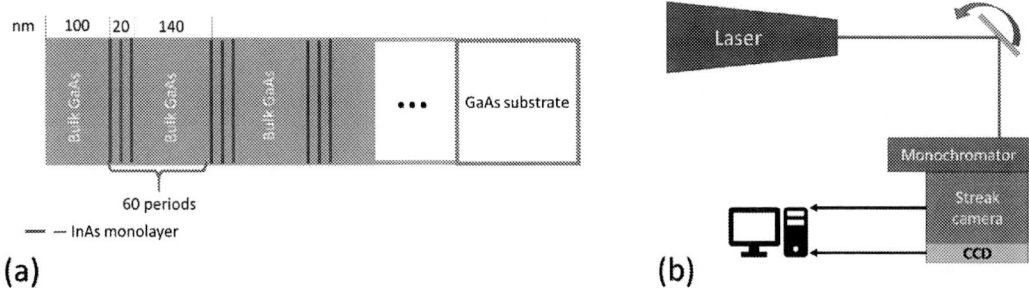

(a)   (b)

Figure 1(a,b). The scheme of **(a)** the structure; **(b)** experimental setup.

The study of the PL spectrum revealed that the structure incorporates four exciton states, as is evident in fig. 2. The highest energy belongs to a bulk GaAs exciton, whereas the triplet of lower-energy states is generated by ML QWs. The splitting of the level is the result of a close disposition of the monolayers, which leads to the holes' wavefunctions being overlapped.

We have then measured the time-resolved spectra of emission at different angles and from various positions of the structure, see next section for details. Overall quality of the results suggests good localization of excitons in ML QWs, which is a main demand for any device aiming at the utilization of quantum properties of semiconductor structures.

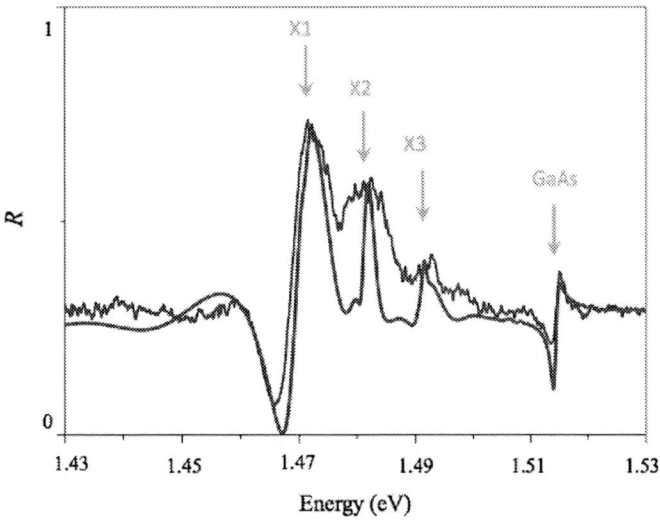

Figure 2. Measured (blue) and calculated (red) reflection spectra. Excitons' peaks are denoted by arrows.

3. Results

The measurement PL spectrum was useful in a sense that it provided us with data to tune a model for later modelling. Indeed, an exciton in a QW can be described by a certain reflection coefficient and a transfer matrix. The parameters of an exciton are its resonance energy, radiative decay rate Γ_0 and non-radiative decay rate γ. The bulk GaAs is of course also easily modelled using the transfer matrix method. This leads us to a procedure of tuning the exciton model, when we can obtain the abovementioned parameters by varying them to match the calculated reflection spectrum and the measured one. The result of this procedure is shown in fig. 2, and the parameters obtained are shown in Table 1.

Table 1. Exciton parameters.

	Resonance energy, eV	Radiative decay Γ_0, meV	Non-radiative decay γ, meV
X1	1.4712	0.36	7.0
X2	1.4820	0.18	7.0
X3	1.4910	0.14	0.6

The reflection spectrum of the structure was then calculated by the transfer matrix method for the whole range of angles using the obtained exciton parameters, and the results are shown in fig. 3(a). Three monolayer QWs produce three horizontal lines of reduced reflection. One can also see, that there are areas of increased and reduced reflection coefficient for angles and frequencies of light in the vicinity of Bragg condition (1).

We have also calculated the dependence of the probability density of spontaneous emission rate on the angle of emission and frequency of light using S-quantization formalism (see [10] for details). Fig. 3(b) shows the dependence of modal Purcell factor (ratio of densities spontaneous emission rates for the emitters placed into the structure and into free space, respectively [10]). It can be seen, that pattern of modal Purcell factor shows peaks for the angles and frequencies satisfying Bragg condition (1). Surprisingly, the pattern of modal Purcell factor does not demonstrate enhanced emission for all the angles and frequencies coupled by Bragg condition which was already pointed out in [1].

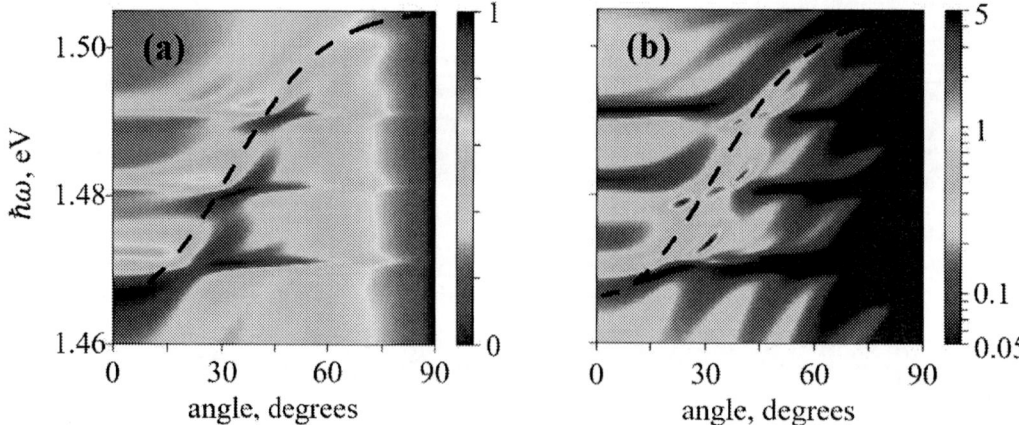

Figure 3(a,b). Modelling of **(a)** the reflection coefficient; **(b)** the modal Purcell factor. The dashed line corresponds to the Bragg condition (1).

We have also measured time-resolved spectra from the surface of the sample $F_s(\omega, t)$ (where emission is affected by the Bragg arrangement of monolayer quantum wells, fig. 4(a)) for different emission angles and from the edge of the sample $F_e(\omega, t)$ as shown in fig. 4(b).

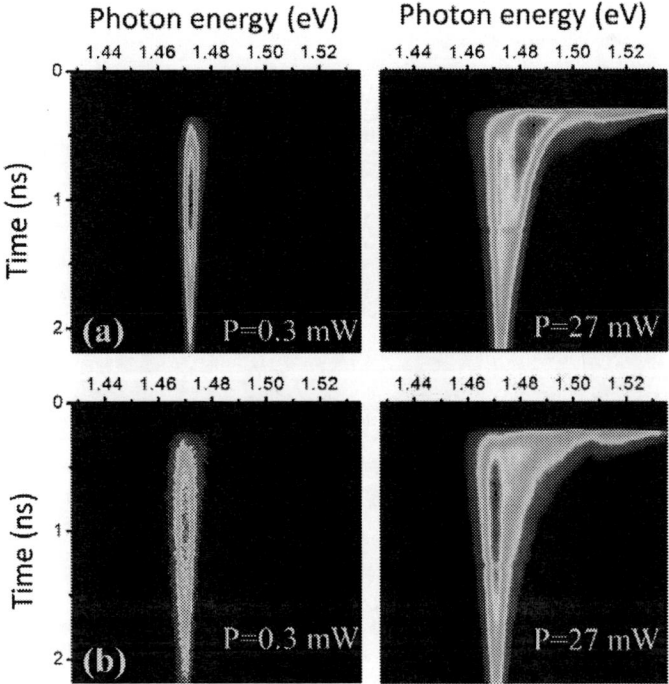

Figure 4(a,b). Time-resolved photoluminescence spectra when measured from **(a)** the surface at emission angle 55 degree; **(b)** the edge of structure for low (left) and high (right) pumping intensities.

A quick look at experimental results satisfies that for the low pumping intensity, there is only one emission line corresponding to ground state exciton for both measurements from the edge and from the surface. For the high pumping intensity additional line appears only for the emission spectrum taken from the surface of the sample (in fact, this line is a superposition of several lines). The picture shown in fig. 3(a) (on the right) corresponds to the emission angle 55 degrees when additional mode has maximal power.

4. Conclusions
We have grown a high-quality ML Bragg structure. The results of the optical study speak for its high quality including uniformity of the periods and high exciton localization. We have modelled and measured time resolved emission pattern from the edge and the surface of InAs monolayer quantum well Bragg structure. The patterns of the emission show that the superradiant mode is only possible when the QWs are excited at once and are coupled. The amplification of the emission shows a vague agreement with the line of Bragg condition, but additional conditions apply.

Acknowledgments
The work was partially supported by the RFBR grant No. 18-32-00801 and by the grant of Minobrnauka № 16.9789.2017/BCh.

References

[1] Ivchenko E L, Nesvizhskii A I and Jorda S 1994 *Superlattices Microstruct* **16**(1) 17.

[2] Chaldyshev V V et al. 2011 *Appl. Phys. Lett.* **98** 073112.

[3] Dicke R H 1954 *Processes Phys. Rev. American Physical Society* **93**(1) 99.

[4] Chaldyshev V V et al. 2011 *Appl. Phys. Lett.* **99** 251103

[5] Askitopoulos A et al. 2011 *Phys. Rev. Lett.* **106** 076401.

[6] Goldberg D et al. 2009 *Nature Photon.* **3** 662.

[7] Pozina G et al. 2015 *Sci. Rep.* **5** 14911.

[8] Pozina G et al. 2017 *physica status solidi (b)* **254**(4) 1600402.

[9] Schaarschmidt M et al. 2004 *Phys. Rev. B* **70** 233302.

[10] Kaliteevski M A, Mazlin V A, Ivanov K A, Gubaidullin A R 2015 *Optics and Spectroscopy* **119**(5) 832.

SPBOPEN 2018

IOP Publishing

Multiple frequency Bloch oscillations in natural superlattices

K A Ivanov[1,2], E I Girshova[2], E D Kolykhalova[3], M A Kaliteevski[1]

[1]ITMO University, St. Petersburg, 197101, Russia
[2]St. Petersburg Academic University, St. Petersburg 194021, Russia
[3]Ioffe Institute, St. Petersburg 194021, Russia

Abstract. Numerical modelling of electron wavefunctions in a one-dimensional superlattice was conducted. The obtained series of wavefunctions were used to calculate matrix elements of dipole transitions and light emission probabilities. The transitions between distant Stark states were shown to be of high probability together with conventional transitions between neighbor states dubbed Bloch oscillations. Distant transitions have emission probabilities with the same dependence on the applied field and somewhat smaller probabilities.

1. Introduction

The phenomenon of Bloch oscillations was predicted quite a long time ago [1] and it revolves around the fact that in a Stark ladder of electronic states resulting from the electric field F application to a periodic structure with period d, the interlevel transitions are possible with a Bloch frequency

$$\omega = eFd/\hbar \qquad (1)$$

In a nutshell, the Bloch oscillations are produced by an electron whose movement is induced by the applied field. Since in a periodic potential the dispersion curve of an electron has a bend, where its effective mass becomes negative, the electron changes its direction of movement, and hence the oscillations.

It is thus essential for the Bloch oscillations to be manifested that an electron energy dispersion is of minizone (sine-like) type. This is true for some crystals in certain directions, but in a crystal such transitions are hardly observable. The difficulty is a consequence of too wide a minizone and hence too large an electric field needed to be applied, breaking the crystal.

However, the superlattices (SLs) as suggested by [2] are attractive for this. The fact that a minizone on these artificial crystals is narrow, the needed electric fields are modest and lead to a Bloch frequency in a terahertz (THz) range. The THz radiation in various semiconductor structures received quite a bit of attention recently [3,4].

One of interesting materials with a potential in THz technology is silicon carbide (SiC). Being a natural crystal, it nevertheless exhibits some properties of a very high-quality superlattice. Its hexagonal polymorphs are well modelled by a layered structure [5]. Indeed, there are recent observations of THz emission from hexagonal SiC with applied electrical bias [6].

As for the phenomenon of the Bloch oscillations in a general sense, there was previously a consensus about its harmonic nature. That is, both in quasi-classical and quantum approaches [7-9] it was predicted that emission or quantum transitions are occurring only on the sole Bloch frequency. The purpose of this work is to demonstrate that emission is not confined to the sole Bloch frequency but is also possible on multiples of the Bloch frequency. The main focus of this work as to application to the real life is SiC and modelling of its properties.

Content from this work may be used under the terms of the Creative Commons Attribution 3.0 licence. Any further distribution of this work must maintain attribution to the author(s) and the title of the work, journal citation and DOI.

Published under licence by IOP Publishing Ltd

2. Model

As was mentioned before, a hexagonal polymorph of SiC (of which 2H, 4H, 6H and 8H are studied in this work) can be considered to be a superlattice. Indeed, as fig. 1a,c shows, the general structure of SiC

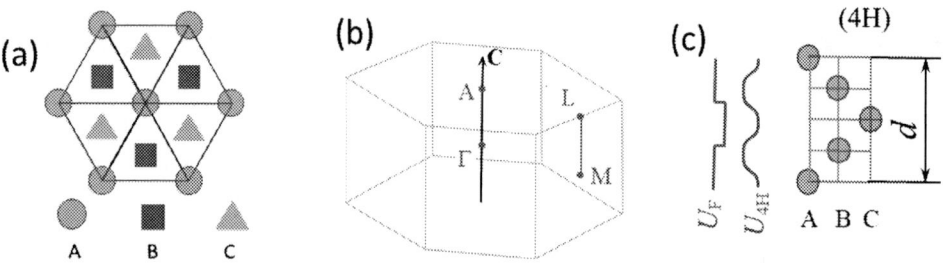

Figure 1(a-b). (a) Densely packed hexagonal lattice; **(b)** the Brillouin zone in a hexagonal lattice; **(c)** arrangement of atomic layers in 4H-SiC.

in its hexagonal form is a densely packed one. The main unit is a layer of silicon-carbon atom pairs that are arranged in triangular fashion. As the layers are stacked one upon another, there are two positions in which atoms could go in relation to the previous two layers. Thus, the layers can be denoted as A, B or

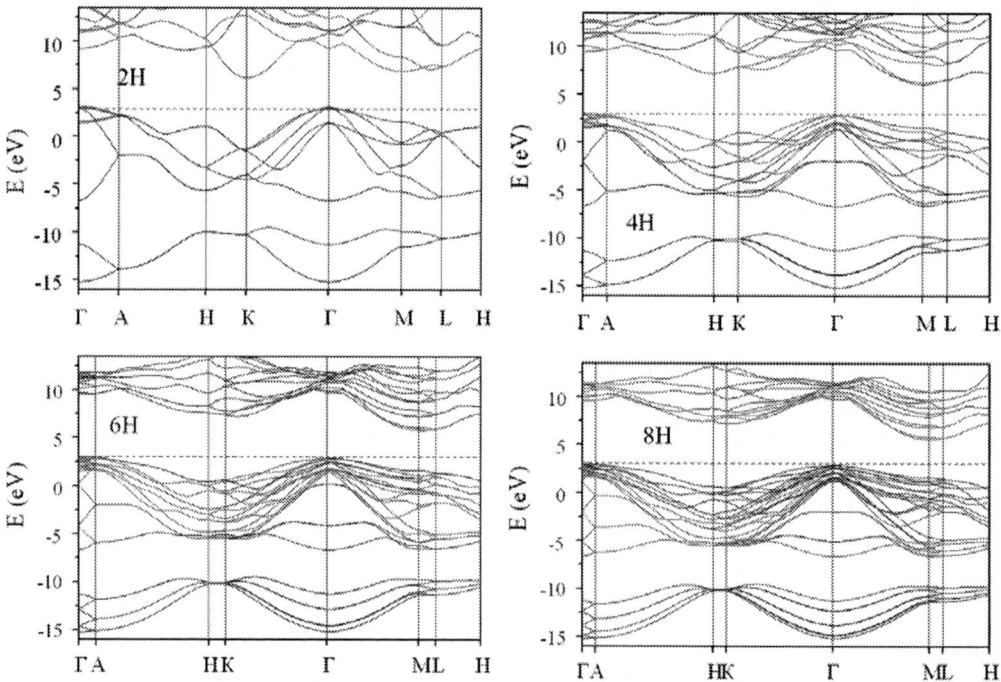

Figure 2. Band structure for different hexagonal SiC polymorphs, modelled by DFT. Fermi level denoted by dashed line.

C. Every layer has to be surrounded by layers of names different to itself to achieve dense packing (i.e. B-C-B or C-B-A). If these two layers are themselves mutually different (as in C-B-A), the layer B has a hexagonal type of near symmetry and lower energy, thus is a quantum well. In the other case the layer is of cubic symmetry and presents a quantum barrier [10-12].

To prove that such a model is indeed plausible we have conducted energy dispersion calculations using the DFT method [13] for 2H, 4H, 6H and 8H polymorphs. The results presented in fig. 2 are to be interpreted in a following way. First, in all the polymorphs except 2H the M-L dispersion branch has the lowest energy in the valence band, which means that the electron transport models should consider this branch. Second, the nature of this branch (except in 2H) is very much like that of an ideal minizone dispersion in a SL.

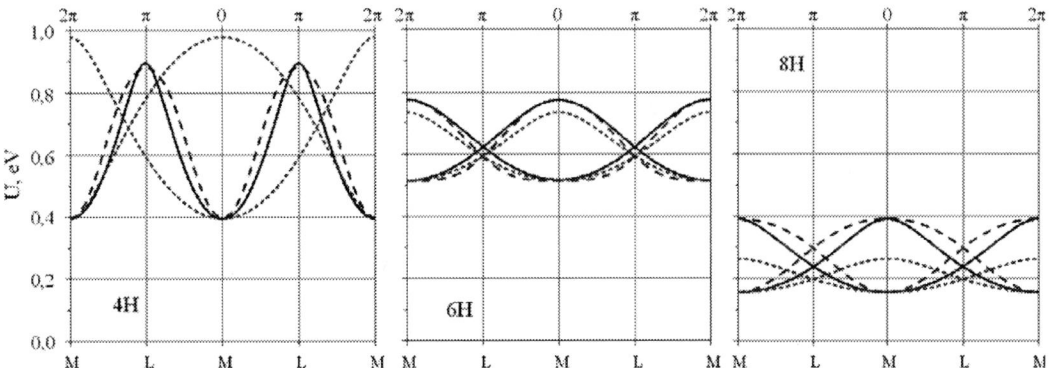

Figure 3. Dispersion in SiC along M-L line near the bottom of the valence band. Solid line – model potential, long dash – DFT calculation, short dash – dispersion from [5].

We went on by selecting for 4H-, 6H- and 8H-SiC such a model potential as to resemble the one calculated by DFT as closely as possible. The results are shown in fig. 3 and table 1. Each potential period is a sum of widths of the barrier and the well and is equal to a lattice constant of a corresponding SiC polymorph in the direction perpendicular to the layers (along the line going through Si-C pair). We have also somewhat improved the findings of [5] by providing a much closer resemblance of the model dispersion to the DFT-calculated one.

Table 1. Model potential parameters.

	Barrier height, eV	Barrier width, nm	Potential period, nm	Electron mass, m_0
4H	1.9	0.095	0.5	1.75
6H	2.35	0.25	0.75	1.0
8H	0.79	0.25	1.0	1.0

The obtained periodic model potential can now be used to conduct calculations of quantum properties of the material under electrical bias. The calculation utilised the transfer matrix approach. The reason for this is that one cannot use DFT when an electric field is applied for the problems of large enough scale. Not only is the calculation time unrealistic, but a parasitic dipole moment often emerges and has to be corrected. The transfer matrix method, on the other hand, isn't general and needs a one-dimensional model, but is very quick and can be generalized for the case of an applied electrical field. It helps to observe that the wavefunction and its flow (proportional to its derivative) are bound to be conserved at the boundary between two layers with different (but constant) potential energy. This leads to construction of a 2x2 transfer matrix for all the boundaries and layers. These matrices depend on the potential energy and the energy of an electron via the electron wavevector $k = \sqrt{2m(E-U)}/\hbar$. Since a wavefunction must take a form of a decaying exponent outside of the structure, one can obtain an equation on E. This provides a method to calculate eigen energies of a certain structure. The slope in potential energy arising from an application of an electrical field can be approximated by enough steps (a piece-wise constant function). In an agreement with quantum theory we have found out that in a biased periodic SL the energy levels are equally spaces and form a Stark ladder of translationally symmetrical states. The spatial width s of the state is proportional to the minizone width and reverse proportional to the applied field:

$$s = W/eF \tag{2}$$

This is shown in fig. 4.

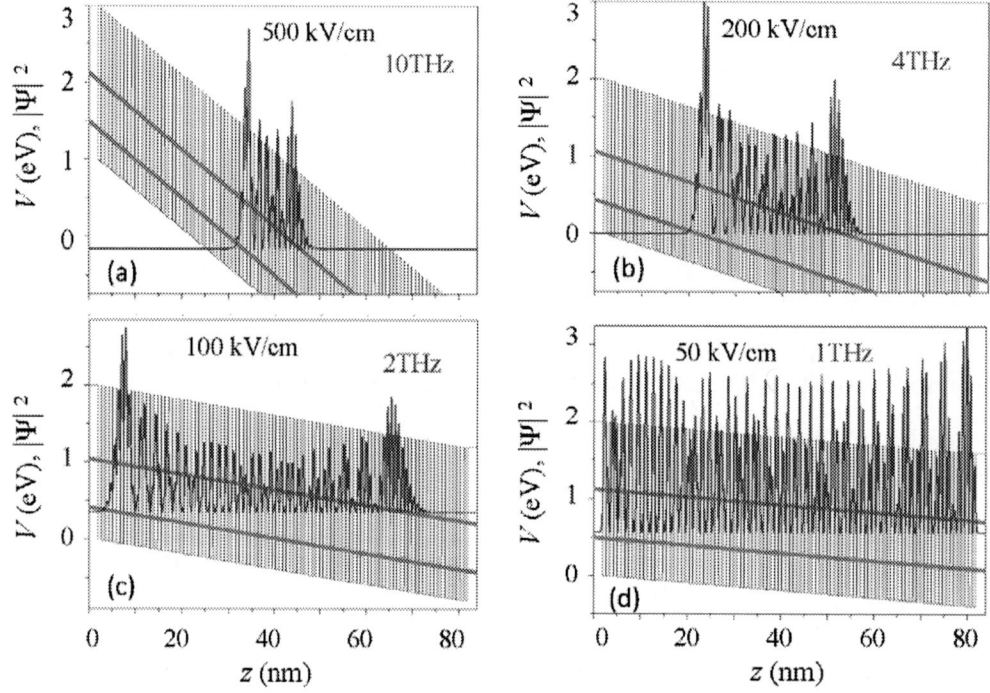

Figure 4(a-d). Spatial distribution of probability density (wavefunction module squared) for an electron in a SL with different bias field applied: **(a)** 500 kV/cm; **(b)** 200 kV/cm; **(c)** 100 kV/cm; **(d)** 50 kV/cm.

With energies and transfer matrices at hand we can calculate wavefunctions $\langle f|$ and $|i\rangle$ of any dipole transition and subsequently a dipole matrix element and the emission probability for any frequency, provided that this frequency is a difference between the energies of Stark states (that is, a multiple of Bloch frequency (1)):

$$P = \frac{\alpha|\langle f|x|i\rangle|^2 n^2 \omega^3}{\pi c^2} \tag{3}$$

For the single Bloch frequency, the matrix element can be estimated [9] to be $s/4$ and the estimate of the transitional probability is

$$P = \frac{\alpha}{16\pi}\left(\frac{nWd}{c\hbar}\right)^2 \omega \tag{4}$$

3. Results

When calculated, it became evident that there exists a possibility for transitions at the frequencies of 2ω and 3ω, together with conventional Bloch transitions at ω [14]. Importantly, the probabilities of such transitions are proportional to the applied electrical field as is the probability of Bloch oscillations. The value of the probability falls with every multiple by an order of magnitude, that is, at 2ω we are getting a ten times weaker emission than at ω, and the same is true for 3ω versus 2ω. The results are summarized in fig. 5. Still, the estimate (4) holds for the single frequency transition which is shown by a dashed line and speaks for the correctness of the chosen method.

Importantly, this behaviour needs a SL with size bigger than an effective state size s. Otherwise, the transition is weaker due to weaker localisation but exhibits a superlinear growth against the growth of applied field (evident in fig. 1 for the part with smaller frequencies).

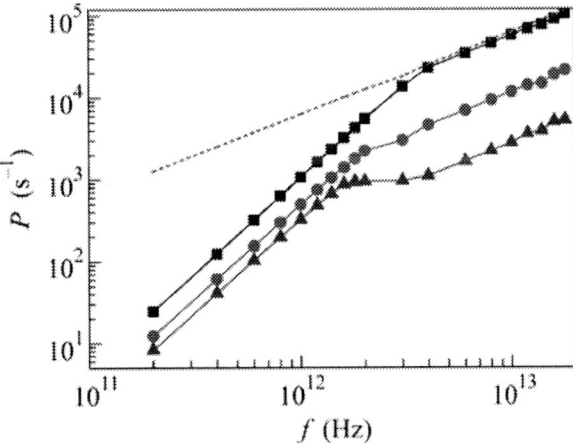

Figure 5. Dependence of transition probabilities for single (squares), double (circles) and triple (triangles) Bloch frequencies on frequency. The dashed line corresponds to (4).

4. Conclusions

We have studied Bloch oscillations in electrically biased and provided a theoretical proof that they might be accompanied by the emission of light on multiples of the Bloch frequency. We have also studied the energy dispersion in hexagonal SiC and provided a model for studying quantum effects in it. The advantages of the model are the speed of calculations and the sustained applicability under symmetry-breaking conditions such as an applied electric field.

Acknowledgements

The work was partially supported by the RFBR grant No. 18-32-00801.

References

[1] Bloch F 1928 *Z. Phys.* **52** 17.
[2] Esaki L, Tsu R 1970 *IBM J. Res. Dev.* **14** 61.
[3] Savenko I G, Shelykh I A, Kaliteevski M A 2011 *Phys. Rev. Lett.* **107** 027401.
[4] Gallant A J et al. 2007 *App. Phys. Lett.* **91** 161115.
[5] Dubrovskiy G B, Lepneva A A 1977 *Sov. Phys. Solid State* **19**(5) 1252.
[6] Sankin V I et al. 2012 *Appl. Phys. Lett.* **100** 111109.
[7] Laref A, Laref S 2008 *Phys. Status Solidi B* **245** 89.

[8] Luban M 1985 *J. Math. Phys.* **26** 2386.
[9] Bouchard A M, Luban M 1995 *Phys. Rev. B* **52** 5105.
[10] Ramsdell L S 1947 *American Mineralogist* **32** 64.
[11] Verma A R, Krishna P 1966 *Polymorphism and Polytypism in Crystals* (New York: Wiley).
[12] Sankin V I, Stolichnov I A 1998 *Superlattices and Microstruct.* **23** 999.
[13] Clark S J et al. 2005 *Zeitschr. fuer Kristallographie* **220**(5-6), 567.
[14] Ivanov K A, Petrov A G, Kaliteevski M A, Gallant A J 2015 *JETP Lett.* **102**(12) 796.

SPBOPEN 2018

IOP Publishing

Enhancement of spontaneous emission probability in a disordered media

K M Morozov[1,2]*, K A Ivanov[2], A R Gubaydullin[1,2], M A Kaliteevski[3]

[1]St. Petersburg Academic University, St. Petersburg, 194021 Russia
[2]ITMO University, St. Petersburg, 197101 Russia
[3]Ioffe Institute, St. Petersburg, 194021 Russia

*E-mail: const.morozov@mail.ru

Abstract. Was provided statistical analysis of Purcell effect dependence on emission frequency and emitter position in disordered structure. Fact that the presence of disorder in one-dimensional photonic crystals could lead to a modification of spontaneous emission rate in the frequency region corresponding to a photonic band gap (PBG) was demonstrated. Was shown, that amount of disorder regulates the two different regimes of the Purcell effect. In case of weak disorder Purcell enhancement of spontaneous emission occurs at PBG edges, and in case of strong disorder at PBG centre.

1. Introduction

Interplay of the Bragg interference and random scattering of light in disordered photonic crystal (PC) structures gives rise to a wide range of fascinating optical phenomena, such as Anderson localization of light [1,2] modification of band structure [3,4] and emission properties [5]. As known, localized states appear in the frequency region corresponding to photonic band gap (PBG) of disordered PC. For those states the local value of the electric field could be very high, they act as a Fabry-Perot resonances and lead to increasing transmission through the structure [6]. Strong spatial variation of the electromagnetic field for the localized photonic states in disordered photonic crystals can also lead to a modification of the spontaneous emission rate (Purcell effect) [7,8]

The rate of spontaneous emission can be estimated by a number of various methods. To get statistical examination of the problem (averaging results over the ensemble), it needed to be fast. The approach developed in [9] requires less computational power but gives correct result only in case of centre symmetrical structures. Here, for the analysis of Purcell effect in disordered PC, we will use generalization of that approach, called S-quantization method.

The aim of present paper is to provide statistical analysis to get understanding how disorder changes the picture of spontaneous emission probability in one-dimensional photonic crystal structures by using S-quantization formalism [10,11].

Content from this work may be used under the terms of the Creative Commons Attribution 3.0 licence. Any further distribution of this work must maintain attribution to the author(s) and the title of the work, journal citation and DOI.

Published under licence by IOP Publishing Ltd

2. Results and discussion

Was considered the situation of one-dimensional photonic crystal (200 layered Bragg structure), based on two layers (A and B) with the same length d and different refractive indexes (n_A and n_B) with presence of disorder. Refractive index profile of disordered structure (figure 1) was defined by relation:

$$n_{A,B}^{dis} = n_0 \pm g + n_0 \delta P, \tag{1}$$

where first two terms describe ideal periodic structure situation ($n_{A,B}^{Ideal} = n_0 \pm g$), and term $n_0 \delta P$ define the amount of disorder in structure by a fluctuation parameter δ (parameter P takes values in range [-0.5, 0.5]).

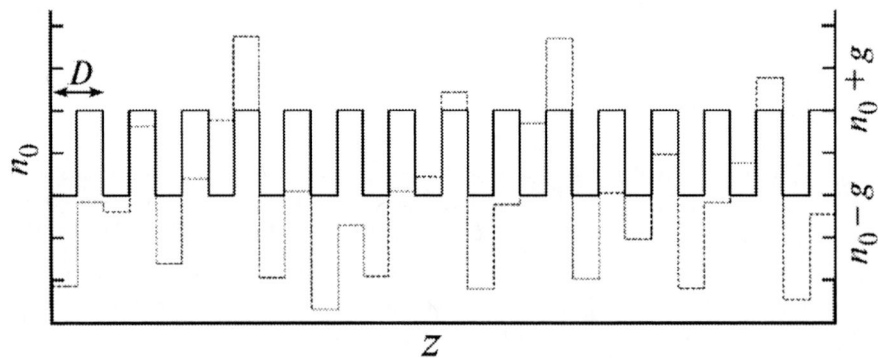

Figure 1. Parameters of the structure. Refractive index profiles in the ideal and disordered structures are shown by solid and dashed lines, respectively.

In this consideration, the main property of periodic structure (optical length) would depend only on parameter δ :

$$D_{opt}^{dis} = n_A^{dis} d_A + n_B^{dis} d_B = d(n_0 + g + n_0 \delta P + n_0 - g + n_0 \delta P) = 2n_0 d(1 + \delta P) \tag{2}$$

For zero disorder case ($\delta = 0$), PBG centered at Bragg frequency

$$\omega_0 = \pi c / (n_0 D), \tag{3}$$

and a relative PBG width is

$$\Delta\omega / \omega_0 = 4g / (\pi n_0). \tag{4}$$

Hereafter we will use the average refractive index $n_0 = 2.0$ and modulation $g = 0.025$, providing relative width of the PBG $\Delta\omega / \omega_0 \approx 0.016$.

The localized states can appear in the PBG then the disorder level rise to value δ_{th} (threshold level of disorder):

$$\delta_{th} = \sqrt{2/\pi}\sqrt{\Delta\omega/\omega_0}. \tag{5}$$

For the parameters used in the modelling $\delta_{th} \approx 0.75$. For the values of the disorder parameter above $\delta = 2\delta_{th} \approx 0.15$ the effects, related to an existence of PBG disappear and the system can be considered completely disordered. In that case, was choose the different parameters of fluctuation parameter δ (0.07, 0.1 and 0.15) to investigate transition between different types of spontaneous emission amplification and suppression in several frequency areas.

To analyze how the presence of disorder changes the known picture of Purcell effect in one dimensional PC S-quantization formalism was chosen. S-quantization is the undemanding to computational power and fast numerical approach to calculate the spontaneous emission rate in arbitrary non-uniform dielectric media [10,11]. Approach gives value of the modal Purcell factor, number that describes how change spontaneous emission rate of single emitter placed in structure for a specific position in structure and emission frequency. Also, to get more complete picture of phenomenon, the results of calculation was averaged over the ensemble (10^4) of structures.

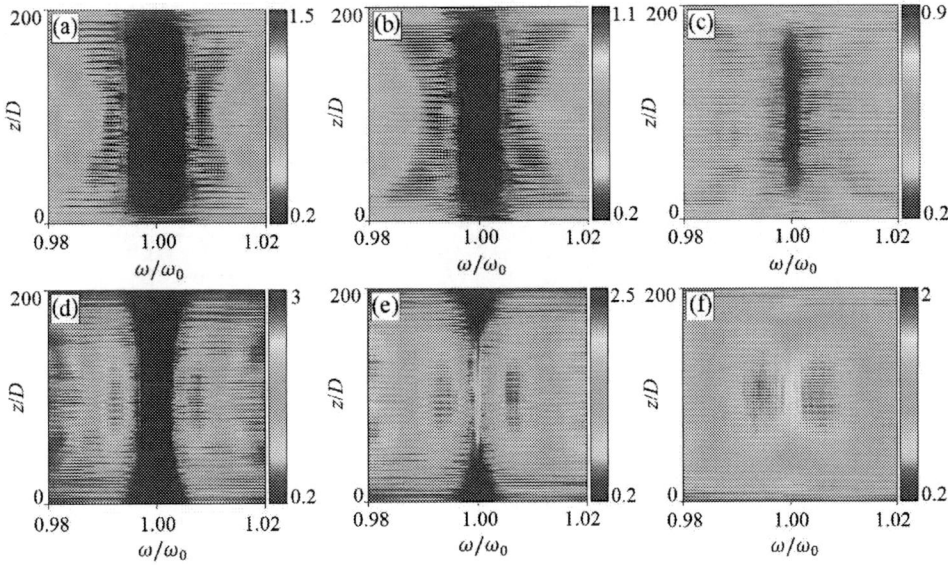

Figure 2. (color online) On the top panels: dependence of the modal Purcell factor on the frequency and position of the dipole source placed inside the disordered structure and averaged over an ensemble of 10^4 structures with δ=0.07 (a) 0.1 (b) and 0.15 (c). On the bottom panels (d, e, f) show the dependence of the standard deviation σ corresponding to the averaged Purcell factor, shown on (a, b, c) respectively, on the frequency and the dipole position.

Figure 2 shows the pattern of the Purcell coefficient and its average over 10^4 configurations. When the disorder parameter equals its threshold value $\delta = 0.07$, there are two ranges related to an area of the reduced Purcell coefficient corresponding to the photonic band gap and to a widened area of enhanced emission corresponding to the edge state, respectively. Figures 3 c-f illustrates the evolution of the pattern of the Purcell coefficient with increasing disorder parameter δ. It can be seen, that though the disorder increases lead to the shrinking of PBG for the Purcell coefficient, PBG in the pattern of the Purcell coefficient still remains noticeable for both values of disorder $\delta = 0.1$ and $\delta = 0.15$. At the same time, in the pattern of standard deviation of Purcell coefficient (figure 2 d,e,f), PBG disappears when δ reaches the value 0.1.

Figure 3 shows dependences of modal Purcell factor on the frequency, in case of dipole position in structures that gives the maximum enhancement of the spontaneous emission. Peaks, that appeared in PBG zone are related to the high localized states. It can be seen, with increase of disorder, maximum value of Purcell factor increases too. That profile of modal Purcell factor is very close to microcavity eigenstate electric field profile, the narrow peak with exponentially decayed tails. Figure 4 demonstrates the averaged modal Purcell factor changing with different value of structures in set. It can be seen here, why the number structures in ensemble was choose as 10^4. Case at quantity 10^3 doesn't give a full picture of phenomenon, especially at high levels of disorder (more than threshold value of δ).

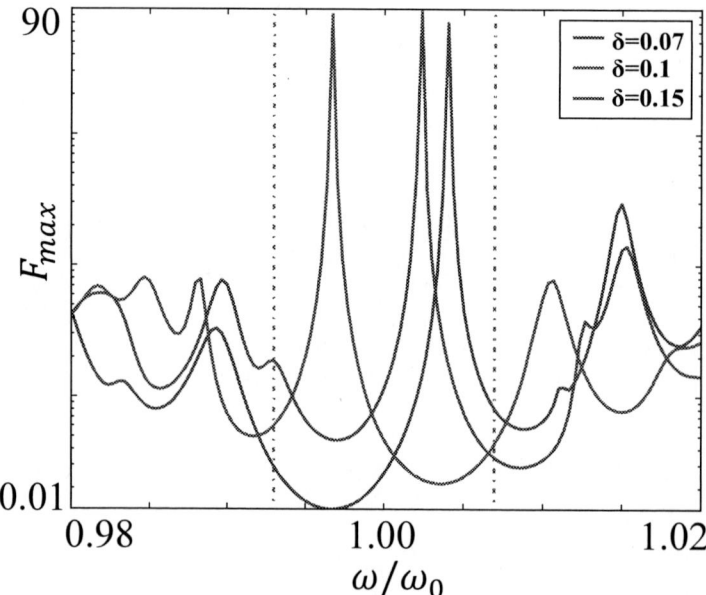

Figure 3. (color online) Dependence of the modal Purcell factor on the frequency, which illustrates the maximum enhancement of the spontaneous emission achieved on the localized state appearing in the PBG due to disorder $\delta=0.07$ (blue), 0.1 (green), 0.15 (red). Purple dashed line shows the edges of PBG in ideal PC structure.

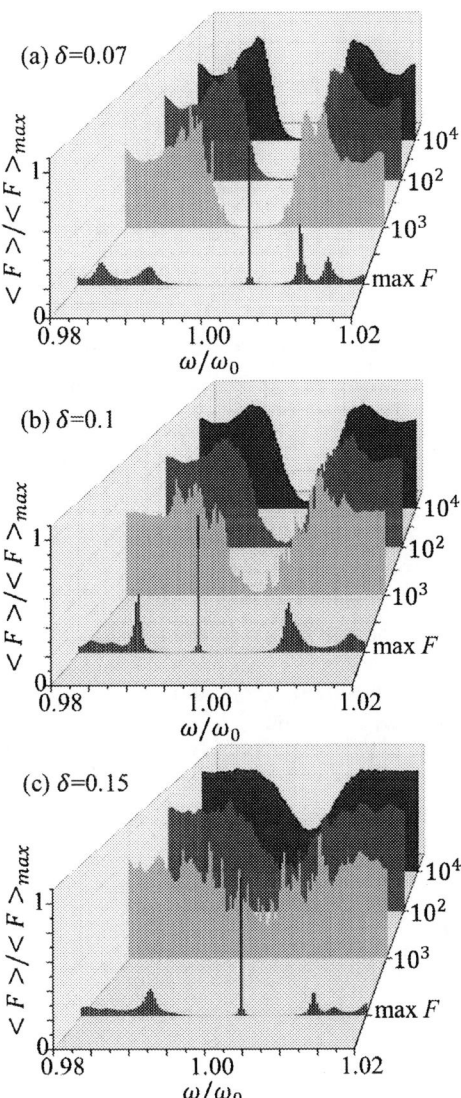

Figure 4. (color online) Normalized dependence of the modal Purcell factor on the frequency, averaged over an ensemble of 10^2 (green), 10^3 (red), 10^4 (dark red) structures with disorder fluctuation parameter δ=0.07 (a), 0.1 (b), and 0.15 (c). Blue profile shows the dependence for single disordered structure, which provides the maximum value of the modal Purcell factor, obtained in the ensemble of 10^4 structures, and illustrates the maximum enhancement of the spontaneous emission achieved on the localized state appearing in the PBG.

3. Conclusions

Calculation of spontaneous emission rate for emitter placed in disordered photonic crystal structures using S-quantization formalism was carried out. Statistical analysis of averaged modal Purcell factor with different amount of disorder was done. Was demonstrated, how variation disorder influences on appearing of eigenstates in PBG. Was shown, that depend on the rate of disorder in system two different cases of Purcell effect take place.

Acknowledgements

The work is supported by Ministry of Higher Education and Science of the Russian Federation (project № 3.9787.2017/8.9).

References

[1] Limonov M F, De La Rue R M 2016 *Optical Properties of Photonic Structures: Interplay of Order and Disorder* (CRC Press).
[2] Lagendijk A et al. 2009 *Phys. Today* **62**, 24–29.
[3] Kaliteevski M A, Beggs D M, Brand S, et.al. 2001 *Phys. Rev. B* **91**(1), 109-118.
[4] Kaliteevski M A, Abram R A, Brand S, Nikolaev V V Opt. Spect. **91**, 109-118 (2001).
[5] Romanov S G, et.al. 2004 *Phys. Status Solidi* **1**, 1522-1530.
[6] Deych L I, Zaslavsky D, Lisyansky A A 1998 *Phys. Rev. Lett.* **81,** 5390–5393.

[7] Purcell E M 1946 Phys. Rev. **69**, 681.
[8] Petrov E P, Bogomolov V N, Kalosha I I, Gaponenko S V 1998 *Phys. Rev. Lett.* **81,** 77–80.
[9] De Martini F, Marrocco M, Mataloni P, Crescentini L, Loudon R 1991 *Phys. Rev. A* **43**, 2480–2497.
[10] Kaliteevski M A, Mazlin V A, Ivanov K A, Gubaidullin A R 2015 Opt. Spect. **119**, 832-837.
[11] Kaliteevski M A, Gubaidullin A R, Ivanov K A, Mazlin V A 2016 Opt. Spect. **121**, 71-81.

SPBOPEN 2018

IOP Publishing

IOP Conf. Series: Journal of Physics: Conf. Series **1124** (2018) 081048   doi:10.1088/1742-6596/1124/8/081048

Room temperature lasing in injection microdisks with InGaAsN/GaAs quantum well active region

E I Moiseev[1], M V Maximov[1], A M Nadtochiy[1], N V Kryzhanovskaya[1], D A Sannikov[2,5], T Yagafarov[2], M Kulagina[3], T Niemi[4], R Isoaho[4], M Guina[4] A E Zhukov[1]

[1]St Petersburg Academic University, 194021 St Petersburg, Russia
[2]Skolkovo Institute of Science and Technology, 143026 Moscow, Russia
[3] Ioffe Institute, 194021 St Petersburg, Russia
[4]Optoelectronics Research Centre, Tampere University of Technology, 33720 Tampere, Finland
[5] The Lebedev Physical Institute of the Russian Academy of Sciences, 119333 Moscow, Russia.

Abstract. Injection microdisk lasers based on three InGaAsN/GaAs quantum wells with different diameters of the resonator were fabricated and studied. Room temperature lasing at 1.2 μm is demonstrated for the first time. Dependence of the threshold current on the diameter is discussed.

1. Introduction

As the level of integration of electronic chips increases, the number of integrated transistors per unit area becomes increasingly large, and electrical interconnection technologies using electrons as signal transmission carriers face serious challenges [1, 2]. For these reasons, many researchers have turned their attention to optoelectronic devices. Microdisk (MD) lasers are prospective candidates as laser sources for optoelectronic circuits since high quality factors can be achieved even in resonators of a few micrometers in diameter [3,4] at a low lasing threshold. MD lasers have a smaller footprint, lower power consumption, and better high-speed modulation characteristics, as compared with Fabry–Pérot cavity lasers [5]. Moreover, semiconductor MDs attract widespread attention due to their unique fundamental properties, as well as exploitability as ultrasmall modulators, detectors, and sensors. Using quantum dots (QDs) as the active region in such lasers have advantages of low threshold and high thermal stability of characteristics [6]. However, the optical gain of the ground-state optical transition, which can be achieved with QDs, is limited due to finite number of the QDs. This may lead to lasing via the excited states of the QDs or completely prevent lasing. The quantum well (QW) active region provides a higher gain compare to the QDs. Thus the use of QWs may help to work out the problem of gain saturation when the laser`s size are scaled down. Meanwhile, the wavelength of InGaAs QWs is limited to about 1.1 μm. Longer wavelengths, which are more preferable for optical interconnect, can be achieved with nitrogen-containing QWs. Previously, we have demonstrated InGaAsN injection MD lasers capable of lasing up to 170K [7]. In the present work, room temperature operation of such microlasers is demonstrated for the first time.

Content from this work may be used under the terms of the Creative Commons Attribution 3.0 licence. Any further distribution of this work must maintain attribution to the author(s) and the title of the work, journal citation and DOI.

Published under licence by IOP Publishing Ltd

2. Experiment

The structures studied were grown by molecular beam epitaxy on GaAs (100) substrate. The structure consists of a GaAs waveguide with three $Ga_{0.7}In_{0.3}N_{0.02}As_{0.98}$ QWs separated with 10-nm-thick GaAs layers and have ground-state emission peak around 1.2 μm at room temperature. The thickness of the waveguide with the QW active region is 400 nm, the thickness of $Al_{0.25}Ga_{0.75}As$ cladding layers with gradient doping was 2 μm on both sides of the active region. The structure represents a semiconductor p-i-n diode, with the p-side on top, and the substrate as the n-side. A sketch of the layers sequence is shown in figure 1. The microdisks with a diameter of the resonators ranging from 11 to 31 μm were formed by means of photolithography and chemical plasma etching. SEM image of the MD array is shown in figure 2. The etching depth was about 4 μm. AgMn/NiAu (AuGe/Ni/Au) metallization was used to form ohmic contacts to the p+ GaAs cap layer and the n+ substrate, respectively. The structures were investigated under pulse injection pumping at room temperature. Electroluminescence was excited with 0.05-μs-long injection pulses with 10 kHz repetition rate. Injection current was limited to 300 mA. A piezoelectrically adjustable objective (Olympus LMPlan IR objective ×10 with NA = 0.25) was used to collect the microelectroluminescence signal (μEL) from the microlasers. The collected μEL was then analyzed with a 1000 mm monochromator Horiba FHR coupled with a cooled InGaAs CCD array.

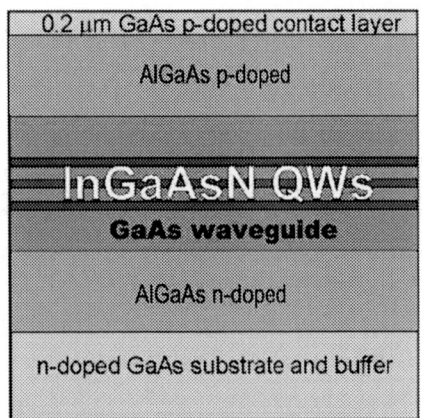

Figure 1. Schematic illustration of the heterostructure layers sequence.

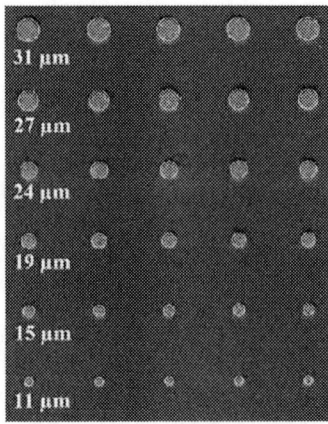

Figure 2. SEM image of the MD array.

3. Results

At low injection currents, a broad spontaneous emission is observed (figure 3). The ground-state peak is centered around 1.2 μm with full width at half maximum of more than 60 nm. μEL spectra were obtained for all investigated diameters. The spectra of the microdisk laser with diameter D=11 μm at different pump current are shown in Figure 4. The narrow lines corresponding to whispering gallery modes (WGM) of the resonator is observed on the long-wavelength side of the InGaAsN QWs spectra for all the lasers. FWHM ($\Delta\lambda$) was 0.05 nm. The Q-factor was estimated as $\lambda/\Delta\lambda$ to be 24'000.

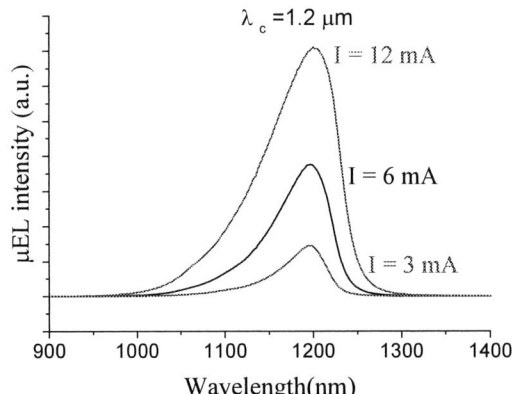

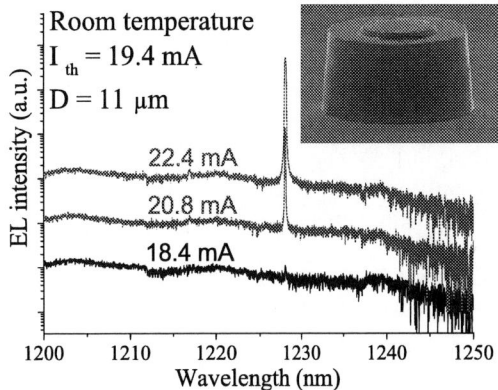

Figure 3. Emission spectra obtained at 300 K at different pump current.

Figure 4. The spectra of the microdisk laser (D=11 μm) at different pump current. The spectra are shifted in the vertical direction for clarity. *Inset: SEM image of the microdisk laser.*

Figure 5 shows the dependence of the integrated emission intensities of the lasing WGM line at room temperature on the pump current for the microdisks of different diameters. The threshold current is 19.4 and 65 mA for microlasers with diameters of 11 and 27 μm, respectively.

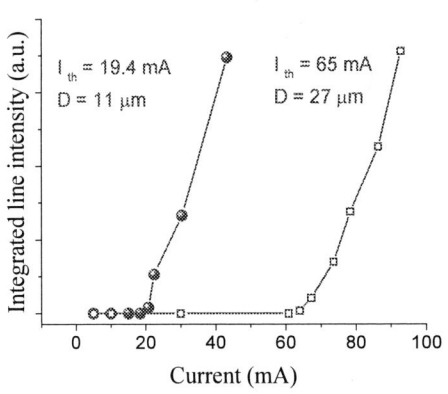

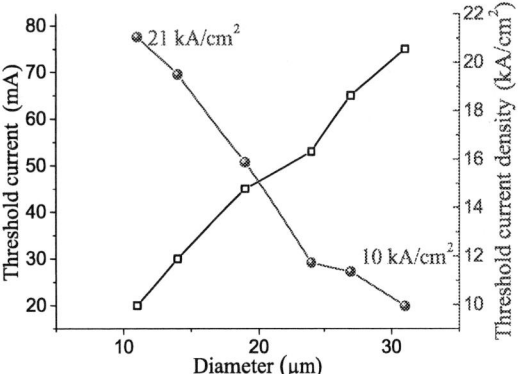

Figure 5. Current dependence of integrated line intensity for 11 μm and 27μm in diameter MDs.

Figure 6. Threshold current (solid symbols) and threshold current density (open symbols) as a function of MD diameter.

The threshold current density J_{th} was estimated for all diameters of resonator (figure 6). When the diameter of MDs decreases from 31 to 10.5 μm, the minimal value of J_{th} grows from 11 kA/cm^{-2} up to 21 kA/cm^{-2}. The growth of J_{th} with a decrease in D for MD lasers is probably caused by non-radiative recombination at the MD sidewalls. The surface passivation procedure will be probably helpful to reduce the threshold current.

To the best of our knowledge, this is the first demonstration of room-temperature lasing in N-containing microdisk lasers under injection excitation.

Acknowledgments

The work is supported by the Skolkovo Foundation (grant agreement for Russian educational and scientific organisation no.6 dd. 30.12.2015)" and Skolkovo Institute of Science and Technology (General agreement no. 3663-MRA dd. 25.12.2017).

References

[1] H. Ron, K.W. Mai, M.A. Horowitz, The future of wires, Proc IEEE, 89 (4) (2001), pp. 490-504

[2] J.A. Davis, et al., Interconnect limits on gigascale integration (GSI) in the 21st century Proceedings of the IEEE, 89 (3) (2001), pp. 305-324

[3] Srinivasan K, Borselli M, Painter O, Stintz A and Krishna S 2006 Opt. Express 14 1094

[4] M V Maximov et al. (2014). Ultrasmall microdisk and microring lasers based on InAs/InGaAs/GaAs quantum dots. Nanoscale research letters, 9(1), 657.

[5] X. M. Lv, Y. Z. Huang, Y. D. Yang, L. X. Zou, H. Long, B. W. Liu, J. L. Xiao, and Y. Du, "Influences of carrier diffusion and radial mode field pattern on high speed characteristics for microring lasers," Appl. Phys. Lett. 104(16), 161101 (2014)

[6] Y. Arakawa, H. Sakaki, Appl. Phys. Lett. 40 939 (1982).

[7] N. V. Kryzhanovskaya, E. I. Moiseev, Yu. S. Polubavkina, F. I. Zubov, M. V. Maximov, A. A. Lipovskii, M. M. Kulagina, S. I. Troshkov, V.-M. Korpijarvi, T. Niemi, R. Isoaho, M. Guina, M. V. Lebedev, T. V. Lvova, and A. E. Zhukov, J. Appl. Phys. 120, 233103 (2016).

AUTHOR INDEX

Abdullin, F A 378
Abdullin, Kh. A......77
Afinogenov, B. I. 408
Agafonov, I Y 444
Ageev, O A 1, 149, 158, 162, 964, 1087, 1141
Aglikov, A......542
Akasov, R. A. 277
Akhmerov, Yu 499
Akimov, A......761
Aksenov, E T 306, 311
Al-Alwani, A J 91, 257, 1018, 1136
Alekhina, O V 1096
Alekseenko, A P 272
Alekseev, A A 984
Alekseev, P A 202, 1038, 1121
Alekseev, P 1083
Alekseev, V 956, 972
Alexeev, A N 1087
Alexeyev, C N 578
Alexeyeva, M C 578
Aleynik, A S 852
Almaev, A V 895
Alonso, E T 905
Aluyev, A 499
Alyabyeva, N I 87
An, P 765, 769, 784, 788
Anikeeva, V E 387
Anokhin, A......659
Anoshkin, I 602
Antipova, O 315
Antonov, I N 1106
Antonov, V I 1151
Arsenov, P V 1171
Artamonov, D V......1043
Arteev, D S 458
Artyukhov, I 968
Arustamyan, D A 915
Arzamastsev, S......956
Ashurov, M S 586
Avdeev, S P 118, 158
Averkin, S 1156
Avilov, V I......87
Avilova, M M 1183
Babich, E S 642, 678
Babichev, A V 486
Babkin, A V 1110, 1115
Babkina, Anastasiia N......856
Bagayev, S N 630

Bagshaw, J. T......537
Baidus, N V......171, 520
Bakarov, A K 848
Bakulin, E 998
Balakirev, S V 1, 81, 112, 462, 1087
Baldycheva, A 905
Balezin, M 813
Baranov, A I......507, 515
Baranov, M A 272
Baryshev, A M 874
Beginin, E N......900
Begletsova, N 1018, 1026
Bektemirova, M 870
Belanova, A 659
Belenkov, E A 6, 43, 49
Belenkov, M E 43
Belikov, A. I. 557
Belorus, A O 262
Belov, A I 30
Belyakov, M 813
Benemanskaya, G V 722
Benimetskiy, F A 1121
Berdnikov, Y 181
Berdnikov, Yu 292
Berezovskaya, T N 980
Bert, N A......122, 127
Berzhansky, V 838, 844, 870
Besedina, N A 292
Bezpaly, A D 651
Bezuglaya, E. V. 417
Bikberdina, N 288
Blokhin, S A 582
Bobrova, D 288
Bogatov, N. M. 96, 915
Bogdanov, A......233, 320, 808, 813, 817
Bogomazova, N......1126
Boikov, I K 476
Bokova, A 838
Bolshakov, A 154, 197, 436, 449, 532, 542, 976
Bolshedvorskii, S 761
Bondarev, A D 176
Borodin, B R 1121
Boronenko, M 288
Boronin, A I 107
Borshchev, O V 593
Bougrov, V......486, 1013
Bouravleuv, A D 202, 980
Breuer, S......352

Bukatin, A	315, 333, 343, 582	
Bulyanitsa, A	283	
Burakov, A E	1096, 1110, 1115	
Burakova, I V	1096, 1110, 1115	
Burtsev, A A	948	
Butt, M A	551, 563	
Buyanova, E S	574	
Bykova, E M	87	
Carapezzi, S	1165	
Chebotarev, S N	68, 915	
Chekhonin, I A	630	
Chekhonin, M A	630	
Chelny, A	499	
Cherkasov, K V	942	
Chernikov, E V	895	
Chernov, V M	43	
Chernova, R K	1026	
Chernyakov, A E	458	
Chernyshev, A	659	
Chikalova-Luzina, O P	328	
Chipouline, A	597	
Chu, Y H	1146	
Chumakov, A	1018, 1026, 1136	
Cirlin, G	20, 202, 1083	
Clemente, I	149, 1156	
Craciun, M F	905	
Danilina, E. M.	96	
Danilov, V V	348	
Danilov, Yu	865	
Danilushkin, A V	948	
Danishevskaya, Y	870	
Davydov, R V	1151, 1160	
Davydov, V V	63, 227, 365, 382, 403, 467	
Davydova, T A	895	
Degtyarev, S A	551	
Deligeorgis, G.	1070	
Demchenko, P.	602, 905	
Dementev, P A	722	
Demidov, A A	928, 952	
Demidov, E	860	
Demina, P. A.	277	
Denisenko, M. V.	673	
Denisov, S A	171	
Derevyannikova, E A	107	
Dermeneva, M.	1013	
Didenko, S	499	
Dikareva, N V	520	
Dimakis, E	181	
Divochiy, A V	689, 699	
Dolzhikova, A	956	
Dorokhin, M	171, 860, 865	
Dostanko, A P	149	

Dragunova, A S	444	
Drakin, A Yu	928, 952	
Dubrovskii, V G	142, 181, 292, 352	
Dudelev, V V	486	
Dukelskii, K V	827	
Dunaeva, T	972	
Dunaevskiy, M	1038, 1083, 1106, 1121	
Dvoreckaia, L N	154	
Dvoretckaia, L N	197, 976	
Dzyubanenko, S	998	
Efimov, A A	1132, 1171	
Efimov, Yu P	743	
Egorkin, V I	1074	
Egorov, A Yu	486	
Egorov, I	1049	
Egorov, V S	630	
Egorova, M N	984	
Elezov, M. S.	611	
Eliáš, M	421	
Eliseev, S A	743	
Elizarov, V	718	
Emelyanov, A	233, 320	
Emelyanov, V M	503	
Enns, Y. B.	878	
Enns, Y.	1008	
Eno, N. A.	537	
Eremenko, M M	1, 81, 112	
Ermak, S V	747	
Ermakov, A	1018	
Eseev, M K	1078	
Evreinova, N V	252	
Evseenkov, A S	937	
Evstrapov, A A	343	
Evstrapov, A	283	
Evstropiev, S K	827	
Evstropov, V V	360, 481	
Evstropyev, K S	827	
Evtikhiev, V P	621	
Ezhov, A A	586	
Faradzhev, S P	994	
Faradzheva, M P	994	
Fatkhutdinova, L I	747	
Fedorov, G E	797	
Fedorov, G F	779	
Fedorov, S. N.	472	
Fedorov, V V	39, 154, 197, 449	
Fedoseev, M	35	
Fedotov, A A	133	
Ferrari, S	756, 765, 769, 784, 788	
Fetisova, M V	444, 582	
Filatov, D O	1106	
Filatov, N A	333, 343	

Filimonov, E D	503, 524
Filippov, S V	1165
Firsov, D A	486
Fomchenkov, S A	568, 616, 710, 739
Fomin, A	53, 968, 1049, 1053, 1057
Fomin, E V	176
Fomina, M	35, 53, 968
Frizyuk, K S	647
Frolov, I V	440
Gabitov, I	597
Gajdoš, A	392, 421
Galkovsky, T	1126
Galunin, E V	1115
Gangrskaia, E	642, 823
Gareev, K G	252
Garmash, V I	1074
Gavrilov, S A	1102
Gavrus, I V	68
Gayduchenko, I A	779, 797
Generalova, A. N.	277
Gerasimov, V A	221
Gets, D	453
Girshova, E I	1193
Gladkov, A	315
Gladyshev, A G	486
Gladyshev, S. A.	817
Glagolev, S F	63
Glechik, D	844
Glukhovskoy, E	91, 257, 1018, 1026, 1136
Golikov, A	664, 765, 784, 788
Goloshchapov, D	72, 243
Golosov, D A	149
Goltsman, G	611, 664, 756, 761, 765, 769, 779, 784, 788, 797
Golubkov, P E	378, 1043
Golubok, A O	436, 976
Gomon, D	602
Goncharova, L M	68
Gorbyak, V V	727
Gordeev, N Yu	372, 427, 547
Gorokh, G	57, 1126
Gorshkov, O N	1106
Goryacheva, I Yu	1061
Goryacheva, V A	257
Goshev, A A	1078
Grebenchukov, A N	793, 905
Grebenikova, N M	403
Greshnyakov, V A	6
Gridin, G E	616
Grigoryev, M N	1179
Grishkanich, A	718
Gromov, D G	1102

Gubaidullin, A R	1188
Gubaydullin, A R	736, 803, 1199
Gudkov, V A	1074
Gudovskikh, A	39, 398, 431, 495, 507, 528
Guina, M	1205
Guryev, V I	267
Gusev, E Yu	118, 158
Guseva, Y A	408
Guseva, Yu A	356, 444
Gvozdev, D A	212
Helm, M	181
Hennrich, F	779
Ierusalimsky, N. V.	277
Il'In, O I	133
Il'Ina, M V	919
Ilkiv, I V	202
Inyushov, A	705
Iorsh, I. V.	808
Ippolitov, Y A	243
Ishteev, A	453
Isoaho, R	1205
Ivanov, K A	803, 1188, 1193, 1199
Ivanov, S V	192
Ivanov, V V	1132, 1171
Ivonin, M	149
Javadzade, J	761
Jean, T	142
Jityaeva, J Y	158
Jmerik, V N	192
Kaasik, V	668, 823
Kadykova, Yu	956
Kaftyreva, L A	267
Kahl, O	756
Kaidashev, E	659
Kalganova, S	956, 968, 972
Kaliteevski, M A	736, 803, 1188, 1193, 1199
Kalyuzhnyy, N	137, 360, 372, 427, 481, 491, 528
Kamalov, R V	15
Kapitanova, P V	813, 817
Kapitonov, A N	984
Kapitonov, Yu V	743
Kapustin, S N	1078
Karachinsky, L Ya	486
Karakchieva, O	832
Karavainikov, A	838
Kardash, T Yu	107
Karenkikh, O G	1087
Karzanov, V	860
Katin, K. P.	1070
Kaydashev, V	659
Kazakin, A. N.	878
Kazakov, V	636

Kazakova, T A 586
Kazantsev, V 315
Khaydukov, E. V. 277
Khochenkov, D. A. 277
Khodzitsky, M K 793, 905
Khodzitsky, M. 602
Khonina, S N 551, 563, 616
Khorin, P A 568
Khrebtov, A I. 20, 348
Khripunov, A 248
Khromov, V S 328
Kiesewetter, D 248
Kinev, N V 874
Kirilenko, A 154
Kirilenko, D A 122, 202
Kivshar, Y. S. 817
Kleider, J-P 515
Kleimanov, R. 1008
Klenov, N. V. 673
Klimenko, V 233, 348
Klimin, V S 81, 149, 162, 462, 960, 964, 1141
Klimonsky, S O 586
Klunnikova, Y V 11
Knyaginin, D A 952
Knyazev, N 233
Kolesnikov, I E 684
Kolesnikova, A S 91, 1136
Kolodeznyi, E S 486
Kolomiytsev, A S 87
Kolosko, A G 1165
Kolpakov, V 315
Kolykhalova, E D 1193
Komarevtsev, I. 1008
Komrakova, S 761
Konakov, S. 323, 998
Kondrashin, V I. 378
Konkov, O I 994
Konopel'Ko, L 718
Konshin, A A 919
Kopyev, V V 933
Korenev, V V 352
Koriazhkina, M N 1106
Korikov, C C 311, 731
Korneev, A 664, 756, 761, 765, 769, 784, 788
Kornev, A A 582
Korolev, D S 30, 1106
Korolev, D V 252
Korotkov, A 998
Korsakov, A S 752
Koryakin, A A 166, 181, 206
Koshelets, V P 874
Koshelev, K. L. 808, 817

Koshelev, O A 192
Koshkin, A Yu 467
Koshuro, V 35, 1057, 1066
Kostrin, D K 221
Kotliar, K P 176
Kotlyar, K P 202, 356, 444, 495, 511, 582
Kotlyar, K. 408
Kots, I N 960, 1141
Koubisy, M S I 574
Kourova, N. 499
Kovalskiy, V A 171
Kovalyuk, V 664, 756, 761, 765, 769, 784, 788
Kozlowski, A V 1003
Krivenkov, R Yu 625
Kryzhanovskaya, N V ... 20, 356, 408, 444, 528, 582, 695, 1205
Kudrin, A. 860
Kudryashov, D 398, 431, 495, 507, 511, 528, 542
Kukushkin, S A 24, 166, 206
Kulagina, A S 348, 827
Kulagina, M 547, 1205
Kulichkova, V A 239
Küppers, F 597
Kupriyanov, A N 910
Kurnosov, D A 1110, 1115
Kurochkin, A S 486
Kurtenkov, R. V. 472
Kuzhelev, M V 133
Kuznetsov, Y 860, 865
Kuznetsova, I A 884
Kuznetsova, O B 306
Kuznetsova, Yu A 413, 574
Lapin, B P 578
Lavrentyev, V 972
Lazarenko, A A 100
Lazarenko, P I 1102
Lebedev, A A 202, 1121
Lebedev, E A 1102
Lebedev, S P 202, 1121
Lenshin, A. 72, 262
Lesnikov, V 860
Levin, D D 337
Levina, S A 503, 524
Levitskii, I V 621
Lewes-Malandrakis, G 756
Liashenko, T. 453
Lihachev, A I 923
Limonov, M. F. 817
Lioubtchenko, D 602
Lipovskii, A 408, 444, 582, 642, 668, 678, 695, 823
Lisitsyn, S A 462
Lobino, M 664
Loboda, V 998

Lobov, S A 296
Logachev, V 499
Loganchuk, S M.. 915
Logunov, S E 467
Lozovenko, A A 57
Lozovskii, V N... 68
Lubenchenko, A V 689
Lugovitskaya, T N 257
Lugovskoy, N... 844
Lukashchuk, A 597
Lukashenko, T 283
Lukashev, E P 212
Lutetskiy, A V 486
Lyanguzov, N 659
Lyashko, S 838
Macku, R 392, 421
Makarenko, D. P. 77
Makarov, S 453, 542, 607
Makarov, V A 296
Makeev, M O 942
Makevnina, V V 217
Maksimova, N K 895
Malekizandi, M... 597
Malyar, I V 1003
Malysheva, E 865
Malyshkin, V G 382
Malyukov, S P 11
Mamaeva, T A... 436
Mamonkina, A N... 747
Manshina, A A 684
Markin, A V 1061
Markina, N E 1061
Martemyanov, N A... 15
Martynov, A V 1043
Mashkovtsev, M A 413
Maslov, A. D. 417
Maslov, M. M. 1070
Masyukov, M S 793
Matyushkin, Y E 797
Maximov, M V.....352, 356, 372, 408, 427, 444, 547, 582, 1205
Mekhov, I B 630
Mekhrengin, M V 267
Melezhik, A V 1096, 1115
Meshkov, S A 942
Meshkovskii, I K 267
Mezentsov, S. 1057
Mezentzov, S 1175
Mezhenny, M 499
Miakonkikh, A 149, 1156
Mikhailova, T 838, 1179
Mikhaylin, I A 1, 81, 112
Mikhaylov, A N 30, 1106

Mikhaylov, A O 1038
Mikhaylovskaya, Z A 574
Mikheev, G M 625
Mikheev, K G 625
Mingaleva, A E 714
Minkov, K N.. 1171
Mintairov, M A.. 360, 481
Mintairov, S A 137, 360, 372, 427, 481, 528
Mintairov, S. 491
Mitin, D M 532
Mitrofanov, M I 621
Mizerov, A 72, 980
Mizerov, M N 621
Mkrtchyan, E S 1096, 1115
Mogileva, T N 625
Mohamed, A A A 915
Moiseenko, I M 655
Moiseev, E I 356, 408, 444, 582, 1205
Mokhov, D V 980
Monastyrenko, A 431
Morozov, I A.. 495, 507, 515
Morozov, K M.. 736, 803, 1199
Morozov, M Y 655
Morozov, P V 689
Morozov, S V 20
Moshnikov, V A 262
Moskaletz, O 636
Moskotin, M V.. 779, 797
Motúz, R 421
Mozharov, A........................ 154, 197, 449, 507, 532, 542, 976
Mukhin, I........................... 154, 197, 449, 532, 542, 976
Mukhin, S... 436
Mukhina, I.. 315
Murashev, V 499
Murashkevich, A 1126
Muravijova, D. 1013
Myasoedova, T N 1179
Myazin, N S 227
Mylnikov, I L 1032
Mynbaeva, M. 1013
Nadtochiy, A M 137, 1205
Naidenko, N A.. 960
Nashchekin, A V 176, 923
Nasibulin, A G... 408, 532
Naumisheva, E B 252
Nebel, C 756
Nebojsa, A 421
Nechaev, D V 192
Nedviga, A 870
Nekipelov, S V 714
Nekorkin, S M 520
Nemtseva, S Yu 1102

Nenasheva, E A..813
Nepomnyashchaya, E K...............................301, 731
Neskorniuk, V...597
Neskoromnaya, E A...............................1096, 1110
Nevedomskiy, V N.....................................127, 137
Nevedomsky, V N...486
Nevolin, V K...337
Nezhentsev, A V..1074
Niemi, T..1205
Nikiforov, K...103
Nikitina, E.....................................72, 100, 980, 1188
Nikolaev, D N...621
Nikolaev, K O...378
Nikolaev, V G...630
Nikolaeva, A A..647
Nikolskaya, A A..30
Nikonorov, N V...827, 856
Novikov, A S...408
Novikov, I I...486
Novopashin, S A...107
Nozdriukhin, D V..333, 343
Nudga, A...589
Nunn, N...625
Obraztsova, E D..797
Oleinik, V L...933
Olekhno, N A...647
Osipov, A V...24
Ovchinnikov, G..597
Pakharukov, Yu V...990
Palkanov, P...1053
Panchak, A..491
Panchenko, I V..1146
Pankin, D V...684
Papež, N...392, 421
Papshev, V...53
Parafin, A...865
Parulin, R A..574
Paschenko, V Z..212
Pashchenko, A. S...96
Pastukhov, A I..262
Pavlov, D A..30, 1106
Pavlov, P. A...537
Pavlov, S I..176
Payusov, A S...372, 427, 547
Pecherskaya, E A...378, 1043
Pelipenko, M I..1146
Perin, A..651, 705, 774
Perkov, S A...320
Permyakov, D V..436
Pernice, W.............................756, 765, 769, 784, 788
Petrenyov, I A...15
Petrov, A A..365

Petrov, M I..647
Petrov, S I...1087
Petrov, V. V..77, 1183
Petrova, J..890
Petrova, O V...714
Petukhov, V A..337
Phyo, Kyaw Z...557
Pigareva, Y...315
Pikhtin, N A..486
Pimashkin, A..315
Pinaev, N..398
Pirogov, E V...100
Pligovka, A...1126
Plotnikova, Y A...243
Podoskin, A A...1038
Pogorelaya, D A..267, 852
Pogosov, A G..848
Pokhabov, D A..848
Poletaev, D...589
Polikarpov, Yu A...337
Polischuk, O V..655
Poltavtsev, S V...743
Polyakova, V V..1141
Ponomarenko, S A..593
Ponomaryova, N S..928
Popov, E O..1165
Popov, V V..655
Porfirev, A P...616, 739
Poshivalova, E..1092
Postnov, V N...252
Pozharov, M V...1136
Pozina, G..1188
Poznyak, A A...57
Prikhodko, A V..994
Prokhodtsov, A...788
Prokopov, A...838, 844
Proskuryakov, V...1175
Protas, N V..1132, 1171
Prudaev, I A..933
Prudkovskiy, V. S...1070
Puech, P...1070
Punegov, V I...187
Pustozerov, A V..774
Putintseva, M V..311
Pyatishev, E...878, 1008
Pyatkov, F..779
Qassime, M M..257
Rabinovich, O...499
Radaev, O A..440
Rajanna, P M...532
Raskhodchikov, A V..695
Raskhodchikov, D V..668

Raskhodchikov, D 823
Raudik, S A 532
Reddy, Andra N. K 563
Redkov, A V 24, 206
Reduto, I 444, 582, 642, 823
Reshetov, I V 668
Reznik, A 248
Reznik, R 20, 511, 1083
Rezvan, A A 162, 462, 960, 964, 1141
Rodin, A 398
Rodionov, I 968, 1049, 1053, 1057, 1092, 1175
Romanov, N R 699
Romashkin, A V 337
Rud', V. Y. 403, 1160, 1160
Rudakov, K I 874
Rudenko, K 149, 1156
Rudnitskaya, G 283
Rudyk, N N 133
Russo, S 905
Ruzankina, J 718
Ryabova, M A 1106
Ryazanov, R M 1102
Rybalka, S B 928, 952
Rybalko, D A 547
Rybin, M G 797
Rybin, M. V. 817
Rychkov, V N 413
Rykov, A V 171, 520
Ryzhuk, R V 1087
Rzhevskiy, S S 684
Sadreev, A F 813
Sadrieva, Z 808, 813, 817
Safargaliev, R F 990
Sakhadzhi, G. V. 948
Salii, R A 137, 360
Salimgareev, D D 752
Samartsev, I V 520
Samosvat, D M 328
Samsonenko, Yu B 20
Samusev, A K 436
Samusev, K. B. 817
Sannikov, D A 1205
Santer, S 1018
Sapunov, G A 154, 449
Saranin, D 453
Satanin, A. M. 673
Savchenko, E A 301, 306
Savchuk, A 499
Savelyev, A V 352, 476
Savelyev, A. G. 277
Savelyev, D A 710
Savenko, O V 884

Sayenko, A V 11
Scherbak, S A 356, 444, 582, 678, 695
Scherbatenko, M. L. 611
Schneider, H 181
Sedykh, S V 928
Selenina, A V 239
Selifonova, E I 1026
Selivanov, L M 221
Semenova, O I 387
Semyonov, V V 747
Seredin, P 72, 243, 262
Sergeev, V A 440
Serin, A A 372, 427, 547
Sevastyanov, E Yu 895
Sevostyanov, V P 91, 1026, 1136
Shabanov, V E 365
Shabiev, F K 990
Shabunina, E I 458
Shaman, Yu P 1102
Shamshin, M O 296
Shandarov, V M 651
Shandyba, N A 1146
Shaposhnikov, A 838
Shapovalov, D V 365
Sharaevskaya, A Yu 900
Sharov, V 1083
Shchelkunov, A 1049
Shenderova, O A 625
Shengurov, V G 171
Shenina, M E 1106
Shepeleva, Y V 1043
Shernyakov, Yu M 356, 372, 427, 547
Shevyrin, A A 848
Shinkarenko, O 1018, 1136
Shipovskaya, A B 257
Shipulya, N D 323
Shishkina, O A 928
Shklyaev, A A 848
Shkoldin, V A 436
Shmakov, S. 233, 320, 348
Shmidt, N M 458
Shmygalev, A S 752
Sholina, N. V. 277
Shomysov, N N 714
Shostka, N 832
Shubina, K Y 495, 980
Shugurov, K U 449
Shumikhin, K. V. 948
Shumilin, A 53, 968
Shurkaeva, I V 689
Shvarts, M. 360, 481, 491, 503, 524
Sibirev, N V 181

Sidorov, A I 727
Simon, V A 221
Sitnikova, A A 122, 154
Sivkov, D V 187, 714
Sivkov, V N 714
Skandakov, R N 714
Skaptsov, A 53
Skarvada, P 392
Skibina, Y S 1061
Skorotetcky, M S 593
Škvarenina, L 392, 421
Skvortsov, A N 301
Slavinskaya, E M 107
Slipchenko, S O 486, 1038
Smagulova, S A 984
Smirnov, E 664
Smirnov, K J 63, 403
Smirnov, K V 689, 699
Smirnov, V A 1146
Smirnova, T E 895, 933
Smolovik, M A 852
Snigirev, L A 122
Sobola, D 392, 421
Sobolev, Dmitrii I 856
Sobolev, M S 100
Sochilina, A. V. 277
Sofronov, A N 486
Sokolenko, B 589, 832
Sokolovskii, A S 181, 292
Sokolovskii, G S 486
Sokura, L A 127
Solodovnik, M S 1, 81, 112, 462, 1087
Solomin, E G 919
Solovev, I A 743
Soshenko, V 761
Soshnikov, I P 176, 511
Sosnin, D 542
Spivak, Yu M 262
Stadnichenko, A I 107
Starkov, A S 1032
Starkov, I A 1032
Starnikova, A. P. 77
Stetsyura, S V 1003
Stolpner, A 248
Strigalev, V E 852
Stroganov, B V 743
Struchkov, N S 337
Stupin, D D 923
Sun, Y 607
Surin, N M 593
Surnin, K A 176
Sushkov, A A 30

Svirina, A 233
Svirko, Yu 823
Sych, D. V. 611
Sychev, S. K. 808
Tagantsev, D K 668
Talnishnikh, N A 458
Tauchnitz, T. 181
Terekhov, V I 752
Tereshchenko, O E 387
Terterov, I 233
Terukov, E I 1160
Tetelbaum, D I 30
Tikhomirov, V G 408, 937
Timoshenko, I V 574
Timoshnev, S N 722
Tingaev, M I 49
Titova, N 779
Tkachev, A G 1110
Tolstunov, M I 11
Tomilin, A N 239
Tomilin, S 838
Tominov, R V 87, 1146
Tomskaya, A E 984
Touel, L 915
Trigorly, S 972
Trushnikov, I 705
Tsimokha, A S 239
Tsvetkova, O Yu 91, 1136
Tsymbalov, A 890
Tupik, A 283
Tushavin, G V 63
Tutova, M S 1022
Uhov, A A 221
Uskov, A V 436, 976
Uvarov, A V 39, 511, 515
Uvarov, I V 910
Vaganov, M 636
Vakhtomin, Yu B 689, 699
Vakulov, Z 149, 1141
Varlamov, D 283
Varnavskaya, A V 68
Vasilev, A 542
Vasileva, A A 684
Vasiliev, A A 436
Vasiliev, V. V. 472
Vasin, S V 440
Vasinkina, E 956, 972
Ved, M 171, 865
Velichko, E N 272, 301, 311
Vergeles, P S 171
Vetter, A 756
Vikulin, D V 578

Vlasov, A A 852
Vokhmintsev, A S 15
Volik, D P 149
Volkova, N A 827
Volodin, A 659
Vongsvivut, J 243
Vorobjev, L E 486
Vorobyov, V 761
Voronov, B 664, 779
Voyko, A 53, 968
Voytitsky, V 589
Vozianova, A V 793
Voznyuk, G V 621
Vyatkina, S P 413
Vysochinskaya, V 233
Weinstein, I A 15
Yagafarov, T 1205
Yagovkina, M 1013
Yakimov, E B 171
Yakimova, R. 1070
Yakovenko, N A 68
Yakovlev, G E 515
Yashina, N Yu 91
Yavorsky, M A 578
Yugay, A 998
Yulmetova, O S 1022
Yulmetova, R F 1022
Yushkanov, A A 884
Yusupova, E A 382
Zabalueva, Z A 731
Zaikovskii, A V 107
Zaitsev, A D 905
Zakharevich, A 53, 1061
Zakhidov, A 453
Zakhlebayeva, A 1126
Zalyotov, D V 365
Zamburg, E G 149, 1146
Zatsepin, A F 413, 574
Zavadskiy, S M 149
Zdoroveyshchev, A 171, 520, 865
Zegrya, A G 328
Zegrya, G G 328
Zelentsov, K S 507, 528
Zemlyakov, V E 1074
Zemlyanoy, V K 1132
Zhdanov, E Yu 848
Zhevlakov, A 718
Zhilkin, B P 752
Zhukov, A E ...20, 352, 356, 372, 408, 427, 444, 582, 695, 1205
Zhukov, M V 436
Zhuravleva, N 248
Zinchenko, T O 378, 1043

Zolotov, P 664, 689, 699
Zolotukhin, D 72
Zolotukhin, P 659
Zorin, V N 252
Zubkova, E 765, 769, 784, 788
Zubov, F I 372
Zuev, D A 607
Zvonkov, B N 520
Zybin, A A 458

Institute of Physics
Dirac House, Temple Back
Bristol BS1 6BE UK

ISSN: 1742-6588
ISBN 978-1-5108-8099-3

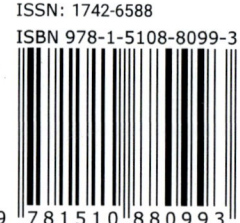

9 781510 880993